LDA	lithium diisopropylamide
LDMAN	lithium 1-(dimethylamino)naphthalenide
LHMDS	= LiHMDS
LICA	lithium isopropylcyclohexylamide
LiHMDS	lithium hexamethyldisilazide
LiTMP	lithium 2,2,6,6-tetramethylpiperidide
LTMP	= LiTMP
LTA	lead tetraacetate
lut	lutidine
m-CPBA	*m*-chloroperbenzoic acid
MA	maleic anhydride
MAD	methylaluminum bis(2,6-di-*t*-butyl-4-methylphenoxide)
MAT	methylaluminum bis(2,4,6-tri-*t*-butylphenoxide)
Me	methyl
MEK	methyl ethyl ketone
MEM	(2-methoxyethoxy)methyl
MIC	methyl isocyanate
MMPP	magnesium monoperoxyphthalate
MOM	methoxymethyl
MoOPH	oxodiperoxomolybdenum(pyridine)-(hexamethylphosphoric triamide)
mp	melting point
MPM	= PMB
Ms	mesyl (methanesulfonyl)
MS	mass spectrometry; molecular sieves
MTBE	methyl *t*-butyl ether
MTM	methylthiomethyl
MVK	methyl vinyl ketone
n	refractive index
NaHDMS	sodium hexamethyldisilazide
Naph	naphthyl
NBA	*N*-bromoacetamide
nbd	norbornadiene (bicyclo[2.2.1]hepta-2,5-diene)
NBS	*N*-bromosuccinimide
NCS	*N*-chlorosuccinimide
NIS	*N*-iodosuccinimide
NMO	*N*-methylmorpholine *N*-oxide
NMP	*N*-methyl-2-pyrrolidinone
NMR	nuclear magnetic resonance
NORPHOS	bis(diphenylphosphino)bicyclo[2.2.1]-hept-5-ene
Np	= Naph
PCC	pyridinium chlorochromate
PDC	pyridinium dichromate
Pent	*n*-pentyl
Ph	phenyl
phen	1,10-phenanthroline
Phth	phthaloyl
Piv	pivaloyl
PMB	*p*-methoxybenzyl

PMDTA	*N,N,N′,N″,N″*-pentamethyldiethylene-triamine
PPA	polyphosphoric acid
PPE	polyphosphate ester
PPTS	pyridinium *p*-toluenesulfonate
Pr	*n*-propyl
PTC	phase-transfer catalyst/catalysis
PTSA	*p*-toluenesulfonic acid
py	pyridine
RAMP	(*R*)-1-amino-2-(methoxymethyl)pyrrolidine
rt	room temperature
salen	bis(salicylidene)ethylenediamine
SAMP	(*S*)-1-amino-2-(methoxymethyl)pyrrolidine
SET	single-electron transfer
Sia	siamyl (3-methyl-2-butyl)
TASF	tris(diethylamino)sulfonium difluorotrimethylsilicate
TBAB	tetrabutylammonium bromide
TBAF	tetrabutylammonium fluoride
TBAD	= DBAD
TBAI	tetrabutylammonium iodide
TBAP	tetrabutylammonium perruthenate
TBDMS	*t*-butyldimethylsilyl
TBDPS	*t*-butyldiphenylsilyl
TBHP	*t*-butyl hydroperoxide
TBS	= TBDMS
TCNE	tetracyanoethylene
TCNQ	7,7,8,8-tetracyanoquinodimethane
TEA	triethylamine
TEBA	triethylbenzylammonium chloride
TEBAC	= TEBA
TEMPO	2,2,6,6-tetramethylpiperidinoxyl
TES	triethylsilyl
Tf	triflyl (trifluoromethanesulfonyl)
TFA	trifluoroacetic acid
TFAA	trifluoroacetic anhydride
THF	tetrahydrofuran
THP	tetrahydropyran; tetrahydropyranyl
Thx	thexyl (2,3-dimethyl-2-butyl)
TIPS	triisopropylsilyl
TMANO	trimethylamine *N*-oxide
TMEDA	*N,N,N′,N′*-tetramethylethylenediamine
TMG	1,1,3,3-tetramethylguanidine
TMS	trimethylsilyl
Tol	*p*-tolyl
TPAP	tetrapropylammonium perruthenate
TBHP	*t*-butyl hydroperoxide
TPP	tetraphenylporphyrin
Tr	trityl (triphenylmethyl)
Ts	tosyl (*p*-toluenesulfonyl)
TTN	thallium(III) nitrate
UHP	urea–hydrogen peroxide complex
Z	= Cbz

Handbook of Reagents
for Organic Synthesis

Reagents for Silicon-Mediated Organic Synthesis

OTHER TITLES IN THIS COLLECTION

e-EROS

For access to information on all the reagents covered in the
Handbooks of Reagents for Organic Synthesis, and many more,
subscribe to e-EROS on the Wiley Online Library website.
A database is available with over 200 new entries and updates every
year. It is fully searchable by structure, substructure and reaction
type and allows sophisticated full text searches.
http://onlinelibrary.wiley.com/book/10.1002/047084289X

Handbook of Reagents
for Organic Synthesis

Reagents for Silicon-Mediated Organic Synthesis

Edited by

Philip L. Fuchs
Purdue University, West Lafayette, IN, USA

A John Wiley & Sons, Ltd , Publication

This edition first published 2011
© 2011 John Wiley & Sons Ltd

Registered office
John Wiley & Sons Ltd, The Atrium, Southern Gate, Chichester, West Sussex, PO19 8SQ,
United Kingdom

For details of our global editorial offices, for customer services and for information about how
to apply for permission to reuse the copyright material in this book please see our website at
www.wiley.com.

Library of Congress Cataloging-in-Publication Data

Handbook of reagents for organic synthesis.
  p.cm
  Includes bibliographical references.
  Contents: [1] Reagents, auxiliaries and catalysts for C–C bond
  formation / edited by Robert M. Coates and Scott E. Denmark
  [2] Oxidizing and reducing agents / edited by Steven D. Burke and
  Riek L. Danheiser [3] Acidic and basic reagents / edited by
  Hans J. Reich and James H. Rigby [4] Activating agents and
  protecting groups / edited by Anthony J. Pearson and William R. Roush
  [5] Chiral reagents for asymmetric synthesis / edited by Leo A. Paquette
  [6] Reagents for high-throughput solid-phase and solution-phase organic
  synthesis / edited by Peter Wipf [7] Reagents for glycoside, nucleotide
  and peptide synthesis / edited by David Crich [8] Reagents for direct
  functionalization of C–H bonds/edited by Philip L. Fuchs [9] Fluorine-
  Containing Reagents/edited by Leo A. Paquette [10] Catalyst Components
  for Coupling Reactions / edited by Gary A. Molander [11] Reagents for
  Radical and Radical Ion Chemistry/edited by David Crich [12] Sulfur-Containing
  Reagents / edited by Leo A. Paquette [13] Reagents for Silicon-Mediated Organic
  Synthesis / edited by Philip L. Fuchs

  ISBN 0-471-97924-4 (v. 1)   ISBN 0-471-97926-0 (v. 2)
  ISBN 0-471-97925-2 (v. 3)   ISBN 0-471-97927-9 (v. 4)
  ISBN 0-470-85625-4 (v. 5)   ISBN 0-470-86298-X (v. 6)
  ISBN 0-470-02304-X (v. 7)   ISBN 0-470-01022-3 (v. 8)
  ISBN 978-0-470-02177-4 (v. 9) ISBN 978-0-470-51811-3 (v.10)
  ISBN 978-0-470-06536-5 (v. 11) ISBN 978-0-470-74872-5 (v.12)
  ISBN 978-0-470-71023-4 (v. 13)

  1. Chemical tests and reagents.   2. Organic compounds-Synthesis.
  QD77.H37 1999   98-53088
  547'.2 dc 21   CIP

A catalogue record for this book is available from the British Library.

ISBN 13: 978-0-470-71023-4

Set in 9½/11½ pt Times Roman by Thomson Press (India) Ltd., New Delhi.
Printed in Singapore by Markono Print Media Pte Ltd.

This volume is dedicated to Professor Leo Paquette who conceived and guided the EROS project and to Ms. Louise Portsmouth who makes its evolution possible on a daily basis.

Contents

Preface

The eight-volume *Encyclopedia of Reagents for Organic Synthesis* (*EROS*), authored and edited by experts in the field, and published in 1995, had the goal of providing an authoritative multivolume reference work describing the properties and reactions of approximately 3000 reagents. With the coming of the Internet age and the continued introduction of new reagents to the field as well as new uses for old reagents, the electronic sequel, *e-EROS*, was introduced in 2002 and now contains in excess of 4000 reagents, catalysts, and building blocks making it an extremely valuable reference work. At the request of the community, the second edition of the encyclopedia, *EROS-II*, was published in March 2009 and contains the entire collection of reagents at the time of publication in a 14-volume set.

While the comprehensive nature of *EROS* and *EROS-II* and the continually expanding *e-EROS* render them invaluable as reference works, their very size limits their practicability in a laboratory environment. For this reason, a series of inexpensive one-volume *Handbooks of Reagents for Organic Synthesis* (*HROS*), each focused on a specific subset of reagents, was introduced by the original editors of *EROS* in 1999:

Reagents, Auxiliaries, and Catalysts for C–C Bond Formation
Edited by Robert M. Coates and Scott E. Denmark

Oxidizing and Reducing Agents
Edited by Steven D. Burke and Rick L. Danheiser

Acidic and Basic Reagents
Edited by Hans J. Reich and James H. Rigby

Activating Agents and Protecting Groups
Edited by Anthony J. Pearson and William R. Roush

This series has continued over the last several years with the publication of a further series of *HROS* volumes, each edited by a current or past member of the *e-EROS* editorial board:

Chiral Reagents for Asymmetric Synthesis
Edited by Leo A. Paquette

Reagents for High-Throughput Solid-Phase and Solution-Phase Organic Synthesis
Edited by Peter Wipf

Reagents for Glycoside, Nucleotide, and Peptide Synthesis
Edited by David Crich

Reagents for Direct Functionalization of C–H Bonds
Edited by Philip L. Fuchs

Fluorine-Containing Reagents
Edited by Leo A. Paquette

Catalyst Components for Coupling Reactions
Edited by Gary A. Molander

Reagents for Radical and Radical Ion Chemistry
Edited by David Crich

Sulfur-Containing Reagents
Edited by Leo A. Paquette

This series now continues with the present volume entitled *Reagents for Silicon-Mediated Organic Synthesis*, edited by Philip Fuchs, long-standing member of the online *e-EROS* Editorial Board. This 13th volume in the *HROS* series, like its predecessors, is intended to be an affordable, practicable compilation of reagents arranged around a central theme that it is hoped will be found at arm's reach from synthetic chemists worldwide. The reagents have been selected to give broad relevance to the volume, within the limits defined by the subject matter of silicon-mediated reagents. We have enjoyed putting this volume together and hope that our colleagues will find it just as enjoyable and useful to read and consult.

David Crich
Centre Scientifique de Gif-sur-Yvette
Institut de Chimie des Substances Naturelles
Gif-sur-Yvette, France

Introduction

As the second most abundant element in the earth's crust (26–28%),[1] silicon is vital to the construction industry as concrete and cement. The manufacture of silicon-based integrated circuits continues to evolve our computer-rich lifestyle. Most high-purity silicon is refined by conversion to Cl_3SiH or Cl_4Si and distilled followed by reduction to elemental Si. These and other low-cost, high-purity materials serve as primary feedstocks for the manufacture of the diverse collection of silicon reagents used in organic synthesis.

The aim of this handbook was to collect the most important organosilicon reagents together into a single volume applicable for the use by the bench chemist. The majority of reagents in this volume are either newly commissioned or recently updated and are not in the *Encyclopedia of Reagents for Organic Synthesis*, second edition. A new feature in this handbook is the reagent finder. This is an alphabetically organized lookup table that is arranged by the organic functionality and specific structure of the silicon atom to which it is bound. It is hoped that visual inspection of this table will both allow easy access to a specific reagent and stimulate creative design of new reagents and silicon-based strategies.

The subsequent section is a short overview of this editor's perspective with regard to evolution of silicon's impact in the field of organic synthesis. In particular, the continued evolution of oxygen–silicon bonds with respect to alcohol protecting groups and silyl enol ethers is only tangentially addressed. The three key areas of greatest synthetic impact are as follows: (1) The use of *silicon as a temporary tether* for organic synthesis by unifying a reactive pair of functional groups and taking advantage of their template-based intramolecular cyclization. (2) A second conceptual growth area is the specific use of silane functionality as a hetero *t*-butyl group, often colloquially referred to as the *use of silicon as a fat proton*. In effect, the large stearic demand of the organosilane group is used to both direct reaction specificity and provide a latent point of attachment that can be either protiodesilylated (reduced) or converted to an alcohol by the Kumada–Fleming–Tamao (KFT) oxidation. (3) A final area briefly highlighted is the use of the *Brook rearrangement as an anion relay stratagem.*

Silicon Bond Formation/Cleavage. The vast majority of applications of silicon intermediates in organic synthesis involve temporary installation and removal of silicon atoms. While there are many well-established examples of classical, backside, S_N2 displacements at tetravalent silicon, it is useful for the purpose of this handbook to artificially consider that all silicon bond formations and cleavages occur via siliconate [Si(V)] intermediates. This key reaction extends across the periodic table especially including groups 4, 5, 6, and 7. Well-known examples include the often used alcohol, phenol, and carboxylic acid protections (and selective deprotections) via the formation of silicon–oxygen derivatives (eq 1).[2]

Y(TMS–Y)	BDE Si–Y (kcal mol^{-1})
H	95
Me	94
SiMe$_3$	79
F	158
Cl	117
Br	102
I	82
—OH	133
—OMe	123
O–H	118
O–Me	96
—NMe$_2$	97

As can be easily appreciated from the table of bond dissociation energies,[3] reactions involving formation of silicon–oxygen bonds from silyl iodides, bromides, and chlorides are exothermic. Of equal significance, reactions that result in formation of the silicon–fluorine bond (the strongest single bond known) enable cleavage of virtually all other bonds to silicon.

Silicon–element bonds that are allylic or benzylic undergo especially facile nucleophile-mediated cleavage reactions by virtue of their ~10–11 kcal mol^{-1} weaker bond (**4–6**, eq 1).

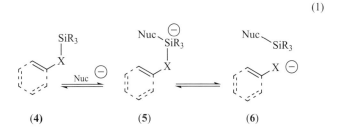

$$(1)$$

Silyl Enol Ethers. Until the early 1960s, functionalization of ketones such as **7** and other enolizable carbonyl compounds was accomplished by either acid- or base-catalyzed electrophilic alkylation of the α-carbon via generation of enol **8** or enolate **9** intermediates (eq 2). Disadvantages of the enol **8** included low equilibrium population, resulting in slow reactions combined with lack of monoselectivity. While the enolate **9** can be efficiently generated in near-quantitative yield, its high reactivity and basicity often led to unwanted side reactions, especially in cases where ketone/enolate equilibrium reactions are involved. By comparison, the easily prepared and purified silyl enol ethers **10** have vastly expanded the synthetic toolbox of the organic chemist. Hindered silyl enol ethers such as **10** ($R_3Si = t\text{-BuMe}_2Si$) enjoy sufficient hydrolytic stability that they can be routinely analyzed and purified using silica gel chromatography.

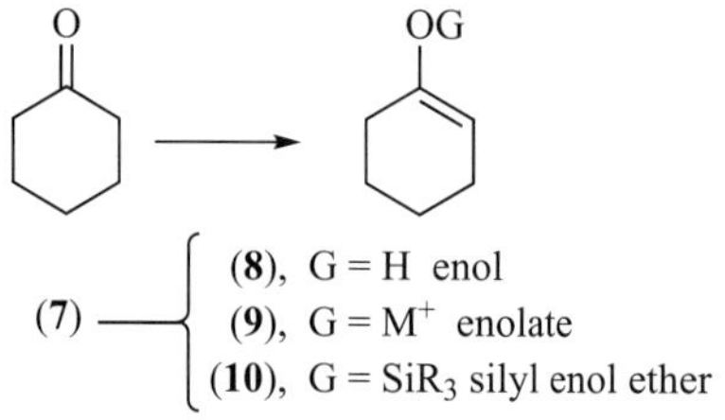

(2)

(**7**)
- (**8**), G = H   enol
- (**9**), G = M$^+$ enolate
- (**10**), G = SiR$_3$ silyl enol ether

While effective methods for regiospecific conversion of silyl enol ethers such as **10** to enolate **9** (eq 3) further extended synthetic options using these basic intermediates, the observation that electrophilic reactions of silyl enol ethers (Friedel–Crafts-type chemistry via the intermediacy of siloxonium ion **11**) were highly general enabled a new array of synthetic opportunities for carbonyl functionalization, often under essentially neutral conditions.[4]

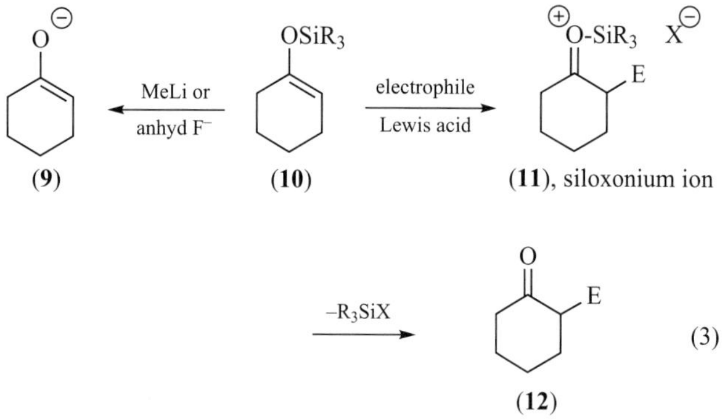

(**9**)   (**10**)   (**11**), siloxonium ion

(3)

(**12**)

Vinyl siloxonium ion **14** also serves as an important intermediate en route to conjugate adducts of unsaturated carbonyl compounds. For example, treatment of cyclohexenone (**13**) with reactive silyl electrophiles affords γ-functionalized silyl enol ethers **15** and **16** suitable for subsequent synthetic transformations (eq 4).[5]

The *temporary silicon tether* (TST)[6] strategy has been updated (2010) by an excellent review focusing upon metal-mediated reactions.[7] The inception of this strategy is attributable to Nishiyama and Itoh who reported the radical cyclization of acyclic bromomethyl silyl ethers to siloxanes and their subsequent oxidation to 1,3-diols.[8] Shortly thereafter, the group of Gilbert

Stork published a series of seminal papers that evolved this concept to the alcohol-directed hydroxymethylation and methylation of decalin and hydrindane systems.[9]

As can be seen in eq 5, treatment of bromomethyl silyl ether **17** with tributyl tin hydride generates an α-alkoxylsilylmethyl radical that suffers intramolecular 5-*exo-trig* addition to the proximal olefin to provide *cis*-fused siloxane **18**, which can be either converted to 1,3-diol **19**, using the KFT oxidation, or protiodesilylated to *syn*-methyl alcohol **20** using fluoride.[9] An important experimental contribution has been provided by the Roush group who describe conditions for protiodesilylation of *unactivated* $C(sp^3)$ phenyldimethylsilyl PDMS groups to CH–bonds.[10]

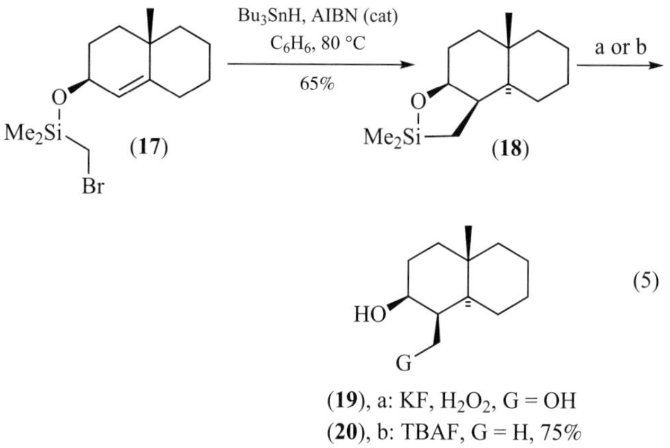

(5)

(**19**), a: KF, H_2O_2, G = OH
(**20**), b: TBAF, G = H, 75%

More recently, the Kigoshi group featured the use of a *bridged* six-membered siloxane stereocontrol element, also established by intramolecular addition of the α-alkoxysilylmethyl radical, yielding ketoester **21**. After conversion to **22**, intramolecular spiroalkylation of the keto enolate kinetically set the stereochemistry at carbons a and b. Methallylation (at c) to give **23** as a single isomer was followed by ring-closing metathesis (RCM), which established the basic carbon framework of the ingenol ring system (eq 6).[11]

The α-halomethyl silyl ether intermediate has also been effectively applied in the context of anion chemistry.[12] Treatment of iodide (or bromide) **25** with 2 equiv of t-BuLi rapidly affords an α-silyllithium species that smoothly undergoes intramolecular addition to the carbonyl group affording bridged bicyclic [3.3.1] alcohol **26** in excellent yield (eq 7). Equally notable is the observation that formation of the ketone enolate of **25** provides the bridged bicyclic [3.2.2] system **27** via intramolecular alkylation.

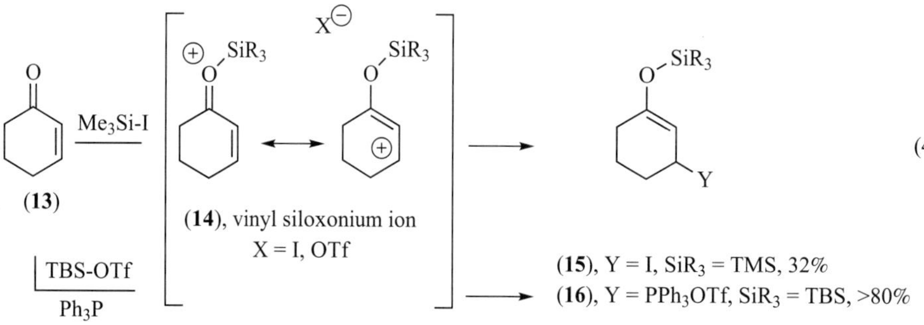

(**13**)

(**14**), vinyl siloxonium ion
X = I, OTf

TBS-OTf
Ph$_3$P

(4)

(**15**), Y = I, SiR$_3$ = TMS, 32%
(**16**), Y = PPh$_3$OTf, SiR$_3$ = TBS, >80%

A list of General Abbreviations appears on the front Endpapers

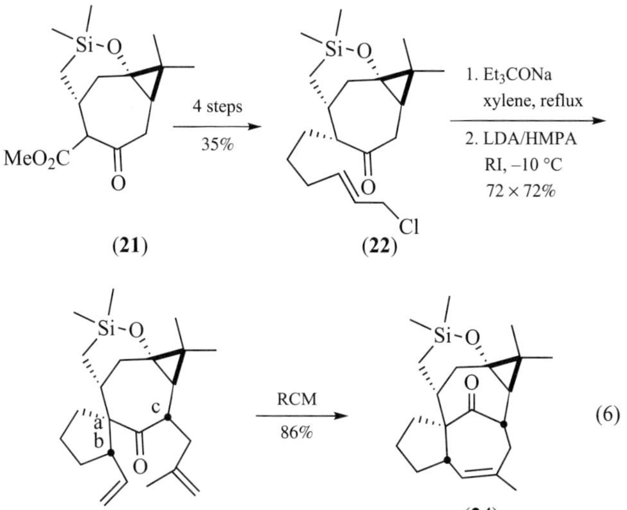

(21)   4 steps   35%   **(22)**

1. Et₃CONa
   xylene, reflux
2. LDA/HMPA
   RI, −10 °C
   72 × 72%

RCM   86%   (6)

(23)   **(24)**

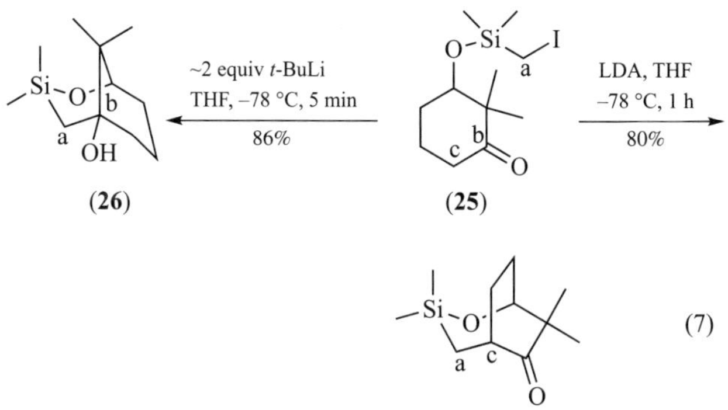

~2 equiv *t*-BuLi
THF, −78 °C, 5 min
86%

(26)   **(25)**

LDA, THF
−78 °C, 1 h
80%

(7)

(27)

The TST strategy has been the subject of diverse applications. For example, the Shea research group employed the concept of type II intramolecular Diels–Alder chemistry in conjunction with a *temporary silicon tether* and a chiral 1,2-diol auxiliary to effect diastereomeric transformation of **28** to compounds **29** and **30** in a 1:6 ratio (eq 8).[13]

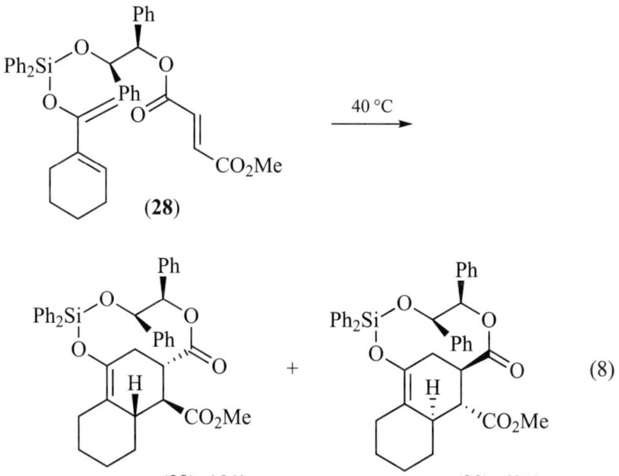

(28)   40 °C

+   (8)

(29), 10%   **(30)**, 62%

Titanocene-mediated RCM using the Takeda thioacetal strategy provides another opportunity to use the silyl ether tether to

direct intramolecular bond formation. In this example, easily prepared substrate **31** was smoothly cyclized to siloxane **33** (eq 9). Subsequent KFT oxidation provided the 1,5-diol with complete retention of the Z-olefin (not shown).[14]

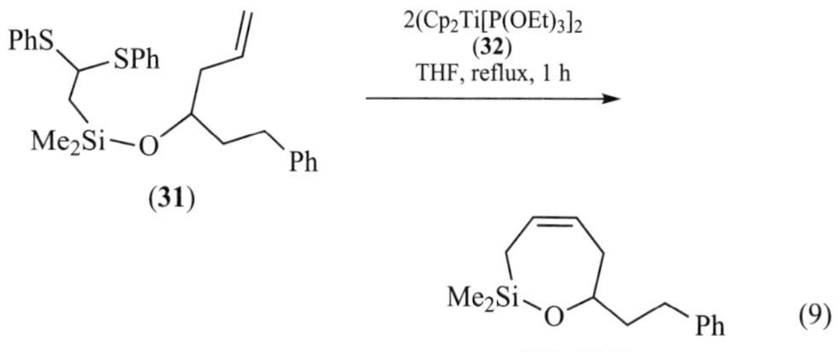

(31)   2(Cp₂Ti[P(OEt)₃]₂)
(32)
THF, reflux, 1 h

(9)

(33), 80%

A striking synergistic application of the TST in combination with RCM involves the catalytic enantiospecific desymmetrization of *meso* silyl ether **34**. Subjecting **34** to enantiopure Mo catalyst **35** generates siloxane **36** with near-perfect yield, enantiomeric excess, and atom economy (eq 10).[15]

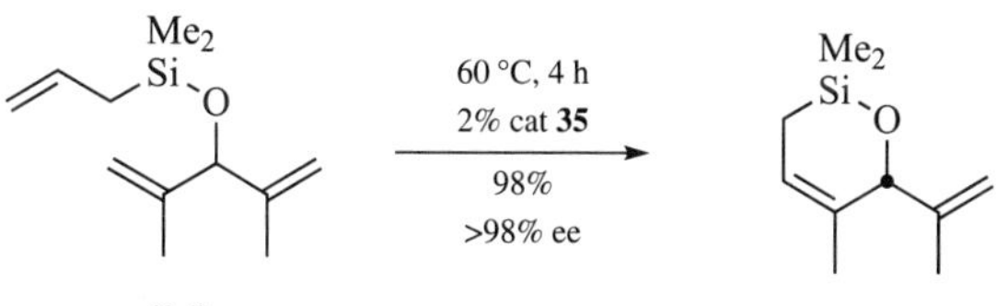

(34)   60 °C, 4 h
2% cat **35**
98%
>98% ee   **(36)**

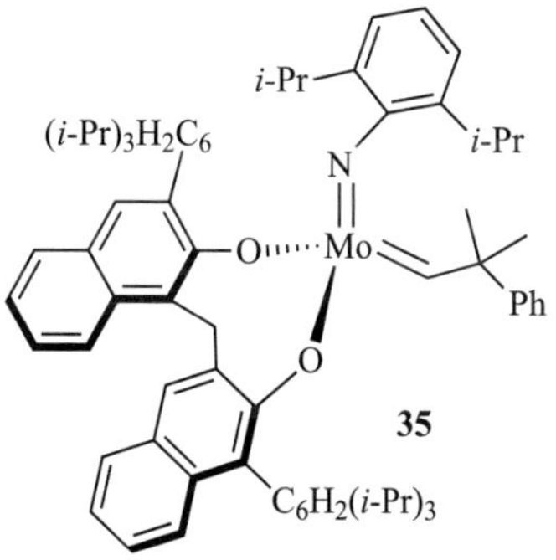

(10)

35

A final TST example involved insightful conformational control as a design element in the total synthesis of spirofungin A by Marjanovic and Kozmin. The key spiroketalization involved deprotection of the benzyl protecting groups of **38** with concomitant hydrogenation of the two olefins to provide a keto-diol conformationally constrained by the bissilyl ether tether. In situ cyclization delivered the desired 1,3-diaxially encumbered spiroketal **39** as a single diastereomer in near-quantitative yield (eq 11).[16]

The propensity of R₃Si to exert a large steric effect (as a 'hetero-*t*-butyl group') in combination with its ability to form pentavalent intermediates and serve as a latent hydroxyl group via KFT oxidation provides exceptional opportunities for the synthetic organic chemist.[17] This feature is parallel to that exhibited by boron compounds in that hydroboration–oxidation enables the borane functional group to be oxidized to a hydroxyl group with retention of stereochemistry. The carbon–silicon bond is far more resistant to oxidation and hydrolysis than the carbon–boron bond and therefore can be carried through many steps of a synthetic sequence.

Avoid Skin Contact with All Reagents

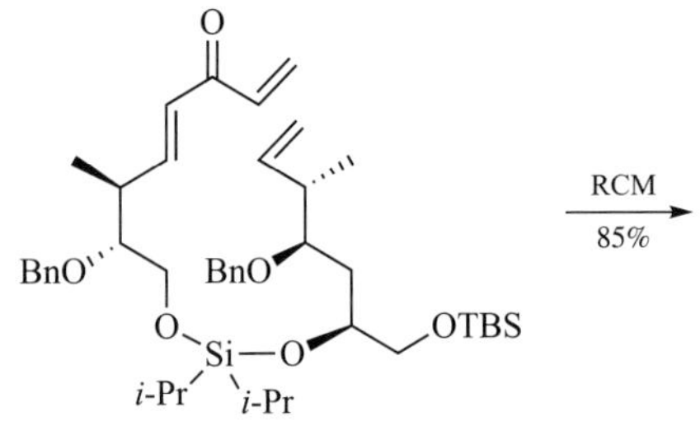

RCM
85%

(37)

H₂, Pd/C
98%

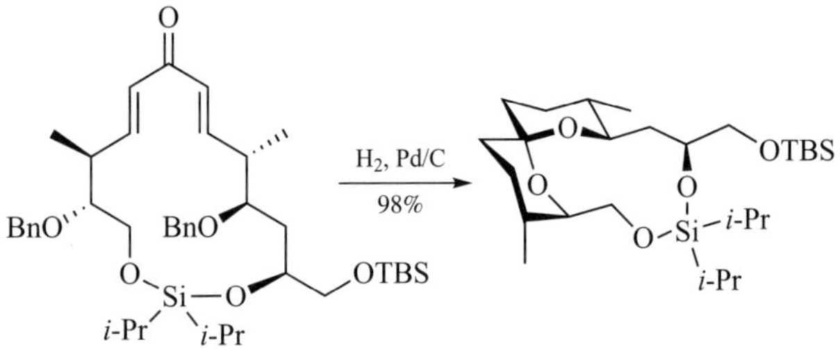

(38) (39) (11)

Although in more casual settings, many chemists emphasize silicon's steric effect by referring to 'silicon as a fat proton', attempts to find this term in the computer searchable database of organic chemistry reveal few instances of this usage.[18]

A good example of this effect is the palladium (0)-catalyzed methylation of allyl phosphate **40**.[19] In this instance, the palladium π-allyl intermediate is regio- and stereospecifically alkylated with inversion, affording vinylsilane **41** via methyl delivery from the metal (eq 12).

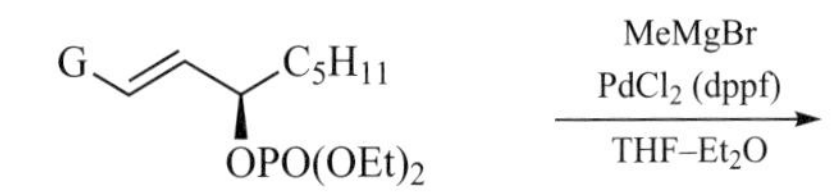

$$G \diagup C_5H_{11}$$
OPO(OEt)₂

MeMgBr
PdCl₂ (dppf)
THF–Et₂O

(**40**), G = TMS, 99% ee (*R*)
(**43**), G = *t*-Bu, rac

$$G \diagup C_5H_{11} \quad + \quad G \diagdown C_5H_{11}$$
Me   Me

(12)

(**41**), G = TMS, 89%, 98% ee (*S*)   (**42**), G = TMS, 0%
(**44**), G = *t*-Bu, rac, 8%   (**45**), G = *t*-Bu, rac, 17%

Control via a distal silicon group is seen in a pair of intramolecular Diels–Alder reactions provided by Wilson. Cyclization of alcohol **47** affords **49** as a pair of secondary alcohol diastereomers in near-equal amounts,[20] while the TMS derivative **48** generates alcohol **50** essentially as a single isomer (eq 13).[21]

A nice example of the steric effect of the TMS group is seen in the triethylborohydride reduction of ketones **51** and **52**.[22] In the former case, the 'small' vinyl group allows hydride to approach from the α-face of the ketone to provide *syn*-1,3-diol **53**. In sharp contrast, similar reduction of bulky vinylsilane **52** affords *anti*-diol **54**, which was protiodesilylated to *anti*-1,3-diol **55** for direct comparison with *syn*-diol **53** (eq 14).

While the above examples show the use of the TMS group to maximal advantage, the requirement of a proximal alcohol for eventual protiodesilylitic removal (or oxidation) has prompted

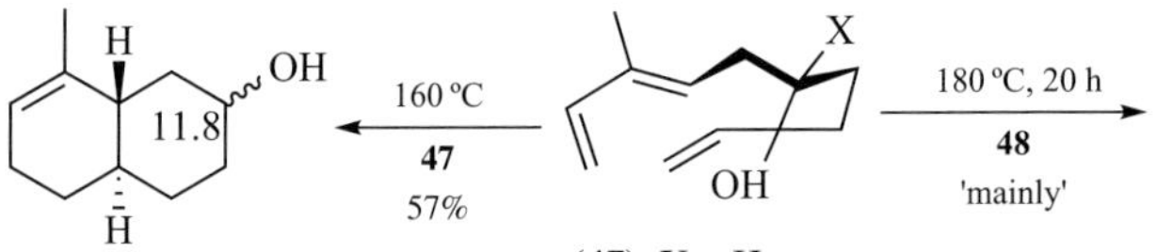

160 °C
47
57%

11.8

(49)

180 °C, 20 h
48
'mainly'

(**47**), X = H
(**48**), X = TMS

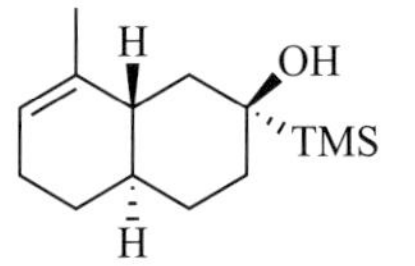

(13)

(50)

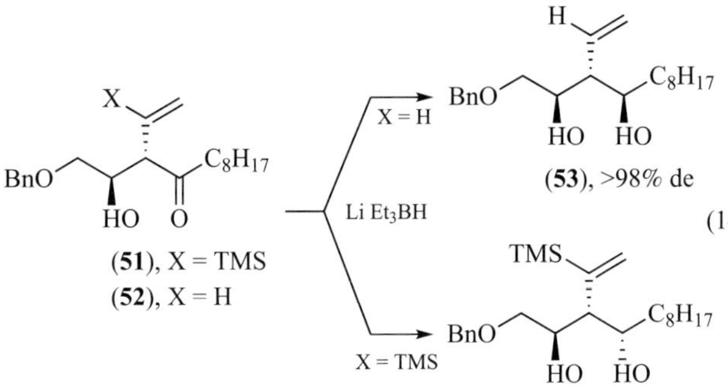

X = H

BnO   C₈H₁₇
   HO   HO

(53), >98% de

Li Et₃BH

(51), X = TMS
(52), X = H

X = TMS

BnO   C₈H₁₇
   HO   HO

cat KH/HMPA { (**54**), X = TMS
              (**55**), X = H, >98% de

(14)

widespread application of the phenyldimethylsilyl (PDMS) group as a more versatile silicon entity.[23] For example, PDMS-Li is readily converted to cyanocuprate **58**, which undergoes ready 1,4-addition to lactone **57**, generating an enolate that suffers subsequent methylation, thus giving **59** bearing two new stereocenters. In addition to introduction of the valuable UV detection feature, the PDMS moiety is a routinely employed surrogate for the hydroxyl group, as seen in the conversion of **59** to lactone **60** by application of the KFT oxidation (eq 15).[24]

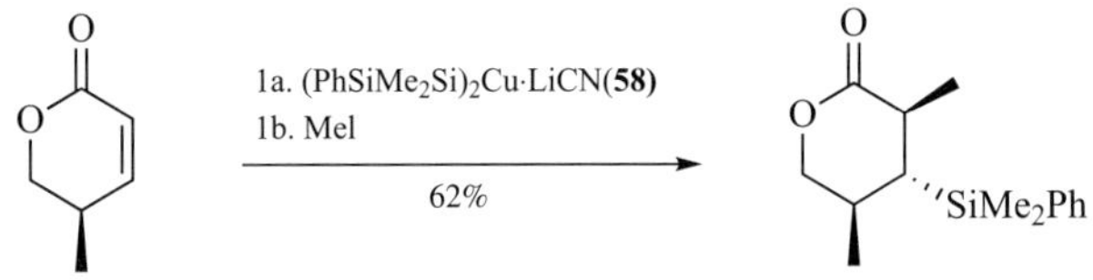

(57)

1a. (PhSiMe₂Si)₂Cu·LiCN(**58**)
1b. MeI
62%

(59)
SiMe₂Ph

Br₂, MeCO₃H, NaOAc
75%

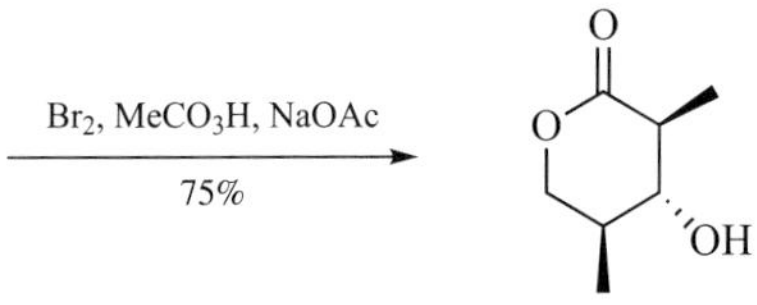

(15)

(60)
OH

Allyl silanes are particularly effective at kinetic regiocontrol. Landais has shown that intramolecular oxyselenylation of alcohol **61** undergoes cyclization exclusively via conformation **61a** to deliver tetrahydrofuran **62** to the exclusion of isomer **63** (eq 16).[25]

A further extension of the allylic stereocontrol of the PDMS moiety is dramatically seen in the radical chain cyclization of substrates **64** and **68** as studied by the Landais group (eq 17).[26]

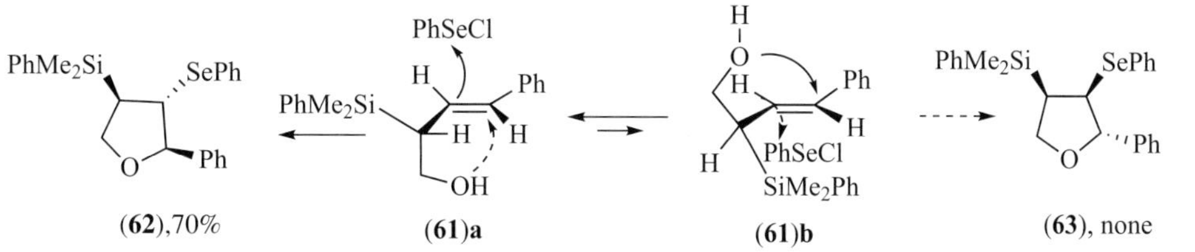

PhSeCl

PhMe$_2$Si ,,,SePh

(62), 70%   (61)a   (61)b   (63), none   (16)

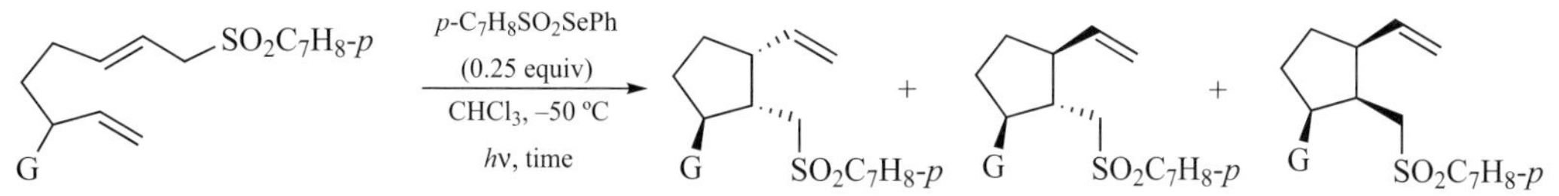

p-C$_7$H$_8$SO$_2$SePh
(0.25 equiv)
CHCl$_3$, –50 °C
$h\nu$, time

SO$_2$C$_7$H$_8$-p SO$_2$C$_7$H$_8$-p SO$_2$C$_7$H$_8$-p SO$_2$C$_7$H$_8$-p (17)

(64), G = PDMS 0.5 h 85% (65), G = PDMS (99 parts) (66), G = PDMS (1 parts) (67), G = PDMS (none)
(68), G = OH   2 h 63%   (69), G = OH (27 parts)   (70), G = OH (none)   (71), G = OH (73 parts)

Roush has provided a particularly noteworthy example showing the synthetic adaptability of stereorich cyclohexylsilane 73. Minimally protected dibenzyl pentaol 73 undergoes acid-mediated Peterson elimination[27] to 74, base-mediated Peterson reaction to 76, and KFT oxidation to inositol derivative 75 (eq 18).[28]

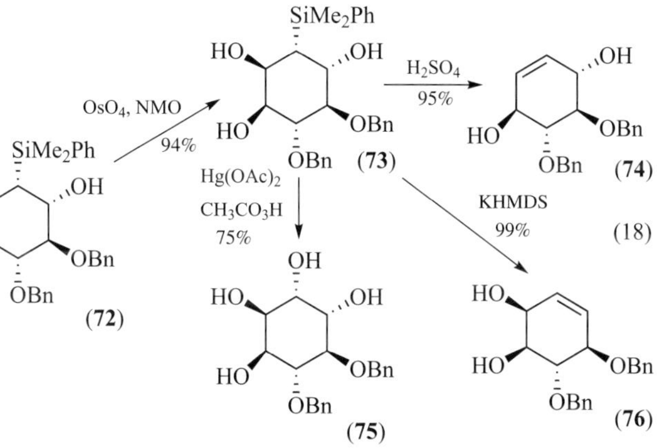

OsO$_4$, NMO
94%

SiMe$_2$Ph
,,,OH
OBn
OBn
(72)

H$_2$SO$_4$
95%

Hg(OAc)$_2$
CH$_3$CO$_3$H
75%

(73)

KHMDS
99%

(74)

(75)

(76)   (18)

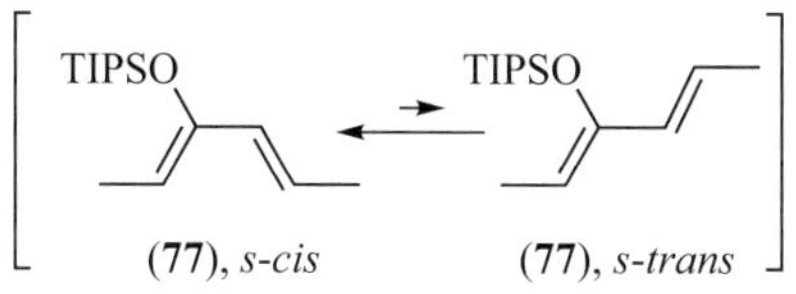

TIPSO   TIPSO

(77), s-cis   (77), s-trans

Taking the steric effect to its next extreme, the triisopropylsilyl (TIPS) group has seen some very interesting applications.[29] Yamamoto has shown that TIPS silyl ether 77 is at least 100 times more reactive than the corresponding TMS and 11 times more reactive than the TBS silyl ethers in the enantiospecific copper-catalyzed hetero-Diels–Alder reaction of heteroaryl nitroso compound 78. The argument is made that the steric demands of the OTIPS group provide a higher concentration of the requisite s-cis conformation of 77 (eq 19).[30]

Another spectacular result attributable to the TIPS group is seen in the carefully engineered oxazaborolidine 81-catalyzed (0.05 equiv) catecholborane (82) reduction of methyl acetylenic ketones 83 and 84. In this study, Helal, Magriotis, and Corey conclusively demonstrated that the distal acetylenic TIPS moiety was responsible for catalyzed delivery of the activated hydride essentially only from the indicated ketone lone pair (eq 20).[31]

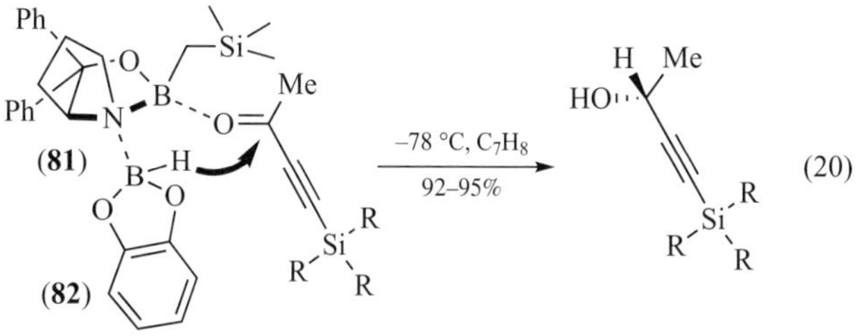

(81)   (82)   –78 °C, C$_7$H$_8$
92–95%   (20)

(83), R = Me   (85), R = Me, (R/S) = 14:1
(84), R = i-Pr   (86), R = i-Pr, (R/S) = 49:1

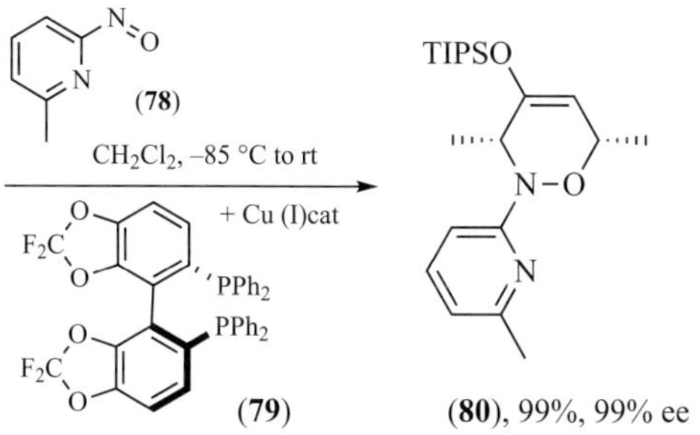

(78)

CH$_2$Cl$_2$, –85 °C to rt
+ Cu (I)cat

(79)   TIPSO

(80), 99%, 99% ee   (19)

TIPS functionality substantially improves the stability of C-2 silylated oxazoles. While C-2 TMS and TBS oxazoles are substantially degraded during aqueous workup or silica chromatography, TIPS oxazole 89 can be handled in routine fashion under nonacidic conditions.[32] Metalation of 89 gives C-4 lithio oxazole 90 that may be captured with most typical electrophiles, including TMSCl, to provide 92 in 89% yield. Subsequent acid treatment effects cleavage of the C-2 TIPS group, giving TMS-oxazole 94 in 90% yield (eq 21).

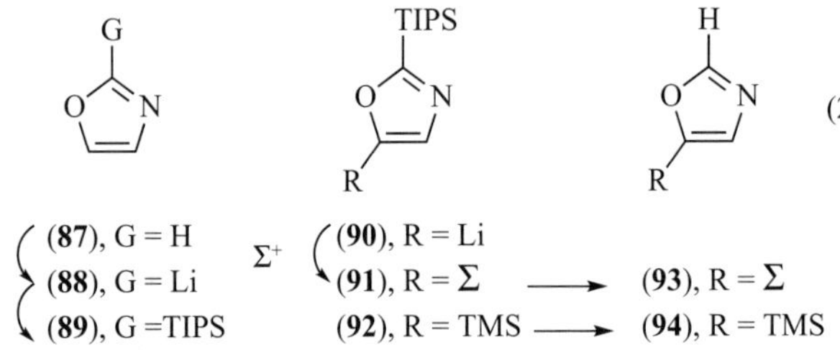

(21)

$$(87), G = H$$
$$(88), G = Li \quad \Sigma^+$$
$$(89), G = TIPS$$

$$(90), R = Li$$
$$(91), R = \Sigma \longrightarrow (93), R = \Sigma$$
$$(92), R = TMS \longrightarrow (94), R = TMS$$

Molecules from the chiral pool are central features in all asymmetric syntheses. In particular, those derived from α-amino acids are exploited with great frequency because of the value derived from the 1,2-N,O functionality. Soderquist has provided a major improvement to the synthesis of the sensitive α-amino aldehyde moiety by employing TIPS chemistry.[33] Bissilylation of the parent acids 95 gives TIPS esters 96 that can be smoothly reduced to the *ideally stable* oxazaborolidines 97 that are conveniently hydrolyzed to α-amino aldehydes 98 with essentially complete retention of enantiopurity. TIPS amino aldehydes 98 are the only known α-amino aldehydes that *can be distilled without decomposition or racemization* (eq 22).[33]

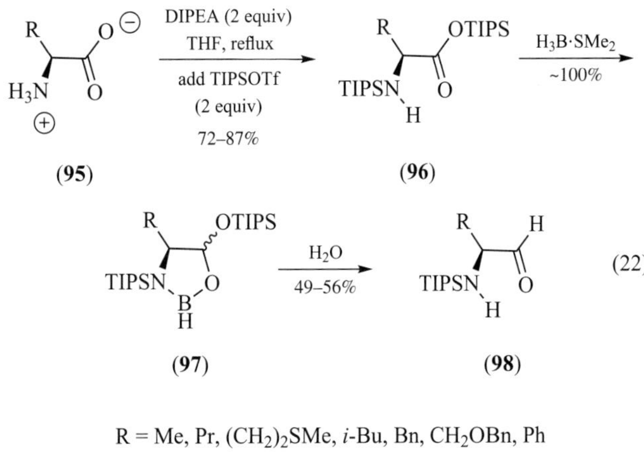

(23)

(99)   (100)   (101)

It is now commonly accepted to refer to the entire family of such carbanion-generating processes as Brook rearrangements, with those involving five- and six-membered intermediates enabling especially powerful synthetic processes.[36] Specifically, the group of Amos Smith has leveraged the chemistry of α-silylated dithiane anion additions to highly functionalized epoxides to a major strategy-level triply convergent process. Addition of metalated dithiane 102 to epoxide 103 affords intermediate 104 that upon warming in the presence of epoxide 105 and HMPA undergoes Brook rearrangement to 106 with concomitant addition to epoxide 105, leading to spongistatin segment 107. The one-pot sequence proceeded in 69% yield and was conducted on a multigram scale (eq 24).[37]

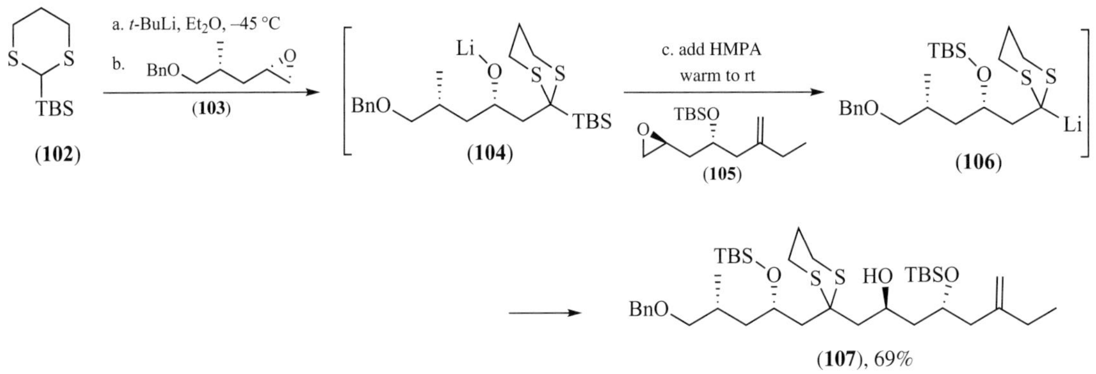

(24)

(22)

R = Me, Pr, (CH$_2$)$_2$SMe, i-Bu, Bn, CH$_2$OBn, Ph

The final focal point of silicon chemistry to be mentioned in this introduction deals with the continuing evolution of the Brook rearrangement.[34] At its inception, the Brook rearrangement involved the intramolecular C $\rightarrow$ O silicon migration of α-oxidosilane 99 to an α-silyloxy anion 101 via the intermediacy of pentavalent siliconate 100 (eq 23).[35]

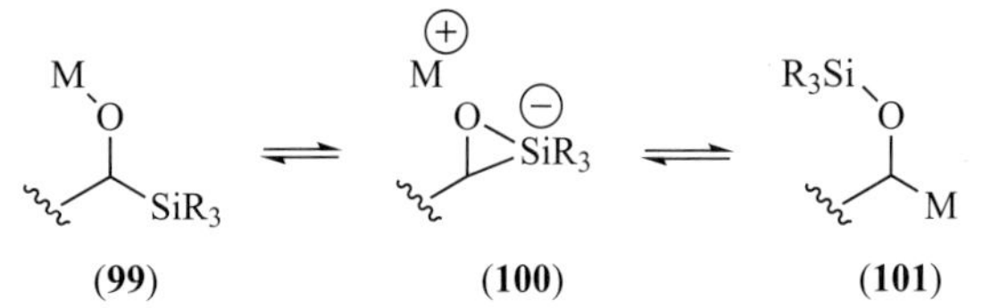

1. Lutgens, F. K; Tarbuck, E. J. *Essentials of Geology*, 7th ed.; Prentice Hall, 2000.

2. (a) Nelson, T. D.; Crouch, R. D. *Synthesis* 1996, 1031. (b) Wuts, P. M.; Green, T. W. *Green's Protective Groups in Organic Synthesis*, 4th ed.; Wiley–Interscience, 2007.

3. (a) Walsh, R., *Acc. Chem. Res.* 1981, *14*, 246. (b) Becerra, R.; Walsh, R. *The Chemistry of Organic Silicon Compounds*; Rappoport, Z.; Apeloig, Y., Eds., Wiley, 1998. Vol. 2, Chapter 4, p 153.

4. (a) Rasmussen, J. K., *Synthesis* 1977, *77*, 91. (b) Colvin, E. W., *Chem. Rev.* 1978, *78*, 15. (c) Fleming, I., *Chem. Soc. Rev.* 1981, *10*, 83. (d) Brownbridge, P., *Synthesis* 1983, *83*, 1. (e) Brownbridge, P., *Synthesis* 1983, *83*, 85. (f) Kuwajima, I.; Nakamura, E., *Acc. Chem. Res.* 1985, *18*, 181.

5. (a) Itoh, K.; Nakanishi, S.; Otsuji, Y., *Chem. Lett.* 1987, 2103. (b) Kim, S.; Lee, P. H., *Tetrahedron Lett.* 1988, *29*, 5413. (c) Kim, S.; Lee, P. H.; Kim, S. S., *Bull. Korean Chem. Soc.* 1989, *10*, 218.

6. Gauthier, D. R. Jr.; Zandi, K. S.; Shea, K. J., *Tetrahedron* 1998, *54*, 2289.

7. Bracegirdle, S.; Anderson, E. A., *Chem. Soc. Rev.* 2010, *39*, 4114.

8. Nishiyama, H.; Kitajima, T.; Matsumoto, M.; Itoh, K., *J. Org. Chem.* 1984, *49*, 2298.

9. (a) Stork, G.; Kahn, M., *J. Am. Chem. Soc.* **1985**, *107*, 500. (b) Stork, G.; Sofia, M. J., *J. Am. Chem. Soc.* **1986**, *108*, 6826; See also the review: (c) Stork, G., *Bull. Chem. Soc. Jpn.* **1988**, *61*, 149, which describes mixed acetal and silyl ether directed radical reactions.

10. Heitzman, C. L.; Lambert, W. T.; Mertz, E.; Shotwell, J. B.; Tinsley, J. M.; Va, P.; Roush, W. R., *Org. Lett.* **2005**, *7*, 2405.

11. Hayakawa, I.; Asuma, Y.; Ohyoshi, T.; Aoki, K.; Kigoshi, H., *Tetrahedron Lett.* **2007**, *48*, 6221.

12. Iwamoto, M.; Miyano, M.; Utsugi, M.; Kawada; H.; Nakada, M., *Tetrahedron Lett.* **2004**, *45*, 8647.

13. Shea, K. J.; Gauthier, D. R., Jr., *Tetrahedron Lett.* **1994**, *35*, 7311.

14. (a) Fujiwara, T.; Yanai, K.; Shimane, K.; Takamori, M.; Takeda, T., *Eur. J. Org. Chem.* **2001**, 155. For a review, see: (b) Takeda, T.; Fujiwara, T., *Rev. Heteratom Chem.* **1999**, *21*, 93.

15. Aeilts, S. L.; Cefalo, D. R.; Bonitatebus, P. J., Jr.; Houser, J. H.; Hoveyda, A. H.; Schrock, R. R., *Angew. Chem., Int. Ed.* **2001**, *40*, 1452.

16. Marjanovic, J.; Kozmin, S. A., *Angew. Chem., Int. Ed.* **2007**, *46*, 8854.

17. (a) Hwu, J.R.; Wang, N., *Chem. Rev.* **1989**, *89*, 1599. (b) Fleming, I.; Barbero, A.; Walter, D., *Chem. Rev.* **1997**, *97*, 2063.

18. A nonscientific e-mail survey of about 20 silicon experts trying to attribute this colorful phrase to a specific chemist has led this editor to learn that both Barry Trost and Ian Fleming remember using it in seminars in the 1970s. In addition, Phil Magnus responded that his colleagues Richard Jones and Don Tilly both believe that the late Colin Eaborn was the originator of the phrase *fat proton for TMS*, and Magnus further remembered hearing it used at a 1976 Eaborn seminar at Ohio State. Significantly, Jih Ru Hwu uses the title 'The trimethylsilyl cationic species as a bulky proton' in a study of dioxolanation (Hwu, J. R.; Wetzel, J. M., *J. Org. Chem.* **1985**, *50*, 3946; see also Ref. 17a). The only 'fat proton' link found in the literature appears to be a silicon section heading in Hans Reich's Wisconsin Chem 842 notes (www.chem.wisc.edu/areas/reich/.../_chem842-05-**silicon**.htm).

19. Urabe, H.; Inami, H.; Sato, F., *J. Chem. Soc., Chem. Commun.* **1993**, 1595.

20. Wilson, S. R.; Mao, D. T., *J. Am. Chem. Soc.* **1978**, *100*, 6289.

21. Wilson, S. R.; Hague, M. S.; Misra, R. N., *J. Org. Chem.* **1982**, *47*, 747.

22. (a) Suzuki, K.; Katayama, E.; Tsuchihashi, G.-I., *Tetrahedron Lett.* **1984**, *25*, 1817. (b) Suzuki, K.; Miyazawa, M.; Masato Shimazaki, M.; Tsuchihashi, G.-I., *Tetrahedron Lett.* **1986**, *27*, 6237.

23. (a) Fleming, I.; Barbero, A.; Walter, D., *Chem. Rev.* **1997**, *97*, 2063. (b) Jones, G. R.; Landais, Y., *Tetrahedron* **1996**, *52*, 7599. (c) Fleming, I.; Roberts, R. S.; Smith, S. C., *Tetrahedron Lett.* **1996**, *37*, 9395. (d) Fleming, I.; Henning, R.; Parker, D. C.; Plaut, H. E.; Sanderson, P. E. J., *J. Chem. Soc., Perkin Trans 1* **1995**, 317.

24. Fleming, I.; Reddy, N. L.; Takaki, K.; Ware, A. C., *Chem. Commun.* **1987**, 1472.

25. (a) Landais, Y.; Planchenault, D.; Weber, V., *Tetrahedron Lett.* **1995**, *36*, 2987. (b) Andrey, O.; Ducry, L.; Landais, Y.; Planchenault, D.; Weber, V., *Tetrahedron* **1997**, *53*, 4339.

26. James, P.; Landais, Y., *Org. Lett.* **2004**, *6*, 325.

27. Peterson reaction: (a) Ager, D. J., *Synthesis* **1984**, *84*, 384. (b) Ager, D. J., *Org. Reac.* **1990**, *38*, 4.

28. Heo, J.-N.; Holson, E. B; Roush, W. R., *Org. Lett.* **2003**, *5*, 1697.

29. Rucker, C., *Chem. Rev.* **1995**, *95*, 1009.

30. Yamamoto, Y.; Yamamoto, H., *Angew. Chem., Int. Ed.* **2005**, *44*, 7082.

31. Helal. C. J.; Magriotis, P. A.; Corey, E. J., *J. Am. Chem. Soc.* **1996**, *118*, 10938.

32. Miller, R. A.; Smith, R. M.; Marcune, B., *J. Org. Chem.* **2005**, *70*, 9074.

33. Soto-Cairoli, B; Justo de Pomar, J.; Soderquist, J. A., *Org. Lett.* **2008**, *10*, 333.

34. Brook, A. G., *J. Organomet. Chem.* **1986**, *300*, 21.

35. (a) Moser, W. H., *Tetrahedron* **2001**, *57*, 2065. (b) Smith, A. B., III; Adams, C. M., *Acc. Chem. Res.* **2004**, *37*, 365. For the most recent review on the Brook rearrangement in tandem with subsequent bond formation, see the 2008 Leighton group meeting by Laura Schacherer, http://www.columbia.edu/cu/chemistry/groups/leighton/gm/20080201-LNS-TandemBrook.pdf.

36. Smith, A. B., III; Xian, M.; Kim, W.-S.; Kim, D.-S., *J. Am. Chem. Soc.* **2006**, *128*, 12368.

37. Smith, A. B., III; Doughty, V. A.; Sfouggatakis, C.; Bennett, C. S.; Koyanagi, J.; Takeuchi, M., *Org. Lett.* **2002**, *4*, 783.

Philip L. Fuchs
Purdue University, West Lafayette, IN, USA

Reagent Finder Table

Si group(s)	Si–X bond(s)	Reagent name	CAS number	Page no.
TMS, TMS, TMS	Al	Tris(trimethylsilyl)aluminum Tris(trimethylsilyl)aluminum·Ethyl Ether Complex	65343-66-0 75441-10-0	745
PDMS	Al(Et)$_2$	Diethyl[dimethyl(phenyl)silyl]aluminum	86014-18-8	237
PDMS	BOCMe$_2$CMe$_2$O	1,3,2-Dioxaborolane, 2-(Dimethylphenylsilyl)-4,4,5,5-tetramethyl-, 1,3,2-Dioxaborolane, 4,4,5,5-Tetramethyl-2-(methyldiphenylsilyl)-, 1,3,2-Dioxaborolane, 4,4,5,5-Tetramethyl-2-(triphenylsilyl)-	185990-03-8	270
TMS	Br	Bromotrimethylsilane	2857-97-8	92
TMS	See footnote [a] for structure	(+)-9-(1S,2S-Pseudoephedrinyl)-(10R)-(trimethylsilyl)-9-borabicyclo[3.3.2]decane 9-((1R,2R)-Pseudoephedrinyl)-(10S)-(trimethylsilyl)-9-borabicyclo[3.3.2]decane	848701-34-8 848946-18-9	453
TMS	C(=CH$_2$)COCH$_3$	3-Trimethylsilyl-3-buten-2-one	43209-86-5	586
TMS	C(CH$_2$=CH)=C=O	(Trimethylsilyl)vinylketene	75232-81-4	725
TMS	C(CH$_2$CH$_2$)SPh	1-(Phenylthio)-1-(trimethylsilyl)cyclopropane	74379-74-1	418
TMS	C(CH$_3$)=CHCH$_2$I	(E)-1-Iodo-3-trimethylsilyl-2-butene	52815-00-6, 52685-51-5	336
TMS	C(CO$_2$Me)=CH$_2$	Methyl 2-Trimethylsilylacrylate	18269-31-3	396
TMS	C(OMe)=C=CH$_2$	1-Trimethylsilyl-1-methoxyallene	77129-88-5	656
TMS	C≡C–CH=CHBr (E)	(E)-1-Bromo-4-trimethylsilyl-1-buten-3-yne	107646-62-8	100
TMS	C≡C–CH$_2$Cu	3-Trimethylsilyl-2-propynylcopper(I)	55630-32-5	711
TIPS	C≡C–H	(Triisopropylsilyl)acetylene	89343-06-6	550
TMS	C≡C–I	(Iodoethynyl)trimethylsilane	18163-47-8	321
TMS	C≡C–OEt	Ethoxy(trimethylsilyl)acetylene	1000-62-0	289
TIPS	C≡CCH$_2$	1,3-Bis(triisopropylsilyl)propyne	82192-59-4	56
TIPS	C≡CCH$_2$Li	3-Lithio-1-triisopropylsilyl-1-propyne	82192-58-3	347

[a]

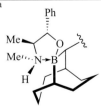

Si group(s)	Si–X bond(s)	Reagent name	CAS number	Page no.
TMS	C≡CCH$_2$OTHP	3-Tetrahydropyranyloxy-1-trimethylsilyl-1-propyne	36551-06-1	487
TMS	C≡CCHMeOH	3-Butyn-2-ol, 4-(Trimethylsilyl)-	6999-19-5	148
TMS	C≡CCu	(Trimethylsilyl)ethynylcopper(I)	53210-13-2	637
TMS	C≡CH	Trimethylsilylacetylene	1066-54-2	569
TIPS	C≡CSO$_2$CF$_3$	Triisopropylsilylethynyl Triflone	196789-82-9	558
TMS	C$_3$H$_2$NS	2-(Trimethylsilyl)thiazole	79265-30-8	712
TMS	C$_3$H$_3$N$_2$	N-(Trimethylsilyl)imidazole	18156-74-6	640
TMS	C$_4$H$_3$O	2-Trimethylsilyloxyfuran	61550-02-5	684
Me$_2$SiH	C$_5$H$_4$N	2-Pyridyldimethylsilane	21032-48-4	457
PhSiH$_2$	C$_6$H$_4$-o-CH$_2$NMe$_2$	(2-Dimethylaminomethylphenyl)phenylsilane	129552-42-7	253
Cl(i-Pr)$_2$Si	C$_6$H$_4$-p-Poly	Diisopropyl Chlorosilane, Polymer-supported	No CAS	240
TMS	C$_6$H$_4$(ortho-OSO$_2$CF$_3$)	2-(Trimethylsilyl)phenyl Triflate	88284-48-4	697
Me$_2$Si	C$_6$H$_4$CH2O-o	2,1-Benzoxasilole, 1,3-Dihydro-1,1-dimethyl-	321903-29-1	43
TMS	CBr=CH$_2$	(1-Bromovinyl)trimethylsilane	13683-41-5	109
TMS	CF$_2$SePh	Benzene, [[Difluoro(trimethylsilyl)methyl]seleno]	868761-01-7	33
TMS	CF$_2$SPh	Benzene, [[Difluoro(trimethylsilyl)methyl]thio]-	536975-49-2	36
TMS	CF$_3$	Trifluoromethyltrimethylsilane	81290-20-2	539
TMS, TMS, TMS	CH	Tris(trimethylsilyl)methane	1068-69-5	746
TMS, TMS	CH–NH$_2$	Methanamine, 1,1-Bis(trimethylsilyl)	134340-00-4	369
TMS	CH(C$_8$H$_{12}$B)–CH=CH$_2$	9-[1-(Trimethylsilyl)-2(E)-butenyl]-9-borabicyclo[3.3.1]nonane	100701-75-5	588
TMS	CH(CH$_2$OH)$_2$	2-Trimethylsilyl-1,3-propanediol	189066-36-2	701
TMS	CH(SCH$_2$CH$_2$CH$_2$S)	2-Trimethylsilyl-1,3-dithiane	13411-42-2	604
TBS	CH(SCH$_2$CH$_2$CH$_2$S)	2-t-Butyldimethylsilyl-1,3-dithiane	95452-06-5	122
TMS	CH=C(OLi)OEt	Ethyl Lithio(trimethylsilyl)acetate	54886-62-3	293
TMS	CH=C=CH$_2$	(Trimethylsilyl)allene	14657-22-8	580
TMS	CH=C=O	Trimethylsilylketene	4071-85-6	645
TMS	CH=CH–CH=CH$_2$ (E)	(E)-1-Trimethylsilyl-1,3-butadiene	71504-26-2	585
TMS	CH=CH$_2$	Vinyltrimethylsilane	754-05-2	755
ClMe$_2$Si	CH=CH$_2$	Chlorodimethylvinylsilane	1719-58-0	157
TMS, TMS	CH=CH$_2$CH$_2$(E) CH=CH$_2$CH$_2$ (Z) CH=CH$_2$CH$_2$ (E + Z)	Silane, 1,1′-(1-Propene-1,3-diyl)bis[1,1,1-trimethyl-(E), Silane, 1,1′-(1-Propene-1,3-diyl)bis[1,1,1-trimethyl-(Z), Silane, 1,1′-(1-Propene-1,3-diyl)bis[1,1,1-trimethyl-(E + Z)	52152-48-4 227612-82-0 34774-00-9	464

Si group(s)	Si–X bond(s)	Reagent name	CAS number	Page no.
TMS	CH=CHSePh (*E*)	Benzene, [[(1*E*)-2-(Trimethylsilyl)ethenyl]seleno]	130267-88-8	39
TMS	CH=CHSn(*n*-Bu)$_3$ (*E*)	(*E*)-1-Tri-*n*-butylstannyl-2-trimethylsilylethylene	58207-97-9	499
TMS	CH=CHSnMe$_3$ (*E*)	(*E*)-1-Trimethylsilyl-2-trimethylstannylethylene	65801-56-1	724
TMS	CH=CHSO$_2$Ph (*E*)	Benzene, [[(1*E*)-2-trimethylsilyl)ethenyl]sulfonyl]-	64489-06-1	42
TMS	CH$_2$–C(CH$_2$CO$_2$Me)=CH$_2$	3-[(Trimethylsilyl)methyl]-3-butenoic Acid Methyl Ester	70639-89-3	662
TMS	CH$_2$–C(CH$_2$I)=CH$_2$	3-Iodo-2-trimethylsilylmethyl-1-propene	80121-73-9	338
TMS	CH$_2$–C(CH$_2$OAc)=CH$_2$	3-Acetoxy-2-trimethylsilylmethyl-1-propene	72047-94-0	1
TMS	CH$_2$–C≡C–H	3-Trimethylsilyl-1-propyne	13361-64-3	704
TMS	CH$_2$–CBr=CH$_2$	2-Bromo-3-trimethylsilyl-1-propene	81790-10-5	106
TMS	CH$_2$–CH=CH$_2$	Allyltrimethylsilane	762-72-1	14
Cl$_3$Si	CH$_2$–CH=CH$_2$	Allyl Trichlorosilane	107-37-9	10
ClMe$_2$Si	CH$_2$Br	(Bromomethyl)chlorodimethylsilane	16532-02-8	85
TMS	CH$_2$C(CH=CH$_2$)=CH$_2$	2-Trimethylsilylmethyl-1,3-butadiene	70901-64-3	660
TMS	CH$_2$C(CH$_2$OH)=CH$_2$	2-Trimethylsilylmethyl-2-propen-1-ol	81302-80-9	675
TMS	CH$_2$C(CH$_2$SnBu$_3$)=CH$_2$	Trimethyl{2-[(tributylstannyl)methyl]-2-propenyl}silane	164662-96-8	731
ClSi	CH$_2$CH=CH$_2$, CH$_2$CH=CH$_2$, CH$_2$CH=CH$_2$	Silane, Chlorotri-2-propen-1-yl-	17865-20-2	463
TMS	CH$_2$CH=CHCH=CH$_2$ (*E*)	5-Trimethylsilyl-1,3-pentadiene	72952-73-9	695
Me$_2$NSiMe$_2$, Me$_2$NSiMe$_2$	CH$_2$CH$_2$	1,4-Bis(dimethylamino)-1,1,4,4-tetramethyldisilethylene	91166-50-6	50
ClMe$_2$Si, ClMe$_2$Si	CH$_2$CH$_2$	1,2-Bis(chlorodimethylsilyl)ethane	13528-93-3	48
Me$_2$Si	CH$_2$CH$_2$CH$_2$	1,1-Dimethylsilacyclobutane	2295-12-7	261
ClMeSi	CH$_2$CH$_2$CH$_2$	1-Chloro-1-methylsiletane	2351-34-0	166
ADMS	CH$_2$CH$_2$Cl	Allyl(2-chloroethyl)dimethylsilane	104107-85-9	9
TMS	CH$_2$CH$_2$SO$_2$Ph	(2-Phenylsulfonylethyl)trimethylsilane	73476-18-3	409
PDMS	CH$_2$Cl	(Chloromethyl)dimethylphenylsilane	1631-83-0	162
TMS	CH$_2$CN	Trimethylsilylacetonitrile	18293-53-3	564
TMS	CH$_2$CO$_2$Et	Ethyl Trimethylsilylacetate	4071-88-9	297
TMS	CH$_2$CO$_2$H	Trimethylsilylacetic Acid	2345-38-2	562
MDPS	CH$_2$CO$_2$R	Ethyl(Methyldiphenylsilyl)acetate Methyl(Methyldiphenylsilyl)acetate *i*-Propyl(Methyldiphenylsilyl)acetate *t*-Butyl(Methyldiphenylsilyl)acetate	13950-57-7 89266-73-9 87776-13-4 77772-21-5	295

Si group(s)	Si–X bond(s)	Reagent name	CAS number	Page no.
TMS	CH$_2$CO$_2$t-Bu	t-Butyl Trimethylsilylacetate	41108-81-0	147
TMS	CH$_2$COCH3	Trimethylsilylacetone	5908-40-7	563
TMS	CH$_2$I	(Iodomethyl)trimethylsilane	4206-67-1	323
TMS	CH$_2$K CH$_2$Na	Trimethylsilylmethylpotassium Trimethylsilylmethylsodium	53127-82-5 53127-81-4	674
TMS	CH$_2$Li	Trimethylsilylmethyllithium	1822-00-0	664
TMS	CH$_2$MgCl	Trimethylsilylmethylmagnesium Chloride	13170-43-9	666
TMS	CH$_2$N$_3$	Trimethylsilylmethyl Azide	87576-94-1	657
TMS	CH$_2$N$_3$C$_6$H$_4$	1-[(Trimethylsilyl)methyl]-1H-benzotriazole	122296-00-8	658
TMS	CH$_2$NBn(CH$_2$OMe)	N-Benzyl-N-(methoxymethyl)-N-trimethylsilylmethylamine	93102-05-7	44
TMS	CH$_2$SPh	(Phenylthiomethyl)trimethylsilane	17873-08-4	412
Cl$_3$Si	CH$_3$	Methyltrichlorosilane	75-79-6	389
TMS	CHBr$_2$	(Dibromomethyl)trimethylsilane	2612-42-2	197
TMS, TMS	CHCl	Silane, 1,1′-(Chloromethylene)bis[1,1,1]-trimethyl-	5926-35-2	461
TMS	CHLi–CH$_2$=CH$_2$	Trimethylsilylallyllithium	67965-38-2	582
(t-Bu)$_2$Si	CHMeCHMe (trans)	trans-1,1-Di-t-butyl-2,3-dimethylsilirane	116888-87-0	207
MDPS	CHMeCO$_2$Et	Ethyl 2-(Methyldiphenylsilyl)propanoate	77772-22-6	296
TES	CHMeHC=NC$_6$H$_{11}$-c	(α-Triethylsilyl)propionaldehyde Cyclohexylimine	119711-55-6	516
TMS	CHN$_2$	Trimethylsilyldiazomethane	18107-18-1	590
TMS	CHOTMSCH=CH$_2$	3-Trimethylsilyl-3-trimethylsilyloxy-1-propene	66662-17-7	723
TMS, TMS	CHSPh	Phenylthiobis(trimethylsilyl)methane	62761-90-4	411
ClSi(i-Pr)$_2$	H	Chlorodiisopropylsilane	2227-29-4	151
TIPS	Cl	Triisopropylsilyl Chloride	13154-24-0	554
TDMS	Cl	Dimethylthexylsilyl Chloride	67373-56-2	263
TBS	Cl	t-Butyldimethylchlorosilane	18162-48-6	110
TBDPS	Cl	t-Butyldiphenylchlorosilane	58479-61-1	135
Ph$_3$Si	Cl	Chlorotriphenylsilane	76-86-8	182
Me$_2$SiH	Cl	Chlorodimethylsilane	1066-35-9	156
MDPS	Cl	Methyldiphenylchlorosilane	144-79-6	381
C$_{21}$H$_{21}$ClSi	Cl	Tribenzylchlorosilane	18740-59-5	496
Me$_2$Si	Cl, Cl	Dichlorodimethylsilane	75-78-5	228
(t-Bu)$_2$Si	Cl, Cl	Di-t-butyldichlorosilane	18395-90-9	201
(i-Pr)$_2$Si	Cl, Cl	Dichlorodiisopropylsilane	7751-38-4	222

A list of General Abbreviations appears on the front Endpapers

Si group(s)	Si–X bond(s)	Reagent name	CAS number	Page no.
TMS	Cl	Chlorotrimethylsilane	75-77-4	170
TMS	CH=CHLi (E)	(E)-2-(Trimethylsilyl)vinyllithium	55339-31-6	729
TMS	CLi=CH$_2$	1-(Trimethylsilyl)vinyllithium	51666-94-5	726
TMS	CLiN$_2$	Diazo(trimethylsilyl)methyllithium	84645-45-4	191
TMS	CMe=C=CH$_2$	1-Methyl-1-(trimethylsilyl)allene	74542-82-8	397
TMS	CMe=CHCH$_2$Cl (E)	(E)-(3-Chloro-1-methyl-1-propenyl)-trimethylsilane	116399-78-1	166
TMS	CMe$_2$–CH=CH$_2$	(α,α-Dimethylallyl)trimethylsilane	67707-64-6	252
TMS	Me$_2$CSPh	2-(Phenylthio)-2-(trimethylsilyl)propane	89656-96-2	420
TMS	CMgBr=CH$_2$·CuI	1-(Trimethylsilyl)vinylmagnesium Bromide–Copper(I) Iodide	49750-22-3, 7681-65-4	730
TMS	CN	Cyanotrimethylsilane	7677-24-9	184
TBS	CN	t-Butyldimethylsilyl Cyanide	56522-24-8	120
Et$_2$Si(CH$_2$)$_4$-C$_6$H$_4$-p-Poly	CN	Trialkylsilyl Cyanide, Polymer-supported	766-24-9P	494
Me$_2$Si	CN, CN	Dicyanodimethylsilane	5158-09-8	234
TMS	CSPh=CH$_2$	1-Phenylthio-1-trimethylsilylethylene	62762-20-3	419
TMS	Cu	Trimethylsilylcopper	91899-54-6	589
PDMS, PDMS	CuLi	Lithium Bis[dimethyl(phenyl)silyl]cuprate	75583-57-2	350
PDMS, PDMS	CuCN·Li2	Dilithium cyanobis(dimethylphenylsilyl)-cuprate	110769-32-9	248
Ph$_2$Si	F, F	Diphenyldifluorosilane	312-40-3	276
TES	H	Triethylsilane	617-86-7	506
Ph$_3$Si	H	Triphenylsilane	789-25-3	733
PDMS	H	Dimethyl(phenyl)silane	766-77-8	254
ClPh$_2$Si	H	Chlorodiphenylsilane	1631-83-0	162
Cl(t-Bu)$_2$Si	H	Di-t-butylchlorosilane	56310-18-0	198
(EtO)$_3$Si	H	Triethoxysilane	998-30-1	501
F$_6$Si	H, H	Fluorosilicic Acid	16961-83-4	301
PhSi	H, H, H	Phenylsilane–Cesium Fluoride	694-53-1	408
TMS	HC(SMe)$_2$	Bis(methylthio)(trimethylsilyl)methane	37891-79-5	52
TMS	HC(ZnCl)–CH=CH$_2$	Allyltrimethylsilylzinc Chloride	89822-47-9	24
TMS, TMS	HC=CH–CH=CH (E,E)	(1E,3E)-1,4-Bis(trimethylsilyl)-1,3-butadiene	22430-47-3	73
TMS	HC=CHCH$_2$OAc (E) HC=CHCH$_2$OAc (Z) HC=CHCH$_2$OAc ($E + Z$)	3-Trimethylsilyl-2-propen-1-yl Acetate (E) 3-Trimethylsilyl-2-propen-1-yl Acetate (Z) 3-Trimethylsilyl-2-propen-1-yl Acetate ($E + Z$ mix)	86422-21-1 86422-22-2 80401-14-5	703

Si group(s)	Si–X bond(s)	Reagent name	CAS number	Page no.
2-PyrMe$_2$Si, 2-PyrMe$_2$Si	HCLi	[Bis(2-pyridyldimethylsilyl)methyl]lithium	356062-49-2	54
TMS	I	Iodotrimethylsilane	16029-98-4	325
Cl$_3$Si	I	Trichloroiodosilane	13465-85-5	500
Me$_2$Si	I, I	Diiododimethylsilane	15576-81-5	238
TMS	K	Trimethylsilylpotassium	56859-17-7	700
TMS, TMS	MeN–O	*N*-Methyl-*N*,*O*-bis(trimethylsilyl)-hydroxylamine	22737-33-3	380
TMS	Li	Trimethylsilyllithium	18000-27-6	650
PDMS	Li	Dimethylphenylsilyllithium	3839-31-4	255
PDMS	Li·CuCN	Lithium Cyano(dimethylphenylsilyl)cuprate	75583-56-1	353
TMS, TMS	N–K	Potassium Hexamethyldisilazide	40949-94-8	432
TMS, TMS	N–Li	Lithium Hexamethyldisilazide	4039-32-1	356
TMS, TMS	N–Na	Sodium Hexamethyldisilazane	1070-89-9	468
TMS	N(Et)$_2$	Trimethylsilyldiethylamine	996-50-9	598
Ph$_3$SiF$_2$	N(*n*-Bu)$_4$	Tetrabutylammonium Difluorotriphenylsilicate	163931-61-1	479
TMS	N(SO$_2$CF$_3$)$_2$	Methanesulfonamide, 1,1,1-Trifluoro-*N*[(trifluoromethyl)sulfonyl]-*N*(trimethylsilyl)-	82113-66-4	372
TMS	N=C=O	Isocyanatotrimethylsilane	1118-01-1	339
TMS	N$_3$	Azidotrimethylsilane	4648-54-8	26
TMS, TMS	NCH$_2$OMe	(Methoxymethyl)bis(trimethylsilyl)amine	88211-44-3	374
TMS, TMS	NCHO	Formamide, *N*,*N*-bis(trimethylsilyl)-	15500-60-4	302
Me$_2$Si	NEt$_2$, NEt$_2$	*N*,*N*-Diethylaminodimethylsilane	13686-66-3	237
TMS, TMS	NH	Hexamethyldisilazane	999-97-3	317
Me$_2$SiH	NH	1,1,3,3-Tetramethyldisilazane	15933-59-2	488
TBS, TBS	NH–HN	1,2-Bis(*t*-butyldimethylsilyl)hydrazine	10000-20-1	46
F$_6$Si	NH$_4$, NH$_4$	Ammonium Hexafluorosilicate	16919-19-0	25
TMS	NLi(Bn)	Lithium *N*-benzyltrimethylsilylamide	113709-50-5	349
TBS	NMeCOCF$_3$	*N*-(*t*-Butyldimethylsilyl)-*N*-methyltrifluoroacetamide	77377-52-7	125
TMS, TMS	NMgBr	Bromomagnesium Hexamethyldisilazide	50916-70-6	85
TMS, TMS	O	Hexamethyldisiloxane	107-46-0	313
HMeSi	O	Polymethylhydrosiloxane	9004-73-3	427
Cl(*i*-Pr)$_2$Si	O	1,3-Dichloro-1,1,3,3-tetraisopropyldisiloxane	69304-37-6	230
Me$_2$Si	(O, O)$_3$	Hexamethylcyclotrisiloxane	541-05-9	310

A list of General Abbreviations appears on the front Endpapers

Si group(s)	Si–X bond(s)	Reagent name	CAS number	Page no.
H$_2$C=CHSiMe	(O, O)$_4$	2,4,6,8-Tetraethenyl-2,4,6,8-tetramethylcyclotetrasiloxane	2554-06-5	484
TMSCH$_2$CH$_2$	See footnote [b] for structure	2-(Trimethylsilyl)ethyl 3-nitro-1H-1,2,4-triazole-1-carboxylate	1001067-09-9	636
TMS, TMS	O–O	Bis(trimethylsilyl) Peroxide	5796-98-5	79
Me$_2$Si	O, O	PDMS Thimbles	42557-10-8	403
TMSCH$_2$CH$_2$	O$_2$C–Imid	2-(Trimethylsilyl)ethoxycarbonylimidazole	81616-20-8	627
TMS	O$_2$CCF$_2$SO$_2$F	Trimethylsilyl Fluorosulfonyldifluoroacetate	120801-75-4	638
TMSCH$_2$CH$_2$	O$_2$CCl	2-(Trimethylsilyl)ethyl Chloroformate	20160-60-5	634
TMS	O$_2$N=CH$_2$	Trimethylsilyl Methanenitronate	51146-35-1	653
TMS	OC(=CH$_2$)CH=CH$_2$	2-Trimethylsilyloxy-1,3-butadiene	38053-91-7	681
TMS	OC(CH$_2$CH$_2$)OEt	1-Ethoxy-1-(trimethylsiloxy)cyclopropane	27374-25-0	285
TMS, TMS	OC(CH$_3$)=N	N,O-Bis(trimethylsilyl)acetamide	10416-59-8	58
TMS, TMS	OC(OEt)=C(OEt)O	1,2-Diethoxy-1,2-bis(trimethylsilyloxy)ethylene	90054-58-3	235
TBS	OC(OMe)=CH$_2$	Ketene t-butyldimethylsilyl Methyl Acetal	91390-62-4	341
Cl$_3$Si	OC(OMe)=CH$_2$	Methyl Trichlorosilyl Ketene Acetal (1)	4519-10-2	393
Me$_2$Si	OC(OMe)=CMe$_2$, OC(OMe)=CMe$_2$	Bis(1-methoxy-2-methyl-1-propenyloxy)dimethylsilane	86934-32-9	51
TBS	OC(t-BuS)=CH$_2$	1-t-Butylthio-1-t-butyldimethyl-silyloxyethylene	121534-52-9	138
TBS	OC(t-BuS)=CHCH$_3$	1-t-Butylthio-1-t-butyldimethylsilyloxypropene	102146-11-2 97250-84-5 97250-83-4	141
TES	OCH(CH=CH$_2$)$_2$	3-Triethylsilyloxy-1,4-pentadiene	62418-65-9	515
TMS	OCH=CH–CH=CH$_2$	1-Trimethylsilyloxy-1,3-butadiene	6651-43-0	677
TMS	OCH=CHBr (Z)	(Z)-2-Bromo-1-(trimethylsilyloxy)ethylene	64556-66-7	105
TMS	OCH=CHLi (Z)	(Z)-2-(Trimethylsilyloxy)vinyllithium	78108-48-2	693
TMSCH$_2$CH$_2$	OCH$_2$Cl	2-(Trimethylsilyl)ethoxymethyl Chloride	76513-69-4	628
TMSCH$_2$CH$_2$	OCH$_2$P(Ph)$_3$Cl	[2-(Trimethylsilyl)ethoxymethyl]-triphenylphosphonium Chloride	82495-75-8	633
TMS	OCMe$_2$COC$_2$H$_5$	2-Methyl-2-(trimethylsilyloxy)-3-pentanone	72507-50-7	400
TMS	OCOMe=C(CH$_3$)$_2$	1-Methoxy-2-methyl-1-(trimethylsilyloxy)propene	31469-15-5	376
TMSCH$_2$CH$_2$	OH	2-(Trimethylsilyl)ethanol	2916-68-9	622
TMS, TMS	ONH	N,O-Bis(trimethylsilyl)hydroxylamine	22737-37-7	75
TBS	ONH$_2$	O-(t-Butyldimethylsilyl)hydroxylamine	41879-39-4	124

[b]

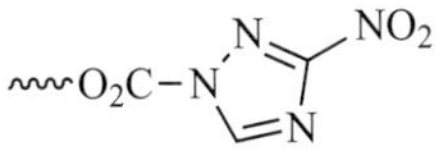

$\sim\!\!\sim\!\!\sim\!O_2C-N$⟨triazole ring⟩$NO_2$

Si group(s)	Si–X bond(s)	Reagent name	CAS number	Page no.
TES	OOOH	Triethylsilyl Hydrotrioxide	101631-06-5	514
TIPS	OSO_2CF_3	Triisopropylsilyl Trifluoromethanesulfonate	80522-42-5	559
TMS	OSO_2CF_3	Trimethylsilyl Trifluoromethanesulfonate	27607-77-8	518
TES	OSO_2CF_3	Triethylsilyl Trifluoromethanesulfonate	79271-56-0	517
TDMS	OSO_2CF_3	Dimethylthexylsilyl Trifluoromethanesulfonate	103588-79-0	265
TBS	OSO_2CF_3	t-Butyldimethylsilyl Trifluoromethanesulfonate	69739-34-0	127
$Cl(t$-$Bu)_2Si$	OSO_2CF_3	Trifluoromethanesulfonic Acid, Chlorobis (1,1-Dimethylethyl)silyl Ester	134436-16-1	537
$(t$-$Bu)_2Si$	OSO_2CF_3, OSO_2CF_3	Di-t-butylsilyl Bis(trifluoromethanesulfonate)	85272-35-7	209
$(i$-$Pr)_2Si$	OSO_2CF_3, OSO_2CF_3	Diisopropylsilyl Bis(trifluoromethanesulfonate)	85272-30-6	241
TMS, TMS, TMS	P	Phosphine, Tris(trimethylsilyl)-	15573-38-3	422
TBS	$P(Ph)_2$	(t-Butyldimethylsilyl)diphenylphosphine	678187-53-6	121
TMS, TMS	S	Bis(trimethylsilyl) Sulfide	3385-94-2	83
$TMSCH_2CH_2$	SH	2-(Trimethylsilyl)ethanethiol	18143-30-1	618
TIPS	SH	Triisopropylsilanethiol	156275-96-6	546
$(t$-$BuO)_3Si$	SH	Tri(t-butoxy)silanethiol	690-52-8	497
TMS	SMe	(Methylthio)trimethylsilane	3908-52-2	388
F_2SiMe_3	$S(NMe_2)_3$	Tris(dimethylamino)sulfonium Difluorotrimethylsilicate	59218-87-0, 59201-86-4	739
TMS	SPh	(Phenylthio)trimethylsilane	4551-15-9	414
$TMSCH_2CH_2$	SO_2Cl	β-Trimethylsilylethanesulfonyl Chloride	106018-85-3	611
$TMSCH_2CH_2$	$SO_2N=IPh$	[N-(2-(Trimethylsilyl)ethanesulfonyl)imino]-phenyliodane	236122-13-7	617
$TMSCH_2CH_2$	SO_2NH_2	2-(Trimethylsilyl)ethanesulfonamide	125486-96-6	609
$TMSCH_2CH_2$, $TMSCH_2CH_2$	$(SO_2)_2NH$	Bis(β-trimethylsilylethanesulfonyl)imide	548462-13-1	74
TMS, TMS	Se	Bis(trimethylsilyl) Selenide	4099-46-1	82
TMS	SePh	Phenyl Trimethylsilyl Selenide	33861-17-5	421
Cl_3Si	$SiCl_3$	Hexachlorodisilane	13465-77-5	309
TMS, TMS, TMS	SiH	Tris(trimethylsilyl)silane	1873-77-4	747
Ph_2SiH	$SiHPh_2$	1,1,2,2-Tetraphenyldisilane	16343-18-3	490
TMS	$Sn(Bu)_3$	Trimethylsilyltributylstannane	17955-46-3	719
TMS	$SiMe_3$	Disilane, 1,1,1,2,2,2-hexamethyl-	1450-14-2	278
TMS	$SiPhCl_2$	1,1-Dichloro-2,2,2-trimethyl-1-phenyldisilane	57519-88-7	233
$TMSCH_2$, $TMSCH_2$	Zn	Bis[(trimethylsilyl)methyl]zinc	41924-26-9	76

A list of General Abbreviations appears on the front Endpapers

Short Note on InChIs and InChIKeys

The IUPAC International Chemical Identifier (InChI^{TM}) and its compressed form, the InChIKey, are strings of letters representing organic chemical structures that allow for structure searching with a wide range of online search engines and databases such as Google and PubChem. While they are obviously an important development for online reference works, such as *Encyclopedia of Reagents for Organic Synthesis* (*e-EROS*), readers of this volume may be surprised to find printed InChI and InChIKey information for each of the reagents.

We introduced InChI and InChIKey to e-EROS in autumn 2009, including the strings in all HTML and PDF files. While we wanted to make sure that all users of *e-EROS*, the second print edition of *EROS*, and all derivative handbooks would find the same information, we appreciate that the strings will be of little use to the readers of the print editions, unless they treat them simply as reminders that *e-EROS* now offers the convenience of InChIs and InChIKeys, allowing the online users to make best use of their browsers and perform searches in a wide range of media.

If you would like to know more about InChIs and InChIKeys, please go to the *e-EROS* website: http://onlinelibrary.wiley.com/book/10.1002/047084289X and click on the InChI and InChIKey link.

3-Acetoxy-2-trimethylsilylmethyl-1-propene[1]

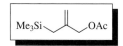

[72047-94-0] $C_9H_{18}O_2Si$ (MW 186.33)

InChI = 1S/C9H18O2Si/c1-8(6-11-9(2)10)7-12(3,4)5/h1,
    6-7H2,2-5H3

InChIKey = IKQWABMHZKGCLX-UHFFFAOYSA-N

(precursor for palladium trimethylenemethane cycloaddition reactions)

Physical Data: bp 68–70 °C/6.5 mmHg;[2] 60–61 °C/2.5 mmHg.[3]
Solubility: sol most organic solvents.
Preparative Methods: commercially available but expensive. Several preparations of (1)[2–4] or the precursor alcohol (2)[5,6] have been reported. The simplest involves the direct silylation of methallyl alcohol (eq 1).[4,5]

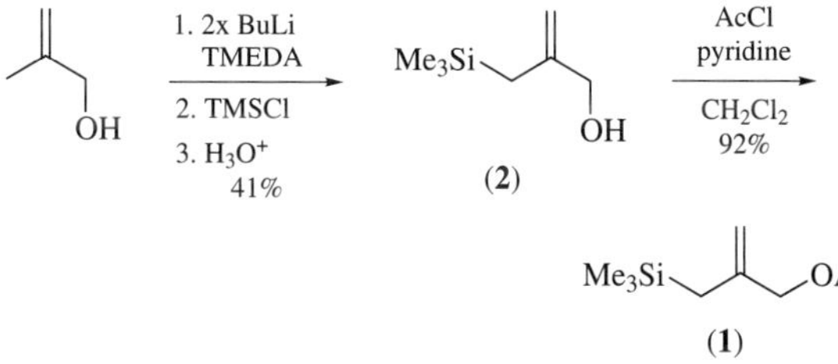

Handling, Storage, and Precautions: volatile, irritant, flammable.

Original Commentary

Daniel A. Singleton
Texas A&M University, College Station, TX, USA

Introduction. The development by Trost and co-workers of palladium–trimethylenemethane (TMM) cycloadditions employing the title reagent (1) and related reagents was a seminal advance in ring-construction methodology. The generality and versatility of these reactions is illustrated below by their use in [3 + 2] cyclo-additions to form both cyclopentanes and heterocycles, [3 + 4] and [3 + 6] cycloadditions, and applications in total synthesis.

Methylenecyclopentanes. Many lines of evidence indicate that in the presence of catalytic Pd[0], (1) forms the reactive intermediate palladium–TMM complex (3).[7–9] This complex appears to be zwitterionic in character, but its reactivity is governed mainly by its nucleophilicity/basicity. Thus the TMM

reactions work best and almost exclusively with electron-deficient alkenes. It appears that in most cases the cycloaddition is stepwise, with the initial Michael-like reaction followed by nucleophilic attack on a cationic π-allylpalladium complex (eq 2). The ion-pair nature of the intermediate (4) is thought to be the source of some unusual stereochemical effects which give the reactions some of the characteristics of concerted cycloadditions.[10,11]

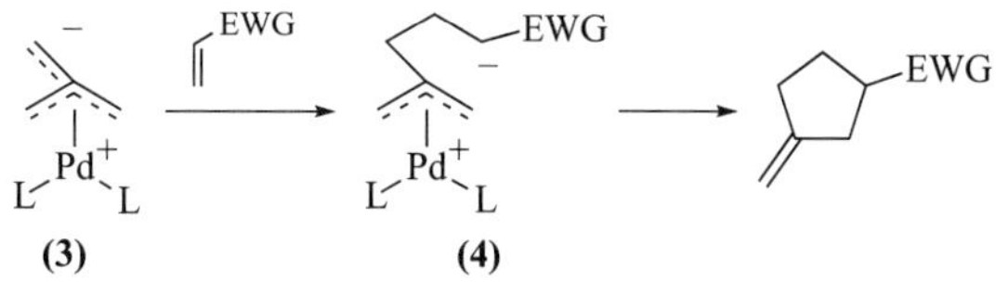

The corresponding methylenecyclopentanes are readily formed from a wide variety of activated alkenes, such as unsaturated esters, lactones, and nitriles, enones, vinyl sulfones, and nitro-alkenes.[4] The breadth of utility of this reaction is exemplified below by its use as key steps in syntheses of albene (eq 3),[12] brefeldin A (eq 4),[13] hirsutene (eq 5),[14] the core spirocarbocyclic rings of the ginkgolides (eq 6),[15] cephalotaxine (eq 7),[16] and kem-panes (eq 8).[17] The use of doubly activated alkenes, as in eq 8, is of particular advantage in the annulation of 2-cycloalkenones.[17b]

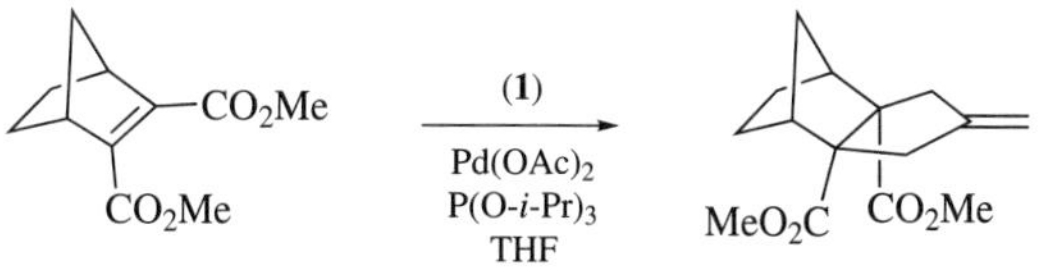

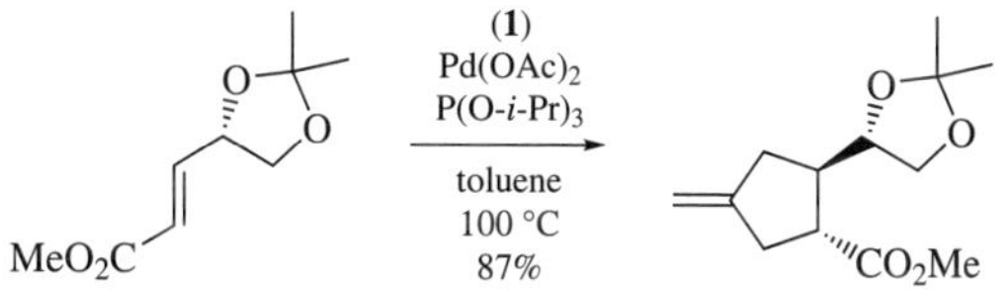

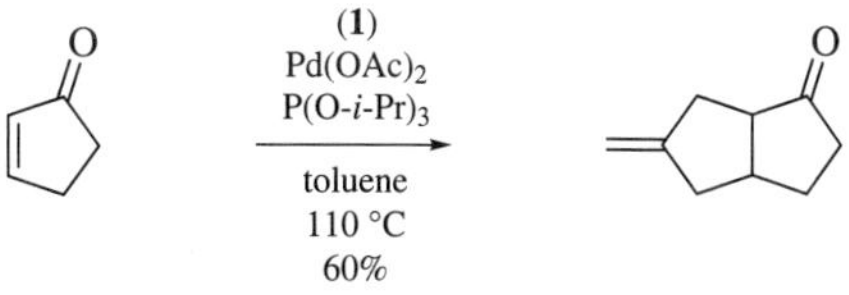

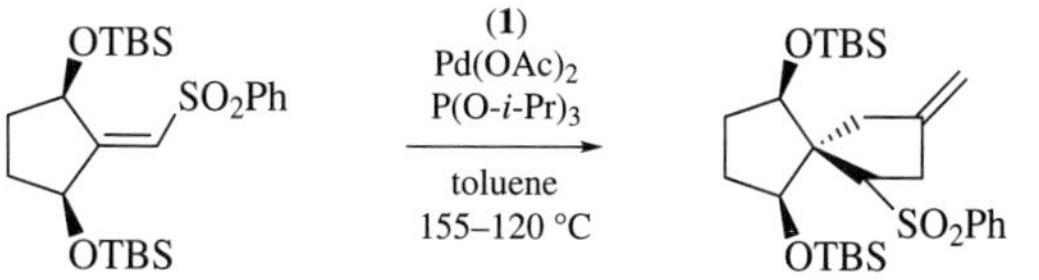

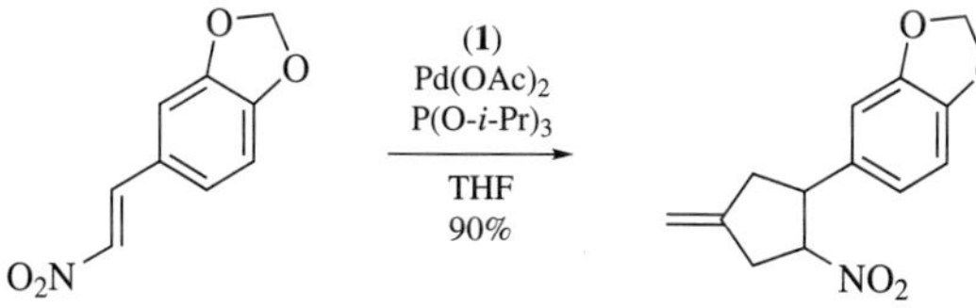

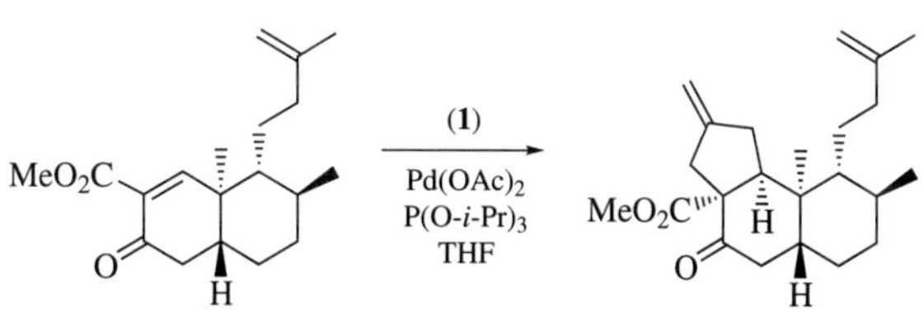

The adducts from vinyl sulfones are readily converted to cyclo-pentenones.[18] Vinyl sulfoxides are also acceptors for Pd–TMM cycloadditions. The use of optically pure vinyl sulfoxides allows for reasonable asymmetric induction (eq 9).[19] Outstanding asymmetric induction has been observed using chiral alkylideneoxazepadiones (eq 10).[20]

$$(9)$$

$$(10)$$

Methylenecyclopentenes. Although Pd–TMM cycloadditions with alkynes are unsuccessful, the formation of methylenecyclopentenes from alkynes can be accomplished via the cyclopentadiene adduct (eq 11).[21]

$$(11)$$

Heterocycles. Aldehydes can be converted to methylenetetrahydrofurans using tributyltin acetate or trimethyltin acetate as a cocatalyst (eq 12).[22] Most ketones are completely unreactive to Pd–TMM chemistry, but some oxacyclohexanones are readily annulated (eq 13), apparently due to activation of the carbonyl by the ring oxygen(s).[23] In reactions of α,β-unsaturated aldehydes and ketones, the competition between conjugate (alkene) and 1,2 (carbonyl) cycloadditions often favors reaction of the alkene. The cocatalyst tris(acetylacetonato)indium can be used to favor the 1,2-cycloaddition.[24]

$$(12)$$

$$(13)$$

Methylenepyrrolidines can be formed in Pd–TMM reactions with imines (eq 14). These reactions have also been accomplished using nickel catalysts.[25]

$$(14)$$

[4 + 3], [6 + 3], and [3 + 3] Cycloadditions. As suggested by the stepwise mechanism of eq 2, appropriately chosen dienes and trienes should undergo [4 + 3] and [6 + 3] cycloadditions in Pd–TMM reactions. Thus pyrones readily undergo [4 + 3] cycloadditions (eq 15),[26] and tropones undergo [6 + 3] cycloadditions (eq 16).[27] The main key to obtaining the higher order cycloadditions appears to be the use of all *s-cis* dienes and trienes.[28]

$$(15)$$

$$(16)$$

In a reaction that illustrates the nucleophilic character of the Pd–TMM intermediate (**3**), activated aziridines can undergo a [3 + 3] cycloaddition (eq 17),[29] presumably via initial nucleophilic ring opening of the aziridine by the Pd–TMM.

$$(17)$$

First Update

Māris Turks & Pierre Vogel
Swiss Federal Institute of Technology (EPFL), Lausanne, Switzerland

Preparative Methods. As discussed above, one of the simplest methods of preparation of 3-acetoxy-2-trimethylsilylmethyl-1-propene (**1**) is the direct silylation of the *C,O*-dianion derived from methallyl alcohol, followed by *O*-desilylation and subsequent acetylation (eq 1).[5,30] However, because of laboratory safety concerns (use of large quantities of 10 M *n*-BuLi) one might choose another more "user-friendly" route such as the copper-mediated conversion of 2-chloromethylallyl chloride to 2-chloromethylallyl trichlorosilane (**5**), followed by methylation[31] and nucleophilic displacement of the chloride by an acetate (eq 18).[2] Moreover, the intermediate 3-chloro-2-trimethylsilylmethyl-1-propene (**6**) is also a useful reagent in organic synthesis and in many aspects has similar reactivity to **1**. Alternatives to **1** include the carbonate derivatives **7**.

More recently, Organ and Murray have proposed a convenient, one pot synthesis of **1**, starting from 2-chloroprop-2-en-1-ol (eq 19).[32,33]

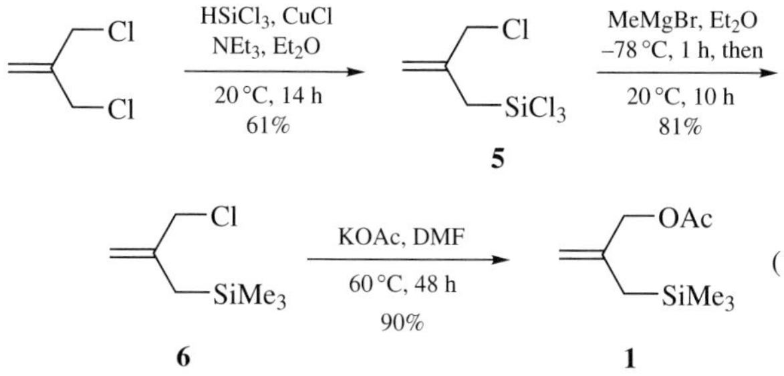

HSiCl₃, CuCl
NEt₃, Et₂O

20 °C, 14 h
61%

5

MeMgBr, Et₂O
–78 °C, 1 h, then

20 °C, 10 h
81%

KOAc, DMF
60 °C, 48 h
90%

(18)

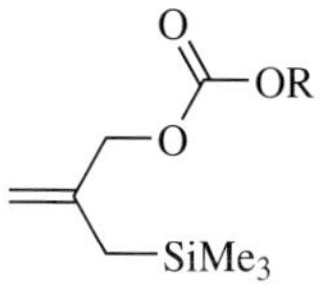

6   **1**

7

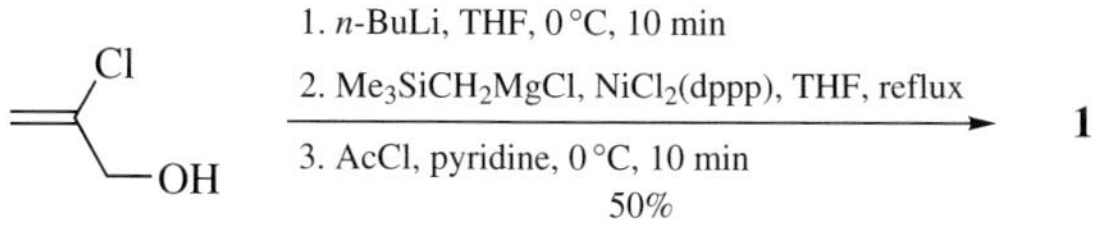

1. *n*-BuLi, THF, 0 °C, 10 min
2. Me₃SiCH₂MgCl, NiCl₂(dppp), THF, reflux
3. AcCl, pyridine, 0 °C, 10 min
50%

1   (19)

Reactions. The chemistry of Pd-TMM [3 + 2], [4 + 3], [6 + 3], and [3 + 3] cycloadditions has been extensively reviewed,[1,34] therefore, only recent advances will be covered here. Additionally, the use of **1** as an nucleophilic allylating agent is covered.

Pd-TMM Cycloadditions.

[3 + 2] Cycloadditions. Today there is enough evidence that **1** in the presence of Pd(0) forms Pd-TMM complexes of type **3**.[8,9] Up to now, a stepwise mechanism (path a, intermediate **9**) was favored.[10,11] However, a concerted mechanism, involving transition state **10**, has found strong support recently (eq 20).[35]

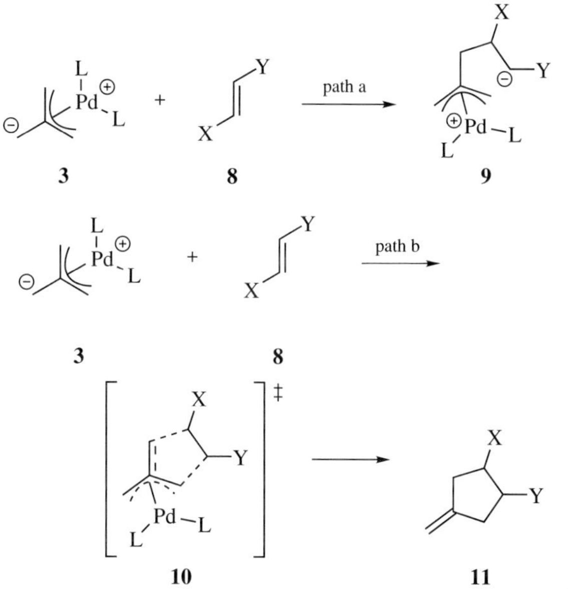

3   +   **8**   path a   **9**

3   +   **8**   path b

10   **11**   (20)

Recent applications include the use of **1** in the total synthesis of (+)-Sulcatine G (eq 21),[36] Bayer drug 36-7620 (**13**) (eq 22), and of (+)-Brefeldin A (eq 23).[37]

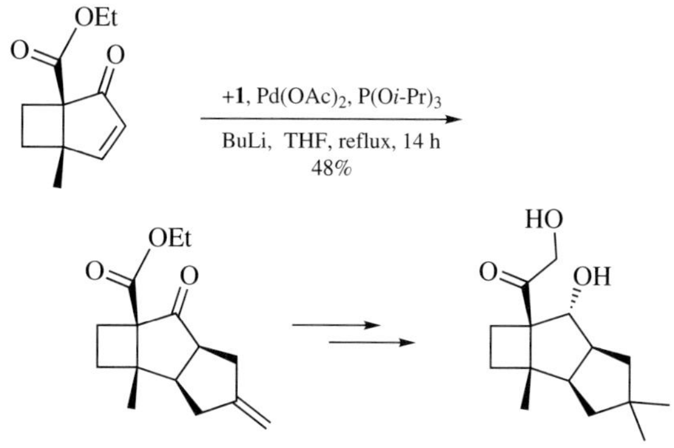

+1, Pd(OAc)₂, P(O*i*-Pr)₃

BuLi, THF, reflux, 14 h
48%

(21)

(+)-Sulcatine G

+1, Pd(OAc)₂, P(O*i*-Pr)₃
toluene, reflux, 91%
dr > 98:2

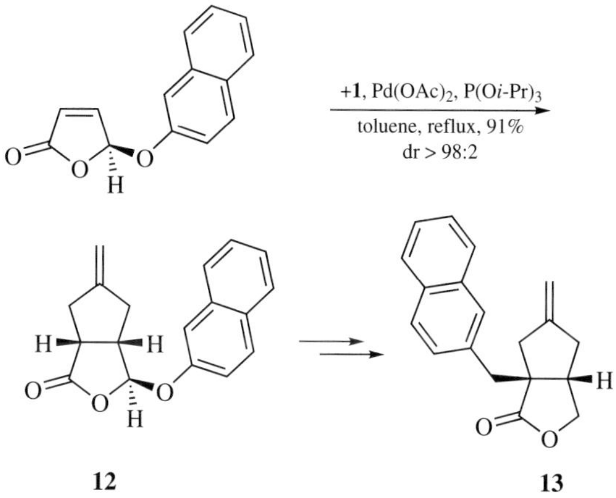

(22)

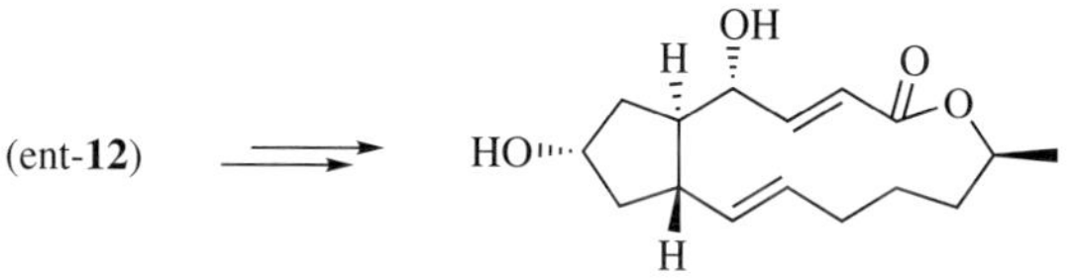

12   **13**

(ent-**12**)   (23)

(+)-Brefeldin A

Cycloadditions have been very successful in generating derivatives of C₆₀, an electron deficient species. Thus Pd-TMM undergoes a smooth cycloaddition with C₆₀ (eq 24).[38]

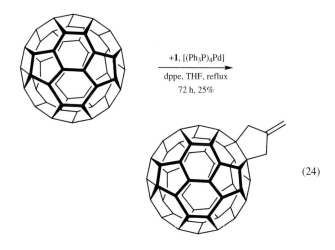

+1, [(Ph₃P)₄Pd]
dppe, THF, reflux
72 h, 25%

(24)

Precursors in the synthesis of β,β'-fused metallocenoporphyrins have been prepared in an one-pot operation that consists of a TMM-cycloaddition and subsequent elimination of HNO₂ (eq 25).[39]

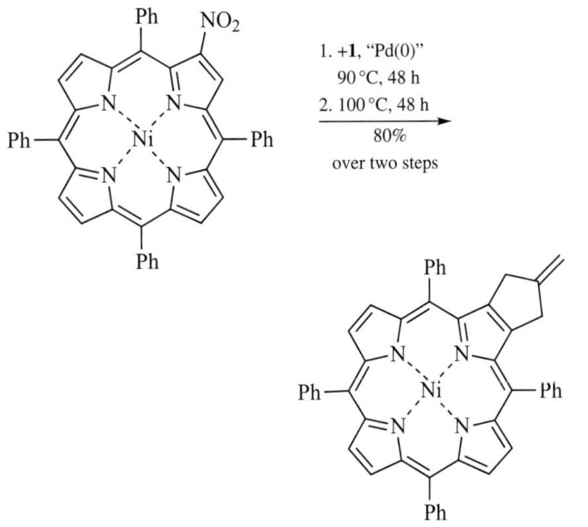

$$ (25) $$

As shown earlier, triisopropyl phosphite has turned out to be a far better ligand than triphenylphosphine *en route* to 5,5-fused proline surrogates (eq 26).[40]

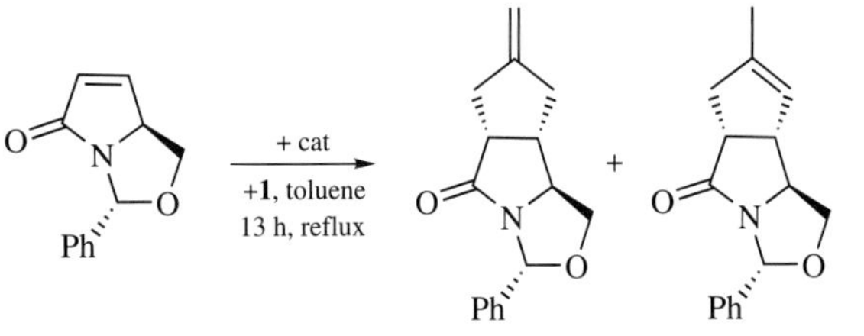

$$ (26) $$

cat:	[(PPh₃)₄Pd]:	9%	8%
	[(*i*-PrO)₃P)₄Pd]:	80%	not detected

[3 + 2] Cycloadditions of TMM to aldehydes allow the preparation of methylene tetrahydrofurans. The challenge was to overcome the poor nucleophilicity of the intermediate alkoxide, resulting from the Pd-TMM nucleophilic addition to the aldehyde. The latter must attack intramolecularly the π-allyl moiety to generate the cycloadduct and liberate the palladium catalyst for the next cycle. The solution is to add trialkyltin acetates (Me₃SnOAc or Bu₃SnOAc) to the reaction mixture. This leads to the formation of stannyl ethers that react readily with π-allylpalladium cation intermediates. Me₃SnOAc gives better results than other tin derivatives and only 5–10 mol % is necessary (eq 27).[29]

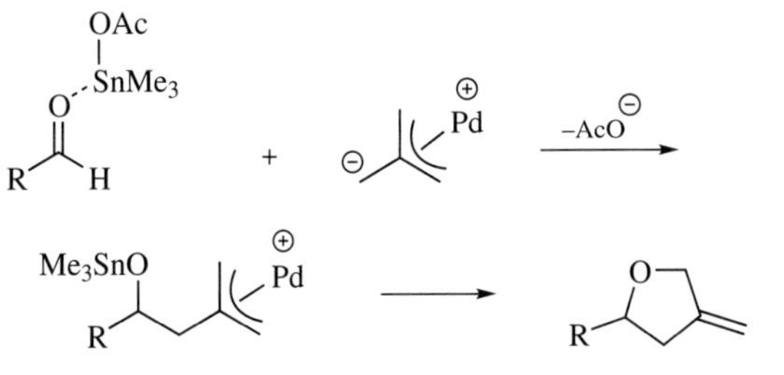

$$ (27) $$

Tin additives have only a minor effect on directing the TMM-Pd 1,2-additions to α,β-unsaturated aldehydes. The use of a more electropositive and oxyphilic cocatalyst such as In(acac)₃ favors 1,2-additions over 1,4-additions (eq 28).[24,41]

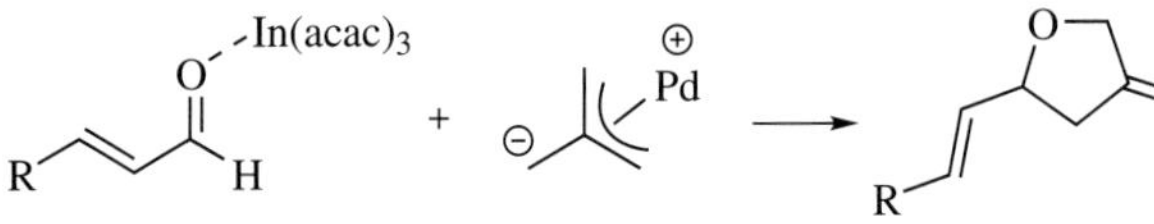

$$ (28) $$

Baldwin and co-workers have successfully used this strategy in their total synthesis of (±)-Aureothin and (±)-*N*-Acetylaureothin (eq 29).[42]

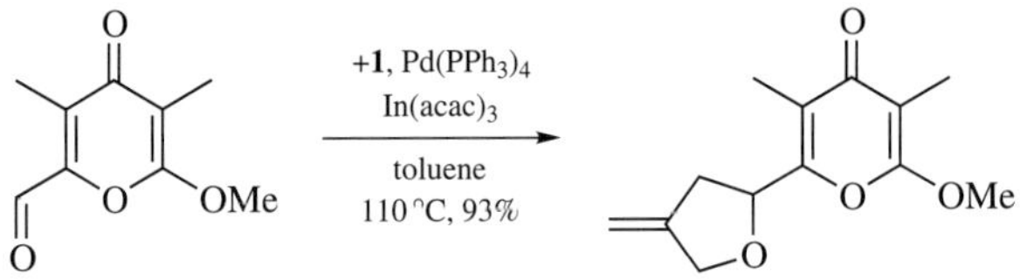

$$ (29) $$

[3 + 2] Cycloadditions of TMM-Pd to imines were initiated by Kemmitt's group,[25] and used by Trost et al. (eq 30).[43]

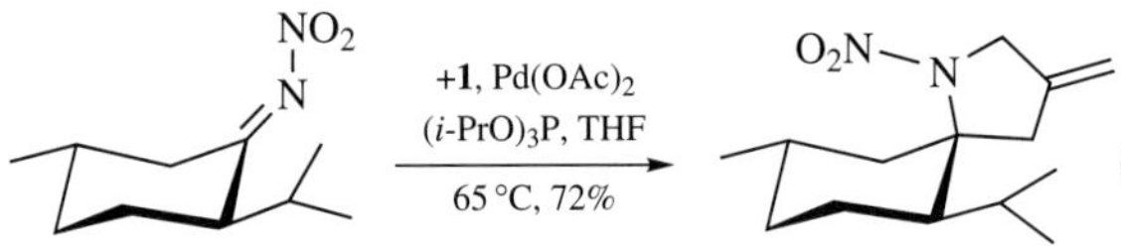

$$ (30) $$

[4 + 3] and [6 + 3] Cycloadditions. Seven-membered ring systems can be obtained from Pd-TMM cycloadditions to electron-deficient 1,3-dienes. Logically, 1,3-dienes restricted to the *cisoid* form are more reactive (eq 31, path a),[28] whereas their *transoid* form mostly provides five-membered rings (eq 31, path b).

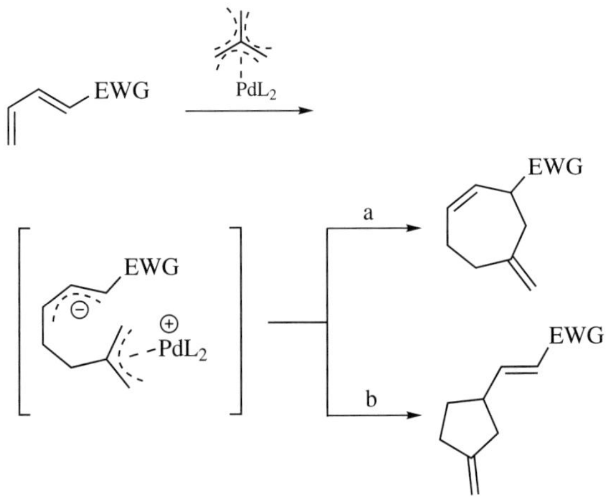

$$ (31) $$

Thus, conformationally restricted 1,2-dimethylenecyclopentane (**14**) provided [4 + 3] product **15** (eq 32),[28] whereas the flexible 2-vinyl cyclopentenone (**16**) gave the [3 + 2] product **17** (eq 33).[44] In the later case, a low yield of [4 + 3] adduct was obtained when carbonate analog **7** was used in the place of **1**.

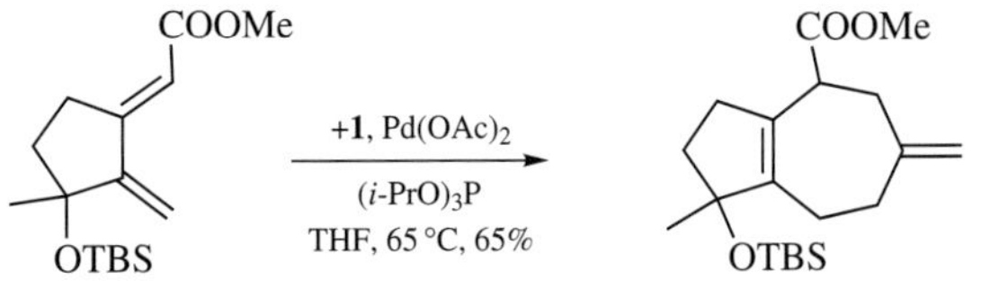

$$ (32) $$

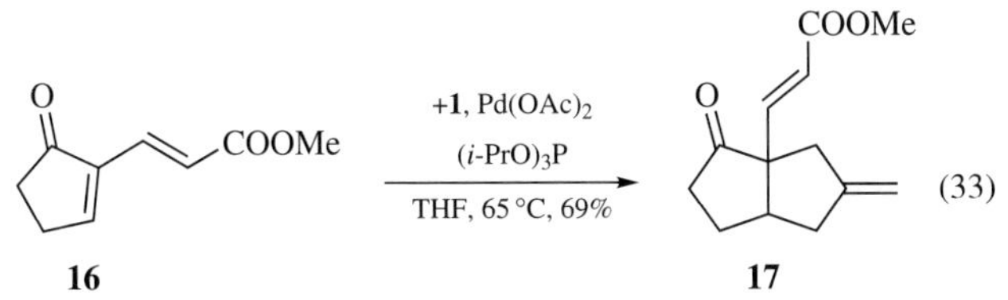

(33)

16   **17**

1-Aza-1,3-dienes undergo [4 + 3] cycloaddition with Pd-TMM providing azepane derivatives (eq 34).[43]

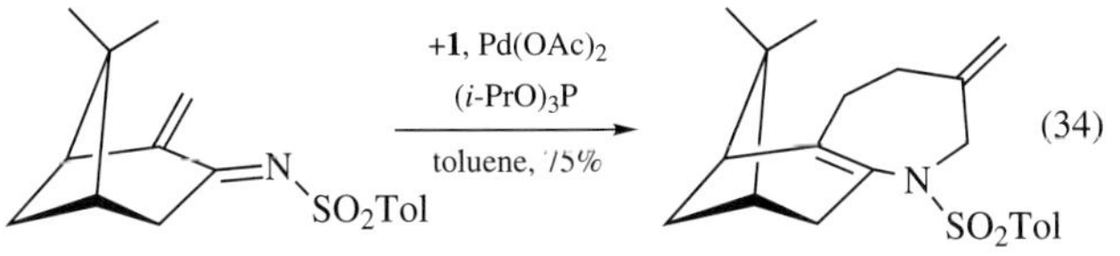

(34)

In theory, conjugated trienes can undergo [6 + 3]-cycloaddition with Pd-TMM to give nine-membered ring compounds. So far, all the known examples involve tropone derivatives as starting materials. Their cyclic structure and reduced conformational flexibility promote the [6 + 3]-cycloaddition at the expense of the competing [2 + 3] and [4 + 3]-cycloadditions. A recent example includes the addition of a methyl-substituted analog of **1** to 2-phenyltropone.[45]

[3 + 3] Cycloadditions. Harrity and co-workers have explored Pd-TMM reactions with activated aziridines, which generate [3 + 3]-cycloadducts. They have used this reaction in their synthesis of (−)-Castoramine and other *Nuphar* alkaloids.[46] In this particular case, the reaction of **18** gave only a modest yield of the desired product **19** along with other substances arising form aziridine ring opening (eq 35).

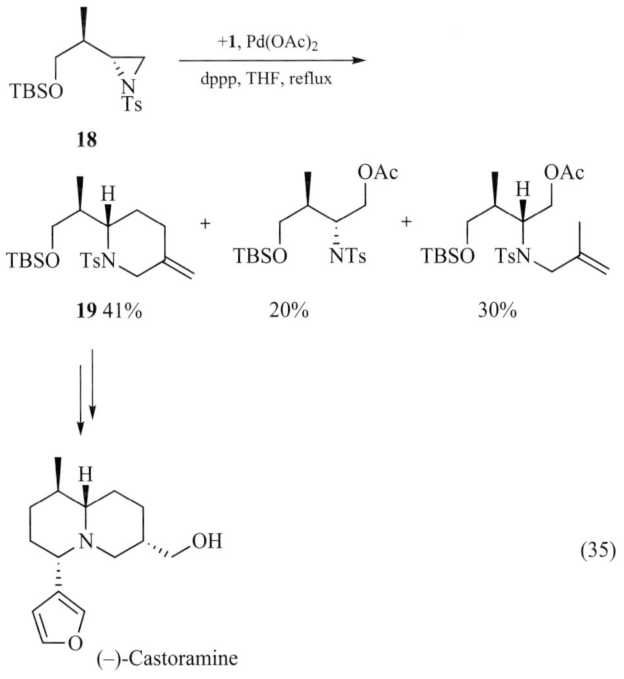

18

19 41%   20%   30%

(35)

(−)-Castoramine

The same group has applied the [3 + 3]-cycloadditions of Pd-TMM to aziridines in the preparation of functionalized piperidines (eq 36).[47]

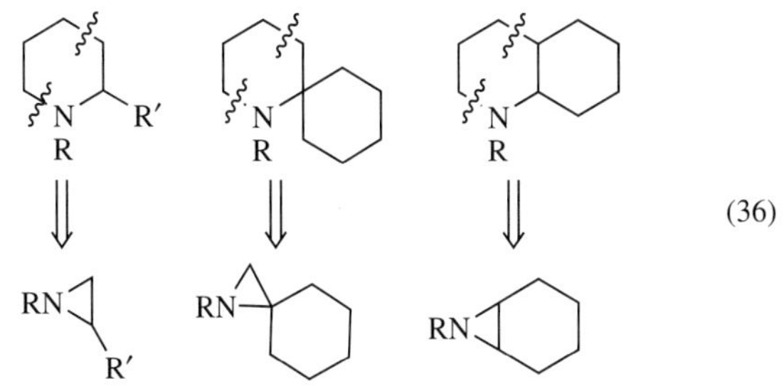

(36)

Hayashi and co-workers have reported Pd-TMM additions to azomethine imines (eq 37) and nitrones (eq 38).[48]

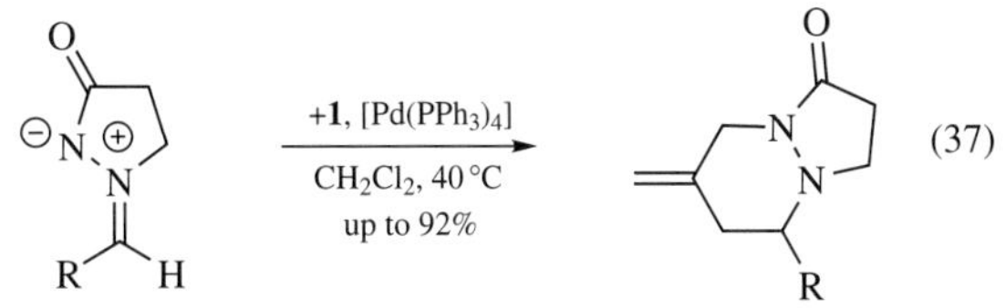

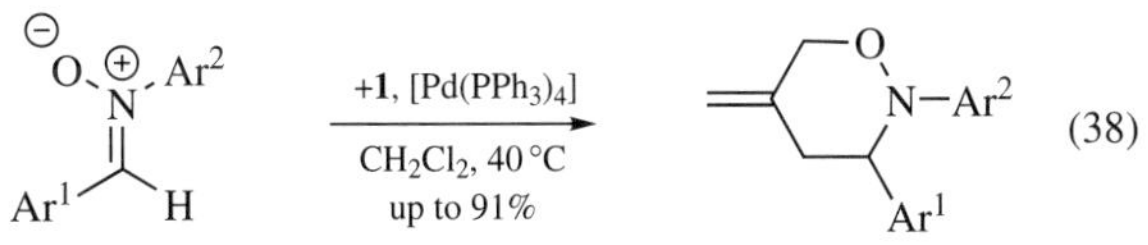

(37)

(38)

Allylation of Electrophilic Carbon Moieties.

Allylation of Aldehydes. Reiser's group has developed a stereoselective synthesis of substituted cyclopropyl carbaldehydes and has demonstrated their diastereoselective Hosomi–Sakurai allylations with several allylsilanes, including **1** (eq 39).[49]

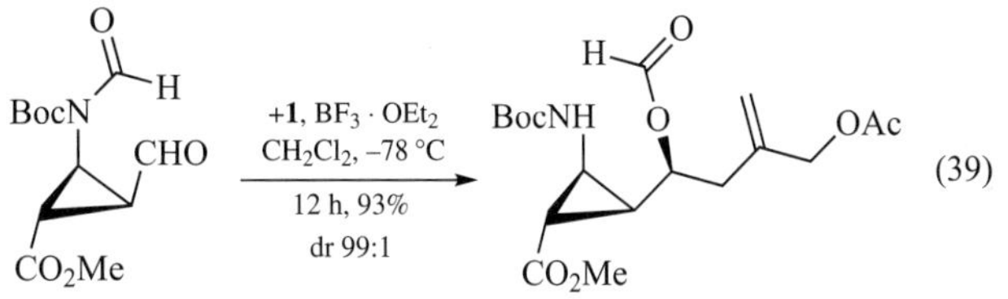

(39)

The same group also reported the synthesis of Paraconic acids, using a Hosomi–Sakurai allylation with **1** combined with a retro-aldol/lactonization cascade (eq 40).[50]

Differently substituted allylsilanes, among them **1**, have been used in the solid phase synthesis of naturally occurring phthalides (eq 41).[51]

Allylation of O,O-Acetals and N,O-Acetals. Nakata and co-workers have used **1** for the *C*-glycosidation of **20** on the way to the total synthesis of Brevetoxin-B (eq 42).[52]

Acetal allylation with **1** has been used in the combinatorial synthesis of selective glucocorticoid modulators (eq 43).[53] Similarly, allylation of **21** with **1** provided 4-allylazetidin-2-one (**23**) (eq 44), (no yield given).[54]

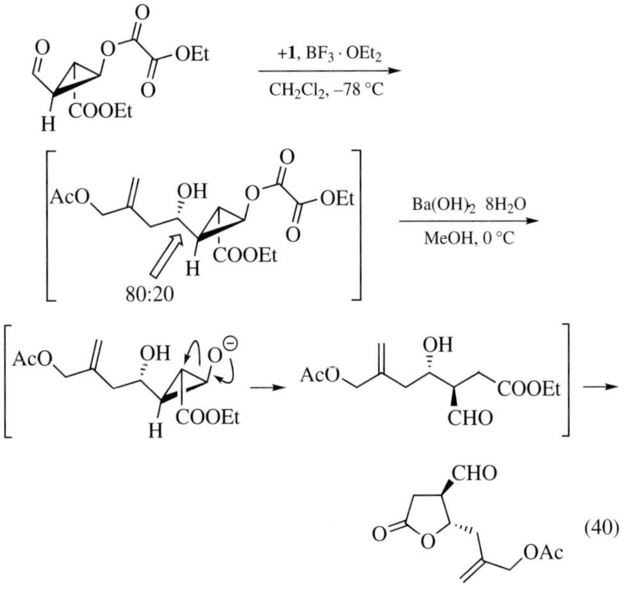

$+\mathbf{1}$, BF$_3$·OEt$_2$

CH$_2$Cl$_2$, –78 °C

Ba(OH)$_2$ 8H$_2$O

MeOH, 0 °C

80:20

$trans/cis = 80:20$
73%

(40)

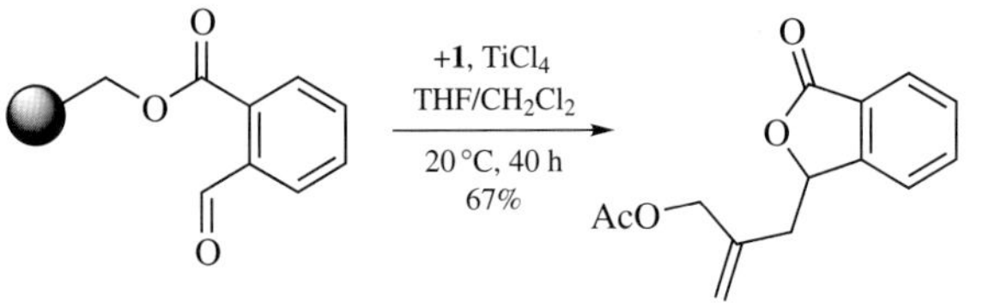

$+\mathbf{1}$, TiCl$_4$

THF/CH$_2$Cl$_2$

20 °C, 40 h
67%

(41)

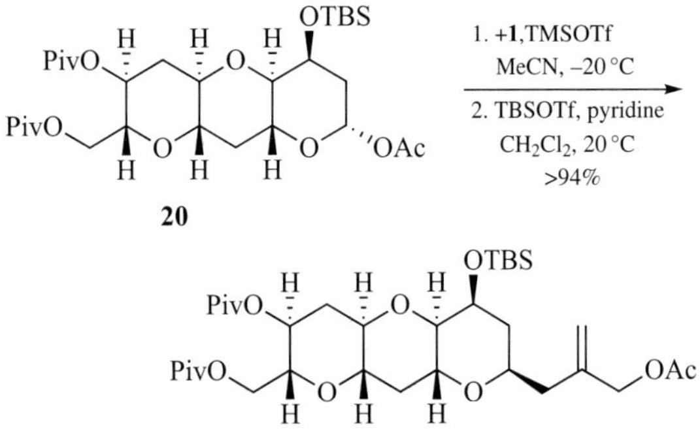

1. $+\mathbf{1}$, TMSOTf
MeCN, –20 °C

2. TBSOTf, pyridine
CH$_2$Cl$_2$, 20 °C
>94%

20

(42)

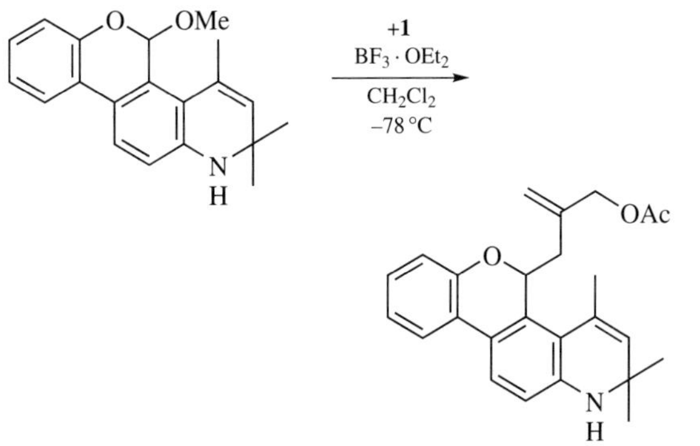

$+\mathbf{1}$
BF$_3$·OEt$_2$
CH$_2$Cl$_2$
–78 °C

(43)

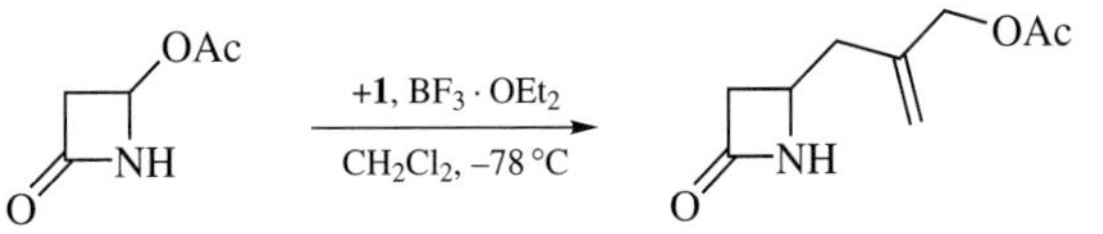

$+\mathbf{1}$, BF$_3$·OEt$_2$

CH$_2$Cl$_2$, –78 °C

(44)

21   22

The Steckhan group has developed an electrochemical synthesis of N,O-acetals and has used them in the allylation reactions with **1** (eq 45).[55]

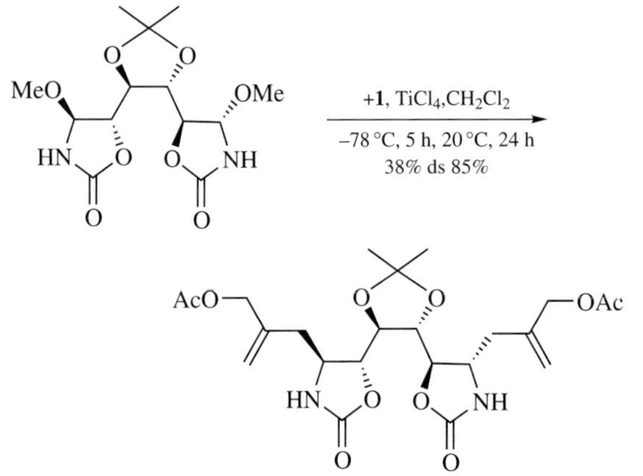

$+\mathbf{1}$, TiCl$_4$, CH$_2$Cl$_2$

–78 °C, 5 h, 20 °C, 24 h
38% ds 85%

(45)

Allylation of Allylic Electrophiles. Allylsilane **1** has been used in the Ferrier C-glycosylation of tri-O-acetyl glucal **23** (eq 46) and acetamidate **24** (eq 47).[56,57]

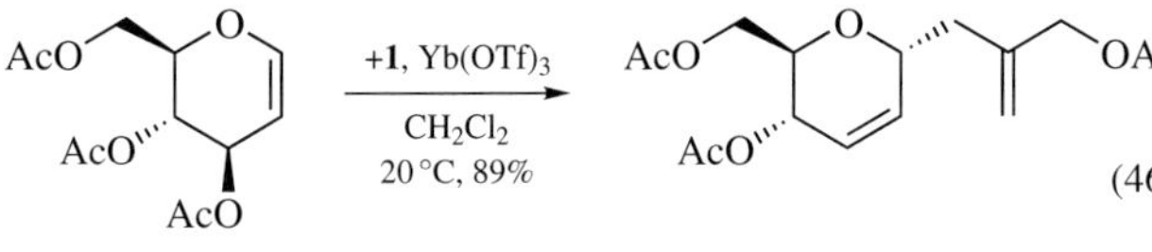

$+\mathbf{1}$, Yb(OTf)$_3$

CH$_2$Cl$_2$
20 °C, 89%

23

(46)

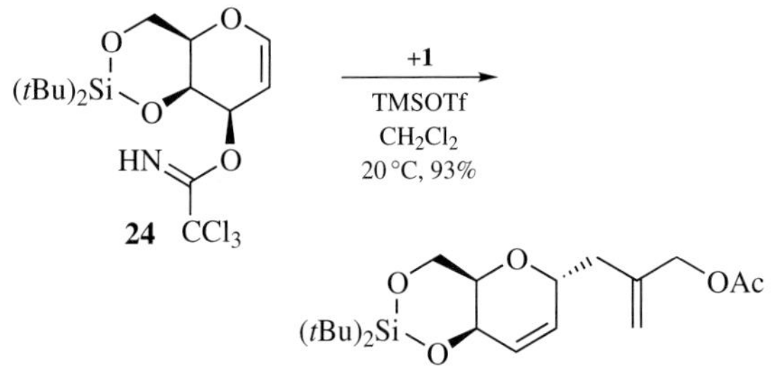

$+\mathbf{1}$

TMSOTf
CH$_2$Cl$_2$
20 °C, 93%

24   CCl$_3$

(47)

Vogel and co-workers have used **1** as double chain elongation synthon in their synthesis of long chain polypropionate fragments (eq 48).[58] Their approach is based on an one-pot reaction cascade which involves hetero-Diels–Alder addition of electron-rich 1-alkoxy-3-acyloxy-1,2-dienes (e.g., **25**) to sulfur dioxide, subsequent ionization into zwitterionic intermediates (e.g., **26**), then reaction of the latter with allylsilanes. The final step of this sequence is the retro-ene desulfinylation of an intermediate allylsilyl sulfinate (e.g., **27**) to produce useful synthetic intermediates containing two more stereogenic centers (e.g., **28**).

The rich chemistry of the resulting allyl acetates (e.g., **28**) has been further demonstrated by their transformation into allylsilanes.[58] Compound **27** has been used as an allylpalladium precursor that undergoes intramolecular sulfonylation to produce tetrahydro-2H-thiocine (**29**) (eq 49). Alternatively, **27** was reacted with N-chlorosuccimide and benzylamine to generate the corresponding sulfonyl chloride that was reacted in situ with benzylamine to give a separable 1:1 mixture of diastereomeric sulfonamides **30**. Both sulfonamides underwent intramolecular Pd-catalyzed allylation giving thiazonine derivatives **31** (eq 49).[59]

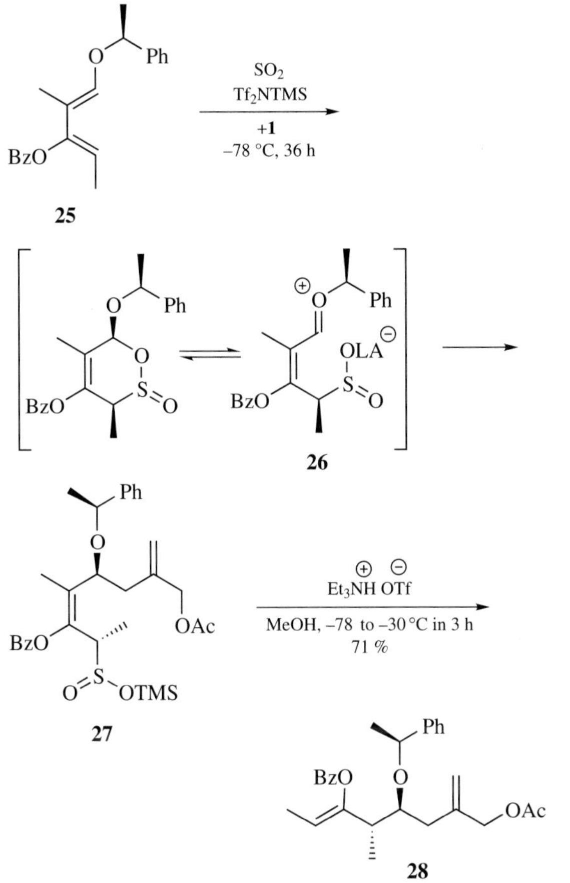

(48)

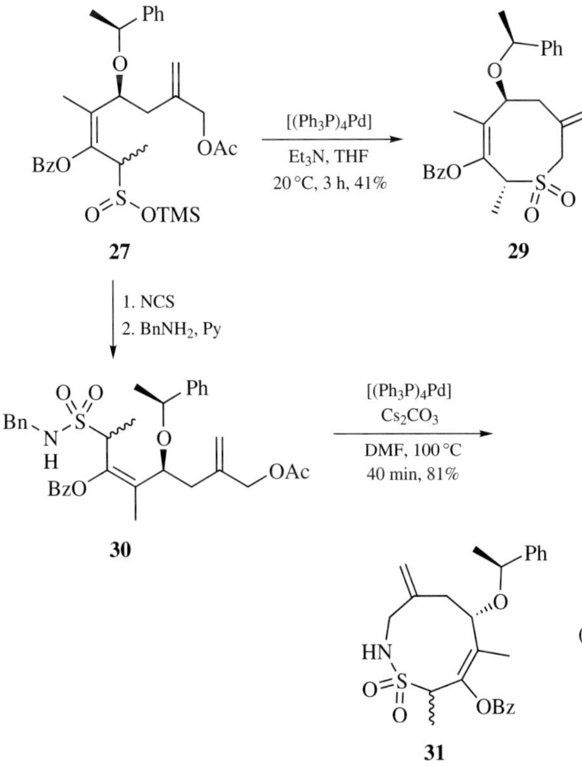

(49)

(50)

64:36

Miscellaneous Reactions. Voelter and co-workers have observed the exclusive formation of the unexpected alkylation product **33**, along with recovered starting material in their attempt to promote a [3 + 2] cycloaddition of **32** with Pd-TMM (eq 51).[61] They speculated that either enolate formation or an electron-transfer process disrupt the expected [3 + 2] cycloaddition of **32** with Pd-TMM.

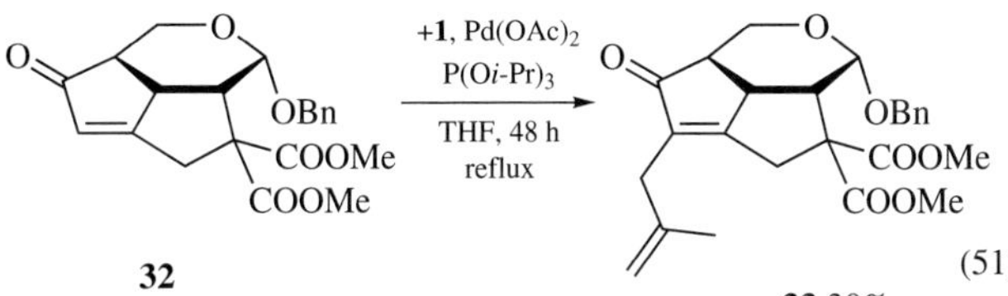

(51)

33 30%
(90% based on rec. st. mat.)

Kim et al. have developed a method for *Si-N* exchange in allyl-silane systems, including **1** (eq 52).[62]

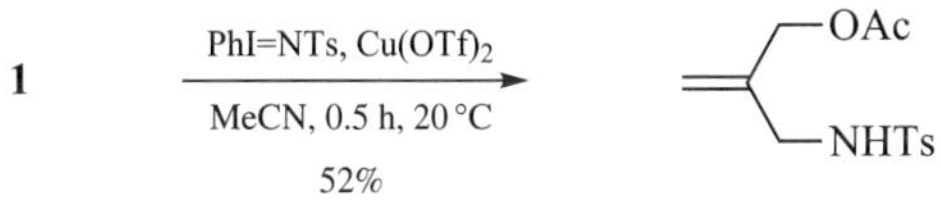

(52)

After the seminal discovery of Trost, the reactivity of other metal complexes than those of palladium has been screened against **1**. Thus, with **1** [Os(CO)$_2$(PPh$_3$)$_2$] gives a stable TMM complex **34**, albeit in low yield (eq 53).[63]

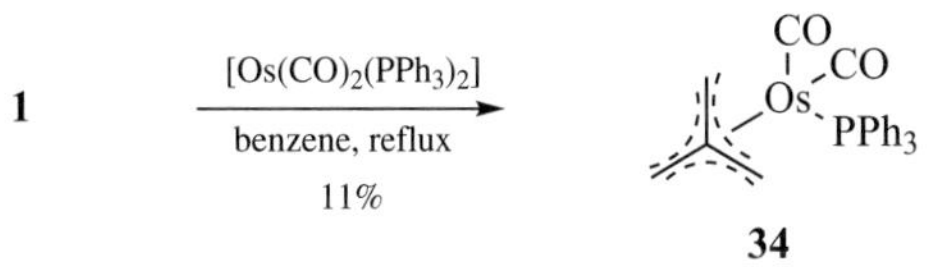

(53)

With [IrCl(CO)(PPh$_3$)$_2$], **1** reacts to give the stable TMM-complex **35** in low yield. In the presence of NaPF$_6$, complex **36** is obtained in 60% yield (eq 54).[64,65]

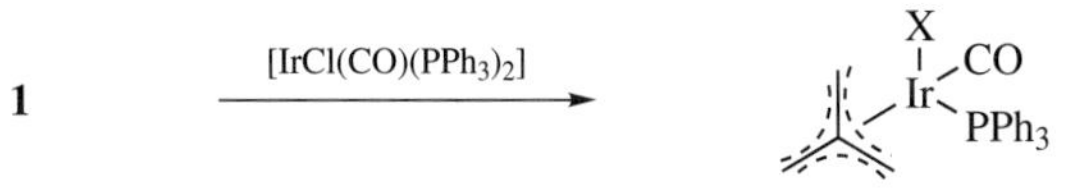

(54)

toluene, 12 h, 110 °C: X = Cl, 11%   X = Cl; **35**
toluene, NaPF$_6$, 12 h, 110 °C: X = PF$_6$, 60%   X = PF$_6$: **36**

Mo, Ru, Os, Rh, Ir, and Pt complexes also react with **1** to give the corresponding TMM-complexes in moderate yields. However, better yields are obtained with the methanesulfonate analog of **1**.[57,58]

Allylsilane **1** was also used in the Lewis acid-mediated Nicholas reaction of 3-acetoxycyclohept-1-en-4-yne(hexacarbonyl) dicobalt (eq 50).[60]

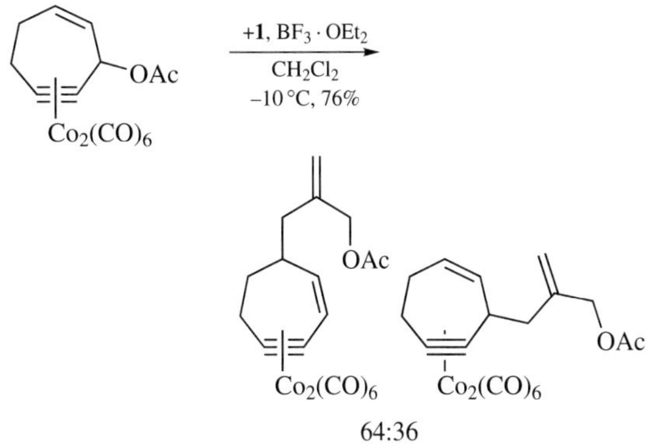

Interestingly, when **1** was treated with ruthenium complex **37** the reaction stopped at the level of stable η^3-allyl complex **38**. No formation of the corresponding trimethylenemethane-ruthenium complex was observed (eq 55).[66]

Related Reagents.   Related reagents include carbonate[43,67] and sulfonate[63–65] derivatives of alcohol **2**, other halide derivatives of **6**, [60,68–70] and corresponding trialkyltin[22,71] derivatives.

X = OC(O)R, OC(O)R, O(SO₂)R, Cl, Br, I
M = Si, Sn

TMM-transition metal complexes can also be obtained from methylidenecyclopropane.[72] Direct methylidenecyclopropane additions to carbonyl compounds can be catalyzed by Lewis acids.[73,74]

1. Trost, B. M., *Angew. Chem., Int. Ed. Engl.* **1986**, *25*, 1.

2. Trost, B. M.; Buch, M.; Miller, M. L., *J. Org. Chem.* **1988**, *53*, 4887.

3. Agnel, G.; Malacria, M., *Synthesis* **1989**, 687.

4. Trost, B. M.; Chan, D. M. T., *J. Am. Chem. Soc.* **1983**, *105*, 2315. Trost, B. M.; Chan, D. M. T., *J. Am. Chem. Soc.* **1979**, *101*, 6429.

5. Trost, B. M.; Chan, D. M. T.; Nanninga, T. N., *Org. Synth.* **1984**, *62*, 58.

6. Knapp, S.; O'Connor, U.; Mobilio, D., *Tetrahedron Lett.* **1980**, *21*, 4557.

7. Trost, B. M.; Chan, D. M. T., *J. Am. Chem. Soc.* **1979**, *101*, 6432.

8. Trost, B. M.; Chan, D. M. T., *J. Am. Chem. Soc.* **1980**, *102*, 6359.

9. Trost, B. M.; Chan, D. M. T., *J. Am. Chem. Soc.* **1983**, *105*, 2326.

10. Trost, B. M.; Miller, M. L., *J. Am. Chem. Soc.* **1988**, *110*, 3687.

11. Trost, B. M.; Mignani, S. M., *Tetrahedron Lett.* **1986**, *27*, 4137.

12. Trost, B. M.; Renaut, P., *J. Am. Chem. Soc.* **1982**, *104*, 6668.

13. Trost, B. M.; Lynch, J.; Renaut, P.; Steinman, D. H., *J. Am. Chem. Soc.* **1986**, *108*, 284.

14. Cossy, J.; Belotti, D.; Pete, J. P., *Tetrahedron Lett.* **1990**, *46*, 1859. Cossy, J.; Belotti, D.; Pete, J. P., *Tetrahedron Lett.* **1987**, *28*, 4547.

15. Trost, B. M.; Acemoglu, M., *Tetrahedron Lett.* **1989**, *30*, 1495.

16. Ishibashi, H.; Okano, M.; Tamaki, H.; Maruyama, K.; Yakura, T.; Ikeda, M., *J. Chem. Soc., Chem. Commun.* **1990**, 1436.

17. (a) Paquette, L. A.; Sauer, D. R.; Cleary, D. G.; Kinsella, M. A.; Blackwell, C. M.; Anderson, L. G., *J. Am. Chem. Soc.* **1992**, *114*, 7375. (b) Cleary, D. G.; Paquette, L. A., *Synth. Commun.* **1987**, *17*, 497.

18. Trost, B. M.; Seoane, P.; Mignani, S.; Acemoglu, M., *J. Am. Chem. Soc.* **1989**, *111*, 7487.

19. Chaigne, F.; Gotteland, J. P.; Malacria, M., *Tetrahedron Lett.* **1989**, *30*, 1803.

20. Trost, B. M.; Yang, B.; Miller, M. L., *J. Am. Chem. Soc.* **1989**, *111*, 6482.

21. Trost, B. M.; Balkovec, J. M.; Angle, S. R., *Tetrahedron Lett.* **1986**, *27*, 1445.

22. Trost, B. M.; King, S. A.; Schmidt, T., *J. Am. Chem. Soc.* **1989**, *111*, 5902. Trost, B. M.; King, S. A., *Tetrahedron Lett.* **1986**, *27*, 5971.

23. Trost, B. M.; King, S. A.; Nanninga, T. N., *Chem. Lett.* **1987**, 15.

24. Trost, B. M.; Sharma, S.; Schmidt, T., *J. Am. Chem. Soc.* **1992**, *114*, 7903.

25. Jones, M. D.; Kemmitt, R. D. W., *J. Chem. Soc., Chem. Commun.* **1986**, 1201.

26. Trost, B. M.; Schneider, S., *Angew. Chem.* **1989**, *101*, 215.

27. Trost, B. M.; Seoane, P. R., *J. Am. Chem. Soc.* **1987**, *109*, 615.

28. Trost, B. M.; MacPherson, D. T., *J. Am. Chem. Soc.* **1987**, *109*, 3483.

29. Bambal, R. B.; Kemmitt, R. D. W., *J. Organomet. Chem.* **1989**, *362*, C18.

30. Trost, B. M.; Chan, D. M. T.; Nanninga, T. N., *Org. Synth., Coll. Vol. 7*, **1990**, 266.

31. The use of freshly prepared MeMgI (Mg + MeI) instead of MeMgBr gives equally good results: Turks, M.; Vogel, P., unpublished results.

32. Organ, M. G.; Murray, A. P., *J. Org. Chem.* **1997**, *62*, 1523.

33. Sakar, T. K. In *Science of Synthesis*; Fleming, I., Ed.; Georg Thieme Verlag: Stuttgart, New York, 2002; Vol. 4, p 837.

34. (a) Chan, D. M. T.; In *Cycloaddition Reaction in Organic Synthesis*; Kobayashi, S.; Jørgensen, K. A., Eds., Wiley-VCH: Weinheim, 2001; p 57. (b) Chan, D. M. T.; In *Comprehensive Organic Synthesis*, Trost, B. M.; Fleming, I.; Paquette, L. E., Eds.; Pergamon: Oxford, 1991; Vol. 5, p 271. (c) Lautens, M.; Klute, W.; Tam, W., *Chem. Rev.* **1996**, *96*, 49. (d) Ojima, I.; Tzamarioudaki, M.; Li, Z.; Donovan, R. J., *Chem. Rev.* **1996**, *96*, 635. (e) Trost, B. M., *Pure Appl. Chem.* **1988**, *60*, 1615.

35. Singleton, D. A.; Schulmeier, B. E., *J. Am. Chem. Soc.* **1999**, *121*, 9313.

36. Taber, D. F.; Frankowski, K. J., *J. Org. Chem.* **2005**, *70*, 6417.

37. (a) Trost, B. M.; Crawley, M. L., *J. Am. Chem. Soc.* **2002**, *124*, 9328. (b) Trost, B. M.; Crawley, M. L., *Chem. Eur. J.* **2004**, *10*, 2237.

38. Shiu, L.-L.; Lin, T.-I.; Peng, S.-M.; Her, G.-R.; Ju, D. D.; Lin, S.-K.; Hwang, J.-H.; Mou, C.-Y.; Luh, T.-Y., *J. Chem. Soc., Chem. Commun.* **1994**, 647.

39. Wang, H. J. H.; Jaquinod, L.; Nurco, D. J.; Vicente, M. G. H.; Smith, K. M., *Chem. Commun.* **2001**, 2646.

40. Jao, E.; Bogen, S.; Saksena, A. K.; Girijavallabhan, V., *Tetrahedron Lett.* **2003**, *44*, 5033.

41. Trost, B. M.; Sharma, S.; Schmidt, T., *Tetrahedron Lett.* **1993**, *34*, 7183.

42. (a) Jacobsen, M. F.; Moses, J. E.; Adlington, R. M.; Baldwin, J. E., *Org. Lett.* **2005**, *7*, 641. (b) Jacobsen, M. F.; Moses, J. E.; Adlington, R. M.; Baldwin, J. E., *Tetrahedron* **2006**, *62*, 1675.

43. Trost, B. M.; Marrs, C. M., *J. Am. Chem. Soc.* **1993**, *115*, 6636.

44. Trost, B. M.; Yamazaki, S., *Chem. Lett.* **1994**, 2245.

45. Trost, B. M.; Parquette, J. R.; Marquart, A. L., *J. Am. Chem. Soc.* **1995**, *117*, 3284.

46. Goodenough, K. M.; Moran, W. J.; Raubo, P.; Harrity, J. P. A., *J. Org. Chem.* **2005**, *70*, 207.

47. Hedley, S. J.; Moran, W. J.; Price, D. A.; Harrity, J. P., *J. Org. Chem.* **2003**, *68*, 4286.

48. Shintani, R.; Hayashi, T., *J. Am. Chem. Soc.* **2006**, *128*, 6330.

49. Böhm, C.; Schinnerl, M.; Bubert, C.; Zabel, M.; Labahn, T.; Parisini, E.; Reiser, O., *Eur. J. Org. Chem.* **2000**, 2955

50. Chhor, R. B.; Nosse, B.; Sörgel, S.; Böhm, C.; Seitz, M.; Reiser, O., *Chem. Eur. J.* **2003**, *9*, 260.

51. Knepper, K.; Ziegert, R. E.; Bräse, S., *Tetrahedron* **2004**, *60*, 8591.

52. (a) Matsuo, G.; Kawamura, K.; Hori, N.; Matsukura, H.; Nakata, T., *J. Am. Chem. Soc.* **2004**, *126* 14374 (b) Matsukura, H.; Hori, N.; Matsuo, G.; Nakata, T., *Tetrahedron Lett.* **2000**, *41*, 7681.

53. Elmore, S. W.; Pratt, J. K.; Coghlan, M. J.; Mao, Y.; Green, B. E.; Anderson, D. D.; Stashko, M. A.; Lin, C. W.; Falls, D.; Nakane, M.; Miller, L.; Tyree, C. M.; Miner, J. N.; Lane, B., *Bioorg. Med. Chem. Lett.* **2004**, *14*, 1721.

54. Oumoch, S.; Rousseau, G., *Bioorg. Med. Chem. Lett.* **1994**, *4*, 2841.

55. (a) Zelgert, M.; Nieger, M.; Lennartz, M.; Steckhan, E., *Tetrahedron* **2002**, *58*, 2641. (b) Lennartz, M.; Sadakane, M.; Steckhan, E., *Tetrahedron* **1999**, *55*, 14407.

56. Takhi, M.; Abdel Rahman, A. A.-H.; Schmidt, R. R., *Tetrahedron Lett.* **2001**, *42*, 4053.

57. Abdel-Rahman, A. A.-H.; Takhi, M.; El Sayed, H.; El, A.; Schmidt, R. R., *J. Carbohydrate Chem.* **2002**, *21*, 113.

58. Turks, M.; Fonquerne, F.; Vogel, P., *Org. Lett.* **2004**, *6*, 1053.

59. Bouchez, L. C.; Turks, M.; Dubbaka, S. R.; Fonquerne, F.; Craita, C.; Laclef, S.; Vogel, P., *Tetrahedron* **2005**, *61*, 11473.

60. DiMartino, J.; Green, J. R., *Tetrahedron* **2006**, *62*, 1402.

61. Naz, N.; Al-Tel, T. H.; Al-Abed, Y.; Voelter, W.; Ficker, R.; Hiller, W., *J. Org. Chem.* **1996**, *61*, 3205.

62. Kim, D. Y.; Choi, J. S.; Rhie, D. Y.; Chang, S. K.; Kim, I. K., *Synth. Commun.* **1997**, *27*, 2753.

63. Jones, M. D.; Kemmitt, R. D. W.; Platt, A. W. G.; Russell, D. R.; Sherry, L. J. S., *J. Chem. Soc., Chem. Commun.* **1984**, 673.

64. Jones, M. D.; Kemmitt, R. D. W.; Platt, A. W. G., *J. Chem. Soc., Dalton Trans.* **1986**, 1411.

65. Jones, M. D.; Kemmitt, R. D. W., *J. Chem. Soc., Chem. Commun.* **1985**, 811.

66. Kondo, H.; Kageyama, A.; Yamaguchi, Y.; Haga, M.-a.; Kirchner, K.; Nagashima, H., *Bull. Chem. Soc. Jpn.* **2001**, *74*, 1927.

67. Holzapfel, C. W.; van der Merwe, T. L., *Tetrahedron Lett.* **1996**, *37*, 2307.

68. Chloro derivative: Lygo, B.; Slack, D.; Wilson, C., *Tetrahedron Lett.* **2005**, *46*, 6629.

69. Bromo derivative: Ryter, K.; Livinghouse, T., *J. Org. Chem.* **1997**, *62*, 4842.

70. Iodo derivative: Zhang, W.; Carter, R. G., *Org. Lett.* **2005**, *7*, 4209.

71. Sibi, M. P.; Hasegawa, H., *Org. Lett.* **2002**, *4*, 3347.

72. (a) Suzuki, T.; Fujimoto, H., *Inorg. Chem.* **2000**, *39*, 1113. (b) Allen, S. R.; Barnes, S. G.; Green, M.; Moran, G.; Trollope, L.; Murrall, N. W.; Welch, A. J.; Sharaiha, D. M., *J. Chem. Soc., Dalton Trans.* **1984**, *6*, 1157.

73. Miura, K.; Takasumi, M.; Hondo, T.; Saito, T.; Hosomi, A., *Tetrahedron Lett.* **1997**, *38*, 4587.

74. For a reviews see: (a) Nakamura, E.; Yamago, S., *Acc. Chem. Res.* **2002**, *35*, 867. (b) Nakamura, I.; Yamamoto, Y., *Adv. Synth. Cat.* **2002**, *344*, 111.

Allyl(2-chloroethyl)dimethylsilane

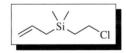

[104107-85-9] $C_7H_{15}ClSi$ (MW 162.71)

InChI = 1S/C7H15ClSi/c1-4-6-9(2,3)7-5-8/h4H,1,5-7H2,2-3H3

InChIKey = QEQUOBOEEJIONU-UHFFFAOYSA-N

(a useful β-hydroxyethyl carbanion equivalent and silacycle precursor)

Physical Data: bp 70 °C/40 mm.

Solubility: soluble in most organic solvents.

Form Supplied in: pale yellow oil; not commercially available.

Handling, Storage, and Precaution: prone to hydrolysis. Toxicological properties not known; assumed to be toxic.

Preparation. The title reagent **4** was first obtained via a multistep sequence starting from chloro(chloromethyl)dimethylsilane (**1**).[1] Reaction of (**1**) with allylmagnesium bromide gives chloride **2**, which is then homologated in two steps. Formation of the Grignard reagent from **2** and reaction with paraformaldehyde leads to alcohol **3**, and further reaction with PPh₃/CCl₄ gives the desired product **4** (eq 1).

$$
\begin{array}{ccccc}
\text{1} & \xrightarrow[\text{Et}_2\text{O}]{\text{allylMgBr}} & \text{2} & \xrightarrow[\text{2. (CH}_2\text{O)}_n]{\text{1. Mg, Et}_2\text{O}} & \\
& 40\% & & 72\% &
\end{array}
$$

$$
\text{3} \xrightarrow[64\%]{\text{PPh}_3, \text{CCl}_4} \text{4} \tag{1}
$$

Silane **4** is more easily prepared in gram quantities in a one-pot procedure[2] by sequential addition of 1 equiv of methyllithium and then 1 equiv of allylmagnesium bromide to a solution of commercially available (2-chloroethyl)dichloromethylsilane (**5**) in THF (eq 2). Reduced pressure distillation then gives **4** in high yield.

$$
\text{5} \xrightarrow[\text{2. allylMgBr} \atop 81\%]{\text{1. MeLi, THF}} \text{4} \tag{2}
$$

As a β-Hydroxyethyl Carbanion Equivalent.[2] The title compound can act efficiently as a β-hydroxyethyl carbanion equivalent. Transformation of compound **4** into the corresponding Grignard reagent, i.e., **6** (eq 3) followed by reaction with aldehydes and ketones gives adducts **7** (eq 4). Tamao–Fleming oxidation of **7** then affords 1,3-diols (**8**) in good yields.

$$
\text{4} \xrightarrow{\text{Mg, THF}} \text{6} \tag{3}
$$

$$
\text{6} \xrightarrow[\text{THF} \atop 50-81\%]{\text{R}_1\text{R}_2\text{C}=\text{O}} \text{7} \xrightarrow[\text{THF/MeOH} \atop 58-84\%]{\substack{\text{1. KHF}_2, \text{TFA, CHCl}_3 \\ \text{2. H}_2\text{O}_2, \text{NaHCO}_3}} \text{8} \tag{4}
$$

R_1/R_2 = alkyl, phenyl, H, cyclic

Grignard reagent **6** can be converted into a cuprate reagent by treatment with CuBr·Me₂S (DMS) complex. This cuprate can then undergo conjugate addition to enones such as 2-cyclohexenone in the presence of TMSCl to give adduct **9**, which upon Tamao–Fleming oxidation gives the β-hydroxyethyl ketone (**10**) (eq 5).

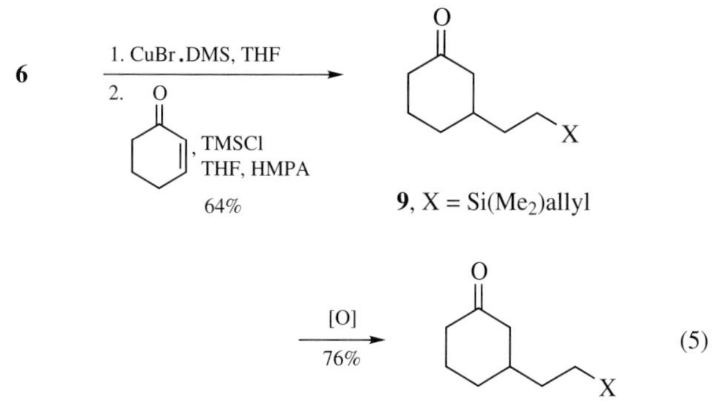

$$6 \xrightarrow[\substack{2. \\ \text{TMSCl} \\ \text{THF, HMPA} \\ 64\%}]{\text{1. CuBr·DMS, THF}}$$

9, X = Si(Me$_2$)allyl

$$\xrightarrow[76\%]{[O]} \quad (5)$$

10, X = OH

Treatment of Grignard reagent **6** with CuI, followed by addition of an epoxide, results in regioselective ring-opening to give adducts **11**. Subsequent Tamao–Fleming oxidation of **11** leads to 1,4-diols (**12**) in good yields (eq 6).

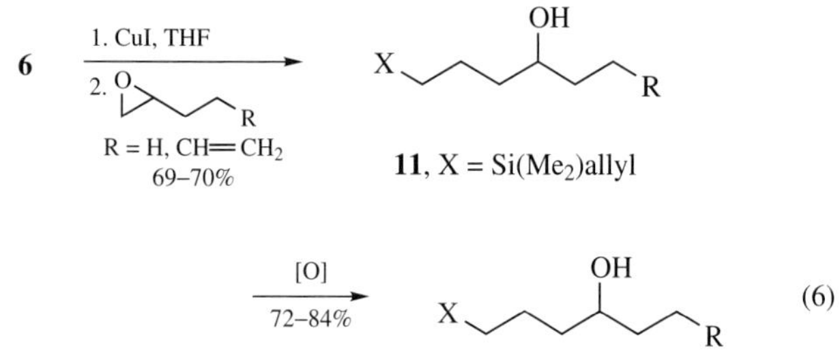

$$6 \xrightarrow[\substack{2. \\ R = H, CH=CH_2 \\ 69-70\%}]{\text{1. CuI, THF}}$$

11, X = Si(Me$_2$)allyl

$$\xrightarrow[72-84\%]{[O]} \quad (6)$$

12, X = OH

Use in Free Radical Chemistry. Chloride **4** has been used in free radical cyclizations.[1,3] Treatment of chloride **4** with tri-*n*-butyltin hydride (TBTH)/AIBN in benzene at reflux gave a high yield of reduced material **13** together with some of the 6-*endo* cyclization product **14**, a minor product (eq 7). The radical cyclization was shown to be irreversible. This example represents a deviation from the usual preference for 5-*exo* radical cyclizations.

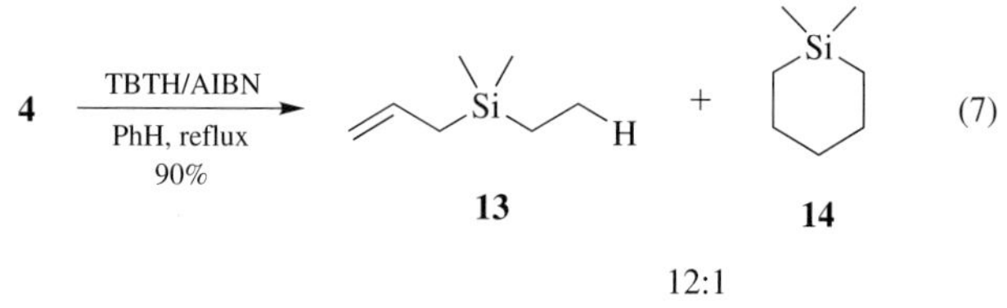

$$4 \xrightarrow[\substack{PhH, reflux \\ 90\%}]{TBTH/AIBN} \quad \mathbf{13} \quad + \quad \mathbf{14} \quad (7)$$

12:1

Synthesis of Precursors for Photochemical and Cationic Cyclizations. Allyl(2-chloroethyl)dimethylsilane (**4**) was converted to thiol **15** by nucleophilic displacement with thioacetic acid and treatment of the resultant thioacetate with ammonia. UV irradiation of **15** gave exclusively the 7-endo cyclization product **16** (eq 8).[4]

Amine **17** was prepared from **4** by reaction with aniline at 80 °C. Treatment of **17** with mercury(II) acetate followed by borohydride reduction gave the azasilacycle (**18**) in modest yield (eq 9).[5] This result is in accord with the β-carbocation stabilizing effect of a silicon atom. Synthon **4** is thus a useful building block for the preparation of silicon-containing heterocycles.

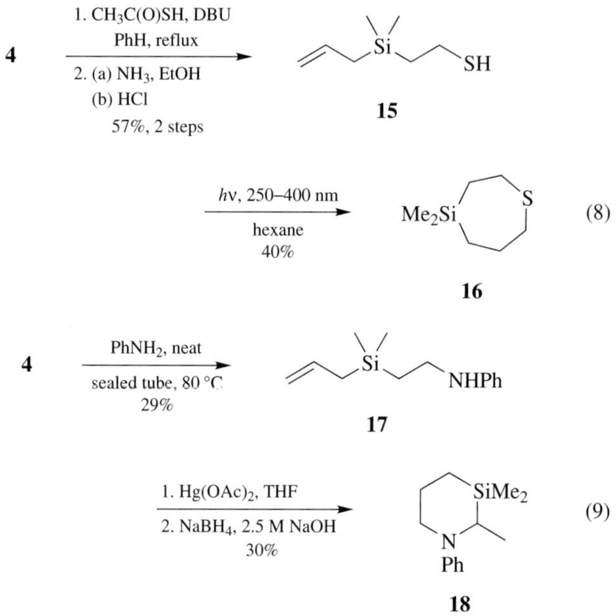

$$4 \xrightarrow[\substack{2. (a) NH_3, EtOH \\ (b) HCl \\ 57\%, 2 steps}]{\text{1. CH}_3\text{C(O)SH, DBU} \\ \text{PhH, reflux}}$$

15

$$\xrightarrow[\substack{\text{hexane} \\ 40\%}]{h\nu, 250-400 \text{ nm}} \quad (8)$$

16

$$4 \xrightarrow[\substack{\text{sealed tube, 80 °C} \\ 29\%}]{\text{PhNH}_2, \text{ neat}}$$

17

$$\xrightarrow[\substack{2. \text{NaBH}_4, 2.5 \text{ M NaOH} \\ 30\%}]{\text{1. Hg(OAc)}_2, \text{ THF}} \quad (9)$$

18

1. Saigo, K.; Tateishi, K.; Adachi, H.; Saotome, Y., *J. Org. Chem.* **1988**, *53*, 1572.
2. Klos, A. M.; Heintzelman, G. R.; Weinreb, S. M., *J. Org. Chem.* **1997**, *62*, 3758.
3. Wilt, J. W., *Tetrahedron* **1985**, *41*, 3979.
4. Kirpichenko, S. V.; Tolstikova, L. L.; Suslova, E. N.; Voronkov, M. G., *Tetrahedron Lett.* **1993**, *34*, 3889.
5. Kirpichenko, S. V.; Abrosimova, A. T.; Albanov, A. I.; Voronkov, M. G., *Russ. J. Gen. Chem.* **2001**, *71*, 1874.

Steven M. Weinreb & Magnus W. P. Bebbington
The Pennsylvania State University, University Park, PA, USA

Allyl Trichlorosilane

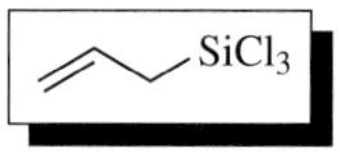

[107-37-9]   C$_3$H$_5$Cl$_3$Si   (MW 175.52)
InChI = 1S/C3H5Cl3Si/c1-2-3-7(4,5)6/h2H,1,3H2
InChIKey = HKFSBKQQYCMCKO-UHFFFAOYSA-N

(allylation of aldehydes[1-3])

Alternate Name: 2-propen-1-yl trichlorosilane.
Physical Data: bp 116 °C/750 Torr; n_D 1.4450; *d* 1.211.[4]
Solubility: soluble in MeCN, THF, CH$_2$Cl$_2$, toluene, benzene, etc.; not compatible with protic solvents, acetone, and other ketones. Dipolar aprotic solvents (DMF, DMSO, HMPA, etc.) coordinate Si, thereby increasing the reactivity of C$_3$H$_5$SiCl$_3$ with electrophiles.
Preparative Methods: allyl trichlorosilane is conveniently prepared from allyl chloride, Cl$_3$SiH, CuCl, and (*i*-Pr)$_2$EtN (or Et$_3$N) in ether at 20 °C,[1,5] or from allyl chloride and copper-silicon powder at 250 °C.[6] Other methods include the reaction of allyl chloride with SiCl$_4$ in the presence of CuCl and Et$_3$N[7]

or in the presence of Cp_2Ni and HMPA at $90\,^\circ C$,[8] the flash pyrolysis of allyl chloride with Si_2Cl_2 in PhCl at $500\,^\circ C$,[9] the reaction of allylMgBr with $SiCl_4$,[10] coupling of allene with Cl_3SiH, catalyzed by $(Ph_3P)_4Pd$ ($120\,^\circ C$, 5 h),[11] and the elimination of HCl from $Cl(CH_2)_3SiCl_3$ using, e.g., quinoline as base.[12] The analogous (E)-crotyl trichlorosilane can be synthesized from (E)-crotyl chloride, Cl_3SiH, CuCl, and (i-Pr)$_2$EtN in ether at $20\,^\circ C$,[1,5,13,14] or from (E)-crotyl chloride, $SiCl_4$, and Cp_2Ni in HMPA at $90\,^\circ C$.[8] The reaction of (E)-crotyl chloride with Cl_3SiH and Bu_4PCl at $150\,^\circ C$ has also been reported.[15] (Z)-crotyl trichlorosilane, on the other hand, was prepared via the 1,4-addition of Cl_3SiH to butadiene, catalyzed by $(Ph_4P)_4Pd$ (at $20\,^\circ C$ or $-78\,^\circ C$)[11,13c,16] or by $(PhCN)_2PdCl_2$;[17] this hydrosilylation can also be catalyzed by Ni[18] or proceed uncatalyzed at $500-600\,^\circ C$.[19] Analogous trifluoro and tribromo derivatives $C_3H_5SiX_3$ (X = F and Br) have also been described.[20,21]

Handling, Storage, and Precaution: corrosive, flammable, moisture sensitive, reacts with water and protic solvents. Keep tightly closed, store in a cool, dry, and dark place.

Reactions at Silicon with Heteroatom Nucleophiles. Alcohols, such as MeOH, EtOH, etc., and t-BuOOH replace chlorides at Si, producing the corresponding $allylSi(OR)_3$.[22] In analogy, NaSCN gives $allylSi(SCN)_3$,[23] and AcONa yields $allylSi(OAc)_3$,[24] as a result of triple nucleophilic displacement; the latter product can also be obtained on reaction of $allylSiCl_3$ with Ac_2O.[24] With water, formation of cyclic oligomers has been observed.[25] Treatment with SbF_3 is known to convert $allylSiCl_3$ into $allylSiF_3$.[20]

With diols, such as diisopropyl tartrate,[26] BINOL,[26] $PhCH(OH)CH_2OH$,[26] and pinacol,[27] the reaction of $allylSiCl_3$ can be controlled to generate the corresponding tetracoordinate Si species by replacement of two chlorides (eq 1).[26] With less bulky diols, such as diisopropyl tartrate, the reaction can be extended to produce pentacoordinate Si(V) species (eq 1);[26] $allylSiF_3$ reacts in a similar way.[26b] Bis-lithium catecholate affords the pentacoordinate species.[17] Vicinal amino alcohols and diamines, namely pseudoephedrine[27b,28] and trans-1,2-diaminocyclohexane,[27b,29] generate tetracoordinated species (eq 2).

tris-substitution (eq 3).[6,10a,22a,30] Aryllithiums react in a similar fashion, affording trisubstituted products,[22c] whereas monosubstitution can be attained with steroid-derived lithium acetylides.[31]

$$\text{allyl–}SiCl_3 \xrightarrow{\text{RMgX}} \text{allyl–}SiCl_2R \xrightarrow{\text{RMgX}}$$
$$\text{allyl–}SiClR_2 \xrightarrow{\text{RMgX}} \text{allyl–}SiR_3 \quad (3)$$

Reactions at the Double Bond. The double bond of $allylSiCl_3$ undergoes standard addition reactions, such as hydrohalogenation (with HCl, HBr, and HI) under Markovnikov control.[32] Chlorination (with Cl_2)[33] and hydrogenation (on "SiH/Ni" catalyst)[34] also proceed uneventfully. Addition of carbenes, generated from Cl_3SiCCl_3 and CHF_2Cl, respectively, results in the formation of the expected cyclopropane products.[35] Hydrosilylation with Cl_3SiH and $(ClCH_2)SiMe_2H$, respectively,[36] and the photochemical addition of Ph_2PH[37] have also been reported. Finally, $allylSiCl_3$ even undergoes Friedel-Crafts reaction with benzene in the presence of $AlCl_3$.[38]

Catalytic Allylation of Aldehydes. While $allylSiMe_3$ is known to react with aldehydes upon activation of the carbonyl group by a Lewis acid, the reaction of $allylSiCl_3$ requires activation by a Lewis base that coordinates to the silicon (eq 4).[1–3,5,17,39,40] The latter reaction is believed to proceed through a closed transition state, which controls diastereoselectivity in the case of (E)- and (Z)-crotylSiCl_3, and creates an opportunity for the transfer of chiral information if a chiral Lewis base is employed. Provided the Lewis base dissociates from the silicon with sufficient rate (eq 4), it can act as a catalyst (rather than a stoichiometric reagent). Typical Lewis bases that promote the allylation reaction are the common dipolar aprotic solvents, such as DMF,[1,40] DMSO,[1,2] and HMPA,[2,41] and other substances possessing a strongly Lewis basic oxygen, such as formamides[42] (also on solid support[1,5,43]), urea derivatives,[44] catecholates,[17] and their chiral modifications;[45] the allylation with $allylSiF_3$ is also promoted by fluoride.[17,39b,d] Unlike aldehydes, ketones are normally inert under catalytic conditions.

$$(1)$$

$$(2)$$

Reactions at Silicon with Carbon Nucleophiles. The reaction of $allylSiCl_3$ with Grignard reagents RMgX (R = Me, alkyl, Ar; X = Br, Cl) can be controlled to afford products of mono-, bis-, and

$$(4)$$

Chiral Lewis-basic catalysts, in particular phosphoramides (**1** and **2**),[2,13c,16,46,47] pyridine *N,N*-bisoxides (**3** and **4**),[48,49] and pyridine *N*-monoxides (**5**, **6** and **7,8**),[3,50] exhibit very high enantioselectivities for the allylation of aromatic, heteroaromatic, and cinnamyl-type aldehydes with allyl, (*E*)- and (*Z*)-crotyl, and prenyl trichlorosilanes. Chiral formamides,[42] pyridine-oxazolines,[51] urea derivatives,[44] and sulfoxides[52] are generally less enantioselective and effective only in stoichiometric quantities. The reaction is much less efficient with aliphatic aldehydes, which require stoichiometric conditions (vide infra). However, α,β-unsaturated aldehydes do react readily and give 1,2-addition products.[1,3,40,50]

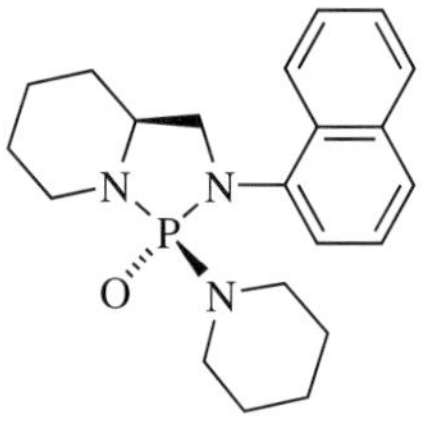

1
($\leq 85\%$ ee)[13b,47]

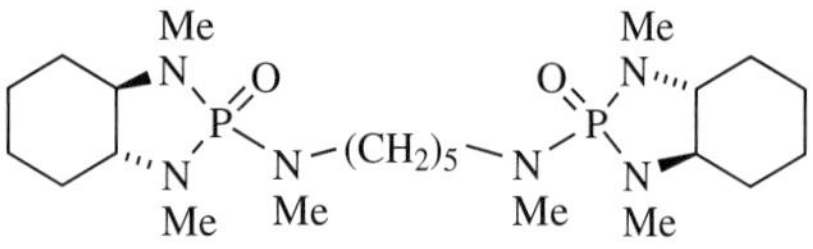

2
($\leq 87\%$ ee)[46]

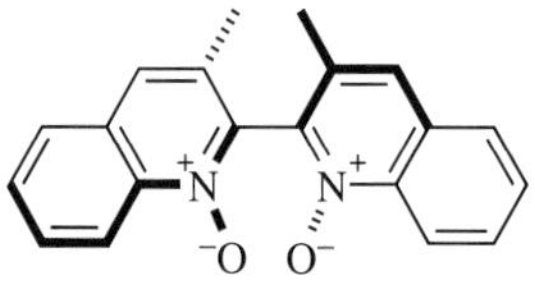

3
(88% ee)[48]

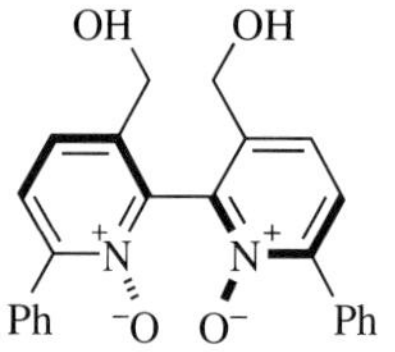

4
(88% ee)[49]

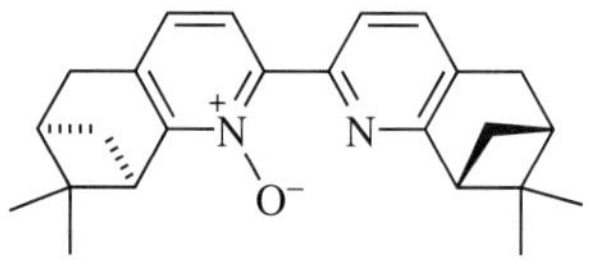

5
($\leq 90\%$ ee)[3,50a]

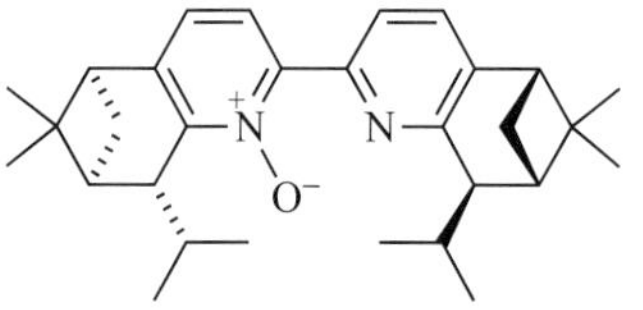

6
($\leq 97\%$ ee)[3]

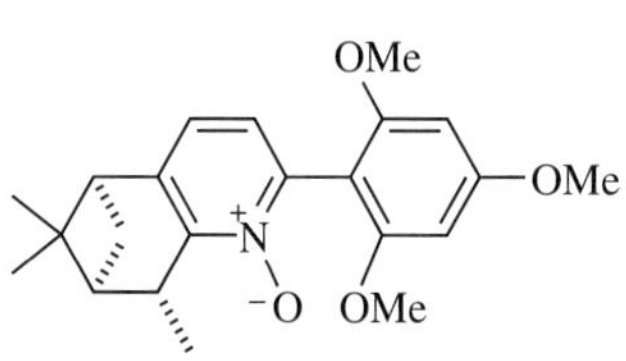

7
($\leq 96\%$ ee)[50d]

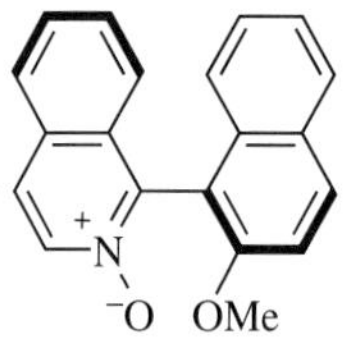

8
($\leq 96\%$ ee)[50c]

Stoichiometric Allylation of Aldehydes and Ketones. As discussed above, catalytic allylation is generally confined to highly reactive aldehydes (aromatic, heteroaromatic, and cinnamyl-type). However, if a stoichiometric activator is employed, e.g., DMF or HMPA as additive or solvent, the substrate range broadens to aliphatic aldehydes and α-keto acids $RCOCO_2H$ (R = aryl, alkyl).[41,42] In the case of chiral aldehydes, such as *N*-protected alaninal, diastereoselectivity attained in DMF is modest ($\sim$3:1).[53] AllylSiF$_3$ can transfer the allyl group to α-hydroxy ketones and β-diketones on heating with Et$_3$N in THF.[54]

Chiral chelates, derived from allylSiCl$_3$ and pseudoephedrine (eq 2) or *trans*-1,2-diaminocyclohexane, react with aromatic and aliphatic aldehydes to produce the corresponding homoallylic alcohols with 78–96% ee (eq 5).[27,29]

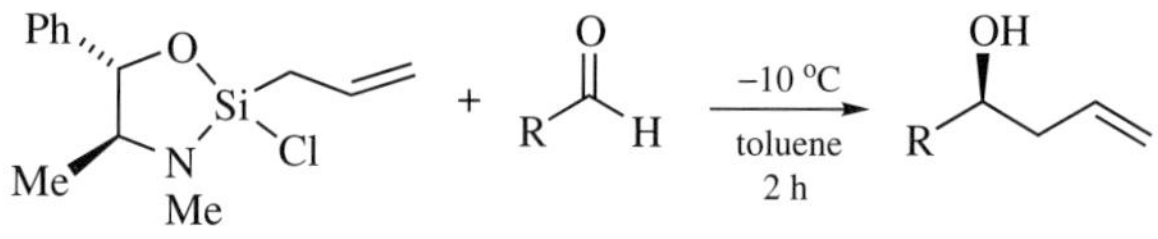

Allylation of Imines. *N*-Benzoyl hydrazones, derived from aromatic and aliphatic aldehydes, undergo allylation with allylSiCl$_3$ in DMF at 20 °C[55] or in CH$_2$Cl$_2$ with excess of DMSO as activator at −78 °C (eq 6).[56] The hydrazones derived from arylalkyl ketones have also been reported to react.[55] *N*-Aryl aldimines are allylated in DMF at 0 °C.[57]

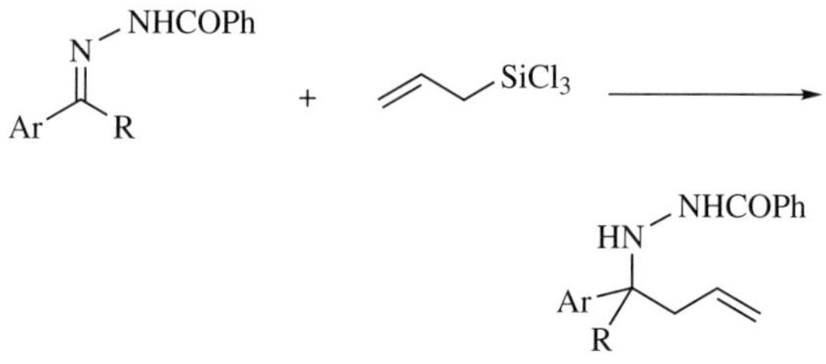

Stoichiometric asymmetric allylation of *N*-benzoyl hydrazones derived from aromatic and heteroaromatic aldehydes can be attained with pseudoephedrine chelates derived from allyl and (*E*)- and (*Z*)-crotyl trichlorosilanes with excellent diastereo- and enantioselectivity (eq 7); note the opposite relative configuration at the vicinal centers of the product compared to that obtained in the aldehyde allylation (eq 4), resulting from different orientation in the transition state.[28]

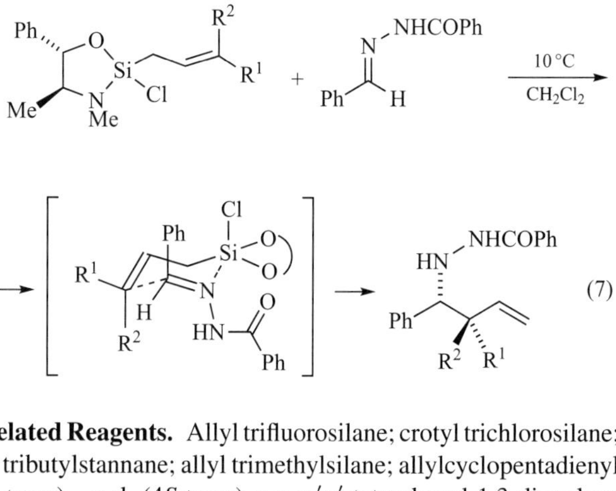

Related Reagents. Allyl trifluorosilane; crotyl trichlorosilane; allyl tributylstannane; allyl trimethylsilane; allylcyclopentadienyl [(4*R,trans*)- and (4*S,trans*)-$\alpha,\alpha,\alpha',\alpha'$-tetraphenyl-1,3-dioxolane -4,5-dimethanolato-*O,O'*]titanium; *B*-allyldiisocaranylborane; diisopropyl 2-allyl-1,3,2-dioxaborolane-4,5-dicarboxylate; diisopropyl 2-crotyl-1,3,2dioxaborolane-4,5-dicarboxylate; *B*-allyldiisopinocampheylborane; *B*-crotyldiisopinocampheylborane.

1. (a) Kobayashi, S.; Nishio, K., *Tetrahedron Lett.* **1993**, *34*, 3453. (b) Kobayashi, S.; Nishio, K., *J. Org. Chem.* **1994**, *59*, 6620. (c) Kobayashi, S.; Nishio, K., *Synthesis* **1994**, 457.

2. Denmark, S. E.; Coe, D. M.; Pratt, N. E.; Griedel, B. D., *J. Org. Chem.* **1994**, *59*, 6161.

3. Malkov, A. V.; Bell, M.; Orsini, M.; Pernazza, D.; Massa, A.; Herrmann, P.; Meghani, P.; Kočovský, P., *J. Org. Chem.* **2003**, *68*, 9659.

4. Aldrich Catalogue.

5. (a) Kobayashi, S.; Nishio, K., *Chem. Lett.* **1994**, 1773. (b) Nakajima, M.; Saito, M.; Hashimoto, S., *Chem. Pharm. Bull.* **2000**, *48*, 306.

6. (a) Hurd, D. T., *J. Am. Chem. Soc.* **1945**, *67*, 1813. (b) Hurd, D. T., *Inorg. Synth.* **1950**, *3*, 59. (c) U.S. Pat. 2420912 (to Gen. Electric Co.) (1945).

7. Hagen, G.; Mayr, H., *J. Am. Chem. Soc.* **1991**, *113*, 4954.

8. (a) Calas, R.; Dunogues, J.; Deleris, G.; Duffaut, N., *J. Organomet. Chem.* **1982**, *225*, 117. (b) Lefort, M.; Simmonet, C.; Birot, M.; Deleris, G.; Dunogues, J.; Calas, R., *Tetrahedron Lett.* **1980**, *21*, 1857.

9. (a) Chernyshev, E. A.; Komalenkova, N. G.; Yakovleva, G. N.; Bykovchenko, V. G.; Khromykh, N. N.; Bochkarev, V. N., *Russ. J. Gen. Chem. (Engl.Transl.)* **1997**, *67*, 894. (b) *Zh. Obshch. Khim.* **1997**, *67*, 955.

10. (a) Scott, R. E. Frisch, K. C., *J. Am. Chem. Soc.* **1951**, *73*, 2599. (b) Henry, C.; Brook, M. A., *Tetrahedron* **1994**, *50*, 11379.

11. Tsuji, J.; Hara, M.; Ohno, K., *Tetrahedron* **1974**, *30*, 2143.

12. (a) Mironov, V. F.; Nepomnina, V. V., *Bull. Acad. Sci. USSR, Div. Chem. Sci. (Engl. Transl.)* **1960**, 1983. (b) *Izv. Akad. Nauk SSSR, Ser. Khim.* **1960**, 2140.

13. (a) Aoki, S.; Mikami, K.; Terada, M.; Nakai, T., *Tetrahedron* **1993**, *49*, 1783. (b) D'Aniello, F.; Falorni, M.; Mann, A.; Taddei, M., *Tetrahedron: Asymmetry* **1996**, *7*, 1217. (c) Iseki, K.; Kuroki, Y.; Takahashi, M.; Kishimoto, S.; Kobayashi, Y., *Tetrahedron* **1997**, *53*, 3513. (d) Shibato, A.; Itagaki, Y.; Tayama, E.; Hokke, Y.; Asao, N.; Maruoka, K., *Tetrahedron* **2000**, *56*, 5373.

14. Kira, M.; Hino, T.; Sakurai, H., *Tetrahedron Lett.* **1989**, *30*, 1099.

15. Cho, Y.-S.; Kang, S.-H.; Han, J. S.; Yoo, B. R.; Jung, I. N., *J. Am. Chem. Soc.* **2001**, *123*, 5584.

16. (a) Wadamoto, M.; Ozasa, N.; Yanagisawa, A.; Yamamoto, H., *J. Org. Chem.* **2003**, *68*, 5593. (b) Iseki, K.; Kuroki, Y.; Takahashi, M.; Kobayashi, Y., *Tetrahedron Lett.* **1996**, *37*, 5149.

17. Kira, M.; Sato, K.; Sakurai, H., *J. Am. Chem. Soc.* **1988**, *110*, 4599.

18. Čapka, M.; Hetflejš, J., *Collect. Czech. Chem. Commun.* **1975**, *40*, 2073 and 3020.

19. Heinicke, J.; Kirst, I.; Gehrhus, B.; Tzschach, A., *Z. Chem.* **1988**, *28*, 261.

20. (a) Mironov, V. F., *Bull. Acad. Sci. USSR, Div. Chem. Sci. (Engl.Transl.)* **1962**, 1797. (b) *Izv. Akad. Nauk SSSR, Ser. Khim.* **1962**, 1884.

21. (a) Chernyshev, E. A., *Bull. Acad. Sci. USSR, Div. Chem. Sci. (Engl.Transl.)* **1961**, 2032. (b) *Izv. Akad. Nauk SSSR, Ser. Khim.* **1961**, 2173.

22. (a) Burkhard, C. A., *J. Am. Chem. Soc.* **1950**, *72*, 1078. (b) Pola, J.; Jakoubková, Chvalovský, V., *Collect. Czech. Chem. Commun.* **1978**, *43*, 3391. (c) Horvath, R. F.; Chan, T. H., *J. Org. Chem.* **1989**, *54*, 317. (d) Mathias, L. J.; Carothers, T. W., *J. Am. Chem. Soc.* **1991**, *113*, 4043. (e) Losev, V. B.; Fridmann, G. E., *J. Gen. Chem. USSR (Engl.Transl.)* **1963**, *33*, 891. (f) *Zh. Obshch. Khim.* **1963**, *33*, 905. (g) Fan, Y. L.; Shaw, R. G., *J. Org. Chem.* **1973**, *38*, 2410.

23. Glowacki, G. R.; Post, H. W., *J. Org. Chem.* **1962**, *27*, 634.

24. (a) U.S. Pat. 2537073 (to Montclair Research Corp. and Ellis-Foster Co.) (1946). (b) Frisch, K. C.; Goodwin, P. A.; Scott, R. E., *J. Am. Chem. Soc.* **1952**, *74*, 4584.

25. (a) Podberezskaya, N. V.; Baidina, I. A.; Alekseev, V. I.; Borisov, S. V.; Martynova, T. N., *J. Struct. Chem. (Engl.Transl.)* **1981**, 737. (b) *Zh. Strukt. Khim.* **1981**, 116. (c) Martynova, T. N.; Korchkov, V. P.; Semyannikov, P. P., *J. Organomet. Chem.* **1983**, *258*, 277. (d) Hendan, B. J.; Marsmann, H. C., *J. Organomet. Chem.* **1994**, *483*, 33. (e) Lücke, S.; Stoppek-Langner, K.; Kuchinke, J.; Krebs, B., *J. Organomet. Chem.* **1999**, *584*, 11.

26. (a) Wang, Z.; Wang, D.; Sui, X., *J. Chem. Soc., Chem. Commun.* **1996**, 2261. (b) Zhang, L. C.; Sakurai, H.; Kira, M., *Chem. Lett.* **1997**, 129. (c) Wang, D.; Wang, Z. G.; Wang, M. W.; Chen, Y. J.; Liu, L.; Zhu, Y., *Tetrahedron: Asymmetry* **1999**, *10*, 327.

27. (a) Wang, X.; Meng, Q.; Nation, A. J.; Leighton, J. L., *J. Am. Chem. Soc.* **2002**, *124*, 10672. (b) Kinnaird, J. W. A.; Ng, P. Y.; Kubota, K.; Wang, X.; Leighton, J. L., *J. Am. Chem. Soc* **2002**, *124*, 7920.

28. Berger, R.; Rabbat, P. M. A.; Leighton, J. L., *J. Am. Chem. Soc.* **2003**, *125*, 9596.

29. Kubota, K.; Leighton, J. L., *Angew. Chem., Int. Ed.* **2003**, *42*, 946.

30. (a) *Dokl. Akad. Nauk SSSR* **1962**, *142*, 1316. (b) Ahmad, I.; Falck-Pedersen, M. L.; Undheim, K., *J. Organomet. Chem.* **2001**, *625*, 160.

31. Peters, R. H.; Crowe, D. F.; Tanabe, M.; Avery, M. A.; Chong, W. K. M., *J. Med. Chem.* **1987**, *30*, 646.

32. (a) Mironov, V. F., *Izv. Akad. Nauk SSSR, Ser. Khim.* **1959**, 1862. (b) Sheludyakov, V. D.; Zhun', V. I.; Loginov, S. V., *J. Gen. Chem.USSR (Engl.Transl.)* **1985**, *55*, 1202. (c) *Zh. Obshch. Khim.* **1985**, *55*, 1345.

33. Leasure, J. K.; Speier, J. L., *J. Med. Chem.* **1966**, *9*, 949.

34. Picard, J.-P.; Dunogues, J.; Elyusufi, A.; Lefort, M., *Synth. Commun.* **1984**, *14*, 95.

35. (a) Müller, R.; Müller, W., *Chem. Ber.* **1965**, *98*, 2916. (b) Sheludyakov, V. D.; Zhun', V. I.; Loginov, S. V.; Shcherbinin, V. V.; Turkel'taub, G. N., *J. Gen. Chem. USSR (Engl.Transl.)* **1984**, *54*, 1883. (c) *Zh. Obshch. Khim.* **1984**, *54*, 2108.

36. (a) Krieble, R. H.; Burkhard, C. A., *J. Am. Chem. Soc.* **1947**, *69*, 2689. (b) Zhun', V. I.; Tsvetkov, A. L.; Bochkarev, V. N.; Slyusarenko, T. F.; Turkel'taub, G. N.; Sheludyakov, V. D., *J. Gen. Chem. USSR (Engl.Transl.)* **1989**, *59*, 344. (c) *Zh. Obshch. Khim.* **1989**, *59*, 390. (d) Babich, E. D.; Karel'skii, V. N.; Vdovin, V. M.; Nametkin, N. S., *Bull. Acad. Sci. USSR, Div. Chem. Sci. (Engl.Transl.)* **1968**, *6*, 1312. (e) *Izv. Akad. Nauk SSSR, Ser. Khim.* **1968**, 1389.

37. Holmes-Smith, R. D.; Osei, R. D.; Stobart, S. R., *J. Chem. Soc. Perkin Trans.1.eps* **1983**, 861.

38. Tamao, K.; Yoshida, J.; Akita, M.; Sugihara, Y.; Iwahara, T.; Kumada, M., *Bull. Chem. Soc. Jpn.* **1982**, *55*, 255.

39. (a) Kira, M.; Kobayashi, M.; Sakurai, H., *Tetrahedron Lett.* **1987**, *28*, 4081. (b) Sakurai, H., *Synlett* **1989**, 1. (c) Kira, M.; Sato, K.; Sakurai, H., *J. Am. Chem. Soc.* **1990**, *112*, 257. (d) Kira, M.; Zhang, L. C.; Kabuto, C.; Sakurai, H., *Organometallics* **1998**, *17*, 887.

40. Short, J. D.; Attenoux, S.; Berrisford, D. J., *Tetrahedron Lett.* **1997**, *38*, 2351.

41. Wang, Z.; Xu, G.; Wang, D.; Pierce, M. E.; Confalone, P. N., *Tetrahedron Lett.* **2000**, *41*, 4523.

42. (a) Iseki, K.; Mizuno, S.; Kuroki, Y.; Kobayashi, Y., *Tetrahedron Lett.* **1998**, *39*, 2767. (b) Iseki, K.; Mizuno, S.; Kuroki, Y.; Kobayashi, Y., *Tetrahedron* **1999**, *55*, 977.

43. Ogawa, C.; Sugiura, M.; Kobayashi, S., *Chem. Commun.* **2003**, 192.

44. Chataigner, I.; Piarulli, U.; Gennari, C., *Tetrahedron Lett.* **1999**, *40*, 3633.

45. (a) Denmark, S. E.; Fu, J., *Chem. Rev.* **2003**, *103*, 2763. (b) Malkov, A. V.; Kočovský, P., *Curr. Org. Chem.* **2003**, *7*, 1737. (c) Chelucci, G.; Murineddu, G.; Pinna, G. A., *Tetrahedron: Asymmetry* **2004**, *15*, 1373.

46. (a) Denmark, S. E.; Fu, J., *J. Am. Chem. Soc.* **2000**, *122*, 12021. (b) Denmark, S. E.; Fu, J., *J. Am. Chem. Soc.* **2001**, *123*, 9488.

47. Hellwig, J.; Belser, T.; Müller, J. F. K., *Tetrahedron Lett.* **2001**, *42*, 5417.

48. Nakajima, M.; Saito, M.; Shiro, M.; Hashimoto, S., *J. Am. Chem. Soc.* **1998**, *120*, 6419.

49. (a) Shimada, T.; Kina, A.; Ikeda, S.; Hayashi, T., *Org. Lett.* **2002**, *4*, 2799. (b) Shimada, T.; Kina, A.; Hayashi, T., *J. Org. Chem.* **2003**, *68*, 6329.

50. (a) Malkov, A. V.; Orsini, M.; Pernazza, D.; Muir, K. W.; Langer, V.; Meghani, P.; Kočovský, P., *Org. Lett.* **2002**, *4*, 1047. (b) Malkov, A. V.; Bell, M.; Vassieu, M.; Bugatti, V.; Kočovský, P., *J. Mol. Catal. A* **2003**, *196*, 179. (c) Malkov, A. V.; Dufková, L.; Farrugia, L.; Kočovský, P., *Angew. Chem., Int. Ed.* **2003**, *42*, 3674. (d) Malkov, A. V.; Bell, M.; Castelluzzo, F.; Kočovský, P., *Org. Lett.* **2005**, *7*, 3219.

51. Angell, R. M.; Barrett, A. G. M.; Braddock, D. C.; Swallow, S.; Vickery, B. D., *Chem. Commun.* **1997**, 919.

52. (a) Massa, A.; Malkov, A. V.; Kočovský, P.; Scettri, A., *Tetrahedron Lett.* **2003**, *44*, 7179. (b) Rowlands, G. J.; Barnes, W. K., *Chem. Commun.* **2003**, 2712.

53. (a) Gryko, D.; Urbanczyk-Lipkowska, Z.; Jurczak, J., *Tetrahedron* **1997**, *53*, 13373. (b) Gryko, D.; Jurczak, J., *Helv. Chim. Acta* **2000**, *83*. 2705.

54. (a) Sato, K.; Kira, M.; Sakurai, H., *J. Am. Chem. Soc.* **1989**, *111*, 6429. (b) Kira, M.; Sato, K.; Sekimoto, K.; Gewald, R.; Sakurai, H., *Chem. Lett.* **1995**, 281. (c) Gewald, R.; Kira, M.; Sakurai, H., *Synthesis* **1996** 111. (d) Chemler, S. R.; Roush, W. R., *Tetrahedron Lett.* **1999**, *40*, 4643. (e) Chemler, S. R.; Roush, W. R., *J. Org. Chem.* **2003**, *68*, 1319.

55. (a) Kobayashi, S.; Hirabayashi, R., *J. Am. Chem. Soc.* **1999**, *121*, 6942. (b) Hirabayashi, R.; Ogawa, C.; Sugiura, M.; Kobayashi, S., *J. Am. Chem. Soc.* **2001**, *123*, 9493.

56. Kobayashi, S.; Ogawa, C.; Konishi, H.; Sugiura, M., *J. Am. Chem. Soc.* **2003**, *125*, 6610.

57. Sugiura, M.; Robvieux, F.; Kobayashi, S., *Synlett* **2003**, 1749.

Pavel Kočovský
University of Glasgow, Glasgow, UK

Allyltrimethylsilane[1]

[762-72-1]   C$_6$H$_{14}$Si   (MW 114.26)

InChI = 1S/C6H14Si/c1-5-6-7(2,3)4/h5H,1,6H2,2-4H3

InChIKey = HYWCXWRMUZYRPH-UHFFFAOYSA-N

(carbon nucleophile for the introduction of allyl groups by Lewis acid-catalyzed or fluoride ion-catalyzed reaction with acid chlorides, aldehydes, ketones, iminium ions, enones, and similar carbon electrophiles)

Physical Data: bp 85–86 °C, *d* 0.717 g cm^{-3}.

Solubility: freely sol all organic solvents.

Form Supplied in: colorless liquid. Methods for the synthesis of allylsilanes in general have been reviewed.[3]

Analysis of Reagent Purity: δ 5.74 (1H, ddt, *J* 16.9, 10.2 and 8), 4.81 (1H, dd, *J* 16.9 and 2), 4.79 (1H, dd, *J* 10.2 and 2), 1.49 (2H, d, *J* 8) and −0.003 (9H, s).[2]

Handling, Storage, and Precautions: inflammable; the vapor is irritating to the skin, eyes, and mucous membranes.

Original Commentary

Ian Fleming
Cambridge University, Cambridge, UK

As a Carbon Nucleophile in Lewis Acid-Catalyzed Reactions. Allyltrimethylsilane is an alkene some 10^5 times more nucleophilic than propene, as judged by its reactions with diarylmethyl cations.[4] It reacts with a variety of cationic carbon electrophiles, usually prepared by coordination of a Lewis acid to a functional group, but also by chemical or electrochemical oxidation,[5] or by irradiation in the presence of 9,10-dicyanoanthracene.[6] The electrophile attacks the terminal alkenic

carbon to give an intermediate cation, and the silyl group is lost to create a double bond at the other terminus. Among the more straightforward electrophiles are acid chlorides (eq 1),[7] aldehydes and ketones (eq 2),[8] their acetals (eq 3),[9] and the related alkoxyalkyl halides (eq 4),[10] iminium ions, and acyliminium ions (eqs 5 and 6),[11,12] and tertiary and allylic or benzylic alkyl halides (eqs 7 and 8).[13,14]

The reaction with acetals does not always need the acetal itself to be synthesized: it can be made in the same flask by mixing the allylsilane, the aldehyde, the silyl ether of the alcohol, and a catalytic amount of an acid such as diphenylboryl trifluoromethanesulfonate,[15] trimethylsilyl trifluoromethanesulfonate,[16] or fluorosulfuric acid.[17]

The further reaction of the double bond in the first-formed product is an occasional complication, as in the formation of tetrahydropyrans from aldehydes when aluminum chloride is the Lewis acid (eq 9),[18] and of piperidines from primary amines and formaldehyde (eq 10).[11]

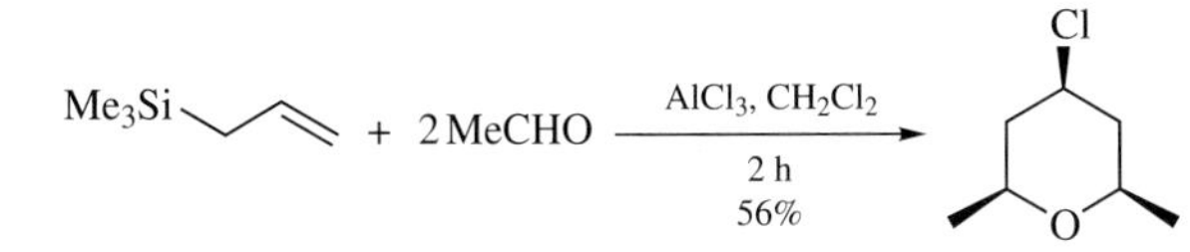

Me$_3$Si + 2 MeCHO $\xrightarrow[\substack{2\text{ h}\\56\%}]{\text{AlCl}_3,\ \text{CH}_2\text{Cl}_2}$ (9)

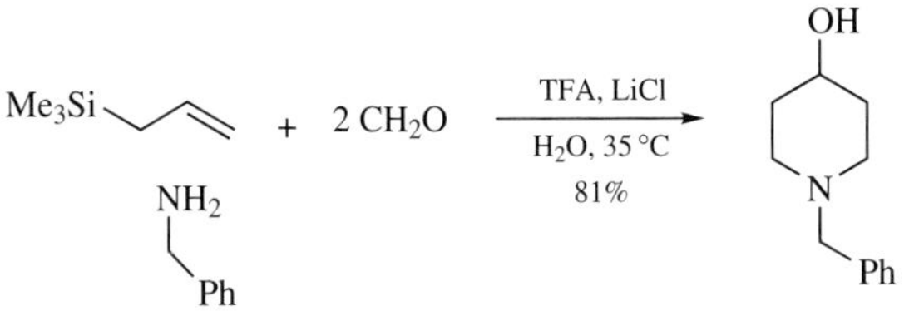

Me$_3$Si + 2 CH$_2$O $\xrightarrow[\substack{\text{H}_2\text{O, 35 °C}\\81\%}]{\text{TFA, LiCl}}$ (10)

α,β-Unsaturated esters, aldehydes, ketones, and nitriles generally react in Michael fashion in what is called a Sakurai reaction (eq 11).[19] The intermediate silyl enol ether may be treated with a second electrophile to set up two C–C bonds in one operation (eq 12).[20] Occasionally the silyl group is not lost from the intermediate cation but migrates instead to give a cyclopentannulation byproduct (eq 13).[21]

α,β-Unsaturated nitro compounds initially give nitronic acid derivatives, which can be reduced directly with titanium(III) chloride to give the corresponding ketone (eq 14).[22]

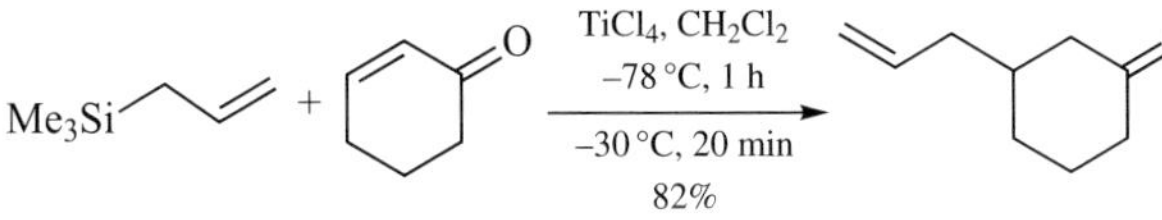

Me$_3$Si + $\xrightarrow[\substack{-30\text{ °C, 20 min}\\82\%}]{\text{TiCl}_4,\ \text{CH}_2\text{Cl}_2\\-78\text{ °C, 1 h}}$ (11)

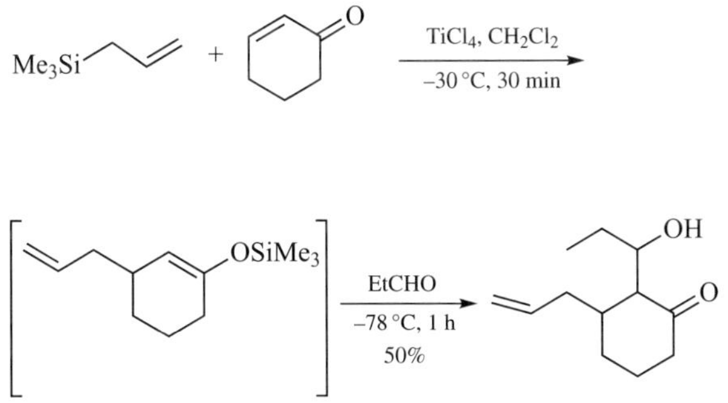

Me$_3$Si + $\xrightarrow[-30\text{ °C, 30 min}]{\text{TiCl}_4,\ \text{CH}_2\text{Cl}_2}$

$\xrightarrow[\substack{-78\text{ °C, 1 h}\\50\%}]{\text{EtCHO}}$ (12)

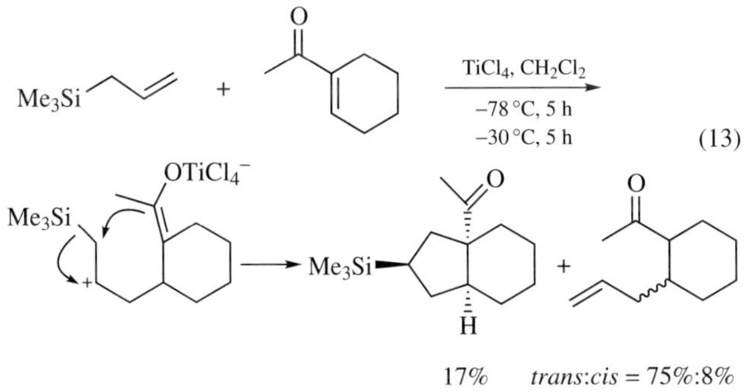

Me$_3$Si + $\xrightarrow[\substack{-78\text{ °C, 5 h}\\-30\text{ °C, 5 h}}]{\text{TiCl}_4,\ \text{CH}_2\text{Cl}_2}$ (13)

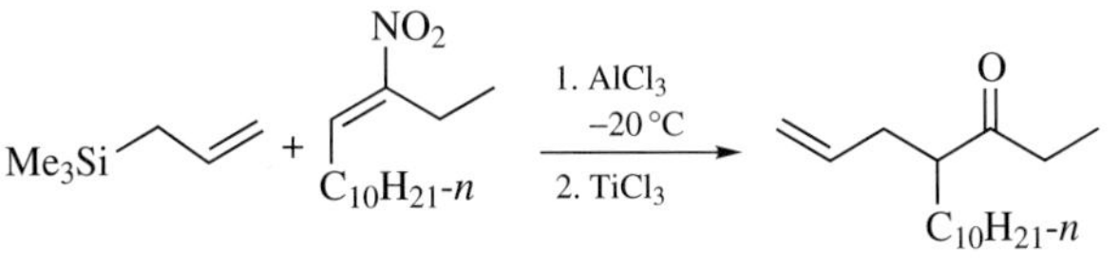

17%   trans:cis = 75%:8%

Me$_3$Si + $\underset{\text{C}_{10}\text{H}_{21}\text{-}n}{\overset{\text{NO}_2}{}}$ $\xrightarrow[\substack{-20\text{ °C}\\2.\ \text{TiCl}_3}]{1.\ \text{AlCl}_3}$ (14)

Some electrophiles require a separate activation step, as in the stereospecific reaction of alkenes with benzenesulfenyl chloride (eq 15).[23] The intermediate that actually reacts with the allylsilane is presumably the episulfonium ion. There is a corresponding reaction of epioxonium ions derived from 2-bromoethyl ethers.[24] Intramolecular hydrosilylation of an ester generates an acetal, which reacts with allyltrimethylsilane in the usual way and with high stereocontrol to make an anti 1,3-diol derivative (eq 16).[25]

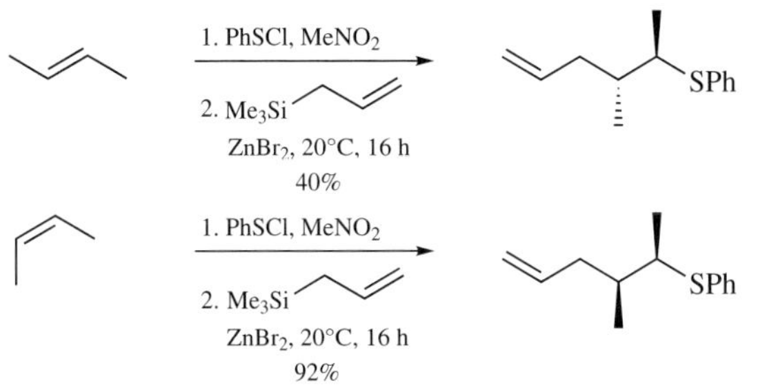

$\xrightarrow[\substack{2.\ \text{Me}_3\text{Si}\\\text{ZnBr}_2,\ 20°\text{C, 16 h}\\40\%}]{1.\ \text{PhSCl, MeNO}_2}$

$\xrightarrow[\substack{2.\ \text{Me}_3\text{Si}\\\text{ZnBr}_2,\ 20°\text{C, 16 h}\\92\%}]{1.\ \text{PhSCl, MeNO}_2}$ (15)

$\xrightarrow[\substack{\text{CH}_2\text{Cl}_2\\-30\text{ °C, 50 min}\\78\%}]{\text{Me}_3\text{Si}\\\text{TfOH}_2^+\text{–B(OTf)}_4}$ (16)

anti:syn = 20:1

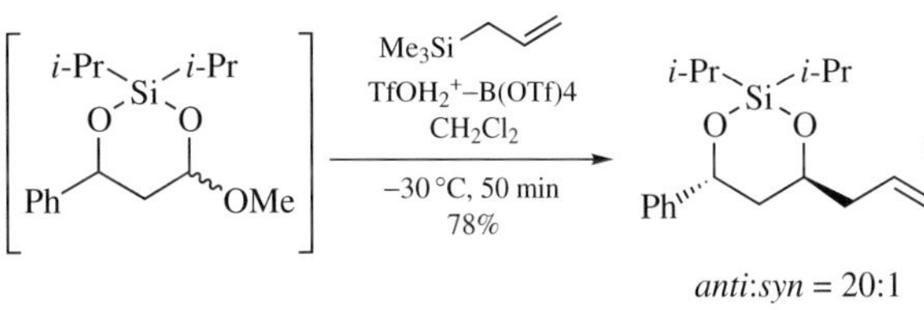

The reactions with α- or β-oxygenated aldehydes can be made to give high levels of stereocontrol in either sense by choosing a chelating or nonchelating Lewis acid (eq 17).[26]

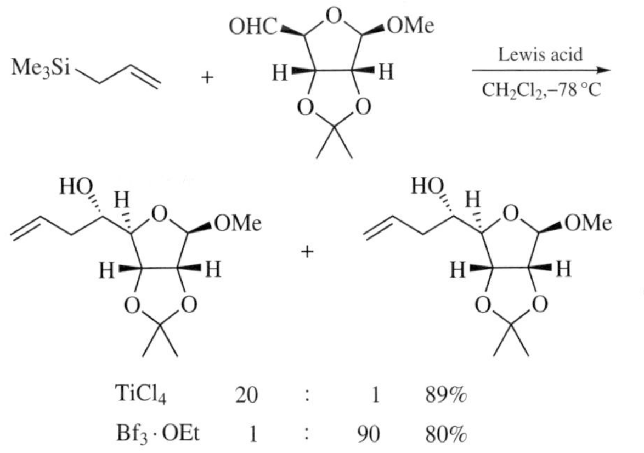

Me$_3$Si + $\xrightarrow[\text{CH}_2\text{Cl}_2,\ -78\text{ °C}]{\text{Lewis acid}}$ (17)

TiCl$_4$	20	:	1	89%
Bf$_3$·OEt	1	:	90	80%

Sugar acetals (eq 18)[27] and glycal 3-acetates (eq 19)[28] and related compounds react stereoselectively in favor of axial attack at the anomeric carbon.

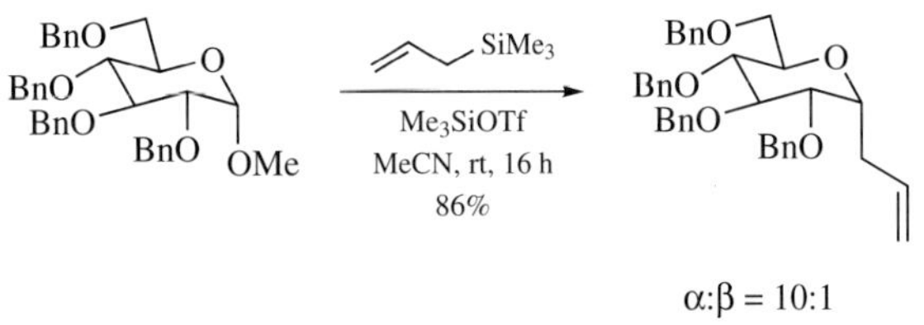

$\xrightarrow[\substack{\text{Me}_3\text{SiOTf}\\\text{MeCN, rt, 16 h}\\86\%}]{\text{SiMe}_3}$ (18)

α:β = 10:1

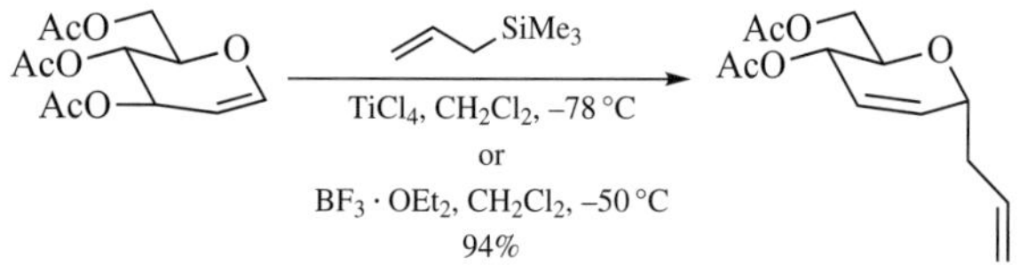

$$\text{(19)}$$

$$\alpha{:}\beta = 16{:}1$$

The choice of Lewis acid in all these reactions is often important in getting the best results, one Lewis acid being much the best in several cases. Most are used in molar amounts, but some, especially iodotrimethylsilane, trimethylsilyl trifluoromethanesulfonate, and the more powerful trifluoromethanesulfoxonium tetrakis(trifluoromethanesulfonyl)boronate.[29] [TfOH$_2{}^+$B(OTf)$_4{}^-$], have the advantage that they can be used in catalytic quantities. In addition to the Lewis acids illustrated, the following less obvious Lewis acids have also been used with one or more of the electrophiles: triphenylmethyllithium,[30] titanocene ditriflate,[31] lithium perchlorate,[32] tin(IV) chloride in the presence of either tin(II) trifluoromethanesulfonate[33] or zinc chloride,[34] antimony(V) chloride,[35] chlorotrimethylsilane combined with indium chloride,[36] ethylaluminum dichloride,[37] and diphenylboryl trifluoromethanesulfonate[38] Among carbon electrophiles not illustrated are quinones,[39] cyclopropanedicarboxylic esters,[40] oxetanes,[41] nitriles,[42] dithioacetals,[43] diselenoacetals,[44] the intermediate sulfur-stabilized cation from a Pummerer rearrangement,[45] and propargyl ethers with[46] or without[47] octacarbonyldicobalt complexation.

As a Carbon Nucleophile in Uncatalyzed Reactions. Some electrophiles do not need Lewis acids, being already cationic and electrophilic enough to react with allyltrimethylsilane. Examples are the dithianyl cation (eq 20),[48] the tricarbonyl(cyclohexadienyl) iron cation (eq 21),[49] (π-allyl) tetracarbonyliron cations,[50] and chlorosulfonyl isocyanate (CSI).[51] Other reagents react directly by cycloaddition, but need further steps to achieve an overall electrophilic substitution, as in the reactions with nitrones (eq 22).[52]

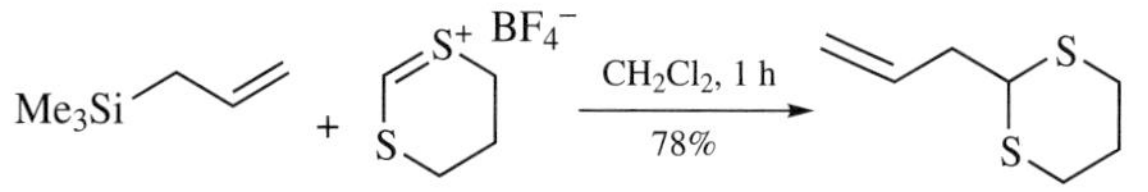

$$\text{(20)}$$

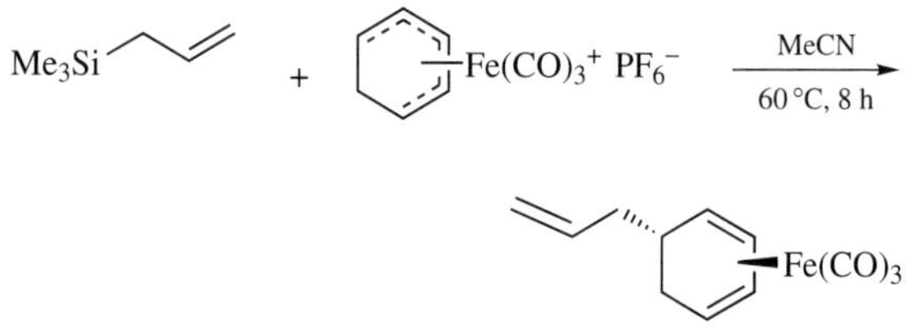

$$\text{(21)}$$

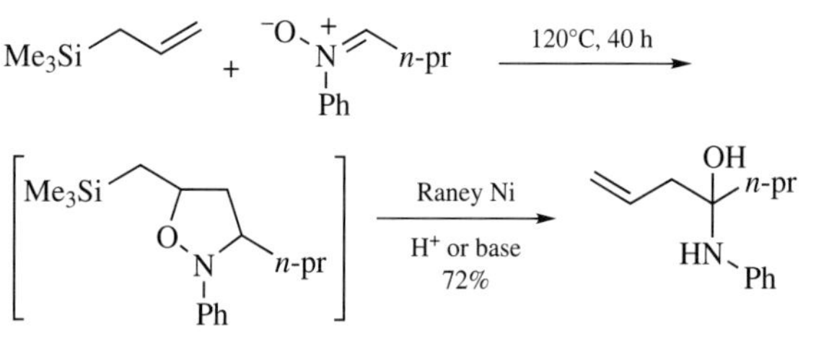

$$\text{(22)}$$

Although heteroatom (N-, P-, O-, S-, Se-, and halogen-based) electrophiles react with allylsilanes, the products with

allyltrimethylsilane itself are usually too simple for this to be an important synthetic method. An exception perhaps is the reaction with palladium(II) chloride, which gives the π-allylpalladium chloride cation.[53]

As a Carbon Nucleophile in Fluoride Ion-Catalyzed Reactions. The reactions with aldehydes, ketones (eq 23),[54] and α,β-unsaturated esters (eq 24)[55] can also be catalyzed by fluoride ion, usually introduced as tetra-n-butylammonium fluoride (TBAF), or other silicophilic ions such as alkoxide. These reactions produce silyl ether intermediates, which are usually hydrolyzed before workup. The stereochemistry of attack on chiral ketones can sometimes be different for the Lewis acid- and fluoride ion-catalyzed reactions.[56] In addition some electrophiles only react in the fluoride-catalyzed reactions, as with the addition to trinitrobenzene giving an allyl Meisenheimer 'complex'.[57]

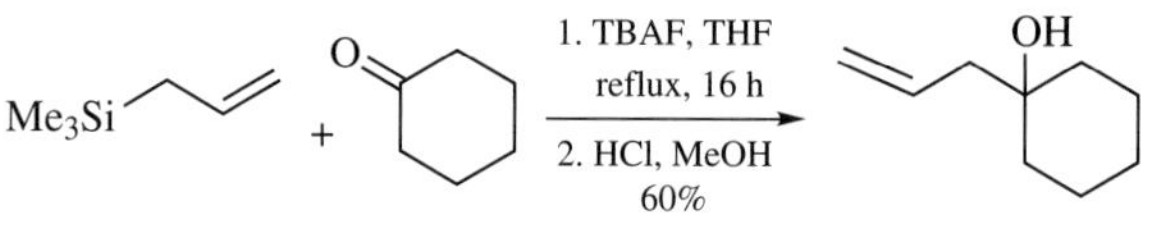

$$\text{(23)}$$

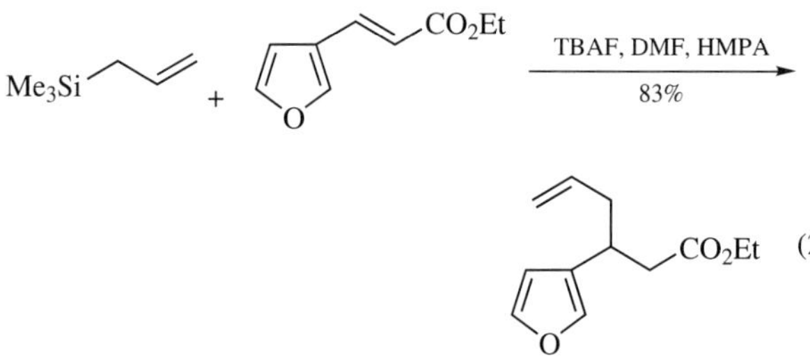

$$\text{(24)}$$

Other Reactions. Allyltrimethylsilane reacts with some highly electrophilic alkenes, carbonyl compounds, azo compounds, and singlet oxygen to a greater or lesser extent in ene reactions that do not involve the loss of the silyl group, and hence give vinylsilanes in a solvent-dependent reaction (eq 25).[58]

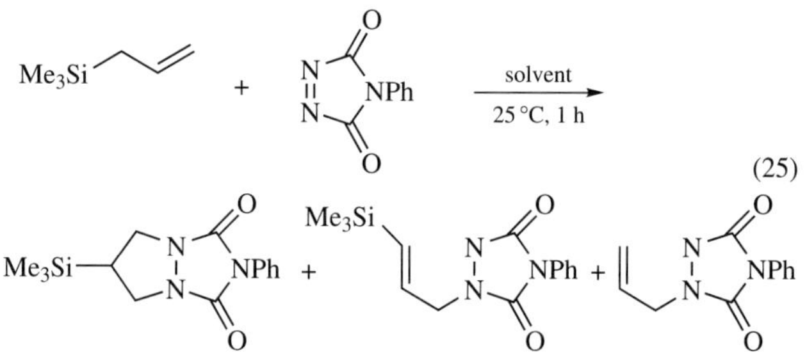

$$\text{(25)}$$

benzene	18	:	73	:	9
MeCN	6	:	28	:	67

Hydroalumination (and hydroboration) take place regioselectively to place the aluminum (or boron) atom at the terminus, creating a 3-trimethylsilylpropyl nucleophile.[59] Radicals attack allyltrimethylsilane at the terminus, and the intermediate radical reacts further without the trimethylsilyl group being expelled, as the corresponding germanium, tin, and lead groups are.[60] Treatment with strong bases gives trimethylsilylallyl–metal compounds.

First Update

Naoki Asao

Tohoku University, Sendai, Japan

As a Carbon Nucleophile in Lewis Acid-catalyzed Reactions.[61]

Allylation Reaction.

Reaction with Aldehydes and Ketones. The allylation of aldehydes with allyltrimethylsilane, leading to homoallyl alcohols, has seen some important developments in recent years. It has now been shown that this reaction proceeds with only a catalytic amount of certain Lewis acids, such as $Me_3SiB(OTf)_4$,[62] $Me_3SiB(OTf)_3Cl$,[63] $Sc(OTf)_3$,[64] $BiBr_3$,[65] Me_2AlNTf_2,[66] $YbCl_3$,[67] Me_3SiNTf_2,[68] $FeCl_3$,[69] $MeC(CH_2NHTf)_3$-Me_3Al,[70] and GaI_3,[71] when traditionally it requires stoichiometric or greater amounts of Lewis acid (eq 2). A combination of $InCl_3$ and R_3SiCl provides a strong Lewis acid catalyst, which is also suitable for the allylation of aldehydes and ketones, such as benzoin.[72] $HN(SO_2F)_2$ works well as a Brønsted acid catalyst for the present reaction,[73] although allyltrimethylsilane decomposes in the presence of strong protic acids.[74] Polystyrene-bound tetrafluorophenylbis(triflyl)methane[75] and dealuminated zeolite-Y, exchanged with rare earth metals (RE-Y),[76] are some other useful heterogeneous catalysts. In addition, iodine also efficiently catalyzes the allylation of both aromatic and aliphatic aldehydes with allyltrimethylsilane to afford the corresponding homoallyl alcohols.[77] On the other hand, 1,6-heptadiene derivatives can be obtained through the bisallylation of carbonyl derivatives.[67,72b,78]

Reaction with Acetals. Allylation reactions of acetals or acylals with allyltrimethylsilane are catalyzed by $TMSN(SO_2F)_2$,[79] $Sc(ClO_4)_3$,[80] $TMSNTf_2$,[81] $BiBr_3$,[65] $InCl_3$,[82] $Sc(OTf)_3$,[64c] $Bi(OTf)_3$,[83] and $FeCl_3$,[84] to give homoallyl ethers (eq 3). A combination of $AlBr_3$, Me_3Al, and $CuBr$ works well as the mixed Lewis acid system.[85] The chemoselective allylation of acetals in ionic liquids, such as butylmethylimidazolium hexafluorophosphate and butylmethylimidazolium triflate, is also possible with the use of catalytic TMS triflate.[86] Microwave-assisted allylation proceeds smoothly in the presence of CuBr.[87] Allylation of acetals also proceeds smoothly when using sulfur dioxide as a Lewis acidic solvent.[88]

Reaction with Conjugate Enones. The conjugate addition of allyltrimethylsilane to conjugate enones can be catalyzed by $TMSNTf_2$, which is formed in situ from $HNTf_2$ and allyltrimethylsilane (eq 11).[89] $InCl_3$ in the presence of excess amount of $TMSCl$,[90] and elemental iodine are also suitable catalysts for this reaction.[91]

Reaction with Acyl Chlorides. In the presence of catalytic amounts of $BiCl_3$-$3NaI$ or $BiCl_3$-$1.5ZnI_2$, acylation of allyltrimethylsilane by a variety of acyl chlorides proceeds to give β,γ-unsaturated ketones (eq 1).[92] On the other hand, aryl chloroformate reacts with allyltrimethylsilane in the presence of a stoichiometric amount of $AlCl_3$ to give the β,γ-unsaturated *O*-aryl esters.[93]

Reaction with Hydrates of α-Keto Aldehydes and Glyoxylates. Catalytic allylations of the hydrates of α-keto aldehydes and glyoxylates with allyltrimethylsilane can be performed by using $Yb(OTf)_3$ and $Sc(OTf)_3$ (eq 26).[94] TMSOMs, which is formed in situ from methansulfonic acid and allyltrimethylsilane, is also a viable catalyst for this reaction.[95]

$$(26)$$

Reaction with Allylic Halides, Alcohols, and their Derivatives. Allylation of allyl and propargyl trimethylsilyl ethers as well as benzyl and propargylic alcohol derivatives proceeds in the presence of a catalytic amount of Lewis acids, such as $ZnCl_2$,[96] $TMS(OTf)$,[97] and $B(C_6F_5)_3$.[98] Direct substitutions of the hydroxyl group of allylic, benzylic, and propargylic alcohols are catalyzed by $HN(SO_2F)_2$,[99] a rhenium-oxo complex,[100] and $InCl_3$ (eq 27).[101] A combination of chlorodimethylsilane and allyltrimethylsilane effectively promotes the deoxygenative allylation of aromatic ketones in the presence of a catalytic amount of an indium compound, such as indium trihalide or metallic indium (eq 28).[102] Allylation of cyclic allylic acetates with allyltrimethylsilane can be catalyzed by molecular iodine.[103]

$$(27)$$

$$(28)$$

Reaction with Aldimines. A combination of $InCl_3$ and R_3SiCl is a suitable catalyst system for allylation of aldimines (eq 29).[72] The allylation of a variety of in situ generated protected aldimines using aldehydes, carbamates, and allyltrimethylsilane in a three-component reaction proceeds in the presence of a stoichiometric amount of BF_3·OEt_2 or $NCS/SnCl_2$,[104] or a catalytic amount of $TrClO_4$, $Bi(OTf)_3$, and I_2 (eq 30).[105-107] On the other hand, BF_3·OEt_2 can be employed in the synthesis of a library of homoallylic amines by the reactions involving a solid phase-bound carbamate.[108] Lewis acids also promote the allylation of quinolines derivatives (eq 31).[109]

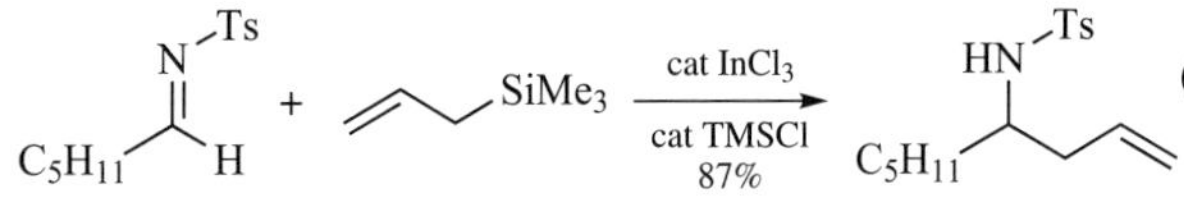

(29)

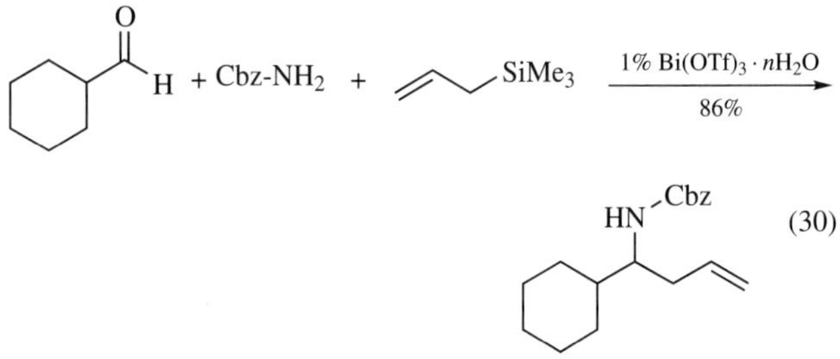

(30)

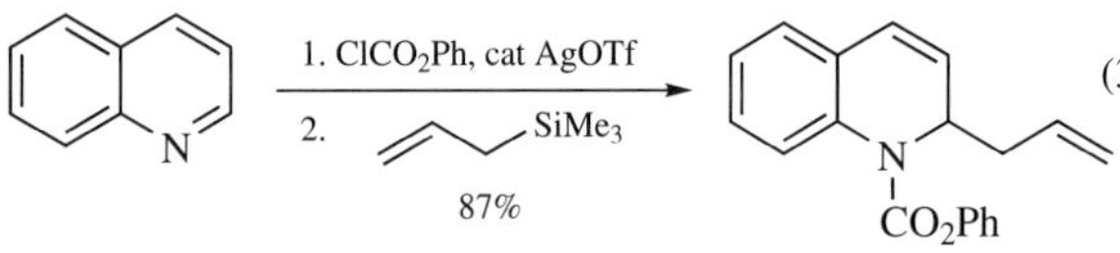

(31)

Reaction with Alkenes and Alkynes. The $AlCl_3$-catalyzed reactions of allyltrimethylsilane with terminal alkenes or cycloalkenes proceed in the *trans* fashion to give the addition products regioselectively (eq 32).[110] Likewise, the allylsilylation of unactivated alkynes are catalyzed by Lewis acids, such as $AlCl_3$, $EtAlCl_2$-TMSCl, and $HfCl_4$, and the *trans* addition products are obtained regio- and stereoselectively (eq 33).[111,112] On the other hand, allylgallation of alkynes proceeds when the reaction is carried out by using a stoichiometric amount of $GaCl_3$ (eq 34).[113]

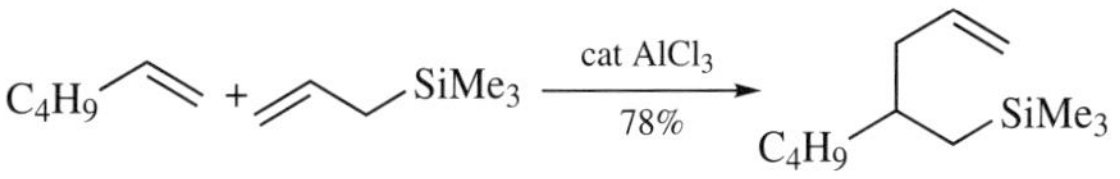

(32)

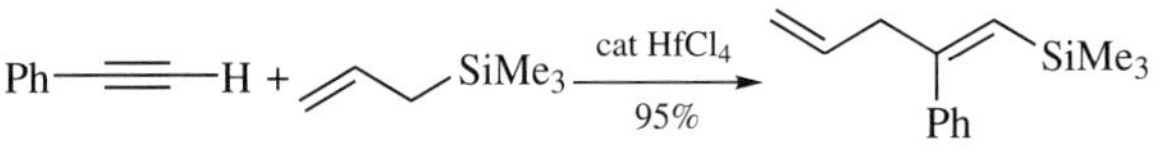

(33)

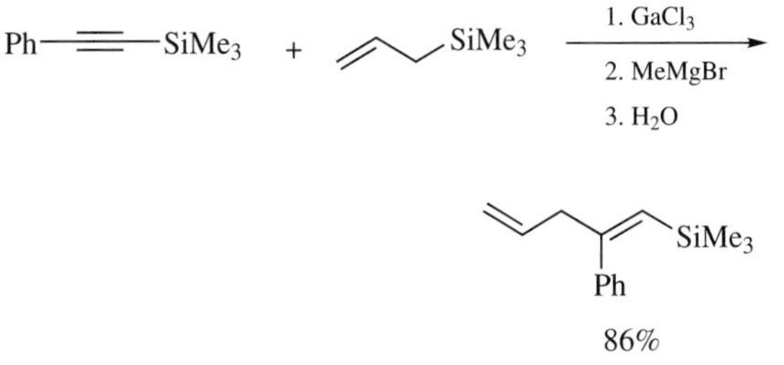

(34)

Stereoselective Allylation Reaction. The enantioselective allylation of aldehydes with allyltrimethylsilane is now possible with good control. One example is the use of a chiral acyloxy borane (CAB) catalyst.[114] Besides that, chiral $Ti(OiPr)_2X_2$-BINOL[115] and TiF_4-BINOL catalysts[116] are effective for the enantioselective allylation of glyoxylates and aldehydes, respectively (eq 35). Chiral homoallylamines can be prepared from the reaction of imines and allyltrimethylsilane using chiral π-allylpalladium complexes with TBAF (eq 36)[117] or an (S)-Tol-BINAP-CuPF$_6$ system.[118] Free radical allyl transfers from allyltrimethylsilane provides another method for this enantioselective C–C bond formation. Promoted by chiral Lewis acids, both

enantiomers can be prepared using the same chiral ligands by choosing the appropriate metal (eq 37).[119]

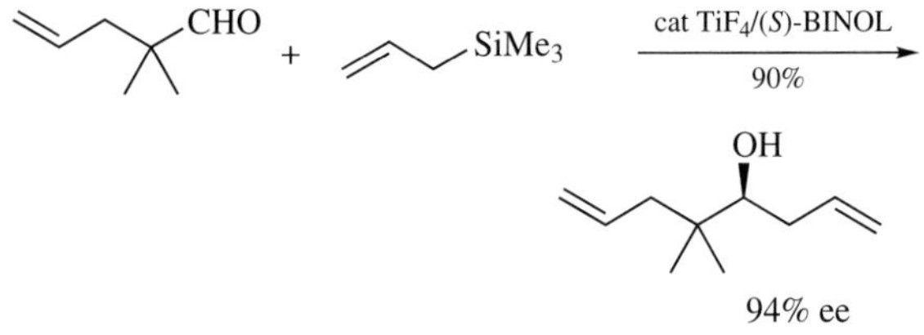

(35)

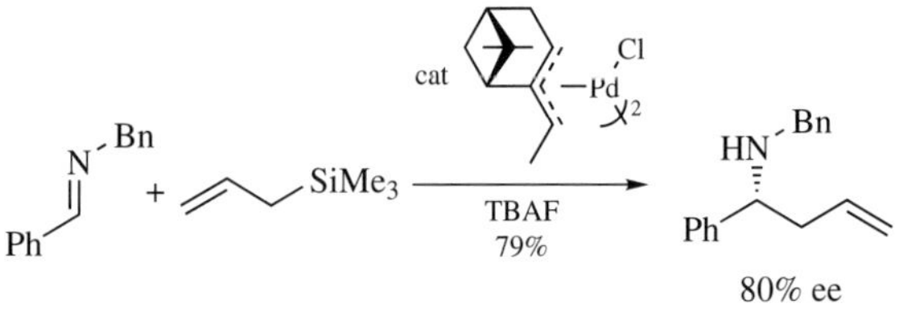

(36)

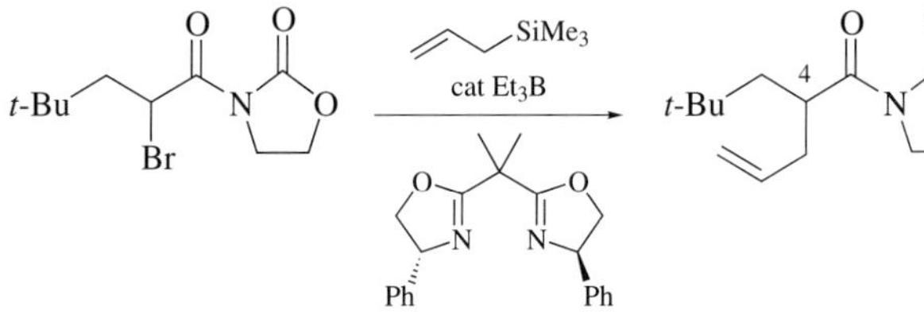

(37)

4R:4S

$Zn(OTf)_2$	88%	95:5
MgI_2	86%	16:84

The asymmetric allylations of aldehydes with allyltrimethylsilane using chiral auxiliaries are also useful. The use of norpseudoephedrine derivatives and a catalytic amount of TMSOTf is an example that proceeds well.[120] This procedure can be applied for the enantioselective synthesis of tertiary homoallyllic alcohols using simple ketones (eq 38).[121] In addition, a number of diastereoselective allylations through acetal-based oxonium ions has been reported.[61,122] High diastereoselectivities are also observed for the radial allylation of β-alkoxy esters initiated by Et_3B, which proceeds under bidentate chelation-controlled conditions to give α-allylated products (eq 39).[123] Chiral N-acylhydrazones are also found to be effective as well for asymmetric allylation and this reaction proceeds in the presence of $In(OTf)_3$ and tetrabutylammonium triphenyldifluorosilicate (TBAT) to give homoallyl hydrazine derivatives (eq 40).[124] Similarly, asymmetric conjugate addition of allyltrimethylsilane to α,β-unsaturated N-acylamides employing chiral auxiliaries were demonstrated.[125] The allylation of arylidenetetralone-$Cr(CO)_3$ complex produces the less common *endo*-adduct exclusively (eq 41).[126]

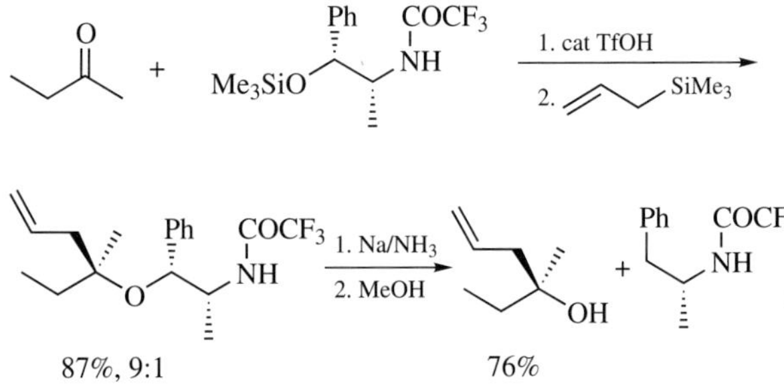

(38)

87%, 9:1   76%

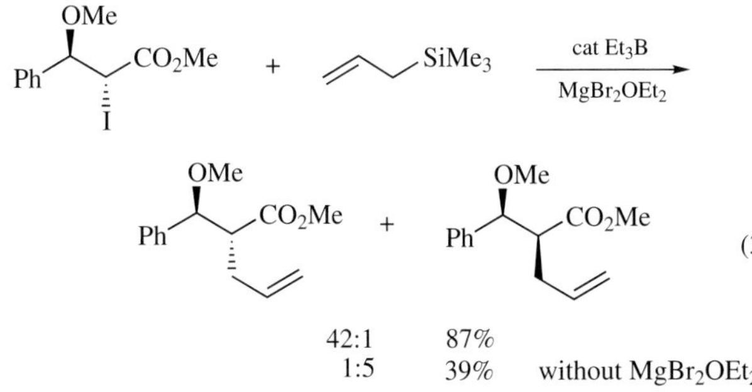

(39)

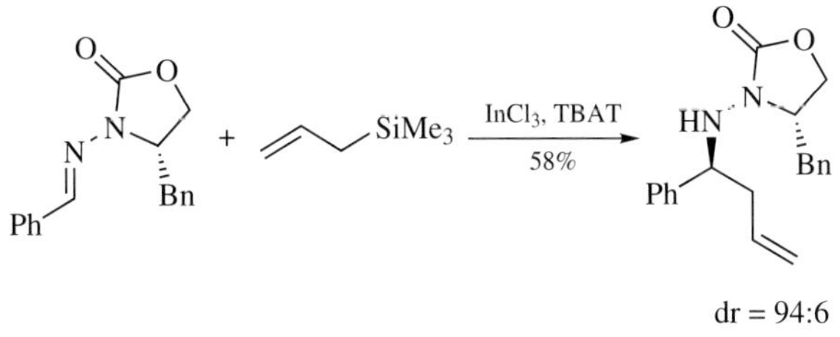

42:1   87%
1:5   39%   without MgBr$_2$OEt$_2$

(40)

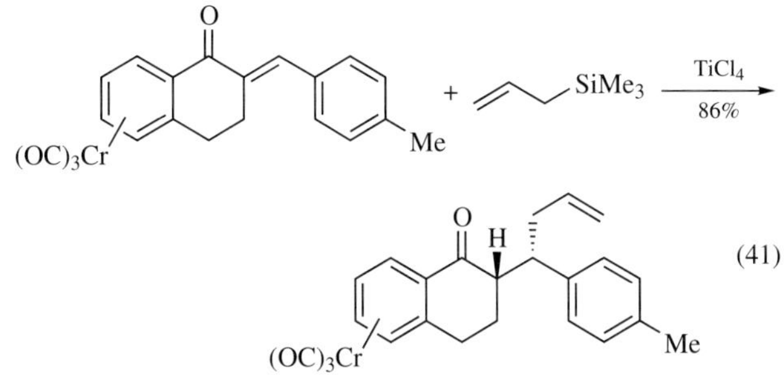

dr = 94:6

(41)

(43)

>98% de, 95% ee

(44)

2:3

(45)

(46)

49%   43%

(47)

(48)

Annulation. Allyltrimethylsilane undergoes [2 + 2] addition reactions with methyl propiolate in the presence of AlCl$_3$ or EtAlCl$_2$ (eq 42).[127] It can also undergo an asymmetric [3 + 2] cycloaddition with α-keto esters by using L-quebrachitol as a chiral auxiliary (eq 43).[128] Allyltrimethylsilane also undergoes ring-expansion reactions with cyclopropane derivatives to give cyclopentane derivatives in the presence of TiCl$_4$ (eq 44).[129] In the presence of AlCl$_3$ catalyst, cycloaddition between allyltrimethylsilane and conjugated dienes also takes place (eq 45).[130] Functionalized bicyclo[2.2.1]heptanones can be obtained together with cyclopentanones from simple dienone and allyltrimethylsilane through the domino electrocyclization/[3 + 2] cycloaddition (eq 46).[131] Other annulation partners, α,β-epoxy alcohol or aziridine derivatives, give the corresponding tetrahydrofuranes or pyrrolidines, respectively, with allyltrimethylsilane.[132] A chiral nitrone undergoes 1,3-dipolar cycloaddition with allyltrimethylsilane under high pressure or in the presence of BF$_3$·OEt$_2$ to give the isoxazoline derivatives stereoselectively, which can be converted to the enantiomerically pure allylglycine (eq 47).[133] BF$_3$OEt$_2$ also mediates the formal [4 + 2] cycloaddition of triethylsilyl-protected β-hydroxy aldehydes with allyltrimethylsilane to afford tetrahydropyrans (eq 48).[134] On the other hand, tetrahydroquinolines can be prepared from the SnCl$_4$-mediated reaction of N-aryl-1H-benzotriazoyl-1-methanamines with allyltrimethylsilane.[135]

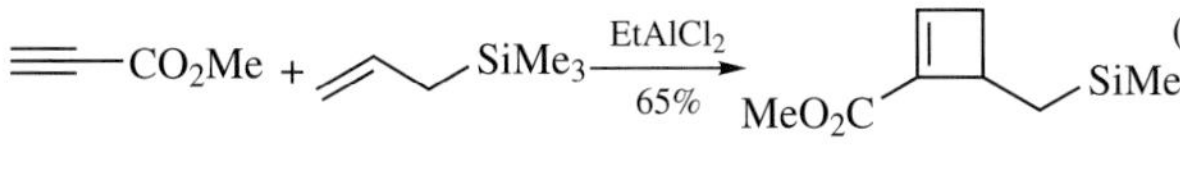

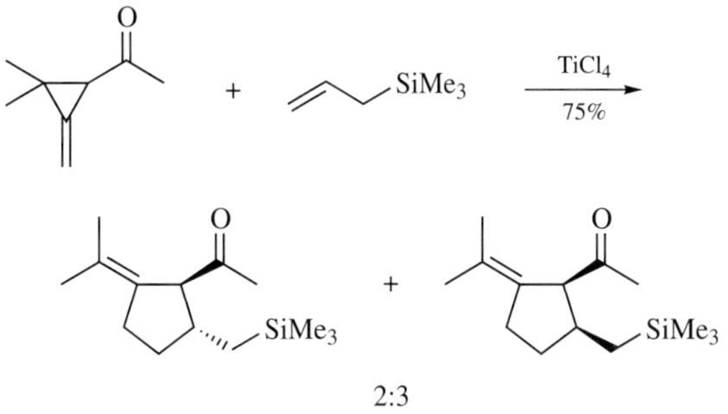

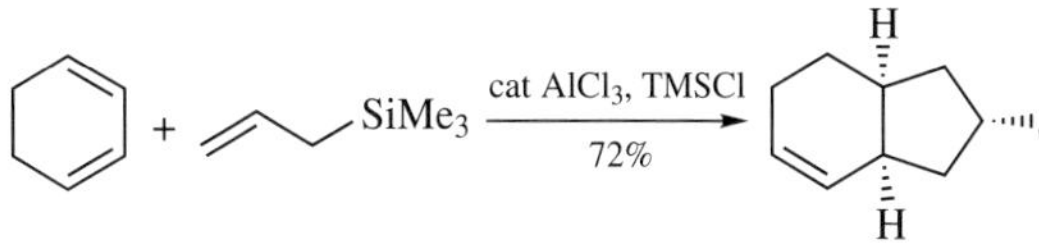

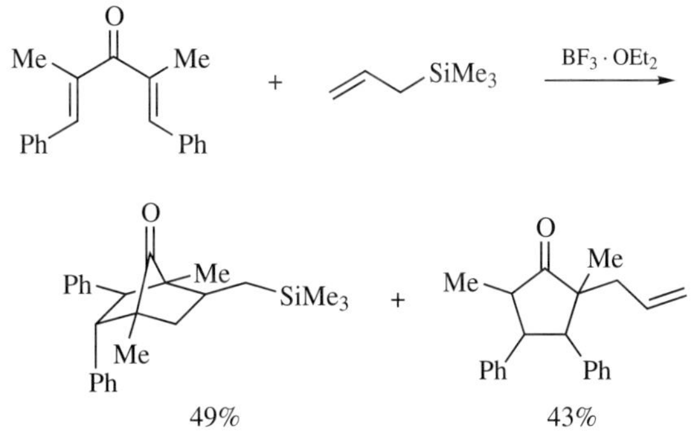

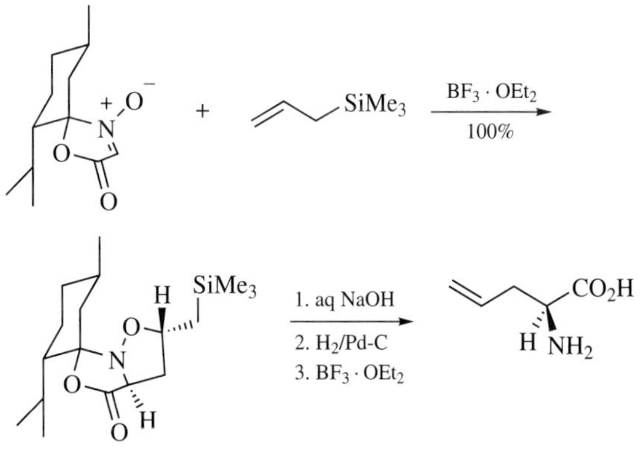

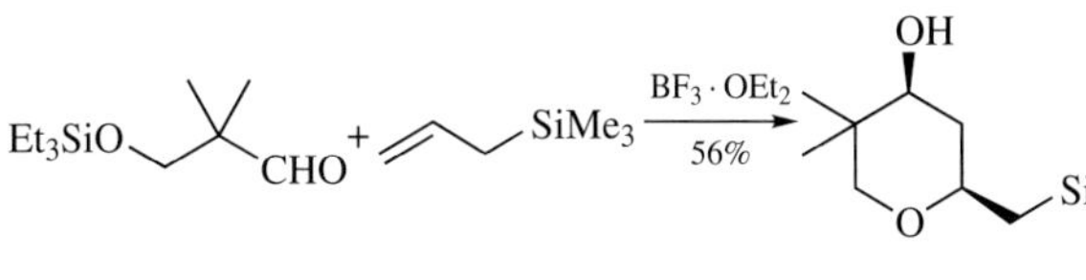

(42)

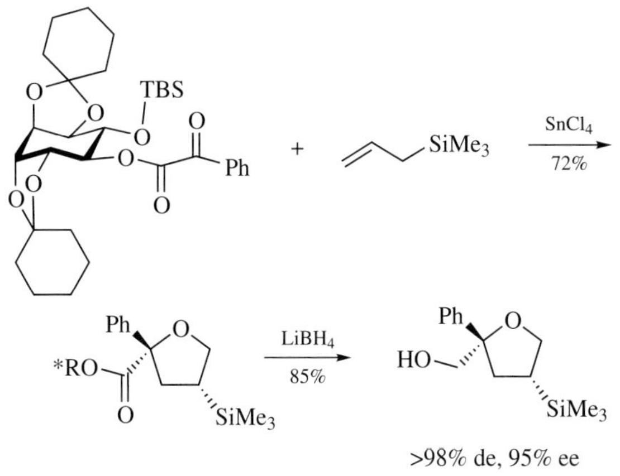

Other Reactions. Allyltrimethylsilane reacts with epoxides in the presence of TiCl$_4$ and tetrabutylammonium iodide to provide alkenes with high stereoselectivity through the formation of iodohydrin intermediates (eq 49).[136] TiCl$_4$ can also mediate the crossed-aldol reaction between two different aldehydes. This proceeds via a titanium enolate, derived from an α-iodo aldehyde and allyltrimethylsilane in the presence of TiCl$_4$, to afford β-hydroxy aldehydes (eq 50).[137] In addition, TiCl$_4$ can mediate the reaction of allyltrimethylsilane with α-haloacylsilanes to provide α-silyl-β',γ'-unsaturated ketones.[138,139] This method can be applied to the three-component coupling reaction of α-chloroacylsilanes, allyltrimethylsilane, and carbonyl compounds (eq 51).[138] Ketones in the presence of MnO$_2$ and acetic acid react with allyltrimethylsilane to give the corresponding hydrogen atom transfer adducts (eq 52).[140] Treatment of 1,3-dicarbonyl compounds, such as 1,3-diketones, β-keto esters, and malonates, with allyltrimethylsilane in acetonitrile gives the allylated products in the presence of CAN or ceric tetra-n-butylammonium nitrate (CTAN). However, replacement of CAN with manganese acetate and acetic acid or CTAN in dichloromethane affords dihydrofuran derivatives (eq 53).[141] The oxidative allylations of silyl enol ethers (eq 54) as well as 1,3-dicarbonyl compounds proceed using oxovanadium compounds.[142] Besides that, allyltrimethylsilane also undergoes coupling reactions with benzylic silanes based on the oxovanadium-induced one-electron oxidative desilylation (eq 55).[143] Monoperoxy acetals (eq 56),[144] selenoacetals,[145] O/Se-heteroacetals,[146] and tris(phenylchalcogeno)methanes[147] can be allylated in the presence of a stoichiometric amount of Lewis acids. On the other hand, Lewis acid-mediated reaction of alkoxyhydroperoxides or ozonides with allyltrimethylsilane gives 1,2-dioxolanes (eq 57).[148] Reaction of triarylbismuth difluorides with allyltrimethylsilane in the presence of BF$_3$·OEt$_2$ generates an allyltriarylbismuthonium compound, which is reactive toward arenes and heteroatom nucleophiles to give the corresponding allylated products (eq 58).[149] Allylated benzene compounds can be also obtained from the lithium perchlorate-mediated reaction of quinones with allyltrimethylsilane (eq 59).[39,150] Treatment of *ortho*-(alkynyl)benzaldehydes with IPy$_2$BF$_4$ and HBF$_4$, and then with allyltrimethylsilane affords isochromene derivatives (eq 60).[151] Cu(OTf)$_2$ catalyzes the allylic amination of allyltrimethylsilane with [N-(p-toluenesulfonyl)imino]phenyliodinane (eq 61).[152] NbCl$_5$-promoted reaction of aldehydes with allyltrimethylsilane proceeds to give cyclopropanes derived from 2 equiv of allylsilane and one of aldehydes (eq 62).[153]

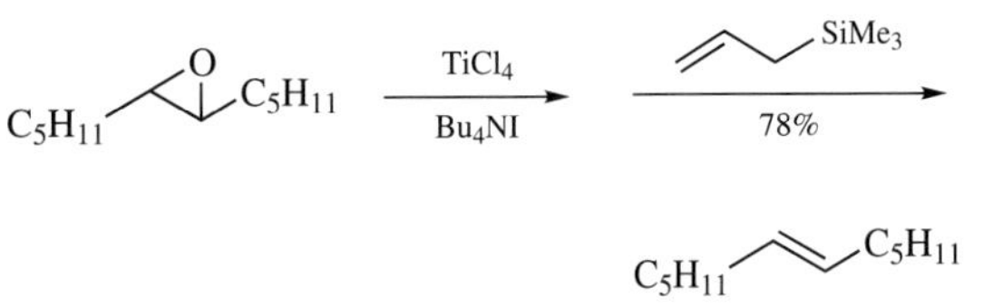

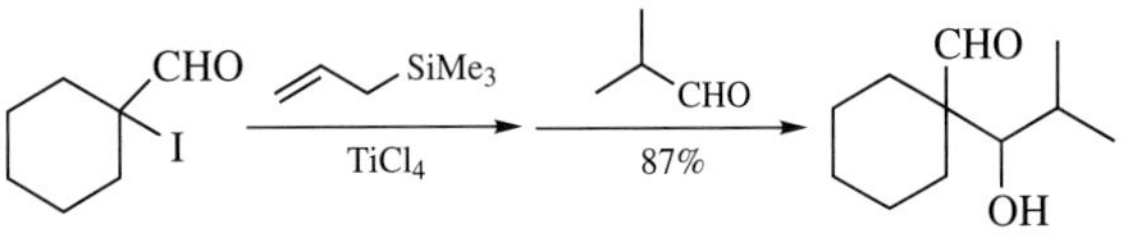

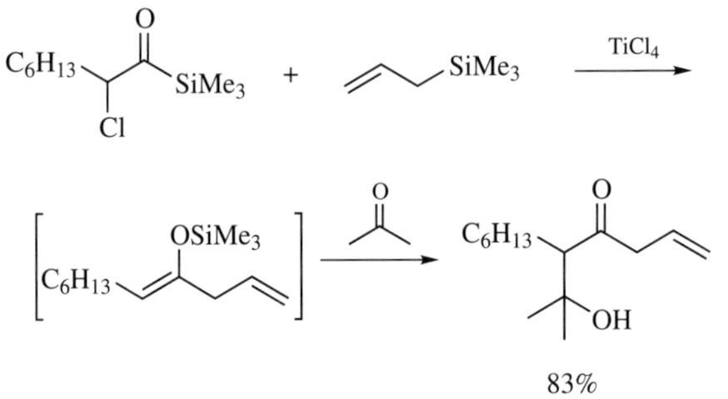

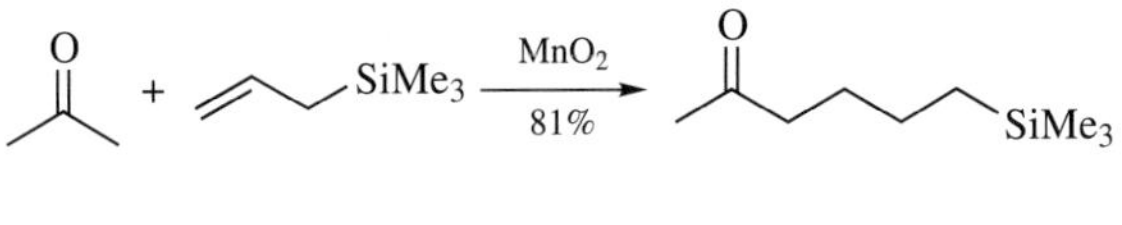

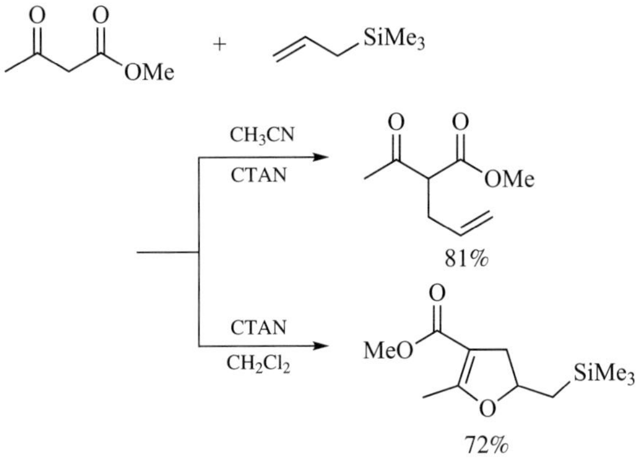

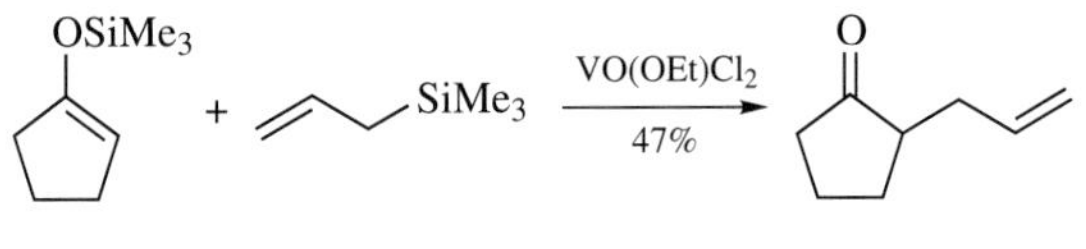

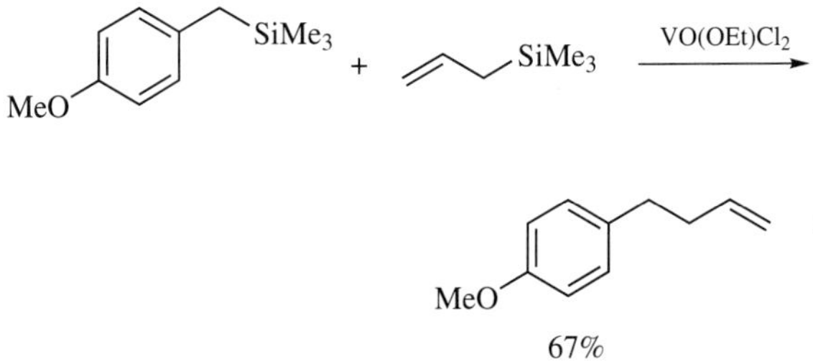

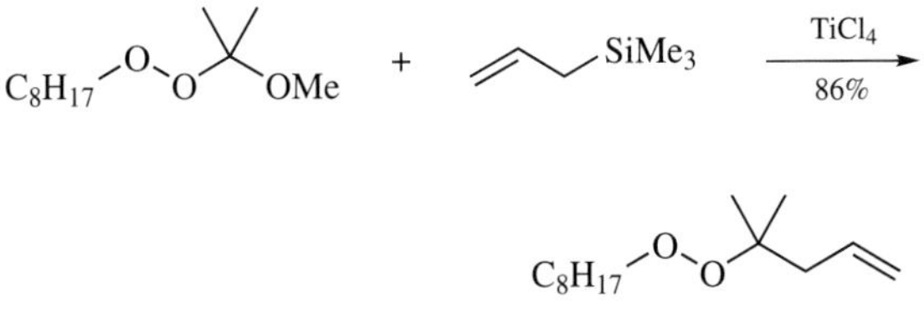

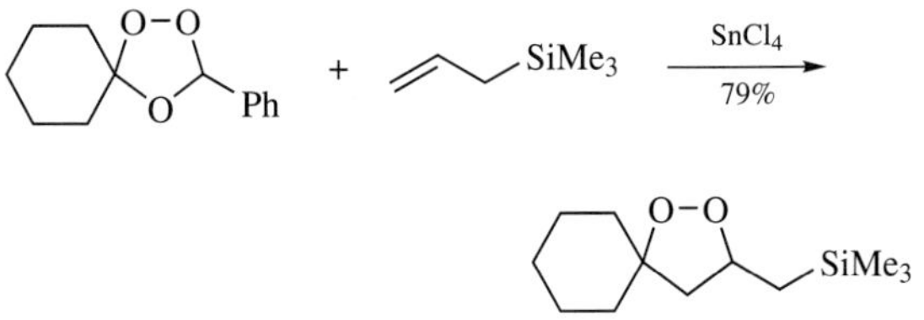

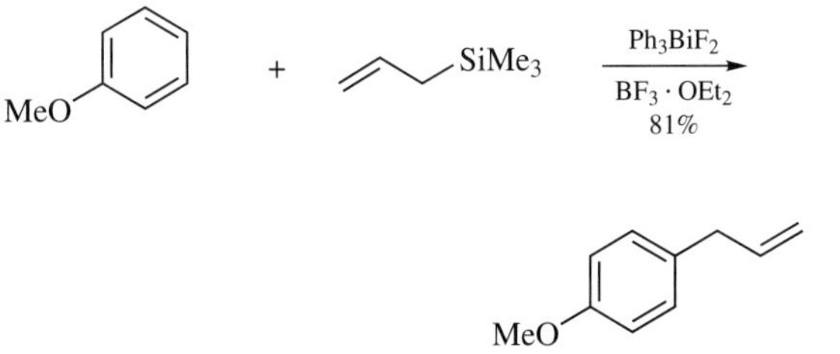

$$o:m:p = 15:0:85 \quad (58)$$

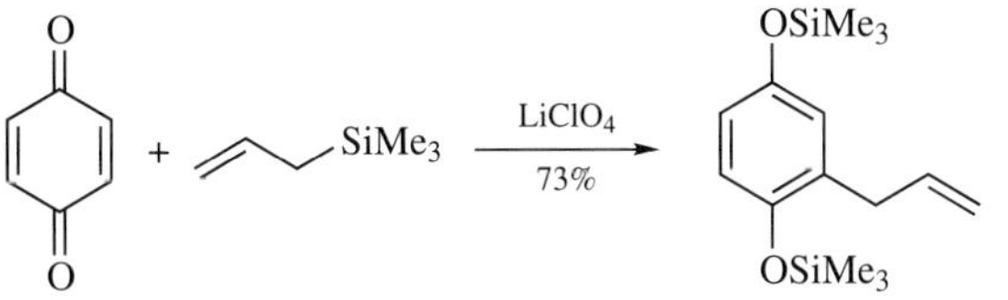

$$(59)$$

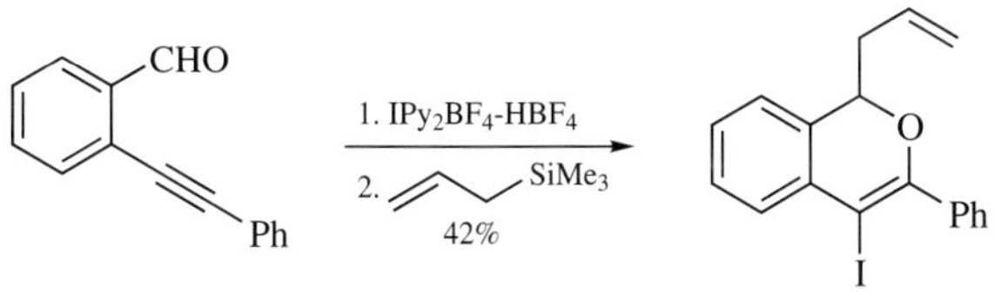

$$(60)$$

$$PhI=NTs \;+\; \diagup\!\!\!\diagdown SiMe_3 \xrightarrow[78\%]{cat\ Cu(OTf)_2} \diagup\!\!\!\diagdown NHTs \quad (61)$$

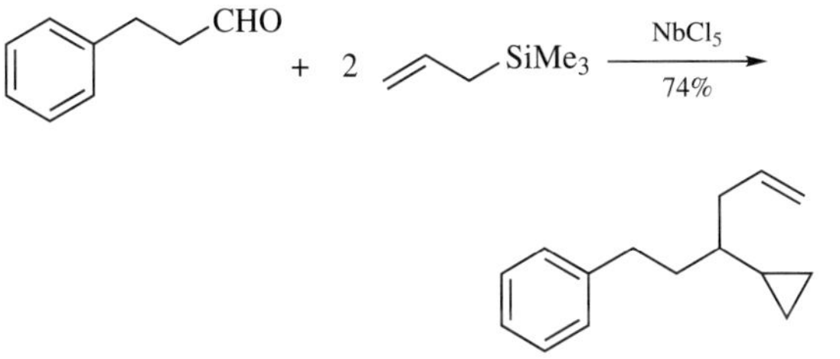

$$(62)$$

As a Carbon Nucleophile in Lewis Base-catalyzed Reactions. Allylation of alkyl iodides with allyltrimethylsilane proceeds in the presence of phosphazenium fluoride.[154] Tetrabutylammonium triphenyldifluorosilicate (TBAT) is useful for allylation of aldehydes, ketones, imines, and alkyl halides with allyltrimethylsilane (eq 63).[155] Similarly, TBAHF$_2$ is an effective catalyst for allylation of aldehydes.[156] The homoallylamines are synthesized from allyltrimethylsilane and imines with a catalytic amount of TBAF (eq 64).[157] The reactions of thioketones as well as sulfines with allyltrimethylsilane can be mediated by TBAF to give allylic sulfides and allyl sulfoxides, respectively.[158] Besides fluoride ion, 2,8,9-triisopropyl-2,5,8,9-tetra-aza-1-phosphabicyclo[3.3.3]undecane promotes the allylation of aldehydes with allyltrimethylsilane as a Lewis base catalyst (eq 65).[159]

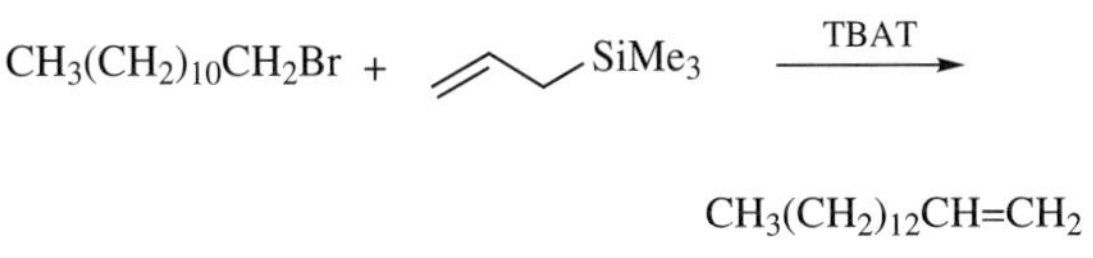

$$CH_3(CH_2)_{10}CH_2Br \;+\; \diagup\!\!\!\diagdown SiMe_3 \xrightarrow{TBAT}$$

$$CH_3(CH_2)_{12}CH=CH_2 \quad (63)$$
$$87\%$$

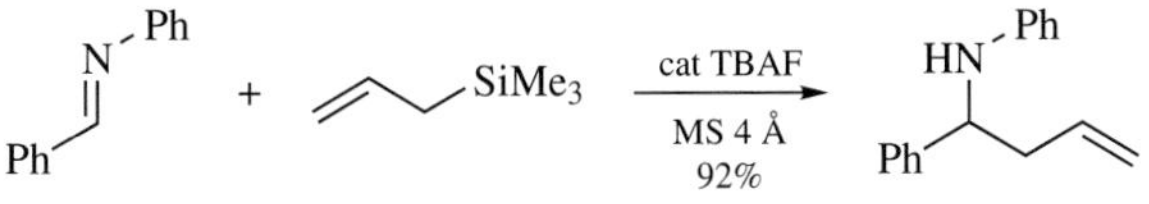

$$(64)$$

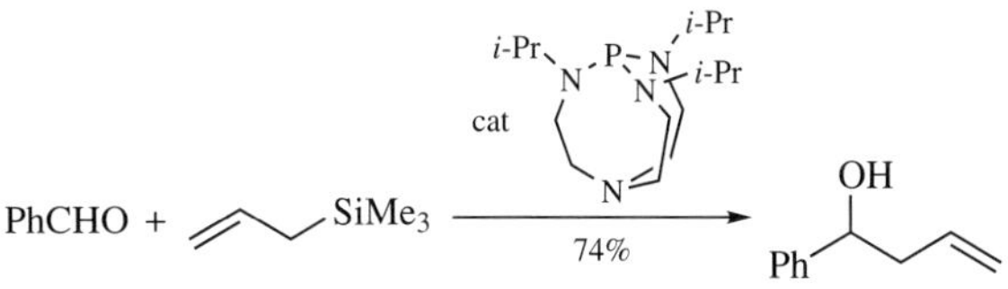

$$(65)$$

Other Reactions.[160] *N*-Triflyloxy amides undergo ionization in refluxing isopropanol to give *N*-acyliminium ions, which react with allyltrimethylsilane to give homoallyl amides (eq 66).[161] Iminium ions are also required in the reaction of *N,N*-dialkylmethyleneammonium hexachloroantimonates with allyltrimethylsilane, which gives alkylideneammonium ions via ene reactions. Treatment of the resulting ammonium ions with NaBH$_4$ gives the corresponding amines (eq 67).[162] Lithium dienolates of α,β-unsaturated acylsilanes can be conveniently obtained by the reaction of (1-silylallyl)lithiums with carbon monoxide via completely α-selective C–C bond formation (eq 68).[163] The direct oxidative carbon-carbon bond formation of carbamates with allyltrimethylsilane also proceeds using the "cation pool" method and a "cation flow" system (eq 69).[164] Irradiating a mixture of electron-deficient alkenes and allyltrimethylsilane with phenanthrene affords allylation products diastereoselectively (eq 70).[165]

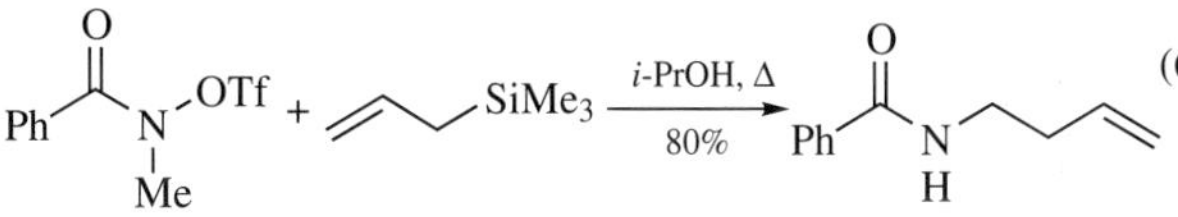

$$(66)$$

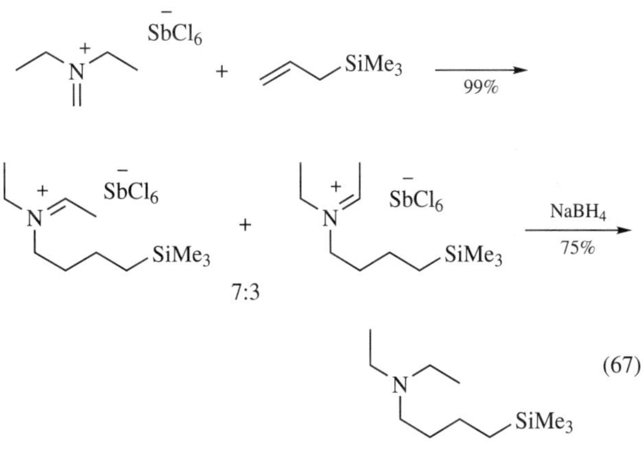

$$7:3 \qquad (67)$$

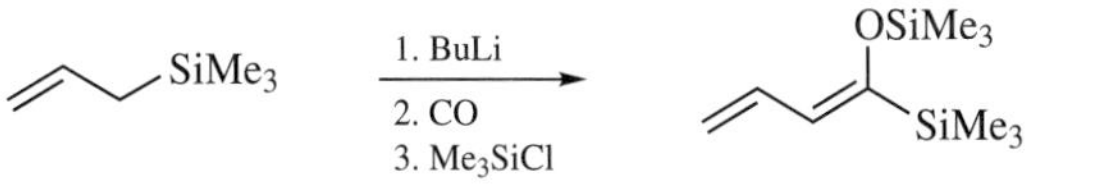

$$(68)$$
$$89\%$$
$$E/Z = 96:4$$

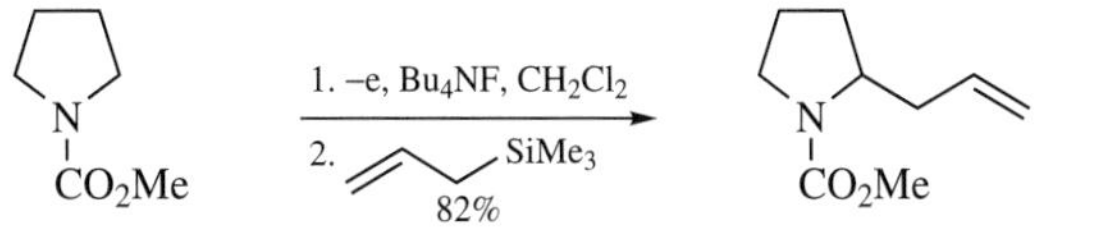

$$(69)$$
$$82\%$$

(70)

97:3

Related Reagents. Allyltriisopropylsilane; allyltrichlorosilane; crotyltrimethylsilane; methallyltrimethylsilane; allyltributylstannane.

1. Fleming, I.; Dunogués, J.; Smithers, R., *Org. React.* **1989**, *37*, 57.

2. Delmulle, L.; van der Kelen, G. P., *J. Mol. Struct.* **1980**, *66*, 315.

3. Sarkar, T. K., *Synthesis* **1990**, 969 and 1101.

4. (a) Hagen, G.; Mayr, H., *J. Am. Chem. Soc.* **1991**, *113*, 4954. (b) Review: Mayr, H., *Angew. Chem., Int. Ed. Engl.* **1990**, *29*, 1371.

5. Yoshida, J.-I.; Murata, T.; Matsunaga, S.-I.; Tsuyoshi, M.; Shiozawa, S.; Isoe, S., *Rev. Heteroatom Chem.* **1991**, *5*, 193.

6. Pandey, G.; Rani, K. S.; Lakshmaiah, G., *Tetrahedron* **1992**, *33*, 5107.

7. Sasaki, T.; Nakanishi, A.; Ohno, M., *J. Org. Chem.* **1982**, *47*, 3219.

8. Mukaiyama, T.; Nagaoka, H.; Murakami, M.; Oshima, M., *Chem. Lett.* **1985**, 977.

9. (a) Hosomi, A.; Endo, M.; Sakurai, H., *Chem. Lett.* **1976**, 941. (b) Review: Mukaiyama, T.; Murakami, M., *Synthesis* **1987**, 1043.

10. Sakurai, H.; Sakata, Y.; Hosomi, A., *Chem. Lett.* **1983**, 409.

11. Larsen, S. D.; Grieco, P. A.; Fobare, W. F., *J. Am. Chem. Soc.* **1986**, *108*, 3512.

12. Kraus, G. A.; Neuenschwander, K., *J. Chem. Soc., Chem. Commun.* **1982**, 134.

13. Kraus, G. A.; Hon, Y.-S., *J. Am. Chem. Soc.* **1985**, *107*, 4341.

14. Hosomi, A.; Imai, T.; Endo, M.; Sakurai, H., *J. Organomet. Chem.* **1985**, *285*, 95.

15. Mukaiyama, T.; Ohshima, M.; Miyoshi, N., *Chem. Lett.* **1987**, 1121.

16. Mekhalfia, A.; Marko, I. E., *Tetrahedron Lett.* **1991**, *32*, 4779.

17. Lipshutz, B. H.; Burgess-Henry, J.; Roth, G. P., *Tetrahedron Lett.* **1993**, *34*, 995.

18. Coppi, L. Ricci, A.; Taddei, M., *Tetrahedron Lett.* **1987**, *28*, 973.

19. Hosomi, A.; Sakurai, H., *J. Am. Chem. Soc.* **1977**, *99*, 1673.

20. Hosomi, A.; Hashimoto, H.; Kobayashi, H.; Sakurai, H., *Chem. Lett.* **1979**, 245.

21. Pardo, R.; Zahra, J.-P.; Santelli, M., *Tetrahedron Lett.* **1979**, *20*, 4557, with products reformulated in the light of Knölker, H. J.; Jones, P. G.; Pannek, J.-B., *Synlett* **1990**, 429 and Danheiser, R. L.; Dixon, B. R.; Gleason, R. W., *J. Org. Chem.* **1992**, *57*, 6094 and references therein.

22. Ochiai, M.; Arimoto, M.; Fujita, E., *Tetrahedron Lett.* **1981**, *22*, 1115.

23. Alexander, R. P.; Paterson, I., *Tetrahedron Lett.* **1983**, *24*, 5911.

24. Nishiyama, H.; Naritomi, T.; Sakuta, K.; Itoh, K., *J. Org. Chem.* **1983**, *48*, 1557.

25. Davis, A. P.; Hegarty, S. C., *J. Am. Chem. Soc.* **1992**, *114*, 2745.

26. Danishefsky, S. J.; De Ninno, M. P.; Phillips, G. B.; Zelle, R. L.; Lartey, P. A., *Tetrahedron Lett.* **1986**, *42*, 2809.

27. Hosomi, A.; Sakata, Y.; Sakurai, H., *Tetrahedron Lett.* **1984**, *25*, 2383.

28. Danishefsky, S.; Kerwin, J. F., *J. Org. Chem.* **1982**, *47*, 3803.

29. Davis, A. P.; Jaspars, M., *J. Chem. Soc., Perkin Trans. 1* **1992**, 2111.

30. Hayashi, M; Mukaiyama, T., *Chem. Lett.* **1987**, 289.

31. Hollis, T. K.; Robinson, N. P.; Whelan, J.; Bosnich, B., *Tetrahedron Lett.* **1993**, *34*, 4309.

32. Pearson, W. H.; Schkeryantz, J. M., *J. Org. Chem.* **1992**, *57*, 2986.

33. Mukaiyama, T.; Shimpuku, T.; Takashima, T.; Kobayashi, S., *Chem. Lett.* **1989**, 145.

34. Hayashi, M.; Inubushi, A.; Mukaiyama, T., *Bull. Chem. Soc. Jpn.* **1988**, *61*, 4037.

35. Mukaiyama, T.; Takenoshita, H.; Yamada, M.; Soga, T., *Chem. Lett.* **1990**, 1259.

36. Mukaiyama, T.; Ohno, T.; Nishimura, T.; Han, J. S.; Kobayashi, S., *Chem. Lett.* **1990**, 2239.

37. Simpkins, N. S., *Tetrahedron Lett.* **1991**, *47*, 323.

38. Mukaiyama, T.; Ohshima, M.; Miyoshi, N., *Chem. Lett.* **1987**, 1121.

39. (a) Hosomi, A.; Sakurai, H., *Tetrahedron Lett.* **1977**, 4041. (b) Hosomi, A.; Sakurai, H., *Tetrahedron Lett.* **1978**, 2589.

40. Bambal, R.; Kemmitt, R. D. W., *J. Chem. Soc., Chem. Commun.* **1988**, 734.

41. Carr, S. A.; Weber, W. P., *J. Org. Chem.* **1985**, *50*, 2782.

42. Hamana, H.; Sugasawa, T., *Chem. Lett.* **1985**, 921.

43. Mori, I.; Bartlett, P. A.; Heathcock, C. H., *J. Am. Chem. Soc.* **1987**, *109*, 7199.

44. Hermans, B.; Hevesi, L., *Tetrahedron Lett.* **1990**, *31*, 4363.

45. Mori, I.; Bartlett, P. A.; Heathcock, C. H., *J. Org. Chem.* **1990**, *55*, 5966.

46. Schreiber, S. L.; Klimas, M. T.; Sammakia, T., *J. Am. Chem. Soc.* **1987**, *109*, 5749.

47. Hayashi, M.; Inubushi, A.; Mukaiyama, T., *Chem. Lett.* **1987**, 1975.

48. Hallberg, A.; Westerlund, C., *Chem. Lett.* **1982**, 1993.

49. Kelly, L. F.; Narula, A. S.; Birch, A. J., *Tetrahedron Lett.* **1980**, *21*, 871.

50. Li, Z.; Nicholas, K. M., *J. Organomet. Chem.* **1991**, *402*, 105.

51. (a) Grignon-Dubois, M.; Pillot, J.-P.; Dunogués, J.; Duffaut, N.; Calas, R.; Henner, B., *J. Organomet. Chem.* **1977**, *124*, 135. (b) Colvin, E. W.; Montieth, M., *J. Chem. Soc., Chem. Commun.* **1990**, 1230.

52. Hosomi, A.; Shoji, H.; Sakurai, H., *Chem. Lett.* **1985**, 1049.

53. (a) Kliegman, J. M., *J. Organomet. Chem.* **1971**, *29*, 73. (b) Yamamoto, K.; Shinohara, K.; Ohuchi, T.; Kumada, M., *Tetrahedron Lett.* **1974**, 1153.

54. Hosomi, A.; Shirahata, A.; Sakurai, H., *Tetrahedron Lett.* **1978**, 3043.

55. Majetich, G.; Casares, A. M.; Chapman, D.; Behnke, M., *J. Org. Chem.* **1986**, *51*, 1745.

56. Taniguchi, M.; Oshima, K.; Utimoto, K., *Chem. Lett.* **1992**, 2135.

57. Artamkina, G. A.; Kovalenko, S. V.; Beletskaya, I. P.; Reutov, O. A., *J. Organomet. Chem.* **1987**, *329*, 139.

58. (a) Ohashi, S.; Ruch, W. E.; Butler, G. B., *J. Org. Chem.* **1981**, *46*, 614. (b) Review: Dubac, J.; Laporterie, A., *Chem. Rev.* **1987**, *87*, 319.

59. Maruoka, K.; Sano, H.; Shinoda, K.; Nakai, S.; Yamamoto, H., *J. Am. Chem. Soc.* **1986**, *108*, 6036.

60. Light, J. P.; Ridenour, M.; Beard, L.; Hershberger, J. W., *J. Organomet. Chem.* **1987**, *326*, 17.

61. (a) Yamamoto, Y.; Asao, N., *Chem. Rev.* **1993**, *93*, 2207. (b) Masse, C. E.; Panek, J. S., *Chem. Rev.* **1995**, *95*, 1293. (c) Langkopf, E.; Schinzer, D., *Chem. Rev.* **1995**, *95*, 1375. (d) Santelli, M.; Pons, J. M., In *Lewis Acids and Selectivity in Organic Synthesis*; CRC Press: Boca Raton, 1996. (e) Fleming, I.; Barbero, A.; Walter, D., *Chem. Rev.* **1997**, *97*, 2063. (f) Luh, T. Y.; Liu, S. T., In *The Chemistry of Organic Silicon Compounds*; Rappaport, Z.; Apeloig, Y., Eds.; Wiley: Chichester, UK, 1998; p 1793. (g) Brook, M. A., *Silicon in Organic, Organometallic and Polymer Chemistry*; Wiley Interscience: New York, 2000. (h) Sarkar, T. K., In *Science of Synthesis*; Houben-Weyl Methods of Molecular Transformations, Georg Thieme Verlag: Stuttgart, 2002, Vol. 4, 837. (i) Denmark, S.; Fu, J., *Chem. Rev.* **2003**, *103*, 2763. (j) Chabaud, L.; James, P.; Landais, Y., *Eur. J. Org. Chem.* **2004**, 3173.

62. Davis, A. P.; Jaspars, M., *Angew. Chem., Int. Ed.* **1992**, *31*, 470.
63. Davis, A. P.; Muir, J. E.; Plunkett, S. J., *Tetrahedron Lett.* **1996**, *37*, 9401.
64. (a) Aggarwal, V. K.; Vennall, G. P., *Tetrahedron Lett.* **1996**, *37*, 3745. (b) Aggarwal, V. K.; Vennall, G. P., *Synthesis* **1998**, 1822. (c) Yadav, J. S.; Reddy, B. V. S.; Srihari, P., *Synlett* **2001**, 673.
65. Komatsu, N.; Uda, M.; Suzuki, H., *Tetrahedron Lett.* **1997**, *38*, 7215.
66. Marx, A.; Yamamoto, H., *Angew. Chem., Int. Ed.* **2000**, *39*, 178.
67. Fang, X.; Watkin, J. G.; Warner, B. P., *Tetrahedron Lett.* **2000**, *41*, 447.
68. Ishihara, K.; Hiraiwa, Y.; Yamamoto, H., *Synlett* **2001**, 1851.
69. Watahiki, T.; Oriyama, T., *Tetrahedron Lett.* **2002**, *43*, 8959.
70. Kanai, M.; Kuramochi, A.; Shibasaki, M., *Synthesis* **2002**, 1956.
71. Sun, P.; Xian, Y.; Xiao, Y., *J. Chem. Res.* **2004**, 216.
72. (a) Onishi, Y.; Ito, T.; Yasuda, M.; Baba, A., *Eur. J. Org. Chem.* **2002**, 1578. (b) Onishi, Y.; Ito, T.; Yasuda, M.; Baba, A., *Tetrahedron Lett.* **2002**, *58*, 8227.
73. Kaur, G.; Manju, K.; Trehan, S., *Chem. Commun.* **1996**, 581.
74. Chan, T. H.; Fleming, I., *Synthesis* **1979**, 761.
75. Ishihara, K.; Hasegawa, A.; Yamamoto, H., *Angew. Chem., Int. Ed.* **2001**, *40*, 4077.
76. Sasidharan, M.; Tatsumi, T., *Chem. Lett.* **2003**, *32*, 624.
77. Yadav, J. S.; Chand, P. K.; Anjaneyulu, S., *Tetrahedron Lett.* **2002**, *43*, 3783.
78. (a) Monti, H.; Afshari, M.; Léandri, G., *J. Organomet. Chem.* **1995**, *486*, 69. (b) Ishii, A.; Kotera, O.; Saeki, T.; Mikami, K., *Synlett* **1997**, 1145. (c) Michaut, M.; Santelli, M.; Parrain, J. L., *J. Organomet. Chem.* **2000**, *606*, 93.
79. Trehan, A.; Vij, A.; Walia, M.; Kaur, G.; Verma, R. D.; Trehan, S., *Tetrahedron Lett.* **1993**, *34*, 7335.
80. Hachiya, I.; Kobayashi, S., *Tetrahedron Lett.* **1994**, *35*, 3319.
81. Ishii, A.; Kotera, O.; Saeki, T.; Mikami, K., *Synlett* **1997**, 1145.
82. Yadav, J. S.; Reddy, B. V. S.; Madhuri, Ch.; Sabitha, G., *Chem. Lett.* **2001**, 18.
83. Wieland, L. C.; Zerth, H. M.; Mohan, R. S., *Tetrahedron Lett.* **2002**, *43*, 4597.
84. Watahiki, T.; Akabane, Y.; Mori, S.; Oriyama, T., *Org. Lett.* **2003**, *5*, 3045.
85. Jung, M. E.; Maderna, A., *Tetrahedron Lett.* **2004**, *45*, 5301.
86. Zerth, H. M.; Leonard, N. M.; Mohan, R. S., *Org. Lett.* **2003**, *5*, 55.
87. Jung, M. E.; Maderna, A., *J. Org. Chem.* **2004**, *69*, 7755.
88. Mayr, H.; Gorath, G.; Bauer, B., *Angew. Chem., Int. Ed.* **1994**, *33*, 788.
89. Kuhnert, N.; Peverley, J.; Robertson, J., *Tetrahedron Lett.* **1998**, *39*, 3215.
90. Lee, P. H.; Lee, K.; Sung, S.; Chang, S., *J. Org. Chem.* **2001**, *66*, 8646.
91. Yadav, J. S.; Reddy, B. V. S.; Sadasiv, K.; Satheesh, G., *Tetrahedron Lett.* **2002**, *43*, 9695.
92. Le Roux, C.; Dubac, J., *Organometallics* **1996**, *15*, 4646.
93. (a) Olah, G. A.; VanVliet, D. S.; Wang, Q.; Prakash, G. K. S., *Synthesis* **1995**, 159. (b) Mayr, H.; Gabriel, A. O.; Schumacher, R., *Liebigs Ann.* **1995**, 1583.
94. (a) Yang, Y.; Wang, M.; Wang, D., *Chem Commun.* **1997**, 1651. (b) Wang, M. W.; Chen, Y. J.; Liu, L.; Wang, D.; Liu, X. L., *J. Chem. Res. (S)* **2000**, 80.
95. (a) Wang, M. W.; Chen, Y. J.; Wang, D., *Synlett* **2000**, 385. (b) Wang, M. W.; Chen, Y. J.; Wang, D., *Heteroatom Chem.* **2001**, *12*, 534.
96. Yokozawa, T.; Furuhashi, K.; Natsume, H., *Tetrahedron Lett.* **1995**, *36*, 5243.
97. (a) Ishikawa, T.; Okano, M.; Aikawa, T.; Saito, S., *J. Org. Chem.* **2001**, *66*, 4635. (b) Ishikawa, T.; Aikawa, T.; Mori, Y.; Saito, S., *Org. Lett.* **2004**, *6*, 1369.
98. (a) Rubin, M.; Gevorgyan, V., *Org. Lett.* **2001**, *3*, 2705. (b) Schwier, T.; Rubin, M.; Gevorgyan, V., *Org. Lett.* **2004**, *6*, 1999.
99. Kaur, G.; Kaushik, M.; Trehan, S., *Tetrahedron Lett.* **1997**, *38*, 2521.
100. Luzung, M. R.; Toste, F. D., *J. Am. Chem. Soc.* **2003**, *125*, 15760.
101. Yasuda, M.; Saito, T.; Ueba, M.; Baba, A., *Angew. Chem., Int. Ed.* **2004**, *43*, 1414.
102. Yasuda, M.; Onishi, Y.; Ito, T.; Baba, A., *Tetrahedron Lett.* **2000**, *41*, 2425.
103. Yadav, J. S.; Reddy, B. V. S.; Rao, K. V.; Raj, K. S.; Rao, P. P.; Prasad, A. R.; Gunasekar, D., *Tetrahedron Lett.* **2004**, *45*, 6505.
104. (a) Veenstra, S. J.; Schmid, P., *Tetrahedron Lett.* **1997**, *38*, 997. (b) Masuyama, Y.; Tosa, J.; Kurusu, Y., *Chem. Commun.* **1999**, 1075.
105. Niimi, L.; Serita, K.; Hiraoka, S.; Yokozawa, T., *Tetrahedron Lett.* **2000**, *41*, 7075.
106. Ollevier, T.; Ba, T., *Tetrahedron Lett.* **2003**, *44*, 9003.
107. Phukan, P., *J. Org. Chem.* **2004**, *69*, 4005.
108. (a) Meester, W. J. N.; Rutjes, F. P. J. T.; Hermkens, P. H. H.; Hiemstra, H., *Tetrahedron Lett.* **1999**, *40*, 1601. (b) Meester, W. J. N.; van Maarseveen, J. H.; Kirchsteiger, K.; Hermkens, P. H. H.; Schoemaker, H. E.; Hiemstra, H.; Rutjes, F. P. J. T., *ARKIVOC* **2004**, 122.
109. Yamaguchi, R.; Nakayasu, T.; Hatano, B.; Nagura, T.; Kozima, S.; Fujita, K., *Tetrahedron Lett.* **2001**, *57*, 109.
110. (a) Yeon, S. H.; Lee, B. W.; Yoo, B. R.; Suk, M. Y.; Jung, I. N., *Organometallics* **1995**, *14*, 2361. (b) Choi, G. M.; Yoo, B. R.; Lee, H. J.; Lee, K. B.; Jung, I. N., *Organometallics* **1998**, *17*, 2409.
111. (a) Yeon, S. H.; Han, J. S.; Hong, E.; Do, Y.; Jung, I. N., *J. Organomet. Chem.* **1995**, *499*, 159. (b) Jung, I. N.; Yoo, B. R., *Synlett* **1999**, 519.
112. (a) Asao, N.; Yoshikawa, E.; Yamamoto, Y., *J. Org. Chem.* **1996**, *61*, 4874. (b) Yoshikawa, E.; Gevorgyan, V.; Asao, N.; Yamamoto, Y., *J. Am. Chem. Soc.* **1997**, *119*, 6781. (c) Asao, N.; Yamamoto, Y., *Bull. Chem. Soc. Jpn.* **2000**, *73*, 1071.
113. Yamaguchi, M.; Sotokawa, T.; Hirama, M., *Chem. Commun.* **1997**, 743.
114. Ishihara, K.; Mouri, M.; Gao, Q.; Maruyama, T.; Furuta, K.; Yamamoto, H., *J. Am. Chem. Soc.* **1993**, *115*, 11490.
115. Aoki, S.; Mikami, K.; Terada, M.; Nakai, T., *Tetrahedron Lett.* **1993**, *49*, 1783.
116. (a) Gauthier, D. R., Jr.; Carreira, E. M., *Angew. Chem., Int. Ed.* **1996**, *35*, 2363. (b) Bode, J. W.; Gauthier, D. R., Jr.; Carreira, E. M., *Chem. Commun.* **2001**, 2560.
117. Nakamura, K.; Nakamura, H.; Yamamoto, Y., *J. Org. Chem.* **1999**, *64*, 2614.
118. Fang, X.; Johannsen, M.; Yao, S.; Gathergood, N.; Hazell, R. G.; Jørgensen, K. A., *J. Org. Chem.* **1999**, *64*, 4844.
119. Porter, N. A.; Wu, J. H.; Zhang, G.; Reed, A. D., *J. Org. Chem.* **1997**, *62*, 6702.
120. (a) Tietze, L. F.; Dölle, A.; Schiemann, K., *Angew. Chem., Int. Ed.* **1992**, *31*, 1372. (b) Tietze, L. F.; Schiemann, K.; Wegner, C.; Wulff, C., *Chem. Eur. J.* **1996**, *2*, 1164. (c) Tietze, L. F.; Wulff, C.; Wegner, C.; Schuffenhauer, A.; Schiemann, K., *J. Am. Chem. Soc.* **1998**, *120*, 4276.
121. (a) Tietze, L. F.; Schiemann, K.; Wegner, C., *J. Am. Chem. Soc.* **1995**, *117*, 5851. (b) Tietze, L. F.; Wegner, C.; Wulff, C., *Synlett* **1996**, 471. (c) Tietze, L. F.; Wegner, C.; Wulff, C., *Eur. J. Org. Chem.* **1998**, 1639. (d) Tietze, L. F.; Schiemann, K.; Wegner, C.; Wulff, C., *Chem. Eur. J.* **1998**, *4*, 1862. (e) Tietze, L. F.; Weigand, B.; Völkel, L.; Wulff, C.; Bittner, C., *Chem. Eur. J.* **2001**, *7*, 161.
122. (a) Kudo, K.; Hashimoto, Y.; Sukegawa, M.; Hasegawa, M.; Saigo, K., *J. Org. Chem.* **1993**, *58*, 579. (b) Andrus, M. B.; Lepore, S. D., *Tetrahedron Lett.* **1995**, *36*, 9149. (c) Maeda, K.; Shinokubo, H.; Oshima, K., *J. Org. Chem.* **1997**, *62*, 6429. (d) Manju, K.; Trehan, S., *Chem. Commun.* **1999**, 1929. (e) Cossrow, J.; Rychnovsky, S. D., *Org. Lett.* **2002**, *4*, 147. (f) Powell, S. A.; Tenenbaum, J. M.; Woerpel, K. A., *J. Am. Chem. Soc.* **2002**, *124*, 12648.
123. (a) Guindon, Y.; Guérin, B.; Chabot, C.; Ogilvie, W., *J. Am. Chem. Soc.* **1996**, *118*, 12528. (b) Guindon, Y.; Guérin, B.; Rancourt, J.; Chabot, C.; Mackintosh, N.; Ogilvie, W. W., *Pure. Appl. Chem.* **1996**, *68*, 89.

124. Friestad, G. K.; Ding, H., *Angew. Chem., Int. Ed.* **2001**, *40*, 4491.

125. (a) Wu, M. J.; Wu, C. C.; Lee, P. C., *Tetrahedron Lett.* **1992**, *33*, 2547. (b) Wu, M. J.; Yeh, J. Y., *Tetrahedron Lett.* **1994**, *50*, 1073.

126. Sur, S.; Ganesh, S.; Pal, D.; Puranik, V. G.; Chakrabarti, P.; Sarkar, A., *J. Org. Chem.* **1996**, *61*, 8362.

127. (a) Snider, B. B.; Rodini, D. J.; Conn, R. S.; Sealfon, S., *J. Am. Chem. Soc.* **1979**, *101*, 5283. (b) Monti, H.; Audran, G.; Léandri, G.; Monti, J. P., *Tetrahedron Lett.* **1994**, *35*, 3073. (c) Dang, H. S.; Davies, A. G., *J. Organomet. Chem.* **1998**, *553*, 67.

128. (a) Akiyama, T.; Ishikawa, K.; Ozaki, S., *Chem. Lett.* **1994**, 627. (b) Akiyama, T.; Yasusa, T.; Ishikawa, K.; Ozaki, S., *Tetrahedron Lett.* **1994**, *35*, 8401.

129. (a) Monti, H.; Rizzotto, D.; Léandri, G., *Tetrahedron Lett.* **1998**, *54*, 6725. (b) Sugita, Y.; Yamadoi, S.; Hosoya, H.; Yokoe, I., *Chem. Pharm. Bull.* **2001**, *49*, 657.

130. Choi, G. M.; Yeon, S. H.; Jin, J.; Yoo, B. R.; Jung, I. N., *Organometallics* **1997**, *16*, 5158.

131. Giese, S.; Kastrup, L.; Stiens, D.; West, F. G., *Angew. Chem., Int. Ed.* **2000**, *39*, 1970.

132. (a) Sugita, Y.; Kimura, Y.; Yokoe, I., *Tetrahedron Lett.* **1999**, *40*, 5877. (b) Yadav, J. S.; Reddy, B. V. S.; Pandey, S. K.; Srihari, P.; Prathap, I., *Tetrahedron Lett.* **2001**, *42*, 9089.

133. (a) Katagiri, N.; Sato, H.; Kurimoto, A.; Okada, M.; Yamada, A.; Kaneko, C., *J. Org. Chem.* **1994**, *59*, 8101. (b) Katagiri, N.; Okada, M.; Kaneko, C.; Furuya, T., *Tetrahedron Lett.* **1996**, *37*, 1801. (c) Katagiri, N.; Okada, M.; Morishita, Y.; Kaneko, C., *Tetrahedron Lett.* **1997**, *53*, 5725.

134. Angle, S. R.; El-Said, N. A., *J. Am. Chem. Soc.* **1999**, *121*, 10211.

135. Katritzky, A. R.; Cui, X.; Long, Q., *J. Heterocycl. Chem.* **1999**, *36*, 371.

136. Yachi, K.; Maeda, K.; Shinokubo, H.; Oshima, K., *Tetrahedron Lett.* **1997**, *38*, 5161.

137. Maeda, K.; Shinokubo, H.; Oshima, K., *J. Org. Chem.* **1998**, *63*, 4558.

138. Horiuchi, Y.; Oshima, K.; Utimoto, K., *J. Org. Chem.* **1996**, *61*, 4483.

139. Chung, W. J.; Welch, J. T., *J. Fluorine Chem.* **2004**, *125*, 543.

140. (a) Hwu, J. R.; Chen, B. L.; Shiao, S. S., *J. Org. Chem.* **1995**, *60*, 2448. (b) Hwu, J. R.; King, K. Y.; Wu, I. F.; Hakimelahi, G. H., *Tetrahedron Lett.* **1998**, *39*, 3721.

141. (a) Hwu, J. R.; Chen, C. N.; Shiao, S. S., *J. Org. Chem.* **1995**, *60*, 856. (b) Zhang, Y.; Raines, A. J.; Flowers, R. A, II, *Org. Lett.* **2003**, *5*, 2363. (c) Zhang, Y.; Raines, A. J.; Flowers, R. A., II, *J. Org. Chem.* **2004**, *69*, 6267.

142. (a) Hirao, T.; Fujii, T.; Ohshiro, Y., *Tetrahedron Lett.* **1994**, *50*, 10207. (b) Hirao, T.; Sakaguchi, M.; Ishikawa, T.; Ikeda, I., *Synth. Commun.* **1995**, *25*, 2579.

143. Hirao, T.; Fujii, T.; Ohshiro, Y., *Tetrahedron Lett.* **1994**, *35*, 8005.

144. Dussault, P. H.; Lee, I. Q., *J. Am. Chem. Soc.* **1993**, *115*, 6458.

145. Hermans, B.; Hevesi, L., *Bull. Soc. Chim. Belg.* **1994**, *103*, 257.

146. Hermans, B.; Hevesi, L., *J. Org. Chem.* **1995**, *60*, 6141.

147. Silveira, C. C.; Fiorin, G. L.; Braga, A. L., *Tetrahedron Lett.* **1996**, *37*, 6085.

148. (a) Dussault, P. H.; Zope, U., *Tetrahedron Lett.* **1995**, *36*, 3655. (b) Dussault, P. H.; Liu, X., *Tetrahedron Lett.* **1999**, *40*, 6553.

149. Matano, Y.; Yoshimune, M.; Suzuki, H., *Tetrahedron Lett.* **1995**, *36*, 7475.

150. Ipaktschi, J.; Heydari, A., *Angew. Chem., Int. Ed.* **1992**, *31*, 313.

151. Barluenga, J.; Vázquez-Villa, H.; Ballesteros, A.; Gnzález, J. M., *J. Am. Chem. Soc.* **2003**, *125*, 9028.

152. (a) Kim, D. Y.; Choi, J. S.; Rhie, D. Y.; Chang, S. K.; Kim, I. K., *Synth. Commun.* **1997**, *27*, 2753. (b) Kim, D. Y.; Kim, H. S.; Park, E. J.; Mang, J. Y.; Lee, K., *Bull. Korean Chem. Soc.* **2001**, *22*, 315.

153. Maeta, H.; Nagasawa, T.; Handa, Y.; Takei, T.; Osamura, Y.; Suzuki, K., *Tetrahedron Lett.* **1995**, *36*, 899.

154. Schwesinger, R.; Link, R.; Thiele, G.; Rotter, H.; Honert, D.; Limbach, H. H.; Männle, F., *Angew. Chem., Int. Ed. Engl.* **1991**, *30*, 1372.

155. Pilcher, A. S.; DeShong, P., *J. Org. Chem.* **1996**, *61*, 6901.

156. Mori, A.; Fujita, A.; Ikegashira, K.; Nishihara, Y.; Hiyama, T., *Synlett* **1997**, 693.

157. Wang, D. K.; Zhou, Y. G.; Tang, Y.; Hou, X. L.; Dai, L. X., *J. Org. Chem.* **1999**, *64*, 4233.

158. (a) Capperucci, A.; Ferrara, M. C.; Degl'Innocenti, A.; Bonini, B. F.; Mazzanti, G.; Zani, P.; Ricci, A., *Synlett* **1992**, 880. (b) Capperucci, A.; Degl'Innocenti, A.; Leriverend, C.; Metzner, P., *J. Org. Chem.* **1996**, *61*, 7174.

159. Wang, Z.; Kisanga, P.; Verkade, J. G., *J. Org. Chem.* **1999**, *64*, 6459.

160. Chan, T. H.; Wang, D., *Chem. Rev.* **1995**, *95*, 1279.

161. Hoffman, R. V.; Nayyar, N. K.; Shankweiler, J. M.; Klinekole, B. W., III, *Tetrahedron Lett.* **1994**, *35*, 3231.

162. Ofial, A. R.; Mayr, H., *J. Org. Chem.* **1996**, *61*, 5823.

163. Ryu, I.; Yamamoto, H.; Sonoda, N.; Murai, S., *Organometallics* **1996**, *15*, 5459.

164. (a) Yoshida, J.; Suga, S.; Suzuki, S.; Kinomura, N.; Yamamoto, A.; Fujiwara, K., *J. Am. Chem. Soc.* **1999**, *121*, 9546. (b) Suga, S.; Okajima, M.; Fujiwara, K.; Yoshida, J., *J. Am. Chem. Soc.* **2001**, *123*, 7941.

165. (a) Hayamizu, T.; Maeda, H.; Ikeda, M.; Mizuno, K., *Tetrahedron Lett.* **2001**, *42*, 2361. (b) Hayamizu, T.; Maeda, H.; Mizuno, K., *J. Org. Chem.* **2004**, *69*, 4997.

Allyltrimethylsilylzinc Chloride

[89822-47-9]   C6H13ClSiZn   (MW 214.09)

InChI = 1S/C6H13Si.ClH.Zn/c1-5-6-7(2,3)4;;/h5-6H,1H2, 2-4H3;1H;/q;;+1/p-1

InChIKey = XLNRPGZRGSEEKG-UHFFFAOYSA-M

(allylmetal reagent; useful for synthesis of spiro-γ-lactones[1])

Preparative Methods: deprotonation of allyltrimethylsilane with *n*-butyllithium in THF at −78 °C or *s*-butyllithium in THF/TMEDA, followed by the addition of zinc chloride.[1,2] The reagent is prepared in situ and used directly.

Spiro-γ-lactone Synthesis via Reactions with Carbonyl Compounds. Although lithiated allyltrimethylsilane adds to simple ketones with good regioselectivity (eq 1),[2] its addition to 17-ketosteroids produces a mixture of α and γ adducts. Transmetalation of allyltrimethylsilyllithium with ZnCl2 leads to a reagent (allyltrimethylsilylzinc chloride) which adds with complete regioselectivity to this type of hindered ketone (eq 2).[1] The resulting silylated homoallylic alcohols can be converted by a simple succession of steps to valuable steroidal γ-lactones. It is interesting to note that the 4-hydroxybutenylsilane obtained by the addition of the allyltrimethylsilyl anion to adamantan-2-one undergoes some unusual transformations[3] in the presence of *N*-bromosuccinimide, leading to the oxetane (**1**) via a 4-*exo*-tet ring opening[4] of the intermediate bromonium ion (eq 3). The extent of α and γ selectivity can be fine-tuned by changing the metal, and very high α-regioselectivity is observed using magnesium as the counterion.[5]

Similarly, the addition of triethylaluminum to the lithium reagent leads to a high α-regioselectivity (eq 4).[6] A stereocontrolled synthesis of *anti*-homoallylic alcohols is possible by preparing the corresponding allylic boronic ester and subsequent reaction with an aldehyde. A stereoselective elimination allows the preparation of either (*E*)- or (*Z*)-dienes (eq 5).[7] The isomeric reagent (**1**) is a useful reagent for the preparation of 2-substituted polyfunctional allylic silanes (eq 6).[8]

6. (a) Yamamoto, H.; Saito, Y.; Maruyama, K., *J. Chem. Soc., Chem. Commun.* **1982**, 1326; (b) Yamamoto, Y.; Saito, Y.; Maruyama, K., *Tetrahedron Lett.* **1982**, *23*, 4597; (c) Yamamoto, Y.; Yatagai, H.; Saito, Y.; Maruyama, K., *J. Org. Chem.* **1984**, *49*, 1096.

7. (a) Tsai, D. J. S.; Matteson, D. S., *Tetrahedron Lett.* **1981**, *22*, 2751; (b) Sato, F.; Suzuki, Y.; Sato, M., *Tetrahedron Lett.* **1982**, *23*, 4589.

8. Minato, A.; Suzuki, K.; Tamao, K.; Kumada, M., *Tetrahedron Lett.* **1984**, *25*, 83.

Paul Knochel
Philipps-Universität Marburg, Marburg, Germany

Ammonium Hexafluorosilicate

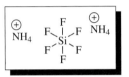

[16919-19-0] $H_8F_6N_2Si$ (MW 178.15)

InChI = 1S/F6Si.2H3N/c1-7(2,3,4,5)6;;/h;2*1H3/q-2;;/p+2

InChIKey = ITHIMUMYFVCXSL-UHFFFAOYSA-P

Physical Data: d 2.01 g cm^{-3}; decomposes prior to melting; odorless.

Solubility: water solubility 18 g/100 g water.

Form Supplied in: white crystalline solid, widely available.

Handling, Storage, and Precautions: store in plastic/poly airtight containers, in a cool, dry place, away from acids and oxidizers. Unsuitable for glass or ceramic containers. Stable compound, decomposes upon heating to HF, ammonia, and nitrogen oxide (NO$_x$) gases. Handle with care in well-ventilated area. Irritant to skin and mucus membranes, however, acute and chronic toxicity of this substance is not fully known.

Epoxide Ring Opening. Recently, Shimizu and Nakahara reported the epoxide ring opening[1] (eq 1) to generate fluorohydrin (**2**). Using only 47% HF provided **2** in low chemical yield (34%); however, the addition of ammonium hexafluorosilicate (AHFS) increased the yield to 46%, and, further, the addition of CsF also increased the yield, up to 67%.

$$\text{(1)}$$

AHFS has also been employed in the opening of epoxides of type **3**.[1] Epoxide **3** is transformed (eq 2) to the fluorohydrin with selectivity for the *syn*-isomer (**4**). In eq 2 without the CsF additive, the *syn*-selectivity was 7:1. However, in the presence of CsF the selectivity was 20:1 in favor of the *syn*-diasteromer *syn* **4**.

$$\text{(2)}$$

1. Ehlinger, E.; Magnus, P., *Tetrahedron Lett.* **1980**, *21*, 11.

2. Ayalon-Chass, D.; Ehlinger, E.; Magnus, P., *J. Chem. Soc., Chem. Commun.* **1977**, 772.

3. Ehlinger, E.; Magnus, P., *J. Chem. Soc., Chem. Commun.* **1979**, 421.

4. Baldwin, J. E., *J. Chem. Soc., Chem. Commun.* **1976**, 734.

5. Lau, P. W. K.; Chan, T. H., *Tetrahedron Lett.* **1978**, 2383.

Trisubstituted epoxides and trisubstituted epoxysilanes (eqs 3 and 4, respectively) have also been examined.[1] However, these substrates required the addition of $(^i Pr)_2 NEt$ to prevent rearrangement of the epoxide.[2–4]

$$\text{(3)}$$

$$\text{(4)}$$

Dealumination of Zeolites. AHFS has been mainly used for the preparation of different materials, mainly inorganic substances. AHFS has been used for the very effective dealumination of zeolites[5,6] to prepare a number of different materials. These materials include silica nanoboxes[7] and activated zeolite catalysts for the oxidation of benzene to phenol with NO_2.[8]

1. Shimizu, M.; Nakahara, Y., *J. Fluorine Chem.* **1999**, *99*, 95.
2. Richborn, B.; Gerkin, R., *J. Am. Chem. Soc.* **1971**, *93*, 1963.
3. Coxon, J. N.; Harshorn, M. P.; Rae, W. J., *Tetrahedron* **1970**, *26*, 1091.
4. Hart, H.; Lerner, L. R., *J. Org. Chem.* **1967**, *72*, 2669.
5. Le Van Mao, R.; Vo, N. T. C.; Denes, G.; Le, T. S., *J. Porous Mater.* **1995**, *1*, 175.
6. Le Van Mao, R.; Lavigne, J. A.; Sjiariel, B.; Langford, C. H., *J. Mater. Chem.*; **1993**, *3, 6*, 679.
7. Lu, L.; Le Van Mao, R.; Al-Yassir, N.; Muntasar, A.; Vu, N. T., *Catal. Lett.* *105*(3–4), 139.
8. Kollmer, F.; Hausmann, H.; Hoelderich, W. F., *J. Catal.* **2004**, *227*, 408.

Christopher G. McDaniel & Jon R. Parquette
The Ohio State University, Columbus, OH, USA

Azidotrimethylsilane[1]

[4648-54-8] $C_3H_9N_3Si$ (MW 115.21)
InChI = 1S/C3H9N3Si/c1-7(2,3)6-5-4/h1-3H3
InChIKey = SEDZOYHHAIAQIW-UHFFFAOYSA-N

(azidation of organic halides, acetals, and esters; α- and β-siloxy azides from carbonyl compounds and epoxides; preparation of heterocyclic compounds; effective substitute for hydrazoic acid)

Alternative Name: trimethylsilyl azide; TMSA.
Physical Data: bp 95–96 °C; n_D^{20} 1.416; d 0.868 g cm^{-3}; fp 23 °C; IR ν_{max} 2100 cm^{-1}; ^{1}H NMR (CDCl$_3$) δ = 0.22.

Solubility: sol most organic solvents.
Form Supplied in: colorless liquid.
Preparative Methods: several methods for the synthesis of this azide have been reported.[1] The procedure involving aluminum chloride is not recommended, since an explosive product is formed.[2] Azidotrimethylsilane is now commercially available, and a representative synthetic procedure is as follows. A mixture of sodium azide and chlorotrimethylsilane is refluxed in di-*n*-butyl ether for 2 days and the azide is safely distilled directly from the reaction vessel. Purer compound (99% content) is obtained by redistillation of the product. Several improved conditions have been reported for the preparation of this azide.[20–22] In these procedures, trimethylsilyl chloride is reacted with sodium azide either neat or in a high boiling point solvent, such as a mixture of silicone oil and polyethylene glycol. Distillation of the crude product usually provides trimethylsilyl azide (TMSA) in high purity (97.9%) and yield (97%).[22]
Handling, Storage, and Precautions: the azide, which decomposes at 500 °C,[1b] is more stable thermally than most organic azides and can be stored in a refrigerator for more than one year. It is moisture sensitive, and should be handled with care since it reacts rapidly with water to release toxic hydrazoic acid. Handle in a fume hood. When inhaled or absorbed, TMSA can release hydrazoic acid, which is a potent vasodilator, in the body. A recent report has shown a synergetic effect between TMSA and some blood pressure medications; therefore, people taking those medications should be extremely careful when dealing with this reagent.[23]

Original Commentary

Kozaburo Nishiyama
Tokai University, Shizuoka, Japan

Many applications of TMSA in organic synthesis have been reported but only representative examples are described herein.

Substitution Reactions. Benzyl, allyl, and substituted alkyl halides are converted to the corresponding azides in 60–100% yields via reactions with TMSA under neutral conditions in a nonaqueous solvent (eq 1).[3] By using tin(IV) chloride as a catalyst, secondary and tertiary cyclic and polycyclic halides are similarly transformed into the corresponding azides in 50–92% yields (eq 1).[4]

$$Me_3SiN_3 + RX \longrightarrow RN_3 + Me_3SiX \qquad (1)$$

$$R = ArCH_2, CH_2CH=CH_2, CH_2CN, CH_2CO_2Et,$$
$$c\text{-}C_5H_9, c\text{-}C_6H_{11}, \text{1-, 2-, 3-, and 4-adamantyl}$$
$$X = Cl, Br$$

Acyl halides react with TMSA to give trimethylsilyl halides and the corresponding isocyanates in a range of 83–98% yields, which are rearranged from the acyl azides via the Curtius degradation (eq 2).[5] However, a reaction of aroyl chlorides with TMSA in the presence of zinc iodide at 0 °C gives the corresponding azides in 85–96% yields (eq 3).[6]

$$Me_3SiN_3 + RCOX \longrightarrow RN{=}C{=}O + Me_3SiX \quad (2)$$

R = c-C_6H_{11}, Ph, p-MeC_6H_4, $CH_2{=}CH(CH_2)_8$–
$MeOCOCH_2CH_2$, phthaloyl, etc.

$$Me_3SiN_3 + ArCOX \xrightarrow[0\,°C]{ZnI_2} ArCON_3 + Me_3SiX \quad (3)$$

The azide reacts with orthoesters at reflux temperature and acetals in the presence of $SnCl_4$ at −78 °C to give the corresponding azides in 40–70% yields (eq 4).[7]

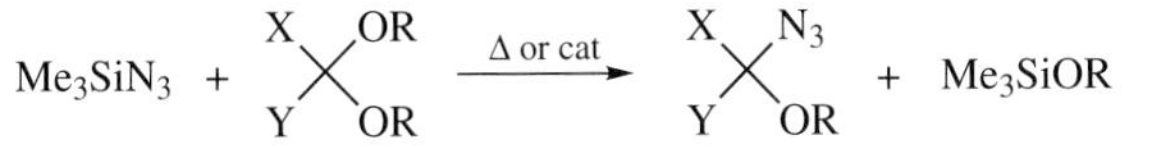
Me3SiN3 +
Δ or cat
+ Me3SiOR

(4)

X= alkyl, aryl, alkoxy

α-Siloxy Azides. TMSA reacts with carbonyl compounds to give α-siloxy azides in the presence of zinc chloride or tin(II) chloride. With aldehydes, α-siloxy azides are obtained in 70–80% yields,[8,9] but the yields of these compounds are very low with most ketones.[9] Carbonyl compounds including ketones are stepwisely or directly converted into *gem*-diazides (30–80%), tetrazoles (40–100%), and nitriles (60–100%), depending on the reaction conditions.[9] The reaction sequences are summarized in Scheme 1.

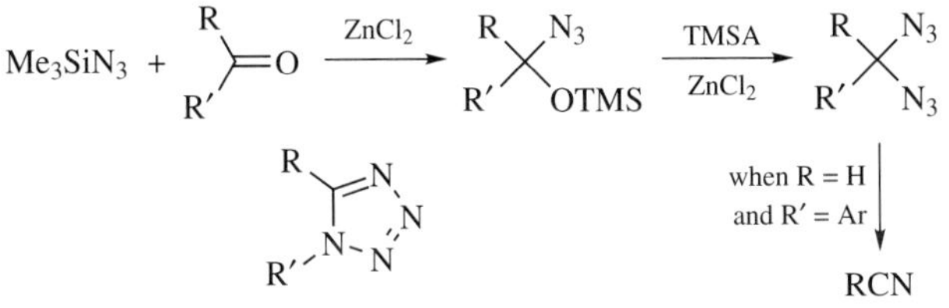
Me3SiN3 +
ZnCl2
TMSA
ZnCl2
when R = H
and R′ = Ar
RCN

Scheme 1

β-Siloxy Azides. The azide reacts with epoxides in the presence of zinc chloride,[8b] titanium and vanadium complexes,[10] zinc tartrate,[11] and aluminum isopropoxide[12] to give β-siloxy azides in 80–100% yields (eq 5), which are precursors of β-amino alcohols. With cyclohexene oxide, the *trans* isomer is formed exclusively, regardless of the catalyst used.[8b,10] However, the regio- and chemoselectivity of the reactions of other epoxides depend upon the catalyst.[10–12]

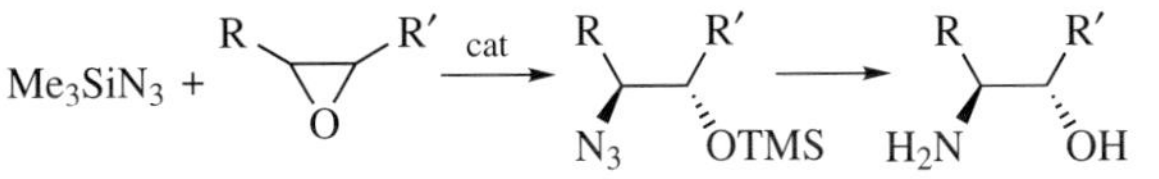
Me3SiN3 +
cat
OTMS
OH

(5)

R = Me, R′ = H; R = $CH_2{=}CH(CH_2)_2$, R′ = H;
R, R′ = -$(CH_2)_n$- (n = 3–5); etc.

Application of this procedure for the regio- and stereoselective synthesis of vinyl azides in the presence of boron trifluoride etherate complex is summarized in eq 6.[13]

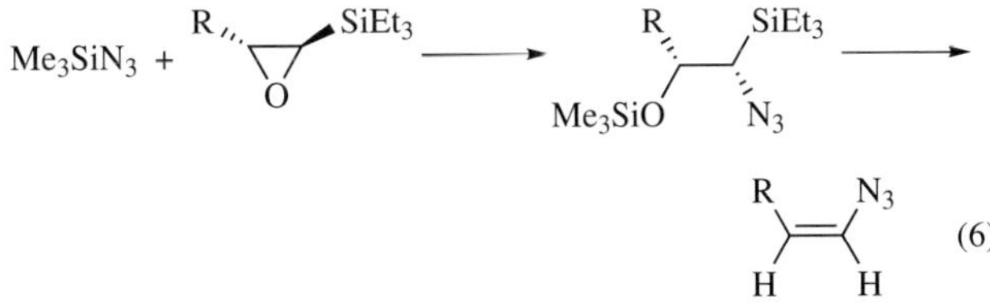
Me3SiN3 +
SiEt3
Me3SiO

(6)

ω-Siloxy Isocyanates. Acyclic anhydrides react with TMSA in a manner similar to acid halides to give equal amounts of trimethylsilyl esters and isocyanates (eq 7).[5b,14] Similarly, cyclic anhydrides react with the azide to give ω-trimethylsiloxycarbonyl alk(en)yl isocyanates (eq 8)[14a,15] which are further transformed into 1,3-oxazine-2,6-dione derivatives (70–90%).[14b]

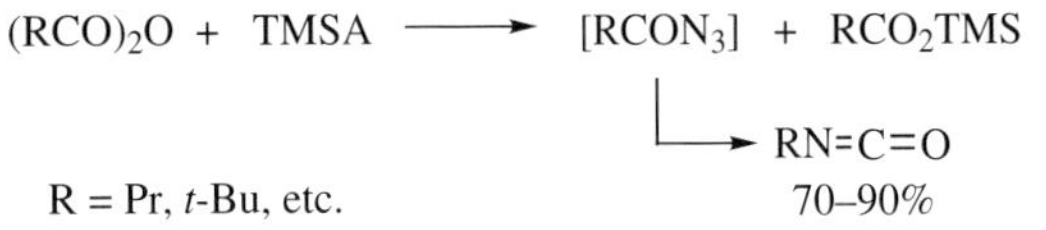
(RCO)2O + TMSA
[RCON3] + RCO2TMS
RN=C=O
70–90%

(7)

R = Pr, t-Bu, etc.

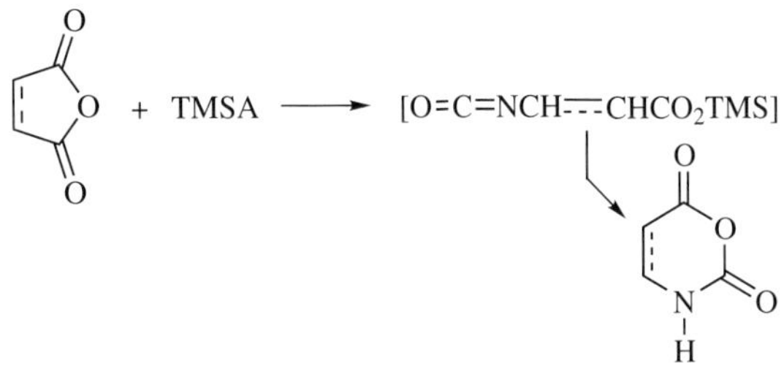
+ TMSA
[O=C=NCH=CHCO2TMS]

(8)

[3 + 2] Cycloaddition. TMSA, like other alkyl azides, reacts with alkynes and C-hetero multiple bonds to give [3 + 2] cycloaddition products, triazoles and tetrazoles.[1] For example, the reaction of TMSA with 2-butyne gives 4,5-dimethyl-2-trimethylsilyl-1,2,3-triazole in 78–87% yields (eq 9),[16] and TMSA and cyanoferrocene react to give 5-ferrocenyl-2-trimethylsilyltetrazole in 75% yield (eq 10).[17]

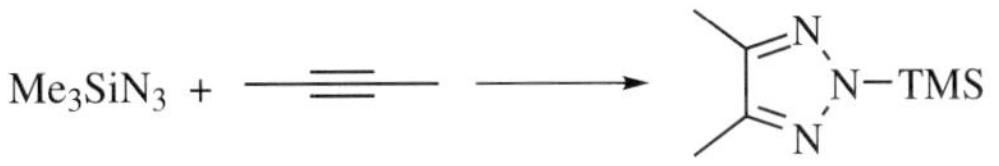
Me3SiN3 +
N–TMS

(9)

Me3SiN3 + FcC≡N
Fc
N–TMS

(10)

Fc = ferrocenyl

Lead and Phenyliodoso Azides. When lead(IV) acetate and phenyliodoso derivatives are used as a catalyst for azidation of alkenic and aromatic compounds (eq 11),[1,18] lead azides and phenyliodoso azides are formed as intermediates.

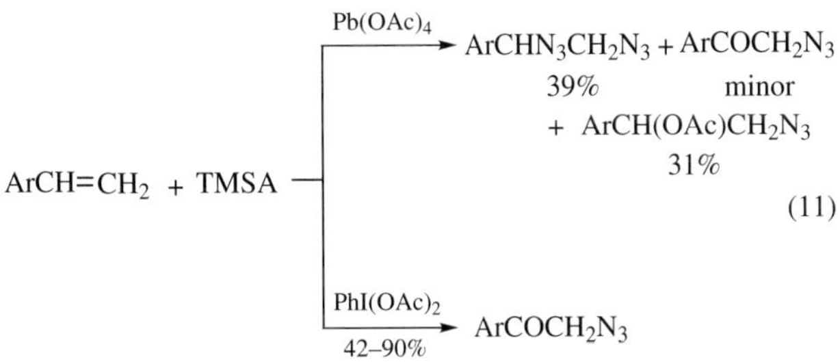
Pb(OAc)4
ArCHN3CH2N3 + ArCOCH2N3
39%
minor
+ ArCH(OAc)CH2N3
31%
ArCH=CH2 + TMSA
PhI(OAc)2
42–90%
ArCOCH2N3

(11)

Miscellaneous. A combination of the azide and triflic acid is a highly efficient electrophilic aromatic amination reagent system.[19]

First Update

Cheng Wang

Merck & Co., Inc., West Point, PA, USA

Substitution Reactions. The scope of substitution reactions possible with TMSA has been expanded beyond activated alkyl halides. Primary and secondary alkyl halides, phosphates, or tosylates are converted to the corresponding alkyl azides in 91–100% yields via reaction with TMSA induced by the fluoride anion (eq 12).[24,25]

$$Me_3SiN_3 \quad + \quad R-X \quad \xrightarrow[\text{THF, rt to } >70\,°C]{\text{TBAF}} \quad RN_3 \qquad (12)$$

R = primary, secondary alkyl
X = Cl, Br, I, OP(O)(OEt)$_2$, OTs

Allyl chlorides and allyl esters react with TMSA in the presence of lithium perchlorate[26] or a catalytic amount of a palladium catalyst to give the corresponding allyl azides (eq 13).[27] This transformation can also be used as a deprotection method for allyl esters.[28]

$$Me_3SiN_3 \quad + \quad \text{(alkene with } R^1, R^2, R^3, X) \quad \xrightarrow{\text{(a) or (b)}} \quad \text{(alkene with } N_3) \qquad (13)$$

(a) LiCLO$_4$ in ether; (b) Pd(0) (cat.), THF

X = Cl, OAc, OCO$_2$Et

β-Siloxy Azides and β-Amino Azides.

Epoxide Ring-opening Reactions. Ring-opening reactions of epoxides continue to be a major application of TMSA. Besides introduction of newer catalysts, e.g., tetraphenylstilbonium hydroxide[29] and lanthanide(III) triflates,[30] asymmetric ring opening (ARO) of epoxides has become a major focus of the research in this area. The palladium-catalyzed stereospecific azide substitution reaction of α,β-unsaturated γ,δ-epoxides with TMSA has been shown to yield azido alcohols with double inversion of configuration.[31] Reactions catalyzed by (salen)Cr(III) complexes (**1**) provide high enantioselectivity, e.g., opening of a variety of *meso*-epoxides with TMSA using catalyst **1** gave β-silyloxy or hydroxyl azides in 65–90% yields with 82–98% enantiomeric excess (eq 14).[32,33] On the contrary, reactions using chiral titanium and aluminum complexes, while viable, were shown to provide only modest enantioselectivity.[34–36]

A kinetic resolution strategy can be utilized for ring opening of racemic monosubstituted or 2,2-disubstituted[37] epoxides. As described in the literature,[32,37–39] substoichiometric amounts of TMSA (0.5–0.7 equiv) can be used with (salen)Cr(III)N$_3$ complexes as a catalyst to give enantiopure β-silyloxy azides in 74–98% yields with 89–98% enantiomeric excess, or the enantiopure (97–98% ee) unreacted starting epoxides (eq 15). Besides the chiral Cr(III) catalyst systems, β-cyclodextrin is also used as a catalyst in enantioselective ring opening of epoxides with TMSA under mild conditions.[40] β-Silyloxyl azides resulting from above asymmetric ring-opening reactions eventually can be reduced to

1,2-amino alcohols. This methodology has been used in several total syntheses.[36,41–43]

$$ \text{(14)} $$

Y = CH$_2$, O, NFmoc, NCOCF$_3$, CH=CH; C=O, etc.
n = 1, 2.

$$ R\text{-epoxide} \xrightarrow[\text{TMSN}_3]{(S,S)\ \textbf{1b}} \text{OTMS product} + \text{epoxide} \qquad (15) $$

R = Me, Et, *i*-Pr, ClCH$_2$, Bn, Ph, CH$_2$CN, etc.

Aziridine Ring-opening Reactions. TMSA also reacts with aziridines under neutral conditions[44] or in the presence of Lewis acids or Lewis bases.[45] The resulting β-azido amines can be used as precursors for 1,2-diamines (eq 16).[46–49]

$$ Me_3SiN_3 \quad + \quad \text{(aziridine with } P, R^1, R^2) \quad \longrightarrow \quad \text{(HN-P, N}_3) \quad \longrightarrow \qquad $$

$$ \text{(H}_2\text{N, NH}_2\text{, } R^1, R^2) \qquad (16) $$

P = Ts, Bn, 4-nitrobenzoyl, 3,5-dinitrobenzoyl

Similar to the ring-opening reaction of epoxides, Cr(III) complexes[50] (**2**) were reported to catalyze the enantioselective ring-opening reaction of *meso*-aziridines in 80–95% yields with 83–94% enantiomeric excess (eq 17). Other catalyst systems, including rare earth alkoxides with chiral ligands, have also proved efficient in this transformation.[49]

$$ Me_3SiN_3 \quad + \qquad \xrightarrow[\text{acetone, 4 Å sieves}]{\textbf{2}} \qquad \qquad (17) $$

R = Me, -(CH$_2$)$_4$-, -CH$_2$OCH$_2$-, -(CH$_2$)$_3$-, -CH$_2$CH=CHCH$_2$-

[3 + 2] Cycloaddition. Recently, additional examples of the use of TMSA in the synthesis of triazoles and tetrazoles have been reported.

Triazoles. TMSA can be utilized to form triazoles by [2 + 3] cycloaddition using alkynes[51,52] or nitro olefins[53] (eq 18) as coupling reagents. To achieve the corresponding intramolecular transformation in a tandem fashion, the precursor is usually obtained through either substitution[54] or ring opening[55] with TMSA, and then is used directly in the cycloaddition.

$$ (18) $$

R = various substituents,   X = CN, CO_2Et

Tetrazoles. Aryl nitriles and alkyl nitriles are converted into the corresponding tetrazoles in the presence of dibutyltin oxide[56,57] or trimethylaluminum[58] in moderate to good yields. Amides[59] or thioamides[60] also react with TMSA to form tetrazoles. Reaction of α,β-unsaturated ketones with TMSA can be catalyzed by TMSOTf in dichloromethane.[61] A tandem strategy for fused tetrazole formation has also been reported;[62] a [3 + 2] cycloaddition between a nitrile and TMSA in presence of dibutyltin oxide is followed by an intramolecular allylation to form a fused tetrazole in 60–82% yields.

Multicomponent Coupling Reactions.

Three-component Coupling Reactions. An aldehyde and an isocyanide can participate in the modified Passerini three-component coupling reaction with TMSA in dichloromethane to form a tetrazole in moderate to high yields.[63] In the Ugi-type three-component coupling reaction, TMSA reacts with an enamine and 1-isocyanomethylbenzotriazole (BetMIC) in methanol to provide a tetrazole in 24–98% yields.[64] Allyl methyl carbonate (AMC) is powerful partner for TMSA in multicomponent coupling reactions. For example, allyl aryl cyanamides with a wide variety of functional groups are obtained in excellent yields through the palladium-catalyzed three-component coupling reaction of aryl isocyanides, AMC, and TMSA (eq 19).[65,66] The reaction of AMC and TMSA with a nitrile, in the presence of a catalytic amount of $Pd_2(dba)_3$·$CHCl_3$ and a phosphine ligand, was shown to selectively produce 2-allyltetrazoles in good to excellent yields.[67]

$$ (19) $$

R = 4-MeO, 3-MeO, 2-MeO, 4-CN, 4-Cl, etc.

Similarly, the palladium-catalyzed three-component coupling reaction of activated alkynes, with TMSA and AMC, results in the regiospecific formation of 2-allyl-1,2,3-triazoles in moderate to good yields (eq 20).[68] The analogous three-component coupling reaction of monosubstituted alkynes lacking an electron-

withdrawing group requires a Pd–Cu bimetallic catalyst system, with the regiospecificity of the allyl substituent in the resulting triazoles dictated by the choice of either triphenylphosphite or triphenylphosphine as the ligand (eqs 21 and 22).[69–71]

$$ (20) $$

$$ (21) $$

$$ (22) $$

Four-component Coupling Reactions. TMSA is one of the components in the Ugi-five–center-four-component reaction (U-5C-4-CR). The other components for this reaction are an amine, an isocyanide, and an oxo-component, such as aldehyde or ketone. The reaction is usually carried out in methanol and provides tetrazoles in moderate to good yields.[72,73]

Oxidative Azidations. The combination of TMSA with chromium trioxide (CrO_3) is an excellent reagent for oxidation of alkyl and aryl aldehydes to the corresponding acyl azides in 64–100% yields.[74] Olefins are converted directly to α-azidoketones via oxidative azidation using the same reagent combination in good yields.[75] Another useful reagent combination leading to oxidative azidation is iodosobenzene (PhIO) and TMSA, which converts allyltrimethylsilanes into allyl azides in the presence of boron trifluoride etherate in 53–82% yields.[76] With this reagent combination, N,N-dimethylanilines are transformed into N-(azidomethyl)-N-methylanilines in excellent yields (>95%).[77] Using the same reagent, amides, carbamates, and ureas are oxidized to their α-azido derivatives in 25–82% yield,[78,79] as are the cyclic sulfides.[80] Another application of this combination is direct β-azido functionalization of triisopropylsilyl enol ethers by rapid addition of TMSA to a suspension of iodosobezene and the enol at −18 to −15 °C.[81] Adding a catalytic amount of the persistent radical TEMPO and lowering the reaction temperature alters the reaction outcome to complete double azidation (eq 23).[82] Yet another application of the PhIO/TMSA combination is a one-step homogeneous azidophenylselenylation of glycols,[83] which is more efficient than the stepwise procedure using TMSA/trimethylsilylnitrate[84] for the similar transformation.

$$(23)$$

Miscellaneous.

Conjugated Additions. Conjugated nitro olefins react with TMSA in the presence of TiCl$_4$ in dichloromethane at room temperature forming the corresponding α-azido hydroximoyl chloride in 71–87% yields.[85] Aliphatic nitro compounds, following activation with trimethylsilyl bromide to form the N,N-bis(siloxy) enamines, react with excess amounts of TMSA in dichloromethane in the presence of catalytic amount of triethylamine to give α-azido oximes in 89–100% yields.[86] Other conjugated systems, such as α,β-unsaturated carbonyl compounds, react with TMSA in the presence of acetic acid and catalytic amount of tertiary amines to form β-azido ketones in good yields.[87]

TMSA-mediated Schmidt Rearrangements. TMSA can be used in place of the explosive hydrazoic acid (HN$_3$) in the TMSA-mediated Schmidt Rearrangement. Thus, certain ketones are converted into the corresponding fused lactams by reaction with TMSA using trifluoroacetic acid (TFA)[88] or chloroform[89] as a solvent, with the regiospecificity of lactam formation being substrate dependent (eq 24).[88,89]

$$(24)$$

For example, $R^1 = R^2 = R^3 = H$, when $n = 1$, **a/b** = 4.5/1

$n = 0$, only **a**

Synthesis of Primary Amines. Primary amines are synthesized in moderate to good yields by reaction of TMSA with trialkylboranes (eq 25).[90]

$$(25)$$

O-Trimethylsilylation. Primary, allylic, benzylic secondary, hindered secondary, tertiary alcohols, and phenols are efficiently converted into the corresponding trimethylsilyl ethers when treated with neat TMSA and, when necessary, in the presence of a catalytic amount of tetrabutylammonium bromide (TBABr) in over 91% yields.[91] This reaction can be also used to achieve selective protection of primary and secondary alcohols in the presence of tertiary ones.

Related Reagents. Hydrazoic Acid; Sodium Azide; Tri-n-butyltin Azide.

1. (a) Weber, W. P. *Silicon Reagents for Organic Synthesis*; Springer: New York, 1983; p 40 and references cited therein; (b) Peterson, W. R. Jr, *Rev. Silicon, Germanium, Tin, Lead Compounds* **1974**, *1*, 193.

2. West, R.; Zigler, S., *Chem. Eng. News* **1984**, *62*(34), 4.

3. Nishiyama, K.; Karigomi, H., *Chem. Lett.* **1982**, 1477.

4. Prakash, G. K. S.; Stephenson, M. A.; Shih, J. G.; Olah, G. A., *J. Org. Chem.* **1986**, *51*,

5. (a) Kricheldorf, H. R., *Synthesis* **1972**, 551; (b) Washburne, S. S.; Peterson, W. R. Jr, *Synth. Commun.* **1972**, *2*, 227.

6. Prakash, G. K. S.; Iyer, P. S.; Arvanaghi, M.; Olah, G. A., *J. Org. Chem.* **1983**, *48*, 3358.

7. (a) Hartmann, W.; Heine, H.-G., *Tetrahedron Lett.* **1979**, 513. (b) Kirchmeyer, S.; Mertens, A.; Olah, G. A., *Synthesis* **1983**, 500. (c) Moriarty, R. M.; Hou, K.-C., *Synthesis* **1984**, 683.

8. (a) Birkofer, L.; Müller, F.; Kaiser, W., *Tetrahedron Lett.* **1967**, 2781. (b) Birkofer, L.; Kaiser, W., *Justus Liebigs Ann. Chem.* **1975**, 266.

9. (a) Nishiyama, K.; Watanabe, A., *Chem. Lett.* **1984**, 455. (b) Nishiyama, K.; Oba, M.; Wantanabe, A., *Tetrahedron* **1987**, *43*, 693. (c) Nishiyama, K.; Yamaguchi, T., *Synthesis* **1988**, 106.

10. Blandy, C.; Choukroun, R.; Gervais, D., *Tetrahedron Lett.* **1983**, *24*, 4189.

11. Yamashita, H., *Chem. Lett.* **1987**, 525.

12. Emziane, M.; Lhoste, P.; Sinou, D., *Synthesis* **1988**, 541.

13. Tomoda, S.; Matsumoto, Y.; Takeuchi, Y.; Nomura, Y., *Bull. Chem. Soc. Jpn.* **1986**, *59*, 3283.

14. (a) Washburne, S. S.; Peterson, W. R. Jr; Berman, D. A., *J. Org. Chem.* **1972**, *37*, 1738; (b) Kricheldorf, H. R., *Chem. Ber.* **1972**, *105*, 3958.

15. Kricheldorf, H. R.; Regel, W., *Chem. Ber.* **1973**, *106*, 3753.

16. Birkofer, L.; Wegner, P., *Chem. Ber.* **1966**, *99*, 2512.

17. Washburne, S. S.; Peterson, W. R. Jr, *J. Organomet. Chem.* **1970**, *21*, 427.

18. (a) Zbiral, E.; Kischa, K., *Tetrahedron Lett.* **1969**, 1167. (b) Ehrenfreund, J.; Zbiral, E., *Tetrahedron* **1972**, *28*, 1697.

19. Olah, G. A.; Ernst, T. D., *J. Org. Chem.* **1989**, *54*, 1204.

20. Sukata, K., *J. Org. Chem.* **1988**, *53*, 4867.

21. Fukunaga, T.; Oguro, K.; Mitsui, O.; Ota, R. JP A2 10 045 769, 19 980 217.

22. Kofukuda, T. WO A1 2 006 038 329, 20 060 413.

23. TerBeek, K. J., *Chem. Eng. News* **1998**, *76*, 6.

24. Ito, M.; Koyakumaru, K. i.; Ohta, T.; Takaya, H., *Synthesis* **1995**, 376.

25. Soli, E. D.; DeShong, P., *J. Org. Chem.* **1999**, *64*, 9724.

26. Pearson, W. H.; Schkeryantz, J. M., *J. Org. Chem.* **1992**, *57*, 2986.

27. Safi, M.; Fahrang, R.; Sinou, D., *Tetrahedron Lett.* **1990**, *31*, 527.

28. Shapiro, G.; Buechler, D., *Tetrahedron Lett.* **1994**, *35*, 5421.

29. Fujiwara, M.; Tanaka, M.; Baba, A.; Ando, H.; Souma, Y., *Tetrahedron Lett.* **1995**, *36*, 4849.

30. Crotti, P.; Di Bussolo, V.; Favero, L.; Macchia, F.; Pineschi, M., *Tetrahedron Lett.* **1996**, *37*, 1675.

31. Miyashita, M.; Mizutani, T.; Tadano, G.; Iwata, Y.; Miyazawa, M.; Tanino, K., *Angew. Chem., Int. Ed.* **2005**, *44*, 5094.

32. Martinez, L. E.; Leighton, J. L.; Carsten, D. H.; Jacobsen, E. N., *J. Am. Chem. Soc.* **1995**, *117*, 5897.

33. Leighton, J. L.; Jacobsen, E. N., *J. Org. Chem.* **1996**, *61*, 389.

34. Hayashi, M.; Kohmura, K.; Oguni, N., *Synlett* **1991**, 774.

35. Adolfsson, H.; Moberg, C., *Tetrahedron: Asymmetry* **1995**, *6*, 2023.

36. Schneider, U.; Pannecoucke, X.; Quirion, J. C., *Synlett* **2005**, 1853.

37. Lebel, H.; Jacobsen, E. N., *Tetrahedron Lett.* **1999**, *40*, 7303.

38. Larrow, J. F.; Schaus, S. E.; Jacobsen, E. N., *J. Am. Chem. Soc.* **1996**, *118*, 7420.

39. Schaus, S. E.; Jacobsen, E. N., *Tetrahedron Lett.* **1996**, *37*, 7937.

40. Kamal, A.; Arifuddin, M.; Rao, M. V., *Tetrahedron: Asymmetry* **1999**, *10*, 4261.

41. Wu, M. H.; Jacobesen, E. N., *Tetrahedron Lett.* **1997**, *38*, 1693.

42. Di Bussolo, V.; Caselli, M.; Romano, M. R.; Pineschi, M.; Crotti, P., *J. Org. Chem.* **2004**, *69*, 8702.

43. Alcaide, B.; Biurrun, C.; Martinez, A.; Plumet, J., *Tetrahedron Lett.* **1995**, *36*, 5417.

44. Wu, J.; Sun, X.; Xia, H. G., *Eur. J. Org. Chem.* **2005**, 4769.

45. Minakata, S.; Okada, Y.; Oderaotoshi, Y.; Komatsu, M., *Org. Lett.* **2005**, *7*, 3509.

46. Leung, W. H.; Yu, M. T.; Wu, M. C.; Yeung, L. L., *Tetrahedron Lett.* **1996**, *37*, 891.

47. Wu, J.; Hou, X. L.; Dai, L. X., *J. Org. Chem.* **2000**, *65*, 1344.

48. Chandrasekhar, M.; Sekar, G.; Singh, V. K., *Tetrahedron Lett.* **2000**, *41*, 10079.

49. Fukuta, Y.; Mita, T.; Fukuda, N.; Kanai, M.; Shibasaki, M., *J. Am. Chem. Soc.* **2006**, *128*, 6312.

50. Li, Z.; Fernandez, M.; Jacobsen, E. N., *Org. Lett.* **1999**, *1*, 1611.

51. Coats, S. J.; Link, J. S.; Gauthier, D.; Hlasta, D. J., *Org. Lett.* **2005**, *7*, 1469.

52. Demina, M. M.; Novopashin, P. S.; Sarapulova, G. I.; Larina, L. I.; Smolin, A. S.; Fundamenskii, V. S.; Kashaev, A. A.; Medvedeva, A. S., *Russ. J. Org. Chem.* **2004**, *40*, 1804.

53. Amantini, D.; Fringuelli, F.; Piermatti, O.; Pizzo, F.; Zunino, E.; Vaccaro, L., *J. Org. Chem.* **2005**, *70*, 6526.

54. Yanai, H.; Taguchi, T., *Tetrahedron Lett.* **2005**, *46*, 8639.

55. Kim, M. S.; Yoon, H. J.; Lee, B. K.; Kwon, J. H.; Lee, W. K.; Kim, Y.; Ha, H. J., *Synlett* **2005**, 2187.

56. Schulz, M. J.; Coats, S. J.; Hlasta, D. J., *Org. Lett.* **2004**, *6*, 3265.

57. Lukyanov, S. M.; Bliznets, I. V.; Shorshnev, S. V.; Aleksandrov, G. G.; Stepanov, A. E.; Vasil'ev, A. A., *Tetrahedron* **2006**, *62*, 1849.

58. Huff, B. E.; Staszak, M. A., *Tetrahedron Lett.* **1993**, *34*, 8011.

59. Duncia, J. V.; Pierce, M. E.; Santella, J. B., III, *J. Org. Chem.* **1991**, *56*, 2395.

60. Lehnhoff, S.; Ugi, I., *Heterocycles* **1995**, *40*, 801.

61. Magnus, P.; Taylor, G. M., *J. Chem. Soc., Perkin Trans. 1* **1991**, 2657.

62. Ek, F.; Manner, S.; Wistrand, L. G.; Frejd, T., *J. Org. Chem.* **2004**, *69*, 1346.

63. Nixey, T.; Hulme, C., *Tetrahedron Lett.* **2002**, *43*, 6833.

64. Doemling, A.; Beck, B.; Magnin-Lachaux, M., *Tetrahedron Lett.* **2006**, *47*, 4289.

65. Kamijo, S.; Jin, T.; Yamamoto, Y., *J. Am. Chem. Soc.* **2001**, *123*, 9453.

66. Kamijo, S.; Yamamoto, Y., *J. Am. Chem. Soc.* **2002**, *124*, 11940.

67. Kamijo, S.; Jin, T.; Yamamoto, Y., *J. Org. Chem.* **2002**, *67*, 7413.

68. Kamijo, S.; Jin, T.; Huo, Z.; Yamamoto, Y., *Tetrahedron Lett.* **2002**, *43*, 9707.

69. Kamijo, S.; Jin, T.; Huo, Z.; Yamamoto, Y., *J. Am. Chem. Soc.* **2003**, *125*, 7786.

70. Kamijo, S.; Jin, T.; Yamamoto, Y., *Tetrahedron Lett.* **2004**, *45*, 689.

71. Kamijo, S.; Jin, T.; Huo, Z.; Yamamoto, Y., *J. Org. Chem.* **2004**, *69*, 2386.

72. Umkehrer, M.; Kolb, J.; Burdack, C.; Ross, G.; Hiller, W., *Tetrahedron Lett.* **2004**, *45*, 6421.

73. Mayer, J.; Umkehrer, M.; Kalinski, C.; Ross, G.; Kolb, J.; Burdack, C.; Hiller, W., *Tetrahedron Lett.* **2005**, *46*, 7393.

74. Lee, J. G.; Kwak, K. H., *Tetrahedron Lett.* **1992**, *33*, 3165.

75. Reddy, M. V. R.; Kumareswaran, R.; Vankar, Y. D., *Tetrahedron Lett.* **1995**, *36*, 6751.

76. Arimoto, M.; Yamaguchi, H.; Fujita, E.; Ochiai, M.; Nagao, Y., *Tetrahedron Lett.* **1987**, *28*, 6289.

77. Magnus, P.; Lacour, J.; Weber, W., *J. Am. Chem. Soc.* **1993**, *115*, 9347.

78. Magnus, P.; Hulme, C.; Weber, W., *J. Am. Chem. Soc.* **1994**, *116*, 4501.

79. Magnus, P.; Hulme, C., *Tetrahedron Lett.* **1994**, *35*, 8097.

80. Tohma, H.; Egi, M.; Ohtsubo, M.; Watanabe, H.; Takizawa, S.; Kita, Y., *Chem. Commun.* **1998**, 173.

81. Magnus, P.; Lacour, J., *J. Am. Chem. Soc.* **1992**, *114*, 767.

82. Magnus, P.; Roe, M. B., *Tetrahedron Lett.* **1996**, *37*, 303.

83. Mironov, Y. V.; Sherman, A. A.; Nifantiev, N. E., *Tetrahedron Lett.* **2004**, *45*, 9107.

84. Reddy, B. G.; Madhusudanan, K. P.; Vankar, Y. D., *J. Org. Chem.* **2004**, *69*, 2630.

85. Kumaran, G.; Kulkarni, G. H., *Synth. Commun.* **1995**, *25*, 3735.

86. Sukhorokov, A. Y.; Bliznets, I. V.; Lesiv, A. V.; Khomutova, Y. A.; Strelenko, Y. A.; Ioffe, S. L., *Synthesis* **2005**, 1077.

87. Guerin, D. J.; Horstmann, T. E.; Miller, S. J., *Org. Lett.* **1999**, *1*, 1107.

88. Kaye, P. T.; Mphahlele, M. J., *Synth. Commun.* **1995**, *25*, 1495.

89. Pozgan, F.; Polanc, S.; Kocevar, M., *Heterocycles* **2002**, *56*, 379.

90. Kabalka, G. W.; Goudgaon, N. M.; Liang, Y., *Synth. Commun.* **1988**, *18*, 1363.

91. Amantini, D.; Fringuelli, F.; Pizzo, F.; Vaccaro, L., *J. Org. Chem.* **2001**, *66*, 6734.

B

Benzene, [[Difluoro(trimethylsilyl)-methyl]seleno]*

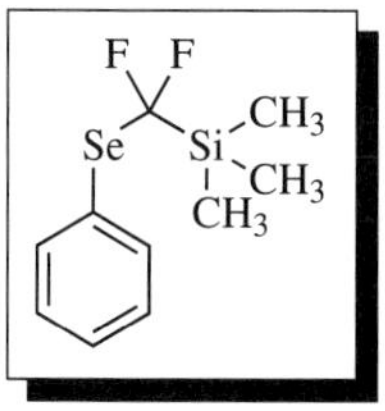

[868761-01-7] $C_{10}H_{14}F_2SeSi$ (MW 279.26)

InChI = 1S/C10H14F2SeSi/c1-14(2,3)10(11,12)13-9-7-5-4-6-8-9/h4-8H,1-3H3

InChIKey = KOWXABGSSCWOHZ-UHFFFAOYSA-N

(reagent used as fluorine-containing building block for the introduction of gem-difluoromethylene group into organic molecules)

Alternative Name: [difluoro(phenylseleno)methyl]trimethylsilane.

Preparation.   [[Difluoro(trimethylsilyl)methyl]seleno]-benzene (PhSeCF$_2$TMS) was prepared by the reaction of [(bromodifluoromethyl)seleno]benzene (PhSeCF$_2$Br), magnesium, and chlorotrimethylsilane (TMSCl) (eq 1).[1] The sequence of adding raw materials had a distinct influence on the formation of PhSeCF$_2$TMS. The proper procedure was to treat PhSeCF$_2$Br with Mg turnings in THF at room temperature, followed by slow addition of TMSCl via a syringe.

$$\text{PhSeCF}_2\text{Br} \xrightarrow[\text{2. TMSCl}]{\text{1. Mg, THF, rt}} \underset{81\%}{\text{PhSeCF}_2\text{TMS}} \quad (1)$$

PhSeCF$_2$Br could be prepared by treatment of selenophenol (PhSeH) with sodium hydride, followed by sequential addition of NaBH$_4$ and CF$_2$Br$_2$ (eq 2).[1] The addition of NaBH$_4$ could prevent the generation of PhSeSePh or convert the resulting PhSeSePh into PhSe$^-$, which could react with CF$_2$Br$_2$ to give PhSeCF$_2$Br. Thus, it was found that PhSeSePh was also be used as starting material for the preparation of PhSeCF$_2$Br (eq 3).[1]

$$\text{PhSeH} \xrightarrow[\substack{\text{2. NaBH}_4,\ 0\ ^\circ\text{C to rt} \\ \text{3. CF}_2\text{Br}_2,\ -70\ ^\circ\text{C}}]{\text{1. NaH, THF, 0 }^\circ\text{C}} \underset{86\%}{\text{PhSeCF}_2\text{Br}} \quad (2)$$

---

*Article previously published in the electronic Encyclopedia of Reagents for Organic Synthesis as [[Difluoro(trimethylsilyl)methyl]seleno]benzene. DOI: 10.1002/047084289X. rn01324.

$$\text{PhSeSePh} \xrightarrow[\text{2. CF}_2\text{Br}_2,\ -70\ ^\circ\text{C}]{\text{1. NaBH}_4,\ \text{THF, 0 }^\circ\text{C to rt}} \underset{61\%}{\text{PhSeCF}_2\text{Br}} \quad (3)$$

gem-Difluoromethylation of Aldehydes and Ketones in the Presence of Tetrabutylammonium Fluoride (TBAF). The reaction of PhSeCF$_2$TMS with aldehydes in the presence of TBAF gave the corresponding α,α-difluoro-β-hydroxy selenides in moderate to high yields (eq 4).[1] It was reported the presence of a small amount of water in TBAF did not pose any serious problem in the trifluoromethylation of aldehydes and ketones with CF$_3$TMS, and the use of anhydrous TBAF did not offer any advantage.[2] However, the presence of a small amount of water in TBAF lowered the yield of the desired alcohols and led to PhSeCF$_2$H as a by-product. This phenomenon was attributed to the electron-donating property of PhSe moiety, which resulted in increased nucleophilicity of PhSeCF$_2^-$ relative to CF$_3^-$ anion. Addition of molecular sieves into the reaction mixture successfully overcame this problem.

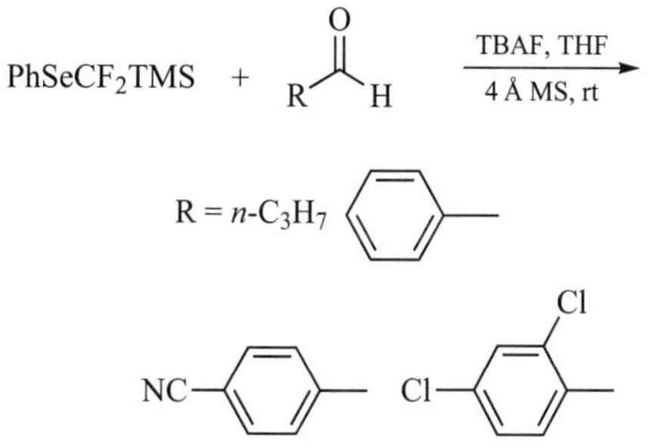

The *gem*-difluoromethylation of ketones under the same reaction conditions resulted in somewhat low yields (eqs 5 and 6).[1]

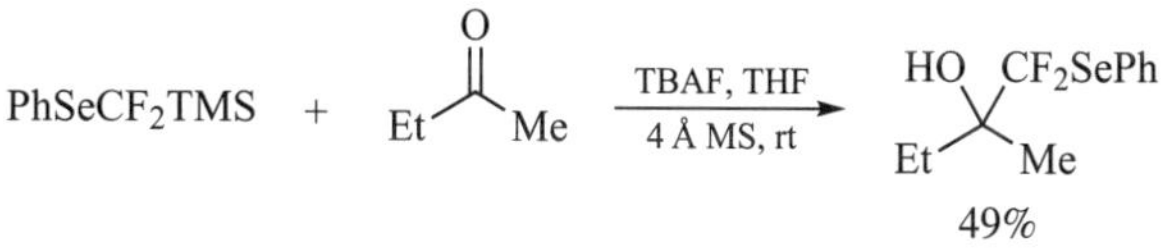

gem-Difluoromethylation of Aldehydes in the Presence of Lewis Acid. Trifluoromethylation and difluoromethylation reactions using nucleophilic catalysts are widely studied. However, there are only a few reports of successful catalytic trifluoromethylation and difluoromethylation employing Lewis acids. Shibata and co-workers described the first Lewis acid-catalyzed trifluoromethylation and difluoromethylation reactions of aldehydes.[3] When Cu(OAc)$_2$/dppe was used as the catalyst in DMF, treatment of aldehydes with 2.5 equiv of PhSeCF$_2$TMS gave the corresponding difluoro(phenylselenyl)methyl adducts in high yields (eq 7).[3]

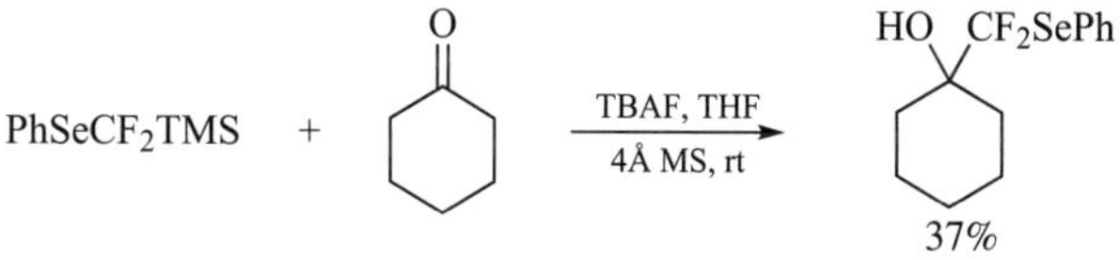

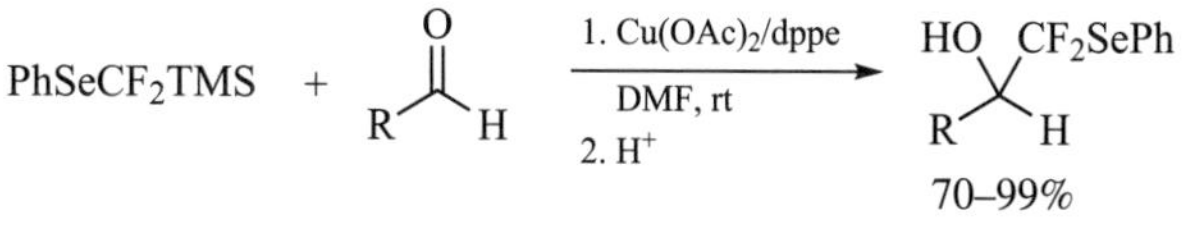

R = 2-naphtyl, Ph, 4-MeOC$_6$H$_4$
  4-NO$_2$C$_6$H$_4$, (E)-CH = CHPh

Avoid Skin Contact with All Reagents

Diastereoselective *gem*-Difluoromethylation of Aldehydes.
Asymmetric induction in the nucleophilic difluoromethylation
by the sulfinyl group as a chiral auxiliary was developed.[4]
Difluoromethylation of 1-[(2,4,4-triisopropylphenyl)sulfinyl]2-
naphthaldehyde with PhSeCF$_2$TMS at $-94\,^\circ$C gave an adduct
in high yield and high diastereoselectivity (eq 8).[4]

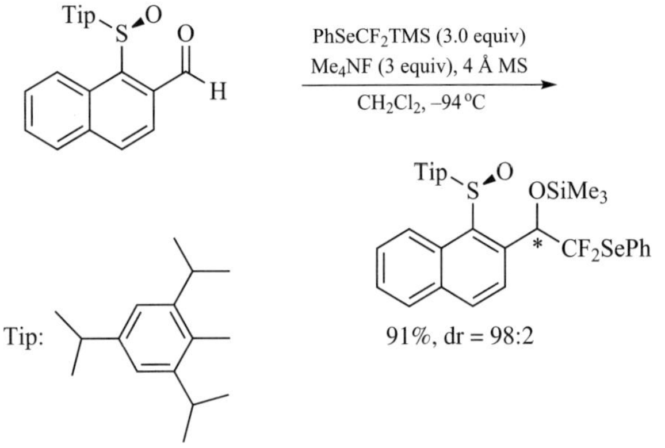

$$\text{(8)}$$

$$91\%, \text{dr} = 98{:}2$$

The addition of PhSeCF$_2$TMS to carbohydrate-derived α-chiral
aldehydes provided the corresponding difluoro(phenylselenyl)-
methyl adducts with acceptable diastereoselectivities (eqs 9–15).[5]
Tetramethylammonium fluoride (TMAF) appeared as the appropriate mediator for these reactions (eqs 9–13 and eq 15). The
benzoyl-protected aldehyde required the use of tetrabutylammonium difluorotriphenylsilicate (TBAT) as a mediator to give the
addition product only in low yield (eq 14). The two-step procedure (the addition of TBAF in the second step) was used to convert the OH/OTMS mixture initially obtained to the free alcohols
(eqs 9 and 13–15). For benzyl- and MOM-protected aldehydes, a
warm-up to room temperature was sufficient to afford alcohols in
high yields (eqs 10–12).

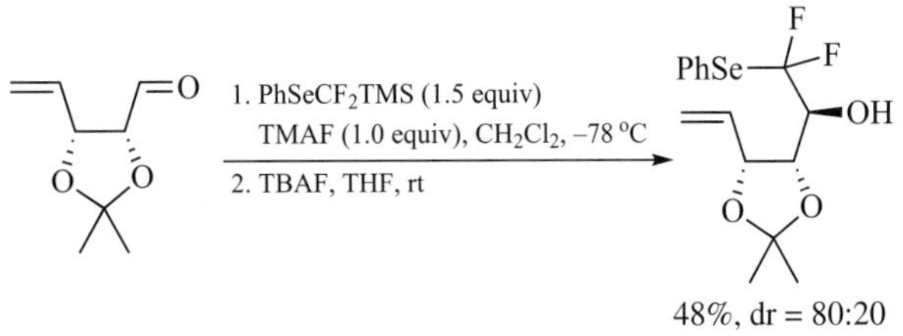

$$48\%, \text{dr} = 80{:}20 \quad \text{(9)}$$

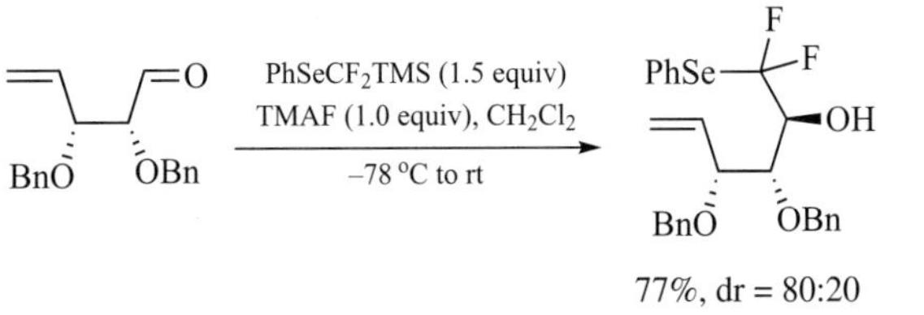

$$77\%, \text{dr} = 80{:}20 \quad \text{(10)}$$

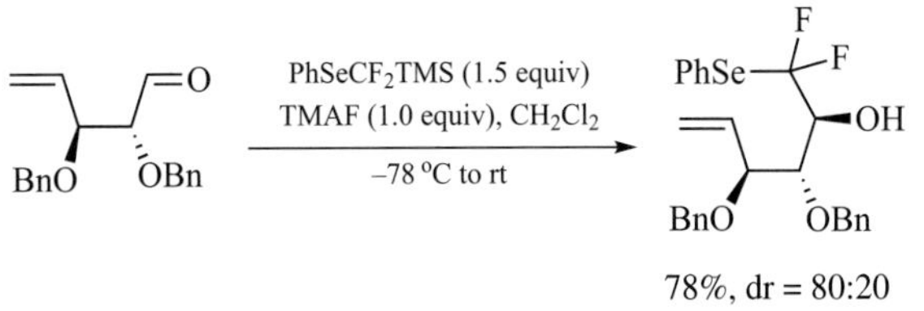

$$78\%, \text{dr} = 80{:}20 \quad \text{(11)}$$

$$75\%, \text{dr} = 80{:}20$$

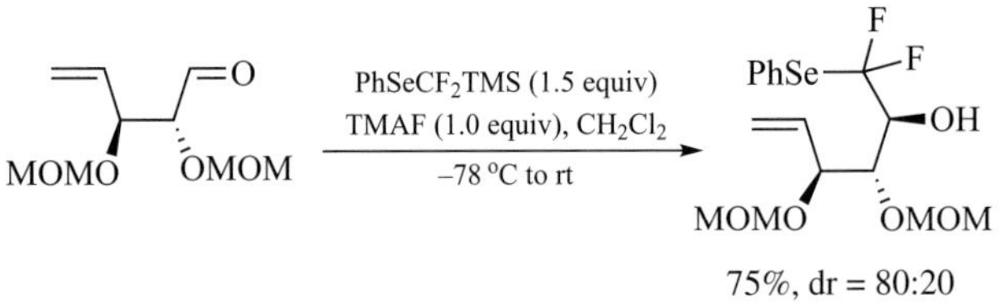

$$\text{(12)}$$

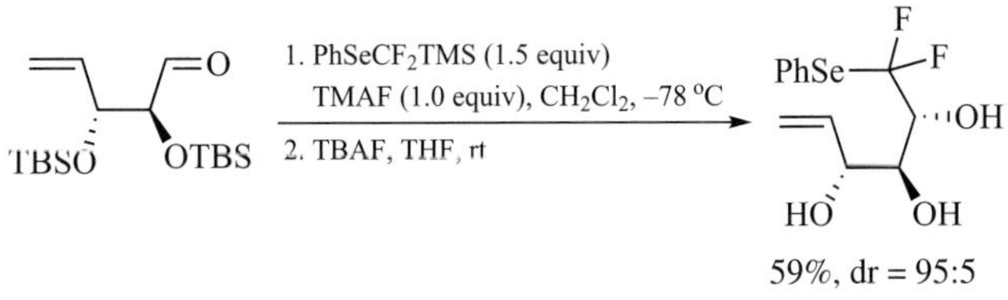

$$59\%, \text{dr} = 95{:}5$$

$$\text{(13)}$$

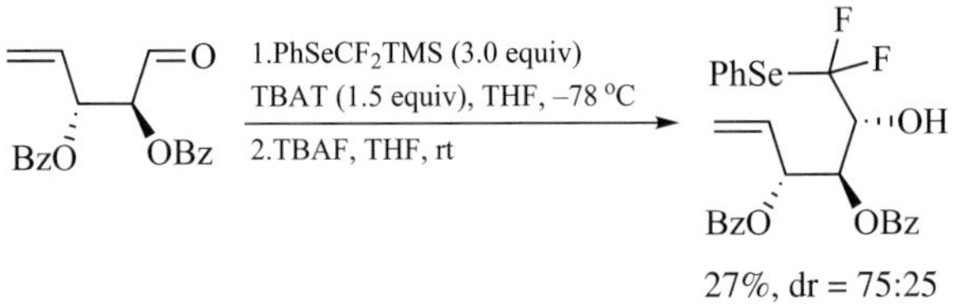

$$27\%, \text{dr} = 75{:}25 \quad \text{(14)}$$

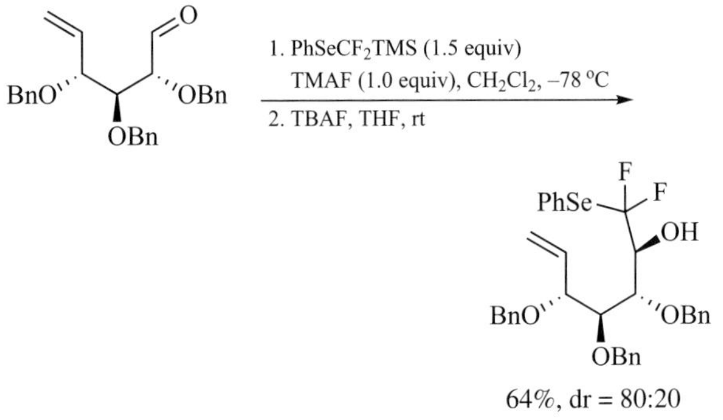

$$64\%, \text{dr} = 80{:}20 \quad \text{(15)}$$

**Diastereoselective *gem*-Difluoromethylation of *t*-
Butanesulfinylimines.** The addition of PhSeCF$_2$TMS to
carbohydrate-derived *t*-butanesulfinylimines proceeded
smoothly using TBAT as a promoter. The matched diastereoselectivity (dr >98:2) (eq 16) and mismatched case (dr = 83:17)
were observed (eq 17).[5b]

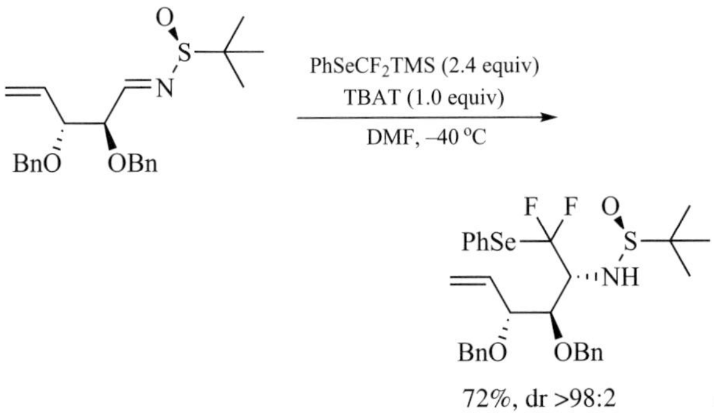

$$72\%, \text{dr} >98{:}2 \quad \text{(16)}$$

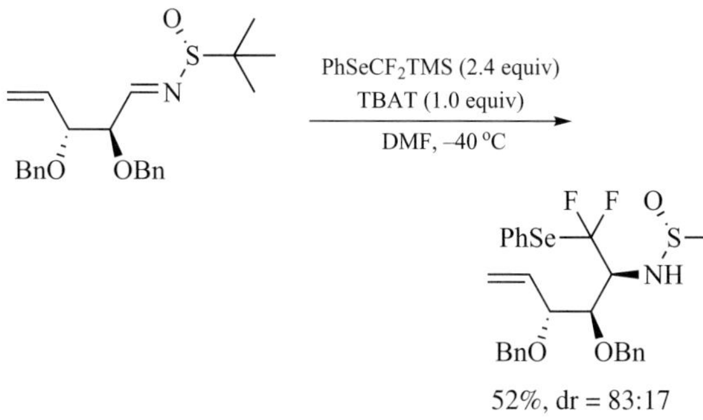

(17)

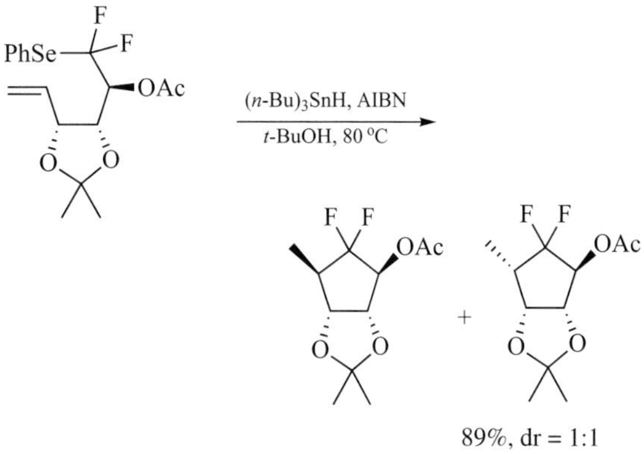

52%, dr = 83:17

(21)

89%, dr = 1:1

Transformation of Difluoro(phenylselenyl)methyl Alcohols (Amine). Reduction of the phenylseleno group to a hydrogen can be readily and directly achieved by treatment with Bu₃SnH/AIBN (eq 18)[1,3] or by treatment with NaBH₄/InCl₃ (eq 19).[6]

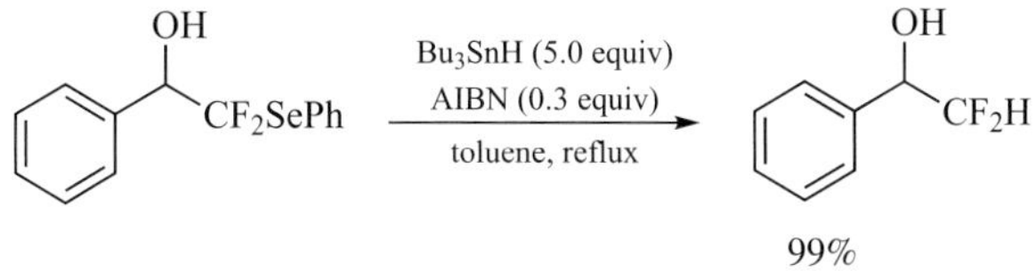

99%   (18)

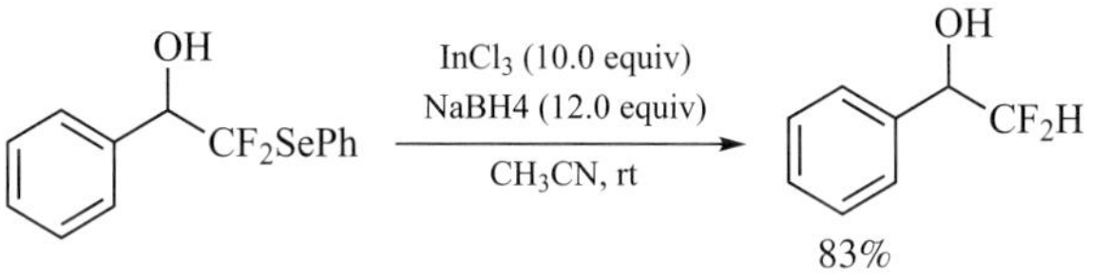

83%   (19)

Allylation of difluoro(phenylselenyl)methyl alcohol with allyl bromide in the presence of KOH provided the difluoro-(phenylselenyl)methyl-containing cyclization precursor. The reductive radical cyclization of this precursor proceeded smoothly in the presence of InCl₃ and NaBH₄ (eq 20).[6]

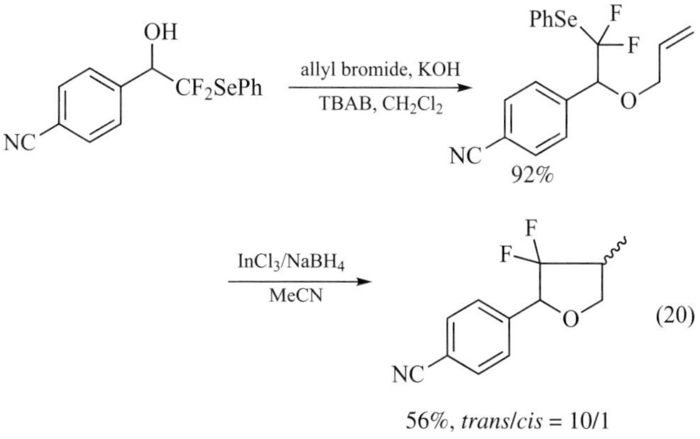

92%

(20)

56%, *trans/cis* = 10/1

The difluoro(phenylselenyl)methyl adduct obtained from the addition of PhSeCF₂TMS to carbohydrate-derived α-chiral aldehyde was acetylated and engaged in reductive radical cyclization to give 5-deoxypentofuranose analogs (eq 21).[5]

The reductive radical cyclization of adduct of the addition of PhSeCF₂TMS to carbohydrate-derived *t*-butanesulfinylimine also afforded 5-deoxypentofuranose analog in good yield (eq 22).[5b]

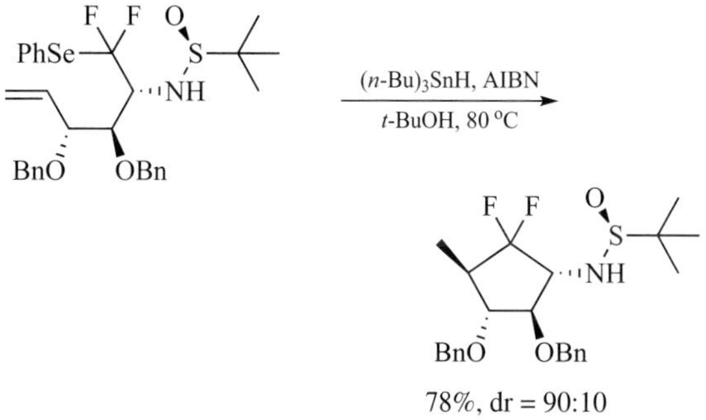

(22)

78%, dr = 90:10

1. Qin, Y. Y.; Qiu, X. L.; Yang, Y. Y.; Meng, W. D.; Qing, F. L., *J. Org. Chem.* **2005**, *75*, 9040.

2. (a) Krishnamurti, R.; Bellew, D. R.; Prakash, G. K. S., *J. Org. Chem.* **1991**, *56*, 984. (b) Singh, R. P.; Shreeve, J. M., *Tetrahedron* **2000**, *56*, 7613.

3. Mizuta, S.; Shibata, N.; Ogawa, S.; Fujimoto, S.; Nakamura, S.; Toru, T., *Chem. Commun.* **2006**, 2575.

4. Sugimoto, H.; Nakamura, S.; Shibata, Y.; Shibata, N.; Toru, T., *Tetrahedron Lett.* **2006**, *47*, 1337.

5. (a) Fourriere, G.; Lalot, J.; Hijfte, N. V.; Quirion, J. C.; Leclerc, E., *Tetrahedron Lett.* **2009**, *50*, 7048. (b) Fourriere, G.; Hijfte, N. V.; Lalot, J.; Dutech, G.; Fragnet, B.; Coadou, G.; Quirion, J. C.; Leclerc, E., *Tetrahedron* **2010**, *66*, 3963.

6. Qin, Y. Y.; Yang, Y. Y.; Qiu, X. L.; Qing, F. L., *Synthesis* **2006**, 1475.

Feng-Ling Qing
Shanghai Institute of Organic Chemistry, Shanghai, China

Benzene, [[Difluoro(trimethylsilyl)-methyl]thio]-

S
SiMe3
F F

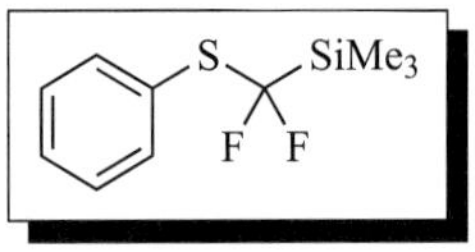

[536975-49-2] $C_{10}H_{14}F_2SSi$ (MW 232.37)

InChI = 1S/C10H14F2SSi/c1-14(2,3)10(11,12)13-9-7-5-4-6-8-9/h4-8H,1-3H3

InChIKey = NNNLERGMVGBYSD-UHFFFAOYSA-N

(nucleophilic (phenylthio)difluoromethylating reagent)

Alternate Names: α,α-difluoro-α-phenylsulfanyl-α-trimethylsilylmethane; $PhSCF_2SiMe_3$; [difluoro(phenylthio)methyl]-trimethylsilane.

Physical Data: colorless liquid; bp 86–87°C/4 mmHg.

Solubility: soluble in most organic solvents, for example hexane, tetrahydrofuran (THF), CH_2Cl_2, dimethylformamide (DMF), and so on.

Preparative Methods: [difluoro(phenylthio)methyl]trimethylsilane ($PhSCF_2SiMe_3$) was prepared for the first time by Prakash et al.,[1] using the Barbier coupling reaction of bromodifluoromethylphenyl sulfide,[2] prepared from dibromodifluoromethane and sodium benzenethiolate,[3] magnesium metal, and chlorotrimethylsilane (TMSCl) in DMF (eq 1).

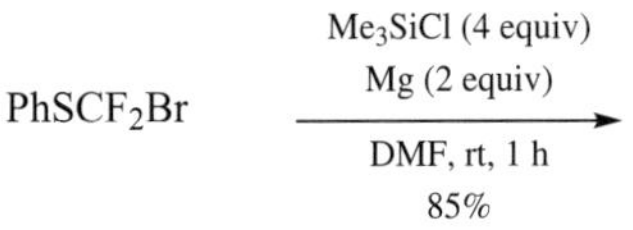

PhSCF2Br
Me3SiCl (4 equiv)
Mg (2 equiv)
DMF, rt, 1 h
85%
PhSCF2SiMe3

(1)

It was recently reported by Langlois and Médebielle[4] that THF was efficiently used as solvent and was preferred to DMF to facilitate the workup (eq 2).

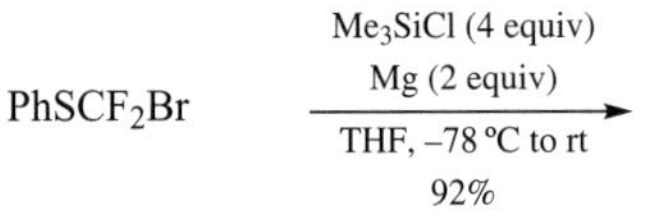

PhSCF2Br
Me3SiCl (4 equiv)
Mg (2 equiv)
THF, –78 °C to rt
92%
PhSCF2SiMe3

(2)

Handling, Storage, and Precautions: the material should be handled in a well-ventilated fume hood and can be stored in a refrigerator for months without any appreciable decomposition.

General. Organofluorine compounds have received remarkable interest in recent years due to their wide-ranging biological effects. The development of general synthetic routes to such compounds and the use of new fluorinated compounds as building blocks are of great importance. Of particular interest is the selective incorporation of the *gem*-difluoromethylene group 'CF_2' into organic molecules. The report on the synthesis of $PhSCF_2SiMe_3$ was published in 2003 by Prakash et al.,[1] and the reagent has become one of the most versatile and efficient nucleophilic (phenylthio)difluoromethylating reagents.

Addition to Aldehydes and Ketones. Fluoride-induced nucleophilic (phenylthio)difluoromethylation of $PhSCF_2SiMe_3$ has been extensively investigated. Prakash et al.[5] and Pohmakotr et al.[6] have independently reported their studies on fluoride-catalyzed nucleophilic difluoro(phenylthio)methylation of aldehydes and ketones with $PhSCF_2SiMe_3$. This methodology efficiently transfers the '$PhSCF_2$' group into both enolizable and non-enolizable aldehydes and ketones. Prakash et al. employed 10 mol % of tetrabutylammonium triphenyldifluorosilicate (TBAT) as the anhydrous fluoride source and obtained the corresponding alcohols, after TMS deprotection, in good to excellent yields (72–91%). The resulting $PhSCF_2$-containing alcohols can be further transformed into difluoromethyl alcohols via oxidation–desulfonylation (eq 3).[5]

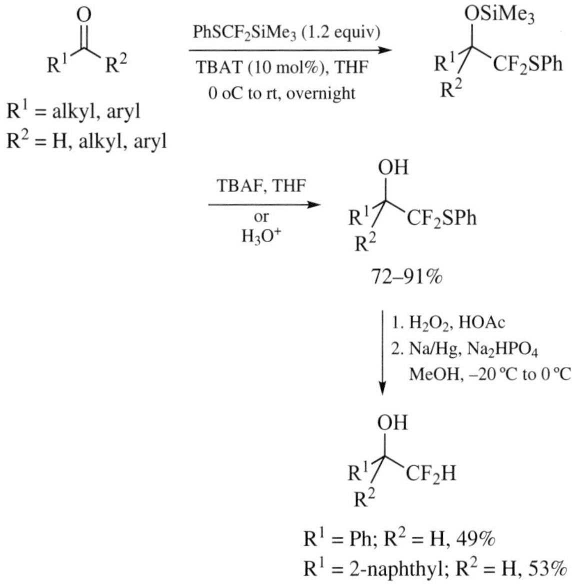

O
R1 R2
R1 = alkyl, aryl
R2 = H, alkyl, aryl
PhSCF2SiMe3 (1.2 equiv)
TBAT (10 mol%), THF
0 oC to rt, overnight
OSiMe3
R1 CF2SPh
R2
TBAF, THF
or
H3O+
OH
R1 CF2SPh
R2
72–91%
1. H2O2, HOAc
2. Na/Hg, Na2HPO4
MeOH, –20 °C to 0 °C
OH
R1 CF2H
R2
R1 = Ph; R2 = H, 49%
R1 = 2-naphthyl; R2 = H, 53%

(3)

Similarly, the reaction of $PhSCF_2SiMe_3$ with aldehydes and ketones using 10 mol % of anhydrous tetra-*n*-butylammonium fluoride (TBAF) reported by Pohmakotr et al. also provided the corresponding adducts, both carbinols and silyl ethers, in moderate to excellent yields (39–92%). The silyl ethers could be converted into the corresponding carbinols in quantitative yields by using KF in acetonitrile/THF (eq 4). Subsequent oxidation of the carbinol sulfides to sulfoxides followed by flash vacuum pyrolysis (FVP) gave the corresponding *gem*-difluoroalkenes in 41–82% yields (eq 5).[6]

O
R1 R2
R1 = alkyl, aryl
R2 = H, alkyl, aryl
PhSCF2SiMe3 (1 equiv)
TBAF (10 mol %), THF
–78 °C to rt, overnight
OSiMe3
R1 CF2SPh
R2
+
OH
R1 CF2SPh
R2
39–92%
(combined yield)
KF, CH3CN/THF
rt, overnight

(4)

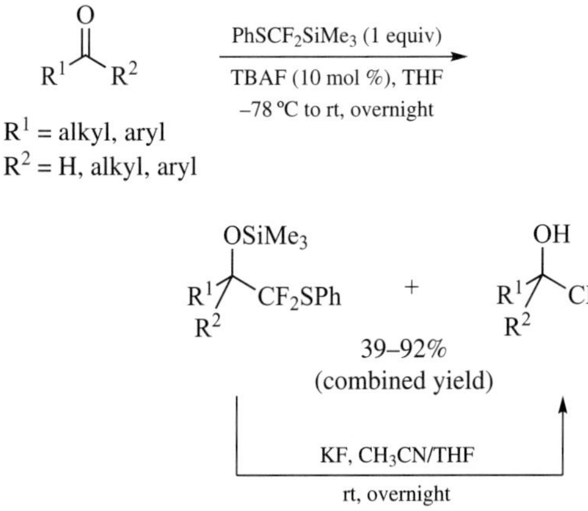

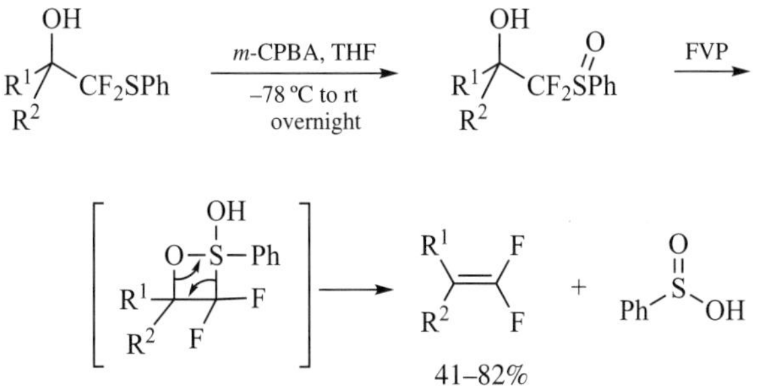

$$ (5) $$

41–82%

The first Lewis acid-catalyzed difluoromethylation of $PhSCF_2$ $SiMe_3$ with an aldehyde was published by Shibata and Toru in 2006.[7] The reaction of p-nitrobenzaldehyde with $PhSCF_2SiMe_3$ in the presence of $Cu(OAc)_2$/1,2-bis(diphenylphosphino)ethane (dppe) as the catalyst system in DMF gave the corresponding adduct in 60% yield (eq 6).

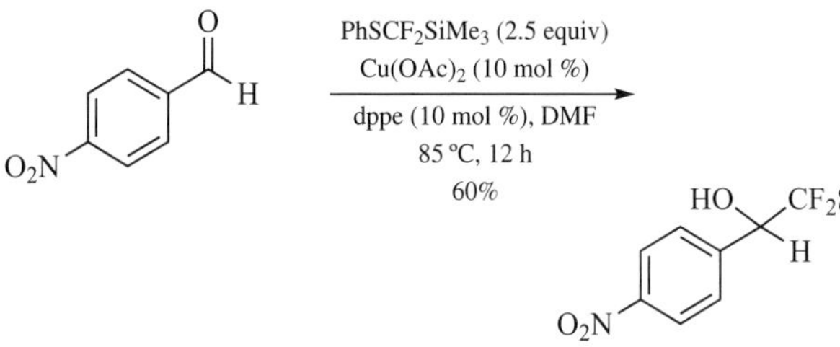

$$ (6) $$

Addition to Ester. Prakash et al. disclosed the reaction between $PhSCF_2SiMe_3$ and methyl benzoate using different solvents at $-78\,^{\circ}C$ to room temperature to yield the corresponding ketone in 28–41% yield (eq 7).[5]

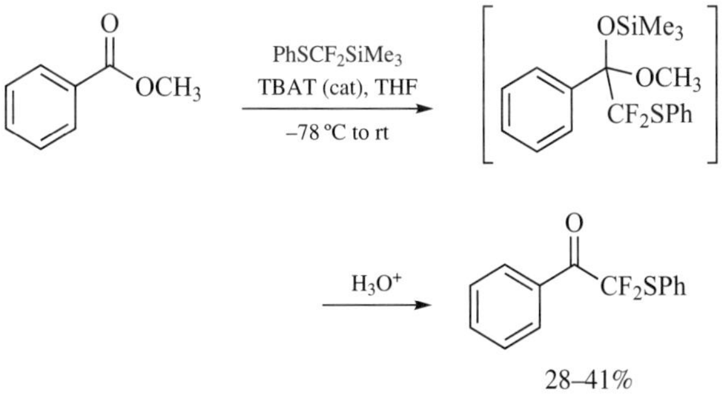

$$ (7) $$

28–41%

Addition to Disulfide. When excess potassium t-butoxide was used as a promoter, $PhSCF_2SiMe_3$ reacted with diphenyl disulfide to give the corresponding dithioacetal in 85% yield (eq 8).[5]

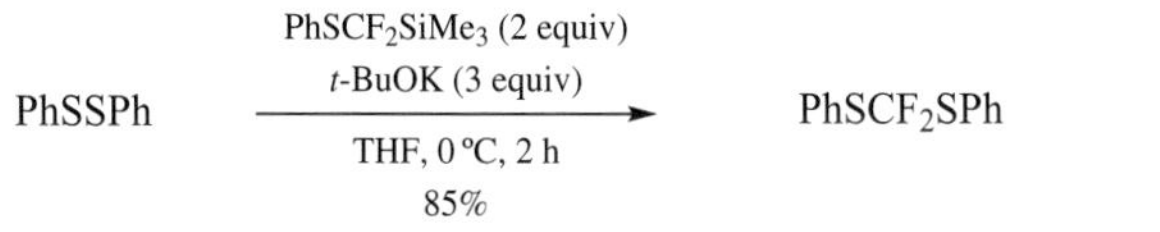

PhSSPh $\xrightarrow[\substack{\text{THF, 0 °C, 2 h} \\ 85\%}]{\substack{PhSCF_2SiMe_3 \ (2 \ \text{equiv}) \\ t\text{-BuOK (3 equiv)}}}$ $PhSCF_2SPh$ $\quad$ (8)

Addition to Imines. An elegant diastereoselective fluoride-induced nucleophilic (phenylthio)difluoromethylation of (R)-(N-t-butylsulfinyl)imines with $PhSCF_2SiMe_3$ was reported by Li and Hu.[8] The reaction afforded the corresponding products in 30–89%

yields and with high diastereoselectivity (dr up to 99:1). The resulting $PhSCF_2$-containing sulfinamides were transformed into chiral 2,4-*trans*-disubstituted 3,3-difluoropyrrolidines through an intramolecular radical cyclization (eq 9).

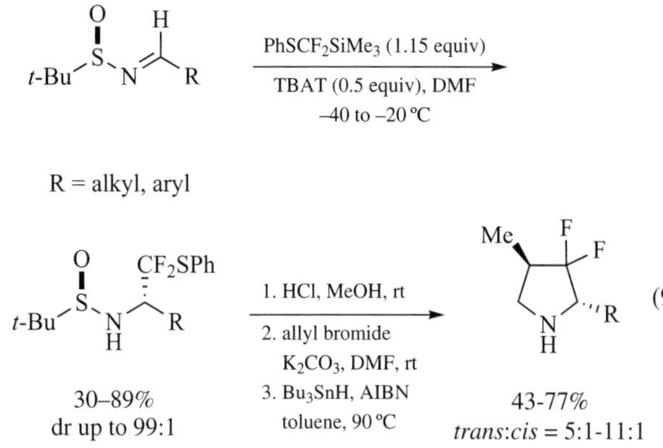

$$ (9) $$

43-77%

trans:cis = 5:1-11:1

Addition to α-Ketoesters. Chemoselective fluoride-catalyzed nucleophilic difluoro(phenylthio)methylation of $PhSCF_2$ $SiMe_3$ to α-ketoesters was demonstrated.[9] The alkylation reaction preferentially took place at the keto group with the ester group untouched yielding the corresponding α-hydroxy esters in good to excellent yields (77–98%) (eq 10). The phenylthio group was removed under standard reaction conditions (Bu_3SnH, AIBN, toluene, reflux). When 2,3-butanedione was employed as a substrate, an α-hydroxy ketone was obtained in 80% yield (eq 11).

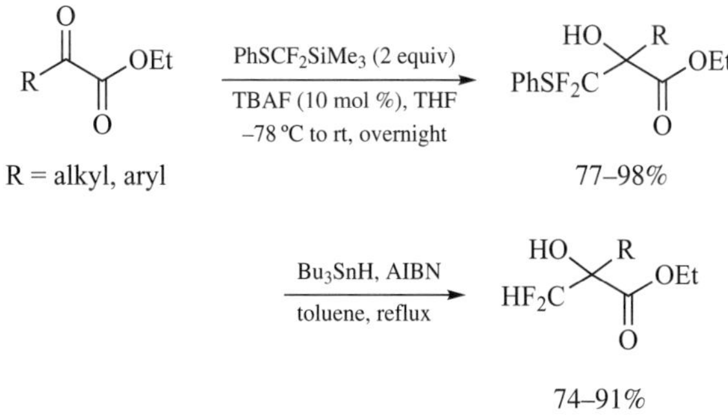

$$ (10) $$

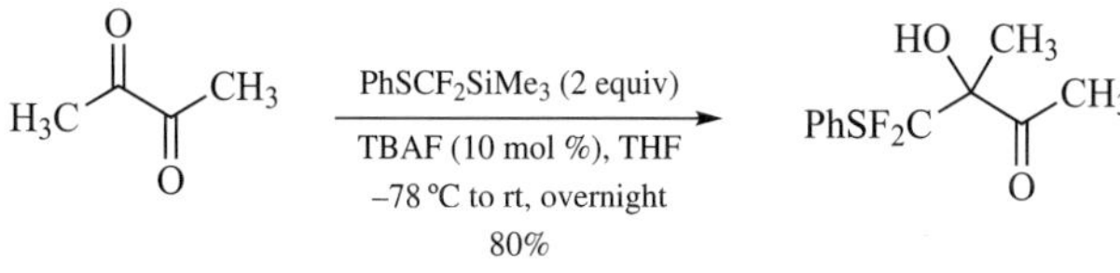

77–98%

74–91%

$$ (11) $$

80%

Addition to β-Ketoester and β-Diketone. Similarly, the reactions of $PhSCF_2SiMe_3$ with highly enolizable ethyl acetoacetate and 2,4-pentanedione also provided the corresponding β-hydroxy ester and β-hydroxy ketone, respectively, albeit in low yields (eqs 12 and 13).[9]

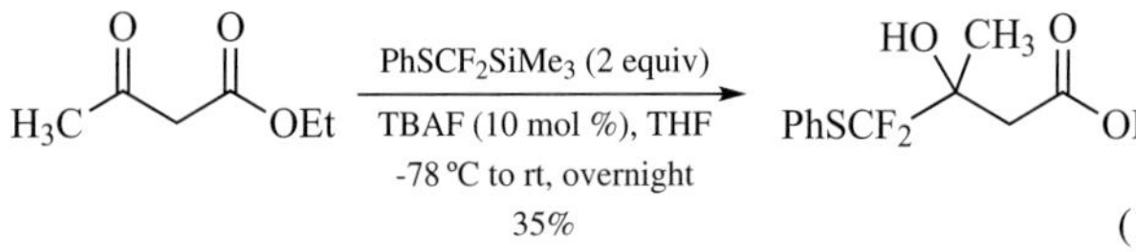

$$(12)$$

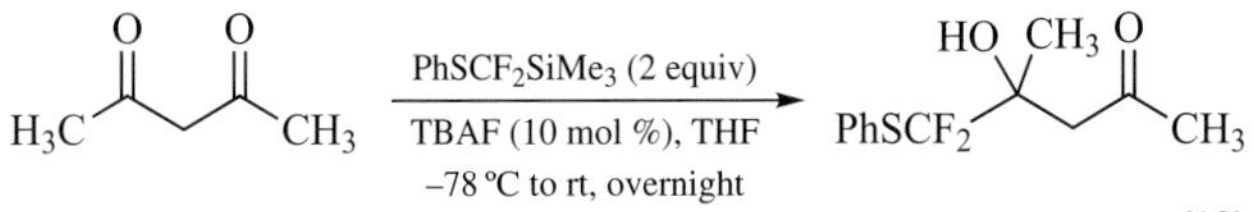

$$(13)$$

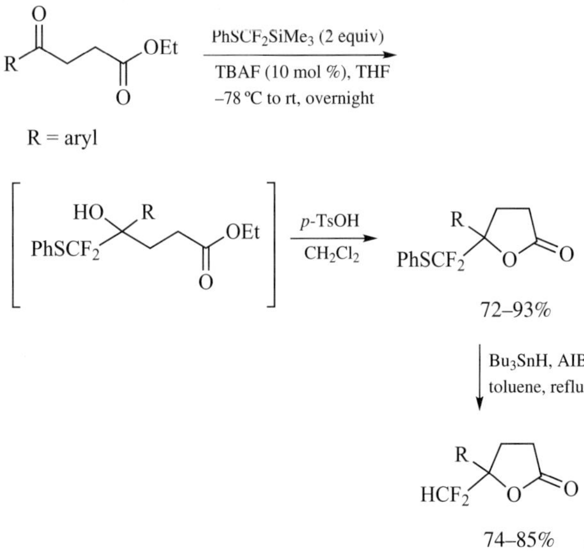

72–93%

Bu$_3$SnH, AIBN
toluene, reflux

74–85%   (14)

Addition to γ-Ketoesters. PhSCF$_2$SiMe$_3$ was also reported to undergo chemoselective addition to the keto group of the γ-ketoesters (eq 14).[9] The corresponding adducts were not isolated but were further treated with acid to furnish the lactones in good yields (72–93%). Upon treatment with Bu$_3$SnH/AIBN, the corresponding γ-butyrolactones containing *gem*-difluoro moieties were obtained.

Addition to *N*-Substituted Cyclic Imides. Recently, PhSCF$_2$ SiMe$_3$ was found, for the first time, to undergo fluoride-catalyzed nucleophilic difluoro(phenylthio)methylation reaction to cyclic imides, affording the corresponding adducts in moderate to good yields (45–95%).[10] Reductive cleavage of the phenylthio group of the *N*-alkylated adducts with Bu$_3$SnH/AIBN yielded *gem*-difluoromethylated products in 66–86% yields. Under similar reduction conditions, *N*-alkenylated and *N*-alkynylated adducts gave the corresponding *gem*-difluoromethylenated 1-azabicyclic compounds. The *trans*- and *E*-isomers were obtained, respectively, as the major isomers (eq 15).

Fluoroalkylation with Alkyl Halides. Fluoride-mediated nucleophilic fluoroalkylation of PhSCF$_2$SiMe$_3$ with alkyl iodides and bromides was reported by Li and Hu.[11] The reaction proceeded well with primary alkyl iodides and bromides in DME solvent when CsF and 15-crown-5 were used as the fluoride source and additive, respectively (eq 16).

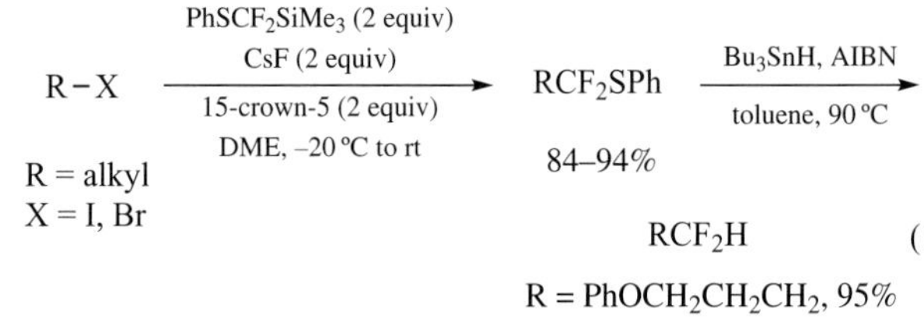

$$(16)$$

R = PhOCH$_2$CH$_2$CH$_2$, 95%

1. Prakash, G. K. S.; Hu, J.; Olah, G. A., *J. Org. Chem.* **2003**, *68*, 4457.

2. (a) Pohmakotr, M.; Ieawsuwan, W.; Tuchinda, P.; Kongsaeree, P.; Prabpai, S.; Reutrakul, V., *Org. Lett.* **2004**, *6*, 4547. (b) Reutrakul, V.; Thongpaisanwong, T.; Tuchinda, P.; Kuhakarn, C.; Pohmakotr, M., *J. Org. Chem.* **2004**, *69*, 6913. See also Ref. 1, 5, and 6.

3. For the preparation of bromodifluoromethylphenyl sulfide, see also (a) Suda, M.; Hino, C., *Tetrahedron Lett.* **1981**, *22*, 1997. (b) Burton, D. J.; Wiemers, D. M., *J. Fluorine Chem.* **1981**, *18*, 573. (c) Rico, I.; Wakselman, C., *Tetrahedron Lett.* **1981**, *22*, 323. (d) Rico, I.; Wakselman, C., *Tetrahedron* **1981**, *37*, 4209.

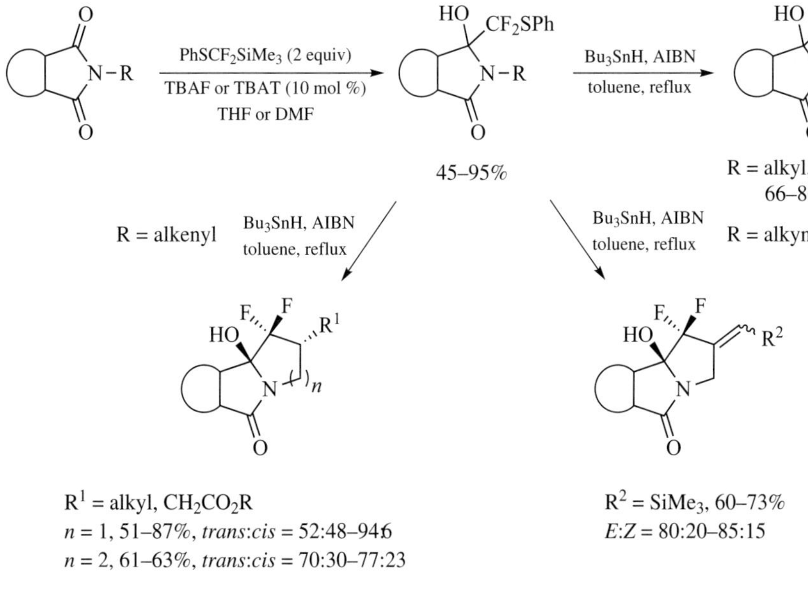

45–95%

R = alkyl, CO$_2$Et
66–86%

R = alkenyl

R = alkynyl

$$(15)$$

R^1 = alkyl, CH$_2$CO$_2$R
n = 1, 51–87%, *trans:cis* = 52:48–94:6
n = 2, 61–63%, *trans:cis* = 70:30–77:23

R^2 = SiMe$_3$, 60–73%
E:Z = 80:20–85:15

Equation (9)

BuOt—C(=O)—C(=CH2)—C(=O)—O'Bu + PhSe—CH=CH—R → (ZnBr₂, CH₂Cl₂, −78 °C to −30 °C, 3 h; then Et₃N, H₂O)

R	Yield %	Yield %
TMS	52	42
TES	63	27
TIPS	11	43

$$(9)$$

Equation (10)

RO—C(=O)—C(=CH2)—C(=O)—OR + PhSe—CH=CH—R₁ → (ZnI₂, CH₂Cl₂, −30 °C to 10 °C, 3 h; then Et₃N, H₂O)

R = (−)-menthyl

R₁	Yield (%)	LiAlH₄	Diol (% ee)	Yield (%)	LiAlH₄	Diol (% ee)
TMS	64		81	35		68
TES	35		79	52		84

$$(10)$$

Equation (11)

EtO—C(=O)—C(=CH—C(=O)Ph)—C(=O)—OEt + PhSe—CH=CH—TMS → (ZnBr₂, CH₂Cl₂, −30 °C, 3 h; then Et₃N, H₂O, 74%)

$$(11)$$

Equation (12)

EtO₂C, 'BuO₂C cyclopropane with TMS and SePh → MeO—, HO₂C cyclopropane with CO₂Et and CO₂H

$$(12)$$

carbon of the acrylate were used to justify the stereochemical outcome.

Equation (13)

(MeO)₂P(=O)—C(=CH2)—C(=O)—OMe + PhSe—CH=CH—TMS → (SnCl₄, CH₂Cl₂, −78 °C, 3 h; then Et₃N, H₂O, 96%)

$$(13)$$

Equation (14)

p-Tol—S(=O)₂—C(=CH2)—C(=O)—OMe + PhSe—CH=CH—TMS → (SnCl₄, CH₂Cl₂, −78 °C, 3 h; then Et₃N, H₂O, 56%)

$$(14)$$

Cyclobutanation via [2 + 2] Cycloaddition.[11] The [2 + 2] cycloaddition of **1** with dimethyl 1,1-dicyanoethene-2,2-dicarboxylate to afford densely functionalized cyclobutanes (eq 15) was achieved with tin tetrachloride (or zinc bromide). The preferential formation of the cyclobutane was attributed to destabilization of the proposed episelenonium ion, favoring the enol-*exo* 4-*endo-tet* mode of cyclization. The regioselectivity of the cycloaddition depends on the coordination ability of the Lewis acid and the resultant stability of acid–substrate complex.

Equation (15)

MeO—C(=O)—C(=C(CN)₂)—C(=O)—OMe + PhSe—CH=CH—TMS → (CH₂Cl₂, SnCl₂, −78 °C, 1 h; or ZnBr₂, rt, 5 h; then Et₃N, H₂O)

Lewis acid	Yield %	Yield %
SnCl₂	71	0
ZnBr₂	0	75

$$(15)$$

Heteroatom Analogs of Methylenemalonate Esters. Phosphorus-substituted cyclopropanes can be prepared by the [2 + 1]-cycloaddition of **1** with 2-phosphonoacrylates (eq 13).[9] Sulfone-substituted cyclopropanes were also prepared (eq 14).[10] Secondary orbital interactions between the selenium and carbonyl

1. Yamazaki, S.; Hama, M.; Yamabe, S., *Tetrahedron Lett.* **1990**, *31*, 2917.
2. Ogawa, A.; Obayashi, R.; Sekiguchi, M.; Masawaki, T.; Kambe, N.; Sonoda, N., *Tetrahedron Lett.* **1992**, *33*, 1329.
3. Yamazaki, S.; Katoh, S.; Yamabe, S., *J. Org. Chem.* **1992**, *57*, 4.
4. Yamazaki, S.; Tanaka, M.; Yamaguchi, A.; Yamabe, S., *J. Am. Chem. Soc.* **1994**, *116*, 2356.
5. Yamazaki, S.; Tanaka, M.; Yamabe, S., *J. Org. Chem.* **1996**, *61*, 4046.

6. Yamazaki, S.; Tanaka, M.; Inoue, T.; Morimoto, N.; Kugamai, H.; Yamamoto, K., *J. Org. Chem.* **1995**, *60*, 6546.

7. Yamazaki, S.; Kataoka, H.; Yamabe, S., *J. Org. Chem.* **1999**, *64*, 2367.

8. Yamazaki, S.; Kumagai, H.; Takada, T.; Yamabe, S.; Yamamoto, K., *J. Org. Chem.* **1997**, *62*, 2968.

9. Yamazaki, S.; Imanishi, T.; Moriguchi, Y.; Takada, T., *Tetrahedron Lett.* **1997**, *38*, 6397.

10. Yamazaki, S.; Yanase, Y.; Tanigawa, E.; Yamabe, S.; Tamura, H., *J. Org. Chem.* **1999**, *64*, 9521.

11. Yamazaki, S.; Kumagai, H.; Yamabe, S.; Yamamoto, K., *J. Org. Chem.* **1998**, *63*, 3371.

Matthew G. Donahue
Queensborough Community College, Bayside, NY, USA

Benzene, [[(1E)-2-(trimethylsilyl)-ethenyl]sulfonyl]*

[64489-06-1] $C_{11}H_{16}O_2SSi$ (MW 240.43)

InChI = 1S/C11H16O2SSi/c1-15(2,3)10-9-14(12,13)11-7-5-4-6-8-11/h4-10H,1-3H3/b10-9+

InChIKey = OAFHXIYOGFKWQE-MDZDMXLPSA-N

(dienophile equivalent to a variety of alkynes in Diels–Alder cycloadditions;[2] used to prepare α-substituted allylsilanes[3])

Physical Data: mp 59–60 °C (petroleum ether).[4]
Solubility: sol common organic solvents.
Form Supplied in: colorless solid; not commercially available.
Preparative Methods: prepared by hydrogenation of the corresponding readily available acetylene in 48% yield (this method is also amenable to the synthesis of the deutero derivative)[2] or by phenylsulfonyl chloride addition to trimethylsilylethylene (78%) followed by triethylamine-mediated dehydrochlorination (97%);[2] selenosulfonation, followed by hydrogen peroxide oxidation, can also be used (84%);[5] an alternative method of preparation involves dehydrochlorination of 1-phenylsulfonyl-1-chloro-2-trimethylsilylethane (PhSO$_2$CHClCH$_2$SiMe$_3$) with 1,8-diazabicyclo[5.4.0]undec-7-ene (quantitative);[4] can also be prepared by treating the lithium salt of the anion of methylthiomethyl phenyl sulfone (PhSO$_2$CH$_2$SMe) with (iodomethyl)trimethylsilane to give PhSO$_2$CH(SMe)CH$_2$SiMe$_3$, followed by oxidation to the sulfoxide and elimination of methanesulfenic acid (60%).[3b]
Handling, Storage, and Precautions: potential alkylating agent; use in a fume hood.

Cycloadditions. The dienophilic properties of (*E*)-phenylsulfonyl-2-trimethylsilylethylene allow the preparation of adducts with reactive dienes such as cyclopentadiene and

anthracene.[2] The adducts are smoothly converted to alkenes upon treatment with fluoride ion, establishing the equivalence of the title reagent to acetylene Alkylation of the α-sulfonyl carbanion can precede the elimination such that synthetic equivalents to HC≡CH, HC≡CD, and RC≡CH are available. The use of this reagent is highlighted by the synthesis of several functionalized dibenzobarrelenes (eq 1).[2] The equivalency to DC≡CD and RC≡CD is illustrated by the preparation of deuterated derivatives.

(1)

The somewhat low reactivity of (*E*)-phenylsulfonyl-2-trimethylsilylethylene in Diels–Alder reactions is probably due to steric hindrance exerted by both substituents and by the poor activation imparted by the silyl group. This drawback is partially offset by the effective elimination to the alkene performed under very mild conditions with fluoride ion. The low dienophilic reactivity of the title reagent is evident in the reaction with isodicyclopentadiene, for which it was demonstrated that only the isomer arising from the [1,5]-hydrogen sigmatropic shift was captured by dienophiles of low reactivity. Highly reactive dienophiles react with the 'symmetric structure', which is obviously a less reactive diene (eq 2).[6]

(2)

Addition Reactions. The utility of (*E*)-phenylsulfonyl-2-trimethylsilylethylene in the synthesis of α-substituted allylsilanes[3] is exemplified in eq 3 for γ-hydroxyvinylsilanes,[3a] and in eq 4 in the preparation of isoprenoid structures.[3b] In these reactions the reagent functions as a Michael acceptor, but α-lithiation may compete with less nucleophilic bases such as butyllithium.[3b]

---

*Article previously published in the electronic Encyclopedia of Reagents for Organic Synthesis as (*E*)-1-Phenylsulfonyl-2-trimethylsilylethylene. DOI: 10.1002/047084289X. rp119.

A list of General Abbreviations appears on the front Endpapers

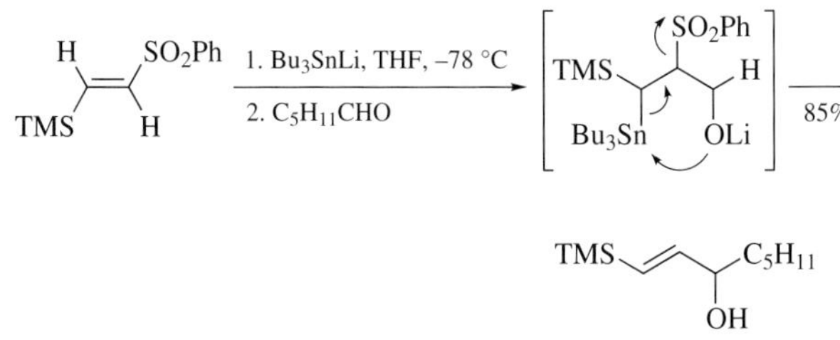

The left column contains reaction schemes:

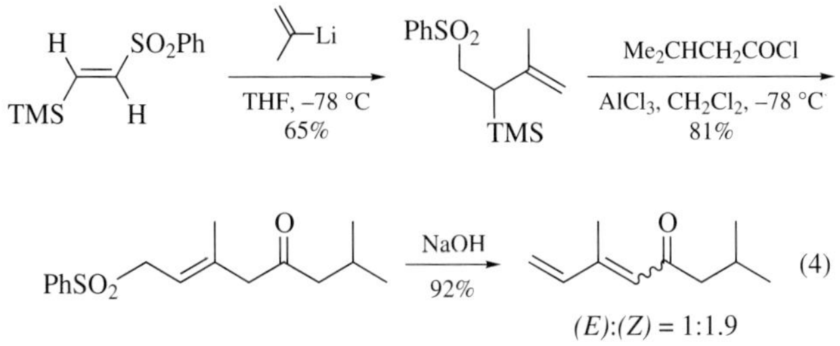

Equation (3): reaction of TMS/SO2Ph alkene with 1. Bu3SnLi, THF, −78 °C; 2. C5H11CHO giving intermediate (85%) then TMS─/═─C5H11 with OH.

Equation (4): reaction with isopropenyllithium (THF, −78 °C, 65%), then Me2CHCH2COCl, AlCl3, CH2Cl2, −78 °C, 81%, then NaOH, 92%, (E):(Z) = 1:1.9

The availability of analogous reagents bearing different atoms in the place of silicon, such as tin,[3] boron,[7a] and chlorine[7b] as well as the alkynic homologs, is notable.[8] Finally, the sulfide related to the title reagent merits mention.[9]

1. Block, E.; Aslam, M., *Tetrahedron* **1988**, *44*, 281.
2. (a) Paquette, L. A.; Williams, R. V., *Tetrahedron Lett.* **1981**, *22*, 4643. (b) Carr, R. V. C.; Williams, R. V.; Paquette, L. A., *J. Org. Chem.* **1983**, *48*, 4976. (c) Paquette, L. A.; Bay, E., *J. Am. Chem. Soc.* **1984**, *106*, 6693.
3. (a) Ochiai, M.; Ukita, T.; Fujita, E., *Tetrahedron Lett.* **1983**, *24*, 4025. (b) Ochiai, M.; Kenzo, S.; Fujita, E.; Tada, S., *Chem. Pharm. Bull.* **1983**, *31*, 3346. (c) Ochiai, M.; Kenzo, S.; Fujita, E., *Chem. Pharm. Bull.* **1984**, *32*, 3686.
4. Hsiao, C.-N.; Shechter, H., *J. Org. Chem.* **1988**, *53*, 2688.
5. (a) Paquette, L. A.; Crouse, G. D., *J. Org. Chem.* **1983**, *48*, 141. (b) Lin, H.-S.; Coghlan, M. J.; Paquette, L. A., *Org. Synth.* **1988**, *67*, 157.
6. Paquette, L. A.; Williams, R. V.; Carr, R. V. C.; Charumilind, P.; Blount, J. F., *J. Org. Chem.* **1982**, *47*, 4566.
7. (a) Martinez-Fresneda, P.; Vaultier, M., *Tetrahedron Lett.* **1989**, *30*, 2929. (b) Montanari, F., *Gazz. Chim. Ital.* **1956**, *86*, 406.
8. (a) Williams, R. V.; Sung, C.-L. A., *J. Chem. Soc., Chem. Commun.* **1987**, 590. (b) Padwa, A.; Wannamaker, M. W., *J. Chem. Soc., Chem. Commun.* **1987**, 1742. (c) Djeghaba, Z.; Jousseaume, B.; Ratier, M.; Duboudin, J.-G., *J. Organomet. Chem.* **1986**, *304*, 115.
9. Magnus, P.; Quagliato, D., *J. Org. Chem.* **1985**, *50*, 1621.

Ottorino De Lucchi
Università di Venezia, Italy

Giovanna Delogu
Istituto CNR, IATCAPA, Sassari, Italy

2,1-Benzoxasilole, 1,3-dihydro-1,1-dimethyl-*[1]

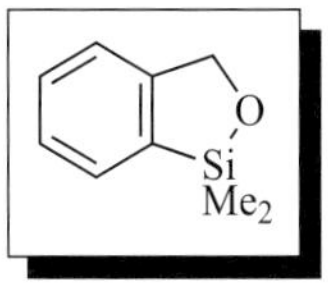

[321903-29-1]   C9H12OSi   (MW 164.28)

InChI = 1S/C9H12OSi/c1-11(2)9-6-4-3-5-8(9)7-10-11/h3-6H, 7H2,1-2H3

InChIKey = GLZOQCBCUKBUJG-UHFFFAOYSA-N

(silylation reagent, cross-coupling, carbonyl addition reaction)

Physical Data: bp 45 °C/2.0 mmHg.

Solubility: generally sol most organic solvents.

Form Supplied in: neat colorless liquid.

Analysis of Reagent Purity: purity is analyzed by ^{1}H NMR (CDCl$_3$): 0.40 (s, 6H), 5.16 (s, 2H), 7.23 (dd, J = 7.5, 0.7 Hz, 1H), 7.28–7.33 (m, 1H), 7.37–7.42 (m, 1H), 7.59 (dd, J = 7.1, 0.4 Hz, 1H).

Preparative Methods: by the halogen–lithium exchange reaction of 2-(2-tetrahydro-2*H*-pyranoxymethyl)bromobenzene followed by treatment with chlorodimethylsilane and subsequently with a catalytic amount of *p*-toluenesulfonic acid in MeOH (eq 1).[2] The reagent is isolated by distillation under reduced pressure. Large quantities can be prepared and stored in a refrigerator.

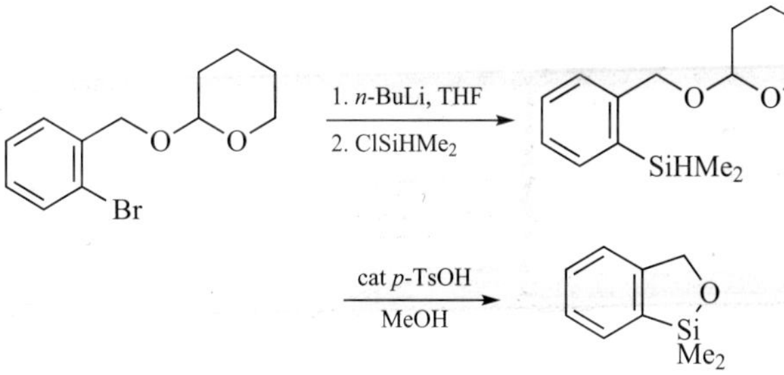

Scheme (eq 1): 2-(THP-oxymethyl)bromobenzene with 1. *n*-BuLi, THF; 2. ClSiHMe2 gives the silane with SiHMe2; then cat *p*-TsOH, MeOH gives the benzoxasilole product.

Purification: distillation at reduced pressure.

Handling, Storage, and Precautions: stable toward moisture, heat, and air; exposure to acid and base should be avoided.

Reaction with Alkenyl and Aryl Nucleophiles. This reagent reacts with various alkenyl- and arylmetallic reagents such as Grignard reagents and organolithium reagents through cleavage/formation of the Si–O/Si–C bonds to give alkenyl- or aryl[2-(hydroxymethyl)phenyl]dimethylsilanes in good yields after conventional aqueous workup (eq 2).[2]

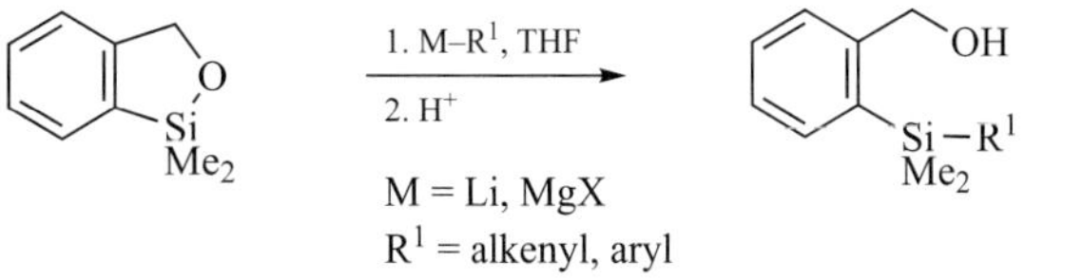

Scheme (eq 2): benzoxasilole with 1. M–R^1, THF; 2. H$^+$ gives ortho-(hydroxymethyl)phenyl-Si(Me2)R^1

M = Li, MgX
R^1 = alkenyl, aryl

*Article previously published in the electronic Encyclopedia of Reagents for Organic Synthesis as 1,3-Dihydro-1,1-dimethyl-2,1-benzoxasilole. DOI: 10.1002/047084289X. rn01341.

The resulting alkenyl- and arylsilane reagents undergo the palladium-catalyzed cross-coupling reactions under mild reaction conditions using K_2CO_3 as a base (eq 3).[2] Use of a copper cocatalyst is beneficial for the cross-coupling reaction of the arylsilanes. Distillation of a crude product obtained by a gram-scale cross-coupling reaction under reduced pressure allows recovery of the title compound.

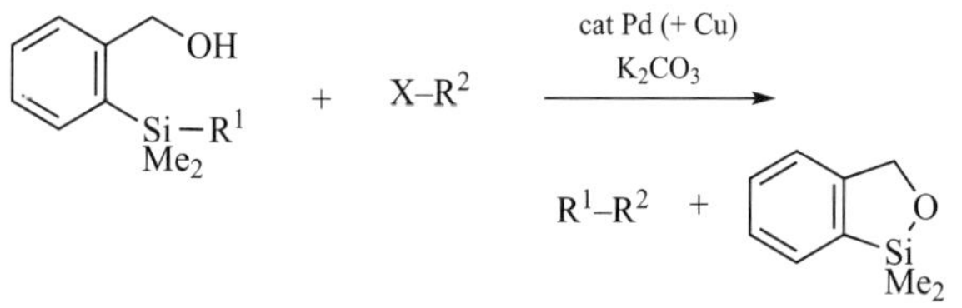

$$X = OCO_2Me, OCO_2t\text{-}Bu, Cl, Br, I$$
$$R^1 = alkenyl, aryl$$
$$R^2 = allyl, benzyl, alkenyl, aryl$$

The organosilicon reagents also participate in the rhodium-catalyzed reaction with α,β-unsaturated carbonyl compounds,[3] imines,[4] and alkynes (eq 4).[5] The use of optically active diene ligands allows the rhodium-catalyzed transformation to proceed in an enantioselective manner.[3,4] The reaction can be performed on a gram scale, and the cyclic silyl ether can also be recovered by distillation of a crude product.

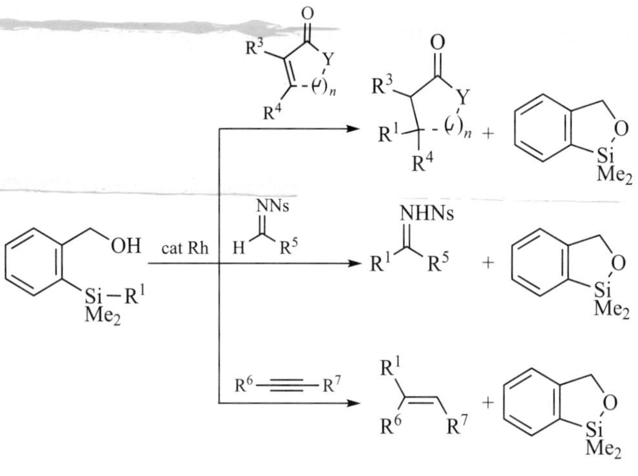

$$R^1 = alkenyl, aryl$$
$$R^3–R^7 = alkyl, alkenyl, aryl, silyl \quad (4)$$

Reaction with Lithium Aluminum Hydride. Treatment of the reagent with $LiAlH_4$ followed by acetylation of the resulting metal alkoxide with acetyl chloride in situ leads to dimethyl[2-(2-acetoxymethyl)phenyl]silane (eq 5).[2c] The hydrosilane thus obtained undergoes the platinum-catalyzed hydrosilylation of alkynes to give alkenyl[2-(hydroxymethyl)phenyl]dimethyl-silanes, which can be used for the above transformations, upon deacetylation under basic conditions.

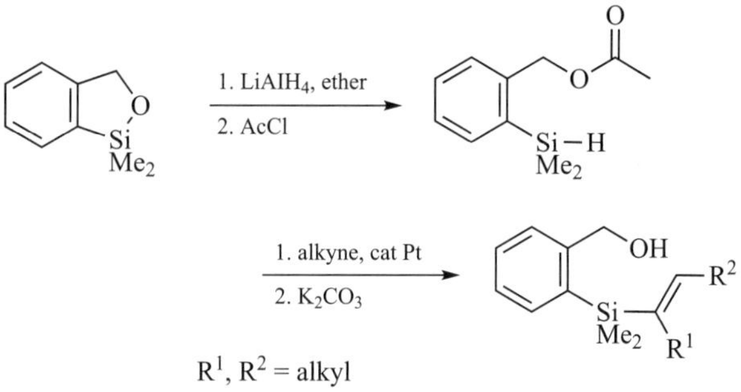

$$R^1, R^2 = alkyl \quad (5)$$

1. Nakao, Y.; Sahoo, A. K.; Imanaka, H.; Yada, A.; Hiyama, T., *Pure Appl. Chem.* **2006**, *78*, 435.
2. (a) Nakao, Y.; Imanaka, H.; Sahoo, A. K.; Yada, A.; Hiyama, T., *J. Am. Chem. Soc.* **2005**, *127*, 6952. (b) Nakao, Y.; Sahoo, A. K.; Yada, A.; Chen, J.; Hiyama, T., *Sci. Technol. Adv. Mater.* **2006**, *7*, 536. (c) Nakao, Y.; Imanaka, H.; Chen, J.; Yada, A.; Hiyama, T., *J. Organomet. Chem.* **2007**, *692*, 585. (d) Nakao, Y.; Ebata, S.; Chen, J.; Imanaka, H.; Hiyama, T., *Chem. Lett.* **2007**, *36*, 606. (e) Chen, J.; Tanaka, M.; Takeda, M.; Sahoo, A. K.; Yada, A.; Nakao, Y.; Hiyama, T., *Bull. Chem. Soc. Jpn.* **2010**, *83*, 554.
3. (a) Nakao, Y.; Chen, J.; Imanaka, H.; Hiyama, T.; Ichikawa, Y.; Duan, W.-L.; Shintani, R.; Hayashi, T., *J. Am. Chem. Soc.* **2007**, *129*, 9137. (b) Shintani, R.; Ichikawa, Y.; Hayashi, T.; Chen, J.; Nakao, Y.; Hiyama, T., *Org. Lett.* **2007**, *9*, 4643.
4. Nakao, Y.; Takeda, M.; Chen, J.; Hiyama, T.; Ichikawa, Y.; Shintani, R.; Hayashi, T., *Chem. Lett.* **2008**, *37*, 290.
5. Nakao, Y.; Takeda, M.; Chen, J.; Hiyama, T., *Synlett* **2008**, 774.

Yoshiaki Nakao
Kyoto University, Kyoto, Japan

Tamejiro Hiyama
Chuo University, Tokyo, Japan

N-Benzyl-*N*-(methoxymethyl)-*N*-trimethylsilylmethylamine[1]

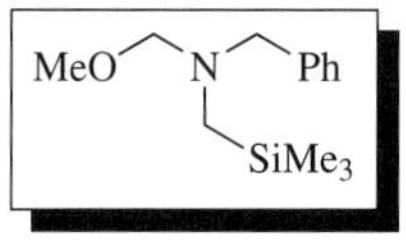

[93102-05-7] $C_{13}H_{23}NOSi$ (MW 237.42)

InChI = 1S/C13H23NOSi/c1-15-11-14(12-16(2,3)4)10-13-8-6-5-7-9-13/h5-9H,10-12H2,1-4H3

InChIKey = RPZAAFUKDPKTKP-UHFFFAOYSA-N

(nonstabilized azomethine ylide precursor;[1] reacts with alkenes to give pyrrolidines;[2] alkynes give 3-pyrrolines;[2] carbonyl and thiocarbonyl groups afford 1,3-oxazolidines and 1,3-thiazolidines, respectively[3])

Physical Data: bp 77–80 °C/0.5 mmHg.

Preparative Methods: most conveniently prepared by treatment of benzylamine with chloromethyltrimethylsilane followed by formaldehyde and methanol.[4] Access to higher ether homologs is achieved by replacing methanol with the appropriate alcohol.[5] An alternate procedure involves alkylation of lithium *N*-benzyltrimethylsilylmethylamide with methoxymethyl chloride.[2]

Purification: distillation under reduced pressure, although good yields can be obtained with undistilled reagent.

Handling, Storage, and Precautions: the reagent should be handled in a well ventilated fume hood.

1,3-Dipolar Cycloadditions. *N*-Benzyl-*N*-(methoxymethyl)-*N*-trimethylsilylmethylamine (**1**) is a valuable reagent for in situ generation of the *N*-benzyl azomethine ylide (**2**). It is generally preferred over alternative silylmethylamine precursors[6–8] because of ease of handling and use. The ylide (**2**) is most conveniently generated from (**1**) using a catalytic amount of trifluoroacetic acid as described by Achiwa.[2] Alternative catalysts include LiF,[3,4] TBAF,[7] Me$_3$SiOTf–CsF,[5] or Me$_3$SiI–CsF.[5] Mechanistic studies provide evidence that the reactive intermediate generated from (**1**) with either CF$_3$CO$_2$H or F$^-$ is a 1,3-dipolar species.[7,8] Reaction of (**2**) with alkenes provides an efficient convergent route to pyrrolidine derivatives. Alkynes afford 3-pyrrolines[2] which can be converted into pyrroles.[6a] The ylide (**2**) reacts most readily with electron deficient alkenes and alkynes since this pairing results in a narrow dipole HOMO–dipolarophile LUMO energy gap.[9] Examples of suitable dipolarophiles include unsaturated esters,[2–5] ketones,[2] imides,[2,4] nitriles,[4] and sulfones.[3,4] Cycloaddition occurs with complete *cis* stereospecificity[2–4] (eq 1) which is consistent with a concerted mechanism. Dipolarophiles containing an endocyclic double bond afford fused bicyclic pyrrolidines,[10] whereas substrates with an exocyclic double bond provide access to spirocyclic systems.[11]

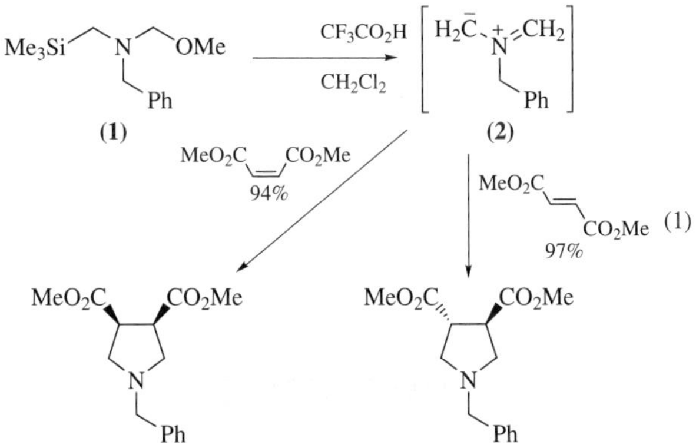

Styrenes bearing electron withdrawing aromatic substituents such as CN and NO$_2$ give high yields of 3-arylpyrrolidines (eq 2).[12] When electron withdrawing groups are absent yields tend to be poor. Vinylpyridines afford 3-pyridylpyrrolidines.[12] The reactivity of alkenes can be enhanced by the introduction of a CF$_3$ substituent. For example, α-trifluoromethylstyrene gives a high yield of cycloadduct whereas α-methylstyrene is unreactive.[13]

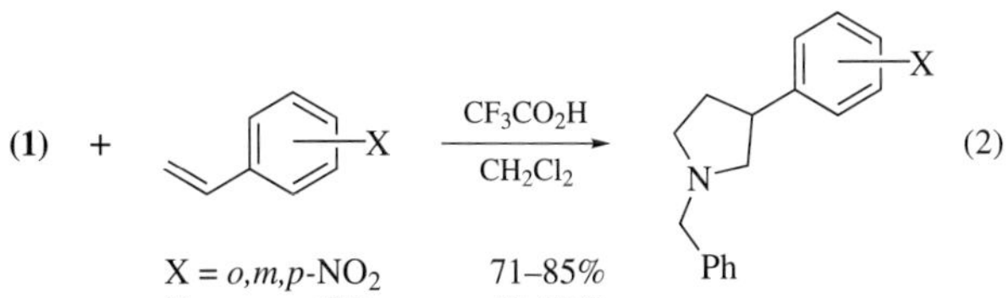

X = *o,m,p*-NO$_2$	71–85%	
X = *o,m,p*-CN	58–76%	
X = *p*-OMe	20%	
X = H	20%	

Cycloaddition reactions with aldehydes and ketones afford 1,3-oxazolidines while thioketones give 1,3-thiazolidines (eq 3).[3] Adducts derived from reaction at both the carbonyl and alkenic double bond have been observed with an α,β-unsaturated aldehyde.[3]

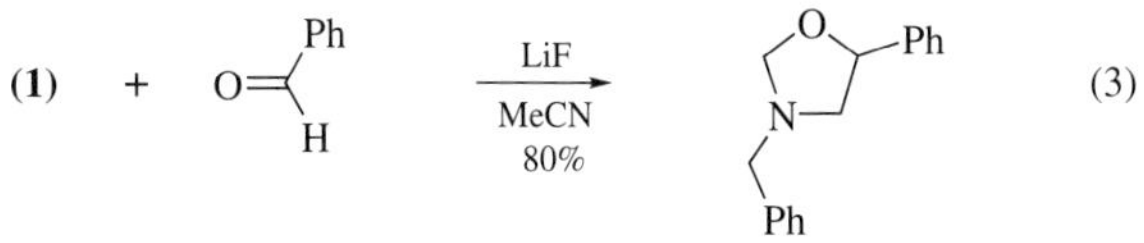

Diastereoselective 1,3-Dipolar Cycloadditions. Several examples of high diastereofacial selectivity with homochiral dipolarophiles have been reported. Cycloaddition of (**1**) with the cyclic dipolarophile (**3**) occurs with complete π-facial selectivity as a result of addition from the side opposite the bulky silyloxymethyl group (eq 4).[14] The key step in an asymmetric synthesis of (*S*)-(−)-cucurbitine involves cycloaddition of (**1**) with the α,β-dehydrolactone (**4**) to give the pyrrolidine (**5**) as a single diastereomer (eq 5).[15]

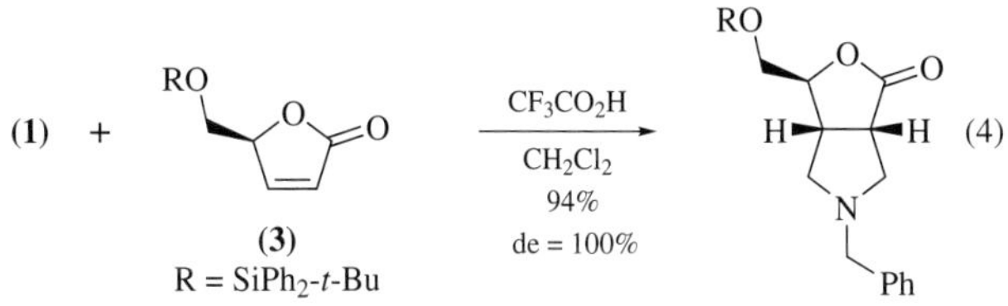

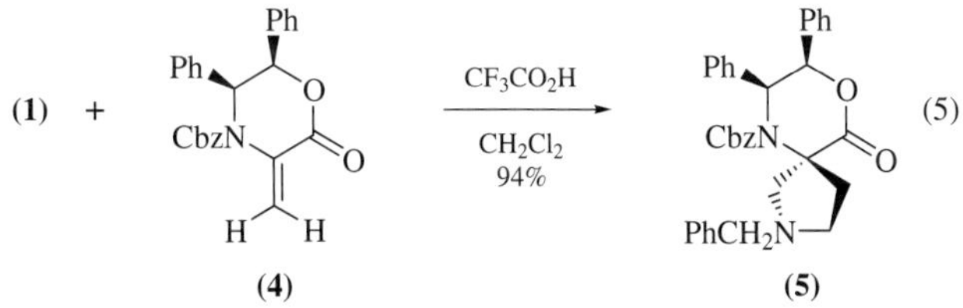

Double asymmetric induction employing a chiral *N*-α-methylbenzyl analog of (**1**) has been described.[16]

---

1. (a) Terao, Y.; Aono, M.; Achiwa, K., *Heterocycles* **1988**, *27*, 981. (b) Tsuge, O.; Kanemasa, S., *Adv. Heterocycl. Chem.* **1989**, *45*, 231.

2. Terao, Y.; Kotaki, H; Imai, N.; Achiwa, K., *Chem. Pharm. Bull.* **1985**, *33*, 2762.

3. Padwa, A.; Dent, W., *J. Org. Chem.* **1987**, *52*, 235.

4. Padwa, A.; Dent, W., *Org. Synth., Coll. Vol.* **1992**, *8*, 231.

5. Hosomi, A.; Sakata, Y.; Sakurai, H., *Chem. Lett.* **1984**, 1117.

6. (a) Padwa, A.; Chen, Y. Y.; Dent, W.; Nimmesgern, H., *J. Org. Chem.* **1985**, *50*, 4006. (b) Pandey, G.; Lakshmaiah, G.; Kumaraswamy, G., *J. Chem. Soc., Chem. Commun.* **1992**, 1313.

7. Terao, Y.; Kotaki, H.; Imai, N.; Achiwa, K., *Chem. Pharm. Bull.* **1985**, *33*, 896.

8. Terao, Y.; Imai, N.; Achiwa, K., *Chem. Pharm. Bull.* **1987**, *35*, 1596.

9. Houk, K. N.; Sims, J.; Watts, C. R.; Luskus, L. J., *J. Am. Chem. Soc.* **1973**, *95*, 7301.
10. Orlek, B. S.; Wadsworth, H.; Wyman, P.; Hadley, M. S., *Tetrahedron Lett.* **1991**, *32*, 1241.
11. Orlek, B. S.; Wadsworth, H.; Wyman, P.; King, F. D., *Tetrahedron Lett.* **1991**, *32*, 1245.
12. Laborde, E., *Tetrahedron Lett.* **1992**, *33*, 6607.
13. Begue, J. P.; Bonnet-Delpon, D.; Lequeux, T., *Tetrahedron Lett.* **1993**, *34*, 3279.
14. Wee, A. G. H., *J. Chem. Soc., Perkin Trans. 1* **1989**, 1363.
15. Williams, R. M.; Fegley, G. J., *Tetrahedron Lett.* **1992**, *33*, 6755.
16. Fray, A. H.; Meyers, A. I., *Tetrahedron Lett.* **1992**, *33*, 3575.

Barry S. Orlek

SmithKline Beecham Pharmaceuticals, Harlow, UK

1,2-Bis(*t*-butyldimethylsilyl)hydrazine

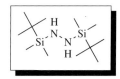

[10000-20-1] $C_{12}H_{32}N_2Si_2$ (MW 260.57)
InChI = 1S/C12H32N2Si2/c1-11(2,3)15(7,8)13-14-16(9,10)12(4,5)6/h13-14H,1-10H3
InChIKey = AIZSLKRCSRDHKC-UHFFFAOYSA-N

(source of latent hydrazine used in the preparation of stable alkyl and aryl *N*-silyl hydrazones, *gem*-dihalides, vinyl halides, and room temperature Wolff–Kishner reductions)

Alternative Names: BTBSH.
Physical Data: bp 118–119 °C/9 Torr;[1] 55–65 °C/0.05 Torr;[2] n^{25}_D 1.4456.
Form Supplied in: colorless liquid.
Preparative Methods: prepared by reacting *t*-butyldimethylchlorosilane with anhydrous hydrazine at 70 °C (eq 1), with or without solvent.

$$\text{TBDMSCl} + \text{NH}_2\text{NH}_2 \xrightarrow[\text{Et}_2\text{O, }\Delta]{\text{neat, 70 °C} \atop \text{or}} \quad (1)$$

2.0 equiv
1   **2**

Purification: fractional distillation.
Handling, Storage, and Precaution: this reagent should be viewed as a latent source of hydrazine; all safety precautions relating to the handling of hydrazine should be considered applicable to this reagent.

Reaction with Aldehydes and Ketones. Classical preparations of aryl and alkyl hydrazones with hydrazine lead to mixtures of hydrazones and azines (eq 2).

$$\xrightarrow{\text{N}_2\text{H}_2} \quad + \quad (2)$$

3   **4**   **5**

This problem can be surmounted by the use of BTBSH, which affords the *N*-silyl hydrazones cleanly and in high yield (eq 3).[2] Thus, *N*-silylated aryl and alkyl hydrazones are prepared by reaction of 1,2-bis(*t*-butyldimethylsilyl)hydrazine with an aldehyde or ketone in the presence of a catalytic amount of Sc(OTf)₃ in dichloromethane (eq 3).[2] A number of Lewis acid catalysts were explored, with Sc(OTf)₃ being the most effective, giving the hydrazones in yields of up to 95% as mixtures of the *syn*- and *anti*-isomers. In the absence of Lewis acid catalysis, the reaction does not proceed. Unlike the simple hydrazine-derived hydrazones, the resulting *N*-silyl hydrazones are stable, isolable, and capable of being stored for extended periods of time.

$$\xrightarrow[\text{0.1–1.0 mol % Sc(OTf)}_3]{\text{BTBSB, CH}_2\text{Cl}_2} \quad (3)$$

3   **6**
                              85–95%

Wide ranges of substrates were successfully converted to the *N*-silyl hydrazones under these conditions. Functional groups such as ketals (**7**), lactones (**8**), ethers (**9**), and basic amines (**10**) are compatible with the reaction conditions. The only caveat is that free phenolic and hydroxy groups are obligatorily silylated.[2]

7   **8**

9   **10**

Wolff–Kishner Reduction. A particularly useful reaction of the *N*-silyl hydrazones is their use as substrates in the Wolff–Kishner reaction, where reduction of the corresponding carbonyl compound can be accomplished at relatively modest temperatures.[2] This is in marked contrast to the more traditional Wolff–Kishner reaction, and its Huang–Minlon modification, which require high temperatures and the use of potassium hydroxide.[3]

Room temperature conditions for the Wolff–Kishner reaction, previously reported by Cram and co-workers,[4] were adapted by Myers and Furrow to the current *N*-silyl hydrazone methodology (eq 4).

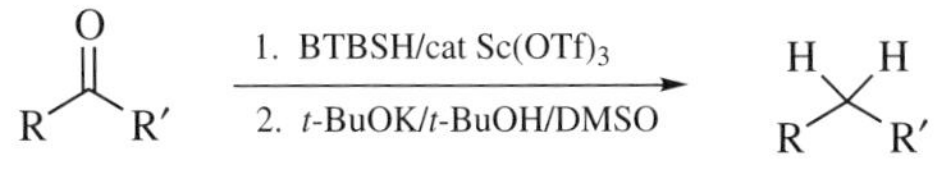

(4)

The Myers' two-step, single-flask procedure differs from the Cram procedure in that the requisite hydrazones are more reliably made and do not require isolation. The reduction of simple aliphatic hydrazones generally takes place at ambient temperature, whereas sterically hindered hydrazones required some heating (100 °C) for the reaction to proceed at a reasonable rate. Complex structures such as the steroid hecogenin (**12**, eq 5), and other tertiary hydrazones,[5] are efficiently reduced in high yield. The morphine derivative naloxone (**14**, eq 6) underwent the expected Kishner–Leonard elimination[6] under the standard conditions.[2]

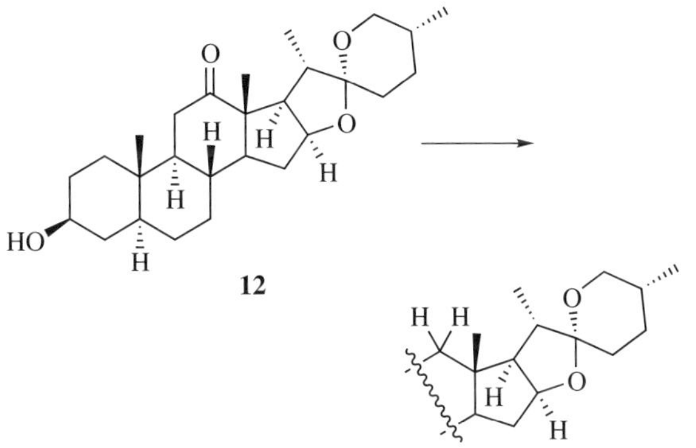

(5)

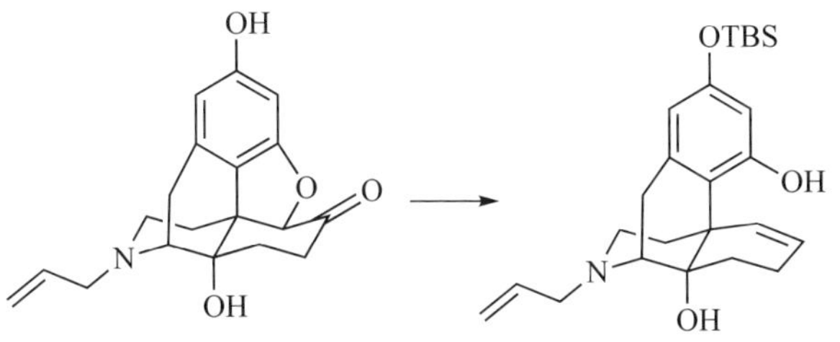

(6)

Preparation of Esters. The *N*-silyl hydrazones react with (difluoroiodo)benzene in the presence of chloropyridine to form an intermediate diazo-species, which participates in typical esterification reactions with carboxylic acids (eq 7).[7]

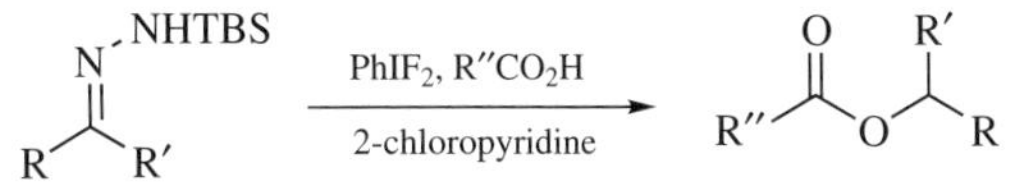

(7)

The reaction conditions are relatively mild and resistant to adventitious moisture; the authors noting that the intentional addition of as much as 5 equiv water had no affect on yield. Also of note is that phenolic and alcoholic functional groups present on

the carboxylic acid substrate were inert toward the diazo precursor and as a consequence required no protection. Sterically hindered carboxylic acids such as that found in podocarpic acid (**18**) were converted to the corresponding benzyl ester, although 9 equiv of the TBSH derivative were needed to drive this particular reaction to completion (eq 8).

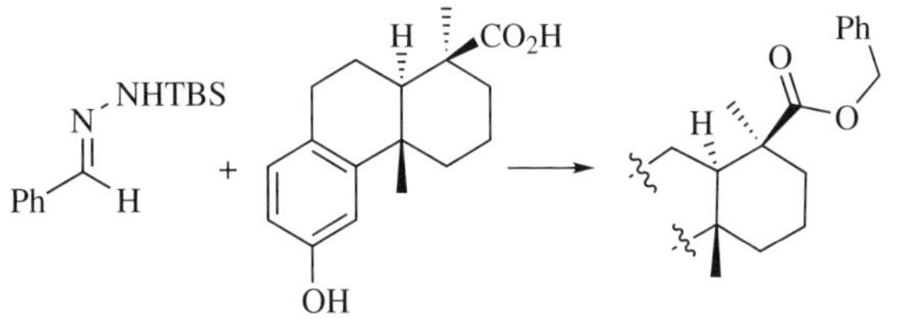

(8)

The esterification of *o*-nitrobenzoic acid with the *N*-silyl hyrdazone derivative (**8**) proceeds in good yield without detriment to the lactone (eq 9).

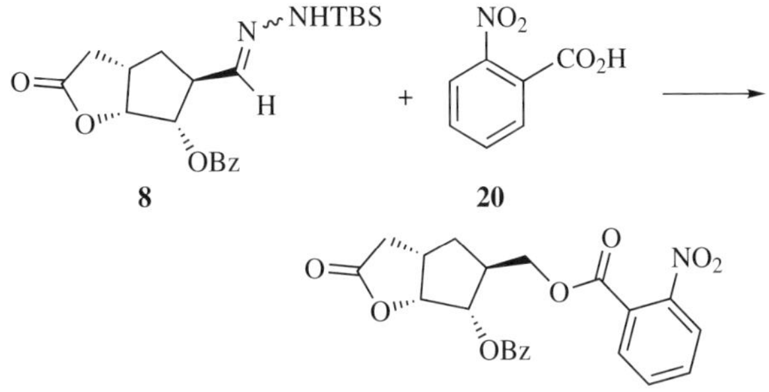

(9)

Preparation of *gem*-Dihalides. *gem*-Dichlorides (**23**) and *gem*-dibromides (**24**) are conveniently prepared from aldehydes and ketones using BTBSH and a copper(II) halide (eq 10).[2]

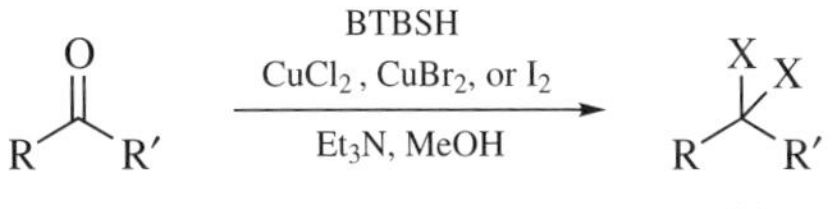

(10)

In contrast to other methods for affecting this transformation, the reaction conditions are mild and tolerant of a variety of acid and base labile functional groups. *gem*-Diiodides (**25**) were also prepared in a similar manner using molecular iodine in place of the copper halide.

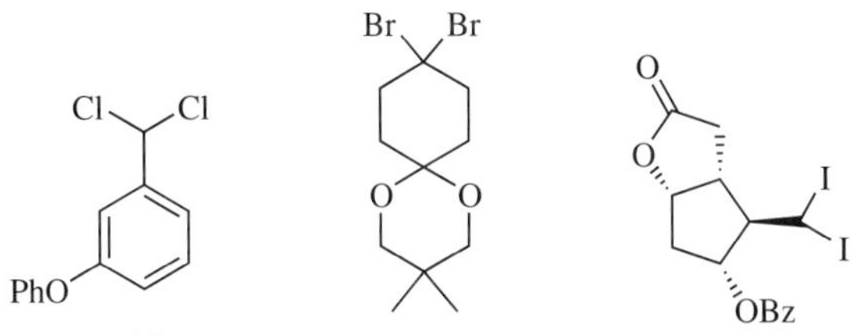

Preparation of Vinyl Bromides and Iodides. Vinyl bromides can be prepared from ketones in moderate to excellent yield via the *N*-silyl hydrazone derivatives (eq 11).[2] The reaction proceeds at ambient temperature and is applicable to sterically encumbered ketones.

$$R \overset{\underset{\displaystyle N \cdot NHTBS}{\|}}{\underset{\displaystyle R'}{C}} \xrightarrow[\text{CH}_2\text{Cl}_2, \ 0-23\,°\text{C}]{\text{Br}_2 \text{ or } \text{I}_2, \text{ BTMG}} R' \overset{X}{\diagdown} R \quad (11)$$

26   **27**

The optimal conditions for this reaction employ the freshly prepared silyl hydrazone, from which residual t-butlydimethyl-silanol has been removed under vacuum, and its slow addition to molecular bromine or iodine and base in dichloromethane. Following the original work of Barton with simple hydrazones,[8] optimal yields were obtained when N''-t-butyl-N,N,N',N'-tetrame-thylguanidine (BTMG) was used as base (eqs 12 and 13).

$$\text{(12)}$$

7   **28**, X = Br, 65%
                              29, X = I, 67%

$$\text{(13)}$$

9   **30**, X = Br, 84%
                              31, X = I, 82%

Related Reagents. Hydrazine.

1. West, R.; Ishikawa, M.; Bailey, R. E., *J. Am Chem Soc.* **1966**, *88*, 4648.

2. Furrow, M. E.; Myers, A. G., *J. Am Chem Soc.* **2004**, *126*, 5436.

3. Todd, D. *Organic Reactions*; John Wiley & Sons Inc.: New York, 1948; Vol. 4, p. 378.

4. Cram, D. J.; Sahyun, M. R.; Knox, G. R., *J. Am Chem Soc.* **1962**, *84*, 1734.

5. Crich, D.; Xu, H.; Kenig, F., *J. Org. Chem.* **2006**, *71*, 5016.

6. Leonard, N. J.; Gelfand, S. J., *J. Am Chem Soc.* **1955**, *77*, 3269.

7. Furrow, M. E.; Myers, A. G., *J. Am Chem Soc.* **2004**, *126*, 12222.

8. Barton, D. H. R.; Bashiardes, G.; Fourrey, J.-L., *Tetrahedron* **1988**, *44*, 147.

Gary F. Filzen
Pfizer Global Research & Development, Ann Arbor, MI, USA

1,2-Bis(chlorodimethylsilyl)ethane[1]

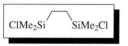

[13528-93-3] $C_6H_{16}Cl_2Si_2$ (MW 215.27)

InChI = 1S/C6H16Cl2Si2/c1-9(2,7)5-6-10(3,4)8/h5-6H2,1-4H3

InChIKey = VGQOKOYKFDUPPJ-UHFFFAOYSA-N

(reagent for the protection of primary amines as their tetramethyl-disilylazacyclopentane or Stabase adducts; used in combination with zinc to form organozinc carbenoids; electrophilic silylating reagent)

Physical Data: mp 36–41 °C; bp 198 °C/734 mmHg.

Form Supplied in: white, waxy solid with a sharp acid odor; 96% pure.

Handling, Storage, and Precautions: the dry solid reacts with moisture to produce hydrogen chloride. Incompatible with strong acids, alcohols, strong bases and strong oxidizers; may be ignited by static electricity; causes severe irritation to the respiratory tract. The reagent therefore should be used in a well ventilated hood.

Protection of Primary Amines. Few protecting groups exist for the N,N-diprotection of primary amines. The tetramethyldisi-lylazacyclopentane or Stabase adduct (**3**) is particularly attractive for this purpose due to its ease of preparation, base stability and facile removal. This cyclic disilane is easily prepared (eq 1) from the commercially available title reagent (**1**).[2] For primary amines with pK_a values between 10 and 11, typical reaction conditions involve the treatment of a dichloromethane solution of the amine with (**1**) in the presence of two equivalents of triethylamine at room temperature. Subsequent workup with aqueous dihydrogen phosphate provides (**3**) in excellent yields.[1] Stabase adducts are quite stable to strongly basic conditions such as n-butyllithium and s-butyllithium (at −25 °C), lithium diisopropylamide, and Grig-nard reagents making it a superb candidate for use in the alkylation of substrates bearing a primary amine. For example, protected amino acid derivatives are easily alkylated (eq 2).[1,3] After formation of the lithium enolate of the protected ethyl glycinate (**4**) under standard conditions, exposure to various alkyl halides or aldehydes yields alkylation products such as (**5**).[1]

$$\text{(1)}$$

(1)   **(2)**   **(3)**

$$\xrightarrow[\text{2. RX}]{\text{1. LDA, THF, }-78\,°\text{C}}$$

49–91%

(4)   **(5)**

RX = MeI, PhCHO, $\diagup\diagdown$Br, Br$\diagup\diagup$———SiMe₃, Br₅⟨benzyl⟩CH₂Br

In addition to its base stability, this protecting group is readily cleaved under acidic conditions; often upon workup. After quenching the metallated dihydropyridine in eq 3 with the Stabase-protected aminoethyl disulfide (**7**), simple in situ treatment with saturated aqueous ammonium chloride liberates the regioisomeric free amines (**8a**) and (**8b**).[4] However, more vigorous conditions are sometimes required. For example, allylic amine (**11**) was produced by exposure of the crude coupling product of (**9**) and (**10**) to an excess of ethanolic HCl at 0 °C (eq 4).[5] Alternatively, Stabase deprotection can be achieved by reaction with ethanolic sodium borohydride, as demonstrated in eq 5. The tertiary alcohol resulting from Grignard addition to aromatic ketone (**12**) was thus deprotected by treatment with NaBH$_4$ in ethanol to afford amino alcohol (**13**).[6]

Cleavage of Stabase adducts reveals a reactive primary amine that can be utilized in further transformations. Several groups have engineered one-pot, multistep processes that exploit both the protecting power and the acid lability of the cyclic disilane to generate amines in situ. An elegant example of such a transformation is illustrated in the general synthesis of 2-substituted pyrrolidines (**17**) which otherwise are difficult to synthesize (eq 6). Addition of the Stabase-protected Grignard (**14**) to an N-methoxy-N-methyl amide is followed by treatment of the reaction mixture with ethanolic HCl. Collapse of the intermediate hemiaminal is followed by Stabase cleavage and concomitant cyclization to the imine (**16**) in a single operation. The substituent R may vary from aliphatic to a wide variety of heterocyclic compounds.[7]

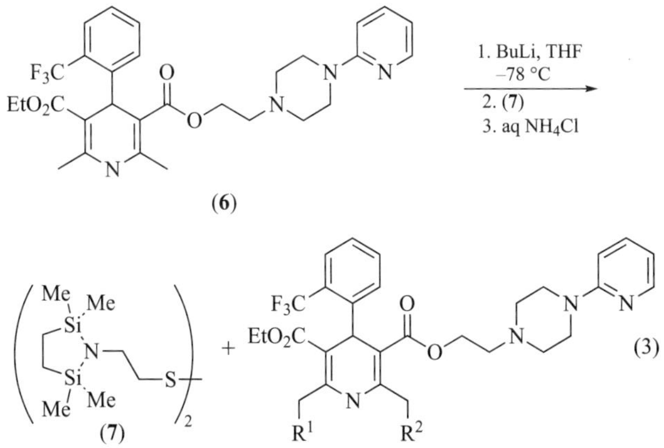

(**6**)

(**8a**) R^1 = H; R^2 = SCH$_2$CH$_2$NH$_2$; 16%

(**8b**) R^1 = SCH$_2$CH$_2$NH$_2$; R^2 = H; 18%

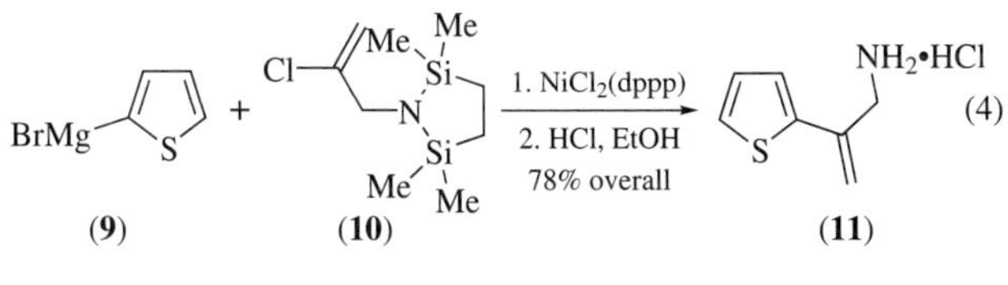

A further extension of this concept is shown in the 6-substituted 3-pyridinol synthesis described in eq 7. Treatment of condensation product (**19**) with 1 M HCl initiates Stabase deprotection that is followed by acid-catalyzed rearrangement of the resulting amino alcohol to pyridine (**20**).[8]

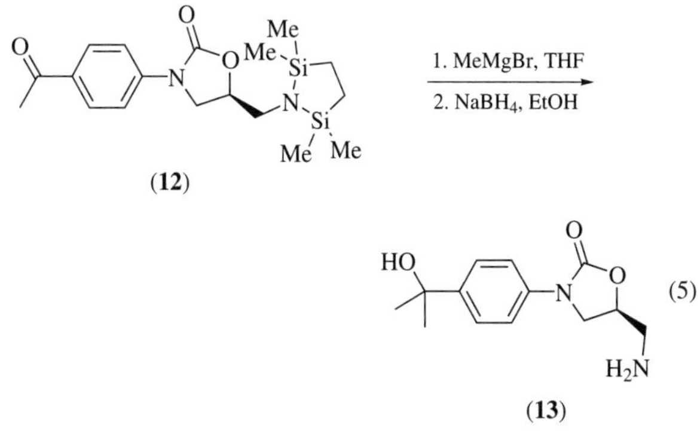

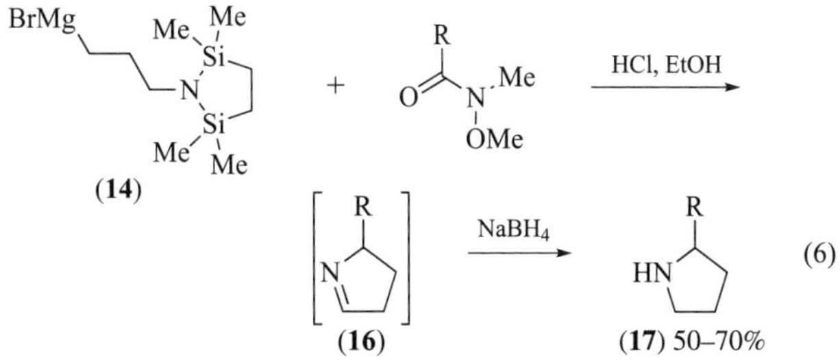

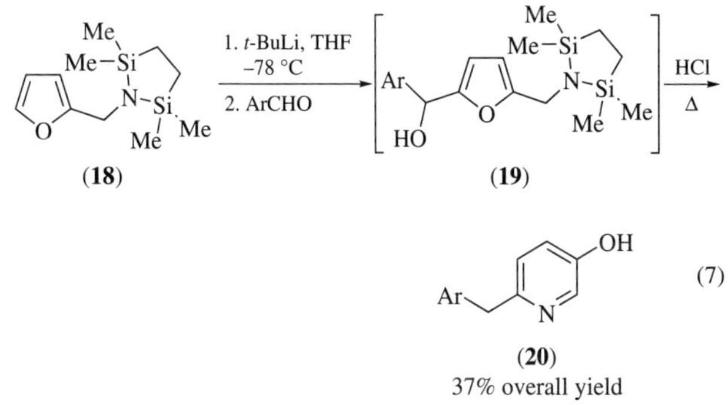

(**12**)

(**13**)   (5)

(**14**)   (**16**)   (**17**) 50–70%   (6)

(**18**)   (**19**)   (20)

(**20**)
37% overall yield   (7)

Formation of Organozinc Carbenoids. When aromatic aldehydes and certain α,β-unsaturated carbonyl compounds are treated with zinc and 1,2-bis(chlorodimethylsilyl)ethane, an organozinc carbenoid results. This two-electron process is postulated to occur as depicted in eq 8. Sequential reaction of the two silicon atoms with the zinc carbonyl complex and subsequent extrusion of the cyclic siloxane (**23**) gives rise to the putative organozinc carbenoid (**24**). When generated in the presence of an alkene, cyclopropanes (**25**) are produced in good yields as in eq 9.[9] When 2 equiv of carbonyl compound are employed, the organozinc carbenoid is trapped by excess carbonyl compound, and the intermediate epoxide is deoxygenated to yield products (**26**) of symmetrical dicarbonyl coupling (eq 10).[10]

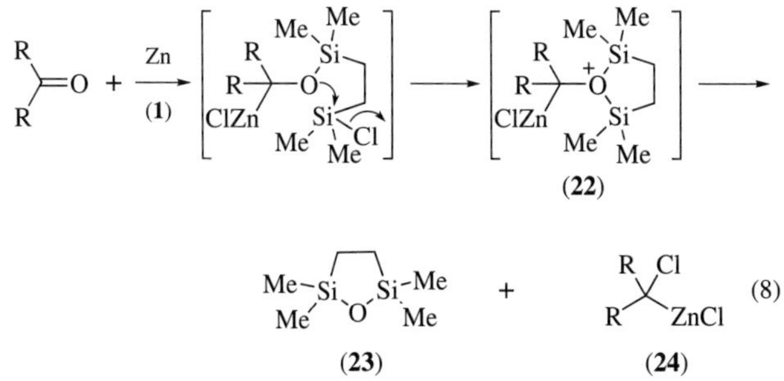

(**1**)   (**22**)   (8)

(**23**)   (**24**)

$$\text{(eq 9)} \quad (25)$$

$$2 \quad \xrightarrow[\text{(1)}]{\text{Zn}} \quad \text{(10)}$$

(26) R = Me, 86%

Electrophilic Silylation Reagent. 1,2-Bis(chlorodimethyl-silyl)ethane is not only an effective amine silylating reagent, but can also be employed in reactions with other anions. For example, when treated with the silver sulfonate salt as in eq 11, the nonafluorobutanesulfonic acid silyl ester (**27**), an extremely powerful silylating reagent, results.[11] Carbanions are also effectively trapped by (**1**) A series of rigid butadiene Diels–Alder precursors such as (**28**) have been prepared in this fashion as outlined in eq 12.[12]

$$C_4F_9SO_2OH \xrightarrow{Ag_2CO_3} C_4F_9SO_2OAg \xrightarrow{(1)} \quad (11)$$

(27)

$$Me_3Sn \quad SnMe_3 \xrightarrow[\text{(1)}]{2\ MeLi}_{60\%} \quad (12)$$

(28)

Related Reagents. 1,4-Bis(dimethylamino)-1,1,4,4-tetramethyldisilethylene.

1. Djuric, S.; Venit, J.; Magnus, P., *Tetrahedron Lett.* **1981**, *22*, 1787.
2. Sakurai, H.; Tominaga, K.; Watanabe, T.; Kumada, M., *Tetrahedron Lett.* **1966**, 5493.
3. (a) Leduc, R.; Bernier, M.; Escher, E., *Helv. Chim. Acta* **1983**, *66*, 960. (b) Cavelier–Frontin, F.; Jacquier, R.; Paladino, J.; Verducci, J., *Tetrahedron* **1991**, *47*, 9807.
4. Poindexter, G. S.; Licause, J. F.; Dolan, P. L.; Foley, M. A.; Combs, C. M., *J. Org. Chem.* **1993**, *58*, 3811.
5. Barger, T. M.; McCowan, J. R.; McCarthy, J. R.; Wagner, E. R., *J. Org. Chem.* **1987**, *52*, 678.
6. Gregory, W. A.; Brittelli, D. R.; Wang, C.-L. J.; Kezar, H. S. III; Carlson, R. K.; Park, C.-H.; Corless, P. F.; Miller, S. J.; Rajagopalan, P.; Wuonola, M. A.; McRipley, R. J.; Eberly, V. S.; Slee, A. M.; Forbes, M., *J. Med. Chem.* **1990**, *33*, 2569.
7. Basha, F. Z.; DeBernardis, J. F., *Tetrahedron Lett.* **1984**, *25*, 5271.
8. Barrett, A. G. M.; Lebold, S. A., *Tetrahedron Lett.* **1987**, *28*, 5791.
9. Motherwell, W. B.; Roberts, L. R., *J. Chem. Soc., Chem. Commun.* **1992**, *21*, 1582.
10. Afonso, C. A. M.; Motherwell, W. B.; O'Shea, D. M.; Roberts, L. R., *Tetrahedron Lett.* **1992**, *33*, 3899.
11. Frasch, M.; Sundermeyer, W.; Waldi, J., *Chem. Ber.* **1992**, *125*, 1763.
12. Reich, H. J.; Reich, I. L.; Yelm, K. E.; Holladay, J. E.; Gschneidner, D., *J. Am. Chem. Soc.* **1993**, *115*, 6625.

Fatima Z. Basha & Steven W. Elmore
Abbott Laboratories, Abbott Park, IL, USA

1,4-Bis(dimethylamino)-1,1,4,4-tetramethyldisilethylene

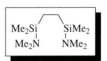

[91166-50-6] $C_{10}H_{28}N_2Si_2$ (MW 232.52)

InChI = 1S/C10H28N2Si2/c1-11(2)13(5,6)9-10-14(7,8)12(3)4/
          h9-10H2,1-8H3

InChIKey = MRAAXSSHMOFDJR-UHFFFAOYSA-N

(reagent for the protection of primary amines with lower pK_a values, such as substituted anilines, as their tetramethyldisilylazacyclopentane or Stabase adduct; precursor of tetramethyldisilylthiacyclopentane, a sulfur transfer reagent)

Physical Data: bp 100–103 °C/13 mmHg; d 0.824 g cm^{-3}.
Preparative Method: the reagent is prepared as a colorless liquid in 90% yield by treating commercially available bis(chlorodimethylsilyl)ethane with 5 equiv of dimethylamine in ether at 0 °C for 5 h.
Handling, Storage, and Precautions: the liquid vapors are heavier than air and may travel a considerable distance to a source of ignition. Vapor–air mixtures are explosive above the flash point of 14 °C. Use in a fume hood.

Protection of Primary Aromatic Amines.[1] The title reagent is used for the preparation of Stabase derivatives of primary amines, and is particularly useful for the protection of primary aromatic amines with pK_a values of 10–11. The protection of substituted anilines is achieved by heating equimolar amounts of the title compound and the anilines in the presence of zinc iodide (0.5 mol %) under a slow stream of nitrogen at 140 °C for 5 h. In many cases the products can be purified by vacuum distillation. Cleavage of the Stabase adducts to the corresponding primary amines can be performed in ether in the presence of 2 equiv of methanol and a small amount of *p*-toluenesulfonic acid monohydrate (0.2 mol %). The tetramethyldisilylazacyclopentane or Stabase adduct (**3**) is a particularly attractive protecting group due to its ease of preparation, base stability, and facile removal (eq 1).

$$\xrightarrow[\text{HNMe}_2]{\text{Et}_2\text{O},\ 0\ ^\circ\text{C}} \quad \xrightarrow[\text{}]{\text{ZnI}_2,\ 140\ ^\circ\text{C}}$$

(2)   **(1)**

$$\quad (1)$$

(3)

The title compound was utilized to protect 2-amino-6-bromo-pyridine (**4**) in 72% yield in a sequence to prepare helicopodant (**7**) (eq 2).[2]

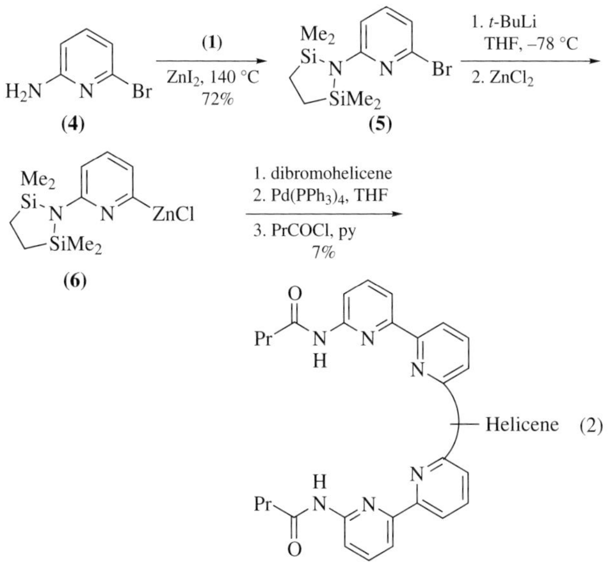

Bis(1-methoxy-2-methyl-1-propenyloxy)-dimethylsilane[1]

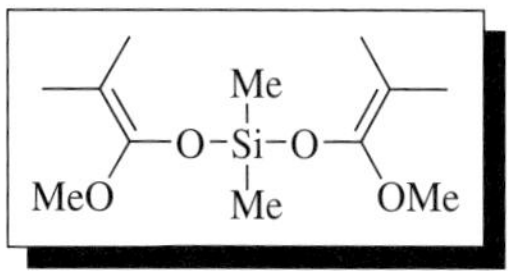

[86934-32-9] $C_{12}H_{24}O_4Si$ (MW 260.40)

InChI = 1S/C12H24O4Si/c1-9(2)11(13-5)15-17(7,8)16-12(14-6)10(3)4/h1-8H3

InChIKey = YIAMKONKFNWPGX-UHFFFAOYSA-N

(very useful bifunctional protecting agent for various types of H-acidic materials under mild conditions)

Physical Data: 89–90 °C/6.5 mmHg.
Form Supplied in: colorless liquid; not commercially available.
Preparative Method: prepared by treating 2 equiv of methyl lithioisobutyrate with 1 equiv of dichlorodimethylsilane at low temperature in dry THF, followed by distillation of the colorless liquid.
Handling, Storage, and Precautions: moisture sensitive; should be stored under nitrogen in a refrigerator.

The Dialkylsilylene Protecting Group. Bifunctional compounds containing alcohol, thiol, acid, and amine units may be protected as cyclic dimethylsilylene derivatives, which are analogous to isopropylidene derivatives, by treatment with the title reagent (**1**).[2] Cleavage of the silylene protective group is easily achieved by hydrolysis or solvolysis. It should be noted that the dimethylsilylene group is highly sensitive to hydrolysis and is unstable to column chromatography.

Silylene derivatives have become very useful in synthetic organic chemistry. For example, the cyclic dimethylsilylene derivative (**2**) of *o*-hydroxyacetophenone is useful for [4 + 2] cycloadditions (eq 1).[3] A glycosyl donor and acceptor have been linked as the dimethylsilylene derivative (**3**) and used for stereoselective glycosylation (eq 2).[4] Also, an efficient and stereocontrolled aldol reaction of silylene enolate (**4**) with aldehydes has been reported (eq 3).[5]

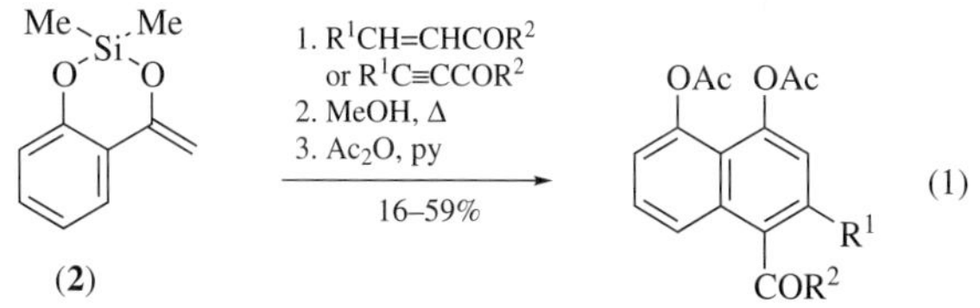

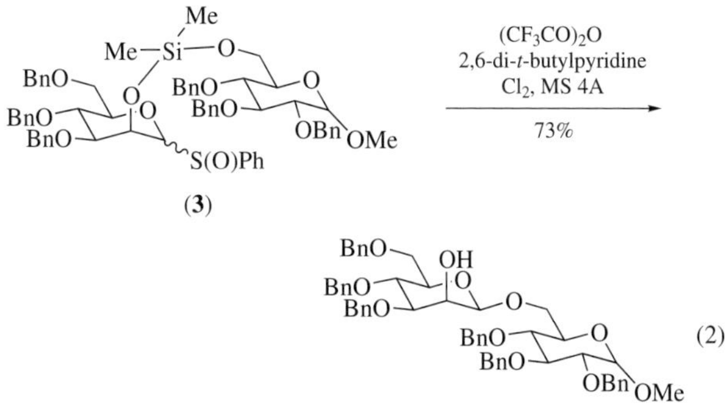

Sulfur Transfer Reagent. The reagent was used to prepare 2,2,5,5-tetramethyl-2,5-disila-1-thiacyclopentane (**8**) which acts as a sulfur transfer reagent. An example of this reaction is shown in eq 3.[3]

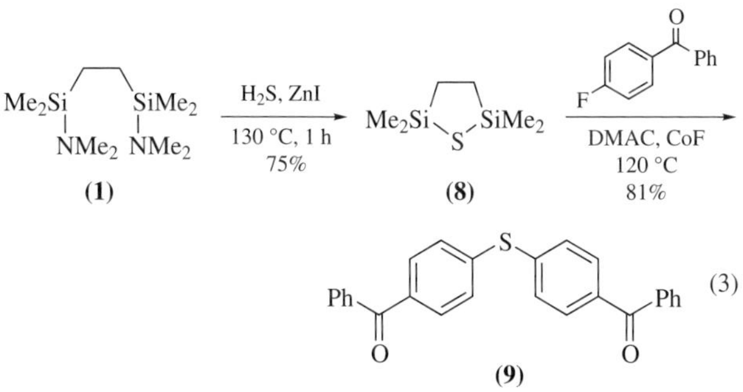

1. Guggenheim, T. L., *Tetrahedron Lett.* **1984**, *25*, 1253.
2. Deshayes, K.; Byoene, R. D.; Chad, I.; Knobler, C. B.; Diederich, F., *J. Org. Chem.* **1991**, *56*, 6787.
3. Guggenheim, T. L., *Tetrahedron Lett.* **1987**, *28*, 6139.

Fatima Z. Basha & Anwer Basha
Abbott Laboratories, Abbott Park, IL, USA

$$\text{(4)} \xrightarrow[\underset{77\%}{23\,^\circ\text{C}}]{\substack{\text{PhCHO}\\ \text{CH}_2\text{Cl}_2}} \qquad \text{(3)}$$

(4)

$$\text{PhCO}_2\text{H} \xrightarrow[96\%]{\substack{\textbf{(1)}\\ \text{CH}_2\text{Cl}_2}} (\text{PhCO}_2)_2\text{SiMe}_2 \qquad (10)$$

Advantages of Bis(1-methoxy-2-methyl-1-propenyloxy)-dimethylsilane. Dichlorosilanes,[6] bis(trifluoromethanesulfonyl)silanes,[7] and dimethoxysilanes[8] have been used for the synthesis of dialkylsilylene derivatives. These methods, however, have some disadvantages: (i) a strong base or acid catalyst is required; (ii) forcing reaction conditions (high temperature for a long period) are sometimes required; (iii) large amounts of side-products such as inorganic salts, or amine or acid salts, are produced that must be separated from the moisture-sensitive silylene derivatives; and (iv) the yields of the products are not always high. Reagent (1), which is much more reactive than other difunctional silyl reagents, allows the ready silylenation of H-acidic materials in the absence of base, and allows isolation of pure products in almost quantitative yields without aqueous workup. The only reaction side-product is methyl isobutyrate, which is volatile.

Bifunctional Protection of H-Acidic Materials.[1,9] The reaction of (1) with H-acidic materials is generally carried out by employing a slight excess of the reagent in an inert solvent such as CH₂Cl₂ or MeCN and usually gives a high yield of the corresponding silylene derivatives. Furthermore, addition of catalytic amounts of dichlorodimethylsilane induces a slightly exothermal reaction, leading to the rapid formation of the silylenes. A wide range of 1,2-, 1,3-, and 1,4-difunctional compounds give the corresponding silylenes in high yields (eqs 4–8). Monofunctional alcohols and carboxylic acids can also be linked via the dimethylsilylene group forming dialkoxy- and diacyloxydimethylsilane, respectively (eqs 9 and 10).

1. Kita, Y.; Yasuda, H.; Sugiyama, Y.; Fukata, F.; Haruta, J.; Tamura, Y., *Tetrahedron Lett.* **1983**, *24*, 1273.

2. Lalonde, M.; Chan, T. H., *Synthesis* **1985**, 817.

3. (a) Kita, Y.; Yasuda, H.; Tamura, O.; Tamura, Y., *Tetrahedron Lett.* **1984**, *25*, 1813. (b) Kita, Y.; Ueno, H.; Kitagaki, S.; Kobayashi, K.; Iio, K.; Akai, S., *J. Chem. Soc., Chem. Commun.* **1994**, 701.

4. Stork, G.; Kim, G., *J. Am. Chem. Soc.* **1992**, *114*, 1087.

5. (a) Myers, A. G.; Widdowson, K. L., *J. Am. Chem. Soc.* **1990**, *112*, 9672. (b) Myers, A. G.; Widdowson, K. L.; Kukkola, P. J., *J. Am. Chem. Soc.* **1992**, *114*, 2765.

6. (a) Spassky, N., *C. R. Hebd. Seances Acad. Sci.* **1960**, *251*, 2371. (b) Wieber, M.; Schmidt, M., *Z. Naturforsch., Tell B* **1963**, *18b*, 846. (c) Wieber, M.; Schmidt, M., *J. Organomet. Chem.* **1963**, *1*, 93. (d) Abel, E. W.; Bush, R. P., *J. Organomet. Chem.* **1965**, *3*, 245. (e) Kober, F.; Ruhl, W. J., *J. Organomet. Chem.* **1975**, *101*, 57. (f) Meyer, H.; Nagorsen, G.; Weiss, A., *Z. Naturforsch., Tell B* **1975**, *30b*, 488. (g) Markiewicz, W. T.; Wiewiorowski, M., *Nucl. Acid Res. Special Publ.* **1978**, *4*, 185. (h) Markiewicz, W. T., *J. Chem. Res. (S)* **1979**, 24; Markiewicz, W. T.; Padyukova, N. Sh.; Samek, Z.; Smrt, J., *Collect. Czech. Chem. Commun.* **1980**, *45*, 1860. (i) van Boeckel, S. A. A.; van Boom, J. H., *Chem. Lett.* **1981**, 581. (j) Djuric, S.; Venit, J.; Magnus, P., *Tetrahedron Lett.* **1981**, *22*, 1787. (k) Cragg, R. H.; Lane, R. D., *J. Organomet. Chem.* **1981**, *212*, 301. (l) Trost, B. M.; Caldwell, C. G., *Tetrahedron Lett.* **1981**, *22*, 4999.

7. Corey, E. J.; Hopkins, P. B., *Tetrahedron Lett.* **1982**, *23*, 4871.

8. Jenner, M. R.; Khan, R., *J. Chem. Soc., Chem. Commun.* **1980**, 50.

9. Brown, R. F. C.; Coulston, K. J.; Eastwood, F. W.; Hill, M. P., *Aust. J. Chem.* **1988**, *41*, 215.

Yasuyuki Kita
Osaka University, Japan

$$\text{(4)} \qquad \xrightarrow[98\%]{\substack{\textbf{(1)}\\ \text{MeCN}}} \qquad (4)$$

$$\text{(5)} \qquad \xrightarrow[94\%]{\substack{\textbf{(1)}\\ \text{THF}}} \qquad (5)$$

$$\xrightarrow[\underset{95\%}{\text{THF}}]{\substack{\textbf{(1)}\\ \text{cat TMSCl}}} \qquad (6)$$

$$\xrightarrow[90\%]{\substack{\textbf{(1)}\\ \text{MeCN}}} \qquad (7)$$

$$\xrightarrow[95\%]{\substack{\textbf{(1)}\\ \text{MeCN}}} \qquad (8)$$

$$\text{PhOH} \xrightarrow[98\%]{\substack{\textbf{(1)}\\ \text{CH}_2\text{Cl}_2}} (\text{PhO})_2\text{SiMe}_2 \qquad (9)$$

Bis(methylthio)(trimethylsilyl)methane

[37891-79-5]   C₆H₁₆S₂Si   (MW 180.40)

InChI = 1S/C6H16S2Si/c1-7-6(8-2)9(3,4)5/h6H,1-5H3

InChIKey = QEPMPXAUMUWNNO-UHFFFAOYSA-N

(reagent used for synthesis of bis(methylthio)ketene acetals from aldehydes and ketones;[1] acyl anion synthon for conjugate additions to enones and enoates;[2] useful for one-carbon homologations[3])

Physical Data: bp 67–70 °C/10 mmHg.

Solubility: insol H₂O; sol organic solvents.

Preparative Methods: prepared from lithiobis(methylthio)methane by silylation with chlorotrimethylsilane (eq 1).[1] Both bis(methylthio)methane and chlorotrimethylsilane are commercially available.

$$(\text{MeS})_2\text{CH}_2 \xrightarrow[\text{2. Me}_3\text{SiCl}]{\substack{\text{1. BuLi, THF}\\ -60\,^\circ\text{C}}} (\text{MeS})_2\text{CHSiMe}_3 \qquad (1)$$

Handling, Storage, and Precautions: use in a fume hood.

Lithiobis(methylthio)(trimethylsilyl)methane. The anion is generated by treatment of the title compound with *n*-butyllithium in THF at −60 °C (eq 2). This species is well-behaved at low temperature. Useful reactions of this anion are discussed in the following sections.

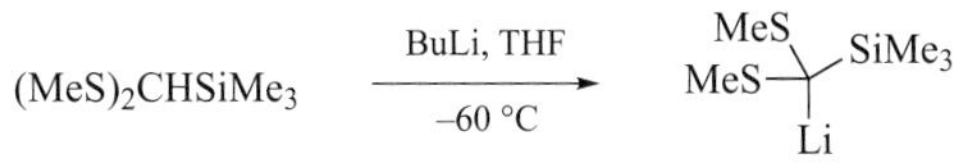

$$(MeS)_2CHSiMe_3 \xrightarrow[-60\,°C]{BuLi,\ THF} \quad \text{(2)}$$

Thioketene Acetals. The lithio reagent readily converts aldehydes and ketones into the corresponding bis(methylthio)ketene acetals via a Peterson alkenation sequence.[1] Aromatic aldehydes give excellent yields under these conditions (eq 3), and certain enolizable ketones such as acetophenone and cyclohexanone also react accordingly (eq 4). Alternative methods for preparing these acetals include a Horner–Wittig process that uses phosphonate derivatives[4] and an alkylative process involving CS_2 and MeI.[5] Relative to these alternatives, the title reagent offers advantages of ease of reagent preparation and control of regioselectivity. The phosphonate reagents are more nucleophilic, however, and avoid competitive deprotonation.

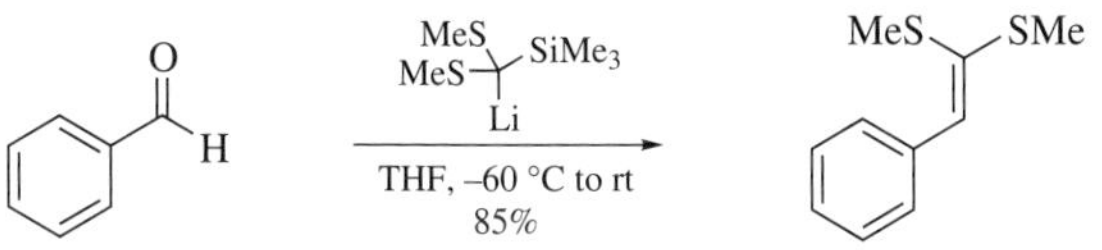

$$\text{(3)}$$

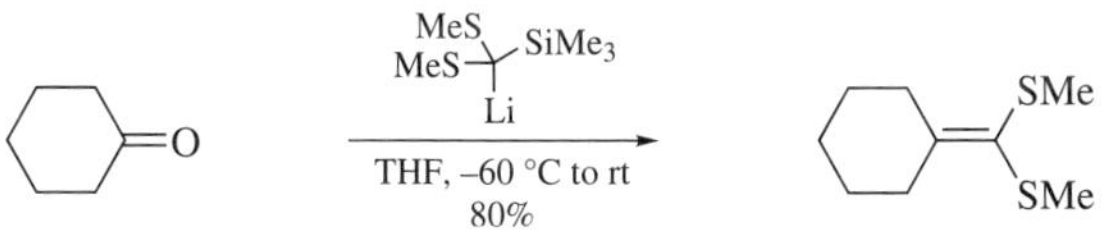

$$\text{(4)}$$

Reaction of the lithio species with α-oxo ketones and aldehydes allows the formation of β,β-disubstituted enones, useful intermediates for the preparation of unsaturated 1,5-diketones (eq 5).[6]

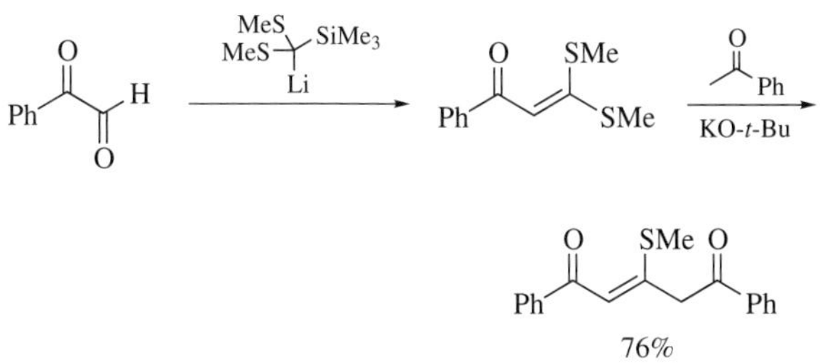

$$\text{(5)}$$

Acyl Anion Conjugate Additions. The lithio reagent readily undergoes 1,4-addition to unsaturated substrates (eq 6), in direct contrast to the corresponding 2-lithio-2-trimethylsilyl-1,3-dithiane, which is a poor Michael donor.[2] The initial Michael adducts can also be alkylated to provide highly functionalized products. Very good levels of diastereoselectivity have been observed in the 1,4-addition and enolate alkylations of cyclic enoates (eq 7)[7] and acyclic enones (eq 8).[8]

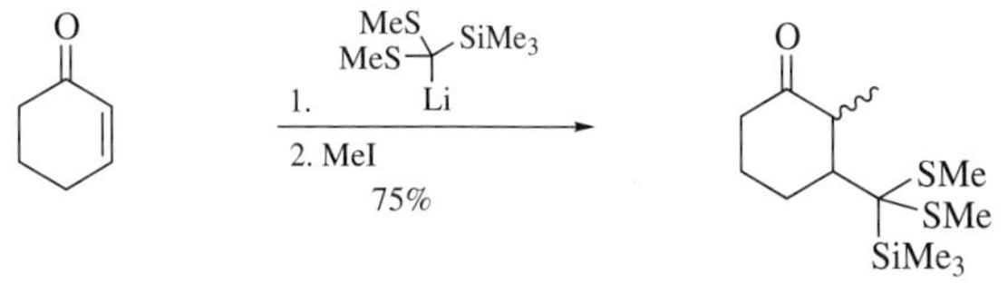

$$\text{(6)}$$

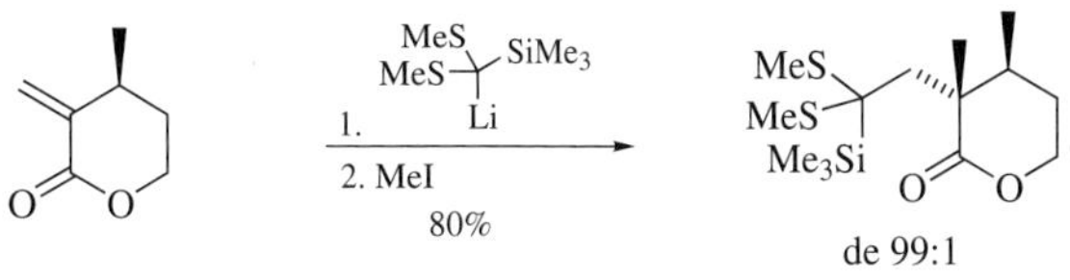

$$\text{(7)}$$

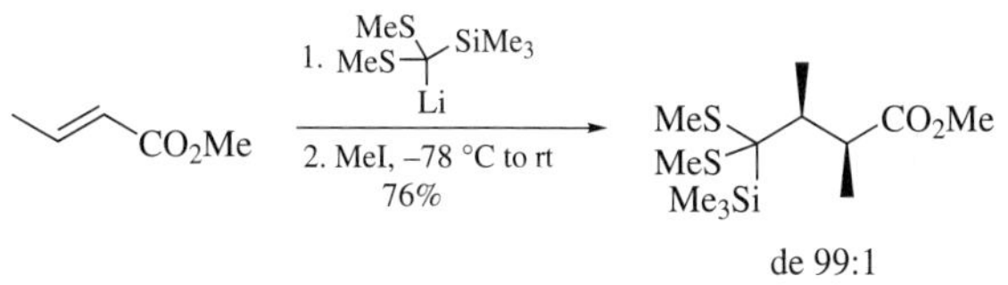

$$\text{(8)}$$

Other effective Michael donors of this type include lithiobis-(phenylthio)(trimethylsilyl)methane,[9] lithio(methoxy)(phenylthio)(trimethylsilyl)methane,[10] and lithio(methylthio)(methylsulfoxido)methane.[11] The isolated yields with lithiobis(phenylthio)(trimethylsilyl)methane are generally superior to those with the title reagent, but the latter compound is more readily available and easier to handle than those listed above.

One-Carbon Homologations.[3] The bis(methylthio)ketene acetals are useful synthetic intermediates because they are readily converted into *S*-methyl thioesters (eq 9)[12] or the corresponding carboxylic acids and esters.[13a]

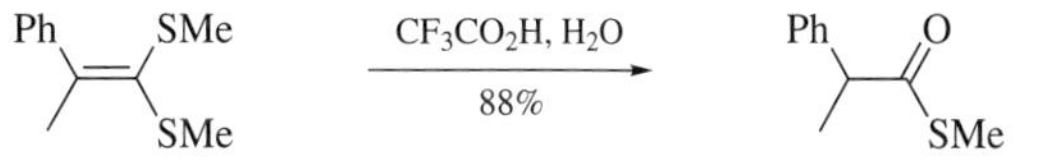

$$\text{(9)}$$

This acetal methodology was used for a 1,3-carbonyl transposition in a synthesis of ($\pm$)-myodesmone (eq 10),[13b] and for a one-carbon homologation en route to vitamin B_{12} (eq 11).[14]

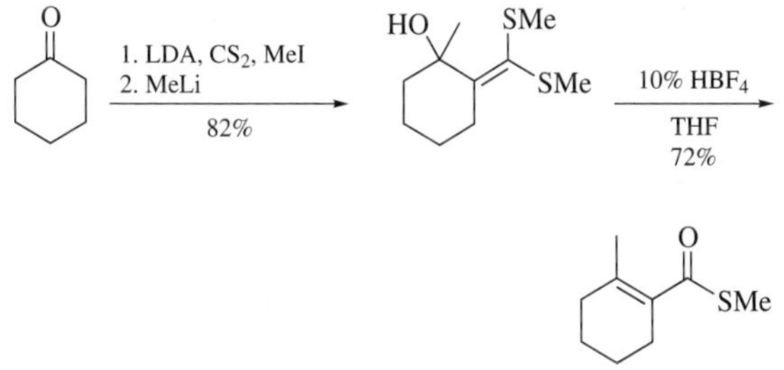

$$\text{(10)}$$

A particularly useful transformation is the fluoride-induced removal of the trimethylsilyl group to provide a mixed *O,S*-acetal that is easily converted into the corresponding aldehyde (eq 12).[10]

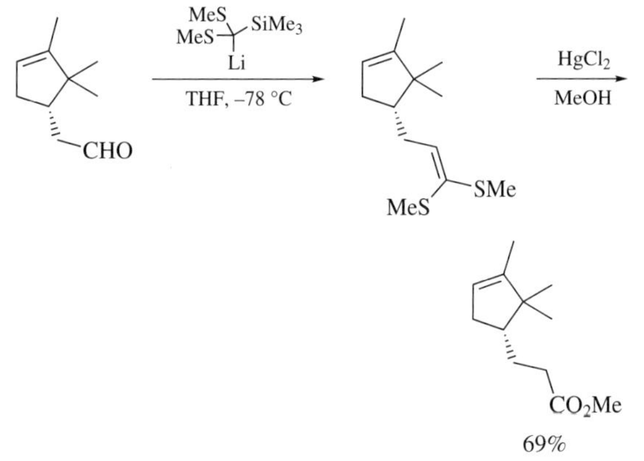

(11)

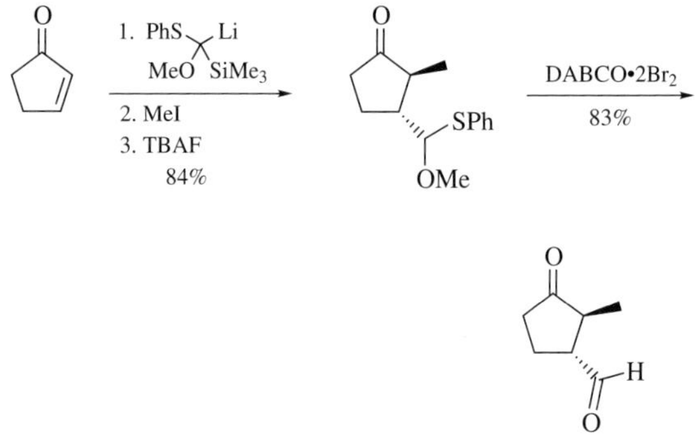

69%

(12)

1. Seebach, D.; Kolb, M.; Gröbel, B.-T., *Chem. Ber.* **1973**, *106*, 2277.

2. Seebach, D.; Bürstinghaus, R., *Angew. Chem., Int. Ed. Engl.* **1975**, *14*, 57.

3. Martin, S. F., *Synthesis* **1979**, 633.

4. (a) Mikolajczyk, M.; Grzejszczak, S.; Zatorski, A., *Tetrahedron Lett.* **1976**, *17*, 2731. (b) Kruse, C. G.; Broekhof, N. L. J. M.; Wijsman, A.; van der Gen, A., *Tetrahedron Lett.* **1977**, *18*, 885.

5. Potts, K. T.; Cipullo, M. J.; Ralli, P.; Theodoridis, G., *J. Org. Chem.* **1982**, *47*, 3027.

6. Potts, K. T.; Cipullo, M. J.; Ralli, P.; Theodoridis, G., *J. Am. Chem. Soc.* **1981**, *103*, 3584.

7. Tomioka, K.; Kawasaki, H.; Yasuda, K.; Koga, K., *J. Am. Chem. Soc.* **1988**, *110*, 3597.

8. Kawasaki, H.; Tomioka, K.; Koga, K., *Tetrahedron Lett.* **1985**, *26*, 3031.

9. Myers, M. R.; Cohen, T., *J. Org. Chem.* **1989**, *54*, 1290.

10. Otera, J.; Niibo, Y.; Nozaki, H., *J. Org. Chem.* **1989**, *54*, 5003.

11. Ogura, K.; Yamashita, M.; Tsuchihashi, G.-I., *Tetrahedron Lett.* **1978**, *19*, 1303.

12. Seebach, D.; Bürstinghaus, R., *Synthesis* **1975**, 461.

13. (a) Myrboh, B.; Ila, H.; Junjappa, H., *J. Org. Chem.* **1983**, *48*, 5327. (b) Dieter, R. K.; Lin, Y. J.; Dieter, J. W., *J. Org. Chem.* **1984**, *49*, 3183.

14. Stevens, R. V.; Chang, J. H.; Lapalme, R.; Schow, S.; Schlageter, M. G.; Shapiro, R.; Weller, H. N., *J. Am. Chem. Soc.* **1983**, *105*, 7719.

John W. Benbow
Lehigh University, Bethlehem, PA, USA

[Bis(2-pyridyldimethylsilyl)methyl]lithium

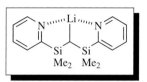

[356062-49-2] $C_{15}H_{21}LiN_2Si_2$ (292.45)

InChI = 1S/C15H21N2Si2.Li/c1-18(2,14-9-5-7-11-16-14)13-19(3,4)15-10-6-8-12-17-15;/h5-13H,1-4H3;

InChIKey = JEMKBBRWRUUGAC-UHFFFAOYSA-N

(reagent used for the preparation of vinylsilanes from carbonyl compounds)

Solubility: soluble in diethyl ether.
Form Supplied in: prepared in situ and used directly.
Analysis of Reagent Purity: NMR.
Preparative Methods: deprotonation of bis(2-pyridyldimethyl-silyl)methane by *n*-butyllithium in dry Et$_2$O at $-78\,°C$ under argon.
Handling, Storage, and Precaution: sensitive to air and moisture.

Introduction. The Peterson-type olefination reaction has emerged as an extremely useful method for the preparation of alkenes from carbonyl compounds.[1,2] Synthetically useful vinyl-silanes can be prepared by using the Peterson-type olefination reaction of bis(silyl)methylmetal with carbonyl compounds.[3–5] Among the various reported bis(silyl)methylmetals, [bis(2-pyri-dyldimethylsilyl)methyl]lithium is an extremely efficient reagent for the stereoselective preparation of vinylsilane.[6]

Preparation of [Bis(2-pyridyldimethylsilyl)methyl]lithium. The synthesis of the starting material bis(2-pyridyldimethylsilyl)methane can be easily accomplished in one-pot by the depro-tonation of 2-pyridyltrimethylsilane with *t*-BuLi followed by reaction with 2-pyridyldimethylsilane (eq 1).[7,8] The subsequent generation of [bis(2-pyridyldimethylsilyl)methyl]lithium is easily accomplished by the deprotonation of bis(2-pyridyldimethylsilyl)methane with *n*-butyllithium in dry diethyl ether (eq 2).[6]

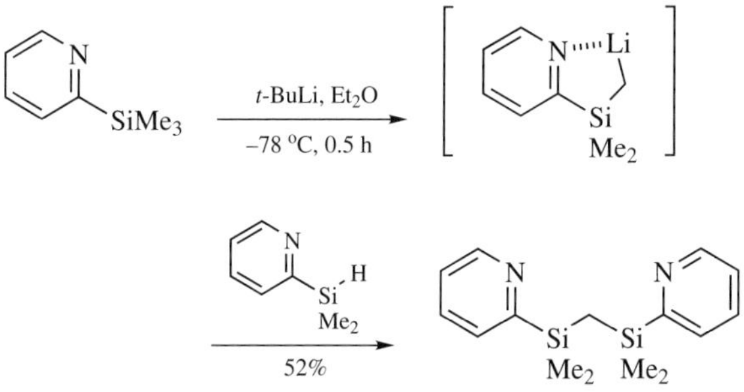

(1)

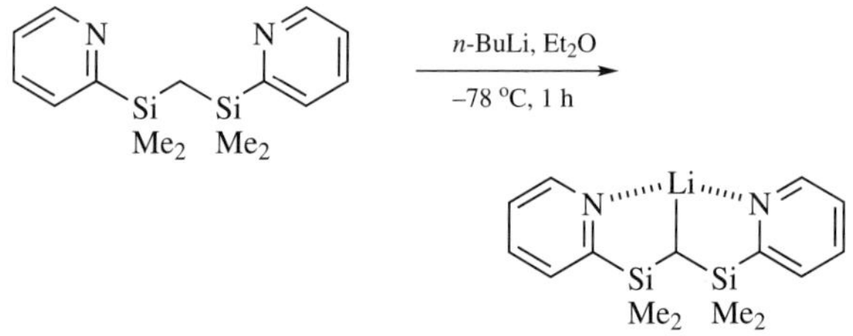

52%

(2)

Table 1 Reactions of [bis(2-pyridyldimethylsilyl)methyl]lithium with carbonyl compounds[6]

Entry	Carbonyl compound	Vinylsilane	Yield (%)
1	Ph–CH₂CH₂–CHO	Ph(CH₂)₂CH=CH–SiMe₂(2-pyridyl)	Quantitative
2	Cyclohexyl–CHO	Cyclohexyl–CH=CH–SiMe₂(2-pyridyl)	Quantitative
3	(CH₃)₃C–CHO	(CH₃)₃C–CH=CH–SiMe₂(2-pyridyl)	Quantitative
4	Ph–CHO	Ph–CH=CH–SiMe₂(2-pyridyl)	90
5	2-Furyl–CHO	2-Furyl–CH=CH–SiMe₂(2-pyridyl)	Quantitative
6	Mesityl–CHO	Mesityl–CH=CH–SiMe₂(2-pyridyl)	94
7	OHC–C₆H₄–CHO (para)	(2-pyridyl)Me₂Si–CH=CH–C₆H₄–CH=CH–SiMe₂(2-pyridyl)	98
8	(CH₃)₂C=O	(CH₃)₂C=CH–SiMe₂(2-pyridyl)	73
9	Ph(CH₃)C=O	Ph(CH₃)C=CH–SiMe₂(2-pyridyl)	56
10	Ph₂C=O	Ph₂C=CH–SiMe₂(2-pyridyl)	84
11	CH₂=CH–CHO	CH₂=CH–CH=CH–SiMe₂(2-pyridyl)	53
12	CH₂=C(CH₃)–CHO	CH₂=C(CH₃)–CH=CH–SiMe₂(2-pyridyl)	Quantitative
13	Ph–CH=CH–CHO	Ph–CH=CH–CH=CH–SiMe₂(2-pyridyl)	79

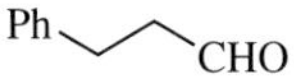

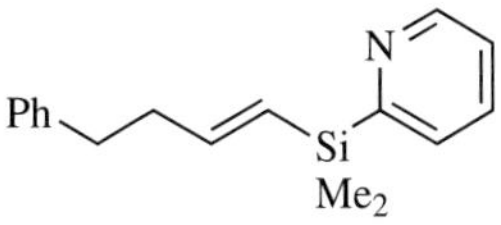

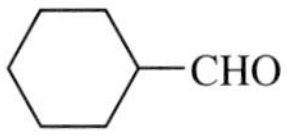

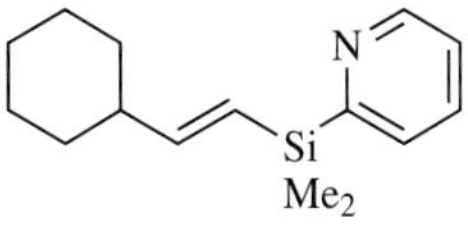

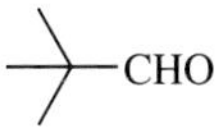

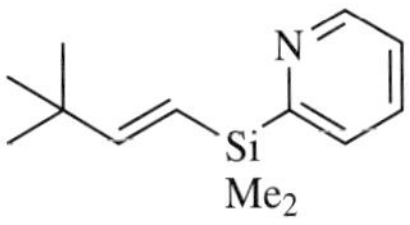

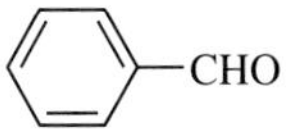

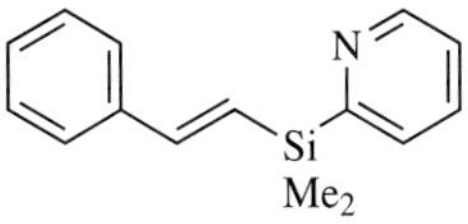

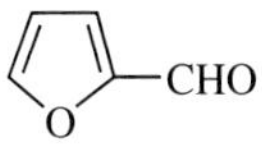

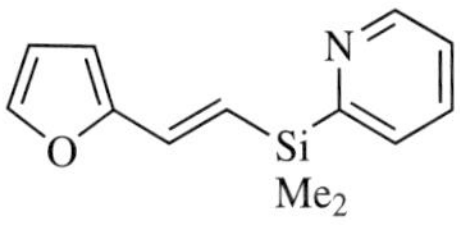

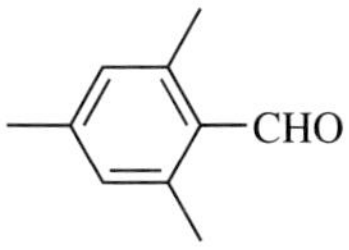

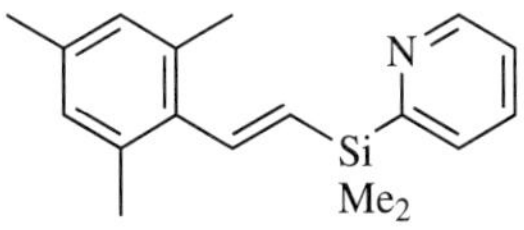

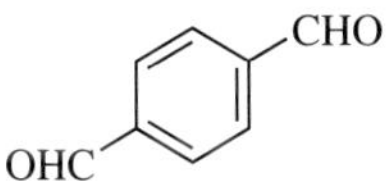

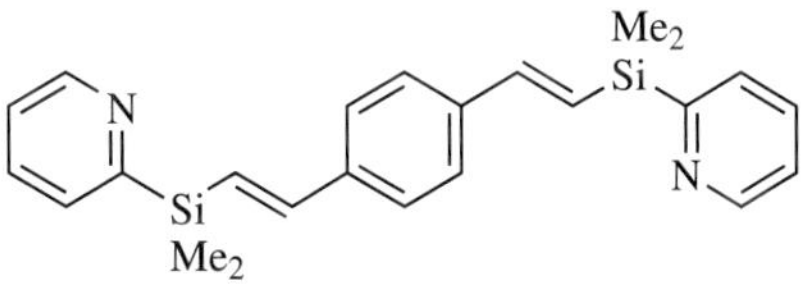

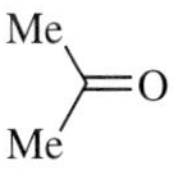

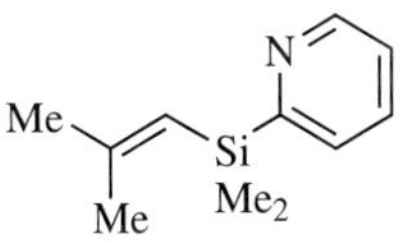

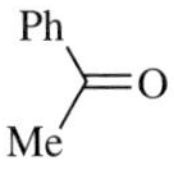

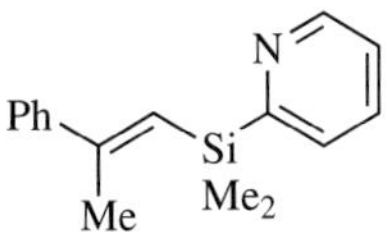

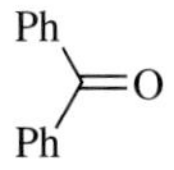

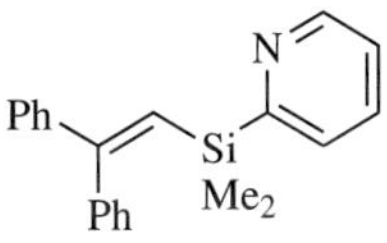

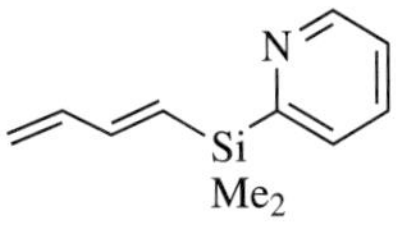

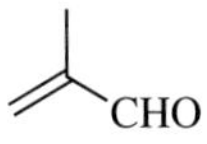

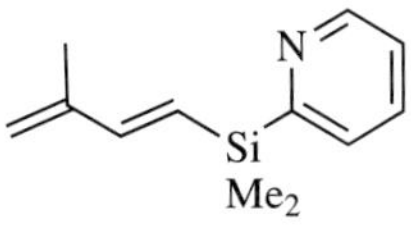

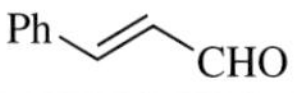

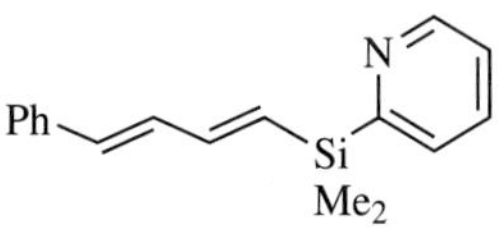

Reaction of [bis(2-pyridyldimethylsilyl)methyl]lithium with Carbonyl Compounds. The Peterson-type reaction of [bis(2-pyridyldimethylsilyl)methyl]lithium with primary, secondary, and tertiary aliphatic and aromatic aldehydes produces the corresponding 2-pyridyl-substituted vinylsilanes in essentially quantitative yields (entries 1–7, Table 1, eq 3).[6] The reaction is also applicable to sterically hindered aldehydes (entries 3 and 6) and di-aldehyde (entry 7). The reactions with ketones give disubstituted vinylsilanes with somewhat lower yields (entries 8–10). The reaction can be applied to the stereoselective synthesis of dienylsilanes as well (entries 11–13). In all cases, the reaction occurs in a complete stereoselective fashion (>99% E).

$$ \qquad (3) $$

Synthetic Transformations of 2-Pyridyl-substituted Vinylsilane. 2-Pyridyl-substituted vinylsilanes can be converted into other vinylsilanes. Subjection of 2-pyridyl-substituted vinylsilanes to potassium fluoride/methanol leads to the formation of methoxy(vinyl)silanes by pyridyl-silyl bond cleavage (eq 4).[9] The resultant methoxysilanes can be further allowed to react with Grignard reagents such as phenylmagnesium bromide to give the corresponding vinylsilanes that are commonly used for various transformations (eq 4).[10]

$$ \qquad (4) $$

More practically, 2-pyridyl-substituted vinylsilanes themselves can be directly subjected to the reactions with various electrophiles. For example, treatment of 2-pyridyl-substituted vinylsilanes with acid chlorides in the presence of aluminum chloride affords the corresponding α,β-unsaturated enones (eq 5).[11] The reaction of 2-pyridyl-substituted vinylsilanes with bromine and subsequent treatment with sodium methoxide affords the corresponding vinyl bromides (eq 6).[11]

$$ \qquad (5) $$

$$ \qquad (6) $$

Related Reagents. *n*-Butyllithium; potassium fluoride; phenylmagnesium bromide; aluminum chloride; bromine; sodium methoxide.

1. Peterson, D. J., *J. Org. Chem.* **1968**, *33*, 780.
2. Ager, D. J., *Org. React.* **1990**, *38*, 1.
3. Gröbel, B. T.; Seebach, D., *Angew. Chem., Int. Ed. Engl.* **1974**, *13*, 83.
4. Hudrlik, P. F.; Agwaramgbo, E. L. O.; Hudrlik, A. M., *J. Org. Chem.* **1989**, *54*, 5613.
5. (a) Sakurai, H.; Nishiwaki, K.; Kira, M., *Tetrahedron Lett.* **1973**, 4193. (b) Hartzell, S. L.; Rathke, M. W., *Tetrahedron Lett.* **1976**, 2737. (c) Sachdev, K., *Tetrahedron Lett.* **1976**, 4041. (d) Carter, M. J.; Fleming, I., *J. Chem. Soc., Chem. Commun.* **1976**, 679. (e) Seyferth, D.; Lefferts, J. L.; Lambert, R. L., Jr, *J. Organomet. Chem.* **1977**, *142*, 39. (f) Seebach, D.; Bürstinghaus, R.; Gröbel, B. T.; Kolb, M., *Liebigs Ann. Chem.* **1977**, 830. (g) Gröbel, B. T.; Seebach, D., *Chem. Ber.* **1977**, *110*, 852. (h) Isobe, M.; Kitamura, M.; Goto, T., *Tetrahedron Lett.* **1979**, 3465. (i) Fleming, I.; Pearce, A., *J. Chem. Soc., Perkin Trans. 1* **1980**, 2485. (j) Sato, Y.; Takeuchi, S., *Synthesis* **1983**, 734. (k) Ager, D. J., *J. Org. Chem.* **1984**, *49*, 168. (l) Takeda, T.; Ando, K.; Mamada, A.; Fujiwara, T., *Chem. Lett.* **1985**, 1149. (m) Ager, D. J.; East, M. B., *J. Org. Chem.* **1986**, *51*, 3983. (n) Inoue, S.; Sato, Y., *Organometallics* **1986**, *5*, 1197. (o) Terao, Y.; Aono M.; Takahashi, I.; Achiwa, K., *Chem. Lett.* **1986**, 2089. (p) Marchand, A. P.; Huang, C.; Kaya, R.; Baker, A. D.; Jemmis, E. D.; Dixon, D. A., *J. Am. Chem. Soc.* **1987**, *109*, 7095. (q) Kira, M.; Hino, T.; Kubota, Y.; Matsuyama, N.; Sakurai, H., *Tetrahedron Lett.* **1988**, *29*, 6939.
6. Itami, K.; Nokami, T.; Yoshida, J., *Org. Lett.* **2000**, *2*, 1299.
7. Itami, K.; Mitsudo, K.; Yoshida, J., *Tetrahedron Lett.* **1999**, *40*, 5533.
8. Itami, K.; Mitsudo, K.; Yoshida, J., *Tetrahedron Lett.* **1999**, *40*, 5537.
9. Itami, K.; Mitsudo, K.; Yoshida, J., *J. Org. Chem.* **1999**, *64*, 8709.
10. (a) Fleming, I.; Dunoguès, J.; Smither, R., *Org. React.* **1989**, *37*, 57. (b) Larson, G. L., *J. Organomet. Chem.* **1992**, *422*, 1. (c) Fleming, I.; Barbero, A.; Walter, D., *Chem. Rev.* **1997**, *97*, 2063.
11. Itami, K.; Mitsudo, K.; Kamei, T.; Koike, T.; Nokami, T.; Yoshida, J., *J. Am. Chem. Soc.* **2000**, *122*, 12013.

Jun-ichi Yoshida & Kenichiro Itami
Kyoto University, Yoshida, Kyoto, Japan

1,3-Bis(triisopropylsilyl)propyne

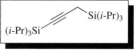

[82192-59-4] $C_{21}H_{44}Si_2$ (MW 352.75)
InChI = 1S/C21H44Si2/c1-16(2)22(17(3)4,18(5)6)14-13-15-23 (19(7)8,20(9)10)21(11)12/h16-21H,14H2,1-12H3
InChIKey = JFDDOGUVACDNTR-UHFFFAOYSA-N

(precursor of a stereoselective bulky C_3 nucleophile, functionalized Peterson reagent[1])

Alternate Name: 1,3-bis(TIPS)propyne.
Physical Data: bp 130–135 °C/0.08 mmHg; d 0.846 g cm^{-3}.
Solubility: both the reagent and its lithio derivative are soluble in ether or THF.
Analysis of Reagent Purity: GC at 210 °C on a glass capillary column coated with OV-17; TLC on silica (hexanes, R_f

0.75). ^{1}H NMR (CDCl$_3$, 80 MHz): $\delta = 1.63$ (s, CH$_2$), 1.3–0.9 (m, 2 TIPS).

Preparative Method: 1,3-bis(TIPS)propyne is prepared in quantitative yield by silylation of 3-lithio-1-triisopropylsilyl-1-propyne with triisopropylsilyl trifluoromethanesulfonate.[2]

Purification: by distillation or chromatography.

Handling, Storage, and Precautions: no special precautions required; the reagent is not changed when stored for several months at ambient temperature. The Li derivative is handled under inert gas using simple syringe equipment.

Generation of Lithio-1,3-bis(triisopropylsilyl)propyne. The main use of the title reagent is as a precursor of the lithium derivative. 1,3-Bis(triisopropylsilyl)propyne is cleanly lithiated by treatment with 1 equiv of *n*-butyllithium in THF at −20 °C (15 min) to what is probably an equilibrium mixture of propargylic and allenic species (**1**) and (**2**) (eq 1).[3] The bulky triisopropylsilyl (TIPS) group serves as a controlling group in the addition of (**1/2**) to electrophiles.

Formation of (Z)-enynes. The lithio derivative reacts with aliphatic aldehydes (RCHO, −78 °C → rt, few h) to form the (Z)-enynes R–CH=CH–C≡C–TIPS (**3**) in good yield (57–79%) and high isomeric purity (Z:E ≈ 20:1, eq 2). The intermediate addition product formed by attack of (**2**) on the aldehyde has *erythro* configuration[2] and undergoes *syn* elimination of silanolate to give (**3**) by the pathway generally observed in Peterson reactions in alkaline media.[4]

The method was used for attachment of the *cis* enyne side chain in a synthesis of gephyrotoxin,[5] and for the preparation of (Z)-non-3-en-1-yne, a building block in the synthesis of prostaglandins and of 11-HETE methyl ester.[6]

Formation of (E)-Enynes. The stereoselectivity of the reaction can be reversed simply by adding 5 equiv hexamethylphosphoric triamide to the solution of (**1/2**) prior to addition of the aldehyde. Under these conditions the reaction is complete within a few seconds at −78 °C, and (E)-enynes (**4**) are isolated in 60–65% yield with high E:Z selectivity (20:1 to 10:1, eq 3). Evidence has been presented that the action of HMPA is exerted in the addition to the aldehyde, not in the elimination step.[2] A *threo* propargylic intermediate is presumably formed by attack of an ion pair similar

to (**1**) onto the aldehyde. The method was used for the introduction of the *trans* enyne side chain in a synthesis of laurenyne.[7]

The stereochemical preferences of these reactions are less pronounced with aromatic aldehydes. For example, the reaction of (**1/2**) with benzaldehyde in THF gives a 2:1 mixture of Z:E enynes (−20 → 0 °C, 1 h, 67%), whereas in the presence of 1 equiv HMPA mostly the (E)-enyne is obtained (1:9 selectivity, −78 °C, 10 s, 78%).

Enynes and Dienynes from Other Bis(silyl)propynes. Corresponding Li reagents containing a less bulky silyl group at the propargylic carbon generally show lower selectivity for (Z)-enynes, e.g. the Li derivatives of 1,3-bis(trimethylsilyl)propyne, 1-trimethylsilyl-3-triethylsilyl-1-propyne, 1-trimethylsilyl-3-*t*-butyldimethylsilyl-1-propyne (TMS–C≡C–CH$_2$–SiR$_3$ with R$_3$ = Me$_3$, Et$_3$, *t*-BuMe$_2$).[8] Selectivity could be improved by using a Mg or Ti counterion instead of Li.[8–10] Cinnamaldehyde, cyclohexanone, and *cis*- and *trans*-hydrindanones[11] likewise underwent the reaction.

1-Trimethylsilyl-3-triisopropylsilyl-1-propyne (TMS–C≡C–CH$_2$–TIPS, **5**) was found to give essentially the same results (yield and stereoselectivity) as bis(TIPS)propyne in two cases of direct comparison (crotonaldehyde, cinnamaldehyde).[12] Scrambling of the silyl groups did not occur. Compound (**5**) is available from TMS–C≡C–CH$_2$Br by reaction with magnesium in the presence of 0.5% mercury(II) chloride followed by silylation with TIPS-OTf (76%). Silane (**5**) can also be prepared from propargyl bromide by reaction with (1) Mg/0.5% HgCl$_2$, (2) TIPS-OTf, (3) *n*-BuLi, (4) TMS-Cl (two-pot procedure, 63% overall).[12]

Using the lithio reagent obtained from (**5**) in THF, (Z)-dienynes were prepared in good yield and stereoselectivity from several α,β-unsaturated aldehydes (eq 4) (Table 1).[12,13] In the corresponding Wittig reactions, the yields and isomeric purity of enynes were far lower. Oxirane and aziridine groups are compatible with the reaction conditions (eq 5) (Table 2).[12]

The preparation of (Z)-enynes by alternative methods (Wittig–Horner and others) is possible;[14–16] and is also possible for (E)-enynes (Wittig).[17–19]

Table 1 TMS-Protected dienynes from α,β-unsaturated aldehydes (eq 4)

R^1	R^2	R^3	Yield (%)	Z:E
H	Me	H	81[a]	4:1[a]
Me	Me	H	89	3.5:1
Me	H	Me	93	34:1
H	–(CH$_2$)$_3$–		80	13:1
H	–(CH$_2$)$_3$–		86[a]	21:1[a]
Ph	–(CH$_2$)$_4$–		74	>20:1
Ph	–(CH$_2$)$_5$–		85	37:1
Ph	–(CH$_2$)$_6$–		60	>20:1
Ph	–(CH$_2$)$_{10}$–		67	9:1
H	Ph	H	82	3.4:1
Ph	Ph	H	83	1.9:1
H	Ph	Me	89	24:1
H	Ph	Ph	91	98:1

[a] In the presence of 1 equiv Ti(O-*i*-Pr)$_4$.

Table 2 TMS-Protected enynes from oxirane (aziridine) carbaldehydes (eq 5)

X	R^1	R^2	R^3	Yield (%)	Z:E
O	H	Ph	H	60	1.5:1
O	Ph	H	Ph	63	1:1.5
O	H	Ph	Ph	75	>20:1
NCH$_2$Ph	Ph	H	H	73	>20:1

Epoxide Opening. The lithio propyne derivative (**1/2**) adds to epoxides in spite of its bulkiness. The TIPS group near the reaction center is useful for preventing secondary reactions in a polyepoxide substrate. Thus while the simple monoadduct (**6a**) from *cis*-benzene trioxide and 3-lithio-1-TIPS-1-propyne was formed as a minor product only (always accompanied by products (**7a**), (**8a**), (**9a**), the monoadduct (**6b**) was isolated in 48% yield (two diastereomers) from the reaction with (**1/2**) along with 10% (**7b**); bisadducts (**8b/9b**) were not formed (eq 6).[20]

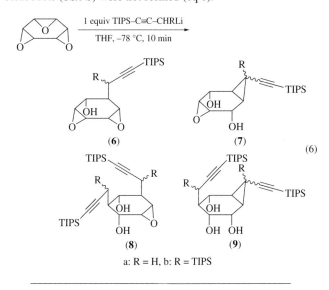

a: R = H, b: R = TIPS

1. *Fieser & Fieser 11*, 63.
2. Corey, E. J.; Rücker, Ch., *Tetrahedron Lett.* **1982**, *23*, 719.
3. Peterson, P. E.; Jensen, B. L., *Tetrahedron Lett.* **1984**, *25*, 5711.
4. (a) Ager, D. J., *Synthesis* **1984**, 384. (b) Ager, D. J., *Org. React.* **1990**, *38*, 1.
5. Overman, L. E.; Lesuisse, D.; Hashimoto, M., *J. Am. Chem. Soc.* **1983**, *105*, 5373.
6. (a) Corey, E. J.; Shimoji, K.; Shih, C., *J. Am. Chem. Soc.* **1984**, *106*, 6425. (b) Corey, E. J.; Shih, C.; Shih, N.-Y.; Shimoji, K., *Tetrahedron Lett.* **1984**, *25*, 5013.
7. Overman, L. E.; Thompson, A. S., *J. Am. Chem. Soc.* **1988**, *110*, 2248.
8. Yamakado, Y.; Ishiguro, M.; Ikeda, N.; Yamamoto, H., *J. Am. Chem. Soc.* **1981**, *103*, 5568.
9. Furuta, K.; Ishiguro, M.; Haruta, R.; Ikeda, N.; Yamamoto, H., *Bull. Chem. Soc. Jpn.* **1984**, *57*, 2768.
10. Palazón, J. M.; Martín, V. S., *Tetrahedron Lett.* **1988**, *29*, 681.
11. Peterson, P. E.; Leffew, R. L. B.; Jensen, B. L., *J. Org. Chem.* **1986**, *51*, 1948.
12. Schulz, D. Diploma Thesis, Universität Freiburg, 1985.
13. (a) Eberbach, W.; Laber, N., *Tetrahedron Lett.* **1992**, *33*, 57. (b) Laber, N. Thesis, Universität Freiburg, 1991.
14. Gao, L.; Murai, A., *Tetrahedron Lett.* **1992**, *33*, 4349.
15. Holmes, A. B.; Pooley, G. R., *Tetrahedron* **1992**, *48*, 7775.
16. Gibson, A. W.; Humphrey, G. R.; Kennedy, D. J.; Wright, S. H. B., *Synthesis* **1991**, 414.
17. (a) Marshall, J. A.; Grote, J.; Shearer, B., *J. Org. Chem.* **1986**, *51*, 1633. (b) Marshall, J. A.; Salovich, J. M.; Shearer, B. G., *J. Org. Chem.* **1990**, *55*, 2398.
18. (a) Luh, T.-Y.; Wong, K.-T., *Synthesis* **1993**, 349. (b) Stadnichuk, M. D.; Voropaeva, T. I., *Russ. Chem. Rev. (Engl. Transl.)* **1992**, *61*, 1091.
19. Wang, K. K.; Wang, Z.; Gu, Y. G., *Tetrahedron Lett.* **1993**, *34*, 8391.
20. Rücker, Ch. unpublished results.

Christoph Rücker
Universität Freiburg, Freiburg, Germany

N,O-Bis(trimethylsilyl)acetamide

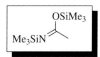

[10416-59-8]   C$_8$H$_{21}$NOSi$_2$ (MW 203.43)

InChI = 1S/C8H21NOSi2/c1-8(9-11(2,3)4)10-12(5,6)7/h1-7H3/b9-8+

InChIKey = SIOVKLKJSOKLIF-CMDGGOBGSA-N

(powerful silylating agent; reacts with a wide range of functional groups)

Alternate Name: BSA.
Physical Data: bp 71–73 °C/35 mmHg.[1,2]
Solubility: very sol most organic solvents.
Form Supplied in: liquid; rapidly contaminated with trimethylsilylacetamide and acetamide if exposed to moist air.
Drying: extreme moisture sensitivity; should be kept in a sealed container.
Preparative Method: made by the reaction of acetamide with a large excess of chlorotrimethylsilane in the presence of triethylamine.[1,2]
Handling, Storage, and Precautions: no special handling and storage procedures; all work must be carried in an efficient fume hood.

Original Commentary

Harry Heaney

Loughborough University of Technology, UK

Bis(trimethylsilyl)acetamide.   Bis(trimethylsilyl)acetamide (BSA) has been formulated as *N,N*-bis(trimethylsilyl)acetamide as well as the imidate form. The NMR spectra of a number of silylamide derivatives (^{13}C, ^{29}Si, ^{14}N, and ^{17}O) have been investigated. Whereas BSA is seen to exist in the imidate form, some analogs, for example bis(trimethylsilyl)formamide, exist in the *N,N*-bis(trimethylsilyl) form, and the product obtained by the interaction of 1,2-bis(chlorodimethylsilyl)ethane with BSA is also clearly in the alternative form.[3] Trimethylsilyl derivatives of hydroxamic acids are formed in reactions using hexamethyl-disilazane and are similar in structure to BSA.[4] This entry will concentrate on examples from the more recent literature.

Protection of Functional Groups.   Early reports of the use of BSA concentrated on the preparation of derivatives of a number of functional group types in order to facilitate chromatographic analysis. The trimethylsilylation of amides, ureas, amino acids and dipeptides, carboxylic acids, enols and hindered phenols were reported under mild reaction conditions.[12] Substituted ureas are converted into their monosilylated derivatives because of the lone electron pair involvement and so *N,N'*-bis(trimethylsilyl)-urea is close in energy to BSA. By the same token, one of the trimethylsilyl residues in BSA is replaced more easily than the second. The conversion of *N*-haloamides into the *N*-haloimidic trimethylsilyl esters, for example *N*-chloroacetamide into *N*-chloroacetimidic trimethylsilyl ester, was probably the first reported use of BSA.[5] Part per billion traces in water of environmentally important phenols and carboxylic acids have been determined by concentration using macroreticular resins and subsequent derivatization using BSA before gas chromatography.[6] It has been claimed that trimethylsilylation of trichothecines prior to gas chromatography is best achieved by using bis(trimethylsilyl)trifluoroacetamide (BSTFA) rather than BSA.[7]

Protection of Hydroxy Groups.   The first example of the use of the trimethylsilyl group for the protection of a sterically hindered hydroxy group involved a 14-hydroxy steroid which was treated with bis(trimethylsilyl)acetamide in DMF at 78 °C.[8] Since that time a wide variety of alcohols have been protected using BSA. The product from the enantioselective hydroboration–oxidation of *N*-benzyloxycarbonyl-1,2,3,6-tetrahydropyridine was converted into its trimethylsilyl ether (TMS ether) in order to allow isolation by gas chromatography.[9] The osmylation of unsaturated ester components of insect pheromones, followed by a reductive workup and silylation with BSA, has been used to allow identification using GC–MS.[10] Bis(trimethylsilylation) using BSA has been shown to be the most convenient method for the protection of a range of biologically important dihydroarene-diols prior to characterization using GC–MS.[11] High yields have been reported with other secondary alcohols using BSA,[12] including the example shown in eq 1.[13] The method has been used to effect the trimethylsilylation of an aldol product,[14] Ninhydrin reacts with allyltrimethylsilane and trimethylsilyl triflate in acetonitrile to give the expected 2,2-bis(trimethylsilyloxy)indan-

1,3-dione. However, using BSA the major product incorporated an acetamido residue.[15] The replacement of an acetoxy group by an amido group (eq 2) presumably occurs via an oxonium species.[16]

$$(1)$$

$$(2)$$

Bis(trimethylsilyl)acetamide has also been used to provide temporary protection of hydroxy and carboxy groups in a number of coupling reactions involved in the preparation of cephalosporins and penicillins, as shown in eq 3.[17] The protection of nucleoside hydroxy groups prior to reduction using tributyltin hydride–AIBN has been reported.[18] The same method has also been used in connection with nucleoside coupling reactions.[19] Transglycosylation catalyzed by trimethylsilyl triflate has also been carried out on trimethylsilyl-protected nucleosides (eq 4), in which the prior protection was effected using BSA.[20] In the case of 6-oxopurineribonucleoside synthesis, the 7β-isomer is first formed and eventually affords the 9β-isomer, presumably by way of the 7,9-diribonucleoside.[21] It has been shown that BSA causes the debromination of some purine nucleosides in the presence of potassium fluoride and a crown ether.[22]

$$(3)$$

$$(4)$$

Ur = uracil-1-yl; Ad = adenin-9-yl

The silylation of tertiary alcohols in good yields has been reported, e.g. di-*t*-hydroxyadamantane.[23] The preparation of 2-methyl-2-trimethylsilyloxynonadecane was achieved in 84% yield using a combination of BSA, chlorotrimethylsilane, and *N*-(trimethylsilyl)imidazole.[24]   2-Methyl-2-trimethylsilyloxy-pentanone, a prototype for a series of compounds that are valuable intermediates used in a study of diastereoselective aldol reactions, can be prepared in good yield using BSA.[25] Even highly hindered silanols react with BSA as indicated in eq 5.[26] It is of interest to note that BSA has been used to reduce the influence of free hydroxy groups on surfaces. Difficult Diels–Alder reactions that require high temperatures and are carried out in sealed glass tubes may suffer from decomposition of the diene on the surface of the glass. Silylation with BSA is then worth investigating.[27] It has also been shown that the charcoal carrier for tungstic acid–tri-*n*-butylchlorostannane, which is used in the catalyzed hydrogen peroxide epoxidation of alkenes, is improved by treatment with BSA. Decomposition of the hydrogen peroxide, catalyzed by surface hydroxy groups, is then less troublesome.[28]

(5)

Formation of Trimethylsilyl Enol Ethers. The silylation of enolizable aldehydes and ketones has been carried out using BSA in HMPA in the presence of very small amounts of metallic sodium. The method is highly regio- and stereoselective, giving the (*Z*)-isomer that results from kinetic control.[29] With more easily enolized compounds the reaction proceeds readily in the absence of other additives. In the case of the bis(trimethylsilylation) of oxindole (eq 6) the effect is evidently related to the gain in resonance energy on silylation at oxygen.[2] Similarly, the conversion of glutaconic anhydride into the silylated α-pyrone (eq 7) was achieved in very high yield using BSA.[30] The product is a very useful Diels–Alder diene. Other highly functionalized dienes have been prepared and used in Diels–Alder syntheses of highly functionalized naturally occurring anthraquinones. The dienes shown in eqs 8 and 9 were used in the syntheses of ceroalbolinic acid[31] and xantholaccaic acid,[32] respectively.

(6)

(7)

(8)

(9)

The protection of phenols was one of the first examples of the use of BSA, including the trimethylsilylation of 2,6-di-*t*-butylphenol.[2] 2,6-Di-*t*-butyl-4-vinylphenol has been protected with

BSA and then polymerized by both cationic and radical initiators before liberating the free phenolic groups by cleavage with acid.[33] A number of calixarene trimethylsilyl ethers have been prepared from the phenols using BSA.[34] Calix[4]arene has been converted into a mixture of tris- and tetrakis(trimethylsilyl) ethers by heating the phenol with BSA in acetonitrile.[35]

Protection of Carboxy Groups. As with hydroxy compounds, the conversion of carboxy groups into their trimethylsilyl derivatives is valuable in connection with GC–MS. The method has been used, for example, in assaying a perhydroindole ACE inhibitor.[36] Amino acid hydrohalides are silylated in almost quantitative yields by BSA in THF at reflux. Other methods require higher temperatures. The free amino acid can be isolated after treatment with water and another important feature is that no racemization was observed.[37] The use of BSA in peptide coupling reactions has been reported,[38] including examples where hydroxy amino acids are involved.[39] A coupling reaction involved in a total synthesis of the immunostimulating peptide FK156 also involved the formation of a peptide TMS ester.[40] The trimethylsilylation of (*E*)-3-bromoacrylic acid proceeds in good yield (eq 10) and has been used in conjunction with a protected alanyl anion in the synthesis of an unsaturated α-amino acid.[41] The bis(trimethylsilyl) ester of squaric acid, formed using BSA (eq 11), has been shown to be in dynamic equilibrium. The trimethylsilyl groups undergo rapid intermolecular migration, as established by crossover experiments.[42]

(10)

(11)

Reactions at Nitrogen. The protection and activation of a number of nitrogen functional groups has been achieved using BSA, for example the silylation of aziridines.[43] Treatment of primary and secondary amines with BSA and then with a chloroformate gives good yields of carbamates.[44] Conversion of the guanine derivative into the bis(trimethylsilyl) derivative and the trimethylsilyl trifluoromethanesulfonate nucleoside coupling reaction (eq 12) was part of a synthesis of AzddMAP.[45] Similarly, the conversion of sulfoximines into their TMS derivatives, prior to deprotonation and reaction with a range of electrophiles, can be carried out in almost quantitative yields using BSA.[46] The preparation of the 1,3,4-oxadiazoline shown in eq 13 was achieved in 60–65% yield.[47]

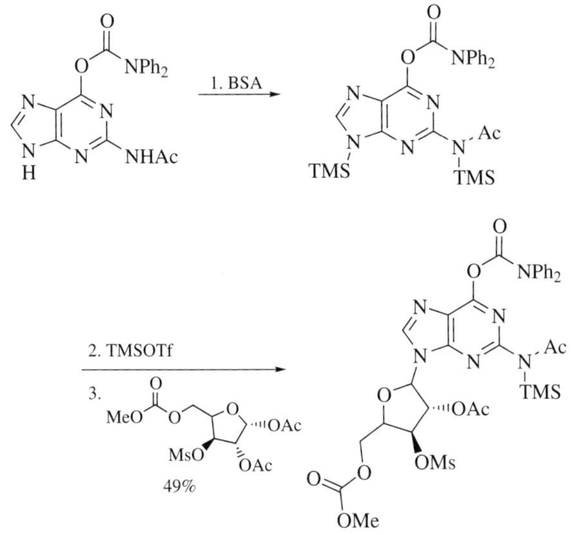

$$(12)$$

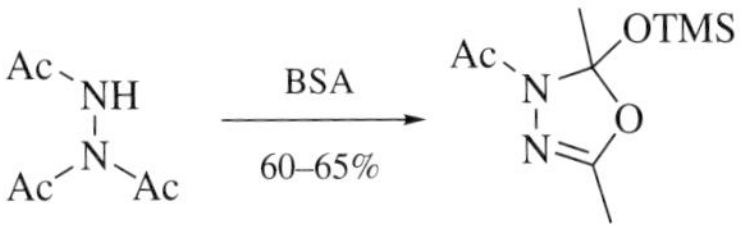

$$(13)$$

The increased volatility on silylation of pyrimido[4,5-*i*] imidazo[4,5-*g*]cinnoline has been investigated in connection with flash vacuum pyrolysis studies.[48] It has been reported that di-hydropyrimidines and dihydro-*s*-triazines were dehydrogenated under vigorous trimethylsilylation conditions.[49] Apparently the reaction does not proceed in the absence of oxygen or in the presence of free radical inhibitors. Aromatic isocyanates react with BSA to give silylated azauracils as shown in eq 14, which are desilyated on treatment with ethanol.[50] We also note that BSA and other bis(trimethylsilyl)amides are converted into ni-triles in high yields by treatment with tetra-*n*-butylammonium fluoride Lewis acids, or iron(II) phthalocyanine,[51] and a range of bis(trimethylsilyl)amides react with phosgene to form acyl isocyanates.[52] Arylsulfonylacetonitriles undergo carbon–carbon coupling reactions when heated in BSA at 120–140 °C.[53] A range of silylnitronates have been prepared using BSA or the trifluoro analog (BSTFA). The products (eq 15) are useful intermediates for the synthesis of isoxazolidines.[54]

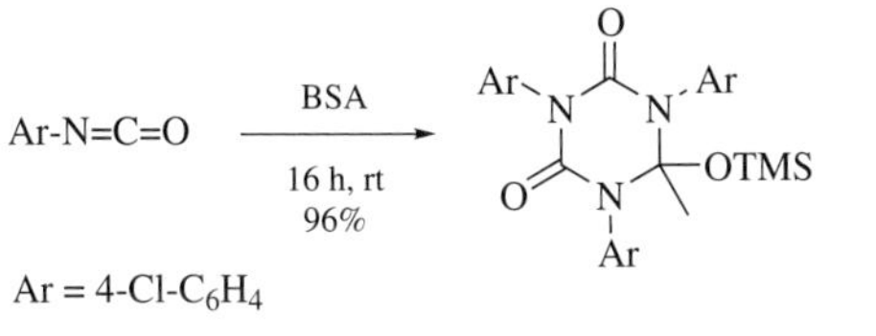

$$(14)$$

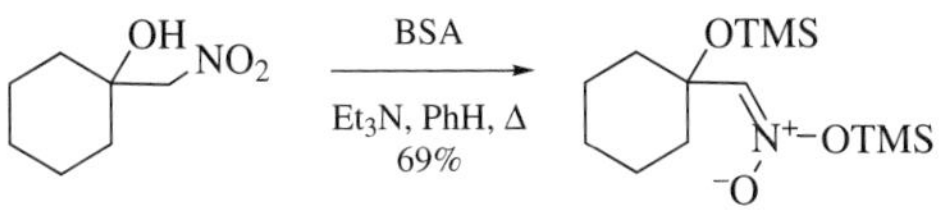

$$Ar = 4\text{-Cl-}C_6H_4$$

$$(15)$$

The Control of Protic Acids.   There are a number of exam-ples where protic acids have been shown to have a deleterious effect on reactions and where partial or complete control is ad-vantageous. The rearrangement of the benzylpenicillin sulfoxide to the cephalosporanic acid derivative (eq 16) gave the best yield using hydrogen bromide and ca. 30 mol % of BSA.[55] Similarly, the preparation of eight-membered oxygen heterocycles that are

present in a number of marine natural products, has been achieved in moderate yields by Lewis acid promoted acetal–alkene cycli-zations in the presence of BSA as shown in eq 17.[56] The same principle has been used to control Friedel–Crafts alkylation reac-tions involving, for example, methyl chloromethoxyacetate and *N*-methylindole.[57] In the absence of BSA a considerable amount of the indole derivative was converted into its dimer while the major product was methyl bis(1-methyl-3-indolyl)acetate. In the presence of BSA the formation of the dimer was suppressed and the initial product, methyl 1-methyl-3-indolylacetate, isolated in a 90% yield. Control in the highly diastereoselective alkylation of 1-methylindole using chiral pyrrole derivatives was achieved in a better yield when using BSA (eq 18).[58]

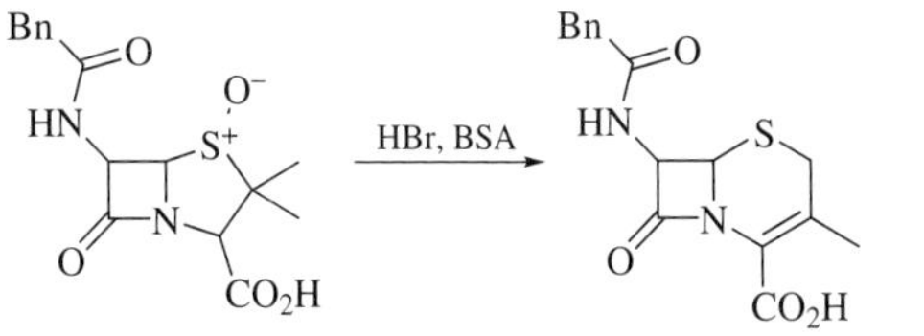

$$(16)$$

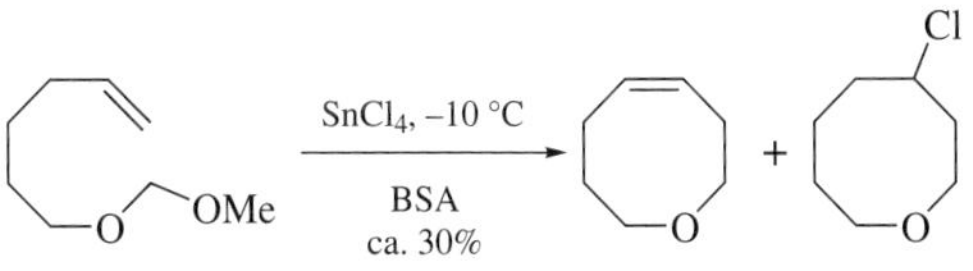

$$(17)$$

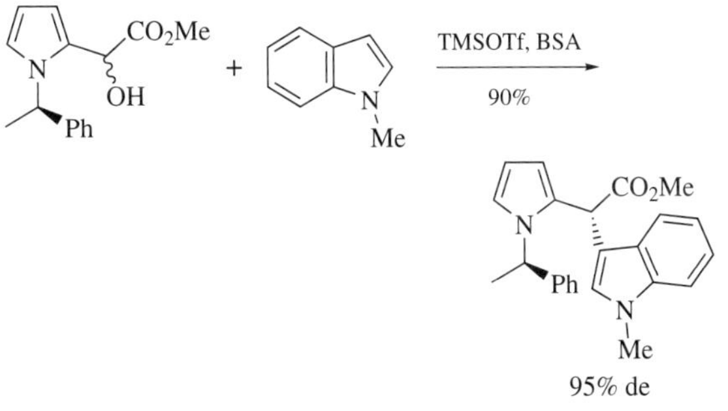

$$(18)$$

Use has also been made of the oxophilicity of silicon, as compared to the azaphilicity of protons, in Mannich re-actions of bis(aminol ethers) with heterocycles such as 2-methylfuran where a proton is liberated during the reaction.[59] The bis(methoxymethyl) derivative from *t*-butylamine gave an iminium salt when treated with hydrogen chloride in ether. In duplicate reactions using 2-methylfuran, one (eq 19) was treated with BSA while the other, in the absence of BSA, was worked up using Hünig's base and gave the expected secondary amine after a hydrolytic work up in 80% yield. Evidently in the first reaction the BSA reacted with the hydrogen chloride produced and gave chlorotrimethylsilane.

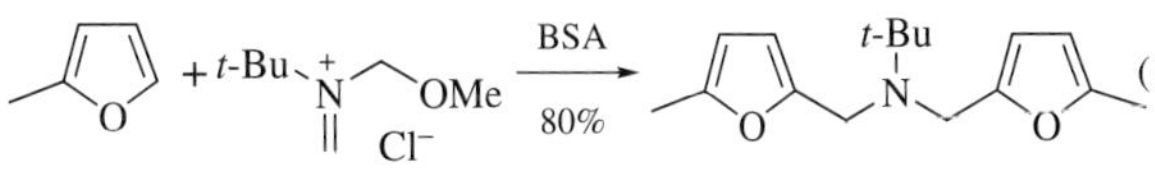

$$(19)$$

The reaction between triflic acid and *N*-trimethylsilyl-1,3-oxa-zolidin-2-one has been shown to afford trimethylsilyl triflate in a high yield, whereas the reaction of BSA with triflic acid gave only a 43% yield of trimethylsilyl triflate.[60] Despite that finding, in situ catalytic reactions involving trimethylsilyl triflate (eq 20)[58b] and trimethylsilyl fluorosulfonate (eq 21)[61] can be carried out in high yields when a catalytic amount of the acid is added to BSA.

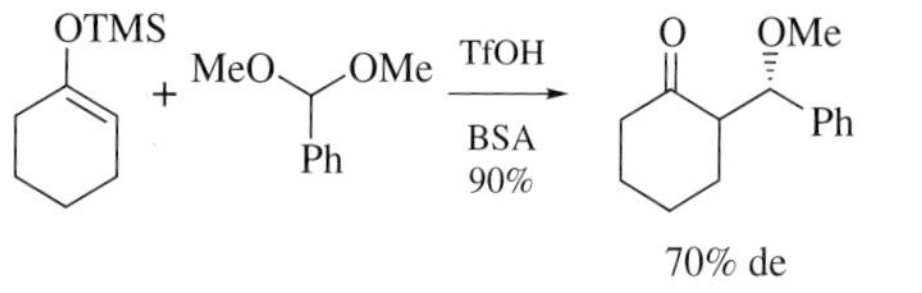

(20)

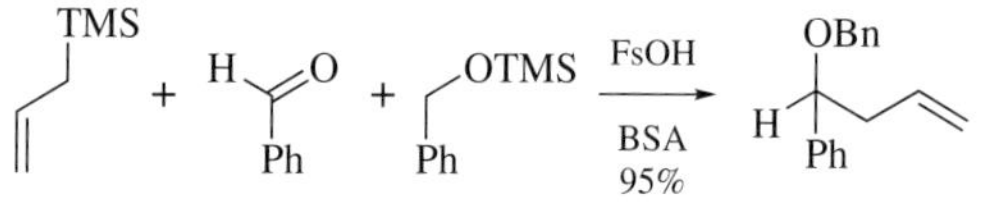

70% de

(21)

Reactions of the Anion from Bis(trimethylsilyl)acetamide. The deprotonation of BSA using lithium diisopropylamide was used to generate the protected acetamide enolate ion for use in the synthesis of the dienone shown in eq 22.[62] It has been shown subsequently that the same product can be prepared by the reaction of the anion, generated from *n*-butyllithium and BSA, with 2,6-dibromo-1,4-benzoquinone.[63] The anion has also been used in reactions with a number of carbonyl compounds, including benzophenone.[64]

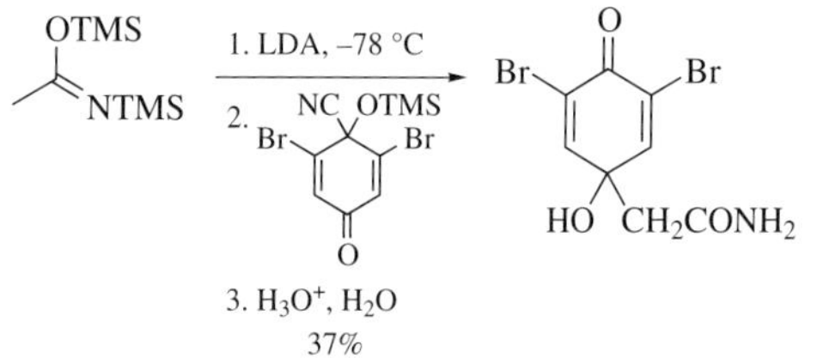

(22)

37%

Reactions of Phosphorus Compounds. A number of papers report the conversion of chlorophosphites, dichlorophosphites, and their thio analogs into trimethylsilyl derivatives.[65] The reaction between bromodifluorophosphine and BSA at −80 °C results in formation of an unstable product with a P–N bond together with bromotrimethylsilane.[66] Nucleoside phosphonates have been converted into the related bis(trimethylsilyl) phosphites by treatment with BSA. Subsequent reaction with an acyl chloride and triethylamine then rapidly affords the corresponding acyl phosphonate.[67] The Michael addition of phosphonous acids and esters by way of intermediate silyl alkyl phosphonites has been used in the preparation of phosphinic acids (eq 23).[68] The coupling of an activated acid with an acylphosphoranylidene in the presence of BSA (eq 24) has been used in the synthesis of the vicinal tricarbonyl portion of the immunosuppressant macrocyclic lactone FK-506.[69] The coupling in that case gave the required product in a 91% yield.

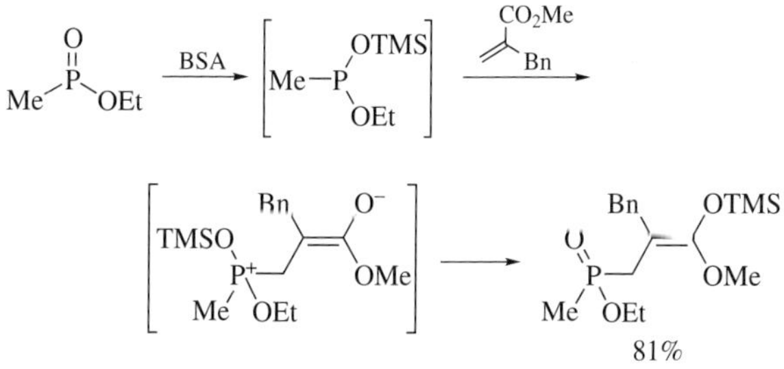

(23)

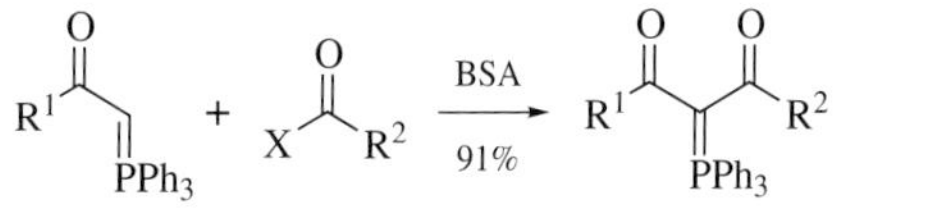

81%

(24)

Metal Catalyzed Reactions of Allylic Esters. hexacarbonylmolybdenum catalyzed elimination of acetic acid in the presence of BSA is effective in a high yield route to conjugated dienes,[70] and BSA has been used as a base in an unexpected palladium catalyzed tandem elimination–cycloaddition reaction.[71] The alkylation of malonic esters with allylic acetates using hexacarbonyl molybdenum and BSA (eq 25) has been reported to give high yields.[72] β-Keto esters are also alkylated using palladium(0) catalysis. In reactions of 2-hydroxymethyl-2-propen-1-ol diacetate with methyl 4-methylcyclohexanone-2-carboxylate, the use of BSA gave the monoalkylated product whereas using DBU resulted in the formation of the bicycloannulation product.[73] The palladium catalyzed alkylation of sodium dimethyl malonate with a silylated allene acetate in the presence of BSA (eq 26) proceeds via the silylated ester.[74] The silylation step is necessary because of the sensitive nature of allenyl acetates. A stereoselective total synthesis of isolobophytolide used BSA to convert an allylic pivalate into the corresponding enol trimethylsilyl derivative and this was then used in the intramolecular allylation of a sulfonyl acetate anion using tetrakis(triphenylphosphine)palladium(0) as catalyst.[75]

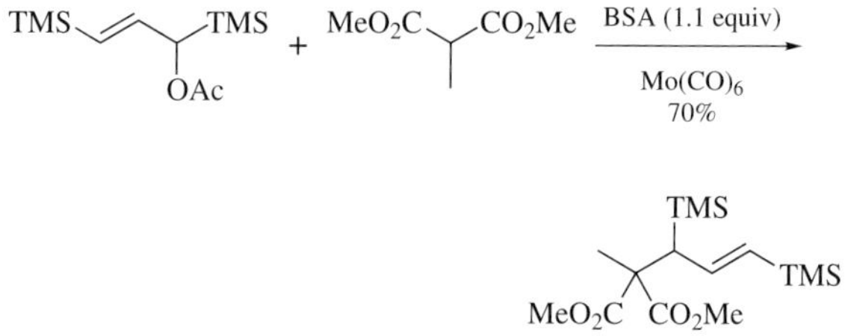

(25)

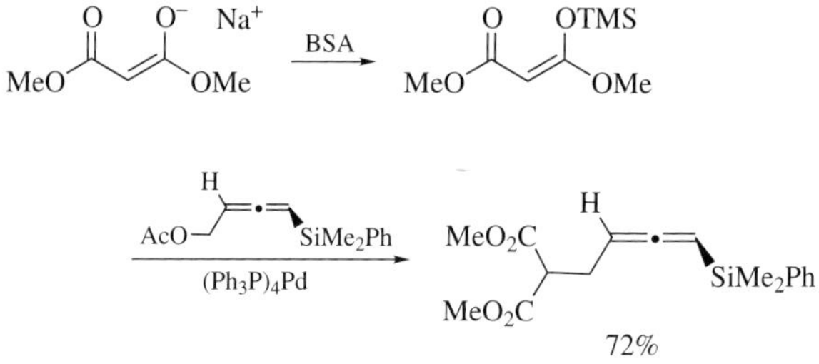

(26)

72%

First Update

Jian Cui

Schering-Plough Research Institute, Union, NJ, USA

Silylation and Activation of Nucleobases. Base glycosidation, using the Vorbrüggen method,[76] and its variations, has been a much exploited synthetic reaction in the formation of many important nucleoside analogs. BSA is used to activate the nucleobases by silylation. The persilylated nucleobases undergo nucleosidations with electrophiles in the presence of Lewis acids, generally tin(IV) chloride or trimethylsilyl trifluoromethanesulfonate,[16,20,45,76,77] while phenylselenenyl chloride,[78] *N*-iodosuccinimide/triflic acid,[79] and iodine[80] have also been used to mediate these reactions. A typical nucleosidation reaction is shown in eq 27. Thymine **1** was treated with BSA in acetonitrile to form the silylated thymine **2** in situ, which is then reacted with protected L-threose **3** in the presence of TMSOTf to give the α-nucleoside-2′,3′-dibenzoate **4** with high diastereoselectivity.[77a]

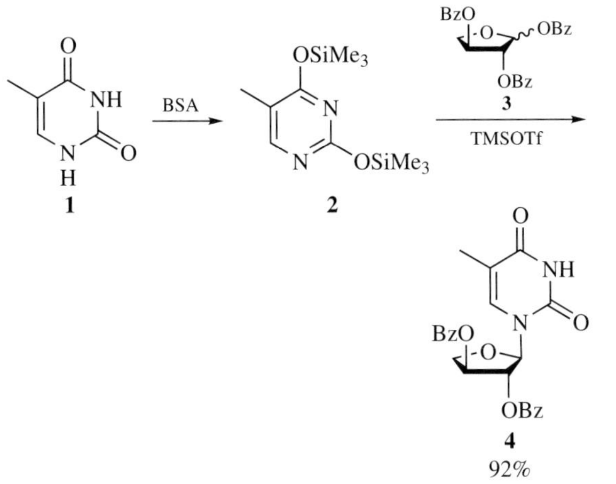

In an alternative one-pot synthesis of *C*-nucleosides **7**, based on a Heck-coupling reaction, BSA was used to solubilize the nucleobase **5** in DMF via in situ *O*-silylation followed by subsequent addition of the glycal (*X* = O) or enamine (*X* = N-CBZ) **6** with Pd(OAc)$_2$ as catalyst (eq 28).[81]

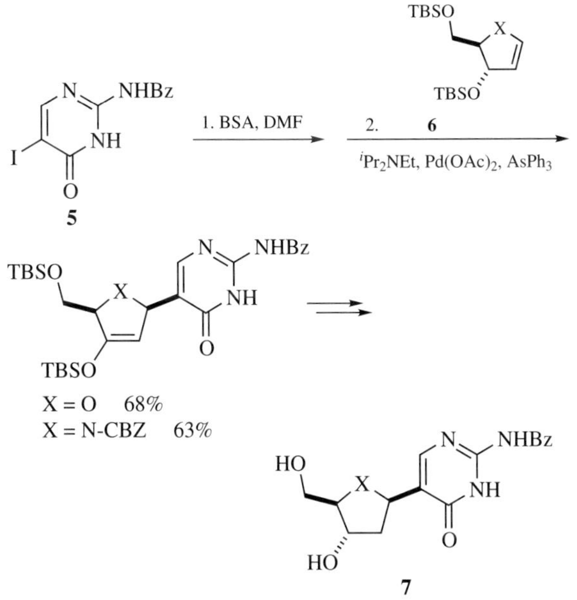

Palladium-catalyzed Allyllation Reactions. Palladium-catalyzed asymmetric allylic substitutions by soft nucleophiles are an extensively studied research area in organic chemistry with huge advances in ligand development and enantiocontrol seen in the past two decades. A typical asymmetric allylic substitution reaction with malonate is shown in eq 29.[82] Diastereo- and enantioselective allylation of substituted nitroalkanes has also been reported.[83] BSA has been used as a standard base in these reactions and functions satisfactorily.

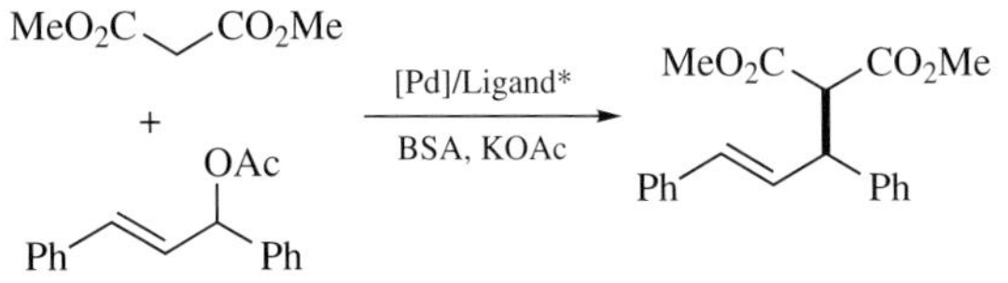

When the reaction conditions were applied to allylic carbonate **8**, the palladium-mediated cyclization occurred to form the 11-membered ketone **9** in 86% yield (eq 30).[84] When the palladiun-catalyzed allylation of propargyl malonate **10** in the presence of BSA was combined with the rhodium-catalyzed Pauson-Khand type reaction, the tandem action of the two catalysts gave an excellent yield (92%) of bicyclopentenone **11** in one-pot (eq 31).[85]

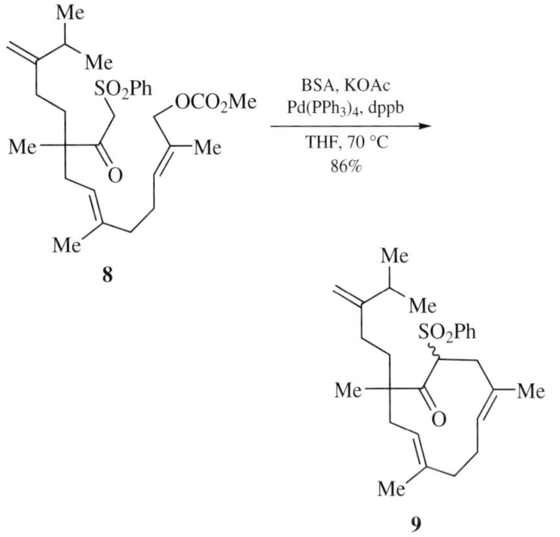

As a powerful silylating agent, BSA reacts with a wide range of functional groups.

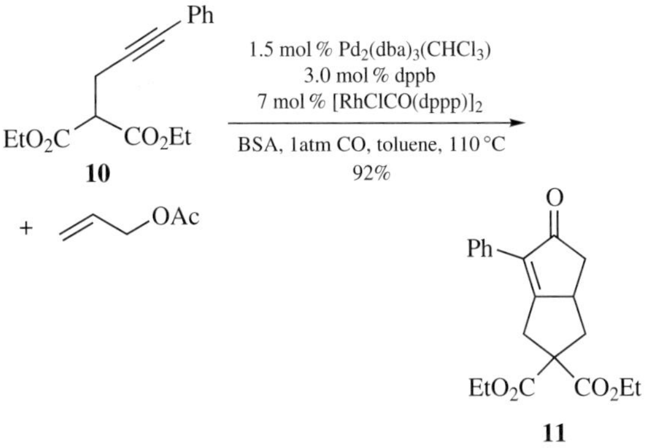

Trimethylsilylation of Hydroxy Groups. BSA was used to trimethylsilylate the hydroxyl group of **12** selectively to form **13** in almost quantitative yield (eq 32).[86] The presence of catalytic amounts (∼0.02 equiv) of tetrabutylammonium fluoride (TBAF) significantly promoted the silylation of alcohols under mild conditions with high chemoselectivity, i.e., TBAF plays a role as a smooth silyl transfer catalyst from nitrogen to the hydroxyl group.

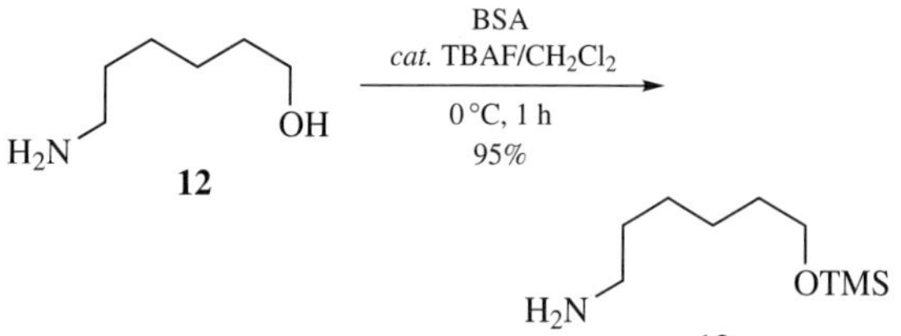

(32)

BSA has been used to isolate and purify a set of unstable intermediary aldol adducts by providing protection of their hydroxyl groups[87] as exemplified in eq 33. Reaction between the triethyl phosphonoacetate **14** and aldehyde **15** utilizing a TiCl₄-Et₃N reagent has been successfully proceeded to give the aldol adduct **16** in excellent yield with high *anti*selectivity. As expected, the adduct and its *O*-acetyl derivative were too unstable for column chromatographic purification. To overcome this problem, a neutral silyl derivatization method using BSA and pyridinium triflate (PyH⁺OTf⁻) was adopted, and the corresponding trimethylsilyl ether **17** was easily isolated, and was stable enough for purification using conventional SiO₂ column chromatography.

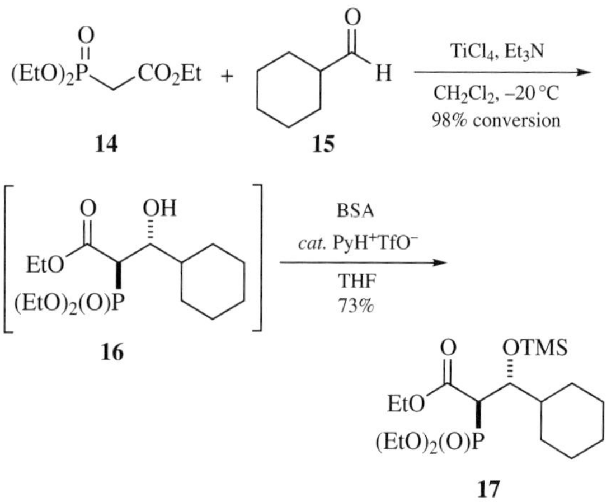

(33)

anti:syn = 93:7

BSA has also been used to trimethylsilylate hydroperoxides.[88] The 1,2-bishydroperoxide **19** (generated from photooxygenation of a solution of 2-phenylnorbornene **18** in acetonitrile containing 30% aqueous hydrogen peroxide) was successfully transformed using BSA into the corresponding bis-trimethylsilylated derivative **20**, which could be isolated by column chromatography on silica gel (eq 34).

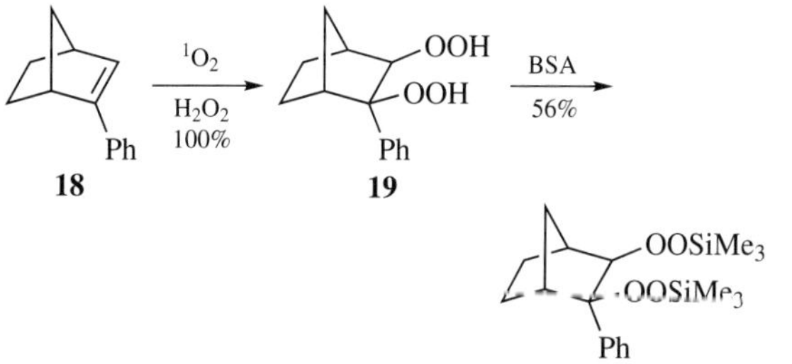

(34)

BSA has been used sometimes to provide in situ protection of hydroxyl groups of reaction products in order to promote desired reactions or prevent the undesired side reactions. In an example, the addition of BSA promoted almost the quantitative 1,3-isomerization of tertiary allylic alcohols **21** to the primary

alcohols **22** (eq 35) via rhenium oxo catalysis.[89] Because silylation by BSA is faster for primary alcohols than for tertiary alcohols, the product was selectively and irreversibly silylated, removing it from the reaction equilibrium. The products can be deprotected and isolated in high yields demonstrating the efficiency of this procedure for the synthesis of allylic alcohols containing either conjugated or nonconjugated trisubstituted alkenes. In another example, BSA has been shown to suppress the formation of abnormal aromatic Claisen rearrangement product **25** from allyl aryl ether **23** by efficiently trapping the incipient normal product **24** as its silyl ether under mild conditions (eq 36).[90] By silylating the incipient phenol **24** before it undergoes the 1,5-hydrogen shift, the otherwise thermodynamically more favored abnormal Claisen rearrangement can be prevented.

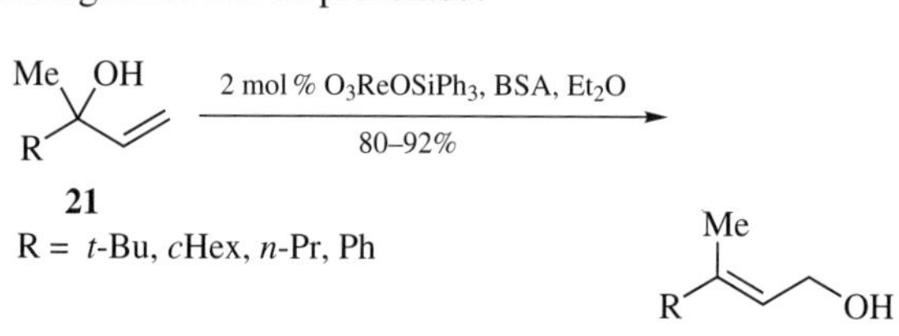

(35)

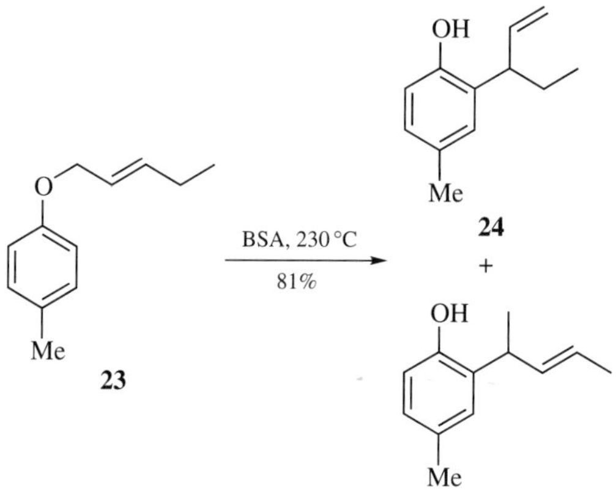

(36)

24:25 > 99:1

1,4-Conjugate Addition Reactions. BSA was found to activate 1,4-conjugate addition of bis[(trimethylstannyl)vinyl)] cuprate reagent **27** to α,β-unsaturated keto-esters without concomitant reduction of the double bond. With the keto ester **26**, a near-quantitative yield (92%) of the stannylvinyl adduct **28** was obtained with no trace of the reduction product **29** (eq 37).[91]

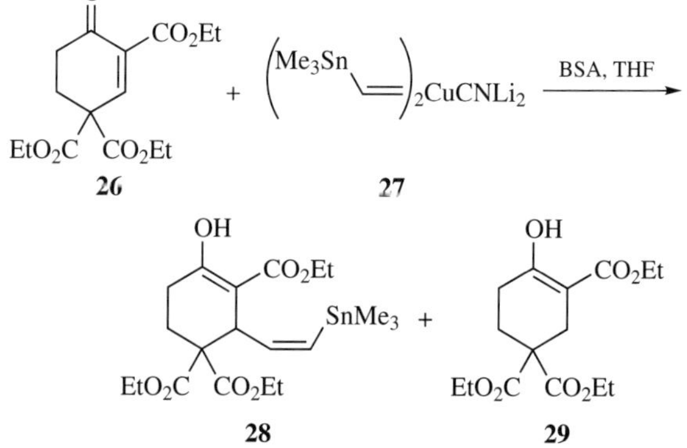

(37)

28   **29**
92%   0%

Formation of Trimethylsilyl Enol Ethers from Enolizable Aldehydes and Ketones. The most common methods for preparing silyl enol ethers use silyl chlorides or silyl triflate/base combinations and need careful attention during workup of the reaction and isolation of the enol ether. Silylations with BSA are generally mild and nearly neutral and do not require the addition of a supplementary base. Ionic liquids have been used for the preparation of silyl enol ethers **31** from aldehydes and ketones **30** with BSA in good yields (eq 38).[92] These new reaction conditions open an important alternative to the use of highly toxic HMPA as solvent.[29,93]

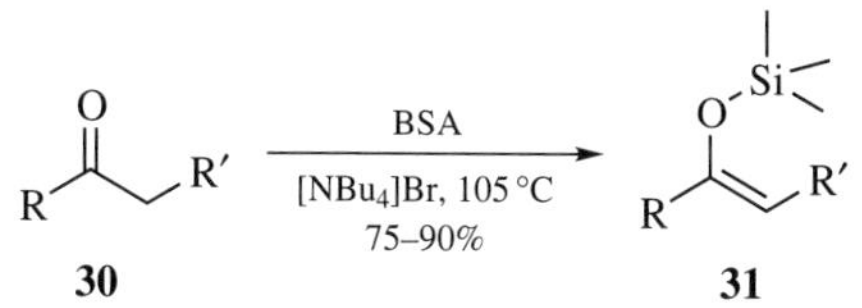

(38)

Claisen Rearrangement. A novel variant of the decarboxylative Claisen rearrangement (dCr) reaction of tosyl acetate esters was reported. In this modified transformation, α-tosyl silylketene acetals, formed in situ from allylic tosylacetates **32** in the presence of stoichiometric or substoichiometric BSA and substoichiometric potassium acetate, undergo [3,3]-sigmatropic rearrangement followed by acetate-induced decarboxylation in situ to provide homoallylic sulfones **33** (eq 39).[94] When allylic acetate esters **34** containing a variety of *N*-arylsulfonyl sulfoximines on the acetyl residue were subjected to the decarboxylative Claisen rearrangement reaction, rearranged products **35** were isolated in generally good yields, and diastereoselectivities up to 82:18 have been obtained (eq 40).[95] Allylic tosyl malonate esters **36** also undergo the decarboxylative Claisen rearrangement reaction using the combination of BSA and potassium acetate under mild conditions (25 °C), thereby providing an effective, regiospecific alternative to the metal-catalyzed allylation of methyl tosyl acetate (eq 41).[96]

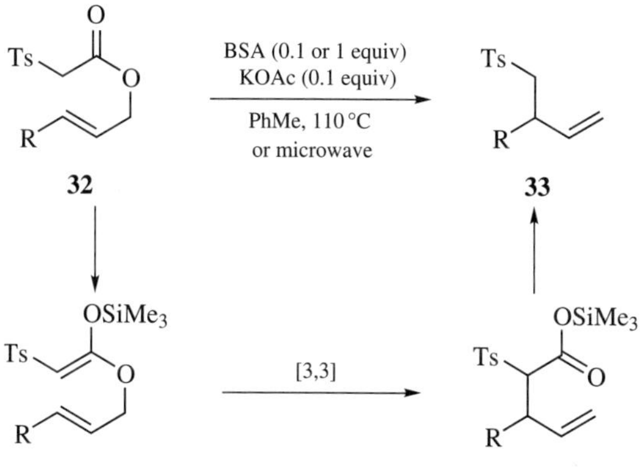

(39)

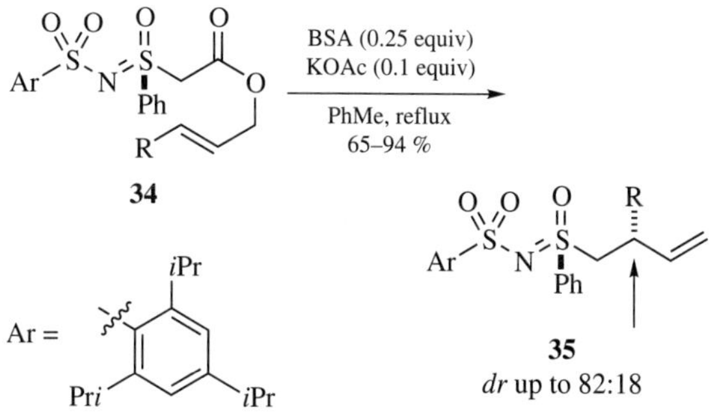

(40)

Formation and Reaction of the Trimethylsilyl Ester of Squaric Acid. The most common method employed in the synthesis of squarate esters uses expensive silver nitrate and goes through the explosive disilver salt of squaric acid. It was conceived that silyl esters of squaric acid **38** may be considered as silyl enol ethers that can undergo Si–O cleavage in the presence of a suitable fluoride source and that *O*-alkylation may be affected using electrophilic halides. Upon preparation of the bis(trimethylsilyl) squarate **39** using BSA and reaction in situ with *p*-methylbenzyl bromide in the presence of a fluoride source, the desired benzyl squarate **40** was obtained in about 50% overall yield (eq 42).[42,97]

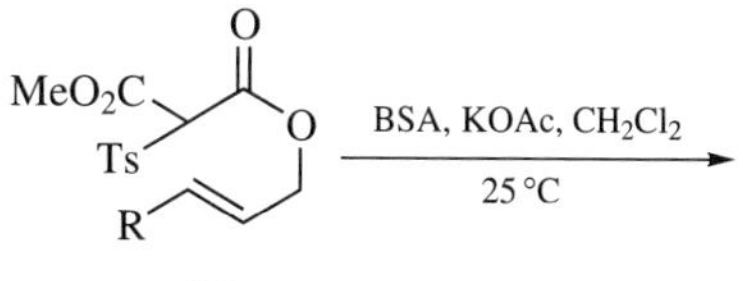

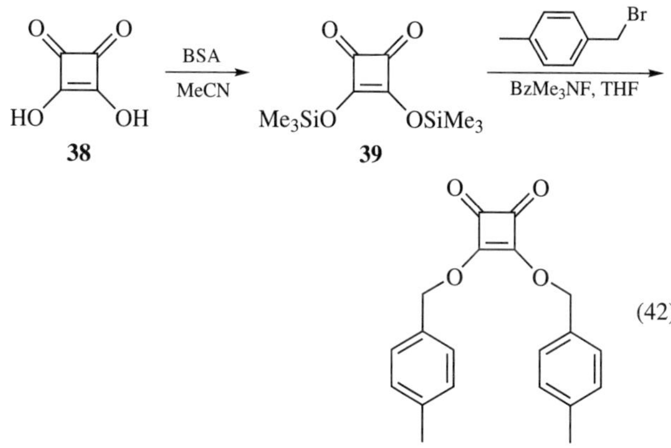

(41)

(42)

Silylation of Boronic Acid and Its Usage in Suzuki-Miyaura Cross-coupling. The anti-MRSA carbapenem, L-742,728, has been prepared in large quantity using the Suzuki-Miyaura cross-coupling as the key reaction. Initial cross-coupling efforts employing the simple fluorenone boronic acid **41** were successful but problematic for scale-up because the product **45** could not be isolated as a crystalline solid without chromatography. In order to solve this problem, boronic acid **41** was fully silylated using BSA in THF. The resulting homogeneous solution of **42** was effective for cross-coupling with vinyl triflate **43**. The TMS-containing product **44** crystallized well from methanol and could be isolated without chromatography (eq 43).[98]

Silylation of Formyl Hydrazone. Addition of the ethyl Grignard reagent to formyl hydrazone **46** via in situ silylation using BSA provided the corresponding formyl hydrazine with excellent diastereoselectivity in favor of the desired *S,S*-diastereoisomer **47** (eq 44).[99]

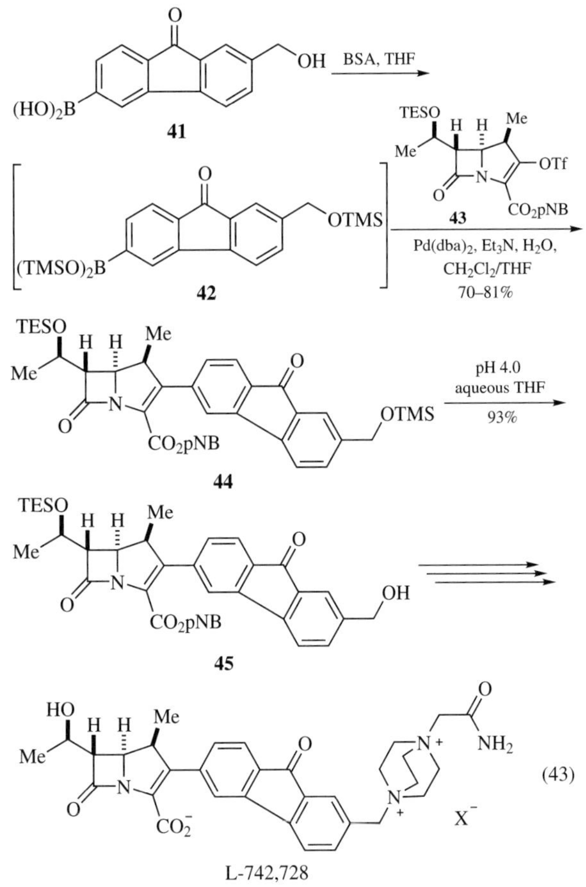

(43)

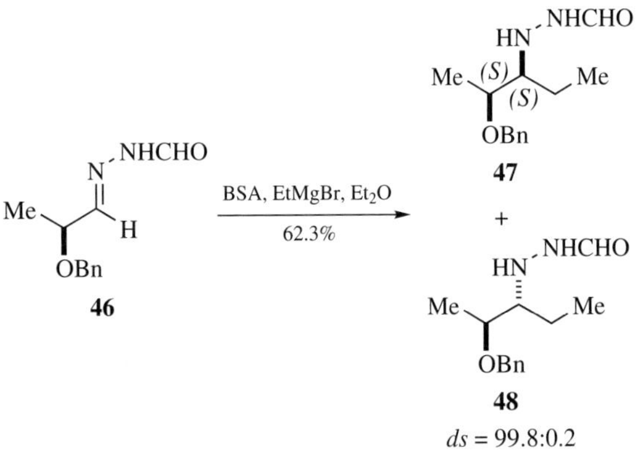

L-742,728

$$ds = 99.8:0.2$$

(44)

(45)

(46)

Xy:

(47)

$\sim 90\%$ (estimated by ^{1}H NMR)

Reactions at Nitrogen. BSA has been used to provide in situ protection of amino groups. As reported in a systematic study of the Pd-catalyzed alkylation of indoles by allylic carbonates, when BSA/Li$_2$CO$_3$ was used as the base system in low-coordinating solvent (CH$_2$Cl$_2$), the desired C-alkylation product **50** was obtained in high regioselectivity from **49** (eq 45).[100] The N-TMS-indole was initially formed in situ followed by subsequent C-alkylation. The initial fast silylation of the indole by BSA might be responsible for preventing the formation of the charged species leading to the kinetic N-product **51**. In another example, imidoyl telluride **52** was transferred to the t-leucine amide **53** in good yield by mercury-mediated oxidative hydrolysis in the presence of BSA (eq 46).[101]

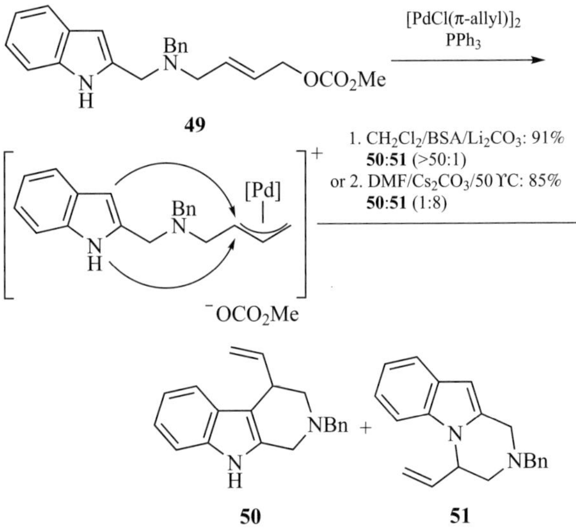

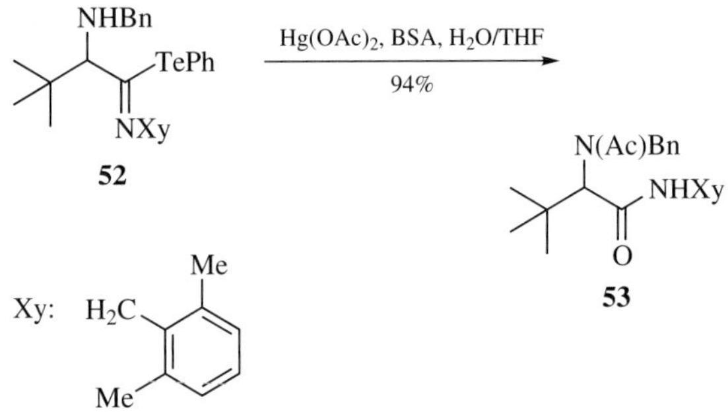

A method that uses BSA and a catalytic amount of methyl iodide for the preparation of enamines from ketones has been reported.[102] A variety of enamines can be prepared by this method, though, in general, pyrrolidine appears to be the most reactive amine. This method was applied to a model system for the synthesis of roseophilin, in which the requisite bridged bicycle **55** was assembled through a bis-allylation protocol on enamine **54**, which was prepared from cyclododecanone (eq 47).[103]

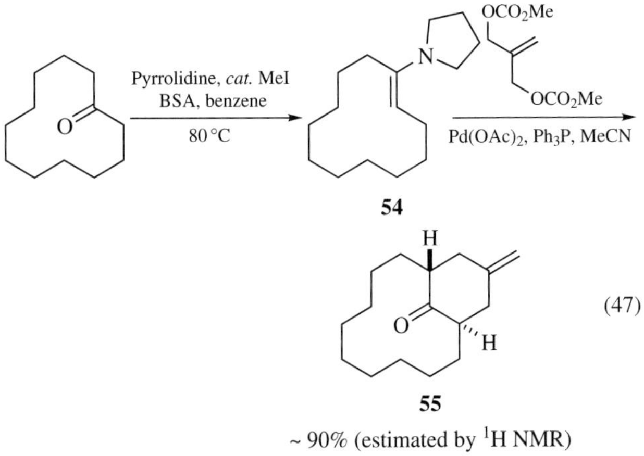

Preparation of Heterocycles. BSA has been used in the solid-phase synthesis of 1,2,3-triazoles via trimethylsilyl-directed 1,3-dipolar cycloaddition reactions of 1-trimethylsilylacetylenes with organic azides (eq 48).[104] When resin **56** was allowed to react with trimethylsilylpropynoic acid in the presence of BSA, high regioselectivity was observed and the trisubstituted triazole **57a** was obtained in near quantitative yield that is greater than 98% purity.

Clearly, BSA protects against desilylation and decarboxylation of the desired triazole system by rapid and complete formation of the silyl ester, which blocks both deleterious pathways. When an excess of ethyl trimethylsilylpropynoate was allowed to react with resin **56** in the presence of BSA and then washed and cleaved with 50% TFA/DCM, a 94% isolated yield of triazole **57b** was obtained after silica gel purification; no 1,4-regioisomer or desilylation were detected under these conditions.

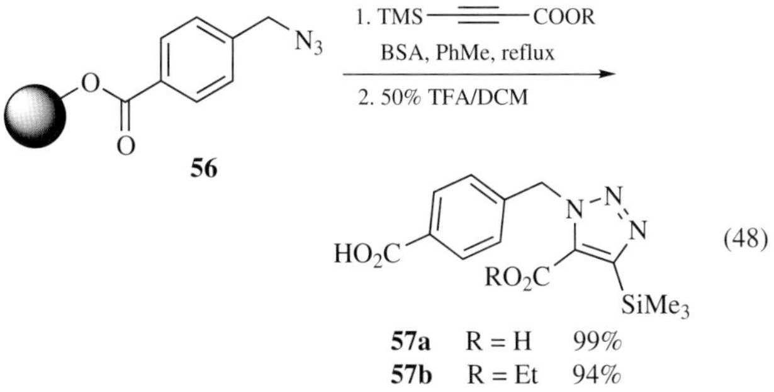

$$(48)$$

57a R = H   99%
57b R = Et   94%

Treatment of **58** with DMF, HMDS, and BSA at 220 °C induced a Dimroth-type rearrangement and formed tricyclic compound **59** in good yield (67%) (eq 49).[105] This transformation takes only 30 min and provides an expedited method for preparation of **59**. In the past, the same compound was obtained by a tedious de-*t*-butylation procedure using 99% formic acid for 48 h.

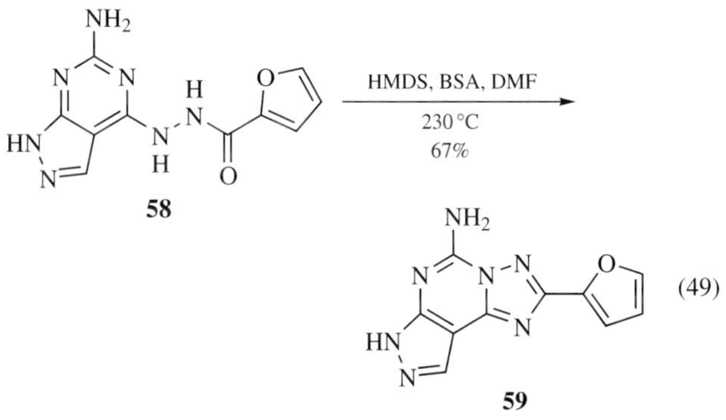

$$(49)$$

Silylation of Nitroarenes. The combination of DBU and BSA mediated double condensation of nitroarenes with cinnamyl-type sulfones smoothly to yield 2-aryl-4-arylsufonyl quinolines and their hetero analogs.[106] For example, nitronaphthalene **60** reacted with cinnamyl phenyl sulfone **61** in the presence of DBU and BSA to form **62** in 82% yield (eq 50). Some nitroarenes were found to react with arylmethylene compounds prone to form benzylic-type carbanions to give some sophisticated fused nitrogen heterocycles in the presence of BSA.[107] Depending on the nitroarene structure and sometimes also on the reaction conditions, the reaction led to the formation of the expected, fused pyridine ring **65** or to the anthranile derivative **67** (eqs 51 and 52). When alkyl/arylthiols were used as nucleophiles to some active bicyclic nitroarenes (**68**) in the presence of DBU and BSA in DMF solution, dithioalkyl/aryl substituted anilines (**69**) were formed in moderate to good yields via displacement of *ortho-* and *para*-hydrogen atoms with simultaneous reduction of the nitro- to amino-group (eq 53).[108]

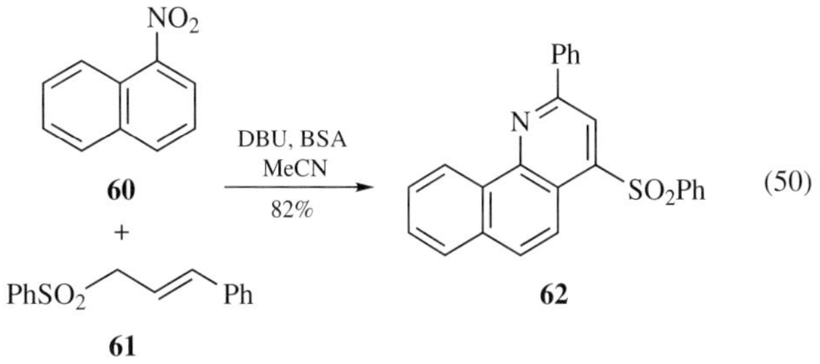

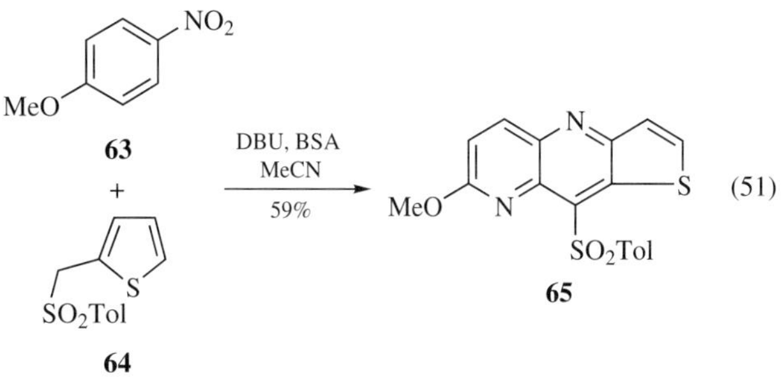

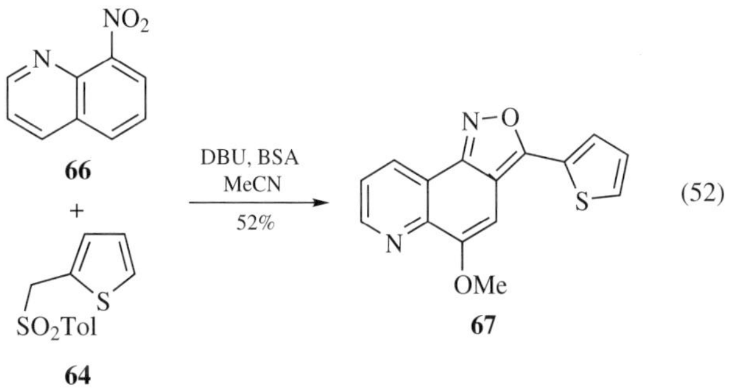

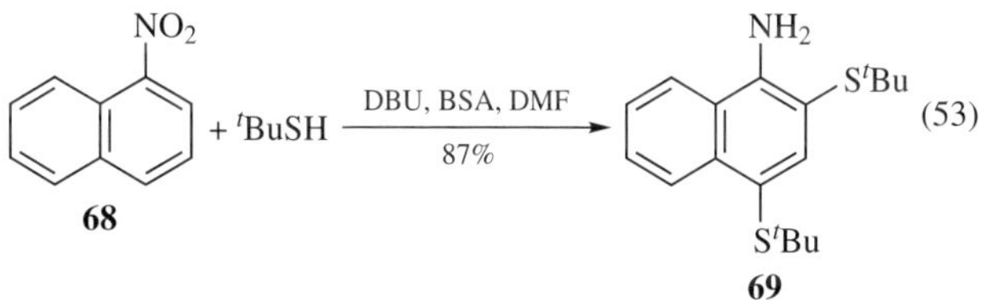

$$(50)$$

$$(51)$$

$$(52)$$

$$(53)$$

Silylation of Aliphatic Nitro Compounds. BSA has been used in silylation of β-substituted aliphatic nitro compounds with subsequent reactions involving the resultant trimethylsilyl nitronates.[54,109] A general method for the synthesis of conjugated enoximes (**74**) via silylation of functionalized aliphatic nitro compounds (**70**) has been reported.[110] The use of BSA triggered a 1,3-*N,C*-elimination of "Me₃SiOH" from the initially formed silylnitronate **71** to give highly reactive nitroso alkene **72**. The subsequent 1,5-migration of a proton followed by silylation of oximino group furnished derivative **73** in 90% yield with desilylation eventually affording the targeted enoxime **74** in 85% yield (eq 54). In another application, β-functionalized nitroso alkene **76**, obtained from methyl β-nitropropionate **75** and BSA, can function as a good heterodienophile in Diels-Alder reactions (eq 55).[111] For example, **76** was trapped by cyclic dienes **77** to give adducts **78** with the corresponding stereoselectivity.

Silylation and Activation of *H*-Phosphonates. *H*-Phosphonate monoesters can be converted into trivalent silyl phosphites ROP(OSiMe₃)₂ using BSA.[112] Silyl phosphites ROP(OSiMe₃)₂ are highly reactive to electrophiles or oxidizing reagents.[67] BSA was used to prepare the bis(silylated) analog **80** of monophosphate

79, a transformation which was not affected by trimethylsilyl chloride and triethylamine. The further reaction with silylated α-ketophosohonate **81** generated AZT-5'-HMBP analog **82** after methanolysis (eq 56).[113] Dinucleoside AZT/d4T boranophosphate **84** was synthesized from **83** through the silylation using BSA, boronation with an amine-borane complex, and hydrolysis in ammonium hydroxide and ethanol (eq 57).[114]

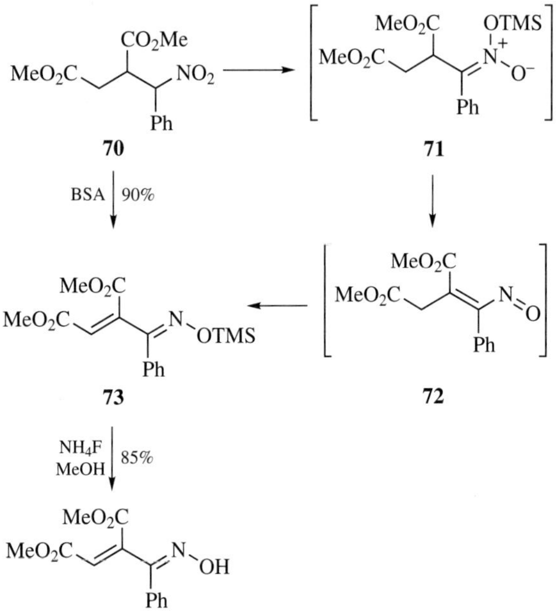

(54)

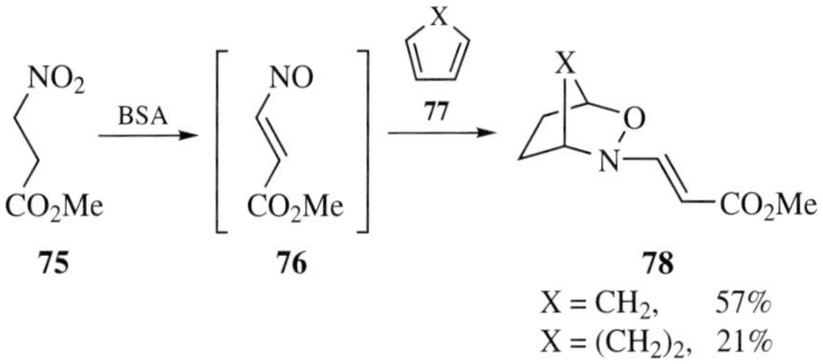

(55)

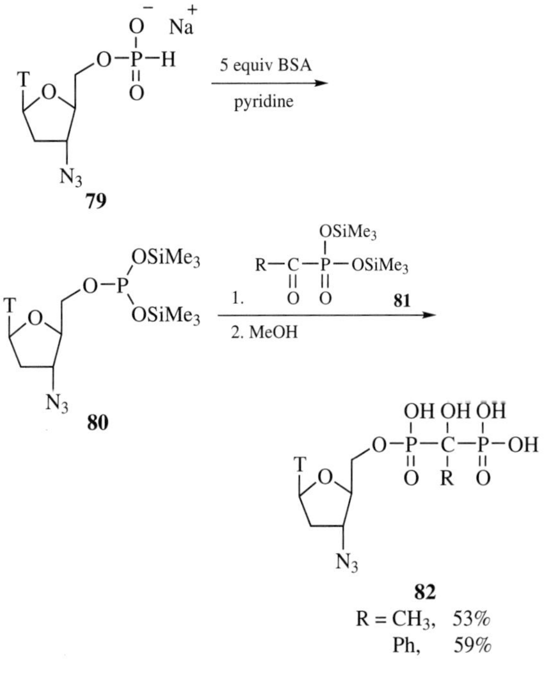

X = CH$_2$,   57%
X = (CH$_2$)$_2$,   21%

(56)

R = CH$_3$,   53%
    Ph,   59%

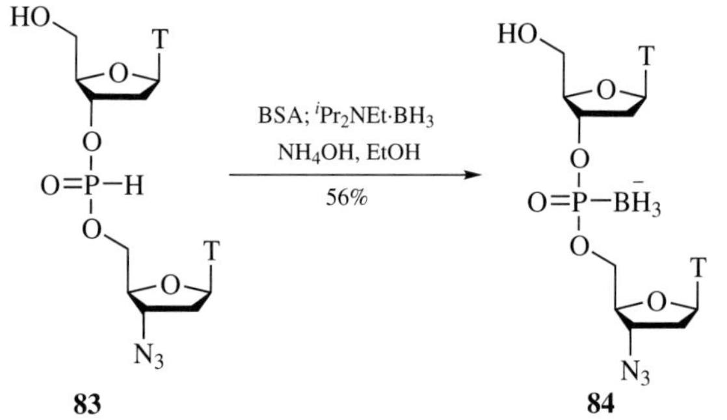

(57)

Oxidation of *H*-phosphonate mono- and diesters was achieved with oxidants and BSA under anhydrous conditions via the corresponding trimethylsilyl phosphites. BSA acts as not only a reagent to convert an *H*-phosphonate to the reactive trimethylsilyl phosphite but it is also a scavenger of all traces of water in this oxidation. 2-(Phenylsulfonyl)-3-(3-nitrophenyl)oxaziridine (PNO) (eq 58),[115] (camphorsulfonyl)oxaziridine (CSO) (eq 59)[116] and bis(trimethylsilyl) peroxide/*cat.* TMSOTf[117] in the presence of BSA were found to be effective for the oxidation of *H*-phosphonate esters.

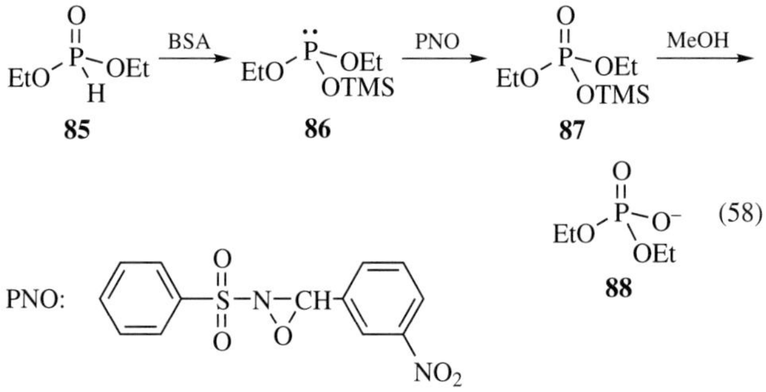

(58)

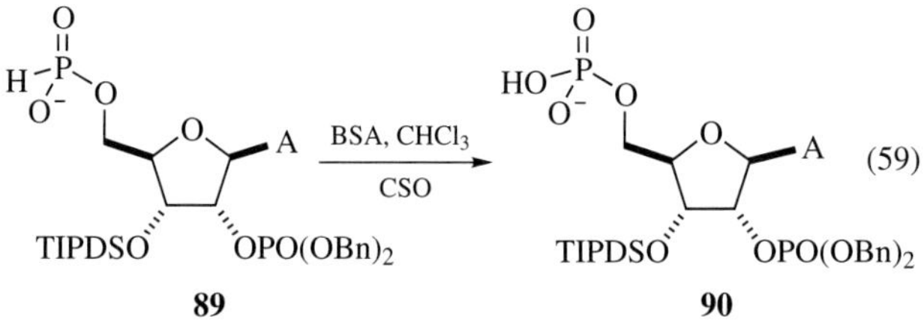

(59)

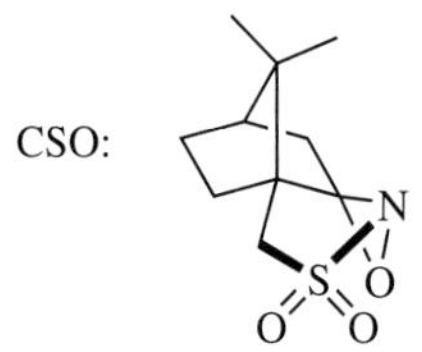

Silylation and Activation of *H*-Phosphinic Acids or Esters. Similar to *H*-Phosphonates, nonnucleophilic *H*-phosphinic acids or esters can be converted into corresponding bis-*O,O*-silylated phosphites RP(OSiMe$_3$)$_2$ using BSA. The resulting nucleophilic P^{III} species RP(OSiMe$_3$)$_2$ are highly reactive to electrophiles and undergo a series of nucleophilic addition reactions.[118] BSA promoted addition of methyl-phosphinic acid benzyl ester **91** to CBZ-protected piperidone and produced the GABA analog **93** after deprotection (eq 60).[119] Phosphinate **96** was prepared through an Arbuzov reaction of *H*-phosphinic acid **94** with benzyl bromide mediated by BSA and subsequent methylation with (trimethyl-silyl)diazomethane (eq 61).[120] BSA-mediated conjugate Michael addition of farnesyl *H*-phosphinate **97** with benzyl acrylate afforded compound **98** (eq 62).[68,119]

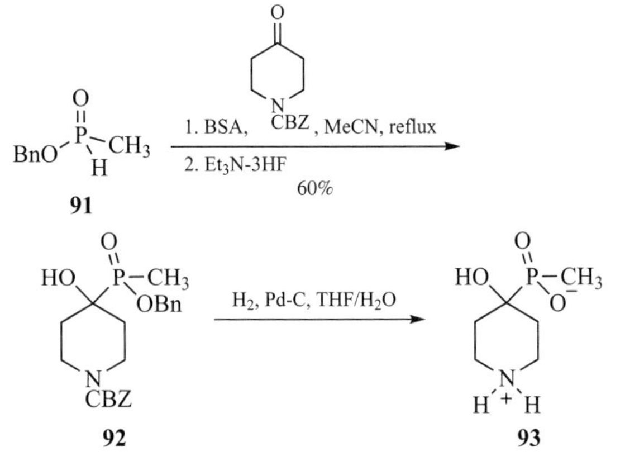

(60)

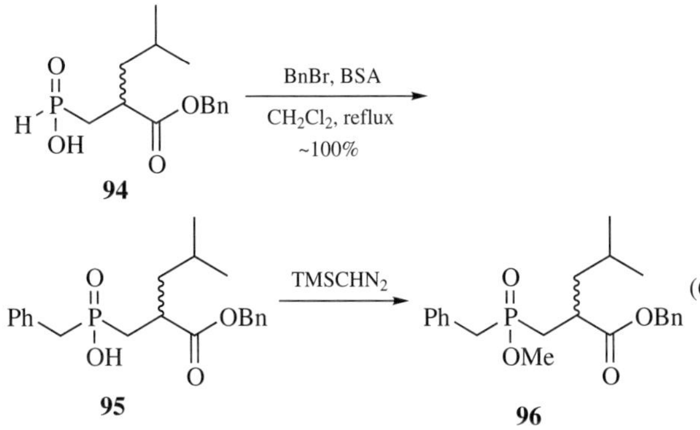

(61)

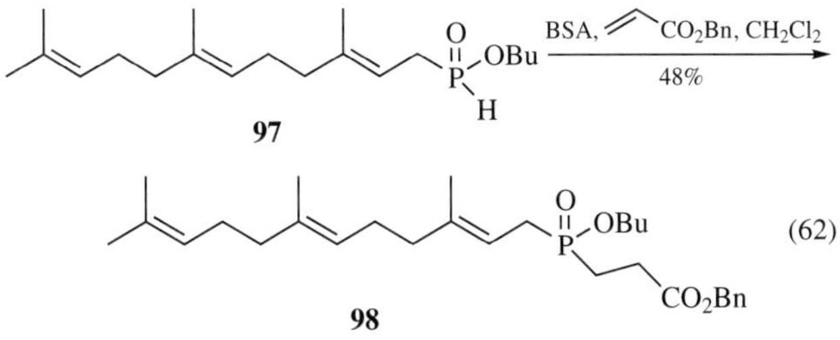

(62)

Deprotection of Phosphates, Phosphonates, and Phosphinates. BSA has been used in removal of alkyl groups from phosphate, phosphonate, and phosphinate esters to generate the corresponding acids. Conversion of phosphinate ester **99** into phosphinic acid **100** was carried out by treatment with BSA and trimethylsilyl iodide (TMSI), which caused simultaneous removal of the carbobenzyloxy group from the terminal proline residue and of the methyl group from the phosphinate ester (eq 63).[121] Same reaction condition can also allow effective removal of ethyl groups from phosphonate diethylesters.[122]

Deprotection of *p*-nitrophenethyl (NPE) phosphonate or phosphate esters is accomplished under basic, nonnucleophilic conditions by treatment with DBU in pyridine with formation of *p*-nitrostyrene. However, this method presents problems for sensitive substrates, since prolonged reaction times are required for deprotection to the dianion. The first elimination step proceeds

more rapidly than the second, since the dianion is much worse as a leaving group than the monoanion **102**. Addition of BSA to the reaction mixture converts the anionic intermediate **102** to the silyl ester **103** and facilitates elimination of the second NPE moiety leading to the bis(trimethylsilyl) ester **104** and then the acid **105** upon hydrolysis (eq 64).[123]

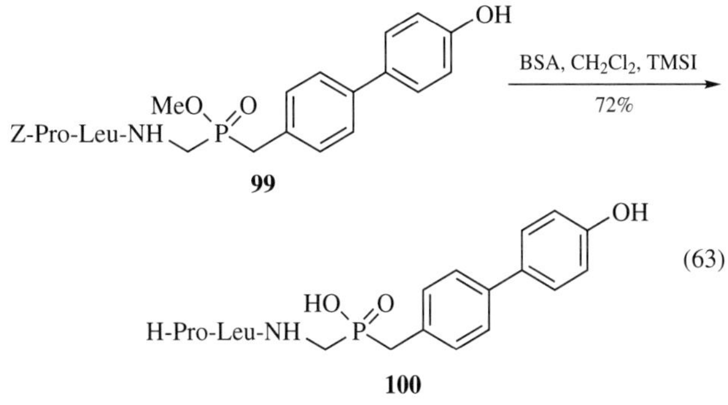

(63)

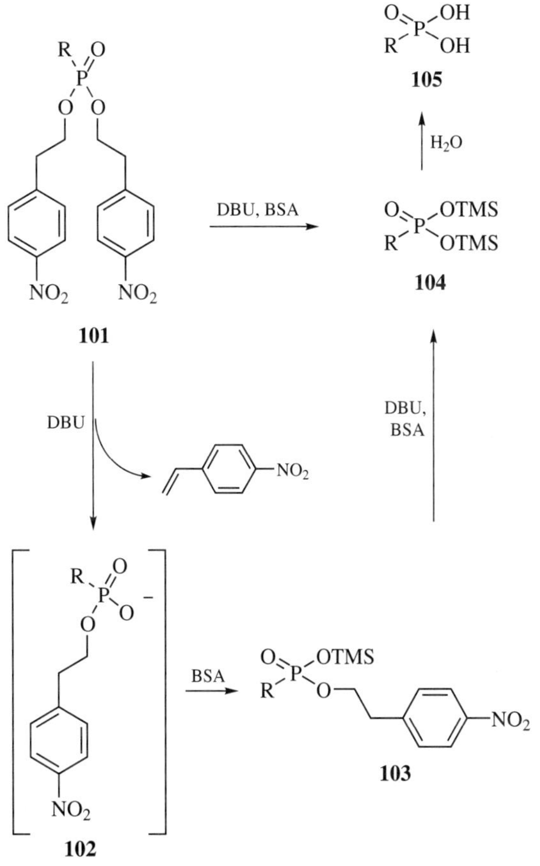

(64)

The combination of base with BSA was also highly effective for the simultaneous and rapid removal of 2-cyanoethyl (CE) protecting group.[124] Alkyl esters of phosphorothioate are extremely stable compared to phosphate esters. Although the commonly used reagents TMSBr or TMSI can remove alkyl esters in phosphates, this reaction fails with phosphorothioates. For this reason, cyanoethyl ester groups are used for the phosphorothioate and phosphoroselenoate syntheses, as these protecting groups can be readily removed under mild basic conditions as demonstrated in eq 65.[125] The CE groups on phosphate **106** were removed prior to deprotection of ethyl esters using *t*-butylamine in the presence of

BSA. The presence of BSA is essential to completely remove all the CE groups, since this reaction is reversible and the reaction is driven to completion by trapping the silyl ethers.

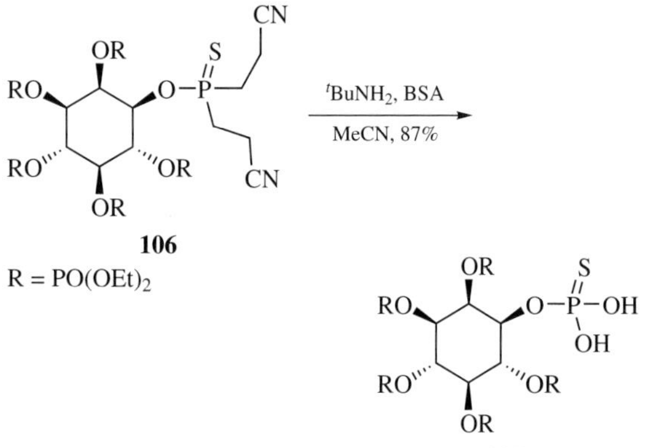

106

R = PO(OEt)$_2$

(65)

107

Silicon-induced Pummerer-type Reaction. As an extension of silicon-induced asymmetric Pummerer-type reaction, chiral, nonracemic sulfoxides were reacted with thiols and BSA in the presence of a catalytic amount of TMSOTf in MeCN at room temperature overnight to give the corresponding chiral, nonracemic *anti*thioacetals with high diastereoselectivity.[126] In the example shown in eq 66, the desired *anti*-**109** was obtained from *syn*-**108** with 11:1 diastereoselctivity over *syn*-**109** in 82% yield. The use of BSA might trigger a 1,3-*S,C*-elimination of "Me$_3$SiOH" from the initially formed silylated-sulfoxide to give the thionium intermediate. The subsequent nucleophilic reaction with thiol through either the cyclic Cram model of the silicon coordinated intermediate or the Cram model of the thionium intermediate generated predominantly the *anti*-isomer.

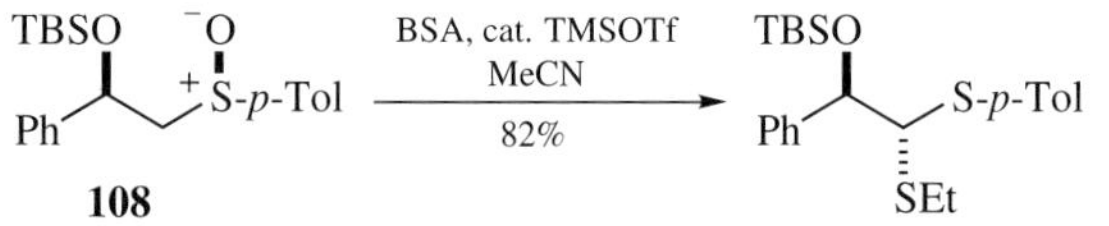

108

+ EtSH

109

dr = 11:1

Preparation of Dts Protecting Group. BSA has been used to introduce the dithiasuccinoyl (Dts) protecting group to amines. The Dts function was developed to meet the requirements of an orthogonally removable N^α-amino protecting group. The Dts group is stable to strong acids and photolysis but is rapidly and specifically removed under mild conditions by thiolysis. Thiocarbamic acid esters **111**, which were prepared from corresponding amines **110**, were suspended in MeCN and treated with BSA to effect trimethylsilylation at the α-carboxyl group and concomitant solubilization. These esters are then treated in situ with (chlorocarbonyl)sulfenyl chloride to achieve conversion to the Dts-protected monomers **112** rapidly in good overall yields (eq 67).[127]

Miscellaneous. The reaction of the 3-keto-4-azasteroid **113** with 2,3-dichloro-5,6-dicyano-1,4-benzoquinone (DDQ) in the presence of BSTFA and catalytic triflic acid in toluene at 25 °C gave the diastereomeric C–C adducts **114**. After thermolysis, a 93% yield of product **115** was observed (eq 68).[128]

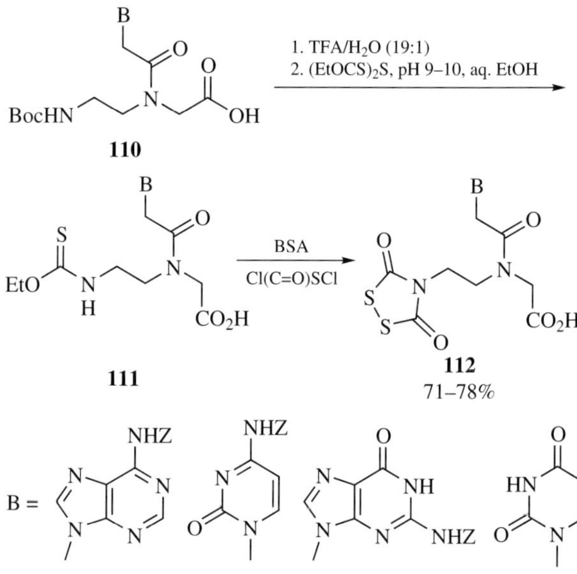

110

111   **112**

71–78%

A(Z)   C(Z)   G(Z)   T

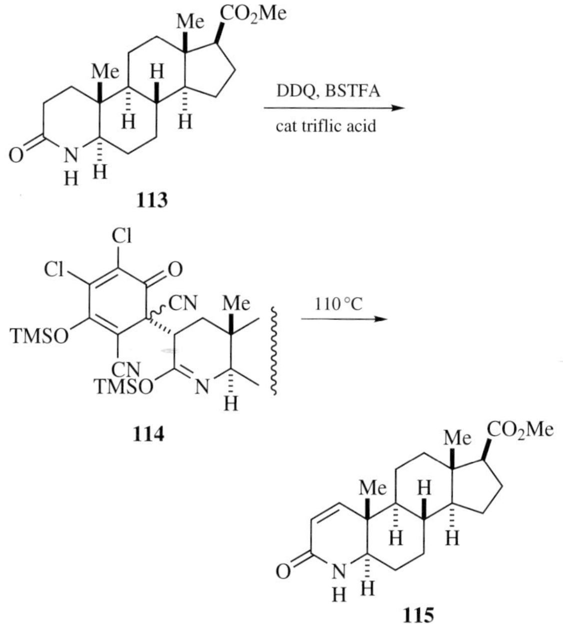

113

114

(68)

115

Related Reagents. Bis(trimethylsilyl)trifluoroacetamide (BSTFA).

1. Birkofer, L.; Ritter, A.; Giessler, W., *Angew. Chem., Int. Ed. Engl.* **1963**, *2*, 96.

2. Klebe, J. F.; Finkbeiner, H.; White, D. M., *J. Am. Chem. Soc.* **1966**, *88*, 3390.

3. (a) Samples, M. S.; Yoder, C. H., *J. Organomet. Chem.* **1987**, *332*, 69. (b) Samples, M. S.; Yoder, C. H.; Schaetter, C. D., *J. Chem. Educ.* **1987**, *64*, 177.

4. Rigaudy, J.; Lytwyn, E.; Wallach, P.; Cuong, N. K., *Tetrahedron Lett.* **1980**, *21*, 3367.

5. Birkofer, L.; Dickopp, H., *Tetrahedron Lett.* **1965**, 4007.

6. Prater, W. A.; Simmons, M. S.; Mancy, K. H., *Anal. Lett.* **1980**, *13*, 205.

7. Kientz, C. E.; Verweij, A., *J. Chromatogr.* **1986**, *355*, 229.

8. Galbraith, M. N.; Horn, D. H. S.; Middleton, E. J.; Hackney, R. J., *J. Chem. Soc., Chem. Commun.* **1968**, 466.

9. Brown, H. C.; Vara Prasad, J. V. N., *J. Am. Chem. Soc.* **1986**, *108*, 2049.

10. Myerson, J.; Haddon, W. F.; Soderstrom, E. L., *Tetrahedron Lett.* **1982**, *23*, 2757.

11. Brooks, C. J. W.; Cole, W. J.; Borthwick, J. H.; Brown, G. M., *J. Chromatogr* **1982**, *239*, 191.

12. Brinker, U. H.; Gomann, K.; Zorn, R., *Angew. Chem., Int. Ed. Engl.* **1983**, *22*, 869.

13. Ohno, M.; Matsuoka, S.; Eguchi, S., *J. Org. Chem.* **1986**, *51*, 4553.

14. Heathcock, C. H.; White, C. T.; Morrison, J. J.; VanDerveer, D., *J. Org. Chem.* **1981**, *46*, 1296.

15. Yalpani, M.; Wilke, G., *Chem. Ber.* **1985**, *118*, 661.

16. Snatzke, G.; Vlahov, *Liebigs Ann. Chem.* **1985**, 439.

17. Böhme, E. H. W.; Bambury, R. E.; Baumann, R. J.; Erickson, R. C.; Harrison, B. L.; Hoffman, P. F.; McCarty, F. J.; Schnettler, R. A.; Vaal, M. J.; Wenstrup, D. L., *J. Med. Chem.* **1980**, *23*, 405.

18. Singh, J.; Wise, D. S.; Townsend, L. B., *Nucleic Acid Chem.* **1991**, 96.

19. Kawasaki, A. M.; Townsend, L. B., *Nucleic Acid Chem.* **1991**, 298.

20. Imazawa, M.; Eckstein, F., *J. Org. Chem.* **1979**, *44*, 2039.

21. Dudycz, L. W.; Wright, G. E., *Nucleosides Nucleotides* **1984**, *3*, 33.

22. Chung, F.-L.; Earl, R. A.; Townsend, L. B., *J. Org. Chem.* **1980**, *45*, 4056.

23. Sasaki, T.; Nakanishi, A.; Ohno, M., *J. Org. Chem.* **1982**, *47*, 3219.

24. Barton, D. H. R.; Crich, D., *J. Chem. Soc., Perkin Trans. 1* **1986**, 1603.

25. (a) Young, S. D.; Buse, C. T.; Heathcock, C. H., *Org. Synth., Coll. Vol.* **1990**, *7*, 381. (b) Bal, B.; Buse, C. T.; Smith, K.; Heathcock, C. H., *Org. Synth., Coll. Vol.* **1990**, *7*, 185.

26. Eaborn, C.; Safa, K. D., *J. Organomet. Chem.* **1982**, *234*, 7.

27. Trost, B. M.; Vladuchick, W. C.; Bridges, A. J., *J. Am. Chem. Soc.* **1980**, *102*, 3548.

28. Itoi, Y.; Inoue, M.; Enomoto, S., *Bull. Chem. Soc. Jpn.* **1985**, *58*, 3193.

29. Dedier, J.; Gerval, P.; Frainnet, E., *J. Organomet. Chem.* **1980**, *185*, 183.

30. Kozikowski, A. P.; Schmiesing, R., *Tetrahedron Lett.* **1978**, 4241.

31. Cameron, D. W.; Conn, C.; Feutrill, G. I., *Aust. J. Chem.* **1981**, *34*, 1945.

32. (a) Cameron, D. W.; Feutrill, G. I.; Perlmutter, P., *Tetrahedron Lett.* **1981**, *22*, 3273. (b) Cameron, D. W.; Feutrill, G. I.; Perlmutter, P., *Aust. J. Chem.* **1982**, *35*, 1469.

33. Braun, D.; Wittig, W., *Makromol. Chem.* **1980**, *181*, 557.

34. Gutsche, C. D.; Bauer, L. J., *J. Am. Chem. Soc.* **1985**, *107*, 6059.

35. Gutsche, C. D.; Pagoria, P. F., *J. Org. Chem.* **1985**, *50*, 5795.

36. Tsaconas, C.; Devissaguet, M.; Padieu, P., *J. Chromatogr.* **1989**, *488*, 249.

37. Rogozhin, S. V.; Davidovich, Yu. A.; Yurtanov, A. I., *Synthesis* **1975**, 113.

38. (a) Zalipsky, S.; Albericio, F.; Slomczynska, V.; Barany, G., *Int. J. Pept. Protein Res.* **1987**, *30*, 740. (b) Taskaeva, Yu. M.; Shvachkin, Yu. P., *Zh. Obshch. Khim.* **1987**, *57*, 961 (*Chem. Abstr.* **1988**, *108*, 6393a).

39. Rabinovich, A. K.; Krysin, E. P., *Khim. Prir. Soedin.* **1988**, 248 (*Chem. Abstr.*, **1989**, *110*, 8671e).

40. Hemmi, K.; Takeno, H.; Okada, S.; Nakaguchi, O.; Kitaura, Y.; Hashimoto, M., *J. Am. Chem. Soc.* **1981**, *103*, 7026.

41. Bey, P.; Vevert, J. P., *J. Org. Chem.* **1980**, *45*, 3249.

42. Reetz, M. T.; Neumeier, G., *Liebigs Ann. Chem.* **1981**, 1234.

43. Weitzberg, M.; Aizenshtat, Z.; Blum, J., *J. Heterocycl. Chem.* **1981**, *18*, 1513.

44. Raucher, S.; Jones, D. S., *Synth. Commun.* **1985**, *15*, 1025.

45. Almond, M. R.; Collins, J. L.; Reitter, B. E.; Rideout, J. L.; Freeman, G. A.; St Clair, M. H., *Tetrahedron Lett.* **1991**, *32*, 5745.

46. Hwang, K.-J., *J. Org. Chem.* **1986**, *51*, 99.

47. Kalinin, A. V.; Khasapov, B. N.; Apasov, E. T.; Kalikhman, I. D.; Ioffe, S. L., *Izv. Akad. Nauk SSSR, Ser. Khim.* **1984**, 694 (*Chem. Abstr.* **1984**, *101*, 91 045m).

48. d'Alarcao, M.; Bakthavachalam, V.; Leonard, N. J., *J. Org. Chem.* **1985**, *50*, 2456.

49. Kelly, J. A.; Abbasi, M. M.; Beisler, J. A., *Anal. Biochem.* **1980**, *103*, 203.

50. Kantlehner, W.; Haug, E.; Speh, P.; Bräuner, H.-J., *Liebigs Ann. Chem.* **1985**, 65.

51. Rigo, B.; Lespagnol, C.; Pauly, M., *Tetrahedron Lett.* **1986**, *27*, 347.

52. Kozyukov, V. P.; Feoktistov, A. E.; Mironov, V. F., *Zh. Obshch. Khim.* **1983**, *53*, 2155 (*Chem. Abstr.*, *100*, 51 682q).

53. Neplyuev, V. M.; Bazarova, I. M.; Lozinskii, M. O.; Lazukina, L. A., *Zh. Org. Khim.* **1984**, *20*, 1451 (*Chem. Abstr.*, **1985**, *102*, 45 570b).

54. Sharma, S. C.; Torssell, K., *Acta Chem. Scand.* **1979**, *B33*, 379.

55. De Koning, J. J.; Kooreman, H. J.; Tan, H. S.; Verweij, J., *J. Org. Chem.* **1975**, *40*, 1346.

56. Blumenkopf, T. A.; Bratz, M.; Castañeda, A.; Look, G. C.; Overman, L. E.; Rodriguez, D.; Thompson, A. S., *J. Am. Chem. Soc.* **1990**, *112*, 4386.

57. Earle, M. J.; Fairhurst, R. A.; Heaney, H., *Tetrahedron Lett.* **1991**, *32*, 6171.

58. (a) Earle, M. J.; Heaney, H., *Synlett* **1992**, 745. (b) El Gihani, M.; Heaney, H., *Synlett* **1993**, 433.

59. Earle, M. J.; Fairhurst, R. A.; Heaney, H.; Papageorgiou, G.; Wilkins, R. F., *Tetrahedron Lett.* **1990**, *31*, 4229.

60. (a) Ballester, M.; Palomo, A. L., *Synthesis* **1983**, 571. (b) Aizpurua, J. M.; Palomo, C.; Palomo, A. L., *Synlett* **1984**, *62*, 336.

61. El Gihani, M.; Heaney, H., *Synlett* **1993**, 583.

62. Evans, D. A.; Wong, R. Y., *J. Org. Chem.* **1977**, *42*, 350.

63. Fischer, A.; Henderson, G. N., *Tetrahedron Lett.* **1983**, *24*, 131.

64. Morwick, T., *Tetrahedron Lett.* **1980**, *21*, 3227.

65. (a) Kibardina, L. K.; Pudovik, M. A., *Zh. Obshch. Khim.* **1986**, *56*, 1906 (*Chem. Abstr.* **1987**, *107*, 154 387w). (b) Ovchinnikov, V. V.; Safina, Yu. G.; Cherkasov, R. A.; Karataeva, F. Kh.; Pudovik, A. N., *Zh. Obshch. Khim.* **1988**, *58*, 2066 (*Chem. Abstr.* **1989**, *111*, 153 914m). (c) Pudovik, M. A.; Kibardina, L. K.; Pudovik, A. N., *Zh. Obshch. Khim.* **1991**, *61*, 1058 (*Chem. Abstr.* **1992**, *116*, 6633u). (d) Al'fonsov, V. A.; Trusenev, A. G.; Batyeva, E. S.; Pudovik, A. N., *Izv. Akad. Nauk SSSR, Ser. Khim.* **1991**, 2103 (*Chem. Abstr.* **1992**, *116*, 21 139f).

66. Ebsworth, E. A. V.; Rankin, D. W. H.; Steger, W.; Wright, J. G., *J. Chem. Soc., Dalton Trans.* **1980**, 1768.

67. De Vroom, E.; Spierenburg, M. L.; Dreef, C. E.; van der Marel, G. A.; van Boom, J. H., *Recl. Trav. Chim. Pays-Bas* **1987**, *106*, 65.

68. Thottathil, J. K.; Ryono, D. E.; Przybyla, C. A.; Moniot, J. L.; Neubeck, R., *Tetrahedron Lett.* **1984**, *25*, 4741.

69. Wasserman, H. H.; Rotello, V. M.; Williams, D. R.; Benbow, J. W., *J. Org. Chem.* **1989**, *54*, 2785.

70. Trost, B. M.; Lautens, M.; Peterson, B., *Tetrahedron Lett.* **1983**, *24*, 4525.

71. Trost, B. M.; Mignani, S., *J. Org. Chem.* **1986**, *51*, 3435.

72. Trost, B. M.; Brandi, A., *J. Org. Chem.* **1984**, *49*, 4811.

73. Gravel, D.; Benoît, S.; Kumanovic, S.; Sivaramakrishnan, H., *Tetrahedron Lett.* **1992**, *33*, 1407.

74. Trost, B. M.; Tour, J. M., *J. Org. Chem.* **1989**, *54*, 484.

75. Marshall, J. A.; Andrews, R. C.; Lebioda, L., *J. Org. Chem.* **1987**, *52*, 2378.

76. (a) Niedballa, U.; Vorbrüggen, H., *J. Org. Chem.* **1974**, *39*, 3654. (b) Vorbrüggen, H.; Krolikiewicz, K.; Bennua, B., *Chem. Ber.* **1981**, *114*, 1234.

77. (a) Schöning, K.-U.; Scholz, P.; Guntha, S.; Wu, X.; Krishnamurthy, R.; Eschenmoser, A., *Science* **2000**, *290*, 1347. (b) Honcharenko, D.; Varghese, O. P.; Plashkevych, O.; Barman, J.; Chattopadhyaya, J., *J. Org. Chem.* **2006**, *71*, 299.

78. (a) Haraguchi, K.; Shiina, N.; Yoshimura, Y.; Shimada, H.; Hashimoto, K.; Tanaka, H., *Org. Lett.* **2004**, *6*, 2645. (b) Dong, S.; Paquette, L. A., *J. Org. Chem.* **2005**, *70*, 1580.

79. Knapp, S.; Thakur, V. V.; Madduru, M. R.; Malolanarasimhan, K.; Morriello, G. J.; Doss, G. A., *Org. Lett.* **2006**, *8*, 1335.

80. Huang, Y.; Dey, S.; Zhang, X.; Sönnichsen, F.; Garner, P., *J. Am. Chem. Soc.* **2004**, *126*, 4626.

81. (a) Mayer, A.; Leumann, C. J., *Nucleosides Nucleotides* **2003**, *22*, 1919. (b) Mayer, A.; Häberli, A.; Leumann, C. J., *Org. Biomol. Chem.* **2005**, *3*, 1653.

82. (a) Togni, A.; Breutel, C.; Schnyder, A.; Spindler, F.; Landert, H.; Tijani, A., *J. Am. Chem. Soc.* **1994**, *116*, 4062. (b) Gilbertson, S. R.; Xie, D., *Angew. Chem., Int. Ed.* **1999**, *38*, 2750. (c) Evans, D. A.; Campos, K. R.; Tedrow, J. S.; Michael, F. E.; Gagné, M. R., *J. Am. Chem. Soc.* **2000**, *122*, 7905. (d) Akiyama, R.; Kobayashi, S., *Angew. Chem., Int. Ed.* **2001**, *40*, 3469. (e) Jansat, S.; Gómez, M.; Philippot, K.; Muller, G.; Guiu, E.; Claver, C.; Castillón, S.; Chaudret, B., *J. Am. Chem. Soc.* **2004**, *126*, 1592. (f) Nemoto, T.; Matsumoto, T.; Masuda, T.; Hitomi, T.; Hatano, K.; Hamada, Y., *J. Am. Chem. Soc.* **2004**, *126*, 3690. (G) Yoshida, M.; Ohsawa, Y.; Ihara, M., *J. Org. Chem.* **2004**, *69*, 1590. (h) Pàmies, O.; Diéguez, M.; Claver, C., *J. Am. Chem. Soc.* **2005**, *127*, 3646.

83. (a) Trost, B. M.; Surivet, J.-P., *J. Am. Chem. Soc.* **2000**, *122*, 6291. (b) Trost, B. M.; Surivet, J.-P., *Angew. Chem., Int. Ed.* **2000**, *39*, 3122.

84. Hu, T.; Corey, E. J., *Org. Lett.* **2002**, *4*, 2441.

85. (a) Jeong, N.; Seo, S. D.; Shin, J. Y., *J. Am. Chem. Soc.* **2000**, *122*, 10220. (b) Son, S. U.; Park, K. H.; Seo, H.; Chung, Y. K.; Lee, S.-G., *Chem. Commun.* **2001**, *23*, 2440. (c) Son, S. U.; Park, K. H.; Chung, Y. K., *J. Am. Chem. Soc.* **2002**, *124*, 6838.

86. Tanabe, Y.; Murakami, M.; Kitaichi, K.; Yoshida, Y., *Tetrahedron Lett.* **1994**, *35*, 8409.

87. Katayama, M.; Nagase, R.; Mitarai, K.; Misaki, T.; Tanabe, Y., *Synlett* **2006**, *1*, 129.

88. (a) Kim, H-S.; Begum, K.; Ogura, N.; Wataya, Y.; Nonami, Y.; Ito, T.; Masuyama, A.; Nojima, M.; McCullough, K. J., *J. Med. Chem.* **2003**, *46*, 1957. (b) Kim, H.-S.; Tsuchiya, K.; Shibata, Y.; Wataya, Y.; Ushigoe, Y.; Masuyama, A.; Nojima, M.; McCullough, K. J., *J. Chem. Soc., Perkin Trans. 1* **1999**, *13*, 1867.

89. Morrill, C.; Grubbs, R. H., *J. Am. Chem. Soc.* **2005**, *127*, 2842.

90. Fukuyama, T.; Li, T.; Peng, G., *Tetrahedron Lett.* **1994**, *35*, 2145.

91. Pereira, O. Z.; Chan, T.-H., *J. Org. Chem.* **1996**, *61*, 5406.

92. Smietana, M.; Mioskowski, C., *Org. Lett.* **2001**, *3*, 1037.

93. El Gihani, M. T.; Heaney, H., *Synthesis* **1998**, 357.

94. Bourgeois, D.; Craig, D.; King, N. P.; Mountford, D. M., *Angew. Chem., Int. Ed.* **2005**, *44*, 618.

95. Craig, D.; Grellepois, F.; White, A. J. P., *J. Org. Chem.* **2005**, *70*, 6827.

96. Craig, D.; Grellepois, F., *Org. Lett.* **2005**, *7*, 463.

97. Mehta, P. G., *Syn. Commun.* **1994**, *24*, 2497.

98. Yasuda, N.; Huffman, M. A.; Ho, G.-J.; Xavier, L. C.; Yang, C.; Emerson, K. M.; Tsay, F.-R.; Li, Y.; Kress, M. H.; Rieger, D. L.; Karady, S.; Sohar, P.; Abramson, N. L.; DeCamp, A. E.; Mathre, D. J.; Douglas, A. W.; Dolling, U.-H.; Grabowski, E. J. J.; Reider, P. J., *J. Org. Chem.* **1998**, *63*, 5438.

99. Saksena, A. K.; Girijavallabhan, V. M.; Wang, H.; Lovey, R. G.; Guenter, F.; Mergelsberg, I.; Puar, M. S., *Tetrahedron Lett.* **2004**, *45*, 8249.

100. Bandini, M.; Melloni, A.; Umani-Ronchi, A., *Org. Lett.* **2004**, *6*, 3199.

101. Yamago, S.; Miyazoe, H.; Nakayama, T.; Miyoshi, M.; Yoshida, J., *Angew. Chem., Int. Ed.* **2003**, *42*, 117.

102. Yamamoto, Y.; Matui, C., *J. Org. Chem.* **1998**, *63*, 377.

103. Salamone, S. G.; Dudley, G. B., *Org. Lett.* **2005**, *7*, 4443.

104. Coats, S. J.; Link, J. S.; Gauthier, D.; Hlasta, D. J., *Org. Lett.* **2005**, *7*, 1469.

105. Baraldi, P. G.; Bovero, A.; Fruttarolo, F.; Romagnoli, R.; Tabrizi, M. A.; Preti, D.; Varani, K.; Borea, P. A.; Moorman, A. R., *Bioorg. Med. Chem.* **2003**, *11*, 4161.

106. (a) Wróbel, Z., *Tetrahedron* **1998**, *54*, 2607. (b) Wróbel, Z., *Eur. J. Org. Chem.* **2000**, *3*, 521.

107. Wróbel, Z., *Synlett* **2004**, *11*, 1929.

108. Wróbel, Z., *Tetrahedron* **2003**, *59*, 101.

109. Ioffe, S. L.; Lyapkalo, I. M.; Tishkov, A. A.; Danilenko, V. M.; Strelenko, Y. A.; Tartakovsky, V. A., *Tetrahedron* **1997**, *53*, 13085.

110. Danilenko, V. M.; Tishkov, A. A.; Ioffe, S. L.; Lyapkalo, I. M.; Strelenko, Y. A.; Tartakovsky, V. A., *Synthesis* **2002**, *5*, 635.

111. Tishkov, A. A.; Lyapkalo, I. M.; Ioffe, S. L.; Strelenko, Y. A.; Tartakovsky, V. A., *Org. Lett.* **2000**, *2*, 1323.

112. Wada, T.; Mochizuki, A.; Sato, Y.; Sekine, M., *Tetrahedron Lett.* **1998**, *39*, 7123.

113. Migianu, E.; Monteil, M.; Even, P.; Lecouvey, M., *Nucleosides Nucleotides* **2005**, *24*, 121.

114. Lin, C.; Fu, H.; Tu, G.; Zhao, Y., *Synthesis* **2004**, *4*, 509.

115. Wada, T.; Sato, Y.; Honda, F.; Kawahara, S.; Sekine, M., *J. Am. Chem. Soc.* **1997**, *119*, 12710.

116. Dowden, J.; Moreau, C.; Brown, R. S.; Berridge, G.; Galione, A.; Potter, B. V. L., *Angew. Chem., Int. Ed.* **2004**, *43*, 4637.

117. Kato, T.; Hayakawa, Y., *Synlett* **1999**, *11*, 1796.

118. (a) Gautier, A.; Garipova, G.; Salcedo, C.; Balieu, S.; Piettre, S. R., *Angew. Chem., Int. Ed.* **2004**, *43*, 5963. (b) Borloo, M.; Jiao, X.-Y.; Wójtowicz, H.; Rajan, P.; Verbruggen, C.; Augustyns, K.; Hacmcrs, A., *Synthesis* **1995**, *9*, 1074. (c) Bartley, D. M.; Coward, J. K., *J. Org. Chem.* **2005**, *70*, 6757. (d) Reiter, L. A.; Mitchell, P. G.; Martinelli, G. J.; Lopresti-Morrow, L. L.; Yocum, S. A.; Eskra, J. D., *Bioorg. Med. Chem. Lett.* **2003**, *13*, 2331. (e) Raguin, O.; Fournié-Zaluski, M.-C.; Romieu, A.; Pèlegrin, A.; Chatelet, F.; Pélaprat, D.; Barbet, J.; Roques, B. P.; Gruaz-Guyon, A., *Angew. Chem., Int. Ed.* **2005**, *44*, 4058. (f) Ribière, P.; Bravo-Altamirano, K.; Antczak, M. I.; Hawkins, J. D.; Montcham, J.-L., *J. Org. Chem.* **2005**, *70*, 4064.

119. Abrunhosa-Thomas, I.; Ribière, P.; Adcock, A. C.; Montcham, J.-L., *Synthesis* **2006**, *2*, 325.

120. Reiter, L. A.; Jones, B. P., *J. Org. Chem.* **1997**, *62*, 2808.

121. Bianchini, G.; Aschi, M.; Cavicchio, G.; Crucianelli, M.; Preziuso, S.; Gallina, C.; Nastari, A.; Gavuzzo, E.; Mazza, F., *Bioorg. Med. Chem.* **2005**, *13*, 4740.

122. (a) Agamennone, M.; Campestre, C.; Preziuso, S.; Consalvi, V.; Crucianelli, M.; Mazza, F.; Politi, V.; Ragno, R.; Tortorella, P.; Gallina, C., *Eur. J. Med. Chem.* **2005**, *40*, 271. (b) Solas, D.; Hale, R. L.; Patel, D. V., *J. Org. Chem.* **1996**, *61*, 1537.

123. (a) Alberg, D. G.; Lauhon, C. T.; Nyfeler, R.; Fässler, A.; Bartlett, P. A., *J. Am. Chem. Soc.* **1992**, *114*, 3535. (b) An, M.; Maitra, U.; Neidlein, U.; Bartlett, P. A., *J. Am. Chem. Soc.* **2003**, *125*, 12759.

124. (a) Sekine, M.; Tsuruoka, H.; Iimura, S.; Wada, T., *Nat. Prod. Lett.* **1994**, *5*, 41. (b) Sekine, M.; Aoyagi, M.; Ushioda, M.; Ohkubo, A.; Seio, K., *J. Org. Chem.* **2005**, *70*, 8400.

125. Xu, Y.; Liu, X.; Prestwich, G. D., *Tetrahedron Lett.* **2005**, *46*, 8311.

126. Shibata, N.; Fujita, S.; Gyoten, M.; Matsumoto, K.; Kita, Y., *Tetrahedron Lett.* **1995**, *36*, 109.

127. Planas, M.; Bardají, E.; Jensen, K. J.; Barany, G., *J. Org. Chem.* **1999**, *64*, 7281.

128. (a) Bhattacharya, A.; DiMichele, L. M.; Dolling, U.-H.; Douglas, A. W.; Grabowski, E. J. J., *J. Am. Chem. Soc.* **1988**, *110*, 3318. (b) Williams, J. M.; Marchesini, G.; Reamer, R. A.; Dolling, U.-H.; Grabowski, E. J. J., *J. Org. Chem.* **1995**, *60*, 5337.

(1*E*,3*E*)-1,4-Bis(trimethylsilyl)-1,3-butadiene

Me₃Si⌢⌢SiMe₃

[22430-47-3] $C_{10}H_{22}Si_2$ (MW 198.45)

InChI = 1S/C10H22Si2/c1-11(2,3)9-7-8-10-12(4,5)6/h7-10H,
  1-6H3/b9-7+,10-8+

InChIKey = KTDRTOYQXRJJTE-FIFLTTCUSA-N

(building block for the stereodefined construction of conjugated polyenyl chains[1-4])

Physical Data: bp 46–49 °C/15 mmHg; n_D^{20} 1.4679.
Analysis of Reagent Purity: GC/MS, IR, ^{1}H and ^{13}C NMR.
Preparative Methods: coupling of the Grignard reagent derived from (*E*)-2-(trimethylsilyl)-1-bromoethylene with the same bromo derivative leads to the silyldiene in 51% yield;[5] a higher yield (62%) is obtained by reaction of bis(trimethylsilyl)-methyllithium with (*E*)-3-(trimethylsilyl)-2-propenal.[6] Other procedures involve the Pd⁰-catalyzed demercuration of bis[(*E*)-(2-trimethylsilylvinyl)]mercury, prepared from (*E*)-2-(trimethylsilyl)-1-lithioethylene and HgCl₂ (77% yield),[7] or the reaction of (*E*)-1,4-bis(trimethylsilyl)-2-butene with *n*-BuLi and TMEDA and subsequent oxidation with HgCl₂ of the isolated bis-lithiated species, stabilized by complexation with TMEDA (57% yield).[8]
Handling, Storage, and Precautions: conveniently stored at −15 °C for several months, without appreciable isomerization.

Electrophilic Substitutions.   A highly selective substitution of a trimethylsilyl group is readily performed with acyl chlorides in the presence of aluminum chloride to obtain silylated ketones with high retention of configuration (>98%). Dicarbonyl compounds are prepared with a one-pot procedure, by simply adding the second acyl chloride after completion of the first step, without isolation of the monosubstitution product (eq 1).[1]

$$(1)$$

The versatility of this approach is illustrated by the synthesis of polyunsaturated fatty acids of special interest. Ostopanic acid, a cytotoxic fatty acid,[9] is directly obtained by acylation reactions with the appropriate acyl chlorides,[2] whereas β-parinaric acid methyl ester, an interesting fluorescent probe for biological membranes,[10] is prepared by a selective double acylation, followed by reduction and dehydration reactions (eq 2).[3]

Ostopanic acid

Parinaric acid methyl ester

$$(2)$$

Moreover, unsaturated keto silanes and diketones obtained according to eq 1 can be easily reduced to saturated keto silanes or 1,6-dicarbonyl compounds;[11] unsaturated keto silanes can be transformed into enantiomerically enriched hydroxy derivatives by reduction and enzymatic kinetic resolution.[12] Monosilylated unsaturated sulfides, precursors of a variety of dienylsilanes,[13] can be also obtained by chemoselective electrophilic substitutions.[4]

Halogen Addition.   Bromine addition leads to a tetrabromo derivative which can be bromodesilylated to (1*Z*,3*Z*)-1,4-dibromobutadiene (eq 3).[14]

$$(3)$$

Cycloaddition Reactions.   Diels–Alder reactions with maleic anhydride (eq 4)[5,6] lead to cyclic allylsilanes, which can undergo further interesting elaborations.[6]

$$(4)$$

Related Reagents.   (1*E*,3*E*,5*E*)-1,6-Bis(trimethylsilyl)-1,3,5-hexatriene; 2,3-Bis(trimethylstannyl)-1,3-butadiene; (*E*)-1-Trimethylsilyl-1,3-butadiene; 2-Trimethylsilylmethyl-1,3-butadiene.

---

1. Babudri, F.; Fiandanese, V.; Marchese, G.; Naso, F., *J. Chem. Soc., Chem. Commun.* **1991**, 237.

2. Babudri, F.; Fiandanese, V.; Naso, F., *J. Org. Chem.* **1991**, *56*, 6245.

3. Babudri, F.; Fiandanese, V.; Naso, F.; Punzi, A., *Synlett* **1992**, 221.

4. Fiandanese, V.; Mazzone, L., *Tetrahedron Lett.* **1992**, *33*, 7067.

5. Bock, H.; Seidl, H., *J. Am. Chem. Soc.* **1968**, *90*, 5694.

6. Carter, J. M.; Fleming, I.; Percival, A., *J. Chem. Soc., Perkin Trans. 1* **1981**, 2415.

7. Seyferth, D.; Vick, S. C., *J. Organomet. Chem.* **1978**, *144*, 1.

8. Field, D. L.; Gardiner, M. G.; Kennard, C. H. L.; Messerle, B. A.; Raston, C. L., *Organometallics* **1991**, *10*, 3167.

9. Hamburger, M.; Handa, S. S.; Cordell, G. A.; Kinghorn, A. D.; Farnsworth, N. R., *J. Nat. Prod.* **1987**, *50*, 281.

10. Sklar, L. A.; Miljanich, G. P.; Bursten, S. L.; Dratz, E. A., *J. Biol. Chem.* **1979**, *254*, 9583.

11. Fiandanese, V.; Punzi, A.; Ravasio, N., *J. Organomet. Chem.* **1993**, *447*, 311.

12. Fiandanese, V.; Hassan, O.; Naso, F.; Scilimati, A., *Synlett* **1993**, 491.

13. Fiandanese, V.; Marchese, G.; Mascolo, G.; Naso, F.; Ronzini, L., *Tetrahedron Lett.* **1988**, *29*, 3705.

14. Ferede, R.; Noble, M.; Cordes, A. W.; Allison, N. T.; Lay, J. Jr., *J. Organomet. Chem.* **1988**, *339*, 1.

Francesco Naso, Vito A. Fiandanese & Francesco Babudri
CNR Centre MISO, University of Bari, Italy

Bis(β-trimethylsilylethanesulfonyl)imide

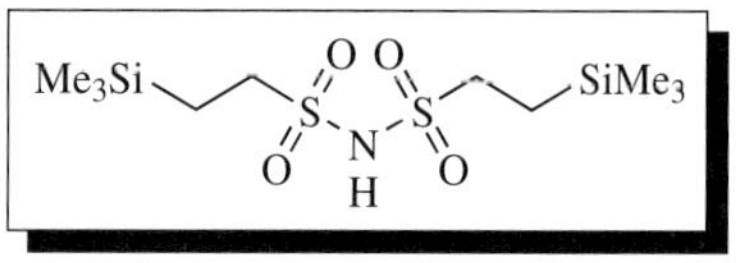

[548462-13-1] $C_{10}H_{27}NO_4S_2Si_2$ (MW 345.64)

InChI = 1S/C10H27NO4S2Si2/c1-18(2,3)9-7-16(12,13)11-17(14,15)8-10-19(4,5)6/h11H,7-10H2,1-6H3

InChIKey = HXORNOSAFNQXMS-UHFFFAOYSA-N

(a useful reagent for the preparation of protected amines)

Physical Data: mp 144–145 °C.

Solubility: soluble in most organic solvents.

Form Supplied in: off-white solid; not commercially available.

Handling, Storage, and Precaution: stable to air and atmospheric moisture. Toxicological properties unknown; assume to be toxic.

Preparation. Compound **2** is easily obtained via a two-step sequence starting from the reaction of commercially available methanesulfonyl chloride and ammonium chloride in the presence of sodium hydroxide (eq 1).[1] The product bis(methanesulfonimide) (**1**) is then treated with 3 equiv of LiHMDS at −78 °C to give a trianion. Subsequent alkylation with 2 equiv of commercially available (iodomethyl)trimethylsilane and warming to room temperature affords the desired sulfonimide **2** in good yield (eq 1).

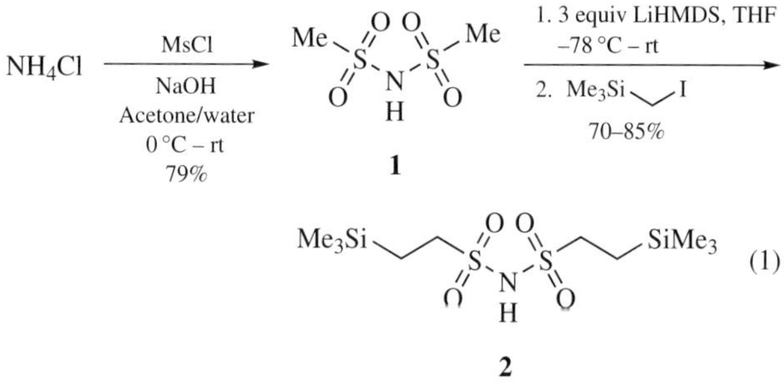

A Useful Reagent for the Synthesis of Protected Amines.[1]
The title compound is sufficiently acidic to act as the nucleophilic partner in Mitsunobu reactions with both primary and secondary alcohols.[1] Secondary alcohols require a higher temperature (80 °C) than primary alcohols (65 °C) in the coupling

process. Treatment of a broad range of substrates with **3** under the appropriate conditions leads to primary amines fully protected as their bis(SES) derivatives **4** (eq 2). These compounds are typically stable to Grignard reagents (e.g., 3 equiv of BnMgCl in THF overnight) and to strong aqueous acids and bases at room temperature (5% aq. NaOH or 1N HCl overnight). However, partial cleavage of one of the SES groups can occur in hot basic alcoholic solution over several hours.

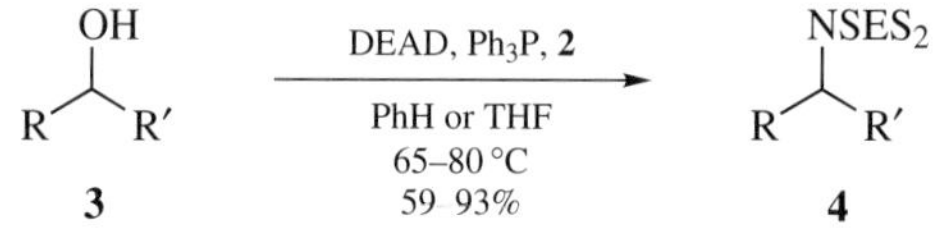

R = alkyl, allyl, or benzyl
R' = H or alkyl

Deprotection Chemistry. Removal of one of the SES groups is readily achieved by heating imide derivatives **4** with CsF in acetonitrile (eq 3) to give monoprotected amines **5**. As yet, a one-pot procedure for direct production of the corresponding primary amine by simultaneous removal of both SES groups has not been developed.

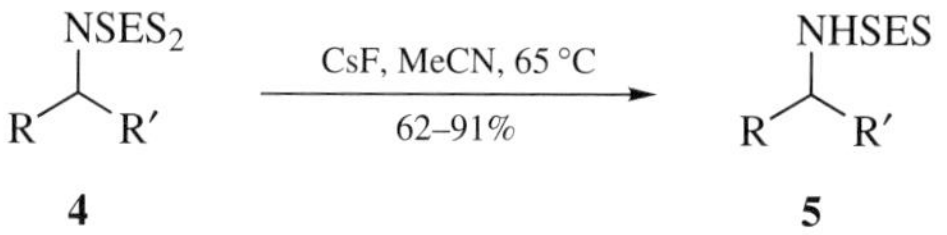

A useful extension of this chemistry is that compounds **4** can be mono-deprotected and alkylated *in a single pot* (eq 4). Treatment of **4** with CsF in hot acetonitrile in the presence of an alkylating agent leads to SES-protected secondary amines **6** in high yields.

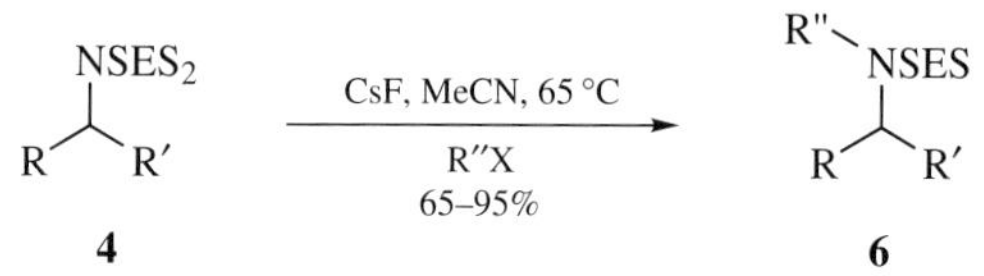

R'' = Bn, Bu, or allyl

Deprotection of **6** can be achieved under the standard conditions for removal of the SES protecting group to give free amines **7** (eq 5).[2]

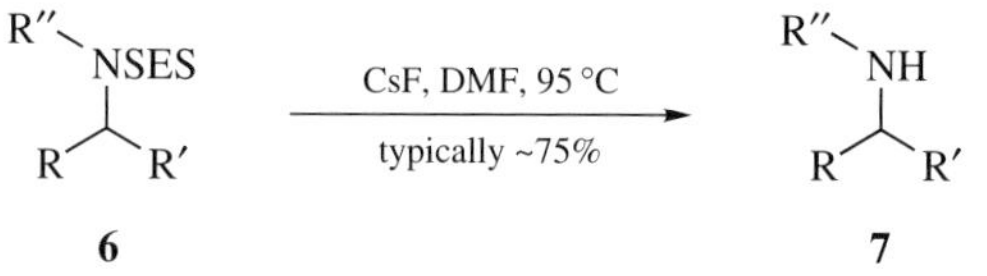

1. Dastrup, D. M.; VanBrunt, M. P.; Weinreb S. M., *J. Org. Chem.* **2003**, *68*, 4112.

2. Weinreb, S. M.; Demko, D. M.; Lessen, T. A.; Demers, J. P., *Tetrahedron Lett.* **1986**, *27*, 2099.

Steven M. Weinreb & Magnus W. P. Bebbington
The Pennsylvania State University, University Park, PA, USA

N,O-Bis(trimethylsilyl)hydroxylamine

[22737-37-7] $C_6H_{19}NOSi_2$ (MW 177.39)

InChI = 1S/C6H19NOSi2/c1-9(2,3)7-8-10(4,5)6/h7H,1-6H3

InChIKey = ZAEUMMRLGAMWKE-UHFFFAOYSA-N

(protected hydroxylamine synthon for the formation of iso-cyanates,[6] hydroxamic acids,[6] *N*-phosphinoylhydroxylamines,[12] and oximes[13])

Physical Data: bp 137–139 °C, 78–80 °C/100 mmHg; *d* 0.830 g cm^{-3}; flash point 28 °C.

Solubility: sol THF, ether, pentane, CH_2Cl_2.

Form Supplied in: liquid; widely available, but expensive.

Analysis of Reagent Purity: ^{1}H NMR (benzene-d_6) δ 0.16 (s, 9, NSiMe$_3$), 0.25 (s, 9, OSiMe$_3$), 4.6 (br s, 1, NH).[1–3]

Preparative Methods: can be prepared in 69% yield by treat-ing dry hydroxylamine with 2 equiv chlorotrimethylsilane and 2 equiv triethylamine.[3] A safer preparation, which avoids the use of the explosive solid, hydroxylamine, uses the reaction of hydroxylamine hydrochloride with excess hexamethyldisi-lazane (71–75% yield).[12] Alternatively, neutralization of hy-droxylamine hydrochloride with ethylenediamine followed by addition of chlorotrimethylsilane can be used.[4]

Handling, Storage, and Precautions: flammable, corrosive liq-uid; hydrolyzed by moist air. Avoid inhalation and prevent con-tact with skin and eyes. The reagent should be stored under argon at 0–4 °C. Use in a fume hood.

Nucleophilic Reactions. *N,O*-Bis(trimethylsilyl)hydroxyl-amine is a protected, lipophilic form of hydroxylamine. It reacts with a variety of electrophiles predominantly by attack on the ni-trogen nucleophilic center. Reaction with acid chlorides (1 equiv) in the presence of triethylamine gives *N,O*-bis(trimethylsilyl)-hydroxamic acids by *N*-acylation.[6] A related reagent, tris(tri-methylsilyl)hydroxylamine, gives the same product in high yields, also by *N*-acylation.[5,6] Hydrolysis gives the free hydroxamic acids, whereas thermal fragmentation affords isocyanates (eq 1).[6]

Excess acetyl chloride (and presumably other acyl chlorides) reacts with *N,O*-bis(trimethylsilyl)hydroxylamine to give *O*-acetyl acetohydroxamate in 88% yield after hydrolytic workup.[7] Isocyanates react exothermically with *N,O*-bis(trimethylsilyl)-hydroxylamine to give *N,O*-bis(trimethylsilyl)-*N*-hydroxyureas

in high yields.[8] Carboxylation of *N,O*-bis(trimethylsilyl)hydro-xylamine with carbon dioxide in THF followed by silyla-tion gives a tris(trimethylsilyl) derivative of *N*-hydroxycarbamic acid.[9] This intermediate rearranges in high yield to afford an *N*-trimethylsilyloxy isocyanate (eq 2).[10]

Reaction of *N,O*-bis(trimethylsilyl)hydroxylamine with diketene affords the explosive *N,O*-bis(trimethylsilyl) derivative of acetoacetylhydroxamic acid. Deprotection and cyclization gives an isoxazole (eq 3) whose dianion serves as a useful β-keto amide synthon.[11]

Phosphorus analogs of hydroxamic acids, prepared from phos-phinic chlorides and *N,O*-bis(trimethylsilyl)hydroxylamine, tend to be unstable. The stable *O*-sulfonyl derivatives of the phos-phinoylhydroxylamines undergo base-induced, Lossen-type rear-rangement to phosphonamidates by aryl group migration from P to N (eq 4).[12]

Deprotonation of *N,O*-bis(trimethylsilyl)hydroxylamine with *n*-butyllithium or potassium hydride at low temperature yields the nitrogen centered anion *N,O*-bis(trimethylsilyl)hydroxyl-amide.[6,13] At higher temperatures the oxyanion *N,N*-bis(tri-methylsilyl)hydroxylamide is formed by rearrangement.[1,14] Each reacts with acyl chlorides chemospecifically by *N*-acylation and *O*-acylation, respectively. The oxyanion also reacts with silyl halides,[1a] methyl iodide,[1] and sulfonyl chlorides[14] chemospecifi-cally on oxygen. In a Peterson-type one-pot reaction, oximes and oxime derivatives can be prepared effectively from the N-anion of *N,O*-bis(trimethylsilyl)hydroxylamine and an aldehyde or ketone (eq 5).[13] Oximes of sterically hindered ketones can be formed in high yields by this procedure.

---

1. (a) West, R.; Boudjouk, P., *J. Am. Chem. Soc.* **1973**, *95*, 3987. (b) West, R.; Boudjouk, P.; Matuszko, A., *J. Am. Chem. Soc.* **1969**, *91*, 5184.

2. Chang, Y. H.; Chiu, F.-T.; Zon, G., *J. Org. Chem.* **1981**, *46*, 342.

3. Wannagat, U.; Smrekar, O., *Monatsh. Chem.* **1969**, *100*, 750 (*Chem. Abstr.* **1969**, *71*, 22 148x).

4. Bottaro, J. C.; Bedford, C. D.; Dodge, A., *Synth. Commun.* **1985**, *15*, 1333.

5. Ando, W.; Tsumake, H., *Synth. Commun.* **1983**, *13*, 1053.

6. King, F. D.; Pike, S.; Walton, D. R. M., *J. Chem. Soc., Chem. Commun.* **1978**, 351.

7. Kozyukov, V. P.; Feoktistov, A. E.; Mironov, V. F., *J. Gen. Chem. USSR (Engl. Transl.)* **1988**, *58*, 1154.

8. Muzovskaya, E. V.; Kozyukov, Vik. P.; Mironov, V. F.; Kozyukov, V. P., *J. Gen. Chem. USSR (Engl. Transl.)* **1989**, *59*, 349.

9. Mironov, V. F.; Sheludyakov, V. D.; Kirilin, A. D., *J. Gen. Chem. USSR (Engl. Transl.)* **1979**, *49*, 817.

10. Sheludyakov, V. D.; Gusev, A. I.; Dmitrieva, A. B.; Los', M. G.; Kirilin, A. D., *J. Gen. Chem. USSR (Engl. Transl.)* **1983**, *53*, 2051.

11. Oster, T. A.; Harris, T. M., *J. Org. Chem.* **1983**, *48*, 4307.

12. (a) Harger, M. J. P.; Shimmin, P. A., *Tetrahedron Lett.* **1991**, *32*, 4769. (b) Fawcett, J.; Harger, M. J. P.; Sreedharan-Menon, R., *J. Chem. Soc., Chem. Commun.* **1992**, 227. (c) Harger, M. J. P.; Shimmin, P. A., *Tetrahedron* **1992**, *48*, 7539.

13. Hoffman, R. V.; Buntain, G. A., *Synthesis* **1987**, 831.

14. King, F. D.; Walton, D. R. M., *Synthesis* **1975**, 788.

M. Catherine Johnson & Robert V. Hoffman
New Mexico State University, Las Cruces, NM, USA

Bis[(trimethylsilyl)methyl]zinc

[41924-26-9] $C_8H_{22}Si_2Zn$ (MW 239.82)

InChI = 1S/2C4H11Si.Zn/c2*1-5(2,3)4;/h2*1H2,2-4H3;

InChIKey = AHQMHEHOLXXTEQ-UHFFFAOYSA-N

(useful organometallic reagent for organic synthesis)

Alternate Name: (TMSM)$_2$Zn.

Physical Data: colorless liquid; mp $-80\,^\circ$C; bp $44\,^\circ$C/1.5 Torr; *d* 1.08 g cm^{-3}.

Solubility: soluble in benzene, dichloromethane, ether, heptane, THF, and toluene.

Preparative Methods: several methods for the preparation of bis[(trimethylsilyl)methyl]zinc have been reported.[1–3] Addition of powdered anhydrous zinc(II) chloride to a solution of [(trimethylsilyl)methyl]magnesium chloride in diethyl ether at $0\,^\circ$C and subsequent stirring for 4 days at room temperature affords the title compound in 90% yield after fractional distillation. A second route consists of the reaction of (iodomethyl)trimethylsilane with zinc/copper couple[1,2] (prepared by reduction of CuO with dihydrogen in the presence of zinc dust) at reflux for 3 h and affords the title compound in 56% yield. Bis[(trimethylsilyl)methyl]zinc can also be prepared quantitatively by reacting bis[(trimethylsilyl)methyl]mercury with an excess of zinc at ambient conditions.[3]

Purity: by distillation at reduced pressure under an inert atmosphere.

Handling, Storage, and Precaution: bis[(trimethylsilyl)methyl]-zinc is spontaneously flammable in air; it also reacts violently with water and should be kept under an inert atmosphere.

Cross-coupling and Nucleophilic Substitution Reactions. Bis[(trimethylsilyl)methyl]zinc has proven to be a useful nucleophile in transition metal-catalyzed cross-coupling reactions with, for example, alkenyl sulfoximines (eq 1).[4,5]

Negishi-type cross-coupling of the title compound with alkenyl bromides, to form allyl silanes, can be achieved in the presence of a palladium catalyst (eq 2).[6]

The reagent also reacts with chiral α-ferrocenyl acetates and diacetates in the presence of Lewis acids to deliver the acetate displacement products in a process that involves retention of configuration and, therefore, delivers the α-trimethylsilylmethylated products in high enantiomeric excess.[7] (TMSM)$_2$Zn reacts with allyl chlorides in the presence of copper(I) catalysts and a chiral ferrocene ligand to give S_N2' products preferentially (eq 3).[8]

Analogous but less efficient transformations are observed with allyl bromides.[9] In a related vein, (TMSM)$_2$Zn reacts in a highly *exo*-selective manner with benzo-7-oxanorbornadiene, in the presence of a palladium catalyst, to give novel dihydronaphthalene derivatives (eq 4).[10]

$$R_1 \diagdown \diagup Cl + Zn(CH_2SiMe_3)_2 \xrightarrow[\substack{THF, -50\,°C \text{ to } -90\,°C \\ 18\,h}]{\substack{\text{(ferrocene ligand) (10 mol \%)} \\ CuBr\cdot Me_2S \text{ (1 mol \%)}}}$$

R_1 = Aryl, c-C_6H_{11},
3-thienyl, (Z)-TIPSOCH$_2$

$$\text{(3)}$$

52% yield (67% ee)
S_N2':S_N2 = 94:6

$$\xrightarrow[\substack{CH_2Cl_2, \text{ rt} \\ 67\%}]{\substack{Zn(CH_2SiMe_3)_2 \text{ (1.5 equiv)} \\ Pd(dppf)Cl_2 \text{ (5 mol \%)}}}$$

$$\text{(4)}$$

Addition to Carbonyl Compounds. (TMSM)$_2$Zn can be used for the preparation of a wide range of mixed organozinc compounds, in which the (trimethylsilyl)methyl (TMSM) moiety acts as a non-transferable group.[11] As a result only a slight excess of the mixed zinc reagent is required in the reaction with aldehydes, a process which affords secondary alcohols in good yields (eq 5).[11,12]

$$(FG\text{-}R)_2Zn \xrightarrow{Zn(CH_2SiMe_3)_2} FG\text{-}R\text{--}Zn(CH_2SiMe_3)$$

$$\xrightarrow[\substack{R'CHO \\ Ti(Oi\text{-}Pr)_4 \text{ (0.6 equiv)}}]{}$$

$$\text{(5)}$$

74–98%
(86–98% ee)

FG-R = Cl(CH$_2$)$_4$, PivO(CH$_2$)$_{3 \text{ or } 4}$, AcO(CH$_2$)$_5$....
R' = Ph, Et....

By using suitably *N*-protected α-, β-, and γ-aminoaldehydes, this protocol allows for the enantioselective synthesis of 1,2-, 1,3-, and 1,4-amino alcohols, respectively.[13]

The same mixed organozinc species participate in Michael-type reactions with a wide range of α,β-unsaturated ketones and aldehydes (eq 6).[11,14] Transition metal catalysts are not required for an effective process.

FG-R-ZnCH$_2$SiMe$_3$ +

$$\xrightarrow[\substack{THF, NMP \\ -30\,°C \text{ to rt}, 12\,h}]{Me_3SiBr}$$

$$\text{(6)}$$

FG-R = alkyl, aryl
X = alkyl, H, OR

Preparation of Transition Metal Complexes. Bis[(trimethylsilyl)methyl]zinc has found wide application as an alkylating reagent in reactions with transition metal halides. Its reaction with TaCl$_5$ allows for the isolation of (Me$_3$SiCH$_2$)$_x$Ta$_{5-x}$ (x = 1,2,3), which is in contrast to the reaction of TaCl$_5$ with Me$_3$SiCH$_2$M (where M = Li or MgX).[2] By reacting the title compound with imido tungsten chlorides, the corresponding alkyl complexes can be isolated, some of which can be converted into tungsten alkylidene complexes [e.g., W(NPh)(CHSiMe$_3$)(CH$_2$SiMe$_3$)$_2$].[15–17] Alkyne complexes of tungsten chlorides react with the title compound to afford W(C$_2$R$_2$)(CH$_2$SiMe$_3$)$_2$Cl in 60–80% yield.[18] Molybdenum alkyl complexes are accessible by related means.[19] Reaction of an osmium glycolate complex with Zn(CH$_2$SiMe$_3$)$_2$ has been reported to afford the octahedral (Me$_3$SiCH$_2$)$_2$OsO$_2$·L$_2$ (L = pyridine).[20,21] A polymeric osmium alkyl complex of the composition [(SiMe$_3$CH$_2$)$_2$OsO$_2$]$_n$ is formed when [PPh$_4$][OsO$_2$Cl$_4$] reacts with the title compound. This product can be depolymerized by treatment with pyridine.[22]

The reaction of bispinacolato osmium(VI) oxide with Zn(CH$_2$SiMe$_3$)$_2$ leads to the replacement of only one pinacolato ligand and formation of the bisalkyl derivative (RO)$_2$Os(O)(CH$_2$SiMe$_3$)$_2$ [(RO)$_2$ = pinacolato].[23] Rhenium trioxo alkyl complexes such as ReO$_3$(CH$_2$SiMe$_3$) can be prepared from Re$_2$O$_7$,[24] while rhenium dioxo trialkyl complexes such as ReO$_2$(CH$_2$CMe$_3$)$_2$(CH$_2$SiMe$_3$) are generated through the reaction of ReO$_2$(CH$_2$CMe$_3$)$_2$X·L (X = Cl, Br; L = pyridine) with the title reagent.[25] A further example of the reaction of a group VII metal complex involves the imido complex Mn(Nt-Bu)$_3$Cl which is transformed into a dimeric Mn(V) complex with bridging imido groups by exposure to the title compound (eq 7).[26]

$$Mn(Nt\text{-}Bu)_3Cl + Zn(CH_2SiMe_3)_2 \xrightarrow[15\%]{-78\,°C \text{ to rt}}$$

$$\text{(7)}$$

Reactions with R–XH (X = S, O, Se). The formation of a monomeric complex, incorporating a tricoordinated zinc with unusual T-shaped geometry, is observed on treatment of bis[(trimethylsilyl)methyl]zinc with 2 equivalents of 2,4,6-tri-t-butylthiophenol in Et$_2$O (70–80% yield)[27] (the production of a polymeric structure is probably suppressed by the steric demands of the aromatic substituents). In contrast, if an equimolar amount of thiol is employed, a trimeric zinc thiolate complex of the formula [ArSZnCH$_2$SiMe$_3$]$_3$ (Ar = 2,4,6-tri-t-butylphenyl or 2,4,6-triisopropylphenyl) is obtained.[28] The analogous zinc–selenium complex is accessible by the same sort of means.[29] The reaction of sterically demanding alcohols with bis[(trimethylsilyl)methyl]zinc gives rise to zinc alkoxy complexes of the general formula [ROZnCH$_2$SiMe$_3$]$_n$, with the value of n depending on the nature of the substituent R (R = 2,6-diisopropylphenyl or 2,4,6-tri-t-butylphenyl: $n = 2$; R = adamantyl: $n = 4$).[28] A more complex range of products is detected if the alcohol incorporates one or more additional donor atoms.[30]

Formation of Zincate Complexes. Zincate compounds such as K(ZnCH$_2$SiMe$_3$)$_3$ are formed in a transmetalation reaction of bis[(trimethylsilyl)methyl]zinc with elemental potassium,[31] or by reaction with potassium salts.[32] Lithium zincates such as Li[Zn(CH$_2$SiMe$_3$)$_2$R] [R = Me, Ph, N(SiMe$_3$)$_2$] can be prepared by reacting the title compound with the corresponding organolithium compounds.[33] Transmetalation with alkaline earth metals such as strontium, barium, and calcium affords zincates of the type illustrated in eq 8.[34,35]

$$\text{M} \ + \ 3\ \text{Zn(CH}_2\text{SiMe}_3)_2 \quad \xrightarrow[79\text{--}93\%]{-\text{Zn}}$$

$$\left[\ \text{Me}_3\text{Si}-\text{Zn}\underset{\text{SiMe}_3}{\overset{\text{SiMe}_3}{<}}\text{M(solvent)}_2\ \right]_2 \quad (8)$$

M = Ca, Sr: solvent = THF
M = Ba: solvent = THF, toluene

Adducts with Nitrogen Heterocycles, Amines, Imines, Phosphines, and Arsines. The addition of nitrogen heterocycles such as 2,2′-bipyridyl or 1,10-phenanthroline to bis[(trimethylsilyl)methyl]zinc leads to the quantitative formation of orange-red to red adducts which show remarkable stability towards oxygen.[2] The complex with N,N,N',N'-tetramethylethylenediamine is less stable.[2]

Bis[(trimethylsilyl)methyl]zinc is reported to react quantitatively with 1,4-diaza-1,3-butadienes to form a 1:1 coordination complex that is stable at room temperature. At temperatures above 35 °C, the complex engages in single-electron transfer reactions.[36,37] The reaction with the secondary phosphine [HP(SiMe$_3$)$_2$] leads to the formation of the dinuclear, heteroleptic complex [Me$_3$SiZnP(SiMe$_3$)$_2$]$_2$ incorporating bridging phosphido ligands.[38] A similar result is observed in the reaction with HAs(SiMe$_3$)$_2$.[39] When the sterically demanding primary amine 2,6-diisopropylaniline is used, the corresponding amido complex is formed.[40]

1.   Moorhouse, S.; Wilkinson, G., *J. Organomet. Chem.* **1973**, *52*, C5.

2.   Moorhouse, S.; Wilkinson, G., *J. Chem. Soc., Dalton Trans.* **1974**, 2187.

3.   Heinekey, D. M.; Stobart, S. R., *Inorg. Chem.* **1978**, *17*, 1463.

4.   Gais, H.-J.; Bulow, G., *Tetrahedron Lett.* **1992**, *33*, 461.

5.   Gais, H.-J.; Bulow, G., *Tetrahedron Lett.* **1992**, *33*, 465.

6.   Kercher, T.; Livinghouse, T., *J. Org. Chem.* **1997**, *62*, 805.

7.   Almena Perea, J. J.; Ireland, T.; Knochel, P., *Tetrahedron Lett.* **1997**, *38*, 5961.

8.   Dübner, F.; Knochel, P., *Angew. Chem., Int. Ed.* **1999**, *38*, 379.

9.   Börner, C.; Gimeno, J.; Gladiali, S.; Goldsmith, P. J.; Ramazzotti, D.; Woodward, S., *Chem. Commun.* **2000**, 2433.

10.   Lautens, M.; Renaud, J.-L.; Hiebert, S., *J. Am. Chem. Soc.* **2000**, *122*, 1804.

11.   Berger, S.; Langer, F.; Lutz, C.; Knochel, P.; Mobley, T. A.; Reddy, C. K., *Angew. Chem., Int. Ed. Engl.* **1997**, *36*, 1496.

12.   Lutz, C.; Knochel, P., *J. Org. Chem.* **1997**, *62*, 7895.

13.   Lutz, C.; Lutz, V.; Knochel, P., *Tetrahedron* **1998**, *54*, 6385.

14.   Jones, P.; Reddy, C. K.; Knochel, P., *Tetrahedron* **1998**, *54*, 1471.

15.   Pedersen, S. F.; Schrock, R. R., *J. Am. Chem. Soc.* **1982**, *104*, 7483.

16.   van der Schaaf, P. A.; Grove, D. M.; Smeets, W. J. J.; Spek, A. L.; van Koten, G., *Organometallics* **1993**, *12*, 3955.

17.   van der Schaaf, P. A.; Abbenhuis, R. A. T. M.; van der Noort, W. P. A.; de Graaf, R.; Grove, D. M.; Smeets, W. J. J.; Spek, A. L.; van Koten, G., *Organometallics* **1994**, *13*, 1433.

18.   Theopold, K. H.; Holmes, S. J.; Schrock, R. R., *Angew. Chem., Int. Ed. Engl.* **1983**, *22*, 1010.

19.   Araujo, J. P.; Wicht, D. K.; Bonitatebus, Jr, P. J.; Schrock, R. R., *Organometallics* **2001**, *20*, 5682.

20.   Herrmann, W. A.; Eder, S. J.; Kiprof, P.; Rypdal, K.; Watzlowik, P., *Angew. Chem., Int. Ed. Engl.* **1990**, *29*, 1445.

21.   Herrmann, W. A.; Eder, S. J.; Kiprof, P., *J. Organomet. Chem.* **1991**, *413*, 27.

22.   LaPointe, A. M.; Schrock, R. R.; Davis, W. M., *J. Am. Chem. Soc.* **1995**, *117*, 4802.

23.   Herrmann, W. A.; Watzlowik, P., *J. Organomet. Chem.* **1992**, *437*, 363.

24.   Herrmann, W. A.; Romao, C. C.; Fischer, R. W.; Kiprof, P.; de Méric de Bellefon, C., *Angew. Chem., Int. Ed. Engl.* **1991**, *30*, 185.

25.   Cai, S.; Hoffman, D. M.; Wierda, D. A., *Organometallics* **1996**, *15*, 1023.

26.   Danopoulos, A. A.; Wilkinson, G.; Sweet, T. K. N.; Hursthouse, M. B., *J. Chem. Soc., Dalton Trans.* **1995**, 205.

27.   Power, P. P.; Shoner, S. C., *Angew. Chem., Int. Ed. Engl.* **1990**, *29*, 1403.

28.   Olmstead, M. M.; Power, P. P.; Shoner, S. C., *J. Am. Chem. Soc.* **1991**, *113*, 3379.

29.   Ruhlandt-Senge, K.; Power, P. P., *Inorg. Chem.* **1993**, *32*, 4505.

30.   van der Schaaf, P. A.; Wissing, E.; Boersma, J.; Smeets, W. J. J.; Spek, A. L.; van Koten, G., *Organometallics* **1993**, *12*, 3624.

31.   Purdy, A. P.; George, C. F., *Organometallics* **1992**, *11*, 1955.

32.   Fabicon, R. M.; Richey, Jr, H. G., *J. Chem. Soc., Dalton Trans.* **2001**, 783.

33.   Westerhausen, M.; Rademacher, B.; Schwarz, W.; Henkel, S., *Z. Naturforsch.* **1994**, *49b*, 199.

34.   Westerhausen, M.; Gückel, C.; Habereder, T.; Vogt, M.; Warchhold, M.; Nöth, H., *Organometallics* **2001**, *20*, 893.

35.   Westerhausen, M.; Gückel, C.; Piotrowski, H.; Vogt, M., *Z. Anorg. Allg. Chem.* **2002**, *628*, 735.

36.   Wissing, E.; van der Linden, S.; Rijnberg, E.; Boersma, J.; Smeets, W. J. J.; Spek, A. L.; van Koten, G., *Organometallics* **1994**, *13*, 2602.

37.   Wissing, E.; Rijnberg, E.; van der Schaaf, P. A.; van Gorp, K.; Boersma, J.; van Koten, G., *Organometallics* **1994**, *13*, 2609.

38.   Rademacher, B.; Schwarz, W.; Westerhausen, M., *Z. Anorg. Allg. Chem.* **1995**, *621*, 287.

39. Rademacher, B.; Schwarz, W.; Westerhausen, M., *Z. Anorg. Allg. Chem.* **1995**, *621*, 1439.

40. Olmsted, M. M.; Grigsby, W. J.; Chacon, D. R.; Hascall, D. R.; Power, P. P., *Inorg. Chim. Acta* **1996**, *251*, 273.

Martin G. Banwell & Jens Renner
The Australian National University, Canberra, ACT, Australia

Bis(trimethylsilyl) Peroxide[1]

$$Me_3Si \diagdown_O \diagup^O \diagdown SiMe_3$$

[5796-98-5] $C_6H_{18}O_2Si_2$ (MW 178.38)
InChI = 1S/C6H18O2Si2/c1-9(2,3)7-8-10(4,5)6/h1-6H3
InChIKey = XPEMYYBBHOILIJ-UHFFFAOYSA-N

(a masked form of 100% hydrogen peroxide;[2] synthon of HO^+ for electrophilic hydroxylation;[3,4] source of Me_3SiO^+ for electrophilic oxidations;[4–6] a versatile oxidant for alcohols, ketones, phosphines, phosphites, and sulfides[1,2,7])

Physical Data: bp 41 °C/30 mmHg;[7] n_D^{20} 1.3970.[1b]
Solubility: highly sol aprotic organic solvents.
Form Supplied in: colorless oil; used as an anhydrous and protected form of hydrogen peroxide; 10% solution in hexane or in CH_2Cl_2.
Analysis of Reagent Purity: ^1H NMR (s) δ 0.18 ppm;[3] ^{29}Si NMR δ 27.2 ppm.[8]
Preparative Methods: obtained in 80–96% yields by reaction of chlorotrimethylsilane with 1,4-diazabicyclo[2.2.2]octane·$(H_2O_2)_2$,[3] hexamethylenetetramine·H_2O_2,[8] or hydrogen peroxide–urea in CH_2Cl_2.[9]
Handling, Storage, and Precautions: thermally stable; can be handled in the pure state and distilled;[8] rearranges at 150–180 °C.[1c]

Original Commentary

Jih Ru Hwu & Buh-Luen Chen
Academia Sinica & National Tsing Hua University, Taiwan, Republic of China

Electrophilic Oxidations. Bis(trimethylsilyl) peroxide (**1**) functions as an electrophilic hydroxylating agent for aliphatic, aromatic, and heteroaromatic anions.[3] Reaction of their lithium or Grignard compounds with (**1**) often affords trimethylsiloxy intermediates, which undergo desilylation with HCl in methanol to give the corresponding alcohols in good yields. In the presence of a catalytic amount of trifluoromethanesulfonic acid, (**1**) reacts with aromatic compounds to produce the corresponding phenols after acidic workup.[10] In these reactions the Me_3SiO^- moiety is considered as a synthon of the hydroxyl cation (i.e. OH^+).

Lithium or Grignard enolates of vinyl compounds react with (**1**) to produce the corresponding α-hydroxy ketones upon acidic workup.[4] Treatment of enolate anions derived from carboxylic acids and amides with (**1**) in THF at rt gives the corresponding α-hydroxy derivatives in 31–58% yields.[11] Furthermore, stereoselective synthesis of silyl enol ethers with retention of configuration is accomplished by oxidation of (*E*)- and (*Z*)-vinyllithiums, prepared from the corresponding bromides and *s*-BuLi, with (**1**) in THF at −78 °C.[6] Heterocyclic silyl enol ethers, such as 3-(trimethylsiloxy)furan and 3-(trimethylsiloxy)thiophene, can be obtained by reaction of (**1**) with 3-lithiofuran and 3-lithiothiophene, respectively.[5] In some reactions involving nucleophiles and (**1**), silylation may compete with siloxylation;[3] the outcome depends upon the counterion.[4]

Oxidative Desulfonylation and Selective Baeyer–Villiger Oxidation. Aliphatic, alicyclic, and benzylic phenyl sulfones react with *n*-BuLi and then with (**1**) in situ to give aldehydes or ketones in 66–91% yields (eq 1).[12] On the other hand, reaction of ketones with (**1**) in the presence of a catalyst, such as trimethylsilyl trifluoromethanesulfonate,[2] tin(IV) chloride, or boron trifluoride etherate,[7] in CH_2Cl_2 gives esters in good to excellent yields (eq 2). This Baeyer–Villiger oxidation proceeds in a regio- and chemoselective manner: the competing epoxidation of a C–C double bond does not occur.

$$RR'C=O + Me_3SiOSiMe_3 + PhSO_2Li \quad (1)$$

$$\text{catalyst} = Me_3SiSO_3CF_3, SnCl_4, BF_3 \cdot OEt_2$$

Oxidation of Alcohols, Sulfur, and Phosphorus Compounds as well as Si–Si and Si–H Bonds. Bis(trimethylsilyl) peroxide acts as an effective oxidant for alcohols in the presence of pyridinium dichromate or $RuCl_2(PPh_3)_3$ complex as the catalyst in CH_2Cl_2.[13] By this method, primary allylic and benzylic alcohols can be selectively oxidized to α-enals in the presence of a secondary alcohol.

Sulfur trioxide reacts with (**1**) in CH_2Cl_2 at −30 °C, affording bis(trimethylsilyl) monoperoxysulfate, an oxidant useful for the Baeyer–Villiger oxidation.[14] In addition, sulfides can be converted to sulfoxides or sulfones by use of (**1**) in benzene at reflux.[1b,15,16] This peroxide can also oxidize phosphines and phosphites to the corresponding oxyphosphoryl derivatives with retention of configuration at the phosphorus center in high yields.[1b,17] Furthermore, it converts the P=S and P=Se functionalities to the P=O group with inversion of configuration at the phosphorus center.[17] The $CF_3SO_3SiMe_3$ or Nafion–$SiMe_3$ catalyzed oxidation of nucleoside phosphites to phosphates under nonaqueous conditions is applied to the solid-phase synthesis of oligonucleotides.[18,19]

Reactions between (**1**) and disilanes containing fluorine atoms or possessing ring strain proceed readily at ambient temperature to generate the corresponding disiloxanes.[20] Oxidation of hydrosilanes (R_3SiH) with (1) gives a mixture of R_3SiOH and $R_3SiOSiMe_3$. The ratio of R_3SiOH to $R_3SiOSiMe_3$ tends to decrease in the order $Et_3SiH > PhMe_2SiH > Me_3SiSiHMe_2$.

Isomerization of Allylic Alcohols and Formation of 1-Halo-1-alkynes. Isomerization of primary and secondary allylic alcohols to tertiary isomers proceeds in CH_2Cl_2 at 25 °C in the presence of a catalyst that is prepared in situ by activation of vanadyl bis(acetylacetonate) or $MoO_2(acac)_2$ with (1) (eq 3).[21] On the other hand, terminal alkynes react with (1) in the presence of copper or zinc halides in THF at -15 °C to afford terminal 1-halo-1-alkynes in 40–85% yields.[22] By the same method, 1-cyano-1-alkynes are obtained in 65% yield by use of copper cyanide.

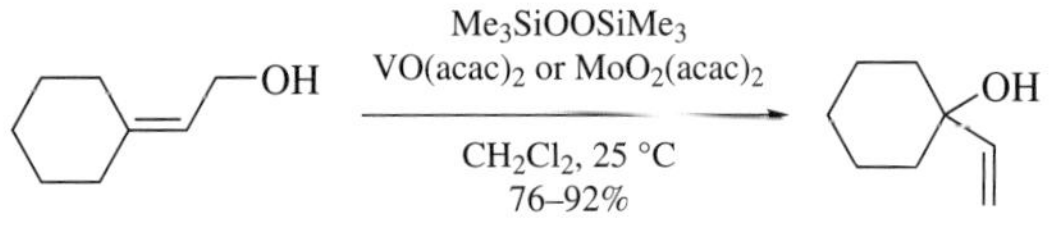

$$\text{(3)}$$

First Update

Santhosh F. Neelamkavil

Schering-Plough Research Institute, Kenilworth, NJ, USA

Enantioselective Synthesis of α-Substituted Ketones. Highly enantio-enriched ketones with a stereogenic center at the α-position have been synthesized by the oxidation of the intermediate chiral cycloalkenyl lithium species with bis(trimethylsilyl) peroxide (eq 4).[23] The lithiated intermediates are readily obtained from the corresponding halides by I/Li exchange (eq 4).[23] These reactions proceed in good yield and enantioselectivity; however, scale-up is an issue due to the sensitive nature of $(TMSO)_2$.

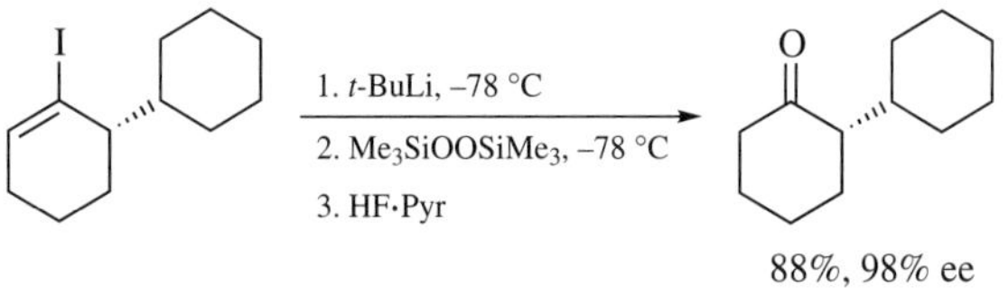

$$\text{(4)}$$

88%, 98% ee

α-Hydroxylation of Lactams. In the total synthesis of the natural product Vindoline, $(TMSO)_2$ was employed for the α-hydroxylation of a key lactam intermediate. This was achieved by generating the lactam enolate which was then treated with $(TMSO)_2$, followed by direct quenching with TIPSOTf, to form the protected α-hydroxylated lactam intermediate (eq 5).[24]

Preparation of Heteroaryl Amides via Sequential S_NAr Substitution and Oxidation. Heteroaryl amides have been prepared from the corresponding dialkylamino acetonitrile derivatives via a sequential base-mediated coupling and oxidation with $(TMSO)_2$. Potential substrates include halogenated pyridine, pyrazine, pyrimidine, pyridazine, triazine, imidazole, pyrazole, oxazole, and thiazole derivatives, in which the halogen α to the nitrogen atom is disposed toward displacement by nucleophiles via an S_NAr substitution process. This one-pot procedure relies on the presence of excess base to regenerate the anion following

substitution, and the direct in situ oxidation of the same, resulting in the collapse of the hydroxylated product with the release of cyanide to provide the heteroaryl amides (eq 6).[25]

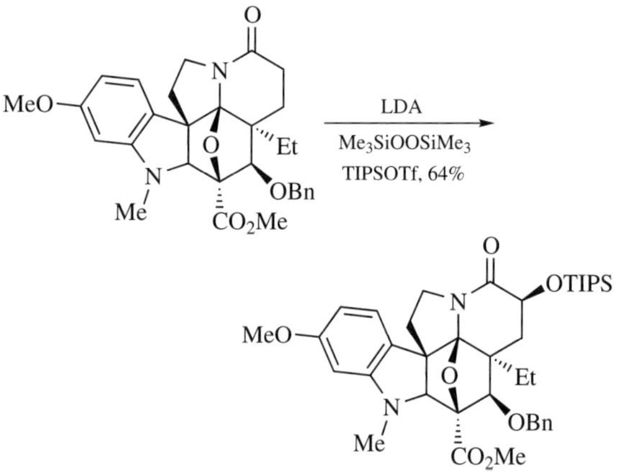

$$\text{(5)}$$

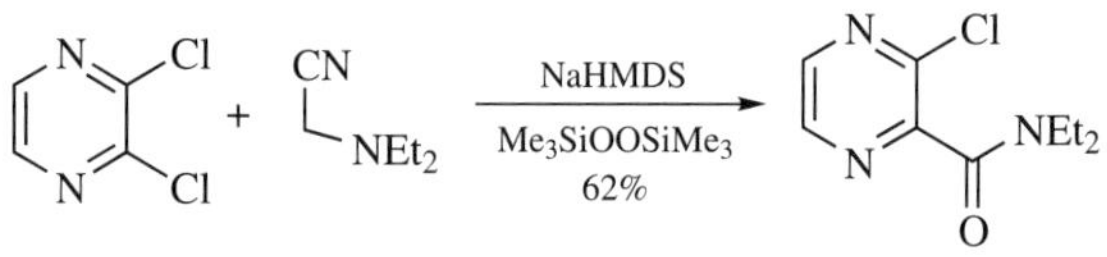

$$\text{(6)}$$

Functionalization of Azaferrocene Catalysts. Chiral azaferrocenes are highly useful in enantioselective acylation as nucleophilic catalysts,[26] and in transition metal-catalyzed asymmetric reactions as chiral ligands.[27] Enantioselective lithiation of an azaferrocene moiety followed by functionalization with $(TMSO)_2$ resulted in a lateral hydroxyl-substituted product with excellent optical purity, but in poor yield (eq 7).[28] The low yield was attributed to the poor reactivity of $(TMSO)_2$ toward the labile azaferrocene substrate.

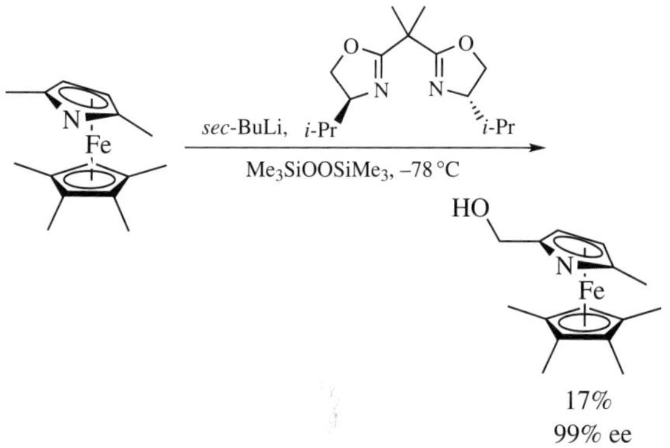

$$\text{(7)}$$

17%
99% ee

Olefin Epoxidation Catalyzed by Oxorhenium Derivatives. Organometallic rhenium species (e.g., CH_3ReO_3) have been recently developed as well-defined catalysts for a variety of processes including olefin epoxidation with aqueous hydrogen peroxide. The major limitation of this system is the acidity of the reaction medium. An improved method was later reported in which the organometallic rhenium species was replaced by cheaper and more readily available inorganic rhenium oxides (e.g., Re_2O_7, $ReO_3(OH)$, and ReO_3). At the same time bis(trimethylsilyl) peroxide replaced aqueous hydrogen peroxide (eq 8). The high epoxidation activity of these inorganic oxorhenium catalysts is thought to be due to the almost anhydrous conditions achieved by using $(TMSO)_2$ as the oxygen atom source, along with the presence of only a trace of a protic agent (e.g. H_2O) to catalyze the slow transfer of the peroxide moiety from silicon to rhenium.[29,30]

$$\text{(8)}$$

with reagents $Me_3SiOOSiMe_3$, Re_2O_7, CH_2Cl_2, 95%

One-pot Synthesis of β-Cyanohydrins from Olefins. *trans-β-Cyanohydrins* can be prepared efficiently by treating the corresponding olefins with bis(trimethylsilyl) peroxide and TMSCN in a reaction promoted by a bulky zirconium catalyst (eq 9).[31] This reaction was much faster and cleaner in the presence of 20 mol % of Ph_3PO. A similar reaction performed in the presence of TMSOAc, albeit with a different catalyst $Zr(O^iPr)_4$, resulted in the formation of *trans-β-acetoxy alcohols*.[32] Chlorohydrins can also be prepared if the corresponding olefins are treated with $SnCl_4$ and TMSCl in the presence of bis(trimethylsilyl) peroxide.[32]

$$\text{(9)}$$

with reagents $Me_3SiOOSiMe_3$, TMSCN, KF, MeOH, 50 °C, 100%

Synthesis of Pyridine-N-Oxides. Methyl isonicotinate was oxidized by $(TMSO)_2$ in the presence of CH_3ReO_3, or other inorganic rhenium derivatives such as ReO_3, Re_2O_7 and $HOReO_3$, to the N-oxide in high yield (95–98% isolated yields). With $ReCl_3$ or $NaOReO_3$ as catalyst, only trace amounts of oxidation products were obtained (eq 10).[33] The importance of a trace amount of water in this oxidation process was confirmed when the reaction was retarded by the presence of molecular sieves. The optimal water content was found to be between a trace and 15 mol %, with higher proportions leading to lower conversion. A commercial 65–70% solution of perrhenic acid in water was found to be a convenient source of both Re and water.

$$\text{(10)}$$

with reagents $HOReO_3$, $Me_3SiOOSiMe_3$, 98%

Selective Functionalization of Saturated Hydrocarbons. Alkanes have been oxidized to ketones using $Fe^{III}(\text{picolinic acid})_3$ in the presence of bis(trimethylsilyl) peroxide in pyridine. Oxidation of cyclohexane gave cyclohexanone as the major product along with a minor amount of cyclohexanol (eq 11). The reaction is catalytic in $Fe^{III}(\text{picolinic acid})_3$ and alcohols can also be converted to ketones with this system.[34]

$$\text{(11)}$$

with reagents $Fe^{III}(\text{picolinic acid})_3$, $Me_3SiOOSiMe_3$; 73% and 19%

Related Reagents. Bis(trimethylsilyl) Peroxide–Vanadyl Bis(acetylacetonate).

1. (a) Brandes, D.; Blaschette, A., *J. Organomet. Chem.* **1973**, *49*, C6. (b) Brandes, D.; Blaschette, A., *J. Organomet. Chem.* **1974**, *73*, 217. (c) Alexandrov, Y. A., *J. Organomet. Chem.* **1982**, *238*, 1. (d) Huang, L.; Hiyama, T., *Yuki Gosei Kagaku Kyokaishi* **1990**, *48*, 1004 (*Chem. Abstr.* **1991**, *114*, 102 079x).
2. Suzuki, M.; Takada, H.; Noyori, R., *J. Org. Chem.* **1982**, *47*, 902.
3. Taddei, M.; Ricci, A., *Synthesis* **1986**, 633.
4. Camici, L.; Dembech, P.; Ricci, A.; Seconi, G.; Taddei, M., *Tetrahedron* **1988**, *44*, 4197.
5. Camici, L.; Ricci, A.; Taddei, M., *Tetrahedron Lett.* **1986**, *27*, 5155.
6. Davis, F. A.; Lal, G. S.; Wei, J., *Tetrahedron Lett.* **1988**, *29*, 4269.
7. Matsubara, S.; Takai, K.; Nozaki, H., *Bull. Chem. Soc. Jpn.* **1983**, *56*, 2029.
8. Babin, P.; Bennetau, B.; Dunogues, J., *Synth. Commun.* **1992**, *22*, 2849.
9. Jackson, W. P., *Synlett* **1990**, *9*, 536.
10. Olah, G. A.; Ernst, T. D., *J. Org. Chem.* **1989**, *54*, 1204.
11. Pohmakotr, M.; Winotai, C., *Synth. Commun.* **1988**, *18*, 2141.
12. Hwu, J. R., *J. Org. Chem.* **1983**, *48*, 4432.
13. Kanemoto, S.; Matsubara, S.; Takai, K.; Oshima, K.; Utimoto, K.; Nozaki, H., *Bull. Chem. Soc. Jpn.* **1988**, *61*, 3607.
14. Adam, W.; Rodriguez, A., *J. Org. Chem.* **1979**, *44*, 4969.
15. Kocienski, P.; Todd, M., *J. Chem. Soc., Chem. Commun.* **1982**, 1078.
16. Curci, R.; Mello, R.; Troisi, L., *Tetrahedron* **1986**, *42*, 877.
17. Kowalski, J.; Wozniak, L.; Chojnowski, J., *Phosphorus Sulfur Silicon* **1987**, *30*, 125.
18. Hayakawa, Y.; Uchiyama, M.; Noyori, R., *Tetrahedron Lett.* **1986**, *27*, 4191.
19. Hayakawa, Y.; Uchiyama, M.; Noyori, R., *Tetrahedron Lett.* **1986**, *27*, 4195.
20. Tamao, K.; Kumada, M.; Takahashi, T., *J. Organomet. Chem.* **1975**, *94*, 367.
21. Matsubara, S.; Okazoe, T.; Oshima, K.; Takai, K.; Nozaki, H., *Bull. Chem. Soc. Jpn.* **1985**, *58*, 844.
22. Casarini, A.; Dembech, P.; Reginato, G.; Ricci, A.; Seconi, G., *Tetrahedron Lett.* **1991**, *32*, 2169.
23. Soorukram, D.; Knochel, P., *Angew. Chem., Int. Ed.* **2006**, *45*, 3686.
24. Choi, Y.; Ishikawa, H.; Velcicky, J.; Elliott, G. I.; Miller, M. M.; Boger, D. L., *Org. Lett.* **2005**, *7*, 4539.
25. Zhang, Z.; Yin, Z.; Kadow, J. F.; Meanwell, N. A.; Wang, T., *J. Org. Chem.* **2004**, *69*, 1360.
26. Ruble, J. C.; Fu, G. C., *J. Org. Chem.* **1996**, *61*, 7230.
27. Dosa, P. I.; Ruble, J. C.; Fu, G. C., *J. Org. Chem.* **1997**, *62*, 444.
28. Fukuda, T.; Imazato, K.; Iwao, M., *Tetrahedron Lett.* **2003**, *44*, 7503.
29. Yudin, A. K.; Sharpless, K. B., *J. Am. Chem. Soc.* **1997**, *119*, 11536.
30. Yudin, A. K.; Chiang, J. P.; Adolfsson, H.; Coperet, C., *J. Org. Chem.* **2001**, *66*, 4713.
31. Yamasaki, S.; Kanai, M.; Shibasaki, M., *J. Am. Chem. Soc.* **2001**, *123*, 1256.
32. Sakurada, I.; Yamasaki, S.; Gottlich, R.; Iida, T.; Kanai, M.; Shibasaki, M., *J. Am. Chem. Soc.* **2000**, *122*, 1245.
33. Coperet, C.; Adolfsson, H.; Chiang, J. P.; Yudin, A. K.; Sharpless, K. B., *Tetrahedron Lett.* **1998**, *39*, 761.
34. Barton, D. H. R.; Chabot, B. M., *Tetrahedron* **1997**, *53*, 487.

Bis(trimethylsilyl) Selenide

$$Me_3Si \overset{Se}{\diagup} \diagdown SiMe_3$$

[4099-46-1] $C_6H_{18}SeSi_2$ (MW 225.34)

InChI = 1S/C6H18SeSi2/c1-8(2,3)7-9(4,5)6/h1-6H3

InChIKey = FKIZDWBGWFWWOV-UHFFFAOYSA-N

(synthesis of unsymmetrical selenides;[6] generation of selenoaldehydes;[7] reduction of sulfoxides, selenoxides, and telluroxides[4])

Physical Data: bp 45–46 °C/5.3 mmHg.

Solubility: sol Et_2O and THF.

Handling, Storage, and Precautions: is an air-sensitive, colorless oil. Because of the strong odor and the release of toxic hydrogen selenide on exposure to moisture, handling of bis(trimethylsilyl) selenide should be performed in a well-ventilated hood. It can be stored at −20 °C in an argon-flushed glass vessel for about a month; after this time, amorphous selenium (red color) is gradually deposited on the vessel.

Synthesis of Bis(trimethylsilyl) Selenide. In the initial studies, bis(trimethylsilyl) selenide was synthesized by the following two methods: silylation of sodium selenide[1] (or lithium selenide[2]) with chlorotrimethylsilane (eq 1) or reaction of bromobenzene, magnesium, and selenium with chlorotrimethylsilane (eq 2).[3]

$$Se \xrightarrow[\text{liq. NH}_3]{\text{Na (or Li)}} \underset{\text{(or Li}_2\text{Se)}}{Na_2Se} \xrightarrow[\text{C}_6\text{H}_6\ (\text{or Et}_2\text{O})]{Me_3SiCl} (Me_3Si)_2Se \quad (1)$$
$$83\%$$

$$PhBr + Mg + Se \xrightarrow{Et_2O} Se(MgBr)_2 + Ph_2Se \xrightarrow[\text{rt, 24 h}]{Me_3SiCl}$$
$$(Me_3Si)_2Se \quad (2)$$
$$39\%$$

However, the former method requires the troublesome manipulation involved in preparing sodium selenide (or lithium selenide) in liquid ammonia[1b] and conducting the silylation in benzene (or diethyl ether). The latter method is inefficient and results in a low yield of the desired disilyl selenide. Thus a one-pot, high-yield procedure based on the lithium triethylborohydride reduction of elemental selenium has been developed, as depicted in eq 3.[4] In this method, both preparation of Li_2Se and silylation with Me_3SiCl are performed in THF. Technically important is the use of selenium shot; if selenium powder is utilized, the yield of Li_2Se decreases sharply. Moreover, addition of small amount of boron trifluoride etherate (1.6 mol %) accelerates the silylation of Li_2Se.[5]

$$Se \xrightarrow[\text{THF}]{LiBEt_3H} Li_2Se \xrightarrow[\text{THF}]{Me_3SiCl} (Me_3Si)_2Se \quad (3)$$
$$95\%$$

Recently it has been reported that the reaction of lithium diselenide (Li_2Se_2) with 2 equiv of chlorotrimethylsilane affords bis(trimethylsilyl) selenide in 73% yield with the concomitant formation of elemental selenium (eq 4).[5]

$$Se \xrightarrow[\text{cat. PhC}\equiv\text{CPh}]{\text{Li, THF}} Li_2Se_2 \xrightarrow[\text{THF}]{Me_3SiCl} (Me_3Si)_2Se + Se \quad (4)$$
$$73\%$$

Synthesis of Unsymmetrical Selenides. Bis(trimethylsilyl) selenide reacts with equimolar amounts of *n*-butyllithium to generate $Me_3SiSeLi$, alkylation of which then provides trimethylsilyl alkyl selenides. Similar treatment of Me_3SiSeR^1 with BuLi / R^2X successfully leads to unsymmetrical selenides in good yields (eq 5).[6] Use of acid chlorides in place of alkyl halides results in the formation of selenoesters.

$$(Me_3Si)_2Se \xrightarrow[\text{2. R}^1\text{X}]{\text{1. BuLi, THF}} [R^1SeSiMe_3] \xrightarrow[\text{2. R}^2\text{X}]{\text{1. BuLi}} R^1SeR^2 \quad (5)$$
$$31–97\%$$

R^1 = alkyl, benzyl, allyl, acyl; R^2 = alkyl, benzyl

Generation and Diels–Alder Reaction of Selenoaldehydes.[7] Direct conversion of aldehydes to selenoaldehydes has been achieved by treatment with bis(trimethylsilyl) selenide in the presence of a catalytic amount of butyllithium (eq 6).[7a]

R = alkyl, aryl
$$45–85\%$$

A mechanistic proposal includes addition of $Me_3SiSeLi$ to aldehydes, followed by elimination of Me_3SiOLi to generate selenoaldehydes, which undergo Diels–Alder reaction with dienes. Aldehydes bearing a conjugate diene unit provide intramolecular Diels–Alder adducts (eq 7).[7b]

endo:exo = 55:45

Reduction of Sulfoxides, Selenoxides, and Telluroxides. Bis(trimethylsilyl) selenide is useful as a two-electron reducing agent for the reduction of sulfoxides, selenoxides, and telluroxides to corresponding sulfides, selenides, and tellurides, respectively (eq 8).[4]

$$\overset{O}{\underset{R^1 \diagdown X \diagup R^2}{\uparrow}} + (Me_3Si)_2Se \xrightarrow[\text{rt, 1 h}]{\text{THF}} R^1XR^2 + (Me_3Si)_2O + Se \quad (8)$$
$$80–100\%$$

X = S, Se, Te; R^1, R^2 = alkyl, aryl

1. (a) Schmidt, M.; Ruf, H., *Z. Anorg. Allg. Chem.* **1963**, *321*, 270. (b) Klemm, W.; Sodomann, H.; Langmesser, P., *Z. Anorg. Allg. Chem.* **1939**, *241*, 281.

2. Drake, J. E.; Glavincevski, B. M.; Henderson, H. E., *Can. J. Chem.* **1978**, *56*, 465.

3. (a) Hooton, K. A.; Allred, A. L., *Inorg. Chem.* **1965**, *4*, 671. (b) Nedugov, A. N.; Pavlova, N. N.; Lapkin, I. I., *Zh. Org. Khim.* **1991**, *27*, 2068.

4. Detty, M. R.; Seidler, M. D., *J. Org. Chem.* **1982**, *47*, 1354.

5. Syper, L.; Mlochowski, J., *Tetrahedron* **1988**, *44*, 6119.

6. Segi, M.; Kato, M.; Nakajima, T.; Suga, S.; Sonoda, N., *Chem. Lett.* **1989**, 1009.

7. (a) Segi, M.; Nakajima, T.; Suga, S.; Murai, S.; Ryu, I.; Ogawa, A.; Sonoda, N., *J. Am. Chem. Soc.* **1988**, *110*, 1976. (b) Segi, M.; Takahashi, M.; Nakajima, T.; Suga, S.; Murai, S.; Sonoda, N., *Tetrahedron Lett.* **1988**, *29*, 6965. (c) Takikawa, Y.; Uwano, A.; Watanabe, H.; Asanuma, M.; Shimada, K., *Tetrahedron Lett.* **1989**, *30*, 6047.

Akiya Ogawa & Noboru Sonoda
Osaka University, Osaka, Japan

Bis(trimethylsilyl) Sulfide[1]

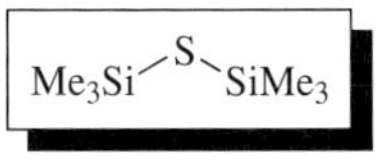

[3385-94-2] $C_6H_{18}SSi_2$ (MW 178.44)

InChI = 1S/C6H18SSi2/c1-8(2,3)7-9(4,5)6/h1-6H3

InChIKey = RLECCBFNWDXKPK-UHFFFAOYSA-N

(reduction of S, Se, and Te oxides;[4] formation of thiocarbonyl derivatives under mild conditions;[7] formation of allylsilanes from ketones or allylic alcohols[13])

Alternate Name: TMS$_2$S.

Physical Data: bp 164 °C; d 0.846 g cm^{-3}.

Solubility: sol diethyl ether, THF, CH$_2$Cl$_2$, MeCN, and aromatic solvents.

Form Supplied in: colorless cloudy liquid, commercially available.

Handling, Storage, and Precautions: avoid strong oxidizing agents. The reagent is moisture sensitive, flammable, and decomposes to CO, CO$_2$, H$_2$S, and sulfur oxides. Stench, mucous membrane (lungs, eyes) and skin irritant. Store in cool dry place free of oxygen. Use in a fume hood.

Original Commentary

Mark A. Matulenko
University of Wisconsin at Madison, WI, USA

Reducing Agent. Bis(trimethylsilyl) sulfide (TMS$_2$S) is used to reduce aromatic nitro groups[2] to amines and the oxides of sulfur, selenium, and tellurium. Conditions for nitro group reduction (eq 1) are forcing; however, yields are good. Reactions of TMS$_2$S with primary aliphatic nitro groups[3] result in the formation of the thiohydroxamic acids (eq 2) which can be isolated or carried further to the nitrile. Secondary alkyl nitro derivatives provide oximes. Sulfoxides are reduced[4] to sulfides, selenoxides to selenides, and telluroxides to tellurides. Conditions are mild and work well on both the aliphatic and aromatic oxides.

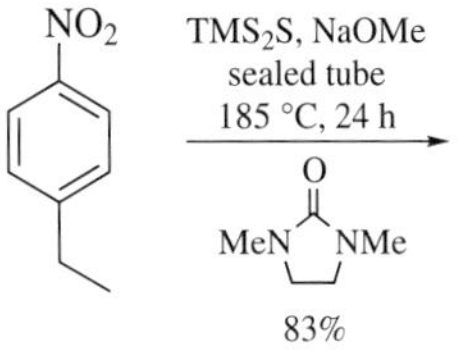

$$(1)$$

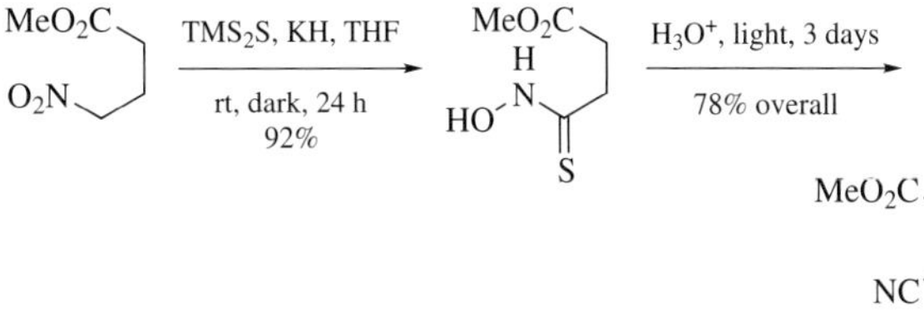

$$(2)$$

Sulfur Transfer Reagent. Alkyl sulfides[5] can be formed from the corresponding alkyl halides using TMS$_2$S. Thioaldehydes,[6] thioketones,[7] thioamides,[8] thioacid anhydrides,[9] thioacylsilanes,[10] divinyl sulfides (eq 4),[11] and thiranes (eq 5)[12] are also produced from TMS$_2$S and their respective carbonyl or alkenic compounds, in combination with the reagents shown.

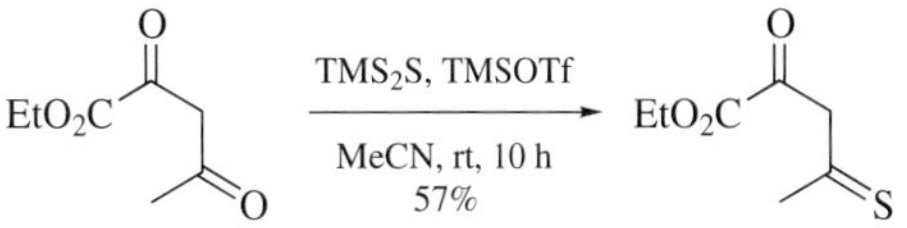

$$(3)$$

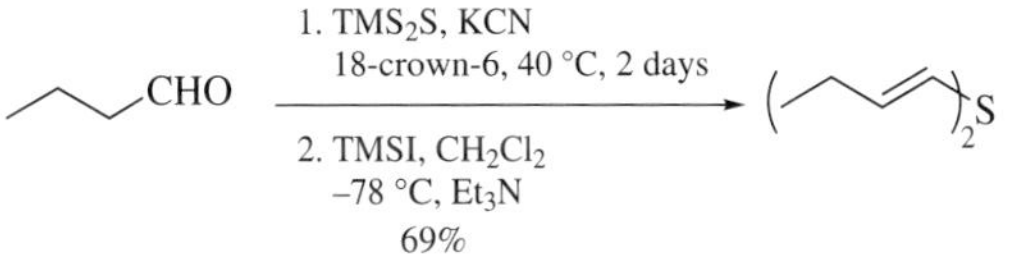

$$(4)$$

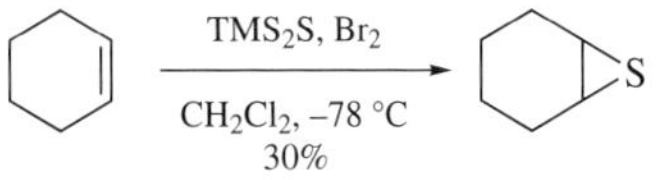

$$(5)$$

Silicon Transfer Reagent. Bis(trimethylsilyl) sulfide can transform[1b] alcohols, acids, and amines into their silylated counterparts. Enol silanes[13] are also formed under the influence of this silyl transfer reagent.

First Update

Alessandro Degl'Innocenti & Antonella Capperucci
Università di Firenze, Firenze, Italy

Reducing Agent. Bis(trimethylsilyl) has recently been reported to reduce selectively aromatic and heteroaromatic azides to amines in mild conditions[14] (eq 6). Application to the synthesis of benzodiazepine antibiotics is reported.[15]

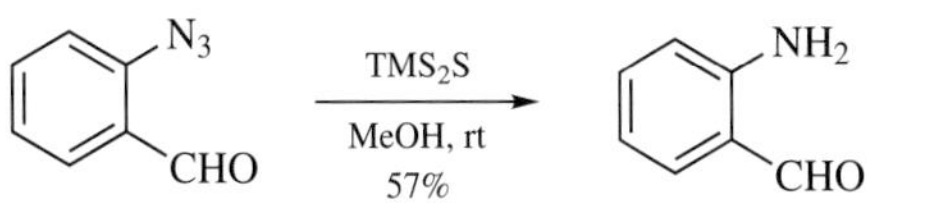

$$(6)$$

Sulfur Transfer Reagent. TMS$_2$S has been shown to be a "counterattack reagent" in multistep chemical transformations such as bis-*O*-demethylation, conversion of primary nitro compounds to thiohydroxamic acids and of secondary nitro compounds to oximes.[16] TMS$_2$S is able to transform regio- and stereoselectively epoxides into β-mercaptoalcohols[17] (eq 7) and alkyl halides into alkyl thiols.[18] Allyl[19] and diallyl sulfides[20] (eq 8) are also produced from TMS$_2$S. Thioamides and thiolactams,[21] peptide thioacids,[22] thioacyl-[23] and thioformylsilanes,[24] α,β-acetylenic thioaldehydes and thioketones,[25] and α,β-unsaturated thioacyl-stannanes[26] are also obtained from TMS$_2$S and suitable reagents and catalysts.[27] Bis(thioacylsilanes)[28] and α,β-ethylenic thioacyl-silanes[29] are produced with TMS$_2$S, but are transformed directly to silylated thiaheterocycles (eq 9) of different ring size and silyl-dithiins (eq 10), respectively.

Nucleoside *H*-Phosphonothioates and -dithioates[30] and thioxophosphines[31] are also generated with TMS$_2$S.

(7)

(8)

(9)

(10)

1. (a) So, J.-H.; Boudjouk, P., *Inorg. Synth.* **1992**, *29*, 30. (b) Mizhiritskii, M. D.; Reikhsfel'd, V. O., *Russ. Chem. Rev. (Engl. Transl.)* **1988**, *57*, 447. (c) Mizhiritskii, M. D.; Yuzhelevskii, Yu. A., *Russ. Chem. Rev. (Engl. Transl.)* **1987**, *56*, 355. (d) Capozzi, G., *Pure Appl. Chem.* **1987**, *59*, 989.

2. Hwu, J. R.; Wong, F. F.; Shiao, M.-J., *J. Org. Chem.* **1992**, *57*, 5254.

3. Tsay, S.-C.; Gani, P.; Hwu, J. R., *J. Chem. Soc., Perkin Trans. 1* **1991**, 1493.

4. (a) Soysa, H. S. D.; Weber, W. P., *Tetrahedron Lett.* **1978**, 235. (b) Detty, M. R.; Seidler, M. D., *J. Org. Chem.* **1982**, *47*, 1354.

5. (a) Ando, W.; Furuhata, T.; Tsumaki, H.; Sekiguchi, A., *Synth. Commun.* **1982**, 627. (b) Steliou, K.; Salama, P.; Corriveau, J., *J. Org. Chem.* **1985**, *50*, 4969.

6. (a) Capperucci, A.; Degl'Innocenti, A.; Ricci, A.; Mordini, A.; Reginato, G., *J. Org. Chem.* **1991**, *56*, 7323. (b) Segi, M.; Nakajima, T.; Suga, S.; Murai, S.; Ryu, I.; Ogawa, A.; Sonoda, N., *J. Am. Chem. Soc.* **1988**, *110*, 1976. (c) Lebedev, E. P.; Mizhiritskii, M. D.; Baburina, V. A.; Zaripov, Sh. I., *J. Gen. Chem. USSR (Engl. Transl.)* **1979**, *49*, 943.

7. (a) Degl'Innocenti, A.; Capperucci, A.; Mordini, A.; Reginato, G.; Ricci, A.; Cerreta, F., *Tetrahedron Lett.* **1993**, *34*, 873. (b) Steliou, K.; Mrani, M., *J. Am. Chem. Soc.* **1982**, *104*, 3104.

8. Lin, P.-Y.; Ku, W. S.; Shiao, M.-J., *Synthesis* **1992**, 1219.

9. Ando, W.; Furuhata, T.; Tsumaki, H.; Sekiguchi, A., *Chem. Lett.* **1982**, 885.

10. Ricci, A.; Degl'Innocenti, A.; Capperucci, A.; Reginato, G., *J. Org. Chem.* **1989**, *54*, 19.

11. Aida, T.; Chan, T. H.; Harpp, D. N., *Tetrahedron Lett.* **1981**, *22*, 1089.

12. Capozzi, F.; Capozzi, G.; Menichetti, S., *Tetrahedron Lett.* **1988**, *29*, 4177.

13. Mizhiritskii, M. D.; Lebedev, E. P.; Fufaeva, A. N., *J. Gen. Chem. USSR (Engl. Transl.)* **1982**, *52*, 1862.

14. (a) Capperucci, A.; Degl'Innocenti, A.; Funicello, M.; Mauriello, G.; Scafato, P.; Spagnolo, P., *J. Org. Chem.* **1995**, *60*, 2254. (b) Capperucci, A.; Degl'Innocenti, A.; Funicello, M.; Scafato, P.; Spagnolo, P., *Synthesis* **1996**, 1185.

15. Kamal, A.; Reddy, K. L.; Reddy, G. S. K.; Reddy, B. S. N., *Tetrahedron Lett.* **2004**, *45*, 3499.

16. Hwu, J. R.; Lin, C.-F.; Tsay, S.-C., *Phosphorus, Sulfur, and Silicon* **2005**, *180*, 1389.

17. Degl'Innocenti, A.; Capperucci, A.; Cerreti, A.; Pollicino, S.; Scapecchi, S.; Malesci, I.; Castagnoli, G., *Synlett* **2005**, 3063.

18. Hu, J.; Fox, M. A., *J. Org. Chem.* **1999**, *64*, 4959.

19. Tsay, S.-C.; Lin, L. C.; Furth, P. A.; Shum, C. C.; King, D. B.; Yu, S. F.; Chen, B.-L.; Hwu, J. R., *Synthesis* **1993**, 329.

20. Tsay, S.-C.; Yep, G. L.; Chen, B.-L.; Lin, L. C.; Hwu, J. R., *Tetrahedron* **1993**, *49*, 8969.

21. Smith, D. C.; Lee, S. W.; Fuchs, P. L., *J. Org. Chem.* **1994**, *59*, 348.

22. Schwabacher, A. W.; Maynard, T. L., *Tetrahedron Lett.* **1993**, *34*, 1269.

23. Degl'Innocenti, A.; Capperucci, A.; Oniciu, D. C.; Katritzky, A. R., *J. Org. Chem.* **2000**, *65*, 9206.

24. Degl'Innocenti, A.; Scafato, P.; Capperucci, A.; Bartoletti, L.; Spezzacatena, C.; Ruzziconi, R., *Synlett* **1997**, 361.

25. Capperucci, A.; Degl'Innocenti, A.; Scafato, P.; Mecca, T.; Mordini, A.; Reginato, G., *Synlett* **1999**, 1739.

26. Degl'Innocenti, A.; Capperucci, A.; Nocentini, T.; Biondi, S.; Fratini, V.; Castagnoli, G.; Malesci, I., *Synlett* **2004**, 2159.

27. (a) Degl'Innocenti, A.; Capperucci, A., *Eur. J. Org. Chem.* **2000**, 2171. (b) Degl'Innocenti, A.; Capperucci, A.; Castagnoli, G.; Malesci, I., *Synlett* **2005**, 1965.

28. Bouillon, J.-P.; Capperucci, A.; Portella, C.; Degl'Innocenti, A., *Tetrahedron Lett.* **2004**, *45*, 87.

29. (a) Capperucci, A.; Degl'Innocenti, A.; Biondi, S.; Nocentini, T.; Rinaudo, G., *Tetrahedron Lett.* **2003**, *44*, 2831. (b) Capperucci, A.; Degl'Innocenti, A.; Nocentini, T.; Biondi, S.; Dini, F., *J. Organomet. Chem.* **2003**, *686*, 363.

30. Cieślak, J.; Jankowska, J.; Stawiński, J.; Kraszewski, A., *J. Org. Chem.* **2000**, *65*, 7049.

31. Kamijo, K.; Toyota, K.; Yoshifuji, M., *Chem. Lett.* **1999**, 567.

Bromomagnesium Hexamethyldisilazide

[50916-70-6] $C_6H_{18}BrMgNSi_2$ (MW 264.59)
InChI = 1S/C6H18NSi2.BrH.Mg/c1-8(2,3)7-9(4,5)6;;/h1-6H3;
1H;/q-1;;+2/p-1
InChIKey = HFIJANVQJMJRRW-UHFFFAOYSA-M

(base for selective formation of magnesium enolates)

Physical Data: white solid from ether of composition
$Mg_2Br_2[N(SiMe_3)_2]_2 \cdot 2Et_2O$, mp 98 °C.
Solubility: sol benzene, cyclohexane, and toluene; very sol ether.
Preparative Method: hexamethyldisilazane is added to ethylmagnesium bromide in ether. Cooling to −20 °C produces white crystals in good yield.[1]

Reaction with Carbonyl Compounds. Although seldom used in comparison with bromomagnesium diisopropylamide, bromomagnesium hexamethyldisilazide has the unique feature of high solubility in organic solvents.

Bromomagnesium hexamethyldisilazide and bromomagnesium diisopropylamide were the most effective of several magnesium bases examined for selective formation of the more substituted silyl enol ether from ketones (eq 1). Less-hindered ketones gave mainly aldol products.[2]

$$\text{eq (1)}$$

96:4

The reagent reacts with hindered esters to give isolable complexes, with no evidence for enolate formation (eq 2).[3]

$$\text{eq (2)}$$

R = Me, *t*-Bu

1:1 complex

---

1. Wannagat, U.; Autzen, H.; Kuckertz, H.; Wismar, H.-J., *Z. Anorg. Allg. Chem.* **1972**, *394*, 254.

2. Krafft, M. E.; Holton, R. A., *Tetrahedron Lett.* **1983**, *24*, 1345.
3. Lochmann, L.; Sŏrm, M., *Collect. Czech. Chem. Commun.* **1973**, *38*, 3449.

Michael Rathke & Robert Elghanian
Michigan State University, East Lansing, MI, USA

(Bromomethyl)chlorodimethylsilane[1]

[16532-02-8] $C_3H_8BrClSi$ (MW 187.54)
InChI = 1S/C3H8BrClSi/c1-6(2,5)3-4/h3H2,1-2H3
InChIKey = CAURZYXCQQWBJO-UHFFFAOYSA-N

(allylic,[2] propargylic,[3] and homoallylic[4] bromomethylsilyl ethers are used for regio- and stereoselective introduction of hydroxymethyl[2–4] or methyl[2c,d] groups through radical cyclization; derivatizing agent for electron capture GC[5])

Physical Data: bp 133–135 °C; bp 130 °C/740 mmHg; d_4^{20} 1.386; n_D^{20} 1.466.
Solubility: sol most organic solvents including benzene, THF, methylene chloride, chloroform, and pyridine.
Form Supplied in: commercially available colorless liquid with 98% (GC) purity.
Analysis of Reagent Purity: GC analysis appears best; the following spectroscopic information may also be used: FTIR (vapor phase at 225 °C): 2978 (w), 1392 (w), 1264 (m), 822 (s) cm^{-1};[6] ^{1}H NMR ($CDCl_3$) δ 0.58 (s, 6H), 2.65 (s, 2H); ^{13}C NMR ($CDCl_3$) δ 0.27, 16.47.
Handling, Storage, and Precautions: moisture-sensitive; toxicological properties have not been thoroughly investigated; symptoms of exposure may include burning sensation, coughing, wheezing, laryngitis, shortness of breath, headache, nausea, and vomiting.[7] Use in a fume hood.

Original Commentary

Masato Koreeda
University of Michigan, Ann Arbor, MI, USA

Use as a Source of α-Silyl Radicals. Treatment of hydroxy groups with $(BrCH_2)Me_2SiCl$ in the presence of triethylamine often with a catalytic amount of 4-dimethylaminopyridine provides the (bromomethyl)dimethylsilyl ether derivatives in high yield. α-Silyl radicals generated from (bromomethyl)dimethylsilyl derivatives of cyclic allylic alcohols undergo highly regio- and stereoselective 5-*exo* radical cyclization reactions to provide the corresponding siloxanes (eq 1).[2,8] These five-membered siloxanes can be efficiently oxidized to 1,3-diols[2a–d,8] by the use of Tamao's conditions[9] or cleaved to β-methyl alcohols with potassium *t*-butoxide–dimethyl sulfoxide (eq 1).[2c,d] Radicals generated from the (bromomethyl)dimethylsilyl ethers of acyclic alcohols also exhibit similar regioselectivity during radical cyclization.[2a] Formation of *trans*-3,4-disubstituted 1-sila-2-oxa-cyclopentanes is generally favored (eq 2).[2a] However, propensity toward increased 6-*endo* cyclization, as is observed in the cyclization of 2-sila-5-hexen-1-yl radicals,[10] can be clearly manifested in certain α-silyl radicals (eq 3).[2a, 11] This 6-*endo* mode of cyclization is observed for α-silyl radicals with appropriate structural bias (eq 4).[12] Detailed studies on the factors that affect these two competing modes of cyclization of allylic α-silyl radicals have resulted in the development of new methodology for the introduction of an angular hydroxymethyl group (eq 5).[13,14]

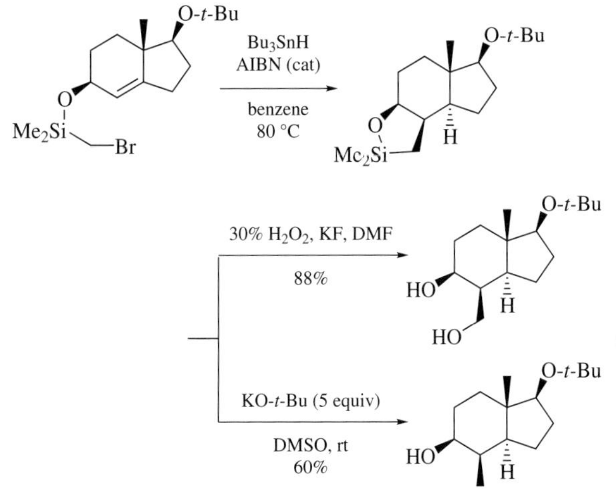

(1)

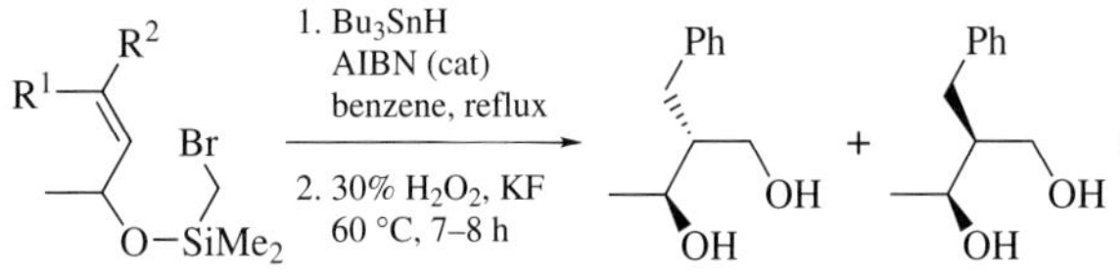

(2)

$$R^1 = H, R^2 = Ph \quad 84:16 \ (85\%)$$
$$R^1 = Ph, R^2 = H \quad 100:0 \ (94\%)$$

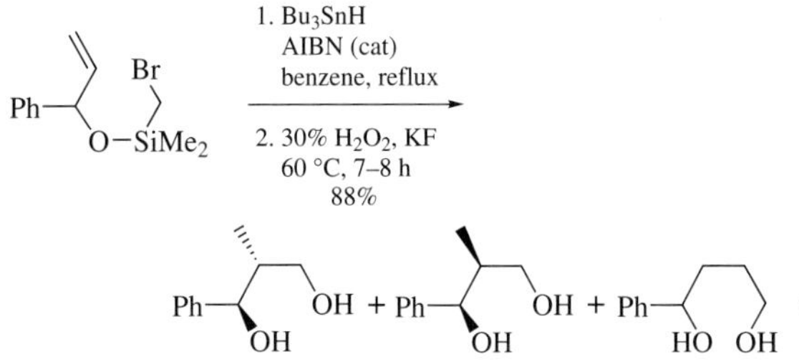

(3)

54:5:41

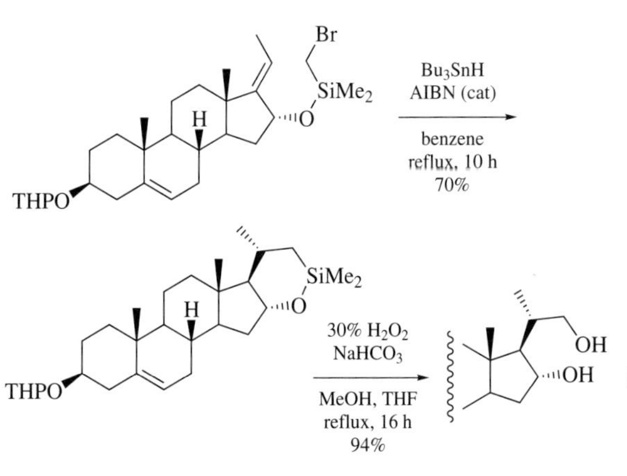

(4)

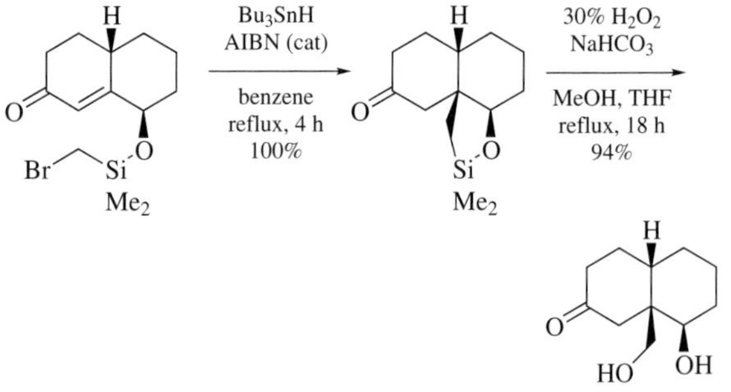

(5)

(Bromomethyl)dimethylsilyl propargylic ethers undergo highly regio-, stereo-, and chemoselective radical cyclization.[3,15] The α-silyl radicals cyclize in a 5-*exo* fashion to provide siloxanes with an exocyclic alkene having *trans* stereochemistry between the newly formed C–C and C–H bonds (eq 6).[15] This stereochemical preference is reversed when a substituent at the distal sp carbon is a TMS or phenyl group. When there are competing sites for addition between alkenic and alkynic carbons by the 5-*exo*-mode of cyclization, these α-silyl radicals preferentially attack an sp (5-*exo*-dig) over sp^2 carbon (5-*exo*-trig).[15]

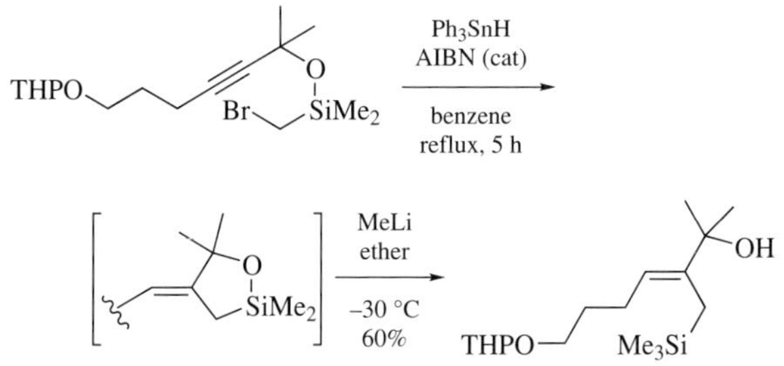

(6)

The mode of cyclization of α-silyl radicals generated from homoallylic (bromomethyl)dimethylsilyl ethers is highly dependent upon the substitution pattern on the distal alkenic carbon.[4] Thus while those with unsubstituted distal alkene carbons provide seven-membered siloxanes (7-*endo*) (eq 7),[4] those having substituted carbons undergo a 6-*exo*-mode of radical cyclization in a highly regio- and stereoselective manner (eq 8).[4] These six-membered siloxanes provide, upon Tamao oxidation,[9] branched chain 1,4-diols (eq 8).[4]

Generation of Chloro(lithiomethyl)dimethylsilane. Treatment of (bromomethyl)chlorodimethylsilane in THF/ether with 1 equiv of *n*-butyllithium in hexanes at −120 °C produces chloro(lithiomethyl)dimethylsilane. This lithiated derivative spontaneously eliminates LiCl to give an intermediate silaethylene species, which undergoes dimerization to afford 1,1,3,3-tetramethyl-1,3-disilacyclobutane (45%), together with a small amount of 1,1,3,3,5,5-hexamethyl-1,3,5-trisilacyclohexane (15%) (eq 9).[16]

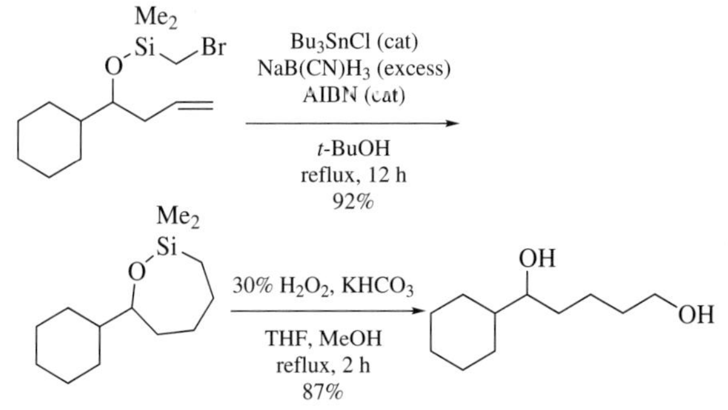

(7)

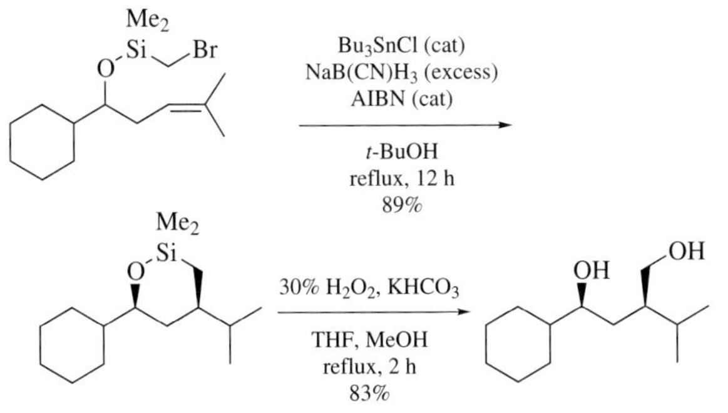

$$ (8) $$

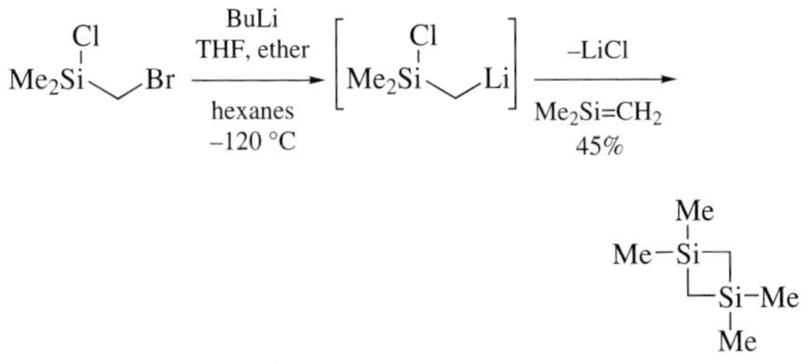

$$ (9) $$

Silazole Synthesis. (Bromomethyl)dimethylsilyl derivatives obtainable from 2-mercaptoimidazoles and -benzimidazoles undergo cyclodehydrohalogenation, in the presence of a slight excess of a proton sponge (1,8-bis(dimethylamino)naphthalene), providing silazoles (eq 10).[17]

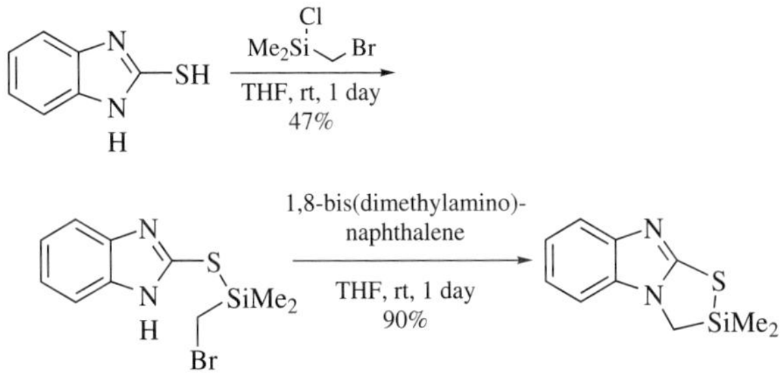

$$ (10) $$

First Update

Anne-Lise Dhimane, Louis Fensterbank, Jean-Philippe Goddard, Emmanuel Lacôte & Max Malacria
Université Pierre et Marie Curie, Paris, France

The conformation of (bromomethyl)chlorodimethylsilane has been studied by vibrational (IR and Raman) spectroscopy and ab initio methods.[18]

Radical Reactions.

Olefins and Alkynes. The increased scope of the silicon tether in radical chemistry, which is still the most important reactivity allowed by the (bromomethyl)dimethylsilane moiety, has been reviewed.[19] Some authors have again illustrated the synthetic interest of Nishiyama–Stork radical cyclizations of allylic (bromomethyl)dimethylsilyl ethers. Jenkins achieved stereoselective cyclizations on a fused cyclopentanol,[20] while Herdewijn prepared pyranosyl nucleosides.[21] Starting from γ-substituted

γ-hydroxy-α-methylene carboxylates, Nagano reported a diastereoselective entry to seven-membered oxacycles.[22]

Because of the relative stability of α-silyl radicals, radical cyclizations of propargylic (bromomethyl)dimethylsilyl ethers usually require slow addition of the reducing agent. Maruoka was able to carry out the same cyclizations without syringe pump addition; upon precomplexation of the substrate with ATPH, a bowl-shaped, aluminum-derived Lewis acid, the reaction proceeded smoothly toward the expected products at low temperature ($-78\,^\circ$C, eq 11).[23] Only trace amounts of the same product were obtained without the Lewis acid. This outcome has been attributed to a conformational restriction introduced by the cavity, which forces the two reacting arms into proximity.

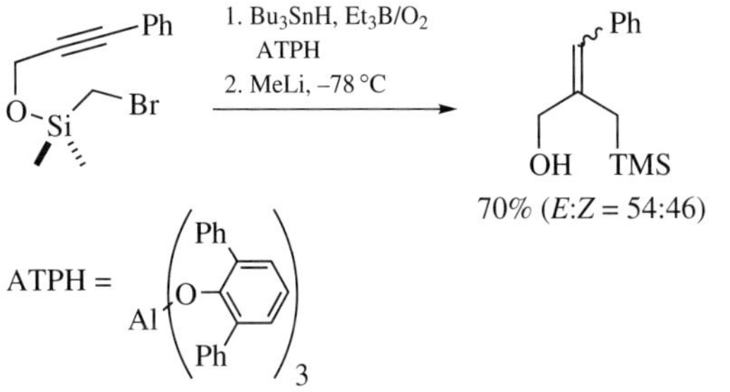

$$ (11) $$

Tanaka's strategy for the preparation of $1'$-C-hydroxymethyl uridine derivatives related to angustmycin antibiotics relied on radical cyclizations from (bromomethyl)dimethylsilyl-tethered $1',2'$-unsaturated uridines.[24] The best yield was obtained with a carbomethoxy substituent (eq 12), but the reaction also worked with a phenoxy substituent, giving access to the $2'$-epimer in the angustmycin family, and thus to analogues for SAR examinations.

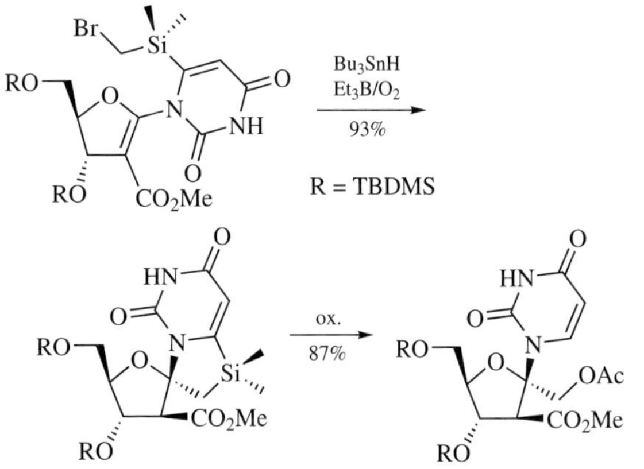

$$ (12) $$

Nitrogen-based Functional Groups. Important efforts have been devoted to (bromomethyl)dimethylsilyl radical reactions featuring nitrogen atoms. Renaud introduced a powerful new terminating step for the Nishiyama–Stork cyclization.[25] After conversion of the silyl ether to the corresponding iodide, cyclization in the presence of phenylsulfonyl azide allowed azidation of the cyclized radical (eq 13).

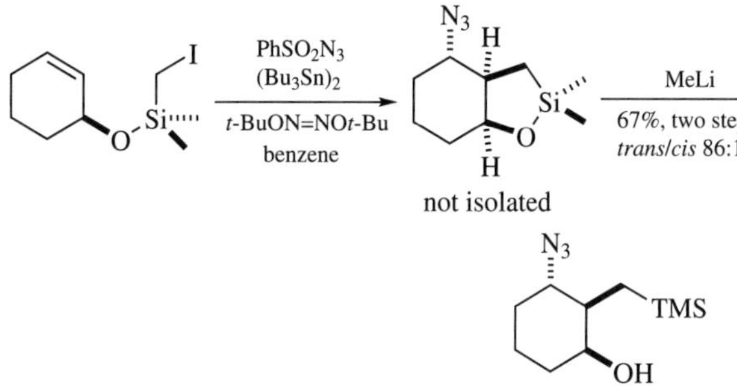

(13)

The Malacria group reported the extension of the silyl tether to nitrogen atoms.[26] Despite the increased fragility of the Si–N bond, the authors could devise a one-pot procedure that led to the formation of amino alcohols from allylamines (eq 14). Different protecting groups could be installed on the nitrogen atom. The reaction could be extended to propargylic amines, albeit in slightly lower yields.

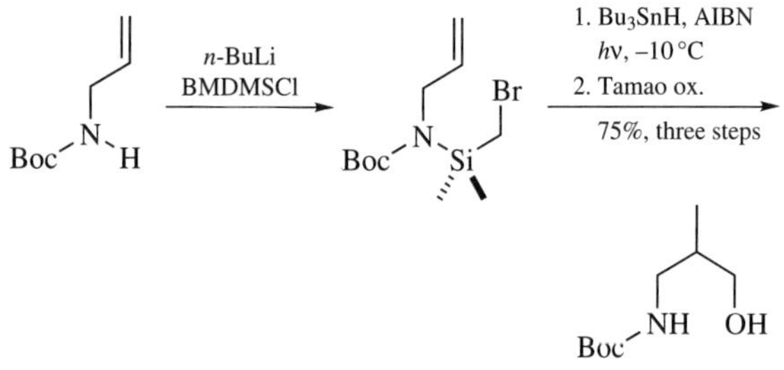

(14)

Finally, Friestad designed the radical addition of BMDMS-derived radicals onto chiral hydrazones.[27] Implementation of this methodology gave him entry to chiral amino alcohols with high diastereoselectivities (eq 15). The silicon tether is an essential tool for the stereochemical outcome, as it allows the reaction to proceed through a temporary cyclic transition state, but eventually leads to formal acyclic control.

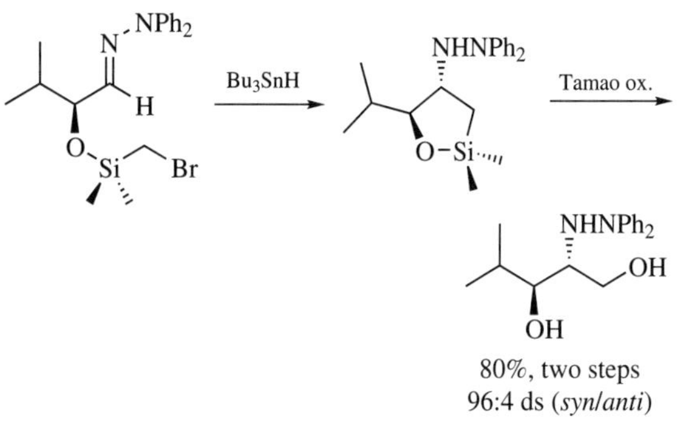

(15)

80%, two steps
96:4 ds (syn/anti)

Radical Cascades. During the last decade, the Malacria group has pursued the design of a large variety of molecular edifices whose assembly relied on the cyclization of a (bromomethyl)dimethylsilyl propargylic ether as an efficient trigger for radical cascades.[19] Indeed, the vinyl radical originating from the efficient and highly regioselective 5-exo-dig cylization of the initial α-silyl radical has proven to constitute a very versatile synthetic tool. Useful carbocyclic cores possessing hydrindene[28] and steroid[29] skeletons have been diastereoselectively built after 5-exo/6-exo cyclization tandem processes.[30] Round-trip strategies, as defined by Curran,[31] could be exploited

in various contexts. Thus, the controlled construction of two stereogenic centers from two sp-hybridized carbon atoms was achieved on the occasion of a highly diastereoselective approach to cyclopentanone derivatives.[32] The key step relies on the back cyclization of translocated radicals generated by activated 1,5-H transfers from the initial vinyl radical.[33] Highly efficient all-carbon 5-endo-trig processes have been devised from (bromomethyl)dimethylsilyl diisopropyl propargylic ethers.[34] These sterically encumbered substrates undergo a diastereoselective translocation of the vinyl radical to an unactivated methyl group of one iso-propyl group.[35] Highly strained bicyclo[3.1.1]heptanes could be obtained through a radical cascade[36] featuring a 5-exo-dig cyclization/1,6-H transfer/6-endo-trig cyclization/4-exo-dig cyclization sequence followed by a final 1,6-H final transfer, which constitutes the driving force of the reversible 4-exo-dig cyclization.[37] Moreover 1,n (n = 4, 5, 6, 7)-H transfers have been incorporated into new radical cascades.[38] Access to enantiomerically pure 1,2,3-triols relied on the previously unexploited 1,4-H transfer.[39] Introduction of a cyclopropyl moiety on the vinyl radical led to a new preparation of allene derivatives. It could also serve as a new radical clock (eq 16).[40]

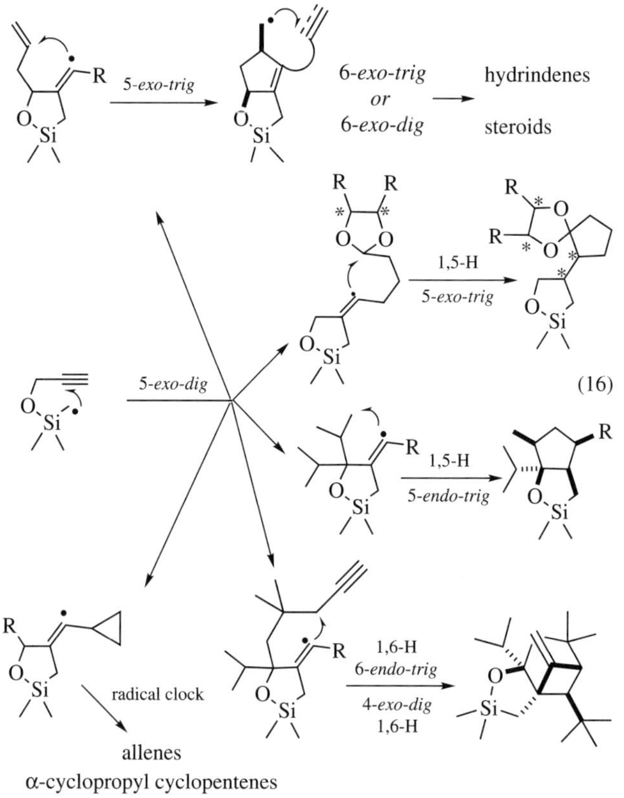

(16)

The diastereoselective construction of polycyclic structures such as linear triquinanes has been explored by the Malacria group from acyclic[41] or macrocyclic precursors.[42] Based on an acyclic (bromomethyl)dimethylsilyl precursor, the outcome of the sequence leading to the triquinane is remarkable; five new C–C bonds, two contiguous quaternary centers, and four new stereogenic centers can be formed with almost complete control (eq 17).[43]

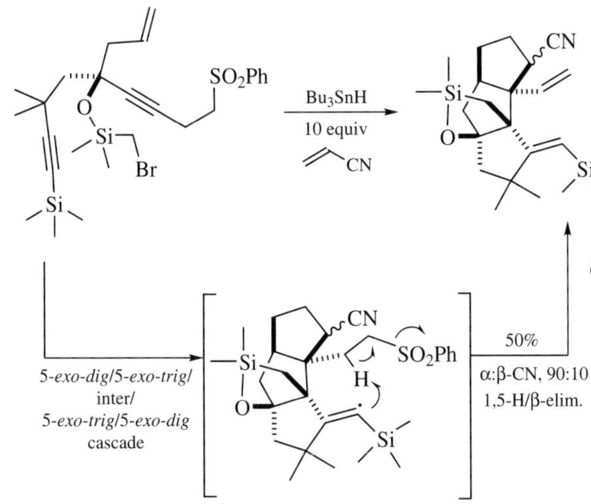
SO2Ph
Bu3SnH
10 equiv
CN
Br
Si
O
5-exo-dig/5-exo-trig/
inter/
5-exo-trig/5-exo-dig
cascade
50%
α:β-CN, 90:10
1,5-H/β-elim.

(17)

In another diastereoselective access to a linear triquinane, a radical transannular cascade strategy from a (bromomethyl) dimethylsilyl ether of a cycloundecadienyne was devised.[44] The vinyl radical, generated from the initial 5-*exo*-dig process, followed a transannular tandem of 5-*exo*/5-*exo* cyclizations (eq 18).

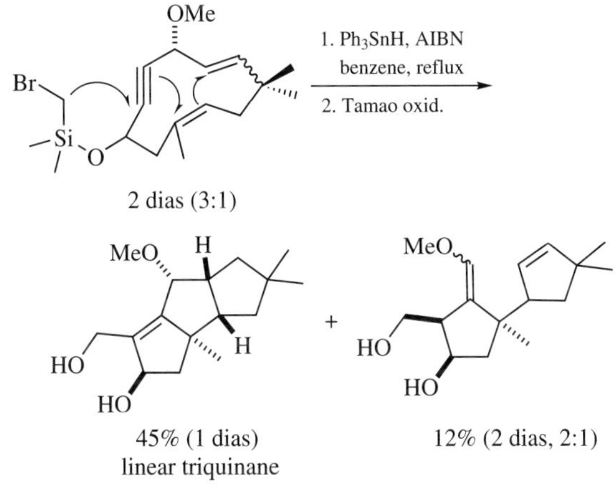
OMe
Br
Si
O
1. Ph3SnH, AIBN
benzene, reflux
2. Tamao oxid.
2 dias (3:1)
MeO
H
HO
HO
H
+
45% (1 dias)
linear triquinane
12% (2 dias, 2:1)

(18)

The unusual angular 4,6,5-tricyclic framework of the protoilludanes was attained by following a similar transannular strategy in which the (bromomethyl)dimethylsilyl ether was switched from one propargylic position to the other on the cycloundecadienyne platform (eq 19).[45] The first total synthesis of *epi*-illudol proceeds through a biomimetic tandem of transannular 4-*exo*/6-*exo* cyclizations of the vinyl radical.

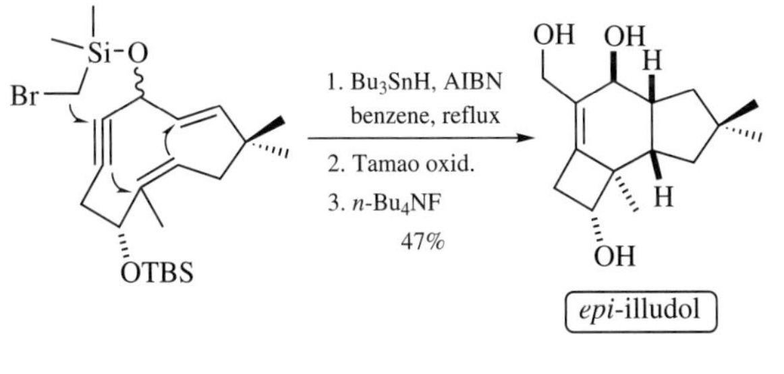
Si-O
Br
OTBS
1. Bu3SnH, AIBN
benzene, reflux
2. Tamao oxid.
3. n-Bu4NF
47%
OH OH
H
H
OH
epi-illudol

(19)

Application in Total Synthesis. Based on a Michael-oriented 6-*endo-trig* cyclization, Corey has reported stereocontrolled syntheses of salinosporamide A and biologically active analogs (eq 20).[46,47] After silylation of an unsaturated γ-lactam, the resulting silyl ether underwent a classical tributyltin hydride radical-chain cyclization to afford a *cis*-fused bicyclic γ-lactam in high yield. The high regioselectivity of the cyclization process allowed the exclusive formation of the 6-*endo-trig* vs. the 5-*exo-trig* cyclization product. The total stereoselectivity (dr > 99:1) of this cyclization can be explained by the *syn*-addition of the radical to the olefinic moiety. Tamao–Fleming oxidation of the cyclic siloxane liberated a 1,4-diol that could be conveyed to the target molecule.

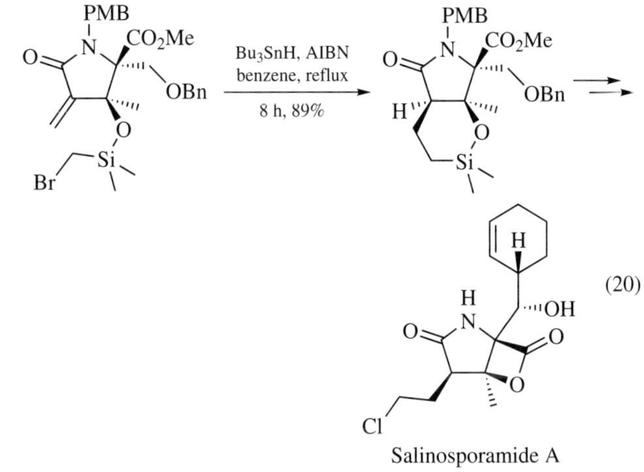
PMB
CO2Me
OBn
Br
Si
Bu3SnH, AIBN
benzene, reflux
8 h, 89%
H
OH
Cl
Salinosporamide A

(20)

Prunet and Férézou reported a hydroxy-directed diastereoselective installation of a methyl group on indalone models and a spiroketal substrate.[48] This process has been studied on different substrates leading to various diastereoselectivities. It has also been applied to the synthesis of the bafilomycin A_1 C_{15}–C_{25} subunit. The (bromomethyl)dimethylsilyl ether was engaged in a radical-mediated cyclization process which corresponds to a radical Michael addition of an α-silyl radical onto an α,β-unsaturated ester. In this case (eq 21), the 5-*exo*-trig cyclization led to the *trans*-diastereoisomer as the major product. Unfortunately, it was not possible to convert this *trans*-isomer into the desired *cis* one. The generation of the methyl group was carried out by fluoride-assisted desilylation in DMF. As shown by Little, the introduction of a hydroxymethyl group can also be accomplished on a more congested system.[49]

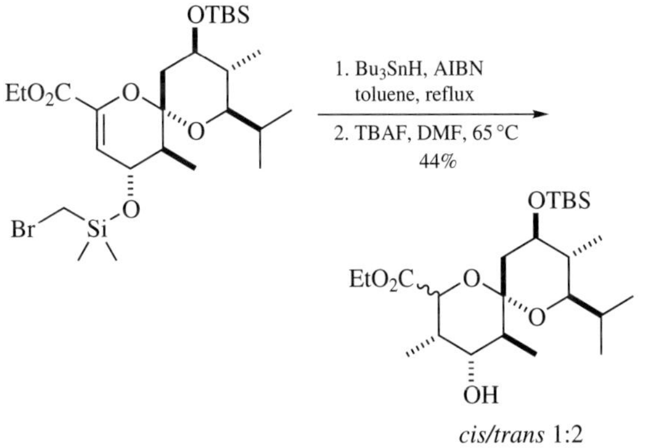
OTBS
EtO2C
Br
Si
1. Bu3SnH, AIBN
toluene, reflux
2. TBAF, DMF, 65 °C
44%
OH
cis/trans 1:2

(21)

The total synthesis of (−)-Lasonolide A has been envisaged by using the radical cyclization of a (bromomethyl)dimethylsilyl ether precursor. Lee proposed the use of such a step at an early stage of the synthesis (eq 22).[50] Upon treatment with tributyltin hydride, the initial α-silyl radical underwent a first 6-*endo*-trig cyclization generating a new carbon-centered radical. This newly formed reactive intermediate proceeded to a second 6-*exo*-trig cyclization to the α,β-unsaturated ester moiety. This radical cascade reaction was highly stereoselective and regioselective, affording the *cis*-fused bicyclic derivative as a single product in high yield. Further synthetic modifications of this skeleton led to the total synthesis of (−)-Lasonolide A.

$$\longrightarrow \quad (\text{−})\text{-Lasonolide A} \quad (22)$$

Another approach to (−)-Lasonolide A has been proposed by Shishido (eq 23).[51] Treatment of a cyclic (bromomethyl)dimethylsilyl allylic ether under radical-mediated tin hydride conditions afforded the 5-*exo-trig* cyclization product as a single product. Due to the unstability of this bicyclic compound, the Tamao–Fleming oxidation was carried out in situ giving the 1,3-diol compound in 59% overall yield. This method allowed the formation of contiguous quaternary and tertiary stereogenic centers with total control of the diastereoselectivity.

$$\longrightarrow \quad (\text{−})\text{-Lasonolide A} \quad (23)$$

A triple diastereoselective radical cyclization of a (bromomethyl)dimethylsilyl allylic ether has been reported by Kende in studies directed toward the total synthesis of taxol.[52] The authors initially planned a cascade process involving a 5-*exo-trig*/8-*exo-dig* cyclization tandem in order to generate the B-ring of the taxane framework. Unfortunately, a 5-*exo-trig*/5-*exo-dig* sequence took place leading to the formation of spirocyclic derivatives. An efficient enantioselective total synthesis of the (−)-epipodophyllotoxine has been proposed by Linker.[53] The crucial step of this approach is the regio- and stereoselective introduction of a hydroxymethylene group as a precursor of the lactone moiety. The use of a (bromomethyl)dimethylsilyl ether allowed the radical Michael addition onto an α,β-unsaturated ester. After Tamao–Fleming oxidation, the exclusive lactonization of the *cis*-isomer afforded the formation of the (−)-epipodophyllotoxine in high overall yield and enantiomeric ratio. A route to the densely functionalized cyclopentane unit of viridenomycin based on a tandem radical cyclization–allylation sequence was disclosed by Pattenden.[54]

Other Applications.

Nucleophilic Addition. The Grignard reagent made from (bromomethyl)chlorodimethyl silane adds to dimethyldichlorosilane to generate the symmetrical $ClMe_2SiCH_2SiMe_2Cl$ adduct that is a useful organosilicon spacer.[55] The hydroxymethyl unit can also be implemented through polar pathways. After metal–halogen exchange with n-BuLi or t-BuLi on the corresponding iodo precursor obtained after substitution of the bromide, the α-silyl carbon acts as a nucleophilic center and adds efficiently to a carbonyl function (eq 24). This reaction can also be accomplished with the bromide precursor and with samarium(II) as single electron transfer agent. The same article shows an umpolung approach relying on an α-silyl electrophilic center that is alkylated by an enolate intermediate.[56]

$$(24)$$

Use as an Electrophile. The chlorosilane function is highly electrophilic and can react with a variety of nucleophiles, for instance with an aryllithium carbanion,[57] to provide silyl derivatives. The (bromomethyl)chlorodimethylsilane can also be utilized as a bis-electrophilic reagent, thanks to the bromomethylene function. Initial substitution of chlorine by an aryllithium[58] or an aryl Grignard[59] followed by the displacement of bromine by a phenoxide anion provided valuable linkages for solid phase synthesis. Based on the same idea, Martin proposed a new synthesis of unsymmetrical C-aryl glycosides.[60] Orthometallation of a furyl moiety followed by silylation sets a diene on a silicon tether. The masked dienophile (a benzyne) is then introduced by O-alkylation. The cycloadduct was then converted to various naphthol derivatives (eq 25).

Alkylation of (bromomethyl)chlorodimethylsilane by the acetylide anion followed by addition of the thioacetate anion opened a new access to a silyl tethered yne-vinylsulfide precursor that found use in new radical cascades.[61] Unsymmetrical ansafluorenyl containing ligands incorporating a CH_2-$SiMe_2$ bridge have been described and result from a dialkylation with fluorenyl carbanions.[62]

(25)

Interestingly, interaction between ureas[63] and 2-quinolones[64] bearing a TMS group on nitrogen with bromomethyl dimethyl chlorosilane allows the generation of pentacoordinate silicon structures that have been characterized by ^{29}Si and ^{13}C NMR. The chlorine analog could be crystallized, permitting X-ray crystallographic analysis. These compounds provide structural information and can be used as model for the nucleophilic substitution reaction at silicon (eq 26).

(26)

+ TMSCl

The electrophilic reactivity of silicon in (bromomethyl)chlorodimethylsilane has also served in a large number of applications concerning the surface modification[65,66] of various silica gel or silicates derivatives.[67] Indeed, haloalkyl-activated silica supports are promising for the preparation of stationary phases (sorbents) with high loadings of immobilized ligands.[68]

1. (a) Curran, D. P., *Synthesis* **1988**, 417. (b) Stork, G., *Bull. Soc. Claim. Fr., Part 2* **1990**, 675.

2. (a) Nishiyama, H.; Kitajima, T.; Matsumoto, M.; Itoh, K., *J. Org. Chem.* **1984**, *49*, 2298. (b) Stork, G.; Kahn, M., *J. Am. Chem. Soc.* **1985**, *107*, 500. (c) Stork, G.; Sofia, M. J., *J. Am. Chem. Soc.* **1986**, *108*, 6826. (d) Stork, G.; Ma, R., *Tetrahedron Lett.* **1989**, *30*, 3609.

3. Magnol, E.; Malacria, M., *Tetrahedron Lett.* **1986**, *27*, 2255.

4. Koreeda, M.; Hamann, L. G., *J. Am. Chem. Soc.* **1990**, *112*, 8175.

5. Poole, C. F.; Zlatkis, A., *J. Chromatogr. Sci.* **1979**, *17*, 115.

6. *Aldrich Library of FT-IR Spectra*; Pouchert, C. J., Ed.; Aldrich: Milwaukee, 1989; Vol. 1(3), p 1634C.

7. *Sigma-Aldrich Library of Chemical Safety Data*; Aldrich: Milwaukee, 1987; Vol. 2, p 548B.

8. (a) Kurek-Tyrlik, A.; Wicha, J.; Snatzke, G., *Tetrahedron Lett.* **1988**, *29*, 4001. (b) Crimmins, M. T.; O'Mahony, R., *J. Org. Chem.* **1989**, *54*, 1157. (c) Bonnert, R. V.; Davies, M. J.; Howarth, J.; Jenkins, P. R., *J. Chem. Soc., Chem. Commun.* **1990**, *148*. (d) Majetich, G.; Song, J.-S.; Ringold, C.; Neumeth, G. A., *Tetrahedron Lett.* **1990**, *31*, 2239. (e) Kurek-Tyrlik, A.; Wicha, J.; Zarecki, A.; Snatzke, G., *J. Org. Chem.* **1990**, *55*, 3484. (f) Majetich, G.; Song, J.-S.; Ringold, C.; Nemeth, G. A.; Newton, M. G., *J. Org. Chem.* **1991**, *56*, 3973. (g) Bonnert, R. V.; Davies, M. J.; Howarth, J.; Jenkins, P. R.; Lawrence, N. J., *J. Chem. Soc., Perkin Trans. 1* **1992**, 27. (h) Pedretti, V.; Mallet, J.-M.; Sinay, P., *Carbohydr. Res.* **1993**, *244*, 247.

9. (a) Tamao, K.; Ishida, N.; Tanaka, T.; Kumada, M., *Organometallics* **1983**, *2*, 1694. (b) Tamao, K.; Ishida, N.; Kumada, M., *J. Org. Chem.* **1983**, *48*, 2120.

10. (a) Wilt, J. W., *J. Am. Chem. Soc.* **1981**, *103*, 5251. (b) Wilt, J. W., *Tetrahedron* **1985**, *41*, 3979. (c) Wilt, J. W.; Lusztyk, J.; Peeran, M.; Ingold, K. U., *J. Am. Chem. Soc.* **1988**, *110*, 281.

11. Lakomy, I.; Scheffold, R., *Helv. Chim. Acta* **1993**, *76*, 804.

12. (a) Koreeda, M.; George, I. A., *J. Am. Chem. Soc.* **1986**, *108*, 8098. (b) Koreeda, M.; George, I. A., *Chem. Lett.* **1990**, 83.

13. Lejeune, J.; Lallemand, J. Y., *Tetrahedron Lett.* **1992**, *33*, 2977.

14. Koreeda, M.; Visger, D. C., *Tetrahedron Lett.* **1992**, *33*, 6603.

15. (a) Agnel, G.; Malacria, M., *Tetrahedron Lett.* **1990**, *31*, 3555. (b) Journet, M.; Magnol, E.; Smadja, W.; Malacria, M., *Synlett* **1991**, 58. (c) Journet, M.; Malacria, M., *J. Org. Chem.* **1992**, *57*, 3085. (d) Journet, M.; Malacria, M., *Tetrahedron Lett.* **1992**, *33*, 1893.

16. Chmielecka, J.; Stanczyk, W., *Synlett* **1990**, 344.

17. Alper, H.; Wolin, M. S., *J. Org. Chem.* **1975**, *40*, 437.

18. Nilsen, A.; Klaeboe, P.; Nielsen, C. J.; Guirgis, G. A.; Aleksa, V., *J. Mol. Struct.* **2000**, *550–551*, 199.

19. Fensterbank, L.; Malacria, M.; Sieburth, S. M., *Synthesis* **1997**, 813.

20. Jenkins, P. R.; Wood, A. J., *Tetrahedron Lett.* **1997**, *38*, 1853.

21. Doboszewski, B.; De Winter, H.; Van Aerschot, A.; Herdewijn, P., *Tetrahedron* **1995**, *51*, 12319.

22. Nagano, H.; Hara, S., *Tetrahedron Lett.* **2004**, *45*, 4329.

23. Ooi, T.; Hokke, Y.; Tayama, E.; Maruoka, K., *Tetrahedron* **2001**, *57*, 135.

24. Ogamino, J.; Mizunuma, H.; Kumamoto, H.; Takeda, S.; Haraguchi, K.; Nakamura, K. T.; Sugiyama, H.; Tanaka, H., *J. Org. Chem.* **2005**, *70*, 1684.

25. Ollivier, C.; Renaud, P., *J. Am. Chem. Soc.* **2001**, *123*, 4717.

26. Blaszykowski, C.; Dhimane, A.-L.; Fensterbank, L.; Malacria, M., *Org. Lett.* **2003**, *5*, 1341.

27. Friestad, G. K.; Massari, S. E., *J. Org. Chem.* **2004**, *69*, 863.

28. Journet, M.; Lacôte, E.; Malacria, M., *J. Chem. Soc., Chem. Commun.* **1994**, 461.

29. Wu, S.; Journet, M.; Malacria, M., *Tetrahedron Lett.* **1994**, *35*, 8601.

30. Wipf, P. F.; Graham, T. H., *J. Org. Chem.* **2003**, *68*, 8798.

31. Haney, B. P.; Curran, D. P., *J. Org. Chem.* **2005**, *65*, 2007.

32. Bogen, S.; Journet, M.; Malacria, M., *Synlett* **1994**, 958.

33. Fensterbank, L.; Dhimane, A. L.; Wu, S.; Lacôte, E.; Bogen, S.; Malacria, M., *Tetrahedron* **1996**, *52*, 11405.

34. Bogen, S.; Malacria, M., *J. Am. Chem. Soc.* **1996**, *118*, 3992.

35. Bogen, S.; Gulea, M.; Fensterbank, L.; Malacria, M., *J. Org. Chem.* **1999**, *64*, 4920.

36. Bogen, S.; Fensterbank, L.; Malacria, M., *J. Am. Chem. Soc.* **1997**, *119*, 5037.

37. Bogen, S.; Fensterbank, L.; Malacria, M., *J. Org. Chem.* **1999**, *64*, 819.

38. Gross, A.; Fensterbank, L.; Bogen, S.; Thouvenot, R.; Malacria, M., *Tetrahedron* **1997**, *53*, 13797.

39. Gulea, M.; Lopez-Romero, J. M.; Fenterbank, L.; Malacria, M., *Org. Lett.* **2000**, *2*, 2591.

40. Mainetti, E.; Fensterbank, L.; Malacria, M., *Synlett* **2002**, 923.

41. Bogen, S.; Devin, P.; Fensterbank, L.; Journet, M.; Lacôte, E.; Malacria, M., *Recent Res. Devel. In Organic Chem.* **1997**, *1*, 385.

42. Aïssa, C.; Dhimane, A. L.; Malacria, M., *Synlett* **2000**, 1585.

43. Devin, P.; Fensterbank, L.; Malacria, M., *J. Org. Chem.* **1998**, *63*, 6764.

44. Dhimane, A. L.; Aïssa, C.; Malacria, M., *Angew. Chem. Int. Ed.* **2002**, *41*, 3284.

45. Rychlet Elliott, M.; Dhimane, A. L.; Malacria, M., *J. Am. Chem. Soc.* **1997**, *119*, 3427.

46. Reddy, L. R.; Saravanan, P.; Corey, E. J., *J. Am. Chem. Soc.* **2004**, *126*, 6230.

47. Reddy, L. R.; Fournier, J.-F.; Reddy, B. V. S.; Corey, E. J., *Org. Lett.* **2005**, *7*, 2699.

48. Poupon, J. C.; Lopez, R.; Prunet, J.; Férézou, J. P., *J. Org. Chem.* **2002**, *67*, 2118.

49. Carroll, G. L.; Allan, A. K.; Schwaebe, M. K.; Little, R. D., *Org. Lett.* **2000**, *2*, 2531.

50. Lee, E.; Song, H. Y.; Kang, J. W.; Kim, D. S.; Jung, C. K.; Joo, J. M., *J. Am. Chem. Soc.* **2002**, *124*, 384.

51. Yoshimura, T.; Bando, T.; Shindo, M.; Shishido, K., *Tetrahedron Lett.* **2004**, *45*, 9241.

52. Kende, A. S.; Journet, M.; Ball, R. G.; Tsou, N. N., *Tetrahedron Lett.* **1996**, *37*, 6295.

53. Engelhardt, U.; Sarkar, A.; Linker, T., *Angew. Chem. Int. Ed.* **2003**, *42*, 2487.

54. Mulholland, N. P.; Pattenden, G., *Tetrahedron Lett.* **2005**, *46*, 937.

55. Ganicz, T.; Stanczyk, W. A.; Bialecka-Florjanczyk, E.; Sledzinska, I., *Polymer* **1999**, *40*, 4733.

56. Iwamoto, M.; Miyano, M.; Utsugi, M.; Kawada, H.; Nakada, M., *Tetrahedron Lett.* **2004**, *45*, 8647.

57. Van Dort, P. C.; Fuchs, P. L., *J. Org. Chem.* **1997**, *62*, 7142.

58. Chenera, B.; Finkelstein, J. A.; Veber, D. F., *J. Am. Chem. Soc.* **1995**, *117*, 11999.

59. Newlander, K. A.; Chenera, B.; Veber, D. F.; Yim, N. C. F.; Moore, M. L., *J. Org. Chem.* **1997**, *62*, 6726.

60. Kaelin, D. E., Jr.; Sparks, S. M.; Plake, H. R.; Martin, S. F., *J. Am. Chem. Soc.* **2003**, *125*, 12994.

61. Journet, M.; Rouillard, A.; Cai, D.; Larsen, R. D., *J. Org. Chem.* **1997**, *62*, 8630.

62. Siedle, A. R.; Newmark, R. A.; Duerr, B. F.; Leung, P. C., *J. Mol. Cat. A: Chemical* **2004**, *214*, 187.

63. Bassindale, A. R.; Glynn, S. J.; Taylor, P. G.; Auner, N.; Herrschaft, B., *J. Organomet. Chem.* **2001**, *619*, 132.

64. Bassindale, A. R.; Parker, D. J.; Taylor, P. G.; Auner, N.; Herrschaft, B., *J. Organomet. Chem.* **2003**, *667*, 66.

65. Liu, Y.-H.; Lin, H.-P.; Mou, C.-Y., *Langmuir* **2004**, *20*, 3231.

66. Cao, C.; Fadeev, A. Y.; McCarthy, T. J., *Langmuir* **2001**, *17*, 757.

67. Hasegawa, I.; Niwa, T.; Takayama, T., *Inorg. Chem. Commun.* **2005**, *8*, 159.

68. Mingalyov, P. G.; Fadeev, A. Yu; *J. Chromatogr. A* **1996**, *719*, 291.

Bromotrimethylsilane[1]

Me$_3$SiBr

[2857-97-8]   C$_3$H$_9$BrSi   (MW 153.09)

InChI = 1S/C3H9BrSi/c1-5(2,3)4/h1-3H3

InChIKey = IYYIVELXUANFED-UHFFFAOYSA-N

(mild and selective reagent for cleavage of lactones, epoxides, acetals, phosphonate esters and certain ethers; effective reagent for formation of silyl enol ethers; can function as brominating agent)

Alternate Name: TMS-Br.

Physical Data: bp 79 °C; d 1.188 g cm^{-3}; n_D^{20} 1.4240; fp 32 °C.

Solubility: sol CCl$_4$, CHCl$_3$, CH$_2$Cl$_2$, ClCH$_2$CH$_2$Cl, MeCN, toluene, hexanes; reactive with THF (ethers), alcohols, and somewhat reactive with EtOAc (esters).

Form Supplied in: colorless liquid, packaged in ampules.

Analysis of Reagent Purity: well characterized by ^{1}H, ^{13}C, and ^{29}Si NMR spectroscopy.

Preparative Methods: although many methods are reported,[1] only a few are provided here: chlorotrimethylsilane undergoes halogen exchange with either magnesium bromide[2] in Et$_2$O or sodium bromide[3] in MeCN, which allows in situ reagent formation (eq 1); alternatively, hexamethyldisilane reacts with bromine in benzene solution or neat, to afford only TMS-Br with no byproducts (eq 2).[4] TMS-Br may also be generated by reaction of hexamethyldisiloxane and aluminum bromide (eq 3).[5] However, it should be noted that the reactivity of in situ generated reagent appears to depend upon the method of preparation.

$$Me_3SiCl \xrightarrow[\substack{or \\ NaBr, MeCN}]{MgBr_2, Et_2O} Me_3SiBr + LiBr \text{ or } NaBr \quad (1)$$

$$Me_3Si\text{-}SiMe_3 + Br_2 \longrightarrow Me_3SiBr \quad (2)$$

$$Me_3SiOSiMe_3 + AlBr_3 \longrightarrow Me_3SiBr \quad (3)$$

Purification: by distillation.

Handling, Storage, and Precaution: extremely sensitive to light, air, and moisture; fumes in air due to hydrolysis (HBr), and becomes discolored upon prolonged storage (free Br$_2$).

Original Commentary

Michael J. Martinelli
Lilly Research Laboratories, Indianapolis, IN, USA

Ester Cleavage.[6] Although esters are readily cleaved with iodotrimethylsilane, reaction of esters with TMS-Br under similar conditions gives somewhat lower yields of silyl esters or acids upon hydrolysis (eq 4). Lactones, however, react with TMS-Br at 100 °C to afford ω-bromocarboxylic acids after hydrolysis of the silyl ester (eq 5).[7]

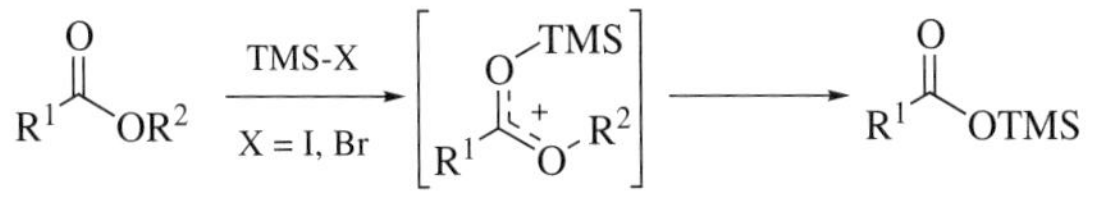

$$R^1\text{—}C(=O)\text{—}OR^2 \xrightarrow[X = I, Br]{TMS\text{-}X} \left[R^1\text{—}\overset{+}{C}(\text{—O—TMS})(\text{O}^-\text{—}R^2) \right] \longrightarrow R^1\text{—}C(=O)\text{—OTMS} \quad (4)$$

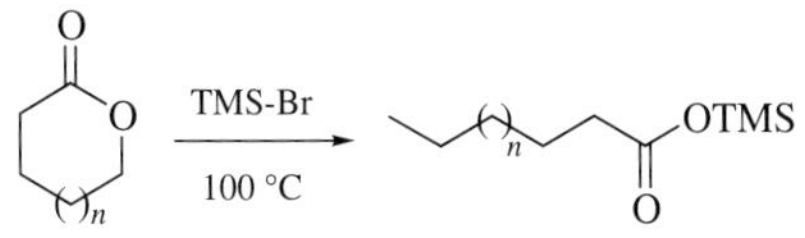

$$(5)$$

Ether Cleavage. THF[8] reacts with TMS-Br, thereby rendering ethereal solvents incompatible with the reagent. Smooth removal of the methoxymethyl (MOM) protecting group can be accomplished with TMS-Br at 0 °C (eq 6),[9] Whereas acetals, THP and silyl ethers are slowly cleaved with TMS-Br, the reagent generated in situ effects selective MOM ether cleavage in the presence of an acetonide.[10] The majority of published ether cleavages have been accomplished with TMS-I, although limited data show that the more vigorous conditions necessary for ethyl ether cleavage also result in bromide formation.[8]

$$RO\text{-}CH_2OMe \xrightarrow[0\,°C]{TMS\text{-}Br} RO\text{-}H \quad (6)$$

Cleavage of Epoxides. Epoxide opening with TMS-Br occurs to provide the primary alkyl bromide at −60 °C (eq 7).[8]

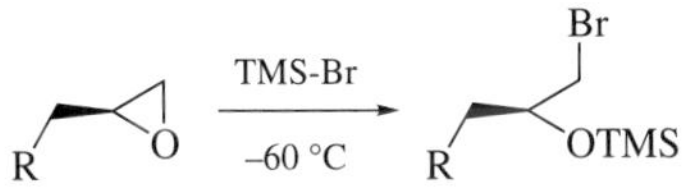

$$(7)$$

Cleavage of Acetals. Acetals can be cleaved by analogy to ethers, providing the parent carbonyl species.[3c, 6b] Glycosyl bromides have been prepared from the corresponding acetate by reaction with TMS-Br in CHCl₃ at rt (eq 8).[11] In conjunction with CoBr₂ and tetra-*n*-butylammonium bromide, TMS-Br converts the glucopyranose to the α-D-glucoside in the presence of an alcohol (eq 9).[12]

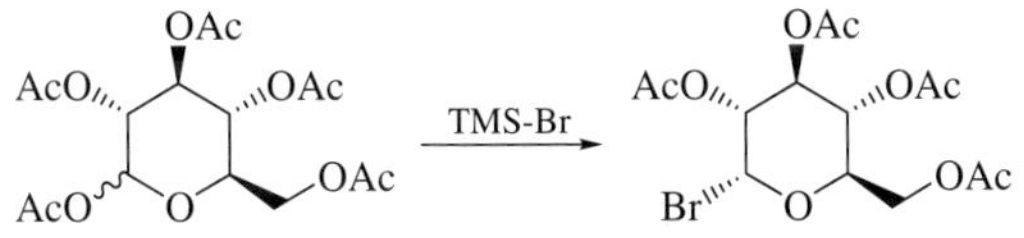

$$(8)$$

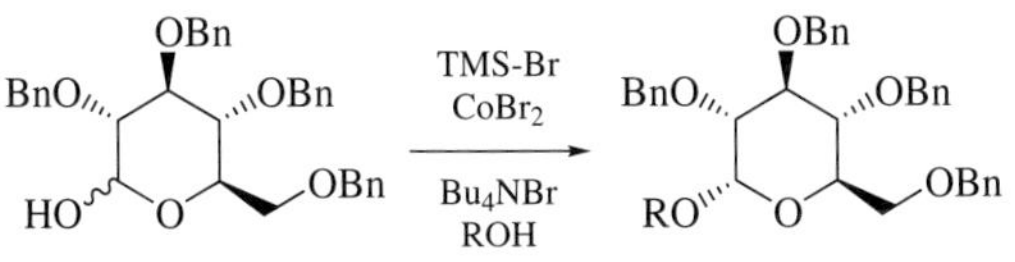

$$(9)$$

An interesting solvent effect was noted in the cleavage of the acetonide moiety in some nucleoside derivatives (eq 10).[13] In CH₂Cl₂, TMS-Br converted the acetonide to the anhydrouridine within 1.5 h, but in MeCN the bromide is formed after 10 min.

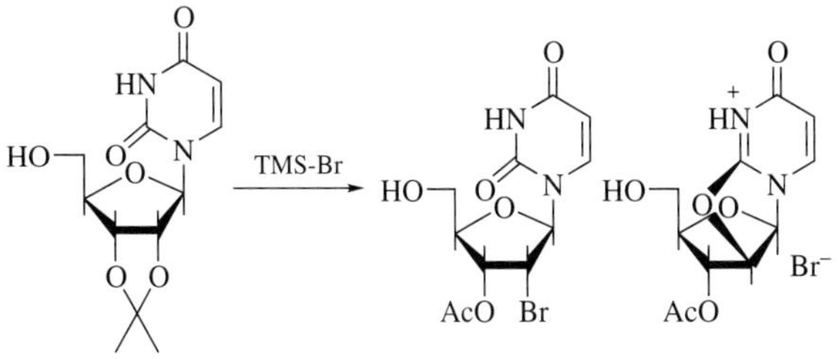

$$(10)$$

Formation of Enol Ethers.[14] Bromotrimethylsilane with triethylamine in DMF is an effective medium for production of thermodynamic (Z) silyl enol ethers (eq 11).

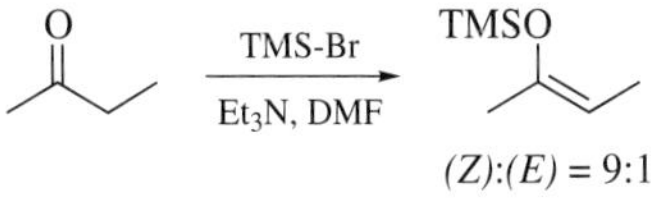

$$(Z){:}(E) = 9{:}1 \qquad (11)$$

Formation of Alkyl Bromides.[6b] Alcohols react with excess TMS-Br (1.5–4 equiv) at 25–50 °C to form the alkyl bromide and hexamethyldisiloxane (eq 12). Benzylic and tertiary alcohols react faster than secondary alcohols.

$$R\text{-}CH_2OH \xrightarrow{TMS\text{-}Br} R\text{-}CH_2Br \quad (12)$$

Reaction with Acid Chlorides.[15] Acid bromides may be prepared from acid chlorides by reaction with TMS-Br (eq 13).

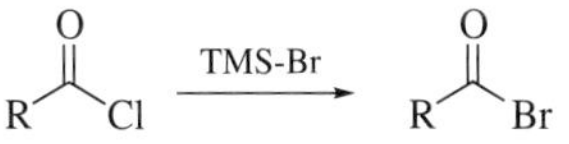

$$R\text{—}C(=O)\text{—}Cl \xrightarrow{TMS\text{-}Br} R\text{—}C(=O)\text{—}Br \quad (13)$$

Cleavage of Phosphonate Esters.[16] Compared to the reactivity of TMS-I with phosphonate and phosphate esters, TMS-Br is more selective and will cleave phosphonate esters even in the presence of carboxylic esters and carbamates. Benzyl ester protecting groups on aryl phosphates are selectively removed with TMS-Br.[17] The reaction of phosphonate esters with TMS-Br proceeds through a mechanism similar to ester cleavage, providing a silyl ester which is subsequently hydrolyzed with MeOH or H₂O (eq 14).

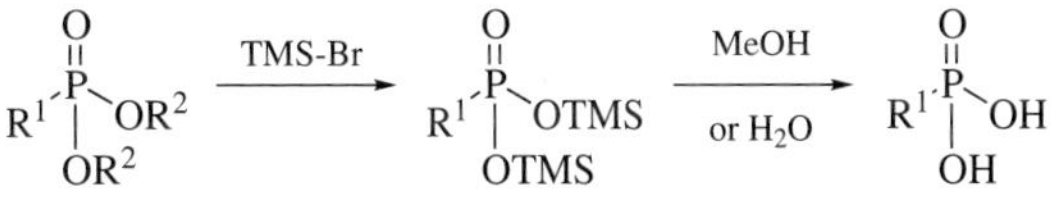

$$(14)$$

Reaction with Amines. Amines react with TMS-Br to form isolable adducts, which react readily with ketones to form enamines under mild conditions (eq 15).[18]

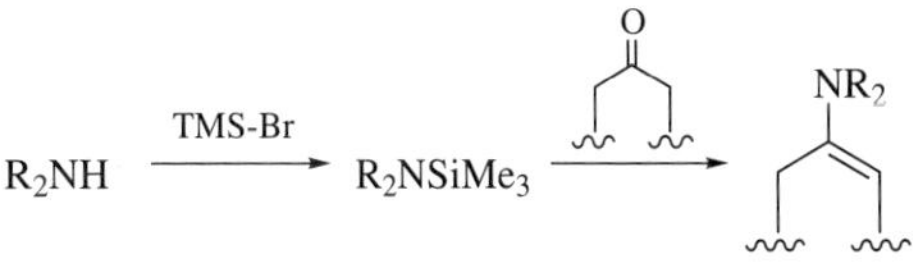

$$R_2NH \xrightarrow{TMS\text{-}Br} R_2NSiMe_3 \longrightarrow \quad (15)$$

Conjugate Addition. α,β-Unsaturated ketones undergo conjugate addition with TMS-Br. Treatment of the intermediate with *p*-toluenesulfonic acid and ethylene glycol provides *β*-bromoethyldioxolanes (eq 16).[19]

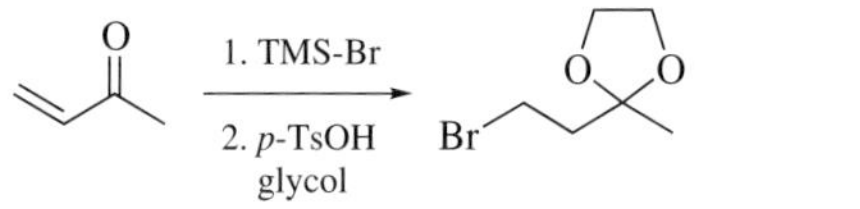

Bromolactonization. *ω*-Unsaturated carboxylic acids react with TMS-Br in the presence of a tertiary amine in DMSO yielding bromolactones, resulting from *cis* addition across the double bond (eq 17).[20]

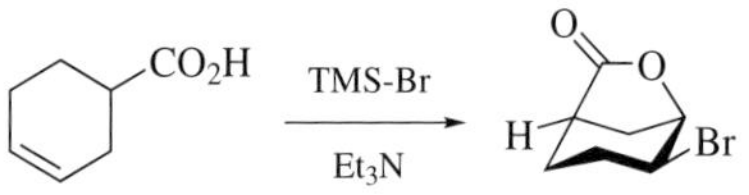

Ylide Formation. *Methylenetriphenylphosphorane* reacts with TMS-Br to provide the corresponding ylide precursor (eq 18).[21]

$$Ph_3P=CH_2 \xrightarrow{\text{TMS-Br}} Ph_3\overset{+}{P}CH_2SiMe_3 \ Br^- \qquad (18)$$

First Update

Scott R. Pollack

Johnson & Johnson Pharmaceutical Research and Development, L.L.C., Raritan, NJ, USA

Reactions with Transition Metals: Forming Carbon-Carbon Bonds. The combination of certain lanthanides and TMS-Br has been found to produce lanthanum halides (LaX_n; $n = 2$ or 3) that are very active reducing reagents (eq 19). So far, the only metals to be used in these reactions are samarium (Sm)[22] and ytterbium (Yb).[23] In addition to TMS-Br, these reactions have been accomplished using Sm/TMS-Cl/NaI under similar conditions with comparable yields.

$$La \ + \ n\,TMS\text{-}Br \xrightarrow{\hspace{2cm}} LaBr_n \qquad (19)$$
$$La = Sm \text{ or } Yb$$

These halides, which can be prepared in situ or prior to the reaction, have been used in carbon-carbon bond forming reactions between carbonyl compounds and/or their derivatives (eqs 20–22). This reaction is similar to the classical pincol coupling. The mechanism of the Yb-promoted reductive couplings is thought to occur via a single electron transfer (SET) from the reductant that then leads to a carbon-based radical, which then undergoes homo- or heterocoupling reactions (eq 23).

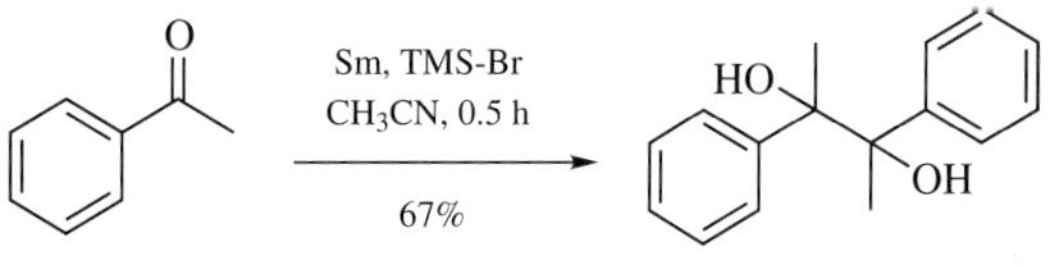

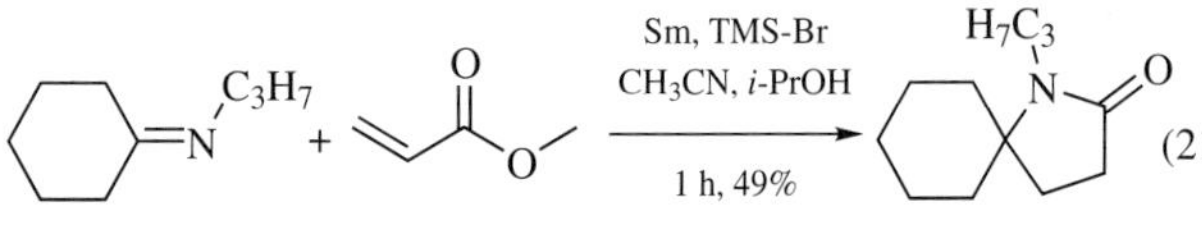

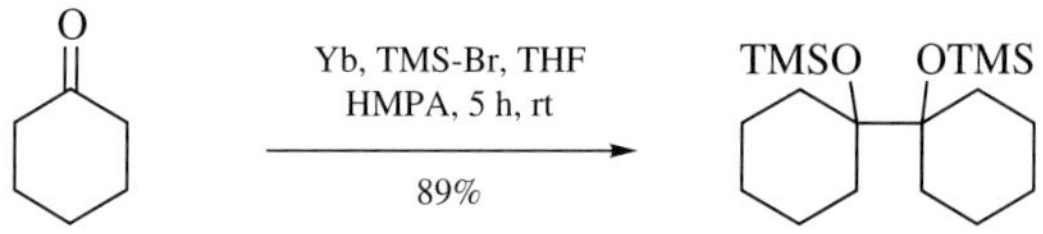

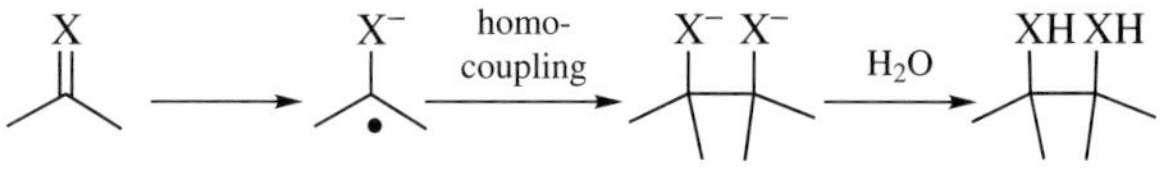

$$X = O$$
$$X = CH\text{-}C(=O)R$$
$$X = N\text{-}R$$

Reaction with Nitroalkanes. An unusual and synthetically versatile transformation has been reported wherein TMS-X (X = Br or OTf) is reacted with nitroalkanes to give *N*-siloxyimines or *N,N*-bis(siloxy)enamines (eq 24), whose structure and stereodynamics have been well investigated.[24,25]

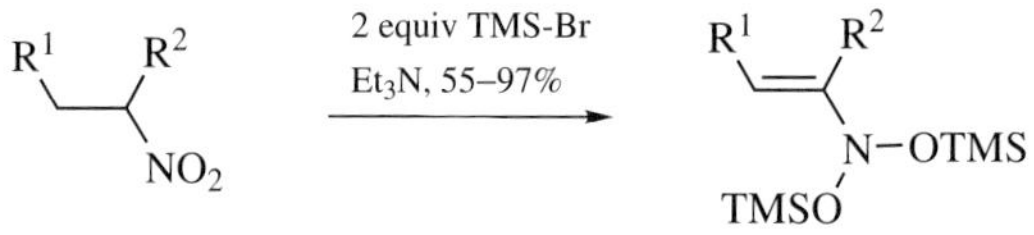

While the *N,N*-bis(siloxy)enanamines have not found much utility, it has been shown that the intermediate monosiloxyimine may be trapped intramolecularly to produced a variety of useful carbo- and heterocycles. For example, as shown in eqs 25 and 26,[26–28] *γ*-nitro esters and ketones may be reacted with TMS-Br to produce an intermediate *N*-siloxyimine. In the case of the diester (eq 25), an additional equivalent of TMS-Br reacts to form a silyl enol ether that then cyclizes to form a *trans*-substituted cyclopropane.

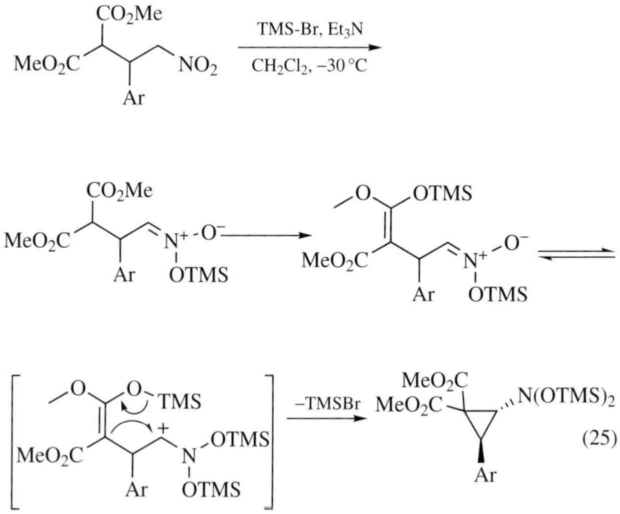

Ar = Ph; yield = 61%
Ar = *p*-MeOPh; yield = 88%

When a ketone derivative is used (eq 26), the presence of a base (Et$_3$N) favors enolization and a dihydrofuran is formed. Once again, the relative stereochemistry is *trans* between the aryl constituent and the amine. Treatment of this product with ammonium fluoride in methanol resulted in elimination of the amine and consequent formation of a 2,4-diarylfuran.

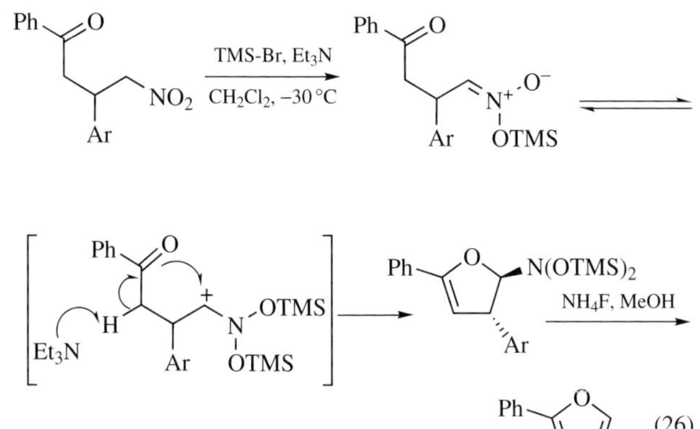

$$Ar = Ph; yield = 83\%$$
$$Ar = p\text{-}ClPh; yield = 87\% \tag{26}$$

In addition to cyclic derivatives, this chemistry has also been used for the preparation of α,β-unsaturated oximes from either nitroalkanes[29] or 1,2-oxazine N-oxides.[30] Equation 27 shows the synthesis of β-functionalized α,β-unsaturated oximes from nitroalkanes.

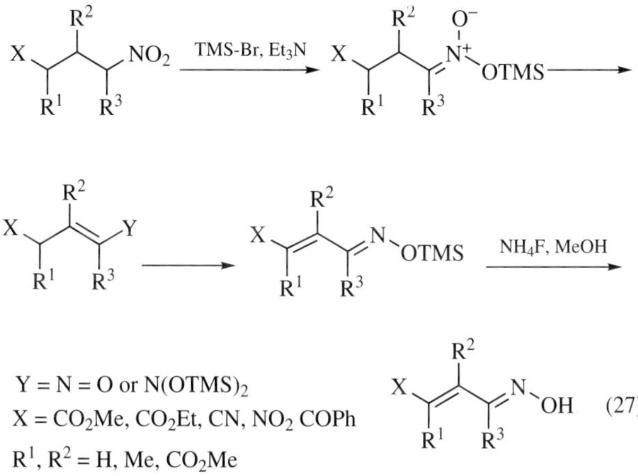

$$Y = N = O \text{ or } N(OTMS)_2$$
$$X = CO_2Me, CO_2Et, CN, NO_2 COPh$$
$$R^1, R^2 = H, Me, CO_2Me$$
$$R^3 = H, Me, Et, Ph, CO_2Me \tag{27}$$

This reaction produces the reported products in yields ranging from 26% to 28% with varying degrees of E/Z selectivity. Another approach to these oximes employs a retro-[4 + 2] fragmentation from the oxazine N-oxides (eq 28). The oxazines themselves were prepared via the hetero Diels-Alder reaction of nitroalkanes and olefins. Upon silylation of the N-oxide, at $-30\,^{\circ}\text{C}$, a rearrangement occurs giving the enamine, which then undergoes a retro Diels-Alder reaction. This leads to expulsion of a ketone and the generation of the unsaturated protected oxime with complete stereocontrol. Finally, the silyl group is removed in a subsequent step to give the oximes in 64–89% yields.

Synthesis of Silylated Thioalkynes in the Preparation of β-Thiolactams.

The derivatization of thioalkynes with TMS-Br (or TMS-Cl) and subsequent reaction with imines provides a method for the synthesis of β-thiolactams (eq 29).[31,32] The yields for the TMS-Br silylation are generally higher than those with TMS-Cl.

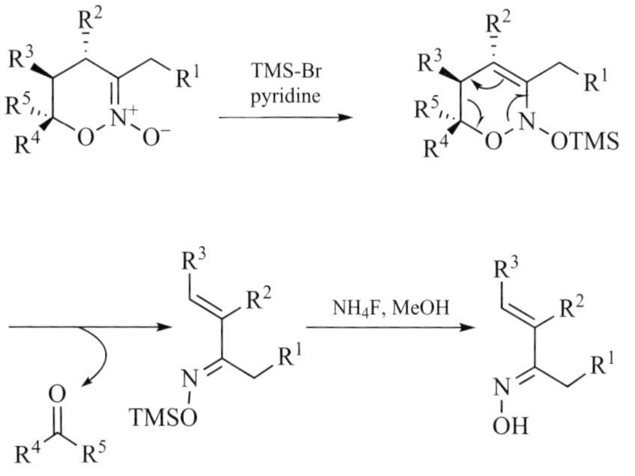

$$R^1 = H, Me; R^2 = Ph, 4\text{-MeOPh}; R^3 = H; R^4 = OMe; R^5 = Me$$

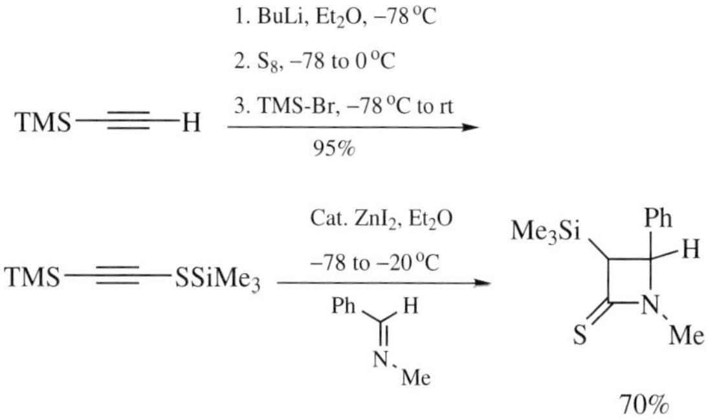

α-Bromination of Alkynyl Ethers and Sulfides.

In addition to being an active silylating agent, TMS-Br has found application as a brominating agent. One such example is the reaction with electron-rich thioalkyl-[33] and alkoxyalkynes.[34,35] The products of these reactions are the synthetically important α-bromo vinyl sulfides and alkoxides. The reaction can also be carried out with TMS-Cl or TMS-I to produce the respective chloro- or iodo-derivatives. Equation 30 shows a typical example for the alkynyl sulfides.

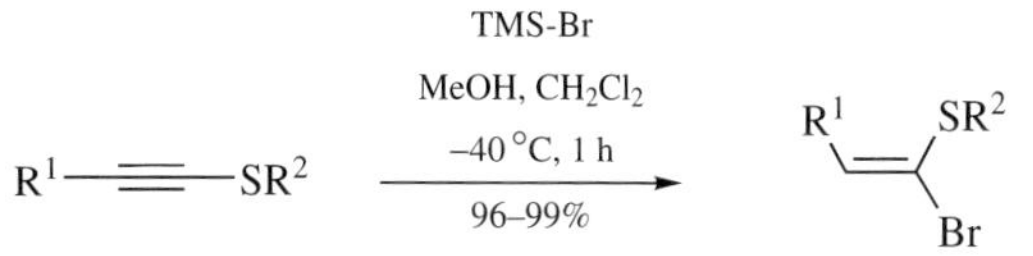

$$R^1 = C_4H_9, C_7H_{15}, PhCH_2CH_2, Cyclohexyl$$
$$R^2 = Me, Ph$$

The reaction uses only a single equivalent of TMS-Br and MeOH, which acts as a trap for the TMS group, to give MeOTMS as a by-product. The reaction is extremely regioselective with E/Z ratios from 11:1 to >33:1. The bromine (or possibly iodine) atom installed in the reaction is easily metallated and reacted with electrophiles (chloroformates, TMS-Cl, etc.) or coupled, under transition metal-catalyzed conditions, with vinyls, CO, or Grignards. Additionally, TMS-Br may also be reacted with alkynyl ethers (eq 31) to produce α-bromo vinyl ethers.

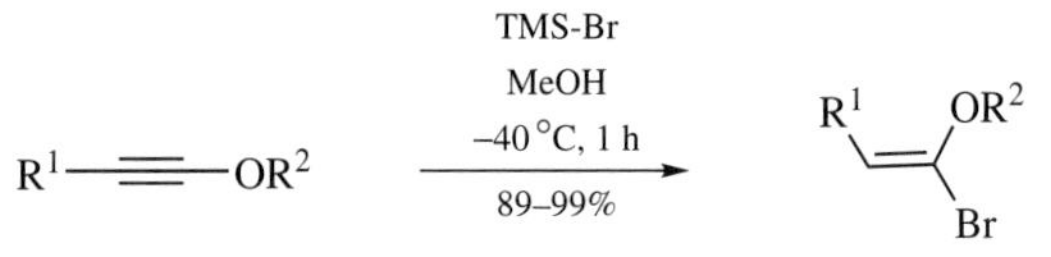

$R^1\!\!=\!\!=\!\!OR^2$ $\xrightarrow[\text{89–99\%}]{\substack{\text{TMS-Br}\\ \text{MeOH}\\ -40\,^{\circ}\text{C, 1 h}}}$ (31)

R^1 = Me, Et, C_6H_{13}, CH_2=$CHCH_2$

R^2 = Menthol, Borneol, Me, Cyclohexyl

Alkyne-Iminium-promoted Cyclizations. Another reaction of TMS-Br with alkynes is the intramolecular alkyne-iminium ion cyclization,[36] which has been used to prepare piperidines of various substitution patterns.

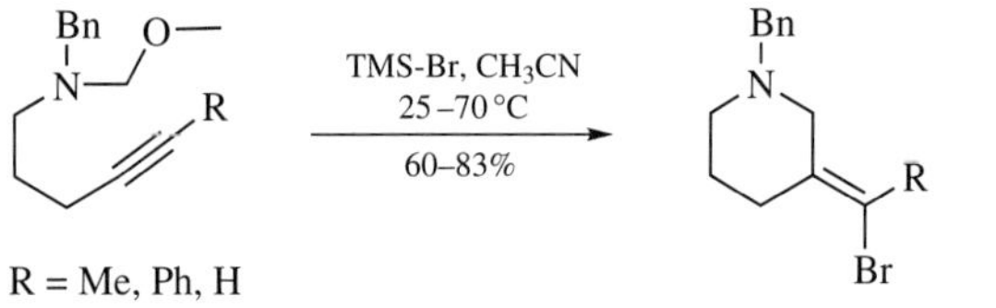

$\xrightarrow[\text{60–83\%}]{\substack{\text{TMS-Br, CH}_3\text{CN}\\ 25\text{–}70\,^{\circ}\text{C}}}$ (32)

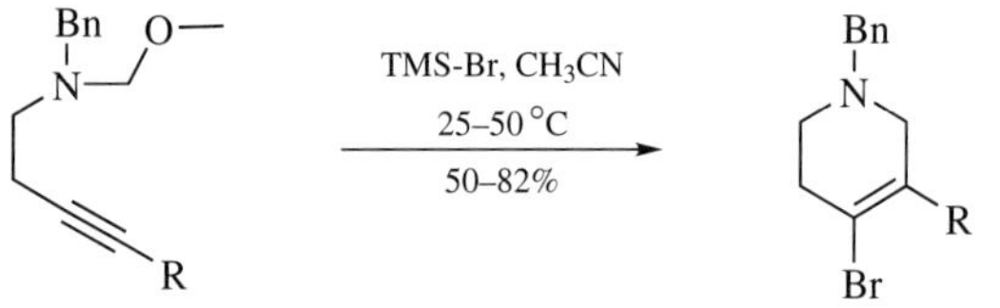

R = Me, Ph, H

$\xrightarrow[\text{50–82\%}]{\substack{\text{TMS-Br, CH}_3\text{CN}\\ 25\text{–}50\,^{\circ}\text{C}}}$ (33)

For example, treatment of the aminals shown in eqs 32 and 33 with TMS-Br leads to piperdines with either *exo-* or *endo-*cyclic unsaturation depending on the tether length. Interestingly, the only reported products were the piperdines; no pyrrolidine-derived compounds were observed even with tethers shorter than two methylene groups.

Bromination on Carbon Adjacent to Aromatic Rings and Carbonyls. As has already been discussed, TMS-Br may be used as a source of bromine. In this case, TMS-Br, when mixed with a second reagent, may be used to introduce a bromine atom at either a benzylic position[37] or alpha to an aldehyde or ketone. In one example of this type selenium(IV) oxyhalides (eq 34) are generated in situ from selenium dioxide and TMS-Br.

$$\text{SeO}_2 \;+\; \text{TMS-Br} \longrightarrow \left[\underset{\text{Br}\,\,\,\,\,\text{Br}}{\overset{\text{O}}{\text{Se}}} \right] \quad (34)$$

Reaction with aldehydes and ketones then leads to selective monobromination (eqs 35 and 36). These reactions are usually carried out in CCl_4, and aldehydes generally react faster than ketones.

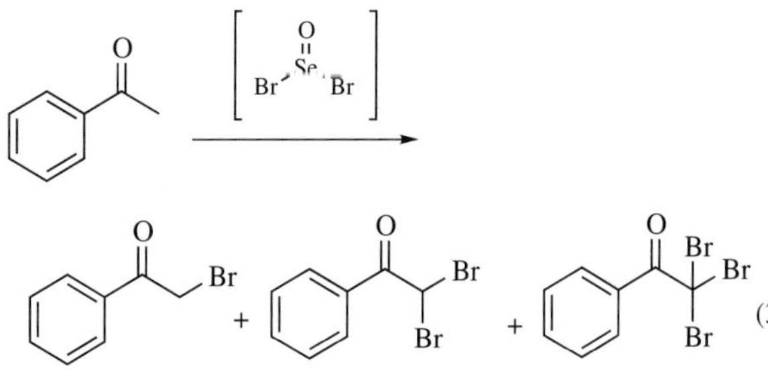

(35)

87% 0% 6%

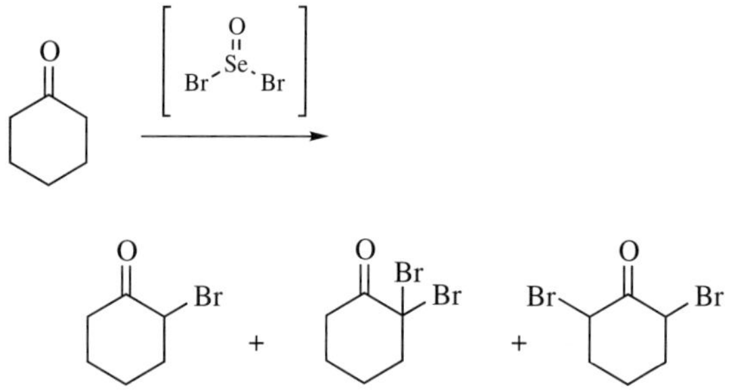

(36)

90% 0% 1%

In order to achieve bromination at benzylic carbons a different set of conditions were used. Specifically, TMS-Br was mixed with sodium bromate in carbon tetrachloride (CCl_4) between room temperature and reflux. An example of this reaction, which is believed to proceed via a radical chain reaction mechanism, is the selective bromination of phenylacetonitrile (eq 37).[38]

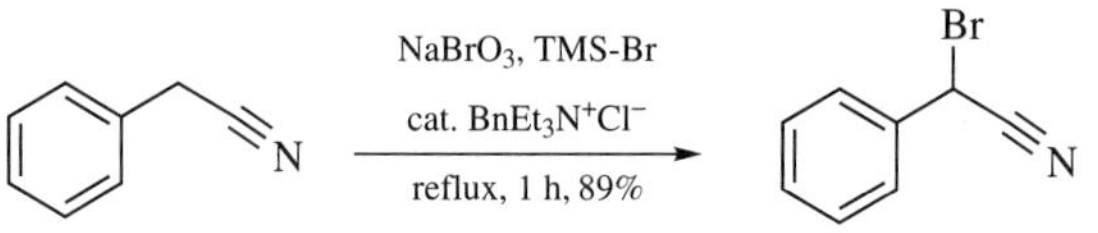

$\xrightarrow[\text{reflux, 1 h, 89\%}]{\substack{\text{NaBrO}_3,\ \text{TMS-Br}\\ \text{cat. BnEt}_3\text{N}^+\text{Cl}^-}}$ (37)

Selective Bromination at an Anomeric Carbon. A further example of the use of TMS-Br to introduce a bromine atom to a molecule is the exchange of an acetoxyl group for bromine at the anomeric carbon of carbohydrates.[39] The pyran depicted in eq 38 is treated with TMS-Br in chloroform at room temperature, and only the axial isomer of the bromide was formed.

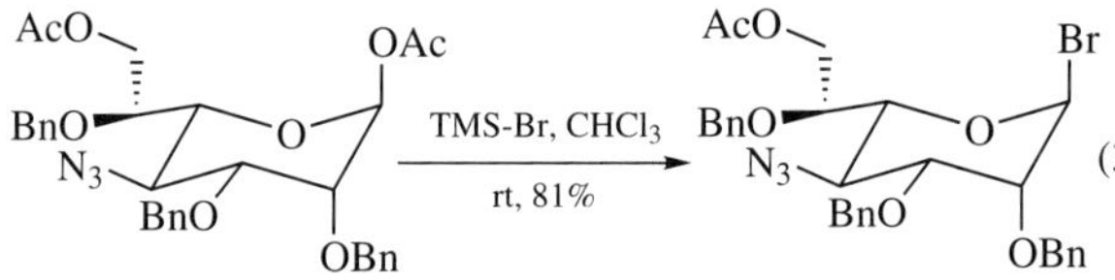

$\xrightarrow[\text{rt, 81\%}]{\text{TMS-Br, CHCl}_3}$ (38)

In Situ Glycosylation System. When TMS-Br is mixed with cobalt(II) bromide, tetrabutylammonium bromide and 4 Å molecular sieves a reagent is generated,[40] which activates anomeric hemiacetals toward glycosylation. Equation 39 shows the coupling of a furanose with a pyranoside. The anomeric selectivity is variable and substrate dependant. This method can be used in the formation of more complex polysaccharides.

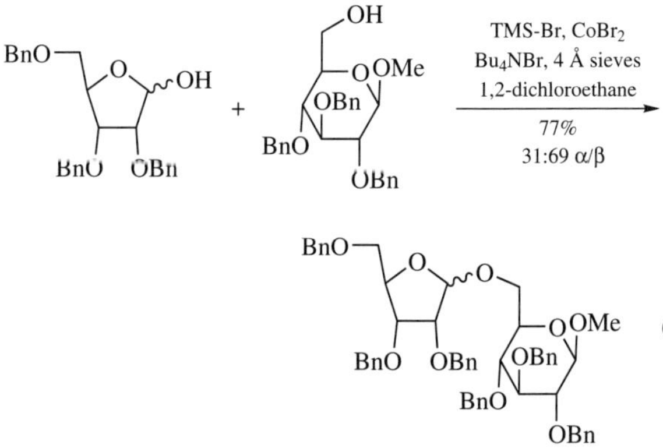

$\xrightarrow[\substack{\text{77\%}\\ 31{:}69\ \alpha/\beta}]{\substack{\text{TMS-Br, CoBr}_2\\ \text{Bu}_4\text{NBr, 4 Å sieves}\\ \text{1,2-dichloroethane}}}$ (39)

Haloacetoxylation: Halogenation with Hypervalent Iodine. In the presence of hypervalent iodine species such as iodosobenzene, TMS-Br can be used to introduce bromine atoms onto electron-rich ring systems.[41] As shown in eqs 40 and 41, bromination may be quickly and efficiently affected in combination with iodobenzene diacetate.

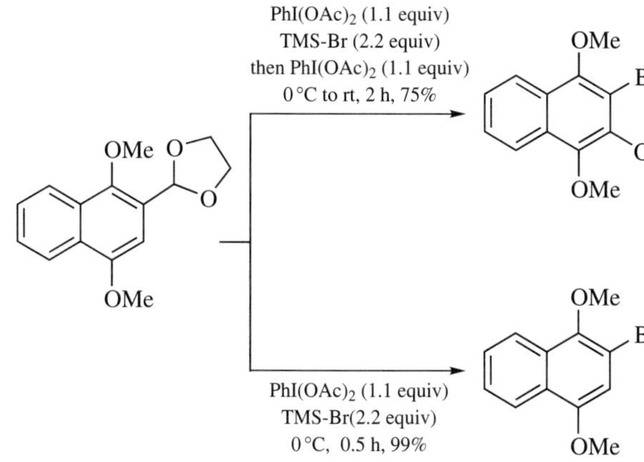
PhI(OAc)2 (1.1 equiv)
TMS-Br (2.2 equiv)
then PhI(OAc)2 (1.1 equiv)
0 °C to rt, 2 h, 75%
PhI(OAc)2 (1.1 equiv)
TMS-Br(2.2 equiv)
0 °C, 0.5 h, 99%
OMe
Br
OAc

(40)

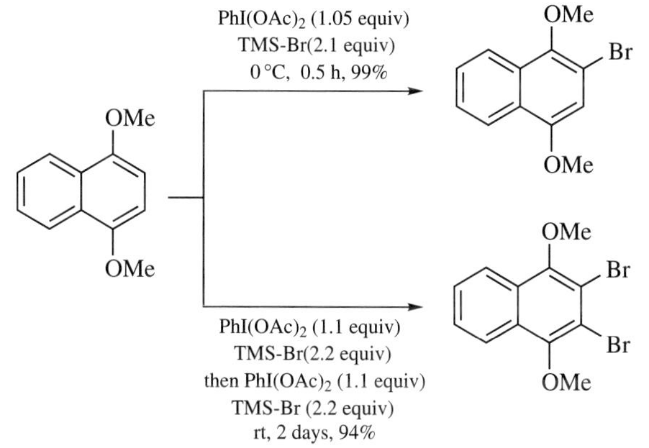
PhI(OAc)2 (1.05 equiv)
TMS-Br(2.1 equiv)
0 °C, 0.5 h, 99%
PhI(OAc)2 (1.1 equiv)
TMS-Br(2.2 equiv)
then PhI(OAc)2 (1.1 equiv)
TMS-Br (2.2 equiv)
rt, 2 days, 94%
OMe
Br

(41)

The bromination is the first step in the reaction; when additional iodosobenzene is present it is followed by an acetoxylation. In the example of eq 41, dibromination may be affected by using longer reaction times and additional equivalents of reagents. The intermediate in the dibromination is the bromoacetoxy derivative from which the acetoxy group is ultimately replaced with a bromine atom.

Bromination of Olefins. Another example of TMS-Br as a source of bromine is its use in the bromination of alkenes (eq 42). This transformation is achieved via the use of tetradecyltrimethylammonium permanganate $[(C_{14}H_{29}(CH_3)_3N)(KMnO_4)]$.[42] This reagent performs a *trans* relative addition of the bromine atoms and is selective for electron-rich olefins over electron-deficient ones. The yields for these reactions are generally high (60–91%).

Synthesis of 1,2,4-Triazolin-3-ones. The use of TMS-Br in the formation of carbo- and heterocycles is well known. A further example of this type of chemistry is the preparation of 1,2,4-triazolin-3-ones. Thus, when an *N*-formylsemicarbazide is treated with TMS-Br in the presence of hexamethyldisilazane (HMDS) and catalytic ammonium sulfate activation takes place followed by cyclization (eq 43).[43] The HMDS acts as both solvent and silylating agent for this reaction. After cyclization is achieved the reaction is completed by loss of bis-TMS ether. The nature of the groups attached to the *N*-4 position can be either alkyl or aryl and steric congestion does not seem to retard this reaction.

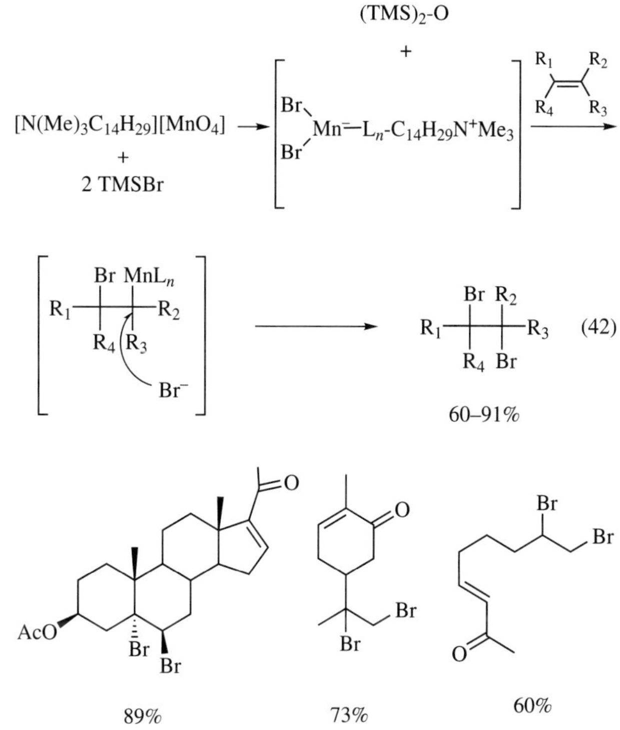
(TMS)2-O
+
[N(Me)3C14H29][MnO4]
+
2 TMSBr
Br
Mn=Ln-C14H29N+Me3
Br
R1 R2
R4 R3
Br MnLn
Br−
Br R2
R1 R3
R4 Br
60–91%
AcO
89%
73%
60%

(42)

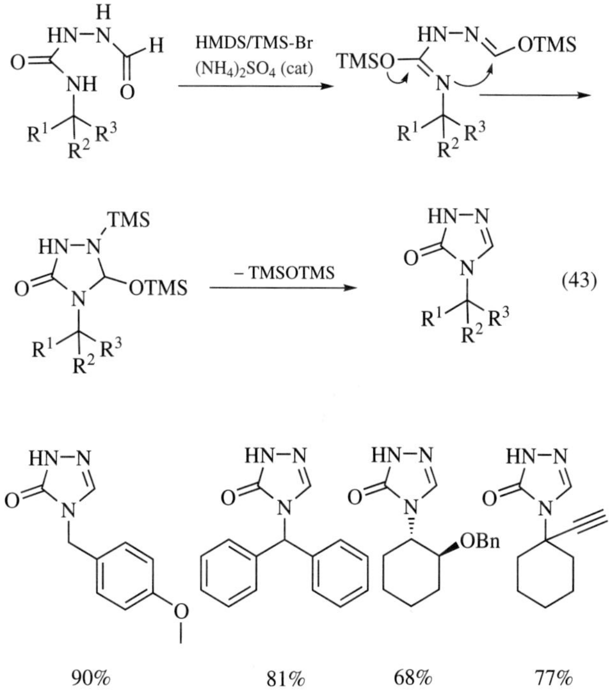
HMDS/TMS-Br
(NH4)2SO4 (cat)
TMSO
OTMS
TMS
- TMSOTMS
OBn
90%
81%
68%
77%

(43)

Chloride Displacement in Pyridines and Other Heterocycles. TMS-Br may also be used as a source of nucleophilic bromide in the displacement of chloride atoms in heterocyclic systems.[44] TMS-I may also be used in this reaction to produce the corresponding iodo compounds. The mechanism of this reaction (eq 44) is believed to involve *N*-acylation to give an activated heteroaryllium species that is susceptible to nucleophilic attack by halide ion. The reaction generally favors a predominance of the bromo (or iodo) products over the chloro compounds owing to the weaker nucleophilicity of chloride anion.

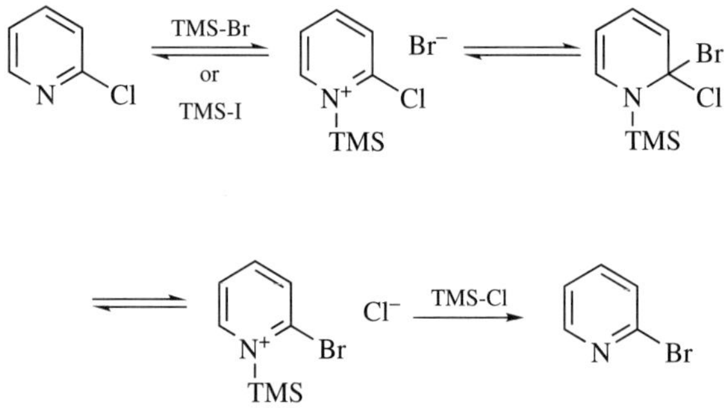

$$(44)$$

As can be seen from the examples (eqs 45–47) the yields for this type of transformation run the gamut from moderate to excellent. Fluoro derivatives could not be made to undergo displacement with either TMS-Br or TMS-I.

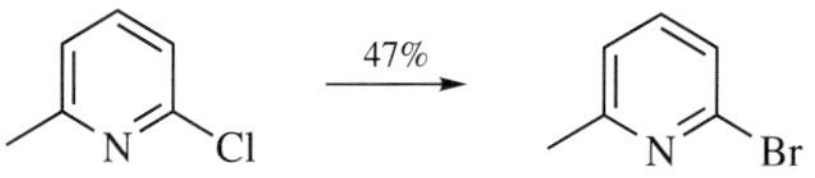

$$(45)$$

$$(46)$$

$$(47)$$

Stereoselective Pummerer Reactions. The Pummerer reaction generally involves the treatment of a sulfoxide containing an α-hydrogen with an acylating agent, such as acid chloride or anhydride, to generate a thionium ion which then undergoes the classical rearrangement. Further research in this area led to the discovery that the active hydrogen may be separated from the sulfoxide sulfur atom by an intervening double bond, including aromatic rings. This so-called vinylogous Pummerer reaction has been shown to take place with the mediation of TMS-Br or TMS-Cl, in a highly stereoselective fashion.[45] Thus, a sulfoxide may be converted to a chiral benzyl alcohol with high enantioselectivity (eq 48). In the unsubstituted toluene derivative, $R^1 = H$, only trimethylsilyl group is transferred to the carbon atom. This is reported to be due to an inversion of the relative reactivity of the two nucleophilic centers (carbon and oxygen) in the molecule upon further substitution of the benzylic carbon. The enantioselectivity in some of the other cases studied, such as derivatives where R^1 is ethyl, benzyl, allyl, etc., was found to be greater than 98%.

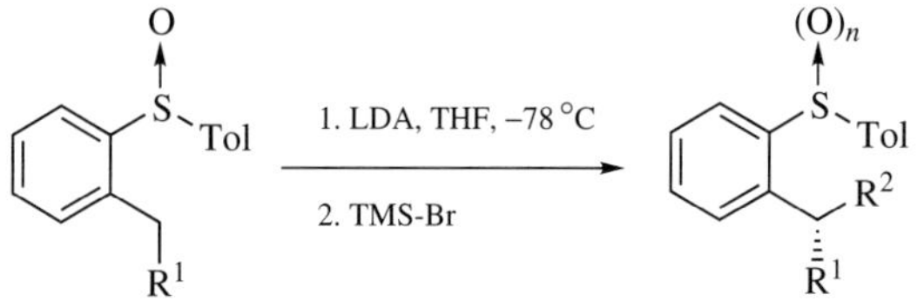

$$(48)$$

$R^1 = H$	$R^1 = CH_3$
$n = 1$	$n = 0$
$R^2 = TMS$	$R^2 = OTMS$
Yield = 63%	Yield = 40%, >98% ee

A mechanism to explain this high selectivity was reported and is shown in eq 49. Upon generation of the carbon-based anion TMS-X is added and reacts preferentially at the sulfoxide oxygen ($R^1 \neq H$). The sulfonium cation then eliminates OTMS to give a conjugated thionium species which then collapses to the benzyl alcohol, wherein the OTMS is reattached to the face it originally occupied.

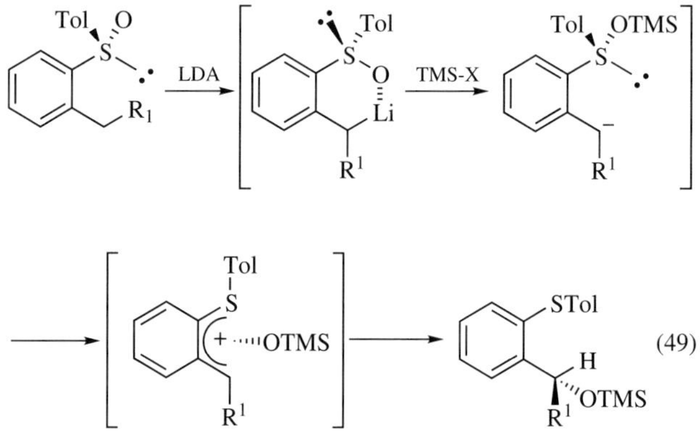

$$(49)$$

Retro-Mannich Reactions for the Synthesis of Pyrroles. The retro-Mannich reaction of tropinones under the influence of certain catalysts, such as TMS-Br affords pyrroles (eq 50).[46] In these reactions the nitrogen atom of the substrates typically is protected as a carbamate or as a sulfonamide. A basic amine is not conducive to this reaction with this type of reagent. The TMS-Br is used catalytically in amounts usually around 10%. The tropinone ketone may be α-substituted with various alkyl groups to give ultimately homologous ketone products.

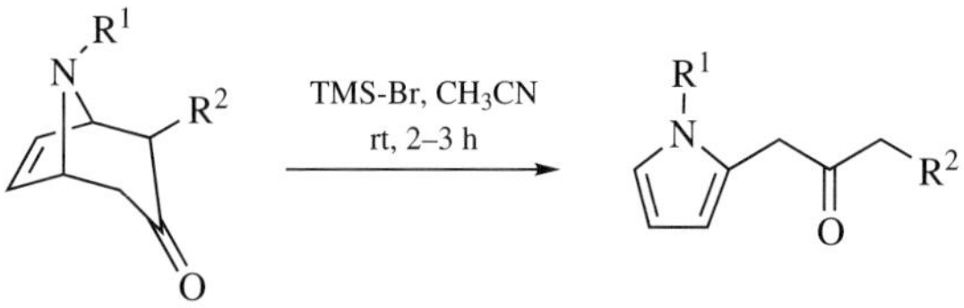

$$(50)$$

$R^1 = CO_2Me$, $R^2 = H$	72%
$R^1 = CO_2Bn$, $R^2 = H$	68%
$R^1 = CO_2Bn$, $R^2 = Et$	52%
$R^1 = Ts$, $R^2 = Et$	64%

The proposed mechanism of this synthetic manipulation (eq 51) begins with polarization of the carbonyl by reaction with TMS-Br, and subsequent retro-Mannich reaction. A subsequent proton loss yields the pyrrole, hydrolyzes the enol ether, and regenerates the TMS-X.

(51)

Oxazoline Fragmentation. As shown in eq 52, a serine-derived oxazoline can be reacted with TMS-Br to produce $N(N)$-protected β-halogeno α-amino esters.[47] This reaction is achieved with TMS-Br which first N-silylates the oxazoline. This renders the β-carbon of the α-amino ester susceptible to nucleophilic attack by the bromide anion. The attack of the halogen anion leads to ring opening of the oxazoline and generation of the N-protected β-bromoalanine derivative.

(52)

THF, 50 °C, 16 h, 57%

or

CH_2Cl_2, 40 °C, 48 h, 85%

The Michaelis-Arbuzov Rearrangement. It has been found that TMS halides, including TMS-Br, may be used to facilitate the Michaelis-Arbuzov reaction (eqs 53 and 54).[48] This transformation is usually high yielding and substantially free from such by-products as phosphine oxides. The reaction of TMS-X with trialkyl phosphites for the preparation of phosphonates is similarly efficient with regards to O to P alkyl group transfer selectivity.

(53)

(54)

R = Et, 85%; R = Ph, 87%

The proposed mechanism for this reaction (eq 55) involves initial coordination of TMS-Br to the oxygen atom. The released bromide anion then attacks a carbon atom liberating an alkyl halide. This alkyl halide then alkylates phosphorous and concomitant desilylation generates the phosphine oxide with an overall transformation of phosphorous(III) to phosphorous(V).

(55)

1. Schmidt, A. H., *Aldrichim. Acta* **1981**, *14*(2), 31.

2. Krüerke, U., *Chem. Ber.* **1962**, *95*, 174.

3. (a) Scheibye, S.; Thomsen, I.; Lawesson, S. O., *Bull. Soc. Chim. Belg.* **1979**, *88*, 1043. (b) Olah, G. A.; Gupta, B. G. B.; Malhotra, R.; Narang, S. C., *J. Org. Chem.* **1980**, *45*, 1638. (c) Schmidt, A. H.; Russ, M., *Chem. Ber.* **1981**, *114*, 1099.

4. Sakurai, H.; Sasaki, K.; Hosomi, A., *Tetrahedron Lett.* **1980**, *21*, 2329.

5. (a) Voronkov, M. G.; Dolgov, B. N.; Dmitrieva, N. A., *Dokl. Akad. Nauk SSSR* **1952**, *84*, 959. (b) Gross, H.; Böck, C.; Costisella, B.; Gloede, J., *J. Prakt. Chem.* **1978**, *320*, 344.

6. (a) Ho, T. L.; Olah, G. A., *Synthesis* **1977**, 417. (b) Jung, M. E.; Hatfield, G. L., *Tetrahedron Lett.* **1978**, 4483.

7. Kricheldorf, H. R., *Angew. Chem., Int. Ed. Engl.* **1979**, *18*, 689.

8. Kricheldorf, H. R.; Mörber, G.; Regel, W., *Synthesis* **1981**, 383.

9. Hanessian, S.; Delorme, D.; Dufresne, Y., *Tetrahedron Lett.* **1984**, *25*, 2515.

10. Woodward, R. B.; Logusch, E.; Nambiar, K. P.; Sakan, K.; Ward, D. E.; Au-Yeung, B.; Balaram, P.; Browne, L. J.; Card, P. J.; Chen, C. H.; Chênevert, R. B.; Fliri, A.; Frobel, K.; Gais, H.-J.; Garratt, D. G.; Hayakawa, K.; Heggie, W.; Hesson, D. P.; Hoppe, D.; Hoppe, I.; Hyatt, J. A.; Ikeda, D.; Jacobi, P. A.; Kim, K. S.; Kobuke, Y.; Kojima, K.; Krowicki, K.; Lee, V. J.; Leutert, T.; Malchenko, S.; Martens, J.; Matthews, R. S.; Ong, B. S.; Press, J. B.; Rajan Babu, T. V.; Rousseau, G.; Sauter, H. M.; Suzuki, M.; Tatsuta, K.; Tolbert, L. M.; Truesdale, E. A.; Uchida, I.; Ueda, Y.; Uyehara, T.; Vasella, A. T.; Vladuchick, W. C.; Wade, P. A.; Williams, R. M.; Wong, H. N.-C., *J. Am. Chem. Soc.* **1981**, *103*, 3213.

11. Gillard, J. W.; Israel, M., *Tetrahedron Lett.* **1981**, *22*, 513.

12. Morishima, N.; Koto, S.; Kusuhara, C.; Zen, S., *Chem. Lett.* **1981**, 427.

13. Logue, M. W., *Carbohydr. Res.* **1975**, *40*, C9.

14. Ahmad, S.; Khan, M. A.; Iqbal, J., *Synth. Commun.* **1988**, *18*, 1679.

15. Schmidt, A. H.; Russ, M.; Grosse, D., *Synthesis* **1981**, 216.

16. (a) McKenna, C. E.; Schmidhauser, J., *J. Chem. Soc., Chem. Commun.* **1979**, 739. (b) Breuer, E.; Safadi, M.; Chorev, M.; Gibson, D., *J. Org. Chem.* **1990**, *55*, 6147.

17. Lazar, S.; Guillaumet, G., *Synth. Commun.* **1992**, *22*, 923.

18. Comi, R.; Franck, R. W.; Reitano, M.; Weinreb, S. M., *Tetrahedron Lett.* **1973**, 3107.

19. Hsung, R. P., *Synth. Commun.* **1990**, *20*, 1175.

20. Iwata, C.; Tanaka, A.; Mizuno, H.; Miyashita, K., *Heterocycles* **1990**, *31*, 987.

21. Seyferth, D.; Grim, S. O., *J. Am. Chem. Soc.* **1961**, *83*, 1610.

22. Akane, N.; Hatano, T.; Kusui, H.; Nishiyama, Y.; Ishii, Y., *J. Org. Chem.* **1994**, *59*, 7902.

23. (a) Taniguchi, Y.; Tatsuhiro, K.; Nakahashi, M.; Takaki, K.; Fujiwara, Y., *Appl. Organomet. Chem.* **1995**, *9*, 491. (b) Taniguchi, Y.; Nakahashi, M.; Kuno, T.; Tsuno, M.; Makioka, Y.; Takaki, K.; Fujiwara, Y., *Tetrahedron Lett.* **1994**, *35*, 4111.

24. Tishkov, A. A.; Dilman, A. D.; Faustov, V.; Birukov, A. A.; Lysenko, K. S.; Belykov, P. A.; Ioffe, S. L.; Strelenko, Y. A.; Antipin, M. Y., *J. Am. Chem. Soc.* **2002**, *124*, 11358.

25. Tishkov, A. A.; Lyapkalo, I. M.; Kozincev, A. V.; Ioffe, S. L.; Strelenko, Y. A.; Tartakovsky, V. A., *Eur. J. Org. Chem.* **2000**, 3229.

26. Tishkov, A. A.; Kozintsev, A. V.; Lyapkalo, I. M.; Ioffe, S. L.; Kachala, V. V.; Strenlenko, Y. A.; Tartakovsky, V. A., *Tetrahedron Lett.* **1999**, *40*, 5075.

27. Smirnov, V. O.; Tishkov, A. A.; Lyapkalo, I. M.; Ioffe, S. L.; Kachala, V. V.; Strelenko, Y. A.; Tartakovsky, V. A., *Russ. Chem. Bull.* **2001**, *50*, 2433.

28. Birin, K. P.; Tishkov, A. A.; Ioffe, S. L.; Strelenko, Y. A.; Tartakovsky, V. A., *Russ. Chem. Bull.* **2003**, *52*, 647.

29. Danilenko, V. M.; Tishkov, A. A.; Ioffe, S. L.; Lyapkalo, I. M.; Strelenko, Y. A.; Tartakovsky, V. A., *Synthesis* **2002**, 635.

30. Tishkov, A. A.; Lesiv, A. V.; Khomutova, Y. A.; Strelenko, Y. A.; Nesterov, I. D.; Antipin, M. Y.; Ioffe, S. L.; Denmark, S. E., *J. Org. Chem.* **2003**, *68*, 9477.

31. Förster, W.-R.; Isecke, R.; Spanka, C.; Schaumann, E., *Synthesis* **1997**, 942.

32. Müller, M.; Förster, W.-R.; Holst, A.; Kingma, A. J.; Shaumann, E.; Adiwidjaja, G., *Chem. Eur. J.* **1996**, *2*, 949.

33. Su, M.; Yu, W.; Jin, Z., *Tetrahedron Lett.* **2001**, *42*, 3771.

34. Yu, W.; Jin, Z., *J. Am. Chem. Soc.* **2000**, *122*, 9840.

35. (a) Yu, W.; Jin, Z., *J. Am. Chem. Soc.* **2002**, *124*, 6576. (b) Yu, W.; Jin, Z., *J. Am. Chem. Soc.* **2001**, *123*, 3369.

36. Murata, Y.; Overman, L. E., *Heterocycles* **1996**, *42*, 549.

37. Lee, J. G.; Park, I. N.; Seo, J. W., *Bull. Kor. Chem. Soc.* **1995**, *16*, 349.

38. Lee, J. G.; Seo, J. W.; Yoon, U. C.; Kang, K.-T., *Bull. Kor. Chem. Soc.* **1995**, *16*, 371.

39. Suzuki, T.; Suzuki, S. T.; Yamada, I.; Koashi, Y.; Yamada, K.; Chida, N., *J. Org. Chem.* **2002**, *67*, 2874.

40. Hirooka, M.; Mori, Y.; Sasaki, A.; Koto, S.; Shinoda, Y.; Morinaga, A., *Bull. Chem. Soc. Jpn.* **2001**, *74*, 1679.

41. Evans, P. A.; Brandt, T. A., *Tetrahedron Lett.* **1996**, *37*, 6443.

42. Hazra, B. G.; Chordia, M. D.; Bahule, B. B.; Pore, V. S.; Basu, S., *J. Chem. Soc., Perkin Trans. 1* **1994**, 1667.

43. Huang, X.; Palani, A.; Xiao, D.; Aslanian, R.; Shih, N.-Y., *Org. Lett.* **2004**, *6*, 4795.

44. Schlosser, M.; Cottet, F., *Eur. J. Org. Chem.* **2002**, 4181.

45. Garcia Ruano, J. L.; Alemán, J.; Aranda, M. T.; Arévalo, M. J.; Padwa, A., *Org. Lett.* **2005**, *7*, 19.

46. Cramer, N.; Juretschke, J.; Laschat, S.; Baro, A.; Frey, W., *Eur. J. Org. Chem.* **2004**, 1397.

47. Laaziri, A.; Uziel, J.; Jugé, S., *Tetrahedron: Asymmetry* **1998**, *9*, 437.

48. Renard, P.-Y.; Vayron, P.; Mioskowski, C., *Org. Lett.* **2003**, *5*, 1661.

(E)-1-Bromo-4-trimethylsilyl-1-buten-3-yne

[107646-62-8]　　　　C$_7$H$_{11}$BrSi　　　　(MW 203.15)

InChI = 1S/C7H11BrSi/c1-9(2,3)7-5-4-6-8/h4,6H,1-3H3/b6-4+

InChIKey = HLZBLUKDUBOQPX-GQCTYLIASA-N

(cross-coupling reagent used as a four-carbon synthon in the synthesis of conjugated oligoenes and oligoenynes)

Physical Data: bp 72–74 °C/15mm-Hg; d 1.076 g cm^{-3}.

Solubility: insoluble in water; soluble in hexane, ether, alcohol, and most other organic solvents.

Analysis of Reagent Purity: ^{1}H NMR (CDCl$_3$) δ 0.19 (s, 9 H), 6.21 (d, J = 14.1 Hz, 1 H), 6.75 (d, J = 14.1Hz, 1 H); ^{13}C NMR (CDCl$_3$) δ −0.325 (3 C), 97.33, 100.91, 117.61, 119.97 ≥ 99% *E* by ^{13}C NMR; IR (neat), 3075, 2168, 2111, 1696, 1577, 1252, 1199, 1062, 845 cm^{-1}.

Preparative Methods: obtained by the Pd(PPh$_3$)$_4$ catalyzed cross-coupling reaction of *(E)*-1-bromo-2-iodoethylene with TMS—≡—ZnBr generated in situ by the reaction of TMS—≡, MeMgBr, and ZnBr$_2$.

Purity: distilled at reduced pressure.

Handling, Storage, and Precaution: the reagent should be stored in the refrigerator.

Preparation.[1] To a solution of (trimethylsilyl)acetylene (2.8 mL, 20 mmol) in THF (30 mL) was added via a syringe MeMgBr (8 mL of 3 M ether solution; 24 mmol). The reaction mixture was stirred at 23 °C for 3 h, and a solution of anhydrous ZnBr$_2$ (5.85 g, 26 mmol) in THF (10 mL) was added at 0 °C. The mixture was stirred at 0 °C for 30 min and added via cannula to a solution of *(E)*-1-bromo-2-iodoethylene (5.12 g, 22 mmol) and Pd(PPh$_3$)$_4$ (0.46 g, 0.02 equiv) in THF (15 mL) (eq 1). The resultant mixture was stirred at 23 °C for 10 h, quenched with aqueous NH$_4$Cl, and extracted with pentane. The pentane extract was washed with aqueous NaHCO$_3$ and brine, dried over MgSO$_4$, and distilled to afford 3.29 g (81% yield) of the title compound (≥ 99% *E* by ^{13}C NMR spectroscopy) as a colorless liquid: bp 72–74 °C (15 mmHg).

$$Br{\nearrow}I + BrZn{=\!=\!=}SiMe_3 \xrightarrow[\text{THF, 23 °C, 10 h}]{2 \text{ mol } \% \text{ Pd(PPh}_3)_4}$$

$$Br{\diagdown}{=\!=\!=}SiMe_3 \quad (1)$$
1

Prior to the development of the procedure described above, there had been no practical route to **1**, although its formation in 5% yield was observed in the reaction of Me$_3$SiC≡CZnCl with a mixture (typically E/Z = 33/67) of *(E)*- and *(Z)*-1,2-dibromoethylenes containing 1 equiv of the *E* isomer relative to Me$_3$SiC≡CZnCl.[2] The stereoisomeric purity was not determined. The major product of the reaction was *(E)*-1,6-bis(trimethylsilyl)-3-hexen-1,5-diyne (55%) with *(Z)*-1,2-dibromoethylene remaining unreacted.[2]

As an Electrophile in the Pd-Catalyzed Cross-Coupling.

Synthesis of Carotenoids and Retinoids. The Pd-catalyzed cross-coupling reactions of **1** with alkenylmetals containing Al, Zn, and Zr have been shown to be high-yielding and selective. The reaction of **1** with β,β-disubstituted alkenylalanes under the conditions of double metal catalysis with Pd and Zn[3] is satisfactory, and the ≥99% stereospecificity level can be maintained in most cases. Thus, this reaction used in conjunction with the Zr-catalyzed carboalumination of terminal alkynes[4] forms the foundation of a highly stereoselective and novel methodology that is fundamentally discrete from the conventional carbonyl olefination-based methodology for the synthesis of carotenoids (Scheme 1).[1] (*E*)-1-Bromo-2-iodoethylene [56798-08-4], one of the starting materials for the synthesis of **1**, also serves as a two-carbon synthon for the central (*E*)-ethylene moiety of carotenoids.

Synthesis of β-carotene (4)

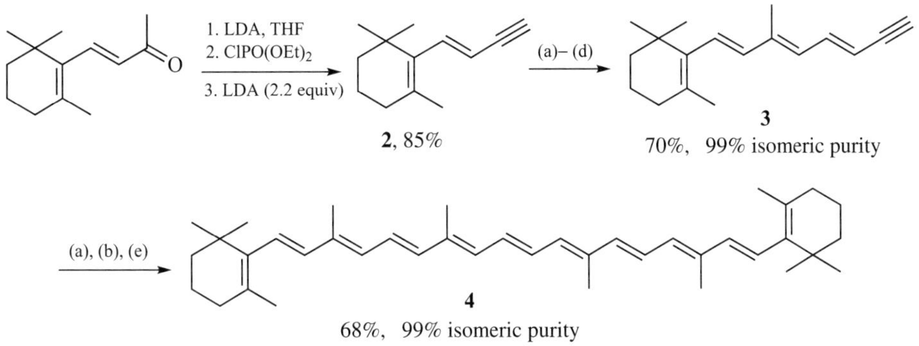

1. LDA, THF
2. ClPO(OEt)2
3. LDA (2.2 equiv)
2, 85%
(a)– (d)
3
70%, 99% isomeric purity
(a), (b), (e)
4
68%, 99% isomeric purity

Synthesis of γ- carotene (7)

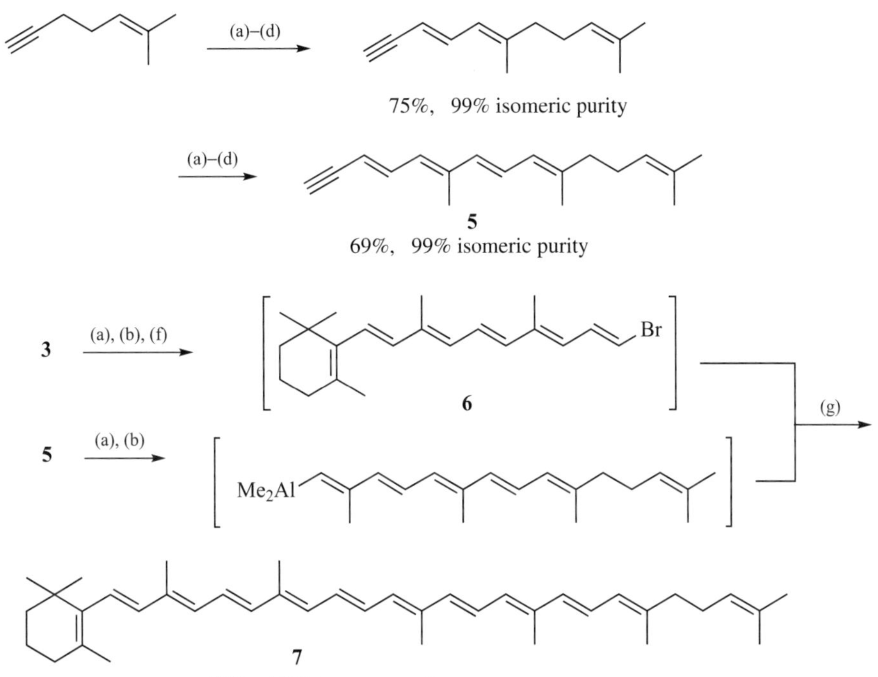

(a)–(d)
75%, 99% isomeric purity
(a)–(d)
5
69%, 99% isomeric purity
3
(a), (b), (f)
Br
6
5
(a), (b)
Me2Al
(g)
7
53%, 99% isomeric purity

Synthesis of vitamin A (8)

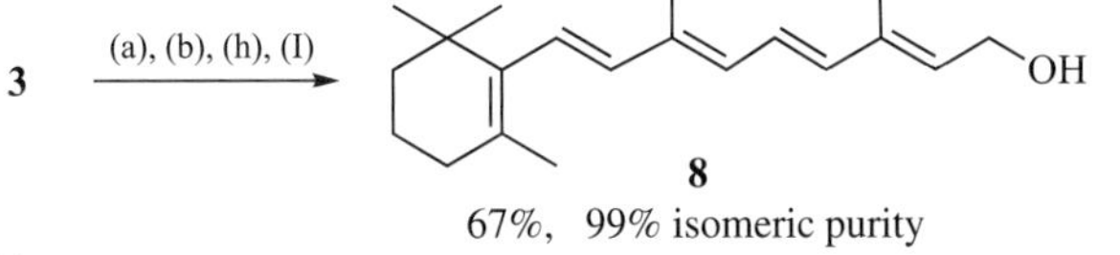

3
(a), (b), (h), (I)
OH
8
67%, 99% isomeric purity

Conditions:

(a) Me_3Al (2 equiv), Cp_2ZrCl_2 (1 equiv), $(CH_2Cl)_2$, 23 °C, 4 h.
(b) Evaporation at 50 °C and < 0.5 mmHg.
(c) BrCH=CHC≡CSiMe$_3$ (1.05 equiv), $ZnCl_2$ (1 equiv) in THF, 2.5 mol % $Pd_2(dba)_3$, 10 mol % TFP [tri(2-furyl)phosphine], DMF, 23 °C, 6 h.
(d) K_2CO_3, MeOH, 23 °C, 3 h.
(e) ICH=CHBr (0.5 equiv), $ZnCl_2$ (1 equiv) in THF, 2.5 mol % $Pd_2(dba)_3$, 10 mol % TFP, DMF, 23 °C, 8 h.
(f) ICH=CHBr (1.05 equiv), $ZnBr_2$ (1 equiv) in THF, 5 mol % $Pd(PPh_3)_4$, DMF, 23 °C, 2 h.
(g) $ZnCl_2$ (1 equiv) in THF, 2.5 mol % $Pd_2(dba)_3$, 10 mol % TFP, DMF , 23 °C, 6 h.
(h) THF, *n*-BuLi (1 equiv), 23 °C, 0.5 h.
(i) $(CH_2O)_n$ (3 equiv), 23 °C, 5 h.

Scheme 1

Synthesis of (all-E)-Oligoenes of the (CH=CH)ₙ Type. The Pd-catalyzed cross-coupling reaction of **1** with (E)-alkenyl-zirconocene derivatives that can be readily generated in situ by hydrozirconation of the corresponding terminal alkynes including oligoenynes proceeds in high yields and ≥98–99% stereo- and regioselectivities (eq 2, Table 1).[5]

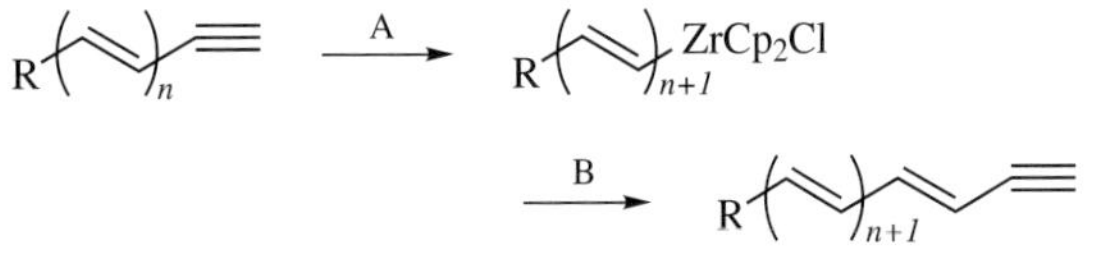

Conditions:

A; HZrCp₂Cl, THF.

B; (a) **1** (1.05 equiv), 5 mol % Cl₂Pd(PPh₃)₂+ 2 i-Bu₂AlH, ZnCl₂, THF.

   (b) K₂CO₃ (1.2 equiv), MeOH.

Oligoenynes containing either an odd or an even number of unsaturated carbon-carbon bonds can be efficiently and selectively prepared without the need for stereoisomeric separation (Scheme 2).[5]

The homologated oligoenyne products can be converted into esters with or without incorporation of one or more additional double bonds, as summarized in Scheme 3.[5]

Hydrozirconation of enynes and oligoenynes with preformed, pure HZrCp₂Cl[6] is high-yielding and both regioselective and stereoselective. The corresponding hydroboration with dialkylboranes proceeds well, but is less clean, producing minor amounts of regioisomers; whereas hydroalumination of enynes and oligoenynes is accompanied by the formation of alkynylalanes in significant amounts. These results make hydrozirconation the method of choice.

Table 1 Efficient and selective four-carbon homologation of oligoenes via hydrozirconation—palladium-catalyzed cross-coupling with (E)-BrCH=CHC≡CSiMe₃[a]

Entry	R—≡ 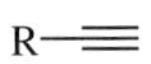	Product R⌇₂ 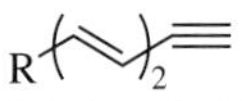	NMR yield[b] (%)	Isolated yield (%)
1	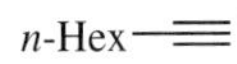 n-Hex—≡[c]	 n-Hex⌇₂	91	85
2	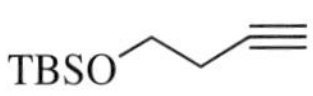 TBSO≡[c]	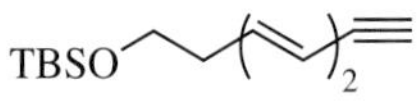 TBSO⌇₂	86	79
3	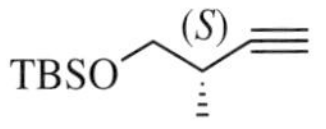 TBSO(S)≡	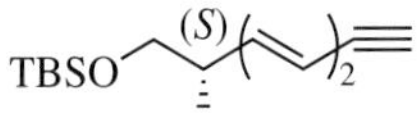 TBSO(S)⌇₂	84[d]	79
4	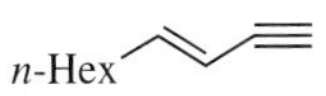 n-Hex≡	 n-Hex⌇₃	86	80
5	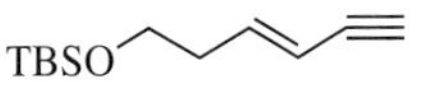 TBSO≡	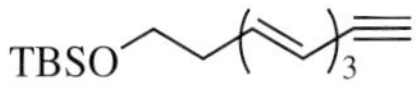 TBSO⌇₃	ND[e]	78
6	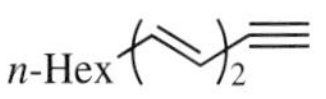 n-Hex⌇₂≡	 n-Hex⌇₄	83	78
7	 TBSO⌇₂≡	 TBSO⌇₄	81	73
8	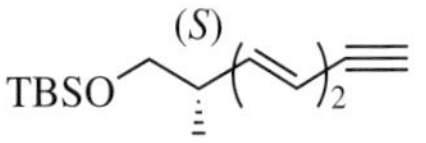 TBSO(S)⌇₂≡	 TBSO(S)⌇₄	79	74
9	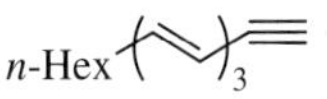 n-Hex⌇₃≡	 n-Hex⌇₅	74[f]	67[g]
10	 n-Hex⌇₄≡	 n-Hex⌇₆	73[f]	NA[h]

[a] Unless otherwise mentioned, the reaction was carried out by hydrozirconation with HZrCp₂Cl in THF at 23 °C for 1 h followed by addition of (E)-BrCH=CHC≡CSiMe₃ (1.05 equiv), 5 mol % Cl₂Pd(PPh₃)₂ + 2DIBAH, and ZnCl₂ (1 equiv) in THF and stirring the resultant mixture for 4 h at 23 °C.

[b] Yield of the Me₃Si derivative determined by NMR analysis.

[c] Hydrozirconation of the alkyne was carried out by using i-BuZrCp₂Cl.

[d] By GLC analysis.

[e] ND stands for not determined.

[f] Isolated yield of the silylated derivative.

[g] Overall isolated yield based on the starting trienyne.

[h] NA stands for clean desilylation with methanolic K₂CO₃ has not been achieved.

A list of General Abbreviations appears on the front Endpapers

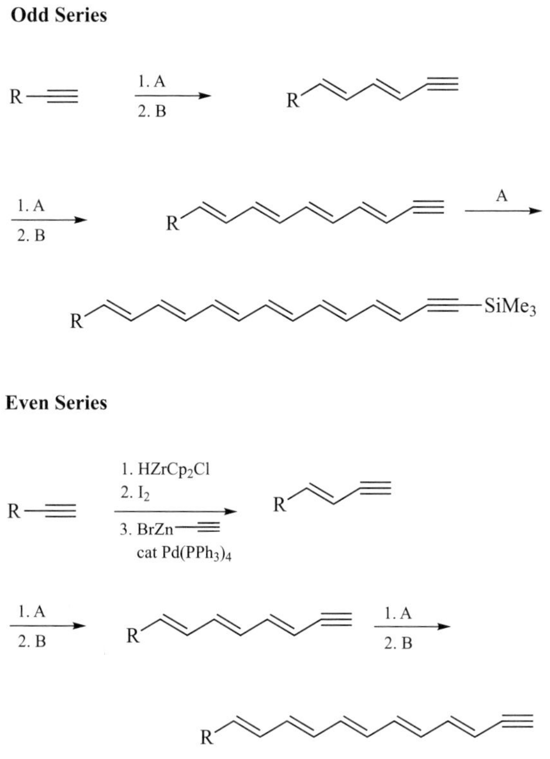

Odd Series

Even Series

Conditions:

A: (a) HZrCp$_2$Cl, THF.

   (b) **1** (1.05 equiv), 5% Cl$_2$Pd(PPh$_3$)$_2$ + 2 *i*-Bu$_2$AlH, ZnCl$_2$, THF.

B: K$_2$CO$_3$ (1.2 equiv), MeOH.

Scheme 2

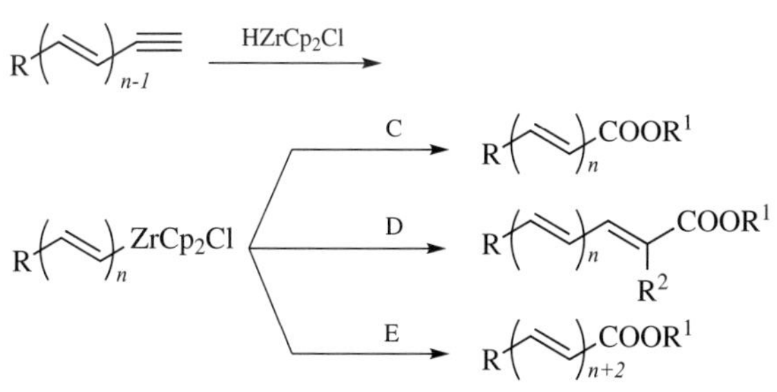

Conditions:

C: (a) Me$_2$AlCl (1.1 equiv).

   (b) ClCOOR^1 (3 equiv).

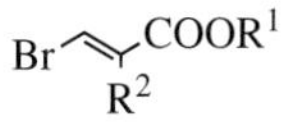

D: Br$\overset{}{\underset{R^2}{\diagup}}$COOR^1 (1.05 equiv), 5 mol % Cl$_2$Pd(PPh$_3$)$_2$,

   10 mol % *i*-Bu$_2$AlH, ZnCl$_2$, THF.

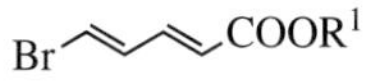

E: Br$\diagup\!\!\diagdown\!\!\diagup$COOR^1 (1.05 equiv), 5 mol % Cl$_2$Pd(PPh$_3$)$_2$,

   10 mol % *i*-Bu$_2$AlH, ZnCl$_2$, THF.

Scheme 3

The hydrozirconation–Pd-catalyzed cross-coupling tandem process described above has been applied to the synthesis of α, ω-difunctional conjugated oligoenes that either have been or

can be used for the synthesis of oligoene macrolide antibiotics (Scheme 4).[5]

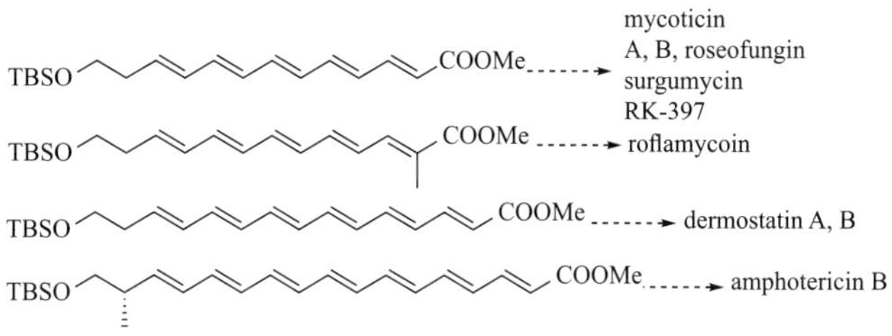

Scheme 4

As a Nucleophile in the Pd-Catalyzed Cross-Coupling. Treatment of **1** with *t*-BuLi (2.1 equiv) in Et$_2$O at −78 °C followed by addition of dry ZnBr$_2$ (1 equiv) in THF at −78 to 0 °C generates in situ the corresponding alkenylzinc reagent **9**. Its Pd-catalyzed cross-coupling reaction provides an alternate and complementary route to conjugated enynes and oligoenynes. Thus, for example, a difunctional trienyne **10** suitable for the synthesis of 6,7-dehydrostipiamide (**11**) has been synthesized from **1** and (*E*)-BrCH=C(Me)COOEt in three steps in 75% overall yield with complete control of regiochemistry and stereochemistry (Scheme 5).

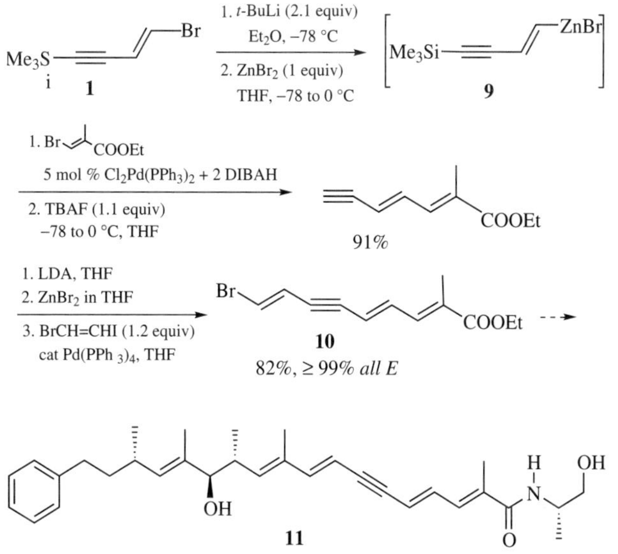

Scheme 5

Particularly attractive is the use of **9** in the trans-selective mono-substitution of 1,1-dibromoalkenes, followed by methylation with MeZnBr to produce chiral alkyl-group-containing dienyne (**12**), also potentially applicable to the synthesis of **11**. The overall yield of **12** based on the starting 1,1-dibromoalkene is 84%, and the crudely isolated product is ≥98% isomerically pure (Scheme 6).

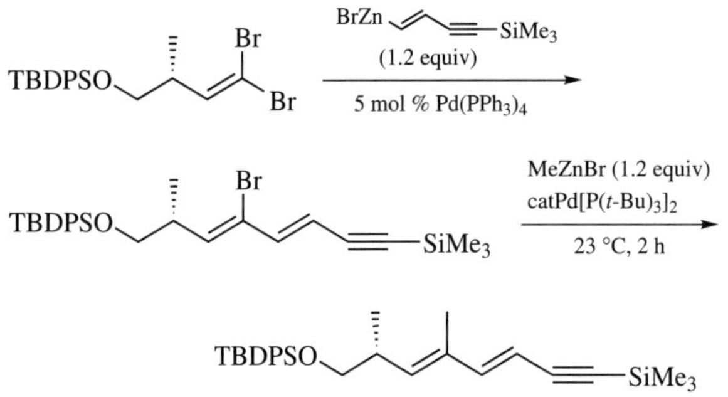

Scheme 6

Related Reagents.

(E)-1-Bromo-4-(t-butyldimethylsilyl)-1-buten-3-yne **(13).** *(CAS no. 259129-34-5).* Before the preparation of **1** by the method herein described, its TBS-protected analog had been prepared in a similar manner and used in the synthesis

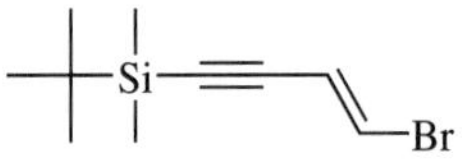

13

of xerulin (Scheme 7).[7] In many cases, **1** and **13** might be expected to be interchangeable. In cases where both are satisfactory, the lower cost of **1** should favor it over **13**.

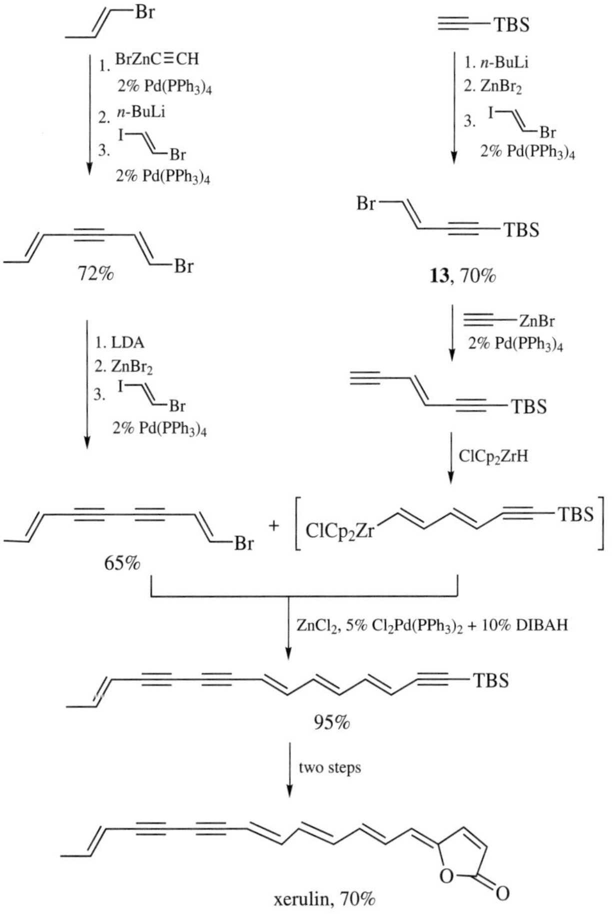

Scheme 7

(Z)-1-Bromo-4-trimethylsilyl-1-buten-3-yne **(14).** There have been a few papers[8,9] reporting the synthesis of the *Z* isomer of **1**. The most reasonable one appears to be that shown in eq 3.[8] Since its cross-coupling and other synthetically useful reactions have not been adequately investigated, its synthetic utility remains largely unknown, although the Sonogashira alkynylation[10] appears to proceed well, as indicated in eq 3.[8]

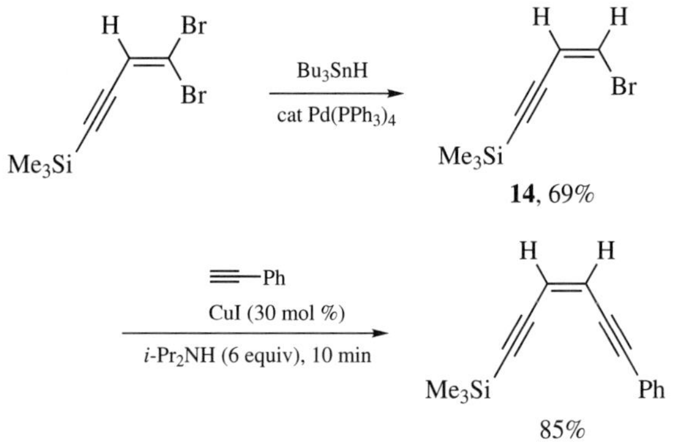

1. Zeng, F.; Negishi, E., *Org. Lett.* **2001**, *3*, 719.

2. (a) Carpita, A.; Rossi, R., *Tetrahedron Lett.* **1986**, *27*, 4351. (b) Andreini, B. P.; Benetti, M.; Carpita, A.; Rossi, R., *Gazz. Chim. Ital.* **1988**, *118*, 469.

3. Negishi, E.; Okukado, N.; King, A. O.; Van Horn, D. E.; Spiegel, B. I., *J. Am. Chem. Soc.* **1978**, *100*, 2254.

4. (a) Van Horn, D. E.; Negishi, E., *J. Am. Chem. Soc.* **1978**, *100*, 2252. (b) Rand, C. L.; Van Horn, D. E.; Moore, M. W.; Negishi, E., *J. Org. Chem.* **1981**, *46*, 4093. (c) Negishi, E.; Van Horn, D. E.; Yoshida, T., *J. Am. Chem. Soc.* **1985**, *107*, 6639.

5. Zeng, F.; Negishi, E., *Org. Lett.* **2002**, *4*, 703.

6. Hart, D. W.; Schwartz, J., *J. Am. Chem. Soc.* **1974**, *96*, 8115.

7. Negishi, E.; Alimardanov, A.; Xu, C., *Org. Lett.* **2000**, *2*, 65.

8. (a) Uenishi, J.; Kawahama, R.; Yonemitsu, O.; Tsuji, J., *J. Org. Chem.* **1996**, *61*, 5716. (b) Uenishi, J.; Kawahama, R.; Yonemitsu, O.; Tsuji, J., *J. Org. Chem.* **1998**, *63*, 8965.

9. Babudri, F.; Fiandanese, V.; Marchese, G.; Punzi, A., *Tetrahedron* **2001**, *57*, 549.

10. (a) Sonogashira, K.; Tohda, Y.; Hagihara, N., *Tetrahedron Lett.* **1975**, 4467. (b) Sonogashira, K.; Yatake, T.; Tohda, Y.; Takahashi, S.; Hagihara, N., *J. Chem. Soc. Chem. Commun.* **1977**, 291. (c) Tohda, Y.; Sonogashira, K.; Hagihara, N., *Synthesis* **1977**, 777. (d) Takahashi, S.; Kuriyama, Y.; Sonogashira, K.; Hagihara, N., *Synthesis* **1980**, 627.

Ei-ichi Negishi & Fanxing Zeng
Purdue University West Lafayette, IN, USA

(Z)-2-Bromo-1-(trimethylsilyloxy)-ethylene

[64556-66-7] $C_5H_{11}BrOSi$ (MW 195.13)

InChI = 1S/C5H11BrOSi/c1-8(2,3)7-5-4-6/h4-5H,1-3H3/b5-4-

InChIKey = KDPKNARRJNETNZ-PLNGDYQASA-N

(reagent used as a precursor of (Z)-(trimethylsilyloxy)vinyl-lithium, α-bromo aldehydes, and α-disilylketene)

Physical Data: bp 48–49 °C/15 mmHg.[1]

Solubility: sol Et$_2$O, THF and most organic solvents; fairly sol cold water.

Analysis of Reagent Purity: ^{1}H NMR (C$_6$D$_6$) (Z) isomer: 6.37 (d; J = 3.8); 5.02 (d; J = 3.8); 0.05 (s). (E) isomer: 6.61 (d; J = 11.4); 5.52 (d; J = 11.4); −0.06 (s).

Preparative Methods: addition of bromine to trimethylsilyl vinyl ether in CH$_2$Cl$_2$ or Et$_2$O at −60 °C, followed by dehydrobromination with triethylamine (50%)[1] or by reaction of trimethylsilyl 2,2-dibromoethyl ether with two equiv of *n*-butyllithium in Et$_2$O from −70 °C to rt (86%)[2] (Z/E:95/5).[3]

Purification: vacuum distillation.

Handling, Storage, and Precaution: must be stored at −20 °C in the absence of moisture (flushing with dry Ar or N$_2$). It turns pale yellow on storage. It is instantaneously hydrolyzed by dilute acidic aqueous solutions.

(Z)-2-Bromo-1-(trimethylsilyloxy)ethylene **1** is used essentially as a precursor of (Z)-2-(trimethylsilyloxy)vinyllithium (eq 1).[3]

$$\text{(1)}$$

Reagent (**1**) has been condensed with allylic alcohols in the presence of a catalytic quantity of a Lewis acid (such as boron trifluoride etherate) to achieve a new synthesis of α,β-unsaturated aldehydes (eq 2).[4] This reaction is analogous to the widely used Müller-Cunradi–Pieroh reaction[5] except that the oxidation state of the cationic species is one unit lower and that of the nucleophilic species is one unit higher. Thus condensation with β-ionol leads to an α-bromo aldehyde which was transformed into β-ionylideneacetaldehyde by dehydrobromination, using 1,8-diazabicyclo[5.4.0]undec-7-ene.[4]

The reaction of (**1**) with lithium diisopropylamide followed by treatment with chlorodimethylsilane was reported to give the disilylketene (**2**) in a modest yield,[6] instead of the expected alkynic ether (**3**) (eq 3).[2] The hydridosilylketene (**2**) is transformed by flash vacuum thermolysis (FVT) to trimethylsilylacetylene by expulsion of dimethylsilanone.[6]

$$\text{(2)}$$

$$\text{(3)}$$

It is noteworthy that compounds homologous with (**1**) couple with Grignard reagents in the presence of nickel–phosphine complexes to produce alkylated and arylated enol ethers (eq 4).[7]

$$\text{(4)}$$

R^1, R^2 = H, alkyl, aryl; R^3 = alkyl, aryl

---

1. (a) Zembayashi, M.; Tamao, K.; Kumada, M., *Synthesis* **1977**, 422. (b) Komarov, N. V.; Lisovin, E. G., *Zh. Obshch. Khim.* **1979**, *49*, 1673 (*Chem. Abstr.* **1979**, *91*, 211 474z).

2. Pirrung, M. C.; Hwu, J. R., *Tetrahedron Lett.* **1983**, *24*, 565.

3. (a) Duhamel, L.; Tombret, F., *J. Org. Chem.* **1981**, *46*, 3741. (b) Duhamel, L.; Tombret, F., Mollier, Y., *J. Organomet. Chem.* **1985**, *280*, 1.

4. (a) Ancel, J. E.; Bienaymé, H.; Duhamel, L.; Duhamel, P.; Fr. Patent 14699; **1991**, Eur. Patent 544588 **1991**, (*Chem. Abstr.* **1993**, *119*, 271 443v). (b) Duhamel, L.; Duhamel, P.; Ancel, J. E., *Tetrahedron Lett.* **1994**, *35*, 1209.

5. Müller-Cunradi, M.; Pieroh, K. US Patent **1939**,2 165 962 (*Chem. Abstr.* **1939**, *33*, 8210²).

6. Barton, T. J.; Groh, B. L., *J. Am. Chem. Soc.* **1985**, *107*, 7221.

7. Tamao, K.; Zembayashi, M.; Kumada, M., *Chem. Lett.* **1976**, 1239.

Lucette Duhamel & Steven A. Kates

University of Rouen, France

2-Bromo-3-trimethylsilyl-1-propene

[81790-10-5]   C$_6$H$_{13}$BrSi   (MW 193.16)

InChI = 1S/C6H13BrSi/c1-6(7)5-8(2,3)4/h1,5H2,2-4H3

InChIKey = LUPQCAARZVEFMT-UHFFFAOYSA-N

(synthon for CH$_2$=C$^-$CH$_2$TMS$^{1-3}$ and CH$_2$=CBrC$^-$H$_2$;^{2} for synthesis of 1-trimethylsilylmethyl-substituted 1,3-butadienes[4])

Alternate Name: 2-bromoallyltrimethylsilane.

Physical Data: bp 46–50 °C/20 mmHg,[3] 64–65 °C/38–39 mmHg.[1]

Solubility: sol alcohol, acetone, ether, THF, pentane; insol water.

Preparative Methods: (1) reaction of 2,3-dibromopropene with lithium (trimethylsilyl)cuprate in HMPA at 0 °C (63–90%);[3,5] (2) reaction of 2,3-dibromopropene with trichlorosilane in the presence of trichlorosilane and copper(I) chloride, followed by treatment with methylmagnesium bromide (63–71%).[1,3,6]

Handling, Storage, and Precaution: no special requirements, except refrigeration in a brown bottle is recommended for long-term storage.

Synthon for CH$_2$=C$^-$CH$_2$TMS. The 1-trimethylsilyl-methylvinyl anion CH$_2$=C(M)CH$_2$TMS (**2**) (M = Li, Mg, Cu, etc.), readily prepared from 2-bromo-3-trimethylsilyl-1-propene (**1**) under typical conditions, allows the introduction of the synthetically useful 1-trimethylsilylmethylvinyl group to a wide variety of substrates. Ring opening of 1-butene oxide with the Grignard reagent (**2**) (M = MgBr) in the presence of copper(I) iodide gives only one regioisomer. Subsequent desilylative oxidation of this allyl alcohol to α-methylene-γ-lactones provides further utility of (**1**) as a 1-hydroxymethylvinyl anion equivalent, i.e. CH$_2$=−C$^-$CH$_2$OH (eq 1).[1] Alternatively, the alcohol from *trans*-2,3-epoxybutane provides a route to the unstable six-membered β,γ-unsaturated lactone (eq 2).[7] The copper-catalyzed 1,4-addition to the typically unreactive mesityl oxide proceeds smoothly. The versatility of the allylsilane moiety is again illustrated in the ethylaluminum dichloride-induced cyclization of the adduct to a tertiary cyclopentanol in high yield (eq 3).[2]

Addition of (**2**) to α,ω-substituted aldehydes and ketones is an efficient entry to inter- and intramolecular [3 + 2] cycloaddition approaches to methylenecyclopentanes.[8] For example, chemoselective addition of the Grignard reagent (**2**) to the aldehyde (**3**), followed by acetylation and treatment with a palladium(0) catalyst, gives a bicyclo[3.3.0]octane (eq 4).[3] Similarly, the ketosulfone (**5**), which derives from the lithium reagent (**2**) (M = Li) and the ketodithiane (**4**), cyclizes to the perhydroindanone system (eq 5).[9] This methodology has also been extended to the synthesis of methylenetetrahydrofurans, e.g. in the formation of the phyllanthocin ring system (eq 6).[10] Lewis acid-mediated cycloadditions, both inter- and intramolecular, have also been developed.[11,12] An intermolecular case which generates the zizaene skeleton is depicted in eq 7.[11]

$$\text{(eq 3)}$$

$$\text{(eq 1)}$$

$$\text{(eq 2)}$$

$$\text{(eq 4)}$$

$$\text{(eq 5)}$$

$$\text{(eq 6)}$$

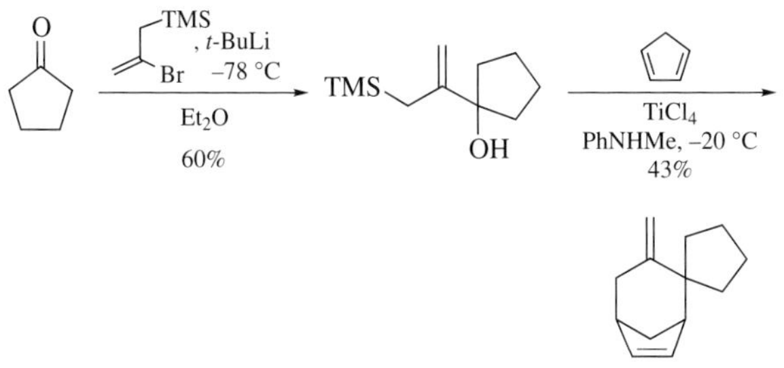

(7)

(10)

(11)

(12)

85%

The tin analog of (**2**) (M = Me₃Sn) undergoes smooth coupling with various aryl bromides and acyl chlorides to produce 2-substituted allylsilanes (eq 8).[13] Interestingly, the organozinc reagent (**2**) (M = ZnCl) has been shown to equilibrate to the allyl isomer (**6**) and react as such with electrophiles to produce vinylsilanes.[14a] However, it is possible in some cases to control the product distribution with the choice of catalyst (eq 9).[14b]

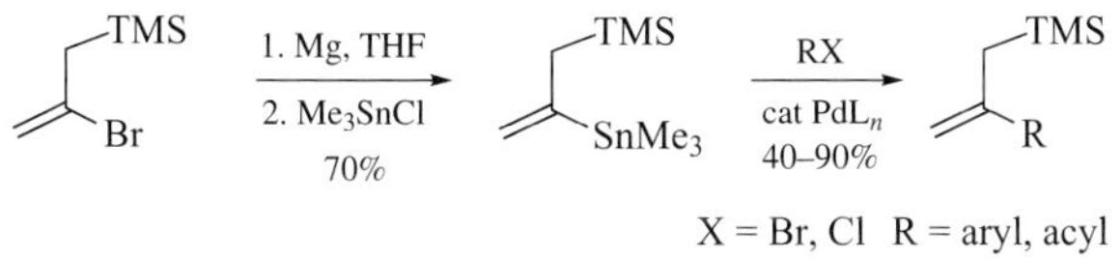

(8)

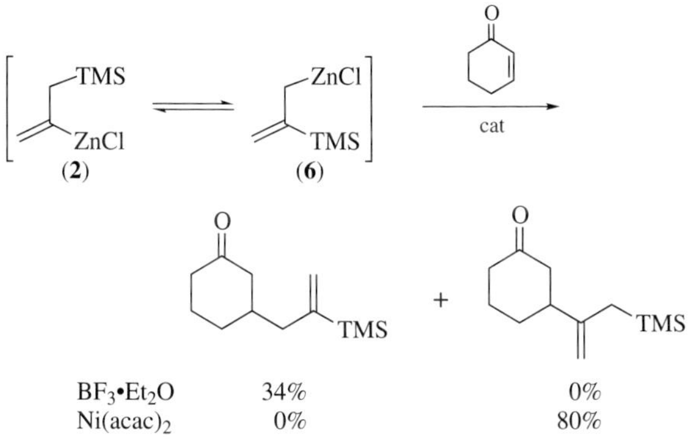

X = Br, Cl   R = aryl, acyl

(9)

BF₃•Et₂O	34%	0%
Ni(acac)₂	0%	80%

Synthon for CH₂=CBrC⁻H₂. Lewis acid-mediated addition of the silyl bromide (**1**) to electrophiles provides a convenient way to introduce the CH₂=CBrCH₂ moiety. In the case of carbonyl substrates, aldehydes react well while aliphatic ketones require higher concentration and aromatic ketones fail to participate. High stereoselectivity is achieved in many cases, e.g. in the addition to aldehyde (**7**). The subsequent conversion of the adduct by carbonylation to a α-methylene-γ-butyrolactone illustrates the synthetic versatility of the vinyl bromide unit (eq 10).[2] Conjugate addition to enones has also been demonstrated.[2,15] In the case of 1-acetylcyclopentene, intramolecular Barbier reaction of the *cis* isomer of the adduct gives an excellent yield of the cyclopentanol (eq 11).[2] Thus by combining the nucleophilic properties of the allylsilane with the ability to effect transmetalation of the vinyl bromide unit, the silyl bromide can serve also as a synthon for CH₂=C⁻–C⁻H₂. Furthermore, the vinyl bromide moiety can be a source of a vinyl radical. The synthesis of a propellane ring skeleton is made possible using this strategy (eq 12).[16]

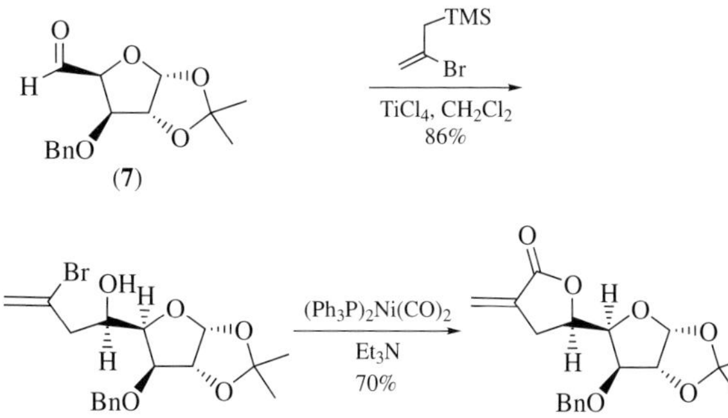

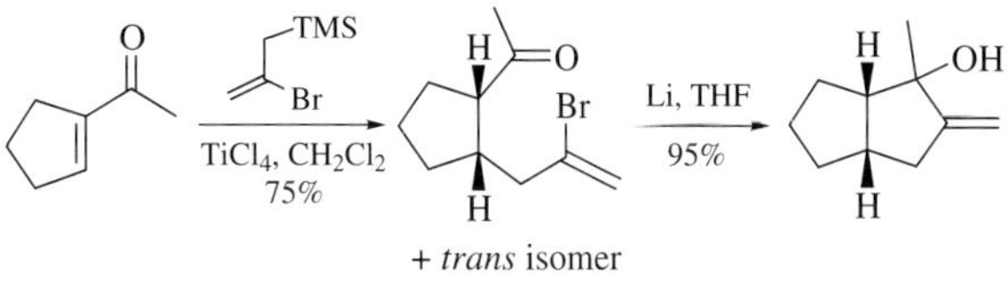

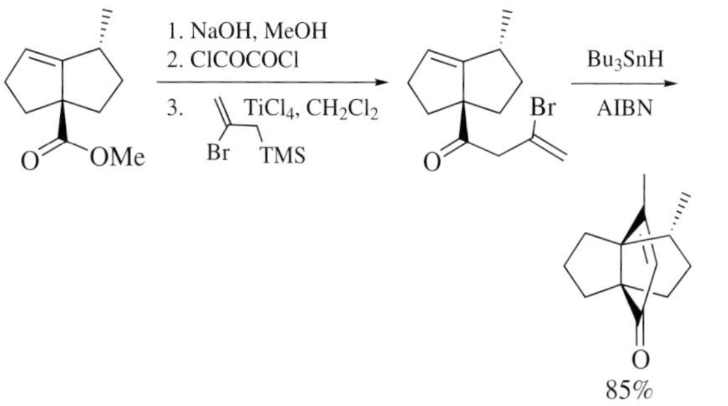

Under Lewis acidic conditions, the silyl bromide (**1**) and α,α′-dimethoxylated amides undergo a [3 + 3] type annulation which is useful for the construction of the piperidine skeleton (eq 13).[17] The silyl bromide (**1**) is also an excellent reagent for the stereoselective introduction of a functionalized allyl group at the anomeric center of the glucosides (eq 14).[6,18]

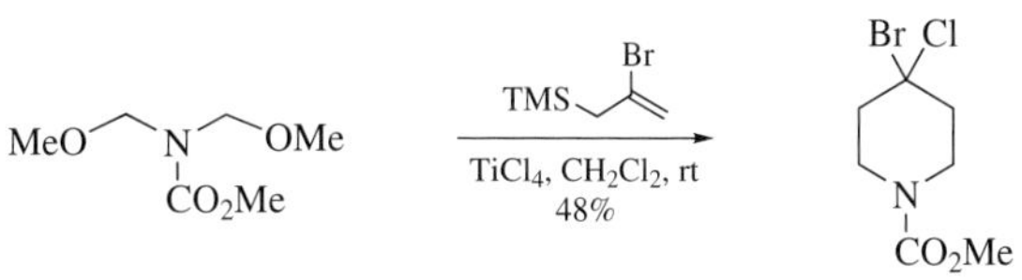

(13)

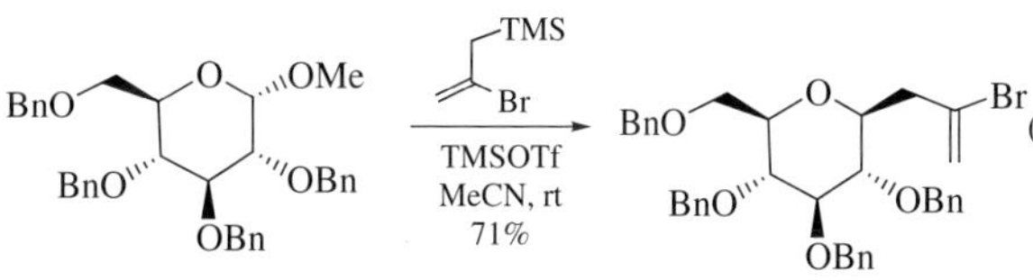

(14)

Coupling Reactions to Dienes and Enynes. The silyl bromide (**1**) participates readily in copper or transition metal-mediated coupling reactions to produce 1,3-butadienes, which are very useful synthetic intermediates. For example, 2,3-bis[(trimethylsilyl)methyl]-1,3-butadiene (**8**), derived from the oxidative dimerization of cuprate (**2**) (M = Cu) is useful for rapid construction of multicyclic systems via tandem Diels–Alder reactions, as depicted in eq 15.[19] The diene 2-dimethylaminomethyl-3-trimethylsilylmethyl-1,3-butadiene (**9**) functions similarly in the synthesis of a [6,7] ring system (eq 16).[20]

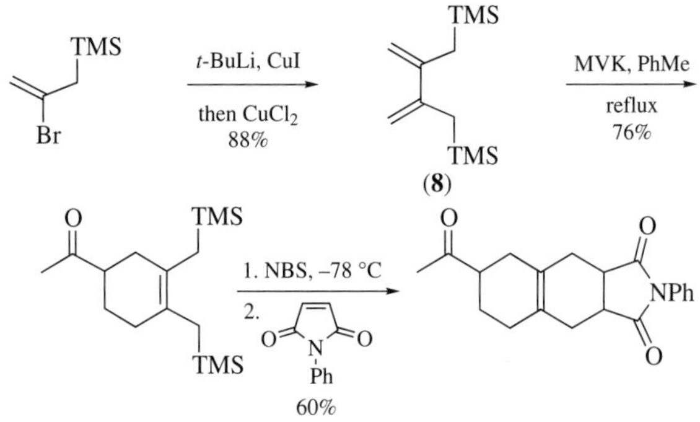

t-BuLi, CuI / then CuCl₂ / 88%

MVK, PhMe / reflux / 76%

(8)

1. NBS, −78 °C
2.

60%

(15)

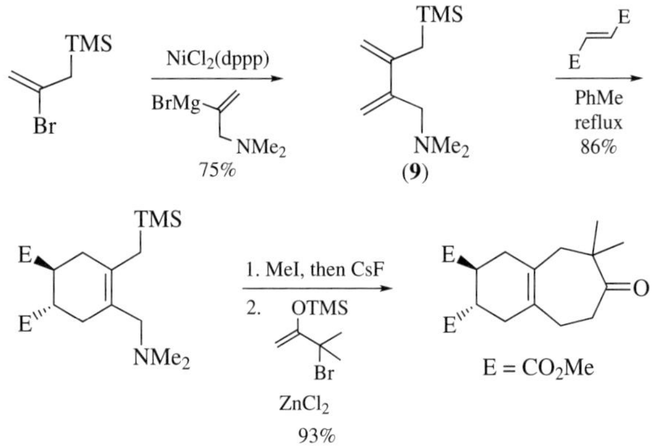

NiCl₂(dppp) / BrMg / NMe₂ / 75%

(9)

PhMe / reflux / 86%

1. MeI, then CsF
2. OTMS / Br / ZnCl₂ / 93%

E = CO₂Me   (16)

(10)

PdCl₂(PPh₃)₂ / *i*-Pr₂NH, THF, rt / 94%

(11)

(12)   (19)

Nickel-catalyzed cross coupling of Grignard reagent (**2**) (M = MgBr) with other vinyl bromides provides an easy access to isoprenylsilanes that display high reactivity and regioselectivity toward various dienophiles, as illustrated in the synthesis of a dihydropyran (eq 17).[4] Carbopalladation of allene with the silyl bromide (**1**) in the presence of a nucleophile also produces dienes. In one case the allylsilane unit is used to generate a cyclohexyl ring (eq 18).[21]

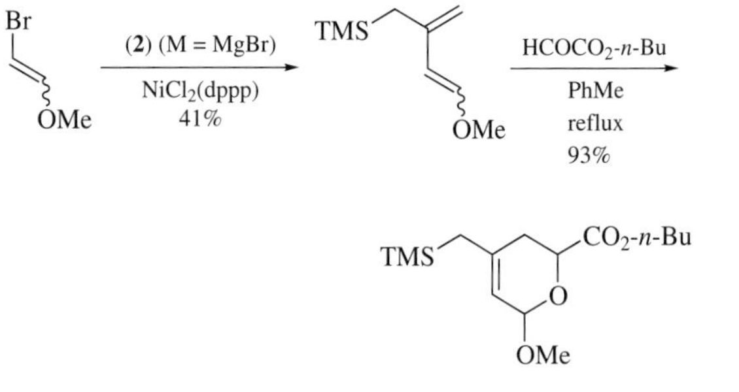

(**2**) (M = MgBr) / NiCl₂(dppp) / 41%

HCOCO₂-*n*-Bu / PhMe / reflux / 93%

(17)

H₂C=C=CH₂ / COMe / Na—C(CO₂Et) , cat PdLₙ / 85%

TiCl₄ / −78 °C, 2 h / 49%

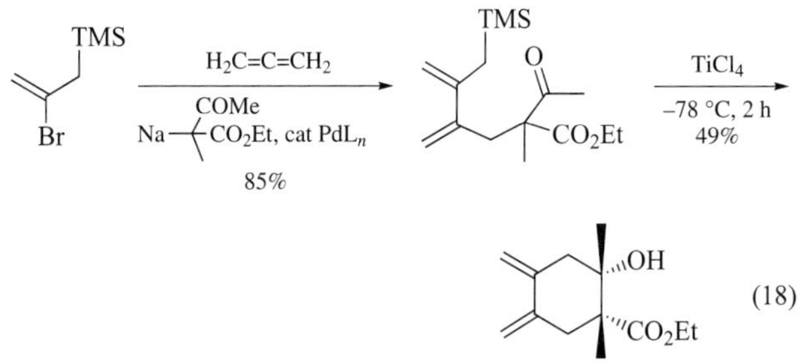

(18)

Palladium-catalyzed coupling of the silyl bromide to the terminal alkyne (**10**) gives the propargylic allylsilane (**11**), which is a key intermediate in the synthesis of a 10-membered cyclodiynol analog (**12**) of the antitumor agent neocarzinostatin (eq 19).[22]

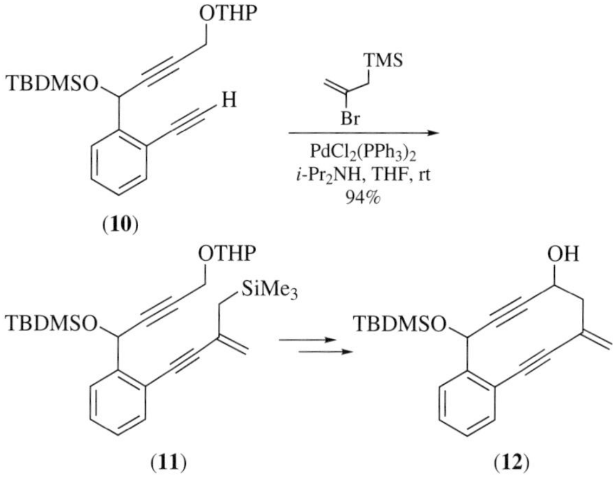

1. Nishiyama, H.; Yokoyama, H.; Narimatsu, S.; Itoh, K., *Tetrahedron Lett.* **1982**, *23*, 1267.
2. Trost, B. M.; Coppola, B. P., *J. Am. Chem. Soc.* **1982**, *104*, 6879.
3. Trost, B. M.; Grese, T. A.; Chan, D. M. T., *J. Am. Chem. Soc.* **1991**, *113*, 7350.
4. Hosomi, A.; Sakata, Y.; Sakurai, H., *Tetrahedron Lett.* **1985**, *26*, 5175.
5. Smith, J. G.; Drozda, S. E.; Petraglia, S. P.; Quinn, N. R.; Rice, E. M.; Taylor, B. S.; Viswanathan, M., *J. Org. Chem.* **1984**, *49*, 4112.
6. Hosomi, A.; Sakata, Y.; Sakurai, H., *Carbohydr. Res.* **1987**, *171*, 223.
7. Isaac, K.; Kocienski, P.; Campbell, S., *J. Chem. Soc., Chem. Commun.* **1983**, 249.
8. For an overview, see: (a) Chan, D. M. T., *Comprehensive Organic Synthesis* **1991**, *5*, 271. (b) Little, R. D., *Comprehensive Organic Synthesis* **1991**, *5*, 239. (c) Trost, B. M., *Angew. Chem., Int. Ed. Engl.* **1986**, *25*, 1. (d) Trost, B. M., *Pure Appl. Chem.* **1988**, *60*, 1615.
9. Trost, B. M.; Grese, T. A., *J. Org. Chem.* **1992**, *57*, 686.
10. Trost, B. M.; Moeller, K. D., *Heterocycles* **1989**, *28*, 321.
11. Hoffmann, H. M. R.; Eggert, U.; Gibbels, U.; Giesel, K.; Koch, O.; Lies, R.; Rabe, J., *Tetrahedron* **1988**, *44*, 3899.
12. (a) Ipaktschi, J.; Lauterbach, G., *Angew. Chem., Int. Ed. Engl.* **1986**, *25*, 354. (b) Collins, M. P.; Drew, M. G. B.; Mann, J.; Finch, H., *J. Chem. Soc., Perkin Trans. 1* **1992**, 3211.
13. Kang, K.-T.; Kim, S. S.; Lee, J. C., *Tetrahedron Lett.* **1991**, *32*, 4341.
14. (a) Eshelby, J. J.; Crowley, P.; Parsons, P. J., *Synlett* **1993**, *277*, 279. (b) Minato, A.; Suzuki, K.; Tamao, K.; Kumada, M., *Tetrahedron Lett.* **1984**, *25*, 83.
15. Ipaktschi, J.; Heydari, A., *Angew. Chem., Int. Ed. Engl.* **1992**, *31*, 313.
16. Jasperse, C. P.; Curran, D. P., *J. Am. Chem. Soc.* **1990**, *112*, 5601.
17. Shono, T.; Matsumura, Y.; Uchida, K.; Kobayashi, H., *J. Org. Chem.* **1985**, *50*, 3243.
18. Hosomi, A.; Sakata, Y.; Sakurai, H., *Tetrahedron Lett.* **1984**, *25*, 2383.
19. Trost, B. M.; Shimizu, M., *J. Am. Chem. Soc.* **1982**, *104*, 4299.
20. Hosomi, A.; Otaka, K.; Sakurai, H., *Tetrahedron Lett.* **1986**, *27*, 2881.
21. Cazes, B.; Colovray, V.; Gore, J., *Tetrahedron Lett.* **1988**, *29*, 627.
22. Suffert, J., *Tetrahedron Lett.* **1990**, *31*, 7437.

Dominic M. T. Chan
DuPont Agricultural Products, Newark, DE, USA

A list of General Abbreviations appears on the front Endpapers

(1-Bromovinyl)trimethylsilane[1]

[13683-41-5] $C_5H_{11}BrSi$ (MW 179.13)

InChI = 1S/C5H11BrSi/c1-5(6)7(2,3)4/h1H2,2-4H3
InChIKey = VVDJVCJVVHHCIB-UHFFFAOYSA-N

(undergoes lithium–halogen exchange in the presence of alkyllithium;[2] used as a dienophile in [4 + 2] cycloaddition reactions;[3] used in palladium-catalyzed reactions;[4] substrate for cyclopropanation[5])

Alternate Names: 1-(trimethylsilyl)vinyl bromide; 1-bromo-1-(trimethylsilyl)ethylene.
Physical Data: bp 124 °C/745 mmHg; d 1.156 g cm^{-3}.
Solubility: completely miscible with THF and Et$_2$O.
Form Supplied in: clear liquid.
Analysis of Reagent Purity: ^{1}H and ^{13}C NMR can be used.
Preparative Methods: prepared, in good yield, from the reaction of vinyltrimethylsilane with bromine at low temperature followed by dehydrohalogenation in the presence of an amine base.[1] Alternative syntheses and reagents, i.e. (1-bromovinyl)-triphenylsilane and (1-bromovinyl)triethylsilane, are also known.[2a]
Handling, Storage, and Precautions: store in a dry area.

Vinyltrimethylsilane Metal.[2]   1-Trimethylsilylvinyl metal species are mostly used as acyl anion equivalents and as hindered vinyl anion substitutes. 1-(trimethylsilyl)vinyllithium is easily accessible from (1-bromovinyl)trimethylsilane by treatment with *n*-butyllithium at −78 °C in ether (eq 1).[2a] Other solvents and alkyllithiums can also be used for the metal–halogen exchange.[2b] The use of (1-chlorovinyl)trimethylsilane or the direct reaction of (1-bromovinyl)trimethylsilane with lithium metal are not proper methods for generating the lithium species.[2c]

$$\underset{Br}{\overset{SiMe_3}{=}} \quad \xrightarrow[-78\,°C]{t\text{-BuLi}} \quad \underset{Li}{\overset{SiMe_3}{=}} \qquad (1)$$

The corresponding Grignard reagent is made by reaction of magnesium with (1-bromovinyl)trimethylsilane in THF (eq 2).[2d] It has the same general uses as the lithium analog (1-(trimethyl-silyl)vinyllithium). Furthermore, it has been shown to react with allylic acetates, in the presence of palladium catalysts, to give the corresponding dienes in moderate to low enantiomeric excess when the catalyst was chiral (eq 3).[6] Generally, Grignard reagents will add to the carbonyl of the acetate group in preference to allylic substitution. The authors described 1-(trimethylsilyl)vinyl-magnesium bromide as a soft stabilized anion to explain this peculiar reactivity.

$$\underset{Br}{\overset{SiMe_3}{=}} \quad \xrightarrow[THF]{Mg} \quad \underset{MgBr}{\overset{SiMe_3}{=}} \qquad (2)$$

$$\underset{MgBr}{\overset{SiMe_3}{=}} + \quad \overset{}{\bigcirc}\!\!-OAc \quad \xrightarrow[THF,\,10\,°C,\,2\,h]{PdCl_2(ProliNOP)} \quad \overset{}{\bigcirc}\!\!\diagup\!\!\underset{SiMe_3}{\diagup} \qquad (3)$$
$$85\% \qquad\qquad 30\%\ ee$$

$$ProliNOP = \quad \overset{O-PPh_2}{\underset{N-PPh_2}{\diagdown}}$$

This Grignard reagent has also been reacted with carbon dioxide[2f] and the cuprate of this Grignard reagent adds to alkynyl compounds and provides a practical synthesis of 1,3-butadienes (eq 4).[2g]

$$\underset{CuMgBr_2}{\overset{SiMe_3}{=}} \quad \xrightarrow[\substack{2.\ H_2O}]{\substack{1.\ \equiv\!\!-OEt \\ THF}} \quad \underset{EtO}{\overset{SiMe_3}{=}\!\!=} \qquad (4)$$
$$66\%$$

[4 + 2] Cycloaddition Reactions.[3]   A number of substituted vinyltrimethylsilanes react with dienes and nitrile oxides in Diels–Alder[6] and 1,3-dipolar cycloadditions.[3] The trimethylsilyl group has a deactivating effect. This reaction with benzonitrile oxide gives 3-phenyl-5-trimethylsilylisoxazole (eq 5).

$$\underset{Br}{\overset{SiMe_3}{=}} + \quad Ph\overset{N}{\underset{Cl}{\diagup}}\!\!\diagdown OH \quad \xrightarrow{Et_3N} \quad Ph\overset{N-O}{\diagdown\!\!\diagdown}\underset{SiMe_3}{} \qquad (5)$$
$$65\%$$

Palladium-Catalyzed Reactions.[4]   Numerous nucleophiles react with (1-bromovinyl)trimethylsilane in the presence of palladium complexes. The bromine has been substituted by phenylthio,[4a] vinyl,[4b] and aryl[4c] groups. This approach gives reasonable yields of the desired products. However, the substitution reactions sometimes lack regiospecificity. For example, a mixture of regioisomers was obtained in eq 6. Two mechanisms have been proposed for the formation of the β-substituted product. One involves an elimination step to give trimethylsilylacetylene as an intermediate, which then undergoes catalyzed additions with the nucleophile at either the α- or β-positions.[4a] The other mechanism involves the formation of a pentacoordinated palladium intermediate, leading to the formation of isomeric products.[4c]

$$\underset{Br}{\overset{SiMe_3}{=}} \quad \xrightarrow[PhS-SnMe_3]{Pd(PPh_3)_4} \quad \underset{SPh}{\overset{SiMe_3}{=}} + \quad PhS\!\!\diagup\!\!\diagdown\!\!SiMe_3 \qquad (6)$$
$$80\% \qquad\qquad 3:1$$

Substrate for Cyclopropanation.[5]   Dichlorocarbene adds to vinyltrimethylsilane, and the adduct is a source of 1-chlorocyclopropene, a reactive dienophile for Diels–Alder reactions.[5b] Dichlorocarbene also reacts with (1-bromovinyl)trimethylsilane[5a] to form (1-bromo-2-dichlorocyclopropyl)trimethylsilane. β-Elimination in the presence of tetra-*n*-butylammonium fluoride gives 1-bromo-2-chlorocyclopropene, another reactive dienophile (eq 7).

$$\underset{Br}{\overset{SiMe_3}{=}} \quad \xrightarrow[\substack{glyme-diglyme \\ (10:1),\,90\,°C}]{Cl_3CO_2Na} \quad \overset{Cl\ \ Cl}{\underset{Br}{\triangle}}\!\!SiMe_3 \quad \xrightarrow{TBAF} \quad \overset{Cl}{\underset{Br}{\triangle}} \qquad (7)$$

Realated Reagents. (2-Bromovinyl)trimethylsilane; 1-(trimethylsilyl)vinyllithium; (*E*)-2-(trimethylsilyl)vinyllithium; vinylmagnesium bromide.

1. Boeckman, R. K. Jr.; Blum, D. M.; Ganem, B.; Halvey, N., *Org. Synth.* **1978**, *58*, 152.

2. (a) Alternative reagents and synthesis. For (1-bromovinyl)triphenyl- and trimethylsilanes see: Chan, T. H.; Mychajlowskij, W.; Ong, B. S.; Harpp, D. N., *J. Org. Chem.* **1978**, *43*, 1526. Gröbel, B. T.; Seebach, D., *Chem. Ber.* **1977**, *110*, 867. For (1-bromovinyl)triethylsilane see: Stork, G.; Ganem, B., *J. Am. Chem. Soc.* **1973**, *95*, 6152 and Ottolenghi, A.; Fridkin, M.; Zilkha, A., *Synlett* **1963**, *41*, 2977. (b) Overman, L. E.; Thompson, A. S., *J. Am. Chem. Soc.* **1988**, *110*, 2248. (c) Cunico, R. F.; Han, Y. K., *J. Organomet. Chem.* **1979**, *174*, 247. Husk, G. R.; Velitchko, A. M., *J. Organomet. Chem.* **1973**, *49*, 85. (d) Vinod, T. K.; Hart, H., *Tetrahedron Lett.* **1988**, *29*, 885. Boeckman, R. K. Jr.; Blum, D. M.; Arthur, S. D., *J. Am. Chem. Soc.* **1979**, *101*, 5060; Huynh, C.; Linstrumelle, G., *Tetrahedron Lett.* **1979**, 1073. (e) Sato, F.; Kobayashi, Y.; Takahashi, O.; Chiba, T.; Takeda, Y.; Kusakabe, M., *J. Chem. Soc., Chem. Commun.* **1985**, 1636. Sato, F.; Takahashi, O.; Kato, T.; Kobayashi, Y., *J. Chem. Soc., Chem. Commun.* **1985**, 1638 and references cited therein. (f) Cooke, M. P. Jr., *J. Org. Chem.* **1987**, *52*, 5729. (g) Foulon, J. P.; Bourgain-Commerçon, M.; Normant, J. F., *Tetrahedron* **1986**, *42*, 1389.

3. (a) Padwa, A.; MacDonald, J. G., *J. Org. Chem.* **1983**, *48*, 3189. (b) Padwa, A.; MacDonald, J. G., *Tetrahedron Lett.* **1982**, *23*, 3219.

4. (a) Carpita, A.; Rossi, R.; Scamuzzi, B., *Tetrahedron Lett.* **1989**, *30*, 2699. (b) Cazes, B.; Colovray, V.; Gore, J., *Tetrahedron Lett.* **1988**, *29*, 627. (c) Ennis, D. S.; Gilchrist, T. L., *Tetrahedron Lett.* **1989**, *30*, 3735. Minato, A.; Suzuki, K.; Tamao, K.; Kumada, M., *Tetrahedron Lett.* **1984**, *25*, 83 and references cited therein.

5. (a) Billups, W. E.; Lin, L. J.; Arney, B. E. Jr.; Rodin, W. A.; Casserly, E. W., *Tetrahedron Lett.* **1984**, *25*, 3935. (b) Chan, T. H.; Massuda, D., *Tetrahedron Lett.* **1975**, 3383.

6. Fotiadu, F.; Cros, P.; Faure, B.; Buono, G., *Tetrahedron Lett.* **1990**, *31*, 77.

Denis Labrecque & Tak-Hang Chan
McGill University, Montréal, Québec, Canada

t-Butyldimethylchlorosilane

$$t\text{-BuMe}_2\text{SiCl}$$

[18162-48-6] $C_6H_{15}ClSi$ (MW 150.72)

InChI = 1S/C6H15ClSi/c1-6(2,3)8(4,5)7/h1-5H3

InChIKey = BCNZYOJHNLTNEZ-UHFFFAOYSA-N

(widely used reagent for the protection of alcohols, amines, carboxylic acids, ketones, amides, thiols, and phenols;[1] useful for regioselective silyl enol ether formation[2] and stereoselective silyl ketene acetal formation[3])

Alternate Names: *t*-butyldimethylsilyl chloride; TBDMSCl; TBSCl.

Physical Data: mp 86–89 °C; bp 125 °C.

Solubility: very sol nearly all common organic solvents such as THF, methylene chloride, and DMF.

Form Supplied in: moist white crystals, commonly available.

Handling, Storage, and Precautions: hygroscopic, store under N_2; harmful if inhaled, swallowed, or absorbed through skin; should be used and weighed out in a fume hood.

Original Commentary

Bret E. Huff
Lilly Research Laboratories, Indianapolis, IN, USA

Protecting Group. The reactions of TBDMSCl closely parallel those of chlorotrimethylsilane (TMSCl). However, TBDMS ethers are about 10^4 more stable toward hydrolysis than the corresponding TMS ethers.[1,4] The hydrolytic stability of the TBDMS group has made it very valuable for the isolation of many silicon-containing molecules. Since its introduction in 1972, the TBDMS protecting group has undoubtedly become the most widely used silicon protecting group in organic chemistry.[1] Alcohols are most commonly protected as their TBDMS ethers by treatment of the alcohol in DMF (2 mL g^{-1}) at rt with 2.5 equiv of imidazole (Im) and 1.2 equiv of TBDMSCl. Alcohol protection in the presence of 4-dimethylaminopyridine (DMAP) allows a greater range of solvents to be used (protection of alcohols in solvents other than DMF using TBDMSCl alone are sluggish) and distinguishes a kinetic preference for protection of primary alcohols in the presence of secondary alcohols.[5] Table 1 outlines conditions for the selective protection of primary and secondary alcohols using TBDMSCl and DMAP as catalyst.

Table 1 Use of DMAP to catalyze the protection of diols with TBDMSI

Amine(s)	Solvent	GC ratio of A:B:C:
DMAP(0.04 equiv), Et$_3$N(1.1 equiv)	CH$_2$Cl$_2$	95:0:5
Im(2.2 equiv)	DMF	59:11:30
Im (0.04 equiv), Et$_3$N(1.1 equiv)	DMF	No reaction
DMAP (0.04 equiv), Et$_3$N(1.1 equiv)	DMF	0:0:96

Alcohol:

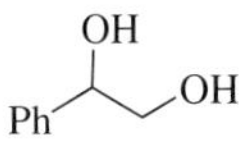

Products:

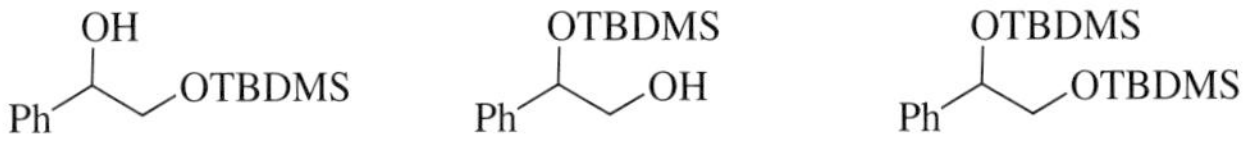

Table 2[6] compares the rate of hydrolysis of several bulky silicon ethers with acid, base, and fluoride; TBDMS ethers are less resistant to hydrolysis than the corresponding TIPS (triisopropylsilyl) and TBDPS (*t*-butyldiphenylsilyl) ethers.

Table 2 Comparison of TBDMS, TIPS, and TBDPS rates of hydrolysis by F$^-$, H$^+$, and OH$^-$ ($t_{1/2}$)

ROSiR$_3$	R = *n*-Bu^a		R = Cy^{a,b}		
	H$^+$	OH$^-$	H$^+$	OH$^-$	F$^-$
TBDMS	<1 min	1h	<4 min	26 h	76 min
TIPS	18 min	14 h	100 min	44 h	137 min
TBDPS	244 min	<4 h	360 min	14 h	–

^{a}H$^+$ refers to 1% HCl in 95% EtOH at 22 °C; OH$^-$ refers to 5% NaOH in 95% EtOH at 90 °C.
^{b}F$^-$ refers to 2 equiv of *n*-Bu$_4$NF in THF at 22 °C.

The TBDMS group is also suitable for the protection of amines, including heterocycles, carboxylic acids, and phenols. Other more reactive reagents are also available for introduction of the TBDMS group including *t*-butyldimethylsilyl trifluoromethane-sulfonate, MTBSA, and TBDMS-imidazole.[1,7]

Anion Trap. TBDMSCl is useful as an anion trapping reagent. For example, TBDMSCl was found to be an efficient trap of the lithio α-phenylthiocyclopropane anion.[8] When dichlorothio-phene was treated with 2 equiv of *n*-butyllithium followed by 2 equiv of TBDMSCl the di-TBDMS-thiophene was isolated.[9] The lithium anions of primary (eq 1) and secondary (eq 2) nitriles were trapped with TBDMSCl to give *C,N*-disilyl- and *N*-silylketenimines in excellent yields.[10]

Silyl stannanes have been prepared by trapping tin anions with TBDMSCl or other silyl chlorides. Alkynes treated with silyl stan-nanes and catalytic tetrakis(triphenylphosphine)palladium(0) give *cis*-silyl stannylalkenes in good yields.[11]

$$\text{(1)}$$

$$\text{(2)}$$

Silyl Ketene Acetals. The lithium enolates of esters may be trapped with TBDMSCl to prepare the corresponding ketene silyl acetals.[3] The resulting TBDMS ketene acetals are more stable than the corresponding TMS ketene acetals and have a greater preference for *O*- vs. *C*-silylation products. When TMSCl was used to trap the enolate of methyl acetate, a 65:35 ratio of *O*- to *C*-silated products was obtained. In addition, *O*-(TMS) silyl ketene acetals are thermally and hydrolytically unstable. However, similar treatment of lithium enolates with TBDMSCl provided the corresponding *O*-(TBDMS) silyl ketene acetals exclusively (eq 3). The *O*-TBDMS ketene acetals generally survive extraction from cold aqueous acid. lithium diisopropylamide was found to be satisfactory for the preparation of the ester enolates. The lower reactivity of TBDMSCl requires that the enol silation be performed at 0 °C with added HMPA.[12]

$$\text{(3)}$$

A detailed study of the formation of (*E*)- and (*Z*)-silyl ketene acetals was recently published.[3d] It was found that the formation of silyl enolates does not correspond to simple kinetic vs. thermo-dynamic formation of the enolates. Formation of the ester enolates occurred under kinetic control and a kinetic resolution accounted for selective formation of (*E*)- and (*Z*)-silyl ketene acetals. Table 3 summarizes some of these results.

Claisen Rearrangement. The first silyl ketene acetal Claisen rearrangement was introduced in 1972 using TMS ketene acetals. Since then, the silyl Claisen rearrangement using TBDMS ketene

Table 3 Effects of reaction condition on (*Z*):(*E*) ratio of TBDMS silyl ketene acetal formation of ethyl propionate

Ester:base ratio	Base	Solvent	Additive	(Z):(E) Silyl ketene acetal ratio	Yield (%)
1:1	LDA	THF	–	6:94	90
1.05:1	LDA	THF	45%DMPU	>98:2	80
1.2:1	LDA	THF	23%HMPA	93:7	65
1:1	LDA	THF	23%HMPA	85:15	80
1:1	LHMDS	THF	23%HMPA	>91:9	85
1.1:1	LHMDS	THF	23%HMPA	>95:5	60

acetals has found widespread use in organic synthesis.[13,14] One advantage of the silyl ketene acetal Claisen rearrangement is that the ketene acetal geometry may be predictably controlled (see above). Two components of the reaction contribute to stereo-control: the geometry of the silyl ketene acetal and the contribution of boat vs. chair transition state. A useful variant of the Claisen rearrangement involves the use of an enantiomerically pure α-silyl secondary alcohol prepared by Brook rearrangement of a TBDMS-protected primary alcohol. In this reaction the stereo-chemistry at the silicon-bearing center is transferred to the Claisen product.[15] The (*E*)-enolate was prepared by treatment of the es-ter with lithium hexamethyldisilazide (eq 4); the (*Z*)-enolate was prepared by treatment of the ester with LDA (eq 5).[15b] As ex-pected, the (*Z*)- and (*E*)-silyl ketene acetals gave the corres-ponding *syn* and *anti* Claisen products in good selectivity. The vinylsilane was hydrolyzed to the alkene using 50% HBF$_4$ in acetonitrile at 55 °C.

$$\text{(4)}$$

$$\text{(5)}$$

The silyl ketene acetals of methyl α-(allyloxy)acetates were found to undergo [3,3]-sigmatropic rearrangement, whereas the corresponding lithium enolates undergo [2,3]-sigmatropic rearrangement.[16] An interesting ring contraction based on the TBDMS silyl ketene acetal Claisen rearrangement has also been reported.[17]

TBDMS Enol Ethers.[18] Enolates trapped with TBDMSCl to prepare the corresponding enol ethers are more stable than the corresponding TMS enol ethers.[19] The potassium enolate of 2-methylcyclohexanone, prepared by addition of potassium hydride to a solution of the ketone and TBDMSCl in THF at $-78\,°C$ followed by warming to rt, gave the thermodynamic enol ether in a 56:44 ratio. In the presence of HMPA, the ratio improves to 98:2 (eq 6). This method works especially well with ketones with a propensity for self-condensation.[20]

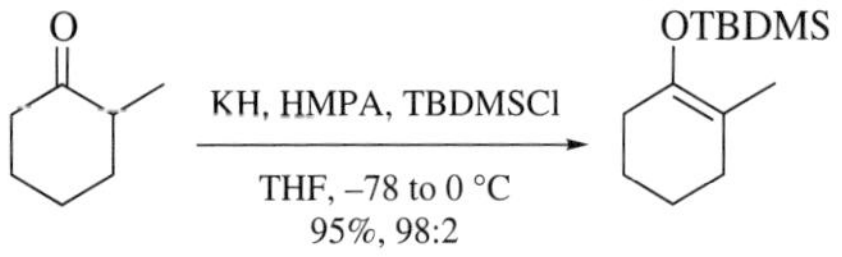

$$(6)$$

Potassium enolates derived from acylfulvalenes were trapped with TBDMSCl but not TMSCl or diphenylmethylsilyl chloride.[21] Interestingly, TBDMSCl was found to be compatible with CpK anion at $-78\,°C$. TBDMS enol ethers have also been used as β-acyl anion equivalents.[22] The TBDMS-silyl enol ethers of diketones (eq 7) and β-keto esters (eq 8) may be prepared by mixing them with TBDMSCl in THF with imidazole.[23] Alcohols may be protected under acidic conditions as their TBDMS ethers by treatment with β-silyl enol ethers in polar solvents.

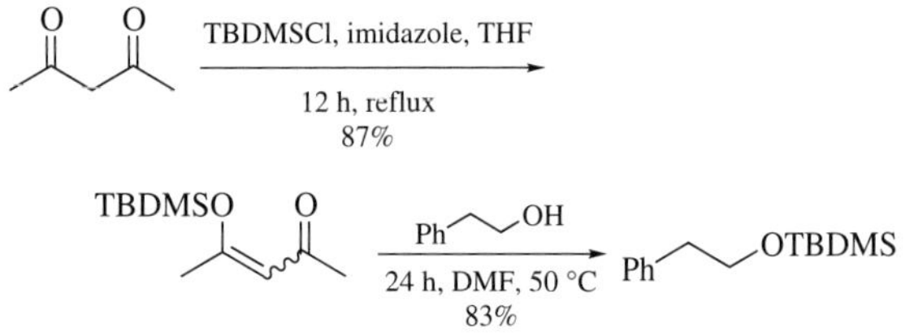

$$(7)$$

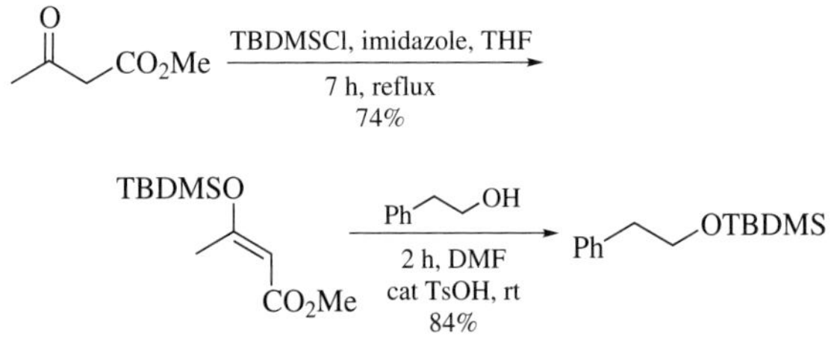

$$(8)$$

Aldol Reaction. The catalyst system TBDMSCl/InCl$_3$ selectively activates aldehydes over acetals for aldol reactions with TBDMS enol ethers.[24] Acetals and aldehydes are activated towards aldol reactions using TMSCl/InCl$_3$ or Et$_3$SiCl/InCl$_3$ as catalysts (eq 9).

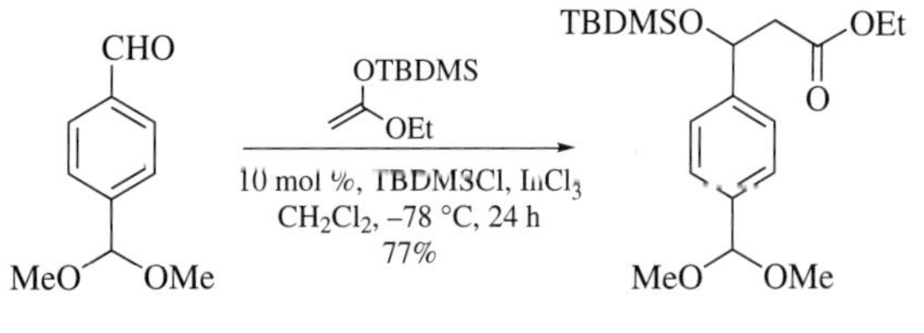

$$(9)$$

TBDMSCl as Cl$^-$ Source. TBDMSCl was used as a source of chloride ion in the Lewis acid-assisted opening of an epoxide.[25] The epoxide was treated with TBDMSCl and triethylamine followed by titanium tetraisopropoxide and additional TBDMSCl to give the *trans* chloride as the major product in 67% yield (eq 10).

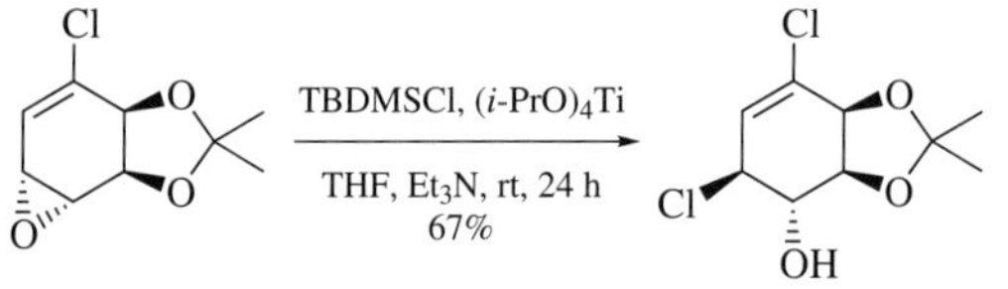

$$(10)$$

TBDMSCl-Assisted Reactions. Nitro aldol (Henry) reactions have been reported to be promoted by TBDMSCl.[26] To a THF solution of tetra-*n*-butylammonium fluoride is added sequentially equimolar amounts of the nitro compound, aldehyde, and Et$_3$N, followed by an excess of TBDMSCl (eq 11). Substitution of TMSCl for TBDMSCl reduces the yield of nitro aldol product. The authors speculate that TBDMSCl is responsible for activation of the aldehyde while *n*-Bu$_4$NF activates the nitro compound. In a related method, primary and secondary nitro alkanes were treated with LDA in THF followed by addition of TBDMSCl to give the corresponding silyl nitronates. The silyl nitronates reacted with a variety of aliphatic and aromatic aldehydes which gave vicinal nitro TBDMS aldol products.[27]

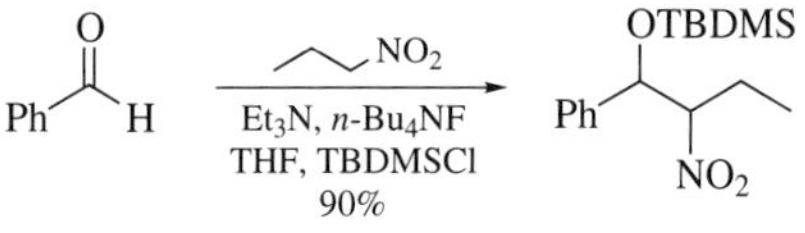

$$(11)$$

Reaction with Nucleophiles. TBDMSCl is the reagent of choice for the preparation of other TBDMS-containing reagents. For example, *t*-butyldimethylsilyl cyanide may be prepared by the reaction of TBDMSCl and potassium cyanide in acetonitrile containing a catalytic amount of zinc iodide.[28] TBDMSCN has also been prepared by treatment of TBDMSCl with KCN and 18-crown-6 in CH$_2$Cl$_2$ at reflux and by treatment of TBDMSCl with lithium cyanide prepared in situ.[29] *t*-butyldimethylsilyl trifluoromethanesulfonate is prepared by treatment of TBDMSCl with trifluoromethanesulfonic acid at $60\,°C$.[30] Other nucleophiles, such as thiolates, also react with TBDMSCl.[31] *t*-butyldimethylsilyl iodide was prepared by treatment of TBDMSCl with sodium iodide in acetonitrile.[32] In contrast to THF cleavage reactions using TMSI, the more stable TBDMS-protected primary alcohol may be isolated from the reaction in eq 12.

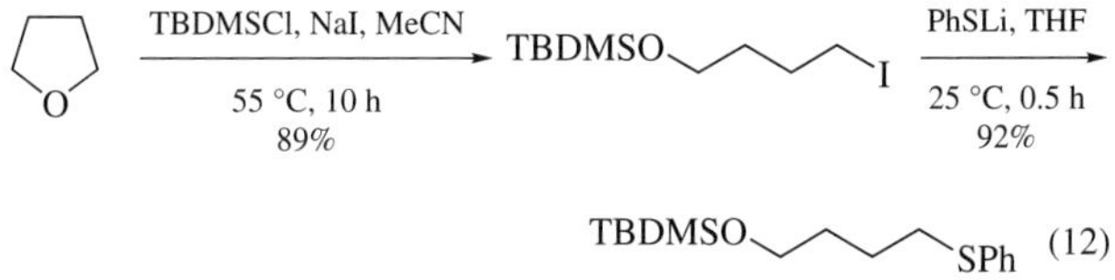

$$(12)$$

Mannich Reaction. The Mannich reaction of *N*-methyl-1,3-oxazolidine with 2-methylfuran was shown to proceed smoothly in the presence of TBDMSCl and catalytic 1,2,4-triazole in 61% yield. Interestingly, this reaction failed with TBDMSOTf due to the destruction of 2-methylfuran. The reaction proceeds with decreased yield (31%) in the absence of triazole. These reaction conditions allowed for the isolation of the TBDMS-protected alcohol (eq 13).[33]

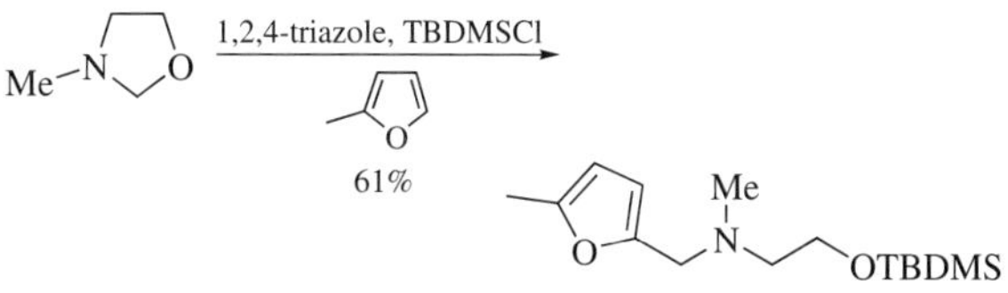

(13)

Acid Chlorides. TBDMS esters, when treated with DMF and oxalyl chloride in methylene chloride at 0 °C, give the corresponding acid chlorides in excellent yields under neutral conditions (eq 14).[34] Similarly, *N*-carboxyamino acid anhydrides were prepared via the intermediacy of an acid chloride prepared from a TBDMS ester (eq 15).[35]

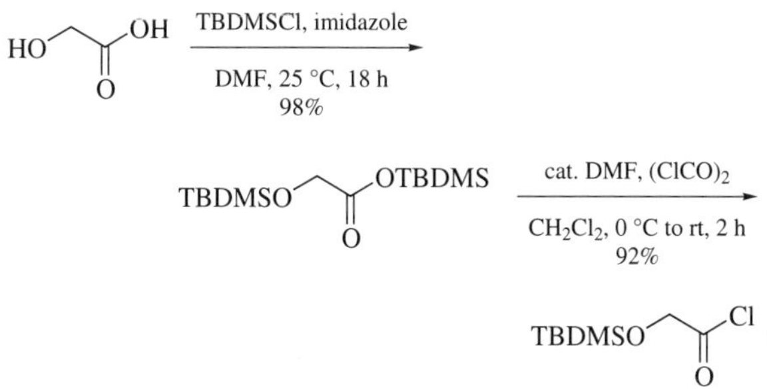

(14)

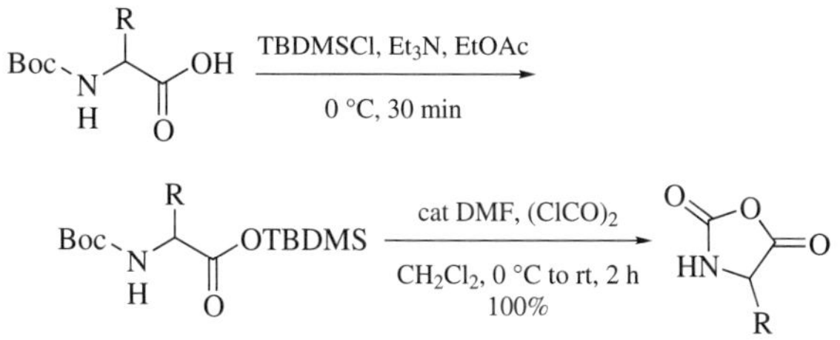

(15)

Conjugate Additions. When a mixture of TBDMSCl and a β-aryl enone (2:1) was added to $Bu_2CuCNLi_2$ at −78 °C, the TBDMS group added 1,4 to the enone to give β-silyl carbonyl compounds (eq 16).[36] *N*-TBDMS silyliminocuprates also add to α,β-unsaturated carbonyl compounds.[37]

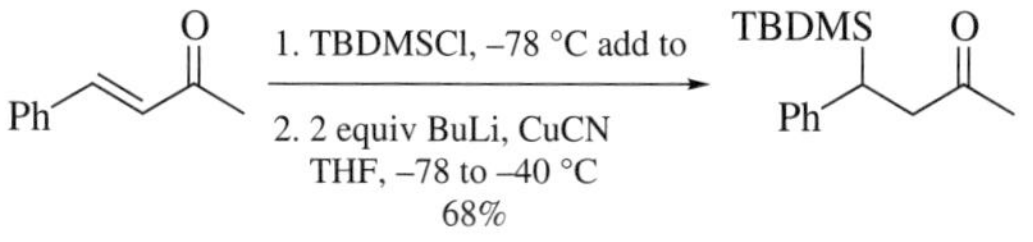

(16)

α-Silyl Aldehydes. Initial attempts to isolate TMS α-silyl aldehydes were unsuccessful due to the lability of the TMS group. However, the α-silyl aldehyde was prepared from the cyclohexyl imine of acetaldehyde by treatment with LDA followed by TBDMSCl (eq 17). Typical of TBDMSCl trapping reactions of imines and hydrazones, *C*-silylation was observed. The imine was hydrolyzed with HOAc in CH_2Cl_2 which gave the α-silyl aldehyde. These compounds, after treatment with organometallic reagents such as ethylmagnesium bromide or ethyllithium, may be eliminated in a Peterson-like manner to give either *cis* or *trans* alkenes (eq 18).[38]

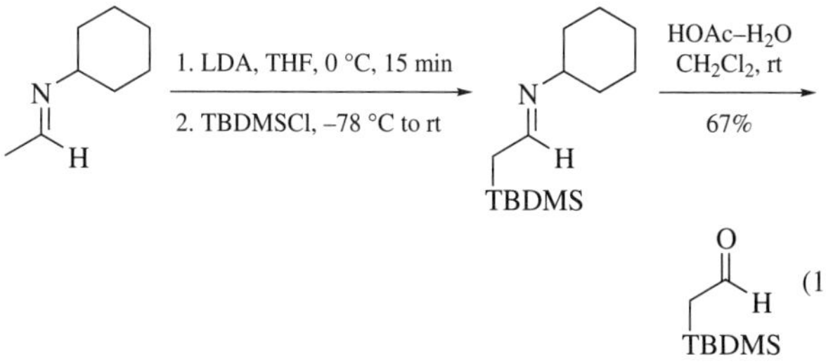

(17)

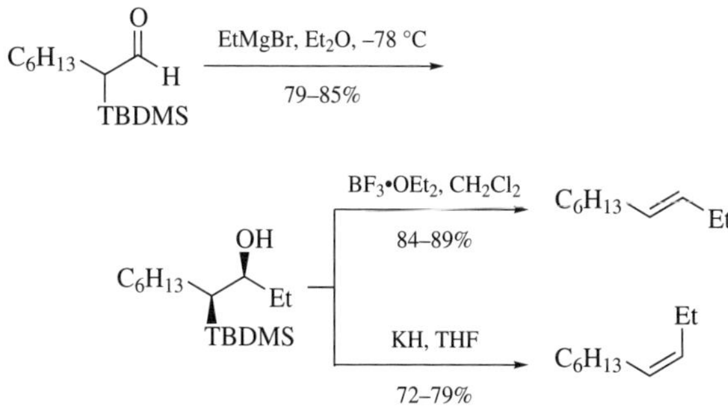

(18)

α-Silyl Ketones. When SAMP or RAMP hydrazones were treated with LDA followed by TBDMSCl, the corresponding α-silyl hydrazones were isolated. Ozonolysis of the hydrazone gave the enantiomerically enriched α-silyl ketones (eq 19).[39] Yields for the overall process are 52–79% for the preparation of α-silyl ketones and 22–42% for the preparation of α-silyl aldehydes. TBDMSOTf may also be used for quench of the SAMP/RAMP hydrazone enolate.

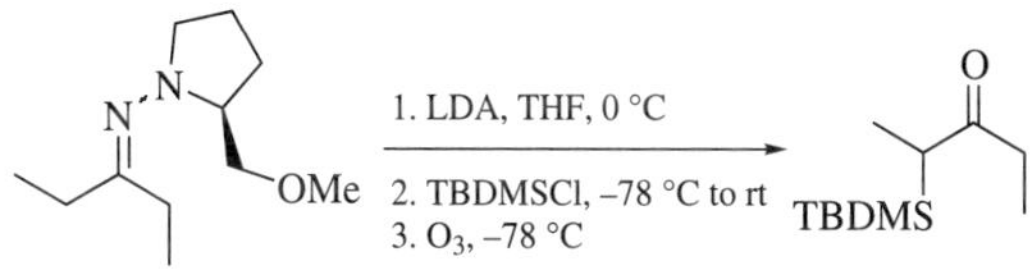

(19)

Acyl Silanes. Although acyl trimethylsilanes are known, they are usually unstable and lead to poor diastereoselectivity in aldol reactions.[40] TBDMS acyl silanes, however, were prepared in 50% yield from 1-methoxy-1-lithiopropene in the presence of TMEDA at rt (eq 20). The lithium enolates of TBDMS acyl silanes were treated with aldehydes to give the corresponding aldol products in reasonable yields.

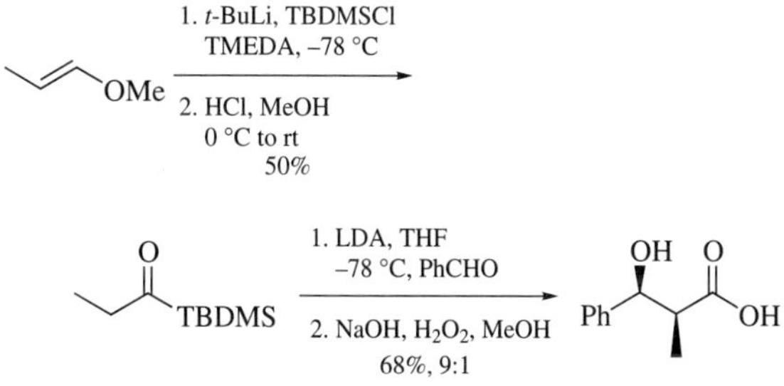

(20)

TBDMS acyl enones were prepared by treatment of an ethoxyethyl (EE)-protected alkoxyallene with *n*-BuLi at −85 °C followed by treatment of resulting anion with TBDMSCl (eq 21). Acid hydrolysis of the OEE group led to the TBDMS acyl enones in good yield.[41]

$$\xrightarrow[\substack{2.\ TBDMSCl,\ HMPA,\ 15\ h \\ 81\%}]{1.\ n\text{-BuLi},\ Et_2O}$$

$$\xrightarrow[\substack{84\%}]{H_2SO_4,\ H_2O\text{–THF}}$$

$$\text{(21)}$$

Modified Amine Base. The regioselectivity of ketone deprotonation was improved by the use of lithium *t*-butyldimethylsilylamide as base.[42] The base was prepared by deprotonation of isopropylamine with *n*-BuLi in THF (eq 22). The resulting anion was quenched with TBDMSCl to give the amine in 70% yield after distillation. Deprotonation of various ketones using this amide base was found to be equally or more selective than LDA. For example, the TBDMS-modified base gave a 62:38 ratio of kinetic to thermodynamic enolate, whereas LDA gave a 34:66 ratio with phenyl acetone.

$$\xrightarrow[\substack{TBDMSCl,\ -78\ °C\ to\ rt}]{n\text{-BuLi},\ THF}$$

$$\xrightarrow[\substack{2.\ -78\ °C,\ TMSCl \\ 2\text{-Me-cyclohexanone} \\ >99:1}]{1.\ n\text{-BuLi},\ -78\ °C\ to\ rt} \quad \text{(22)}$$

N-Formylation. When secondary amines were treated with TBDMSCl, DMAP, and Et₃N in DMF the corresponding *N*-formyl derivatives were formed (eq 23).[43] It was found that the reaction proceeds through a Vilsmeier type reagent formed by the reaction of TBDMSCl and DMF. It is possible that other TBDMS alkylation reactions, such as protection of alcohols in DMF, may proceed through a similar DMF-derived Vilsmeier reagent.

$$\xrightarrow[\substack{40\ °C,\ 10\ h}]{DMF,\ TBDMSCl,\ Et_3N}$$

$$\xrightarrow[\substack{87\%}]{} \quad \text{(23)}$$

N-Silyl Imines. When aldehydes were treated first with tris(trimethylstannyl)amine followed by TBDMSCl, the corresponding *N*-TBDMS imines were isolated in good yields.[44] These silyl imines reacted with ester enolates to give *β*-lactams (eq 24).

$$\xrightarrow[\substack{Et_2O,\ rt,\ 15\ min}]{N(SnMe_3)_3,\ TBDMSCl}$$

$$\xrightarrow[\substack{45\ min \\ 65\%}]{} \quad \text{(24)}$$

First Update

Wenming Zhang

Dupont Crop Protection, Stine-Haskell Research Center, Elkton Road, Newark, DE, USA

Protecting Group. Several alternative protocols have been developed for protecting alcohols as *t*-butyldimethylsilyl ethers. TBDMSCl was found to protect alcohols efficiently when DMSO–hexane or a room temperature ionic liquid, such as 1-*n*-butyl-3-methylimidazolium hexafluorophosphate, was used as solvent.[45] With DMSO–hexane as solvent, neither a base nor an acid catalyst is needed to complete the protection reaction. Thus, this catalyst-free method is especially useful for blocking alcohols with acid- or base-sensitive functional groups. Allylic alcohols and alcohols containing cyclopropane or tetrahydrofuran moieties undergo the same silylation in DMSO–hexanes with no undesired side reactions. *t*-Butyldiphenylsilyl (TBDPS) and triphenylsilyl (TPS) ethers can be formed in a similar way while chlorotriisopropylsilane (TIPSCl) gives the corresponding silyl ether in only low yield.

Simply heating a neat mixture of an alcohol and TBDMSCl with no solvent, at elevated temperature, affords the TBDMS ether in good yield. This method is not only useful for its solvent-free conditions, but also for its enhanced silylation ability. For example, 1,1-dimethyl-2-phenylethanol was not silylated when heated with TBDMSCl in DMF at 120 °C for 5 h. In contrast, the desired silyl ether was obtained in 86% yield under the same conditions, but without adding DMF as solvent (eq 25).[46]

$$Ph\diagup OH \xrightarrow{TBDMSCl,\ 120\ °C,\ 5\ h} Ph\diagup OTBDMS \quad \text{(25)}$$

DMF as solvent:	No reaction
Solvent-free:	86%

Silyl ethers of alcohols and phenols were also prepared efficiently by treatment of the alcohol or phenol with TBDMSCl and a catalytic amount of imidazole or iodine under the solvent-free and microwave irradiation conditions.[47] Under the same microwave conditions, treatment of the silyl ether in methanol and in the presence of catalytic amount of iodine released the parent alcohol in quantitative yield.

It is often critical to the successful synthesis of natural products to achieve selective protection of one hydroxyl group in a polyol. Such selective protection has been the subject of a number of studies. For example, it has been found that *t*-butyldimethylsilylation of *threo*-α,β-dihydroxyphosphonates afforded the silyl ether at the β-position with a high degree of regioselectivity when pyridine or γ-collidine was used as base instead of imidazole (eq 26).[48] The silylation proceeded with a kinetic preference since the reaction with imidazole at lower temperature provided similar selectivity.

TBDMSCl has been used to selectively protect the primary hydroxyl group of 1,2- and 1,3-diols under mild conditions, via dibutylstannylene acetal intermediates. For example, treatment of the dibutylstannylene acetal of 1-phenylethane-1,2-diol with TBDMSCl in chloroform affords the primary ether in excellent yield (eq 27).[49] This selectivity for the primary hydroxyl group is different from acylation or alkylation of dibutylstannylene acetal intermediates, in which the secondary hydroxyl group is preferentially derivatized. The authors attributed this selectivity difference to the steric bulkiness of the TBDMS group.

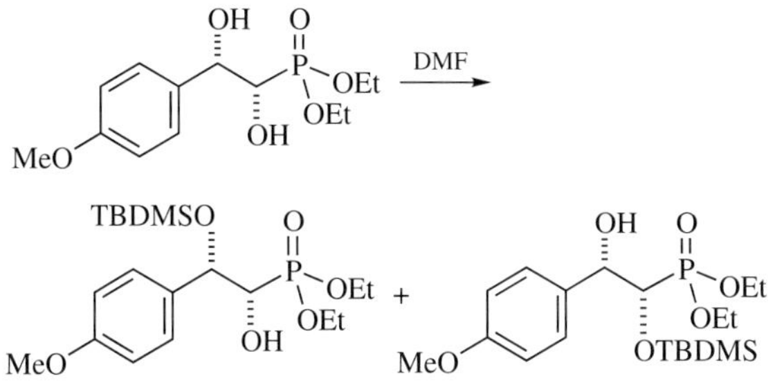

Imidazole, rt	85%	78:22
Imidazole, –20 °C	65%	91:9
Pyridine, rt	83%	97:3

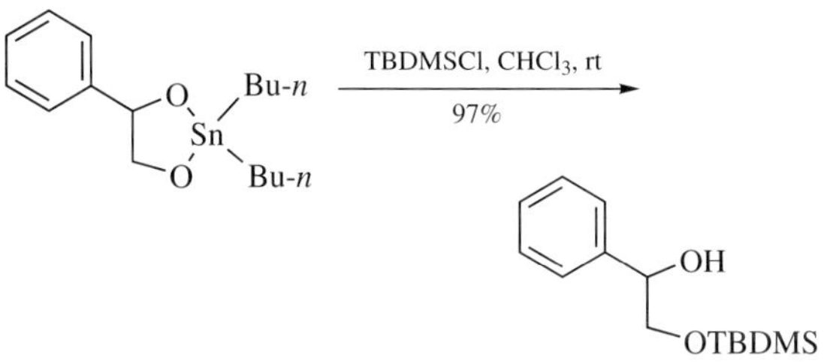

The above kinetically controlled regioselectivity is also reflected in the selective monosilylation of butane-1,2,4-triol. In this case, the sole product obtained is 4-*t*-butyldimethylsiloxy-butane-1,2-diol with 99% yield (eq 28). The authors rationalized that, with the bulky TBDMSCl, the dibutylstannylene acetal rapidly migrates between the 1,2-diol and 1,3-diols and affords the product of kinetic control. This sequence of reactions was also used to selectively block the equatorial alcohol of a *cis*-diol on the pyranoside ring.[50]

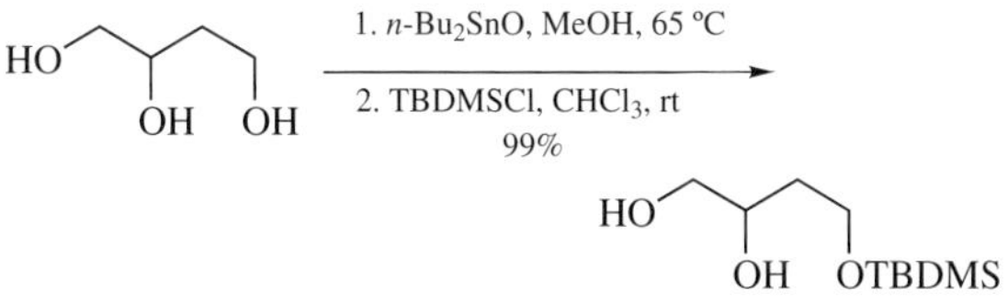

Bases, such as imidazole or TEA, are often employed in TBDMS protection of alcohols. They have been used to release masked alcohols which are subsequently protected as TBDMS ethers. For example, treatment of a 2-pyridone with TBDMSCl in the presence of TEA afforded the di-TBDMS ether in good yield (eq 29).[51]

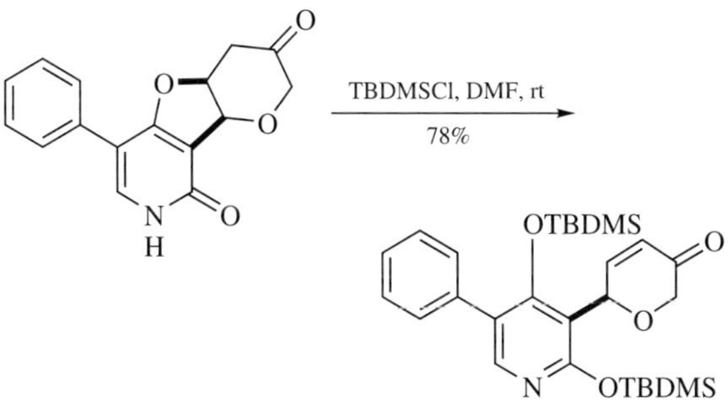

TBDMSCl has been employed to convert benzylidene acetals into TBDMS phenyl ethers. Stirring a mixture of the acetal,

TBDMSCl, and sodium cyanoborohydride in acetonitrile afforded the protected TBDMS ether in 89% yield (eq 30).[52] Interestingly, when either smaller silyl chlorides (TESCl or TMSCl) or larger silyl chlorides (TBDPSCl or TIPSCl) were used instead of TBDMSCl, the free secondary alcohol was obtained as the major product.

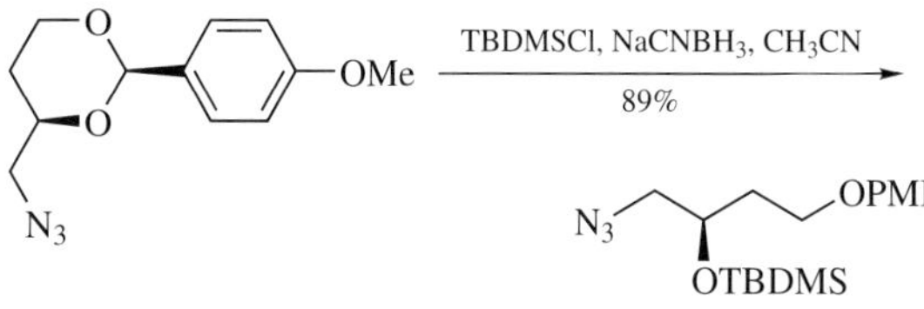

In the presence of a strong base, the TBDMS group of silyl ethers may undergo migration to a proximal hydroxyl group to give regioisomeric silyl ethers. For example, when an inseparable atropisomeric mixture of the biphenyl homoallyl alcohols was treated with potassium *t*-butoxide, the isomer with TBDMS group proximal to the homoallylic hydroxyl group underwent migration, and the nascent hydroxyl group was protected again as a TBDMS ether (eq 31).[53] The other atropisomer remained intact and was separated out due to the now large physical property difference between two compounds.

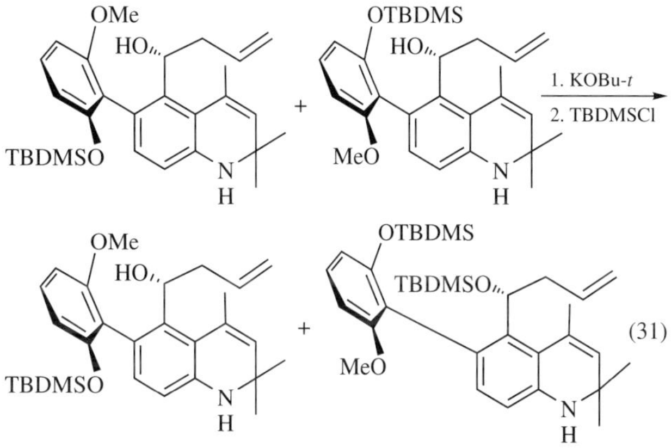

TBDMSCl has been used extensively in protecting hemiacetals and generating mixed acetals.[54] For example, a lactol was reacted with TBDMSCl in the presence of silver oxide to provide a silyl acetal in 93% yield (eq 32).[55] This transformation takes place only for hemiacetals derived from aldehydes, while hemiacetals of ketones (hemiketals) give the siloxy ketones instead.[56]

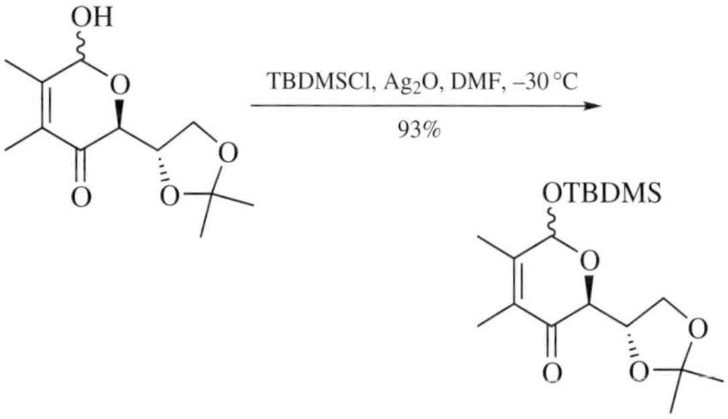

Mixed TBDMS silyl ketals, such as 1-allyloxy-1-siloxycyclo-propanes, have been prepared by TBDMSCl directly trapping intermediates from the reductive cyclization of ω-haloesters. For example, treatment of methallyl iodopropionate with zinc–copper

couple and TBDMSCl in THF affords 1-methallyloxy-1-siloxy-cyclopropane in 66% yield (eq 33).[57]

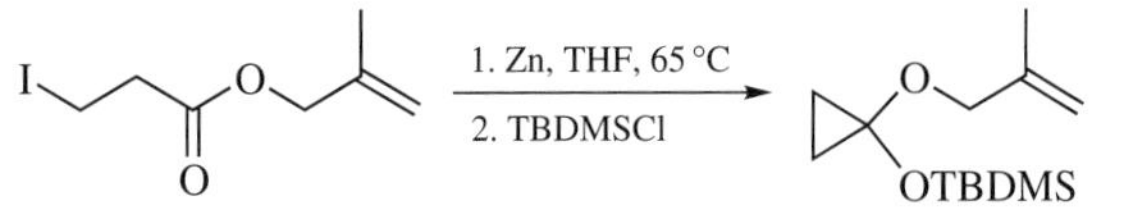

$$\text{(33)}$$

TBDMSCl and other trialkylsilyl chlorides have been used for the "temporary" *N*-protection of secondary amides. Such substitution is very helpful in equilibrating the amide *s-trans* conformation favored by the original secondary amides to the *s-cis* conformation preferred by the tertiary amide. The resulting tertiary amides were applied to reactions demanding a defined conformation, such as radical cyclizations. For example, under the standard radical reaction conditions, a secondary acrylamide furnished only reduction product while the tertiary amide, protected by TBDMS group afforded two radical cyclization products in 65% combined yield (eq 34).[58]

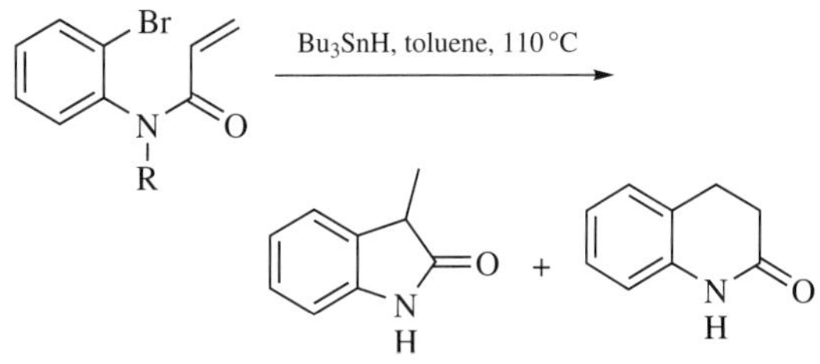

R = H,   Reduction product
R = TBDMS,   65% (5-*exo*: 6-*endo* = 3:1)

In the above radical reaction, the 6-*endo* cyclization product was unexpected and indicates that there exists an equilibrium between the desired *N*-silylation and the undesired *O*-silylation products. For the *O*-silyl imidate, the C–C *s-cis* conformer must be favored over the *s-trans* conformer in order to avoid the steric interaction between vinyl group and the TBDMS group (eq 35). The ^{29}Si NMR spectrum provided further evidence for the silyl group scrambling and indicated a rapid *N*-silyl/*O*-silyl equilibration since a single peak at δ 13.38 ppm was observed which is the averaged value of chemical shifts for an *O*-silyl imidate (δ 19.6 ppm) and an *N*-silyl amide (δ 8.7 ppm).

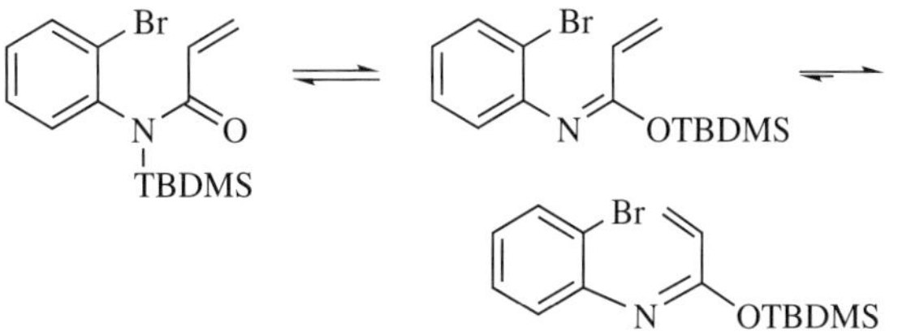

$$\text{(35)}$$

Hydroxylamines have been protected as siloxyamines by treatment with TBDMSCl and imidazole in DMF.[59] Heating is necessary for reaction of hindered hydroxylamines.[60] Aldoximes and ketoximes are even less reactive in this protection process, and higher temperature (130 °C) or stronger bases, such as DBU or LDA, are required to ensure a satisfactory result.[61] Interestingly, dichloromethane seems to be a superior solvent for the efficient protection of oximes as the corresponding *t*-butyldimethyl-siloximes.[62]

Combination of a silylating agent and a tertiary amine (TEA, DBU, or DIPEA) provides a convenient protocol to convert aldehydes and ketones into silyl enol ethers. However, when coupled with mild base, i.e., imidazole (Im), TBDMSCl provides a useful way to protect aldehydes as *O-t*-butyldimethylsilylimidazolyl aminals. This combination of TBDMS and imidazole in a blocking group is especially beneficial in protecting aldehydes which are prone to racemization or aldol reaction. The aminal protecting group showed considerable stability under mild acidic conditions (HOAc-H$_2$O-THF) at 80 °C, while it was readily removed by treatment with 9:1 CH$_3$CN-49% HF (eq 36).[63]

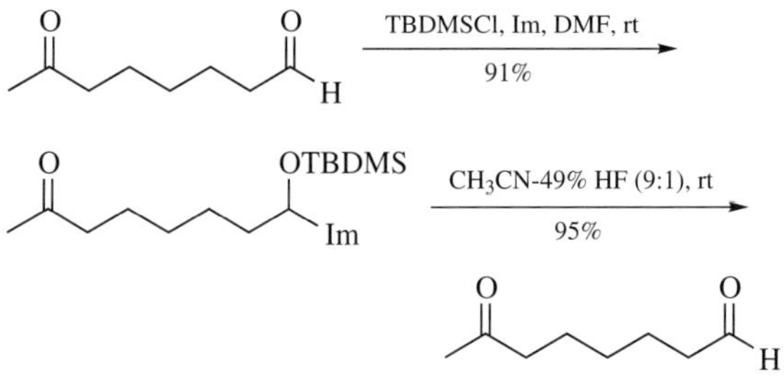

$$\text{(36)}$$

The aminals remained intact when subjected to various other transformation conditions, including the BF$_3$-1,2-ethanedithiol conditions. The parent imidazole aminals react with organolithium reagents similarly to the unprotected aldehyde by first proton abstract of 2-H proton of imidazole and subsequent retro-[1,4]-Brook rearrangement. The 2-substituted imidazoles (e.g., 2-methylimidazole), on the contrary, generate aminals which are stable to organolithium reagents.[64]

Anion Trap. In contrast to TMSCl, 1 equiv of TBDMSCl can be used to trap the α'- and γ-dianions of β-(monoalkylamino) α,β-unsaturated ketones, affording the corresponding α'- and γ-TBDMS β-(monoalkylamino) α,β-unsaturated ketones.[65] Interestingly, upon hydrolysis with 2 N hydrochloric acid, γ-TBDMS ketones were converted into the starting enaminones (eq 37), while the α'-TBDMS ketones were transformed into the corresponding α'-TBDMS 1,3-diketones (eq 38).

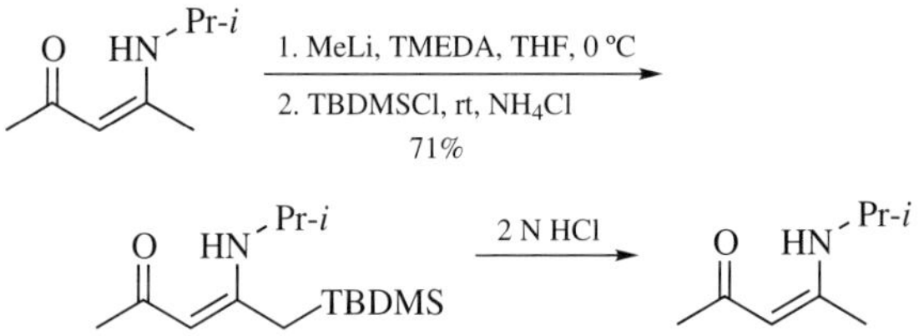

$$\text{(37)}$$

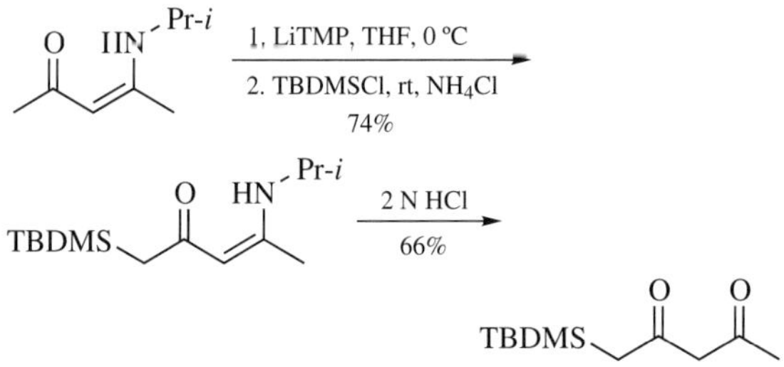

$$\text{(38)}$$

TBDMSCl has also been used to trap phosphides. For example, 2,6-diisopropylphenylphosphide, generated by metallation of the corresponding phosphane, was intercepted by TBDMSCl to give the monosilylated phosphane. The process was repeated to afford the disilylated phosphane with good overall yield (eq 39).[66]

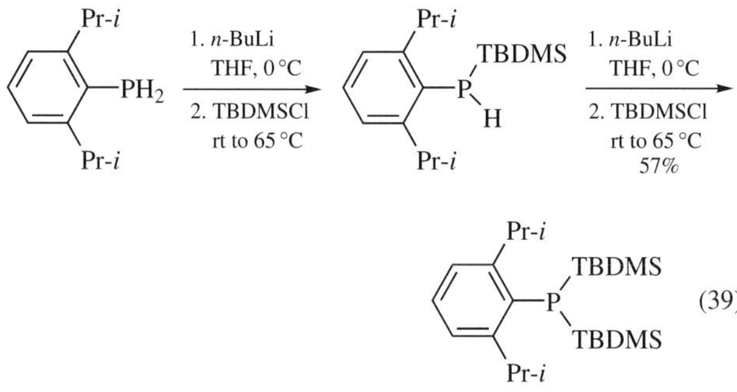

(39)

3-Silylated 1,4-cyclohexadienes were used to replace tin hydride as reducing agents in radical chain reactions.[67] They were prepared by TBDMSCl interception of the regioselective metallation intermediate of cyclohexadienes, and subsequent methylation (eq 40).[68] The 3-TBDMS-2,4-dimethoxy-1,4-cyclohexadiene was found to be a very efficient hydrogen donor reagent in various radical reactions.

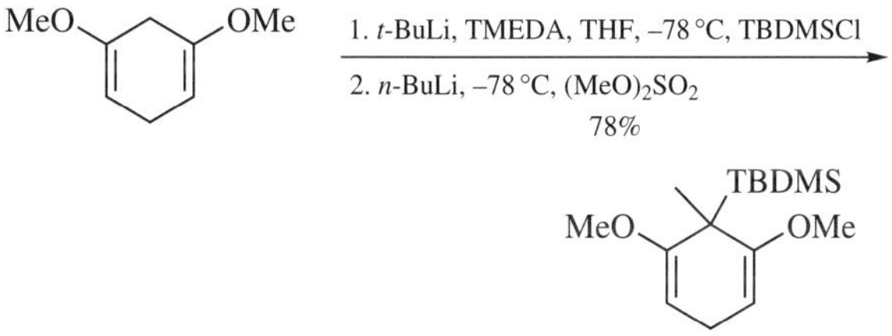

(40)

Claisen Rearrangement. Allyl *N*-phenylimidates were reported to undergo thermal isomerization to form ketene *N,O*-acetals which then transformed into γ,δ-unsaturated anilides readily via Claisen rearrangement (eq 41). Since the intermediate ketene acetals were formed with no stereoselectivity, the relative configuration of the two newly formed stereogenic centers α and β to the anilides was under thermodynamic control. However, the corresponding *N*-silyl ketene, generated by deprotonation and subsequent *N*-silylation of the same imidates, underwent Claisen rearrangement and subsequent desilylation to give the γ,δ-unsaturated anilides with good stereoselectivity for the α and β chiral centers (eq 42).[69] Interestingly, the Claisen rearrangement does not happen until the reaction mixture is heated above 130 °C if the olefin starting material in eq 42 has the Z-configuration.

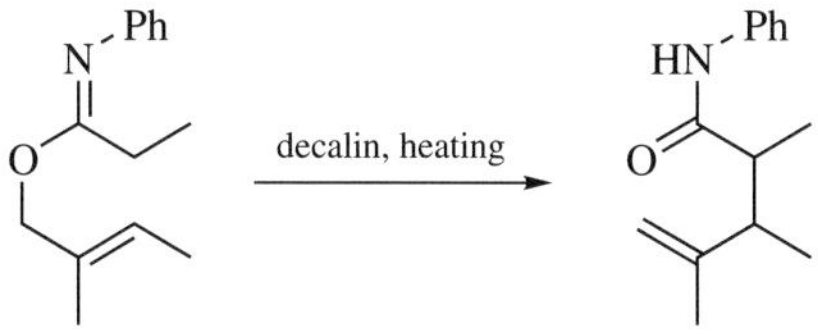

(41)

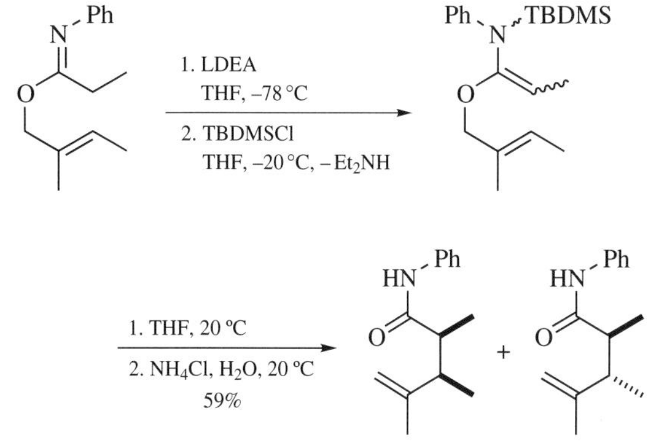

(42)

syn:anti = 98.1:1.9

TBDMS Enol Ethers. After deprotonation with alkyllithiums, *O*-allylic *N,N*-diisopropyl carbamate anions were found to undergo condensation reactions with aldehydes or trialkyltin chlorides and to generate the corresponding vinyl carbamates. These vinyl carbamates have been used to prepare TBDMS silyl enol ethers, in excellent yield, through addition of methyllithium followed by trapping the resultant enolate anion with TBDMSCl (eq 43).[70] The initial Z-configuration of the starting enol carbamates is retained during the formation of the silyl enol ether, this method, thus, providing a nice method for the preparation of Z-configured TBDMS enol ethers.

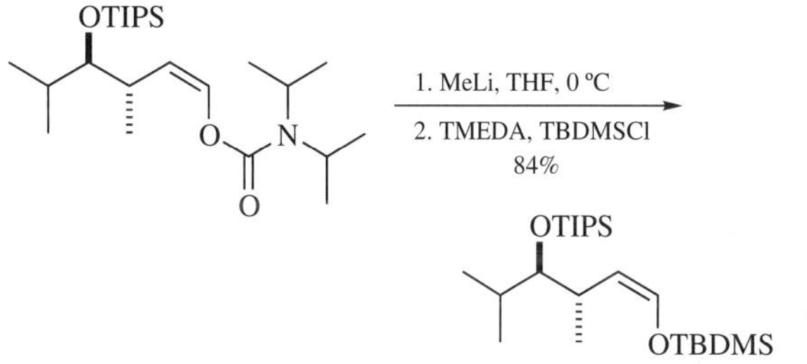

(43)

The Cu(acac)$_2$-catalyzed cycloaddition between a TBDMS enol ether and alkyl diazoacetates provides a useful method for the preparation of siloxycyclopropanecarboxylates (eq 44). Since these siloxycyclopropane derivatives possess a masked ketone group, and serve as important building blocks for the synthesis of the corresponding γ-oxo esters (eq 45).[71]

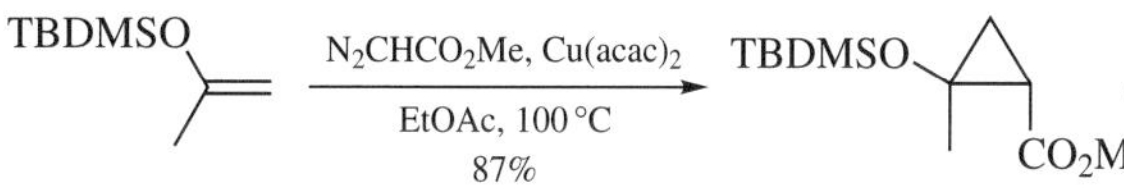

(44)

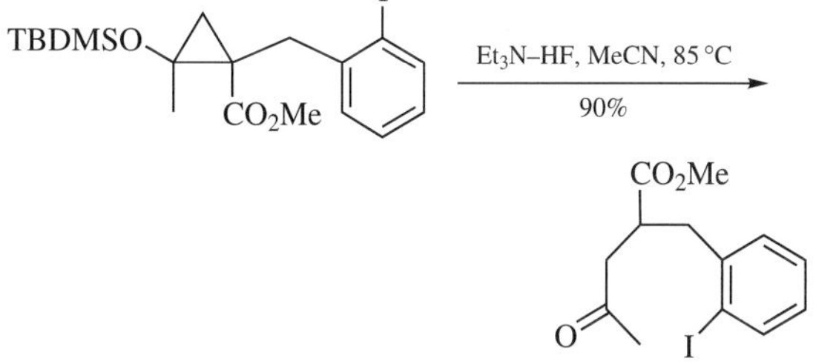

(45)

In the presence of excess LDA, silyl enol ethers of methyl ketones readily underwent elimination to furnish lithiated allenes, which were then trapped by a variety of electrophiles, such as chlorosilanes, chlorostannanes, and ketones (eq 46). Only silyl enol ethers containing the hindered TBDMS or TIPS groups underwent the elimination of silanolates, and good yields of the interception products were obtained.[72]

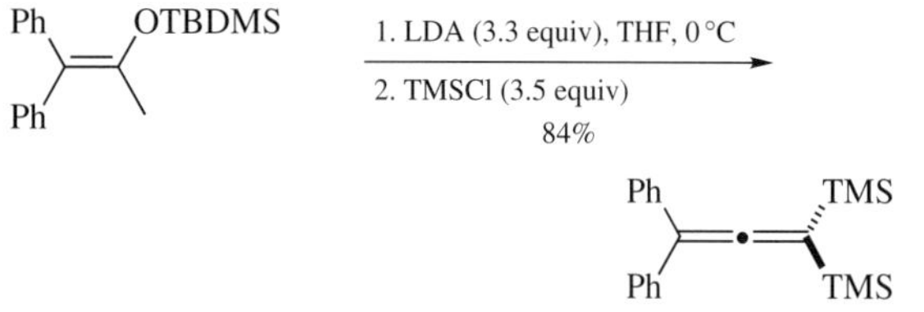

$$(46)$$

TBDMSCl as Cl⁻ Source. During the TBDMS protection of a primary alcohol under the standard conditions (TBDMSCl, DIPEA, cat. DMAP), a primary alcohol tosylate was simultaneously partially converted into the corresponding chloride, with a 70% combined yield (eq 47).[73]

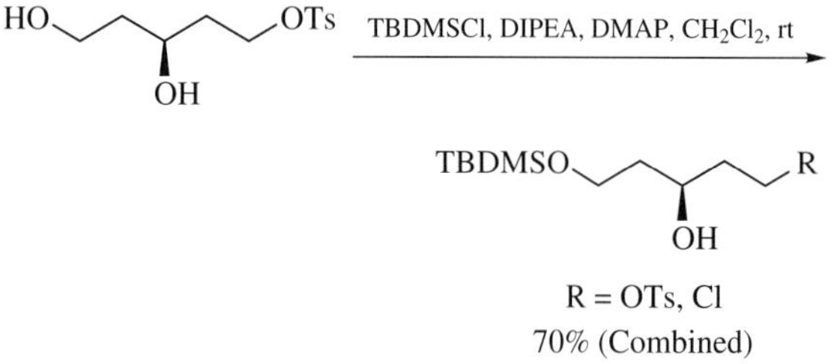

$$(47)$$

R = OTs, Cl
70% (Combined)

TBDMSCl-assisted Reaction. TBDMSCl is suitable for the activation of electrophiles in the conjugate addition of titanium "ate" complexes of ketone enolates to α,β-unsaturated ketones. Under such conditions only a stoichiometric amount of the "ate" complex is required to achieve comparable yields, with no loss of diastereoselectivity.[74]

Ring opening of a cyclohexyl epoxide with dilithium tetra-chlorocuprate (Li₂CuCl₄) gave two alcohol isomers in the ratio of ca. 2:1 and 77% total yield. Addition of TBDMSCl improved both the yield and regioselectivity, with only one product isolated in excellent yield (eq 48).[75]

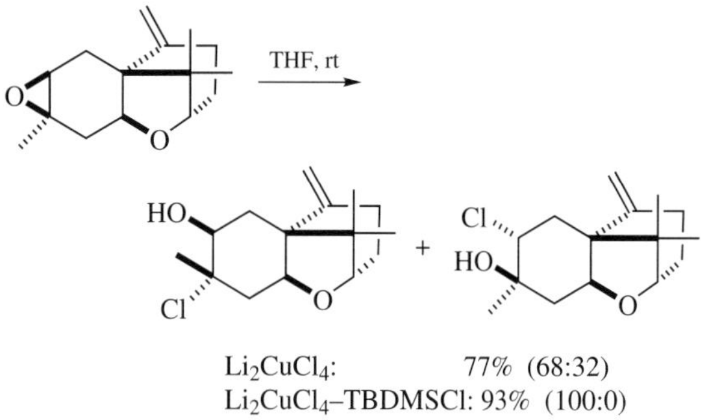

$$(48)$$

Li₂CuCl₄: 77% (68:32)
Li₂CuCl₄–TBDMSCl: 93% (100:0)

N-Silyl Imines. Promoted with one molar equivalent of trialkylsilyl chloride, and in the presence of catalytic amount of cuprous bromide, the reaction of benzonitrile with a Grignard reagent or organolithium reagent generated the corresponding N-trialkylsilylketimines. Isomerization of the TBDMS silyl imines to silylenamine was slower than the corresponding reaction of TMS silyl imines. Reduction of these imines in situ with borane reagent generated the corresponding primary phenylalkyl

amine. Simply quenching the silyl imines with water afforded the aromatic ketones in good yield (eq 49). This sequence with TBDMSCl and cuprous bromide avoided the harsh acidic workup associated with the similar transformations lacking the TBDMS silyl imine intermediate.[76]

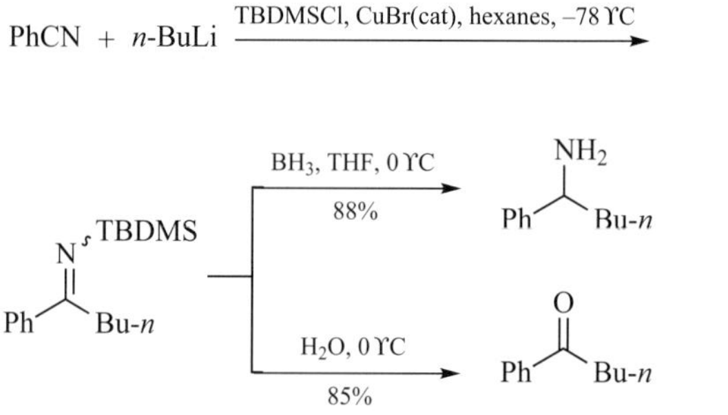

$$(49)$$

N-Containing Heterocycles. Nitroalkenes were reacted with TBDMSCl and DBU to afford N-silyloxyisoazolidines through the intramolecular olefin cycloadditions of the nitronate intermediates.[77] Only one diastereomer with three new consecutive stereogenic centers was isolated in good to excellent yield. Desilylation with TBAF and further spontaneous aerobic oxidation furnished the hydroxymethyl nitro compounds (eq 50). 1-Bromo-nitroalkanes, on the contrary, when treated with TBDMSCl and TEA, underwent the same cycloaddition reaction to produce the nitronates after in situ elimination of TBDMSBr (eq 51).[78]

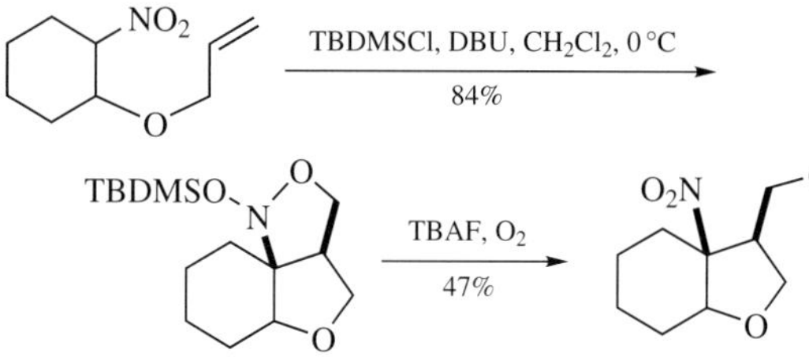

$$(50)$$

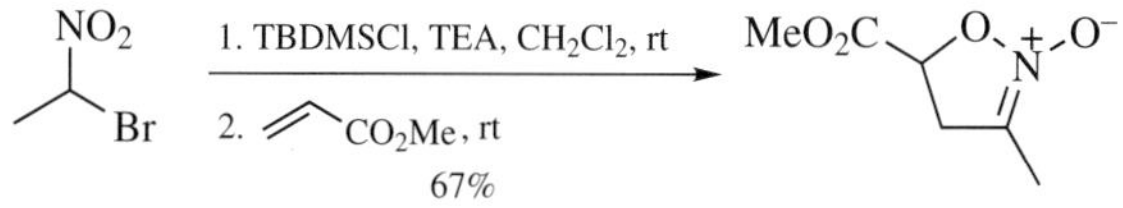

$$(51)$$

Similar intramolecular cycloaddition of TBDMS-protected oximes, tethered with olefin moieties, in the presence of boron trifluoride-ether generated N-nonsubstituted isoxazolidines in moderate to good yield (eq 52).[79] The olefin geometry of the substrate was retained under the reaction conditions and the cycloaddition proceeded with high diastereoselectivity.

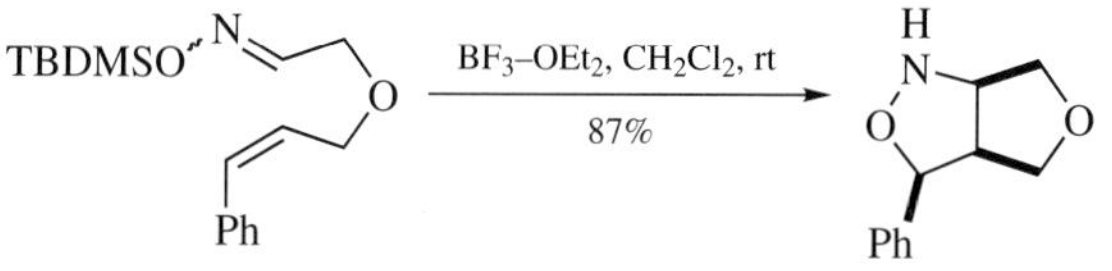

$$(52)$$

1. (a) Corey, E. J.; Venkateswarlu, A., *J. Am. Chem. Soc.* **1972**, *94*, 6190. (b) Lalonde, M.; Chan, T. H., *Synthesis* **1985**, 817. (c) Greene, T. W.; Wuts, P. G. M. *Protective Groups in Organic Synthesis*; Wiley: New York, 1991. (d) Colvin, E. *Silicon in Organic Synthesis*; Butterworths: London, 1981.

2. Stork, G.; Hudrlik, P. F., *J. Am. Chem. Soc.* **1968**, *90*, 4462.

3. (a) Rathke, M. W.; Sullivan, D. F., *Synth. Commun.* **1973**, *3*, 67. (b) Rathke, M. W.; Sullivan, D. F., *Tetrahedron Lett.* **1973**, 1297. (c) An interesting application of TBS silyl ketene acetals as homo-Reformatsky reagents may be found in: Oshino, H.; Nakamura, E.; Kuwajima, I., *J. Org. Chem.* **1985**, *50*, 2802. (d) Ireland, R. E.; Wipf, P.; Armstrong, J. D., III, *J. Org. Chem.* **1991**, *56*, 650.

4. For general papers comparing the stability of silanes containing various alkyl groups on Si see: (a) Sommer, L. H.; Tyler, L. J., *J. Am. Chem. Soc.* **1954**, *76*, 1030. (b) Ackerman, E., *Acta Chem. Scand.* **1956**, *10*, 298; *Acta Chem. Scand.* **1957**, *11*, 373.

5. Hernandez, O.; Chaudhary, S. K., *Tetrahedron Lett.* **1979**, 99.

6. Cunico, R. F.; Bedell, L., *J. Org. Chem.* **1980**, *45*, 4797.

7. Mawhinney, T. P.; Madson, M. A., *J. Org. Chem.* **1982**, *47*, 3336.

8. Wells, G. J.; Yan, T.-H.; Paquette, L. A., *J. Org. Chem.* **1984**, *49*, 3604.

9. Okuda, Y.; Lakshmikantham, M. V.; Cava, M. P., *J. Org. Chem.* **1991**, *56*, 6024.

10. Watt, D. S., *Synth. Commun.* **1974**, *4*, 127.

11. Chenard, B. L.; Van Zyl, C. M., *J. Org. Chem.* **1986**, 3561.

12. Ireland, R. E.; Mueller, R. H., *J. Am. Chem. Soc.* **1972**, *94*, 5897.

13. Ireland, R. E.; Mueller, R. H.; Willard, A. K., *J. Am. Chem. Soc.* **1976**, *98*, 2868.

14. For other examples see (a) Mohammed, A. Y.; Clive, D. L. J., *J. Chem. Soc., Chem. Commun.* **1986**, 588. (b) Kita, Y.; Shibata, N.; Miki, T.; Takemura, Y.; Tamura, O., *J. Chem. Soc., Chem. Commun.* **1990**, 727. (c) Metz, P.; Mues, C., *Synlett* **1990**, 97.

15. (a) Ireland, R. E.; Varney, M. D., *J. Am. Chem. Soc.* **1984**, *106*, 3668. (b) Ireland, R. E.; Daub, J. P., *J. Org. Chem.* **1981**, *46*, 479.

16. Raucher, S.; Gustavson, L. M., *Tetrahedron Lett.* **1986**, *27*, 1557.

17. (a) Abelman, M. M.; Funk, R. L.; Munger, J. D., Jr., *J. Am. Chem. Soc.* **1982**, *104*, 4030. (b) Funk, R. L.; Munger, J. D., Jr., *J. Org. Chem.* **1984**, *49*, 4320.

18. For a review see: Brownbridge, P., *Synthesis* **1983**, 1; *Synthesis* **1983**, 29.

19. (a) Ireland, R. E.; Courtney, L.; Fitzsimmons, B. J., *J. Org. Chem.* **1983**, *48*, 5186. (b) Piers, E.; Burmeister, M. S.; Reissig, H.-U., *Synlett* **1986**, *64*, 180.

20. (a) Orban, J.; Turner, J. V.; Twitchin, B., *Tetrahedron Lett.* **1984**, *25*, 5099. (b) Orban, J.; Turner, J. V., *Tetrahedron Lett.* **1983**, *24*, 2697. (c) Ireland, R. E.; Thompson, W. J.; Mandel, N. S.; Mandel, G. S., *J. Org. Chem.* **1979**, *44*, 3583.

21. McLoughlin, J. I.; Little, R. D., *J. Org. Chem.* **1988**, *53*, 3624.

22. Trimitsis, G.; Beers, S.; Ridella, J.; Carlon, M.; Cullin, D.; High, J.; Brutts, D., *J. Chem. Soc., Chem. Commun.* **1984**, 1088.

23. Veysoglu, T.; Mitscher, L. A., *Tetrahedron Lett.* **1981**, *22*, 1299.

24. Mukaiyama, T.; Ohno, T.; Han, J. S.; Kobayashi, S., *Chem. Lett.* **1991**, 949.

25. Hudlicky, T.; Luna, H.; Olivo, H. F.; Andersen, C.; Nugent, T.; Price, J. D., *J. Chem. Soc., Perkin Trans. 1* **1991**, 2907.

26. Fernández, R.; Gasch, C.; Gómez-Sánchez, A.; Vílchez, J. E., *Tetrahedron Lett.* **1991**, *32*, 3225.

27. Colvin, E. W.; Beck, A. K.; Seebach, D., *Helv. Chim. Acta* **1981**, *64*, 2264.

28. Rawal, V. H.; Rao, J. A.; Cava, M. P., *Tetrahedron Lett.* **1985**, *26*, 4275.

29. (a) Gassman, P. G.; Haberman, L. M., *J. Org. Chem.* **1986**, *51*, 5010. (b) Mai, K.; Patil, G., *J. Org. Chem.* **1986**, *51*, 3545.

30. Corey, E. J.; Cho, H.; Rücker, C.; Hua, D. H., *Tetrahedron Lett.* **1981**, *22*, 3455.

31. Aizpurua, J. M.; Paloma, C., *Tetrahedron Lett.* **1985**, *26*, 475.

32. (a) Nyström, J.-E.; McCanna, T. D.; Helquist, P.; Amouroux, R., *Synthesis* **1988**, 56. (b) Detty, M. R.; Seidler, J. D., *J. Org. Chem.* **1981**, *46*, 1283.

33. Fairhurst, R. A.; Heaney, H.; Papageorgiou, G.; Wilkins, R. F.; Eyley, S. C., *Tetrahedron Lett.* **1989**, *30*, 1433.

34. Wissner, A.; Grudzinskas, C. V., *J. Org. Chem.* **1978**, *43*, 3972.

35. Mobashery, S.; Johnston, M., *J. Org. Chem.* **1985**, *50*, 2200.

36. Amberg, W.; Seebach, D., *Angew. Chem., Int. Ed. Engl.* **1988**, 1718.

37. (a) Murakami, M.; Matsuura, T.; Ito, Y., *Tetrahedron Lett.* **1988**, *29*, 355. (b) Ager, D. J.; Fleming, I.; Patel, S. K., *J. Chem. Soc., Perkin Trans. 1* **1981**, 2520.

38. Hudrlik, P. F.; Kulkarni, A. K., *J. Am. Chem. Soc.* **1981**, *103*, 6251.

39. (a) Lohray, B. B.; Enders, D., *Helv. Chim. Acta* **1989**, *72*, 980. (b) Enders, D.; Lohray, B. B., *Angew. Chem., Int. Ed. Engl.* **1987**, *26*, 351.

40. Schinzer, D., *Synthesis* **1989**, 179.

41. Reich, H. J.; Kelly, M. J.; Olsen, R. E.; Holtan, R. C., *Tetrahedron* **1983**, *39*, 949.

42. Prieto, J. A.; Suarez, J.; Larson, G. L., *Synth. Commun.* **1988**, *18*, 253.

43. Djuric, S. W., *J. Org. Chem.* **1984**, *49*, 1311.

44. Busato, S.; Cainelli, G.; Panunzio, M.; Bandini, E.; Martelli, G.; Spunta, G., *Synlett* **1991**, 243.

45. (a) Watahiki, T.; Matsuzaki, M.; Oriyama, T., *Green Chem.* **2003**, *5*, 82. (b) Xu, Z.-Y.; Xu, D.-Q.; Liu, B.-Y.; Luo, S.-P., *Synth. Commun.* **2003**, *33*, 4143.

46. Hatano, B.; Toyota, S.; Toda, F., *Green Chem.* **2001**, *3*, 140.

47. (a) Saxena, I.; Deka, N.; Sarma, J. C.; Tsuboi, S., *Synth. Commun.* **2003**, *33*, 4005. (b) Bastos, E. L.; Ciscato, L. F. M. L.; Baader, W. J., *Synth. Commun.* **2005**, *35*, 1501.

48. Yokomatsu, T.; Suemune, K.; Yamagishi, T.; Shibuya, S., *Synlett* **1995**, 847.

49. Leigh, D. A.; Martin, R. P.; Smart, J. P.; Truscello, A. M., *Chem. Commun.* **1994**, 1373.

50. Garegg, P. J.; Olsson, L.; Stefan, O., *J. Carbohydr. Chem.* **1993**, *12*, 955.

51. (a) Fürstner, A.; Feyen, F.; Prinz, H.; Waldmann, H., *Angew. Chem., Int. Ed.* **2003**, *42*, 5361. (b) Nhien, A. N. V.; Tomassi, C.; Len, C.; Marco-Contelles, J. L.; Balzarini, J.; Pannecouque, C.; De Clercq, E.; Postel, D., *J. Med. Chem.* **2005**, *48*, 4276.

52. Gustafsson, T.; Schou, M.; Almqvist, F.; Kihlberg, J., *J. Org. Chem.* **2004**, *69*, 8694.

53. Ku, Y.-Y.; Grieme, T.; Raje, P.; Sharma, P.; King, S. A.; Morton, H. E., *J. Am. Chem. Soc.* **2002**, *124*, 4282.

54. (a) Jonke, S.; Liu, K.-g.; Schmidt, R. R., *Chem. Eur. J.* **2006**, *12*, 1274. (b) Takao, K.-I.; Watanabe, G.; Yasui, H.; Tadano, K.-I., *Org. Lett.* **2002**, *4*, 2941.

55. Honda, T.; Tomitsuka, K.; Tsubuki, M., *J. Org. Chem.* **1993**, *58*, 4274.

56. (a) Kraus, G. A.; Bougie, D., *Tetrahedron* **1994**, *50*, 2681. (b) Kraus, G. A.; Bougie, D., *Synlett* **1992**, 279.

57. (a) Yasui, K.; Fugami, K.; Tanaka, S.; Tamaru, Y.; Ii, A.; Yoshida, Z.-I.; Saidi, M. R., *Tetrahedron Lett.* **1992**, *33*, 785. (b) Yasui, K.; Tanaka, S.; Tamaru, Y., *Tetrahedron* **1995**, *51*, 6881.

58. (a) Jones, K.; Wilkinson, J.; Ewin, R., *Tetrahedron Lett.* **1994**, *41*, 7673. (b) Takasu, K.; Nishida, N.; Ihara, M., *Synthesis* **2004**, 2222.

59. Gomes, M. J. S.; Sharma, L.; Prabhakar, S.; Lobo, A. M.; Glória, P. M. C., *Chem. Commun.* **2002**, 746.

60. Liao, Y.; Xie, C.; Lahti, P. M.; Weber, R. T.; Jiang, J. J.; Barr, D. P., *J. Org. Chem.* **1999**, *64*, 5176.

61. (a) Ortiz, M.; Cordero, J. F.; Pinto, S.; Alverio, I., *Synth. Commun.* **1994**, *24*, 409. (b) Kim, K. S.; Hurh, E. Y.; Youn, J. N.; Park, J. I., *J. Org Chem.* **1999**, *64*, 9272.

62. Ortiz-Marciales, M.; De Jesús, M.; Figueroa, D.; Hernández, J.; Vázquez, L.; Vega, R.; Morales, E. M.; López, J. A., *Synth. Commun.* **2003**, *33*, 311.

63. Quan, L. G.; Cha, J. K., *Synlett* **2001**, 1925.

64. Gimisis, T.; Arsenyan, P.; Georganakis, D.; Leondiadis, L., *Synlett* **2003**, 1451.

65. Bartoli, G.; Bosco, M.; Dalpozzo, R.; De Nino, A.; Iantorno, E.; Tagarelli, A.; Palmieri, G., *Tetrahedron* **1996**, *52*, 9179.

66. Boeréaaa, R. T.; Masuda, J. D., *Can. J. Chem.* **2002**, *80*, 1607.

67. Studer, A.; Amrein, S.; Schleth, F.; Schulte, T.; Walton, J. C., *J. Am. Chem. Soc.* **2003**, *125*, 5726.

68. Amrein, S.; Studer, A., *Helv. Chim. Acta* **2002**, *85*, 3559.

69. Metz, P.; Linz, C., *Tetrahedron* **1994**, *50*, 3951.

70. Madec, D.; Henryon, V.; Férézou, J.-P., *Tetrahedron Lett.* **1999**, *40*, 8103.

71. Khan, F. A.; Czerwonka, R.; Reissig, H.-U., *Eur. J. Org. Chem.* **2000**, 3607.

72. Langer, P.; Doring, M.; Seyferth, D.; Gorls, H., *Chem. Eur. J.* **2001**, *7*, 573.

73. Mounéaaa, S.; Niel, G.; Busquet, M.; Eggleston, I.; Jouin, P., *J. Org. Chem.* **1997**, *62*, 3332.

74. Bernardi, A.; Cavicchioli, M.; Scolastico, C., *Tetrahedron* **1993**, *49*, 10913.

75. Miyashita, K.; Yoneda, K.; Akiyama, T.; Koga, Y.; Tanaka, M.; Yoneyama, T.; Iwata, C., *Chem. Pharm. Bull.* **1993**, *41*, 465.

76. Ortiz-Marciales, M.; Tirado, L. M.; Colón, R.; Ufret, M. L.; Figueroa, R.; Lebrón, M.; DeJesús, M.; Martínez, J.; Malavéaaa, T., *Synth. Commun.* **1998**, *28*, 4067.

77. Roger, P.-Y.; Durand, A.-C.; Rodriguez, J.; Dulcère, J.-P., *Org. Lett.* **2004**, *6*, 2027.

78. Kunetsky, R. A.; Dilman, A. D.; Ioffe, S. L.; Struchkova, M. I.; Strelenko, Y. A.; Tartakovsky, V. A., *Org. Lett.* **2003**, *5*, 4907.

79. Tamura, O.; Mitsuya, T.; Huang, X.; Tsutsumi, Y.; Hattori, S.; Ishibashi, H., *J. Org. Chem.* **2005**, *70*, 10720.

t-Butyldimethylsilyl Cyanide

t-BuMe₂SiCN

[56522-24-8] $C_7H_{15}NSi$ (MW 141.29)

InChI = 1S/C7H15NSi/c1-7(2,3)9(4,5)6-8/h1-5H3
InChIKey = CWAKIXKDPQTVTA-UHFFFAOYSA-N

(preparation of stable silylated cyanohydrins;[1] silylating agent;[2] synthesis of isocyanides via oxirane ring opening[3])

Alternate Names: *t*-butylcyanodimethylsilane; TBDMSCN.
Physical Data: bp 163–167 °C/760 mmHg; mp 76–78 °C.
Solubility: sol organic solvents (methylene chloride, THF, chloroform); reacts rapidly with water and protic solvents.
Preparative Methods: conveniently prepared by refluxing *t*-butyldimethylchlorosilane (1 equiv), potassium cyanide (1.3 equiv) and 18-crown-6 (0.31 equiv) in dry methylene chloride under nitrogen.[4] The reagent can also be prepared by the reaction of silver(I) cyanide with *t*-butyldimethylchlorosilane,[2] or stirring at 60 °C for 4 h a mixture of sodium cyanide, Amberlite XAD-4 resin, and *t*-butyldimethylchlorosilane in acetonitrile.[5]
Handling, Storage, and Precautions: must be protected from moisture and handled in a well-ventilated fume hood. Exposure to water releases highly toxic HCN gas.

Cyanosilylation. The Lewis acid-catalyzed reaction of *t*-butyldimethylsilyl cyanide (TBDMSCN) with aldehydes and ketones affords the corresponding silylated cyanohydrins in good yield (eq 1).[1,6,7] These have a greater stability than the silylated cyanohydrins obtained by using cyanotrimethylsilane The

addition of TBDMSCN to sterically hindered ketones proceeds smoothly in the presence of a catalytic amount of zinc iodide or potassium cyanide/18-crown-6 (eqs 2–4).[8] A chelation-controlled stereoselective synthesis of silylated cyanohydrins has also been reported (eq 5).[9]

Silylation. TBDMSCN silylates alcohols, phenols, and carboxylic acids efficiently and in high yield.[2]

Oxirane Ring Opening. Oxiranes undergo ring opening when refluxed with TBDMSCN/zinc iodide in methylene chloride, yielding isocyanides (eq 6).[3] The rate of the oxirane ring opening is slower than the corresponding reactions with trimethylsilyl cyanide/zinc iodide.[10]

1. LaMattina, J. L.; Mularski, C. J., *J. Org. Chem.* **1986**, *51*, 413.
2. Mai, K.; Patil, G., *J. Org. Chem.* **1986**, *51*, 3545.
3. Gassman, P. G.; Haberman, L. M., *J. Org. Chem.* **1986**, *51*, 5010.
4. Hwu, J. R.; Lazar, J. G.; Corless, P. F., *Synthesis* **1984**, 1020.
5. Sukata, K., *Bull. Chem. Soc. Jpn.* **1987**, *60*, 2257.
6. Baker, C. B.; Putt, S. R.; Showalter, H. D. H., *J. Heterocycl. Chem.* **1983**, *20*, 629.
7. Corey, E. J.; Crouse, D. N.; Anderson, J. E., *J. Org. Chem.* **1975**, *40*, 2140.
8. Golinski, M.; Brock, C. P.; Watt, D. S., *J. Org. Chem.* **1993**, *58*, 159.
9. Reetz, M. T.; Kesseler, K.; Jung, A., *Angew. Chem., Int. Ed. Engl.* **1985**, *24*, 989.
10. Gassman, P. G.; Guggenheim, T. L., *J. Am. Chem. Soc.* **1982**, *104*, 5849.

William C. Groutas
Wichita State University, KS, USA

(*t*-Butyldimethylsilyl)diphenyl-phosphine

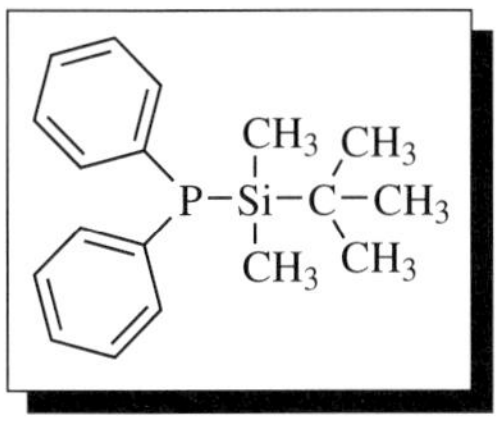

[678187-53-6]   C$_{18}$H$_{25}$PSi   (MW 300.45)

InChI = 1S/C18H25PSi/c1-18(2,3)20(4,5)19(16-12-8-6-9-13-16)
        17-14-10-7-11-15-17/h6-15H,1-5H3

InChIKey = MXOUGJGMEFGDCN-UHFFFAOYSA-N

(reagent used as diphenylphosphino group source in acidic, basic, metal-catalyzed carbon–phosphorus bond formation. Also used as silyl group source for silylation of alcohols)

Physical Data: bp 138 °C (0.4 mmHg); *d* 1.103 gcm^{-3}.

Solubility: soluble in hexane, ether, and most organic solvents.

Form Supplied in: colorless liquid; easily prepared from chlorophosphine and chlorosilane.

Analysis of Reagent Purity: by ^{1}H and ^{31}P NMR.

Preparative Methods: reaction of lithium diphenylphosphide with *t*-butyldimethylchlorosilane in THF, or reaction of chlorodiphenylphosphine and *t*-butyldimethylchlorosilane with magnesium in THF.

Purification: distilled under reduced pressure.

Handling, Storage, and Precautions: sensitive toward moisture and air. The compound is corrosive and liberates diphenylphosphine upon contact with moisture. To help prevent oxidation and hydrolysis, store and handle under nitrogen or preferably under argon. It can be stored for months under argon.

Fluoride-mediated Phosphinations. The silylphosphine serves as a nucleophilic phosphine group source in the presence of TBAF under mild conditions. Alkenes and alkynes with an electron-withdrawing substituent or an anion stabilizing substituent are regioselectively phosphinated via nucleophilic conjugate addition of the diphenylphosphide equivalent generated by the action of fluoride with the reagent, producing the corresponding alkylphosphines and alkenylphosphines, respectively (eqs 1 and 2).[1]

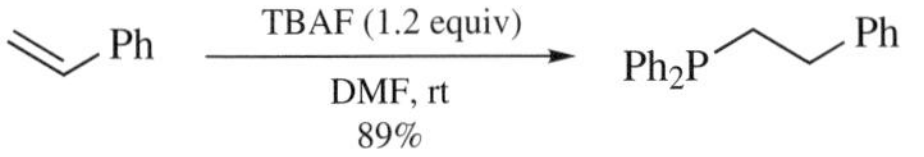

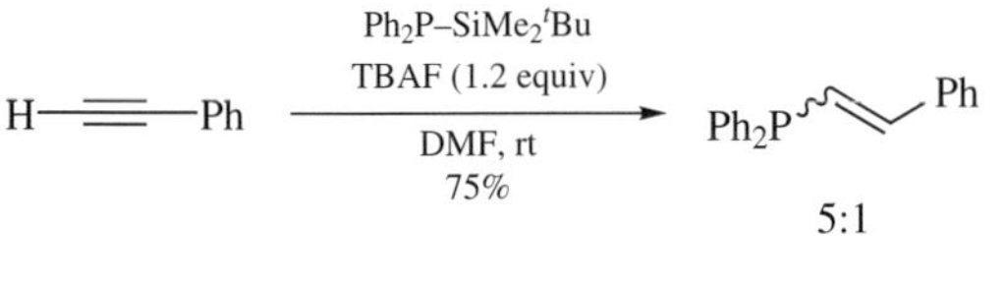

By using an anhydrous fluoride like TASF, the anionic intermediate could be trapped by an aldehyde to give the three-component coupling product (eq 3).[1]

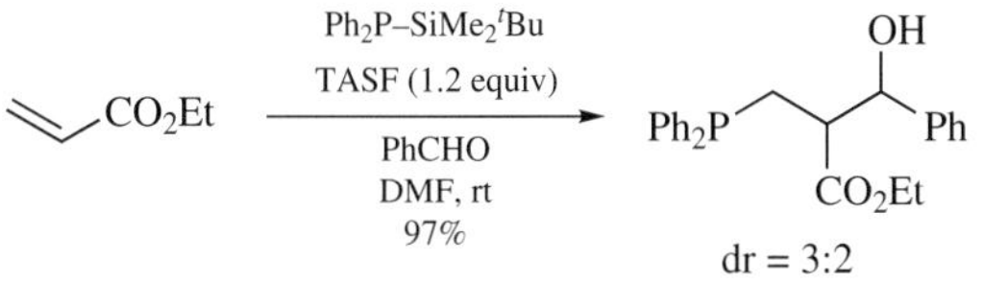

This coupling reaction could also be carried out by using a catalytic amount of CsF in DMF (eq 4). The silyl ether of the adduct is initially formed by catalytic transfer of the silyl group from the silylphosphine to an intermediate cesium alkoxide, which following hydrolysis gave the same coupling product as above (eq 4).[2]

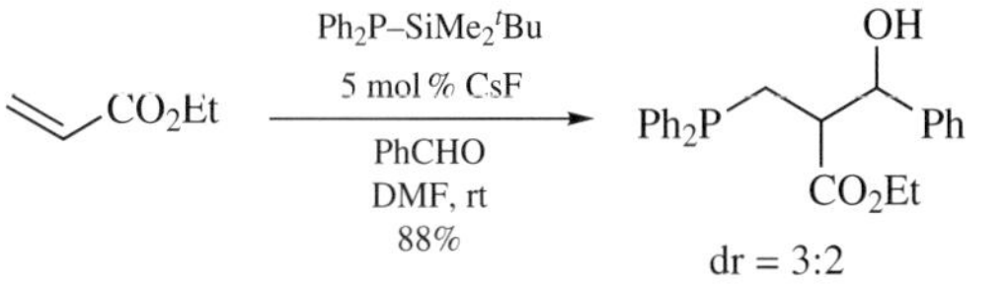

Lewis-acid-catalyzed/mediated Phosphination. The silylphosphine smoothly reacted with electrophiles activated by a Lewis acid. Addition of the silylphosphine to the activated propiolates also proceeded to produce vicinal *syn*-silylphosphinated acrylates (eq 5).[3]

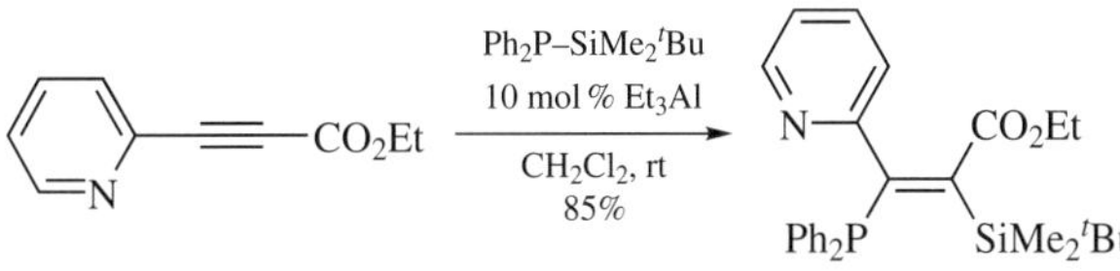

In the presence of Lewis acids such as Et$_3$Al, Et$_2$AlCl, and BF$_3$·OEt$_2$, the silylphosphine reacted with aldehydes and epoxides to give the corresponding α- and β-hydroxyalkylphosphine derivatives, respectively (eqs 6 and 7).[4] The more substituted alkylphosphine was preferentially formed by using an unsymmetrical epoxide. This is in contrast to the basic phosphination of epoxides, which provides mainly less substituted alkylphosphines.[5]

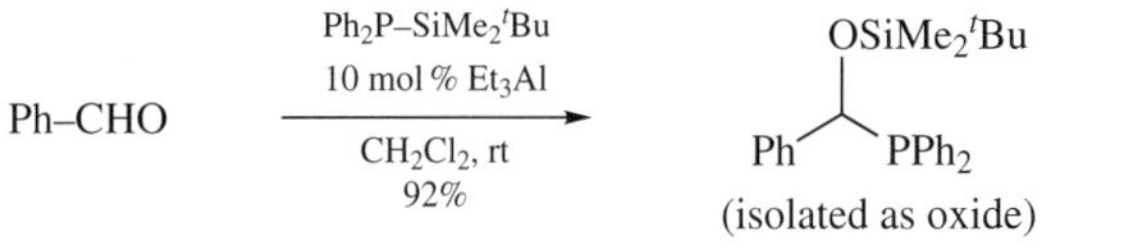

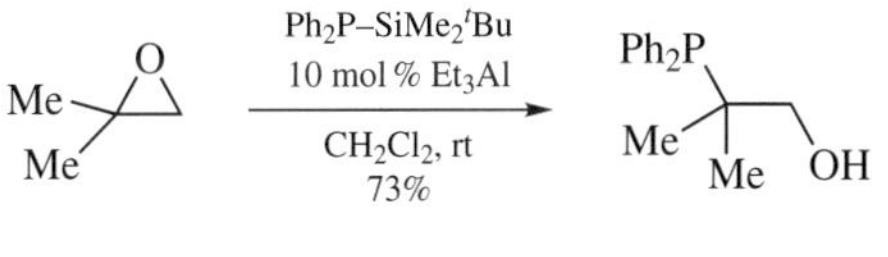

Rhodium-catalyzed Hydrophosphination. In the presence of a cationic rhodium catalyst, the hydrophosphination of alkynes has been achieved with the silylphosphine. The silylphosphine reacts with alkynes to give the alkenylphosphines in a regio- and stereoselective manner. A variety of alkynes, including alkyl and aryl substituted, terminal, and internal alkynes are hydrophosphinated. The phosphine group tends to locate at the less hindered site in the case of alkynes having an alkyl or aryl substituent, whereas the electron-withdrawing substituent on sp-carbon strongly controlled the regioselectivity to the Michael-type reaction (eq 8).[6] The silyl group is completely lost in this case to give the alkenylphosphine. The parent phosphine source, diphenylphosphine, did not react under the same conditions.

$$^{n}Bu\!\!=\!\!=\!\!CO_2Et \xrightarrow[\substack{MeOH-C_6H_6, 80\,°C \\ 81\%}]{\substack{Ph_2P-SiMe_2{}^{t}Bu \\ 5\ mol\ \%\ [Rh(cod)Cl]_2/AgOTf}} \begin{array}{c} ^{n}Bu \quad CO_2Et \\ \diagup\!\!\!\!\diagdown \\ Ph_2P \quad H \end{array} \quad (8)$$

$$E:Z = >99:1$$

Silylation of Alcohols. The silylphosphine serves as a silyl group source for the silylation of alcohols in the presence of DEAD and PPTS.[7] Primary and secondary hydroxy groups are rapidly converted to the TBDMS ether under acidic conditions (eq 9). Other silyl groups could also be introduced by using silylphosphines having the corresponding silyl group.

$$Ph\diagdown\!\!\!\diagup\!\!\!\diagdown\!\!\!OH \xrightarrow[\substack{CH_2Cl_2,\ rt \\ 5\ min \\ 95\%}]{\substack{Ph_2P\ SiMe_2{}^{t}Bu \\ DEAD,\ PPTS}} Ph\diagdown\!\!\!\diagup\!\!\!\diagdown\!\!\!OSiMe_2{}^{t}Bu \quad (9)$$

Related Reagents. Trimethylsilyldiphenylphosphine; triethylsilyldiphenylphosphine; triisopropyldiphenylphosphine; (*t*-butyldimethylsilyl)phenylmethylphosphine; bis(*t*-Butyldimethylsilyl)phenylphosphine; tris(trimethylsilyl)phosphine; trimethylsilyldiethylphosphine.

1. Hayashi, M.; Matsuura, Y.; Watanabe, Y., *Tetrahedron Lett.* **2004**, *45*, 9167.

2. Hayashi, M.; Matsuura, Y.; Watanabe, Y., *Tetrahedron Lett.* **2005**, *46*, 5135.

3. Hayashi, M.; Matsuura, Y.; Kurihara, K.; Maeda, D.; Nishimura, Y.; Morita, E.; Okasaka, M.; Watanabe, Y., *Chem. Lett.* **2007**, *36*, 634.

4. Hayashi, M.; Matsuura, Y.; Yamasaki, T.; Watanabe, Y., *submitted*.

5. Muller, G.; Sainz, D., *J. Organomet. Chem.* **1995**, *495*, 103.

6. Hayashi, M.; Matsuura, Y.; Watanabe, Y., *J. Org. Chem.* **2006**, *78*, 9248.

7. Hayashi, M.; Matsuura, Y.; Watanabe, Y., *Tetrahedron Lett.* **2004**, *45*, 1409.

Minoru Hayashi
Ehime University, Matsuyama, Japan

2-*t*-Butyldimethylsilyl-1,3-dithiane

[95452-06-5] $C_{10}H_{22}S_2Si$ (MW 234.50)
InChI = 1S/C10H22S2Si/c1-10(2,3)13(4,5)9-11-7-6-8-12-9/h9H, 6-8H2,1-5H3
InChIKey = JISLIURJHXAPTN-UHFFFAOYSA-N

(anion undergoes alkylation with electrophiles subsequently providing a general synthesis to acyl silanes;[1] anion undergoes

a one-flask bisalkylation[2] with an epoxide and a second electrophile via 1,4-Brook rearrangement[3])

Physical Data: bp 115 °C/0.03 mm Hg, 124 °C/5 mm Hg, *d* 1.010.
Solubility: insoluble in H$_2$O; soluble in organic solvents.
Form Supplied in: commercially available as colorless liquid.
Preparative Methods: prepared by alkylation of 2-lithio-1,3-dithiane with *t*-butyldimethylsilyl chloride (eq 1).[1a,c,2b,4]

$$\begin{array}{c}S\diagdown\diagup S\end{array} \xrightarrow[\substack{2.\ TBSCl,\ -78\,°C\ to\ -30\,°C}]{\substack{1.\ n\text{-}BuLi,\ THF,\ -78\,°C}} \begin{array}{c}S\diagdown\diagup S \\ TBS\end{array} \quad (1)$$

Handling, Storage, and Precaution: use in a fume hood; unpleasant odor.

2-Lithio-2-*t*-butyldimethylsilyl-1,3-dithiane (1). Reagent **1** is generated from 2-*t*-butyldimethylsilyl-1,3-dithiane by treatment with *n*-butyllithium, *t*-butyllithium, or lithium diisopropylamide (LDA)[1h] in THF at −78 °C followed by warming to approximately −45 °C (eq 2),[2b] although lithiation at higher temperatures is also reported (−30 °C to 0 °C).[1c–i] Hexamethylphosphoramide (HMPA) often is employed as a cosolvent.[1a,2bc] Diethyl ether (Et$_2$O) also can be used as the solvent of choice for lithiation.[2b]

$$\begin{array}{c}S\diagdown\diagup S \\ TBS\end{array} \xrightarrow[\substack{-78\,°C\ to\ -30\,°C}]{\substack{n\text{-}BuLi,\ THF/HMPA}} \begin{array}{c}S\diagdown\diagup S \\ Li \quad TBS\end{array} \quad (2)$$

$$\mathbf{1}$$

Alkylations. Treatment of **1** with various primary alkyl halides provides the corresponding substituted dithianes (eqs 3–5[1a,d,e]). Removal of the dithiane under the Lewis acid conditions, illustrated in eqs 3–5, unmasks the acyl silane for subsequent transformations such as photolysis,[1a] radical reactions,[1d] and heterocyclic synthesis.[1e] Other conditions for removing the dithiane moiety of 2-substituted-2-*t*-butyldimethylsilyl-1,3-dithianes include anodic oxidation,[5] ceric ammonium nitrate (CAN)/NaHCO$_3$ in CH$_3$CN/H$_2$O,[1b,c,f,i] iodomethane/CaCO$_3$ in THF/H$_2$O,[1g,h] and I$_2$/CaCO$_3$ in THF/H$_2$O.[1e] The formyl silane of 2-*t*-butyldimethylsilyl-1,3-dithiane has also been reported.[4]

$$(3)$$

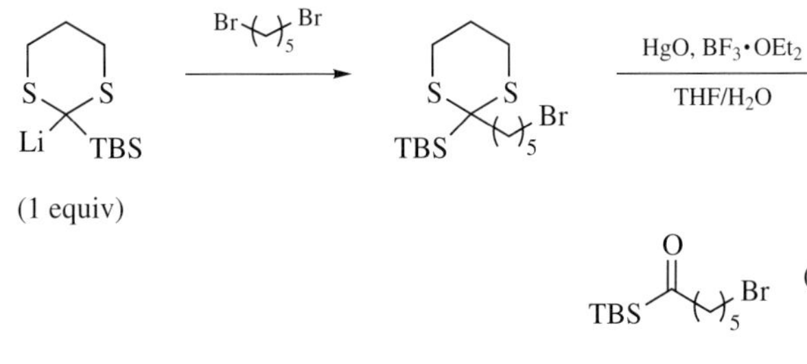

(1 equiv)

70% over two steps (4)

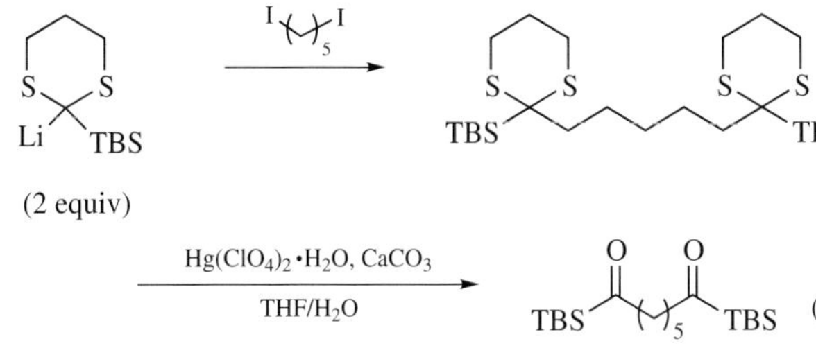

(2 equiv)

71% over two steps (5)

n = 1, 75%   n = 1, 17%
n = 3, 46%   n = 3, 3% (8)

The three-component process with 2-*t*-butyldimethylsilyl-1,3-dithiane has been further optimized to incorporate two different epoxides, employing solvent as a means to control the 1,4-Brook rearrangement (eq 9).[2b] In this protocol, lithiation and alkylation of the first epoxide are conducted in Et_2O rather than THF to prevent premature silyl migration. Introduction of HMPA (or DMPU)[2b] with the second epoxide in Et_2O initiates the Brook process subsequently leading to the desired unsymmetrical, three-component adducts.

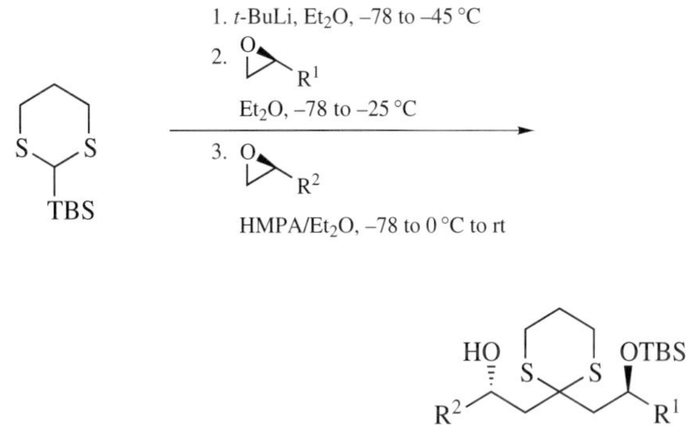

1. *t*-BuLi, Et_2O, –78 to –45 °C
2. R^1 Et_2O, –78 to –25 °C
3. R^2 HMPA/Et_2O, –78 to 0 °C to rt

56–74% (9)

Tandem Bisalkylations. Tandem bisalkylations of the anion generated from either 2-trimethylsilyl-1,3-dithiane or 2-*t*-butyldimethylsilyl-1,3-dithiane with scalemic epoxides proceed via a 1,4-Brook rearrangement yielding pseudo-C_2-symmetrical 1,5-diols (eq 6).[2a] Epoxides bearing an additional nucleofugal site can undergo bisalkylation resulting in carbocycles.[6] For example, lithiation of 2-*t*-butyldimethylsilyl-1,3-dithiane followed by introduction of a bisepoxide derived from D-mannitol gives polyhydroxylated cycloheptanes and cyclohexanes (eq 7).[6b] Interestingly, high regioselectivity is achieved in the second epoxide opening based on the protecting groups of the C3- and C4-diol. Reaction of **1** with epoxy tosylates also yield carbocycles. However, side products are often observed (eq 8).[6a]

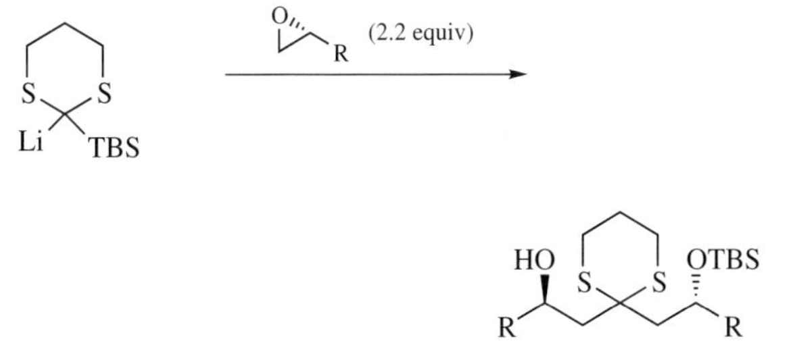

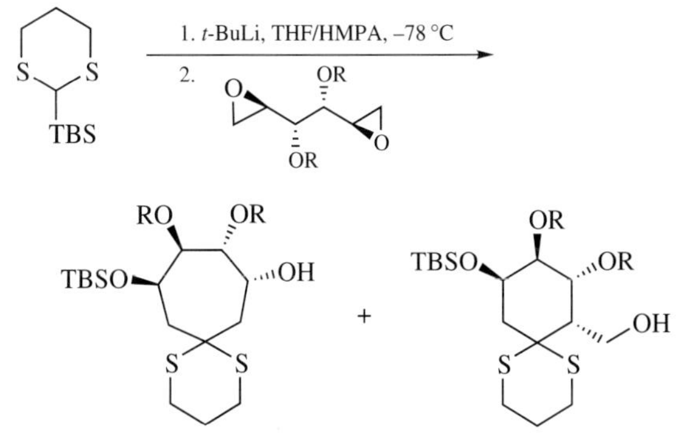

Careful selection of the second electrophile also has led to higher-order multicomponent couplings. Bisalkylation of 1 equiv of either epichlorohydrin or 1,2:4,5-diepoxypentane yields five-component adducts which can be further elaborated to extended 1,3-polyol moieties (eqs 10[2b] and 11[7]).

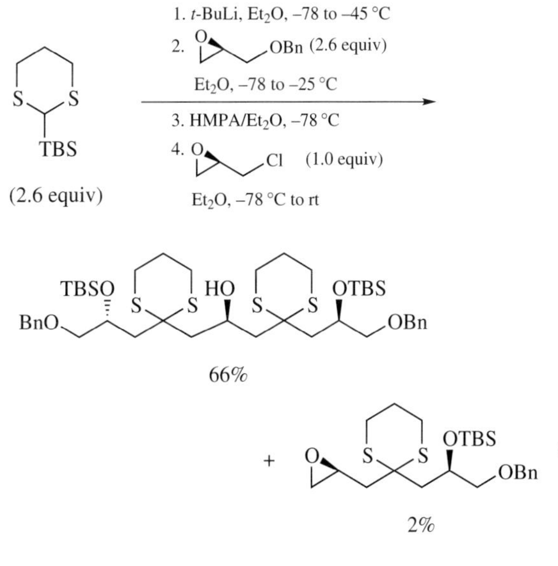

(2.2 equiv)

(6)

1. *t*-BuLi, THF/HMPA, –78 °C
2.

R = Bn, 82%   R = Bn, 10%
R = CMe₂, 24%   R = CMe₂, 67% (7)

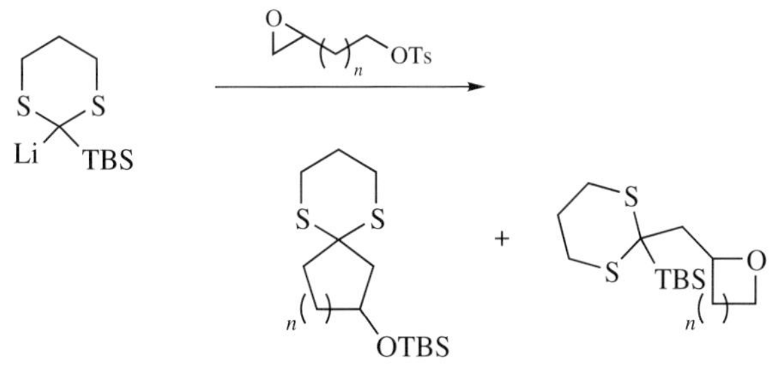

1. *t*-BuLi, Et_2O, –78 to –45 °C
2. OBn (2.6 equiv) Et_2O, –78 to –25 °C
3. HMPA/Et_2O, –78 °C
4. Cl (1.0 equiv) Et_2O, –78 °C to rt

(2.6 equiv)

66%

2% (10)

1. *t*-BuLi, Et$_2$O, −78 to −45 °C

2. (2.3 equiv)
Et$_2$O, −78 to −25 °C

3. HMPA/THF, −78 °C

4. (1.0 equiv)
THF, −78 °C to rt

59%

+ (11)

13%

Related Reagents. 2-Trimethylsilyl-1,3-dithiane; 2-lithio-1,3-dithiane; 1,3-dithiane.

1. (a) Scheller, M. E.; Frei, B., *Helv. Chim. Acta* **1984**, *67*, 1734. (b) Tsai, Y. M.; Nieh, H. C.; Cherng, C. D., *J. Org. Chem.* **1992**, *57*, 7010. (c) Chuang, T.-H.; Fang, J.-M.; Jiaang, W.-T.; Tsai, Y.-M., *J. Org. Chem.* **1996**, *61*, 1794. (d) Chang, S.-Y.; Jiaang, W.-T.; Cherng, C.-D.; Tang, K.-H.; Huang, C.-H.; Tsai, Y.-M., *J. Org. Chem.* **1997**, *62*, 9089. (e) Bouillon, J.-P.; Portella, C., *Eur. J. Org. Chem.* **1999**, 1571. (f) Jiaang, W.-T.; Lin, H.-C.; Tang, K.-H.; Chang, L.-B.; Tsai, Y.-M., *J. Org. Chem.* **1999**, *64*, 618. (g) Bouillon, J.-P.; Saleur, D.; Portella, C., *Synthesis* **2000**, 843. (h) Saleur, D.; Bouillon, J.-P.; Portella, C., *Tetrahedron Lett.* **2000**, *41*, 321. (i) Huang, C.-H.; Chang, S.-Y.; Wang, N.-S.; Tsai, Y.-M., *J. Org. Chem.* **2001**, *66*, 8983.

2. (a) Tietze, L. F.; Geissler, H.; Gewert, J. A.; Jakobi, U., *Synlett* **1994**, 511. (b) Smith, A. B., III; Boldi, A. M., *J. Am. Chem. Soc.* **1997**, *119*, 6925. (c) Smith, A.B., III; Pitram, S. M.; Boldi, A. M.; gawnt, m. f.; Sfoaffatakis, c.; Moser, W.W. *J. Am. Chem. Soc.* **2003**, 47 14935

3. Brook, A. G., *Acc. Chem. Res.* **1974**, *7*, 77.

4. Silverman, R. B.; Lu, X.; Banik, G. M., *J. Org. Chem.* **1992**, *57*, 6617.

5. Suda, K.; Watanabe, J.; Takanami, T., *Tetrahedron Lett.* **1992**, *33*, 1355.

6. (a) Michel, T.; Kirsching, A.; Beier, C.; Braeuer, N.; Schaumann, E.; Adiwidjaja, G., *Liebigs Ann.* **1996**, 1811. (b) Le Merrer, Y.; Gravier-Pelletier, C.; Maton, W.; Numa, M.; Depezay, J.-C., *Synlett* **1999**, 1322.

7. Smith, A. B., III; Pitram, S. M., *Org. Lett.* **1999**, *1*, 2001.

Amos B. Smith, III & Suresh M. Pitram
University of Pennsylvania, Philadelphia, PA, USA

O-(*t*-Butyldimethylsilyl)hydroxylamine

(**1**; R^1 = Me, R^2 = *t*-Bu)

[41879-39-4]   C$_6$H$_{17}$NOSi   (MW 147.29)

InChI = 1S/C6H17NOSi/c1-6(2,3)9(4,5)8-7/h7H2,1-5H3

InChIKey = SSUCKKNRCOFUPT-UHFFFAOYSA-N

(**2**; R^1 = R^2 = Me)

[22737-36-6]   C$_3$H$_{11}$NOSi   (MW 105.21)

InChI = 1S/C3H11NOSi/c1-6(2,3)5-4/h4H2,1-3H3

InChIKey = AEKHNNJSMVVESS-UHFFFAOYSA-N

(**3**; R^1 = Ph, R^2 = *t*-Bu)

[103587-51-5]   C$_{16}$H$_{21}$NOSi   (MW 271.43)

InChI = 1S/C16H21NOSi/c1-16(2,3)19(18-17,14-10-6-4-7-11-14)15-12-8-5-9-13-15/h4-13H,17H2,1-3H3

InChIKey = PWTWOGMXABHJOA-UHFFFAOYSA-N

(preparation of *O*-silyloximes;[1] generation of nitrosoalkenes;[1,2] generation of azadienes[3])

Physical Data: (**1**) mp 62–65 °C; bp 87–90 °C/40 mmHg. (**2**) bp 98–100 °C; *d* 0.860 g cm^{-3}. (**3**) mp 74–75 °C; bp 175 °C/0.6 mmHg.

Solubility: freely sol hexane, chloroform, and most organic solvents; insol water.

Form Supplied in: colorless solid or liquid; commercially available.

Preparative Methods: the reagents are most conveniently prepared by treatment of hydroxylamine hydrochloride with ethylenediamine in dichloromethane, followed by the appropriate chlorosilane.[4] A two-step procedure using ammonia in place of ethylenediamine has also been reported.[1]

Purification: reagents are purified via distillation and/or recrystallization (>97% GC).

Handling, Storage, and Precautions: *O*-silylhydroxylamines should be protected from moisture and handled in a fume hood.

Preparation of *O*-Silyloximes and Oximes. *O*-Silylhydroxylamines are used to prepare *O*-silyloximes via treatment of carbonyl compounds with these reagents in CHCl$_3$ in the presence of 4 Å molecular sieves.[1] Treatment of *O*-silyloximes with fluoride ion yields the desilylated oximes, thus allowing the oximation of acid- or base-sensitive ketones and aldehydes.

A series of 3-alkyl-2-chlorocyclohexanone silyloximes (methyl, ethyl, *i*-propyl) have been prepared and their conformations extensively examined.[5] By analysis of the vicinal interproton coupling, these compounds were shown to exist predominantly in the diaxial chair conformation (**5**) while the precursor ketones prefer the diequatorial conformation (**4**). This observation was corroborated by an X-ray crystal structure of *trans*-3-methyl-2-chlorocyclohexanone *t*-butyldiphenylsilyloxime, which showed that the chair with diaxial substituents is indeed preferred in the solid state. A strong hyperconjugative stabilization of the axial conformation, termed the vinylogous anomeric effect, was proposed to be the origin of this preference.

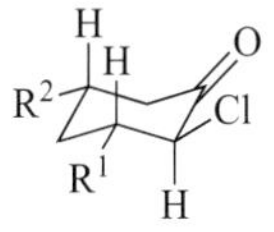

(4) 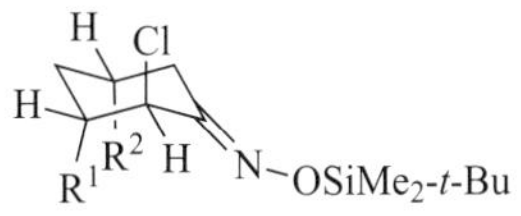   (5)

Generation of Nitrosoalkenes. Treatment of silyloximes of α-halocarbonyl compounds with fluoride ion generates nitrosoalkenes without extraneous nucleophilic species present.[1] Nitrosoalkenes generated from (6) have been used in intramolecular [4 + 2] cycloadditions (eq 1).[2] Electron-rich alkenes are required for concerted cycloadditions as this is an inverse electron-demand cycloaddition. The choice of fluoride source is critical for success in that the highest yields are obtained when the nitrosoalkene is slowly generated via use of sparingly soluble metal fluorides such as cesium fluoride or potassium fluoride.

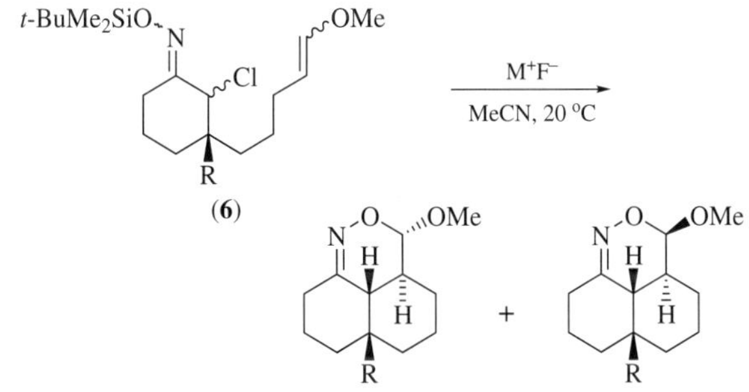

R = H; CsF   70%   13%
R = Me; KF   65%   17%   (1)

Generation of 1-Azadienes. *O*-Silyloximes have also been used to generate 1-azadienes for [4 + 2] cycloadditions.[3] Thus an *O*-silyloxime of an α,β-unsaturated aldehyde can be treated with an acid chloride or chloroformate in the presence of chlorotrimethylsilane and aluminum chloride to give an α-cyanohydroxamic acid derivative, which upon mild thermolysis forms an azadiene (eq 2). These azadienes undergo efficient intermolecular [4 + 2] cycloadditions or, with a tethered alkene, intramolecular cycloadditions.

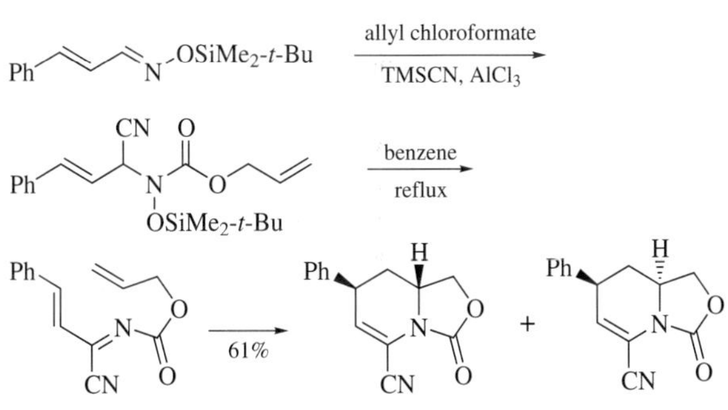

89:11   (2)

Reactions at the Hydroxylamine Nitrogen. These reagents have been used as *O*-protected forms of hydroxylamine for synthetic transformations involving the unprotected nitrogen. Thus *O*-silylhydroxylamines have been used to prepare *N*-hydroxy-β-lactams[6] and substituted dihydro-3-hydroxy-1,2,3-benzotriazines,[7] as well as *N*-alkylated hydroxylamines for use as leukotriene biosynthesis inhibitors.[8]

Related Reagents. *O*-Benzylhydroxylamine Hydrochloride; *N*,*O*-Bis(trimethylsilyl)hydroxylamine; Hydroxylamine; *O*-(Mesitylsulfonyl)hydroxylamine.

---

1. Denmark, S. E.; Dappen, M. S., *J. Org. Chem.* **1984**, *49*, 798.
2. Denmark, S. E.; Dappen, M. S.; Sternberg, J. A., *J. Org. Chem.* **1984**, *49*, 4741.
3. Teng, M.; Fowler, F. W., *J. Org. Chem.* **1990**, *55*, 5646.
4. Bottaro, J. C.; Bedford, C. D.; Dodge, A., *Synth. Commun.* **1985**, *15*, 1333.
5. Denmark, S. E.; Dappen, M. S.; Sear, N. S.; Jacobs, R. T., *J. Am. Chem. Soc.* **1990**, *112*, 3466.
6. Zercher, C. K.; Miller, M. J., *Tetrahedron Lett.* **1989**, *50*, 7009.
7. Jakobsen, M. H.; Buchardt, O.; Holm, A.; Meldal, M., *Synthesis* **1990**, 1008.
8. Stewart, A. O.; Martin, J. G., *J. Org. Chem.* **1989**, *54*, 1221.

Michael S. Dappen
Athena Neurosciences, South San Francisco, CA, USA

Scott E. Denmark
University of Illinois, Urbana, IL, USA

N-(*t*-Butyldimethylsilyl)-*N*-methyltrifluoroacetamide[1]

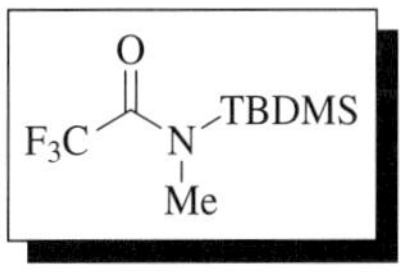

[77377-52-7] $C_9H_{18}F_3NOSi$ (MW 241.33)
InChI = 1S/C9H18F3NOSi/c1-8(2,3)15(5,6)13(4)7(14)9(10,11)12/h1-6H3
InChIKey = QRKUHYFDBWGLHJ-UHFFFAOYSA-N

(excellent *t*-butyldimethylsilylating reagent for alcohols, amines, carboxylates, thiols, and inorganic oxyanions;[1,2] is often used to protect hydroxy and amino groups;[1] derivatizing reagents for GC–MS analysis[3])

Alternate Name: BSMTFA.
Physical Data: bp 168–170 °C; *d* 1.12 g cm^{-3}.[1]
Solubility: sol most aprotic organic solvents.[1]
Form Supplied in: colorless liquid;[1] neat or with 1% *t*-butyldimethylsilyl chloride commercially available.[4,5]
Preparative Method: obtained in 91% yield by reaction of *N*-methyl-2,2,2-trifluoroacetamide (1.0 equiv) in benzene and acetonitrile (1:1 v/v) with NaH (1.0 equiv) and then with *t*-BuMe$_2$SiCl (1.2 equiv) at 4 °C.[1]

Handling, Storage, and Precautions: moisture-sensitive; is easily transferable with a gas-tight syringe.[1]

***t*-Butyldimethylsilylation of Functional Groups.**[1] In the presence of 1% of *t*-butyldimethylchlorosilane as the catalyst, *N*-(*t*-butyldimethylsilyl)-*N*-methyltrifluoroacetamide (BSMTFA) functions as an extremely reactive *t*-butyldimethylsilylating reagent for alcohols, amines, carboxylic acids, and thiols. Silylation is generally completed within 5 min at 25 °C in acetonitrile.[1] This amide is more reactive than *N*-(*t*-butyldimethylsilyl)-*N*-methylacetamide.

In the protection of a hydroxy group, the resultant *t*-butyldimethylsilyl (TBDMS) ethers are stable under the conditions for acetate saponification and hydrogenation.[1] These silyl ethers also remain intact towards the Jones reagent and Wittig reagents. The TBDMS ethers are approx. 10^4 times more stable against hydrolysis than the corresponding trimethylsilyl (TMS) ethers.[6] Selective removal of the TBDMS group can be accomplished by use of dilute acetic acid or tetra-*n*-butylammonium fluoride in THF at 25 °C.

Silylation of ketones by use of BSMTFA occurs in triethylamine and DMF at 40–60 °C to give the corresponding silyl enol ethers in good to excellent yields (eq 1).[7,8] In addition, silyl ether formation takes place in *N*-hydroxysuccinimide (88% yield) and *N*-hydroxypyrrole (99% yield) by use of BSMTFA in THF.[9]

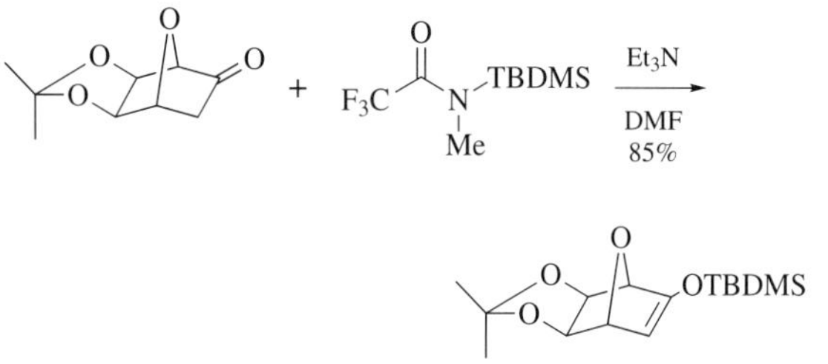

$$(1)$$

TBDMS amines are approx. 100 times more stable than the labile TMS ethers towards solvolysis.[1] Reaction of dibenzyl aspartate with BSMTFA in acetonitrile and then with *t*-BuMgCl in ether affords an *N*-silylated β-lactam in 75% yield (eq 2).[10] On the other hand, *N*-silylation of the sulfoximine PhMeS(=O)=NH proceeds in 92% yield with BSMTFA at 90 °C.[11] The oxime and amide functionalities can also be silylated with BSMTFA (eq 3).[12,13]

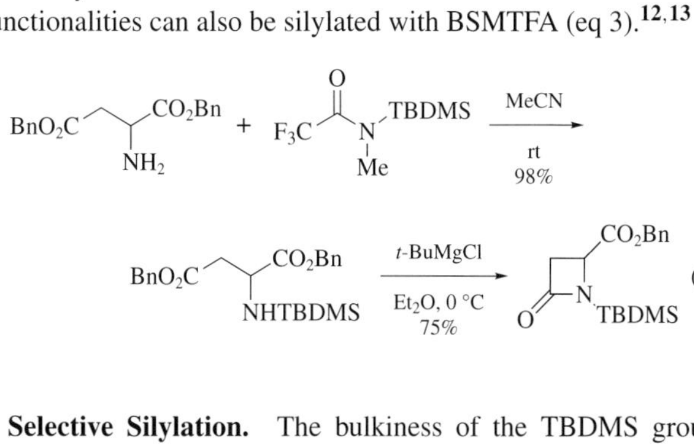

$$(2)$$

Selective Silylation. The bulkiness of the TBDMS group enables BSMTFA to possess greater selectivity than the corresponding TMS-containing agents in silylation. In general, *t*-butyldimethylsilylation of secondary amines, which are sterically more congested, proceeds more slowly than that of alcohols, carboxylates, thiols, and primary amines.[1,4] Reaction of alkyl-diamines with BSMTFA gives disilylated derivatives, in which

each of the two primary amino groups is monosilylated. Trisilylation occurs after a prolonged reaction time.[5] Selective monosilylation of a dihydroxy aromatic compound can be accomplished in 92% yield by use of 1.5 equiv of BSMTFA in a mixture of DMF and THF (eq 4).[14]

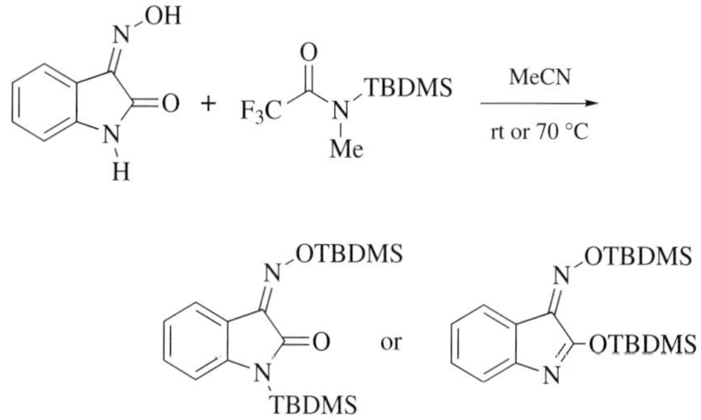

$$(3)$$

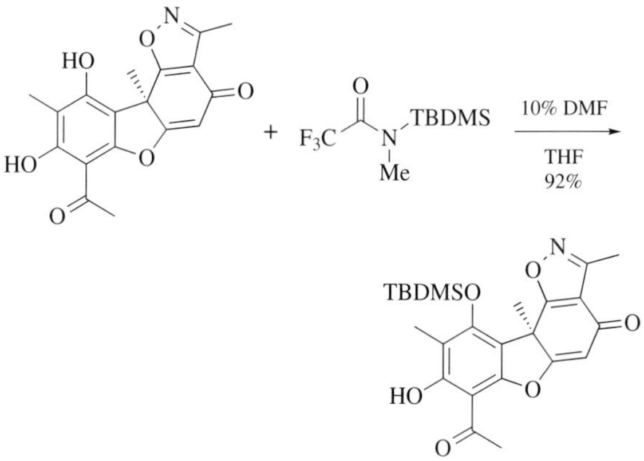

$$(4)$$

Derivatizing Reagent for GC–MS Analyses. For organic compounds containing active protons (e.g. alcohols, amines, carboxylates, and thiols), silylation is often used to improve resolution and peak symmetry, or to decrease adsorption on the column for GC–MS analysis.[3,15–17] Use of BSMTFA as the derivatizing reagent produces a mass spectrum with a prominent peak at $M - 57$ *m/z*, which represents loss of a *t*-butyl group from the molecular ion of the silylated species. In comparison, the trimethylsilylated amino acids are often unstable, easily hydrolyzed, and produce varying GC results.[17]

Silylation of Inorganic Oxyanions.[2,18] The TBDMS derivatives of some inorganic oxyanions can be prepared by reaction of the corresponding free acids or the ammonium salts with BSMTFA in DMF. Those oxyanions include arsenate, arsenite, borate, carbonate, molybdenate, phosphate, phosphite, pyrophosphate, selenate, selenite, sulfate, sulfite, and vanadate. The *t*-butyldimethylsilylated inorganic oxyanions are relatively stable in comparison with the corresponding TMS derivatives, which show significant degradation on GC. Moreover, those *t*-butyldimethylsilylated oxyanions display single, sharp, and symmetrical chromatographic peaks without tailing on GC and produce easily interpretable mass spectra dominated by a characteristic $M - 57$ ion peak.

1. Mawhinney, T. P.; Madson, M. A., *J. Org. Chem.* **1982**, *47*, 3336.

2. Mawhinney, T. P., *J. Chromatogr.* **1983**, *257*, 37.

3. Schwenk, W. F.; Berg, P. J.; Beaufrere, B.; Miles, J. M.; Haymond, M. W., *Anal. Biochem.* **1984**, *141*, 101.

4. Ballard, K. D.; Knapp, D. R.; Oatis, J. E., Jr.; Walle, T., *J. Chromatogr.* **1983**, *277*, 333.

5. Ng, L.-K., *J. Chromatogr.* **1984**, *314*, 455.

6. Corey, E. J.; Venkateswarlu, A., *J. Am. Chem. Soc.* **1972**, *94*, 6190.

7. Auberson, Y.; Vogel, P., *Helv. Chim. Acta* **1989**, *72*, 278.

8. Fattori, D.; De Guchteneere, E.; Vogel, P., *Tetrahedron Lett.* **1989**, *30*, 7415.

9. Keana, J. F. W.; Heo, G. S.; Mann, J. S.; Van Nice, F. L.; Lex, L.; Prabhu, V. S.; Ferguson, G., *J. Org. Chem.* **1988**, *53*, 2268.

10. Baldwin, J. E.; Adlington, R. M.; Gollins, D. W.; Schofield, C. J., *Tetrahedron* **1990**, *46*, 4733.

11. Hwang, K.-J.; Logusch, E. W.; Brannigan, L. H.; Thompson, M. R., *J. Org. Chem.* **1987**, *52*, 3435.

12. Glover, V.; Halket, J. M.; Watkins, P. J.; Clow, A.; Goodwin, B. L.; Sandler, M., *J. Neurochem.* **1988**, *51*, 656.

13. Halket, J. M.; Watkins, P. J.; Przyborowska, A.; Goodwin, B. L.; Clow, A.; Glover, V.; Sandler, M., *J. Chromatogr.* **1991**, *562*, 279.

14. Cooper, A. B.; Wang, J.; Saksena, A. K.; Girijavallabhan, V.; Ganguly, A. K.; Chan, T.-M.; McPhail, A. T., *Tetrahedron* **1992**, *48*, 4757.

15. Wilson, R. T.; Groneck, J. M.; Henry, A. C.; Rowe, L. D., *J. Assoc. Off. Anal. Chem.* **1991**, *74*, 56.

16. Landrum, D. C.; Mawhinney, T. P., *J. Chromatogr.* **1989**, *483*, 21.

17. Mawhinney, T. P.; Robinett, R. S. R.; Atalay, A.; Madson, M. A., *J. Chromatogr.* **1986**, *358*, 231.

18. Mawhinney, T. P., *Anal. Lett.* **1983**, *16*(A2), 159.

t-Butyldimethylsilyl Trifluoromethane-sulfonate[1]

$$\boxed{\text{t-BuMe}_2\text{SiOSO}_2\text{CF}_3}$$

[69739-34-0] $C_7H_{15}F_3O_3SSi$ (MW 264.33)

InChI = 1S/C7H15F3O3SSi/c1-6(2,3)15(4,5)13-14(11,12)7(8,9)10/h1-5H3

InChIKey = WLLIXJBWWFGEHT-UHFFFAOYSA-N

(highly reactive silylating agent and Lewis acid capable of converting primary, secondary, and tertiary alcohols[1b] to the corresponding TBDMS ethers, and converting ketones[2] and lactones[2a,3] into their enol silyl ethers; promoting conjugate addition of alkynylzinc compounds[4] and triphenylphosphine[5] to α,β-enones; activation of chromones in [4 + 2] cycloaddition reactions;[6] rearrangement of allylic tributylstannyl silyl ethers;[7] activation of pyridine rings toward Grignard reagents[8] and trans-alkylation of tertiary amine *N*-oxides;[9] and transformation of *N*-*t*-butoxycarbonyl groups into *N*-alkoxycarbonyl groups[10])

Alternate Name: TBDMS triflate.

Physical Data: bp 60 °C/7 mmHg; colorless oil, d 1.151 g cm^{-3}.

Solubility: sol most organic solvents such as pentane, CH$_2$Cl$_2$, etc.

Analysis of Reagent Purity: ^{1}H NMR (CDCl$_3$) δ 1.00 (s, 9H, *t*-Bu), 0.45 (s, 6H, Me).

Form Supplied in: liquid; widely available.

Preparative Method:[1b] to 24 g (0.16 mol) of *t*-butyldimethyl-chlorosilane at 23 °C under argon is added 14 mL (0.16 mol) of trifluoromethanesulfonic acid dropwise. The solution is heated at 60 °C for 10 h, at which time no further hydrogen chloride evolves (removed through a bubbler). The resulting product is distilled under reduced pressure: 34 g (80% yield) of TBDMS triflate; bp 60 °C/7 mmHg.

Handling, Storage, and Precautions: the material should be stored under argon at 0 °C. The compound has an unpleasant odor and reacts rapidly with water and other protic solvents.

Original Commentary

Duy H. Hua & Jinshan Chen
Kansas State University, Manhattan, KS, USA

Silylation of Alcohols.[1]   Primary, secondary, and tertiary alcohols are silylated by reaction with TBDMS triflate in excellent yields. For instance, treatment of *t*-butanol with 1.5 equiv of TBDMS triflate and 2 equiv of 2,6-lutidine in CH$_2$Cl$_2$ at 25 °C for 10 min gives a 90% yield of (*t*-butoxy)-*t*-butyldimethylsilane.[1b] The following alcohols are similarly silylated in excellent yields (70–90%): 2-phenyl-2-propanol, *endo*-norborneol, *cis*-2,2,4,4-tetramethylcyclobutane-1,3-diol, and 9-*O*-methylmaytansinol (converted to the 3-TBDMS derivative) (eq 1).[1b]

$$\text{R-OH} \xrightarrow[90\%]{\substack{\text{TBDMSOTf} \\ \text{2,6-lutidine, CH}_2\text{Cl}_2}} \text{R-OTBDMS} \qquad (1)$$

Formation of Enol Silyl Ethers.[2,3]   Various sterically hindered ketones have been converted into enol silyl ethers by treatment with 1–2 equiv of TBDMS triflate and 1.5 equiv of triethylamine in CH$_2$Cl$_2$ or 1,2-dichloroethane at rt. A representative example is depicted in eq 2.[2a]

Reactions of chiral β-keto sulfoxides with 1.1 equiv of lithium diisopropylamide in THF at −78 °C followed by 1.2 equiv of TBDMS triflate at −78 °C produce the corresponding (*Z*)-enol silyl ethers (eq 3).[2b]

R^1 = H, R^2 = Me; R^1 = R^2 = Me; R^1 = H, R^2 = Ph

Lactones have also been transformed into silyl ketene acetals upon treatment with TBDMS triflate and triethylamine in CH$_2$Cl$_2$ (eqs 4 and 5).[2a,3] In the case of 8a-vinyl-2-oxooctahydro-2*H*-1,4-benzoxazine, the resulting silyl ketene acetal undergoes Claisen rearrangement to provide the octahydroquinoline (eq 6).[3]

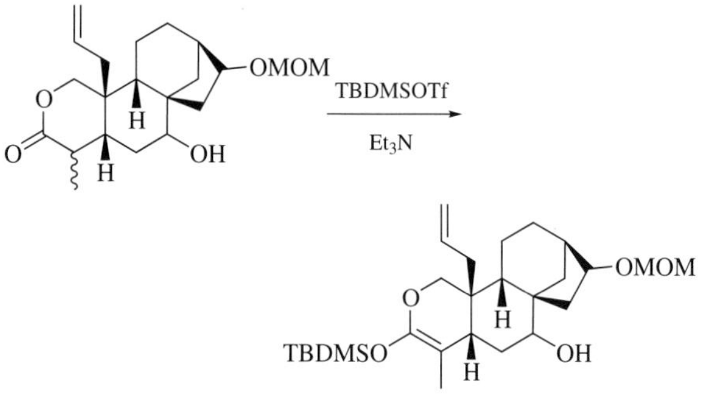

$$ \text{(4)} $$

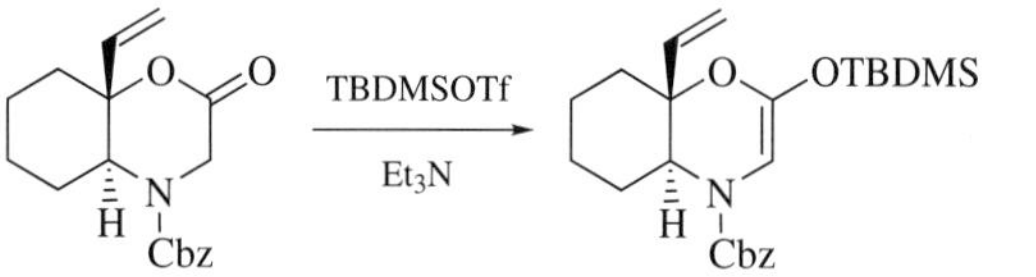

$$ \text{(5)} $$

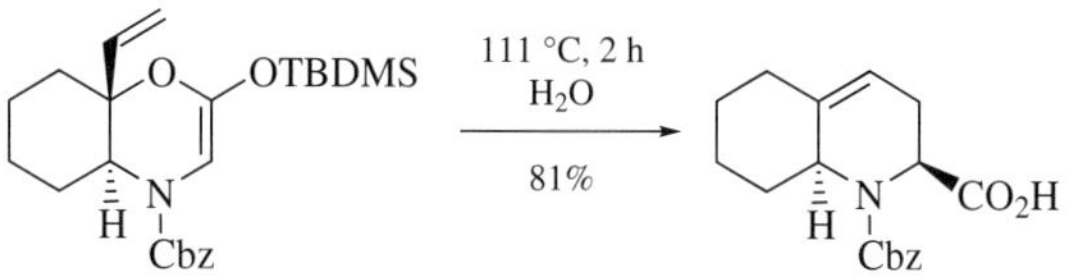

$$ \text{(6)} $$

Conjugate Addition of Alkynylzinc Bromides.[4] Alkynylzinc bromides undergo conjugate addition with α,β-unsaturated ketones in the presence of TBDMS triflate in ether–THF at $-40\,^{\circ}\text{C}$ to give the corresponding 1,4-adducts (54–96% yields). A representative example is illustrated in eq 7.[4] Other trialkylsilyl triflates such as triisopropylsilyl trifluoromethanesulfonate or trimethylsilyl trifluoromethanesulfonate can effectively replace TBDMS triflate.

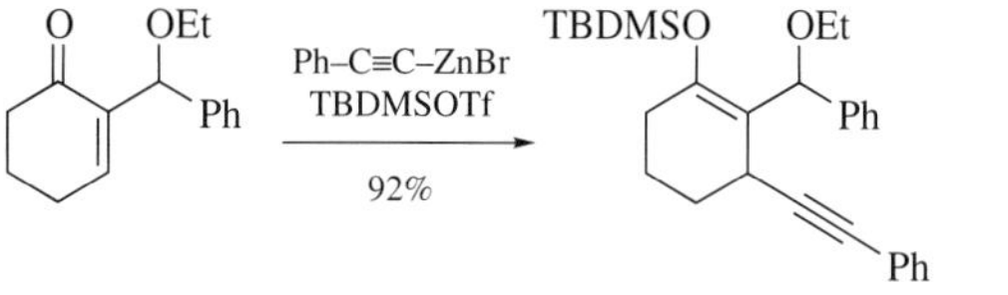

$$ \text{(7)} $$

Phosphoniosilylation.[5] Cyclic enones treated with TBDMS triflate and triphenylphosphine in THF at rt provide the corresponding 1-(3-*t*-butyldimethylsilyloxy-2-cycloalkenyl)triphenylphosphonium triflates (eq 8) which, upon lithiation with *n*-butyllithium followed by Wittig reaction with aldehydes, afford various conjuated dienes.[5]

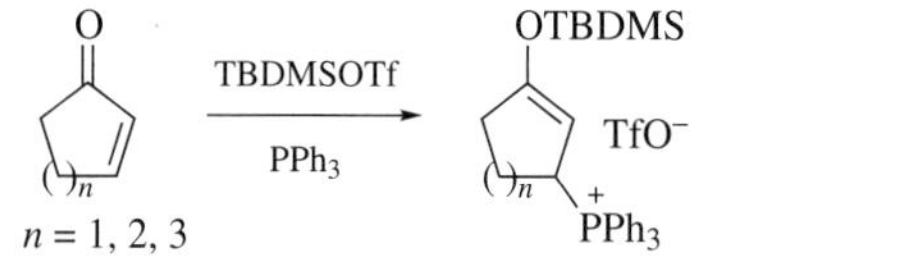

$$ \text{(8)} $$
$$ n = 1, 2, 3 $$

Silylation of Chromones.[6] The preparation of 4-*t*-butyldimethylsilyloxy-1-benzopyrylium triflate is carried out by heating chromone and TBDMS triflate at $80\,^{\circ}\text{C}$ for 1 h (without solvent) under nitrogen (eq 9).[6b] The silylated chromones undergo addition reaction with enol silyl ethers and 2,6-lutidine,[6b] and [4 + 2]-type cycloaddition reactions with α,β-unsaturated ketones in the presence of TBDMS triflate and 2,6-lutidine. An example of the cycloaddition reaction is shown in eq 10.[6a]

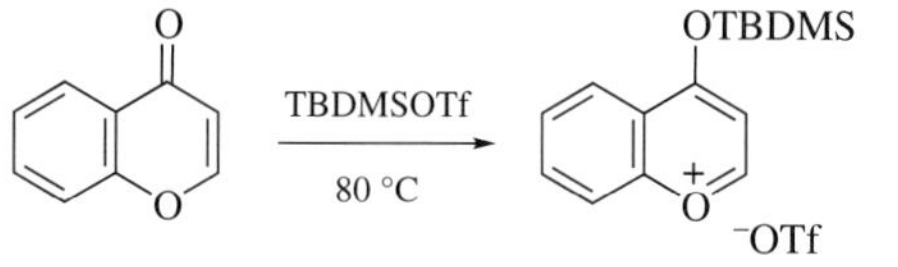

$$ \text{(9)} $$

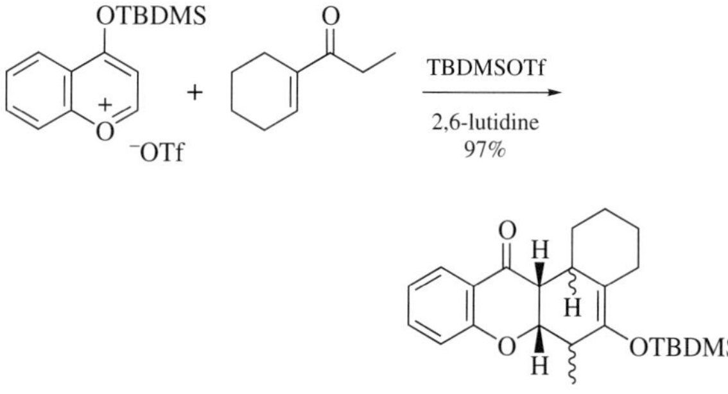

$$ \text{(10)} $$

Rearrangement of Allylic Tributylstannyl Silyl Ethers.[7] α-Silyloxy allylic stannanes are isomerized with TBDMS triflate to (Z)-γ-silyloxy allylic stannanes (eq 11).[7] The resulting allylic stannanes undergo addition reactions with aldehydes in the presence of boron trifluoride etherate to provide the 3-(*t*-butyldimethylsilyloxy)-4-hydroxyalkenes.[7a]

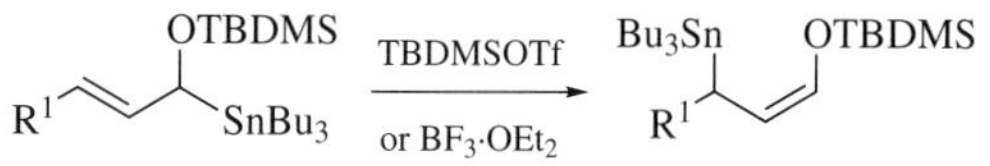

$$ \text{(11)} $$

Activation of Pyridine.[8] N-(*t*-Butyldimethylsilyl)pyridinium triflate, prepared from pyridine and TBDMS triflate in CH_2Cl_2 at rt, undergoes addition reactions with alkyl and aryl Grignard reagents to give 4-substituted pyridines after oxidation with oxygen (eq 12).[8] Only about 1% of the 2-substituted pyridines were formed in the cases studied.

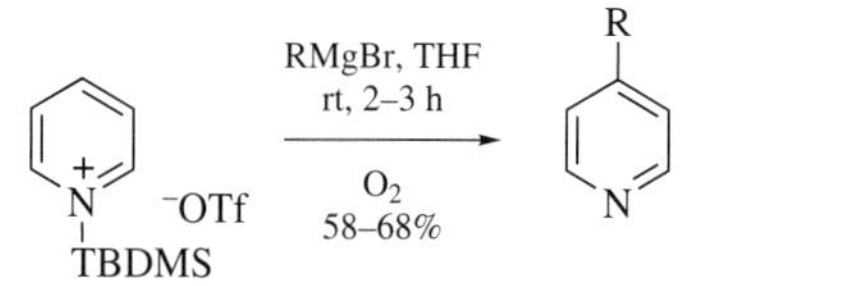

$$ \text{(12)} $$

Transalkylation of Tertiary Amine N-Oxides.[9] N-(*t*-Butyldimethylsilyloxy)-N-methylpiperidinium triflate is quantitatively formed from the reaction of N-methylpiperidine N-oxide (eq 13).[9b] The resulting amine salts derived from various trialkylamine N-oxides undergo transalkylation by treatment with methyllithium in THF at $0\,^{\circ}\text{C}$ followed by alkyl halides and tetra-*n*-butylammonium Fluoride in a sealed tube at $110\,^{\circ}\text{C}$ for 10 h, to afford trisubstituted amines (eq 14).[9a]

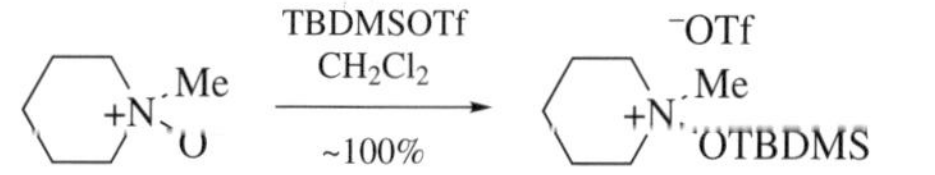

$$ \text{(13)} $$

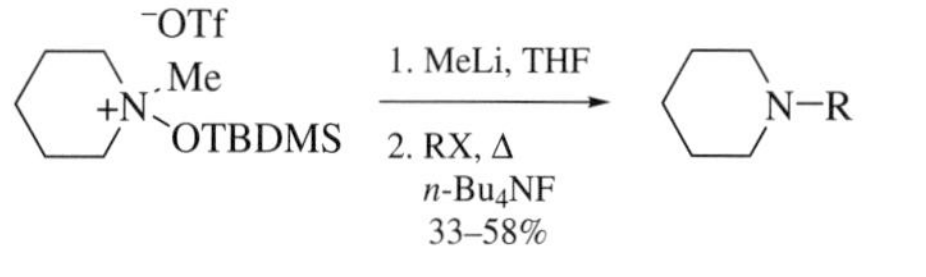

$$ \text{(14)} $$

Interconversion of N-Boc Group into N-Alkoxycarbonyl Group.[10] Treatment of *t*-butyl alkylcarbamates with 1.5 equiv of TBDMS triflate and 2 equiv of 2,6-lutidine in CH_2Cl_2 at rt for about 15 min furnishes the corresponding TBDMS carbamates (eq 15).[10] Desilylation of these silyl carbamates with aqueous fluoride ion gives excellent yields of the corresponding primary amines. The silyl carbamates are also converted into other *N*-alkoxycarbonyl derivatives by treatment with TBAF and alkyl halides in THF at 0 °C (82–88% yields).[10]

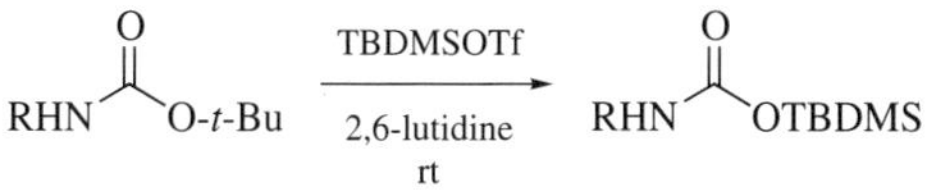

(15)

First Update

Richard S. Grainger & Caterina S. Aricó
University of Birmingham, Edgbaston, Birmingham, UK

Carbonyl Activation. The high oxaphilicity and Lewis acidity of TBDMS triflate activates carbonyl compounds toward nucleophilic attack. Examples of heteroatom- and carbon-based nucleophiles have been reported.

TBDMS triflate has been used in place of the more commonly employed TMS triflate as a catalyst in the Noyori protocol for the formation of cyclic acetals from ketones and 1,2-bis(trimethylsiloxy)ethane (eq 16).[11]

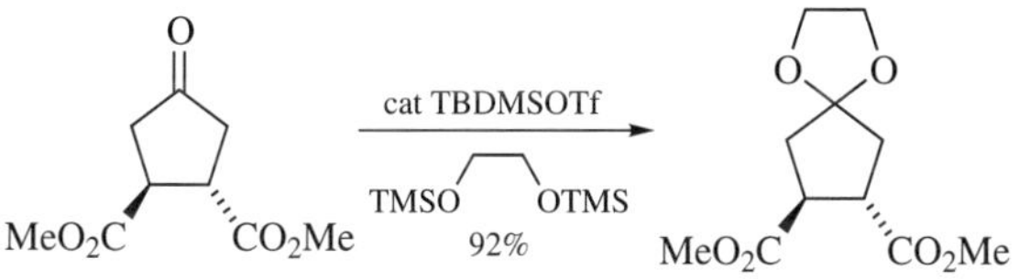

(16)

The intermolecular addition of δ-valerolactam to 2,3-*O*-isopropylidene-D-glyceraldehyde is promoted by a combination of 3 equiv each of TBDMS triflate and *i*-PrNEt$_2$ (eq 17). Addition of nucleobases such as thymine, uracil, and 6-chloropurine is also effective under these conditions, with a similar preference for the *anti*-diastereomer.[12]

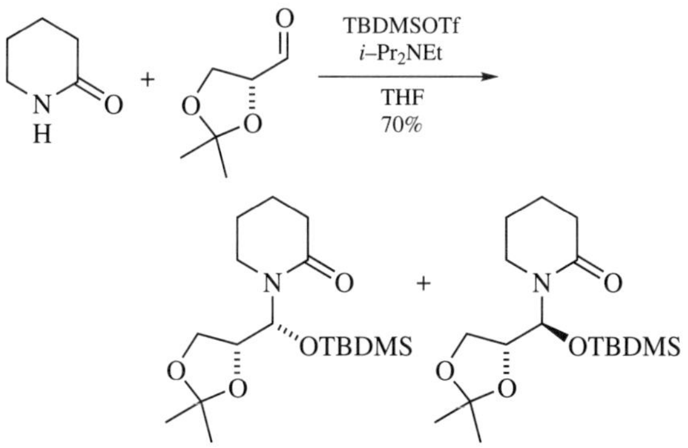

(17)

77:23 *anti:syn*

O-Acyl mandelamides undergo highly stereoselective intramolecular cyclization reactions in the presence of TBDMS triflate to give stable 2-silyloxy-1,3-oxazolidin-4-ones (eq 18).[13]

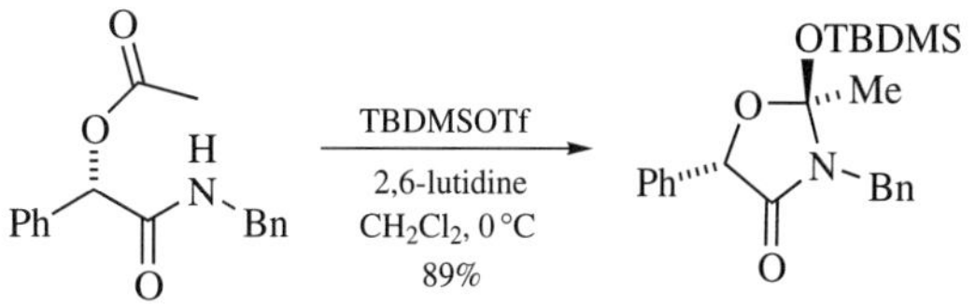

(18)

An intramolecular silylative aldol reaction has been effectively employed as a key step in the synthesis of a variety of carbasugars (eq 19).[14] The TBDMS triflate plays a dual role in this chemistry, forming an enol silane and activating the aldehyde toward nucleophilic attack.

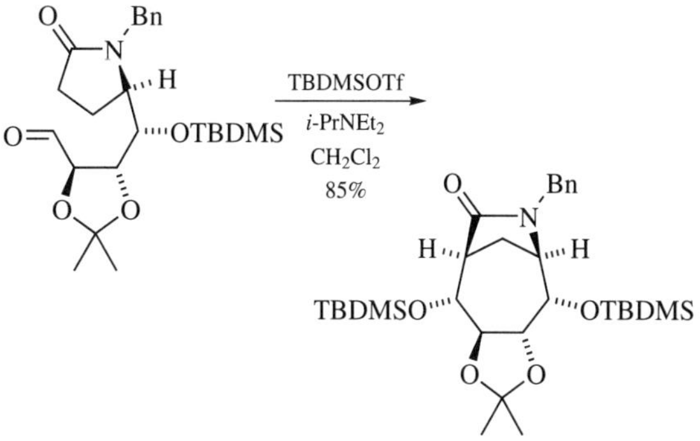

(19)

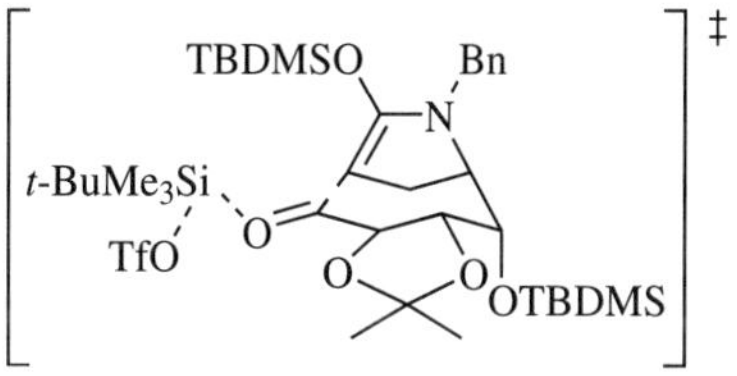

TBDMS triflate-activated carbonyl compounds have also been trapped intramolecularly with carbon–carbon double bonds, such as in the intramolecular Friedel–Crafts alkylation of a furan (eq 20).[15]

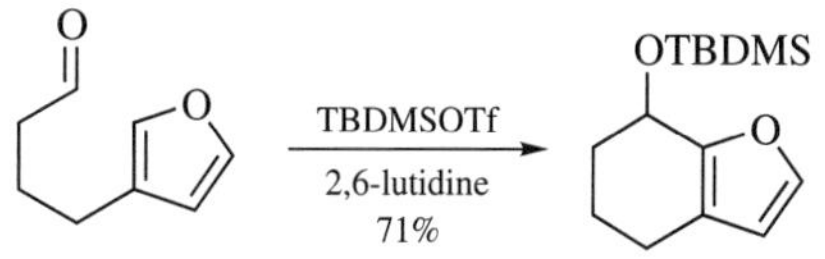

(20)

A highly stereoselective intramolecular Prins double cyclization catalyzed by TBDMS triflate has been reported (eq 21). The more bulky TBDMS triflate was used in preference to TES triflate to discourage competing protection of the tertiary alcohol.[16]

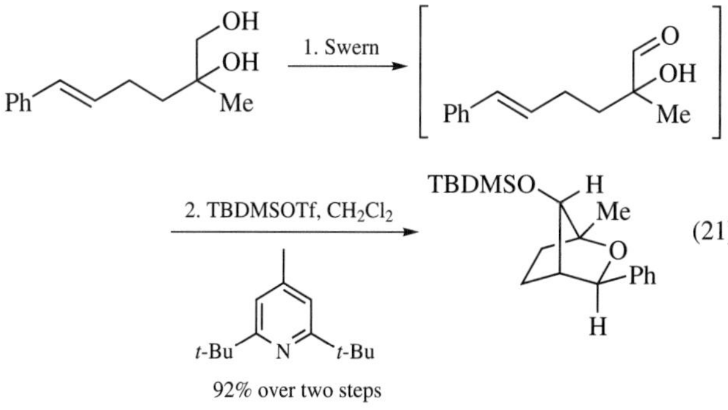

(21)

92% over two steps

Conjugate Addition. TBDMS triflate has been used to promote the conjugate addition of carbon- and heteroatom-based nucleophiles to a range of α,β-unsaturated carbonyl compounds, in both stoichiometric and catalytic quantities. In some cases, the silyl enol ether is isolated, in other cases, it is implied as an intermediate but hydrolyzed either in situ or by addition of an acid or a reagent known to cleave a carbon–silicon bond e.g., TBAF. Examples of carbon–based nucleophiles are shown in eqs 22–26.

TBDMS triflate promotes the addition of organoaluminates, generated in situ from the corresponding organolithium compound and trimethylaluminum, to α,β-unsaturated carbonyl compounds (eq 22).[17] The greater hydrolytic stability of TBDMS enol ethers compared with TMS enol ethers made TBDMS triflate preferable to TMS triflate in this chemistry, although additions to β-disubstituted α,β-unsaturated ketones and α,β-unsaturated esters could not be achieved.

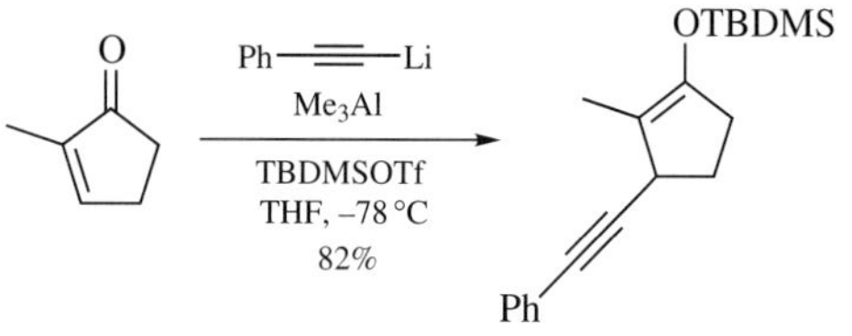

$$(22)$$

The addition of a stoichiometric amount of TBDMS triflate resulted in the best yields and ee values in the asymmetric copper-catalyzed conjugate addition of trimethylaluminum to a cyclohexa-2,5-dienone (eq 23).[18]

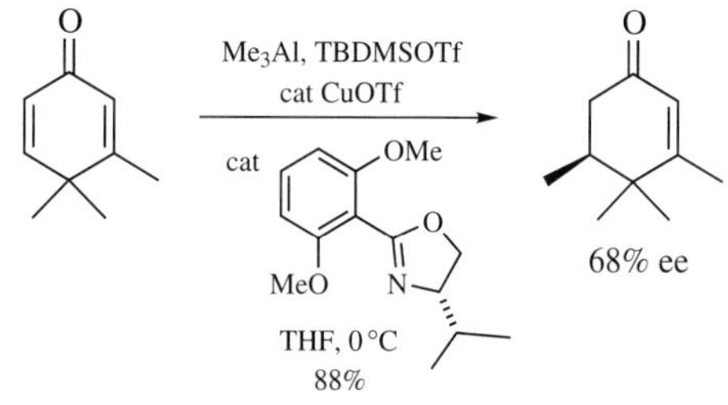

$$(23)$$

TBDMS triflate promotes the conjugate addition of formaldehyde hydrazones to α,β-unsaturated ketones and lactones (eq 24).[19,20] Another bulky silylating agent, dimethyl(1,2,2-trimethylpropyl)silyl (TDS) triflate, has also been used to the same effect with enones.[20]

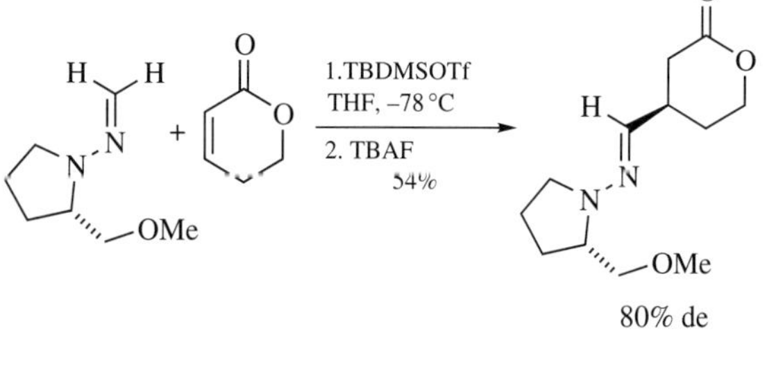

$$(24)$$

TBDMS triflate was the most selective and efficient catalyst for the addition of *t*-butyldimethylsilyl ketene acetal to *N*-tosyl-2-pyridone (eq 25).[21]

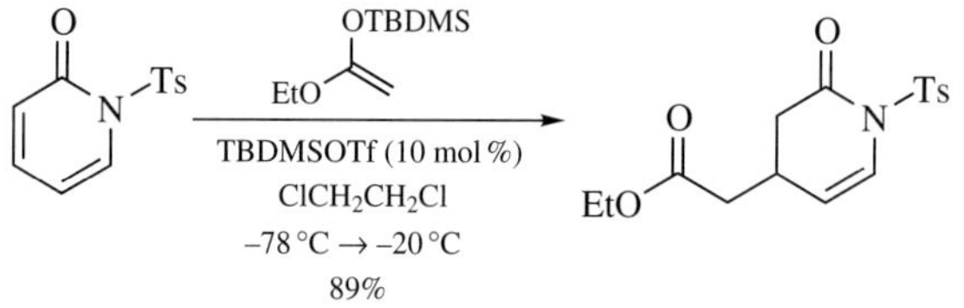

$$(25)$$

The first step in a one-pot formation of a phenanthrene ring system from an enantiomerically pure rhenium-bound naphthalene is a TBDMS triflate-promoted Michael addition to 3-penten-2-one (eq 26). Electron-rich rhenium-bound naphthalenes also undergo TBDMS triflate-promoted conjugate addition reactions to less-activated Michael acceptors such as methyl acrylate, leading to the formal Diels–Alder cycloaddition product of naphthalene with this dienophile.[22]

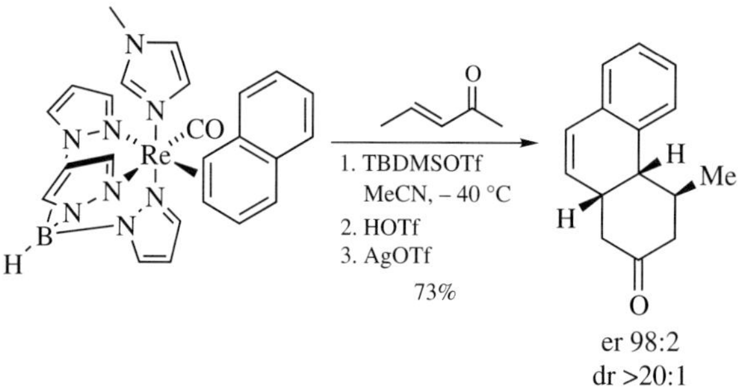

$$(26)$$

The addition of heteroatom nucleophiles to α,β-unsaturated carbonyl compounds has also been promoted using TBDMS triflate. An intermolecular aza-double Michael reaction has been developed as a route to functionalized piperidin-2-ones (eq 27). Due to its relative ease of handling, TBDMS triflate was generally preferred to TMSI–HMDS as an additive in such reactions.[23]

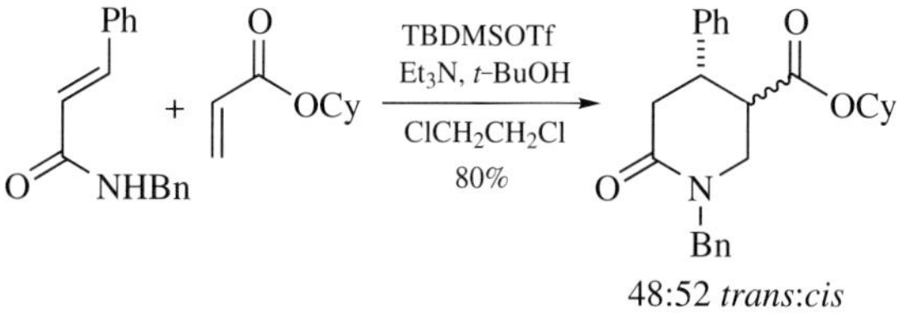

$$(27)$$

The addition of a sulfide and TBDMS triflate to an enone results in the formation of 3-*t*-butyldimethylsilyloxyalk-2-enylsulfonium salts, which have been characterized by low temperature ^{1}H NMR spectroscopy. Although too unstable to be isolated, they undergo nucleophilic displacement reaction with in situ-generated allylindium reagents (eq 28).[24] Such species are also proposed as intermediates in the Morita–Baylis–Hillman reaction of enones with dimethyl acetals (eq 29). The TBDMS triflate may play a dual role in this reaction, both promoting the conjugate addition, and also activating the acetal toward addition of sulfide to create a more active electrophile for the intermolecular addition reaction.[25]

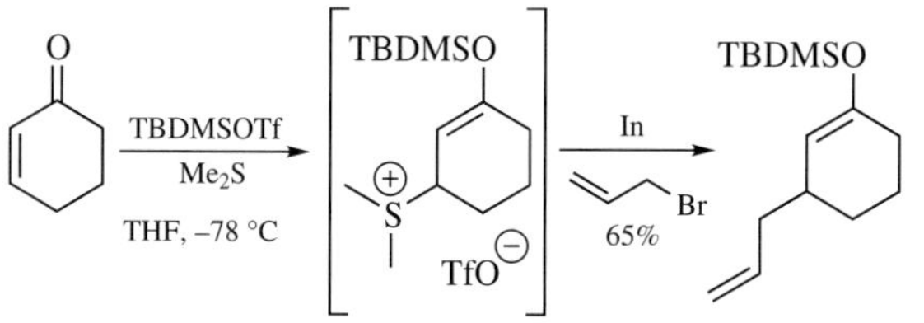

$$(28)$$

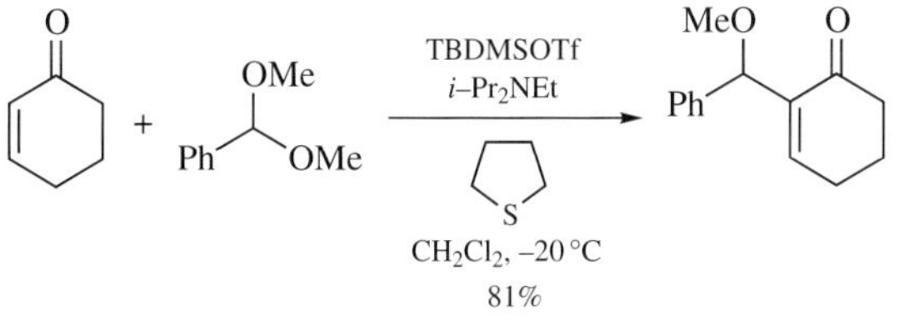

(29)

Acetal Activation.   Activated carbonyl derivatives are also obtained by treatment of acetals with TBDMS triflate. In the case of oxygen nucleophiles, this results in the formation of a new acetal (eqs 30 and 31).[26,27]

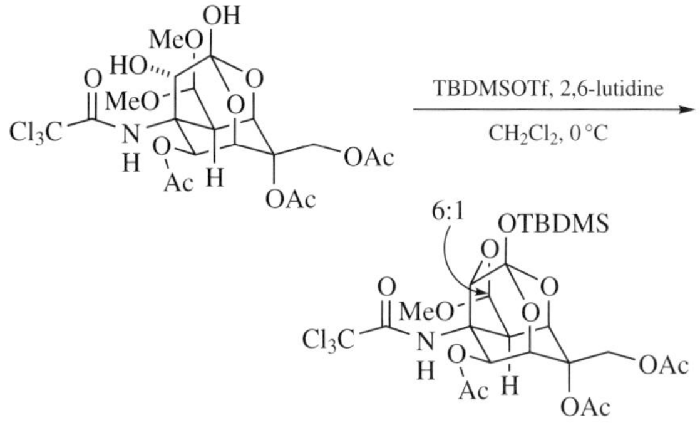

(30)

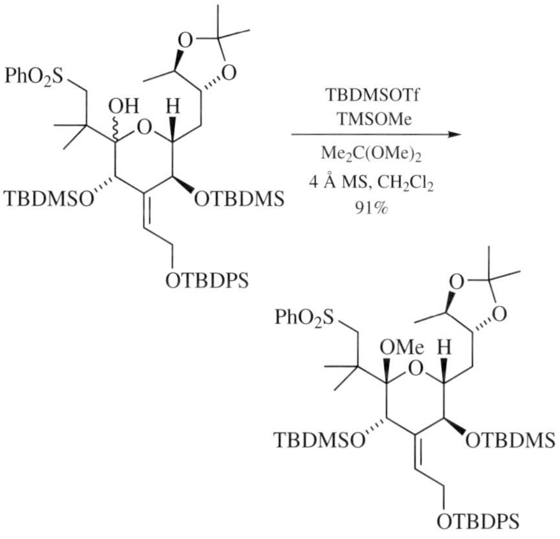

(31)

$\alpha{:}\beta = 1{:}10$

Addition of carbon-based nucleophiles to acetals activated with TBDMS triflate has been reported. In a Mukaiyama aldolization-type reaction, oxygen-, sulfur-, and nitrogen-based heterocyclic silyloxydienes add to chiral electrophilic species derived from the same building blocks (eq 32). All nine possible combinations of nucleophile and electrophile have been achieved.[28]

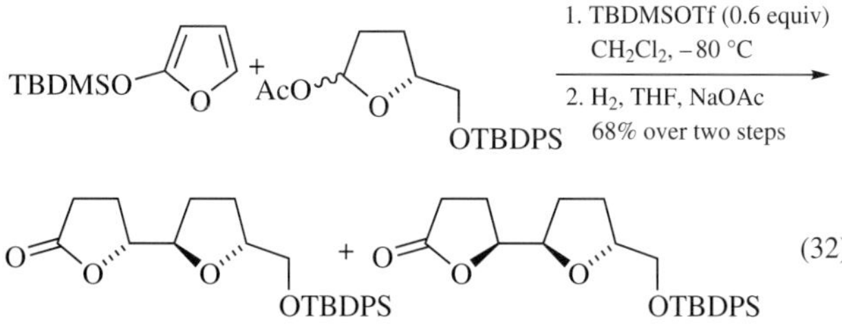

(32)

45:55

Oxonium species derived from the combination of TBDMS triflate and acetals have been used as electrophiles in carbon–carbon bond forming reactions with π-basic transition metal complexes of aromatic rings, such as the rhenium-bound naphthalene in eq 26.[23]

Glycosylation.   Glycosyl trichloroacetimidates undergo glycosidation reactions with a wide range of glycosyl acceptors in a highly stereoselective manner catalyzed by TBDMS triflate at low temperature in CH_2Cl_2 (eq 33).[29,30] Glycosyl acetates[31] and glycosyl phosphates[32] have also been employed as glycosyl donors in TBDMS triflate-mediated glycosylations. The diminished Lewis acidity of TBDMS triflate compared with the more commonly employed TMS triflate can be used advantageously to promote glycosidation reactions of sensitive substrates and minimize formation of unwanted side products.

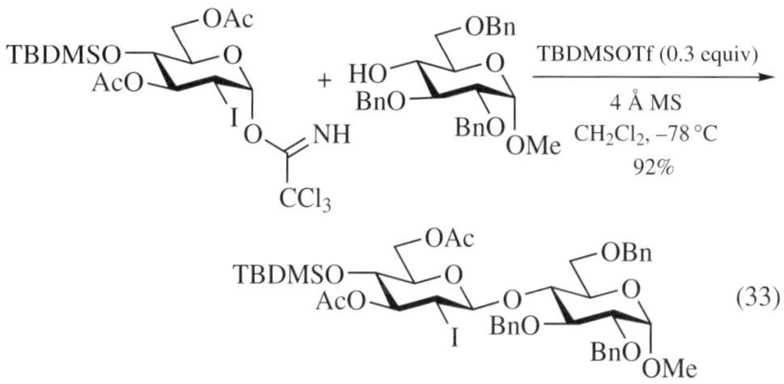

(33)

Epoxide Opening.   Epoxides undergo stereospecific opening reactions when treated with a combination of trimethylaluminum, Et_3N, and TBDMS triflate in CH_2Cl_2 at $-50\,°C$ to give the corresponding alkylation–silylation products in excellent yield (eq 34). TMS triflate and TES triflate are equally effective and have been used in combination with other trialkylaluminum reagents.[33]

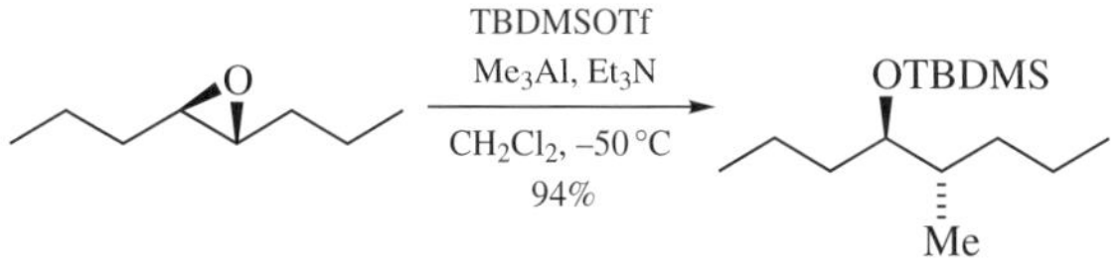

(34)

Treatment of epoxy silyl ethers with TBDMS triflate and base results in a "non-aldol aldol rearrangement" to a silyl enol ether.[34] Although other unhindered silyl triflates can be used, the cleanest reaction mixtures were obtained using TBDMS triflate with TBDMS-protected epoxy alcohols. The resulting silyl enol ether can be used directly in a subsequent Mukaiyama aldol addition reaction (eq 35). In the case of benzaldehyde, *syn*-selectivity is observed. The non-aldol aldol rearrangement is a diastereoselective process, and depending on the relative stereochemistry of the initial epoxy silyl ether and the size of the alkyl group at the protected secondary alcohol, can result in competing migrations and silyl triflate-promoted Payne rearrangements.[35]

Activation of C=N Double Bonds.   TBDMS triflate has been used to promote additions to, and for the isomerization of, a number of systems containing carbon–nitrogen double bonds. TBDMS triflate is the optimal silyl triflate to promote the Mukaiyama-type vinylogous imino-aldol (Mannich-type) addition of 2-[(*t*-butyldimethylsilyl)oxy]furan to the *N*-benzyl imine derived from (2S)-2,3-*O*-isopropylideneglyceraldehyde (eq 36).[36]

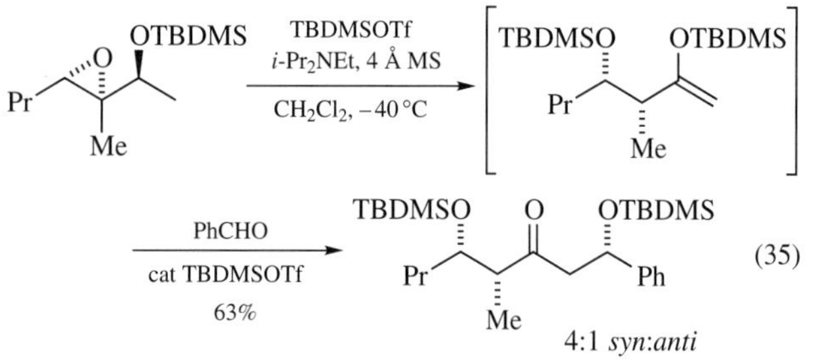

(35)

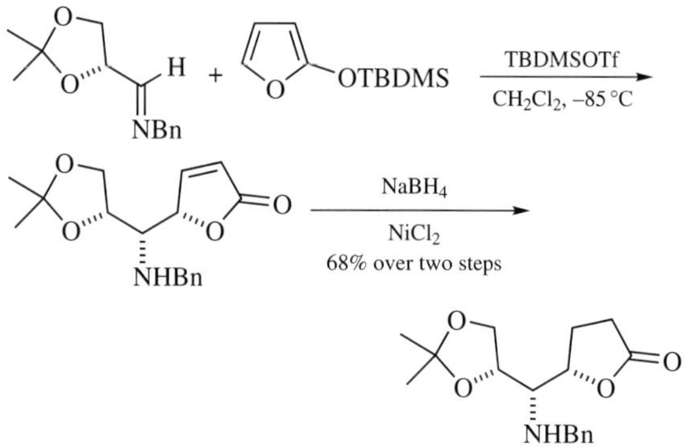

63%

4:1 *syn:anti*

NaBH₄
NiCl₂
68% over two steps

(36)

NHBn

TBDMS triflate is a highly effective promoter for the stereoselective addition of diethylphosphite to a range of chiral *N*-benzyl nitrones (eq 37).[37]

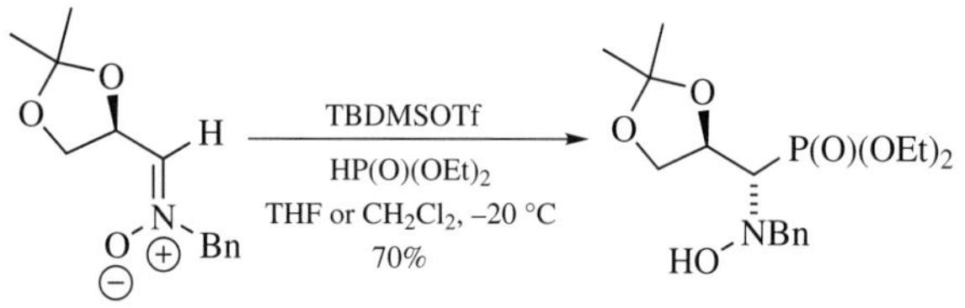

(37)

HP(O)(OEt)₂
THF or CH₂Cl₂, –20 °C
70%

O-Methyl-*O*-*t*-butyldimethylsilyl ketene acetal undergoes a *syn*-selective addition to the same nitrone in the presence of TBDMS triflate (eq 38). *anti*-Selectivity is observed with BF₃·OEt₂.[38] Additions of *O*-methyl-*O*-*t*-butyldimethylsilyl ketene acetal to other chiral nitrones promoted by TBDMS triflate have also been reported.[39]

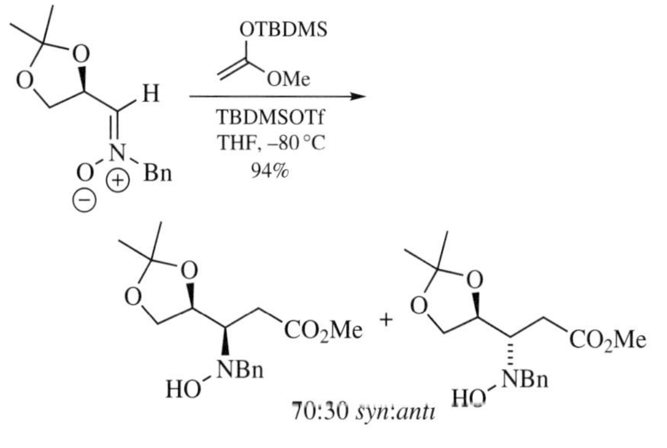

(38)

TBDMSOTf
THF, –80 °C
94%

70:30 *syn:anti*

O-Silylation of 3-alkyl-substituted 1,2-oxazine *N*-oxides (cyclic nitronates) with TBDMS triflate/Et₃N gives α,β-unsaturated nitrosal acetals (nitrosals) which can act as nucleophiles in further bond-forming reactions with oxygen- and nitrogen-substituted carbocations (eq 39). TMS-substituted variants are available in comparable yields using TMSBr/Et₃N, although reaction times are much longer (24 h vs. 1 h for TBDMS triflate). A

TMS-substituted variant has been shown to act as an electrophile toward morpholine, demonstrating that α,β-unsaturated nitrosal acetals can exhibit ambident reactivity.[40]

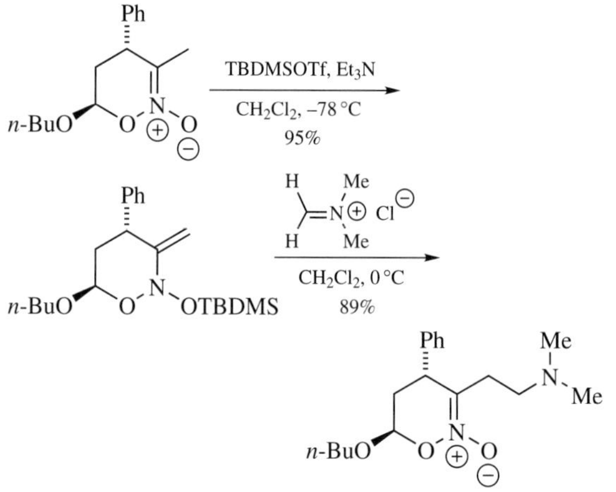

TBDMSOTf, Et₃N
CH₂Cl₂, –78 °C
95%

CH₂Cl₂, 0 °C
89%

(39)

In the absence of Et₃N, the intermediate iminium cation can be captured with nucleophiles such as silyl ketene acetals, silyl enol ethers, enamines, and allyl stannanes (eq 40).[41]

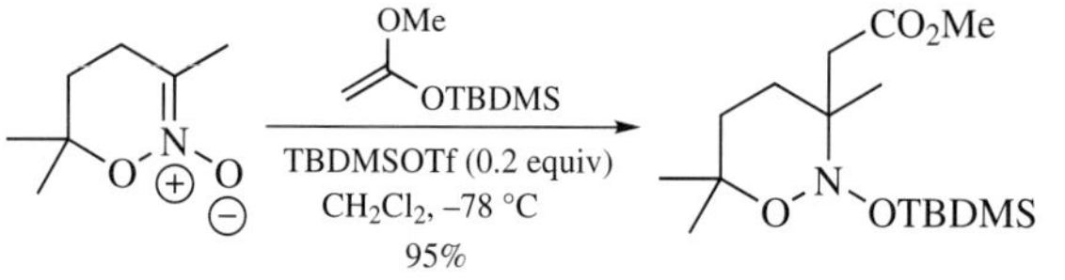

TBDMSOTf (0.2 equiv)
CH₂Cl₂, –78 °C
95%

(40)

N-Silylation. TBDMS is often used as a protecting group for nitrogen, which can, depending on the functional group, be carried through multiple transformations before being removed with a source of fluoride. The combination of TBDMS triflate and a base has been used for the *N*-silylation of indoles,[42] oxazolidin-2-ones,[43] and β-lactams[44] (eqs 41–43).

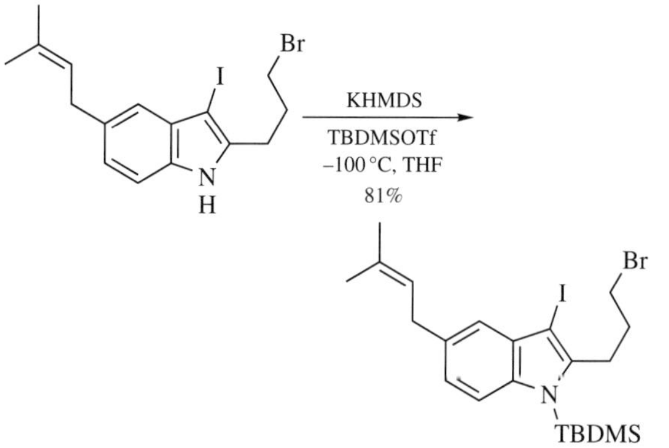

KHMDS
TBDMSOTf
–100 °C, THF
81%

(41)

TBDMSOTf
2,6-lutidine
CH₂Cl₂
55%

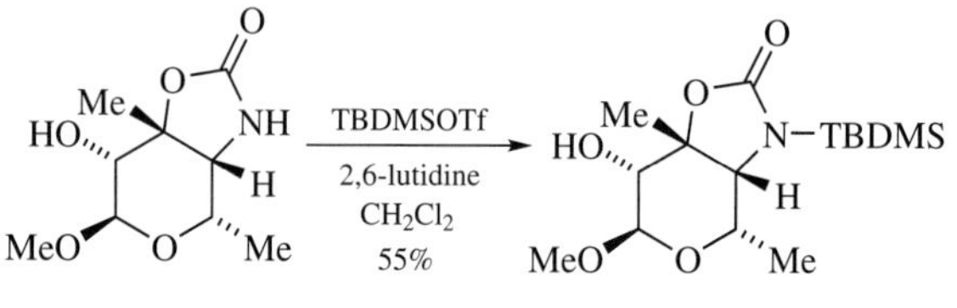

(42)

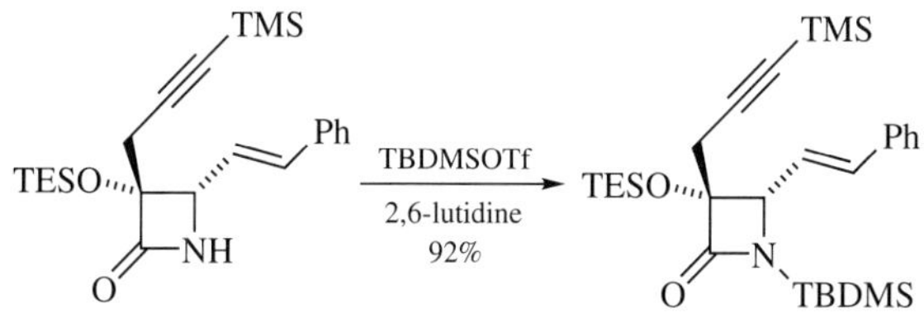

$$(43)$$

$$(46)$$

O-Aryl *N*-isopropylcarbamates are temporarily *N*-protected with TBDMS triflate to form stable intermediates for low-temperature-directed lithiation reactions (eq 44). *N*-Desilylation occurs during aqueous work-up. TBDMS triflate proved superior to TMS triflate for certain biaryl systems.[45]

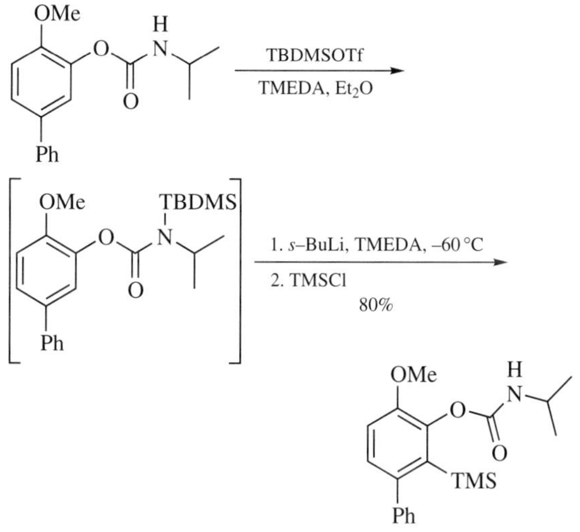

$$(44)$$

Aldehyde tosyl hydrazones are readily *N*-silylated, in quantitative yield, with a combination of TBDMS triflate and Et$_3$N. Although susceptible to hydrolysis on silica gel, these useful synthetic intermediates, which adopt well-defined geometries, can be used directly in carbon–carbon bond-forming processes, for example in reductive coupling reactions with organolithium reagents (eq 45).[46]

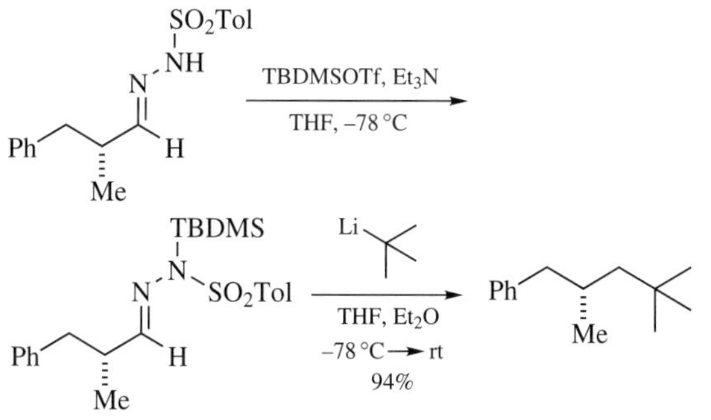

$$(45)$$

C-Silylation.   2-Silyloxazoles can be prepared by deprotonation and quenching the resulting ambient oxazole anion (isocyano enolate) chemoselectively with trialkylsilyl triflates (eq 46). Trialkylsilyl chlorides generally give the product of *O*-silylation.[47]

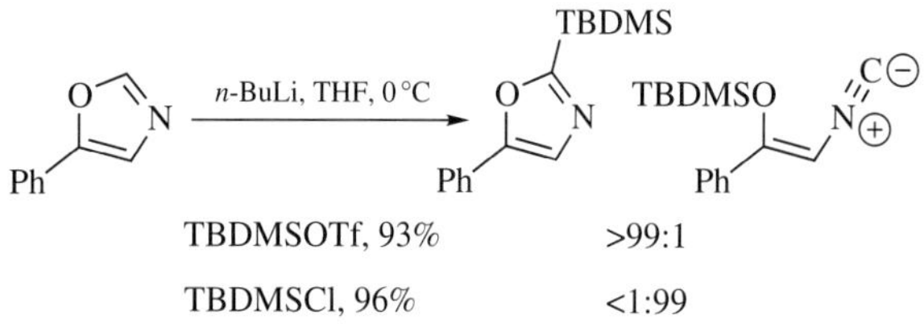

A range of carbanions have been silylated with TBDMS triflate, including allylic,[48] alkynyl,[49] and indenyl systems.[50] Functionalized TBDMS-substituted diazoacetates are prepared by *C*-silylation of unsubstituted diazoacetates using TBDMS triflate in combination with Hünig's base (eq 47).[51] The crude silyl diazoacetate can be directly oxidized without further purification to give silylglyoxylates (eq 48).[52]

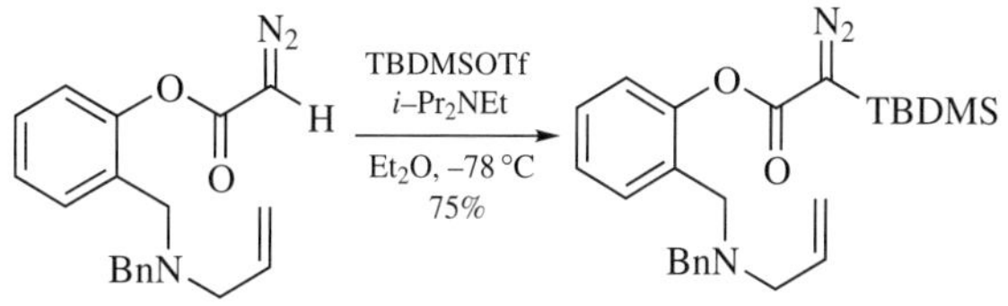

$$(47)$$

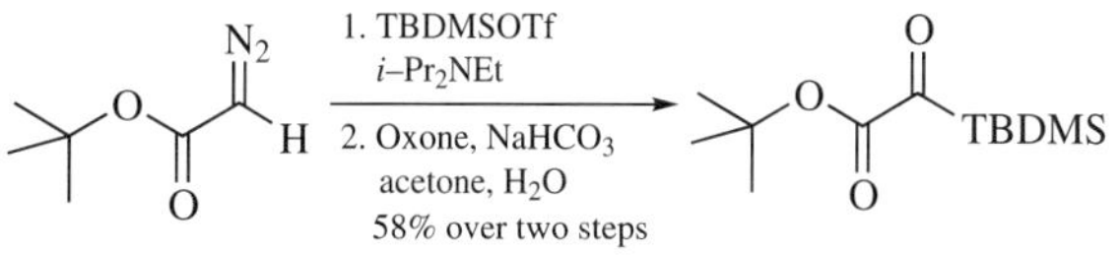

$$(48)$$

Benzannulation of Diaryl-(dichlorocyclopropyl)methanols. Substituted α-arylnaphthalenes are obtained upon treatment of aryl(aryl′)-2,2-dichlorocyclopropylmethanols with Lewis acids (eq 49). Complementary regioselectivity is obtained using silyl triflates and TiCl$_4$, with TBDMS triflate proving superior to TMS triflate in terms of yield and selectivity.[53]

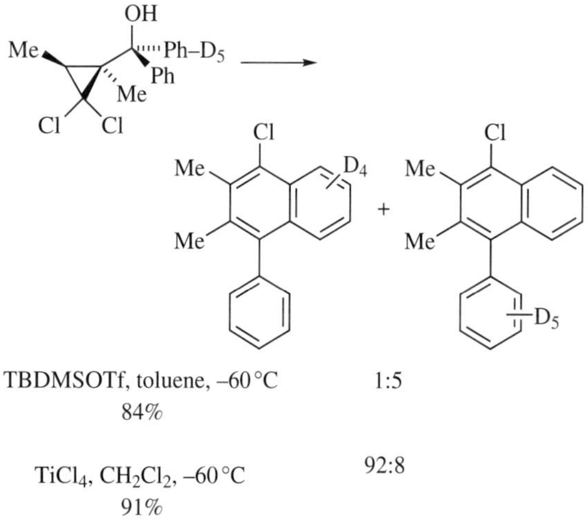

$$(49)$$

SO$_2$ Activation.   At low temperatures, 1-alkoxy-1,3-dienes add to sulfur dioxide activated by TBDMS triflate to generate a zwitterionic intermediate that can be quenched with silyl enol ethers to give β,γ-unsaturated silyl sulfinates. Desilylation and reaction with methyl iodide provide a stereoselective route to polyfunctional sulfones in an overall four-component coupling sequence (eq 50).[54]

$$(50)$$

TBDMS triflate also catalyzes the ene reaction of sulfur dioxide with silyl enol ethers. The silyl sulfinate intermediates can be halogenated to the corresponding sulfonyl halide, and subsequently trapped with an amine or alcohol to give sulfonamides or sulfonate esters, respectively (eq 51).[55]

$$(51)$$

Related Reagents. *t*-Butyldimethylchlorosilane; triethylsilyl trifluoromethanesulfonate; trimethylsilyl trifluoromethanesulfonate; *t*-butyldimethylsilyl iodide; *N-t*-(butyldimethylsilyl)-*N*-methyltrifluoroacetamide; triisopropylsilyl trifluoromethanesulfonate.

1. (a) Stewart, R. F.; Miller, L. L., *J. Am. Chem. Soc.* **1980**, *102*, 4999. (b) Corey, E. J.; Cho, H.; Rucker, C.; Hua, D. H., *Tetrahedron Lett.* **1981**, *22*, 3455. (c) For a review of trialkylsilyl triflates: Emde, H.; Domsch, D.; Feger, H.; Frick, U.; Gotz, A.; Hergott, H. H.; Hofmann, K.; Kober, W.; Krageloh, K.; Oesterle, T.; Steppan, W.; West, W.; Simchen, G., *Synthesis* **1982**, 1.

2. (a) Mander, L. N.; Sethi, S. P., *Tetrahedron Lett.* **1984**, *25*, 5953. (b) Solladie, G.; Mangein, N.; Morreno, I.; Almario, A.; Carreno, M. C.; Garcia-Ruano, J. L., *Tetrahedron Lett.* **1992**, *33*, 4561.

3. Angle, S. R.; Breitenbucker, J. G.; Arnaiz, D. O., *J. Org. Chem.* **1992**, *57*, 5947.

4. Kim, S.; Lee, J. M., *Tetrahedron Lett.* **1990**, *31*, 7627.

5. Kozikowski, A. P.; Jung, S. H., *J. Org. Chem.* **1986**, *51*, 3400.

6. (a) Lee, Y.; Iwasaki, H.; Yamamoto, Y.; Ohkata, K.; Akiba, K., *Heterocycles* **1989**, *29*, 35. (b) Iwasaki, H.; Kume, T.; Yamamoto, Y.; Akiba, K., *Tetrahedron Lett.* **1987**, *28*, 6355.

7. (a) Marshall, J. A.; Welmaker, G. S., *J. Org. Chem.* **1992**, *57*, 7158. (b) Marshall, J. A.; Welmaker, G. S., *Tetrahedron Lett.* **1991**, *32*, 2101. (c) Marshall, J. A.; Welmaker, G. S., *Synlett* **1992**, 537.

8. Akiba, K.; Iseki, Y.; Wata, M., *Tetrahedron Lett.* **1982**, *23*, 3935.

9. (a) Tokitoh, N.; Okazaki, R., *Chem. Lett.* **1984**, 1937. (b) Okazaki, R.; Tokitoh, N., *J. Chem. Soc., Chem. Commun.* **1984**, 192.

10. Sakaitani, M.; Ohfune, Y., *Tetrahedron Lett.* **1985**, *26*, 5543.

11. Semmelhack, M. F.; Jaskowski, M.; Sarpong, R.; Ho, D. M., *Tetrahedron Lett.* **2002**, *43*, 4947.

12. Battistini, L.; Casiraghi, G.; Curti, C.; Rassu, G.; Zambrano, V.; Zanardi, F., *Tetrahedron* **2004**, *60*, 2957.

13. Kamimura, A.; Omata, Y.; Kakehi, A.; Shirai, M., *Tetrahedron* **2002**, *58*, 8763.

14. Rassu, G.; Auzzas, L.; Pinna, L.; Zambrano, V.; Zanardi, F.; Battistini, L.; Gaetani, E.; Curti, C.; Casiraghi, G., *J. Org. Chem.* **2003**, *68*, 5881.

15. Miles, W. H.; Connell, K. B., *Tetrahedron Lett.* **2003**, *44*, 1161.

16. Jung, M. E.; Angelica, S.; D'Amico, D. C., *J. Org. Chem.* **1997**, *62*, 9182.

17. Kim, S.; Park, J. H., *Synlett* **1995**, 163.

18. Takemoto, Y.; Kuraoka, S.; Hamaue, N.; Aoe, K.; Hiramatsu, H.; Iwata, C., *Tetrahedron* **1996**, *52*, 14177.

19. Enders, D.; Vázquez, J.; Raabe, G., *Eur. J. Org. Chem.* **2000**, 893.

20. Díez, E.; Fernández, R.; Gasch, C.; Lassaletta, J. M.; Llera, J. M.; Martín-Zamora, E.; Vázquez, J., *J. Org. Chem.* **1997**, *62*, 5144.

21. Hiroya, K.; Jouka, R.; Katoh, O.; Sakamura, T.; Anzai, M.; Sakamoto, T., *ARKIVOK* **2003**, *(viii)*, 232.

22. Ding, F.; Valahovic, M. T.; Keane, J. M.; Anstey, M. R.; Sabat, M.; Trindle, C. O.; Harman, W. D., *J. Org. Chem.* **2004**, *69*, 2257.

23. Takasu, K.; Nishida, N.; Tomimura, A.; Ihara, M., *J. Org. Chem.* **2005**, *70*, 3957.

24. Lee, K.; Kim, H.; Miura, T.; Kiyota, K.; Kusama, H.; Kim, S.; Iwasawa, N.; Lee, P. H., *J. Am. Chem. Soc.* **2003**, *125*, 9682.

25. Rao, J. S.; Brière, J.-F.; Metzner, P.; Basavaiah, D., *Tetrahedron Lett.* **2006**, *47*, 3553.

26. Nishikawa, T.; Urabe, D.; Isobe, M., *Angew. Chem., Int. Ed.* **2004**, *43*, 4782.

27. Obitsu, T.; Ohmori, K.; Ogawa, Y.; Hosomi, H.; Ohba, S.; Nishiyama, S.; Yamamura, S., *Tetrahedron Lett.* **1998**, *39*, 7349.

28. Zanardi, F.; Battistini, L.; Rassu, G.; Pinna, L.; Mor, M.; Culeddu, N.; Casiraghi, G., *J. Org. Chem.* **1998**, *63*, 1368.

29. Roush, W. R.; Gung, B. W.; Bennett, C. E., *Org. Lett.* **1999**, *1*, 891.

30. Orgueira, H. A.; Bartolozzi, A.; Schell, P.; Litjens, R. E. J. N.; Palmacci, E. R.; Seeberger, P. H., *Chem. Eur. J.* **2003**, *9*, 140.

31. Roush, W. R.; Narayan, S., *Org. Lett.* **1999**, *1*, 899.

32. Plante, O. J.; Palmacci, E. R.; Andrade, R. B.; Seeberger, P. H., *J. Am. Chem. Soc.* **2001**, *123*, 9545.

33. Shanmugam, P.; Miyashita, M., *Org. Lett.* **2003**, *5*, 3265.

34. Jung, M. E.; van den Heuvel, A., *Org. Lett.* **2003**, *5*, 4705.

35. Jung, M. E.; van den Heuvel, A., *Tetrahedron Lett.* **2002**, *43*, 8169.

36. Rassu, G.; Auzzas, L.; Zambrano, V.; Burreddu, P.; Pinna, L.; Battistini, L.; Zanardi, F.; Casiraghi, G., *J. Org. Chem.* **2004**, *69*, 1625.

37. De Risi, C.; Perrone, D.; Dondoni, A.; Pollini, G. P.; Bertolasi, V., *Eur. J. Org. Chem.* **2003**, 1904.

38. Merino, P.; del Alamo, E. M.; Bona, M.; Franco, S.; Merchan, F. L.; Tejero, T.; Vieceli, O., *Tetrahedron Lett.* **2000**, *41*, 9239.

39. Merino, P.; Franco, S.; Merchan, F. L.; Tejero, T., *J. Org. Chem.* **2000**, *65*, 5575.

40. Tishkov, A. A.; Lesiv, A. V.; Khomutova, Y. A.; Strelenko, Y. A.; Nesterov, I. D.; Antipan, M. Yu.; Ioffe, S. L.; Denmark, S. E., *J. Org. Chem.* **2003**, *68*, 9477.

41. Smirnov, V. O.; Ioffe, S. L.; Tishkov, A. A.; Khomutova, Y. A.; Nesterov, I. D.; Antipin, M. Yu.; Smit, W. A.; Tartakovsky, V. A., *J. Org. Chem.* **2004**, *69*, 8485.

42. Smith, A. B., III; Cui, H., *Org. Lett.* **2003**, *5*, 587.

43. Trost, B. M.; Gunzner, J. L., *J. Am. Chem. Soc.* **2001**, *123*, 9449.

44. Banfi, L.; Guanti, G., *Tetrahedron Lett.* **2002**, *43*, 7427.

45. Kauch, M.; Snieckus, V.; Hoppe, D., *J. Org. Chem.* **2005**, *70*, 7149.

46. Myers, A. G.; Movassaghi, M., *J. Am. Chem. Soc.* **1998**, *120*, 8891.

47. Miller, R. A.; Smith, R. M.; Karady, S.; Reamer, R. A., *Tetrahedron Lett.* **2002**, *43*, 935.

48. Myers, A. G.; Sogi, M.; Lewis, M. A.; Arvedson, S. P., *J. Org. Chem.* **2004**, *69*, 2516.

49. Myers, A. G.; Goldberg, S. D., *Angew. Chem., Int. Ed.* **2000**, *39*, 2732.

50. Möller, A. C.; Heyn, R. H.; Blom, R.; Swang, O.; Görbitz, C. H.; Kopf, J., *J. Chem. Soc., Dalton Trans.* **2004**, 1578.

51. Clark, J. S.; Middleton, M. D., *Org. Lett.* **2002**, *4*, 765.

52. Nicewicz, D. A.; Johnson, J. S., *J. Am. Chem. Soc.* **2005**, *127*, 6170.

53. Nishii, Y.; Yoshida, T.; Asano, H.; Wakasugi, K.; Morita, J.-i.; Aso, Y.; Yoshida, E.; Motoyoshiya, J.; Aoyoma, H.; Tanabe, Y., *J. Org. Chem.* **2005**, *70*, 2667.

54. Narkevitch, V.; Megevand, S.; Schenk, K.; Vogel, P., *J. Org. Chem.* **2001**, *66*, 5080.

55. Bouchez, L. C.; Dubbaka, S. R.; Turks, M.; Vogel, P., *J. Org. Chem.* **2004**, *69*, 6413.

t-Butyldiphenylchlorosilane[1]

t-BuPh$_2$SiCl

[58479-61-1]   C$_{16}$H$_{19}$ClSi   (MW 274.86)

InChI = 1S/C16H19ClSi/c1-16(2,3)18(17,14-10-6-4-7-11-14)
15-12-8-5-9-13-15/h4-13H,1-3H3

InChIKey = MHYGQXWCZAYSLJ-UHFFFAOYSA-N

(reagent for the temporary protection of hydroxy groups as their *t*-butyldiphenylsilyl ethers; selectivity can be obtained for primary vs. secondary hydroxy groups; other protection (carbonyl, amine, etc.) is also possible;[2] deprotection is conveniently effected with fluoride ion, strong acid, or base; the TBDPS ether group is compatible with a large number of organic functional group transformations,[1-3] in nucleoside chemistry,[1] as well as in numerous examples in which polyfunctional molecules are chemically manipulated[4])

Alternate Name: TBDPS-Cl.
Physical Data: colorless liquid, bp 93–95 °C/0.015 mmHg; n_D^{20} 1.5680; *d* 1.057 g cm^{-3}.
Solubility: miscible in most organic solvents.
Form Supplied in: colorless liquid 98%.
Preparative Method: a dry 1 L, three-necked round bottomed flask is equipped with a magnetic stirring bar, a 500 mL equalizing dropping funnel fitted with a rubber septum, a reflux condenser, and nitrogen inlet tube. The flask is flushed with nitrogen, then charged with 127 g (0.5 mol) of diphenyldichlorosilane in 300 mL of redistilled pentane. A solution of *t*-butyllithium in pentane (500 mL, 0.55 mol), is transferred under nitrogen pressure to the dropping funnel using a stainless steel, double-tip transfer needle. This solution is slowly added to the contents of the flask and when the addition is complete, the mixture is refluxed 30 h under nitrogen with stirring. The suspension is allowed to cool to rt, the precipitated lithium chloride is rapidly filtered through a pad of Celite, and the latter is washed with 200 mL of pentane. The solvent is removed by evaporation, and the colorless residue is distilled through a short (10 cm), Vigreux column, to give 125–132 g of the colorless title compound.
Handling, Storage, and Precautions: the reagent is stable when protected from moisture and protic solvents. A standard M solution of the reagent can be prepared in anhyd DMF and kept at 0 °C under argon in an amber bottle. Use in a fume hood.

Original Commentary

Stephen Hanessian
University of Montréal, Montréal, Quebéc, Canada

Preparation of *t*-Butyldiphenylsilyl Ethers and Related Transformations. A standard protocol[1] involves the addition of the reagent (1.1 equiv) to a solution of the alcohol (1 equiv) and imidazole (2 equiv) in DMF at 0 °C or at room temperature. When silylation is complete, the solution is diluted with water and ether or dichloromethane and the organic layer is processed in the usual way. In some instances the addition of 4-dimethylaminopyridine (DMAP) can enhance the reaction rate.[5,6] Selective protection of the primary hydroxy group in carbohydrate derivatives has been achieved using this reagent and poly(4-vinylpyridine) in CH$_2$Cl$_2$ or THF in the presence of hexamethylphosphoric triamide.[7] Enol *t*-butyldiphenylsilyl ethers are easily formed by trapping of enolates with the chloride.[8] Amines can be selectively converted into the corresponding primary *t*-butyldiphenylsilylamines essentially under the same conditions.[9] *t*-Butyldiphenylsilyl cyanohydrins are prepared in the presence of potassium cyanide, zinc iodide, and a carbonyl compound.[10]

Selectivity and Compatibility of *O*- and *N*-*t*-Butyldiphenyl Silyl Derivatives. The *O*-*t*-butyldiphenylsilyl ether protective group offers some unique advantages and synthetically useful features compared to other existing counterparts.[1,2] Silylation of a primary hydroxy group takes place in preference to a secondary, giving products that are often crystalline, and detectable on TLC plates under UV light, because of the presence of a strong chromophore. The TBDPS group is compatible with a variety of conditions[1] used in synthetic transformations such as hydrogenolysis (Pd(OH)$_2$, C/H$_2$, etc.), *O*-alkylation (NaH, DMF, halide), de-*O*-acylation (NaOMe, NH$_4$OH, K$_2$CO$_3$/MeOH), mild chemical reduction with hydride reagents, carbon–carbon bond formation with organometallic reagents, transition metal-mediated reactions, Wittig reactions, etc. These reactions were tested during the total synthesis of thromboxane B$_2$[11] for which the TBDPS group was originally designed. Subsequently, numerous related organic transformations have been carried out in the presence of TBDPS ethers in the context of natural product synthesis[3,4] and functional group manipulations.[2] It is distinctly more acid stable than the *O*-trityl, *O*-THP, and *O*-TBDMS groups, and is virtually unaffected under conditions that cause complete cleavage of these groups (50% HCO$_2$H, 2–6 h; 50% aq TFA, 15 min; HBr/AcOH, 0 °C, few min; 80% aq acetic acid, few h).[1] It is also compatible with conditions used for the acid-catalyzed formation and hydrolysis of acetal groups and in the presence of some Lewis acids (e.g. TMSBr,[12] BCl$_3$ · SMe$_2$,[13] Me$_2$AlCl,[8] etc.). The mono-TBDPS derivative of a primary amine[9] is stable to base, as well as to alkylating and acylating reagents. Thus a secondary amine can be acylated in the presence of a primary *t*-butyldiphenylsilylamino group.

Deprotection.[1,2] The *O*-TBDPS group can be cleaved under the following conditions: tetra-*n*-butylammonium fluoride in THF at rt; variations of F$^-$-catalyzed reactions (e.g. F$^-$/AcOH; HF/pyridine;[14] HF/MeCN,[15] etc.); aqueous acids and bases[1] (1 N HCl, 5 N NaOH, aq methanolic NaOH or KOH); ion exchange resins;[7] potassium superoxide, dimethyl sulfoxide, 18-crown-6.[16] Selective cleavage of a TBDPS ether in the presence of a TBDMS ether in carbohydrate derivatives involves treatment with sodium hydride/HMPA at 0 °C.[17] Although the *O*-TBDPS group is stable to most hydride reagents, lithium aluminum hydride reduction of an amide group has resulted in the cleavage[18] of an adjacent *O*-TBDPS ether, possibly via internal assistance. TBDPS-amines are cleaved by 80% AcOH or by HF/pyridine.[9]

First Update

Wenming Zhang
Dupont Crop Protection, Stine-Haskell Research Center, Newark, DE, USA

Protecting Group. Several new methods have been developed for using the reagent to protect primary and secondary alcohols as their TBDPS ethers. In the presence of ammonium nitrate or ammonium perchlorate, reaction between TBDPS-Cl and a primary alcohol, such as benzyl alcohol, in DMF provided excellent yields of the corresponding silyl ethers in just 15 min (eq 1).[19] When silver nitrate was used as promoter, the reactions gave inferior yields under otherwise identical conditions.

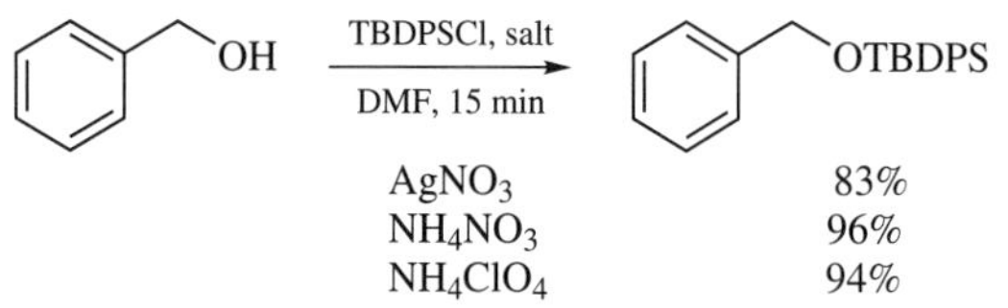

$$\text{(1)}$$

AgNO$_3$	83%
NH$_4$NO$_3$	96%
NH$_4$ClO$_4$	94%

However, when the same protocol was used to protect secondary alcohols, silver nitrate afforded the TBDPS silyl ethers with much shorter reaction times. For example, cyclohexanol was converted to its TBDPS silyl ether in 83% yield in 15 min when silver nitrate was the additive (eq 2). In contrast, it took 24 h for ammonium nitrate or ammonium perchlorate to complete the reaction.

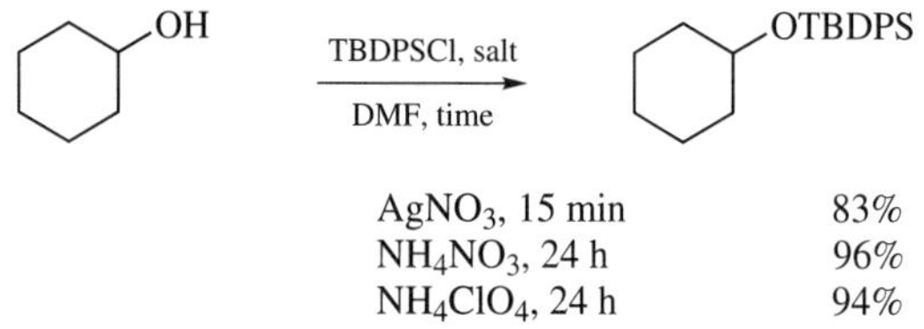

$$\text{(2)}$$

AgNO$_3$, 15 min		83%
NH$_4$NO$_3$, 24 h		96%
NH$_4$ClO$_4$, 24 h		94%

It is proposed that the above reactions are driven by the equilibrium starting with TBDPS-Cl and the inorganic salt, and precipitation of ammonium chloride or silver chloride in DMF (eq 3).

$$\text{TBDPSCl} + \text{NH}_4\text{X} \rightleftharpoons \text{NH}_4\text{Cl} \downarrow + \text{TBDPSX}$$
$$\xrightarrow{\text{ROH}} \text{ROTBDPS} + \text{HX} \qquad \text{(3)}$$

Proazaphosphatrane oxides have been found to catalyze silylation reaction of alcohols to the TBDPS silyl ethers.[20] Thus, benzyl alcohol was converted into benzyl TBDPS ether in 95% yield

when 0.1 equiv of a proazaphosphatrane oxide was added as catalyst (eq 4). Without the catalyst, the reaction provided only 47% of the silyl ether product under the same conditions.

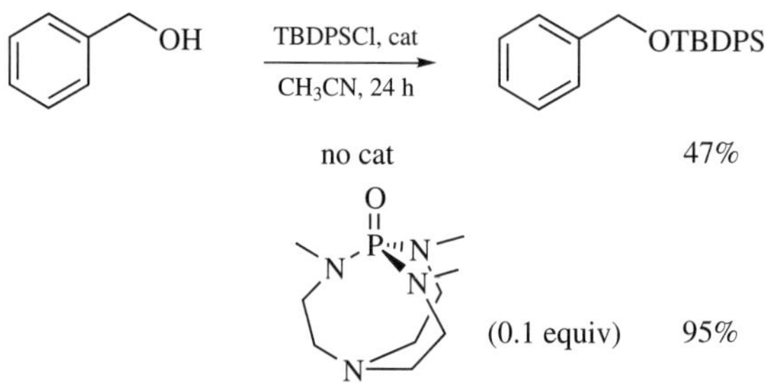

$$\text{(4)}$$

TBDPS-Cl, among other trialkylsilyl chlorides, has been transformed into *O*-trisubstituted silyl benzamide (*Si*-BEZA) which, when combined with pyridinium triflate, proved to be a powerful agent for the silylation of alcohols (eq 5).[21] Even tertiary alcohols, such as 2-methyl-2-pentanol, can be converted into their corresponding TBDPS silyl ethers in good yield (eq 6).

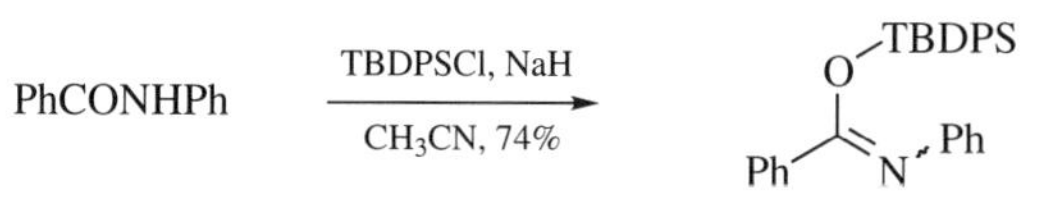

$$\text{(5)}$$

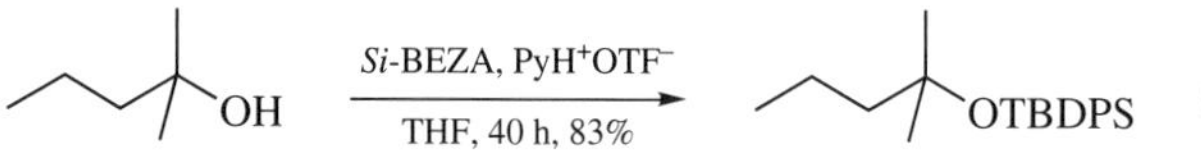

$$\text{(6)}$$

Due to the bulky size of the TBDPS group, TBDPS-Cl has been studied under a number of different circumstances for the selective protection of a single hydroxyl group in various polyols. For example, *t*-butyldiphenylsilylation of *threo*-α,β-dihydroxyphosphonate afforded the silyl ether at the β-position with good regioselectivity (eq 7).[22]

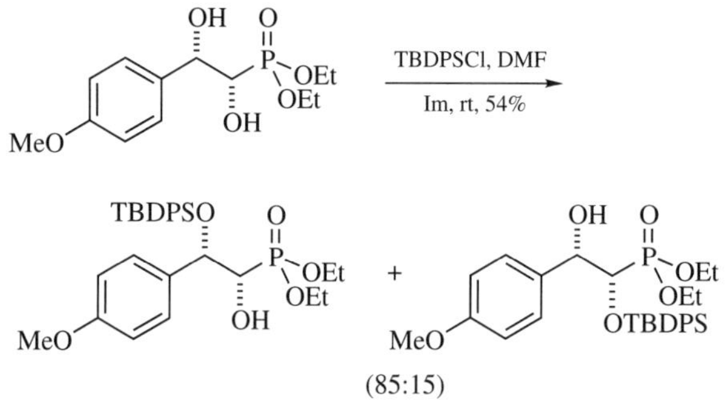

$$\text{(7)}$$

$$(85:15)$$

Selective protection between two different primary hydroxyl groups can also be achieved with TBDPS-Cl. For example, when 2-methyl-1,4-butan-diol was treated with TBDPS-Cl and DBU in DMF at −50 °C for half an hour, it provided two mono-silylated alcohols in good combined yield and excellent selectivity, favoring protection of the less hindered hydroxy group (eq 8).[23]

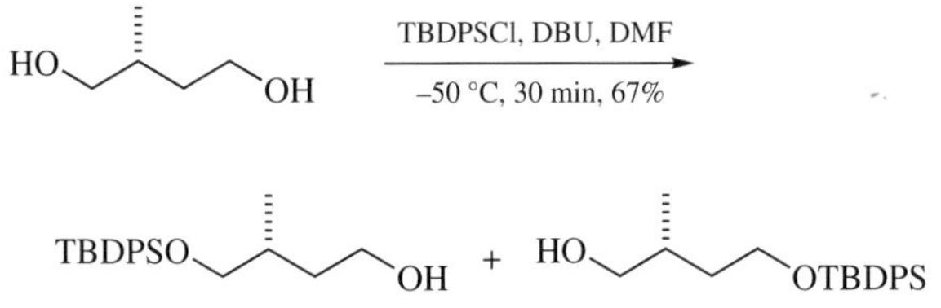

(8)

(1:16)

(10)

Diacetal protected butane tetrol provides another example of selective mono-protection of two primary hydroxyl groups (eq 9).[24] Treatment of this compound with TBDPS-Cl and imidazole in THF gave mainly the axial alcohol in very good yield. In contrast, when the compound was pretreated with sodium hydride, the silylation reaction furnished the equatorial alcohol in excellent yield and good selectivity. Reactions with TBDMS-Cl in place of TBDPS-Cl gave comparable selectivity under the same conditions but gave lower yields. When imidazole was used as base, the preference for equatorial alcohol silylation is due to its higher accessibility. In contrast, the axial alkoxide, formed through deprotonation with sodium hydride, is favored by chelation of the sodium counterion, resulting in the highly selective axial silylation.

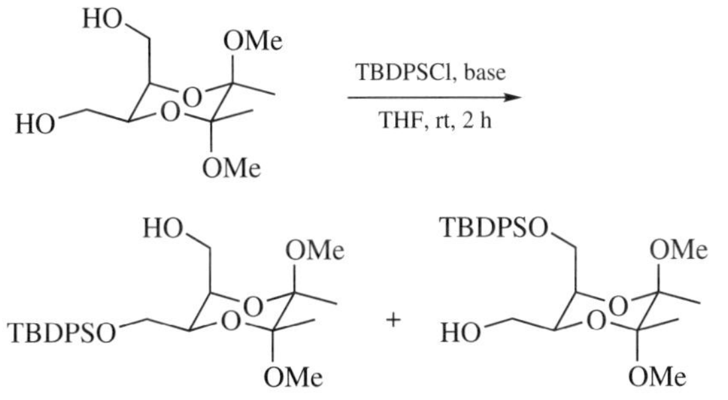

(9)

Im (1.5 equiv), 83%   5:1
NaH (1.0 equiv), 91%   1:9

When TBDPS-Cl is used to react with hemiacetals, it converts hemiacetals into ring-opened silyl ether carbonyl compounds, instead of mixed acetals.[25] Presumably, the sizable TBDPS group presents too much steric hindrance for the formation of the corresponding mixed silyl acetals.

As an efficient protecting group, the use of TBDPS-Cl has also been extended to protect the hydroxy or amino groups in oximes, carboxylic acids, and sulfoximes.[26]

Synthesis of Biaryls. When TBDPS protected 2-bromobenzyl alcohol is subjected to standard radical conditions, the initially formed aryl radical reacts with one phenyl ring of the TBDPS group at its *ipso* position, and the resulting intermediate further transforms into a silyl radical, which then reduced to the corresponding silane.[27] After this phenyl migration, desilylation by methyl lithium eventually provides the biphenyl product (eq 10). Other phenyl silyl ethers of this type can also experience the same aryl migration and produce biaryls.

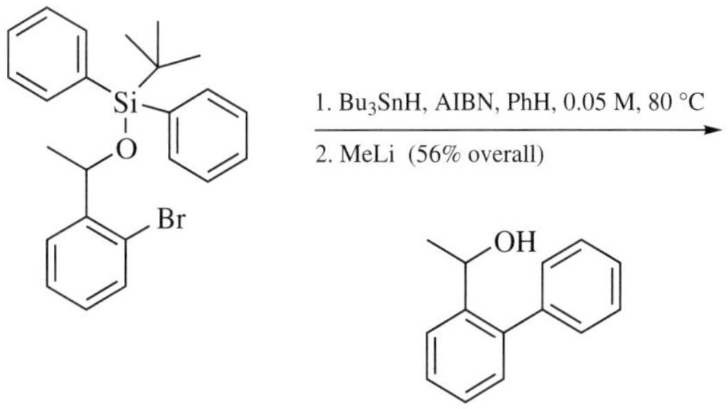

Peterson Olefination. The TBDPS silyl group has been applied in Peterson olefination reactions. The reagent's bulkiness playcd its role twice as shown in the following example, and provided the desired product with high selectivity.[28] α-TBDPS silyl aldehyde can be prepared by a sequence of addition of TBDPS silyl cuprate onto acetylene to form the vinylsilanes, epoxidation of the resulting vinylsilanes, and further acid-catalyzed rearrangement of the epoxide intermediate.[29] When *n*-butyllithium was added to the α-TBDPS aldehyde, it provided the corresponding β-hydroxysilane in very good yield (eq 11). This addition process follows the Felkin-Anh model and offers the *erythro*-β-hydroxysilane exclusively, due to the presence of sterically demanding TBDPS group at the α-position of the aldehyde. When the resulting hydroxysilane was subjected to standard Peterson olefination protocols under either acidic (BF_3) or basic conditions, the desired *Z*- or *E*- disubstituted alkenes were obtained, with excellent yields and very high selectivity, which again is attributed to the bulky TBDPS group (eq 12).

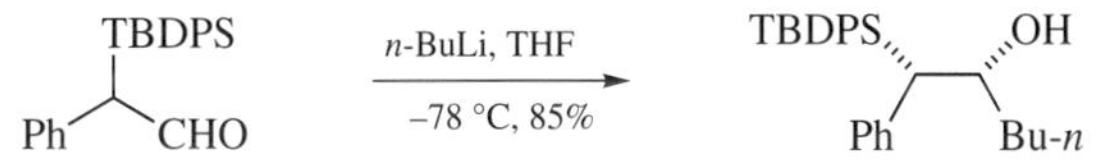

(11)

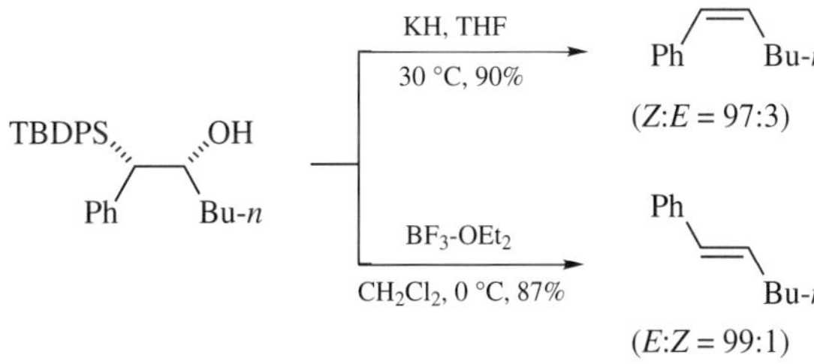

(12)

Retro-Brook Rearrangement. The TBDPS silyl group also found its use in the retro-Brook rearrangement. For example, when the TBDPS silyl ether of trifluoroethanol was treated with excess LDA, it underwent base-promoted elimination of hydrogen fluoride to form 2,2-difluorovinyl TBDPS silyl ether, which was further deprotonated to generate corresponding vinyl anion.[30] This newly generated anion underwent retro-Brook rearrangement to relocate the TBDPS group from oxygen to the anionic carbon, leading to an enolate which, upon treatment with water, furnished the difluoroacetyl TBDPS silane in good yield (eq 13). Various other silyl ethers gave lower yields or no desired product for the same reaction.

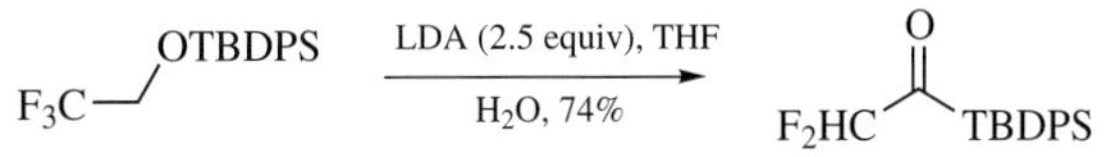

(13)

Pummerer Rearrangement. When a δ-hydroxysulfoxide is treated with TBDPS-Cl and imidazole in DMF, the sulfoxide oxygen coordinates with the TBDPS silicon atom to form the sulfoxonium salt, which further converts into the corresponding sulfonium ion and eliminates a molecule of TBDPS-OH.[31] Intramolecular attack by the δ-hydroxy group furnishes the substituted THF as the sole product in excellent yield (eq 14). In contrast, when the same δ-hydroxysulfoxide is treated with TBDMS-Cl, the reaction offers a mixture of both the rearrangement product and the corresponding TBDMS silyl ether in 2:1 ratio.

| TBDPSCl | 80% | |
| TBDMSCl | 60% | 30% |

1. Hanessian, S.; Lavallée, P., *Can. J. Chem.* **1975**, *53*, 2975; Lavallée, P. Ph.D. Thesis, Université de Montréal, 1977.

2. Greene, T. W.; Wuts, P. G. M. *Protective Groups in Organic Synthesis*, 2nd ed.; Wiley: New York, 1991.

3. Corey, E. J.; Cheng, X.-M. *The Logic of Chemical Synthesis*; Wiley: New York, 1989.

4. Hanessian, S. *The Total Synthesis of Natural Products: The Chiron Approach*; Pergamon: Oxford, 1983.

5. Ireland, R. E.; Obrecht, D. M., *Helv. Chim. Acta* **1986**, *69*, 1273.

6. Chaudhary, S. K.; Hernandez, O., *Tetrahedron Lett.* **1979**, 99.

7. Cardillo, G.; Orena, M.; Sandri, S.; Tomasini, C., *Chem. Ind. (London)* **1983**, 643.

8. Horiguchi, Y.; Suehiro, I.; Sasaki, A.; Kuwajima, I., *Tetrahedron Lett.* **1992**, *34*, 6077.

9. Overman, L. E.; Okazaki, M. E.; Mishra, P., *Tetrahedron Lett.* **1986**, *27*, 4391.

10. Duboudin, F.; Cazeau, P.; Moulines, F.; Laporte, O., *Synthesis* **1982**, 212.

11. Hanessian, S.; Lavallée, P., *Can. J. Chem.* **1981**, *59*, 870.

12. Hanessian, S.; Delorme, D.; Dufresne, Y., *Tetrahedron Lett.* **1984**, *25*, 2515.

13. Congreve, M. S.; Davison, E. C.; Fuhry, M. A. M.; Holmes, A. B.; Payne, A. N.; Robinson, R. A.; Ward, S. E., *Synlett* **1993**, 663.

14. Nicolaou, K. C.; Seitz, S. P.; Pavia, M. R.; Petasis, N. A., *J. Org. Chem.* **1979**, *44*, 4011.

15. Ogawa, Y.; Nunomoto, M.; Shibasaki, M., *J. Org. Chem.* **1986**, *51*, 1625.

16. Torisawa, Y.; Shibasaki, M.; Ikegami, S., *Chem. Pharm. Bull.* **1983**, *31*, 2607.

17. Shekhani, M. S.; Khan, K. M.; Mahmood, K.; Shah, P. M.; Malik, S., *Tetrahedron Lett.* **1990**, *31*, 1669.

18. Rajashekhar, B.; Kaiser, E. T., *J. Org. Chem.* **1985**, *50*, 5480.

19. Hardinger, S. A.; Wijaya, N., *Tetrahedron Lett.* **1993**, *34*, 3821.

20. Liu, X.; Verkade, J. G., *Heteroatom Chem.* **2001**, *12*, 21.

21. Misaki, T.; Kurihara, M.; Tanabe, Y., *Chem. Commun.* **2001**, 2478.

22. Yokomatsu, T.; Suemune, K.; Yamagishi, T.; Shibuya, S., *Synlett.* **1995**, 847.

23. Lautens, M.; Stammers, T. A., *Synthesis* **2002**, 1993.

24. Dixon, D. J.; Foster, A. C.; Ley, S. V.; Reynolds, D. J., *J. Chem. Soc., Perkin Trans. 1* **1999**, 1635.

25. (a) Taishi, T.; Takechi, S.; Mori, S., *Tetrahedron Lett.* **1998**, *39*, 4347. (b) Sibley, R.; Hatoum-Mokdad, H.; Schoenleber, R.; Musza, L.; Stirtan, W.; Marrero, D.; Carley, W.; Xiao, H.; Dumas, J., *Bioorg. Med. Chem. Lett.* **2003**, *13*, 1919.

26. (a) Pyne, S. G.; Dong, Z.; Skelton, B. W.; White, A. H., *J. Chem. Soc., Perkin Trans. 1* **1994**, 2607. (b) Holzapfel, C. W.; Crous, R.; Greyling, H. F.; Verdoorn, G. H., *Heterocycles* **1999**, *51*, 2801. (c) Thyrann, T.; Lightner, D. A., *Tetrahedron* **1996**, *52*, 447.

27. Studer, A.; Bossart, M.; Vasella, T., *Org. Lett.* **2000**, *2*, 985.

28. Barbero, A.; Blanco, Y.; García, C.; Pulido, F. J., *Synthesis* **2000**, 1223.

29. Barbero, A.; Cuadrado, P.; Fleming, I.; González, A. M.; Pulido, F. J.; Sánchez, A., *J. Chem. Soc., Perkin Trans. 1* **1995**, 1525.

30. Higashiya, S.; Chung, W. J.; Lim, D. S.; Ngo, S. C.; Kelly, W. H., IV; Toscano, P. J.; Welch, J. T., *J. Org. Chem.* **2004**, *69*, 6323.

31. Raghavan, S.; Rajender, A.; Rasheed, M. A.; Reddy, S. R., *Tetrahedron Lett.* **2003**, *44*, 8253.

1-*t*-Butylthio-1-*t*-butyldimethyl-silyloxyethylene[1]

[121534-52-9] $C_{12}H_{26}OSSi$ (MW 246.48)

InChI = 1S/C12H26OSSi/c1-10(14-11(2,3)4)13-15(8,9)12(5,6)7/h1H2,2-9H3

InChIKey = CNXNDSMRZZONKQ-UHFFFAOYSA-N

(excellent reagent for Lewis acid-mediated aldol-type reactions,[2] under either stoichiometric[3c–e] or catalytic conditions[3a,b,4–6])

Physical Data: colorless liquid; bp 145 °C/20 mmHg.

Solubility: sol *n*-pentane, diethyl ether, dichloromethane, etc.

Analysis of Reagent Purity: (NMR) ^{1}H 0.20 (s, 6H), 0.96 (s, 9H), 1.39 (s, 9H), 4.68 (m, 2H) ppm.

Preparative Method: the reagent is obtained by reaction of *t*-butyl thioacetate with lithium diisopropylamide in THF/HMPT and subsequent trapping with *t*-butyldimethylchlorosilane (75% yield).[2]

Handling, Storage, and Precautions: the reagent should be stored in the absence of moisture at −15 °C.

Original Commentary

Cesare Gennari
Università di Milano, Milan, Italy

Enantioselective Aldol Additions.[1] The reagent undergoes Lewis acid-catalyzed Mukaiyama-type additions to aldehydes to give β-hydroxy thioesters with good yields and remarkable enantioselectivity (eq 1).[3a,b] Slow addition of the aldehyde (3–20 h) is necessary for high enantioselectivity.[3a]

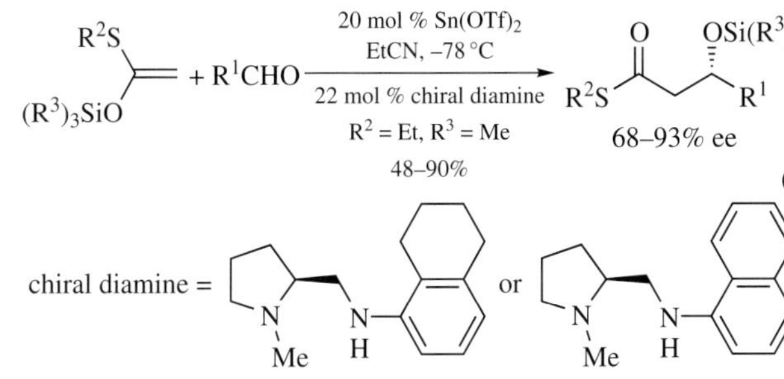

$$R^2S\!\!=\!\!CH_2\ (R^3)_3SiO \quad + \quad R^1CHO \xrightarrow[\substack{22\ mol\%\ chiral\ diamine \\ R^2 = Et,\ R^3 = Me \\ 48\text{--}90\%}]{\substack{20\ mol\%\ Sn(OTf)_2 \\ EtCN,\ -78\,^\circ C}} \quad R^2S\!\!-\!\!CH_2\!\!-\!\!CH(OSi(R^3)_3)R^1 \quad 68\text{--}93\%\ ee \tag{1}$$

chiral diamine =

This reaction is the catalytic version of the stoichiometric reaction outlined in eq 2.[3c,d]

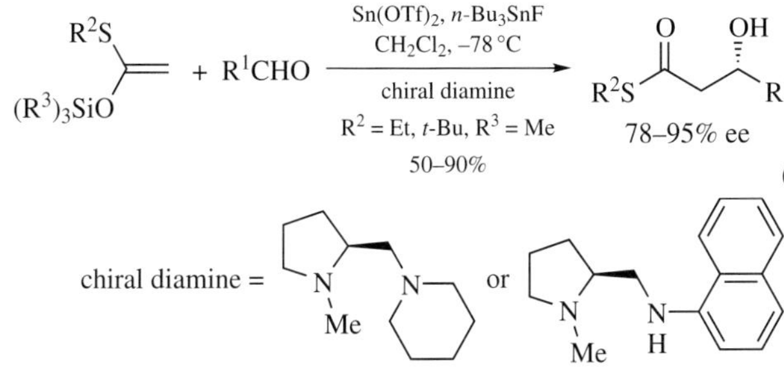

$$R^2S\!\!=\!\!CH_2\ (R^3)_3SiO \quad + \quad R^1CHO \xrightarrow[\substack{chiral\ diamine \\ R^2 = Et,\ t\text{-}Bu,\ R^3 = Me \\ 50\text{--}90\%}]{\substack{Sn(OTf)_2,\ n\text{-}Bu_3SnF \\ CH_2Cl_2,\ -78\,^\circ C}} \quad R^2S\!\!-\!\!CH_2\!\!-\!\!CH(OH)R^1 \quad 78\text{--}95\%\ ee \tag{2}$$

chiral diamine =

Under the same conditions as eq 2, the reagent adds to α-keto esters to give the corresponding 2-substituted malates in good yield (74–81%) and excellent enantiomeric excesses (92–98%).[3e]

Other catalytic asymmetric aldol reactions are shown in eq 3. Slow addition of the aldehyde is necessary for catalyst B[5] but not for A.[4]

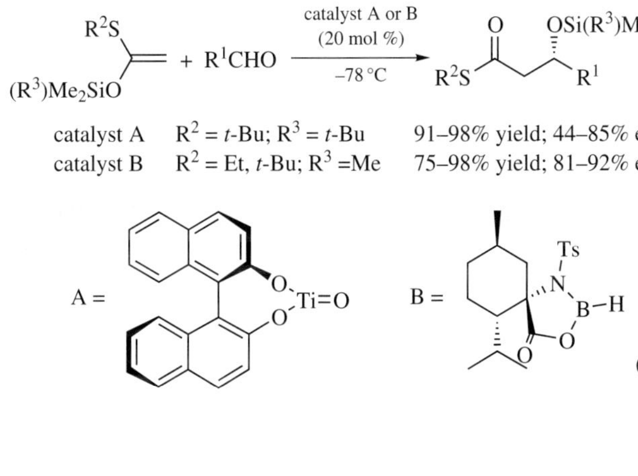

$$R^2S\!\!=\!\!CH_2\ (R^3)Me_2SiO \quad + \quad R^1CHO \xrightarrow[-78\,^\circ C]{\substack{catalyst\ A\ or\ B \\ (20\ mol\%)}} \quad R^2S\!\!-\!\!CH_2\!\!-\!\!CH(OSi(R^3)Me_2)R^1 \tag{3}$$

catalyst A $R^2 = t$-Bu; $R^3 = t$-Bu   91–98% yield; 44–85% ee
catalyst B $R^2 = $ Et, t-Bu; $R^3 =$ Me   75–98% yield; 81–92% ee

A =   B =

Under either the catalytic (eq 1) or the stoichiometric conditions (eq 2), the reagent undergoes addition to chiral aldehydes with complete 'reagent control', i.e. the stereochemistry of the aldol reaction is totally controlled by the chiral catalyst regardless of the inherent diastereofacial preference of the chiral aldehydes (eq 4).[6]

Titanium(IV) chloride and tin(IV) chloride mediate the addition of the title reagent to chiral α-alkoxy aldehydes[2,7] and β-alkoxy aldehydes[2] with complete chelation control (eq 5), whereas the corresponding silyl ketene acetal is unselective.[1,2]

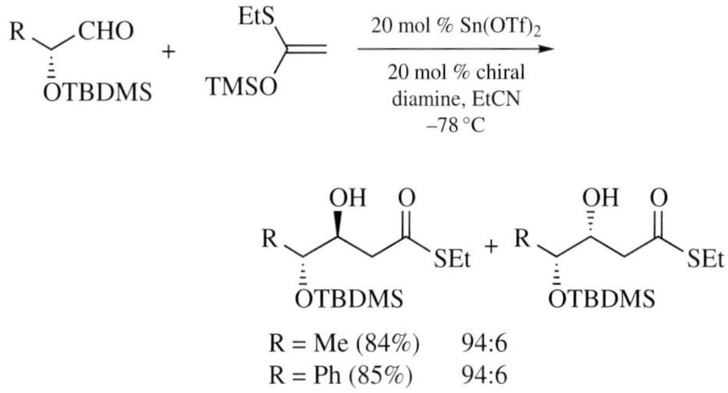

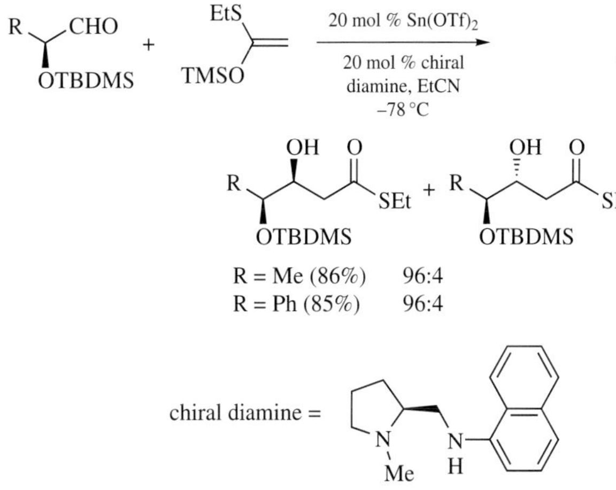

$$R\!\!-\!\!CH(OTBDMS)CHO \quad + \quad EtS\!\!=\!\!CH_2\ TMSO \xrightarrow[\substack{-78\,^\circ C}]{\substack{20\ mol\%\ Sn(OTf)_2 \\ 20\ mol\%\ chiral \\ diamine,\ EtCN}}$$

R = Me (84%)   94:6
R = Ph (85%)   94:6

$$R\!\!-\!\!CH(OTBDMS)CHO \quad + \quad EtS\!\!=\!\!CH_2\ TMSO \xrightarrow[\substack{-78\,^\circ C}]{\substack{20\ mol\%\ Sn(OTf)_2 \\ 20\ mol\%\ chiral \\ diamine,\ EtCN}} \tag{4}$$

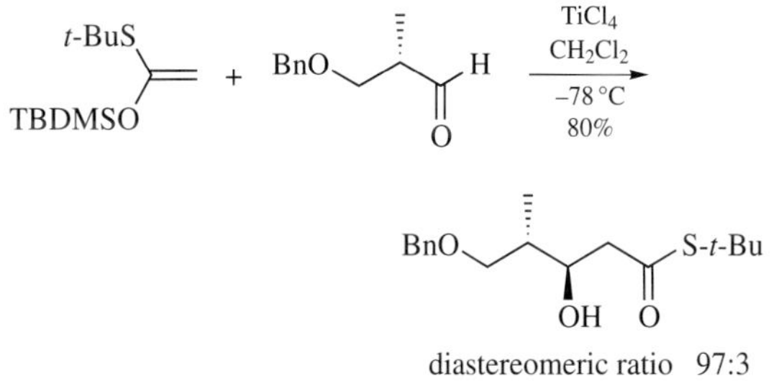

R = Me (86%)   96:4
R = Ph (85%)   96:4

chiral diamine =

$$t\text{-}BuS\!\!=\!\!CH_2\ TBDMSO \quad + \quad BnO\!\!-\!\!CH_2CH(CH_3)CHO \xrightarrow[\substack{-78\,^\circ C \\ 80\%}]{\substack{TiCl_4 \\ CH_2Cl_2}}$$

$$BnO\!\!-\!\!CH_2CH(CH_3)CH(OH)CH_2\!\!-\!\!CO\!\!-\!\!S\text{-}t\text{-}Bu \tag{5}$$

diastereomeric ratio 97:3

Addition to Other Electrophiles. Trimethylsilyl trifluoromethanesulfonate catalyzes the addition to a chiral iminium ion with good yield and complete stereoselectivity (eq 6).[8]

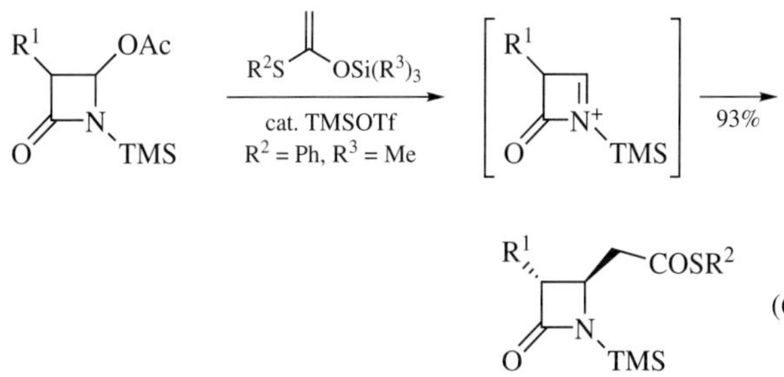

$$\underset{O}{\overset{R^1\ OAc}{\big|}}\!\!N\text{-}TMS \quad \xrightarrow[\substack{R^2 = Ph,\ R^3 = Me}]{\substack{R^2S\!\!=\!\!CH_2\ OSi(R^3)_3 \\ cat.\ TMSOTf}} \quad \left[\ \underset{O}{\overset{R^1}{\big|}}\!\!N^+\text{-}TMS\ \right] \xrightarrow{93\%}$$

$$\underset{O}{\overset{R^1}{\big|}}\!\!CH_2COSR^2 \atop N\text{-}TMS \tag{6}$$

Chlorotrimethylsilane–InCl$_3$ catalyzes the addition to O-trimethylsilyl monothioacetals to afford the corresponding sulfide derivatives in good yield.[9] Titanium tetrachloride mediates the addition of the title reagent to a chiral acetal with high diastereoselectivity (eq 7).[10]

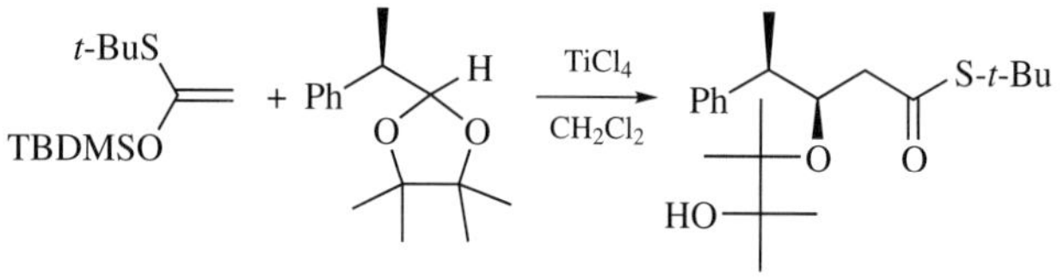

$$racemate \quad (7)$$
diastereomeric ratio 91:9

First Update

Paul R. Blakemore

Oregon State University, Corvallis, OR, USA

Aldol Reactions. The title reagent continues to find application in Mukaiyama-type aldol reactions and excellent stereoselectivity is obtainable in both substrate-controlled and reagent-controlled scenarios. Stannic chloride catalyzed addition to an aldehyde with an α-benzyloxy substituent proceeded with chelation control to afford the 3,4-*syn* product exclusively (eq 8).[11] High levels of diastereocontrol (Felkin–Anh) were also observed in Lewis acid catalyzed additions to α-methylated aldehydes; however, in this case, the triisopropylsilyl congener of the reagent gives significantly higher stereoselectivity.[12] Other simple aldol additions have been explored in a limited fashion from polymer-supported aldehydes.[13]

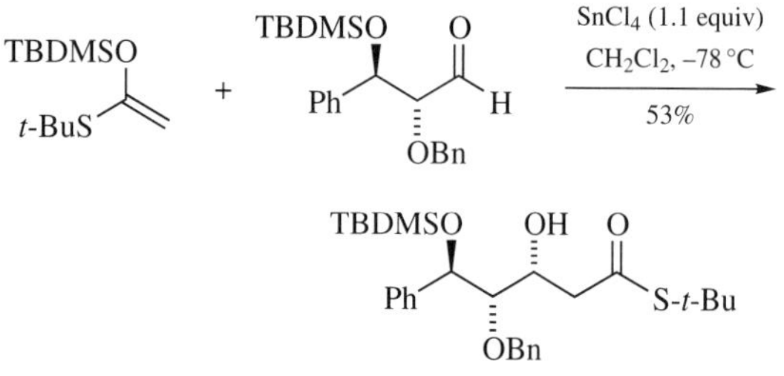

only isolated product

Recently, a novel chiral sulfoxime ligand, the result of a modular design principle by Bolm and coworkers, was demonstrated to provide effective stereoinduction in Cu(II)-catalyzed aldol additions of the reagent to pyruvate esters (eq 9).[14] Comparable results were obtained from the simpler less encumbered reagent 1-*t*-butylthio-1-trimethylsilyloxyethene.

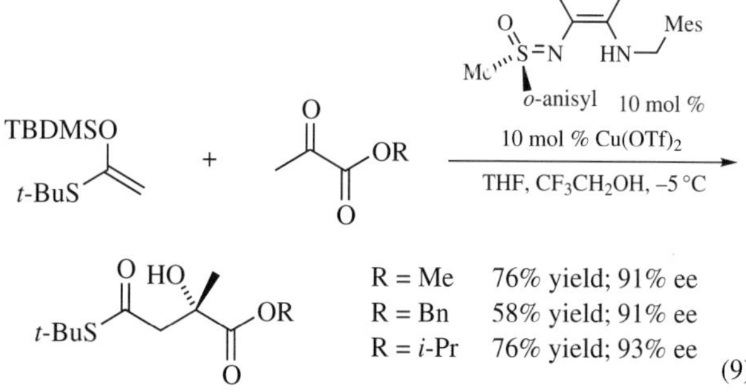

R = Me	76% yield; 91% ee
R = Bn	58% yield; 91% ee
R = *i*-Pr	76% yield; 93% ee

$$(9)$$

Mannich Reactions. Advances in asymmetric Mannich methodology have seen the reagent employed for the catalytic enantioselective synthesis of β-aminothioesters. Addition of the reagent to *N*-(2-thienyl)sulfonyl aldimines catalyzed by a readily prepared planar chiral Cu(I) sulfenylphosphinoferrocene complex gave the expected products with moderate to excellent enantioselectivity (eq 10).[15] The thienyl moiety is essential for acceptable results and free β-aminothioesters may be obtained from the initial addition adducts in high yield (>80%) by deprotection with Mg/MeOH. *N*-Acyl imines derived from glyoxalate esters have also been used as electrophiles in related Cu(I)-catalyzed Mannich processes utilizing chiral 1,2-diamine ligands.[16]

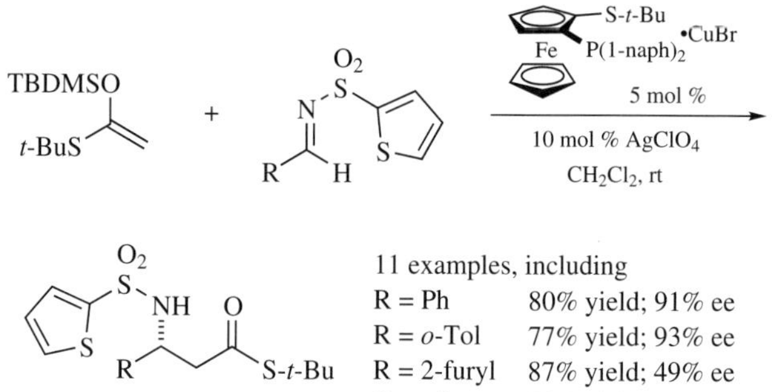

11 examples, including	
R = Ph	80% yield; 91% ee
R = *o*-Tol	77% yield; 93% ee
R = 2-furyl	87% yield; 49% ee
R = *c*-hex	70% yield; 76% ee

$$(10)$$

Michael Reactions. Several reports highlighting utilization of the title reagent in Mukaiyama-type Michael reactions have now appeared. 3,4-Dihydro-α-pyrones are obtained in quantitative yield from the reagent and α,β-unsaturated ketones via a two-step sequence comprising conjugate addition followed by mercuric ion mediated cyclization (eq 11).[17]

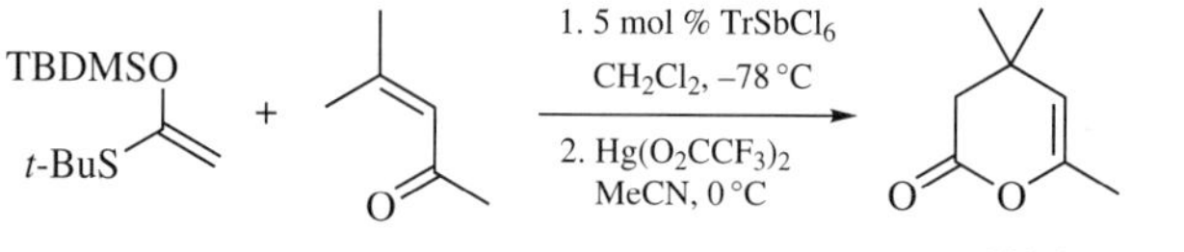

100%

$$(11)$$

Enantioselective 1,4-addition of the reagent to enones catalyzed by chiral oxazaborolidine Lewis acids has been described by Harada and coworkers (eq 12).[18] An undesired '*t*-BuMe$_2$Si$^+$' catalyzed pathway leads to the racemic silyl enol ether by-product: this behavior is suppressed by using 1-*t*-butylthio-1-trimethyl-silyloxyethene as an alternate silyl ketene *S,O*-acetal in the presence of a suitable silyl cation trapping agent (2,6-diisopropyl-phenol). Catalytic enantioselective Michael additions employing the title reagent and a chiral titanium oxide complex have also been described.[19]

RO + Ph + O →

R = TBDMS   6% yield; 83% ee   71% yield; 0% ee

R = TMS*   88% yield; 79% ee   <10% yield

* with 2,6-diisopropylphenol additive   (12)

Related Reagents. For comparison and more details, see also those entries that deal with other silyl ketene acetals, in particular 1-*t*-butylthio-1-*t*-butyldimethylsilyloxypropene, ketene *t*-butyldimethylsilyl methyl acetal, and analogs.

1. Gennari, C., *Comprehensive Organic Synthesis* **1991**, *2*, 629.

2. Gennari, C.; Cozzi, P. G., *Tetrahedron* **1988**, *44*, 5965.

3. (a) Kobayashi, S.; Furuya, M.; Ohtsubo, A.; Mukaiyama, T., *Tetrahedron: Asymmetry* **1991**, *2*, 635. (b) Mukaiyama, T.; Furuya, M.; Ohtsubo, A.; Kobayashi, S., *Chem. Lett.* **1991**, 989. (c) Kobayashi, S.; Mukaiyama, T., *Chem. Lett.* **1989**, 297. (d) Kobayashi, S.; Uchiro, H.; Fujishita, Y.; Shiina, I.; Mukaiyama, T., *J. Am. Chem. Soc.* **1991**, *113*, 4247. (e) Kobayashi, S.; Fujishita, Y.; Mukaiyama, T., *Chem. Lett.* **1989**, 2069.

4. Mukaiyama, T.; Inubushi, A.; Suda, S.; Hara, R.; Kobayashi, S., *Chem. Lett.* **1990**, 1015.

5. Parmee, E. R.; Hong, Y.; Tempkin, O.; Masamune, S., *Tetrahedron Lett.* **1992**, *33*, 1729.

6. Kobayashi, S.; Ohtsubo, A.; Mukaiyama, T., *Chem. Lett.* **1991**, 831.

7. Kim, S.; Salomon, R., *Tetrahedron Lett.* **1989**, *30*, 6279.

8. (a) Yoshida, A.; Tajima, Y.; Takeda, N.; Oida, S., *Tetrahedron Lett.* **1984**, *25*, 2793. (b) *Tetrahedron Lett.* **1985**, *26*, 673.

9. Mukaiyama, T.; Ohno, T.; Nishimura, T.; Han, J. S.; Kobayashi, S., *Chem. Lett.* **1990**, 2239.

10. Mori, I.; Ishihara, K.; Flippin, L. A.; Nozaki, K.; Yamamoto, H.; Bartlett, P. A.; Heathcock, C. H., *J. Org. Chem.* **1990**, *55*, 6107.

11. Mukai, C.; Miyakawa, M.; Hanaoka, M., *J. Chem. Soc., Perkin Trans. 1* **1997**, 913.

12. Davis, A. P.; Plunkett, S. J.; Muir, J. E., *Chem. Commun.* **1998**, 1797.

13. Gennari, C.; Ceccarelli, S.; Piarulli, U.; Aboutayab, K.; Donghi, M.; Paterson, I., *Tetrahedron* **1998**, *54*, 14999.

14. Langner, M.; Rémy, P.; Bolm, C., *Chem. Eur. J.* **2005**, *11*, 6254.

15. González, A. S.; Arrayás, R. G.; Carretero, J. C., *Org. Lett.* **2006**, *8*, 2977.

16. Kobayashi, S.; Matsubara, R.; Nakamura, Y.; Kitagawa, H.; Sugiura, M., *J. Am. Chem. Soc.* **2003**, *125*, 2507.

17. Kobayashi, S.; Moriwaki, M., *Synlett* **1997**, 551.

18. (a) Harada, T.; Iwai, H.; Takatsuki, H.; Fujita, K.; Kubo, M.; Oku, A., *Org. Lett.* **2001**, *3*, 2101. (b) Wang, X.; Adachi, S.; Iwai, H.; Takatsuki, H.; Fujita, K.; Kubo, M.; Oku, A.; Harada, T., *J. Org. Chem.* **2003**, *68*, 10046.

19. Kobayashi, S.; Suda, S.; Yamada, M.; Mukaiyama, T., *Chem. Lett.* **1994**, 97.

1-*t*-Butylthio-1-*t*-butyldimethyl-silyloxypropene[1]

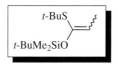

[102146-11-2]   C₁₃H₂₈OSSi   (MW 260.51)

InChI = 1S/C13H28OSSi/c1-10-11(15-12(2,3)4)14-16(8,9)13(5,6)7/h10H,1-9H3

InChIKey = GSIJEPGSYHDQHD-UHFFFAOYSA-N

(E)

[97250-84-5]

InChI = 1S/C13H28OSSi/c1-10-11(15-12(2,3)4)14-16(8,9)13(5,6)7/h10H,1-9H3/b11-10-

InChIKey = GSIJEPGSYHDQHD-KHPPLWFESA-N

(Z)

[97250-83-4]

InChI = 1S/C13H28OSSi/c1-10-11(15-12(2,3)4)14-16(8,9)13(5,6)7/h10H,1-9H3/b11-10+

InChIKey = GSIJEPGSYHDQHD-ZHACJKMWSA-N

(excellent reagent for Lewis acid-mediated aldol-type[2–9] and Michael additions,[17–21] under either stoichiometric or catalytic conditions)

Physical Data: colorless liquid.

Solubility: sol *n*-pentane, diethyl ether, dichloromethane, etc.

Analysis of Reagent Purity: (NMR) ¹H 0.15 (s, 6H), 0.95 (s, 9H), 1.30 (s, 9H), 1.65 (d, 3H, *J* = 6.7 Hz), 5.22 (q, 1H, *J* = 6.7 Hz) ppm (*Z* isomer). ¹H 0.18 (s, 6H), 0.95 (s, 9H), 1.30 (s, 9H), 1.60 (d, 3H, *J* = 6.8 Hz), 5.22 (q, 1H, *J* = 6.8 Hz) ppm (*E* isomer).

Preparative Methods: obtained by reaction of *t*-butyl thiopropionate with lithium diisopropylamide in THF and subsequent trapping with *t*-butyldimethylsilyl trifluoromethanesulfonate (95% *Z* isomer),[2] or with LDA in THF/HMPT and subsequent trapping with *t*-butyldimethylchlorosilane (95% *E* isomer).[2]

Handling, Storage, and Precautions: the reagent should be stored in the absence of moisture at −15 °C.

Original Commentary

Cesare Gennari

Università di Milano, Italy

Enantioselective Aldol Additions.[1] The reagent undergoes Lewis acid-catalyzed Mukaiyama-type additions to aldehydes to give *syn* α-methyl-β-hydroxy thioesters with good yields and remarkable enantioselectivities (eq 1).[3a–d] Slow addition of the aldehyde (3–4.5 h) in propionitrile is necessary for high enantioselectivity.[3b–d]

A similar version of the reaction makes use of 100 mol % tin(II) oxide, 65 mol % trimethylsilyl trifluoromethanesulfonate, and 50 mol % chiral diamine under slow addition conditions (9 h), with slightly worse results (58–82% yield; *syn:anti* = 91:9–98:2;

68–94% ee).[3e] These reactions are the catalytic versions of the stoichiometric reaction outlined in eq 2.[3c,3d,3f–i]

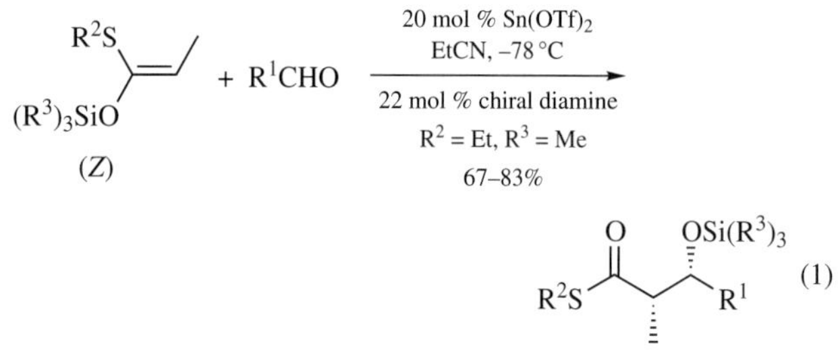

$$syn:anti = 86:14 \text{ to } 100:0 \qquad (1)$$
$$86\text{–}98\% \text{ ee}$$

chiral diamine =

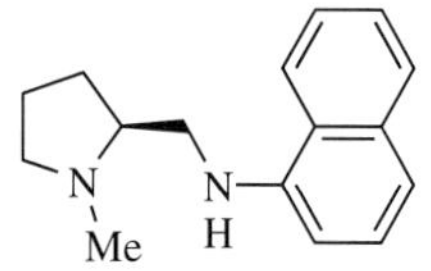

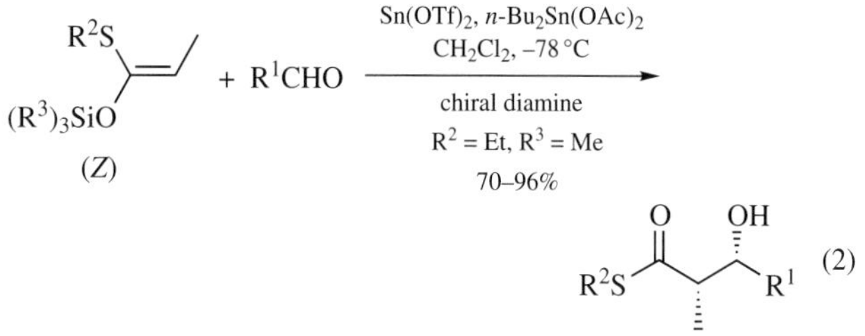

$$syn:anti = 100:0 \qquad (2)$$
$$>98\% \text{ ee}$$

chiral diamine =

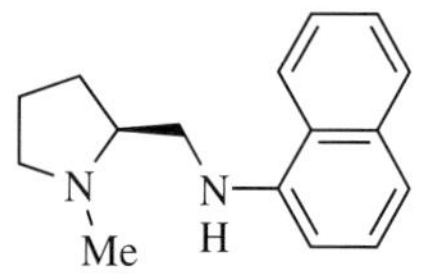

Under similar conditions to eq 2, the reagent adds to α-keto esters to give the corresponding substituted malates in good yield and high stereoselectivity (eq 3).[3j]

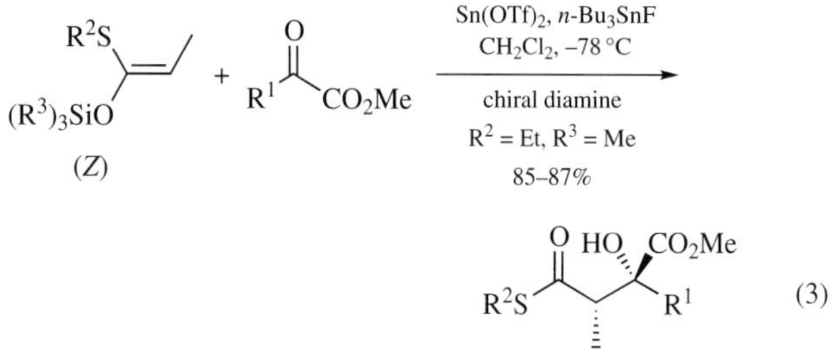

$$syn:anti = 94:6 \text{ to } 96:4 \qquad (3)$$
$$82\text{–}96\% \text{ ee}$$

chiral diamine =

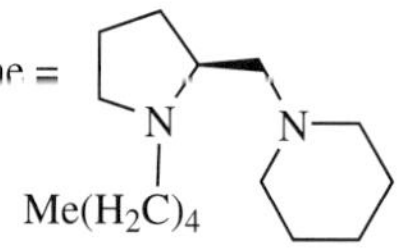

Under similar conditions to eq 3, silyl ketene acetals derived from α-benzyloxy thioesters[4a–c] or α-silyloxy thioesters[4d] add to various aldehydes to give either *anti* (91–99%) or *syn* (88–97%) α,β-dihydroxy thioesters, depending on the reagent used, in good yield (46–93%) and with high enantioselectivity (82–98% ee).

Another catalytic asymmetric aldol reaction is shown in eq 4. Slow addition of the aldehyde is again necessary for high selectivity.[5] Catalyst A is preferred in reactions with aromatic and unsaturated aldehydes, while catalyst B is employed in reactions with primary aldehydes.[5] In contrast to the catalysts mentioned above (eqs 1–3), both A and B give *anti* aldols as the predominant stereoisomers.[5]

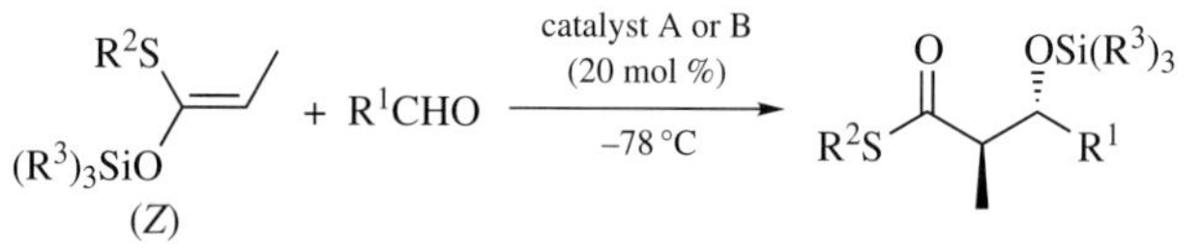

catalyst A	R^2 = Et, *t*-Bu; R^3 = Me	54–94% yield; 60–89% ee
		anti:syn = 66:33 to 94:6
catalyst B	R^2 = Et; R^3 =Me	81–85% yield; 70–82% ee
		anti:syn = 88:12 to 91:9

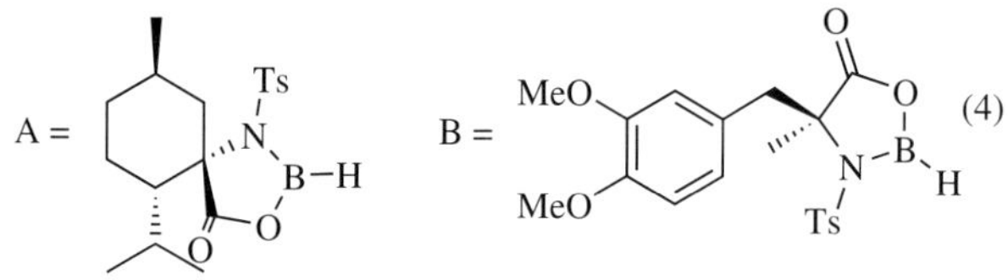

$$(4)$$

Diastereoselective Aldol Additions.[1] Boron trifluoride etherate mediates the addition of both (*E*) and (*Z*) title reagents to various aldehydes to give the aldol products with high *anti* selectivity (*anti:syn* = 91:9–96:4) irrespective of the silyl ketene acetal geometry (eq 5).[2,6] *Syn* selectivity is obtained using the (*E*) isomer via fluoride-mediated addition to benzaldehyde (*syn:anti* = 95:5),[2] or via TMSOTf-catalyzed addition to benzaldehyde dimethyl acetal (*syn:anti* = 78:22).[7a–d] titanium(IV) chloride mediates the addition of both (*E*) and (*Z*) title reagents to propynals to give the aldol products with high *anti* selectivity (*anti:syn* > 95:5) irrespective of the silyl ketene acetal geometry, while high *syn* selectivity (*syn:anti* ≥ 98:2) is obtained in the TiCl$_4$-mediated addition to cobalt-complexed propynals.[7e] High *syn* selectivity is obtained in the uncatalyzed reaction of either (*E*)- or (*Z*)-*O*-silacyclobutyl analogs via a direct silicon group transfer (eq 5).[8]

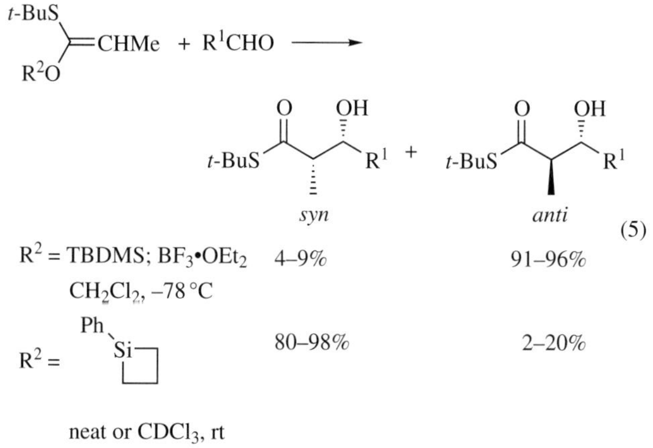

	syn	*anti*	
R^2 = TBDMS; BF$_3$•OEt$_2$	4–9%	91–96%	
CH$_2$Cl$_2$, –78 °C			(5)
R^2 = (Ph-Si-cyclobutyl)	80–98%	2–20%	

neat or CDCl$_3$, rt

Simple stereoselectivity (*syn:anti* ratios) can be controlled in the Sn(OTf)$_2$-mediated addition of the reagent to α-keto esters by varying the silyl ketene acetal geometry (eq 6).[3j]

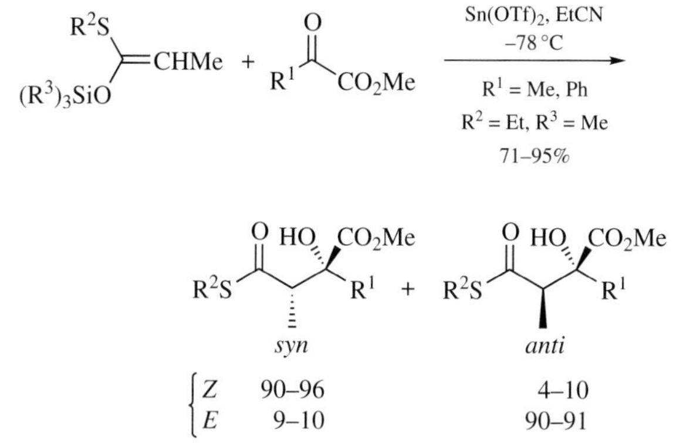

$$\text{(6)}$$

$$
\begin{cases}
Z & 90\text{–}96 \qquad 4\text{–}10 \\
E & 9\text{–}10 \qquad 90\text{–}91
\end{cases}
$$

$$\text{(8)}$$

$$
\begin{array}{ll}
R^2 = \text{MeSCH}_2 & 98:2 \\
R^2 = 2\text{-picolyl} & 2:98
\end{array}
$$

The reagent undergoes Lewis acid-promoted Mukaiyama-type additions[1] to chiral aldehydes with moderate to good stereocontrol (eq 7).[2,6,9a–d] It is remarkable that high diastereoselectivity (>99%) can be obtained using TiCl$_4$ due to 'chelation control', and that these results are independent of the silyl ketene acetal geometry.[2,9b] These reactions have been used for establishing three consecutive stereocenters in the total synthesis of several biologically interesting compounds.[10–13]

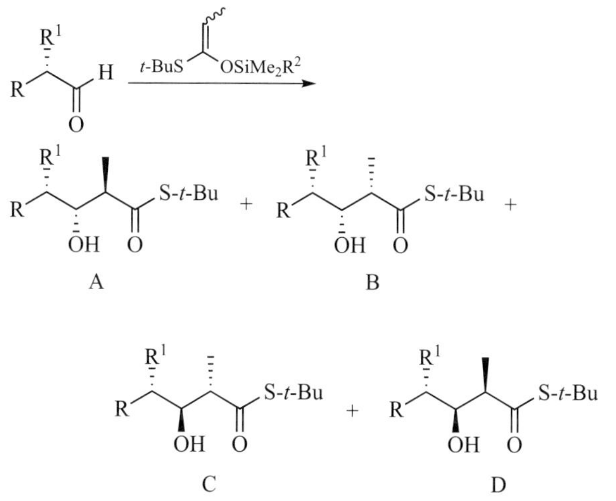

$$\text{(9)}$$

$$
\begin{array}{lll}
E & R = n\text{-Pr}, i\text{-Pr}, \text{Ph} & 6\text{–}25 \qquad 75\text{–}94 \\
Z & R = \text{Ph}, i\text{-Pr} & \text{other isomers} + \\
& & 24 \qquad\qquad 76
\end{array}
$$

$$\text{(7)}$$

Tin(IV) chloride or BF$_3$·OEt$_2$ mediate the addition of the title reagents to norephedrine-derived 2-methoxyoxazolidines; yields and diastereomeric ratios are independent of the silyl ketene acetal geometry (eq 10).[16]

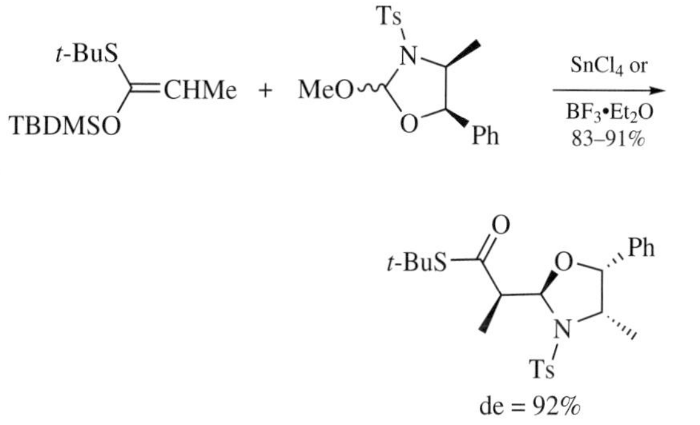

R	R^1	Promoter	(E):(Z)	Yield (%)	Ref.	Ratio A:B:C:D
Ph	Me	BF$_3$·OEt$_2$	95:5	75	2	91:7:2:0
Me	OBn	SnCl$_4$	95:5	89	2, 9a	3:97:0:0
Me	OBn	SnCl$_4$	5:95	90	2, 9a	24:76:0:0
Me	OBn	TBAF	93:7	72	2, 9a	3:16:8:73
Me	OBn	TBAF	10:90	68	2, 9a	3:13:12:72
Me	OBn	BF$_3$·OEt$_2$	93:7	57	2	22:6:60:12
BnOCH$_2$	OBn	SnCl$_4$	95:5	75	2	<5:>95:0:0
BnOCH$_2$	Me	TiCl$_4$	95:5	67	2, 9b	0:0:<1:>99
BnOCH$_2$	Me	TiCl$_4$	5:95	65	2, 9b	0:0:<1:>99
BnOCH$_2$	Me	BF$_3$·OEt$_2$	95:5	75	2, 9b	77:14:9:0
BnOCH$_2$	Me	BF$_3$·OEt$_2$	5:95	71	2, 9b	77:16:7:0
Et	SMe	BF$_3$·OEt$_2$	5:95	83	9c	0:0:91:9
Et	SMe	TiCl$_4$	5:95	81	9c	0:100:0:0

Addition to Other Electrophiles. Zinc chloride catalyzes the addition to a chiral azetinone with good yield. The stereochemical outcome depends on the sulfur substituent (eq 8).[14]

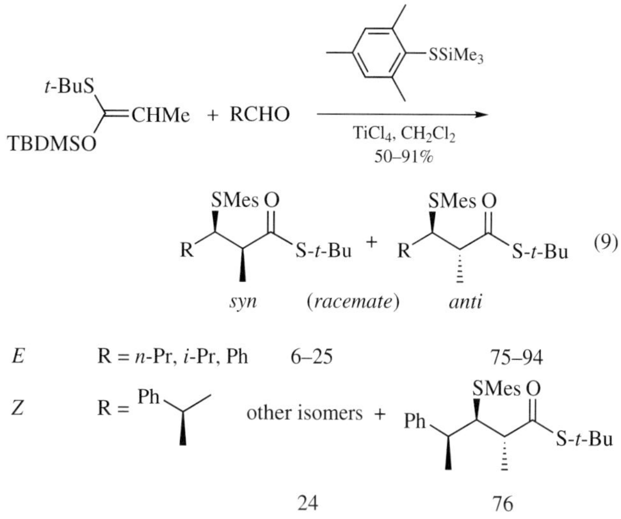

TiCl$_4$ mediates the addition of the title reagents to achiral and chiral thionium ions with good diastereoselectivity (eq 9).[15]

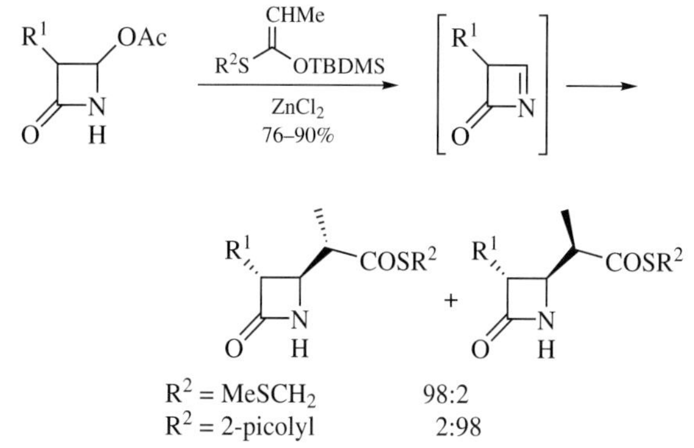

$$\text{(10)}$$

$$de = 92\%$$

Michael Additions. In the presence of a catalytic amount of trityl salts (2–3 mol %), thioester silyl ketene acetals react stereoselectively with acyclic α,β-unsaturated ketones to give the Michael adducts in high yield (eq 11).[17a]

In the case of cyclic ketones, either the *syn* or the *anti* products can be obtained stereoselectively by varying catalyst and substituents (eq 12).[17a,b] For high *syn* selectivity the *t*-butyl thioester group (R^2 = *t*-Bu) is essential, while both the substituent on silicon (R^3) and silyl ketene acetal geometry have little effect on the diastereoselectivity.[17b]

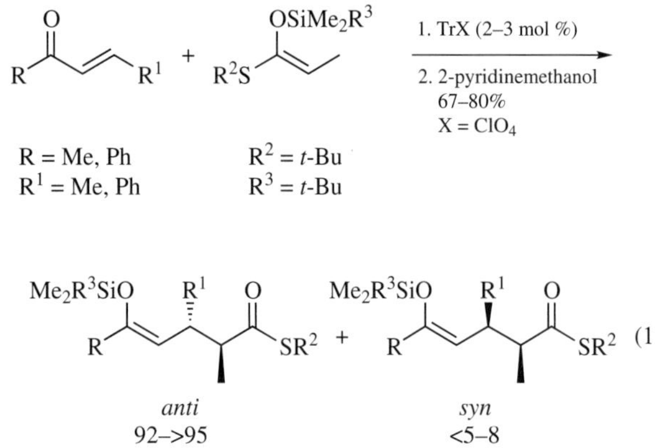

R = Me, Ph R^2 = *t*-Bu
R^1 = Me, Ph R^3 = *t*-Bu

$$anti \quad\quad\quad\quad syn \qquad (11)$$
$$92->95 \quad\quad\quad\quad <5-8$$

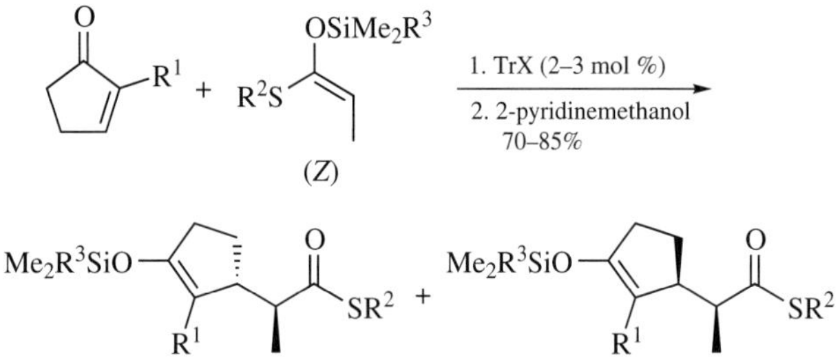

(12)

R^1	R^2	R^3	X	(E):(Z)	anti	syn
H	Et	Ph	$SbCl_6$	100:0	92	8
H	*t*-Bu	Me	OTf	0:100	21	79
Me	*t*-Bu	Me	$SbCl_6$	0:100	<5	>95
SPh	*t*-Bu	*t*-Bu	$SbCl_6$	0:100	<5	>95
CO_2Me	*t*-Bu	*t*-Bu	$SbCl_6$	0:100	10	90

Catalytic amounts (5 mol %) of trityl chloride and tin(II) chloride,[18] or antimony(V) chloride and tin(II) triflate,[19] have also been used to promote the Michael additions, with stereoselectivity in favor of the *anti* isomers and slightly reduced ratios in comparison with the above results. No stereoselectivity was observed with catalytic amounts (10 mol %) of ytterbium(III) trifluoromethanesulfonate.[20]

Using the Michael-aldol protocol, several key intermediates on the way to natural products have been synthesized, e.g. dehydroiridodiol,[17a] aromatin,[17b] and fastigilin-C (eq 13).[21]

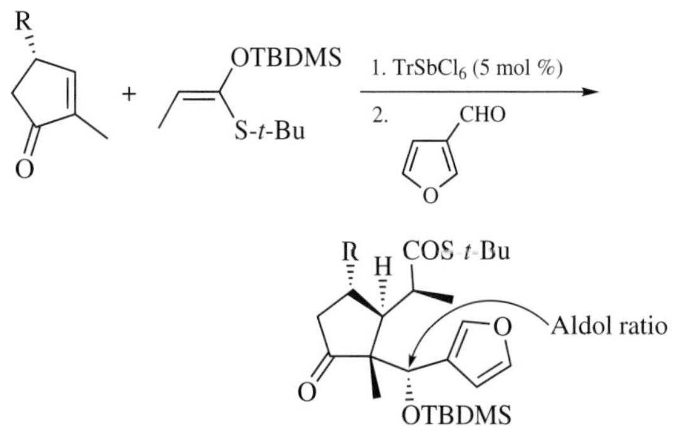

R = H   73%   Aldol ratio = 6:1 (−95 °C, CH_2Cl_2)
R = OMe   81–90%   Aldol ratio = 100:0 (−78 °C, CH_2Cl_2)

First Update

Paul R. Blakemore
Oregon State University, Corvallis, OR, USA

Aldol Reactions. The title reagent and its various congeners continue to find application in Mukaiyama-type aldol reactions as versatile nucleophilic propionate synthons with predictable behavior.[22] For example, the (*E*)-isomer of the reagent produced the expected major isomer in a boron trifluoride etherate mediated addition to a complex aldehyde en route to a C19–C35 subunit of swinholide A (eq 14).[23] Remote stereoinduction in Mukaiyama aldol reactions of the reagent with (2-sulfinylphenyl)acetaldehydes was recently observed; *anti* aldol adducts were obtained preferentially regardless of the geometry of the silyl ketene thioacetal employed.[24]

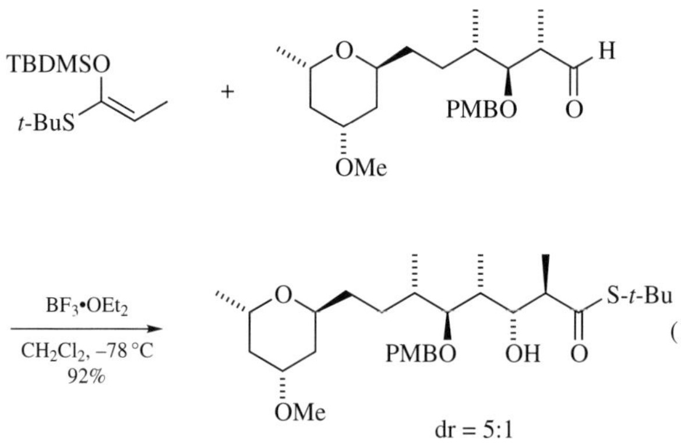

$$(14)$$
dr = 5:1

S-Pyrid-2-yl congeners of the reagent have been employed for the synthesis of β-lactones via a tandem Mukaiyama aldol lactonization (TMAL) process.[25] Herein, choice of Lewis acid promoter (zinc(II) chloride or tin(IV) chloride) determines whether *cis* or *trans* β-lactone products are obtained (eq 15). The analogous Mannich initiated sequence provides β-lactams.[26]

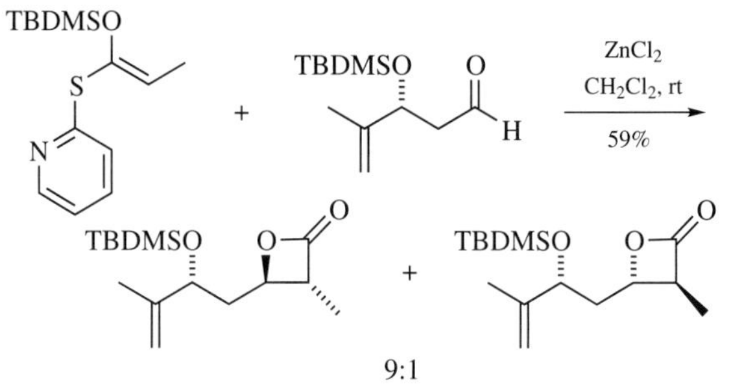

$$(15)$$
9:1

The Evans group has introduced a number of effective chiral Lewis acids (based on Sn(II), Cu(II), Zn(II), and Sc(III) complexes of 'box' and 'pybox' ligand systems) for the catalytic enantioselective addition of silyl ketene thioacetals related to the title reagent to glyoxylate and pyruvate esters (e.g., eq 16).[27] The carbonyl substrates employed must be able to engage in two-point binding for the attainment of high enantioselectivity and comparable results are generally obtained from either geometrical isomer of the ketene thioacetal.

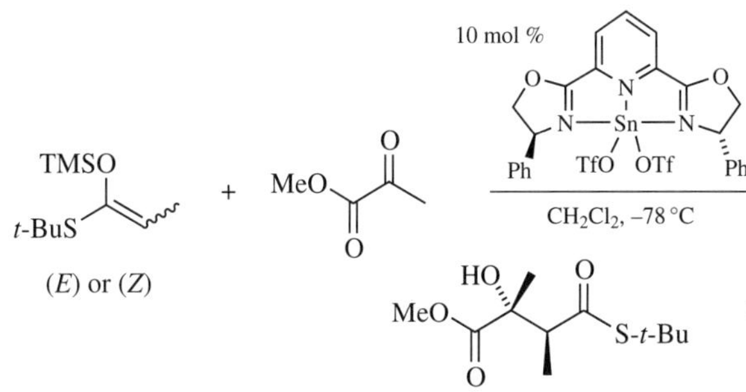

from (Z)   94% yield; *anti:syn* = 99:1; 99% ee
from (E)   84% yield; *anti:syn* = 99:1; 96% ee

anti:syn = 85:15

Mannich Reactions. Silyl ketene thioacetals, including the title reagent, have been demonstrated to be effective nucleophiles in various Mukaiyama-type Mannich processes. For example, Kobayashi and coworkers recently described stereoselective additions of the reagent to *N*-acyl glyoxylate aldimines catalyzed by chiral Cu(II) complexes (eq 17).[28] As may be expected on the basis of comparable aldol reactions, the relative stereochemistry of the product was independent of the geometrical isomer of ketene thioacetal employed: *syn* adducts were obtained preferentially. Higher enantioselectivity was observed from the (Z)-configured reagent and other types of silyl ketene thioacetals (e.g., SEt or OSiEt$_3$) generally gave inferior results. Related enantioselective additions to analogous *N*-acyl hydrazone substrates catalyzed by chiral Zn(II) diamine complexes in water have also been reported by the same group.[29]

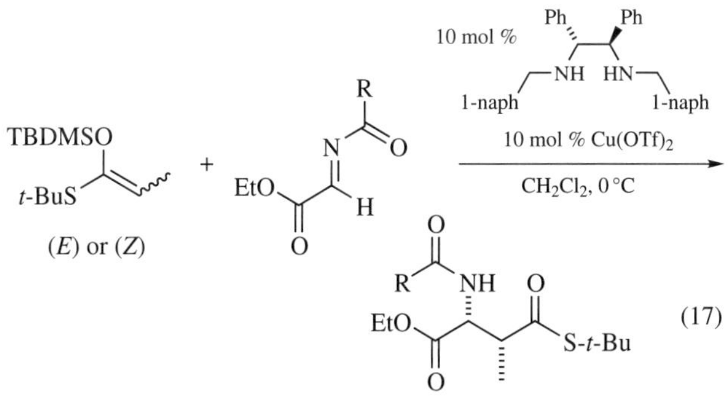

Michael Reactions. Significant advances have been made in the study of Mukaiyama–Michael additions of silyl ketene thioacetals to α,β-unsaturated carbonyl compounds. Diastereoselectivity in titanium(IV) chloride mediated addition of the title reagent to enones was studied in detail by Fukuzumi and coworkers (eq 19).[33] In most cases, *anti* or *syn* configured products could be targeted as desired by respective deployment of either the (E)- or (Z)-isomer of the reagent; however, the sense and magnitude of diastereoselectivity was also enigmatically dependent on enone substituent. The trimethylsilyl congener of the reagent gave consistently inferior results in this particular application. Conjugate addition of the title reagent to simple enones catalyzed by trityl hexachloroantimonate (TrSbCl$_6$) followed by mercuric cation mediated cyclization provides for an efficient synthesis of 3,4-dihydro-α-pyrans.[34] More recently, trimethylsilyl trifluoromethanesulfonate (TMSOTf) and other trialkylsilyl triflates have been shown to afford similar stereoselectivity to TrSbCl$_6$ in related Mukaiyama–Michael reactions.[35]

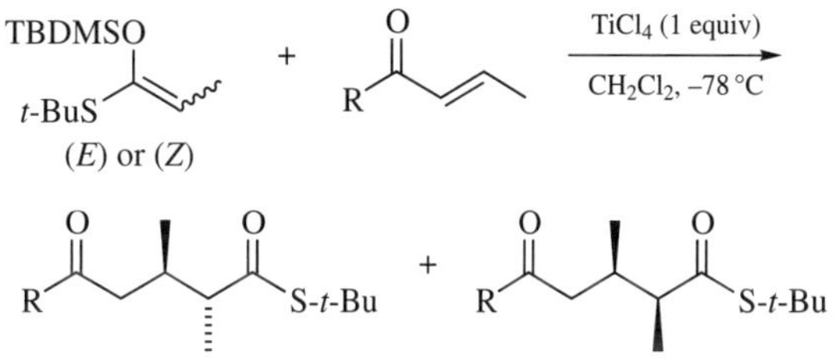

(E):(Z)	R	Yield (%)	*anti*	*syn*
93:7	Me	77	92	8
<1:99	Me	67	60	40
93:7	*t*-Bu	99	95	5
<1:99	*t*-Bu	83	9	91
<93:7	Ph	66	93	7
<1:99	Ph	73	46	54
93:7	p-anis	71	95	5
<1:99	p-anis	82	9	91
93:7	Mes	60	11	89
<1:99	Mes	67	11	89

Less traditional Lewis acid catalysts have also been employed to mediate additions of silyl ketene thioacetals to imines and hydrazones, including scandium(III) triflate[26,30] and bismuth(III) triflate.[31] In a significant recent advance, Mukaiyama and coworkers reported Lewis base catalyzed additions of the trimethylsilyl congener of the title reagent to *N*-sulfonyl aldimines (eq 18).[32] The reaction is tolerant of water and competent Lewis base activators include readily obtained salts such as lithium acetate and lithium benzamide.

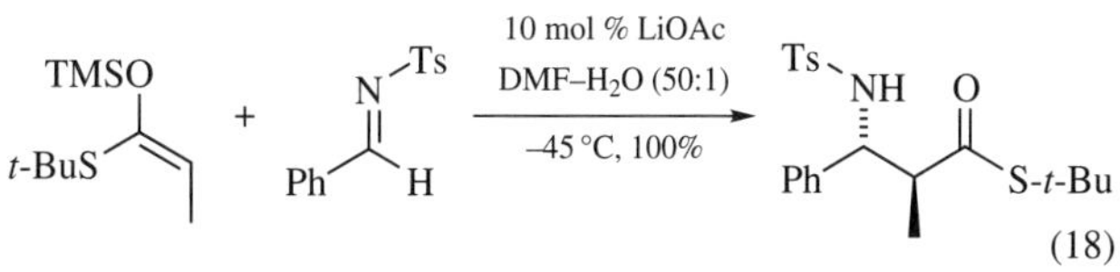

A number of effective enantioselective Mukaiyama–Michael reactions using silyl ketene thioacetals have emerged. For example, the versatile Evans-type cationic Cu(II) bisoxazoline ('box') complexes catalyze the addition of congeners of the title reagent to acryloyl oxazolidinones with excellent enantio- and diastereoselectivity (eq 20).[36] Again, the geometry of the silyl ketene thioacetal correlates to the relative stereochemical configuration of the product, but in this case stereoselectivity is further influenced by the steric demand of the thioalkyl substituent. The conjugate addition acceptor must allow for two-point binding to the catalytically active cationic Cu(II) center for successful results; also, these reactions are greatly accelerated in the presence of 1,1,1,3,3,3-hexafluoro-2-propanol (reaction time reduced by 70% on average) without overly compromising stereoselectivity.[36] Harada and coworkers recently reported a chiral oxazaborolidinone catalyst for the *syn* selective asymmetric addition of silyl ketene thioacetals to enones.[37] The enantioselectivity for these reactions did not exceed 90% ee; however, in principle at least, application of the Harada process is not limited to any special enone structural type.

(20)

	R	Yield (%)	syn:anti	% ee
(Z)	*t*-Bu	73	99:1	99
	Me	90	66:34	90
(E)	*t*-Bu	65	22:78	96
	Me	90	5:95	90

Related Reagents. For comparisons and more details, see also those entries that deal with other silyl ketene acetals, in particular 1-*t*-Butylthio-1-*t*-butyldimethylsilyloxyethylene, ketene *t*-butyldimethylsilyl methyl acetal, and analogs.

1. Gennari, C., *Comprehensive Organic Synthesis* **1991**, *2*, 629.

2. Gennari, C.; Beretta, M. G.; Bernardi, A.; Moro, G.; Scolastico, C.; Todeschini, R., *Tetrahedron* **1986**, *42*, 893.

3. (a) Mukaiyama, T.; Kobayashi, S.; Uchiro, H.; Shiina, I., *Chem. Lett.* **1990**, 129. (b) Kobayashi, S.; Fujishita, Y.; Mukaiyama, T., *Chem. Lett.* **1990**, 1455. (c) Kobayashi, S.; Uchiro, H.; Shiina, I.; Mukaiyama, T., *Tetrahedron* **1993**, *49*, 1761. (d) Mukaiyama, T.; Furuya, M.; Ohtsubo, A.; Kobayashi, S., *Chem. Lett.* **1991**, 989. (e) Mukaiyama, T.; Uchiro, H.; Kobayashi, S., *Chem. Lett.* **1990**, 1147. (f) Mukaiyama, T.; Uchiro, H.; Kobayashi, S., *Chem. Lett.* **1989**, 1757. (g) Mukaiyama, T.; Asanuma, H.; Hachiya, I.; Harada, T.; Kobayashi, S., *Chem. Lett.* **1991**, 1209. (h) Mukaiyama, T.; Uchiro, H.; Kobayashi, S., *Chem. Lett.* **1989**, 1001. (i) Kobayashi, S.; Uchiro, H.; Fujishita, Y.; Shiina, I.; Mukaiyama, T., *J. Am. Chem. Soc.* **1991**, *113*, 4247. (j) Kobayashi, S.; Hachiya, I., *J. Org. Chem.* **1992**, *57*, 1324.

4. (a) Mukaiyama, T.; Uchiro, H.; Shiina, I.; Kobayashi, S., *Chem. Lett.* **1990**, 1019. (b) Mukaiyama, T.; Shiina, I.; Kobayashi, S., *Chem. Lett.* **1990**, 2201. (c) Kobayashi, S.; Onozawa, S.; Mukaiyama, T., *Chem. Lett.* **1992**, 2419. (d) Mukaiyama, T.; Shiina, I.; Kobayashi, S., *Chem. Lett.* **1991**, 1901.

5. Parmee, E. R.; Hong, Y.; Tempkin, O.; Masamune, S., *Tetrahedron Lett.* **1992**, *33*, 1729.

6. Gennari, C.; Bernardi, A.; Cardani, S.; Scolastico, C., *Tetrahedron Lett.* **1985**, *26*, 797.

7. (a) Murata, S.; Suzuki, M.; Noyori, R., *J. Am. Chem. Soc.* **1980**, *102*, 3248. (b) Noyori, R.; Murata, S.; Suzuki, M., *Tetrahedron* **1981**, *37*, 3899. (c) Murata, S.; Noyori, R., *Tetrahedron Lett.* **1982**, *23*, 2601. (d) Murata, S.; Suzuki, M.; Noyori, R., *Tetrahedron* **1988**, *44*, 4259. (e) Mukai, C.; Kataoka, O.; Hanaoka, M., *Tetrahedron Lett.* **1991**, *32*, 7553.

8. Denmark, S. E.; Griedel, B. D.; Coe, D. M., *J. Org. Chem.* **1993**, *58*, 988.

9. (a) Gennari, C.; Bernardi, A.; Poli, G.; Scolastico, C., *Tetrahedron Lett.* **1985**, *26*, 2373. (b) Gennari, C.; Bernardi, A.; Scolastico, C.; Potenza, D., *Tetrahedron Lett.* **1985**, *26*, 4129. (c) Annunziata, R.; Cinquini, M.; Cozzi, F.; Cozzi, P. G.; Consolandi, E., *J. Org. Chem.* **1992**, *57*, 456. (d) Yamazaki, T.; Yamamoto, T.; Kitazume, T., *J. Org. Chem.* **1989**, *54*, 83.

10. Gennari, C.; Cozzi, P. G., *J. Org. Chem.* **1988**, *53*, 4015.

11. Pilli, R. A.; Murta, M. M., *J. Org. Chem.* **1993**, *58*, 338.

12. Dìez-Martin, D.; Kotecha, N. R.; Ley, S. V.; Mantegani, S.; Menéndez, J. C.; Organ, H. M.; White, A. D., *Tetrahedron* **1992**, *48*, 7899.

13. Rosemberg, S. H.; Boyd, S. A.; Mantei, R. A., *Tetrahedron Lett.* **1991**, *32*, 6507.

14. (a) Martel, A.; Daris, J.-P.; Bachand, C.; Corbeil, J.; Menard, M., *Can. J. Chem.* **1988**, *66*, 1537. See also: (b) Shibata, T.; Iino, K.; Tanaka, T.; Hashimoto, T.; Kameyama, Y.; Sugimura, Y., *Tetrahedron Lett.* **1985**, *26*, 4739. (c) Kim, C. U.; Luh, B.; Partyka, R. A., *Tetrahedron Lett.* **1987**, *28*, 507.

15. Mori, I.; Bartlett, P. A.; Heathcock, C. H., *J. Org. Chem.* **1990**, *55*, 5966.

16. (a) Bernardi, A.; Cardani, S.; Carugo, O.; Colombo, L.; Scolastico, C.; Villa, R., *Tetrahedron Lett.* **1990**, *31*, 2779. (b) Bernardi, A.; Cavicchioli, M.; Poli, G.; Scolastico, C.; Sidjimov, A., *Tetrahedron* **1991**, *47*, 7925.

17. (a) Mukaiyama, T.; Tamura, M.; Kobayashi, S., *Chem. Lett.* **1986**, 1817. (b) Mukaiyama, T.; Tamura, M.; Kobayashi, S., *Chem. Lett.* **1987**, 743.

18. Mukaiyama, T.; Kobayashi, S.; Tamura, M.; Sagawa, Y., *Chem. Lett.* **1987**, 491.

19. Kobayashi, S.; Tamura, M.; Mukaiyama, T., *Chem. Lett.* **1988**, 91.

20. Kobayashi, S.; Hachiya, I.; Takahori, T.; Araki, M.; Ishitani, H., *Tetrahedron Lett.* **1992**, *33*, 6815.

21. (a) Tanis, S. P.; McMills, M. C.; Scahill, T. A.; Kloosterman, D. A., *Tetrahedron Lett.* **1990**, *31*, 1977. (b) Tanis, S. P.; Robinson, E. D.; McMills, M. C.; Watt, W., *J. Am. Chem. Soc.* **1992**, *114*, 8349.

22. Kobayashi, S.; Manabe, K.; Ishitani, H.; Matsuo, J.-I., *Sci. Synth.* **2002**, *4*, 317.

23. Keck, G. E.; Lundquist, G. D., *J. Org. Chem.* **1999**, *64*, 4482.

24. García Ruano, J. L.; Fernández-Ibáñez, M. Á.; Maestro, M. C., *Tetrahedron* **2006**, *62*, 12297.

25. (a) Mitchell, T. A.; Romo, D., *J. Org. Chem.* **2007**, *72*, 9053. (b) Wang, Y.; Zhao, C.; Romo, D., *Org. Lett.* **1999**, *1*, 1197. (c) Yang, H. W.; Romo, D., *J. Org. Chem.* **1997**, *62*, 4.

26. Benaglia, M.; Cozzi, F.; Puglisi, A., *Eur. J. Org. Chem.* **2007**, 2865.

27. (a) Evans, D. A.; Masse, C. E.; Wu, J., *Org. Lett.* **2002**, *4*, 3375. (b) Evans, D. A.; Burgey, C. S.; Kozlowski, M. C.; Tregay, S. W., *J. Am. Chem. Soc.* **1999**, *121*, 686. (c) Evans, D. A.; MacMillan, D. M. C.; Campos, K. R., *J. Am. Chem. Soc.* **1997**, *119*, 10859. (d) Evans, D. A.; Kozlowski, M. C.; Tedrow, J. A., *Tetrahedron Lett.* **1996**, *37*, 7481.

28. Kobayashi, S.; Matsubara, R.; Nakamura, Y.; Kitagawa, H.; Sugiura, M., *J. Am. Chem. Soc.* **2003**, *125*, 2507.

29. (a) Hamada, T.; Manabe, K.; Kobayashi, S., *J. Am. Chem. Soc.* **2004**, *126*, 7768. (b) Hamada, T.; Manabe, K.; Kobayashi, S., *Chem. Eur. J.* **2006**, *12*, 1205.

30. Díez, E.; Prieto, A.; Simon, M.; Vázquez, J.; Álvarez, E.; Fernández, R.; Lassaletta, J. M., *Synthesis* **2006**, 540.

31. Ollevier, T.; Nadeau, E., *Org. Biomol. Chem.* **2007**, *5*, 3126.

32. Fujisawa, H.; Takahashi, E.; Mukaiyama, T., *Chem. Eur. J.* **2006**, *12*, 5082.

33. Fujita, Y.; Otera, J.; Fukuzumi, S., *Tetrahedron* **1996**, *52*, 9419.

34. Kobayashi, S.; Moriwaki, M., *Synlett* **1997**, 551.

35. Michalak, K.; Wicha, J., *Pol. J. Chem.* **2004**, *78*, 205.

36. Evans, D. A.; Scheidt, K. A.; Johnston, J. N.; Willis, M. C., *J. Am. Chem. Soc.* **2001**, *123*, 4480.

37. Harada, T.; Yamauchi, T.; Adachi, S., *Synlett* **2005**, 2151.

t-Butyl Trimethylsilylacetate

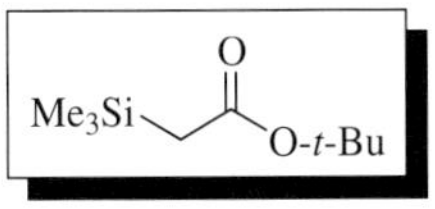

[41108-81-0]   C$_9$H$_{20}$O$_2$Si   (MW 188.34)

InChI = 1S/C9H20O2Si/c1-9(2,3)11-8(10)7-12(4,5)6/h7H2,
1-6H3

InChIKey = HOHBQWITMXOSOW-UHFFFAOYSA-N

(used to prepare lithium enolates which can be alkylated;[1,2] can be alkenated with aldehydes and ketones;[3] can be acylated;[4] can be reacted with Group 14 halides[5])

Physical Data: air stable, colorless liquid; bp 67 °C/13 mmHg; n_D^{20} 1.4184; ^{1}H NMR (CDCl$_3$, 90 MHz) 0.13 (s, 9H), 1.45 (s, 9H), 1.80 (s, 2H); IR 1710 cm^{-1}.

Preparative Methods: an advantage of *t*-butyl trimethylsilylacetate over its methyl and ethyl ester analogs is that it can be prepared by *C*-silylation of the lithium enolate of *t*-butyl acetate at −78 °C in THF (85–90% yield). Under the same conditions the enolates of methyl and ethyl acetate give primarily *O*-silylated products.[3] The reagent has also been prepared by a rapid boron trifluoride etherate catalyzed reaction of trimethylsilylketene with *t*-butyl alcohol.[6]

Lithium Enolate. Most applications require initial formation of the enolate (**1**) (eq 1), which is obtained as a white solid suspended in THF, stable indefinitely at −78 °C.[1]

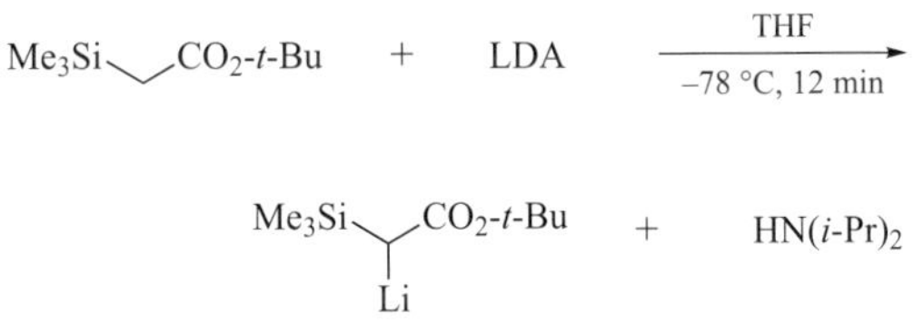

Alkylation Reactions. Enolate (**1**) is readily alkylated with a variety of primary alkyl halides at −70 to 0 °C in THF,[2,7] usually in the presence of TMEDA or DMSO. The alkylation products may be desilylated (eq 2) to provide a two-carbon chain extension which is not prone to dialkylation or elimination side reactions.[7a]

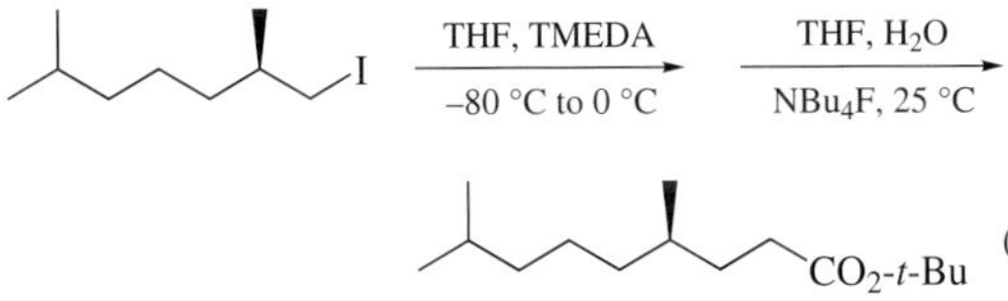

The reagent is spiroalkylated in two steps with 2-methyl-1,4-diiodobutane (eq 3).[7b]

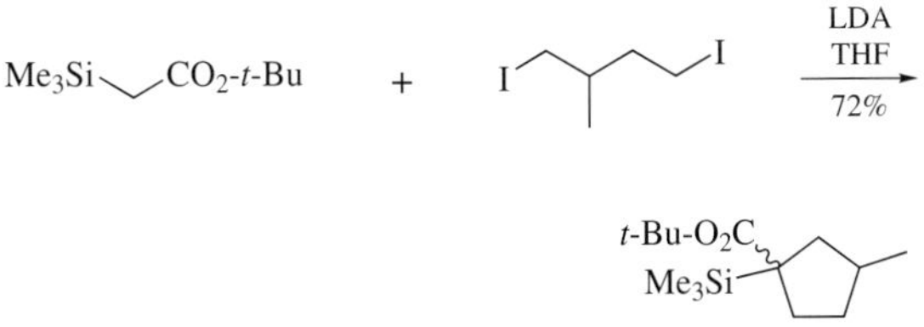

Enolate (**1**) can be vinylated with 2-bromopropene in the presence of a Ni catalyst.[7c]

Peterson Alkenation.[8] Enolate (**1**) reacts with aldehydes or ketones to give α,β-unsaturated esters (eq 4).[1,8] In some cases (**1**) gives better yields and different *E:Z* ratios from the analogous Horner–Emmons reactions with phosphonoacetates.[9]

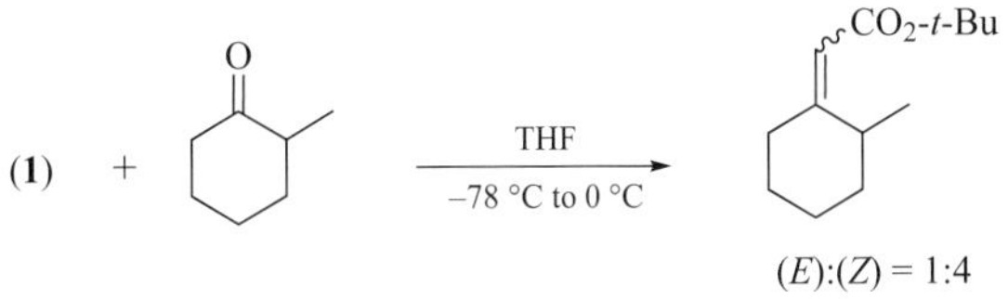

Reaction with Acyl Derivatives. Enolate (**1**) does not react with ethyl acetate or *N,N*-dimethylacetamide, and gives a complicated mixture of products with acetyl chloride. However, (**1**) is acetylated with acetylimidazole to give the lithium enolate of *t*-butyl acetoacetate (eq 5).[4a]

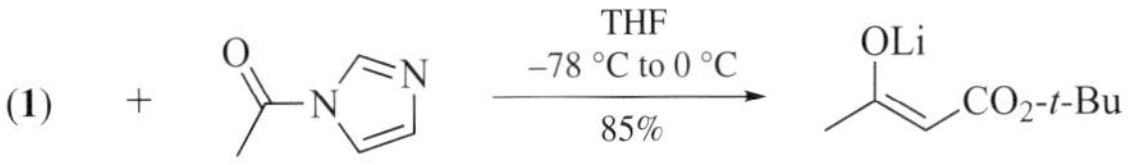

Protonation of the reaction mixtures obtained with a variety of acylimidazoles gives β-keto esters in 50–94% yields. Enolate (**1**) reacts with a variety of lactones to give Peterson alkenation products (eq 6).[4b]

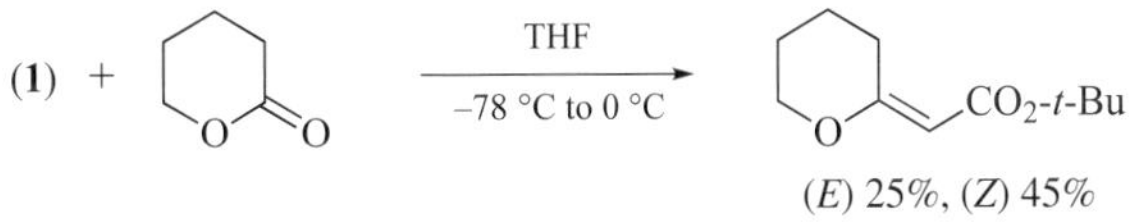

Enolate (**1**) reacts with some aromatic nitriles (but not with benzonitrile) to give *N*-silylenamines.[10]

Reaction with Group 14 Halides. Enolate (**1**) reacts with chlorotrimethylgermane or tri-*n*-butylchlorostannane to give *t*-butyl (trimethylgermyl)(trimethylsilyl)acetate (83% yield)[5a] and *t*-butyl (trimethylsilyl)(tri-*n*-butylstannyl)acetate,[5b] respectively.

Related Reagents. *t*-Butyl α-Lithiobis(trimethylsilyl)acetate; Ethyl Lithioacetate; Ethyl Lithio(trimethylsilyl)acetate; Ethyl Trimethylsilylacetate; Triethyl Phosphonoacetate; Trimethylsilylacetic Acid.

---

1. Hartzell, S. L.; Sullivan, D. F.; Rathke, M. W., *Tetrahedron Lett.* **1974**, *15*, 1403.

2. Hudrlik, P. F.; Peterson, D.; Chou, D., *Synth. Commun.* **1975**, *5*, 359.

3. Rathke, M. W.; Sullivan, D. F., *Synth. Commun.* **1973**, *3*, 67.

4. (a) Hartzell, S. L.; Rathke, M. W., *Tetrahedron Lett.* **1976**, 2757. (b) Saúve, G.; Deslongchamps, P., *Synth. Commun.* **1985**, *15*, 201.

5. (a) Kanemoto, N.; Sato, Y.; Inoue, S., *J. Organomet. Chem.* **1988**, *348*, 25. (b) Ager, D. J.; Cooke, G. E.; East, M. B.; Mole, S. J.; Rampersaud, A.; Webb, V. J., *Organometallics* **1986**, *5*, 1906.

6. Ruden, R. A., *J. Org. Chem.* **1974**, *39*, 3607.

7. (a) Helmchen, G.; Schmierer, R., *Tetrahedron Lett.* **1983**, *24*, 1235. (b) Paquette, L. A.; Maynard, G. D.; Ra, C. S.; Hoppe, M., *J. Org. Chem.* **1989**, *54*, 1408. (c) Chenera, B.; Chuang, C.-P.; Hart, D. J.; Lai, C.-S., *J. Org. Chem.* **1992**, *57*, 2018. (d) Heathcock, C. H.; Piettre, S.; Ruggeri, R. B.; Ragan, J. A.; Kath, J. C., *J. Org. Chem.* **1992**, *57*, 2554.

8. For reviews see: Ager, D. J., *Synthesis* **1984**, 384; *Org. React.* **1990**, *38*, 1.

9. (a) Crimmin, M. J.; O'Hanlon, P. J.; Rogers, N. H., *J. Chem. Soc., Perkin Trans. 1* **1985**, 541. (b) Budhram, R. S.; Palaniswamy, V. A.; Eisenbraun, E. J., *J. Org. Chem.* **1986**, *51*, 1402. (c) Rosen, G. M.; Rauckman, E. J., *Org. Prep. Proced. Int.* **1978**, *10*, 17. (d) Duraisamy, M.; Walborsky, H. M., *J. Am. Chem. Soc.* **1983**, *105*, 3252. (e) Bennett, M.; Gill, G. B.; Pattenden, G.; Shuker, A. J., *Synlett* **1990**, 455.

10. Konakahara, T.; Kurosaki, Y., *J. Chem. Res. (S)* **1989**, 130.

Michael Rathke & Robert Elghanian
Michigan State University, East Lansing, MI, USA

3-Butyn-2-ol, 4-(trimethylsilyl)-*

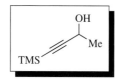

[6999-19-5]   C$_7$H$_{14}$OSi   (MW 142.27)

InChI = 1S/C7H14OSi/c1-7(8)5-6-9(2,3)4/h7-8H,1-4H3

InChIKey = HJJSDJHRTMFJLP-UHFFFAOYSA-N

Alternate Name: TMS-butynol.

Physical Data: bp 83–85 °C (13 mmHg).

Solubility: highly soluble in all standard organic solvents (hexanes, toluene, CH$_2$Cl$_2$, EtOAc, alcohols, ethers). Partially soluble in water.

Form Supplied in: colorless to light yellow liquid; racemic compound is commercially available. Nonracemic material must be prepared.

Handling, Storage, and Precaution: limited data are available for 4-TMS-3-butyn-2-ol. Prudent laboratory practices for handling chemicals should be followed. Use in a fume hood with adequate ventilation.

Preparative Methods: racemic 4-trimethylsilyl-3-butyn-2-ol can be prepared by deprotonation with strong bases (BuLi, LDA, Grignards reagents) of trimethylsilylacetylene followed by addition to acetaldehyde.[1] Deprotonation of 3-butyn-2-ol followed by quenching with excess trimethylsilyl chloride followed by concomitant hydrolysis of the trimethylsilyl ether is generally the most straightforward route.[2] Enzymatic reduction of 4-TMS-3-butyn-2-one has also been used to prepare the reagent using alcohol dehydrogenase.[3]

Preparation of nonracemic 4-TMS-3-butyn-2-ol has been accomplished by asymmetric addition of dimethylzinc to acetaldehyde promoted by TADDOL[4] or addition of a trimethylsilylvinylsulfoxide to acetaldehyde followed by thermal elimination of the sulfoxide.[5] Asymmetric reduction of 4-TMS-3-butyn-2-one using stoichiometric reducing reagents,[6] catalytic transfer hydrodrogenation,[7] and enzymatic reduction with isolated protein or whole cells afford the 4-TMS-3-butyn-2-ol with varying degrees of enantioenrichment.[8] Enzymatic

resolution by esterification of the racemic alcohol is the method of choice for the large-scale preparation.[9]

TMS Allenyl Metal Reagents.   Addition of an enantioenriched TMS allenyl metal reagent derived from the nonracemic mesylate of 4-TMS-3-butyn-2-ol to aldehydes has been shown to be a powerful method to form the corresponding homopropargyl adducts with high diastereoselectivity (eq 1). TMS-3-butyn-2-ol was successfully utilized by Marshall and coworkers with the focus directed on generation of a highly stereoselective, chiral allenylindium reagent formed by transmetallation with an in situ generated allenylpalladium species.[10]

$$ \text{(eq 1)} \tag{1} $$

More recently, Mulzer has reported use of the corresponding allenylsilane derived from 4-TMS-3-butyn-2-ol for use in the synthesis of the C13–C18 fragment of branimycin (eq 2).[11]

$$ \text{(eq 2)} \tag{2} $$

*Article previously published in the electronic Encyclopedia of Reagents for Organic Synthesis as 4-Trimethylsilyl-3-butyn-2-ol. DOI: 10.1002/047084289X. rn01324.

Generation of Optically Active α-Trimethylsilylbutenolides. Sato and coworkers published a highly efficient method to prepare enantiomerically pure α,β-unsaturated butenolides from enantioenriched 4-TMS-3-butyn-2-ol through a Ti-catalyzed hydromagnesiation reaction of the alkyne, carboxylation using CO_2, and a subsequent lactonization (eq 3).[12] This approach was later employed elegantly by Negishi in the synthesis of epolactaene.[13]

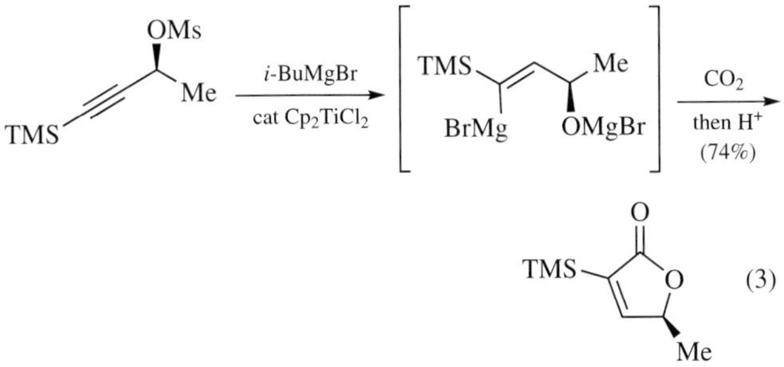

Cross-couplings. Fu and Smith recently reported a Ni-catalyzed asymmetric cross-coupling of racemic 4-TMS-3-butyn-2-ol by initial conversion to the corresponding bromide and subsequent Negishi cross-coupling in the presence of a chiral catalyst system (eq 4).[14] The key carbon–carbon bond was generated with a reported ee of 93% and in very good overall yield.

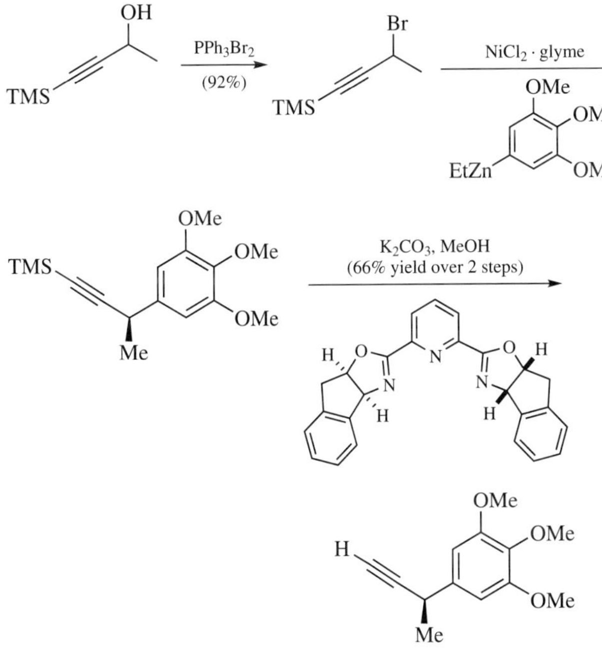

1. Danheiser, R. L.; Carini, D. J.; Fink, D. M.; Basak, A., *Tetrahedron* **1983**, *39*, 935.

2. Pearson, A. J.; Kim, J. B., *Org. Lett.* **2002**, *4*, 2837.

3. Wong, C.; Bradshaw, C. W. U.S. Pat. 5,225,339 (to Scripps Research Institute) (1993).

4. Oguni, N.; Satoh, N.; Fujii, H., *Synlett* **1995**, *10*, 1043.

5. Nakamura, S.; Kusuda, S.; Kawamura, K.; Toru, T., *J. Org. Chem.* **2002**, *67*, 640.

6. Helal, C. J.; Magriotis, P. A.; Corey, E. J., *J. Am. Chem. Soc.* **1996**, *118*, 10938.

7. (a) Matsumura, K.; Hashiguchi, S.; Ikariya, T.; Noyori, R., *J. Am. Chem. Soc.* **1997**, *119*, 8738. (b) Thompson, C. F.; Jamison, T. F.; Jacobsen, E. N., *J. Am. Chem. Soc.* **2001**, *123*, 9974.

8. (a) Zhang, B.-B.; Lou, W.-Y.; Zong, M.-H.; Wu, H., *J. Mol. Catal. B* **2008**, *54*, 122. (b) Schubert, T.; Hummel, W.; Kula, M. R.; Muller, M., *Eur. J. Org. Chem.* **2001**, *22*, 4181.

9. (a) Burgess, K.; Jennings, L. D., *J. Am. Chem. Soc.* **1991**, *113*, 6129. (b) Dickman, D. A.; Ku, Y.-Y.; Morton, H. E.; Chemburkar, S. R.; Patel, H. H.; Thomas, A.; Plata, D. J.; Sawick, D. P., *Tetrahedron: Asymmetry* **1997**, *8*, 1791. (c) Marshall, J. A.; Chobanian, H. R., *Org. Synth.*, **2005**, *82*, 43. (d) Hungerhoff, B.; Sonnenschein, H.; Theil, F., *J. Org. Chem.* **2002**, *67*, 1781.

10. Marshall, J. A.; Chobanian, H. R.; Yanik, M. M., *Org. Lett.* **2001**, *3*, 3369.

11. Felzmann, W.; Castagnolo, D.; Rosenbeiger, D.; Mulzer, J., *J. Org. Chem.* **2007**, *72*, 2182.

12. Ito, T.; Okamoto, S.; Sato, F., *Tetrahedron Lett.* **1990**, *31*, 6399.

13. Tan, Z.; Negishi, E., *Org. Lett.* **2006**, *8*, 2783.

14. Smith, S. W.; Fu, G. C., *J. Am. Chem. Soc.* **2008**, *130*, 12645.

Harry R. Chobanian
Merck & Co., Inc., Rahway, NJ, USA

Mathew M. Yanik
J-Star Research, South Plainfield, NJ, USA

Chlorodiisopropylsilane

[2227-29-4] C$_6$H$_{15}$ClSi (MW 150.73)

InChI = 1S/C6H15ClSi/c1-5(2)8(7)6(3)4/h5-6,8H,1-4H3

InChIKey = CGXYLRTYLIHXEE-UHFFFAOYSA-N

(reagent for the synthesis of *O*-diisopropylsilyl derivatives for intramolecular hydrosilylation reactions; used in a 1,3-*anti*-selective reduction of β-hydroxy ketones,[1–3] and in a mild method for reducing β-hydroxy esters[4])

Physical Data: bp 150–153 °C,[5] bp 54–55 °C/45 mmHg;[6] *d* 0.872 g mL^{-1}.[6]

Solubility: generally sol organic solvents; reacts with alcohols, ammonia,[6] and water.

Form Supplied in: neat liquid, commercially available.

Analysis of Reagent Purity: ^{1}H NMR δ(CDCl$_3$) 4.35 (Si–H).[5] The most likely impurity is tetraisopropyldisiloxane, (*i*-Pr$_2$-SiH)$_2$O, for which the corresponding signal is slightly upfield at 4.28.[7]

Preparative Method: obtained by reaction of trichlorosilane with isopropylmagnesium chloride;[6] the original yield of 45% may be raised to 70–80% by employing conc hydrochloric acid to quench the reaction.[2]

Purification: distillation at atmospheric or reduced pressure.

Handling, Storage, and Precautions: shelf-stable provided moisture is excluded. Corrosive; yields HCl on reaction with water. Use in a fume hood.

Original Commentary

Anthony P. Davis
Trinity College, Dublin, Ireland

1,3-*Anti*-Selective Reduction of β-Hydroxy Ketones. Treatment of β-hydroxy ketones (**1**) with chlorodiisopropylsilane in the presence of pyridine, or (more generally) triethylamine with catalytic 4-dimethylaminopyridine, gives β-diisopropylsilyloxy ketones (**2**) (eq 1).[1,2] These derivatives may be purified by chromatography on silica gel, and are obtained in yields of ca. 70–80%. Under the influence of Lewis acidic catalysts, most usefully tin (IV) chloride, they undergo intramolecular hydrosilylation to give *trans*-siladioxanes (**3**), along with minor amounts of *cis* stereoisomers (**4**) (e.g. R^1, R^2 = *i*-Pr, diastereomer ratio (dr) = 120:1; R^1, R^2 = Bu, dr = 40:1).[1,2] After desilylation, *anti*-diols

(**5**) may be isolated. The overall yields of (**5**) from hydroxy ketones (**1**) are in the region 60–70%.[2]

The hydrosilylation step is presumed to occur via six-membered cyclic transition state (**6**). The stereoselectivity results from (a) the equatorial disposition of R^1, and (b) steric and stereoelectronic effects (including possibly a C–O···Si interaction) which promote the axial orientation for the carbonyl group. The dimethylsilyl derivative (**7**) has been used in similar sequences,[2] but selectivities were found to be lower and optimization was complicated by the hydrolytic instability of Me$_2$Si–O bonds.

The 1,3-*anti* selectivity is maintained in the presence of an α-substituent, i.e. in substrates such as (**8**). Thus diisopropylsilylation/hydrosilylation of both diastereomers of (**8**) gave the 1,3-*trans*-siladioxane (**9**), with dr ≥ 250:1 (eq 2).[3] However, with the more challenging substrates (**10**) the α-position was found to have more influence. Both diastereomers gave principally the 1,2-*trans* product (**11**) (eq 3) but, while *anti*-(**10**) resulted in dr = 97:1, *syn*-(**10**) gave only dr = 5:1.

As part of the above work it was found that the superacid TfOH$_2^+$B(OTf)$_4^-$ may also be used as the hydrosilylation catalyst.[3] In one case it gave superior results to those obtained with SnCl$_4$.

The 1,3-*anti*-selective reduction of β-hydroxy ketones may also be accomplished using tetramethylammonium triacetoxyborohydride,[8] and the samarium-catalyzed intramolecular Tishchenko reduction.[9] The former gives the *anti*-diols directly, although with slightly lower stereoselectivity than the intramolecular hydrosilylation. The latter results in monoacylated products, and gives excellent yields and selectivities.

Intramolecular Reduction of β-Hydroxy Esters. Intramolecular hydrosilylation is also possible within β-diisopropylsilyloxy esters (**13**), constituting an exceptionally mild method for reducing ester groups to the aldehyde oxidation level (eq 4).[4] The derivatives (**13**) may be synthesized from β-hydroxy esters (**12**) as described above for the analogous ketones. Treatment with fluoride ions (but not Lewis acids) induces hydride transfer to give alkoxysiladioxanes (**14**) in excellent yields (≥95%). Although usually performed in dichloromethane, the hydrosilylation may also be accomplished with ethyl acetate as solvent, providing strong evidence for intramolecularity.

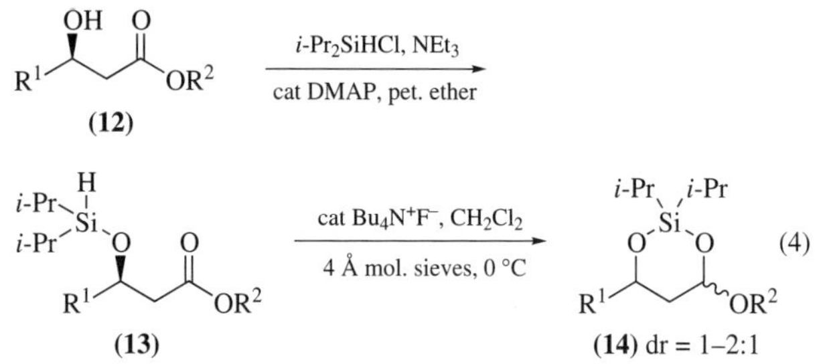

While the reduction in eq 4 is not stereoselective, the stereocontrolled elaboration of (**14**) may be achieved in principle via conformationally biased carbocations (**15**). This has been demonstrated for the allylation of (**14**) with allyltrimethylsilane (eq 5).[4] The major products (**16**) appear to derive from axial attack on (**15**). The most effective catalyst was $TfOH_2^+B(OTf)_4^-$, which could be used at levels of 1–2 mol% and gave yields of 78–85% with dr ≥ 20:1. While it is possible that the superacid acts by protonation of the alkoxy oxygen in (**14**), it is probably more likely that the active species is the supersilylating agent $Me_3Si\,B(OTf)_4$,[10] derived from superacid and allylsilane.

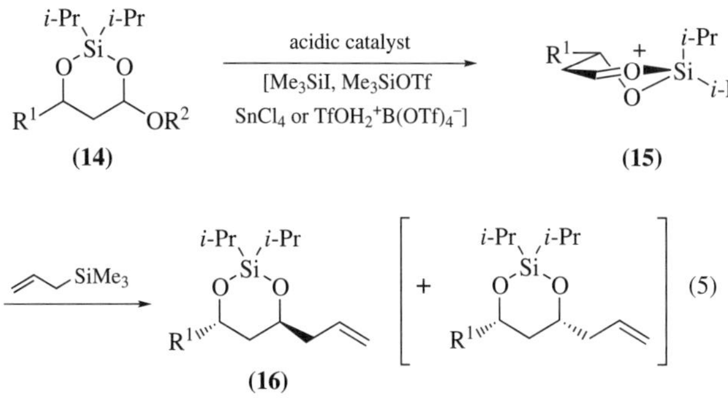

Intramolecular Hydrosilylation/Oxidation of Alkenols. Since the advent of methods for the oxidative degradation of organosilicon compounds, it has been possible to develop 'controlled hydration' methodology based on the hydrosilylation of C=C bonds. A useful variant involves the *O*-silylation of an alkenol with a dialkylsilylating agent R_2SiHX, followed by intramolecular hydrosilylation/oxidation to give a diol product.[11,12] Because of the intramolecular delivery of the Si–H moiety, these reactions often take place with excellent stereoselectivity. While

most of this work has employed Me_2SiH as the 'internal reagent', the use of *i*-Pr_2SiH has been explored occasionally.[13,14] An example where it proved advantageous is the conversion of allylic alcohol (**17**) to *syn*-diol (**18**) (eq 6) in dr 30:1.[14] A related method involving dimethylsilyl groups gave dr of only 2.4:1 with this substrate.[12]

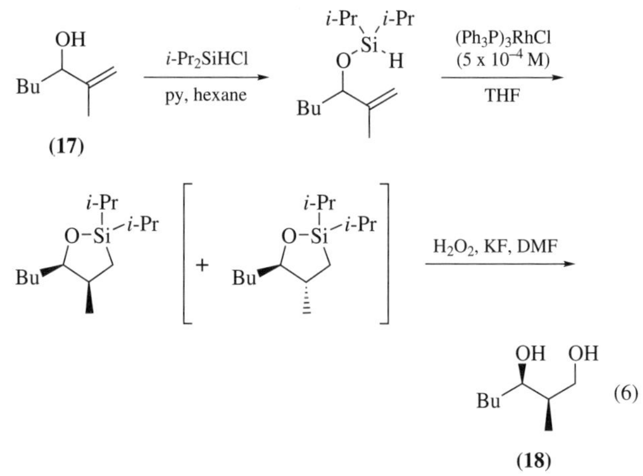

Synthesis of Other Organosilicon Reagents. Aside from the direct applications referred to above, *i*-Pr_2SiHCl serves as an intermediate for the synthesis of certain other reagents. Examples are (a) *i*-$Pr_2Si(OTf)_2$, used for the protection of diols as diisopropylsilylene derivatives,[15] (b) dialkylarylsilanes (**19**) for the introduction of a fluorescent *O*-protecting group,[16] and (c) the disiloxane (**20**), which may be converted to (**21**) (for diol protection)[7] or (**22**) (used in intramolecular ionic hydrogenation).[17]

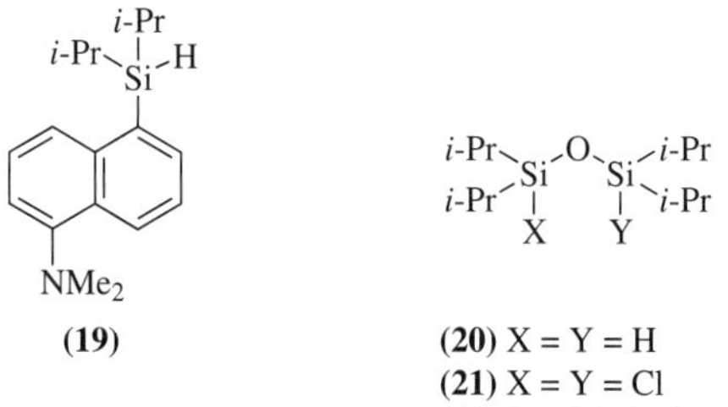

First Update

Daesung Lee & Yi Jin Kim
University of Wisconsin, University Avenue, Madison, WI

Intramolecular Hydrosilylation of Alkynes. Hydrosilylation with chlorodiisopropylsilane has been extended to include alkynes. In a synthetically useful example of an intramolecular variant of alkyne hydrosilylation,[18–21] treatment of a homopropargyl alcohol (**23**) with triethylamine and catalytic DMAP in the presence of chlorodiisopropylsilane affords a silyl ether (**24**) that is suitable for intramolecular hydrosilylation (eq 7).[20] The hydrosilylation occurs in the presence of Speier's catalyst to afford alkylidenesilacyclopentane (**25**) cleanly and in high yield.

(7)

(8)

By switching the catalyst from platinum to ruthenium, the stereoselectivity of the *syn*-hydrosilylation can be reversed to afford the *anti*-hydrosilylation product (**26**) in the presence of catalytic [RuCl$_2$(benzene)]$_2$ (eq 8).[19]

The synthetic utility of this chemistry has been demonstrated through a palladium-catalyzed cross-coupling reaction of alkylidenesilacyclopentanes (**25**) with aryl iodides (eq 9). Treatment of **25** with TBAF and catalytic Pd(dba)$_2$ in the presence of iodobenzene yielded arylated homoallylic alcohol (**27**) in good yield. This reaction was tolerant of a variety of functional groups and occured stereospecifically with respect to the double bond geometry. Unfortunately, cross-coupling between the *anti*-hydrosilylation product (**26**) and iodobenzene was very slow and reaction with other substituted aryl iodides failed to occur. However, this problem can be alleviated by switching from the diisopropylsilacyclopentane to dimethyl variants, which indicates that steric hindrance imposed by the isopropyl groups plays a significant role in the efficiency of the cross-coupling reaction.

(9)

Silyl-substituted Enol Ethers via Direct Metalation and Their Cross-Coupling. In addition to silacyclopentanes and silanols,[22] silyl-substituted enol ethers prepared from chlorodiisopropylsilane have also been shown to undergo cross-coupling reactions with various aryl iodides.[23] Lithiation of 2,3-dihydropyran (**28**) followed by the addition of chlorodiiosopropylsilane yields hydridosilane (**29**),[24] which can directly undergo palladium-catalyzed cross-coupling with aryl iodides in good to excellent yields (eq 10).

(10)

Intramolecular Silylformylation of Alkynyl Hydrosilanes. Intramolecular silylformylation is possible with alkynyl hydrosilanes.[25] Following standard conditions for silyl ether formation between a homopropargyl alcohol (**31**) and chlorodiisopropylsilane using triethylamine in hexane, the silyl ether (**32**) was subjected to [Rh(CN*t*-Bu)$_4$][Co(CO)$_4$] under a carbon monoxide atmosphere in toluene yielding the silylformylation product (**33**) stereoselectively and in good yield (eq 11). The synthetic utility of these silylformylation products is similar to that of the hydrosilylation products. These formylated alkenyl siloxanes have been shown to undergo cross-coupling reactions with aryl iodides under slightly modified conditions.

(11)

Ti(II)-mediated Enyne Cyclization. Five-membered cyclic siloxanes are typically synthesized by intramolecular hydrosilylation or silylformylation of alkynes[25] or via radical additions to alkynyl silanes.[26] An alternative route to these synthetically useful intermediates involves a Ti(II)-mediated cyclization of (silyloxy)enynes (eq 12).[27] Silylation of propynyllithium (**34**) with chlorodiisopropylsilane and subsequent bromination with NBS affords bromodiisopropylpropynylsilane (**35**). Upon silylation of allylic alcohol (**36**), the resulting (silyloxy)enyne (**37**) undergoes Ti(II)-mediated enyne cyclization to afford exclusively the *anti*-diastereoemer of the cyclic siloxane (**38**) in good yield. This reductive enyne coupling offers a new route to the construction of polypropionate subunits and has been successfully implemented in the total syntheses of natural products (−)-dictyostatin and (−)-7-demethylpiericidin A$_1$.[28,29]

Silicon-tethered Ene Reactions. Silicon-tethered cyclization reactions have been widely utilized because of the ease of formation and removal of the temporary silicon tether and the ability of organosilicon compounds to tolerate many commonly used organic transformations. One example involves the ene reaction of α-(prenyl)dialkylsilyloxy aldehydes (eq 13).[30] A 1,2-diol (**39**) can be selectively silylated by treatment with *n*-BuLi and chlorodiisopropylsilane followed by prenyllithium to afford internally silylated siloxane (**40**). Oxidation of the remaining primary alcohol to the corresponding aldehyde leads to the ene cyclization precursor, which upon treatment with Me$_2$AlCl undergoes cyclization to afford a mixture of diastereomeric ene products (**41** and **42**) with the *anti, anti*-isomer **41** predominating. The major product can be oxidized under Tamao–Kumada conditions to afford the triol (**43**),[31] albeit in low yield (17%). Related substrates containing a diphenylsilyl tether (prepared from chlorodiphenysilane) gave much better conversion to the triol after oxidative cleavage (48%).

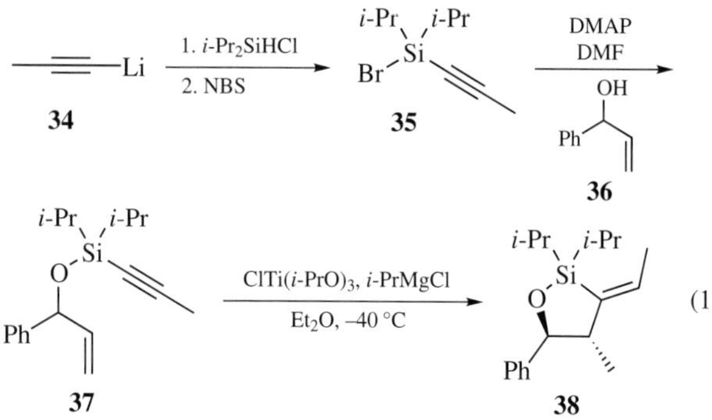

Reductive Etherification of Alkoxydialkylsilanes with Carbonyl Compounds. Chlorordiisopropylsilane along with other silicon hydride reagents are commonly used for the reduction of carbonyl compounds, often in the context of intramolecular reduction of β-dialkylsilyloxy ketones formed from silylation of β-hydroxy ketones.[1,2] One variant of the utilization of the Si–H bond as the hydride source is a reductive etherification of alkoxydialkylsilanes with carbonyl compounds using a Cl(*i*-Pr)$_2$SiBr (BiBr$_3$/*i*-Pr$_2$SiHCl) catalytic system (eq 14),[32] Omission of a catalytic amount of chlorodiisopropylsilane from the BiBr$_3$-mediated coupling results in lower yields even after extended reaction times. Thus, treatment of benzyl alcohol (**44**) under standard conditions (DMAP, Et$_3$N) with chlorodiisopropylsilane affords benzyloxydiisopropylsilane (**45**). On addition of **45** and then benzaldehyde to a mixture of BiBr$_3$ and chlorodiisopropylsilane, formation of dibenzyl ether (**46**), via hydride delivery to an oxonium intermediate, is observed in excellent yield (90%).

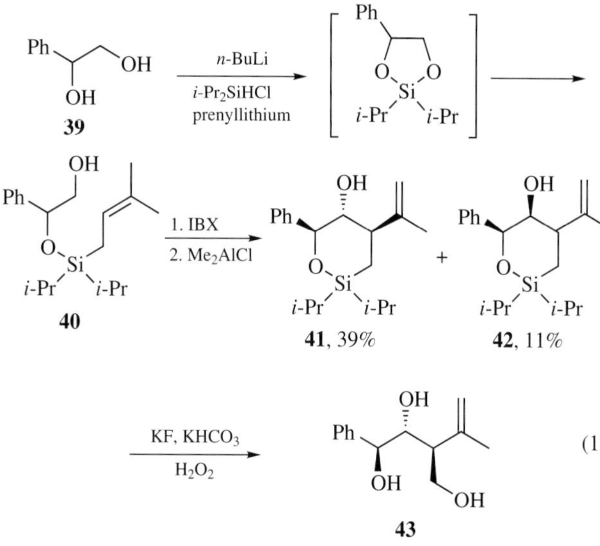

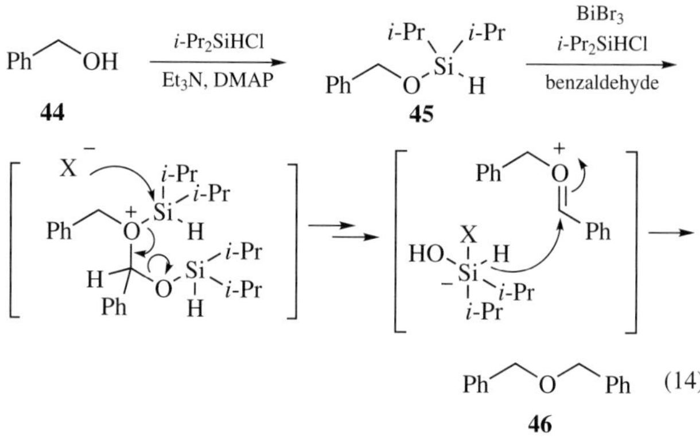

Preparation of Unsymmetrical Silaketals. Silaketals have gained popularity due to the fact that they serve as temporary tethers that convert less efficient and less selective intermolecular transformations into intermolecular ones with increased efficiency and selectivity. Furthermore, silaketals, like many other silicon-derived tethers, are easily formed and removed in organic transformations. A three-step protocol for the preparation of unsymmetrical silaketals without isolation of any of the intermediates formed has been reported using chlorodiisopropylsilane (eq 15).[33] A clear advantage to a one-pot preparation of silaketals from chlorodiisopropylsilane is that problematic separation issues of the intermediates from other organosilicon side products or from the final product are avoided. Initial treatment of alcohol (**47**) with Et$_3$N and DMAP affords siloxane intermediate (**48**), which upon addition of NBS is converted to the bromide (**49**). Finally, this intermediate species can be treated with an alcohol again in the presence of Et$_3$N and DMAP to afford silaketal (**50**) in very good yield without contamination by undesired symmetrical silaketal side products. These conditions were applicable to a range of substrate alcohols and are also amenable to the preparation of silyl ethers and unsymmetrical silanes.[34]

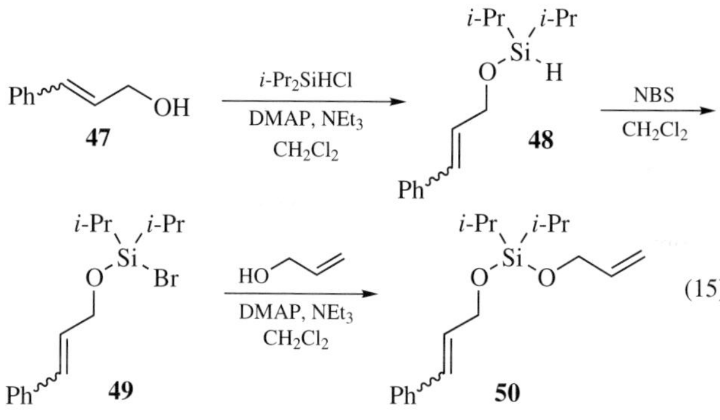

Photochemically Removable Silyl Protecting Groups. A photochemically removable silyl protecting group offers an alternative method to control the release of this useful functional group in protecting group chemistry.[35] *o*-Ethynylphenyl acetate (**51**) was subjected to hydrosilylation with chlorodiisopropylsilane in the presence of catalytic H$_2$PtCl$_6$ to afford chloroalkenylsilane (**52**, eq 16).[36,37] Treatment of this intermediate with lithium dimethylamide followed by addition of 4-*t*-butylcyclohexanol allowed simultaneous protection of the alcohol and removal of the acetate group to afford silyl ether (**53**). Protection of various primary and secondary alcohols can be achieved using this chloroalkenylsilane, although, tertiary alcohols remained unreactive. In addition, a variant of this silylating reagent containing the dimethyl counterpart requires significantly shorter reaction times than those with diisopropyl groups. The photochemical removal of the silyl protecting group can be achieved by irradiating the silyl ether (**53**) at 254 nm in acetonitrile to produce cyclic siloxane (**54**) and 4-*t*-butylcyclohexanol.

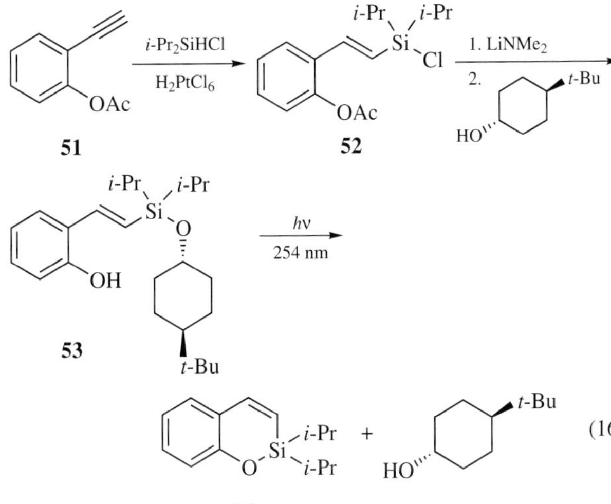

the diphenyl variant of the silanol giving lower values (30% and 89% ee, respectively).

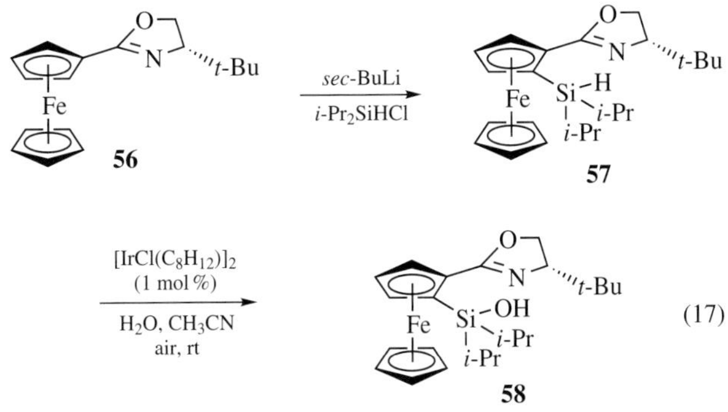

The related photocleavable silylating agent [(2-acetoxy-3-naphthyl)vinyl]-diisopropylsilyl chloride (**55**) can also be used under similar conditions with cleavage from the alcohol on irradiation at 350 nm.

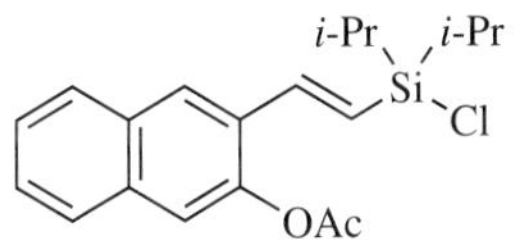

Organosilanols as Catalysts in Asymmetric Aryl Addition Reactions. Although organosilanols have been utilized in organic synthesis, as intermediates for palladium-catalyzed cross-coupling reactions,[22,23] they have not been employed as chiral ligands in catalytic asymmetric reactions until recently. Chlorodiisopropylsilane can be used for the formation of ferrocene-based organosilanols to catalyze asymmetric aryl addition reactions.[38] A chiral ferrocenyl oxazoline (**56**), available in two steps from ferrocene carboxylic acid, can be simply treated with *sec*-BuLi and then with chlorodiisopropylsilane to yield silylated 1,2-disubstituted ferrocene (**57**) in good yield (eq 17). Then the silyl group can be oxidized to give the silanol (**58**) by treatment with catalytic [IrCl(C$_8$H$_{12}$)]$_2$ in the presence of water and air.[39] The synthesis of this silanol can be varied with respect to the substituent on the oxazoline ring and the alkyl groups on the silicon (i.e., by the use of a related reagent chlorodimethylsilane) without significant changes in yield and efficiency.

This chiral organosilanol is capable of catalyzing asymmetric phenyl addition to various substituted benzaldehydes. Thus, treatment of 4-chlorobenzaldehyde with Ph$_2$Zn/Et$_2$Zn in the presence of catalytic organosilanol (**58**) yielded product (**59**) in 82% and 91% ee (eq 18). Changes in ee were noted depending on the dialkyl substituents on the silicon center, with the dimethyl or

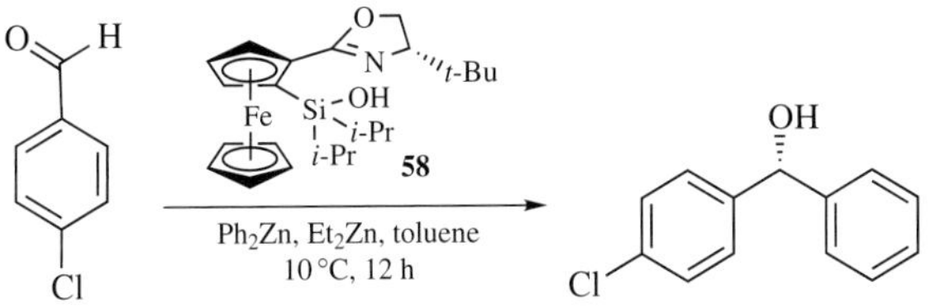

Development of Diisopropylsilyl Solid Supports. A diisopropylsilyl linker-based solid support (**62**) can be synthesized in six steps beginning with silylation of 4-lithioanisole (**60**) with chlorodiisopropylsilane to form silane (**61**, eq 19). Substrate loading can be accomplished by activation of the silane to the silyl triflate followed by addition of the substrate alcohol to form the silyl ether-linked substrate (**61**). Once the small molecule synthesized on the bead is ready for cleavage, treatment with a 5% solution of HF/pyridine in THF accomplishes efficient recovery of the desired small molecule and in high purity.[40,41]

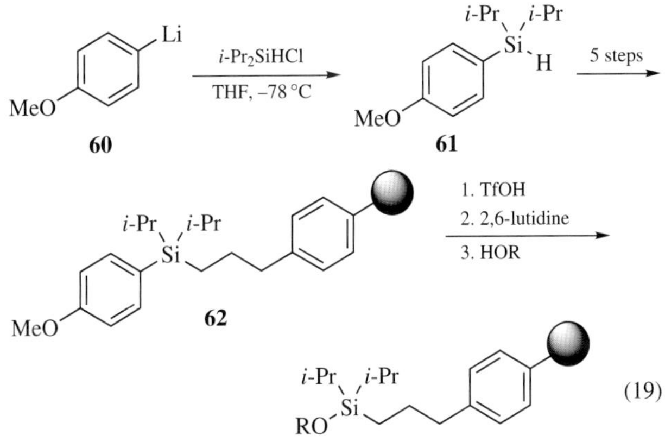

Related Reagents. Chlorodimethylsilane; Tetramethylammonium Triacetoxyborohydride.

1. Anwar, S.; Davis, A. P., *J. Chem. Soc., Chem. Commun.* **1986**, 831.

2. Anwar, S.; Davis, A. P., *Tetrahedron* **1988**, *44*, 3761.

3. Anwar, S.; Bradley, G.; Davis, A. P., *J. Chem. Soc., Perkin Trans. 1* **1991**, 1383.

4. Davis, A. P.; Hegarty, S. C., *J. Am. Chem. Soc.* **1992**, *114*, 2745.

5. Bradley, G. MSc thesis, Univ. of Dublin, 1990.

6. Metras, F.; Valade, J., *Bull. Soc. Claim. Fr.* **1965**, 1423.

7. Markiewicz, W. T., *J. Chem. Res. (S)* **1979**, 24; *J. Chem. Res. (M)*, 0181.

8. Evans, D. A.; Chapman, K. T.; Carreira, E. M., *J. Am. Chem. Soc.* **1988**, *110*, 3560.

9. Evans, D. A.; Hoveyda, A. H., *J. Am. Chem. Soc.* **1990**, *112*, 6447.

10. Davis, A. P.; Jaspars, M., *Angew. Chem., Int. Ed. Engl.* **1992**, *31*, 470.

11. Tamao, K.; Ishida, N.; Tanaka, T.; Kumada, M., *Organometallics* **1983**, *2*, 1694.

12. Tamao, K.; Nakajima, T.; Sumiya, R.; Arai, H.; Higuchi, N.; Ito, Y., *J. Am. Chem. Soc.* **1986**, *108*, 6090.

13. (a) Curtis, N. R.; Holmes, A. B., *Tetrahedron Lett.* **1992**, *33*, 675. (b) Denmark, S. E.; Forbes, D. C., *Tetrahedron Lett.* **1992**, *33*, 5037. (c) Bergens, S. H.; Noheda, P.; Whelan, J.; Bosnich, B., *J. Am. Chem. Soc.* **1992**, *114*, 2121.

14. Anwar, S.; Davis, A. P., *Proc. R. Irish Acad.* **1989**, *89B*, 71.

15. Corey, E. J.; Hopkins, P. B., *Tetrahedron Lett.* **1982**, *23*, 4871.

16. Horner, L.; Mathias, J., *J. Organomet. Chem.* **1985**, *282*, 175.

17. McCombie, S. W.; Ortiz, C.; Cox, B.; Ganguly, A. K., *Synlett* **1993**, 541.

18. Maifeld, S. V.; Tran, M. N.; Lee, D., *Tetrahedron Lett.* **2005**, *46*, 105.

19. Denmark, S. E.; Pan, W., *Org. Lett.* **2002**, *4*, 4163.

20. Denmark, S. E.; Pan, W., *Org. Lett.* **2001**, *3*, 61.

21. Sudo, T.; Asao, N.; Yamamoto, Y., *J. Org. Chem.* **2000**, *65*, 8919.

22. Denmark, S. E.; Wehrli, D., *Org. Lett.* **2000**, *2*, 565.

23. Denmark, S. E.; Neuville, L., *Org. Lett.* **2000**, *2*, 3221.

24. Evans, P. A.; Baum, E. W., *J. Am. Chem. Soc.* **2004**, *126*, 11150.

25. Denmark, S. E.; Kobayashi, T., *J. Org. Chem.* **2003**, *68*, 5153.

26. Tamao, K.; Maeda, K.; Yamaguchi, T.; Ito, Y., *J. Am. Chem. Soc.* **1989**, *111*, 4984.

27. O'Neil, G. W.; Phillips, A. J., *Tetrahedron Lett.* **2004**, *45*, 4253.

28. O'Neil, G. W.; Phillips, A. J., *J. Am. Chem. Soc.* **2006**, *128*, 5340.

29. Keaton, K. A.; Phillips, A. J., *J. Am. Chem. Soc.* **2006**, *128*, 408.

30. Robertson, J.; Hall, M. J.; Stafford, P. M.; Green, S. P., *Org. Biomol. Chem.* **2003**, *1*, 3758.

31. Tamao, K.; Kakui, M.; Akita, T.; Iwahara, R.; Kanatani, J.; Yoshida, J.; Kumada, M., *Tetrahedron* **1983**, *39*, 983.

32. Jiang, X.; Bajwa, J. S.; Slade, J.; Prasad, K.; Pepic, O.; Blacklock, T. J., *Tetrahedron Lett.* **2002**, *43*, 9225.

33. Petit, M.; Chouraqui, G.; Aubert, C.; Malacria, M., *Org. Lett.* **2003**, *5*, 2037.

34. Chouraqui, G.; Petit, M.; Aubert, C.; Malacria, M., *Org. Lett.* **2004**, *6*, 1519.

35. Pillai, V. N. R., *Synthesis* **1980**, 1.

36. Pirrung, M. C.; Lee, Y. R., *J. Org. Chem.* **1993**, *58*, 6961.

37. Pirrung, M. C.; Fallon, L.; Zhu, J.; Lee, Y. R., *J. Am. Chem. Soc.* **2001**, *123*, 3638.

38. Özcubukcu, S.; Schmidt, F.; Bolm, C., *Org. Lett.* **2005**, *7*, 1407.

39. Lee, Y.; Seomoon, D.; Kim, S.; Han, H.; Chang, S.; Lee, P. H., *J. Org. Chem.* **2004**, *69*, 1741.

40. Tallarico, J. A.; Depew, K. M.; Pelish, H. E.; Westwood, N. J.; Lindsley, C. W.; Shair, M. D.; Schrieber, S. L.; Foley, M., *J. Comb. Chem.* **2001**, *3*, 312.

41. Woolard, F. X.; Paetsch, J.; Ellman, J. A., *J. Org. Chem.* **1997**, *62*, 6102.

Chlorodimethylsilane

$[1066-35-9]$ C_2H_7ClSi (MW 94.62)

InChI = 1S/C2H7ClSi/c1-4(2)3/h4H,1-2H3

InChIKey = YGHUUVGIRWMJGE-UHFFFAOYSA-N

(intramolecular hydrosilation of allyl alcohols, homoallyl alcohols, allyl amines, and α-hydroxy ketones for the regio- and stereoselective synthesis of polyols and amino alcohols)

Physical Data: bp 35–36 °C; mp −111 °C; n_D^{20} 1.3830; d 0.852 g cm^{-3}.

Solubility: sol common aprotic organic solvents.

Form Supplied in: neat liquid.

Handling, Storage, and Precautions: moisture sensitive and corrosive. Must be stored under nitrogen in a tightly capped bottle in a freezer and transferred or weighed quickly via syringe. This reagent should be handled in a fume hood.

Intramolecular Hydrosilation of Allyl and Homoallyl Alcohols. For a detailed description of the intramolecular hydrosilation–oxidation sequence, see 1,1,3,3-tetramethyldisilazane. 1,1,3,3-Tetramethyldisilazane and *N,N*-diethylaminodimethylsilane are frequently employed for preparation of hydrodimethylsilyl ethers for the intramolecular hydrosilation of allyl and homoallyl alcohols. Chlorodimethylsilane in combination with a tertiary amine such as triethylamine is another useful reagent for the synthesis of hydrodimethylsilyl ethers, especially for large scale preparations.[1]

Other chlorodiorganosilanes (ClR$_2$SiH) such as chlorodiisopropylsilane (Cl(i-Pr)$_2$SiH),[2] chlorodiphenylsilane (ClPh$_2$SiH),[3] and 1-chloro-1-silacyclohexane ((CH$_2$)$_5$SiClH),[4] are also useful reagents for intramolecular hydrosilation, including catalytic asymmetric intramolecular hydrosilation.[3,4] These bulky hydrosilanes have some advantages over ClMe$_2$SiH in these reactions, since the silylated intermediates are less moisture sensitive, are more easily handled, and can lead to higher stereoselectivity in the alkene hydrosilation step.

Intramolecular Hydrosilation of Allyl Amines. The amino group of an allyl amine is silylated with chlorodimethylsilane, and then the resulting silazane is subjected to standard platinum-catalyzed cyclization and oxidation steps to form a *syn*-2-amino alcohol (eq 1). Exclusive formation of the four-membered cyclic product and the high stereoselectivity should be noted.[5]

Intramolecular Hydrosilation of Carbonyl Groups. Catalytic asymmetric intramolecular hydrosilation of the carbonyl group of an α-hydroxy ketone has been attained by using chlorodimethylsilane as the silylating agent and a rhodium catalyst containing an optically active bidentate ligand (eq 2).[6]

In addition, chlorodiisopropylsilane is a useful reagent for Lewis acid catalyzed intramolecular hydrosilation of β-hydroxy ketones[7] and for fluoride ion induced reduction of esters.[8] It may also be mentioned that chlorodi-*t*-butylsilane is used in highly stereoselective intramolecular ionic hydrogenations of hydroxy alkenes.[9]

$$Ph-\underset{NH_2}{CH}-CH=C(CH_3)_2 \xrightarrow[\text{(repeated)}]{\substack{\text{1. BuLi} \\ \text{2. ClSiMe}_2\text{H}}}$$

$$\underset{HMe_2Si}{Ph}\underset{SiMe_2H}{N}\ C=C(CH_3)_2 \xrightarrow[\text{rt}]{\text{cat Pt[\{(CH}_2\text{=CH)Me}_2\text{Si\}}_2\text{O]}_2}$$

$$\underset{HMe_2Si}{Ph}\underset{N-SiMe_2}{}\!\!\!\!\!\!\!\!\!\! i\text{-Pr} \xrightarrow[\text{76\% overall}]{\substack{30\% \text{ H}_2\text{O}_2 \\ \text{KF, KHCO}_3, \text{THF, MeOH, rt}}} \underset{NH_2}{Ph}\underset{}{\overset{OH}{CH}}-CH(CH_3)_2 \quad (1)$$

trans >99%   *syn* >99%

$$\underset{OH}{\overset{O}{\parallel}}\!\!\!\! \xrightarrow[\text{R}_3\text{N}]{\text{ClSiMe}_2\text{H}}$$

$$\underset{O}{\overset{O}{\parallel}}\!\!\!\! SiMe_2H \xrightarrow[\text{CH}_2\text{Cl}_2, \text{rt, 0.3–24 h}]{\text{cat [Rh(cod)(diphosphine)]}^+\text{TfO}^-}$$

$$\underset{O}{\overset{O-SiMe_2}{}}\!\!\!\! \xrightarrow[\text{MeOH}]{\text{K}_2\text{CO}_3} \underset{}{\overset{OH}{}}\!\!\!\!\text{OH} \quad (2)$$

75–90%
(*R*)-(+)
93% ee

$$\text{diphosphine} = $$

(*R,R*)-*i*-Pr-DuPHOS

1. Tamao, K.; Nakagawa, Y.; Ito, Y., *Org. Synth.*, **1995**, in press.
2. Anwar, S.; Davis, A. P., *Proc. R. Ir. Acad.* **1989**, *89B*, 71.
3. Tamao, K.; Tohma, T.; Inui, N.; Nakayama, O.; Ito, Y., *Tetrahedron Lett.* **1990**, *31*, 50, 7333.
4. Bergens, S. H.; Noheda, P.; Whelan, J.; Bosnich, B., *J. Am. Chem. Soc.* **1992**, *114*, 2121.
5. Tamao, K.; Nakagawa, Y.; Ito, Y., *J. Org. Chem.* **1990**, *55*, 3438; *J. Am. Chem. Soc.* **1992**, *114*, 218.
6. Burk, M. J.; Feaster, J. E., *Tetrahedron Lett.* **1992**, *33*, 2099.
7. Anwar, S.; Bradley, G.; Davis, A. P., *J. Chem. Soc., Perkin Trans. 1* **1991**, 1383.
8. Davis, A. P.; Hegarty, S. C., *J. Am. Chem. Soc.* **1992**, *114*, 2745.
9. McCombie, S. W.; Ortiz, C.; Cox, B.; Ganguly, A. K., *Synlett* **1993**, 541.

Kohei Tamao
Kyoto University, Kyoto, Japan

Chlorodimethylvinylsilane

[1719-58-0]   C$_4$H$_9$ClSi   (MW 120.65)
InChI = 1S/C4H9ClSi/c1-4-6(2,3)5/h4H,1H2,2-3H3
InChIKey = XSDCTSITJJJDPY-UHFFFAOYSA-N

(removable silyl tether for directing regio- and stereochemistry)

Alternate Names: dimethylvinylchlorosilane; vinyldimethyl-chlorosilane.
Physical Data: bp 82–83 °C; *d* 0.874 g mL^{-1} at 25 °C.
Solubility: most aprotic organic solvents, but CH$_2$Cl$_2$ is most commonly used; reacts with H$_2$O and other protic solvents.
Form Supplied in: clear, colorless to yellow to brown liquid.
Handling, Storage, and Precautions: avoid inhalation of vapor; keep away from sources of ignition; store in a cool place in a tightly sealed container away from moisture; air and moisture sensitive.

Lithiation. In the presence of *t*-BuLi in a hydrocarbon solvent or THF, chlorodimethylvinylsilane dimerizes to form 1,3-disilacyclobutanes (**1**) (eq 1).[1] Two different mechanisms are possible depending on the solvent used. In hydrocarbon solvents, the formation of silaethylene **2** occurred before dimerization (eq 1), but when THF was used, the α-lithiosilane **3** was formed as the intermediate in the reaction (eq 2).[2]

$$\text{Me}_2\text{ClSi}\diagup\!\!\!\diagup \xrightarrow[\text{hexane, }-78\,°C]{t\text{-BuLi}} \left[\underset{Me}{\overset{Me}{}}Si=\diagup\ t\text{-Bu}\right] \quad (2)$$

$$\longrightarrow \quad (1)$$

(**1**)

$$\text{Me}_2\text{ClSi}\diagup\!\!\!\diagup \xrightarrow[\text{THF, }-78\,°C]{t\text{-BuLi}} \left[\underset{Me}{\overset{Cl}{Si}}\overset{}{\underset{Li}{C}}\!\!\!\!\! t\text{-Bu}\right] \quad (3)$$

$$\longrightarrow \quad (2)$$

(**1**)

The intermediates were trapped with various dienes in order to confirm the mechanism of the reaction in different solvents. When 2,3-dimethyl-1,3-butadiene was used, the major product formed in hexane was the [2 + 4]-cycloaddition product **4**, but in THF the major product was the coupling product of the lithiated intermediate and the chlorodimethylvinylsilane **5** (eq 3).[2]

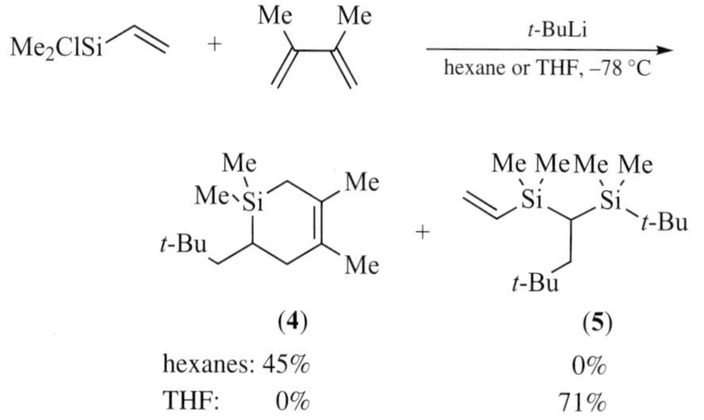

(3)

(4)   **(5)**

hexanes: 45%   0%
THF: 0%   71%

Hydroboration.   Metallacycloalkanones have been prepared by the cyclic hydroboration of dialkenyl silanes. In order to obtain one such dialkenyl silane, chlorodimethylvinylsilane was allylated with allyl magnesium bromide (eq 4).[3] The dialkenyl silane **6** was then hydroborated sequentially with thexylborane, followed by treatment with potassium cyanide, trifluoroacetic anhydride, and finally hydrogen peroxide in sodium hydroxide to produce the silacycloheptanone **7** (eq 4).[3] Cyclic silylketones were then further reacted to form acyclic compounds with chiral centers that are difficult to introduce without the use of the silicon tether.[4–6]

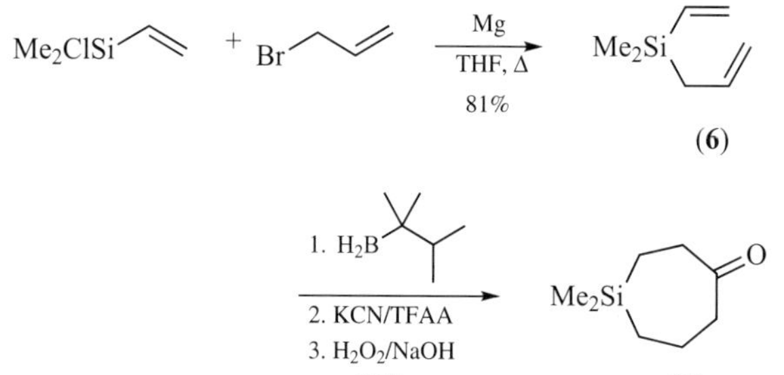

(4)

The hydroboration of vinylsilane has been used to make β-dimethylphenylsilylethyl ester linkers **10** (e.g., R = 4-CHO-C_6H_4) for solid-phase chemistry. In order to generate **8**, polysytrene–divinylbenzene resin was lithiated and quenched with chlorodimethylvinylsilane. Hydroboration followed by an oxidative workup gave the silyl ethanol resin **9** (eq 5), which can be further functionalized and cleaved from the resin with a protodesilylation reaction using TFA.[7,8]

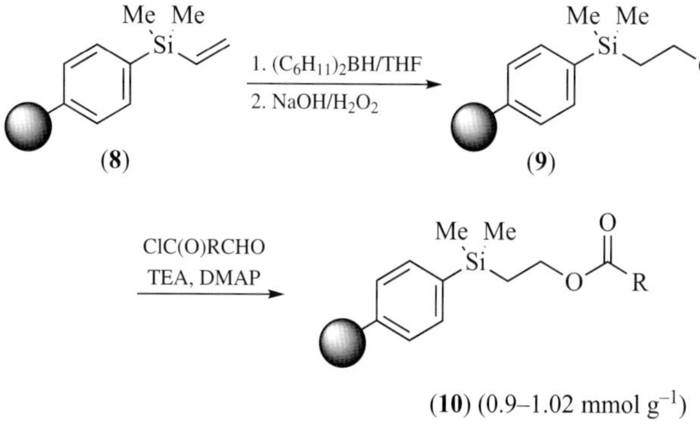

(10) (0.9–1.02 mmol g^{-1})

Hydroboration of a vinylsilane was utilized as a disposable silicon tether twice in the total synthesis of vineomycinone B$_2$ methyl ester.[9] In the first instance, the vinyl silyl functionality was added to the furan ring by a regioselective lithiation followed by addition

of chlorodimethylvinylsilane, generating **11**. The vinyl group of **11** was subjected to hydroboration/oxidation conditions to furnish the alcohol **12** in high yield (eq 6). In the second case, a bromofuran was subjected to a lithium–halogen exchange and quenched with chlorodimethylvinylsilane to form **13**, followed by hydroboration to give the alcohol **14** (eq 7). Both compounds were used in a tandem intramolecular Diels–Alder reaction with the removable silyl linkers directing the regiochemistry of the cycloaddition step.

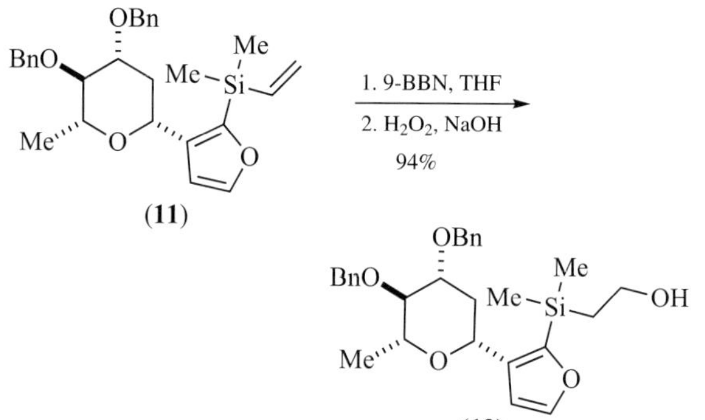

(6)

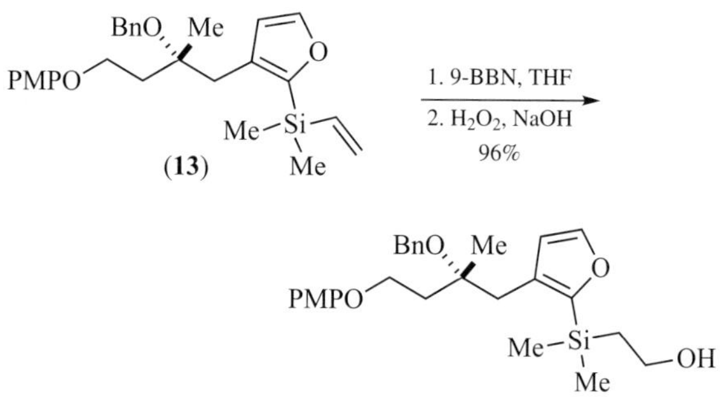

(7)

Radical Reactions.   The dimethylvinylsilyl group has been used in radical addition reactions in total synthesis[10] and as a reaction directing tether for radical cyclizations followed by cleavage of the silyl group.[11] For example, the synthesis of *C*-glycosides was accomplished using this method.[12] The dimethylvinylsilyl group was installed on the C2-OH in **15** using chlorodimethylvinylsilane, and a radical reaction with the SePh functionality was induced with tributyltin hydride and AIBN. The final product **16** was obtained in good yield and stereoselectivity after oxidation and deprotection of the silyl protecting groups, followed by benzyl protection of the alcohols (eq 8). This method has been applied to the synthesis of an IP$_3$ receptor ligand.[13]

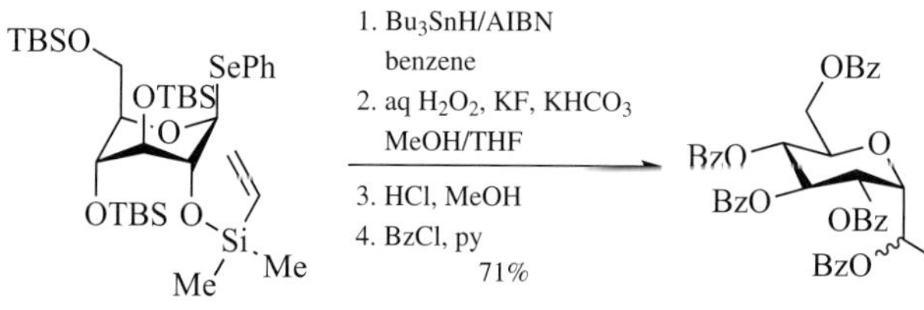

(8)

This chemistry was also applied in the synthesis of unnatural oligodeoxynucleotide building blocks for the synthesis of antisense agents. Dimethylvinylsilylated 4′-phenylselenothymidine **17** was subjected to the same conditions as before (Bu$_3$SnH and

AIBN), followed by oxidation of the silyl-cyclized product to afford the diol **18** in good yield (eq 9).[14]

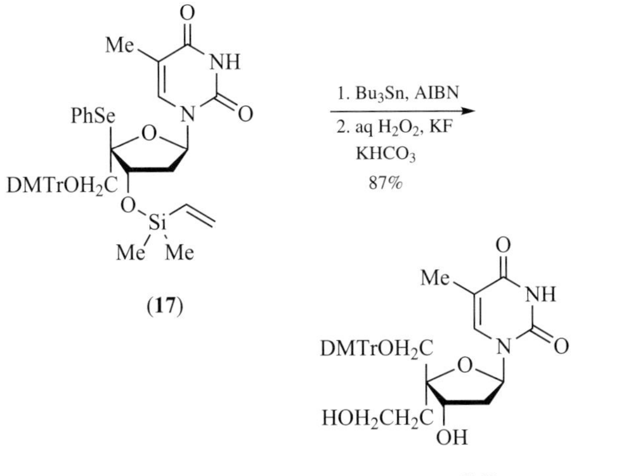

(9)

The key step toward the synthesis of daunosamine was an intramolecular vinyl group transfer of the dimethylvinylsilyl-protected **19** via a radical cyclization reaction, followed by treatment with TBAF to give the product **20** in good yield with excellent diastereoselectivity (eq 10).[15]

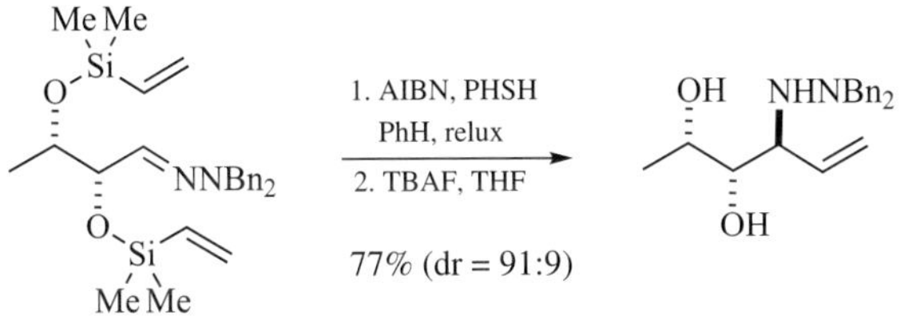

(10)

Vinyl groups have also been transferred to quaternary carbons using the vinyl silicon tether strategy. The vinylsiloxy ethers **21** were made by the reaction of the corresponding alcohols in DMF or DCM with chlorodimethylvinylsilane in the presence of pyridine or imidazole. They were subjected to radical reaction conditions using triethylborane to initiate the intramolecular cyclization followed by acid treatment to form the elimination products **22** favoring the 2,3-*syn*-diastereomer in good yields (eq 11).[16]

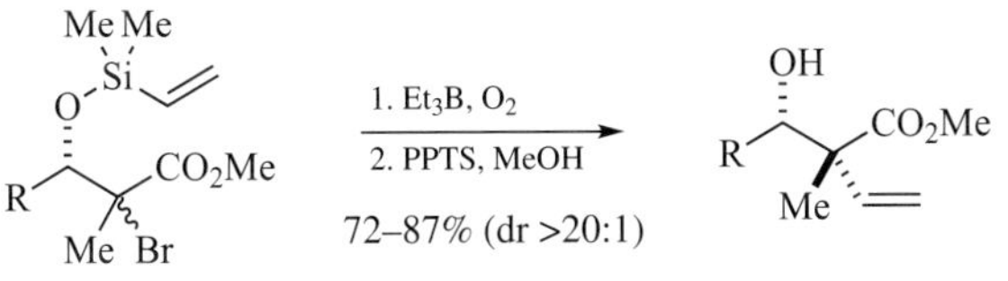

(11)

Cycloaddition Reactions. The dimethylvinylsilyl group has also been successfully used in cycloaddition reactions, including Diels–Alder reactions, to control regio- and stereochemistry.[11,17] For example, the silicon–tethered triene **23**, generated from (2*E*,4*E*)-hexa-2,4-dien-1-ol and chlorodimethylvinylsilane, underwent a thermal Diels–Alder reaction to form the silicon-tethered bicyclic product **24** in good yield (eq 12).[18] Removal of the silicon tether with TBAF produced the product **25** as a

single diastereomer. Alternatively, treatment with TBAF and hydrogen peroxide furnished the corresponding diol **26** as a single regioisomer and in a *cis*-1,2:*trans*-1,2 ratio of 70:30 (eq 13).[18]

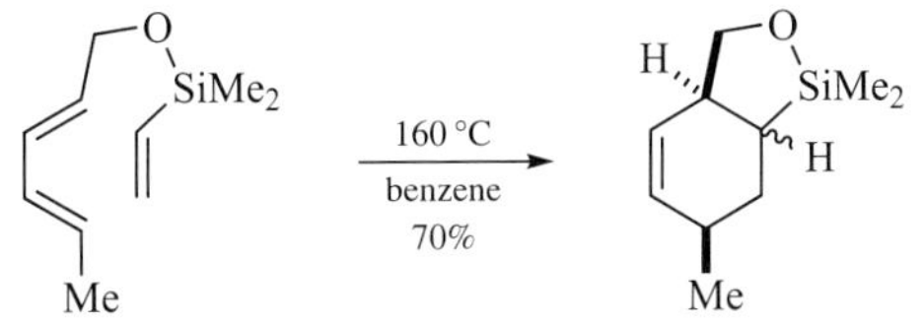

(12)

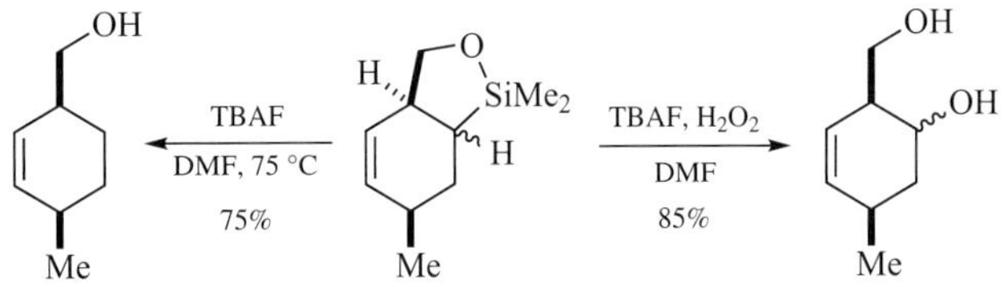

(13)

Sterically demanding dialkylvinylsilanes enabled the control of the *endo/exo* selectivity of the Diels–Alder reaction. Three vinylsilyl-tethered precursors **27** were investigated in an intramolecular Diels–Alder reaction to furnish the *exo* product **28** over the *endo* product **29**. As the size of the R group was increased, the selectivity for **28** increased, with the di-*t*-butyl group giving the highest selectivity (eq 14).[19]

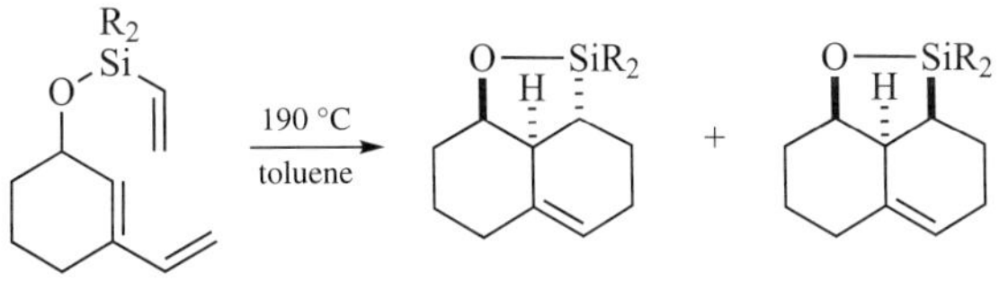

(14)

R = CH$_3$	4:1
Ph	10:1
t-Bu	>20:1

An example of a [5 + 2]-cycloaddition reaction using a silicon tether for regiocontrol has been performed with a pyrone alkene. Here, the alcohol **30** was reacted with dimethylvinylchlorosilane in the presence of triethylamine to afford the silicon-tethered intermediate **31**. Without purification, **31** was heated in toluene to give the cycloaddition product, and this was treated with potassium fluoride and *m*-CPBA in DMF to produce the final diol **32** in a 78% overall yield (eq 15).[20]

Silicon-tethered 1,3-dipolar cycloaddition reactions have been performed to regio- and stereoselectively assemble complex compounds.[21–23] A tandem reaction sequence was conducted by first installing the dimethylvinylsilane on **33** (R^1, R^2 include H, Me, Et, Bu, (CH$_2$)$_5$, *n*-C$_{12}$H$_{25}$, Ph, *anti*-PhCH(OH)) to form intermediate **34** followed by the 1,3-cycloaddition with the cyclic nitronate furnishing **35** in good yield (eq 16).[21]

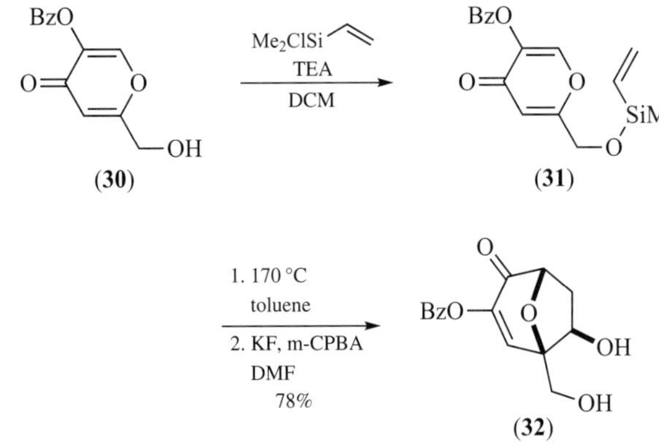

$$ (15) $$

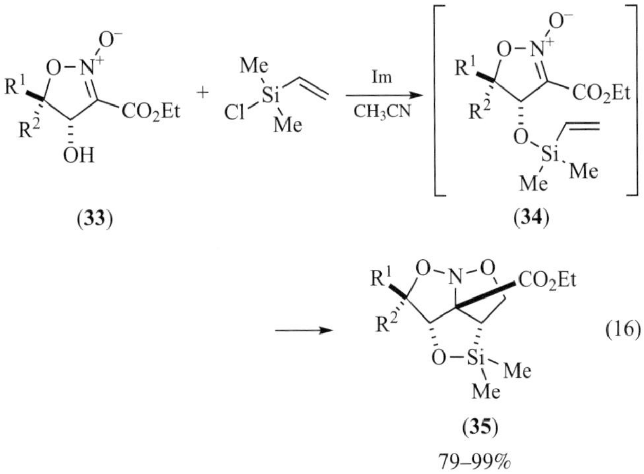

$$ (16) $$

(33) + (34) → (35)

79–99%

Transition Metal-catalyzed Reactions. Chlorodimethylvinyl-silane has been used in the synthesis of silyl-containing Heck reaction precursors.[24] A Heck reaction of aryl or alkenyl iodides with dimethylvinylsilylpyridine (36) using $Pd_2(dba)_3$ and tri-2-furylphosphine (TFP) produced the coupled alkene products 37 (R = Ph, 2-py, 2-thiophene, and others) in high yields and with exclusive E selectivity due to the pyridine directing group (eq 17).[25] The pyridine moiety was also employed as a phase tag, which enabled easy purification via acid–base extraction. The silicon linker was subsequently cleaved by H_2O_2 oxidation.[26]

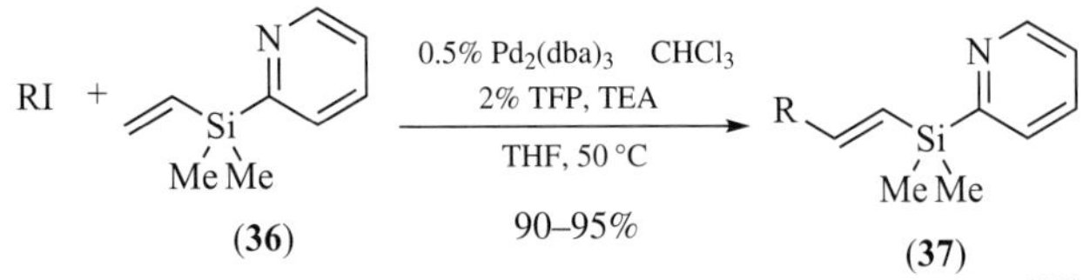

$$ (17) $$

Silicon has also been employed as a tether for directing ring-closing metathesis (RCM) reactions. The RCM precursors 38 were generated by attachment of chlorodimethylvinylsilane to alkenyl alcohols. An olefin metathesis reaction using Schrock's catalyst provided the cyclized products 39 in excellent yields (eq 18).[27] Although Grubbs' catalyst had been successfully employed in the case of allylsiloxy dienes, Schrock's catalyst resulted in better yields in the case of the vinylsiloxy dienes 39.[28] Oxidative cleavage of the silicon tether was then used to provide acyclic hydroxyaldehydes.

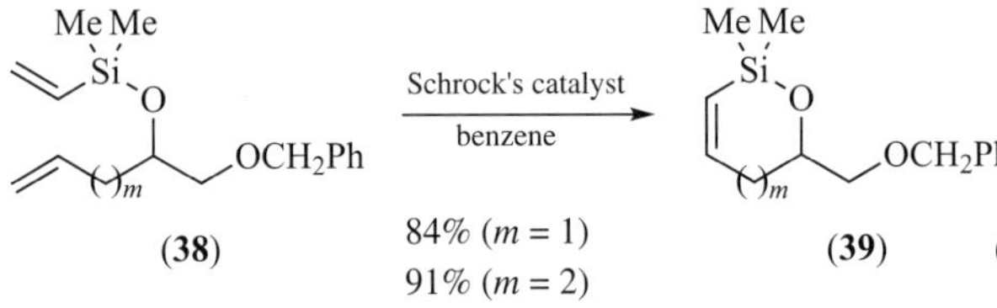

$$ (18) $$

84% (m = 1)
91% (m = 2)

A sequential RCM/silicon-assisted cross-coupling reaction has been developed and was applied to the formation of natural products.[28–33] First, an RCM reaction using Schrock's catalyst, similar to eq 18, is performed to give 40, and this is followed by an intermolecular palladium cross-coupling reaction with various aryl iodides. TBAF is added followed by the $Pd(dba)_2$ catalyst to generate the coupled aryl alkenes 41 (eq 19).[28]

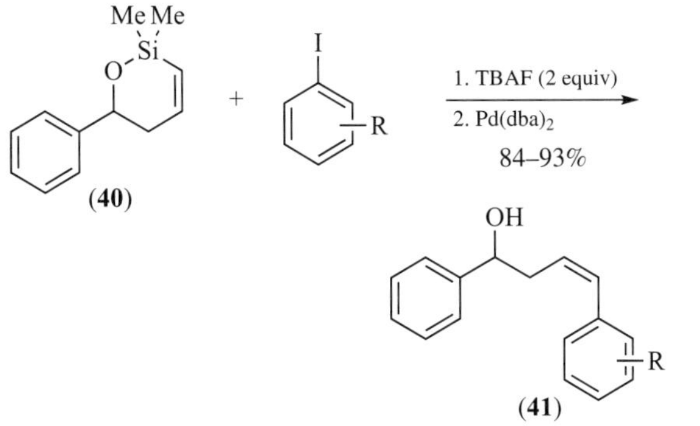

$$ (19) $$

84–93%

The sequence has also been conducted intramolecularly to form medium-sized rings. In order to accomplish this, an iodoalkene was tethered to the starting alcohol, prior to the installation of the vinylsilyl group. The RCM reaction was performed to give 42 followed by the palladium-catalyzed cross-coupling reaction using allyl palladium chloride (APC), generating the 10-membered ring 43 in good yield (eq 20).[29] This methodology has been applied to the synthesis of (+)-brasilenyne and the formation of macrolactones.[30,32,33]

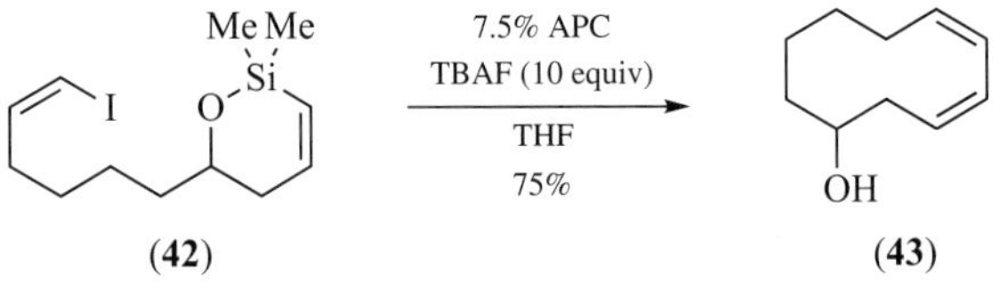

$$ (20) $$

7.5% APC
TBAF (10 equiv)
THF
75%

An *endo*-selective enyne RCM has been developed using a vinylsilane tether. The vinylsilane 44 is reacted with two different Mo catalysts to produce almost exclusively the *endo* product 45 in good yield (eq 21).[34]

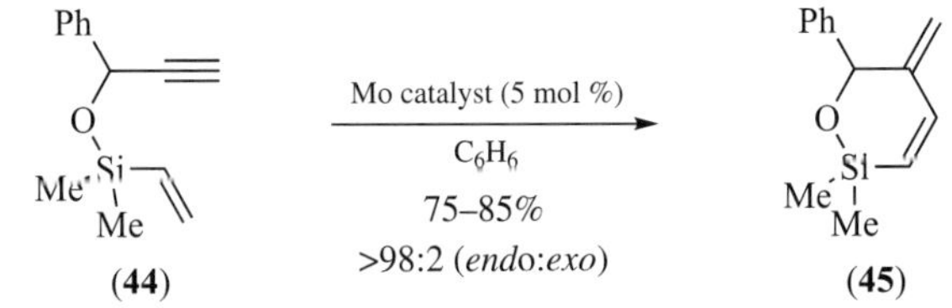

$$ (21) $$

Mo catalyst (5 mol %)
C_6H_6
75–85%
>98:2 (endo:exo)

Titanium-mediated reductive cross-coupling reactions between allylic alcohols and vinylsilanes have been reported to not only occur with high regio- and stereoselectivity, but also afford products complementary to those obtained from Claisen rearrangements.[35] A reaction of the lithiated allylic alcohols 46 (R^1 = PMB, C_5H_9, *t*-Bu, and others; R^2 = Me; R^3 = H, Me) with dimethylvinylcholorsilane was treated with ClTi(O*i*-Pr)$_3$ and C$_5$H$_9$MgCl to form the intermediate **47**, followed by oxidation to afford the alcohols **48** in moderate yield but with almost exclusive *Z* geometry (eq 22). Ti has also been used with vinylsilanes to form cyclopropanols.[36,37]

(22)

Reactions involving cobalt catalysis have been used with the vinylsilyl group to form cyclized products. A Co-catalyzed intramolecular hydroacylation reaction was performed with a vinylsilyl aromatic aldehyde **49** to form the cyclized silylether product **50** in quantitative conversion (eq 23).[38]

(23)

An intramolecular Pauson–Khand reaction of silicon-tethered enynes has been reported as an alternative method to an intermolecular Pauson–Khand reaction with ethylene. The alkynyl vinylsilylether **51** was cyclized with Co$_2$(CO)$_8$ in refluxing acetonitrile with 1% H$_2$O to afford the cyclized enone **52** in good yield (eq 24).[39] The final product **52** was exclusively formed without any silicon-tethered product detected, and five other examples were reported with the same outcome.[39] Other examples of silicon-tethered Pauson–Khand reactions similar to the one in eq 24 have been reported.[17,40]

(24)

1. Jones, P. R.; Lim, T. F. O., *J. Am. Chem. Soc.* **1977**, *99*, 2013.

2. Jones, P. R.; Lim, T. F. O.; Pierce, R. A., *J. Am. Chem. Soc.* **1980**, *102*, 4970.

3. Soderquist, J. A.; Hassner, A., *J. Org. Chem.* **1983**, *48*, 1801.

4. Soderquist, J. A.; Negron, A., *Tetrahedron Lett.* **1998**, *39*, 9397.

5. Sellars, J. D.; Steel, P. G., *Tetrahedron* **2009**, *65*, 5588.

6. Sen, S.; Purushotham, M.; Qi, Y. M.; Sieburth, S. M., *Org. Lett.* **2007**, *9*, 4963.

7. Alonso, C.; Nantz, M. H.; Kurth, M. J., *Tetrahedron Lett.* **2000**, *41*, 5617.

8. Wang, B.; Chen, L.; Kim, K., *Tetrahedron Lett.* **2001**, *42*, 1463.

9. Chen, C. L.; Sparks, S. M.; Martin, S. F., *J. Am. Chem. Soc.* **2006**, *128*, 13696.

10. Merten, J.; Hennig, A.; Schwab, P.; Frohlich, R.; Tokalov, S. V.; Gutzeit, H. O.; Metz, P., *Eur. J. Org. Chem.* **2006**, *5*, 1144.

11. Bols, M.; Skrydstrup, T., *Chem. Rev.* **1995**, *95*, 1253.

12. Yahiro, Y.; Ichikawa, S.; Shuto, S.; Matsuda, A., *Tetrahedron Lett.* **1999**, *40*, 5527.

13. Shuto, S.; Yahiro, Y.; Ichikawa, S.; Matsuda, A., *J. Org. Chem.* **2000**, *65*, 5547.

14. Ueno, Y.; Nagasawa, Y.; Sugimoto, I.; Kojima, N.; Kanazaki, M.; Shuto, S.; Matsuda, A., *J. Org. Chem.* **1998**, *63*, 1660.

15. Friested, G. K.; Jiang, T.; Mathies, A. K., *Org. Lett.* **2007**, *9*, 777.

16. Duplessis, M.; Waltz, M.-E.; Bencheqroun, M.; Cardinal-David, B.; Guindon, Y., *Org. Lett.* **2009**, *11*, 3148.

17. Dobbs, A. P.; Miller, I. J.; Martinovic, S., *Beilstein J. Org. Chem.* **2007**, *3*(21).

18. Stork, G.; Chan, T. Y.; Breault, G. A., *J. Am. Chem. Soc.* **1992**, *114*, 7578.

19. Sieburth, S. M.; Fensterbank, L., *J. Org. Chem.* **1992**, *57*, 5279.

20. Rumbo, A.; Castedo, L.; Mourino, A.; Mascarenas, J. L., *J. Org. Chem.* **1993**, *58*, 5585.

21. Righi, P.; Marotta, E.; Landuzzi, A.; Rosini, G., *J. Am. Chem. Soc.* **1996**, *118*, 9446.

22. Marotta, E.; Righi, P.; Rosini, G., *Tetrahedron Lett.* **1998**, *39*, 1041.

23. Kudoh, T.; Ishikawa, T.; Shimizu, Y.; Saito, S., *Org. Lett.* **2003**, *5*, 3875.

24. Taylor, R. E.; Engelhardt, F. C.; Yuan, H., *Org. Lett.* **1999**, *1*, 1257.

25. Itami, K.; Mitsudo, K.; Kamei, T.; Koike, T.; Nokami, T.; Yoshida, J., *J. Am. Chem. Soc.* **2000**, *122*, 12013.

26. Yoshida, J.; Itami, K.; Mitsudo, K.; Suga, S., *Tetrahedron Lett.* **1999**, *40*, 3403.

27. Chang, S.; Grubbs, R. H., *Tetrahedron Lett.* **1997**, *38*, 4757.

28. Denmark, S. E.; Yang, S.-M., *Org. Lett.* **2001**, *3*, 1749.

29. Denmark, S. E.; Yang, S.-M., *J. Am. Chem. Soc.* **2002**, *124*, 2102.

30. Denmark, S. E.; Yang, S.-M., *J. Am. Chem. Soc.* **2002**, *124*, 15196.

31. Denmark, S. E.; Yang, S.-M., *Tetrahedron* **2004**, *60*, 9695.

32. Denmark, S. E.; Yang, S.-M., *J. Am. Chem. Soc.* **2004**, *126*, 12432.

33. Denmark, S. E.; Muhuhi, J. M., *J. Am. Chem. Soc.* **2010**, *132*, 11768.

34. Lee, Y.-J.; Schrock, R. R.; Hoveyda, A. H., *J. Am. Chem. Soc.* **2009**, *131*, 10652.

35. Belardi, J. K.; Micalizio, G. C., *J. Am. Chem. Soc.* **2008**, *130*, 16870.

36. Mizojiri, R.; Urabe, H.; Sato, F., *Tetrahedron Lett.* **1999**, *40*, 2557.

37. Mizojiri, R.; Urabe, H.; Sato, F., *J. Org. Chem.* **2000**, *65*, 6217.

38. Lenges, C. P.; Brookhart, M., *J. Am. Chem. Soc.* **1997**, *119*, 3165.

39. Reichwein, J. F.; Iacono, S. T.; Patel, M. C.; Pagenkopf, B. L., *Tetrahedron Lett.* **2002**, *43*, 3739.

40. Itami, K.; Mitsudo, K.; Fujita, K.; Ohashi, Y.; Yoshida, J., *J. Am. Chem. Soc.* **2004**, *126*, 11058.

Alexander Deiters & Andrew L. McIver

North Carolina State University, Raleigh, NC, USA

Chlorodiphenylsilane[1]

$$\begin{array}{c} Ph \diagdown \underset{\diagup}{Si} \diagup H \\ Ph \diagup \quad Cl \end{array}$$

[1631-83-0] $C_{12}H_{11}ClSi$ (MW 218.76)

InChI = 1S/C12H11ClSi/c13-14(11-7-3-1-4-8-11)12-9-5-2-6-
10-12/h1-10,14H

InChIKey = VTMOWGGHUAZESJ-UHFFFAOYSA-N

(hydrosilylation;[1] precursor to alkyldiphenyl- and aryldiphenylsilanes[2])

Physical Data: bp 140–145 °C/7 mmHg (99–101 °C/1 mmHg;
83–85 °C/0.4 mmHg); d 1.137 g cm^{-3}; n 1.5845.

Solubility: sol benzene, chloroform, carbon tetrachloride, and
ether.

Form Supplied in: liquid. Commercially available.

Preparative Methods: chlorodiphenylsilane may be prepared in
high yield by treatment of diphenylsilane with triphenylmethyl
chloride in refluxing benzene[3] or with PCl$_5$ at rt in CCl$_4$.[4]

Handling, Storage, and Precautions: moisture sensitive. Reactions are typically conducted under an inert atmosphere. The
compound is corrosive and liberates HCl upon contact with
moisture; may burn exposed skin and can be destructive to eyes,
mucous membranes, and upper respiratory tract. Chemically
incompatible with water, alcohols, and amines.

Hydrosilylation of Alkenes and Alkynes. In the presence of a
catalyst (typically chloroplatinic acid, H$_2$PtCl$_2$), chlorodiphenylsilane hydrosilylates double and triple bonds. As a difunctional
reagent, chlorodiphenylsilane may be used in a reaction with an
organometallic reagent (see the following section) to displace
chlorine prior to the hydrosilylation reaction as in eq 1[5] or hydrosilylation may be conducted first with subsequent reaction at
silicon as illustrated in eq 2.[6] The regiochemistry of the reaction
has been reviewed.[7]

$$\text{(1)}$$

$$\text{(2)}$$

An enantioselective synthesis of chiral diols via intramolecular hydrosilylation using a chiral catalyst followed by oxidative
cleavage of the Si–C bond (with retention of configuration) has

been reported.[8] Intramolecular hydrosilylation/oxidation of allylamines provides a highly regio- and stereoselective synthesis of
2-amino alcohols (eq 3).[9]

$$\text{(3)}$$

Chloride Displacement by Organometallic Reagents.
Organolithium and Grignard reagents react with chlorodiphenylsilane to produce substituted alkyl- or aryldiphenylsilanes (eqs 4
and 5).[2]

$$\text{(4)}$$

$$\text{(5)}$$

Related Reagent. Triethylsilane.

1. (a) Jones, D. N. In *Comprehensive Organic Chemistry*; Barton, D. H. R.,
   Ed.; Pergamon: New York, 1979; Vol 3, p 567. (b) Chalk, A. J., *Trans. N.
   Y. Acad. Sci.* **1970**, *32*, 481.

2. (a) Buynak, J. D.; Strickland, J. B.; Lamb, G. W.; Khasnis, D.; Modi, S.;
   Williams, D.; Zhang, H., *J. Org. Chem.* **1991**, *56*, 7076. (b) Tacke, R.;
   Strecker, M.; Sheldrick, W. S.; Heeg, E.; Berndt, B.; Knapstein, K. M.,
   Chem. Ber. **1980**, *113*, 1962.

3. Cory, J. Y.; West, R., *J. Am. Chem. Soc.* **1963**, *85*, 2430.

4. Mawaziny, S., *J. Chem. Soc. (A)* **1970**, 1641.

5. Merkl, G.; Berr, K. P., *Tetrahedron Lett.* **1992**, *33*, 1601.

6. Brook, A.; Kucera, H. W., *J. Organomet. Chem.* **1975**, *87*, 263.

7. E. W, Colvin, *Silicon in Organic Synthesis*; Butterworths: Boston, 1981;
   p 46, 325.

8. Bergens, S. H.; Noheda, P.; Whelan, J.; Bosnich, B., *J. Am. Chem. Soc.*
   1992, *114*, 2121.

9. Tamao, K.; Kakagawa, Y.; Ito, Y., *J. Org. Chem.* **1990**, *55*, 3438.

Dallas K. Bates
Michigan Technological University, Houghton, MI, USA

(Chloromethyl)dimethylphenylsilane

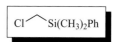

[1833-51-8] $C_9H_{13}ClSi$ (MW 184.74)

InChI = 1S/C9H13ClSi/c1-11(2,8-10)9-6-4-3-5-7-9/h3-7H,8H2,
1-2H3

InChIKey = RJCTVFQQNCNBHG-UHFFFAOYSA-N

(masked hydroxyl group,[7-8,11,26-31] preparation of SMOM-Cl,[4] C-alkylating agent,[5-10] N-alkylating agent,[12-17] O-alkylating agent,[18-22] S-alkylating agent,[23] preparation of Grignard reagents,[27-33] alkene formation,[28e,35] preparation of organolithium reagents[40-45])

Alternate Names: [(chloromethyl)dimethylsilyl]benzene; chloromethylphenyldimethylsilane; dimethylphenylsilylmethyl chloride; phenyldimethyl(chloromethyl)silane.

Physical Data: bp 106–107 °C/15 mmHg; d 1.024 g cm^{-3}

Preparative Methods: can be prepared either by the reaction of phenylmagnesium bromide with chloro(chloromethyl)dimethylsilane in ether at reflux,[1] or by the reaction of phenylmagnesium bromide with chloro(chloromethyl)dimethylsilane in the presence of catalytic (N,N,N',N'-tetramethylethylenediamine)zinc in 1,4-dioxane at 20 °C.[2]

Purification: for most applications, the material may be used as received. For applications requiring rigorous exclusion of impurities and oxygen, the material may be further purified by stirring with phosphorus pentoxide followed by distillation.

Handling, Storage, and Precautions: reagent is stable to normal temperatures and pressures. Causes skin irritation and serious eye irritation. May cause respiratory irritation. Wear suitable personal protective equipment and use in a fume hood.

Introduction. The first reported preparation and use of (chloromethyl)dimethylphenylsilane was in 1949.[3] At the time, the main utility of (chloromethyl)dimethylphenylsilane was as the starting material for conversion to the corresponding Grignard reagent (see below). Since then, it has also been used for heteroatom alkylation, carbon alkylation, and conversion to a variety of organometallic and organolanthanide reagents. The main advantage of this reagent over the closely related (chloromethyl)trimethylsilane is the ability of (chloromethyl)dimethylphenylsilane to undergo a Fleming oxidation, thus allowing (chloromethyl)dimethylphenylsilane to serve as a masked hydroxyl group. This utility has been exploited for both the C-substituted and N-substituted adducts. A major disadvantage of the use of (chloromethyl)dimethylphenylsilane is its propensity to undergo rearrangements under a variety of conditions (see below). (Chloromethyl)dimethylphenylsilane is used as the precursor for the preparation of (phenyldimethylsilyl)methoxymethyl chloride (SMOM-Cl), a hydroxyl protecting group.[4]

C-Alkylation. (Chloromethyl)dimethylphenylsilane (**1**) can be utilized to install a silylmethyl group on carbon via base promoted C-alkylation of terminal alkynes,[5] dihydropyrazines,[6] malonic esters,[7] phenylacetonitriles,[8] sulfoxides,[9] and imines.[10] Although **1** has been used directly in the alkylation, conversion to the corresponding iodide **2**[5,11] via Finkelstein displacement (eq 1) prior to alkylation is sometimes warranted. Except for the malonic esters (eq 2), strongly basic conditions and low temperatures (with slow warming) are generally employed in the transformation (eqs 3 and 4).

(1)

(2)

(3)

(4)

N-Alkylation. (Chloromethyl)dimethylphenylsilane (**1**) has been utilized for the alkylation of nitrogen nucleophiles including amines,[12] α-aminoesters (eq 5),[13] anilides and anilines (eq 6),[14] benzimidazole,[15] potassium cyanate (with in situ treatment with ammonia to afford a nitrosourea),[16] and sodium azide (eq 7).[17] Although **1** has been utilized for N-alkylation of a variety of nitrogen nucleophiles, the reaction conditions tend to be similar and usually involve employing an exogenous base, elevated temperatures, and polar aprotic solvents (e.g., DMF, DMSO, DMPU) to effect the N-alkylation. In some cases, for instance, alkylation of α-aminoester **9**, potassium iodide is incorporated to facilitate the reaction.

(5)

(6)

(7)

O-Alkylation. The most thoroughly investigated alkylation using **1** is with oxygen nucleophiles. Only a limited number of successful O-alkylations exist and are currently limited to sodium acetate,[18] a potassium carboxylate (**15**),[19] and substituted phenols.[20] The conditions used in all cases employ polar aprotic solvents (e.g., DMF, DMSO) and elevated temperatures, and use either **2** directly (eq 8) or sodium iodide or potassium iodide for in situ conversion of **1** to **2** (eq 9).

$$\text{(15)} \xrightarrow[\text{DMSO, 50 °C}]{\text{ICH}_2\text{Si(CH}_3)_2\text{Ph}} \text{(16) (93%)} \qquad (8)$$

$$\text{(17)} \xrightarrow[\text{DMSO, 80 °C}]{\substack{\text{ClCH}_2\text{Si(CH}_3)_2\text{Ph} \\ \text{K}_2\text{CO}_3, \text{KI}}} \text{(18) (80%)} \qquad (9)$$

Although the reaction of (chloromethyl)dimethylphenylsilane and some sodium alkoxides does proceed to provide the O-alkylated products (eq 10), the substitution product ethers **19** are contaminated with variable amounts of products derived from O-silylation with concomitant phenyl migration **20** (i.e., rearrangement products) and/or products derived from expulsion of the chloromethyl group **21** (i.e., cleavage products).[21] If the alkyl ether adducts are desired, it is preferable to first convert (chloromethyl)dimethylphenylsilane to the corresponding alcohol, and then alkylate using the substrate as the electrophilic partner in the transformation. In one instance, it was possible to perform an O-alkylation of a primary alcohol by converting **1** to the corresponding triflate.[18,22]

$$\text{(1)} \xrightarrow[\substack{\text{EtOH} \\ \text{80°C}}]{\text{NaOEt}} \text{EtO}\frown\text{Si(CH}_3)_2\text{Ph} + \text{Ph}\frown\text{Si(CH}_3)_2\text{OEt} + \text{EtO}-\text{Si(CH}_3)_2\text{Ph}$$

$$\text{(19) (42%)} \qquad \text{(20) (32%)} \qquad \text{(21) (16%)} \qquad (10)$$

S-Alkylation. Examples of S-alkylation with (chloromethyl)-dimethylphenylsilane are currently very limited. One example that does exist is the dialkylation of sodium sulfide.[23]

Rearrangements. It should be noted that upon treatment with certain reagents (including alkoxides, see above), (chloromethyl)-dimethylphenylsilane is capable of undergoing rearrangements under radical, electrophilic, and nucleophilic conditions. Thus, it has been observed that upon treatment with ethylmagnesium bromide and cobalt(II) chloride, (chloromethyl)dimethylphenylsilane undergoes a Kharasch–Grignard radical rearrangement to provide three products (eq 11).[24] Furthermore, treatment of **1** with catalytic ethylaluminum dichloride in dichloromethane followed by quenching with methylmagnesium iodide affords the product from migration of the phenyl group from silicon to carbon (eq 12) via an electrophilic rearrangement pathway.[25] Finally, in addition to the rearrangements observed upon treatment with alkoxides, treatment with nucleophilic fluoride sources (e.g., potassium fluoride or cesium fluoride) in the presence of 18-crown-6, (chloromethyl)dimethylphenylsilane rearranges to

afford primarily the product from phenyl migration due to its propensity to bear the negative charge that develops in the transition state of the rearrangement (eq 13).[26]

$$\text{(1)} \xrightarrow[\text{THF, }\Delta]{\substack{\text{EtMgBr} \\ \text{CoCl}_2}} \text{H}_3\text{C}-\text{Si(CH}_3)_2\text{Ph} + \frown\text{Si(CH}_3)_2\text{OEt}$$

$$\text{(22) (29% brsm)} \qquad \text{(23) (6% brsm)}$$
$$+$$
$$\text{Ph(CH}_3)_2\text{SiH}_2 \qquad \text{Si(CH}_3)_2\text{Ph} \qquad (11)$$
$$\text{(24) (64% brsm)}$$

$$\text{(1)} \xrightarrow[\text{CH}_2\text{Cl}_2, \text{rt}]{\text{EtAlCl}_2} \left[\underset{\text{Cl}}{\underset{|}{\text{Ph}\frown\text{Si(CH}_3)_2}} \right] \xrightarrow[\text{CH}_2\text{Cl}_2]{\text{MeMgI}}$$

$$\text{(25)}$$
$$\text{Ph}\frown\text{Si(CH}_3)_3 \qquad (12)$$
$$\text{(26) (71%)}$$

$$\text{(1)} \xrightarrow[\text{PhMe}]{\substack{\text{KF or CsF} \\ \text{18-crown-6}}}$$

$$\underset{\text{F}}{\underset{|}{\text{Ph}\frown\text{Si(CH}_3)_2}} + \underset{\text{F}}{\underset{|}{\text{H}_3\text{C}\frown\text{SiCH}_3\text{Ph}}} \qquad (13)$$

$$\text{(27)} \qquad \text{(16:1)} \qquad \text{(28)}$$

Grignard Reagent. The main utility of (chloromethyl)dimethylphenylsilane is for the introduction of a masked hydroxyl group on carbon. This task is generally accomplished by first converting (chloromethyl)dimethylphenylsilane to the corresponding Grignard **29** (eq 14),[27] followed by 1,2-addition to carbonyls[28] or imines,[29] 1,4-addition to unsaturated carbonyls,[30] nickel-catalyzed coupling with vinyl halides,[31] or S_N2' attack on activated alkenes bearing an allylic carbamate[8] or allylic strained C–O bond (e.g., allylic epoxides, diepoxynaphthalenes).[32] Grignard **29** has also been used to append a silylmethyl group to C60 fullerenes.[33] Once incorporated, the hydroxyl functionality may be uncovered by Fleming oxidation.[34]

Similar to (chloromethyl)trimethylsilane, (chloromethyl)dimethylphenylsilane can be used to prepare alkenes via a Peterson olefination by treating C=O compounds with Grignard **29**. However, due to its propensity for rearrangements under the conditions needed for the elimination step, **1** appears less suited for this task than (chloromethyl)trimethylsilane if the des-silyl terminal alkenes are desired.[35] However, in a parallel transformation, it was found that if the β-silylalcohol **31** is treated with magnesium iodide etherate, then the Julia ring-opened homoallylation adduct **32** could be isolated (eq 15).[28e]

$$Cl\diagup\diagdown Si(CH_3)_2Ph \xrightarrow[\text{Et}_2\text{O, }\Delta]{\text{Mg}^0} ClMg\diagup\diagdown Si(CH_3)_2Ph \qquad (14)$$

(1)   (29)

(30)   (31)

$$\xrightarrow[\substack{\text{Et}_2\text{O/PhH} \\ 0\,^\circ\text{C}}]{\text{MgI}_2 \quad \text{OEt2})n} \qquad (15)$$

(32) (68%, sole diastereomer)

(Chloromethyl)dimethylphenylsilane can also be converted to the corresponding deuterated,[36] organoplatinum,[1] phosphino,[37] silyl,[38] and stannyl,[39] adducts via the Grignard 29.

Organolithium Reagent. Similar to (chloromethyl)trimethylsilane, treatment of 1 with *sec*-butyllithium at low temperature affords the reactive intermediate from Li–H exchange,[40] leaving the chlorine substituent in place. This intermediate has been treated with carbon electrophiles to afford the *C*-alkylated products,[29a, 41] trivalent aluminum reagents, dialkylmagnesium reagents, divalent zinc reagents, and dialkyl cadmium reagents to afford the 1,2-migration products,[42] and cyclic organozirconium reagents to afford the ring-expanded adducts.[43] If the lithiated adduct of (chloromethyl)dimethylphenylsilane arising from Li–Cl exchange is desired, this intermediate can be accessed by treating 1 with lithium metal (50% dispersion in mineral oil) in pentane at reflux in a glove box.[44] This procedure was used to access the corresponding organothorium adduct. The organoscandium adduct has also been prepared via a similar procedure.[45]

Related Reagents.   (Chloromethyl)trimethylsilane; *t*-butyl-(chloromethyl)dimethylsilane.

1. Ankianiec, B. C.; Christou, V.; Hardy, D. T.; Thomson, S. K.; Young, G. B., *J. Am. Chem. Soc.* **1994**, *116*, 9963.

2. (a) Murakami, K.; Yorimitsu, H.; Oshima, K., *J. Org. Chem.* **2009**, *74*, 1415. (b) Murakami, K.; Yorimitsu, H.; Oshima, K.; Panteleev, J.; Lautens, M., *Org. Synth.* **2010**, *87*, 178.

3. Sommer, L. H.; Gold, J. R.; Goldberg, G. M.; Marans, N. S., *J. Am. Chem. Soc.* **1949**, *71*, 1509.

4. Boons, G. J. P. H.; Elie, C. J. J.; van der Marel, G. A.; van Boom, J. H., *Tetrahedron Lett.* **1990**, *31*, 2197.

5. Organ, M. G.; Mallik, D., *Can. J. Chem.* **2006**, *84*, 1259.

6. Tacke, R.; Merget, M.; Bertermann, R.; Bernd, M.; Beckers, T.; Reissmann, T., *Organometallics* **2000**, *19*, 3486.

7. Sommer, L. H.; Goldberg, G. M.; Barnes, G. H.; Stone, L. S., *J. Am. Chem. Soc.* **1954**, *76*, 1609.

8. Itoh, H.; Tanaka, H.; Ohta, H.; Takeshiba, H., *Chem. Pharm. Bull.* **2001**, *49*, 909.

9. Corey, E. J.; Luo, G.; Lin, L. S., *J. Am. Chem. Soc.* **1997**, *119*, 9927.

10. Smitrovich, J. H.; Woerpel, K. A., *J. Am. Chem. Soc.* **1998**, *120*, 12998.

11. Eaborn, C.; Jeffrey, J. C., *J. Chem. Soc.* **1954**, 4266.

12. (a) Noll, J. E.; Speier, J. L.; Daubert, B. F., *J. Am. Chem. Soc.* **1951**, *73*, 3867. (b) Sato, Y.; Toyo'oka, T.; Aoyama, T.; Shirai, H., *J. Org. Chem.* **1976**, *41*, 3559.

13. (a) Sun, H.; Moeller, K. D., *Org. Lett.* **2003**, *5*, 3189. (b) Sun, H.; Martin, C.; Kesselring, D.; Keller, R.; Moeller, K. D., *J. Am. Chem. Soc.* **2006**, *128*, 13761.

14. (a) Gotteland, J.-P.; Delhon, A.; Junquero, D.; Oms, P.; Halazy, S., *Bioorg. Med. Chem. Lett.* **1996**, *6*, 533. (b) Hwu, J. R.; King, K. Y., *Chem. Eur. J.* **2005**, *11*, 3805.

15. Klaehn, J. R.; Luther, T. A.; Orme, C. J.; Jones, M. G.; Wertsching, A. K.; Peterson, E. S., *Macromolecules* **2007**, *40*, 7487.

16. Ninomiya, S.-I.; Liu, F.-Z.; Nakagawa, H.; Kohda, K.; Kawazoe, Y.; Sato, Y., *Chem. Pharm. Bull.* **1986**, *34*, 3273.

17. Hachiya, H.; Kakuta, T.; Takami, M.; Kabe, Y., *J. Organomet. Chem.* **2009**, *694*, 630.

18. (a) Boons, G. J. P. H.; van der Marel, G. A.; van Boom, J. H., *Tetrahedron Lett.* **1990**, *31*, 2197. (b) Simov, B. P.; Wuggenig, F.; Mereiter, K.; Andres, H.; France, J.; Schnelli, P.; Hammerschmidt, F., *J. Am. Chem. Soc.* **2005**, *127*, 13934.

19. Gassmann, S.; Guintchin, B.; Bienz, S., *Organometallics* **2001**, *20*, 1849.

20. (a) Gotteland, J.-P.; Brunel, I.; Gendre, F.; Desire, J.; Delhon, A.; Junquero, D.; Oms, P.; Halazy, S., *J. Med. Chem.* **1995**, *38*, 3207. (b) Mateo, C.; Rernandez-Rivas, C.; Echavarren, A. M.; Cardenas, D. J., *Organometallics* **1997**, *16*, 1997. (c) Mateo, C.; Fernandez-Rivas, C.; Cardenas, D. J.; Echavarren, A. M., *Organometallics* **1998**, *17*, 3661.

21. (a) Eaborn, C.; Jeffrey, J. C., *J. Chem. Soc.* **1957**, 137. (b) Kreeger, R. L.; Menard, P. R.; Sans, E. A.; Shechter, H., *Tetrahedron Lett.* **1985**, *26*, 1115. (c) Sans, E. A.; Shechter, H., *Tetrahedron Lett.* **1985**, *26*, 1119.

22. For preparation of the requisite alcohol for triflate formation, see Ref. 18b.

23. Hosomi, A.; Ogata, K.; Ohkum, M.; Hojo, M., *Synlett*, **1991**, 557.

24. Wilt, J. W.; Kolewe, O.; Kraemer, J. F., *J. Am. Chem. Soc.* **1969**, *91*, 2624.

25. Hudrlik, P. F.; Abdallah, Y. M.; Kulkarni, A. K.; Hudrlik, A. M., *J. Org. Chem.* **1992**, *57*, 6552.

26. (a) Damrauer, R.; Danahey, S. E.; Yost, V. E., *J. Am. Chem. Soc.* **1984**, *106*, 7633. (b) Damrauer, R.; Yost, V. E.; Danahey, S. E.; O'Connell, B. K., *Organometallics* **1985**, *4*, 1779.

27. A typical procedure for preparing the Grignard reagent involves adding an ethereal solution of (chloromethyl)dimethylphenylsilane to magnesium metal and warming under reflux until the magnesium is consumed. See Refs 28–33 for specific procedures.

28. (a) Boons, G. J. P. H.; van der Marel, G. A.; van Boom, J. H., *Tetrahedron Lett.* **1989**, *30*, 229. (b) Boons, G. J. P. H.; Overhand, M.; van der Marel, G. A.; van Boom, J. H., *Carbohydr. Res.* **1989**, *192*, c1. (c) Boons, G. J. P. H.; Overhand, M.; van der Marel, G. A.; van Boom, J. H., *Angew. Chem. Int. Ed., Engl.* **1989**, *28*, 1504. (d) Boons, G. J. P. H.; van der Marel, G. A.; van Boom, J. H., *Tetrahedron Lett.* **1990**, *31*, 2197. (e) Li, W.-D. Z.; Yang, J.-H., *Org. Lett.* **2004**, *6*, 1849. (f) Chiara, J. L.,: Garcia, A.; Sesmilo, E.; Vacas, T., *Org. Lett.* **2006**, *8*, 3935. (g) Rodgen, S. A.; Schaus, S. E., *Angew. Chem., Int. Ed.* **2006**, *45*, 4929.

29. For addition to imines: (a) van Delft, F. L.; de Kort, M.; van der Marel, G. A.; van Boom, J. H., *J. Org. Chem.* **1996**, *61*, 1883. (b) Trost, B. M.; Lee, C., *J. Am. Chem. Soc.* **2001**, *123*, 12191. (c) Ko, C. H.; Jung, D. Y.; Kim, M. K.; Kim, Y. H., *Synlett* **2005**, 304. (d) For addition to pyridine: Krow, G. R.; Lin, G.; Yu, F., *J. Org. Chem.* **2005**, *70*, 590.

30. Comins, D. L.; Libby, A. H.; Al-Awar, R. S.; Foti, C. J., *J. Org. Chem.* **1999**, *64*, 2184.

31. (a) Organ, M. G.; Murray, A. P., *J. Org. Chem.* **1997**, *62*, 1523. (b) Organ, M. G.; Winkle, D. D., *J. Org. Chem.* **1997**, *62*, 1881. (c) Robertson, C. W.; Woerpel, K. A., *J. Org. Chem.* **1999**, *64*, 1434.

32. (a) Jung, M. E.; D'Amico, D. C., *J. Am. Chem. Soc.* **1995**, *117*, 7379. (b) Enev, V. S.; Drescher, M.; Mulzer, J., *Org. Lett.* **2008**, *10*, 413. (c) Gromov, A.; Enev, V.; Mulzer, J., *Org. Lett.* **2009**, *11*, 2884.

33. (a) Matsuo, Y.; Nakamura, E., *Inorg. Chim. Acta* **2006**, *359*, 1979. (b) Matsuo, Y.; Iwashita, A.; Abe, Y.; Li, C.-Z.; Matsuo, K.; Hashiguchi, M.; Nakamura, E. J., *Am. Chem. Soc.* **2008**, *130*, 15429.

34. Fleming, I.; Henning, R.; Parker, D. C.; Plaut, H. E.; Sanderson, P. E. J., *J. Chem. Soc., Perkin Trans. 1* **1995**, 317.

35. Yield for preparation of 3-ethenylfuran using (chloromethyl)dimethylphenylsilane was 25% versus >90% using (chloromethyl)trimethylsilane. See (a) Tanaka, R.; Nakano, K.; Nozaki, K. J., *Org. Chem.* **2007**, *72*, 8671. (b) Hersel, U.; Steck, M.; Seifert, K., *Eur. J. Chem.* **2000**, 1609.

36. Miura, K.; Inoue, G.; Sasagawa, H.; Kinoshita, H.; Ichikawa, J.; Hosomi, A., *Org. Lett.* **2009**, *11*, 5066.

37. Bennett, M. A.; Edward, A. J.; Harper, J. R.; Khimyak, T.; Willis, A. C., *J. Organomet. Chem.* **2001**, *629*, 7.

38. Reich, H. J.; Goldenberg, W. S.; Gudmundsson, B. O.; Sanders, A. W.; Kulicke, K. J.; Simon, K.; Guzei, I. A., *J. Am. Chem. Soc.* **2001**, *123*, 8067.

39. Itami, K.; Kamei, T.; Yoshida, J.-I., *J. Am. Chem. Soc.* **2001**, *123*, 8773.

40. Burfor, C.; Cooke, F.; Roy, G.; Magnus, P., *Tetrahedron* **1983**, *39*, 867.

41. Smitrovich, J. H.; Woerpel, K. A., *J. Org. Chem.* **2000**, *65*, 1601.

42. Negishi, E.-I.; Akiyoshi, K. J., *Am. Chem. Soc.* **1988**, *110*, 646.

43. Dixon, S.; Fillery, S. M.; Kasatkin, A.; Norton, D.; Thomas, E.; Whitby, R. J., *Tetrahedron* **2004**, *60*, 1401.

44. Bruno, J. W.; Smith, G. M.; Marks, T. J.; Fair, C. K.; Schultz, A. J.; Williams, J. M., *J. Am. Chem. Soc.* **1986**, *108*, 40.

45. Emslie, D. J. H.; Piers, W. E.; Parvez, M.; McDonald, R., *Organometallics* **2002**, *21*, 4226.

Valerie A. Cwynar
The Ohio State University, Columbus, OH, USA

(*E*)-(3-Chloro-1-methyl-1-propenyl)-trimethylsilane

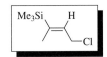

[116399-78-1]   C₇H₁₅ClSi   (MW 162.76)

InChI = 1S/C7H15ClSi/c1-7(5-6-8)9(2,3)4/h5H,6H2,1-4H3/b7-5+

InChIKey = LGIOTKSIJIWLPM-FNORWQNLSA-N

(annulation reagent; alkylative equivalent of 3-buten-2-one; 3-oxobutyl synthon)

Alternate Name: (*E*)-1-chloro-3-trimethylsilyl-2-butene.
Physical Data: bp 92–94 °C/74 mmHg.
Solubility: freely sol organic solvents.
Form Supplied in: not commercially available.
Analysis of Reagent Purity: δ_H(CCl₄) 0.0 (9H, s), 1.67 (3H, d, *J* 1.5 Hz), 3.93 (2H, d, *J* 7 Hz), 5.8 (1H, t of q, *J* 7 and 1.5 Hz).
Preparative Method: see (*E*)-1-iodo-3-trimethylsilyl-2-butene.

Basis and Example of Use. This reagent acts as the alkylative equivalent of 3-buten-2-one (methyl vinyl ketone), permitting regiospecific enolate alkylation under nonequilibrating conditions. As a vinylsilane it can be converted,[1] via the corresponding α,β-epoxy silane, into a 3-oxobutyl moiety, and hence used in an alkylative equivalent of Robinson annulation.[2] Application of such methodology can be seen (eq 1) in the preparation of bicyclodecenones via the alkylation of dihydrocarvone with this reagent.[3] Apart from this application, and its use in combination with potassium iodide[4], it has seen little use so far.

$$(1)$$

1. Stork, G.; Colvine, E., *J. Am. Chem. Soc.* **1971**, *93*, 2080; see also Gröbel, B.-Th.; Seebach, D., *Angew. Chem., Int. Ed. Engl.* **1974**, *13*, 83.
2. Bergmann, E. D.; Ginsburg, D.; Pappo, R., *Org. React.* **1959**, *10*, 179.
3. Noureldin, N., *Egypt. J. Chem.* **1986**, *29*, 25 (*Chem. Abstr.* **1988**, *109*, 129 323f).
4. Jung, M. E., *Tetrahedron* **1976**, *32*, 3.

Ernest W. Colvin
University of Glasgow, Glasgow, UK

1-Chloro-1-methylsiletane

[2351-34-0]   C₄H₉ClSi   (MW 120.65)

InChI = 1S/C4H9ClSi/c1-6(5)3-2-4-6/h2-4H2,1H3

InChIKey = JWKKXHBGWAQZOK-UHFFFAOYSA-N

(synthesis of alkyl, aryl, and heteroatom-substituted siletanes)

Physical Data: bp 103 °C; *d* 0.985 g cm⁻³.
Solubility: soluble in most aprotic organic solvents.
Form Supplied in: colorless liquid.
Analysis of Reagent Purity: elemental analysis, IR, M.S., NMR.
Preparative Methods: siletane **1** can be prepared in large quantities by Wurtz coupling reaction of dichloro-3-chloropropylmethylsilane with Mg.[1] It is commercially available from Gelest and Aldrich chemical company.
Handling, Storage, and Precautions: moisture sensitive and corrosive; store under an inert atmosphere; use in a fume hood.

Introduction. Siletanes (silacyclobutanes) represent a unique class of silicon-based reagents that have enjoyed much success in a number of synthetic applications including aldol reactions, allylations, cross-coupling reactions, carbosilane oxidations, and ring expansions. The manifestation of their strain in enhanced Lewis acidity and in the ring opening to silanols accounts for their utility.[2] 1-Chloro-1-methylsiletane **1** can be used for the preparation of the various siletane derivatives for these applications.

Reaction of 1-Chloro-1-methylsiletane 1 with Nucleophiles. Grignard reagents react efficiently with 1-chloro-1-methylsiletane **1** to provide the corresponding aryl-, alkenyl-, or alkylsiletanes (eqs 1–3).[3–5] On the contrary, reactions with alkyllithiums are limited to the reaction of **1** with a lithium acetylide (eq 4).[6] The reaction of **1** with an intermediate vinylalane obtained by hydroalumination of 1-heptyne gave (E)-1-heptenylmethylsiletane in >99/1 isomeric purity (eq 5).[7] The reaction of **1** with $LiAlD_4$ afforded 1-deuterio-1-methylsiletane (eq 6).[8]

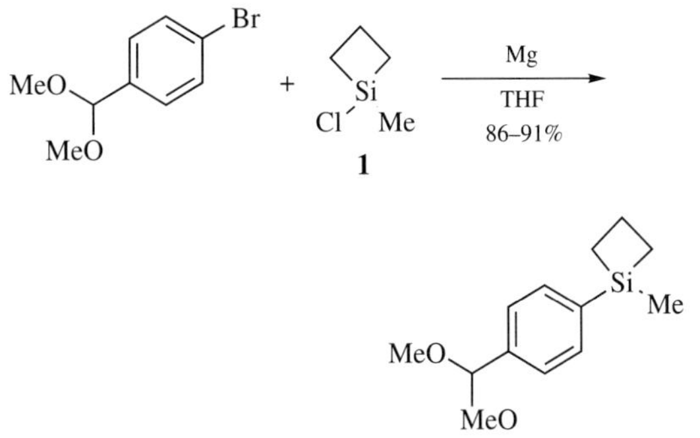

$$(1)$$

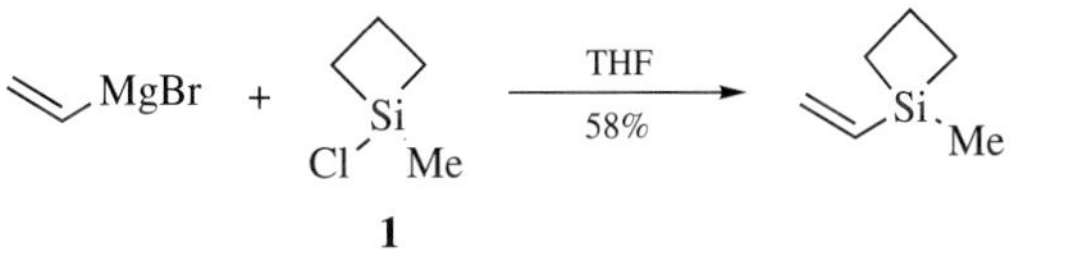

$$(2)$$

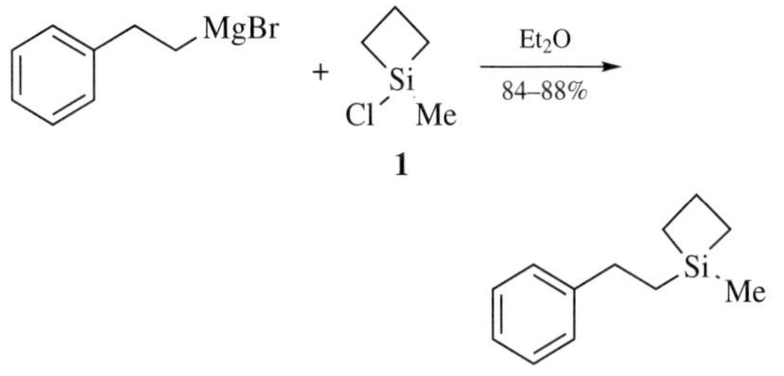

$$(3)$$

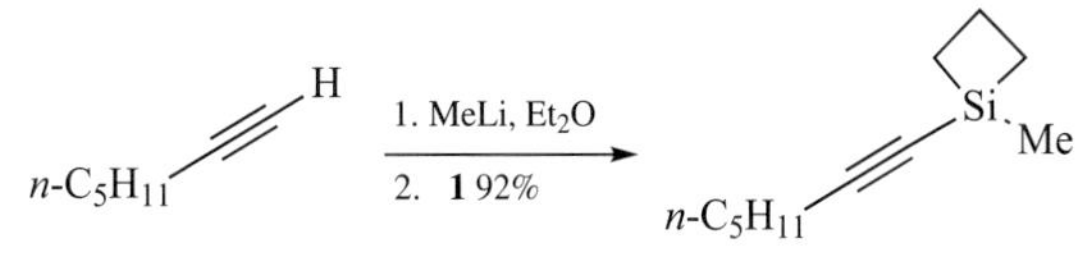

$$(4)$$

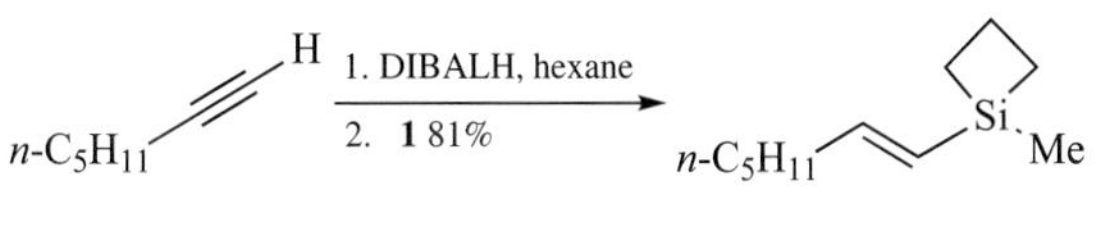

$$(5)$$

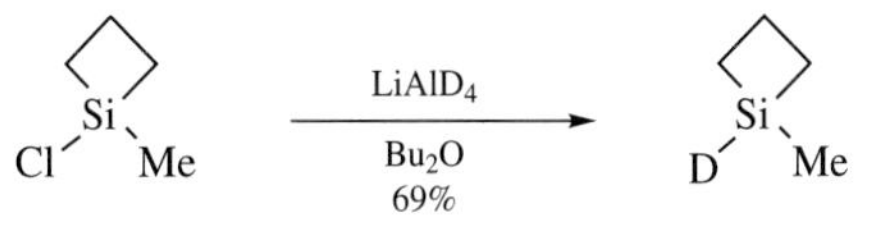

$$(6)$$

Enoxysiletane derivatives are prepared by standard enolization/silation protocols of carbon acids (ketones and esters) with

lithium diisopropylamide (Scheme 1). For the preparation of ketone-derived enoxysilanes (**2** and **3**), the use of lithium tetramethylpiperidide was required to minimize competitive reaction of the amine with the highly reactive 1-chloro-1-methylsiletane.[9a,9b] This hyperreactivity also presented problems in the preparation of enoxysilane derivatives from esters. Under standard silylation conditions, reaction with **1** led to substantial amounts of C-silylation product **5**. This undesirable side reaction could be suppressed by the use of tripiperidinophosphoric triamide (TPPA) as a polar cosolvent for the preparation of **4**.[9b]

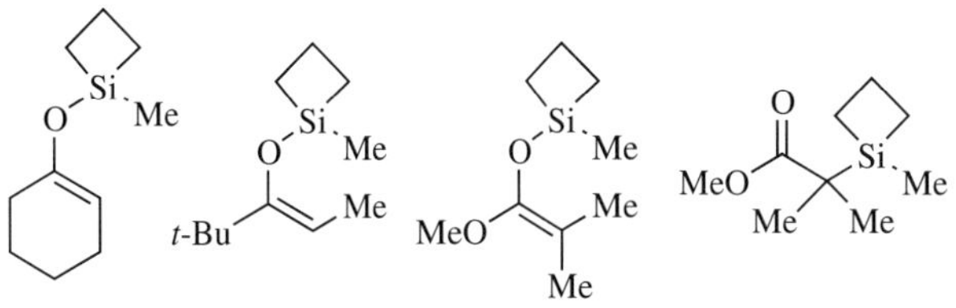

Scheme 1

Finally, chlorosiletane **1** reacts with N-lithiopyrrole to afford 1-pyrrolyl-1-methylsilacyclobutane (eq 7).[10]

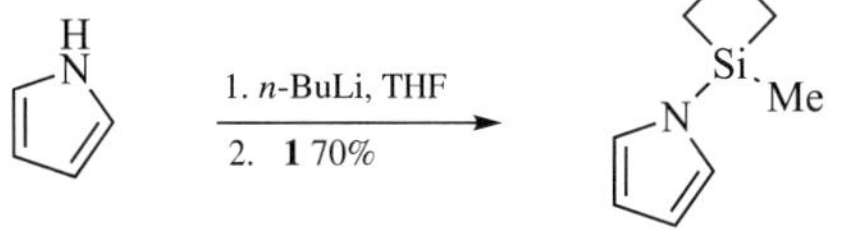

$$(7)$$

Aldol Addition. Enoxy(methyl)siletanes are successfully employed for aldol additions of ketones and esters with aldehydes without the need for catalysis (eq 8).[9] For example, the O-silacyclobutyl ketene acetal **4** undergoes rapid and clean aldol addition with benzaldehyde to afford the corresponding β-silyloxy ester as the only product. In a control experiment, the reaction of O-trimethylsilyl ketene acetal with benzaldehyde shows no sign of reaction under the same reaction conditions after 15 days. Although the reaction of enoxysilane **2** with benzaldehyde (1 M, C_6D_6) was extremely slow and required heating for prolonged periods for completion (100 °C, $t_{1/2}$ 800 min; syn/anti 85/15), the control trimethylsilyl enol ether showed absolutely no sign of reaction under the same reaction condition after 66 h. Thus, a significant enhancement in reactivity for enoxysilacyclobutanes was demonstrated.

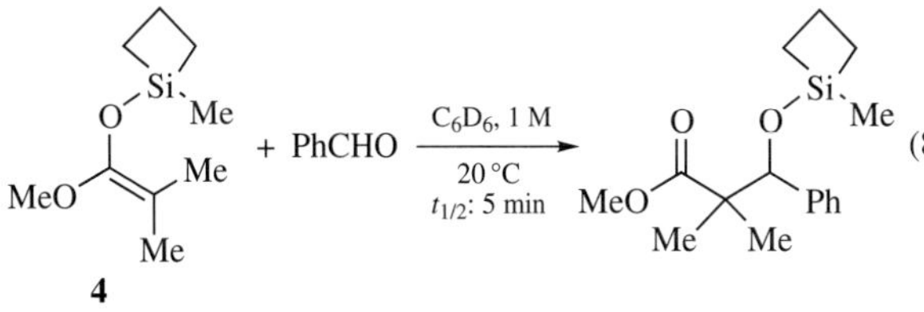

$$(8)$$

The t-butyl derivatives (E)-**6** and (Z)-**6** were employed since the methyl analogs were contaminated by C-silylated esters. The reactions of (E)-**6** with representative aldehydes proceeded smoothly at room temperature to afford, predominantly, the syn aldol products. The fact that an 89/11 E/Z mixture of **6** afforded a >95/5 syn/anti mixture of aldol products most likely arises from the more rapid reaction of the E isomer. Indeed, pure (Z)-**6** reacted sluggishly and with modest anti selectivity (eq 9, Table 1).

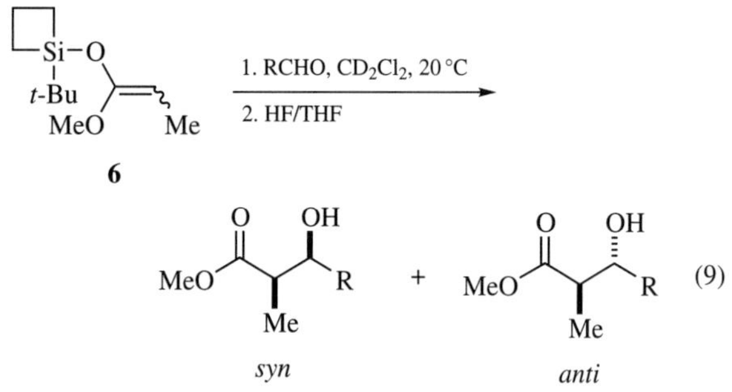

$$\text{(9)}$$

Table 1 Aldol reaction of the O,O-ketene acetals **6**

Entry	6, E/Z	R	$t_{1/2}$ (h)	syn/anti
1	89/11	Ph	2.8	98/2
2	95/5	Ph	2.2	95/5
3	0/100	Ph	28.3	42/58
4	89/11	Cinnamyl	6.7	93/7
5	89/11	n-Pentyl	17.0	93/7
6	89/11	Cyclohexyl	38.3	>99/1

The results from reactions of O-silyl S,O-ketene acetals (E)-**7** and (Z)-**7** are collected in Table 2.[9b] 1-Phenylsiletanes are used for this reaction because 1-t-butylsiletanes did not work in this series. The sulfur analog was slightly less reactive than its oxygen partner. The reactions of (Z)-**7** afforded predominantly the syn isomer albeit less selectively than (E)-**7**. Unlike O,O-ketene acetals, in this series changing ketene acetal geometry did not affect the stereochemical course, but only reduced the syn selectivity (eq 10, Table 2).[9b]

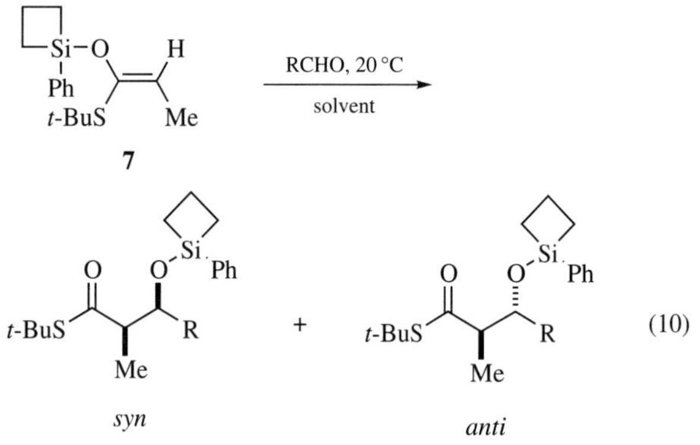

$$\text{(10)}$$

Table 2 Aldol reaction of the S,O-ketene acetals **7**

Entry	7, E/Z	R	Solvent	Time (h)	Convn (%)	syn/anti
1	4/96	Ph	CDCl₃	50.5	84	98/2
2	100/0	Ph	Neat			85/15
3	4/96	Cinnamyl	CDCl₃	51.0	91	70/30
4	4/96	n-Pentyl	CDCl₃	50.5	42	90/10
5	4/96	n-Pentyl	Neat	24.0	68	90/10
6	4/96	Cyclohexyl	CDCl₃	50.0	NR	
7	4/96	Cyclohexyl	Neat	29.0	46	80/20

It was unambiguously established that these reactions proceed by direct silicon group transfer by the double-label, crossover experiment (eq 11). The ketene acetal **4** reacted extremely rapidly

$(1\ \text{M}, \text{C}_6\text{D}_6, 18\,°\text{C}, t_{1/2}\ 4.5\ \text{min})$ demonstrating a significant effect of the silicon substituent.[9b] Deuterium analysis of the product **8** from a 1:1 mixture of **4** and d_6-**4** (96.8% d_6) showed less than 1% crossover (0.54% d_3-**8**). This result strongly suggests involvement of pentacoordinate silicon species.

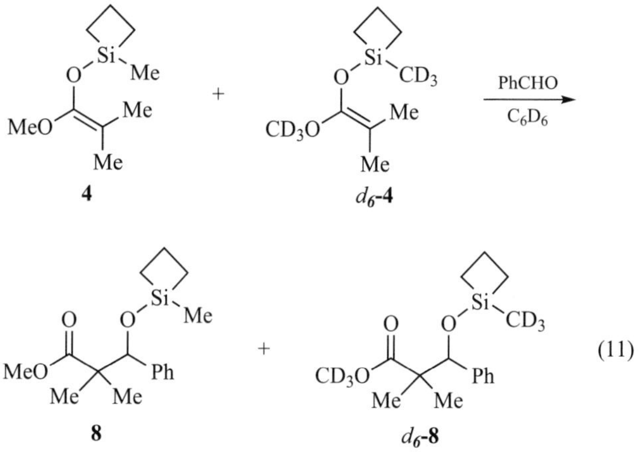

$$\text{(11)}$$

Allylation. Heating a mixture of 1-allyl-1-phenylsilacyclobutane and benzaldehyde at 130 °C for 12 h under argon in a sealed tube provided 1-phenyl-3-buten-1-ol in 85% yield, after workup with aq. 1 M HCl solution followed by silica gel chromatography (eq 12).[11] On the contrary, the use of allyldimethylphenylsilane resulted in the recovery of benzaldehyde and allylsilane, after heating at 160 °C for 24 h.

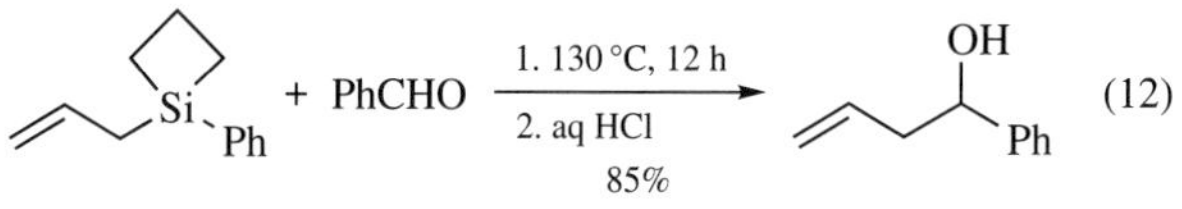

$$\text{(12)}$$

Cross-Coupling. 1-Alkenyl-1-methylsilcyclobutanes represent a special class of alkenyl(trialkyl)silanes that can undergo rapid and high-yielding palladium-catalyzed cross-coupling reactions with various aryl and vinyl iodides.[7] These reactions proceed rapidly in the presence of TBAF and 5 mol % of Pd(dba)₂ in THF at ambient temperature (eq 13). In a control experiment, (E)-heptenyltrimethylsilane failed to give more than traces of the product after 3 days. Initially, the remarkable rate of the cross-coupling was attributed to the enhanced Lewis acidity of the silicon center from strain release during the formation of the pentacoordinate fluorosiliconate.[12] However, subsequent studies showed that the siletane serves as a precursor to the silanol following a rapid, fluoride-promoted ring opening. The silanols are the active species for the cross coupling.[2b,13]

Ring Expansion. In the presence of nucleophiles or carbenoids, silacyclobutanes undergo ring expansion to silacyclopentanes.[6] As an example, the addition of i-PrOLi to 1-[(Z)-1,2-epoxyhexyl]-1-methylsilacyclobutane gives the ring expanded silacyclopentane (eq 14).[6a] In addition, silacyclobutanes also undergo facile, palladium-catalyzed ring expansion with alkynes (eq 15),[14a] acid halides (eq 16)[14b] (or alkyl halides in the presence of carbon monoxide).[14c] In addition, they undergo ring expansion with aldehydes under the action of potassium t-butoxide (eq 17).[14d]

$$R_1 \overset{R_2}{=}\!\!-\!\!\overset{}{\underset{Me}{Si}} + R_3{-}I \xrightarrow[\substack{5\ mol\ \%\ Pd(dba)_2,\ THF \\ rt,\ 10\ min}]{TBAF\ (3.0\ equiv)} R_1 \overset{R_2}{=}\!\!-\!\!R_3 \quad (13)$$

E Siletane Series

$n\text{-}C_5H_{11}$

91% (E/Z 99.9/0.1)

84% (E/Z 99.7/0.3)

$n\text{-}C_5H_{11}$ — OMe

94% (E/Z 99.0/0.9)

73% [$(E, E)/(E, Z)$ 98.0/2.0]

Z Siletane Series

$n\text{-}C_5H_{11}$

90% (E/Z 0.9/99.1)

75%

[$(Z, E)/(Z, Z)+(E, E)$97.2/2.8]

$$\text{(14)} \qquad \xrightarrow[88\%]{LiO i\text{-}Pr}$$

$$\text{SiMe}_2 + \overset{CO_2Me}{\underset{CO_2Me}{|||}} \xrightarrow[\substack{benzene,\ reflux,\ 2\ h \\ 50\%}]{(Ph_3P)_2PdCl_2\ (1\ mol\ \%)} \quad (15)$$

$$\text{SiMe}_2 + \underset{Ph}{\overset{O}{\underset{}{\|}}}\!\!-\!\!Cl \xrightarrow[\substack{Et_3N\ (\ 2.0\ equiv) \\ toluene,\ 80\ °C,\ 4\ h \\ 92\%}]{(Ph_3P)_2PdCl_2\ (4\ mol\ \%)} \quad (16)$$

$$\text{SiMe}_2 + \underset{Ph}{\overset{O}{\underset{}{\|}}}\!\!-\!\!H \xrightarrow[\substack{THF,\ 0\ °C,\ 2\ h \\ 65\%}]{KO t\text{-}Bu\ (10\ mol\ \%)} \quad (17)$$

Oxidation of Carbon-silicon Bonds. 1-Methylsiletanes are viable substrates for the Tamao oxidation of carbon-silicon bonds that combine the high reactivity associated with heteroatom-substituted silanes with the ease of purification and handling of tetraalkylsilanes (eq 18).[5] The siletane is also employed as an oxidation labile protecting group. The carbosilane oxidation triggered the deprotection of 4-siletanylbenzyl ethers to give the corresponding alcohols (eq 19).[15]

$$\text{TBSO} \xrightarrow[\substack{KHCO_3 \\ 82–85\%}]{H_2O_2,\ KF} \text{TBSO} \text{—OH} \quad (18)$$

$$\xrightarrow[\substack{K_2CO_3 \\ 86–97\%}]{H_2O_2,\ KF} R{-}OH \quad (19)$$

Miscellaneous. 1-Methylsiletanes are also used for polymerization,[16] insertion of carbenoid into β C–H bond of silacyclobutane (eq 20),[17] and siletane amphiphiles (eq 21).[18]

$$Me_2Si + N_2\!\!=\!\!CO_2t\text{-}Bu \xrightarrow[CH_2Cl_2,\ rt]{Rh_2(OAc)_4}$$

$$Me_2Si \overset{CO_2t\text{-}Bu}{} \quad (20)$$

$$\underset{Me}{\overset{}{Si}}\!\!-\!\!SnBu_3 \xrightarrow[\substack{2.\ Ph_2CO,\ -78\ °C\ to\ rt \\ 40\%}]{1.\ n\text{-}BuLi,\ THF,\ -78\ °C} \underset{Ph}{\overset{H}{\underset{}{}}}\!\!=\!\!\underset{Ph}{\overset{H}{\underset{}{}}} \quad (21)$$

1. Laane, J., *J. Am. Chem. Soc.* **1967**, *89*, 1144.

2. (a) Denmark, S. E.; Sweis, R. F., *Acc. Chem. Res.* **2002**, *35*, 835. (b) Denmark, S. E.; Wehrli, D.; Choi, J. Y., *Org. Lett.* **2000**, *2*, 2491.

3. House, S. E.; Poon, K. W. C.; Lam, H.; Dudley, G. B., *J. Org. Chem.* **2006**, *71*, 420.

4. Denmark, S. E.; Wang, Z., *Synthesis* **2000**, 999.

5. Sunderhaus, J. D.; Lam, H.; Dudley, G. B., *Org. Lett.* **2003**, *5*, 4571.

6. (a) Matsumoto, K.; Takeyama, Y.; Miura, K.; Oshima, M.; Utimoto, K., *Bull. Chem. Soc. Jpn.* **1995**, *68*, 250. (b) Matsumoto, K.; Aoki, Y.; Oshima, K.; Utimoto, K., *Tetrahedron Lett.* **1993**, *49*, 8487.

7. Denmark, S. E.; Choi, J. Y., *J. Am. Chem. Soc.* **1999**, *121*, 5821.

8. Barton, T. J.; Tillman, N., *J. Am. Chem. Soc.* **1987**, *109*, 6711.

9. (a) Denmark, S. E.; Griedel, B. D.; Coe, D. M., *J. Org. Chem.* **1993**, *58*, 988. (b) Denmark, S. E.; Griedel, B. D.; Coe, D. M.; Schnute, M. E., *J. Am. Chem. Soc.* **1994**, *116*, 7026. (c) Denmark, S. E.; Griedel, B. D.; *J. Org. Chem.* **1994**, *59*, 5136. (d) Myers, A. G.; Kephart, S. E.; Chen, H., *J. Am. Chem. Soc.* **1992**, *114*, 7922.

10. Smith, C. J.; Tsang, M. W. S.; Holmes, A. B.; Danheiser, R. L.; Tester, J. W., *Org. Biomol. Chem.* **2005**, *3*, 3767.

11. Masumoto, K.; Oshima, K.; Utimoto, K., *J. Org. Chem.* **1994**, *59*, 7152.

12. Denmark, S. E.; Jacobs, R. T.; Ho, G.-D.; Wilson, S. R., *Organometallics* **1990**, *9*, 3015.

13. Denmark, S. E.; Sweis, R. F.; Wehrli, D., *J. Am. Chem. Soc.* **2004**, *126*, 4865.

14. (a) Takeyama, Y.; Nozaki, K.; Matsumoto, K.; Oshima, K.; Utimoto, K., *Bull. Chem. Soc. Jpn.* **1991**, *64*, 1461. (b) Tanaka, M.; Yamashita, H.; Tanaka, M., *Organometallics* **1996**, *15*, 1524. (c) Chauhan, B. P. S.; Tanaka, Y.; Yamashita, H.; Tanaka, M., *Chem. Commun.* **1996**, 1207. (d) Takeyama, Y.; Oshima, K.; Utimoto, K., *Tetrahedron Lett.* **1990**, *31*, 6059.

15. Lam, H.; House, S. E.; Dudley, G. B., *Tetrahedron Lett.* **2005**, *46*, 3283.

16. (a) Matsumoto, K.; Hasegawa, H.; Matsuoka, H., *Tetrahedron* **2004**, *60*, 7197. (b) Wu, X.; Grinevich, O.; Neckers, D., *Chem. Mater.* **1999**, *11*, 3687. (c) Nametkin, N. S.; Vdovin, V. M.; Poletaev, V. A.; Zavyalov, V. I., *Dokl. Chem. Akad. Nauk SSSR* **1967**, *175*, 716.

17. Hatanaka, Y.; Watanabe, M.; Onozawa, S.; Tanaka, M.; Sakurai, H., *J. Org. Chem.* **1998**, *63*, 422.

18. Kozytska, M. V.; Dudley, G. M., *Chem. Commun.* **2005**, 3047.

Jun Young Choi

Synta Pharmaceuticals Corp., MA, USA

Scott E. Denmark

University of Illinois at Urbana-Champaign, IL, USA

Chlorotrimethylsilane[1,2]

[75-77-4]   C$_3$H$_9$ClSi   (MW 108.64)

InChI = 1S/C3H9ClSi/c1-5(2,3)4/h1-3H3

InChIKey = IJOOHPMOJXWVHK-UHFFFAOYSA-N

(protection of silyl ethers,[3] transients,[5–7] and silylalkynes;[8] synthesis of silyl esters,[4] silyl enol ethers,[9,10] vinylsilanes,[13] and silylvinylallenes;[15] Boc deprotection;[11] TMSI generation;[12] epoxide cleavage;[14] conjugate addition reactions catalyst[16–18])

Alternate Names: trimethylsilyl chloride; TMSCl.

Physical Data: bp 57 °C; *d* 0.856 g cm^{-3}.

Solubility: sol THF, DMF, CH$_2$Cl$_2$, HMPA.

Form Supplied in: clear, colorless liquid; 98% purity; commercially available.

Analysis of Reagent Purity: bp, NMR.

Purification: distillation over calcium hydride with exclusion of moisture.

Handling, Storage, and Precautions: moisture sensitive and corrosive; store under an inert atmosphere; use in a fume hood.

HCl impurities in TMSCl are not completely removed by treatment with monomeric pyridine bases due to the solubility of their hydrochloride salts in TMSCl. Therefore use of thus treated TMSCl to react with 'hard' organometallic reagents often produces unwanted hydrogen quenched by-products that are difficult to remove from the targeted silylated material. This problem is solved by storing TMSCl for 48h in a septum-sealed bottle as neat reagent over polyvinylpyridine. Any HCl present forms forms a polymer-bound hydrochloride that is insoluble in the reagent. Thus simply removing the clear supernatant TMSCl via a dry syringe insures delivery of acid-free material.[144]

Original Commentary

Ellen M. Leahy

Affymax Research Institute, Palo Alto, CA, USA

Protection of Alcohols as TMS Ethers. The most common method of forming a silyl ether involves the use of TMSCl and a base (eqs 1–3).[3,19–22] Mixtures of TMSCl and **Hexamethyldisilazane** (HMDS) have also been used to form TMS ethers. Primary, secondary, and tertiary alcohols can be silylated in this

manner, depending on the relative amounts of TMS and HMDS (eqs 4–6).[23]

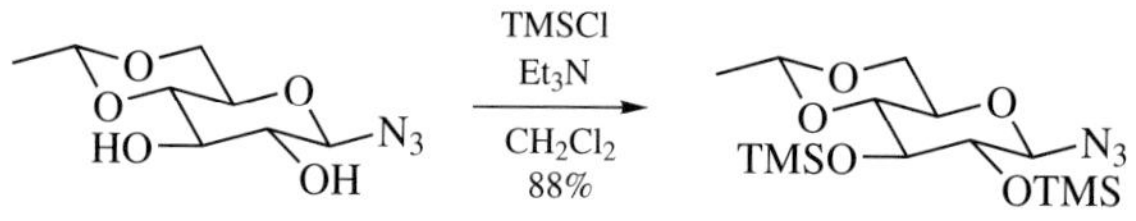

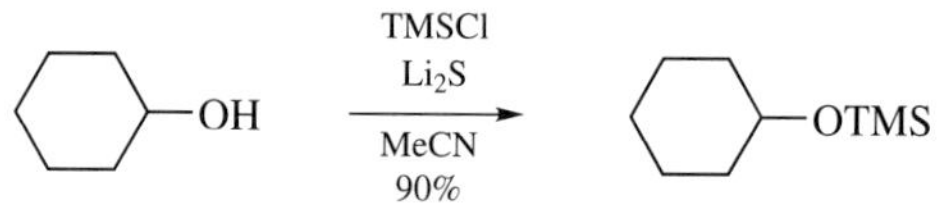

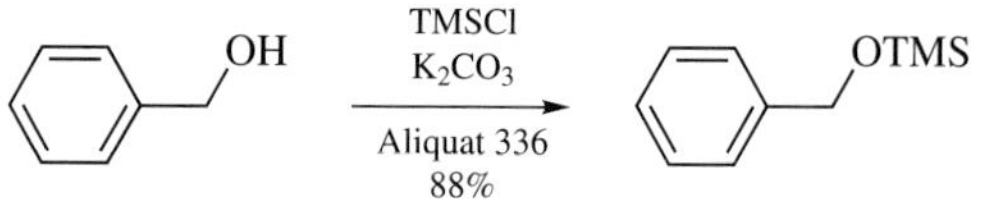

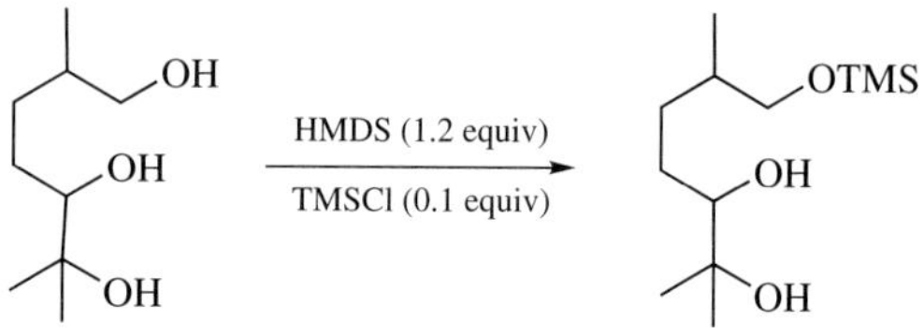

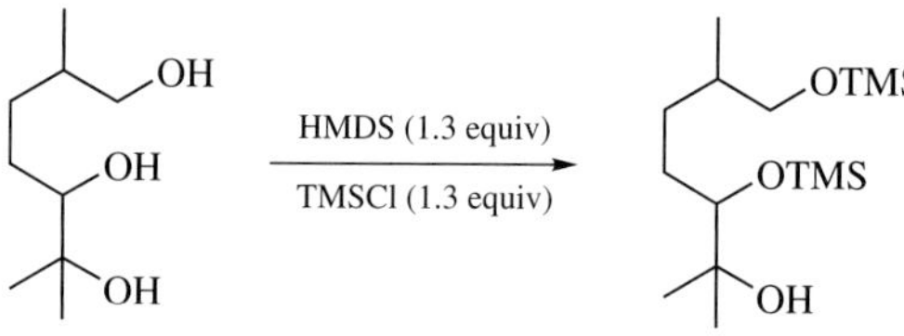

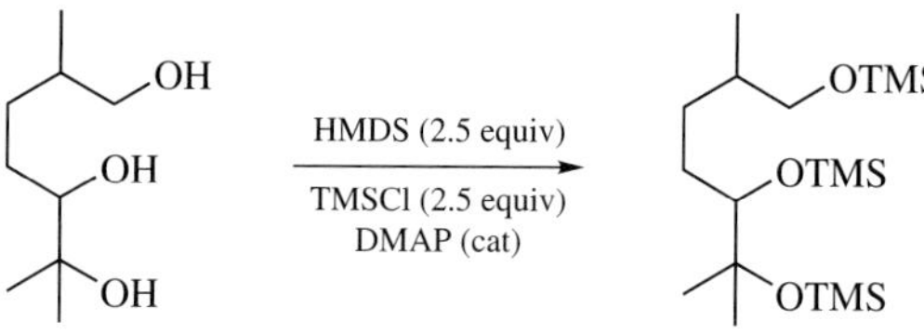

Trimethysilyl ethers can be easily removed under a variety of conditions,[19] including the use of tetra-*n*-butylammonium fluoride (TBAF) (eq 7),[20] citric acid (eq 8),[24] or potassium carbonate in methanol (eq 9).[25] Recently, resins (OH$^-$ and H$^+$ form) have been used to remove phenolic or alcoholic TMS ethers selectively (eq 10).[26]

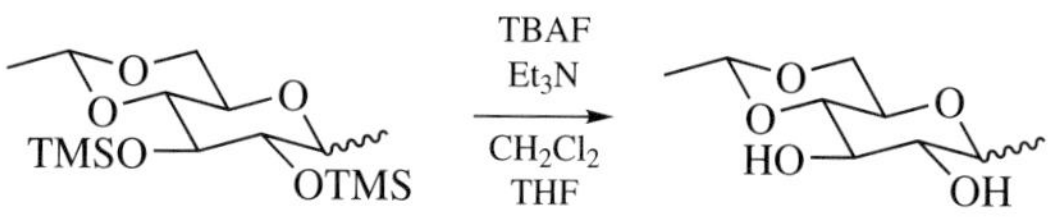

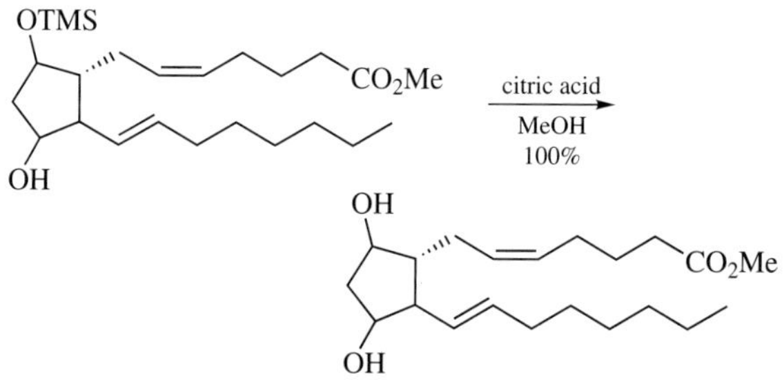

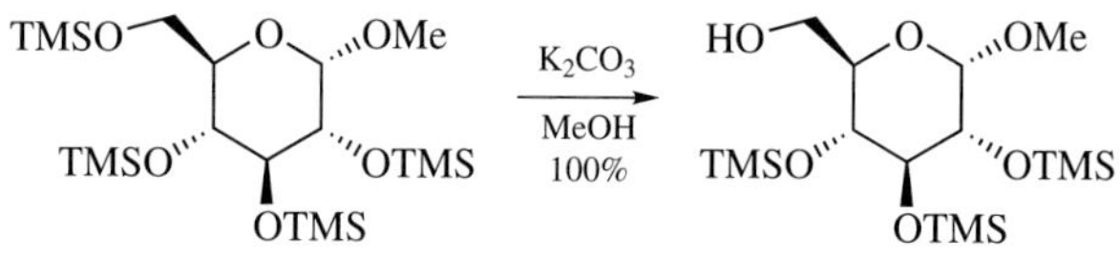

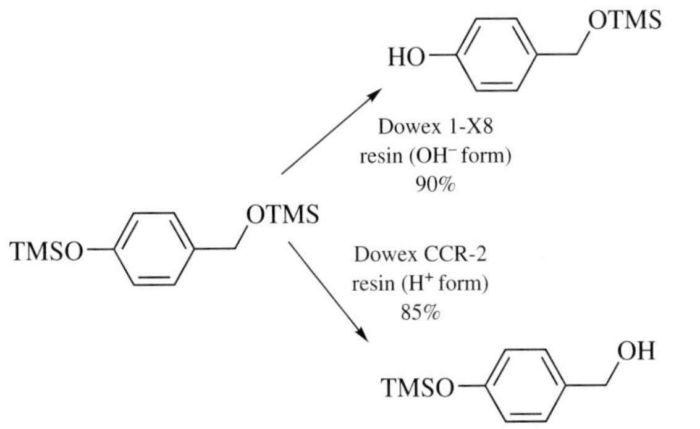

Transient Protection. Silyl ethers can be used for the transient protection of alcohols (eq 11).[27] In this example the hydroxyl groups were silylated to allow tritylation with concomitant desilylation during aqueous workup. The ease of introduction and removal of TMS groups make them well suited for temporary protection.

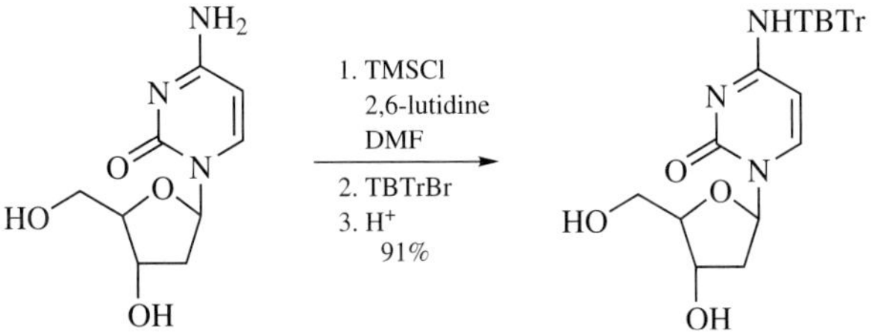

Trimethylsilyl derivatives of amino acids and peptides have been used to improve solubility, protect carboxyl groups, and improve acylation reactions. TMSCl has been used to prepare protected amino acids by forming the *O,N*-bis-trimethylsilylated amino acid, formed in situ, followed by addition of the acylating agent (eq 12).[5] This is a general method which obviates the production of oligomers normally formed using Schotten–Baumann conditions, and which can be applied to a variety of protecting groups.[5]

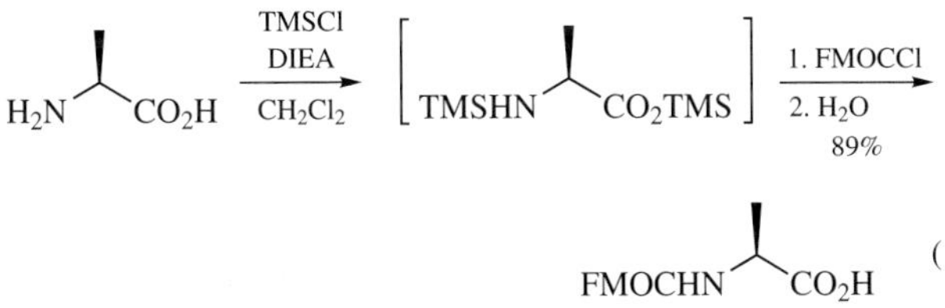

Transient hydroxylamine oxygen protection has been successfully used for the synthesis of *N*-hydroxamides.[6] Hydroxylamines can be silylated with TMSCl in pyridine to yield the *N*-substituted *O*-TMS derivative. Acylation with a mixed anhydride of a protected amino acid followed by workup affords the *N*-substituted hydroxamide (eq 13).[6]

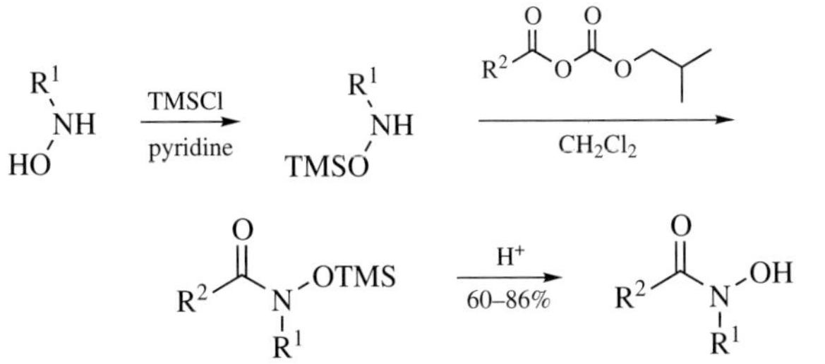

Formation of Silyl Esters. TMS esters can be prepared in good yields by reacting the carboxylic acid with TMSCl in 1,2-dichloroethane (eq 14).[4] This method of carboxyl group protection has been used during hydroboration reactions. The organoborane can be transformed into a variety of different carboxylic acid derivatives (eqs 15 and 16).[7] TMS esters can also be reduced with metal hydrides to form alcohols and aldehydes or hydrolyzed to the starting acid, depending on the reducing agent and reaction conditions.[28]

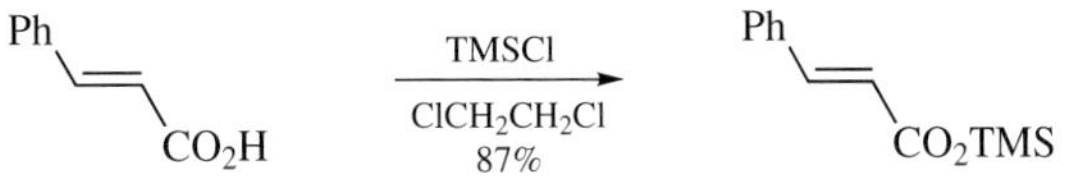

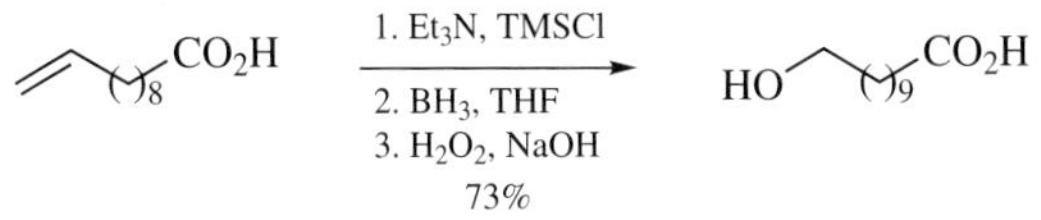

Protection of Terminal Alkynes. Terminal alkynes can be protected as TMS alkynes by reaction with *n*-butyllithium in THF followed by TMSCl (eq 17).[8] A one-pot β-elimination–silylation process (eq 18) can also yield the protected alkyne.

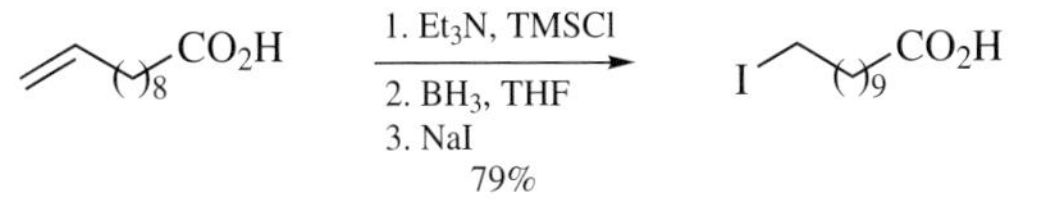

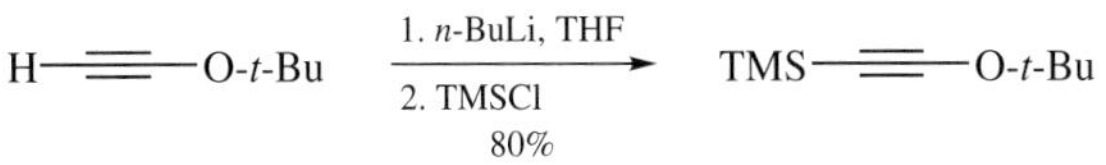

Silyl Enol Ethers. TMS enol ethers of aldehydes and symmetrical ketones are usually formed by reaction of the carbonyl compound with triethylamine and TMSCl in DMF (eq 19), but other bases have been used, including sodium hydride[29] and potassium hydride.[30]

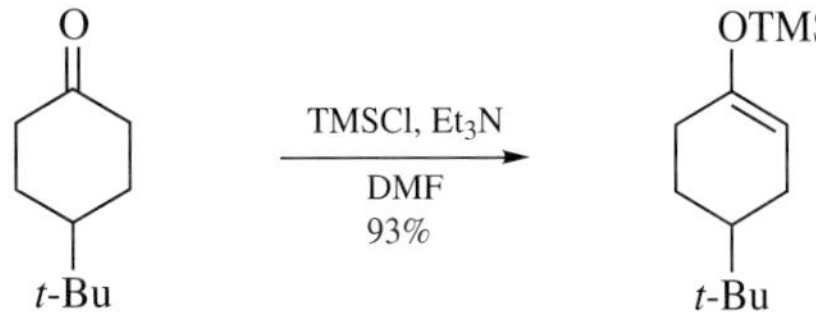

Under the conditions used for the generation of silyl enol ethers of symmetrical ketones, unsymmetrical ketones give mixtures of structurally isomeric enol ethers, with the predominant product being the more substituted enol ether (eq 20).[10] Highly hindered bases, such as lithium diisopropylamide (LDA),[31] favor formation of the kinetic, less substituted silyl enol ether, whereas bromomagnesium diisopropylamide (BMDA)[10] generates the more substituted, thermodynamic silyl enol ether. A combination of TMSCl/sodium iodide has also been used to form silyl enol ethers of simple aldehydes and ketones[32] as well as from α,β-unsaturated aldehydes and ketones.[33] Additionally, treatment of α-halo ketones with zinc, TMSCl, and TMEDA in ether provides a regiospecific method for the preparation of the more substituted enol ether (eq 21).[34]

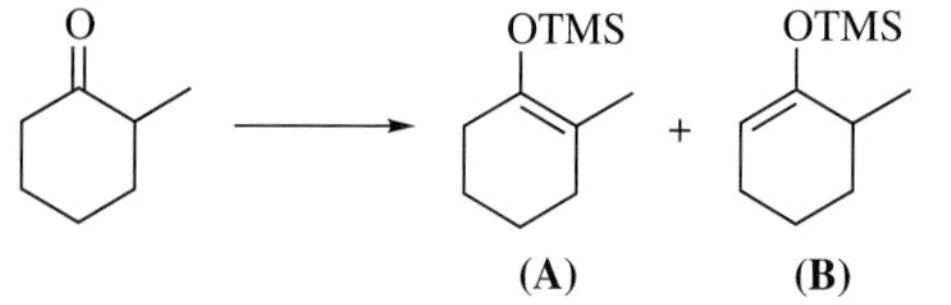

Reagents	Ratio (A):(B)
LDA, DME; TMSCl	1:99
NaH, DME; TMSCl	73:27
Et₃N, TMSCl, DMF	78:22
KH, THF; TMSCl	67:33
TMSCl, NaI, MeCN, Et₃N	90:10
BMDA, TMSCl, Et₃N	97:3

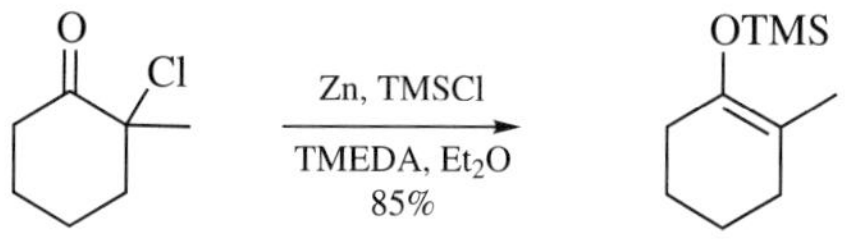

$$(20)$$

$$(21)$$

Mild Deprotection of Boc Protecting Group. The Boc protecting group is used throughout peptide chemistry. Common ways of removing it include the use of 50% trifluoroacetic acid in CH_2Cl_2, trimethylsilyl perchlorate, or iodotrimethylsilane (TMSI).[19] A new method has been developed, using TMSCl–phenol, which enables removal of the Boc group in less than one hour (eq 22).[11] The selectivity between Boc and benzyl groups is high enough to allow for selective deprotection.

$$\text{Boc-Val-OCH}_2\text{-resin} \xrightarrow[\substack{20 \text{ min} \\ 100\%}]{\text{TMSCl, phenol}} \text{Val-OCH}_2\text{-resin} \quad (22)$$

In Situ Generation of Iodotrimethylsilane. Of the published methods used to form TMSI in situ, the most convenient involves the use of TMSCl with NaI in acetonitrile.[12] This method has been used for a variety of synthetic transformations, including cleavage of phosphonate esters (eq 23),[35] conversion of vicinal diols to alkenes (eq 24),[36] and reductive removal of epoxides (eq 25).[37]

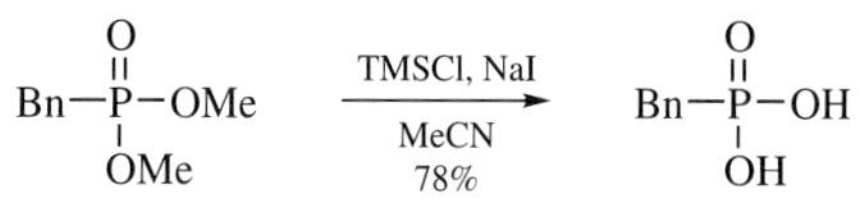

$$(23)$$

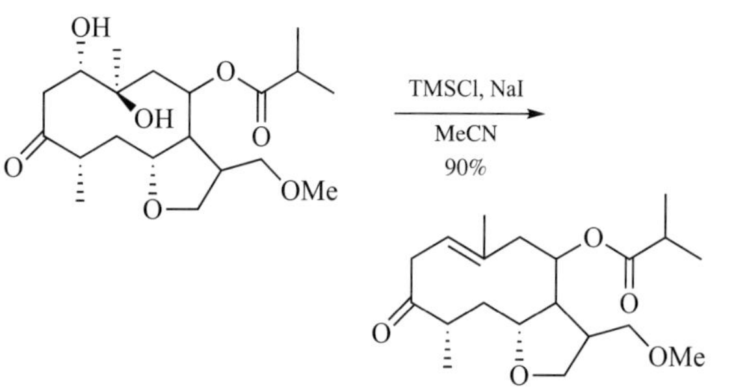

$$(24)$$

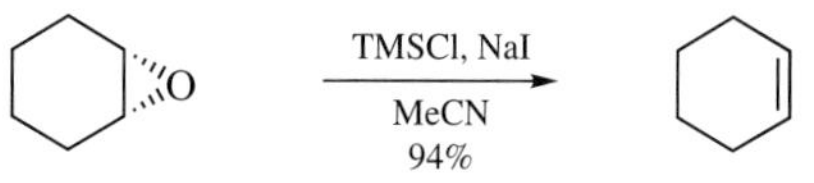

$$(25)$$

Conversion of Ketones to Vinylsilanes. Ketones can be transformed into vinylsilanes via intermediate trapping of the vinyl anion from a Shapiro reaction with TMSCl. Formation of either the tosylhydrazone[38] or benzenesulfonylhydrazone (eq 26)[13,39] followed by reaction with n-butyllithium in TMEDA and TMSCl gives the desired product.

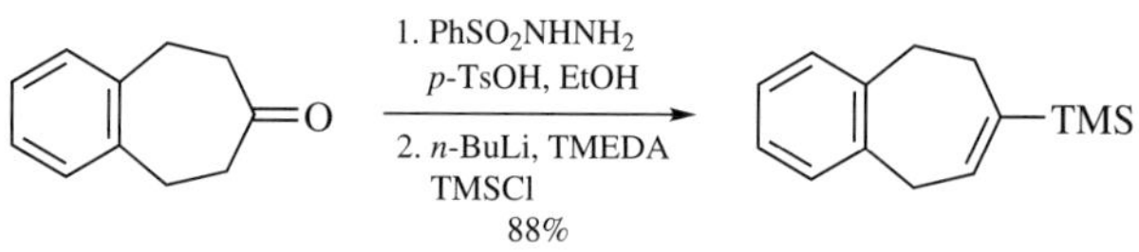

$$(26)$$

Epoxide Cleavage. Epoxides open by reaction with TMSCl in the presence of triphenylphosphine or tetra-n-butylammonium chloride to afford O-protected vicinal chlorohydrins (eq 27).[14]

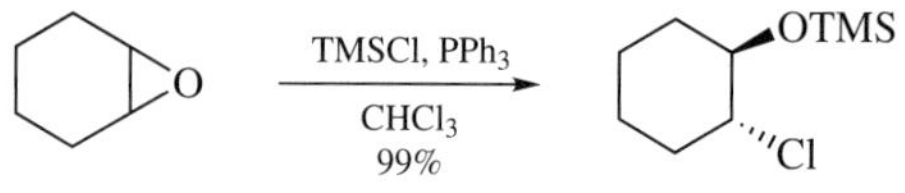

$$(27)$$

Formation of Silylvinylallenes. Enynes couple with TMSCl in the presence of Li/ether or Mg/hexamethylphosphoric triamide to afford silyl-substituted vinylallenes. The vinylallene can be subsequently oxidized to give the silylated cyclopentanone (eq 28).[15]

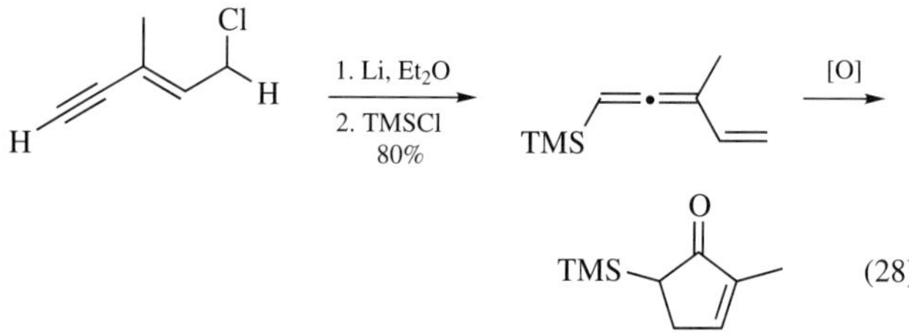

$$(28)$$

Conjugate Addition Reactions. In the presence of TMSCl, cuprates undergo 1,2-addition to aldehydes and ketones to afford silyl enol ethers (eq 29).[16] In the case of a chiral aldehyde, addition of TMSCl follows typical Cram diastereofacial selectivity (eq 30).[16,40]

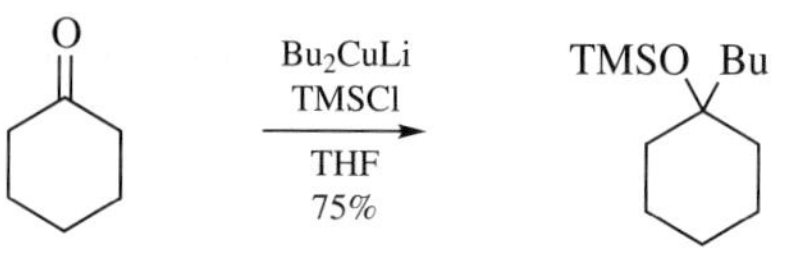

$$(29)$$

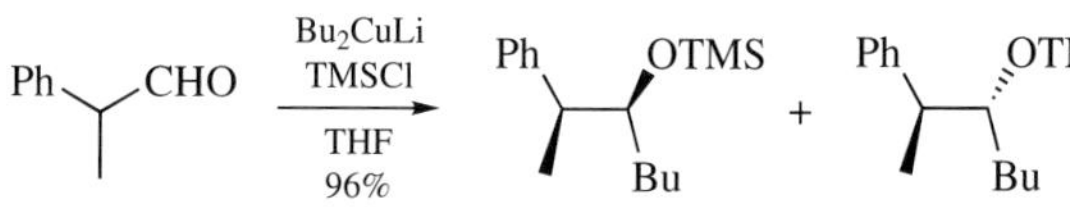

$$(30)$$

Conjugate addition of organocuprates to α,β-unsaturated carbonyl compounds, including ketones, esters, and amides, are accelerated by addition of TMSCl to provide good yields of the 1,4-addition products (eq 31).[17,41,42] The effect of additives such as HMPA, DMAP, and TMEDA have also been examined.[18,43] The role of the TMSCl on 1,2- and 1,4-addition has been explored by several groups, and a recent report has been published by Lipshutz.[40] His results appear to provide evidence that there is an interaction between the cuprate and TMSCl which influences the stereochemical outcome of these reactions.

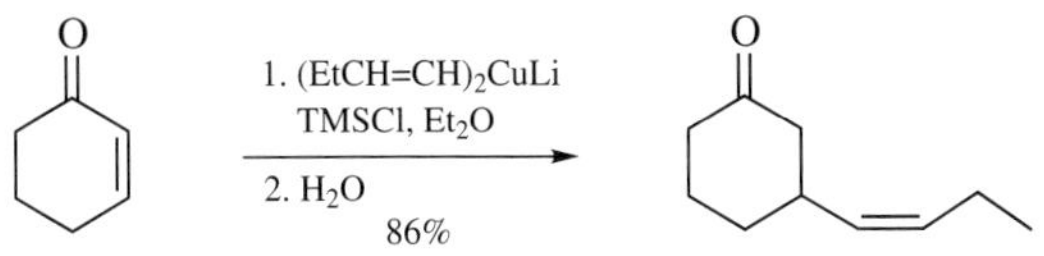

The addition of TMSCl has made 1,4-conjugate addition reactions to α-(nitroalkyl)enones possible despite the presence of the acidic α-nitro protons (eq 32).[44] Copper-catalyzed conjugate addition of Grignard reagents proceeds in high yield in the presence of TMSCl and HMPA (eq 33).[45] In some instances the reaction gives dramatically improved ratios of 1,4-addition to 1,2-addition.

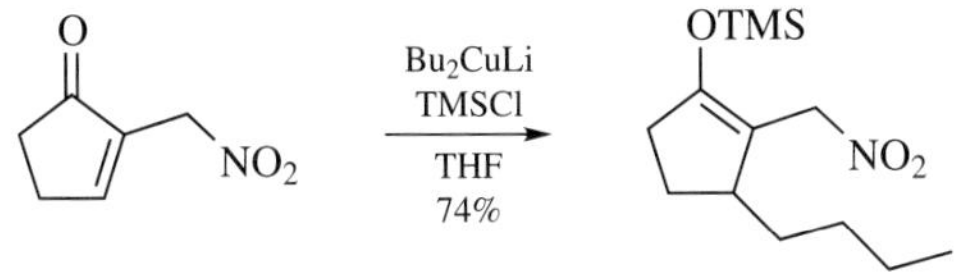

First Update

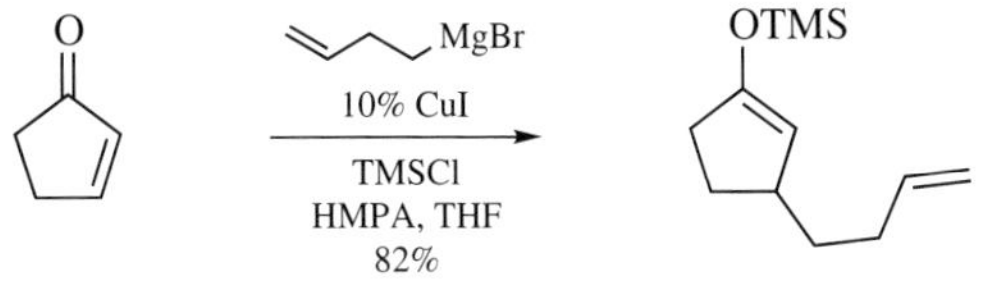

Wenming Zhang

Dupont Crop Protection, Newark, DE, USA

Protection of Alcohols as TMS Ethers. Several new methods have been developed for the protection of alcohols as TMS ethers. For example, TMS silyl ethers of alcohols and phenols can be prepared efficiently by treatment of the alcohol or phenol with TMSCl and catalytic amount of imidazole or iodine under the solvent-free and microwave irradiation conditions.[46] This transformation proved to be reversible. Under the same microwave conditions, treatment of the silyl ether in methanol and in the presence of catalytic amount of iodine releases the parent alcohol in quantitative yield.

In another new method, treatment of aliphatic alcohols with TMSCl in DMF containing magnesium turnings at rt produces the corresponding TMS silyl ethers in good to excellent yields. This protocol provides a viable way to protect sterically hindered alcohols as their TMS ethers. For example, *O*-silylation of ethyl 2-methyl-2-hydroxypropanate generates the desired TMS ether in 91% yield (eq 34).[47]

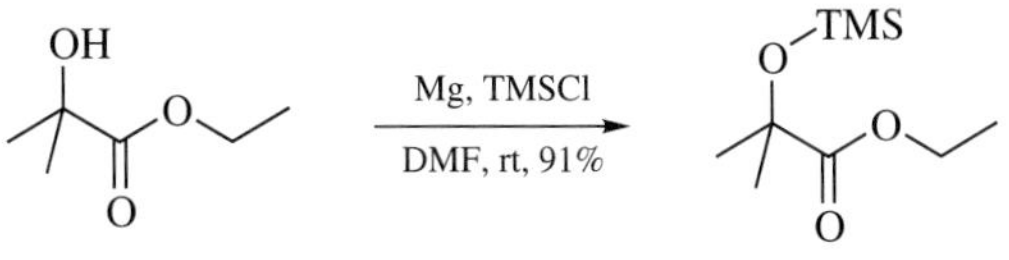

Mercaptans have been protected as trimethylsilylated sulfides by treating the mercaptan first with a strong base such as *n*-hexyllithium and then capturing the sulfide anion with TMSCl.[48] Aldehydes have been converted into *N*-trimethylsilylimines by sequential treatment with LiHMDS and TMSCl.[49]

Silyl Enol Ethers. α,α-Difluorotrimethylsilyl enol ethers can be prepared through a Mg/TMSCl-promoted selective C–F bond cleavage of the corresponding trifluoromethyl ketones. An α,α-difluorinated analog of Danishefsky's diene was prepared in good yield when the ketone was exposed to excessive Mg/TMSCl in DMF at 50 °C for 3 min (eq 35).[50]

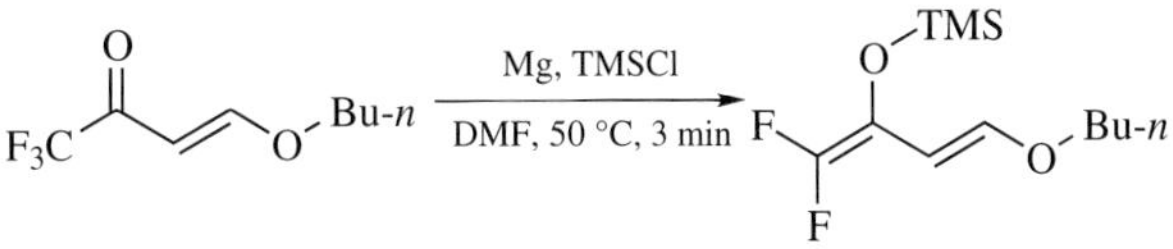

Treatment of a vinyl ketone with TMSCl in THF provides the corresponding 2-chloroenol TMS ether, which can be used to generate the Machenzie complex through reaction with Ni(cod)$_2$, and which can be further applied in several multicomponent assembly reactions.[51]

Epoxide Cleavage. Chlorotrimethylsilane has been employed in combination with various Lewis acids, such as SnCl$_2$,[52] BF$_3 \cdot$ Et$_2$O,[53] and concentrated LiClO$_4$ in ether,[54] to open epoxide rings and generate chlorohydrins or their derivatives. For example, when glycidyl phenyl ether was reacted with TMSCl in the presence of SnCl$_2$ as catalyst, the oxirane ring was cleaved to generate two chlorohydrin regioisomers that were further transformed into corresponding acetates (eq 36). In this particular case, the C$_3$–O bond cleavage is preferred due to stabilizing chelation of the tin(II) cation by the 1-phenoxy oxygen. However, when glycidyl benzoate was treated under the identical conditions, the C$_3$–O bond cleavage product now was obtained as the minor product (eq 37). The major product came from the C$_2$–O bond cleavage, due to the neighboring benzoate group participation and rearrangement.

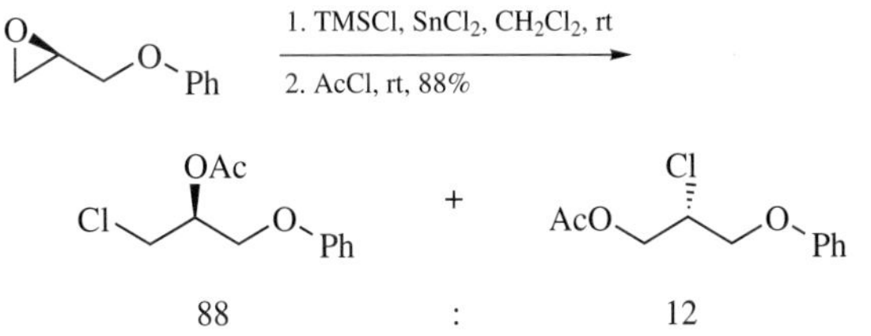

It was demonstrated that the ring opening of expoxides with TMSCl can also be facilitated by nucleophilic catalysts, such as 1,2-ferrocenediylazaphosphinines,[55] phosphaferrocenes, and phosphazirconcenes.[56] For example, in the presence of 5% of 1,2-ferrocenediylazaphosphinine, 1-hexene oxide was converted into 1-chloro-2-hexanol in 97% yield with 100% regioselectivity (eq 38). Various toluenesulfonyl aziridines also undergo a similar ring opening reaction to produce the corresponding chloramine derivatives under mild conditions in DMF.[57] As solvent, DMF also serves as an activator.

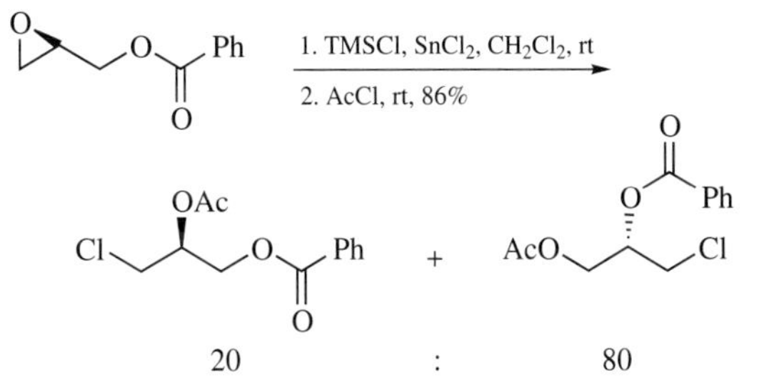

n-Bu $\overset{O}{\triangle}$ H $\xrightarrow[\text{rt, 70 min. 97\%}]{\text{TMSCl, cat}}$

n-Bu $\overset{OH}{\underset{Cl}{\diagup}}$ + n-Bu $\overset{Cl}{\underset{OH}{\diagup}}$ (38)

100 : 0

cat =

Related to the epoxide ring opening, TMSCl also mediates some cyclopropane ring opening reactions. For example, treatment of 1-aceto-2,2-dimethylcyclopropane with TMSCl and sodium chloride in acetonitrile at 55 °C for 24 h generated 5-chloro-5-methyl-2-hexanone in 84% yield (eq 39).[58] When sodium iodide was employed to replace sodium chloride, iodotrimethylsilane generated in situ, and the reaction completed under more facile conditions (rt and 12 h). Interestingly, the dominant product is 5-iodo-4,4-dimethyl-2-pentanone (eq 40), arising from iodide attacking at the less hindered secondary methylene carbon, instead of the quaternary dimethylmethylene carbon.

$\xrightarrow[\text{CH}_3\text{CN, 55 °C, 24 h}]{\text{TMSCl, NaCl}}$

$\overset{O}{\underset{Cl}{\diagdown}}$ + $\overset{O}{\underset{Cl}{\diagdown}}$ (39)

84% : 0%

$\xrightarrow[\text{CH}_3\text{CN, rt, 12 h}]{\text{TMSCl, NaI}}$

$\overset{O}{\underset{I}{\diagdown}}$ + $\overset{O}{\underset{I}{\diagdown}}$ (40)

33% : 47%

Conjugate Addition Reactions. Besides copper salts, a number of other catalyst systems, such as Ni(acac)$_2$/DIBAL,[59] ZnEt$_2$,[60] InCl$_3$,[61] and Pd(PPh$_3$)$_4$/LiCl,[62] catalyze Michael addition of various enones at the β-carbon to form new C–C bonds. In addition to the formation of C–C bonds, FeCl$_3 \cdot 6H_2O$, when combined with a stoichiometric amount of TMSCl, can also catalyze the aza-Michael addition to form new C–N bonds.[63] For example, in the presence of 5% of FeCl$_3 \cdot 6H_2O$ and 1.1 equiv of TMSCl, stirring a mixture of 2-cyclohexenone and ethyl carbamate in dichloromethane provides an 89% yield of the protected β-aminocyclohexanone (eq 41).

$\overset{O}{\diagdown}$ + NH$_2$CO$_2$Et $\xrightarrow[\text{rt, 12 h, 89\%}]{\substack{\text{TMSCl, (1.1equiv)} \\ \text{FeCl}_3 \cdot \text{H}_2\text{O (5\%)}}}$ $\overset{O}{\underset{\text{NHCO}_2\text{Et}}{\diagdown}}$ (41)

Cohen and Liu have examined the yield and selectivity enhancement effects of TMSCl and TMSCl/HMPA on conjugate addition of some stabilized organolithium reagents onto, particularly, easily polymerized α,β-unsaturated carbonyl compounds. The authors proposed that the beneficial effect is due to the prevention of 1,2-addition and polymerization of the α,β-unsaturated carbonyl compounds.[64]

The regiochemistry of 1,2- versus 1,4-addition of lithiated N-Boc-N-(p-methoxyphenyl)benzylamine with 2-cyclohexenone can be ligand controlled.[65] In the absence of a ligand, the reaction provides a 77% yield of the 1,2-addition product allylic alcohol (eq 42). When (−)-sparteine is premixed with n-BuLi before lithiation, the reaction furnishes a 70% yield of the corresponding 1,4-addition product. Furthermore, in the presence of TMSCl, the same reaction now affords 82% yield of the Michael adduct, with excellent diastereoselectivity (>99:1 dr) and enantioselectivity (96:4 er) favoring the (S,S)-cyclohexanone.

(42)

Ligand			
None	77%	5%	
(-)-sparteine	<5%	70%	-
(-)-sparteine + TMSCl	-	-	82%, >99:1 dr, 94:6 er

Related to the copper-catalyzed Michael addition reaction, allyl phosphates furnish substitution products via the *anti* S_N2' mechanism when treated with an appropriate Grignard reagent in the presence of catalytic amount of CuCN and 1 equiv of TMSCl (eq 43).[66]

$\xrightarrow[\text{CuCN, Et}_2\text{O, 0 °C, 73\%}]{\text{MeMgBr, TMSCl}}$

(43)

86 : 4 : 10

Anion Trap. Besides formation of silyl enol ethers, TMSCl has also been applied to trap other oxide anions to form the desired trimethylsilyl derivatives. For example, reaction of ethyl 4-phenylbutanoate with 1,1-dichloroethyllithium, which was generated from 1,1-dichloroethane and LDA, produced exclusively the mixed acetal as expected in 86% yield in the presence of TMSCl (eq 44).[67] Without TMSCl, the reaction gave the corresponding ketone as the final product. However, the yield is rather low and the ketone was obtained in only 16% (eq 45). When the same reaction run with methoxymethyl 4-phenylbutanoate, it always afforded the ketone as the final product, regardless of the presence of TMSCl, although, the yield was improved dramatically from 20% to 73% in the presence of the reagent (eq 46).

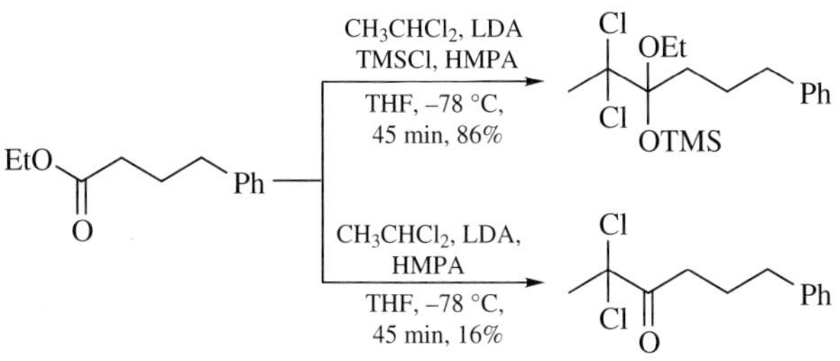

(44)

(45)

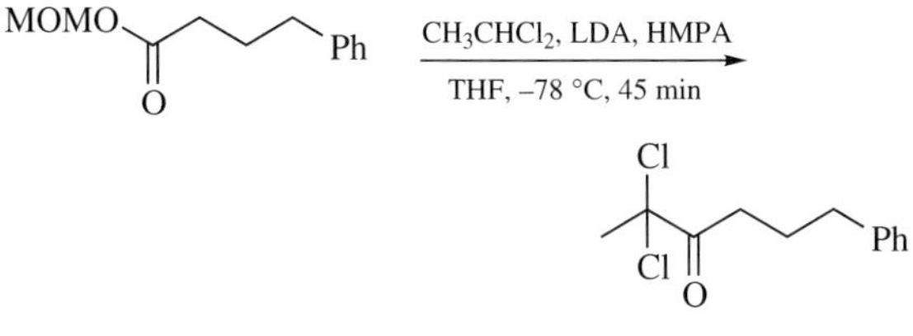

(46)

with TMSCl	73%
without TMSCl	20%

Dehydrating Agent. TMSCl is also a suitable dehydrating agent for scavenging water generated in various reactions. For example, in the esterification of 5-methylpyrazinoic acid with *n*-nonanol, a 44% yield of the desired ester can be distilled out directly from the reaction mixture of the acid, alcohol, and TMSCl (eq 47).[68]

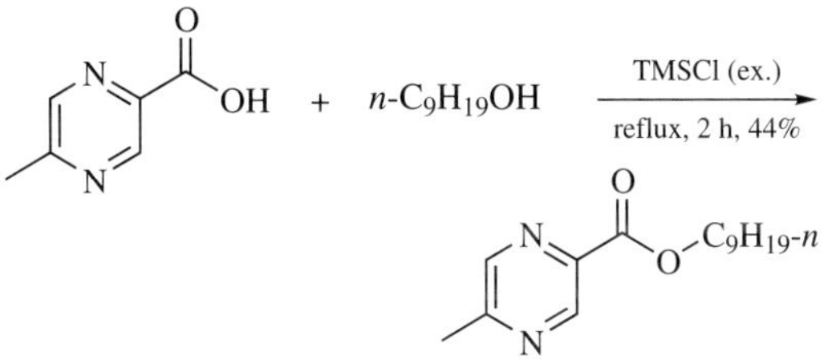

(47)

TMSCl also effects some dehydration cyclization reactions for the formation of heterocyclic compounds. For example, in the presence of TMSCl, boiling of *N*-benzoyl-*α*,*β*-dehydrophenylalanine anilide in DMF leads to isolation of the cyclization product imidazol-5-one in 80% yield (eq 48).[69] Interestingly, the same compound can be prepared by directly heating the corresponding oxazolone and aniline, also in the presence of TMSCl, although in somewhat lower yield (eq 49).

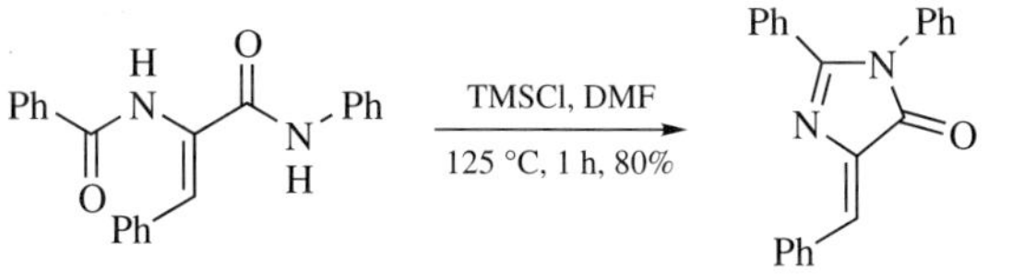

(48)

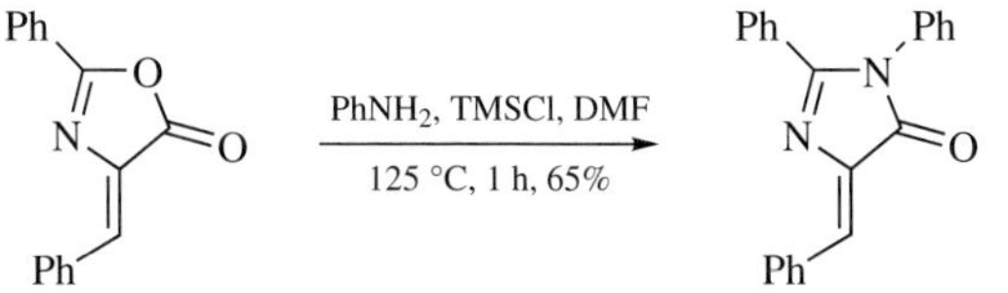

(49)

The combination of TMSCl and triethylamine provides another choice of dehydrating agent. For example, stirring a mixture of 1-nitronaphthyl tolyl sulfone, TMSCl, and triethylamine in DMF at rt for 4 days provides the corresponding anthranil in

72% yield (eq 50).[70] In a similar transformation, treatment of substituted *ortho*-nitroarylethanes with TMSCl and triethylamine in DMF results in formation of 1-hydroxyindoles as the dehydration/cyclization product.[71]

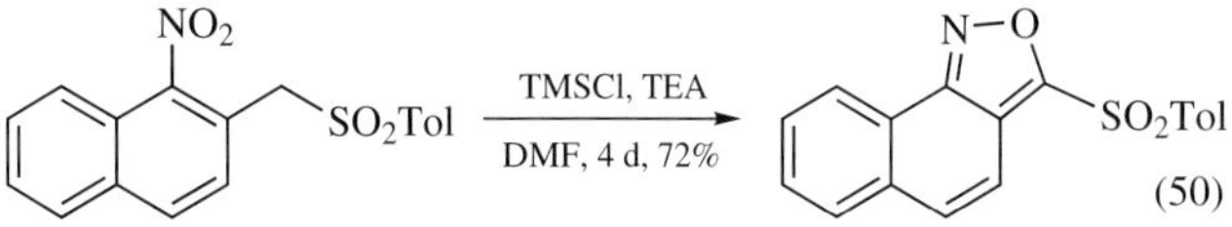

(50)

Chlorination. TMSCl has been used as the chloride source in a variety of substitution reactions. For example, when 1-bromoundecane is heated with stoichiometric amount of TMSCl in DMF at 90 °C and in the presence of imidazole, a Finkelstein-like reaction provides the corresponding 1-chloroundecane in quantitative yield (eq 51).[72]

$$n\text{-}C_{11}H_{23}Br \xrightarrow[\text{90 °C, 1 h, 99%}]{\text{TMSCl, Im, DMF}} n\text{-}C_{11}H_{23}Cl \quad (51)$$

In another example, treatment of an oxazoline ester with TMSCl in THF at reflux leads to oxazoline ring opening and furnishes the corresponding *β*-chloro-amino ester in quantitative yield (eq 52).[73] Treatment with TMSBr and TMSI at rt affords the corresponding *β*-halo-amino esters in quantitative yield.

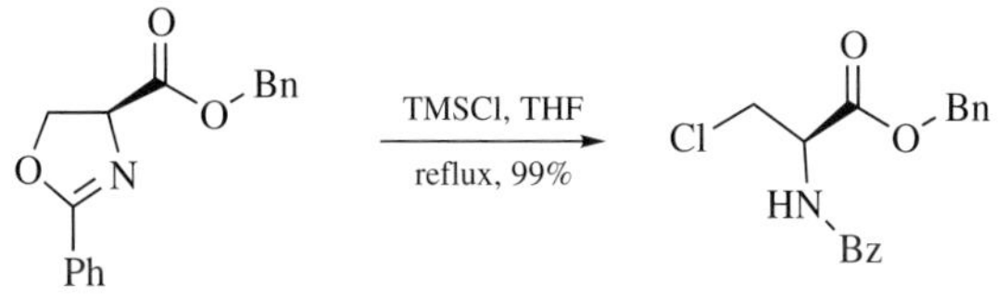

(52)

In a reaction generating *β*-halo-amino carboxylic acid derivatives similar to the ones above, stirring a mixture of a serine amide and TMSCl in acetonitrile at reflux for 8 h furnishes the *β*-chloro-amino amide in 72% yield (eq 53).[74] Treatment with TMSI results in the same transformation, but provides only 20% yield of the desired iodide.

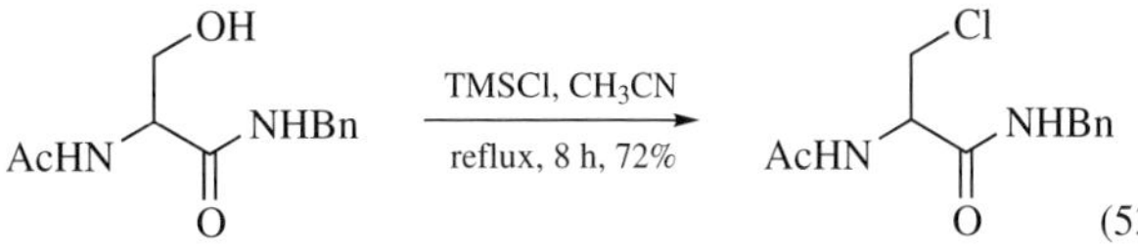

(53)

The above transformation of alcohols into chlorides does not occur for simple alcohols unless a catalytic amount of DMSO is added. Thus, addition of the catalyst into a mixture of 1-propanol and TMSCl (2 equiv) at rt provides 1-chloropropane in 93% yield in only 10 min (eq 54).[75] The reaction also works for other primary and tertiary alcohols, but not for secondary alcohols. The reaction of secondary alcohols with TMSCl and a stoichiometric amount of DMSO follows a procedure similar to that of the Swern oxidation and the corresponding ketones are produced.[76]

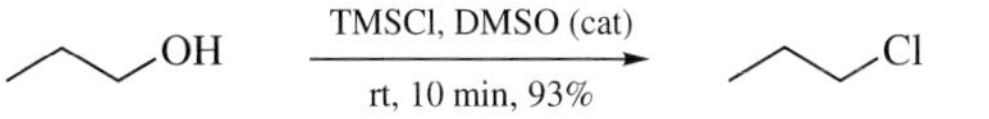

(54)

The TMSCl and DMSO combination can also be utilized to convert thiols into the corresponding disulfides if a stoichiometric amount of DMSO and a catalytic amount of TMSCl are used. For example, addition of 0.05 equiv of TMSCl into a mixture of thiophenol and DMSO in dichloromethane affords the diphenyl disulfide product in 90% yield (eq 55).[77]

$$PhSH \xrightarrow[\text{rt, 30 min, 90\%}]{\text{DMSO, TMSCl (cat)}} PhSSPh \quad (55)$$

TMSCl is also a suitable chlorine source for the electrophilic chlorination reactions. For example, a combination of TMSCl, DMSO, and a catalytic amount of Bu$_4$NBr can be used for the chlorination of isoxazolin-5(4H)-ones (eq 56).[78] In another report, the combination of TMSCl and iodobenzene diacetate effectively converts various flavone derivatives into the corresponding 3-chloroflavones under mild conditions.[79]

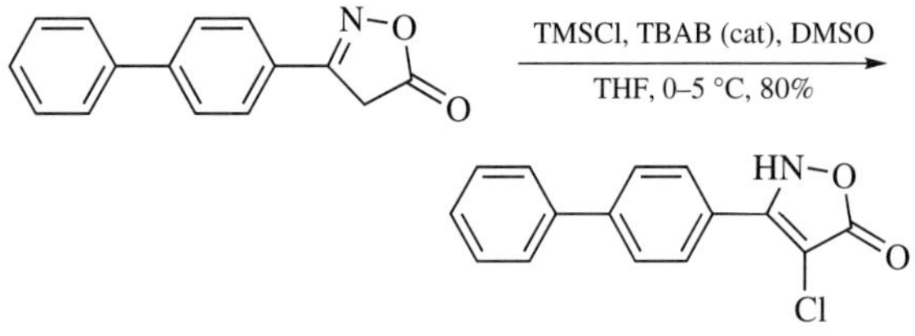

$$(56)$$

TMSCl can be used as chlorine source for hydrochlorination of olefins or acetylenes. For example, treatment of a mixture of $\Delta^{9,10}$-octaline and a substoichiometric amount of water with TMSCl furnishes the *trans*-hydrochlorination product in quantitative yield (eq 57).[80]

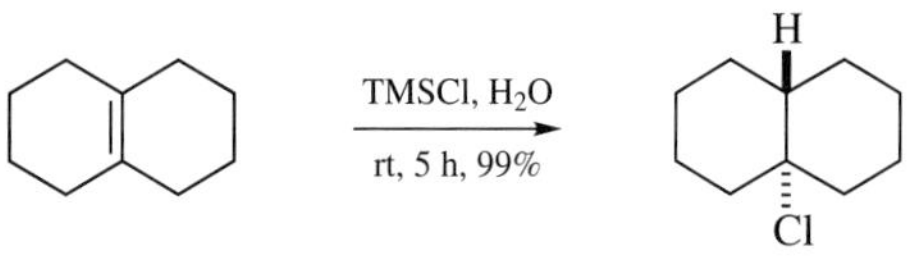

$$(57)$$

In another reaction, in the presence of TMSCl and FeCl$_2$ catalyst, decomposition of propargyloxycarbonyl azide results in the formation of an intramolecular *syn*-aminochlorination product under mild conditions (eq 58).[81]

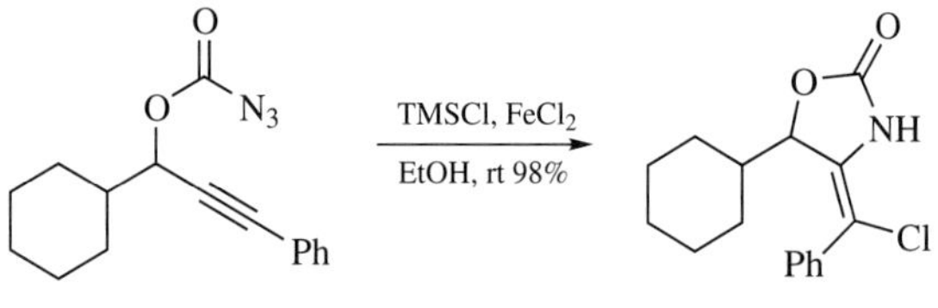

$$(58)$$

H$^+$ Surrogate. TMSCl can function similarly to a Brønsted acid or can generate a Brønsted acid in situ in various reactions. For example, addition of 2 equiv of TMSCl and 2 equiv of water to various nitriles furnishes the corresponding amides in good yield under ambient conditions (eq 59).[82] Furthermore, heating a mixture of 2 equiv of TMSCl and an amide in an alcohol solvent at 40 °C provides the corresponding ester in moderate to good yield (eq 60).[83] Finally, one-pot direct conversion of a nitrile into the corresponding ester can be achieved by heating a mixture of the nitrile and 2 equiv of TMSCl in the desired alcohol solvent at 50 °C (eq 61).[84]

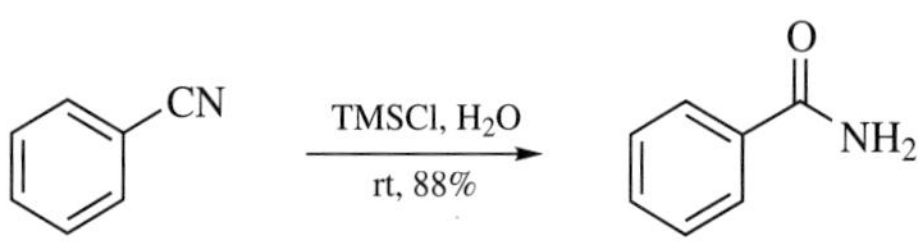

$$(59)$$

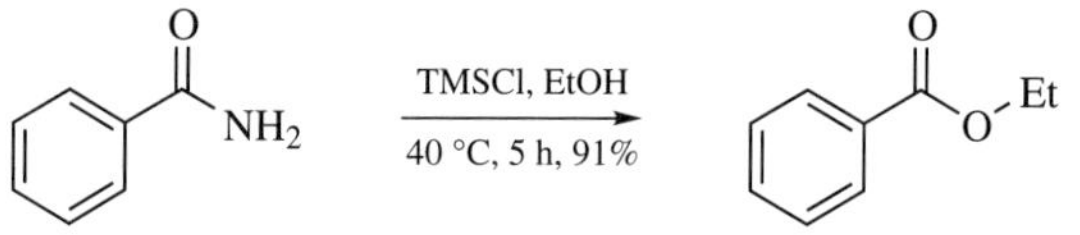

$$(60)$$

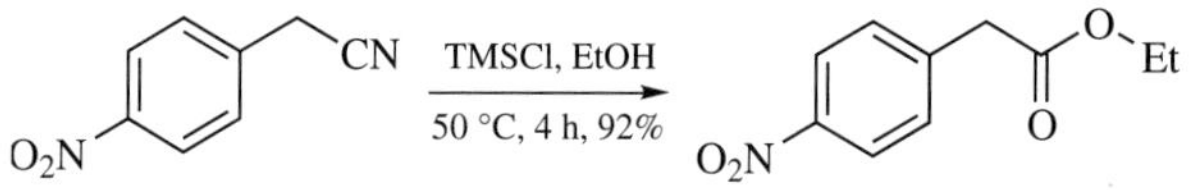

$$(61)$$

TMSCl has also been used as an efficient catalyst in acetalization reactions to protect ketones as dioxalanes[85] or thioacetals.[86] Treatment of 4-methylidene-3,4-dihydro-2H-pyrrole by TMSCl affords the corresponding pyrrole as the sole product (eq 62).[87]

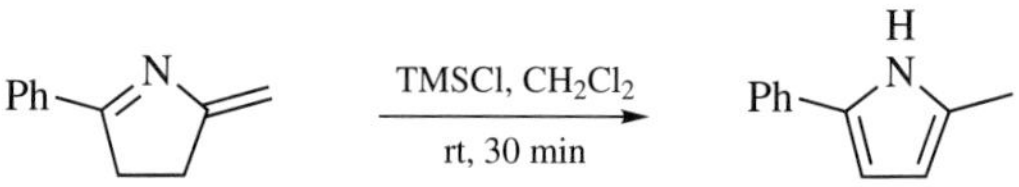

$$(62)$$

A combination of TMSCl and triethylamine has allowed the successful cyclization of several deoxybenzoins to form isoflavones (eq 63), while a number of other protocols under either basic or acidic conditions failed to provide the desired products.[88]

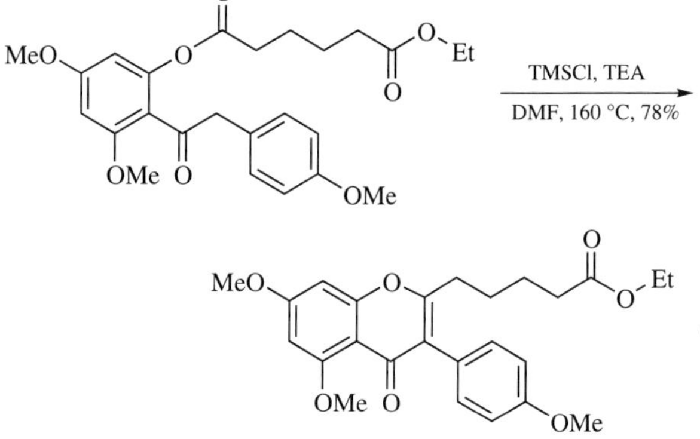

$$(63)$$

Lewis Acid. In addition to the numerous examples of the type throughout this article, several specific examples of TMSCl as a Lewis acid are given here. Addition of diethylzinc to various imines is promoted by different Lewis acids, such as TMSCl, BF$_3$ · Et$_2$O, and ZnCl$_2$, to offer a variety of secondary amines (eq 64).[89] The reaction is simple and effective and, unlike the comparable additions to carbonyl compounds, no amino alcohol is needed.

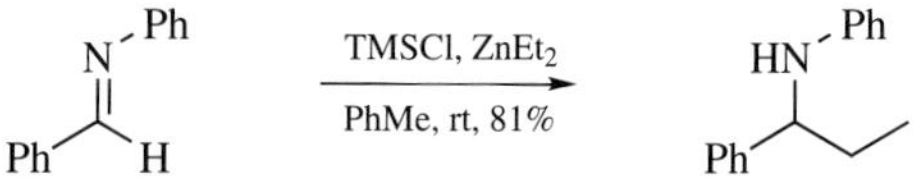

$$(64)$$

TMSCl also promotes the Pummerer reaction. For example, sequential addition LDA and TMSCl into a THF solution of 2-ethylbenzene sulfoxide produces 2-(1-hydroxy)-ethylbenzene thioether in reasonable yield and excellent stereoselectivity (eq 65).[90] Interestingly, treatment of 2-methylbenzene sulfoxide under the identical conditions gives 2-trimethylsilylmethylbenzene sulfoxide (eq 66).

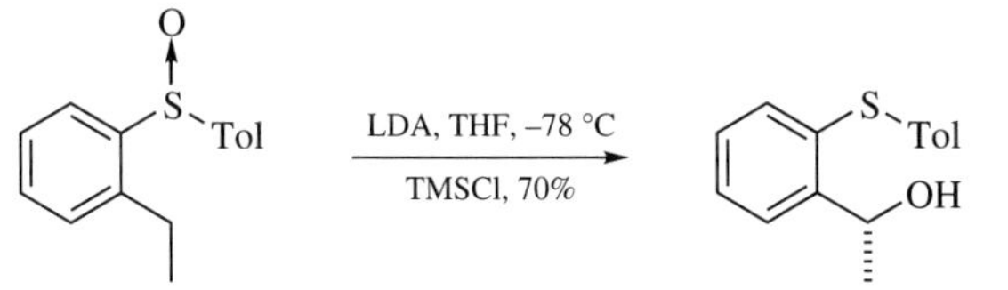

(65)

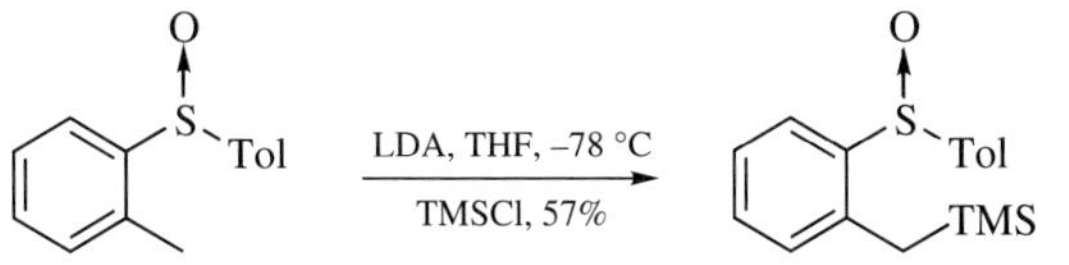

(66)

In another example, addition of phenyl vinyl selenoxide into a mixture of an indanedione, hexamethyldisilazane, and TMSCl in dichloromethane affords the α-trimethylsiloxyselenide in 70% yield (eq 67).[91] The reaction proceeds through a sequence of in situ formation of TMS silyl enol ether, Michael addition onto the phenyl vinyl selenoxide, and seleno Pummerer rearrangement of the resulting selenoxide. Trifluoroacetic anhydride and various other trialkylchlorosilanes give the same product for this reaction, but in much lower yields.

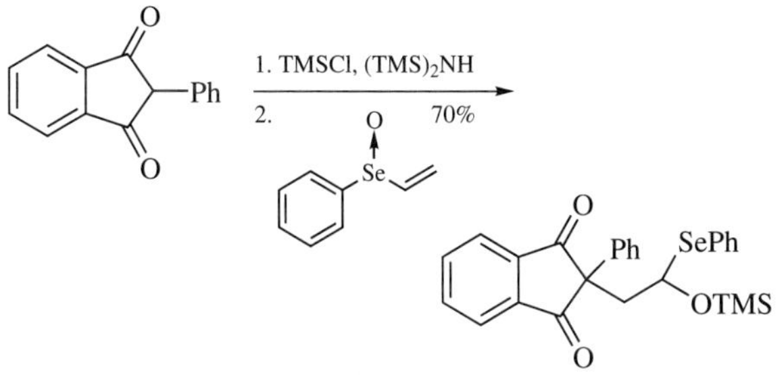

(67)

The reactivity of nitrones is enhanced by addition of TMSCl as well as TESCl and TMSOTf. Thus, stirring a mixture of TMSCl, the benzylnitrone of propanal (1 equiv), and indole (2 equiv) in dichloromethane at rt provides the bis-indole in good yields (eq 68).[92] In contrast, when equimolar amounts of TMSCl, benzylnitrone, and indole are mixed, and pyridine is added into the reaction mixture to trap hydrogen chloride liberated during the reaction, the reaction gives the indole hydroxylamine as the major product (eq 69).

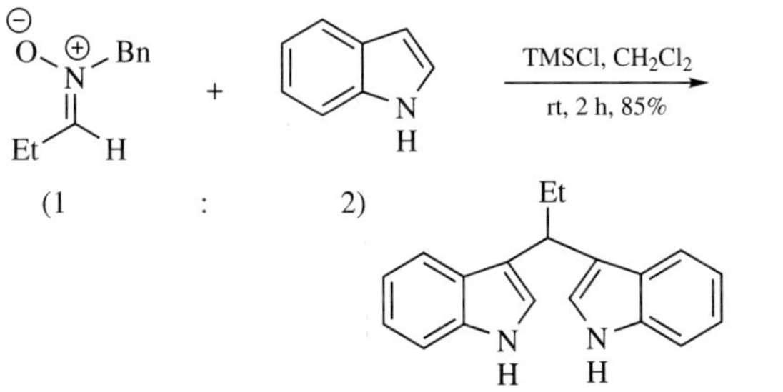

(68)

NaBH$_4$/TMSCl and NaBH$_3$CN/TMSCl. TMSCl has been combined with NaBH$_4$ to alter its reducing ability and applied to the reduction of a number of different functional groups. For example, when this combination was mixed with a succimide in ethanol at 0 °C, the corresponding mono- aminal was isolated in good yield, with no further reduction product detected (eq 70).[93]

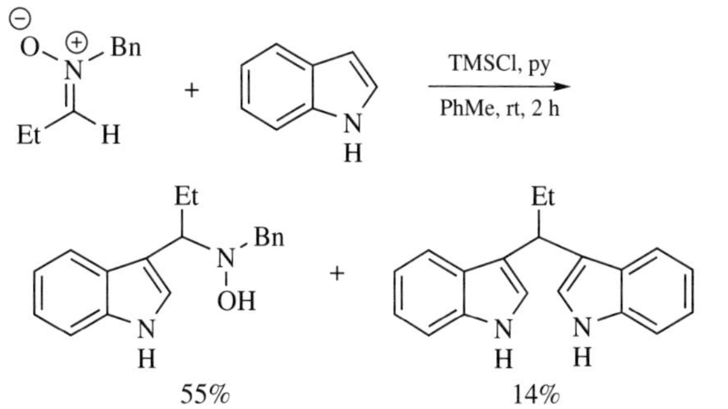

(69)

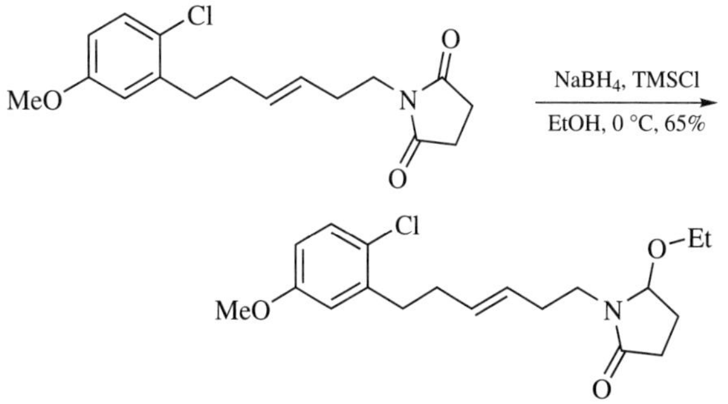

(70)

With the addition of a catalytic amount of (S)-α,α-diphenyl-pyrrolidinemethanol, this reagent combination of NaBH$_4$/TMSCl has been successfully utilized in the enantioselective reduction of various ketones. The chiral alcohols were produced in excellent yields and very high enantiomeric excess (eq 71).[94]

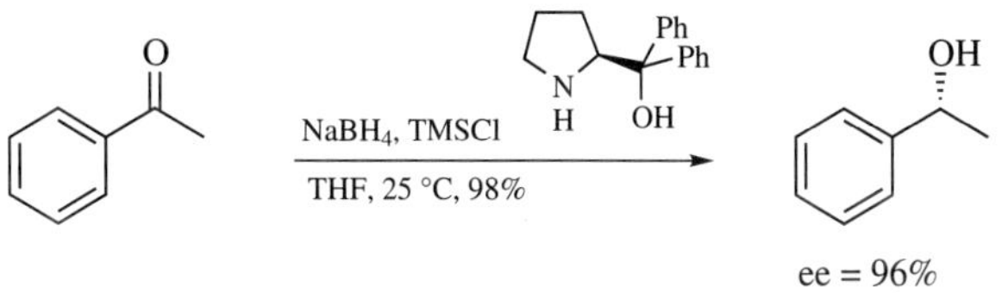

(71)

ee = 96%

In the presence of acetic acid, the NaBH$_4$/TMSCl combination is the reagent of choice for reductive amination of urea and aromatic aldehydes. Either the mono-alkylated urea or the di-alkylated urea can be obtained in good yields, depending on the ratio of urea and the aldehyde (eq 72).[95]

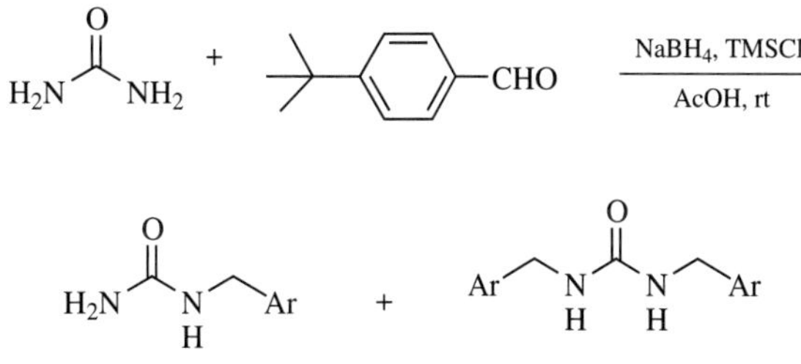

(72)

molar ratio				
0.5	:	1		75%
20	:	1	94%	6%

NaBH$_3$CN/TMSCl is another reducing agent combination that has been studied. With this reagent system, aldehydes, ketones, and acetals attached to benzo[b]furans or activated aromatic rings, such as methoxyphenyl rings, are completely deoxygenated to produce the corresponding alkyl arenes (eq 73).[96] When attached to nonactivated or deactivated aromatic rings, such as chlorophenyl and nitrophenyl rings, those aldehydes, ketones, and acetals groups are reduced to the alcohol or ether stage upon treatment with NaBH$_3$CN/TMSCl (eq 74). Ethyl and methyl benzofuroate esters are completely inactive under the same reaction conditions.

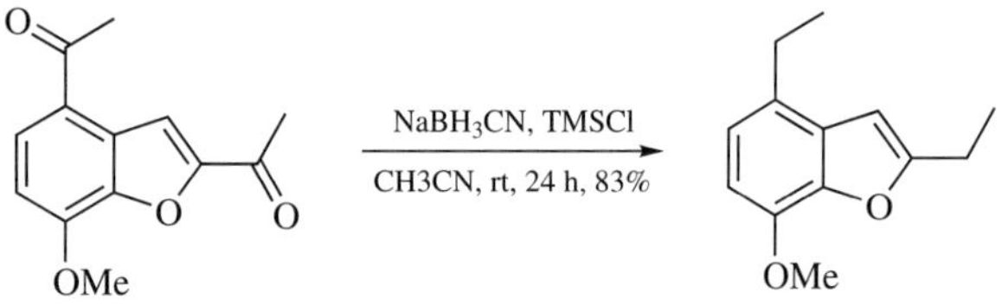

$$\text{(73)}$$

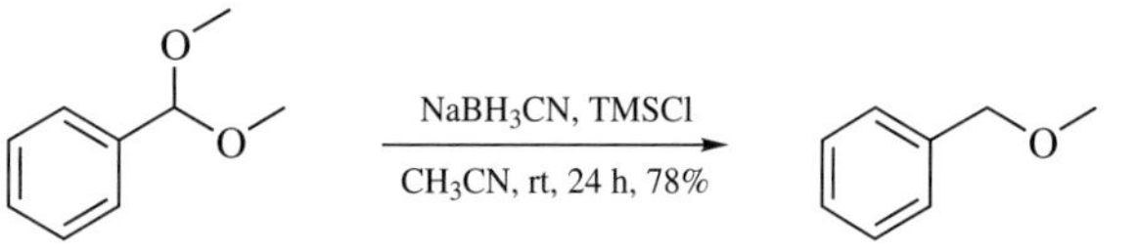

$$\text{(74)}$$

This reduction system also provides an alternative choice for the selective regeneration of alcohols from the corresponding allylic ethers, while nitro, ester, carbamate, and acetal groups maintain intaction under the reaction conditions (eq 75).[97]

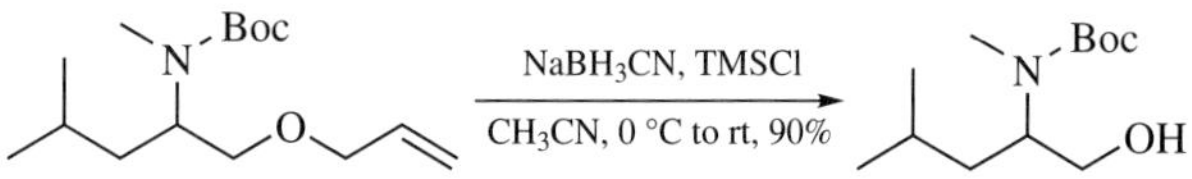

$$\text{(75)}$$

Additive in Various Reduction Systems. TMSCl has been employed in many different reduction systems to improve the reactivity and/or increase the selectivity. The reagent has been used to activate zinc dust[98] or as additive for Reformatskii reactions.[99] While less reactive than Mg/TMSCl combination, the Zn/TMSCl combination proved to be the preferred reagent system for large scale and high concentrated preparation of diamines through the pinacol coupling of two molecules of imines (eq 76).[100]

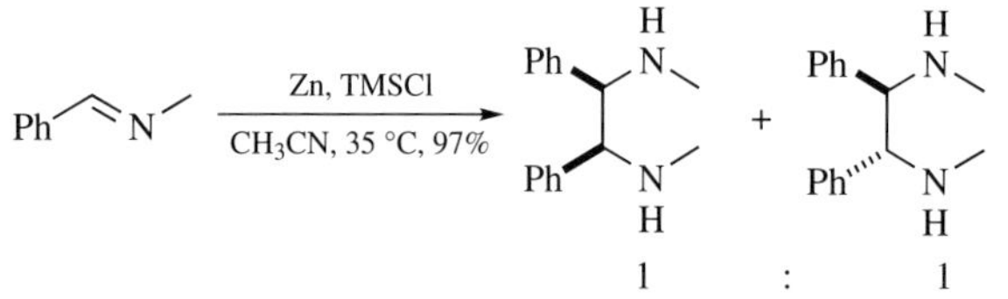

$$\text{(76)}$$

$$1 \quad : \quad 1$$

A catalytic amount of TMSCl can substitute for toxic HgCl$_2$ as an excellent activating agent in the preparation of SmI$_2$, and also in Sm-promoted cyclopropanation of both allylic and α-allenic alcohols.[101] The Sm/TMSCl reduction system has been explored in the pinacol coupling reaction of aromatic carbonyl compounds, reductive dimerization/cyclization of 1,1-dicyanoalkenes,[102] debromination of vicinal dibromides to produce (E)-alkenes, and reductive coupling of sodium thiosulfates to generate disulfides.[103]

TMSCl has been combined with various metals and used in a variety of reactions, such as in combination with Na in acyloin condensation,[104] with Ti or Ti/Zn in the McMurry reaction and deoxygenation of epoxides,[105] and with Mn/PbCl$_2$ in three component coupling of alkyl iodides, electron-deficient olefins, and carbonyl compounds (eq 77).[106] The reagent has also been used in reductive ring opening of sugar-rings,[107] with La/I$_2$/CuI in deoxygenative dimerization of benzylic and allylic alcohols, ethers, and esters (eq 78),[108] with In to selectively reduce β-nitrostyrenes to α-phenyl-α-methoxy oximes,[109] with CHI$_3$/Mn/CrCl$_2$ to convert

aldehydes into the corresponding homologated (E)-1-alkenyltrimethylsilanes,[110] and with CrCl$_2$/H$_2$O to transform terminal ynones and aldehydes into 2,5-disubstituted furans through the Baylis-Hillman type adduct intermediate and its subsequent cyclization (eq 79).[111]

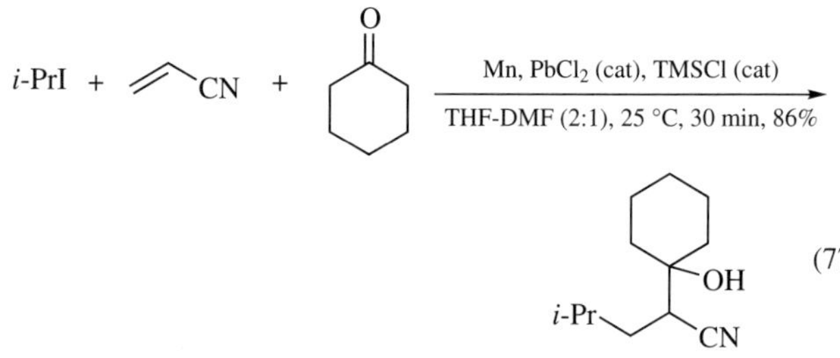

$$\text{(77)}$$

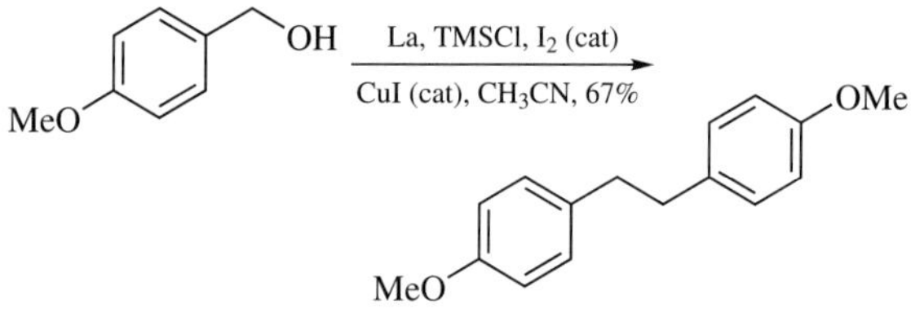

$$\text{(78)}$$

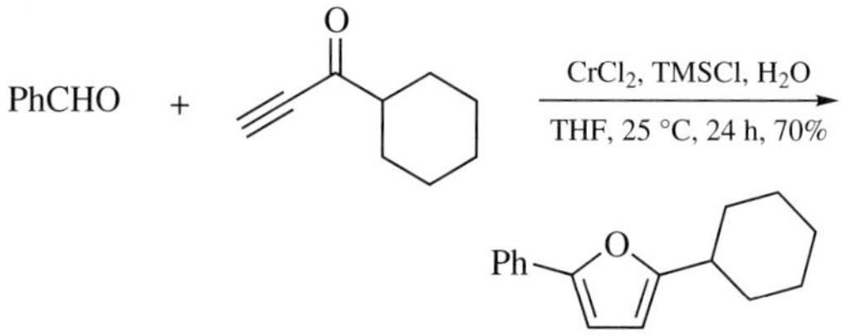

$$\text{(79)}$$

TMSCl, in combination with Mn and a catalytic amount of CrCl$_2$ can mediate the addition of various organic halides and alkenyl triflates to aldehydes to form the corresponding secondary alcohols.[112] In particular, regardless of the configuration of the crotyl bromide, addition of this bromide to various aldehydes always generates *anti*-configured homoallyl alcohols with excellent diastereomeric excess and in good yields (eq 80). It was demonstrated that, as catalyst, CrCl$_3$ is equally efficient and is preferred for practical reasons. Chromocene (Cp$_2$Cr) or CpCrCl$_2 \cdot$ THF further upgrades the number of turnovers at chromium. This reduction system has also been used to produce 3-substituted furans in good yields via reductive annelation of 1,1,1-trichloroethyl propargyl ethers (eq 81).[113]

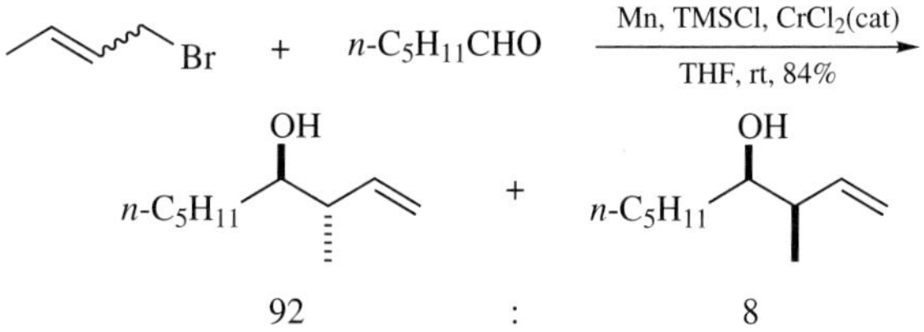

$$\text{(80)}$$

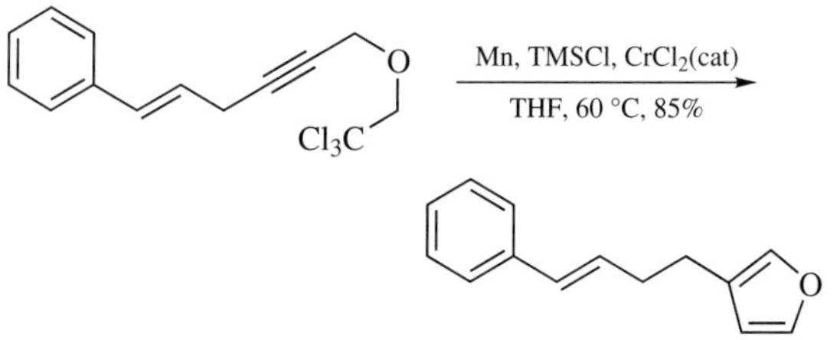

$$92 \quad : \quad 8$$

$$\text{(81)}$$

The Mg/TMSCl combination in DMF is another reducing system to be thoroughly studied. When treated with this reducing system, aromatic carbonyl compounds are converted into α-trimethylsilylalkyl trimethylsilyl ethers, generally, in reasonable to good yield.[114] When bis(chlorodimethylsilyl)ethane or 1,5-dichlorohexamethyltrisiloxane is used instead, the corresponding cyclic silylalkyl silyl ether is formed in moderate to good yield (eq 82).[115]

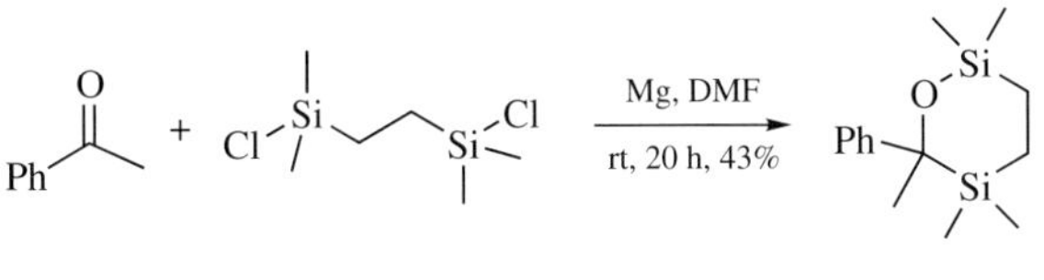

$$(82)$$

Reactions of various activated alkenes with Mg/TMSCl in DMF system have been explored as well. When treated with this reduction mixture, α-substituted arylvinyl sulfones affords (E)-β-substituted styrenes in high stereoselectivity and reasonable yields, through the desulfonation reaction (eq 83).[116] Under the reaction conditions, α,β-unsaturated ketones undergo facile and regioselective reductive dimerization to afford the corresponding bis(silyl enol ethers), that is, 1,6-bis(trimethylsilyloxy)-1,5-dienes (eq 84).[117] In contrast, α,β-unsaturated esters or α,β-epoxy esters were converted into β-trimethylsilyl esters when Mg/TMSCl in HMPA was used instead.[118] In the presence of an electrophile, such as an acid anhydride or acid chloride, α,β-unsaturated carbonyl compounds,[119] vinylphosphorus compounds,[120] and stilbenes and acenaphthylenes all experience a reductive cross-coupling reaction to give β-C-acylation products (eq 85).[121] Under these conditions, the reaction between ethyl β-arylacrylates and an aldehyde furnishes the corresponding γ-hydroxyester as the intermediate product, which is further converted into a β-aryl-γ-lactone as final product (eq 86).[122] In general, the *trans* γ-lactone is obtained as the preferred product.

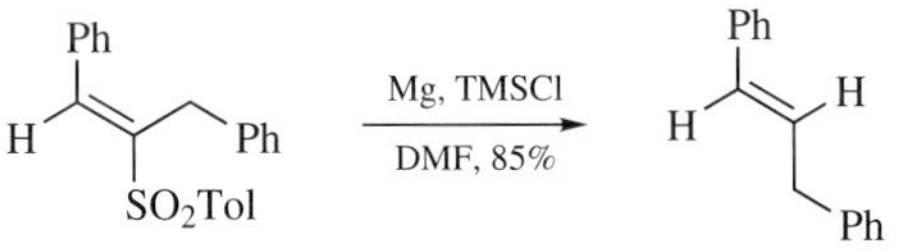

$$(83)$$

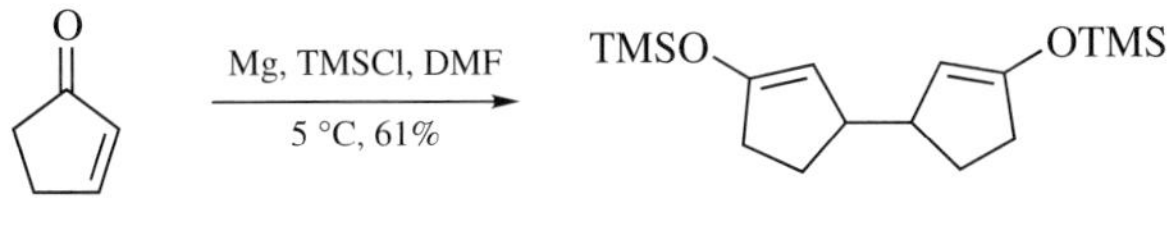

$$(84)$$

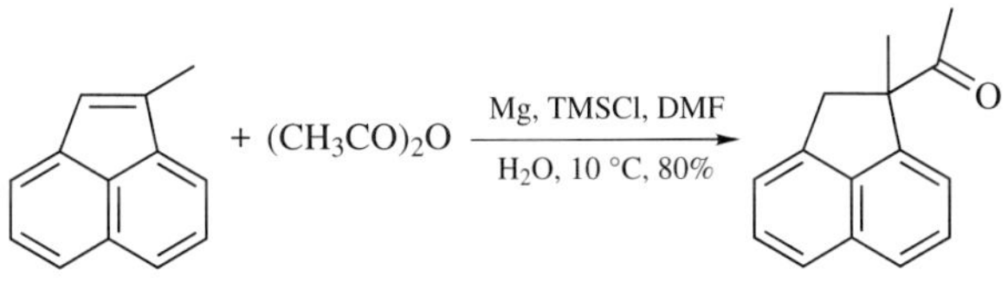

$$(85)$$

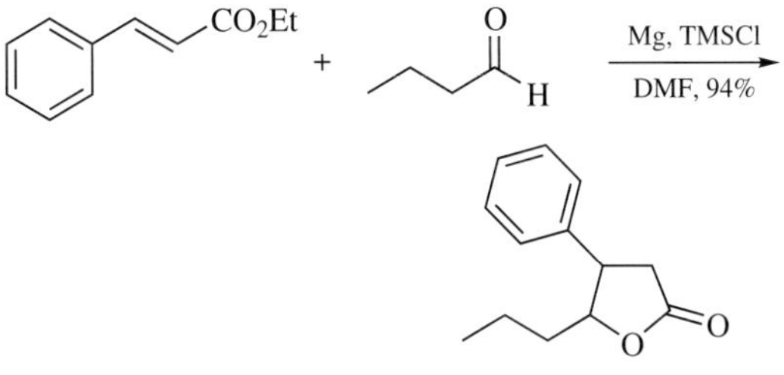

$$(86)$$

cis/trans = 1/1.7

Mg/TMSCl has also been applied to reduce Ti(O-i-Pr)$_4$ and generate a low-valent titanium species, which can mediate allyl

and propargyl ether cleavage reactions to produce the corresponding aliphatic or aromatic alcohols (eq 87).[123] Selective cleavage of a propargyl ether in the presence of an allyl ether was made possible through addition of 2 equiv of ethyl acetate (eq 88).

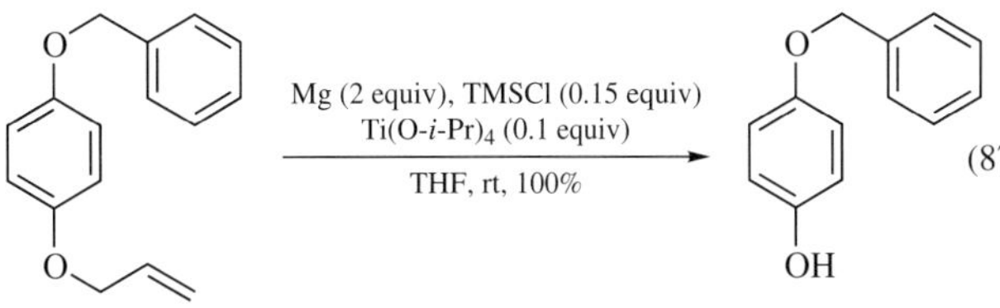

$$(87)$$

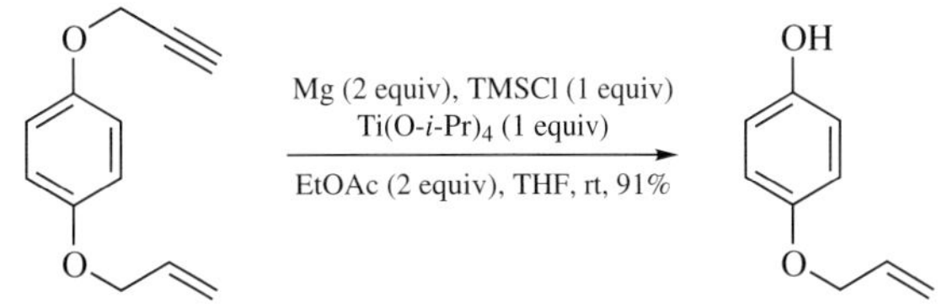

$$(88)$$

Additive in Miscellaneous Reactions. Concurrent addition of TMSCl and an enol triflate into a preformed Boc-protected α-aminoalkyl cuprate results in formation of Boc-protected secondary allylamine in good yield (eq 89).[124]

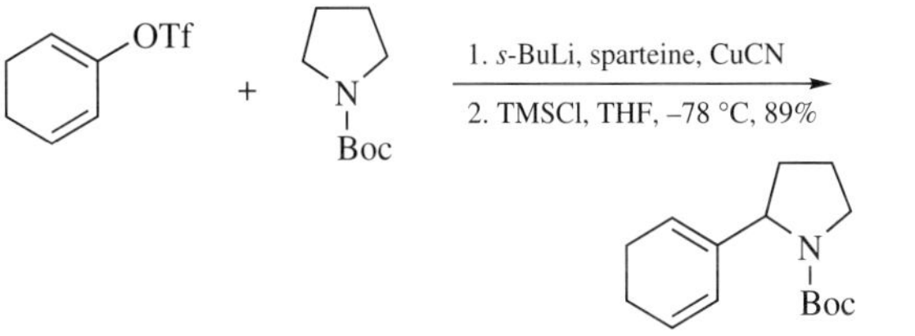

$$(89)$$

Treatment of terminal acetylenes with hydrogen iodide, generated in situ from TMSCl, sodium iodide, and water, provides a highly regioselective synthesis of 2-iodo-1-alkenes. Followed by addition of cuprous cyanide, the protocol offers a convenient one-pot preparation of 2-substituted acrylonitriles in fair to good yield (eq 90).[125] In contrast, when the reaction with arylacetylene proceeds in DMSO and with catalytic amount of sodium iodide, the 3-arylpropynenitrile is obtained as the preferred product.[126]

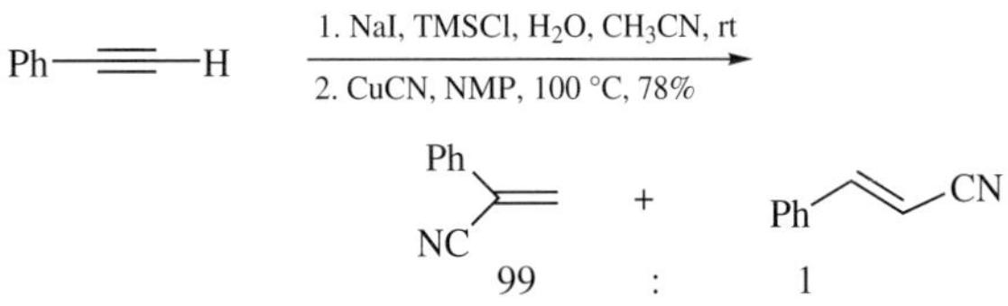

$$(90)$$

In the presence of TMSCl and ZnCl$_2$, Zn(Hg) can convert ortho formates, or acetals into organozinc carbenoid species, which then undergo a variety of reactions, such as cyclopropanation.[127] N-Diethoxymethyl amides function similarly and give amidocyclopropanation reactions,[128] or diastereoselective amidocyclopropanations if a chiral auxiliary is incorporated into the original amide (eq 91).[129]

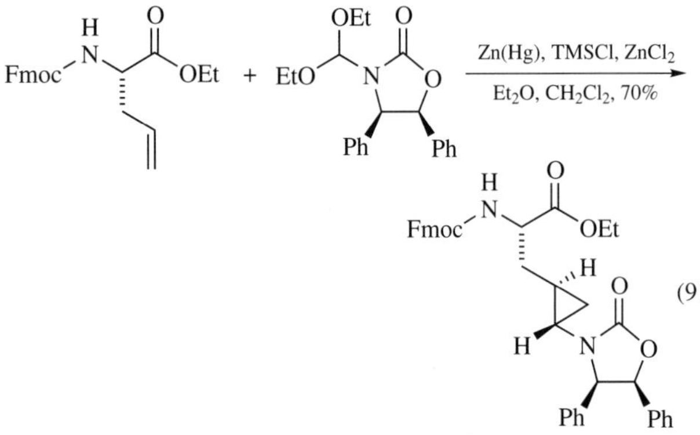

(91)

TMSCl is a suitable additive for some Rh-[130] and FeCl$_2$-[131] catalyzed diazo decomposition reactions. TMSCl itself can also convert β-trimethylsiloxy α-diazocarbonyl compounds into a mixture of α-chloro-β,γ-unsaturated and γ-chloro-α,β-unsaturated carbonyl compounds.[132]

In a number of multicomponent condensation reactions, TMSCl has also been utilized to improve the yield or efficacy of the desired product, or is directly incorporated into the final molecules. Examples include the synthesis of N-aryl-3-arylamino acids from a three-component reaction of phenols, glyoxylates, and anilines,[133] preparation of 2,4,5-trisubstituted oxazoles,[134] or 4-cyanooxazoles,[135] and the three-component Biginelli reaction (eq 92) and Biginelli-like Mannich reaction of carbamates, aldehydes, and ketones.[136]

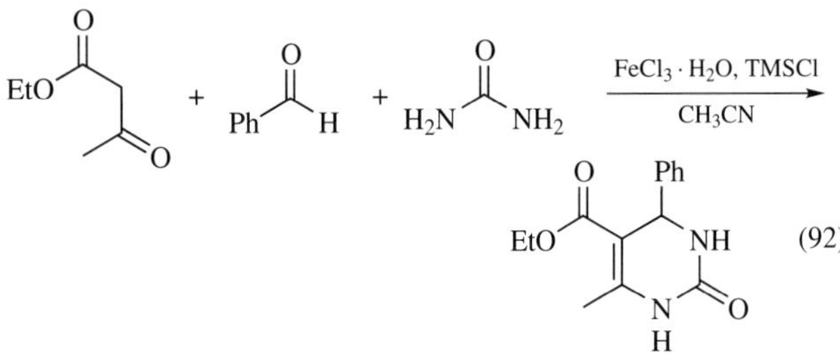

(92)

In the presence of TMSCl, various sulfides are oxidized by KO$_2$ to afford sulfoxides in excellent yields, with little further oxidation to the sulfones (eq 93).[137] A trimethylsilylperoxy radical species is proposed to mediate the reaction. When promoted by a Lewis acid such as SnCl$_4$, (SnO)$_n$, or Zr(O-i-Pr)$_4$, bis(trimethylsilyl) peroxide (BTSP) and TMSCl convert olefins into chlorohydrins (eq 94).[138] Replacing TMSCl with TMS acetate furnishes the corresponding acetoxy alcohols as the final product. Stirring a mixture of allenic zinc reagents, TMSCl, and ZnCl$_2$ under an oxygen atmosphere produces propargyl hydroperoxides regioselectively; further transformation to the corresponding propargyl alcohols is achieved with Zn and hydrochloric acid.[139]

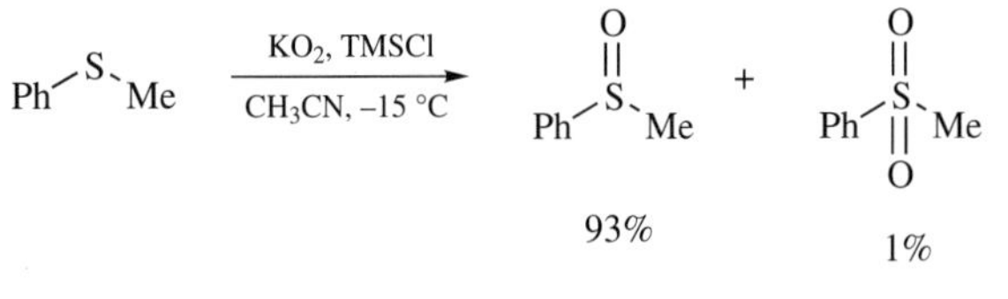

(93)

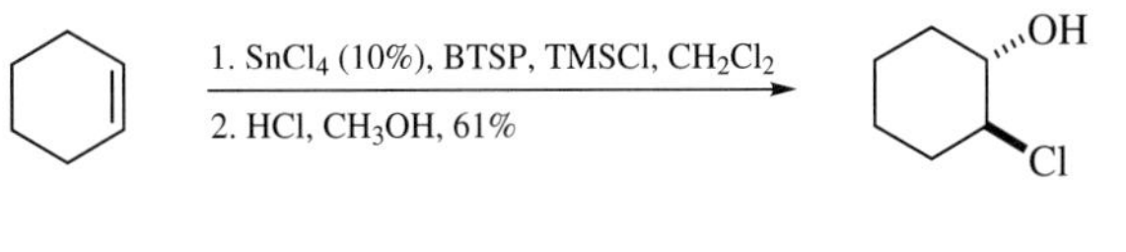

(94)

A mixture of an inorganic nitrate salt and TMSCl promotes the *ipso*-nitration of arylboronic acids to the corresponding nitroarenes in moderate to excellent yields, with high regioselectivity (eq 95).[140] Treatment of various secondary nitro compounds subsequently with 1 equiv of KH and then a catalytic amount of TMSCl furnishes the corresponding ketones in good to excellent yields.[141] In the Lewis acid mediated carboxylation of aromatic compounds with CO$_2$, addition of a large excess of TMSCl significantly improves the yields of the resulting aromatic carboxylic acids.[142]

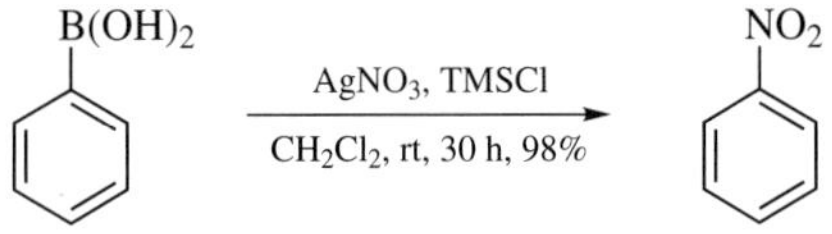

(95)

Upon addition of TMSCl, the yield dramatically increased from 21% to 91% for the palladium-catalyzed intramolecular conversion of methyl 3-oxo-6-heptenonate to 2-carbomethoxycyclohexanone (eq 96).[143] In this particular case, however, hydrogen chloride, generated from hydrolysis of TMSCl with adventitious moisture, was identified as the active promoter. Thus, in a similar cyclization reaction, addition of hydrogen chloride, instead of TMSCl, provided the desired product with comparable yield (eq 97).

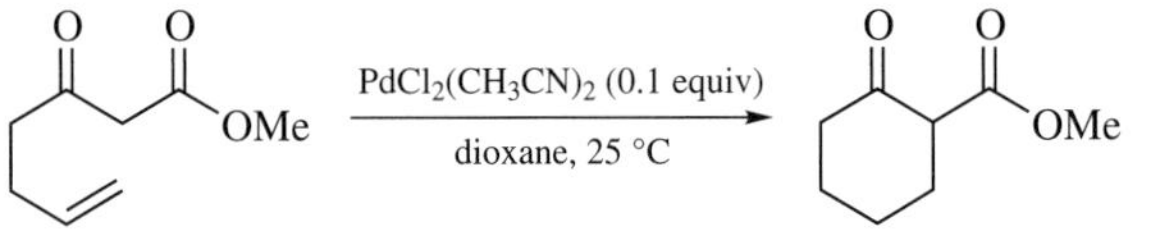

TMSCl	
-	21%
2–3 equiv	91%

(96)

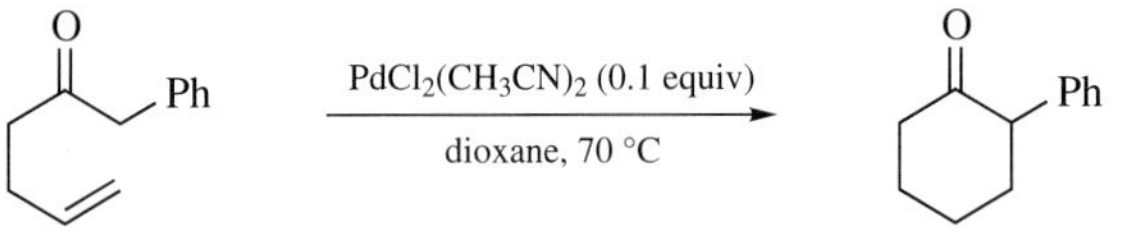

| CuCl$_2$ (1 equiv), TMSCl (2–3 equiv) | 78% |
| CuCl$_2$ (0.3 equiv), HCl (0.1 equiv) | 79% |

(97)

1. Colvin, E. *Silicon in Organic Synthesis*; Butterworths: Boston, 1981.

2. Weber, W. P., *Silicon Reagents for Organic Synthesis*; Springer: New York, 1983.

3. Langer, S. H.; Connell, S.; Wender, I., *J. Org. Chem.* **1958**, *23*, 50.

4. Hergott, H. H.; Simchen, G., *Synthesis* **1980**, 626.

5. Bolin, D. R.; Sytwu, I.-I; Humiec, F.; Meinenhofer, J., *Int. J. Peptide Protein Res.* **1989**, *33*, 353.

6. Nakonieczna, L.; Chimiak, A., *Synthesis* **1987**, 418.

7. Kabalka, G. W.; Bierer, D. E., *Synth. Commun.* **1989**, *19*, 2783.

8. Valenti, E.; Pericàs, M. A.; Serratosa, F., *J. Org. Chem.* **1990**, *55*, 395.

9. House, H. O.; Czuba, L. J.; Gall, M.; Olmstead, H. D., *J. Org. Chem.* **1969**, *34*, 2324.

10. Krafft, M. E.; Holton, R. A., *Tetrahedron Lett.* **1983**, *24*, 1345.

11. Kaiser, E.; Tam, J. P.; Kubiak, T. M.; Merrifield, R. B., *Tetrahedron Lett.* **1988**, *29*, 303.

12. Olah, G. A.; Narang, S. C.; Gupta, B. G. B.; Malhotra, R., *J. Org. Chem.* **1979**, *44*, 1247.

13. Paquette, L. A.; Fristad, W. E.; Dime, D. S.; Bailey, T. R., *J. Org. Chem.* **1980**, *45*, 3017.

14. Andrews, G. C.; Crawford, T. C.; Contillo, L. G., *Tetrahedron Lett.* **1981**, *22*, 3803.

15. Dulcere, J.-P; Grimaldi, J.; Santelli, M., *Tetrahedron Lett.* **1981**, *22*, 3179.

16. Matsuzawa, S.; Isaka, M.; Nakamura, E.; Kuwajima, I., *Tetrahedron Lett.* **1989**, *30*, 1975.

17. Alexakis, A.; Berlan, J.; Besace, Y., *Tetrahedron Lett.* **1986**, *27*, 1047.

18. Horiguchi, Y.; Matsuzawa, S.; Nakamura, E.; Kuwajima, I., *Tetrahedron Lett.* **1986**, *27*, 4025.

19. Green, T. W.; Wuts, P. G. M., *Protective Groups in Organic Synthesis*; Wiley: New York, 1991.

20. Allevi, P.; Anastasia, M.; Ciufereda, P., *Tetrahedron Lett.* **1993**, *34*, 7313.

21. Olah, G. A.; Gupta, B. G. B.; Narang, S. C.; Malhotra, R., *J. Org. Chem.* **1979**, *44*, 4272.

22. Lissel, M.; Weiffen, J., *Synth. Commun.* **1981**, *11*, 545.

23. Cossy, J.; Pale, P., *Tetrahedron Lett.* **1987**, *28*, 6039.

24. Bundy, G. L.; Peterson, D. C., *Tetrahedron Lett.* **1978**, 41.

25. Hurst, D. T.; McInnes, A. G., *Synlett* **1965**, *43*, 2004.

26. Kawazoe, Y.; Nomura, M.; Kondo, Y.; Kohda, K., *Tetrahedron Lett.* **1987**, *28*, 4307.

27. Sekine, M.; Masuda, N.; Hata, T., *Tetrahedron* **1985**, *41*, 5445.

28. Larson, G. L.; Ortiz, M.; Rodrigues de Roca, M., *Synth. Commun.* **1981**, 583.

29. Stork, G.; Hudrlik, P. F., *J. Am. Chem. Soc.* **1968**, *90*, 4462.

30. Negishi, E.; Chatterjee, S., *Tetrahedron Lett.* **1983**, *24*, 1341.

31. Corey, E. J.; Gross, A. W., *Tetrahedron Lett.* **1984**, *25*, 495.

32. Cazeau, P.; Duboudin, F.; Moulines, F.; Babot, O.; Dunogues, J., *Tetrahedron* **1987**, *43*, 2075.

33. Cazeau, P.; Duboudin, F.; Moulines, F.; Babot, O.; Dunogues, J., *Tetrahedron* **1987**, *43*, 2089.

34. Rubottom, G. M.; Mott, R. C.; Krueger, D. S., *Synth. Commun.* **1977**, *7*, 327.

35. Morita, T.; Okamoto, Y.; Sakurai, H., *Tetrahedron Lett.* **1978**, *28*, 2523.

36. Barua, N. C.; Sharma, R. P., *Tetrahedron Lett.* **1982**, *23*, 1365.

37. Caputo, R.; Mangoni, L.; Neri, O.; Palumbo, G., *Tetrahedron Lett.* **1981**, *22*, 3551.

38. Taylor, R. T.; Degenhardt, C. R.; Melega, W. P.; Paquette, L. A., *Tetrahedron Lett.* **1977**, 159.

39. Fristad, W. E.; Bailey, T. R.; Paquette, L. A., *J. Org. Chem.* **1980**, *45*, 3028.

40. Lipschutz, B. H.; Dimock, S. H.; James, B., *J. Am. Chem. Soc.* **1993**, *115*, 9283.

41. Nakamura, E.; Matsuzawa, S.; Horiguchi, Y.; Kuwajima, I., *Tetrahedron Lett.* **1986**, *27*, 4029.

42. Corey, E. J.; Boaz, N. W., *Tetrahedron Lett.* **1985**, *26*, 6015.

43. Johnson, C. R.; Marren, T. J., *Tetrahedron Lett.* **1987**, *28*, 27.

44. Tamura, R.; Tamai, S.; Katayama, H.; Suzuki, H., *Tetrahedron Lett.* **1989**, *30*, 3685.

45. Booker-Milburn, K. I.; Thompson, D. F., *Tetrahedron Lett.* **1993**, *34*, 7291.

46. (a) Saxena, I.; Deka, N.; Sarma, J. C.; Tsuboi, S., *Synth. Commun.* **2003**, *33*, 4005. (b) Bastos, E. L.; Ciscato, L. F. M. L.; Baader, W. J., *Synth. Commun.* **2005**, *35*, 1501.

47. Nishiguchi, I.; Kita, Y.; Watanabe, M.; Ishino, Y.; Ohno, T.; Maekawa, H., *Synlett* **2000**, 1025.

48. Pesti, J. A.; Yin, J.; Chung, J., *Synth. Commun.* **1999**, *29*, 3811.

49. Barluenga, J.; del Pozo, C.; Olano, B., *Synthesis* **1995**, 1529.

50. Amii, H.; Kobayashi, T.; Terasawa, H.; Uneyama, K., *Org. Lett.* **2001**, *3*, 3103.

51. (a) García-Gómez, G.; Moretó, J. M., *Chem. Eur. J.* **2001**, *7*, 1503. (b) García-Gómez, G.; Moretó, J. M., *J. Am. Chem. Soc.* **1999**, *121*, 878.

52. Oriyama, T.; Ishiwata, A.; Hori, Y.; Yatabe, T.; Hasumi, N.; Koga, G., *Synlett* **1995**, 1004.

53. Concellon, J. M.; Suarez, J. R.; del Solar, V.; Llavona, R., *J. Org. Chem.* **2005**, *70*, 10348.

54. Azizi, N.; Saidi, MR., *Tetrahedron Lett.* **2003**, *44*, 7933.

55. Paek, S. K.; Shim, S. C.; Cho, C. S.; Kim, T.-J., *Synlett* **2003**, 849.

56. Wang, L-S.; Hollis, T. K., *Org. Lett.* **2003**, *5*, 2543.

57. Wu, J.; Sun, X.; Xia, H-G., *Eur. J. Org. Chem.* **2005**, 4769.

58. Huang, H.; Forsyth, C. J., *Tetrahedron* **1997**, *53*, 16341.

59. Ikeda, S.; Kondo, K.; Sato, Y., *J. Org. Chem.* **1996**, *61*, 8248.

60. Reddy, C. K.; Devasagayaraj, A.; Knochel, P., *Tetrahedron Lett.* **1996**, *37*, 4495.

61. Lee, P. H.; Lee, K.; Sung, S.; Chang, S., *J. Org. Chem.* **2001**, *66*, 8646.

62. Yuguchi, M.; Tokuda, M.; Orito, K., *J. Org. Chem.* **2004**, *69*, 908.

63. Xu, L.-W.; Xia, C.-G.; Hu, X.-X., *Chem. Commun.* **2003**, 2570.

64. Liu, H.; Cohen, T., *Tetrahedron Lett.* **1995**, *36*, 8925.

65. Park, Y. S.; Weisenburger, G. A.; Beak, P., *J. Am. Chem. Soc.* **1997**, *119*, 10537.

66. Kimura, M.; Yamazaki, T.; Kitazume, T.; Kubota, T., *Org. Lett.* **2004**, *6*, 4651.

67. Shiina, I.; Imai, Y.; Suzuki, M.; Yanagisawa, M.; Mukaiyama, T., *Chem. Lett.* **2000**, 1062.

68. Cynamon, M. H.; Gimi, R.; Gyenes, F.; Sharpe, C. A.; Bergmann, K. E.; Han, H. J.; Gregor, L. B.; Rapolu, R.; Luciano, G.; Welch, J. T., *J. Med. Chem.* **1995**, *38*, 3902.

69. Topuzyan, V. O.; Oganesyan, A. A.; Panosyan, G. A., *Russ. J. Org. Chem.* **2004**, *40*, 1644.

70. Wrobel, Z., *Synthesis* **1997**, 753.

71. Wrobel, Z., *Tetrahedron* **1997**, *53*, 5501.

72. Peyrat, J-F.; Figadere, B.; Cave, A., *Synth. Commun.* **1996**, *26*, 4563.

73. Meyer, F.; Laaziri, A.; Papini, A. M.; Uziel, J.; Juge, S., *Tetrahedron: Asymmetry.* **2003**, *14*, 2229.

74. Choi, D.; Kohn, H., *Tetrahedron Lett.* **1995**, *36*, 7011.

75. Snyder, D. C., *J. Org. Chem.* **1995**, *60*, 2638.

76. Roa-Gutierrez, F.; Liu, H-J., *Bull. Inst. Chem., Acad. Sin.* **2000**, *47*, 19.

77. Karimi, B.; Hazarkhani, H.; Zareyee, D., *Synthesis* **2002**, 2513.

78. Huppe, S.; Rezaei, H.; Zard, S. Z., *Chem. Commun.* **2001**, 1894.

79. Rho, H. S.; Ko, B-S.; Ju, Y.-S., *Synth. Commun.* **2001**, *31*, 2101.

80. Boudjouk, P.; Kim, B-K.; Han, B-H., *Synth. Commun.* **1996**, *26*, 3479.

81. (a) Bach, T.; Schlummer, B.; Harms, K., *Synlett* **2000**, 1330. (b) Danielec, H.; Kluegge, J.; Schlummer, B.; Bach, T., *Synthesis* **2006**, 551.

82. Basu, M. K.; Luo, F-T., *Tetrahedron Lett.* **1998**, *39*, 3005.

83. Xue, C.; Luo, F-T., *J. Chin. Chem. Soc. (Taipei)* **2004**, *51*, 359.

84. (a) Luo, F-T.; Jeevanandam, A., *Tetrahedron Lett.* **1998**, *39*, 9455. (b) Cravotto, G.; Giovenzana, G. B.; Pilati, T.; Sisti, M.; Palmisano, G., *J. Org. Chem.* **2001**, *66*, 8447.

85. Su, X.; Gao, H.; Hang, L.; Li, Z., *Synth. Commun.* **1995**, *25*, 2807.

86. Papernaya, L. K.; Levanova, E. P.; Sukhomazova, E. N.; Albanov, A. I.; Deryagina, E. N., *Russ. J. Org. Chem.* **2003**, *39*, 1533.

87. Tsutsui, H.; Narasaka, K., *Chem. Lett.* **1999**, 45.

88. Pelter, A.; Ward, R. S.; Whalley, J. L., *Synthesis* **1998**, 1793.

89. Hou, X. L.; Zheng, X. L.; Dai, L. X., *Tetrahedron Lett.* **1998**, *39*, 6949.

90. Garcia Ruano, J. L.; Aleman, J.; Aranda, M. T.; Arevalo, M. J.; Padwa, A., *Org. Lett.* **2005**, *7*, 19.

91. Hagiwara, H.; Kafuku, K.; Sakai, H.; Kirita, M.; Hoshi, T.; Suzuki, T., *J. Chem. Soc., Perkin Trans.1* **2000**, 2577.

92. Chalaye-Mauger, H.; Denis, J.-L.; Averbuch-Pouchot, M.-T.; Vallee, Y., *Tetrahedron* **2000**, *56*, 791.

93. Romero, A. G.; Leiby, J. A.; Mizsak, S. A., *J. Org. Chem.* **1996**, *61*, 6974.

94. Jiang, B.; Feng, Y.; Zheng, J., *Tetrahedron Lett.* **2000**, *41*, 10281.

95. Xu, D.; Ciszewski, L.; Li, T.; Repic, O.; Blacklock, T. J., *Tetrahedron Lett.* **1998**, *39*, 1107.

96. Box, V. G. S.; Meleties, P. C., *Tetrahedron Lett.* **1998**, *39*, 7059.

97. Rao, G.; Venkat, R. D. S.; Mohan, G. H.; Iyengar, D. S., *Synth. Commun.* **2000**, *30*, 3565.

98. (a) Deboves, H. J. C.; Montalbetti, C. A. G.; Jackson, R. F. W., *J. Chem. Soc., Perkin Trans. 1* **2001**, 1876. (b) Kim, Y.; Choi, E. T.; Lee, M. H.; Park, Y. S., *Tetrahedron Lett.* **2007**, *48*, 2833.

99. Dartiguelongue, C.; Payan, S.; Duval, O.; Gomes, L. M.; Waigh, R. D., *Bull. Soc. Chim. Fr.* **1997**, *134*, 769.

100. Alexakis, A.; Aujard, I.; Mangeney, P., *Synlett* **1998**, 873.

101. Lautens, M.; Ren, Y., *J. Org. Chem.* **1996**, *61*, 2210.

102. Wang, L.; Zhang, Y., *Tetrahedron* **1998**, *54*, 11129.

103. Xu, X.; Lu, P.; Zhang, Y., *Synth. Commun.* **2000**, *30*, 1917.

104. Matzeit, A.; Schaefer, H. J.; Amatore, C., *Synthesis* **1995**, 1432.

105. Fürstner, A.; Hupperts, A., *J. Am. Chem. Soc.* **1995**, *117*, 4468.

106. Takai, K.; Ueda, T.; Ikeda, N.; Moriwake, T., *J. Org. Chem.* **1996**, *61*, 7990.

107. Tanaka, K.; Yamano, S.; Mitsunobu, O., *Synlett* **2001**, 1620.

108. Nishino, T.; Nishiyama, Y.; Sonoda, N., *Bull. Chem. Soc. Jpn.* **2003**, *76*, 635.

109. Yadav, J. S.; Reddy, B. V. S.; Srinivas, R.; Ramalingam, T., *Synlett* **2000**, 1447.

110. Takai, K.; Hikasa, S.; Ichiguchi, T.; Sumino, N., *Synlett* **1999**, 1769.

111. Takai, K.; Morita, R.; Sakamoto, S., *Synlett* **2001**, 1614.

112. (a) Fürstner, A.; Shi, N., *J. Am. Chem. Soc.* **1996**, *118*, 2533. (b) Fürstner, A.; Shi, N., *J. Am. Chem. Soc.* **1996**, *118*, 12349.

113. Barma, D. K.; Kundu, A.; Baati, R.; Mioskowski, C.; Falck, J. R., *Org. Lett.* **2002**, *4*, 1387.

114. Ishino, Y.; Maekawa, H.; Takeuchi, H.; Sukata, K.; Nishiguchi, I., *Chem. Lett.* **1995**, 829.

115. Uchida, T.; Kita, Y.; Maekawa, H.; Nishiguchi, I., *Tetrahedron* **2006**, *62*, 3103.

116. Nishiguchi, I.; Matsumoto, T.; Kuwahara, T.; Kyoda, M.; Maekawa, H., *Chem. Lett.* **2002**, 478.

117. Maekawa, H.; Sakai, M.; Uchida, T.; Kita, Y.; Nishiguchi, I., *Tetrahedron Lett.* **2004**, *45*, 607.

118. Bolourtchian, M.; Mamaghani, M.; Badrian, A., *Phos., Sulfur, Silicon, Rel. Elem.* **2003**, *178*, 2545.

119. Ohno, T.; Sakai, M.; Ishino, Y.; Shibata, T.; Maekawa, H.; Nishiguchi, I., *Org. Lett.* **2001**, *3*, 3439.

120. Kyoda, M.; Yokoyama, T.; Maekawa, H.; Ohno, T.; Nishiguchi, I., *Synlett* **2001**, 1535.

121. (a) Nishiguchi, I.; Yamamoto, Y.; Sakai, M.; Ohno, T.; Ishino, Y.; Maekawa, H., *Synlett* **2002**, 759. (b) Yamamoto, Y.; Kawano, S.; Maekawa, H.; Nishiguchi, I., *Synlett* **2004**, 30.

122. Ohno, T.; Ishino, Y.; Tsumagari, Y.; Nishiguchi, I., *J. Org. Chem.* **1995**, *60*, 458.

123. Ohkubo, M.; Mochizuki, S.; Sano, T.; Kawaguchi, Y.; Okamoto, S., *Org. Lett.* **2007**, *9*, 773.

124. Dieter, R. K.; Dieter, J. W.; Alexander, C. W.; Bhinderwala, N. S., *J. Org. Chem.* **1996**, *61*, 2930.

125. Luo, F-T.; Ko, S-L.; Chao, D-Y., *Tetrahedron Lett.* **1997**, *38*, 8061.

126. Cheng, Z-Y.; Li, W-J.; He, F.; Zhou, J-M.; Zhu, X-F., *Bioorg. Med. Chem.* **2007**, *15*, 1533.

127. Fletcher, R. J.; Motherwell, W. B.; Popkin, M. E., *Chem. Commun.* **1998**, 2191.

128. Begis, G.; Cladingboel, D.; Motherwell, W. B., *Chem. Commun.* **2003**, 2656.

129. Begis, G.; Sheppard, T. D.; Cladingboel, D. E.; Motherwell, W. B.; Tocher, D. A., *Synthesis* **2005**, 3186.

130. (a) Mori, T.; Oku, A., *Chem. Commun.* **1999**, 1339. (b) Sawada, Y.; Mori, T.; Oku, A., *Chem. Commun.* **2001**, 1086.

131. Bach, T.; Schlummer, B.; Harms, K., *Synlett* **2000**, 1330.

132. Xiao, F.; Zhang, Z.; Zhang, J.; Wang, J., *Tetrahedron Lett.* **2005**, *46*, 8873.

133. Huang, T.; Li, C-J., *Tetrahedron Lett.* **2000**, *41*, 6715.

134. Wang, Q.; Ganem, B., *Tetrahedron Lett.* **2003**, *44*, 6829.

135. Xia, Q.; Ganem, B., *Synthesis* **2002**, 1969.

136. Xu, L.-W.; Wang, Z.-T.; Xia, C.-G.; Li, L.; Zhao, P.-Q., *Helv. Chim. Acta* **2004**, *87*, 2608.

137. Chen, Y.-J.; Huang, Y.-P., *Tetrahedron Lett.* **2000**, *41*, 5233.

138. Sakurada, I.; Yamasaki, S.; Goettlich, R.; Iida, T.; Kanai, M.; Shibasaki, M., *J. Am. Chem. Soc.* **2000**, *122*, 1245.

139. Harada, T.; Kutsuwa, E., *J. Org. Chem.* **2003**, *68*, 6716.

140. Prakash, G. K. S.; Panja, C.; Mathew, T.; Surampudi, V.; Petasis, N. A.; Olah, G. A., *Org. Lett.* **2004**, *6*, 2205.

141. Hwu, J. R.; Josephrajan, T.; Tsay, S-C., *Synthesis* **2006**, 3305.

142. Nemoto, K.; Yoshida, H.; Suzuki, Y.; Morohashi, N.; Hattori, T., *Chem. Lett.* **2006**, 820.

143. Wang, X.; Pei, T.; Han, X.; Widenhoefer, R. A., *Org. Lett.* **2003**, *5*, 2699.

144. Hutchinson, D. K., *Aldrichchima Acta*, **1986**. *19*, 58 (Labnotes).

Chlorotriphenylsilane

$$\boxed{Ph_3SiCl}$$

[76-86-8] $C_{18}H_{15}ClSi$ (MW 294.87)

InChI = 1S/C18H15ClSi/c19-20(16-10-4-1-5-11-16,17-12-6-2-7-13-17)18-14-8-3-9-15-18/h1-15H

InChIKey = MNKYQPOFRKPUAE-UHFFFAOYSA-N

(bulky silylating agent[1-3])

Alternate Name: triphenylsilyl chloride.

Physical Data: mp 92–94 °C; bp 240–243 °C/35 mmHg.

Solubility: readily sol most aprotic organic solvents, although CH_2Cl_2 is most commonly used; reacts with H_2O and other protic solvents.

Form Supplied in: colorless solid.

Handling, Storage, and Precautions: somewhat prone to hydrolysis, and should be handled and stored under an anhydrous, inert atmosphere.

General Considerations. Like other bulky trialkylsilyl groups, the triphenylsilyl group was introduced to serve as a hydrolytically stable protecting group for alcohols. The hydroxyl functionality can be easily derivatized using Ph_3SiCl in the presence of a tertiary amine or imidazole as a base (eqs 1 and 2).[1-3]

A wide variety of Grignard reagents and organolithium complexes participate in reactions with Ph_3SiCl to afford organosilanes. Thus the reaction of allylmagnesium chloride with Ph_3SiCl

gives the allylsilane in high yield (eq 3).[4] However, when hydrosilyl Grignard reagents are employed, reduction to the silane occurs (eq 4).[5] Azaallyl anion[6] and diazolithium salts[7] can each be silylated to give the anticipated silane adducts (eqs 5 and 6). Similarly, arylsilanes[8] and alkynylsilanes[9–12] are produced upon silylation of the appropriate metalated[13] carbanions with Ph_3SiCl (eqs 7–11).

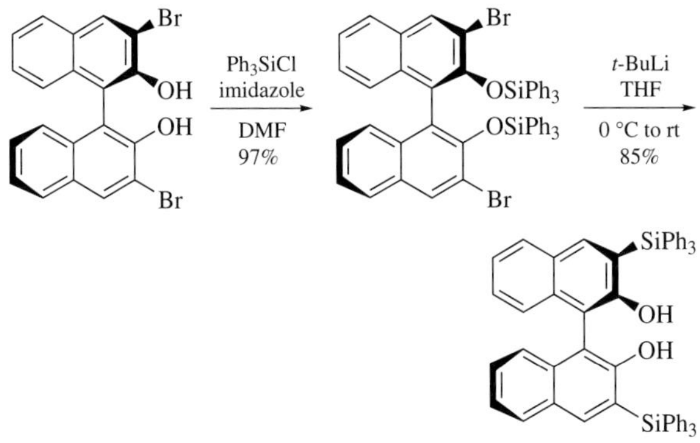

Ph_3SiCl to give the iron silane (eq 12).[14] Silylcuprates can be prepared by lithiation of Ph_3SiCl followed by treatment with copper(I) ion.[15] Addition of the lithium disilylcuprate to aroyl chlorides produces the acyl silane (eq 13),[16] while conjugate addition of the higher order silyl cyanocuprate to an alkynic morpholinium species was utilized for the synthesis of a heterosubstituted allenylsilane (eq 14).[17]

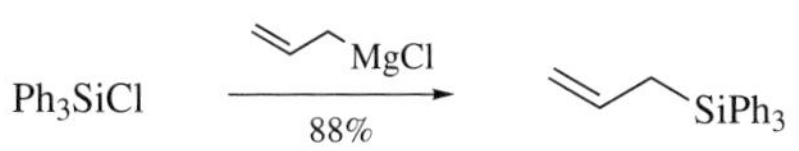

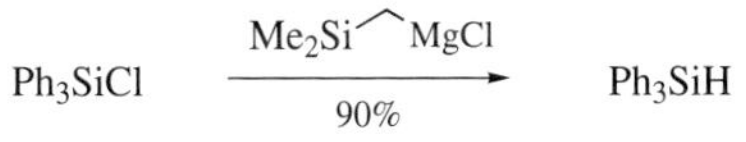

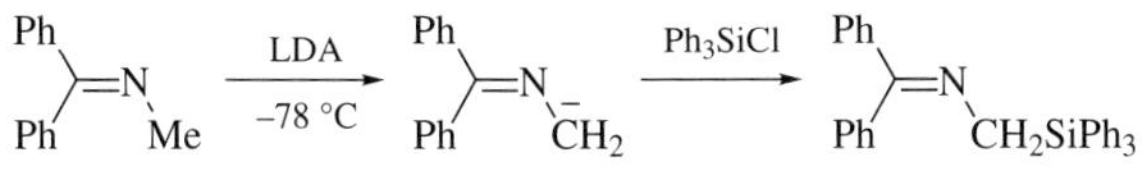

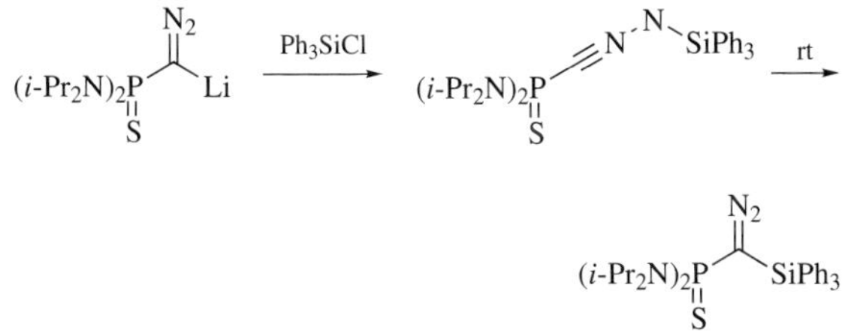

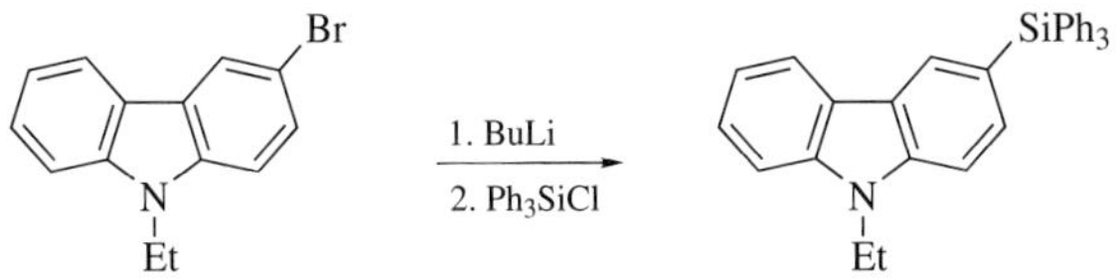

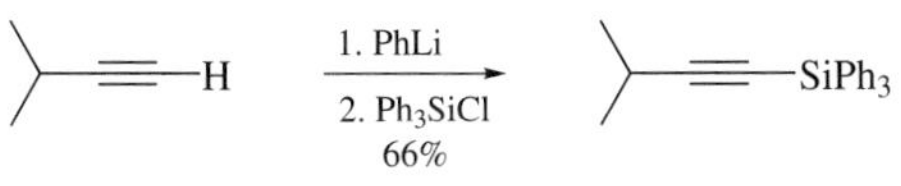

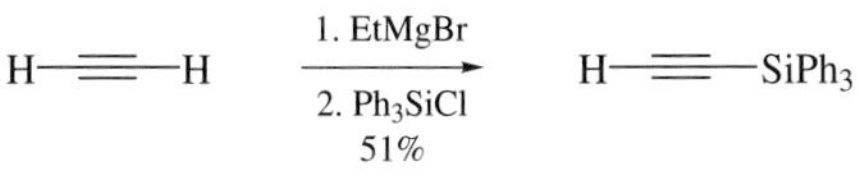

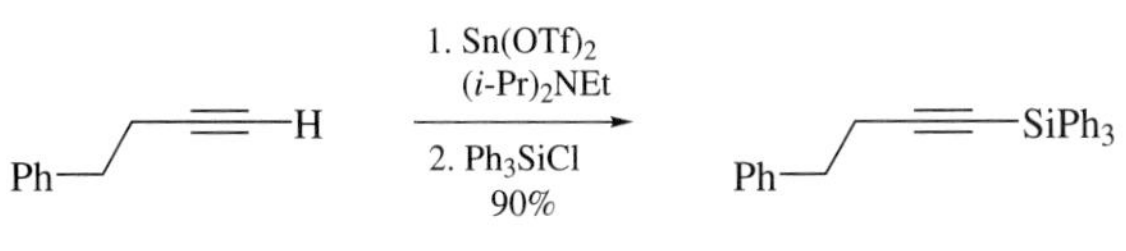

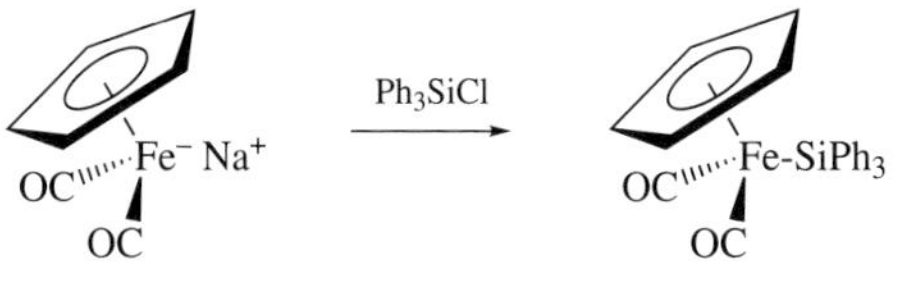

(2)

$$Ph_3SiCl \xrightarrow[88\%]{\diagup\!\!\diagup\ MgCl} \diagup\!\!\diagup\ SiPh_3 \quad (3)$$

$$Ph_3SiCl \xrightarrow[90\%]{Me_2Si\diagup\ MgCl} Ph_3SiH \quad (4)$$

(5)

(6)

(7)

(8)

$$H\!-\!\!\equiv\!\!-\!H \xrightarrow[\substack{2.\ Ph_3SiCl \\ 51\%}]{1.\ EtMgBr} H\!-\!\!\equiv\!\!-\!SiPh_3 \quad (9)$$

(10)

$$Et_2N\!-\!\!\equiv\!\!-\!\!\equiv\!\!-\!Li \xrightarrow{Ph_3SiCl} Et_2N\!-\!\!\equiv\!\!-\!\!\equiv\!\!-\!SiPh_3 \quad (11)$$

Other types of heteroatomic nucleophiles are capable of reacting with Ph_3SiCl to produce the silyl derivative. For example, sodium dicarbonylcyclopentadienylferrate combines with

(12)

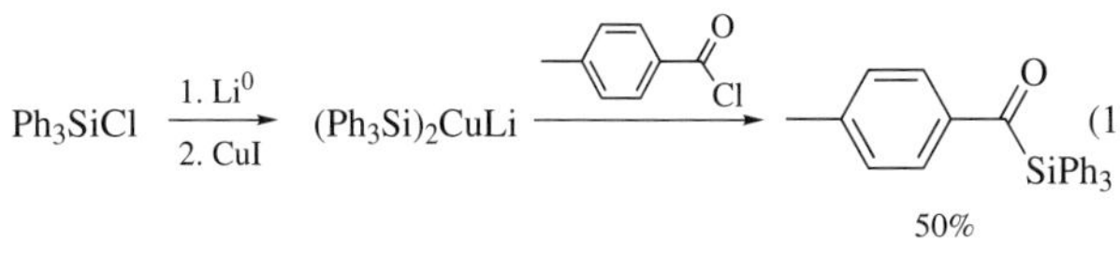

(13)

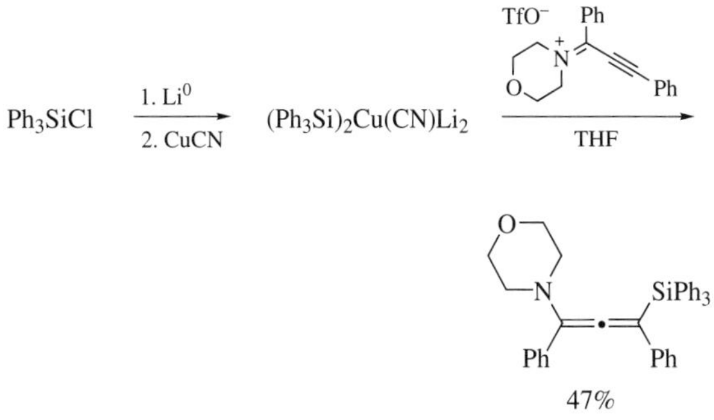

(14)

Related Reagents.   *t*-Butyldimethylchlorosilane; Chlorotriethylsilane; Chlorotrimethylsilane.

1. (a) Maruoka, K.; Hasegawa, M.; Yamamoto, H.; Suzuki, K.; Shimazaki, M.; Tsuchihashi, G.-i., *J. Am. Chem. Soc.* **1986**, *108*, 3827. (b) Maruoka, K.; Sato, J.; Yamamoto, H., *Tetrahedron* **1992**, *48*, 3749.

2. Maruoka, K.; Itoh, T.; Araki, Y.; Shirasaka, T.; Yamamoto, H., *Bull. Chem. Soc. Jpn.* **1988**, *61*, 2975.

3. Cox, P. J.; Wang, W.; Snieckus, V., *Tetrahedron Lett.* **1992**, *33*, 2253.

4. (a) Eisch, J. J.; Gupta, G., *J. Organomet. Chem.* **1979**, *168*, 139. (b) Majetich, G.; Hull, K.; Casares, A. M.; Khetani, V., *J. Org. Chem.* **1991**, *56*, 3958.

5. (a) Jarvie, A. W. P.; Rowley, R. J., *J. Organomet. Chem.* **1973**, *57*, 261. (b) Jarvie, A. W. P.; Rowley, R. J., *J. Organomet. Chem.* **1972**, *34*, C7.

6. Popowski, E.; Hahn, A.; Kelling, H., *J. Organomet. Chem.* **1976**, *110*, 295.

7. Castan, F.; Baceiredo, A.; Bigg, D.; Bertrand, G., *J. Org. Chem.* **1991**, *56*, 1801.

8. Meen, R. H.; Gilman, H., *J. Org. Chem.* **1955**, *20*, 73.

9. (a) Brook, A. G.; Duff, J. M.; Reynolds, W. F., *J. Organomet. Chem.* **1976**, *121*, 293. (b) Gilman, H.; Brook, A. G.; Miller, L. S., *J. Am. Chem. Soc.* **1953**, *75*, 3757. (c) Petrov, A. D.; Shchukovskaya, L. L., *Zh. Obshch. Khim.* **1955**, *25*, 1128.

10. Denmark, S. E.; Habermas, K. L.; Hite, G. A.; Jones, T. K., *Tetrahedron* **1986**, *42*, 2821.

11. Yamaguchi, M.; Hayashi, A.; Minami, T., *J. Org. Chem.* **1991**, *56*, 4091.

12. Fuestel, M.; Himbert, G., *Liebigs Ann. Chem.* **1984**, 586.

13. The alkynyllithium complexes give higher yields of alkynylsilanes than do the alkynyl Grignard reagents; see Fitzmaurice, N. J.; Jackson, W. R.; Perlmutter, P., *J. Organomet. Chem.* **1985**, *285*, 375.

14. Corriu, R. J. P.; Douglas, W. E., *J. Organomet. Chem.* **1973**, *51*, C3.

15. Ager, D. J.; Fleming, I.; Patel, S. K., *J. Chem. Soc., Perkin Trans. 1* **1981**, 2520.

16. (a) Bunyak, J. D.; Strickland, J. B.; Lamb, G. W.; Khasnis, D.; Modi, S.; Williams, D.; Zhang, H., *J. Org. Chem.* **1991**, *56*, 7076. (b) Duffaut, N.; Dunogues, J.; Biran, C.; Calas, R., *J. Organomet. Chem.* **1978**, *161*, C23.

17. Mayer, T.; Maas, G., *Synlett* **1990**, 399.

Edward Turos
State University of New York at Buffalo, NY, USA

Cyanotrimethylsilane[1]

[7677-24-9] C$_4$H$_9$NSi (MW 99.21)

InChI = 1S/C4H9NSi/c1-6(2,3)4-5/h1-3H3

InChIKey = LEIMLDGFXIOXMT-UHFFFAOYSA-N

(agent for the cyanosilylation of saturated and unsaturated aldehydes and ketones;[2] powerful silylating agent;[3] reagent for carbonyl umpolung[4])

Alternate Name: trimethylsilyl cyanide.

Physical Data: mp 11–12 °C; bp 118–119 °C; *d* 0.744 g cm^{-3}.

Solubility: sol organic solvents (CH$_2$Cl$_2$, CHCl$_3$); reacts rapidly with water and protic solvents.

Form Supplied in: colorless liquid; available commercially. Can also be prepared.[1e]

Handling, Storage, and Precautions: flammable liquid which must be stored in the absence of moisture and used in an inert atmosphere. Reagent is highly toxic: contact with water produces hydrogen cyanide.

Original Commentary

William C. Groutas
Wichita State University, KS, USA

Introduction. Cyanotrimethylsilane is a highly versatile reagent that reacts with a multitude of functional groups to yield an array of products and/or highly valuable synthetic intermediates.

Cyanohydrin Trimethylsilyl Ethers and Derivatives. Aldehydes and ketones are readily transformed into the corresponding cyanohydrin trimethylsilyl ethers when treated with cyanotrimethylsilane in the presence of Lewis acids (eq 1),[5,6] triethylamine,[7] or solid bases such as CaF$_2$ or hydroxyapatite.[8] The products can be readily hydrolyzed to the corresponding cyanohydrins. The cyanosilylation of aromatic aldehydes can be achieved with high enantioselectivity in the presence of catalytic amounts of a modified Sharpless catalyst consisting of titanium tetraisopropoxide and L-(+)-diisopropyl tartrate (eq 2).[9] Catalysis with chiral titanium reagents yields aliphatic and aromatic cyanohydrins in high chemical and optical yields

(eq 3).[10] Cyanohydrins can be subsequently transformed into a variety of useful synthetic intermediates (eq 4).[11,12]

$$\text{(1)}$$

$$\text{(2)}$$

81% ee

$$\text{(3)}$$

R = Ph, 79%, 96% ee
R = Bn, 66%, 77% ee
R = phenylethyl, 89%, 89% ee

$$\text{(4)}$$

Conjugate Additions. Cyanotrimethylsilane reacts with α,β-unsaturated ketones in the presence of Lewis acids (aluminum chloride, tin(II) chloride, triethylaluminum) to yield, upon hydrolysis, the corresponding 1,4-addition products (eqs 5 and 6).[13] This methodology is superior to other procedures.[14] By controlling the reaction conditions and the stoichiometry of the reaction, the kinetically controlled 1,2-addition products can also be obtained in high yields (eq 7).[13,15]

$$\text{(5)}$$

95:5

$$\text{(6)}$$

$$\text{(7)}$$

Regioselective cyanosilylation of unsaturated ketones can also be effected efficiently using a solid acid or solid base support.[16] 1,4-Adducts are obtained when a strong solid acid such as Fe^{3+}- or Sn^{4+}-montmorillonite is used, while 1,2-adducts are obtained in the presence of solid bases such as CaO and MgO.

Reactions with Oxiranes, Oxetanes, and Aziridines. Lewis acids, lanthanide salts, and titanium tetraisopropoxide or aluminum isopropoxide catalyze the reactions of cyanotrimethylsilane with oxiranes, oxetanes, and aziridines, yielding ring-opened products. The nature of the products and the regioselectivity of the reaction are primarily dependent on the nature of the Lewis acid, the substitution pattern in the substrate, and the reaction conditions. Monosubstituted oxiranes undergo regiospecific cleavage to form 3-(trimethylsiloxy)nitriles when refluxed with a slight excess of cyanotrimethylsilane in the presence of a catalytic amount of potassium cyanide–18-crown-6 complex (eqs 8–10).[17] The addition of cyanide occurs exclusively at the least-substituted carbon.

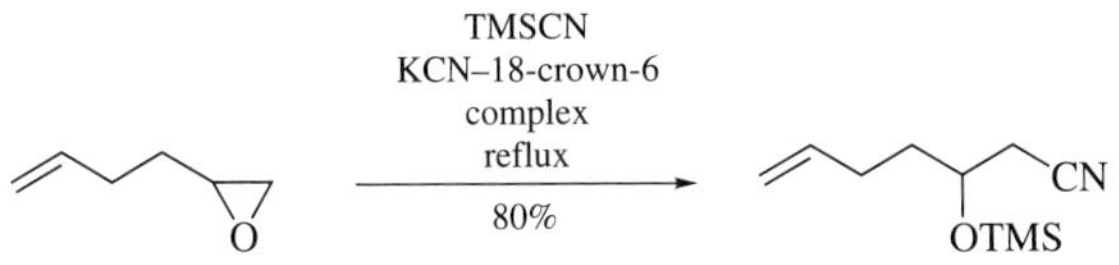

$$\text{(8)}$$

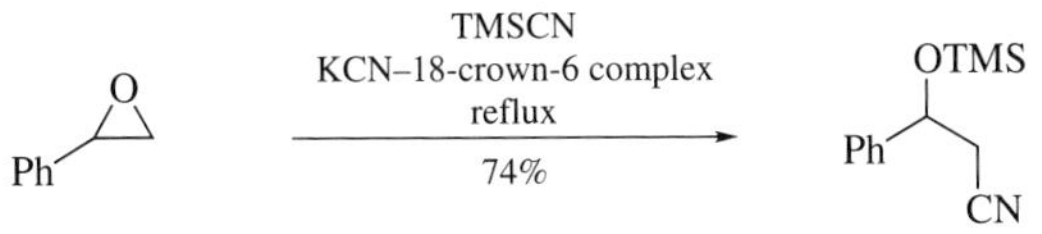

$$\text{(9)}$$

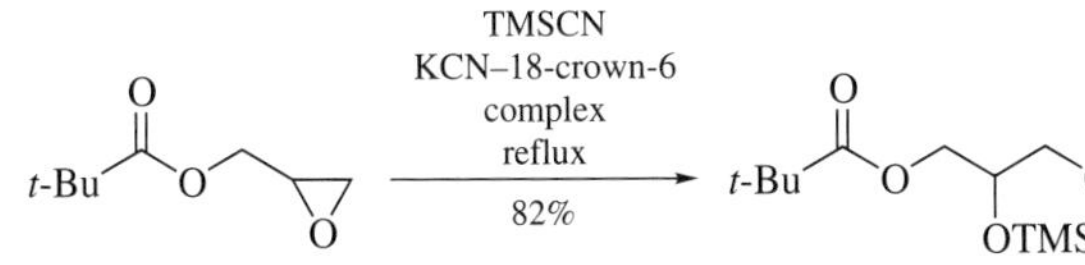

$$\text{(10)}$$

Good yields of 3-(trimethylsiloxy)nitriles are also obtained from the reactions of oxiranes with cyanotrimethylsilane in the presence of lanthanide salts,[18] or when the reaction is catalyzed by AlCl$_3$ or diethylaluminum chloride (eq 11).[19] Ring opening of chiral glycidyl derivatives by Me$_3$SiCN catalyzed by Ti(O-i-Pr)$_4$ or Al(O-i-Pr)$_3$ occurs in a regiospecific and highly stereoselective manner (eq 12).[20]

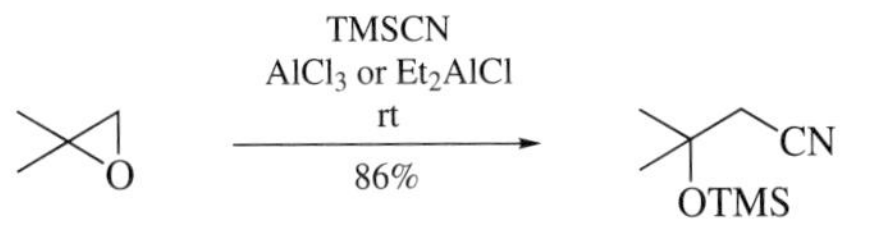

$$\text{(11)}$$

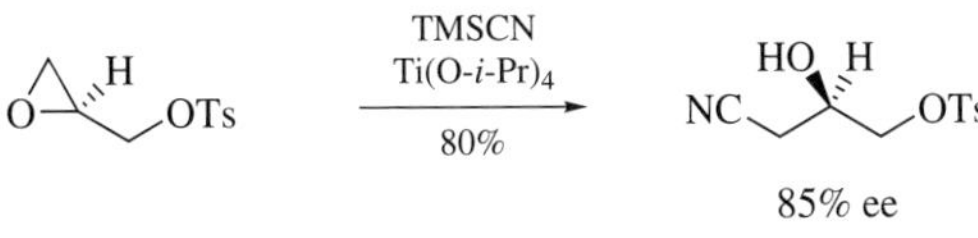

$$\text{(12)}$$

Oxetanes give rise to 4-(trimethylsiloxy)propionitriles (eq 13).[19] Similar observations have been made in the reactions of N-tosylaziridines and cyanotrimethylsilane catalyzed by lanthanum salts (eq 14).[21]

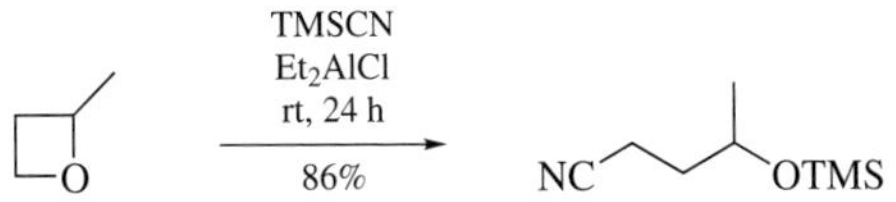

$$\text{(13)}$$

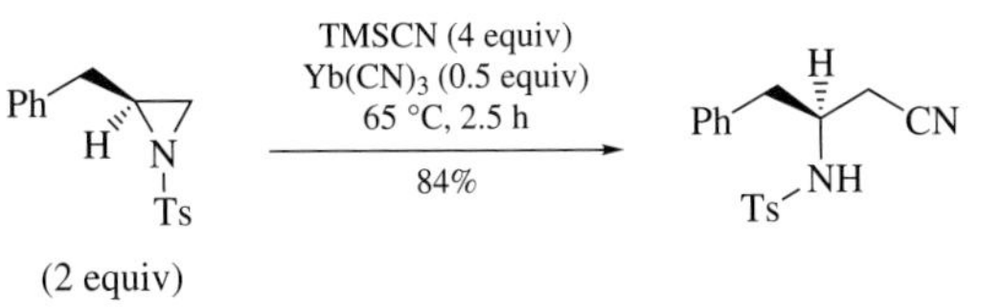

$$\text{(14)}$$

The ambident nature of cyanotrimethylsilane[22] can lead to the formation of nitriles or isocyanides, depending on the nature of the catalyst. For example, cyanotrimethylsilane reactions with epoxides and oxetanes catalyzed by soft Lewis acids give rise to 2-(trimethylsiloxy) isocyanides arising by attack on the more substituted carbon (eqs 15 and 16).[23–25] Milder reaction conditions and better yields of isocyanides can be realized when the reaction of cyanotrimethylsilane with oxiranes is carried out in the presence of Pd(CN)$_2$, SnCl$_2$, or Me$_3$Ga (eq 17).[26] Isocyanides are useful precursors for the synthesis of β-amino alcohols and oxazolines.

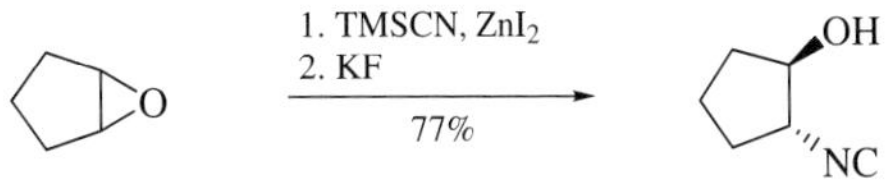

$$\text{(15)}$$

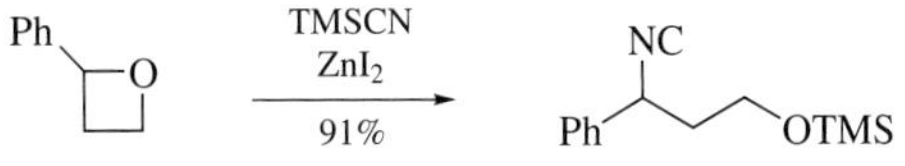

$$\text{(16)}$$

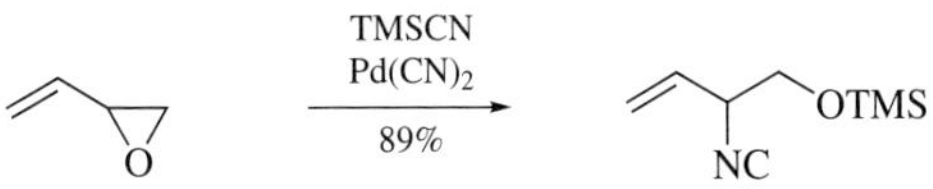

$$\text{(17)}$$

Carbonyl Umpolung. When deprotonated with a strong base, O-(trimethylsilyl) cyanohydrins can function as effective acyl anion equivalents that can be used to convert aldehydes to ketones,[27] and in the synthesis of 1,4-diketones,[28] tricyclic ketones,[29] and the highly sensitive α,β-epoxy ketone functionality (eq 18).[30]

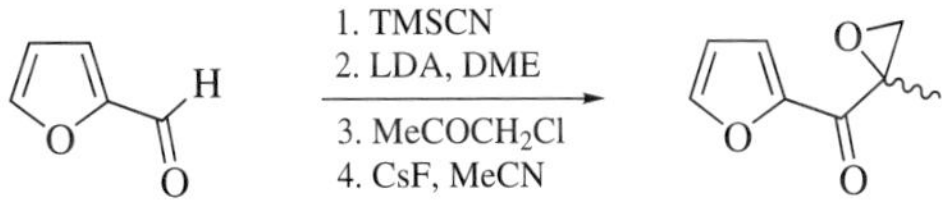

$$\text{(18)}$$

Miscellaneous Transformations. Cyanotrimethylsilane effects the transformation of acyl chlorides to acyl cyanides,[31] α-chloro ethers and α-chloro thioethers to α-cyano ethers[32] and α-cyano thioethers (eq 19),[33] t-butyl chlorides to nitriles (eqs 20 and 21),[34] 1,3,5-trisubstituted hexahydro-1,3,5-triazines to aminoacetonitriles,[35] the cyanation of allylic carbonates and acetates (eqs 22 and 23),[36] and the formation of aryl thiocyanates from aryl sulfonyl chlorides and sulfinates.[37] The reagent has been used effectively in peptide synthesis[38] and in a range of other synthetic applications.[39–46]

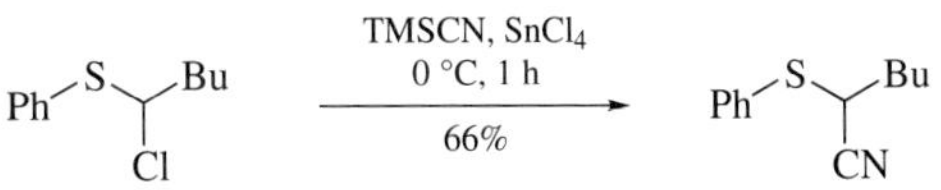

$$\text{(19)}$$

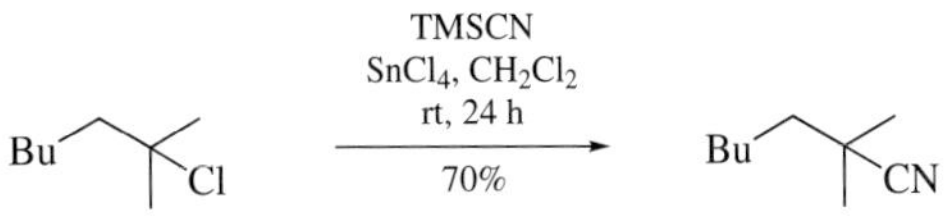

$$\text{(20)}$$

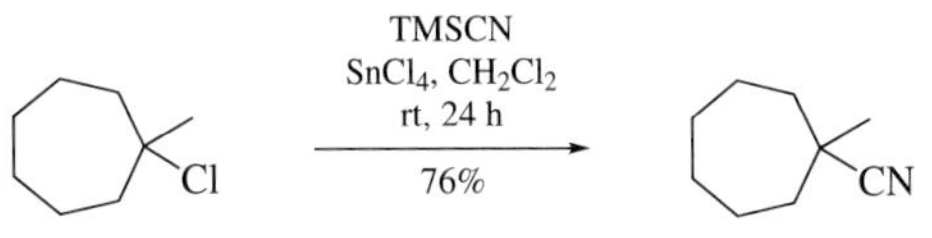

$$\text{(21)}$$

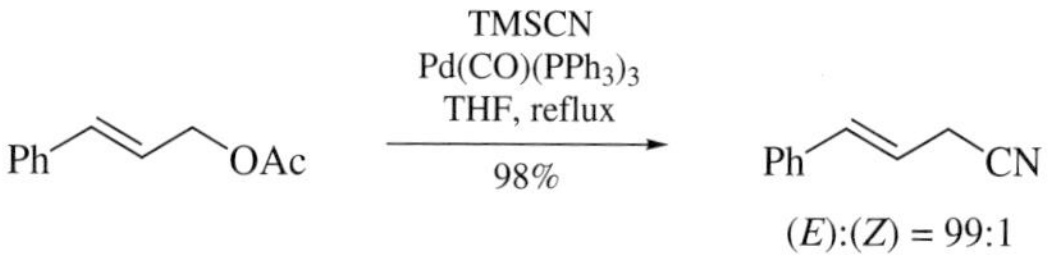

$$Ph\diagdown OAc \xrightarrow[\text{98\%}]{\substack{\text{TMSCN} \\ \text{Pd(CO)(PPh}_3)_3 \\ \text{THF, reflux}}} Ph\diagdown CN \quad (22)$$

$(E):(Z) = 99:1$

$$\diagdown OCO_2Me \xrightarrow[\text{78\%}]{\substack{\text{TMSCN} \\ \text{Pd(PPh}_3)_3 \\ \text{THF, reflux}}} \diagdown CN \quad (23)$$

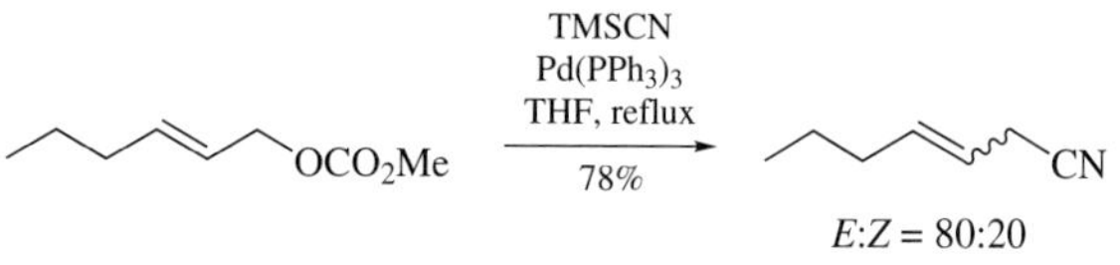

$E:Z = 80:20$

First Update

Zhendong Jin & Heng Zhang
University of Iowa, Iowa City, IA, USA

Addition to Carbonyls, Imines (Strecker-type Reactions), and Heteroaromatic Rings (Reissert-type Reactions). Cyanohydrin trimethylsilyl ethers are of significant synthetic interest as they can be transformed into a variety of multifunctional intermediates. Aldehydes and ketones can be enantioselectively converted to cyanohydrin trimethylsilyl ethers when treated with cyanotrimethylsilane in the presence of a Lewis acid and a chiral ligand. Enantioselective and/or diastereoselective formation of cyanohydrins and their derivatives has been reported and most of these reactions involve chiral ligands and metal catalysts containing Ti (eq 24),[47] Sm (eq 25),[48] and Al (eq 26).[49]

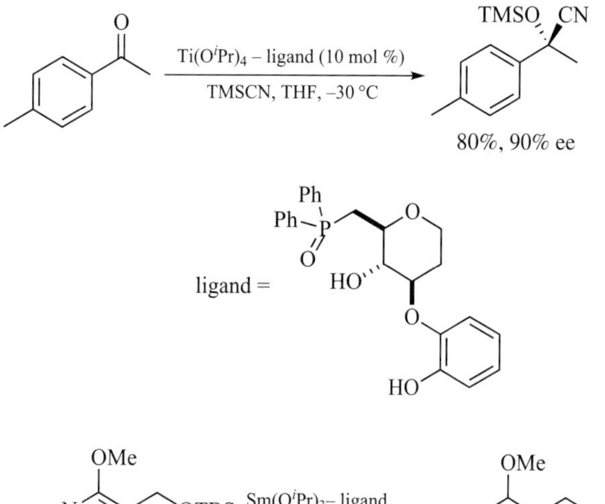

$$(24)$$

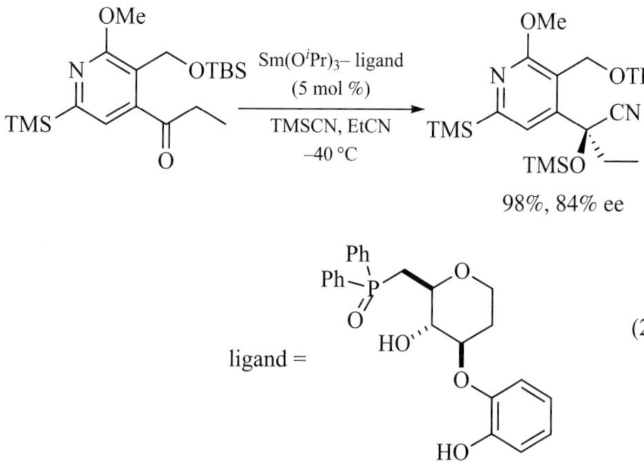

$$(25)$$

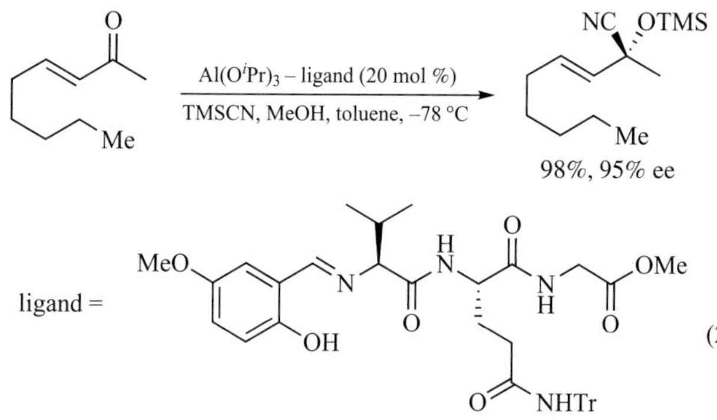

Chiral oxazaborolidinium ion is an excellent catalyst for the cyanosilylation of methyl ketones promoted by cyanotrimethylsilane and diphenylmethyl phosphine oxide (eq 27).[50] Chiral thiourea (eq 28)[51,52] and amino acid salt (eq 29)[53] have been developed and employed as catalysts in the enantioselective synthesis of cyanohydrins and their derivatives.

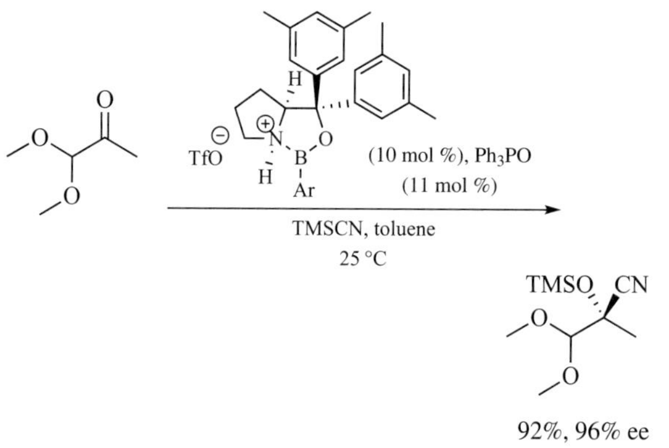

$$(26)$$

$$(27)$$

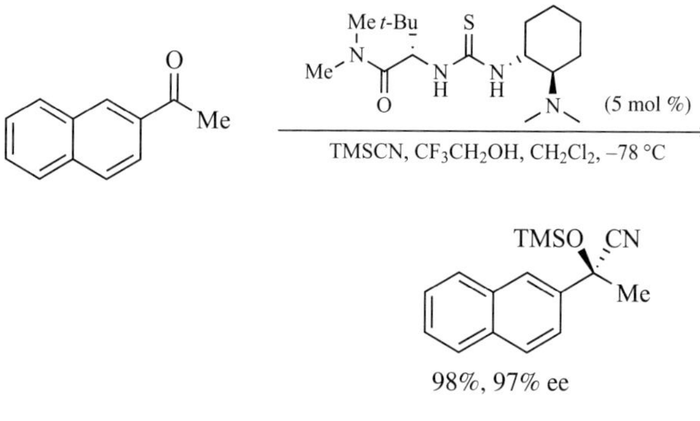

92%, 96% ee

$$(28)$$

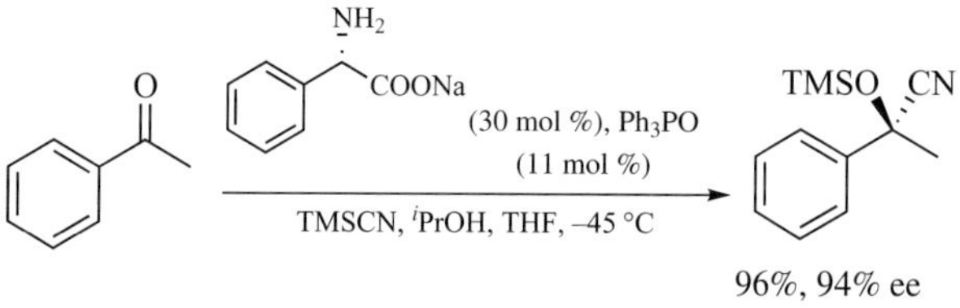

98%, 97% ee

$$(29)$$

96%, 94% ee

Excellent enantioselectivity has also been achieved in the reaction between imines and cyanotrimethylsilane, which provides an effective synthesis of chiral α-amino acids. Ti-catalyzed reactions using a chiral tripeptide Schiff base have been developed to realize such transformations (eq 30).[54] In addition, a combination of 2,2′-biphenol and Cinchonine as chiral ligands has been employed to effect such reactions (eq 31).[55] An enantioselective route was also developed for the conversion of readily prepared and air-stable aliphatic hydrazones to synthetically valuable chiral α-hydrazinonitriles (eq 32).[56]

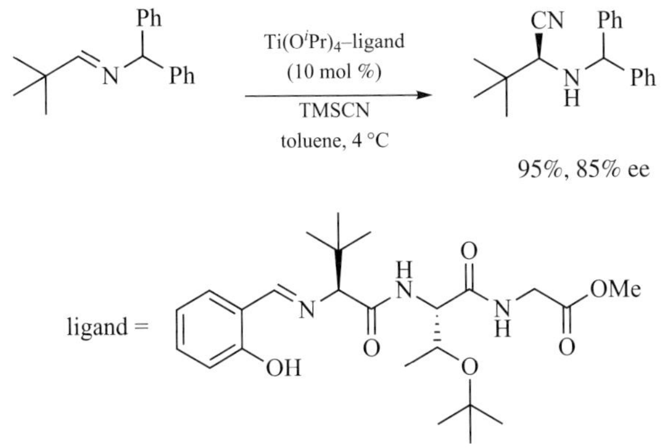

(30)

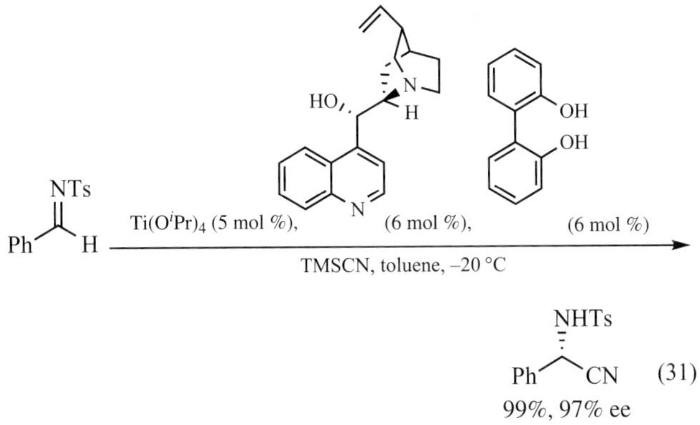

(31)

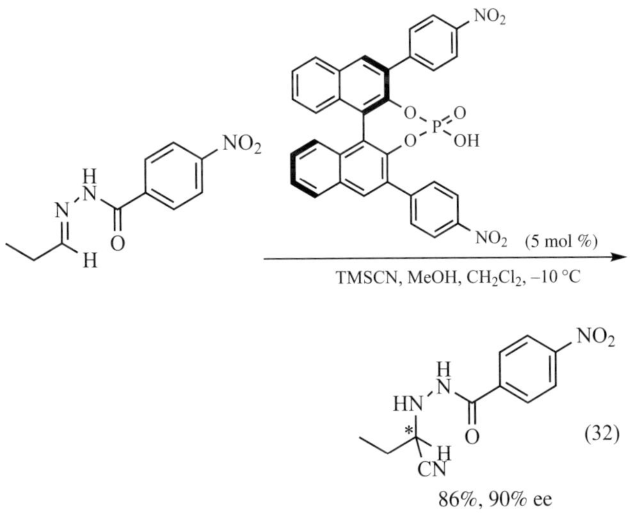

(32)

Enantioselective Reissert reaction using an Al-containing asymmetric bifunctional catalyst has been developed (eq 33).[57] The Reissert reaction catalyzed by an Al-containing chiral catalyst can be used to enantioselectively construct a quaternary stereocenter (eq 34).[58]

Conjugate Addition. Catalytic enantioselective conjugate addition of cyanotrimethylsilane to α,β-unsaturated imides has been reported to afford 1,4-addition products in excellent yield and enantiomeric excess (eq 35).[59] Under the catalysis of a chiral gadolinium complex, cyanotrimethylsilane can undergo facile conjugate addition to enones (eq 36)[60] and α,β-unsaturated

N-acylpyrroles (eq 37)[61] with high enantioselectivity. The conjugate addition of cyanotrimethylsilane to β,β-disubstituted α,β-unsaturated carbonyl compounds catalyzed by a chiral Sr complex provides a convenient route for the asymmetric construction of a quaternary carbon (eq 38).[62]

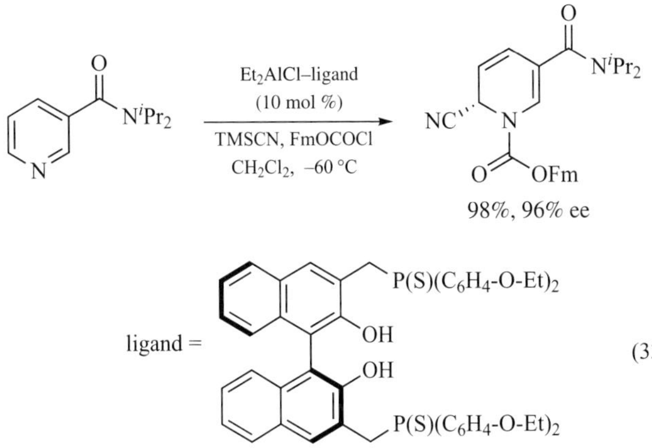

(33)

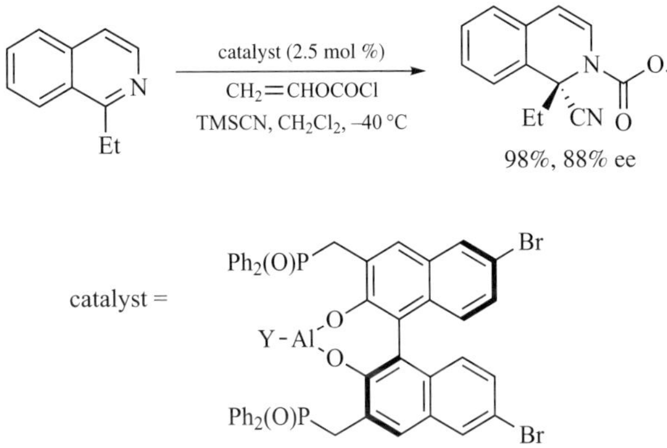

(34)

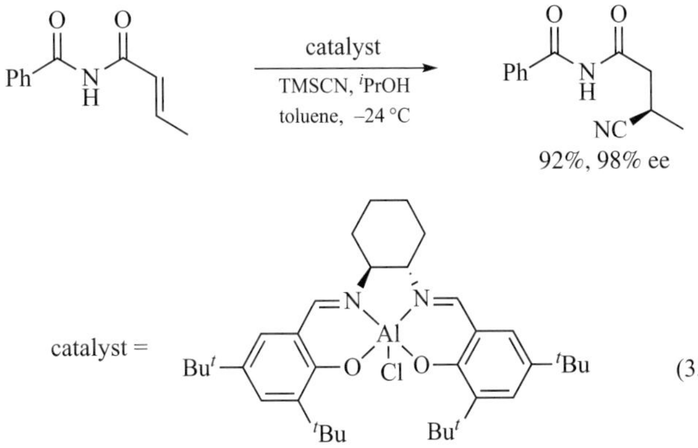

(35)

Reactions with Oxiranes and Aziridines. Under the catalysis of a chiral Yb, Ti, and Gd, oxiranes and aziridines undergo enantioselective ring opening. The asymmetric ring opening of *meso*-epoxides by cyanotrimethylsilane catalyzed by (pybox)YbCl3 complexes yields β-trimethylsilyloxy nitriles with excellent enantioselectivity (eq 39).[63] Aziridines with relatively low activity can also react with cyanotrimethylsilane to provide excellent yield of ring-opening products with excellent enantioselectivity under the catalysis of Gd (eq 40)[64] and Y (eq 41)[65] complexes.

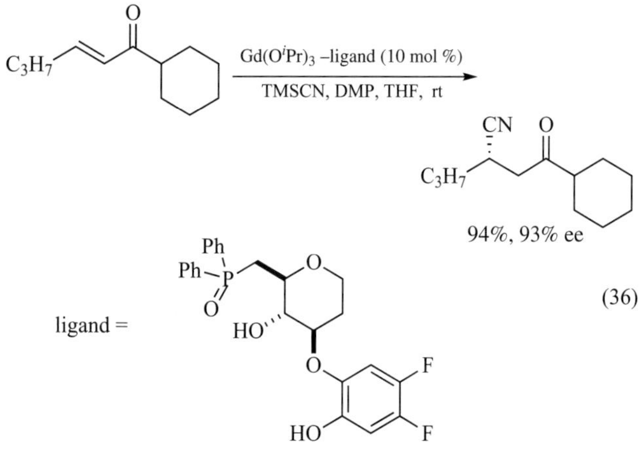

(36)

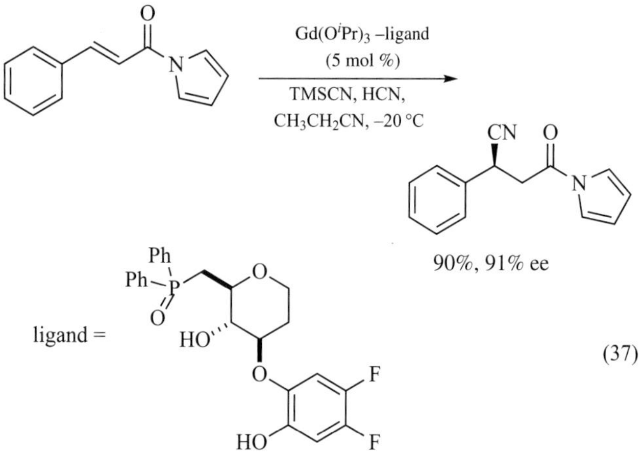

(37)

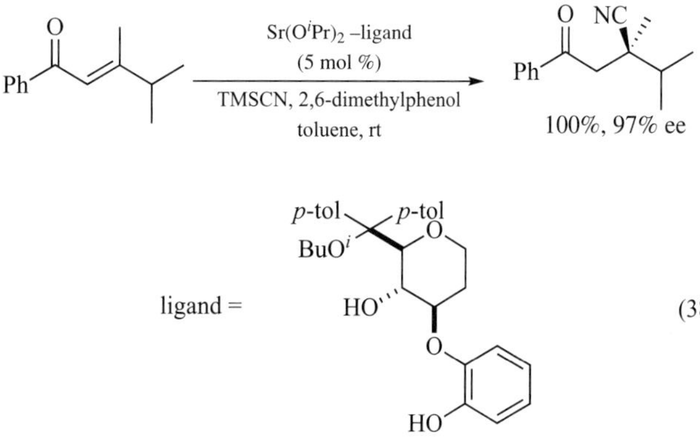

(38)

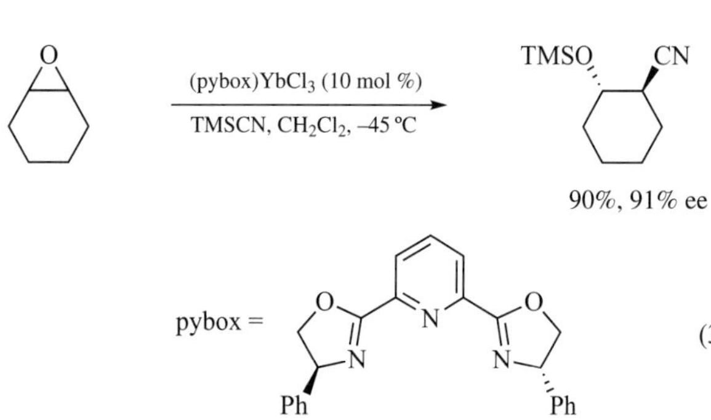

pybox =

(39)

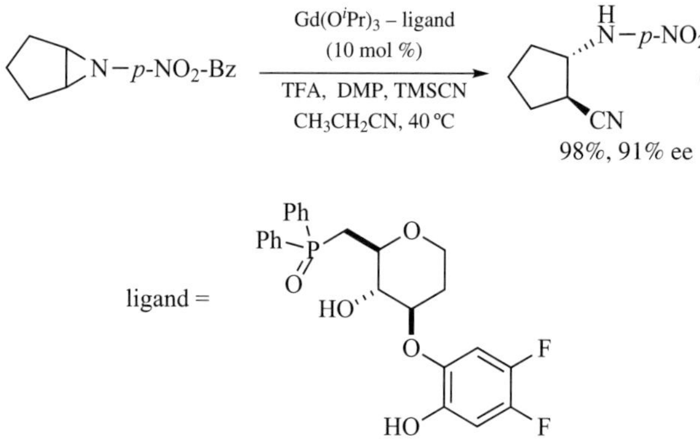

(40)

ligand =

Miscellaneous Transformations. The hypervalent iodine(III) reagent phenyliodine bis(trifluoroacetate) (PIFA) mediates the selective cyanation reaction of a wide range of electron-rich heteroaromatic compounds such as pyrroles, thiophenes, and indoles under mild conditions (eq 42).[66] Nonactivated arylalkenes are effectively converted to tertiary benzylic nitriles in the presence of triflic acid and cyanotrimethylsilane (eq 43).[67]

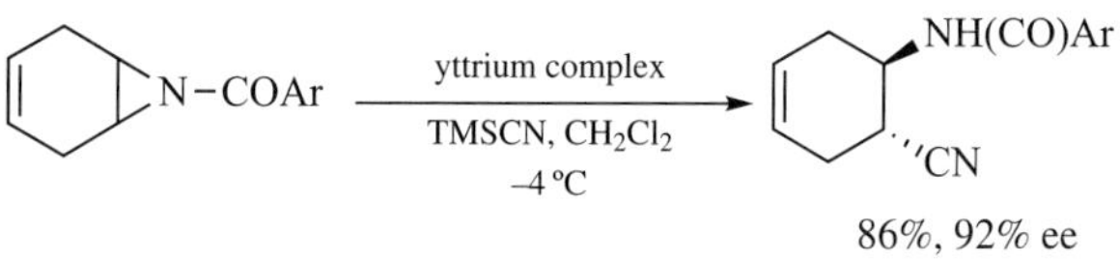

(41)

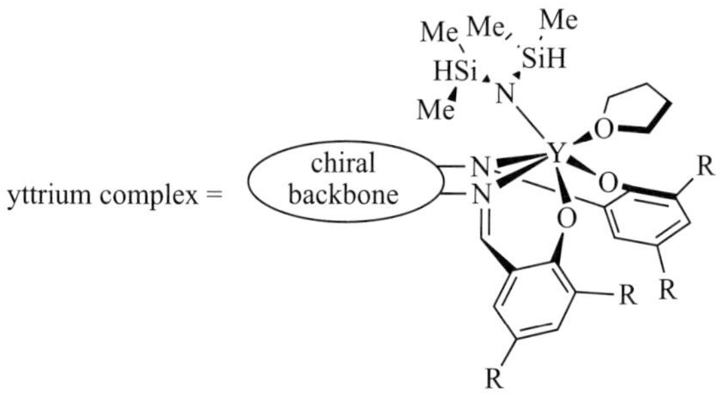

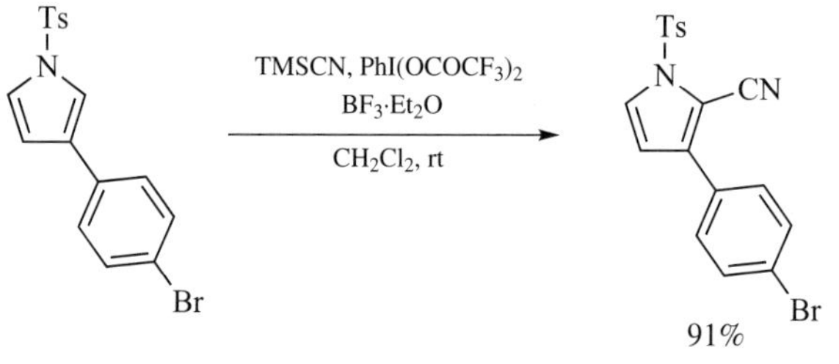

yttrium complex =

(42)

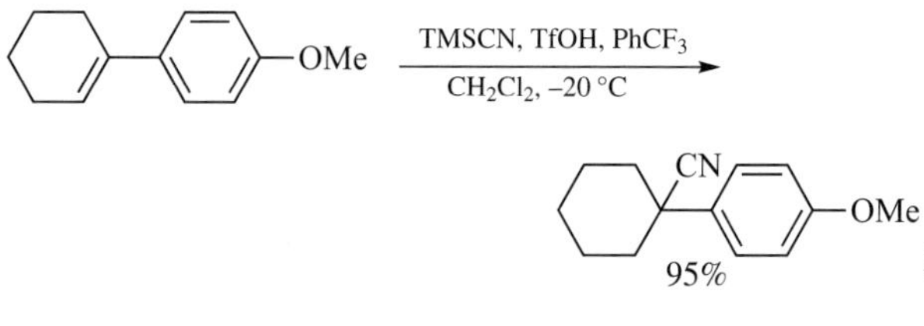

(43)

Convenient and efficient synthesis of α-aryl nitriles has been developed by direct cyanation of alcohols with TMSCN under the catalysis of Lewis acid (eq 44).[68] A variety of benzylic alcohols can be converted to the corresponding nitriles in good to high yields under mild conditions.

OH

TMSCN, InBr$_3$ (10 mol %)

CH$_2$Cl$_2$, rt

CN

(44)

98%

Related Reagents.   *t*-Butyldimethylsilyl Cyanide; Hydrogen Cyanide.

1. (a) Colvin, E. W. *Silicon in Organic Synthesis*; Butterworths: London, 1981. (b) Weber, W. P. *Silicon Reagents for Organic Synthesis*; Springer: Berlin, 1983. (c) *Advances in Silicon Chemistry 1*, Larson, G. Ed.; JAI: Greenwich, CT, 1991. (d) Groutas, W. C.; Felker, D., *Synthesis* **1980**, 861. (e) Livinghouse, T., *Org. Synth., Coll. Vol.* **1990**, *7*, 517.
2. Gassman, P. G.; Talley, J. J., *Tetrahedron Lett.* **1978**, 3773.
3. Mai, K.; Patil, S., *J. Org. Chem.* **1986**, *51*, 3545.
4. Hunig, S., *Chimia* **1982**, *36*, 1.
5. Evans, D. A.; Carroll, G. L.; Truesdale, L. K., *J. Org. Chem.* **1974**, *39*, 914.
6. Lidy, W.; Sundermeyer, W., *Chem. Ber.* **1973**, *106*, 587.
7. Kobayashi, S.; Tsuchiya, Y.; Mukaiyama, T., *Chem. Lett.* **1991**, 537.
8. Onaka, M.; Higuchi, K.; Sugita, K.; Izumi, Y., *Chem. Lett.* **1993**, 1393.
9. Hayashi, M.; Matsuda, T.; Oguni, N., *J. Chem. Soc., Perkin Trans. 1* **1992**, 3135.
10. Minamikawa, H.; Hayakawa, S.; Yamada, T.; Iwasawa, N.; Narasaka, K., *Bull. Chem. Soc. Jpn.* **1988**, *61*, 4379.
11. Somanathan, R.; Aguilar, H. R.; Ventura, G. R., *Synth. Commun.* **1983**, *13*, 273.
12. Confalone, P. N.; Pizzolato, G., *J. Am. Chem. Soc.* **1981**, *103*, 4251.
13. Utimoto, K.; Obayashi, M.; Sishiyama, Y.; Inoue, M.; Nozaki, H., *Tetrahedron Lett.* **1980**, *21*, 3389.
14. Nagata, W.; Yoshioka, M.; Hirai, S., *J. Am. Chem. Soc.* **1972**, *94*, 4635.
15. Evans, D. A.; Truesdale, L. K.; Carroll, G. L., *J. Chem. Soc., Chem. Commun.* **1973**, 55.
16. Higuchi, K.; Onaka, M.; Izumi, Y., *J. Chem. Soc., Chem. Commun.* **1991**, 1035.
17. Sassaman, M. B.; Prakash, G. K. S.; Olah, G. A., *J. Org. Chem.* **1990**, *55*, 2016.
18. Matsubara, S.; Onishi, H.; Utimoto, K., *Tetrahedron Lett.* **1990**, *31*, 6209.
19. Mullis, J. C.; Weber, W. P., *J. Org. Chem.* **1982**, *47*, 2873.
20. Sutowardoyo, K. I.; Sinou, D., *Tetrahedron: Asymmetry* **1991**, *2*, 437.
21. Matsubara, S.; Kodama, T.; Utimoto, K., *Tetrahedron Lett.* **1990**, *31*, 6379.
22. Seckar, J. A.; Thayer, J. S., *Inorg. Chem.* **1976**, *15*, 501.
23. Gassman, P. G.; Guggenheim, T. L., *J. Am. Chem. Soc.* **1982**, *104*, 5849.
24. Gassman, P. G.; Haberman, L. M., *Tetrahedron Lett.* **1985**, *26*, 4971.
25. Spessard, G. O.; Ritter, A. R.; Johnson, D. M.; Montgomery, A. M., *Tetrahedron Lett.* **1983**, *24*, 655.
26. Imi, K.; Yanagihara, N.; Utimoto, K., *J. Org. Chem.* **1987**, *52*, 1013.
27. (a) Hunig, S.; Wehner, W., *Synthesis* **1975**, 180. (b) Deuchert, K.; Hertenstein, U.; Hunig, S., *Synthesis* **1973**, 777. (c) Hertenstein, U.; Hunig, S.; Oller, M., *Synthesis* **1976**, 416.
28. Hunig, S.; Wehner, G., *Chem. Ber.* **1980**, *113*, 324.
29. Fischer, K.; Hunig, S., *J. Org. Chem.* **1987**, *52*, 564.
30. Hunig, S.; Marschner, C., *Chem. Ber.* **1990**, *123*, 107.
31. Herrmann, K.; Simchen, G., *Synthesis* **1979**, 204.
32. Schwindeman, J. A.; Magnus, P. D., *Tetrahedron Lett.* **1981**, *22*, 4925.
33. Fortes, C. C.; Okino, E. A., *Synth. Commun.* **1990**, *20*, 1943.
34. Reetz, M. T.; Chatziiosifidis, I.; Kunzer, H.; Muller-Starke, H., *Tetrahedron* **1983**, *39*, 961.
35. Ha, H.-J.; Nam, G.-S.; Park, K. P., *Synth. Commun.* **1991**, *21*, 155.
36. Tsuji, Y.; Yamada, J.; Tanaka, S., *J. Org. Chem.* **1993**, *58*, 16.
37. Kagabu, S.; Maehara, M.; Sawahara, K.; Saito, K., *J. Chem. Soc., Chem. Commun.* **1988**, 1485.
38. Anteunis, M. J. O.; Becu, C.; Becu, F., *Bull. Soc. Chim. Belg.* **1987**, *96*, 133.
39. Molander, G. A.; Haar, J. P., *J. Am. Chem. Soc.* **1991**, *113*, 3608.
40. Linderman, R. J.; Chen, K., *Tetrahedron Lett.* **1992**, *33*, 6767.
41. Utimoto, K.; Wakabayashi, Y.; Horiie, T.; Inoue, M.; Shishiyama, Y.; Obayashi, M.; Nozaki, H., *Tetrahedron* **1983**, *39*, 967.
42. Zimmer, H.; Reissig, H.-U.; Lindner, H. J., *Liebigs Ann. Chem.* **1992**, 621.
43. Mukaiyama, T.; Soga, T.; Takenoshita, H., *Chem. Lett.* **1989**, 997.
44. Fuchiyami, T.; Ichikawa, S.; Konno, A., *Chem. Lett.* **1989**, 1987.
45. Krepski, L. R.; Lynch, L. E.; Heilmann, S. M.; Rasmussen, J. K., *Tetrahedron Lett.* **1985**, *26*, 981.
46. Mai, K.; Patil, G., *Tetrahedron Lett.* **1984**, *25*, 4583.
47. Hamashima, Y.; Kanai, M.; Shibasaki, M., *J. Am. Chem. Soc.* **2000**, *122*, 7412.
48. Yabu, K.; Masumoto, S.; Yamasaki, S.; Hamashima, Y.; Kanai, M.; Du, W.; Curran, D. P.; Shibasaki, M., *J. Am. Chem. Soc.* **2001**, *123*, 9908.
49. Deng, H.; Isler, M. P.; Snapper, M. L.; Hoveyda, A. H., *Angew. Chem., Int. Ed.* **2002**, *41*, 1009.
50. Ryu, D. H.; Corey, E. J., *J. Am. Chem. Soc.* **2005**, *127*, 5384.
51. Fuerst, D. E.; Jacobsen, E. N., *J. Am. Chem. Soc.* **2005**, *127*, 8964.
52. Zuend, S. J.; Jacobsen, E. N., *J. Am. Chem. Soc.* **2007**, *129*, 15872.
53. Liu, X.; Qin, B.; Zhou, X.; He, B.; Feng, X., *J. Am. Chem. Soc.* **2005**, *127*, 12224.
54. Krueger, C. A.; Kuntz, K. W.; Dzierba, C. D.; Wirschun, W. G.; Gleason, J. D.; Snapper, M. L.; Hoveyda, A. H., *J. Am. Chem. Soc.* **1999**, *121*, 4284.
55. Wang, J.; Hu, X.; Jiang, J.; Gou, S.; Huang, X.; Liu, X.; Feng, X., *Angew. Chem., Int. Ed.* **2007**, *46*, 8468.
56. Zamfir, A.; Tsogoeva, S. B., *Org. Lett.* **2010**, *12*, 188.
57. Ichikawa, E.; Suzuki, M.; Yabu, K.; Albert, M.; Kanai, M.; Shibasaki, M., *J. Am. Chem. Soc.* **2004**, *126*, 11808.
58. Funabashi, K.; Ratni, H.; Kanai, M.; Shibasaki, M., *J. Am. Chem. Soc.* **2001**, *123*, 10784.
59. Sammis, G. M.; Jacobsen, E. N., *J. Am. Chem. Soc.* **2003**, *125*, 4442.
60. Tanaka, Y.; Kanai, M.; Shibasaki, M., *J. Am. Chem. Soc.* **2008**, *130*, 6072.
61. Mita, T.; Sasaki, K.; Kanai, M.; Shibasaki, M., *J. Am. Chem. Soc.* **2005**, *127*, 514.
62. Tanaka, Y.; Kanai, M.; Shibasaki, M., *J. Am. Chem. Soc.* **2010**, *132*, 8862.
63. Schaus, S. E.; Jacobsen, E. N., *Org. Lett.* **2000**, *2*, 1001.
64. Mita, T.; Fujimori, I.; Wada, R.; Wen, J.; Kanai, M.; Shibasaki, M., *J. Am. Chem. Soc.* **2005**, *127*, 11252.
65. Wu, B.; Gallucci, J. C.; Parquette, J. R.; RajanBabu, T. V., *Angew. Chem. Int. Ed.* **2009**, *48*, 1126.
66. Dohi, T.; Morimoto, K.; Kiyono, Y.; Tohma, H.; Kita, Y., *Org. Lett.* **2005**, *7*, 537.
67. Yanagisawa, A.; Nezu, T.; Mohri, S., *Org. Lett.* **2009**, *11*, 5286.
68. Chen, G.; Wang, Z.; Wu, J.; Ding, K., *Org. Lett.* **2008**, *10*, 4573.

Diazo(trimethylsilyl)methyllithium[1]

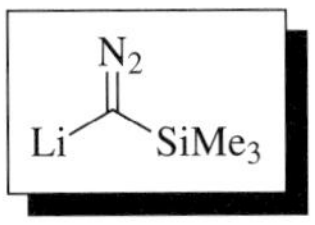

[84645-45-4] $C_4H_9LiN_2Si$ (MW 120.16)

InChI = 1S/C4H9N2Si.Li/c1-7(2,3)4-6-5;/h1-3H3;

InChIKey = WNCZNZOHHBHKSX-UHFFFAOYSA-N

(nucleophilic diazo species for introduction of one carbon; used in homologation and alkenylation of ketones,[2,4,5] preparation of diazo ketones,[6] acylsilanes,[7] vinylsilanes,[8] and heterocycles[9,10])

Alternate Names: lithiotrimethylsilyldiazomethane; trimethyl-silyldiazomethyllithium.

Solubility: sol THF/hexane mixtures in which it is generated; precipitates out of ether/hexane.

Preparative Methods: prepared in situ by lithiation of trimethylsi-lyldiazomethane using *n*-butyllithium; prepared in situ by lithiation of trimethylsilyldiazomethane (TMSCHN$_2$) using butyllithium, lithium diisopropylamide (LDA), or lithium 2,2,6,6-tetramethylpiperidide (LTMP). The lithium salt is easily converted to the corresponding magnesium bromide salt (eq 1).

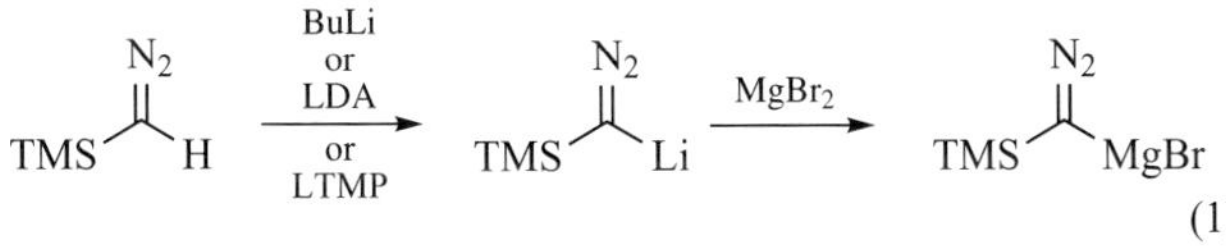

Handling, Storage, and Precautions: handle as for other organo-lithium reagents; stable in solution below 0 °C. Use in a fume hood; TMSCHN$_2$ should be regarded as toxic and all operations must be carried out in a well-ventilated fume hood and all skin contact should be avoided.[12]

Original Commentary

Christopher J. Moody
Loughborough University of Technology, Loughborough, UK

Homologation Reactions of Ketones. When generated in pentane/THF at −100 °C, diazo(trimethylsilyl)methyllithium re-acts with ketones (three examples) and benzaldehyde to give the unstable 1-diazo-1-trimethylsilyl-2-alkanols (eq 2).[2] On gentle warming, the diazo alcohols rapidly lose N$_2$ to give trimethylsilyl epoxides. Since such epoxides readily give aldehydes on treatment

with acid,[3] the method represents an overall homologation of a ke-tone to the next higher aldehyde.

When diaryl ketones are used, diarylalkynes are formed in good yield (eq 3).[4] The reaction presumably proceeds via the diazoalkene followed by aryl migration with loss of nitrogen.

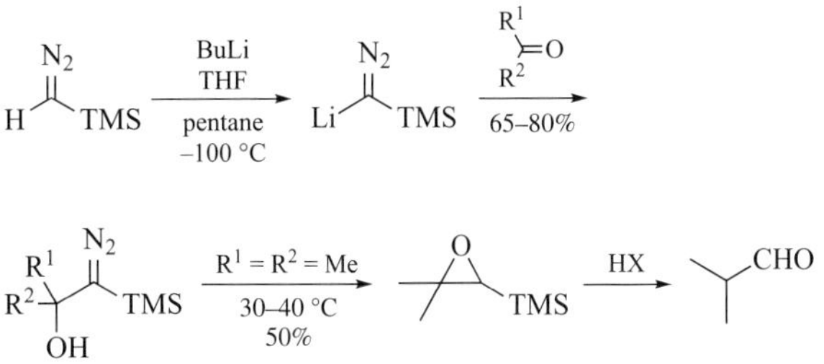

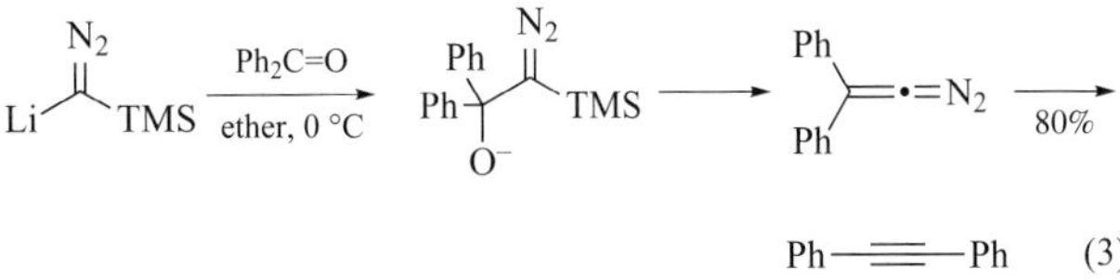

In contrast to the above, a more recent study, using slightly different conditions, reports that whereas aldehydes give terminal alkynes (eq 4), ketones give cyclopentene derivatives, presumably by intramolecular C–H insertion of intermediate alkylidenecar-benes (eq 5).[5] No detectable amounts of diazo alcohols or epoxysi-lanes were produced.

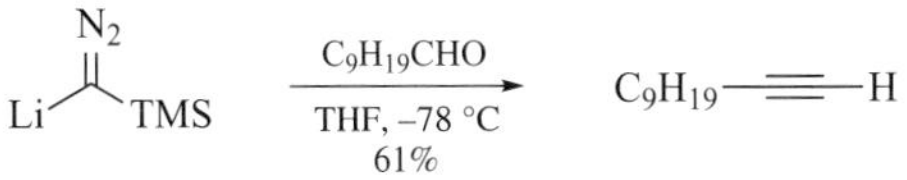

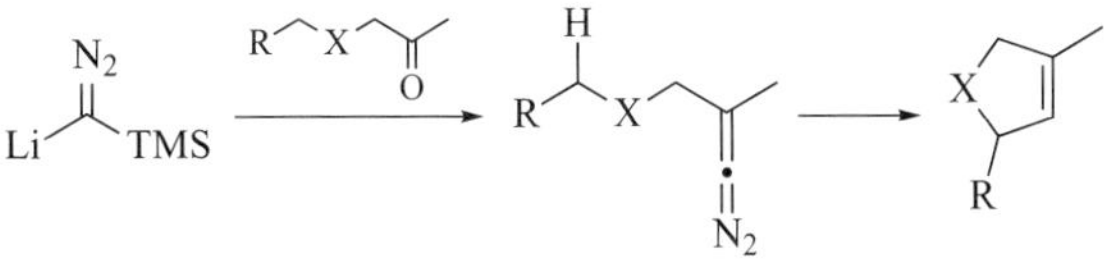

Preparation of Diazo Ketones. Diazo(trimethylsilyl)methyl-lithium reacts with lactones to give the product of ring opening, the diazo alcohol, although the silyl group is lost during workup (eq 6).[6] The diazo alcohol can subsequently be cyclized to the corresponding oxepane by treatment with dirhodium(II) tetra-acetate.

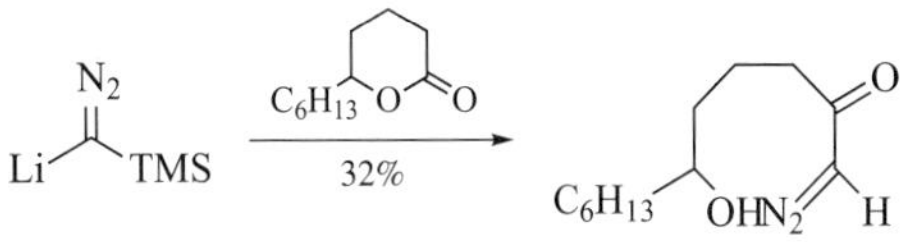

Preparation of Acylsilanes. Diazo(trimethylsilyl)methyl-lithium is alkylated by a range of alkyl halides RX (R = alkyl, benzyl) to give the corresponding diazosilanes, which are ox-idized to the acylsilanes in reasonable yield by treatment with *m*-chloroperbenzoic acid (eq 7).[7]

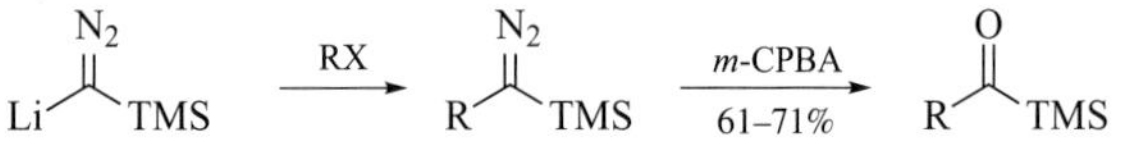

(7)

Preparation of Vinylsilanes ((E)-Trimethylsilylalkenes).
Treatment of trimethylsilyldiazoalkanes, obtained by alkylation of diazo(trimethylsilyl)methyllithium with RCH_2X, with copper(I) chloride gives excellent yields of (E)-1-trimethylsilylalkenes (eq 8).[8]

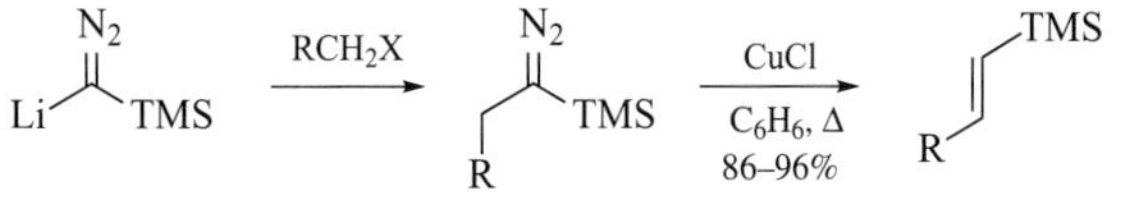

(8)

Preparation of Heterocycles.
On reaction with methyl esters, diazo(trimethylsilyl)methyllithium (2 equiv) gives 2-substituted 5-trimethylsilyltetrazoles in 50–90% yield (eq 9).[9a] Similarly, reaction with nitriles gives 1,2,3-triazoles in high yield (eq 10).[9b] Reaction with $(Me_3Si)_2C=PCl$ leads to 1,2,4-diazaphospholes (eq 11).[10]

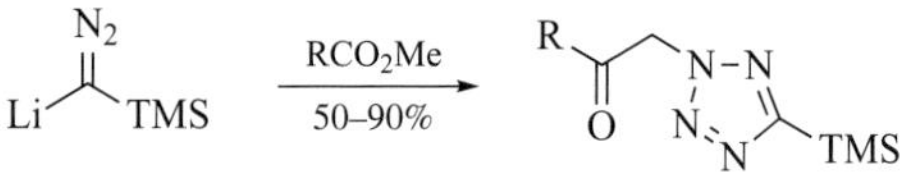

(9)

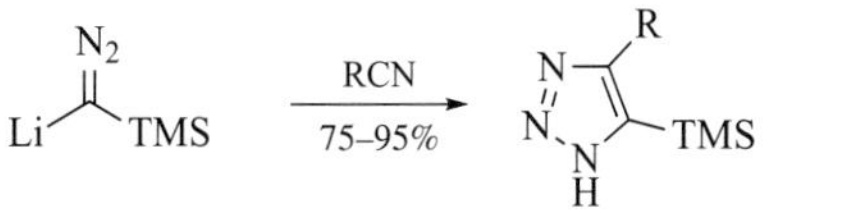

(10)

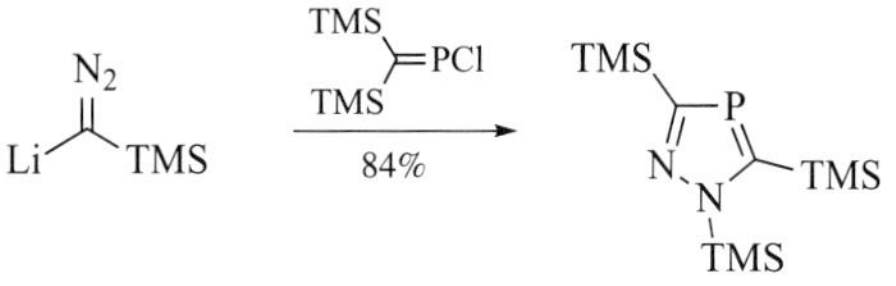

(11)

First Update

Takayuki Shioiri & Toyohiko Aoyama
Nagoya City University, Nagoya, Japan

Reactions with Carbonyl Compounds and Utilization of 2-Diazo-2-(trimethylsilyl)ethanols.
Diazo(trimethylsilyl)-methyllithium $(TMSC(Li)N_2)$ reacts with aldehydes and ketones to give lithium 2-diazo-2-(trimethylsilyl)ethoxides by nucleophilic addition,[2] which produce alkylidene carbenes by expulsion of TMSOLi and nitrogen (eq 12).

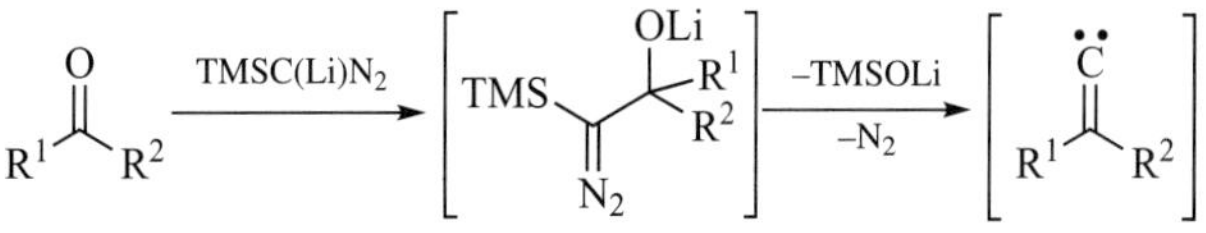

(12)

2-Diazo-2-(trimethylsilyl)ethanols can be almost quantitatively obtained from carbonyl compounds by the reaction with diazo-(trimethylsilyl)methylmagnesium bromide $(TMSC(MgBr)N_2)$

and can undergo [3 + 2]cycloaddition reaction with ethyl propiolate or dimethyl acetylenedicarboxylate to give di- and trisubstituted pyrazoles (eq 13).[13]

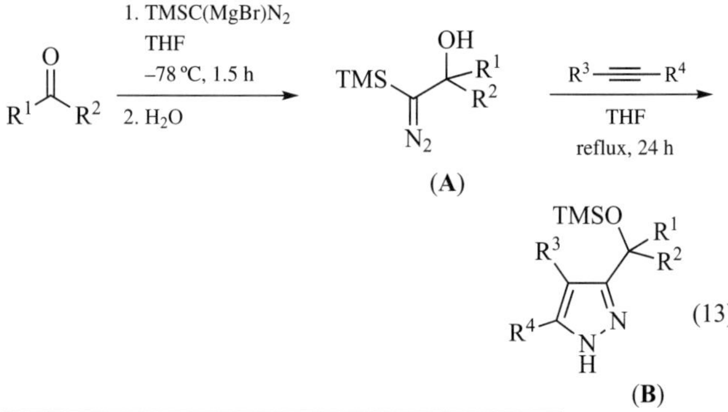

(13)

R^1	R^2	R^3	R^4	Yield (%)	
				A	**B**
$PhCH_2CH_2$	Me	CO_2Me	CO_2Me	quant	83
$PhCH_2CH_2$	Me	H	CO_2Et	quant	92
$4-MeOC_6H_4$	H	H	CO_2Et	quant	78

Analogously, 2-diazo-2-(trimethylsilyl)ethanols react with benzynes from o-(trimethylsilyl)aryltriflates to furnish 3-substituted indazoles, [3 + 2] cycloadducts (eq 14).[14]

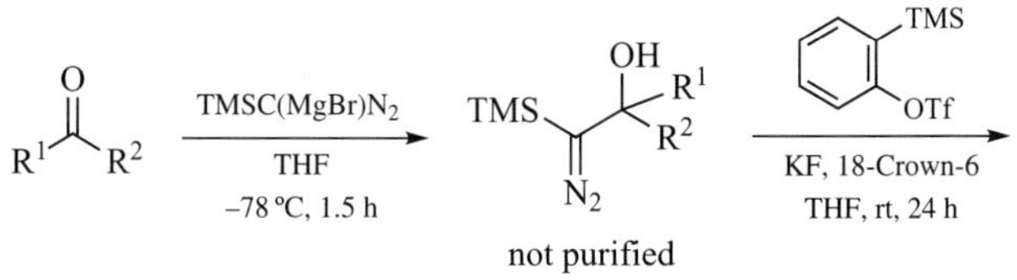

(14)

R^1	R^2	Yield (%)
$PhCH_2CH_2$	Me	93
$4-MeOC_6H_4$	Ph	74
2-Thienyl	Me	82

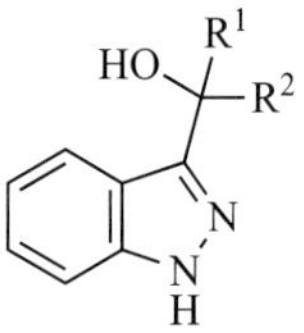

Reaction of α-ketoesters with $TMSC(MgBr)N_2$ followed by in situ treatment of the resulting magnesium 2-diazo-2-(trimethylsilyl)ethoxides with pivalic acid produces α-substituted β-trimethylsilyl-α,β-epoxyesters in Z-selective manner (eq 15).[15]

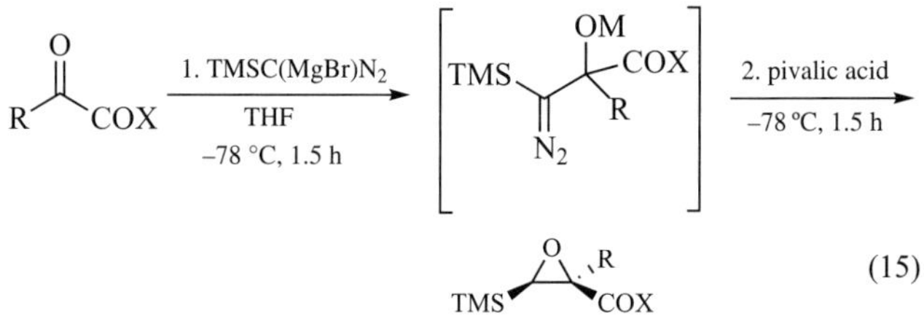

(15)

R	X	Yield (%)	Z/E
$PhCH_2CH_2$	OEt	88	96/4
$PhCH_2CH_2$	NPr^i_2	80	>99/1
Pr^i	OEt	83	94/6

Ligand-induced control of C–H versus aliphatic C–C migration reaction of Rh carbenoids has been observed. Thus, treatment of lithium 2-diazo-2-(trimethylsilyl)ethoxides with TMSCl affords the corresponding trimethylsilyl derivatives, which produce the silyl enol ethers **A** (1,2-C–H migration products) with high Z-selectivity as the major products utilizing $Rh_2(OAc)_4$.[16,17] In contrast, the use of $Rh_2(tfa)_4$ in place of $Rh_2(OAc)_4$ gives the silyl enol ethers **B** (1,2-C-C migration products) with high Z- selectivity (eq 16).[17]

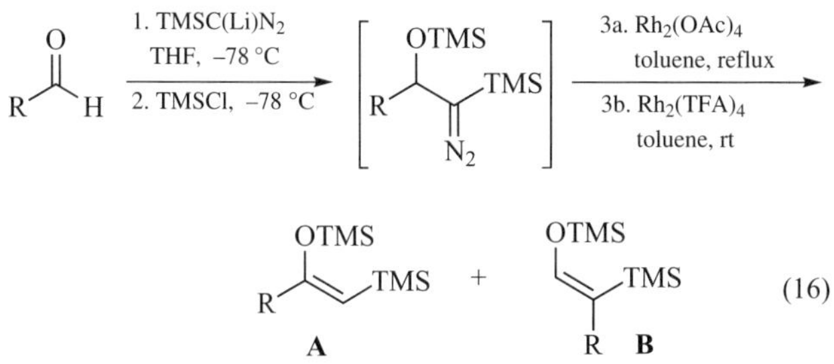

$$ \text{(16)} $$

R	Method	A (Z/E)	B (Z/E)
Bu	3a	69 (93/7)	18 (81/19)
	3b	—	95 (82/18)
Ph(CH$_2$)$_2$	3a	91 (97/3)	9 (84/16)
	3b	—	100 (90/10)

Reactions with Carbonyl Compounds Followed by the Colvin Rearrangement. Alkylidene carbenes formed from lithium 2-diazo-2-(trimethylsilyl)ethoxides as shown in eq 12 undergo the Colvin rearrangement to furnish the acetylenic compounds.[4,18] Utilizing this methodology, a polyyne formation was carried out (eq 17).[19] This is the first example of polyyne formation using the Colvin rearrangement.

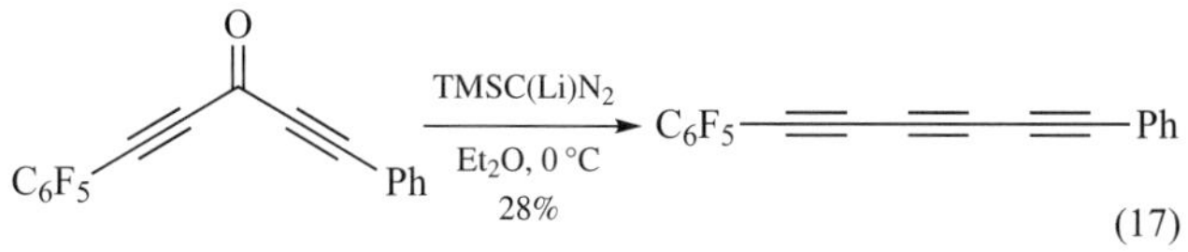

$$ \text{(17)} $$

Acetylenic compounds derived from aryl and heteroaryl aldehydes were further transformed into functionalized acetylenes in a one-pot reaction (eq 18).[20] In the Colvin conversion of aryl or heteroaryl (oxo) acetates to aryl or heteroaryl propiolates, TMSC-(MgBr)N$_2$ was found to be more efficient than its lithium salt (eq 19).[21]

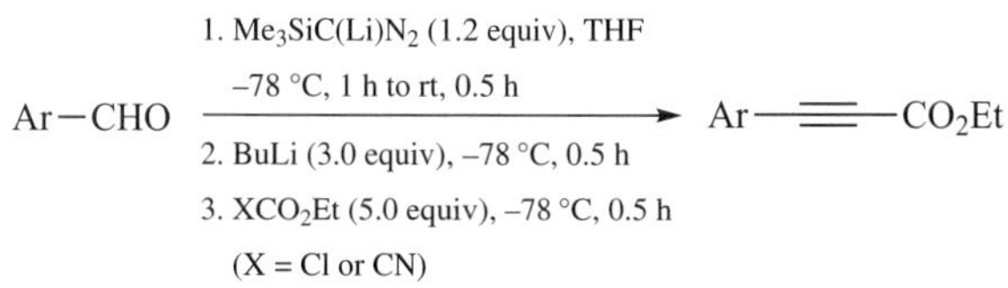

Ar	Yield (%)
4-MeC$_6$H$_4$	69
4-MeOC$_6$H$_4$	79
2-MeOC$_6$H$_4$	72

$$ \text{(18)} $$

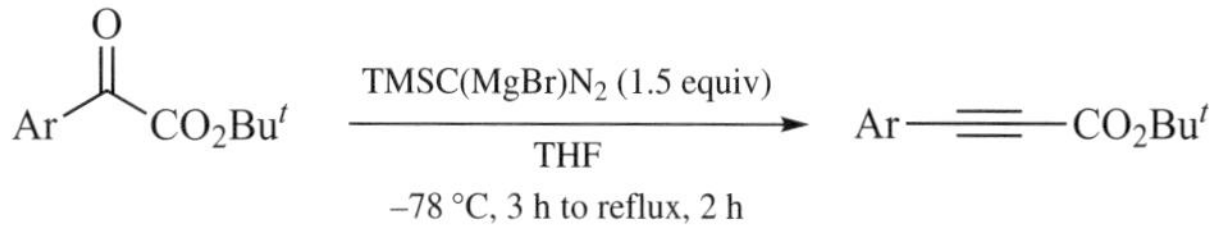

Ar	Yield (%)
4-MeOC$_6$H$_4$	64
2-naphthyl	61
2-furyl	52

$$ \text{(19)} $$

Reaction of *o*-acyl-*N*-pivaloylanilines with TMSC(Li)N$_2$ efficiently undergoes the Colvin rearrangement to give *o*-alkynyl-*N*-pivaloylanilines via alkylidene carbene intermediates (eq 20),[22] while *o*-acyl-*N*-tosylanilines smoothly react with TMSC(Li)N$_2$ to give 3-substituted *N*-tosylindoles in good to high yields through an N–H insertion (eq 21).[23] After the reaction of *N*-tosylanilines with TMSC(Li)N$_2$ in the latter example, addition of Bu^{t}Li and then electrophiles such as benzaldehyde gives 2,3-disubstituted *N*-tosylindoles (eq 22).[24]

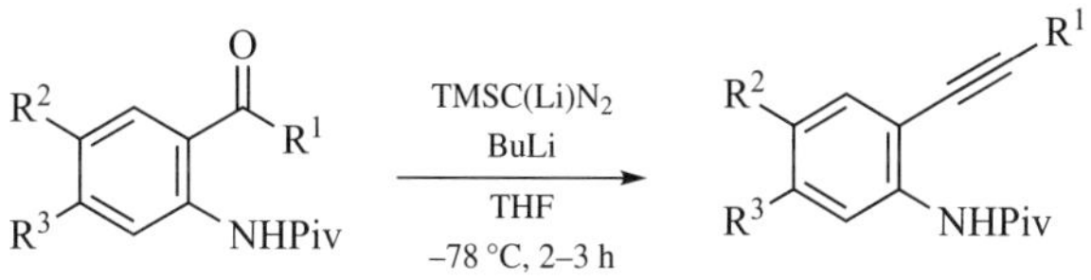

$$ \text{(20)} $$

R^1	R^2	R^3	Yield (%)
2-Furyl	H	H	84
Pr^i	Me	H	82
Pr^i	-CH=CH-CH=CH-		63

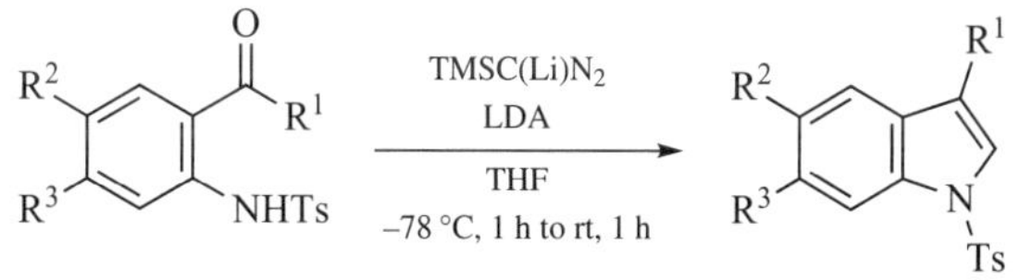

$$ \text{(21)} $$

R^1	R^2	R^3	Yield (%)
Ph	H	H	91
Ph	Cl	H	87
Me	-OCH$_2$O-		78

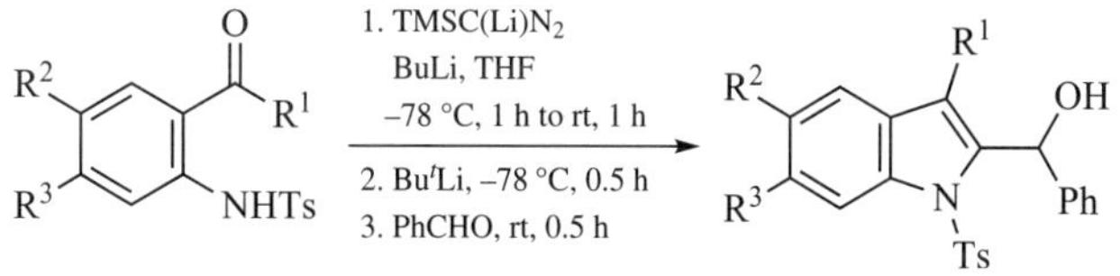

$$ \text{(22)} $$

R^1	R^2	R^3	Yield (%)
Ph	H	H	77
Bu	H	H	79
Me	-OCH$_2$O-		69

o-Triisopropylsiloxyaryl ketones and aldehydes smoothly react with TMSC(Li)N$_2$ to produce *o*-triisopropylsiloxyarylacetylenes through the Colvin rearrangement. The acetylenes are easily converted to benzofurans by treatment with tetrabutylammonium

R^1	R^2	Yield (%)
H	H	69
H	3-MeO	74
Me	H	78

R^1	R^2	Yield (%)
H	H	56
H	3-MeO	75
Me	H	71

R^1	R^2	R^3	Yield (%)
H	3-MeO	Ph	86
Me	H	Ph	81
Me	H	2-Pyridyl	73

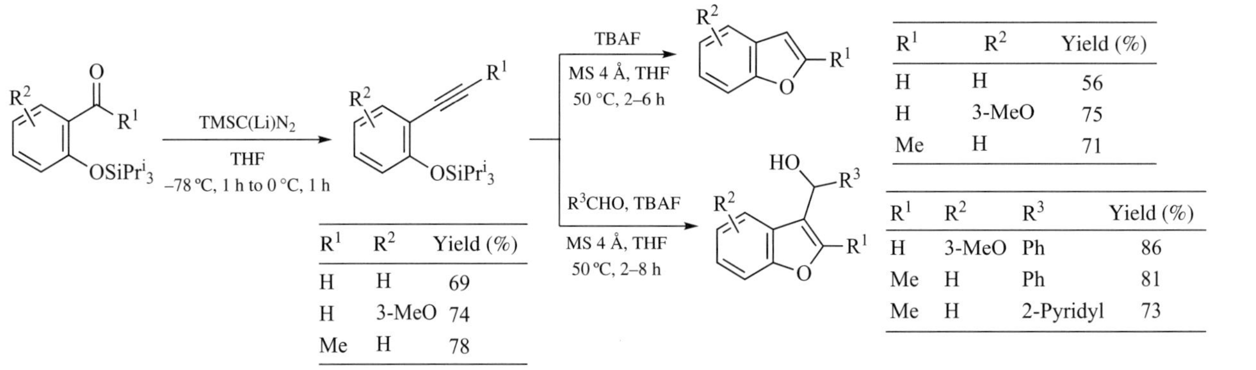

(23)

fluoride (TBAF). 3-Benzofuranmethanols are also obtained when the reaction is conducted in the presence of carbonyl compounds (eq 23).[25]

Reactions with Carbonyl Compounds Followed by Insertion Reactions. *N*-Substituted *β*-aminoketones react with TMSC(Li)N_2 to give 2-pyrroline derivatives via the alkylidene carbene intermediates by intramolecular N–H insertion, analogously to the reaction of *o*-acyl-*N*-tosylanilines with TMSC(Li)N_2 (eq 21). Dehydrogenation of the 2-pyrrolines with active MnO_2 (CMD, chemical manganese dioxide, produced for battery manufacture) affords the corresponding pyrroles (eq 24).[26]

R^1	R^2	R^3	Yield (%) A	Yield (%) B
Me	Me	H	81	59
Et	Me	H	83	68
Et	H	Me	66	71

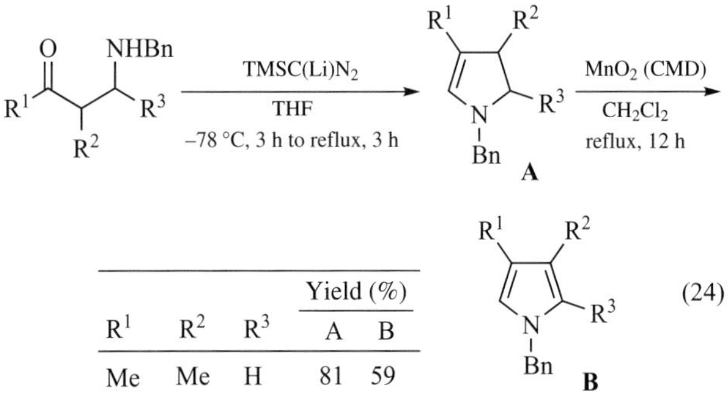

(24)

Interestingly, when the reaction of *N*-benzyl-*β*-aminoketones with TMSC(Li)N_2 is carried out at low temperature, intramolecular *N*-alkylation via initially formed diazoalkoxides preferentially occurs to give 3-substituted-3-(trimethylsiloxy)pyrrolidines as the major products (eq 25).[27]

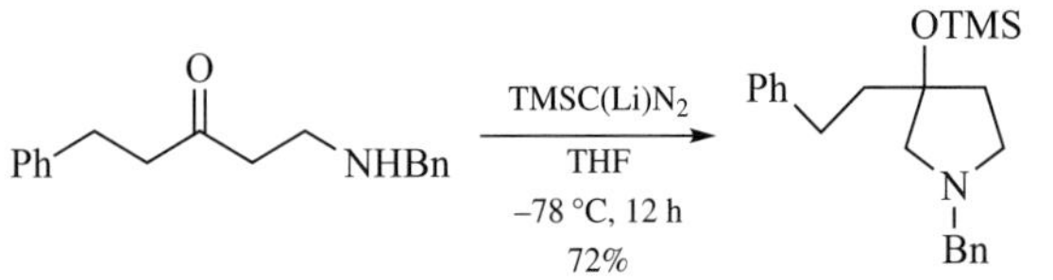

(25)

2-Unsubstituted benzofuran-3-carboxylates are formed as the major products by the reaction of *t*-butyl (*o*-methoxyphenyl)(oxo)acetates with TMSC(MgBr)N_2 (eq 26).[28] Alkylidene carbenes are first formed from *α*-ketoesters, which cyclize to oxonium ylides. Elimination of methylene affords benzofurans.

R^1	R^2	Yield (%) A	Yield (%) B
H	H	55	16
Me	H	42	15
H	MeO	62	9

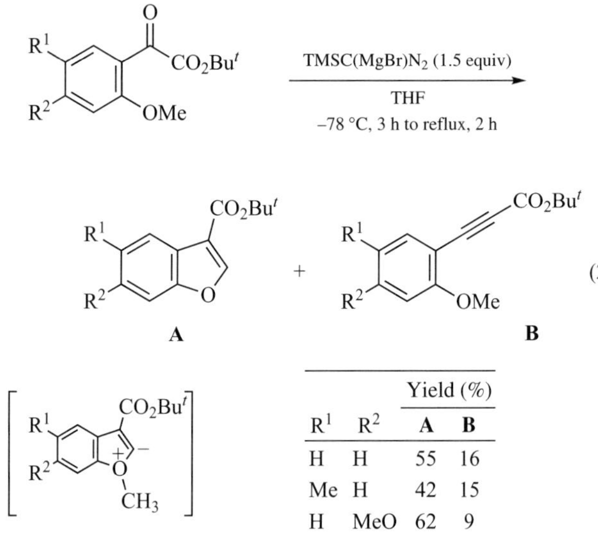

(26)

TMSC(Li)N_2 reacts with *α*-oxoketene dithioacetals to furnish thiophenes via sulfonium ylides, homologous ketones, and homologous alkynes as a function of substrate structure (eq 27).[29]

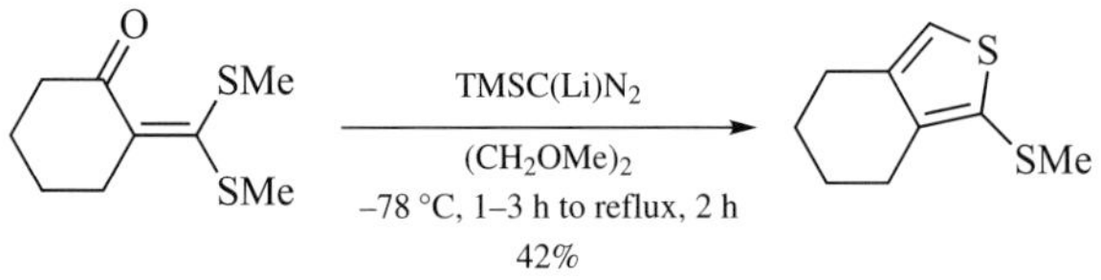

(27)

A cyclopropenation involving selective insertion of alkylidene carbenes into the C*α*–Si bond of *α*-silyl ketones has been achieved by the reaction of TMSC(Li)N_2 with *α*-silyl ketones that are prepared by the indium chloride-catalyzed reaction between $TMSCHN_2$ and aldehydes (eq 28).[30] In the case of *α*-silyl-*α*′-alkoxy ketones, the C*γ*–H insertion competitively occurs and a mixture of cyclopropenes (C*α*–Si insertion products) and dihydrofurans (C*γ*–H insertion products) are formed in preference of the former (eq 29).

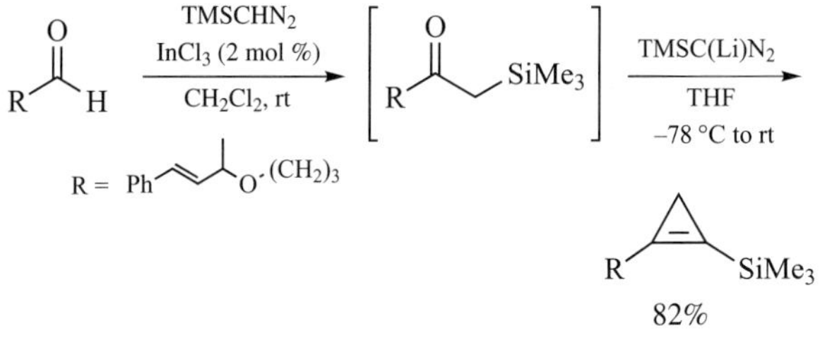

(28)

R = Et, (CH$_2$)$_4$

$$84\% \ (\mathbf{A/B} = 2.4/1) \tag{29}$$

Azulene-1-carboxylic esters are synthesized in two steps: (1) reaction of TMSC(Li)N$_2$ with ethyl or *t*-butyl 4-aryl-2-oxo-butanoates, and (2) dehydrogenation of the resulting 2,3-dihydroazulenes with active MnO$_2$ (CMD) (eq 30).[31]

$$\tag{30}$$

	Yield (%)	
R	**A**	**B**
H	60	52
Me	72	64
MeO	76	71

Reaction of linear compounds containing a ketone and diene functional groups with TMSC(Li)N$_2$ produces alkylidene carbenes that undergo intramolecular cyclopropanation followed by formation of trimethylenemethane diyls and then [3 + 2]cycloaddition, giving linearly fused triquinanes (eq 31).[32]

[3 + 2]Cycloaddition. [3 + 2]Cycloaddition reaction of TMSC(Li)N$_2$ with benzynes, generated from halobenzenes, gives the corresponding 3-trimethylsilylindazoles (eq 32).[33]

$$\tag{31}$$

$$\tag{32}$$

R^1	R^2	Yield (%)	**A/B**
MeO	H	61	73/27
H	MeO	75	50/50
H	F$_3$C	74	50/50

White phosphorus, P$_4$, reacts with TMSC(Li)N$_2$ to give 1,2,3,4-diazadiphospholide anion, which results from a formal [3 + 2] cycloaddition between [P=P] and the diazomethyl anion. The protonation of the pholide anion with TFA affords 2*H*-1,2,3,4-diazadiphosphole (eq 33).[34]

$$\tag{33}$$

Transformation of Benzocyclobutenones into 2,3-Benzodiazepines. Benzocyclobutenones react with TMSC(Li)N$_2$ to give 2,3-benzodiazepines via 2-diazo-2-(trimethylsilyl)ethoxide anions and *o*-quinodimethane intermediates (eq 34).[35]

Diazoketone Formation. Reaction of carbamate derivatives of pyroglutamic acid esters with TMSC(Li)N$_2$ below −100 °C gives the corresponding substituted 6-diazo-5-oxo-norleucine esters (eq 35).[36]

$$\tag{34}$$

R^1	R^2	Yield (%)
H	H	70
H	MeO	79
MeO	MeO	82

$$\text{(35)}$$

R^1	R^2	Yield (%)
Et	Fmoc	72
Bn	Boc	71
Bu^t	Boc	75

The anion prepared from $TMSCHN_2$ and $NaN(TMS)_2$ is effective to convert the *N*-carboxylated β-lactams to the analogous diazoketones.[37]

Reaction of TMSC(Li)N₂ with Halides of Boron, Phosphorus, Sulfur, and Tin. $TMSC(Li)N_2$ undergoes the reaction with halides of boron,[38] phosphorus,[39] and sulfur[40] to give the corresponding diazo trimethylsilylmethyl derivatives (eqs 36–38), while tin(II) chloride affords the carbodiimide (eq 39).[41]

$$\text{(36)}$$

$$\text{(37)}$$

R^1	R^2	Yield (%)
Pr^i	H	92
Bu^t	H	90
Bu^t	Ph	89

$$\text{(38)}$$

R	Yield (%)
Ph	51
Me	59

$$\text{(39)}$$

Ar = 2,6-$Pr^i_2C_6H_3$

Related Reagents. Triisopropylsilyldiazomethane can be similarly converted into its lithium derivative.[11] Diazo compounds containing tin (R_3SnCHN_2) and germanium (R_3GeCHN_2) are also known,[1] but have not been used in synthesis. Other lithiated diazo compounds are covered elsewhere, as are other lithiated silanes.

1. For a review on organometallic derivatives of diazo compounds, see: Kruglaya, O. A.; Vyazankin, N. S., *Russ. Chem. Rev. (Engl. Transl.)* **1980**, *49*, 357.
2. Schöllkopf, U.; Scholz, H.-U., *Synthesis* **1976**, 271.
3. Stork, G.; Ganem, B., *J. Am. Chem. Soc.* **1973**, *95*, 6152.
4. Colvin, E. W.; Hamill, B. J., *J. Chem. Soc., Perkin Trans. 1* **1977**, 869.
5. Ohira, S.; Okai, K.; Moritani, T., *J. Chem. Soc., Chem. Commun.* **1992**, 721.
6. Davies, M. J.; Moody, C. J.; Taylor, R. J., *J. Chem. Soc., Perkin Trans. 1* **1991**, 1.
7. Aoyama, T.; Shioiri, T., *Tetrahedron Lett.* **1986**, *27*, 2005.
8. Aoyama, T.; Shioiri, T., *Tetrahedron Lett.* **1988**, *29*, 6295.
9. (a) Aoyama, T.; Shioiri, T., *Chem. Pharm. Bull.* **1982**, *30*, 3450. (b) Aoyama, T.; Shioiri, T., *Chem. Pharm. Bull.* **1982**, *30*, 3849.
10. Schnurr, W.; Regitz, M., *Synthesis* **1989**, 511.
11. Arthur, M. P.; Goodwin, H. P.; Baceiredo, A.; Dillon, K. B.; Bertrand, G., *Organometallics* **1991**, *10*, 3205.
12. (a) Shioiri, T.; Aoyama, T.; Mori, S., *Org. Synth., Coll. Vol.* **1993**, *8*, 612. (b) Barnhart, R.; Dale, D. J.; Ironside, M. D.; Vogt, P. F., *Org. Process Res. Dev.* **2009**, *13*, 1388.
13. Hari, Y.; Tsuchida, S.; Sone, R.; Aoyama, T., *Synthesis* **2007**, 3371.
14. Hari, Y.; Sone, R.; Aoyama, T., *Org. Biomol. Chem.* **2009**, *7*, 2804.
15. Hari, Y.; Tsuchida, S.; Aoyama, T., *Tetrahedron Lett.* **2006**, *47*, 1977.
16. Aggarwal, V. K.; Sheldon, C. G.; MacDonald, G. J.; Martin, W. P., *J. Am. Chem. Soc.* **2002**, *124*, 10300.
17. Vitale, M.; Lecourt, T.; Sheldon, C. G.; Aggarwal, V. K., *J. Am. Chem. Soc.* **2006**, *128*, 2524.
18. Miwa, K.; Aoyama, T.; Shioiri, T., *Synlett* **1994**, 107.
19. Kendall, J.; McDonald, R.; Ferguson, M. J.; Tykwinski, R. R., *Org. Lett.* **2008**, *10*, 2163.
20. Hari, Y.; Date, K.; Aoyama, T., *Heterocycles* **2007**, *74*, 545.
21. Hari, Y.; Date, K.; Kondo, R.; Aoyama, T., *Tetrahedron Lett.* **2008**, *49*, 4965.
22. Hari, Y.; Kanie, T.; Aoyama, T., *Tetrahedron Lett.* **2006**, *47*, 1137. See also, Taber, D. F.; Plepys, R. A., *Tetrahedron Lett.* **2005**, *46*, 6045.
23. Miyagi, T.; Hari, Y.; Aoyama, T., *Tetrahedron Lett.* **2004**, *45*, 6303.
24. Hari, Y.; Kanie, T.; Miyagi, T.; Aoyama, T., *Synthesis* **2006**, 1249.
25. (a) Ito, Y.; Aoyama, T.; Shioiri, T., *Synlett* **1997**, 1163. (b) Sakai, A.; Aoyama, T.; Shioiri, T., *Tetrahedron Lett.* **1999**, *40*, 4211. (c) Sakai, A.; Aoyama, T.; Shioiri, T., *Heterocycles* **2000**, *52*, 643.
26. Yagi, T.; Aoyama, T.; Shioiri, T., *Synlett* **1997**, 1063.
27. Hari, Y.; Yokoyama, T.; Aoyama, T., *Heterocycles* **2010**, *80*, 679.
28. Hari, Y.; Kondo, R.; Date, K.; Aoyama, T., *Tetrahedron* **2009**, *65*, 8708.
29. Miyabe, R.; Shioiri, T.; Aoyama, T., *Heterocycles* **2002**, *57*, 1313.
30. Li, J.; Sun, C.; Lee, D., *J. Am. Chem. Soc.* **2010**, *132*, 6640.
31. Hari, Y.; Tanaka, S.; Takuma, Y.; Aoyama, T., *Synlett* **2003**, 2151.
32. Lee, H.-Y.; Kim, W.-Y.; Lee, S., *Tetrahedron Lett.* **2007**, *48*, 1407.
33. Shoji, Y.; Hari, Y.; Aoyama, T., *Tetrahedron Lett.* **2004**, *45*, 1769.
34. Charrier, C.; Maigrot, N.; Ricard, L.; Floch, P. L.; Mathey, F., *Angew. Chem., Int. Ed. Engl.* **1996**, *35*, 2133.
35. Matsuya, Y.; Ohsawa, N.; Nemoto, H., *J. Am. Chem. Soc.* **2006**, *128*, 13072.

36. (a) Coutts, I. G. C.; Saint, R. E., *Tetrahedron Lett.* **1998**, *39*, 3243. (b) Coutts, I. G. C.; Saint, R. E.; Saint, S. L.; Chambers-Asman, D. M., *Synthesis* **2001**, 247.

37. Ha, D.-C.; Kang, S.; Chung, C.-M.; Lim, H.-K., *Tetrahedron Lett.* **1998**, *39*, 7541.

38. Weber, L.; Wartig, H. B.; Stammler, H.-G.; Neumann, B., *Organometallics* **2001**, *20*, 5248.

39. (a) Krysiak, J.; Lyon, C.; Baceiredo, A.; Gornitzka, H.; Mikolajczyk, M.; Bertrand, G., *Chem. Eur. J.* **2004**, *10*, 1982. (b) Illa, O.; Bagan, X.; Baceiredo, A.; Branchadell, V.; Ortuno, R. M., *Tetrahedron: Asymmetry* **2008**, *19*, 2353.

40. Wagner, T.; Lange, J.; Grote, D.; Sander, W.; Schaumann, E.; Adiwidjaja, G.; Adam, A.; Kopf, J., *Eur. J. Org. Chem.* **2009**, 5198.

41. Jana, A.; Roesky, H. W.; Schulzke, C.; Samuel, P. P., *Inorg. Chem.* **2010**, *49*, 3461.

(Dibromomethyl)trimethylsilane

$$Me_2SiCHBr_2$$

[2612-42-2] $C_4H_{10}Br_2Si$ (MW 246.03)

InChI = 1S/C4H10Br2Si/c1-7(2,3)4(5)6/h4H,1-3H3

InChIKey = SIDSNGIDUFIXBW-UHFFFAOYSA-N

(reagent for the preparation of vinylsilanes)

Alternate Name: trimethyl(dibromomethyl)silane.
Physical Data: bp 49.5–51.5 °C/12 mmHg; d 1.519 g cm^{-3}.[1b]
Solubility: sol most organic solvents.
Preparative Method: to a dry, 1.0 L three-neck flask, equipped with a mechanical stirrer, a pentane thermometer, a nitrogen inlet tube, and a pressure-equalizing addition funnel, was added 62.0 g (0.245 mol) of bromoform and 300 mL of dry THF. The reaction was cooled to −90 °C and 0.25 mol of isopropylmagnesium chloride in 200 mL of THF was added during 30 min at such a rate that the temperature was kept at −80 °C or below. The reaction was stirred at −90 °C for 15 min and then 0.25 mol of chlorotrimethylsilane in 100 mL of THF was added at such a rate that the temperature did not exceed −80 °C. After 1 h, the reaction was allowed to warm to rt. The reaction mixture was hydrolyzed with saturated NH$_4$Cl solution until large lumps of salt cake formed. The organic layer was poured off, dried over anhydrous Na$_2$SO$_4$, and distilled at atmospheric pressure to remove most of the THF. A trap-to-trap distillation of the residue at 0.1 mmHg was followed by distillation through an 11 in. Vigreux column. The fraction boiling at 49.5–51.5 °C/12 mmHg was collected, affording 47.9 g (79%) of (dibromomethyl)trimethylsilane.[1a]

Lithium–Halogen Exchange. Reaction of (dibromomethyl)trimethylsilane with *n*-butyllithium at −110 °C results in the formation of trimethylsilylbromomethyllithium. The silane and the *n*-BuLi are added simultaneously in order to suppress side reactions. Treatment of the intermediate lithium reagent with chlorotrimethylsilane (eq 1) or mercury(II) bromide (eq 2) affords bis(trimethylsilyl)bromomethane or bis(trimethylsilylbromomethyl)mercury, respectively. If the lithium reagent is

allowed to warm in the presence of cyclohexene, the only reaction observed is an alkylation with the *n*-butyl bromide formed in the exchange reaction. The alkylated product is formed in 89% yield.[1a]

$$TMSCHBr_2 \xrightarrow[\substack{2.\ TMSCl \\ 52\%}]{1.\ BuLi,\ -110\ °C} (TMS)_2CHBr \qquad (1)$$

$$TMSCHBr_2 \xrightarrow[\substack{2.\ HgBr_2 \\ 44\%}]{1.\ BuLi,\ -110\ °C} (TMSCHBr)_2Hg \qquad (2)$$

In an alternative procedure, the lithium reagent was obtained by reaction with lithium metal. Subsequent reaction with a trialkylborane (eq 3) followed by oxidation provided access to α-hydroxysilanes. The yields are good for simple trialkylboranes but are diminished as the alkyl groups become more hindered.[2]

$$TMSCHBr_2 \xrightarrow[\substack{2.\ (C_5H_{11})_3B \\ 3.\ H_2O_2,\ OH^- \\ 79\%}]{1.\ 2\ equiv\ Li,\ THF} TMSCH(OH)C_5H_{11} \qquad (3)$$

Deprotonation. (Dibromomethyl)trimethylsilane can be deprotonated with lithium diisopropylamide at −78 °C. Reaction of the intermediate anion with *n*-butyl iodide (eq 4) furnished the alkylated product in 93% yield.[3,4] If the anion was allowed to warm to 20 °C in the absence of an electrophile, the bis(trimethylsilyl)ethylene (eq 5) was produced.[4]

$$TMSCHBr_2 \xrightarrow[\substack{2.\ BuI \\ 93\%}]{1.\ LDA,\ -78\ °C} TMSCBr_2Bu \qquad (4)$$

$$TMSCHBr_2 \xrightarrow[\substack{2.\ 20\ °C \\ 50\%}]{1.\ LDA,\ -78\ °C} TMSBrC=CBrTMS \qquad (5)$$

Vinylsilanes. Two reports have appeared on the formation of the 1,1-di-Grignard reagent using magnesium amalgam. In the first report, the di-Grignard, prepared indirectly from the dizinc reagent, was combined with cyclohexanone to produce the vinylsilane in 40% yield.[5] In the second, satisfactory results were obtained only with the non-enolizable benzophenone (eq 6). Cyclohexanone afforded only 13% of the vinylsilane. In addition, the di-Grignard was treated with chlorotrimethylstannane to furnish bis(trimethylstannyl)trimethylsilylmethane in 94% yield.[6]

$$TMSCHBr_2 \xrightarrow[\substack{2.\ Ph_2CO \\ 80\%}]{1.\ Mg(Hg)} \underset{Ph \qquad Ph}{\overset{TMS}{\diagup\!\!\diagdown}} \qquad (6)$$

The *gem*-dichromium reagent was prepared by reduction of the dibromide with chromium(II) chloride. This reagent will stereoselectively produce (*E*)-alkenylsilanes in excellent yield. The reagent can chemoselectively react with an aldehyde (eq 7) in the presence of a ketone.[7]

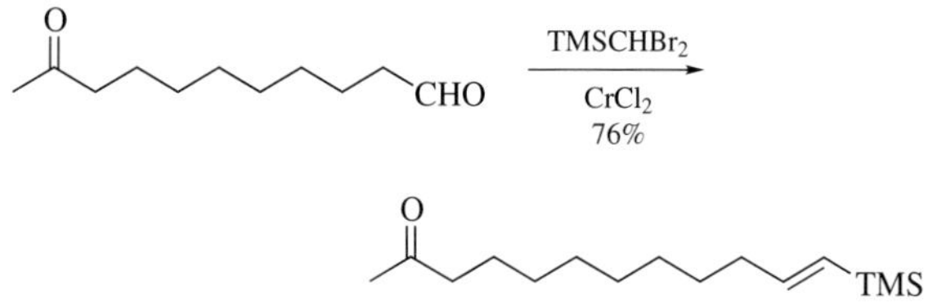

$$\text{(7)}$$

Esters (eq 8) and thioesters (eq 9) can be silylmethylenated by the action of the dibromide with low-valent titanium prepared from titanium(IV) chloride and zinc. The reaction produces β-hetero-substituted vinylsilanes and was (Z) selective.[8]

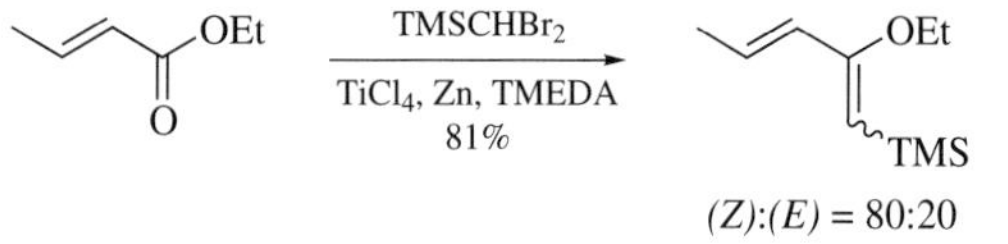

$$\text{(8)}$$

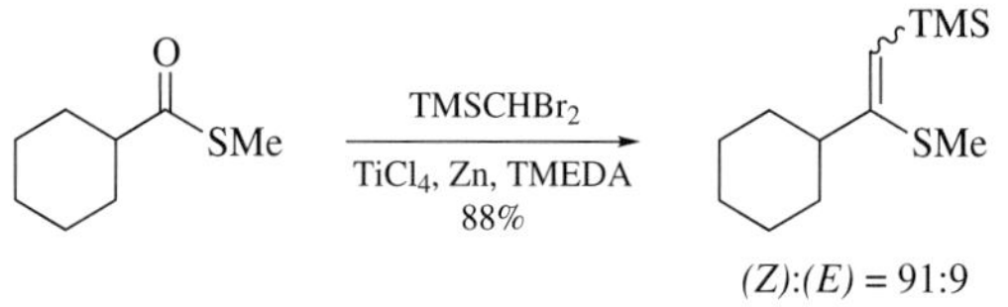

$$(Z){:}(E) = 80{:}20$$

$$\text{(9)}$$

$$(Z){:}(E) = 91{:}9$$

Miscellaneous. The kinetics of the cleavage of the dibromomethyl group by ammonia buffer have been reported.[9]

1. (a) Seyferth, D.; Lambert, R. L.; Hanson, E. M., *J. Organomet. Chem.* **1970**, *24*, 647. (b) Villieras, J., *Bull. Soc. Claim. Fr., Part 2* **1967**, 1520.

2. (a) Rosario, O.; Oliva, A.; Larson, G. L., *J. Organomet. Chem.* **1978**, *146*, C8. (b) Larson, G. L.; Argüelles, R.; Rosario, O.; Sandoval, S., *J. Organomet. Chem.* **1980**, *198*, 15.

3. Villieras, J.; Bacquet, C.; Masure, D.; Normant, J. F., *J. Organomet. Chem.* **1973**, *50*, C7.

4. Villieras, J.; Bacquet, C.; Normant, J. F., *Bull. Soc. Claim. Fr., Part 2* **1975**, 1797.

5. Martel, B.; Varache, M., *J. Organomet. Chem.* **1972**, *40*, C53.

6. van de Heisteeg, B. J. J.; Schat, G.; Tinga, M. A. G. M.; Akkerman, O. S.; Bickelhaupt, F., *Tetrahedron Lett.* **1986**, *27*, 6123.

7. Takai, K.; Kataoka, Y.; Okazoe, T.; Utimoto, K., *Tetrahedron Lett.* **1987**, *28*, 1443.

8. Takai, K.; Tezuka, M.; Kataoka, T.; Utimoto, K., *Synlett* **1989**, 27.

9. (a) Chojnowski, J.; Stanczyk, W., *J. Organomet. Chem.* **1974**, *73*, 41. (b) Chojnowski, J.; Stanczyk, W., *J. Organomet. Chem.* **1975**, *99*, 359.

Michael J. Taschner
The University of Akron, OH, USA

Di-*t*-butylchlorosilane

$$(t\text{-Bu})_2\text{SiClH}$$

[56310-18-0]   C$_8$H$_{19}$ClSi   (MW 178.775)

InChI = 1S/C8H19ClSi/c1-7(2,3)10(9)8(4,5)6/h10H,1-6H3

InChIKey = OGWXFZNXPZTBST-UHFFFAOYSA-N

Physical Data: bp 82–85 °C/45 mmHg; flash point 39 °C (closed cup); *d* 0.880 g cm^{-3}.

Form Supplied in: clear, colorless liquid.

Preparative Methods: can be prepared by the chlorination of di-*t*-butylsilane or by treatment of silicon tetrachloride with *t*-BuLi.[1,2]

Purification: distillation.

Handling, Storage, and Precautions: moisture sensitive; avoid strong oxidizing agents and strong bases; flammable; causes burns; avoid inhalation of vapor or mist; use in a fume hood.

Intramolecular Reducing Agent. Considerable study of alkoxysilanes as intramolecular hydrogen transfer agents has been undertaken.[3,4] The di-*t*-butyloxysilanes, prepared from di-*t*-butylchlorosilane and the requisite alcohol, are particularly effective and are generally stable to column chromatography.

Treatment of a γ-iodoallylic alcohol with NaH and *t*-Bu$_2$SiHCl afforded the corresponding di-*t*-butyloxysilanes which were exposed to UV irradiation in the presence of 10% hexabutylditin in the so-called unimolecular chain transfer (UMCT) reaction of silicon hydrides to afford the reduced alkene (eq 1).[5]

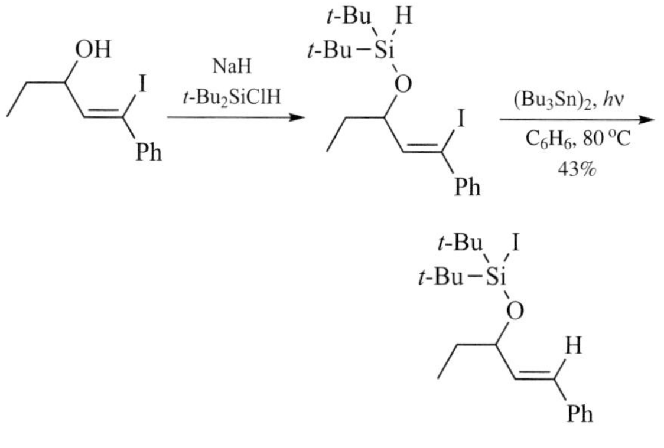

$$\text{(1)}$$

These alkoxysilanes are useful for making five-membered rings *via* a 5-*endo*-trigonal radical cyclization from a selenide precursor (eq 2).[6,7] It was observed that the (di-*t*-butyl)silyl group may be difficult to cleave in the absence of an adjacent oxygen functionality.

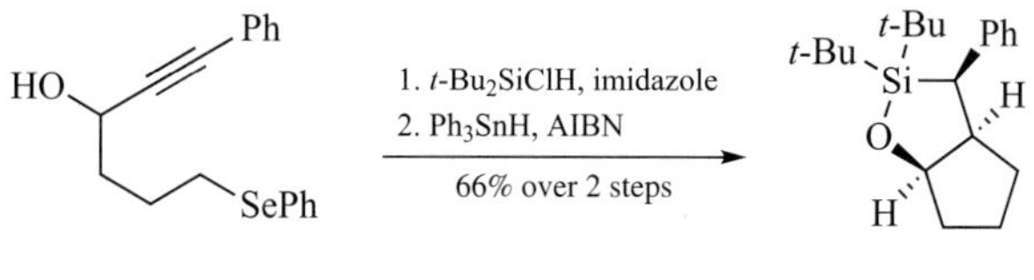

$$\text{(2)}$$

This approach has been further developed to encompass polycyclic compounds.[8] The reaction proceeds by a sequential 5-*exo*-digonal cyclization, followed by 1,5-hydrogen transfer, and then a final 5-*endo*-trigonal cyclization. This methodology has been applied to the synthesis of optically pure products when a single enantiomer of the chiral alcohol is used as the initial substrate (eq 3).[9] In addition, alkyl iodides can be used in place of the selenide precursors.[5,10]

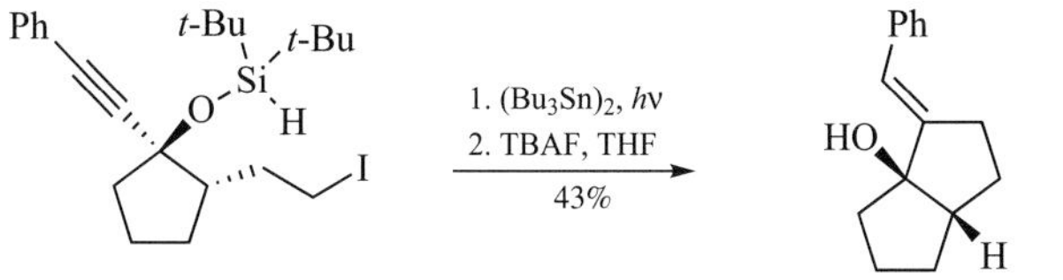

(3)

The (di-*t*-butyl)silyl group has also proved effective for the conversion of *cis*-2,5-disubstituted THF derivatives to the corresponding *trans*-2,5-disubstituted rings.[11] Activation of the hydroxyl group followed by a 1,2-hydride shift generates the oxonium ion at the C2 position. The di-*t*-butyloxysilane then delivers the hydride stereospecifically to form the *trans*-disubstituted product (eq 4). This motif is found in many natural products and a similar approach has been successfully applied to the synthesis of (+)-sylvaticin.[12]

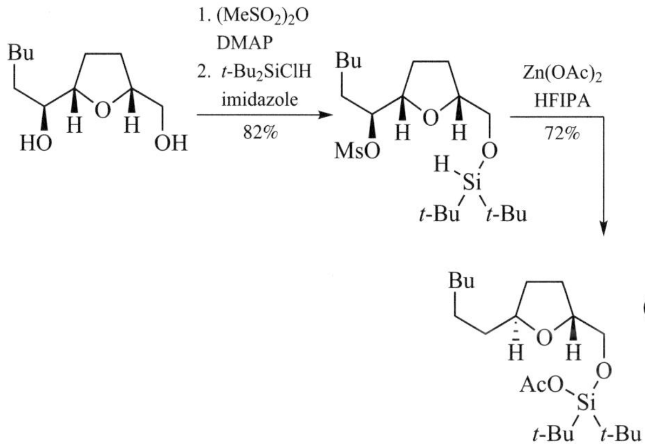

(4)

Selective Protecting Group. Silyl groups have been widely employed as protecting groups for alcohols. Where a choice exists, typical silylation conditions lead to selective protection of the less hindered hydroxyl group. In contrast, di-*t*-butylchlorosilane can be used for the one-pot silylation of the internal hydroxyl group of a 1,2-alkanediol.[13] The observed selectivity arises from the kinetically controlled ring cleavage of the cyclic silyl ether intermediate where lithium complexes preferentially at the less hindered oxygen. Selectivity was noted to increase when N,N,N',N'-tetramethylethylenediamine (TMEDA) was present in the reaction mixture (eq 5).

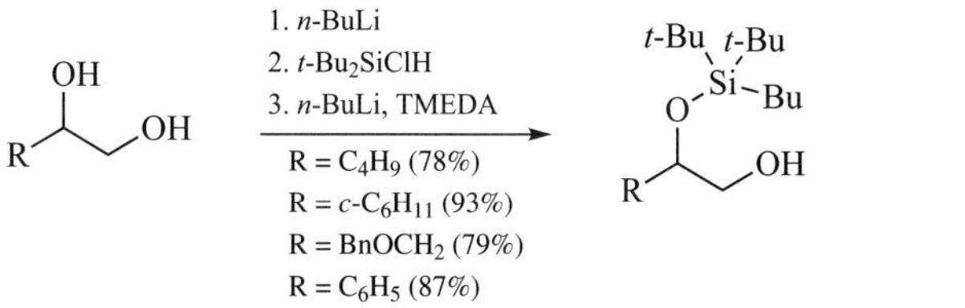

(5)

Silyl ether protecting groups are of interest in the synthesis of compounds containing vinyl ether groups, such as in the plasmalogens, where other protecting group strategies invariably lead to decomposition (eq 6).[14]

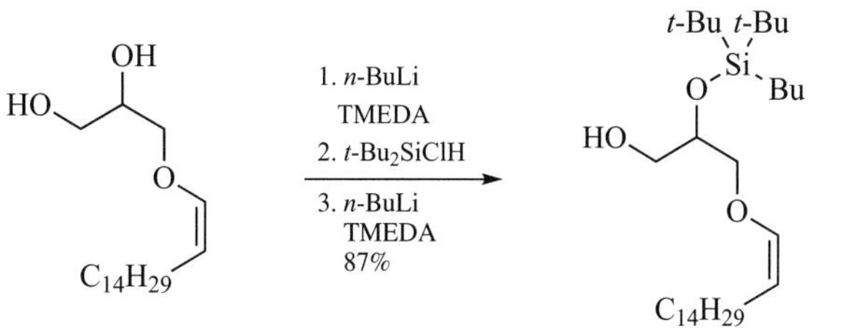

(6)

Protection of Diols. Di-*t*-buylchlorosilane has been used in the preparation of di-*t*-butylsilyl ditriflate, a highly effective reagent for the protection of a wide range of 1,2-, 1,3-, and 1,4-diols under mild conditions (eq 7).[15]

$$t\text{-Bu}_2\text{SiHCl} + 2\,\text{CF}_3\text{SO}_3\text{H} \longrightarrow t\text{-Bu}_2\text{Si(OSO}_2\text{CF}_3)_2 + \text{H}_2 + \text{HCl}$$

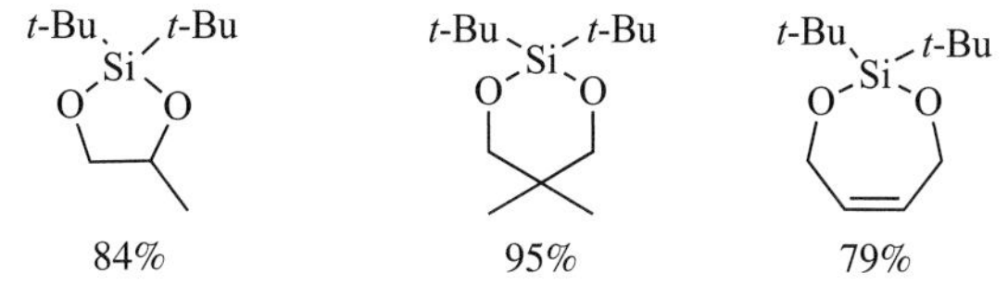

(7)

Si–H Insertion Reactions. Di-*t*-butylchlorosilane has found application in the rhodium-catalyzed Si–H insertion of carbenoids, formed by the decomposition of α-diazoesters (eq 8).[16] The chlorosilanes generated can be readily converted to alkoxysilanes by treatment with an alcohol and a base. In a study of a range of chlorosilanes by Landais, it was found that the bulky di-*t*-butylchlorosilane was the most reactive.[17,18]

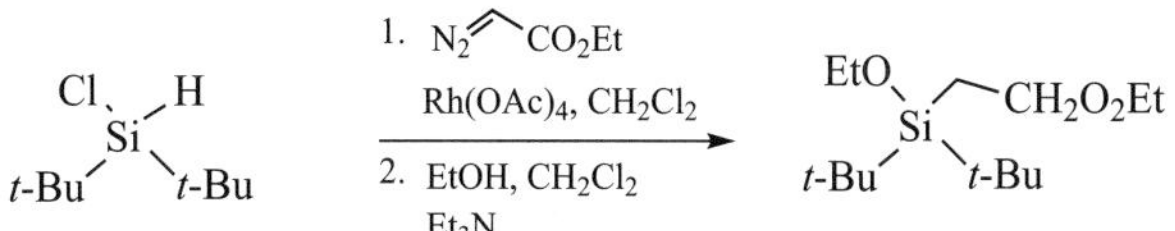

(8)

62%
(isolated with 20% of silanol:ethyl-2-(di-*tert*-butylhydroxysilyl)ethanoate)

Preparation of Alkynylsilanes. Alkynylsilanes are versatile synthetic intermediates which display interesting reactivity.[19,20] Such groups allow for the introduction of allene substituents at the C3 position of 2,3-epoxyalcohols.[21] The regioselectivity depends on the configuration of the epoxide moiety with *cis*-epoxides proceeding by a 5-*exo*-type cyclization while *trans*-epoxides undergo a 6-*endo* cyclization. The allenylsilane intermediates are readily converted to the corresponding allenes with TBAF in 1-methyl-2-pyrrolidine (NMP) (eq 9).

Hydrosilylation of Alkynes. Di-*t*-butylchlorosilane has been utilized in the preparation of (*E*)-di-*t*-butyl-(1-heptenyl)-silanol by hydrosilylation of 1-heptyne and hydrolysis of the intermediate chlorosilanes (eq 10).[22]

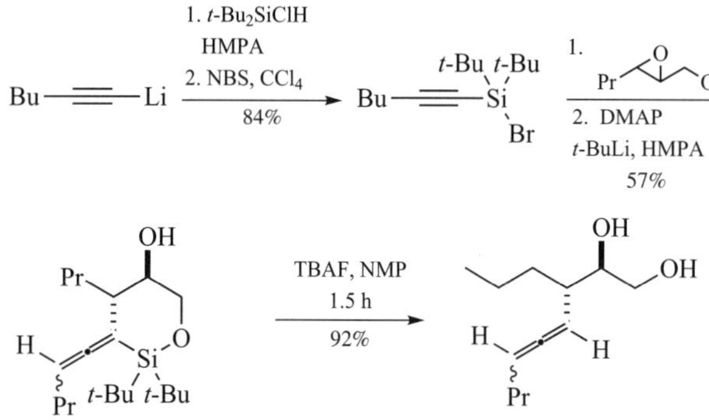

$$\text{Bu}\!\!-\!\!\equiv\!\!-\!\!\text{Li} \xrightarrow[\substack{\text{2. NBS, CCl}_4 \\ 84\%}]{\substack{1.\ t\text{-Bu}_2\text{SiClH} \\ \text{HMPA}}} \text{Bu}\!\!-\!\!\equiv\!\!-\!\!\underset{\underset{\text{Br}}{t\text{-Bu}}}{\overset{t\text{-Bu}}{\text{Si}}} \xrightarrow[\substack{\text{2. DMAP} \\ t\text{-BuLi, HMPA} \\ 57\%}]{1.}$$

(9)

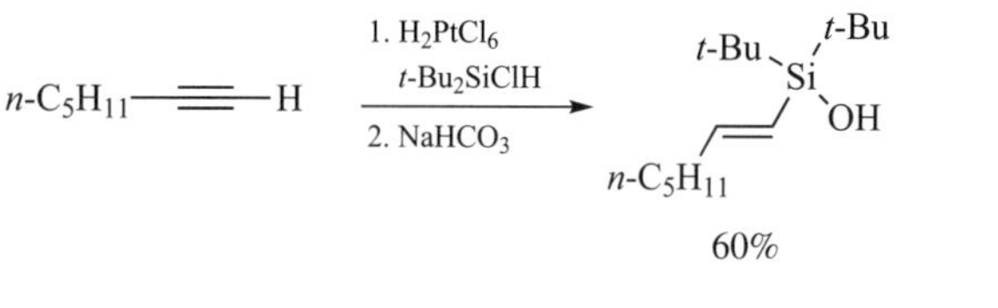

(10)

Internal Controlling Groups. Di-*t*-butylsilyl ethers are capable of controlling the regioselectivity and stereoselectivity of certain reactions. The ability of silyl groups to stabilize carbocations at the *β*-position has been exploited in the diastereoselective synthesis of cyclic polyols.[23] The di-*t*-butylsilyl ether group undergoes regio- and stereoselective migration in the presence of lithium and 4,4'-di-*t*-butylbiphenyl (DBB) followed by cyclization under basic conditions to form the allylsilane (eq 11). Epoxidation of the double bond occurs primarily on the face opposite to the bulky *t*-butyl group. SiO$_2$-induced ring opening leads to a *β*-silyl cationic intermediate and stereoselective introduction of the hydroxyl group *via* neighboring group participation (eq 11).

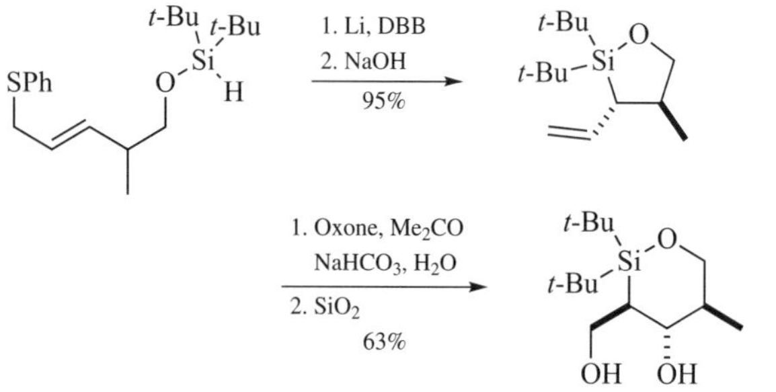

(11)

By linking reactive dienes and dienophiles with a silaketal tether, the course of the intramolecular Diels–Alder reaction can be controlled with a high degree of stereoselectivity and with regiochemistry opposite to that predicted by bond polarization models.[24] The cyclization proceeds in a 'head-to-tail' manner and the methyl group on the diene strongly favors an *endo* cyclization due to steric factors imposed in the transition state (eq 12). The di-*t*-butylsilyl group was found to be more thermally stable than related alkylsilyls.

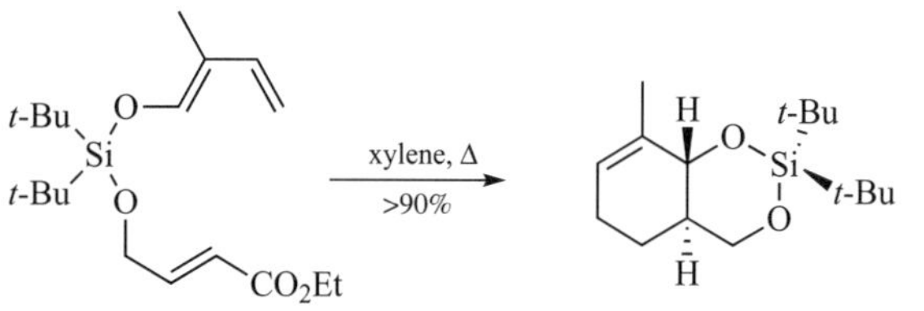

(12)

Applications in PET Imaging. Di-*t*-buylchlorosilane has been utilized in the synthesis of silicon-based building blocks for ^{18}F-radiolabeling of peptides for application in PET imaging.[25] Nucleophilic substitution of di-*t*-butylchlorosilane with {4-[2-(tetrahydro-2*H*-pyran-2-yloxy)ethyl]phenyl}lithium proceeded in 74% yield (eq 13), the product of which was further modified to afford the ^{18}F-radiolabeled compound.

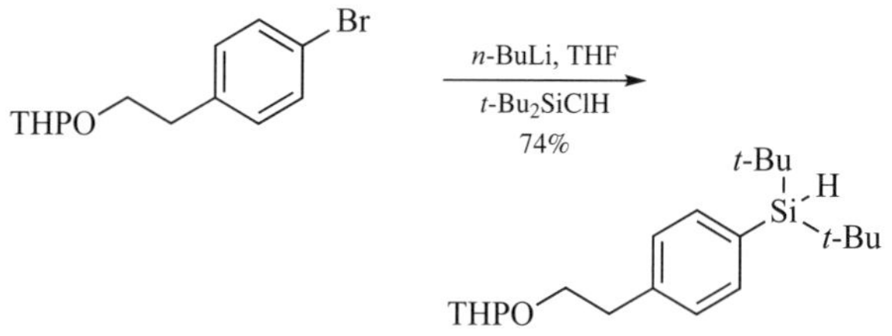

(13)

The lithiated derivative of di-*t*-butylchlorosilane has been utilized in the synthesis of 3'-silylated thymidine derivatives for application in PET imaging.[26]

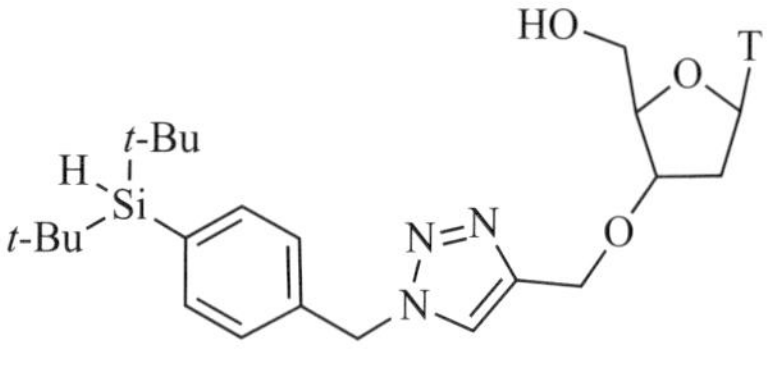

Formation of Organometallic Complexes. Di-*t*-buylchlorosilane has been used in the preparation of a range of organometallic complexes, many through an oxidative addition process to iridium, manganese, and molybdenum.[27–30]

It has also been utilized in the synthesis of aluminum amides (eq 14) and metal siloxanes containing Si–O–Sn linkages (eq 15).[31,32]

$$t\text{-Bu}_2\text{SiClH} \xrightarrow[\text{Et}_2\text{O}]{\text{NH}_3} t\text{-Bu}_2\text{Si(H)NH}_2 \xrightarrow[\substack{\text{hexane} \\ \text{reflux}}]{\text{Ali-Bu}_3}$$

$$1/2[i\text{-Bu}_2\text{AlN(H)Si(H)}t\text{-Bu}_2]_2$$

(14)

70%

$$t\text{-Bu}_2\text{SiClH} \ + \ (t\text{-Bu}_2\text{SnO})_3 \xrightarrow[\text{reflux}]{\text{CDCl}_3}$$

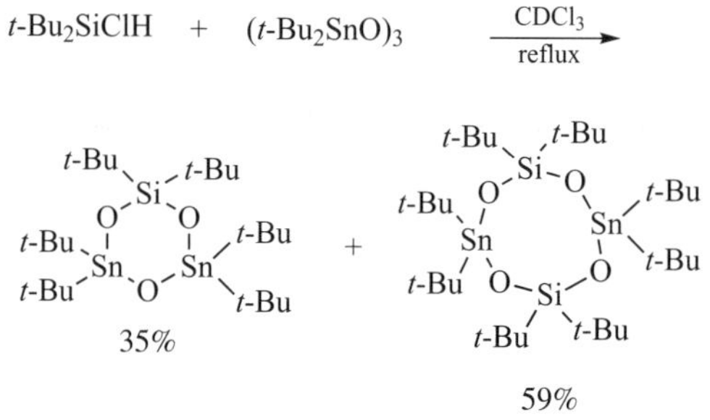

(15)

35%   59%

1. Dexheimer, E. M.; Spialter, L., *Tetrahedron Lett.* **1975**, 1771.
2. Doyle, M. P.; West, C. T., *J. Am. Chem. Soc.* **1975**, *97*, 3777.
3. Curran, D. P.; Xu, J.; Lazzarini, E., *J. Am. Chem. Soc.* **1995**, *117*, 6603.
4. Curran, D. P.; Xu, J.; Lazzarini, E., *J. Chem. Soc., Perkin Trans. 1* **1995**, 3049.
5. Martinez-Grau, A.; Curran, D. P., *Tetrahedron* **1997**, *53*, 5679.
6. Clive, D. L. J.; Cantin, M., *J. Chem. Soc., Chem. Commun.* **1995**, 319.
7. Clive, D. L.; Yang, W., *Chem. Commun. (Cambridge)* **1996**, 1605.

8. Clive, D. L. J.; Yang, W.; MacDonald, A. C.; Wang, Z.; Cantin, M., *J. Org. Chem.* **2001**, *66*, 1966.

9. Clive, D. L. J.; Ardelean, E.-S., *J. Org. Chem.* **2001**, *66*, 4841.

10. Martinez-Grau, A.; Curran, D. P., *J. Org. Chem.* **1995**, *60*, 8332.

11. Donohoe, T. J.; Williams, O.; Churchill, G. H., *Angew. Chem., Int. Ed.* **2008**, *47*, 2869.

12. Donohoe, T. J.; Harris, R. M.; Williams, O.; Hargaden, G. C.; Burrows, J.; Parker, J., *J. Am. Chem. Soc.* **2009**, *131*, 12854.

13. Tanino, K.; Shimizu, T.; Kuwahara, M.; Kuwajima, I., *J. Org. Chem.* **1998**, *63*, 2422.

14. Van, d. B. J.; Shin, J.; Thompson, D. H., *J. Org. Chem.* **2007**, *72*, 5005.

15. Corey, E. J.; Hopkins, P. B., *Tetrahedron Lett.* **1982**, *23*, 4871.

16. Andrey, O.; Landais, Y.; Planchenault, D., *Tetrahedron Lett.* **1993**, *34*, 2927.

17. Andrey, O.; Landais, Y.; Planchenault, D.; Weber, V., *Tetrahedron* **1995**, *51*, 12083.

18. Landais, Y.; Parra-Rapado, L.; Planchenault, D.; Weber, V., *Tetrahedron Lett.* **1997**, *38*, 229.

19. Molander, G. A.; Romero, J. A. C.; Corrette, C. P., *J. Organomet. Chem.* **2002**, *647*, 225.

20. Mukherjee, S.; Kontokosta, D.; Patil, A.; Rallapalli, S.; Lee, D., *J. Org. Chem.* **2009**, *74*, 9206.

21. Tanino, K.; Honda, Y.; Miyashita, M., *Tetrahedron Lett.* **2000**, *41*, 9281.

22. Denmark, S. E.; Neuville, L.; Christy, M. E. L.; Tymonko, S. A., *J. Org. Chem.* **2006**, *71*, 8500.

23. Tanino, K.; Yoshitani, N.; Moriyama, F.; Kuwajima, I., *J. Org. Chem.* **1997**, *62*, 4206.

24. Gillard, J. W.; Fortin, R.; Grimm, E. L.; Maillard, M.; Tjepkema, M.; Bernstein, M. A.; Glaser, R., *Tetrahedron Lett.* **1991**, *32*, 1145.

25. Mu, L.; Hohne, A.; Schubiger, P. A.; Ametamey, S. M.; Graham, K.; Cyr, J. E.; Dinkelborg, L.; Stellfeld, T.; Srinivasan, A.; Voigtmann, U.; Klar, U., *Angew. Chem., Int. Ed.* **2008**, *47*, 4922.

26. James, D.; Escudier, J.-M.; Amigues, E.; Schulz, J.; Vitry, C.; Bordenave, T.; Szlosek-Pinaud, M.; Fouquet, E., *Tetrahedron Lett.* **2010**, *51*, 1230.

27. Handwerker, H.; Beruda, H.; Kleine, M.; Zybill, C., *Organometallics* **1992**, *11*, 3542.

28. Koloski, T. S.; Pestana, D. C.; Carroll, P. J.; Berry, D. H., *Organometallics* **1994**, *13*, 489.

29. Zarate, E. A.; Kennedy, V. O.; McCune, J. A.; Simons, R. S.; Tessier, C. A., *Organometallics* **1995**, *14*, 1802.

30. Driess, M.; Pritzkow, H.; Rell, S.; Winkler, U., *Organometallics* **1996**, *15*, 1845.

31. Choquette, D. M.; Timm, M. J.; Hobbs, J. L.; Rahim, M. M.; Ahmed, K. J.; Planalp, R. P., *Organometallics* **1992**, *11*, 529.

32. Beckmann, J.; Mahieu, B.; Nigge, W.; Schollmeyer, D.; Schuermann, M.; Jurkschat, K., *Organometallics* **1998**, *17*, 5697.

Gráinne C. Hargaden
Dublin Institute of Technology, Dublin, Ireland

Timothy P. O'Sullivan
University College Cork, Cork, Ireland

Di-*t*-butyldichlorosilane

[18395-90-9]   C$_8$H$_{18}$Cl$_2$Si   (MW 213.22)

InChI = 1S/C8H18Cl2Si/c1-7(2,3)11(9,10)8(4,5)6/h1-6H3

InChIKey = PDYPRPVKBUOHDH-UHFFFAOYSA-N

(reagent for the protection of diols; used as a silylene precursor)

Physical Data: mp −15 °C; bp 190 °C/729 mmHg; *d* 1.009 g cm^{-3}.

Solubility: sol most common organic solvents.

Form Supplied in: liquid.

Preparative Methods: can be conveniently prepared by chlorination of di-*t*-butylsilane (CCl$_4$/PdCl$_2$ (cat), 85%)[1,2] but various other methods of preparation have been reported.[3–6]

Purification: distillation.

Handling, Storage, and Precautions: moisture sensitive; reacts with hydroxylic solvents, corrosive; lachrymator. Use in a fume hood.

Original Commentary

Snorri T. Sigurdsson & Paul B. Hopkins
University of Washington, Seattle, WA, USA

Protection of Alcohols. The presence of the bulky *t*-butyl groups in di-*t*-butyldichlorosilane has been found to increase the Si–C bond lengths slightly and to widen the CSiC bond angles by 11.1° relative to dichlorodimethylsilane as determined by electron diffraction and molecular mechanics calculations.[7]

The di-*t*-butylsilylene protecting group for diols was introduced by Trost and Caldwell[4] and used in a total synthesis of deoxypillaromycinone.[5] It is introduced by the reaction of di-*t*-butyldichlorosilane with a 1,2- or 1,3-diol in acetonitrile in the presence of triethylamine and 1-hydroxybenzotriazole (HOBt) at 45–90 °C (eq 1). For a related, highly reactive reagent see di-*t*-butylsilyl Bis(trifluoromethanesulfonate).

$$ \text{(1)} $$

When at least one of the hydroxy groups is phenolic or primary, reactions proceed smoothly at 25–65 °C, but when both hydroxy groups are secondary, more forcing conditions (95 °C, sealed tube) are required. In all cases, yields range from 64% to 85%. No examples have been reported with tertiary alcohols. pyridinium poly(hydrogen fluoride) is used for deprotection and under those conditions a *t*-butyldimethylsilyl ether and a β-hydroxy ketone are unaffected (eq 2).

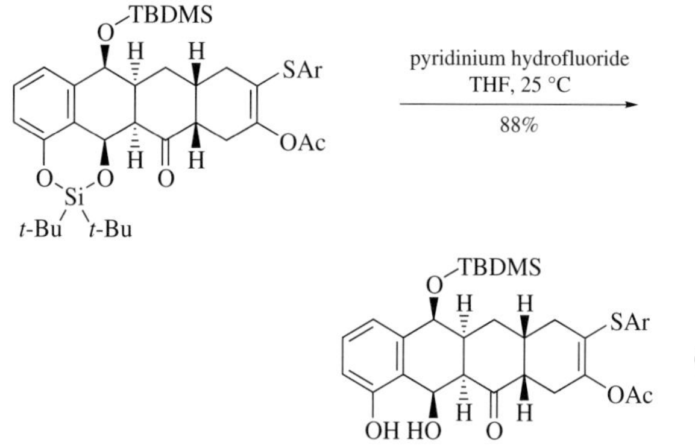

$$(2)$$

This protecting group has seen limited use but has been applied in ribonucleoside chemistry.[8–10] The reagent reacts slowly with nucleosides in the presence of imidazole/DMF but formation of the more reactive di-*t*-butylsilyl dinitrate in situ, followed by addition of cytidine, results in a protected 3′,5′-silylene derivative in excellent yield (90%) (eq 3). The protecting group can be removed conveniently with tributylamine hydrofluoride.[11]

A one-pot procedure has been reported for the selective protection of a secondary alcohol over a primary alcohol in a 1,3-diol in 2′-deoxynucleosides (eq 4).[12,13]

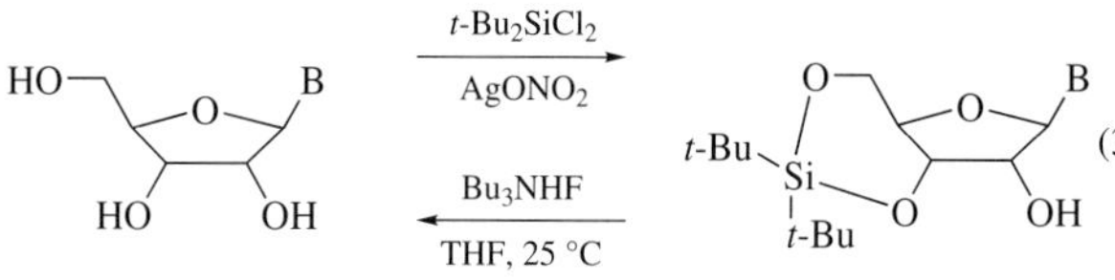

$$(3)$$

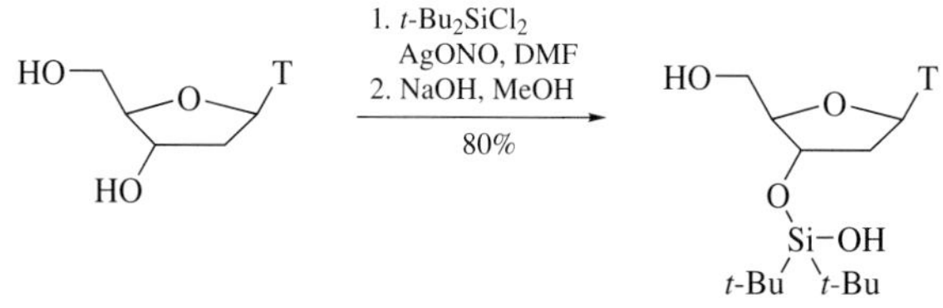

$$(4)$$

Derivatization of Diols and Hydroxy Acids. Di-*t*-butyldichlorosilane has been used to derivatize α-hydroxy acids, β-hydroxy acids, alkylsalicyclic acids, anthranilic acid, catechols, and 1,2- and 1,3-diols for analysis by gas chromatography–electron impact mass spectrometry (eq 5).[14,15] These derivatives are useful for separation. The major fragmentation is that of Si–C bonds. The 1,2-diol in the antibiotic sorangicin has also been derivatized with di-*t*-butyldichlorosilane.[16]

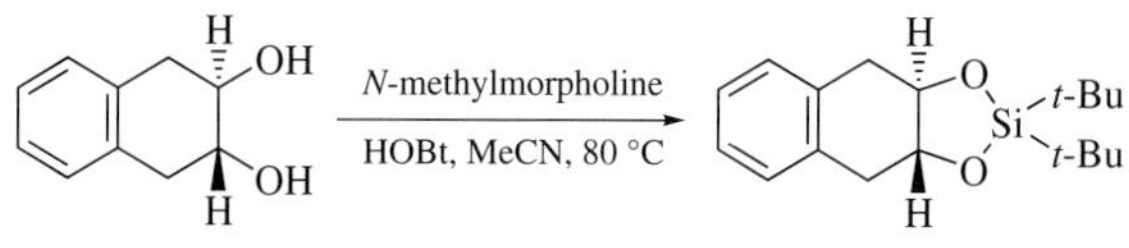

$$(5)$$

m/z 304 (M^+), 247 ($M^+ -$ Bu), 205 (B, 247 $-$ C$_3$H$_6$)

Silylene Precursor. Di-*t*-butyldichlorosilane can be reduced using lithium/THF to give a putative silylene derivative that will

react with double bonds to give 1,1-silirane derivatives[17–19] or react with triethylsilane to give an Si–H insertion product (eq 6).[17]

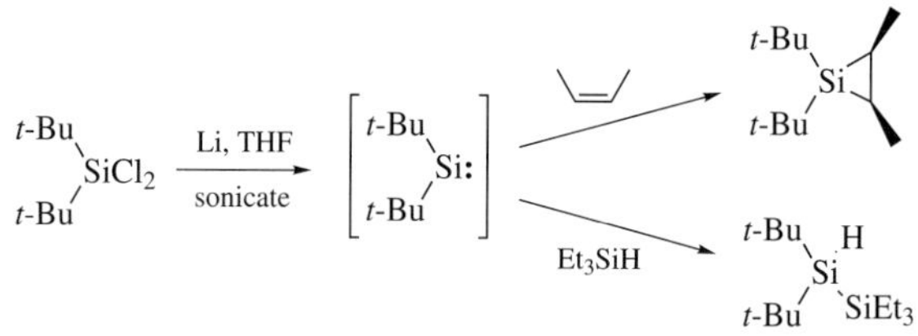

$$(6)$$

Reaction of di-*t*-butyldichlorosilane with lithium naphthalenide in DME gives compound (**1**) which upon irradiation gives, in addition to naphthalene, tetra-*t*-butyldisilene that subsequently reacts with 2,3-dimethylbutadiene to give a Diels–Alder adduct along with a product arising from an ene reaction (eq 7).[20]

Reduction of di-*t*-butyldichlorosilane with lithium in THF at 0 °C in the presence of an excess of dichlorodimethylsilane gives compound (**2**).[21] Reaction of di-*t*-butyldichlorosilane with lithium and 1,4-diaza-1,3-butadienes gives 1,3-diaza-2-sila-4-cyclopentenes (eq 8).[22]

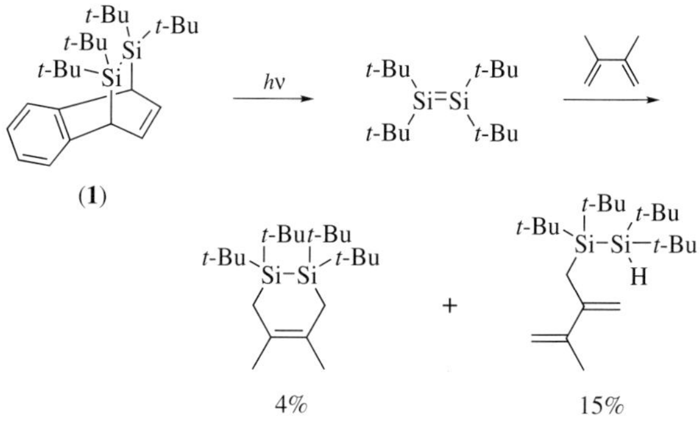

$$(7)$$

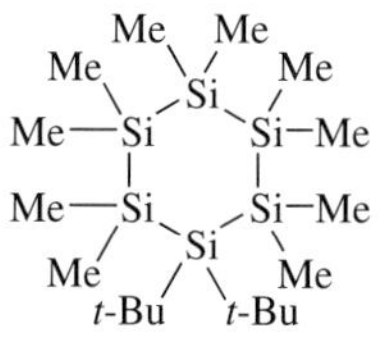

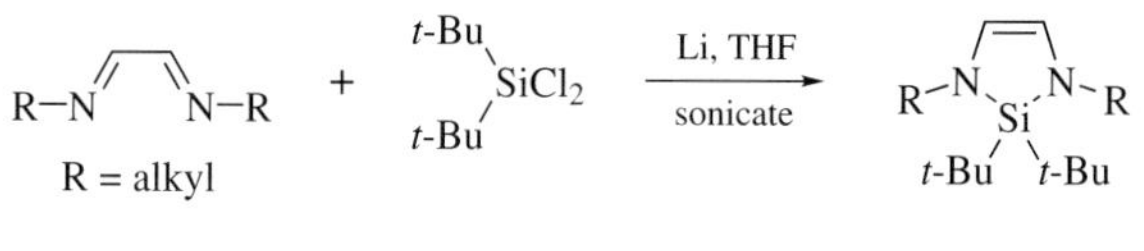

$$(8)$$

Other Substitution Reactions. The chlorine atoms in di-*t*-butyldichlorosilane can be replaced with various nucleophiles. Di-*t*-butyldifluorosilane has been prepared by using SbF$_3$,[23,24] ZnF$_2$,[25] or (NH$_4$)$_2$SiF$_6$.[26,27] Reaction of di-*t*-butyldichlorosilane with lithium aluminum hydride gives di-*t*-butylsilane[28] and reaction with sodium azide gives di-*t*-butyldiazidosilane; upon irradiation this gives a putative di-*t*-butylsilylene which undergoes various reactions depending on the conditions.[29] Reactions of di-*t*-butyldichlorosilane with LiPHMe[30] and LiPH$_2$[31] yield (**3**) and (**4**), respectively.

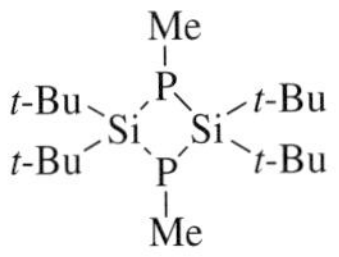

(3)

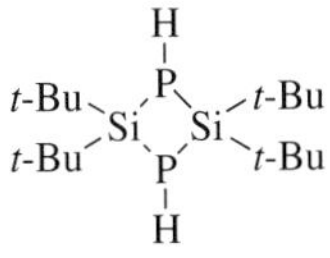

(4)

Treatment of di-*t*-butyldichlorosilane with trimethylsilyl-lithium gives di-(*t*-butyl)bis(trimethylsilyl)silane.[32]

First Update

Jitendra Belani

Infinity Pharmaceuticals, Cambridge, MA, USA

Applications in Silylene Transfer Reactions. Tremendous progress has been made in understanding silylene transfer reactions in the past 10 years. Significant contributions have been made by Woerpel and coworkers who have not only extended synthesis of silacyclopropanes from functionalized, chiral alkenes[33] but also provided new insights into mechanistic details of silylene transfer reactions.[34,35] The authors investigated influence of oxygen functionality and steric interactions on the diastereoselectivity of silacyclopropanation. Silacyclopropanation of cyclohexene by reduction of *t*-Bu$_2$SiCl$_2$ provided **5** in 71% yield and has been done on 50 g scale (eq 9). The silacyclopropane **5** can itself be used as a silylene transfer agent.[33] Silacyclopropanation of 3-methyl-1-cyclohexene by reduction of *t*-Bu$_2$SiCl$_2$ gave silacycle **6** with 92:8 diastereoselectivity (eq 9). Increasing the size of the allylic substituent from methyl to isopropyl improved the diastereoselectivity to 96:4. Silacyclopropanations of homoallylic ethers (**8a** and **8b**) afforded *trans*-silacyclopropanes in excellent diastereoselectivities that were attributed to approach of the lithium silylenoid intermediate **9** from the less hindered side of the olefin (eq 10). The ether functionality in such substrates does not direct silacyclopropanation. In contrast to substituted cyclohexenes that gave mixtures of silacycles under thermal transfer conditions, quantitative silylene transfer occurred in case of functionalized cyclopentenes with high diastereoselectivity (eq 11).[33]

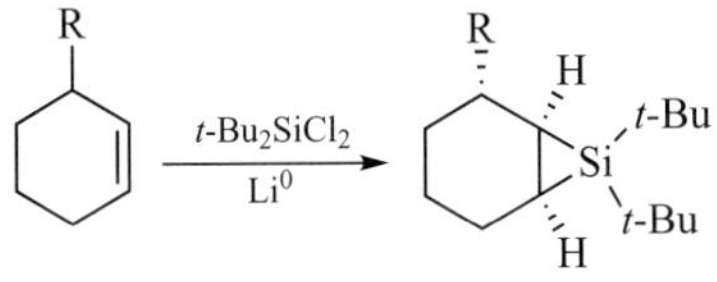

(9)

Diastereoselective silacyclopropanation of 5-*endo*-substituted norbornene **13** and 1,1-disubstituted alkenes **14** and **15** by lithium reduction of *t*-Bu$_2$SiCl$_2$ was also reported (Table 1). In the case of norbornene **13**, the selectivity arises due to *exo*-approach of the silylene intermediate to the olefin. However, in case of 1,1-disubstituted alkenes, minimization of 1,2-allylic strain accounts for the observed acyclic stereocontrol. Synthetic utility of such diastereoselective silacyclopropanation was also demonstrated by

elaboration of the resulting silanes to afford complex polyoxygenated structures (eq 12).[33]

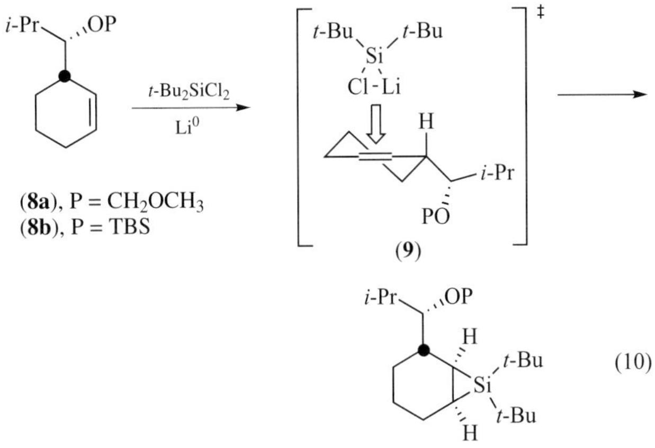

(**8a**), P = CH$_2$OCH$_3$
(**8b**), P = TBS

(9)

(10)

(**10a**), P = 54%, 96:4 ds
(**10b**), P = 46%, 92:8 ds

97%
(96:4 ds)

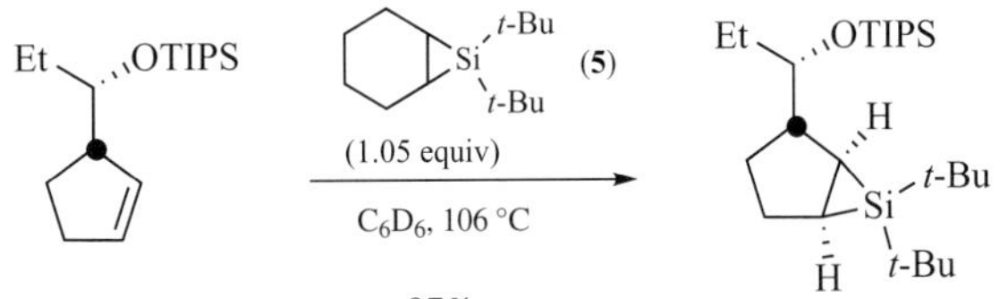

(11)

(11)

(12)

Table 1 Diastereoselective silacyclopropanations of norbornenes and 1,1-disubstituted alkenes using lithium reduction of *t*-Bu$_2$SiCl$_2$

Alkene	ds	Silacycle	Yield (%)
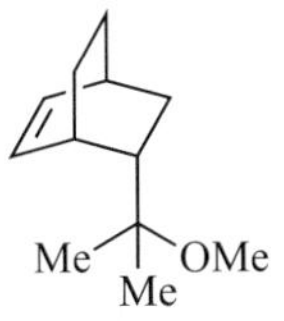 (13)	>99:1	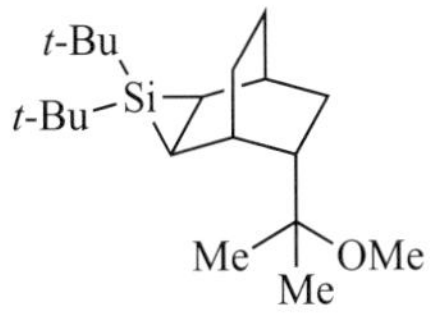 (16)	67
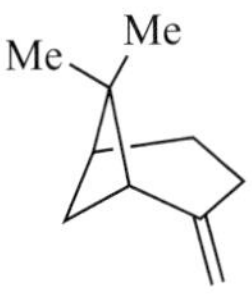 (14)	98:2	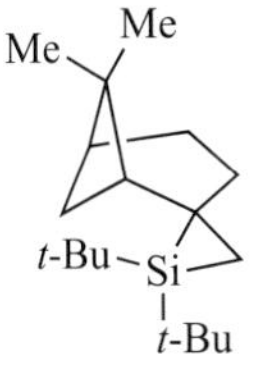 (17)	71
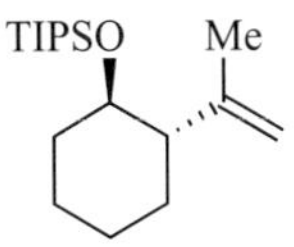 (15)	97:3	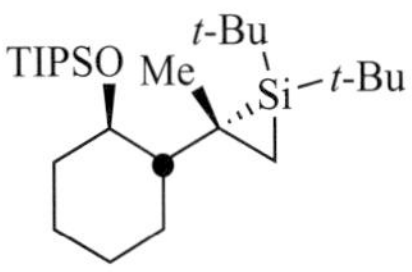 (18)	61

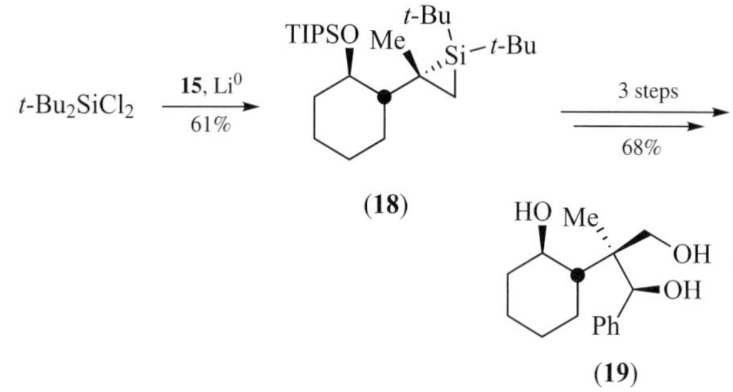

(18)

(19)

Howard and Woerpel have also reported silylene transfer to α-ketoesters using dimethylsilirane **21** derived from *t*-Bu$_2$SiCl$_2$ and isobutene.[36] This results in direct formation of silalactones via 6π-electrocyclization and subsequent Ireland–Claisen rearrangement. The silalactones underwent hydrolysis upon treatment with HF to provide the desired α-hydroxy acids (eq 13).[36]

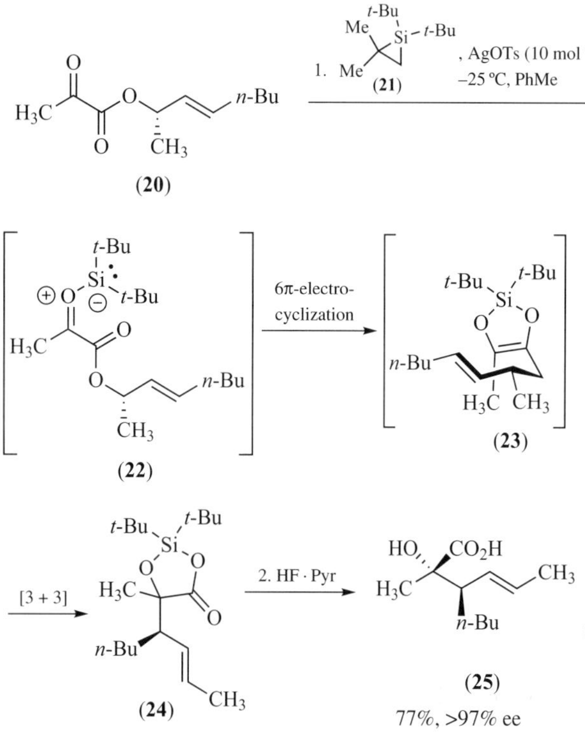

(20)

(22)   (23)

(24)

(25)

77%, >97% ee

Protection of Alcohols. Di-*t*-butyldichlorosilane has been used repeatedly for protection of diols to provide conformationally constrained ring systems, thereby assisting in stereoselective synthesis of fragments. Ito and coworkers reported enhanced selectivity for β-(1,2-*cis*)-glycosylation of the furanose ring using 3,5-*O*-tetraisopropyldisiloxanylidene (TIPDS)- and di-*t*-butylsilyl-protected thioglycosides as donors. Compared to protection using di-*t*-butylsilyl group, the conformation restraint introduced by the eight-membered ring 3,5-*O*-protection resulted in an enhanced β-selectivity of 20:1 (eqs 14 and 15) . This enhanced selectivity was then used to demonstrate synthesis of heptaarabinofuranoside.[37]

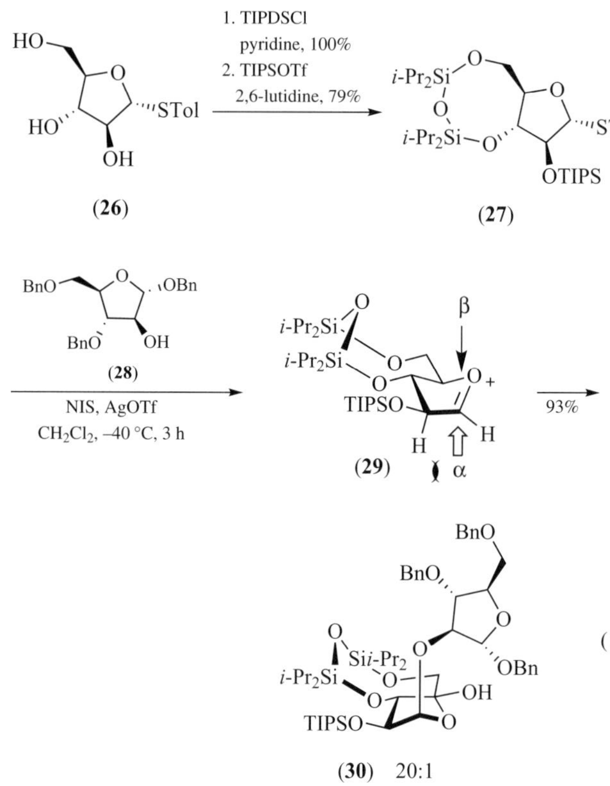

(26)   (27)

(28)

NIS, AgOTf
CH$_2$Cl$_2$, –40 °C, 3 h

(29)   93%

(30)   20:1   (14)

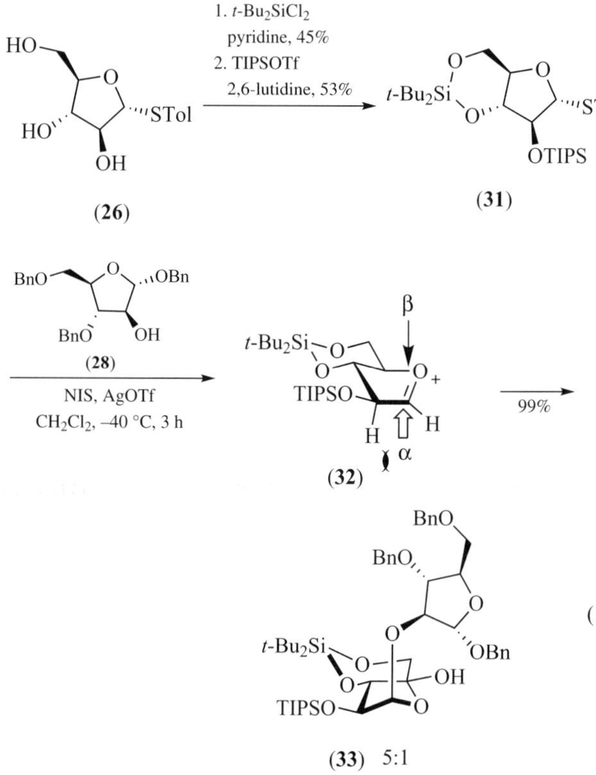

(26)

1. *t*-Bu$_2$SiCl$_2$
pyridine, 45%
2. TIPSOTf
2,6-lutidine, 53%

(31)

(28)

NIS, AgOTf
CH$_2$Cl$_2$, –40 °C, 3 h

(32)   99%

(33)   5:1   (15)

Mishra and Misra have employed *t*-Bu$_2$SiCl$_2$ to prepare di(*t*-butyl)[(pyren-1-yl)methoxy]silyl chloride (TBMPSCl) that was subsequently used to protect 5′-hydroxy group of deoxyribonucleosides. The resultant 5′-*O*-protected nucleosides were converted into 3′-[(2-cyanoethyl)-*N*,*N*-diisopropyl]phosphoramidites and subsequently 2-mer, 4-mer, 13-mer, and 26-mer sequences

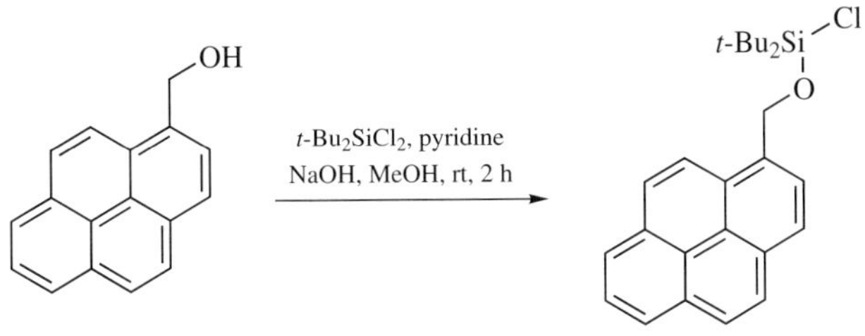

(**34**) (**35**)
TBMPSCl

(16)

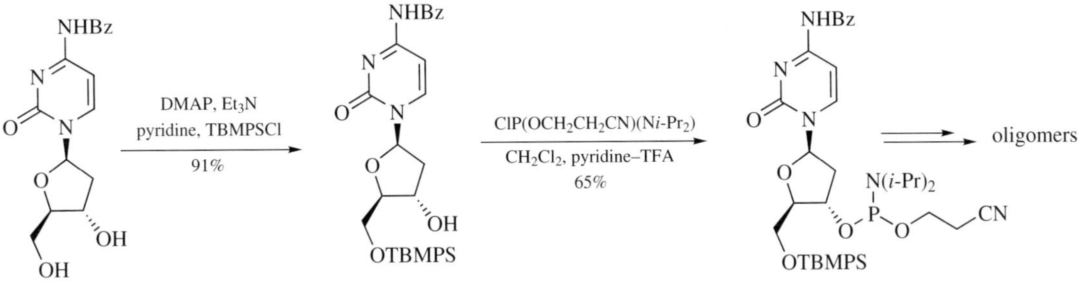

(**36**) (**37**) (**38**)

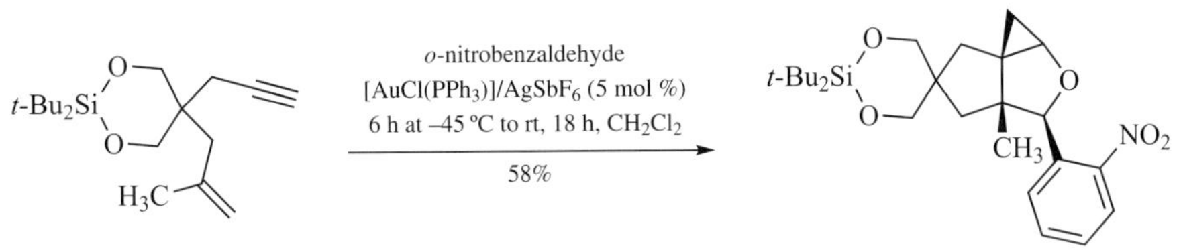

(17)

(**39**) (**40**)

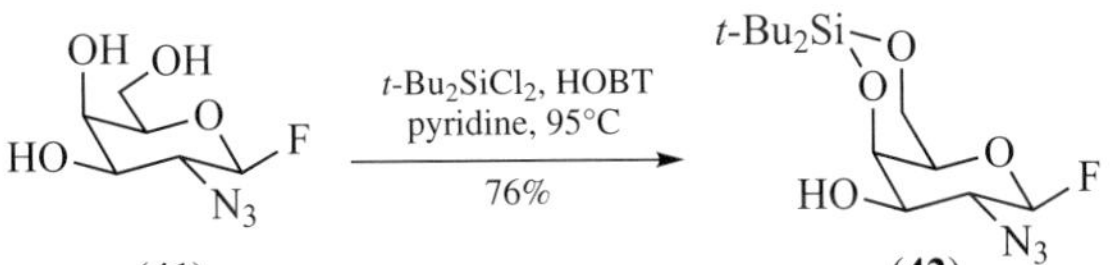

(18)

(**41**) (**42**)

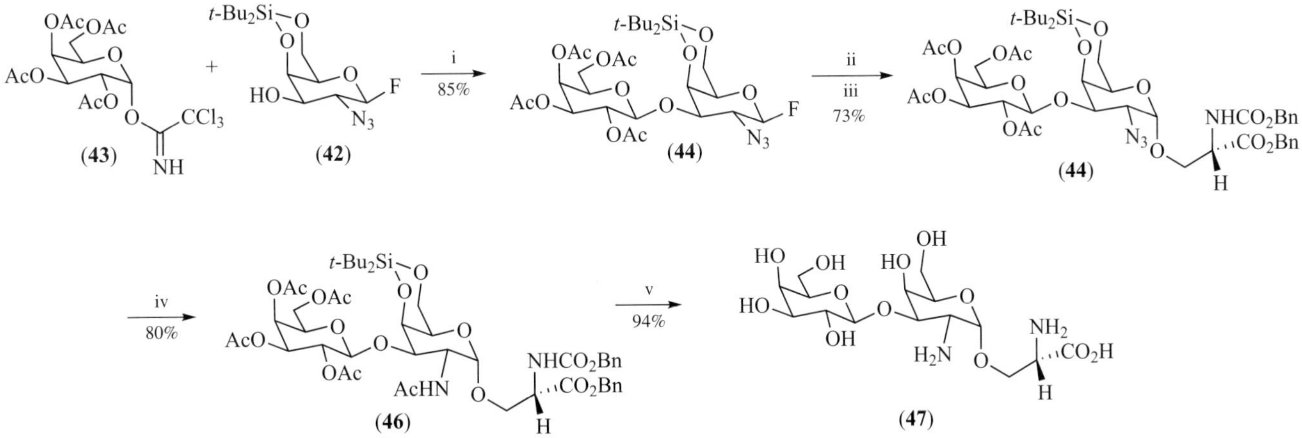

(**43**) (**42**) (**44**) (**44**) (**46**) (**47**)

(i) TMSOTf, CH_2Cl_2, –20 °C, 3 h; (ii) Cp_2ZrCl_2, $AgClO_4$, CH_2Cl_2, 25 °C, 2 h
(iii) *N*-benzyloxycarbonyl L-serine benzyl ester, Cp_2ZrCl_2, $AgClO_4$, CH_2Cl_2, 25 °C, 2 h; (iv) AcSH, pyridine, 16 h
(v) **a.** Et_3N–3HF, THF, 0–25 °C, 4 h, **b.** 0.1 N NaOH aq, MeOH, 0 °C, 20 min, **c.** Pd-C, MeOH, H_2, 1.5 h

(48) → DMF, K$_2$CO$_3$, 63% → **(50)**   reagent **(49)**

1. (Boc)$_2$O, THF, reflux
2. TBAF, THF, rt
91%

(51) →
1. DHP, PPTS, DCM, rt
2. MsCl, Et$_3$N, DCM, rt
58% → **(52)**

1. [^{18}F]HF/K$_2$CO$_3$/K222, DMSO
2. aq HCl →

^{18}F ... OH   **(53)**   (19)

were synthesized, purified, and characterized using the fluorescent properties of these oligomers (eq 16).[38]

In a separate communication, Helmchen and coworkers showed that di-*t*-butylsilyl-protected 1,3-diol-enyne **39** underwent a diastereoselective gold(I)-catalyzed cycloaddition of *o*-nitobenzaldehyde yielding 2-oxabicyclo-[3.1.0]hexane **40** in 58% yield (eq 17).[39]

Nishimura and coworkers reported use of di-*t*-butyldichlorosilane to protect diols in an effort to find an alternative protecting group in oligosaccharide synthesis. The resultant 4,6-cyclic di-*t*-butylsilylenediyl ether was easily formed and was found to be a versatile group that allowed regioselective manipulations of hydroxyl groups in sugar residues. The cyclic silyl ether was selectively removed by treating with fluoride under mild conditions. Accordingly, 2-azido-2-deoxy-*β*-D-galactopyranosyl fluoride **41** was efficiently converted to glycoconjugate **47** (eq 18).[40]

Kung and coworkers have described preparation of ^{18}F-labeled stilbenes as PET imaging agents for detecting *β*-amyloid plaques in the brain. The authors used di-*t*-butyldichlorosilane to protect 2-hydroxypropyl-1,3-diol that was subsequently converted to the bromide **49**. Alkylation of phenol **48** with the bromide **49** provided the ether **50** that was transformed into the desired ^{18}F-labeled stilbene (eq 19).[41]

1. Watanabe, H.; Ohkawa, T.; Muraoka, T.; Nagai, Y., *Chem. Lett.* **1981**, 1321.

2. Watanabe, H.; Muraoka, T.; Kageyama, M.; Yoshizumi, K.; Nagai, Y., *Organometallics* **1984**, *3*, 141.

3. Doyle, M. P.; West, C. T., *J. Am. Chem. Soc.* **1975**, *97*, 3777.

4. Trost, B. M.; Caldwell, C. G., *Tetrahedron Lett.* **1981**, *22*, 4999.

5. Trost, B. M.; Caldwell, C. G.; Murayama, E.; Heissler, D., *J. Org. Chem.* **1983**, *48*, 3252.

6. Graalmann, O.; Klingebiel, U., *J. Organomet. Chem.* **1984**, *275*, C1.

7. Forsyth, G. A.; Rankin, D. W. H., *J. Mol. Struct.* **1990**, *222*, 467.

8. Furusawa, K.; Ueno, K.; Katsura, T., *Chem. Lett.* **1990**, 97.

9. Furusawa, K.; Katsura, T., *Tetrahedron Lett.* **1985**, *26*, 887.

10. Furusawa, K.; Katsura, T.; Sakai, T.; Tsuda, K., *Nucleic Acids Symp. Ser.* **1984**, *15*, 41.

11. Furusawa, K., *Chem. Lett.* **1989**, 509.

12. Furusawa, K.; Sakai, T.; Tsuda, K., *Jpn. Kokai Tokkyo Koho* **1988**, 913.

13. Furusawa, K.; Sakai, T.; Tsuda, K., *Kenkyu Hokou - Sen'i Kobunshi Zairyo Kenkyusho* **1988**, *158*, 121.

14. Brooks, C. J. W.; Cole, W. J.; Barrett, G. M., *J. Chromatogr.* **1984**, *315*, 119.

15. Brooks, C. J. W.; Cole, W. J., *Analyst* **1985**, *110*, 587.

16. Hoefle, G.; Jansen, R.; Schummer, D. Ger. Offen. 3 930 950, 1991.

17. Boudjouk, P.; Samaraweera, U.; Sooriyakumaran, R.; Chrusciel, J.; Anderson, K. R., *Angew. Chem., Int. Ed. Engl.* **1988**, *27*, 1355.

18. Weidenbruch, M.; Lesch, A.; Marsmann, H., *J. Organomet. Chem.* **1990**, *385*, C47.

19. Boudjouk, P.; Black, E.; Kumarathasan, R., *Organometallics* **1991**, *10*, 2095.

20. Masamune, S.; Murakami, S.; Tobita, H., *Organometallics* **1983**, *2*, 1464.

21. Helmer, B. J.; West, R., *J. Organomet. Chem.* **1982**, *236*, 21.

22. Weidenbruch, M.; Lesch, A.; Peters, K., *J. Organomet. Chem.* **1991**, *407*, 31.

23. Weidenbruch, M.; Peter, W., *Angew. Chem., Int. Ed. Engl.* **1975**, *14*, 642.

24. Weidenbruch, M.; Pesel, H.; Peter, W; Steichen, R., *J. Organomet. Chem.* **1977**, *141*, 9.

25. Rempfer, B.; Oberhammer, H.; Auner, N., *J. Am. Chem. Soc.* **1986**, *108*, 3893.

26. Damrauer, R.; Simon, R. A., *Organometallics* **1988**, *7*, 1161.

27. Auner, N., *Z. Anorg. Allg. Chem.* **1988**, *558*, 87.

28. Triplett, K.; Curtis, M. D., *J. Organomet. Chem.* **1976**, *107*, 23.

29. Welsh, K. M.; Michl, J.; West, R., *J. Am. Chem. Soc.* **1988**, *110*, 6689.

30. Fritz, G.; Uhlmann, R., *Z. Anorg. Allg. Chem.* **1978**, *442*, 95.

31. Fritz, G.; Biastoch, R., *Z. Anorg. Allg. Chem.* **1986**, *535*, 63.

32. Becker, G.; Hartmann, H. M.; Muench, A.; Riffel, H., *Z. Anorg. Allg. Chem.* **1985**, *530*, 29.

33. Driver, T. G.; Franz, A. K.; Woerpel, K. A., *J. Am. Chem. Soc.* **2002**, *124*, 6524.

34. Driver, T. G.; Woerpel, K. A., *J. Am. Chem. Soc.* **2004**, *126*, 9993.

35. Driver, T. G.; Woerpel, K. A., *J. Am. Chem. Soc.* **2003**, *125*, 10659.

36. Howard, B. E.; Woerpel, K. A., *Org. Lett.* **2007**, *9*, 4651.

37. Ishiwata, A.; Akao, H.; Ito, Y., *Org. Lett.* **2006**, *8*, 5525.

38. Mishra, R.; Misra, K., *Chem. Lett.* **2007**, *36*, 768.

39. Schelwies, M.; Moser, M.; Dempwolff, A. L; Rominger, F.; Helmchen, G., *Chem. Eur. J.* **2009**, *15*, 10888.

40. Kumagai, D.; Miyazaki, M.; Nishimura, S.-I., *Tetrahedron Lett.* **2001**, *42*, 1953.

41. Zhang, W.; Oya, S.; Kung, M.; Hou, C.; Maier, D. L.; Kung, H. F., *J. Med. Chem.* **2005**, *48*, 5980.

trans-1,1-Di-*t*-butyl-2,3-dimethylsilirane

[116888-87-0]   C$_{12}$H$_{26}$Si   (MW 198.21)

InChI = 1S/C12H26Si/c1-9-10(2)13(9,11(3,4)5)12(6,7)8/
h9-10H,1-8H3/t9-,10-/m1/s1

InChIKey = PZPSUANGKSNRSG-NXEZZACHSA-N

(substituted silirane that participates in thermal and metal-catalyzed insertion reactions)

Physical Data: liquid, bp 45 °C (2 mm Hg).

Solubility: soluble in common organic solvents (i.e., benzene, toluene, CH$_2$Cl$_2$).

Analysis of Reagent Purity: ^{1}H-NMR, gas chromatography.

Preparative Methods: prepared by silacyclopropanation of (*E*)-2-butene.

Purity: purification was accomplished by distillation at reduced pressure.

Handling, Storage, and Precaution: sensitive to atmospheric oxidation and moisture. Operations should be performed in rigorously dried glassware using degassed solvents.

Introduction. Silacyclopropanes are strained ring systems that undergo several diverse stereoselective carbon-carbon bond-forming reactions, typically involving one- or two-atom insertions of substrates such as aldehydes, formamides, isocyanates, and alkynes (eq 1). Early seminal studies of Seyferth[1] and Ando[2] established the ability of hexamethylsilirane to undergo two-atom insertion reactions. Recently, Woerpel extended this reactivity to several new stereoselective carbon-carbon bond-forming processes, greatly expanding the utility of these reactions.[3] Although thermolysis or photolysis of silacyclopropane generates free silylene, which can react with alkynes and alkenes,[4] most insertion processes do not involve free silylene. Rather, diradical intermediates have been postulated for the thermal insertions.[1g] Further

oxidative transformations of the insertion products provide functionalized products useful in synthetic applications.

$$(1)$$

Synthesis. Silacyclopropanes such as *trans*-1,1-di-*t*-butyl-2,3-dimethylsilirane (**1**) are most accessible via the reaction of silylene[5] or silylenoids[6] with (*E*)-2-butene. Accordingly, exposure of di-*t*-butyldichlorosilane to lithium metal in the presence of (*E*)-2-butene provides silirane **1** (eq 2).[6] However, generating silylenoid intermediates requires elevated temperatures and highly reducing conditions, severely limiting substrate generality. Because free silylene can also be generated thermally or photochemically from silacyclopropanes, silylene can be transferred thermally from a silacyclopropane to an alkene. In particular, silver triflate-catalyzed silylene transfer using silacyclopropane **2** proceeds at low temperatures and short reaction times with mono- and disubstituted alkenes and exhibits broad functional group compatibility.[7]

$$(2)$$

Aldehyde Insertions. Seyferth reported the first example of the thermal insertion of aldehydes into the C–Si bond of hexamethyldisilirane affording oxasilacyclopentane (eq 3).[1f] To study the scope and mechanism of this process, Woerpel made recourse to the di-*t*-butylsilacyclopropanes, developed by Boudjouk,[5,6,8] because silylenes bearing bulky substituents undergo smooth additions to olefins, in contrast to siliranes having small substituents, and can be obtained as single diastereomers. Accordingly, Woerpel observed that silirane **1** reacts with benzaldehyde at 100 °C providing a mixture of four diastereomers (**3a–3d**) in a ratio 75:7:8:10 (**3a:3b:3c:3d**) in 52% yield (eq 4).[9] Insertion occurs at the more substituted C–Si bond of the silirane when unsymmetrical silacyclopropanes are used (eq 5). The addition of KO*t*-Bu/18-C-6 lowers the required temperature to 22 °C and reverses the stereoselectivity, providing **3a–3d** in the ratio 3:1:13:83. The authors propose a mechanism involving alkoxide attack at the silicon atom affording a pentacoordinate siliconate species to explain the rate increase. Although this insertion process was limited to aromatic aldehydes, the use of a copper(II) catalyst permits α,β-unsaturated aldehydes to undergo insertion at the silirane under mild conditions (eq 5). In this case, the authors suggest that the mechanism involves an initial transmetallation of silicon to copper.[10] Zinc-catalyzed insertions occur even with aliphatic aldehydes, similarly affording the product of insertion at the least hindered C–Si bond.[11]

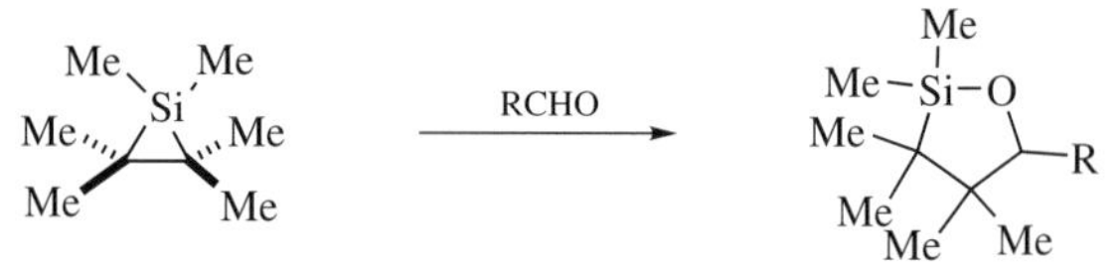

(3)

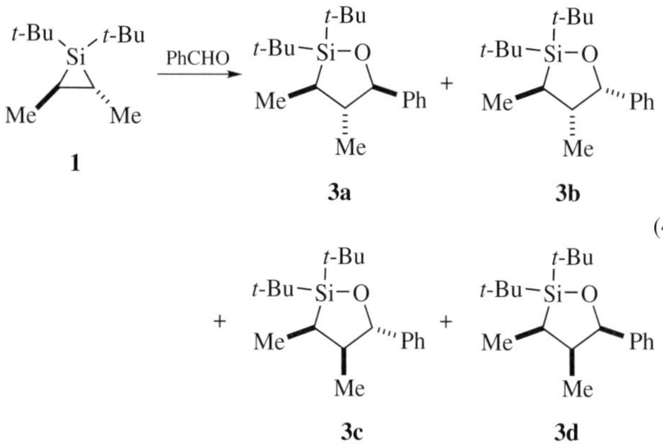

(4)

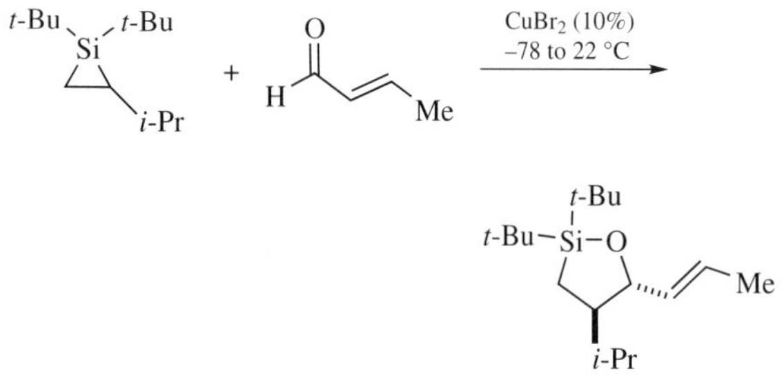

(5)

Formamide Insertions. Silirane **1** also reacts with formamides at elevated temperatures providing *N,O*-acetal products such as oxasilacyclopentane (**4**), with extremely high stereoselectivity relative to that observed with aldehydes (eq 6).[12] However, the formamide insertion was not catalyzed by Lewis bases such as fluoride or *t*-butoxide anion. Coordination of the amide carbonyl oxygen to the silicon atom of silacyclopropane, which is more electrophilic due to the strained silirane ring, increases the nucleophilicity of the silane and accounts for the greater reactivity of the formamides over esters and alkyl aldehydes. The high stereoselectivity of this process is consistent with a greater degree of structural preorganization present in this transition state. The use of a copper(II) catalyst permits the reaction to take place at or below room temperature with improved regioselectivity.[10]

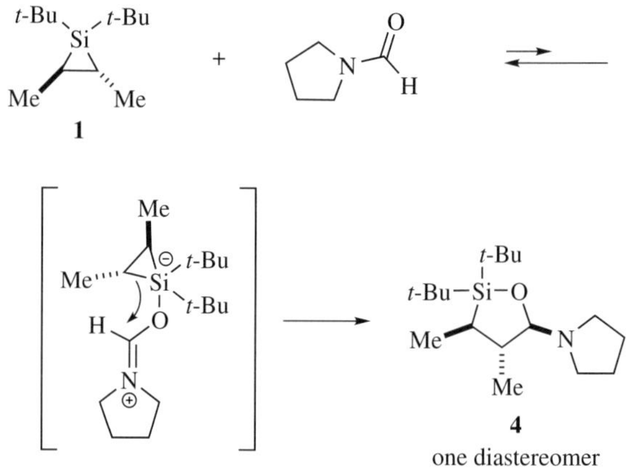

(6)

one diastereomer

One-Atom Insertions. Siliranes also undergo one-atom insertion reactions with elemental sulfur, oxygen, or selenium.[8] Similarly, the insertion of aryl or alkyl isocyanates provides iminosilacyclobutanes in high yield as single stereoisomers (eq 7).[13] The reaction proceeds with the retention of silacyclopropane configuration and the insertion occurs at the more substituted C–Si bond when unsymmetrical siliranes are employed, similar to the regioselectivity observed for aldehydes and formamides.

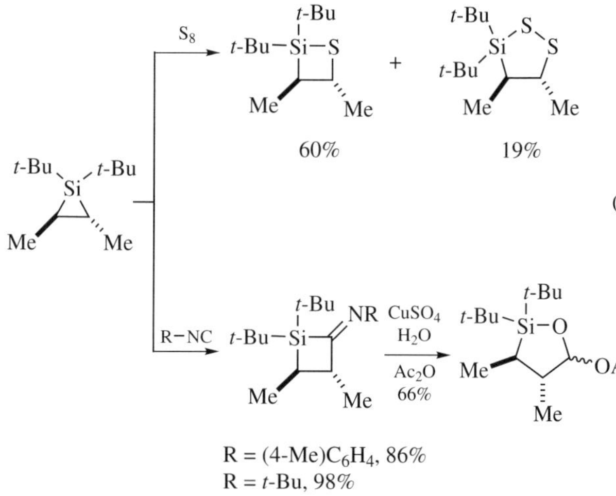

(7)

R = (4-Me)C$_6$H$_4$, 86%
R = *t*-Bu, 98%

Alkyne Insertions. Alkynes undergo insertion reactions with silacyclopropanes in the presence of palladium catalysts.[14] For example, **1**-*cis* reacts with terminal acetylenes such as phenylacetylene in the presence of PdCl$_2$(PPh$_3$)$_2$ at 22 °C to give silole **5**, the product of reductive coupling of two alkynes, and insertion product **6** (eq 8). In contrast, **1**-*trans* affords only 1% of the corresponding insertion product giving instead *trans*-2-butene as the major product. The mechanism of this process is thought to occur via palladacyclobutane **7**, which is produced by oxidative addition of the Pd species into the C–Si bond of the silacyclopropane.[15] Internal alkynes react with silacyclopropanes to give silacyclopropenes via silylene transfer rather than silole products (eq 9).[16]

Oxasilacyclopentane Functionalization. The synthetic utility of these insertion products is revealed via oxidation of the C–Si bond of the oxasilacyclopentane products to afford oxygenated compounds.[9] Accordingly, oxidation of **8** with *t*-BuOOH in the presence of a strong base such as Bu$_4$NF affords diol **9** (eq 10).

These oxidation products hold great promise in the development of new synthons for natural product synthesis.[17]

1. (a) Seyferth, D.; Annarelli, D. C., *J. Am. Chem. Soc.* **1975**, *97*(24), 7162. (b) Seyferth, D.; Annarelli, D. C., *J. Am. Chem. Soc.* **1975**, *97*(8), 2273. (c) Seyferth, D.; Annarelli, D. C.; Vick, S. C.; Duncan, D. P., *J. Organomet. Chem.* **1980**, *201*(1), 179. (d) Seyferth, D.; Escudie, J.; Shannon, M. L.; Satge, J., *J. Organomet. Chem.* **1980**, *198*(3), C51. (e) Seyferth, D.; Annarelli, D. C.; Vick, S. C., *J. Organomet. Chem.* **1984**, *272*(2), 123. (f) Seyferth, D.; Goldman, E. W.; Escudie, J., *J. Organomet. Chem.* **1984**, *271*(1–3), 337. (g) Seyferth, D.; Vick, S. C.; Shannon, M. L., *Organometallics* **1984**, *3*(12), 1897. (h) Seyferth, D.; Duncan, D. P.; Shannon, M. L.; Goldman, E. W., *Organometallics* **1984**, *3*(4), 574.

2. Saso, H.; Ando, W.; Ueno, K., *Tetrahedron* **1989**, *45*(7), 1929.

3. Franz, A. K.; Woerpel, K. A., *Acc. Chem. Res.* **2000**, *33*(11), 813.

4. Ostendorf, D.; Kirmajer, L.; Saak, W.; Marsmann, H.; Weidenbruch, M., *Eur. J. Inorg. Chem.* **1999**, *12*, 2301.

5. Boudjouk, P.; Black, E.; Kumarathasan, R., *Organometallics* **1991**, *10*(7), 2095.

6. Boudjouk, P.; Samaraweera, U.; Sooriyakumaran, R.; Chrusciel, J.; Anderson, K. R., *Angew. Chem.* **1988**, *100*(10), 1406.

7. Cirakovic, J.; Driver, T. G.; Woerpel, K. A., *J. Am. Chem. Soc.* **2002**, *124*(32), 9370.

8. (a) Boudjouk, P.; Black, E.; Kumarathasan, R.; Samaraweera, U.; Castellino, S.; Oliver, J. P.; Kampf, I. W., *Organometallics* **1994**, *13*(9), 3715. (b) Boudjouk, P.; Black, E.; Samaraweera, U., *Inorg. Synth.* **1997**, *31*, 81.

9. Bodnar, P. M.; Palmer, W. S.; Shaw, J. T.; Smitrovich, J. H.; Sonnenberg, J. D.; Presley, A. L.; Woerpel, K. A., *J. Am. Chem. Soc.* **1995**, *117*(42), 10575.

10. Franz, A. K.; Woerpel, K. A., *J. Am. Chem. Soc.* **1999**, *121*(5), 949.

11. Franz, A. K.; Woerpel, K. A., *Angew. Chem., Int. Ed.* **2000**, *39*(23), 4295–4299.

12. Shaw, J. T.; Woerpel, K. A., *J. Org. Chem.* **1997**, *62*(3), 442.

13. Nguyen, P. T.; Palmer, W. S.; Woerpel, K. A., *J. Org. Chem.* **1999**, *64*(6), 1843.

14. Palmer, W. S.; Woerpel, K. A., *Organometallics* **1997**, *16*(6), 1097.

15. Palmer, W. S.; Woerpel, K. A., *Organometallics* **2001**, *20*(17), 3691.

16. Palmer, W. S.; Woerpel, K. A., *Organometallics* **1997**, *16*(22), 4824.

17. (a) Shaw, J. T.; Woerpel, K. A., *Tetrahedron* **1997**, *53*(48), 16597. (b) Tenenbaum, J. M.; Woerpel, K. A., *Org. Lett.* **2003**, *5*(23), 4325. (c) Powell, S. A.; Tenenbaum, J. M.; Woerpel, K. A., *J. Am. Chem. Soc.* **2002**, *124*(43), 12648.

Jon R. Parquette
The Ohio State University, Columbus, Ohio, USA

Di-*t*-butylsilyl Bis(trifluoromethanesulfonate)

[85272-31-7] $C_{10}H_{18}F_6O_6S_2Si$ (MW 440.44)

InChI = 1S/C10H18F6O6S2Si/c1-7(2,3)25(8(4,5)6,21-23(17,18)9(11,12)13)22-24(19,20)10(14,15)16/h1-6H3

InChIKey = HUHKPYLEVGCJTG-UHFFFAOYSA-N

(reagent for the protection of diols)

Physical Data: bp 73–75 °C/0.35 mmHg; *d* 1.208 g cm^{-3}.
Solubility: sol most common organic solvents.
Form Supplied in: liquid.
Preparative Method: by the treatment of di-*t*-butylchlorosilane with trifluoromethanesulfonic acid, followed by distillation (71% yield).[1]
Purification: distillation.
Handling, Storage, and Precautions: moisture sensitive; reacts with hydroxylic solvents; corrosive.

Original Commentary

Snorri T. Sigurdsson & Paul B. Hopkins
University of Washington, Seattle, WA, USA

Protection of Alcohols. Di-*t*-butylsilyl bis(trifluoromethanesulfonate) is a reagent for the selective protection of polyhydroxy compounds. This reagent reacts with 1,2-, 1,3-, and 1,4-diols under mild conditions to give the corresponding dialkylsilylene derivatives in high yield (0–50 °C, 79–96%). Deprotection is conveniently achieved by using aqueous hydrofluoric acid in acetonitrile (eq 1).

Unlike di-*t*-butyldichlorosilane, this reagent reacts with hindered alcohols. Even pinacol reacts to give the silylene derivative (100 °C, 24 h, 70%). Di-*t*-butylsilylene derivatives of 1,2-diols are more reactive than those of 1,3- and 1,4-diols and undergo rapid hydrolysis (5 min) in THF/H$_2$O at pH 10, while the 1,3- and 1,4-derivatives are unaffected at pH 4–10 (22 °C) for several hours. This protecting group is stable under the conditions of PDC oxidation of alcohols (CH$_2$Cl$_2$, 25 °C, 27 h) and tosylation of alcohols (pyridine, 25 °C, 27 h).

The reagent has seen limited use for the protection of alcohols but has been used to protect nucleosides (eq 2).[2–5] The procedure consists of sequential addition of the ditriflate and triethylamine to the nucleoside in DMF. The choice of solvent is critical.

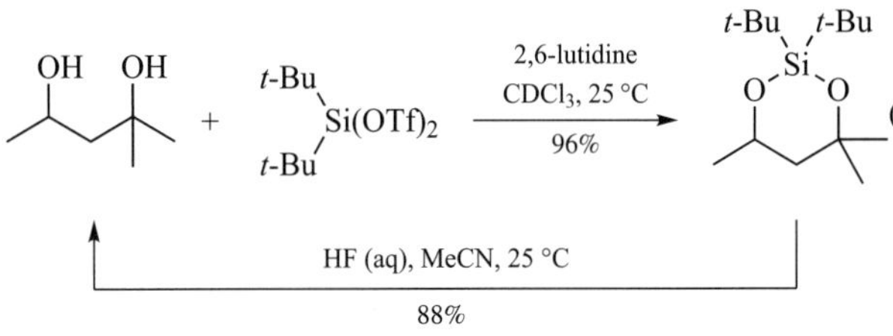

(1)

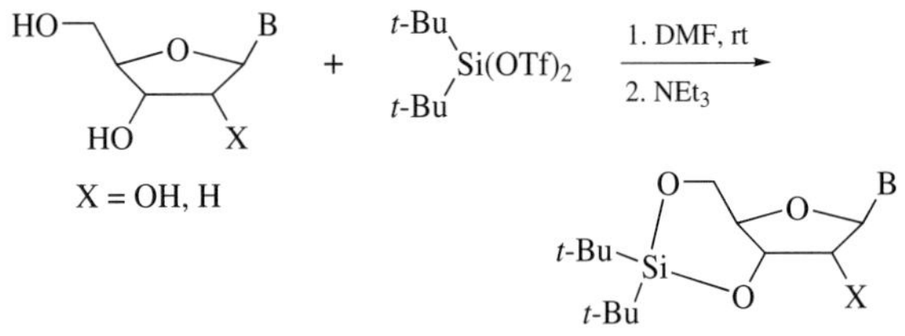

The ribonucleosides of uracil, adenine, and guanine give the protected derivative in 94–95% yield.[2] Cytidine gives a low yield of the desired product under these conditions. Subsequent studies suggested that O^2 of cytosine participates in the reaction. Addition of trifluoromethanesulfonic acid or silver(I) trifluoromethanesulfonate at 0 °C prior to addition of the silylating agent results in a 99% yield of the desired derivative.[3] The derivatives are acid sensitive, presumably due to the proximity of the 2′-hydroxy group. Acetylation, tetrahydropyranylation, methoxytetrahydropyranylation, and silylation of the 2′-hydroxy group are accomplished without affecting the dialkylsilylene protecting group. The 2′-deoxyribonucleosides, including 2′-deoxycytidine, can also be prepared by the aforementioned procedure (yields 90–99%). These cyclic silylene derivatives of nucleosides can be deprotected conveniently using tributylamine hydrofluoride in THF (5 min, 1 M, rt, 20 equiv).[4] A one-pot procedure has been reported for simultaneously protecting the 2′-, 3′-, and 5′-hydroxys of a ribonucleoside, which utilizes the acid generated upon silylating the 3′- and 5′-hydroxys for catalyzing the formation of a THP acetal at the 2′-position (eq 3).[5]

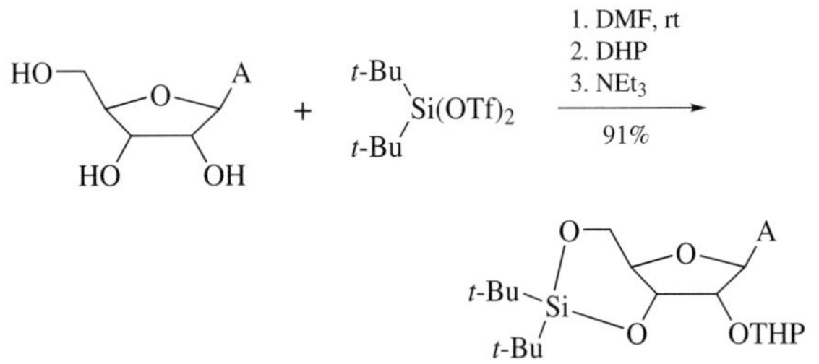

Derivatization of Alcohols. Di-*t*-butylsilyl bis(trifluoromethanesulfonate) has been used to derivatize hindered diols, to give derivatives such as (**1**), for analysis by gas chromatography–electron impact mass spectrometry.[6] The major fragmentation is that of the Si–C bonds.

m/z 366 (M^+), 309 (M^+ – Bu), 191 (B)

(**1**)

Reagent in Enantioselective Additions. In a study of enantioselective conjugate addition to cyclohexanone it was found that the presence of HMPA and various silyl reagents markedly increases the enantioselectivity (eq 4).[7] Di-*t*-butylsilyl bis(trifluoromethanesulfonate) gives a 67% yield and 40% ee but *t*-butyldiphenylchlorosilane gives a 97% yield and 78% ee.

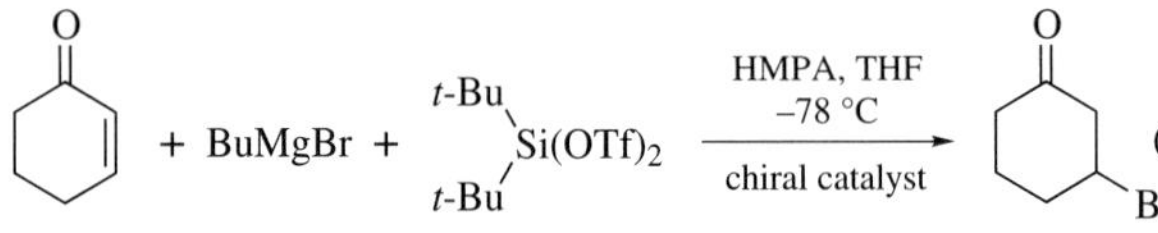

(4)

Other Substitution Reactions. An extremely hindered silyl reagent, tri-*t*-butylsilyl trifluoromethanesulfonate, was prepared from di-*t*-butylsilyl bis(trifluoromethanesulfonate) and *t*-butyllithium (eq 5).[8] This reagent might find use in the protection of alcohols.

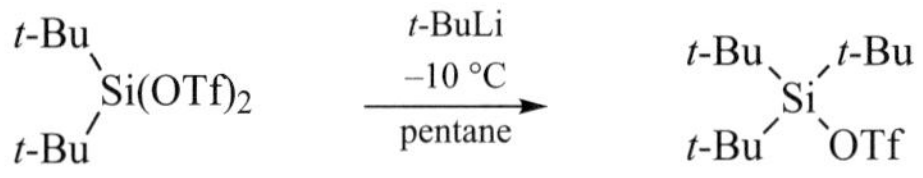

(5)

In conjunction with the study of alkyl-substituted silyl triflates, (**2**) and (**3**) have been prepared from the corresponding alkynyllithium reagents and di-*t*-butylsilyl bis(trifluoromethanesulfonate).[9]

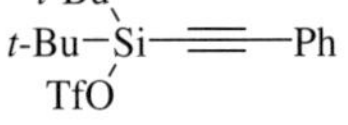

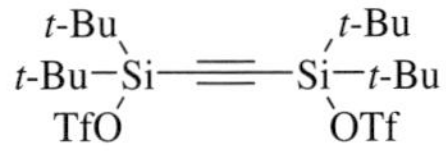

The preparation of other derivatives of di-*t*-butylsilyl bis(trifluoromethanesulfonate) using germanium[10] and phosphorus[11] nucleophiles has been reported and provides bifunctional silanes such as (**4**) and (**5**).

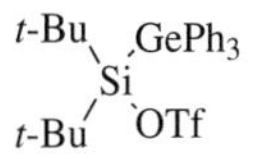

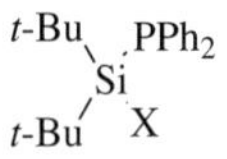

X = –C≡CPh, –SiPh₃, –N(TMS)₂, –PHPh, –O-*i*-Pr, –SPh

A compound closely related to di-*t*-butylsilyl bis(trifluoromethanesulfonate) is di-*t*-butylchlorosilyl trifluoromethanesulfonate, which has been used to tether two structurally different alcohol derivatives in order to effect an intramolecular Diels–Alder reaction (eq 6).[12]

t-Bu, *t*-Bu Si with O groups, 160–190 °C, xylene, >90%, CO$_2$Et

$$+ \quad (6)$$

CO$_2$Et   CO$_2$Et

>99:<1

First Update

Jitendra Belani

Infinity Pharmaceuticals, Cambridge, MA, USA

In the past 10 years, there has been a dramatic increase in use of di-*t*-butylsilyl bis(trifluoromethanesulfonate) as a protecting group for diols to provide the required orthogonality or to impart selectivity in various reactions both in synthesis of natural products and in carbohydrate and oligonucleotide syntheses.

Applications in Total Synthesis of Natural Products and Development of New Methods. Paterson et al. strategically used di-*t*-butylsilyl bis(trifluoromethanesulfonate) to avoid differential protection of C$_{21}$ and C$_{23}$ hydroxyl groups in their efforts to synthesize the C$_{19}$–C$_{32}$ fragment of swinholide A (eq 7).[13a] The elaboration of the fragment to the natural product was reported in the subsequent communications[13b–d] and the authors were able to remove the silylene protecting group using HF·pyridine. The authors have also reported similar application of di-*t*-butylsilyl bis(trifluoromethanesulfonate) to protect 1,3-diols in total synthesis of concanamycin A.[14]

t-Bu$_2$Si(OTf)$_2$, 2,6-lutidine
CH$_2$Cl$_2$, 20 °C, 16 h
86%

OMe   OH   OH   OBn

(6)

swinholide A

(7)

(7)

The first total synthesis of concanamycin F by Toshima et al. also utilized di-*t*-butylsilyl bis(trifluoromethanesulfonate) for protection of 1,3-diols.[15] The key aldol reaction between **11** and **12** was best achieved using PhBCl$_2$ and *i*-Pr$_2$NEt in CH$_2$Cl$_2$ at −78 °C to provide the desired aldol **13** as the sole isomer in 84% yield. The selective deprotection of di-*t*-butylsilylene was then achieved using HF·pyridine, which resulted in concomitant

formation of the hemiacetal **15**. Removal of the diethylisopropylsilyl group using TBAF provided concanamycin F (eq 8).

In another report, Panek and Jain used chiral allylsilane methodology for construction of C$_1$–C$_{17}$ polypropionate fragment of rutamycin B and oligomycin C.[16] The aldehyde partner used for the sequence consisted of a diol protected as a di-*t*-butylsilylene **18**. The reaction proceeded in the presence of TiCl$_4$ in excellent diastereoselectivity and yield (eq 9).

Di Fabio et al. reported synthesis of 3-alkoxy-substituted trinems from commercially available 4-acetoxy-3-[(*R*)-l-*t*-butyldimethylsilyloxyethyl]-2-azetidinone.[17] Epoxide **20** was treated with ceric ammonium nitrate in acetonitrile to provide the intermediate nitrate ester **21** in 55% yield. Simultaneous protection of the secondary alcohol and the amide was accomplished using di-*t*-butylsilyl bis(trifluoromethanesulfonate) to provide the tricyclic β-lactam **22**. Removal of the nitro group by catalytic hydrogenation provided the desired secondary alcohol that was alkylated with allyl bromide to provide **23** in quantitative yield (eq 10).[17]

A successful approach to construction of the tricyclic core common to hamigeran terpenes was demonstrated using an intramolecular Pauson–Khand reaction.[18] For an effective cyclization, the authors report that it was necessary to tether the olefin-containing moiety to the aromatic framework to reduce its conformational mobility using the di-*t*-butylsilylene protecting group (eq 11).

Stoltz and coworkers reported a highly selective catalytic reductive isomerization reaction using 10% Pd/C and hydrogen in MeOH. During their explorations of the total synthesis of (+)-dragmacidin F, olefin isomerization using the carbonate **27** (eq 12) was developed.[19] The authors provided evidence that the method does not proceed via a stepwise reduction/elimination sequence or a π-allylpalladium intermediate. Replacement of the carbonate by a dioxasilyl linkage (**29**), however, did not result in isomerization, only diastereoselective reduction of the *exo*-olefin was observed (eq 12).[19]

Hillaert and Van Calenbergh have reported synthesis of (*S*)-3-(hydroxymethyl)butane-1,2,4-triol, a versatile, chiral building block that can be transformed into useful compounds such as (*S,S*)-4-(hydroxymethyl)pyrrolidine-3-ol and oxetanocin A.[20] The authors reported a short synthesis of the protected triol **33** from the epoxide **31** in six steps. The use of di-*t*-butylsilylene as the protecting group provided the desired differentiation of the hydroxyl groups, which was necessary for their successful investigation into the stereoselective synthesis of *N*-homoceramides (eq 13).[20]

Synthesis of polypropionate marine natural product (+)-membrenone C and its 7-*epi*-isomer has been reported using a key desymmetrization technique to create five contiguous chiral centers from bicyclic precursor **36**.[21] The diol was protected using di-*t*-butylsilyl bis(trifluoromethanesulfonate) and further elaborated into the natural product and its epimer (eq 14).[21] In a separate communication, Perkins et al. utilized di-*t*-butylsilyl bis(trifluoromethanesulfonate) for synthesis of a model system en route to the polypropionate natural products auripyrones A and B.[22]

Danishefsky and coworkers reported studies toward the total synthesis of tetrahydroisoquinoline alkaloid, ecteinascidin.[23] The synthesis required a pentasubstituted E-ring system where they utilized the di-*t*-butylsilylene protecting group in the sequence to prepare the amino acid **42** (eq 15). In addition to above

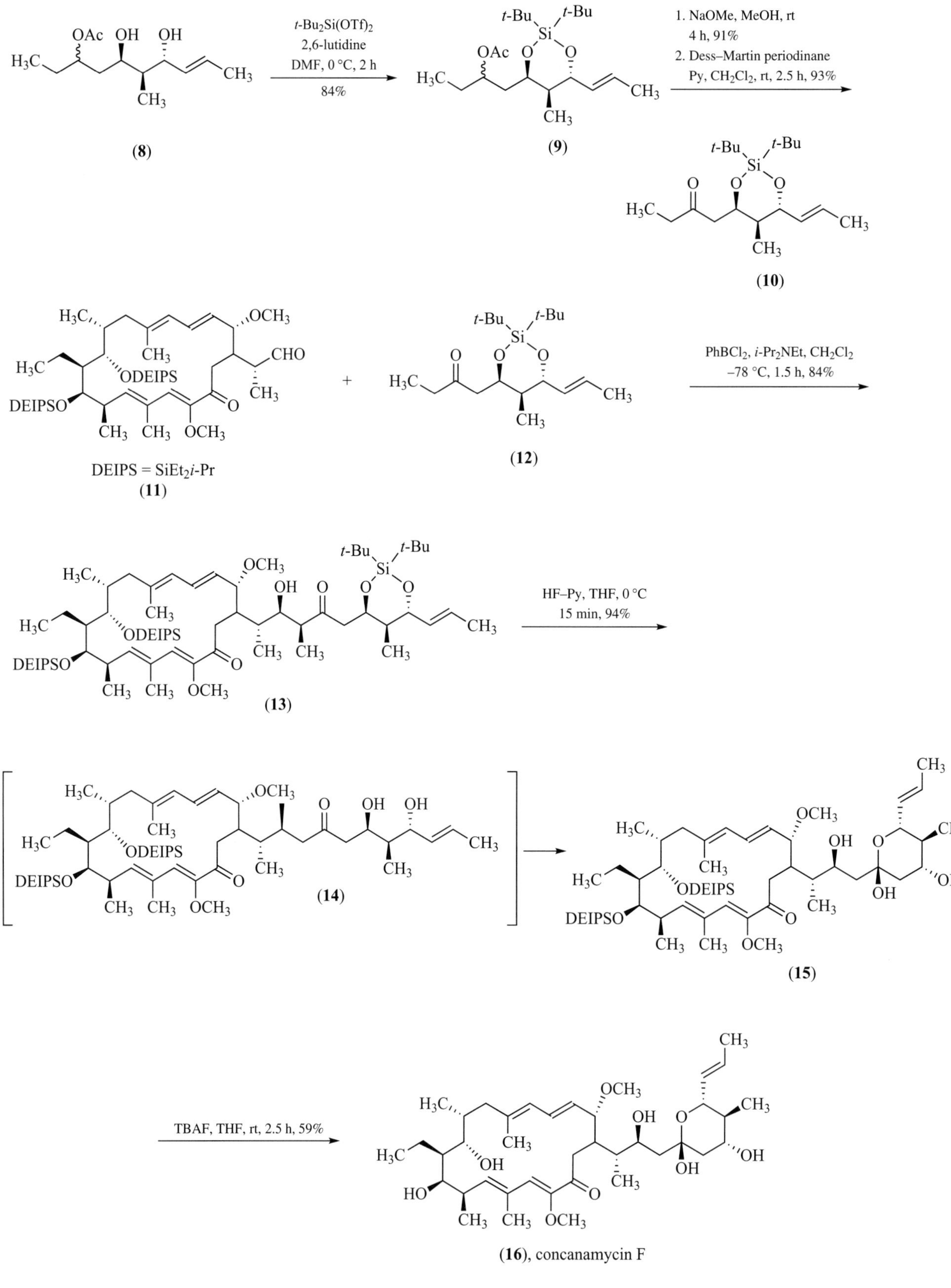

(8)

t-Bu$_2$Si(OTf)$_2$
2,6-lutidine
DMF, 0 °C, 2 h
84%

(9)

1. NaOMe, MeOH, rt
   4 h, 91%
2. Dess–Martin periodinane
   Py, CH$_2$Cl$_2$, rt, 2.5 h, 93%

(10)

DEIPS = SiEt$_2$$i$-Pr
(11)

+

(12)

PhBCl$_2$, i-Pr$_2$NEt, CH$_2$Cl$_2$
−78 °C, 1.5 h, 84%

(13)

HF–Py, THF, 0 °C
15 min, 94%

(14)

(15)

TBAF, THF, rt, 2.5 h, 59%

(16), concanamycin F

(8)

examples, protection of 1,3-diol using di-*t*-butylsilyl bis(trifluoromethanesulfonate) has been reported in the synthesis of brasilinolides,[24] polyol fragments of ansamycin antibiotics,[25] (±)-*epi*-stegobinone,[26] 24-demethylbafilomycin C1,[27] peloruside A,[28] bafilomycin A1,[29] premisakinolide A,[30] (+)-papulacandin D,[31] and other polyether natural products such as maitotoxin,[32] yessotoxin,[33] and gambieric acids.[34]

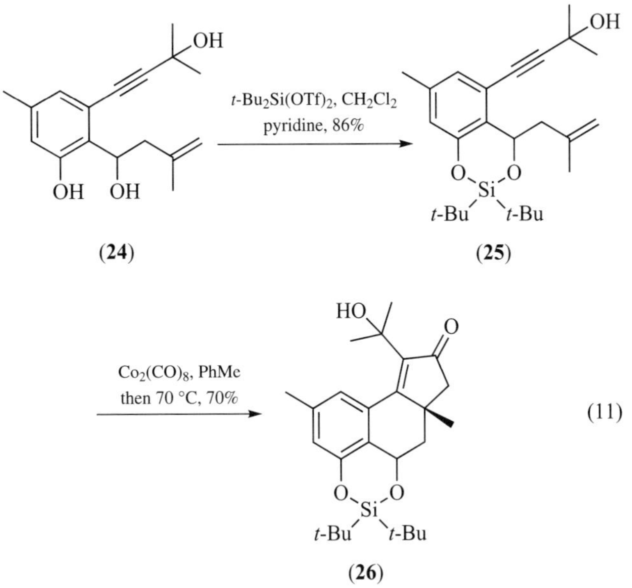
OH
OH OH
(24)
t-Bu2Si(OTf)2, CH2Cl2
pyridine, 86%
OH
O Si O
t-Bu t-Bu
(25)
HO
O
Co2(CO)8, PhMe
then 70 °C, 70%
O Si O
t-Bu t-Bu
(26)

(11)

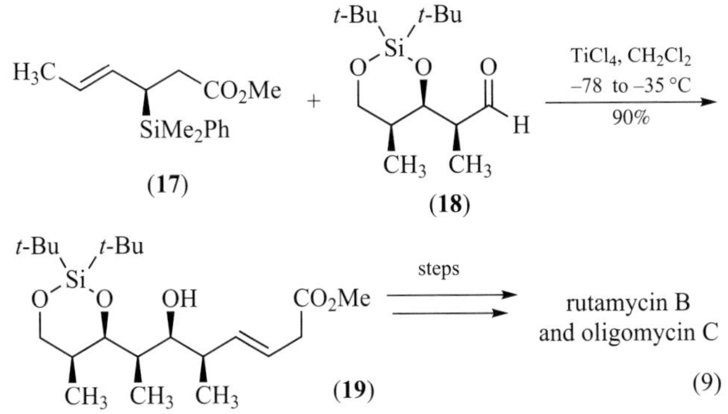
H3C
CO2Me
SiMe2Ph
(17)
t-Bu t-Bu
Si
O O O
H
CH3 CH3
(18)
TiCl4, CH2Cl2
−78 to −35 °C
90%
t-Bu t-Bu
Si
O O OH CO2Me
CH3 CH3 CH3
(19)
steps
rutamycin B
and oligomycin C

(9)

diastereoselection: >30:1 *syn*:*anti*

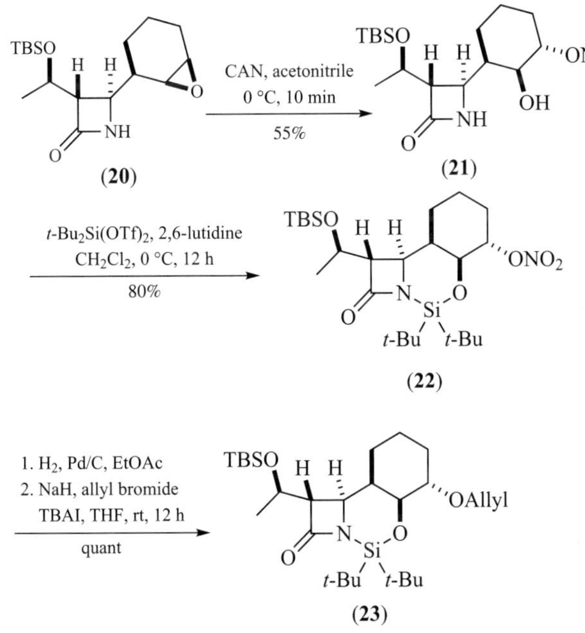
TBSO
H H
O
NH
O
(20)
CAN, acetonitrile
0 °C, 10 min
55%
TBSO
H H
ONO2
OH
NH
O
(21)
t-Bu2Si(OTf)2, 2,6-lutidine
CH2Cl2, 0 °C, 12 h
80%
TBSO H H
ONO2
N Si O
t-Bu t-Bu
(22)
1. H2, Pd/C, EtOAc
2. NaH, allyl bromide
TBAI, THF, rt, 12 h
quant
TBSO H H
OAllyl
N Si O
t-Bu t-Bu
(23)

(10)

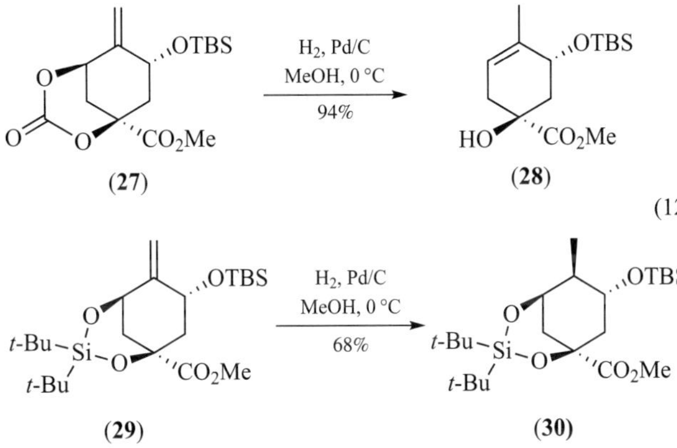
OTBS
O
O O
CO2Me
(27)
H2, Pd/C
MeOH, 0 °C
94%
OTBS
HO
CO2Me
(28)
OTBS
t-Bu
Si O
t-Bu
CO2Me
(29)
H2, Pd/C
MeOH, 0 °C
68%
OTBS
t-Bu
Si O
t-Bu
CO2Me
(30)

(12)

A mixed di-*t*-butylsilylene has been prepared from (*S*)-5-hexen-2-ol and prochiral 1,4-pentadiene-3-ol for synthesis of (2*S*,7*S*)-dibutyroxynonane, the sex pheromone of *Sitodiplosis mosellana*. The intermediate **43** was then subjected to ring closing metathesis to provide the diene **44** in 70% yield over two steps (eq 16). Deprotection of the nine-membered silylene was achieved using TBAF under refluxing condition in the presence of molecular sieves. Reduction with H_2/PtO_2 and diacetylation yielded the desired (2*S*,7*S*)-dibutyroxynonane in 22% overall yield.[35]

Kita and coworkers reported a novel method to prepare 2,2′-substituted biphenyl compounds using phenyliodine(III) bis(trifluoroacetate) (PIFA), a hypervalent iodine reagent.[36] The substrate for the coupling was first prepared by reacting 1 equiv of di-*t*-butylsilyl bis(trifluoromethanesulfonate) with 2 equiv of the phenol **46**. Di-*t*-butylsilylene **47** then underwent an intramolecular cyclization in the presence of $BF_3 \cdot OEt_2$ to provide the desired tricyclic compound **48**. Removal of silylene ether was accomplished by TBAF in excellent yields. This sequence worked well for both 2,2′-disubstituted symmetrical and unsymmetrical biphenyls (eq 17).[36]

In a separate communication, Kita and coworkers also reported an enantiodivergent synthesis of an ABCDE ring analog of the antitumor antibiotic fredericamycin A via an intramolecular [4 + 2] cycloaddition. Late-stage oxidations were performed on the di-*t*-butylsilylene for protection of phenolic hydroxyl groups. Syntheses of both (*R*)- and (*S*)-enantiomers were reported (eq 18).[37]

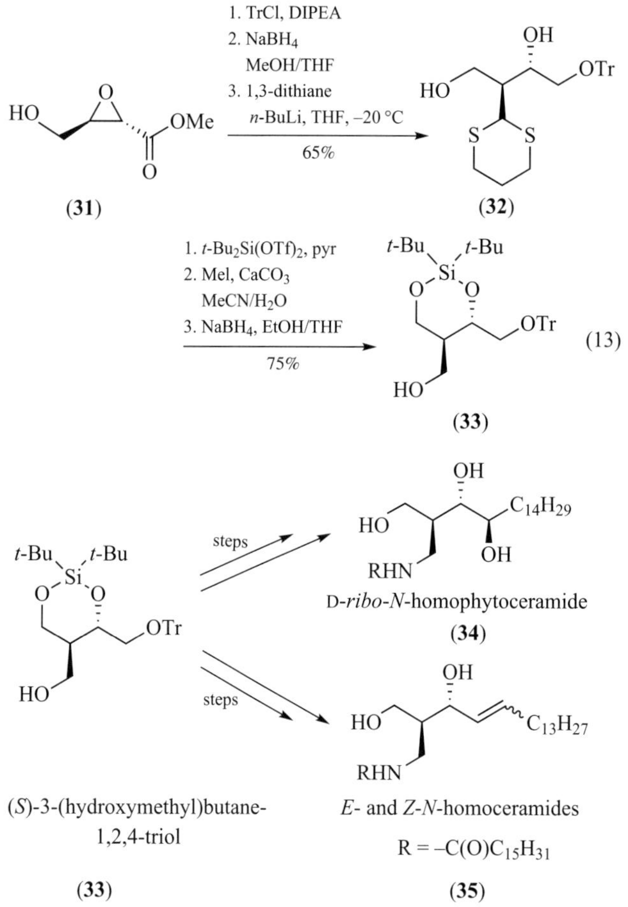

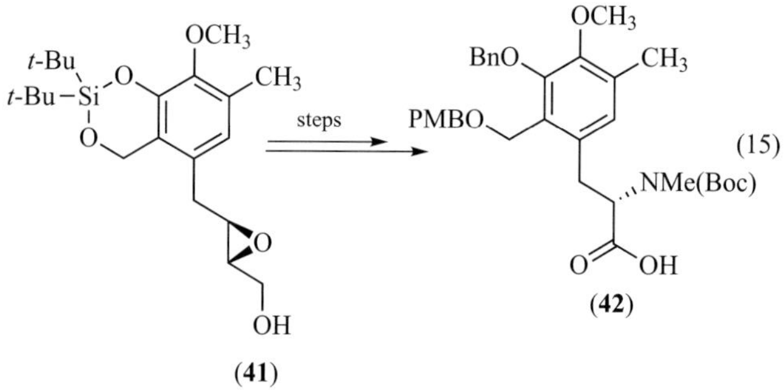

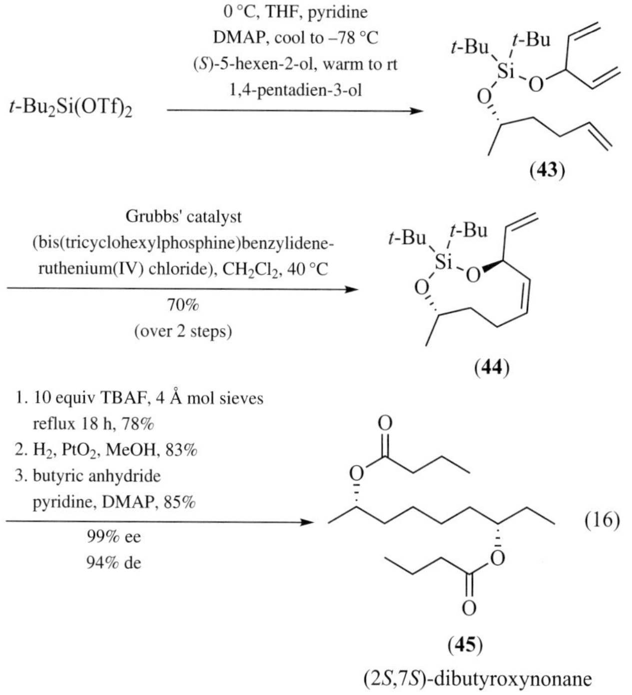

(31) → (32)

1. TrCl, DIPEA
2. NaBH₄ MeOH/THF
3. 1,3-dithiane *n*-BuLi, THF, −20 °C
65%

(31) → (33)

1. *t*-Bu₂Si(OTf)₂, pyr
2. MeI, CaCO₃ MeCN/H₂O
3. NaBH₄, EtOH/THF
75% (13)

(33) → (34) D-*ribo*-*N*-homophytoceramide

(33) → (35) *E*- and *Z*-*N*-homoceramides
R = −C(O)C₁₅H₃₁

(*S*)-3-(hydroxymethyl)butane-1,2,4-triol (33)

(41) → (42) steps (15)

t-Bu₂Si(OTf)₂ → (43)
0 °C, THF, pyridine
DMAP, cool to −78 °C
(*S*)-5-hexen-2-ol, warm to rt
1,4-pentadien-3-ol

(43) → (44)
Grubbs' catalyst
(bis(tricyclohexylphosphine)benzylidene-ruthenium(IV) chloride), CH₂Cl₂, 40 °C
70% (over 2 steps)

(44) → (45)
1. 10 equiv TBAF, 4 Å mol sieves reflux 18 h, 78%
2. H₂, PtO₂, MeOH, 83%
3. butyric anhydride pyridine, DMAP, 85%
99% ee
94% de
(16)

(2*S*,7*S*)-dibutyroxynonane

In a study on copper-catalyzed enantioselective intramolecular aminooxygenation of alkenes, Chemler and coworkers showed that this reaction allows synthesis of chiral indolines and pyrrolidines.[38] Accordingly, the protected amino-diol **53** undergoes aminooxygenation in the presence of copper triflate, bisoxazoline ligands, and molecular oxygen to provide the desired spirodecane **54** in 87% yield and 89% ee (eq 19).[38]

Craig and coworkers have reported stereocontrolled polyol synthesis via C–H insertion reactions of silicon-tethered diazoacetates.[39] Menthol was treated with di-*t*-butylsilyl bis(trifluoromethanesulfonate) and the product was condensed with ethyl diazoacetate to provide the precursor **57a** to the C–H insertion reaction (eq 20). The rhodium(II) octanoate-catalyzed

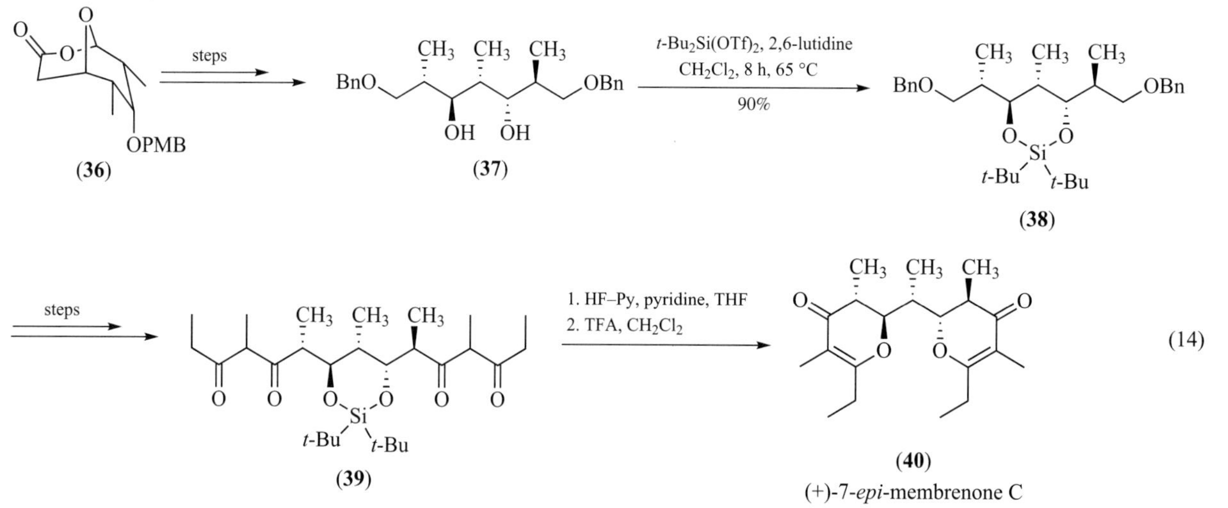

(36) → (37) steps

(37) → (38)
t-Bu₂Si(OTf)₂, 2,6-lutidine
CH₂Cl₂, 8 h, 65 °C
90%

(38) → (39) steps

(39) → (40)
1. HF–Py, pyridine, THF
2. TFA, CH₂Cl₂
(14)

(+)-7-*epi*-membrenone C

decomposition of the diazoacetate **57a**, however, did not provide the desired C–H insertion product, although the reaction was successful with diisopropylsilyl bis(trifluoromethanesulfonate) (eq 20).

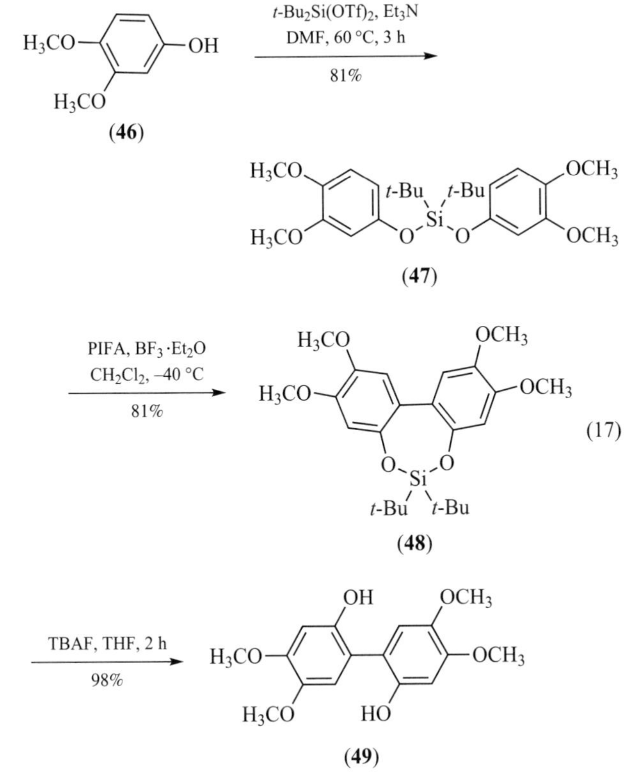

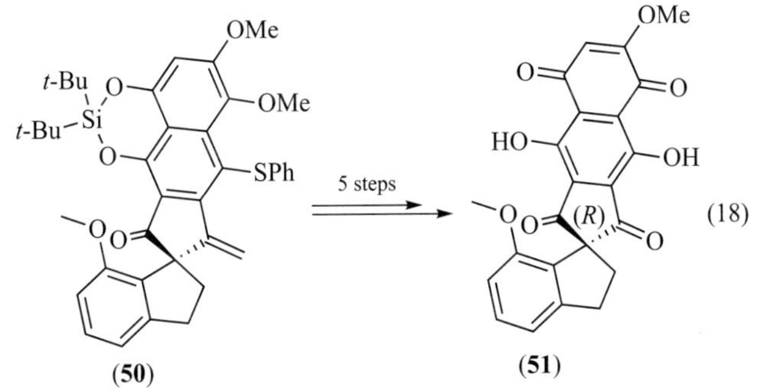

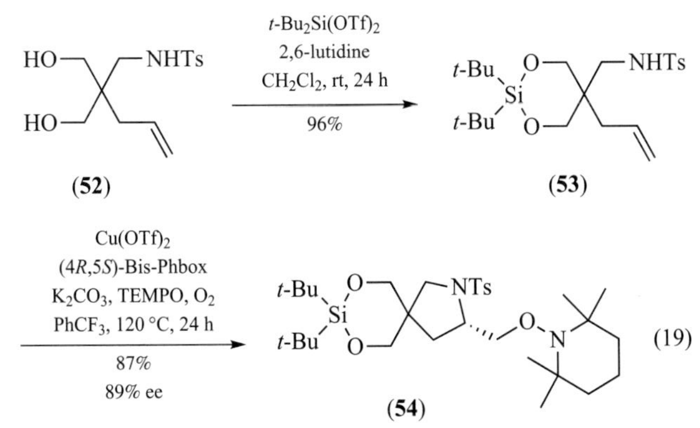

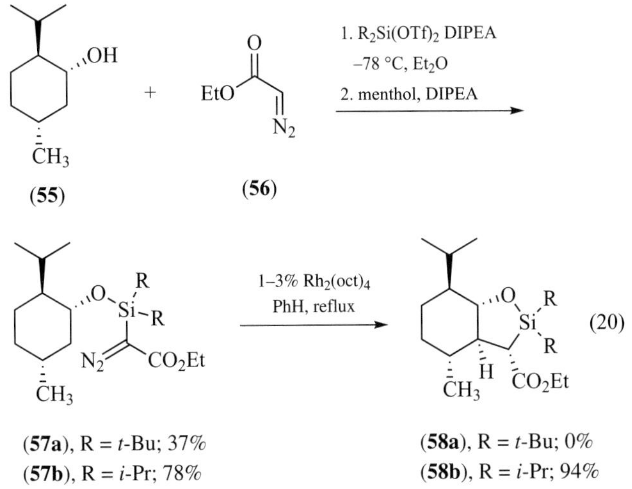

(46)

(47)

(17)

(48)

(49)

(50)   (51)   (18)

(52)   (53)

(54)

(55)   (56)

(57a), R = *t*-Bu; 37%   (58a), R = *t*-Bu; 0%
(57b), R = *i*-Pr; 78%   (58b), R = *i*-Pr; 94%

In a study about stereoselective allylation of chiral monoperoxides, Ahmed and Dussault focused on induction from 2-, 3-, and 4-substituted monoperoxyacetals.[40] It was observed that neighboring iodo-, alkoxy-, acetoxy-, and silyl, groups imparted useful levels of diastereoselection in the Lewis acid-mediated allylation of monoperoxyacetals. While investigating 1,3-stereoinduction, the homoallylic alcohol **59** was ozonized and the resultant hydroperoxy alcohol was silylated using di-*t*-butylsilyl bis(trifluoromethanesulfonate) to provide 3-sila-1,2,4-trioxepane **60**. Unfortunately, the substrate provided only 5% of allylated silatrioxepane under SnCl$_4$-mediated allylation (eq 21).[40]

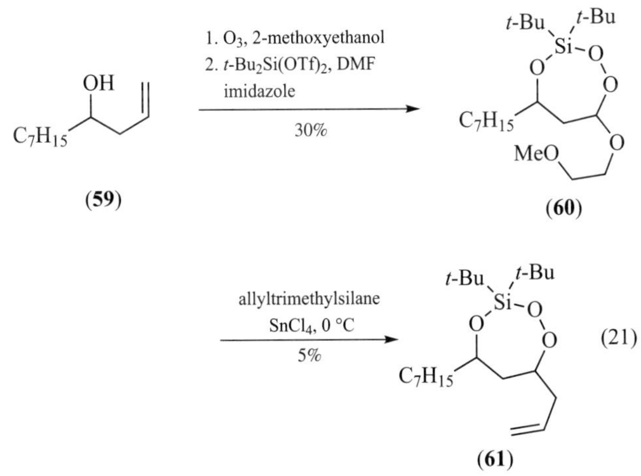

(59)   (60)

(21)

(61)

Applications in Carbohydrate and Oligonucleotide Syntheses. In efforts to synthesize aza-C-disaccharides, Baudat and Vogel used a key cross-aldolization of aldehyde **62** and (−)-(1S,4R,5R,6R)-6-chloro-5-(phenylseleno)-7-oxabicyclo-[2.2.1] heptan-2-one ((−)-**63**) to provide the product alcohol **64** stereoselectively in 70% yield.[41] The stereochemistry of the aldol product was confirmed by reduction of the ketone followed by protection of the diol using di-*t*-butylsilyl bis(trifluoromethanesulfonate) providing a mixture of rotamers **65** resulting from benzyl carbamate. Treatment of this mixture with *m*-CPBA provided the vinyl chloride **66** that displayed typical ^{1}H NMR vicinal coupling constants, thus confirming the stereochemistry of *exo-anti* aldol (+)-**64** (eq 22). A NOE experiment further confirmed the structure of (+)-**66**.[41]

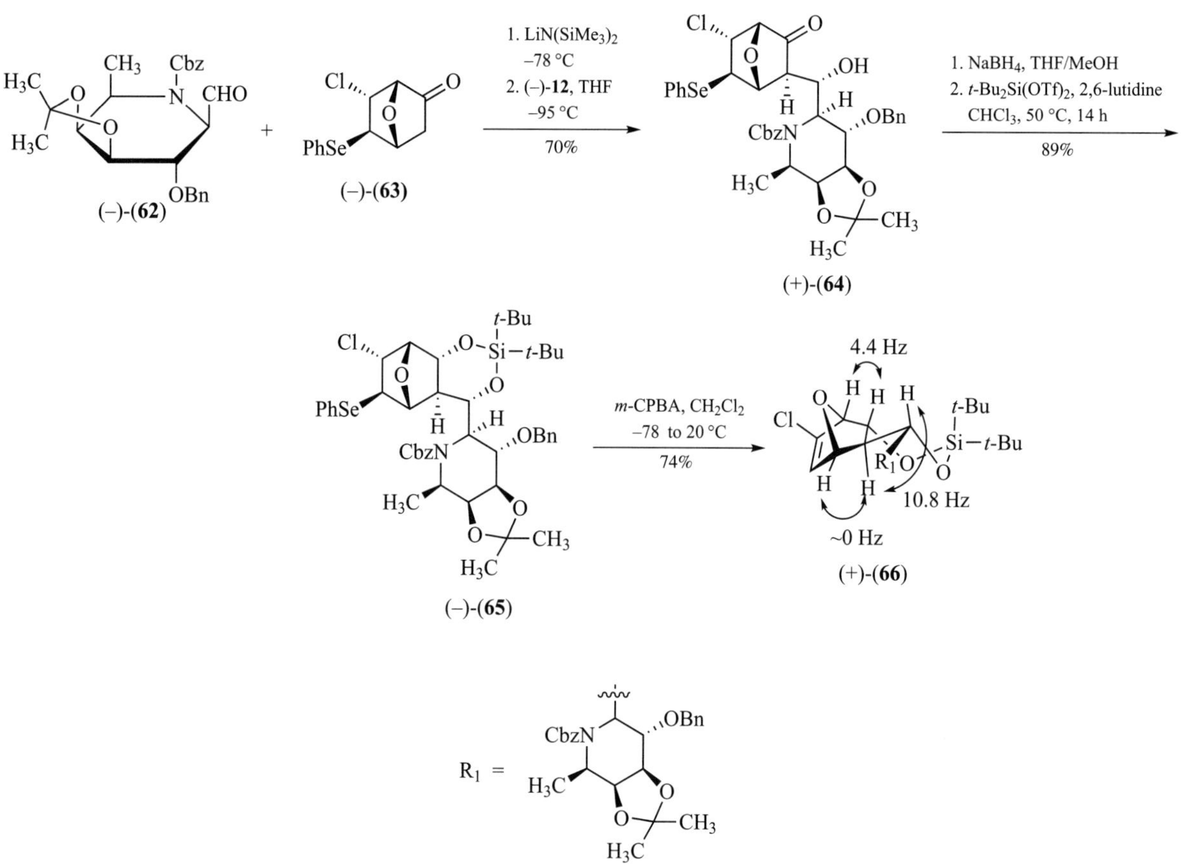

$$(22)$$

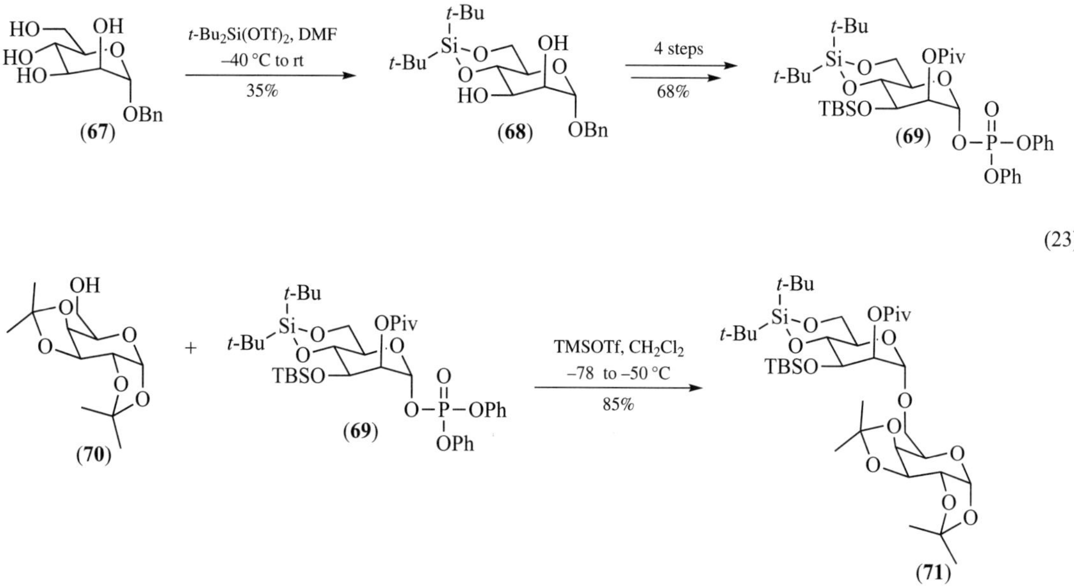

$$(23)$$

While investigating glycosyl phosphates for selective synthesis of α- and β-glycosidic linkages using conformationally restrained mannosyl phosphate **69**, Seeberger and coworkers observed only the desired α-isomer **71** by virtue of the participating pivaloyl group as expected (eq 23). These reactions were high yielding, fast, and completely selective.[42]

In a study of synthesis and conformational analysis of arabino-furanosyl oligosaccharide analogs in which one ring is locked into either the E$_3$ or °E conformation, Houseknecht and Lowary used di-*t*-butylsilyl bis(trifluoromethanesulfonate) to protect the 1,4-diol. The di-*t*-butylsilylene **73** is stable under a variety of reaction conditions and can be easily removed under mild HF–pyridine conditions (eq 24).[43]

Electrophilic glycosidation of 3,5-*O*-(di-*t*-butylsilylene)-4-thioglycal **80** has been reported to exclusively provide the β-anomer of 4′-thionucleosides irrespective of the nucleobase employed. The face selectivity of the approach to 1′,2′-double bond by the incoming electrophile can be controlled by changing

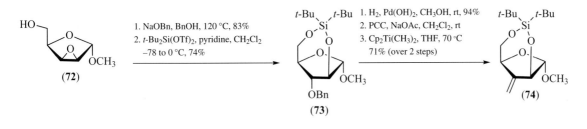

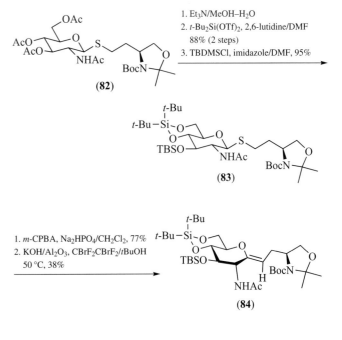

the protecting group of the 3′- and 5′-hydroxyl groups. Hence, approach to the α-face using NIS or PhSeCl increased in the order of 3′,5′-*O*-(di-*t*-butylsilylene):DTBS > 3′,5′-*O*-(1,1,3,3-tetraisopropyldisilox-ane-1,3-diyl):TIPDS > 3′,5′-bis-*O*-(*t*-butyldimethylsilyl):TBDMS (eq 25).[44]

Ohnishiy and Ichikawa reported a stereoselective synthesis of a C-glycoside analog of *N*-Fmoc-serine β-*N*-acetylglucosaminide (**85**) using a key Ramberg–Bäcklund rearrangement.[45] The di-*t*-butylsilylene protecting group provided the desired stability under the strongly basic conditions of the rearrangement (eq 26).

Entry	Electrophile	Base	Yield (%)
81a	PhSeCl	uracil-1-yl	88
81b	NIS	uracil-1-yl	73
81c	PhSeCl	thymin-1-yl	62
81d	PhSeCl	cytosin-1-yl	85

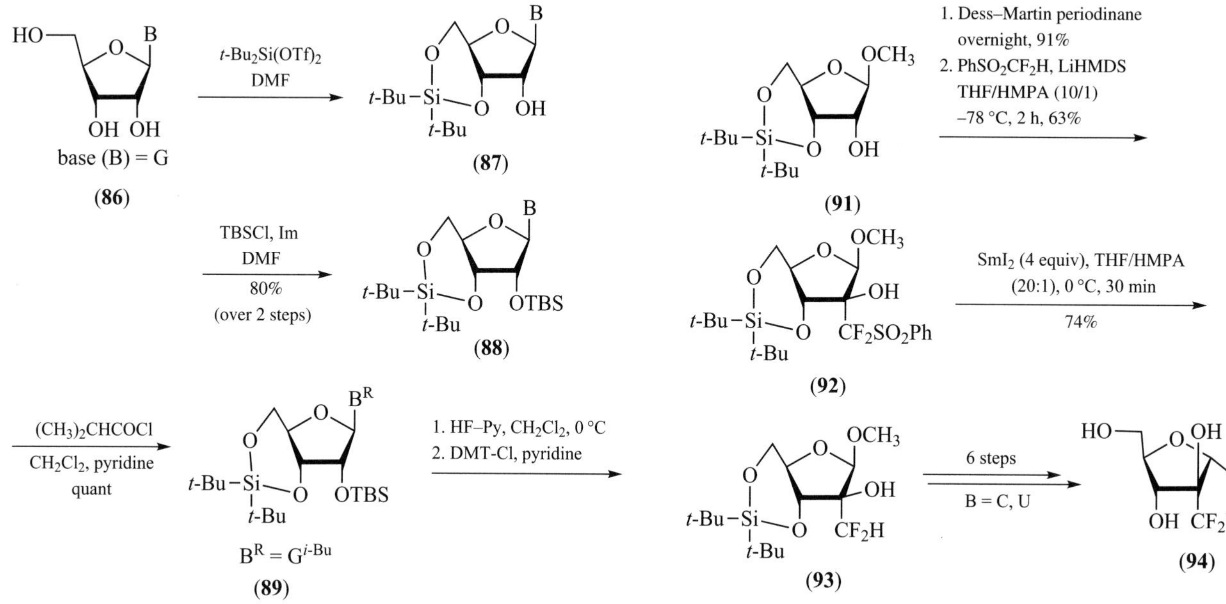

In a report on synthesis of 5′-*O*-DMT-*N*-acyl-2′-*O*-TBDMS-protected nucleoside precursors for phosphoramidite RNA synthesis, Serebryany and Beigelman used di-*t*-butylsilyl bis(trifluoromethanesulfonate) to protect 3′- and 5′-hydroxyl functions of a nucleoside.[46] The silylated derivatives were obtained in high yields and were crystalline and easy to purify.

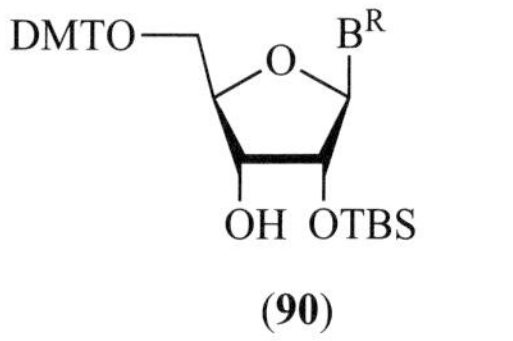

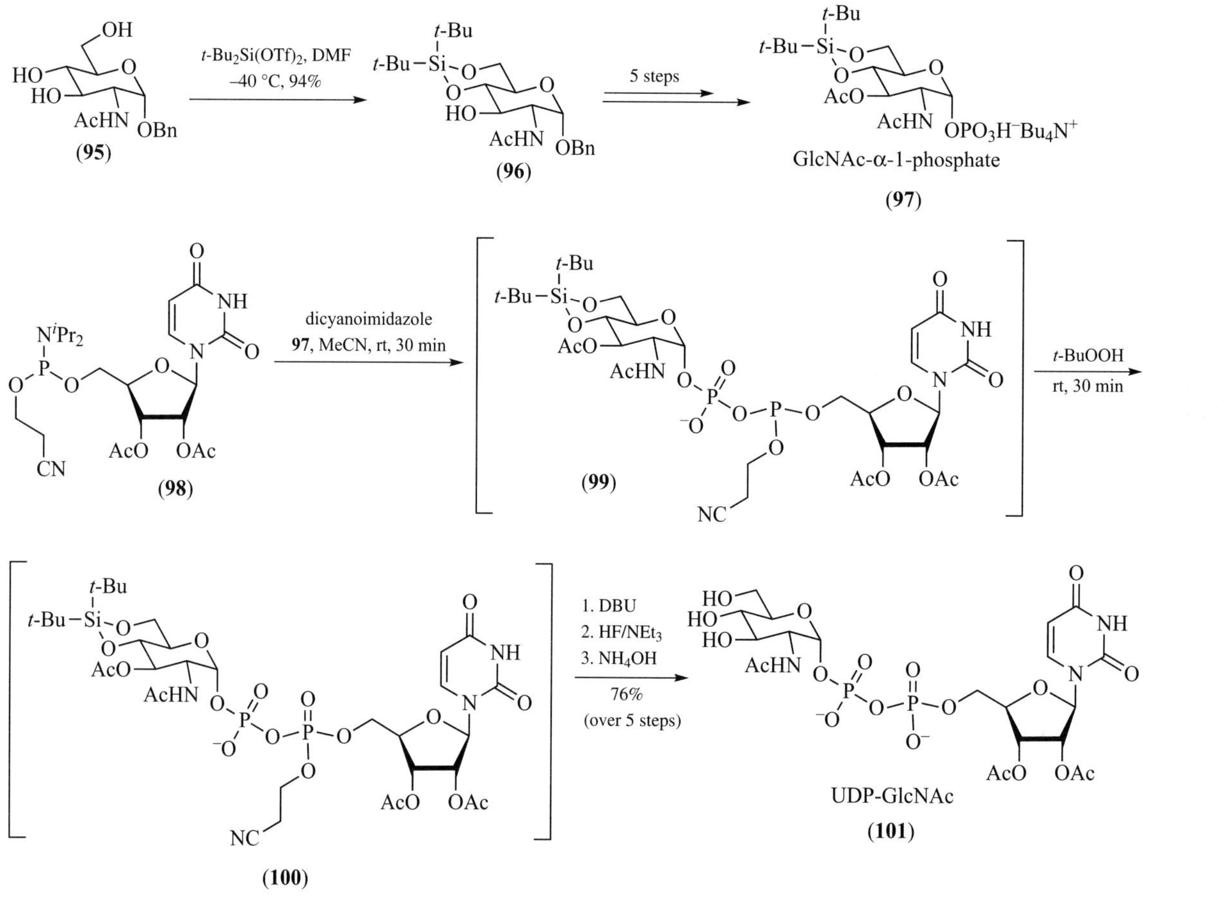

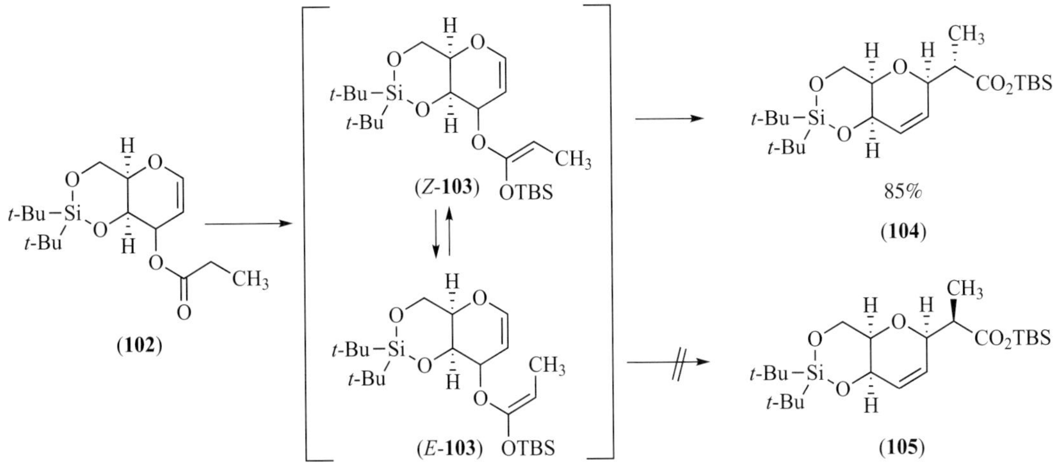

(30)

Acylation of the N_2-position of the guanosine derivative provided crystalline compound **89** in quantitative yield. Deprotection of the silylene **89** using mild HF–pyridine followed by protection of the primary alcohol with dimethoxytrityl chloride in pyridine gave the desired compound **90** in about 60% yield over five steps (eq 27).[46]

Synthesis of 2'-*C*-difluoromethylribonucleosides has been separately reported where simultaneous protection of the 3- and 5-hydroxyl groups of the methyl D-ribose with di-*t*-butylsilyl bis(trifluoromethanesulfonate) afforded the silylene **91**.[47] Dess–Martin periodinane oxidation of the C-2 hydroxyl group followed by nucleophilic addition of difluoromethyl phenyl sulfone in the presence of lithium hexamethyldisilazane converted **91** exclusively to sulfone **92** in 57% yield. Reduction of the sulfone using SmI$_2$, coupling with persilylated bases, and a sequence of deprotection steps provided the desired 2'-*C*-difluoromethylribonucleosides (eq 28).[47]

Van der Marel and coworkers reported a novel method to prepare pyrophosphates by coupling of a sugar phosphate and a nucleoside phosphoramidite.[48] Benzyl-2-acetamido-2-deoxy-α-D-glucoside **95** was first protected using di-*t*-butylsilyl bis(trifluoromethanesulfonate) to improve solubility of GlcNAc derivative. In four additional steps, the silyl ketal was converted into the tetrabutylammonium salt of GlcNAc-α-1-phosphate **97**. Uridine phosphoramidite **98** was then coupled with GlcNAc-α-1-phosphate **97** in the presence of dicyanoimidazole. The reaction was monitored using ^{31}P NMR spectroscopy. Upon complete disappearance of amidite **99**, the mixture was treated with *t*-butylperoxide to provide diastereomeric cyanoethyl-protected pyrophosphate **100**. The cyanoethyl group was then removed by treatment with anhydrous DBU, followed by treatment with HF/Et$_3$N to deprotect the di-*t*-butylsilylene group, and finally ammonium hydroxide was used to hydrolyze the acetate protecting groups to provide UDP-*N*-acetylglucosamine **101** in 76% yield (eq 29).[48]

Kishi and coworkers reported a protocol to improve the overall stereoselectivity of the Ireland–Claisen rearrangement for pyranoid and furanoid glucals.[49] To prepare building blocks for marine natural product halichondrins, the authors reported synthesis of silylene **102** in two steps from D-galactal. Under the conditions reported by Ireland (LHMDS, TBSCl, HMPA, THF, −78 °C), **102** was converted to the corresponding *O*-silyl ketene acetal **103**, which was estimated as a 7.3:1 mixture of *Z*-**103** and *E*-**103** as monitored using ^{1}H NMR analysis. Upon heating at 80 °C in

benzene for 1 day, this mixture furnished the carboxylate **104** as a *single diastereomer* in >85% yield, along with **102** and *E*-**103** in ca. 12% combined yield. The authors further provided experimental evidence that Claisen rearrangement took place through *Z*-**103** and that the Me stereochemistry of **104** indicated that the Claisen rearrangement proceeded exclusively via the boatlike transition state (eq 30).[49]

A practical approach for the stereoselective introduction of β-arabinofuranosides has been developed by locking an arabinosyl donor in a conformation in which nucleophilic attack from the β-face is favored.[50] This was achieved by using di-*t*-butylsilyl bis(trifluoromethanesulfonate) to protect the C-5 and C-3 hydroxyl groups. This resulted in C-5 and O-3 in a pseudoequatorial orientation, resulting in a perfect chair conformation of the protecting group. The nucleophilic attack from the

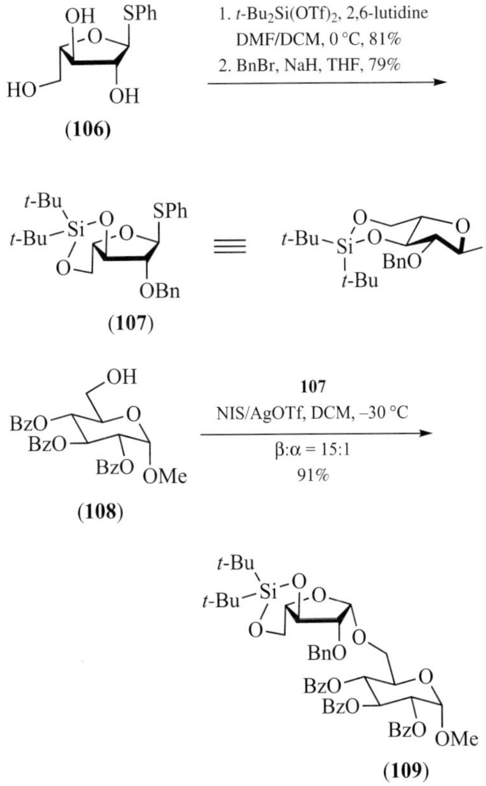

(31)

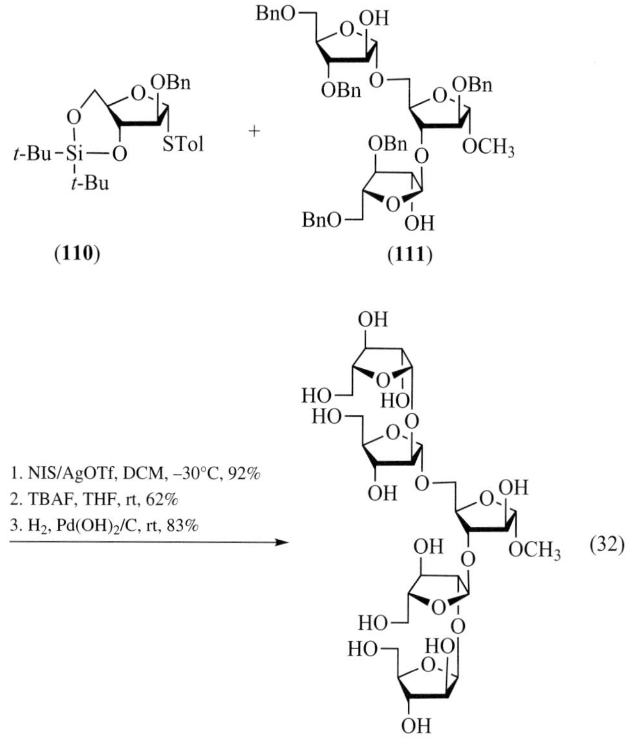

(110)   **(111)**

1. NIS/AgOTf, DCM, −30°C, 92%
2. TBAF, THF, rt, 62%
3. H₂, Pd(OH)₂/C, rt, 83%

(112)

α-face is disfavored due to unfavorable steric interactions with H-2. The glycosyl donor **107** was prepared in two convenient steps from commercially available thioglycoside **106**. The coupling of the conformationally constrained glycosyl donor **107** with the glycosyl acceptor **108** in the presence of the powerful thiophilic promoter system *N*-iodosuccinimide/silver triflate (NIS/AgOTf) in DCM at −30 °C gave disaccharide **109** with excellent β-selectivity (β:α = 15:1) in 91% yield (eq 31).[50]

Another β-selective glycosylation was reported by Lowary and coworkers who intended to study ligand specificity of CS-35, a monoclonal antibody that recognizes mycobacterial lipo-arabinomannan.[51] Selective glycosylation followed by two-step deprotection of the hydroxyl groups provided the desired pentasaccharide **112** in 51% overall yield (eq 32). Subsequently, Crich et al. studied the importance of the activation method for the observed selectivity in glycosylation using 3,5-*O*-(di-*t*-butylsilylene)-2-*O*-benzylarabinothiofuranosides as glycosyl donors for the synthesis of β-arabinofuranosides.[52]

A novel six-step synthesis of N_2-dimethylaminomethylene-2'-*O*-methylguanosine has been reported.[53] The synthesis utilizes di-*t*-butylsilylene protection for 3'- and 5'-hydroxyl groups. The arenesulfonylation at the O_6 position of 3'- and 5'-*O*-protected guanosine without 2'-*O*-protection was found to be completely selective. Compound **117** is a useful intermediate for oligonucleotide construction (eq 33). A similar application of di-*t*-butylsilylene protecting group was originally reported for the synthesis of N_2-isobutyryl-2'-*O*-methyl guanosine.[54]

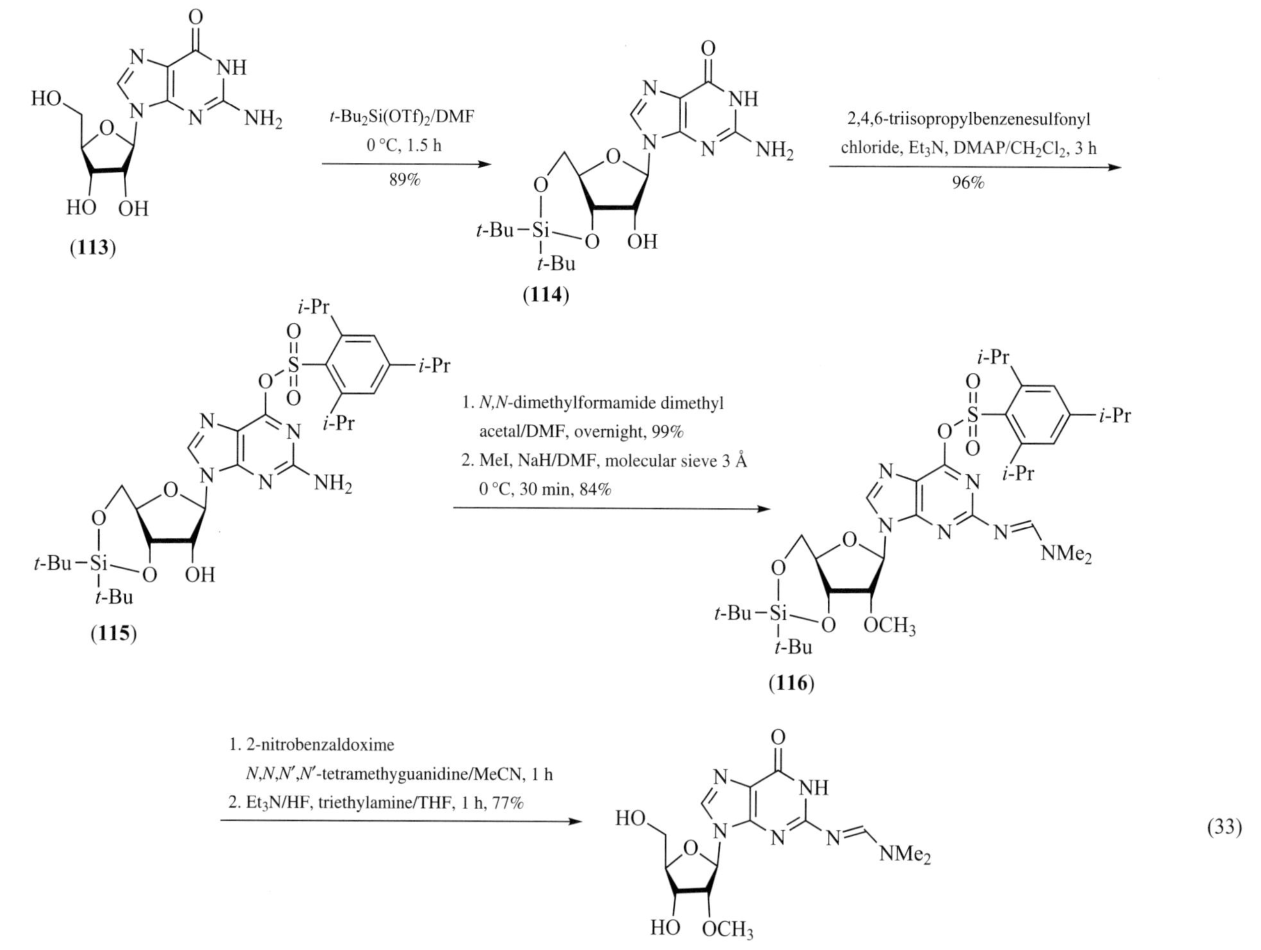

(113)

t-Bu₂Si(OTf)₂/DMF
0 °C, 1.5 h
89%

(114)

2,4,6-triisopropylbenzenesulfonyl
chloride, Et₃N, DMAP/CH₂Cl₂, 3 h
96%

(115)

1. *N,N*-dimethylformamide dimethyl
acetal/DMF, overnight, 99%
2. MeI, NaH/DMF, molecular sieve 3 Å
0 °C, 30 min, 84%

(116)

1. 2-nitrobenzaldoxime
N,N,N′,N′-tetramethyguanidine/MeCN, 1 h
2. Et₃N/HF, triethylamine/THF, 1 h, 77%

(33)

(117)

Sabatino and Damha recently reported synthesis, characterization, and properties of oxepane nucleic acids.[55] These sugar phosphate oligomers have the pentofuranose ring of DNA and RNA replaced with a seven-membered sugar ring. The oxepane nucleoside monomers were prepared from the ring expansion reaction of a cyclopropanated glycal, **118**, and their conversion into phosphoramidite derivatives. Properties of these oxepane nucleic acids were then compared to naturally occurring DNA (eq 34).

Regioselective Monodeprotection. The monodeprotection of di-*t*-butylsilylene ethers prepared from substituted 1,3-pentanediols and 2,4-hexanediols has been achieved with BF$_3$·SMe$_2$.[56] The reaction is highly selective and provides access to 1,3-diols silylated at the sterically more hindered position. This is consistent with coordination of boron to the sterically more accessible oxygen prior to intramolecular delivery of fluoride. The reaction conditions for deprotection are compatible with esters, allyl ethers, and TIPS ethers. The resulting secondary di-*t*-butylfluorosilyl ethers are stable to various conditions including low pH aqueous solutions and silica gel chromatography; the di-*t*-butylfluorosilyl ethers are readily cleaved with HF–pyridine. An example is provided below (eq 35).

1. Corey, E. J.; Hopkins, P. B., *Tetrahedron Lett.* **1982**, *23*, 4871.
2. Furusawa, K.; Ueno, K.; Katsura, T., *Chem. Lett.* **1990**, 97.
3. Furusawa, K., *Chem. Express* **1991**, *6*, 763.
4. Furusawa, K., *Chem. Lett.* **1989**, 509.
5. Furusawa, K.; Sakai, T., *Jpn. Kokai Tokkyo Koho* **1989**, 629.
6. Brooks, C. J. W.; Cole, W. J., *Analyst (London)* **1985**, *110*, 587.
7. Ahn, K.-H.; Klassen, R. B.; Lippard, S. J., *Organometallics* **1990**, *9*, 3178.
8. Uhlig, W., *Chem. Ber.* **1992**, *125*, 47.
9. Uhlig, W., *Z. Anorg. Allg. Chem.* **1991**, *603*, 109.
10. Uhlig, W., *J. Organomet. Chem.* **1991**, *421*, 189.
11. Uhlig, W., *Z. Anorg. Allg. Chem.* **1991**, *601*, 125.
12. Gillard, J. W.; Fortin, R.; Grimm, E. L.; Maillard, M.; Tjepkema, M.; Bernstein, M. A.; Glaser, R., *Tetrahedron Lett.* **1991**, *32*, 1145.
13. (a) Paterson, I.; Cumming, J. G.; Ward, R. A.; Lamboley, S., *Tetrahedron* **1995**, *51*, 9393. (b) Paterson, I.; Smith, J.; Ward, R. A., *Tetrahedron* **1995**, *51*, 9413. (c) Paterson, I.; Ward, R. A.; Smith, J. D.; Cumming, J. G.; Yeung, K., *Tetrahedron* **1995**, *51*, 9437. (d) Paterson, I.; Yeung, K.; Ward, R. A.; Smith, J. A.; Cumming, J. G.; Lamholey, S., *Tetrahedron* **1995**, *51*, 9467.
14. Paterson, I.; McLeodl, M. D., *Tetrahedron Lett.* **1997**, *38*, 4183.
15. Toshima, K.; Jyojima, K.; Miyamoto, N.; Katohno, M.; Nakata, M.; Matsumura, S., *J. Org. Chem.* **2001**, *66*, 1708.
16. Panek, J. S.; Jain, N. F., *J. Org. Chem.* **2001**, *66*, 2747.
17. Di Fabio, R.; Rossi, T.; Thomas, R. J., *Tetrahedron Lett.* **1997**, *38*, 3587.
18. Madu, C. E.; Lovely, C. J., *Org. Lett.* **2007**, *9*, 4697.
19. Caspi, D. D.; Garg, N. K.; Stoltz, B. M., *Org. Lett.* **2005**, *7*, 2513.
20. Hillaert, U.; Van Calenbergh, S., *Org. Lett.* **2005**, *7*, 5769.
21. Yadav, J. S.; Srinivas, R.; Sathaiah, K., *Tetrahedron Lett.* **2006**, *47*, 1603.
22. Perkins, M. V.; Sampson, R. A.; Joannou, J.; Taylor, M. R., *Tetrahedron Lett.* **2006**, *47*, 3791.
23. Chan, C.; Zheng, S.; Zhou, B.; Guo, J.; Heid, R. H.; Wright, B. J. D.; Danishefsky, S. J., *Angew. Chem., Int. Ed.* **2006**, *45*, 1749.
24. Paterson, I.; Mühlthau, F. A.; Cordier, C. J.; Housden, M. P.; Burton, P. M.; Loiseleur, O., *Org. Lett.* **2009**, *11*, 353.
25. Obringer, M.; Barbarotto, M.; Choppin, S.; Colobert, F., *Org. Lett.* **2009**, *11*, 3542.
26. Calad, S. A.; Cirakovic, J.; Woerpel, K. A., *J. Org. Chem.* **2007**, *72*, 1027.
27. Guan, Y.; Wu, J.; Sun, L.; Dai, W., *J. Org. Chem.* **2007**, *72*, 4953.
28. Engers, D. W.; Bassindale, M. J.; Pagenkopf, B. L., *Org. Lett.* **2004**, *6*, 663.
29. Quéron, E.; Lett, R., *Tetrahedron Lett.* **2004**, *45*, 4533.
30. Nakamura, R.; Tanino, K.; Miyashita, M., *Org. Lett.* **2005**, *7*, 2929.
31. Denmark, S. E.; Kobayashi, T.; Regens, C. S., *Tetrahedron* **2010**, *66*, 4745.
32. Nicolaou, K. C.; Gelin, C. F.; Seo, J. H.; Huang, Z.; Umezawa, T., *J. Am. Chem. Soc.* **2010**, *132*, 9900.
33. Watanabe, K.; Suzuki, M.; Murata, M.; Oishi, T., *Tetrahedron Lett.* **2005**, *46*, 3991.
34. Clark, J. S.; Kimber, M. C.; Robertson, J.; McErlean, C. S.; Wilson, C., *Angew. Chem., Int. Ed.* **2005**, *44*, 6157.
35. Hooper, A. M.; Dufour, S.; Willaert, S.; Pouvreau, S.; Pickett, J. A., *Tetrahedron Lett.* **2007**, *48*, 5991.
36. Takada, T.; Arisawa, M.; Gyoten, M.; Hamada, R.; Tohma, H.; Kita, Y., *J. Org. Chem.* **1998**, *63*, 7698.
37. Akai, S.; Tsujino, T.; Fukuda, N.; Iio, K.; Takeda, Y.; Kawaguchi, K.; Naka, T.; Higuchi, K.; Kita, Y., *Org. Lett.* **2001**, *3*, 4015.
38. Fuller, P. H.; Kim, J.; Chemler, S. R., *J. Am. Chem. Soc.* **2008**, *130*, 17638.
39. Kablean, S. N.; Marsden, S. P.; Craig, A. M., *Tetrahedron Lett.* **1998**, *39*, 5109.
40. Ahmed, A.; Dussault, P. H., *Tetrahedron* **2005**, *61*, 4657.
41. Baudat, A.; Vogel, P., *J. Org. Chem.* **1997**, *62*, 6252.
42. Plante, O. J.; Palmacci, E. R.; Seeberger, P. H., *Org. Lett.* **2000**, *2*, 3841.
43. Houseknecht, J. B.; Lowary, T. L., *J. Org. Chem.* **2002**, *67*, 4150.
44. Haraguchi, K.; Takahashi, H.; Shiina, N.; Horii, C.; Yoshimura, Y.; Nishikawa, A.; Sasakura, E.; Nakamura, K. T.; Tanaka, H., *J. Org. Chem.* **2002**, *67*, 5919.

45. Ohnishiy, Y.; Ichikawa, Y., *Bioorg. Med. Chem. Lett.* **2002**, *12*, 997.

46. Serebryany, V.; Beigelman, L., *Tetrahedron Lett.* **2002**, *43*, 1983.

47. Ye, J.; Liao, X.; Piccirilli, J. A., *J. Org. Chem.* **2005**, *70*, 7902.

48. Gold, H.; Van Delft, P.; Meeuwenoord, N.; Codée, J. D.; Filippov, D. V.; Eggink, G.; Overkleeft, H. S.; Van der Marel, G. A., *J. Org. Chem.* **2008**, *73*, 9458.

49. Chen, C.; Namba, K.; Kishi, Y., *Org. Lett.* **2009**, *11*, 409.

50. Zhu, X.; Kawatkar, S.; Rao, Y.; Boons, G., *J. Am. Chem. Soc.* **2006**, *128*, 11948.

51. Rademacher, C.; Shoemaker, G. K.; Kim, H.;, Zheng, R. B.; Taha, H.; Liu, C.; Nacario, R. C.; Schriemer, D. C.; Klassen, J. S.; Peters, T.; Lowary, T. L., *J. Am. Chem. Soc.* **2007**, *129*, 10489.

52. Crich, D.; Pedersen, C. M.; Bowers, A. A.; Wink, D. J., *J. Org. Chem.* **2007**, *72*, 1553.

53. Mukobata, T.; Ochi, Y.; Ito, Y.; Wada, S.; Urata, H., *Bioorg. Med. Chem. Lett.* **2010**, *20*, 129.

54. Zlatev, I.; Vasseur, J.; Morvan, F., *Tetrahedron* **2007**, *63*, 11174.

55. Sabatino, D.; Damha, M. J., *J. Am. Chem. Soc.* **2007**, *129*, 8259.

56. Yu, M.; Pagenkopf, B. L., *J. Org. Chem.* **2002**, *67*, 4553.

Dichlorodiisopropylsilane

$$i\text{-}Pr_2SiCl_2$$

[7751-38-4] $C_6H_{14}Cl_2Si$ (MW 185.17)

InChI = 1S/C6H14Cl2Si/c1-5(2)9(7,8)6(3)4/h5-6H,1-4H3

InChIKey = GSENNYNYEKCQGA-UHFFFAOYSA-N

(nucleoside 3′,5′-hydroxy protection; component in Peterson alkenation reaction)

Physical Data: bp 50–55 °C/5 mmHg.[1,2]

Solubility: sol most organic solvents.

Preparative Method: conveniently prepared by the reaction of diisopropylsilane in carbon tetrachloride with palladium(II) chloride at 140 °C for 8 h in a steel bomb. Washing the residue with ether and removing the ether by rotary evaporation yields the product which is purified by distillation to yield a clear colorless oil.

Handling, Storage, and Precautions: special care should be used when handling since the reagent is corrosive.

Original Commentary

Janet Wisniewski Grissom

University of Utah, Salt Lake City, UT, USA

Nucleoside-protecting Group. The use of silicon groups as protection for hydroxy groups has been widely exploited.[3] One potentially useful way to protect a nucleoside is by using a protecting group which spans the 3′- and 5′-hydroxy groups. One of the useful groups which can be used as a bifunctional protecting group in nucleosides is the diisopropylsilyl group. Reaction of the nucleoside with dichlorodiisopropylsilane and imidazole in DMF gives the protected nucleoside (eq 1).[4] This method has not been

widely used; the more common 3′,5′-hydroxy-bridging protecting group is a tetraisopropyldisiloxy group.[5]

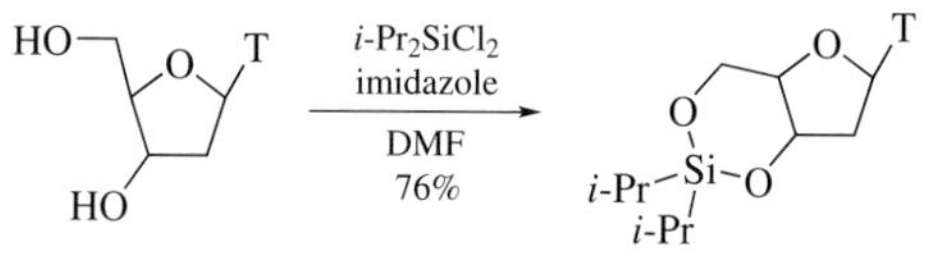

$$(1)$$

Nucleosides are generally linked together through phosphate linkages. Studies are being directed toward synthesizing alternative linkages that will impart different solubilities to oligonucleotides. A diisopropylsilyl protecting group has also been used as a phosphate linker isostere (eq 2).[1]

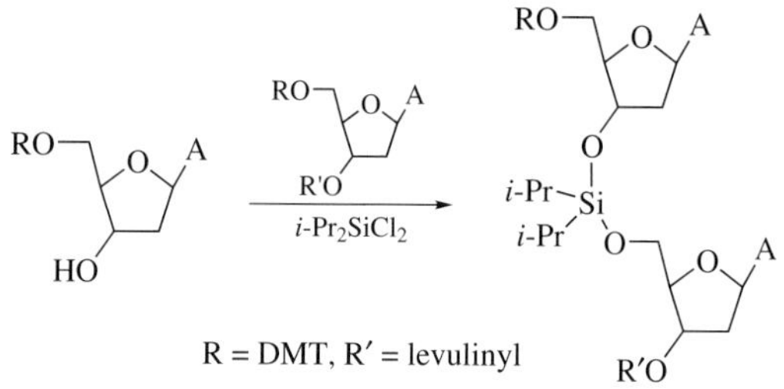

R = DMT, R′ = levulinyl

$$(2)$$

A diisopropylsilyl group has also been used as a linker in a radical cyclization reaction to form eight-membered rings (eq 3).[6]

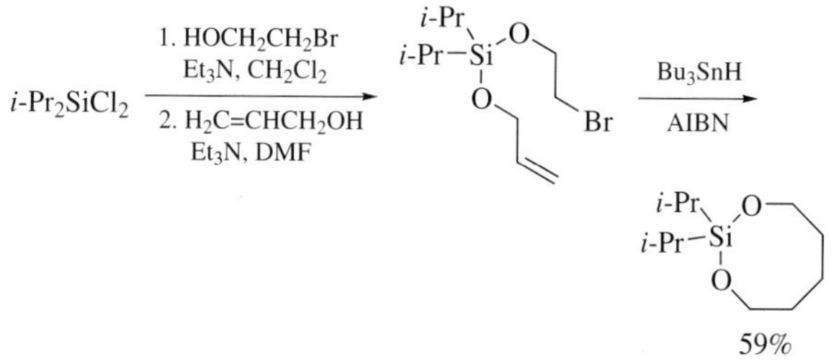

$$(3)$$

Modified Peterson Alkenation. The Peterson alkenation is a useful method for the alkenation of carbonyl compounds. Dichlorodiisopropylsilane has been used in a Peterson alkenation modification (eq 4).[7]

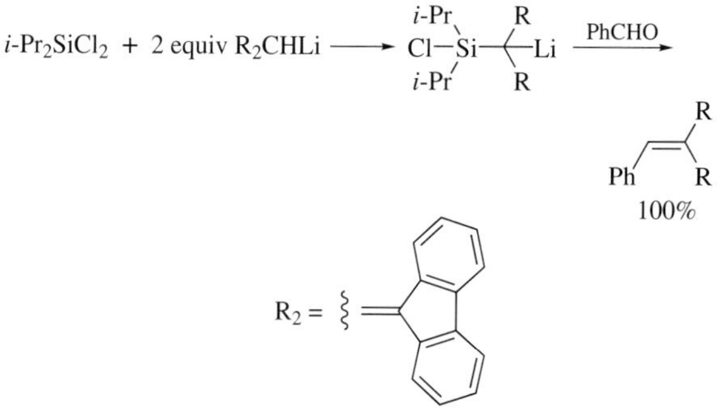

$$(4)$$

First Update

Shiyue Fang
Michigan Technological University, Houghton, MI, USA

Donald E. Bergstrom
Purdue University, West Lafayette, IN, USA

Introduction. In comparison to silyl groups containing methyl or ethyl substituents, those containing isopropyl groups are considerably more stable to chemical reagents including acid, base, and fluoride. This frequently allows one to prepare and manipulate silyl compounds with fewer complications from adventitious side reactions and provides greater control in both introduction and removal of the silyl group. The two chlorine atoms in i-Pr$_2$SiCl$_2$ can be displaced by carbon, oxygen, silicon, phosphorus, and other nucleophiles. Of these, oxygen nucleophiles are most common. The formation of silyl diethers using various silylation reagents including i-Pr$_2$SiCl$_2$ has been reviewed.[8] Most of these silyl diether formation reactions were used for protecting diols and to form cleavable linkages between two molecules or between a molecule and solid support in solid-phase synthesis, which will be discussed in the following sections.

Building Block in Organometallic Compounds. i-Pr$_2$SiCl$_2$ has been used in the synthesis of organosilicon compounds by displacement of either one or two chlorides with a carbon nucleophile. The resulting i-Pr$_2$Si, or less commonly i-Pr$_2$SiCl, building block becomes an integral part of the target compound. Metalated terminal alkynes have been shown to react with i-Pr$_2$SiCl$_2$.[9–12] For example, treating **1** with 2 equiv n-BuLi followed by i-Pr$_2$SiCl$_2$ gave the strained planar silacyclyne **2** in 32% yield along with its dimer (13%, not shown, eq 5).[9]

Reactive carbon nucleophiles also include metalated metallocene and (arene)metal complexes.[13–15] Equation 6 shows an example of the former; treating the suspension of dilithoferrocene–tmeda complex in Et$_2$O with i-Pr$_2$SiCl$_2$ gave **3** in 34% yield.[15] An example of the latter is shown in eq 7; complex **4** was metalated with 2 equiv n-BuLi in the presence of N,N,N',N',N'-pentamethyldiethylenetriamine (pmdta) in pentane, the dilithiated intermediate was isolated and then treated with i-Pr$_2$SiCl$_2$ in heptane to give complex **5**.[14]

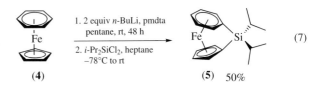

Metalated arenes were also used to react with i-Pr$_2$SiCl$_2$; the dichloroarene **6** was first metalated with activated magnesium metal in THF and then treated with i-Pr$_2$SiCl$_2$ to give the silaanthracene **7** in 32% yield (eq 8).[16] Metalated sp^3 hybridized carbons are additional suitable nucleophiles; for example, the i-Pr$_2$Si group was used to link two 1,3-benzodithiole molecules.[17] The same group was used to link two phosphines via two methylene groups; the products were tested as bidentate ligands in transition metal-mediated catalysis.[18]

Metalated silicon,[19–22] phosphorus,[23,24] and osmium[25] atoms were reacted with i-Pr$_2$SiCl$_2$ to form silicon–silicon, phosphorus–silicon, and osmium–silicon bonds, respectively. For example, the diionic **8** was reacted with i-Pr$_2$SiCl$_2$ in benzene at reflux temperature to give the cyclopentasilane **9** in 35% yield (eq 9).[19]

Because the Si–Cl bond is sensitive to moisture, compounds containing i-Pr$_2$SiCl are less common. However, with careful manipulation, such compounds were characterized by NMR, MS, elemental analysis, and single-crystal X-ray diffraction analysis.[26–28] For example, the bis(benzene)molybdenum complex was metalated with 2 equiv t-BuLi to give the diionic intermediate **10**, which was subjected to react with i-Pr$_2$SiCl$_2$ to give complex **11**; the yield of the latter step is 37% (eq 10).[28]

Diol- and Bis(hydroxyamino)-protecting Group. i-Pr$_2$SiCl$_2$ is a good choice for protecting 1,2-, 1,3-, and potentially 1,n-(with $n > 3$) diols;[29] it was also used for protecting the bis(hydroxyamino) functionality.[30,31] Compared to Me$_2$SiCl$_2$, i-Pr$_2$SiCl$_2$ gives much more robust protection due to the bulkiness of the two isopropyl groups, which slow hydrolysis of the Si–O bonds; compared to $t\text{-Bu}_2\text{SiCl}_2$, the reaction between $i\text{-Pr}_2\text{SiCl}_2$ and a diol is smoother and usually gives higher yield; in addition, diols protected with $i\text{-Pr}_2\text{SiCl}_2$ are easier to deprotect than those protected with $t\text{-Bu}_2\text{SiCl}_2$. Equation 11 shows an example of 1,3-diol protection using $i\text{-Pr}_2\text{SiCl}_2$; stirring compound **12** with $i\text{-Pr}_2\text{SiCl}_2$ in the presence of imidazole gave **13** in excellent yield. The $i\text{-Pr}_2\text{Si}$ group was later removed by HF-pyridine in THF at $0\,^\circ\text{C}$ for 3 h.[29]

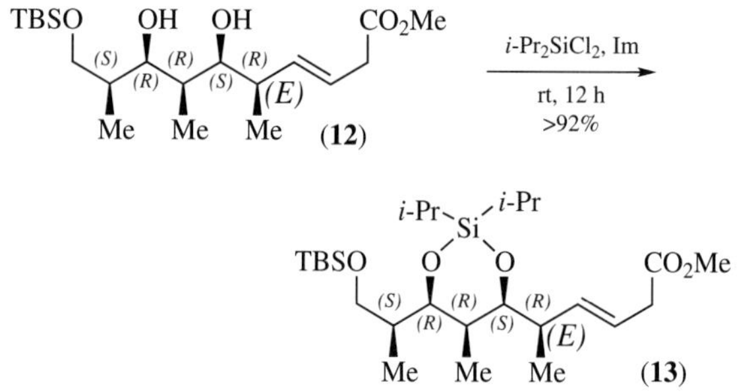

$$(11)$$

1,2-Bis(hydroxyamino) compounds can be protected using $i\text{-Pr}_2\text{SiCl}_2$ as illustrated in eq 12; the protection was achieved by first neutralizing **14** with Et_3N, followed by treating with $i\text{-Pr}_2\text{SiCl}_2$, imidazole, and DMAP to afford **15**. After acylation of the two nitrogen atoms, the $i\text{-Pr}_2\text{Si}$ group was removed by 3 N HCl in methanol at room temperature for 30 min.[31]

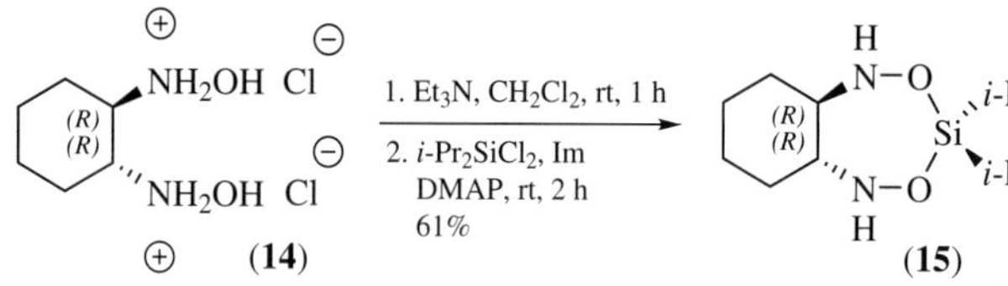

$$(12)$$

Linker for Solid-phase Synthesis. Because the Si–O bond is easy to form, is stable under many conditions, and can be cleaved by fluoride and acid, $i\text{-Pr}_2\text{SiCl}_2$ is used as a cleavable linker in solid-phase synthesis.[32–40] Linkage types between a solid support and a substrate include solid support–O-Si-O-,[33,34,36] solid support–C-Si-O-,[35,37–39] and solid support–O-Si-C-.[32,40] When the first type is used, the substrate is released after cleavage as an alcohol. For example, alcohol **16** was silylated with $i\text{-Pr}_2\text{SiCl}_2$ in DMF and imidazole, and then affixed to a hydroxymethylpolystyrene resin (**17**, $0.87\ \text{mmol g}^{-1}$ hydroxyl loading) in two cycles (each cycle, 6 equiv silylated **16**, rt, 36 h) giving **18** with a high loading value ($0.75\ \text{mmol g}^{-1}$, eq 13).[34] After solid-phase synthesis, the substrate was cleaved from the support by HF–pyridine complex or TBAF.[34]

The solid support–C-Si-O- linkage enjoys greater stability, but is more complicated to create. Upon cleavage, this type of linkage was reported to release the substrate as an alcohol,[37,38] a phenol,[35] and an amine.[39] Equation 14 shows an example in which an alcohol is released upon cleavage. Polystyrene (**19**) was treated with n-BuLi in the presence of tmeda in cyclohexane; after removal of the solvent, the metalated intermediate was reacted with $i\text{-Pr}_2\text{SiCl}_2$. Alcohol **20** was then loaded in CH_2Cl_2 in the presence of imidazole to give **21** after manipulation on solid support, the alcohol

was released with fluoride (5 equiv TBAF, 1.7 equiv AcOH, THF, 48 h, $25\,^\circ\text{C}$).[37]

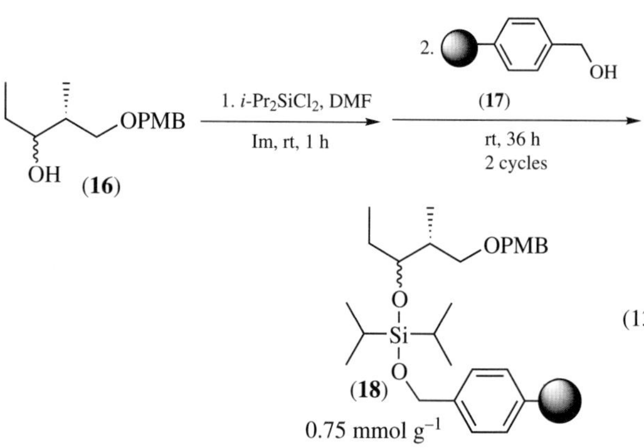

$$(13)$$

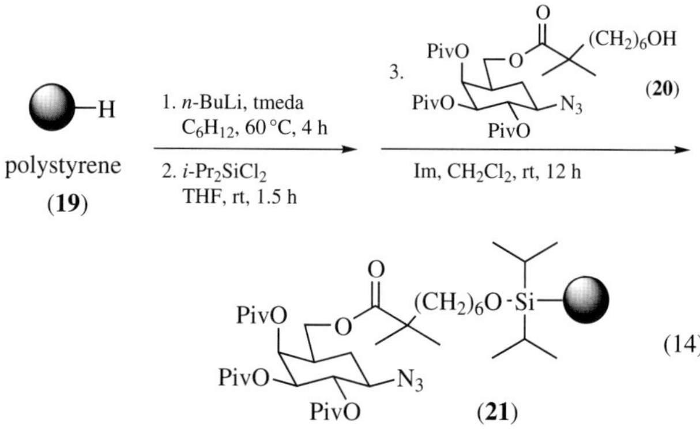

$$(14)$$

Equation 15 shows an example in which an amine is released upon cleavage. *para*-Bromopolystyrene (**22**) was lithiated in THF at $60\,^\circ\text{C}$ and then silylated with $i\text{-Pr}_2\text{SiCl}_2$. The resulting chlorodiisopropylsilyl support was then allowed to react with **23** to give urethane **24**. After peptide synthesis on the support, the product was cleaved by treatment with aqueous HF.[39]

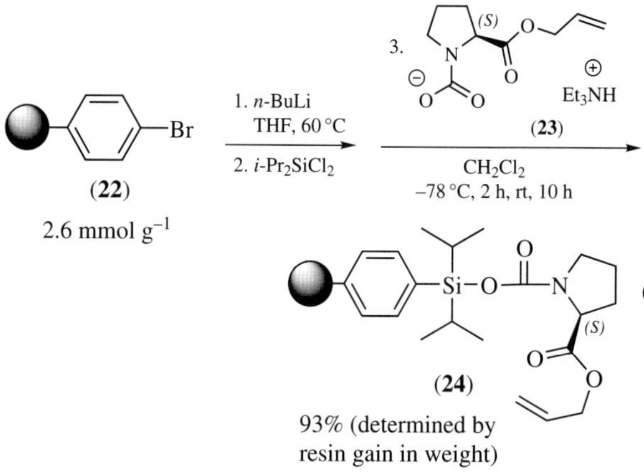

$$(15)$$

93% (determined by resin gain in weight)

The third type of linkage (solid support–O-Si-C-) is called a traceless linker; after cleavage, the substrate is released in an unmodified form, which means that the silicon atom is replaced with a hydrogen atom. An example is shown in eq 16. The substrate **25** was lithiated and then silylated with $i\text{-Pr}_2\text{SiCl}_2$ to give a chlorodiisopropylsilyl intermediate in 71% yield; this was next attached to 4-hydroxylmethyl polystyrene (**26**) to give **27**. The ultimate product was cleaved by treating with TBAF in DMF at 65 °C for 2 h.[40]

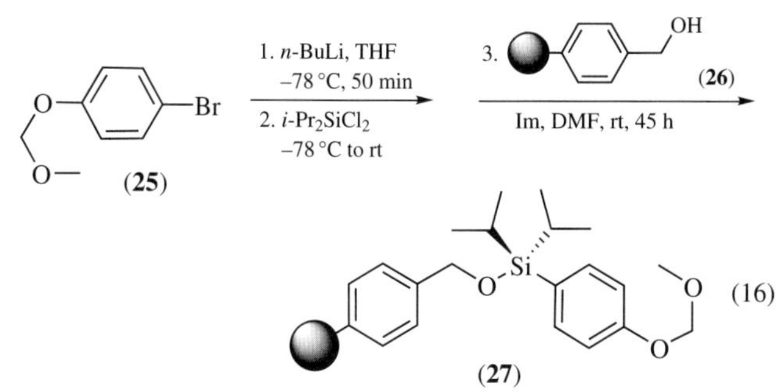

(25)   (26)   (27)

Linker for Intramolecular Reactions. i-Pr$_2$SiCl$_2$ has been widely used for linking two reactants together to convert an intermolecular reaction into an intramolecular reaction. In most cases, a silyl diether (in the form of -O-Si-O-) is employed to link the two molecules.[41–47] Equation 17 shows an example for converting an intermolecular alkene metathesis reaction to an intramolecular one.[43] The racemic **28** was treated with 10 equiv i-Pr$_2$SiCl$_2$ in the presence of imidazole; excess i-Pr$_2$SiCl$_2$ was removed and the intermediate reacted with **29** to give **30** in 86% yield. The alkene metathesis reaction was catalyzed with Grubbs' first generation ruthenium catalyst giving **31** with a diastereoselectivity of cis/$trans$ 20:1 in 82% combined yield. The temporary silyl linker was removed by treating with aqueous HF to give the corresponding diol in 99% yield.[43]

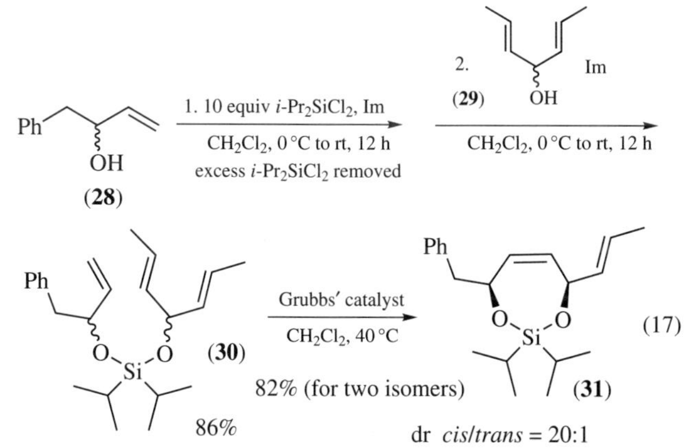

(28)   (29)   (30)   (31)   (17)

Although the use of a large excess of i-Pr$_2$SiCl$_2$ in derivatization of the first of the two alcohols in preparation of unsymmetrical silaketals is frequently successful, as illustrated by the preparation of **30** in eq 17, there is an alternative if difficulties are encountered. The first alcohol coupling can be done with the reagent chlorodiisopropylsilane and the resulting alkoxydiisopropylsilane oxidized with NBS to the alkoxybromodiisopropylsilane, which can then be coupled to a second alcohol with displacement of bromide to yield the unsymmetrical silaketal.[48,49]

Much larger cyclic systems containing the diisopropylsilyl group have been obtained by alkyne metathesis. The example shown in eq 18 describes the formation of a molecular triangle (not shown) and the rhomboid **33** involving the reactions between i-Pr$_2$SiCl$_2$ and oxygen nucleophiles to form silyl diether **32**, alkynylation of **32**, and subsequent Mo(CO)$_6$-catalyzed alkyne metathesis.[50]

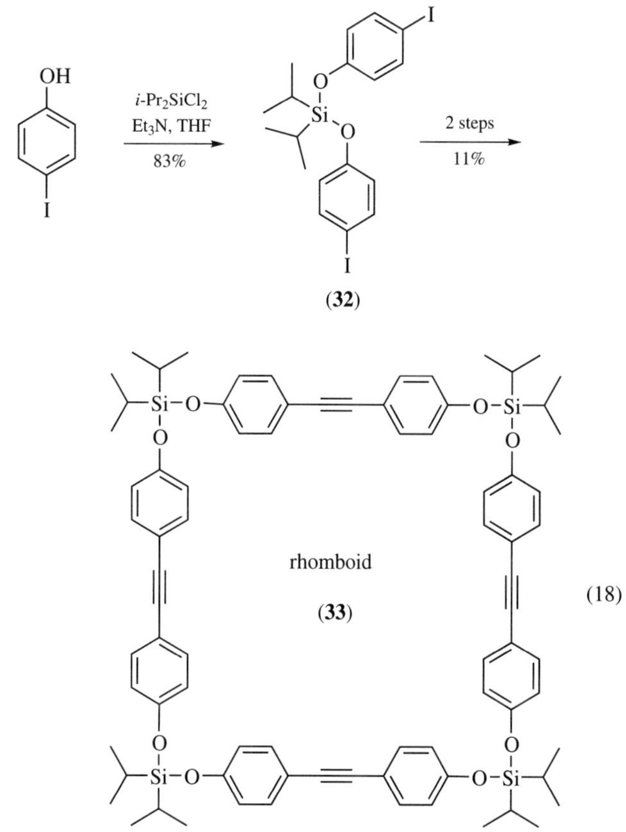

(32)   (18)

rhomboid
(33)

The silyl linkage can also be in the form of -O-Si-C-.[51,52] or -C-Si-C-.[53] One such example is shown in eq 19 Compound **36** was prepared by treating **34** with i-Pr$_2$SiCl$_2$, followed by enol **35**. Subjection of **36** to radical cyclization gave product **37** in 50% yield.[52]

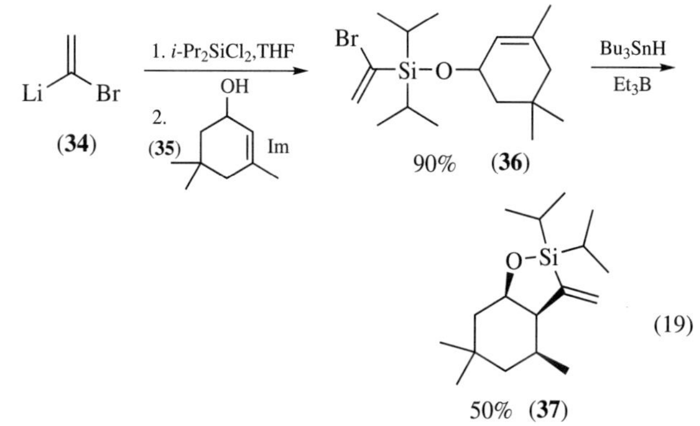

(34)   (35)   (36)   (19)   (37)

Similarly, i-Pr$_2$SiCl$_2$ has been used as a reagent to create an allylsilane that could be coupled in an intramolecular Pauson–Khand reaction, as illustrated in eq 20. Ring closure was successful when a six-membered ring was formed, but not for seven-membered ring formation.[54]

Finally, the i-Pr$_2$Si linker was proposed to render intermolecular glycosidic bond formation reactions to intramolecular ones in oligosaccharide synthesis[46] and has been used to facilitate [2 + 2] cycloaddition reactions.[42,44] A particularly useful example of a [2 + 2] photochemical cycloaddition facilitated by the i-Pr$_2$Si linker is the synthesis of the cis-syn thymidine photodimer, as illustrated in eq 21.[55] After suitable modification, the photodimer can be incorporated into oligonucleotides by automated phospho-ramidite synthesis.

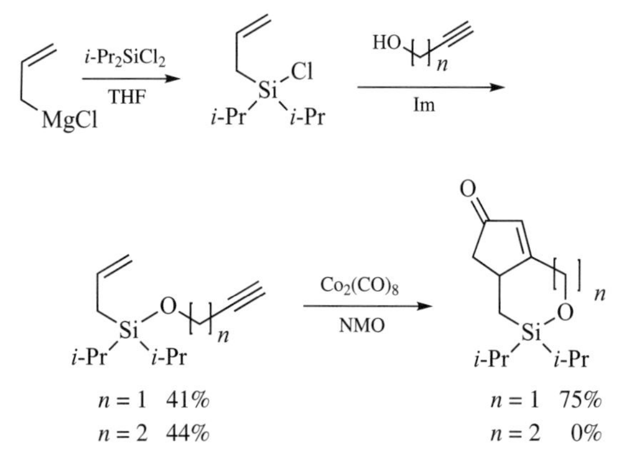

$$ (20) $$

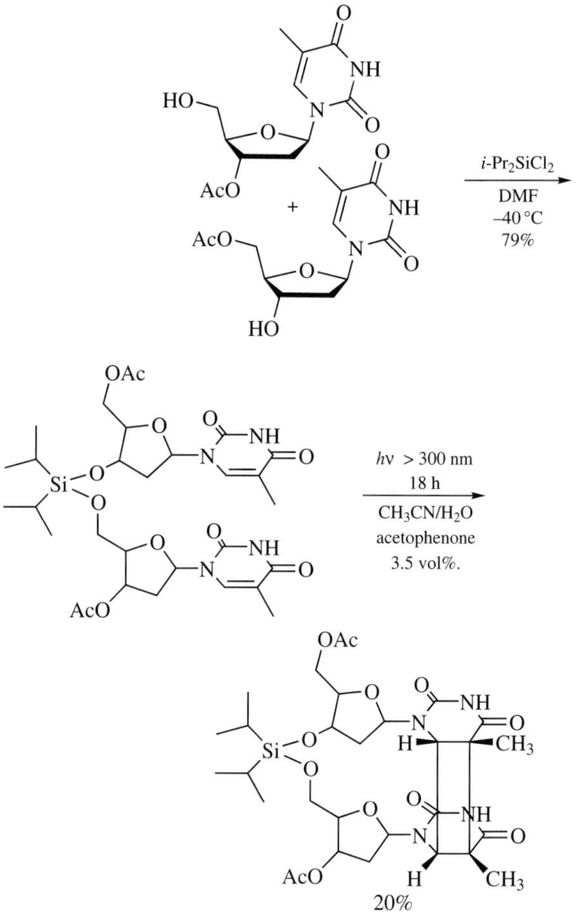

$$ n = 1 \quad 41\% \qquad n = 1 \quad 75\% $$
$$ n = 2 \quad 44\% \qquad n = 2 \quad 0\% $$

Linker for Oligonucleotide Purification. i-Pr$_2$SiCl$_2$ was used as a linking reagent in synthetic oligonucleotide purification in several reports.[56–59] For example, the highly hydrophobic tocopherol **38** was silylated with i-Pr$_2$SiCl$_2$ and then reacted with 1,6-hexanediol to give **39** (44%), which was phosphinylated with 2-cyanoethyl N,N-diisopropylchlorophosphoramidite and incorporated onto the 5′-end of oligodeoxyribonucleotides on a solid-phase synthesizer. After cleavage and deprotection, the modified oligonucleotide **40** containing a hydrophobic group was obtained (eq 22). Failure sequences are major impurities in oligonucleotide synthesis, which are usually difficult to remove and do not contain such a hydrophobic group; therefore, the full-length sequence **40** was easily separated from the failure ones by reverse-phase column chromatography.[58]

In another example, the phosphoramidite **41** with a biotin moiety linked through a silyl diether linkage was coupled to the 5′-end of oligonucleotide on a solid support (eq 23). After synthesis, cleavage, and deprotection, the oligonucleotide **42** that contained

$$ (21) $$

20%

a biotin was obtained along with failure sequences devoid of biotin. Incubation of the crude oligonucleotides with avidin-coated beads attached **42** to the beads through the noncovalent interaction between biotin and avidin, and the failure sequences were readily removed by washing. Unmodified pure oligonucleotide was obtained in high yield by treating the beads with HF–pyridine.[57] Oligonucleotides with a 5′-phosphate group and short synthetic RNAs were purified in a similar fashion.[56,59]

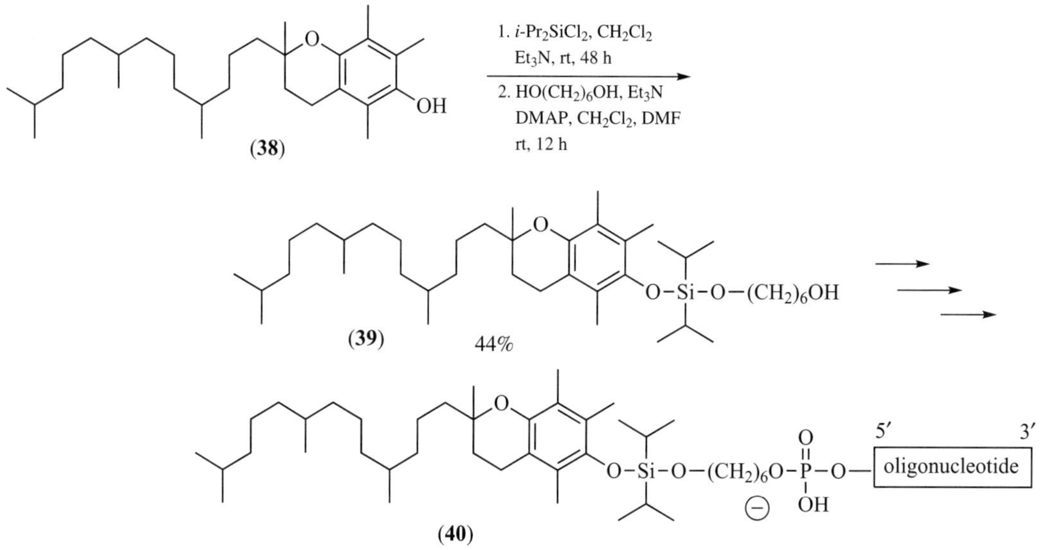

$$ (22) $$

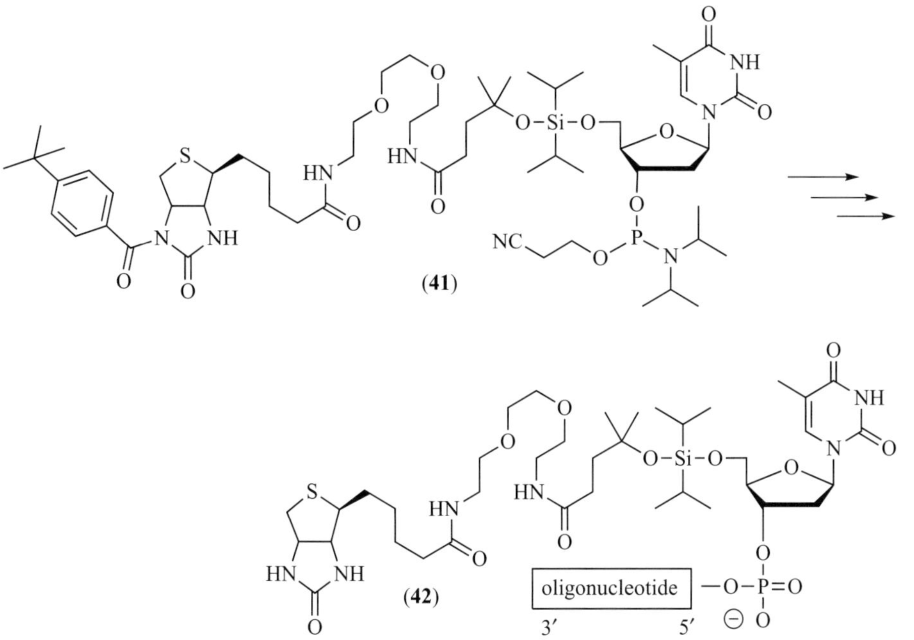

(41)

(23)

(42)

Miscellaneous. *i*-Pr$_2$SiCl$_2$ was used in the preparation of the silanol **43**. The hydroxyl group on the silicon atom in **43** was proposed to direct the molecule to dimerize to form the [4 + 4] product **44** (eq 24).[60]

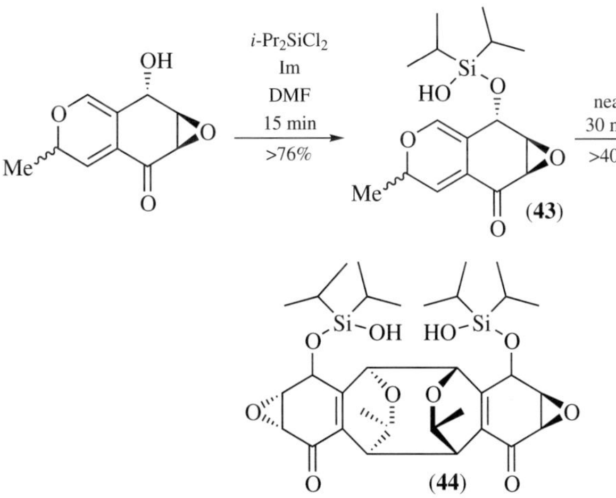

(43)

(24)

It is well accepted that silicon atom can stabilize a carbon cation at its β position. Equation 25 shows an example that uses this phenomenon for the formation of multiple bonds, which involves using *i*-Pr$_2$SiCl$_2$ to prepare the substrate. *i*-Pr$_2$SiCl$_2$ was first treated with PhMgBr, the intermediate was next lithiated at the silicon atom with lithium metal, and then allowed to react with **45** to give **46**. After unmasking the two ketone groups, **46** was treated with a Lewis acid to give **47**. A silicon-stabilized β carbon cation was believed to effect the multiple ring formation reactions (eq 25).[61]

Equation 26 shows an example in which a Si–C bond, formed involving using *i*-Pr$_2$SiCl$_2$, is oxidized to give an alcohol product. *i*-Pr$_2$SiCl$_2$ was treated sequentially with PhLi, *n*-BuLi, and CH$_2$BrCl to give **48**, which was converted to **49** in several steps.

Compound **49** was oxidized to the alcohol **50** by treating with *t*-BuO$_2$H under basic conditions in DMF at elevated temperature.[62]

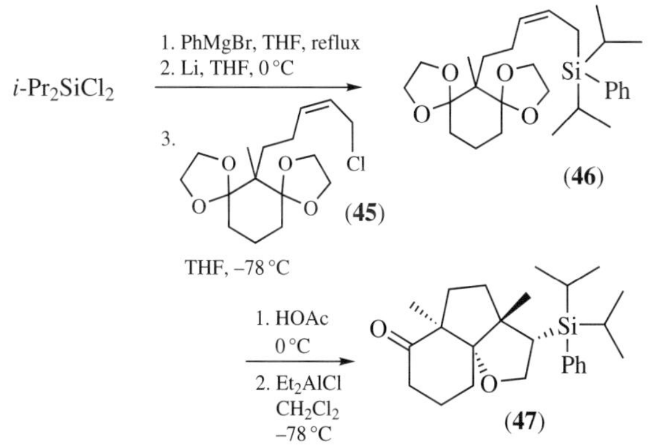

(25)

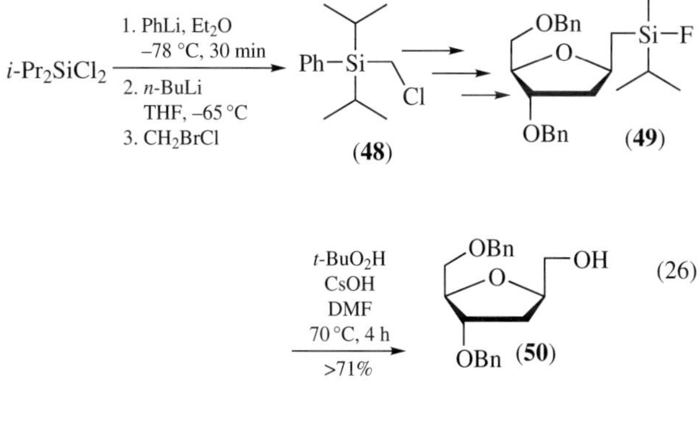

(49)

(26)

Related Reagents. Di-*t*-butyldichlorosilane; Di-*t*-butylmethylsilyl Trifluoromethanesulfonate; Di-*t*-butylsilyl Bis(trifluoromethanesulfonate); Dichlorodimethylsilane; Diisopropylsilyl Bis(trifluoromethanesulfonate).

1. Cormier, J. F.; Ogilvie, K. K., *Nucleic Acid Res.* **1988**, *16*, 4583.

2. Hurd, C. D.; Yarnall, W. A., *J. Am. Chem. Soc.* **1949**, *71*, 755.

3. For reviews on silylations see:(a) Klebe, J. F., *Acc. Chem. Res.* **1970**, *3*, 299. (b) Cooper, B. E., *Chem. Ind. (London)* **1978**, 794.

4. Furusawa, K.; Katsura, T., *Tetrahedron Lett.* **1985**, *26*, 887.

5. Markiewicz, W. T., *J. Chem. Res. (S)* **1979**, 24.

6. Hutchinson, J. H.; Daynard, T. S.; Gilliard, J. W., *Tetrahedron Lett.* **1991**, *32*, 573.

7. Couret, C.; Escudie, J.; Delpon-Lacaze, G.; Satge, J., *J. Organomet. Chem.* **1992**, *440*, 233.

8. Skrydstrup, T., *Sci. Synth.* **2002**, 269.

9. Guo, L.; Hrabusa J. M.; Tessier C. A.; Youngs W. J.; Lattimer R., *J. Organomet. Chem.* **1999**, *578*, 43.

10. Unno, M.; Negishi, K.; Matsumoto, H., *Chem. Lett.* **2001**, 340.

11. Guo, L.; Bradshaw J. D.; Mcconville D. B.; Tessier C. A.; Youngs W. J., *Organometallics* **1997**, *16*, 1685.

12. Unno, M.; Saito, T.; Matsumoto, H., *Bull. Chem. Soc. Jpn.* **2001**, *74*, 2407.

13. Bartole-Scott, A.; Braunschweig, H.; Kupfer, T.; Lutz, M.; Manners, I.; Nguyen T. L.; Radacki, K.; Seeler, F., *Chem. Eur. J.* **2006**, *12*, 1266.

14. Braunschweig, H.; Kupfer, T.; Radacki, K., *Angew. Chem., Int. Ed.* **2007**, *46*, 1630.

15. Bourke S. C.; Maclachlan M. J.; Lough A. J.; Manners, I., *Chem. Eur. J.* **2005**, *11*, 1989.

16. Bedard T. C.; Corey J. Y.; Lange L. D.; Rath N. P., *J. Organomet. Chem.* **1991**, *401*, 261.

17. Li H. Q.; Nishiwaki, K.; Itami, K.; Yoshida, J., *Bull. Chem. Soc. Jpn.* **2001**, *74*, 1717.

18. Heath, H.; Wolfe, B.; Livinghouse, T.; Bae S. K., *Synthesis* **2001**, 2341.

19. Tanaka, R.; Unno, M.; Matsumoto, H., *Chem. Lett.* **1999**, 595.

20. Watanabe, H.; Suzuki, H.; Takahashi, S.; Ohyama, K.; Sekiguchi, Y.; Ohmori, H.; Nishiyama, M.; Sugo, M.; Yoshikawa, M.; Hirai, N.; Kanuma, Y.; Adachi, T.; Makino, M.; Sakata, K.; Kobayashi, K.; Kudo, T.; Matsuyama, H.; Kamigata, N.; Kobayashi, M.; Kijima, M.; Shirakawa, H.; Honda, K.; Goto, M., *Eur. J. Inorg. Chem.* **2002**, 1772.

21. Fischer, R.; Konopa, T.; Baumgartner, J.; Marschner, C., *Organometallics* **2004**, *23*, 1899.

22. Tobita, H.; Kurita, H.; Ogino, H., *Organometallics* **1998**, *17*, 2850.

23. Van Hanisch, C.; Rubner, O., *Eur. J. Inorg. Chem.* **2006**, 1657.

24. Von Hanisch, C., *Eur. J. Inorg. Chem.* **2003**, 2955.

25. Woo L. K.; Smith D. A.; Young V. G., *Organometallics* **1991**, *10*, 3977.

26. Quintard, D.; Keller, M.; Breit, B., *Synthesis* **2004**, 905.

27. Kyushin, S.; Ikarugi, M.; Goto, M.; Hiratsuka, H.; Matsumoto, H., *Organometallics* **1996**, *15*, 1067.

28. Braunschweig, H.; Buggisch, N.; Englert, U.; Homberger, M.; Kupfer, T.; Leusser, D.; Lutz, M.; Radacki, K., *J. Am. Chem. Soc.* **2007**, *129*, 4840.

29. Panek J. S.; Jain N. F., *J. Org. Chem.* **2001**, *66*, 2747.

30. Zhang, W.; Yamamoto, H., *J. Am. Chem. Soc.* **2007**, *129*, 286.

31. Barlan A. U.; Zhang, W.; Yamamoto, H., *Tetrahedron* **2007**, *63*, 6075.

32. Briehn C. A.; Kirschbaum, T.; Bauerle, P., *J. Org. Chem.* **2000**, *65*, 352.

33. Jesberger, M.; Jaunzems, J.; Jung, A.; Jas, G.; Schonberger, A.; Kirschning, A., *Synlett* **2000**, 1289.

34. Paterson, I.; Temal-Laib, T., *Org. Lett.* **2002**, *4*, 2473.

35. Heinze, K., *Chem. Eur. J.* **2001**, *7*, 2922.

36. Paterson, I.; Gottschling, D.; Menche, D., *Chem. Commun.* **2005**, 3568.

37. Zech, G.; Kunz, H., *Chem. Eur. J.* **2004**, *10*, 4136.

38. Reggelin, M.; Brenig, V.; Welcker, R., *Tetrahedron Lett.* **1998**, *39*, 4801.

39. Lipshutz B. H.; Shin, Y., *J. Tetrahedron Lett.* **2001**, *42*, 5629.

40. Boehm, T. L.; Showalter, H. D. H., *J. Org. Chem.* **1996**, *61*, 6498.

41. Stocker, B. L.; Teesdale-Spittle, P.; Hoberg, J. O., *Eur. J. Org. Chem.* **2004**, 330.

42. Bookermilburn, K. I.; Gulten, S.; Sharpe, A., *Chem. Commun.* **1997**, 1385.

43. Evans, P. A.; Cui, B.; Buffone, G. P., *Angew. Chem., Int. Ed.* **2003**, *42*, 1734.

44. Gulten, S.; Sharpe, A.; Baker, J. R.; Booker-Milburn, K. I., *Tetrahedron* **2007**, *63*, 3659.

45. Bradford, C. L.; Fleming, S. A.; Ward, S. C., *Tetrahedron Lett.* **1995**, *36*, 4189.

46. Colombier, C.; Skrydstrup, T.; Beau, J. M., *Tetrahedron Lett.* **1994**, *35*, 8167.

47. Evans, P. A.; Cui, J.; Gharpure, S. J.; Polosukhin, A.; Zhang, H. R., *J. Am. Chem. Soc.* **2003**, *125*, 14702.

48. Petit, M.; Chouraqui, G.; Aubert, C.; Malacria, M., *Org. Lett.* **2003**, *5*, 2037.

49. Cordier, C.; Morton, D.; Leach, S.; Woodhall, T.; O'leary-Steele, C.; Warriner, S.; Nelson, A., *Org. Biomol. Chem.* **2008**, *6*, 1734.

50. Pschirer, N. G.; Fu, W.; Adams, R. D.; Bunz, U. H. F., *Chem. Commun.* **2000**, 87.

51. Denmark, S. E.; Hurd, A. R.; Sacha, H. J., *J. Org. Chem.* **1997**, *62*, 1668.

52. Novikov, Y. Y.; Sampson, P., *J. Org. Chem.* **2005**, *70*, 10247.

53. Rowlands, G. J.; Singleton, J., *J. Chem. Res.* **2004**, 247.

54. Dobbs, A. P.; Miller, I. J.; Martinovic, S., *Beilstein J. Org. Chem.* **2007**, *3*, doi: 10.1186/1860-5397-1183-1121.

55. Kundu, L. M.; Burgdorf, L. T.; Kleiner, O.; Batschauer, A.; Carell, T., *Chembiochem* **2002**, *3*, 1053.

56. Fang, S. Y.; Bergstrom, D. E., *Bioconjug. Chem.* **2003**, *14*, 80.

57. Fang, S. Y.; Bergstrom, D. E., *Nucleic Acids Res.* **2003**, *31*, 708.

58. Sproat, B. S.; Rupp, T.; Menhardt, N.; Keane, D.; Beijer, B., *Nucleic Acids Res.* **1999**, *27*, 1950.

59. Fang, S. Y.; Bergstrom, D. E., *Tetrahedron Lett.* **2004**, *45*, 7987.

60. Li, C. M.; Porco, J. A., *J. Org. Chem.* **2005**, *70*, 6053.

61. Van Delft, F. L.; Valentijn, A. R. P. M.; Van Der Marel, G. A.; Van Boom, J. H., *J. Carbohydr. Chem.* **1999**, *18*, 165.

62. Schinzer, D.; Muller, N.; Fischer, A. K.; Prieb, J. W., *Synlett* **2000**, 1265.

Dichlorodimethylsilane[1]

[75-78-5] $C_2H_6Cl_2Si$ (MW 129.06)

InChI = 1S/C2H6Cl2Si/c1-5(2,3)4/h1-2H3

InChIKey = LIKFHECYJZWXFJ-UHFFFAOYSA-N

(additive for pinacol cyclization;[2] protecting group for diols[3] and carbonyl compounds;[4] precursor for a wide variety of silicon-based reagents)

Physical Data: mp −76 °C; bp 70–71 °C; *d* 1.064 g cm⁻³.

Solubility: sol chlorinated solvents and ethereal solvents; reacts with protic solvents.

Form Supplied in: liquid form from laboratory to commercial scale (typical purities >95%).

Purification: can be purified by distillation.

Handling, Storage, and Precautions: reacts with water, amines, alcohols, amides, and other protic species with the evolution of hydrogen chloride. Use in a fume hood.

Pinacol Reaction. Dichlorodimethylsilane (**1**) allows clean pinacol cyclization of a keto aldehyde to occur without competition from an aldol reaction (eq 1).[2]

$$\text{(1)}$$

Protection. The success of the pinacol cyclization is due to the formation of a 'siliconide', the silicon equivalent of an acetonide. Reaction of a diol with (**1**) in the presence of base also provides these cyclic silicon compounds (eq 2).[3,5] However, the labile dimethylsiliconide does not afford a widely applicable form of protection for this functional group, although it has found some use with pericyclic reactions.[6]

$$\text{(2)}$$

Enolates derived from aldehydes can also form bisenol ethers with (**1**), but the reaction fails with aldehydes containing four or fewer carbons.[7] Ketone lithium enolates provide bisenol ethers in good yields when treated with (**1**) These bisenol ethers can then be employed in further reactions such as intramolecular aldol reactions.[8] The formation of bisenol ethers by treatment of ketones with the silyl dichloride (**1**) in the presence of triethylamine and sodium iodide has been advocated. However, the yields can still be variable and the use of N,N-diethylaminodimethylchlorosilane is suggested as a superior silylating agent.[4] The lithium enolate derived from pinacolone reacts with an equivalent of (**1**) to provide the monoenol silyl chloride that can then be treated with a wide variety of oxygen and nitrogen based nucleophiles.[9] Chiral enol ethers are also available through the use of the dichloride (**1**) as the silicon source (eq 3).[10]

$$\text{(3)}$$

Silanes. Dichlorodimethylsilane (**1**) is a precursor to a wide variety of silicon reagents. For example, reaction with an organometallic reagent, such as phenyl- or t-butyllithium, results in the corresponding silyl chloride (eq 4).[11] Many of these silyl chlorides have found application for the protection of functional groups.[12]

$$\text{Me}_2\text{SiCl}_2 \xrightarrow{\text{RLi}} \text{RMe}_2\text{SiCl} \qquad (4)$$

The presence of two leaving groups allows for the introduction of two different nucleophilic species.[13] In addition to carbon, heteroatom nucleophiles have been used.[14] A wide variety of heterocyclic derivatives are available when difunctional, nucleophilic compounds are treated with (**1**) (eq 5).[15] The synthetic potential of all of these derivatives is as wide as organosilicon chemistry.[16]

$$\text{(5)}$$

The dichloride (**1**) has also found widespread application for the formation of polysiloxanes and other high molecular weight silanes.[17] In addition to water, the chlorine atoms can be displaced by a wide variety of nucleophiles to provide some useful reagents, such as dimethyldiacetoxysilane.[18]

A controlled reaction of dichloride (**1**) with lithium in THF affords dodecamethylsilane,[19] a reagent that has been used for the preparation of silyl enol ethers through photolytic cleavage.[20] Silanes and polysilanes are readily available from the addition of an organometallic species to (**1**) Thus the bisvinylsilane (**2**) can be prepared from (**1**) (eq 6).[21]

$$\text{(6)}$$

Related Reagents. Di-t-butyldichlorosilane; Di-t-butylsilyl Bis(trifluoromethanesulfonate).

1. Colvin, E. W. *Silicon in Organic Synthesis*; Butterworths: London, 1981.
2. Corey, E. J.; Carney, R. L., *J. Am. Chem. Soc.* **1971**, *93*, 7318.
3. Kelly, R. W., *Tetrahedron Lett.* **1969**, 967.
4. Rathke, M. W.; Weipert, P. D., *Synth. Commun.* **1991**, *21*, 1337.
5. Cragg, R. H.; Lane, R. D., *J. Organomet. Chem.* **1985**, *289*, 23.
6. (a) Jenneskens, L. W.; Kostermans, G. B. M.; Harmannus, J. B.; De Wolf, W. H.; Bickelhaupt, F., *J. Chem. Soc., Perkin Trans. 1* **1985**, 2119. (b) Kita, Y.; Okunaka, R.; Honda, T.; Shindo, M.; Tamura, O., *Tetrahedron Lett.* **1989**, *30*, 3995.
7. Fataftah, Z. A.; Ibrahim, M. R.; Abdel-Rahman, H. N., *Bull. Chem. Soc. Jpn.* **1991**, *64*, 671.
8. Fataftah, Z. A.; Ibrahim, M. R.; Abu-Agil, M. S., *Tetrahedron Lett.* **1986**, *27*, 4067.
9. Walkup, R. D., *Tetrahedron Lett.* **1987**, *28*, 511.
10. (a) Kaye, P. T.; Learmonth, R. A., *Synth. Commun.* **1989**, *19*, 2337. (b) Walkup, R. D.; Obeyesekere, N. H., *J. Org. Chem.* **1988**, *53*, 920.
11. (a) Benkeser, R. A.; Foster, D. J., *J. Am. Chem. Soc.* **1952**, *74*, 5314. (b) Sommer, L. H.; Tyler, L. J., *J. Am. Chem. Soc.* **1954**, *54*, 1030.
12. Corey, E. J.; Venkateswarlu, A., *J. Am. Chem. Soc.* **1972**, *94*, 6190.
13. (a) Gillard, J. W.; Fortin, R.; Morton, H. E.; Yoakim, C.; Quesnelle, C. A.; Daignault, S.; Guindon, Y., *J. Org. Chem.* **1988**, *53*, 2602. (b) Barluenga, J.; Foubelo, F.; Gonzalez, R.; Fananas, F. J.; Yus, M., *J. Chem. Soc., Chem. Commun.* **1991**, 1001.

14. (a) Wei, Z. Y.; Wang, D.; Li, J. S.; Chan, T. H., *J. Org. Chem.* **1989**, *54*, 5768. (b) Tamao, K.; Nakajo, E.; Ito, Y., *Tetrahedron* **1988**, *44*, 3997. (c) Wei, Z. Y.; Li, J. S.; Wang, D.; Chan, T. H., *Tetrahedron Lett.* **1987**, *28*, 3441.

15. (a) Yamamoto, Y.; Takeda, Y.; Akiba, K., *Tetrahedron Lett.* **1989**, *30*, 725; Nifant'ev, I. E.; Yarnykh, V. L.; Borzov, M. V.; Mazurchik, B. A.; Mstislavskii, V. I.; Roznyatovskii, V. A.; Ustynyuk, Y. A., *Organometallics* **1991**, *10*, 3739. (b) Roesky, H. W.; Meller-Rehbein, B.; Noltemeyer, M., *Z. Naturforsch., Teil B* **1991**, *46*, 1053. (c) Barluenga, J.; Tomas, M.; Ballesteros, A.; Gotor, V., *Synthesis* **1987**, 489.

16. Tamao, K.; Nakajo, E.; Ito, Y., *J. Org. Chem.* **1987**, 52.

17. Pawlenko, S. *Organosilicon Chemistry*; de Gruyter: Berlin, 1986.

18. (a) Kelly, R. W., *J. Chromatogr.* **1969**, *43*, 229. (b) Matyjaszewski, K.; Chen, Y. L., *J. Organomet. Chem.* **1988**, *340*, 7.

19. (a) Gilman, H.; Tomasi, R. A., *J. Org. Chem.* **1963**, *28*, 1651. (b) Krasnova, T. L.; Mudrova, N. A.; Bochkarev, V. N.; Kisin, A. V., *Zh. Obshch. Khim.* **1985**, *55*, 1528. (c) Cen, S. M.; Katti, A.; Blinka, T. A.; West, R., *Synthesis* **1985**, 684.

20. (a) Ishikawa, M.; Kumada, M., *J. Organomet. Chem.* **1972**, *42*, 325. (b) Ando, W.; Ikeno, M., *Chem. Lett.* **1978**, 609.

21. Nativi, C.; Perrotta, E.; Ricci, A.; Taddei, M., *Tetrahedron Lett.* **1991**, *32*, 2265.

David J. Ager
The NutraSweet Company, Mount Prospect, IL, USA

1,3-Dichloro-1,1,3,3-tetraisopropyl-disiloxane

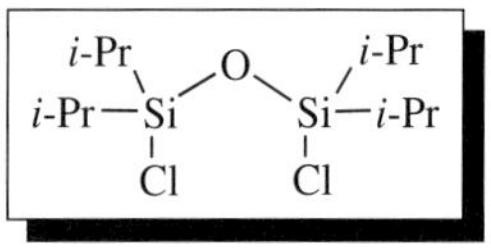

[69304-37-6] $C_{12}H_{28}Cl_2OSi_2$ (MW 315.43)

InChI = 1S/C12H28Cl2OSi2/c1-9(2)16(13,10(3)4)15-17(14,11(5)6)12(7)8/h9-12H,1-8H3

InChIKey = DDYAZDRFUVZBMM-UHFFFAOYSA-N

(simultaneous protection of the 3′- and 5′-hydroxyl functions of furanose nucleosides;[1,2] protecting group for carbohydrates;[3,4] protecting group reagent for open-chain polyhydroxy compounds;[5] preparation of cyclic bridged peptides[6])

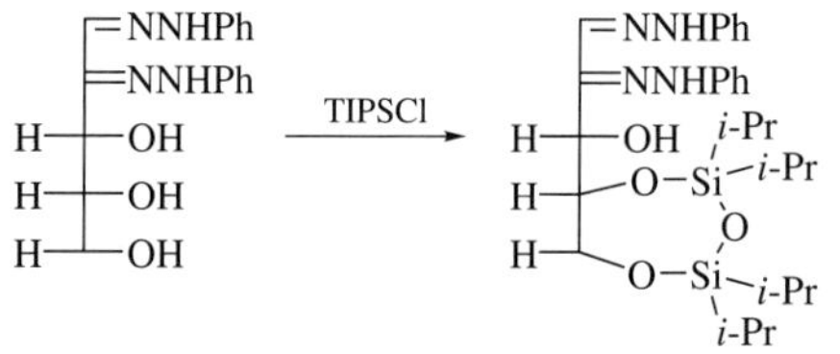

Alternate Names: TIPSCl, TIPDSCl$_2$.
Physical Data: bp 70 °C/0.5 mmHg; *d* 0.986 g cm^{-3}.
Solubility: insoluble in water, 0.3M TsOH in dioxane, 10% TFA in chloroform, 5M NH$_3$ in dioxane–water (4:1), or isobutyl-amine–methanol (1:9), or tertiary amines (Et$_3$N, pyridine).
Form Supplied in: liquid; commercially available.
Preparative Method: can be prepared starting from trichloro-silane by treatment with *i*-PrMgCl, H$_2$O, and acetyl chloride.[7] Alternatively, it can be prepared by the reaction of 1,1,3,3-tetraisopropyldisiloxane with CCl$_4$ in the presence of catalytic PdCl$_2$.[8] This method is also suitable for its in situ preparation.
Handling, Storage, and Precautions: the liquid is corrosive and insoluble in water.

Original Commentary

Joel Slade
Ciba-Geigy Corp., Summit, NJ, USA

Protection of the 3′- and 5′-Hydroxy Functions. The reagent design was based upon the fact that triisopropylsilyl chloride reacts 1000 times faster with primary alcohols than secondary alcohols. Thus in the case of 3′,5′-dihydroxy nucleosides the reagent initially silylates at the 5′-position. This is then followed by intramolecular reaction with the secondary alcohol at the 3′-position to give the doubly protected derivative (eq 1).[1,2,9]

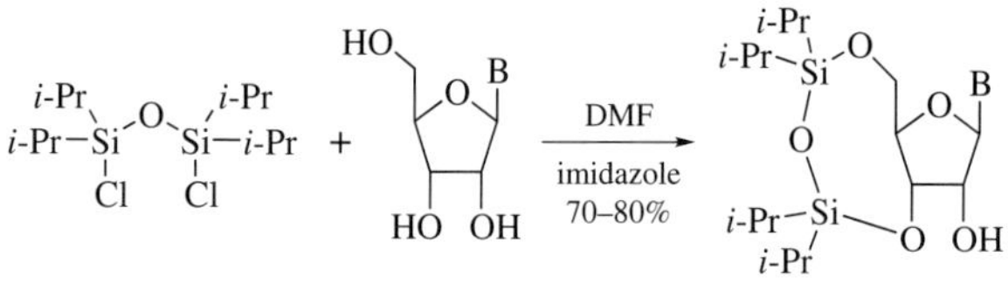

Rendering this protecting group of further utility is the fact that, in DMF under acidic conditions, silyl migration occurs affording the 2′,3′-disilyl derivative (eq 2).[3] This migration has been employed in the synthesis of carbohydrates (eq 3).[4,10]

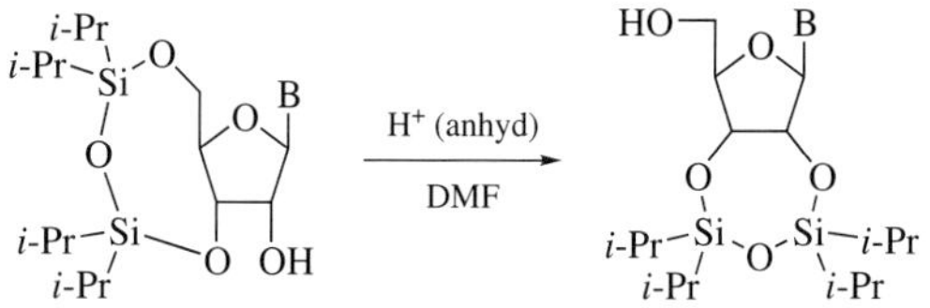

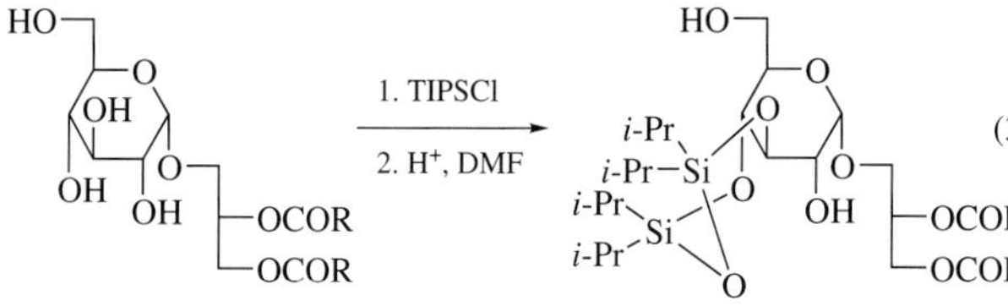

Protection of Open-Chain Polyhydroxy Compounds. The utility of 1,3-dichloro-1,1,3,3-tetraisopropyldisiloxane was extended to include open-chain polyhydroxy compounds. Glycerol reacted to form the eight-membered ring (eq 4) while the D-erythropentose derivative gave only the seven-membered ring (eq 5).[5] This strategy was used as a key step in the synthesis of the steroid 19-norcanrenone.[11]

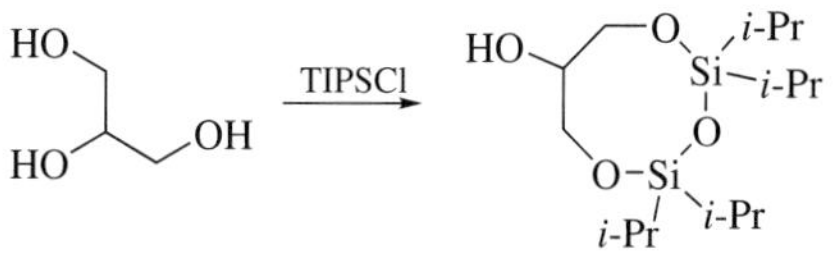

Preparation of Cyclic Bridged Peptides. As part of a study aimed at improving the transport of peptides by derivatization with silyl reagents, the feasibility of restricting the conformational mobility of the enkephalin peptide backbone by bridging two serine residues with a disiloxane bridge was investigated. TIPSCl was chosen due to its specificity for primary alcohols as well as its ability to bridge secondary alcohols if present in the correct spatial environment.[6] Thus the solution phase syntheses of protected [Ser[2],Ser[5]]- and [D-Ser[2],Ser[5]]enkephalins were carried out followed by the reaction of the two free hydroxy groups in each with TIPSCl to afford the bridged disiloxane structures.

First Update

Chryssostomos Chatgilialoglu, Maria Luisa Navacchia & Tamara Perchyonok
ISOF, Consiglio Nazionale delle Ricerche, Bologna, Italy

Protection of the 3′- and 5′-Hydroxy Functions in Nucleosides. The synthesis of 3′,5′-O-(TIPDS)-2′-deoxycytidine,[12] 3′,5′-O-(TIPDS)-2′-deoxyadenosine,[12] and 3′,5′-O-(TIPDS)-2′-deoxyinosine[13] starting from the corresponding 2′-deoxyribonucleosides has been reported in pyridine using equimolar or excess of TIPDSCl$_2$ at room temperature (eq 6).

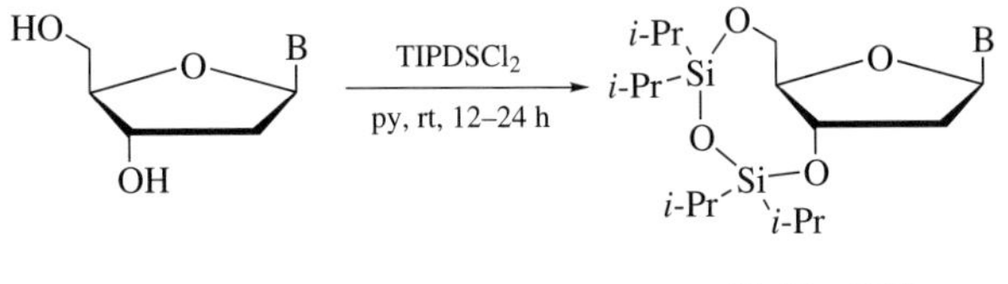

$$\text{Yields} > 90\%$$

The 3′,5′-O-(TIPDS) protection is also reported in the ribonucleosides under similar experimental conditions. For example, the base-modified derivatives **1**, **2**, and **3** are obtained in 61%, 91%, and 57% yield, respectively.[14] In all these cases, the regioselective protection of 3′ and 5′ positions is followed by 2′-deoxygenation reaction by the Barton–McCombie procedure, and it results in the preparation of 2′-deoxyribo analogs (eq 7).

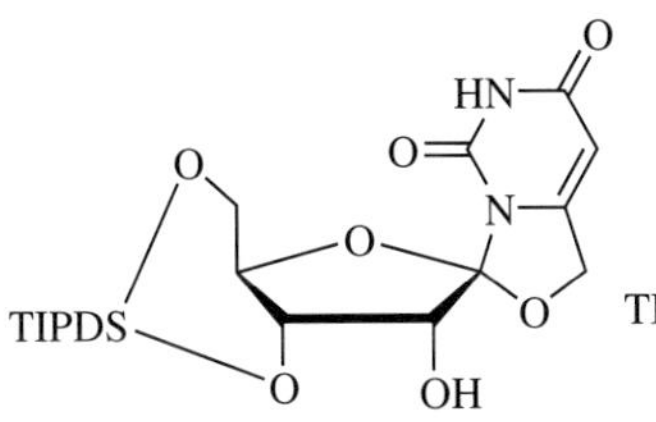

1

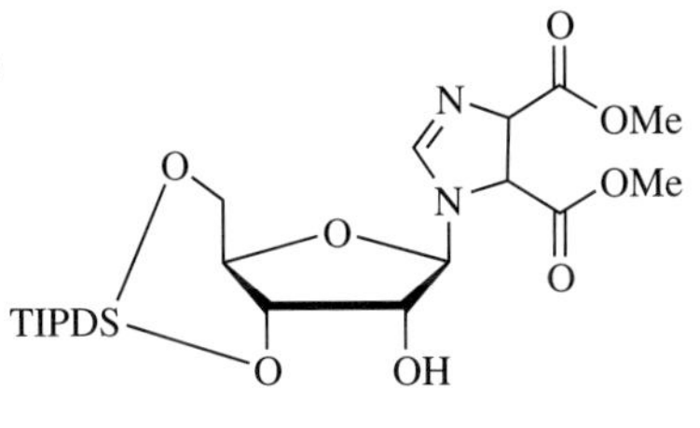

2

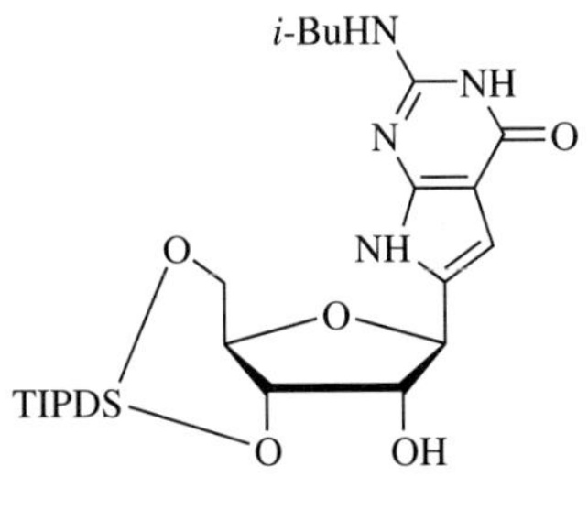

3

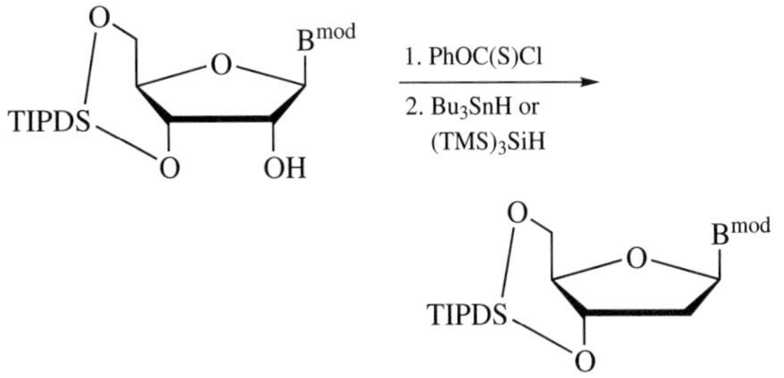

(7)

Other conditions for 3′,5′-O-(TIPDS) protection in nucleosides using pyridine-DMF mixture or imidazole in DMF are reported.[15] The 3′,5′-O-(TIPDS) protection of several 2′-deoxyribo- and ribonucleosides was also accomplished in yields > 90% by preparing TIPDSCl$_2$ in situ from the corresponding silicon hydride (eq 8).[8] In the case of psicofuranosyluracil derivative, where two primary hydroxy groups are present in the molecule, the two silytated products are formed in a 1:1 ratio (eq 9).[8] The TIPDSCl$_2$ reagent is commercially available, albeit expensive. The in situ preparation is straightforward and economically convenient and it can be easily extended to a range of new commercially unavailable halosilanes.[16] For example, 5′-O-(TIPDS) protection is labile under strong alkaline conditions that are generally required for the subsequent 2′-O-alkylation and an analogous protective group methylene-bis(diisopropylsilyl chloride) is prepared using this approach.[17]

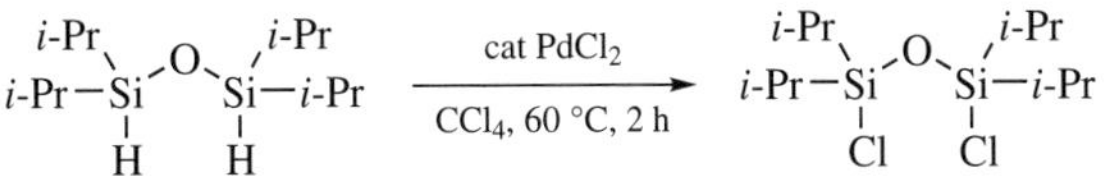

(8)

$$85\%$$

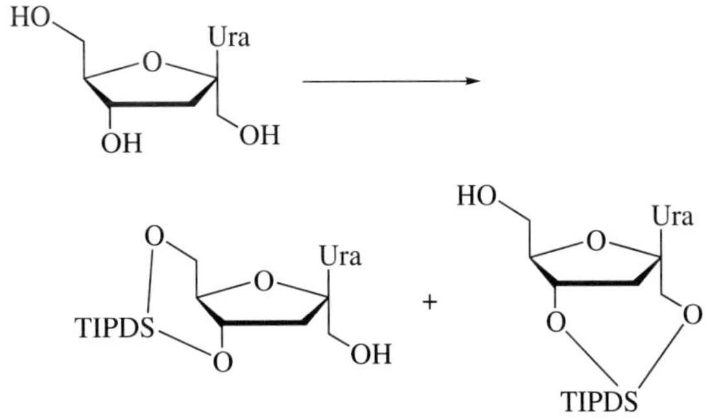

(9)

Protection of Carbohydrates. Protected carbohydrates (**4** and **5**), having one free hydroxyl and a phenylseleno group on the adjacent carbon, were obtained by treatment of commercially available tetra-O-acetyl-β-D-ribofuranose with PhSeH/BF$_3$·Et$_2$O, followed by hydrolysis and silylation with TIPDSCl$_2$ in pyridine (reported yields are 77–80%).[18] As expected, the silyl chloride reacts much faster with primary than secondary alcohols, regiospecifically affording the desired protection.

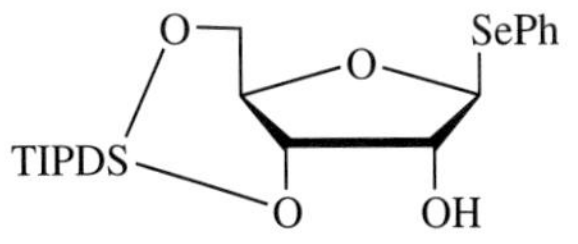

4

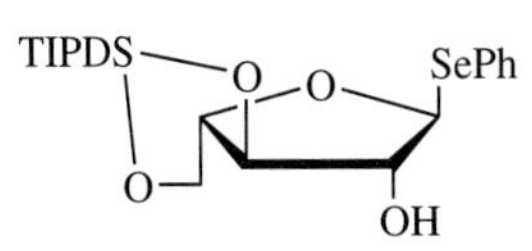

5

The 4,6-O-(TIPDS) protection of methyl β-D-glucopyranoside (**6**) was obtained in a 66% yield by reaction with 1.5 equiv of TIPDSCl$_2$ in dry pyridine. This compound was incorporated in a disaccharide taking advantage of the bulkiness of the protecting group, which strongly decreased the reactivity of the hydroxyl group at C3.[19] On the other hand, the 4,6-O-(TIPDS) protection of methyl 1-thio-α-D-mannopyranoside (**7**) was obtained by reaction of TIPDSCl$_2$ and imidazole in CH$_3$CN in a 90% yield.[20]

Tri-O-acetyl-D-glucal was converted to **8** via methanolysis of the acetate and successive protection of the resulting triol by reaction with TIPDSCl$_2$ and pyridine. Rendering this protecting group of further utility, treatment of **8** with NaH in THF/DMF (2/1, v/v) at 0 °C for 1 h and then with MPMCl at 0–25 °C afforded the derivative **9**, via silyl migration (eq 10). Both **8** and **9** have been used as starting materials for the synthesis of Ciguatoxin fragments.[21]

$$(10)$$

An α:β = 3:1 inseparable mixture of 1-O-benzyl-D-fucose anomers was treated with TIPDSCl$_2$ in DMF in the presence of imidazole and DMAP, to provide a separable mixture of α- and β-disiloxane derivatives in 59% and 20% yield, respectively (eq 11).[22]

$$(11)$$

The protection of all four equatorial hydroxy groups of D-*chiro*-inositol by TIPDS was accomplished in a single step and 76% yield with 2.1 equiv of TIPSCl$_2$ in DMF in the presence of imidazole and 4-(dimethylamino)pyridine (DMAP) (eq 12).[23] When pyridine is used as the base, the reaction yield was very low. The analogous reaction of 3-O-methyl-D-*chiro*-inositol afforded the 4,5-O-(TIPDS) monoprotection, although the yield was found to decrease to 24%.[24]

$$(12)$$

Protection of Open-chain Polyhydroxy Compounds. An example of TIPDS protection using other solvents (CH$_2$Cl$_2$) and imidazole is provided in good yield (86%) during the synthesis of a precursor (eq 13) for a Diels-Alder strategy for preparation of biologically active CP-molecules.[25]

$$(13)$$

Protection of the Proximal Hydroxy Groups of Calix[4]-arenes. When p-t-butylcalix[4]arines and its sulfur-bridged analog were treated with an excess of TIPSCl$_2$ in DMF in the presence of imidazole at room temperature, proximally O,O'-bridged derivatives were obtained in yields >90%.[26] It is interesting to note that the reaction gave neither detectable amounts of O,O''-bridged derivatives nor intermolecularly bridged oligomers without relying on a high dilution technique. It is suggested that the short chain length of three atoms, with aid of four isopropyl substituents, prevented such unwanted bridging.

1. (a) Markiewicz, W., *J. Chem. Res. (S)* **1979**, 24. (b) Markiewicz, W., *J. Chem. Res. (M)* **1979**, 181.

2. Markiewicz, W.; Padyukova, N. Sh.; Samek, Z.; Smrt, J., *Collect. Czech. Chem. Commun.* **1980**, *45*, 1860.

3. Verdegaal, C. H. M. N.; Jansse, P. L.; de Rooij, J. F. M.; van Boom, J. H., *Tetrahedron Lett.* **1980**, *21*, 1571.

4. van Boeckel, C. A. A.; van Boom, J. H., *Tetrahedron Lett.* **1980**, *21*, 3705.

5. Markiewicz, W. Samek, Z; Smrt, J., *Tetrahedron Lett.* **1980**, *21*, 4523.

6. Davies, J. S.; Tremer, E. J., *J. Chem. Soc., Perkin Trans. 1* **1987**, 1107.

7. Zhang, H. X.; Guibe, F.; Balavoine, G., *Synth. Commun.* **1987**, *17*, 1299.

8. Ferreri, C.; Costantino, C.; Romeo, R.; Chatgilialoglu, C., *Tetrahedron Lett.* **1999**, *40*, 1197.

9. (a) Schaumberg, J. P.; Hokanson, G. C.; French, J. C.; Smal, E.; Baker, D. C., *J. Org. Chem.* **1985**, *50*, 1651. (b) Tatsuoka, T.; Imao, K.; Suzuki, K., *Heterocycles* **1986**, *24*, 617. (c) Robins, M. J.; Wilson, J. S.; Hansske, F., *J. Am. Chem. Soc.* **1983**, *105*, 4059. (d) Robins, M. J.; Wilson, J. S.; Sawyer, L.; James, M. N. G., *Can. J. Chem.* **1983**, *61*, 1911. (e) Hagen, M. D.; Scalfi-Happ, C.; Happ, E.; Chladek, S., *J. Org. Chem.* **1988**, *53*, 5040.

10. (a) Ziegler, T.; Eckhardt, E.; Neumann, K.; Birault, V., *Synthesis* **1992**, 1013. (b) Thiem, J.; Duckstein, V.; Prahst, A.; Matzke, M., *Liebigs Ann. Chem.* **1987**, 289. (c) Oltvoort, J. J.; Kloosterman, M.; van Boom, J. H., *Recl. Trav. Chim. Pays-Bas* **1983**, *102*, 501.

11. Nemoto, H.; Fujita, S.; Nagai, M.; Fukumoto, K.; Kametani, T., *J. Am. Chem. Soc.* **1988**, *110*, 2931.

12. Sierzchala, A. B.; Dellinger, D. J.; Betley, J. R.; Wyrzykiewicz, T. K.; Yamada, C. M.; Caruthers, M. H., *J. Am. Chem. Soc.* **2003**, *125*, 13427.

13. Kowalczyk, A.; Harris, C. M.; Harris, T. M., *Chem. Res. Toxicol.* **2001**, *14*, 746.

14. (a) Chatgilialoglu, C.; Gimisis, T.; Spada, G. P., *Chem. Eur. J.* **1999**, *5*, 2866. (b) Zhang, N.; Chen, H.-M.; Koch, V.; Schmitz, H.; Liao, C.-L.;

Bretner, M.; Bhadti, V. S.; Fattom, A. I.; Naso, R. B.; Hosmane, R. S.; Borowski, P., *J. Med. Chem.* **2003**, *46*, 4149. (c) Seela, F.; Debelak, H., *J. Org. Chem.* **2001**, *66*, 3303.

15. (a) Noronha, A. M.; Wilds, C. J.; Lok, C.-N.; Viazovkina, K.; Arion, D.; Parniak, M. A.; Damha, M. J., *Biochemistry* **2000**, *39*, 7050. (b) Chan, J. H.; Chamberlain, S. D.; Biron, K. K.; Davis, M. G.; Harvey, R. J.; Selleseth, D. W.; Dornsife, R. E.; Dark, E. H.; Frick, L. W.; Townsend, L. B.; Drach, J. C.; Koszalka, G. W., *Nucleosides, Nucleotides & Nucleic Acids* **2000**, *19*, 101. (c) Nechev, L. V.; Kozekov, I. D.; Brock, A. K.; Rizzo, C. J.; Harris, T. M., *Chem. Res. Toxicol.* **2002**, *15*, 607.

16. Jin, S.; Miduturu, C. V.; McKinney, D. C.; Silverman, S. K., *J. Org. Chem.* **2005**, *70*, 4284.

17. Wen, K.; Chow, S.; Sanghvi, Y. S.; Theodorakis, E. A., *J. Org. Chem.* **2002**, *67*, 7887.

18. Zhang, J.; Clive, D. L. J., *J. Org. Chem.* **1999**, *64*, 770.

19. Roën, A.; Padrón, J. I.; Vázquez, J. T., *J. Org. Chem.* **2003**, *68*, 4615.

20. Ohnishi, Y.; Ando, H.; Kawai, T.; Nakahara, Y.; Ito, Y., *Carbohydr. Res.* **2000**, *328*, 263.

21. (a) Maeda, K.; Oishi, T.; Oguri, H.; Hirama, M., *Chem. Commun.* **1999**, 1063. (b) Nagumo, Y.; Oguri, H.; Shindo, Y.; Sasaki, S.; Oishi, T.; Hirama, M.; Tomioka, Y.; Mizugaki, M.; Tsumuraya, T., *Bioorg. Med. Chem. Lett.* **2001**, *11*, 2037.

22. Myers, A. G.; Glatthar, R.; Hammond, M.; Harrington, P. M.; Kuo, E. Y.; Liang, J.; Schaus, S. E.; Wu, Y.; Xiang, J.-N., *J. Am. Chem. Soc.* **2002**, *124*, 5380.

23. Martín-Lomas, M.; Flores-Mosquera, M.; Khiar, N., *Eur. J. Org. Chem.* **2000**, 1539.

24. Bonilla, J. B.; Muñoz-Ponce, J.; Nieto, P. M.; Cid, M. B.; Khiar, N.; Martín-Lomas, M., *Eur. J. Org. Chem.* **2002**, 889.

25. Nicolau, K. C.; Jung, J.; Yoon, W. H.; Fong, K. C.; Choi, H.-S.; He, Y.; Zhong, Y.-L.; Baran, P. S., *J. Am. Chem. Soc.* **2002**, *124*, 2183.

26. Nazumi, F.; Hattori, T.; Morohashi, N.; Matsumura, N.; Yamabuki, W.; Kameyama, H.; Miyano, S., *Org. Biomol. Chem.* **2004**, *2*, 890.

1,1-Dichloro-2,2,2-trimethyl-1-phenyl-disilane

[57519-88-7] $C_9H_{14}Cl_2Si_2$ (MW 249.29)

InChI = 1S/C9H14Cl2Si2/c1-12(2,3)13(10,11)9-7-5-4-6-8-9/h4-8H,1-3H3

InChIKey = ZLKKOKIRGMWLGY-UHFFFAOYSA-N

(silylating agent for π-allylpalladium complexes[1] and palladium-catalyzed allylic silation, giving allylsilanes;[2] for asymmetric 1,4-disilylation of α,β-unsaturated ketones, giving optically active β-hydroxy ketones after oxidation[3])

Physical Data: bp 50–60 °C/0.5 mmHg.
Solubility: sol THF, benzene.
Form Supplied in: oil, with some 1-chloro-2,2,2-trimethyl-1-phenyldisilane.[2]
Analysis of Reagent Purity: NMR, MS.
Preparative Method: the disilane is prepared in 75% yield by selective chlorodephenylation of 2,2,2-trimethyl-1,1,1-triphenyldisilane with hydrogen chloride in the presence of a catalytic amount of aluminum chloride in benzene.[2,4]
Purification: fractional distillation.

Handling, Storage, and Precautions: the disilane reacts with air and water, and should be kept sealed to preclude decomposition. The reagent is a potential acidic corrosive. It should be handled in a well ventilated hood.

1,4-Disilylation of α,β-Unsaturated Ketones. In the presence of palladium–phosphine catalysts such as tetrakis(triphenylphosphine)palladium(0), the unsymmetrically substituted disilane $PhCl_2SiSiMe_3$ reacts with α,β-unsaturated ketones in refluxing benzene to give γ-(phenyldichlorosilyl) silyl enol ethers, which can be converted into β-(phenyldiethoxysilyl) ketones by treatment with ethanol and triethylamine (eq 1).[3] Oxidation of the phenyldiethoxysilyl group with hydrogen peroxide in the presence of fluoride gives β-hydroxy ketones in high yields. γ-(Phenyldimethylsilyl) lithium enolates, generated by treatment of the γ-(phenyldichlorosilyl) silyl enol ethers with methyllithium, undergo alkylation with alkyl halides to give β-(phenyldimethylsilyl)-α-alkyl ketones with high *anti* selectivity (eq 2). These β-(phenyldimethylsilyl) ketones are subjected to fluorodephenylation followed by oxidation to give *anti*-β-hydroxy-α-alkyl ketones. The 1,4-addition of the disilane is enantioselective when catalyzed by $PdCl_2[(R)$-BINAP], thereby providing optically active *anti*-β-hydroxy-α-alkyl ketones with 66–92% ee (Table 1).[3a]

$$\text{Pd catalyst} = \text{Pd(PPh}_3)_4, \text{PdCl}_2[(R)\text{-BINAP}]$$

Silylation of π-Allylpalladium Complexes and Palladium-Catalyzed Allylic Silylation. The disilane is a highly reactive reagent for the silylation of π-allylpalladium complexes, producing phenyldichloro(allyl)silanes in high yields.[1] The retention of configuration observed in the silylation of an optically active π-allylpalladium complex indicates that the silylation occurs by way of a π-allyl(phenyldichlorosilyl)palladium intermediate (eq 3). Palladium-catalyzed allylic silylation of allylic chlorides is also effected with the disilane reagent. In the presence of 1 mol% of palladium catalyst coordinated with triphenylphosphine or 1,1'-bis(diphenylphosphino)ferrocene (dppf), the silylation gives allylsilanes in high yields (eq 4).[2] The functionalized allylsilanes may be oxidized to allylic alcohols.

Table 1 β-Hydroxy ketones prepared via the asymmetric disilylation of enones

R^1	R^2	R^3	Yield of β-hydroxy ketone (%)	% ee
Me	Ph	H	90	87
Me	Ph	Me	70	85
Me	4-MeOC$_6$H$_4$	H	81	92
i-Pr	Ph	H	100	86
Ph	Me	H	83	78
Ph	Me	Me	78	45
Me	Me	H	74	69
Me	Me	Bn	74	66

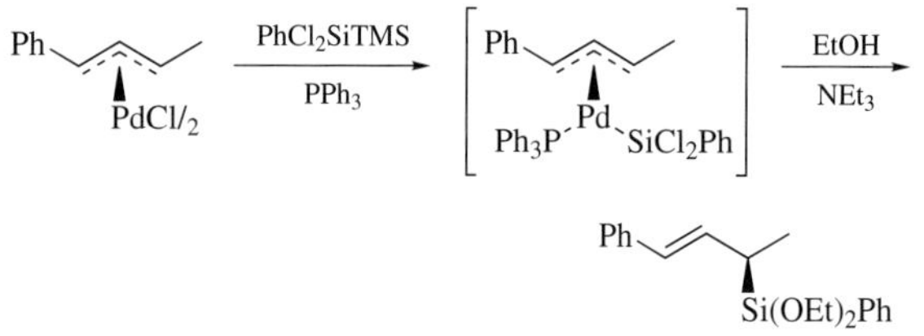

$$(3)$$

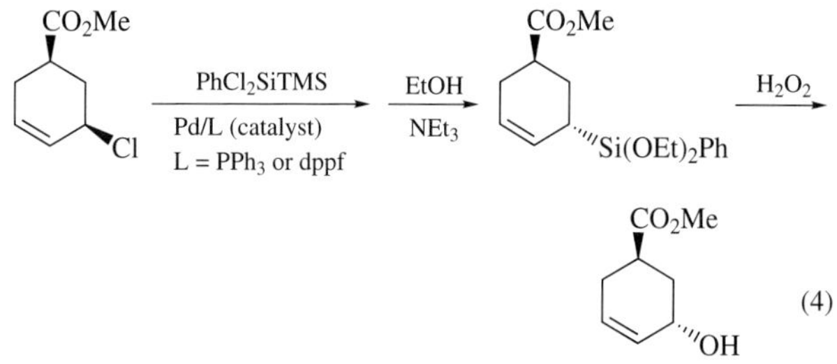

$$(4)$$

1. Hayashi, T.; Yamamoto, A.; Iwata, T.; Ito, Y., *J. Chem. Soc., Chem. Commun.* **1987**, 398.

2. Matsumoto, Y.; Ohno, A.; Hayashi, T., *Organometallics* **1993**, *12*, 4051.

3. (a) Hayashi, T.; Matsumoto, Y.; Ito, Y., *J. Am. Chem. Soc.* **1988**, *110*, 5579. (b) Hayashi, T.; Matsumoto, Y.; Ito, Y., *Tetrahedron Lett.* **1988**, *29*, 4147. (c) Matsumoto, Y.; Hayashi, T.; Ito, Y., *Tetrahedron* **1994**, *50*, 335.

4. Hengge, E.; Bauer, G.; Brandstätter, E.; Kollmann, G., *Monatsh. Chem.* **1975**, *106*, 887.

Tamio Hayashi
Hokkaido University, Sapporo, Japan

Wesley K. M. Chong
Agouron Pharmaceuticals, San Diego, CA, USA

Dicyanodimethylsilane

Me$_2$Si(CN)$_2$

[5158-09-8]   C$_4$H$_6$N$_2$Si   (MW 110.19)

InChI = 1S/C4H6N2Si/c1-7(2,3-5)4-6/h1-2H3

InChIKey = XRRLRPCAMVHBDK-UHFFFAOYSA-N

(protecting reagent of β-diketones[1,2] and β-hydroxy ketones,[1,3] diols, and alcohols[4])

Physical Data: mp 85–87 °C; bp 108–109 °C/54 mmHg.

Solubility: sol dichloromethane, chloroform; slightly sol ether.

Preparative Methods: dicyanodimethylsilane is conveniently prepared by the reaction of dibromodimethylsilane[5] or dichlorodimethylsilane[6] with silver(I) cyanide lithium cyanide,[4] sodium cyanide,[7] and potassium cyanide[7] can also be used in place of silver cyanide. A method which uses hydrogen cyanide as a source of cyano groups[8] and a procedure which utilizes transcyanation with cyanotrimethylsilane have also been reported.[9] Dicyanodimethylsilane is obtained as colorless crystals via these methods.

Handling, Storage, and Precautions: dicyanodimethylsilane reacts instantly with water to form highly toxic hydrogen cyanide and siloxane oligomers. Therefore it should be handled carefully in a fume hood. If possible, it is recommended that CH$_2$Cl$_2$ solutions of this reagent be transferred by using syringe techniques.

Reaction with β-Diketones. While enol silylations and cyanosilylations of β-diketones with cyanotrimethylsilane often give mixtures of products, the reactions with dicyanodimethylsilane give 5-cyano-2,6-dioxa-1-silacyclohex-3-enes in good yields (eq 1).[1] For enolizable β-diketones, the concurrent silylation and cyanosilylation takes place immediately without catalysts. On the other hand, zinc iodide catalysis is effective for reactions with less easily enolizable β-diketones (eq 2). Methyl 2,4-dioxopentanoate also reacts with dicyanodimethylsilane to give the corresponding cyclic product.[2] Treatment of the adducts in methanol (rt, overnight) furnishes the parent 1,3-diketones in good yields. Taking advantage of the stepwise nature of the methanolysis, a formal cyanosilylation can be achieved by treating the adduct with 1 equiv of methanol in dichloromethane at rt (eq 3).[1]

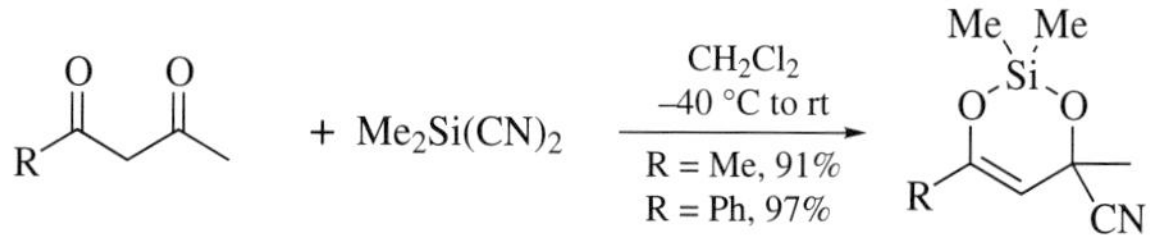

$$(1)$$

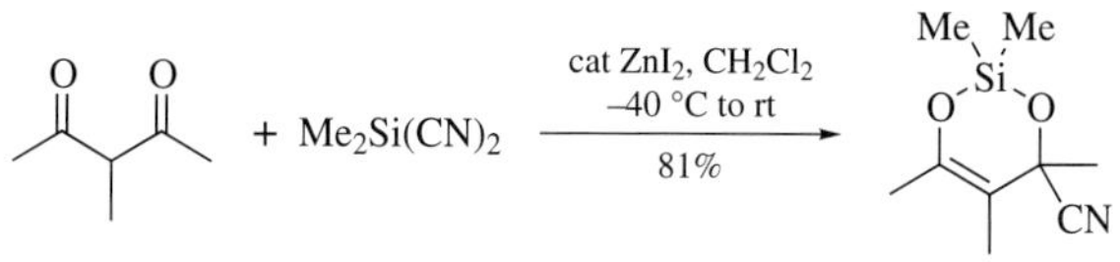

$$(2)$$

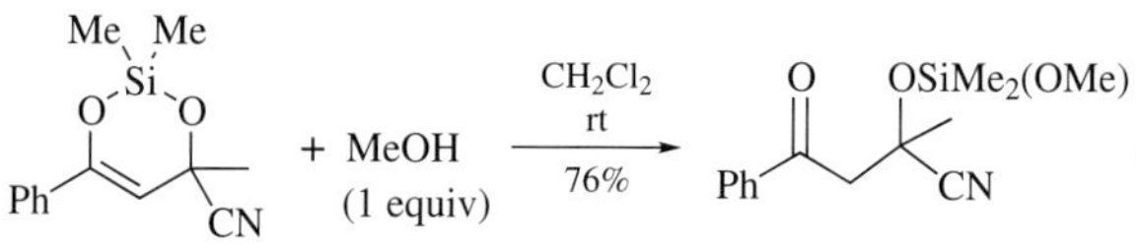

$$(3)$$

Reaction with β-Hydroxy Ketones and Alcohols. The reaction of dicyanodimethylsilane with diacetone alcohol occurs according to eq 4.[1] The reaction, which is initiated at −78 °C, proceeds by way of the indicated cyanodimethylsilyl ether. When warmed to rt, the intermediate silyl ether cyclizes to give a mixture of the cyanosilylation product and the cyclic enol silyl ether. The cyanosilylation of secondary β-hydroxy ketones is highly diastereoselective (eq 5).[3] After desilylation, the cyanohydrins are obtained with high *syn* selectivity (de > 95%). The stereochemical result is rationalized by assuming a silyl-bridged chair-type transition state. Dicyanodimethylsilane reacts readily with alcohols, with liberation of hydrogen cyanide, to give bis(alkoxy)silanes in high yields (eq 6).[4]

$$\text{(4)} \quad 53\% \qquad 36\%$$

$$\text{(5)}$$

$$2\,PhCH_2OH \;+\; Me_2Si(CN)_2 \xrightarrow[91\%]{\substack{neat \\ 25\,°C,\,5\,min}} Me_2Si(OCH_2Ph)_2 \quad (6)$$

Dicyanodiethylsilane, Dicyanodiphenylsilane, and Dicyanomethylphenylsilane. Dicyanodimethylsilane, dicyanodiethylsilane,[7,10] dicyanodiphenylsilane,[7,11] and dicyanomethylphenylsilane[12] are a family of related $R_2Si(CN)_2$ reagents that can be used as protecting groups for bifunctional compounds. The silyl-protection of 1,2- and 1,3-diols with these dicyanosilanes is well documented, affording the corresponding five- and six-membered ring compounds (eqs 7 and 8).[4] The reaction of dicyanomethylphenylsilane with ketones is reported to give biscyanosilylated products, but the yield is modest due to competitive aldol condensations.[12]

$$\text{(7)}$$

$$\text{(8)}$$

Related Reagents. Di-*t*-butyldichlorosilane; Di-*t*-butylsilyl Bis(trifluoromethanesulfonate); Dichlorodimethylsilane.

1. Ryu, I.; Murai, S.; Shinonaga, A.; Horiike, T.; Sonoda, N., *J. Org. Chem.* **1978**, *43*, 780.
2. Foley, L. H., *J. Org. Chem.* **1985**, *50*, 5204.
3. Batra, M. S.; Brunet, E., *Tetrahedron Lett.* **1993**, *34*, 711.
4. Mai, K.; Patil, G., *J. Org. Chem.* **1986**, *51*, 3545.
5. McBride, J. J., Jr.; Beachell, H. C., *J. Am. Chem. Soc.* **1952**, *74*, 5247.
6. Ryu, I.; Murai, S.; Horiike, T.; Shinonaga, A.; Sonoda, N., *Synthesis* **1978**, 154.
7. Sukata, K., *Bull. Chem. Soc. Jpn.* **1987**, *60*, 2257.
8. Hundeck, J., *Z. Anorg. Allg. Chem.* **1966**, *345*, 23.
9. Becu, C.; Anteunis, M. J. O., *Bull. Soc. Chim. Belg.* **1987**, *96*, 115.
10. Earborn, C., *J. Chem. Soc* **1949**, 2755.
11. Johns, I. B.; DiPiertro, H. R., *J. Org. Chem.* **1964**, *29*, 1970.
12. Neef, H., *J. Prakt. Chem.* **1974**, *316*, 817.

Ilhyong Ryu
Osaka University, Osaka, Japan

1,2-Diethoxy-1,2-bis(trimethylsilyloxy)-ethylene[1]

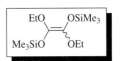

[90054-58-3] $C_{12}H_{28}O_4Si_2$ (MW 292.52)

InChI = 1S/C12H28O4Si2/c1-9-13-11(15-17(3,4)5)12(14-10-2)16-18(6,7)8/h9-10H2,1-8H3

InChIKey = WYMQBNKGHLKCCU-UHFFFAOYSA-N

(synthetic equivalent of the acyl anion $^-C(O)CO_2Et$; useful in $ZnCl_2$-mediated nucleophilic additions to aldehydes and ketones,[1] Michael additions to enones,[1] and substitution reactions of S_N1-active alkyl halides[1] and carboxylic acid chlorides[2])

Physical Data: bp 68–70 °C/5 mmHg.
Solubility: sol CH_2Cl_2, toluene, THF.
Preparative Method: can be prepared in one step from diethyl oxalate using the Rühlmann version[2] of the acyloin condensation (eq 1).[3] Although a yield of 64% of isolated product was reported,[3] this could not be reproduced in a careful series of experiments.[1,4] Rather, yields of 21–25% are consistently obtained (0.1–1.0 molar scale).[1,4] The product consists of a (*Z/E*) isomeric mixture which is of no consequence for synthetic applications.

$$\text{(1)}$$

Handling, Storage, and Precautions: somewhat air and moisture sensitive and should therefore be handled and stored under an atmosphere of an inert gas such as N_2 and in nonprotic solvents. Use in a fume hood.

Reactions Forming Carbon–Carbon Bonds. The reagent described herein has the structural features of an *O*-silyl ketene acetal, and it therefore undergoes the typical reactions known for this class of compounds, especially Lewis acid-mediated C–C bond-forming processes.[5] Since the reagent is extremely electron rich[6] and therefore very reactive, catalytic amounts of mild Lewis acids such as ZnX_2 suffice.[1,4] The nucleophile which reacts with electrophilic organic substrates is a special case of an acyl anion equivalent (synthon) (**1**).

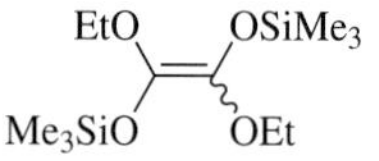

Group transfer-type[7] of Mukaiyama aldol addition[8] to aldehydes or ketones affords the isolable *O*-silylated aldol adducts, which can be hydrolyzed under acidic conditions to provide α-keto esters in high yield (eq 2).[1] In the case of aldehydes the initial products are *O*-silyl-protected reductones.[1]

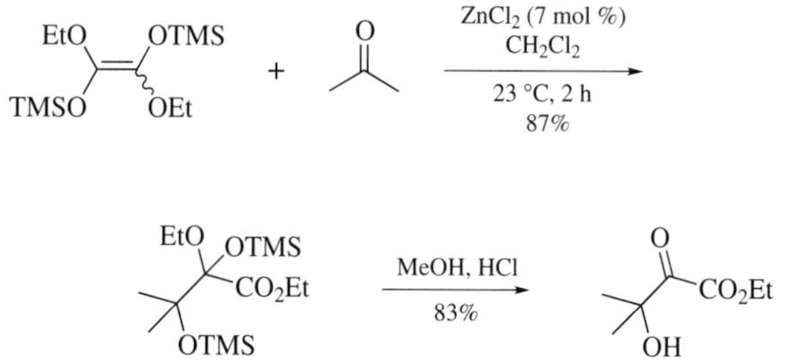

Group transfer-type of Michael additions to α,β-enones are just as smooth, with acidic workup affording α,δ-diketo esters (eq 3).[1,4]

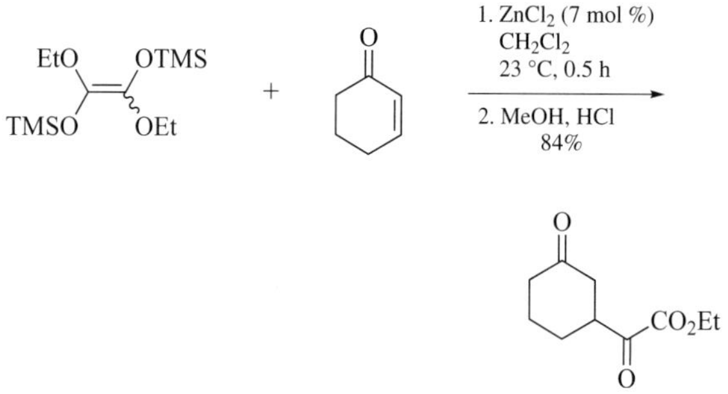

S_N1-active alkyl halides require longer reaction times (typically 3–6 h),[1,4] and the yields are lower. *t*-Butyl chloride and adamantyl bromide have been used successfully, but the generality of the procedure needs to be demonstrated (eq 4). In contrast, acylation reactions using a wide variety of carboxylic acid chlorides have been carried out (eq 5).[9,10] The initial products are protected forms of α,β-diketo esters which can be isolated in yields of 72–93%. Deprotection is accomplished using MeOH/HCl.[9,10] The products are vicinal tricarbonyl compounds having the usual characteristic chemical properties such as rapid formation of hydrated derivatives.[11] The protected forms can be used directly in the synthesis of heterocycles.[9,10]

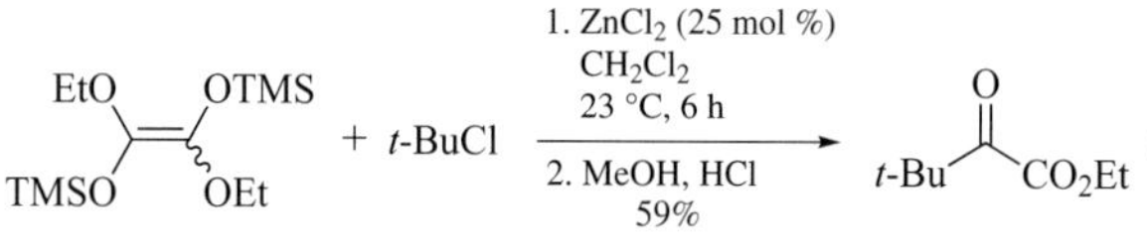

R = Me, 72%; R = Ph, 82%;
R = *o*-Me-C$_6$H$_4$, 91%;
R = *i*-Pr, 93%; R = *t*-Bu, 88%

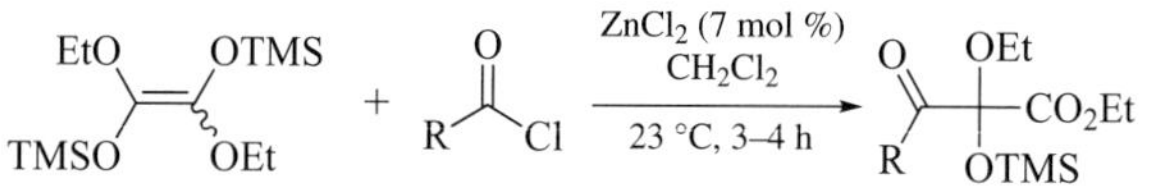

In summary, the title compound, readily available from diethyl oxalate, is a synthetically useful synthon in various C–C bond-forming reactions, requiring only catalytic amounts of mild Lewis acids such as zinc chloride. Alternatives to such an acyl anion equivalent include dithiane derivatives of glyoxylic acid esters, which have not been tested as widely and which require some synthetic effort in the final deprotection step.[12]

Related Reagents. Ethyl Diethoxyacetate; Glyoxylic Acid Diethyl Dithioacetal; Ketene Bis(trimethylsilyl) Acetal; Ketene Diethyl Acetal; 1-Methoxy-2-trimethylsilyl-1-(trimethylsilyloxy)ethylene Methyl Glyoxylate; 8-Phenylmenthyl Glyoxylate; Tetramethoxyethylene; Tris(trimethylsilyloxy)ethylene.

1. Reetz, M. T.; Heimbach, H.; Schwellnus, K., *Tetrahedron Lett.* **1984**, *25*, 511.
2. Rühlmann, K., *Synthesis* **1971**, 236.
3. Kuo, Y. N.; Chen, F.; Ainsworth, C.; Bloomfield, J. J., *J. Chem. Soc., Chem. Commun.* **1971**, 136.
4. Heimbach, H. Dissertation, Universität Marburg, 1982.
5. Reviews of the chemistry of enol silanes: (a) Mukaiyama, T., *Org. React.* **1982**, *28*, 203. (b) Brownbridge, P., *Synthesis* **1983**, 1 and 85. (c) Gennari, C., *Comprehensive Organic Synthesis* **1991**, *2*, 629. (d) Review of Lewis acid-mediated α-alkylation of carbonyl compounds: Reetz, M. T., *Angew. Chem.* **1982**, *94*, 97; *Angew. Chem., Int. Ed. Engl.* **1982**, *21*, 96.
6. Review of tetramethoxyethylene and other electron-rich alkenes: Hoffmann, R. W., *Angew. Chem.* **1968**, *80*, 823; *Angew. Chem., Int. Ed. Engl.* **1968**, *7*, 754.
7. Sogah, D. Y.; Webster, O. W., *Macromolecules* **1986**, *19*, 1775.
8. Mukaiyama, T., *Angew. Chem.* **1977**, *89*, 858; *Angew. Chem., Int. Ed. Engl.* **1977**, *16*, 817.
9. Reetz, M. T.; Kyung, S.-H., *Tetrahedron Lett.* **1985**, *26*, 6333.
10. Kyung, S.-H. Dissertation, Universität Marburg, 1985.
11. (a) Rubin, M. B., *Chem. Rev.* **1975**, *75*, 177. (b) Wassermann, H. H.; Han, W. T., *Tetrahedron Lett.* **1984**, *25*, 3743.
12. (a) Eliel, E. L.; Hartman, A. A., *J. Org. Chem.* **1972**, *37*, 505. (b) Lissel, M., *Synth. Commun.* **1981**, *11*, 343. (c) Lever, O. W. Jr., *Tetrahedron* **1976**, *32*, 1943. (d) Hase, T. A.; Koskimies, J. K., *Aldrichim. Acta* **1982**, *15*, 35.

Manfred T. Reetz
Max-Planck-Institut für Kohlenforschung, Mulheim an der Ruhr, Germany

N,N-Diethylaminodimethylsilane

$$\boxed{Et_2NSiHMe_2}$$

[13686-66-3] $C_6H_{17}NSi$ (MW 131.29)

InChI = 1S/C6H17NSi/c1-5-7(6-2)8(3)4/h8H,5-6H2,1-4H3

InChIKey = KYIHWAKEVGKFOA-UHFFFAOYSA-N

(reagent for intramolecular hydrosilation of allyl alcohols[1] and intermolecular hydrosilation of 1,7-diynes[2])

Physical Data: bp 112 °C/760 mmHg; n_D^{20} 1.4087; d_4^{20} 0.7505 g cm^{-3}.

Solubility: sol common aprotic organic solvents.

Preparative Method: by treatment of commercially available chlorodimethylsilane (ClMe$_2$SiH) and diethylamine in the presence of triethylamine in ether, followed by filtration and distillation.

Handling, Storage, and Precautions: moisture sensitive and corrosive. Must be stored under nitrogen in a tightly capped bottle in a freezer and transferred or weighed quickly via syringe. Use in a fume hood.

Intramolecular Hydrosilation Agent. For a detailed description of the intramolecular hydrosilation procedure, see 1,1,3,3-tetramethyldisilazane. *N,N*-Diethylaminodimethylsilane is a useful reagent for the conversion of hydroxy groups of allyl or homoallyl alcohols into hydrodimethylsilyl ethers for use in intramolecular hydrosilation reactions. In some cases, *N,N*-diethylaminodimethylsilane gives superior results compared to the more commonly employed 1,1,3,3-tetramethyldisilazane (eq 1).[1]

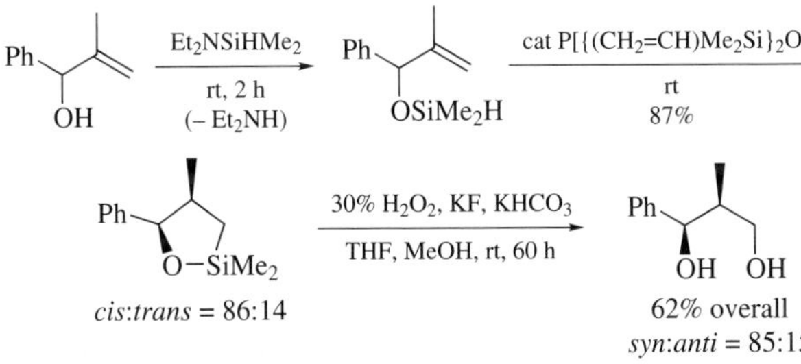

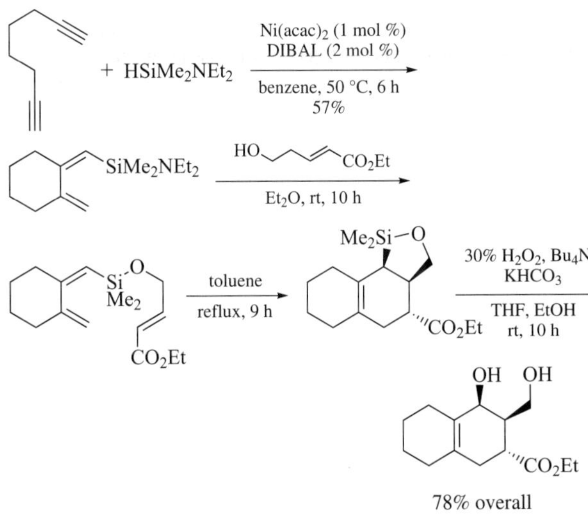

Hydrosilation Agent. Another application of this difunctional hydrosilane is in the nickel-catalyzed hydrosilation of 1,7-octadiyne, which affords a 1,2-dialkylidenecyclohexane with a (Z)-vinylsilane moiety. This exocyclic silyl diene is treated with an allyl alcohol to give a silicon-tethered triene that undergoes intramolecular Diels–Alder reaction and subsequent oxidative cleavage of the silicon–carbon bond to afford a bicyclic diol (eq 2).[2]

1. Tamao, K.; Nakagawa, Y.; Ito, Y., *Organometallics* **1993**, *12*, 2297.
2. Tamao, K.; Kobayashi, K.; Ito, Y., *J. Am. Chem. Soc.* **1989**, *111*, 6478.

Kohei Tamao
Kyoto University, Kyoto, Japan

Diethyl[dimethyl(phenyl)silyl]aluminum

$$\boxed{PhMe_2SiAlEt_2}$$

[86014-18-8] $C_{12}H_{21}AlSi$ (MW 220.36)

InChI = 1S/C8H11Si.2C2H5.Al/c1-9(2)8-6-4-3-5-7-8;2*1-2;/h3-7H,1-2H3;2*1H2,2H3;

InChIKey = VJCFOIBHDDYFAM-UHFFFAOYSA-N

(can silylaluminate terminal alkynes;[1] can cross couple with alkenyl, allyl, and aryl phosphates;[3] can deoxygenate epoxides[5])

Physical Data: see diethylaluminum chloride and dimethylphenylsilyllithium.

Solubility: commonly prepared and used in THF–hexane.

Preparative Method: prepared in situ by adding PhMe$_2$SiLi in THF to a hexane solution of Et$_2$AlCl at 0 °C.[1]

Handling, Storage, and Precautions: see diethylaluminum chloride and dimethylphenylsilyllithium.

Silylalumination of Alkynes and Alkenes. The reagent adds to 1-alkynes in a regio- and stereoselective (*cis* addition) manner. Transition metal catalysts alter the regioselectivity (eq 1).[1]

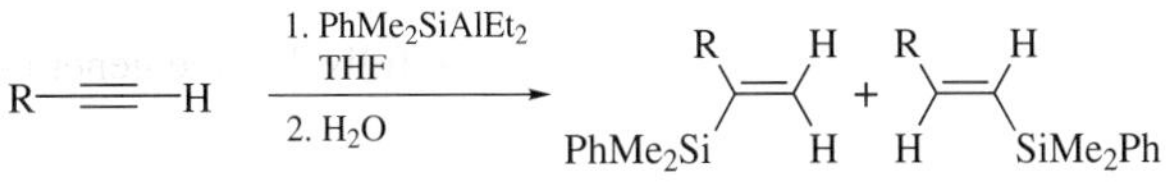

In addition reactions with allenes, PhMe$_2$SiAlEt$_2$ gives, albeit in low yield, allylsilanes as the major product, while contrasting regioselectivity is observed in the Cu-catalyzed silylmagnesiation by a related silylmagnesium reagent, PhMe$_2$SiMgMe (eq 2).[2]

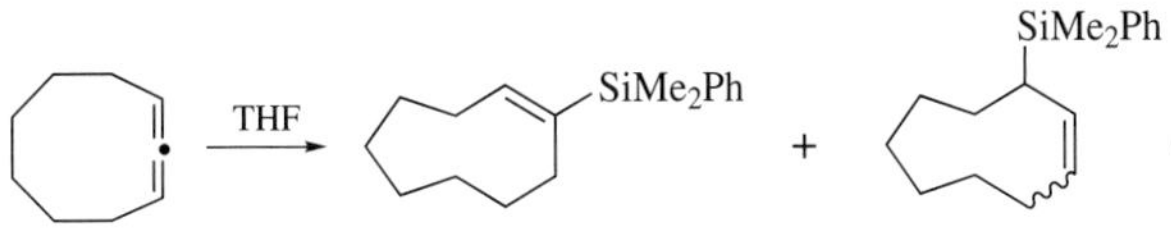

Avoid Skin Contact with All Reagents

Coupling with Allylic and Enol Phosphates. The reagent converts allylic phosphates into allylsilanes (eq 3).[3] Reaction with (2-vinyl-1,1-cyclopropanedicarboxylate gives the 1,7-homo-conjugate addition product (eq 4).[4] In the presence of a Pd catalyst, PhMe$_2$SiAlEt$_2$ converts enol phosphates into vinylsilanes (eq 5).[3] Higher yields are usually obtained with PhMe$_2$SiMgMe.

$$\text{(3)}$$

70%   15%

$$\text{(4)}$$

$(E):(Z) = 81:19$

$$\text{(5)}$$

Deoxygenation of Epoxides. The reagent deoxygenates epoxides in a stereospecific manner, although the reaction is limited to aryl-substituted epoxides (eq 6). A related reagent, [n-Bu$_3$SnAlMe$_3$]$^-$Li$^+$, has broader applicability.[5]

$$\text{(6)}$$

71%

1. Hayami, H.; Sato, M.; Kanemoto, S.; Morizawa, Y.; Oshima, K.; Nozaki, H., *J. Am. Chem. Soc.* **1983**, *105*, 4491.

2. Morizawa, Y.; Oda, H.; Oshima, K.; Nozaki, H., *Tetrahedron Lett.* **1984**, *25*, 1163.

3. Okuda, Y.; Sato, M.; Oshima, K.; Nozaki, H., *Tetrahedron Lett.* **1983**, *24*, 2015.

4. Fugami, K.; Oshima, K.; Utimoto, K.; Nozaki, H., *Bull. Chem. Soc. Jpn.* **1987**, *60*, 2509.

5. Matsubara, S.; Nonaka, T.; Okuda, Y.; Kanemoto, S.; Oshima, K.; Nozaki, H., *Bull. Chem. Soc. Jpn.* **1985**, *58*, 1480.

Keisuke Suzuki & Tetsuya Nagasawa
Keio University, Yokohama, Japan

Diiododimethylsilane

Me$_2$SiI$_2$

[15576-81-5]   C$_2$H$_6$I$_2$Si   (MW 311.96)
InChI = 1S/C2H6I2Si/c1-5(2,3)4/h1-2H3
InChIKey = UYZARHCMSBEPFF-UHFFFAOYSA-N

(reducing agent; preparation of *N*-(iododimethylsilyl)trialkyl-phosphinimines, bis(2,2′-bipyridine) complexes of silicon, di-μ-iodo-bis(tetracarbonyltungsten); chemistry of 7-silanorborna-dienes and dimethylsilylene)

Alternate Name: dimethylsilyl diiodide.
Physical Data: bp 170 °C; *d* 2.203 g cm^{-3}.
Solubility: sol most organic solvents.
Preparative Method: 8.2 g (21 mmol) of boron triiodide and 7.4 g (57 mmol) of dichlorodimethylsilane are refluxed for 3 days at 120 °C; the crude product is then isolated by adding copper powder until the color disappears followed by fractional distillation in a spinning band column apparatus (65% yield).[1]
Handling, Storage, and Precautions: reacts vigorously with water, but without spattering; highly toxic, flammable liquid; use in a fume hood.

Reduction of α-Aryl Alkanols. This reaction can be accomplished in CH$_2$Cl$_2$ at rt with 1 equiv of diiododimethylsilane (eq 1).[2]

$$\text{Ar}{-}\overset{R^1}{\underset{R^2}{\vert}}{-}\text{OH} + \text{Me}_2\text{SiI}_2 \xrightarrow[\text{rt}]{\text{CH}_2\text{Cl}_2} \text{Ar}{-}\overset{R^1}{\underset{R^2}{\vert}}{-}\text{H} + (\text{Me}_2\text{SiO})_n + \text{I}_2 \quad (1)$$

Ar	R^1	R^2	Yield (%)
Ph	Ph	Ph	94
Ph	Ph	Me	100
Ph	Me	Me	50
Ph	Ph	H	100
Ph	PhCO	H	45
			72
			79

The reaction is far milder and more effective than comparative reductions of alcohols with aluminum chloride–palladium and lithium–ammonia. This method also permits reduction in the presence of other functional groups. A neighboring β-keto group, for example, is not reduced by diiododimethylsilane. The reaction is rapid, and reduction occurs best with secondary and tertiary α-aryl alkanols to give the corresponding alkanes in good yields.

Reductive Condensation of Ketones. Diiododimethylsilane reacts with various ketones in the presence of zinc to give the corresponding reductive condensation products in good yield (eq 2).[3]

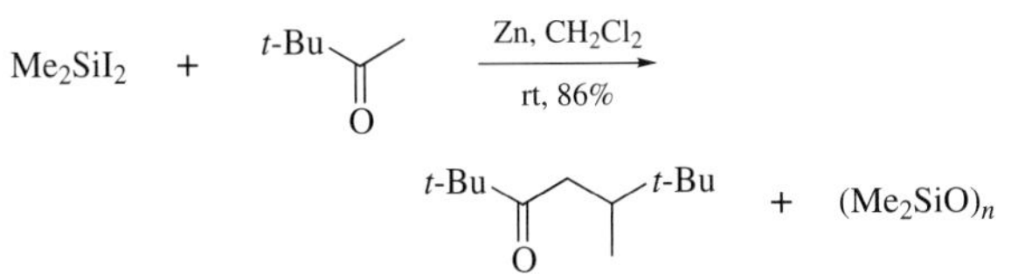

$$\text{Me}_2\text{SiI}_2 \quad + \quad \underset{\text{O}}{t\text{-Bu}} \quad \xrightarrow[\text{rt, 86\%}]{\text{Zn, CH}_2\text{Cl}_2}$$

$$\underset{\text{O}}{t\text{-Bu}} \qquad t\text{-Bu} \quad + \quad (\text{Me}_2\text{SiO})_n \quad (2)$$

Reductive condensation products have been obtained from this reaction for t-BuCOMe, MeCOMe, EtCOEt, and cyclohexanone. The reaction of Me_2SiI_2 with Ph_2CO in the presence of zinc gave benzopinacolone.

Preparation of N-(Iododimethylsilyl)trialkylphosphinimines.

N-(Iododimethylsilyl)trialkylphosphinimines are accessible by transsilylation of the corresponding N-trimethylsilyl compounds or by the reaction of bis(trialkylphosphinimino)-dimethylsilane with diiododimethylsilane (eq 3).[4]

$$R_3P\underset{N}{\overset{\text{Me Me}}{\diagdown\text{Si}\diagdown}}_N PR_3 \quad + \quad \text{Me}_2\text{SiI}_2 \quad \longrightarrow \quad 2 \; R_3P\underset{N}{\overset{\text{Me Me}}{\diagdown\text{Si}\diagdown}}_I \quad (3)$$

In a subsequent reaction, most of the compounds undergo dimerization to form dicationic four-membered systems (eq 4). Only compounds with bulky groups at the phosphorus atom are stable as monomers.

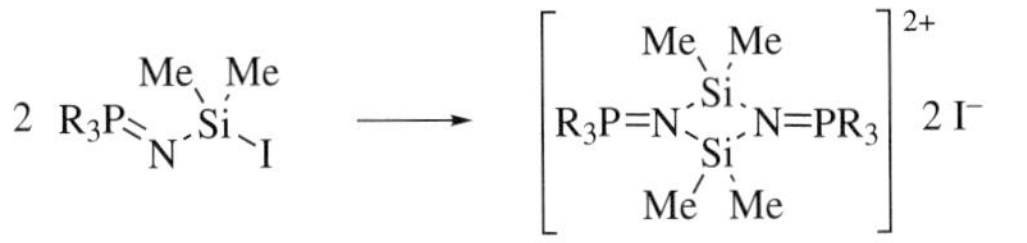

$$2 \; R_3P\underset{N}{\overset{\text{Me Me}}{\diagdown\text{Si}\diagdown}}_I \quad \longrightarrow \quad \left[R_3P=N\underset{\text{Me Me}}{\overset{\text{Me Me}}{\diagdown\text{Si}\diagup}}N=PR_3 \right]^{2+} 2\,I^- \quad (4)$$

Preparation of Bis(2,2′-bipyridine) Complexes of Silicon.

A bis(2,2′-bipyridine) complex of silicon was synthesized by the direct reaction of 2,2′-bipyridine with diiododimethylsilane in chloroform (eq 5).[5] An ionic compound is formed. In solution it contains the *cis*-octahedral $[\text{SiMe}_2(\text{bipy})]^{2+}$ cation (**1**).

$$\text{Me}_2\text{SiI}_2 \quad + \quad 2\text{bipy} \quad \xrightarrow{\text{CHCl}_3} \quad \text{SiMe}_2\text{I}_2 \cdot 2\text{bipy} \quad (5)$$

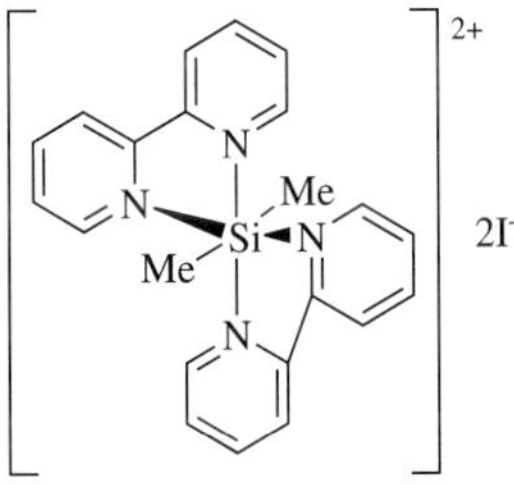

(**1**)

Preparation of Di-μ-iodobis(tetracarbonyltungsten).

Hexacarbonyltungsten reacts with diiododimethylsilane to give $[(\text{CO})_4\text{WI}]_2$ (eq 6).[6] The dinuclear complex (**2**) is diamagnetic. The iodine atoms serve as bridging ligands between metal atoms.

$$\text{W(CO)}_6 \quad \xrightarrow[h\nu]{\text{Me}_2\text{SiI}_2} \quad [(\text{CO})_4\text{WI}]_2 \quad (6)$$

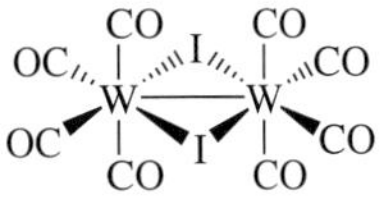

(**2**)

Chemistry of 7-Silanorbornadienes and Dimethylsilylene, Me₂Si.

7-Silanorbornadienes react rapidly with halogens at $-20\,°\text{C}$ to $+20\,°\text{C}$ to yield Me_2SiX_2 (X = Cl, Br, I) and the naphthalene or benzene derivatives (eq 7).[7]

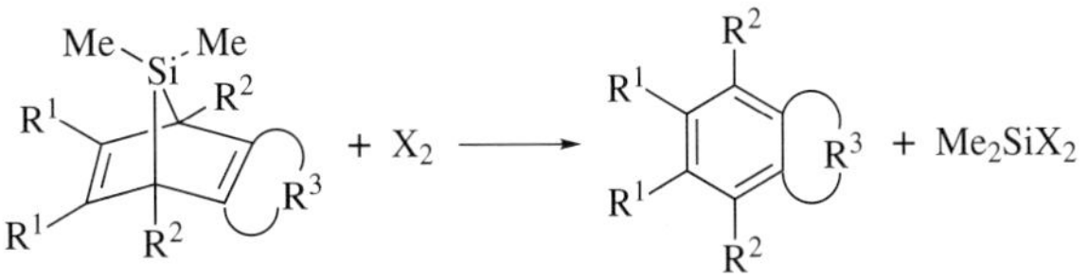

R^1	R^2	R^3
H	H	benzo
H	Ph	cycloocteno
Ph	Me	cycloocteno
Ph	Ph	benzo

Halogen can be abstracted from the C–Hal bond in benzyl iodide (or bromide), leading to Me_2SiX_2 and toluene (eq 8).[7]

$$2 \; \text{PhCH}_2\text{X} \quad \xrightarrow[\text{X = Br, I}]{\text{Me}_2\text{Si}} \quad 2 \; \text{PhMe} \quad + \quad \text{Me}_2\text{SiX}_2 \quad (8)$$

High-Purity Alkali Metal Halides.

Alkali metal halides are refined by reaction in the gas phase with alkyl-, aryl-, or alkoxyhalosilanes.[8] This brings about an exchange between the Si-bound halogen and the anion of the contaminant, while the C-containing radical has a reducing effect. Diiododimethylsilane is used to purify Sodium Iodide in this manner.

Related Reagent. Diiodosilane

1. Wolfsberger, W.; Schmidbaur, H., *J. Organomet. Chem.* **1971**, *28*, 301.
2. Ando, W.; Ikeno, M., *Tetrahedron Lett.* **1979**, 4941.
3. Ando, W.; Ikeno, M., *Chem. Lett.* **1980**, 1255.
4. Wolfsberger, W., *J. Organomet. Chem.* **1979**, *173*, 277.
5. Kummer, D.; Gaisser, K. E.; Seshadri, T., *Chem. Ber.* **1977**, *110*, 1950.
6. Schmid, G.; Boese, R.; Welz, E., *Chem. Ber.* **1975**, *108*, 260.
7. Appler, H.; Neumann, W. P., *J. Organomet. Chem.* **1986**, *314*, 247.
8. Lebl, M., *Chem. Abstr.* **1972**, *77*, 37 185e.

Harowin O'Dowd, Laurent Deloux & Morris Srebnik
University of Toledo, OH, USA

Diisopropyl Chlorosilane, Polymer-supported

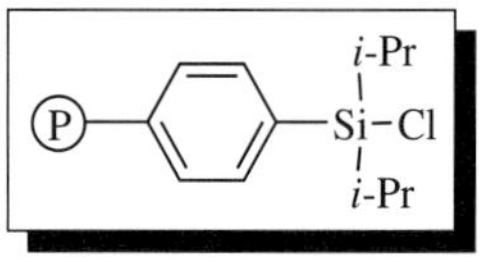

(reagent used as a linker for solid-phase organic synthesis)

Alternate Name: (diisopropylchlorosilyl)polystyrene.
Preparative Methods: lithiated polystyrene can be prepared via direct lithiation of cross-linked polystyrene with (1:1) BuLi and TMEDA at 65 °C in cyclohexane (eq 1).[1] Subsequent addition of diisopropyl-dichlorosilane in benzene followed by washing with excess benzene and drying in vacuo provides the desired functionalized polymer. Loading can be determined by hydrolysis and acid-base titration.[2]

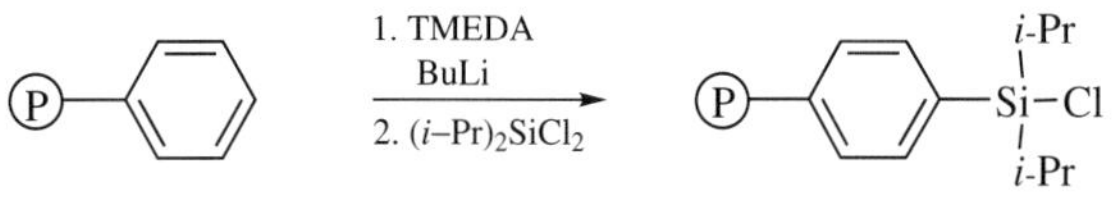

$$\textcircled{P} = \text{polystyrene polymer}$$

Handling, Storage, and Precaution: reagent is water sensitive and cannot be stored for prolonged periods.

Rasta-(diisopropylchlorosilyl) TEMPO Methylpolystyrene Resin. Silyl styrene monomers were prepared from *p*-bromostyrene via lithium-halogen exchange with BuLi followed by quenching with diisopropylchlorosilane (eq 2).[3] Addition of a large molar excess of monomer to a suspension of TEMPO-methylpolystyrene resin and heating to 130 °C for 20 h in a capped vial under nitrogen gave a solid. Washing of the solid alternately with methylene chloride and methanol and drying in vacuo gave Rasta-(diisopropylsilyl)-TEMPO methylpolystyrene resin (eq 3).[3]

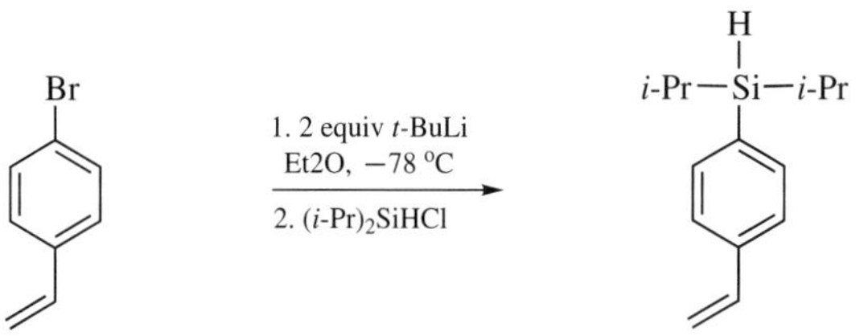

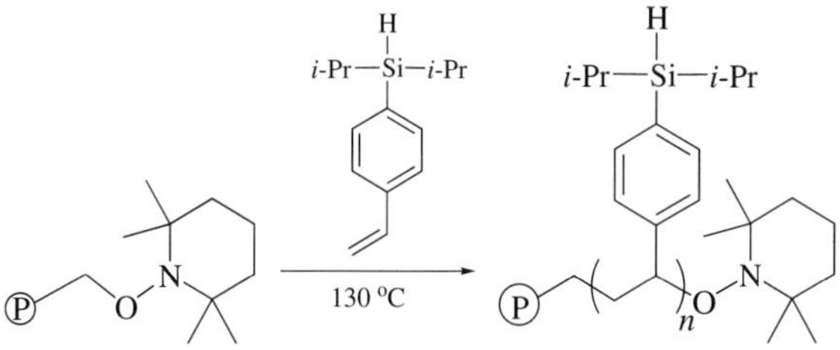

Subsequent treatment with 1,3-dichloro-5,5-dimethyl-hydantoin in THF provides the Rasta-(diisopropylchlorosilyl) TEMPO methylpolystyrene resin (eq 4).[3]

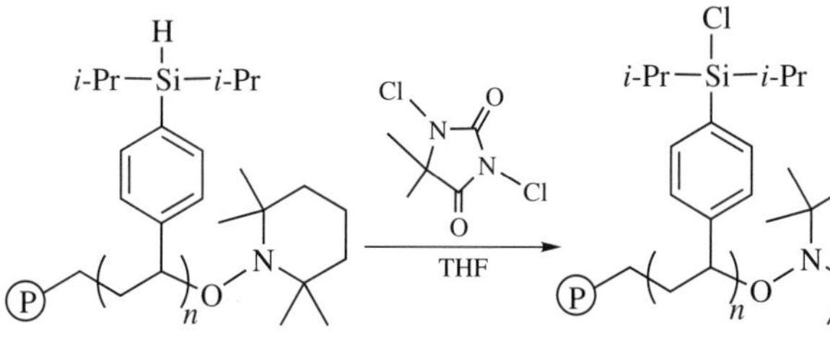

Silyl Linkers for Solid-Phase Organic Synthesis. Silyl ether linkers are advantageous for solid-phase organic synthesis due to their ease of formation and cleavage, provided that the synthetic strategy does not require other silyl protecting groups or strongly acidic conditions.[4] (Diisopropylchlorosilyl)polystyrene can be used as a silyl ether linker by treatment of an alcohol with the reagent resin in the presence of a base such as pyridine or diisopropylethylamine (eq 5). Cleavage of the silyl ether linker can be effected cleanly with TBAF in methylene chloride or tetrahydrofuran.

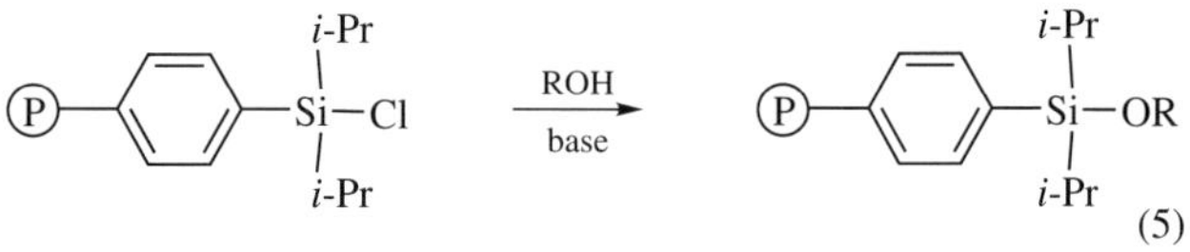

Diisopropylsilyl linkers on polystyrene resins have been used in solid-phase synthesis of oligosaccharides and glycoconjugates via the glycal assembly method (eq 6).[1,5,6]

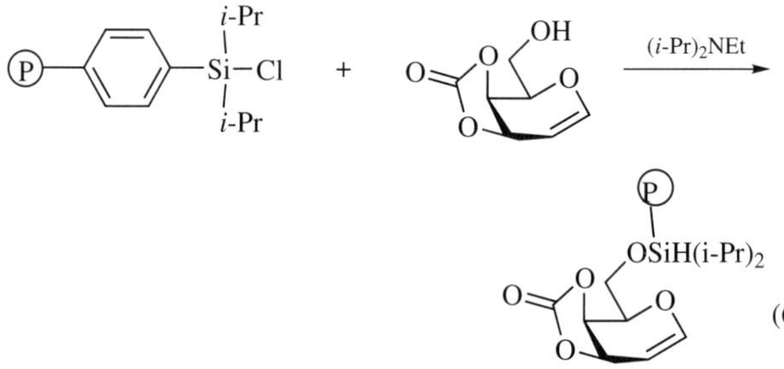

Diisopropylsilyl linkers are more stable to glycosylation conditions than their diphenyl counterparts. Rasta-(diisopropylchlorosilyl) TEMPO methylpolystyrene resins have been demonstrated as viable polymeric supports (eq 7),[2] however, the increased complexity of their preparation from commercially available starting materials has thus far precluded broader general application.

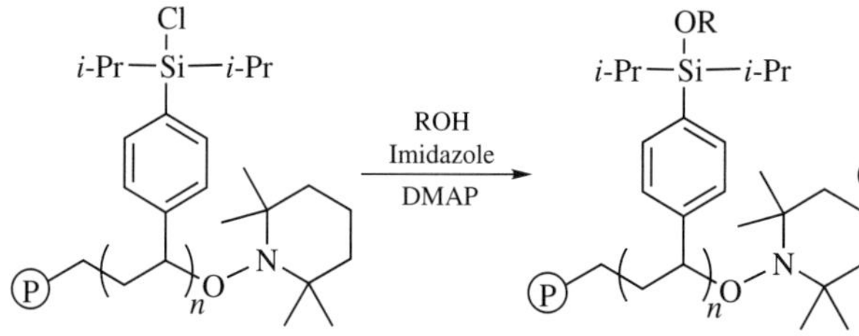

Related Reagents. Polymer-bound diphenyl chlorosilane, trityl chloride resin; Merrifield resin.

1. Zheng, C.; Seeberger, P. H.; Danishefsky, S. J., *J. Org. Chem.* **1998**, *63*, 1126.
2. Chan, T.-H.; Huang, W.-Q., *J. Chem. Soc., Chem Commun.* **1985**, 909.
3. Lindsley, C. W.; Hodges, J. C.; Filzen, G. F.; Watson, B. M.; Geyer, A. G., *J. Comb. Chem.* **2000**, *2*, 550.
4. Seeberger, P. H.; Haase, W.-C., *Chem. Rev.* **2000**, *100*, 4349.
5. Seeberger, P. H.; Danishefsky, S. J., *Acc. Chem. Res.* **1998**, *31*, 685.
6. Savin, K. A.; Woo, J. C. G.; Danishefsky, S. J., *J. Org. Chem.* **1999**, *64*, 4183.

Diana K. Hunt & Peter H. Seeberger
Massachusetts Institute of Technology, Cambridge, MA, USA

Diisopropylsilyl Bis(trifluoromethane-sulfonate)

$$i\text{-Pr}_2\text{Si}(\text{OSO}_2\text{CF}_3)_2$$

[85272-30-6] $C_8H_{14}F_6O_6S_2Si$ (MW 412.39)

InChI = 1S/C8H14F6O6S2Si/c1-5(2)23(6(3)4,19-21(15,16)7(9,10)11)20-22(17,18)8(12,13)14/h5-6H,1-4H3

InChIKey = BHZPNBNSRVSTBW-UHFFFAOYSA-N

(bifunctional silyl protecting group for 1,2-, 1,3-, and 1,4-diols)

Physical Data: bp 85–86 °C/2 mmHg; *d* 1.396 g cm^{-3}; fp 110 °C.

Preparative Method: conveniently prepared by the slow addition of 2 equiv of trifluoromethanesulfonic acid to a stirring mixture of chlorodiisopropylsilane, followed by direct distillation of the product from the reaction flask to yield the product as a pale yellow oil.[1]

Handling, Storage, and Precautions: store in the refrigerator under an inert atmosphere of argon or nitrogen. Special care should be used when handling since diisopropysilyl bis(trifluoromethanesulfonate) has been classified as a corrosive.

Original Commentary

Janet Wisniewski Grissom
University of Utah, Salt Lake City, UT, USA

The use of silyl protecting groups in organic synthesis has been limited to monofunctional silicon compounds. Their use has been enhanced by the fact that they may be put on and removed under mild conditions. Diisopropylsilyl bis(trifluoromethanesulfonate) and its analog, di-*t*-butylsilyl bis(trifluoromethanesulfonate), go one step further in protecting group technology. The two electrophilic sites on silicon allow for the simultaneous protection of 1,2-, 1,3-, and 1,4-diols, and in some cases form a rigid system which allows for asymmetric stereocontrol in subsequent organic transformations.[1]

(**1**) 83%   (**2**) 88%   (**3**) quant

Polyfunctional silicon protecting groups have been employed by researchers before the use of silicon ditriflates. Trost and co-workers found in the development of methodology towards the synthesis of erythronolide A that di-*t*-butyldichlorosilane was relatively unreactive under standard silylation conditions and required forcing conditions (65–95 °C, 1-hydroxybenzotriazole, MeCN, or DMF) for its employment.[2] These two factors make dichlorosilanes limited in their applicability. Silicon ditriflates will react readily with many diols in the presence of 3 equiv of 2,6-lutidine in chloroform at temperatures ranging from 0–25 °C. The generality of this polyfunctional protecting group is illustrated by the formation of the silyl-protected diols (**1**), (**2**), and (**3**).

Corey and Link utilized diisopropylsilyl bis(trifluoromethanesulfonate) as a silylating agent for catechols in the first enantioselective synthesis of pure (*R*)- or (*S*)-isoproterenol (eq 1).[3]

(1)

First Update

Jitendra D. Belani
Infinity Pharmaceuticals, Cambridge, MA, USA

Applications in Synthesis of Oligonucleotides. Diisopropylsilyl bis(trifluoromethanesulfonate) has more recently been used to prepare diisopropylsilyl-linked oligonucleotide analogs.[4] Silylation of 5′-*O*-(dimethoxytrityl)-2′-deoxynucleosides using diisopropylsilyl bis(trifluoromethanesulfonate) and the hindered base 2,6-di-*t*-butyl-4-methylpyridine (Dtbp) provided quantitative yield of 3′-*O*-diisopropylsilanols (eq 2). Hence, silylation of 5′-*O*-(dimethoxytrityl)thymidine, N^6-benzoyl-2′-deoxy-5′-*O*-(dimethoxytrityl)adenosine, N^4-benzoyl-2′-deoxy-5′-*O*-(dimethoxytrityl)cytidine, and N^2-isobutyryl-2′-deoxy-5′-*O*-(dimethoxytrityl)guanosine gave the desired products **5a–d** in excellent yields.

Coupling of products **5a–d** with unprotected thymidine resulted in formation of 3′,5′-linked dinucleosides **6a–d** selectively, and the dimer **7** from reaction with secondary hydroxyl group of thymidine was not observed (eq 3). Subsequent silylation and coupling with another molecule of thymidine gave 3′,5′-linked trinucleoside **8** in 76% yield (eq 4). A tetrathymidylate oligomer was prepared by 3′-O-silylation of **8** followed by coupling with thymidine. Although no self-condensation was observed, only 30% of the product **9** was obtained after purification by preparative HPLC.

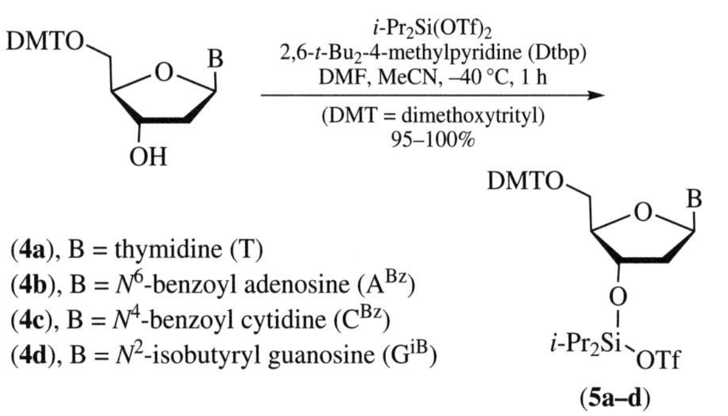

(4a), B = thymidine (T)
(4b), B = N^6-benzoyl adenosine (A^{Bz})
(4c), B = N^4-benzoyl cytidine (C^{Bz})
(4d), B = N^2-isobutyryl guanosine (G^{iB})

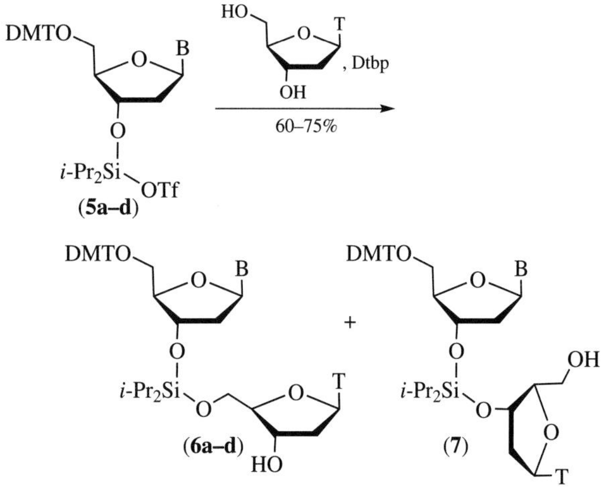

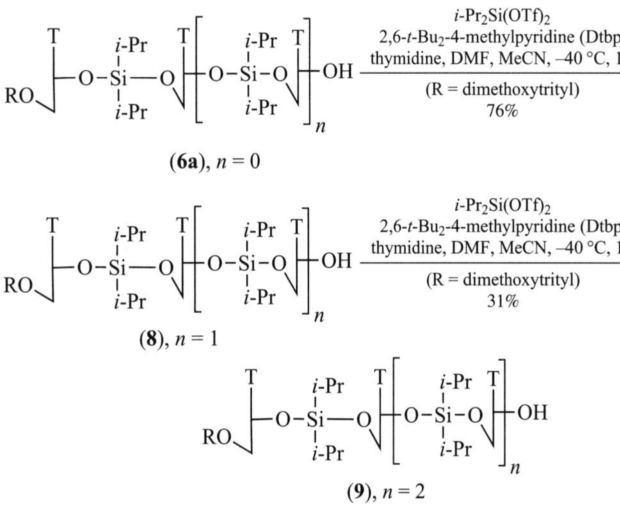

In addition to facile preparation of these oligomers, the authors report that detritylation can be achieved in high yields using 3% trichloroacetic acid, indicating good acid stability of the diisopropylsilyl linkage (eq 5). Solution-phase synthesis of higher oligomers (>5 nucleosides) was tedious and authors reported solid-phase automated synthesis of strands of varying length up to a 10-mer of thymidines containing the all-silicon backbone (eq 6). After the synthesis, the controlled pore glass (CPG) solid support

was cleaved with aqueous ammonium hydroxide solution with good recoveries and the final compound was isolated by preparative reversed phase HPLC.[4]

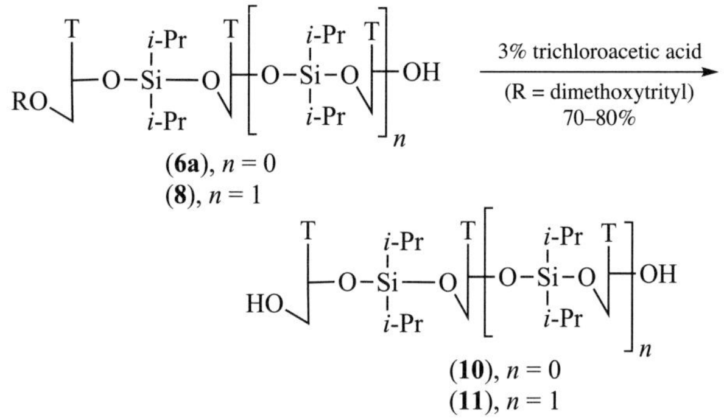

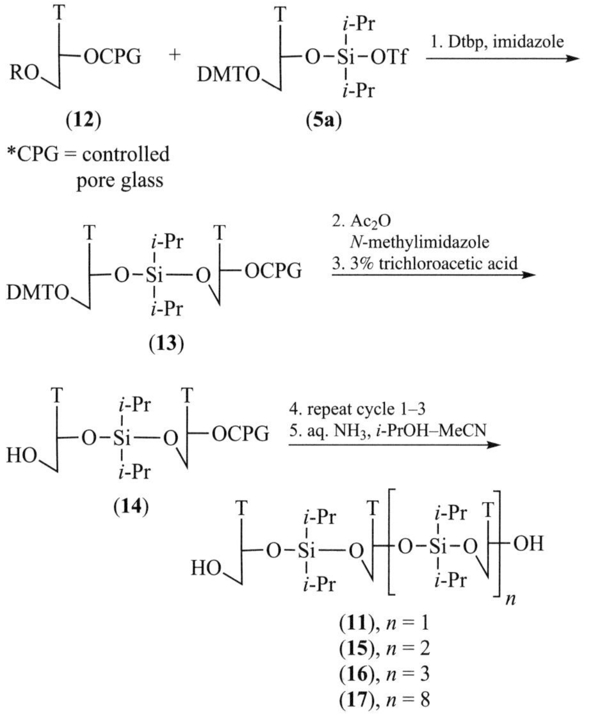

(6a), n = 0
(8), n = 1

(10), n = 0
(11), n = 1

(12)   **(5a)**

*CPG = controlled pore glass

(13)

(14)

(11), n = 1
(15), n = 2
(16), n = 3
(17), n = 8

To further study the physiochemical properties of the diisopropylsilyl linkage as the phosphodiester mimic, the authors prepared phosphodiester-linked deoxyribonucleotide oligomers containing single or multiple units of the silyl-linked di- and trithymidylates. The building blocks were prepared by treating the silyl-linked thymidine dimer (T-Si-T) **6a** and the trimer (T-Si-T-Si-T) **8** with 2-cyanoethyl N,N-diisopropylphosphoramidochloridite (eq 7) and the products were subsequently incorporated into 11-mer oligonucleotides.[4]

(20) 5′ TTT TTT TTTSiT T-3′
(21) 5′ TTT TTT TTSiT TT-3′
(22) 5′ TTT TTTSiT TTT T-3′
(23) 5′ TTT TTT TTSiTSiT T-3′
(24) 5′ TTT TTSiTSiT TTT T-3′
(25) 5′ TTT TTT TTT TT-3′

The oligomers (**20–24**) were subjected to a nuclease stability assay against the control T_{11} (**25**) and it was observed that the exonuclease cleaved the phosphodiester bonds in the 3′–5′ direction but stopped at the siloxane link. The authors also report the thermodynamic melting (T_m) data, UV absorbance, and solubility of the oligomers.[4]

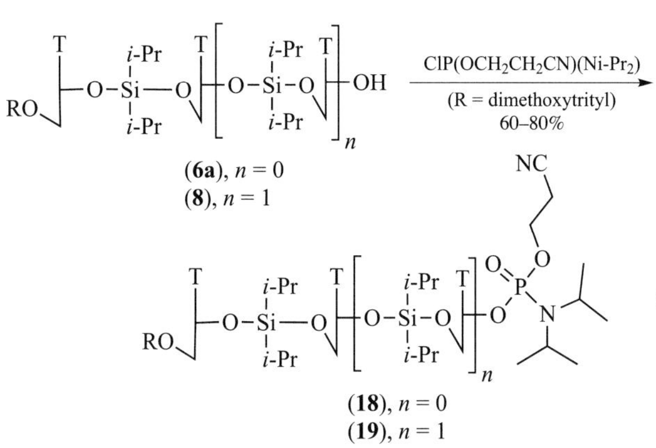

(6a), $n = 0$
(8), $n = 1$

(18), $n = 0$ (7)
(19), $n = 1$

In a separate communication, Sekine and coworkers reported a new synthesis of oligodeoxyribonucleotides containing 4-N-alkoxycarbonyldeoxycytidine derivatives.[5] A variety of oligo-deoxyribonucleotides incorporating the acyl- and alkoxyacyl groups were prepared. One example is shown in eq 8. First, poly-styrene-type ArgoPore resins having a benzyloxy (diisopropyl)-silyl linker (**27**) were prepared. Zinc bromide was used as the detritylating agent to avoid the partial loss of silyl linker that occurred when the traditional acidic conditions were applied. Chain elongation was followed by release of the DNA chain from the resin by treatment with TBAF under neutral condi-tions. The T_m experiments showed that incorporation of 4-N-alkoxycarbonyldeoxycytidines into the DNA strand resulted in higher hybridization affinity with the complementary DNA strands when compared to 4-N-acyldeoxycytidines.[5]

The diisopropylsilyl linker has also been used to prepare a cleavable substrate containing molecular beacons for the quan-tification of DNA photolyase activity.[6] Accordingly, 5′- and 3′-acetyl-protected thymidines **32** and **33** were linked with diiso-propylsilyl bis(trifluoromethanesulfonate) to provide the dinu-cleotide **34** (eq 9). Irradiation of the dimer **34** using ultravio-let light ($h\nu > 300$ nm) gave *cis–syn* thymidine dimer **35** in 20% yield. Removal of the acetate protecting groups using anhy-drous methanolic ammonia solution followed by protection of the primary alcohol using dimethoxytrityl chloride provided the de-sired dimer **36** in 61% yield. Formation of the phosphoramidite **37** was followed by chain elongation using standard procedures. The cleavage of the oligonucleotide from solid support and re-moval of the silyl linkage was achieved using ammonium hydrox-ide in ethanol at 55 °C. The final molecular beacon **39** was puri-fied by reversed-phase HPLC and was used to characterize DNA photolyase activity.[6]

Preparation of Substrates for C–H Insertion Reactions.
Craig and coworkers have reported stereocontrolled polyol synthesis via C–H insertion reactions of silicon-tethered diazo-acetates.[7] Menthol was treated with diisopropylsilyl bis(trifluoro-methanesulfonate) and the product was condensed with ethyl diazoacetate to provide the precursor **43** to the C–H insertion reac-tion (eq 10). The rhodium(II) octanoate-catalyzed decomposition of the diazoacetate **43** gave siloxane **44** as a single stereoisomer. Reduction of the ester, followed by oxidative cleavage of the C–Si bond, gave the desired polyol **46** in 30% yield, along with the elim-

ination product **47** in 38% yield over three steps. The sequence was also reported with steroids lanosterol and cholesterol (eq 11). The observed selectivity with cholesterol (C2 vs. C4) in the inser-tion reaction was ascribed to the deactivation of the allylic posi-tions toward C–H insertion or to the lack of favorable 'equatorial'

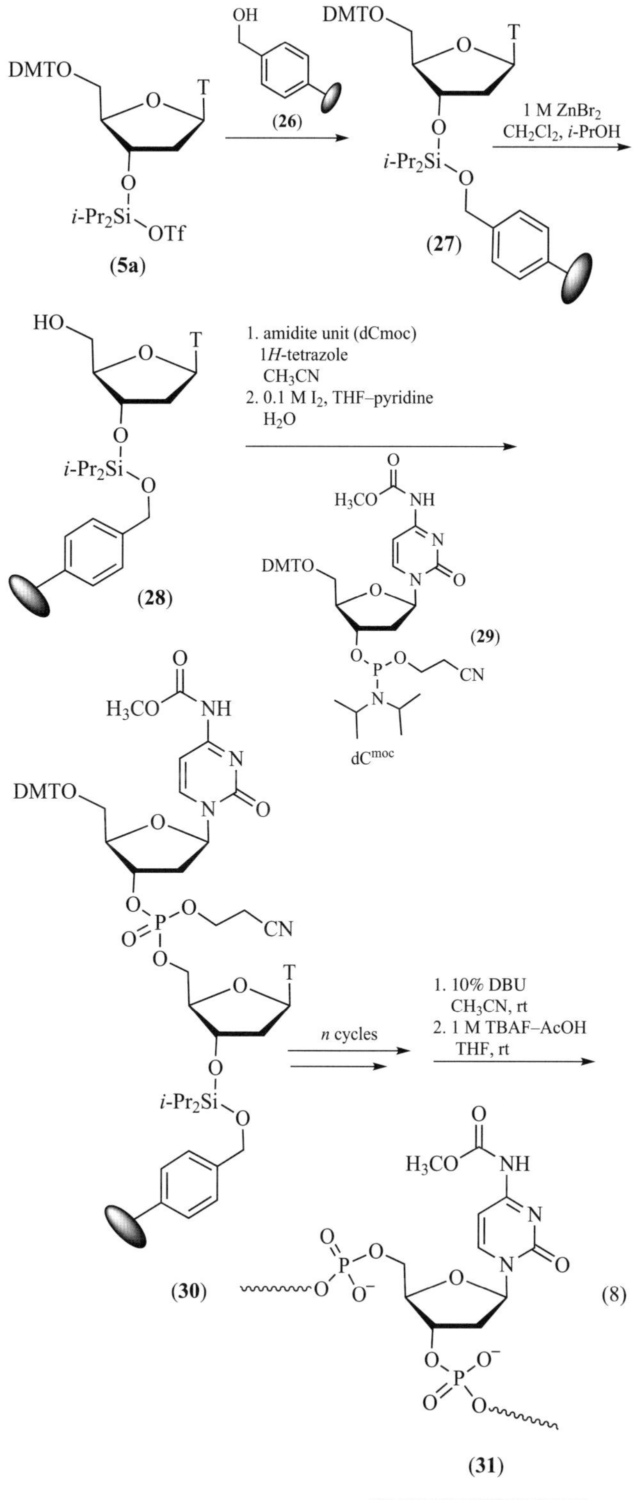

(5a)

(26)

(27)

(28)

1. amidite unit (dCmoc)
   1H-tetrazole
   CH$_3$CN
2. 0.1 M I$_2$, THF–pyridine
   H$_2$O

(29)

dC^{moc}

(30)

n cycles

1. 10% DBU
   CH$_3$CN, rt
2. 1 M TBAF–AcOH
   THF, rt

(31) (8)

d(TTTTTTC^{moc}TTTTTT)

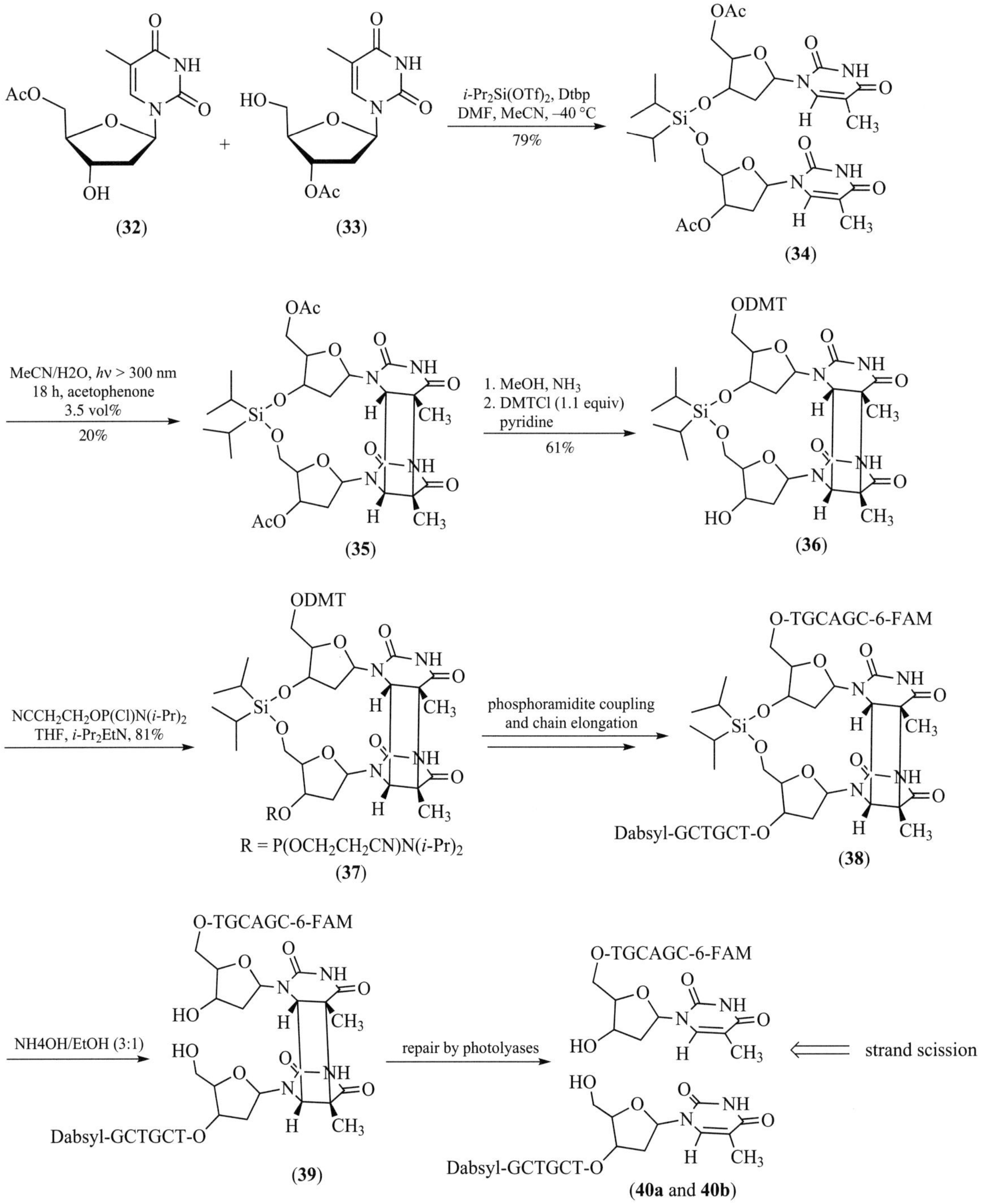

$$i\text{-Pr}_2\text{Si(OTf)}_2, \text{Dtbp}$$
$$\text{DMF, MeCN, } -40\,°\text{C}$$
$$79\%$$

(32) **(33)** **(34)**

MeCN/H2O, $h\nu > 300$ nm
18 h, acetophenone
3.5 vol%
20%

(35)

1. MeOH, NH₃
2. DMTCl (1.1 equiv)
   pyridine
   61%

(36)

NCCH₂CH₂OP(Cl)N(i-Pr)₂
THF, i-Pr₂EtN, 81%

R = P(OCH₂CH₂CN)N(i-Pr)₂
(37)

phosphoramidite coupling
and chain elongation

O-TGCAGC-6-FAM

Dabsyl-GCTGCT-O
(38)

NH4OH/EtOH (3:1)

O-TGCAGC-6-FAM

Dabsyl-GCTGCT-O
(39)

repair by photolyases

O-TGCAGC-6-FAM

Dabsyl-GCTGCT-O
(40a and 40b)

⟸ strand scission

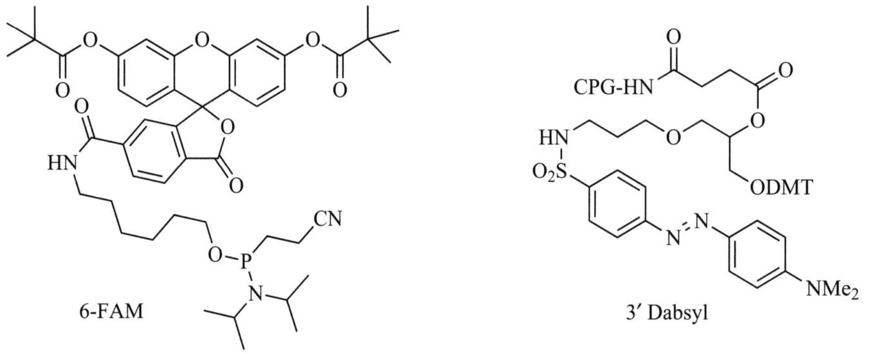

6-FAM

3′ Dabsyl

(9)

disposition of the C4–H bond for attack by the carbenoid due to the presence of the neighboring sp^2 center. In addition, the authors also report formation of silyl diazoacetates **52** and **53** from prochiral alcohols and their conversion to the respective polyols (eq 12).[7]

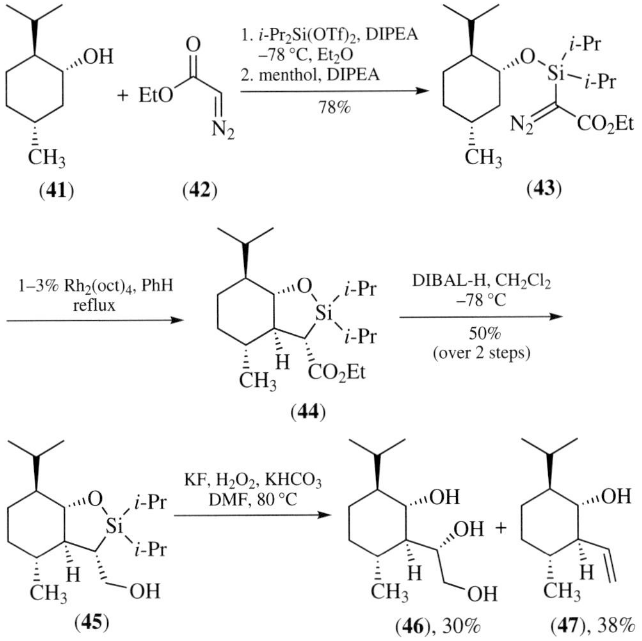

(**41**)   (**42**)   (**43**)

(**44**)

(**45**)   (**46**), 30%   (**47**), 38%

(10)

Maas and coworkers have similarly reported preparation of α-oxysilyl-α-diazoacetates utilizing aliphatic, allyl, and homoallyl alcohols as nucleophiles (eq 13).[8] Photochemical extrusion from (isopropyloxy)silyl-diazoacetates using 254 nm light provided modest yields of tetrahydro-1,2-oxasiloles (**61–63**) (eq 14).

The carbenes generated from thermal decomposition of allyloxy-silyldiazoacetates at temperatures ≥140 °C provided 2,5-dihydro-1,2-oxasiloles **66** and **67** as major products (eq 15). The authors propose that dihydro-oxasiloles **66** and **67** are generated from bicyclic pyrazolines upon thermal extrusion of N$_2$ followed by isomerization of biradicals (eq 16).

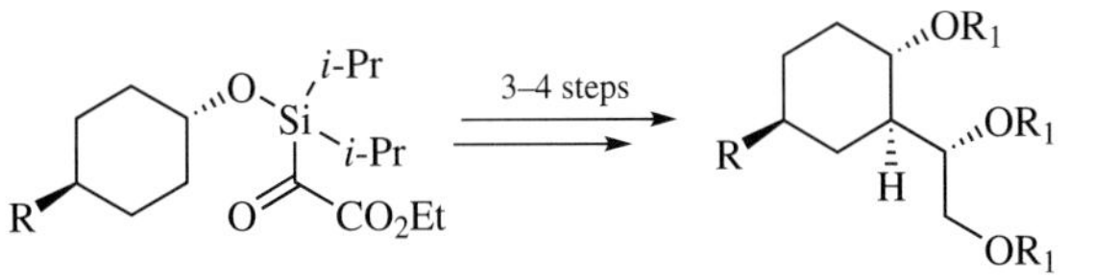

(**52**), R = H   (**54**), R = H, R$_1$ = Ac; 7% over 4 steps
(**53**), R = t-Bu   (**55**), R = t-Bu, R$_1$ = H; 10% over 3 steps

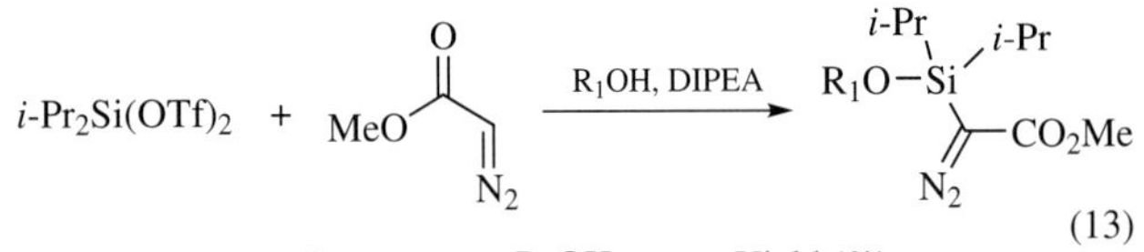

(13)

Product	R$_1$OH	Yield (%)
56	EtOH	62
57	PrOH	65
58	⟍⟍⟍OH	69
59	⟍⟍⟍⟍OH	95
60	MeO$_2$C⟍⟍⟍OH	61

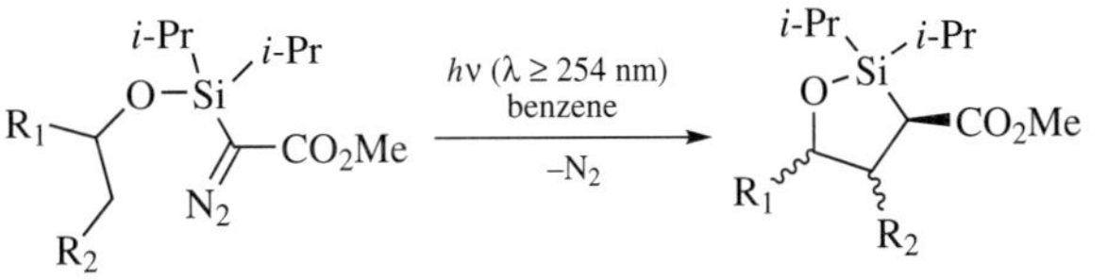

(14)

Product	R$_1$	R$_2$	Yield (%)
61	H	H	39
62	H	Me	62
63	Me	H	51

Decomposition using catalytic CuOTf, however, gave products **68** and **69** resulting from intramolecular cyclopropanation in excellent yields (eq 15). Similarly, [(3-butenyl)oxysilyl]diazoacetate **73** underwent intramolecular cyclopropanation to give silabicyclohexane system **74** under thermal conditions in 58% yield (eq 17).[9]

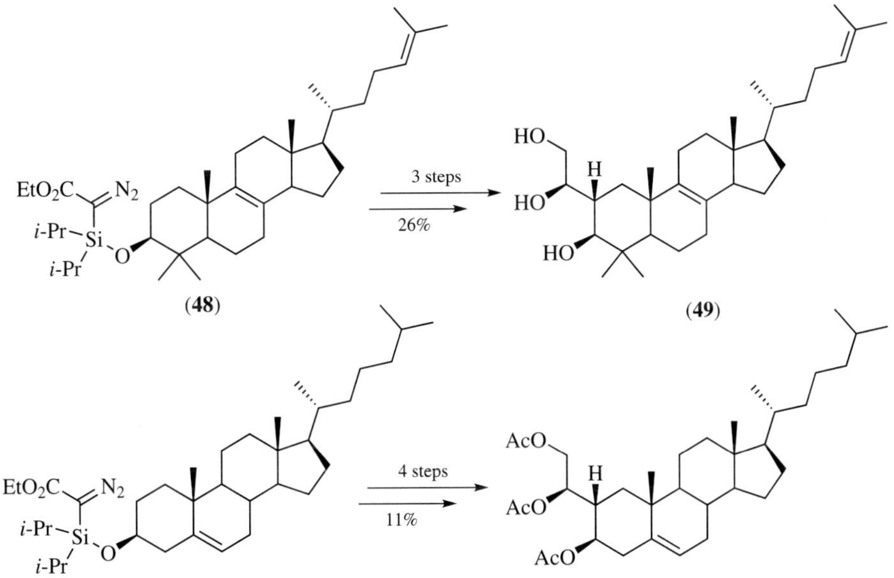

(**48**)   (**49**)

(11)

(**50**)   (**51**)

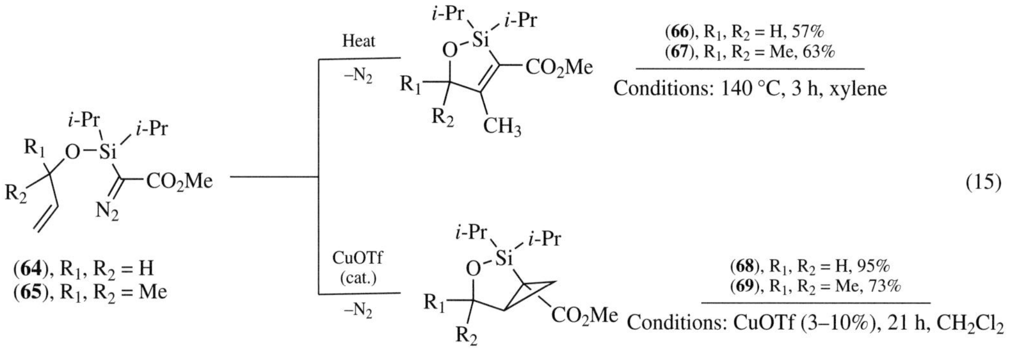
Heat
$-N_2$
CuOTf (cat.)
$-N_2$
(64), R_1, R_2 = H
(65), R_1, R_2 = Me
(66), R_1, R_2 = H, 57%
(67), R_1, R_2 = Me, 63%
Conditions: 140 °C, 3 h, xylene
(68), R_1, R_2 = H, 95%
(69), R_1, R_2 = Me, 73%
Conditions: CuOTf (3–10%), 21 h, CH_2Cl_2

(15)

The electron-deficient alkene **75** did not favor intramolecular [3 + 2] cycloaddition and instead gave rearranged diazoacetate **76** in 64% yield upon heating at 125 °C. Copper(I) triflate-catalyzed decomposition of **75** gave a mixture of compounds and only 14% of cyclopropane system **77** (eq 18) was isolated. Finally, the authors showed that desilylation of these silabicyclic compounds can be achieved using cesium fluoride in wet THF to provide *trans*-1,2-cyclopropanes (eq 19).[9]

Maas and coworkers have also reported preparation of α-[(2-alkynyloxy]silyl-α-diazoacetates (**80–83**) for screening copper- and rhodium-mediated carbenoid reactions (eq 20).[10] Treatment of diazoacetate **80** with catalytic amounts of copper(I) triflate gave a mixture of compounds from which the nitrogen-free oxasilaheterocycles **85** and **86** were isolated in low yields (eq 21). Maas and coworkers have prepared novel diazoacetates bearing an aminosilyl substituent that show remarkable thermal stability.[11] In yet another communication, the authors report synthesis of (diazomethyl)-silyl functionalized Fisher-type carbene complexes (eq 22).[12]

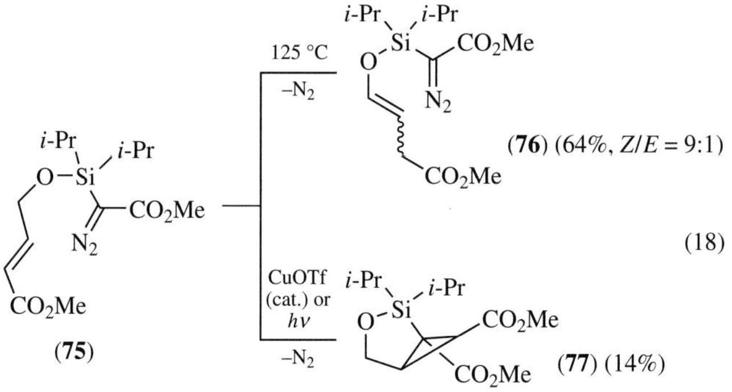
125 °C
$-N_2$
CuOTf (cat.) or hν
$-N_2$
(75)
(76) (64%, Z/E = 9:1)
(77) (14%)

(18)

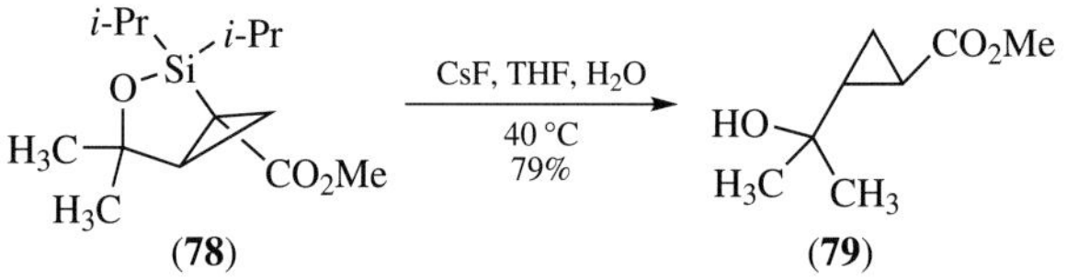
CsF, THF, H_2O
40 °C
79%
(78)
(79)

(19)

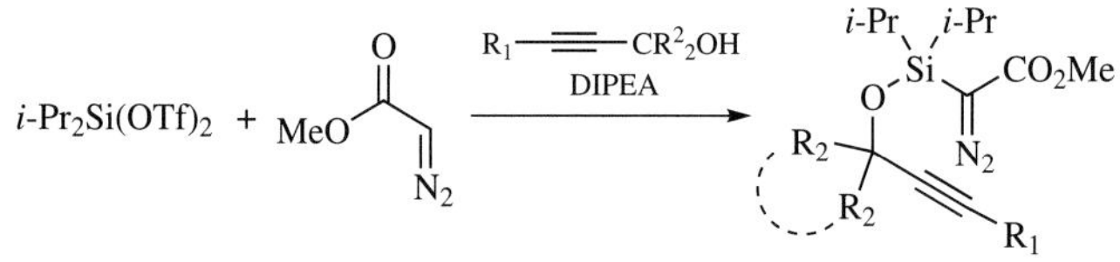
i-$Pr_2Si(OTf)_2$ +
R_1—≡—CR^2_2OH
DIPEA

(20)

Product	R_1	R_2	Yield (%)
80	H	H	67
81	H	Me	73
82	Me	H	81
83	H	-$(CH_2)_5$-	76

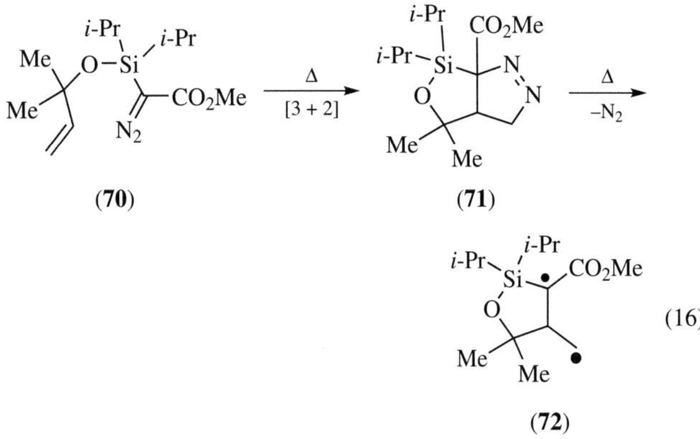
Δ
[3 + 2]
Δ
$-N_2$
(70)
(71)
(72)

(16)

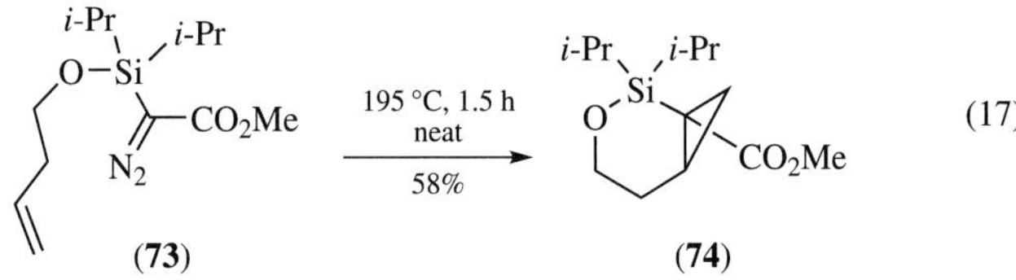
195 °C, 1.5 h
neat
58%
(73)
(74)

(17)

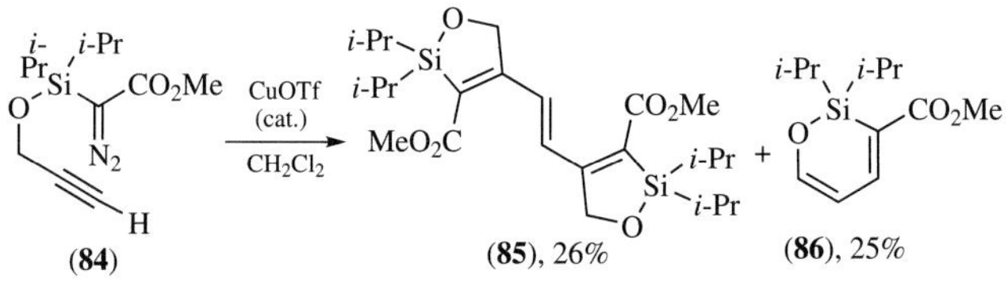
CuOTf (cat.)
CH_2Cl_2
(84)
(85), 26%
(86), 25%

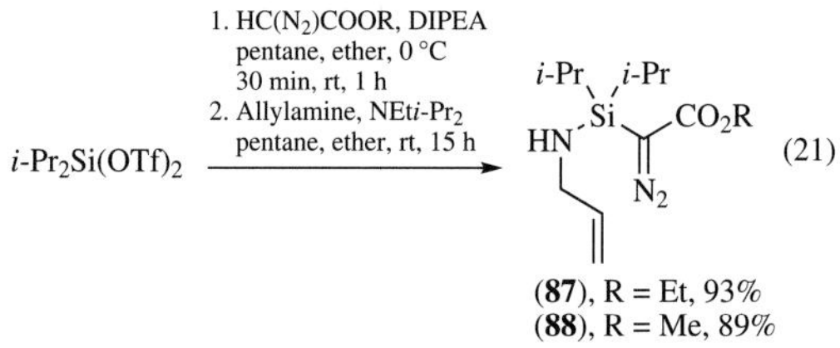
1. $HC(N_2)COOR$, DIPEA
pentane, ether, 0 °C
30 min, rt, 1 h
2. Allylamine, NEti-Pr_2
pentane, ether, rt, 15 h
i-$Pr_2Si(OTf)_2$
(87), R = Et, 93%
(88), R = Me, 89%

(21)

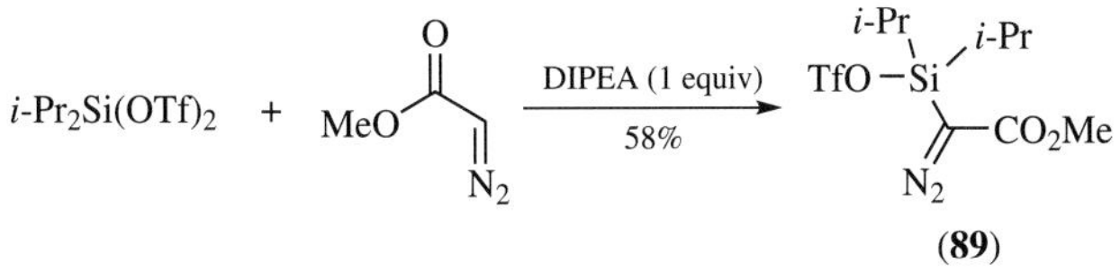

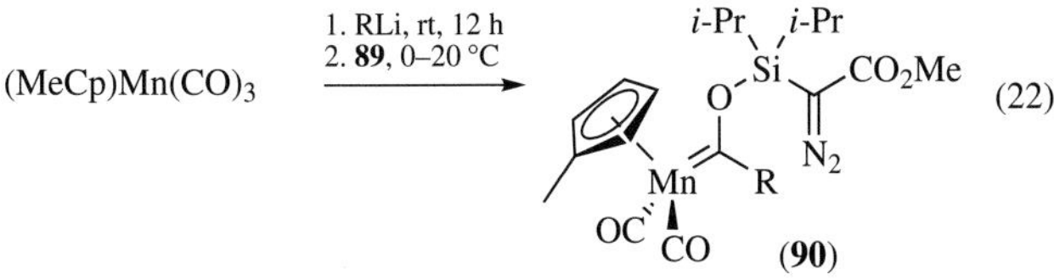

i-Pr$_2$Si(OTf)$_2$ + (MeO, O, N$_2$ structure) $\xrightarrow[\text{58\%}]{\text{DIPEA (1 equiv)}}$ (TfO—Si(i-Pr)$_2$ structure with CO$_2$Me, N$_2$)   (89)

(MeCp)Mn(CO)$_3$ $\xrightarrow[\text{2. 89, 0–20 °C}]{\text{1. RLi, rt, 12 h}}$ (90)   (22)

R	Me	Ph	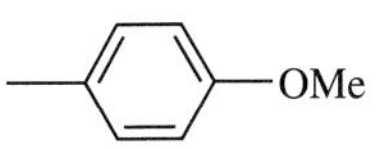 ⬡—OMe	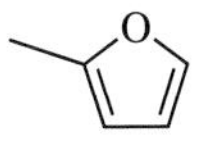 (furan)	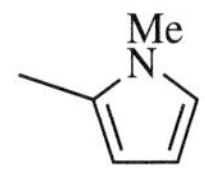 (N-Me pyrrole)
Yield (%)	61	80	69	23	72

Protection of Diols: Applications in Total Synthesis.
Danishefsky and coworkers have used diisopropylsilyl bis(trifluoromethanesulfonate) to protect 1,3-diol **91** in the synthesis of oxadecalin core of phomactin A.[13] The protecting group enforced a rigid conformation about the pyran ring, whereby –OTBDPS and Me groups were forced to adopt axial positions resulting in observed complete facial stereoselectivity in the subsequent Diels–Alder reaction (eq 23).

In efforts toward the synthesis of auripyrone A and B, Perkins and coworkers have utilized diisopropylsilyl bis(trifluoromethanesulfonate) to protect 1,3-diol of intermediate **96**, which was subsequently converted into dihydropyrone **99** in eight steps (eq 24).[14]

Mukaiyama reported an asymmetric synthesis of ABC ring system of 8-demethyltaxoids from optically active eight-membered ring compound **100**. The 1,3-diol in **101** was protected

using diisopropylsilyl bis(trifluoromethanesulfonate) in the presence of pyridine to give silylene compound in 95% yield (eq 25). Alkylation of **102** with methyllithium provided the desired C-1 protected hydroxy group **103** and this compound was subsequently transformed to the desired diol **104** in three steps.[15]

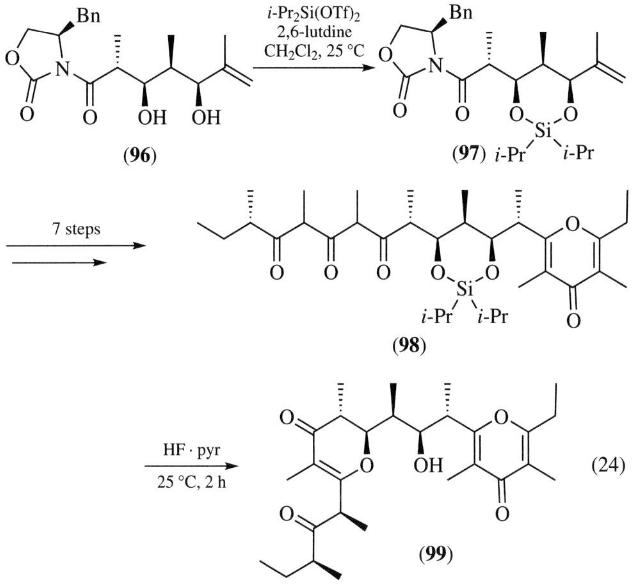

(96) $\xrightarrow[\text{CH}_2\text{Cl}_2, \text{ 25 °C}]{\substack{i\text{-Pr}_2\text{Si(OTf)}_2 \\ 2,6\text{-lutidine}}}$ (97)

$\xrightarrow{\text{7 steps}}$ (98)

$\xrightarrow[\text{25 °C, 2 h}]{\text{HF · pyr}}$ (99)   (24)

Related Reagents. Di-t-butyldichlorosilane; Di-t-butylmethylsilyl Trifluoromethanesulfonate; Di-t-butylsilyl Bis(trifluoromethanesulfonate); Dichlorodimethylsilane; Dichlorodiisopropylsilane.

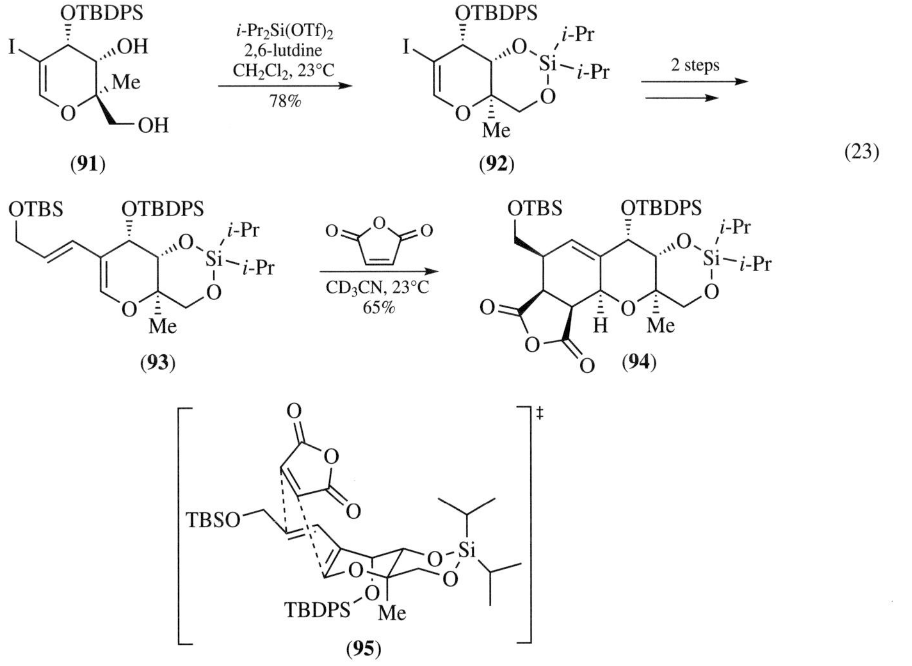

(91) $\xrightarrow[\text{78\%}]{\substack{i\text{-Pr}_2\text{Si(OTf)}_2 \\ 2,6\text{-lutidine} \\ \text{CH}_2\text{Cl}_2, \text{ 23°C}}}$ (92) $\xrightarrow{\text{2 steps}}$ (23)

(93) $\xrightarrow[\text{65\%}]{\text{CD}_3\text{CN, 23°C}}$ (94)

(95)

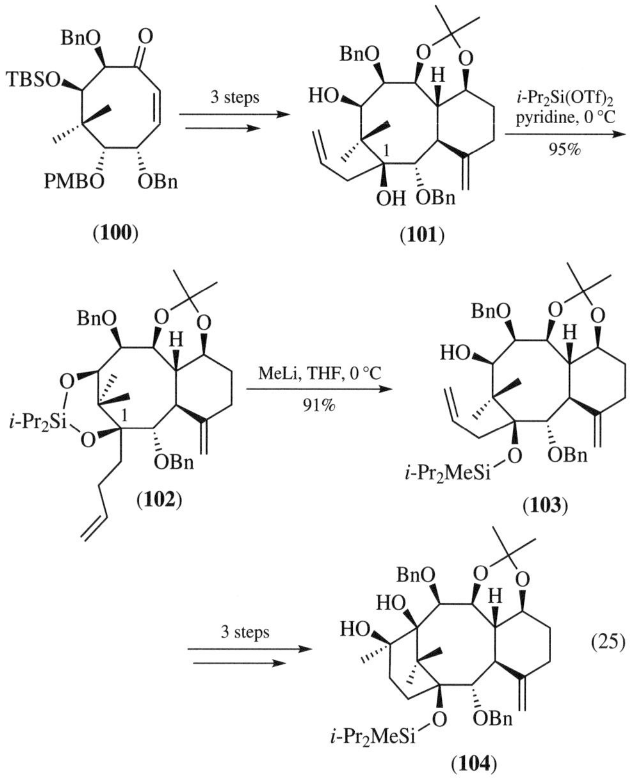

(100)   (101)

(102)   (103)

(104)

1. Corey, E. J.; Hopkins, P. B., *Tetrahedron Lett.* **1982**, *23*, 4871.
2. Trost, B. M.; Caldwell, C. G., *Tetrahedron Lett.* **1981**, *22*, 4999.
3. Corey, E. J.; Link, J. O., *Tetrahedron Lett.* **1990**, *31*, 601.
4. Saha, A. K.; Sardaro, M.; Waychunas, C.; Delecki, D.; Kruse, L. I.; Kutny, R.; Cavanaugh, P.; Yawman, A.; Upson, D. A., *J. Org. Chem.* **1993**, *58*, 7827.
5. Kobori, A.; Miyata, K.; Ushioda, M.; Seio, K.; Sekine, M., *J. Org. Chem.* **2002**, *67*, 476.
6. Kundu, L. M.; Burgdorf, L. T.; Kleiner, O.; Batschauer, A.; Carell, T., *ChemBioChem* **2002**, *3*, 1053.
7. Kablean, S. N.; Marsden, S. P.; Craig, A. M., *Tetrahedron Lett.* **1998**, *39*, 5109.
8. Fronda, A.; Krebs, F.; Daucher, B.; Werle, T.; Maas, G., *J. Organomet. Chem.* **1992**, *424*, 253.
9. Maas, G.; Krebs, F.; Werle, T.; Gettwert, V.; Striegler, R., *Eur. J. Org. Chem.* **1999**, *8*, 1939.
10. Gettwert, V.; Krebs, F.; Maas, G., *Eur. J. Org. Chem.* **1999**, *5*, 1213.
11. Maas, G.; Bender, S., *Synthesis* **1999**, 1175.
12. Maas, G.; Mayer, D., *J. Organomet. Chem.* **2001**, *617–618*, 339.
13. Chemler, S. R.; Iserloh, U.; Danishefsky, S., *J. Org. Lett.* **2001**, *3*, 2949.
14. Perkins, M. V.; Sampson, R. A.; Joannou, J.; Taylor, M. R., *Tetrahedron Lett.* **2006**, *47*, 3791.
15. Shiina, I.; Nishimura, T.; Ohkawa, N.; Sakoh, H.; Nishimura, K.; Saitoh, K.; Mukaiyama, T., *Chem. Lett.* **1997**, 419.

Dilithium Cyanobis(dimethylphenylsilyl)cuprate

$$(Me_2PhSi)_2Cu(CN)Li_2$$

[110769-32-9] $C_{17}H_{22}CuLi_2NSi_2$ (MW 373.97)

InChI = 1/2C8H11Si.CN.Cu.2Li/c2*1-9(2)8-6-4-3-5-7-8;1-2;;;/
h2*3-7H,1-2H3;;;;/q;;;-2;2*+1/rC17H22CuNSi2.2Li/
c1-20(2,16-11-7-5-8-12-16)18(15-19)21(3,4)17-13-9-6-
10-14-17;;/h5-14H,1-4H3;;/q-2;2*+1

InChIKey = ADZKFQKKOLYIFV-CHDOJITCAZ

(higher-order copper reagent for transfer of a dimethylphenylsilyl group to enones, allenes, alkynes, epoxides, and alkyl halides)

Physical Data: relatively stable reagent at rt; usually prepared in situ at 0 °C.

Solubility: sol THF (−78 °C and above).

Preparative Methods: prepared[1] by adding 0.5 equiv of dry copper(I) cyanide to dimethylphenylsilyllithium in THF (eq 1), which in turn is generated from chlorodimethylphenylsilane and lithium metal. Analogous bis(trimethylsilyl)cyanocuprates are prepared in a similar manner.

$$2\,Me_2PhSi\text{-}Li \xrightarrow[\text{THF, 0 °C}]{CuCN} (Me_2PhSi)_2Cu(CN)Li_2 \qquad (1)$$

(1)

More detail for the preparation of this and related reagents may be found in Fleming, I. In *Organocopper Reagents: A Practical Approach*; Taylor, R. J. K., Ed.; OUP: Oxford, 1994; Chapter 12, pp 257–292.

Handling, Storage, and Precautions: cannot be stored or isolated. Care must be taken to avoid oxygen and moisture in the reaction system. Use in a fume hood.

Original Commentary

Joseph P. Marino & David P. Holub
The University of Michigan, Ann Arbor, MI, USA

Structure and Properties of the Reagent. Low-temperature ^{29}Si, ^{13}C, and ^{1}H NMR spectroscopic techniques have been used to probe the nature of these higher-order cuprates.[2] These studies showed conclusively the facile dissociation–reassociation of ligands on copper. In the presence of methyllithium, dimethylphenylsilyllithium, and copper(I) cyanide, a mixed metallocuprate is formed (eq 2). The metallocuprate (**2**) preferentially transfers the dimethylphenylsilyl group to enones and 1-alkynes.

$$(Me_2PhSi)_2Cu(CN)Li_2 + MeLi \rightleftharpoons (Me_2PhSi)(Me)Cu(CN)Li_2 \ (2)$$

(2)

Conjugate Additions to Enones. In general, the reactivities of lower-order and higher-order cyano silylcuprates are very similar toward α,β-unsaturated ketones in ether (eq 3).[2] In some cases, as with alkynic esters, a higher-order reagent gives higher yields and greater stereoselectivity at −78 °C. When the bis-(trimethylsilyl)cyanocuprate (**3**) was added to ethyl 2-butynoate, a high yield of only the (*E*)-isomer was obtained (eq 4).[3]

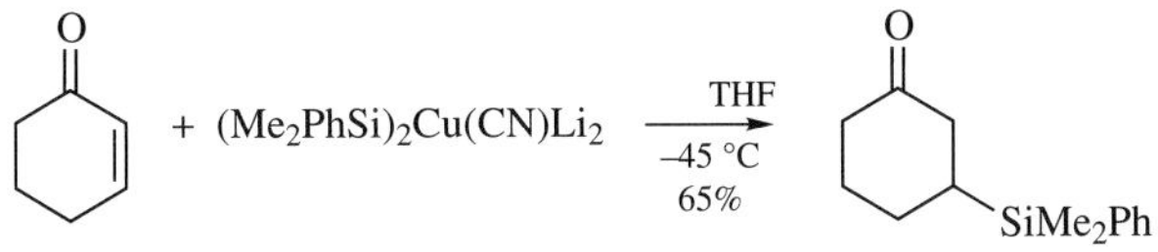

(3)

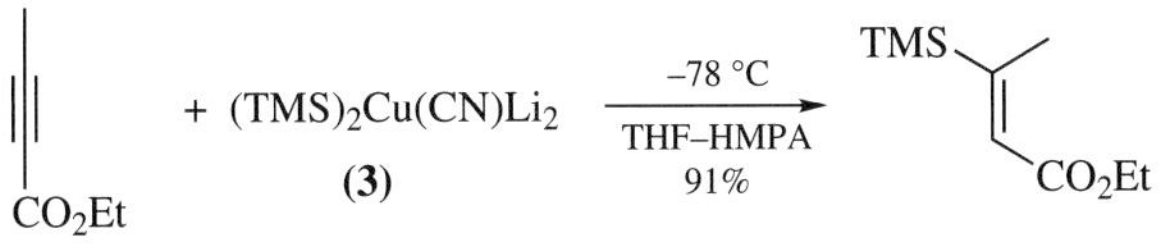

(4)

With cyclohexenone, the corresponding lower-order cuprate gave a 58% product yield and, in the case of eq 4, a mixture of (*E*/*Z*) adducts was observed with the corresponding lower-order cuprate. Interestingly, the mixed (dimethylphenylsilyl)(methyl)-cyanocuprate (**2**) gave the highest yield (94%) of adduct with cyclohexenone.

Silylcupration of Alkynes.

Higher-order silyl cuprates such as (**1**) add regioselectively to terminal alkynes in a *cis* fashion, and the intermediate vinylcopper species (**4**) may be trapped with electrophiles to give 2,2-disubstituted vinylsilanes (**5**) (eq 5).[1,2] Corresponding reactions with the lower-order cuprate of reagent (**1**) give low yields of regioisomeric vinylsilanes upon protonation.[4]

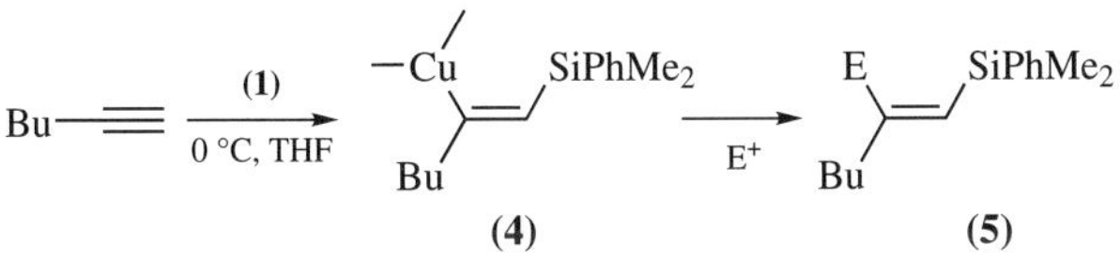

(5)

Typical electrophiles (E$^+$) that have been employed are ammonium chloride for protonation (94% yield), iodomethane (71%), cyclohexenone (54%), and propylene oxide (89%). An example of trapping a vinylcopper species with an ethyl acrylate precursor is shown in eq 6 for the synthesis of the vinylsilane (**6**).[4,5]

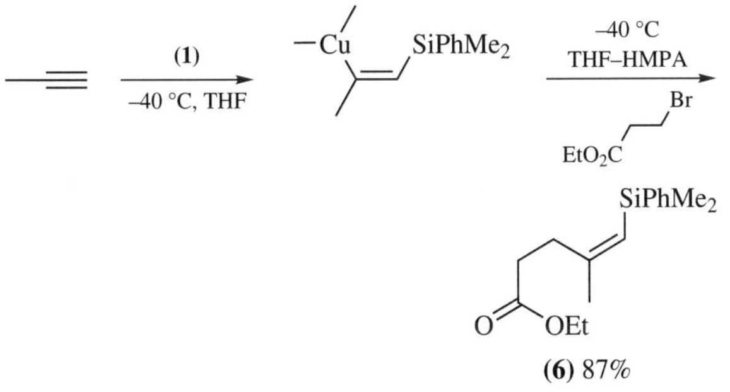

(6)

(**6**) 87%

Silylcupration of Allenes.

As a type of activated alkene, allenes undergo silylcupration with reagent (**1**) to give either vinyl- or allylsilanes.[6] Simple alkylallenes react quickly at −78 °C to give allylsilanes cleanly (eq 7). In this case the silyl group is transferred to the less substituted carbon of the allene.

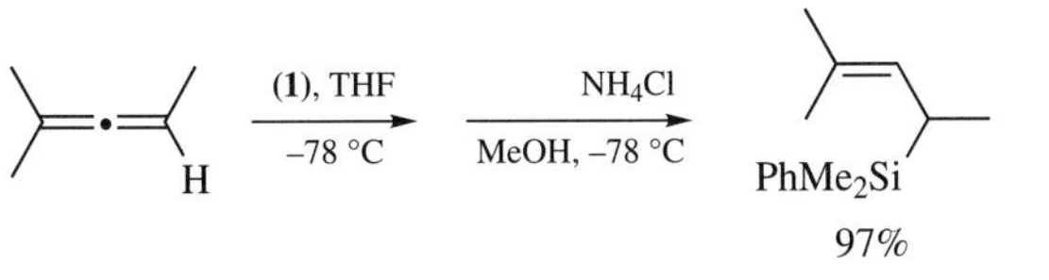

(7)

With allene itself or terminal allenes the regioselectivity is reversed and vinylsilanes are the major products.[7,8] In these cases an intermediate allylcuprate (**7**) is generated which can be quenched with electrophiles such as chlorine, acetyl chloride, or a proton to give products (**8**), (**9**) and (**10**) respectively (eq 8).

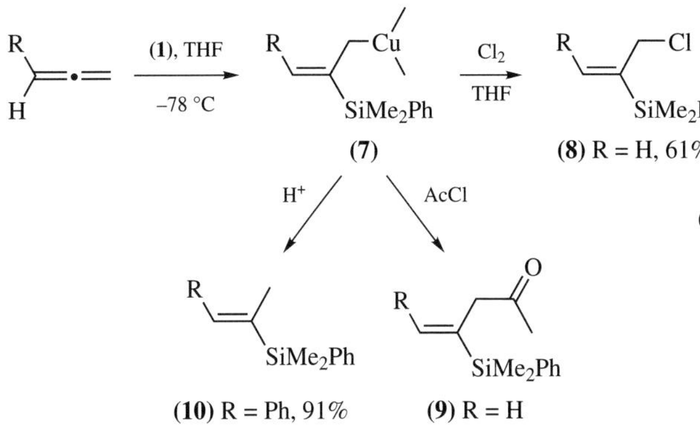

(8)

(**10**) R = Ph, 91%   (**9**) R = H

Intermediate (**7**) (R = H) behaves abnormally when reacted with cyclohexenone and iodine. In the former case, 1,2-addition, and not conjugate addition, prevails; in the latter case, an unusual rearrangement occurs to form the 2-iodoallylsilane (**11**) (eq 9).

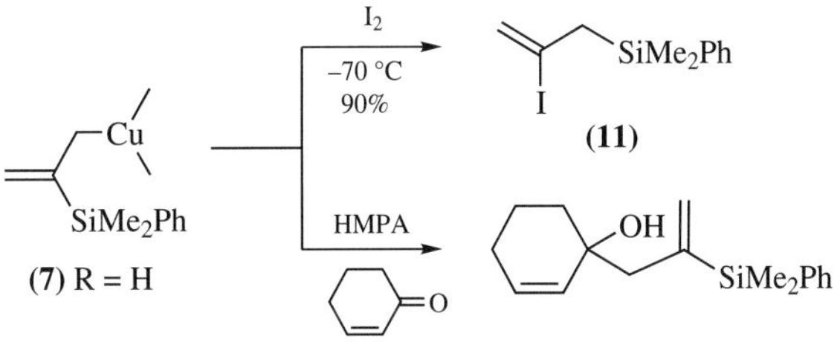

(9)

(**7**) R = H

Previously, the reactions of reagents (**1**) and (**3**) with 3,3-dimethylallene were reported to give the allylsilane adduct (**12**) initially which, upon trapping with cyclohexenone and acetyl chloride, produced the expected adducts (**13**) and (**14**), respectively (eq 10).[9]

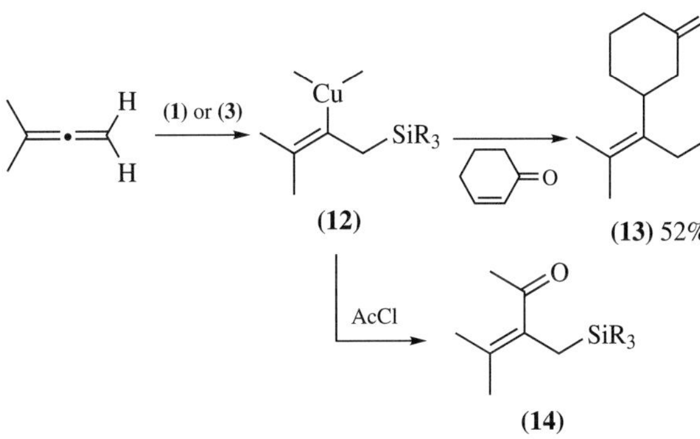

(10)

(**14**)

Silylcuprations of Strained Alkenes.

The reactions between oxabicyclo[3.2.1]octenones and reagent (**1**) and its lower-order analog have been studied.[10] Silylcupration of 8-oxabicyclo[3.2.1] oct-6-en-3-one (**15**) with (**1**) is rapid and is followed by a slower step involving trapping onto the carbonyl group at C-3 to give a ring-closed cyclobutanol product (**16**) along with some of the protonated ketone (**17**) (eq 11).

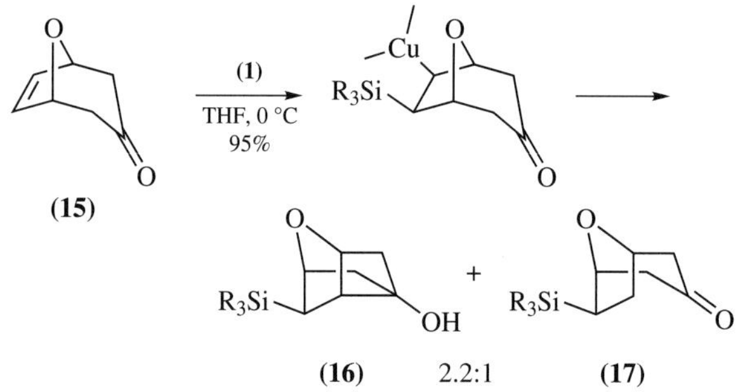

(11)

(15) → (16) 2.2:1 (17)

Silylcupration of α-methyl substituted derivatives of (15) produced intermediates (18) which undergo such slow ring closure onto the ketone that (18) could be trapped with electrophiles (MeI or H$^+$). In an unusual oxidation process, treatment of intermediate (18) (R^1 = R^2 = Me) with dry silica gel produced the β-hydroxysilane (19) in quantitative yield (eq 12). Product (19) easily undergoes a base-catalyzed Peterson alkenation to the starting alkene.

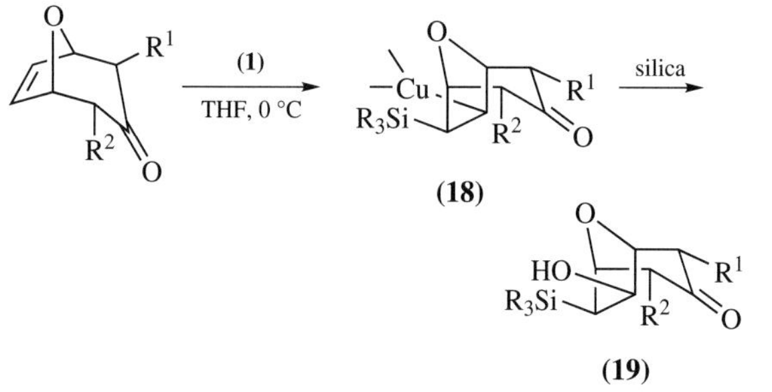

(12)

In summary, the higher-order silyl cuprates (1) and (3) tend to be more reactive and more nucleophilic than their lower-order counterparts in silylcuprations and S$_N$2-like displacements. Both classes of reagents are fairly stable up to ambient temperatures.

First Update

Ian Fleming
University of Cambridge, Cambridge, UK

Copper(I) cyanide is not the only salt used to make silylcuprate reagents, any more than dimethylphenylsilyl is the only silyl group. The cyanide has the advantage that it is not hygroscopic, but in general it appears to make little difference to the reactivity or selectivity of the reagent whether the cyanide, the bromide, or the iodide is used. In most cases, direct comparison of one reagent with another has not been made, with the exception of silylcupration to terminal acetylenes, for which the cyanide-derived reagent is superior in its regioselectivity. Other cuprate reagents are included in this update, to avoid having separate entries for each.

Conjugate Addition to α,β-Unsaturated Carbonyl Compounds Other than Ketones. In addition to α,β-unsaturated ketones, the silylcuprate reagent also reacts with α,β-unsaturated aldehydes, esters (eq 13), amides, and nitriles,[11] and with vinylsulfoxides[12] without needing Lewis acid catalysis as carbon-based cuprates do. With esters, the intermediate enolates, which have the E geometry (20), may be used directly in highly diastereoselective

reactions with alkyl halides and aldehydes.[13,14] The β-hydroxy esters like (21) can be used in the synthesis of allylsilanes by decarboxylative elimination.[15]

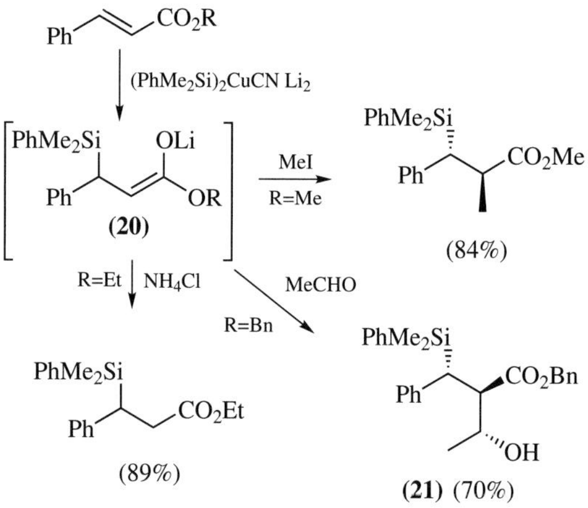

(13)

α,β-Unsaturated carbonyl compounds attached to a chiral auxiliary, like Koga's lactam or Oppolzer's sultam, give products with a stereogenic center carrying a silyl group having high levels of enantiomeric purity.[16,17] When a stereogenic center is at the γ-position of an α,β-unsaturated ester, stereocontrol is poor with the cuprate, and the **Zincate** is to be preferred.[18]

Silylcupration of Alkynes. A few nonterminal unsymmetrical alkynes have been shown to undergo regioselective silylcupration. Control has been achieved by intramolecular protonation (eq 14)[19] and by having different environments that allow diastereofacial chelation (eq 15), where the vinylsilane was subjected immediately to iododesilylation.[20] When one side of the acetylene has a methyl group and the other a branched carbon, complete regioselectivity is achieved via attachment by the silyl group at the less hindered side.[21]

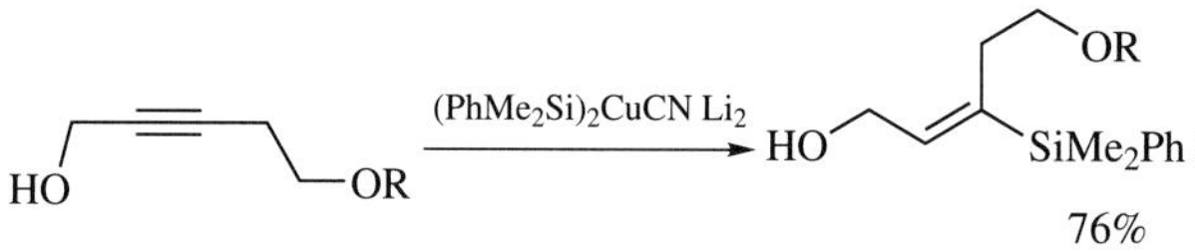

(14)

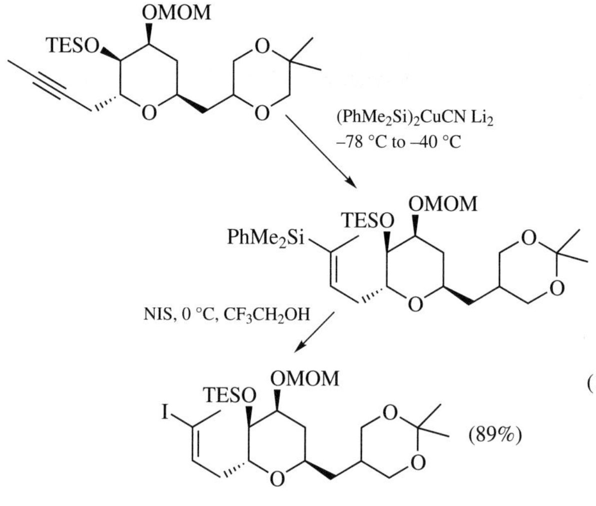

76%

(15)

(89%)

Reaction with Allylic Acetates. The silylcuprate reacts with allylic acetates directly to give allylsilanes (eqs 16–18). Bis secondary allylic acetates are apt to form both regioisomers, but some control of the regiochemistry is possible, using a *cis* double bond, which encourages reaction with allylic shift, especially when the silyl group is delivered to the less-hindered end of the allylic system (eq 16).[22] An alternative protocol, assembling a mixed silylcuprate on a carbamate group, is even better at controlling the regioselectivity, usually giving complete allylic shift (eq 17).[22,23] Tertiary acetates in contrast are well behaved regiochemically, giving only the product with the silyl group at the less-substituted end of the allylic system (eq 18).[23,24] Allylsilanes have many uses as carbon nucleophiles in organic synthesis.[25] A similar reaction takes place with allylic epoxides.[26]

The acetate reaction (eq 16) and the carbamate alternative (eq 17) are complementary in their stereochemistry, the former taking place stereospecifically *anti*[22,23,27] and the latter stereospecifically *syn*.[22]

(16)
(51%) 82 : 18

(17)
(68%)

(18)
(85%)

Other Reactions. Acid chlorides,[28,29] and *N*-acylimidazoles[30] react with the silylcuprate reagent to give acylsilanes. *N*-Acylpyridinium salts react with the silylcuprate reagent to give the product of addition of the silyl group at the 4-position.[31] Propargyl acetates,[32] mesylates[33] sulfides,[34] and a propargylic ether[35] react with the silylcuprate reagent to give allenylsilanes. Other substrates that have been found to react with the silylcuprate are allyl chlorides,[36] alkyl halides,[37] epoxides,[17,38] halopyrazoles,[39] a vinyl iodide,[28] and an aminomethyl acetate.[40]

Related Reagents. Dimethylphenylsilyllithium; Dimethylphenylsilyl(methyl)magnesium; Lithium Cyano(dimethylphenylsilyl)cuprate.

1. Fleming, I.; Roessler, F., *J. Chem. Soc., Chem. Commun.* **1980**, 276.

2. Sharma, S.; Oehlschlager, A. C., *J. Org. Chem.* **1991**, *56*, 770.

3. Audia, J. E.; Marshall, J. A., *Synth. Commun.* **1983**, *13*, 531.

4. Hoeman, M. Ph. D. Thesis, The University of Michigan, 1993.

5. Fleming, I.; Newton, T. W.; Roessler, F., *J. Chem. Soc., Perkin Trans. 1* **1981**, 2527.

6. Fleming, I.; Rowley, M.; Cuadrado, P.; González-Nogal, A.M.; Pulido, F.J., *Tetrahedron* **1989**, *45*, 413.

7. Singh, S. M.; Oehlschlager, A. C., *Can. J. Chem.* **1991**, *69*, 1872.

8. Fleming, I.; Rowley, M., *Tetrahedron* **1989**, *45*, 413.

9. Cuadrado, P.; González, A. M.; Pulido, F. J.; Fleming, I., *Tetrahedron Lett.* **1988**, *29*, 1825.

10. Lautens, M.; Belter, R. K.; Lough, A. J., *J. Org. Chem.* **1992**, *57*, 422.

11. Ager, D. J.; Fleming, I.; Patel, S. K., *J. Chem. Soc., Perkin Trans. 1* **1981**, 2520.

12. Takaki, K.; Maeda, T.; Ishikawa, M., *J. Org. Chem.* **1989**, *54*, 58.

13. Crump, R. A.; Fleming, I.; Hill, J. H. M.; Parker, D.; Reddy, N. L.; Waterson, D., *J. Chem. Soc., Perkin Trans. 1* **1992**, 3277.

14. Fleming, I.; Kilburn, J. D., *J. Chem. Soc., Perkin Trans. 1* **1992**, 3295.

15. Fleming I.; Gil, S.; Sarkar, A. K.; Schmidlin, T., *J. Chem. Soc., Perkin Trans. 1* **1992**, 3351.

16. Fleming, I.; Kindon, N. D., *J. Chem. Soc., Chem. Commun.* **1987**, 1177.

17. Oppolzer, W.; Mills, R. J.; Pachinger, W.; Stevenson, T., *Helv. Chim. Acta* **1986**, *69*, 1542.

18. Krief, A.; Dumont, W.; Baillieul, D., *Tetrahedron Lett.* **2005**, *46*, 8033.

19. Shi, B.; Hawryluk, N. A.; Snider, B. B., *J. Org. Chem.* **2003**, *68*, 1030.

20. (a) Betzer, J.-F.; Pancrazi, A., *Synlett* **1998**, 1129. (b) Zakarian, A.; Batch, A.; Holton, R. A., *J. Am. Chem. Soc.* **2003**, *125*, 7822.

21. Archibald, S. C.; Barden, D. J.; Bazin, J. F. Y.; Fleming, I.; Foster, C. F.; Mandal, A. K.; Mandal, A. K.; Parker, D.; Takaki, K.; Ware, A. C.; Williams A. R. B.; Zwicky, A. B., *OBC* **2004**, *2*, 1051.

22. Fleming, I.; Higgins, D.; Lawrence, N. J.; Thomas, A. P., *J. Chem. Soc., Perkin Trans. 1* **1992**, 3331.

23. Fleming, I.; Terrett, N. K., *J. Organomet. Chem.* **1984**, *264*, 99.

24. Fleming, I.; Marchi, D., *Synthesis* **1981**, 560.

25. Fleming, I.; Dunoguès, J.; Smithers, R., *Org. React.* **1989**, *37*, 575.

26. Merten, J.; Hennig, A.; Schwab, P.; Fröhlich, R.; Tokalov, S. V.; Gutzeit, H. O.; Metz, P., *Eur. J. Org. Chem.* **2006**, 1144.

27. (a) Schmidtmann, E. S.; Oestreich, M., *Chem. Commun.* **2006**, 3643. (b) Spino, C.; Beaulieu, C.; Lafreniere, J., *J. Org. Chem.* **2000**, *65*, 7091.

28. Duffaut, N.; Dunogès, J.; Biran, C.; Calas, R.; Gerval, J., *J. Organomet. Chem.* **1978**, *161*, C23.

29. (a) Brook, A. G.; Harris, J. W.; Lennon, J.; Sheikh, M. E., *J. Am. Chem. Soc.* **1979**, *101*, 83. (b) Bonini, B. F.; Comes-Franchini, M.; Mazzanti, G.; Passamonti, U.; Ricci, A.; Zani, P., *Synthesis* **1995**, 92. (c) Bonini, B. F.; Busi, F.; de Laet, R. C.; Mazzanti, G.; Thuring, J.-W. J. F.; Zani, P.; Zwanenburg, B., *J. Chem. Soc., Perkin Trans. 1* **1993**, 1011.

30. (a) Bonini, B. F.; Comes-Franchini, M.; Fochi, M.; Laboroi, F.; Mazzanti, G.; Ricci, A.; Varchi, G., *J. Org. Chem.* **1999**, *64*, 8008. (b) Bonini, B. F.; Comes-Franchini, M.; Fochi, M.; Mazzanti, G.; Ricci, A.; Varchi, G., *Polyhedron* **2000**, *19*, 529.

31. Hösl, C. E.; Wanner, K. T., *Heterocycles* **1998**, *48*, 2653.

32. Fleming, I.; Takaki, K.; Thomas, A., *J. Chem. Soc., Perkin Trans. 1* **1987**, 2269.

33. Sakae, A.; Shintaro, H.; Saito, K.; Kibayashi, C., *J. Org. Chem.* **2002**, *67*, 5517.

34. Casarini, A.; Jousseaume, B.; Lazzari, D.; Porciatti, E.; Reginato, G.; Ricci, A.; Seconi, G., *Synlett* **1992**, 981.

35. O'Sullivan, P. T.; Buhr, W.; Fuhry, M. A.; Harrison, J. R.; Davies, J. E.; Feeder, N.; Marshall, D. R.; Burton, J. W.; Holmes, A. B., *J. Am. Chem. Soc.* **2004**. *126*, 2194.

36. Smith, J. G.; Drozda, S. E.; Petraglia, S. P.; Quinn, N. R.; Rice, E. M.; Taylor, B. S.; Viswanathan, M., *J. Org. Chem.* **1984**, *49*, 4112.

37. (a) Singer, R. D.; Oehlschlager, A. C., *J. Org. Chem.* **1991**, *56*, 3510. (b) Fürstner, A.; Weidmann, H., *J. Organomet. Chem.* **1988**, *354*, 15. (c) Sibi, M. P.; Harris, B. J.; Shay, J. J.; Hajra, S., *Tetrahedron* **1998**, *54*, 7221.

38. Lipshutz, B. H.; Reuter, D. C.; Ellsworth, E. L., *J. Org. Chem.* **1989**, *54*, 4975.

39. Calle, M.; Cuadrado, P.; González-Nogal, A. M.; Valero, R., *Synthesis* **2001**, 1949.

40. Nativi, C.; Ricci, A.; Taddei, M., *Tetrahedron Lett.* **1990**, *31*, 2637.

(α,α-Dimethylallyl)trimethylsilane[1]

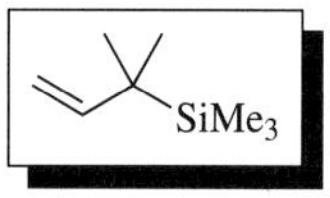

[67707-64-6]   C$_8$H$_{18}$Si   (MW 142.32)

InChI = 1S/C8H18Si/c1-7-8(2,3)9(4,5)6/h7H,1H2,2-6H3

InChIKey = UVAINNXKNVZOJU-UHFFFAOYSA-N

(regiospecific prenylation of acetals;[2] precursor of substituted tetrahydrofurans via reactions with carbonyl compounds[2])

Solubility: sol common organic solvents.

Analysis of Reagent Purity: ^{1}H NMR.

Preparative Methods: treatment of trichlorosilane with 3-methylbut-2-en-1-ylmagnesium chloride followed by methylmagnesium bromide (2 equiv) provides (α,α-dimethyl-allyl)dimethylsilane (bp 108 °C) in 76% yield with 99% isomeric purity. The hydrosilane is converted to the ethoxysi-lane quantitatively by the reaction with ethanol catalyzed by hydrogen hexachloroplatinate(IV), and then fluorinated in situ with tetrafluoroboric acid to give the fluorosilane in 87% yield. The latter is methylated with a small excess of methyllithium to give (α,α-dimethylallyl)trimethylsilane in 96% yield.[2] Alternatively, (α,α-dimethylallyl)trimethylsilane can be prepared from 3-methylbut-1-yn-3-ol in six steps.[3]

Purification: distillation.

Handling, Storage, and Precautions: can be stored in a glass bottle.

Prenylation of Acetals. (α,α-Dimethylallyl)trimethylsilane reacts with acetals in the presence of a Lewis acid to give the corresponding homoallyl alcohols with complete regioselectivity (eqs 1 and 2). The reaction with carbonyl compounds often leads to the formation of tetrahydrofuran derivatives (eq 3), presumably through the formation of chloride.[2]

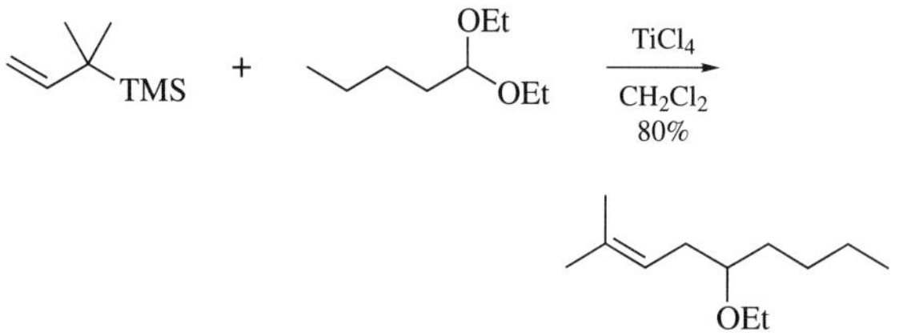

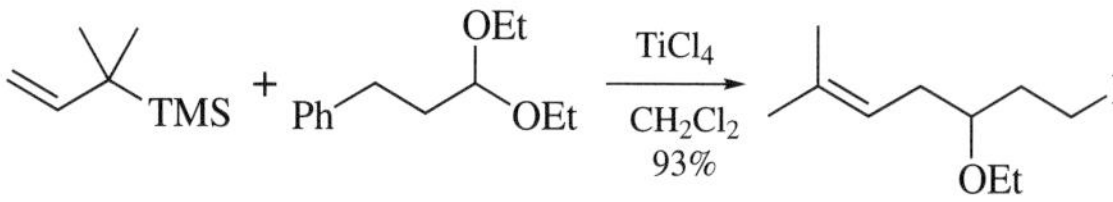

gives a cycloaddition product where the trimethylsilyl group undergoes a 1,2-shift (eq 4).[4]

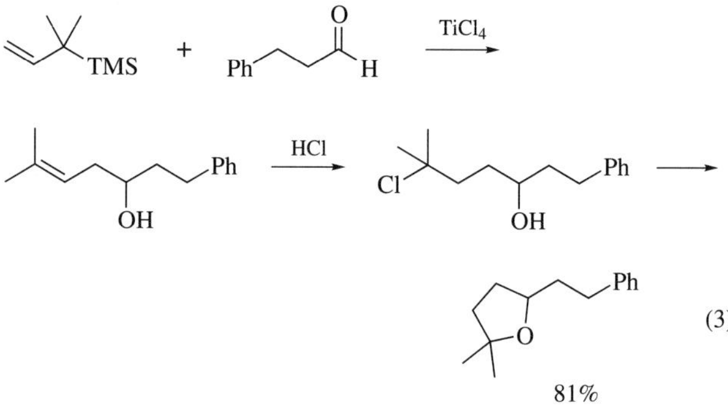

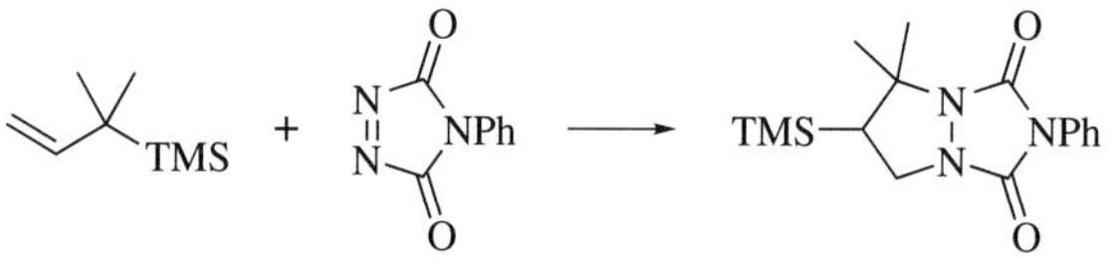

Photochemical Prenylation. Photolysis of (α,α-dimethyl-allyl)trimethylsilane in the presence of 1-methyl-2-phenyl-1-pyrrolinium perchlorate leads to the formation of a (3,3-dimethyl-prop-2-en-1-yl)pyrrolidine as the sole regioisomer (eq 5).[5]

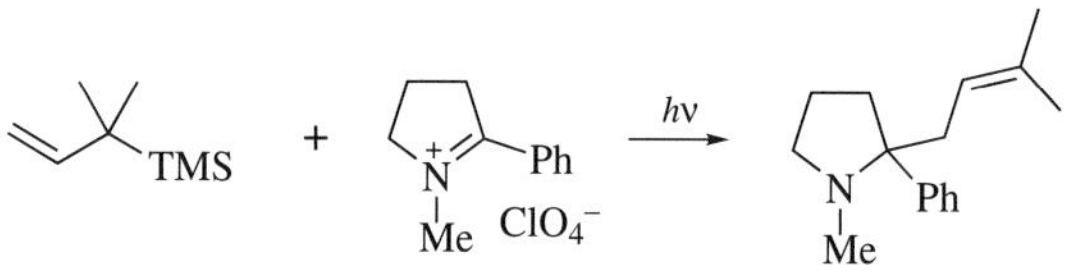

Isomerization. (α,α-Dimethylallyl)trimethylsilane can undergo isomerization to (γ,γ-dimethylallyl)trimethylsilane with fluoride ion catalysis.[6]

---

1. Fieser, M., *Fieser & Fieser* **1980**, *8*, 181.

2. Hosomi, A.; Sakurai, H., *Tetrahedron Lett.* **1978**, 2589.

3. Bennetau, B.; Pillot, J.-P.; Dunogues, J.; Calas, R., *J. Chem. Soc., Chem. Commun.* **1981**, 1094.

4. Cai, J.; Davies, A. G., *J. Chem. Soc., Perkin Trans. 2* **1992**, 1743.

5. Ohga, K.; Yoon, U. C.; Mariano, P., *J. Org. Chem.* **1984**, *49*, 213.

6. Hosomi, A.; Shirahata, A.; Sakurai, H., *Chem. Lett.* **1978**, 901.

Hideki Sakurai
Tohoku University, Sendai, Japan

Cycloaddition with Silyl Shift. The reaction of (α,α-dimethyl-allyl)trimethylsilane with 4-Phenyl-1,2,4-triazoline-3,5-dione

(2-Dimethylaminomethylphenyl)-phenylsilane

[129552-42-7]   C₁₅H₁₉NSi   (MW 241.41)

InChI = 1S/C15H19NSi/c1-16(2)12-13-8-6-7-11-15(13)17-14-9-4-3-5-10-14/h3-11H,12,17H2,1-2H3

InChIKey = MZFJXCUWXJGJTB-UHFFFAOYSA-N

(hydrosilylating agent; selective reducing agent for carbonyls, acyl chlorides, carboxylic acids, and heterocumulenes under neutral conditions)

Physical Data: bp 115 °C/0.05 mmHg.

Solubility: sol ether, THF, alkanes, aromatics, chlorinated solvents.

Preparative Methods: to 67 g (0.5 mol) of *N,N*-dimethylbenzylamine in 300 mL of anhydrous ether are added dropwise under nitrogen 200 mL (0.5 mol) of *n*-butyllithium (2.5 M in hexane). After being stirred for 48 h at 25 °C, the yellow suspension is transferred into a dropping funnel and added at 0 °C to 54 g (0.5 mol) of phenylsilane in 200 mL of ether. The mixture is stirred for 16 h, then filtered over celite. Solvents are removed under vacuum, and the residue is diluted in pentane, filtered again and fractionally distilled to give 82 g (68%) of 2-(dimethylaminomethylphenyl)phenylsilane (**1**).[1]

Related aminoarylsilanes are similarly prepared[2] from 1-(dimethylaminomethyl)naphthalene and 1-(dimethylamino)-naphthalene, giving [8-(dimethylaminomethyl)-1-naphthyl]-phenylsilane (**2**) and [8-(dimethylamino-1-naphthyl]phenyl-silane (**3**), respectively.

General Discussion. The reducing properties of silicon hydrides are enhanced by intramolecular pentacoordination.[1,3] Solvolysis is observed with alcohols and phenols, and addition to carbonyls occurs without the requirement of external catalysts. Some previous reports concerned the reduction of aroyl chlorides to aldehydes using organosilanes under catalytic conditions with Pd/C[4] or tributyltin hydride with catalytic Pd⁰,[5] but incompatible functional groups on the substrate must be avoided. Such problems are overcome with organosilicon hydrides activated by intramolecular coordination of an aminoaryl group. These reagents react with acyl chlorides in carbon tetrachloride, affording chlorosilanes and aldehydes (eq 1).[6] The reactions are fast and quantitative. The aldehydes are either isolated by direct distillation from the crude product mixture, or trapped after filtration over wet silica. The organosilanes can be recycled after separation and treatment with lithium aluminum hydride. Halogen, methoxy, and nitroaryl substituents, heteroaryl groups, and alkenes remain unchanged. The method is also suitable for the conversion of dicarboxylic acid chlorides into dialdehydes.

Hydrosilylation of isocyanates and isothiocyanates (eq 2) is readily achieved under mild conditions. Treatment of the adducts through the one-pot addition of acyl chlorides gives *N*-acylform-amides and *N*-acylthioformamides.[7] *N,N′*-Dialkyl *C*-acylami-dines result from the acylation[8] of *C*-silylamidines, obtained (eq 3) by hydrosilylation of alkylcarbodiimides with (**2**).

$$(1)$$

$$(2)$$

X = O, 85%; X = S, 70%

$$(3)$$

Pentacoordinated hydrosilanes react with excess aryl isocyanates to give isocyanurates.[9] Carboxylic acids are directly reduced to aldehydes in a one-pot process[10] through the thermal decomposition of pentacoordinated silyl carboxylates (eq 4). The aldehydes are extracted from the crude product mixture by distillation, or separated from the trisiloxane by column chromatography over Florisil. The reaction is selective, since fluoro, nitro, cyano, methoxy and heteroaryl substituents do not react with the silane. The present method also permits the reduction of α,β-unsaturated acids. The efficiency follows the order (**2**) > (**1**) > (**3**).

$$(4)$$

Related Reagents. Triethylsilane; Triethylsilane–Trifluoroacetic Acid

1. Arya, P.; Corriu, R. J. P.; Gupta, K.; Lanneau, G. F.; Yu, Z., *J. Organomet. Chem.* **1990**, *399*, 11.

2. Boyer, J.; Brelière, C.; Care, F.; Corriu, R. J. P.; Kpoton, A.; Poirier, M.; Royo, G.; Young, J. C., *J. Chem. Soc., Dalton Trans.* **1989**, 43.

3. Boyer, J.; Brelière, C.; Corriu, R. J. P.; Kpoton, A.; Poirier, M.; Royo, G., *J. Organomet. Chem.* **1986**, *311*, C39.

4. (a) Jenkins, J. W.; Post, H. W., *J. Org. Chem.* **1950**, *15*, 556. (b) Citron, J. D., *J. Org. Chem.* **1969**, *34*, 1977. (c) Dent, S. P.; Eaborn, C.; Pidcok, A., *J. Chem. Soc. (C)* **1970**, 1703.

5. (a) Four, P.; Guibe, F., *J. Org. Chem.* **1981**, *43*, 4439. (b) Guibe, F.; Four, P.; Rivière, H., *J. Chem. Soc. (C)* **1980**, 432.

6. Corriu, R. J. P.; Lanneau, G. F.; Perrot, M., *Tetrahedron Lett.* **1988**, *29*, 1271.

7. Corriu, R. J. P.; Lanneau, G. F.; Perrot-Petta, M.; Mehta, V. D., *Tetrahedron Lett.* **1990**, *31*, 2585.

8. Corriu, R. J. P.; Lanneau, G. F.; Perrot-Petta, M., *Synthesis* **1991**, *11*, 954.

9. Corriu, R. J. P.; Lanneau, G. F.; Mehta, V. D., *Heteroatom Chem.* **1991**, *2*, 461.

10. Corriu, R. J. P.; Lanneau, G. F.; Perrot, M., *Tetrahedron Lett.* **1987**, *28*, 3941.

Robert J. P. Corriu & Gérard F. Lanneau
Université Montpellier II, Montpellier, France

Dimethyl(phenyl)silane

[766-77-8] $C_8H_{12}Si$ (MW 136.29)

InChI = 1S/C8H12Si/c1-9(2)8-6-4-3-5-7-8/h3-7,9H,1-2H3

InChIKey = ZISUALSZTAEPJH-UHFFFAOYSA-N

(hydrosilylating reagent; reductant in combination with acid or F⁻)

Physical Data: bp 157 °C/744 mmHg; *d* 0.889 g cm⁻³.

Solubility: sol common organic solvent such as chloroform, 1,2-dichloroethane, benzene, ether, acetone, dioxane, THF; insol H_2O.

Form Supplied in: oil, commercially available.

Analysis of Reagent Purity: ¹H NMR (CDCl₃–Me₄Si) δ 7.6–7.4 (m, 2 H), 7.4–7.2 (m, 3 H), 4.42 (septet, *J* = 3.6 Hz, 1 H), 0.34 (d, *J* = 3.6 Hz, 6 H). ¹³C and ²⁹Si NMR have also been reported.[2]

Preparative Method: see Benkeser and Foster.[1]

Purification: distillation.

Handling, Storage, and Precautions: moisture sensitive; evolves H_2 when added to wet piperidine.

Diastereoselective Reduction of Carbonyl Compounds.[3]
Dimethyl(phenyl)silane reduces aldehyde and ketone carbonyls with the aid of fluoride ion or acid. α-Acylpropionamides, 1-aminoethyl ketones, and 1-alkoxyethyl ketones are readily converted into the corresponding β-hydroxy amides, α-amino alcohols, and α-alkoxy alcohols, respectively. The stereoselectivity is complementary and generally high: *erythro* (or *syn*) isomers are obtained with trifluoroacetic acid (TFA), whereas *threo* (or *anti*) isomers are obtained with fluoride ion activator (eq 1). The *erythro* selectivity in the acid-promoted carbonyl reduction is ascribed to a proton-bridged Cram's cyclic transition state. On the other hand, the *threo* selectivity in the fluoride-mediated reduction is explained in terms of the Felkin–Anh type model, wherein a penta- or hexacoordinated fluorosilicate is involved. No epimerization at the chiral center is observed during the reaction.

Hydrosilylation.[4] With a transition metal catalyst, the hydrosilane adds to carbon–carbon double or triple bonds to give alkylsilanes or alkenylsilanes. Hydrosilylation of deoxydiisopropylidene-*arabino*-hexenopyranose (**1**) catalyzed by [Rh(norbornadiene)Cl]₂ (bis(bicyclo[2.2.1]hepta-2,5-diene)-dichlorodirhodium) gives 6-deoxy-6-phenyldimethylsilyldiisopropylidenealtrose (**2**), which is desilylated by tetra-*n*-butylammonium fluoride to yield 6-deoxydiisopropylidenealtrose (**3**) (eq 2).[5]

A list of General Abbreviations appears on the front Endpapers

$$
\begin{array}{c}
\text{PhMe}_2\text{SiH} \\
\text{TFA, 0 °C} \\
\text{66–98\%}
\end{array}
\longrightarrow
\begin{array}{c}
\text{OH} \\
R\!\!-\!\!X \\
86\text{–}98\% \text{ de}
\end{array}
$$

$$
\begin{array}{c}
\text{PhMe}_2\text{SiH} \\
\text{Bu}_4\text{N}^+ \text{F}^- \\
\text{HMPA, 0 °C to rt} \\
77\text{–}95\%
\end{array}
\longrightarrow
\begin{array}{c}
\text{OH} \\
R\!\!-\!\!X \\
74\text{–}98\% \text{ de}
\end{array}
\qquad (1)
$$

$$
X = \overset{O}{\underset{}{\text{C}}}-N\!\!\diagdown, \; NMe_2,\, O\text{-}t\text{-Bu},\, OAc,\, OBz,\, OTHP
$$

OEE, NHSO₂Ph, etc.

$$(1) \xrightarrow[\text{(L = norbornadiene)}]{\substack{\text{PhMe}_2\text{SiH} \\ (\text{RhLCl})_2 \\ 70\%}} (2) \xrightarrow[\substack{\text{HMPA} \\ 94\%}]{\text{Bu}_4\text{N}^+ \text{F}^-} (3) \qquad (2)$$

In the presence of a Pt complex like bis(η-divinyltetramethyl-disiloxane)tri-*t*-butylphosphineplatinum(0) (**4**), phenyldimethyl-silane reacts with 1-butyn-3-ol to give an 8:1 mixture of alkenylsilanes (**5** and **6**) (eq 3). The major product 1-dimethyl-phenylsilyl-1-buten-3-ol (**5**) is separated and resolved by lipase-catalyzed esterification.[6]

$$(4) \; \xrightarrow[\substack{t\text{-Bu}_3\text{P-Pt} \\ 96\%}]{\text{PhMe}_2\text{SiH}} (5) \; + \; (6) \qquad 8:1 \qquad (3)$$

Hydrosilylation of 1,4-bis(trimethylsilyl)-1,3-butadiyne (**7**) gives 1,4-bis(trimethylsilyl)-1,3-bis(dimethylphenylsilyl)-1,2-butadiene (**8**) (chlorotris(triphenylphosphine)rhodium(I), 100 °C) or 1,4-bis(trimethylsilyl)-2-(dimethylphenylsilyl)-1-buten-3-yne (**9**) (Pt(PPh₃)₄, 100 °C) (eq 4) depending on the catalyst.[7] The enyne (**9**) is considered to be a precursor of the allene (**8**).

$$
\text{TMS}\!\!\equiv\!\!\equiv\!\!\text{TMS}
$$

$$(7)$$

$$\downarrow \text{PhMe}_2\text{SiH}$$

$$
\underset{\substack{\text{RhCl(PPh}_3)_3 \\ 100 °C, 2 h}}{86\%} \qquad \underset{\substack{\text{Pt(PPh}_3)_4 \\ 100 °C, 2 h}}{94\%} \qquad (4)
$$

$$(8) \qquad (9)$$

Transition metal-catalyzed hydrosilylation of α,β-unsaturated ketones and aldehydes with PhMe₂SiH proceeds in a 1,4-fashion to give silyl vinyl ethers, which are hydrolyzed to give ketones and aldehydes. Asymmetric 1,4-reduction has been studied using a chiral transition metal catalyst, but with little success (eq 5).[8]

$$\text{Ph} \underset{O}{\overset{}{\diagdown}} \quad \xrightarrow[\text{2. H}^+]{\underset{\text{chiral cat}}{\text{1. PhMe}_2\text{SiH}}} \quad \text{Ph} \underset{O}{\overset{}{\diagdown}} \quad (5)$$

16% ee

chiral cat = $[\text{Rh}\{(R)\text{-PMePhCH}_2\text{Ph}\}\text{H}_2\text{S}_2]^+ \text{ClO}_4^-$

Dimethyl (Z,Z)-2,4-hexadienedioate is hydrosilylated with $\text{PhMe}_2\text{SiH}/\text{RhCl}(\text{PPh}_3)_3$ to give the 1,6-hydrosilylation product, methyl (3E)-6-methoxy-6-(phenyldimethylsilyloxy)-3,5-hexadienoate (eq 6).[9]

$$\text{MeO}_2\text{C} \underset{}{\overset{}{\diagdown}} \text{CO}_2\text{Me} \quad \xrightarrow[\text{PhH, 80 °C}]{\underset{\text{RhCl(PPh}_3)_3}{\text{PhMe}_2\text{SiH}}} \quad$$

$$\text{MeO}_2\text{C} \underset{1 \quad 2 \quad 4}{\overset{3 \quad 5}{\diagup\diagdown}} \underset{6}{\overset{}{}}\text{OMe} \quad (6)$$
$$\text{OSiMe}_2\text{Ph}$$

quantitative

Formylsilylation. Under a carbon monoxide atmosphere (1–4 MPa), silylformylation of alkynes occurs in the reaction of alkynes and PhMe_2SiH to give β-silyl enals in the presence of a rhodium catalyst (eq 7).[10] Terminal alkynes exclusively give products having the SiMe_2Ph group at the terminal carbon. Depending on the structure of the alkyne, the isomer ratio (E/Z) varies from 0:100 to 100:0. Hydroxyl groups survive these conditions; however, in the presence of a tertiary amine, alkynols are converted into α-(dimethylphenylsilylmethylene) β-lactones (eq 8).[11]

$$\text{R}^1 \underset{}{=\!=\!=} \text{R}^2 \quad \xrightarrow[\underset{43–99\%}{\text{PhH, 100 °C, 2 h}}]{\underset{\text{Rh}_4(\text{CO})_{12} \text{ or Co}_2\text{Rh}_2(\text{CO})_{12}}{\text{PhMe}_2\text{SiH, CO (1–3 MPa)}}} \quad \underset{\text{OHC}}{\overset{\text{R}^1 \quad \text{R}^2}{\diagup}}\!\!\!\underset{\text{SiMe}_2\text{Ph}}{} \quad (7)$$

$$\text{HO} \underset{}{\overset{}{\diagdown}}\!=\!= \quad \xrightarrow[\underset{43–86\%}{\text{PhH, 100 °C, 2 h}}]{\underset{\text{DBU, Rh}_4(\text{CO})_{12}}{\text{PhMe}_2\text{SiH, CO (1.5–4 MPa)}}} \quad \underset{O}{\overset{\text{SiMe}_2\text{Ph}}{}} \quad (8)$$

Silylformylation of N-propargyl tosylamides gives normal β-silyl enal products, whereas the same reaction in the presence of 1,8-diazabicyclo[5.4.0]undec-7-ene or triethylamine affords α-(dimethylphenylsilyl)methylene β-lactams (eq 9).[12] Propargylamines are transformed to 2-(dimethylphenylsilyl)methyl-2-propenals (eq 10).[13]

$$\text{TosNH} \underset{\text{R}^1}{\overset{\text{R}^2}{\diagdown}}\!=\!= \quad \xrightarrow[\underset{63–81\%}{\text{Rh}_4(\text{CO})_{12}, \text{PhH, 25 °C}}]{\text{PhMe}_2\text{SiH, CO (2 MPa)}} \quad \underset{\text{OHC}}{\overset{\text{R}^1 \quad \text{R}^2}{}}\text{TosNH}\!\!-\!\!\underset{}{\overset{}{}}\!\!\text{H} \quad \underset{\text{SiMe}_2\text{Ph}}{}$$

$$(9)$$

$$\xrightarrow[\underset{28–81\%}{\text{Rh}_4(\text{CO})_{12}, \text{PhH, 100 °C}}]{\underset{\text{DBU or NEt}_3}{\text{PhMe}_2\text{SiH, CO (3 MPa)}}} \quad \underset{\text{Tos}}{\overset{\text{R}^2}{\underset{N}{\diagup}}}\!\!\!\underset{O}{\overset{\text{R}^1 \quad \text{SiMe}_2\text{Ph}}{}}$$

$$\text{R}^3\text{R}^4\text{N} \underset{\text{R}^2}{\overset{\text{R}^1}{\diagdown}}\!=\!= \quad \xrightarrow[\underset{20–94\%}{\text{100 °C, 3 h}}]{\underset{\text{Rh}_4(\text{CO})_{12}, \text{PhH}}{\text{PhMe}_2\text{SiH, CO (2 MPa)}}} \quad \underset{\text{CHO}}{\overset{\text{R}^1}{\text{R}^2}}\!\!\!\underset{\text{SiMe}_2\text{Ph}}{} \quad (10)$$

1. Benkeser, R. A.; Foster, D. J., *J. Am. Chem. Soc.* **1952**, *74*, 5314.

2. Olah, G. A.; Hunadi, R. J., *J. Am. Chem. Soc.* **1980**, *102*, 6989.

3. Fujita, M.; Hiyama, T., *J. Am. Chem. Soc.* **1985**, *107*, 8294; **1984**, *106*, 4629.

4. Hiyama, T.; Kusumoto, T., *Comprehensive Organic Synthesis* **1991**, *8*, 763.

5. Pegram, J. J.; Anderson, C. B., *Carbohydr. Res.* **1988**, *184*, 276.

6. Panek, J. S.; Yang, M.; Solomon, J. S., *J. Org. Chem.* **1993**, *58*, 1003; Heneghan, M.; Procter, G., *Synlett* **1992**, 489; Ward, R. A.; Procter, G., *Tetrahedron Lett.* **1992**, *33*, 3363; Lewis, L. N.; Sy, K. G.; Bryant, G. L., Jr.; Donahue, P. E., *Organometallics* **1991**, *10*, 3750.

7. Kusumoto, T.; Ando, K.; Hiyama, T., *Bull. Chem. Soc. Jpn.* **1992**, *65*, 1280. Kusumoto, T.; Hiyama, T., *Rev. Heteroatom. Chem.* **1994**, *11*, 143.

8. Ojima, I.; Hirai, K., *Asymmetric Synth.* **1985**, *5*, 102.

9. Yamamoto, K.; Tabei, T., *J. Organomet. Chem.* **1992**, *428*, C1.

10. Ojima, I.; Ingallina, P.; Donovan, R. J.; Clos, N., *Organometallics* **1991**, *10*, 38; Matsuda, I.; Ogiso, A.; Sato, S.; Izumi, Y., *J. Am. Chem. Soc.* **1989**, *111*, 2332.

11. Matsuda, I.; Ogiso, A.; Sato, S., *J. Am. Chem. Soc.* **1990**, *112*, 6120.

12. Matsuda, I.; Sakakibara, J.; Nagashima, H., *Tetrahedron Lett.* **1991**, *32*, 7431.

13. Matsuda, I.; Sakakibara, J.; Inoue, H.; Nagashima, H., *Tetrahedron Lett.* **1992**, *33*, 5799.

Tamejiro Hiyama & Manabu Kuroboshi
Tokyo Institute of Technology, Yokohama, Japan

Dimethylphenylsilyllithium[1]

$$\boxed{\text{PhMe}_2\text{SiLi}}$$

[3839-31-4] $\qquad$ C$_8$H$_{11}$LiSi $\qquad$ (MW 142.22)

InChI = 1S/C8H11Si.Li/c1-9(2)8-6-4-3-5-7-8;/h3-7H,1-2H3;

InChIKey = RVCQHCLPKCVLFY-UHFFFAOYSA-N

(reagent for addition to carbonyl compounds;[2] deoxygenation reactions;[3,4] in presence of copper(I) salts, an effective reagent for conjugate addition to α,β-unsaturated compounds,[5] conjugate displacement of tertiary allylic acetates,[6,7] and silylcupration of alkynes[8] and allenes[9])

Physical Data: NMR studies have shown the reagent is monomeric in ether.[10c]

Solubility: sol THF, Et$_2$O.

Preparative Methods: readily prepared from the reaction of PhMe_2SiCl with 2 equiv of Li metal in THF or from the reaction of $\text{PhMe}_2\text{SiSiMe}_2\text{Ph}$ with Li metal in THF.[10a,b] Typical reaction procedure involves stirring chlorodimethylphenylsilane with lithium in THF at 0 °C under N$_2$ for 4–36 h; the chlorosilane is converted into tetramethyldiphenyldisilane in the first 10 min, and the remainder of the time is needed to cleave the disilane.[8a,11] The supernatant reagent solution is calibrated using standard methods and used without further purification for reactions.[8a,11] It is sometimes advantageous to activate the Li metal prior to use by ultrasound irradiation of a hexane suspension,[12a] and to use argon as the inert gas to prevent formation of lithium nitrides.[12b]

Handling, Storage, and Precautions: the reagent solution in THF can be stored under argon or nitrogen at −20 °C for several weeks without significant degradation.[8a] Lithium is a highly flammable solid and must be handled and stored in an inert atmosphere (preferably argon).

Original Commentary

Anil S. Guram
Massachusetts Institute of Technology, Cambridge, MA, USA

Grant A. Krafft
Abbott Laboratories, Abbott Park, IL, USA

Addition to Carbonyl Compounds. PhMe$_2$SiLi in THF undergoes addition to aldehydes and ketones under mild conditions to afford α-silyl alcohols upon aqueous workup. The α-silyl alcohols are versatile synthetic intermediates which can be readily manipulated into a variety of products. Consequently, these reactions have been widely utilized in organic syntheses,[13–16] for example in the synthesis of cytochalasin D cycloundecanone ring system (eq 1),[13] and in the synthesis of (±)-pregn-4-en-20-one.[13] Enones react with PhMe$_2$SiLi predominantly via 1,2-addition to afford α-silyl alcohols which can be converted to silyl enol ethers via Brook rearrangement (eq 2).[16]

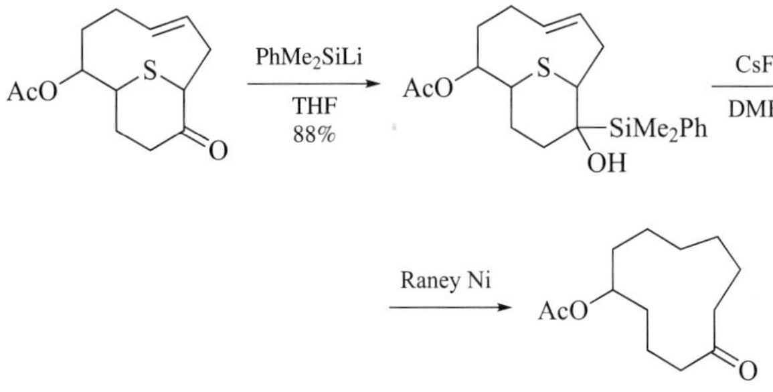

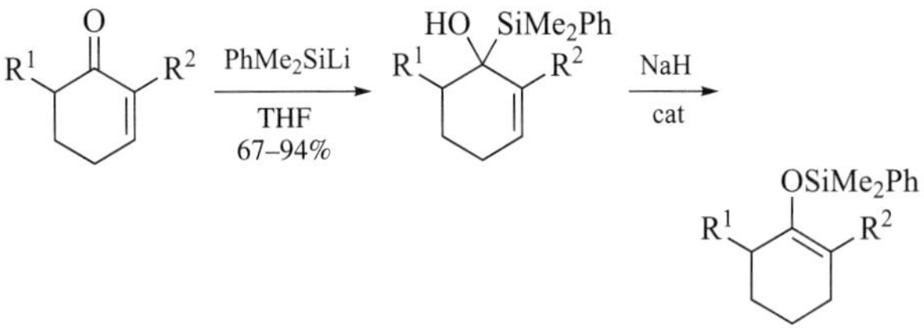

Addition of PhMe$_2$SiLi to α-phenylthio ketones directly affords silyl enol ethers with very high (Z)-stereoselectivity (eq 3).[17]

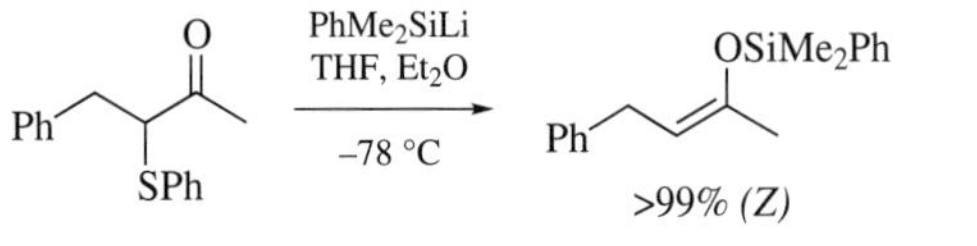

Addition of PhMe$_2$SiLi to aldehydes and subsequent O-sulfonylation of the generated alkoxides with p-toluenesulfonyl chloride affords α-silyl tosylates in good yields.[2] Conversion of these intermediates into α-silyl selenocyanates followed by fluoride-induced cyanide elimination reaction permitted the synthesis and study of reactivity of previously unknown selenoaldehydes (eq 4).[2]

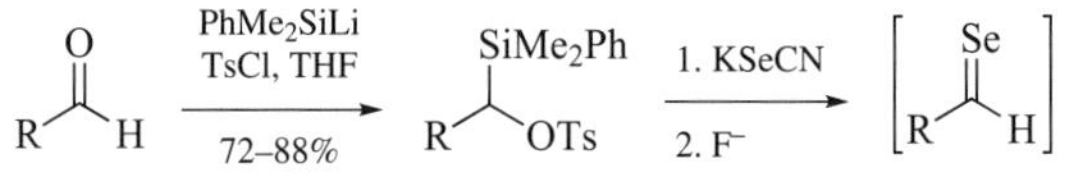

Deoxygenation Reactions. PhMe$_2$SiLi reacts with oxabicyclo[2.2.1]heptenes in Et$_2$O at 0 °C to afford 1,3-cyclohexadienes (eq 5).[3] 6–8 equiv of PhMe$_2$SiLi are required for complete consumption of the starting bicyclic compound. The excess stoichiometry sometimes leads to further reaction of the 1,3-cyclowhexadiene product with the unreacted reagent. Use of the silylcuprate reagent PhMe$_2$SiCu·LiCN in THF at 0 °C alleviates this problem.[3]

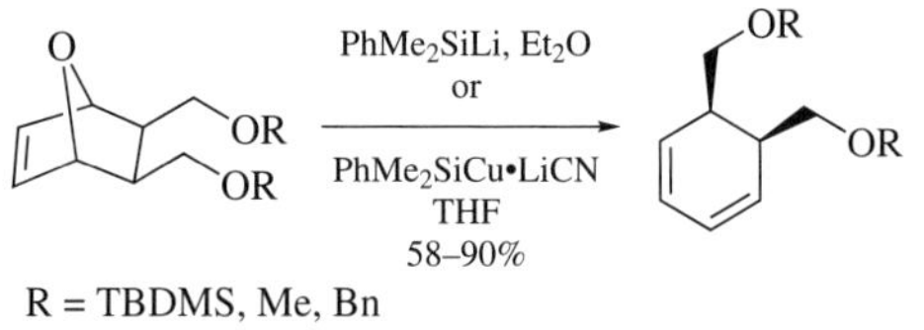

Epoxides are stereospecifically deoxygenated to alkenes with inversion of stereochemistry upon treatment with PhMe$_2$SiLi in THF.[4] *trans*-Stilbene is obtained in 83% yield and >99% stereoselectivity from the reaction of oxide of *cis*-stilbene with PhMe$_2$SiLi. Similarly, *trans*-stilbene oxide is converted to *cis*-stilbene.

Reactions via Cuprates. Dimethylphenylsilylcuprates [(PhMe$_2$Si)$_2$CuLi·LiX] (X = CN or halide), generated in situ from the reaction of 2 equiv of PhMe$_2$SiLi with copper(I) salts (copper(I) cyanide, copper(I) iodide), are popular reagents for the incorporation of PhMe$_2$Si unit into organic molecules.[5,9,18] In general, these reagents react with substrates under milder conditions and tolerate polar functional groups better than PhMe$_2$SiLi. However, these reagents suffer from one major drawback in that not all the Si anions bound to copper are transferred to the substrate. Consequently, reaction workup affords nonvolatile byproducts which complicate product isolation.[5–9] This problem can be avoided by using the mixed cuprate (PhMe$_2$Si)MeCuLi, prepared from the reaction of 1 equiv of PhMe$_2$SiLi and 1 equiv of MeLi with CuCN.[19] The mixed cuprate specifically transfers the silyl group (not the methyl group) to the substrate and exhibits reactivity comparable to that of lithium bis[dimethyl(phenyl)silyl] cuprate Synthetically useful reactions of dimethylphenylsilyl-cuprates include addition to saturated aldehydes, 1,4-addition to α,β-unsaturated enones, conjugate displacement of tertiary allylic acetates, and silylcupration of alkynes and allenes.

The addition of (PhMe$_2$Si)$_2$CuLi to aldehydes has been utilized in the elegant synthesis of isocarbacyclin (eq 6).[20]

$$R^1 = (CH_2)_3CO_2Me$$
$$R^2 = pentyl$$
$$R^3 = TBDMS$$

(6)

$(PhMe_2Si)_2CuLi$ undergoes 1,4-(conjugate) addition to a variety of cyclic and acyclic α,β-unsaturated enones to afford saturated β-silyl carbonyl compounds (eq 7).[5,21–23] In acyclic systems, enones can be regenerated from the β-silyl carbonyl compounds by bromination (bromine–CCl_4) followed by desilyl-bromination (sodium fluoride–EtOH).[23] In cyclic systems, treatment of the β-silyl carbonyl compounds with copper(II) bromide affords the enones directly (eq 7).[21] This overall 'silylation–desilylation' protocol not only represents a method for the protection of α,β-unsaturation but can also be used advantageously by alkylating the intermediate enolate of conjugate silyl addition prior to desilylation (eq 7). This methodology has been successfully applied in the synthesis of carvone and dihydrojasmone.[5a] The alkylation of intermediate enolates is highly diastereoselective, favoring the isomer in which the silyl group and the alkyl group are in *anti* orientation (eq 8). The silyl functionality of the β-silyl carbonyl compounds also can be converted into a hydroxy group with retention of configuration (eq 8).[5b]

(7)

(8)

$(PhMe_2Si)_2CuLi$ reacts with allylic acetates via stereospecific (*anti*) conjugate displacement of the acetate group to afford allylsilanes in excellent yields.[7,24] Similar stereospecificity is observed in reactions with propargyl acetates.[7a]

$(PhMe_2Si)_2CuLi$ reacts with alkynes via *syn* addition of the silyl group and copper to the carbon–carbon triple bond.[8] In reactions with terminal alkynes, the silyl group becomes attached predominantly to the terminal carbon. The acidic hydrogen of

terminal alkynes does not get abstracted under the reaction conditions. The vinylcuprate intermediates react with electrophiles including alkyl/acyl halides, iodine, epoxides, and enones to regio/stereoselectively afford a variety of vinylsilanes.[8]

Silylcupration of allene with $(PhMe_2Si)_2CuLi$ followed by treatment of the intermediate cuprate with H^+, carbon electrophiles, and Cl_2 affords the corresponding vinylsilanes (eq 9).[9] However, reaction of the intermediate cuprate with I_2 affords an allylsilane, from which other more highly substituted allylsilanes are readily obtainable (eq 9). In reaction with substituted allenes, the product ratio (vinylsilane vs. allylsilane) is dependent upon the degree of substitution.[9]

(9)

Related Reactions. In contrast to $PhMe_2SiLi$, related R_3SiLi reagents either are appreciably less stable, may require generation in toxic solvents (hexamethylphosphoric triamide), and can be less nucleophilic, favoring deprotonation or electron transfer reactions. Nevertheless, complementary synthetic organic transformations have been described for these reagents,[25] for example the 1,4-addition of trimethylsilyllithium to enones[26] and the reduction of isocyanates to isocyanides with $t\text{-}BuPh_2SiLi$.[27] The wide synthetic utility of $PhMe_2SiLi$ and $(PhMe_2Si)_2CuLi$ in organic synthesis has also encouraged the development and use of related $PhMe_2Si^-$ derivatives of Al,[28] Mg,[29] Mn,[29] and Zn.[28a, 30]

First Update

Ian Fleming
University of Cambridge, Cambridge, UK

Addition to Carbonyl Compounds. The intermediate α-silyl alcohols produced by the addition of $PhMe_2SiLi$ to aldehydes can be oxidized (Swern) to acylsilanes,[31] or, with lead tetraacetate, to mixed siloxy- acetoxy-acetals.[32] The Brook rearrangements of the α-silyl alcohols have many uses,[33] including, the overall reduction of acyloin silyl ethers to regiocontrolled silyl enol ethers.[34] The corresponding 1,1-disilyl alcoholates produced by the addition of 2 equiv of $PhMe_2SiLi$ to esters and acid chlorides can be oxidized (PDC on the alcohol or CCl_4 on the alcoholate) to acylsilanes (eq 10).[35] The same intermediates can undergo Brook rearrangement to create an acyl anion equivalent (eq 11).[36] In certain cases, the corresponding alcohols can be dehydrated without or with rearrangement to create vinylsilanes (eqs 12 and 13).[37]

$$R^1\text{-C(=O)-X} \xrightarrow{\text{2 equiv PhMe}_2\text{SiLi}}$$

$$R^1\text{-C(O}^-)(\text{SiMe}_2\text{Ph})_2 \xrightarrow[\substack{\text{or CCl}_4 \\ 65\text{–}84\%}]{\substack{1.\ H^+ \\ 2.\ PDC}} R^1\text{-C(=O)-SiMe}_2\text{Ph} \quad (10)$$

$$R^1\text{-C(=O)-X} \xrightarrow{\text{2 equiv PhMe}_2\text{SiLi}}$$

$$R^1\text{-C(O}^-)(\text{SiMe}_2\text{Ph})_2 \xrightarrow[\substack{R^2 = \text{Me, allyl, Bn} \\ 41\text{–}98\%}]{R^2X} R^1\text{-C(OH)(SiMe}_2\text{Ph})\text{R}^2 \quad (11)$$

$$R^1\text{-C(=O)(R}^2)\text{-X} \xrightarrow[2.\ H_3O^+]{1.\ 2\ \text{equiv PhMe}_2\text{SiLi}}$$

$$R^1\text{-C(OH)(SiMe}_2\text{Ph})_2\text{R}^2 \xrightarrow[\substack{\text{Py} \\ (R^2 = H)\ 41\text{–}68\%}]{\text{SOCl}_2} R^1\text{-CH=CH-SiMe}_2\text{Ph} \quad (12)$$

$$R^1\text{-C(=O)(R}^2)\text{-X} \xrightarrow[2.\ H_3O^+]{1.\ 2\ \text{equiv PhMe}_2\text{SiLi}}$$

$$R^1\text{-C(OH)(SiMe}_2\text{Ph})_2\text{R}^2 \xrightarrow[\substack{\text{Py, SO}_2 \\ (R^2 \neq H)\ 58\text{–}70\%}]{\text{SOCl}_2} R^1\text{-CH=C(SiMe}_2\text{Ph})\text{R}^2 \quad (13)$$

The addition of the PhMe$_2$SiLi reagent to *N,N*-dimethylamides can be used to make acylsilanes if the reaction is carried out and quenched at −78 °C (eq 14).[35] Alternatively, if the mixture is allowed to warm to −20 °C or above before quenching, the product is an enediamine (eq 15),[38] while if 2 equiv of PhMe$_2$SiLi are used, the product is an α-silylamine (eq 16).[38] These latter reactions probably take place through an intermediate with carbenoid character produced by a Brook rearrangement of the tetrahedral intermediate. Although these three reactions are fairly general, many amides, even those with comparatively unreactive functional groups, give a surprising variety of other products under these conditions,[38] as do nitriles.[39]

$$R\text{-C(=O)-NMe}_2 \xrightarrow[2.\ H^+,\ -78\,°C]{1.\ \text{PhMe}_2\text{SiLi},\ -78\,°C} R\text{-C(=O)-SiMe}_2\text{Ph} \quad (14)$$
$$69\text{–}91\%$$

$$R\text{-C(=O)-NMe}_2 \xrightarrow[\substack{2.\ \geq-20\,°C \\ 3.\ H^+,\ \geq-20\,°C}]{1.\ \text{PhMe}_2\text{SiLi},\ -78\,°C} \text{Me}_2\text{N-C(R)=C(R)-NMe}_2 \quad (15)$$
$$46\text{–}81\%$$

$$R\text{-C(=O)-NMe}_2 \xrightarrow[2.\ H^+,\ \geq-20\,°C]{1.\ 2\ \text{PhMe}_2\text{SiLi}} R\text{-CH(NMe}_2)(\text{SiMe}_2\text{Ph}) \quad (16)$$
$$35\text{–}96\%$$

The corresponding reaction with *N,N*-dimethylthioamides gives the enethiol tautomer of a thioacylsilane.[40] α-Silylamines can also be made by addition of PhMe$_2$SiLi to pyrazolium salts,[41] and to iminium ions, created in situ from aldehydes and *N*-trimethylsilylamines.[42]

Conjugate Addition Reactions. The cuprate and zincate reagents {lithium bis[dimethyl(phenyl)silyl]cuprate (75583-57-2) and lithium dimethyl[dimethyl(phenyl)silyl]-zincate (100112-20-7)} are usually used in conjugate addition reactions, but PhMe$_2$SiLi can be used with catalytic amounts of dimethylzinc (eq 17).[43] More resistant substrates may employ the zincate produced in situ, with catalytic amounts of copper(I) additives, and aided by the cocatalyst Sc(OTf)$_3$ (eq 18).[44]

$$\xrightarrow[\substack{\text{Me}_2\text{Zn} \\ (10\ \text{mol \%}) \\ 69\%}]{\text{PhMe}_2\text{SiLi}} \quad (17)$$

$$\xrightarrow[\substack{\text{Me}_2\text{CuCN Li}_2\ (3\ \text{mol \%}) \\ \text{Sc(OTf)}_3\ (5\ \text{mol \%}) \\ -78\,°C,\ 3\ \text{min} \\ 81\%}]{\text{PhMe}_2\text{SiLi, Me}_2\text{Zn}} \quad (18)$$

In addition, the PhMe$_2$SiLi reagent itself undergoes conjugate addition with suitable substrates, such as naphthyl isoxazolidine (eq 19),[45] the vinylogous carbamate (eq 20),[46] acetophenone coordinated to a bulky Lewis acid (eq 21),[47] and the allylic epoxide (eq 22).[48]

$$\xrightarrow[2.\ \text{MeI}\ 79\%]{1.\ \text{PhMe}_2\text{SiLi}} \quad (19)$$

$$\text{Me}_2\text{N-CH=CH-CO}_2\text{Et} \xrightarrow[2.\ \text{MeI}\ 96\%]{1.\ \text{PhMe}_2\text{SiLi}} \text{PhMe}_2\text{Si-CH=CH-CO}_2\text{Et} \quad (20)$$

$$\xrightarrow[2.\ \text{MeI}\ 51\%]{1.\ \text{PhMe}_2\text{SiLi}} \quad (21)$$

$$\xrightarrow[89\%]{3.5\ \text{equiv PhMe}_2\text{SiLi}} \quad (22)$$

Nucleophilic Attack on Other Electrophiles. PhMe$_2$SiLi reacts with carbenes like isonitriles to make ynediamines (eq 23)[49] and with carbon monoxide to make an enediolate (eq 24).[49]

$$R-N\equiv C: \xrightarrow[\text{53–70\%}]{PhMe_2SiLi} \underset{PhMe_2Si}{\overset{R}{N}}\equiv\underset{SiMe_2Ph}{\overset{R}{N}} \quad (23)$$

$$O=C: \xrightarrow[\text{2. Me}_3SiCl]{\text{1. PhMe}_2SiLi} \underset{Me_3SiO}{\overset{PhMe_2Si}{\diagdown}}=\underset{OSiMe_3}{\overset{SiMe_2Ph}{\diagup}} \quad (24)$$
"moderate yield"

PhMe$_2$SiLi cleaves silyl enol ethers to give the lithium enolate more rapidly than does methyllithium, allowing t-butyldimethylsilyl enol ethers to be purifed, and then converted into the lithium enolate at low temperature.[34] It reacts with the N-sulfonamides of pyrroles, indoles, and secondary amines, but not primary amines, to remove the sulfonyl group (eq 25), but it opens N-sulfonylaziridines by attack on carbon.[50] It adds to iron-coordinated cyclohexadiene (eq 26) and to chromium-coordinated benzene rings (eq 27).[51]

$$\text{(eq 25)} \quad (25)$$

$$\text{(eq 26)} \quad (26)$$

$$\text{(eq 27)} \quad (27)$$

In addition to the reactions with copper and zinc halides, PhMe$_2$SiLi reacts readily with metal halides in general, and with metals carrying less powerful nucleofugal groups, providing, for example, a simple preparation of silylboranes from alkoxyboranes.[52] Some silylboranes can be cleaved by photolysis to give silyl radicals,[52b] others with methyllithium to give back the silyllithium reagent, enabling the reagent to be prepared in a hydrocarbon solvent.[53] Reaction with Cp$_2$ZrCl$_2$ gives a zirconium-coordinated silylene, which adds to diarylalkynes by insertion into the Si–Zr bond and with dialkylalkynes with insertion into the Zr–C bond (eq 28).[54] It reacts with PtCl$_2$ to give a silyl-platinum reagent, which also adds to arylalkynes.[55]

$$Cp_2ZrCl_2 \xrightarrow{PhMe_2SiLi} \quad (28)$$

Reactions via Copper Reagents and Cuprates. The conjugate addition reactions of the copper-derived reagents prepared from PhMe$_2$SiLi, their addition reactions with alkynes and allenes, and their use in displacements of allyl halides and esters are covered in the relevant sections.

Reactions via Zincates. Similarly the reactions of the zinc reagents derived from PhMe$_2$SiLi are covered in the relevant section.

Related Reagents. Dimethylphenylsilyl(methyl)magnesium; lithium bis[dimethyl(phenyl)silyl]cuprate; trimethylsilylcopper; trimethylsilyllithium.

Triphenylsilylpotassium was intensively studied by Gilman before he made the silyllithium reagent, and he showed that it reacted with a variety of electrophilic substances in addition to carbonyl compounds and epoxides, such as silyl halides, pyridine, azo- and azoxybenzene, thiocyanates, and benzonitrile, as well as with far less electrophilic substances like stilbene, anthracene, diphenyl sulfide, and diphenyl sulfone.[56]

Trimethylsilylsodium, prepared by treating hexamethyldisilane with sodium methoxide in HMPA, reacts with halobenzenes to give trimethylsilyl benzene, and reacts with benzene itself to give the same product.[57] Several silyllithium reagents add one and 2 equiv to C$_{60}$, but the product from the parent PhMe$_2$SiLi is not well characterized.[58]

Dimethyl(o-methoxyphenyl)silyllithium cannot be made from the silyl chloride in the usual way because lithium does not cleave the disilane. However, the reagent can be made by treating the disilane with methyllithium in THF-HMPA. This reagent, and several analogous reagents made in the same way, has an aryl group significantly more easily removed than the phenyl group when it is being used in the Tamao-Fleming reaction.[59] Other reagents which introduce silyl groups susceptible to the Tamao-Fleming reaction in the presence of functional groups like alkenes, and which can be made into cuprate reagents for conjugate addition, are diethylamino(diphenyl)silyllithium[60] and 2-methylbut-2-enyl(diphenyl)silyllithium,[61] both of which can be made from the silyl chloride in the usual way. Dimesitylsilyllithium is a stable reagent that can be used in conjugate addition reactions by way of a copper-catalyzed reaction of the zincate, and the products are susceptible to Tamao-Fleming oxidation.[62]

The chiral silyllithium reagent methyl(naphthyl)(phenyl)silyllithium is configurationally stable at and below room temperature, and reacts with methyl(naphthyl)(phenyl)silyl chloride with retention at the nucleophilic silyl group and inversion at the electrophilic silyl group.[63]

1. (a) Colvin, E. W., *Silicon in Organic Synthesis*; Butterworths: London, 1981; p 134. (b) Fleming, I., *Chem. Soc. Rev.* **1981**, *10*, 83.

2. (a) Krafft, G. A.; Meinke, P. T., *J. Am. Chem. Soc.* **1986**, *108*, 1314. (b) Krafft, G. A.; Meinke, P. T., *J. Am. Chem. Soc.* **1988**, *110*, 8671.

3. Lautens, M.; Ma, S.; Belter, R. K.; Chiu, P.; Leschziner, A., *J. Org. Chem.* **1992**, *57*, 4065.

4. Reetz, M. T.; Plachky, M., *Synthesis* **1976**, 199.

5. (a) Ager, D. J.; Fleming, I.; Patel, S. K., *J. Chem. Soc., Perkin Trans. 1* **1981**, 2520. (b) Fleming, I.; Henning, R.; Plaut, H., *J. Chem. Soc., Chem. Commun.* **1984**, 29.

6. (a) Fleming, I.; Marchi, D., Jr., *Synthesis* **1981**, 560. (b) Fleming, I.; Higgins, D., *J. Chem. Soc., Perkin Trans. 1* **1989**, 206.

7. (a) Fleming, I.; Terrett, N. K., *J. Organomet. Chem.* **1984**, *264*, 99. (b) Fleming, I.; Terrett, N. K., *Tetrahedron Lett.* **1984**, *25*, 5103.

8. (a) Fleming, I.; Newton, T. W.; Roessler, F., *J. Chem. Soc., Perkin Trans. 1* **1981**, 2527. (b) Fleming, I.; Roessler, F., *J. Chem. Soc., Chem. Commun.* **1980**, 276.

9. (a) Fleming, I.; Landais, Y.; Raithby, P. R., *J. Chem. Soc., Perkin Trans. 1* **1991**, 715. (b) Fleming, I.; Rowley, M.; Cuadrado, P.; Gonzalez-Nogal, A. M.; Pulido, F. J., *Tetrahedron* **1989**, *45*, 413.

10. (a) George, M. V.; Peterson, D. J.; Gilman, H., *J. Am. Chem. Soc.* **1960**, *82*, 403. (b) Gilman, H.; Lichtenwalter, G. D., *J. Am. Chem. Soc.* **1958**, *80*, 608. (c) Edlund, U.; Lejon, T.; Venkatachalan, T. K.; Buncel, E.; *J. Am. Chem. Soc.* **1985**, *107*, 6408.

11. Fleming, I.; Roberts, R. S.; Smith, S. C., *J. Chem. Soc., Perkin Trans. 1* **1998**, 1209.

12. (a) Asao, K.; Iio, H.; Tokoroyama, T., *Synthesis* **1990**, 382. (b) Meinke, P. T. Ph.D. Thesis, Syracuse University, August 1987, p 35.

13. Vedejs, E.; Arnost, M. J.; Eustache, J. M.; Krafft, G. A., *J. Org. Chem.* **1982**, *47*, 4384.

14. Bishop, P. M.; Pearson, J. R.; Sutherland, J. K., *J. Chem. Soc., Chem. Commun.* **1983**, 123.

15. (a) Barrett, A. G. M.; Hill, J. M.; Wallace, E. M., *J. Org. Chem.* **1992**, *57*, 386. (b) Burke, S. D.; Saunders, J. O.; Oplinger, J. A.; Murtiashaw, C. W., *Tetrahedron Lett.* **1985**, *26*, 1131.

16. Koreeda, M.; Koo, S., *Tetrahedron Lett.* **1990**, *31*, 831.

17. Reich, H. J.; Holtan, R. C.; Bolm, C., *J. Am. Chem. Soc.* **1990**, *112*, 5609. Also see ref. 12.

18. Proposed structures of $(PhMe_2Si)_nCuLi_{n-1} \cdot LiX$ have been based solely on the stoichiometries of the precursors used in their generation. NMR studies dealing with the composition of these reagents have been recently reported: Singer, R. D.; Oehlschlager, A. C., *J. Org. Chem.* **1991**, *56*, 3510 and references therein.

19. Fleming, I.; Newton, T. W., *J. Chem. Soc., Perkin Trans. 1* **1984**, 1805.

20. Suzuki, M.; Koyano, H.; Noyori, R., *J. Org. Chem.* **1987**, *52*, 5583.

21. Ager, D. J.; Fleming, I., *J. Chem. Soc., Chem. Commun.* **1978**, 177.

22. For Michael addition of $(PhMe_2Si)_2CuLi$ to α,β-unsaturated sulfoxide, see Takaki, K.; Maeda, T.; Ishikawa, M., *J. Org. Chem.* **1989**, *54*, 58.

23. Fleming, I.; Goldhill, J., *J. Chem. Soc., Chem. Commun.* **1978**, 176.

24. Related displacements of F, OTHP, and Cl with $PhMe_2SiLi$ have been described:(a) Hiyama, T.; Obayashi, M.; Sawahata, M., *Tetrahedron Lett.* **1983**, *24*, 4113. (b) Ishii, T.; Kawamura, N.; Matsubara, S.; Utimoto, K.; Kozima, S.; Hitomi, T., *J. Org. Chem.* **1987**, *52*, 4416. (c) Fleming, I.; Sanderson, P. E. J.; Terret, N. K., *Synthesis* **1992**, 69.

25. Magnus, P. D.; Sarkar, T.; Djuric, S. In *Comprehensive Organometallic Chemistry*; Wilkinson, G., Ed.; Pergamon: Oxford, 1982; Vol. 7, p 608.

26. Still, W. C., *J. Org. Chem.* **1976**, *41*, 3063.

27. (a) Baldwin, J. E.; Bottaro, J. C.; Riordan, P. D.; Derome, A. E., *J. Chem. Soc., Chem. Commun.* **1982**, 942. For preparation of t-BuPh$_2$SiLi, see: (b) Cuadrado, P.; Gonzalez, A. M.; Gonzales, B.; Pulido, F. J., *Synth. Commun.*, **1989**, *19*, 275.

28. (a) Wakamatsu, K.; Nonaka, T.; Okuda, Y.; Tückmantel, W.; Oshima, K.; Utimoto, K.; Nozaki, H., *Tetrahedron* **1986**, *42*, 4427. (b) Trost, B. M.; Tour, J. M., *J. Org. Chem.* **1989**, *54*, 484.

29. Fugami, K.; Hibino, J.; Nakatsukasa, S.; Matsubara, S.; Oshima, K.; Utimoto, K.; Nozaki, H., *Tetrahedron* **1988**, *44*, 4277 and references therein.

30. Okuda, Y.; Wakamatsu, K.; Tückmantel, W.; Oshima, K.; Nozaki, H., *Tetrahedron Lett.* **1985**, *26*, 4629.

31. Takeda, K.; Nakajima, A.; Takeda, M.; Okamoto, Y.; Sato, T.; Yoshii, E.; Koizumi, T.; Shiro, M., *J. Am. Chem. Soc.* **1998**, *120*, 4947.

32. Paredes, M. D.; Alonso, R., *J. Org. Chem.* **2000**, *65*, 2292.

33. Moser, W. H., *Tetrahedron* **2001**, *57*, 2065.

34. Fleming, I.; Roberts, R. S.; Smith, S. C., *J. Chem. Soc., Perkin Trans. 1,* **1998**, 1215.

35. Fleming, I.; Ghosh, U., *J. Chem. Soc., Perkin Trans. 1* **1994**, 257.

36. Fleming, I.; Lawrence, A. J.; Richardson, R. D.; Surry, D. S.; West, M. C., *Helv. Chim. Acta* **2002**, *85*, 3349.

37. (a) Chénedé, A.; Fleming, I.; Salmon, R.; West, M. C., *J. Organomet. Chem.* **2003**, *686*, 84. (b) Chénedé, A.; Abd. Rahman, N.; Fleming, I., *Tetrahedron Lett.* **1997**, *38*, 2381.

38. Buswell, M.; Fleming, I.; Ghosh, U.; Mack, S.; Russell, M. G.; Clark, B. P., *Org. Biomol. Chem.* **2004**, *2*, 3006.

39. Fleming, I.; Solay, M.; Stolwijk, F., *J. Organomet. Chem.* **1996**, *521*, 121.

40. Buswell, M.; Fleming, I., *ARKIVOC*, **2002**, *2002*(vii), 46.

41. Cuadrado, P.; Gonzalez-Nogal, A. M., *Tetrahedron Lett.* **1998**, *39*, 1449.

42. Naimi-Jamal, M. R.; Mojtahedi, M. M.; Ipaktschi, J.; Saidi, M. R., *J. Chem. Soc., Perkin Trans. 1* **1999**, 3709.

43. MacLean, B. L.; Henningar, K. A.; Kells, K. W.; Singer, R. D., *Tetrahedron Lett.* **1997**, *38*, 7313.

44. Lipshutz, B. H.; Sclafani, J. A.; Takanami, T., *J. Am. Chem. Soc.* **1998**, *120*, 4021.

45. (a) Hulme, A. N.; Henry, S. S.; Meyers, A. I., *J. Org. Chem.* **1995**, *60*, 1265. (b) Reynolds, A. J.; Scott, A. J.; Turner, C. I., Sherburn, M. S., *J. Am. Chem. Soc.* **2003**, *125*, 12108.

46. Fleming, I.; Marangon, E.; Roni, C.; Russell, M. G.; Taliansky Chamudis, S., *Can. J. Chem.* **2004**, *82*, 325.

47. Saito, S.; Shimada, K.; Yamamoto, H.; Martínez de Marigorta, E.; Fleming, I., *Chem. Commun.* **1997**, 1299.

48. Clive, D. L. J.; Zhang, C., *J. Org. Chem.* **1995**, *60*, 1413.

49. Ito, Y.; Matsuura, T.; Nishimura, S.; Ishikawa, M., *Tetrahedron Lett.* **1986**, *27*, 3261.

50. Fleming, I.; Frackenpohl, J.; Ila, H., *J. Chem. Soc., Perkin Trans. 1* **1998**, 1229.

51. Yeh, M.-C. P.; Sheu, P.-Y.; Ho, J.-X.; Chiang, Y.-L.; Chiu, D.-Y.; Narasimha Rao, U., *J. Organomet. Chem.* **2003**, *675*, 13.

52. (a) Buynak, J. D.; Geng, B., *Organometallics* **1995**, *14*, 3112. (b) Matsumoto, A.; Ito, Y., *J. Org. Chem.* **2000**, *65*, 5707. (c) Suginome, M.; Matsuda, T.; Ito, Y., *Organometallics* **2000**, *19*, 4647.

53. Kawachi, A.; Minamimoto, T.; Tamao, K., *Chem. Lett.* **2001**, 1216.

54. (a) Kuroda, S.; Dekura, F.; Sato, Y.; Mori, M., *J. Am. Chem. Soc.* **2001**, *123*, 4139. (b) Kuroda, S.; Sato, Y.; Mori, M., *J. Organomet. Chem.* **2000**, *611*, 304.

55. Ozawa, F.; Kamite, J., *Organometallics* **1998**, *17*, 5630.

56. (a) Wittenberg, D.; Gilman, H., *Q. Rev., Chem. Soc.* **1959**, *13*, 116. (b) Eisch, J. J., *Organometallics* **2002**, *21*, 5439.

57. Postigo, A.; Rossi, R. A., *Org. Lett.* **2001**, *3*, 1197.

58. Kusukawa, T.; Ando, W., *J. Organomet. Chem.* **1998**, *561*, 109.

59. Lee, T. W.; Corey, E. J., *Org. Lett.* **2001**, *3*, 3337.

60. Tamao, K.; Kawashi, A.; Ito, Y., *J. Am. Chem. Soc.* **1992**, *114*, 3989.

61. (a) Fleming, I.; Winter, S. B. D., *J. Chem. Soc., Perkin Trans. 1* **1998**, 2687. (b) Fleming, I.; Lee, D., *J. Chem. Soc., Perkin Trans. 1* **1998**, 2701.

62. Powell, S. A.; Tenenbaum, J. M.; Woerpel, K. A., *J. Am. Chem. Soc.* **2002**, *124*, 12648.

63. (a) Oh, H.-S.; Imae, I.; Kawakami, Y.; Raj, S. S. S.; Yamane, T., *J. Organomet. Chem.*, **2003**, *685*, 35. (b) Kawachi, A.; Maeda, H.; Tamao, K., *Organometallics* **2002**, *21*, 1319. (c) Strohmann, C.; Hörnig, J.; Auer, D., *Chem. Commun.* **2002**, 766.

1,1-Dimethylsilacyclobutane

[2295-12-7]   C$_5$H$_{12}$Si   (MW 100.23)

InChI = 1S/C5H12Si/c1-6(2)4-3-5-6/h3-5H2,1-2H3

InChIKey = YQQFFTNDQFUNHB-UHFFFAOYSA-N

(reactive silicon compound capable of undergoing various insertion and ring-opening reactions)

Physical Data: bp 79–80 °C.

Form Supplied in: colorless oil, commercially available.

Preparative Methods: 1,1-dichlorosilacyclobutane is treated with methylmagnesium halides or methyllithium in THF or Et$_2$O. Exposure of chloro(3-chloropropyl)dimethylsilane to activated magnesium in THF or Et$_2$O also gives the titled compound.

Handling, Storage, and Precautions: the title compound should be used only in a well-ventilated fume hood. Keep away from sources of ignition. Do not breathe gas and contact with skin and eyes should be avoided. Highly flammable.

Ring-opening Reactions with Carbon and Oxygen Nucleophiles. 1,1-Dimethylsilacyclobutane is a highly strained organosilicon compound such that ring-opening reactions with various nucleophiles take place. The reactions with acetic acid and water provide Me$_2$(*n*-Pr)SiOAc and Me$_2$(*n*-Pr)SiOH, respectively.[1] In contrast, nitrogen nucleophiles such as diethylamine do not cause the ring-opening reaction. Treatment with phenyllithium followed by protonolysis affords Me$_2$(*n*-Pr)PhSi (eq 1).[2]

$$\text{SiMe}_2 \quad \xrightarrow{\text{PhLi}} \quad \xrightarrow{\text{H}^+} \quad \diagup\!\!\diagdown\!\!\diagup\text{SiMe}_2\text{Ph} \quad (1)$$

75%

Precursor of Silene. 1,1-Dimethylsilacyclobutane is pyrolyzed at ca. 600 °C to generate 1,1-dimethylsilene. The highly reactive silene rapidly undergoes dimerization to provide 1,3-disila-1,1,3,3-tetramethylbutane.[3] C–C[4] and C–O[5] multiple bonds are able to trap the silene. The copyrolysis

of 1,1-dimethylsilacyclobutane and propyne furnishes 1,1,3-trimethyl-1-silacyclo-2-butene albeit the yield is low. Upon exposure of benzophenone to the pyrolytic conditions of 1,1-dimethylsilacyclobutane, Wittig-type olefination occurs to afford 1,1-diphenylethene in good yield (eq 2).

$$\text{SiMe}_2 + \underset{\text{Ph}}{\overset{O}{\|}}\underset{\text{Ph}}{} \xrightarrow{600\,°C} \underset{\text{Ph}}{\overset{}{}}\!\!=\!\!\underset{\text{Ph}}{} \quad \text{via} \left[=\!\text{Si}\!\!\overset{Me}{\underset{Me}{}}\right] \quad (2)$$

73%

Base-induced Ring Enlargement. Lithium carbenoids induce ring enlargement of 1,1-dimethylsilacyclobutane into 1,1-dimethyl-2-halo-1-silacyclopentanes.[6] Treatment of 1,1-dimethylsilacyclobutane with CHI$_2$Li in THF at −78 °C gives 1,1-dimethyl-2-iodo-1-silacyclopentane, which is a useful synthetic intermediate. The reaction proceeds through the formation of pentacoordinated silicate followed by 1,2-migration of the methylene group (eq 3).

$$\text{SiMe}_2 \xrightarrow[\text{THF, }-78\,°C]{\text{CHI}_2\text{Li}} \left[\begin{array}{c} \text{Si}\overset{I}{\underset{Me_2}{\diagup}} I \\ \end{array} Li^+ \right] \xrightarrow[-\text{LiI}]{\text{1,2-migration}}$$

$$\underset{\substack{Si \\ Me_2}}{\diagdown}\!\!-I \quad (3)$$

83%

On the contrary, a catalytic amount of *t*-BuOK promotes the reaction with benzaldehyde to afford the corresponding oxasilacyclohexane (eq 4).[7]

$$\text{SiMe}_2 + \underset{\text{Ph}}{\overset{O}{\|}}\underset{\text{H}}{} \xrightarrow[\text{THF, 0 °C}]{10\text{ mol }\%\ t\text{-BuOK}} \underset{\substack{Si \\ Me_2}}{\overset{Ph}{\diagup}}\text{O} \quad (4)$$

65%

Catalytic Reactions under Platinum Catalysis. Phosphine-free platinum complexes catalyze the ring-opening polymerization of 1,1-dimethylsilacyclobutane (eq 5).[8] In contrast, with phosphine as a ligand for platinum catalysts, dimerization of 1,1-dimethylsilacyclobutane proceeds to produce 1,5-disila-1,1,5,5-tetramethylcyclooctane (eq 6).[9] The reaction intermediate is a five-membered silaplatinacycle generated by the oxidative addition of 1,1-dimethylsilacyclobutane to the zerovalent platinum.

$$\text{SiMe}_2 \xrightarrow{\text{cat Pt}} \left[\begin{array}{c} Pt \\ SiMe_2 \end{array} \right] \xrightarrow[\text{phosphine}]{\text{without}} \underset{Me_2}{\overset{}{\left(Si\diagdown\!\!\diagup\right)_n}} \quad (5)$$

$$\text{SiMe}_2 \xrightarrow{\text{cat Pt}} \left[\begin{array}{c} Pt \\ SiMe_2 \end{array} \right] \xrightarrow[\text{phosphine}]{\text{with}} Me_2Si\diagdown\!\!\diagup\!\!\diagdown SiMe_2 \quad (6)$$

Catalytic Reactions under Palladium Catalysis. PdCl$_2$(PPh$_3$)$_2$ catalyzes [4 + 2] cycloaddition of 1,1-dimethylsilacyclobutane with dimethyl acetylenedicarboxylate to give the

Avoid Skin Contact with All Reagents

corresponding cyclic vinylsilane (eq 7).[10,11] The formation of the product proceeds as follows. Initial oxidative addition of 1,1-dimethylsilacyclobutane to palladium followed by insertion of a carbon–carbon triple bond of dimethyl acetylenedicarboxylate to the Pd–Si bond provides seven-membered silapalladacycle. Subsequent reductive elimination furnishes the product and regenerates palladium.

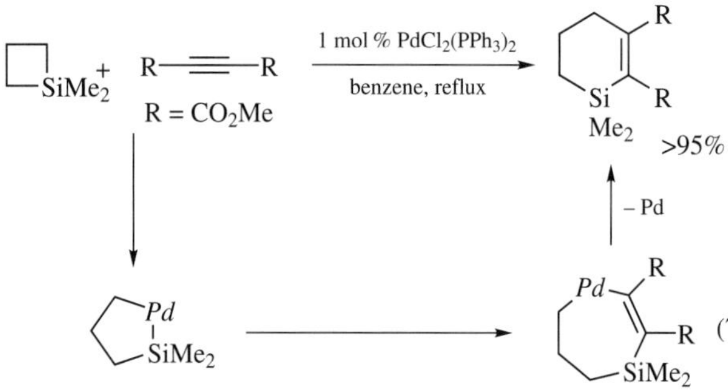

A phenyl-substituted cyclic silyl enolate is synthesized from 1,1-dimethylsilacyclobutane and benzoyl chloride in the presence of a stoichiometric amount of triethylamine and a catalytic amount of PdCl$_2$(PhCN)$_2$ (eq 8).[12] In the case of aliphatic acid chlorides, the use of diisopropylamine instead of triethylamine gives better results. A platinum complex, Pt(CH$_2$=CH$_2$)(PPh$_3$)$_2$, also mediates the reaction although its performance is inferior to that of palladium complexes. Acid anhydrides[13] or combinations of organic halides and carbon monoxide[14] are alternatives to acid chlorides. In the case of acid anhydrides, the amine is not necessary for the reaction.

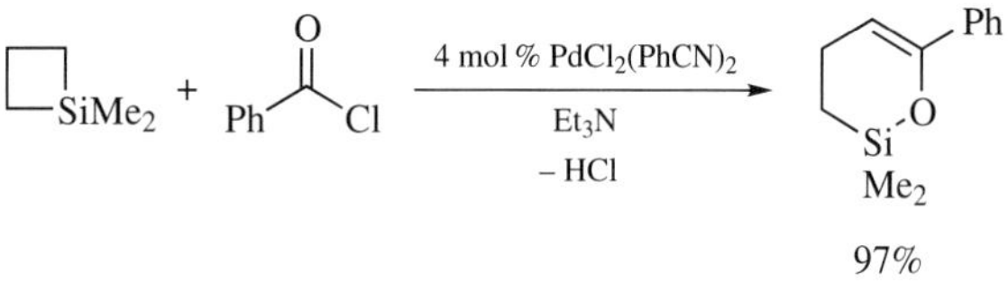

Palladium complexes are known to catalyze cross-metathesis between the C–Si bond of 1,1-dimethylsilacyclobutane and the Si–Si bond of 1,2-disila-1,1,2,2-tetramethylcyclopentane (eq 9).[15]

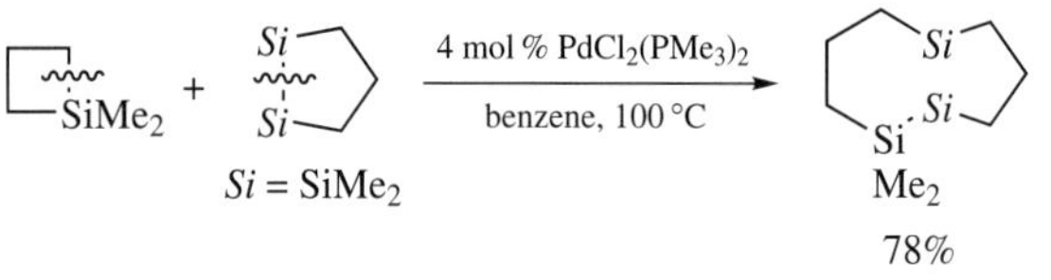

Catalytic Reactions under Rhodium Catalysis. The β-C–H bond of 1,1-dimethylsilacyclobutane undergoes insertion of an ethoxycarbonyl-substituted rhodium carbenoid in a highly regioselective fashion (eq 10).[16] The reaction provides a facile access to functionalized 1,1-dimethylsilacyclobutanes.

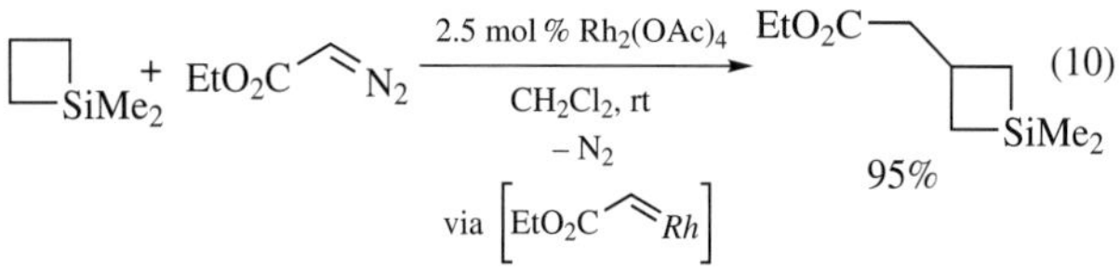

Catalytic Reactions under Nickel Catalysis. Nickel catalyst systems allow 1,1-dimethylsilacyclobutane to be a reagent for hydrosilane-free reductive silylation of aldehydes.[17] Treatment of benzaldehyde with 1,1-dimethylsilacyclobutane in the presence of 10 mol % of Ni(cod)$_2$ and 20 mol % of PPh$_2$Me provides allylbenzyloxydimethylsilane in good yield (eq 11). Divalent nickel precatalysts such as NiCl$_2$ and Ni(acac)$_2$ show no catalytic activity.

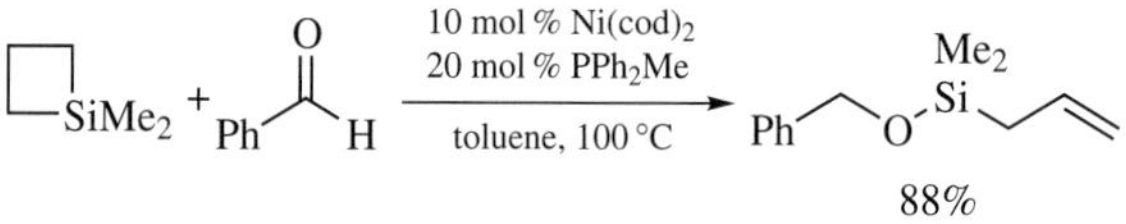

Related Reagents. 1,1-Diphenylsilacyclobutane; 1,1-Dimethyl-2,3-benzosilacyclobutene.

1. Nametkin, N. S.; Vdovin, V. M.; Grinberg, P. L., *Dokl. Akad. Nauk SSSR* **1964**, *155*, 849.
2. Nametkin, N. S.; Vdovin, V. M.; Grinberg, P. L.; Babich, E. D., *Dokl. Akad. Nauk SSSR* **1965**, *161*, 358.
3. Nametkin, N. S.; Vdovin, V. M.; Gusel'nikov, L. E.; Zav'yalov, V. I., *Izv. Akad. Nauk SSSR, Ser. Khim.* **1966**, 358.
4. Conlin, R. T.; Kwak, Y.-W.; Huffaker, H. B., *Organometallics* **1983**, *2*, 343.
5. Golino, C. M.; Bush, R. D.; Roark, D. N.; Sommer, L. H., *J. Organomet. Chem.* **1974**, *66*, 29.
6. Matsumoto, K.; Aoki, Y.; Oshima, K.; Utimoto, K., *Tetrahedron* **1993**, *49*, 8487.
7. Takeyama, Y.; Oshima, K.; Utimoto, K., *Tetrahedron Lett.* **1990**, *31*, 6059.
8. Weyneberg, D. R.; Nelson, L. E., *J. Org. Chem.* **1965**, *30*, 2618.
9. Yamashita, H.; Tanaka, M.; Honda, K., *J. Am. Chem. Soc.* **1995**, *117*, 8873.
10. Sakurai, H.; Imai, T., *Chem. Lett.* **1975**, 891.
11. Takeyama, Y.; Nozaki, K.; Matsumoto, K.; Oshima, K.; Utimoto, K., *Bull. Chem. Soc. Jpn.* **1991**, *64*, 1461.
12. Tanaka, Y.; Yamashita, H.; Tanaka, M., *Organometallics* **1996**, *15*, 1524.
13. Tanaka, Y.; Yamashita, M., *Appl. Organomet. Chem.* **2002**, *16*, 51.
14. Chauhan, B. P. S.; Tanaka, Y.; Yamashita, H.; Tanaka, M., *Chem. Commun.* **1996**, 1207.
15. Reddy, N. P.; Hayashi, T.; Tanaka, M., *Chem. Commun.* **1996**, 1865.
16. Hatanaka, Y.; Watanabe, M.; Onozawa, S.; Tanaka, M.; Sakurai, H., *J. Org. Chem.* **1998**, *63*, 422.
17. Hirano, K.; Yorimitsu, H.; Oshima, K., *Org. Lett.* **2006**, *8*, 483.

Koji Hirano, Hideki Yorimitsu & Koichiro Oshima
Kyoto University, Kyoto, Japan

Dimethylthexylsilyl Chloride

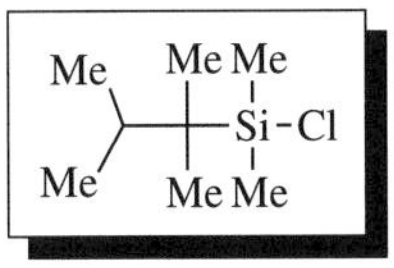

[67373-56-2] $C_8H_{19}ClSi$ (MW 178.78)

InChI = 1S/C8H19ClSi/c1-7(2)8(3,4)10(5,6)9/h7H,1-6H3

InChIKey = KIGALSBMRYYLFJ-UHFFFAOYSA-N

(silylating reagent for the protection of 1° and 2° alcohols and *N*-silylation of amines and amides; used in the preparation of thexyldimethylsilyl triflate; formation of α-silyl hydrazones; formylation of polyfluoroalkyl halides)

Alternate Names: TDS-Cl; chlorodimethylthexylsilane; dimethyl-thexylchlorosilane.

Physical Data: mp 14–15 °C; bp 55–56 °C/10 mmHg;[1] *d* 0.909 g cm^{-3}.

Solubility: sol in THF, DMF, CH_2Cl_2.

Form Supplied in: colorless liquid,[1] commercially available (95% purity).

Analysis of Reagent Purity: ^1H NMR.[1]

Preparative Method: to a mixture of dimethylchlorosilane (10 mL, 92 mmol) and $AlCl_3$ (0.680 g, 5 mmol) at 25 °C was added 2,3-dimethyl-2-butene (11 mL, 91 mmol). After stirring for 4 h at ambient temperature, the mixture was filtered and distilled to yield 15.2 g of TDS-Cl (93%).[1]

Handling, Storage, and Precautions: moisture sensitive, should be stored under inert atmosphere; corrosive, use in fume hood.

Silylating Agent for Protection of Alcohols. Thexyldi-methylsilyl chloride (TDS-Cl) was first introduced as a more easily prepared alternative to the more commonly used *t*-butyl-dimethylsilyl chloride for the protection of alcohols.[1,2] Direct silylation of 1° and 2° alcohols occurs upon treatment with TDS-Cl and imidazole in DMF as illustrated by the protection of 1-butanol (**1**) and cyclohexanol (**3**) to form the corresponding silyl ethers (eqs 1 and 2).[2] Phenols can also be protected using TDS-Cl and imidazole.[3,4] Typical solvents for the formation of silyl ethers using TDS-Cl include DMF,[2] THF,[5] or CH_2Cl_2.[6]

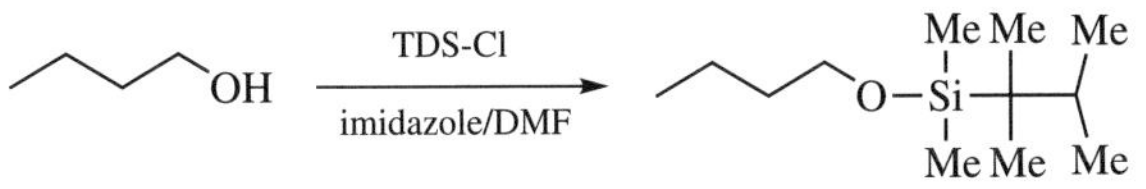

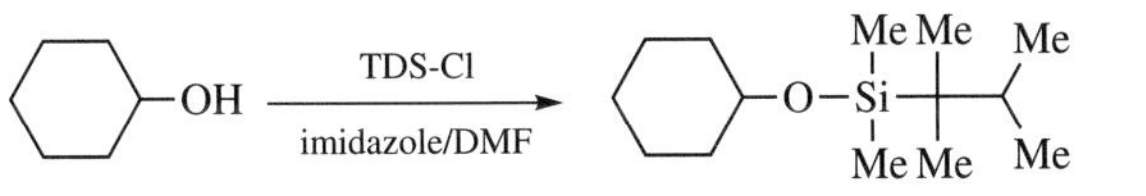

Alternatively, an alcohol can be treated sequentially with *n*-BuLi in THF at −78 °C and TDS-Cl to yield the desired silyl ether.[7]

Silyl protection of 3° alcohols, however, requires the use of thexyldimethylsilyl triflate (TDS-OTf, **5**), which is also prepared from TDS-Cl (eq 3).[2] Subsequent treatment of 2-methyl-2-butanol

(**6**) with TDS-OTf in triethylamine or 2,6-lutidine resulted in the formation of silyl ether **7** in 82% yield (eq 4).[2] The lack of reactivity of TDS-Cl toward 3° alcohols explains the selective protection of a 2° alcohol in the presence of a 3° alcohol in the synthesis of a steroidal side chain.[8]

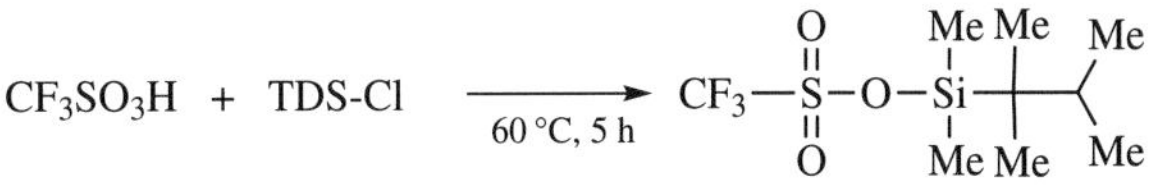

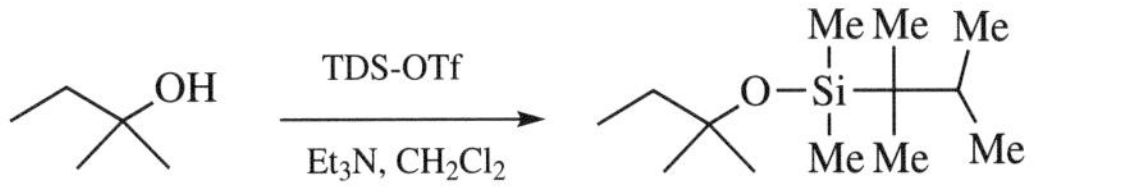

Due to the increased steric bulk, thexyldimethylsilyl ethers undergo deprotection approximately two to three times more slowly than *t*-butyldimethylsilyl ethers under acidic and basic conditions.[2] Similarly, TDS ethers derived from 1° alcohols undergo deprotection more rapidly than 2° TDS ethers.[2,9]

The steric bulk of the thexyldimethylsilyl group allows selective protection of less sterically hindered alcohols in the presence of other alcohols. For example, partially protected sugar **8** was selectively protected at the 1° alcohol to yield 1° silyl ether **9** (eq 5).[5] In this instance, selectivity is obviously due to differing steric constraints on the alcohols.

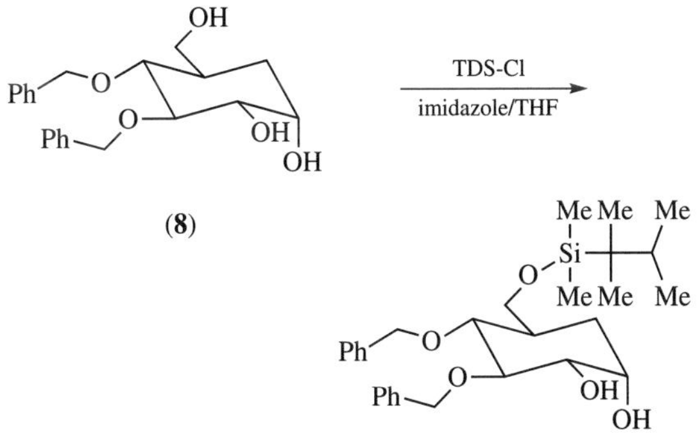

In an example of a 2° alcohol being protected in the presence of another 2° alcohol, diol **10** was selectively protected with thexyldimethylsilyl chloride to produce the monosilyl ether **11** (eq 6).[10] Other examples similarly employed thexyldimethyl-silyl chloride as the reagent of choice for selective protection reactions.[11–13]

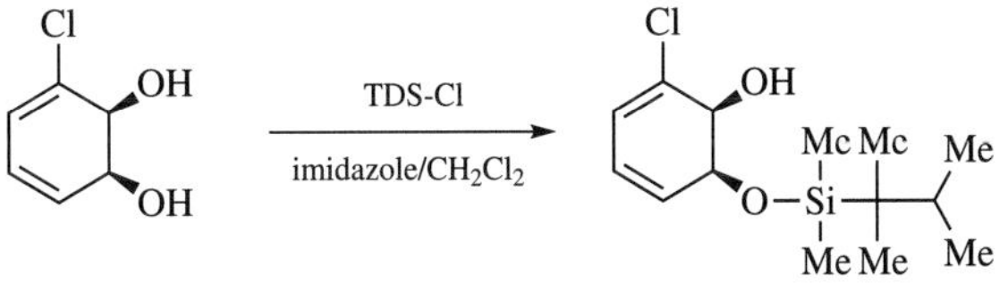

Thexyldimethylsilyl chloride has also been used to produce a new protecting group, thexyldimethylsiloxymethyl chloride, via a two-step synthetic process.[14] Ethylthiomethanol **12** was converted to the TDS ether followed by treatment with sulfuryl chloride to produce thexyldimethylsiloxymethyl chloride **13** (eq 7). Siloxymethyl chloride **13** was used to protect 1°, 2°, and 3° alcohols and phenols, and the protecting group was easily removed by treatment with TBAF in THF or Et₄NF in CH₃CN.[14]

$$EtS \diagdown OH \xrightarrow[\substack{Et_3N,\ DMAP \\ CH_2Cl_2}]{TDS\text{-}Cl} EtS \diagdown OTDS \xrightarrow{SO_2Cl_2}$$

(12)

$$Cl \diagdown O-\underset{\substack{| \\ Me}}{\overset{\substack{Me \\ |}}{Si}}-\underset{\substack{| \\ Me}}{\overset{\substack{Me \\ |}}{C}}-\underset{\substack{| \\ Me}}{\overset{\substack{Me \\ |}}{C}}H \quad (7)$$

(13), 95%

Protection of Other Functional Groups. Other functional groups can also be protected using thexyldimethylsilyl chloride. Both amines and amides undergo N-silylation upon treatment with thexyldimethylsilyl chloride and triethylamine.[2] N-Silyl amines undergo facile hydrolysis and, unlike N-silyl amides, cannot be exposed to aqueous workup.[2] The half-life of a lactam protected with t-butyldimethylsilyl (**14**) and thexyldimethylsilyl (**15**) groups in aqueous acid shows the enhanced stability afforded by the latter protecting group.[2]

(14), $t_{1/2}$ = 40 min   **(15)**, $t_{1/2}$ = 90 min

In a related example, S-methyl-S-phenylsulfoximine (**16**) was treated with thexyldimethylsilyl chloride and pyridine that, after aqueous workup, yielded N-silylated sulfoximine (**17**) in 79% yield (eq 8).[15]

$$\underset{\substack{| \\ Ph}}{\overset{\substack{Me \\ |}}{O=S=NH}} \xrightarrow[pyridine]{TDS\text{-}Cl} \underset{\substack{| \\ Ph}}{\overset{\substack{Me \\ |}}{O=S=N-Si}} \quad (8)$$

(16)   **(17)**, 79%

Carboxylic acids can also be converted to silyl esters using thexyldimethylsilyl chloride and triethylamine in DMF,[2] diethyl ether,[2] or CH₂Cl₂.[16] The resultant silyl ester can be hydrolyzed using triethylamine in aqueous acetone.[16]

Neither ketones nor thiols are reactive with thexyldimethylsilyl chloride as the silylating agent.[2] Ketones require the more reactive thexyldimethylsilyl triflate and triethylamine to form silyl enol ethers.[2,17] But thiols can be converted into the corresponding thiolates, which then undergo silylation with thexyldimethylsilyl chloride to form alkyl silyl thioethers.[2]

Preparation of α-Silyl Ketones. SAMP- and RAMP-hydrazones of methyl ketones have been converted into α-silyl ketones[18] using thexyldimethylsilyl chloride.[19] For example, the SAMP-hydrazone of 2-butanone **18** was deprotonated with LiTMP, and the intermediate anion was treated with thexyldimethylsilyl chloride to form the α-silyl derivative **19** in high yield (eq 9).[19] When other dialkyl ketones were used, thexyldimethylsilyl triflate was required for silylation to occur, but poor regioselectivity was observed.[20]

(18)

(19), 95%

Typically, silyl groups in these compounds act as directing groups in asymmetric synthesis, and removal of both the hydrazone and silyl group may be desired. Ozonolysis was employed to remove the hydrazone, yielding the α-silyl ketone.[18,20] However, in the synthesis of a diketoterpenoid isolated from *Hyrtios erectus*, both the hydrazone and the silyl group were removed by refluxing in aqueous oxalic acid.[19]

Formylation of Polyfluoroalkanes. Thexyldimethylsilyl chloride has been used in a formylation reaction via addition of fluorine-containing organozinc intermediates to DMF.[21] When organozinc complex **20** was heated in the presence of thexyldimethylsilyl chloride and DMF, silyl-protected hemiaminal **21** was isolated in 67% yield (eq 10).[21]

(20)

(21), 67%

Development of a one-pot version of the reaction allowed silyl-protected hemiaminals to be prepared at low temperatures. For example, perfluoroheptyl iodide **22** was treated with zinc, thexyldimethylsilyl chloride, and DMF at 0 °C for 2 h to yield the corresponding hemiaminal **23** in 80% yield (eq 11).[21] Although isolable, the silyl-protected hemiaminals are subject to acidic hydrolysis, producing the corresponding aldehyde **24** (eq 11).[21]

$$CF_3(CF_2)_5—CF_2—I \quad + \quad DMF \quad \xrightarrow[\text{0 °C to rt}]{\text{TDS-Cl, Zn}}$$

(22)

$$CF_3(CF_2)_5—CF_2—\underset{\underset{NMe_2}{|}}{CH} \quad \xrightarrow[\text{90 °C}]{\text{H}_2\text{SO}_4} \quad CF_3(CF_2)_5—CF_2—\overset{\overset{O}{\|}}{CH}$$

(23), 80%   **(24)**, 75%

(11)

This two-step process thus represents the formylation of alkyl halides to yield aldehydes. Polyfluorinated aldehydes had proven difficult to access via other methods.[21]

1. Oertle, K.; Wetter, H., *Tetrahedron Lett.* **1985**, *26*, 5511.
2. Wetter, H.; Oertle, K., *Tetrahedron Lett.* **1985**, *26*, 5515.
3. Crisp, G. T.; Turner, P. D., *Tetrahedron* **2000**, *56*, 8335.
4. Messmore, B. W.; Hulvat, J. F.; Sone, E. D.; Stupp, S. I., *J. Am. Chem. Soc.* **2004**, *126*, 14452.
5. Toma, L.; Legnani, L.; Rencurosi, A.; Poletti, L.; Lay, L.; Russo, G., *Org. Biomol. Chem.* **2009**, *7*, 3734.
6. Lipshutz, B. H.; Tirado, R., *J. Org. Chem.* **1994**, *59*, 8307.
7. Nantz, M. H.; Hitchcock, S. R.; Sutton, S. C.; Smith, M. D., *Organometallics* **1993**, *12*, 5012.
8. Maehr, H.; Uskokovic, M. R.; Adorini, L.; Reddy, G. S., *J. Med. Chem.* **2004**, *47*, 6476.
9. Brandänge, S.; Leijonmarck, H.; Minassie, T., *Acta Chem. Scand.* **1997**, *51*, 953.
10. McKibben, B. P.; Barnosky, G. S.; Hudlicky, T., *Synlett* **1995**, 806.
11. Hudlicky, T.; Seoane, G.; Pettus, T., *J. Org. Chem.* **1989**, *54*, 4239.
12. Pitzer, K.; Hudlicky, T., *Synlett* **1995**, 803.
13. Fonseca, G.; Seoane, G. A., *Tetrahedron: Asymmetry* **2005**, *16*, 1393.
14. Gundersen, L.-L.; Benneche, T.; Undheim, K., *Acta. Chem. Scand* **1989**, *43*, 706.
15. Pearson, A. J.; Blystone, S. L.; Nar, H.; Pinkerton, A. A.; Roden, B. A.; Yoon, J., *J. Am. Chem. Soc.* **1989**, *111*, 134.
16. Abel, S.; Faber, D.; Hüter, O.; Giese, B., *Synthesis* **1999**, 188.
17. Snider, B. B.; Kwon, T., *J. Org. Chem.* **1992**, *57*, 2399.
18. Enders, D.; Adam, J.; Klein, D.; Otten, T., *Synlett* **2000**, 1371.
19. Enders, D.; Schüßeler, T., *Synthesis* **2002**, 2280.
20. Enders, D.; Lohray, B. B., *Angew. Chem., Int. Ed. Engl.* **1987**, *26*, 351.
21. Lang, R. W., *Helv. Chim. Acta* **1988**, *71*, 369.

R. David Crouch
Dickinson College, Carlisle, PA, USA

Dimethylthexylsilyl Trifluoromethanesulfonate

[103588-79-0] $C_9H_{19}F_3O_3SSi$ (MW 292.39)

InChI = 1S/C9H19F3O3SSi/c1-7(2)8(3,4)17(5,6)15-16(13,14)9(10,11)12/h7H,1-6H3

InChIKey = HUCZOYCXPWONIR-UHFFFAOYSA-N

(silylation reagent used to introduce the dimethylthexylsilyl (TDS) group; complexation reagent to activate electrophiles)

Alternate Names: TDSOTf; TDS triflate; TDMSOTf; dimethylthexylsilyl triflate; thexyldimethylsilyl triflate; dimethyl(1,1,2-trimethylpropyl)silyl trifluoromethanesulfonate; dimethyl(1,1,2-trimethylpropyl)silyl-1,1,1-trifluoromethanesulfonate; dimethyl-(2,3-dimethyl-2-butyl)silyl trifluoromethanesulfonate; thexyldimethylsilyl trifluoromethanesulfonate; trifluoromethanesulfonic acid thexyldimethylsilyl ester.

Physical Data: clear light yellow liquid, bp 40–42 °C/0.15 Torr,[1] 53–54 °C/0.6 mmHg; d = 1.16 g mL^{-1} at 20 °C; n_D^{20} 1.405.

Solubility: CH$_2$Cl$_2$, CHCl$_3$, Et$_2$O, THF (can result in ring opening of THF), DMF; reacts with alcohols and water.

Preparative Method: dimethylthexylsilyl chloride and trifluoromethanesulfonic acid are heated to 60 °C for 4–5 h and the resulting silyl ester is isolated by distillation.[1]

Handling, Storage, and Precautions: corrosive; reacts violently with water; unstable to air and moisture; stable upon storage under inert gas in cool place; causes burns; skin/eye contact and inhalation should be avoided; no toxicity data are available but the material is extremely destructive to the tissue of the mucous membranes and upper respiratory tract.

General. Dimethylthexylsilyl trifluoromethanesulfonate (TDSOTf) is generally used as a bulky silylation reagent.[2] The commercially available (≤97%) TDSOTf is more reactive than the corresponding dimethylthexylsilyl chloride (TDSCl) to introduce the sterically demanding TDS functional group on oxygen or carbon atoms. The TDS group is slightly more hindered and has a higher stability than the TBDMS group with the disadvantage that the NMR spectrum is not as simple as in the case when the TBDMS group is applied. In general, more sterically demanding groups on silicon make the silyl group less reactive and more difficult to introduce (Me$_3$Si > Ph$_2$MeSi > Et$_3$Si > i-PrMe$_2$Si > t-BuMe$_2$Si > t-BuPh$_2$Si > t-HexMe$_2$Si > i-Pr$_3$Si). The leaving group on silicon also determines the reactivity for R$_3$Si–X (X = CN > OTf > I > Br > Cl). The use of TDSOTf is somewhat limited by its accessibility and ease of preparation compared to the readily accessible and cheaper TDSCl that has the advantage of being a liquid useful for handling large quantities of material. The relative stabilities of the silyl-protected functional groups are roughly the same as the relative rates for their introduction. In general terms, the relative stabilities of the silyl-protected functional groups will follow the following order: i-Pr$_3$Si > t-HexMe$_2$Si > t-BuPh$_2$Si > t-BuMe$_2$Si > i-PrMe$_2$Si > Et$_3$Si > Ph$_2$MeSi > Me$_3$Si.[3] The methods for cleavage of the TDS group are the same as the methods used to cleave the TBDMS group, but require longer reaction times (2–3 times) due to its increased steric bulk.[1] Corey also reported that tetrafluorosilane readily cleaves TBDMS ethers, whereas TDS ethers remain essentially intact.[4] Moreover, TDS-protected primary alcohols are more rapidly deprotected than secondary alcohols under acidic conditions or upon treatment with fluoride.[1]

O-Silylation of Alcohols. The introduction of the sterically impeded TDS group on an alcohol typically requires the presence of a catalyst such as a hindered tertiary amine or pyridine base. The first use of TDSOTf, prepared by Wetter and Oertle from the less reactive silylation reagent TDSCl, involved the O-silylation of tertiary alcohols (eq 1) in the presence of 2,6-lutidine or Et$_3$N as base in CH$_2$Cl$_2$.[1] Under similar reaction conditions, Franck-Neumann et al. silylated a number of olefinic secondary alcohols with TDSOTf in high yield (eq 2),[5,6] for example, en route toward the total synthesis of the sesquiterpene 2,3-dihydroilludin M.[7,8] To protect the apparently shielded hydroxyl group in functionalized 4-hydroxypentanoates, the very reactive TDSOTf in the presence of collidine had to be used as well (eq 3).[9] Also a diol has been converted to the corresponding bis(silyl) ether in 95% yield upon treatment with TDSOTf and 2,6-lutidine in CH$_2$Cl$_2$ at room temperature overnight (eq 4).[10] The use of TDSOTf as an alternative reagent to silylate alcohols that are unreactive toward TDSCl is not always successful as illustrated by the decomposition of hydroxylactones in the presence of TDSOTf.[11]

$$(1)$$

82%

$$(2)$$

98%

86% $$(3)$$

$$(4)$$

95%

A TDS protecting group can also be introduced on alcohols in the presence of other silyl ether functions and selectively elaborated in later stages of synthetic sequences. The secondary

hydroxyl group of diphenylmethylsilyl (DPMS) ethers can be protected by reaction with TDSOTf in the absence of a base in THF at $-10\,°$C for 5–10 min (eq 5).[12] The use of TDSCl or TBM-SCl in DMF and imidazole at $25\,°$C resulted in exchange of both silyl protecting groups. The TDS ether function was selected for its stability toward BuLi, in contrast to the DPMS group, and the latter reactivity was exploited in subsequent transformations to cyclododeca-2,8-diyne-1,7-dione.

$$(5)$$

no yield reported

A process for the selective synthesis of $2'$-O-TDS uridine has been developed involving treatment of uridine with di-t-butyl-silylbis(trifluoromethanesulfonate) and pyridine in DMF, followed by reaction with TDSOTf and final addition of hydrofluoric acid in THF (eq 6).[13]

$$(6)$$

78%

The TDS ether is a valuable and versatile protecting group to regioselectively prepare cellulose derivatives. Depending on the state of dispersity of cellulose in the reaction mixture, two different types of O-TDS-protected cellulose derivatives have been prepared.[14,15] In a mixture of ammonia and N-methylpyrrolidinone (NMP), swollen cellulose is converted with TDSCl to 6-O-TDS cellulose exclusively with in situ formed TDS-NH$_2$ acting as the silylating reagent. In contrast, silylation of cellulose dissolved in N,N-dimethyl acetamide (DMA)/LiCl with TDSCl in the presence of a base (pyridine, imidazole) afforded 2,6-di-O-TDS cellulose with degree of substitution (DS) of 2. The lesser reactive hydroxyl group at position 3 could then be regioselectively further modified. Although TDSOTf might be a useful alternative reagent to silylate cellulose, no such use has been reported to date.

Enantiomerically pure 1,6-disubstituted 1,5-dienes with an aldol syn-substitution pattern, which undergo highly stereoselective Cope rearrangements, have been synthesized in 60–70% yield by asymmetric aldol reaction of the boron dienolates of chiral imides with α,β-unsaturated aldehydes followed by silylation with TDSOTf and chromatographic purification (eq 7).[16] The free aldol products readily undergo retro-aldol cleavage and thermal decomposition, making them unsuitable for Cope rearrangement.

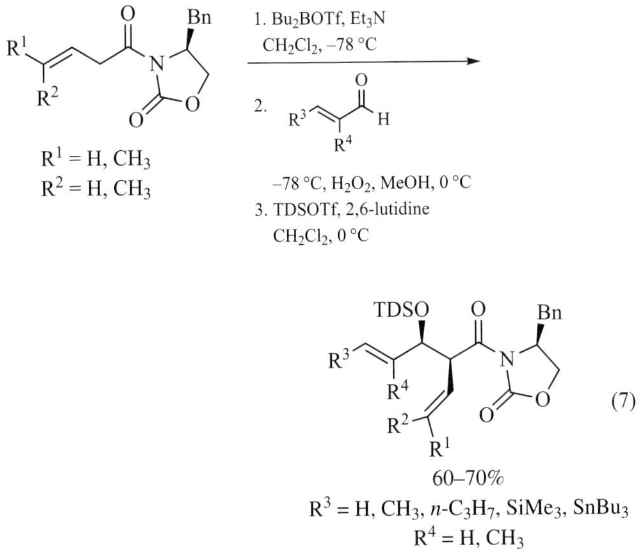

R^1 = H, CH$_3$
R^2 = H, CH$_3$

1. Bu$_2$BOTf, Et$_3$N
   CH$_2$Cl$_2$, −78 °C

2. (acrolein derivative with R^3, R^4)

−78 °C, H$_2$O$_2$, MeOH, 0 °C

3. TDSOTf, 2,6-lutidine
   CH$_2$Cl$_2$, 0 °C

$$(7)$$

60–70%
R^3 = H, CH$_3$, n-C$_3$H$_7$, SiMe$_3$, SnBu$_3$
R^4 = H, CH$_3$

Silyl enol ethers have been prepared in situ by reaction of 2,2-dialkyl-2,3-dihydro-4H-pyran-4-ones with TDSOTf and 2,6-di(t-butyl)pyridine in CHCl$_3$ at room temperature to 60 °C (eq 10).[18] The in situ prepared electron-rich dienes were used in carbonyl–alkyne exchange reactions with electron-poor alkynes to yield highly substituted silylated phenols that could be purified by flash chromatography. The latter aromatic compounds could be hydrolyzed to the corresponding phenols.

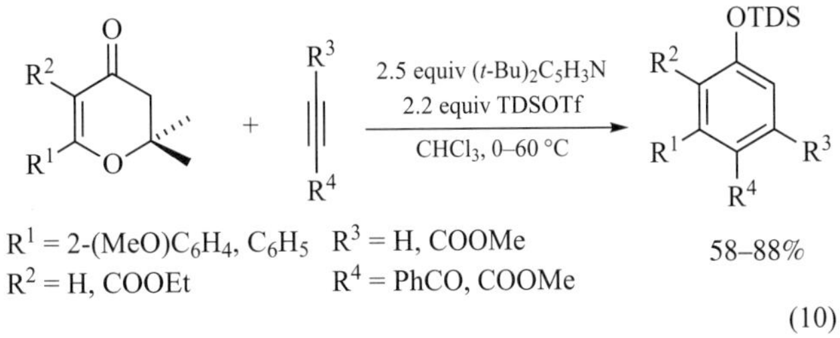

2.5 equiv (t-Bu)$_2$C$_5$H$_3$N
2.2 equiv TDSOTf
CHCl$_3$, 0–60 °C

R^1 = 2-(MeO)C$_6$H$_4$, C$_6$H$_5$ R^3 = H, COOMe
R^2 = H, COOEt R^4 = PhCO, COOMe

58–88%

$$(10)$$

Synthesis of Silyl Enol Ethers. Wetter and Oertle also described the first application of TDSOTf in the conversion of 3,3-dimethylbutan-2-one to the corresponding silyl enol ether in the presence of Et$_3$N or 2,6-lutidine in CH$_2$Cl$_2$ (eq 8).[1] The latter conditions have been used in the efficient synthesis of the TDS enol ether of the t-butyl ketone derived from the treatment of citronellal with t-BuLi and Jones oxidation (eq 9).[17] Oxidative cyclization of the TDS enol ether with CAN and excess NaHCO$_3$ provided 42% of the corresponding cyclic ketone. The less hindered TBDMS enol ether gave only 20% of cyclic ketone besides 20% of the parent ketone resulting from significant hydrolysis under the oxidative cyclization conditions.

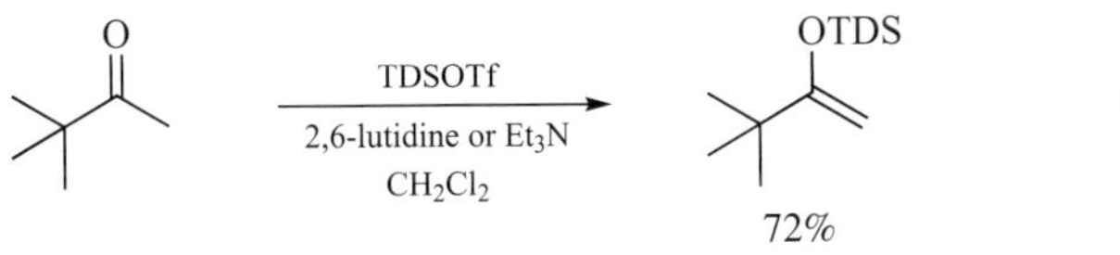

The 4-silyloxyquinolinium triflate was easily prepared in situ by Beifuss and Ledderhose, via reaction of N-benzyloxycarbonyl-protected 4-quinolone with TDSOTf at 20 °C for 1 h, and reacted with phenylmagnesium bromide in CH$_2$Cl$_2$/THF in the presence of 2,6-lutidine to regioselectively form the corresponding 2-substituted silyl enol ether in 54% yield after column chromatography (eq 11).[19] Somewhat surprisingly, the Pd/C-catalyzed hydrogenation of the silyl enol ether gave the corresponding N-deprotected 2-phenyl-4-quinolone directly in 94% yield. This methodology could also be performed with triisopropyl- and t-butyldimethylsilyl enol ethers.

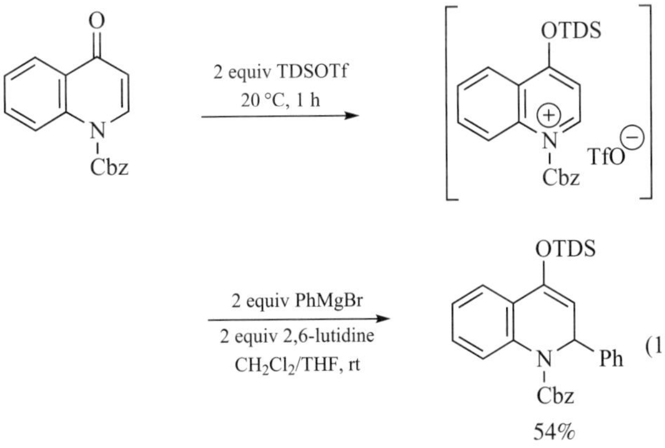

TDSOTf
2,6-lutidine or Et$_3$N
CH$_2$Cl$_2$

$$(8)$$

72%

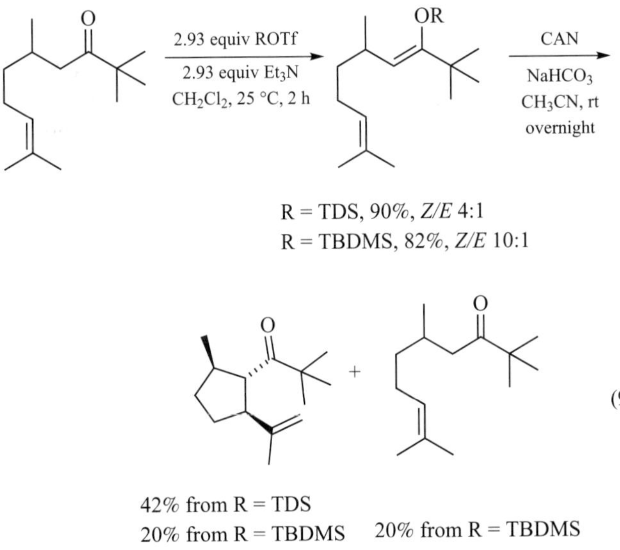

2.93 equiv ROTf
2.93 equiv Et$_3$N
CH$_2$Cl$_2$, 25 °C, 2 h

CAN
NaHCO$_3$
CH$_3$CN, rt
overnight

R = TDS, 90%, Z/E 4:1
R = TBDMS, 82%, Z/E 10:1

$$(9)$$

42% from R = TDS
20% from R = TBDMS   20% from R = TBDMS

2 equiv TDSOTf
20 °C, 1 h

2 equiv PhMgBr
2 equiv 2,6-lutidine
CH$_2$Cl$_2$/THF, rt

$$(11)$$

54%

Besides the use of TDSOTf as a silylation reagent, several reactions have been reported in which TDSOTf functions also as an activating reagent, similar to other Lewis acids, to enhance the reactivity of electrophiles. The reaction of the 4-silyloxypyrylium triflate, prepared in situ from pyran-4-one and TDSOTf, and the 2-silyloxybuta-1,3-diene, generated in situ from 4-phenylbut-3-en-2-one and TDSOTf in the presence of 2,6-lutidine, led to the corresponding tetrahydro-2H-chromene in the presence of excess TDSOTf with 90% yield (eq 12).[20] In this case, somewhat better yields and E/Z selectivity were obtained upon using TBDMSOTf or TESOTf instead of TDSOTf.

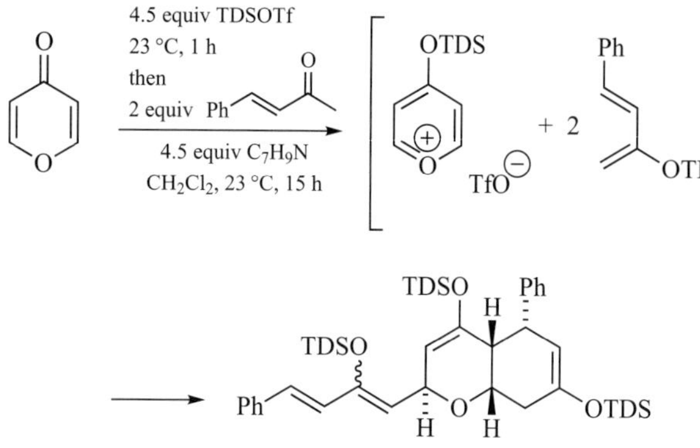

$$90\%, \; E/Z \; 3.8{:}1.0 \qquad (12)$$

Trapping of a deprotonated six-membered vinylogous ester with TDSOTf in THF gave a TDS dienolate that underwent a Dieckmann condensation by addition of a second equivalent of TDSOTf or TiCl₄ at low temperature in 62 and 50% yield, respectively (eq 13).[21] Similarly, in situ trapping of the enolate derived from Michael addition of a deprotonated five-membered vinylogous ester with acrylate gave a TDS enol ether that formed a bicyclooctene in 57% yield by addition of EtAlCl₂ at low temperature (eq 14).[21]

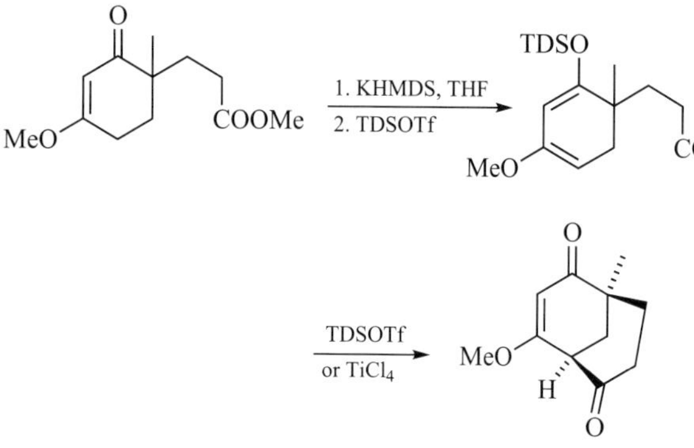

$$50\text{–}62\% \qquad (13)$$

$$57\% \qquad (14)$$

The TDS enol ether of a keto soraphenic acid, in situ derived via silylation with TDSOTf in the presence of Et₃N in CH₂Cl₂ and hydrolysis of the simultaneously formed silyl ester, was selectively cleaved by ozonization at −40 °C to afford the corresponding

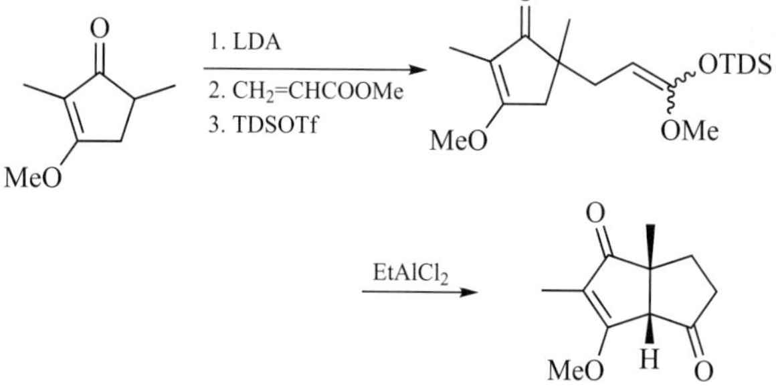

aldehyde carboxylic acid (eq 15).[22] The latter aldehyde was further used for the synthesis of ring-contracted soraphen analogs.

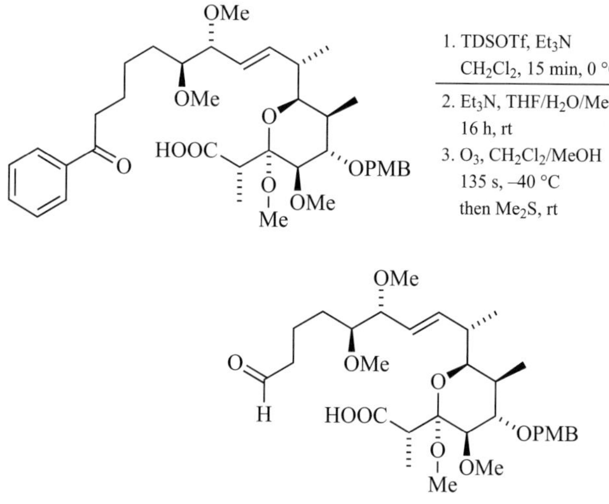

$$26\% \qquad (15)$$

The TDSOTf-promoted regioselective addition of formaldehyde SAMP-hydrazone to prochiral cyclic and acyclic conjugated enones afforded the corresponding silyl enol ethers at −78 °C in THF or deprotected ketones upon treatment with TBAF in good yields and with excellent de's (85 to ≥98%) (eq 16).[23] The use of the promoting bulky trialkylsilyl (TBDMS or TDS) triflate to form a reactive complexed ketone was essential as in the absence of a Lewis acid or in the presence of other Lewis acids (ZnCl₂, TiCl₄, BF₃·Et₂O, etc.) no successful reaction could be achieved. The large steric hindrance around the promoting trialkylsilyl triflate makes the complexation of the small carbonyl oxygen on the ketone much easier than that of the dialkyl-substituted nitrogen on the hydrazone moiety.[24] The latter complexation results in decomposition of the hydrazone and explains the incomplete conversion with sterically hindered substrates, such as 4-t-butylcyclohexenone and 4,4-dimethyl-3-cyclohexenone. The chelation of the carbonyl group with the bulky trialkylsilyl groups also induces high regioselectivity in the reaction, resulting in the absence of 1,2-adducts.

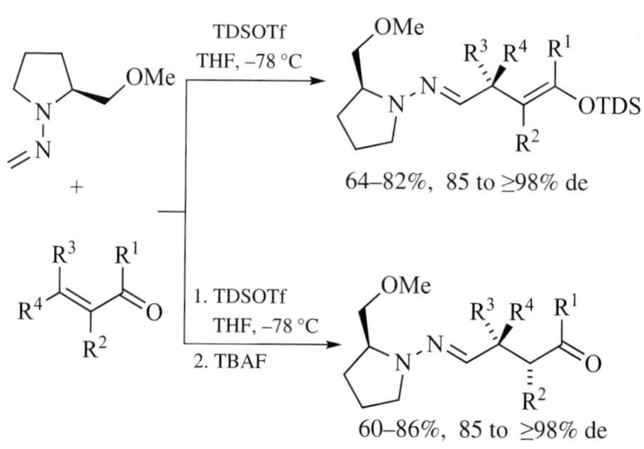

$$(16)$$

$$64\text{–}82\%, \; 85 \text{ to } {\geq}98\% \text{ de}$$

$$60\text{–}86\%, \; 85 \text{ to } {\geq}98\% \text{ de}$$

R¹–R³ = (CH₂)₂, (CH₂)₂C(Me)₂, (CH₂)₃; R¹ = Ph, Me
R² = H, Me; R³ = H; R⁴ = H, Me, Ph

The use of TDSOTf (or other acids such as BF$_3$·OEt$_2$, *p*-TsOH, ZnBr$_2$, ZnCl$_2$) as a catalyst to improve the addition of formaldehyde dialkylhydrazones to sugar-derived aldehydes resulted in undesired hydrazones formed via a direct and retro-[2 + 2]-cycloaddition process between the formaldehyde dialkyl-hydrazones and the sugar aldehyde.[25]

Synthesis of Siloxybutylated Aldehydes and Ketones. During the course of their investigations on the enantioselective synthesis of α-silylated aldehydes and ketones (see below),[26] Enders et al. observed that lithiated SAMP-/RAMP- or SADP-hydrazones reacted with the solvent tetrahydrofuran at −78 °C in the presence of various trialkylsilyl trifluoromethanesulfonates, such as TDSOTf, to give high yields of α-siloxybutylated hydrazones (eq 17).[27] The hydrazones were oxidatively cleaved by ozone at −78 °C to afford α-trialkylsiloxybutylated aldehydes and ketones in excellent yields and enantiomeric purities (≥95% ee, 81%).

R = H, Me
R^1 = Ph, Bn, Et, H
R^2 = Ph, Me, Bn
R^1–R^2 = -(CH$_2$)$_3$-, -(CH$_2$)$_4$-

54–95%
>95% de, 81%
(R^3 = *t*Bu, *t*Hex)

84–94%   (17)
>95% ee, 81%

Asymmetric Synthesis of α-Dimethylthexylsilyl Ketones. Enders et al.[26] established the enantioselective synthesis of α-dimethylthexylsilyl ketones (≥96% ee) via *C*-silylation of lithiated SAMP-hydrazones with TDSOTf in diethyl ether and subsequent oxidative and racemization-free cleavage of the silylhydrazones with ozone (eq 18). The use of THF in the silylation step had to be avoided due to competitive TDSOTf-mediated ring opening of tetrahydrofuran (see above).[27] The α-silyl ketones were obtained in 68–79% overall yield after chromatographic separation from the *S*-nitrosamine, which allowed recycling of the chiral auxiliary. The α-silyl ketones are useful compounds for further carbon–carbon bond forming reactions such as regio- and stereoselective aldol and Mannich reactions or carbon–heteroatom bond forming reactions in which the trialkylsilyl substituent is used as a 'traceless' directing group.[28] The use of this method for the preparation of α-silyl dialkylketones proved problematic due to poor regioselectivity of the α-silylation of the corresponding dialkyl ketone hydrazones. However, an alternative route involving α-silylation of acetaldehyde/methyl ketone hydrazones with chlorosilanes followed by stereoselective α-alkylation with alkyl halides was developed as well.[26b]

1. 1.05 equiv LDA, Et$_2$O
   0 °C, 4 h
2. 1 equiv TDSOTf
   −78 °C, 2 h; then rt, 10 h
3. O$_3$, *n*-pentane,−78 °C

R = Et, Ph

68–79%, ≥96% ee

(18)

Similarly, the TDS group was also introduced by deprotonation of the SAMP-hydrazone of 2,2-dimethyl-1,3-dioxan-5-one with *t*-BuLi at −78 °C followed by silylation with TDSOTf and oxidative cleavage of the chiral auxiliary with ozone to afford the corresponding 4-silylated 2,2-dimethyl-1,3-dioxan-5-one after chromatography (eq 19).[29] In a similar way, the TBDMS derivative of the ketone could be prepared as well, whereas dimethylisopropyl- and dimethylphenylsilylated derivatives were unstable at room temperature. The presence of the 4-dimethylthexylsilyl substituent as an auxiliary group in the 2,2-dimethyl-1,3-dioxan-5-one allowed further application in highly *anti* diastereoselective boron-mediated aldol reactions with aldehydes. After removal of the silyl group via the triethylamine–trihydrofluoride complex, protected oxopolyols were obtained in good yields and excellent diastereomeric and enantiomeric excesses (96–98% de, ee).

1. SAMP, C$_6$H$_6$
2. *t*-BuLi, TDSOTf
   Et$_2$O

95%, >96% de, ee

1. O$_3$, CH$_2$Cl$_2$ (84%)
2. *c*-Hex$_2$BCl, EtMe$_2$N
   Et$_2$O; RCHO (78–93%)
3. Et$_3$N · 3HF, THF (64–79%)

96 to ≥98% de, ee
R = *n*-Pr, *i*-Pr, H$_2$C=C(Me),
BnOCH$_2$, BnO(CH$_2$)$_2$, Ph, *c*-Hex

(19)

Synthesis of Benzyl 2-Silyl-2-diazoacetates. Bolm et al. synthesized benzyl 2-dimethylthexylsilyl-2-diazoacetate from benzyl 2-diazoacetate with TDSOTf in the presence of ethyldiisopropylamine in Et$_2$O in 86% yield after column chromatography (eq 20).[30] In contrast to other benzyl 2-silyl-2-diazoacetates, an attempt to prepare the corresponding benzyl 2-dimethylthexyl-silyl-2-oxoacetate via rhodium-catalyzed reaction of benzyl 2-dimethylthexylsilyl-2-diazoacetate with propylene oxide failed because of steric hindrance, forming a silacyclobutane via intramolecular C–H insertion reaction.

TDSOTf
Et(*i*-Pr)$_2$N
Et$_2$O, −78 °C
rt, overnight

86%

(20)

Synthesis of Propargylsilanes. A propargylsilane with a bulky group on silicon has been prepared in 34% yield from reaction of the Grignard reagent derived from propargyl bromide with TDSOTf (eq 21).[31] Less reactive TBDMSCl did not react with the Grignard reagent even under reflux conditions. The obtained dimethylpropargylthexylsilane reacted with *N*-alkoxycarbonyliminium ions to give stabilized β-silyl vinylic cations.

$$\text{eq (21)} \qquad 34\%$$

1. Wetter, H.; Oertle, K., *Tetrahedron Lett.* **1985**, *26*, 5515.

2. Wuts, P. G. M.; Greene, T. W. *Greene's Protective Groups in Organic Synthesis*, 4th ed.; John Wiley & Sons, Inc.: Hoboken, NJ, 2007.

3. Nelson, T. D.; Crouch, R. D., *Synthesis* **1996**, 1031.

4. Corey, E. J.; Yi, K. Y., *Tetrahedron Lett.* **1992**, *33*, 2289.

5. Franck-Neumann, M.; Miesch, M.; Gross, L., *Tetrahedron Lett.* **1992**, *33*, 3879.

6. Franck-Neumann, M.; Miesch, M.; Lacroix, E., *Tetrahedron Lett.* **1989**, *30*, 3533.

7. Franck-Neumann, M.; Miesch, M.; Barth, F., *Tetrahedron Lett.* **1989**, *30*, 3537.

8. Franck-Neumann, M.; Miesch, M.; Barth, F., *Tetrahedron* **1993**, *49*, 1409.

9. Allmendinger, T.; Rihs, G.; Wetter, H., *Helv. Chim. Acta* **1988**, *71*, 395.

10. Mattes, H.; Benezra, C., *J. Org. Chem.* **1988**, *53*, 2732.

11. Lalic, G.; Galonic, D.; Matovic, R.; Saičić, R. N., *J. Serb. Chem. Soc.* **2002**, *67*, 221.

12. Bodenmann, B.; Keese, R., *Tetrahedron Lett.* **1993**, *34*, 1467.

13. Furusawa, K. U.S. Pat. 6,175,005 (2001); *Chem. Abstr.* **2000**, *132*, 180826.

14. Koschella, A.; Fenn, D.; Nicolas, I.; Heinze, T., *Macromol. Symp.* **2006**, *244*, 59.

15. Heinze, T., *Macromol. Symp.* **2009**, *280*, 15.

16. Schneider, C.; Rehfeuter, M., *Tetrahedron* **1997**, *53*, 133.

17. Snider, B. B.; Kwon, T., *J. Org. Chem.* **1992**, *57*, 2399.

18. Obrecht, D., *Helv. Chim. Acta* **1991**, *74*, 27.

19. Beifuss, U.; Ledderhose, S., *Synlett* **1997**, 313.

20. Beifuss, U.; Goldenstein, K.; Döring, F.; Lehmann, C.; Noltemeyer, M., *Angew. Chem., Int. Ed.* **2001**, *40*, 568.

21. Schinzer, D.; Kalesse, M., *Tetrahedron Lett.* **1991**, *32*, 4691.

22. Schummer, D.; Jahn, T.; Höfle, G., *Liebigs Ann.* **1995**, 803.

23. Lassaletta, J. M.; Fernández, R.; Martín-Zamora, E.; Díez, E., *J. Am. Chem. Soc.* **1996**, *118*, 7002.

24. Díez, E.; Fernández, R.; Gasch, C.; Lassaletta, J. M.; Llera, J. M.; Martín-Zamora, E.; Vázquez, J., *J. Org. Chem.* **1997**, *62*, 5144.

25. Lassaletta, J. M.; Fernández, R.; Martín-Zamora, E.; Pareja, C., *Tetrahedron Lett.* **1996**, *37*, 5787.

26. (a) Enders, D.; Lohray, B. B., *Angew. Chem., Int. Ed. Engl.* **1987**, *26*, 351. (b) Enders, D.; Lohray, B. B.; Burkamp, F.; Bhushan, V.; Hett, R., *Liebigs Ann.* **1996**, 189.

27. Lohray, B. B.; Enders, D., *Synthesis* **1993**, 1092.

28. Enders, D.; Adam, J.; Klein, D.; Otten, T., *Synlett* **2000**, 1371.

29. Enders, D.; Prokopenko, O. F.; Raabe, G.; Runsink, J., *Synthesis* **1996**, 1095.

30. Bolm, C.; Kasyan, A.; Heider, P.; Saladin, S.; Drauz, K.; Günther, K.; Wagner, C., *Org. Lett.* **2002**, *4*, 2265.

31. Esch, P. M.; Hiemstra, H.; Speckamp, W. N., *Tetrahedron* **1992**, *48*, 3445.

Dieter Enders
RWTH Aachen University, Aachen, Germany

Sven Mangelinckx
Ghent University, Ghent, Belgium

1,3,2-Dioxaborolane, 2-(dimethylphenylsilyl)-4,4,5,5-tetramethyl-, 1,3,2-Dioxaborolane, 4,4,5,5-tetramethyl-2-(methyldiphenylsilyl)-, 1,3,2-Dioxaborolane, 4,4,5,5-tetramethyl-2-(triphenylsilyl)-*

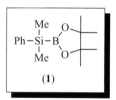

(1)

[185990-03-8]   C$_{14}$H$_{23}$BO$_2$Si   (MW 262.23)

InChI = 1S/C14H23BO2Si/c1-13(2)14(3,4)17-15(16-13)18(5,6)12-10-8-7-9-11-12/h7-11H,1-6H3

InChIKey = ARMSAQNLTKGMGM-UHFFFAOYSA-N

(silaboration and silylation reagent for unsaturated organic compounds; a derivative having diethylamino group on the silicon atom for use as a precursor for silylene generation)

Physical Data: bp 97–99° C/0.1 mmHg.
Solubility: sol common organic solvents.
Analysis of Reagent Purity: ^{1}H and ^{11}B NMR, GC.
Preparative Methods: see below.
Purification: distillation.
Handling, Storage, and Precautions: the reagent can be handled in air, but should be stored under an inert atmosphere.

Preparation and Derivatization of Silylboronic Esters. 2-(Dimethylphenylsilyl)-4,4,5,5-tetramethyl-1,3,2-dioxaborolane (**1**) is prepared by reaction of dimethylphenylsilyllithium with 4,4,5,5-tetramethyl-1,3,2-dioxaborolane (pinacolborane, 2 equiv) or 2-isopropoxy-4,4,5,5-tetramethyl-1,3,2-dioxaborolane (2 equiv) (eq 1).[2] Silylboronic esters **2** and **3a** are also synthesized via similar routes using methyldiphenylsilyllithium and [(diethylamino)diphenylsilyl]lithium, respectively (eq 2). Diethylamino-substituted silylborane **3a** is converted to a chlorine-substituted **4a** by reaction with dry hydrogen chloride (eq 3). Silylborane **1** is

*Article previously published in the electronic Encyclopedia of Reagents for Organic Synthesis as 2-(Dimethylphenylsilyl)-4,4,5,5-tetramethyl-1,3,2-dioxaborolane. DOI: 10.1002/047084289X. rn01328.

converted to 2-(chlorodimethylsilyl)-4,4,5,5-tetramethyl-1,3,2-dioxaborolane (**4b**) in 78% yield by treatment with dry hydrogen chloride in the presence of a catalytic amount (10 mol%) of aluminum chloride (eq 4).[3] The chloro group in **4b** is subjected to substitution by nucleophiles to give silylboronic esters bearing alkoxy, fluoro, and dialkylamino groups (**3b**).

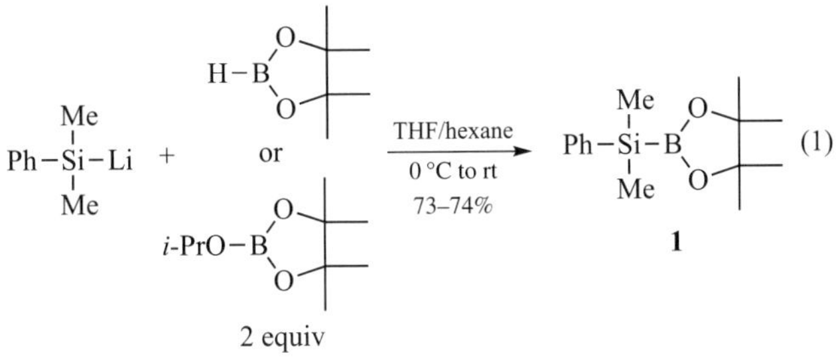

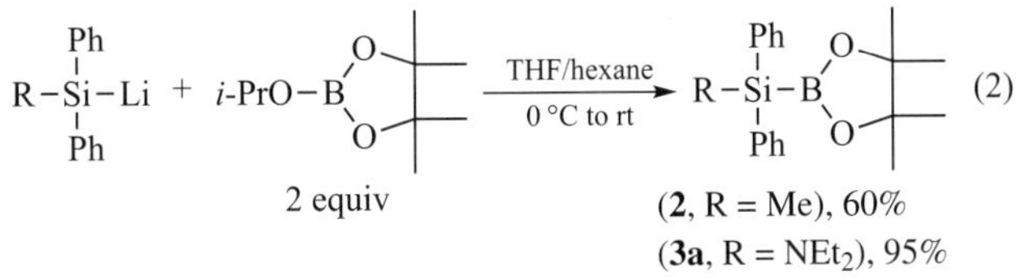

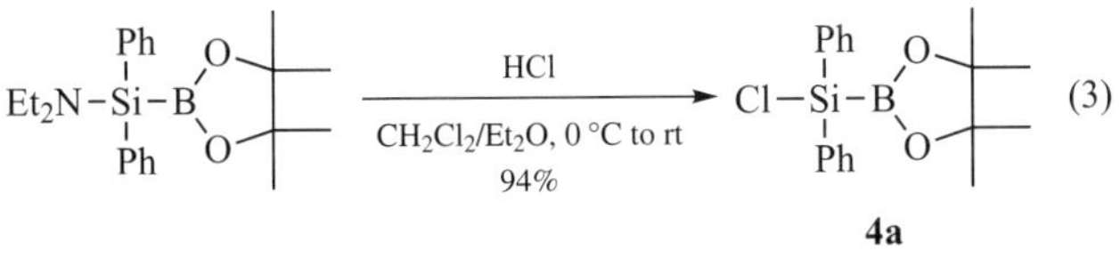

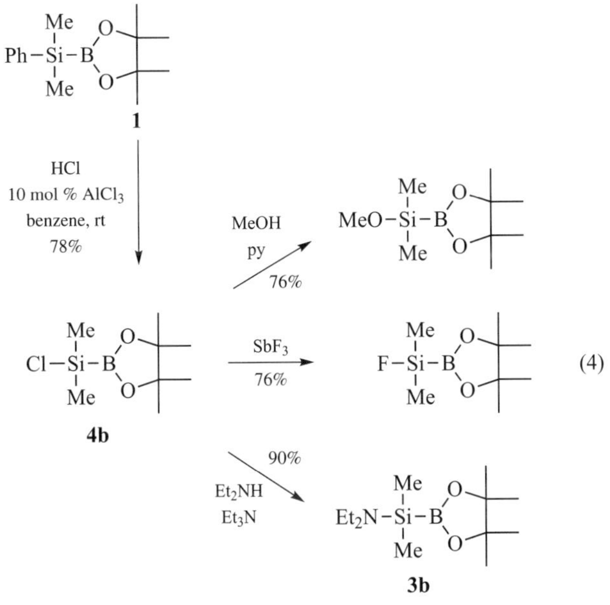

$$R{\equiv\!\!\!=} \xrightarrow[\text{toluene, 110 °C} \atop \text{up to 94%}]{\text{Me}_2\text{PhSi–B(pin) (1)} \atop \text{cat Pd(OAc)}_2/t\text{-BuCH}_2\text{CMe}_2\text{NC}} \quad \underset{\textbf{5}}{\overset{\text{Me}_2\text{PhSi}}{R}{\diagdown}\text{B(pin)}} \quad (5)$$

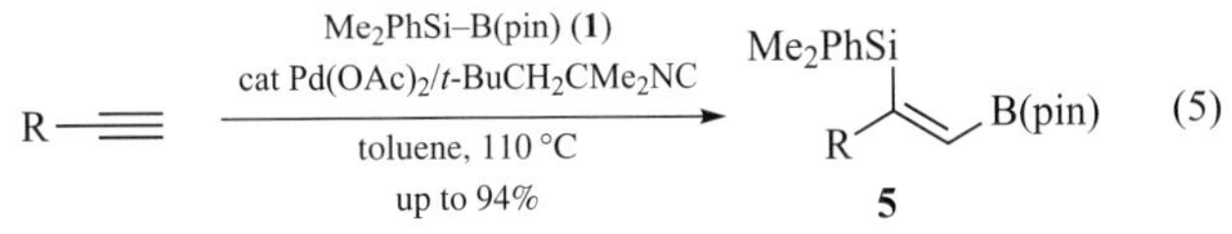

The stereoselectivity in silaboration of terminal alkynes is switchable, when chlorine-substituted **4b** is used in the reaction with subsequent treatment with isopropyl alcohol in the presence of pyridine (eq 6).[5] The normal Z-isomer is formed when the silaboration is carried out using a slight excess of alkyne (**4b**:alkyne = 1.0:1.2), whereas the silaboration using an excess of **4b** (**4b**:alkyne = 1.2:1.0) results in stereoselective formation of the E-isomer. This switch of stereoselectivity is ascribed to Z-to-E isomerization of initially formed Z isomer in the workup stage, in which a palladium catalyst for E–Z isomerization catalyst is generated from i-PrOH and **4b** used in excess.

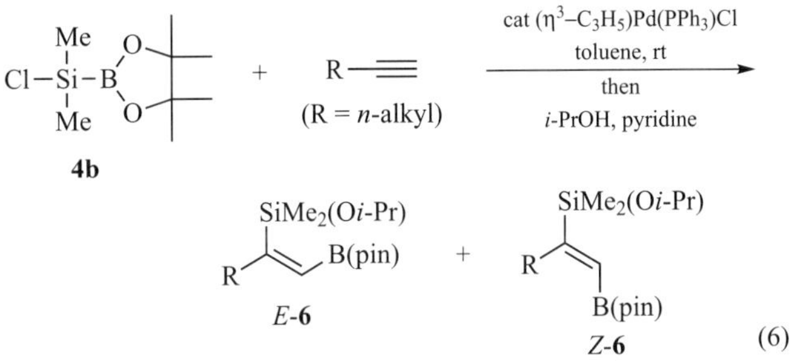

(1)

(2)

(**2**, R = Me), 60%
(**3a**, R = NEt₂), 95%

(3)

4a

(4)

3b

(6)

excess of alkyne	87–93% (E:Z = >99:1)
excess of **4b**	81–92% (E:Z = 11:89–7:93)

The regioselectivity in silaboration of terminal alkynes is switched by the phosphine ligand on palladium catalysts (eq 7).[6] In contrast to the normal regioselectivity (B-to-terminal) with PPh₃, selective formation of the inverse (Si-to-terminal) regioisomer **7**, in which the silyl groups are attached to the terminal positions, takes place when the reaction is carried out with P(t-Bu)₂(biphenyl-2-yl) as a ligand.

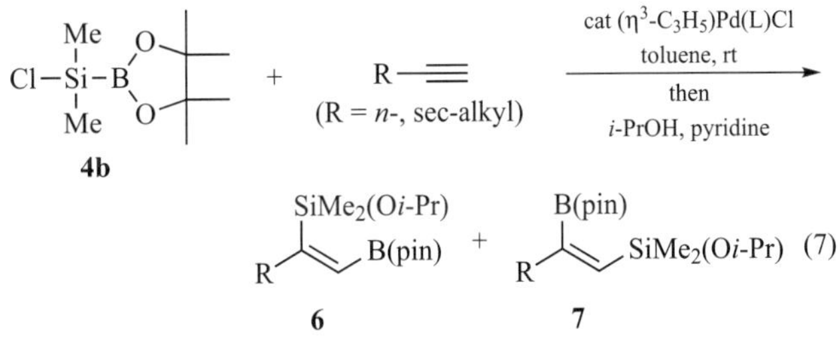

(7)

L = PPh₃	87–93% (**6**:**7** = >99:1)
L = P(t-Bu)₂(biphenyl-2-yl)	62–99% (**6**:**7** = 5:95–2:98)

Silaboration of Unsaturated Organic Compounds. In the presence of palladium or platinum catalysts, **1** reacts with alkynes to give cis-1,2-adducts.[4] Palladium complexes bearing isocyanide or phosphorus ligands such as t-BuCH₂CMe₂NC, P(OEt)₃, and PPh₃ are effective for regioselective silaboration of terminal alkynes, which affords (Z)-1-boryl-2-silyl-1-alkenes **5** (eq 5). Alkynes having various functional groups such as Cl, CN, OTHP, and OH are applicable to this silaboration.

Silaboration of terminal alkenes with **1** takes place under reflux in dioxane in the presence of platinum catalysts such as Pt(PPh₃)₄ and Pt(CH₂=CH₂)(PPh₃)₂ (eq 8).[7] 1,2-Silaboration products in which the silyl group is attached to the terminal position are formed as major products. In the reaction of aliphatic alkenes, 1-boryl-1-silylalkanes **9** are also formed as minor isomers.

$$R = n\text{-}C_6H_{13} \qquad 53\% \ (\mathbf{8}:\mathbf{9} = 87:13)$$
$$R = 4\text{-}MeOC_6H_4 \qquad 53\% \ (\mathbf{8}:\mathbf{9} = >99:1)$$

Intramolecular silaboration of silylboronic esters **10**, which are prepared by reaction of **4a** with homoallylic alcohols (eq 9),[8] gives cyclic silyl ethers with high regioselectivity (eq 10).[8] A platinum catalyst is effective for this silaboration. Stereochemical course of the cyclization in the reaction of secondary homoallylic substrates is highly dependent on the phosphorus ligand on the platinum catalyst. *trans* cyclic products (*trans*-**11**) are selectively formed when PCyPh$_2$ is used as ligand, whereas a catalyst bearing P[O(2,4-*t*-Bu$_2$C$_6$H$_3$)]$_3$ gives *cis* products. In contrast to the intramolecular silaboration of **10**, intramolecular exchange of the boryl group and a hydrogen atom on the double bond takes place in the reaction of **12**, which has a geminally disubstituted C=C bond, leading to stereoselective formation of β,β-disubstituted alkenylboronic ester **13** (eq 11).[9]

Silylboronic ester **1** reacts with allenes in the presence of a palladium catalyst bearing isocyanide or phosphorus ligands such as 4-ethyl-2,6,7-trioxa-1-phosphabicyclo[2.2.2]octane (etpo) and

PPh$_3$ (eq 12).[10,11] The reaction proceeds with selective introduction of the boryl group to the central carbon atom of the allene to give β-borylallylsilanes. In the palladium isocyanide complex-catalyzed silaboration of terminal allenes, regioselectivity of the silaboration depends on the electronic nature of the substituents of the allenes. Silaboration takes place at the internal C=C bond of allenes having electron-donating substituents such as alkyl, 4-methoxyphenyl, and methoxy groups, giving 2-boryl-3-silyl-1-alkenes **14**. In contrast, **1** preferably adds to terminal C=C bond to afford 2-boryl-1-silyl-2-alkenes **15** in the silaboration of electron-deficient allenes such as those having 4-trifluorophenyl and perfluoroalkyl groups. When the reaction is carried out in the presence of a catalytic amount of organic iodide with Pd(dba)$_2$, introduction of a silyl group to the central carbon atom becomes the major reaction pathway to give β-silylallylborane **16** (eq 13).[12]

$$R = PhCH_2CH_2 \qquad 99\% \ (\mathbf{14}:\mathbf{15} = 100:0)$$
$$R = n\text{-}C_6F_{13} \qquad 94\% \ (\mathbf{14}:\mathbf{15} = 0:100)$$

Enantioselective silaboration of terminal allenes is achieved using **2** and a palladium catalyst bearing chiral monodentate phosphine ligand (*R*)-**17**, giving the corresponding β-borylallylsilanes with up to 93% ee (eq 14).[13] Use of enantiopure silylboranes bearing pinanediol instead of pinacol in **1** results in higher enantiofacial selectivities up to 96% de.[14]

Silylboronic ester **1** adds to 1,3-dienes in the presence of platinum and nickel catalysts to give 1,4-silaboration products. A nickel(0) catalyst generated in situ from reaction of $Ni(acac)_2$ with DIBAL is effective for the silaboration of butadiene, isoprene, and 2,3-dimethylbutadiene to give (Z)-1-boryl-4-silyl-2-butenes in high yields with complete Z selectivity (eq 15).[15] Silaboration of cyclic 1,3-dienes such as 1,3-cyclohexadiene and 1,3-cycloheptadiene is also catalyzed by nickel catalyst bearing $PCyPh_2$ as a ligand, giving *cis*-1,4-adducts with high stereoselectivities (eq 16).[15] Silaboration of 1,3-cyclohexadiene with **1** in the presence of a platinum catalyst bearing chiral phosphoramidite (S)-**18** gives enantioenriched silaboration products **19** with 82% ee (eq 17).[16] Nickel-catalyzed reaction of **1** with acyclic 1,4-disubstituted 1,3-butadienes gives dienylboranes and allylsilanes via dehydrogenative borylation along with 1,4-hydrosilylation (eq 18).[17] Platinum-catalyzed 1,4-silaboration of conjugated enynes having bulky substituents at the alkynyl carbons gives allenylmethylborane derivatives (eq 19).[18]

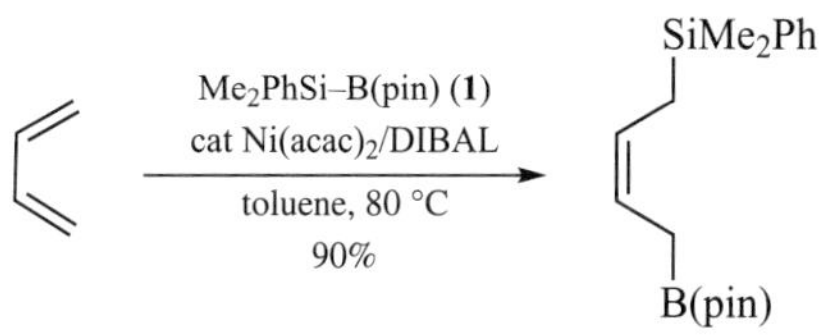

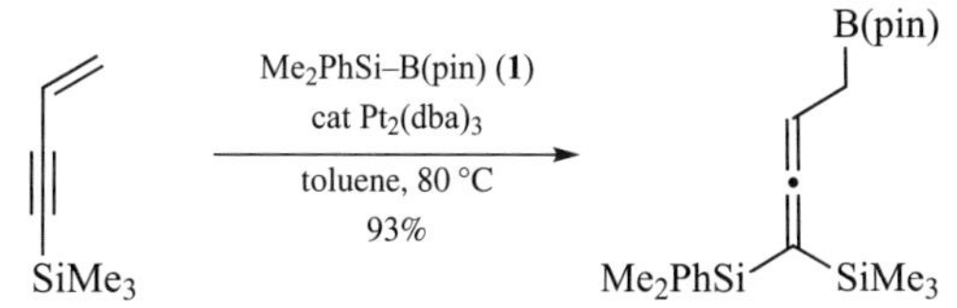

$$(19)$$

Silaborative C–C Bond Forming Reactions. In the presence of a nickel(0) catalyst generated in situ by the reaction of $Ni(acac)_2$ with DIBAH, **1** reacts with 2 equiv of alkynes to give 1-boryl-4-silyl-1,3-butadienes (eq 20).[19]

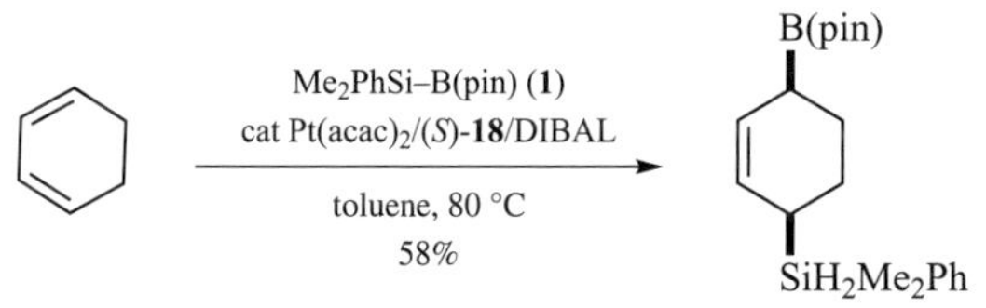

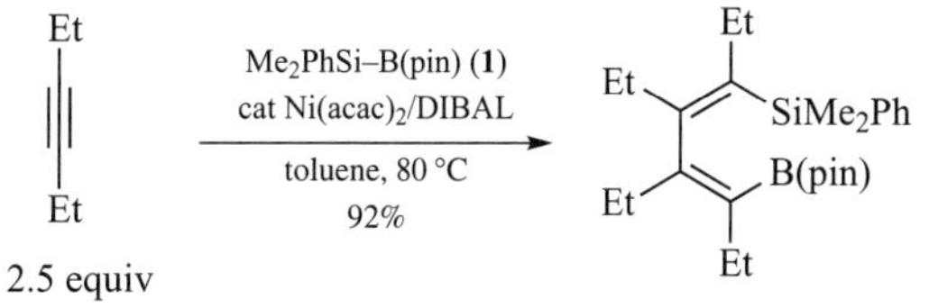

$$(20)$$

Silaborative cyclization of 1,6-enyne using **1** takes place in the presence of a palladium catalyst bearing *N*-heterocyclic carbene as a ligand (eq 21).[20] The cyclization proceeds with regioselective introduction of the boryl group into the terminal carbon atom of the C–C triple bond. A similar enyne cyclization has also been achieved with Pd/P(*t*-Bu)₂(biphenyl-2-yl) catalyst.[6]

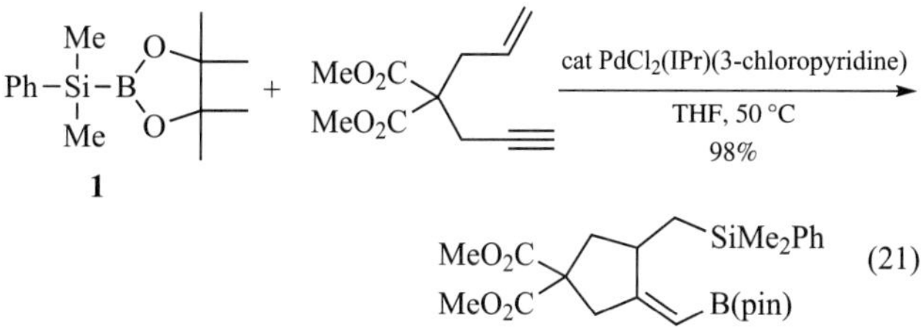

$$(21)$$

Three-component coupling of **1** with 1,3-dienes and aldehydes proceeds in the presence of a platinum catalyst, giving boryl-substituted homoallylic alcohol derivatives in good yields with high diastereoselectivities (eq 22).[21] This reaction is not considered to be a stepwise silaboration/allylation sequence, but catalytic allylplatination of the aldehyde.

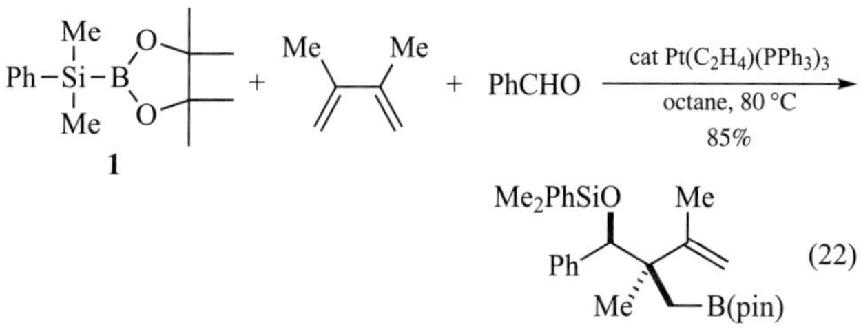

$$(22)$$

dr 96:4

$$(15)$$

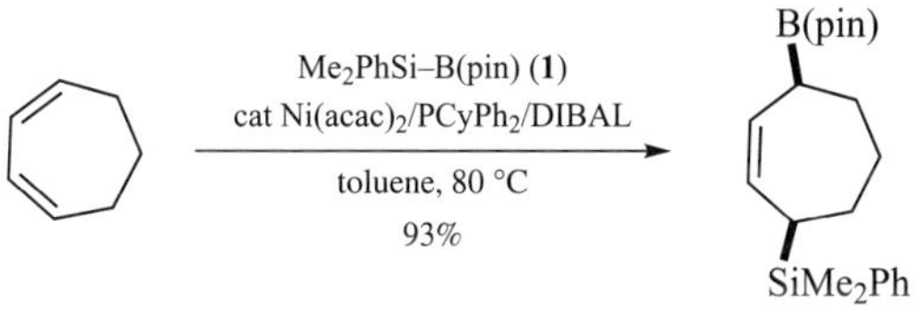

$$(16)$$

cis:trans = >99:1

$$(17)$$

19
82% ee

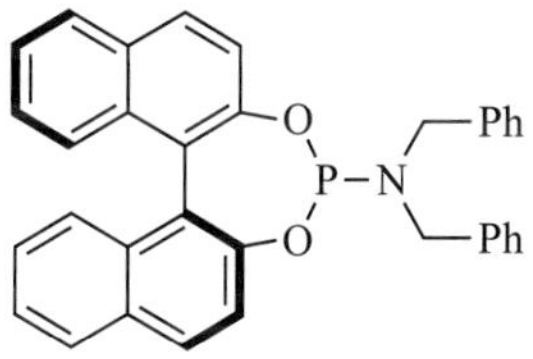

S-**18**

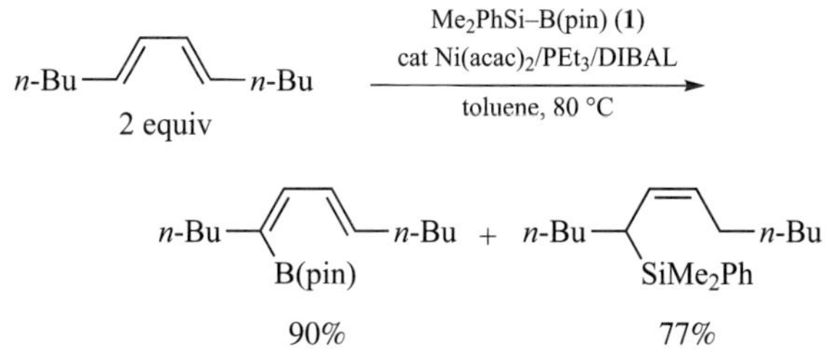

$$(18)$$

90%   77%

Silaborative C–C Bond Cleaving Reactions. Silaboration of methylenecyclopropanes with **1** proceeds in the presence of palladium and platinum catalysts via cleavage of the C–C bond in the cyclopropane ring.[22] E- and Z-alkenylboron products are selectively formed with $Pd(OAc)_2$/t-$BuCH_2CMe_2NC$ (E:Z = 83:17) and $Pt(C_2H_4)(PPh_3)_2$ (E:Z = 17:83), respectively, via cleavage of proximal C–C bond in reactions of benzylidenecyclopropane with **1** (eq 23). In the reaction of cyclohexylidenecyclopropane, proximal C–C bond cleavage takes place selectively with a platinum catalyst to afford alkenylborane **20**, whereas a palladium catalyst promotes distal C–C bond cleavage, resulting in selective formation of allylborane **19** (eq 24). Selective 1,3-introduction of boryl and silyl groups is observed in the reaction of cyclohexane-fused methylenecyclopropanes **21** in the presence of a palladium catalyst. In sharp contrast, use of $Pt(C_2H_4)(PPh_3)_2$ catalyst results in 1,4-introduction of the two groups to give **23** selectively (eq 25).

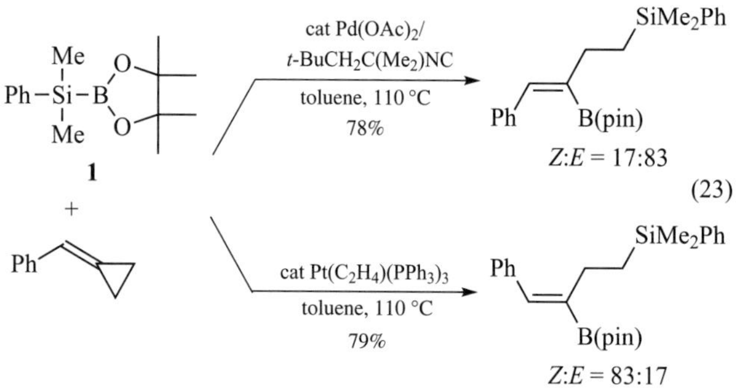

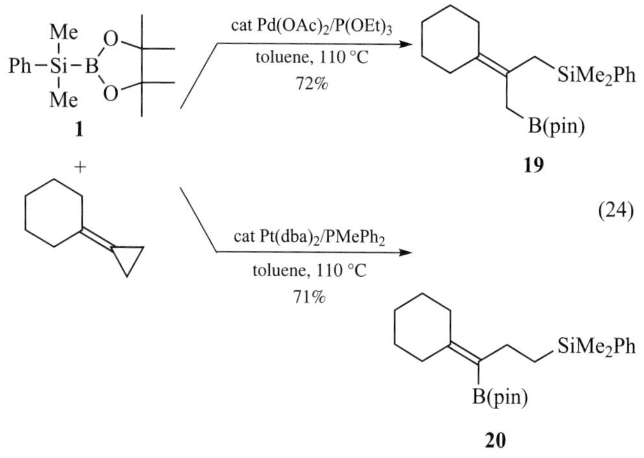

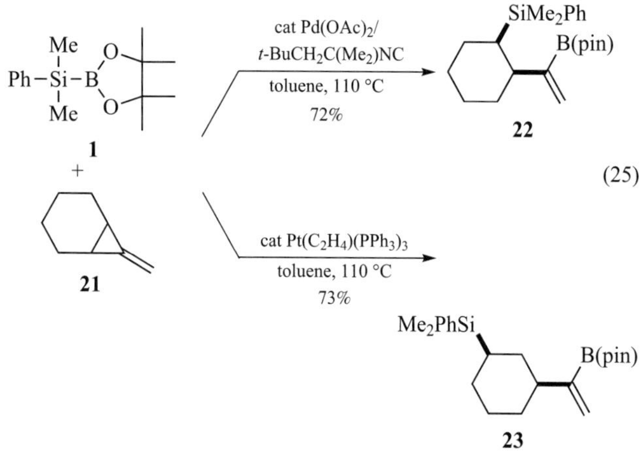

(23)

(24)

(25)

Enantioselective silaborative C–C cleavage of *meso*-methylenecyclopropanes **24** with **2** is catalyzed by a palladium catalyst bearing a chiral monodentate phosphine ligand (*R*)-**17** (eq 26).[23] The reaction proceeds via selective cleavage of one of the two enantiotopic C–C bonds to give alkenylboronic ester **25** with 91% ee.

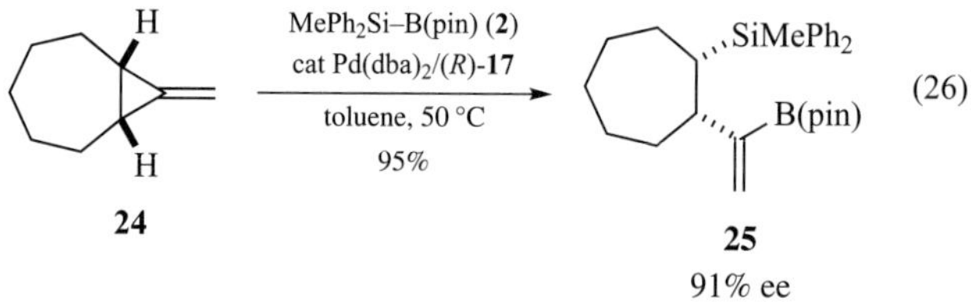

(26)

Kinetic resolution of racemic 1-alkyl-2-methylenecyclopropanes **26** is achieved by silaborative C–C cleavage using **2** with a palladium catalyst bearing chiral phosphoramidite (*S*,*S*,*S*)-**27** (eq 27).[24] The resolution can be classified into parallel kinetic resolution, in which the slower reacting enantiomer is converted into a constitutional isomer of the major enantioenriched products.

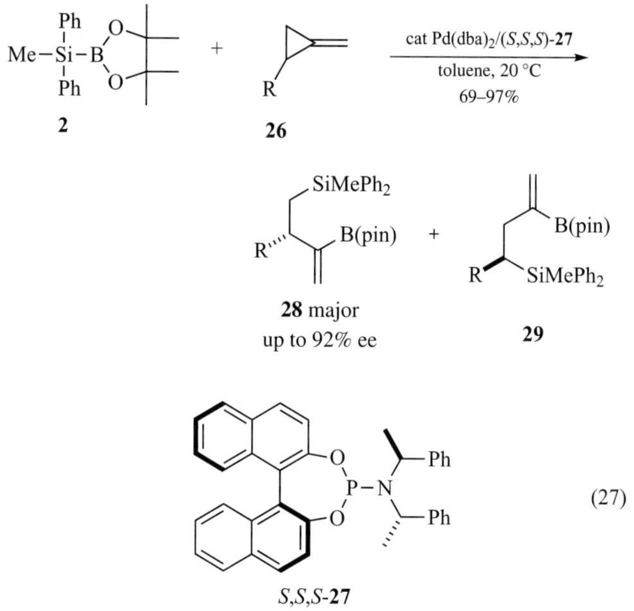

(27)

Nickel-catalyzed reaction of **1** with vinylcyclopropanes proceeds via cleavage of the cyclopropane ring to give formal 1,5-silaboration products **30** (eq 28).[25] Similar silaborative C–C cleavage takes place in the reaction of **1** with vinylcyclobutanes, affording allylic silanes **31** with regioselective introduction of the boron atom via cleavage of the proximal C–C bond in the cyclobutane ring (eq 29).

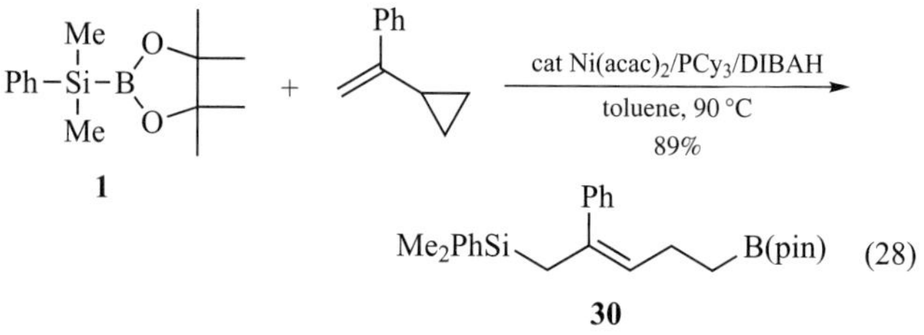

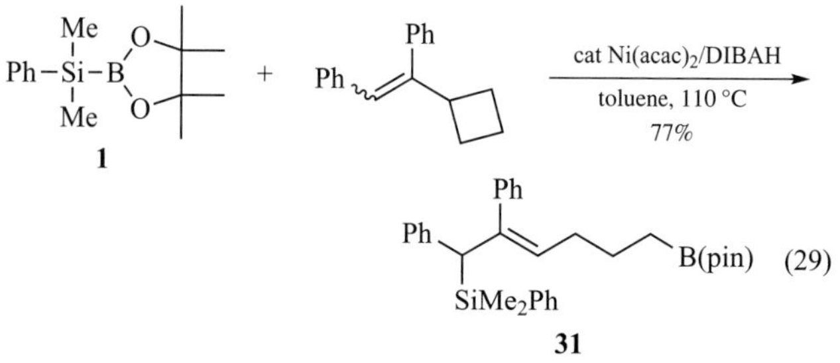

(28)

(29)

Silyl Anion Equivalent. Silylboronic ester **1** reacts as a silyl anion equivalent in the presence of transition metal catalysts. Cyclic and acyclic α,β-unsaturated carbonyl compounds serve as good acceptors of the silyl groups in conjugate addition of **1** catalyzed by rhodium and copper complexes, giving β-silylcarbonyl compounds (eq 30).[26] The silylation takes place with high enantioselectivity when Rh/(S)-BINAP or Cu/chiral NHC catalysts are used. Three-component coupling of **1**, α,β-unsaturated carbonyl compounds, and aldehydes affords β-hydroxyketone stereoselectively in the presence of a copper catalyst (eq 31).[27] The copper enolate **32** is presumed as an intermediate of the reaction.

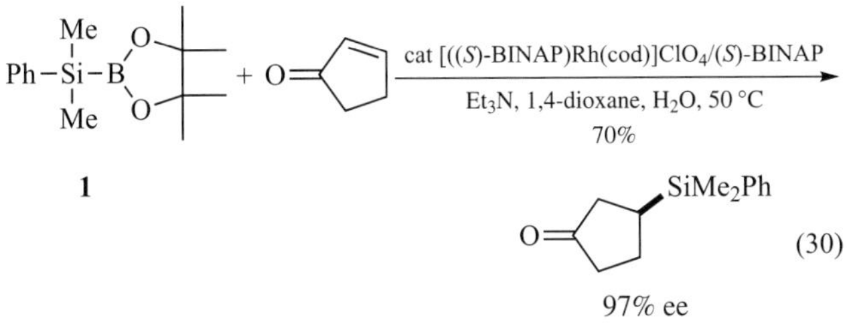

$$ (30) $$

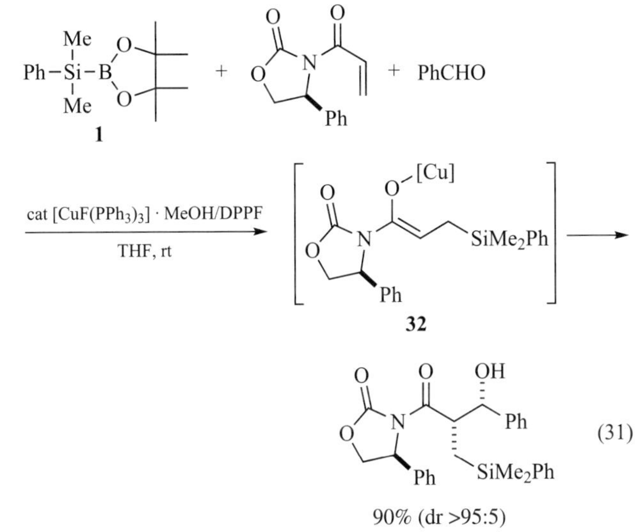

97% ee

$$ (31) $$

90% (dr >95:5)

The silylboronic ester **1** reacts with propargylic carbonate **33** in the presence of a rhodium catalyst to give allenylsilane **34** via S_N2' reaction (eq 32).[28] A high level of chirality transfer is attained when the enantioenriched substrate is used. Allylic chlorides also undergo S_N2' reaction with **1** using a copper catalyst, giving branched allylic silane (eq 33).[29]

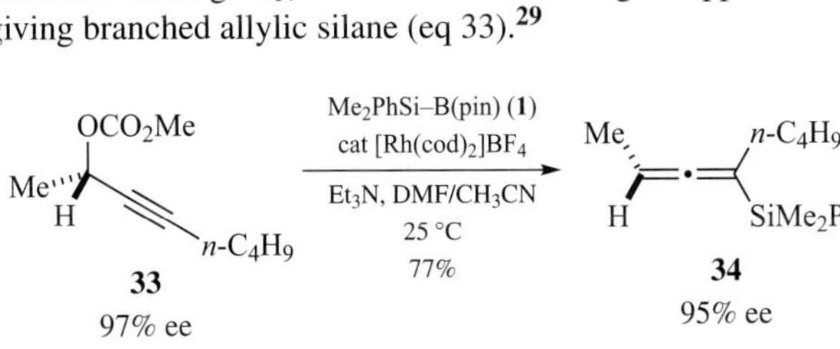

$$ (32) $$

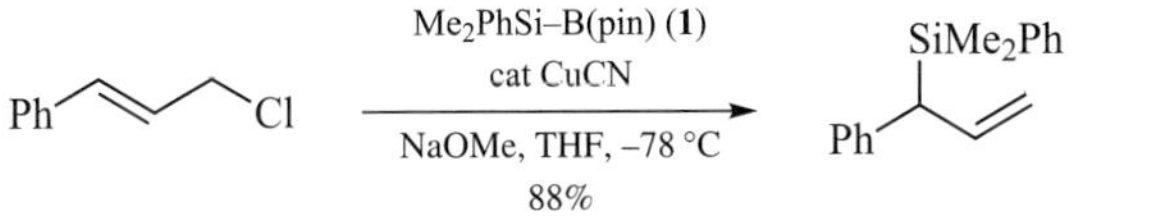

$$ (33) $$

Silylative cyclization of isocyanide **35** takes place with **1** in the presence of a copper catalyst, giving 2-silylindole **37** (eq 34).[30]

It is presumed that the reaction proceeds via formation of imidoyl-copper intermediate **36**.

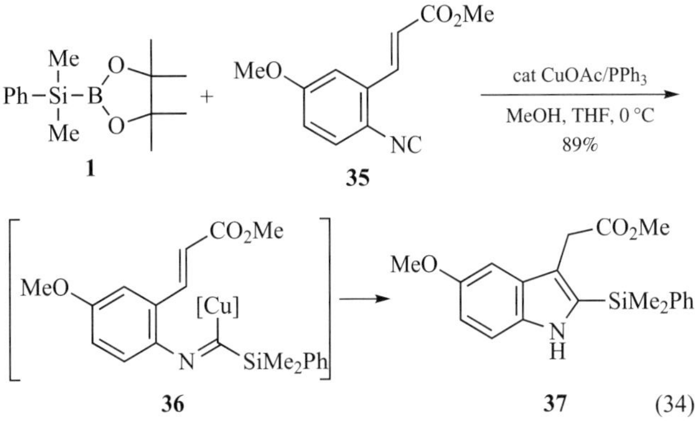

$$ (34) $$

Silylene Equivalent. Silylboronic esters **3** bearing a dialkyl-amino group on the silicon atoms react as silylene equivalents in the palladium-catalyzed reaction with unsaturated hydrocarbons. The reaction of **3b** with terminal alkynes (2 equiv) proceeds smoothly in the presence of Pd/PPh₃ catalyst to give 2,4-disubstituted siloles **38** with formation of aminoborane **39** (eq 35).[31] The diethylamino group on the silicon atom is critically important for the reaction, while no silole formation is found in reactions using other silylboronic esters such as **1** and **3**, which carry aryl and chloro groups on the silicon atoms. A wide variety of 1,3-dienes are also reactive with **3** as silylene equivalent (eq 36).[32] Silacyclopent-3-enes **40** are formed in high yields via stereospecific silylene-1,3-diene [4 + 1]-cycloaddition in the presence of Pd/PMePh₂ catalyst.

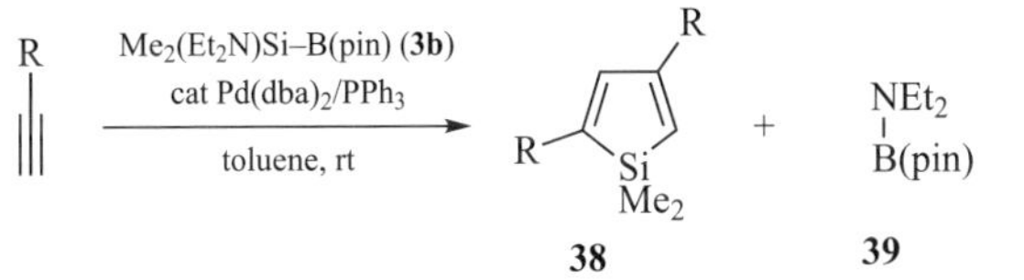

$$ (35) $$

R = Ph	92%
R = 4-methoxyphenyl	96%
R = 1-naphthyl	75%

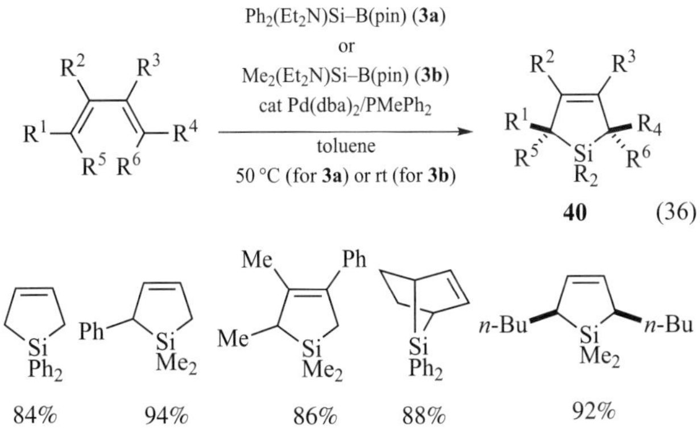

$$ (36) $$

84%   94%   86%   88%   92%

Reaction with Carbenoids and Isocyanides. Silylboronic ester **1** undergoes nucleophilic attack of carbenes and carbenoids, which is followed by subsequent 1,2-migration of the silyl group to give α-silylated organoboron compounds.[33] Reaction of **1** with various lithium carbenoids such as 1-chloro-1-lithio-1-alkenes (eq 37), 1-chloro-1-lithio-2-alkenes (eq 38), and 3-chloro-1-lithio-1-alkynes (eq 39) affords the corresponding α-silylated organoboron compounds in good yields.[34]

$$\text{(37)}$$

Me$_2$PhSi–B(pin) **(1)**
THF/Et$_2$O
$-110\,°C$ to rt
84%

$$\text{(38)}$$

Me$_2$PhSi–B(pin) **(1)**
THF/Et$_2$O
$-98\,°C$ to rt
82%

$$\text{(39)}$$

Me$_2$PhSi–B(pin) **(1)**
THF/Et$_2$O
$-110\,°C$ to rt
70%

Isocyanides, which have carbene-like divalent carbon atoms, also insert into Si–B bond of **1** in the absence of catalysts to give (boryl)(silyl)iminomethanes, which are isolable after conversion to borane-imine complex (eq 40).[35]

$$t\text{-Bu−NC} \qquad \text{(40)}$$

1. Me$_2$PhSi–B(pin) **(1)**
   neat, 80 °C
2. BH$_3$–THF, 0 °C
   80%

1. (a) Suginome, M.; Matsuda, T.; Ohmura, T.; Seki, A.; Murakami, M. In *Comprehensive Organometallic Chemistry III*; Crabtree, R.; Mingos, M., Eds.; Ojima, I., Vol. Ed.; Elsevier, 2007; Vol. 10, p 725. Beletskaya, I.; Moberg, C., *Chem. Rev.* **2006**, *106*, 2320. Burks, H. E.; Morken, J. P., *Chem. Commun.* **2007**, 4717. Ohmura, T.; Suginome, M., *Bull. Chem. Soc. Jpn.* **2009**, *82*, 29.

2. Suginome, M.; Matsuda, T.; Ito, Y., *Organometallics* **2000**, *19*, 4647.

3. Ohmura, T.; Masuda, K.; Furukawa, H.; Suginome, M., *Organometallics* **2007**, *26*, 1291.

4. (a) Suginome, M.; Nakamura, H.; Ito, Y., *J. Chem. Soc., Chem. Commun.* **1996**, 2777. Suginome, M.; Matsuda, T.; Nakamura, H.; Ito, Y., *Tetrahedron* **1999**, *55*, 8787.

5. Ohmura, T.; Oshima, K.; Suginome, M., *Chem. Commun.* **2008**, 1416.

6. Ohmura, T.; Oshima, K.; Taniguchi, H.; Suginome, M., *J. Am. Chem. Soc.* **2010**, *132*, 12194.

7. Suginome, M.; Nakamura, H.; Ito, Y., *Angew. Chem., Int. Ed. Engl.* **1997**, *36*, 2516.

8. Ohmura, T.; Fururkawa, H.; Suginome, M., *J. Am. Chem. Soc.* **2006**, *128*, 13366.

9. Ohmura, T.; Takasaki, Y.; Furukawa, H.; Suginome, M., *Angew. Chem., Int. Ed.* **2006**, *48*, 2372.

10. (a) Suginome, M.; Ohmori, Y.; Ito, Y., *Synlett* **1999**, 1567. Suginome, M.; Ohmori, Y.; Ito, Y., *J. Organomet. Chem.* **2000**, *611*, 403.

11. Onozawa, S.; Hatanaka, Y.; Tanaka, M., *Chem. Commun.* **1999**, 1863.

12. Chang, K.-J.; Rayabarapu, D.-K.; Yang, F.-Y.; Cheng, C.-H., *J. Am. Chem. Soc.* **2004**, *127*, 126.

13. Ohmura, T.; Taniguchi, H.; Suginome, M., *J. Am. Chem. Soc.* **2006**, *128*, 13682.

14. (a) Suginome, M.; Ohmura, T.; Miyake, Y.; Mitani, S.; Ito, Y.; Murakami, M., *J. Am. Chem. Soc.* **2003**, *125*, 11174. (b) Ohmura, T.; Suginome, M., *Org. Lett.* **2006**, *8*, 2503.

15. Suginome, M.; Matsuda, T.; Yoshimoto, T.; Ito, Y., *Org. Lett.* **1999**, *1*, 1567.

16. (a) Gerdin, M.; Moberg, C., *Adv. Synth. Catal.* **2005**, *347*, 749. (b) Gerdin, M.; Penhoat, M.; Zalubovskis, R.; Pétermann, C.; Moberg, C., *J. Organomet. Chem.* **2008**, *693*, 3519.

17. Gerdin, M.; Moberg, C., *Org. Lett.* **2006**, *8*, 2929.

18. Lüken, C.; Moberg, C., *Org. Lett.* **2008**, *10*, 2505.

19. Suginome, M.; Matsuda, T.; Ito, Y., *Organometallics* **1998**, *17*, 5233.

20. Gerdin, M.; Nadakudity, S. K.; Worch, C.; Moberg, C., *Adv. Synth. Catal.* **2010**, *352*, 2559.

21. Suginome, M.; Nakamura, H.; Matsuda, T.; Ito, Y., *J. Am. Chem. Soc.* **1998**, *120*, 4248.

22. Suginome, M.; Matsuda, T.; Ito, Y., *J. Am. Chem. Soc.* **2000**, *122*, 11015.

23. Ohmura, T.; Taniguchi, H.; Kondo, Y.; Suginome, M., *J. Am. Chem. Soc.* **2007**, *129*, 3518.

24. Ohmura, T.; Taniguchi, H.; Suginome, M., *Org. Lett.* **2009**, *11*, 2880.

25. Suginome, M.; Matsuda, T.; Yoshimoto, T.; Ito, Y., *Organometallics* **2002**, *21*, 1537.

26. (a) Walter, C.; Auer, G.; Oestreich, M., *Angew. Chem., Int. Ed.* **2006**, *45*, 5675. Walter, C.; Oestreich, M., *Angew. Chem., Int. Ed.* **2008**, *47*, 3818. (c) Lee, K.-S.; Hoveyda, A. H., *J. Am. Chem. Soc.* **2010**, *132*, 2898.

27. Welle, A.; Petrignet, J.; Tinant, B.; Wouters, J.; Riant, O., *Chem. Eur. J.* **2010**, *16*, 10980.

28. Ohmiya, H.; Ito, H.; Sawamura, M., *Org. Lett.* **2009**, *11*, 5618.

29. Vyas, D.; Oestreich, M., *Angew. Chem., Int. Ed.* **2010**, *49*, 8513.

30. Tobisu, M.; Fujihara, H.; Koh, K.; Chatani, N., *J. Org. Chem.* **2010**, *75*, 4841.

31. Ohmura, T.; Masuda, K.; Suginome, M., *J. Am. Chem. Soc.* **2008**, *130*, 1526.

32. Ohmura, T.; Masuda, K.; Takase, I.; Suginome, M., *J. Am. Chem. Soc.* **2009**, *131*, 16624.

33. Buynak, J. D.; Geng, B., *Organometallics* **1995**, *14*, 3112.

34. (a) Shimizu, M.; Kitagawa, H.; Kurahashi, T.; Hiyama, T., *Angew. Chem., Int. Ed.* **2001**, *40*, 4283. Kurahashi, T.; Hata, T.; Masai, H.; Kitagawa, H.; Shimizu, M.; Hiyama, T., *Tetrahedron* **2002**, *58*, 6381. Shimizu, M.; Kurahashi, T.; Kitagawa, H.; Hiyama, T., *Org. Lett.* **2003**, *5*, 225.

35. Suginome, M.; Fukuda, T.; Nakamura, H.; Ito, Y., *Organometallics* **2000**, *5*, 719.

Toshimichi Ohmura & Michinori Suginome
Kyoto University, Kyoto, Japan

Diphenyldifluorosilane

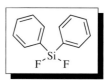

[312-40-3]   C$_{12}$H$_{10}$F$_2$Si   (MW 220.30)

InChI = 1S/C12H10F2Si/c13-15(14,11-7-3-1-4-8-11)12-9-5-2-6-10-12/h1-10H

InChIKey = BOMPXIHODLVNMC-UHFFFAOYSA-N

(a moisture insensitive equivalent of diphenyldichlorosilane that is more reactive toward nucleophiles. Source of phenyl group in the Hiyama coupling reaction. More stable to electrophilic aromatic substitution than diphenyldichlorosilane.)

Physical Data: bp 246–247 °C, bp 156 °C 50 mmHg, bp 66–70 °C 0.5 mmHg; d 1.155 g cm^{-1}.

Solubility: soluble in ether.

Form Supplied in: colorless liquid.

Preparative Method: commercially available. Readily prepared by oxidation of diphenylsilane with CuF$_2$,[1] or by treatment of diphenyldichlorosilane with HF,[2] ZnF$_2$,[3,4] NaBF$_4$,[5,6] or (NH$_4$)$_2$SiF$_6$.[7] Rate enhancements of the chlorine–fluorine exchange have been achieved with ultrasound and addition of water.[8] Diethoxydiphenylsilane and related structures are converted to diphenyldifluorosilane with 48% HF(aq).[9,10]

Purification: distillation.

Handling, Storage, and Precautions: this substance will etch glass at elevated temperatures. Although it is stable to neutral and acidic aqueous conditions, it is a possible source of hydrofluoric acid (see safety precautions).

Nucleophilic Substitution. Eaborn reported that fluorosilanes are more reactive toward organometallic reagents than the corresponding chlorosilanes, and therefore more suited for preparation of sterically hindered organosilanes.[11] Two examples of this enhanced reactivity are shown for the Tsi anion (eq 1)[12] and a dianion (eq 2).[13] The use of less-hindered reagents in stoichiometric amounts still allow for the sequential introduction of nucleophiles (eq 3).[14]

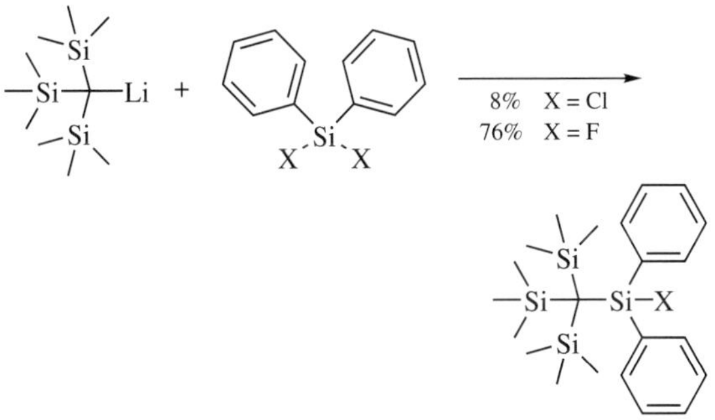

$$(1)$$

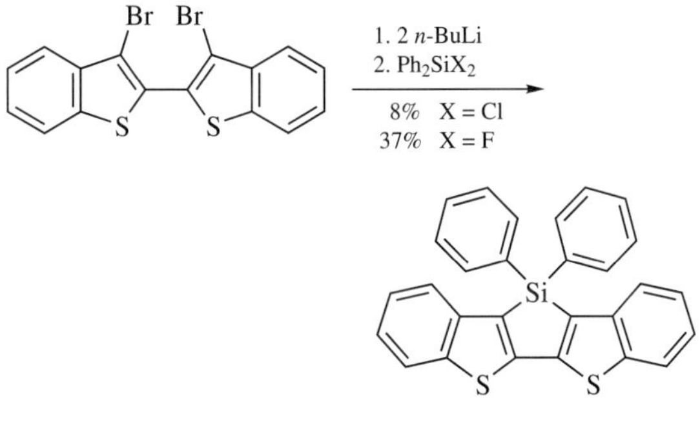

$$(2)$$

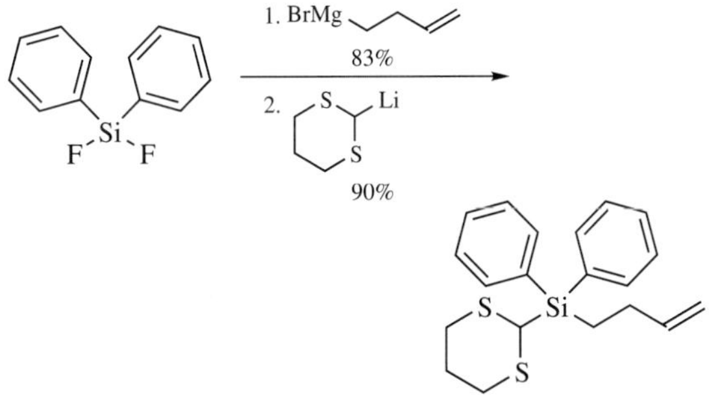

$$(3)$$

Hiyama Coupling. Diphenyldifluorosilane is an effective phenyl transfer reagent for the palladium-catalyzed coupling with aryl iodides (eq 4).[15] It can also be used to prepare biphenyl in high yield when ethyl dibromosenecioate is used as an oxidant (eq 5).[16]

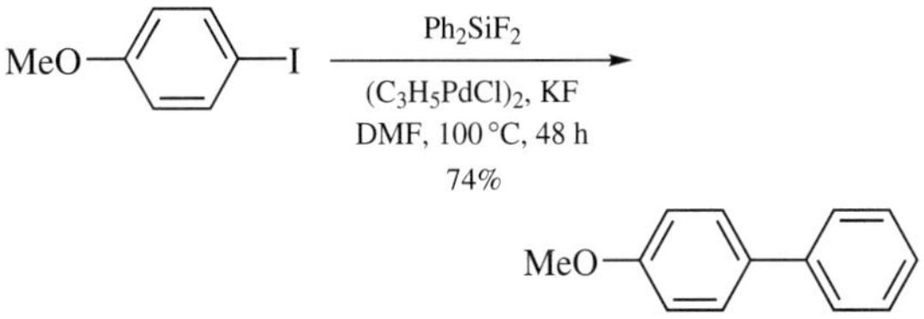

$$(4)$$

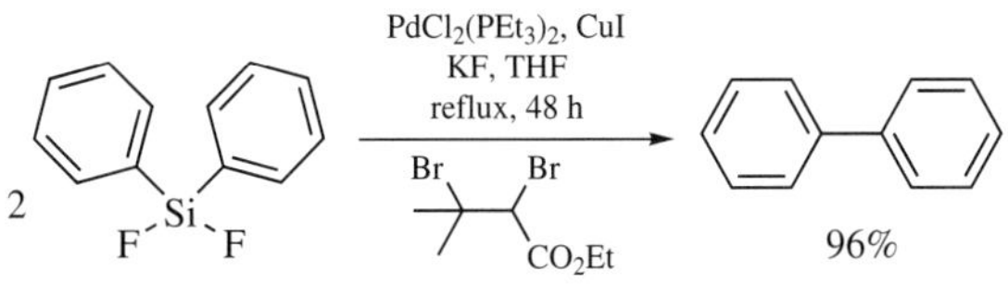

$$(5)$$

Electrophilic Aromatic Substitution. The fluorine substitution of the silicon suppresses the normal ipso attack of electrophiles of arylsilanes, allowing nitration of the ring (eq 6).[17] When the corresponding dichlorosilane is used, significant decomposition is observed under similar conditions.

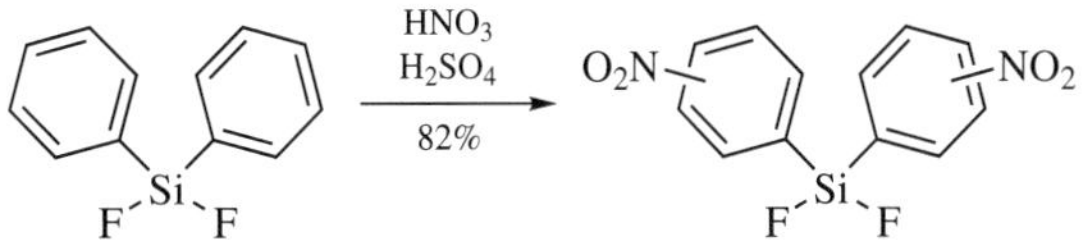

$$(6)$$

---

1. Yoshida, J.-i.; Tsujishima, H.; Nakano, K.; Teramoto, T.; Nishiwaki, K.; Isoe, S., *Organometallics* **1995**, *14*, 567.
2. Pearlson, W. H.; Brice, T. J.; Simons, J. H., *J. Am. Chem. Soc.* **1945**, *67*, 1769.
3. Curran, C.; Witucki, R. M.; McCusker, P. A., *J. Am. Chem. Soc.* **1950**, *72*, 4471.
4. Eméleus, H. J.; Wilkins, C. J., *J. Chem. Soc.* **1944**, 454.
5. Farooq, O.; Tiers, G. V. D., *J. Org. Chem.* **1994**, *59*, 2122.
6. Farooq, O., *J. Chem. Soc., Perkin Trans. 1* **1998**, 661.
7. Damrauer, R.; Simon, R. A.; Kanner, B., *Organometallics* **1988**, *7*, 1161.
8. Lickiss, P. D.; Lucas, R., *J. Organomet. Chem.* **1996**, *510*, 167.
9. Marans, N. S.; Sommer, L. H.; Whitmore, F. C., *J. Am. Chem. Soc.* **1951**, *73*, 5127.
10. Eaborn, C., *J. Chem. Soc.* **1952**, 2846.
11. Eaborn, C., *J. Chem. Soc.* **1952**, 2840.
12. Dua, S. S.; Eaborn, C.; Happer, D. A. R.; Hopper, S. P.; Safa, K. D.; Walton, D. R. M., *J. Organomet. Chem.* **1979**, *178*, 75.
13. Ohshita, J.; Lee, K. H.; Kimura, K.; Kunai, A., *Organometallics* **2004**, *23*, 5622.
14. Mutahi, M.; Nittoli, T.; Guo, L.; Sieburth, S. Mc. N., *J. Am. Chem. Soc.* **2002**, *124*, 7363.
15. Hatanaka, Y.; Goda, K.; Okahara, Y.; Hiyama, T., *Tetrahedron* **1994**, *50*, 8301.
16. Yamaguchi, S.; Ohno, S.; Tamao, K., *Synlett* **1997**, 1199.
17. Bailey, D. L.; Pike, R. M., *Nitroarylfluorosilanes*, US Pat. 3020302 (1962).

Jin Kyung Kim & Scott McN. Sieburth
Temple University, Philadelphia, PA, USA

Disilane, 1,1,1,2,2,2-hexamethyl-

[1450-14-2] $C_6H_{18}Si_2$ (MW 146.42)

InChI = 1S/C6H18Si2/c1-7(2,3)8(4,5)6/h1-6H3

InChIKey = NEXSMEBSBIABKL-UHFFFAOYSA-N

(silylating or reducing reagent in combination with a Pd catalyst
or a nucleophile)

Physical Data: bp 112–114 °C; d 0.729 g cm^{-3}.

Solubility: sol common organic solvents; insol H_2O.

Form Supplied in: oil; commercially available.

Analysis of Reagent Purity: ^{1}H NMR (CDCl$_3$–Me$_4$Si)
δ 0.045 (s).

Preparative Methods: by standard synthesis.[1]

Purification: grossly impure sample (25% impurities) is purified by repeated spinning band distillation. This lowered the
impurity level to 500 ppm. The main impurity is identified as
1-hydroxypentamethyldisilane.

Handling, Storage, and Precautions: stable to H_2O and air;
flammable liquid; irritant. Readily oxidized by oxidants such
as halogen and peroxide. Use in a fume hood.

Original Commentary

Tamejiro Hiyama & Manabu Kuroboshi

Tokyo Institute of Technology, Yokohama, Japan

Trimethylsilyl Anion. Treatment of hexamethyldisilane (1)
with an alkyllithium, metal alcoholate, or fluoride ion generates
the trimethylsilyl anion (2) (eq 1).[2] Reaction of (2) with aryl
halides gives trimethylsilylarenes in 63–80% yields along with
reduced products (ArH, 4–27%).[3] tetra-*n*-butylammonium
Fluoride (TBAF) catalyzed reaction of Me$_6$Si$_2$ with butadienes gives 1,4-disilyl-2-butenes (15–81% yield) with high
(*E*) selectivity.[4]

$$Me_3Si\text{–}SiMe_3 \xrightarrow[\text{or F}^-]{\substack{RLi \\ ROM}} Me_3Si^- M^+ \quad (1)$$
$$\textbf{(1)} \qquad\qquad \textbf{(2)}\ M = Li, Na, K, Bu_4N$$

Aliphatic aldehydes react with Me$_6$Si$_2$/TBAF in HMPA to
afford 1-trimethylsilyl-1-alkanols (3), whereas aryl aldehydes
give pinacols (4) under the same conditions (eq 2).[5]

$$\textbf{(3)} \qquad\qquad\qquad\qquad \textbf{(4)}$$

The trimethylsilyl anion adds to α,β-unsaturated ketones in a
1,4-manner to give β-silyl ketones. This reaction has been applied
to the synthesis of carbacyclin.[6]

α-Diketones (5) react with Me$_6$Si$_2$ in the presence of
PdCl$_2$(PMe$_3$)$_2$ or Pt$_2$(dba)$_3$–P(OCH$_2$)$_3$CEt to give the corresponding 1,2-bis(trimethylsilyloxy)ethylenes (6) (eq 3), while

α-keto esters (7) provide pinacol 2,3-bis(trimethylsilyl) ethers (8)
(eq 4).[7] The regio- and stereoselectivities are moderate (62:38–
87:13).

$$\textbf{(5)} \qquad\qquad\qquad\qquad \textbf{(6)}$$

$$\textbf{(7)} \qquad\qquad\qquad\qquad \textbf{(8)}$$

Allyl alcohols are converted into allylsilanes (11) with
n-butyllithium and Me$_6$Si$_2$ (eq 5). Here, the in situ generated
lithium allyl alcoholates (9) first react with Me$_6$Si$_2$ to produce
trimethylsilyllithium and trimethylsilyl allyl ethers (10), which
subsequently react in an S_N2' manner to give the products.
Similarly, the lithium alcoholates generated from α,β-unsaturated
aldehydes (or ketones) and an alkyllithium, or alternatively from
ketones and vinyllithium, can be converted into the corresponding
allylsilanes.[8]

$$\textbf{(9)} \qquad\qquad \textbf{(10)} \qquad\qquad \textbf{(11)}$$

**Palladium-Catalyzed Coupling Reaction with Organo-
halides.** In the presence of tetrakis(triphenylphosphine)-
palladium(0), vinyl halides (12) are converted to the
corresponding vinylsilanes (13) with the aid of Me$_6$Si$_2$ and tris-
(dimethylamino)sulfonium difluorotrimethylsilicate (TASF) in
53–92% yields with high chemoselectivity and stereospecificity
(eq 6).[9] Similarly, aryl-, benzyl-, and allylsilanes are obtained
from the corresponding halides in 57–100, 8–48, and 52–93%
yield, respectively.[10] Allyl esters also are transformed by
Pd-catalyzed reaction of Me$_6$Si$_2$ into allylsilanes.[11]

$$\textbf{(12)} \qquad\qquad\qquad\qquad \textbf{(13)}\ R = \text{vinyl, allyl} \\ \text{benzyl, aryl}$$

Intermolecular Disilylation of Alkynes. Intermolecular
disilylation of 1-alkynes (14) are promoted using palladium(II)
acetate–1,1,3,3-tetramethylbutyl isocyanide at reflux[12] or
Pd(dba)$_2$–P(OCH$_2$)$_3$CEt at 120 °C[13] to give (*Z*)-1,2-
bis(trimethylsilyl)alkenes (15) in yields up to 91% (eq 7).
Internal alkynes are unreactive. The latter catalyst system

allows 1-alkynes to insert between the Si–Si bonds of poly[*p*-(disilanylene)phenylene] and poly(dimethylsilylene).

$$R\text{—}\!\!\!\equiv\!\!\!\text{—} \quad \xrightarrow[\substack{\text{toluene, reflux or} \\ \text{Pd}_2(\text{dba})_3, 2 \text{ equiv } \text{P(OCH}_2)_3\text{CEt} \\ 120\,^\circ\text{C} \\ 81\text{–}91\%}]{\substack{\text{Me}_3\text{Si–SiMe}_3 \\ \text{Pd(OAc)}_2, t\text{-BuC(Me)}_2\text{NC}}} \quad \underset{(15)}{\underset{R}{\text{Me}_3\text{Si}\diagdown\diagup\text{SiMe}_3}} \quad (7)$$

(14)

Silyl Vinyl Ethers. Enolizable ketones (**16**) are readily transformed to silyl vinyl ethers (**17**) by the action of disilane and a catalytic amount of sodium in HMPA (eq 8).[14] The regioselectivity of the silyl vinyl ether formation depends on the structure of the ketones.

$$\underset{(16)}{\overset{O}{\diagdown\!\!\!\diagup\!\!\!\diagdown}} \quad \xrightarrow[\substack{\text{cat Na, HMPA} \\ 90\,^\circ\text{C}}]{\text{Me}_3\text{Si–SiMe}_3} \quad \underset{(17)\ 35\text{–}45\%}{\overset{\text{OSiMe}_3}{\diagup\!\!\!\diagdown}} \quad + \quad \text{HSiMe}_3 \quad (8)$$

Reductive Coupling of Acyl Halides. Electron-deficient aromatic acid chlorides (Ar$_W$-COCl; **18**) are converted into biphenyls (**19**) by the dichlorobis(triphenylphosphine)palladium(II) catalyzed reaction of Me$_6$Si$_2$ in mesitylene at 165 °C (eq 9).[15] Arylsilanes are produced as byproducts, but these are switched to major products with (ClMe$_2$Si)$_2$.[16] In contrast, aromatic acid chlorides (Ar$_D$-COCl; **20**) bearing an electron-donating group give acylsilanes (**21**) upon catalysis with bis(benzonitrile)dichloropalladium(II)–triphenylphosphine (eq 10).[17]

$$\underset{(18)}{\text{Ar}_W\text{–COCl}} \quad \xrightarrow[\substack{\text{mesitylene, 165 °C} \\ (-2\ \text{Me}_3\text{SiCl}, -2\ \text{CO}) \\ 76\%}]{\text{Me}_6\text{Si}_2,\ \text{PdCl}_2(\text{PPh}_3)_2} \quad \underset{(19)}{\text{Ar}_W\text{–Ar}_W} \quad (9)$$

$$\text{Ar}_W = \text{—}$$

$$\underset{(20)}{\text{Ar}_D\text{–COCl}} \quad \xrightarrow[\substack{\text{neat, 125 °C} \\ (-\text{Me}_3\text{SiCl}) \\ \sim 50\%}]{\text{Me}_6\text{Si}_2,\ \text{PdCl}_2(\text{PhCN})_2,\ \text{PPh}_3} \quad \underset{(21)}{\overset{O}{\underset{\text{Ar}_D}{\diagup\!\!\diagdown}}\text{SiMe}_3} \quad (10)$$

Reduction of α-Halo Carbonyl Compounds and α-Halo Nitriles. A halogen substituent on the carbon next to a carbonyl or cyano group is reduced with Me$_6$Si$_2$ in the presence of Pd(PPh$_3$)$_4$ catalyst (eq 11). In the case of α-halo ketones, oxy-π-allyl(trimethylsilyl)palladium intermediates are proposed.[18]

$$\underset{\substack{(22)\ W = R^2\text{CO, CN} \\ X = \text{Cl, Br}}}{\overset{R^1}{\underset{X}{\diagdown\!\!\diagup\!\!\diagdown}}W} \quad \xrightarrow[\substack{\text{Pd(PPh}_3)_4 \\ \text{PhH, 100 °C} \\ 36\text{–}100\%}]{\text{Me}_6\text{Si}_2} \quad \underset{(23)}{\overset{R^1}{\underset{H}{\diagdown\!\!\diagup\!\!\diagdown}}W} \quad (11)$$

Palladium-Catalyzed Insertion of Isocyanides into Si–Si Linkage.[19] A mixture of Me$_6$Si$_2$, aryl isocyanide (**24**), a catalytic amount of palladium(II) acetate and toluene is heated to reflux to give disilylformaldehyde imines (**25**) (eq 12). This reaction is applied to the insertion of the isocyanide unit into all the Si–Si linkages of (tetradecamethyl)hexasilane (**26**) to give oligomeric silylimines (**27**) (eq 13).

$$\underset{(24)}{\text{Ar–NC}} \quad \xrightarrow[\substack{\text{Pd(OAc)}_2 \\ \text{toluene, reflux} \\ 66\text{–}81\%}]{\text{Me}_3\text{Si–SiMe}_3} \quad \underset{(25)}{\overset{\text{NAr}}{\underset{\text{Me}_3\text{Si}\ \ \ \text{SiMe}_3}{\diagup\!\!\!\diagdown}}} \quad (12)$$

$$\underset{(26)}{\text{Me}\diagdown\overset{\displaystyle \underset{\text{Me}_2}{\overset{\text{Me}_2}{\text{Si}}}}{}\diagdown\!\!\diagup\cdots} \quad \xrightarrow[\substack{\text{Pd(OAc)}_2 \\ \text{DMF, 70 °C} \\ 28\%}]{\text{Ar–NC}}$$

$$\underset{\substack{(27)\ \text{Ar} = 2,6\text{-xylyl}}}{\text{Me}\diagdown\!\!\diagup\cdots\diagdown\text{Me}} \quad (13)$$

Alkyl Bromides from Alcohols. Alkyl bromides (**29**) are obtained in high yields by the reaction of the corresponding alcohols (**28**) with Me$_6$Si$_2$ and pyridinium hydrobromide perbromide (eq 14).[20] The substitution reaction is very fast for tertiary, allylic, and benzylic alcohols, but very slow for primary and secondary alcohols. Inversion of configuration is observed with secondary alcohols.

$$\underset{(28)}{\text{R–OH}} \quad \xrightarrow[\substack{\text{CHCl}_3,\ \text{rt} \\ 79\text{–}100\%}]{\text{Me}_6\text{Si}_2,\ \text{py}^+\text{Br}_3^-} \quad \underset{(29)}{\text{R–Br}} \quad (14)$$

Reductive Silylation of *p*-Quinones. Me$_6$Si$_2$ in combination with a catalytic amount of iodine converts *p*-quinones (**30**) into 1,4-bis(trimethylsiloxy)arenes (**31**) in almost quantitative yields (eq 15).[21] The products are a protected form of quinones, which are regenerated by oxidation with pyridinium chlorochromate (PCC).[22]

$$\underset{(30)}{\overset{O}{\underset{R}{\diagup\!\!\!\diagdown}}} O \quad \underset[\substack{\text{PCC, CH}_2\text{Cl}_2 \\ \text{rt, 2 h} \\ 51\text{–}94\%}]{\overset{\substack{\text{Me}_6\text{Si}_2,\ \text{I}_2,\ \text{PhH} \\ 60\,^\circ\text{C, 7 h} \\ 86\text{–}100\%}}{\rightleftharpoons}} \quad \underset{(31)}{\text{Me}_3\text{SiO}\text{—}\!\!\diagup\!\!\!\!\diagdown\!\!\text{—OSiMe}_3} \quad (15)$$

Reductive Coupling of *gem*-Dichloroalkanes.[23] The reaction of *gem*-dichloroalkanes (**32**) with Me$_6$Si$_2$ proceeds smoothly in the presence of a catalytic amount of Pd(PPh$_3$)$_4$ to give dimerized alkenes (**33**) in fairly good yields (eq 16). The (*E/Z*) ratio of the products is moderate. R^1–CCl$_2$–SiMe$_3$ also reacts with Me$_6$Si$_2$ in the presence of Pd catalyst to yield (*E*)-R^1(Me$_3$Si)C= CR^1(SiMe$_3$) quantitatively.

$$\underset{(32)}{\overset{\text{Cl}\ \ \ \text{Cl}}{\underset{R^1\ \ \ \ R^2}{\diagdown\!\!\diagup}}} \quad \xrightarrow[\substack{\text{mesitylene, 130 °C} \\ 79\text{–}100\%}]{\substack{\text{Me}_6\text{Si}_2\ (2\ \text{equiv}) \\ \text{Pd(PPh}_3)_4\ (1\ \text{mol \%})}} \quad \underset{(33)}{\overset{R^1\ \ \ \ R^1}{\underset{R^2\ \ \ \ R^2}{\diagdown\!\!\diagup}}} \quad (16)$$

First Update

Jih Ru Hwu, Suvendu Sekhar Dey & Ming-Hua Hsu
National Tsing Hua University, Hsinchu, Taiwan

'One-flask' Method. Hexamethyldisilane ($Me_3SiSiMe_3$) can be used to bring about various organic functional group transformations using the 'one-flask' method. These transformations are expedient and more efficient than other classical procedures. Accordingly, direct synthesis of ketene dithioacetals (**35**) and 2-trimethylsilyl-1,3-dithianes (**36**) in good to excellent yields is achieved from 1,3-dithiane (**34**) (eqs 17 and 18). These reactions are applicable to substrates including aldehydes, ketones, enones, nitroalkenes, alkyl, allyl, and benzyl bromides.[24]

$$
\begin{array}{c}
\text{(34)} \xrightarrow[\substack{\text{3. } R^1R^2CO, -78 \text{ to } 25 \text{ °C} \\ \text{4. } NH_4Cl \\ 65-94\%}]{\substack{\text{1. BuLi, THF, } -25 \text{ °C} \\ \text{2. } Me_3SiSiMe_3, HMPA, -25 \text{ °C}}} \text{(35)}
\end{array} \quad (17)
$$

$$
\begin{array}{c}
\text{(34)} \xrightarrow[\substack{\text{3. } RCH_2X, -78 \text{ to } 25 \text{ °C} \\ \text{4. } NH_4Cl \\ 54-98\%}]{\substack{\text{1. BuLi, THF, } -25 \text{ °C} \\ \text{2. } Me_3SiSiMe_3, HMPA, -25 \text{ °C}}} \text{(36)}
\end{array} \quad (18)
$$

Similarly, secondary nitroalkanes (**37**) can be reduced to ketoximes (**38**) by treatment of the corresponding nitronate anions with hexamethyldisilane through a 1,2-elimination process (eq 19). The conversion of nitrones (**39**) to imines (**40**) is also performed by using trimethylsilyllithium, which is prepared in situ from hexamethyldisilane and MeLi (eq 20). Under the same conditions, heterocyclic *N*-oxides (**41**) can be deoxygenated to the corresponding *N*-heterocycles (**42**) in 73–86% yields (eq 21).[25]

$$
\begin{array}{c}
\text{(37)} \xrightarrow[\substack{\text{2. 10\% aq HCl} \\ 40-73\%}]{\text{1. } n\text{-BuLi, } Me_3SiSiMe_3, \text{ THF, } 0-25 \text{ °C}} \text{(38)}
\end{array} \quad (19)
$$

$$
\begin{array}{c}
\text{(39)} \xrightarrow[\substack{\text{2. } H_2O \\ 82-88\%}]{\substack{\text{1. } Me_3SiSiMe_3, MeLi \\ HMPA, THF, -78 \text{ to } 25 \text{ °C}}} \text{(40)}
\end{array} \quad (20)
$$

$$
\begin{array}{c}
\text{(41)} \xrightarrow[\substack{\text{2. } H_2O \\ 73-86\%}]{\substack{\text{1. } Me_3SiSiMe_3, MeLi, HMPA \\ \text{THF, } -78 \text{ to } 25 \text{ °C}}} \text{(42)}
\end{array} \quad (21)
$$

Reductive Deuteration of Multiple Bonds. A combination of hexamethyldisilane and deuterium oxide (10 equiv) works as a deuterium-transfer reagent for alkynes (**43**). This reaction requires a catalytic amount of a palladium complex and selectively gives *E*-1,2-dideuterioalkenes (**44**) (eq 22).[26] Use of H_2O to replace D_2O successfully produces the hydrogenation product in excellent

yields. Moreover, use of the same solvent at 60 °C causes reduction of enones to produce α,β-dideuterioketones.[27]

$$
\begin{array}{c}
R^1 \!\!=\!\!\!=\!\!R^2 \xrightarrow[\substack{5 \text{ mol \% } [PdCl(\eta^3-C_3H_5)]_2 \\ 10 \text{ mol \% } PPh_3, DMA, 80 \text{ °C} \\ 64-99\%}]{Me_3SiSiMe_3, D_2O}
\begin{array}{c} R^1 \quad D \\ \diagup\!\!\!\diagdown \\ D \quad R^2 \end{array}
\end{array} \quad (22)
$$

(43) → **(44)**

$$E:Z = {>}98{:}2$$

R^1 = Ph, 4-$CF_3C_6H_4$, 4-$MeOCOC_6H_4$, $MeOC_6H_4$, 4-MeC_6H_4, 2-MeC_6H_4, 4-*n*-$C_5H_{11}C_6H_4$
R^2 = Ph, 4-$CF_3C_6H_4$, 2-MeC_6H_4, *t*-Bu, Me

Oxidation of Substrates. In addition to acting as a reducing agent, hexamethyldisilane may play a crucial role in the oxidation of organic compounds. For example, treatment of hydrazines (**45**) with hexamethyldisilane in the presence of 1.0 equiv of potassium hydride gives the corresponding 2-tetrazenes (**46**) in fair to good yields (eq 23). In this reaction, hexamethyldisilane acts as an oxidizing agent.[28] Nevertheless, the outcome of this reaction changed dramatically when HMPA is added as the cosolvent: Persilylations occur to all N–H groups in the parent and substituted hydrazines upon their treatment with hexamethyldisilane and potassium hydride (0.30 equiv) in a mixture of THF and HMPA.[29]

$$
\begin{array}{c}
\begin{array}{c} R^1 \\ \diagdown \\ N\!-\!NH_2 \\ \diagup \\ R^2 \end{array} \xrightarrow[\substack{\text{2. } Me_3SiSiMe_3 \\ \text{3. } H_2O \\ 26-67\%}]{\text{1. KH, THF, 0 °C}}
\begin{array}{c} R^1R^2N \\ \diagdown \\ N\!\!=\!\!N \\ \diagup \\ NR^2R^1 \end{array}
\end{array} \quad (23)
$$

(45) → **(46)**

R^1 = Ph, -$(CH_2)_5$-, -$(CHMeCH_2CH_2CH_2CHMe)$-, -$CH_2CH_2OCH_2CH_2$-, -$(CH_2)_6$-
R^2 = Me, -$(CH_2)_5$-, -$(CHMeCH_2CH_2CH_2CHMe)$-, -$CH_2CH_2OCH_2CH_2$-, -$(CH_2)_6$-

Hexamethyldisilane can also act as an oxidant in the conversion of benzyl alcohols to carbonyl compounds. Under alkaline conditions, reaction of α-cyclopropylbenzyl alcohol[30] or 3-methoxybenzyl alcohol[31] with hexamethyldisilane generates (γ-trimethylsilyl)butyrophenone or 3-methoxybenzaldehyde, respectively.

Aroylarylation. Internal alkynes (**48**) effectively undergo 1,2-addition of aroyl and aryl groups upon treatment with aroyl chlorides (**47**) in the presence of [RhCl(cod)]$_2$ and PPh$_3$, using hexamethyldisilane as a reducing agent. The corresponding 1,3-diaryl-2-propen-1-ones (**49**) are generated in good yields (eq 24).[32] The reaction proceeds for relatively reactive alkenes, such as norbornenes. However, similar treatment of phenylacetylene with aroyl chlorides brings about aroylsilylation giving 1-aryl-2-phenyl-3-trimethylsilyl-2-propene-1-ones in 13–44% yields.[32]

Silylation of Aromatic C–H and C–CN Bonds. The [RhCl(cod)]$_2$ catalyst is also used to perform regioselective silylation of the *ortho* carbon–hydrogen bond in 2-arylpyridines (**50**) with the aid of hexamethyldisilane (eq 25).[33] The reaction of

2-(2-methylphenyl)pyridine with 2.0 equiv of hexamethyldisilane in the presence of 5.0 mol % [RhCl(cod)]$_2$ in o-xylene at 130 °C gives 2-[2-methyl-6-(trimethylsilyl)phenyl]pyridine in 86% yield. A variety of substituents, including alkoxy, amino, fluoro, and ester groups, are compatible with this catalysis. When substrates contain two *ortho* C–H bonds, monosilylation occurs selectively by utilization of the 3-methyl-2-pyridyl moiety as a directing group.[33]

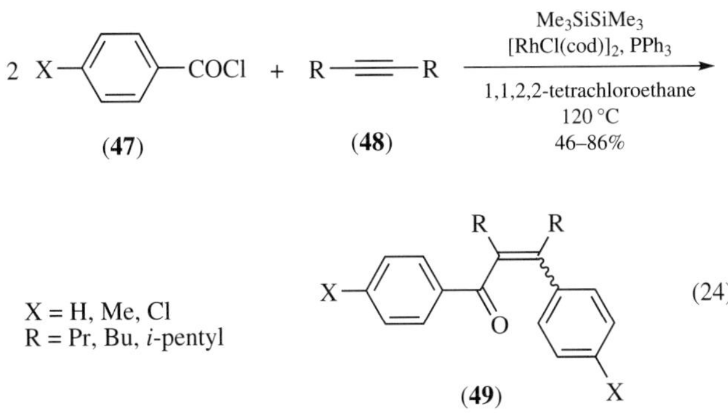

$$X = H, Me, Cl \qquad (24)$$
$$R = Pr, Bu, i\text{-pentyl}$$

(49)

$$Z/E = 89/11 \text{ to } 68/32$$

(51) major   **(52)** minor

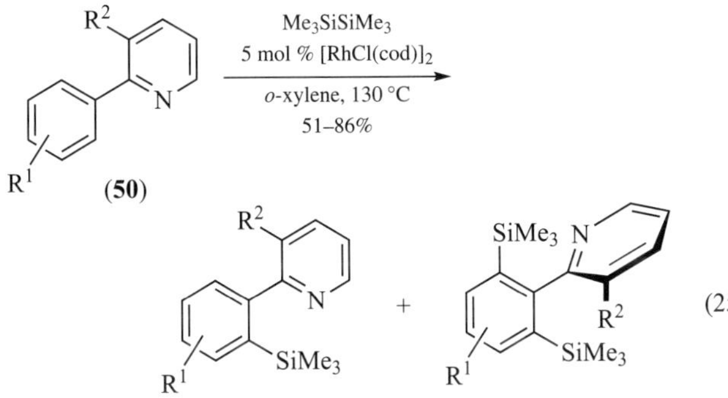

$$R^1 = H, Me, OMe, OiPr, NMe_2, COOEt, CF_3$$
$$R^2 = H, Me$$

Aryl, alkenyl, allyl, and benzyl cyanides (**53**) bearing a variety of functional groups can be silylated with hexamethyldisilane through the cleavage of unreactive C–CN and Si–Si bonds (eq 26).[34] Among the Rh(I) catalysts, [RhCl(cod)]$_2$ is the most effective for this reaction. Enamines (**55**) may be produced as minor products (0–18%) from benzyl derivatives.[35]

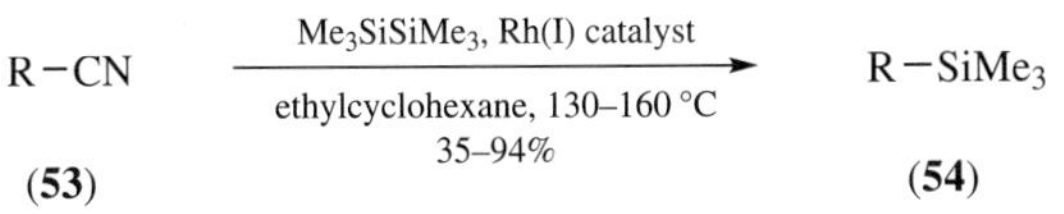

$$R-CN \xrightarrow[\substack{\text{ethylcyclohexane, 130–160 °C} \\ 35\text{–}94\%}]{\text{Me}_3\text{SiSiMe}_3, \text{ Rh(I) catalyst}} R-SiMe_3 \quad (26)$$

(53)   **(54)**

R = aryl,
  alkenyl
  benzyl derivatives

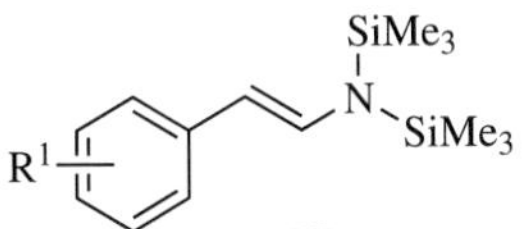

(55)

The scope of this reaction is extended to intramolecular arylation involving a C–CN bond cleavage to produce tricyclic heterocycles (**57**) (eq 27). In this process, cocatalysts such as InCl$_3$, P(O-^{i}Pr)$_3$, and P(4-CF$_3$C$_6$H$_4$)$_3$ are used.[35]

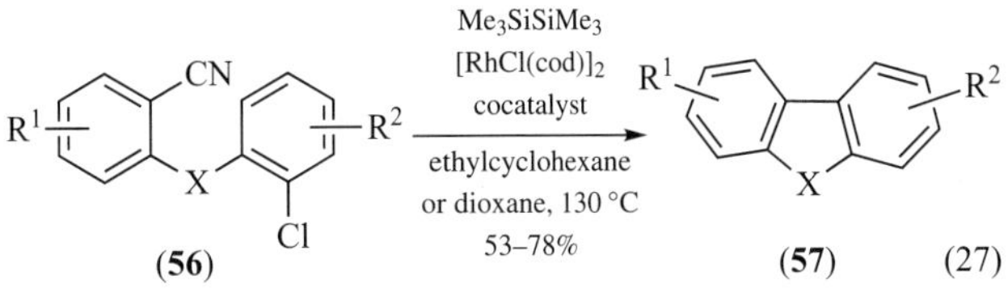

$$(27)$$

X = O, N
R^1 = H, OMe
R^2 = H, OMe, CF$_3$

1,4-Carbosilylation. Hexamethyldisilane is used along with Pd(dba)$_2$ catalyst to accomplish various silylations. The first example is a decarbonylative coupling of acid chlorides (**58**) with 1,3-dienes (**59**). Allylic silanes (**60**) are generated with high regio- and stereoselectivity (eq 28).[36] Butadiene gives E-1,4-isomer solely, whereas stereoselectivity is modest for isoprene and 2,3-dimethyl-1,3-diene. Increase of steric hindrance in acid chlorides suppresses decarbonylation.[37] For example, adamantane-1-carboxylic acid chloride and t-butylacetyl chloride afford selectively the product containing only the acyl functionality. Pivaloyl chloride gives a mixture of decarbonyl- and carbonyl-containing products.

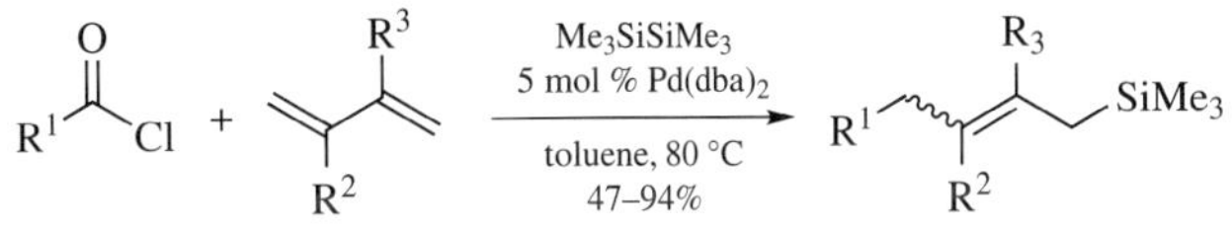

(58)   **(59)**   **(60)**

R^1 = C$_6$H$_5$, 4-Br-C$_6$H$_5$, 4-Cl-C$_6$H$_5$, 4-Me-C$_6$H$_5$,
  4-NO$_2$-C$_6$H$_5$, 2-C$_4$H$_3$O, 5-Br-2-C$_4$H$_2$O, C$_6$H$_5$CH=CH
R^2, R^3 = H, Me   (28)

Dimerization–Double Silylation. The second application of the mixture containing hexamethyldisilane (0.50 mmol) and Pd(dba)$_2$ is to convert 1,3-dienes (**61**) (3.0 mmol) in DMF or dioxane to α,ω-disilylated E-1,4 head-to-head dimer (**62**) exclusively (eq 29).[38] No crossover silylation takes place. Accordingly, carbocyclization–disilylation product (**64**) is generated from bisdienes (**63**) (eq 30) with high regioselectivity and modest stereoselectivity.[39]

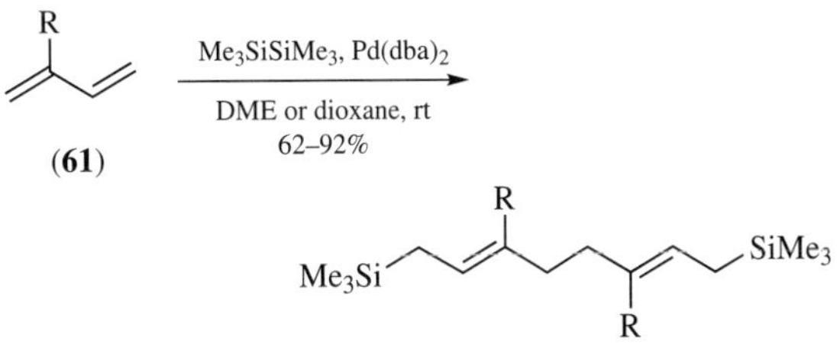

(61)

(62)   (29)

R = H, Me, Ph, OSiMe$_3$

EtO₂C compound **(63)** with Me₃SiSiMe₃, Pd(dba)₂, DME or dioxane, rt, 96% gives **(64)** (30)

2-Acylallylsilane Formation. The third way to use Pd(dba)₂ along with hexamethyldisilane is to acylsilylate allenes (**66**) with acid chlorides or chloroformates in acetonitrile (eq 31).[40] The corresponding allylsilanes (**67**) are generated with high regio- and stereoselectivity.

$$X-COCl\ (65)\ +\ allene\ (66)\ \xrightarrow[\text{CH}_3\text{CN, 80 °C}]{\text{Me}_3\text{SiSiMe}_3\ 5\text{ mol % P(dba)}_2}\ (67)\quad (31)$$

X = alkyl, aryl   47–88%   E/Z = 1/99

X = –OR   R¹ = R² = Me   66–70%   —

Conjugate Silylation of α,β-Enones. Use of (CuOTf)₂·C₆H₆ as a catalyst can cleave the Si–Si bond in hexamethyldisilane to generate a silyl nucleophile.[41] Its 1,4-addition onto α,β-enones, such as cyclo-2-hexen-1-one, 4-benzylbut-3-ene-2-one, and 1,3-dibenzylprop-2-ene-1-one, in aprotic polar solvents such as 1,3-dimethyl-2-imidazolidinone or DMF at 100 °C produces the corresponding β-trimethylsilyl ketones in 76–96% yields.

Silylation of Acid Chlorides. Reaction of hexamethyldisilane with MeLi and CuCN at −23 °C generates (Me₃Si)₂CuLi. It reacts in situ with aliphatic, aromatic, heteroaromatic, and sterically hindered acyl chlorides (**68**) to produce the corresponding acylsilanes (**69**) in fair to good yields (eq 32).[42]

$$RCOCl\ (68)\ \xrightarrow[\substack{\text{2. CuCN, THF, } -23\,°\text{C}\\ \text{3. NH}_4\text{Cl}}]{\substack{\text{1. Me}_3\text{SiSiMe}_3\text{, MeLi, HMPA}\\ \text{THF, } -78\text{ to }25\,°\text{C}}}\ RCOSiMe_3\ (69)\quad (32)$$

39–87%

Reactions of aroyl chlorides with neat hexamethyldisilane at 110 °C are effectively catalyzed by [(η³-C₃H₅)PdCl]₂ with triethylphosphite to give benzoyltrimethylsilane in satisfactory yields (14–71%).[43] Aryltrimethylsilanes are also formed in a few cases. This reaction is applicable to a variety of *ortho-*, *meta-*, and *para*-substituted aroyl chlorides. Under these conditions, phenyl-glyoxalyl chloride produces benzoyltrimethylsilane as the sole product in 52% yield through a facile decarbonylation process.[43]

Silylation of Aryl Halides. In the presence of HMPA and a catalytic amount of Pd(PPh₃)₄, hexamethyldisilane reacts with various electron-deficient aryl and heteroaryl bromides, such as pyridine and quinoline, to give the corresponding aryl and heteroaryl silanes in 50–95% yields.[44] Moreover, diphenylphos-phinophenolate (PO) acts as an effective ligand in the palladium-catalyzed silylation of aryl bromides and iodides (**70**) that possess electron-withdrawing or -donating substituents to give the corresponding trimethylsilylarenes (**71**) (eq 33).[45]

$$X\text{-arene (70)}\ \xrightarrow[\substack{\text{NaOH, 100 °C}\\ \text{THF:H}_2\text{O (1:1) or}\\ \text{TBAB, THF:H}_2\text{O (1:1)}}]{\substack{\text{Me}_3\text{SiSiMe}_3\\ [\text{PdCl}(\pi\text{-C}_3\text{H}_5)]_2/\text{ligand}\\ (5\text{ mol % of Pd, ligand/Pd = 2})}}\ \text{Me}_3\text{Si-arene (71)}\quad (33)$$

65–92%

X = Br, I

R = H, 4-MeC₆H₄, 4-MeOC₆H₄, 4-NH₂C₆H₄, 4-EtOCOC₆H₄, 4-CF₃C₆H₄, 4-HCOC₆H₄, 4-PhC₆H₄

ligand = 2-HO-C₆H₄-PPh₂, 2-MeO-C₆H₄-PPh₂, C₆H₅-PPh₂, 4-HO-C₆H₄-PPh₂, 2-NH₂-C₆H₄-PPh₂

Palladium-catalyzed silylation of aryl chlorides (**72**) is also developed (eq 34),[46] which involves the use of phosphine ligand (**73**) and KF in dioxane. While displaying broad functional group tolerance, the conditions are compatible with numerous electron-rich and -neutral aryl chlorides. Furthermore, use of the ligand (**74**) with LiOAc in DMF was effective for a range of electron-deficient aryl chlorides bearing base-sensitive groups such as ketones, esters, nitriles, and nitro groups.[46]

$$Cl\text{-arene (72)}\ \xrightarrow[\substack{\textbf{73}, \text{KF, dioxane}\\ \text{or}\\ \textbf{74}, \text{LiOAc, DMF}}]{\substack{\text{Pd}_2(\text{dba})_3, \text{H}_2\text{O}\\ \text{Me}_3\text{SiSiMe}_3, 100\,°\text{C}}}\ \text{Me}_3\text{Si-arene (71)}\quad (34)$$

66–91%

R = H, *n*-Bu, OMe, SMe, NH₂, NMe₂, OH, Ac

(73)   **(74)**

The reaction of functionalized fluoroarenes, such as fluoro-acetophenones and (fluorophenyl)oxazolines, with hexamethyl-disilane in the presence of a catalytic amount of [Rh(cod)₂]BF₄ in toluene results in a site-selective Si–F exchange to give the corresponding (trimethylsilyl)fluoroarenes.[47] Being carried out in a stainless steel pressure vial at 130 °C, the yields of silylated products increase as a function of the increasing number of fluoro groups on the aromatic rings.

Silylation of Alcohols and Acetates. The combination of hexamethyldisilane and $[PdCl(\eta^3\text{-}C_3H_5)]_2PPh_3$ in a catalytic amount acts as an effective silylating agent for alcohols (**75**) (eq 35).[48] The most striking feature lies in its high atom economy by generating H_2 gas as the sole by-product. Primary and secondary alcohols are also conveniently silylated by hexamethyldisilane in the presence of TBAF under mild conditions at room temperature.[49]

$$
\text{ROH} \xrightarrow[\substack{\text{DMA, 80 °C} \\ 90\text{–}95\%}]{\substack{Me_3SiSiMe_3 \\ 5\text{ mol }\% [PdCl(\eta^3\text{-}C_3H_5)]_2 \\ 10\text{ mol }\% PPh_3}} \text{ROSiMe}_3 \qquad (35)
$$

(**75**)   (**76**)

R = primary, secondary, and tertiary alkyl

Application of Pd(dba)$_2$ as the catalyst along with LiCl allows hexamethyldisilane to silylate various allylic acetates (**77**) to afford the corresponding more stable allylic silanes (**78**) in high yields (eq 36).[50] Silylation of allylic trifluoroacetates gives the corresponding allylsilanes in good yields at room temperature without the need of the chloride catalyst.[51]

(**77**)   (**78**)

R = alkyl, aryl   E/Z = 53/47 to 96/4

Moreover, the cross-couplings between hexamethyldisilane and Baylis–Hillman acetate (**79**) can also be catalyzed by use of Pd$_2$(dba)$_3$. Accordingly, 3-substituted-2-carbonylallylsilanes (**80**) are produced in high yields with high regio- and stereoselectivity (eq 37).[52]

(**79**)   (**80**)

Z:E = 100:0 to 92:8

Generator of Iodotrimethylsilane. The mixture of hexamethyldisilane and iodine is used as an effective generator of iodotrimethylsilane and gives no by-products.[53] By monitoring the ratio of hexamethyldisilane to iodine, this mixture can be used for the mild cleavage of esters, carbamates, ethers, and sulfoxides to produce the corresponding carboxylic acids, amines, alcohols, and sulfides individually in excellent yields. Alcohols can be equally well transformed into corresponding alkyl iodides with these reagents. Unprotected reducing sugars (**81**) can be transformed to thioglycosides (**82**) with excellent anomeric stereoselectivity in very good yields (eq 38).[54]

(**81**)   (**82**)

1. Wilson, G. R.; Smith, A. G., *J. Org. Chem.* **1961**, *26*, 557.

2. Fujita, M.; Hiyama, T., *Yuki Gosei Kagaku Kyokai Shi* **1984**, *42*, 293, (*Chem. Abstr.* **1984**, *101*, 55 130).

3. Shippey, M. A.; Dervan, P. B., *J. Org. Chem.* **1977**, *42*, 2654.

4. Hiyama, T.; Obayashi, M.; Mori, I.; Nozaki, H., *J. Org. Chem.* **1983**, *48*, 912.

5. Sagami Chemical Research Center. Jpn. Patent 59 42 331 (*Chem. Abstr.* **1984**, *101*, 38 214); Jpn. Patent Patent 59 42 391 (*Chem. Abstr.* **1984**, *101*, 72 944).

6. Shibasaki, M.; Fukasawa, H.; Ikegami, S., *Tetrahedron Lett.* **1983**, *24*, 3497.

7. Yamashita, H.; Reddy, N. P.; Tanaka, M., *Chem. Lett.* **1993**, 315.

8. Hwu, J. R.; Lin, L. C.; Liaw, B. R., *J. Am. Chem. Soc.* **1988**, *110*, 7252.

9. Hatanaka, Y.; Hiyama, T., *Tetrahedron Lett.* **1987**, *28*, 4715.

10. Eaborn, C.; Griffiths, R. W.; Pidcock, A., *J. Organomet. Chem.* **1982**, *225*, 331. Matsumoto, H.; Yako, T.; Nagashima, S.; Motegi, T.; Nagai, Y., *J. Organomet. Chem.* **1978**, *148*, 97.

11. Urata, H.; Suzuki, H.; Moro-oka, Y.; Ikawa, T., *Bull. Chem. Soc. Jpn.* **1984**, *57*, 607.

12. Ito, Y.; Suginome, M.; Murakami, M., *J. Org. Chem.* **1991**, *56*, 1948.

13. Yamashita, H.; Catellani, M.; Tanaka, M., *Chem. Lett.* **1991**, 241.

14. Gerval, P.; Frainnet, E., *J. Organomet. Chem.* **1978**, *153*, 137.

15. Rich, J. D.; Krafft, T. E.; McDermott, P. J.; Chang, T. C. T., *Eur. Pat. Appl.* 339 455 (*Chem. Abstr.* **1990**, *112*, 235 164). Krafft, T. E.; Rich, J. D.; McDermott, P. J., *J. Org. Chem.* **1990**, *55*, 5430.

16. Rich, J. D.; Krafft, T. E., *Organometallics* **1990**, *9*, 2040. Rich, J. D., *Organometallics* **1989**, *8*, 2609.

17. Rich, J. D., *J. Am. Chem. Soc.* **1989**, *111*, 5886.

18. Urata, H.; Suzuki, H.; Moro-oka, Y.; Ikawa, T., *J. Organomet. Chem.* **1982**, *234*, 367.

19. Ito, Y.; Suginome, M.; Matsuura, T.; Murakami, M., *J. Am. Chem. Soc.* **1991**, *113*, 8899.

20. Olah, G. A.; Gupta, B. G. B.; Malhotra, R.; Narang, S. C., *J. Org. Chem.* **1980**, *45*, 1638.

21. Matsumoto, H.; Koike, S.; Matsubara, I.; Nakano, T.; Nagai, Y., *Chem. Lett.* **1982**, 533.

22. Willis, J. P.; Gogins, K. A. Z.; Miller, L. L., *J. Org. Chem.* **1981**, *46*, 3215.

23. Matsumoto, H.; Arai, T.; Takahashi, M.; Ashizawa, T.; Nakano, T.; Nagai, Y., *Bull. Chem. Soc. Jpn.* **1983**, *56*, 3009.

24. Hwu, J. R.; Lee, T.; Gilbert, B. A., *J. Chem. Soc., Perkin Trans. 1* **1992**, 3219.

25. Hwu, J. R.; Tseng, W. N.; Patel, H. V.; Wong, F. F.; Horng, D.-N.; Liaw, B. R.; Lin, L. C., *J. Org. Chem.* **1999**, *64*, 2211.

26. Otsuka, H.; Shirakawa, E.; Hayashi, T., *Chem. Commun.* **2005**, 5885.

27. Shirakawa, E.; Otsuka, H.; Hayashi, T., *Chem. Commun.* **2007**, 1819.

28. Hwu, J. R.; Wang, N.; Yung, R. T., *J. Org. Chem.* **1989**, *54*, 1070.

29. Hwu, J. R.; Wang, N., *Tetrahedron* **1988**, *44*, 4181.

30. Hwu, J. R., *J. Chem. Soc., Chem. Commun.* **1985**, 452.

31. Hwu, J. R.; Tsay, S.-C.; Wang, N.; Hakimelahi, G. H., *Organometallics* **1994**, *13*, 2461.

32. Kokubo, K.; Matsumasa, K.; Miura, M.; Nomura, M., *J. Organomet. Chem.* **1998**, *560*, 217.

33. Tobisu, M.; Ano, Y.; Chatani, N., *Chem. Asian J.* **2008**, *3*, 1585.

34. Tobisu, M.; Ano, Y.; Chatani, N., *J. Am. Chem. Soc.* **2006**, *128*, 8152.

35. Tobisu, M.; Ano, Y.; Chatani, N., *J. Am. Chem. Soc.* **2008**, *130*, 15982.

36. Obora, Y.; Tsuji, Y.; Kawamura, T., *J. Am. Chem. Soc.* **1993**, *115*, 10414.

37. Obora, Y.; Tsuji, Y.; Kawamura, T., *J. Am. Chem. Soc.* **1995**, *117*, 9814.

38. Obora, Y.; Tsuji, Y.; Kawamura, T., *Organometallics* **1993**, *12*, 2853.

39. Obora, Y.; Tsuji, Y.; Kakehi, T.; Kobayashi, M.; Shinkai, Y.; Ebihara, M.; Kawamura, T., *J. Chem. Soc., Perkin Trans. 1* **1995**, 599.

40. Yang, F.-Y.; Shanmugasundaram, M.; Chuang, S.-Y.; Ku, P.-J.; Wu, M.-Y.; Cheng, C.-H., *J. Am. Chem. Soc.* **2003**, *125*, 12576.

41. Ito, H.; Ishizuka, T.; Tateiwa, J.; Sonoda, M.; Hosomi, A., *J. Am. Chem. Soc.* **1998**, *120*, 11196.

42. Capperucci, A.; Degl'Innocenti, A.; Faggi, C.; Ricci, A., *J. Org. Chem.* **1988**, *53*, 3612.

43. Yamamoto, K.; Hayashi, A.; Suzuki, S.; Tsuji, J., *Organometallics* **1987**, *6*, 974.

44. Babin, P.; Bennetau, B.; Theurig, M.; Dunoguès, J., *J. Organomet. Chem.* **1993**, *446*, 135.

45. Shirakawa, E.; Kurahashi, T.; Yoshida, H.; Hiyama, T., *Chem. Commun.* **2000**, 1895.

46. McNeill, E.; Barder, T. E.; Buchwald, S. L., *Org. Lett.* **2007**, *9*, 3785.

47. Ishii, Y.; Chatani, N.; Yorimitsu, S.; Murai, S., *Chem. Lett.* **1998**, 157.

48. Shirakawa, E.; Hironaka, K.; Otsuka, H.; Hayashi, T., *Chem. Commun.* **2006**, 3927.

49. Tanabe, Y.; Okumura, H.; Maeda, A.; Murakami, M., *Tetrahedron Lett.* **1994**, *35*, 8413.

50. Tsuji, Y.; Kajita, S.; Isobe, S.; Funato, M., *J. Org. Chem.* **1993**, *58*, 3607.

51. Tsuji, Y.; Funato, M.; Ozawa, M.; Ogiyama, H.; Kajita, S.; Kawamura, T., *J. Org. Chem.* **1996**, *61*, 5779.

52. Kabalka, G. W.; Venkataiah, B.; Dong, G., *Organometallics* **2005**, *24*, 762.

53. Olah, G. A.; Narang, S. C.; Gupta, B. G. B.; Malhotra, R., *Angew. Chem., Int. Ed. Engl.* **1979**, *18*, 612.

54. Mukhopadhyay, B.; Kartha, K. P. R.; Russell, D. A.; Field, R. A., *J. Org. Chem.* **2004**, *69*, 7758.

E

1-Ethoxy-1-(trimethylsilyloxy)cyclopropane

[27374-25-0] $C_8H_{18}O_2Si$ (MW 174.32)

InChI = 1S/C8H18O2Si/c1-5-9-8(6-7-8)10-11(2,3)4/h5-7H2,1-4H3

InChIKey = BZMMRNKDONDVIB-UHFFFAOYSA-N

(preparation of 3-metallopropionates;[1] metal homoenolate precursor;[2] γ-hydroxy esters; cyclopentenones; 3-aminopropionates; cyclopropylamine formation; 1-aminocyclopropanecarboxylic acids and 1-aminocyclopropanephosphonic acids; β- and γ-amino acids)

Physical Data: bp 50–53 °C/22 mmHg.
Solubility: insol H_2O.
Form Supplied in: colorless liquid.
Analysis of Reagent Purity: GLC, NMR.
Preparative Methods: for the synthesis of the parent and the 2-monoalkyl-substituted compounds, reduction of ethyl 3-chloropropionate with sodium–potassium alloy in the presence of chlorotrimethylsilane in ether.[3] A recent modification using ultrasound irradiation is much more convenient and more widely applicable.[4] Other substituted derivatives are prepared by cyclopropanation of alkyl silyl ketene acetals with the Furukawa reagent (diiodomethane/diethylzinc).[5]
Purification: distillation under reduced pressure.
Handling, Storage, and Precautions: moisture sensitive, yet, once purified by distillation, is stable for a long period of time in a tightly capped bottle at room temperature.

Original Commentary

Eiichi Nakamura
Tokyo Institute of Technology, Japan

Stoichiometric Precursor of Metal Homoenolates. The reaction of 1-alkoxy-1-trimethylsilyloxycyclopropane with a variety of Lewis acidic metal chlorides affords the 3-metallated propionate esters in good to excellent yield (see below).[1,6,7] For instance, the reaction of 1-ethoxy-1-trimethylsilyloxycyclopropane with one equivalent of tin(IV) chloride gives a 3-stannylpropionate, which further reacts with another equivalent of the cyclopropane to give a dialkylated tin compound (eq 1).

The reaction of the siloxycyclopropane with titanium(IV) chloride produces the titanium homoenolate (3-titaniopropionate) in good yield; this, however, is relatively unreactive (eq 2).[8] Addition of one equivalent of $Ti(OR')_4$ generates a more reactive $RTiCl_2OR'$ species, which smoothly reacts with carbonyl compounds below room temperature.[9] The γ-hydroxy ester adducts are useful synthetic intermediates and serve as precursors to γ-lactones and cyclopropanecarboxylates.[10] A useful variation involves the use of the cyclopropanecarboxylate ester as a functionalized homoenolate precursor to obtain levulinic acid derivatives (eq 3).[11]

$$\text{(1)}$$

$$\text{(2)}$$

$$\text{(3)}$$

The zinc homoenolate prepared by the treatment of the siloxycyclopropane with zinc chloride is a versatile synthetic reagent (eq 4).[12] Reduction of 3-iodopropionate with activated zinc also produces a zinc homoenolate species.[13]

Treatment of the silyloxycyclopropane with $ZnCl_2$ followed by addition of an enone, HMPA, THF, and a catalytic amount of a Cu^I salt results in quantitative formation of a conjugate adduct as an enol silyl ether (eq 4). The chlorosilane, a byproduct, is essential for the conjugate addition of the copper homoenolate.[14,15] boron trifluoride etherate promotes the copper-catalyzed conjugate addition reaction with a different stereochemical outcome.[16] A useful application of the conjugate addition reaction is a [3+2] synthesis of cyclopentenones, wherein the homoenolate acts as a 1,3-dipole equivalent (eq 5).[17]

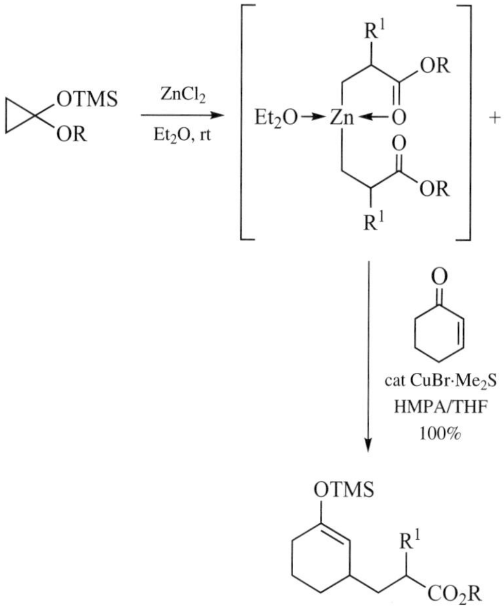

(4)

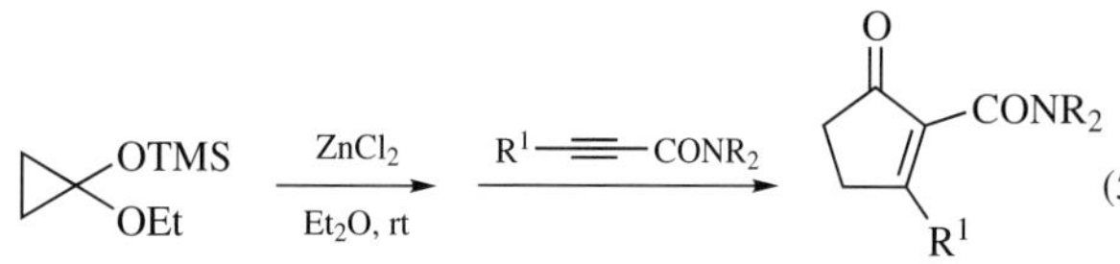

(5)

The zinc homoenolate undergoes copper-catalyzed allylation with allylic chlorides. The reaction is not only extremely S_N2' regioselective but stereoselective for δ-chiral allylic chlorides.[18] Arylation and vinylation of the zinc homoenolates proceed in the presence of a palladium–phosphine complex.[19] Similarly, palladium-catalyzed acylation reaction gives γ-keto esters (eq 6).

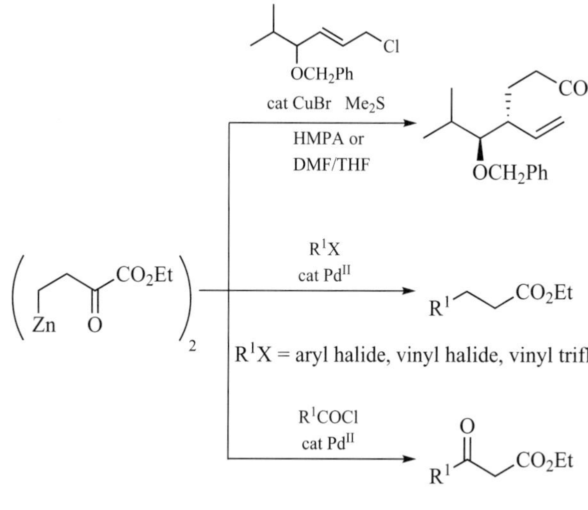

(6)

R^1X = aryl halide, vinyl halide, vinyl triflate

Catalytic Generation of Homoenolate Reactive Species.
Homoaldol reaction between the siloxycyclopropane and an aldehyde with a catalytic amount of zinc iodide in methylene

chloride affords a γ silyloxy ester (eq 7).[18,20] Arylation[21] and acylation[22,23] of the silyloxycyclopropanes in the presence of a palladium catalyst take place via direct attack of an aryl- or acylpalladium intermediate on the C–C bond of the cyclopropane (eqs 8 and 9). The reaction is applicable not only to ester synthesis but also to ketone and aldehyde synthesis. Heating a chloroform solution of the silyloxycyclopropane in the presence of a palladium–phosphine catalyst under 1 atm carbon monoxide produces a γ-keto pimelate (eq 10).[24]

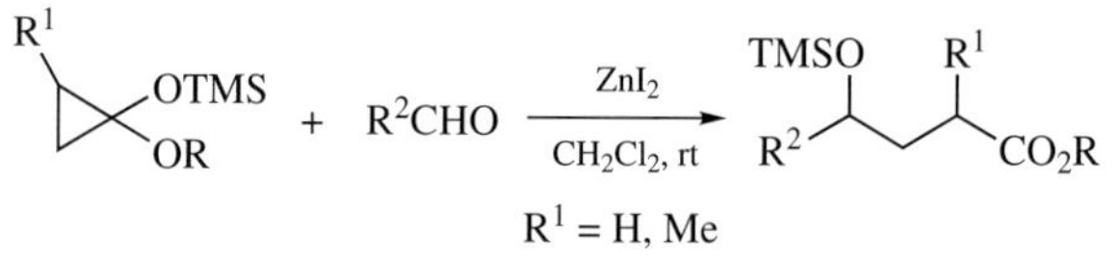

(7)

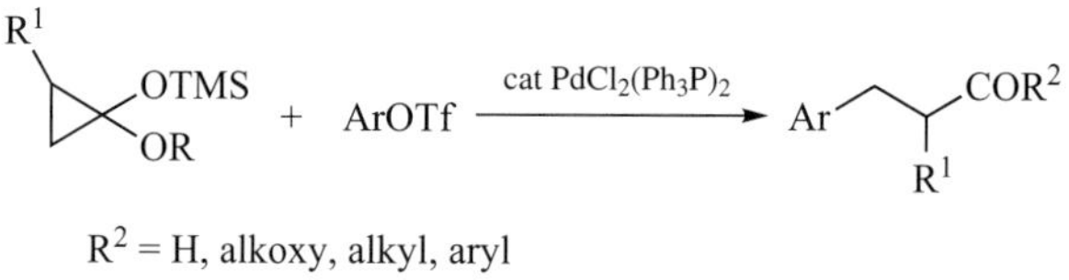

R^1 = H, Me

(8)

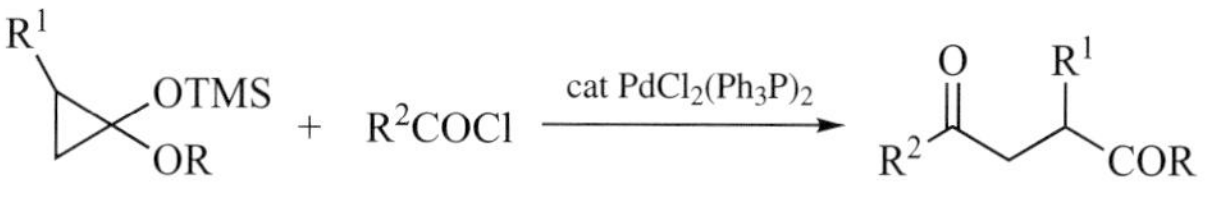

R^2 = H, alkoxy, alkyl, aryl

(9)

R = alkoxy, alkyl, aryl

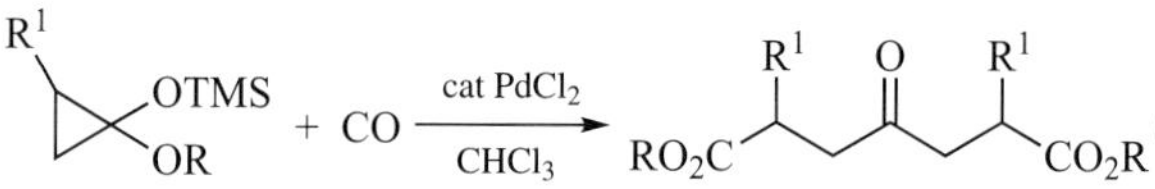

(10)

Precursor of Lithiocyclopropane. Bromination of the silyloxycyclopropane with phosphorus(III) bromide produces 1-bromo-1-ethoxycyclopropane. Successive treatment of the bromide with t-butyllithium and an enal affords a cyclopropylcarbinol, which undergoes acid-catalyzed ring enlargement to give 2-vinylcyclobutanone (eq 11).[25]

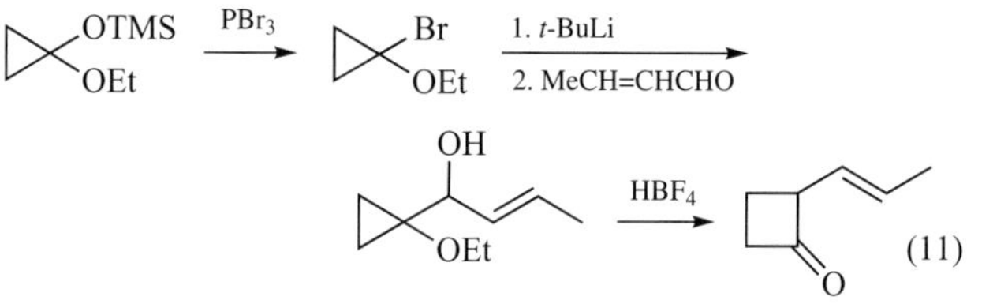

(11)

Reactions with Azidoformates. Photolysis of an acetonitrile solution of the cyclopropane and ethyl azidoformate at rt gives a C–H insertion product (eq 12).[26] However, thermolysis of the same mixture in DMSO gives a 3-aminopropionate by insertion of nitrene into the cyclopropane ring (eq 13).[27]

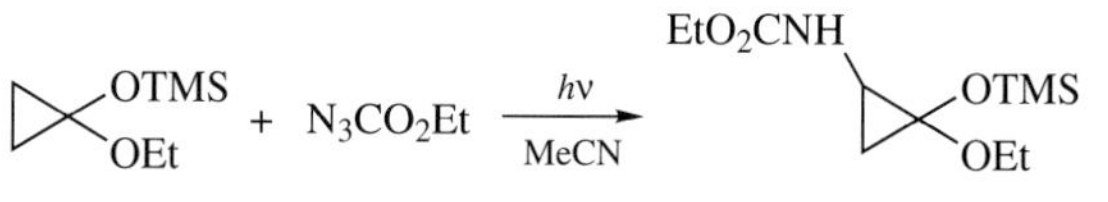

(12)

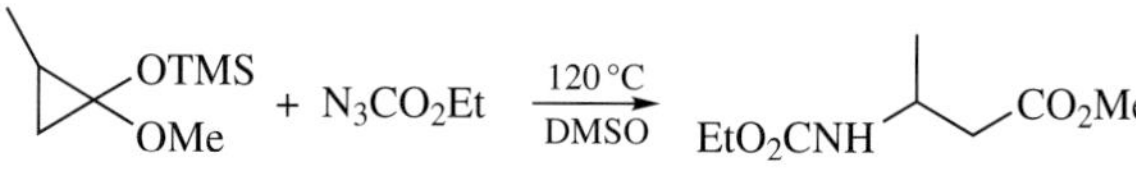

(13)

Cyclopropanone Hemiacetals and Their Use. Mild alcoholysis of the silyloxycyclopropane gives a cyclopropanone hemiacetal. This compound serves as a stable equivalent of unstable cyclopropanones.[3] Treatment with two equivalents of alkynylmagnesium bromide gives a 1-ethynyl-1-hydroxycyclopropane (eq 14).[28]

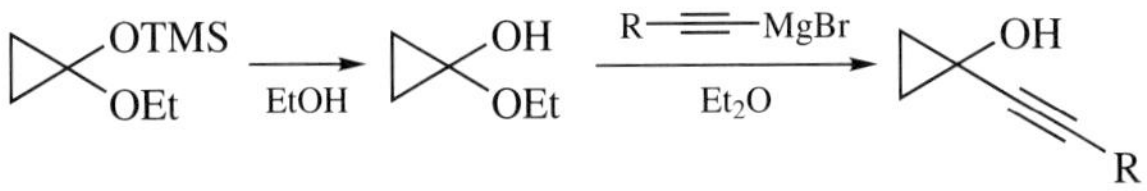

$$\text{(14)}$$

The cyclopropanol also serves as a source of homoenolate radical species. Treatment of a mixture of the cyclopropanol and an enol silyl ether with manganese(III) 2-pyridinecarboxylate in DMF gives a 1,5-dicarbonyl compound (eq 15).[29]

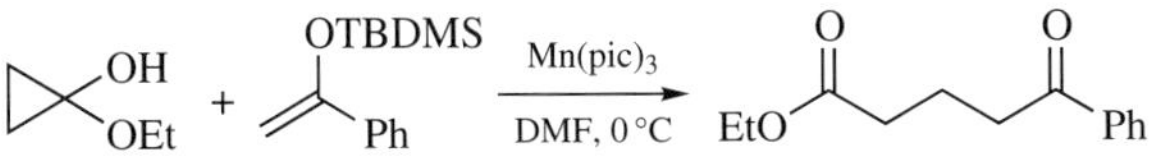

$$\text{(15)}$$

Strecker amino acid synthesis starting with the cyclopropanone hemiacetal provides a enantioselective route to a cyclopropane amino acid (eq 16).[30]

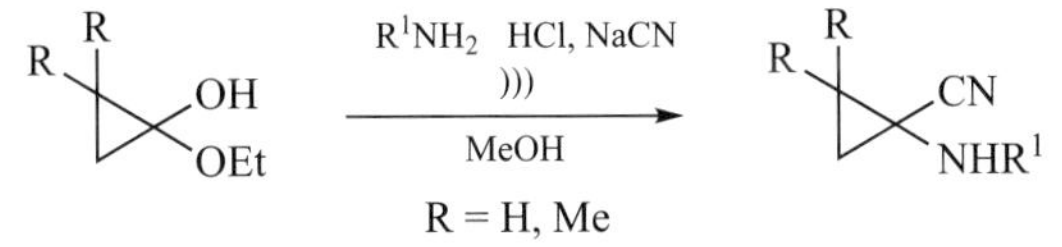

$$\text{(16)}$$

$$R = H, Me$$

First Update

Antoine Fadel

Université Paris-Sud, Orsay, France

The 1-Ethoxy-1-(trimethylsilyloxy)cyclopropane[31] is widely used in various reactions. Cyclization of optically pure β-halo esters gives cyclopropanone acetals enantiomerically pure at C-2 and a 1:1 diastereomeric mixture at C-1 (eq 17).[4,32]

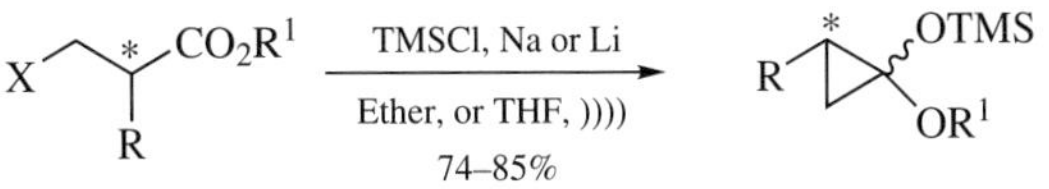

$$\text{(17)}$$

$$X = Cl, Br$$

$$R = H, alkyl; R^1 = Me, Et$$

Homoenolate Reactivity. Since the previous e-EROS report, a number of examples have been described using the cyclopropanone acetals. Thus, the zinc homoenolate, known to undergo a highly regioselective and stereoselective S_N2' allylation reaction (eq 6),[18] is used in the synthesis of moenomycin analogues.[33] The activated titanium homoenolate reacts with aldehydes or ketones to give γ-hydroxy esters that serve as precursors to γ-lactones.[34]

Organozinc Cuprate to Nitroolefins. The copper-catalyzed conjugate addition of functionalized diorganozinc cuprates to nitroolefins leads to synthetically versatile nitro compounds in good yields (eq 18).[35]

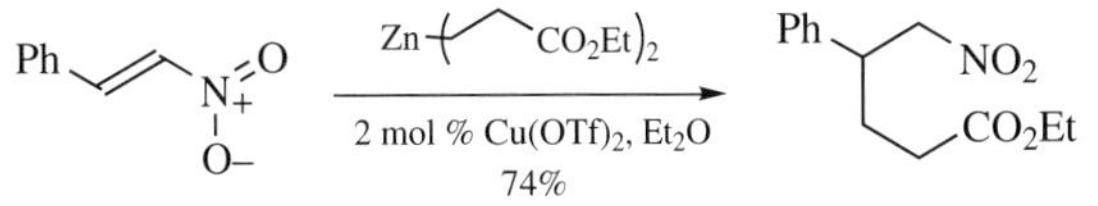

$$\text{(18)}$$

Acylation and Carbonylation. The zinc homoenolate reacts more rapidly with acyl halides than with α-β-enones to give 4-keto esters in good yield (X = Cl) (eq 6).[12,19,36] The Ni-catalyzed coupling of acyl fluoride and zinc homoenolate provides γ-keto esters.[37] The same protocols with catalytic $PdCl_2(Ph_3P)_2$[23] have also been extended to amino acid thioesters giving 4-keto esters in high yield, but suffer complete racemization. Fortunately, the use of phthalic anhydride as thiolate scavenger effectively preserved the enantiopurity of α-amino ketone (eq 19).[38]

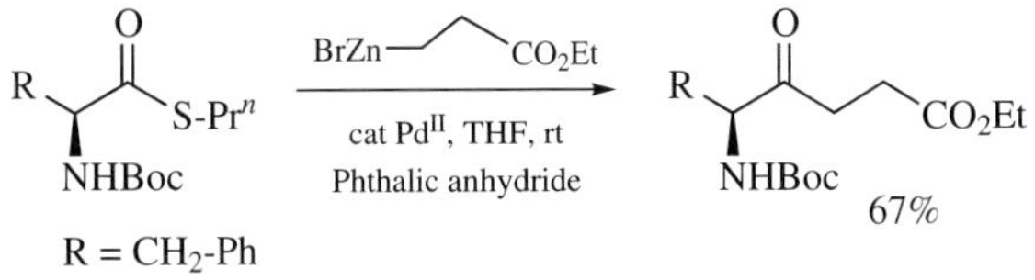

$$\text{(19)}$$

$$R = CH_2-Ph$$

Carbonylative coupling of 1-ethoxy-1-trimethylsilyloxycyclopropane with diphenyliodonium tetrafluoroborate is accomplished in the presence of $Pd(OAc)_2$ and DME, under 1 atm carbon monoxide, to provide γ-keto esters (eq 20).[39]

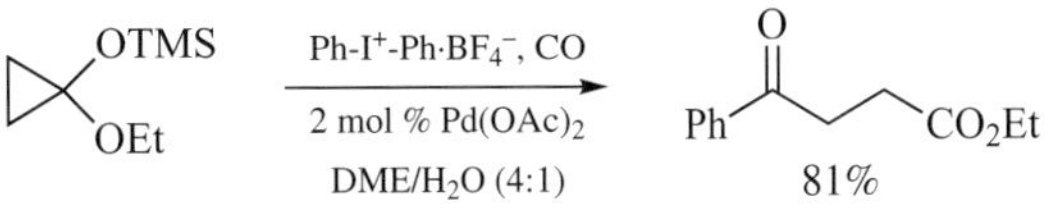

$$\text{(20)}$$

Uses of Transformation Products from Silyloxycyclopropanes.

Reaction with Imines and Arylazo Tosylates. The Lewis acid $Cu(OTf)_2$ is suitable for preparing the homoenolate and activating the imine to produce γ-amino acids or γ-lactams via a homo-Mannich reaction. An asymmetric catalytic version of this reaction is achieved using enantiopure bisoxazolidine (eq 21).[40]

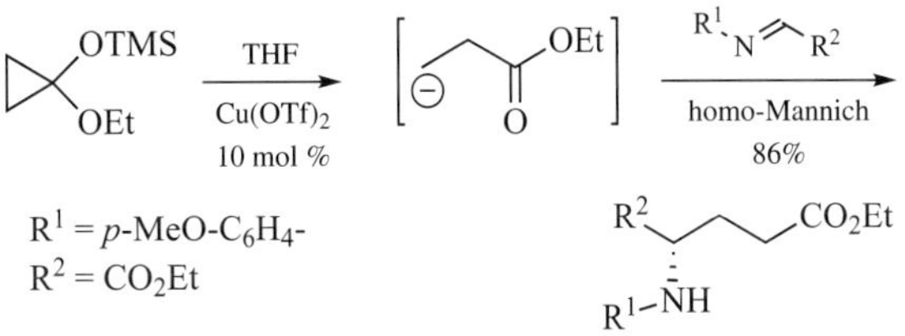

$$\text{(21)}$$

$$R^1 = p\text{-MeO-}C_6H_4-$$

$$R^2 = CO_2Et$$

β-Amino acids are also prepared by adding zinc homoenolate to an arylazo tosylate, followed by a Raney nickel reduction of N-N hydrazine bond in ethanol at reflux.[41]

The conjugate addition–cyclization (formal [3 + 2] cycloaddition) of zinc homoenolates to acetylenic esters and amides provides easy access to highly functionalized cyclopentenones (see eq 5).[42] Applications for the synthesis of ginkgolide B[43] and lubiminol[44] are reported.[45]

Generation of Homoenolate Radicals. The homoenolate radical produced by Cerium(IV) Ammonium Nitrate (CAN) undergoes a fast oxidative addition to electron-rich alkenes such as ethyl vinyl ether and allyltrimethylsilane. Oxidative addition of the homoenolate radical in acetonitrile to cyclic enones in the presence of allyltrimethylsilane gives 2,3-disubstituted cycloalkanones (eq 22).[46] Photochemical generation of the homoenolate radical has also been described.[47]

$$\text{(22)}$$

35–48%

Cyclopropylamine Formation. The cyclopropylation of a variety of amines is easily accomplished by reductive amination with 1-ethoxy-1-trimethylsilyloxycyclopropane (eq 23).[48] This mild and general one-pot cyclopropylation method is useful in medicinal and synthetic organic chemistry to enhance the biological activity of compounds bearing amine functions.[49]

$$\text{(23)}$$

63–91%

R = H, alkyl; R^1 = Ph, alkyl
R-N-R^1 = piperidines

Use of Cyclopropanone Hemiacetals. Heating cyclopropanone hemiacetal at 100 °C in an aqueous buffer provides the cyclopropanone hydrate.[50] It also serves as a source of homoenolate radical species with a catalytic amount of AgNO$_3$ (see, eq 15).[51]

Formation of Amino Acid and Phosphonic Acid Analogues. Interestingly, cyclopropanone hemiacetals provide a rapid way to prepare α-aminocyclopropane-carboxylic acids (ACCs) and phosphonic acids analogues (ACPs). Thus, asymmetric Strecker reaction of hemiacetal with an amine in acidic sodium cyanide selectively affords *cis*-α-aminocyclopropane-carbonitriles, precursors of ACC amino acid derivatives (eq 24).[30,52]

$$\text{(24)}$$

ds 90:10

R = H, Me, Et
R^* = benzyl derivatives

Cyclopropanone acetal presumably undergoes deprotection, condensation with amine, and addition of phosphite to afford the cyclopropane aminophosphonates, which finally give *trans*-α-aminocyclopropanephosphonic acids (eq 25).[53,54]

$$\text{(25)}$$

ds 88–100:12–0

R = H, Me, Et, Bn, iPr, tBu
R^* = benzyl derivatives

Related Reagents. Cyclopropanone; 1-ethoxy-1-hydroxy-cyclopropane; 1-(tetrahydro-2H-pyranyloxy)-cyclopropane-carbaldehyde.

1. Nakamura, E.; Shimada, J.; Kuwajima, I., *Organometallics* **1985**, *4*, 641.

2. Kuwajima, I.; Nakamura, E., *Comprehensive Organic Synthesis* **1991**, *2*, Chapter 1.14.

3. Salaün, J.; Marguerite, J., *Org. Synth.* **1985**, *63*, 147.

4. Fadel, A.; Canet, J.-L.; Salaün, J., *Synlett* **1990**, 89.

5. Rousseau, G.; Slougui, N., *Tetrahedron Lett.* **1983**, *24*, 1251.

6. Murakami, M.; Inouye, M.; Suginome, M.; Ito, Y., *Bull. Chem. Soc. Jpn.* **1988**, *51*, 3649.

7. Ryu, I.; Murai, S.; Sonoda, N., *J. Org. Chem.* **1986**, *51*, 2389.

8. Nakamura, E.; Kuwajima, I., *J. Am. Chem. Soc.* **1983**, *105*, 651.

9. Nakamura, E.; Oshino, H.; Kuwajima, I., *J. Am. Chem. Soc.* **1986**, *108*, 3745.

10. Nakamura, E.; Kuwajima, I., *J. Am. Chem. Soc.* **1985**, *107*, 2138.

11. Reissig, H.-U., *Top. Curr. Chem.* **1988**, *144*, 73.

12. Nakamura, E.; Aoki, S.; Sekiya, K.; Oshino, H.; Kuwajima, I., *J. Am. Chem. Soc.* **1987**, *109*, 8056.

13. Tamaru, Y.; Ochiai, H.; Nakamura, T.; Tsubaki, K.; Yoshida, Z.-I., *Tetrahedron Lett.* **1985**, *26*, 5559. Tamaru, Y.; Ochiai, H.; Nakamura, T.; Yoshida, Z.-I., *Tetrahedron Lett.* **1986**, *27*, 955. Yeh, M. C. P.; Knochel, P., *Tetrahedron Lett.* **1988**, *29*, 2395. Tamaru, Y.; Ochiai, H.; Nakamura, T.; Yoshida, Z.-I., *Angew. Chem., Int. Ed. Engl.* **1987**, *26*, 1157.

14. Nakamura, E.; Kuwajima, I., *J. Am. Chem. Soc.* **1984**, *106*, 3368. Nakamura, E.; Kuwajima, I., *Org. Synth.* **1987**, *66*, 43.

15. (a) Corey, E. J.; Boaz, N. W., *Tetrahedron Lett.* **1985**, *26*, 6015, 6019–6021. (b) Alexakis, A.; Berlan, J.; Besace, Y., *Tetrahedron Lett.* **1986**, *27*, 1047. (c) Horiguchi, Y.; Matsuzawa, S.; Nakamura, E.; Kuwajima, I., *Tetrahedron Lett.* **1986**, *27*, 4025. (d) Nakamura, E.; Matsuzawa, S.; Horiguchi, Y.; Kuwajima, I., *Tetrahedron Lett.* **1986**, *27*, 4029. (e) Matsuzawa, S.; Horiguchi, Y.; Nakamura, E.; Kuwajima, I., *Tetrahedron* **1989**, *45*, 349. (f) Nakamura, E. In *Organocopper Reagents*; Taylor, R. J. K., Ed.; Oxford: Oxford University Press, 1994; Chapter 6.

16. Horiguchi, Y.; Nakamura, E.; Kuwajiama, I., *J. Am. Chem. Soc.* **1989**, *111*, 6257.

17. Crimmins, M. T.; Nantermet, P. G.; Wesley, B.; Vallin, I. M.; Watson, P. S.; McKerlie, L. A.; Reinhold, T. L.; Cheung, A. W.-H.; Stetson, K. A.; Dedopoulou, D.; Gray, J. L., *J. Org. Chem.* **1993**, *58*, 1038.

18. Nakamura, E.; Sekiya, K.; Arai, M.; Aoki, S., *J. Am. Chem. Soc.* **1989**, *111*, 3091.

19. Aoki, S.; Fujimura, T.; Nakamura, E.; Kuwajima, I., *Tetrahedron Lett.* **1989**, *30*, 6541.

20. Gore, V. G.; Mahendra, D.; Chordia, D.; Narasimhan, S., *Tetrahedron* **1990**, *46*, 2483.

21. Aoki, S.; Fujimura, T.; Nakamura, E.; Kuwajima, I., *J. Am. Chem. Soc.* **1988**, *110*, 3296.

22. Fujimura, T.; Aoki, S.; Nakamura, E., *J. Org. Chem.* **1991**, *56*, 2810.

23. Aoki, S.; Nakamura, E., *Tetrahedron* **1991**, *47*, 3935.

24. Aoki, S.; Nakamura, E.; Kuwajima, I., *Tetrahedron Lett.* **1988**, *29*, 1541.

25. Gadwood, R. C.; Rubino, M. R.; Nagarajan, S. C.; Michel, S. T., *J. Org. Chem.* **1985**, *50*, 3255.

26. Mitani, M.; Tachizawa, O.; Takeuchi, H.; Koyama, K., *Chem. Lett.* **1987**, 1029.

27. Mitani, M.; Tachizawa, O.; Takeuchi, H.; Koyama, K., *J. Org. Chem.* **1989**, *54*, 5397.

28. Salaün, J., *J. Org. Chem.* **1976**, *41*, 1237.

29. Iwasawa, N.; Hayakawa, S.; Isobe, K.; Narasaka, K., *Chem. Lett.* **1991**, 1193.

30. Fadel, A., *Tetrahedron* **1991**, *47*, 6265. Fadel, A., *Synlett* **1993**, 503.

31. 1-Ethoxy-1-trimethylsilyloxycyclopropane is now commercially available.

32. Nakamura, E.; Sekiya, K.; Kuwajima, I., *Tetrahedron Lett.* **1987**, *28*, 337.

33. Eichelberger, U.; Neundorf, I.; Hennig, L.; Findeisen, M.; Geisa, S.; Müller, D.; Welzel, P., *Tetrahedron* **2002**, *58*, 545.

34. (a) Tekenouchi, K.; Sogawa, R.; Manabe, K.; Saitoh, H.; Gao, Q.; Miura, D.; Ishizuka, S. J., *Steroid. Biochem. Mol. Biol.* **2004**, *89–90*, 31. (b) Martin, E. O.; Gleason, J. L., *Org. Lett.* **1999**, *1*, 1643.

35. (a) Rimkus, A.; Sewald, N., *Org. Lett.* **2002**, *4*, 3289. (b) Rimkus, A.; Sewald, N., *Synthesis* **2004**, 135.

36. Wang, J.; Scott, A. I., *Tetrahedron Lett.* **1997**, *38*, 739.

37. Zhang, Y.; Rovis, T., *J. Am. Chem. Soc.* **2004**, *126*, 15964.

38. Li, B.; Buzon, R. A.; Chiu, C. K.-F.; Colgan, S. T.; Jorgensen, M. L.; Kasthurikrishnan, N., *Tetrahedron Lett.* **2004**, *45*, 6887.

39. Kang, S.-K.; Yamaguchi, T.; Ho, P.-S.; Kim, W.-Y.; Yoon, S.-K., *Tetrahedron Lett.* **1997**, *38*, 1947.

40. (a) Abbas, M.; Neuhaus, C.; Krebs, B.; Westerman, B., *Synlett* **2005**, 473. (b) Wissing, E.; Kaupp, M.; Boersma, J.; Spek, A. L.; Van Koten, G., *Organometallics* **1994**, *13*, 2349.

41. Sinha, P.; Kofink, C.; Knochel, P., *Org. Lett.* **2006**, *8*, 3741.

42. Crimmins, M. T.; Nantermet, P. G., *J. Org. Chem.* **1990**, *55*, 4235.

43. (a) Crimmins, M. T.; Jung, D. K.; Gray, J. L., *J. Am. Chem. Soc.* **1993**, *115*, 3146. (b) Crimmins, M. T.; Pace, J. M.; Nantermet, P. G.; Kim-Meade, A. S.; Thomas, J. B.; Watterson, S. H.; Wagman, A. S., *J. Am. Chem. Soc.* **1999**, *121*, 10249. (c) Crimmins, M. T.; Pace, J. M.; Nantermet, P. G.; Kim-Meade, A. S.; Thomas, J. B.; Watterson, S. H.; Wagman, A. S., *J. Am. Chem. Soc.* **2000**, *122*, 8453.

44. (a) Crimmins, M. T.; Wang, Z.; McKerlie, L. A., *Tetrahedron Lett.* **1996**, *37*, 8703. (b) Crimmins, M. T.; Wang, Z.; McKerlie, L. A., *J. Am. Chem. Soc.* **1998**, *120*, 1747.

45. (a) Crimmins, M. T.; Hauser, B., *Org. Lett.* **2000**, *2*, 281. (b) Li, C.-C.; Liang, S.; Zhang, X.-H.; Xie, Z.-X.; Chen, J.-H.; Wu, Y.-D.; Yang, Z., *Org. Lett.* **2005**, *7*, 3709.

46. Paolobelli, A. B.; Ruzziconi, R., *J. Org. Chem.* **1996**, *61*, 6434.

47. Mizuno, K.; Nishiyama, T.; Takahashi, N.; Inoue, H., *Tetrahedron Lett.* **1996**, *37*, 2975.

48. (a) Gillaspy, M. L.; Lefker, B. A.; Had, W. A.; Hoover, D.; J., *Tetrahedron Lett.* **1995**, *36*, 7399. (b) Yoshida, Y.; Umezo, K.; Hamada, Y.; Atsumi, N.; Tabuchi, F., *Synlett* **2003**, 2139.

49. Becker, D. P.; Villamil, C. I.; Barta, T. E.; Bedell, L. J.; Boehm, T. L.; DeCrescenzo, G. A.; Freskos, J. N.; Getman, D. P.; Hockerman, S.; Heintz, R.; Howard, S. C.; Li, M. H.; McDonald, J. J.; Carron, C. P.; Funckes-Shippy, C. L.; Mehta, P. P.; Munie, G. E.; Swearingen, C. A., *J. Med. Chem.* **2005**, *48*, 6713.

50. Shaffer, C. L.; Harriman, S.; Koen, Y. M.; Hanzlik, R. P., *J. Am. Chem. Soc.* **2002**, *124*, 8268.

51. Chiba, S.; Cao, Z.; El Bialy, S. A. A.; Narasaka, K., *Chem. Lett.* **2006**, *35*, 18.

52. Fadel, A.; Khesrani, A., *Tetrahedron: Asymmetry* **1998**, *9*, 305.

53. Fadel, A., *J. Org. Chem.* **1999**, *64*, 4953.

54. (a) Fadel, A.; Tesson, N., *Eur. J. Org. Chem.* **2000**, 2153. (b) Fadel, A.; Tesson, N., *Tetrahedron: Asymmetry* **2000**, *11*, 2023. (c) Tesson, N.; Dorigneux, B.; Fadel, A., *Tetrahedron: Asymmetry* **2002**, *13*, 2267.

Ethoxy(trimethylsilyl)acetylene

$$EtO-C{\equiv}C-SiMe_3$$

[1000-62-0] $C_7H_{14}OSi$ (MW 142.27)

InChI = 1S/C7H14OSi/c1-5-8-6-7-9(2,3)4/h5H2,1-4H3

InChIKey = FKMCADCEOYUAFV-UHFFFAOYSA-N

(mild agent for conversion of carboxylic acids into carboxylic anhydrides;[1] dehydrative condensation agent for carboxylic acids and amines and alcohols;[2] trimethylsilylketene precursor;[3] 2π component in cycloaddition reactions[4])

Physical Data: bp 57 °C/34 mmHg; *d* 0.828 g cm^{-3}.

Solubility: insol H_2O; freely sol most common solvents.

Form Supplied in: colorless liquid; widely available.

Preparative Methods: by silylation of ethoxyacetylene.[1,3,5]

Handling, Storage, and Precautions: storable in a refrigerator for years without any polymerization or decomposition (one of the features that favors the use of this reagent over ethoxyacetylene for many applications).

Original Commentary

Yasuyuki Kita

Osaka University, Osaka, Japan

Carboxylic Anhydrides. Dehydration of carboxylic acids or dicarboxylic acids to the corresponding anhydrides is effected by using ethoxy(trimethylsilyl)acetylene (**1**). In general, a carboxylic acid is treated with 1.5 equiv of (**1**) in CH_2Cl_2, $ClCH_2CH_2Cl$, or MeCN at rt to 60 °C for 2 h to 1 d followed by concentration of the reaction mixture to give a pure anhydride in an almost quantitative yield (eq 1).[1] The only byproduct of this reaction is the neutral and volatile ethyl (trimethylsilyl)acetate. Use of ethoxyacetylene for this purpose has been limited because of its instability to heat, high volatility, and low reactivity with some carboxylic acids.[6] The reagent (**1**) resolves these problems and is especially effective for preparation of the carboxylic anhydrides with acid-sensitive functional groups such as (**2**)[1] and (**3**)[7] Suitably functionalized homophthalic anhydrides (**2**) and (**4**)[8] and their heteroaromatic derivatives (**5**)[9] are key intermediates for total synthesis of natural anthracyclines and their heteroaromatic analogs.

(2) 99%   (3) 100%

(4) 99%   (5) 100%

Dehydrative Condensation of Carboxylic Acids with Amines and Alcohols. In the presence of a mercuric catalyst, (1) causes dehydrative condensation of carboxylic acids with H-acidic materials such as amines and alcohols to give the corresponding amides, esters, lactones, and peptides in high yields.[2]

Preparation of (Trimethylsilyl)ketene. Heating of (1) at $120\,°C$ is a convenient method for preparation of trimethylsilylketene (6) (eq 2).[3] This ketene is an easy-handling, storable, and distillable monomeric liquid, and is useful for acylation of sterically hindered alcohols and amines (eq 3). Other useful applications of (6) are represented in the preparation of α-silyl acetates,[10] α-silyl ketones,[11] α,β-unsaturated esters,[12] and silylallenes.[17]

Cycloaddition Reactions. Dichloroketene, N-sulfonylimines, diazomethane, and aroylketenes add to (1) to provide good yields of a cyclobutenone,[4] 2-azetines,[13] a pyrazole,[14] and 4-pyrones,[15] respectively. Cooligomerization of three molecules of (1) with one molecule of carbon dioxide proceeds in the presence of Ni[0] catalyst, giving a 2-pyrone derivative in 90% yield.[16]

First Update

Nanyan Fu
Fuzhou University, Fuzhou, Fujian, China

Thomas T. Tidwell
University of Toronto, Toronto, Ontario, Canada

Formation of Carboxylic Anhydrides. The use of ethoxy(trimethylsilyl)acetylene as a mild dehydrating agent was utilized by Kita and coworkers[18–20] for the efficient formation of homophthalic anhydrides from diacids that were used to prepare peri-hydroxy aromatic compounds that formed the key frameworks of certain polycyclic antibiotics. The general procedure utilized treating the diacid with 2 equiv of ethoxy(trimethylsilyl)acetylene in dichloromethane at room temperature for several hours followed by concentration of the reaction mixture to give the pure homophthalic anhydride in almost quantitative yield, followed by [4 + 2] cycloaddition with

dienophiles (eq 4).[19] This synthetic strategy has been used in the total synthesis of natural products such as fredericamycin A,[21] lactonamycin,[22] and (±)-dynemicin A[23] as well as nonnatural products such as a mimic of chromomycin A_3 as a DNA binding ligand.[24]

The application of ethoxy(trimethylsilyl)acetylene as a dehydrating agent was also extended to polyanhydride synthesis (eq 5). In contrast to traditional methods, this was carried out at low temperature ($20–40\,°C$) by the electrophilic addition–elimination reaction of ethoxy(trimethylsilyl)acetylene, avoiding decomposition due to heating of sensitive monomers and polymers.[25]

Dehydrative Condensation of Carboxylic Acids with Amines and Alcohols. For reversible fluorescent labeling of amino groups, ethoxy(trimethylsilyl)acetylene was used as a dehydrating reagent to couple dansylaminomethylmaleic acid (DAM) with benzylamine through a maleic anhydride intermediate (eq 6).[26]

Preparation of Trimethylsilylketene. The convenient preparation of trimethylsilylketene by thermolysis of ethoxy-(trimethylsilyl)acetylene at $120\,°C$ (eq 7)[27] has been studied by means of ab initio and semiempirical computations.[28]

Cycloaddition Reactions. Ethoxy(trimethylsilyl)acetylene reacted with photochemically generated bisketene 7 at $-25\,°C$ to give the spirocyclopropenylbutenolide 8 and the unstable [4 + 2] cycloaddition product 9 in 8 and 64% yields, respectively. Upon chromatography, 9 tended to undergo desilylation to 10, which was also rather unstable (eq 8). A mechanism involving attack of the alkyne at the carbonyl carbon of one ketenyl group followed by cyclization was suggested.[29]

Ethoxy(trimethylsilyl)acetylene reacted by a [2 + 2 + 2] cycloaddition with diyne 11 in the presence of 10 mol% $Rh(cod)_2$-BF_4 and 10 mol % rac-BINAP in dichloromethane (0.021 M) at ambient temperature to give aryl ether derivative 12 in 50% yield, thus providing a valuable synthetic route to highly substituted benzene derivatives (eq 9).[30]

Electron-rich alkynes, including ethoxy(trimethylsilyl)-acetylene, reacted with chromium alkenylcarbene complexes at $-10\,°C$ in acetonitrile in the presence of a stoichiometric amount of $Ni(cod)_2$ with warming to $20\,°C$ over 2 h to afford moderate yields (40–49%) of cyclopentenones (eq 10). This completely stereoselective nickel(0)-mediated [3 + 2] cyclization reaction of chromium alkenyl(methoxy)carbene complexes with both electron-withdrawing and electron-donating substituted alkynes provides a general synthesis of substituted 2-cyclopentenone derivatives, important synthons for the construction of more complex molecules.[31]

(5)

(6)

(7)

Oxidations. Alkynes of high nucleophilicity such as ethoxy-(trimethylsilyl)acetylene react with electrophilic O_3 to give vicinal dicarbonyl derivatives. In contrast to alkylated or arylated acetylenes, neither products of complete C–C cleavage nor peroxidic materials were detected as primary products. Ethoxy-(trimethylsilyl)acetylene reacted with ozone to yield a mixture of ethyl 2-oxo-2-(trimethyl)acetate (**13**) and ethyl trimethylsilyl oxalate (**14**) (eq 11). The mechanism of this reaction was also discussed.[32]

$$ TMS \equiv OEt \xrightarrow[\text{CH}_2\text{Cl}_2,\ -78\ °C]{\text{O}_3} $$

$$ (11) $$

84% **(13)**   16% **(14)**

(7)

8% **(8)**   64% **(9)**

(10)

(11)

Rh(cod)$_2$BF$_4$, (±)-BINAP

CH$_2$Cl$_2$, rt
50%

(12)

$$ (9) $$

R = Ph, p-MeOC$_6$H$_4$, 2-furyl, t-Bu

1. Ni(cod)$_2$, MeCN
2. SiO$_2$

$$ (10) $$

40–49%

1. Kita, Y.; Akai, S.; Ajimura, N.; Yoshigi, M.; Tsugoshi, T.; Yasuda, H.; Tamura, Y., *J. Org. Chem.* **1986**, *51*, 4150.

2. Kita, Y.; Akai, S.; Yamamoto, M.; Taniguchi, M.; Tamura, Y., *Synthesis* **1989**, 334.

3. Ruden, R. A., *J. Org. Chem.* **1974**, *39*, 3607.

4. Danheiser, R. L.; Sard, H., *Tetrahedron Lett.* **1983**, *24*, 23.

5. Shchukovskaya, L. L.; Pal'chik, R. I., *Izv. Akad. Nauk SSSR, Ser. Khim.* **1964**, 2228.

6. Eglinton, G.; Jones, E. R. H.; Shaw, B. L.; Whiting, M. C., *J. Chem. Soc* **1954**, 1860.

7. Kita, Y.; Okunaka, R.; Honda, T.; Shindo, M.; Taniguchi, M.; Kondo, M.; Sasho, M., *J. Org. Chem.* **1991**, *56*, 119.

8. Tamura, Y.; Sasho, M.; Ohe, H.; Akai, S.; Kita, Y., *Tetrahedron Lett.* **1985**, *26*, 1549. Tamura, Y.; Akai, S.; Kishimoto, H.; Kirihara, M.; Sasho, M.; Kita, Y., *Tetrahedron Lett.* **1987**, *28*, 4583.

9. Kita, Y.; Kirihara, M.; Sekihachi, J.; Okunaka, R.; Sasho, M.; Mohri, S.; Honda, T.; Akai, S.; Tamura, Y.; Shimooka, K., *Chem. Pharm. Bull.* **1990**, *38*, 1836.

10. Kita, Y.; Sekihachi, J.; Hayashi, Y.; Da, Y.-Z.; Yamamoto, M.; Akai, S., *J. Org. Chem.* **1990**, *55*, 1108.

11. Kita, Y.; Matsuda, S.; Kitagaki, S.; Tsuzuki, Y.; Akai, S., *Synlett* **1991**, 401.

12. Akai, S.; Tsuzuki, Y.; Matsuda, S.; Kitagaki, S.; Kita, Y., *J. Chem. Soc., Perkin Trans. 1* **1992**, 2813.

13. Zaitseva, G. S.; Novikova, O. P.; Livantsova, L. I.; Petrosyan, V. S.; Baukov, Yu. I., *Zh. Obshch. Khim.* **1991**, *61*, 1389.

14. Kostyuk, A. S.; Knyaz'kov, K. A.; Ponomarev, S. V.; Lutsenko, I. F., *Zh. Obshch. Khim.* **1985**, *55*, 2088.

15. Kolesnikova, O. N.; Livantsova, L. I.; Shurov, S. N.; Zaitseva, G. S.; Andreichikov, Yu. S., *Zh. Obshch. Khim.* **1990**, *60*, 467.

16. Tsuda, T.; Hasegawa, N.; Saegusa, T., *J. Chem. Soc., Chem. Commun.* **1990**, 945.

17. Kita, Y.; Tsuzuki, Y.; Kitagaki, S.; Akai, S., *Chem. Pharm. Bull.* **1994**, *42*, 233.

18. Kita, Y.; Iio, K.; Okajima, A.; Takeda, Y.; Kawaguchi, K.; Whelan, B. A.; Akai, S., *Synlett* **1998**, 292.

19. Iio, K.; Ramesh, N. G.; Okajima, A.; Higuchi, K.; Fujioka, H.; Akai, S.; Kita, Y., *J. Org. Chem.* **2000**, *65*, 89.

20. Iio, K.; Okajima, A.; Takeda, Y.; Kawaguchi, K.; Whelan, B. A.; Akai, S.; Kita, Y., *ARKIVOC* **2003**, *viii*, 144

21. Kita, Y.; Higuchi, K.; Yoshida, Y.; Iio, K.; Kitagaki, S.; Ueda, K.; Akai, S.; Fujioka, H., *J. Am. Chem. Soc.* **2001**, *123*, 3214.

22. Cox, C. D.; Siu, T.; Danishefsky, S. J., *Angew. Chem., Int. Ed.* **2003**, *42*, 5625.

23. Shair, M. D.; Yoon, T.; Danishefsky, S. J., *Angew. Chem., Int. Ed. Engl.* **1995**, *34*, 1721.

24. Imoto, S.; Haruta, Y.; Watanabe, K.; Sasaki, S., *Bioorg. Med. Chem. Lett.* **2004**, *14*, 4855.

25. Qian, H.; Mathiowitz, E., *Macromolecules* **2007**, *40*, 7748.

26. Sakata, K.; Hamase, K.; Sasaki, S.; Maeda, M.; Zaitsu, K., *Anal. Sci.* **1999**, *15*, 1095.

27. Shchukovskaya, L. L.; Pal'chilk, R. I.; Lazarev, A. N., *Dokl. Akad. Nauk. SSSR* **1965**, *164*, 357.

28. Oblin, M.; Rajzmann, M.; Pons, J.-M., *Tetrahedron* **1997**, *53*, 8165.

29. Colomvakos, J. D.; Egle, I.; Ma, J.; Pole, D. L.; Tidwell, T. T.; Warkentin, J., *J. Org. Chem.* **1996**, *61*, 9522.

30. Clayden, J.; Moran, W. J., *Org. Biomol. Chem.* **2007**, *5*, 1028.

31. Barluenga, J.; Barrio, P.; Riesgo, L.; López, L. A.; Tomás, M., *J. Am. Chem. Soc.* **2007**, *129*, 14422.

32. Schank, K.; Beck, H.; Werner, F., *Helv. Chim. Acta* **2000**, *83*, 1611.

Ethyl Lithio(trimethylsilyl)acetate[1]

[54886-62-3]   C$_7$H$_{15}$LiO$_2$Si   (MW 166.25)

InChI = 1S/C7H16O2Si.Li/c1-5-9-7(8)6-10(2,3)4;/h6,8H,5H2,
1-4H3;/q;+1/p-1

InChIKey = SCBUGTRKOCPYLZ-UHFFFAOYSA-M

(reacts with carbonyl compounds to give α,β-unsaturated esters[2])

Solubility: sol ethereal solvents; reacts with protic solvents.

Preparative Methods: available by reaction of ethyl trimethylsilylacetate with lithium diisopropylamide, lithium isopropylcyclohexylamide, or lithium dicyclohexylamide in THF at low temperature.[1b,2]

Handling, Storage, and Precautions: as with any other organometallic agent, the reagent should be used under an inert atmosphere.

Peterson Alkenation. The use of a Peterson approach for the preparation of α,β-unsaturated esters can hold a number of advantages over a Wittig reaction or one of its derivatives.[2] In addition to an easily removed byproduct, hexamethyldisiloxane, silyl esters can provide the desired α,β-unsaturated esters in moderate to good yields when a Wittig approach fails.[1a,3] Stereoselection can also be observed. In some cases, this can be increased by use of a magnesium enolate and acidic workup.[1a,4] While the analogous Wittig approach requires the use of a stabilized ylide and, hence, usually results in the formation of just the (*E*) isomer (or the thermodynamically more stable isomer) of the α,β-unsaturated ester product, the Peterson reaction can either provide a mixture of the isomeric product alkenes, or the (*Z*) isomer if selection is seen.[5] The stereochemical differences can be interpreted in terms of the Wittig reaction being under thermodynamic control, while the Peterson method is kinetically controlled. Thus stereochemical control during the addition of the silicon reagent can result in stereoselective formation of the α,β-unsaturated ester (eq 1).[6]

Alkylations. Alkylation of the title enolate (**1**) with alkyl iodides in the presence of HMPA allows for the preparation of higher α-silyl esters.[7]

Ester enolate (**1**) can be alkylated by vinyl halides in the presence of a nickel catalyst to afford α-trimethylsilyl-β,γ-unsaturated esters. These products react with a wide variety of electrophiles in the presence of a Lewis acid to provide γ-subsituted α,β-unsaturated esters (eq 2).[8]

Other Electrophiles. In addition to carbonyl compounds, ester enolate (**1**) also reacts with other electrophiles. With nitrones, the product is dependent upon the structure of the nitrone: α,N-dialkyl nitrones provide alkenes, while α-aryl-N-alkyl nitrones or α,N-diaryl nitrones usually give aziridines.[9] With the phenylhydrazone of a 1,2-dicarbonyl compound, reaction with (**1**) provides a convenient preparation of 3(2*H*)-pyridazinones (eq 3).[10]

Analogs. The ester enolates derived from methyl or *t*-butyl trimethylsilylacetate react in an analogous manner to ethyl ester (**1**) with carbonyl compounds.[1a,3,11] However, with the *t*-butyl ester the carbonyl reactant has to be an aldehyde as steric problems result in enolization of ketonic substrates.[12] As with the ethyl ester, addition of the enolate to the carbonyl substrate may allow for stereochemical control of the resultant α,β-unsaturated ester geometry.[5c,13] In addition, the groups attached to silicon can be modified without substantial changes to the reactivity with carbonyl compounds.[14]

It is also possible to introduce additional functionality into the α-silyl ester. Thus *t*-butyl (trimethylsilyl)chloroacetate reacts with carbonyl compounds after ester enolate formation with LDA to form *t*-butyl α-chloro-α,β-unsaturated esters, although the elimination of the silyl moiety may have to be encouraged by the use of thionyl chloride,[3,15] as it also is with the α-bromo analog.[16] A second silyl group, with its additional bulk, can allow for high stereoselection, although the outcome does depend on the metal counterion used in the enolate.[12,17]

Analogs of the ester enolate derived from higher carboxylic acids can also be prepared by the addition of an organometallic agent to methyl 2-trimethylsilylacrylate (**2**). The resultant ester enolate from the Michael addition can be used for a subsequent Peterson alkenation reaction when reacted with a carbonyl compound (eq 4).[18]

Avoid Skin Contact with All Reagents

$$(4)$$

(2)

M = Li, MgX or Cu•MgX

The sterically demanding t-butyl ester enolate reacts with acylimidazoles to provide β-keto esters (eq 5).[19]

$$(5)$$

The ester enolates derived from the methyl or t-butyl esters react with lactones to yield vinyl ethers as a mixture of isomers (eq 6).[20]

$$(6)$$

R = Me or t-Bu

With cyclopentenone derivatives, 1,4-addition is observed for the ester enolates of methyl and t-butyl trimethylsilylacetate,[21] although 1,2-addition occurs with acyclic conjugated enals.[11,12] With a steroidal cyclopentenone substrate, both 1,2- and 1,4-addition were observed.[22] Conjugate addition is observed for the methyl ester with chiral vinyl sulfoxides. High enantioselectivity can be attained (eq 7).[23]

$$(7)$$

As with the ethyl ester, other ester derivatives can be alkylated through their lithium enolates.[24] Use of the menthyl ester provides a route to chiral silanes (eq 8).[25]

$$(8)$$

Related Reagents. t-Butyl α-Lithiobis(trimethylsilyl)acetate; t-Butyl Trimethylsilylacetate; Ethyl Bromozincacetate; Ethyl Lithioacetate; Ethyl Trimethylsilylacetate; Ketene Bis(trimethylsilyl) Acetal; Ketene t-Butyldimethylsilyl Methyl Acetal; 1-Methoxy-2-trimethylsilyl-1-(trimethylsilyloxy)ethylene; Methyl (Methyldiphenylsilyl)acetate; Methyl 2-Trimethylsilylacrylate; Triethyl Phosphonoacetate; Trimethylsilylacetic Acid.

1. (a) Ager, D. J., *Org. React.* **1990**, *38*, 1. (b) Ager, D. J., *Synthesis* **1984**, 384; (c) Fleming, I. In *Comprehensive Organic Chemistry*; Barton, D. H. R.; Ollis, W. D., Eds.; Pergamon: Oxford, 1979; Vol. 3, p 541. (d) Birkofer, L.; Stuhl, O., *Top. Curr. Chem.* **1980**, *88*, 33.
2. Shimoji, K.; Taguchi, H.; Oshima, K.; Yamamoto, H.; Nozaki, H., *J. Am. Chem. Soc.* **1974**, *96*, 1620.
3. Crimmin, M. J.; O'Hanlon, P. J.; Rogers, N. H., *J. Chem. Soc., Perkin Trans. 1* **1985**, 541.
4. Larchevêque, M.; Debal, A., *J. Chem. Soc., Chem. Commun.* **1981**, 877.
5. (a) Černý, I.; Pouzar, V.; Drašar, P.; Tureček, F.; Havel, M., *Collect. Czech. Chem. Commun.* **1986**, *51*, 128. (b) Novák, L.; Rohály, J.; Poppe, L.; Hornyánszky, G.; Kolonits, P.; Zelei, I.; Fehér, I.; Fekete, J.; Szabó, E.; Záhorszky, U.; Jávor, A.; Szántay, C., *Liebigs Ann. Chem.* **1992**, 145. (c) Larson, G. L.; Prieto, J. A.; Hernández, A., *Tetrahedron Lett.* **1981**, *22*, 1575. (d) Strekowski, L.; Visnick, M.; Battiste, M. A., *Tetrahedron Lett.* **1984**, *25*, 5603. (e) Szychowski, J.; MacLean, D. B., *Can. J. Chem.* **1979**, *57*, 1631.
6. Pak, H.; Dickson, J. K.; Fraser-Reid, B., *J. Org. Chem.* **1989**, *54*, 5357. Drian, C. L.; Greene, A. E., *J. Am. Chem. Soc.* **1982**, *104*, 5473. Greene, A. E.; Drian, C. L.; Crabbé, P., *J. Org. Chem.* **1980**, *45*, 2713.
7. Cunico, R. F., *J. Org. Chem.* **1990**, *55*, 4474.
8. Albaugh-Robertson, P.; Katzenellenbogen, J. A., *J. Org. Chem.* **1983**, *48*, 5288.
9. Tsuge, O.; Sone, K.; Urano, S.; Matsuda, K., *J. Org. Chem.* **1982**, *47*, 5171.
10. Patel, H. V.; Vyas, K. A.; Pandey, S. P.; Tavares, F.; Fernandes, P. S., *Synth. Commun.* **1991**, *21*, 1935.
11. Tulshian, D. B.; Fraser-Reid, B., *J. Am. Chem. Soc.* **1981**, *103*, 474.
12. Hartzell, S. L.; Rathke, M. W., *Tetrahedron Lett.* **1976**, 2737.
13. Larcheveque, M.; Legueut, C.; Debal, A.; Lallemand, J. Y., *Tetrahedron Lett.* **1981**, *21*, 1595.
14. Larson, G. L.; Quiroz, F.; Suárez, J., *Synth. Commun.* **1983**, *13*, 833.
15. Chan, T. H.; Moreland, M., *Tetrahedron Lett.* **1978**, 515.
16. Zapata, A.; Ferrer, G., F., *Synth. Commun.* **1986**, *16*, 1611.
17. Boeckman, R. K.; Chinn, R. L., *Tetrahedron Lett.* **1985**, *26*, 5005.
18. Tsuge, O.; Kanemasa, S.; Ninomiya, Y., *Chem. Lett.* **1984**, 1993.
19. Hartzell, S. L.; Rathke, M. W., *Tetrahedron Lett.* **1976**, 2757.
20. Sauvé, G.; Deslongchamps, P., *Synth. Commun.* **1985**, *15*, 201; Takahashi, A.; Kirio, Y.; Sodeoka, M.; Sasai, H.; Shibasaki, M., *J. Am. Chem. Soc.* **1989**, *111*, 643.
21. Nishiyama, H.; Sakuta, K.; Itoh, K., *Tetrahedron Lett.* **1984**, *25*, 2487. Oppolzer, W.; Guo, M.; Baettig, K., *Helv. Chim. Acta* **1983**, *66*, 2140.
22. Wicha, J.; Kabat, M. M., *J. Chem. Soc., Perkin Trans. 1* **1985**, 1601.
23. Posner, G. H.; Weitzberg, M.; Hamill, T. G.; Asirvatham, E.; Cun-Heng, H.; Clardy, J., *Tetrahedron* **1986**, *42*, 2919.
24. Hudrlik, P. F.; Peterson, D.; Chou, D., *Synth. Commun.* **1975**, *5*, 359; Paquette, L. A.; Maynard, G. D.; Ra, C. S.; Hoppe, M., *J. Org. Chem.* **1989**, *54*, 1408.
25. Paquette, L. A.; Gilday, J. P.; Ra, C. S.; Hoppe, M., *J. Org. Chem.* **1988**, *53*, 704. Gilday, J. P.; Gallucci, J. C.; Paquette, L. A., *J. Org. Chem.* **1989**, *54*, 1399.

David J. Ager
The NutraSweet Company, Mount Prospect, IL, USA

Ethyl (Methyldiphenylsilyl)acetate[1]

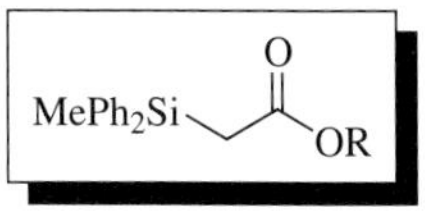

(R = Et)
[13950-57-7] $C_{17}H_{20}O_2Si$ (MW 284.46)
InChI = 1S/C17H20O2Si/c1-3-19-17(18)14-20(2,15-10-6-4-7-
    11-15)16-12-8-5-9-13-16/h4-13H,3,14H2,1-2H3
InChIKey = MFIAJOGCRUYZQO-UHFFFAOYSA-N
(R = Me)
[89266-73-9] $C_{16}H_{18}O_2Si$ (MW 270.43)
InChI = 1S/C16H18O2Si/c1-18-16(17)13-19(2,14-9-5-3-6-10-
    14)15-11-7-4-8-12-15/h3-12H,13H2,1-2H3
InChIKey = PDQUXEJAGLPTDH-UHFFFAOYSA-N
(R = *i*-Pr)
[87776-13-4] $C_{18}H_{22}O_2Si$ (MW 298.49)
InChI = 1S/C18H22O2Si/c1-15(2)20-18(19)14-21(3,16-10-6-4-
    7-11-16)17-12-8-5-9-13-17/h4-13,15H,14H2,1-3H3
InChIKey = BTVMSVMHNBLMQQ-UHFFFAOYSA-N
(R = *t*-Bu)
[77772-21-5] $C_{19}H_{24}O_2Si$ (MW 312.52)
InChI = 1S/C19H24O2Si/c1-19(2,3)21-18(20)15-22(4,16-11-
    7-5-8-12-16)17-13-9-6-10-14-17/h5-14H,15H2,1-4H3
InChIKey = CBTKWYFEEQGNBF-UHFFFAOYSA-N

(vinyl 1,1-dication synthetic equivalent in reactions with Grignard reagents;[2] reagent for Peterson synthesis of α,β-unsaturated esters[3])

Physical Data: clear to light yellow liquids. R = Et, n_D^{25} 1.5381; R = *i*-Pr, n_D^{26} 1.5398; R = *t*-Bu, n_D^{24} 1.5324.
Preparative Method: whereas the direct silylation of the lithium enolate of an ester normally results in the formation of a mixture of the α-silyl ester and the corresponding silyl ketene acetal, the same reaction with methyldiphenylchlorosilane gives exclusively the α-methyldiphenylsilyl ester.[2d,4] This direct C-silylation is the best general route to α-silyl esters.
Purification: best purified by silica gel chromatography. The esters can be distilled through a suitable short path apparatus.
Handling, Storage, and Precautions: the esters desilylate in acid or alkaline medium. They desilylate very slowly with water or alcohols under neutral conditions.

Alkene Synthesis. The first synthetically successful conversion of an α-silyl ester to an alkene was accomplished by the reaction of ethyl trimethylsilylacetate with 2 equiv of the Grignard reagent formed from high purity magnesium and a primary or aryl halide, followed by elimination of the β-hydroxy silane formed (eq 1).[2a] Although the same reaction has never been reported for ethyl α-(methyldiphenylsilyl)acetate, use has been made of the ready C-methyldiphenylsilylation of the lithium enolates of esters to form ethyl α-alkyl-α-(methyldiphenylsilyl)acetates.[4a] These α-silyl esters react with primary or aryl Grignard reagents to provide, after elimination, trisubstituted alkenes (eqs 2–5).[2c,d] Improved results are often obtained when the first addition is carried out with a Grignard reagent and the second with an organolithium reagent. The use of a Grignard reagent followed by an

organolithium reagent allows the preparation of mixed trisubstituted alkenes (eq 5). The stereoselectivity is excellent when the elimination step is accomplished under basic conditions, but only moderate when carried out under acidic conditions.

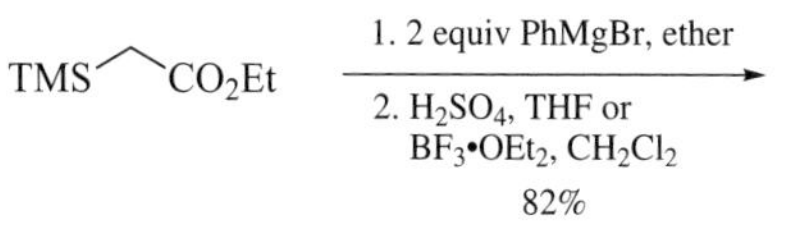

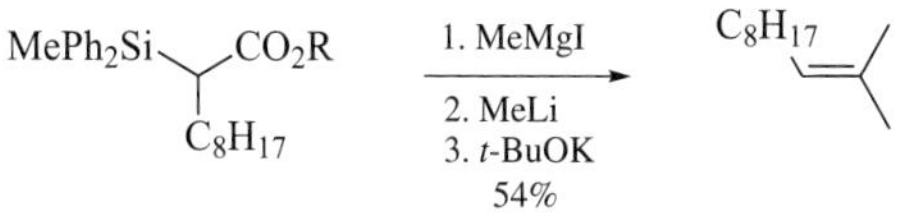

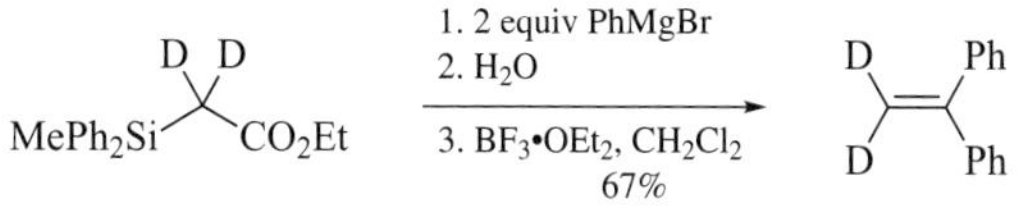

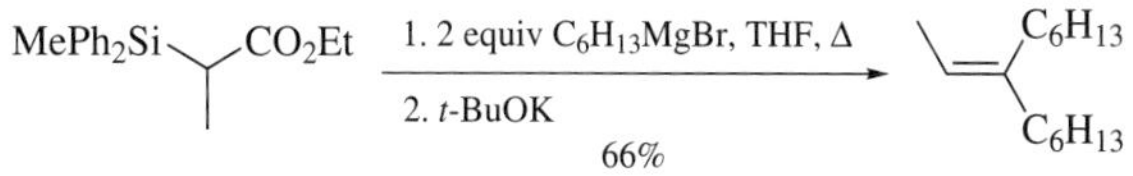

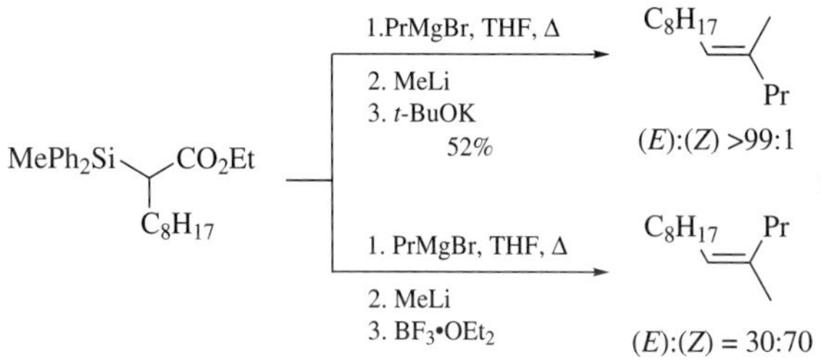

Tetrasubstituted alkenes can be prepared in an analogous fashion from α,α-dialkyl-α-(methyldiphenylsilyl)acetate, prepared by alkylation of the corresponding α-alkyl-α-silyl ester (eqs 6 and 7).[4b] These reactions are, however, strongly influenced by the steric demands of the highly substituted α-carbon and therefore tend to give ketones even with an excess of the Grignard reagent. Alkenes are obtained only with the Grignard/organolithium reagent combination, and even this sequence sometimes succumbs to steric strain. No reports of the use of cerium reagents in these transformations have appeared.

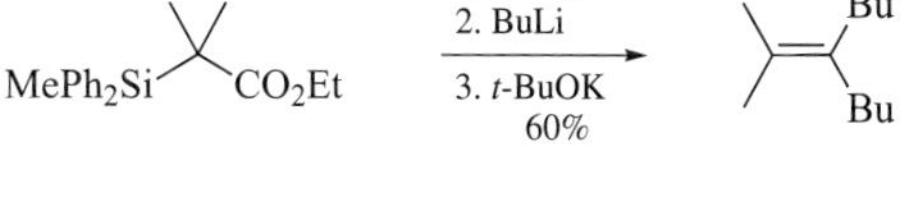

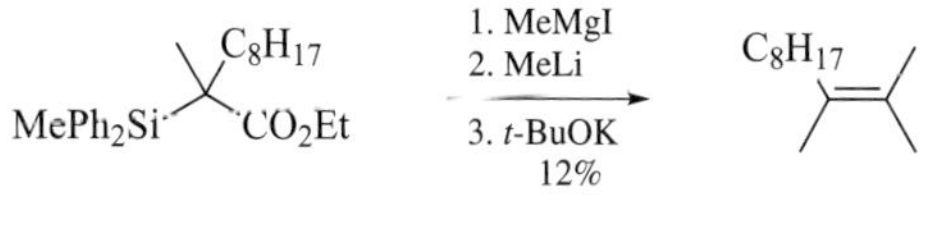

Synthesis of α,β-Unsaturated Esters. Yamamoto and coworkers[3c] and others[3d,e] have shown that α-trimethylsilyl acetates effect Peterson alkenation of ketones and aldehydes. The lithium enolates of α-(methyldiphenylsilyl) esters react similarly with ketones and aldehydes to give α,β-unsaturated esters (eq 8).[3a,b] In the case of α-silyl acetates, no particular advantage of either the trimethylsilyl or methyldiphenylsilyl group over the other is apparent. In the case of α-substituted α-silylacetates, the methyldiphenylsilyl group has the distinct advantage of being directly prepared by methyldiphenylsilylation of the lithium enolate of the ester.[1b] The (E)/(Z) stereoselectivity of the α,β-unsaturated ester synthesis shows no correlation with the steric bulk of the alcohol portion of the α-(methyldiphenylsilyl)acetates. However, the (E)/(Z) ratio is affected by the temperature and mode of addition. It appears that the elimination step in the process leads to a mixture of diastereomeric alkenes.[3a,b]

$$\text{MePh}_2\text{Si} \diagup \text{CO}_2\text{R} \xrightarrow[\text{2. } i\text{-PrCHO}]{\text{1. LDA, THF, }-78\,°\text{C}} \quad \overset{i\text{-Pr}}{\diagdown} \diagup_{\text{CO}_2\text{R}} \qquad (8)$$

R = Et, i-Pr, i-Bu, (–)-menthyl

A Michael addition of the silyl enolate was employed in a short synthesis of (±)-methyl jasmonate from cyclopentenone (eqs 9 and 10).[5] This convergent scheme was carried out in three steps: conjugate addition of methyl α-(methyldiphenylsilyl)lithioacetate to cyclopentenone, alkylation of the resulting enolate with (Z)-1-bromopent-2-ene, and desilylation with potassium fluoride (±)-ethyl jasmonate was prepared in a similar fashion. In the conjugate addition step, the α-(methyldiphenylsilyl)ester gave superior results to those obtained with the α-trimethylsilyl esters.

$$\text{MePh}_2\text{Si} \diagup \text{CO}_2\text{Me} \xrightarrow[\substack{\text{2. cyclopentenone} \\ \text{3. Br} \diagup \diagdown \text{Et}}]{\text{1. BuLi, THF, }-78\,°\text{C}} \qquad (9)$$

$$\xrightarrow{\text{KF, MeOH}} \qquad (10)$$

Related Reagents. *t*-Butyl Trimethylsilylacetate; *N,N*-Dimethyl-2-(trimethylsilyl)acetamide; Ethyl 2-(Methyldiphenylsilyl)propanoate; Ethyl Trimethylsilylacetate; Trimethylsilylacetic Acid; Trimethylsilylacetone.

1. Larson, G. L., *Pure Appl. Chem.* **1990**, *62*, 2021.
2. (a) Larson, G. L.; Hernández, D., *Tetrahedron Lett.* **1982**, *23*, 1035. (b) Hernandez, D. Doctoral Dissertation, University of Puerto Rico, 1984. (c) Hernández, D; Larson, G. L., *J. Org. Chem.* **1984**, *49*, 4285. (d) Cruz de Maldonado, V.; Larson, G. L., *Synth. Commun.* **1983**, *13*, 1163. (e) Larson, G. L.; Lopez-Cepero, I. M.; Mieles, L. R., *Org. Synth., Coll. Vol.* **1993**, *8*, 474.
3. (a) Larson, G. L.; Fernandez de Kaifer, C.; Seda, R.; Torres, L. E.; Ramirez, J. R., *J. Org. Chem.* **1984**, *49*, 3385. (b) Larson, G. L.; Quiroz, F.; Suárez, J., *Synth. Commun.* **1983**, *13*, 833. For related papers on the use of α-trimethylsilylacetates in the synthesis of α,β-unsaturated esters, see: (c) Shimoji, K.; Taguchi, H.; Oshima, K.; Yamamoto, H.; Nozak, H., *J. Am. Chem. Soc.* **1974**, *96*, 1620. (d) Hartzell, S. L.; Sullivan, D. F.; Rathke, M. W., *Tetrahedron Lett.* **1974**, 1403. (e) Larcheveque, M.; Debal, A., *J. Chem. Soc., Chem. Commun.* **1981**, 877.
4. (a) Larson, G. L.; Fuentes, L. M., *J. Am. Chem. Soc.* **1981**, *103*, 2418. (b) Larson, G. L.; Cruz de Maldonado, V.; Fuentes, L. M.; Torres, L. E., *J. Org. Chem.* **1988**, *53*, 633.
5. Oppolzer, W.; Modao, G.; Baettig, K., *Helv. Chim. Acta* **1983**, *66*, 2140.

Gerald L. Larson
Hüls America, Piscataway, NJ, USA

Ethyl 2-(Methyldiphenylsilyl)propanoate[1]

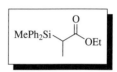

[77772-22-6]   C$_{18}$H$_{22}$O$_2$Si   (MW 298.49)

InChI = 1S/C18H22O2Si/c1-4-20-18(19)15(2)21(3,16-11-7-5-8-12-16)17-13-9-6-10-14-17/h5-15H,4H2,1-3H3

InChIKey = KJOXBWOJYZDXAD-UHFFFAOYSA-N

(precursor to 1,1-disubstituted 1-propenes;[1c,2] reacts with primary Grignard reagents to give 2-(methyldiphenylsilyl)-3-alkanones,[3] which in turn lead to regioselectively derived methyldiphenylsilyl enol ethers[4] or to 3-alkanones; reagent for the synthesis of α-methyl-α,β-unsaturated esters[5])

Alternate Name: ethyl 2-(methyldiphenylsilyl)propionate.
Physical Data: clear to pale yellow liquid; n_D^{24} 1.5407.
Preparative Method: whereas the direct silylation of the lithium enolate of an ester normally results in the formation of a mixture of the α-silyl ester and the corresponding silyl ketene acetal, the reaction of lithium ester enolates with methyldiphenylchlorosilane gives exclusively the α-methyldiphenylsilyl ester.[1b,c] This direct *C*-silylation is the best general route to α-silyl esters.
Purification: can be purified by silica gel chromatography, eluting with ethyl acetate/hexane (2:98 v/v), or by short path distillation.
Handling, Storage, and Precautions: this compound and other α-silyl carbonyl compounds are hydrolyzed under acidic or basic conditions to give the parent desilylated esters.

1,1-Disubstituted 1-Propenes. The reaction of ethyl (methyldiphenylsilyl)propanoate (**1**) with primary or aryl Grignard reagents results in the formation of the α-silyl ketone or a 1,1-disubstituted 1-propene.[1c,2] For example, the reaction with phenylmagnesium bromide in refluxing THF for several hours leads to 1,1-diphenyl-1-propene in good yield (eq 1). In general, the more sterically demanding the Grignard reagent, the lower the yield of the alkene. The synthesis of alkenes from other more sterically hindered α-(methyldiphenylsilyl) esters requires a sequential reaction of the ester with a Grignard reagent followed by an organolithium reagent (eq 2).

$$\text{MePh}_2\text{Si} \diagdown_{\text{OEt}}^{\text{O}} \xrightarrow[\text{2. } t\text{-BuOK}]{\substack{\text{1. 2 equiv PhMgBr, THF, }\Delta \\ \text{85\%}}} \quad \diagdown_{\text{Ph}}^{\text{Ph}} \qquad (1)$$

$$\text{MePh}_2\text{Si}-\overset{\text{O}}{\overset{\|}{\underset{\text{C}_8\text{H}_{17}}{\text{C}}}}-\text{OEt} \xrightarrow[\substack{3.\ t\text{-BuOK} \\ 54\%}]{\substack{1.\ \text{MeMgI} \\ 2.\ \text{MeLi}}} \text{C}_8\text{H}_{17}\diagup\diagdown \qquad (2)$$

$$\text{MePh}_2\text{Si}-\overset{\text{O}}{\overset{\|}{\text{C}}}-\text{OEt} \xrightarrow[\substack{3.\ \text{TMSCl} \\ 65\%}]{\substack{1.\ \text{LDA, THF, }-78\,^\circ\text{C} \\ 2.\ \text{acrolein}}} \diagup\diagup\diagdown\text{CO}_2\text{Et} \qquad (8)$$
$$(Z){:}(E) = 77{:}23$$

α-Silyl Ketones, Silyl Enol Ethers, and Ketones. The reaction of (**1**) with Grignard reagents leads to the corresponding 2-(methyldiphenylsilyl)-3-alkanone (eq 3).[3] This reaction is favored when the acid portion of the ester is larger than propionyl. The reaction proceeds through formation of a magnesium enolate on the side of the α-silyl ketone opposite to that of the bulky methyldiphenylsilyl group (eq 4). The direct conversion of α-(methyldiphenylsilyl) esters to ketones can be accomplished by treatment of the ester with a Grignard reagent followed by protiodesilylation with potassium fluoride/methanol (eq 5).

Related Reagents. *t*-Butyl Trimethylsilylacetate; *N,N*-Dimethyl-2-(trimethylsilyl)acetamide; Ethyl (Methyldiphenylsilyl)acetate; Ethyl Trimethylsilylacetate; Trimethylsilylacetic Acid; Trimethylsilylacetone.

1. (a) Larson, G. L., *Pure Appl. Chem.* **1990**, *62*, 2021. (b) Larson, G. L.; Fuentes, L. M., *J. Am. Chem. Soc.* **1981**, *103*, 2418. (c) Larson, G. L.; Lopez-Cepero, I. M.; Mieles, L. R., *Org. Synth., Coll. Vol.* **1993**, *8*, 474.
2. Hernández, D; Larson, G. L., *J. Org. Chem.* **1984**, *49*, 4285.
3. Larson, G. L.; Cruz de Maldonado, V.; Fuentes, L. M.; Torres, L. E., *J. Org. Chem.* **1988**, *53*, 633.
4. Larson, G. L.; Berrios, R.; Prieto, J. A., *Tetrahedron Lett.* **1989**, *30*, 283.
5. Larson, G. L.; Fernandez de Kaifer, C.; Seda, R.; Torres, L. E.; Ramirez, J. R., *J. Org. Chem.* **1984**, *49*, 3385.

Gerald L. Larson
Hüls America, Piscataway, NJ, USA

$$\text{MePh}_2\text{Si}-\overset{\text{O}}{\overset{\|}{\underset{\text{C}_8\text{H}_{17}}{\text{C}}}}-\text{OEt} \xrightarrow[\substack{2.\ \text{moist ether} \\ 93\%}]{1.\ \text{PrMgBr}} \text{MePh}_2\text{Si}-\overset{\text{O}}{\overset{\|}{\underset{\text{C}_8\text{H}_{17}}{\text{C}}}}-\text{Pr} \qquad (3)$$

Ethyl Trimethylsilylacetate[1]

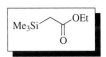

$$[4071\text{-}88\text{-}9] \qquad\qquad \text{C}_7\text{H}_{16}\text{O}_2\text{Si} \qquad\qquad (\text{MW } 160.29)$$
InChI = 1S/C7H16O2Si/c1-5-9-7(8)6-10(2,3)4/h5-6H2,1-4H3
InChIKey = QQFBQBDINHJDMN-UHFFFAOYSA-N

$$\text{MePh}_2\text{Si}-\overset{\text{O}}{\overset{\|}{\underset{\text{C}_8\text{H}_{17}}{\text{C}}}}-\text{OEt} \xrightarrow{\text{PrMgBr}} \left[\text{MePh}_2\text{Si}-\overset{\text{OMgBr}}{\overset{|}{\underset{\text{C}_8\text{H}_{17}}{\text{C}}}}\diagdown\diagup \right] \xrightarrow[\substack{\text{HMPA} \\ 87\%}]{\text{MeI}}$$

$$\text{MePh}_2\text{Si}-\overset{\text{O}}{\overset{\|}{\underset{\text{C}_8\text{H}_{17}}{\text{C}}}}\diagdown\overset{}{\underset{}{}}\diagup\diagdown \qquad (4)$$

(silylating agent;[2] source of an ethyl acetate anion equivalent[3])

Alternate Name: ETSA.
Physical Data: bp 157–158 °C; *d* 0.876 g cm^{-3}.
Solubility: sol ethereal and chlorinated solvents; reacts with protic solvents.
Form Supplied in: liquid (>98%); commercially available.
Preparative Methods: available by a Reformatsky reaction from ethyl bromoacetate,[4] and by reaction of trimethylsilylmethylmagnesium chloride with ethyl chloroformate.[5] An alternative approach requires the treatment of ethyl acetate with triphenylmethylsodium followed by chlorotrimethylsilane[6] The use of a nitrogen base with ethyl acetate in THF followed by reaction with chlorotrimethylsilane results in a mixture of *C*- and *O*-silylation. The use of HMPA as additive in the reaction medium increases the amount of *O*-silylation to 90%.[7] Similar methods can be used to prepare analogs.
Handling, Storage, and Precautions: reacts with protic solvents to give the desilylated product.

$$\text{MePh}_2\text{Si}-\overset{\text{O}}{\overset{\|}{\text{C}}}-\text{OEt} \xrightarrow[\substack{2.\ \text{KF, MeOH} \\ 71\%}]{1.\ \text{C}_9\text{H}_{19}\text{MgBr, THF, }\Delta} \text{Et}-\overset{\text{O}}{\overset{\|}{\text{C}}}-\text{C}_9\text{H}_{19} \qquad (5)$$

The α-(methyldiphenylsilyl) ketones isomerize to the corresponding silyl enol ether regioselectively simply by mild thermolysis at about 150 °C (eq 6). This reaction can also be done in a stereoselective fashion to give the (*Z*)-enol silyl ether when the thermolysis is carried out in acetonitrile (eq 7).[4]

$$\text{MePh}_2\text{Si}-\overset{\text{O}}{\overset{\|}{\underset{\text{C}_8\text{H}_{17}}{\text{C}}}}-\text{Pr} \xrightarrow[\substack{\text{neat} \\ 100\%}]{140\text{-}160\,^\circ\text{C}} \overset{\text{C}_8\text{H}_{17}}{\underset{\text{Pr}}{}}\diagdown\diagup\overset{\text{OSiPh}_2\text{Me}}{} \qquad (6)$$
$$(Z){:}(E) = 70{:}30$$

$$\text{MePh}_2\text{Si}-\overset{\text{O}}{\overset{\|}{\underset{\text{C}_8\text{H}_{17}}{\text{C}}}}-\text{Pr} \xrightarrow[\substack{\text{CD}_3\text{CN} \\ 100\%}]{140\text{-}160\,^\circ\text{C}} \overset{\text{C}_8\text{H}_{17}}{\underset{\text{Pr}}{}}\diagdown\diagup\overset{\text{OSiPh}_2\text{Me}}{} \qquad (7)$$
$$(Z){:}(E) >99{:}1$$

α-Methyl-α,β-unsaturated Esters. Deprotonation of (**1**) and related compounds followed by reaction of the resulting enolate with aldehydes provides α-methyl(alkyl)-α,β-unsaturated esters in good yields.[5] The (*Z*) diastereomer predominates, with (*Z*)/(*E*) ratios between 90:10 and 79:21. Acrolein reacts in a similar manner, undergoing 1,2-alkenation to give a 77:23 mixture of (*Z*)- and (*E*)-ethyl 2-methyl-2,4-pentadienoates (eq 8).

Ethyl trimethylsilylacetate (**1**) is reactive to nucleophiles and readily undergoes desilylation reactions with acid or alkali, ethanol, and bromine (eq 1).[5,8]

$$\text{TMS} \diagup \text{CO}_2\text{Et} \longrightarrow \text{X} \diagup \text{CO}_2\text{Et} + \text{TMSY} \qquad (1)$$

(1)

H_3O^+	X = H	Y = OTMS
HO^-	= H	= OTMS
HCl	= H	= Cl
Br_2	= Br	= Br
EtOH	= H	= OEt

Silylation Reactions. In the presence of a catalytic amount of tetra-*n*-butylammonium fluoride (TBAF), ester (**1**) is a very efficient silylating agent for a wide variety of substrates including carbonyl compounds, alcohols, phenols, carboxylic acids, and alkynes.[2,9] With unsymmetrical ketones the kinetic enol ether is the preferred product.[2,10] Indeed, the use of (**1**) can provide superior selectivity to the use of hindered bases for enolate formation.[11] This is illustrated with a β,γ-epoxy ester which provides an entry to γ-keto-α,β-unsaturated esters (eq 2).[12] These silylation reactions of (**1**) can also be catalyzed by TBAF supported on silica.[13]

$$ (2) $$

Alkylation Reactions. The use of fluoride ion or a base catalyst allows reaction of the silyl ester (**1**) with electrophilic substrates (eq 3). Some of these reactions are similar to those of the ester enolate derived from ethyl acetate.[3,14,15]

$$ (3) $$

With conjugated enones a carbon–carbon bond is formed by a Michael addition with concomitant formation of the proximal silyl enol ether (eq 4).[16] In the presence of a Lewis acid, (**1**) undergoes conjugate additions with α,β-unsaturated carbonyl compounds (eq 5).[17]

$$ (4) $$

$$ (5) $$

Nitroaromatic compounds react with (**1**) in the presence of potassium fluoride to provide an anionic σ-complex.[18] Subsequent oxidation with 2,3-dichloro-5,6-dicyano-1,4-benzoquinone then provides ethyl arylacetates.[19] With an areneiron(I) complex, the alkylcarbonyl substituent from (**1**) is introduced on to the ring.[20]

Electrophilic Reactions. Silylacetate (**1**) reacts with 2 equiv of a Grignard reagent to provide 1,1-disubstituted alkenes in good to excellent yield (eq 6), but this synthesis is limited to Grignard reagents that are not sterically demanding.[21]

$$ (6) $$

Analogs. The methyl ester (**2**) undergoes similar reactions to the ethyl ester (**1**),[22] as do various silyl analogs.[23] These reactions also include silylations with methyl trimethylsilylacetate in the presence of fluoride ion.[24,25]

Reaction of the lithium enolate of (**1**) followed by reaction with chlorotrimethylsilane provides ketene acetal (**3**). This enol reacts with aldehydes in the presence of a Lewis acid to provide (Z)-α,β-unsaturated esters (eq 7).[26] With conjugated enones, (**3**) undergoes conjugate addition when Titanium(IV) Chloride is used as catalyst (cf. eq 5).[27]

$$ (7) $$

The use of ethyl diphenylmethylsilylacetate (**4**; R^1 = H) provides for some alternative methodology as the larger silyl group allows for the selective addition of only 1 equiv of Grignard reagent to afford the β-silyl ketone (eq 8).[28] The chemistry can also be extended to other carboxylic acid analogs. A number of approaches are then available for the preparation of ketones.[22,29,30]

$$ (8) $$

In addition to these enolate reactions, β-silyl ketones can be thermally isomerized to silyl enol ethers by a regioselective rearrangement (eq 9),[22,31] or treated with another Grignard reagent to provide alkenes through Peterson alkenation reactions.[32]

$$ (9) $$

Many of the reactions of ethyl trimethylsilylacetate and its analogues involve the ester enolate; these are discussed in Ethyl Lithio(trimethylsilyl)acetate.

Related Reagents. *t*-Butyl α-Lithiobis(trimethylsilyl)acetate; *t*-Butyl Trimethylsilylacetate; Dilithioacetate; Ethyl Bromozincacetate; Ethyl Lithioacetate; Ethyl Lithio(trimethylsilyl)acetate; Ketene Bis(trimethylsilyl) Acetal; Ketene *t*-Butyldimethylsilyl Methyl Acetal; 1-Methoxy-2-trimethylsilyl-1-(trimethylsilyloxy)-ethylene; Methyl (Methyldiphenylsilyl)acetate; Trimethylsilyl-acetic Acid.

1. Fleming, I. In *Comprehensive Organic Chemistry*; Barton, D. H. R.; Ollis, W. D., Eds.; Pergamon: Oxford, 1979; Vol. 3; p 541.

2. Nakamura, E.; Murofushi, T.; Shimizu, M.; Kuwajima, I., *J. Am. Chem. Soc.* **1976**, *98*, 2346.

3. Latouche, R.; Texier-Boullet, F.; Hamelin, J., *Tetrahedron Lett.* **1991**, *32*, 1179.

4. Fessenden, R. J.; Fessenden, J. S., *J. Org. Chem.* **1967**, *32*, 3535.

5. Gold, J. R.; Sommer, L. H.; Whitmore, F. C., *J. Am. Chem. Soc.* **1948**, *70*, 2874.

6. Hance, C. R.; Hauser, C. R., *J. Am. Chem. Soc.* **1953**, *75*, 994.

7. Rathke, M. W.; Sullivan, D. F., *Synth. Commun.* **1973**, *3*, 67.

8. Birkofer, L.; Ritter, A., *Angew. Chem., Int. Ed. Engl.* **1965**, *4*, 417.

9. Nakamura, E.; Hashimoto, K.; Kuwajima, I., *Bull. Chem. Soc. Jpn.* **1981**, *54*, 805.

10. Nakamura, E.; Shimizu, M.; Kuwajima, I., *Tetrahedron Lett.* **1976**, 1699.

11. Crimmins, M. T.; Mascarella, S. W.; Bredon, L. D., *Tetrahedron Lett.* **1985**, *26*, 997.

12. Fujisawa, T.; Takeuchi, M.; Sato, T., *Chem. Lett.* **1982**, 1795.

13. Gambacorta, A.; Turchetta, S.; Botta, M., *Synth. Commun.* **1989**, *19*, 2441.

14. Birkofer, L.; Ritter, A.; Wieden, H., *Chem. Ber.* **1962**, *95*, 971 (*Chem. Abstr.* **1962**, *57*, 4690f).

15. Csuk, R.; Glanzer, B. I., *J. Carbohydr. Chem.* **1990**, *9*, 809.

16. RajanBabu, T. V., *J. Org. Chem.* **1984**, *49*, 2083.

17. Matsuda, I.; Murata, S.; Izumi, Y., *J. Org. Chem.* **1980**, *45*, 237.

18. Artamkina, G. A.; Kovalenko, S. V.; Beletskaya, I. P.; Reutov, O. A., *J. Organomet. Chem.* **1987**, *329*, 139.

19. RajanBabu, T. V.; Chenard, B. L.; Petti, M. A., *J. Org. Chem.* **1986**, *51*, 1704.

20. Cambie, R. C.; Coulson, S. A.; Mackay, L. G.; Janssen, S. J.; Rutledge, P. S.; Woodgate, P. D., *J. Organomet. Chem.* **1991**, *409*, 385.

21. Larson, G. L.; Hernández, D., *Tetrahedron Lett.* **1982**, *23*, 1035.

22. Larson, G. L.; Hernández, D.; Montes de Lopez-Cepero, I.; Torres, L. E., *J. Org. Chem.* **1985**, *50*, 5260.

23. Cruz de Maldonado, V.; Larson, G. L., *Synth. Commun.* **1983**, *13*, 1163.

24. Paquette, L. A.; Sugimura, T., *J. Am. Chem. Soc.* **1986**, *108*, 3841.

25. Sugimura, T.; Paquette, L. A., *J. Am. Chem. Soc.* **1987**, *109*, 3017.

26. Matsuda, I.; Izumi, Y., *Tetrahedron Lett.* **1981**, *22*, 1805.

27. Matsuda, I., *J. Organomet. Chem.* **1987**, *321*, 307.

28. Larson, G. L.; Montes de López-Cepero, I.; Torres, L. E., *Tetrahedron Lett.* **1984**, *25*, 1673.

29. Larson, G. L.; Fuentes, L. M., *J. Am. Chem. Soc.* **1981**, *103*, 2418.

30. Larson, G. L.; Quiroz, F.; Suárez, J., *Synth. Commun.* **1983**, *13*, 833.

31. Brook, A. G., *Acc. Chem. Res.* **1974**, *7*, 77.

32. Hernández, D.; Larson, G. L., *J. Org. Chem.* **1984**, *49*, 4285.

David J. Ager
The NutraSweet Company, Mount Prospect, IL, USA

Fluorosilicic Acid

[16961-83-4]   H$_2$F$_6$Si   (MW 144.09)

InChI = 1S/F6Si/c1-7(2,3,4,5)6/q-2/p+2

InChIKey = OHORFAFFMDIQRR-UHFFFAOYSA-P

(a fluoride source with both protic and Lewis acid properties providing efficient cleavage of silicon–oxygen bonds, e.g. silyl ether deprotection)

Physical Data: d 1.220 g cm^{-3} for a 25% aq solution.

Form Supplied in: 25% aq solution, clear and colorless. Upon dehydration, this reagent decomposes giving HF and SiF$_4$.

Handling, Storage, and Precautions: fluorosilicic acid is toxic and very corrosive. Like HF, this compound attacks glass and therefore must be stored in plastic containers. Use the same precautions as for aq HF solutions. Use in a fume hood.

Desilylation Reactions. Fluorosilicic acid is a superior reagent for Si–O bond cleavage applications.[1] One of the common reagents used for desilylation is hydrofluoric acid; however, significant problems can accompany its use. Standard protocols utilizing HF specify a large excess of the reagent, leading to low pH conditions which can decompose acid-sensitive substrates.[2] Hydrofluoric acid also lacks selectivity in removing silyl ethers. For example, a substrate which incorporates both a *t*-butyldimethylsilyl (TBDMS) ether and a triisopropylsilyl (TIPS) ether will undergo cleavage of both protecting groups with little difference in cleavage rates. Of course, this is not problematic if the goal is to remove both groups; however, it is often desirable to remove one group selectively while retaining the other.

For the deprotection of silyl ethers, aq H$_2$SiF$_6$ is superior to aq HF. Fluorosilicic acid is a more potent cleaving agent than HF, allowing its use in stoichiometric or even catalytic quantities (eqs 1 and 2). The lower acid concentrations result in milder reaction conditions that are compatible with several acid labile moieties (see below).

Fluorosilicic acid also has the unique ability to differentiate effectively between TBDMS and TIPS ethers, as seen in the competitive deprotection experiment in eq 3. The presence of bulky cosolvents, such as *t*-butanol, which serve as ligands to silicon, further enhance selectivity at the expense of reaction rate. The reaction time can be reduced by increasing the amount of H$_2$SiF$_6$; however, the accompanying increase in acid concentration precludes the use of acid-labile substrates.

The optimal compromise between selectivity, reaction rate, and acid concentration is achieved by using a 90:10 acetonitrile–*t*-butanol solvent system. Under these conditions, selectivity is only slightly degraded and the reaction rate is relatively fast, so the amount of reagent can be reduced to 0.25 equiv (eq 4). Due to the lower acid concentration, certain acid labile groups are tolerated. In competitive deprotection experiments, Bn–O–TBDMS is deprotected in the presence of another acid labile protecting group (an example is shown in eq 5 for THP): MEM and MOM groups are 100% retained; THP (86%) and benzylidine derivatives (77%) are partially retained. An acetonide derivative is retained only to the extent of 16%.

The effects of increasing the steric bulk of the substrate at carbon were also probed. Compounds (**1**)–(**4**) (eq 6) were paired and used in competitive deprotection reactions, providing the results reported in Table 1. Excellent selectivity was observed in the deprotection of primary TBDMS vs. tertiary TBDMS, and secondary TBDMS vs. tertiary TBDMS derivatives. On the other hand, selectivity was only fair for primary TBDMS vs. secondary TBDMS, and secondary TBDMS vs. secondary TIPS ethers. As observed before, *t*-butanol provided greater selectivity than *t*-BuOH–MeCN solvent mixtures.

	R^1	R^2	R^3
(**1**)	H	H	TBDMS
(**2**)	H	Me	TBDMS
(**3**)	H	Me	TIPS
(**4**)	Me	Me	TBDMS

Table 1 Study of substrate steric effects

Compd. A	Compd. B	Solvent[a]	Temp. (°C)	mmol H_2SiF_6	Time (h)	% A[b] (±2%)	% B[b] (±2%)	Selectivity
(1)	**(2)**	*t*-Butanol	25	0.500	3.3	4	62	58
(1)	**(2)**	90:10 MeCN–*t*-butanol	0	0.125	3.0	4	52	48
(1)	**(4)**	90:10 MeCN–*t*-butanol	0	0.125	3.0	0.3	100	100
(2)	**(4)**	90:10 MeCN–*t*-butanol	0	0.250	8.0	2	99	97
(1)	**(3)**	90:10 MeCN–*t*-butanol	0	0.250	24	2	65	63

[a] In each experiment, 0.5 mmol of compound A and compound B were dissolved in 5.0 mL of solvent.

[b] Yields determined by GC analysis.

1. (a) Pilcher, A. S.; Hill, D. K.; Shimshock, S. J.; Waltermire, R. E.; DeShong, P., *J. Org. Chem.* **1992**, *57*, 2492. (b) Pilcher, A. S.; DeShong, P., *J. Org. Chem.* **1993**, *58*, 5130.
2. (a) Mascarenas, J. L.; Mourino, A.; Castedo, L., *J. Org. Chem.* **1986**, *51*, 1269. (b) Newton, R. F.; Reynolds, D. P.; Finch, M. A. W.; Kelly, D. R.; Roberts, S. M., *Tetrahedron Lett.* **1979**, 3981.

Anthony S. Pilcher & Philip DeShong
University of Maryland, College Park, MD, USA

Formamide, *N-N*,bis(trimethylsilyl)-[1]

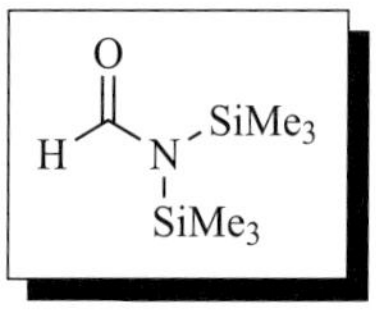

[15500-60-4] $C_7H_{19}NOSi_2$ (MW 189.40)
InChI = 1S/C7H19NOSi2/c1-10(2,3)8(7-9)11(4,5)6/h7H,1-6H3
InChIKey = FRCLVJBDJNNNGI-UHFFFAOYSA-N

(source of nucleophilic formamide; *N*-silylaldimine synthon)[1,2]

Alternate Names: bis(TMS)formamide; BSF.
Physical Data: 81–82 °C/20 mmHg;[3] 154 °C/760 mmHg; mp 16–17 °C; n_D^{20} 1.4388.[4]
Solubility: benzene, CCl_4, CH_2Cl_2, $CHCl_3$.
Form Supplied in: commercially available as neat oil.
Analysis of Reagent Purity: by ^1H NMR.
Preparative Method: chlorotrimethylsilane was added at room temperature to a solution of formamide and triethylamine in dry benzene and the mixture was heated at reflux for 1 h. The reaction mixture was filtered, and the filtrate was evaporated to give a crude oil. This was purified by high-vacuum distillation to afford *N,N*-bis(trimethylsilyl)formamide (BSF) as an oil in 75% yield (eq 1).[3–5]

(1)

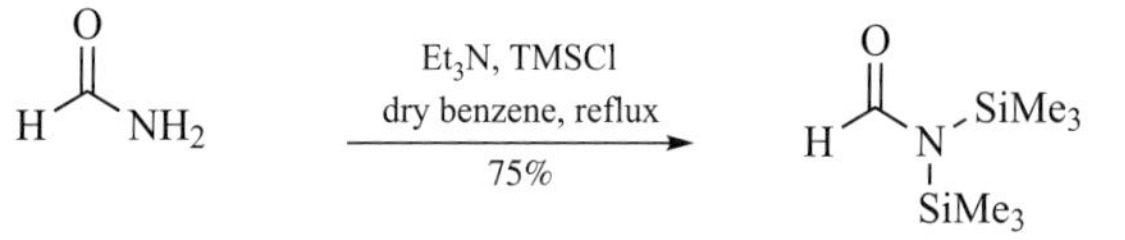

Handling, Storage, and Precautions: stable at room temperature; hydrolyzes in water.[4]

*Article previously published in the electronic Encyclopedia of Reagents for Orgenic Synthesis as N-N,bis(trimethylsilyl)formamide. DOI: 10.1002/047084289X. rn01018.

N,N-Bis(trimethylsilyl)formamide (BSF) exists in the amide form with both silyl groups attached to the nitrogen atom. This is in contrast to most other bis(TMS)amides (e.g., bis(TMS)-acetamide), which favor the *N,O*-bis(TMS)imidate isomer.[2,5–7] BSF behaves as an *N*-formamido nucleophile, reacting with the carbonyl group of aldehydes and activated ketones to give *N*-formyl-*O*-trimethylsilyl-*N,O*-acetals. Similarly, reaction with imines (or their precursors) can in some cases give *N*-formyl-*N,N*-acetals. BSF undergoes reactions with a range of other electrophiles including acid chlorides, chloroformate esters, and isocyanates to give a range of interesting products. When treated with organolithium nucleophiles, attack occurs at the carbonyl function, to provide *N*-silylaldimines after elimination of a silyloxy unit. Such imines are reported to undergo further reactions to give heterocycles. There are also reports of BSF acting as a silylating agent, to give silyl enol ethers, although it is not commonly used for this application.

Silylation Reactions. Treatment of some enolizable aldehydes and ketones with BSF has been reported to provide the corresponding silylenol ethers (eq 2).[1–3,8] BSF is reported to undergo sequential hydrolysis of the TMS functions to yield formamide and hexamethyldisiloxane.[4]

R^1 = Me, Ph, R^2 = OEt; R^1 = $MeOCOCH_2$, R^2 = OMe
R^1 = Ph, R^2 = Me: 60–80%

(2)

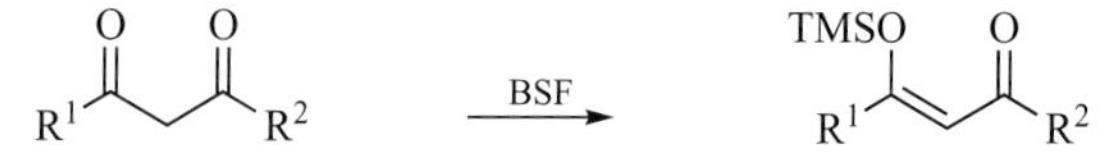

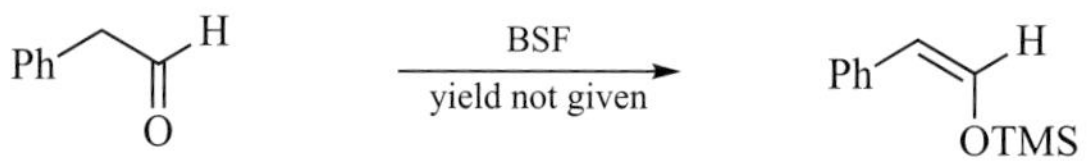

Reactions with Aldehydes and Ketones. BSF can react with aldehydes to give *N*-formyl-*O*-trimethylsilyl-*N*,*O*-acetals. The reactions can be performed through treatment of the aldehyde with a moderate excess (<2 equiv) of BSF, usually in refluxing CHCl$_3$ or CCl$_4$ (eq 3).[3,8] However, poor yields are obtained with hindered aldehydes and prolonged reaction times are sometimes needed. In other cases (e.g., phenylacetaldehyde), only the undesired silylenol ether is obtained.[3,8] The reaction can be greatly accelerated using TMSOTf as a Lewis acid catalyst, providing good yields of adducts even with aldehydes that perform poorly in the uncatalyzed reaction (e.g., hindered or readily enolized aldehydes) (eq 4).[8] A problem with the catalyzed reactions is that the initial product readily undergoes further reaction with a second molecule of aldehyde and a large excess of BSF can be required to provide the monoadduct in reasonable yield (eq 5).[9] Overreaction is less pronounced with hindered aldehydes, thus lower amounts of BSF can be employed to obtain monoadducts.

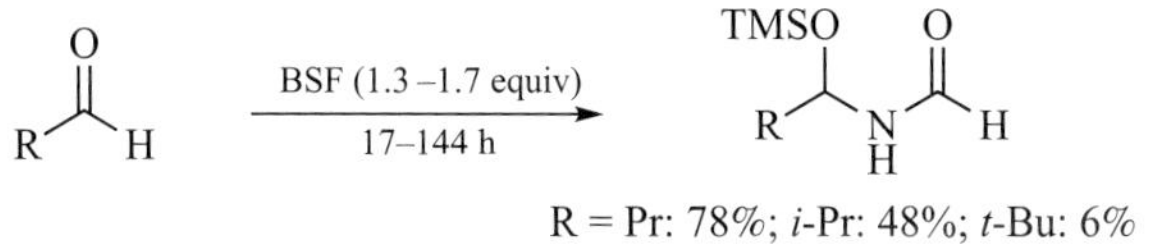

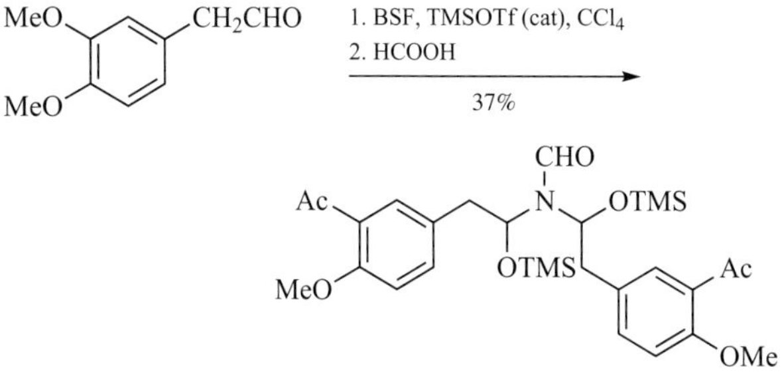

(5)

The reactivity of the initial monoadduct can be used advantageously, as it can undergo a second reaction with a different aldehyde to give unsymmetrical bisadducts (eqs 6 and 7).[9]

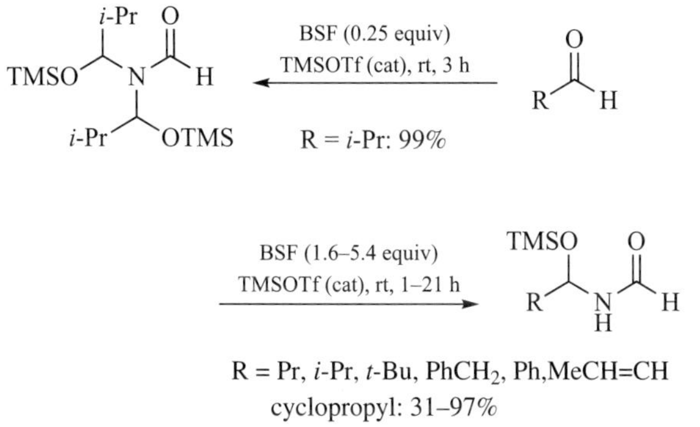

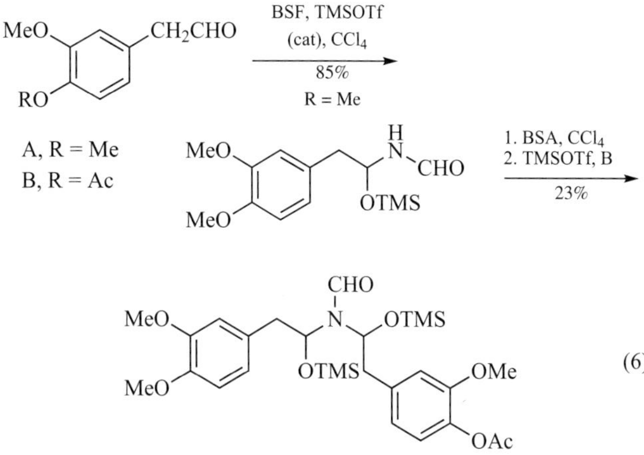

(6)

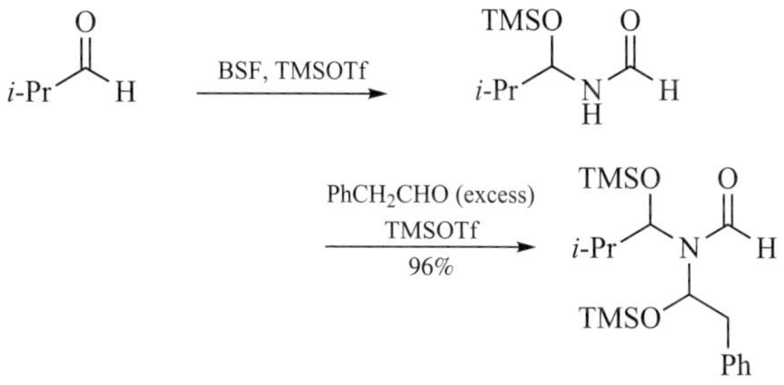

(7)

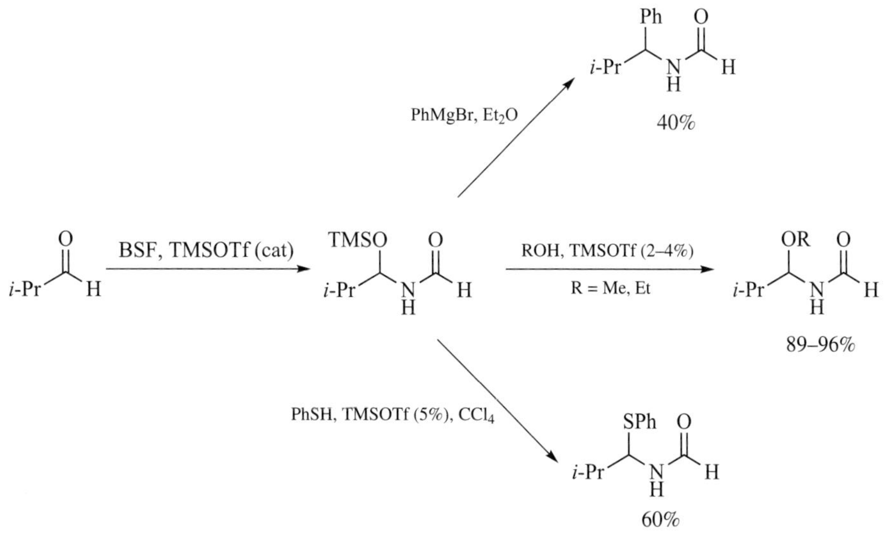

(3)

R = Pr: 78%; *i*-Pr: 48%; *t*-Bu: 6%

R = *i*-Pr: 99%

(4)

R = Pr, *i*-Pr, *t*-Bu, PhCH$_2$, Ph, MeCH=CH
cyclopropyl: 31–97%

(8)

The reactive structure of the monoadduct has been exploited in a number of different ways as reactions with a range of nucleophiles including Grignard reagents, alcohols, and thiophenol have been reported (eq 8).[8]

Intramolecular reactions have also been reported. For example, after monoaddition of BSF to tetracyclic aldehyde (**1**), treatment with TfOH provides a ring-expanded tetracyclic *N*-acylimine presumably via the mechanism shown (eq 9).[3] A direct one-pot synthesis of argemonine alkaloid precursor, *N*-formylpavine, was achieved via reaction of BSF with 3,4-dimethoxyphenylacetaldehyde, providing a bisadduct, followed by acid-catalyzed cyclization (eq 10).[9] The reaction is thought to proceed via the bisadduct, which then undergoes acid-mediated acyliminium ion formation, followed by attack of the aryl ring.

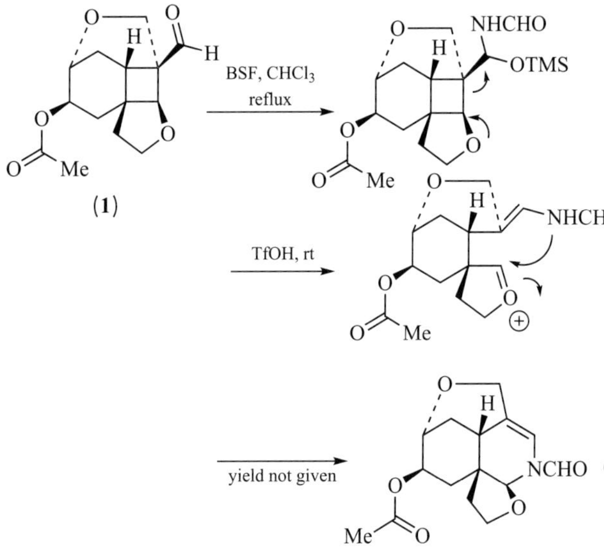

Aldehydes that are masked as acetals have also been reported to react with BSF to give mono- and bisadducts (eq 11).[3] Ketones are generally much less reactive toward BSF, usually requiring longer reaction times and providing lower yields. However, as one would expect, activated ketones such as methyl pyruvate provide better results (eq 12).[3]

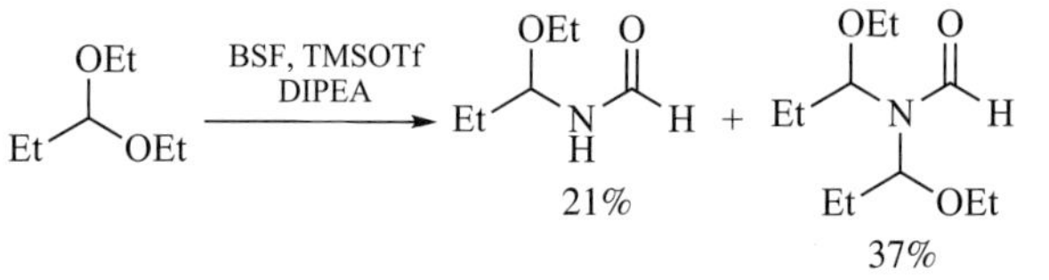

(1)

(9)

yield not given

(11)

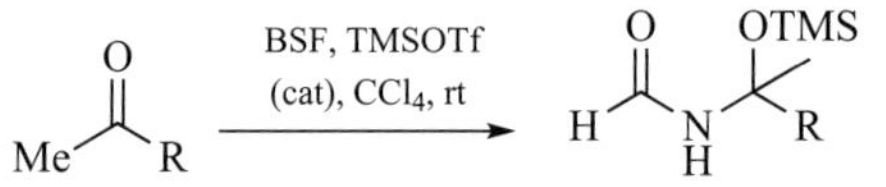

R = Me: 2.9 equiv BSF, 168 h, 38%
CO_2Me: 3.5 equiv BSF, 21 h, 83%

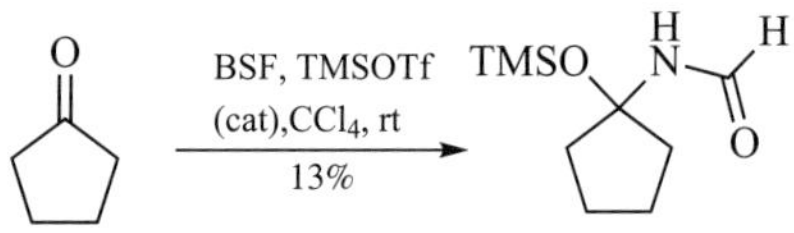

(12)

13%

Reactions with Imines. In some cases, BSF can react with imines to provide *N*,*N*-acetal derivatives. For example, treatment of the quinone methide-substituted imine **3** (R = *t*-Bu, X = OAc, formed via lead(IV) oxide-mediated oxidation of Schiff base **2**) provides the corresponding *N*-formyl-*N*-imino-*N*,*N*-acetal **4**, with concurrent rearomatization of the quinone unit (eq 13).[10] However, when this procedure was adapted for multigram scale, the yield of desired product was very low, with 9% epimer formation. Changing of the oxidant to DDQ allowed smoother reactions

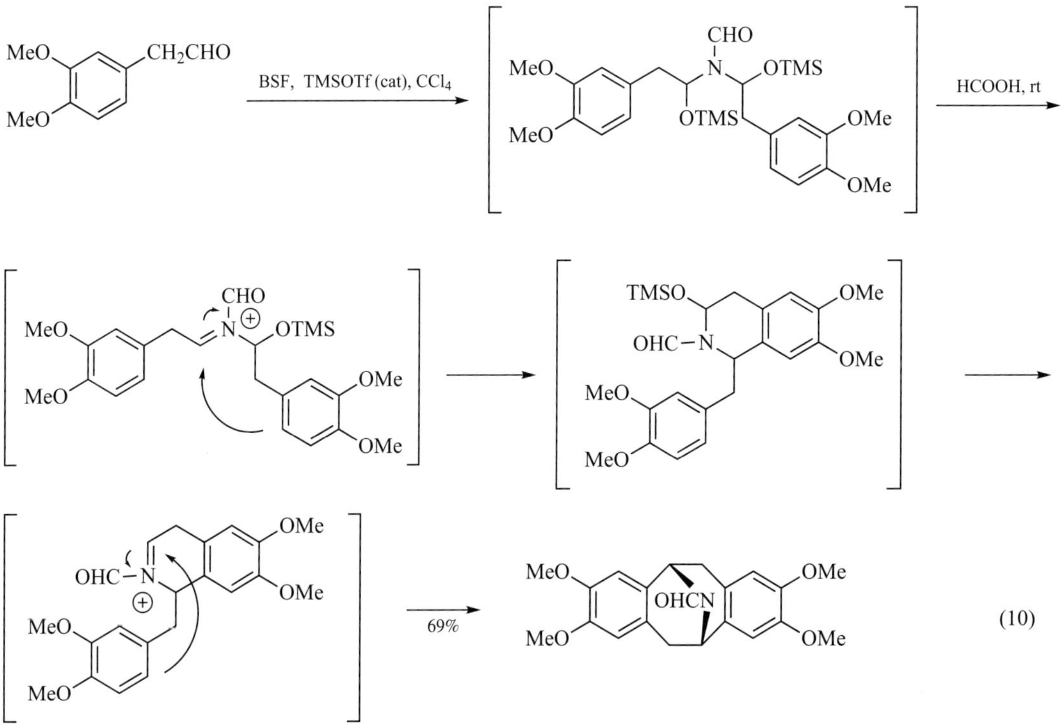

(10)

69%

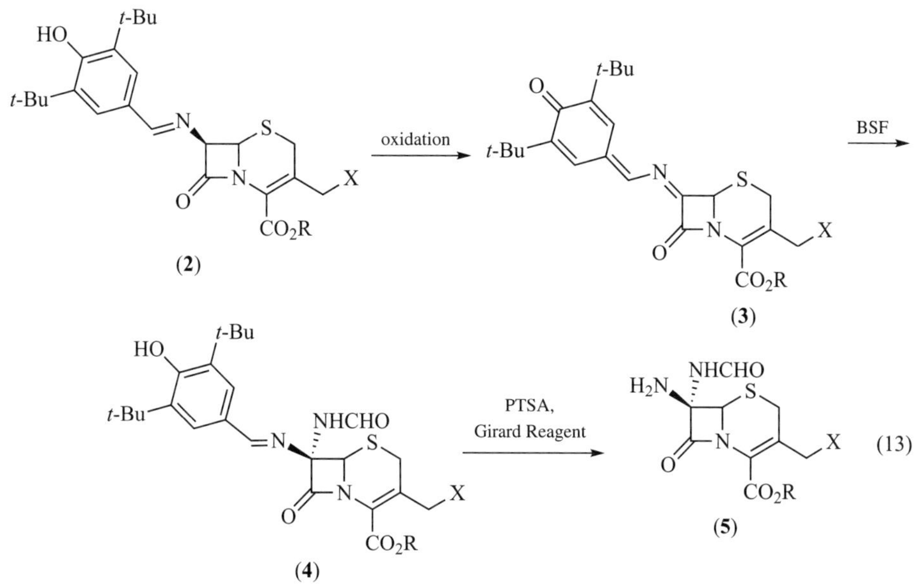

R = *t*-Bu, X = OAc; oxidation conditions: PbO$_2$, PhMe, rt; 58% **4** from **2**

R = PMB, X = Cl; oxidation conditions: DDQ, DCM, rt; 67% **5** from **2**

R = TMS, X = OAc; oxidation conditions: DDQ, DCM, rt; 46% **5** (R = H) from **2** on 1 kg scale

(e.g., formation of **4**, R = PMB, X = Cl) and allowed the one-pot synthesis of 1 kg of 7(α)-formamidocephalosporanate derivative **5** (R = H, X = OAc) in an overall yield of 46%, starting with a simple cephalosporanate substrate **2** (R = TMS). Notably, no epimers were observed.

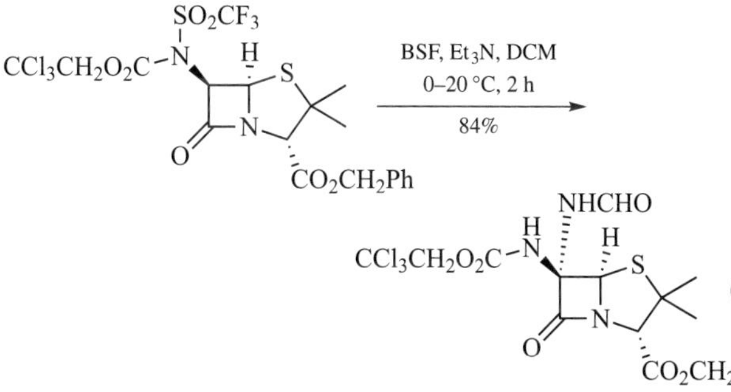

In related reactions, 6α-formamidopenicillins and 7α-formamidocephalosporins[11,12] can be prepared by the reaction of BSF with penicillin and cephalosporin imines prepared from activated amine substrates under basic conditions (eqs 14 and 15).[13] Similarly, BSF reacts with 6α-succinimidooxypencillin **6** to provide the corresponding 6-formamido-6-phenoxyacetamidopencillanate **7** in good yield (eq 16).[14]

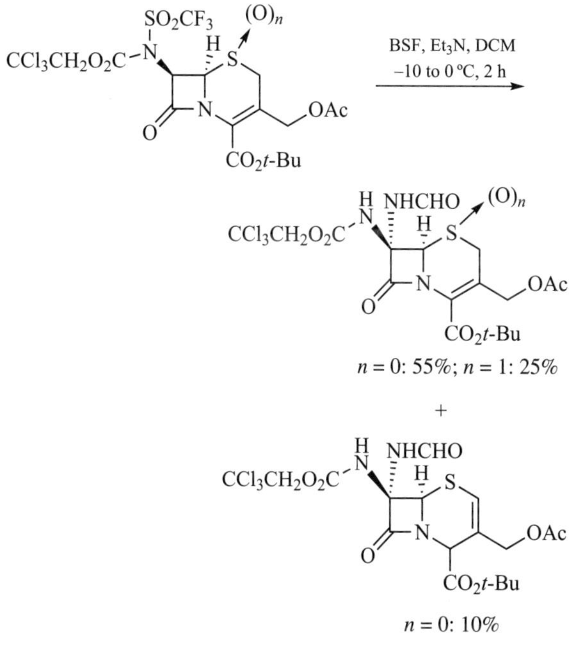

Reaction with Acyl Chlorides, Isocyanates, and Related Reactions. BSF is readily acylated with acid chlorides and chloroformate esters to provide *N*-TMS-*N*-formylamides and carbamates, respectively, which readily undergo protodesilylation (eq 17).[2,15] BSF reacts with 2 equiv of an alkyl or aryl isocyanate to give 1,3,5-triazine-2,4-diones in good to excellent yields (eq 18).[2] In the case of orthocarboxamides, *N*-formylformamidines can be obtained (eq 19).[2]

Reaction with Organometallic Reagents. BSF reacts with organolithium reagents to give aldimines. With nucleophiles such as phenyllithium, nonenolizable stable *N*-trimethylsilylaldimines can be obtained (eq 20).[16] With alkyllithium species, enolizable *N*-silylaldimines can be prepared. Such intermediates have been used in situ for the generation of heterocycles. For example, substituted azetidin-2-ones have been prepared through the reaction of BSF with alkyl- or aryllithium species followed by further reaction of the product with the lithium enolate of a carboxylic ester (eq 21).[16] In another example, *N*-unsubstituted-(4- and 4,5-substituted)-imidazoles can be prepared in moderate yield via a mild, efficient one-pot synthesis from BSF and an alkyllithium reagent followed by treatment with a toluenesulfonylalkyl iso-cyanate anion (eq 22).[17] Use of Grignard reagents in place of organolithium species has also been demonstrated.[17]

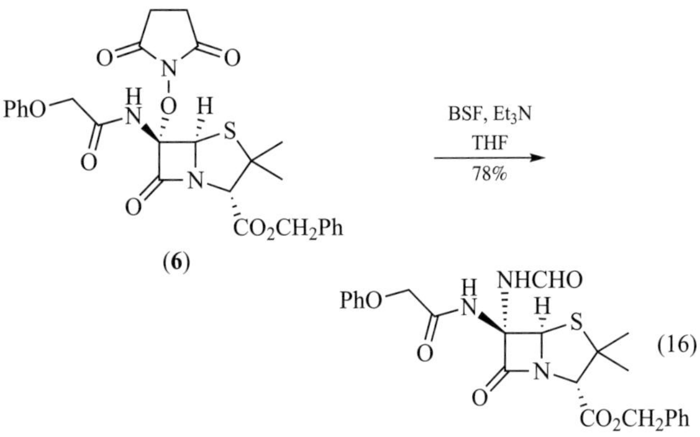

(6)

(16)

(7)

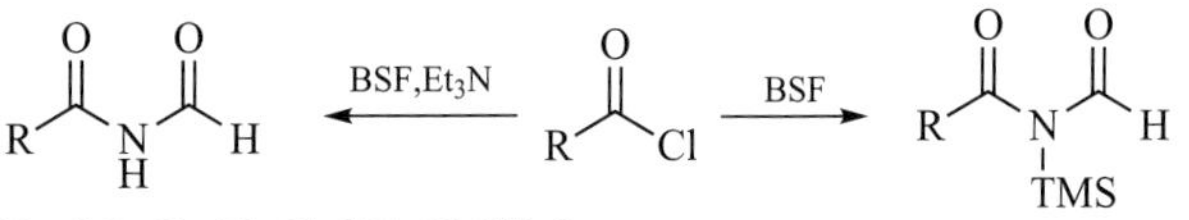

R = Me, Et, Ph, PhCH$_2$, PhCH$_2$O
PhCH$_2$CH$_2$: 59–90%

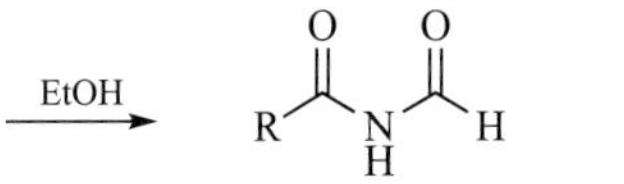

(17)

R = Ph, Me, MeO: 61–71%

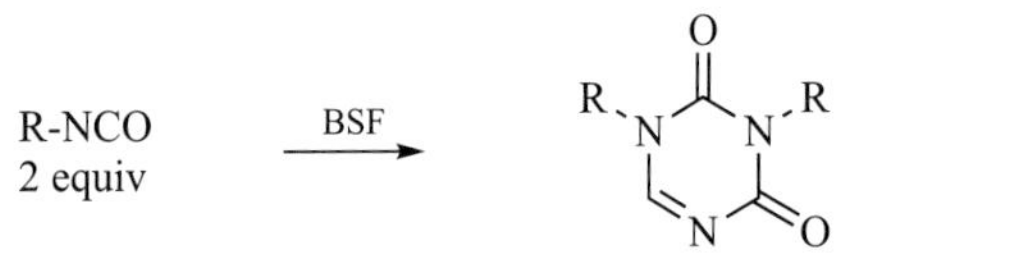

(18)

R = Me, Pr, Bu, cyclohexyl, Ph, 4-MeC$_6$H$_4$, 3-MeC$_6$H$_4$
4-ClC$_6$H$_4$, 3-ClC$_6$H$_4$, 2,6-Me$_2$C$_6$H$_3$, 3,4-Cl$_2$C$_6$H$_3$: 70–99%

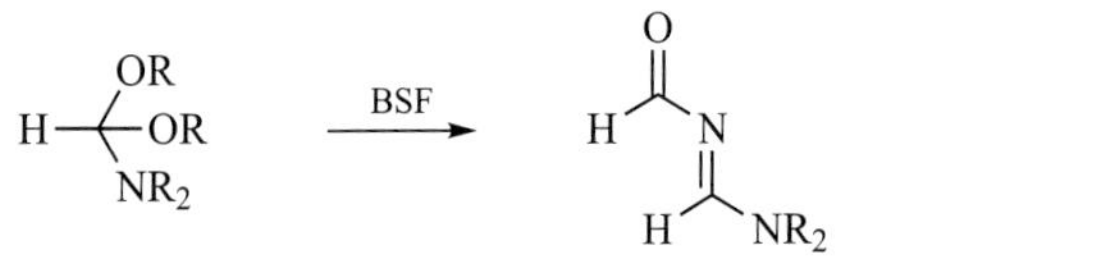

(19)

R = Me; NR$_2$ = NMe$_2$, NEt$_2$, NBu$_2$, pyrrolidinyl, piperidinyl: 74–82%
R = Et; NR$_2$ = NMe$_2$: 80%

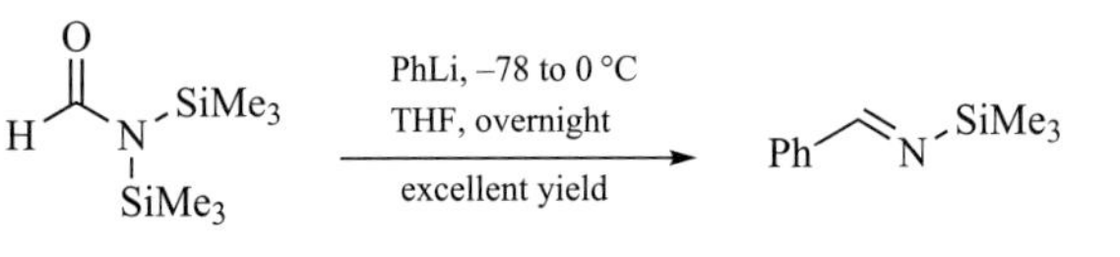

(20)

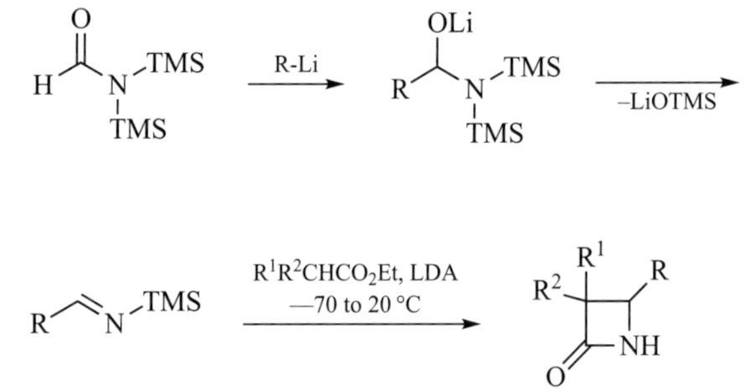

(21)

R = Bu; R^1 = R^2 = Me; R^1 = *i*-Pr, R^2 = H; R^1 = MeCH$_2$, R^2 = H: 93–99%
R = Me, Ph, *s*-Bu, PhCH$_2$OCH$_2$, MeOCH$_2$; R^1 = R^2 = Me: 58–94%
R = *s*-Bu; R^1 = *i*-Pr, R^2 = H: 46%
R = CH$_2$=C(OMe); R^1 = MeCH$_2$, R^2 = H: 46% (1.4:1 *syn:anti*)

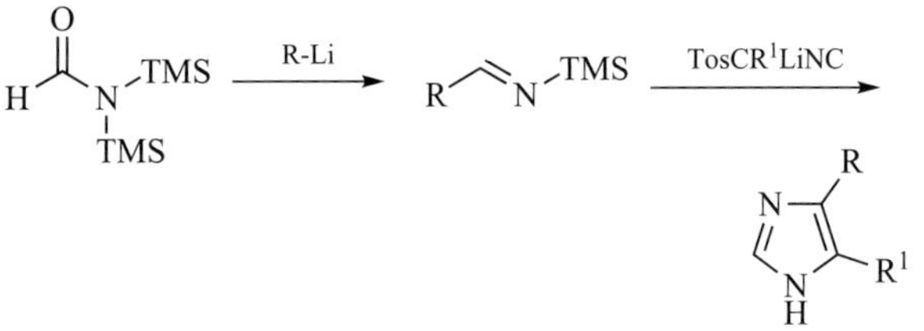

(22)

R = Me, Bu, Ph; R^1 = H: 23–55%
R = Bu; R^1 = Me, PhCH$_2$: 51–66%

In another reaction with a carbon-based nucleophile, BSF reacts with activated methylene compounds such as cyanoacetate esters and malononitrile to give *β*-aminoacrylonitriles (eq 23).[1]

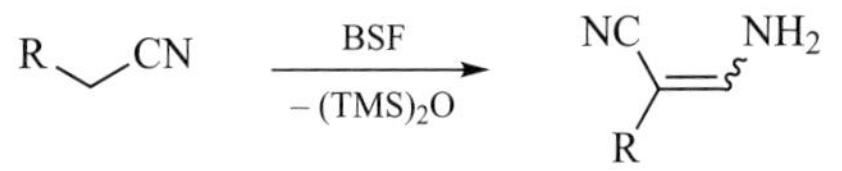

(23)

R = CN (56%)
R = CO$_2$Me (65%)
R = CO$_2$Et (65%)

1. Kantlehner, W.; Kugel, W.; Bredereck, H., *Chem. Ber.* **1972**, *105*, 2264.

2. Kantlehner, W.; Fischer, P.; Kuge, W.; Mohring, E.; Bredereck, H., *Liebigs Ann. Chem.* **1978**, 512.

3. Johnson, A. P.; Luke, R. W. A.; Steele, R. W.; Boa, A. N., *J. Chem. Soc., Perkin Trans. 1* **1996**, 883.

4. Schirawski, G.; Wannagat, U., *Monatshe. Chem.*, **1969**, *100*, 1901.

5. Yoder, C. H.; Copenhafer, W. C.; Dubeshter, B., *J. Am. Chem. Soc.* **1974**, *96*, 4283.

6. Samples, M. S.; Yoder, C. H., *J. Organomet. Chem.* **1987**, *332*, 69.

7. Yoder, C. H.; Bonelli, D., *Inorg. Nucl. Chem. Lett.* **1972**, *8*, 1027.

8. Johnson, A. P.; Luke, R. W. A.; Steele, R. W., *J. Chem. Soc., Chem. Commun.* **1986**, 1658.

9. Johnson, A. P.; Luke, R. W. A.; Singh, G.; Boa, A. N., *J. Chem. Soc., Perkin Trans. 1*, **1996**, 907.

10. Berry, P. B.; Brown, A. C.; Hanson, J. C.; Kaura, A. C.; Milner, P. H.; Moores, C. J.; Quick, J. K.; Saunders, R. N.; Southgate, R.; Whittall, N., *Tetrahedron Lett.* **1991**, *32*, 2683.

11. Ponsford, R. J.; Basker, M. J.; Burton, G.; Guest, A. W.; Harrington, F. P.; Milner, P. H.; Pearson, M. J.; Smale, T. C.; Stachulski, A. V., *Recent Advances in the Chemistry of β-Lactam Antibiotics*. Special Publication No. 52; Cambridge: Royal Society of Chemistry, 1985, p 32.

12. Milner, P. H.; Guest, A. W.; Harrington, F. P.; Ponsford, R. J.; Smale, T. C.; Stachulski, A. V., *J. Chem. Soc., Chem. Commun.* **1984**, 1335.

13. Pearson, M. J., *Tetrahedron Lett.* **1985**, *26*, 377.

14. Milner, P. H.; Stachulski, A. V., *J. Chem. Soc. Perkin Trans. 1* **1991**, 2343.

15. Klinge, M.; Cheng, H.; Zabriskie, T. M.; Vederas, J. C., *J. Chem. Soc., Chem. Commun.* **1994**, 1379.

16. Uyehara, T.; Suzuki, I.; Yamamoto, Y., *Tetrahedron Lett.* **1989**, *32*, 4275.

17. Shih, N. Y., *Tetrahedron Lett.* **1993**, *34*, 595.

Vijayanand Chandrasekaran
Christiana Albertina University of Kiel, Kiel, Germany

Steven J. Collier
Codexis Laboratories Singapore Pte Ltd, Singapore

Hexachlorodisilane[1]

$$Cl_3Si\!-\!SiCl_3$$

[13465-77-5] Cl_6Si_2 (MW 268.88)

InChI = 1S/Cl6Si2/c1-7(2,3)8(4,5)6
InChIKey = LXEXBJXDGVGRAR-UHFFFAOYSA-N

(deoxygenation and desulfurization of phosphine oxides, phosphine sulfides, and amine oxides; reducing agent for nitro groups and sulfur diimides)

Physical Data: bp 144–145.5 °C; *d* 1.562 g cm^{-3}.
Solubility: sol CHCl$_3$, CH$_2$Cl$_2$, benzene, THF; reacts violently, producing toxic fumes, with H$_2$O, alcohols.
Form Supplied in: neat as clear colorless liquid.
Handling, Storage, and Precautions: neat liquid and solutions react violently with water and protic solvents to produce HCl gas. Avoid contact with strong bases. Hexachlorodisilane is corrosive and should be handled and stored in a tightly sealed vessel under N$_2$. Store cool. Vapor ignites in air when heated. Avoid prolonged and repeated exposure. Use in a fume hood.

Deoxygenation and Desulfurization. Hexachlorodisilane has been shown to effect the stereospecific reduction of phosphine oxides to phosphines with inversion of configuration (eq 1),[2] although short reaction times are required to prevent chemical racemization of the phosphine products under the reaction conditions. This protocol complements trichlorosilane which, under appropriate conditions, reduces phosphine oxides with retention of configuration.[2,3] Reductions of bridged bicyclic phosphine oxides[4] and cyclic halophospholene oxides[5] by hexachlorodisilane have also been reported.

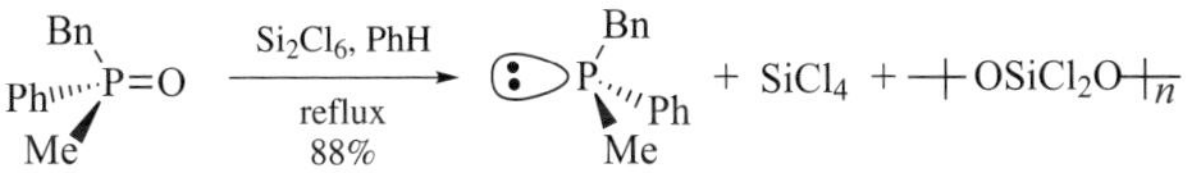

$$\text{eq (1)}$$

In contrast to the deoxygenation of phosphine oxides by hexachlorodisilane, desulfurization of phosphine sulfides by this reagent occurs stereospecifically with *retention* of configuration (eq 2).[6]

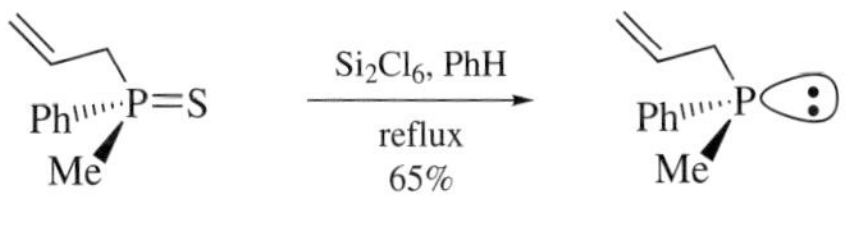

$$\text{eq (2)}$$

Quinuclidine *N*-oxide was reduced by hexachlorodisilane (eq 3).[2] Likewise, azo oxide,[7] azo dioxide,[8,9] and sulfoxide deoxygenations[2] by the reagent were also reported. Treatment of aryl

nitro compounds with hexachlorodisilane afforded products which suggested the possibility of a nitrene intermediate being involved (eq 4).[10] Other disilanes were evaluated in the study, including hexamethyl- and hexaphenyldisilane, pentamethylphenoxy- and pentamethylphenyldisilane, as well as novel benzodisilacyclopentane (**1**).

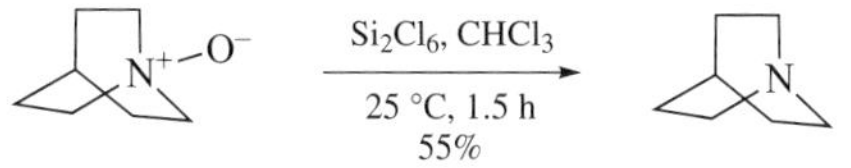

$$\text{eq (3)}$$

(1)

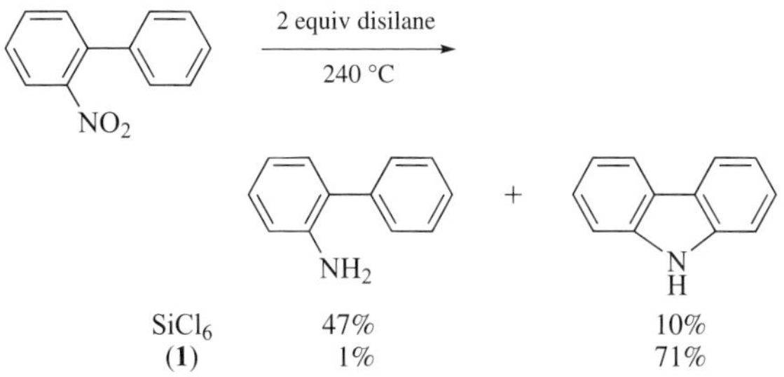

$$\text{eq (4)}$$

	SiCl$_6$	
	(**1**)	
	47%	10%
	1%	71%

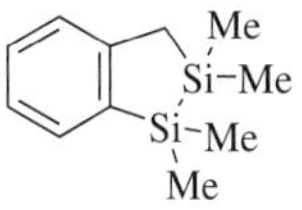

Quinoxaline 1,4-dioxide derivatives were reduced by hexachlorodisilane to give the corresponding quinoxalines (eq 5).[11] Also evaluated in the study were the reducing agents iodotrimethylsilane, trifluoroacetic anhydride–sodium iodide, and titanium(IV) chloride–zinc dust. Hexachlorodisilane, for its convenience and efficiency, was recommended by the authors as the reagent of choice for this transformation.

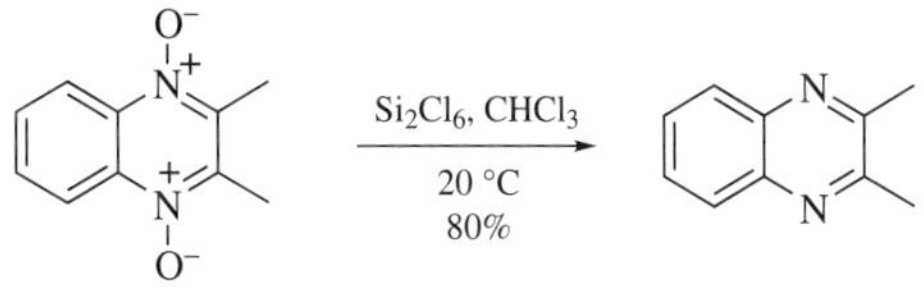

$$\text{eq (5)}$$

Dehydroxylation occurred upon treatment of the alkoxide of a 1-hydroxyimidazole derivative with hexachlorodisilane (eq 6).[12] In the same report it was observed that nitrones are deoxygenated upon treatment with hexachlorodisilane, while C=N reduction occurs to yield hydroxylamines with trichlorosilane (eq 7).

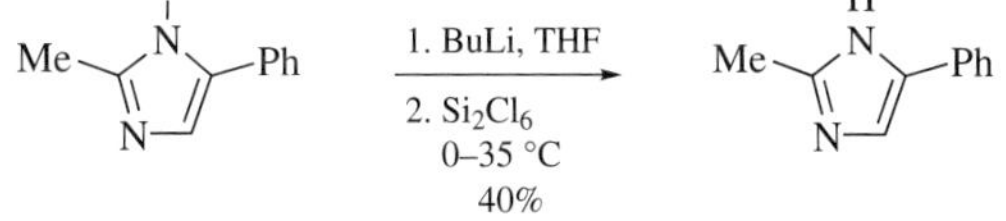

$$\text{eq (6)}$$

$$\text{eq (7)}$$

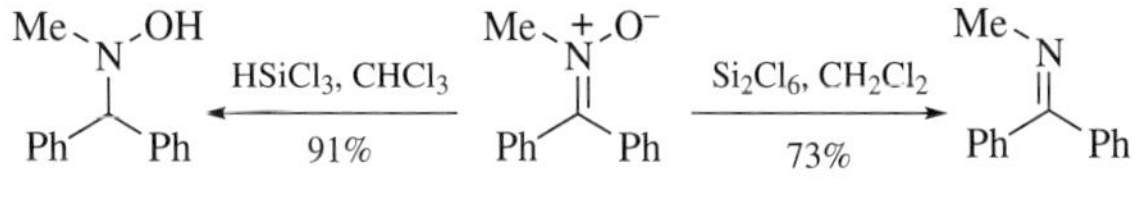

(Trichlorosilyl)phosphane Synthesis. Chlorophosphanes or (trimethylsilyl)phosphanes are converted to trichlorosilylphosphanes upon treatment with hexachlorodisilane (eq 8).[14]

$$i\text{-Pr} \diagdown \atop i\text{-Pr} \diagup P{-}X \quad \xrightarrow[\substack{20\,°C \\ 57-70\%}]{Si_2Cl_6} \quad i\text{-Pr} \diagdown \atop i\text{-Pr} \diagup P{-}SiCl_3 \;+\; SiCl_4 \qquad (8)$$

X = Cl, TMS

Sulfur Diimide Reduction. *N*-Trichlorosilyldiaminosulfanes were prepared upon treatment of sulfur diimides with hexachlorodisilane (eq 9).[13] Likewise, the corresponding *N*-dichloromethylsilyldiaminosulfanes resulted by substituting 1,2-dimethyl-1,1,2,2-tetrachlorodisilane for hexachlorodisilane.

$$t\text{-Bu} \diagdown_{N} {\diagup}^{S} {\diagdown}_{N} {\diagup} t\text{-Bu} \quad \xrightarrow[\substack{20\,°C \\ X = Cl,\, Me}]{Si_2Cl_4X_2} \quad t\text{-Bu} \diagdown_{N} {\diagup}^{S} {\diagdown}_{N} {\diagup} t\text{-Bu} \atop XCl_2Si \quad SiCl_2X \qquad (9)$$

Related Reagent. Trichlorosilane.

1. (a) *Silicon Chemistry*, Corey, J. Y.; Corey, E. R.; Gaspar, P. P., Eds.; Ellis Horwood: Chichester, 1987. (b) *Organosilicon Compounds*, Eaborn, C., Ed.; Butterworth: London, 1960.
2. Naumann, K.; Zon, G.; Mislow, K., *J. Am. Chem. Soc.* **1969**, *91*, 7012.
3. Horner, L.; Balzer, W. D., *Tetrahedron Lett.* **1965**, 1157.
4. Katz, T. J.; Carnahan, J. C., Jr.; Clarke, G. M.; Acton, N., *J. Am. Chem. Soc.* **1970**, *92*, 734.
5. Myers, D. K.; Quin, L. D., *J. Org. Chem.* **1971**, *36*, 1285.
6. Zon, G.; DeBruin, K. E.; Naumann, K.; Mislow, K., *J. Am. Chem. Soc.* **1969**, *91*, 7023.
7. Synder, J. P.; Lee, L.; Bandurco, V. T.; Yu, C. Y.; Boyd, R. J., *J. Am. Chem. Soc.* **1972**, *94*, 3260.
8. Greene, F. D.; Gilbert, K. E., *J. Org. Chem.* **1975**, *40*, 1409.
9. Snyder, J. P.; Heyman, M. L.; Suciu, E. N., *J. Org. Chem.* **1975**, *40*, 1395.
10. Tsui, F.-P.; Vogel, T. M.; Zon, G., *J. Org. Chem.* **1975**, *40*, 761.
11. Homaidan, F. R.; Issidorides, C. H., *Heterocycles* **1981**, *16*, 411.
12. Hortmann, A. G.; Koo, J.-Y.; Yu, C.-C., *J. Org. Chem.* **1978**, *43*, 2289.
13. Herberhold, M.; Frank, S. M.; Wrackmeyer, B.; Bormann, H.; Simon, A., *Chem. Ber.* **1990**, *123*, 75.
14. Martens, R.; du Mont, W.-W., *Chem. Ber.* **1992**, *125*, 657.

David P. Sebesta
University of Colorado, Boulder, CO, USA

Hexamethylcyclotrisiloxane

[541-05-9] $C_6H_{18}O_3Si_3$ (MW 222.46)

InChI = 1S/C6H18O3Si3/c1-10(2)7-11(3,4)9-12(5,6)8-10/h1-6H3

InChIKey = HTDJPCNNEPUOOQ-UHFFFAOYSA-N

(reagent for the preparation of dimethylsilanols,[1] trapping of in-situ-generated silanones,[2] silylenes,[3] silyl azides,[4] silanethiones,[5] and derivatives)

Physical Data: mp 64–64.5 °C; bp 133–134 °C.

Solubility: soluble in most organic solvents.

Form Supplied in: white, crystalline solid.

Purification: sublimation: 55 °C at 55 mmHg.

Handling, Storage, and Precautions: hexamethylcyclotrisiloxane is a stable, crystalline solid that can be stored in a bottle without precaution for air and moisture exposure.

Introduction. Hexamethylcyclotrisiloxane, (**1**), (D₃) is the simplest member of a series of cyclic oligodimethylsiloxanes, primarily used as synthetic equivalents for the reactive intermediate dimethylsilanone (**2**). This characteristic is responsible for its extensive application in polymer chemistry.[6] However, its use in synthetic chemistry can be traced to a 1970 report in which Lee and co-workers attempted to induce the anionic polymerization of D₃ with *n*-butyllithium[1]. To their surprise, rather than the expected[7] PDMS (polydimethylsilicone) polymer, they obtained an 82% yield of *n*-butyldimethylsilanol. This reaction has been shown to be general for alkyl-, alkenyl-, aryl-, and heteroaryllithium reagents, resulting in the production of many organic dimethylsilanols. Hexamethylcyclotrisiloxane has also been exploited as an efficient trapping reagent for a variety of short-lived intermediates, including silanones and silyenes. The ability of D₃ to intercept these reactive species has been attributed to its rigid, planar structure,[8] and the resultant release of strain from ring expansion.

$$\left[\begin{array}{c} Me \diagdown \atop Si {=} O \\ | \\ Me \end{array} \right]$$
2

Preparation of Substituted Dimethylsilanols: General Considerations. In 1993, Sieburth and co-workers reported the preparation and use of dimethyl- and diphenylsilanols as coordinating, directing groups for the lithiation of adjacent aryl C–H bonds.[9] Since that time, the use of silanols has increased dramatically, with applications ranging from coordination chemistry to natural product synthesis.[10–12] Considering the increased awareness of silanols, together with the facility of their preparation, there are remarkably few substituted dimethylsilanols reported as reagents in organic synthesis. Despite their limited number, the utility of dimethylsilanols as synthetic reagents has increased because of their role as effective donors in palladium-catalyzed, cross-coupling reactions with aryl- and alkenyl halides and pseudohalides.[13] Many of the organodimethylsilanols used for this process have been prepared from the combination of D₃ with organolithium reagents to afford substituted dimethylsilanols in good to excellent yields (eq 1). There are no examples that report the combination of other organometallic species, such as Grignard reagents, with D₃ to provide dimethylsilanols. However, Sieburth and co-workers describe the reaction of *n*-butylmagnesium bromide with hexaphenylcyclotrisiloxane to

$$R{\diagup}^{Li} \xrightarrow{D_3} \left[\begin{array}{c} Me \\ R{\diagdown}Si{\diagup}Me \\ O^-Li^+ \end{array} \right] \xrightarrow[\text{(workup)}]{H_2O} \begin{array}{c} Me \\ R{\diagdown}Si{\diagup}Me \\ OH \end{array} \qquad (1)$$

R = 1° alkyl-, alkenyl-,
     alkynyl-, aryl-, heteroaryl-

provide *n*-butyldiphenylsilanol in 27% yield.[14] In comparison, the combination of *n*-butyllithium and hexaphenylcyclotrisiloxane under the same conditions afforded the 91% yield of the desired silanol product.

Although D$_3$ is less reactive than other silicon electrophiles, such as dichlorodimethylsilane and chlorodimethylsilane, it is more useful because it affords fewer by-products and is less prone to undesirable reaction pathways. One of the major problems associated with the preparation and isolation of silanols is their tendency to dehydratively dimerize to form disiloxanes.[15] This dimerization has been shown to be rapid under either strongly basic or acidic conditions.[16] Hydrolysis of the chlorosilane resulting from the reaction of an organolithium reagent with dichlorodimethylsilane generates significant amounts of HCl and therefore requires careful monitoring of the pH of the reaction solution to prevent disiloxane formation. In contrast, the neutralization of the lithium silanolate generated in the reaction with D$_3$ is less problematic. Another advantage of D$_3$ is the selective formation of the organodimethylsilanol whereas with, e.g., chlorodimethylsilane or dimethyldichlorosilane, double addition to provide a tetraorganosilane is a common problem.[17]

Organolithium Reagents from Halogen-metal Exchange. Although the reaction of polydimethylsiloxane and methyllithium to provide lithium trimethylsilanolate had been known for some time,[18] Lee and co-workers were the first to use a defined cyclic dimethylsiloxane, D$_3$, with *n*-butyllithium to prepare *n*-butyldimethylsilanol.[1] Later, Sieburth combined D$_3$ with an aryllithium reagent, generated by lithium–halogen exchange, to prepare an aryldimethylsilanol (eq 2).[9] In each of these reactions the 3/1 stoichiometry of the organolithium reagent to D$_3$ is used as all three dimethylsilanone units are available.

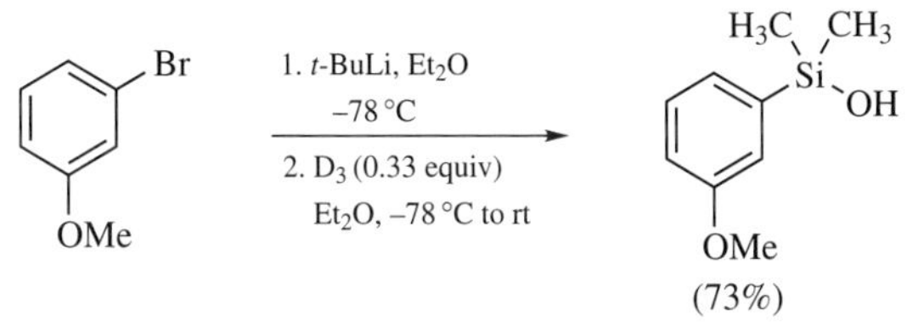

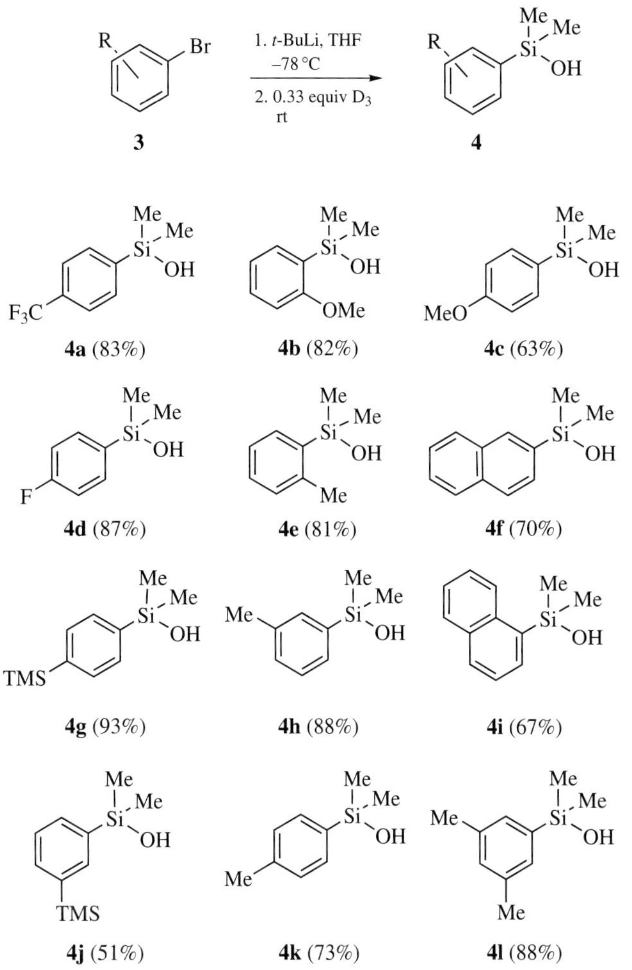

Scheme 1 Substituted aryldimethylsilanols

(2)

Many different aryldimethylsilanols have been prepared from the corresponding aryl bromides. Electron-poor (**4a, d**), electron-rich (**4b, c**), and moderately encumbered (**4e, i, l**) aryl bromides are excellent substrates for the silanol preparation (Scheme 1).[19–22] Trimethylsilyl-substituted aryl bromides are also competent and provide dimethylsilanols **4g** and **4j**. These aryldimethylsilanols are useful as donors in the palladium-catalyzed Heck-type[20] and cross-coupling reactions.[19,21]

Similarly, alkenyllithium reagents derived from lithium–halogen exchange of the corresponding bromides react with D$_3$ to provide alkenylsilanols in good to excellent yields. Geometrically defined, 1-alkenyldimethylsilanols (Scheme 2, **6a, b**)[23–25] have been prepared, as well as the *E*- and *Z*-β-styrylsilanols (**6d**).[23,26] In addition, the dimethylsilanol can be installed in the 2-position of the alkene (**6c, e**).[23,27] Alkenyldimethylsilanols incorporating heteroatom functionality are prepared in comparable yields (**6f**).[25]

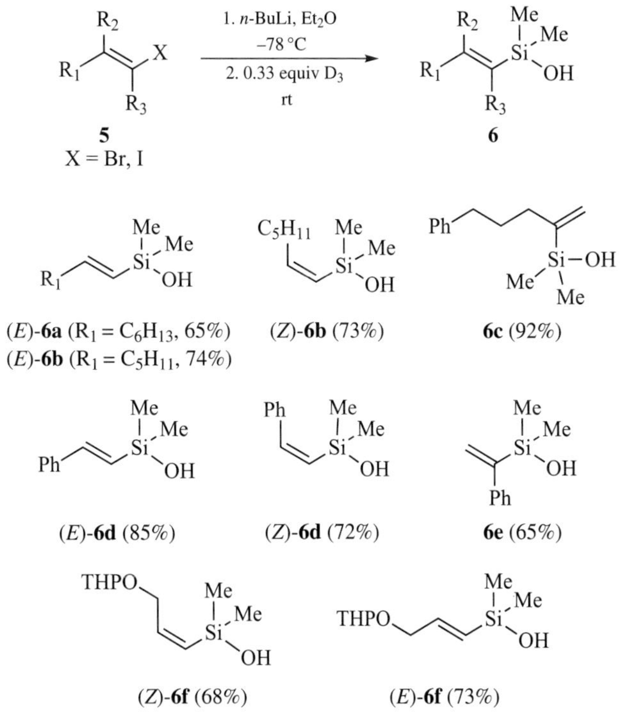

Scheme 2 Simple alkenyldimethylsilanols

A series of tri- and tetrasubstituted alkenyldimethylsilanols were prepared to study the substituent effects on the stereochemical outcome of a subsequent cross-coupling reaction.[28] The products were obtained in high geometrical purity by the combination of D$_3$ and the geometrically defined alkenyllithium reagents generated by bromine–lithium exchange with t-butyllithium (Scheme 3).

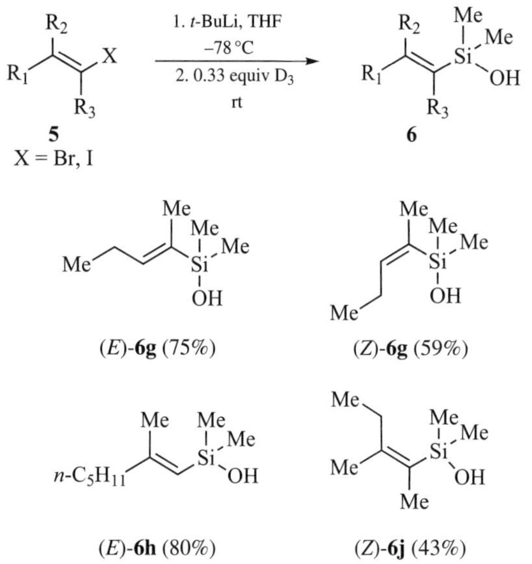

Scheme 3   Highly substituted alkenyldimethylsilanols

Organolithium Reagents from Direct Deprotonation.   Organolithium reagents prepared via lithium–halogen exchange require either installation of the halide or commercially available, halogenated substrates. Alternatively, substrates that can be deprotonated directly with reliable site selectivity provide organolithium reagents that have also been used in combination with D$_3$ to provide dimethylsilanols. Such carbon acidic substrates include alkynes[23] and heterocycles, which have produced the corresponding dimethylsilanols in good yields (Scheme 4).[23,29,30] The preparation and use of indolyl-2-dimethylsilanol (**10**) is noteworthy because it is considerably more stable than the corresponding boronic acid.[29]

Miscellaneous Reactions.   Highly reactive silicon intermediates, such as dimethylsilanone and dimethylsilylene can be trapped with D$_3$. The facility with which D$_3$ intercepts such reactive intermediates is ascribed to the relief of strain in the planar cyclotrisiloxane ring. Upon insertion of a silylene or silanone, the ring expands from six- to eight-membered and alleviates the nonbonding interactions among the methyl groups. For example, the thermolysis of silacyclobutanes such as **19** generates a dimethylsilylene (**20**) that can be trapped with D$_3$ to form silacycle **21**, which is expanded by one Si–C unit (eq 3).[3]

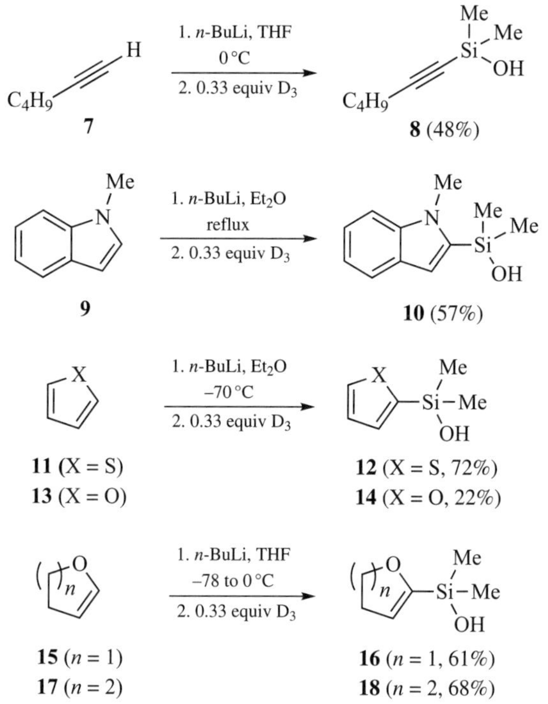

Scheme 4   Substituted dimethylsilanols derived from direct lithiation

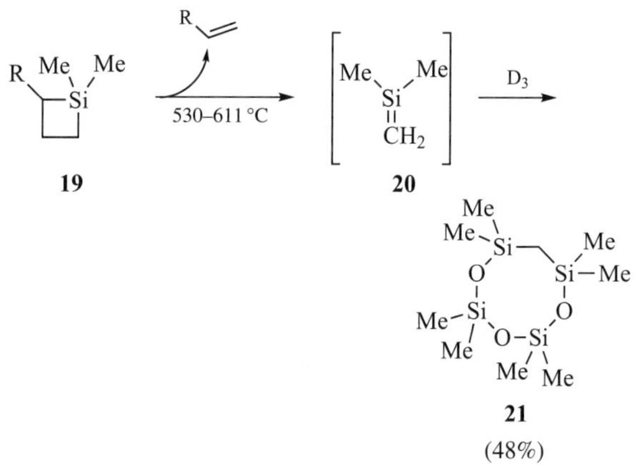

Similarly, dimethylsilyl azide (**22**),[4] dimethylsilene (**23**),[31] dimethylsilanethione (**24**),[5] and diethylsilaselanone (**25**)[32] have also been trapped using D$_3$ (Scheme 5).

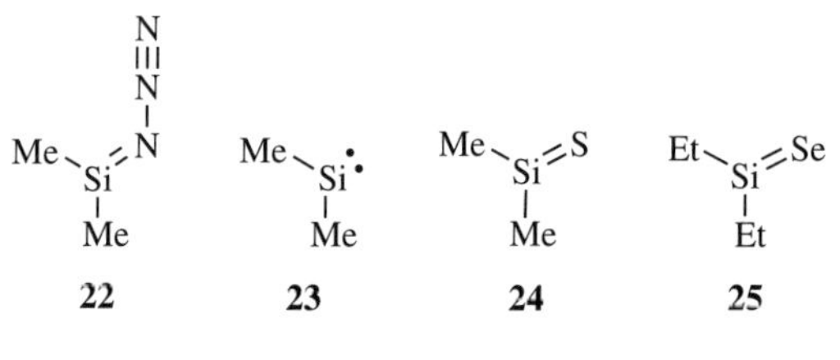

Scheme 5   Highly reactive unsaturated silicon species

1. Frye, C. L.; Salinger, R. M.; Fearon, F. W. G.; Klosowski, J. M.; DeYoung, T., *J. Org. Chem.* **1970**, *35*, 1308.

2. Soysa, H. S. D.; Okinoshima, H.; Weber, W. H., *J. Organomet. Chem.* **1977**, *133*, C17.

3. (a) Golino, C. M.; Bush, R. D.; On, P.; Sommer, L. H., *J. Am. Chem. Soc.* **1975**, *97*, 1957. (b) Okinoshima, H.; Weber, W. P., *J. Organomet. Chem.* **1978**, *150*, C25.

4. (a) Parker, D. R.; Sommer, L. H., *J. Am. Chem. Soc.* **1976**, *98*, 618. (b) Kazuora, S. A.; Weber, W. P., *J. Organomet. Chem.* **1984**, *271*, 47.

5. Soysa, H. S. D.; Weber, W. P., *J. Organomet. Chem.* **1979**, *165*, C1.

6. Toskas, G.; Moreau, M.; Sigwalt, P., *Macromol. Symp.* **2006**, *240*, 68 and references cited within.

7. Lee, C. L.; Frye, C. L.; Johannson, G. K., *Polymer Preprints* **1969**, *10*, 1361.

8. Aggarwal, E. H.; Bauer, S. H., *J. Chem. Phys.* **1950**, *18*, 42.

9. Sieburth, S. M.; Fensterbank, L., *J. Org. Chem.* **1993**, *58*, 6314.

10. Lickiss, P. D., *Adv. Inorg. Chem.* **1995**, *42*, 147.

11. Chandrasekhar, V.; Boomishankar, R.; Nagendran, S., *Chem Rev.* **2004**, *104*, 5847.

12. Li, C.; Porco, J. A. Jr., *J. Am. Chem. Soc.* **2004**, *126*, 1310.

13. (a) Denmark, S. E.; Sweis, R. F. In *Metal-Catalyzed Cross-Coupling Reactions*; de Meijere, A.; Diedrich, F., Eds.; Wiley-VCH: Weinheim, Germany, 2004; Vol 1, Chapter 4. (b) Denmark, S. E.; Ober, M. H., *Aldrichimica Acta* **2003**, *36*, 75 and references cited within.

14. Sieburth, S. M.; Mu, W., *J. Org. Chem.* **1993**, *58*, 7584.

15. Akerman, E., *Acta Chem. Scand.* **1956**, *10*, 298.

16. Pohl, E. R.; Osterholtz, F. D. In *Silanes, Surfaces and Interfaces*; Leyden, D. E., Ed.; Gordon and Breach: New York, 1986; Vol. 1, p 481.

17. Novikov, Y. Y.; Sampson, P., *J. Org. Chem.* **2005**, *70*, 10247.

18. Rudisch, I.; Schmidt, M., *Angew. Chem., Int. Ed. Engl.* **1963**, *2*, 328.

19. Hirabayashi, K.; Kawashima, J.; Nishihara, Y.; Mori, A.; Hiyama, T., *Org. Lett.* **1999**, *1*, 299.

20. Hirabayashi, K.; Ando, J.; Kawashima, J.; Nishihara, Y.; Mori, A.; Hiyama, T., *Bull. Chem. Soc. Jpn.* **2000**, *73*, 1409.

21. Denmark, S. E.; Ober, M. H., *Adv. Syn. Cat.* **2004**, *346*, 1703.

22. Yoshida, H.; Yamaryo, Y.; Oshita, J.; Kunai, A., *Chem. Commun.* **2003**, 1510.

23. Hirabayashi, K.; Takahisa, E.; Nishihara, Y.; Mori, A.; Hiyama, T., *Bull. Chem. Soc. Jpn.* **1998**, *71*, 2409.

24. (a) Denmark, S. E.; Wehrli, D., *Org. Lett.* **2000**, *2*, 565. (b) Denmark, S. E.; Wang, Z., *Org. Synth.* **2005**, *81*, 42.

25. Denmark, S. E.; Pan, W., *J. Organomet. Chem.* **2002**, *653*, 98.

26. Hirabayashi, K.; Mori, A.; Hiyama, T., *Tetrahedron Lett.* **1997**, *38*, 461.

27. Anderson, J. C.; Munday, R. H., *J. Org. Chem.* **2004**, *69*, 8971.

28. Denmark, S. E.; Kallemeyn, J. M., *J. Am. Chem. Soc.* **2006**, *128*, 15958.

29. Denmark, S. E.; Baird, J. D., *Org. Lett.* **2004**, *6*, 3649.

30. Denmark, S. E.; Baird, J. D., *Org. Lett.* **2006**, *8*, 793.

31. (a) Seyferth, D.; Annarelli, D. C.; Duncan, D. P., *Organometallics* **1982**, *1*, 1288. (b) Sommer, L. H.; McLick, J., *J. Organomet. Chem.* **1975**, *101*, 171.

32. Thompson, D. P.; Boudjouk, P., *J. Chem. Soc., Chem. Comm.* **1987**, 1466.

Scott E. Denmark & Christopher R. Butler
University of Illinois, Urbana, IL, USA

Hexamethyldisiloxane

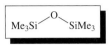

[107-46-0] $C_6H_{18}OSi_2$ (MW 162.42)

InChI = 1S/C6H18OSi2/c1-8(2,3)7-9(4,5)6/h1-6H3

InChIKey = UQEAIHBTYFGYIE-UHFFFAOYSA-N

(silylating agent for carboxylic acids[1,2] and alcohols;[3] used in preparation of aroyl chlorides;[4] precursor for a variety of trimethylsilyl derivatives[5])

Alternate Names: bis(trimethylsilyl) ether; bis(trimethylsilyl) oxide.

Physical Data: mp $-59\,°C$; bp $101\,°C/760$ mmHg; $d\ 0.764\ \mathrm{g\,cm^{-3}}$.

Solubility: sol organic solvents such as toluene, chloroform.

Form Supplied in: colorless liquid; available in NMR (99.5%+) and technical (98 or 99%) grade; available in research and commercial quantities.

Analysis of Reagent Purity: ^{1}H NMR 0.49 ppm; ^{13}C NMR 1.94 ppm (in CDCl$_3$).

Handling, Storage, and Precautions: skin irritant. Incompatible with strong acids, strong bases, and oxidizing agents. Use in a fume hood.

Original Commentary

Joerg Pfeifer
Eli Lilly and Company, Lafayette, IN, USA

Silylation of Carboxylic Acids. Alkanoic and aromatic carboxylic acids can be readily silylated with hexamethyldisiloxane.[1] The reaction is carried out under acid catalysis (usually sulfuric acid) and azeotropic removal of water (eq 1). The rate of reaction is strongly influenced by the acidity of the carboxylic acid and increases with increased acidity. This method avoids the use of an amine and thus simplifies isolation of the product because it does not have to be separated from an amine hydrochloride. A limitation is the need for a solvent which allows for azeotropic removal of water.

$$\text{(eq 1)}$$

An alternate silylation procedure employs zinc chloride as the catalyst and uses the carboxylic acid anhydride as starting material for the preparation of the silyl ester (eq 2).[2] A limited number of examples (benzoic and acetic anhydrides) of this reaction have been published. Distillation is the preferred method for isolation of the product.

$$\text{(eq 2)}$$

Silylation of Alcohols. pyridinium *p*-toluenesulfonate can be used as the catalyst in the silylation of primary and secondary alcohols and phenols (eq 3).[3] Even alcohols sensitive to acid-catalyzed rearrangements can be successfully silylated under these conditions. Continuous removal of the water generated during the reaction is necessary and is achieved via the use of a Soxhlet extractor filled with 4Å molecular sieves.

$$\text{(3)}$$

A special case for the preparation of a phenyl silyl ether involves ultrasound irradiation of benzenediazonium fluoroborate in the presence of hexamethyldisiloxane.[6]

Preparation of Aroyl Chlorides. Substituted benzylidene chlorides (trichloromethylarenes) can be converted to the corresponding substituted benzoyl chlorides (aroyl chlorides) with hexamethyldisiloxane in the presence of a catalytic amount of iron(III) chloride (eq 4).[4] These mild reaction conditions offer a distinct advantage over procedures which employ inorganic oxides and high temperatures to achieve the transformation.[7]

$$\text{(4)}$$

Preparation of Other Trimethylsilyl Derivatives. Hexamethyldisiloxane has been used as an inexpensive starting material for the preparation of numerous synthetically useful trimethylsilyl derivatives. Examples include fluorotrimethylsilane,[5a,b] bromotrimethylsilane,[5c] and iodotrimethylsilane,[5d,e] as well as 'polyphosphoric acid trimethylsilyl ester' (PPSE).[5f]

First Update

Thomas J. Curphey
Dartmouth Medical School, Hanover, NH, USA

Silylation. Mukaiyama et al. have reported that Lewis-acid-catalyzed glycosylation reactions of 1-hydroxy sugars with various nucleophiles are markedly accelerated by the incorporation of 10% HMDO in the reaction mixture (eq 5).[8] Both ribosidyl and pyranosyl sugars respond well to the procedure. Yields with various alcohols as nucleophiles exceed 90%. Electron-rich carbon compounds such as 1,3,5-trimethoxybenzene and carbomethoxy ketene methyl trimethylsilyl acetal may also be used as nucleophiles but afford somewhat lower yields. The glycosylation reactions of eq 5 proceed with high stereoselectivity to produce predominantly either the α- or β-glycoside. HMDO is thought to function by silylation of the nucleophile.

$$\text{1-Hydroxy sugar + Nucleophile} \xrightarrow[\text{Drierite/MeNO}_2]{\overset{\text{catalyst}}{\underset{}{\text{HMDO (10 mol \%)}}}} \text{Glycoside} \quad \text{(5)}$$

Polymethylhydrosiloxane (PMHS, $Me_3SiO(HSi(MeO)_nSiMe_3$, where $n \sim 35$) is an economical reagent for ionic hydrogenations. However, its use in this application can be complicated by the formation of insoluble oligomers that cause problems on workup and have prevented applications on a commercial scale. Lim et al. have reported that incorporation of HMDO into the PMHS reaction mixture leads to a much more tractable silicon by-product, greatly facilitating large-scale synthesis of estradiol (eq 6).[9] HMDO is thought to function as a capping reagent to prevent gel-forming cross-linking of the silicone by-products.

$$\text{(6)}$$

The combination of HMDO and trimethylsilyl chloride has been used to silylate a family of silicate/aluminosilicate mesoporous molecular sieves.[10]

Thionation. Thionation, the conversion of C=O to C=S, has long been a standard method for the preparation of a variety of thiono derivatives.[11] Phosphorus pentasulfide (P_4S_{10}), a classic reagent for bringing about this transformation, suffers from a number of problems. These include low yields and the formation of reaction mixtures containing large amounts of intractable by-products from which isolation of the desired product can be difficult. In recent years, Lawesson's reagent (**1**, LR) has displaced P_4S_{10} as the reagent of choice for many thionations, as LR generally produces higher yields and cleaner reaction mixtures than P_4S_{10}.[12] However, aside from its high cost, LR has the major disadvantage that by-products derived from the reagent itself cannot, in general, be removed by any extractive procedure and must be separated by column chromatography on silica gel. The high equivalent weight of LR means that relatively large columns must be used, and the procedure becomes unwieldy for any but small-scale reactions.

1

In a recent series of publications, it has been reported that P_4S_{10} in combination with hexamethyldisiloxane (P_4S_{10}/HMDO, Curphey's Reagent, CR) is a superior reagent for the conversion of a variety of carbonyl compounds to their corresponding thiono analogs (eq 7).[13–16] Functional groups that respond to this treatment include esters, lactones, amides, lactams, and ketones. In the presence of elemental sulfur, 3-oxoesters are converted to dithiolethiones, a promising class of cancer chemoprotective agents (eq 8).[14,15]

$$R_1-\underset{\underset{O}{\|}}{C}-R_2 \quad \xrightarrow{\text{P}_4\text{S}_{10}/\text{HMDO}} \quad R_1-\underset{\underset{S}{\|}}{C}-R_2 \qquad (7)$$

$$\underset{R_2}{\underset{|}{R_1}}\overset{O}{\underset{\|}{C}}\overset{O}{\underset{\|}{C}}\text{OEt} \quad \xrightarrow[S_8]{\text{P}_4\text{S}_{10}/\text{HMDO}} \qquad (8)$$

In a detailed study involving nearly 40 carbonyl compounds of diverse structure and reactivity, thionation by CR, by LR, and in some cases by P_4S_{10} alone was examined and compared.[15] For almost all the substrates examined, the $\text{P}_4\text{S}_{10}/$HMDO combination gave yields comparable to, or greater than those obtained with LR or with P_4S_{10} alone. Except for a few isolated examples, yields with P_4S_{10} alone were markedly inferior to those obtained with CR or with LR. The most generally useful solvents for thionations by CR were found to be the aromatic hydrocarbons, such as toluene or xylene, although in a few instances, use of other solvents like dichloromethane or acetonitrile was advantageous.

In cases where comparable yields were obtained with $\text{P}_4\text{S}_{10}/$HMDO and LR, CR offered three advantages. First, it is less costly, a significant consideration when large amounts of thiono derivative are to be prepared. Second, the progress of thionation reactions with CR may be judged by simple visual inspection, with the insoluble P_4S_{10} being converted to soluble species as the reaction proceeds. Thionation with LR generally yields insoluble by-products, which means that the progress of the reaction is best determined by periodic chromatographic analysis of the reaction mixture.

The third and perhaps most beneficial advantage of the $\text{P}_4\text{S}_{10}/$HMDO combination is the ease with which phosphorus-containing by-products may be removed. This may be done by hydrolysis under mild conditions, for example, by 1–2-h exposure to phosphate buffer at room temperature. Alternatively, product cleanup may be effected by passage of the crude reaction product through silica gel, with which the phosphorus-containing by-products generated from the $\text{P}_4\text{S}_{10}/$HMDO reagent appear to react.[15] The ability to remove by-products by simple hydrolysis without the need for extensive chromatography as required for LR greatly facilitates large-scale preparations. It has the further advantage that interference in the purification process by reagent-derived by-products is avoided. A good example of the latter advantage is shown by thionation of δ-valerolactone (eq 9).

$$\xrightarrow{\text{LR or CR}} \qquad (9)$$

Lawesson had reported that this lactone failed to give the corresponding thionolactone with his reagent, although a reagent-derived by-product indicating that thionation had occurred was isolated.[17] Reexamination of this thionation by HPLC showed that the desired lactone was in fact formed in 71% chromatographic yield with LR and in 82% yield with CR.[15] Subsequent examination by TLC of the LR reaction mixture showed two overlapping spots, one of which corresponded in R_f to authentic thionolactone and the other, presumably, to a reagent-derived by-product. The anticipated difficulty of separating these two substances by silica gel column chromatography may explain the failure of Lawesson to isolate the thionolactone. In contrast, the thionolactone was readily isolated in 65% yield from the reaction with CR.[15]

It has been proposed that the overall reaction for thionations by the $\text{P}_4\text{S}_{10}/$HMDO combination follows eq 10.[15] This conclusion was reached by analysis of the amounts of reagents consumed under the limiting conditions of a large excess of HMDO, and by NMR determination of the nature and amounts of phosphorus-containing products.

$$6\,R_1-\underset{\underset{O}{\|}}{C}-R_2 \;+\; \text{P}_4\text{S}_{10} \;+\; 5\,(\text{Me}_3\text{Si})_2\text{O} \longrightarrow 6\,R_1-\underset{\underset{S}{\|}}{C}-R_2$$
$$+\; 2\,(\text{Me}_3\text{SiO})_3\text{P=S} \;+\; (\text{Me}_3\text{SiO})_2\underset{\underset{S}{\|}}{P}-\text{O}-\underset{\underset{S}{\|}}{P}(\text{OSiMe}_3)_2 \qquad (10)$$

2

Based on the stoichiometry of eq 10 and the fact that HMDO reacts with P_4O_{10} but not with P_4S_{10} under the thionation conditions, it was proposed that HMDO operates in CR by trapping electrophilic polyphosphates generated as thionation proceeds. It was hypothesized that these polyphosphates are responsible for the side-reactions which lead to lower yields when P_4S_{10} alone is used as the thionating reagent. The failure of pyrophosphate **2** to react with HMDO, as shown in eq 10, led to the suggestion that the culprit polyphosphates are tri- and/or higher polyphosphates.[15]

The detailed studies by Curphey of stoichiometry, solvent, temperature, and reaction time, for thionations by the $\text{P}_4\text{S}_{10}/$HMDO reagent,[15] should greatly facilitate the proper choice of conditions by chemists who wish to employ this reagent.

It has been reported that thionations with CR can be effected in the absence of solvent by the use of microwave radiation.[18] Yields appear to be comparable to or in some cases better than those obtained in solution, while reaction times are very much shorter. Difficulties with charring during the reaction were overcome by using intermittent irradiation accompanied by periodic remixing.

Use of a microwave protocol for the conversion of the formal P–O double bond to the P–S double bond using CR has been reported for synthesis of thiophosphoramidodichloridates (eq 11) and thiophosphoramidate diesters (eq 12).[19] In two cases, reactions using conventional heating were examined and shown to proceed at a slower rate and in lower yield than under microwave irradiation. However, the conditions for reaction using conventional heating, in particular the solvent employed, if any, the reaction temperature, and the ultimate yields obtained were not specified, making it difficult to evaluate these claims.[20]

$$\underset{R_2}{\underset{\diagdown}{R_1}}N-\underset{\underset{Cl}{|}}{\overset{\overset{O}{\|}}{P}}-Cl \quad \xrightarrow[\text{MW (900 W)}]{\text{P}_4\text{S}_{10}/\text{HMDO}} \quad \underset{R_2}{\underset{\diagdown}{R_1}}N-\underset{\underset{Cl}{|}}{\overset{\overset{S}{\|}}{P}}-Cl \qquad (11)$$

$$\underset{R_1\text{O}}{\underset{\diagdown}{R_1\text{O}}}\overset{\overset{O}{\|}}{P}-\text{NHR}_2 \quad \xrightarrow[\text{MW (900 W)}]{\text{P}_4\text{S}_{10}/\text{HMDO}} \quad \underset{R_1\text{O}}{\underset{\diagdown}{R_1\text{O}}}\overset{\overset{S}{\|}}{P}-\text{NHR}_2 \qquad (12)$$

The thionation of five-, six-, and seven-membered ring lactones by the combination of LR and HMDO under microwave irradiation has been reported.[21] The reactions were very fast, being completed in a matter of minutes. In initial trials using γ-dodecalactone as substrate (eq 13) and microwave radiation, thionation by LR alone, by CR, and by the combination of LR/HMDO was compared.

$$\text{(13)}$$

R = n-C$_8$H$_{17}$ **3**   **4**

Both LR alone and CR gave similar mixtures of lactones **3** and **4**, with the proportion of dithiolactone **4** in the mixture comprising 13–14%. When HMDO was incorporated into the LR reaction mixture, the proportion of **4** was reduced to 1%. Under these conditions **3** may be isolated from the reaction mixture in 85% yield. Clearly, the addition of HMDO to the LR reaction mixture increased the selectivity of the thionating reagent, with an accompanying increase in the overall yield of thionolactone **3**. Whether the effect of HMDO in this case arose from a cause similar to that with CR is not known. The formation of a bis(trimethylsilyl)-anisylthiophosphonate is alluded to by the authors,[21] but the references cited[14,15,22] either do not report formation of such a species from LR,[14,15] or are clearly erroneous.[22] It was further reported that when the reaction of eq 13 was conducted with LR/HMDO and conventional heating (solvent, if any, not specified), the reaction was very much slower than with microwave irradiation, but the yields and selectivity with respect to **3** versus **4** were similar.

After completion of the pilot studies with γ-dodecalactone, thionation of a series of γ-lactones was examined and found in most cases to give the corresponding thionolactone with good yield and selectivity.[21] Yields obtained with the one δ-lactone and the one ε-lactone examined were less satisfactory (44% and 15%, respectively).

In considering the various alternatives described above for carrying out thionations, it should be pointed out that claims of economic or environmental benefits derived from conducting these reactions by microwave heating in the absence of solvent are questionable. The amount of solvent used in reactions employing conventional heating, which are normally run at close to molar concentration,[15] is usually trivial compared to that required for isolation and purification of the product. However, reactions runs with microwave irradiation are clearly much faster than with conventional heating and may sometimes give better yields.

Oxidation. Reaction of chromic anhydride with HMDO in halogenated solvents leads to dissolution of the solid with formation of bis(trimethylsilyl)chromate (eq 14).

$$CrO_3 + (Me_3Si)_2O \xrightarrow{\text{CH}_2\text{Cl}_2 \text{ or CCl}_4} (Me_3SiO)_2CrO_2 \quad \text{(14)}$$

Addition of either silica gel or Montmorillite K-10 affords supported reagents (BTSC/SiO$_2$ and BTSC/M K-10, respectively) which are reported to be useful for the oxidation of alcohols.[23,24] Aliphatic and benzylic primary alcohols are oxidized cleanly to aldehydes by these reagents without overoxidation to carboxylic acids, while secondary alcohols afford good yields of ketones. Allylic alcohols are oxidized to the corresponding α,β-unsaturated aldehydes and ketones. However, with BTSC/SiO$_2$ some carbon–carbon double bond cleavage is reported for oxidation of cinnamyl alcohol to cinnamaldehyde.[23] No cleavage is reported to occur when BTSC/M K-10 is used as the oxidant in this reaction.[24]

BTSC/SiO$_2$ is used as the oxidant in an efficient one-pot conversion of aromatic aldehydes to aroyl cyanides (eq 15).[23] BTSC/M K-10 has been employed for direct oxidative deprotection of TMS ethers to aldehydes and ketones (eq 16).[25]

$$\text{ArCHO} \xrightarrow{\text{TMSCN}} \left[\text{Ar} \begin{array}{c} \text{OTMS} \\ \text{CN} \end{array} \right] \xrightarrow[\text{CH}_2\text{Cl}_2]{\text{BTSC/SiO}_2} \text{Ar} \begin{array}{c} \text{O} \\ \text{CN} \end{array} \quad \text{(15)}$$

$$\begin{array}{c} R_1 \\ R_2 \end{array} \begin{array}{c} \text{H} \\ \text{OTMS} \end{array} \xrightarrow[\text{CH}_2\text{Cl}_2]{\text{BTSC/M K-10}} \begin{array}{c} R_1 \\ R_2 \end{array} \text{O} \quad \text{(16)}$$

Oxygen Nucleophile. Uchiro et al. have reported hydrolysis of a thioglycoside bond under very mild conditions using the combination of tetrabutylammonium periodate as oxidant, trityl tetrakis(pentafluorophenyl)borate as Lewis acid catalyst, and HMDO in anhydrous acetonitrile at 0 °C (eq 17).[26] The reaction also proceeded well in the absence of HMDO, using aqueous triflic or perchloric acid as catalyst in place of the trityl derivative. The intersaccharidic bond of a disaccharide was found to be stable under these conditions, with no anomerization being observed.

$$\text{(17)}$$

HMDO serves as an oxygen nucleophile in the conversion of a phosphazene to its corresponding oxide derivative (eq 18)[27] and in the substitution reactions of two phosphoryl derivatives (eq 19).[28]

$$\text{(18)}$$

$$\text{POFX}_2 + (Me_3Si)_2O \xrightarrow{-Me_3SiX} \begin{array}{c} Me_3SiO \\ X \end{array} \begin{array}{c} \text{O} \\ \text{F} \end{array} \quad \text{(19)}$$

55% When X = Cl
53% When X = Br

Reaction of methyl lithium with HMDO in THF generates solutions of anhydrous Me$_3$SiOLi, a synthetically useful form of protected LiOH. For example, this reagent was recently reported to react with CoBr$_2$ to form a Co(OSiMe$_3$)$_2$ derivative possessing in the solid state a dimeric structure in which the two cobalt atoms are joined by silanolato bridges (eq 20).[29]

$$(Me_3Si)_2O \xrightarrow[\text{THF}]{\text{MeLi}} [Me_3SiOLi] \xrightarrow[60\%]{\text{CoBr}_2} \quad \text{(20)}$$

Carbon Nucleophile. It has long been known that HMDO can be metalated with *t*-BuLi in hydrocarbon solvent (eq 21).[30] More recently, Lyszak et al. have isolated lithio derivative **5** as a stable white solid and used it in the preparation of novel zirconium compounds (eqs 22 and 23).[31]

$$(Me_3Si)_2O \xrightarrow[\substack{\text{pentane} \\ \text{or hexane} \\ 64.6\%}]{t\text{-BuLi}} LiCH_2SiMe_2OSiMe_3 \qquad (21)$$
$$\mathbf{5}$$

$$Cp_2ZrCl_2 \xrightarrow[\substack{Et_2O \\ 71.2\%}]{\mathbf{5}} Cp_2Zr(CH_2SiMe_2OSiMe_3)_2 \qquad (22)$$

$$Cp^*ZrCl_3 \xrightarrow[\substack{Et_2O \\ 18.8\%}]{\mathbf{5}} Cp^*ZrCl_2(CH_2SiMe_2OSiMe_3) \qquad (23)$$

Related Reagents. *N,O*-Bis(trimethylsilyl)acetamide; *N,N'*-Bis(trimethylsilyl)urea; Chlorotrimethylsilane; Hexamethyldisilazane; *N*-(Trimethylsilyl)imidazole.

1. Matsumoto, H.; Hoshino, Y.; Nakabayashi, J.; Nakano, T.; Nagai, Y., *Chem. Lett.* **1980**, 1475.

2. Valade, M. J., *C. R. Hebd. Seances Acad. Sci.* **1958**, *246*, 952.

3. Pinnick, H. W.; Bal, B. S.; Lajis, N. H., *Tetrahedron Lett.* **1978**, 4261.

4. Nakano, T.; Ohkawa, K.; Matsumoto, H.; Nagai, Y., *J. Chem. Soc., Chem. Commun.* **1977**, 808.

5. (a) Schmidt, M.; Schmidbaur, H., *Chem. Ber.* **1961**, *94*, 2446. (b) Pray, B. O.; Sommer, L. H.; Goldberg, G. M.; Kerr, G. T.; Di Giorgio, P. A.; Whitmore, F. C., *J. Am. Chem. Soc.* **1948**, *70*, 433. (c) Aizpurua, J. M.; Palomo, C., *Nouv. J. Chim.* **1984**, *8*, 51. (d) Jung, M. E.; Lyster, M. A., *Org. Synth., Coll. Vol.* **1988**, *6*, 353. (e) Voronkov, M. G.; Turkin, Y. A.; Mirskov, R. G.; Kuzimina, E. E.; Rakhlin, V. I., *Zh. Obshch. Khim.* **1989**, *59*, 2641 (*Chem. Abstr.* **1990**, *112*, 118 935t). (f) Yamamoto, K.; Watanabe, H., *Chem. Lett.* **1982**, 1225.

6. Olah, G. A.; Wu, A., *Synthesis* **1991**, 204.

7. (a) Schreyer, R. C., *J. Am. Chem. Soc.* **1958**, *80*, 3483. (b) Rondestvedt, C. S., Jr., *J. Org. Chem.* **1976**, *41*, 3569, 3574, 3577.

8. Mukaiyama, T.; Matsubara, K.; Hora, M., *Synthesis* **1994**, 1368.

9. Lim, C.; Evenson, G. N.; Perrault, W. R.; Pearlman, B. A., *Tetrahedron Lett.* **2006**, *47*, 6417.

10. Beck, J. S.; Vartuli, J. C.; Roth, W. J.; Leonowicz, M. E.; Kresge, C. T.; Schmitt, K. D.; Chu, C. T. W.; Olson, D. H.; Sheppard, E. W.; McCullen, S. B.; Higgins, J. B.; Schlenker, J. L., *J. Am. Chem. Soc.* **1992**, *114*, 10834.

11. For a recent review of thionation reagents see: Polshettiwar, V.; Kaushik, M. P., *J. Sulfur Chem.* **2006**, *27*, 353.

12. For a review see: Cava, M. P.; Levinson, M. I., *Tetrahedron* **1985**, *41*, 5061.

13. Curphey, T. J., *Tetrahedron Lett.* **2002**, *43*, 371.

14. Curphey, T. J., *Tetrahedron Lett.* **2000**, *41*, 9963.

15. Curphey, T. J., *J. Org. Chem.* **2002**, *67*, 6461.

16. Curphey, T. J., U. S. Pat. 7, 012, 148; 2006.

17. Scheibye, S.; Kristensen, J.; Lawesson, S. O., *Tetrahedron* **1979**, *35*, 1339.

18. Polshettiwar, V.; Nivsarkar, M.; Pardashani, D.; Kaushik, M. P., *J. Chem. Res.* **2004**, 474.

19. Nivsarkar, M.; Gupta, A. K.; Kaushik, M. P., *Tetrahedron Lett.* **2004**, *45*, 6863.

20. The authors appear to claim that the transformations of eqs 11 and 12 have been effected by LR. However the reference cited (Maier, L., *Helv. Chim. Acta.* **1963**, *46*, 2026) predates the discovery of the utility of LR for thionations, and deals, *inter alia*, with the reaction between elemental sulfur and halophosphines. There also appears to be a problem with the endnote of reference 19 entitled "Typical procedure." We are instructed to place 1.5 mmol of HMDO (approximately 0.3 ml), 0.8 mmol of P$_4$S$_{10}$ (0.356 g), and 1 mmol of substrate (a solid or high boiling liquid) into a 250 ml conical flask, mix it thoroughly, then irradiate it. The mechanical problem of thoroughly mixing such a thick slurry should be apparent. Moreover, once irradiation is started and the flask contents exceed the bp of HMDO (101 °C) all of this substance should be in the vapor phase (whether the flask is closed or not is not stated).

21. Filippi, J. J.; Fernandez, X.; Lizzani-Cuvelier, L.; Loiseau, A. M., *Tetrahedron Lett.* **2003**, *44*, 6647.

22. Sauer, R. O., *J. Am. Chem. Soc.* **1944**, *22*, 1707. This paper is entitled "Derivatives of Methylchlorosilanes. I. Trimethylsilanol and its Simple Ethers."

23. Lee, J. G.; Lee, J. A.; Sohn, S., *Synth. Commun.* **1996**, *26*, 543.

24. Heravi, M. M.; Ajami, D.; Tabar-Heydar, K., *Monatsh. Chem.* **1998**, *129*, 1305.

25. Heravi, M. M.; Ajami, D.; Tabar-Heydar, K.; Mojtahedi, M. M., *J. Chem. Res.* **1998**, 620.

26. Uchiro, H.; Wakiyama, Y.; Mukaiyama, T., *Chem. Lett.* **1998**, *27*, 567.

27. Kingston, M.; Lork, E.; Mews, R., *J. Fluorine Chem.* **2004**, *125*, 681.

28. Rovnanik, P.; Cernik, M., *J. Fluorine Chem.* **2004**, *125*, 83.

29. Asadi, A.; Eaborn, C.; Hill, M. S.; Hitchcock, P. B.; Smith, J. D., *J. Organomet. Chem.* **2005**, *690*, 944.

30. Frye, C. L.; Salinger, R. M.; Fearon, F. W. G.; Klosowski, J. M.; DeYoung, T., *J. Org. Chem.* **1970**, *35*, 1308.

31. Lyszak, E. L.; O'Brien, J. P.; Kort, D. A.; Hendges, S. K.; Redding, R. N.; Bush, T. L.; Hermen, M. S.; Renkema, K. B.; Silver, M. E.; Huffman, J. C., *Organometallics.* **1993**, *12*, 338.

Hexamethyldisilazane

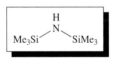

[999-97-3]   C$_6$H$_{19}$NSi$_2$ (MW 161.44)

InChI = 1S/C6H19NSi2/c1-8(2,3)7-9(4,5)6/h7H,1-6H3

InChIKey = FFUAGWLWBBFQJT-UHFFFAOYSA-N

(selective silylating reagent;[1] aminating reagent; nonnucleophilic base[2])

Alternate Name: HMDS.

Physical Data: bp 125 °C; *d* 0.765 g cm^{-3}.

Solubility: sol acetone, benzene, ethyl ether, heptane, perchloroethylene.

Form Supplied in: clear colorless liquid; widely available.

Purification: may contain trimethylsilanol or hexamethyldisiloxane; purified by distillation at ambient pressures.

Handling, Storage, and Precautions: may decompose on exposure to moist air or water, otherwise stable under normal temperatures and pressures. Harmful if swallowed, inhaled, or absorbed through skin. Fire hazard when exposed to heat, flames, or oxidizers. Use in a fume hood.

Silylation. Alcohols,[3] amines,[3] and thiols[4] can be trimethylsilylated by reaction with hexamethyldisilazane (HMDS). Ammonia is the only byproduct and is normally removed by distillation over the course of the reaction. Hydrochloride salts, which are typically encountered in silylation reactions employing chlorosilanes, are avoided, thereby obviating the need to handle large amounts of precipitates. Heating alcohols with hexamethyldisilazane to reflux is often sufficient to transfer the trimethylsilyl group (eq 1).[5] Completion of the reaction is indicated by either a change in the reflux temperature (generally a rise) or by the cessation of ammonia evolution.

$$HN(TMS)_2 + 2 ROH \xrightarrow{\Delta} 2 ROTMS + NH_3 \quad (1)$$

Silylation with HMDS is most commonly carried out with acid catalysis.[5] The addition of substoichiometric amounts of chlorotrimethylsilane (TMSCl) to the reaction mixtures has been found to be a convenient method for catalysis of the silylation reaction.[5,6] The catalytically active species is presumed to be hydrogen chloride, which is liberated upon reaction of the chlorosilane with the substrate. Alternatively, protic salts such as ammonium sulfate can be employed as the catalyst.[7] Addition of catalytic lithium iodide in combination with TMSCl leads to even greater reaction rates.[8] Anilines can be monosilylated by heating with excess HMDS (3 equiv) and catalytic TMSCl and catalytic LiI (eq 2). Silylation occurs without added LiI; however, the reaction is much faster in the presence of iodide, presumably due to the in situ formation of a catalytic amount of the more reactive iodotrimethylsilane.

$$(2)$$

R = H, alkyl, halogen

Hexamethyldisilazane is the reagent of choice for the direct trimethylsilylation of amino acids, for which TMSCl cannot be used due to the amphoteric nature of the substrate.[9] Silylation of glutamic acid with excess hexamethyldisilazane and catalytic TMSCl in either refluxing xylene or acetonitrile followed by dilution with alcohol (methanol or ethanol) yields the derived lactam in good yield (eq 3).[10]

$$(3)$$

The efficiency of HMDS-mediated silylations can be markedly improved by conducting reactions in polar aprotic solvents. For example, treatment of methylene chloride solutions of primary alcohols or carboxylic acids at ambient temperatures with HMDS (0.5–1 equiv) in the presence of catalytic amounts of TMSCl (0.1 equiv) gives the corresponding silyl ether and the trimethylsilyl ester, respectively (eq 4).[1] N-Silylation of secondary amines occurs in preference to primary alcohols when treated with 1 equiv of HMDS and 0.1 equiv TMSCl (eq 5). The silylation of secondary amines cannot be effected in the absence of solvent.[5] Secondary and tertiary alcohols can also be silylated at ambient temperatures in dichloromethane with HMDS and TMSCl mixtures; however, stoichiometric quantities of the silyl chloride are required.

Catalysis by 4-dimethylaminopyridine (DMAP) is necessary for the preparation of tertiary silyl ethers.

$$(4)$$

$$(5)$$

DMF is a useful solvent for HMDS-induced silylation reactions, and reaction rates 10–20 times greater than those carried out in pyridine have been reported.[11] DMSO is also an excellent solvent; however, a cosolvent such as 1,4-dioxane is required to provide miscibility with HMDS.[12]

Imidazole (ImH) catalyzes the silylation reaction of primary, secondary, and tertiary alkanethiols with hexamethyldisilazane.[12] The mechanism is proposed to involve the intermediacy of N-(trimethylsilyl)imidazole (ImTMS), since its preparation from hexamethyldisilazane and imidazole to yield 1-(trimethylsilyl)-imidazole is rapid.[13] The imidazole-catalyzed reactions of hexamethyldisilazane, however, are more efficient than the silylation reactions effected by ImTMS (eq 6 vs. eq 7) due to reversibility of the latter. Imidazole also catalyzes the reaction of HMDS with hydrogen sulfide, which provides a convenient preparation of hexamethyldisilathiane, a reagent which has found utility in sulfur transfer reactions.[14]

$$(6)$$

$$(7)$$

Silyl Enol Ethers. Silylation of 1,3-dicarbonyl compounds can be accomplished in excellent yield by heating enolizable 1,3-dicarbonyl compounds with excess HMDS (3 equiv) and catalytic imidazole (eq 8).[15]

$$(8)$$

In combination with TMSI, hexamethyldisilazane is useful in the preparation of thermodynamically favored enol ethers (eq 9).[16] Reactions are carried out at rt or below and are complete within 3 h.

$$(9)$$

Related thermodynamic enolization control has been observed using metallated hexamethyldisilazide to give the more substituted bromomagnesium ketone enolates.[17] Metallation reactions of HMDS to yield Li, K, and Na derivatives are well known and the resulting nonnucleophilic bases have found extensive applications in organic synthesis.[2]

Amination Reactions. Hexamethyldisilazane is a useful synthon for ammonia in amination reactions. Preparation of primary amides by the reaction of acyl chlorides and gaseous Ammonia, for example, is not an efficient process. Treatment of a variety of acyl halides with HMDS in dichloromethane gives, after hydrolysis, the corresponding primary amide (eq 10).[18] Omitting the hydrolysis step allows isolation of the corresponding monosilyl amide.[19]

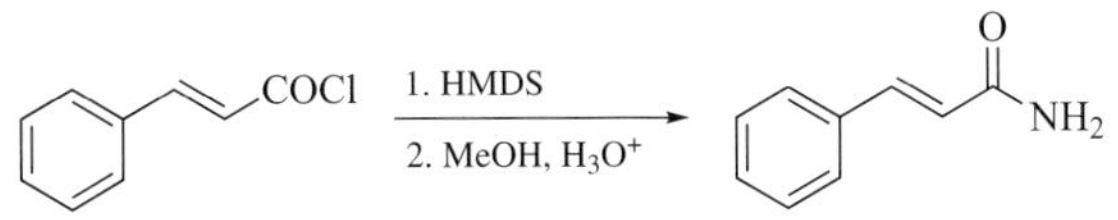

Reductive aminations of ketones with HMDS to yield α-branched primary amines can be effected in the presence of titanium(IV) chloride (eq 11).[20] The reaction is successful for sterically hindered ketones even though HMDS is a bulky amine and a poor nucleophile. The use of ammonia is precluded in these reactions since it forms an insoluble complex with $TiCl_4$.

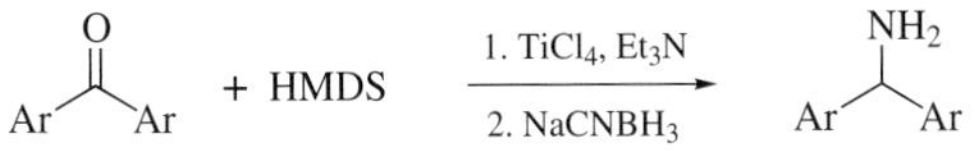

The reaction of phenols with diphenylseleninic anhydride and hexamethyldisilazane gives the corresponding phenylselenoimines (eq 12).[21] The products thus obtained can be converted to the aminophenol or reductively acetylated using zinc and acetic anhydride The use of ammonia or tris(trimethylsilyl)amine in place of HMDS gives only trace amounts of the selenoimines.

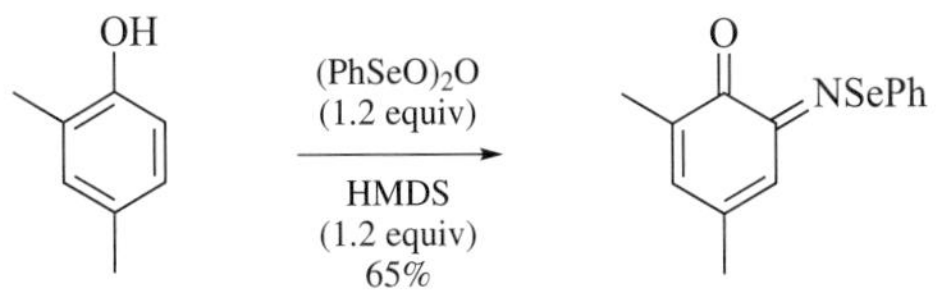

1. Cossy, J.; Pale, P., *Tetrahedron Lett.* **1987**, *28*, 6039.
2. Colvin, E. W. *Silicon in Organic Synthesis*, Butterworths: London, 1981.
3. Speier, J. L., *J. Am. Chem. Soc.* **1952**, *74*, 1003.
4. Bassindale, A. R.; Walton, D. R. M., *J. Organomet. Chem.* **1970**, *25*, 389.
5. Langer, S. H.; Connell, S.; Wender, I., *J. Org. Chem.* **1958**, *23*, 50.
6. Sweeley, C. C.; Bentley, R.; Makita, M.; Wells, W. W., *J. Am. Chem. Soc.* **1963**, *85*, 2497.
7. Speier, J. L.; Zimmerman, R.; Webster, J., *J. Am. Chem. Soc.* **1956**, *78*, 2278.
8. Smith, A. B., III; Visnick, M.; Haseltine, J. N.; Sprengeler, P. A., *Tetrahedron* **1986**, *42*, 2957.
9. Birkofer, L.; Ritter, A., *Angew. Chem., Int. Ed. Engl.* **1965**, *4*, 417.
10. Pellegata, R.; Pinza, M.; Pifferi, G., *Synthesis* **1978**, 614.
11. Kawai, S.; Tamura, Z., *Chem. Pharm. Bull.* **1967**, *15*, 1493.
12. Glass, R. S., *J. Organomet. Chem.* **1973**, *61*, 83.
13. Birkofer, L.; Richter, P.; Ritter, A., *Chem. Ber.* **1960**, *93*, 2804.
14. Harpp, D. N.; Steliou, K., *Synthesis* **1976**, 721.
15. Torkelson, S.; Ainsworth, C., *Synthesis* **1976**, 722.
16. (a) Hoeger, C. A.; Okamura, W. H., *J. Am. Chem. Soc.* **1985**, *107*, 268. (b) Miller, R. D.; McKean, D. R., *Synthesis* **1979**, 730.
17. Kraft, M. E.; Holton, R. A., *Tetrahedron Lett.* **1983**, *24*, 1345.
18. Pellegata, R.; Italia, A.; Villa, M., *Synthesis* **1985**, 517.
19. Bowser, J. R.; Williams, P. J.; Kuvz, K., *J. Org. Chem.* **1983**, *48*, 4111.
20. Barney, C. L.; Huber, E. W.; McCarthy, J. R., *Tetrahedron Lett.* **1990**, *31*, 5547.
21. Barton, D. H. R.; Brewster, A. G.; Ley, S. V.; Rosenfeld, M. N., *J. Chem. Soc., Chem. Commun.* **1977**, 147.

Benjamin A. Anderson
Lilly Research Laboratories, Indianapolis, IN, USA

I

(Iodoethynyl)trimethylsilane

[18163-47-8] C_5H_9ISi (MW 224.13)

InChI = 1S/C5H9ISi/c1-7(2,3)5-4-6/h1-3H3

InChIKey = HNIRHTRSZDSMOF-UHFFFAOYSA-N

(alkyne coupling reagent predominantly used in conjunction with organocopper species or palladium catalysis)

Alternate Name: 1-iodo-2-(trimethylsilyl)acetylene.
Physical Data: bp 96 °C, 55 °C/20 mmHg; n_D^{20} 1.5110; *d* 1.46 g cm^{-3}, fp 48 °C.
Solubility: sol organics (THF, Et$_2$O, pyridine, DCM, CHCl$_3$, DMF).
Analysis of Reagent Purity: ^{1}H NMR, ^{13}C NMR, GC, HPLC.
Form Supplied in: colorless liquid; commercially available.
Preparative Method: all involve iodination of trimethylsilyl-acetylenes. Direct iodination of metalated (lithium or magnesium) trimethylsilylacetylene has been largely supplanted due to the modest yields obtained.[1] Utilizing trimethylsilyl-acetylene and bis(trimethylsilyl) peroxide in the presence of copper(I) iodide gives good yields.[2] Substituting zinc iodide requires using the lithium acetylide.[3] Copper(I) iodide (0.05 equiv.) in the presence of iodine, sodium carbonate, and a phase-transfer catalyst gives good yields without having to use *n*-butyllithium or a Grignard reagent to first deprotonate the trimethylsilylacetylene.[4] The most popular procedure uses bis(trimethylsilyl)acetylene, treating it with iodine monochloride in dichloromethane at 25 °C to obtain excellent yields of (iodoethynyl)trimethylsilane.[5,6]
Handling, Storage, and Precautions: (iodoethynyl)trimethyl-silane is harmful if swallowed, inhaled, or absorbed through skin; use only in a fume hood with suitable personal protective equipment.

Original Commentary

Arthur G. Romero
The Upjohn Company, Kalamazoo, MI, USA

Introduction. (Iodoethynyl)trimethylsilane has been largely confined to aryl- and vinylcopper coupling, where it serves well since the reverse procedure, i.e. coupling with alkynylcopper reagents, is not straightforward due to the lack of reactivity of alkynyl moieties bound to copper.

Cross Coupling with Organocopper Reagents. Vinylcopper species, prepared by the copper-catalyzed addition of ethylmagnesium bromide to monosubstituted alkynes, react with (iodo-ethynyl)trimethylsilane (eq 1) in the presence of 1.2–2 equiv of TMEDA to afford the enyne in good yield.[7] In the absence of TMEDA, iodine exchange to the vinyl group occurs instead of alkynyl–vinyl coupling. In a similar fashion,[8] addition of lithium tributylstannyl(cyano)cuprate across the alkyne gave a vinyl-cuprate which required the addition of lead(IV) acetate–copper(II) acetate to couple successfully with (iodoethynyl)tri-methylsilane. In the absence of zinc(II) chloride, very low yields of the desired product are obtained.

$$\text{(eq 1)} \qquad (1)$$

Allenylcopper species, prepared by *n*-butyllithium deprotona-tion of substituted allenes followed by addition to copper(I) bro-mide, have been coupled with (iodoethynyl)trimethylsilane to give allenyne products (eq 2).[9]

$$\text{(eq 2)} \qquad (2)$$

$$R^1 = t\text{-Bu}, R^2 = H; R^1 = R^2 = Me$$

Arylcopper(I) adducts, prepared by adding an aryllithium or Grignard reagent to a slight excess of copper(I) bromide in ethereal solvent, give rise to arylalkynes when treated with (iodo-ethynyl)trimethylsilane (eq 3).[10] Yields are modest to good, with phenylcopper giving 64%. It was noted that substitution of a catalytic amount of copper(I) bromide, with either phenyl-lithium or phenylmagnesium bromide, or using lithium diphenyl-cuprate, gives iodine transfer to the phenyl group, resulting in the copper acetylide which undergoes coupling with the reactive (iodoethynyl)trimethylsilane to give bis(trimethylsilyl)butadiyne. Although yields are modest, the above sequence is complementary to the Stephens–Castro coupling where aryl iodides are heated in the presence of relatively unreactive copper acetylides to obtain arylalkynes.

$$\text{PhM} \xrightarrow[\text{2. I—≡—TMS}]{\text{1. CuBr (1.1 equiv), Et}_2\text{O}} \text{Ph—≡—TMS} \qquad (3)$$
$$64\%$$
$$M = \text{Li, MgBr}$$

Both (bromo- and (iodoethynyl)trimethylsilane undergo base-induced desilylation, followed by decomposition, when subjected to Cadiot–Chodkiewicz coupling conditions. Switching to the more stable (bromoethynyl)triethylsilane affords good yields of the terminal diynes after protodesilylation (eq 4).[11]

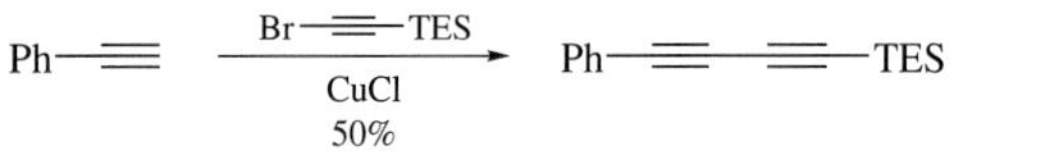

Ph—≡— $\xrightarrow[\substack{CuCl \\ 50\%}]{Br—≡—TES}$ Ph—≡—≡—TES   (4)

Palladium-Catalyzed Coupling. Palladium catalysis (eq 5) has been successfully applied to the coupling of (iodoethynyl)trimethylsilane and acrylate esters. palladium(II) acetate (0.004 equiv.), reduced in situ to palladium(0), in the presence of the alkyne, methyl acrylate, carbonate base, and a phase-transfer salt at 25 °C, gave a 40% yield of the coupled product.[12]

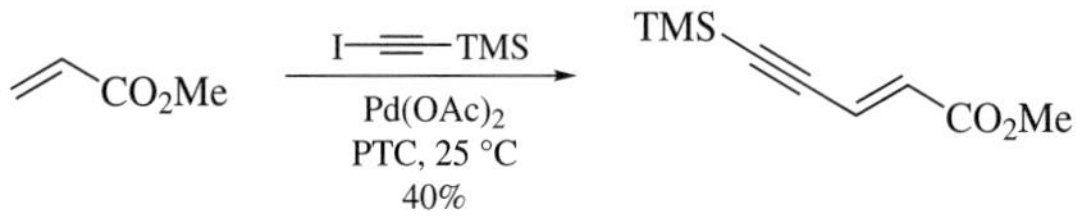

≡CO_2Me $\xrightarrow[\substack{Pd(OAc)_2 \\ PTC, 25\ °C \\ 40\%}]{I—≡—TMS}$ TMS\≡\/\CO_2Me   (5)

At some point, cleavage of the alkynyl–silane bond is generally desired, which can be accomplished quickly and almost quantitatively by treatment with alkali in methanol at 25 °C.[13]

First Update

Craig R. Smith
The Ohio State University, Columbus, OH, USA

Cross-coupling with Organocopper Reagents. Hexasilylbicyclo[2.2.2]octan-1-yl copper, prepared from the treatment of hexasilylbicyclo[2.2.2]octan-1-yl lithium with copper(I) iodide in THF, reacts with (iodoethynyl)trimethylsilane (eq 6) in the presence of pyridine to afford the alkynyl-substituted polysilacage compound.[14]

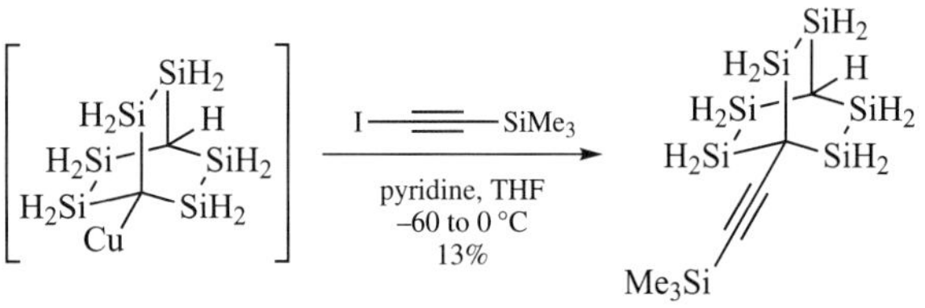

(6)

Copper-promoted aminations have recently been reported as a general strategy for the N-alkynylation of carbamates, sulfonates, and chiral oxazolidinones and imidazolidinones (eq 7).[15] A variety of substituted ynamides can be synthesized via deprotonation of amides with KHMDS followed by reaction with copper(I) iodide and an alkynylbromide or iodide.[16,17]

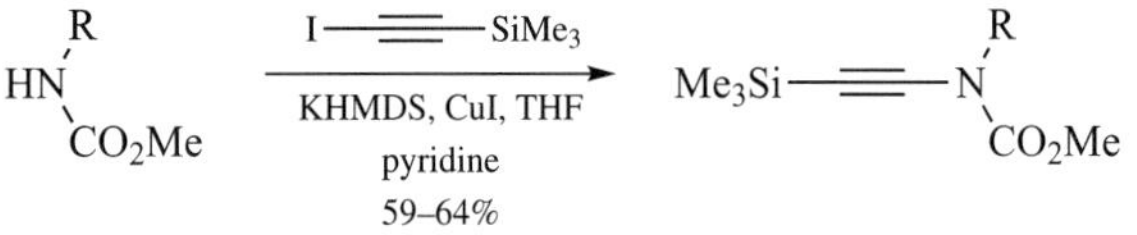

$\underset{\substack{HN \\ |\ \\ CO_2Me}}{R}$ $\xrightarrow[\substack{KHMDS, CuI, THF \\ pyridine \\ 59–64\%}]{I—≡—SiMe_3}$ $Me_3Si—≡—\underset{\substack{| \\ CO_2Me}}{\overset{R}{N}}$ (7)

R = Me, allyl, CH_2CH_2Ph

Palladium-catalyzed Coupling. Palladium catalysis has been successfully applied to the coupling of (iodoethynyl)trimethylsilane and terminal alkynes.[18] Diacetylenes were synthesized via a palladium-catalyzed sp–sp carbon cross-coupling reaction, utilizing Sonogashira–Hagihara conditions. A catalyst system comprised of *trans*-dichlorobis(triphenylphosphine)palladium(II) and

copper(I) iodide in the presence of (iodoethynyl)trimethylsilane, terminal alkyne, and diisopropylamine in THF at 0 °C gave a 48% yield of the desired product (eq 8).[19]

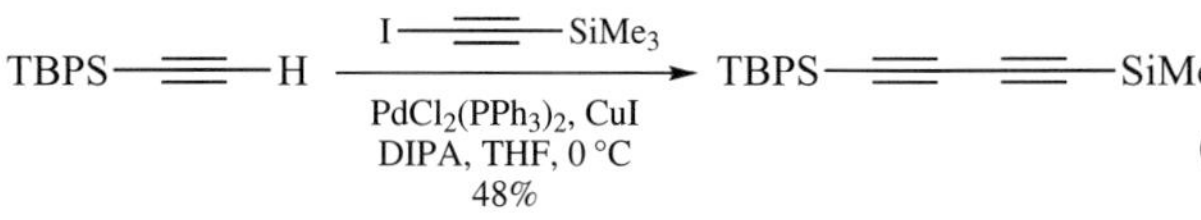

TBPS—≡—H $\xrightarrow[\substack{PdCl_2(PPh_3)_2,\ CuI \\ DIPA,\ THF,\ 0\ °C \\ 48\%}]{I—≡—SiMe_3}$ TBPS—≡—≡—SiMe_3   (8)

The palladium-mediated cross-coupling of organozinc reagents, derived from the asymmetric carboalumination of allyl alcohol, with (iodoethynyl)trimethylsilane (eq 9), has also been reported en route to callystatin A.[20]

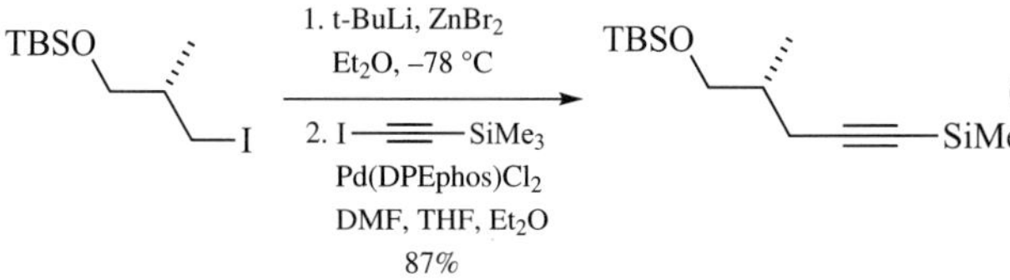

(9)

Miscellaneous Metal-mediated Reactions. The gallium-catalyzed *ortho*-ethynylation of phenols has been achieved with chloro- or (iodoethynyl)trimethylsilane and a variety of phenols. Gallium(III) chloride (0.10 equiv), n-BuLi (0.30 equiv), and 2,6-di-t-butyl-4-methylpyridine (0.10 equiv) in the presence of haloalkyne and phenol in chlorobenzene at 120 °C gave 81–90% yield of a variety of o-(trimethylsilylethynyl)phenols (eq 10).[21]

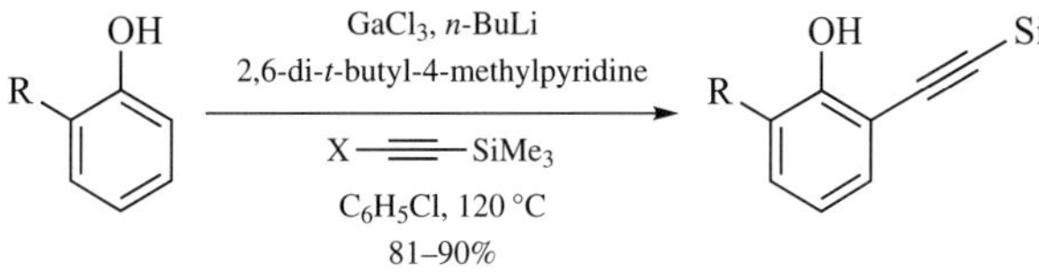

(10)

R = H, Me, *t*-Bu, MeO, TBSO, Cl, Br
X = Cl, Br, I

Recently, two indium-mediated reactions that produce C-glycosides have been disclosed. An efficient Ferrier-type alkynylation of glycals and alkynylindium reagents, derived from iodoacetylenes under Barbier conditions, was disclosed.[22] The Ferrier-type alkynylation proceeds in moderate to good yields (34–94%) affording the α-2,3-unsaturated-alkynyl-C-glycosides (eq 11) with high stereoselectivity (84–95%). The indium-mediated alkynylation of carbohydrates bearing anomeric acetates for the synthesis of C-glycosides has also been disclosed. The creation of a C–C bond from easily accessible carbohydrate precursors was achieved under Barbier conditions, in which carbohydrates bearing anomeric acetates and iodoacetylenes were treated with indium metal in refluxing dichloromethane (eq 12), affording the desired C-glycosides in good yields (44–96%).[23]

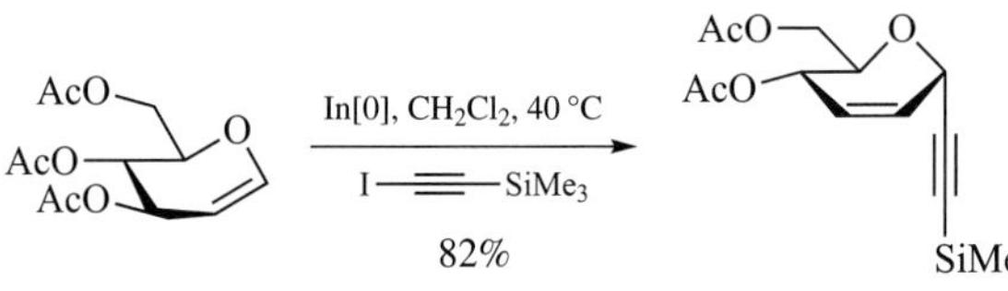

(11)

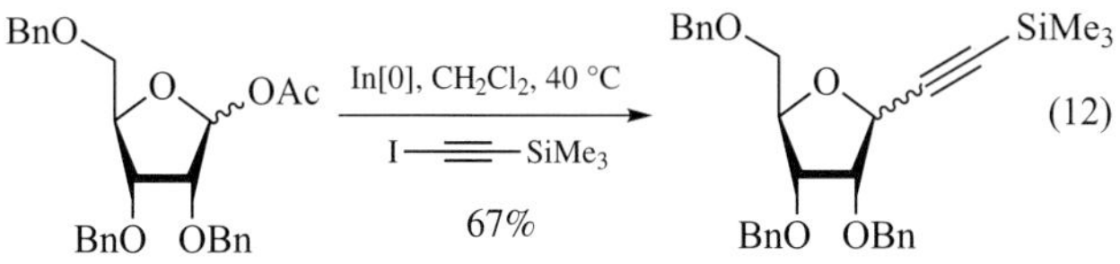

The reaction scheme at top left shows:

BnO—(furanose ring)—OAc, with BnO and OBn substituents, reacting under In[0], CH$_2$Cl$_2$, 40 °C with I—≡—SiMe$_3$, 67%, to give BnO—(furanose ring with C≡C—SiMe$_3$), BnO OBn (12)

Related Reagent. Trimethylsilylacetylene.

1. Buchert, H.; Zeil, W., *Spectrochim. Acta* **1962**, *18*, 1043.

2. Casarini, A.; Dembech, P.; Reginato, G.; Ricci. A.; Seconi, G., *Tetrahedron Lett.* **1991**, *32*, 2169.

3. Ricci, A.; Taddei, M.; Dembech, P.; Guerrini, A.; Seconi, G., *Synthesis* **1989**, 461.

4. Jeffery, T., *J. Chem. Soc., Chem. Commun.* **1988**, 909.

5. Walton, D. R. M.; Waugh, F., *J. Organomet. Chem.* **1972**, *37*, 45.

6. Walton, D. R. M.; Webb, M. J., *J. Organomet. Chem.* **1972**, *37*, 41.

7. Normant, J. F.; Commercon, A.; Villieras, J., *Tetrahedron Lett.* **1975**, 1465.

8. (a) Magriotis, P. L.; Scott, M. E.; Kim, K. D., *Tetrahedron Lett.* **1991**, *32*, 6085. (b) Westmijze, H.; Ruitenberg, K.; Meijer, J.; Vermeer, P., *Tetrahedron Lett.* **1982**, *23*, 2797.

9. Ruitenberg, K.; Meijer, J.; Bullee, R. J.; Vermeer, P., *J. Organomet. Chem.* **1981**, *217*, 267.

10. Oliver, R.; Walton, D. R. M., *Tetrahedron Lett.* **1972**, 5209. Luteyn, J. M.; Spronck, H. J. W.; Salemink, C. A., *Recl. Trav. Chim. Pays-Bas* **1978**, *97*, 187.

11. Eastmond, R.; Walton, D. R. M., *Tetrahedron* **1972**, *28*, 4591.

12. Jeffery, T., *Synthesis* **1987**, 70.

13. Eaborn, C.; Walton, D. R. M., *J. Organomet. Chem.* **1965**, *4*, 217.

14. Shimizu, M.; Mizukoshi, H.; Hiyama, T., *Synthesis* **2004**, 1363.

15. Dunetz, J. R.; Danheiser, R. L., *Org. Lett.* **2003**, *5*, 4011.

16. Dunetz, J. R.; Danheiser, R. L., *J. Am. Chem. Soc.* **2005**, *127*, 5776.

17. (a) Kohnen, A. L.; Mak, X. Y.; Lam, T. Y.; Dunetz, J. R.; Danheiser, R. L., *Tetrahedron* **2006**, *62*, 3815. (b) Kohnen, A. L.; Dunetz, J. R.; Danheiser, R. L., *Org. Synth.* **2007**, *84*, 88.

18. Jahnke, E.; Weiss, J.; Neuhaus, S.; Hoheisel, T. N.; Frauenrath, H., *Chem. Eur. J.* **2009**, *15*, 388.

19. Chalifoux, W. A.; Ferguson, M. J.; Tykwinski, R. R., *Eur. J. Org. Chem.* **2007**, 1001.

20. Liang, B.; Novak, T.; Tan, Z.; Negishi, E., *J. Am. Chem. Soc.* **2006**, *128*, 2770.

21. Kobayashi, K.; Arisawa, M.; Yamaguchi, M., *J. Am. Chem. Soc.* **2002**, *124*, 8528.

22. Lubin-Germain, N.; Hallonet, A.; Huguenot, F.; Palmier, S.; Uziel, J.; Augé, J., *Org. Lett.* **2007**, *9*, 3679.

23. Lubin-Germain, N.; Baltaze, J.-P.; Coste, A.; Hallonet, A.; Lauréano, H.; Legrave, G.; Uziel, J.; Augé, J., *Org. Lett.* **2008**, *10*, 725.

(Iodomethyl)trimethylsilane[1]

Me$_3$SiCH$_2$I

[4206-67-1]   C$_4$H$_{11}$ISi   (MW 214.14)

InChI = 1S/C4H11ISi/c1-6(2,3)4-5/h4H2,1-3H3

InChIKey = VZNYXGQMDSRJAL-UHFFFAOYSA-N

(electrophile for the preparation of allylsilanes[2] and propargyl-silanes;[3] forms carbon alkylation adducts useful for alkene synthesis;[4] alkylation adducts are frequently a source of fluoride-induced reactive intermediates, forming nitrogen[5] and sulfur[6] alkylation adducts that function as ylide precursors; readily undergoes metal–halogen exchange to generate a reagent for Peterson methylenation[7])

Alternate Name: trimethylsilylmethyl iodide.

Physical Data: bp 140–142 °C; d 1.442 g cm^{-3}.

Form Supplied in: colorless liquid; widely available.

Handling, Storage, and Precautions: highly flammable liquid; corrosive; irritant; should be stored cold, protected from light, and stabilized with copper metal. Handle in a fume hood.

Allylsilane and Propargylsilane Synthesis. (Iodomethyl)-trimethylsilane has been used to prepare allylsilanes, which are typically utilized in Lewis acid or fluoride ion-promoted reactions with carbonyl compounds to provide homoallylic alcohols. Treatment of Me$_3$SiCH$_2$I with methyltriphenylphosphonium bromide and base produces an intermediate phosphonium salt which can be further utilized to prepare allylsilanes from carbonyl compounds.[2a,b] This procedure has been modified so that this series of transformations can be done as a one-pot procedure (eq 1).[2c]

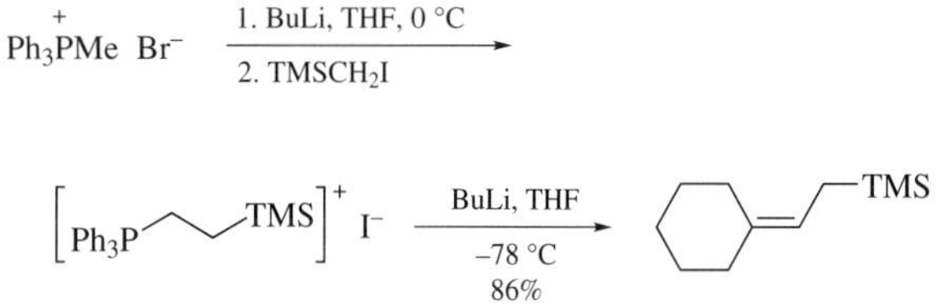

Ph$_3$PMe$^+$ Br$^-$ →(1. BuLi, THF, 0 °C / 2. TMSCH$_2$I)→

[Ph$_3$P—⌒—TMS]$^+$ I$^-$ →(BuLi, THF, –78 °C, 86%)→ (cyclohexylidene)—⌒—TMS   (1)

More recently, it has been demonstrated that activated vinyl-cuprates can be directly added to Me$_3$SiCH$_2$I to give functionalized allylsilanes in moderate yields (eqs 2–4).[8]

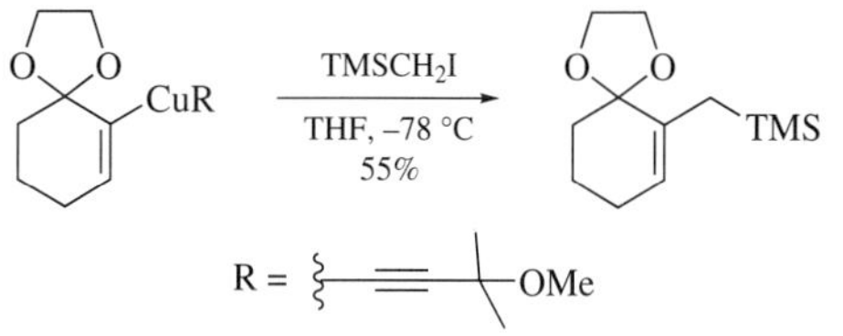

(1,4-dioxaspiro ring)—CuR →(TMSCH$_2$I, THF, –78 °C, 55%)→ (1,4-dioxaspiro ring)—CH$_2$—TMS   (2)

R = ≡—OMe

Ph—CH=CH—CuR →(TMSCH$_2$I, THF, –78 °C, 71%)→ Ph—CH=CH—TMS   (3)

R = ≡—OMe

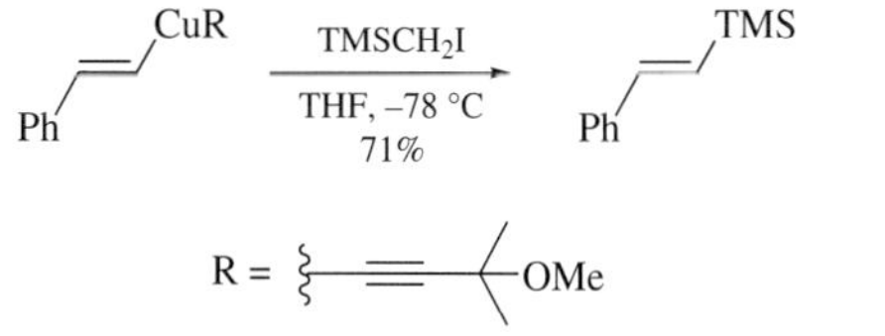

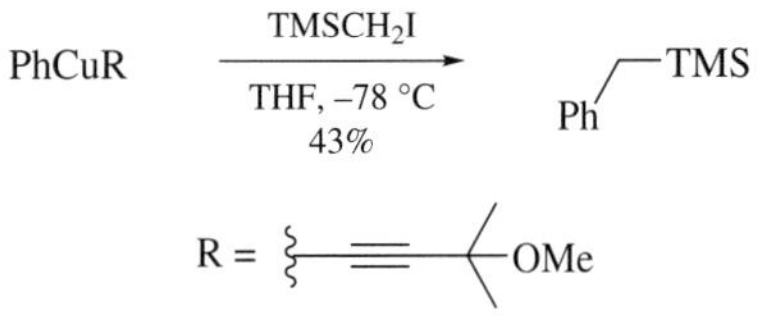

$$PhCuR \xrightarrow[\substack{THF, -78\ ^\circ C \\ 43\%}]{TMSCH_2I} \underset{Ph}{\diagup}\!\!\!-TMS \quad (4)$$

$$R = \text{—}\!\!\equiv\!\!\diagdown\!\!-OMe$$

The synthesis of propargylsilanes has also been thoroughly explored, and these useful reagents are readily prepared by the addition of substituted lithioacetylides to Me_3SiCH_2I.[3] Propargylsilanes have subsequently been used in titanium(IV) chloride-promoted reactions with acetals to yield functionalized α-allenylsilanes or functionalized alkenes.

Alkene Synthesis. α-Alkylation of ketones with Me_3SiCH_2I, followed by a bromination/elimination sequence, gives rise to α-methylene ketones (eq 5).[4a]

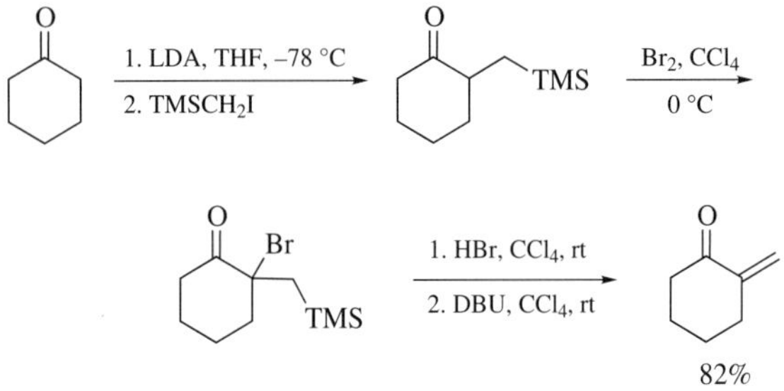

$$(5)$$

$$82\%$$

α-Alkylation of sulfones with Me_3SiCH_2I, followed by fluoride-induced elimination of the resultant β-silyl sulfones, cleanly produces alkenes, and this method has been applied to the synthesis of terminal alkenes, 1,4-dienes (by sequential dialkylation with allyl bromide and Me_3SiCH_2I), 1,3-dienes (eq 6), and *exo*-methylene compounds (eq 7).[4b] The intermediate β-silyl sulfones function as a synthon for $CH_2=CH^-$.

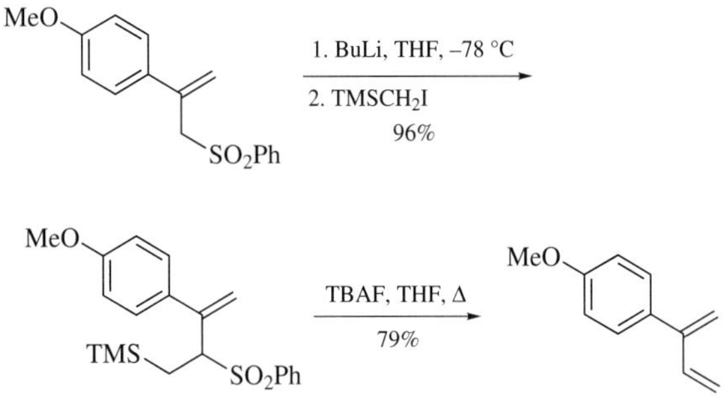

$$(6)$$

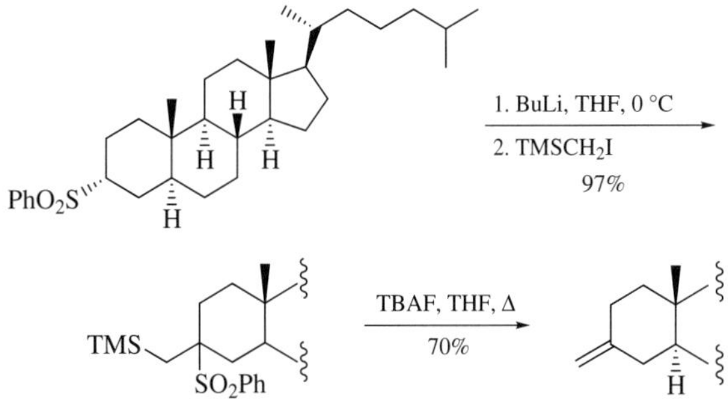

$$(7)$$

Generation of Nitrogen and Sulfur Ylides. *N*-Trimethylsilylmethyl amides, prepared by the alkylation of the deprotonated amide with Me_3SiCH_2I, provide convenient access to nonstabilized azomethine ylide precursors. This method involving fluoride-promoted generation of a nonstabilized imidate ylide for 1,3-dipolar cycloaddition has been applied to the efficient construction of a key intermediate for the synthesis of the pyrrolizidine alkaloids retronecine and indicine by in situ trapping of the ylide intermediate with methyl acrylate (eq 8).[5] This more versatile desilylation method of ylide generation and trapping with dipolarophiles demonstrates applicability to a wider range of synthetic targets than those derived from α-deprotonation of iminium salts, where an often unwanted anion-stabilizing substituent is required.

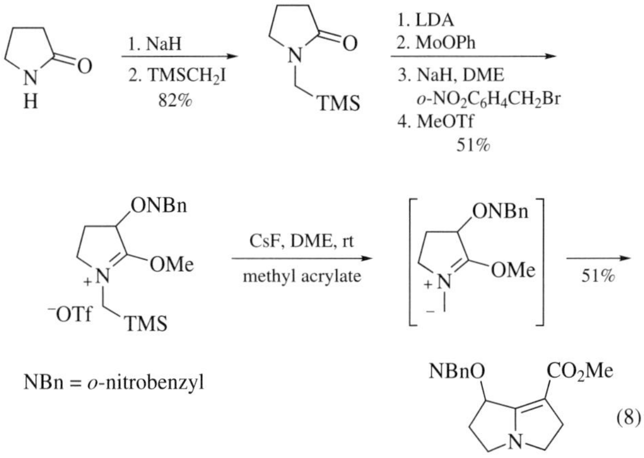

$$NBn = \text{o-nitrobenzyl}$$

$$(8)$$

Similarly, the cesium fluoride-induced desilylation of α-trimethylsilylbenzylsulfonium alkyl triflate salts produces sulfur ylides, which rapidly equilibrate in DME solution to the thermodynamically more stable ylide prior to reaction with aromatic aldehydes to produce predominantly *trans*-diaryl epoxides in high yields.[6a] In the absence of an aldehyde trapping agent, Sommelet–Hauser rearrangements occur.

Related Reagents. (Chloromethyl)trimethylsilane; Trimethylsilylmethyllithium; Trimethylsilylmethylmagnesium Chloride; Trimethylsilyl Trifluoromethanesulfonate.

1. There are currently no published reviews on the use of this reagent. For a review of (chloromethyl)trimethylsilane, see Anderson, R., *Synthesis* **1985**, 717.
2. (a) Seyferth, D.; Wursthorn, K. R.; Mammarella, R. E., *J. Org. Chem.* **1977**, *42*, 3104. (b) Seyferth, D.; Wursthorn, K. R.; Lim, T. F. O.; Sepelak, D. J., *J. Organomet. Chem.* **1979**, *181*, 293. (c) Fleming, I.; Paterson, I., *Synthesis* **1979**, 446. (d) Iio, H.; Ishii, M.; Tsukamoto, M.; Tokoroyama, T., *Tetrahedron Lett.* **1988**, *29*, 5965. (e) Chakraborty, R.; Simpkins, N. S., *Tetrahedron* **1991**, *47*, 7689.
3. (a) Pornet, J.; Kolani, N., *Tetrahedron Lett.* **1981**, *22*, 3609. (b) Pornet, J.; Randrianoelina, B.; Miginiac, L., *Tetrahedron Lett.* **1984**, *25*, 651. (c) Pornet, J.; Damour, D.; Miginiac, L., *Tetrahedron* **1986**, *42*, 2017. (d) Pornet, J., *J. Organomet. Chem.* **1988**, *340*, 273.
4. (a) Fleming, I.; Goldhill, J., *J. Chem. Soc., Perkin Trans. 1* **1980**, 1493. (b) Kocienski, P. J., *Tetrahedron Lett.* **1979**, *20*, 2649.
5. Vedejs, E.; Larsen, S.; West, F. G., *J. Org. Chem.* **1985**, *50*, 2170.
6. (a) Padwa, A.; Gasdaska, J. R., *Tetrahedron* **1988**, *44*, 4147. (b) Vedejs, E.; Martinez, G. R., *J. Am. Chem. Soc.* **1979**, *101*, 6452.

7. (a) Ager, D., *Synthesis* **1984**, 384. (b) Chan, T. H.; Chang, E., *J. Org. Chem.* **1974**, *39*, 3264. (c) Peterson, D. J., *J. Org. Chem.* **1968**, *33*, 780.

8. Majetich, G.; Leigh, A. J., *Tetrahedron Lett.* **1991**, *32*, 609.

Todd K. Jones & Lawrence G. Hamann
Ligand Pharmaceuticals, San Diego, CA, USA

Iodotrimethylsilane[1]

[16029-98-4]   C$_3$H$_9$ISi   (MW 200.11)

InChI = 1S/C3H9ISi/c1-5(2,3)4/h1-3H3

InChIKey = CSRZQMIRAZTJOY-UHFFFAOYSA-N

(a versatile reagent for the mild dealkylation of ethers, carboxylic esters, lactones, carbamates, acetals, phosphonate and phosphate esters; cleavage of epoxides, cyclopropyl ketones; conversion of vinyl phosphates to vinyl iodides; neutral nucleophilic reagent for halogen exchange reactions, carbonyl and conjugate addition reactions; use as a trimethylsilylating agent for formation of enol ethers, silyl imino esters, and *N*-silylenamines, alkyl, alkenyl and alkynyl silanes; Lewis acid catalyst for acetal formation, α-alkoxymethylation of ketones, for reactions of acetals with silyl enol ethers and allylsilanes; reducing agent for epoxides, enediones, α-ketols, sulfoxides, and sulfonyl halides; dehydrating agent for oximes)

Alternate Names: TMS-I; TMSI; trimethylsilyl iodide.

Physical Data: bp 106–109 °C; *d* 1.406 g cm^{-3}; n_D^{20} 1.4710; fp −31 °C.

Solubility: sol in CCl$_4$, CHCl$_3$, CH$_2$Cl$_2$, ClCH$_2$CH$_2$Cl, MeCN, PhMe, hexanes; reactive with THF (ethers), alcohols, and EtOAc (esters).

Form Supplied in: clear colorless liquid, packaged in ampules, stabilized with copper; widely available.

Analysis of Reagent Purity: easily characterized by ^{1}H, ^{13}C, or ^{29}Si NMR spectroscopy.

Preparative Methods: although more than 20 methods have been reported[1] for the preparation of TMS-I, only a few are summarized here. chlorotrimethylsilane undergoes halogen exchange with either lithium iodide[2] in CHCl$_3$ or sodium iodide[3] in MeCN, which allows in situ reagent formation (eq 1). Alternatively, hexamethyldisilane reacts with iodine at 25–61 °C to afford TMS-I with no byproducts (eq 2).[4]

$$ \text{TMSCl} \xrightarrow[\text{or}]{\substack{\text{LiI, CHCl}_3 \\ \\ \text{NaI, MeCN}}} \text{TMSI} \quad + \quad \text{LiCl or NaCl} \quad (1) $$

$$ \text{TMS–TMS} + \text{I}_2 \xrightarrow{25\text{–}61\,°C} \text{TMSI} \quad (2) $$

Several other methods for in situ generation of the reagent have been described.[5,6] It should be noted, however, that the reactivity of in situ generated reagent appears to depend upon the method of preparation.

Purity: by distillation from copper powder.

Handling, Storage, and Precaution: extremely sensitive to light, air, and moisture, it fumes in air due to hydrolysis (HI), and becomes discolored upon prolonged storage due to generation of I$_2$. It is flammable and should be stored under N$_2$ with a small piece of copper wire. It should be handled in a well ventilated fume hood and contact with eyes and skin should be avoided.

Original Commentary

Michael E. Jung
University of California, Los Angeles, CA, USA

Michael J. Martinelli
Lilly Research Laboratories, Indianapolis, IN, USA

Use as a Nucleophilic Reagent in Bond Cleavage Reactions.

Ether Cleavage.[5,7] The first broad use of TMS-I was for dealkylation reactions of a wide variety of compounds containing oxygen–carbon bonds, as developed independently by the groups of Jung and Olah. Simple ethers initially afford the trimethylsilyl ether and the alkyl iodide, with further reaction giving the two iodides (eq 3).[7,8] This process occurs under neutral conditions, and is generally very efficient as long as precautions to avoid hydrolysis by adventitious water are taken. Since the silyl ether can be quantitatively hydrolyzed to the alcohol, this reagent permits the use of simple ethers, e.g. methyl ethers, as protective groups in synthesis. The rate of cleavage of alkyl groups is: tertiary ≈ benzylic ≈ allylic methyl > secondary > primary. Benzyl and *t*-butyl ethers are cleaved nearly instantaneously at low temperature with TMS-I. Cyclic ethers afford the iodo silyl ethers and then the diiodide, e.g. THF gives 4-iodobutyl silyl ether and then 1,4-diiodobutane in excellent yield.[7,8] Alcohols and silyl ethers are rapidly converted into the iodides as well.[8a,9] Alkynic ethers produce the trimethylsilylketene via dealkylative rearrangement.[4b] Phenolic ethers afford the phenols after workup.[5,7,10] In general, ethers are cleaved faster than esters. Selective cleavage of methyl aryl ethers in the presence of other oxygenated functionality has also been accomplished in quinoline.[11] γ-Alkoxyl enones undergo deoxygenation with excess TMS-I (2 equiv), with the first step being conjugate addition of TMS-I.[12]

$$ R^1\text{–O–}R^2 \xrightarrow{\text{TMSI}} R^2\text{I} + R^1\text{OTMS} \xrightarrow{\text{TMSI}} R^1\text{I} \quad (3) $$
$$ \downarrow\text{H}_2\text{O} \qquad\qquad \nearrow\text{TMSI} $$
$$ R^1\text{OH} $$

Cleavage of Epoxides. Reaction of epoxides with 1 equiv of TMS-I gives the vicinal silyloxy iodide.[8e] With 2 equiv of TMS-I, however, epoxides are deoxygenated to afford the corresponding alkene (eq 4).[13a,b] However, allylic alcohols are efficiently prepared by reaction of the intermediate iodosilane with base.[13c,d] Furthermore, acyclic 2-ene-1,4-diols react with TMS-I to undergo dehydration, affording the corresponding diene.[13e]

Ester Dealkylation.[14]   Among the widest uses for TMS-I involves the mild cleavage of carboxylic esters under neutral conditions. The ester is treated with TMS-I to form an initial oxonium intermediate which suffers attack by iodide (eq 5). The trimethylsilyl ester is cleaved with H_2O during workup. Although the reaction is general and efficient, it is possible to accomplish selective cleavage according to the reactivity trend: benzyl, *t*-butyl > methyl, ethyl, *i*-propyl. Neutral transesterification is also possible via the silyl ester intermediate.[15] Aryl esters are not cleaved by TMS-I, however, since the mechanism involves displacement of R^2 by I^-. Upon prolonged exposure (75 °C, 3 d) of simple esters to excess TMS-I (2.5 equiv), the corresponding acid iodides are formed.[14b,16] β-Keto esters undergo decarboalkoxylation when treated with TMS-I.[17] An interesting rearrangement reaction provides α-methylene lactones from 1-(dimethylaminomethyl)cyclopropanecarboxylates (eq 6).[18]

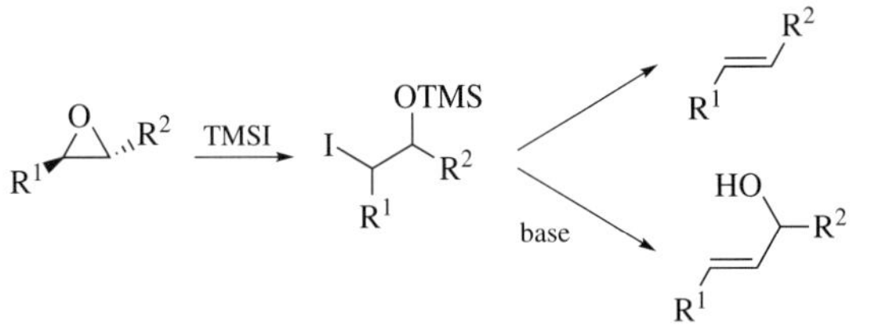

$$\text{(4)}$$

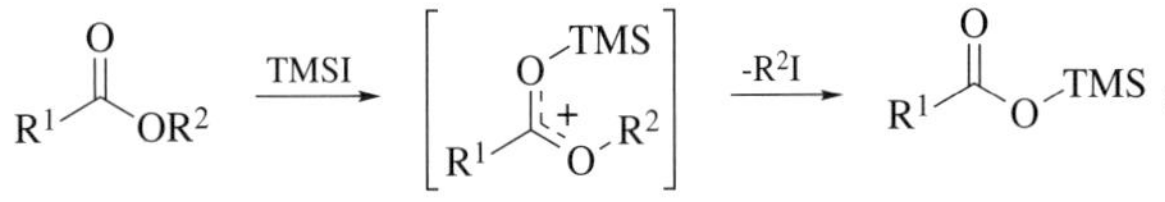

$$\text{(5)}$$

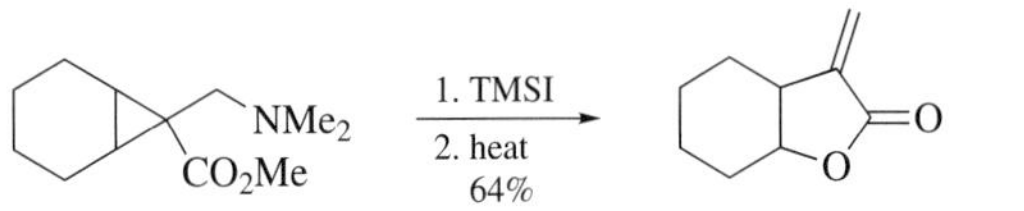

$$\text{(6)}$$

Lactone Cleavage.[14,19]   Analogous to esters, lactones are also efficiently cleaved with TMS-I to provide ω-iodocarboxylic acids, which may be further functionalized to afford bifunctional building blocks for organic synthesis (eq 7). Diketene reacts with TMS-I to provide a new reagent for acetoacylation.[20]

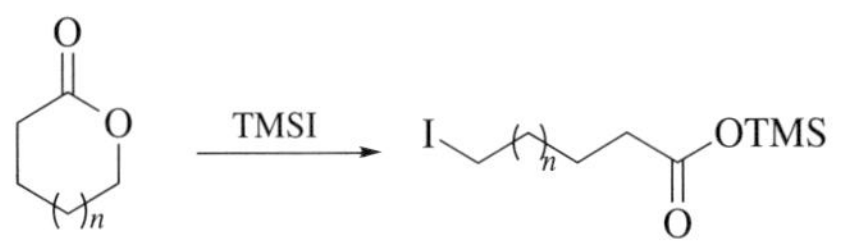

$$\text{(7)}$$

Cleavage of Carbamates.[21]   Since strongly acidic conditions are typically required for the deprotection of carbamates, use of TMS-I provides a very mild alternative. Benzyl and *t*-butyl carbamates are readily cleaved at rt,[22] whereas complete cleavage of methyl or ethyl carbamates may require higher temperatures (reflux). The intermediate silyl carbamate is decomposed by the addition of methanol or water (eq 8). Since amides are stable to TMS-I-promoted hydrolysis,[7a] this procedure can be used to deprotect carbamates of amino acids and peptides.[21d]

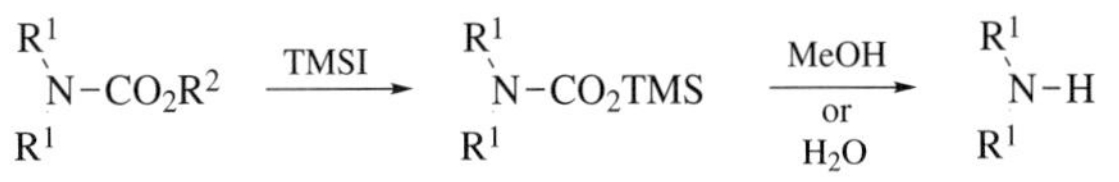

$$\text{(8)}$$

A recent example used TMS-I to deprotect three different protecting groups (carbamate, ester, and orthoester) in the same molecule in excellent yield (eq 9).[23]

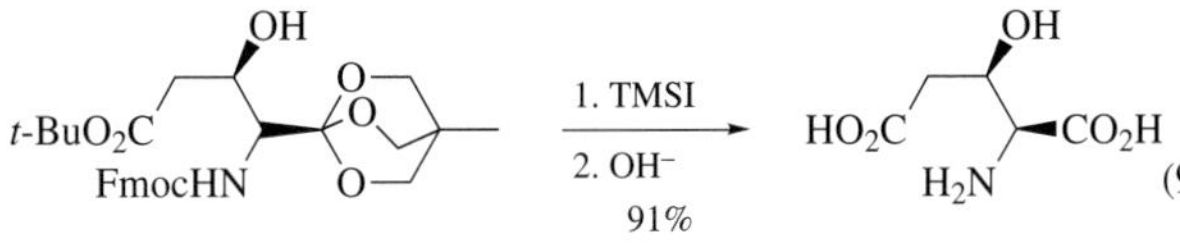

$$\text{(9)}$$

Cleavage of Acetals.[24]   Acetals can be cleaved in analogy to ethers, providing a newly functionalized product (eq 10), or simply the parent ketone (eq 11). Glycals have also been converted to the iodopyrans with TMS-I,[25] and glycosidation reactions have been conducted with this reagent.[26]

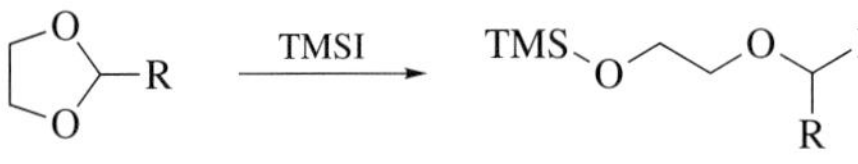

$$\text{(10)}$$

$$\text{(11)}$$

Orthoesters are converted into esters with TMS-I. The dimethyl acetal of formaldehyde, methylal, affords iodomethyl methyl ether in good yield (eq 12)[27a] (in the presence of alcohols, MOM ethers are formed).[27b] α-Acyloxy ethers also furnish the iodo ethers,[28] e.g. the protected β-acetyl ribofuranoside gave the α-iodide which was used in the synthesis of various nucleosides in good yield (eq 13).[28b] Aminals are similarly converted into immonium salts, e.g. Eschenmoser's reagent, dimethyl(methylene)ammonium iodide, in good yield.[29]

$$\text{(12)}$$

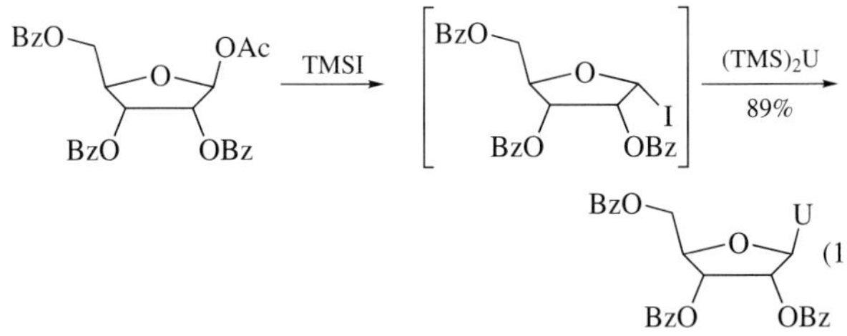

$$\text{(13)}$$

Cleavage of Phosphonate and Phosphate Esters.[30]   Phosphonate and phosphate esters are cleaved even more readily with TMS-I than carboxylic esters. The reaction of phosphonate esters proceeds via the silyl ester, which is subsequently hydrolyzed with MeOH or H_2O (eq 14).

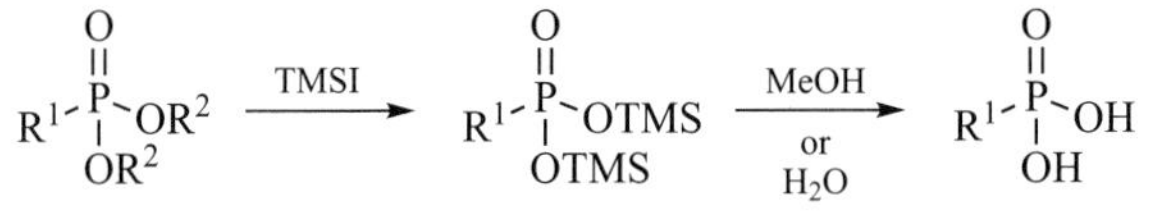

$$\text{(14)}$$

Conversion of Vinyl Phosphates to Vinyl Iodides.[31]   Ketones can be converted to the corresponding vinyl phosphates which react with TMS-I (3 equiv) at rt to afford vinyl iodides (eq 15).

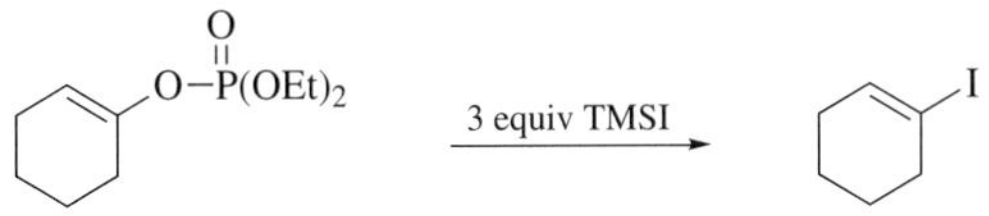

$$(15)$$

Cleavage of Cyclopropyl Ketones.[32] Cyclopropyl ketones undergo ring opening with TMS-I, via the silyl enol ether (eq 16). Cyclobutanones react analogously under these conditions.[33]

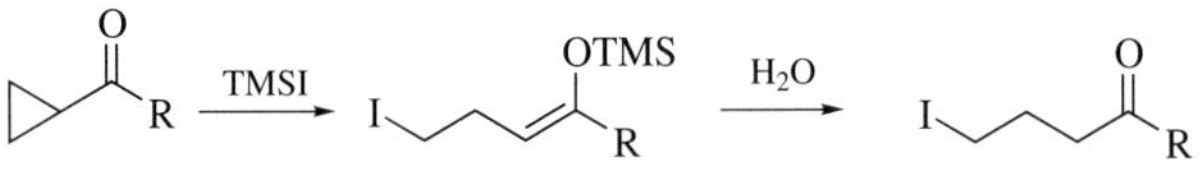

$$(16)$$

Halogen Exchange Reactions.[34] Halogen exchange can be accomplished with reactive alkyl halides, such as benzyl chloride or benzyl bromide, and even with certain alkyl fluorides, by using TMS-I in the presence of $(n\text{-}Bu)_4NCl$ as catalyst (eq 17).

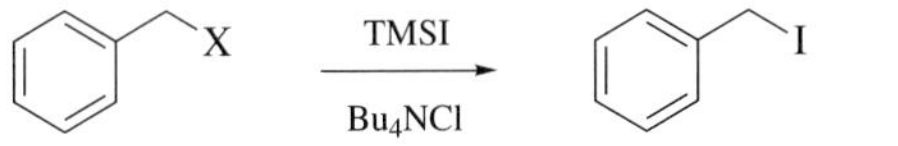

$$X = Br, Cl \qquad (17)$$

Use of TMS-I in Nucleophilic Addition Reactions.

Carbonyl Addition Reactions.[35] α-Iodo trimethylsilyl ethers are produced in the reaction of aldehydes and TMS-I (eq 18). These compounds may react further to provide the diiodo derivative or may be used in subsequent synthesis.

$$RCHO \xrightarrow{\text{TMSI}} \quad (18)$$

An example of a reaction of an iodohydrin silyl ether with a cuprate reagent is summarized in eq 19.[36] An interesting reaction of TMS-I with phenylacetaldehydes gives a quantitative yield of the oxygen-bridged dibenzocyclooctadiene, which was then converted in a few steps to the natural product isopavine (eq 20).[35,37]

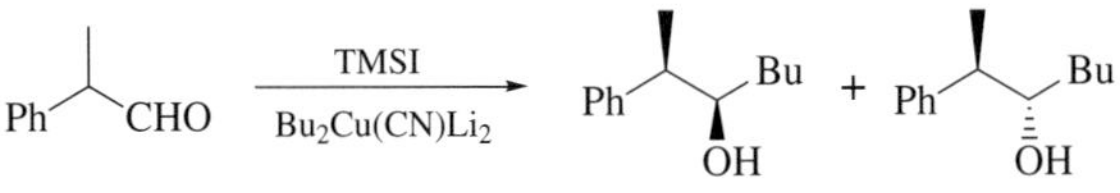

$$7:1 \qquad (19)$$

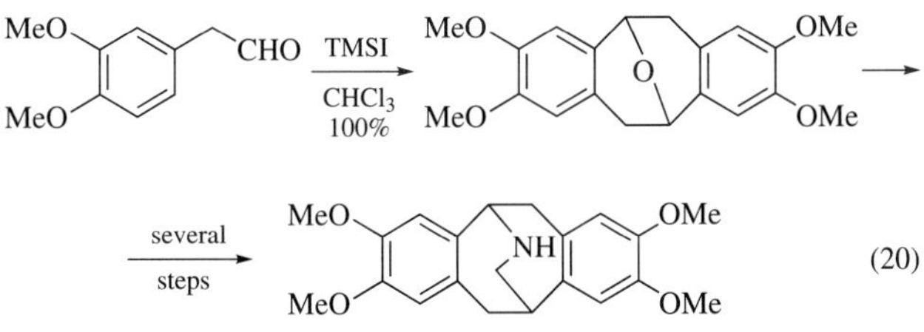

$$(20)$$

β-Iodo ketones have been produced from reactions of TMS-I and ketones with α-hydrogens.[38] This reaction presumably involves a TMS-I catalyzed aldol reaction followed by 1,4-addition of iodide.

Conjugate Addition Reactions.[39] α,β-Unsaturated ketones undergo conjugate addition with TMS-I to afford the β-iodo

adducts in high yield (eq 21). The reaction also works well with the corresponding alkynic substrate.[40]

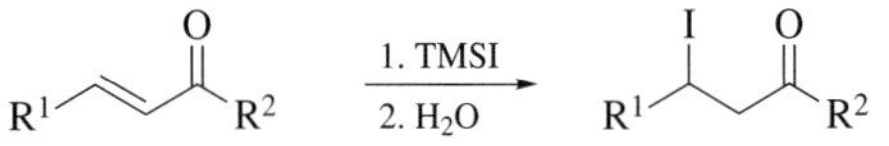

$$(21)$$

TMS-I has also been extensively utilized in conjunction with organocopper reagents to effect highly stereoselective conjugate additions of alkyl nucleophiles.[41]

Use of TMS-I as a Silylating Agent.

Formation of Silyl Enol Ethers.[42] TMS-I in combination with triethylamine is a reactive silylating reagent for the formation of silyl enol ethers from ketones (eq 22). TMS-I with hexamethyldisilazane has also been used as an effective silylation agent, affording the thermodynamic silyl enol ethers. For example, 2-methylcyclohexanone gives a 90:10 mixture in favor of the tetra-substituted enol ether product.[42a] The reaction of TMS-I with 1,3-diketones is a convenient route to 1,3-bis(trimethylsiloxy)-1,3-dienes.[42c]

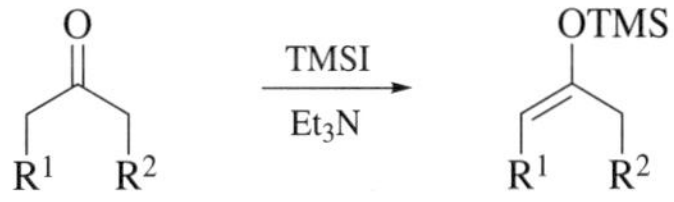

$$(22)$$

In an analogous process, TMS-I reacts with lactams in the presence of Et_3N to yield silyl imino ethers (eq 23).[43a]

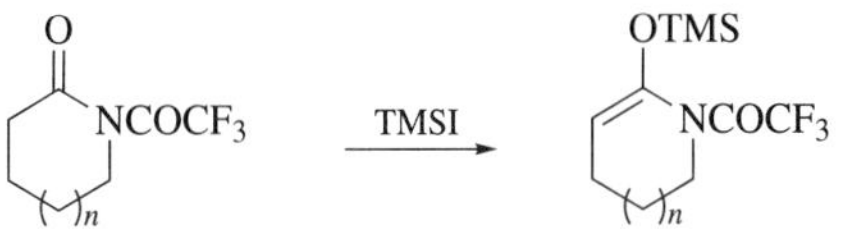

$$(23)$$

Halogenation of Lactams.[43b] Selective and high yielding iodination and bromination of lactams occurs with iodine or bromine, respectively, in the presence of TMS-I and a tertiary amine base (eq 24). The proposed reaction mechanism involves intermediacy of the silyl imino ether.

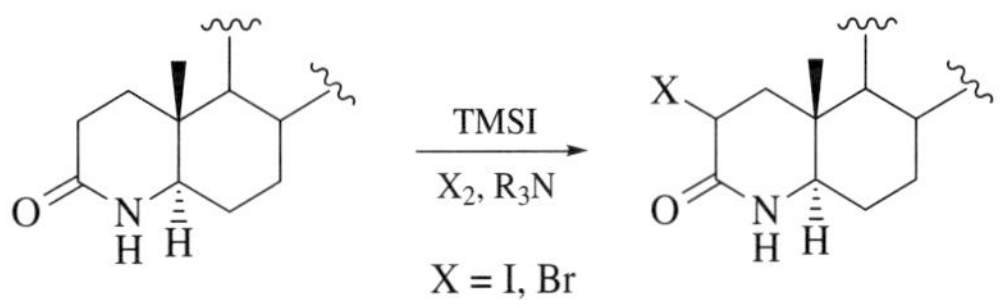

$$X = I, Br \qquad (24)$$

Reaction with Carbanions.[44] TMS-I has seen limited use in the silylation of carbanions, with different regioselectivity compared to other silylating reagents in the example provided in eq 25.

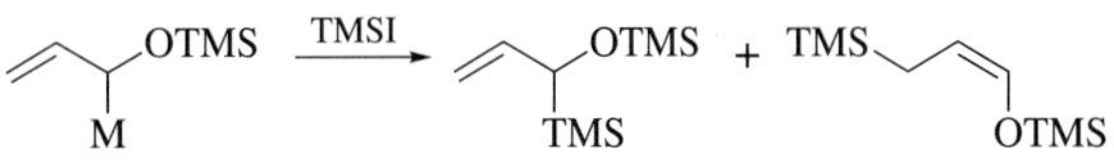

$$85:15 \qquad (25)$$

Silylation of Alkynes and Alkenes.[45] A Heck-type reaction of TMS-I with alkenes in the presence of Pd^0 and Et_3N affords alkenyltrimethylsilanes (eq 26).

$$Ar-CH=CH_2 \xrightarrow[\substack{Pd^0 \\ Et_3N}]{TMSI} Ar-CH=CH-TMS \quad (26)$$

Oxidative addition of TMS-I to alkynes can also be accomplished with a three-component coupling reaction to provide the enyne product (eq 27).

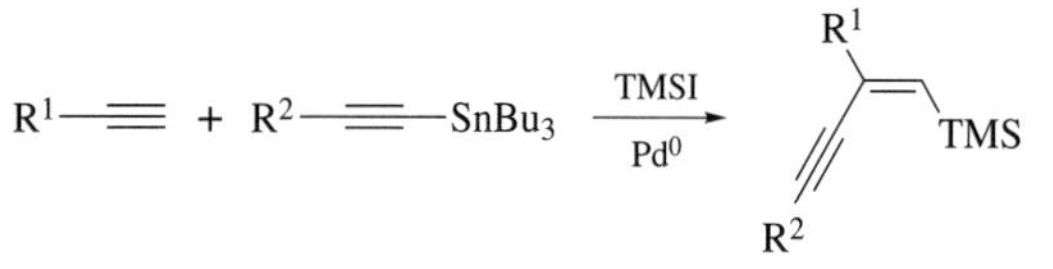

$$R^1\!\!\!\equiv + R^2\!\!\!\equiv\!\!-SnBu_3 \xrightarrow[Pd^0]{TMSI} \quad (27)$$

Use of TMS-I as a Lewis Acid.

Acetalization Catalyst.[46] TMS-I used in conjunction with $(MeO)_4Si$ is an effective catalyst for acetal formation (eq 28).

$$PhCHO \xrightarrow[(MeO)_4Si]{TMSI} PhCH(OMe)_2 \quad (28)$$

Catalyst for α-Alkoxymethylation of Ketones. Silyl enol ethers react with α-chloro ethers in the presence of TMS-I to afford α-alkoxymethyl ketones (eq 29).[47]

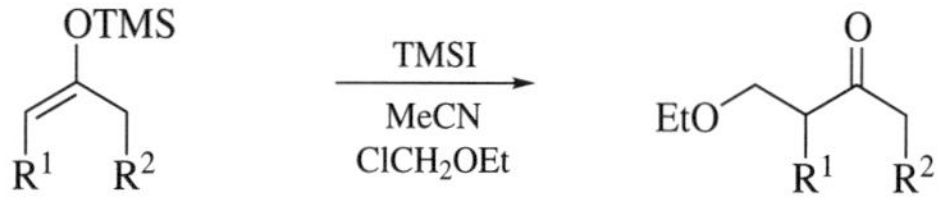

$$\xrightarrow[\substack{MeCN \\ ClCH_2OEt}]{TMSI} \quad (29)$$

Catalyst for Reactions of Acetals with Silyl Enol Ethers and Allylsilanes. TMS-I catalyzes the condensation of silyl enol ethers with various acetals (eq 30)[48] and imines,[49] and of allylsilanes with acetals.[50]

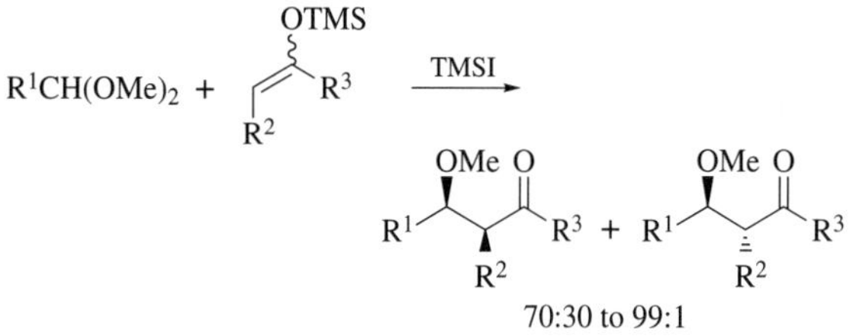

$$R^1CH(OMe)_2 + \quad \xrightarrow{TMSI}$$

70:30 to 99:1

Use of TMS-I as a Reducing Agent. TMS-I reduces enediones to 1,4-diketones,[51] while both epoxides and 1,2-diols are reduced to the alkenes.[13a,b,52] The Diels–Alder products of benzynes and furans are converted in high yield to the corresponding naphthalene (or higher aromatic derivative) with TMS-I (eq 31).[53]

$$\xrightarrow{TMSI} \quad (31)$$

Styrenes and benzylic alcohols are reduced to the alkanes with TMS-I (presumably via formation of HI).[54] Ketones produce the symmetrical ethers when treated with trimethylsilane as a reducing agent in the presence of catalytic TMS-I.[55]

Reduction of α-Ketols.[56,57] Carbonyl compounds containing α-hydroxy, α-acetoxy, or α-halo groups react with excess TMS-I to give the parent ketone. α-Hydroxy ketone reductions proceed via the iodide, which is then reduced with iodide ion to form the parent ketone (eq 32).

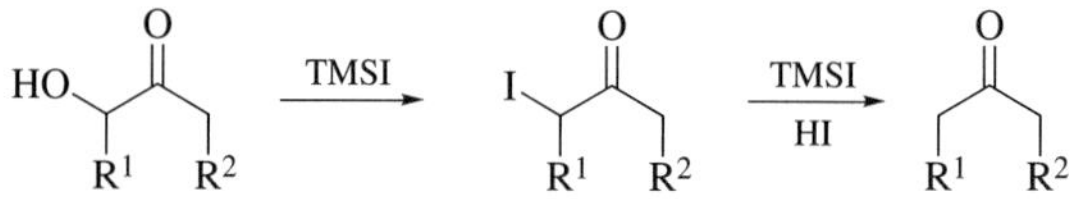

$$\xrightarrow{TMSI} \quad \xrightarrow[HI]{TMSI} \quad (32)$$

Sulfoxide Deoxygenation.[58] The reduction of sulfoxides occurs under very mild conditions with TMS-I to afford the corresponding sulfide and iodine (eq 33). Addition of I_2 to the reaction mixture accelerates the second step. The deoxygenation occurs faster in pyridine solution than the reactions with a methyl ester or alcohol.[59]

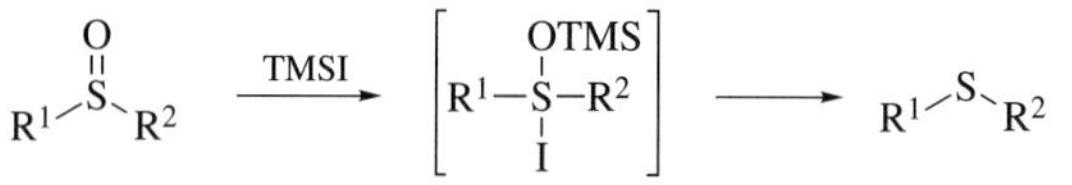

$$R^1\!\!\diagdown\!\!{}^S\!\!\diagup\!\!R^2 \xrightarrow{TMSI} \left[R^1\!-\!\underset{I}{\overset{OTMS}{\underset{|}{\overset{|}{S}}}}\!-\!R^2 \right] \longrightarrow R^1\!\!\diagdown\!\!{}^S\!\!\diagup\!\!R^2 \quad (33)$$

Pummerer reactions of sulfoxides can be accomplished in the presence of TMS-I and an amine base, leading to vinyl sulfides.[60] An efficient synthesis of dithioles was accomplished with TMS-I and Hünig's base (diisopropylethylamine) (eq 34).[61]

$$\xrightarrow[DIPEA]{TMSI} \quad (34)$$

Reaction with Sulfonyl Halides.[62] Arylsulfonyl halides undergo reductive dimerization to form the corresponding disulfides (eq 35). Alkylsulfonyl halides, however, undergo this process under somewhat more vigorous conditions. Although sulfones generally do not react with TMS-I, certain cyclic sulfones are cleaved in a manner analogous to lactones.[63]

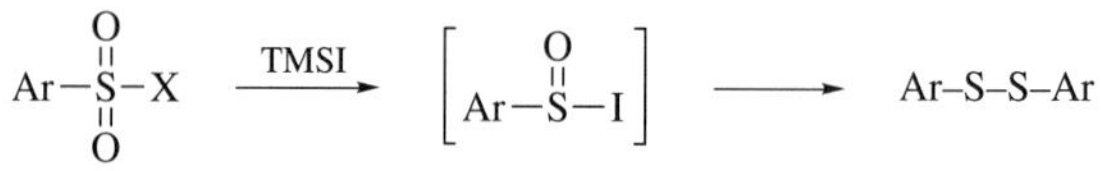

$$Ar\!-\!\underset{O}{\overset{O}{\underset{\|}{\overset{\|}{S}}}}\!-\!X \xrightarrow{TMSI} \left[Ar\!-\!\overset{O}{\underset{\|}{\overset{\|}{S}}}\!-\!I \right] \longrightarrow Ar\!-\!S\!-\!S\!-\!Ar \quad (35)$$

Other Reactions of TMS-I.

Reaction with Phosphine Oxides.[64] Phosphine oxides react with TMS-I to form stable adducts (eq 36). These O-silylated products can undergo further thermolytic reactions such as alkyl group cleavage.

$$R_3P{=}O \xrightarrow{TMSI} R_3\overset{+}{P}{-}O{-}TMS \ I^- \quad (36)$$

Chlorophosphines undergo halogen exchange reactions with TMS-I.[65]

Reaction with Imines.   Imines react with TMS-I to form *N*-silylenamines, in a process analogous to the formation of silyl enol ethers from ketones.[66]

Reaction with Oximes.[67]   Oximes are activated for dehydration (aldoximes, with hexamethylsilazane) or Beckmann rearrangement (ketoximes) with TMS-I (eq 37).

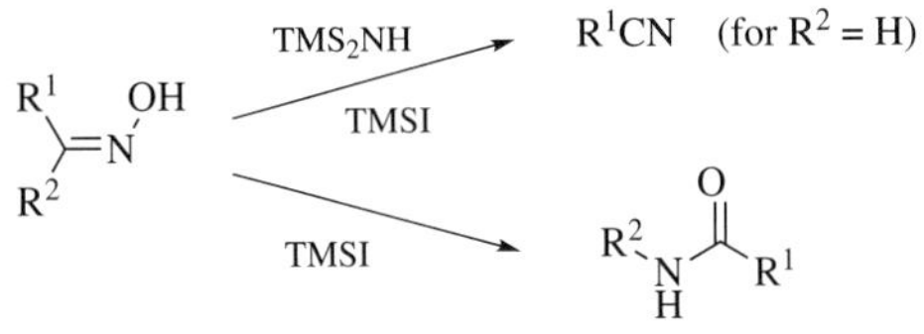

(37)

Reactions with Nitro and Nitroso Compounds.[68]   Primary nitro derivatives react with TMS-I to form the oximino intermediate via deoxygenation, which then undergoes dehydration as discussed for the oximes (eq 38). Secondary nitro compounds afford the silyl oxime ethers, and tertiary nitro compounds afford the corresponding iodide. Nitroalkenes, however, react with TMS-I at 0 °C to afford the ketone as the major product (eq 39).[69]

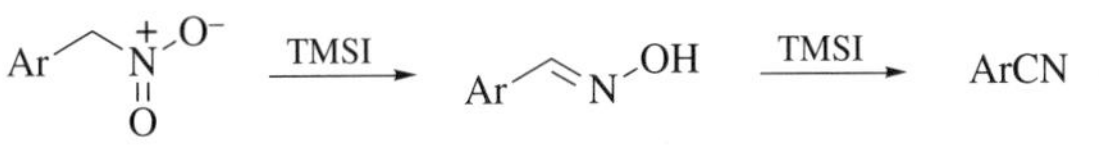

$$Ar\overset{+}{N}\text{---}O^- \xrightarrow{TMSI} Ar\text{---}N\text{---}OH \xrightarrow{TMSI} ArCN \quad (38)$$

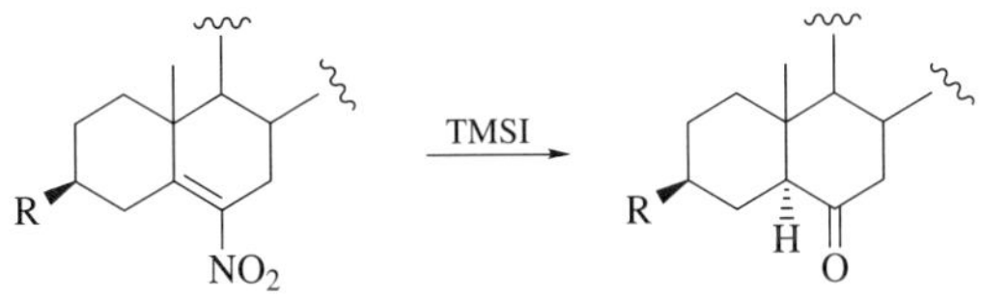

(39)

An interesting analogy to this dehydration process is found in the reductive fragmentation of a bromoisoxazoline with TMS-I, which yields the nitrile (eq 40).[70]

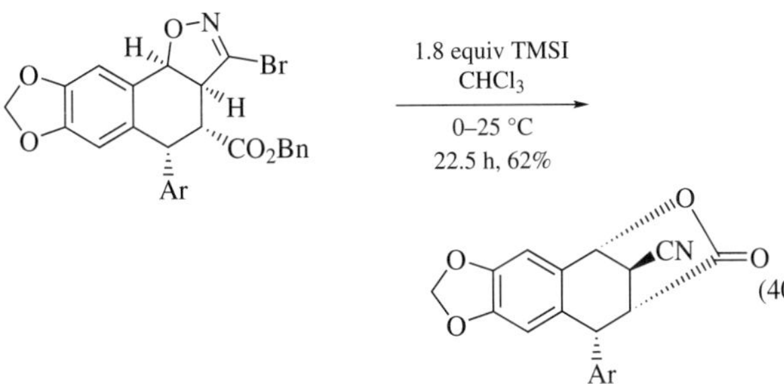

1.8 equiv TMSI
CHCl₃

0–25 °C
22.5 h, 62%

(40)

Rearrangement Reactions.   An interesting rearrangement occurs on treatment of a β-alkoxy ketone with TMS-I which effects dealkylation and retro-aldol reaction to give the eight-membered diketone after reductive dehalogenation (eq 41).[71] Tertiary allylic silyl ethers α to epoxides undergo a stereocontrolled rearrangement to give the β-hydroxy ketones on treatment with catalytic TMS-I (eq 42).[72]

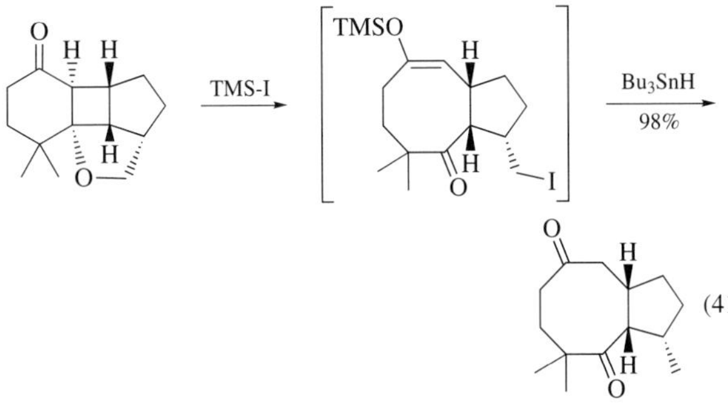

(41)

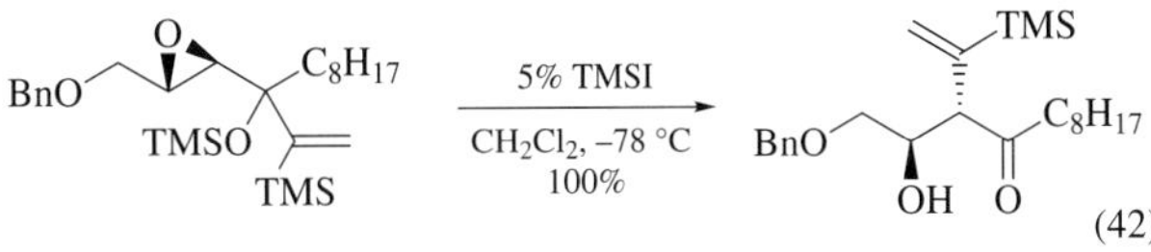

5% TMSI
CH₂Cl₂, –78 °C
100%

(42)

First Update

George A. Olah, G. K. Surya Prakash & Jinbo Hu
University of Southern California, Los Angeles, CA, USA

Selective Bond Cleavage Reactions.

Ether Cleavage.   Iodotrimethylsilane (TMSI) continues to be a versatile ether-cleaving agent in the past decade, and it has been widely applied to complex molecules with high chemoselectivity. Demethylation of 8-methoxy-[2,2]metacyclophanes with TMSI can be accomplished in excellent yields.[73] Treatment of isosorbide and isomannide with TMSI [in situ generated from chlorotrimethylsilane (TMSCl) and sodium iodide] in acetonitrile in the presence of acetone induces the cleavage of only one of the two rings and provides chiral trisubstituted tetrahydrofurans (eq 43).[74] The formation of cyclic sulfonium ions by TMSI-mediated intramolecular displacement of hydroxide or methoxide by sulfide has led to ring contraction reactions from thiepanes to thiolanes (eq 44).[75] The cyclization is especially favored with secondary and tertiary alcohols or ethers, and with an aliphatic more than an aromatic sulfide function.[75]

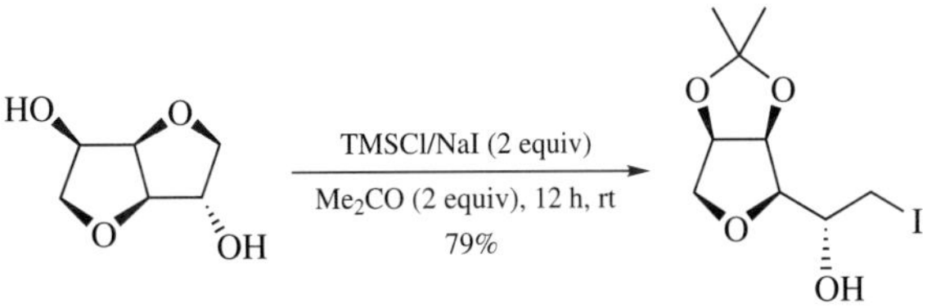

TMSCl/NaI (2 equiv)
Me₂CO (2 equiv), 12 h, rt
79%

(43)

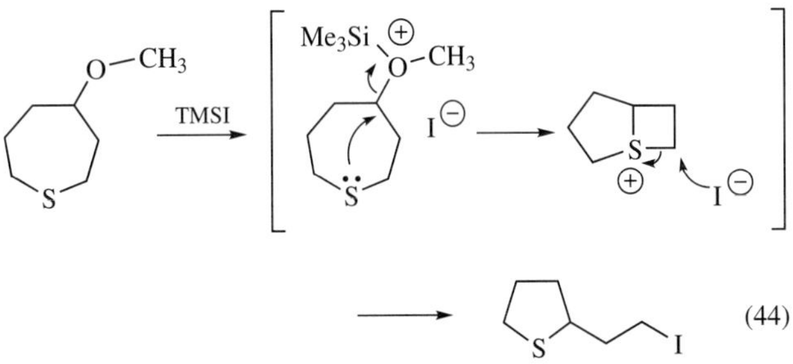

(44)

Ester and Lactone Cleavage. TMSI has been combined with triphenylphosphine (TPP) (in dichloromethane solution) as a more stable, milder, and more selective ester-cleaving agent compared with TMSI itself.[76] TPP increases the stability of TMSI as well as the selectivity by decreasing its reactivity and plays a significant role in preventing side reaction by scavenging the reactive alkyl iodides, generated from the cleavage of ester compounds, to give the corresponding phosphonium iodide salts that are inactive under the reaction conditions.[76] *p*-Methoxybenzyl and diphenylmethyl esters can be easily converted into the corresponding carboxylic acids using TMSI/TPP in dichloromethane at room temperature in good yields (eq 45).[76]

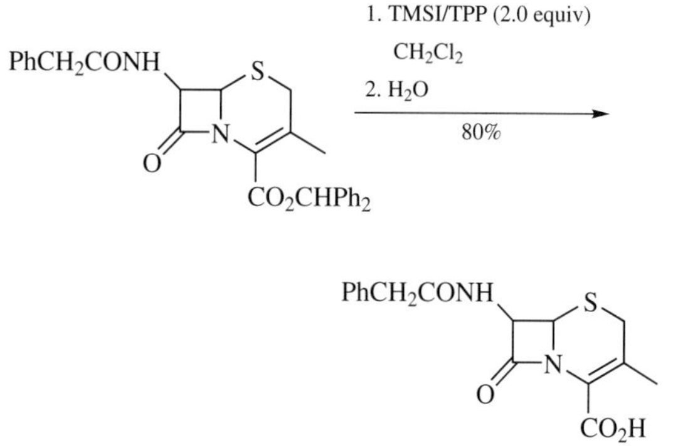

(45)

An unusual sugar lactone cleavage reaction, followed by an intramolecular rearrangement, leads to the formation of primary iodides with the same configuration. The lactone is proposed to be opened by an iodide anion, with inversion of configuration at C-5. The formation of acetoxonium ion then occurs from secondary iodide, with a second inversion at C-5. The acetoxonium ion is then opened regioselectively by an iodide ion, leading to the primary iodide (eq 46).[77]

Cleavage of Carbamates. TMSI has been used as a deblocking agent for the benzyloxycarbonyl group (Cbz) in the synthesis of the nicotinic receptor tracer 5-IA-85380 precursor (eq 47).[78] The TMSI-mediated selective carbamate cleavage can be achieved to afford amino-tin compounds without removing the stannyl moiety, which becomes the key step in the multiple-step synthesis.[78] TMSI has also been applied to the selective cleavage in chiral *N*-substituted 4-phenyl-2-oxazolidinones, and this method allows a more versatile use of 4-phenyl-2-oxazolidinone as a chiral auxiliary and *N*-protective group in the synthesis of carbacephems.[79]

Cleavage of Acetals. The action of TMSI on 7-phenyl-6-alkynal dimethylacetals gives the oxonium ion intermediates, which undergo an intramolecular electrophilic reaction with the carbon-carbon triple bond to afford 2-(1-iodobenzylidene)cyclohexyl methyl ether (eq 48).[80] TMSI has also been used in a one-pot conversion of tetraacetal tetraoxa-cages to aza-cages in alkyl nitriles at room temperature via the ring expansion intermediates, which was interpreted to involve a Ritter-type of reaction mechanism (eq 49).[81] Interestingly, the reaction of tetraacetal tetraoxa-cages with TMSCl and NaI in nitriles at room temperature gives the amido-cages.[81]

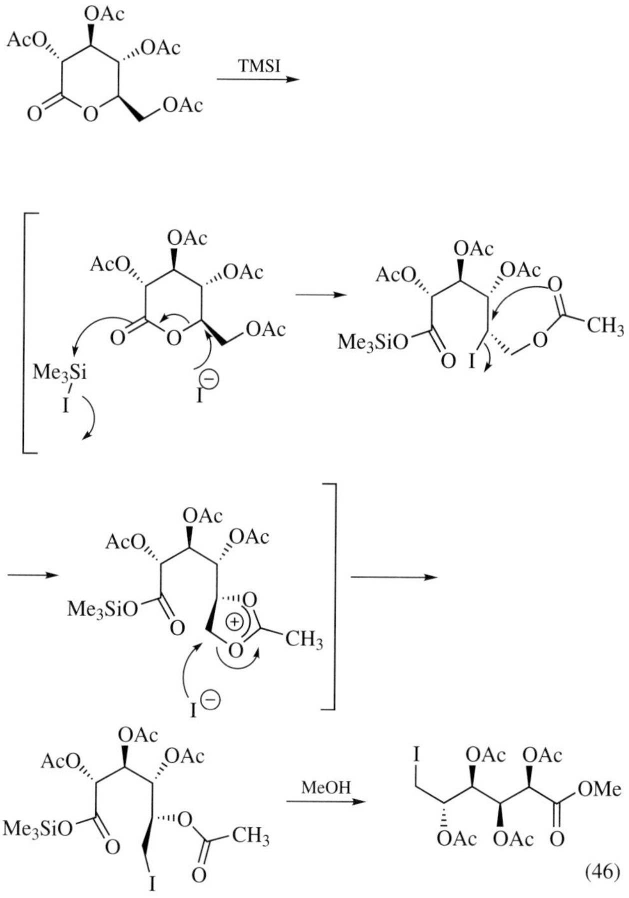

(46)

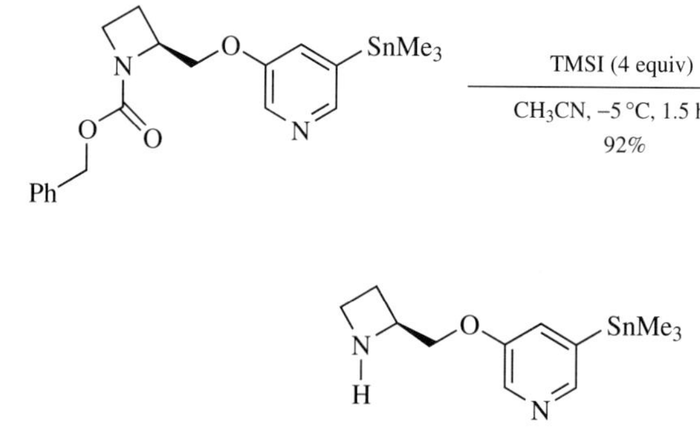

(47)

Cleavage of Phosphorous(III) Esters Leading to a Michaelis-Arbuzov Rearrangement. A new and catalytic version of the Michaelis-Arbuzov rearrangement has been reported by directly forming an alkyl halide through the action of trimethylsilyl halide (TMSX, X = I, Br) on phosphorous(III) esters (eq 50).[82] This rearrangement occurs at temperatures from 20 to 80 °C, and only a catalytic amount (5 mol%) of TMSI (or TMSBr) is needed. Unlike the usual Arbuzov rearrangement, alkyl halides are not required for this type of direct and easy-to-handle TMSX-catalyzed Michaelis-Arbuzov-like rearrangement.[82]

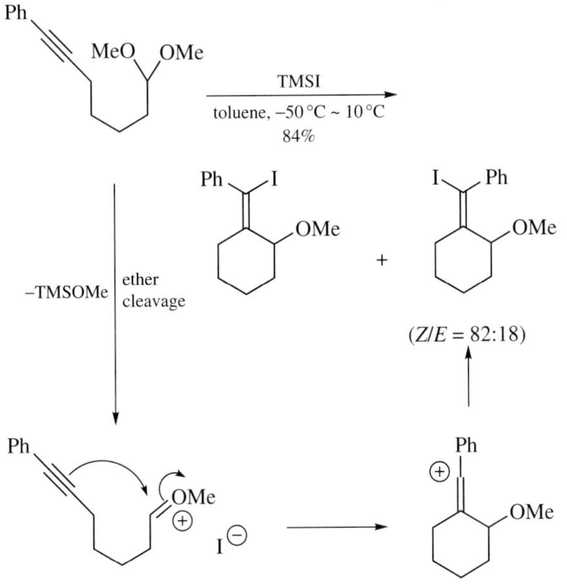

Use of TMSI as a Lewis-Acidic Activation Agent.

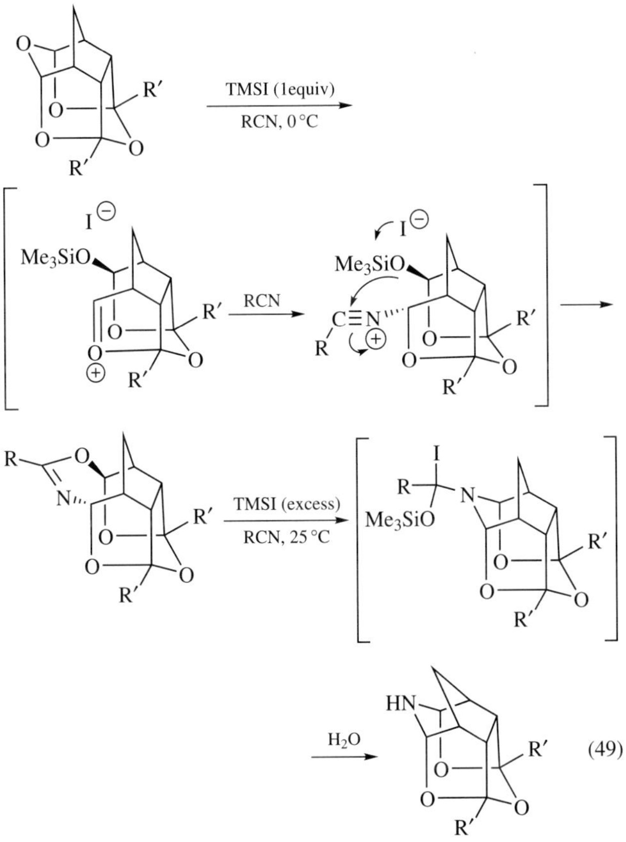

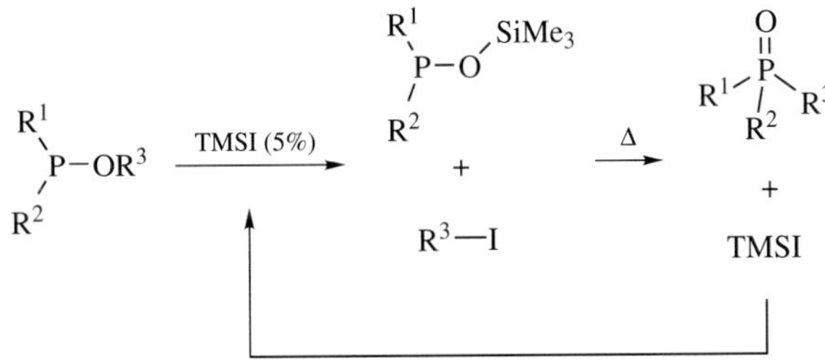

For Biginelli Reaction. TMSI (in situ generated from TMSCl and NaI) is an excellent promoter for the one-pot synthesis of dihydropyrimidinones via the Biginelli reaction (eq 51), which involves the condensation of an aldehyde, a β-ketoester and urea (or thiourea).[83] The traditional Biginelli reaction commonly proceeds under strongly acidic conditions, and this protocol often suffers from low yields particularly in case of substituted aromatic or aliphatic aldehydes. However, when TMSI is applied as a promoter, the reaction usually affords excellent yields of dihydropyrimidinones even at ambient temperature (eq 51).[83]

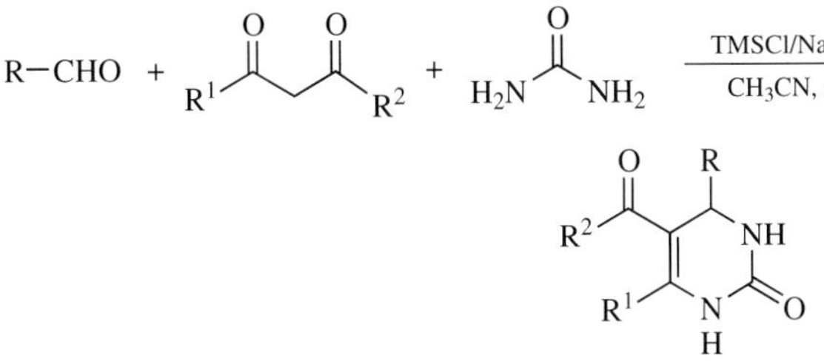

$$R-CHO + R^1 \overset{O}{\underset{}{\diagdown}} \overset{O}{\underset{}{\diagup}} R^2 + H_2N \overset{O}{\underset{}{\diagdown}} NH_2 \xrightarrow[CH_3CN]{TMSCl/NaI} \tag{51}$$

For the Synthesis of Hantzsch 1,4-Dihydropyridines. TMSI (in situ generated) has been used in the efficient synthesis of various substituted Hantzsch 1,4-dihydropyridines using both the classical and modified Hantzsch procedures at room temperature in acetonitrile (eq 52).[84] The usage of TMSI enables the reaction to proceed smoothly with good to excellent yields of products.[84]

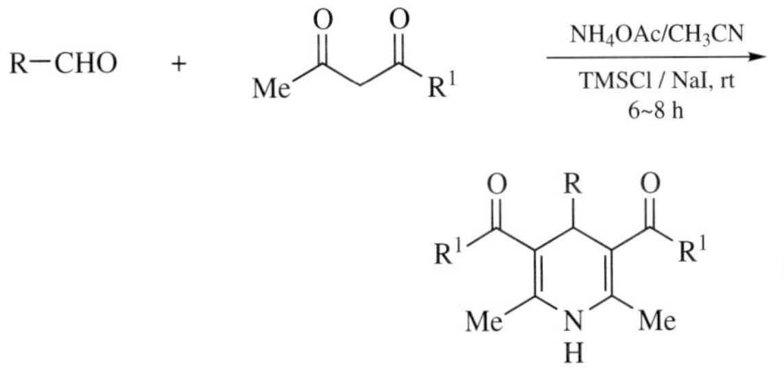

$$R-CHO + Me \overset{O}{\underset{}{\diagdown}} \overset{O}{\underset{}{\diagup}} R^1 \xrightarrow[\substack{TMSCl / NaI, rt \\ 6\sim8\,h}]{NH_4OAc/CH_3CN} \tag{52}$$

For Aldol and Related Reactions. The TMSI/(TMS)$_2$NH combination can be used for the synthesis of polycyclic cyclobutane derivatives by tandem intramolecular Michael-aldol reaction.[85] TMSI-induced diastereoselective synthesis of tetrahydropyranones by a tandem Knoevenagel-Michael reaction, has also been developed.[86] More recently, the facile synthesis of α,α'-bis(substituted benzylidene)cycloalkanones has been reported, using TMSI (in situ generated) mediated cross-aldol condensations (eq 53).[87]

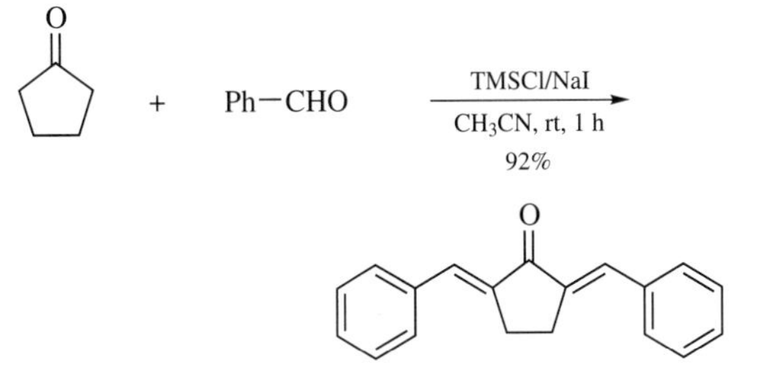

$$\overset{O}{\diagup} + Ph-CHO \xrightarrow[\substack{CH_3CN, rt, 1\,h \\ 92\%}]{TMSCl/NaI} \tag{53}$$

(48) $(Z/E = 82:18)$

(49)

(50)

For the Synthesis of N-Substituted Phthalimides/Naphthalimides. *N*-Substituted phthalimides and naphthalimides can be synthesized in good to excellent yields, employing TMSI (in situ generated) from corresponding azides and anhydrides under mild conditions (eq 54).[88]

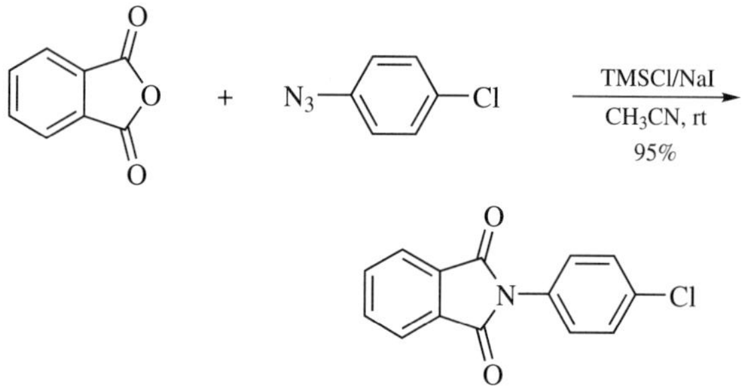

$$ (54) $$

Use of TMSI as a Reducing Agent.

Selective Reduction of α,α-Diaryl Alcohols. TMSI has been utilized as a reducing agent for the rapid and highly selective reduction of α,α-diaryl alcohols to the corresponding alkanes.[89] The reaction proceeds particularly well for electron-rich substrates, which may be associated with the proposed intermediacy of an aryl-stabilized benzylic carbocation at the reduction site.[90] The moderately electron-deficient benzylic alcohols can also be selectively reduced to analogous toluenes, and the reaction condition tolerates other reduction-sensitive functional groups such as ketone, aldehyde, nitrile, and nitro groups (eq 55).[90] The preparation of biarylmethanes, involving benzylation via tandem Grignard reaction-TMSI-mediated reduction, has also been reported.[91]

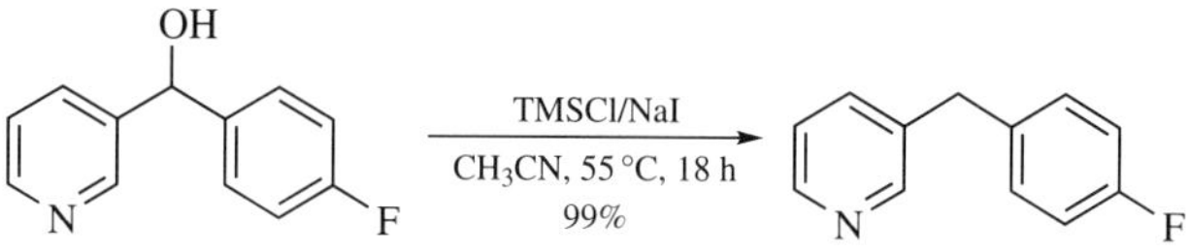

$$ (55) $$

Reduction of Azides to Amines. In situ generated TMSI has been found to be a useful reducing agent for the reduction of azides to amines (eq 56).[92] The reaction is carried out under extremely mild and neutral conditions, and a number of aryl, alkyl, and aroyl azides are suitable for this transformation. This methodology has also been applied to the synthesis of pyrrolo[2,1-c][1,4]benzodiazepines via reductive cyclization of ω-azido carbonyl compounds.[93]

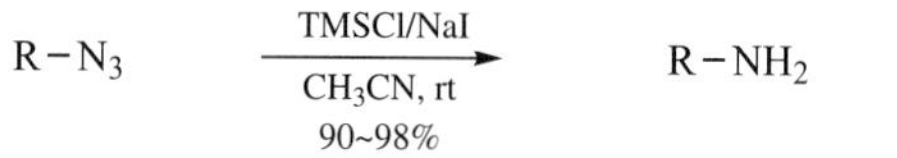

$$ R-N_3 \xrightarrow[\substack{CH_3CN, \, rt \\ 90\sim98\%}]{TMSCl/NaI} R-NH_2 \qquad (56) $$

(R = alkyl, aryl, and aroyl)

Reductive Cleavage of Phthalides and Sulfonamides. 3-Arylphthalides can be readily cleaved reductively by means of TMSI (in situ generated) to give corresponding 2-benzylbenzoic acids (eq 57) or 2-(2-thienylmethyl)benzoic acids.[94]

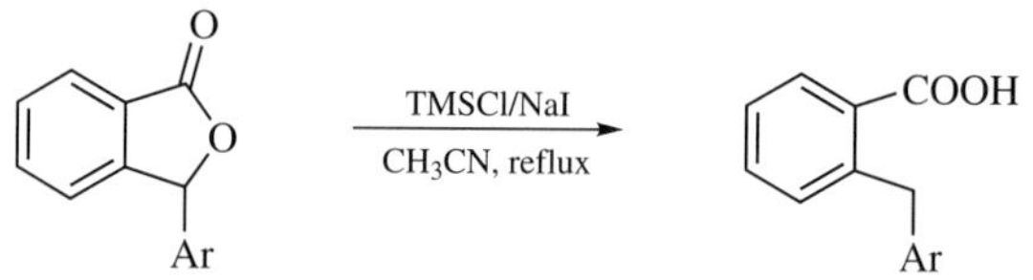

$$ (57) $$

TMSI, in situ generated from TMSCl and NaI, is also a robust reagent for the deprotection of sulfonamides (eq 58).[95] The reductive desulfonylation usually proceeds in good yields with 1.5 equiv of TMSI in acetonitrile under reflux for 3~4 hours. The mild reaction conditions employed in this deprotection method allow the selective deprotection of sulfonamides in the presence of *N*-alkyl and *N*-benzyl groups.[95]

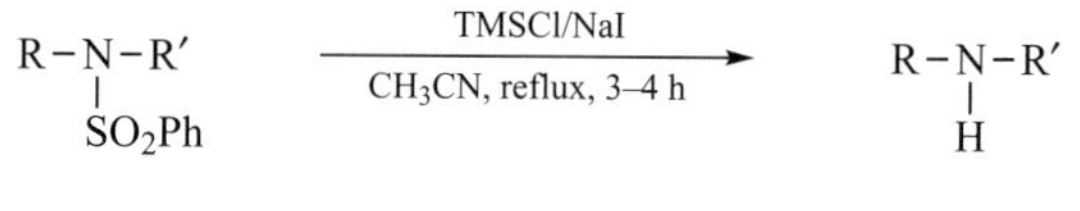

$$ \underset{\underset{SO_2Ph}{|}}{R-N-R'} \xrightarrow[CH_3CN, \, reflux, \, 3-4 \, h]{TMSCl/NaI} \underset{\underset{H}{|}}{R-N-R'} \quad (58) $$

Reductive Cleavage of Heteroaryl C-Halogen Bonds. Regioselective reductive dehalogenation of heterocyclic antibiotic compounds, such as pyrrolnitrins, halo-uridines and pyrimidines, has been successfully accomplished (eq 59).[96a] A single-electron transfer (SET) mechanism was proposed for this type of dehalogenation (eq 59).[96a] TMSI-mediated reductive C-2 dechlorination of some 5-allyl(allenyl)-2,5-dichloro-3-dialkylamino-4,4-dimethoxy-2-cyclopentenones has also been successful.[96b]

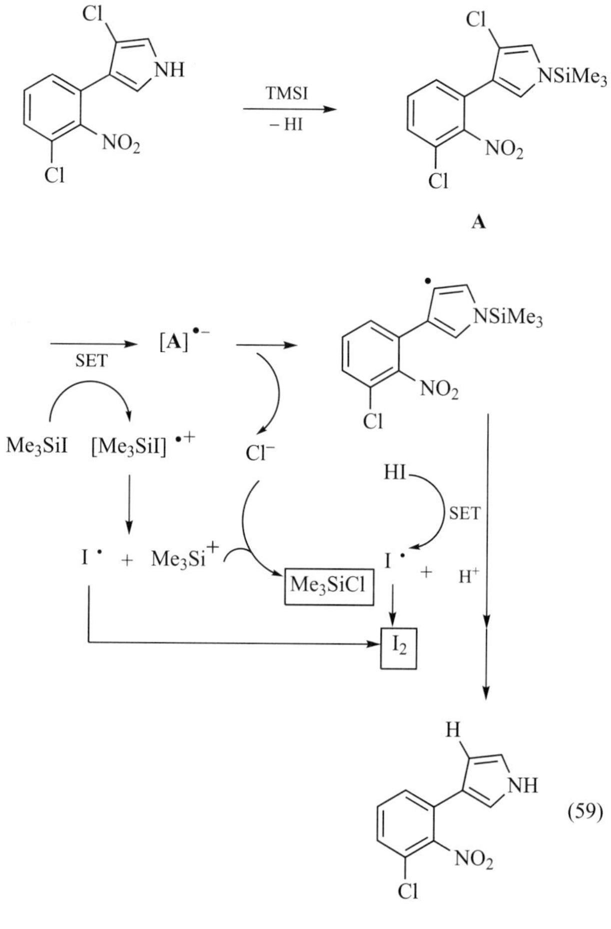

$$ (59) $$

Use of TMSI in Conjunction with Organometallic Reagents.

In Conjugate Monoorganocopper Addition to α,β-unsaturated Carbonyl Compounds.

TMSI has been demonstrated to efficiently promote the reaction of conjugate 1,4-additions of monoorganocopper compounds into a variety of α,β-unsaturated carbonyl compounds, such as cyclic and acyclic enones, β-alkoxy enones, enoates, and lactones, often at −78 °C (eq 60).[97] The RCu(LiI)-TMSI reagent gives a good economy of group transfer with good to excellent yields of conjugate adducts. Lithium iodide, present from preparation of the organocopper compounds, increases the rate of the reaction and is a favorable component.[97] The mechanistic study of the role of TMSI in the conjugate addition of butylcopper-TMSI to α-enones shows that a direct silylation of an intermediate π-complex by TMSI is most likely (eq 60).[98] The conjugate additions of MeCu, PhCu, and n-BuCu to the chiral enoylimides in the presence of TMSI and LiI in THF give the adducts in excellent yields and high diastereoselectivity (80 ∼ 93% de).[99] In the presence of TMSI and LiI in THF, the otherwise unreactive copper acetylides can add to enones present as s-trans conformers to provide good yields of the silyl enol ethers of β-acetylido carbonyl compounds.[100] Similaly, TMSI-promoted diastereoselective conjugate additions of monoorganocuprates Li[RCuI] to different α,β-unsaturated N-acyl oxazolidinones with high yields and diastereomeric ratios have also been reported.[101]

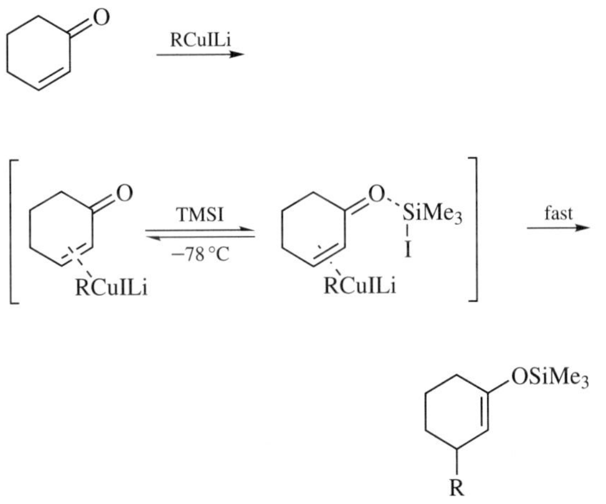

(60)

Pd-Catalyzed Coupling Reactions.

The reaction of terminal acetylenes with TMSI and organozinc reagents (or organostannanes) in the presence of Pd(PPh$_3$)$_4$ results in addition of the trimethylsilyl group of TMSI and an alkyl group of the organozinc reagent (or alkynyl group of organostannane) to the acetylenes to give vinylsilanes (eq 61).[102] This catalytic reaction involves an oxidative addition of the Si-I bond in TMSI to Pd(0) leading to a silylpalladium(II) species, and silylpalladation of an acetylene with the Si-Pd species followed by coupling with organozinc reagents (or organostannanes).[102]

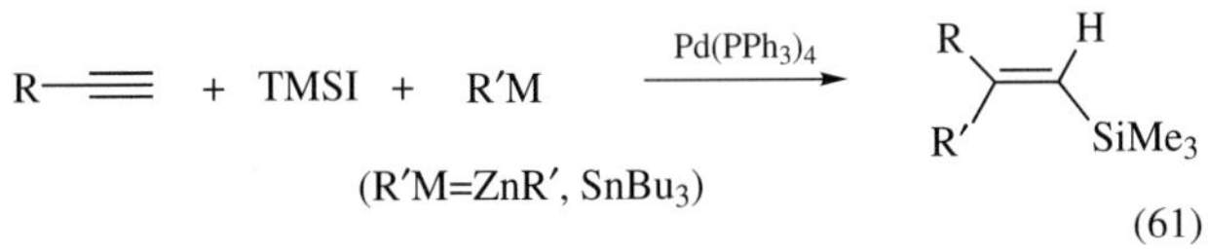

(R′M=ZnR′, SnBu$_3$)   (61)

TMSI as Iodination Agent in Organometallic Complexes.

TMSI has been applied as an iodinating agent in organometallic complexes.[103–105] For example, reaction of the P,N-chelated dimethylplatinum complexes with TMSI stereoselectively gives the corresponding methyl iodo complexes in which only the methyl group trans to the phosphorus atom is exchanged (eq 62).[103] TMSI was also used as a halogen-exchange reagent in the thorium(IV) complex to form Th-I bond.[104] The application of TMSI as a Cl-I exchange reagent in transition metal coordination chemistry, has also been studied.[105]

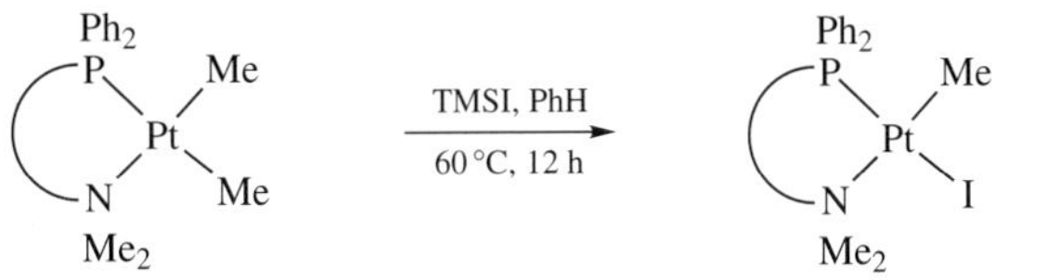

(62)

Other Reactions of TMSI.

Reaction with α,β-Unsaturated Sulfoxides.

The reaction of TMSI with α,β-unsaturated sulfoxides in chloroform at ambient temperature is a mild, efficient, and general method for the preparation of carbonyl compounds (eq 63).[106] The proposed reaction mechanism is shown in eq 63.[106a] Formation of a strong oxygen-silicon bond is followed by reduction of the sulfur function and oxidation of iodide to iodine, the latter precipitating in chloroform. The trimethylsiloxy anion attacks the unsaturated carbon linked to the sulfur function, which leaves the substrate, allowing the formation of the silyl enol ether species. Finally, hydrolysis converts the silyl enol ether into the carbonyl compound.[106a]

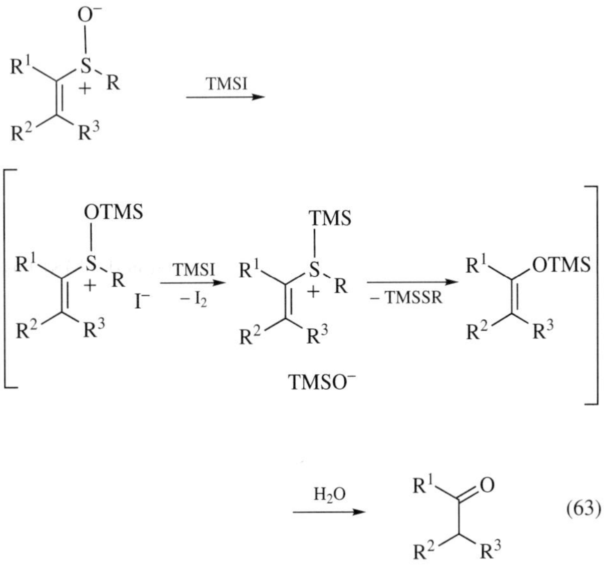

(63)

Iodination of β-Hydroxy Amino Acid Derivatives.

TMSI has been used as an iodinating agent to convert β-hydroxy amino acid derivatives into corresponding β-iodo amino acid derivatives (eq 64).[107] The low yields for the reaction have been attributed in part to the sensitivity of the β-iodo products to the reflux conditions. Higher yields can be obtained when TMSCl and TMSBr are used as chlorinating and brominating agents.[107]

$$(64)$$

1. (a) Olah, G. A.; Prakash, G. K.; Krishnamurti, R., *Adv. Silicon Chem.* **1991**, *1*, 1. (b) Lee, S. D.; Chung, I. N., *Hwahak Kwa Kongop Ui Chinbo* **1984**, *24*, 735. (c) Olah, G. A.; Narang, S. C., *Tetrahedron* **1982**, *38*, 2225. (d) Hosomi, A., *Yuki Gosei Kagaku Kyokai Shi* **1982**, *40*, 545. (e) Ohnishi, S.; Yamamoto, Y., *Annu. Rep. Tohoku Coll. Pharm.* **1981**, *28*, 1. (f) Schmidt, A. H., *Aldrichim. Acta* **1981**, *14*, 31. (g) Groutas, W. C.; Felker, D., *Synthesis* **1980**, *11*, 86. (h) Schmidt, A. H., *Chem.-Ztg.* **1980**, *104* (9), 253.

2. (a) Lissel, M.; Drechsler, K., *Synthesis* **1983**, 459. (b) Machida, Y.; Nomoto, S.; Saito, I., *Synth. Commun.* **1979**, *9*, 97.

3. (a) Schmidt, A. H.; Russ, M., *Chem.-Ztg.* **1978**, *102*, 26, 65. (b) Olah, G. A.; Narang, S. C.; Gupta, B. G. B., *Synthesis* **1979**, 61. (c) Morita, T.; Okamoto, Y.; Sakurai, H., *Tetrahedron Lett.* **1978**, 2523; *J. Chem. Soc., Chem. Commun.* **1978**, 874.

4. (a) Kumada, M.; Shiiman, K.; Yamaguchi, M., *Kogyo Kagaku Zasshi* **1954**, *57*, 230. (b) Sakurai, H.; Shirahata, A.; Sasaki, K.; Hosomi, A., *Synthesis* **1979**, 740.

5. Ho, T. L.; Olah, G. A., *Synthesis* **1977**, 417.

6. (a) Jung, M. E.; Lyster, M. A., *Org. Synth., Coll. Vol.* **1988**, *6*, 353. (b) Jung, M. E.; Blumenkopf, T. A., *Tetrahedron Lett.* **1978**, 3657.

7. (a) Jung, M. E.; Lyster, M. A., *J. Org. Chem.* **1977**, *42*, 3761. (b) Voronkov, M. G.; Dubinskaya, E. I.; Pavlov, S. F.; Gorokhova, V. G., *Izv. Akad. Nauk SSSR, Ser. Khim.* **1976**, 2355.

8. (a) Olah, G. A.; Narang, S. C.; Gupta, B. G. B.; Malhotra, R., *J. Org. Chem.* **1979**, *44*, 1247. (b) Voronkov, M. G.; Dubinskaya, E. J., *J. Organomet. Chem.* **1991**, *410*, 13. (c) Voronkov, M. G.; Puzanova, V. E.; Pavlov, S. F.; Dubinskaya, E. J., *Bull. Acad. Sci. USSR, Div. Chem. Sci.* **1975**, *14*, 377. (d) Voronkov, M. G.; Dubinskaya, E. J.; Pavlov, S. F.; Gorokhova, V. G., *Bull. Acad. Sci. USSR, Div. Chem. Sci.* **1976**, *25*, 2198. (e) Voronkov, M. G.; Komarov, V. G.; Albanov, A. I.; Dubinskaya, E. J., *Bull. Acad. Sci. USSR, Div. Chem. Sci.* **1978**, *27*, 2347. (f) Hirst, G. C.; Johnson, T. O., Jr.; Overman, L. E., *J. Am. Chem. Soc.* **1993**, *115*, 2992.

9. (a) Jung, M. E.; Ornstein, P. L., *Tetrahedron Lett.* **1977**, 2659. (b) Voronkov, M. G.; Pavlov, S. F.; Dubinskaya, E. J., *Dokl. Akad. Nauk SSSR* **1976**, *227*, 607 (Eng. p. 218); *Bull. Acad. Sci. USSR, Div. Chem. Sci.* **1975**, *24*, 579.

10. (a) Casnati, A.; Arduini, A.; Ghidini, E.; Pochini, A.; Ungaro, R., *Tetrahedron* **1991**, *47*, 2221. (b) Silverman, R. B.; Radak, R. E.; Hacker, N. P., *J. Org. Chem.* **1979**, *44*, 4970. (c) Vickery, E. H.; Pahler, L. F.; Eisenbraun, E. J., *J. Org. Chem.* **1979**, *44*, 4444. (d) Brasme, B.; Fischer, J. C.; Wartel, M., *Can. J. Chem.* **1977**, *57*, 1720. (e) Rosen, B. J.; Weber, W. P., *J. Org. Chem.* **1977**, *42*, 3463.

11. Minamikawa, J.; Brossi, A., *Tetrahedron Lett.* **1978**, 3085.

12. Hartman, D. A.; Curley, Jr., R. W., *Tetrahedron Lett.* **1989**, *30*, 645.

13. (a) Denis, J. N.; Magnane, R. M.; van Eenoo, M.; Krief, A., *Nouv. J. Chim.* **1979**, *3*, 705. (b) Detty, M. R.; Seidler, M. D., *Tetrahedron Lett.* **1982**, *23*, 2543. (c) Sakurai, H.; Sasaki, K.; Hosomi, A., *Tetrahedron Lett.* **1980**, *21*, 2329. (d) Kraus, G. A.; Frazier, K., *J. Org. Chem.* **1980**, *45*, 2579. (e) Hill, R. K.; Pendalwar, S. L.; Kielbasinski, K.; Baevsky, M. F.; Nugara, P. N., *Synth. Commun.* **1990**, *20*, 1877.

14. (a) Ho, T. L.; Olah, G. A., *Angew. Chem., Int. Ed. Engl.* **1976**, *15*, 774. (b) Jung, M. E.; Lyster, M. A., *J. Am. Chem. Soc.* **1977**, *99*, 968. (c) Schmidt, A. H.; Russ, M., *Chem.-Ztg.* **1979**, *103*, 183, 285. (d) See also refs. 8a, 9a.

15. Olah, G. A.; Narang, S. C.; Salem, G. F.; Gupta, B. G. B., *Synthesis* **1981**, 142.

16. Acyl iodides are also available from acid chlorides and TMS-I: Schmidt, A. N.; Russ, M.; Grosse, D., *Synthesis* **1981**, 216.

17. (a) Ho, T. L., *Synth. Commun.* **1979**, *9*, 233. (b) Sekiguchi, A.; Kabe, Y.; Ando, W., *Tetrahedron Lett.* **1979**, 871.

18. Hiyama, T.; Saimoto, H.; Nishio, K.; Shinoda, M.; Yamamoto, H.; Nozaki, H., *Tetrahedron Lett.* **1979**, 2043.

19. Kricheldorf, H. R., *Angew. Chem., Int. Ed. Engl.* **1979**, *18*, 689.

20. Yamamoto, Y.; Ohnishi, S.; Azuma, Y., *Chem. Pharm. Bull.* **1982**, *30*, 3505.

21. (a) Jung, M. E.; Lyster, M. A., *J. Chem. Soc., Chem. Commun.* **1978**, 315. (b) Rawal, V. H.; Michoud, C.; Monestel, R. F., *J. Am. Chem. Soc.* **1993**, *115*, 3030. (c) Wender, P. A.; Schaus, J. M.; White, A. W., *J. Am. Chem. Soc.* **1980**, *102*, 6157. (d) Lott, R. S.; Chauhan, V. S.; Stammer, C. H., *J. Chem. Soc., Chem. Commun.* **1979**, 495. (e) Vogel, E.; Altenbach, H. J.; Drossard, J. M.; Schmickler, H.; Stegelmeier, H., *Angew. Chem., Int. Ed. Engl.* **1980**, *19*, 1016.

22. Olah, G. A.; Narang, S. C.; Gupta, B. G. B.; Malhotra, R., *Angew. Chem., Int. Ed. Engl.* **1979**, *18*, 612.

23. Blaskovich, M. A.; Lajoie, G. A., *J. Am. Chem. Soc.* **1993**, *115*, 5021.

24. (a) Jung, M. E.; Andrus, W. A.; Ornstein, P. L., *Tetrahedron Lett.* **1977**, 4175. (b) Bryant, J. D.; Keyser, G. E.; Barrio, J. R., *J. Org. Chem.* **1979**, *44*, 3733. (c) Muchmore, D. C.; Dahlquist, F. W., *Biochem. Biophys. Res. Commun.* **1979**, *86*, 599.

25. Chan, T. H.; Lee, S. D., *Tetrahedron Lett.* **1983**, *24*, 1225.

26. Kobylinskaya, V. I.; Dashevskaya, T. A.; Shalamai, A. S.; Levitskaya, Z. V., *Zh. Obshch. Khim.* **1992**, *62*, 1115.

27. (a) Jung, M. E.; Mazurek, M. A.; Lim, R. M., *Synthesis* **1978**, 588. (b) Olah, G. A.; Husain, A.; Narang, S. C., *Synthesis* **1983**, 896.

28. (a) Thiem, J.; Meyer, B., *Ber. Dtsch. Chem. Ges./Chem. Ber.* **1980**, *113*, 3075. (b) Tocik, Z.; Earl, R. A.; Beranek, J., *Nucl. Acids Res.* **1980**, *8*, 4755.

29. Bryson, T. A.; Bonitz, G. H.; Reichel, C. J.; Dardis, R. E., *J. Org. Chem.* **1980**, *45*, 524.

30. (a) Zygmunt, J.; Kafarski, P.; Mastalerz, P., *Synthesis* **1978**, 609. (b) Blackburn, G. M.; Ingleson, D., *J. Chem. Soc., Chem. Commun.* **1978**, 870. (c) Blackburn, G. M.; Ingleson, D., *J. Chem. Soc., Perkin Trans. 1* **1980**, 1150.

31. Lee, K.; Wiemer, D. F., *Tetrahedron Lett.* **1993**, *34*, 2433.

32. (a) Miller, R. D.; McKean, D. R., *J. Org. Chem.* **1981**, *46*, 2412. (b) Giacomini, E.; Loreto, M. A.; Pellacani, L.; Tardella, P. A., *J. Org. Chem.* **1980**, *45*, 519. (c) Dieter, R. K.; Pounds, S., *J. Org. Chem.* **1982**, *47*, 3174.

33. (a) Miller, R. D.; McKean, D. R., *Tetrahedron Lett.* **1980**, *21*, 2639. (b) Crimmins, M. T.; Mascarella, S. W., *J. Am. Chem. Soc.* **1986**, *108*, 3435.

34. (a) Friedrich, E. C.; Abma, C. B.; Vartanian, P. F., *J. Organomet. Chem.* **1980**, *187*, 203. (b) Friedrich, E. C.; DeLucca, G., *J. Organomet. Chem.* **1982**, *226*, 143.

35. Jung, M. E.; Mossman, A. B.; Lyster, M. A., *J. Org. Chem.* **1978**, *43*, 3698.

36. (a) Jung, M. E.; Lewis, P. K., *Synth. Commun.* **1983**, *13*, 213. (b) Lipshutz, B. H.; Ellsworth, E. L.; Siahaan, T. J.; Shirazi, A., *Tetrahedron Lett.* **1988**, *29*, 6677.

37. Jung, M. E.; Miller, S. J., *J. Am. Chem. Soc.* **1981**, *103*, 1984.

38. Schmidt, A. H.; Russ, M., *Chem.-Ztg.* **1979**, *103*, 183, 285.

39. (a) Miller, R. D.; McKean, D. R., *Tetrahedron Lett.* **1979**, 2305. (b) Larson, G. L.; Klesse, R., *J. Org. Chem.* **1985**, *50*, 3627.

40. Taniguchi, M.; Kobayashi, S.; Nakagawa, M.; Hino, T.; Kishi, Y., *Tetrahedron Lett.* **1986**, *27*, 4763.

41. (a) Corey, E. J.; Boaz, N. W., *Tetrahedron Lett.* **1985**, *26*, 6015, 6019. (b) Bergdahl, M.; Nilsson, M.; Olsson, T.; Stern, K., *Tetrahedron* **1991**, *47*, 9691, and references cited therein.

42. (a) Miller, R. D.; McKean, D. R., *Synthesis* **1979**, 730. (b) Hergott, H. H.; Simchen, G., *Liebigs Ann. Chem.* **1980**, 1718. (c) Babot, O.; Cazeau, P.; Duboudin, F., *J. Organomet. Chem.* **1987**, *326*, C57.

43. (a) Kramarova, E. P.; Shipov, A. G.; Artamkina, O. B.; Barukov, Y. I., *Zh. Obshch. Khim.* **1984**, *54*, 1921. (b) King, A. O.; Anderson, R. K.; Shuman, R. F.; Karady, S.; Abramson, N. L.; Douglas, A. W., *J. Org. Chem.* **1993**, *58*, 3384.

44. (a) Lau, P. W. K.; Chan, T. H., *J. Organomet. Chem.* **1979**, *179*, C24. (b) Wilson, S. R.; Phillips, L. R.; Natalie, K. J., Jr., *J. Am. Chem. Soc.* **1979**, *101*, 3340.

45. (a) Yamashita, H.; Kobayashi, T.; Hayashi, T.; Tanaka, M., *Chem. Lett.* **1991**, 761. (b) Chatani, N.; Amishiro, N.; Murai, S., *J. Am. Chem. Soc.* **1991**, *113*, 7778.

46. Sakurai, H.; Sasaki, K.; Hayashi, J.; Hosomi, A., *J. Org. Chem.* **1984**, *49*, 2808.

47. Hosomi, A.; Sakata, Y.; Sakurai, H., *Chem. Lett.* **1983**, 405.

48. Sakurai, H.; Sasaki, K.; Hosomi, A., *Bull. Chem. Soc. Jpn.* **1983**, *56*, 3195.

49. Mukaiyama, T.; Akamatsu, H.; Han, J. S., *Chem. Lett.* **1990**, 889.

50. Sakurai, H.; Sasaki, K.; Hosomi, A., *Tetrahedron Lett.* **1981**, *22*, 745.

51. Vankar, Y. D.; Kumaravel, G.; Mukherjee, N.; Rao, C. T., *Synth. Commun.* **1987**, *17*, 181.

52. Sarma, J. C.; Barua, N. C.; Sharma, R. P.; Barua, J. N., *Tetrahedron* **1983**, *39*, 2843.

53. Jung, K.-Y.; Koreeda, M., *J. Org. Chem.* **1989**, *54*, 5667.

54. Ghera, E.; Maurya, R.; Hassner, A., *Tetrahedron Lett.* **1989**, *30*, 4741.

55. Sassaman, M. B.; Prakash, G. K.; Olah, G. A., *Tetrahedron* **1988**, *44*, 3771.

56. (a) Ho, T.-L., *Synth. Commun.* **1979**, *9*, 665. (b) Sarma, D. N.; Sarma, J. C.; Barua, N. C.; Sharma, R. P., *J. Chem. Soc., Chem. Commun.* **1984**, 813. (c) Nagaoka, M.; Kunitama, Y.; Numazawa, M., *J. Org. Chem.* **1991**, *56*, 334. (d) Numazawa, M.; Nagaoka, M.; Kunitama, Y., *Chem. Pharm. Bull.* **1986**, *34*, 3722; *J. Chem. Soc., Chem. Commun.* **1984**, 31. (e) Hartman, D. A.; Curley, R. W., Jr., *Tetrahedron Lett.* **1989**, *30*, 645. (f) Cherbas, P.; Trainor, D. A.; Stonard, R. J.; Nakanishi, K., *J. Chem. Soc., Chem. Commun.* **1982**, 1307.

57. Olah, G. A.; Arvanaghi, M.; Vankar, Y. D., *J. Org. Chem.* **1980**, *45*, 3531.

58. (a) Olah, G. A.; Gupta, B. G. B.; Narang, S. C., *Synthesis* **1977**, 583. (b) Pitlik, J.; Sztaricskai, F., *Synth. Commun.* **1991**, *21*, 1769.

59. Nicolaou, K. C.; Barnette, W. E.; Magolda, R. L., *J. Am. Chem. Soc.* **1978**, *100*, 2567.

60. Miller, R. D.; McKean, D. R., *Tetrahedron Lett.* **1983**, *24*, 2619.

61. Schaumann, E.; Winter-Extra, S.; Kummert, K.; Scheiblich, S., *Synthesis* **1990**, 271.

62. Olah, G. A.; Narang, S. C.; Field, L. D.; Salem, G. F., *J. Org. Chem.* **1980**, *45*, 4792.

63. Shipov, A. G.; Baukov, Y. I., *Zh. Obshch. Khim.* **1984**, *54*, 1842.

64. (a) Beattie, I. R.; Parrett, F. W., *J. Chem. Soc. (A)* **1966**, 1784. (b) Livanstov, M. V.; Proskurnina, M. V.; Prischenko, A. A.; Lutsenko, I. F., *Zh. Obshch. Khim.* **1984**, *54*, 2504.

65. Kabachnik, M. M.; Prischenko, A. A.; Novikova, Z. S.; Lutsenko, I. F., *Zh. Obshch. Khim.* **1979**, *49*, 1446.

66. Kibardin, A. M.; Gryaznova, T. V.; Gryaznov, P. I.; Pudovik, A. N., *J. Gen. Chem. USSR (Engl. Transl.)* **1991**, *61*, 1969.

67. (a) Jung, M. E.; Long-Mei, Z., *Tetrahedron Lett.* **1983**, *24*, 4533. (b) Godleski, S. A.; Heacock, D. J., *J. Org. Chem.* **1982**, *47*, 4820.

68. Olah, G. A.; Narang, S. C.; Field, L. D.; Fung, A. P., *J. Org. Chem.* **1983**, *48*, 2766.

69. Singhal, G. M.; Das, N. B.; Sharma, R. P., *J. Chem. Soc., Chem. Commun.* **1989**, 1470.

70. Haber, A., *Tetrahedron Lett.* **1989**, *30*, 5537.

71. Inouye, Y.; Shirai, M.; Michino, T.; Kakisawa, H., *Bull. Chem. Soc. Jpn.* **1993**, *66*, 324.

72. Suzuki, K.; Miyazawa, M.; Tsuchihashi, G., *Tetrahedron Lett.* **1987**, *28*, 3515.

73. Yamato, T.; Matsumoto, J.; Tashiro, M., *J. Chem. Research (S)* **1994**, 246.

74. Ejjiyar, S.; Saluzzo, C.; Amouroux, R.; Massoui, M., *Tetrahedron Lett.* **1997**, *38*, 1575.

75. Cere, V.; Pollicino, S.; Fava, A., *Tetrahedron* **1996**, *52*, 5989.

76. Cha, K. H.; Kang, T. W.; Cho, D. O.; Lee, H.-W.; Shin, J.; Jin, K. Y.; Kim, K.-W.; Kim, J.-W.; Hong, C.-I., *Synth. Commun.* **1999**, 3533.

77. Heck, M. P.; Monthiller, S.; Mioskowski, C.; Guidot, J. P.; Gall, T. L., *Tetrahedron Lett.* **1994**, *35*, 5445.

78. Brenner, E.; Baldwin, R. M.; Tamagnan, G., *Tetrahedron Lett.* **2004**, *45*, 3607.

79. Fisher, J. W.; Dunigan, J. M.; Hatfield, L. D.; Hoying, R. C.; Ray, J. E.; Thomas, K. L., *Tetrahedron Lett.* **1993**, *34*, 4755.

80. Takami, K.; Yorimitsu, H.; Shinokubo, H.; Matsubara, S.; Oshima, K., *Synlett* **2001**, 293.

81. (a) Wu, H.-J.; Chern, J.-H., *Tetrahedron Lett.* **1997**, *38*, 2887. (b) Chern, J.-H.; Wu, H.-J., *Tetrahedron* **1998**, *54*, 5967.

82. Renard, P.-Y.; Vayron, P.; Mioskowski, C., *Org. Lett.* **2003**, *5*, 1661.

83. Sabitha, G.; Reddy, G. S. K. K.; Reddy, C. S.; Yadav, J. S., *Synlett* **2003**, 858.

84. Sabitha, G.; Reddy, G. S. K. K.; Reddy, C. S.; Yadav, J. S., *Tetrahedron Lett.* **2003**, 4129.

85. Ihara, M.; Taniguchi, T.; Makita, K.; Takano, M.; Ohnishi, M.; Taniguchi, N.; Fukumoto, K.; Kabuto, C., *J. Am. Chem. Soc.* **1993**, *115*, 8107.

86. Sabitha, G.; Reddy, G. S. K. K.; Reddy, Rajkumar, M.; C. S.; Yadav, J. S.; Ramakrishna, K. V. S.; Kunwar, A. C., *Tetrahedron Lett.* **2003**, 7455.

87. Sabitha, G.; Reddy, G. S.; Reddy, K. K.; Reddy, C. S.; Yadav, J. S., *Synthesis* **2004**, 263.

88. Kamal, A.; Laxman, E.; Laxman, N.; Rao, N. V., *Tetrahedron Lett.* **1998**, *39*, 8733.

89. Perry, P. J.; Pavlidis, V. H.; Coutts, I. G. C., *Synth. Commun.* **1996**, *26*, 101.

90. Cain, G. A.; Holler, E. R., *Chem. Commun.* **2001**, 1168.

91. Stoner, E. J.; Cothron, D. A.; Balmer, M. K.; Roden, B. A., *Tetrahedron* **1995**, *51*, 11043.

92. Kamal, A.; Rao, N. V.; Laxman, E., *Tetrahedron Lett.* **1997**, *38*, 6945.

93. Kamal, A.; Laxman, E.; Laxman, N.; Rao, N. V., *Bioorg. Med. Chem. Lett.* **2000**, *10*, 2311.

94. Sabitha, G.; Yadav, J. S., *Synth. Commun.* **1998**, *28*, 3065.

95. Sabitha, G.; Reddy, B. V. S.; Abraham, S.; Yadav, J. S., *Tetrahedron Lett.* **1999**, *40*, 1569.

96. (a) Sako, M.; Kihara, T.; Okada, K.; Ohtani, Y.; Kawamoto, H., *J. Org. Chem.* **2001**, *66*, 3610. (b) Akbutina, F. A.; Torosyan, S. A.; Miftakhov, M. S., *Russ. J. Org. Chem.* **2000**, *36*, 1265.

97. (a) Bergdahl, M.; Eriksson, M.; Nilsson, M.; Olsson, T., *J. Org. Chem.* **1993**, *58*, 7238. (b) Eriksson, M.; Nilsson, M.; Olsson, T., *Synlett* **1994**, 271.

98. Eriksson, M.; Johansson, A.; Nilsson, M.; Olsson, T., *J. Am. Chem. Soc.* **1996**, *118*, 10904.

99. Bergdahl, M.; Iliefski, T.; Nilsson, M.; Olsson, T., *Tetrahedron Lett.* **1995**, *36*, 3227.

100. Eriksson, M.; Iliefski, T.; Nilsson, M.; Olsson, T., *J. Org. Chem.* **1997**, *62*, 182.

101. Pollock, P.; Dambacher, J.; Anness, R.; Bergdahl, M., *Tetrahedron Lett.* **2002**, *43*, 3693.

102. (a) Chatani, N.; Amishiro, N.; Morii, T.; Yamashita, T.; Murai, S., *J. Org. Chem.* **1995**, *60*, 1834. (b) Chatani, N.; Amishiro, N.; Murai, S., *J. Am. Chem. Soc.* **1991**, *113*, 7778.

103. Pfeiffer, J.; Kickelbick, G.; Schubert, U., *Organometallics* **2000**, *19*, 62.

104. Rabinovich, D.; Bott, S. G.; Nielsen, J. B.; Abney, K. D., *Inorg. Chim. Acta* **1998**, *274*, 232.

105. Leigh, G. J.; Sanders, J. R.; Hitchcock, P. B.; Fernandes, J. S.; Togrou, M., *Inorg. Chim. Acta* **2002**, *330*, 197.

106. (a) Aversa, M.; Barattucci, A.; Bonaccorsi, P.; Giannetto, P., *Tetrahedron* **2002**, *58*, 10145. (b) Aversa, M.; Barattucci, A.; Bonaccorsi, P.; Bruno, G.; Giannetto, P.; Policicchio, M., *Tetrahedron Lett.* **2000**, *41*, 4441.

107. Choi, D.; Kohn, H., *Tetrahedron Lett.* **1995**, *39*, 7011.

(*E*)-1-Iodo-3-trimethylsilyl-2-butene[1]

(*E*)

[52815-00-6]   C$_7$H$_{15}$ISi   (MW 254.21)

InChI = 1S/C7H15ISi/c1-7(5-6-8)9(2,3)4/h5H,6H2,1-4H3

InChIKey = NWNYZMZBMPWCQQ-UHFFFAOYSA-N

[52685-51-5]   C$_7$H$_{15}$ISi   (MW 254.21)

InChI = 1S/C7H15ISi/c1-7(5-6-8)9(2,3)4/h5H,6H2,1-4H3/
b7-5+

InChIKey = NWNYZMZBMPWCQQ-FNORWQNLSA-N

(annulation reagent; alkylative equivalent of 3-buten-2-one; 3-ketobutyl synthon)

Alternate Name: (3-iodo-1-methyl-1-propenyl)trimethylsilane.
Physical Data: bp 60 °C/20 mmHg.
Solubility: freely sol organic solvents.
Form Supplied in: not commercially available.
Preparative Method: the first reported preparation[2] of this reagent lacked full experimental details. The preparation given (eq 1) is a compilation of the work of several groups; the various steps have been selected to give optimum yields. 2-Propyn-1-ol is converted[3] into its 3-(trimethylsilyl) derivative (**1**), which is then reduced[4] using sodium bis(2-methoxyethoxy)aluminum

hydride (Red-Al). The intermediate vinylaluminum is reacted with iodine to produce the (*Z*)-iodoalkene (**2**). Treatment of this with lithium dimethylcuprate gives the allylic alcohol (**3**);.[5] transformation into (*E*)-(3-chloro-1-methyl-1-propenyl)trimethylsilane (**4**) followed by halogen exchange[6] with sodium iodide then gives reagent (**5**).

An alternative, high-yielding synthesis of (**3**) (eq 2) involves the use of HMPA, but is otherwise straightforward.[7]

$$\text{TMS—TMS} \xrightarrow[\text{2. CuCN}]{\text{1. LiMe, HMPA}} (\text{TMS})_2\text{CuLi·LiCN} \xrightarrow{\equiv\!-\text{CO}_2\text{Et}}$$

$$\underset{\text{CO}_2\text{Et}}{\overset{\text{TMS}}{\diagup\!\diagdown}} \xrightarrow[\text{90\% overall}]{i\text{-Bu}_2\text{AlH}} \underset{\text{OH}}{\overset{\text{TMS}}{\diagup\!\diagdown}} \quad (2)$$

Enolate Alkylation. One of the most useful applications of the Michael reaction[8] is Robinson annulation, using 3-buten-2-one as electrophilic acceptor. However, the Michael reaction, being reversible, is prone to side reactions, and Robinson annulation itself is of only significant utility in relatively simple cases. Hence there has been considerable effort[9] devoted to devising alkylative equivalents of 3-buten-2-one, with the goal of achieving regiospecific enolate alkylation under nonequilibrating conditions. Allylic halides possess sufficiently high reactivity for such alkylation. Useful reagents which have been devised include 1,3-dichloro-2-butene (Wichterle's reagent),[10] Stork's halomethylisoxazoles,[11] and Stotter's γ-iodotiglate.[12] However, these alternatives can all present problems either before or during further transformation to liberate the 3-ketoalkyl chain prior to final cyclization. For example, the Wichterle sequence requires the use of concentrated sulfuric acid to hydrolyze the vinyl chloride moiety.

The use of halomethyl vinylsilanes would appear to offer significant promise. As allylic halides, they should be sufficiently reactive to permit regiospecific alkylation; as vinylsilanes, they can be converted, via α,β-epoxysilanes, into carbonyl compounds.[13] The conditions for this last transformation, with simple α,β-epoxysilanes, are rather vigorous, requiring hot methanol/sulfuric acid.

Examples of Use. Iodide (**5**) reacts with lithium enolates,[9,14] generated in a variety of ways, including kinetic generation from enol acetates (eq 3), regiospecific production by lithium–ammonia reduction of enones, and similar regiospecific production by lithium dimethylcuprate addition to enones (eq 4). Enamines can also be used in a sequence which demonstrates the possibility of reducing enones without affecting the vinylsilane (eq 5); the reagent used in this case was chloride (**4**) in the presence of potassium iodide, i.e. via the in situ generated iodide (**5**).

$$\underset{\text{OH}}{\equiv} \xrightarrow[\substack{\text{2. Me}_3\text{SiCl} \\ \text{3. dil H}_2\text{SO}_4 \\ 91\text{–}94\%}]{\text{1. EtMgBr, THF}} \underset{\text{OH}}{\overset{\text{TMS}}{\equiv}} \xrightarrow[\substack{\text{2. I}_2, 67\%}]{\text{1. Red-Al}}$$
(1)

$$\underset{I \quad \diagup\!\diagdown \text{OH}}{\overset{\text{TMS}}{}} \xrightarrow[\substack{\text{Et}_2\text{O} \\ 85\%}]{\text{LiCuMe}_2\text{·LiCN}} \underset{\diagup\!\diagdown \text{OH}}{\overset{\text{TMS}}{}} \xrightarrow[60\%]{\text{Ph}_3\text{P, CCl}_4}$$
(2)   **(3)**

$$\underset{\diagup\!\diagdown \text{Cl}}{\overset{\text{TMS}}{}} \xrightarrow[\substack{\text{butanone} \\ 79\%}]{\text{NaI}} \underset{\diagup\!\diagdown \text{I}}{\overset{\text{TMS}}{}} \quad (1)$$
(4)   **(5)**

$$\underset{\text{OAc}}{} \xrightarrow[\substack{\text{2. (5)} \\ 91\%}]{\text{1. 2 equiv LiMe}} \underset{\text{O} \quad \text{TMS}}{} \quad (3)$$

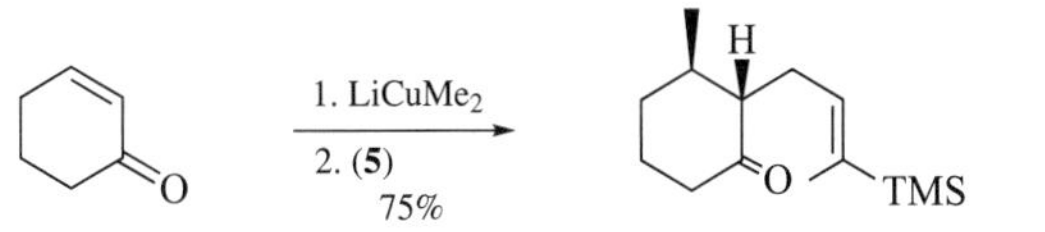

(4)

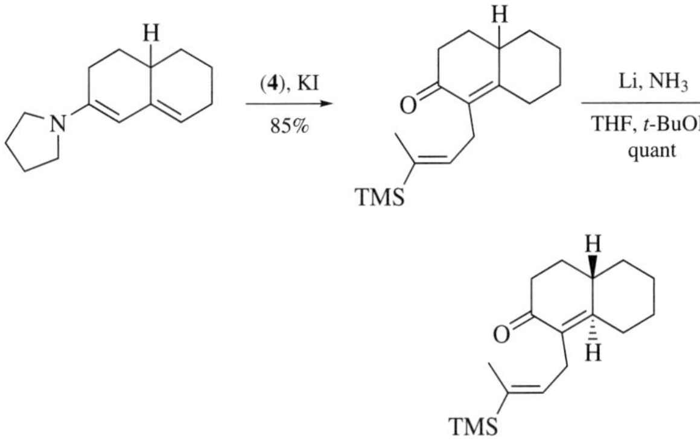

(5)

Release of the latent carbonyl function in these vinylsilanes can be achieved[9,14] under much milder conditions than for simple epoxysilanes, which require hot methanol/sulfuric acid. Here, treatment of the epoxide, formed using *m*-chloroperbenzoic acid, with formic acid for 30 s led cleanly to the dione (eq 6); alternatively, use of a slight excess of *m*-CPBA and a reaction time of 4 h achieved the same conversion. The extreme and contrasting ease of this transformation has been ascribed to nucleophilic participation of the carbonyl group in epoxide opening (eq 7).

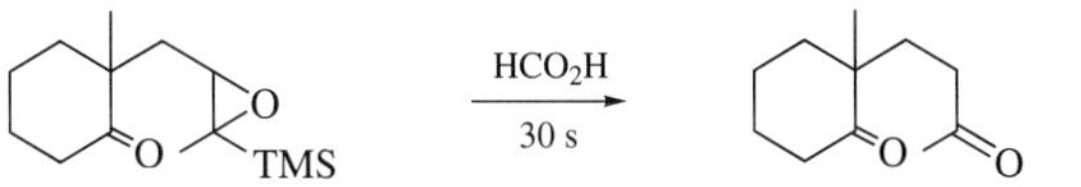

(6)

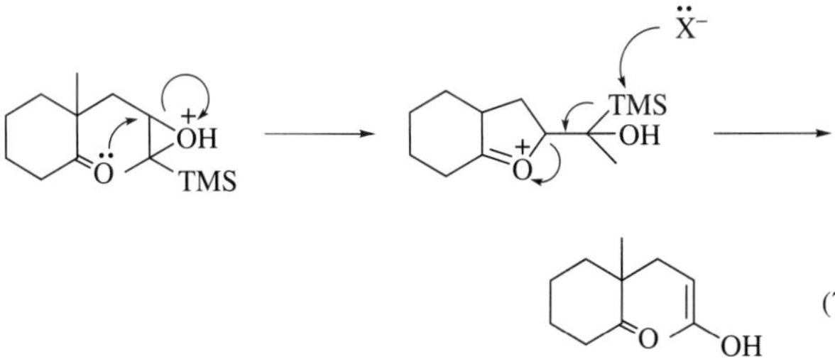

(7)

The diones so formed can be cyclized under basic conditions to the corresponding enones (eqs 8 and 9).

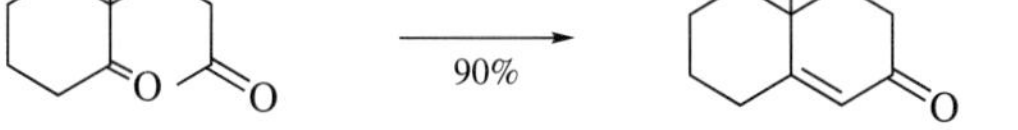

(8)

(9)

Iodide (5) has also been used[15] in a regiocontrolled route (eq 10) to 4,5-disubstituted 2-(and 3-)cyclohexenones. Alkylation of the anion of sulfone (6) with (5) gave (7), subsequent transformations leading sequentially to enones (8) and (9). No further manipulation of these enones has been reported.

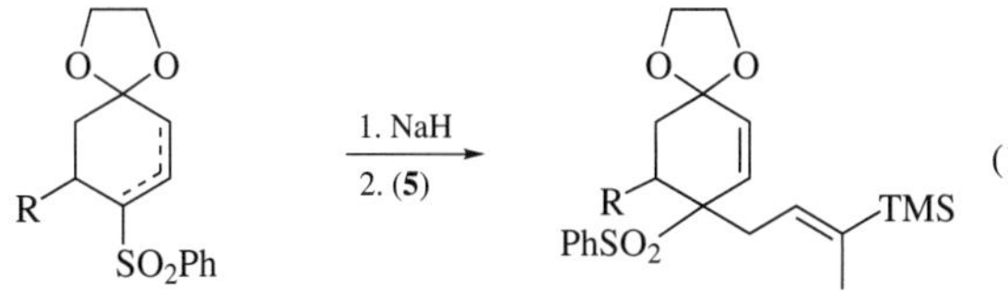

(10)

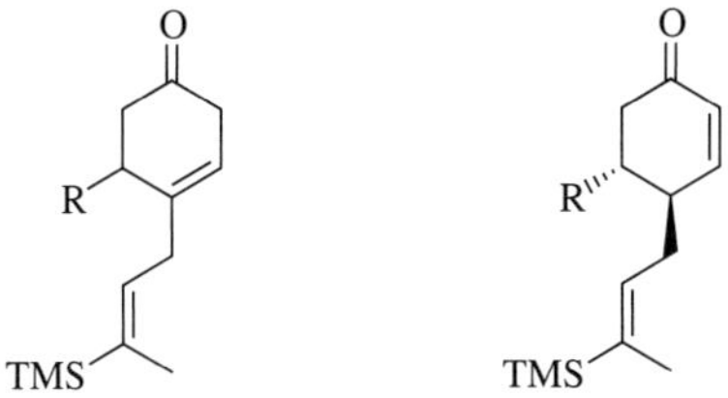

(8)   (9)

A different application[16] of reagent (5) can be seen in the total synthesis of artemisinin (10). Alkylation of the kinetic enolate of (11), derived from (−)-isopulegol, with (5) gave an 6:1 mixture of epimeric products from which the major isomer (12) was isolated in 62% yield. Further transformations led to vinylsilane (13). Liberation of the latent carbonyl group to provide ketone (14) was again achieved under very mild conditions, although here the possibility of carbonyl participation would seem less likely. Further steps then led to artemisinin (10).

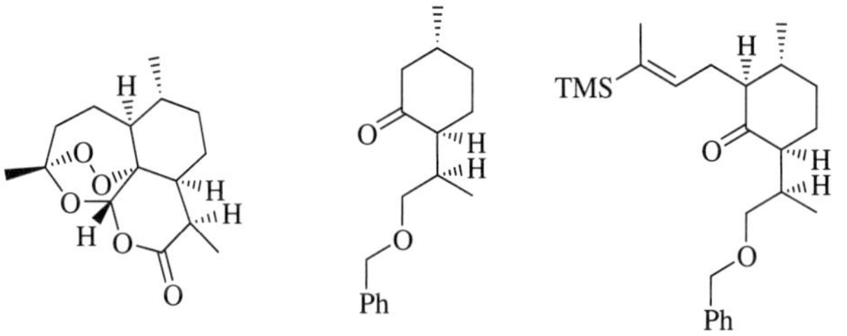

(10)   (11)   (12)

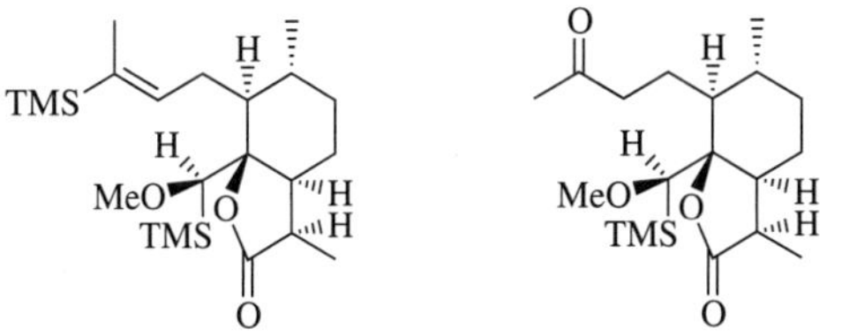

(13)   (14)

Additional examples of the use of this reagent are found in an incorrectly named e-EROS article (3-Iodo-1-methyl-propenyl)-trimethyl-silane (Stork–Jung Vinylsilane).[17]

1. *Fieser & Fieser* **1975**, *5*, 355.

2. Stork, G.; Jung, M. E.; Colvin, E; Noel, Y., *J. Am. Chem. Soc.* **1974**, *96*, 3684.

3. Denmark, S. E.; Jones, T., *Org. Synth.* **1986**, *64*, 182.

4. Denmark, S. E.; Habermas, K. L.; Hite, G. A., *Helv. Chim. Acta* **1988**, *71*, 168.

5. For alternative preparations of this alcohol, see (a) Altnau, G.; Rösch, L.; Bohlmann, F.; Lonitz, M., *Tetrahedron Lett.* **1980**, *21*, 4069. (b) Sato, F; Watanabe, H.; Tanaka, Y.; Sato, M., *J. Chem. Soc., Chem. Commun.* **1982**, 1126; see also Ref. 7.

6. Jung, M. E., cited in Gawley, R. E., *Synthesis* **1976**, 777.

7. Audia, J. E.; Marshall, J. A., *Synth. Commun.* **1983**, *13*, 531.

8. Bergmann, E. D.; Ginsburg, D.; Pappo, R., *Org. React.* **1959**, *10*, 179.

9. Jung, M. E., *Tetrahedron* **1976**, *32*, 3.

10. (a) Wichterle, O.; Prochazka, J.; Hofmann, J., *Collect. Czech. Chem. Commun.* **1948**, *13*, 300. (b) Review: House, H. O. *Modern Synthetic Reactions*, 2nd ed; Benjamin: Menlo Park, CA, 1972; p 611.

11. (a) Stork, G.; Danishefsky, S.; Ohashi, M., *J. Am. Chem. Soc.* **1967**, *89*, 5459. (b) Stork, G.; McMurry, J. E., *J. Am. Chem. Soc.* **1967**, *89*, 5463.

12. Stotter, P. L.; Hill, K. A., *J. Am. Chem. Soc.* **1974**, *96*, 6524.

13. (a) Stork, G.; Colvin, E., *J. Am. Chem. Soc.* **1971**, *93*, 2080. (b) See also Gröbel, B.-Th.; Seebach, D., *Angew. Chem., Int. Ed. Engl.* **1974**, *13*, 83.

14. Stork, G.; Jung, M. E., *J. Am. Chem. Soc.* **1974**, *96*, 3682.

15. (a) Paquette, L. A.; Kinney, W. A., *Tetrahedron Lett.* **1982**, *23*, 5127. (b) Kinney, W. A.; Crouse, G. D.; Paquette, L. A., *J. Org. Chem.* **1983**, *48*, 4986.

16. Schmid, G.; Hofheinz, W., *J. Am. Chem. Soc.* **1983**, *105*, 624.

17. Lam, H., *e-EROS*, **2009**, DOI: 10.1002/047084289X.rn01018

Ernest W. Colvin
University of Glasgow, Glasgow, UK

3-Iodo-2-trimethylsilylmethyl-1-propene

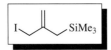

[80121-73-9]   C_7H_15ISi   (MW 254.21)

$InChI = 1S/C7H15ISi/c1-7(5-8)6-9(2,3)4/h1,5-6H2,2-4H3$

$InChIKey = FLDGWNPNLFEXIQ-UHFFFAOYSA-N$

(a bifunctional conjunctive reagent that serves as a dipolar trimethylenemethane synthon;[1,6] treatment with SnF_2 affords a dianionic trimethylenemethane synthon[9])

Physical Data: bp 25 °C/0.5 mmHg.[2]

Solubility: wide range of organic solvents.

Preparative Method: from the corresponding mesylate upon treatment with sodium iodide in acetone.[2] The mesylate can be prepared from the commercially available alcohol.[3]

Purification: distillation in the dark (25 °C/0.5 mmHg, trap at −78 °C).[2]

Handling, Storage, and Precautions: can be stored in the dark at −20 °C for up to 3 months without significant deterioration.[2]

Dipolar Synthon for Trimethylenemethane. The most common use of this reagent is as a 1,3-dipolar synthon for trimethylenemethane (TMM) (eq 1).[1]

$$I\diagdown\diagup{SiMe_3} \equiv \diagup\diagdown + \quad -$$

(1)

The dipolar character is unmasked in two separate synthetic steps. In the first, the synthon reacts at the positive end upon treatment with a nucleophile such as an enolate or metalloenamine.[2]

In the second step, the nucleophilic end is unveiled, usually by treatment with fluoride ion. This nucleophile attacks the carbonyl or imine, completing the sequence. A similar reagent to the iodide, also a 1,3-dipolar synthon for TMM, is derived by treatment of the corresponding acetate with Pd^0.[1] The resulting 1,3-dipolar synthon differs from the iodide in several ways:

1) one synthetic step is required to complete addition of the activated reagent;

2) the reaction simulates a cycloaddition, with an electron-deficient alkene as a reaction partner, rather than a carbonyl or imine;

3) the nucleophilic end of the synthon reacts first, in contrast to the iodide.

Reagent (**1**) has been used successfully with several classes of substrates.[1,4,5] For example, ketone (**2**) was used in a synthesis employing the dipolar synthon as a means of annulating the third ring of the coriolin skeleton (eq 2).[4]

$$
\text{(2)} \quad \xrightarrow[\begin{array}{c}\text{1. KH, DME, (1)}\\\text{2. m-CPBA}\\\text{3. F}^-\text{, THF}\\39\%\end{array}]{} \quad \text{(2)}
$$

Enone alkylation using (**1**), followed by Michael addition of the activated silane constitutes a second class of reactions employing this synthon.[6] As shown in eq 3, the mode of addition in systems containing a second conjugated double bond can be controlled by changing the method used to activate the silane toward nucleophilic attack. Use of a Lewis acid favors 1,6-addition leading to seven-membered rings,[6b,7] while treatment with fluoride ion favors 1,4-addition and formation of five-membered rings.[6b]

$$
\xleftarrow[55\%]{\text{EtAlCl}_2} \quad \text{TMS} \quad \xrightarrow[45\%]{\text{F}^-} \quad \text{(3)}
$$

Finally, imines can be used in the synthesis of nitrogen-containing heterocycles.[8] One particularly interesting method involves a sequential photoexcitation–electron-transfer desilylation method for generating a diradical species capable of forming a spirocyclic product (eq 4).[8c] Notably, the ionic cyclization method involving fluoride ion leads, in this case, to mixtures of products including only small amounts (ca. 7%) of the desired spirocycle.[8c]

$$
\xrightarrow[\begin{array}{c}h\nu\\\text{SET}\\45\text{–}88\%\end{array}]{} \quad \text{(4)}
$$

Dianionic Synthon for Trimethylenemethane.[9] Reagent (**1**), upon metalation with tin(II) fluoride can be used as a dianionic synthon for trimethylenemethane (eq 5).[10] This versatile synthon, in conjunction with bis-electrophiles such as (**4**), has been used to synthesize rings that range in size from five to eight members. An intramolecular hemiacetalization reaction followed by cyclization of the allylsilane makes possible the formation of seven- and eight-membered rings (eq 6).[9b]

(5)

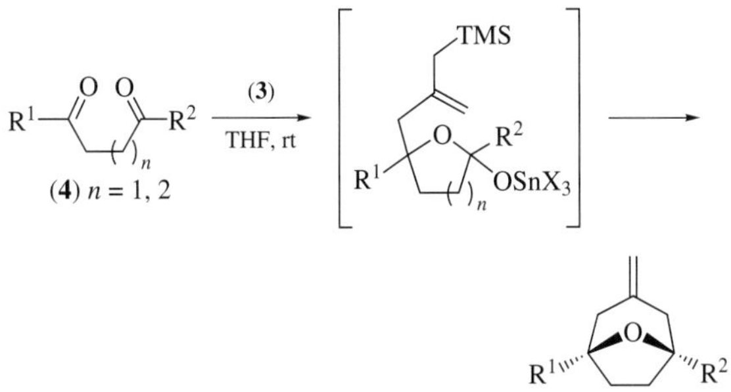

(6)

Related Reagent.   3-Acetoxy-2-trimethylsilylmethyl-1-propene.

1. Review of TMM and equivalents: Trost, B. M., *Angew. Chem., Int. Ed. Engl.* **1986**, *25*, 1.
2. Trost, B. M.; Curran, D. P., *Tetrahedron Lett.* **1981**, *22*, 5023.
3. The alcohol can be prepared in two steps from 2-methyl-2-propen-1-ol: Trost, B. M.; Chan, D. M. T.; Nanninga, T., *Org. Synth.* **1984**, *62*, 58.
4. Trost, B. M.; Curran, D. P., *J. Am. Chem. Soc.* **1981**, *103*, 7380.
5. Posner, G. H.; Asirvatham, E.; Hamill, T. G., *J. Org. Chem.* **1990**, *55*, 2132.
6. (a) Majetich, G.; Desmond, Jr., R. W.; Soria, J. J., *J. Org. Chem.* **1986**, *51*, 1753. (b) Majetich, G; Hull, K.; Defauw, J.; Desmond, R., *Tetrahedron Lett.* **1985**, *26*, 2747. (c) Majetich, G.; Desmond, R.; Casares, A. M., *Tetrahedron Lett.* **1983**, *24*, 1913.
7. Majetich, G.; Leigh, A. J.; Condon, S., *Tetrahedron Lett.* **1991**, *32*, 605.
8. (a) Gelas-Mialhe, Y.; Gramain, J.-C.; Hajouji, H.; Remuson, R., *Heterocycles* **1992**, *34*, 37. (b) Bell, T. W.; Hu, L.-Y., *Tetrahedron Lett.* **1988**, *29*, 4819. (c) Ahmed-Schofield, R.; Mariano, P. S., *J. Org. Chem.* **1985**, *50*, 5667.
9. (a) Molander, G. A.; Shubert, D. C., *J. Am. Chem. Soc.* **1987**, *109*, 576. (b) Molander, G. A.; Shubert, D. C., *J. Am. Chem. Soc.* **1987**, *109*, 6877. (c) Molander, G. A.; Shubert, D. C., *J. Am. Chem. Soc.* **1986**, *108*, 4683.
10. For other dianion TMM synthons, see: (a) Molander, G. A.; Shubert, D. C., *Tetrahedron Lett.* **1986**, *27*, 787. (b) Jones, M. D.; Kemmitt, R. D. W., *J. Chem. Soc., Chem. Commun.* **1985**, 811.

Therese M. Bregant & R. Daniel Little
University of California, Santa Barbara, CA, USA

Isocyanatotrimethylsilane

$$Me_3SiN=C=O$$

[1118-01-1] C_4H_9NOSi (MW 115.23)

InChI = 1S/C4H9NOSi/c1-7(2,3)5-4-6/h1-3H3

InChIKey = NIZHERJWXFHGGU-UHFFFAOYSA-N

(preparation of primary amides from Grignard reagents;[1–3] synthesis of trifluoroacetyl isocyanate;[4] synthesis of cyclic imides[5])

Alternate Name: trimethylsilyl isocyanate.
Physical Data: colorless oil; bp 90–92 °C; fp −2 °C; *d* 0.851 g cm^{-3}.
Solubility: sol ethers, *o*-dichlorobenzene, etc.
Form Supplied in: commercially available.
Preparative Method: from chlorotrimethylsilane and silver isocyanate.[6]
Handling, Storage, and Precautions: harmful vapor; irritant; flammable liquid; moisture sensitive; use in a fume hood.

Amidation.   Grignard reagents are carboxamidated by isocyanatotrimethylsilane to afford primary amides, presumably by way of the *N*-silyl amides (eqs 1 and 2).[1–3] The use of dioxane as an additive to the ether or THF solvent eliminates the formation of silylated Grignard products.[1] Isocyanatotriphenylsilane reacts similarly with phenyllithium and phenylmagnesium bromide to give benzamide (77–80%).[7] Reaction of heterocyclic benzyllithium reagents with isothiocyanatotrimethylsilane affords primary thioamides in moderate yield (eq 2).[2] *N*-Hydroxyurea has been prepared from *O*-silylhydroxylamine and isocyanatotrimethylsilane.[4]

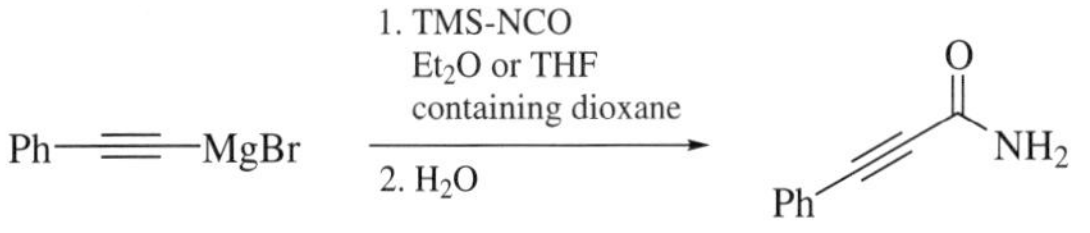

(1)

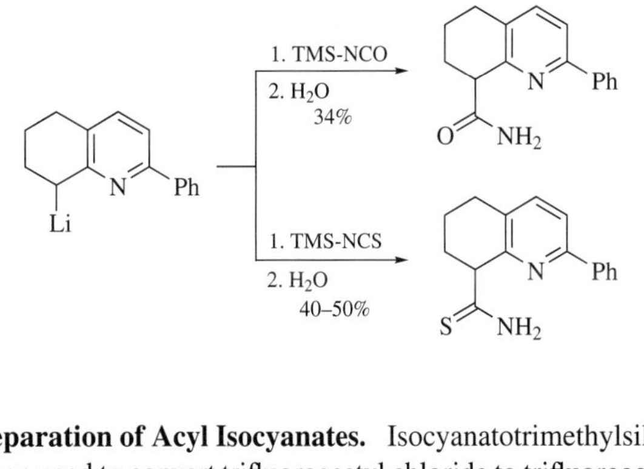

(2)

Preparation of Acyl Isocyanates. Isocyanatotrimethylsilane has been used to convert trifluoroacetyl chloride to trifluoroacetyl isocyanate.[5] In the reaction with 1-naphthalenecarbonyl chloride, cyclization of the intermediate acyl isocyanate provides the imide through an intramolecular Friedel–Crafts reaction (eq 3).[8]

Related Reagents. Benzenesulfonyl Isocyanate; 3-(2-Benzyloxyacetyl)thiazolidine-2-thione; Chlorosulfonyl Isocyanate; Isocyanic Acid; *p*-Toluenesulfonyl Isocyanate; Trichloroacetyl Isocyanate.

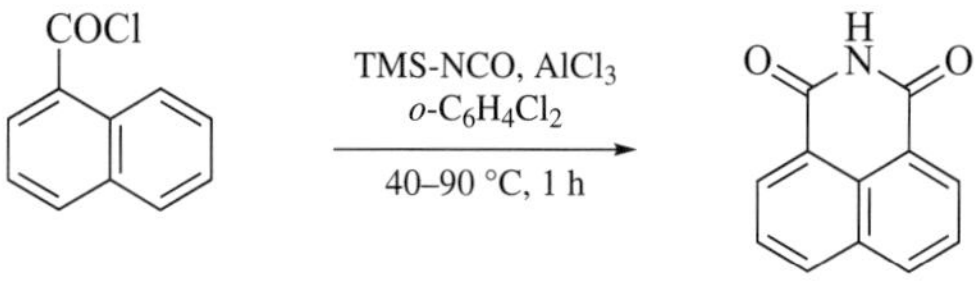

$$\text{(3)}$$

TMS-NCO, AlCl$_3$
o-C$_6$H$_4$Cl$_2$
40–90 °C, 1 h

1. Parker, K. A.; Gibbons, E. G., *Tetrahedron Lett.* **1975**, 981.

2. Curran, A. C. W.; Shepherd, R. G., *J. Chem. Soc., Perkin Trans. 1* **1976**, 983.

3. For a review on carboxamidation of organometallic reagents, see Screttas, C. G.; Steele, B. R., *Org. Prep. Proced. Int.* **1990**, *22*, 269.

4. Muzovskaya, E. V.; Kozyukov, V. P.; Mironov, V. F.; Kozyukov, V. P., *J. Org. Chem. USSR (Engl. Transl.)* **1989**, *59*, 349.

5. Kiemstedt, W.; Sundermeyer, W., *Chem. Ber.* **1982**, *115*, 919.

6. Forbes, G. S.; Anderson, H. H., *J. Am. Chem. Soc.* **1948**, *70*, 1222.

7. Gilman, H.; Hofferth, B.; Melvin, H. W., *J. Am. Chem. Soc.* **1950**, *72*, 3045.

8. Frenzel, R.; Domschke, G.; Mayer, R., *J. Prakt. Chem.* **1991**, *333*, 805.

Kathlyn A. Parker & David Taveras
Brown University, Providence, RI, USA

Ketene *t*-Butyldimethylsilyl Methyl Acetal[1]

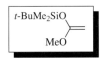

[91390-62-4] $C_9H_{20}O_2Si$ (MW 188.34)

InChI = 1S/C9H20O2Si/c1-8(10-5)11-12(6,7)9(2,3)4/
      h1H2,2-7H3

InChIKey = UVCCWXJGWMGZAB-UHFFFAOYSA-N

(silylation of a variety of substrates;[3] promotion of Pummerer reaction of sulfoxides;[4–8] Lewis acid mediated aldol-type[8–22] and Michael additions,[31–38] under either stoichiometric or catalytic conditions)

Physical Data: colorless liquid; bp 76–76.5 °C/24 mmHg.
Solubility: sol *n*-pentane, diethyl ether, dichloromethane, etc.
Analysis of Reagent Purity: (NMR) ^{1}H 0.14 (s, 6H), 0.93 (s, 9H), 2.95 (d, 1H, $J = 3.0$ Hz), 3.10 (d, 1H, $J = 3.0$ Hz), 3.49 (s, 3H) ppm.
Preparative Method: obtained by reaction of methyl acetate with lithium diisopropylamide in THF/HMPA and subsequent trapping with *t*-butyldimethylchlorosilane (72% yield).[2]
Handling, Storage, and Precautions: should be stored in the absence of moisture at −15 °C.

Silylation. The reagent silylates a variety of substrates (alcohols, acids, thiols, phenols, imides) under mild conditions (a catalytic amount of *p*-toluenesulfonic acid is occasionally added) with excellent yields (91–100%).[3]

Pummerer Reaction. The reagent transforms sulfoxides in the presence of catalytic amounts of zinc iodide into the corresponding α-silyloxy sulfides (eq 1).[4] Vinyl sulfoxides undergo a Michael–Pummerer type reaction to give γ-silyloxy-γ-phenylthio esters (eq 2).[5]

As an extention of this reaction, the reagent promotes the intramolecular Pummerer-type rearrangement of ω-carbamoyl

sulfoxides to give α-thio lactams.[6] The reaction proves particularly useful in the field of β-lactam synthesis (eq 3).[7,8a]

$$(3)$$

Addition to C=X Double Bonds (X = N, O).[1b] The reagent undergoes Lewis acid catalyzed Mukaiyama-type additions[1b] to azetinones, or their corresponding iminium ions, to give *trans*-azetidin-2-one esters with good yields and excellent stereoselectivity (eq 4).[7b,8,9]

$$(4)$$

79%
racemate, *trans*:*cis* = 94:6

In the presence of a diphosphonium salt (7 mol %),[10a] trimethylsilyl trifluoromethanesulfonate (10 mol %),[10b] or iron(III) nitrate–K10 montmorillonite clay,[10c] the reagent adds to imines or *N*-tosyliminium ions to give the corresponding β-amino esters in good yield.

ZnI_2-catalyzed additions of the reagent to chiral α,β-dialkoxy nitrones (eq 5; R^1 = H) proceed with good yield (86–100%) and high diastereoselectivity (ca. 90:10) in favor of the *syn* isomer (R^1 = H, R^2 = CH_2Ph, R^3 = Me). The *anti* isomer is obtained (ca. 90:10) by increasing the steric hindrance of R^2 and R^3 (R^2 = $CHPh_2$, R^3 = *t*-Bu).[11a,11b] Addition to a different nitrone (eq 5; R^1 = Me) gives the *anti* isomer (R^1 = Me, R^2 = CHMePh, R^3 = Me) in quantitative yield and 100% diastereofacial selectivity. This material has been further elaborated to *N*-benzoyl-L-daunosamine.[11c,11d]

$$(5)$$

New catalysts have been recently developed for promoting the aldol-type addition of acetate-derived silyl ketene acetals with high efficiency: 10-methylacridinium perchlorate (5 mol %),[12a] cationic mono- and dinuclear iron complexes (5 mol %),[12b] *t*-butyldimethylsilyl chloride–indium(III) chloride (10 mol %),[12c] [1,2-benzenediolato(2−)-*O*,*O*′]oxotitanium (20 mol %),[12d] phosphonium salts (7 mol %),[12e] and trityl salts (5–20 mol %).[1]

The reagent undergoes Lewis acid promoted Mukaiyama-type additions[1b] to chiral aldehydes with moderate to good stereocontrol (eq 6).[13] It is remarkable that high 'chelation control' can be obtained by using a catalytic amount (3 mol %) of lithium perchlorate.[13e,13f]

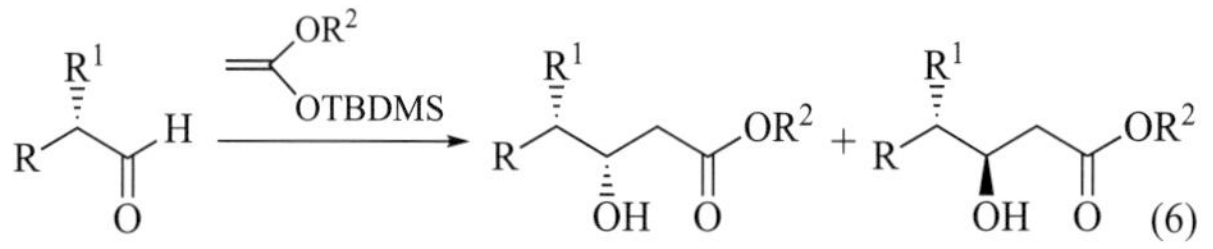

(6)

R	R^1	Promoter	R^2	Ratio	Yield (%)
Ph	Me	$BF_3 \cdot OEt_2$	*t*-Bu	97:3	81[13a]
Ph	Me	10 kbar, 50 °C	Me	71:29	73[13b]
Me	OBn	$SnCl_4$	*t*-Bu	65:35	65[13c,d]
Me	OBn	cat $LiClO_4$	Me	92:8	84[13e,f]
$TBDMSOCH_2$	OBn	$SnCl_4$	Me	>98:<2	90[13g,h]
$CyCH_2$	$NHCO_2$-*i*-Pr	$TiCl_4$	Me	96:4	95[13k,l]

With other substrates and under different conditions (eqs 7 and 8), 'non-chelated' products are obtained with excellent selectivity.[13f,14]

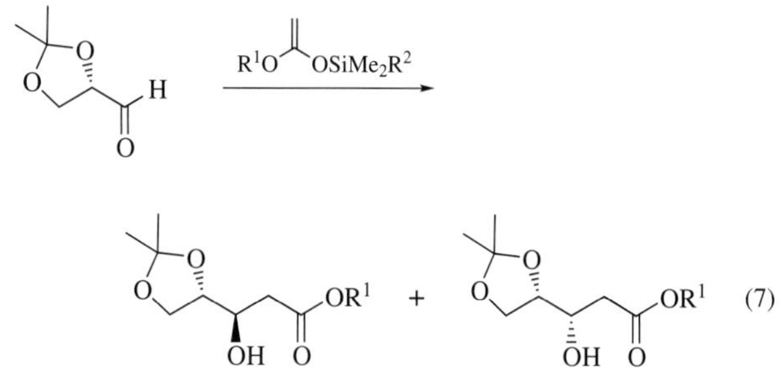

(7)

R^1 = Me, R^2 = *t*-Bu   cat ZnI_2 96:4   67%[14a,b]
R^1 = Et, R^2 = Me   cat $Eu(dppm)_3$ 95:5   100%[14c]

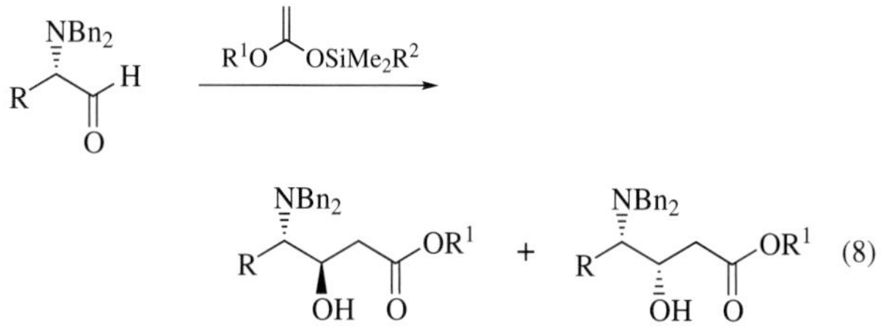

(8)

R^1 = Bn, R^2 = Me $EtAlCl_2$ >99:<1   94%[14d]
R^1 = Me, R^2 = *t*-Bu   cat $LiClO_4$ in CH_2Cl_2 >98:<2   58%[13f]

In the field of *C*-glycoside synthesis, selective β-glycosylation is realized via neighboring group participation of a 2α-acyl group.[15a] In the case of 2-deoxy sugars the neighboring participation of a group at the 3α-position is exploited for selective formation of the β-anomer (β:α = 91:9) (eq 9).[15b]

Enantioselective Aldol-Type Additions. Highly enantioselective aldol-type reactions are successfully carried out by the combined use of a chiral diamine-coordinated tin(II) triflate and tributyltin fluoride (eq 10).[16,17] A catalytic amount of chiral bis(sulfonamido)zinc(II), easily prepared from diethylzinc and chiral sulfonamides, promotes the aldol addition in high yield and

good enantiomeric excess (72–93%) only with chloral and bromal (CX_3CHO).[18]

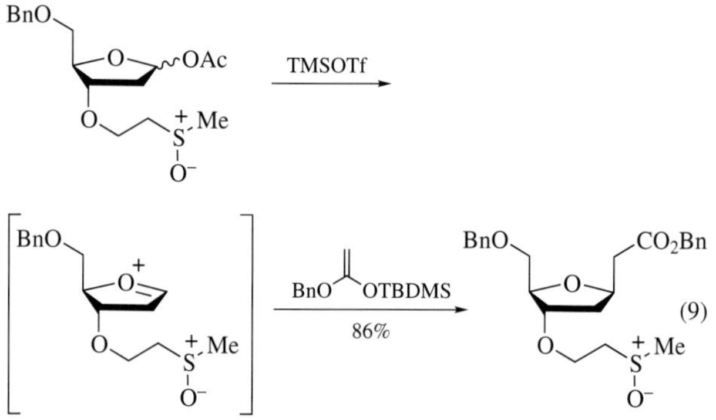

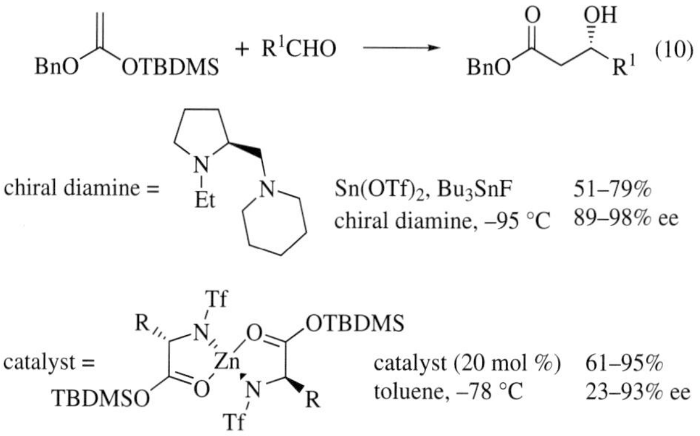

(9)

(10)

chiral diamine =

$Sn(OTf)_2$, Bu_3SnF 51–79%
chiral diamine, –95 °C   89–98% ee

catalyst =

catalyst (20 mol %)   61–95%
toluene, –78 °C   23–93% ee

Chiral borane complexes (20 mol %) catalyze the aldol-type addition to achiral aldehydes in good to excellent yield and enantiomeric excess (eq 11).[19–21]

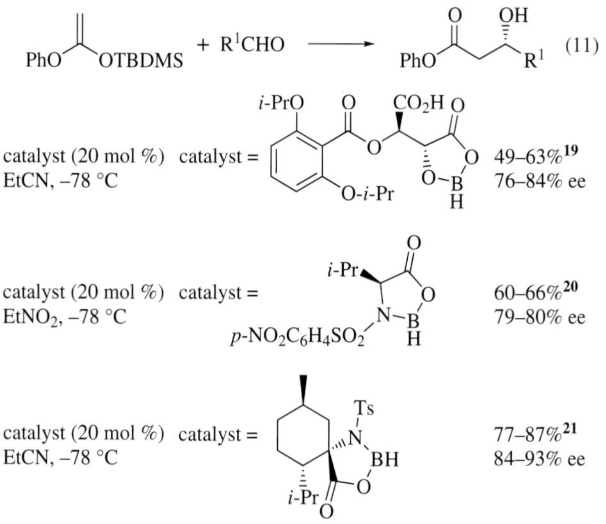

(11)

catalyst (20 mol %)   catalyst =   49–63%[19]
EtCN, –78 °C   76–84% ee

catalyst (20 mol %)   catalyst =   60–66%[20]
$EtNO_2$, –78 °C   79–80% ee

catalyst (20 mol %)   catalyst =   77–87%[21]
EtCN, –78 °C   84–93% ee

A chiral boron reagent, derived from equimolar amounts of (*R*)- or (*S*)-binaphthol and triphenyl borate, promotes the condensation of chiral imines with *t*-butyl acetate silyl ketene acetal in high diastereomeric excess (eq 12).[22]

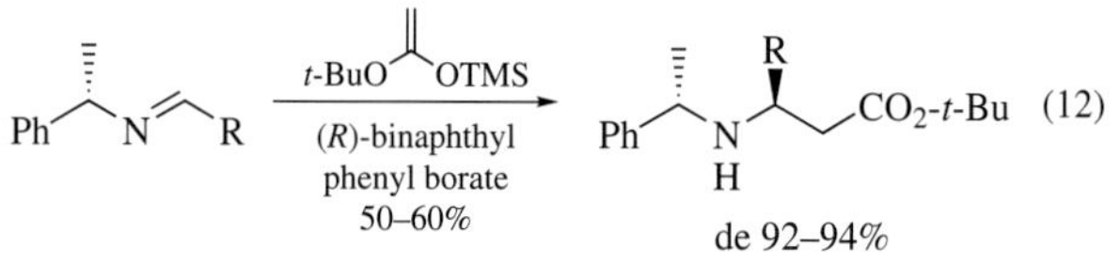

(12)

de 92–94%

Addition to Various Electrophiles.   Various Lewis acids promote the addition of the reagent to an allylic acetate, following a carbon-Ferrier rearrangement pathway.[23] Titanium(IV) chloride promotes the addition of the reagent to 2,2-dialkoxycyclopropanecarboxylic esters to give 3-alkoxy-2-cyclopentenones (eq 13).[24]

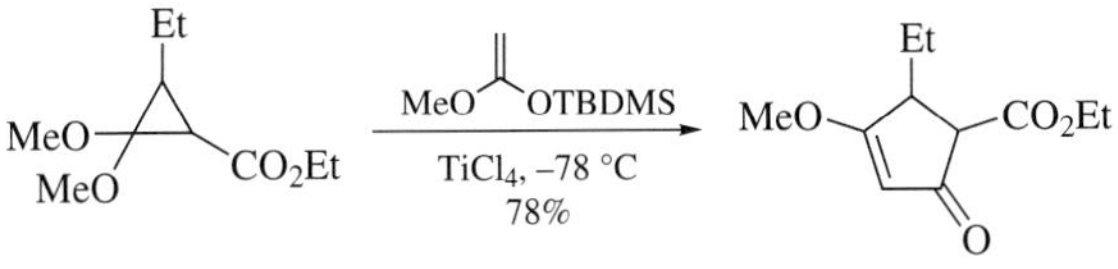

(13)

1,3-Dioxolan-2-ylium cations, derived from aldehyde ethylene acetals and trityl cation, react with the reagent to give the corresponding β-keto esters.[25] Montmorillonite K10 catalyzes the addition of the reagent to pyridine derivatives with electron withdrawing groups to give *N*-silyldihydropyridines.[26] The ketene silyl acetal of ethyl acetate reacts with a chiral bromide to give the corresponding *syn*-lactone in 64% yield via a direct S_N2-type displacement and inversion of stereochemistry (eq 14).[27]

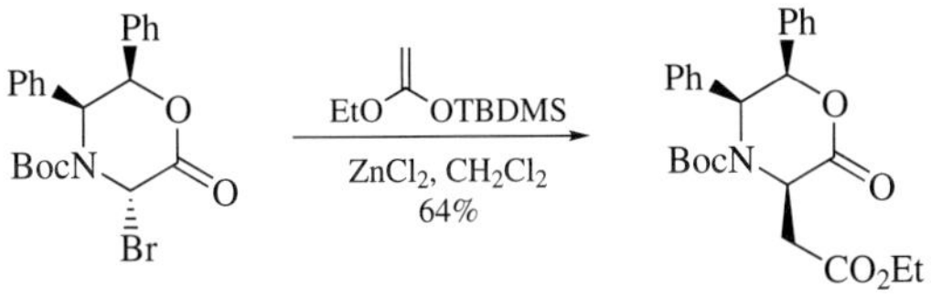

(14)

syn:anti = 45:1

Six-membered chiral acetals, derived from aliphatic aldehydes, undergo aldol-type coupling reactions with silyl ketene acetals in the presence of $TiCl_4$ with high diastereoselectivity (eq 15).[28] This procedure, in combination with oxidative destructive elimination of the chiral auxiliary, has been applied to the preparation of (*R*)-(+)-α-lipoic acid[28a] and mevinolin analogs.[28b]

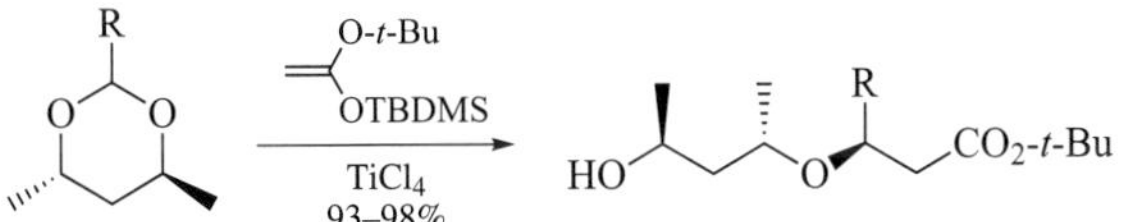

(15)

dr 97–98:2–3

Addition to chiral, bicyclic acetals has been exploited in an approach to the synthesis of the tetrahydropyran subunit of the polyether nigerin.[29] The particular acetal generated by the diisobutylaluminum hydride reduction of aliphatic esters undergoes aldol addition in good yields (eq 16).[30]

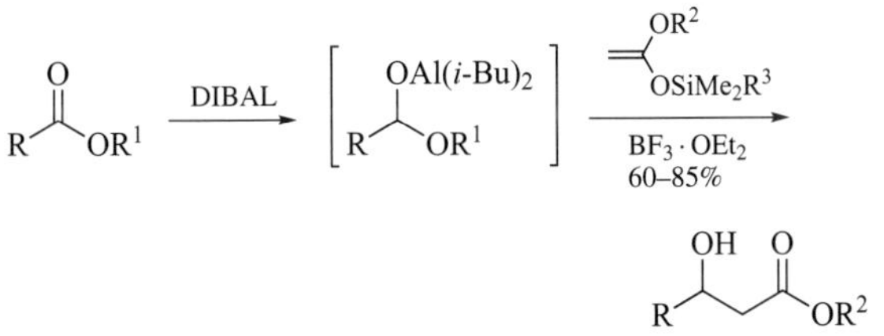

(16)

Michael Addition.   The reagent undergoes Michael addition to α,β-enones in acetonitrile in the absence of a Lewis acid to afford the corresponding *O*-silylated Michael adducts in high yield. These silyl enol ethers undergo site-specific reaction with a variety

of electrophiles (eq 17).[31a,31b] Inability to repeat this procedure led to the discovery that the 'noncatalyzed' Michael reaction is due to traces of phosphorus compounds introduced by drying acetonitrile with P_4O_{10}. The new catalyst system, formed from P_4O_{10} in acetonitrile, was found to be highly effective with a variety of substrates.[31c]

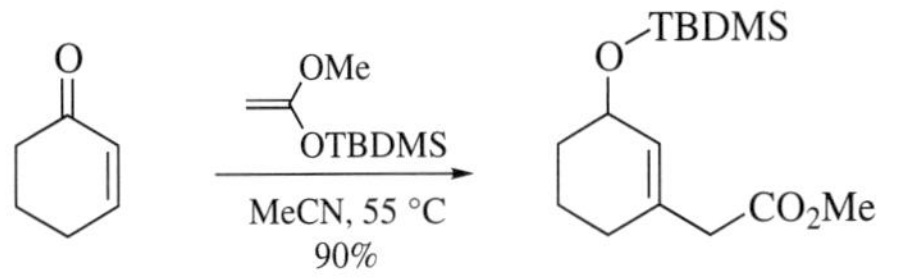

(17)

In those instances where the thermal Michael reaction is sluggish due to sterically hindered substrates, the use of high pressure (15 kbar, 20 °C),[32a] or of $LiClO_4$ (3 mol % in CH_2Cl_2[13f] or 1.0–2.5 M in Et_2O[32b]) prove extremely advantageous (eq 18). The lithium perchlorate-catalyzed Michael reaction can be carried out on α,β-unsaturated δ-lactones and on sterically demanding β,β-disubstituted unsaturated carbonyl systems in high yield and under mild conditions.[32b]

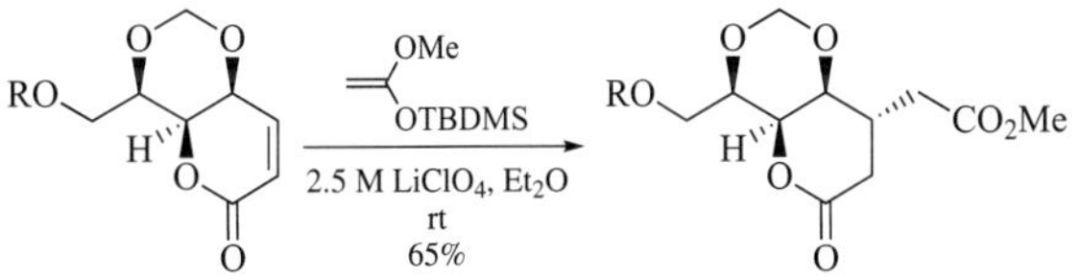

(18)

Michael addition of the reagent to enoates and enones occurs at low temperature (-50 to $-78\,°C$) in the presence of catalytic amounts of various Lewis acids.[33] A catalytic amount of triphenylmethyl perchlorate (5 mol %) effectively catalyzes the tandem Michael reaction of ethyl acetate-derived silyl ketene acetal to α,β-unsaturated ketones and the sequential aldol addition to aldehydes with high stereoselectivity.[34] HgI_2 mediates the Michael addition to chiral enones, followed by Lewis acid-mediated addition to aldehydes. The Michael-aldol protocol has been used for the stereoselective synthesis of key intermediates on the way to prostaglandins, compactin, and ML-236A (eq 19).[35]

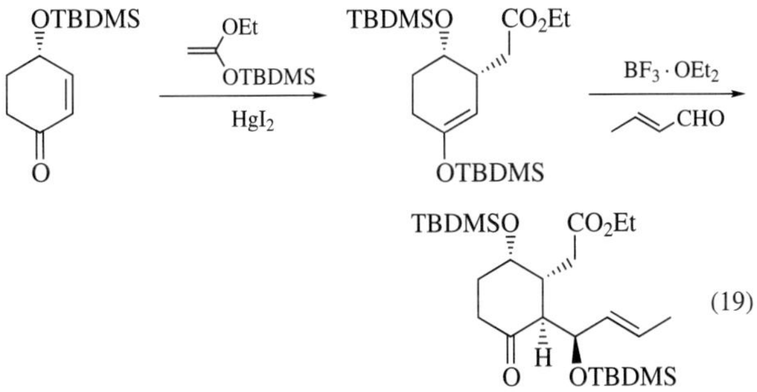

(19)

(single diastereomer)

The mechanism of the $TiCl_4$-mediated Michael addition of silyl ketene acetals has been investigated, and criteria for suppressing the electron transfer process have been devised.[36] Chiral enones show good to excellent diastereofacial preference in $TiCl_4$-mediated reactions with silyl ketene acetals (eq 20).[37]

ZnI$_2$-mediated multiple Michael additions to bis-enoates proceed in good yield and with modest stereocontrol (eq 21).[38]

Related Reagents. 1-*t*-Butylthio-1-*t*-butyldimethylsilyloxy-propene; 1-*t*-Butylthio-1-*t*-butyldimethylsilyloxyethylene.

1. (a) Fieser, L. F.; Fieser, M., *Fieser & Fieser* **1984**, *11*, 279. (b) Gennari, C., *Comprehensive Organic Synthesis* **1991**, *2*, 629.

2. (a) Ainsworth, C.; Chen, F.; Kuo, Y.-N., *J. Organomet. Chem.* **1972**, *46*, 59. (b) Kita, Y.; Segawa, J.; Haruta, J.; Fujii, T.; Tamura, Y., *Tetrahedron Lett.* **1980**, *21*, 3779.

3. Kita, Y.; Haruta, J.; Fujii, T.; Segawa, J.; Tamura, Y., *Synthesis* **1981**, 451.

4. (a) Kita, Y.; Yasuda, H.; Tamura, O.; Itoh, F.; Tamura, Y., *Tetrahedron Lett.* **1984**, *25*, 4681. (b) Kita, Y.; Tamura, O.; Yasuda, H.; Itoh, F.; Tamura, Y., *Chem. Pharm. Bull.* **1985**, *33*, 4235.

5. Kita, Y.; Tamura, O.; Itoh, F.; Yasuda, H.; Miki, T.; Tamura, Y., *Chem. Pharm. Bull.* **1987**, *35*, 562.

6. (a) Kita, Y.; Tamura, O.; Miki, T.; Tamura, Y., *Tetrahedron Lett.* **1987**, *28*, 6479. (b) Kita, Y.; Tamura, O.; Shibata, N.; Miki, T., *Chem. Pharm. Bull.* **1990**, *38*, 1473.

7. (a) Kita, Y.; Tamura, O.; Miki, T.; Tono, H.; Shibata, N.; Tamura, Y., *Tetrahedron Lett.* **1989**, *30*, 729. (b) Kita, Y.; Tamura, O.; Shibata, N.; Miki, T., *J. Chem. Soc., Perkin Trans. 1* **1989**, 1862.

8. (a) Kita, Y.; Shibata, N.; Miki, T.; Takemura, Y.; Tamura, O., *J. Chem. Soc., Chem. Commun.* **1990**, 727. (b) Kita, Y.; Shibata, N.; Tamura, O.; Miki, T., *Chem. Pharm. Bull.* **1991**, *39*, 2225.

9. (a) Chiba, T.; Nakai, T., *Chem. Lett.* **1987**, 2187. (b) Chiba, T.; Nagatsuma, M.; Nakai, T., *Chem. Lett.* **1985**, 1343. (c) Yoshida, A.; Tajima, Y.; Takeda, N.; Oida, S., *Tetrahedron Lett.* **1984**, *25*, 2793. (d) Tajima, Y.; Yoshida, A.; Takeda, N.; Oida, S., *Tetrahedron Lett.* **1985**, *26*, 673. (e) Murahashi, S.-I.; Saito, T.; Naota, T.; Kumobayashi, H.; Akutagawa, S., *Tetrahedron Lett.* **1991**, *32*, 5991.

10. (a) Mukaiyama, T.; Kashiwagi, K.; Matsui, S., *Chem. Lett.* **1989**, 1397. (b) Åhman, J.; Somfai, P., *Tetrahedron* **1992**, *43*, 9537. (c) Onaka, M.; Ohno, R.; Yanagiya, N.; Izumi, Y., *Synlett* **1993**, 141.

11. (a) Kita, Y.; Tamura, O.; Itoh, F.; Kishino, H.; Miki, T.; Kohno, M.; Tamura, Y., *J. Chem. Soc., Chem. Commun.* **1988**, 761. (b) Kita, Y.; Tamura, O.; Itoh, F.; Kishino, H.; Miki, T.; Kohno, M.; Tamura, Y., *Chem. Pharm. Bull.* **1989**, *37*, 2002. (c) Kita, Y.; Itoh, F.; Tamura, O.;

Yan Ke, Y. Y.; Tamura, Y., *Tetrahedron Lett.* **1987**, *28*, 1431. (d) Kita, Y.; Itoh, F.; Tamura, O.; Yan Ke, Y. Y.; Miki, T.; Tamura, Y., *Chem. Pharm. Bull.* **1989**, *37*, 1446.

12. (a) Otera, J.; Wakahara, Y.; Kamei, H.; Sato, T.; Nozaki, H.; Fukuzumi, S., *Tetrahedron Lett.* **1991**, *32*, 2405. (b) Bach, T.; Fox, D. N. A.; Reetz, M. T., *J. Chem. Soc., Chem. Commun.* **1992**, 1634. (c) Mukaiyama, T.; Ohno, T.; Han, J. S.; Kobayashi, S., *Chem. Lett.* **1991**, 949. (d) Hara, R.; Mukaiyama, T., *Chem. Lett.* **1989**, 1909. (e) Mukaiyama, T.; Matsui, S.; Kashiwagi, K., *Chem. Lett.* **1989**, 993. (f) Kobayashi, S.; Matsui, S.; Mukaiyama, T., *Chem. Lett.* **1988**, 1491. (g) Mukaiyama, T.; Leon, P.; Kobayashi, S., *Chem. Lett.* **1988**, 1495. (h) Homma, K.; Mukaiyama, T., *Chem. Lett.* **1990**, 161. (i) Homma, K.; Takenoshita, H.; Mukaiyama, T., *Bull. Chem. Soc. Jpn.* **1990**, *63*, 1898.

13. (a) Heathcock, C. H.; Flippin, L. A., *J. Am. Chem. Soc.* **1983**, *105*, 1667. (b) Yamamoto, Y.; Maruyama, K.; Matsumoto, K., *Tetrahedron Lett.* **1984**, *25*, 1075. (c) Heathcock, C. H.; Davidsen, S. K.; Hug, K. T.; Flippin, L. A., *J. Org. Chem.* **1986**, *51*, 3027. (d) Heathcock, C. H.; Hug, K. T.; Flippin, L. A., *Tetrahedron Lett.* **1984**, *25*, 5973. (e) Reetz, M. T.; Raguse, B.; Marth, C. F.; Hügel, H. M.; Bach, T.; Fox, D. N. A., *Tetrahedron* **1992**, *48*, 5731. (f) Reetz, M. T.; Fox, D. N. A., *Tetrahedron Lett.* **1993**, *34*, 1119. (g) Reetz, M. T.; Kesseler, K., *J. Org. Chem.* **1985**, *50*, 5434. (h) Reetz, M. T., *Pure Appl. Chem.* **1985**, *57*, 1781. (i) Rama Rao, A. V.; Chakraborty, T. K.; Purandare, A. V., *Tetrahedron Lett.* **1990**, *31*, 1443. (j) Barrett, A. G. M.; Raynham, T. M., *Tetrahedron Lett.* **1987**, *28*, 5615. (k) Takemoto, Y.; Matsumoto, T.; Ito, Y.; Terashima, S., *Tetrahedron Lett.* **1990**, *31*, 217. (l) Takemoto, Y.; Matsumoto, T.; Ito, Y.; Terashima, S., *Chem. Pharm. Bull.* **1991**, *39*, 2425. (m) Gennari, C.; Cozzi, P. G., *Tetrahedron* **1988**, *44*, 5965. (n) Shirai, F.; Nakai, T., *Chem. Lett.* **1989**, 445. (o) Yamazaki, T.; Yamamoto, T.; Kitazume, T., *J. Org. Chem.* **1989**, *54*, 83.

14. (a) Kita, Y.; Yasuda, H.; Tamura, O.; Itoh, F.; Yuan Ke, Y.; Tamura, Y., *Tetrahedron Lett.* **1985**, *26*, 5777. (b) Kita, Y.; Tamura, O.; Itoh, F.; Yasuda, H.; Kishino, H.; Yuan Ke, Y.; Tamura, Y., *J. Org. Chem.* **1988**, *53*, 554. (c) Mikami, K.; Terada, M.; Nakai, T., *Tetrahedron: Asymmetry* **1991**, *2*, 993. (d) Mikami, K.; Kaneko, M.; Loh, T.-P.; Terada, M.; Nakai, T., *Tetrahedron Lett.* **1990**, *31*, 3909. (e) Reetz, M. T.; Schmitz, A.; Holdgrün, X., *Tetrahedron Lett.* **1989**, *30*, 5421. (f) Annunziata, R.; Cinquini, M.; Cozzi, F.; Cozzi, P. G.; Consolandi, E., *J. Org. Chem.* **1992**, *57*, 456.

15. (a) Yokoyama, Y. S.; Elmoghayar, M. R. H.; Kuwajima, I., *Tetrahedron Lett.* **1982**, *23*, 2673. (b) Narasaka, K.; Ichikawa, Y.; Kubota, H., *Chem. Lett.* **1987**, 2139.

16. Mukaiyama, T.; Kobayashi, S.; Sano, T., *Tetrahedron* **1990**, *46*, 4653.

17. Kobayashi, S.; Sano, T.; Mukaiyama, T., *Chem. Lett.* **1989**, 1319.

18. Mukaiyama, T.; Takashima, T.; Kusaka, H.; Shimpuku, T., *Chem. Lett.* **1990**, 1777.

19. Furuta, K.; Maruyama, T.; Yamamoto, H., *Synlett* **1991**, 439.

20. (a) Kiyooka, S.; Kaneko, Y.; Komura, M.; Matsuo, H.; Nakano, M., *J. Org. Chem.* **1991**, *56*, 2276. (b) Kiyooka, S.; Kaneko, Y.; Kume, K., *Tetrahedron Lett.* **1992**, *33*, 4927.

21. Parmee, E. R.; Hong, Y.; Tempkin, O.; Masamune, S., *Tetrahedron Lett.* **1992**, *33*, 1729.

22. Hattori, K.; Miyata, M.; Yamamoto, H., *J. Am. Chem. Soc.* **1993**, *115*, 1151.

23. (a) Paterson, I.; Smith, J. D., *J. Org. Chem.* **1992**, *57*, 3261. (b) Kozikowski, A. P.; Park, P., *J. Org. Chem.* **1990**, *55*, 4668.

24. Saigo, K.; Shimada, S.; Shibasaki, T.; Hasegawa, M., *Chem. Lett.* **1990**, 1093.

25. Hayashi, Y.; Wariishi, K.; Mukaiyama, T., *Chem. Lett.* **1987**, 1243.

26. Onaka, M.; Ohno, R.; Izumi, Y., *Tetrahedron Lett.* **1989**, *30*, 747.

27. Williams, R. M.; Sinclair, P. J.; Zhai, D.; Chen, D., *J. Am. Chem. Soc.* **1988**, *110*, 1547.

28. (a) Elliott, J. D.; Steele, J.; Johnson, W. S., *Tetrahedron Lett.* **1985**, *26*, 2535. (b) Johnson, W. S.; Kelson, A. B.; Elliott, J. D., *Tetrahedron Lett.* **1988**, *29*, 3757.

29. Holmes, C. P.; Bartlett, P. A., *J. Org. Chem.* **1989**, *54*, 98.

30. Kiyooka, S.; Shirouchi, M., *J. Org. Chem.* **1992**, *57*, 1.

31. (a) Kita, Y.; Segawa, J.; Haruta, J.; Fujii, T.; Tamura, Y., *Tetrahedron Lett.* **1980**, *21*, 3779. (b) Kita, Y.; Segawa, J.; Haruta, J.; Yasuda, H.; Tamura, Y., *J. Chem. Soc., Perkin Trans. 1* **1982**, 1099. (c) Berl, V.; Helmchen, G.; Preston, S., *Tetrahedron Lett.* **1994**, *35*, 233.

32. (a) Bunce, R. A.; Schlecht, M. F.; Dauben, W. G.; Heathcock, C. H., *Tetrahedron Lett.* **1983**, *24*, 4943. (b) Grieco, P. A.; Cooke, R. J.; Henry, K. J.; VanderRoest, J. M., *Tetrahedron Lett.* **1991**, *32*, 4665.

33. (a) Kawai, M.; Onaka, M.; Izumi, Y., *Bull. Chem. Soc. Jpn.* **1988**, *61*, 2157. (b) Onaka, M.; Mimura, T.; Ohno, R.; Izumi, Y., *Tetrahedron Lett.* **1989**, *30*, 6341. (c) Mukaiyama, T.; Hara, R., *Chem. Lett.* **1989**, 1171. (d) Minowa, N.; Mukaiyama, T., *Chem. Lett.* **1987**, 1719. (e) Hashimoto, Y.; Sugumi, H.; Okauchi, T.; Mukaiyama, T., *Chem. Lett.* **1987**, 1691.

34. Kobayashi, S.; Mukaiyama, T., *Chem. Lett.* **1986**, 1805; *Heterocycles* **1987**, *25*, 205.

35. (a) Danishefsky, S. J.; Cabal, M. P.; Chow, K., *J. Am. Chem. Soc.* **1989**, *111*, 3456. (b) Danishefsky, S. J.; Simoneau, B., *J. Am. Chem. Soc.* **1989**, *111*, 2599. (c) Danishefsky, S. J.; Simoneau, B., *Pure Appl. Chem.* **1988**, *60*, 1555. (d) Chow, K.; Danishefsky, S. J., *J. Org. Chem.* **1989**, *54*, 6016. (e) Audia, J. E.; Boisvert, L.; Patten, A. D.; Villalobos, A.; Danishefsky, S. J., *J. Org. Chem.* **1989**, *54*, 3738.

36. (a) Sato, T.; Wakahara, Y.; Otera, J.; Nozaki, H.; Fukuzumi, S., *J. Am. Chem. Soc.* **1991**, *113*, 4028. (b) Otera, J.; Fujita, Y.; Sato, T.; Nozaki, H.; Fukuzumi, S.; Fujita, M., *J. Org. Chem.* **1992**, *57*, 5054.

37. Heathcock, C. H.; Uehling, D. E., *J. Org. Chem.* **1986**, *51*, 279.

38. Klimko, P. G.; Singleton, D. A., *J. Org. Chem.* **1992**, *57*, 1733.

Cesare Gennari

Università di Milano, Milano, Italy

L

3-Lithio-1-triisopropylsilyl-1-propyne

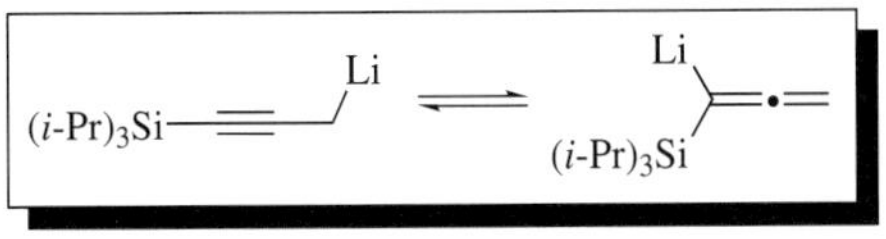

[82192-58-3] $C_{12}H_{23}LiSi$ (MW 202.38)

InChI = 1S/C12H23Si.Li/c1-8-9-13(10(2)3,11(4)5)12(6)7;/
h10-12H,1H2,2-7H3;

InChIKey = IBUTXKYKTSRDRC-UHFFFAOYSA-N

(functionalized nucleophilic C_3 building block; forms C–C bonds with many carbon electrophiles[1])

Alternate Name: 3-triisopropylsilyl-2-propynyllithium.
Physical Data: (precursor) bp 100–101 °C/5 mmHg; d 0.813 g cm^{-3}.
Solubility: good sol ether, THF.

Introduction. The title reagent (**1**) was conceived as an improved version of TMS–C≡C–CH$_2$Li.[2,3] The bulky triisopropylsilyl (TIPS) group provides efficient screening of the carbon atom to which it is attached, and (**1**) therefore usually forms propargylic rather than allenic products.[2,4] The TIPS group is much more inert in basic and nucleophilic reaction mixtures than the widely used TMS group. At the same time, TIPS (in contrast to tri-*t*-butylsilyl) is commercially available, easily introduced, and readily cleaved (F$^-$) from the product.

Reagent (**1**) is made by deprotonation of 1-TIPS-propyne[2] using *n*-butyllithium in ether, better without than with TMEDA added,[5] or *t*-butyllithium in ether/pentane, or preferably *n*-BuLi in THF (−20 °C, 15 min). 1-TIPS-propyne can be prepared from propyne by silylation of either its lithio derivative using TIPS triflate[2] or of the bromomagnesium derivative using TIPS chloride;[6] it is also commercially available.

Homopropargylic Alcohols. Reagent (**1**) smoothly reacts with aldehydes and ketones (eqs 1–3).[2,7]

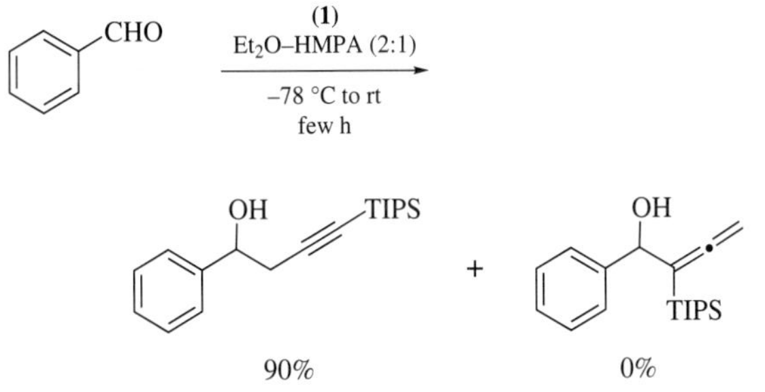

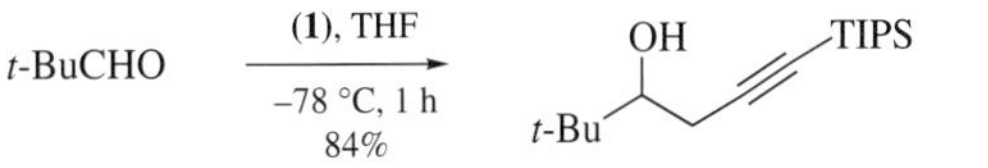

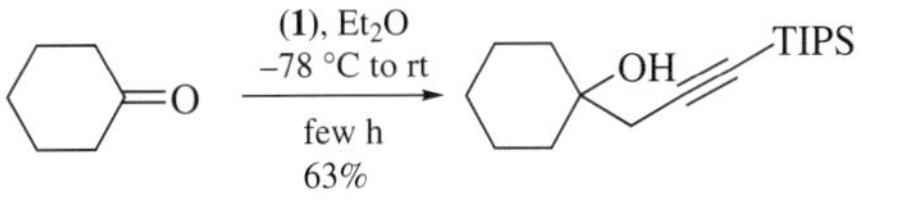

The reaction in eq 1, when performed in ether, gave rise to a 2:1 mixture of the same products. The TMS analog gave none of the corresponding adducts at all in ether/HMPA; in ether it produced both allenic and propargylic major products.

Conjugate Addition. Reagent (**1**) in THF adds in 1,2-fashion to α,β-unsaturated ketones, whereas in THF/HMPA, 1,4-addition is observed as result of kinetic control (eq 4).[2] The hindered α,β-unsaturated ketone isophorone gave smooth 1,2-addition in THF (89%), while in THF/HMPA, 1,2- and 1,4-addition occurred (1:1) in low conversion. 1,2-Addition of (**1**) to acrolein afforded a building block for bongkrekic acid.[8a] 1,4-Addition of (**1**) to η^4-(dienyl)Fe(CO)$_3$ substituted alkylidene malonates is highly stereoselective.[8b]

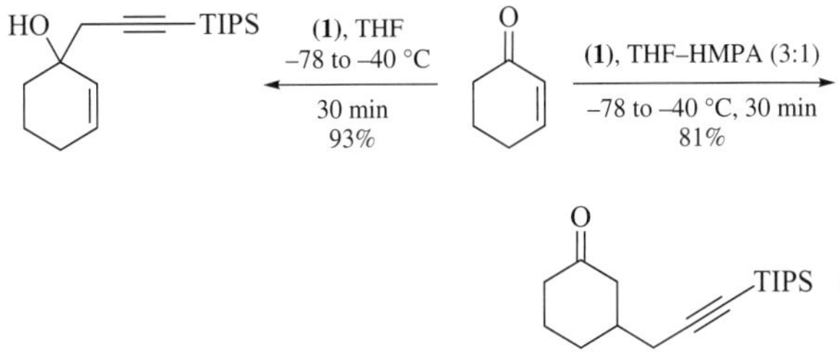

Reactions of similar organoalkynyl/allenyl silanes containing various metals with carbonyl compounds to give propargylic products have been reported.[9–11] Conjugate addition of allenyl stannanes to α,β-unsaturated carbonyl compounds to give propargylic products can be effected in the presence of titanium(IV) chloride.[12]

Chain Elongation of Activated Halides or Phospates. Reagent (**1**) displaces halide ion from benzyl and allyl halides (eqs 5 and 6).[2] The TMS analog reacted less selectively and gave lower yields.[13]

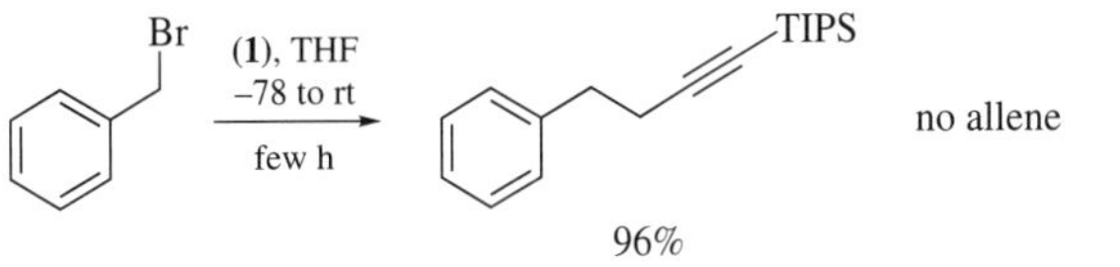

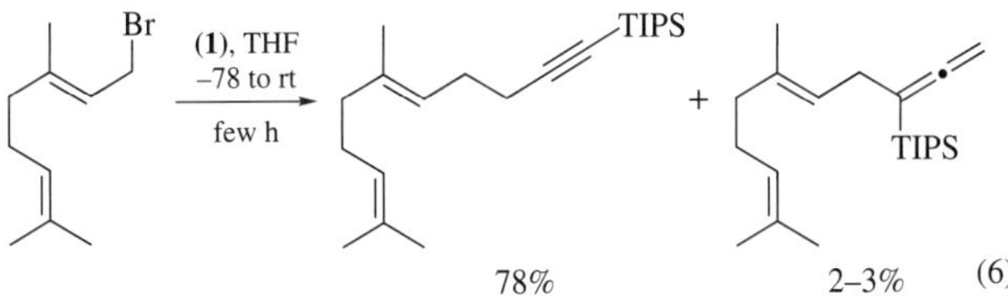

This type of coupling was routinely used by Marshall (allyl chlorides or phosphates) in the synthesis of cembranoids, employing a cuprate made from the bromomagnesium analog TIPS–C≡C–CH$_2$MgBr (**2**), which is prepared from TIPS–C≡C–CH$_2$Br.[14] As a rule, with these allylic electrophiles, S$_N$2′

reaction is not observed. An OH function does not interfere if protected as its bromomagnesium salt,[14g] silyl ether,[14c,d] or even acetate.[14e,g]

Epoxide Opening and Cyclopropane Formation. Reagent (1) cleanly opens epoxides (eq 7; S_N2).[2]

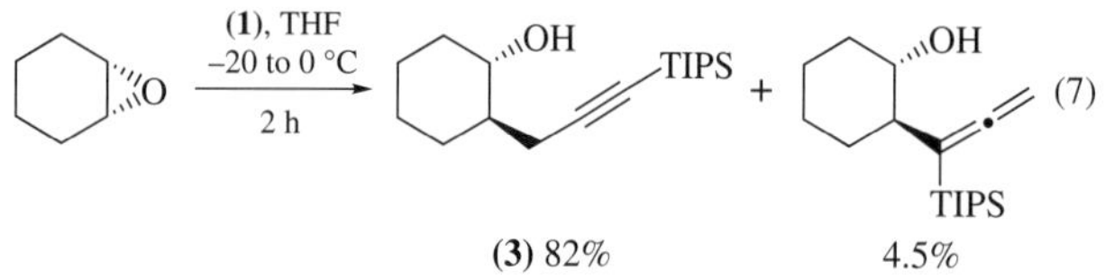

$$(7)$$

(3) 82%   4.5%

In vinyl epoxides, both exclusive S_N2 and exclusive S_N2' reactions were observed for similar reagents, depending on the substrate (eqs 8 and 9).[5,14a,b] In eq 9, the TMS analog gives a mixture of propargylic and allenic products.

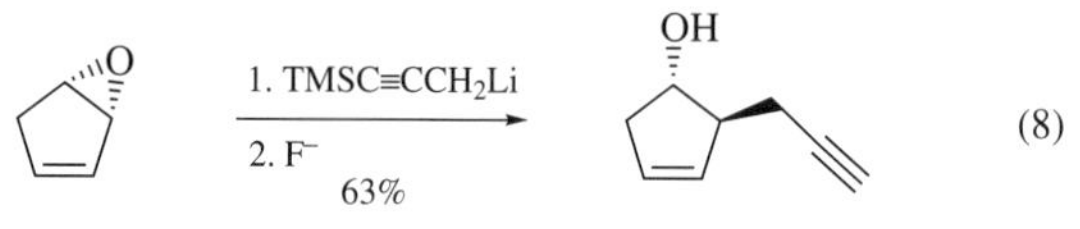

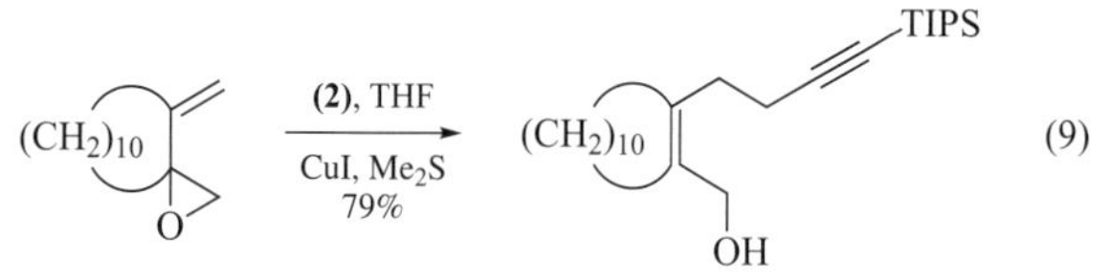

$$(8)$$

$$(9)$$

The products of epoxide opening by (1) can be transformed into cyclopropanes (eq 10). The TMS analog gave 75% in the cyclization step.

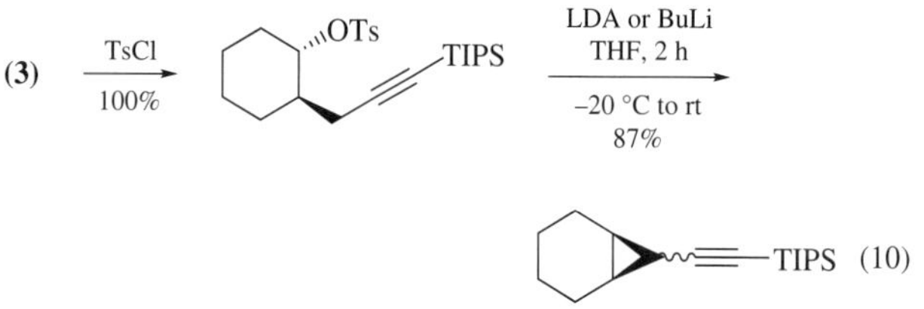

$$(10)$$

This method was triply used in the preparation of a carbocyclic *cis*-tris-σ-homobenzene (eq 11).[15]

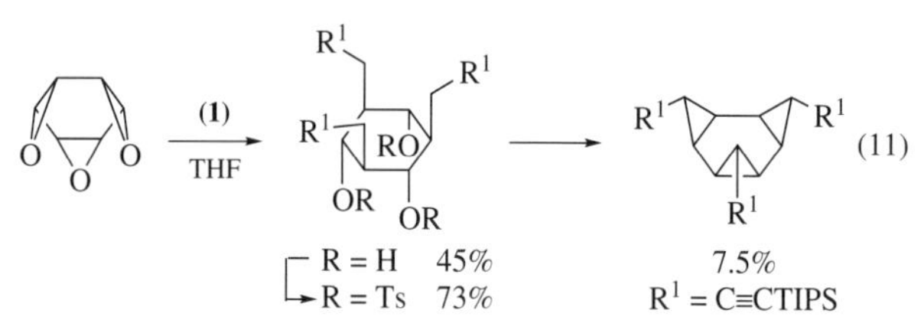

$$(11)$$

R = H   45%   7.5%
R = Ts   73% R^1 = C≡CTIPS

The triple epoxide opening gave higher yields (up to 67%) when organocuprates were used.[16,17] The TMS analog of this tris adduct could not be obtained, due to uncontrolled desilylation reactions.

Transformations of the C≡C–TIPS Group. The C≡C–TIPS group is cleaved by the action of various fluoride reagents, but not by the $AgNO_3/KCN$ procedure employed for cleavage of C≡C–TMS. Corey and Rücker describe other useful transformations.[2] The deprotected propargyl group may be further degraded to an acetic acid moiety by decarboxylative oxidation using

$RuO_2/NaIO_4$ in $CCl_4/MeCN/H_2O$, as in the preparation of *cis*-tris-σ-homobenzenes (eq 12).[17]

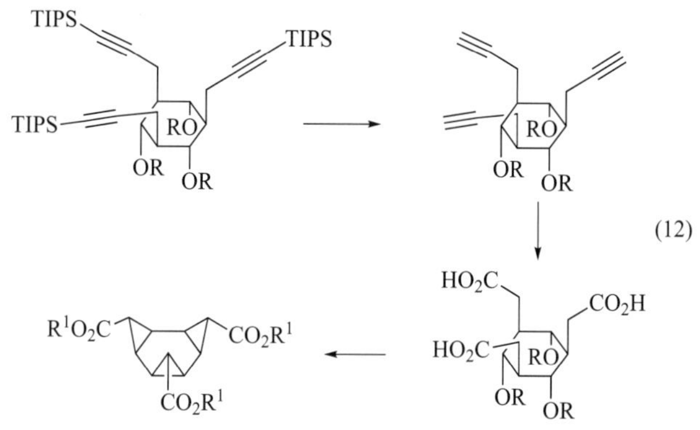

$$(12)$$

Oxidation of the products by ruthenium(VIII) oxide with the TIPS group still in place gives a mixture of carboxylic acid and an α-keto acyl silane (silyl α-diketone; eq 13),[7] an otherwise rare class of compounds;[18] compare the oxidation of disubstituted alkynes to 1,2-diketones by RuO_4[19] and the osmium tetroxide–*t*-butyl hydroperoxide oxidation of TMS-alkynes.[20]

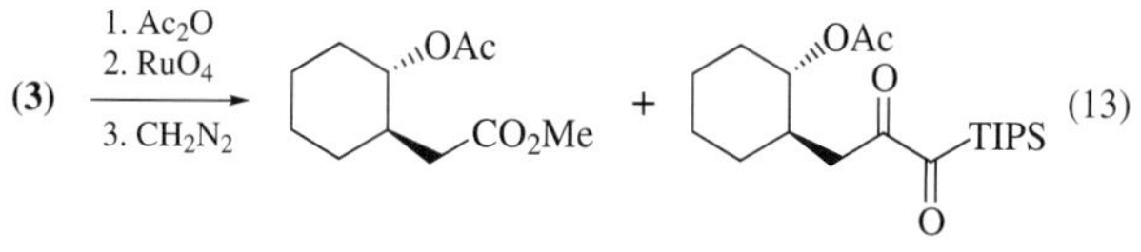

$$(13)$$

TMS-alkynes are oxidized at the terminal carbon to carboxylic acids by hydroboration/oxidation (dicyclohexylborane/NaOH, H_2O_2). This does not work with TIPS-alkynes.[14b] Instead, TIPS-alkynes are cleanly monohydroborated at the internal carbon by 9-borabicyclo[3.3.1]nonane dimer to give (*Z*)-β-borylvinyl-silanes.[6] These can be oxidized in high yields to α-silyl ketones, or cross coupled with a bromide R^1Br (R^1 = aryl, benzyl, dimethyl-vinyl) in the presence of NaOH and tetrakis(triphenylphos-phine)palladium(0) to give β,β-disubstituted vinylsilanes (Suzuki reaction; eq 14).[21a] The same nucleophilic substituted vinylsilane can be added to an aromatic aldehyde to provide access to (*E*)-3-silyl allyl alcohols.[21b]

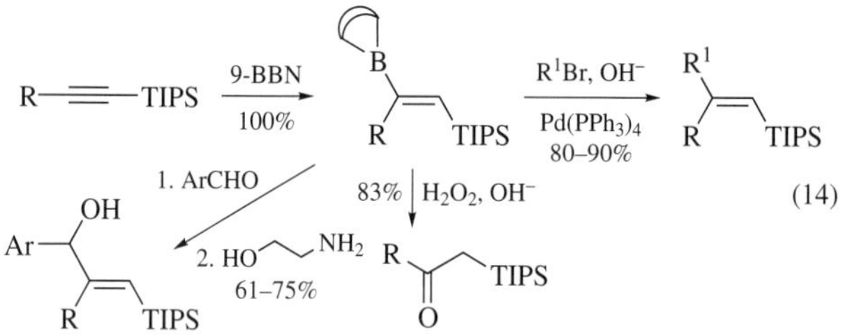

$$(14)$$

Propargyl Ketones. The cyanocuprate derived from (1) reacts with esters to form propargyl ketones, even in the presence of epoxide functionality (eq 15).[16]

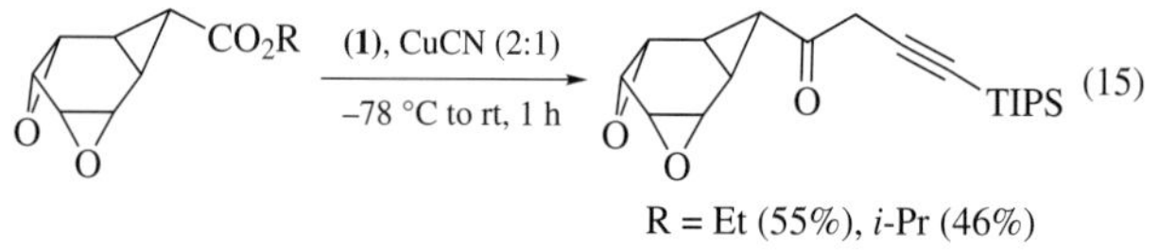

$$(15)$$

R = Et (55%), *i*-Pr (46%)

Bis(silyl)propynes. Reaction of (**1**) with silylating agents affords 1,3-bis(silyl)propynes, valuable Peterson reagents (eq 16).[2]

Isomeric Species. The isomeric substituted lithium acetylide LiC≡C–CH$_2$TIPS (**4**), from 3-TIPS-propyne,[22] was silylated to 1-TMS-3-TIPS-propyne.[23] Compound (**4**) isomerizes to (**1**) in THF/HMPA solution at rt, as shown by isolation of pure 1-TIPS-propyne after quenching a solution of (**4**) with H$_2$O.[7]

1. *Fieser & Fieser* **1984**, *11*, 566.
2. Corey, E. J.; Rücker, Ch., *Tetrahedron Lett.* **1982**, *23*, 719.
3. Stowell, J. C., *Chem. Rev.* **1984**, *84*, 409.
4. Furuta, K.; Ishiguro, M.; Haruta, R.; Ikeda, N.; Yamamoto, H., *Bull. Chem. Soc. Jpn.* **1984**, *57*, 2768.
5. Stork, G.; Kowalski, C.; Garcia, G., *J. Am. Chem. Soc.* **1975**, *97*, 3258.
6. Soderquist, J. A.; Colberg, J. C.; Del Valle, L., *J. Am. Chem. Soc.* **1989**, *111*, 4873.
7. Rücker, Ch. unpublished results.
8. (a) Corey, E. J.; Tramontano, A., *J. Am. Chem. Soc.* **1984**, *106*, 462. (b) Wada, C. K.; Roush, W. R., *Tetrahedron Lett.* **1994**, *35*, 7351.
9. (a) Mesnard, D.; Miginiac, L., *J. Organomet. Chem.* **1990**, *397*, 127. (b) Suzuki, M.; Morita, Y.; Noyori, R., *J. Org. Chem.* **1990**, *55*, 441.
10. Zhang, L.-J.; Mo, X.-S.; Huang, J.-L.; Huang, Y.-Z., *Tetrahedron Lett.* **1993**, *34*, 1621.
11. (a) Danheiser, R. L.; Carini, D. J.; Kwasigroch, C. A., *J. Org. Chem.* **1986**, *51*, 3870. (b) Brown, H. C.; Khire, U. R.; Racherla, U. S., *Tetrahedron Lett.* **1993**, *34*, 15.
12. Haruta, J.; Nishi, K.; Matsuda, S.; Akai, S.; Tamura, Y.; Kita, Y., *J. Org. Chem.* **1990**, *55*, 4853.
13. (a) Corey, E. J.; Kirst, H. A., *Tetrahedron Lett.* **1968**, 5041. (b) Kirst, H. A. Ph. D. Thesis, Harvard University, 1971.
14. (a) Marshall, J. A.; Peterson, J. C.; Lebioda, L., *J. Am. Chem. Soc.* **1983**, *105*, 6515. (b) Marshall, J. A.; Peterson, J. C.; Lebioda, L., *J. Am. Chem. Soc.* **1984**, *106*, 6006. (c) Marshall, J. A.; Jenson, T. M.; DeHoff, B. S., *J. Org. Chem.* **1987**, *52*, 3860. (d) Marshall, J. A.; Lebreton, J.; DeHoff, B. S.; Jenson, T. M., *J. Org. Chem.* **1987**, *52*, 3883. (e) Marshall, J. A.; DeHoff, B. S.; Crooks, S. L., *Tetrahedron Lett.* **1987**, *28*, 527. (f) Marshall, J. A.; Lebreton, J.; DeHoff, B. S.; Jenson, T. M., *Tetrahedron Lett.* **1987**, *28*, 723. (g) Marshall, J. A.; Crooks, S. L.; DeHoff, B. S., *J. Org. Chem.* **1988**, *53*, 1616. (h) Marshall, J. A.; Gung, W. Y., *Tetrahedron Lett.* **1989**, *30*, 309. (i) Marshall, J. A.; Andersen, M. W., *J. Org. Chem.* **1992**, *57*, 2766.
15. Rücker, Ch.; Prinzbach, H., *Tetrahedron Lett.* **1983**, *24*, 4099.
16. Braschwitz, W.-D. Ph. D. Thesis, Universität Freiburg, 1990.
17. Braschwitz, W.-D.; Otten, T.; Rücker, Ch.; Fritz, H.; Prinzbach, H., *Angew. Chem.* **1989**, *101*, 1383 (*Angew. Chem., Int. Ed. Engl.* **1989**, *28*, 1348).
18. Reich, H. J.; Kelly, M. J.; Olson, R. E.; Holtan, R. C., *Tetrahedron* **1983**, *39*, 949.
19. Gopal, H.; Gordon, A. J., *Tetrahedron Lett.* **1971**, 2941.
20. Page, P. C. B.; Rosenthal, S., *Tetrahedron Lett.* **1986**, *27*, 1947.
21. (a) Soderquist, J. A.; Colberg, J. C., *Synlett* **1989**, 25. (b) Soderquist, J. A.; Vaquer, J., *Tetrahedron Lett.* **1990**, *31*, 4545.
22. Danheiser, R. L.; Dixon, B. R.; Gleason, R. W., *J. Org. Chem.* **1992**, *57*, 6094.
23. Schulz, D. Diplom Thesis, Universität Freiburg, 1985.

Christoph Rücker
Universität Freiburg, Freiburg, Germany

Lithium *N*-Benzyltrimethylsilylamide

LiN(SiMe$_3$)CH$_2$Ph

[113709-50-5]   C$_{10}$H$_{16}$LiNSi   (MW 185.31)

InChI = 1S/C10H16NSi.Li/c1-12(2,3)11-9-10-7-5-4-6-8-10;/h4-8H,9H2,1-3H3;/q-1;+1

InChIKey = XMNOZVJUCZKWTB-UHFFFAOYSA-N

(less-basic lithium amide capable of regioselective conjugate addition to α,β-unsaturated esters;[1] amide cuprates have increased selectivity and reactivity toward enoates[2])

Alternate Name: LSA.

Solubility: sol THF.

Preparative Method: to a solution of freshly distilled *N*-benzyltrimethylsilylamine (0.2 mL, 1.0 mmol)[3] in THF at −78 °C is slowly added a solution of *n*-butyllithium (0.61 mL, 1.64 M in hexane, 1.0 mmol) under Ar; a pale yellow solution of lithium *N*-benzyltrimethylsilylamide in THF is obtained after stirring for several min at −78 °C.[1]

Handling, Storage, and Precautions: is moisture and air-sensitive, and should thus be handled under inert atmosphere; prepare just prior to use; use in a fume hood.

Conjugate Additions. Lithium dialkylamides (e.g. lithium diisopropylamide, LDA) with weak nucleophilicity are commonly used as strong bases for deprotonation. However, lithium *N*-benzyltrimethylsilylamide acts as a nucleophile with a basicity weaker than LDA. For example, the reaction of LSA with (*E*)-crotonates affords the conjugate adducts, β-amino esters, in high yields without formation of 1,2-adducts (amides) or products arising from deprotonation at the γ-position of the enoates (eq 1).[1] On the other hand, the reaction of LDA with (*E*)-methyl crotonate gives a mixture of the conjugate adduct and deprotonation product (eq 2).[1,4] The reaction of lithium benzylamide gives the 1,2-adduct as a major product. The (*E*) geometry of α,β-unsaturated esters is essential for conjugate addition, since the treatment of (*Z*)-methyl 2-decenoate with LSA gives the deconjugated ester, methyl 3-decenoate, in high yield.[5] Since the N–Si bond is cleaved easily under weak acidic conditions, purification of the conjugate adduct (see eq 1) by silica gel column chromatography provides desilylated benzylamino esters.

$$R^1 \diagdown CO_2R^2 \xrightarrow{\text{LiN}(i\text{-Pr})_2}$$

$$\underset{\text{CO}_2\text{R}^2}{R^1 \diagdown} \quad + \quad R^1 \diagdown CO_2R^2 \quad (2)$$

The conjugate addition of LSA to methyl crotonate followed by trapping of the resulting enolate with chlorotrimethylsilane affords the (Z)-ketene silyl acetal nearly exclusively (Z:E = 99:1) (eq 3). On the other hand, if the conjugate addition reaction is quenched with methanol, deprotonation of the resulting β-amino ester with LDA followed by treatment with Me_3SiCl gives the (E)-ketene silyl acetal with very high stereoselectivity (Z:E = 2:98) (eq 4). Accordingly, the stereodivergent synthesis of (Z)- or (E)-enolates of β-amino esters is achieved via the direct conjugate addition of LSA or the two-step procedure, respectively.[6]

$$\text{LiN(TMS)CH}_2\text{Ph} + \diagup CO_2Me \xrightarrow{\text{TMSCl}} \quad (3)$$

$$\text{LiN(TMS)CH}_2\text{Ph} + \diagup CO_2Me \xrightarrow[\substack{\text{2. LDA \\ 3. TMSCl}}]{\text{1. MeOH}} \quad (4)$$

The conjugate addition of LSA to enoates followed by trapping with alkyl halides gives α-alkyl-β-amino esters, which are converted to α-alkylated α,β-unsaturated esters upon deamination with silica gel.[1] The reaction of the enolates with aldehydes produces β-amino-β'-hydroxy esters, which are converted to β-lactams having a 1-hydroxyalkyl group at the C-3 position (eq 5). Thus the β-lactam skeleton is prepared stereoselectively and expeditiously via a three-component coupling reaction.[6] The amide cuprate reagents, prepared from 2 equiv LSA and copper(I) cyanide, react effectively with α,β,γ,δ-unsaturated esters to give the corresponding β-amino esters. Based upon this procedure, asymmetric syntheses of β-lactams have been accomplished (eq 6).[2]

$$\text{LiN(TMS)CH}_2\text{Ph} + R^1 \diagdown CO_2Me \xrightarrow{R^2CHO}$$

$$\xrightarrow[\substack{\text{2. PPh}_3, \text{(pyS)}_2}]{\text{1. KOH}} \quad (5)$$

$$\xrightarrow[\substack{\text{2. MeCHO \\ 3. TBDMSCl}}]{\text{1. [Bn(TMS)N]}_2\text{Cu(CN)Li}_2}$$

$$\longrightarrow \quad (6)$$

$X^*_N = (-)$-bornanesultam

The reaction of LSA with ω-halo-α,β-unsaturated esters produces cyclization products.[5] In the case of α,β,χ,ψ-unsaturated

dioic acid esters, tandem conjugate additions occur stereoselectively; this cyclization strategy has been applied to the synthesis of racemic cyclopentane monoterpenes.[5]

1. (a) Uyehara, T.; Asao, N.; Yamamoto, Y., *J. Chem. Soc., Chem. Commun.* **1987**, 1410. (b) Asao, N.; Uyehara, T.; Yamamoto, Y., *Tetrahedron* **1988**, *44*, 4173.
2. Yamamoto, Y.; Asao, N.; Uyehara, T., *J. Am. Chem. Soc.* **1992**, *114*, 5427.
3. Diekman, J.; Thomson, J. B.; Djerassi, C., *J. Org. Chem.* **1967**, *32*, 3904.
4. (a) Rathke, M. W.; Sullivan, D., *Tetrahedron Lett.* **1972**, 4249. (b) Herrmann, J. L.; Kieczykowski, G. R.; Schlessinger, R. H., *Tetrahedron Lett.* **1973**, 2433.
5. (a) Uyehara, T.; Shida, N.; Yamamoto, Y., *J. Chem. Soc., Chem. Commun.* **1989**, 113. (b) Uyehara, T.; Shida, N.; Yamamoto, Y., *J. Org. Chem.* **1992**, *57*, 3139.
6. (a) Uyehara, T.; Asao, N.; Yamamoto, Y., *J. Chem. Soc., Chem. Commun.* **1989**, 753. (b) Asao, N.; Uyehara, T.; Yamamoto, Y., *Tetrahedron* **1990**, *46*, 4563.

Naoki Asao & Yoshinori Yamamoto
Tohoku University, Sendai, Japan

Lithium Bis[dimethyl(phenyl)silyl]cuprate[1]

$(\text{PhMe}_2\text{Si})_2\text{CuLi}$

[75583-57-2] $C_{17}H_{22}CuLiSi_2$ (MW 341.01)

InChI = 1S/2C8H11Si.Cu.Li/c2*1-9(2)8-6-4-3-5-7-8;;/h2*3-7H,1-2H3;;

InChIKey = ZPEQCKOXMDRTHF-UHFFFAOYSA-N

(dimethyl(phenyl)silyl nucleophile for making Si–C bonds by reaction with α,β-unsaturated carbonyl compounds, alkynes, allenes, and allylic acetates)

Physical Data: typically a 0.6M, dark, reddish-brown solution in THF.

Analysis of Reagent Purity: the silyllithium solution can be double-titrated for active reagent using allyl bromide, but the cuprate is usually used without further checks; NMR (of the cuprate made with CuCN in THF-d_8): ^1H δ 0.09; ^{13}C δ 5.1; ^{29}Si δ −24.4; ^7Li δ −3.33.[2]

Preparative Method: the silyl cuprate is prepared[1] from the corresponding silyllithium reagent (dimethylphenylsilyllithium); commercially available chlorodimethylphenylsilane is stirred with lithium shot, wire, or powder under Ar or N_2 in THF at 0 °C for 4–12 h; the silyllithium solution may also be prepared, free of halide ion, by cleaving tetramethyldiphenyldisilane with lithium and ultrasound irradiation;[2] the silyllithium solution (2 equiv), after assay, is transferred by syringe on to anhyd copper(I) iodide, copper(I) bromide, or copper(I) cyanide (1 equiv), kept under argon or nitrogen at 0 °C, stirred at this temperature for 20 min, and then used immediately.

Handling, Storage, and Precautions: must be kept free of O_2 and H_2O; while somewhat more stable thermally than alkyl cuprates, surviving for a few hours at $0\,^\circ$C, it is best used immediately after its preparation; the copper salts, and especially copper(I) cyanide, are toxic; the solutions should therefore be handled in a fume hood wearing impermeable gloves, and the aqueous washings disposed of appropriately, immediately after use.

Introduction. Because dimethyl(phenyl)silyllithium is much easier to prepare than trimethylsilyllithium, the most commonly used silyl cuprate reagent is derived from this silyl group. The reagent can be prepared using CuI, CuBr·SMe$_2$, or CuCN. The three reagents appear to be very similar in their reactivity, except for the higher regioselectivity of the cyanide-derived reagent with terminal alkynes. The dimethyl(phenyl)silyl group in products like allyl- and vinylsilanes appears to impart very similar reactivity to that imparted by the trimethylsilyl group, and it has an advantage over the trimethylsilyl group in that the presence of the phenyl group allows the dimethyl(phenyl)silyl group to be converted into a hydroxyl group with retention of configuration at carbon (eq 1). This transformation requires first a reaction with an electrophile, such as a proton,[3] bromine, or the mercury(II) cation,[4] to remove the phenyl ring and place a nucleofugal group X on the silicon atom. This step is followed by treatment either with peracid or with hydrogen peroxide and a base. The two steps may be combined in one pot;[4] bromine itself does not have to be used, since the peracetic acid oxidizes bromide ion to bromine in situ.

$$ \text{(1)} $$

This capacity of the dimethyl(phenyl)silyl group cannot be drawn upon, however, when there is a C=C double bond in the molecule; no matter which electrophile is used, it attacks the double bond more rapidly than it removes the phenyl ring from the silyl group. This limitation has been overcome using the corresponding diethylamino(diphenyl)silyl-[5] and 2-methylbut-2-enyl(diphenyl)silylcuprate[6] reagents.

A mixed cuprate, [dimethyl(phenyl)silyl]methyl(or-butyl)-cuprate,[7] containing one silyl and one alkyl group, has some advantages. Only the silyl group is transferred to the substrate, and hence only one silyl group is needed, and the byproduct of the silyl-cupration step, methane or butane, is volatile. Yields in silyl-cupration reactions carried out with only 1:1 stoichiometry are apt not to be quite so good, however. Other silyl copper reagents and cuprates with more specific or limited applications are (t-butyldimethylsilyl)butylcuprate,[8] triphenylsilylcopper,[9] the bis(t-butyldiphenylsilyl)cuprate,[10] and the bis[tris(trimethylsilyl)silyl]cuprate.[11]

Reaction with α,β-Unsaturated Carbonyl Compounds. Although silyllithium reagents add kinetically at the β-position of α,β-unsaturated ketones,[12] the reactions are better with the cuprate when the enone is hindered.[13] The cuprate, unlike the lithium reagent, also reacts with α,β-unsaturated aldehydes, esters (eq 2), amides, and nitriles[13] and with vinyl sulfoxides.[14] With esters, the intermediate enolates, which have the (E) geometry (1), may be used directly in highly stereoselective reactions

with alkyl halides and aldehydes.[15,16] The β-hydroxy esters, such as (2), can be used in the synthesis of allylsilanes.[17]

$$ \text{(2)} $$

α,β-Unsaturated carbonyl compounds attached to a homochiral auxiliary, such as Koga's lactam or Oppolzer's sultam, give products with a stereogenic center carrying a silyl group having high levels of enantiomeric purity.[18,19]

Reaction with Alkynes. The silyl cuprate reacts with alkynes by *syn* stereospecific metallo-metallation (eq 3). Provided that the cuprate is derived from copper cyanide, the regioselectivity with terminal alkynes is highly in favor of the isomer with the silyl group on the terminus. The intermediate vinyl cuprate (3) reacts with many substrates, familiar in carbon-based cuprate chemistry, to give overall *syn* addition of a silyl group and an electrophile to the alkyne.[20] A curious feature of this reaction is that the intermediate (3), although uncharacterized, has the stoichiometry of a mixed silicon–carbon cuprate, and yet it transfers the carbon-based group to most substrates, in contrast to the behavior of mixed silyl alkyl cuprates.

$$ \text{(3)} $$

E = H (94%), I (88%), CO$_2$H (69%), Ac (72%), Me (71%)

OH (64%),   (54%)

Disubstituted alkynes also react, and, if the two substituents are well differentiated sterically, as with a methyl group on one side and a branched chain on the other, the regioselectivity is highly in favor of the silyl group appearing at the less hindered end.

Reaction with Allenes. Allenes react with the silyl cuprate at low temperature. The regiochemistry with allene itself places the silyl group on the central carbon atom and the added electrophile at the terminus (eq 4).[21] One surprising exception to this rule is with iodine as the electrophile, when the product (5) has the opposite regiochemistry even from that of the reaction with chlorine. Since the iodide (5) can be converted into a lithium reagent [and a cuprate that is not identical with 4], it is possible to achieve overall either regiochemistry in additions to allene. Monosubstituted allenes give mixtures of regioisomers,[21,22] and disubstituted and trisubstituted allenes give largely allylsilanes whatever electrophile is used. The metallo-metallation step is stereospecifically *syn*.[23]

(4)

(5) 90%

Reaction with Allylic Acetates. Silyl cuprates react with allylic acetates to give allylsilanes directly (eqs 5–7). Allylic acetates that are secondary at both ends are apt to give both regioisomers. Some control of the regiochemistry is possible, however, by using a *cis* double bond; this encourages reaction with allylic shift, especially when the silyl group is delivered to the less-hindered end of the allylic system (eq 5).[24] An alternative protocol, assembling a mixed silyl cuprate on a carbamate group, is even better in controlling the regioselectivity, usually giving complete allylic shift (eq 6).[24,25] Tertiary acetates, on the other hand, are very well behaved regiochemically, giving only the product with the silyl group at the less-substituted end of the allylic system (eq 7).[25,26] Allylsilanes have many uses as carbon nucleophiles in organic synthesis.[27]

The acetate reaction (eq 5) and the carbamate alternative (eq 6) are complementary in their stereochemistry, the former taking place stereospecifically *anti* and the latter stereospecifically *syn*.[24,25]

Other Reactions. Other substrates that have been found to react with silyl-copper reagents and with silyl cuprates are allyl chlorides,[28] an alkyl bromide,[29] epoxides,[8,22] acid chlorides,[9,30] propargyl acetates[31] and sulfides,[32] a vinyl iodide,[9] an aminomethyl acetate,[33] ethyl tetrolate,[34] and some strained allylic ethers.[35]

1. Fleming, I. In *Organocopper Reagents*; Taylor, R. J. K., Ed.; OUP: Oxford, 1995; p. 257.

2. Sharma, S.; Oehlschlager, A. C., *Tetrahedron* **1989**, *45*, 557; *J. Org. Chem.* **1991**, *56*, 770.

3. Fleming, I.; Henning, R.; Plaut, H., *J. Chem. Soc., Chem. Commun.* **1984**, 29.

4. Fleming, I.; Sanderson, P. E. J., *Tetrahedron Lett.* **1987**, *28*, 4229.

5. Tamao, K.; Kawachi, A.; Ito, Y., *J. Am. Chem. Soc.* **1992**, *114*, 3989.

6. Fleming, I.; Winter, S. B. D., *Tetrahedron Lett.* **1993**, *34*, 7287.

7. Fleming, I.; Newton, T. W., *J. Chem. Soc., Perkin Trans. 1* **1984**, 1805.

8. Lipshutz, B. H.; Reuter, D. C.; Ellsworth, E. L., *J. Org. Chem.* **1989**, *54*, 4975.

9. Duffaut, N.; Dunoguès, J.; Biran, C.; Calas, R.; Gerval, J., *J. Organomet. Chem.* **1978**, *161*, C23.

10. Cuadrado, P.; Gonzalez, A. M.; Gonzalez, B.; Pulido, F. J., *Synth. Commun.* **1989**, *19*, 275.

11. Chen, H.-M.; Oliver, J. P., *J. Organomet. Chem.* **1986**, *316*, 255.

12. (a) Still, W. C., *J. Org. Chem.* **1976**, *41*, 3063. (b) Still, W. C.; Mitra, A., *Tetrahedron Lett.* **1978**, 2659.

13. Ager, D. J.; Fleming, I.; Patel, S. K., *J. Chem. Soc., Perkin Trans. 1* **1981**, 2520.

14. Takaki, K.; Maeda, T.; Ishikawa, M., *J. Org. Chem.* **1989**, *54*, 58.

15. Crump, R. A.; Fleming, I.; Hill, J. H. M.; Parker, D.; Reddy, N. L.; Waterson, D., *J. Chem. Soc., Perkin Trans. 1* **1992**, 3277.

16. Fleming, I.; Kilburn, J. D., *J. Chem. Soc., Perkin Trans. 1* **1992**, 3295.

17. Fleming, I.; Gil, S.; Sarkar, A. K.; Schmidlin, T., *J. Chem. Soc., Perkin Trans. 1* **1992**, 3351.

18. Fleming, I.; Kindon, N. D., *J. Chem. Soc., Chem. Commun.* **1987**, 1177.

19. Oppolzer, W.; Mills, R. J.; Pachinger, W.; Stevenson, T., *Helv. Chim. Acta* **1986**, *69*, 1542.

20. Fleming, I.; Newton, T. W.; Roessler, F., *J. Chem. Soc., Perkin Trans. 1* **1981**, 2527.

21. (a) Fleming, I.; Rowley, M.; Cuadrado, P.; González-Nogal, A. M.; Pulido, F. J., *Tetrahedron* **1989**, *45*, 413. (b) Morizawa, Y.; Oda, H.; Oshima, K.; Nozaki, H., *Tetrahedron Lett.* **1984**, *25*, 1163.

22. Singh, S. M.; Oehlschlager, A. C., *Can. J. Chem.* **1991**, *69*, 1872.

23. Fleming, I.; Landais, Y.; Raithby, P. R., *J. Chem. Soc., Perkin Trans. 1* **1991**, 715.

24. Fleming, I.; Higgins, D.; Lawrence, N. J.; Thomas, A. P., *J. Chem. Soc., Perkin Trans. 1* **1992**, 3331.

25. Fleming, I.; Terrett, N. K., *J. Organomet. Chem.* **1984**, *264*, 99.

26. Fleming, I.; Marchi, D., *Synthesis* **1981**, 560.

27. Fleming, I.; Dunoguès, J.; Smithers, R., *Org. React.* **1989**, *37*, 575.

28. Smith, J. G.; Drozda, S. E.; Petraglia, S. P.; Quinn, N. R.; Rice, E. M.; Taylor, B. S.; Viswanathan, M., *J. Org. Chem.* **1984**, *49*, 4112.

29. (a) Singer, R. D.; Oehlschlager, A. C., *J. Org. Chem.* **1991**, *56*, 3510. (b) Fürstner, A.; Weidmann, H., *J. Organomet. Chem.* **1988**, *354*, 15.

30. Brook, A. G.; Harris, J. W.; Lennon, J.; Sheikh, M. E., *J. Am. Chem. Soc.* **1979**, *101*, 83.

31. Fleming, I.; Takaki, K.; Thomas, A., *J. Chem. Soc., Perkin Trans. 1* **1987**, 2269.

32. Casarini, A.; Jousseaume, B.; Lazzari, D.; Porciatti, E.; Reginato, G.; Ricci, A.; Seconi, G., *Synlett* **1992**, 981.

33. Nativi, C.; Ricci, A.; Taddei, M., *Tetrahedron Lett.* **1990**, *31*, 2637.

34. Audia, J. E.; Marshall, J. A., *Synth. Commun.* **1983**, *13*, 531.

35. (a) Lautens, M.; Belter, R. K.; Lough, A. J., *J. Org. Chem.* **1992**, *57*, 422.
(b) Lautens, M.; Ma, S.; Belter, R. K.; Chiu, P.; Leschziner, A., *J. Org. Chem.* **1992**, *57*, 4065.

Ian Fleming
Cambridge University, Cambridge, UK

Lithium Cyano(dimethylphenylsilyl)cuprate

[75583-56-1] $C_9H_{11}CuLiNSi$ (MW 231.79)
InChI = 1S/C8H11Si.CN.Cu.Li/c1-9(2)8-6-4-3-5-7-8;1-2;;/
h3-7H,1-2H3;;;/q;;-1;+1
InChIKey = ZDBAFMCAWBYEBB-UHFFFAOYSA-N

(organometallic reagent for the silylcupration of enones, alkynes, and strained alkenes)

Solubility: sol ether, THF from −78 to 25 °C.
Preparative Methods: the title reagent (**1**), as well as lithium cyano (trimethylsilyl)cuprate *[81802-36-0]*, can be prepared from the corresponding silyllithium species and 1 equiv copper(I) cyanide in ether or THF at −78 to −5 °C;[1,2] the precursor silyllithium (**2**) is generated by the reaction of chlorodimethylphenylsilane with lithium metal in THF or ether (eq 1). More detail for the preparation of this and related reagents may be found in Fleming, I. In *Organocopper Reagents: A Practical Approach*; Taylor, R. J. K., Ed.; OUP: Oxford, 1994; Chapter 12, pp 257–292.

$$Me_2PhSiCl \xrightarrow[THF, -5\,°C]{Li^0} \underset{(\mathbf{2})}{Me_2PhSiLi} \xrightarrow{CuCN} \underset{(\mathbf{1})}{Me_2PhSiCuCNLi} \quad (1)$$

Handling, Storage, and Precautions: thermally stable up to ambient temperatures; oxygen- and moisture-sensitive; use in a fume hood.

Original Commentary

Joseph P. Marino & David P. Holub
University of Michigan, Ann Arbor, MI, USA

Structure and Properties. Low-temperature ^{29}Si, ^{13}C, and ^{1}H NMR spectroscopic techniques have been used to probe the structure of (**1**),[2] which is referred to as a lower-order cuprate. When 2 equiv of the silyllithium (**2**) are combined with 1 equiv copper(I) cyanide, a new reagent, referred to as a higher-order cuprate, is generated.[3] Evidence for the different cuprate structures comes from separate signals in the ^{29}Si and ^{13}C spectra.[2]

Conjugate Additions to Enones. Reagent (**1**) effectively adds the dimethylphenylsilyl group to α,β-unsaturated ketones, esters,

and alkynic analogs. In a typical case, (**1**) adds to cyclohexenone in ether at −78 °C in 58% yield (eq 2).[2]

$$\quad (2)$$

The analogous trimethylsilylcuprate (**3**) adds to ethyl-2-butynoate at 0 °C to give a mixture of stereoisomers, while the higher-order species reacts at −78 °C to yield a single geometrical isomer (eq 3).[4]

$$\quad (3)$$

The lower-order silyl cuprates are apparently less reactive than the higher-order cuprates in conjugate addition reactions. Reagent (**1**) has been added in an addition–elimination sequence to an alkylidenemalonate system as a route to vinylsilanes (eq 4).[5]

$$\quad (4)$$

This reaction was a prelude to a more elaborate scheme for producing more substituted vinylsilanes. The product of eq 4 further reacted with a geminal diorganometallic species to yield an (*E*)/(*Z*) (86:14) mixture of a vinylsilane (eq 5).

$$\quad (5)$$

Alkylations with Allyl Halides. The enhanced reactivity of allyl halides is reflected in the reactions of both (**1**) and (**3**). Thus (**1**) reacts with 2-*t*-butylsulfonylallyl bromide at low temperatures (eq 6).[6]

$$\quad (6)$$

Reagent (**3**) can be employed in the preparation of the 2-bromo-allylsilane (eq 7).[7]

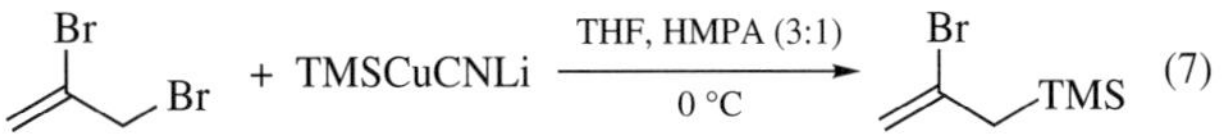

$$\text{(7)}$$

Silylcuprations of Unsaturated Systems. Carbocupration of alkynes is a major synthetic route to substituted alkenes. With (**1**), additions to terminal alkynes usually lead to regioisomeric vinylsilanes (eq 8).[2,8] In contrast to the lower-order cuprates, the corresponding higher-order cuprates generally are regioselective for the generation of the 1-silyl-substituted vinylsilanes (98:2 ratio).

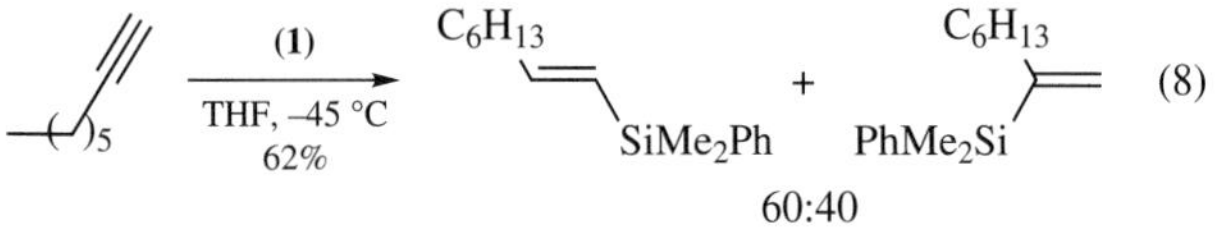

$$\text{(8)}$$

The regioselectivity of the addition of the title reagent (**1**) to 1,2-allenes, such as 1,2-undecadiene, is much more favorable for silylation at the 1-carbon atom. In fact, the major regio- and stereoisomer produced is the (Z)-1-dimethylphenylsilyl-2-undecene (eq 9).[9]

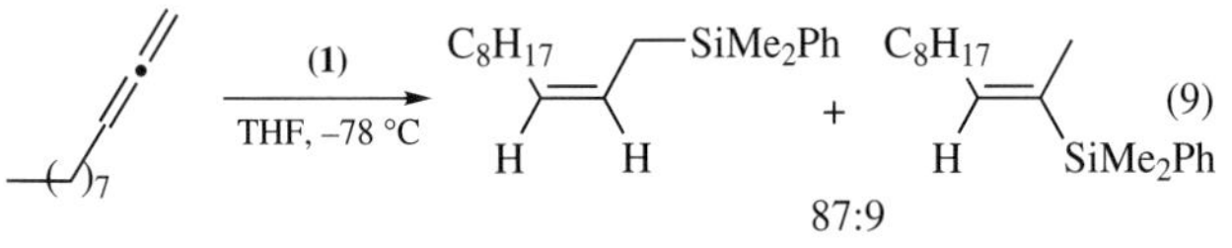

$$\text{(9)}$$

In general, (**1**) does not add to unactivated alkenes. However, strained bicyclic alkenes do undergo silylcupration with the reagent. The [2.2.1] bicyclic ethers (**4**) react with (**1**) at 0 °C, ultimately giving substituted 1,3-cyclohexadienes (**7**) via intermediates (**5**) and (**6**). The latter intermediate undergoes a Peterson alkenation to give (**7**) (eq 10).[10]

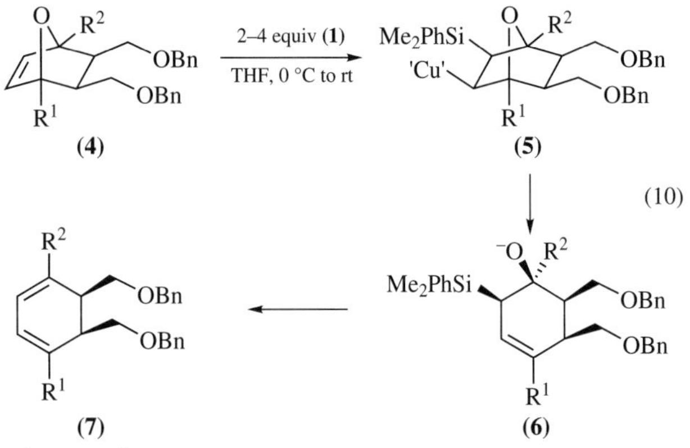

$$\text{(10)}$$

R^1 = Me, R^2 = H (81%)
R^1 = Me, R^2 = Me (83%)

Silylcupration of the oxabicyclic ketone (**8**) takes a different course, in that the bridged oxygen system is not cleaved. Instead, the initial *exo* adduct (**9**) produces a cyclobutanol product (**10**) exclusively (eq 11).[11]

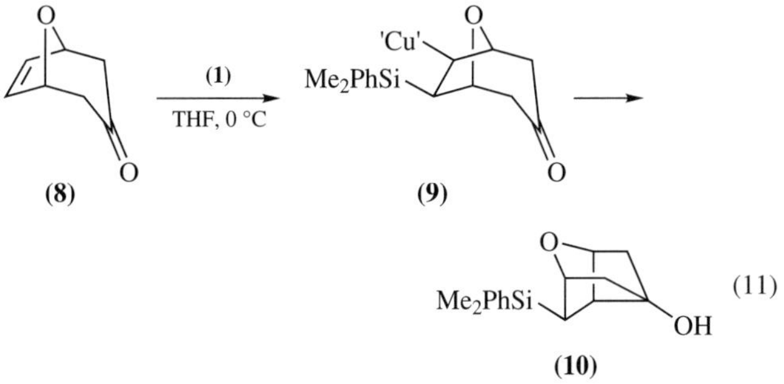

$$\text{(11)}$$

In the same [3.2.1] bicyclic ether system with methyl substitution α to the carbonyl, (**1**) adds to the double bond without cyclobutanol formation (eq 12). With the monomethylated ketone (**11**) an 8:1 mixture of regioisomers is produced, with (**12**) predominating.

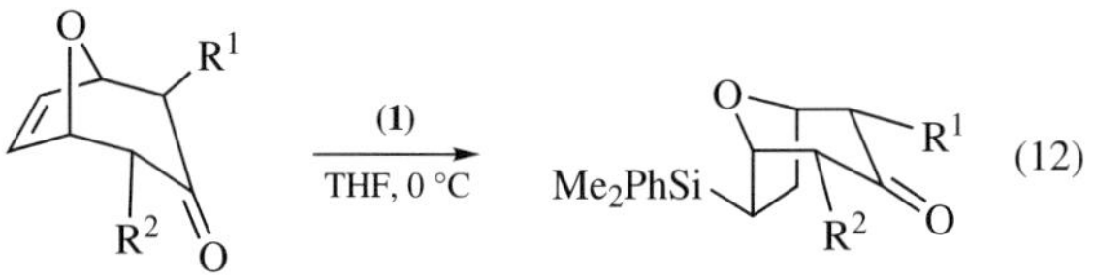

$$\text{(12)}$$

(**11**) R^1 = Me, R^2 = H   (**12**) R^1 = Me, R^2 = H

By allowing the above reaction to proceed for extended periods of time, the corresponding cyclobutanol product derived from (**11**) is favored, while with the dimethyl ketone (**13**) ($R^1 = R^2$ = Me), no cyclobutanols are formed. Intermediate cuprates (**9**) derived from higher-order cuprate additions may be trapped with iodomethane, acid, or hydroxyl groups (eq 13).

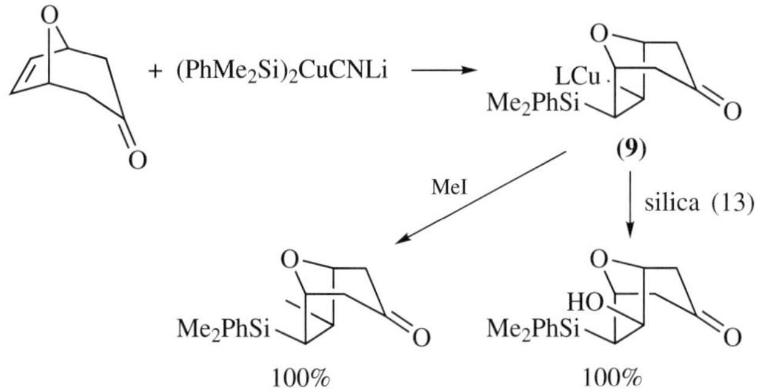

$$\text{(13)}$$

In summary, the lower-order cyanotrialkylsilylcuprates (**1**) and (**3**) react as do most carbon-based cuprates in conjugate additions and carbocupration processes. These reagents are less basic than higher-order cyanocuprates, and less reactive with some less electrophilic partners.

First Update

Ian Fleming
University of Cambridge, Cambridge, UK

Copper(I) cyanide is not the only salt used to make silylcuprate reagents, any more than dimethylphenylsilyl is the only silyl group. The cyanide has the advantage that it is not hygroscopic, but in general it appears to make little difference to the reactivity or selectivity of the reagent whether the cyanide or the iodide is used. In most cases, direct comparison of one reagent with another has not been made. Other cuprate reagents are included in this update, rather than having separate entries for each. Dilithium cyanobis(dimethylphenylsilyl)cuprate shows many similar reactions to the 1:1 reagent, and is sometimes superior in giving higher yields based on the substrate. On the other hand, the 1:1 reagent is more efficient in silicon and has some distinctive chemistry.

Conjugate Addition to Enones. The presence of dimethylsulfide improves the yield of conjugate addition of the 1:1 lithium iodo(dimethylphenylsilyl)cuprate to α,β-unsaturated aldehydes, ketones, esters, and amides, allowing conjugate additions comparable to or better than those using the 2:1 reagent, dilithium cyanobis(dimethylphenylsilyl)cuprate.[12] The lithium cyano-(dimethylphenylsilyl)cuprate may even be made in situ by adding copper(I) cyanide to the flask containing the silyllithium reagent and the α,β-unsaturated ketone at low temperature.[13]

Silylcupration of Alkynes. Silylcupration of alkynes may be made catalytic in copper by treating the silyllithium reagent with methylmagnesium iodide, followed by adding a catalytic amount of copper(I) cyanide. The regiochemistry with terminal acetylenes is unimpaired, and the intermediate vinyl copper species reacts efficiently with allyl diphenyl phosphate.[14]

Silylcupration of Allenes. The 1:1 silylcuprate (**1**) reacts with allene and mono-substituted allenes to give an intermediate allyl-silane-vinylcopper species (**14**),[15] in contrast to the reaction with the 2:1 reagent, which gives a vinylsilane-allylcopper intermediate. The intermediate is regiochemically stable at $-40\,^{\circ}$C, but rearranges at $0\,^{\circ}$C. The copper reagent (**14**) reacts with protons, primary alkyl iodides, ethylene oxide, acid chlorides, and α,β-unsaturated aldehydes and ketones, and with allyl diphenyl phosphate, all of which react without rearrangement (eq 14). The intermediate copper species (**14**) also reacts with vinyl bromides with catalysis by tetrakis(triphenylphosphine)palladium to give 1,3-dienes.[16]

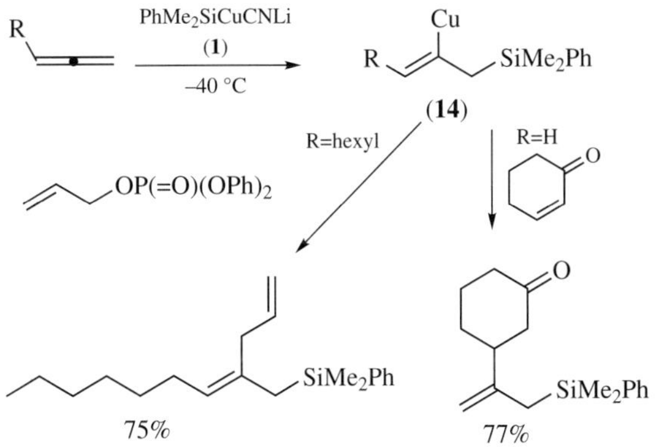

(14)

In contrast, α,β-unsaturated nitriles, catalyzed by boron trifluoride, selectively attack the rearranged, vinylsilane-allylcopper species (**15**), but the intermediate eniminium species (**16**) is then selectively attacked by the original adduct (**14**) (eq 15), indicating that the regioisomeric adducts (**14**) and (**15**) in the silylcupration are interconverting.[17]

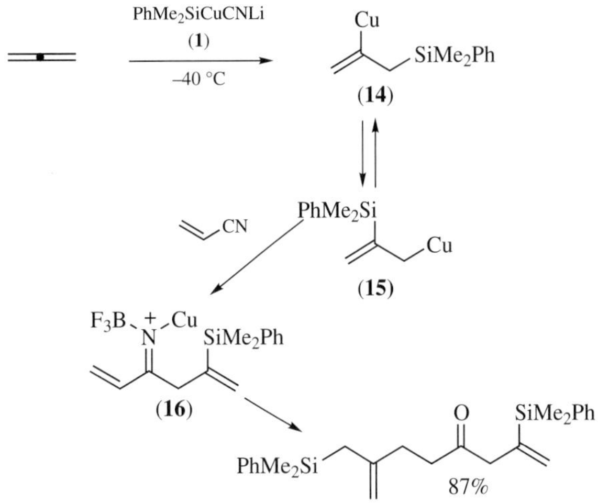

(15)

Silylcupration of Styrenes and Dienes. Styrene reacts with the silylcuprate (**1**), but more slowly than with allenes and acetylenes. Nevertheless, the intermediate copper species can be trapped with electrophiles like allyl diphenyl phosphate (eq 16).[18]

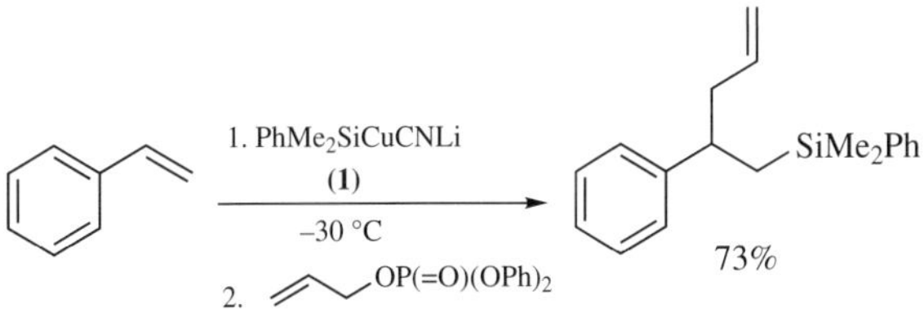

(16)

Dienes also react with the silylcuprate (**1**), but the intermediate copper species is inherently a mixture of regioisomers, summarized as (**17**). With reactive electrophiles, it gives mixture of regioisomers such as (**18**) and (**19**), but, with the less reactive electrophile allyl diphenyl phosphate it gives only regioisomer (**20**) (eq 17).[19]

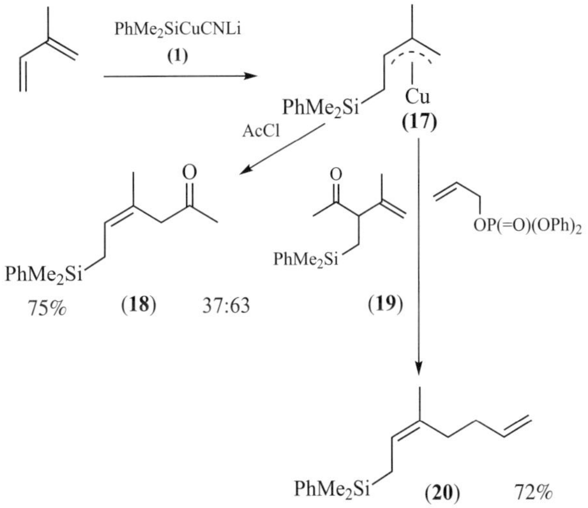

(17)

Other Reactions. The 1:1 silylcuprate, prepared from the iodide rather than the cyanide, added to diisobutylaluminum hydride, makes a more chemoselective reagent for conjugate reduction of the Hajos–Parish diketone than did the earlier t-butylcopper additive (eq 18).[20]

[Reaction scheme with DIBAH + PhMe₂SiCu, then Br₂, giving 70% product] (18)

Related Reagents. Dilithium Cyanobis(dimethylphenylsilyl)cuprate; Dimethylphenylsilyllithium; Dimethylphenylsilyl(methyl)magnesium.

1. (a) Fleming, I.; Marchi, D., *Synthesis* **1981**, 560. (b) Fleming, I.; Terrett, N. K., *J. Organomet. Chem.* **1984**, *264*, 99.
2. Sharma, S.; Oehlschlager, A. C., *J. Org. Chem.* **1991**, *56*, 770.
3. Fleming, I.; Rowley, M., *Tetrahedron* **1989**, *45*, 413.
4. Audia, J. E.; Marshall, J. A., *Synth. Commun.* **1983**, *13*, 531.
5. Tucker, C. E.; Rao, S. A.; Knochel, P., *J. Org. Chem.* **1990**, *55*, 5446.
6. Auvray, P.; Knochel, P.; Normant, J. F., *Tetrahedron* **1988**, *44*, 4495.
7. Trost, B. M.; Chan, D. M. T., *J. Am. Chem. Soc.* **1982**, *104*, 3733.
8. Fleming, I.; Newton, T.W.; Roessler, F., *J. Chem. Soc., Perkin Trans. 1* **1981**, 2527.
9. Singh, S. M.; Oehlschlager, A. C., *Can. J. Chem.* **1991**, *69*, 1872.
10. Lautens, M.; Ma, S.; Belter, R. K.; Chiu, P.; Leschziner, A., *J. Org. Chem.* **1992**, *57*, 4065.
11. Lautens, M.; Belter, R. K.; Lough, A. J., *J. Org. Chem.* **1992**, *57*, 422.
12. Dambacher, J.; Bergdahl, M., *J. Org. Chem.* **2005**, *70*, 580.
13. Hwu, J. R.; Chen, B.-L.; Lin, C.-F.; Murr, B. L., *J. Organomet. Chem.* **2003**, *686*, 198.
14. Liepins, V.; Karlstöm, A. S. E.; Bäckvall, J.-E., *J. Org. Chem.* **2002**, *67*, 2136.
15. (a) Barbero, A.; García, C.; Pulido, F. J., *Tetrahedron* **2000**, *56*, 2739. (b) Barbero, A.; Castreno, P.; Pulido, F. J., *Org. Lett.* **2003**, *5*, 4045. (c) Liepins, V.; Karlström, A. S. E.; Bäckvall, J.-E., *Org. Lett.* **2000**, *2*, 1237.
16. Barbero, A.; García, C.; Pulido, F. J., *Synlett* **2001**, 824.
17. Barbero, A.; Blanco, Y.; Pulido, F. J., *Chem. Commun.* **2001**, 1606.
18. Liepins, V.; Bäckvall, J.-E., *Chem. Commun.* **2001**, 265.
19. Liepins, V.; Bäckvall, J.-E., *Eur. J. Org. Chem.* **2002**, 3527.
20. Daniewski, A. R.; Liu, W., *J. Org. Chem.* **2001**, *66*, 626.

Lithium Hexamethyldisilazide

$[4039\text{-}32\text{-}1]$ $C_6H_{18}LiNSi_2$ (MW 167.37)

InChI = 1S/C6H18NSi2.Li/c1-8(2,3)7-9(4,5)6;/h1-6H3;/q-1;+1

InChIKey = YNESATAKKCNGOF-UHFFFAOYSA-N

(strong nonnucleophilic base)

Alternate Name: LHMDS; lithium bis(trimethylsilyl)amide.
Physical Data: distillable low-melting solid; mp 70–72 °C, bp 115 °C/1 mm Hg.[2] LHMDS is a cyclic trimer in the solid state,[3]

whereas in benzene solution it exists in a monomer–dimer equilibrium.[4] LHMDS exists as a tetramer-dimer mixture in hydrocarbons and as a dimer-monomer mixture in THF and ether. Treatment of LHMDS with trialkylamines increases the monomer concentration, whereas the use of diamines affords exclusively the corresponding chelated monomer.[24] LHMDS is less soluble, less basic, more stable, and much less sensitive to air compared to lithium diisopropylamide. pK_a 29.5 (THF, 27 °C).[1]

Solubility: soluble in most nonpolar solvents, e.g. aromatic hydrocarbons, hexanes, THF.

Form Supplied in: colorless crystalline solid, 1 M solution in THF or hexanes, 1.3 M solution in THF, 1 M solution in THF/cyclohexane.

Preparative Methods: conveniently prepared by the reaction of hexamethyldisilazane with *n*-butyllithium in hexane. For most uses the hexane is then evaporated and replaced with THF.[5]

Handling, Storage, and Precaution: a flammable, moisture sensitive solid; stable in a nitrogen atmosphere. Use in a fume hood.

Original Commentary

Matthew Gray & Victor Snieckus
University of Waterloo, Ontario, Canada

Ketone Enolates. A high yielding synthesis of 6-aryl-4,6-dioxohexanoic acids, precursors to antiinflammatory agents, is achieved using LHMDS (eq 1).[6] This process is applicable to large scale and involves relatively high reaction temperatures. The use of LDA gives reduced yields and small amounts of a diisopropylamide byproduct.

[Reaction scheme]
1. LHMDS, THF, –20 °C
2. succinic anhydride
78–95%
(1)

Ar = 4-ClC₆H₄, 2-thienyl, 3-pyridyl, 4-MeC₆H₄

Ester Enolates. Enantiomerically pure amino acids may ultimately be prepared via stereospecific ester enolate generation using an oxazolidine chiral auxiliary (eq 2).[7] Moderate diastereoselectivity is observed using potassium hexamethyldisilazide.

[Reaction scheme]
1. LHMDS, THF, –78 °C
2. trisyl azide
3. AcOH, rt
61–85%
(2)

trisyl = 2,4,6-triisopropylbenzenesulfonyl

Lithio ethyl acetate is prepared in quantitative yield by reaction of LHMDS with ethyl acetate in THF at –78 °C.[8a] Reaction with carbonyl compounds leads to condensation products in high yield (eqs 3 and 4).[8] No racemization of the α-silyloxy esters occurs (eq 4).

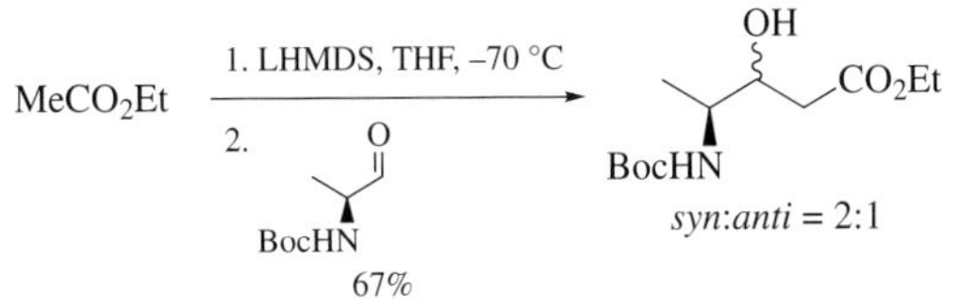

$$MeCO_2Et \xrightarrow[\substack{2.}]{\text{1. LHMDS, THF, }-70\,^\circ C}} \quad (3)$$

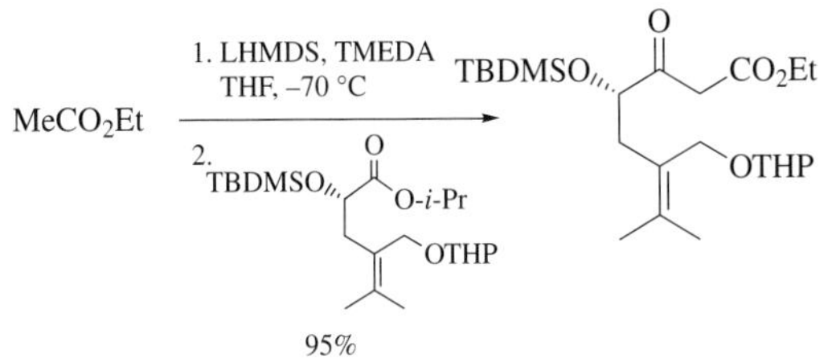

syn:anti = 2:1

67%

$$MeCO_2Et \xrightarrow[\substack{2.}]{\text{1. LHMDS, TMEDA THF, }-70\,^\circ C}} \quad (4)$$

95%

Kinetic Enolates. LHMDS is the recommended base for the generation of kinetic enolates. The resulting enolates are more regiostable than those generated with the corresponding sodium base, sodium hexamethyldisilazide. Thus reaction of Δ^4-3-keto steroids with LHMDS yields 2,4-dienolate ions which can be methylated at C-2 or trapped as 2,4-dienolsilyl ethers (eq 5).[9] Use of potassium t-butoxide/t-BuOH produces the thermodynamically more stable 3,5-dienolate. Acid-catalyzed conditions yields the 3,5-enol ether. Enolates generated with LHMDS may serve as ketone protecting groups during metal hydride reductions (eq 6).[10]

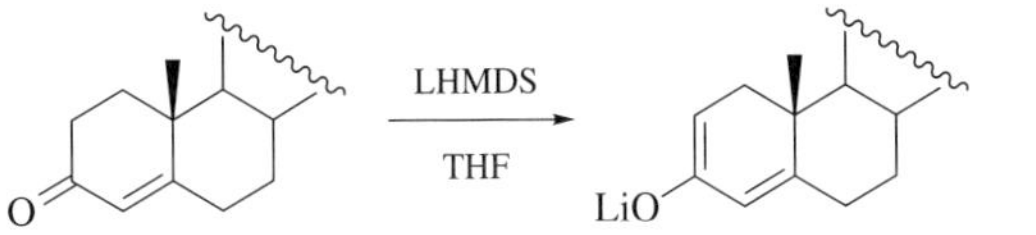

$$\xrightarrow[\text{THF}]{\text{LHMDS}} \quad (5)$$

$$\xrightarrow[\substack{2.\ LiAlH_4 \\ 3.\ NH_3}]{\text{1. LHMDS, THF}} \quad (6)$$

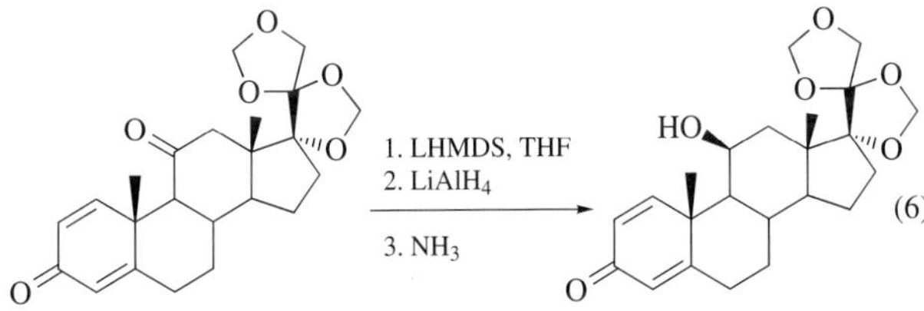

LHMDS has also been used in directed aldol condensations. The compatibility of the base with a silyl ether moiety is of note in the synthesis of ($\pm$)-[6]-gingerol (eq 7).[11]

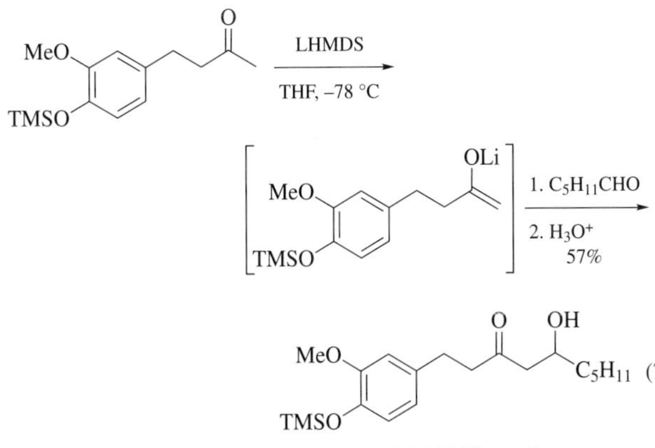

$$\xrightarrow[\text{THF, }-78\,^\circ C]{\text{LHMDS}}$$

$$\xrightarrow[\substack{2.\ H_3O^+ \\ 57\%}]{\text{1. }C_5H_{11}CHO} \quad (7)$$

($\pm$)-[6]-Gingerol

Darzens Condensation. The Darzens reaction invariably fails with aldehydes due to competing base-catalyzed self-condensation reactions.[12] With LHMDS as base, even acetaldehyde provides the desired glycidic ester products in high yield (eq 8).[13]

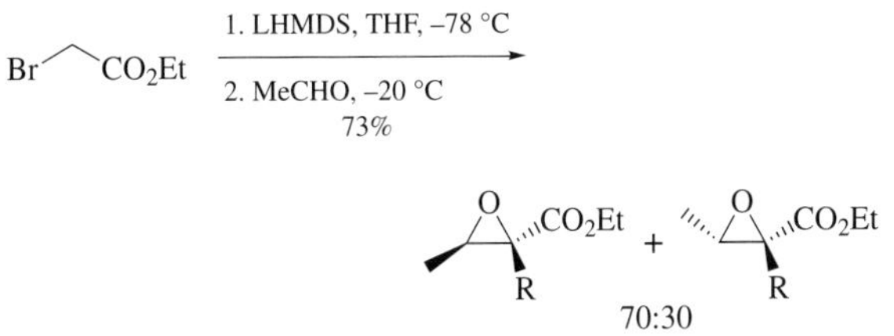

$$Br\diagup CO_2Et \xrightarrow[\substack{2.\ MeCHO, -20\,^\circ C \\ 73\%}]{\text{1. LHMDS, THF, }-78\,^\circ C}}$$

$$+ \quad (8)$$

70:30

Intramolecular Cyclizations. LHMDS-mediated intramolecular cyclizations have been demonstrated (eq 9).[14] The choice of counter cation has a dramatic effect on the stereochemistry of the cyclization.

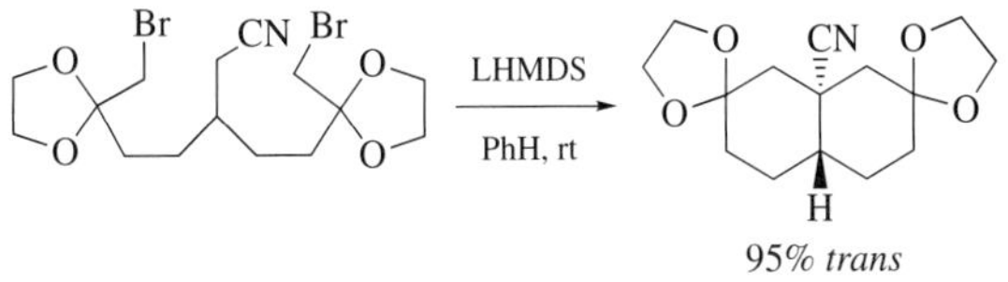

$$\xrightarrow[\text{PhH, rt}]{\text{LHMDS}} \quad (9)$$

95% trans

Ester Enolate Claisen Rearrangement. LHMDS is comparable to LDA for the stereoselective Ireland–Claisen rearrangement of ester enolates (eq 10).[15]

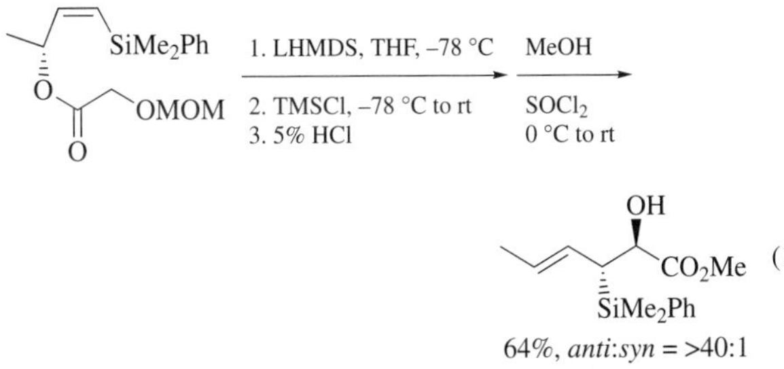

$$\xrightarrow[\substack{2.\ TMSCl, -78\,^\circ C\ to\ rt \\ 3.\ 5\%\ HCl}]{\text{1. LHMDS, THF, }-78\,^\circ C}} \xrightarrow[\substack{SOCl_2 \\ 0\,^\circ C\ to\ rt}]{\text{MeOH}}$$

$$\quad (10)$$

64%, anti:syn = >40:1

Intramolecular Double Michael Addition. LHMDS-mediated sequential Michael reactions constitute the key component of a total synthesis of the diterpene alkaloid atisine (eq 11).[16]

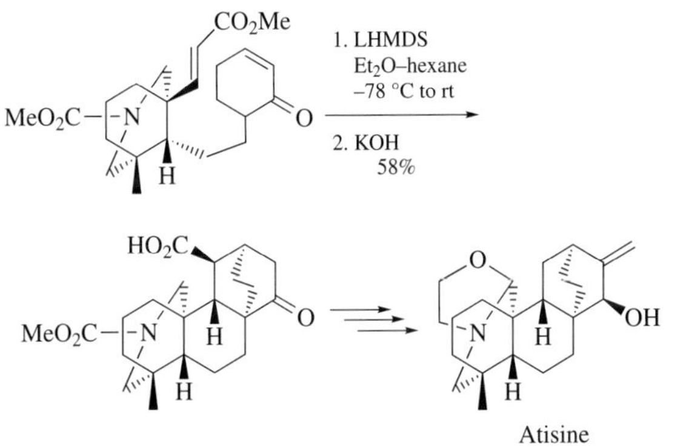

$$\xrightarrow[\substack{2.\ KOH \\ 58\%}]{\substack{\text{1. LHMDS} \\ Et_2O-hexane \\ -78\,^\circ C\ to\ rt}}$$

$$\quad (11)$$

Atisine

Synthesis of Primary Amines. *N,N*-Bis(trimethylsilyl)-methoxymethylamine, formally a $^+CH_2NH_2$ equivalent, is obtained in high yield by treating chloromethyl methyl ether with LHMDS. Treatment of the bis-silylamine with organometallic reagents followed by mild solvolysis gives primary amines in good to excellent yield (eq 12).[17]

$$ClCH_2OMe \xrightarrow[\substack{THF, 0\,°C \\ 86\%}]{LHMDS} (TMS)_2NCH_2OMe \xrightarrow[\text{rt or reflux}]{RMgX, Et_2O}$$

$$(TMS)_2NCH_2R \xrightarrow{HO^-} RCH_2NH_2 \quad (12)$$
$$50–92\%$$

N-Trimethylsilylaldimines. Aldehydes, even enolizable ones, undergo Peterson reactions with LHDMS to give *N*-trimethylsilylaldimines (eq 13),[18] which are valuable intermediates for a variety of systems, including primary amines,[18a,19] β-lactams,[18] and α-methylene-γ-lactams (eq 14).[20] Extension of this chemistry to include α-keto ester substrates allows for the preparation of α-amino esters.[21]

$$RCH=O + LHMDS \xrightarrow{THF} RCH=NTMS \quad (13)$$

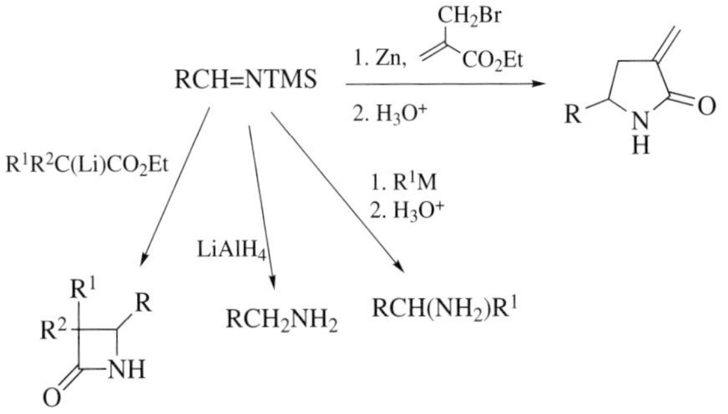

$$(14)$$

N,N-Bis(trimethylsilyl)aminomethyl Acetylide. The reaction of LHMDS with propargyl bromide constitutes a straightforward route to a γ-amino lithium acetylide, a useful precursor to a wide variety of unsaturated protected primary amines (eq 15).[22]

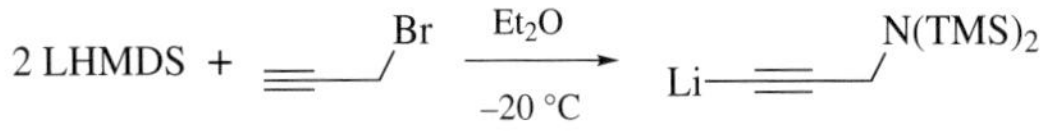

$$(15)$$

For example, reaction with aromatic aldehydes gives α-alkynyl amino alcohols, which can be trapped as their silyl ethers. Base-catalyzed isomerization to an allenic isomer followed by hydrolysis and concomitant cyclization affords 2-substituted pyrroles (eq 16).[22]

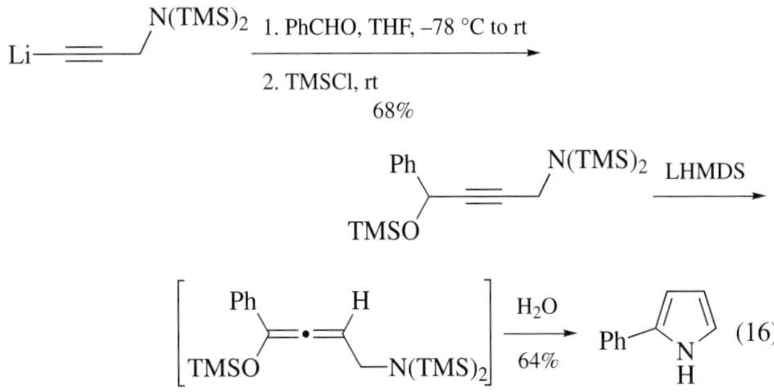

$$(16)$$

β-Ketosilanes. β-Ketosilanes may be prepared from α-bromo ketones using LHMDS to generate intermediate silyl enol ethers followed by metal–halogen exchange (eq 17).[23] They undergo facile rearrangement to silyl enol ethers and are also substrates, after carbonyl reduction, for overall Peterson alkenation.

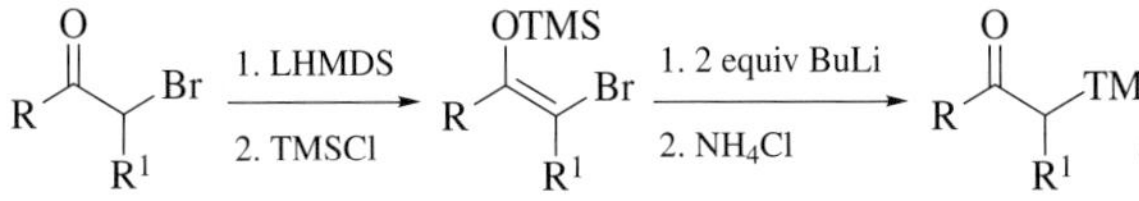

$$(17)$$

First Update

Professor Hélène Lebel
Université de Montréal, Montréal, Canada

Synthesis of Sulfinamides. The nucleophilic substitution of $(1R,2S,5R)$-$(-)$-menthyl (S)-*p*-toluenesulfinate with LHMDS leads to the synthesis of (S)-$(+)$-*p*-toluenesulfinamide in 72% yield with complete stereochemical inversion (eq 18).[25,26] (S)-$(+)$-*p*-Toluenesulfinamide is a versatile intermediate for the preparation of various chiral sulfinimines that are used for the synthesis of chiral amine derivatives via diastereoselective addition of nucleophiles.[27]

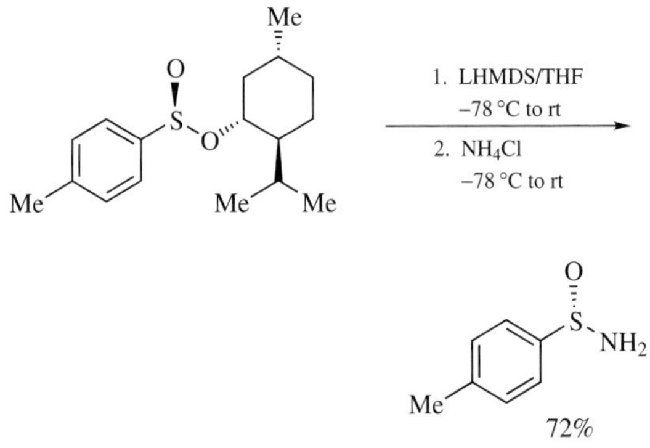

$$(18)$$

Transition Metal-Catalyzed Coupling Reactions: Synthesis of Aryl Amines. LHMDS was first used as a base to deprotonate alkylamines in the palladium-catalyzed synthesis of arylamines.[28,29] This base is mild enough such that the alkylamide is generated at the transition metal center. Earlier examples reported that the reaction of an aryl bromide with cyclohexylamine and LHMDS in the presence of 5 mol % of (tri-*o*-tolylphosphine)$_2$PdCl$_2$ in toluene at 100 °C produced the desired arylamine in 89% yield after 2 h (eq 19).[30]

In the presence of more bulky and electron-rich phosphines, such as 2-dicyclohexylphosphinobiphenyl, palladium-catalyzed C-N bond-forming reactions proceed under milder reaction conditions. While NaO*t*-Bu, LiO*t*-Bu, or K$_3$PO$_4$ are typically used in the palladium-catalyzed synthesis of aryl amines, a better functional group compatibility was observed in the presence of LHMDS.[31] Indeed, aryl halides containing phenol, alcohol, amide, or enolizable keto groups could be converted into the corresponding anilines in good to excellent yields (eq 20).

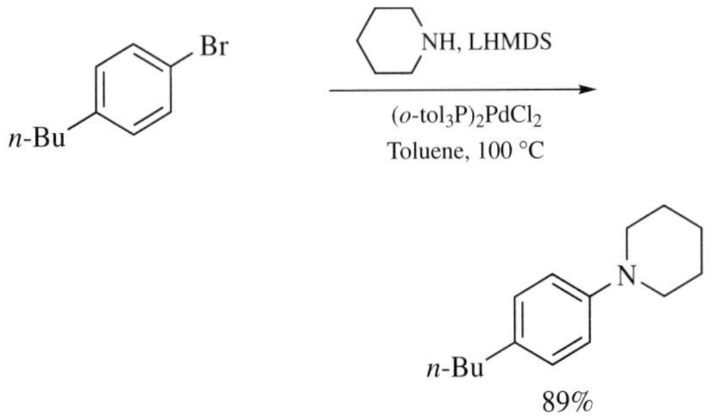

(19)

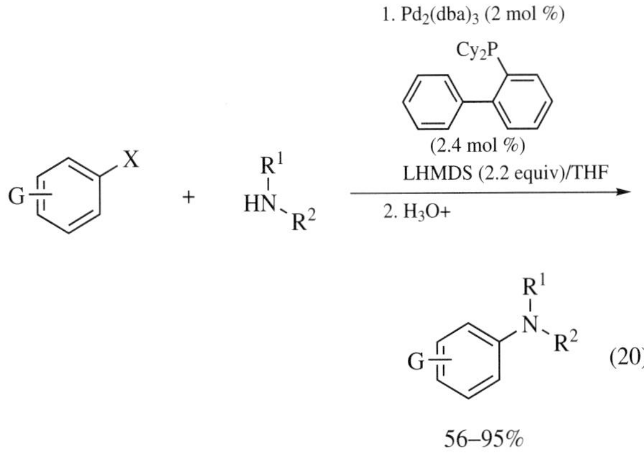

89%

(20)

56–95%

X = Cl; G = 3-CH$_2$CH$_2$OH, 3-NHAc

X = Br; G = 4-(CH$_2$)$_2$OH, 4-O(CH$_2$)$_2$OH, 4-OH,
3-OH, 4-NHAc, 3-C(O)Me, 4-(CH$_2$)$_2$C(O)Me

The use of LHMDS as an ammonia equivalent has also been demonstrated in the palladium-catalyzed synthesis of aryl amines from aryl halides (eq 21).[32] The optimal catalyst is generated from a 1:1 ratio of Pd(dba)$_2$ and P(t-Bu)$_3$, and the loading can be as little as 0.2 mol %. As silylamines underwent hydrolysis during chromatography, the corresponding anilines were isolated after addition of acid and neutralization. In general, the reaction of m- and p-substituted aryl chlorides and bromides provided high yields of the desired products. In contrast, the reaction did not occur with substrates bearing $ortho$-substituents. The reaction conditions are compatible with many electron-donating and electron-withdrawing groups, such as ester, trifluoromethyl, and halide substituents. However, the reaction does not proceed in the presence of cyano and nitro groups, neither with substrates possessing enolizable hydrogen, such as haloacetophenones.

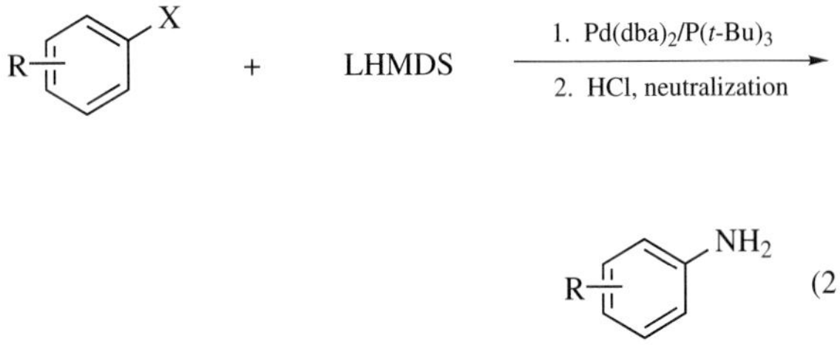

(21)

X = Cl; R = 4-CO$_2$Me, H, 4-F, 4-Me, 4-n-Bu, 4-OMe, 3-OMe;
or 2-Cl-pyridine; 64–99%

X = Br; R = 4-CO$_2$Me, H, 4-F, 4-Me, 4-OMe, 4-t-Bu, 4-n-Bu, 4-CF$_3$,
4-NMe$_2$, 4-Ph, 4-OPh, 3-OMe, 3-CF$_3$; or 2-Br-pyridine; 75–99%

A protocol that utilized a mixture of aminotriphenylsilane and LHMDS has been developed to overcome the limitation of the $ortho$-substituted aryl halides (eq 22).[33]

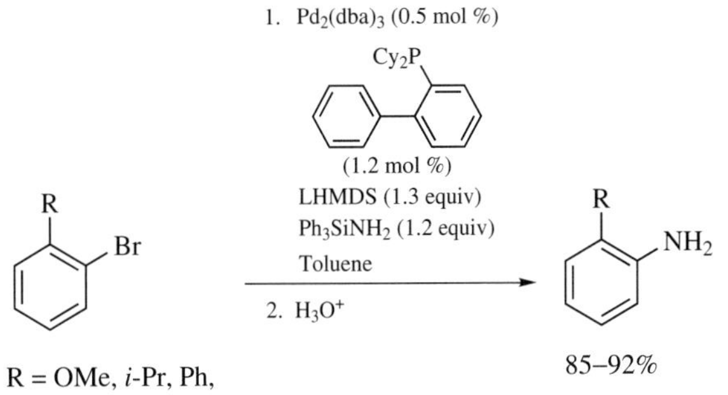

(22)

R = OMe, i-Pr, Ph,
Cl, vinyl

85–92%

Presumably, the two different silyl amides are in equilibrium and the less sterically hindered triphenylsilylamide reacts preferentially. The desired $ortho$-anilines could be isolated in excellent yield after deprotection with an aqueous acid.

Palladium-Catalyzed α-Arylation of Carbonyl Compounds and Nitriles. A variety of bases have been used in the palladium-catalyzed α-arylation of carbonyl derivatives.[34,35] The pKa of the carbonyl moiety determines the choice of the base. For instance, α-arylation with ester derivatives requires the use of strong bases such as LHMDS.[36] For example, the arylation of t-butyl acetate was first disclosed with LHMDS (eq 23).[37]

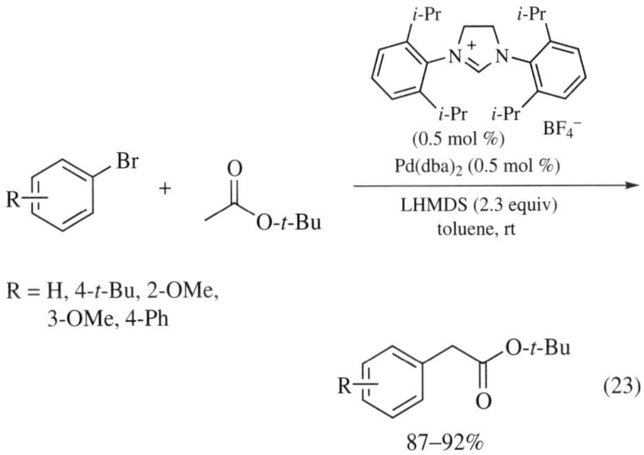

R = H, 4-t-Bu, 2-OMe,
3-OMe, 4-Ph

(23)

87–92%

Although LHMDS was also used with methyl isobutyrate, better yields were obtained for the same reaction with t-butyl propionate and NaHMDS at room temperature. The α-arylation with t-butyl propionate and LHMDS proceeds only at a higher temperature (80 °C).[38] It was shown later that LiNCy$_2$ could also be very effective for this reaction.[39] More sensitive substrates, such as α-imino esters, required the use of milder base as decomposition was observed with LHMDS.[40] Finally, α-arylation of nitrile derivatives was also reported using LHMDS (eq 24).[41]

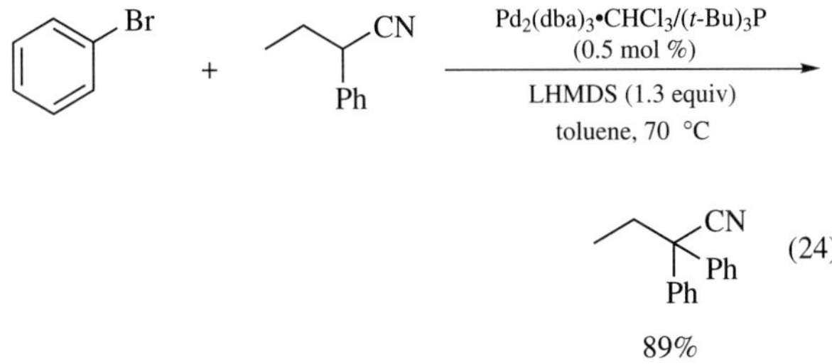

(24)

89%

Transition Metal-Catalyzed Allylic Alkylation. Chelated amino acid ester enolates were found to be suitable nucleophiles for palladium-catalyzed allylic alkylations (eq 25).[42] They were conveniently prepared by deprotonation of a glycine derivative with LHMDS followed by transmetallation with zinc chloride. The palladium-catalyzed allylic alkylation then takes place in the presence of allyl carbonates to produce the desired *anti* amino acid derivative.[43]

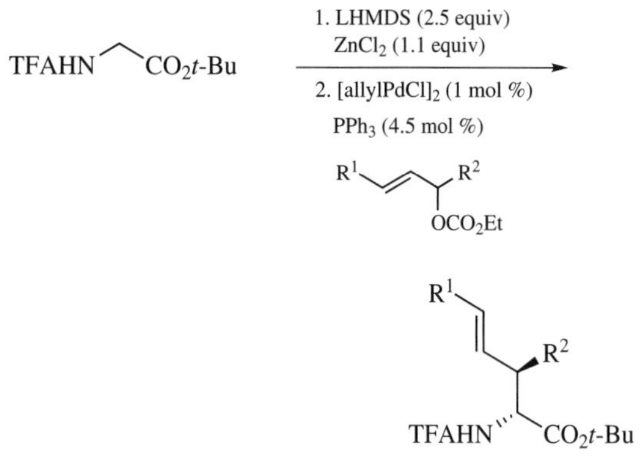

$$(25)$$

58–76%
74–92% ds

The *anti*-product was also formed preferentially when a mixture of LHMDS and an iminoester was used in iridium-catalyzed allylic substitution (eq 26).[44] In contrast, other sodium or potassium bases led to the formation of the *syn*-diastereomer as the major product. Here again, the formation of a Z-chelated enolate has been proposed to account for the observed diastereoselectivities.

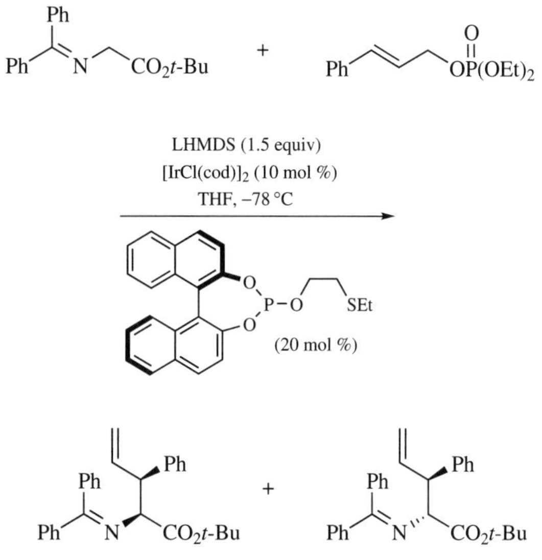

$$(26)$$

56% ee   92% ee

12:88
82%

The deprotonation of azalactones with LHMDS in the molybdenum-catalyzed asymmetric allylic alkylation was reported to lead to quaternary amino acids with excellent yields, diastereoselectivities, and enantioselectivities (eq 27).[45]

LHMDS was also used in the rhodium-catalyzed allylic amination and proved superior to both NaHMDS and KHMDS in terms of reaction rate and regioselectivity (eq 28).[46,47]

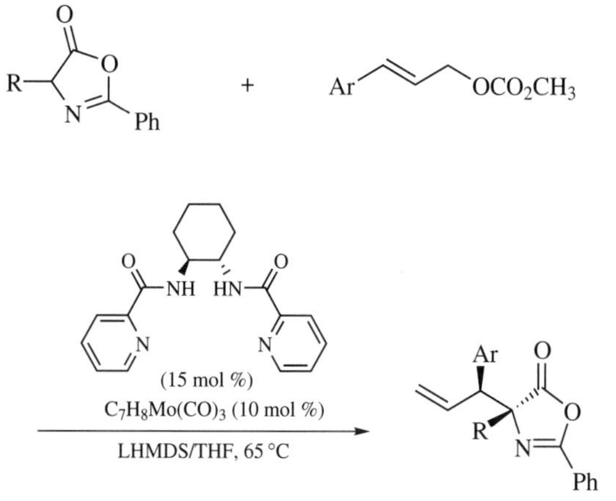

$$(27)$$

76–92%
96–>98% ds
85–99% ee

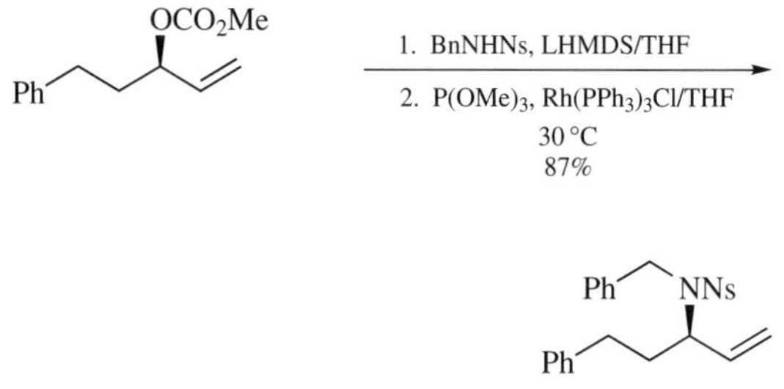

$$(28)$$

99% ee
Regioselectivity = 19:1

Copper(I) alkoxides were mandatory to achieve rhodium-catalyzed allylic etherifications with aliphatic alkoxide derivatives (eq 29).[48] A one-pot procedure that involves the treatment of the corresponding alcohol with LHMDS to generate the lithium enolate, followed by the addition of a copper salt, allowed the preparation of the requisite copper alkoxide.

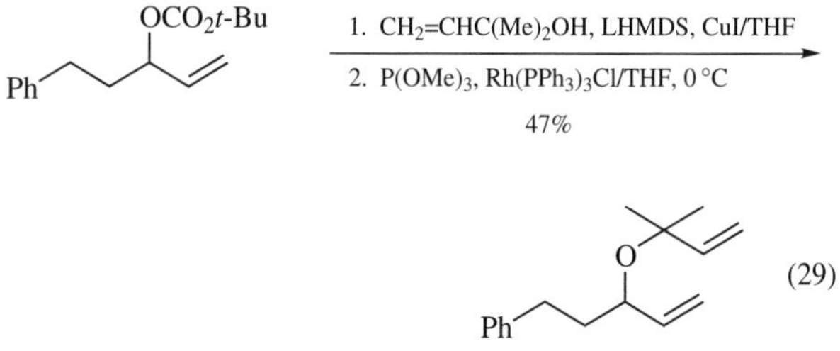

$$(29)$$

Regioselectivity = 47:1

Electrophilic Amination of Alkoxides. Generation of the lithium alkoxide of benzyl alcohol with LHMDS in THF at −78 °C, followed by the addition of oxaziridine **1** and slow warming to ambient temperature, delivered an essentially quantitative yield of the desired *O*-benzyl-*N*-Boc-hydroxylamine (eq 30).[49]

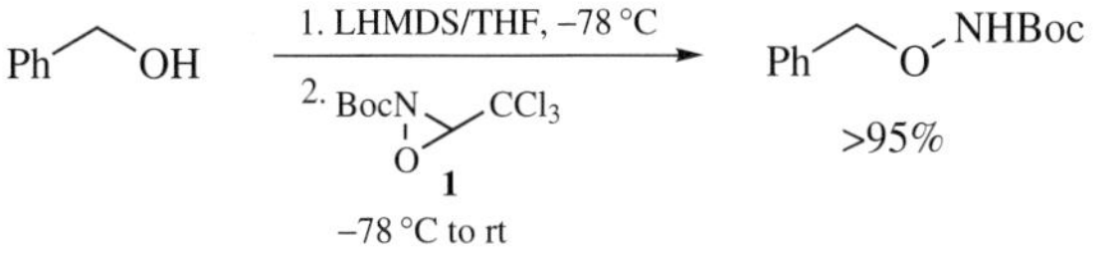

Ph OH
1. LHMDS/THF, −78 °C
2. BocN CCl3
O
1
−78 °C to rt
Ph O NHBoc
>95%

(30)

Synthesis of 1-Bromoalkynes. There are a few examples of the use of LHMDS in the dehydrobromination of 1,1-dibromoalkenes leading to 1-bromoalkynes (eqs 31 and 32),[50,51] although this reaction is usually performed with NaHMDS.[52] In comparison, stronger bases such as butyllithium lead to the corresponding acetylide anion via a metal-halogen process.[53]

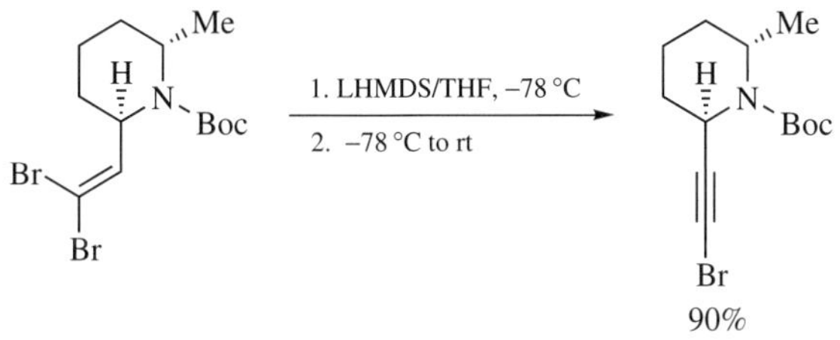

Me
H
N
Boc
Br
Br
1. LHMDS/THF, −78 °C
2. −78 °C to rt
Me
H
N
Boc
Br
90%

(31)

Cl
Br
Br
Cl
N
CO2Et
SEM
LHMDS/THF, 0 °C
Cl
Br
Cl
N
CO2Et
SEM
87%

(32)

Synthesis of Conjugated Enediynes. LHMDS was used in a carbenoid coupling-elimination strategy to synthesize linear and cyclic enediynes from various propargylic halides.[54] The stereoselectivity in the linear series is controlled by substitution at the propargylic center: primary propargylic bromides favored the *Z*-isomer (eq 33),[55] whereas secondary propargylic bromides produced exclusively the *E*-isomer (eq 34).[56] Conversely, only the *Z*-isomer was obtained for cyclic enediynes.

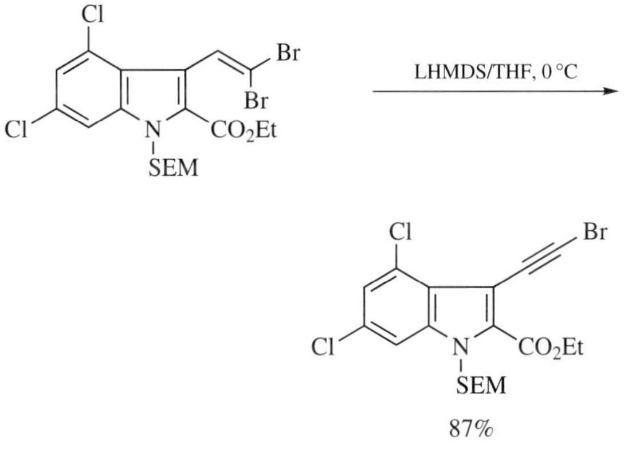

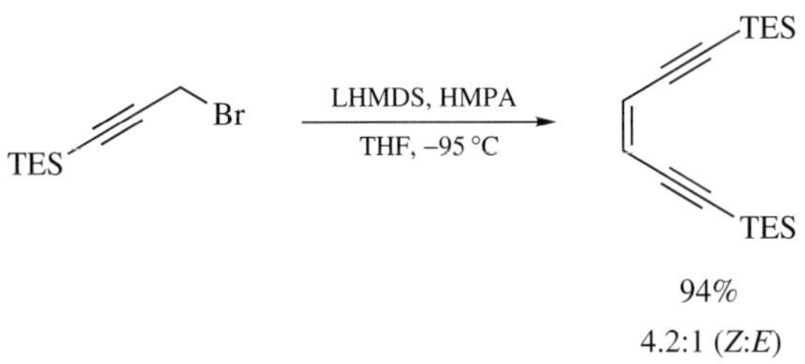

TES
Br
LHMDS, HMPA
THF, −95 °C
TES
TES
94%
4.2:1 (Z:E)

(33)

TIPS
Br
LHMDS, HMPA
THF, −80 °C
TIPS
TIPS
87%
1:100 (Z:E)

(34)

Deprotonation of Bis(oxazolines). It has been recently shown that LHMDS can efficiently replace LDA for the deprotonation/alkylation of the bis(oxazoline), derived from aminoindanol, with 1,2-dibromoethane (eq 35).[57]

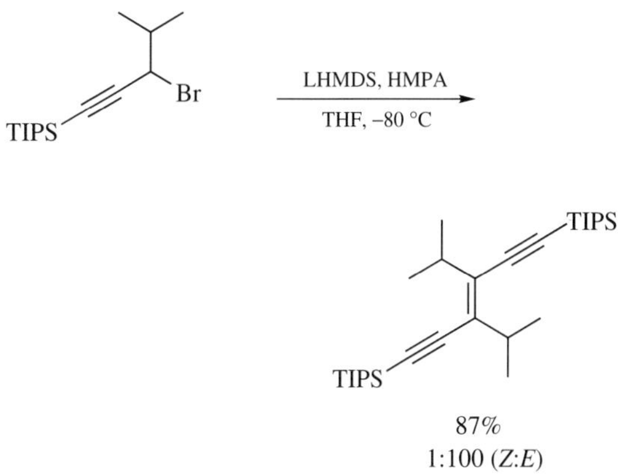

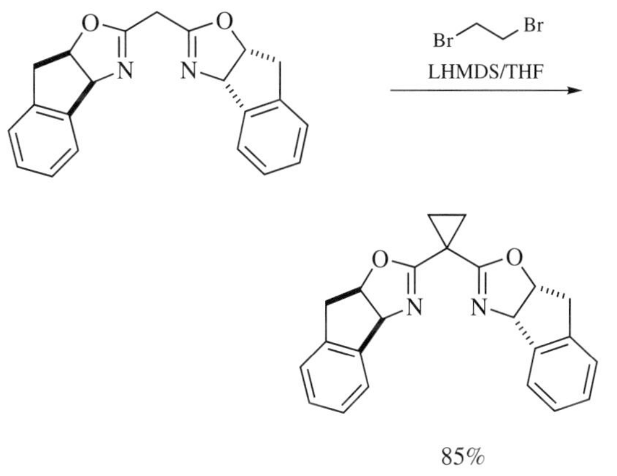

O
N
O
N
Br
Br
LHMDS/THF
O
N
O
N
85%

(35)

Deprotonation of Tosylhydrazones. The deprotonation of tosylhydrazones with LHMDS provides the corresponding lithium salts, which can be further decomposed into the diazo intermediates. The addition of late transition metal complexes leads to the formation of metal carbenoid species which undergo various reactions, such as cyclopropanation, aziridination, epoxidation, and C-H insertion. For instance, the lithium salt of tosylhydrazone **2**, prepared from LHMDS, is reacted with an imine or an alkene in the presence of rhodium(II) acetate and a chiral sulfide to give respectively, the corresponding aziridine or cyclopropane derivatives (eqs 36 and 37).[58,59] Under similar reaction conditions, the sodium salt prepared from NHMDS works equally well.

However, in the case of reaction with aldehydes, which led to the formation of epoxides, it was established that the sodium salt provided higher yields and selectivity than the lithium salt (eq 38).[60]

Such a cation effect has not been observed for the ruthenium–porphyrin-catalyzed intramolecular carbenoid C–H insertion of tosylhydrazones. The lithium and the sodium salts provided equally good results, although the use of LHMDS has been preferred by the researchers (eq 39).[61]

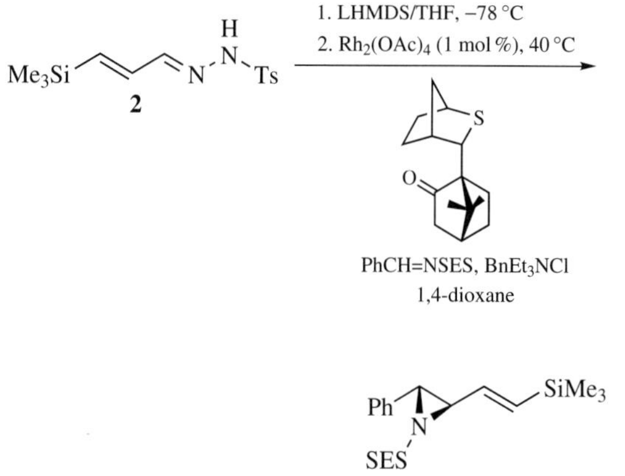

1. LHMDS/THF, −78 °C
2. Rh$_2$(OAc)$_4$ (1 mol %), 40 °C

PhCH=NSES, BnEt$_3$NCl
1,4-dioxane

$$\text{Ph} \quad \overset{\text{N}}{\underset{\text{SES}}{\triangle}} \quad \text{SiMe}_3 \qquad (36)$$

82%
78:22 (*trans:cis*)
94% ee (*trans*)
73% ee (*cis*)

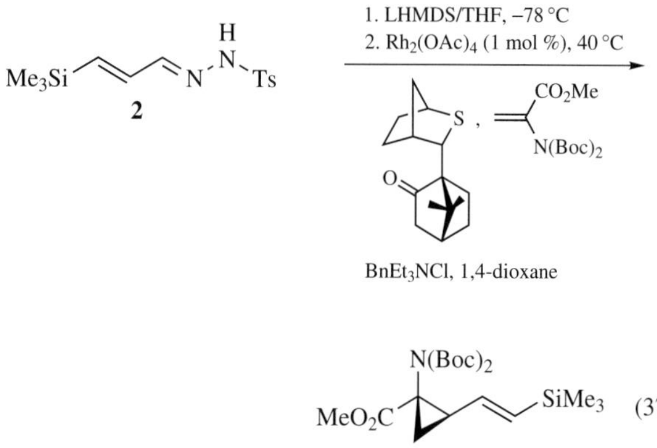

1. LHMDS/THF, −78 °C
2. Rh$_2$(OAc)$_4$ (1 mol %), 40 °C

BnEt$_3$NCl, 1,4-dioxane

$$\text{MeO}_2\text{C} \quad \overset{\text{N(Boc)}_2}{\underset{}{\triangle}} \quad \text{SiMe}_3 \qquad (37)$$

65%
1:6 (*trans:cis*)
75% ee (*cis*)

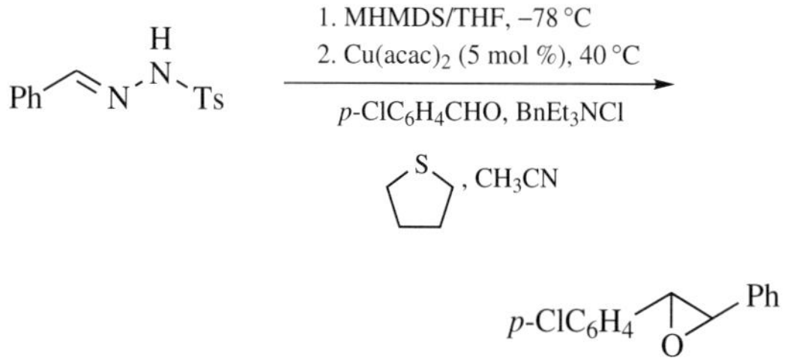

1. MHMDS/THF, −78 °C
2. Cu(acac)$_2$ (5 mol %), 40 °C

p-ClC$_6$H$_4$CHO, BnEt$_3$NCl

$\langle S \rangle$, CH$_3$CN

$$p\text{-ClC}_6\text{H}_4 \quad \overset{\text{O}}{\triangle} \quad \text{Ph} \qquad (38)$$

M = Li, 54%
M = Na, 73%

1. LHMDS/THF, −78 °C
2. [Ru^{II}(TTP)(CO)] (1 mol %), 110 °C
n-Bu$_4$NBr, Toluene, MS 4Å

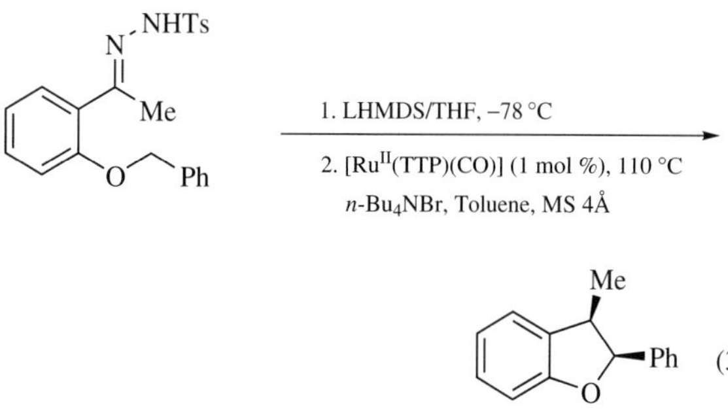

$$\text{(39)}$$

89%
95:5 (*cis:trans*)

Generation of Phosphorus Ylides and Phosphonate Anions.

LHMDS has been utilized for the deprotonation of a variety of phosphonium salts to generate the corresponding ylides. The reactivity and the selectivity of the subsequent Wittig reaction with a carbonyl compound is often strongly influenced by the type of base used for the deprotonation step. Indeed, a systematic investigation showed the strong dependence of the isomeric ratio with regards to the base used.[62] While high *E*-selectivities were observed with oxido ylide generated from phosphonium salt **3** and LHMDS (eq 40), simple alkyl substituted phosphorus ylides obtained from **4** provided low *E*-selectivities when reacted with benzaldehyde (eq 41).[63]

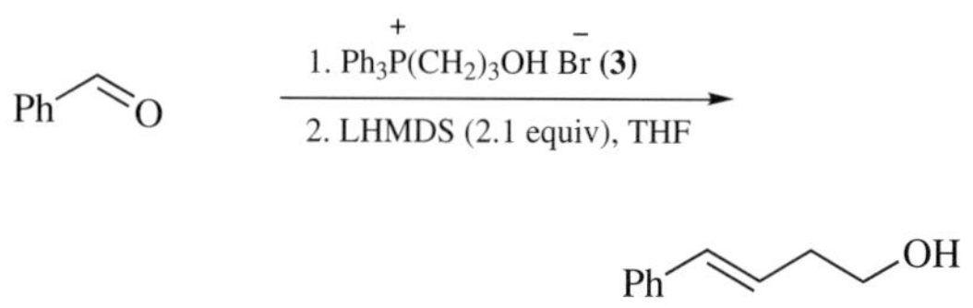

$$\text{Ph} \overset{\text{O}}{\diagdown} \quad \xrightarrow[\text{2. LHMDS (2.1 equiv), THF}]{\text{1. Ph}_3\text{P(CH}_2)_3\text{OH Br (3)}} \quad \text{Ph} \diagup\diagdown\diagup \text{OH} \qquad (40)$$

80%
96:4 (*E:Z*)

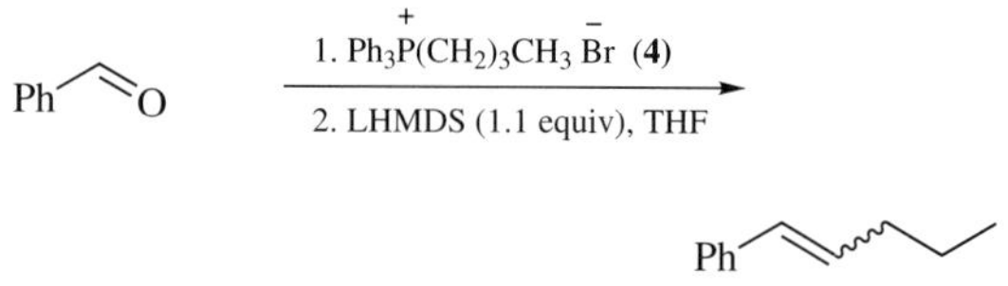

$$\text{Ph} \overset{\text{O}}{\diagdown} \quad \xrightarrow[\text{2. LHMDS (1.1 equiv), THF}]{\text{1. Ph}_3\text{P(CH}_2)_3\text{CH}_3 \text{ Br (4)}} \quad \text{Ph} \diagup\diagdown\diagup\diagdown \qquad (41)$$

35:65 (*E:Z*)

Good *Z*-selectivities were observed for the reaction of aldehydes with carboxy ylides prepared from the corresponding phosphonium salt and LHMDS. This strategy has been extensively used in the total synthesis of prostaglandins,[64] monohydroxy-eicosatetraenoic acids (HETE),[65,66] and leukotrienes.[67,68] For instance, the Wittig condensation between aldehyde **5** and carboxy phosphonium salt **6** in the presence of LHMDS provided 49% yield of the TBDPS-protected 10(*S*)-HETE methyl ester **7** (eq 42).[69]

Similar reaction conditions were used in the synthesis of L-α-aminosuberic acid. The Wittig condensation of chiral aldehyde **8** with carboxy phosphonium salt **9** in the presence of LHMDS leads to the formation of the corresponding optically pure *Z*-alkene (eq 43).[70] In contrast, the utilization of dimsyl sodium base at room temperature resulted in a racemic product mixture.

LHMDS was also used in the synthesis of vinyl serine derivatives, as shown in eqs 44[71] and 45.[72] Here again, the stereochemical integrity was preserved and the best selectivities were observed using a lithium base.

In addition, LHMDS has been successfully used in the deprotonation of other functionalized phosphonium salts, such as oxazolidinone methyl **10** (eq 46),[73] *N*-methylformamide **11** (eq 47),[74] and thiazolylmethyl **12** (eq 48).[75]

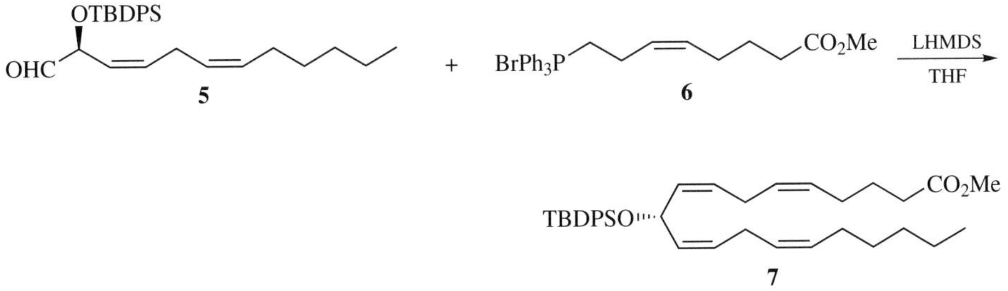

OTBDPS

OHC

5

+ BrPh$_3$P

CO$_2$Me

6

$\xrightarrow[\text{THF}]{\text{LHMDS}}$

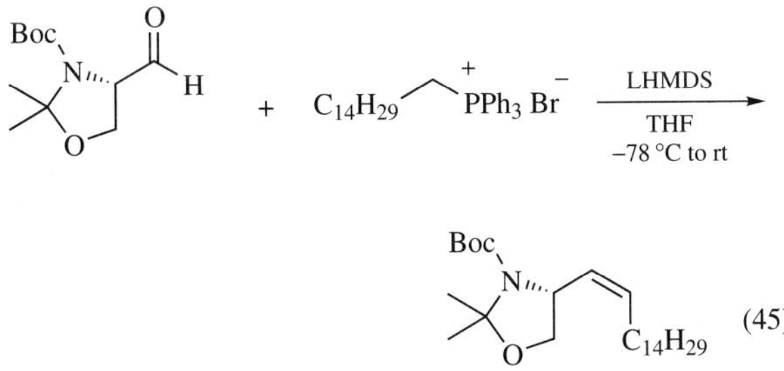

TBDPSO

CO$_2$Me

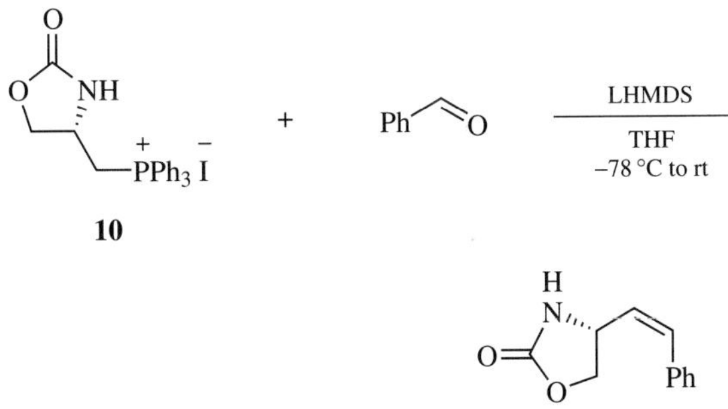

7

(42)

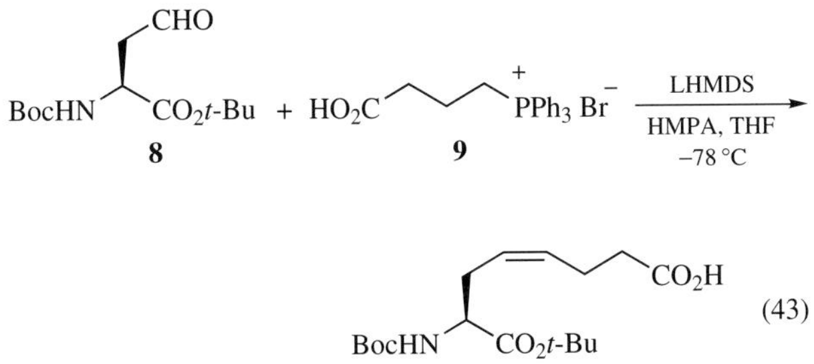

CHO

BocHN

CO$_2$t-Bu

8

+ HO$_2$C

PPh$_3$ Br$^-$

9

$\xrightarrow[\substack{\text{HMPA, THF} \\ -78\,°C}]{\text{LHMDS}}$

CO$_2$H

BocHN CO$_2$t-Bu

(43)

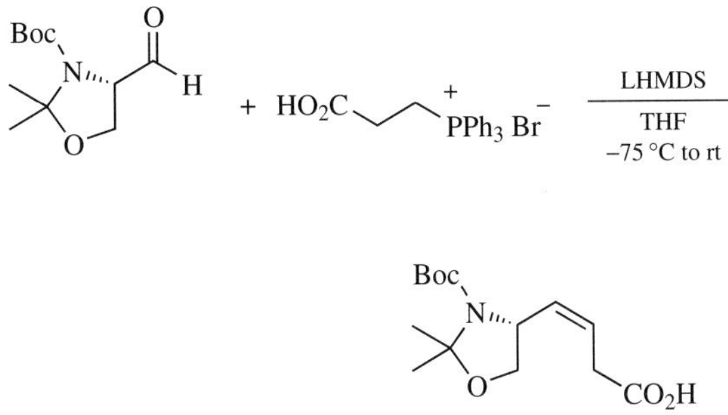

Cl$^-$ Ph$_3$P$^+$

Me
N
CHO

11

+ Ph

CHO

$\xrightarrow[\substack{\text{THF} \\ -78\,°C \text{ to rt}}]{\text{LHMDS}}$

Ph

Me
N
CHO

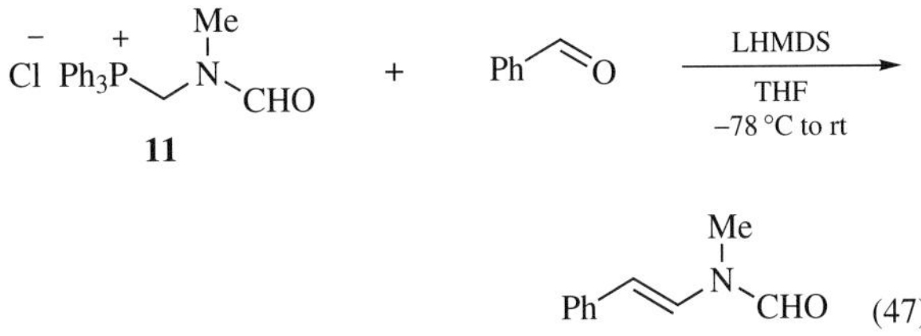

(47)

88%
10:1 (*E:Z*)

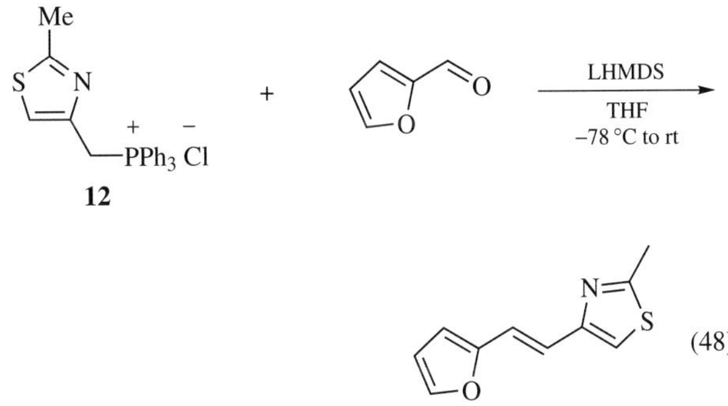

Boc
N
O
H

+ HO$_2$C

PPh$_3$ Br$^-$

$\xrightarrow[\substack{\text{THF} \\ -75\,°C \text{ to rt}}]{\text{LHMDS}}$

Boc
N
O

CO$_2$H

(44)

73%
>98:1 (*Z:E*)

Me
S
N

PPh$_3$ Cl$^-$

12

+

O
CHO

$\xrightarrow[\substack{\text{THF} \\ -78\,°C \text{ to rt}}]{\text{LHMDS}}$

O

N
S

(48)

61%
84:16 (*E:Z*)

Boc
N
O
H

+ C$_{14}$H$_{29}$

PPh$_3$ Br$^-$

$\xrightarrow[\substack{\text{THF} \\ -78\,°C \text{ to rt}}]{\text{LHMDS}}$

Boc
N
O

C$_{14}$H$_{29}$

(45)

66%
90:10 (*Z:E*)

The latter reagent has been slightly modified to be used in the preparation of a vinyl thiazolium unit, en route to the total synthesis of epothilones (eq 49).[76]

Other recent uses of LHMDS as base for the deprotonation of phosphonium salts include the final coupling in the total synthesis of (+)-spongistatin[77] and the assembly of the subunit C31-C40 of pectenotoxins.[78]

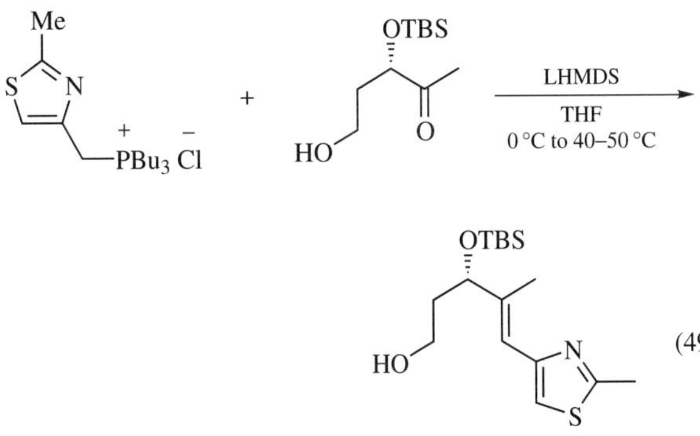

Me
S
N

PBu$_3$ Cl$^-$

+

HO

OTBS
O

$\xrightarrow[\substack{\text{THF} \\ 0\,°C \text{ to } 40–50\,°C}]{\text{LHMDS}}$

HO

OTBS

N
S

Me

(49)

77%
>99:1 (*E:Z*)

O
O
NH

PPh$_3$ I$^-$

10

+ Ph

CHO

$\xrightarrow[\substack{\text{THF} \\ -78\,°C \text{ to rt}}]{\text{LHMDS}}$

O
O
N
H

Ph

(46)

88%
>99:1 (*Z:E*)

Finally, LHMDS has also been the base of choice for the deprotonation of more exotic phosphonium salts, such as allylic phosphonium salt **13** (eq 50), nitrogen phosphonium **14** (eq 51), and pinacolboratomethylphosphonium **15** (eq 52).

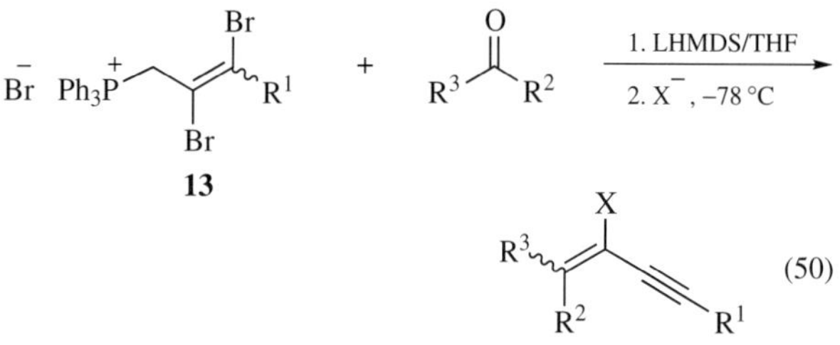

$$(50)$$

38–62%

$$(51)$$

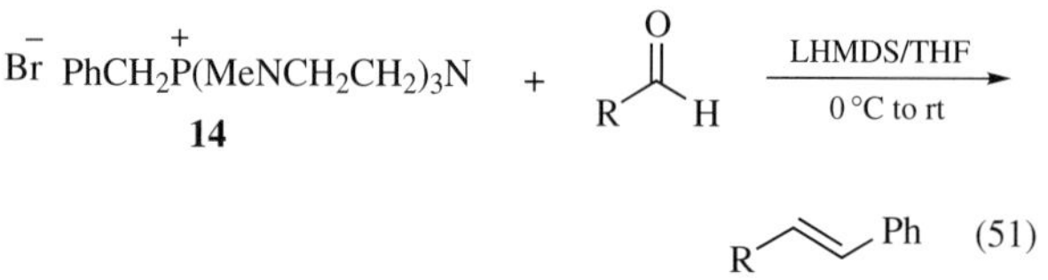

88–89%
>99:1 (E:Z)

$$(52)$$

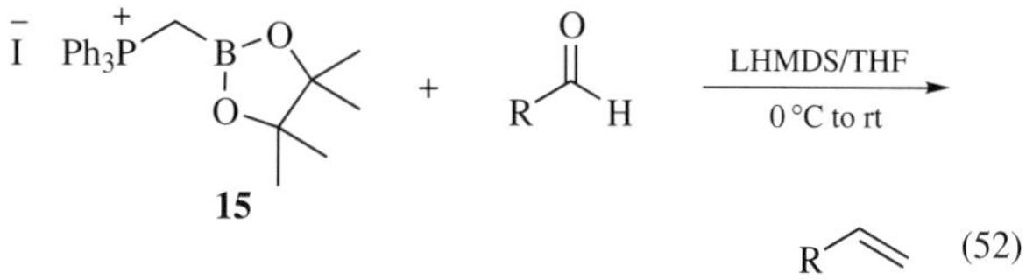

99% conv.

The formation of phosphonate anions has also been achieved using LHMDS to generate Horner–Wadsworth–Emmons reagents. For instance, the final coupling step in the synthesis of leukotriene B_4 involving phosphonate **16** and aldehyde **17**, in the presence of LHMDS, leads to *E*-alkene **18** in 77% yield (eq 53).[79]

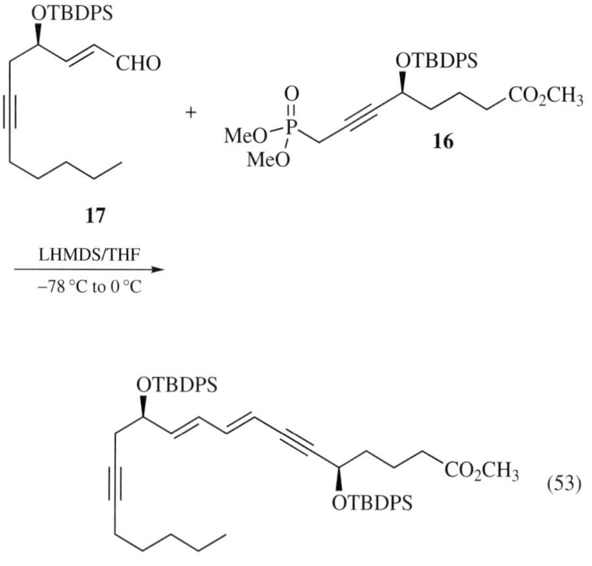

$$(53)$$

18

LHMDS was identified as the base of choice for the in situ formation of diiodophosphonates, which upon reaction with carbonyl compounds provided the corresponding diiodoalkene (eq 54).[80]

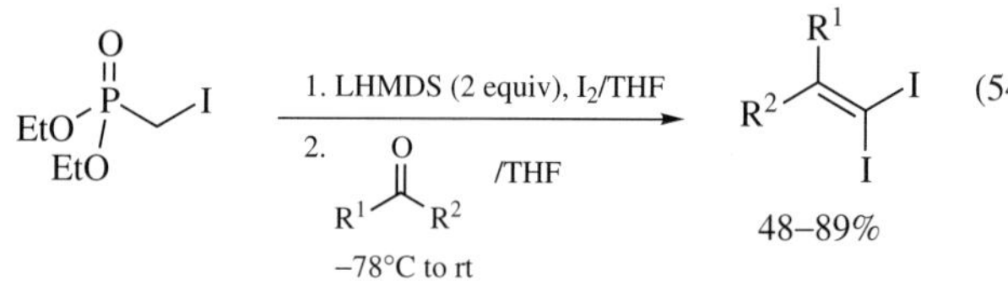

$$(54)$$

48–89%

Other uses of phosphonate anions prepared from deprotonation with LHMDS include the acylation reaction for the synthesis of (diphenylphosphono)acetic acid esters (eq 55)[81] as well as the aza-Darzens reaction leading to aziridine 2-phosphonates (eq 56).[82] In both cases, LHMDS proved to be equal or superior to other lithium bases.

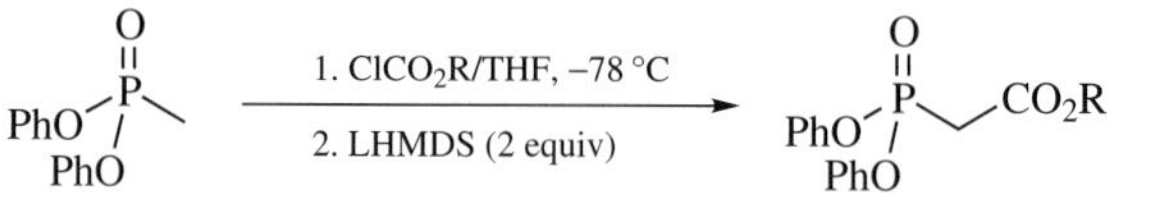

$$(55)$$

71–89%

$$(56)$$

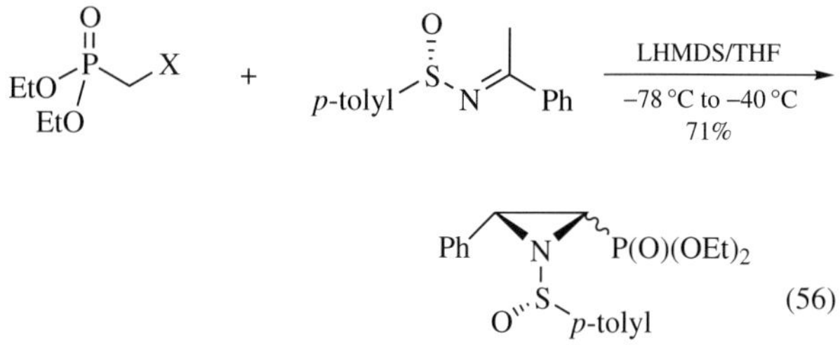

81:19
(*syn:anti*)

Generation of Sulfur Ylide: Julia Olefination and Related Reactions. The deprotonation at a position α to a sulfone group is effected using a strong base, typically *n*-butyllithium or LDA, to give an anion which then reacts with an aldehyde or a ketone leading to the corresponding β-hydroxy sulfone. There are, however, a number of examples where LHMDS was used as a base. For instance, sulfone **19** was deprotonated with LHMDS and then reacted with aldehyde **20** to produce β-hydroxy sulfone **21** (eq 57).[83]

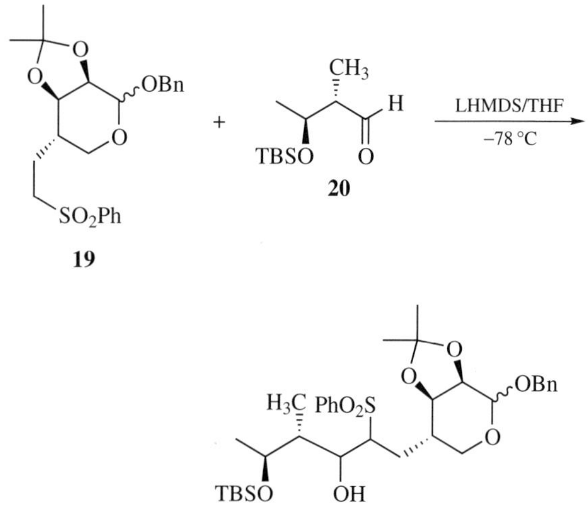

$$(57)$$

21

LHMDS has also been used for the synthesis of fluorosulfone anion **22** which upon condensation with carbonyl derivatives led to the formation of fluorovinyl sulfone (eq 58).[84]

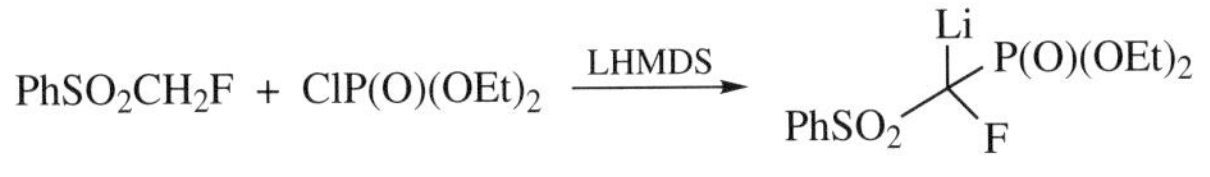

PhSO$_2$CH$_2$F + ClP(O)(OEt)$_2$ $\xrightarrow{\text{LHMDS}}$

22

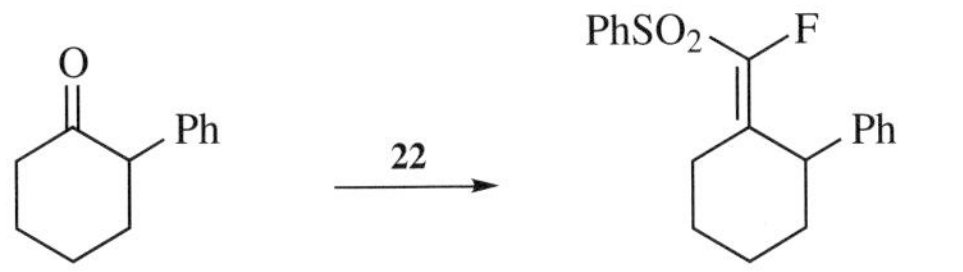

(58)

The highest yield and selectivity for the formation of a simple diene from a benzothiazoyl sulfone was observed when using LHMDS in dichloromethane (eq 60).[88]

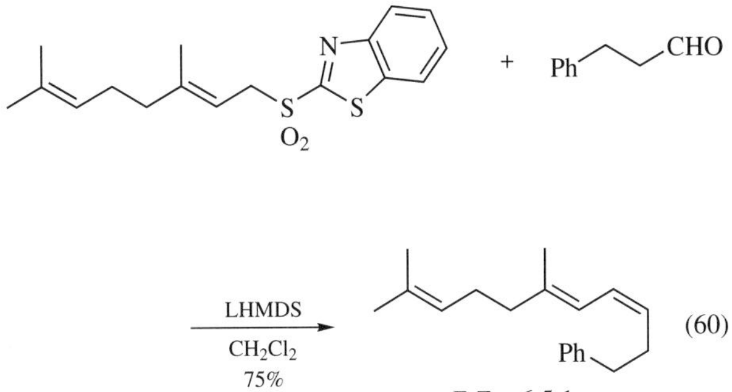

(60)

$E:Z = 6.5:1$

More recently, the 'modified' Julia olefination, which employed certain heteroaryl sulfones instead of the traditional phenyl sulfones, has emerged as a powerful tool for alkene synthesis.[85] Although the reaction was first reported with LDA, bases such as LHMDS, NaHMDS, and KHMDS are now commonly used.[86] In addition, solvent as well as base counter-cation have been shown to markedly affect the stereochemical outcome of the olefination reaction. LHMDS has been selected as the base of choice in some systems, particularly the one that led to the formation of dienes and trienes. Indeed, triene **25** was obtained in 68% yield from sulfone **23** and aldehyde **24** in the presence of LHMDS (eq 59).[87]

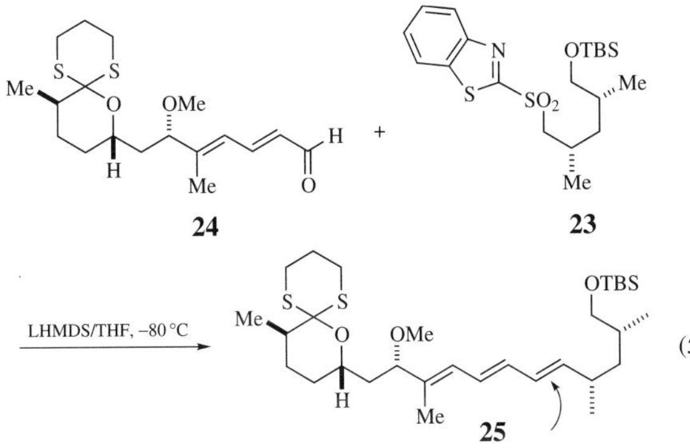

24   **23**

(59)

25

$E:Z = 19:1$

1-Phenyl-1*H*-tetrazol-5-yl sulfones were introduced as another alternative to phenyl sulfones, which led to alkenes in a one-pot procedure from aldehydes. Earlier results with simple systems showed that KHMDS provided higher selectivity than LHMDS.[89] However, in a more complex system, it was observed that *E*-vinyl cyclopropane (**28**) could be synthesized exclusively from aldehyde **26** and sulfone **27** in the presence of LHMDS in DMF and HMPA (eq 61).[90] In contrast, the *Z*-isomer was mainly formed in the presence of NaHMDS or KHMDS.

Generation of Arsonium Ylides. The use of LHMDS as base to deprotonate arsonium salts has been reported. Lithium was found to be the counter-ion that provides the highest selectivity (eq 62).[91]

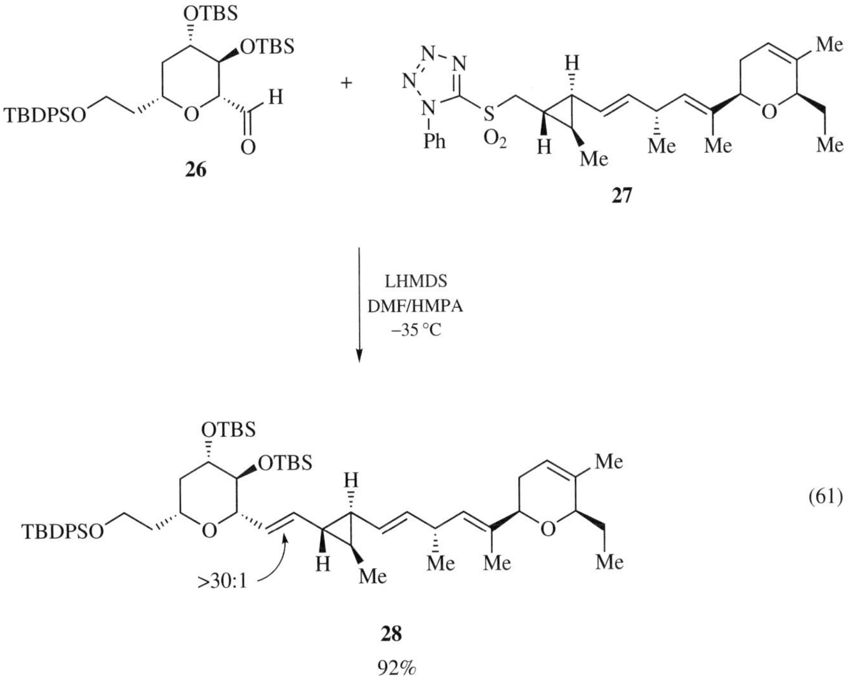

26   **27**

LHMDS
DMF/HMPA
−35 °C

(61)

28

92%

96%, 70:30 dr

Related Reagents. Lithium diethylamide; lithium diisopropylamide; lithium piperidide; lithium pyrrolidide; lithium 2,2,6,6-tetramethylpiperidide.

1. Wannagat, U.; Nierderprüm, H., *Ber. Dtsch. Chem. Ges.* **1961**, *94*, 1540.

2. (a) Mootz, D.; Zinnius, A.; Böttcher, B., *Angew. Chem., Int. Ed. Engl.* **1969**, *8*, 378. (b) Rogers, R. D.; Atwood, J. L.; Grüning, R., *J. Organomet. Chem.* **1978**, *157*, 229. For further structural information see (c) Lappert, M. F.; Power, P. P.; Sanger, A. R.; Srivastava, R. C., *Metal and Metalloid Amides*; Wiley: New York, 1980.

3. Kimura, B. Y.; Brown, T. L., *J. Organomet. Chem.* **1971**, *26*, 57.

4. Fraser, R. R.; Mansour, T. S., *J. Org. Chem.* **1984**, *49*, 3442.

5. (a) Rathke, M. W., *Org. Synth., Coll. Vol.* **1988**, *6*, 598. (b) Amonoo-Neizer, E. H.; Shaw, R. A.; Skovlin, D. O.; Smith, B. C., *Inorg. Synth.* **1966**, *8*, 19.

6. Murray, W.; Wachter, M.; Barton, D.; Forero-Kelly, Y., *Synthesis* **1991**, 18.

7. Es-Sayed, M.; Gratkowski, C.; Krass, N.; Meyers, A. I.; de Meijere, A., *Synlett* **1992**, 962.

8. (a) Rathke, M. W., *J. Am. Chem. Soc.* **1970**, *92*, 3222. (b) Mori, K.; Matsuda, H., *Justus Liebigs Ann. Chem.* **1992**, 131. (c) Pettersson, L.; Magnusson, G.; Frejd, T., *Acta Chem. Scand.* **1993**, *47*, 196.

9. Tanabe, M.; Crowe, D. F., *J. Chem. Soc., Chem. Commun.* **1973**, 564.

10. Barton, D. H. R.; Hesse, R. H.; Pechet, M. M.; Wiltshire, C., *Chem. Commun.* **1972**, 1017.

11. Denniff, P.; Whiting, D. A., *J. Chem. Soc., Chem. Commun.* **1976**, 712.

12. (a) Newman, M. S.; Magerlein, B. J., *Org. React.* **1949**, *5*, 413. (b) Morrison, J. D.; Mosher, H. S., *Asymmetric Organic Reactions*; Prentice–Hall: New York, 1971.

13. Borch, R. F., *Tetrahedron Lett.* **1972**, 3761.

14. (a) Stork, G.; Gardner, J. O.; Boeckman, Jr., R. K.; Parker, K. A., *J. Am. Chem. Soc.* **1973**, *95*, 2014. (b) Stork, G.; Boeckman, Jr., R. K., *J. Am. Chem. Soc.* **1973**, *95*, 2016. (c) Stork, G.; Cohen, J. F., *J. Am. Chem. Soc.* **1974**, *96*, 5270.

15. (a) Ireland, R. E.; Daub, J. P., *J. Org. Chem.* **1981**, *46*, 479. (b) Fujisawa, T.; Maehata, E.; Kohama, H.; Sato, T., *Chem. Lett.* **1985**, 1457. (c) Sato, T.; Tsunekawa, H.; Kohama, H.; Fujisawa, T., *Chem. Lett.* **1986**, 1553. (d) Ireland, R. E.; Wipf, P.; Armstrong, III, J. D., *J. Org. Chem.* **1991**, *56*, 650. (e) Panek, J. S.; Clark, T. D., *J. Org. Chem.* **1992**, *57*, 4323.

16. Ihara, M.; Suzuki, M.; Fukumoto, K.; Kabuto, C., *J. Am. Chem. Soc.* **1990**, *112*, 1164.

17. (a) Morimoto, T.; Takahashi, T.; Sekiya, M., *J. Chem. Soc., Chem. Commun.* **1984**, 794. For other closely related examples of amine synthesis see: (b) King, F. D.; Walton, D. R. M., *J. Chem. Soc., Chem. Commun.* **1974**, 256. (c) Murai, T.; Yamamoto, M.; Kondo, S.; Kato, S., *J. Org. Chem.* **1993**, *58*, 7440.

18. (a) Hart, D. J.; Kanai, K.-i.; Thomas, D. G.; Yang, T.-K., *J. Org. Chem.* **1983**, *48*, 289. (b) Cainelli, G.; Giacomini, D.; Panunzio, M.; Martelli, G.; Spunta, G., *Tetrahedron Lett.* **1987**, *28*, 5369. (c) Andreoli, P.; Billi, L.; Cainelli, G.; Panunzio, M.; Bandini, E.; Martelli, G.; Spunta, G., *Tetrahedron* **1991**, *47*, 9061 and references cited therein. (d) Colvin, E. W., *Silicon Reagents in Organic Synthesis*; Academic: London, 1988; p 73 and references cited therein.

19. Leboutet, L.; Courtois, G.; Miginiac, L., *J. Organomet. Chem.* **1991**, *420*, 155.

20. El Alami, N.; Belaud, C.; Villieras, J., *Synth. Commun.* **1988**, *18*, 2073.

21. Matsuda, Y.; Tanimoto, S.; Okamoto, T.; Ali, S. M., *J. Chem. Soc., Perkin Trans. 1* **1989**, 279.

22. Corriu, R. J. P.; Huynh, V.; Iqbal, J.; Moreau, J. J. E.; Vernhet, C., *Tetrahedron* **1992**, *48*, 6231.

23. (a) Sampson, P.; Hammond, G. B.; Weimer, D. F., *J. Org. Chem.* **1986**, *51*, 4342. (b) Kowalski, C. J.; O'Dowd, M. L.; Burke, M. C.; Fields, K. W., *J. Am. Chem. Soc.* **1980**, *102*, 5411.

24. Lucht, B. L.; Collum, D. B., *Acc. Chem. Res.* **1999**, *32*, 1035.

25. Davis, F. A.; Reddy, R. E.; Szewczyk, J. M.; Reddy, G. V.; Portonovo, P. S.; Zhang, H. M.; Fanelli, D.; Reddy, R. T.; Zhou, P.; Carroll, P. J., *J. Org. Chem.* **1997**, *62*, 2555.

26. Fanelli, D. L.; Szewczyk, J. M.; Zhang, Y.; Reddy, G. V.; Burns, D. M.; Davis, F. A., *Org. Synth.* **2000**, *77*, 50.

27. Davis, F. A.; Reddy, R. E.; Szewczyk, J. M., *J. Org. Chem.* **1995**, *60*, 7037.

28. Hartwig, J. F., *Angew. Chem., Int. Ed.* **1998**, *37*, 2047.

29. Muci, A. R.; Buchwald, S. L., *Top. Curr. Chem.* **2002**, *219*, 131.

30. Hartwig, J. F.; Louie, J., *Tetrahedron Lett.* **1995**, *36*, 3609.

31. Harris, M. C.; Huang, X. H.; Buchwald, S. L., *Org. Lett.* **2002**, *4*, 2885.

32. Lee, S.; Jorgensen, M.; Hartwig, J. F., *Org. Lett.* **2001**, *3*, 2729.

33. Huang, X.; Buchwald, S. L., *Org. Lett.* **2001**, *3*, 3417.

34. Culkin, D. A.; Hartwig, J. F., *Acc. Chem. Res.* **2003**, *36*, 234.

35. Miura, M.; Nomura, M., *Top. Curr. Chem.* **2002**, *219*, 211.

36. Lloyd-Jones, G. C., *Angew. Chem., Int. Ed.* **2002**, *41*, 953.

37. Lee, S.; Beare, N. A.; Hartwig, J. F., *J. Am. Chem. Soc.* **2001**, *123*, 8410.

38. Moradi, W. A.; Buchwald, S. L., *J. Am. Chem. Soc.* **2001**, *123*, 7996.

39. Jorgensen, M.; Lee, S.; Liu, X. X.; Wolkowski, J. P.; Hartwig, J. F., *J. Am. Chem. Soc.* **2002**, *124*, 12557.

40. Gaertzen, O.; Buchwald, S. L., *J. Org. Chem.* **2002**, *67*, 465.

41. Culkin, D. A.; Hartwig, J. F., *J. Am. Chem. Soc.* **2002**, *124*, 9330.

42. Kazmaier, U., *Curr. Org. Chem.* **2003**, *7*, 317.

43. Weiss, T. D.; Helmchen, G.; Kazmaier, U., *Chem. Commun.* **2002**, 1270.

44. Kanayama, T.; Yoshida, K.; Miyabe, H.; Takemoto, Y., *Angew. Chem., Int. Ed.* **2003**, *42*, 2054.

45. Trost, B. M.; Dogra, K., *J. Am. Chem. Soc.* **2002**, *124*, 7256.

46. Evans, P. A.; Robinson, J. E.; Nelson, J. D., *J. Am. Chem. Soc.* **1999**, *121*, 6761.

47. Evans, P. A.; Robinson, J. E.; Moffett, K. K., *Org. Lett.* **2001**, *3*, 3269.

48. Evans, P. A.; Leahy, D. K., *J. Am. Chem. Soc.* **2002**, *124*, 7882.

49. Foot, O. F.; Knight, D. W., *Chem. Commun.* **2000**, 975.

50. Wong, L. S. M.; Sharp, L. A.; Xavier, N. M. C.; Turner, P.; Sherburn, M. S., *Org. Lett.* **2002**, *4*, 1955.

51. Giacobbe, S. A.; Di Fabio, R.; Baraldi, D.; Cugola, A.; Donati, D., *Synth. Commun.* **1999**, *29*, 3125.

52. Grandjean, D.; Pale, P.; Chuche, J., *Tetrahedron Lett.* **1994**, *35*, 3529.

53. Corey, E. J.; Fuchs, P. L., *Tetrahedron Lett.* **1972**, 3769.

54. Jones, G. B.; Wright, J. M.; Plourde, G. W.; Hynd, G.; Huber, R. S.; Mathews, J. E., *J. Am. Chem. Soc.* **2000**, *122*, 1937.

55. Wright, J. M.; Jones, G. B., *Tetrahedron Lett.* **1999**, *40*, 7605.

56. Hynd, G.; Jones, G. B.; Plourde, G. W.; Wright, J. M., *Tetrahedron Lett.* **1999**, *40*, 4481.

A list of General Abbreviations appears on the front Endpapers

57. Barnes, D. M.; Ji, J. G.; Fickes, M. G.; Fitzgerald, M. A.; King, S. A.; Morton, H. E.; Plagge, F. A.; Preskill, M.; Wagaw, S. H.; Wittenberger, S. J.; Zhang, J., *J. Am. Chem. Soc.* **2002**, *124*, 13097.

58. Aggarwal, V. K.; Alonso, E.; Fang, G. Y.; Ferrar, M.; Hynd, G.; Porcelloni, M., *Angew. Chem., Int. Ed.* **2001**, *40*, 1433.

59. Aggarwal, V. K.; de Vicente, J.; Bonnert, R. V., *Org. Lett.* **2001**, *3*, 2785.

60. Aggarwal, V. K.; Alonso, E.; Bae, I.; Hynd, G.; Lydon, K. M.; Palmer, M. J.; Patel, M.; Porcelloni, M.; Richardson, J.; Stenson, R. A.; Studley, J. R.; Vasse, J.-L.; Winn, C. L., *J. Am. Chem. Soc.* **2003**, *125*, 10926.

61. Cheung, W. H.; Zheng, S. L.; Yu, W. Y.; Zhou, G. C.; Che, C. M., *Org. Lett.* **2003**, *5*, 2535.

62. Maryanoff, B. E.; Reitz, A. B.; Duhl-Emswiler, B. A., *J. Am. Chem. Soc.* **1985**, *107*, 217.

63. Maryanoff, B. E.; Duhl-Emswiler, B. A., *Tetrahedron Lett.* **1981**, *22*, 4185.

64. Hanessian, S.; Guindon, Y.; Lavallee, P.; Dextraze, P., *Carbohydr. Res.* **1985**, *141*, 221.

65. Leblanc, Y.; Fitzsimmons, B. J.; Adams, J.; Perez, F.; Rokach, J., *J. Org. Chem.* **1986**, *51*, 789.

66. Saniere, M.; Le Merrer, Y.; Barbe, B.; Koscielniak, T.; Depezay, J. C., *Tetrahedron* **1989**, *45*, 7317.

67. Delorme, D.; Girard, Y.; Rokach, J., *J. Org. Chem.* **1989**, *54*, 3635.

68. Morris, J.; Wishka, D. G., *Tetrahedron Lett.* **1988**, *29*, 143.

69. Yeola, S. N.; Saleh, S. A.; Brash, A. R.; Prakash, C.; Taber, D. F.; Blair, I. A., *J. Org. Chem.* **1996**, *61*, 838.

70. Wernic, D.; DiMaio, J.; Adams, J., *J. Org. Chem.* **1989**, *54*, 4224.

71. Beaulieu, P. L.; Duceppe, J. S.; Johnson, C., *J. Org. Chem.* **1991**, *56*, 4196.

72. Azuma, H.; Tamagaki, S.; Ogino, K., *J. Org. Chem.* **2000**, *65*, 3538.

73. Sibi, M. P.; Renhowe, P. A., *Tetrahedron Lett.* **1990**, *31*, 7407.

74. Paterson, I.; Cowden, C.; Watson, C., *Synlett* **1996**, 209.

75. Williams, D. R.; Brooks, D. A.; Moore, J. L.; Stewart, A. O., *Tetrahedron Lett.* **1996**, *37*, 983.

76. Mulzer, J.; Mantoulidis, A.; Oehler, E., *J. Org. Chem.* **2000**, *65*, 7456.

77. Smith, A. B.; Zhu, W. Y.; Shirakami, S.; Sfouggatakis, C.; Doughty, V. A.; Bennett, C. S.; Sakamoto, Y., *Org. Lett.* **2003**, *5*, 761.

78. Evans, D. A.; Rajapakse, H. A.; Chiu, A.; Stenkamp, D., *Angew. Chem., Int. Ed.* **2002**, *41*, 4573.

79. Kerdesky, F. A. J.; Schmidt, S. P.; Brooks, D. W., *J. Org. Chem.* **1993**, *58*, 3516.

80. Bonnet, B.; Le Gallic, Y.; Ple, G.; Duhamel, L., *Synthesis* **1993**, 1071.

81. Ando, K., *J. Org. Chem.* **1999**, *64*, 8406.

82. Davis, F. A.; Wu, Y. Z.; Yan, H. X.; McCoull, W.; Prasad, K. R., *J. Org. Chem.* **2003**, *68*, 2410.

83. Keck, G. E.; Kachensky, D. F.; Enholm, E. J., *J. Org. Chem.* **1985**, *50*, 4317.

84. McCarthy, J. R.; Matthews, D. P.; Paolini, J. P., *Org. Synth.* **1995**, *72*, 216.

85. Blakemore, P. R., *J. Chem. Soc., Perkin Trans. 1* **2002**, 2563.

86. Baudin, J. B.; Hareau, G.; Julia, S. A.; Lorne, R.; Ruel, O., *Bull. Soc. Chim. Fr.* **1993**, *130*, 856.

87. Bellingham, R.; Jarowicki, K.; Kocienski, P.; Martin, V., *Synthesis* **1996**, 285.

88. Charette, A. B.; Berthelette, C.; St-Martin, D., *Tetrahedron Lett.* **2001**, *42*, 5149.

89. Blakemore, P. R.; Cole, W. J.; Kocienski, P. J.; Morley, A., *Synlett* **1998**, 26.

90. Liu, P.; Jacobsen, E. N., *J. Am. Chem. Soc.* **2001**, *123*, 10772.

91. Dai, W. M.; Lau, C. W., *Tetrahedron Lett.* **2001**, *42*, 2541.

M

Methanamine, 1,1-Bis(trimethylsilyl)[1]

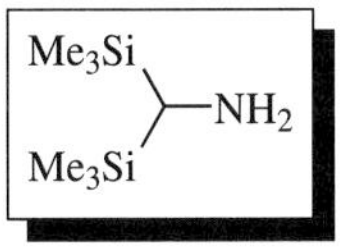

[134340-00-4]   C$_7$H$_{21}$NSi$_2$ (MW 175.42)

InChI = 1S/C7H21NSi2/c1-9(2,3)7(8)10(4,5)6/h7H,8H2,1-6H3

InChIKey = RTSDKFPTCHQXCH-UHFFFAOYSA-N

(reagent used for the preparation of *N*-bis(trimethylsilyl)methyl amines, imines, as a synthetic equivalent of *N*-methyl, *N*-(trimethylsilyl)methyl carbanion, and [*N*-CH] dianion synthons, as an *N*-protecting group)

Alternate Names: bis(trimethylsilyl)methylamine, BSMA.

Physical Data: colorless liquid, bp 70 °C at 25 mm Hg; its hydrochloride salt, mp 191 °C.

Solubility: sol ether, alcohol, dichloromethane, and most organic solvents; hydrochloride salt sol in H$_2$O, alcohol, DMF, DMSO.

Analysis of Reagent Purity: ^{1}H, ^{13}C, ^{29}Si NMR, elemental analysis, EIMS.

Preparative Methods: prepared by lithium metal mediated reductive silylation of trimethylsilyl cyanide.[2]

Purification: by fractional distillation at 25 mm Hg through a 120 mm Vigreux column.

Handling, Storage, and Precautions: best stored under nitrogen or argon in the refrigerator.

Peterson Olefination Involving Tertiary *N*-[Bis(trimethylsilyl)methyl] Amides. Tertiary amide derivatives of bis(trimethylsilyl)methylamine undergo monodesilylation, when treated with tetrabutylammonium fluoride, to form the corresponding *N*-(trimethylsilylmethyl) α-carbanion. Inter- and intramolecular Peterson olefinations of this α-carbanion with aldehydes or ketones leads to acyclic and cyclic enamides.[3] For example, the amide (**1**) reacted efficiently with the aldehyde (**2**) to afford the enamide (**3**, *E* : *Z* = 1:1) in 71% yield (eq 1). Enolizable aldehydes and ketones were unsuitable reaction partners.

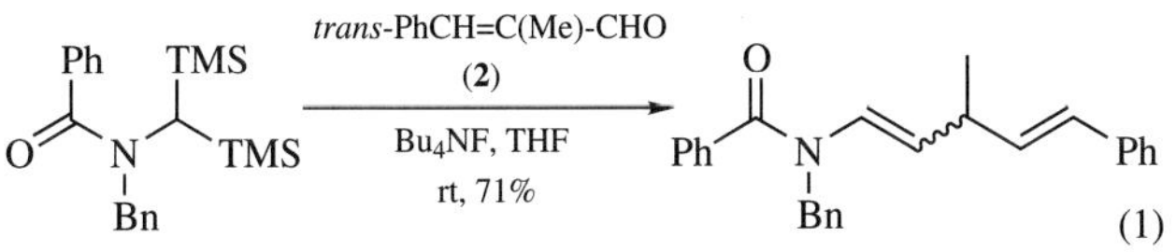

Intramolecular Peterson olefination leads to dihydroisoquinoline derivatives in modest yields. Interestingly, unlike the intermolecular reaction, the enolizable ketone (**4**) reacted to give the dihydroquinoline (**5**) in 58% yield (eq 2).

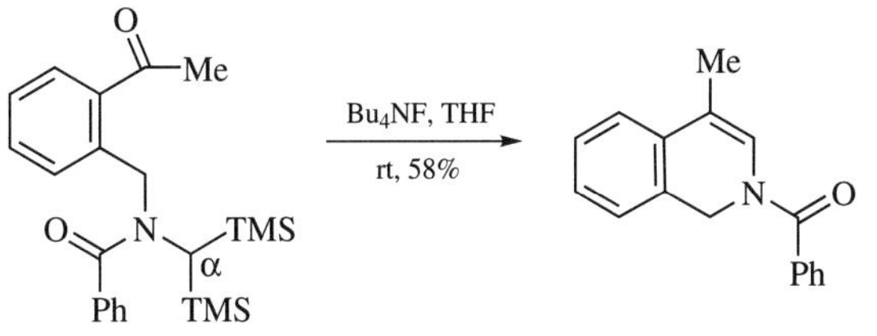

Reactions Involving *N*-Bis(trimethylsilyl)methyl Imine Derivatives. Bis(trimethylsilyl)methylamine condenses with aldehydes and ketones to form stable imines, which are useful intermediates in a variety of synthetic transformations.[4] Thus, various *N*-bis(trimethylsilyl)methyl-1-aza-1,3-dienes underwent thermally mediated, *trans*-stereoselective C→N silyl migration to furnish 1,2-bis(trimethylsilyl)-2,3-dihydropyrroles in good yields (eq 3). These products can be *N*-desilylated to afford 2-trimethylsilyl-3,4-dihydropyrroles (68–93%), which can be reduced with lithium aluminum hydride to afford pyrrolidine derivatives.

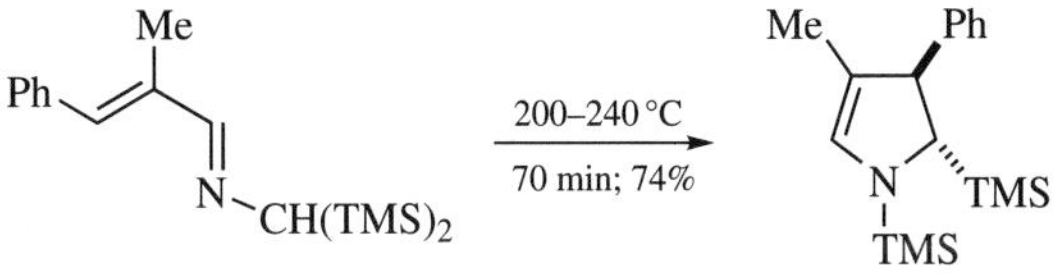

N-[Bis(trimethylsilyl)methyl] imines react efficiently with ketenes, generated in situ via base-mediated dehydrochlorination of the acid chloride, to form β-lactam products in good overall yields.[5]

A variety of aldehydes including formaldehyde were used to form the imines. β-Lactam formation was found to be *cis*-diastereoselective (>98:2 to 88:12, *cis*:*trans* ratio) when the chiral oxazolidinone derived acid chloride (**6**) was used (eq 4). The use of *N*-[bis(trimethylsilyl)methyl] β-lactams in synthesis, such as in the preparation of Type-II β-turn dipeptides have been reported.[6]

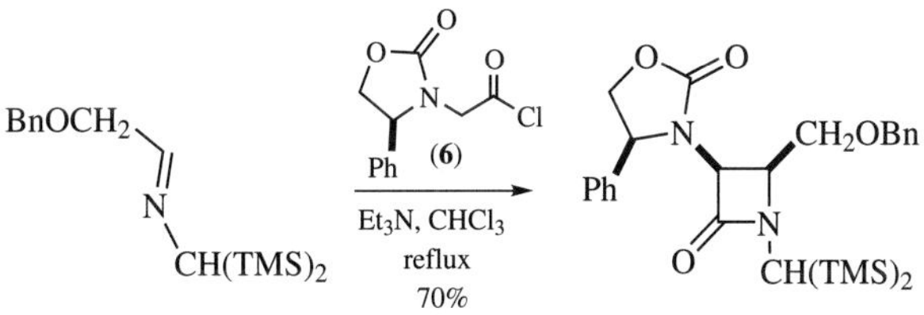

cis:*trans* = 98:2

The 2-azaallyl-carbanion of *N*-[bis(trimethylsilyl)methyl] imines can also be generated; the regioselectivity of the carbanion quenching was found to be dependent upon the nature of the base and electrophile. The steric size of the *N*-[bis(trimethylsilyl)methyl] group was also found to play a role in influencing regioselectivity of the reaction especially when bulky electrophiles are used.

$$\text{(7)} \quad \xrightarrow[\substack{-78\,^\circ\text{C} \\ 2.\ \text{“E}^+\text{”, THF} \\ -50\,^\circ\text{C} \\ 60\text{--}65\%}]{1.\ \text{MeLi, THF}} \quad \text{(8)} \quad \text{or} \quad \text{(9)} \tag{5}$$

Thus, treatment of imine (**7**, R = Ph) with methyl lithium (MeLi) and then methyl iodide gave a 63% yield of (**8**, R = Ph, E = Me). With trimethylsilyl chloride (TMSCl) as the trapping agent, product (**9**, R = Ph, E = TMS) was obtained in 60% (eq 5).[7] Similarly, deprotonation of (**7**, R = *trans*-EtCH=CHCH₂) with MeLi followed by butylation (*n*-BuI), provided a 65% yield of (**8**, R = *trans*-EtCH=CHCH₂, E = *n*-Bu), whereas with TMSCl, (**9**, R = *trans*-EtCH=CHCH₂, E = TMS) was formed in 65% yield.

The reaction of *N*-[bis(trimethylsilyl)methyl]-1-aza-1,3-butadienes with cuprates produced 1,2- and 1,4-conjugate addition products (eq 6).[8] With Gilman reagents, LiCuR₂ (e.g., R = *n*-Bu or Ph), 1,4-conjugate addition products (**10**, R = *n*-Bu; 70% or Ph; 66%) were obtained. Higher order cyanocuprates, Li₂Cu(CN)R₂, possessing heteroaryl ligands were found to lead to 1,2-conjugate addition products of type (**11**).

$$\text{(eq 6)} \tag{6}$$

For example, when R = 2-thienyl, a 55% yield of (**11**); with R = 2-furyl, compound (**11**, R = 2-furyl) was obtained in 45% yield along with a 10% yield of the 1,4-addition product (**10**, R = 2-furyl).

1,2-Nucleophilic addition of lithium alkynides to *N*-[bis(trimethylsilyl)methyl] aldimines mediated by boron trifluoride etherate afforded *N*-[bis(trimethylsilyl)methyl] propargylamines in moderate to good yields (eq 7).[9] For example, imine (**12**, R = *n*-Pr) reacted with lithium phenylethynide to give propargylamine (**13**, R = *n*-Pr, R' = Ph) in 81% yield.

$$\xrightarrow[\substack{\text{THF, } -78\,^\circ\text{C} \\ \text{MeOH, K}_2\text{CO}_3 \\ 46\text{--}85\%}]{\text{R'}\!\!\equiv\!\!\equiv\!\!-\text{Li} \\ \text{BF}_3\cdot\text{OEt}_2} \tag{7}$$

Functionalized aldimines (**12**, R = MOMOCH₂) upon reaction with lithium trimethylsilylethynide gave a 67% yield of (**13**, R = MOMOCH₂, R' = H). Arylaldimines were unreactive under the reaction conditions and enolizable aldimines gave lower yields.

C–H Insertion Reaction of *N*-[Bis(trimethylsilyl)methyl] Diazoamides.

N-[Bis(trimethylsilyl)methyl] diazoamides undergo efficient intramolecular C–H insertion reaction, catalyzed by rhodium(II) catalysts to form γ- and/or β-lactams (eq 8).[10]

Rhodium(II)-carbenoid C–H insertion into the methine C–H bond of the bis(trimethylsilyl)methyl unit was not observed. Further, the bis(trimethylsilyl)methyl group also served as an effective site-selective control element in the rhodium(II)-carbenoid reaction of diazoamides.

$$\xrightarrow[\substack{\text{CH}_2\text{Cl}_2, \text{ reflux} \\ 72\%}]{\text{Rh}_2(\text{OAc})_4 \\ (2\ \text{mol \%})} \tag{8}$$

Thus, exposure of diazoamide (**14**) to rhodium(II) tetraacetate gave a 72% yield of γ-lactam (**15**) and the corresponding β-lactam was not detected.

The *N*-[bis(trimethylsilyl)methyl] moiety is also useful in intramolecular C–H insertion reactions catalyzed by chiral rhodium(II) carboxylates and carboxamidates (eq 9).[11] Overall good regio- and chemoselectivity was observed, favoring the formation of γ-lactam products. However, these selectivities were also found to be dependent on the type of chiral rhodium(II) catalysts. The degree of asymmetric induction at the newly formed C-4 stereocenter was determined only for γ-lactam products.

$$\xrightarrow[\substack{\text{CH}_2\text{Cl}_2 \\ \text{reflux} \\ 73\text{--}92\%}]{\text{Rh}_2\text{L}^*_4 \\ (2\ \text{mol \%})} \tag{9}$$

For example, upon treatment of diazoamide (**16**, R = H, R¹ = 4-MeOC₆H₄) with 2 mol % of dirhodium(II) tetrakis[methyl 4(*R*)-2-oxazolidinone-4-carboxylate] [Rh₂(4*R*-MEOX)₄] afforded the γ-lactam (**17**) in 83% yield (69% ee). With (**16**, R = CO₂Me, R¹ = Ph) and dirhodium(II) tetrakis[(*S*)-*N*-phthaloyl-*t*-leucinate] [Rh₂(*S*-PTTL)₄], the γ-lactam (**17**, 57% yield, 65% ee) was obtained along with 23% yield of the β-lactam (**18**).

Intramolecular C–H insertion with *meso*-dioxanyl diazoamide, catalyzed by Rh₂(4*R*-MEOX)₄, led in high yield and ee to the desired bicyclic lactam (eq 10).[12] The *N*-benzhydryl diazoamide afforded the desired lactam in low yield (17%), but also resulted in the formation of products from rhodium(II)-carbenoid attack at the *N*-benzhydryl group.

$$\xrightarrow[\substack{\text{CH}_2\text{Cl}_2, \text{ reflux} \\ 95\%, 90\% \text{ ee}}]{\text{Rh}_2(4R\text{-MEOX})_4 \\ (1\ \text{mol \%})} \tag{10}$$

N-[Bis(trimethylsilyl)methyl] Group as an N-Protecting Group. The N-[bis(trimethylsilyl)methyl] group also serves as an N-protecting group in amides and lactams. The N-[bis(trimethylsilyl)methyl] moiety is readily removed by oxidation using ceric(IV) ammonium nitrate (CAN) in aqueous acetonitrile. The initially formed N-formyl amide or lactam is readily deformylated by stirring in methanolic sodium or potassium carbonate (eqs 11, 12, and 13).[11-13]

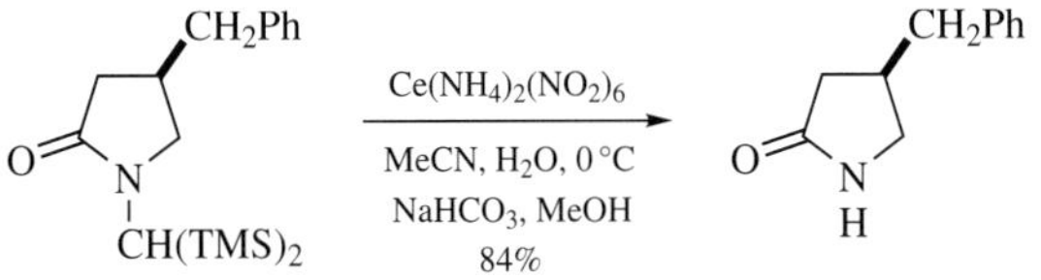

(11)

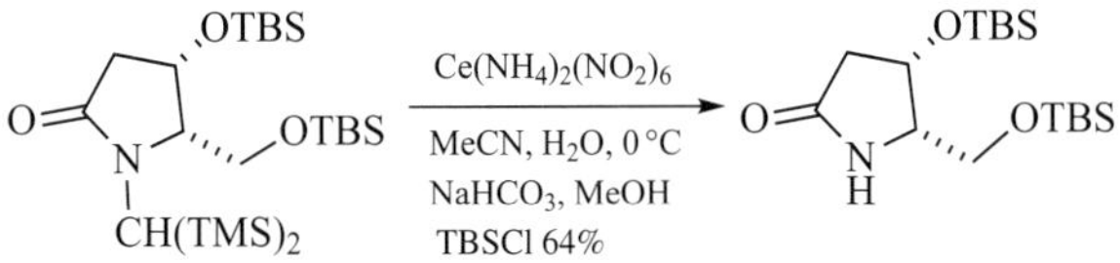

(12)

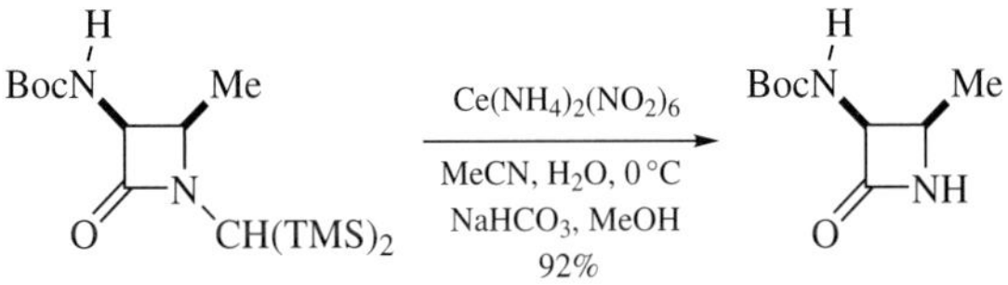

(13)

N-[Bis(trimethylsilyl)methyl] Group as a Synthetic Equivalent for the N-Methyl and N-(Trimethylsilyl)methyl Moieties. The N-[bis(trimethylsilyl)methyl] group in amides and lactams can be di- and monodesilylated using tetrabutylammonium fluoride in THF (eq 14)[6] and cesium fluoride in acetonitrile (eq 15)[6] to furnish the corresponding N-methyl or N-(trimethylsilyl)methyl amides and lactams.

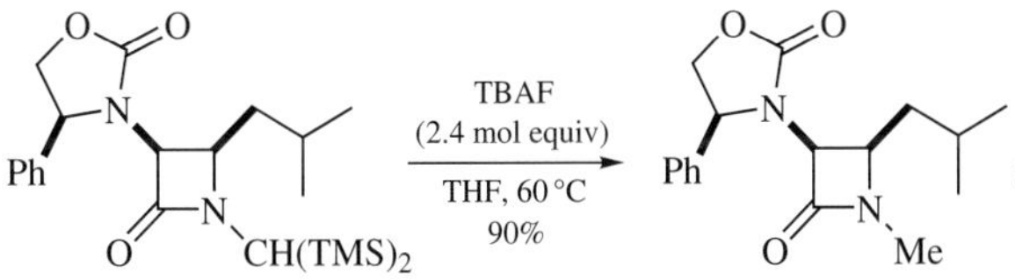

(14)

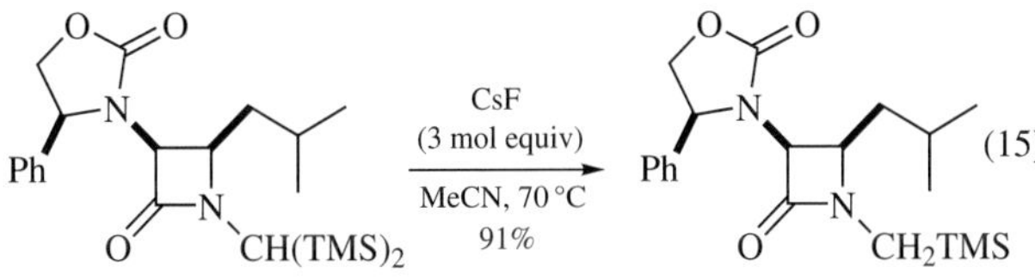

(15)

Alternatively, di- and monodesilylation can be realized using potassium fluoride dihydrate in aqueous DMSO (eq 16)[11] and sodium chloride in aqueous DMSO (eq 17).[11]

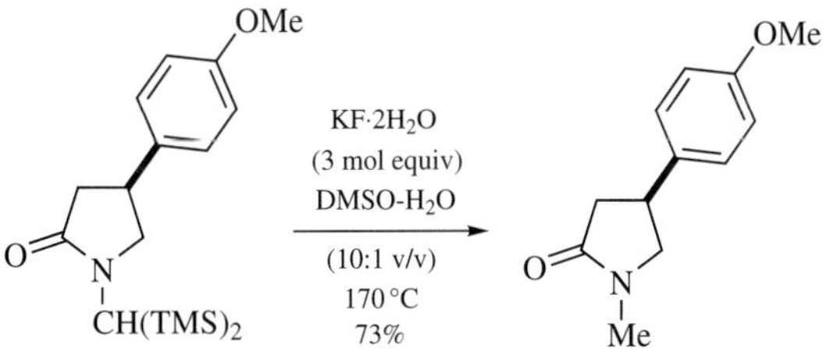

(16)

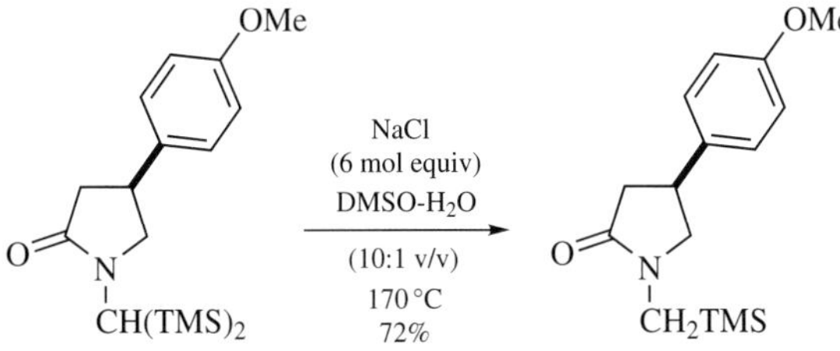

(17)

Trimethylsilyl-stabilized Azomethine Ylides. Some studies were aimed at using N-[bis(trimethylsilyl)methyl] amides and thioamides as precursors for azomethine ylides (eq 18).[14] Amides and thioamides were treated with methyl triflate to form the corresponding methyl imidate or thioimidate. Fluoride mediated monodesilylation generated the corresponding azomethine ylides, which were trapped with methyl acrylate. The 1,3-dipolar adducts were not isolated but were oxidized by 2,3-dichloro-5,6-dicyano-p-benzoquinone (DDQ) to furnish the pyrrole derivatives. With the amide (**19**, X = O), (**20**) and (**21**) were obtained in 25% and 13% yield, respectively. Only pyrrole (**20**) was obtained, in 41% yield, in the case of the thioamide (**19**, X = S).

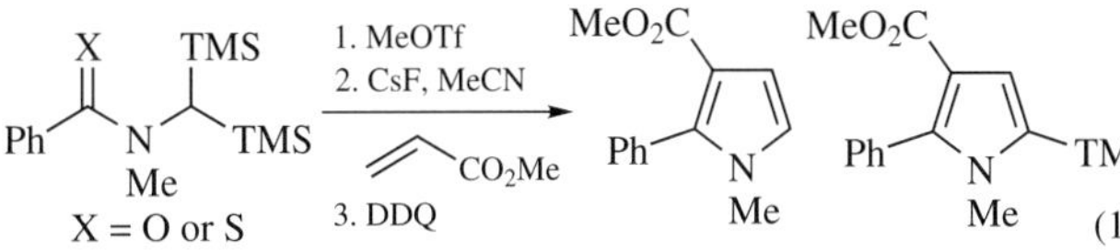

(18)

Synthesis of Amino Alcohols and Diamines. Ring opening of epoxides with bis(trimethylsilyl)methylamine gave amino alcohols. Overall good to excellent regioselectivity was observed; β-amino alcohols being favored over α-amino alcohols (eq 19).[15] In refluxing ethanol, the β-amino alcohol was obtained in 72% yield along with a minor amount (8%) of the regioisomeric α-amino alcohol.

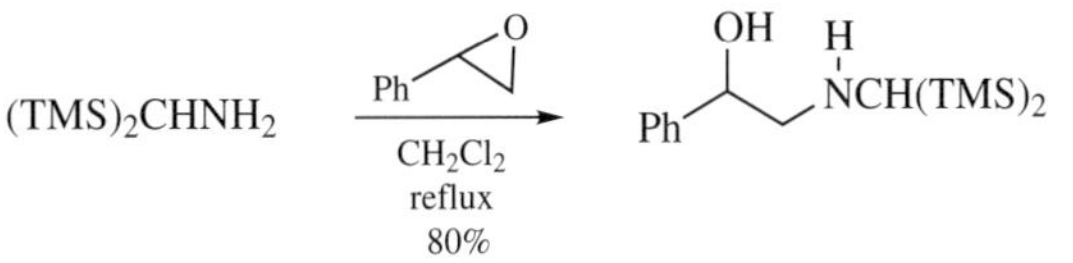

(19)

Epichlorohydrin also reacted with bis(trimethylsilyl)methylamine to give azetidine-3-ol in good yield (eq 20).[15]

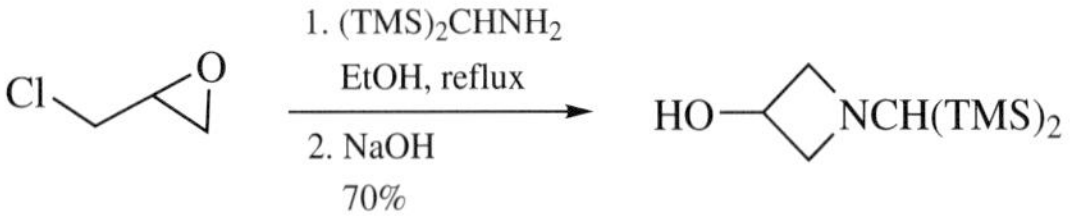

(20)

1,2- Diamines possessing N-[bis(trimethylsilyl)methyl moieties were prepared from 1,2-diimines, which in turn were obtained via condensation of bis(trimethylsilyl)methylamine with 1,2-diketones. Subsequent reduction of the diimines with lithium aluminum hydride gave the 1,2-diamine products (eq 21).[16]

Avoid Skin Contact with All Reagents

(21)

1,3-Diamines were also prepared via Michael addition reaction of bis(trimethylsilyl)methylamine with acrylonitrile followed by reduction of the nitrile group with lithium aluminum hydride (eq 22).[16]

(22)

Related Reagents. Acrylonitrile; boron trifluoride etherate; ceric(IV) ammonium nitrate; 2,3-dichloro-5,6-dicyano-*p*-benzoquinone; dirhodium(II) tetrakis[methyl 4(*R*)-2-oxazolidinone-4-carboxylate]; dirhodium(II) tetrakis[(*S*)-*N*-phthaloyl-*t*-leucinate]; DMSO; epichlorohydrin; lithium aluminum hydride; methyl acrylate; methyl lithium; methyl triflate; tetrabutylammonium fluoride; trimethylsilyl chloride; rhodium(II) tetraacetate.

1. Picard, J.-P., *Can. J. Chem.* **2000**, *78*, 1363.

2. Picard, J.-P.; Grelier, S.; Constantieux, T.; Dunogues, J.; Aizpurua, J. M.; Palomo, C.; Petraud, M.; Barbe, B., *Organometallics* **1993**, *12*, 1378.

3. Palomo, C.; Aizpurua, J. M.; Legido, M.; Picard, J. P.; Dunogues, J.; Constantieux, T., *Tetrahedron Lett.* **1992**, *33*, 3903.

4. Palomo, C.; Aizpurua, J. M.; Garcia, J. M.; Legido, M., *J. Chem. Soc., Chem. Commun.* **1991**, 524.

5. Palomo, C.; Aizpurua, J. M.; Legido, M.; Mielgo, A.; Galarza, R., *Chem. Eur. J.* **1997**, *3*, 1432.

6. Palomo, C.; Aizpurua, J. M.; Ganboa, I.; Benito, A.; Cuerdo, L.; Fratila, R. M.; Jimenez, A.; Loinaz, I.; Miranda, J. I.; Pytlewska, K. R.; Micle, A.; Linden, A., *Org. Lett.* **2004**, *6*, 4443.

7. Ricci, A.; Guerrini, A.; Seconi, G.; Mordini, A.; Constantieux, T.; Picard, J. P.; Aizpurua, J. M.; Palomo, C., *Synlett* **1994**, 955.

8. Bonini, B. F.; Fochi, M.; Franchini, M. C.; Mazanti, G.; Ricci, A.; Picard, J. P.; Dunogues, J.; Aizpurua, J. M.; Palomo, C., *Synlett* **1997**, 1321.

9. Wee, A. G. H.; Zhang, B., *Tetrahedron Lett.* **2007**, *48*, 4135.

10. Wee, A. G. H.; Duncan, S. C., *J. Org. Chem.* **2005**, *70*, 8372.

11. Wee, A. G. H.; Duncan, S. C.; Fan, G. J., *Tetrahedron: Asymmetry* **2006**, *17*, 297.

12. Doyle, M. P.; Hu, W.; Wee, A. G. H.; Wang, Z.; Duncan, S. C.; Wang, Z., *Org. Lett.* **2003**, *5*, 407.

13. Palomo, C.; Aizpurua, J. M.; Legido, M.; Galarza, R.; Deya, P. M.; Dunogues, J.; Picard, J. P.; Ricci, A.; Seconi, G., *Angew. Chem., Int. Ed. Engl.* **1996**, *35*, 1239.

14. Cuevas, J.; Patil, P.; Snieckus, V., *Tetrahedron Lett.* **1989**, *30*, 5841.

15. Constantieux, T.; Grelier, S.; Picard, J. P., *Synlett* **1998**, 510.

16. Picard, J. P.; Fortis, F.; Grelier, S., *J. Organomet. Chem.* **2003**, *686*, 306.

Andrew G. Wee & Bao Zhang
University of Regina, Regina, Saskatchewan, Canada

Methanesulfonamide, 1,1,1-Trifluoro-*N*[(trifluoromethyl)sulfonyl]-*N*(trimethylsilyl)-*

[82113-66-4] $C_5H_9F_6NO_4S_2Si$ (MW 353.34)

InChI = 1S/C5H9F6NO4S2Si/c1-19(2,3)12(17(13,14)4(6,7)8)18(15,16)5(9,10)11/h1-3H3

InChIKey = MLIRNWUYOYIGBZ-UHFFFAOYSA-N

(Lewis acid catalyst for a variety of organic transformations)

Alternate Names: TMSNTf$_2$.

Physical Data: colorless liquid, bp: 115 °C (2 Torr).

Solubility: soluble in most common organic solvents, commonly used in dichloromethane, toluene, diethyl ether.

Form Supplied in: not commercially available.

Preparative Methods: easily prepared in situ through protodesilation by mixing commercially available allyltrimethylsilane (or any TMS enol ether) and Tf$_2$NH. Originally prepared by mixing Tf$_2$NH with trimethylsilane at low temperature or mixing AgNTf$_2$ with TMSCl.

Handling, Storage, and Precaution: flammable, corrosive, very hygroscopic.

Synthesis. The original synthesis of TMSNTf$_2$ by DesMarteau involved mixing of Tf$_2$NH and TMSH, liberating H$_2$.[1] Schreeve then used metathesis with AgNTf$_2$ and TMSCl to generate TMSNTf$_2$.[2] Currently, TMSNTf$_2$ is commonly prepared in situ by mixing allyl-TMS and TMS-enol ethers or TMS-ketene acetals in a 1:1 ratio with the commercially available Tf$_2$NH (eq 1).[3]

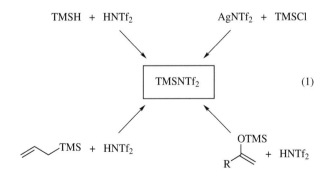

(1)

Friedel–Crafts Alkylation. One of the first uses for this highly Lewis acidic reagent in organic synthesis was found by Mikami in the Friedel–Crafts alkylation reaction of anisole (eq 2).[4] Here the first example of TMSNTf$_2$ outperforming TMSOTf was found.

*Article previously published in the electronic Encyclopedia of Reagents for Organic Synthesis as (Bis(trifluoromethylsulfonyl)amino)trimethylsilane. DOI: 10.1002/047084289X.rn00809.

Cyclization Reactions. Ghosez further explored the reactivity of this reagent and realized success in the Diels–Alder reaction of cyclic and acyclic dienes (eq 3).[3,5] Although TfOH is reported to be a stronger Brønsted acid than Tf_2NH, $TMSNTf_2$ was found to have a higher Lewis acidity than TMSOTf and has since been found to promote a variety of reactions in which TMSOTf fails.

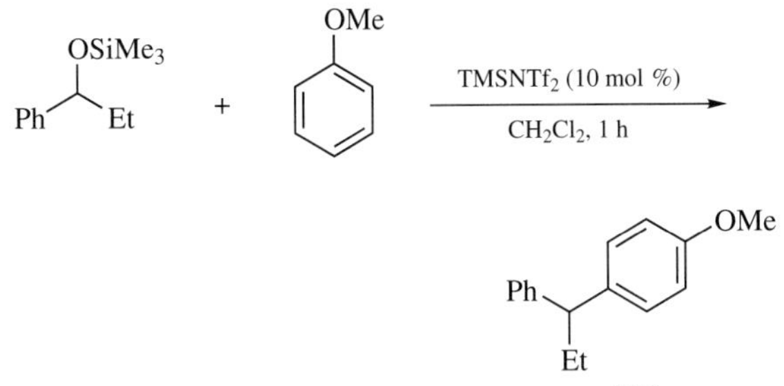

$$
\begin{array}{c}
\underset{Ph}{\overset{OSiMe_3}{\diagdown}}\!\!\!-\!Et \;+\; \text{(anisole)} \xrightarrow[\text{CH}_2\text{Cl}_2,\,1\text{ h}]{\text{TMSNTf}_2\,(10\text{ mol}\%)}
\end{array}
$$

(2)

95%

$$
\text{(cyclopentadiene)} + \overset{O}{\underset{}{\diagup}}\!\!\text{OMe} \xrightarrow[\text{toluene}]{\text{catalyst (10 mol \%)}}
$$

(3)

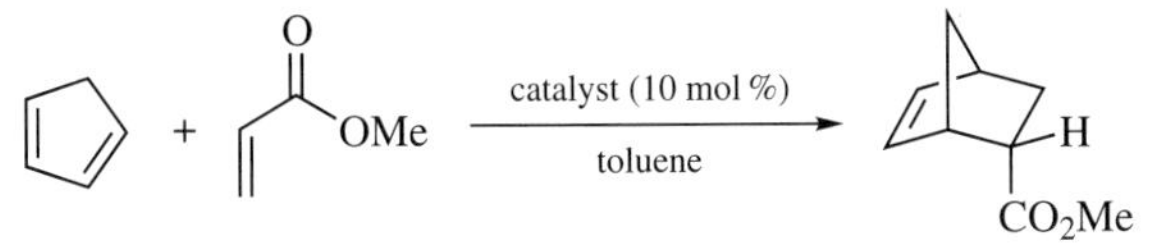

TMSNTf$_2$: 83% yield, 24/1 *endo/exo*
TMSOTf: <5% yield, 13.3/1 *endo/exo*

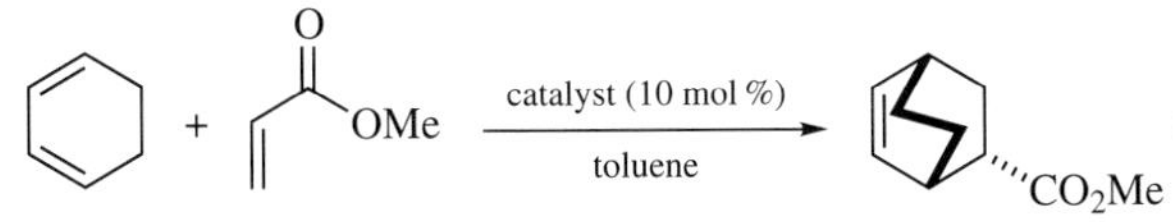

$$
\text{(cyclohexadiene)} + \overset{O}{\underset{}{\diagup}}\!\!\text{OMe} \xrightarrow[\text{toluene}]{\text{catalyst (10 mol \%)}}
$$

(3)

TMSNTf$_2$: 92% yield, 49/1 *endo/exo*
TMSOTf: nr

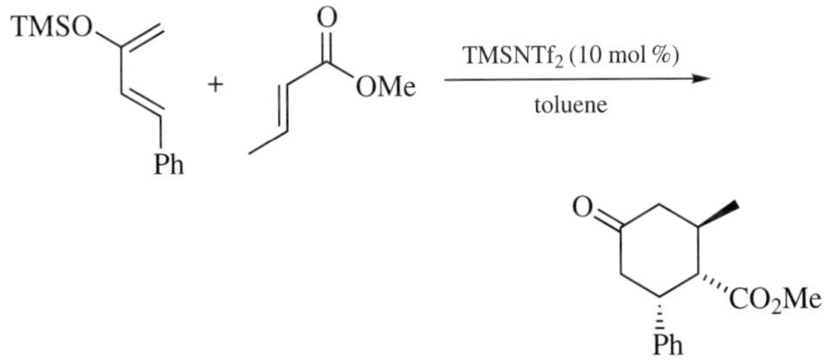

$$
\underset{Ph}{\overset{TMSO}{\diagdown}} + \overset{O}{\underset{}{\diagup}}\!\!\text{OMe} \xrightarrow[\text{toluene}]{\text{TMSNTf}_2\,(10\text{ mol}\%)}
$$

79% yield
9/1 *endo/exo*

TMSNTf$_2$ [as well as other trialkylsilyl-bis(trifluoromethanesulfonyl)imides] was also found to promote a [2 + 2] cyclization of allylsilanes with acrylates (eq 4).[6] Acrylonitrile and methyl propiolate were found to be suitable [2 + 2] reaction partners. Many examples from this report use Tf_2NH as the catalyst initiator and with the common trend, TfOH was unable to promote this cyclization reaction.

Allylation. The use of allyl-TMS and Tf_2NH to generate $TMSNTf_2$ in situ has also been applied to the allylation reaction. Robertson reported the highly regioselective and diastereoselective Sakurai reaction to cyclopentenones (eq 5).[7] Again, $TMSNTf_2$ was required for reactivity whereas TfOH or standard conditions (TiCl$_4$, 24 h) gave no Sakurai product. This system was even able to promote the 1,4-addition to more difficult α,β, unsaturated esters (eq 5).

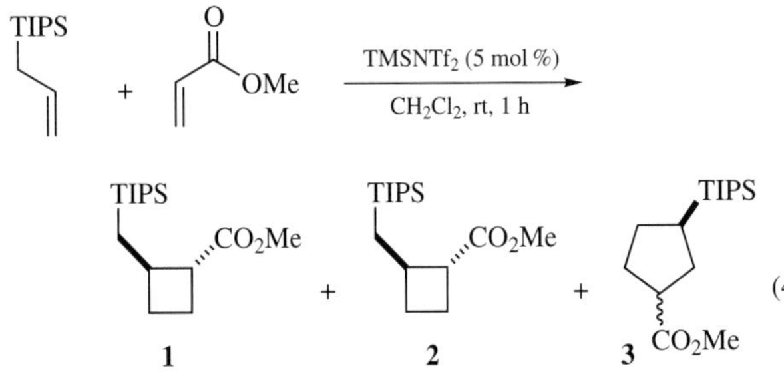

$$
\text{TIPS} + \overset{O}{\underset{}{\diagup}}\!\!\text{OMe} \xrightarrow[\text{CH}_2\text{Cl}_2,\,\text{rt, 1 h}]{\text{TMSNTf}_2\,(5\text{ mol}\%)}
$$

1 + **2** + **3** (4)

93%
75/8/17 (**1/2/3**)

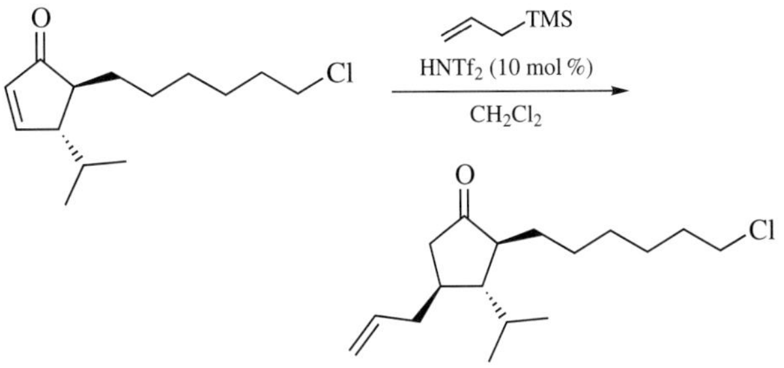

$$
\xrightarrow[\text{CH}_2\text{Cl}_2]{\overset{\text{TMS}}{\diagup}\!\!\diagup} \quad \text{HNTf}_2\,(10\text{ mol}\%)
$$

97% yield
de >98%

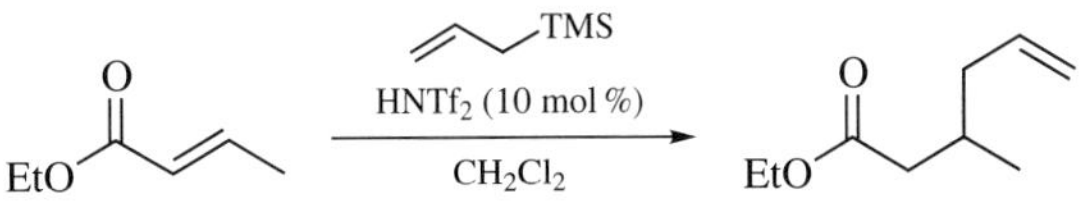

$$
\text{EtO}\overset{O}{\diagup}\!\!\diagup \xrightarrow[\text{CH}_2\text{Cl}_2]{\text{HNTf}_2\,(10\text{ mol}\%)} \text{EtO}\overset{O}{\diagup}
$$

(5)

55%

$$
\xrightarrow[\text{CH}_2\text{Cl}_2]{\text{HNTf}_2\,(10\text{ mol}\%)}
$$

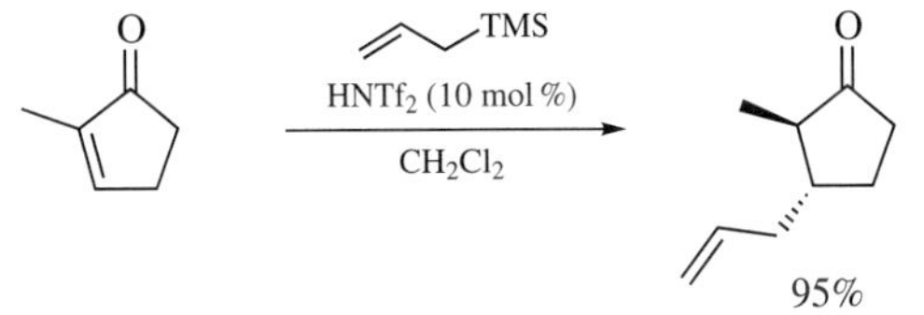

95%

The allylation of ketones and aldehydes was accomplished by Yamamoto,[8] where preformation of $TMSNTf_2$ was done in diethyl ether at $-78\,^{\circ}$C for 15 min. Slow addition of the carbonyl compound followed by acidic hydrolysis gave the allylation products in high yield (eq 6).

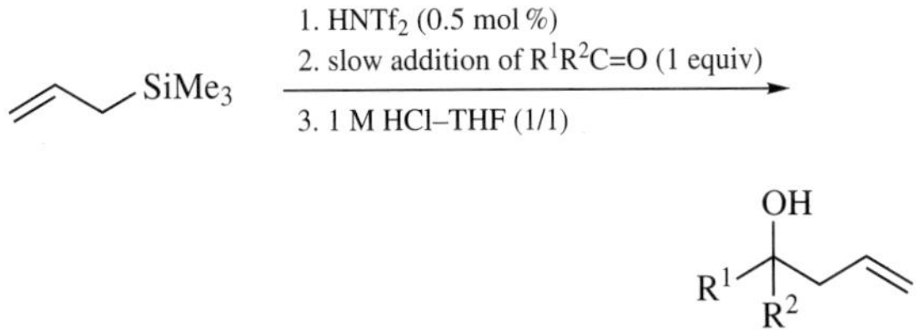

$$
\overset{}{\diagup}\!\!\diagup\text{SiMe}_3 \xrightarrow[\substack{\text{2. slow addition of } R^1R^2C=O\,(1\text{ equiv}) \\ \text{3. 1 M HCl–THF (1/1)}}]{\text{1. HNTf}_2\,(0.5\text{ mol}\%)} \underset{R^2}{\overset{OH}{\underset{R^1}{\diagup}}}
$$

(6)

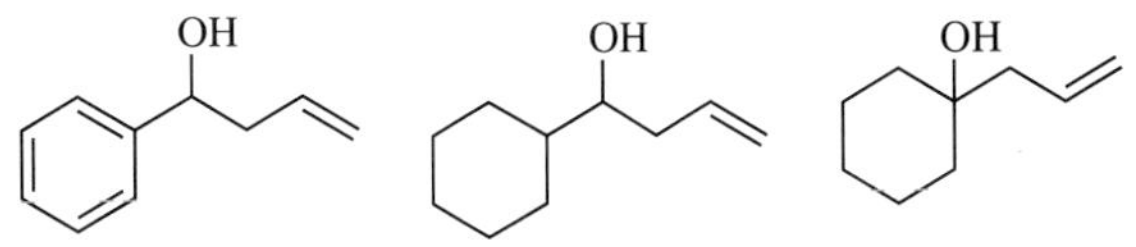

98% yield

92% yield

91% yield

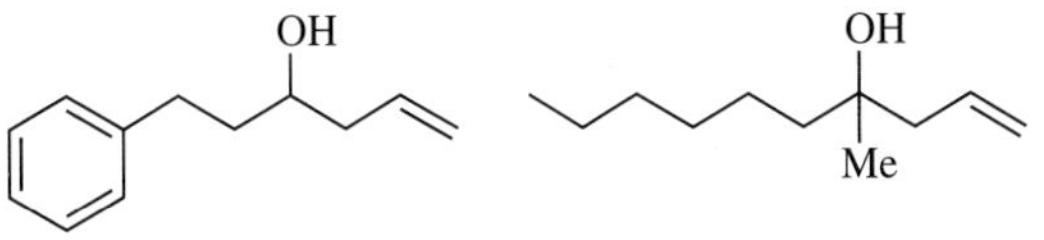

90% yield

89% yield

Avoid Skin Contact with All Reagents

Aldol and Aldol-type Reactions.   Yamamoto also used this procedure for the aldol reaction between TMS enol ethers with aldehydes and ketones (eq 7).[8]

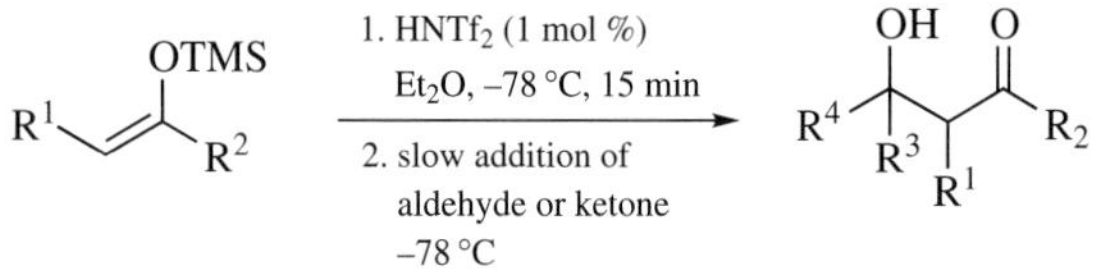

$$R^1 \underset{R^2}{\overset{OTMS}{\diagup\!\!\!\diagdown}} \xrightarrow[\substack{\text{2. slow addition of}\\ \text{aldehyde or ketone} \\ -78\,°C}]{\substack{\text{1. HNTf}_2 \text{ (1 mol \%)} \\ \text{Et}_2\text{O, } -78\,°C, 15\,\text{min}}} \underset{\underset{R^1}{R^3}}{\overset{OH\quad O}{R^4}} R_2 \qquad (7)$$

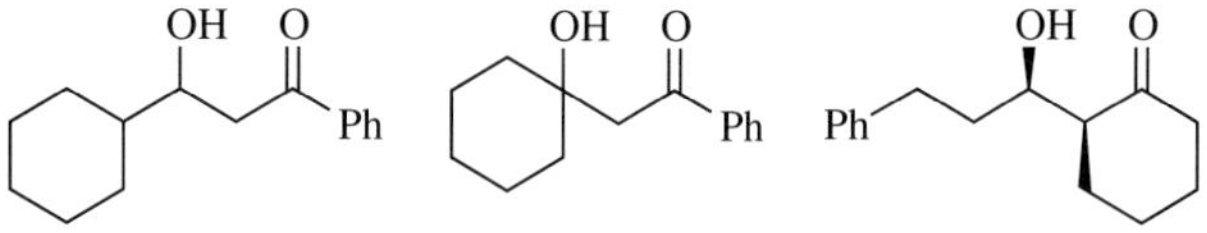

87%

87%

92%
70/30 *syn/anti*

The use of TMS-enol ethers, and -ketene acetals with the TMSNTf$_2$ catalyst was also shown to be quite effective for the α-amido alkylation (eq 8).[9] High diastereoselectivty was obtained in many cases, whereas TfOH was again unable to promote the reaction.

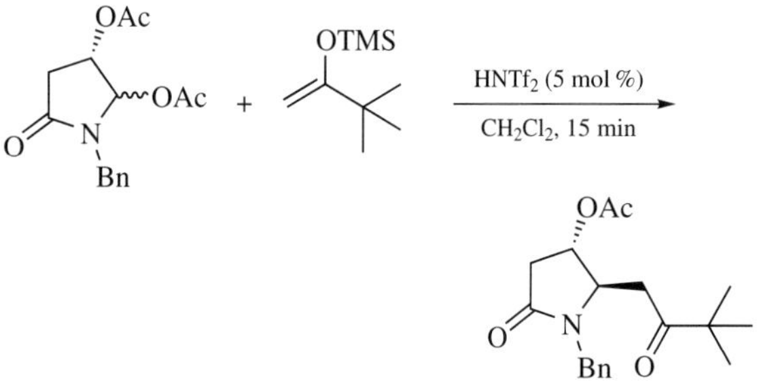

$$\xrightarrow[\text{CH}_2\text{Cl}_2, 15\,\text{min}]{\text{HNTf}_2 \text{ (5 mol \%)}} \qquad (8)$$

90%
95/5 *trans/cis*

$$\xrightarrow[\text{CH}_2\text{Cl}_2, 15\,\text{min}]{\text{HNTf}_2 \text{ (5 mol \%)}}$$

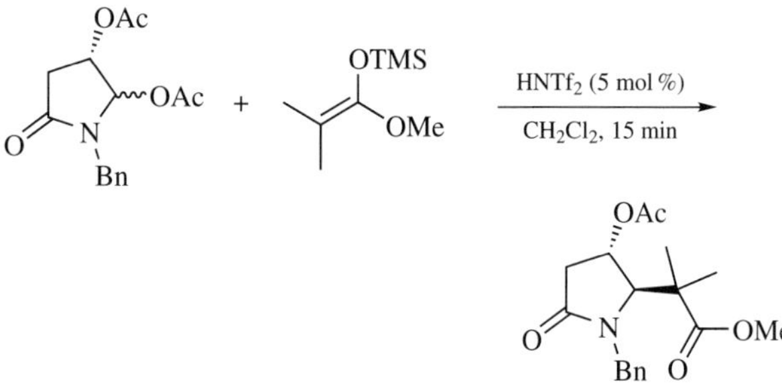

86%
99/1 *trans/cis*

Tetrahydropyran Formation.   The use of an appropriately substituted allylsilane with a variety of electrophiles with 10% TMSNTf$_2$ gave the cyclized tetrahydropyran compounds (eq 9). For this reaction, TMSNTf$_2$ *outperformed* a variety of Lewis acids including: TMSOTf, TiCl$_4$, TiCl$_2$(O*i*-Pr)$_2$, SnCl$_4$, BF$_3$·OEt$_2$, and EtAlCl$_2$.[10]

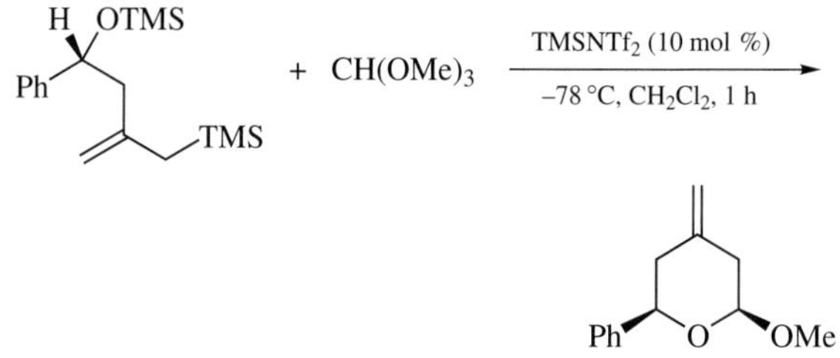

$$+ \text{CH(OMe)}_3 \xrightarrow[-78\,°C, \text{CH}_2\text{Cl}_2, 1\,\text{h}]{\text{TMSNTf}_2 \text{ (10 mol \%)}} \qquad (9)$$

83%
dr 11:1

1. Foropoulos, J.; Jr, DesMarteau, D. D., *Inorg. Chem.* **1984**, *23*, 3720.
2. Vij, A.; Zheng, Y. Y.; Kirchmeier, R. L.; Shreeve, J. M., *Inorg. Chem.* **1994**, *33*, 3281.
3. Mathieu, B.; Ghosez, L., *Tetrahedron* **2002**, *58*, 8219.
4. Ishii, A.; Kotera, O.; Saeki, T.; Mikami, K., *Synlett* **1997**, 1145.
5. Mathieu, B.; Ghosez, L., *Tetrahedron Lett.* **1997**, *38*, 5497.
6. Takasu, K.; Hosokawa, N.; Inanaga, K.; Ihara, M., *Tetrahedron Lett.* **2006**, *47*, 6053.
7. Kuhnert, N.; Peverley, J.; Robertson, J., *Tetrahedron Lett.* **1998**, *39*, 3215.
8. Ishihara, K.; Hiraiwa, Y.; Yamamoto, H., *Synlett* **2001**, 1851.
9. Othman, R. B.; Bousquet, T.; Othman, M.; Dalla, V., *Org. Lett.* **2005**, *7*, 5335.
10. Yu, C.-M.; Lee, J.-Y.; So, B.; Hong, J., *Angew. Chem., Int. Ed.* **2002**, *41*, 161.

Matthew Brian Boxer
The University of Chicago, Chicago, IL, USA

(Methoxymethyl)bis(trimethylsilyl)amine

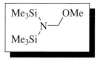

[88211-44-3]   C$_8$H$_{23}$NOSi$_2$ (MW 205.50)
InChI = 1S/C8H23NOSi2/c1-10-8-9(11(2,3)4)12(5,6)7/h8H2, 1-7H3
InChIKey = AYYQEQCMWOPLLS-UHFFFAOYSA-N

(electrophilic *N,N*-bis(trimethylsilyl)aminomethylating agent for a variety of molecules; after desilylation, the overall transformation achieves primary aminomethylation[1,2])

Physical Data: bp 94–95 °C/83 mmHg; *d* 0.873 g cm^{-3}; n_D^{20} 1.431–1.434.
Solubility: sol organic solvents such as hexane, dichloromethane, ether, alcohol; stable in neutral or basic solvents; unstable in protic solvents.
Form Supplied in: colorless liquid.
Analysis of Reagent Purity: ^{1}H NMR δ (CDCl$_3$): 0.14 (18H, s, 2 × (CH$_3$)$_3$Si), 3.17 (3H, s CH$_3$O), 4.82 (2H, s, CH$_2$); ^{13}C NMR δ (CDCl$_3$): 1.90 (q), 53.4 (q), 80.9 (t).
Handling, Storage, and Precautions: fairly stable at rt; for long storage, keep at 0–10 °C in the absence of moisture.

General Considerations. This reagent, conveniently prepared[3,4b] by the reaction of lithium hexamethyldisilazide or sodium hexamethyldisilazide with chloromethyl methyl ether in THF, is utilized for introduction of a primary aminomethyl unit into a variety of molecules by a two-step procedure involving *N,N*-bis(trimethylsilyl)aminomethylation followed by solvolytic removal of the trimethylsilyl groups. The reagent can be regarded as a synthetic equivalent for $^+CH_2NH_2$, and it functions as a reagent for achieving Mannich-type transformations[5] which are rarely possible to afford primary amines. The reagent reacts successfully with Grignard and organolithium compounds,[3] ketene silyl (or bissilyl) acetals,[4] silyl sulfides, and silyl phosphites.[6] While the *N,N*-bis(trimethylsilyl)aminomethylated products themselves are useful as *N*-protected primary amines,[7] they are easily desilylated to the corresponding primary aminomethyl compounds in protic media.

Primary Aminomethylation of Organometallic Compounds.[3] Grignard compounds react with the reagent in equimolar proportions to produce aminomethylated products in the *N*-silyl-protected form. Alkyl, allyl, and aryl Grignard reagents react smoothly in Et_2O at rt usually in high yield, whereas alkynyl Grignard reagents react under reflux in THF. Alkyllithium compounds and aryllithium compounds (prepared in situ by lithiation of arenes with *n*-BuLi) do not react with the reagent unless an equimolar amount of anhydrous magnesium bromide (best prepared from 1,2-dibromoethane and Mg in Et_2O) or $MgCl_2$ is present. Organoaluminum compounds derived from allyl and propargyl bromides have also been shown to undergo *N,N*-bis-(trimethylsilyl)aminomethylation with this reagent (eqs 1–6).[8]

$$C_6H_{13}Br \xrightarrow[\substack{2.\ (TMS)_2NCH_2OMe \\ 76\%}]{1.\ Mg,\ Et_2O} C_7H_{15}N(TMS)_2 \quad (1)$$

(2)

(3)

(4)

(5)

(6)

Solvolytic removal of the *N*-trimethylsilyl groups in the products leads to the corresponding primary amines and is easily accomplished in refluxing MeOH or EtOH in the presence of a catalytic amount of acid such as TsOH or HCl. Silica gel is also capable of catalyzing the solvolysis.

Primary Aminomethylation of Carboxylic Acids and Esters.[4] The reaction of ketene silyl acetals with the reagent followed by desilylation provides a general method for α-aminomethylation of carboxylic acids and esters. Ketene silyl acetals and bissilyl acetals, prepared from carboxylic esters and acids, respectively, by treatment with lithium diisopropylamide (LDA) and chlorotrimethylsilane, react smoothly with the reagent in CH_2Cl_2 at rt in the presence of a catalytic amount (0.01 mol equiv) of trimethylsilyl trifluoromethanesulfonate to afford *N,N*-bis(trimethylsilyl)-β-aminocarboxylic esters, generally in high yields. The triflate-catalyzed reaction involves electrophilic attack of $[(Me_3Si)_2N=CH_2]^+TfO^-$ with release of Me_3SiOMe (eqs 7–9).

(7)

$$R^1, R^2 = H\ or\ alkyl,\ R^3 = alkyl\ or\ H,\ R^4 = alkyl\ or\ TMS$$

(8)

(9)

The alkyl and trimethylsilyl esters of *N,N*-bis(trimethylsilyl)-β-aminocarboxylic acids are desilylated to the corresponding β-amino acids by treatment with catalytic amounts of acid in aqueous solvents. Selective removal of the *N*-trimethylsilyl groups of the alkyl esters can be performed with AcOH–aq. THF to give the corresponding β-amino acid esters.

***N,N*-Bis(trimethylsilyl)aminomethylation of Heteroatoms.[6]** Electrophilic introduction of a *N,N*-bis(trimethylsilyl)aminomethyl group with this reagent is possible at sulfur and phosphorus atoms by using silyl sulfides and silyl phosphites. The reaction of the reagent with phenyl- or alkylthiotrimethylsilanes and trimethylsilyl dithiocarbamates smoothly proceeds on heating at 60 °C in the presence of zinc bromide (0.05 mol equiv) to give *N,N*-bis(trimethylsilyl)aminomethylated sulfur compounds. $PhSCH_2N(SiMe_3)_2$ behaves similarly to the title reagent and is especially effective for *N,N*-bis(trimethylsilyl)aminomethylation of cuprate complexes.[2]

The reaction of the reagent with trimethylsilyl phosphite proceeds well at 60 °C in the presence of tin(II) chloride (0.05 mol equiv) to give *N,N*-bis(trimethylsilyl)aminomethylphosphonates in good yields (eqs 10–12).

$$\text{PhSTMS} \xrightarrow[\substack{\text{ZnBr}_2 \text{ (5 mol\%)} \\ \Delta \\ 98\%}]{(\text{TMS})_2\text{NCH}_2\text{OMe}} \text{PhSCH}_2\text{N(TMS)}_2 \quad (10)$$

$$\text{(piperidine-N-C(=S)-STMS)} \xrightarrow[\substack{\text{ZnBr}_2 \text{ (5 mol\%)} \\ \Delta \\ 94\%}]{(\text{TMS})_2\text{NCH}_2\text{OMe}} \text{(piperidine-N-C(=S)-SCH}_2\text{N(TMS)}_2) \quad (11)$$

$$(\text{EtO})_2\text{POTMS} \xrightarrow[\substack{\text{SnCl}_2 \text{ (5 mol\%)} \\ \Delta \\ 90\%}]{(\text{TMS})_2\text{NCH}_2\text{OMe}} (\text{EtO})_2\overset{\text{O}}{\underset{}{\text{P}}}\text{CH}_2\text{N(TMS)}_2 \quad (12)$$

N,N-Bis(trimethylsilyl)aminomethylphosphonates are desilylated to give aminomethylphosphonates, whereas *N,N*-bis(trimethylsilyl)aminomethylated sulfur compounds are not convertible to primary aminomethyl sulfides because of their instability.

1. Fieser, M., *Fieser & Fieser* **1986**, *12*, 62.

2. Morimoto, T. *Advances in Pharmaceutical Sciences*; The Research Foundation for Pharmaceutical Sciences: Japan, 1987; Vol. 3, p 153.

3. (a) Morimoto, T.; Takahashi, T.; Sekiya, M., *J. Chem. Soc.* **1984**, 794. (b) Bestmann, H. J.; Wölfel, G., *Angew. Chem., Int. Ed. Engl.* **1984**, *23*, 53. (c) Bestmann, H. J.; Wölfel, G.; Mederer, K., *Synthesis* **1987**, 848. (d) McCarthy, J. R.; Charlotte, L. B.; Matthews, D. P.; Bargar T. M., *Tetrahedron Lett.* **1987**, *28*, 2207.

4. (a) Okano, K.; Morimoto, T.; Sekiya, M., *J. Chem. Soc.* **1984**, 883. (b) Okano, K.; Morimoto, T.; Sekiya, M., *Chem. Pharm. Bull.* **1985**, *33*, 2228.

5. (a) Blicke, F. F., *Org. React.* **1942**, *1*, 303. (b) Tramontini, M., *Synthesis* **1973**, 703.

6. Morimoto, T.; Aono, M.; Sekiya, M., *J. Chem. Soc.* **1984**, 1055.

7. Morimoto, T.; Sekiya, M., *Chem. Lett.* **1985**, 1371.

8. Courtois, G.; Mesnard, D.; Dugue, B.; Miginiac, L., *Bull. Soc. Claim. Fr., Part 2* **1987**, 93.

Toshiaki Morimoto & Minoru Sekiya
University of Shizuoka, Shizuoka, Japan

1-Methoxy-2-methyl-1-(trimethylsilyloxy)propene[1]

[31469-15-5]   C$_8$H$_{18}$O$_2$Si   (MW 174.35)

InChI = 1S/C8H18O2Si/c1-7(2)8(9-3)10-11(4,5)6/h1-6H3

InChIKey = JNOGVQJEBGEKMG-UHFFFAOYSA-N

(functional equivalent of enolate of methyl isobutyrate; ester enolate surrogate in electrophilic reactions including alkylation,[2] aldol reaction,[3–5] Michael reaction,[5a, 6–8] initiator for group transfer polymerization of acrylates,[9] nitroarylation,[10] oxidation,[11] dimerization,[12] and cycloadditions[13–15])

Alternate Name: dimethylketene methyl trimethylsilyl acetal.
Physical Data: bp 35 °C/15 mmHg; *d* 0.858 g cm^{-3}.

Solubility: freely sol organic solvents.
Form Supplied in: clear colorless liquid; 95% pure.
Analysis of Reagent Purity: GC and NMR; likely impurities are methyl isobutyrate and hexamethyldisiloxane.
Handling, Storage, and Precautions: irritant; flammable; moisture sensitive; use in a fume hood.

General Considerations. Ketene alkyl silyl acetals are functional equivalents of enolates of esters that can be readily prepared and stored. The title compound is a prototypical ketene silyl acetal (KSA) that can been prepared by either of the two most commonly employed methods: (a) deprotonation of the α-hydrogen of an ester followed by silylation (eq 1),[16] and (b) metal-catalyzed hydrosilylation of α,β-unsaturated esters (eq 2).[17]

$$\text{(isobutyrate-CO}_2\text{Me)} \xrightarrow[\substack{\text{2. TMSCl} \\ 0\,°\text{C to rt} \\ 95\%}]{\text{1. LDA, 0 °C}} \text{(TMSO,OMe ketene acetal)} \xrightarrow[\substack{\text{reflux, 40 h} \\ 97\%}]{\text{Na, TMSCl}} \text{(CO}_2\text{Me, Cl)} \quad (1)$$

(**1**)

$$\text{(methacrylate-CO}_2\text{Me)} + \text{TMSH} + \underset{\substack{(0.000015 \\ \text{equiv})}}{\text{RhCl}_3 \cdot \text{H}_2\text{O}} \xrightarrow[\substack{1\% \text{ 2,6-di-}t\text{-Bu-4-Me-phenol} \\ 85\%}]{38 \text{ to } 50\,°\text{C, 26 h}} (\mathbf{1}) \quad (2)$$

Most of the reactions of KSAs are characterized by the highly nucleophilic nature of this class of compounds. Under many reaction conditions they also serve as silylating agents, often facilitating product formation via the trapping of an unstable intermediate.[10]

Electrophilic Reactions.

Alkylation. Ketene silyl acetals readily undergo Lewis acid-mediated alkylation (eq 3) with alkylating agents that form stable carbenium ions (tertiary, benzylic, allylic, those carrying an α-oxygen, sulfur, or nitrogen).[2] The reaction can be extended to primary alkylation by use of α-chloroalkyl phenyl sulfides which are valuable intermediates for not only alkylation but also alkylidenation.[18] Examples of electrophiles (see Table 1) include alkyl glycosides,[19] *N*-alkylhexahydro-1,3,5-triazine,[20] α-acetoxy[21] or α-sulfonyl amides,[22] cationic η^5-dienyliron complexes,[23] electron-deficient pyridines,[24] and elemental fluorine.[25] Addition to dimethyl acetals[5] give products which are protected aldol (eq 4) derivatives. Dibutyltin ditriflate can be used to activate acetals selectively in these type of reactions.[26]

$$(\mathbf{1}) + \underset{R^2}{\overset{R^1}{\text{>}}}{+}{-}Y \xrightarrow{\text{Lewis acid}} \underset{R^2}{\overset{\text{MeO}_2\text{C} \quad Y}{\underset{R^1}{\text{><}}}} \quad (3)$$

(Y = cation-stabilizing group)

Table 1 Prototypical reactions of ketene alkyl silyl acetals

Electrophile	Reaction conditions	Product	Yield (%)
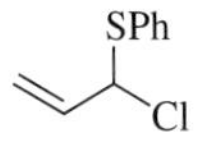	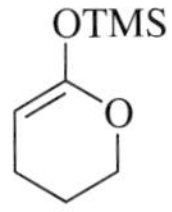 , 0.02 equiv ZnBr$_2$, CH$_2$Cl$_2$, 20 h, rt	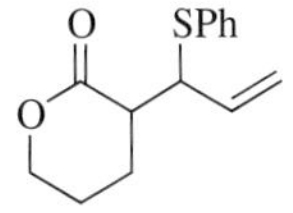	76
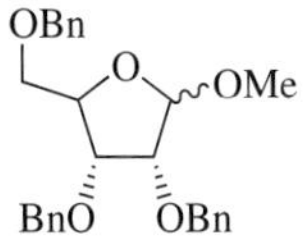	(**1**), ZnCl$_2$, TMSBr, CH$_2$Cl$_2$, rt	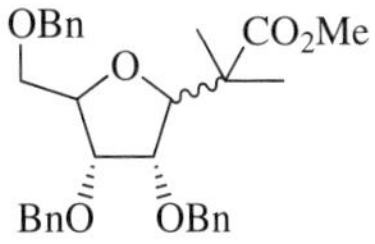	63 $\alpha{:}\beta = 5{:}1$
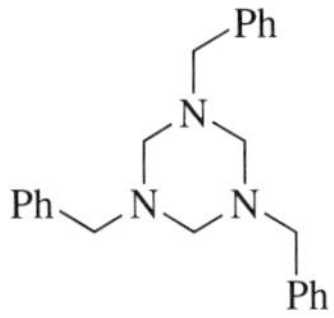	(**1**), 0.01 equiv CF$_3$SO$_3$H, CH$_2$Cl$_2$, rt, 3 h	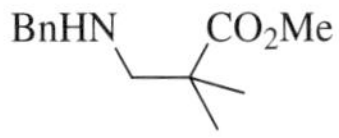	83
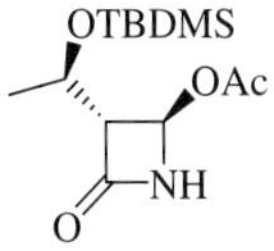	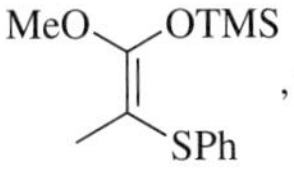 , 0.1 equiv ZnI$_2$, 1 h, MeCN, rt	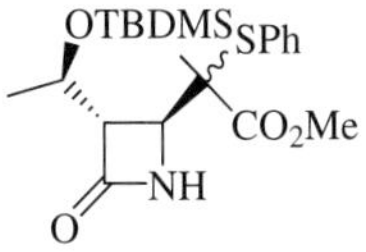	95
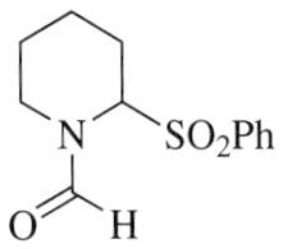	(**1**), AlCl$_3$, −78°C to rt, CH$_2$Cl$_2$, 30 min	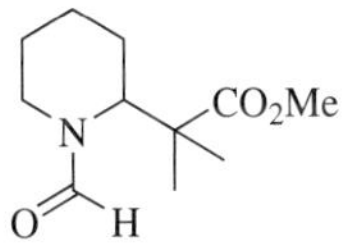	89
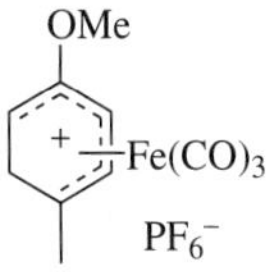	1. (**1**), CH$_2$Cl$_2$, MeCN, δ; 2. Me$_3$NO, PhH, 55 °C	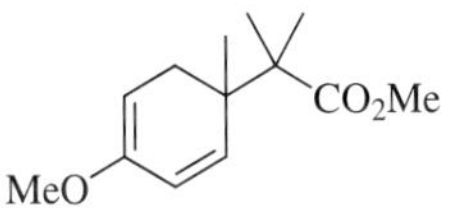	–
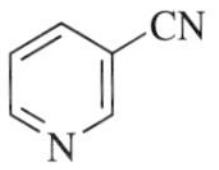	(**1**), Fe montmorillonite, CH$_2$Cl$_2$, rt, 2 h	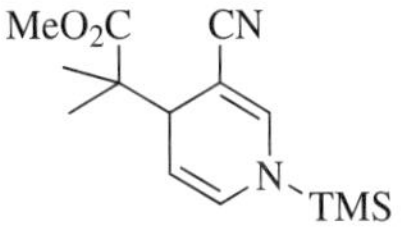	94
5% F2/N2	(**1**), Cl$_3$CF, −78 °C	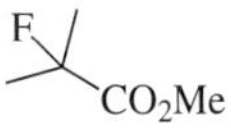	76

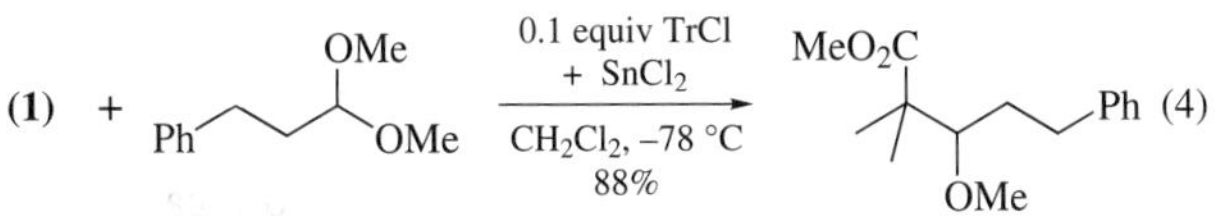

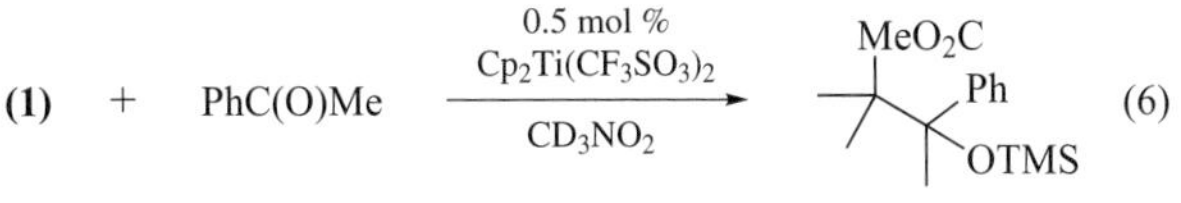

Aldol Reactions: Addition to Aldehydes and Imines. Since its discovery, the Mukaiyama aldol reaction[27] has attracted considerable attention and several improvements in reaction conditions have been reported. Most useful catalysts for this reaction appear to be recently reported lanthanide triflates (eq 5),[3] bis(cyclopentadienyl)titanium bis(trifluoromethanesulfonate), or Cp$_2$Zr(OTf)$_2$·THF.[28] The metallocene salt also catalyzes additions to ketones (eq 6). This reaction can also be carried out under essentially neutral conditions by warming (70 °C) a stoichiometric mixture of the aldehyde and the KSA in acetonitrile (eq 7).[8,29] When an optically active aldehyde is used, a slightly better stereochemical control is noticed under catalysis of zinc iodide.[29]

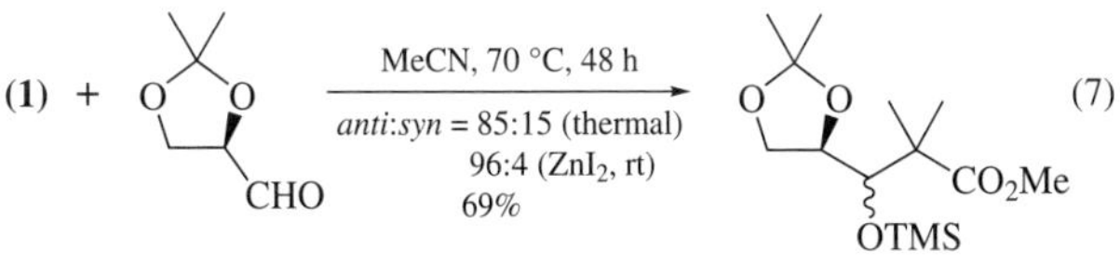

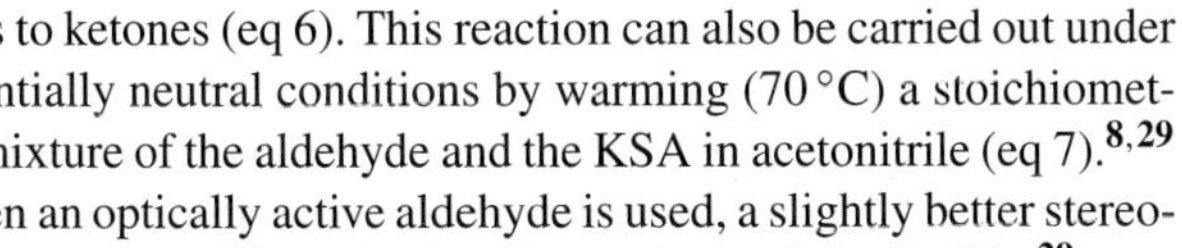

In the presence of an amino acid-derived boronate (e.g. **2**)[30] or a diamine–tin(II), complex,[31] as Lewis acids, optically active aldol products are obtained in good yields (eq 8). In the addition of KSAs to iminies, a diphosphonium ditriflate[32] (eq 9) or an acidic montmorillonite clay[33] has been claimed to give better results than the originally reported Lewis acids (titanium(IV) chloride[34] and trimethylsilyl trifluoromethanesulfonate[35]). The products of this reaction are valuable intermediates for the synthesis of β-lactams.[36] Two excellent reviews covering this area have recently appeared.[4,37]

$$4\text{-}NO_2C_6H_4SO_2N\text{-}BH$$

(2)

$$PhCHO + \underset{\text{TMSO}\quad\text{OEt}}{} \xrightarrow[\substack{\text{EtNO}_2, 1\text{ h}, -78\,°C \\ 92\%}]{0.2\text{ equiv }(\mathbf{2})} \underset{\text{OH}}{\overset{EtO_2C}{}}Ph \quad (8)$$

90% ee

$$(\mathbf{1}) + \underset{Ph}{\overset{Ph}{N}} \xrightarrow[\substack{\text{CH}_2\text{Cl}_2, -78\,°C \\ 2.\ \text{NaHCO}_3 \\ (\text{with Fe–montmorillonite} \\ 0\,°C, 92\%) \\ 94\%}]{1.\ 0.07\text{ equiv Bu}_3\overset{+}{P}\text{-O-}\overset{+}{P}Bu_3\ 2OTf^-} \underset{Ph}{\overset{CO_2Me}{NHPh}} \quad (9)$$

Mukaiyama–Michael Reactions. 1,4-Addition of ketene alkyl silyl acetals to α,β-unsaturated carbonyl compounds (eq 10) is promoted by a variety of Lewis acids[38] (for example, $TiCl_4$, $Ti(OR)_4$, $SnCl_4$, trityl perchlorate, lanthanide salts, Al–montmorillonite clay), or Lewis bases such as fluoride ion (eq 11),[7] or quaternary ammonium carboxylates.[39] Lanthanide salts are particularly effective catalysts,[40] and in the case of ytterbium(III) trifluoromethanesulfonate, the catalyst can be recovered (eq 10).[38]

$$(\mathbf{1}) + \underset{Ph}{\overset{O}{}}Ph \xrightarrow[\substack{\text{CH}_2\text{Cl}_2,\ rt \\ 94\%,\ \text{cat recovered}}]{\substack{Yb(OTf)_3 \\ (0.01\text{ equiv})}} \underset{Ph}{\overset{\text{TMSO}}{}}\underset{Ph}{\overset{CO_2Me}{}} \quad (10)$$

$$(\mathbf{1}) + \underset{}{\overset{O}{}} + \xrightarrow[\substack{(0.04\text{ equiv}) \\ \text{or} \\ \text{TMSCl, rt} \\ \text{MeNO}_2}]{\text{TASF(Me)}^*, -78\,°C} \xrightarrow{67\%} \underset{CO_2Me}{\overset{OTMS}{}} \quad (11)$$

$* = (Me_2N)_3S^+ \ (Me_3SiF_2)^-$

This reaction can also be carried out thermally,[7,8] under neutral conditions. The thermal reaction is the most convenient when the conjugate adduct is to be trapped as the regiochemically pure silyl enolate. In the case of the disubstituted KSAs such as the title compound, this addition reaction is sluggish, and may require electrophilic catalysis (eq 11) (TMSCl,[7] 5.0 M LiClO$_4$/Et$_2$O[41]) or high pressure.[42] Sequential Michael additions to α,β-unsaturated esters lead to polymers (group transfer polymerization) whose molecular weight and end group functionality can be controlled (eq 12).[9] Mechanistic studies indicate an associative intramolecular silicon transfer process via (**2**), with concomitant C–C bond formation during the polymer growth. Other Michael-type reactions include additions to acetylenecarboxylic esters (eq 13),[15,43] and nitroalkenes (eq 14).[44] Stereochemical[45] and mechanistic (electron transfer?)[46] studies on the Mukaiyama–Michael reaction have been reported and ways of improving the diastereoselectivity have been prescribed.[6]

$$(\mathbf{1}) + \underset{(MMA)}{\overset{CO_2Me}{}} \xrightarrow{F^- \text{ or } HF_2^-} \underset{CO_2Me}{\overset{\text{TMSO}\quad\text{OMe}}{}} \xrightarrow[\sim 100\%]{n(\text{MMA})}$$

$$\underset{\text{OTMS}}{\overset{MeO_2C\ \ MeO_2C\ \ OMe}{}}_n \qquad \underset{\text{poly}}{\overset{MeO}{}} \quad (12)$$

MMA = methyl methacrylate

(3)

$Si^* = SiR_3F$

poly = polymer chain

$$(\mathbf{1}) + \underset{}{\overset{CO_2Et}{}} \xrightarrow[\text{CH}_2\text{Cl}_2]{TiCl_4, -78\,°C} \underset{CO_2Et}{\overset{CO_2Me}{TiCl_3}} \xrightarrow[86\%]{NBS} \underset{Br\ \ CO_2Et}{\overset{CO_2Me}{}} \quad (13)$$

$(E:Z) = 99:1$

$$(\mathbf{1}) + \underset{}{\overset{NO_2}{}} \xrightarrow[\text{CH}_2\text{Cl}_2, -78\,°C \text{ to } 0\,°C]{1\text{ equiv TiCl}_4, Ti(O\text{-}i\text{-}Pr)_4\ (1:1)} \underset{}{\overset{MeO_2C}{}}\overset{O}{} \quad (14)$$

Nitroarylation. In the presence of stoichiometric amounts of tris(dimethylamino)sulfonium difluorotrimethylsilicate, (**1**) undergoes addition to aromatic nitro compounds to give dihydroaromatic nitro intermediates (eq 15), which are easily oxidized to α-nitroaryl carbonyl compounds by 2,3-dichloro-5,6-dicyano-1,4-benzoquinone or bromine.[10] These compounds are potentially useful for the synthesis of oxindoles and aryl acetic acids.

$$(\mathbf{1}) + \underset{Cl}{\overset{NO_2}{}}OMe \xrightarrow[\substack{\text{THF}, -40\,°C \\ 2.\ \text{DDQ or Br}_2}]{1.\ 1\text{ equiv TASF(Me)}} \underset{Cl}{\overset{NO_2\ \ CO_2Me}{}}MeO \quad (15)$$

Oxidation of Ketene Silyl Acetals. Singlet oxygen oxidizes KSAs resulting in a cleavage of C–C bonds (eq 16),[47] the reaction presumably proceeding through a silyl peroxide intermediate. Lead(IV) acetate[11] and *m*-chloroperbenzoic acid[48] oxidize KSAs to α-acetoxy (or hydroxy) carboxylic acid derivatives. Oxidative dimerization of ketene silyl acetals by $TiCl_4$ (eq 17)[12] can be understood in terms of single-electron transfer to the Lewis acids.[49]

$$\underset{OR}{\overset{\text{TBDMSO}}{}}\underset{Ar}{\overset{N}{}} \xrightarrow[\substack{\text{THF, pentane} \\ 56\%}]{^1O_2,\ 0\text{–}5\,°C} \underset{MeO}{\overset{O}{}}\underset{OMe}{\overset{N\ \ OMe}{}} \quad (16)$$

$$2\ \underset{\text{OTMS}}{\overset{OMe}{}} \xrightarrow[\substack{rt,\ 1\text{ h} \\ 80\%}]{TiCl_4,\ \text{CH}_2\text{Cl}_2} \underset{CO_2Me}{\overset{CO_2Me}{}} \quad (17)$$

Cycloaddition Reactions of Ketene Silyl Acetals. In the presence of zirconium(IV) chloride, KSA adds to ethyl propiolate in a [2 + 2] fashion (eq 18).[15] Other cycloaddition reactions include cycloaddition to imines (eq 19),[13] addition of ethyl azidoformate,[14] chlorovinylcarbene,[50] and (ethoxycarbonyl)nitrene.[51] Synthesis of β-lactams from KSAs and imines have attracted considerable attention and the subject has been reviewed recently.[4,37] Cyclopropanation of KSAs with Et_2Zn/CH_2I_2 to provide cyclopropanone acetals is also known (eq 20).[52]

$$(\mathbf{1}) \quad + \quad HC\equiv CCO_2Et \xrightarrow[CH_2Cl_2,\ 18\ h]{cat\ ZrCl_4,\ rt} \quad (18)$$

$$(\mathbf{1}) \quad + \quad \xrightarrow[\substack{-78\ ^\circ C\ to\ rt \\ 24\ h \\ 81\%}]{TiCl_4} \quad (19)$$

~98% de

$$\quad + \quad CH_2I_2 \quad + \quad Et_2Zn \xrightarrow[70\%]{Et_2O,\ 2\ h} \quad (20)$$

Related Reagents. *t*-Butyl α-Lithioisobutyrate; Dilithioacetate; Ethyl Lithioacetate; Ketene Bis(trimethylsilyl) Acetal; Ketene *t*-Butyldimethylsilyl Methyl Acetal; 1-Methoxy-1-(trimethylsilyloxy)propene; 1-Methoxy-2-trimethylsilyl-1-(trimethylsilyloxy)ethylene; Tris(trimethylsilyloxy)ethylene.

1. (a) Colvin, E. W. In *The Chemistry of the Metal–Carbon Bond*; Hartley, F. R., Ed.; Wiley: Chichester, UK, 1987; Vol. 4, p 539. (b) Weber, W. P. *Silicon Reagents for Organic Synthesis*; Springer: Berlin, 1983. (c) Mukaiyama, T., *Org. React.* **1982**, *28*, 203. (d) Brownbridge, P., *Synthesis* **1983**, 1. (e) Brownbridge, P., *Synthesis* **1983**, 85.

2. (a) Reetz, M. T., *Angew. Chem., Int. Ed. Engl.* **1982**, *21*, 96. (b) Reetz, M. T.; Schwellnus, K.; Hubner, F.; Massa, W.; Schmidt, R. E., *Chem. Ber.* **1983**, *116*, 3708.

3. Kobayashi, S.; Hachiya, I.; Takahori, T., *Synthesis* **1993**, 371. For earlier work on related lanthanides, see Ref. 6 of this article. See also: van de Weghe, P.; Collin, J., *Tetrahedron Lett.* **1993**, *34*, 3881. Mikami, K.; Terada, M.; Nakai, T., *J. Chem. Soc., Chem. Commun.* **1993**, 343.

4. Gennari, C., *Comprehensive Organic Synthesis* **1991**, *2*, 629.

5. (a) Mukaiyama, T.; Kobayashi, S., *J. Organomet. Chem.* **1990**, *382*, 39. For the use of other Lewis acids in aldol-type reactions see: (b) Montmorillonite clay (Al^{3+}): Kawai, M.; Onaka, M.; Izumi, Y., *Bull. Chem. Soc. Jpn.* **1988**, *61*, 1237. (c) TMSOTf: Murata, S.; Suzuki, M.; Noyori, R., *Tetrahedron* **1988**, *44*, 4259. (d) $Bu_2Sn(OTf)_2$: Sato, T.; Otera, J.; Nozaki, H., *J. Am. Chem. Soc.* **1990**, *112*, 901.

6. Otera, J.; Fujita, Y.; Sato, T.; Nozaki, H., *J. Org. Chem.* **1992**, *57*, 5054.

7. RajanBabu, T. V., *J. Org. Chem.* **1984**, *49*, 2083. See also: Refs. 5(a) and 39.

8. (a) Kita, Y.; Segawa, J.; Haruta, J.; Fuji, T.; Yasuda, H.; Tamura, Y., *J. Chem. Soc., Perkin Trans. 1* **1982**, 1099. For the use of nitromethane as a solvent, see Ref. 7. (b) A recent report (Berl, V.; Helmchen, G.; Preston, S., *Tetrahedron Lett.* **1994**, *35*, 233) questions the role of the solvent (acetonitrile) and suggests phosphorus-containing impurities as responsible for the catalysis of this reaction. Indeed, these workers find that P_4O_{10} is a good catalyst for the reaction In light of this observation, the thermal aldol reaction, shown in eq 7, may also have to be reinvestigated.

9. Webster, O. W.; Hertler, W. R.; Sogah, D. Y.; Farnham, W. B.; RajanBabu, T. V., *J. Am. Chem. Soc.* **1983**, *105*, 5706; *J. Macromol. Sci., Chem.* **1984**, *A21*, 943.

10. (a) RajanBabu, T. V.; Reddy, G. S.; Fukunaga, T., *J. Am. Chem. Soc.* **1985**, *107*, 5473. (b) RajanBabu, T. V.; Chenard, B. L.; Petti, M. A., *J. Org. Chem.* **1986**, *51*, 1704.

11. Rubottom, G. M.; Gruber, J. M.; Marrero, R.; Juve, H. D., Jr.; Kim, C. W., *J. Org. Chem.* **1983**, *48*, 4940.

12. (a) Inaba, S.; Ojima, I., *Tetrahedron Lett.* **1977**, 2009. (b) Hirai, K.; Ojima, I., *Tetrahedron Lett.* **1983**, *24*, 785.

13. Ojima, I.; Inaba, S., *Tetrahedron Lett.* **1980**, *21*, 2081. For applications involving KSAs from *N*-methylephedrine esters, see Gennari, C.; Schimperna, G.; Venturini, I., *Tetrahedron* **1988**, *44*, 4221.

14. Cipollone, A.; Loretto, M. A.; Pellacani, L.; Tardella, P. A., *J. Org. Chem.* **1987**, *52*, 2584.

15. Quendo, A.; Rousseau, G., *Tetrahedron Lett.* **1988**, *29*, 6443.

16. Ainsworth, C.; Chen, F.; Kuo, Y., *J. Organomet. Chem.* **1972**, *46*, 59. KSAs can also be prepared from α-halo esters using Na and TMSCl: Schulz, W. J., Jr.; Speier, J. L., *Synthesis* **1989**, 163.

17. (a) Revis, A.; Hilty, T. K., *J. Org. Chem.* **1990**, *55*, 2972. (b) See also: Yoshii, E.; Kobayashi, Y.; Koizumi, T.; Oribe, T., *Chem. Pharm. Bull.* **1974**, *22*, 2767.

18. Paterson, I., *Tetrahedron* **1988**, *44*, 4207.

19. Reetz, M. T.; Muller-Starke, H., *Liebigs Ann. Chem.* **1983**, 1726.

20. Ikeda, K.; Achiwa, K.; Sekiya, M., *Tetrahedron Lett.* **1983**, *24*, 913.

21. Kita, Y.; Shibata, N.; Tohjo, T.; Yoshida, N., *J. Chem. Soc., Perkin Trans. 1* **1992**, 1795.

22. Brown, D. S.; Hansson, T.; Ley, S. V., *Synlett* **1990**, *1*, 48. See also: (a) Brown, D. S.; Bruno, M.; Ley, S. V., *Heterocycles* **1989**, *28* (Special Issue No. 2), 773. (b) Ley, S. V.; Lygo, B.; Sternfeld, F.; Wonnacott, A., *Tetrahedron* **1986**, *42*, 4333.

23. Pearson, A. J.; O'Brien, M. K., *Tetrahedron Lett.* **1988**, *29*, 869.

24. Onaka, M.; Ohno, R.; Izumi, Y., *Tetrahedron Lett.* **1989**, *30*, 747.

25. Purrington, S. T.; Woodard, D. L., *J. Org. Chem.* **1990**, *55*, 3423.

26. Sato, T.; Otera, J.; Nozaki, H., *J. Am. Chem. Soc.* **1990**, *112*, 901.

27. Mukaiyama, T., *Org. React.* **1982**, *28*, 203.

28. Hollis, T. K.; Robinson, N. P.; Bosnich, B., *Tetrahedron Lett.* **1992**, *33*, 6423. See also: Hong, Y.; Norris, D. J.; Collins, S., *J. Org. Chem.* **1993**, *58*, 3591. Hollis, T. K.; Bosnich, B., *J. Am. Chem. Soc.* **1995**, *117*, 4750.

29. Kita, Y.; Tamura, O.; Itoh, F.; Yasuda, H.; Kishino, H.; Ke, Y. Y.; Tamura, Y., *J. Org. Chem.* **1988**, *53*, 554. The rate of this aldol reaction is accelerated by incorporating the silicon in a four-membered ring: Myers, A. G.; Kephart, S. E.; Chen, H., *J. Am. Chem. Soc.* **1992**, *114*, 7922.

30. Kiyooka, S.; Kaneko, Y.; Kume, K., *Tetrahedron Lett.* **1992**, *33*, 4927.

31. Mukaiyama, T.; Kobayashi, S.; Sano, T., *Tetrahedron* **1990**, *46*, 4653.

32. Mukaiyama, T.; Kashiwagi, K.; Matsui, S., *Chem. Lett.* **1989**, 1397.

33. Onaka, M.; Ohno, R.; Yanagiya, N.; Izumi, Y., *Synlett* **1993**, 141.

34. Ojima, I.; Inaba, S.; Yoshida, K., *Tetrahedron Lett.* **1977**, 3643.

35. Guanti, G.; Narisano, E.; Banfi, L., *Tetrahedron Lett.* **1987**, *28*, 4331.

36. Colvin, E. W.; McGarry, D.; Nugent, M. J., *Tetrahedron* **1988**, *44*, 4157. For the corresponding Li enolate version see: Ha, D.-C.; Hart, D. J.; Yang, T.-K., *J. Am. Chem. Soc.* **1984**, *106*, 4819.

37. Kleinman, E. F., *Comprehensive Organic Synthesis* **1991**, *2*, 893.

38. Kobayashi, S.; Hachiya, I.; Takahori, T.; Araki, M.; Ishitani, H., *Tetrahedron Lett.* **1992**, *33*, 6815.

39. Klimko, P. G.; Singleton, D. A., *J. Org. Chem.* **1992**, *57*, 1733. For the use of Montmorillonite clay, see: Kawai, M.; Onaka, M.; Izumi, Y., *Bull. Chem. Soc. Jpn.* **1988**, *61*, 2157.

40. van de Weghe, P.; Collin, J., *Tetrahedron Lett.* **1993**, *34*, 3881.

41. Grieco, P. A.; Cooke, R. J.; Henry, K. J.; VanderRoest, J. M., *Tetrahedron Lett.* **1991**, *32*, 4665.

42. Yamamoto, Y.; Maruyama, K.; Matsumoto, K., *Tetrahedron Lett.* **1984**, *25*, 1075. See also Bunce, R. A.; Schlecht, M. F.; Dauben, W. G.; Heathcock, C. H., *Tetrahedron Lett.* **1983**, *24*, 4943.

43. Kelly, T. R.; Ghoshal, M., *J. Am. Chem. Soc.* **1985**, *107*, 3879. See also for montmorillonite catalysis Onaka, M.; Mimura, T.; Ohno, R.; Izumi, Y., *Tetrahedron Lett.* **1989**, *30*, 6341.

44. Miyashita, M.; Yanami, T.; Kumazawa, T.; Yoshikoshi, A., *J. Am. Chem. Soc.* **1984**, *106*, 2149.

45. Heathcock, C. H.; Norman, M. H.; Uehling, D. E., *J. Am. Chem. Soc.* **1985**, *107*, 2797.

46. Sato, T.; Wakahara, Y.; Otera, J.; Nozaki, H.; Fukuzumi, S., *J. Am. Chem. Soc.* **1991**, *113*, 4028.

47. Wasserman, H. H.; Lipshutz, B. H.; Wu, J. S., *Heterocycles* **1977**, *7*, 321. See also: Adam, W.; del Fierro, J., *J. Org. Chem.* **1978**, *43*, 1159.

48. Rubottom, G. M.; Marrero, R., *Synth. Commun.* **1981**, *11*, 505.

49. Fukuzumi, S.; Fujita, M.; Otera, M.; Fujita, Y., *J. Am. Chem. Soc.* **1992**, *114*, 10271.

50. Slougui, N.; Rousseau, G., *Tetrahedron Lett.* **1987**, *28*, 1651.

51. Loreto, M. A.; Pellacani, L.; Tardella, P. A., *J. Chem. Res. (S)* **1988**, 304.

52. Rousseau, G.; Slougui, N., *Tetrahedron Lett.* **1983**, *24*, 1251.

T. V. (Babu) RajanBabu
The Ohio State University, Columbus, OH, USA

N-Methyl-*N*,*O*-bis(trimethylsilyl)-hydroxylamine

[22737-33-3] $C_7H_{21}NOSi_2$ (MW 191.47)

InChI = 1S/C7H21NOSi2/c1-8(10(2,3)4)9-11(5,6)7/h1-7H3
InChIKey = LXNSJZVQLMUDMT-UHFFFAOYSA-N

(bis-trimethylsilylated *N*-methylhydroxylamine;[1–3] can convert aldehydes and ketones to *N*-methyl nitrones with regio- and chemoselectivity;[3] source of methylnitrene by pyrolysis;[4,5] source of the trimethylsilyl(methyl)aminyl radical by photolysis[6])

Physical Data: bp 40–41 °C/10 mmHg;[2–8] n_D^{20} 1.4146.[2]
Solubility: sol benzene, toluene, CH_2Cl_2, $CHCl_3$.[2]
Form Supplied in: colorless liquid;[2–9] commercially available.
Preparative Method: obtained in 52% yield by reaction of *N*-methylhydroxylamine with chlorotrimethylsilane (2.0 equiv) and triethylamine (3.3 equiv) in ether at 25 °C.[1,2]
Handling, Storage, and Precautions: hydrolyzes readily upon exposure to moisture; stored without decomposition for several months in a sealed bottle at −10 to −15 °C;[2] relatively thermally stable, having $t_{1/2}$ of 46.5 h at 140 °C;[4] dec at 180–190 °C.[10]

N-Methyl Nitrone Formation. Aldehydes or ketones react with a stoichiometric amount of $Me_3SiN(Me)OSiMe_3$ **(1)** in benzene at 50 °C for 24 h to give *N*-methyl nitrones in good to excellent yields (eq 1).[1] For deactivated carbonyl compounds, such as *p*-nitrobenzaldehyde, *p*-(dimethylamino)benzaldehyde, and 2-furaldehyde, addition of trimethylsilyl trifluoromethanesulfonate (0.03–0.04 equiv) as a catalyst is necessary for the formation of *N*-methyl nitrones; otherwise accumulation of hemiaminal intermediates **(2)** occur. Sequential intra- and intermolecular [3 + 2]

cycloadditions can be carried out in situ for substrates containing a C=C or a C=N double bond (eqs 2 and 3).

$$R^1R^2C=O + MeN(OTMS)(TMS) \longrightarrow \left[\begin{array}{c} TMSO \\ R^1 \\ R^2 \end{array} \quad \begin{array}{c} NMe \\ \\ OTMS \end{array} \right] \longrightarrow$$

        (1)   **(2)**

$$\begin{array}{c} R^1 \\ R^2 \end{array} C=\overset{+}{N}\overset{O^-}{\underset{Me}{}} + TMSOTMS \qquad (1)$$

$$\text{CHO} \xrightarrow[\substack{1.\ (\mathbf{1}),\ 50\ °C \\ 2.\ 80\ °C}]{78\%} \qquad (2)$$

$$\text{cyclohexanone} \xrightarrow[\substack{1.\ (\mathbf{1}),\ 50\ °C \\ 2.\ PhN=C=O,\ rt}]{76\%} \qquad (3)$$

Regio- and Chemoselective Nitrone Formation. The trimethylsilyl cationic (Me_3Si^+) moiety in $Me_3SiN(Me)OSiMe_3$ serves as a 'bulky proton',[11] which enables this reagent to react selectively with carbonyl groups having different steric environments. For example, regioselective nitrone formation occurs exclusively at the C-3, instead of the sterically more crowded C-20, carbonyl group in 5α-pregnane-3,20-dione in benzene at reflux (eq 4). Also, $Me_3SiN(Me)OSiMe_3$ preferentially reacts with trimethylacetaldehyde in the presence of acetone to afford the corresponding nitrone. The selectivity offered by $Me_3SiN(Me)OSiMe_3$ is higher than that offered by MeNHOH·HCl (14:1 versus 6:1). $Me_3SiN(Me)OSiMe_3$ reacts preferentially with aldehyde functionality in the presence of chlorides. This is evidenced by the reaction of $Me_3SiN(Me)OSiMe_3$ with 5-chloropentanal (1.0 equiv) in CH_2Cl_2 at reflux to give a nitrone in 79% yield without any substitution product being formed.

$$\xrightarrow[71\%]{\substack{(\mathbf{1}) \\ PhH,\ \Delta}} \qquad (4)$$

Protection of the Carbonyl Group. For carbonyl compounds containing a second reducible functionality, selective reduction of the latter can be accomplished by protection of the carbonyl group in its hemiaminal form by use of $Me_3SiN(Me)OSiMe_3$. An example is shown in eq 5, in which 4-cyanobenzaldehyde is first treated with 1.3 equiv of $Me_3SiN(Me)OSiMe_3$ to give a hemiaminal intermediate. In situ, 1.2 equiv of a reducing agent are added to afford terephthalaldehyde in 80–93% yields after acidic workup. The applicable reducing agents include $NaBH_4$, Super-Hydride, L-Selectride, K-Selectride, LS-Selectride, BH_3·THF,

9-BBN, DIBAL, Red-Al, and Bu_3SnH.[2] This method provides a way to carry out protection–reduction–deprotection in one flask.

$$NEC-\text{—}-CHO \xrightarrow[CH_2Cl_2]{(1)} \left[NEC-\text{—}-\underset{OTMS}{\overset{OTMS}{\underset{|}{\overset{|}{C}}}}\text{N—Me} \right] \xrightarrow{[H]}$$

$$\left[\bar{N}\equiv\text{—}-\underset{OTMS}{\overset{OTMS}{\underset{|}{\overset{|}{C}}}}\text{N—Me} \right] \xrightarrow{H_3O^+} OHC-\text{—}-CHO \quad (5)$$

$$80\text{–}93\%$$

Methylnitrene Formation. Pyrolysis of $Me_3SiN(Me)$-$OSiMe_3$ at 150 °C gives methylnitrene (MeN:) through α-deoxysilylation.[4,5] The nitrene intermediate can be trapped by $(MeO)_2MeSiH$ to give the insertion product $(MeO)_2MeSiNHMe$.

Silylalkylaminyl Radical Formation. Irradiation of $Me_3SiN(Me)OSiMe_3$ with 260–340 nm UV light gives the trimethylsilyl(methyl)aminyl radical $(Me(Me_3Si)N\cdot)$.[6] This radical is more reactive than the $Me_2N\cdot$ radical and can be trapped by use of trialkyl phosphites to afford phosphoranyl radicals, such as $\cdot P(OEt)_3N(Me)SiMe_3$.

1. Robl, J. A.; Hwu, J. R., *J. Org. Chem.* **1985**, *50*, 5913.
2. Hwu, J. R.; Robl, J. A.; Wang, N.; Anderson, D. A.; Ku, J.; Chen, E., *J. Chem. Soc., Perkin Trans. 1* **1989**, 1823.
3. Hwu, J. R.; Khoudary, K. P.; Tsay, S.-C., *J. Organomet. Chem.* **1990**, *399*, C13.
4. Chang, Y. H.; Chiu, F.-T.; Zon, G., *J. Org. Chem.* **1981**, *46*, 342.
5. Tsui, F. P.; Vogel, T. M.; Zon, G., *J. Am. Chem. Soc.* **1974**, *96*, 7144.
6. Brand, J. C.; Roberts, B. P.; Winter, J. N., *J. Chem. Soc., Perkin Trans. 2* **1983**, 261.
7. Smrekar, O.; Wannagat, U., *Monatsh. Chem.* **1969**, *100*, 760.
8. West, R.; Boudjouk, P., *J. Am. Chem. Soc.* **1973**, *95*, 3987.
9. West, R.; Boudjouk, P.; Matuszko, A., *J. Am. Chem. Soc.* **1969**, *91*, 5184.
10. West, R.; Nowakowski, P.; Boudjouk, P., *J. Am. Chem. Soc.* **1976**, *98*, 5620.
11. Hwu, J. R.; Wetzel, J. M., *J. Org. Chem.* **1985**, *50*, 3946.

Jih Ru Hwu & Shwu-Chen Tsay
National Tsing Hua University, Taiwan, Republic of China

Methyldiphenylchlorosilane

$$\boxed{MePh_2SiCl}$$

[144-79-6] $C_{13}H_{13}ClSi$ (MW 232.80)

InChI = 1S/C13H13ClSi/c1-15(14,12-8-4-2-5-9-12)13-10-6-3-7-11-13/h2-11H,1H3

InChIKey = OJZNZOXALZKPEA-UHFFFAOYSA-N

(protecting group for alcohols; reacts with carboxylic acids and sulfoximes; C-silylates lithium ester enolates)

Physical Data: bp 295 °C/760 mmHg; d 1.128 g cm^{-3}; n_D^{20}-1.5742.

Form Supplied in: clear or transparent, pale yellow liquid.

Analysis of Reagent Purity: gas chromatography on a nonpolar column such as SE-30.

Solubility: sol most organic solvents; reacts with protic solvents such as alcohols, acids, amines, water.

Purification: vacuum distillation is preferred method.

Handling, Storage, and Precautions: chlorosilanes react with water and even moisture in the air to generate HCl and siloxanes, care should be taken to minimize contact with air. Use in a fume hood.

Original Commentary

Gerald L. Larson
Hüls America, Piscataway, NJ, USA

Protecting Group for Alcohols. The reaction of methyldiphenylchlorosilane with alcohols in the presence of imidazole provides the methyldiphenylsilyl-protected alcohol. The protection of primary, secondary, and tertiary alcohols proceeds with high yields. The methyldiphenylsilyl-protected alcohols are intermediate in stability between those of the trimethylsilyl- and *t*-butyldimethylsilyl-protected ones.[1] It has been shown that, in general, the phenyl-substituted silyl groups are more readily removed by base, where electronic factors are more important, than the trialkylsilyl protecting groups. On the other hand, with acid hydrolysis steric factors play a more significant role and the phenyl-substituted silyl groups are more stable than the trimethylsilyl group, but less stable than the more sterically hindered triethylsilyl and other trialkylsilyl groups.[2] The methyldiphenylsilyl ethers can be oxidized by the chromium reagents dipyridine chromium(VI) oxide, pyridinium chlorochromate, and pyridinium dichromate (eq 1).[1]

$$\text{OSiPh}_2\text{Me} \xrightarrow[\substack{\text{or PCC} \\ \text{or PDC} \\ 95\text{–}100\%}]{\text{CrO}_3, \text{py}} \overset{O}{\underset{}{\text{}}}\text{—H} \quad (1)$$

A primary alcohol was readily methyldiphenylsilylated by the reaction of methyldiphenylchlorosilane in the presence of an amine. In eq 2 the amine is also the reagent being silylated.[3]

$$MePh_2SiCl + 2\ HOCH_2CH_2NR_2 \longrightarrow$$

$$MePh_2SiOCH_2CH_2NR_2 + HOCH_2CH_2NR_2\cdot HCl \quad (2)$$

$$R = Me\ (99\%),\ Et\ (95\%)$$

The methyldiphenylsilyl esters of carboxylic acids can be prepared by the reaction of the chlorosilane with the acid in the presence of triethylamine or with the potassium salt of the acid (eqs 3 and 4).[4]

$$MePh_2SiCl + RCO_2H \xrightarrow{Et_3N} RCO_2SiPh_2Me + Et_3N \cdot HCl \quad (3)$$

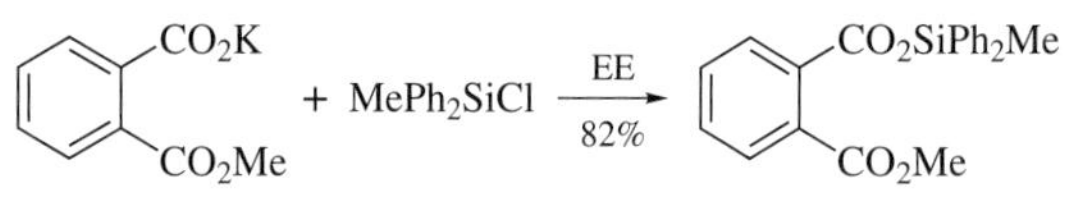

Sulfoximes can be silylated with methyldiphenylchlorosilane. The resulting methyldiphenylsilylated sulfoximes can be deprotonated at the methyl group and the anion condensed with aldehydes. The diastereospecificity is greater with the methyldiphenylsilyl group than with the *t*-butyldimethylsilyl group (eqs 5 and 6).[5]

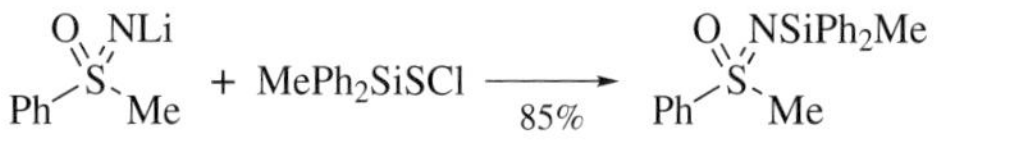

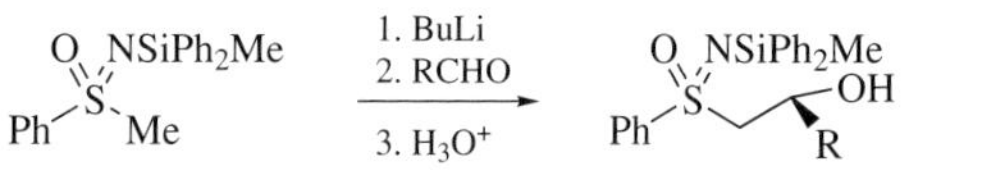

***C*-Silylation of Lithium Ester Enolates.** A very useful asset of methyldiphenylchlorosilane is its ability to directly provide α-(methyldiphenylsilyl) esters (eq 7).[6] These α-(methyldiphenylsilyl) esters have themselves proved to be precursors to alkenes, ketones, α-(methyldiphenylsilyl) ketones, and enol silyl ethers and have been used in the synthesis of some natural products.[6b] An elegant use of the methyldiphenylsilyl group in synthesis (eq 8) is that reported by Villanueva and Prieto,[7] who methyldiphenylsilylated the protected ethyl 4-oxolevulinate to give the α-silyl ester (**1**) This was then treated with 3-phenylpropylmagnesium bromide to give ketone (**2**), a reaction made possible by the presence of the bulky methyldiphenylsilyl group. This ketone was then regioselectively deprotonated on the side opposite to the methyldiphenylsilyl group and the resulting enolate trimethylsilylated to give enol silyl ether (**3**), which was subjected to an intramolecular Mukaiyama reaction to give cyclopentanone (**4**). Deprotonation and reaction with formaldehyde produces a methylenomycin B derivative, (**5**).

Other Methyldiphenylsilylations. Methyldiphenylchlorosilane was used to form 6-silyluracils and uridines.[8] The yields were better than those obtained with chlorotrimethylsilane (eq 9). The methyldiphenylsilylation of the lithium anion of 3-triethylsilyloxy-1,4-pentadiene occurs to place the methyldiphenylsilyl group on the 1-position, providing, after deprotection, the β-methyldiphenylsilyl ketone (**6**) (eq 10).[9]

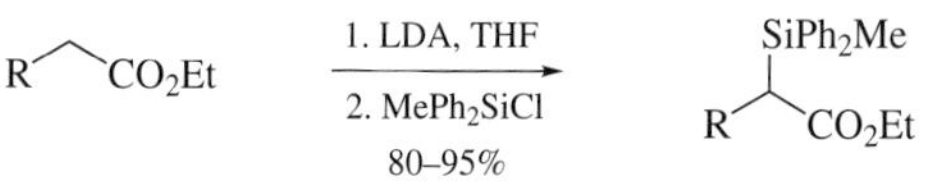

First Update

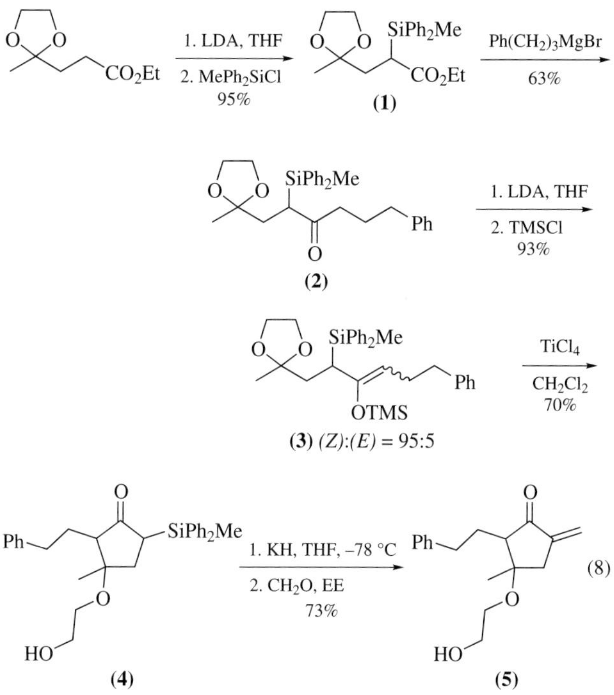

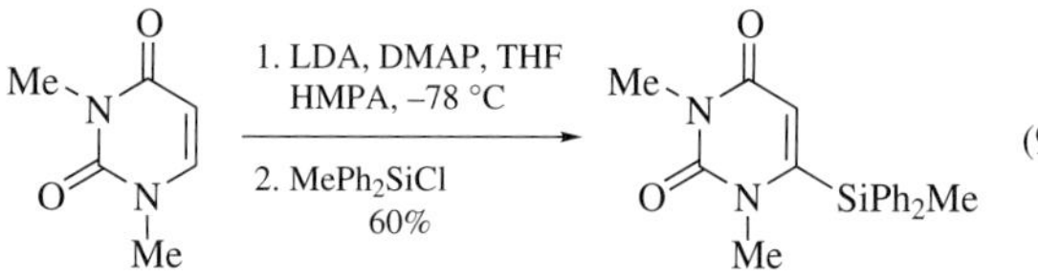

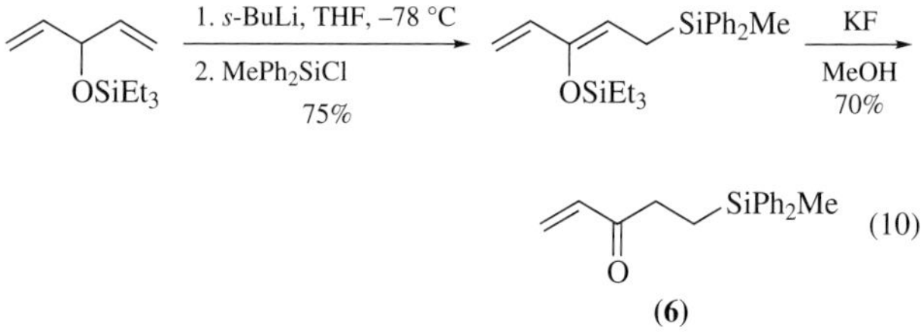

Andreas Luxenburger
Industrial Research Limited, Carbohydrate Chemistry, Lower Hutt, New Zealand

Protecting Group. Methyldiphenylchlorosilane has been used to protect a variety of additional alcohols.[10,11] In a study aimed at the stereo-controlled preparation of α-chiral crotylsilanes, Behrens et al. investigated methyldiphenylsilyl as a protecting group (eq 11).[11] The synthesis began with methyldiphenylsilyl protection of 1-phenyl-3-butyn-1-ol (**7**). Lithiation of the product and subsequent reaction with acetaldehyde gave the acetylenic alcohols, which were subjected to Lindlar partial hydrogenation to afford the *cis*-configured allyl alcohols (**8**). Mukaiyama redox condensation then yielded the phenylthio analogs (**9**) and (**10**) as a 3.5:1 *cis/iso*-mixtures. Retro-[1,4]-Brook rearrangement applied to this mixture was initiated by lithium naphthalenide and resulted in a 59:41 mixture of the *anti*- and *syn, trans*-alcohols (**11**) and (**12**). After oxidation of PhS-containing impurities, the double bonds were hydrogenated over Pd/C and Raney Ni, respectively (eq 12). Finally, Fleming-Tamao oxidation and basic methanolysis yielded the *anti*- and *syn*-diols (**13**) and (**14**). Better stereoselectivities were achieved when the bulkier *t*-BuPh$_2$Si group was used.

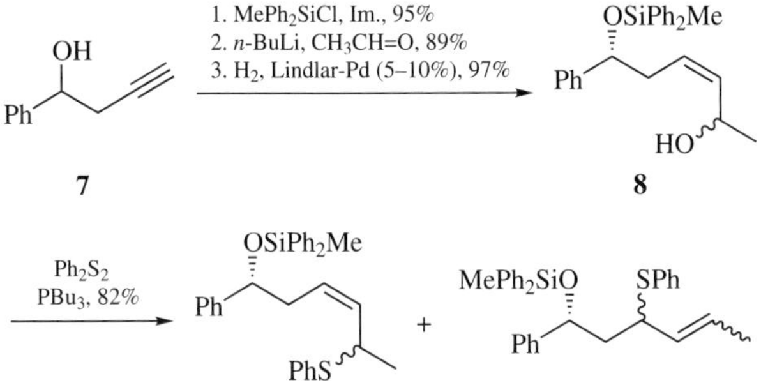

The selective cleavage of specific silyl ethers, including methyldiphenylsilyl ethers, in the presence of other silyl ethers has been reviewed by Nelson and Crouch.[12] In addition to the commonly used silyl deprotection methods, which operate either under acidic reaction conditions or in the presence of a fluoride source, Behloul et al. described a desilylation procedure, dependent on the use of lithium powder and a catalytic amount of naphthalene, which is selective for silyl protecting groups that contain at least one phenyl group. The method can be applied to silyl ethers derived from saturated aliphatic or aromatic alcohols, while allylic and benzylic analogs suffer from C–O bond cleavage.[13] The methyldiphenylsilyl group has also been used to protect the amido function of lactams in the synthesis of benzthiodiazepinones.[14]

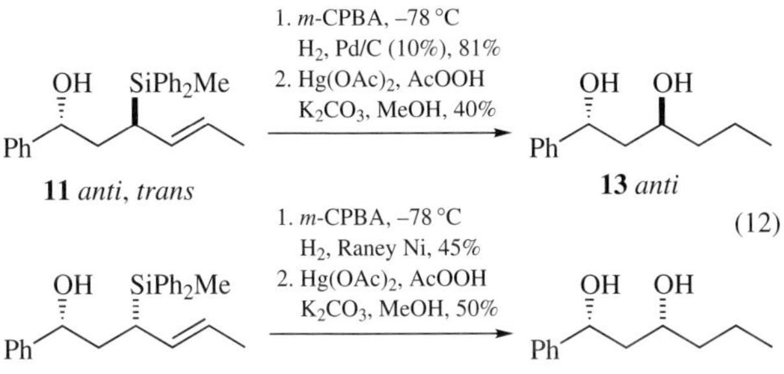

***C*-Silylation of Li-Ester Enolates.** Further applications of α-diphenylmethylsilyl carboxylates have been reported.[15–17] Gais and co-workers employed α-methyldiphenylsilyl acetate (**16**), prepared from (−)-8-phenylmenthol acetate (**15**), by lithiation with LDA and reaction with methyldiphenylchlorosilane, in an asymmetric Peterson reaction with the bicyclo[3.3.0]octan-3-one derivative (**17**) to yield a 89:11 mixture of the (*Z*)- and (*E*)-esters (**18**) and (**19**) About 50% of the starting material was recovered in the unoptimized procedure (eq 13).[15]

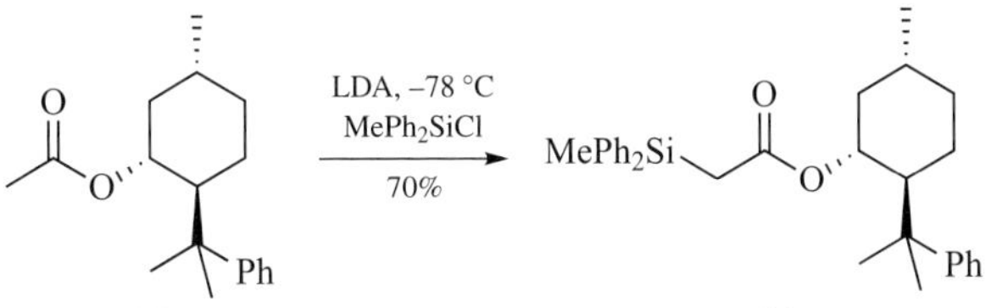

In another example, α-silyl ester (**20**) underwent Michael addition to cyclopentenone after deprotonation with LDA (eq 14).[17] The intermediate α-silylated ester enolate of (**20**) served as a softer nucleophile compared to its unsubstituted analog favoring 1,4- over 1,2-addition. After quenching with allyl iodide the *trans*-cyclopentanone (**21**) was obtained. Desilylation with potassium fluoride provided the diene (**22**).

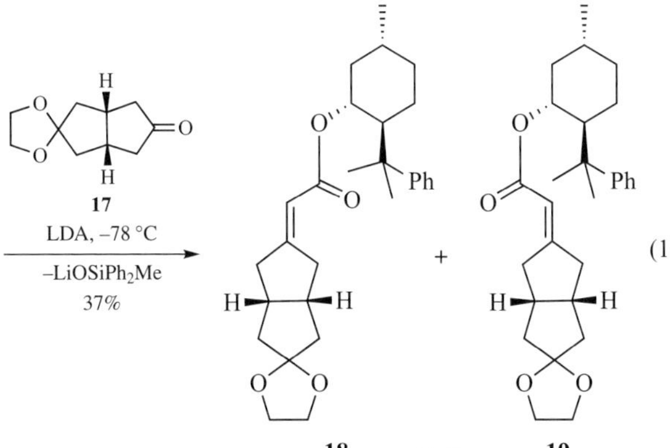

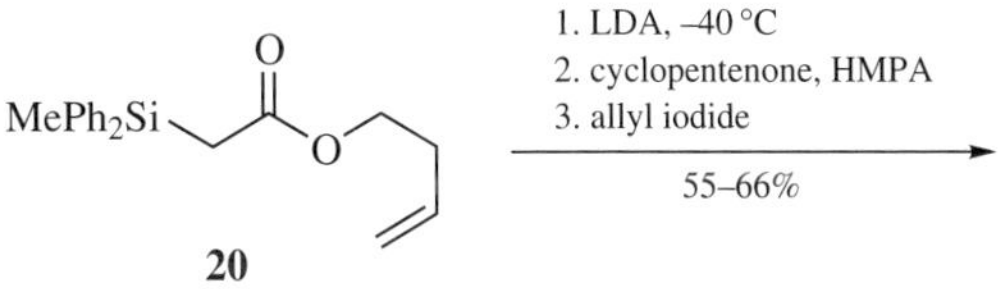

***C*-Silylation of Carbamates.** Methyldiphenylchlorosilane has also found utility in the silylation of carbamates. In the case of the neryl *N*,*N*-diisopropylcarbamate (**23**), enantioselective deprotonation was achieved upon treatment with *n*-BuLi/(−)-sparteine, and reaction with methyldiphenylchlorosilane provided the corresponding (−)-α-silyl alkenyl carbamate.[10] Decarbomylation with DIBAL-H and subsequent reaction of the derived α-silyl acohol with phenyl isocyanate furnished the crystalline (−)-*N*-phenylurethane (**24**) (eq 15).

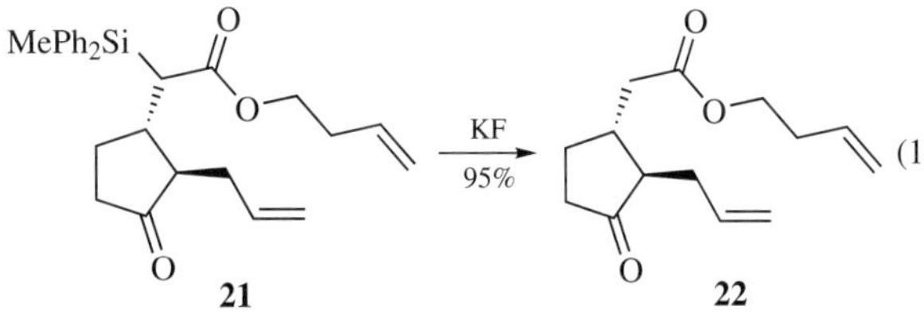

In an effort to prepare chiral 1,2-diamines, Coldham et al. started with the asymmetric deprotonation of *N*-Boc-imidazolidine (**25**) using *sec*-BuLi/(−)-sparteine. The deprotonation occurred exclusively at C-5 and trapping with methyldiphenylchlorosilane gave the 5-methyldiphenylsilylated imidazolidine (**26**) in high optical purity (ee 88%) (eq 16).[11] Ethanolysis with malonic acid as catalyst led to the selective formation of the acyclic *N*-Boc-diamine, and TFA-mediated cleavage of the Boc group afforded the chiral diamine (**27**). The synthesis of the corresponding 1,3-diamine by the same approach was also described.

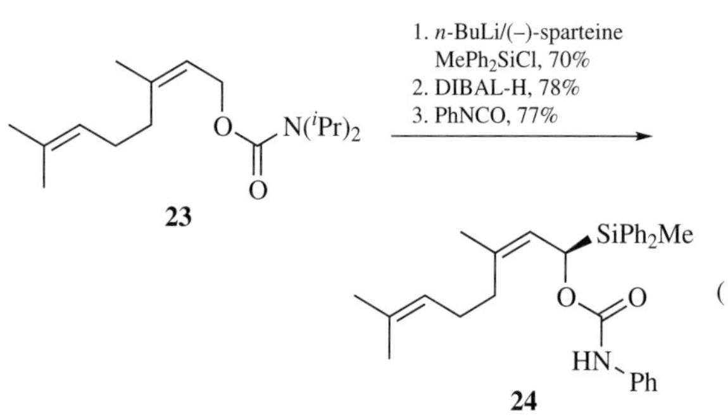

$$ \tag{15} $$

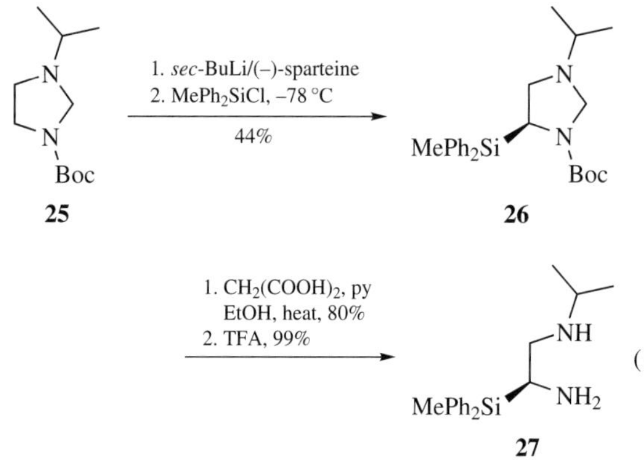

$$ \tag{16} $$

Acylsilanes. Acylsilanes have emerged as a very attractive class of compounds that feature unique chemical properties as well as a wide range of applications in organic chemistry. Particularly notable is their use as aldehyde or ester equivalents.[20] Moreover, acylsilanes are excellent radical acceptors (e.g. eq 24), while α,β-unsaturated acylsilanes are useful Michael acceptors (eq 20).

Synthetic routes to acyl(methyldiphenyl)silanes have mainly been based on methodology developed by Brook and Corey. According to Brook's synthesis of aryl silyl ketones[21], benzylmagnesium chloride or bromide, for example, were silylated with methyldiphenylchlorosilane to afford benzylmethyldiphenylsilane (**28**) that, on dibromination of the benzyl methylene group and subsequent hydrolysis, gave the benzoylsilane (**29**) (eq 17).[22]

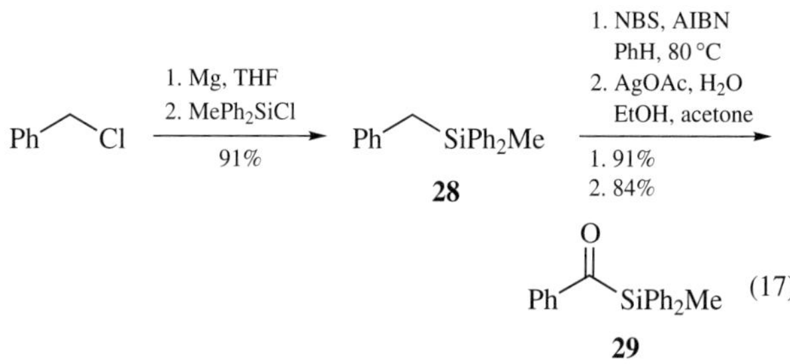

$$ \tag{17} $$

Another general approach to acyl(methyldiphenyl)silanes starts from 1,3-dithiane (**30**), which, on lithiation with n-BuLi and subsequent treatment with methyldiphenylchlorosilane forms 2-methyldiphenylsilyl-1,3-dithiane (**31a**). This may be alkylated following a second deprotonation to yield the corresponding substituted 1,3-dithianes (**31b**). Hydrolysis of the thioacetals to afford the corresponding acylsilanes (**32**) was conducted with

HgO/BF$_3$·OEt$_2$. Ce(NH$_4$)$_2$(NO$_3$)$_6$ (CAN) or PhI(CF$_3$CO$_2$)$_2$ may also be used (eq 18).[23]

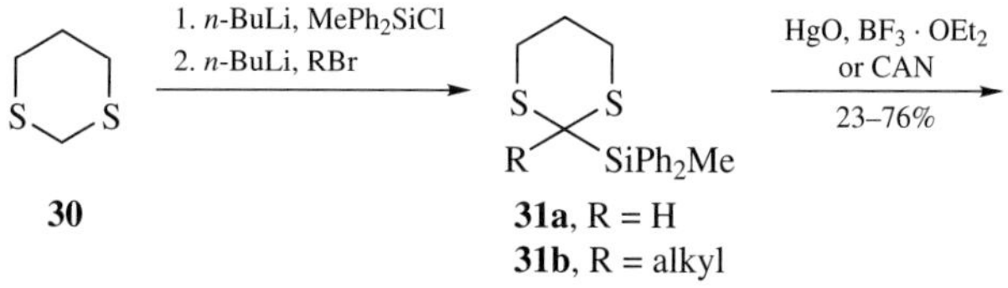

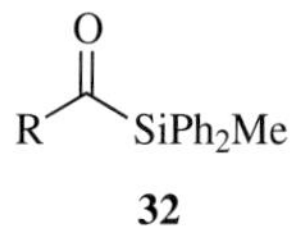

$$ \tag{18} $$

Kondo et al. reported on the synthesis of acyl(methyldiphenyl)silanes starting from dichlorobis(methyldiphenylsilyl)methane (**33**) prepared by LDA-mediated deprotonation of dichloromethane and subsequent reaction with methyldiphenylchlorosilane (eq 19).[24] Halogen-metal exchange, followed by reaction with alkyl or aryl Grignard reagent in the presence of a copper salt, such as copper cyanide, provides 1,1-disilylalkylcopper intermediates (**34**), which undergo oxidation to the corresponding acylsilanes (**35**) when exposed to air.

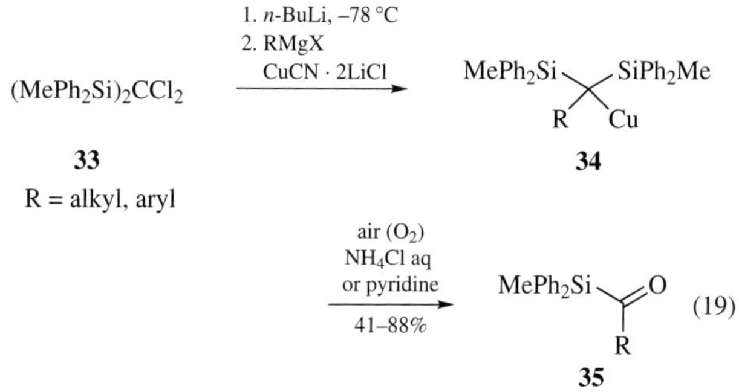

$$ \tag{19} $$

An example of the preparation of an α,β-unsaturated acyl-(methyldiphenyl)silane starting from *trans*-cinnamaldehyde (**36**) was published by Tsai and Sieh (eq 20).[25a] Conversion of the enal into the corresponding dithiane derivative followed by methyldiphenylsilylation and subsequent cleavage of the thioacetal yielded the α,β-unsaturated acylsilane (**37**). Since α,β-unsaturated acylsilanes are potent Michael acceptors, this product underwent 1,4-addition in good yield to give adducts such as compound (**38**) when added to a lower order cuprate at low temperatures. No products of 1,2-addition were observed.

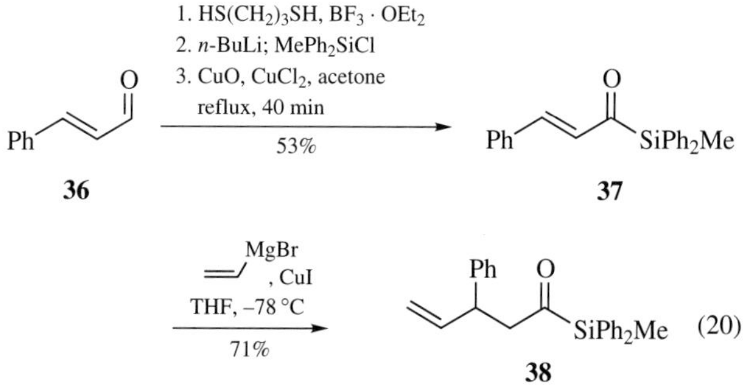

$$ \tag{20} $$

In a study of regioselectivity in [3 + 4] and [3 + 5] annulation addition reactions of bis(trimethylsilyl) enol ethers, for example

(**41**), with carbonyl-containing acylsilanes, Molander and Siedem synthesized a variety of 1,4- and 1,5-acyl(methyldiphenyl)lsilanes (eq 21).[26] The methyldiphenylsilyl group was employed as previous experiments targeting the synthesis of the more unstable acyltrimethylsilane dicarbonyl counterparts revealed that the envisaged compounds could not all be isolated. In the example illustrated in eq 21, 2-methyldiphenylsilyl-1,3-dithiane was alkylated with 2-(3-chloropropyl)-2-methyl-1,3-dioxolane to give compound (**39**) Sequential hydrolysis of the ketal and dithiane moieties yielded 1-(methyldiphenylsilyl)hexane-1,5-dione (**40**). TMSOTf-mediated annulation with bis(trimethylsilyl) enol ether (**41**) gave the bicyclic ether (**42**), decarboxylation and desilylation of which furnished the bicylic ketone (**43**).

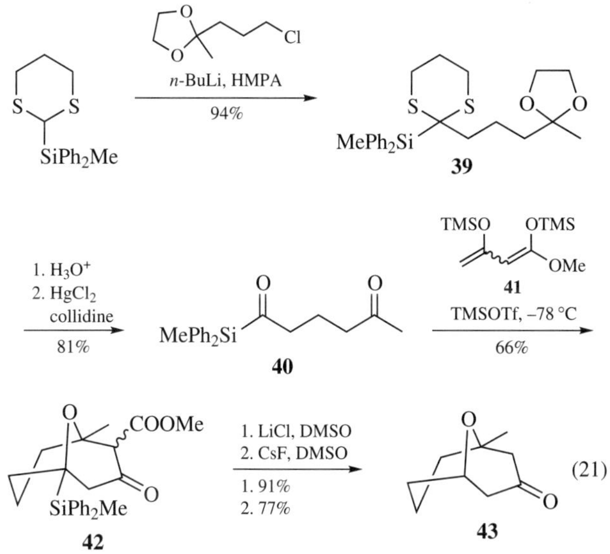

The instability of aldehydo-acylsilane (**46**) required a different approach for its preparation (eq 22). Thus 1-(methyldiphenylsilyl)propan-1-one (**44**) was converted into the corresponding *N,N*-dimethylhydrazone that was alkylated with 3-bromo-1,1-dimethoxypropane to give compound (**45**). Hydrolysis with Amberlyst-15® as catalyst afforded the target aldehyde (**46**).

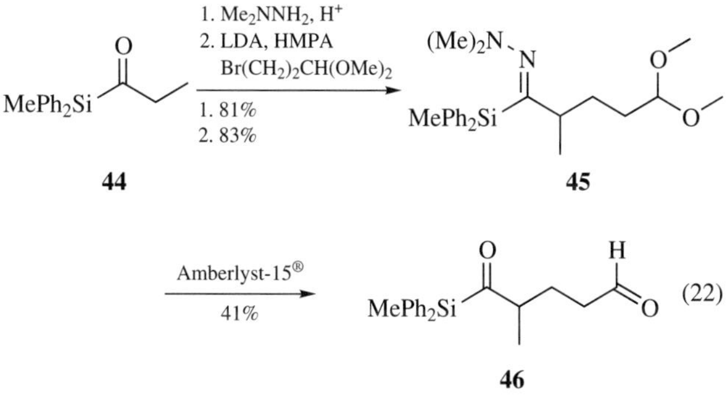

The synthesis of symmetrical and unsymmetrical bis-(acylsilanes) has also been reported.[27,23] The former category of compounds are easily prepared from the symmetrical bis-(dithianes) by lithiation, methyldiphenylsilylation, and hydrolysis. The synthesis of unsymmetrical analogs has been effected by the stepwise treatment of dihaloalkanes with different lithiated 2-silyl-1,3-dithianes (eq 23). Deprotection of the products (e.g. compound (**47**)) with Hg(ClO₄)₂/CaCO₃ or I₂/CaCO₃

provided the unsymmetrical 1,5-dicarbonyl products that may be cyclized under acid catalysis to 2,6-bis(trialkylsilyl)-4*H*-pyrans (e.g. compound (**48**)). 1,4-Bis(acylsilanes) can be similarly converted to 2,5-bis(trialkylsilyl)furans.

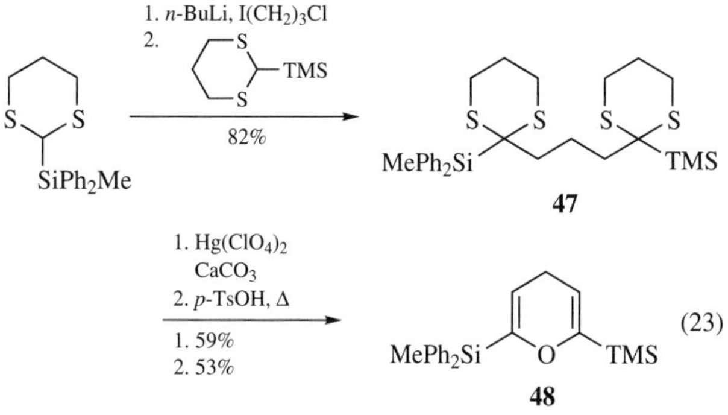

1,5-Bis(acyl(methyldiphenyl)silanes) can be cyclized by a free radical procedure to the corresponding 2,6-bis(methyldiphenylsilyl)-4,5-dihydro-6*H*-pyrans by treatment with tributyltin hydride and a catalytic amount of AIBN (eq 24).[23]

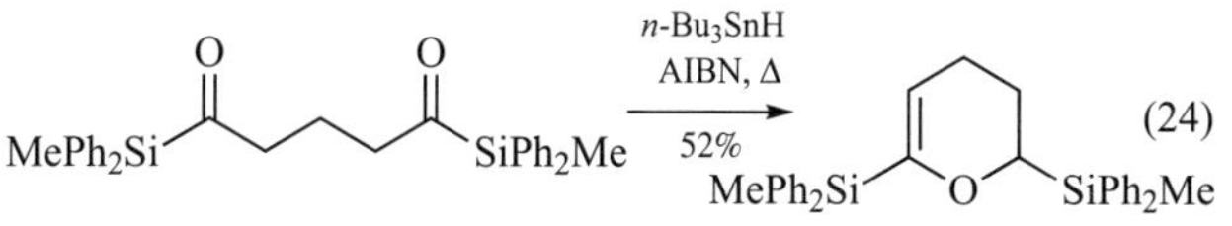

Likewise, 2-silyldihydropyrans and furans can be prepared by heating the corresponding bromo- or chloroacylsilanes in *N*-methyl-2-pyrrolidinone (NMP) in the presence of potassium iodide as shown in eqs 25 and 26, respectively.[28] Treatment of unsaturated acylsilanes with either iodine or phenylselenenyl bromide generates the cyclized iodides or selenides (eqs 27 and 28).[28]

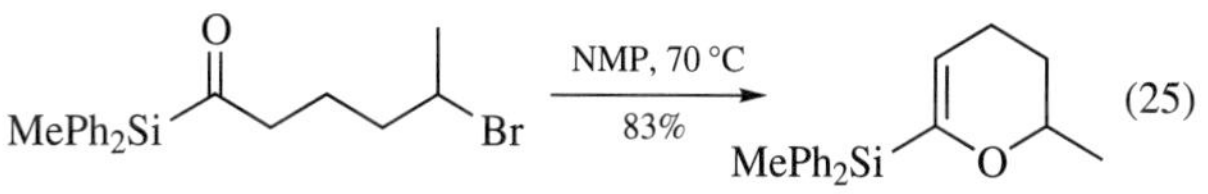

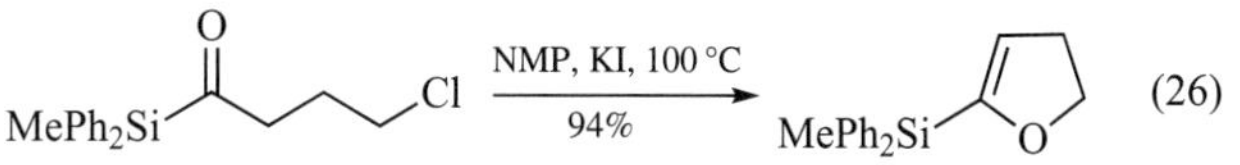

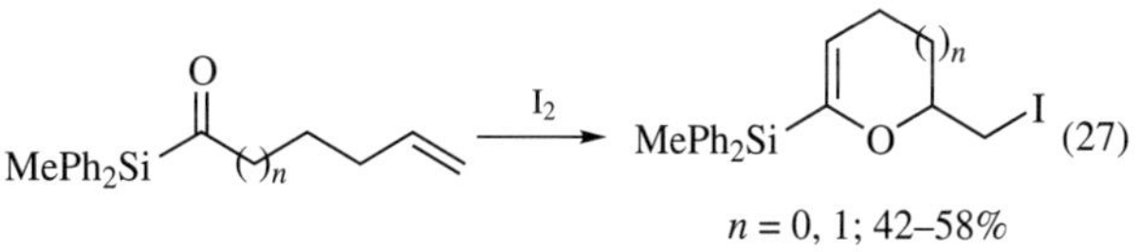

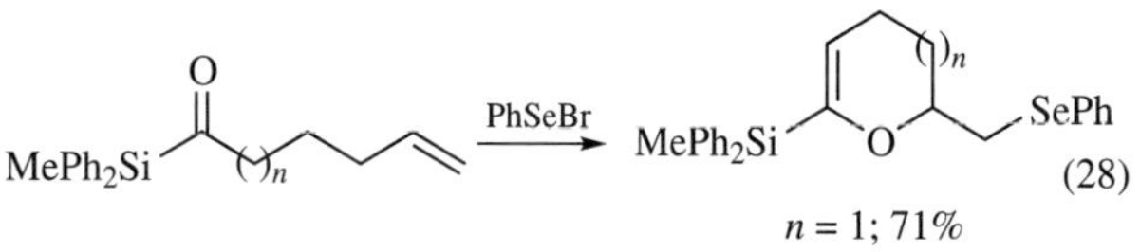

Intramolecular radical cyclizations of acylsilanes containing the methyldiphenylsilyl moiety were investigated by the Tsai group (eq 29).[29] As indicated in eq 29 products of radical addition to the carbonyl group involve silyl migration, whereas attack at the alkene center gives simple adducts.

MePh₂Si—...—Br $\xrightarrow{\begin{array}{c} n\text{-Bu}_3\text{SnH, AIBN} \\ C_6H_6,\ 1\ h,\ \Delta \end{array}}$

(29)

$n = 1$:	78%; cis/trans 1.1:1	12%; cis/trans 1:2
$n = 2$:	24%; cis/trans 3.8:1	46%; cis/trans 1:2.7

Reduction of the carbonyl group of acyl(methyldiphenyl)silanes to the corresponding α-silyl alcohols can be achieved by using DIBAL-H, and enantioselective reduction has been realized by employing Noyori's chiral catalyst.[22]

Other Methyldiphenylsilylations. Oshima and co-workers have synthesized dibromobis(methyldiphenylsilyl)methane (**49**) and dibromo(methyldiphenylsilyl)methane (**50**) from tetra- and dibromomethanes, respectively (eqs 30 and 31).[30] The preparation of dichlorobis(methyldiphenylsilyl)methane from dichloromethane is also reported.[24]

$$CBr_4\ +\ Ph_2MeSiCl\ \xrightarrow[75\%]{n\text{-BuLi, }-78\,°C}\ (MePh_2Si)_2CBr_2 \quad (30)$$
49

$$CH_2Br_2\ +\ Ph_2MeSiCl\ \xrightarrow[85\%]{LDA,\ -78\,°C}\ MePh_2Si\underset{Br\quad Br}{\diagdown}H \quad (31)$$
50

Dibromo-compound (**50**) may be deprotonated with LDA and either alkylated with for example methyl iodide to give compound (**51**), as shown in eq 32, or treated with different chlorosilanes to yield the unsymmetrically substituted dibromodisilyl-methane (**52**) that may be further converted to the 1,1-disilylethene (**53**) upon monomethylation with lithium trimethylmagnesate and base-promoted dehydrobromination (eq 33).[30]

$$(MePh_2Si)CHBr_2\ \xrightarrow[\text{MeI, }90\%]{LDA,\ -78\,°C}\ MePh_2Si\underset{Br\quad Br}{\diagdown}Me \quad (32)$$
50 → **51**

$$(MePh_2Si)CHBr_2\ \xrightarrow[\substack{-78\,°C \\ TMSCl,\ 85\%}]{LDA}\ MePh_2Si\underset{Br\quad Br}{\diagdown}TMS \quad (—)$$
50 → **52**

$$\xrightarrow[\substack{1.\ 93\% \\ 2.\ 98\%}]{\substack{1.\ Me_3MgLi,\ -78\,°C \\ 2.\ DBU,\ 90\,°C}}\ MePh_2Si\diagup TMS \quad (33)$$
53

Reaction of dibromo(methyldiphenylsilyl)methane (**50**) with tributylmagnesate (n-Bu₃MgLi) in the presence of catalytic copper(I) cyanide affords the α-silylpentylmagnesium intermediate (**54**) that may be alkylated or acylated (eq 34). Intermediate (**54**) can also be trapped with for example methyl vinyl ketone as Michael-acceptor to afford 5-(methyldiphenylsilyl)-2-nonanone (**55**) (eq 35).[31]

$$(MePh_2Si)CHBr_2\ \xrightarrow{\substack{1.\ n\text{-Bu}_3MgLi \\ 2.\ CuCN\cdot 2LiCl}}\ MePh_2Si\overset{}{\underset{n\text{-Bu}}{\diagdown}}Mgn\text{-Bu}\ \xrightarrow[50-79\%]{electrophile}$$
50 → **54**

Electrophile: CH₂=CHCH₂Br, PhCOCl

$$MePh_2Si\overset{}{\underset{n\text{-Bu}}{\diagdown}}E \quad (34)$$

$$\textbf{54}\ \xrightarrow[79\%]{TMSCl}\ MePh_2Si\underset{n\text{-Bu}}{\diagdown}\cdots\overset{O}{\diagup} \quad (35)$$
55

Allyl and vinyl halides can be methyldiphenylsilylated.[32–34] The conversion of the allyl chloride (**56**) into the corresponding allyl(methyldiphenyl)silane (**57**) was achieved by treatment with (methyldiphenylsilyl)lithium, prepared from methyldiphenyl-chlorosilane and metallic lithium, at low temperatures (eq 36).[32] Vinyl silanes can readily be obtained from the corresponding vinyl halides via lithiation and subsequent reaction with methyldiphenyl-lchlorosilane (eq 37).[33,34] The vinyl silane (**58**), shown in eq 37 was subjected further to a rhodium-catalyzed acylation providing the α,β-unsaturated ketone (**59**) in good yield.[34] When a triph-enylsilyl analog of (**58**) was used longer reaction times were required and the yield dropped to 21%. Higher yields were obtained and shorter reaction times were necessary when the trimethylsilyl or dimethylphenylsilyl groups were used.

BnO—...—Cl $\xrightarrow[61-75\%]{MePh_2SiLi,\ -78\,°C}$
56

BnO—...—SiPh₂Me (36)
57

$\xrightarrow{\substack{tert\text{-BuLi, }-100\,°C \\ MePh_2SiCl,\ -78\,°C}}$ ...—SiPh₂Me
58

$\xrightarrow[71\%]{\substack{[RhCl(CO_2)_2]_2\ (5\ mol\ \%) \\ 1,4\text{-dioxane, }(AcO)_2O \\ 90\,°C,\ 30\ h}}$ (37)
59

Methyldiphenylsilylation can also be achieved by the 1,4-addition of methyldiphenylsilyl cuprate, generated from methyl-diphenylsilyllithium and copper(I) iodide, to a Michael-acceptor, such as methyl p-methylcinnamate (eq 38).[35]

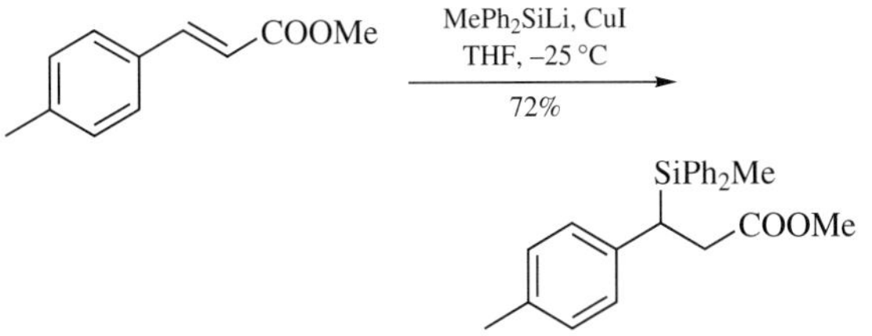

$$(38)$$

The methyldiphenylsilylations of cyclopentadiene[36] and indene[37] have also been reported. Starting from cyclopentadienyl methyldiphenylsilane (**60**), Landais and Para-Rapado developed a stereo-controlled and flexible route to carba-sugars (eq 39).[36]

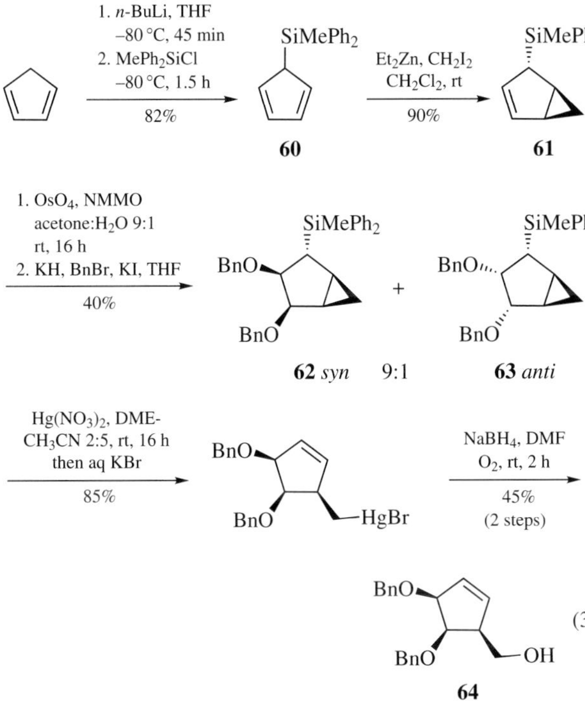

$$(39)$$

Simmons-Smith cyclopropanation of compound (**60**) led to the exclusive formation of the *anti*-isomer (**61**). Subsequent *cis*-dihydroxylation of the remaining double bond followed by *O*-benzylation of the hydroxyl groups gave rise to a 9:1 mixture of the isomers (**62**) and (**63**) having the oxygenated substituents *syn* and *anti* to the methylene group, respectively. Good diastereo-control was achieved with compound (**60**) and the Me$_2$PhSi analog, but when larger silyl groups were used, the reaction was less selective. In addition, a base-catalyzed Peterson elimination occurred with the minor isomer as a side reaction during protection of the diol and caused a moderate overall yield in this reaction step. Mercury(II)-induced desilylation of compounds (**62**) and (**63**) ultimately furnished the homoallylic alcohol (**64**), which is a synthetic precursor of carba-hexofuranoses.

Alkynes can also be silylated with methyldiphenylchlorosilane.[37,39] In an example described by the Barrett group, methyldiphenylchlorosilane was introduced as a bulky silyl protecting group able to selectively disable terminal alkenes toward ring-closing-metathesis (RCM) and hydrogenation reactions. By double deprotonation of propargyl alcohol, followed by reaction with methyldiphenylchlorosilane and acid-mediated mono-deprotection, alkyne (**65**) was made in 95% yield (eq 40).

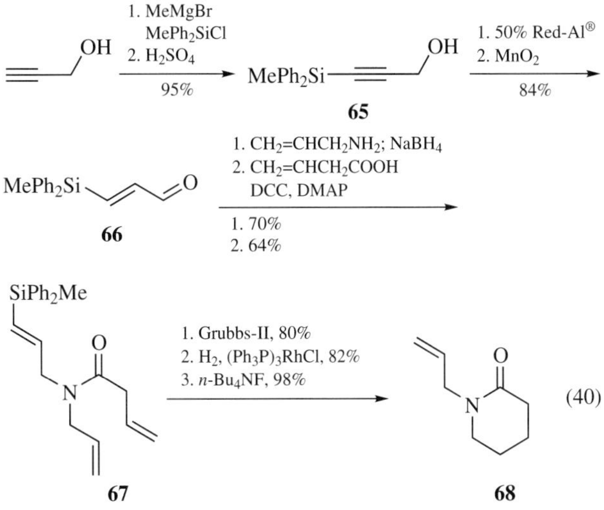

$$(40)$$

Stereoselective reduction by Red-Al® of compound (**65**), and subsequent oxidation of the resulting allylic alcohol, gave the α,β-unsaturated aldehyde (**66**) exclusively as the *trans*-isomer. Reductive amination with allylamine, followed by acylation with 3-butenoic acid in the presence of DCC and DMAP of the derived secondary allyl amine, led to amide (**67**), the triene substrate for the projected ring-closing-metathesis reaction. Use of Grubbs' second generation metathesis catalyst resulted in ring-closure between the two nonsilylated unsaturated *N*-substituents groups and gave the corresponding lactam. Selective hydrogenation of the remaining endocyclic double bond using Wilkinson's catalyst followed by cleavage of the methyldiphenylsilyl protecting group with tetrabutylammonium fluoride afforded the *N*-allyl saturated lactam (**68**). By extending this methodology, the authors prepared lactams having up to a 16-membered ring.

Terao et al. have developed a simple synthetic route to bis-(allylsilanes) (eq 41).[40] When 1,3-butadiene was treated with methyldiphenylchlorosilane and a Grignard reagent, such as phenylmagnesium bromide, in the presence of catalytic Pd(acac)$_2$, the diene (**69**) was formed stereo- and regioselectively in good yield. Complete stereo- and regioselectivity was also achieved for other phenyl- or allyl-substituted chorosilanes.

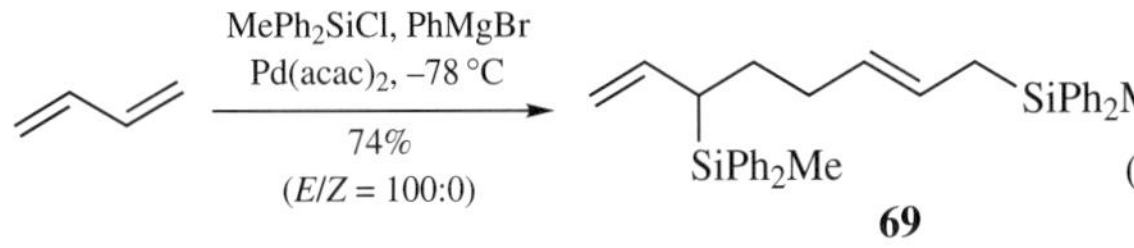

$$(41)$$

Related Reagents. *t*-Butyldimethylchlorosilane; *t*-Butyldi-phenylchlorosilane; Ethyl 2-(Methyldiphenylsilyl)propanoate.

1. Denmark, S. E.; Hammer, R. P.; Weber, E. J.; Habermas, K. L., *J. Org. Chem.* **1987**, *52*, 165.

2. Sommer, L. H. *Stereochemistry, Mechanism and Silicon*, McGraw–Hill: New York, 1965, 127.

3. (a) Tacke, R.; Wannagat, U., *Monatsh. Chem.* **1975**, *106*, 1005. (b) Friedrich, G.; Bartsch, R.; Ruehlmann, K., *Pharmazie* **1977**, *32*, 394.

4. (a) Verweij, J.; Ger. Patent. 2 166 561 (*Chem. Abstr.* **1974**, *81*, 152 216k). (b) Rutherford, K. G.; Seidewand, R. J., *Can. J. Chem.* **1975**, *53*, 67.

5. Hwang, H. J.; Logusch, E. W.; Brannigan, L. H.; Thompson, M. R., *J. Org. Chem.* **1987**, *52*, 3435.

6. (a) Larson, G. L.; Fuentes, L. M., *J. Am. Chem. Soc.* **1981**, *103*, 2418. (b) Larson, G. L., *Pure Appl. Chem.* **1990**, *62*, 2021.

7. Villanueva, O.; Prieto, J. A., *J. Org. Chem.* **1993**, *58*, 2718.

8. Ikehira, H.; Matsuura, T.; Saito, I., *Tetrahedron Lett.* **1984**, *25*, 3325.

9. Kim, S.; Emeric, G.; Fuchs, P. L., *J. Org. Chem.* **1992**, *57*, 7362.

10. (a) Duvold, T.; Jørgensen, A.; Andersen, N. R.; Henriksen, A. S.; Sørensen, M. D.; Björkling, F., *Bioorg. Med. Chem. Lett.* **2002**, *12*, 3569. (b) Duvold, T.; Sørensen, M. D.; Björkling, F.; Henriksen, A. S.; Rastrup-Andersen, N., *J. Med. Chem.* **2001**, *44*, 3125. (c) Matile, S.; Berova, N.; Nakanishi, K.; Fleischhauer, J.; Woody, R. W., *J. Am. Chem. Soc.* **1996**, *118*, 5198. (d) Thommen, M.; Veretenov, A. L.; Guidetti-Grept, R.; Keese, R., *Helv. Chim. Acta* **1996**, *79*, 461. (e) Lewis, R. T.; Motherwell, W. B.; Shipman, M.; Slawin, A. M. Z.; Williams, D. J., *Tetrahedron* **1995**, *51*, 3289.

11. Behrens, K.; Kneisel, B. O.; Noltemeyer, M.; Brückner, R., *Liebigs Ann. Chem.* **1995**, 385.

12. Nelson, T. D.; Crouch, R. D., *Synthesis* **1996**, 1031.

13. Behloul, C.; Guijarro, D.; Yus, M., *Tetrahedron* **2005**, *61*, 6908.

14. Hamilton, H. W.; Nishiguchi, G.; Hagen, S. E.; Domagala, J. D.; Weber, P. C.; Gracheck, S.; Boulware, S. L.; Nordby, E. C.; Cho, H.; Nakamura, T.; Ikeda, S.; Watanabe, W., *Bioorg. Med. Chem. Lett.* **2002**, *12*, 2981.

15. Chao, H.-G.; Bernatowicz, M. S.; Reiss, P. D.; Matsueda, G. R., *J. Org. Chem.* **1994**, *59*, 6687.

16. Gais, H.-J.; Schmiedl, G.; Ossenkamp, R. K. L., *Liebigs Ann. Chem.* **1997**, 2419.

17. (a) Fürstner, A.; Müller, T., *Synlett* **1997**, 1010. (b) Oppolzer, W.; Guo, M.; Baettig, K., *Helv. Chim. Acta* **1983**, *66*, 2140.

18. Zeng, W.; Fröhlich, R.; Hoppe, D., *Tetrahedron* **2005**, *61*, 3281.

19. (a) Ashweek, N. J.; Coldham, I.; Haxell, T. F. N.; Howard, S., *Org. Biomol. Chem.* **2003**, *1*, 1532. (b) Coldham, I.; Copley, R. C. B.; Haxell, T. F. N.; Howard, S., *Org. Lett.* **2001**, *3*, 3799.

20. (a) Patrocínio, A. F.; Moran, P. J. S., *J. Braz. Chem. Soc.* **2001**, *12*, 7. (b) Bonini, B. F.; Comes-Franchini, M.; Fochi, M.; Mazzanti, G.; Ricci, A., *J. Organomet. Chem.* **1998**, *567*, 181. (c) Cirillo, P. F.; Panek, J. S., *Org. Prep. Proced. Int.* **1992**, *24*, 553. (d) Bulman Page, P. C.; Klair, S. S.; Rosenthal, S., *Chem. Soc. Rev.* **1990**, *19*, 147. (e) Ricci, A.; Degl'Innocenti, A., *Synthesis* **1989**, 647. (f) Fleming, I.; Barbero, A.; Walter, D., *Chem. Rev.* **1997**, *97*, 2063.

21. Brook, A. G., *J. Am. Chem. Soc.* **1957**, *79*, 4373.

22. Huckins, J. R.; Rychnovsky, S. D., *J. Org. Chem.* **2003**, *68*, 10135.

23. Chuang, T.-H.; Fang, J.-M.; Jiaang, W.-T.; Tsai, Y.-M., *J. Org. Chem.* **1996**, *61*, 1794.

24. (a) Kondo, J.; Inoue, A.; Ito, Y.; Shinokubo, H.; Oshima, K., *Tetrahedron* **2005**, *61*, 3361. (b) Inoue, A.; Kondo, J.; Shinokubo, H.; Oshima, K., *J. Am. Chem. Soc.* **2001**, *123*, 11109. (c) Kondo, J.; Shinokubo, H.; Oshima, K., *Org. Lett.* **2006**, *8*, 1185.

25. (a) Tsai, Y.-M.; Sieh, J.-A., *J. Chin. Chem. Soc.* **1999**, *46*, 825. (b) Danheiser, R. L.; Fink, D. M., *Tetrahedron Lett.* **1985**, *26*, 2509.

26. Molander, G. A.; Siedem, C. S., *J. Org. Chem.* **1995**, *60*, 130.

27. Bouillon, J.-P.; Saleur, D.; Portella, C., *Synthesis* **2000**, 843.

28. (a) Tsai, Y.-M.; Cherng, C.-D.; Nieh, H.-C.; Sieh, J.-A., *Tetrahedron* **1999**, *55*, 14587. (b) Tsai, Y.-M.; Nieh, H.-C.; Cherng, C.-D., *J. Org. Chem.* **1992**, *57*, 7010.

29. Jiaang, W.-T.; Lin, H.-C.; Tang, K.-H.; Chang, L.-B.; Tsai, Y.-M., *J. Org. Chem.* **1999**, *64*, 618.

30. Inoue, A.; Kondo, J.; Shinokubo, H.; Oshima, K., *Chem. Lett.* **2001**, 956.

31. Inoue, A.; Kondo, J.; Shinokubo, H.; Oshima, K., *Chem. Eur. J.* **2002**, *8*, 1730.

32. Asao, K.; Iio, H.; Tokoroyama, T., *Synthesis* **1990**, 382.

33. Barnhart, R. W.; Wang, X.; Noheda, P.; Bergens, S. H.; Whelan, J.; Bosnich, B., *J. Am. Chem. Soc.* **1994**, *116*, 1821.

34. Yamane, M.; Uera, K.; Narasaka, K., *Bull. Chem. Soc. Jpn.* **2005**, *78*, 477.

35. (a) Ramage, R.; Barron, C. A.; Bielecki, S.; Holden, R.; Thomas, D. W., *Tetrahedron* **1992**, *48*, 499. (b) Ramage, R.; Barron, C. A.; Bielecki, S.; Thomas, D. W., *Tetrahedron Lett.* **1987**, *28*, 4105.

36. Landais, Y.; Parra-Rapado, R., *Eur. J. Org. Chem.* **2000**, 401.

37. Möller, A. C.; Heyn, R. H.; Bolm, R.; Swang, O.; Görbitz, C. H.; Kopf, J., *J. Chem. Soc., Dalton Trans.* **2004**, 1578.

38. (a) Trost, B. M.; Machacek, M.; Schnaderbeck, M. J., *Org. Lett.* **2000**, *12*, 1761. (b) Miura, K.; Hondo, T.; Okajima, S.; Nakagawa, T.; Takahashi, T.; Hosomi, A., *J. Org. Chem.* **2002**, *67*, 6082. (c) Shibata, T.; Yamashita, K.; Takagi, K.; Ohta, T.; Soai, K., *Tetrahedron* **2000**, *56*, 9259. (d) Clive, D. L. J.; Cole, D. C., *J. Chem. Soc., Perkin Trans. 1* **1991**, 3263. (e) Himbert, G.; Henn, L., *Liebigs Ann. Chem.* **1987**, 771.

39. Schultz-Fademrecht, C.; Deshmukh, P. H.; Malagu, K.; Procopiou, P. A.; Barrett, A. G. M., *Tetrahedron* **2004**, *60*, 7515.

40. Terao, J.; Oda, A.; Kambe, N., *Org. Lett.* **2004**, *6*, 3341.

(Methylthio)trimethylsilane

[3908-52-2]   C$_4$H$_{12}$SSi   (MW 120.32)

InChI = 1S/C4H12SSi/c1-5-6(2,3)4/h1-4H3

InChIKey = JDYCDPFQWXHUDL-UHFFFAOYSA-N

(thioacetalization under neutral conditions;[1,2] cleavage of ethers[3,4])

Physical Data: bp 110–114 °C; *d* 0.848 g cm^{-3}.

Solubility: sol organic solvents (CH$_2$Cl$_2$, benzene, acetonitrile, ether).

Form Supplied in: colorless liquid; commercially available.

Preparative Method: prepared as described by Evans et al.[14]

Handling, Storage, and Precautions: the reagent is sensitive to moisture and incompatible with strong oxidizing agents. Should be used in a well-ventilated fume hood.

Carbonyl Protection. The Lewis acid-catalyzed reaction of aldehydes and ketones with (methylthio)trimethylsilane gives rise to the corresponding thioacetals (eq 1).[1,2] Selective thioacetalization can be achieved with this reagent (eq 2).[1] The reaction with α,β-unsaturated aldehydes and ketones leads to the formation of the 1,4-addition product.

$$\text{(eq 1)}$$

$$\text{(eq 2)}$$

Cleavage of Acyclic and Cyclic Ethers. Acyclic ethers are readily cleaved by (methylthio)trimethylsilane (eq 3).[3] In contrast to iodotrimethylsilane, esters are unaffected under the reaction conditions. The reagent is also compatible with acetals and amides.[4] The zinc chloride-catalyzed reaction of oxetanes with (methylthio)trimethylsilane leads to the formation of ring-opened products (eq 4).[5] No reaction takes place in the absence of zinc chloride.

$$
\text{(3)}
$$

$$
\text{(4)}
$$

Vinyl Sulfides. A general route to vinyl sulfides has been achieved through the reaction of thiosilanes with silyl enol ethers in the presence of boron trifluoride etherate (eq 5).[6]

$$
\text{(5)}
$$

Miscellaneous Transformations. The reaction of Me_3SiSMe with glycopyranoses leads to the stereoselective formation of 1-thioglycosides (eq 6).[7,8] When primary nitro compounds are treated sequentially with n-butyllithium followed by Me_3SiSMe, the corresponding thiohydroximates are produced in good yields (eq 7).[9] 4-Sulfinylazetidin-2-ones undergo a novel substitution reaction with silylated heteronucleophiles in the presence of a catalytic amount of zinc iodide (eq 8).[10,11] When propenoyltrimethylsilane is treated with Me_3SiSMe, the synthetically useful silyl enol ethers of acylsilanes are formed (eq 9).[12] Orthothioesters are formed when esters are treated with Me_3SSiMe and aluminum chloride.[13]

$$
\text{(6)}
$$

$$
\text{(7)}
$$

$$
\text{(8)}
$$

$$
\text{(9)}
$$

Related Reagent. Methanethiol.

1. Evans, D. A.; Truesdale, L. K.; Grimm, K. G.; Nesbitt, S. L., *J. Am. Chem. Soc.* **1977**, *99*, 5009.
2. Groutas, W. C.; Felker, D. S., *Synthesis* **1980**, 861.
3. Hanessian, S.; Guindon, Y., *Tetrahedron Lett.* **1980**, *21*, 2305.
4. Guindon, Y.; Young, R. N.; Frenette, R., *Synth. Commun.* **1981**, *11*, 391.
5. Firgo, H. A.; Weber, W. P., *J. Organomet. Chem.* **1981**, *222*, 201.
6. Degl'Innocenti, A.; Ulivi, P.; Capperucci, A.; Mordini, A.; Reginato, G.; Ricci, A., *Synlett* **1992**, 499.
7. Pozsgay, V.; Jennings, H. J., *Tetrahedron Lett.* **1987**, *28*, 1375.
8. (a) Hasegawa, A.; Ohki, H.; Nagahama, T.; Ishida, H.; Kiso, M., *Carbohydr. Res.* **1991**, *212*, 277: (b) Kameyama, A.; Ishida, H.; Kiso, M.; Hasegawa, A., *Carbohydr. Res.* **1990**, *200*, 269.
9. Hwu, J. R.; Tsay, S.-C.; *Tetrahedron* **1990**, *46*, 7413.
10. Kita, Y.; Shibata, N.; Yoshida, N.; Tohjo, T., *Tetrahedron Lett.* **1991**, *32*, 2375.
11. de Vries, J. G.; Hauser, G.; Sigmund, G., *Heterocycles* **1985**, *23*, 1081.
12. Ricci, A.; Degl'Innocenti, A.; Borselli, G.; Reginato, G., *Tetrahedron Lett.* **1987**, *28*, 4093.
13. Matthews, D. P.; Whitten, J. P.; McCarthy, J. R., *Tetrahedron Lett.* **1986**, *27*, 4861.
14. Evans, D. A.; Grimm, E. G.; Truesdale, L. K., *J. Am. Chem. Soc.* **1975**, *97*, 3229.

William C. Groutas
Wichita State University, KS, USA

Methyltrichlorosilane[1]

[75-79-6] $\quad$ CH_3Cl_3Si $\quad$ (MW 149.48)

InChI = 1S/CH3Cl3Si/c1-5(2,3)4/h1H3

InChIKey = JLUFWMXJHAVVNN-UHFFFAOYSA-N

(precursor to organosilicon compounds;[1] silylating agent;[1] Lewis acid[2])

Physical Data: bp 66 °C; d 1.273 g cm^{-3}.

Solubility: sol methylene chloride.

Form Supplied in: liquid; commercially available.

Purification: can be purified by distillation.

Handling, Storage, and Precautions: is corrosive and moisture sensitive. It should be handled in an anhydrous atmosphere in a fume hood.

Original Commentary

George A. Olah, G. K. Surya Prakash, Qi Wang & Xing-ya Li
University of Southern California, Los Angeles, CA, USA

Organosilicon Compounds. $MeSiCl_3$ is an important precursor to organosilicon compounds.[1] It reacts with carbanions to form corresponding alkyl(aryl) methylsilanes.[3] Grignard and organolithium reagents are the most frequently used carbanion sources in this transformation. Generally, the organolithiums are preferred over Grignard reagents, especially in the preparation of tetraorganosilanes. However, it was reported that the reaction using Grignard reagents was facilitated by a catalytic amount of cyanide or thiocyanate ions.[4] Stepwise substitution on MeSiCl$_3$ can be accomplished by introducing carbanions sequentially, yielding dichloro-, monochloro-, and tetraorganosilanes, respectively (eq 1). This makes it possible to design a silane with a specific steric bulkiness, a feature often required for the monochlorosilanes to serve as protecting groups.[5]

$$ MeSiCl_3 \xrightarrow{R^-M^+} \underset{Cl}{\overset{Me}{Cl-\underset{|}{\overset{|}{Si}}-R}} \xrightarrow{(R^1)^-M^+} \underset{R^1}{\overset{Me}{Cl-\underset{|}{\overset{|}{Si}}-R}} \xrightarrow{(R^2)^-M^+} \underset{R^1}{\overset{Me}{R^2-\underset{|}{\overset{|}{Si}}-R}} \quad (1) $$

MeSiCl$_3$ has been used as a coupling agent in the preparation of the multidentate ligand (*RRR*)-siliphos {(*RRR*)-MeSi[CH$_2$P(*t*-Bu)Ph]$_3$} (eq 2).[6] The preparation of this optically pure tristertiary phosphine was achieved by deprotonation of optically pure (*R*)-P(BH$_3$)Me(*t*-Bu)Ph, followed by silylation of the resulting carbanion with MeSiCl$_3$, and the subsequent removal of the BH$_3$ protecting group with morpholine The function of MeSiCl$_3$ as a coupling agent has also been widely used for the synthesis of star-branched polymers.[7] Cyclic silanes are obtained by treating MeSiCl$_3$ with dicarbanions and tricarbanions, albeit in lower yields (eqs 3 and 4).[8] Treatment of CFCl$_3$/(Et$_2$N)$_3$P with MeSiCl$_3$ led to the formation of CFCl$_2$SiCl$_2$Me.[9]

$$ (R)\text{-MeP(BH}_3)(t\text{-Bu})Ph \xrightarrow[\substack{2.\ MeSiCl_3 \\ 3.\ morpholine \\ 43\%}]{1.\ TMEDA,\ BuLi} \quad (2) $$

$$ N[(CH_2)_3MgCl]_3 \xrightarrow[\substack{THF,\ C_6H_6 \\ 14\%}]{MeSiCl_3} \quad (3) $$

$$ \xrightarrow[\substack{2.\ MeSiCl_3 \\ 2\%}]{1.\ Mg,\ THF} \quad (4) $$

Substitutions of MeSiCl$_3$ can also be performed under radical conditions or with transition metal catalysis.[1,10] The radical reactions can be initiated either by higher temperature or photolysis. However, the selectivity of the reaction is generally poor.[1] Better results were obtained with transition metal-catalyzed reactions. It was reported that MeSiCl$_3$ reacted with terminal alkynes in the presence of a catalytic amount of copper(I) chloride to form alkynyldichlorosilanes (eq 5).[11]

$$ RC{\equiv}CH + MeSiCl_3 \xrightarrow[\substack{150\ °C \\ 60-80\%}]{CuCl,\ Et_3N} R{\equiv}SiCl_2Me \quad (5) $$

Besides *C*-silylation, MeSiCl$_3$ is also capable of silylating heteroatoms such as, O, N, and S.[1] It reacts with alcohols,

oximes, alkoxysilanes, and esters to form corresponding silylated products.[12] In fact, certain poly- or oligoorganosiloxanes were prepared from MeSiCl$_3$ and hydroxy-containing compounds.[13] When aldehydes and ketones are used, tris(alkenyloxy)silanes are obtained in good yields (eq 6).[14] In the case of 1,3-diketones, organosilicon(IV) β-diketonate complexes are formed (eq 7).[15] Amines are silylated in a similar fashion.[16]

$$ \underset{R}{\overset{R^1}{>}}{=}\!O + MeSiCl_3 \xrightarrow[\substack{MeCN \\ 30-88\%}]{NaI,\ Et_3N} \left(\underset{R}{\overset{R^1}{>}}{=}\!\underset{R^2}{\overset{O{-}SiMe}{}}\right)_3 \quad (6) $$

$$ \underset{R}{\overset{O\quad O}{\bigvee}}R^1 \xrightarrow[\substack{ether \\ 44-89\%}]{MeSiCl_3} \left(\underset{R^1}{\overset{R}{}}\right)_2\overset{Me}{\underset{Cl}{Si}} \quad (7) $$

Reactions as a Lewis Acid. MeSiCl$_3$ is an effective Lewis acid in the condensation of carboxylic acids with alcohols and amines.[17] Good yields of esters and amides are obtained. MeSiCl$_3$ reacts with epoxides to form β-chloroalkoxysilanes in good yields.[18] It also promotes the addition of allylstannanes to aldehydes (eq 8).[19] Compared to the traditional Lewis acids, MeSiCl$_3$ is relatively mild and less moisture sensitive. The reaction is very selective. When both allylstannanes and allylsilanes are present in the reaction, aldehydes react only with allylstannanes. It was reported that acetylacetone was condensed to pyrylium chloride by using MeSiCl$_3$.[20]

$$ \underset{R}{\overset{O}{\bigvee}} + \diagdown\!\!\diagup\!\!SnBu_3 \xrightarrow[\substack{THF,\ C_6H_6 \\ \sim100\%}]{MeSiCl_3} \underset{R}{\overset{OSiCl_2Me}{\diagdown\!\!\diagup\!\!\diagdown}} \quad (8) $$

MeSiCl$_3$ cleaves the ether linkage of *gem*-amino ethers to form dialkyl(methylene)ammonium chloride (Mannich-type salts) in excellent yields (eq 9).[21a] This procedure is much more convenient, efficient, economical, and less dangerous compared to earlier methods. The Mannich salts were also prepared in situ for electrophilic substitution of nucleophilic aromatic compounds (eq 10).[21b,c]

$$ \underset{R^1}{\overset{R^2O}{\diagdown}}{N}{-}R \xrightarrow[\substack{MeCN \\ 85-98\%}]{MeSiCl_3} {=}\underset{R^1}{\overset{R}{N^+}}\ Cl^- \quad (9) $$

$$ \underset{Me}{\overset{\text{(pyrrole)}}{N}} \xrightarrow[\substack{EtOCH_2N(i\text{-}Pr)_2 \\ 42\%}]{MeSiCl_3} \underset{Me}{\overset{\text{(pyrrole)}}{N}}{-}CH_2CH_2N(i\text{-}Pr)_2 \quad (10) $$

In conjunction with CuCl, MeSiCl$_3$ promotes conjugate addition of both organomagnesium and manganese(II) reagents to α,β-ethylenic esters (eq 11).[22] These are much higher yield reactions compared to conjugate addition of copper or cuprate reagents to the substrates. The exact role of MeSiCl$_3$ in the reactions is not clear.

$$R^2 \underset{R^2}{\overset{R^1}{\diagdown}} CO_2Et \xrightarrow[\text{MeSiCl}_3, \text{THF}]{\text{RMCl, CuCl}} R^2 \underset{R}{\overset{R^1}{\diagdown}} CO_2Et \qquad (11)$$

$$M = Mg, Mn^{II}$$

MeSiCl$_3$ has been found useful as a deprotecting agent in peptide chemistry.[23] It also promotes, in conjunction with diphenyl sulfoxide, formation of disulfide bonds in peptides (eq 12).[24]

$$2 \underset{H_2N}{\overset{SR}{\diagdown}} CO_2H \xrightarrow[\text{sulfoxide}]{\text{MeSiCl}_3} \underset{H_2N}{\overset{S \text{——} S}{\diagdown}} \underset{H_2N}{\overset{}{\diagdown}} CO_2H \quad \underset{H_2N}{\overset{}{\diagdown}} CO_2H \qquad (12)$$

MeSiCl$_3$/sodium iodide[2] is a surrogate of iodotrimethylsilane,[25] a widely used silicon Lewis acid, which has been used for various organic transformations. Compared to Me$_3$SiI, MeSiCl$_3$/NaI is milder, and thus it is more regioselective when used in ether cleavage (eq 13). Methyl ethers (ROMe) are cleaved affording alcohols as sole products, provided that the alkyl groups are either primary or secondary. Benzyl, trityl, and tetrahydropyranyl ethers are also cleaved regioselectively at ambient temperature to give quantitative yields of alcohols. If a tertiary alkyl group is present in the ethers, an alkyl iodide is then obtained. Esters and lactones are similarly cleaved to carboxylic acids. Acetals are converted to carbonyl compounds by MeSiCl$_3$/NaI (eq 14).

$$R^1OR \xrightarrow[\text{2. H}_2\text{O}]{\text{1. MeSiCl}_3, \text{NaI}} R^1OH \qquad (13)$$

$$R^1 = \text{primary and secondary alkyl group};$$
$$R = \text{Me, benzyl, trityl, tetrahydropyranyl}$$

$$\underset{R^1}{\overset{R}{\diagdown}} \underset{OMe}{\overset{OMe}{\diagup}} \xrightarrow[\substack{\text{2. H}_2\text{O} \\ 70\text{–}96\%}]{\text{1. MeSiCl}_3, \text{NaI}} \underset{R^1}{\overset{R}{\diagdown}} {=}O \qquad (14)$$

MeSiCl$_3$/NaI is a good iodinating agent, which converts alcohols to corresponding iodides (eq 15). Primary alcohols react very slowly even under reflux conditions, giving low yields. However, moderate and excellent yields of alkyl iodides can be obtained with secondary, tertiary, and benzylic alcohols.

$$ROH \xrightarrow[\substack{\text{NaI} \\ 20\text{–}96\%}]{\text{MeSiCl}_3} RI \qquad (15)$$

$$R = \text{alkyl}$$

MeSiCl$_3$/NaI was found to dehalogenate α-halo ketones reductively (eq 16). The conversion of 5-cyano-7-oxabicyclo[2.2.1]-hept-2-ene to 2-furanpropanenitrile was also reported (eq 17).[26]

$$\underset{Br}{\overset{R^1}{\diagdown}} \underset{R}{\overset{O}{\diagup}} \xrightarrow[\substack{\text{NaI} \\ 60\text{–}98\%}]{\text{MeSiCl}_3} \underset{}{\overset{R^1}{\diagdown}} \underset{R}{\overset{O}{\diagup}} \qquad (16)$$

$$\xrightarrow[\substack{\text{NaI} \\ 94\%}]{\text{MeSiCl}_3} \qquad (17)$$

First Update

Margaret A. Brimble

The University of Auckland, Auckland, New Zealand

Organosilicon Compounds. MeSiCl$_3$ continues to be an important precursor to *C*-silylated organosilicon compounds. Tetrahedral (triaryl)methylsilanes prepared via lithiation of halogenated aromatic compounds followed by reaction with methyltrichlorosilane provide a silicon-based building block to access 3D self-assembled structures containing a tetrahedral silicon in a key shape-defining position (eq 18).[27] MeSiCl$_3$ is also used to access a triphenolic silane used to prepare a cage-shaped silicon tethered borate (eq 19).[28]

$$3 \xrightarrow[\text{2 h, then MeSiCl}_3]{\text{BuLi, Et}_2\text{O, }-90 \text{ to } 0\,^\circ\text{C}} \qquad (18)$$

no yield given

$$3 \xrightarrow[\substack{\text{1. }^t\text{BuLi, TMEDA} \\ \text{2. MeSiCl}_3 \\ \text{3. LiAlH}_4}]{} \qquad (19)$$

no yield given

Three generations of bithiophenesilane monodendrons and dendrimers are synthesized from methyl(2,2′-bithien-5-yl)dichlorosilane that is obtained by dropwise addition of 2,2′-bithien-5-ylmagnesium bromide to MeSiCl$_3$ (eq 20).[29]

$$\xrightarrow[\substack{\text{1. Mg, THF} \\ \text{2. MeSiCl}_3}]{} \qquad (20)$$

no yield given

The kinetics of the Grignard reaction of butylmagnesium chloride and phenylmagnesium bromide with MeSiCl$_3$ was investigated in diethyl ether and diethyl ether–toluene mixtures establishing that there was no significant difference in the two solvents.[30] A later study of the kinetics of the reactions of phenylmagnesium chloride and bromide and diphenylmagnesium with MeSiCl$_3$ in THF and THF–hydrocarbon mixtures established that the reaction proceeds much faster in THF than in diethyl ether or THF–hydrocarbon mixtures.[31]

Silylation of heteroatoms using MeSiCl$_3$ continues to extend the range and use of organosilicon compounds in synthesis. Chiral bidentate silyl derivatizing reagents have been developed for the GC analysis of aliphatic 1,3-diols. These reagents react with the diols to form cyclic siloxanes, thereby enabling the determination of their enantiomeric composition. The chiral silyl reagents are readily prepared by reaction of MeSiCl$_3$ with 1 equiv of (*S*)-[1-(2-bromophenyl)ethoxy]dichloromethylsilane or (−)-menthol (eq 21).[32]

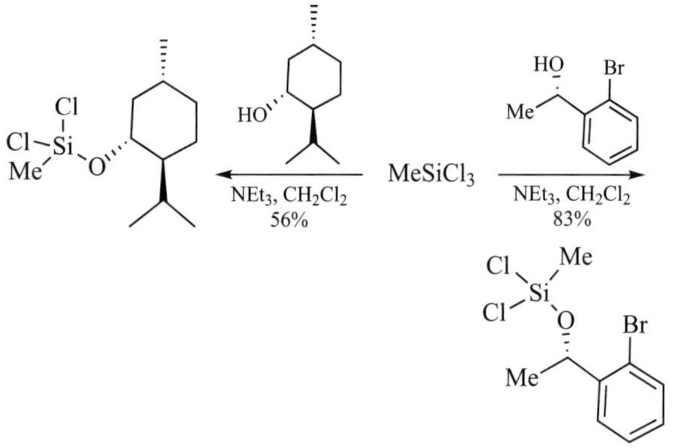

(21)

The synthesis of pentacoordinate silicon complexes with enamine-functionalized salen ligands is easily performed by the addition of $MeSiCl_3$ and diethylamine to a salen ligand (eq 22).[33]

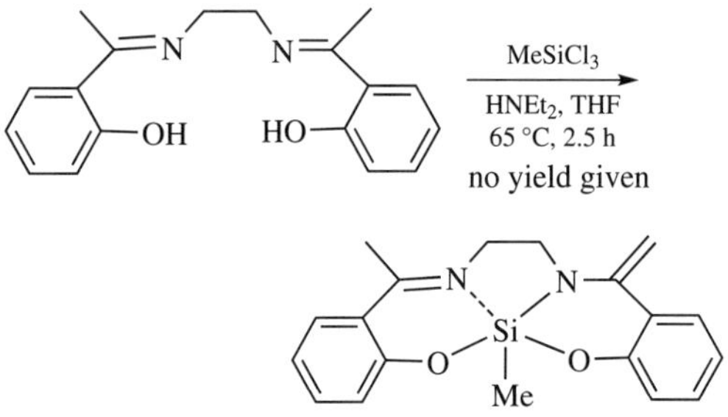

(22)

Various imidazolylsilane derivatives are synthesized by a *trans*-silylation procedure between 1-(trimethyl)imidazoles and $MeSiCl_3$ (eq 23).[34]

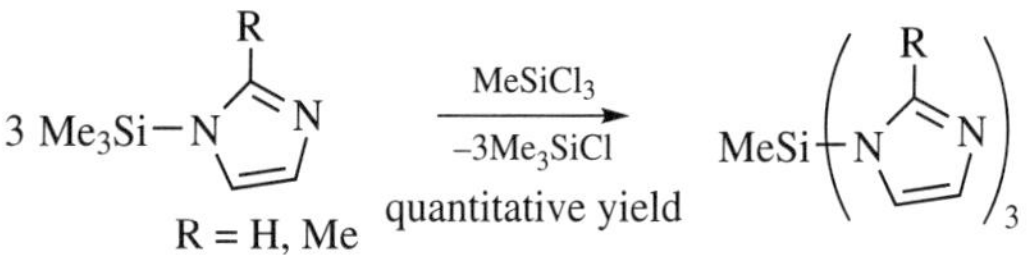

(23)

Reactions as a Lewis Acid. $MeSiCl_3$ is a useful Lewis acid to effect an Ireland–Claisen rearrangement to assemble the limonoid framework of the insect antifeedant azadirachtin.[35] The synergistic effect of boron trifluoride etherate and $MeSiCl_3$ effects an efficient crossed imino pinacol coupling reaction affording 1,2-diamines in good yield in the presence of a zinc–copper couple (eq 24).[36] Double Mannich reaction of cyclic β-ketoesters with bis(aminol) ethers using $MeSiCl_3$ as activator allows rapid assembly of azabicyclo[3.3.1]nonanes and azabicyclo[3.2.1]octanes useful for the synthesis of alkaloids related to methyllycaconitine (eq 25).[37]

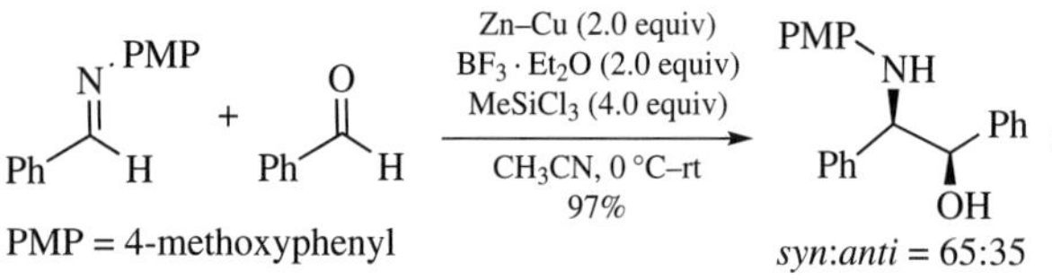

PMP = 4-methoxyphenyl

(24)

syn:anti = 65:35

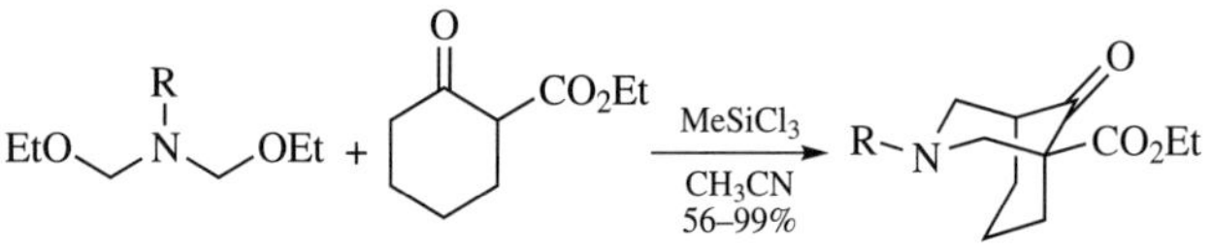

R = Et, ^{i}Pr, ^{n}Bu, ^{t}Bu, Cy, 2-phenylethyl, 3-phenylpropyl

(25)

Oligosilanes and Polysilanes. Polysilanes comprising catenated silicon–silicon bonds in their backbone exhibit unique electronic properties due to σ-conjugation through the silicon framework. Reaction of a 1:3 mixture of $MeSiCl_3$ and Me_3SiCl with lithium metal in THF at ambient temperature affords a polysilane that can be further transformed to a permethylbicyclo[2.2.2]-octasilane (eq 26).[38] Stable polysilanols constructed of oligosilane frameworks can be conceived of as silicon analogs of polyalcohols. Treatment of $Ph(Me_3Si)_2Si-K\cdot3THF$ with $MeSiCl_3$ provides an organosilane bearing an easily exchangeable chloro function that is readily converted to a silanol upon hydrolysis (eq 27).[39]

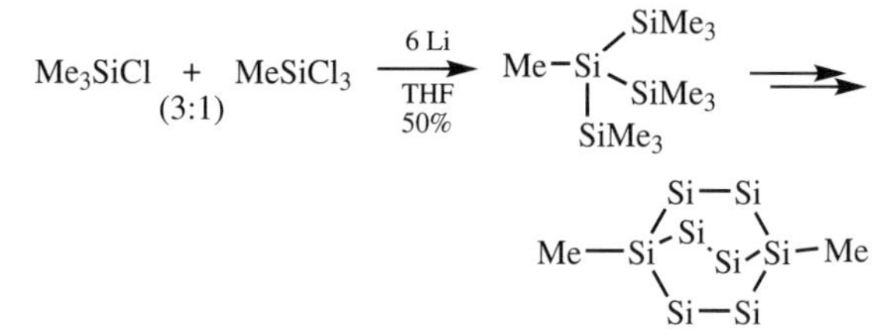

(26)

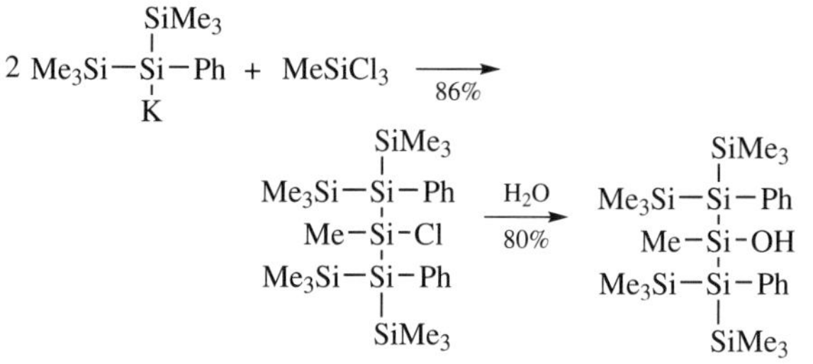

(27)

1. Voorhoeve, R. J. H. *Organohalosilanes, Precursors to Silicones*; Elsevier: Amsterdam, 1967.

2. (a) Olah, G. A.; Husain, A.; Singh, B. P.; Mehrotra, A. K., *J. Org. Chem.* **1983**, *48*, 3667. (b) Olah, G. A.; Husain, A.; Gupta, B. G. B.; Narang, S. C., *Angew. Chem., Int. Ed. Engl.* **1981**, *20*, 690.

3. (a) Rubinazztajn, S.; Zeldin, M.; Fife, W. K., *Synth. React. Inorg. Met. Org. Chem.* **1990**, *20*, 495. (b) Van den Ancker, T.; Jolly, B. S.; Lappert, M. F.; Raston, C. L.; Skelton, B. W.; White, A. H., *J. Chem. Soc., Chem. Commun.* **1990**, 1006. (c) Chen, G. J.; Tamborski, C., *J. Organomet. Chem.* **1985**, *293*, 313. (d) Tamborski, C.; Chen, G. J.; Anderson, D. R.; Snyder, Jr., C. E., *Ind. Eng. Chem., Prod. Res. Dev.* **1983**, *22*, 172. (e) Erchak, N. P.; Ashmane, A. R.; Popelis, Y. Y.; Lukevits, E., *J. Gen. Chem. USSR (Engl. Transl.)* **1983**, *53*, 334 (*Chem. Abstr.* **1983**, *99*, 53 874p). (f) Shi, B. C.; Jutzi, P., *Chem. J. Chin. Univ.* **1991**, *12*, 1338 (*Chem. Abstr.* **1992**, *117*, 212 575a). (g) Wada, M.; Wakamori, H.; Hiraiwa, A.; Erabi, T., *Bull. Chem. Soc. Jpn.* **1992**, *65*, 1389.

4. Lennon, P. J.; Mack, D. P.; Thompson, Q. E., *Organometallics* **1989**, *8*, 1121.

5. Toshima, K.; Tatsuta, K.; Kinoshita, M., *Bull. Chem. Soc. Jpn.* **1988**, *61*, 2369.

6. Ward, T. R.; Venanzi, L. M.; Albinati, A.; Lianza, F.; Gerfin, T.; Gramlich, V.; Tombo, G. M. R., *Helv. Chim. Acta* **1991**, *74*, 983.

7. For example, see: (a) Storey, R. F.; George, S. E.; Nelson, M. E., *Macromolecules* **1991**, *24*, 2920. (b) Jenkins, D. K., *Polymer* **1985**, *26*, 147. (c) Hadjichristidis, N.; Roovers, J., *Polymer* **1985**, *26*, 1087.

8. (a) Luchina, N.; Ciobanu, A.; Bostan, M., *Rev. Roum. Chem.* **1981**, *26*, 1479. (b) Boudjouk, P.; Sooriyakumaran, R.; Kapfer, C. A., *J. Organomet. Chem.* **1985**, *281*, C21. (c) Jurkschat, K.; Mugge, C.; Schmidt, J.; Tzschach, A., *J. Organomet. Chem.* **1985**, *287*, C1.

9. Josten, R.; Ruppert, I., *J. Organomet. Chem.* **1987**, *329*, 313.

10. Son, V. V.; Ivashchenko, S. P.; Son, T. V., *J. Gen. Chem. USSR (Engl. Transl.)* **1990**, *60*, 624 (*Chem. Abstr.* **1990**, *113*, 78 480c).

11. Deleris, G.; Dunogues, J.; Calas, R.; Lapouyade, P., *J. Organomet. Chem.* **1974**, *80*, C45.

12. (a) Fomin, V. A.; Etlis, V. S.; Petrukhin, I. V., *J. Gen. Chem. USSR (Engl. Transl.)* **1983**, *53*, 711 (*Chem. Abstr.* **1983**, *99*, 105 322d). (b) Ryasin, G. V.; Fedotov, I. S.; Luk'yanova, I. A.; Mironov, V. F., *J. Appl. Chem. USSR* **1974**, *47*, 2683 (*Chem. Abstr.* **1975**, *82*, 43 513e). (c) Voronkov, M. G.; Kuznetsova, G. A.; Baryshok, V. P., *J. Gen. Chem. USSR (Engl. Transl.)* **1983**, *53*, 1512 (*Chem. Abstr.* **1983**, *99*, 212 600q). (d) Mbah, G.; Speier, J. L., *J. Organomet. Chem.* **1984**, *271*, 77.

13. For example, see: (a) Martynova, T. N.; Chupakhina, T. I., *J. Organomet. Chem.* **1988**, *345*, 11. (b) Rebrov, E. A.; Muzafarov, A. M.; Papkov, V. S.; Zhdanov, A. A., *Dokl. Akad. Nauk SSSR* **1989**, *309*, 376 (*Chem. Abstr.* **1990**, *112*, 235 956m). (c) Andrianov, K. A.; Chernyavskii, A. I.; Makarova, N. N., *Izv. Akad. Nauk SSSR, Ser. Khim.* **1979**, *1835* (*Chem. Abstr.* **1980**, *92*, 42 027u).

14. (a) Shchepin, V. V.; Lapkin, I. I., *J. Gen. Chem. USSR (Engl. Transl.)* **1981**, *51*, 1957 (*Chem. Abstr.* **1982**, *96*, 52 374b). (b) Rochin, C.; Babot, O.; Moulines, F.; Duboudin, F., *J. Organomet. Chem.* **1984**, *273*, C7. (c) Rochin, C.; Babot, O.; Duboudin, F., *J. Organomet. Chem.* **1985**, *281*, C24.

15. (a) Schott, V. G.; Golz, K., *Z. Anorg. Allg. Chem.* **1973**, *399*, 7. (b) Serpone, N.; Hersh, K. A., *J. Organomet. Chem.* **1975**, *84*, 177.

16. (a) Issleib, K.; Kuhne, U.; Krech, F., *Phosphorus Sulfur/Phosphorus Sulfur Silicon* **1985**, *21*, 367. (b) tom Dieck, H.; Zettlitzer, M., *Chem. Ber.* **1987**, *120*, 795. (c) Brauer, D. J.; Burger, H.; Liewald, G. R.; Wilke, J., *J. Organomet. Chem.* **1985**, *287*, 305. (d) Grobe, J.; Voulgarakis, N., *Z. Anorg. Allg. Chem.* **1984**, *517*, 125. (e) Grobe, J.; Hildebrandt, W.; Martin, R.; Walter, A., *Z. Anorg. Allg. Chem.* **1991**, *592*, 121.

17. (a) Nakao, R.; Oka, K.; Fukumoto, T., *Bull. Chem. Soc. Jpn.* **1981**, *54*, 1267. (b) Akpoyraz, M., *Doga, Ser. C* **1980**, *4*(2), 1 (*Chem. Abstr.* **1982**, *96*, 103 600g).

18. Viktorova, I. P.; Gladkikh, A. F.; Viktorov, O. F., *J. Gen. Chem. USSR (Engl. Transl.)* **1976**, *46*, 1947 (*Chem. Abstr.* **1977**, *86*, 155 731p).

19. Marshall, R. L.; Young, D. J., *Tetrahedron Lett.* **1992**, *33*, 1365.

20. Serpone, N.; Ignacz, T. F., *Gazz. Chim. Ital.* **1985**, *115*, 419.

21. (a) Rochin, C.; Babot, O.; Dunogues, J.; Duboudin, F., *Synthesis* **1986**, 228. (b) Heaney, H.; Papageorgiou, G.; Wilkins, R. F., *J. Chem. Soc., Chem. Commun.* **1988**, 1161. (c) Fairhurst, R. A.; Heaney, H.; Papageorgiou, G.; Wilkins, R. F.; Eyley, S. C., *Tetrahedron Lett.* **1989**, *30*, 1433.

22. (a) Cahiez, G.; Alami, M., *Tetrahedron Lett.* **1990**, *31*, 7423. (b) Cahiez, G.; Alami, M., *Tetrahedron Lett.* **1990**, *31*, 7425.

23. Kiso, Y.; Yoshida, M.; Fujisaki, T.; Mimoto, T.; Kimura, T.; Shimokura, M. *Proc. 24th Symp. Peptide Chem.*; Protein Research Foundation: Osaka, **1986**; p 205 (*Chem. Abstr.* **1988**, *108*, 112 924j).

24. (a) Akaji, K.; Tatsumi, T.; Yoshida, M.; Kimura, T.; Fujiwara, Y.; Kiso, Y., *J. Am. Chem. Soc.* **1992**, *114*, 4137. (b) Akaji, K.; Tatsumi, T.; Yoshida, M.; Kimura, T.; Fujiwara, Y.; Kiso, Y., *J. Chem. Soc., Chem. Commun.* **1991**, 167.

25. Olah, G. A.; Prakash, G. K. S.; Krishnamurti, R. In *Advances in Silicon Chemistry*; Larson, G. L. Ed; JAI Press: Greenwich, CT, **1991**; *1*, p 1.

26. Kibayashi, T.; Ishii, Y.; Ogawa, M., *Bull. Chem. Soc. Jpn.* **1985**, *58*, 3627.

27. Kryschenko, Y. K.; Seidel, S. R.; Muddiman, D. C.; Nepomuceno, A. I.; Stang, P. J., *J. Am. Chem. Soc.* **2003**, *125*, 9647.

28. Yasuda, Y.; Yoshioka, S.; Nakajima, H.; Chiba, K.; Baba, A., *Org. Lett.* **2008**, *10*, 929.

29. Luponosov, Y. N.; Ponomarenko, S. A.; Surin, N. M.; Muzafarov, A. M., *Org. Lett.* **2008**, *10*, 2753.

30. Tuulmets, A.; Nguyen, B. T.; Panov, D.; Sassian, M.; Jarv, J., *J. Org. Chem.* **2003**, *68*, 9933.

31. Tuulmets, A.; Nguyen, B. T.; Panov, D., *J. Org. Chem.* **2004**, *69*, 5071.

32. Arsene, C.; Schulz, S., *Org. Lett.* **2002**, *4*, 2869.

33. Wagler, J.; Bohme, U.; Roewer, G., *Angew. Chem., Int. Ed.* **2002**, *41*, 1732.

34. Tozawa, T.; Yamane, Y.; Mukaiyama, T., *Chem. Lett.* **2005**, *34*, 734.

35. Fukuzaki, T.; Kobayashi, S.; Hibi, T.; Ikuma, Y.; Ishihara, J.; Kanoh, N.; Murai, A., *Org. Lett.* **2002**, *4*, 2877.

36. (a) Shimizu, M.; Suzuki, I.; Makino, H., *Synlett* **2003**, 1635. (b) Shimizu, M.; Iwata, A.; Makino, H., *Synlett* **2002**, 1538.

37. Brocke, C.; Brimble, M. A.; Lin, D. S.-H.; McLeod, M. D., *Synlett* **2004**, 2359.

38. Setaka, W.; Hamada, N.; Kira, M., *Chem. Lett.* **2004**, *33*, 626.

39. Krempner, C.; Kopf, J.; Mamat, C.; Reinke, H.; Spannenberg, A., *Angew. Chem., Int. Ed.* **2004**, *43*, 5406.

Methyl Trichlorosilyl Ketene Acetal (1)

[4519-10-2] $C_3H_5Cl_3O_2Si$ (MW 207.52)

InChI = 1S/C3H5Cl3O2Si/c1-3(7-2)8-9(4,5)6/h1H2,2H3

InChIKey = LMVJWNCKICXPHV-UHFFFAOYSA-N

(reagent for the enantioselective acetate aldol additions to aldehydes[1] and ketones[2])

Physical Data: bp 25 °C (5 mmHg), $d_{20} = 1.3144$ g dm^{-3}.

Solubility: soluble in halogenated organic solvents such as methylene chloride, chloroform, carbon tetrachloride; sparingly soluble in hydrocarbon and ethereal solvents.

Analysis of Reagent Purity: ^{1}H NMR and elemental analysis.

Preparative Methods: two different routes have been developed for the preparation of methyl trichlorosilyl ketene acetal. One route is through bis(tributyltin) oxide-catalyzed transmetalation of methyl tributylstannyl acetate (**2**) with silicon tetrachloride (eq 1).[3] When the reaction is complete, excess silicon tetrachloride is removed and the product is purified by fractional distillation under reduced pressure.

$$\underset{\textbf{2}}{Bu_3Sn\!\!-\!\!\overset{O}{\overset{\|}{C}}\!\!-\!\!OMe} \quad \xrightarrow[\substack{(Bu_3Sn)_2O,\ 10\ mol\ \% \\ rt}]{\substack{SiCl_4 \\ (6\ equiv)}} \quad \underset{\textbf{1}}{H_2C\!\!=\!\!\overset{OSiCl_3}{\overset{|}{C}}\!\!-\!\!OMe} \quad (1)$$

Alternatively, methyl trimethylsilylacetate (**3**) can be converted to **1** cleanly when trichlorosilyl triflate is used (eq 2).[3] This route has the advantage of not using toxic tin compounds. However, due to the similar boiling point, a complete removal of trimethylsilyl trifluoromethanesulfonate by-product (**4**) from **1** by fractional distillation is difficult. This method is therefore suitable only for use when **4** does not interfere with the subsequent reaction of **1**.

$$\underset{\textbf{3}}{TMS\!\!-\!\!\overset{O}{\overset{\|}{C}}\!\!-\!\!OMe} \quad \xrightarrow[rt]{Cl_3SiOTf} \quad \underset{\textbf{1}}{H_2C\!\!=\!\!\overset{OSiCl_3}{\overset{|}{C}}\!\!-\!\!OMe} \ +\ \underset{\textbf{4}}{TMSOTf} \quad (2)$$

Handling, Storage, and Precautions: methyl trichlorosilyl ketene acetal is a thermally unstable and hydrolytically labile compound, which is best used immediately after its preparation. The thermal rearrangement to the *C*-silyl isomer occurs rapidly at 70 °C[4] and proceeds steadily albeit slowly at room temperature. If its storage is unavoidable, it must be kept at the lowest possible temperature. Its hydrolytic decomposition is instantaneous and anhydrous conditions are required for all the manipulations. This reagent is volatile and corrosive. Good ventilation and personal protection equipment are required for its handling.

Introduction. The preparation of **1** was first reported by Baukov and co-workers in 1966.[4] Thirty years later, Denmark et al. demonstrated its unique reactivity as an aldolization reagent.[1] Under catalysis by Lewis bases, a variety of aldehydes and ketones react smoothly with **1** at room temperature to afford the aldol addition products in high yields. Catalytic, asymmetric aldol additions of **1** have also been demonstrated with both aldehyde and ketone substrates. However, the level of enantioselection is highly variable and dependent on the substrate. Synthetically useful enantioselectivities have been achieved with aldehydes using chiral phosphoramides and with aryl ketones using a chiral bipyridine bis-*N*-oxide catalyst.

To obtain the highest yields and selectivities when employing **1**, it is essential that the reaction be conducted in a noncoordinating solvent such as methylene chloride. The work-up process involves hydrolytic cleavage of three Si–Cl bonds. As 3 equiv of hydrogen chloride are liberated during the aqueous work-up, the addition of the reaction mixture to the aqueous buffer solution is always preferred to avoid high local concentration of acid in case the aldol product is acid sensitive.

Aldol Additions to Aldehydes. Methyl trichlorosilyl ketene acetal is highly reactive toward aldehydes. The aldol additions of **1** to a number of aldehydes proceed to completion cleanly within minutes at −80 °C.[1] Only sterically congested pivalaldehyde reacts slowly enough to be followed spectroscopically.[5] The trichlorosilyl aldolates (**5**) are readily converted to the aldol addition products (**6**) following an aqueous work-up (eq 3). The preparative utility of **1** is demonstrated by the excellent yields with a wide range of substrates (Table 1). Branched and highly enolizable aldehydes are shown to be compatible with the unpromoted reaction (entries 2 and 5). With a potential Michael acceptor such as cinnamaldehyde, no conjugate addition products are observed and only aldol addition product are formed in high yield (entry 3).

Table 1 Aldol additions of 1 with aldehydes

Entry	R	Yield (%)
1	Phenyl	99
2	Benzyl	94
3	(*E*)-Cinnamyl	89
4	2-Phenethyl	96
5	Cyclohexyl	96
6	*t*-Butyl	99

are used.[1] However, the development of enantioselective addition of **1** to aldehydes is challenging because of the competing achiral background reaction.[5] A synthetically useful asymmetric variant of this reaction is still lacking.

Aldol Additions to Ketones. Traditionally, cerium enolates[6] or the Reformatsky-type reaction[7] have been employed to achieve high-yielding aldol additions to enolizable ketones. In this regard, methyl trichlorosilyl ketene acetal provides a reliable alternative for the synthesis of tertiary β-hydroxy esters. In the absence of a Lewis base promoter, the aldol additions of **1** to ketones are too slow to be synthetically useful.[2] On the contrary, with pyridine *N*-oxide as catalyst, methyl trichlorosilyl ketene acetal reacts smoothly with nearly all classes of ketones (**7**) (Scheme 1).[8] Good yields of the tertiary alcohol products (**8**) are obtained (eq 4), table 2 from aromatic (entries 1–2 and 4–6), heteroaromatic (entry 3), olefinic (entries 7–8), acetylenic (entries 9–10), and aliphatic (entries 11–14) ketones. The only poorly performing substrate is 2-tetralone (**7o**), which affords a 45% yield of the addition product and returns 45% of unreacted starting material, most likely from competitive enolization.

Scheme 1

Appreciable levels of enantioselection in the addition of **1** to aldehydes are achievable when chiral phosphoramide promoters

Table 2 Pyridine *N*-oxide-catalyzed aldol additions of **1** with ketones

Entry	Ketone	Time (h)	Yield (%)
1	**7a**	2	94
2	**7b**	16	91
3	**7c**	2	91
4	**7d**	2	87
5	**7e**	4	90
6	**7f**	2	94
7	**7g**	2	92
8	**7h**	2	71
9	**7i**	2	92
10	**7j**	2	72
11	**7k**	8	83
12	**7l**	2	87
13	**7m**	2	94
14	**7n**	8	93
15	**7o**	2	45

Table 3 Chiral bipyridine bis-*N*-oxide-catalyzed aldol additions of **1** with ketones

Entry	Ketone	Yield (%)	er
1	**7a**	96	91.2/8.7
2	**7b**	89	77.9/22.1
3	**7c**	87	74.3/25.7
4	**7d**	90	90.1/9.9
5	**7e**	91	88.0/12.0
6	**7f**	94	83.9/16.1
7	**7g**	87	55.4/44.6
8	**7h**	86	53.9/46.1
9	**7i**	94	67.6/32.4
10	**7j**	89	93.1/6.9
11	**7k**	84	66.1/33.9
12	**7l**	84	60.1/39.9
13	**7m**	91	66.0/34.0
14	**7n**	87	71.5/28.5

Modest to good enantioselectivities are achieved for the asymmetric aldol addition of **1** with ketones when bipyridine bis-*N*-oxide (**9**) possessing both central and axial chirality is used.[8,9] For all the nonactivated ketones, the bis-*N*-oxide-catalyzed reactions proceed cleanly and afford enantiomerically enriched products in good yields (eq 5, Table 3). The enantioselectivity, however, is highly dependent on the structure of the ketone. The best performing substrates are aromatic ketones (entries 1–6 and entry 10). The highest enantioselectivity (er 93.1/6.9, entry 10) is with substrates wherein the sizes of the two flanking groups are the most different. For these substrates, the chiral bis-*N*-oxide-catalyzed additions of **1** provide a valuable alternative to the traditional auxiliary-based methods.

additions to benzaldehyde occur only at significantly higher temperature than the other analogs.

Scheme 2

$$1 + \overset{O}{\underset{R^1}{\overset{\|}{C}}}R^2 \xrightarrow[\text{2. sat. aq. NaHCO}_3]{\substack{\text{1. bis } N\text{-oxide } \mathbf{9} \\ 10\ \text{mol}\% \\ \text{CH}_2\text{Cl}_2\ -20\ ^\circ\text{C}}} \text{MeO}\underset{R^2}{\overset{O}{\underset{\|}{C}}}\underset{R^2}{\overset{OH}{\underset{R^1}{C}}} \quad (5)$$

The copper(I) fluoride–phosphine-catalyzed aldol additions of trimethylsilyl ketene acetals to ketones give generally higher enantioselectivities.[10]

Related Electrophilic Silyl Ketene Acetal Reagents. A variety of other silyl ketene acetals (**10–14**) that possess different silyl groups (Scheme 2) are prepared by the transmetalation reaction of methyl tributylstannyl acetate (**3**) with the corresponding chlorosilane.[5] Because of the modulated electrophilicity of the silicon moieties, the rates of the unpromoted reactions with both benzaldehyde and pivalaldehyde follow the order **10** ≥ **1** > **11** > **12**. For ketene acetals **13** and **14**, the two alkyl substituents at the silicon center dramatically attenuate reactivity and the aldol

1. Denmark, S. E.; Winter, S. B. D.; Su, X.; Wong, K-T., *J. Am. Chem. Soc.* **1996**, *118*, 7404.
2. Denmark, S. E.; Fan, Y., *J. Am. Chem. Soc.* **2002**, *124*, 4233.
3. Denmark, S. E.; Stavenger, R. A.; Winter, S. B. D.; Wong, K.-T.; Barsanti, P. A., *J. Org. Chem.* **1998**, *63*, 9517.
4. Burlachenko, G. S.; Khasapov, B. N.; Petrovskaya, L. I.; Baukov, Yu. I.; Lutsenko, I. F., *J. Gen. Chem. USSR (Engl. Transl.)* **1966**, *36*, 532.
5. Denmark, S. E.; Winter, S. B. D., *Synlett*, **1997**, 1087.
6. Xiao, S.; Liu, H.-J., *Synth. Commun.* **1994**, *24*, 2485.
7. Ojida, A.; Yamano, T.; Taya, N.; Tasaka, A., *Org. Lett.* **2002**, *4*, 3051.
8. Denmark, S. E.; Fan, Y.; Eastgate, M. D., *J. Org. Chem.* **2005**, *70*, 5235.
9. Denmark, S. E.; Fan, Y., *Tetrahedron: Asymmetry* **2006**, *17*, 687.
10. Oisaki, K.; Zhao, D.; Kanai, M.; Shibasaki, M., *J. Am. Chem. Soc.* **2006**, *128*, 7164.

Scott E. Denmark
University of Illinois, Urbana, IL, USA

Yu Fan
Bristol Myers Squibb Co., New Brunswick, NJ, USA

Methyl 2-Trimethylsilylacrylate

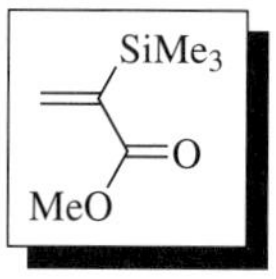

[18269-31-3]   C$_7$H$_{14}$O$_2$Si   (MW 158.30)

InChI = 1S/C7H14O2Si/c1-6(7(8)9-2)10(3,4)5/h1H2,2-5H3
InChIKey = XYEVJASMLWYHGL-UHFFFAOYSA-N

(reactive, α-silyl Michael acceptor for conjugate addition of carbanions, enolate anions, organolithium reagents, and Grignard reagents; affords 3-cyclopentenecarboxylates and cyclohexanecarboxylates by [3 + 2] and [4 + 2] annulation methods; provides α-silyl esters and vinylsilanes by organolithium conjugate additions[1,3,5])

Physical Data: bp 75 °C/50 mmHg, 152 °C/690 mmHg.
Solubility: sol Et$_2$O, THF, EtOH, and most organic solvents.
Analysis of Reagent Purity: IR, NMR, GC.
Preparative Methods: prepared[2a,b] by the reaction of (*E*)-2-(trimethylsilyl)vinyllithium with methyl chloroformate or by carbonation of the corresponding Grignard reagent prepared from (1-bromovinyl)trimethylsilane,[2c] giving 2-trimethylsilylacrylic acid. The methyl ester is prepared from the acid by direct esterification with absolute methanol in the presence of mineral acid, by reaction with diazomethane at low temperature, or by treatment with BF$_3$·MeOH complex. For preparation of various trialkylsilylacrylic acid esters[2d] and *t*-butyl 2-trimethylsilylcrotonate,[8] see the cited references.
Handling, Storage, and Precautions: should be stored in a refrigerator and addition of a few crystals of ionol is recommended in order to prevent polymerization. Use in a fume hood.

Cyclopentenes by [3 + 2] Annulation. Conjugate addition of allyllithium reagents bearing stabilizing groups in the α- and β-positions to methyl 2-trimethylsilylacrylate leads to an anionic [3 + 2] cyclization,[3a,b] generating the trimethylsilyl-substituted 3-cyclopentenecarboxylate in 57% yield (eq 1). α-Unsubstituted Michael acceptors do not give good yields of the cyclopentene product.

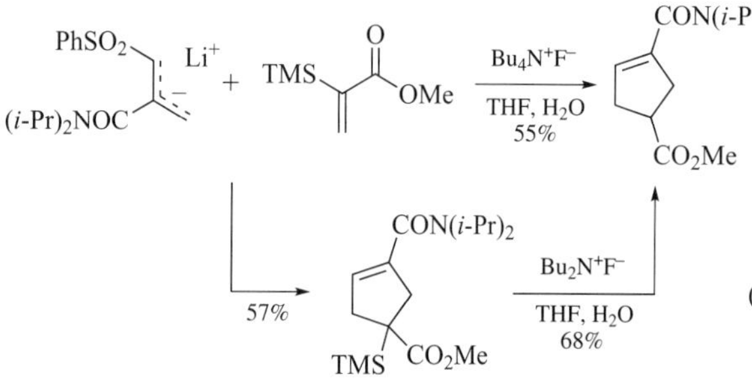

Cyclohexanecarboxylates by [4 + 2] Annulations. The kinetic enolate of 1-acetylcyclohexene on reaction with methyl 2-trimethylsilylacrylate undergoes a two-fold Michael addition,

giving rise to 5-substituted 1-decalones (eq 2), one of which was converted to (±)-khusitone[4a] and (±)-khusilal.[4b]

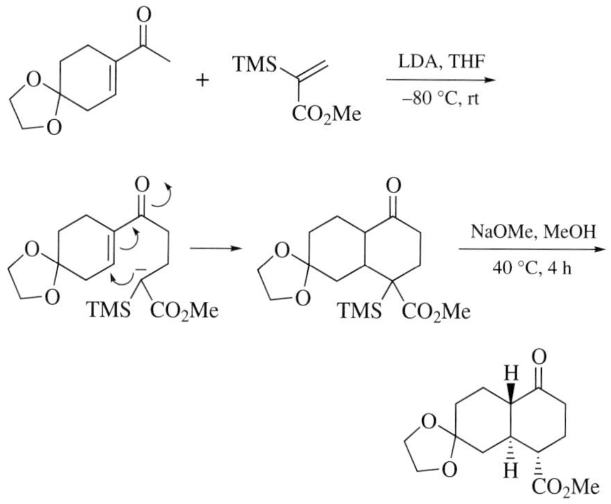

In another annulation sequence,[5] a functionalized aryllithium generated chemoselectively from (2-bromo-4,5-dimethoxyphenyl)ethyl bromide at −100 °C undergoes conjugate addition with methyl 2-trimethylsilylacrylate, and the adduct cyclizes by intramolecular alkylation to give the tetrahydronaphthalenecarboxylate in 80% yield (eq 3).

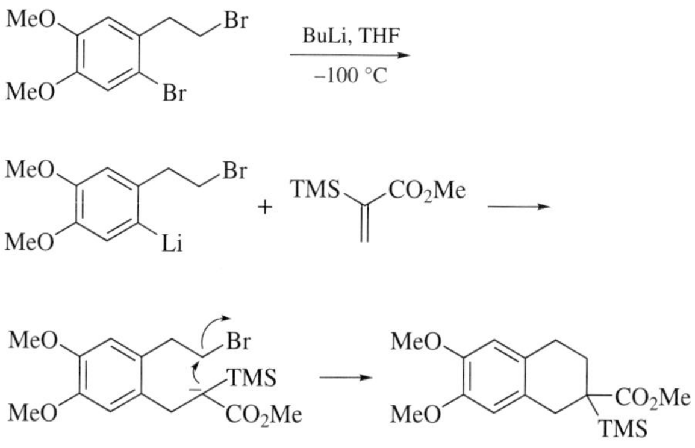

α-Silyl Esters. Conjugate addition of organometallic reagents (magnesium or lithium) to methyl 2-trimethylsilylacrylate gives either 1:1 or 1:2 adduct[6c] anions (eq 4), depending upon the reaction conditions and reactivity of donor molecules. These anions undergo either alkylation[6b] with a suitable alkylation agent or add to another electrophile[6a] such as ketones, aldehydes, or α,β-unsaturated aldehydes.

Synthesis of Pyrrolidones. 2-Azaallyl anions[7] derived from aryl-substituted imines undergo highly diastereoselective Michael addition to methyl 2-trimethylsilylcrotonate. The Michael adducts are converted to pyrrolidones through hydrolytic cyclization (eq 5).

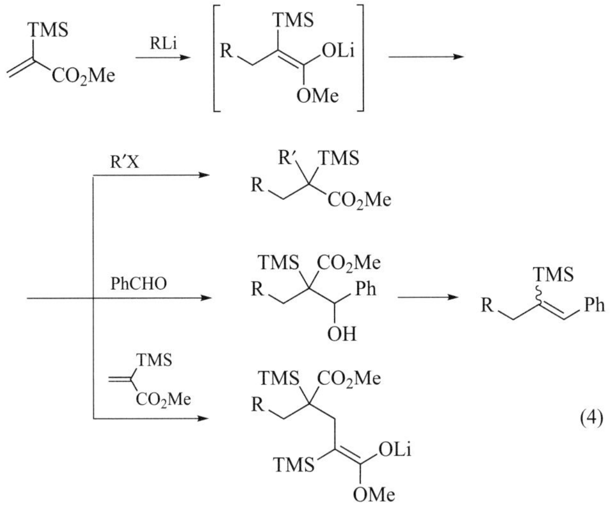

(4)

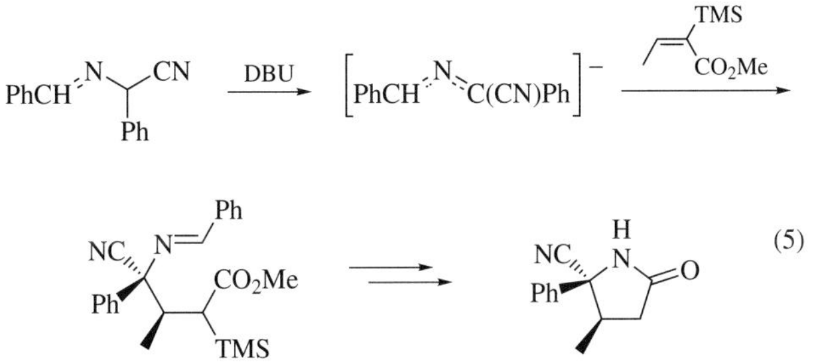

(5)

Related Reagents. Ethyl Acrylate; 1-(Trimethylsilyl)vinyllithium; Methyl 2-Methylthioacrylate; Trimethyl 2-Phosphonoacrylate; 3-Trimethylsilyl-3-buten-2-one; Vinyltrimethylsilane.

1. *Silicon in Organic Synthesis*; Colvin, E., Ed.; Butterworths: London, 1981.
2. (a) Ottolenghi, A.; Fridkin, M.; Zilkha, A., *Can. J. Chem.* **1963**, *41*, 2977. (b) Cunico, R. F.; Lee, H. M.; Herbach, J., *J. Organomet. Chem.* **1973**, *52*, C7. (c) Boeckman, R. K.; Blum, D. M.; Ganem, B.; Halvey, N., *Org. Synth.* **1974**, *58*, 152. (d) Eryoma, J.; Matsumoto, S.; Furukawa, J.; Tsutsumi, K., *Chem. Abstr.* **1993**, *107*, 176 222r.
3. (a) Peter, B.; Burg, D. A., *J. Org. Chem.* **1989**, *54*, 1647. (b) Peter, B.; Burg, D. A., *Tetrahedron Lett.* **1986**, 5911.
4. (a) Hagiwara, H.; Akama, T.; Okano, A.; Uda, H., *Chem. Lett.* **1988**, 1793. (b) Hagiwara, H.; Akama, T.; Uda, H., *Chem. Lett.* **1989**, 2067.
5. Narula, A. P. S.; Schuster, D. I., *Tetrahedron Lett.* **1981**, 3707.
6. (a) Tanaka, J.; Kanemasa, S.; Ninomiya, Y.; Tsuge, O., *Bull. Chem. Soc. Jpn.* **1990**, *63*, 466. (b) Tanaka, J.; Kanemasa, S.; Ninomiya, Y.; Tsuge, O., *Bull. Chem. Soc. Jpn.* **1990**, *63*, 476. (c) Tanaka, J.; Kanemasa, S.; Kobayashi. Tsuge, O., *Chem. Lett.* **1989**, 1453.
7. Tsuge, O.; Kazunori, V.; Kanemasa, S.; Kiyotaka, Y., *Bull. Chem. Soc. Jpn.* **1987**, *60*, 3347.
8. Hartzell, S. L.; Rathke, M. W., *Tetrahedron Lett.* **1976**, 2737.

Anubhav P. S. Narula
International Flavors Fragrances, Union Beach, NJ, USA

1-Methyl-1-(trimethylsilyl)allene[1]

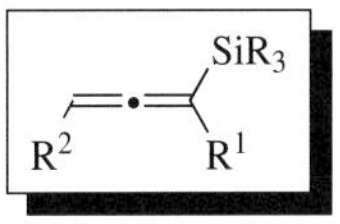

(**1a**; $R_3 = Me_3$, $R^1 = Me$, $R^2 = H$)[4b,9]
[74542-82-8] $C_7H_{14}Si$ (MW 126.30)
InChI = 1S/C7H14Si/c1-6-7(2)8(3,4)5/h1H2,2-5H3
InChIKey = BARFIQOXZWRCGW-UHFFFAOYSA-N

(**1b**; $R_3 = t\text{-}BuMe_2$, $R^1 = Me$, $R^2 = H$)[7]
[99035-24-2] $C_{10}H_{20}Si$ (MW 168.39)
InChI = 1S/C10H20Si/c1-8-9(2)11(6,7)10(3,4)5/h1H2,2-7H3
InChIKey = ZDJOEJQCVQKQAK-UHFFFAOYSA-N

(**1c**; $R_3 = (i\text{-}Pr)_3$, $R^1 = Me$, $R^2 = H$)[7]
[120789-53-9] $C_{13}H_{26}Si$ (MW 210.48)
InChI = 1S/C13H26Si/c1-9-13(8)14(10(2)3,11(4)5)12(6)7/h10-12H,1H2,2-8H3
InChIKey = ZLYUYKJQRQJIBU-UHFFFAOYSA-N

(**1d**; $R_3 = Me_3$, $R^1 = H$, $R^2 = Me$)[4b]
[74542-82-8] $C_7H_{14}Si$ (MW 126.30)
InChI = 1S/C7H14Si/c1-5-6-7-8(2,3)4/h5,7H,1-4H3
InChIKey = SBPLHANFNZGLJQ-UHFFFAOYSA-N

(propargylic anion equivalents;[2,3] three-carbon synthons for [3 + 2] annulations leading to five-membered compounds including cyclopentenes,[4] dihydrofurans,[5] pyrrolines,[5] isoxazoles,[6] furans,[7] and azulenes[8])

Physical Data: (**1a**) bp 111 °C; bp 54–56 °C/90 mmHg; (**1b**) bp 62–65 °C/30 mmHg; (**1d**) bp 107–110 °C.

Solubility: sol CH_2Cl_2, benzene, THF, Et_2O, and most organic solvents.

Form Supplied in: colorless liquid; not commercially available.

Analysis of Reagent Purity: (**1a**) IR (neat) 2955, 2910, 2900, 2860, 1935, 1440, 1400, 1250, 935, 880, 830, 805, 750, and 685 cm^{-1}; ^1H NMR (250 MHz, $CDCl_3$) δ 0.08 (s, 9H), 1.67 (t, 3H, $J = 3.3$), and 4.25 (q, 2H, $J = 3.3$); ^{13}C NMR (67.9 MHz, $CDCl_3$) δ −2.1, 15.1, 67.3, 89.1, and 209.1.[4b,9]

Preparative Methods: 1-methyl-1-(trialkylsilyl)allenes can be conveniently prepared by the method of Vermeer.[9,10] Silyl-substituted propargyl mesylates thus undergo S_N2' displacement by the organocopper reagent generated from methylmagnesium chloride, copper(I) bromide, and lithium bromide 1-Methyl-1-(trimethylsilyl)allene is produced in 52% yield from commercially available (trimethylsilyl)propargyl alcohol in this fashion (eq 1).[4b,9] The *t*-butyldimethylsilyl and triisopropylsilyl analogs are synthesized by the same method in 90% and 58% yield, respectively.[7] Propargyl alcohols bearing these and other trialkylsilyl groups can be prepared by treatment of propargyl alcohol with *n*-butyllithium and the appropriate trialkylsilyl chloride.[7] Allenylsilanes bearing other C-1 substituents can be prepared in an analogous manner by using the appropriate Grignard reagents.

$$ \text{(1a)} \tag{1} $$

1-Methyl-1-(trialkylsilyl)allenes can be alkylated at C-3 with a variety of alkyl halides by treatment of the allenylsilane with *n*-butyllithium and the desired alkyl halide (e.g. eq 2).[7] In addition to ethyl bromide, alkylating agents such as *n*-heptyl bromide,[6] 1,2-dibromobutane,[7] and 5-bromo-1-pentene[8] have been employed in this reaction.

$$ \text{(1b)} \tag{2} $$

3-Methyl-1-(trimethylsilyl)allene (**1d**) is prepared by the direct silylation of the lithium derivative of 1,2-butadiene.[4b] Sequential treatment of a THF solution of 1,2-butadiene with 1.0 equiv of lithium 2,2,6,6-tetramethylpiperidide (−78 °C, 3 h) and 1.05 equiv of chlorotrimethylsilane (−78 °C to 25 °C, 12 h) affords (**1d**) in 41% yield after distillation (eq 3).

$$ \text{(1d)} \tag{3} $$

Purification: allenes (**1a**), (**1b**), and (**1c**) are purified by distillation at reduced pressure or by column chromatography. Allene (**1d**) is distilled at atmospheric pressure. The allenylsilanes obtained by the Vermeer method typically contain up to 7–8% of the trialkylsilyl-1-butyne isomer produced by S_N2 reaction. This mixture can be used directly in most subsequent reactions without further purification. If desired, however, the alkynyl contaminant can be selectively removed by treatment of the mixture with silver(I) nitrate in (10:1) methanol–water at room temperature for one hour.[9] 1-Methyl-1-(trimethylsilyl)-allene is obtained in 79% yield after pentane extraction and distillation.

Handling, Storage, and Precautions: 1-methyl-1-(trimethylsilyl)allene is stable indefinitely when stored under nitrogen in the refrigerator.

Propargylic Anion Equivalents. Due to the silicon β-effect, allenylsilanes react with electrophiles at the 3-position in a fashion analogous to the behavior of allyl- and propargylsilanes.[11] Allenylsilanes can thus function as propargylic anion equivalents. Particularly important is the reaction of (trimethylsilyl)allenes with aldehydes and ketones to provide a regiocontrolled route to homopropargylic alcohols of a variety of substitution types. Allenylsilanes substituted at the 1-position undergo the addition to carbonyl compounds in the presence of titanium(IV) chloride to afford homopropargylic alcohols directly (eq 4).[2]

$$ \text{(1a)} + \qquad \tag{4} $$

In contrast, allenylsilanes lacking a substituent at C-1 react with carbonyl compounds to produce a mixture of the desired homopropargylic alcohols and (trimethylsilyl)vinyl chlorides. This initial product can be converted to the desired alkyne using the method of Cunico and Dexheimer:[12] exposure of the crude mixture of allenylsilane adducts to 2.5 equiv of potassium fluoride in DMSO furnishes the homopropargylic alcohols in good yield (eq 5).[2]

$$ \text{(1d)} + \qquad \tag{5} $$

A number of methods have been reported for the preparation of homopropargylic alcohols.[13] Alcohols having the substitution pattern represented in structure **3**, for example, can be prepared using 3-alkyl-substituted allenyltitanium,[14] -alanate,[15] and -zinc compounds.[16] Unfortunately, these methods are not applicable to the synthesis of type **4** products. Zweifel has shown that allenyldialkylboranes (generated via the reaction of lithium chloropropargylide with trialkylboranes) combine with aldehydes (but not ketones) to produce type **4** homopropargylic alcohols.[17] The preparation of type **2** homopropargylic alcohols is discussed in the article on (trimethylsilyl)allene.

$$ \text{(2)} \qquad \text{(3)} \qquad \text{(4)} $$

Santelli has demonstrated that allenylsilanes without a C-1 substituent undergo conjugate addition to α,β-unsaturated acyl cyanides to give δ,ε-alkynic acyl cyanides.[3]

Three-Carbon Synthon for [3 + 2] Annulations. Danheiser and co-workers have exploited allenylsilanes as the three-carbon components in a [3 + 2] annulation strategy for the synthesis of a variety of five-membered carbocycles and heterocycles. The pathway by which a typical annulation proceeds is shown in eq 6. Reaction of the 2-carbon component (the 'allenophile') at C-3 of the allenylsilane is followed by rapid rearrangement of the silicon-stabilized vinyl cation. Ring closure then affords the five-membered product.

$$ \tag{6} $$

Synthesis of Five-Membered Carbocycles. 1-Substituted allenylsilanes react with α,β-unsaturated carbonyl compounds in the presence of titanium tetrachloride to produce cyclopentenes.[4] For example, carvone and 1-methyl-1-(trimethylsilyl)allene react smoothly to give a *cis*-fused bicyclic system (eq 7).[4d]

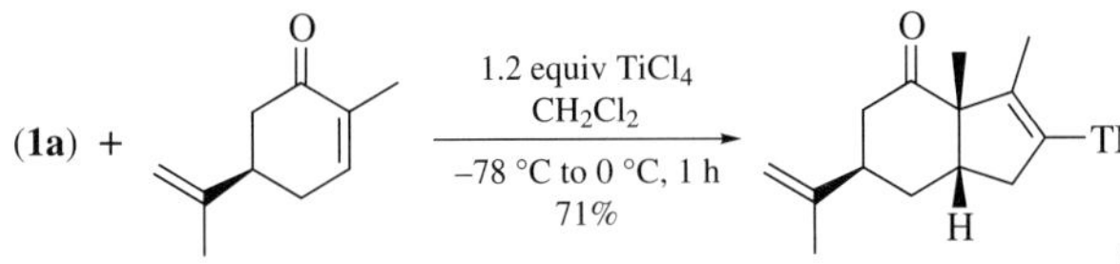

(7)

As illustrated above, the reaction proceeds with a strong preference for suprafacial addition of the allene to the allenophile, thus permitting the stereocontrolled synthesis of a variety of mono- and polycyclic systems. Both cyclic and acyclic enones participate in the reaction. Spiro-fused products are obtained from α-alkylidene ketone substrates (eq 8).[4b]

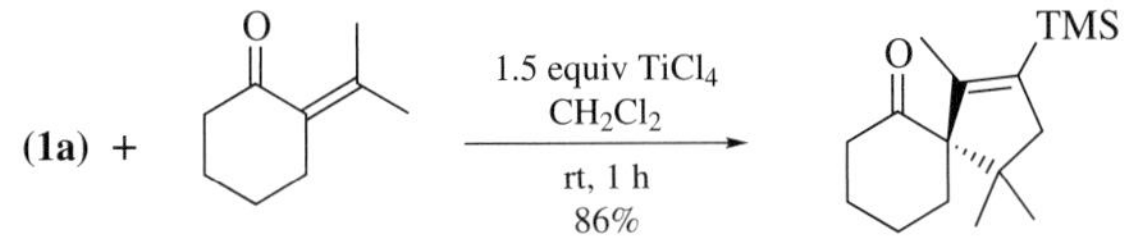

(8)

When α,β-unsaturated acyl silanes are employed, the type of product formed varies depending on the trialkylsilyl substituent of the acyl silane: five-membered carbocycles are produced from reaction with *t*-butyldimethylsilyl derivatives, whereas six-membered carbocycles are obtained from trimethylsilyl compounds (eq 9).[4c]

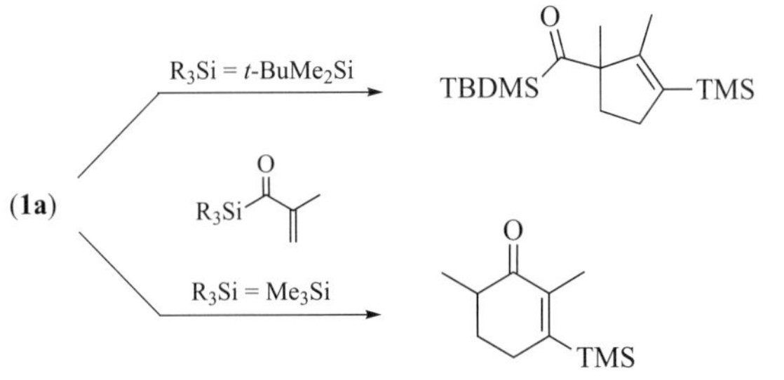

(9)

Allenylsilanes lacking a C-1 alkyl substituent do not function efficiently as three-carbon synthons in the [3 + 2] annulation. This phenomenon is attributable to the relative instability of the terminal vinyl cation intermediate required according to the proposed mechanism for the annulations (eq 6). Fully substituted five-membered rings result from annulations employing allenylsilanes substituted at both C-1 and C-3.[4]

Synthesis of 1,3-Dihydrofurans. (*t*-Butyldimethylsilyl)-allenes combine with aldehydes to produce dihydrofurans (eq 10).[5] In a typical reaction, the aldehyde and 1.1 equiv of titanium tetrachloride are premixed at $-78\,^{\circ}$C in methylene chloride for 10 min. The allenylsilane (1.2 equiv) is then added, and the reaction mixture is stirred in the cold for 15–45 min.

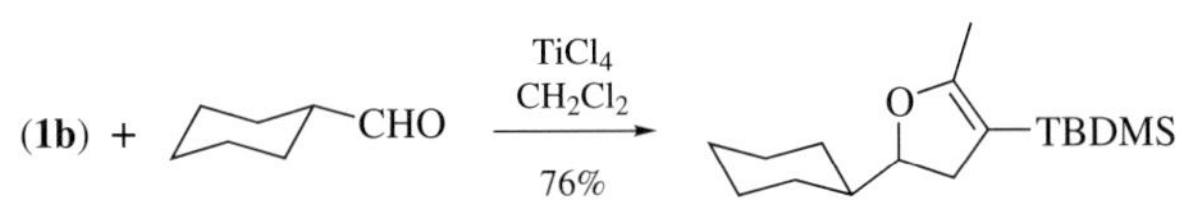

(10)

In reactions of a C-3 substituted allenylsilane with achiral aldehydes, *cis*-substituted dihydrofurans are the predominant products (eq 11).[5]

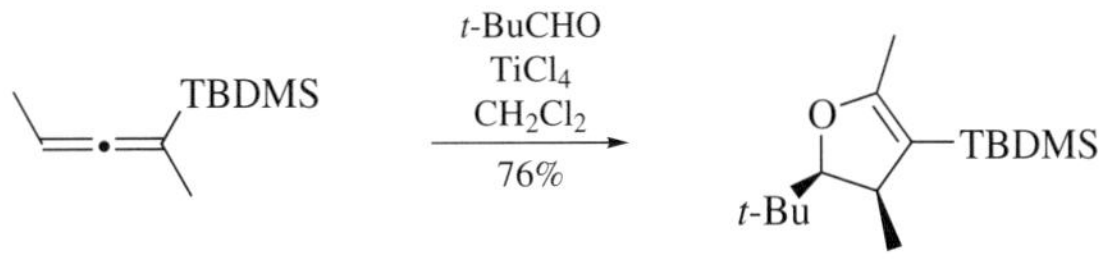

(11)

(Trimethylsilyl)allenes are unsuitable for this [3 + 2] annulation, as the intermediate carbocations undergo chloride-initiated desilylation to produce alkynic byproducts. This unwelcome reaction pathway is suppressed when the bulkier (*t*-butyldimethylsilyl)allenes are employed.

Synthesis of Pyrrolizinones. Cyclic *N*-acyl imine derivatives combine with (*t*-butyldimethylsilyl)allenes to afford nitrogen heterocycles (eq 12).[5] The *N*-acyliminium ions are generated from ethoxypyrrolidinones in the presence of titanium tetrachloride.

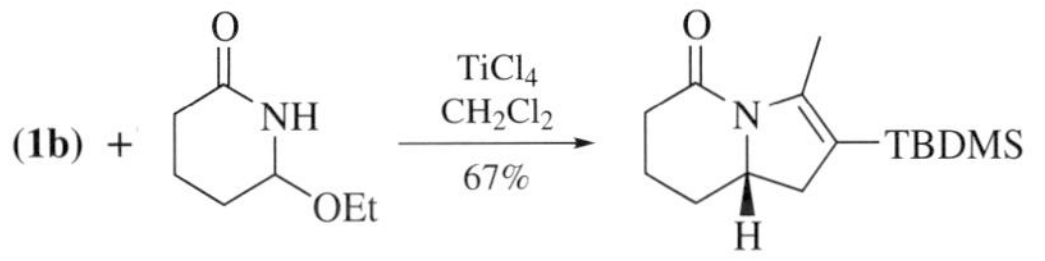

(12)

Synthesis of Furans and Isoxazoles. Electrophilic species of the general form $Y\equiv X^+$ serve as 'heteroallenophiles', combining with allenylsilanes in a regiocontrolled [3 + 2] annulation method. As illustrated in the mechanism shown in eq 13, addition of the heteroallenophile at C-3 of the allenylsilane produces a vinyl cation stabilized by hyperconjugative interaction with the adjacent carbon–silicon σ-bond. A 1,2-trialkylsilyl shift then occurs to generate an isomeric vinyl cation, which is intercepted by nucleophilic X. Elimination of H^+ furnishes the aromatic heterocycle.

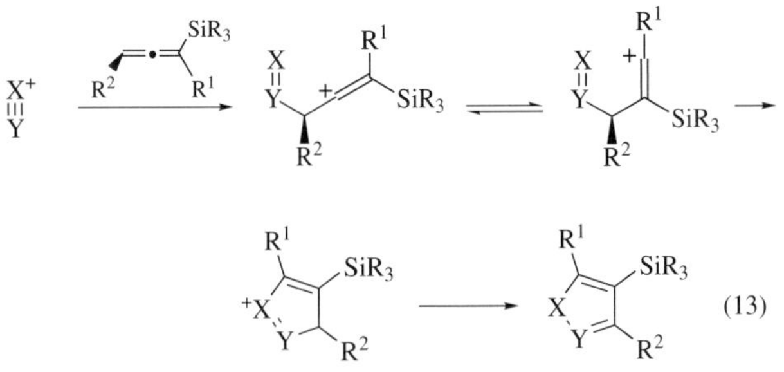

(13)

Isoxazoles are synthesized when the heteroallenophile is nitrosonium ion. Thus, reaction of commercially available nitrosonium tetrafluoroborate with allenylsilanes in acetonitrile at $-30\,^{\circ}$C affords silyl-substituted isoxazoles in good yield (eq 14).[6]

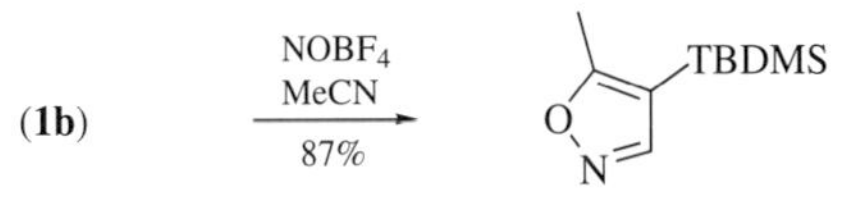

(14)

In a variation of the above procedure, (trimethylsilyl)allenes are employed in a one-pot procedure that produces 5-substituted and 3,5-disubstituted isoxazoles lacking the 4-silyl substituent (eq 15). Desilylation is encouraged by addition of water and warming the reaction mixture to 65–70 °C after the initial annulation.[6] Alternatively, addition of electrophilic reagents to the reaction mixture leads to isoxazoles with C-4 substituents such as Br, COMe, etc.

$$\text{(15)}$$

The heteroaromatic strategy is extended to the synthesis of furans when acylium ions are employed as the heteroallenophile (eq 16).[7] Acylium ions are generated in situ via the reaction of acyl chlorides and aluminum chloride. Typically, the allenylsilane is added to a solution of 1.0 equiv each of AlCl₃ and the acyl chloride in methylene chloride at −20 °C. The reaction is complete in 1 h at −20 °C.

$$\text{(16)}$$

(1c)

Intramolecular [3 + 2] annulation affords bicyclic furans (eq 17).[7]

$$\text{(17)}$$

(*t*-Butyldimethylsilyl)- and (triisopropylsilyl)allenes are superior to their trimethylsilyl counterparts for this annulation, presumably due to the ability of the larger trialkylsilyl groups to suppress undesirable desilylation reactions. In annulations involving allenylsilanes which lack C-3 substituents, the bulkier triisopropylsilyl derivatives are superior to *t*-butyldimethylsilyl analogs.

Synthesis of Azulenes.   Reaction of tropylium cations with allenylsilanes produces substituted azulenes.[8] Typically, commercially available tropylium tetrafluoroborate (2 equiv) is employed. The second equivalent dehydrogenates the dihydroazulene intermediate to produce the aromatic product. Poly(4-vinylpyridine) (poly (4-VP)) or methyltrimethoxysilane is used to scavenge the HBF₄ produced in the reaction.

$$\text{(18)}$$

The azulene synthesis proceeds best with 1,3-dialkyl (*t*-butyldimethylsilyl)allenes. (Trimethylsilyl)allenes desilylate to generate propargyl-substituted cycloheptatrienes as significant byproducts. As observed in the other [3 + 2] annulations discussed already, allenylsilanes lacking C-1 alkyl substituents do not participate in the reaction.

Synthesis of Silylalkynes via Ene Reactions.   (Trimethylsilyl)allenes undergo ene reactions with 4-phenyl-1,2,4-triazoline-3,5-dione and other reactive enophiles to give silylalkenes.[18]

1. Review: Panek, J. S., *Comprehensive Organic Synthesis* **1991**, *2*, 579.
2. (a) Danheiser, R. L.; Carini, D. J., *J. Org. Chem.* **1980**, *45*, 3925. (b) Danheiser, R. L.; Carini, D. J.; Kwasigroch, C. A., *J. Org. Chem.* **1986**, *51*, 3870.
3. (a) Jellal, A.; Santelli, M., *Tetrahedron Lett.* **1980**, *21*, 4487. (b) Santelli, M.; Abed, D. E.; Jellal, A., *J. Org. Chem.* **1986**, *51*, 1199.
4. (a) Danheiser, R. L.; Carini, D. J.; Basak, A., *J. Am. Chem. Soc.* **1981**, *103*, 1604. (b) Danheiser, R. L.; Carini, D. J.; Fink, D. M.; Basak, A., *Tetrahedron* **1983**, *39*, 935. (c) Danheiser, R. L.; Fink, D. M., *Tetrahedron Lett.* **1985**, *26*, 2513. (d) Danheiser, R. L.; Fink, D. M.; Tsai, Y.-M., *Org. Synth.* **1988**, *66*, 8.
5. Danheiser, R. L.; Kwasigroch, C. A.; Tsai, Y.-M., *J. Am. Chem. Soc.* **1985**, *107*, 7233.
6. Danheiser, R. L.; Becker, D. A., *Heterocycles* **1987**, *25*, 277.
7. Danheiser, R. L.; Stoner, E. J.; Koyama, H.; Yamashita, D. S.; Klade, C. A., *J. Am. Chem. Soc.* **1989**, *111*, 4407.
8. Becker, D. A.; Danheiser, R. L., *J. Am. Chem. Soc.* **1989**, *111*, 329.
9. Danheiser, R. L.; Tsai, Y.-M.; Fink, D. M., *Org. Synth.* **1988**, *66*, 1.
10. Westmijze, H.; Vermeer, P., *Synthesis* **1979**, 390.
11. For reviews on allylsilanes and related systems, see Ref. 1 and: Fleming, I.; Dunogues, J.; Smithers, R., *Org. React.* **1989**, *37*, 57; Fleming, I., *Comprehensive Organic Synthesis* **1991**, *2*, 563.
12. Cunico, R. F.; Dexheimer, E. M., *J. Am. Chem. Soc.* **1972**, *94*, 2868.
13. For reviews of the chemistry of propargylic anion equivalents, see: (a) Yamamoto, H., *Comprehensive Organic Synthesis* **1991**, *2*, 81. (b) Epsztein, R. In *Comprehensive Carbanion Chemistry*; Buncel, E.; Durst, T., Eds.; Elsevier: Amsterdam, 1984; Part B, p 107. (c) Moreau, J.-L. In *The Chemistry of Ketenes, Allenes, and Related Compounds*; Patai, S., Ed.; Wiley: New York, 1978, p 343.
14. (a) Furuta, K.; Ishiguro, M.; Haruta, R.; Ikeda, N.; Yamamoto, H., *Bull. Chem. Soc. Jpn.* **1984**, *57*, 2768. (b) Ishiguro, M.; Ikeda, N.; Yamamoto, H., *J. Org. Chem.* **1982**, *47*, 2225.
15. Hahn, G.; Zweifel, G., *Synthesis* **1983**, 883.
16. Zweifel, G.; Hahn, G., *J. Org. Chem.* **1984**, *49*, 4565.
17. Zweifel, G.; Backlund, S. J.; Leung, T., *J. Am. Chem. Soc.* **1978**, *100*, 5561.
18. Laporterie, A.; Dubac, J.; Manuel, G.; Deleris, G.; Kowalski, J.; Dunogues, J.; Calas, R., *Tetrahedron* **1978**, *34*, 2669.

Katherine L. Lee & Rick L. Danheiser
Massachusetts Institute of Technology, Cambridge, MA, USA

2-Methyl-2-(trimethylsilyloxy)-3-pentanone[1]

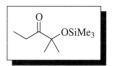

[72507-50-7]   C₉H₂₀O₂Si   (MW 188.38)

InChI = 1S/C9H20O2Si/c1-7-8(10)9(2,3)11-12(4,5)6/h7H2, 1-6H3

InChIKey = BLYRAXGLJVABLF-UHFFFAOYSA-N

(lithium enolate used in the synthesis of *syn-β*-hydroxy acids, aldehydes, and ketones)

Physical Data: bp 71–75 °C/15 mmHg.

Preparative Methods: prepared in four steps from propionaldehyde in 14–33% overall yield;[2] in three steps from acetone

cyanohydrin in 40% overall yield;[3] and in two steps from 3-hydroxy-3-methyl-2-butanone in 55% overall yield.[4]
Handling, Storage, and Precautions: use in a fume hood.

Stereoselection in Reactions of Lithium Enolates. The (Z)-lithium enolate (**2**), obtained from the reaction of 2-methyl-2-trimethylsilyloxypentan-3-one (**1**) with lithium diisopropylamide, reacts with aldehydes to afford *syn*-β-hydroxy ketones (**3**) exclusively (eq 1).[2,3,5] The synthetic utility of (**3**) is demonstrated by conversion to β-hydroxy acids (**4**), β-hydroxy aldehydes (**5**), and other β-hydroxy ketones by straightforward procedures (eqs 2–3 and eq 4).[2–6] Table 1 illustrates examples of condensations of (**2**) with simple aldehydes and subsequent conversions to (**4**), (**5**), and (**6**).

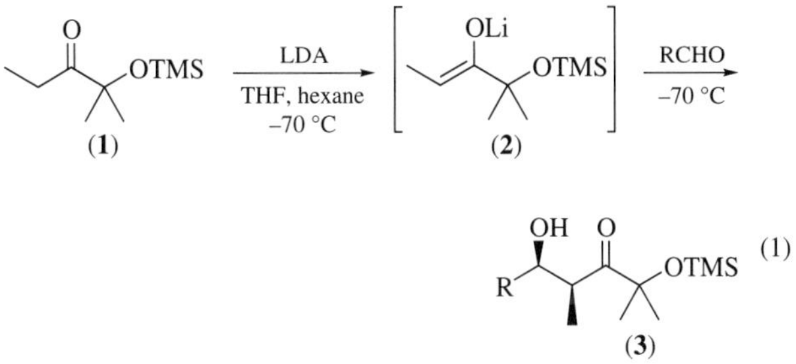

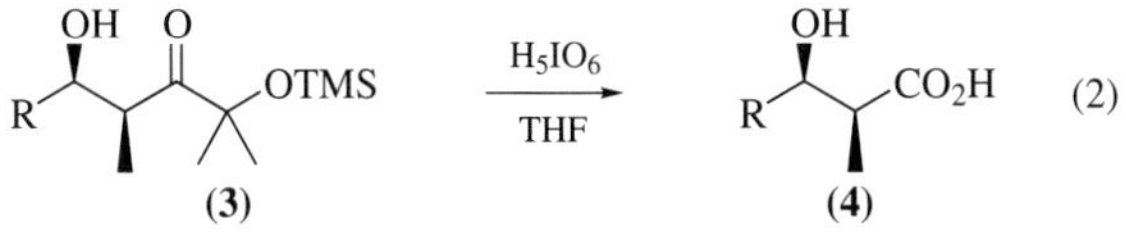

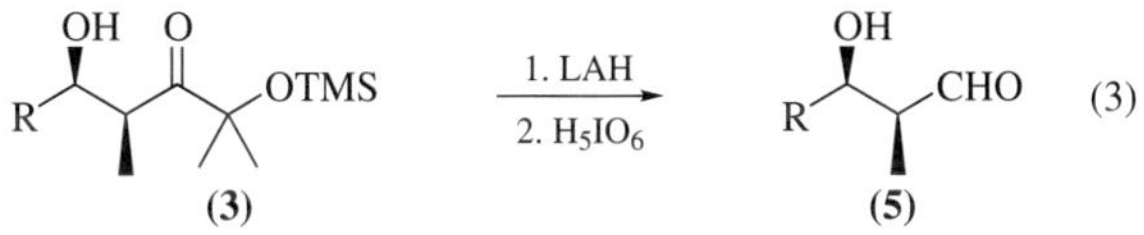

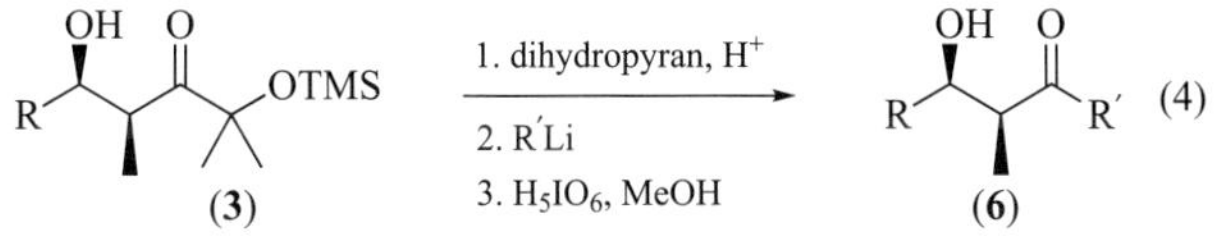

The reaction of (**2**) with aldehydes containing additional functional groups and stereocenters has been reported (eq 5, Table 2).[2,3,9,10] In all instances the stereoselection is high in favor of aldol product of *syn* relative stereochemistry at the α- and β-positions (**7** and **8**), but the degree of stereoselection at the β- and γ-positions is highly variable. Only 2,4-dimethylpent-3-enal exhibits a high level of diastereoselection in formation of the product predicted by Cram's rule (**7**); this product has been used to prepare the C-1 to C-7 fragment of erythronolide A.[10] Additional examples of aldol condensations of (**2**) with complex aldehydes are given in (eq 6).[9] When R and R′ are H, the ratio of (**9**):(**10**) is 5.7:1. When R is Et and R′ is Me, the ratio of (**9**):(**10**) is 3:1. It is noteworthy that significantly higher levels of diastereocontrol are achieved in the synthesis of compounds with the same relative stereochemistry as (**7**) (R = Ph, X = Me; 98:2) and (**9**) (R = Et, R′ = Me; >98:2) using propionate equivalents (**11**)[8a] and (**12**),[8b] respectively.

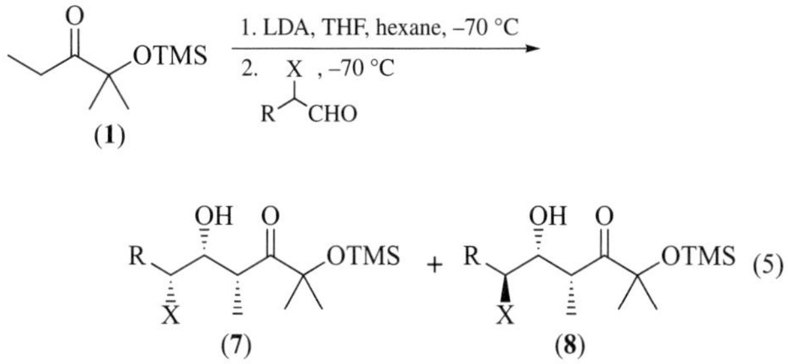

Table 2 Ratios of diastereomeric product (**7**) and (**8**) in reactions of (**2**) with aldehydes containing stereocenters at the α-position (eq 5)

R	X	(**7**):(**8**)	Yield (%)	Ref.
Ph	Me	4:1	51[a]	2,3
Me₂C=CH	Me	15:1	42	7a
BnOCH₂[b]	Me[b]	1:2	63	7a
t-BuPh₂SiOCH₂[c]	Me[b]	1:3.7	57	7a
AcOCH₂[b]	Me[b]	1:3.4	42	7a
MeCH(OSiMe₂-t-Bu)[c]	Me[c]	1:2.6	d	7b
Me	OSiMe₂-t-Bu	1:2	41	7c
Me	OBn	1:2	51	7c
Me	OCH₂OBn	1:3.5	e	7c

[a] Yield of pure (**7**).

[b] (S)-Enantiomer.

[c] (2S,3S)-Enantiomer.

[d] The crude product was converted to the corresponding δ-lactone in 55% overall yield.

[e] Yield not reported.

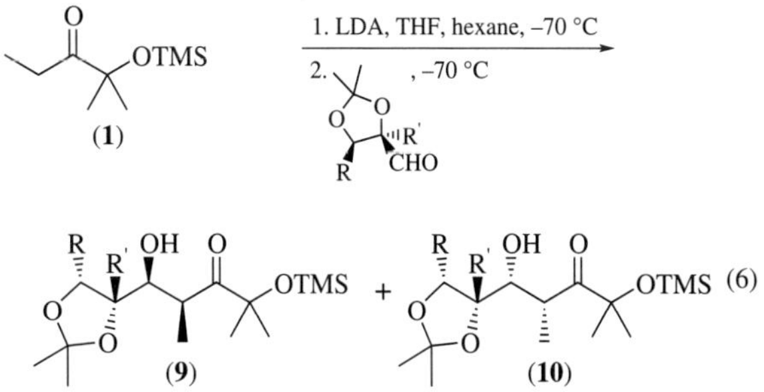

Table 1 Aldol condensations of enolate (**2**) with aldehydes (RCHO) and subsequent conversions of (**3**) to *syn*-β-hydroxy acids (**4**), aldehydes (**5**) and ketones (**6**) (eqs 1–4)

R	Yield of (**3**) (%)	Yield of (**4**) (%)	Yield of (**5**) (%)	Yield of (**6**) (%)	Ref.
Et	82	–	75[a]	–	5a
n-C₅H₁₁	43	–	–	–	2
i-Pr	61	77–83	–	–	2, 3
Bn	51	76	–	–	2, 3
Ph	78	87	87 (R′ = Me)	–	2, 3, 6
			65 (R′ = Et)		
			88 (R′ = n-Bu)		
2-Thienyl	88	80[b]	–	–	5h
CH₂=CH	80	100	–	–	5c
trans-MeCH=CH	97	c	–	–	5c
CH₂=C(Me)	99	c	–	–	5c

[a] Isolated as an acetate ester.

[b] Isolated as a methyl ester after esterification with diazomethane.

[c] Yield of this conversion was not reported.

(11)   (12)

Aldol reactions of lithium enolates of higher homologs of propionate equivalent (1) have been studied in the context of natural product syntheses and yield similar levels of stereoselection.[9] The amino sugars (±)-ristosamine and (±)-megalosamine and the nucleus of crassin acetate were prepared using this methodology.[9] Aldol reactions of (2) generated from chiral bases or in the presence of chiral solvents yield products of low to moderate levels of optical purity.[10] Asymmetric aldol reactions are best accomplished using other methods.[1] Michael additions of (2) have been shown to be highly stereoselective in some instances.[11]

Stereoselection in Reactions of Other Enolates and Enol Derivatives. A few examples of reactions of boron, magnesium, and tin enolates of (1) and Lewis acid-catalyzed reactions of enol silanes derived from (1) have been reported.[12–15] The boron enolate formed by reaction of (1) with di-n-butylboryl trifluoromethanesulfonate and diisopropylethylamine yields *syn* aldol products.[12] The magnesium enolate formed from (2) by exchange with magnesium bromide yields a higher ratio (5:1) of chelation to nonchelation controlled product than (2) in reaction with (R)-3-benzyloxymethoxy-2-methylpropionaldehyde (see Table 2).[13] Excess tin enolate of (1) reacts with β-lactam (13) to give (14) and its 1α-epimer in a 95:5 ratio.[14] In contrast to the enolates, enol silane (15) reacts with benzaldehyde to afford primarily *anti* aldol product (16) with boron trifluoride etherate (16:3 (R = Ph) = 5:1) or titanium(IV) chloride (16:3 (R = Ph) = 9:1) as a catalyst.[15a] Reaction of (15) with (S)-2-benzyloxypropionaldehyde catalyzed by tin(IV) chloride gave a single isomer (17) that possessed the same relative stereochemistry as the minor isomer in the reaction with (2) (see Table 2).[15b]

(13)   (14)

(15)   (16)   (17)

Related Reagents. (S)-4-Benzyl-2-oxazolidinone; 10,2-Camphorsultam; (R)-(+)-t-Butyl 2-(p-Tolylsulfinyl)propionate; 10-Dicyclohexylsulfonamidoisoborneol; Diisopinocampheylboron Trifluoromethanesulfonate; (R,R)-2,5-Dimethylborolane; 2,6-Dimethylphenyl Propionate; Ethyl 2-(Methyldiphenylsilyl)-propanoate; α-Methyltoluene-2,α-sultam; 3-Propionylthiazolidine-2-thione; (R,R)-1,2-Diphenyl-1,2-diaminoethane N,N'-Bis-[3,5-bis(trifluoromethyl)benzenesulfonamide]; 1,1,2-Triphenyl-1,2-ethanediol.

1. (a) Heathcock, C. H. *In Modern Synthetic Methods*; Scheffold, R., Ed.; VCH: Weinheim, 1992. (b) Heathcock, C. H., *Comprehensive Organic Synthesis* 1991, Vol. 2, Chapter 1.6. (c) Kim, B. M.; Williams, S. F.; Masamune, S., *Comprehensive Organic Synthesis* 1991, Vol. 2, Chapter 1.7. (d) Heathcock, C. H. *Asymmetric Synthesis*; Academic: New York, 1984; Vol. 3. (e) Heathcock, C. H. *Comprehensive Carbanion Chemistry*; Buncel, E.; Durst, T. Eds.; Elsevier: Amsterdam, 1984; Vol. 2. (f) Evans, D. A.; Nelson, J. V.; Taber, T. R., *Top. Stereochem.* 1982, 13, 1.

2. Young, S. D.; Buse, C. T.; Heathcock, C. H., *Org. Synth., Coll. Vol.* 1990, 7, 381.

3. (a) Heathcock, C. H.; Buse, C. T.; Kleschick, W. A.; Pirrung, M. C.; Sohn, J. E.; Lampe, J., *J. Org. Chem.* 1980, 45, 1066. (b) Bal, B.; Buse, C. T.; Smith, K.; Heathcock, C. H., *Org. Synth., Coll. Vol.* 1990, 7, 185.

4. Smith, A. B., III; Levenberg, P. A.; Jerris, P. J.; Scarborough, R. M., Jr.; Wovkulich, P. M., *J. Am. Chem. Soc.* 1981, 103, 1501.

5. (a) Pilli, R. A.; Murta, M. M., *Synth. Commun.* 1988, 18, 981. (b) Kusakabe, M.; Sato, F., *J. Org. Chem.* 1989, 54, 3486. (c) Heathcock, C. H.; Finkelstein, B. L.; Jarvi, E. T.; Radel, P. A.; Hadley, C. R., *J. Org. Chem.* 1988, 53, 1922.

6. White, C. T.; Heathcock, C. H., *J. Org. Chem.* 1981, 46, 191.

7. (a) Heathcock, C. H.; Young, S. D.; Hagen, J. P.; Pilli, R.; Badertscher, U., *J. Org. Chem.* 1985, 50, 2095. (b) Brooks, D. W.; Kellogg, R. P., *Tetrahedron Lett.* 1982, 23, 4991. (c) Heathcock, C. H.; Young, S. D.; Hagen, J. P.; Pirrung, M. C.; White, C. T.; Van Derveer, D., *J. Org. Chem.* 1980, 45, 3846.

8. (a) Mori, I.; Ishihara, K.; Heathcock, C. H., *J. Org. Chem.* 1990, 55, 1114. (b) Heathcock, C. H.; Pirrung, M. C.; Buse, C. T.; Hagen, J. P.; Young, S. D.; Sohn, J. E., *J. Am. Chem. Soc.* 1979, 101, 7077.

9. (a) Heathcock, C. H.; Montgomery, S. H., *Tetrahedron Lett.* 1983, 24, 4637. (b) Dauben, W. G.; Saugier, R. K.; Fleischhauer, I., *J. Org. Chem.* 1985, 50, 3767.

10. (a) Shioiri, T.; Ando, A., *Tetrahedron* 1989, 45, 4969. (b) Heathcock, C. H.; White, C. T.; Morrison, J. J.; Van Derveer, D., *J. Org. Chem.* 1981, 46, 1296.

11. Oare, D. A.; Heathcock, C. H., *J. Org. Chem.* 1990, 55, 157.

12. Danda, H.; Hansen, M. M.; Heathcock, C. H., *J. Org. Chem.* 1990, 55, 173.

13. Collum, D. B.; McDonald, J. H., III; Still, W. C., *J. Am. Chem. Soc.* 1980, 102, 2118.

14. Shirai, F.; Nakai, T., *J. Org. Chem.* 1987, 52, 5491.

15. (a) Heathcock, C. H.; Davidsen, S. K.; Hug, K. T.; Flippin, L. A., *J. Org. Chem.* 1986, 51, 3027. (b) Reetz, M. T.; Kesseler, K.; Jung, A., *Tetrahedron* 1984, 40, 4327.

William A. Kleschick
DowElanco, Indianapolis, IN, USA

P

PDMS Thimbles*

[42557-10-8] C_2H_6OSi (MW 74.02)

InChI = 1S/C2H6OSi/c1-4(2)3/h1-2H3

InChIKey = SEUDSDUUJXTXSV-UHFFFAOYSA-N

Alternate Names: PDMS; dimethicone; polymerized siloxane; silicone; Sylgard 184; trimethylsiloxy-terminated polydimethylsiloxane.

Physical Data: colorless free-flowing liquid; average molecular weight is supplier dependent; d 0.98 g mL^{-1}

Solubility: most chlorinated, hydrocarbon, or ethereal solvents. Insoluble in MeOH, EtOH, glycerol, and water. Limited solubility in DMF and DMSO.

Form Supplied in: PDMS thimbles are fabricated from commercially available Sylgard 184 – a two-component prepolymer of PDMS. Typical impurities include residual Pt catalyst. This polymer has a high concentration of silica nanoparticles.

Preparative Methods: reaction between dimethylchlorosiloxane and water yields PDMS. The anionic polymerization of hexamethylcyclotrisiloxane yields PDMS. PDMS thimbles are fabricated by curing Sylgard 184 in an oven.

Handling, Storage, and Precautions: PDMS thimbles may be stored for over a year under ambient conditions. No specific precautions are necessary.

Preparation of PDMS Thimbles. The thimbles are readily fabricated from commercially available PDMS kits (Sylgard 184) from Dow Corning. The kit is a two-component mixture that is mixed in a 10:1 ratio to yield a viscous prepolymer of PDMS. A Pt catalyst found in one component adds a terminal [Si]–H bond across a [Si]–CH=CH$_2$ to yield cross-links in the polymer matrix of the form [Si]–CH$_2$CH$_2$–[Si]. The cross-linking chemistry is robust even when the two components are mixed at ratios distant from 10:1. Importantly, amines inhibit the cross-linking reaction through coordination to the Pt catalyst.

Thimbles were fabricated by first forming a monolayer of trichloro(1*H*,1*H*,2*H*,2*H*-perfluorooctyl)silane on a glass vial with dimensions of several centimeters.[1] The vials were dipped into the PDMS mixture, cured in an oven at 65 °C for 30 min, and then dipped into the PDMS mixture one more time before curing for

several hours at 65 °C. Two coats of PDMS led to reproducible thicknesses for the PDMS walls of 100 μm. The thimbles were removed from the glass vials by swelling in hexanes, which caused the PDMS to delaminate from the surface of the vials. Alternatively, metal rods were also used in the same fashion.[2]

Differential Flux of Reagents and Catalysts Through PDMS Thimbles. Flux of a molecule through a polymeric membrane depends on several factors, but two of the most important are the partition coefficient of a molecule from the solvent into the polymer and the rate of diffusion of a molecule in the polymer matrix. Simply, a molecule must be soluble in PDMS and possess a reasonable rate of diffusion within PDMS to flux through it. Numerous reviews cover these concepts with mathematical rigor.[3–7]

Most organic molecules flux through PDMS thimbles with reasonable rates. Molecules with carboxylic acids, alcohols, amines, aromatic rings, nitro groups, esters, olefins, epoxides, and ketones readily flux through the walls of PDMS thimbles.[1,2,8–16] For instance, when the molecules shown in eq 1 were dissolved in CH$_2$Cl$_2$ on the interior of PDMS thimbles with walls approximately 100 μm thick, they readily fluxed to CH$_2$Cl$_2$ on the exterior of the thimbles. The time it took for the ratio of the concentration for each molecule on the exterior to the interior of the thimble to reach 0.80 is shown. Clearly even large molecules such as cholesterol can flux through PDMS walls at reasonable rates.

$$\text{(1)}$$

(120 min) (35 min) (70 min) (240 min)

The flux was also dependent on the thickness of the walls of the thimbles, solvents, and temperature. Briefly, as the thickness of walls increased or the temperature decreased, the flux decreased. The solvent was a critical parameter and the flux increased with the ability of solvents to swell PDMS. To quantify how well solvents swelled PDMS, Whitesides and coworkers measured the lengths of PDMS rods before and after swelling in various solvents.[14] Apolar, aprotic solvents swelled PDMS well, but polar protic solvents were poor solvents for PDMS. For instance, PDMS rods were swollen in both CH$_2$Cl$_2$ and DMF and increased in length. The ratios of the lengths of the rods after swelling to their original, preswollen lengths were 1.22 for CH$_2$Cl$_2$ and 1.02 for DMF. When the flux of *p*-nitrobenzaldehyde through PDMS thimbles with walls approximately 100 μm thick was measured, the flux was much faster when CH$_2$Cl$_2$ was used as a solvent than when DMF was used.[10] Specifically, in experiments where *p*-nitrobenzaldehyde was initially added to the interior of the thimbles, the time it took for the ratio of its concentration on the exterior to the interior of the thimble to reach 0.8 was approximately 70 min when CH$_2$Cl$_2$ was the solvent and 80 h when DMF was

*Article previously published in the electronic Encyclopedia of Reagents for Organic Synthesis as Polydimethylsiloxane Thimbles DOI: 10.1002/047084289X.rn01331

the solvent.

Molecules that possessed very low flux through the walls of PDMS thimbles included ionic liquids, p-toluenesulfonic acid, and polystyrene.[10,16] Others have shown that ionic liquids have no solubility within PDMS and our results support this. For instance, the ionic liquid shown in eq 2 did not partition into PDMS.[17] In addition, p-toluenesulfonic acid did not flux through PDMS when DMF was used as the solvent. The negligible flux was probably due to its deprotonation and the fact that ionic molecules have little to no solubility in PDMS. Finally, it was well known that polymers do not flux through polymeric membranes. When polystyrene (MW = 18 000 g mol^{-1}) was added to the interior of a thimble with CH_2Cl_2 on the interior and exterior, it did not flux through the walls of the thimble even after 3 days.[10]

$$(2)$$

In summary, a wide variety of organic molecules with various functional groups flux through PDMS thimbles, but care should be taken to consider the solvent, thickness of the walls, temperature, and the presence of any ionic functional group.

Site Isolation of Water from LiAlH₄. Water was successfully added to the interior of a PDMS thimble and site isolated from LiAlH₄ that was added to the exterior to complete a two-step, one-pot reaction as shown in eq 3.[1] In this reaction, 74 equiv of water was added for every 1.25 equiv of LiAlH₄, so control reactions where water and LiAlH₄ were not site isolated by the walls of a PDMS thimble failed to yield the product.

$$(3)$$

Two methods were studied to demonstrate how to think about using PDMS thimbles. In the first method, all of the solvents and reagents were added and allowed to react as shown in eq 4. The cyclic acetal, water, PTSA, and an organic solvent were added to the interior of a PDMS thimble and LiAlH₄ and an organic solvent were added to the exterior. This method was successful when hexanes were used as the solvent (quantitative conversions and 89% isolated yield), but significant amounts of the ketone were isolated when CH_2Cl_2, THF, or mixtures of hexanes with CH_2Cl_2 or THF were used as the organic solvents. The limitation in this method was that the THF and CH_2Cl_2 swelled the PDMS thimbles and allowed some water to flux through the walls of the thimble and quenched some of the LiAlH₄ before it had reacted with all of the ketone. When hexanes were used as the organic solvent on the interior and exterior of the thimbles, the flux of water through the walls of the thimbles was lower relative to when THF or CH_2Cl_2 was used.

$$(4)$$

In the second method, the reaction on the interior of the thimble was allowed to go to completion with no solvent or LiAlH₄ on the exterior (eq 5). Only after the reaction on the interior was complete was the organic solvent and LiAlH₄ added to the exterior of the thimble. These reactions proceeded to quantitative conversions and high yields for solvent mixtures of 1:3 CH_2Cl_2:hexanes, 1:3 THF:hexanes, and 1:1 THF:hexanes (v/v). These reactions were successful in solvent mixtures that failed for the method shown in eq 4. The reason for this success was that the cyclic acetal was not allowed to flux through the walls of the thimbles until it had all been converted to ketone. The flux of the cyclic acetal to the exterior of the thimbles shown in eq 4 slowed its conversion to the ketone (the cyclic acetal would flux to the exterior and had to flux back to the interior to yield the ketone) such that the flux of water to the exterior to quench LiAlH₄ was competitive with the deprotection of the acetal. By converting all of the cyclic acetal to ketone as shown in eq 5 prior to the addition of solvent and LiAlH₄ to the exterior of the thimble, the success of the reduction depended only on the fast flux through PDMS of the ketone compared to the slower flux for water.[15,16]

$$(5)$$

Site Isolation of Water from Grignard Reagents. The same two-step, one-pot cascade reaction was utilized for the reaction shown in eq 6.[1] The acetal, water, an organic solvent (typically hexanes, diethyl ether, or a mixture of both), and PTSA were added to the interior of a PDMS thimble. When the first reaction was judged to be complete, an organic solvent and either Grignard reagent, butyl lithium, or an organocuprate were added to the exterior of the thimble. These reactions used 74 equiv of water for every 5–8 equiv of nucleophile and every 1 equiv of cyclic acetal. Thus, the reactions failed without PDMS to site isolate water from the nucleophile. Reaction times for the first step ranged from 3.5 to 12 h and for the second step they ranged from 11 to 22 h. These reaction times were not optimized. Isolated yields of 77–93% were obtained.

The use of butyl lithium was problematic because it rapidly reacted with the walls of the PDMS thimbles such that holes formed and water leaked out to quench any remaining butyl lithium.

Site Isolation of the Grubbs' Second-generation Catalyst from *m*-CPBA.   One concept that was critical for site isolation of organometallic catalysts was that big molecules generally have slower flux through membranes than small molecules. This concept was critical for molecules such as the Grubbs' catalysts, which possessed high molecular weights and cross-sectional areas (eq 7). This discussion will address how to site isolate organometallic and inorganic catalysts.

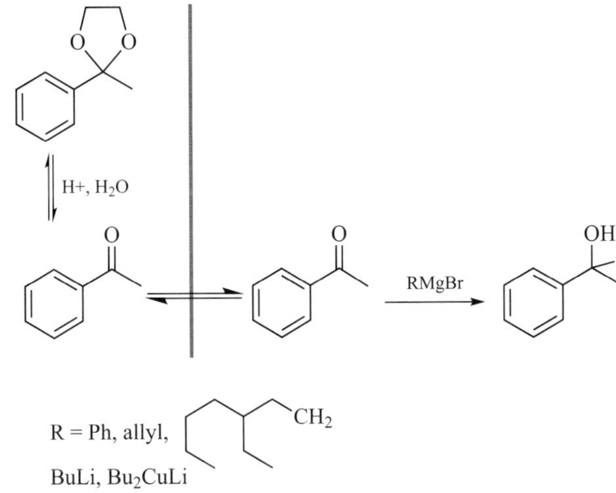

R = Ph, allyl,
BuLi, Bu$_2$CuLi

(6)

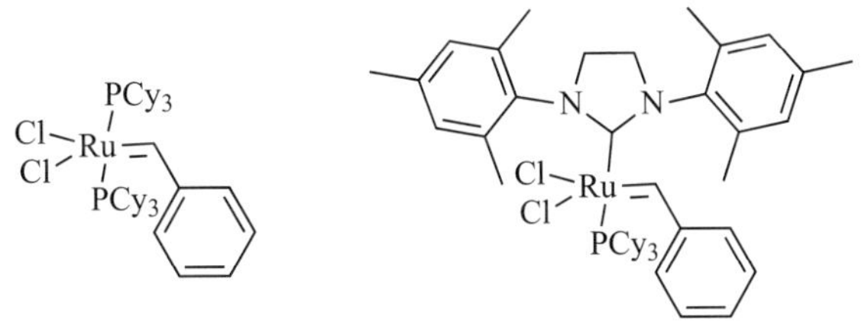

(7)

The Grubbs' second-generation catalyst was site isolated within PDMS thimbles because its flux was low due to its size and because it was insoluble in the solvent found on the exterior of the thimble.[2,13] Both factors were critically important to site isolate the catalyst on the interior of the PDMS thimbles. When the catalyst was added to the interior of a PDMS thimble with CH$_2$Cl$_2$ on both the interior and the exterior, the catalyst fluxed to the exterior within minutes. When the solvent on the interior and the exterior of a thimble was changed to a 1:1 (v/v) mixture of CH$_2$Cl$_2$:1-butyl-3-methylimidazolium hexafluorophosphate [BMIM][PF$_6$], 14% of the ruthenium in the system was found on the exterior of the thimble after 16 h. The best site isolation occurred when the solvent on the interior of the thimble was CH$_2$Cl$_2$:[BMIM][PF$_6$] and the solvent on the exterior was a 1:1 mixture of MeOH:H$_2$O. With these solvents, only 0.4% of the ruthenium in the system was found on the exterior after 16 h. The reason for the success was that the Grubbs' catalyst was not soluble in 1:1 MeOH:H$_2$O but it was soluble in the CH$_2$Cl$_2$:[BMIM][PF$_6$] solvent mixture on the

interior of the thimble. Although the Grubbs' catalyst could flux through the thimble, it was insoluble in the solvent on the exterior of the thimble, so it did not partition into the exterior. This method may be a general method to site isolate catalysts using PDMS membranes.

This method was used to complete a series of ring-closing and cross-metathesis reactions (eqs 8 and 9). The Grubbs' second-generation catalyst and CH$_2$Cl$_2$:[BMIM][PF$_6$] were added to the interior of a thimble and the starting material and MeOH:H$_2$O were added to the exterior. The starting materials fluxed to the interior of the thimble to react while the Grubbs' catalyst remained on the interior. Yields ranged from 69 to 93% with reaction times of 2.5–26 h.

X = C(CO$_2$Et)$_2$, N-Ts

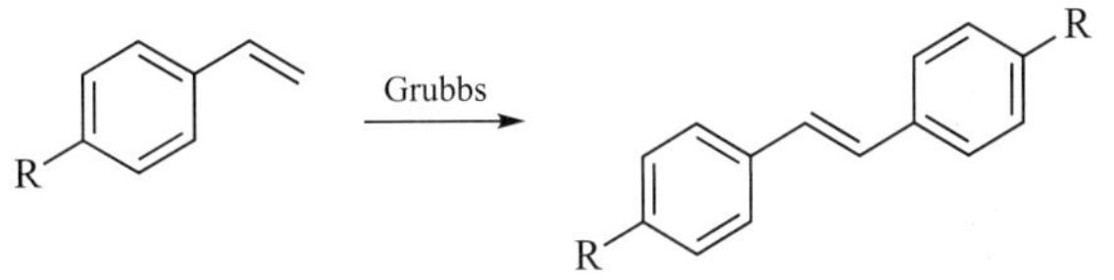

(8)

R = H, OAc, CH$_3$

(9)

The catalyst could be recycled for five cycles at quantitative conversions and with yields from 66 to 82% as shown in eq 10. The catalyst was added to the interior of a thimble with CH$_2$Cl$_2$:[BMIM][PF$_6$] and the malonate. After the reaction was complete, the product was extracted to the exterior of the thimble with MeOH. The MeOH on the exterior was removed from the reaction vessel and new malonate was added to the interior of the thimble to react with the catalyst.

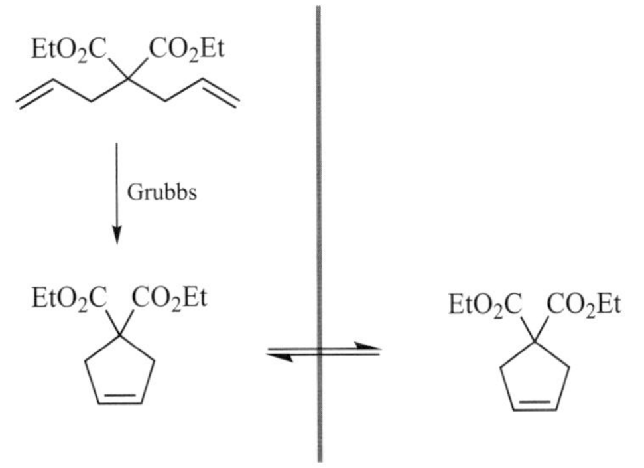

(10)

The Grubbs' catalyst was successfully site isolated from *m*-CPBA using PDMS thimbles as shown in eqs 11 and 12. In control reactions, it was shown that at ratios of *m*-CPBA to Grubbs' catalyst of 3000:1, the Grubbs' catalyst catalytically decomposed *m*-CPBA in a very rapid reaction such that the epoxidation of olefins proceeded to give low yields (<22%). The Grubbs' second-generation catalyst was added to the interior of a thimble with CH$_2$Cl$_2$:[BMIM][PF$_6$] and a substrate. After the metathesis reaction was complete, *m*-CPBA and MeOH:H$_2$O were added to the exterior to complete the epoxidation. Isolated yields of 67–83% were obtained for these substrates.

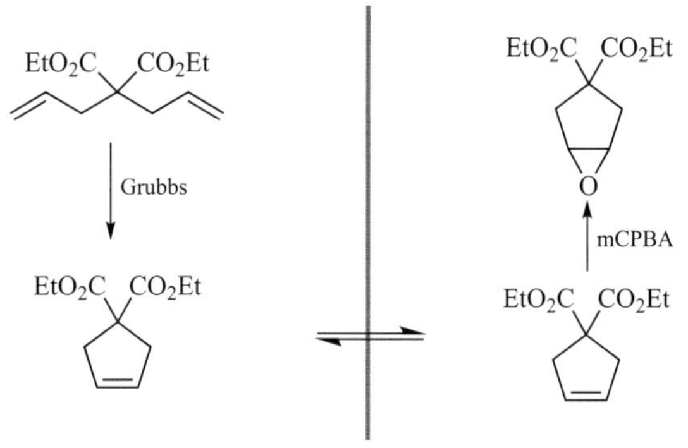

(11)

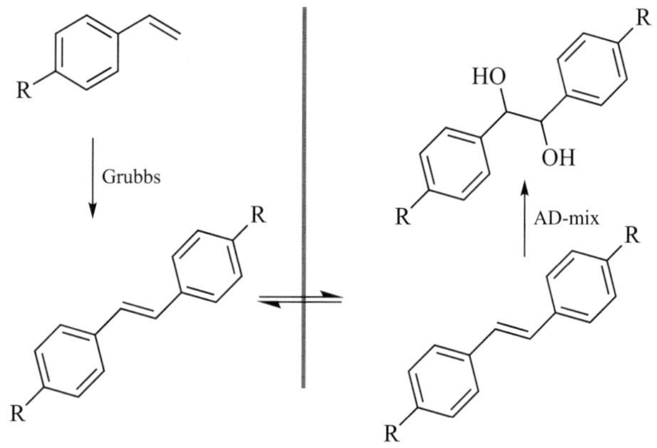

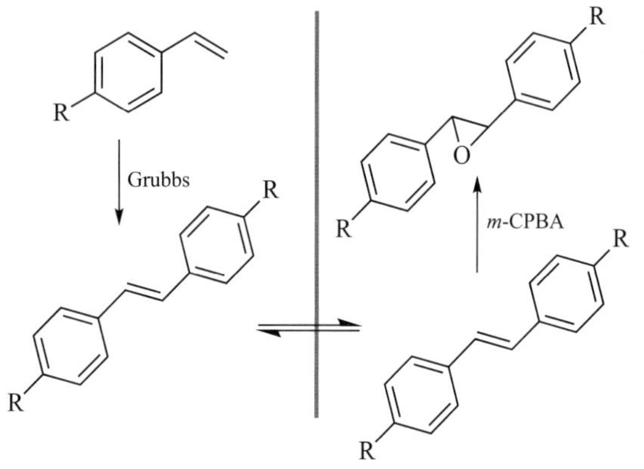

(12)

R = H, OAc, CH$_2$Cl$_2$, OMe, CH$_3$

(14)

R = H, OMe, CH$_3$

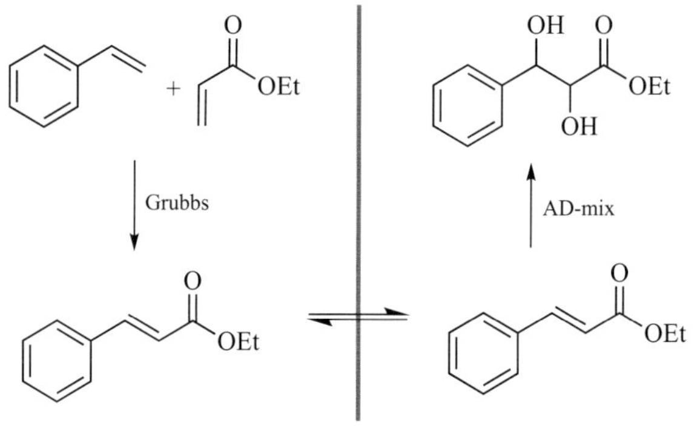

(15)

The Grubbs' catalyst was recycled in a cascade sequence as shown in eq 11. The Grubbs' catalyst was recycled as described before, but in these reactions the extracted product was epoxidized with *m*-CPBA after removal from the reaction vessel containing the PDMS thimble. The isolated yields of the product ranged from 60 to 86% through five cycles.

Site Isolation of the Grubbs' Catalyst from the Sharpless Dihydroxylation Catalyst. The Grubbs' catalyst was site isolated from AD-mix to complete cascade reactions as shown in eqs 13–15.[8] The Grubbs' catalyst was added to the interior of a thimble with CH$_2$Cl$_2$:[BMIM][PF$_6$] and the metathesis reaction was run to completion. Next, the AD-mix was added to the exterior of a thimble with 1:1 (v/v) *t*-BuOH:H$_2$O or 1:2:3 (v/v/v) [BMIM][PF$_6$]:H$_2$O:acetone. The product of the metathesis reaction fluxed to the exterior and reacted to yield the diol.

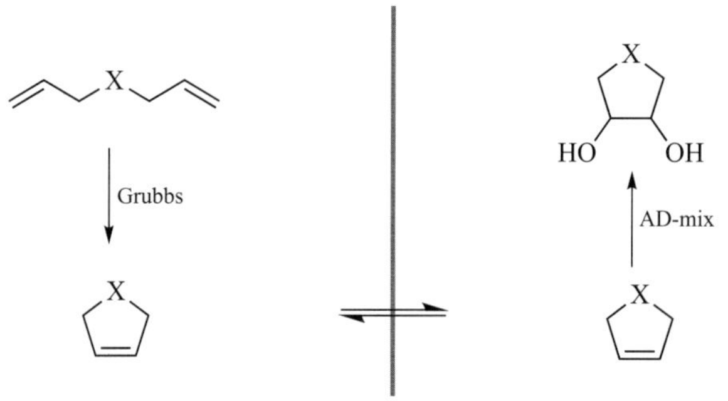

The Grubbs' catalyst was site isolated using the thimbles and solvents described here. Less than 1% fluxed to the exterior even after 24 h. In contrast, Os(VIII) fluxed to the interior of the thimble but this had no apparent impact on the yields and enantiomeric excesses. An achiral product was also isolated to demonstrate that the two-step process was successful for a cyclic substrate. The yields for the reactions ranged from 61 to 95% and the enantiomeric excesses were 84–98%. The enantiomeric excesses matched those reported by others for identical substrates in reactions completed in glass vessels.

This method was extended to substrates with intermediates that were challenging to isolate.[8] For instance, a metathesis–dihydroxylation cascade reaction was completed on a challenging substrate (eq 16). The metathesis reaction of diallylamine yielded a molecule with an estimated boiling point of 55 °C; this molecule was not isolated but rather converted to the diol and isolated as a high-boiling product (80% isolated yield). It was not necessary to isolate the volatile intermediate in this sequence.

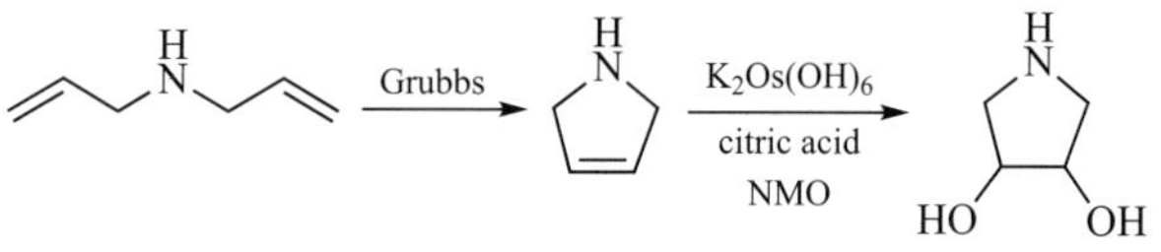

(13)

X = C(CO$_2$Et), CHOBn

(16)

A second challenging metathesis–dihydroxylation cascade reaction was completed (eq 17). Diallyl sulfide (boiling point: 138 °C) was reacted with the Grubbs' catalyst to yield a cyclic intermediate with an estimated boiling point of 90 °C. The intermediate had an unpleasant odor, but it was not isolated in this sequence. Rather, the formation of the diol and oxidation of the sulfur were completed in situ to yield a product that was essentially odorless (79% isolated yield). Three reactions were completed in one pot with this substrate.

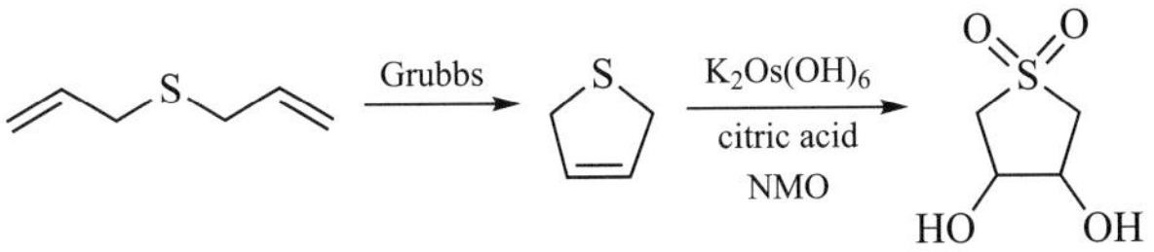

(17)

Site Isolation of Polymeric Catalysts. A common method to site isolate catalysts from the products of a reaction is to attach the catalyst to a polymer that can be precipitated at the end of a reaction.[18–21] Catalysts attached to a polymer are generally not site isolated from other catalysts attached to a different polymer such that only one catalyst can be added to a reaction mixture. The use of PDMS thimbles solves this problem.[10]

It is well known that polymers do not flux through polymeric membranes. In experiments with PDMS thimbles and CH_2Cl_2 as the solvent, the polymers shown in eqs 18 and 19 did not flux through the membranes even after 3 days. No evidence of the polymers was found on the exterior of the thimbles, which can be contrasted with the rapid flux of small molecules under identical conditions (eq 1).

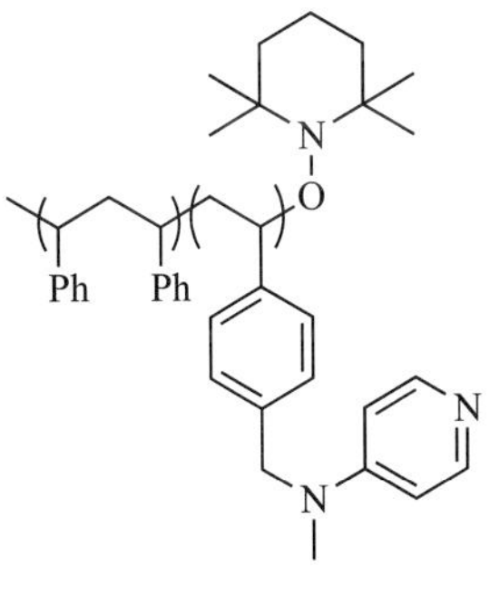

(18)

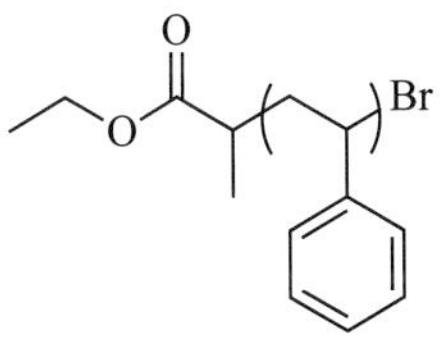

(19)

A series of cascade reactions with an acid catalyst on the interior of a PDMS thimble and a base catalyst on the exterior were completed (eq 20). In these reactions, the acetal was added to the interior of a thimble with an acid catalyst and methyl acrylate was added to the exterior with a basic catalyst. The solvent on the interior and exterior of a thimble was 7:1 (v/v) DMF:H_2O. The reactions were heated to 70 °C for 72 h and then product was isolated.

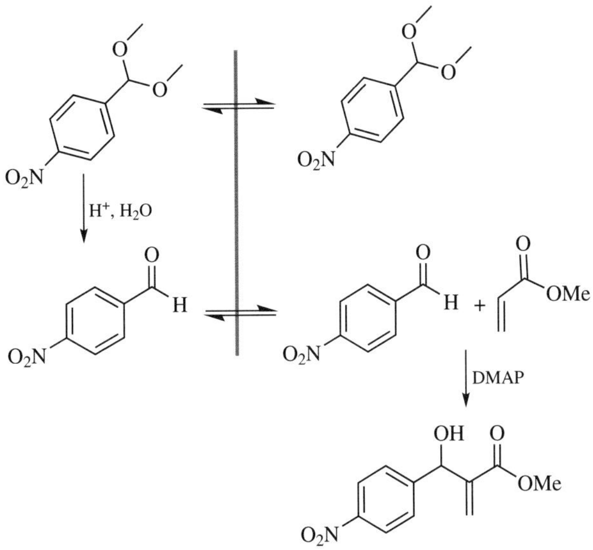

(20)

Striking differences in yields were observed in reactions run with or without PDMS thimbles. The acid catalyst was either PTSA or commercially available beads with PTSA bonded to the backbone (eq 21). The basic catalyst was either DMAP, commercially available beads with DMAP bonded to the backbone (eq 22), or a linear polymer (eq 18). Six reactions were completed with either PTSA or the beads with PTSA and each of the three different bases. In reactions with PDMS thimbles to site isolate the acid from the base, the isolated yields of the final product were 71–93%. The lowest yields were when DMAP was the basic catalyst. This result was probably due to the flux of DMAP through the thimbles that quenched some of the acid on the interior of the thimble.

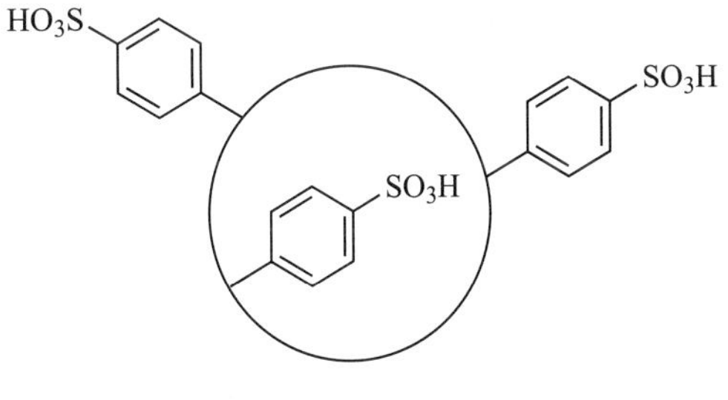

(21)

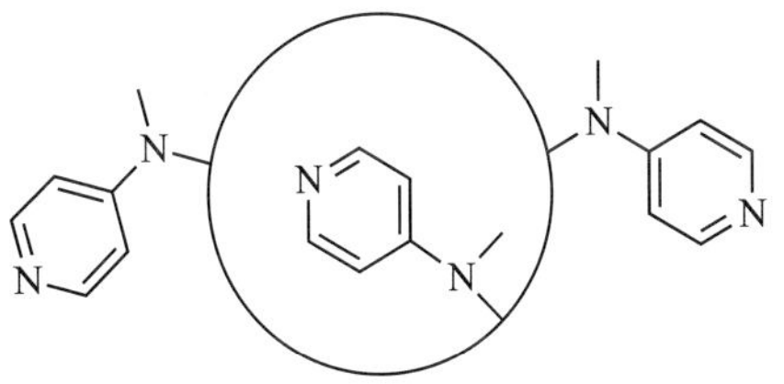

(22)

In five of the six reactions without thimbles to site isolate the acid and base catalysts, the isolated yields of the product ranged from 0 to 15%. Only in the reaction with polymeric beads with PTSA and beads with DMAP did the isolated yield reach 50%. In summary, the PDMS thimbles successfully site isolated polymeric catalysts from one another and this approach should work for other polymeric catalysts.

Site Isolation of PdCl2. Due to the importance of Pd in organic synthesis, the site isolation of $PdCl_2$ was studied.[9] The

Wacker–Tsuji oxidation of olefins (eq 23) and Pd-mediated homocoupling of aryl boronic acids (eq 24) were completed on the interior of PDMS thimbles. In each of these reactions, the catalyst was $PdCl_2$ and a polar protic solvent was used on the interior of the thimbles. The products were fluxed to the exterior of the thimbles by the addition of CH_2Cl_2 or hexanes to the exterior. Isolated yields were 56–93%.

$$R \diagdown \xrightarrow[\text{}]{\text{PdCl}_2,\ \text{CuCl},\ \text{O}_2} R \diagdown \begin{matrix} O \\ \| \\ \end{matrix} \qquad (23)$$

R = Ph

$$R \diagdown \diagdown -B(OH)_2 \xrightarrow[\text{Na}_2\text{CO}_3]{\text{PdCl}_2} \left(R \diagdown \diagdown \right)_2 \qquad (24)$$

$$R = H,\ \overset{O}{\overset{\|}{C}H},\ Br$$

In each of these reactions, the Pd was site isolated on the interior of the thimbles. To characterize the concentration of Pd in the products, the reactions were run to completion and the product was extracted to the exterior. The solvent on the exterior of the thimble was isolated and removed. The concentration of Pd in the product was found by ICP-MS. Between 0.225 and < 0.002% of the Pd that was added to the interior of the thimble was found in the products on the exterior of the thimbles. In other words, >99.998% and no less than 99.775% of the Pd remained on the interior of the thimbles even after the product was extracted to the exterior.

The reason for the low flux of $PdCl_2$ was that it was soluble only in polar protic solvents. Its solubility in PDMS was negligible, so it did not flux through the walls of the thimbles. When phosphines were added to the reaction mixture to coordinate to $PdCl_2$ and render it less polar, the phosphines and Pd readily fluxed through the walls of the PDMS thimbles. This method demonstrates that catalysts that do not partition into PDMS will remain site isolated by it.

Final Notes. Small molecules, organometallic catalysts, inorganic catalysts, and polymeric catalysts were all successfully site isolated at levels up to >99.998% using PDMS thimbles. When site isolating catalysts or reagents, the most important parameters to consider are the solubilities and rates of diffusion within the swollen PDMS matrix and the solubilities in the organic solvents on the interior and exterior of thimbles.

1. Runge, M. B.; Mwangi, M. T.; Miller, A. L.; Perring, M.; Bowden, N. B., *Angew. Chem., Int. Ed.*, **2008**, *47*, 935.

2. Mwangi, M. T.; Runge, M. B.; Hoak, K. M.; Schulz, M. D.; Bowden, N. B., *Chem. Eur. J.* **2008**, *14*, 6780.

3. Crank, J. *The Mathematics of Diffusion*; Clarendon Press: Oxford, 1970.

4. Shah, M. R.; Noble, R. D.; Clough, D. E., *J. Member. Sci.* **2007**, *287*, 111.

5. Watson, J. M.; Zhang, G. S.; Payne, P. A., *J. Member. Sci.* **1992**, *73*, 55.

6. Tamai, Y.; Tanaka, H.; Nakanishi, K., *Macromolecules* **1994**, *27*, 4498.

7. Tamai, Y.; Tanaka, H.; Nakanishi, K., *Macromolecules* **1995**, *28*, 2544.

8. Mwangi, M. T.; Schulz, M. D.; Bowden, N. B., *Org. Lett.* **2009**, *11*, 33.

9. Miller, A. L.; Bowden, N. B., *J. Org. Chem.* **2009**, *74*, 4834.

10. Miller, A. L.; Bowden, N. B., *Adv. Mater.* **2008** *20*, 4195.

11. Miller, A. L.; Bowden, N. B., *Chem. Commun.* **2007**, 2051.

12. Runge, M. B.; Mwangi, M. T.; Bowden, N. B., *J. Organomet. Chem.* **2006**, *691*, 5278.

13. Mwangi, M. T.; Runge, M. B.; Bowden, N. B., *J. Am. Chem. Soc.* **2006**, *128*, 14434.

14. Lee, J. N.; Park, C.; Whitesides, G. M., *Anal. Chem.* **2003**, *75*, 6544.

15. Balmer, T. E.; Schmid, H.; Stutz, R.; Delamarche, E.; Michel, B.; Spencer, N. D.; Wolf, H., *Langmuir* **2005**, *21*, 622.

16. Banerjee, S.; Asrey, R.; Saxena, C.; Vyas, K.; Bhattacharya, A., *J. Appl. Polym. Sci.* **1997**, *65*, 1789.

17. Schafer, T.; Di Paolo, R. E.; Franco, R.; Crespo, J. G., *Chem. Commun.* **2005**, 2594.

18. Madhaven, N.; Weck, M., *Adv. Synth. Catal.* **2008**, *350*, 419.

19. Harned, A. M.; He, H. S.; Toy, P. H.; Flynn, D. L.; Hanson, P. R., *J. Am. Chem. Soc.* **2005**, *127*, 52.

20. Ryo, A.; Shu, K., *Chem. Rev.* **2009**, *109*, 594.

21. Bergbreiter, D. E.; Tian, J.; Hongfa, C., *Chem. Rev.* **2009**, *109*, 530.

Tyler R. Long & Ned B. Bowden
University of Iowa, Iowa City, IA, USA

Phenylsilane–Cesium Fluoride

$$\boxed{PhSiH_3 \!-\! CsF}$$

(PhSiH_3)

[694-53-1]	C_6H_8Si	(MW 108.23)

InChI = 1S/C6H8Si/c7-6-4-2-1-3-5-6/h1-5H,7H3
InChIKey = PARWUHTVGZSQPD-UHFFFAOYSA-N

(CsF)

[13400-13-0]	CsF	(MW 151.91)

InChI = 1S/Cs.FH/h;1H/q+1;/p-1
InChIKey = XJHCXCQVJFPJIK-UHFFFAOYSA-M

(mild nucleophilic non-Lewis acidic hydride donor[1] used under aprotic conditions)

Physical Data: phenylsilane: bp 120 °C; d 0.877 g cm^{-3}. Cesium fluoride: mp 683 °C.

Solubility: phenylsilane: sol THF, MeCN. Cesium fluoride: sol H_2O; slightly sol THF, MeCN.

Form Supplied in: phenylsilane is a clear liquid; cesium fluoride is a white solid; both widely available.

Drying: anhydrous cesium fluoride is prepared by flame drying under vacuum, taking care not to fuse the salt.

Handling, Storage, and Precautions: no particular requirements.

4-Oxazoline Formation. The $PhSiH_3$–cesium fluoride reagent[1,2] is a nonnucleophilic hydride donor without Lewis acid character with characteristics similar to the diphenylsilane–cesium fluoride,[2] α-NaphSiH$_3$–CsF,[2] and $(RO)_3SiH$–CsF[2] reagents on which it is based. In aprotic solvents, it has been used to reduce 4-oxazolium salts to 4-oxazolines.[1] The 4-oxazolines open at -40 °C to azomethine ylides, which are trapped with dipolarophiles to give 3-pyrrolines (**2**) (eq 1). For isolation purposes the 3-pyrrolines are oxidized to pyrroles (**3**). The mildness of the reagent allows for the selective reduction of the oxazolium salt without further reduction of the 4-oxazoline, azomethine ylide, or dipolarophile which are present at different stages of the reaction. Of the reducing agents tried for (**1**), PhSiH$_3$ gave the highest yield (95%), although Ph$_2$SiH$_2$ (93%) worked nearly as well.[1b] Other reducing agents gave much poorer yields. Propiolates, acrylates, *N*-phenylmaleimide, and vinyl sulfone dipolarophiles have been used in this [3 + 2] cycloaddition without dipolarophile reduction.

Dipolarophiles such as ethyl acrylate exhibit regiochemical preference for '*meta*' acyl or carboxyl groups (eq 2).[1] With one exception (entry 6), all cases are consistent with FMO approximation in which the largest dipole HOMO coefficient is at the acceptor-substituted carbon (Table 1).[1b]

Table 1 Regiochemistry of acrylate trapping (eq 2)

| Entry | Group | | Isolated yield (%) | |
	R^1	R^2	(4)	(5)
1	Ph	Ph	55	9
2	Ph	OEt	63	10
3	Ph	Me	87	0
4	Me	Ph	40	20
5	Me	OEt	61	0
6	H	Ph	0	67
7	H	OEt	47	0
8	Ph	H	57	0

A comparison of methods to generate azomethine ylides via 4-oxazoline ring opening vs. aziridine thermolysis has shown that the oxazoline method can produce the same azomethine dipole under kinetic control as the aziridine method under thermodynamic

control.[3] Since the 4-oxazoline method allows improved regiochemical selectivity in many cases, the synthesis of 4-oxazolines from PhSiH$_3$–CsF-induced reduction of oxazolium salts possesses greater potential for azomethine ylide applications in stereocontrolled synthesis.

Intramolecular trapping of an azomethine ylide generated from an oxazolium salt has been accomplished (eq 3).[4] This approach allows a new entry into the mitosene skeleton.

Related Reagents. Diphenylsilane–Cesium Fluoride.

1. (a) Vedejs, E.; Grissom, J. W., *J. Am. Chem. Soc.* **1986**, *108*, 6433. (b) Vedejs, E.; Grissom, J. W., *J. Am. Chem. Soc.* **1988**, *110*, 3238.
2. Boyer, J.; Corriu, R. J. P; Perz, R.; Reye, C., *J. Organomet. Chem.* **1979**, *172*, 143 (*Chem. Abstr.* **1979**, *91*, 108 039n).
3. Vedejs, E.; Grissom, J. W., *J. Org. Chem.* **1988**, *53*, 1882.
4. Vedejs, E.; Piotrowski, D. W., *J. Org. Chem.* **1993**, *58*, 1341.

Thomas J. Fleck
The Upjohn Company, Kalamazoo, MI, USA

(2-Phenylsulfonylethyl)trimethylsilane

[73476-18-3]   C$_{11}$H$_{18}$O$_2$SSi   (MW 242.45)
InChI = 1S/C11H18O2SSi/c1-15(2,3)10-9-14(12,13)11-7-5-4-6-8-11/h4-8H,9-10H2,1-3H3
InChIKey = IVPQHMQOYMNETK-UHFFFAOYSA-N

(reagent for the synthesis of mono- and 1,1-disubstituted alkenes via sulfone metalation, alkylation, and fluoride-induced elimination[1])

Physical Data: mp 52 °C.
Solubility: sol all common ethereal, halocarbon, and hydrocarbon solvents.
Form Supplied in: solid; commercially available.
Preparative Methods: (2-phenylsulfonylethyl)trimethylsilane (**2**) is prepared by radical addition of thiophenol to vinyltrimethylsilane to give (2-phenylthioethyl)trimethylsilane (**1**), which is then oxidized with hydrogen peroxide (eq 1).[2,3]

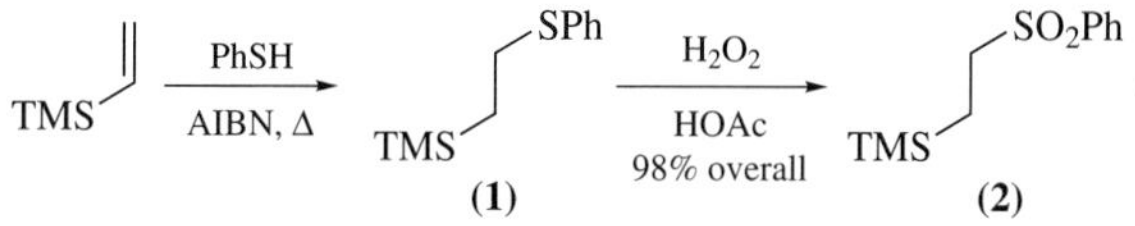

$$TMS \xrightarrow[\text{AIBN, }\Delta]{\text{PhSH}} TMS{-}SPh \text{ (1)} \xrightarrow[\substack{\text{HOAc}\\ \text{98\% overall}}]{\text{H}_2\text{O}_2} TMS{-}SO_2Ph \text{ (2)} \quad (1)$$

$$TMS{-}SO_2Ph \text{ (3)} \xrightarrow[\substack{180\,^\circ C,\ 7\,d\\ 98\%}]{\text{anthracene}} \text{[} TMS, SO_2Ph \text{]} \xrightarrow[84\%]{\text{TBAF}}$$

$$\quad (4)$$

General Considerations. A sequence involving metalation, alkylation, and fluoride-induced elimination of benzenesulfinate allows the conversion of (**2**) to a terminal alkene. An analogous sequence involving a double alkylation of (**2**) provides a 1,1-disubstituted alkene (eq 2).[4,5] The lithio derivative of (**2**) has also been used to prepare cyclopropylidene derivatives,[6] homo-allylic alcohols,[7] and allyl silanes via the Julia alkenation.[3]

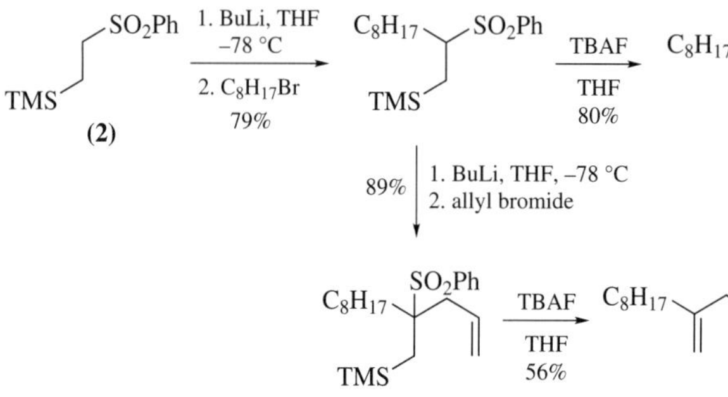

$$TMS{-}SO_2Ph \text{ (2)} \xrightarrow[\substack{2.\ C_8H_{17}Br\\ 79\%}]{1.\ \text{BuLi, THF }-78\,^\circ C} C_8H_{17}{-}SO_2Ph\ (TMS) \xrightarrow[\substack{\text{THF}\\ 80\%}]{\text{TBAF}} C_8H_{17}$$

$$89\% \downarrow \substack{1.\ \text{BuLi, THF, }-78\,^\circ C\\ 2.\ \text{allyl bromide}} \quad (2)$$

$$C_8H_{17}{-}SO_2Ph\ (TMS) \xrightarrow[\substack{\text{THF}\\ 56\%}]{\text{TBAF}} C_8H_{17} \quad$$

Alternative Routes to Substituted (2-Phenylsulfonylethyl)-trimethylsilanes. Although the fluoride-induced elimination is a reliable and efficient method for synthesizing alkenes, the requisite silyl sulfone precursors are often better prepared by methods which avoid the use of (**2**). For example, eq 3 illustrates an alternative synthesis of 2,2-dialkyl-2-(phenylsulfonylethyl)trimethylsilanes involving alkylation of a lithio sulfone with (iodomethyl)-trimethylsilane.[8,9]

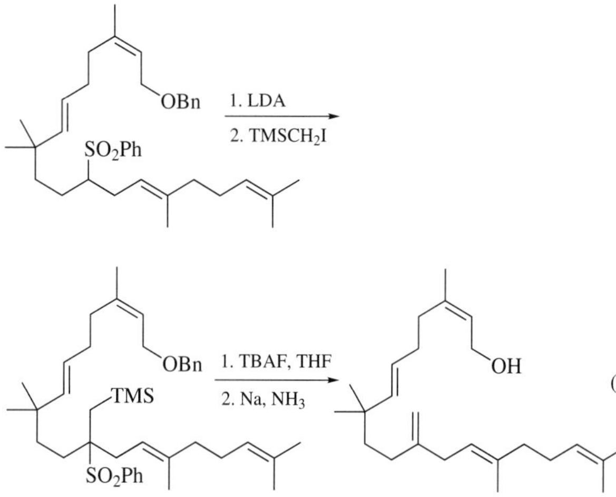

$$\xrightarrow[\text{2. TMSCH}_2\text{I}]{\text{1. LDA}}$$

$$\xrightarrow[\text{2. Na, NH}_3]{\text{1. TBAF, THF}} \quad (3)$$

2-(Phenylsulfonylvinyl)trimethylsilane (**3**) is a useful acetylene equivalent for use in Diels–Alder reactions (eq 4).[10,11]

Phenylsulfonylethylene derivatives undergo conjugate addition reactions. Thus (**3**) reacts with unsaturated Grignard reagents and organocuprates,[12] while the substituted sulfone (**5**) undergoes conjugate addition by the enolate derivative (**4**) to give a bicyclomycin intermediate (eq 5).[13]

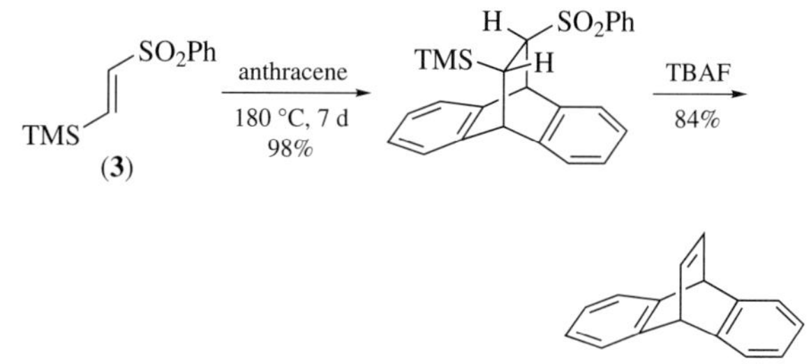

$$\text{(4)} + \text{(5)} \xrightarrow[50\%]{-40\,^\circ C}$$

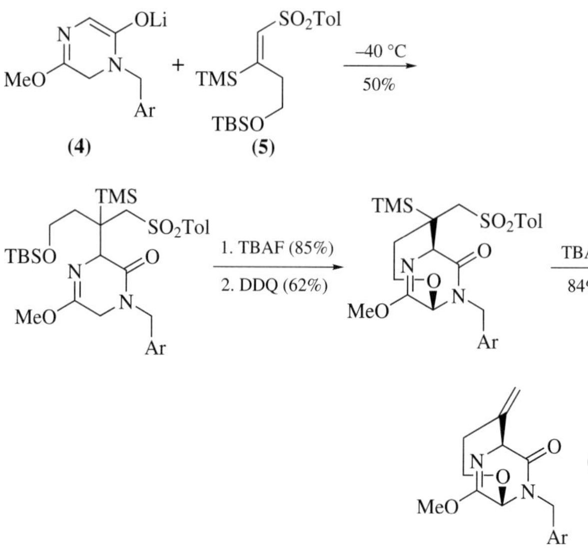

$$\xrightarrow[\text{2. DDQ (62\%)}]{1.\ \text{TBAF (85\%)}} \xrightarrow[84\%]{\text{TBAF}}$$

$$\quad (5)$$

Related Reagents. 1,1-Bis(phenylsulfonyl)ethylene; Phenyl-sulfonylethylene; α-Phenylsulfonylethyllithium; 1-Phenyl-sulfonyl-2-(trimethylsilyl)acetylene; (E)-1-Phenylsulfonyl-2-trimethylsilylethylene; Phenylsulfonyl(trimethylsilyl)methane.

1. Kocienski, P. J., Comprehensive Organic Synthesis **1991**, 6, 1002.
2. Hsiao, C.-N.; Shechter, H., Tetrahedron Lett. **1982**, 23, 1963.
3. Hsiao, C.-N.; Shechter, H., J. Org. Chem. **1988**, 53, 2688.
4. Kocienski, P. J., Tetrahedron Lett. **1979**, 20, 2649.
5. Nájera, C.; Sansano, J. M., Tetrahedron Lett. **1993**, 34, 3781.
6. Hsiao, C.-N.; Hannick, S. M., Tetrahedron Lett. **1990**, 31, 6609.
7. Kabat, M. M.; Wicha, J., Tetrahedron Lett. **1991**, 32, 1073.
8. Kocienski, P. J., J. Org. Chem. **1980**, 45, 2037.
9. Kocienski, P. J.; Todd, M., J. Chem. Soc., Perkin Trans. 1 **1983**, 1777, 1783.
10. Paquette, L. A.; Williams, R. V., Tetrahedron Lett. **1981**, 22, 4643.
11. Carr, R. V. C.; Williams, R. V.; Paquette, L. A., J. Org. Chem. **1983**, 48, 4976.
12. Eisch, J. J.; Behrooz, M.; Dua, S. K., J. Organomet. Chem. **1985**, 285, 121.
13. Dawson, I. M.; Gregory, J. A.; Herbert, R. B.; Sammes, P. G., J. Chem. Soc., Chem. Commun. **1986**, 620.

Georges Hareau & Philip Kocienski
Southampton University, Southampton, UK

Phenylthiobis(trimethylsilyl)methane[1]

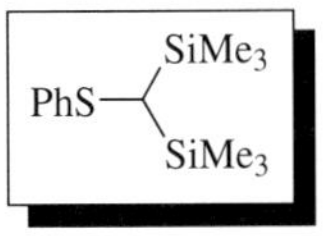

[62761-90-4] $C_{13}H_{24}SSi_2$ (MW 268.62)

InChI = 1S/C13H24SSi2/c1-15(2,3)13(16(4,5)6)14-12-10-8-7-
9-11-12/h7-11,13H,1-6H3

InChIKey = QHCXDRRVCOVYHN-UHFFFAOYSA-N

(formation of vinylsilanes and as a methoxycarbonyl anion equivalent)

Physical Data: bp 88 °C/0.16 mmHg.
Solubility: sol most common organic solvents.
Analysis of Reagent Purity: ^1H NMR.
Preparative Methods: synthesized from phenylthiotrimethyl-silylmethane.[2]
Handling, Storage, and Precautions: use in a fume hood.

Vinylsilane Preparation.   The most general utility of the title reagent (1) involves Peterson alkenation reactions. Deprotonation of this reagent, followed by addition to carbonyl-containing substrates, affords phenylthiovinylsilanes (eq 1).[3]

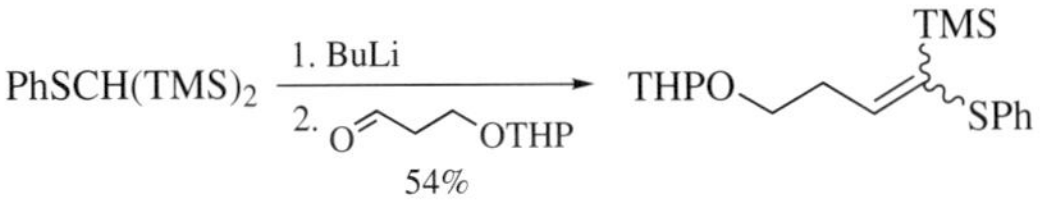

The reduction of the carbon–sulfur bond in (1) with lithium naphthalenide also produces an alkenation reagent (eq 2).[4] This protocol is limited, however, to use with nonenolizable ketones and aldehydes. Lithiobis(trimethylsilyl)methane, the active intermediate in this reaction, can also be produced through the deprotonation of bis(trimethylsilyl)methane.

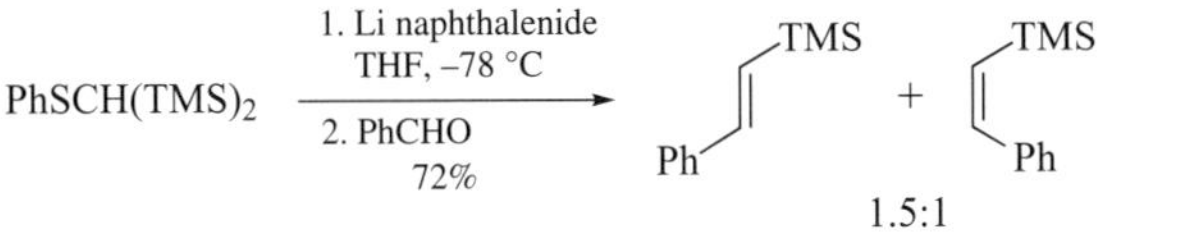

When (1) is treated with a fluoride source, a mild alkenation reagent is produced (eq 3).[5] The efficient transformation of enolizable ketones and aldehydes under these conditions, as opposed to the lithium naphthalenide reduction process, aptly demonstrates this fact.

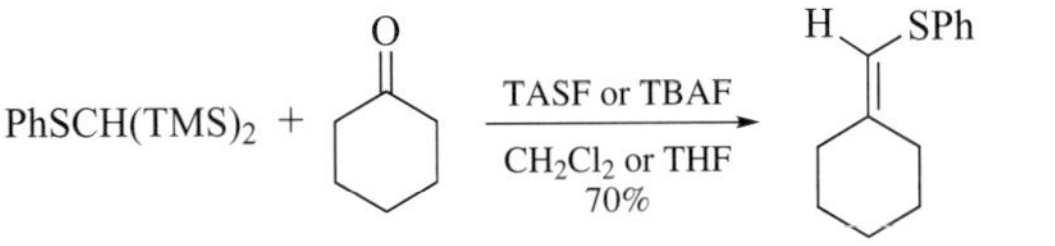

Conjugate Addition.   The lithium anion of (1) adds exclusively in Michael fashion to cyclohexenone (eq 4).[6] In comparison, lithiobis(phenylthio)methane predominantly adds to the carbonyl, providing a 2:1 mixture of the possible regioisomers. As with many other 1,4-conjugate additions, substitution at the alkene terminus alters the reaction regioselectivity (eq 5).[6]

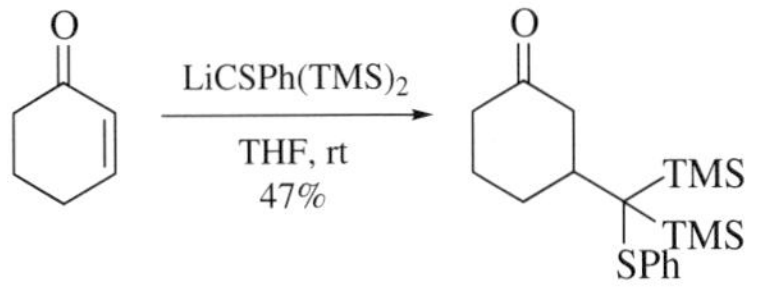

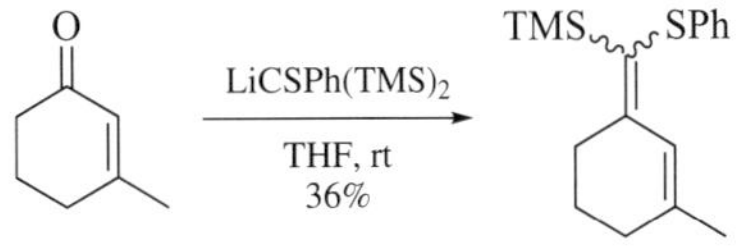

Homo-Peterson Alkenation.   Styrene oxide reacts with lithiophenylthiobis(trimethylsilyl)methane to afford cyclopropane-containing products (eq 6).[7] This reaction is limited, due to the complexity of its mechanism: the alkenation reagent must serve to generate both alkenic and carbenic species. For this reason, only styrene oxide and trimethylsilyloxirane undergo this transformation.

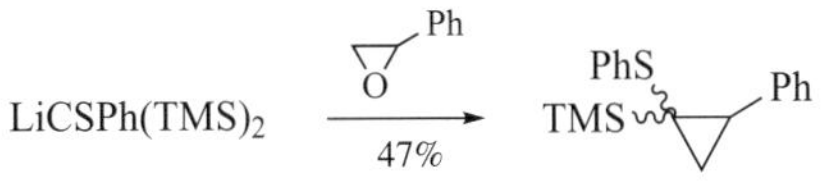

Methoxycarbonyl Anion Equivalent.   The alkylation of lithiophenylthiobis(trimethylsilyl)methane with an alkyl halide, followed by electrochemical oxidation of the resulting thiobis-(silane), provides a homologated methyl ester (eq 7).[8] The electrochemical oxidation, due to its mild nature, can be tolerated by a wide variety of functional groups.

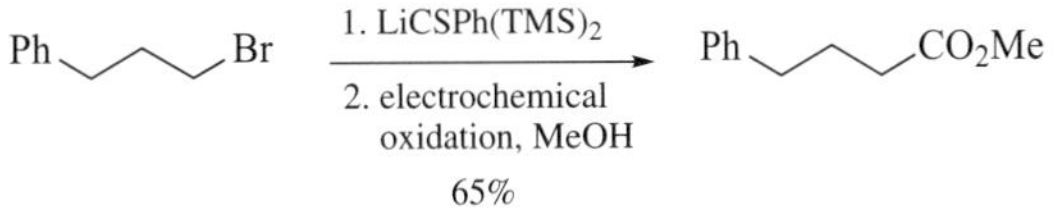

1. For a review of organosulfur–silicon compounds, see: Block, E.; Aslam, M., *Tetrahedron* **1988**, *44*, 281.
2. Grobel, B.-T.; Seebach, D., *Chem. Ber.* **1977**, *110*, 852.
3. Hart, D. J.; Tsai, Y.-M., *J. Am. Chem. Soc.* **1984**, *106*, 8209.
4. Ager, D. J., *J. Org. Chem.* **1984**, *49*, 168.
5. Palomo, C.; Aizpurua, J. M.; Garcia, J. M.; Ganboa, I.; Cossio, F. P.; Lecea, B.; Lopez, C., *J. Org. Chem.* **1990**, *55*, 2498.
6. Ager, D. J.; East, M. B., *J. Org. Chem.* **1986**, *51*, 3983.
7. Schaumann, E.; Friese, C., *Tetrahedron Lett.* **1989**, *30*, 7033.
8. Yoshida, J.; Matsunaga, S.; Murata, T.; Isoe, S., *Tetrahedron* **1991**, *47*, 615.

Jeffrey A. McKinney
Zeneca Pharmaceuticals, Wilmington, DE, USA

(Phenylthiomethyl)trimethylsilane

$$PhS \diagdown SiMe_3$$

[17873-08-4] $C_{10}H_{16}SSi$ (MW 196.42)
InChI = 1S/C10H16SSi/c1-12(2,3)9-11-10-7-5-4-6-8-10/h4-8H,
    9H2,1-3H3
InChIKey = UOQDIMYVQSHALM-UHFFFAOYSA-N

(lithio derivative is a one-carbon homologating agent leading to
    aldehydes via sila-Pummerer rearrangement,[1] vinyl sulfides via
    Peterson alkenation,[2] and vinylsilanes from silyl epoxides[3])

Physical Data: bp 158–159 °C/52 mmHg; d^{20} 0.967 g cm^{-3}; n_D^{20}
    1.5390.
Solubility: sol most common organic solvents.
Form Supplied in: oil; commercially available.
Analysis of Reagent Purity: by GC or ^{1}H NMR analysis.
Preparative Methods: by reaction of phenylthiomethyllithium
    with chlorotrimethylsilane.[4]
Purification: by distillation under reduced pressure.
Handling, Storage, and Precautions: irritant; foul smelling; use
    in a fume hood.

Original Commentary

Akira Hosomi & Makoto Hojo
University of Tsukuba, UK

Georges Hareau & Philip Kocienski
Southampton University

Formylation Reactions. Metalation of the title reagent (**1**)
with *n*-butyllithium gives the fairly stable anion phenylthio-
(trimethylsilyl)methyllithium (**2**), which is a formyl anion equivalent as illustrated by the sequence in eq 1.[1] Anion (**2**) reacts
with primary alkyl bromides and iodides to yield 1-phenylthio-
1-trimethylsilylalkanes (**3**) in high yields. These intermediates
are easily oxidized by *m*-chloroperbenzoic acid to the corresponding sulfoxides. Subsequent heating promotes a sila-
Pummerer rearrangement[5] to produce aldehydes after hydrolysis.[1]
Reagent (**1**) is superior to 1,3-dithiane for the synthesis of aldehydes from alkyl halides, in view of the greater ease of hydrolysis of
(**3**) Application of this procedure to the synthesis of unsymmetrical ketones has also been reported.[6]

The zinc bromide catalyzed alkylation of enol silanes by
α-halo-α-(phenylthio)methyltrimethylsilanes (eq 2) provides an
efficient method for electrophilic formylation (i.e. an umpolung

of the transformation in eq 1) which is useful for the synthesis
of β-ketoaldehydes.[7] The sequence can also be diverted for the
synthesis of β-trimethylsilylenones (eq 2).[8]

Phenylthiobis(trimethylsilyl)methane (**4**), prepared by the
reaction of (**2**) with Me$_3$SiCl, serves as a synthetic equivalent
of LiCO$_2$Me.[9] Metalation of (**4**) with BuLi in the presence
of N,N,N',N'-tetramethylethylenediamine gives phenylthiobis-
(trimethylsilyl)methyllithium (**5**); subsequent alkylation and
electrochemical oxidation achieves chain homologation to form
an ester (eq 3). Alternatively, alkylation of (**5**) with a terminal oxirane followed by cycloelimination in situ generates 1-phenylthio-
1-trimethylsilylcyclopropanes.[10] Anion (**2**) itself reacts with
oxiranes to give adducts which can also be converted to 1-phenyl-
thio-1-trimethylsilylcyclopropanes, but two intermediary steps
are required to accomplish cycloelimination (eq 4).[11]

Synthesis of Vinyl Sulfides.[2] Anion (**2**) adds to aldehydes,
ketones, and enones, in the last-named case in either 1,2- or 1,4-
manner depending on the reaction conditions.[12] In the case of 1,2-
addition the β-hydroxysilane adduct usually cannot be isolated
owing to rapid elimination (Peterson alkenation) to vinyl sulfides
(eq 5), which can be hydrolyzed to the corresponding aldehydes.[2]
The vinyl sulfides are not formed stereoselectively unless there
are proximate bulky substituents.[13] Peterson alkenation products
have also been observed in the reaction of (**2**) with amides, ureas,
and carbonates.[14]

The reaction of (2) with esters provides α-phenylthio-α-tri-methylsilyl ketones, which can be easily converted to α-phenylthio ketones (eq 6).[15] β-Hydroxy sulfides are obtained from the reaction of (1) with carbonyl compounds in the presence of a catalytic amount of fluoride ion (eq 7).[16]

$$(2) \xrightarrow[\text{TMEDA}]{\text{RCO}_2\text{R}'} \text{PhS} \underset{\text{TMS}}{\overset{O}{\|}} R \xrightarrow{\text{SiO}_2} \text{PhS} \overset{O}{\|} R \quad (6)$$

$$(1) \xrightarrow[\text{THF, rt}]{\text{TBAF (cat.)}} \text{PhS} \overset{OH}{\underset{R^2}{\|}} R^1 \quad (7)$$

Synthesis of Vinylsilanes.[3] Lithiated reagent (2) reacts with α-trimethylsilyloxiranes to yield vinylsilanes (eq 8). The reaction involves nucleophilic opening of the oxirane ring by (2), followed by 1,2-elimination of a silyl and a phenylthio group.[3]

$$\text{TMS} \xrightarrow[\text{2. TBAF}]{\text{1. (2)}} \text{TMS} \overset{OH}{\diagup} \text{OBn} \quad (8)$$
69% trans:cis = 2:1

Vinylsilanes are also prepared from (2) as illustrated in eq 9.[17] Treatment of (2) with bis(trimethylsilyl) peroxide provides (6) which reacts with phosphonium ylides in the presence of tetra-n-butylammonium fluoride to give mixtures of vinylsilanes favoring the (Z)-isomer.

$$(2) \xrightarrow{\text{TMSO-OTMS}} \underset{\text{TMS}}{\overset{\text{PhS}}{\diagup}}\text{OTMS} \xrightarrow[\text{50\%}]{\text{Ph}\diagup\text{PPh}_3 \atop 15\% \text{TBAF}} \text{Ph}\diagup\diagup\text{TMS} \quad (9)$$
(6) (Z):(E) = 9:1

Other Reactions. In addition to haloalkanes, carbonyl derivatives, and oxiranes, anion (2) reacts with a wide range of other electrophiles including N-chlorosuccinimide,[18] chlorotrimethylsilane,[8] Bu₃SnCl,[15] diphenyl disulfide, and benzenesulfenyl chloride,[15] and trialkylboranes.[19]

Related Reagents. Methoxy(phenylthio)(trimethylsilyl)-methane (7) is prepared by metalation–silylation of phenylthiomethyl methyl ether.[20] [Methoxy(phenylthio)(trimethylsilyl)-methyl]lithium (8), prepared from (7) by metalation with s-butyllithium in the presence of TMEDA, undergoes clean and efficient Peterson alkenation with aldehydes to give (E)-ketene-O,S-acetals stereoselectively, which can then be hydrolyzed to thioesters (eq 10).[21]

$$R\overset{\text{TMS}}{\underset{\text{OMe}}{\mid}}\text{SPh} \xrightarrow[\text{96\%}]{i\text{-PrCHO, THF} \atop -78\,°C \text{ to rt, 12 h}} \diagup\overset{\text{SPh}}{\underset{\text{OMe}}{}} \xrightarrow[\text{90\%}]{\text{TMSCl, NaI} \atop \text{MeCN, rt, 5 min}}$$
(7) R = H
(8) R = Li ← s-BuLi, TMEDA, THF, −78 °C, 2.5 h
$$\diagup\overset{\text{SPh}}{\underset{O}{}} \quad (10)$$

The selenium analog of (2) is prepared by deprotonation of (phenylselenomethyl)trimethylsilane with lithium diisopropyl-amide or lithium 2,2,6,6-tetramethylpiperidide (eq 11). The anion (9) thus formed can be alkylated in good yields, and the alkylated product can be oxidized to the corresponding selenoxide at low temperature, but selenoxide elimination leading to a vinyl-silane competes with the sila-Pummerer rearrangement.[22,23]

$$\underset{\text{Li}}{\overset{\text{TMS}}{\mid}}\text{SePh} \xrightarrow{\text{PhCH}_2\text{Br}} \underset{\text{Ph}}{\overset{\text{TMS}}{\mid}}\text{SePh} \xrightarrow[\text{2. CCl}_4, \Delta]{1.\ m\text{-CPBA}} \underset{\text{Ph}}{\overset{\text{TMS}}{\diagup}} + \underset{\text{Ph}}{\overset{O}{\diagup}} \quad (11)$$
(9) 47% 24%

See also 2-lithio-1,3-dithiane, phenylthiomethyllithium, and bis(phenylthio)methane.

First Update

Craig R. Smith
The Ohio State University, Columbus, OH, USA

Recently, (phenylthiomethyl)trimethylsilane (1), or the anion (2) thereof, has been used to synthesize vinylsilanes from α-trimethylsilyloxiranes,[24] in Peterson-type olefination reactions to generate vinyl sulfides,[25] and also as a radical precursor under electrochemical conditions.[26] Addition reactions of (phenylthiomethyl)trimethylsilyl carbene (10), prepared via two independent methods, to olefins have also been disclosed.[27]

Synthesis of Vinyl Sulfides. Treatment of furanoses with anion 2 induced the ring-opening reaction via addition to the aldehyde of the omega hydroxy aldehyde in equilibrium followed by the rapid elimination of the β-hydroxysilane which afforded vinyl sulfides (eq 12) in moderate yields (50–61%) with poor stereoselectivity (E/Z 2:3).[25] Alternatively, modified Horner–Wittig reactions afforded the desired products in much higher yields (72–100%) with improved selectivity for the E-stereoisomer (E/Z, as high as 17:1).

$$\underset{\text{BnO OBn}}{\overset{\text{BnO}}{\diagup}}\text{OH} \xrightarrow[\text{PhS}\diagup\text{SiMe}_3 \atop 50\%]{\text{BuLi, THF} \atop -78-0\,°C} \underset{\text{BnO}}{\overset{\text{BnO}}{\diagup}}\text{OH SPh} \quad (12)$$
(E/Z, 2:3)

Other Reactions. Catalytic electro-initiated coupling reactions of N-acyliminium ions and arylthiomethylsilanes have been disclosed, which relied upon anodic oxidation of arylthiomethyl-silane leading to facile cleavage the C–Si bond, generating aryl-thiomethyl radicals.[26a] After trapping of the initial thioanisyl radical with an N-acyliminium ion, a catalytic radical chain process ensues affording the desired coupling products (eq 13) in variable yields (14–90%).[26b]

$$\underset{\text{CO}_2\text{Me}}{\overset{\text{F}_4\text{B}^-}{}} + \diagup\overset{\text{SPh}}{\underset{\text{SiMe}_3}{}} \xrightarrow[\text{CH}_2\text{Cl}_2 \atop 46\%]{\text{electrochemical} \atop \text{initiation}} \underset{\text{CO}_2\text{Me}}{\overset{}{}}\diagup\text{SPh} \quad (13)$$

(Phenylthiomethyl)trimethylsilyl carbene (**10**) has been generated via two independent methods, either from the diazo compound through copper catalysis (eq 14),[27] or from the chloro-(phenylthiomethyl)silane by base-induced α-elimination. The generation of carbene **10** was verified by [2 + 1] cycloadditions with olefins affording cyclopropanes (eq 15) in low to moderate yields (12–61%).[27]

(14)

(15)

Anion **2** reacts with primary chloroalkanes to afford 1-phenylthio-1-trimethylsilylalkanes in high yields. Mixed P–S ligands for the copolymerization of olefins and carbon monoxide have been prepared by treating *o*-(diphenylphosphino)benzyl chloride with anion **2** in THF affording the desired ligand in 80% isolated yield (eq 16).[28]

(16)

1. (a) Kocienski, P. J., *Tetrahedron Lett.* **1980**, *21*, 1559. (b) Ager, D. J.; Cookson, R. C., *Tetrahedron Lett.* **1980**, *21*, 1677. (c) Ager, D. J., *J. Chem. Soc., Perkin Trans. 1* **1983**, 1131. (d) Homologs of the title compound are also known: Ager, D. J., *Tetrahedron Lett.* **1983**, *24*, 95.

2. (a) Ager, D. J., *J. Chem. Soc., Perkin Trans. 1* **1986**, 183. (b) Horiguchi, Y.; Furukawa, T.; Kuwajima, I., *J. Am. Chem. Soc.* **1989**, *111*, 8277.

3. (a) Kobayashi, Y.; Ito, T.; Yamakawa, I.; Urabe, H.; Sato, F., *Synlett* **1991**, 813. (b) For another route to vinylsilanes from **1**, see Ogura, F.; Otsubo, T.; Ohira, N., *Synthesis* **1983**, 1006.

4. (a) Gilman, H.; Webb, F. J., *J. Am. Chem. Soc.* **1940**, *62*, 987. (b) Cooper, G. D., *J. Am. Chem. Soc.* **1954**, *76*, 3713. (c) Carey, F. A.; Court, A. S., *J. Org. Chem.* **1972**, *37*, 939.

5. (a) Brook, A. G., *Acc. Chem. Res.* **1974**, *7*, 77. (b) Vedejs, E.; Mullins, M., *Tetrahedron Lett.* **1975**, 2017.

6. (a) Ager, D. J., *J. Chem. Soc., Chem. Commun.* **1984**, 486. (b) Ager, D. J., *J. Chem. Soc., Perkin Trans. 1* **1986**, 195.

7. (a) Ager, D. J., *Tetrahedron Lett.* **1983**, *24*, 419. (b) Paterson, I., *Tetrahedron* **1988**, *44*, 4207.

8. Fleming, I.; Perry, D. A., *Tetrahedron* **1981**, *37*, 4027.

9. Yoshida, J.-i.; Isoe, S., *Chem. Lett.* **1987**, 631.

10. Schaumann, E.; Friese, C., *Tetrahedron Lett.* **1989**, *30*, 7033.

11. Cohen, T.; Sherbine, J. P.; Mendelson, S. A.; Myers, M., *Tetrahedron Lett.* **1985**, *26*, 2965.

12. Ager, D. J.; East, M. B., *J. Org. Chem.* **1986**, *51*, 3983.

13. Prieto, J. A.; Larson, G. L.; Gonzalez, P., *Synth. Commun.* **1989**, *19*, 2773.

14. Agawa, T.; Ishikawa, M.; Komatsu, M.; Ohshiro, Y., *Bull. Chem. Soc. Jpn.* **1982**, *55*, 1205.

15. Ager, D. J., *Tetrahedron Lett.* **1981**, *22*, 2803.

16. (a) Kitteringham, J.; Mitchell, M. B., *Tetrahedron Lett.* **1988**, *29*, 3319. (b) Hosomi, A.; Ogata, K.; Hoashi, K.; Kohra, S.; Tominaga, Y., *Chem. Pharm. Bull.* **1988**, *36*, 3736.

17. Dembech, P.; Guerrini, A.; Ricci, A.; Seconi, G.; Taddei, M., *Tetrahedron* **1990**, *46*, 2999.

18. (a) Yamamoto, I.; Okuda, K.; Nagai, S.; Motoyoshiya, J.; Gotoh, H.; Matsuzaki, K., *J. Chem. Soc., Perkin Trans. 1* **1984**, 435. (b) Ishibashi, H.; Nakatani, H.; Maruyama, K.; Minami, K.; Ikeda, M., *J. Chem. Soc., Chem. Commun.* **1987**, 1443. (c) Ishibashi, H.; Nakatani, H.; Umei, Y.; Yamamoto, W.; Ikeda, M., *J. Chem. Soc., Perkin Trans. 1* **1987**, 589.

19. Larson, G. L.; Argüelles, R.; Rosario, O.; Sandoval, S., *J. Organomet. Chem.* **1980**, *198*, 15.

20. De Groot, A.; Jansen, B. J. M., *Synth. Commun.* **1983**, *13*, 985.

21. Hackett, S.; Livinghouse, T., *J. Org. Chem.* **1986**, *51*, 879.

22. Reich, H. J.; Shah, S. K., *J. Org. Chem.* **1977**, *42*, 1773.

23. Reich, H. J.; Chow, F.; Shah, S. K., *J. Am. Chem. Soc.* **1979**, *101*, 6638.

24. Lange, J.; Schaumann, E., *Eur. J. Org. Chem.* **2009**, 4674.

25. (a) Arnés, X.; Díaz, Y.; Castillón, S., *Synlett* **2003**, 2143. (b) Rodríquez, M. Á.; Boutureira, O.; Arnés, X.; Matheu, M. I.; Díaz, Y.; Castillón, S., *J. Org. Chem.* **2005**, *70*, 10297.

26. (a) Kira, M.; Nakazawa, H., *Chem. Lett.* **1986**, 497. (b) Suga, S.; Shimizu, I.; Ashikari, Y.; Mizuno, Y.; Maruyama, T.; Yoshida, J.-I., *Chem. Lett.* **2008**, *37*, 1008.

27. Wagner, T.; Lange, J.; Grote, D.; Sander, W.; Schaumann, E.; Adiwidjaja, G.; Adam, A.; Kopf, J., *Eur. J. Org. Chem.* **2009**, 5198.

28. Sakakibara, K.; Nozaki, K., *Bull. Chem. Soc. Jpn.* **2009**, *82*, 1006.

(Phenylthio)trimethylsilane

PhSSiMe₃

[4551-15-9]   C₉H₁₄SSi   (MW 182.39)

InChI = 1S/C9H14SSi/c1-11(2,3)10-9-7-5-4-6-8-9/h4-8H,1-3H3

InChIKey = VJMQFIRIMMSSRW-UHFFFAOYSA-N

(selective carbonyl protection;[1,2] preparation of thioesters;[3] cleavage of ethers[4])

Physical Data: bp 93–99 °C/12 mmHg; *d* 0.967 g cm⁻³.

Solubility: sol organic solvents; incompatible with strong acids and strong oxidizing agents.

Preparative Method: commercially available, but can be prepared.[5]

Handling, Storage, and Precautions: moisture sensitive; use in a fume hood.

Original Commentary

William C. Groutas

Wichita State University, KS, USA

Thioacetalization. In the presence of an acid catalyst (zinc iodide, aluminum chloride, anhyd hydrogen chloride, titanium(IV) chloride), (phenylthio)trimethylsilane reacts with aldehydes and ketones at rt or below to form the corresponding thioacetals (eqs 1 and 2).[1–6] With α,β-unsaturated aldehydes and ketones, initiation of the reaction with potassium cyanide–18-crown-6 leads to exclusive formation of the 1,4-addition product (eq 3).[1]

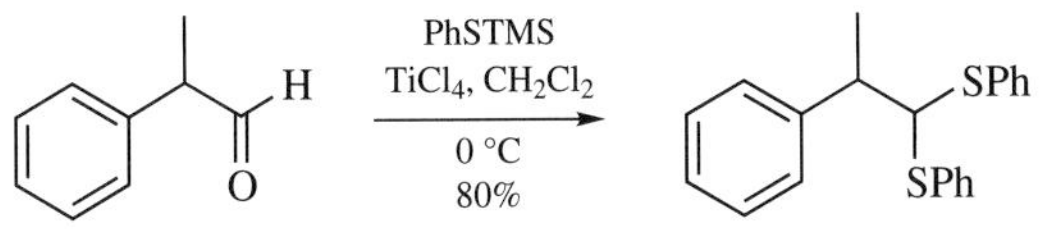

(1)

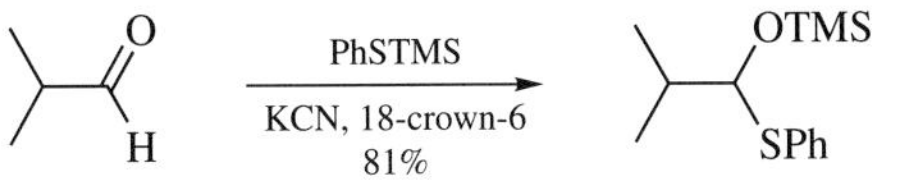

(2)

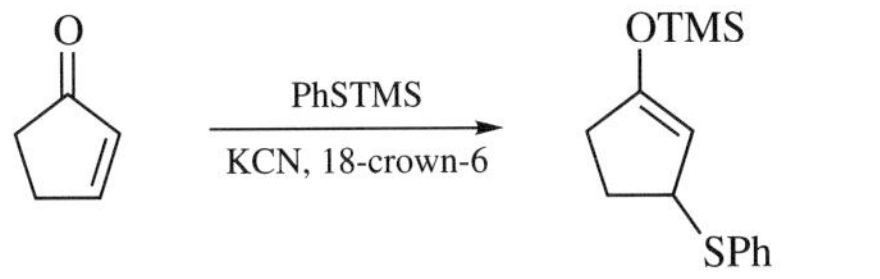

(3)

Cleavage of Ethers. Methyl and benzyl ethers can be readily cleaved under mild conditions (eq 4).[4] The reagent also effects the stepwise and selective dealkylation of phosphotriesters.[7]

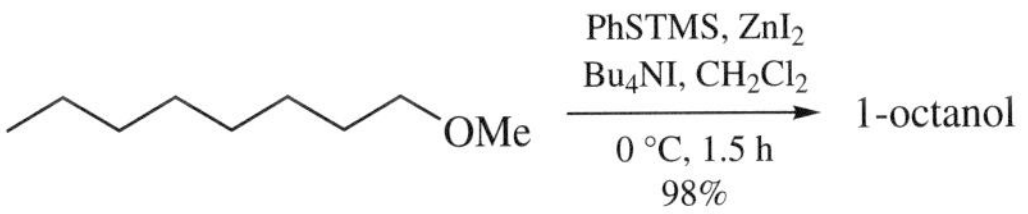

(4)

Miscellaneous Transformations. (Phenylthio)trimethylsilane effects the direct conversion of glycosides into 1-thioglycosides,[8,9] esters into thioesters,[3] epoxides into β-(trimethylsilyloxy) thioethers and β-hydroxy ethers (eq 5),[10] sulfoxides into sulfides,[11] chalcone-derived acetals into α-alkoxy-α,β-unsaturated nitriles,[12] and silyl enol ethers into vinyl sulfides (eq 6).[13] The AlCl$_3$-catalyzed reaction of (phenylthio)trimethylsilane with electrophilic cyclopropanes leads to the formation of ring-opened products (eq 7).[14]

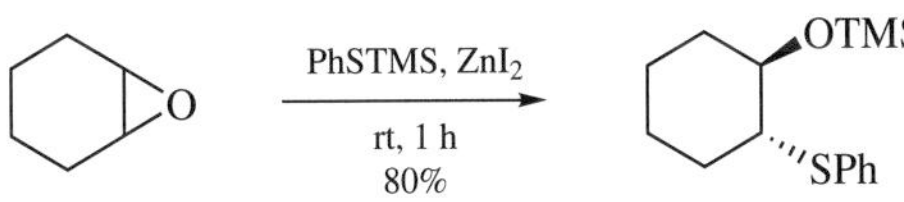

(5)

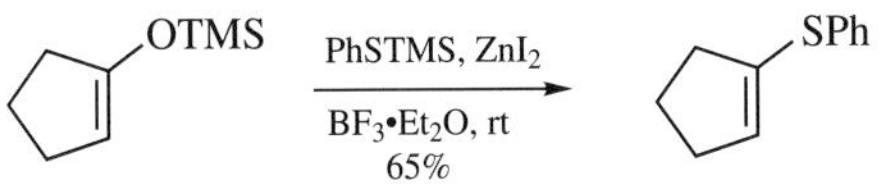

(6)

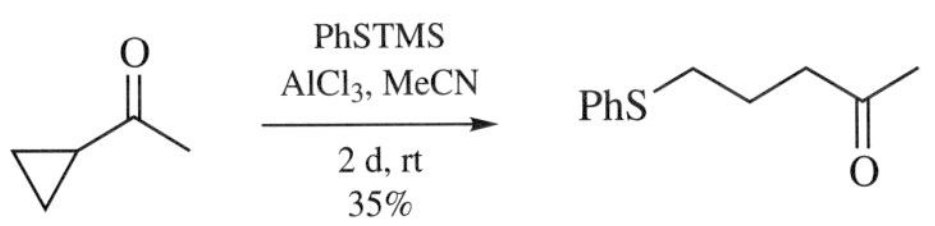

(7)

First Update

Larissa B. Krasnova
The Scripps Research Institute, La Jolla, CA, USA

Andrei K. Yudin
University of Toronto, Toronto, ON, Canada

Hemiacetal Formation and Related Reactions. A chemoselective functionalization of acyclic acetals in the presence of cyclic acetals can be achieved using stoichiometric amounts of TMSOTf as a Lewis acid (eq 8).[15] Contamination with triflic acid, which is a product of TMSOTf hydrolysis, can lead to the formation of undesired by-products. Addition of 2,6-di-*t*-butyl-4-methylpyridine (DTBMP) was shown to be beneficial for the reaction yield. It is believed to trap triflic acid without reducing the Lewis acidity of TMSOTf (eq 9).[16,19]

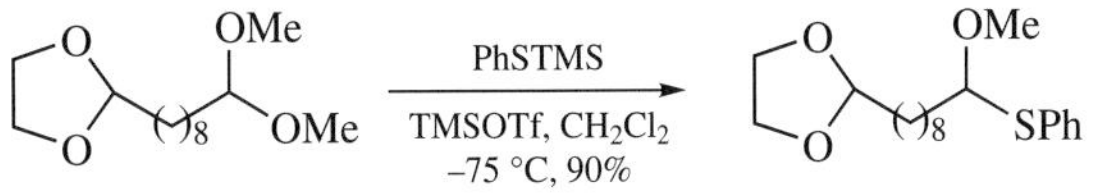

(8)

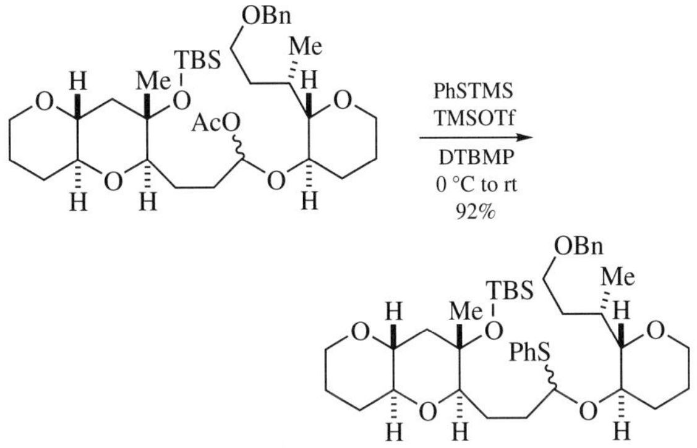

(9)

The preparation of hemithioacetals using (phenylthio)trimethylsilane in the presence of stoichiometric amounts of a Lewis acid has become a routine procedure in carbohydrate synthesis. The resulting thioglucosides are used as coupling reagents in polysaccharide synthesis. The standard reaction conditions for displacement of the anomeric acetoxy group with (phenylthio)trimethylsilane involve TMSOTf,[18] BF$_3$ · OEt$_2$,[19] or SnCl$_4$[20,21] at ambient temperature, and ZnI$_2$,[22–24] which usually requires heating (eq 10). Simultaneous addition of ZnI$_2$ and PhSTMS was found to be important in order to avoid substitution of 6-O-acyl group by thiophenyl nucleophile.[24] Acetal modification using only catalytic amounts of Lewis acid (zinc(II) chloride) has been reported (eq 11).[25]

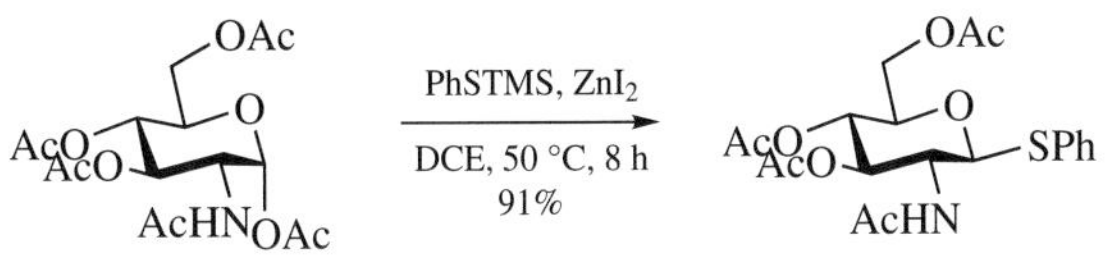

(10)

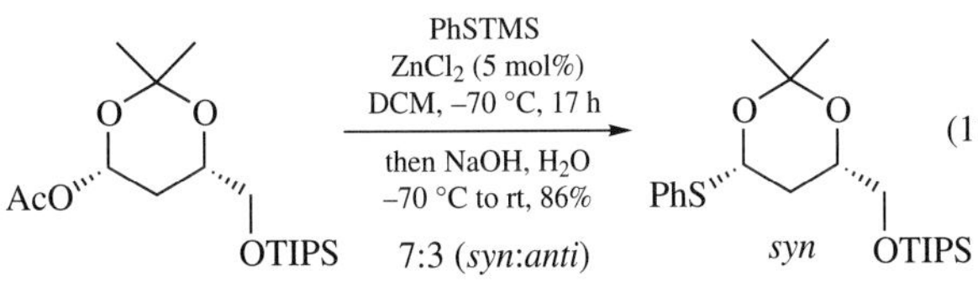

(11)

Displacement of the anomeric benzoyl group by (phenylthio)trimethylsilane proceeds under milder conditions than displacement of the acetoxy group. Zinc(II) iodide induces generation of an oxonium intermediate at room temperature,[26] whereas TMSOTf promotes the displacement at 0 °C.[27,28] Substitution of the anomeric *p*-nitrobenzoyl group with (phenylthio)trimethylsilane occurs in the presence of TMSOTf and gives product within 1 h (eq 12).[28]

$$H_3C \xrightarrow[\text{RO}]{\text{OR}} \xrightarrow[\text{NHCOOAll}]{} \xrightarrow[\substack{\text{DCM, 0 °C} \\ \text{1h, 95\%}}]{\substack{\text{PhSTMS} \\ \text{TfOTMS}}} H_3C \xrightarrow[\text{RO}]{\text{SPh}} \xrightarrow[\text{NHCOOAll}]{} \quad (12)$$

$$R = \text{—C(=O)—C}_6\text{H}_4\text{—NO}_2$$

In cases when reaction proceeds through oxonium[29,30] or iminium ion[31] intermediates with the methoxy as a leaving group, phenylthioacetal installation is generally carried out using PhSTMS/ZnI$_2$/Bu$_4$NI combination of reagents. Phenylthioacetal can be subsequently reduced with lithium di-t-butylbiphenylide (LiDBB) to give an organometallic compound, which in turn can be coupled with various electrophiles (eq 13).[32] Preparation of phenylthioacetal from the galactotrioside via formal displacement of p-methoxyphenoxy group at the α-position has been shown.[33]

$$\text{(13)}$$

1. PhSTMS, ZnI$_2$, Bu$_4$NI, DCE, 85 °C, 70%;
2. LiDBB, THF, −78 °C then

THF, −78 °C, 82%
48:26:14:12 dr

Phenylthioacetals can also be obtained from 1,6-anhydroglucopuranoses using PhSTMS/ZnI$_2$ system (eq 14).[34–39] An intermediate silylated alcohol derivative is not stable in the presence of water and is converted to the corresponding alcohol during K$_2$CO$_3$/MeOH work up[34–36] or upon treatment with TBAF,[37] 70% acetic acid,[38] or TFA.[39]

$$\text{(14)}$$

PhSTMS, ZnI$_2$, 53%   K$_2$CO$_3$, MeOH, 85%

= R

When hemiacetal formation is followed by reduction, sulfides can be obtained from aldehydes in a one-pot manner under mild reaction conditions (eq 15).[40] Vinyl sulfides can be directly prepared from silyl enol ethers upon treatment with PhSTMS in the presence of equimolar amounts of BF$_3$·OEt$_2$ (eqs 16 and 17).[41,42] Ketones under the same reaction conditions give only trace amount of vinyl sulfides.

$$\text{Ph—CHO} \xrightarrow[\substack{\text{Et}_3\text{SiH, CH}_2\text{Cl}_2, \text{ 0 °C to rt} \\ 82\%}]{\text{PhSTMS, TMSCl-InCl}_3 \text{ (cat)}} \text{Ph—CH(SPh)—H} \quad (15)$$

$$\text{(16)}$$

PhSTMS, BF$_3$·OEt$_2$, rt, 58%

$$\text{(17)}$$

PhSTMS, BF$_3$·OEt$_2$, rt, 87%

3/2 (E/Z)

Nucleophilic Displacement. PhTMS-BF$_3$·OEt$_2$ system has been shown to be useful in the transformation of allylic alcohols to allylic sulfides (eq 18).[43] Preparation of unsymmetrical diaryl sulfides can be achieved by reaction of arenediazonium tetrafluoroborates with PhSTMS (eq 19).[44] In some cases, addition of cupric sulfide increases the yield of the diaryl sulfides. The use of (phenylthio)trimethylsilane as a coupling partner in palladium catalyzed reactions with allyl carbonates (eq 20) and aryl iodide (eq 21) has been explored.[45,46]

$$\text{(18)}$$

PhSTMS, BF$_3$·OEt$_2$, CH$_2$Cl$_2$, rt, 92%

$$\text{(19)}$$

PhSTMS, DMF −10 °C to rt then Na$_2$S (aq.) 64%

$$\text{(20)}$$

PhSTMS
Pd$_2$(dba)$_3$CHCl$_3$ (2.5 mol %)
dppp (3 mol %), THF, rt,
(i-PrO)$_3$P (30 mol %), N$_2$
70%

$$\text{Ph-I} \xrightarrow[\text{DMF, 50 °C, K}_3\text{PO}_4, 78\%]{\text{PhSTMS, PdCl}_2\text{(dppf) (3 mol \%)}} \text{Ph-S-Ph} \quad (21)$$

Silicon activation of (phenylthio)trimethylsilane in the presence of TBAF allows epoxide ring opening reactions to proceed without addition of external Lewis acids (eq 22).[47] Epoxide opening in the presence of Bu$_4$NCl has also been documented.[48] Taking advantage of the catalyst employed for the enantioselective epoxidation of α,β-unsaturated amides, one-pot sequential asymmetric epoxidation-regioselective epoxide opening process has been developed (eq 23).[49] After desilylation, the corresponding β-phenylthio-α-hydroxyamides are obtained with high regio- and stereoselectivities. Nucleophilic ring opening of vinyl epoxides with (phenylthio)trimethylsilane in the presence of butyl lithium takes place from the less hindered side, whereas ring opening in the presence of zinc(II) iodide as catalyst gives rearranged product (eq 24).[50] Ring opening of the thiiranium ions generated from 2,3-epoxisulfides proceeds slowly at low temperature to give products as single stereo- and regioisomers (eq 25).[51]

$$\text{(22)}$$

PhSTMS
TBAF (20 mol %)
69%

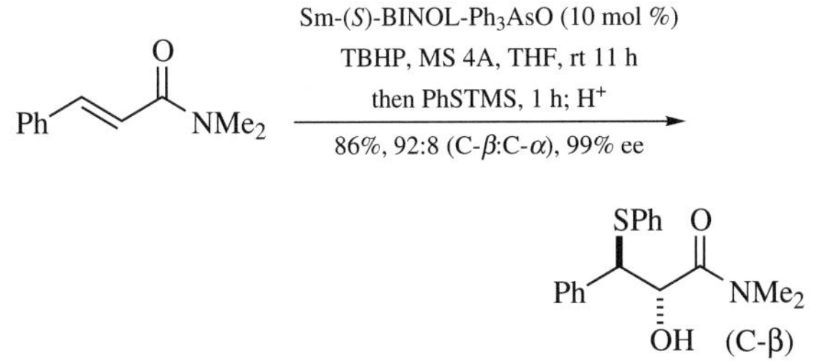

$$
\text{(23)}
$$

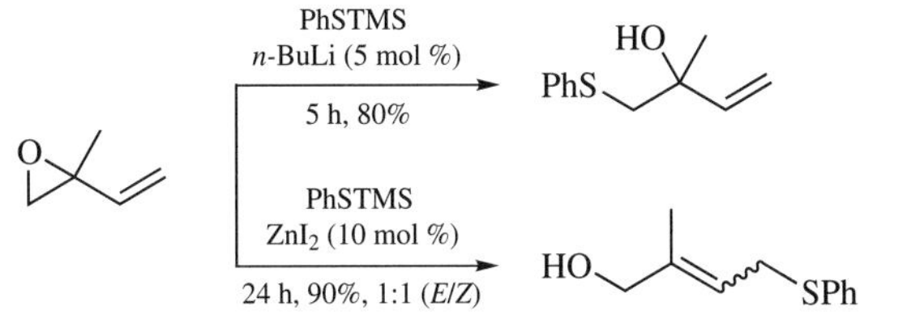

$$
\text{(24)}
$$

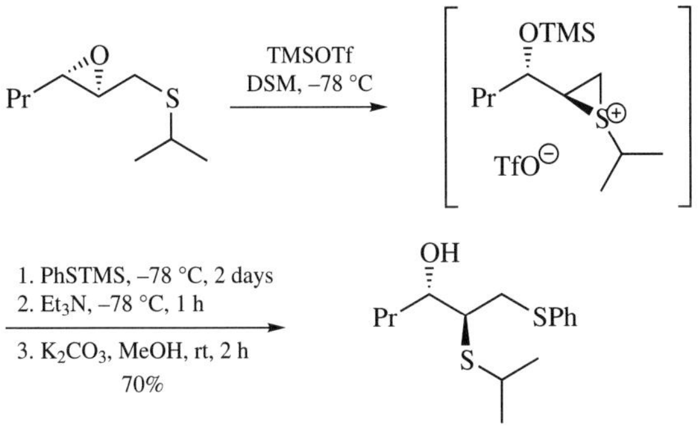

$$
\text{(25)}
$$

Synthesis of *S*-phenyl carbothioates from the corresponding silylated acids can be carried out with PhSTMS under Lewis acid catalysis (eq 26).[52,53] The choice of Lewis acid has direct impact on the chemoselectivity of the reaction.

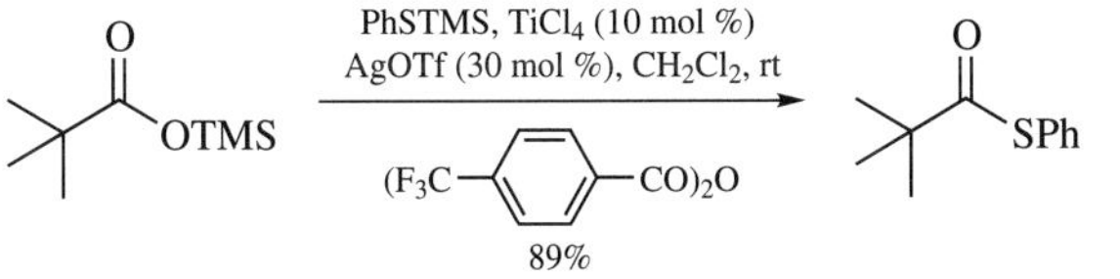

$$
\text{(26)}
$$

(Phenylthio)trimethylsilane has been used for the methyl and benzyl deprotection of alcohols.[54] In the following example, removal of the benzyl group was shown to be problematic under a variety of conditions (hydrogenation, oxidation with DDQ, AcOH, and HCl treatment, reaction with TMSI) (eq 27). However, debenzylation can be performed at ambient temperature using (phenylthio)trimethylsilane in the presence of ZnI$_2$ and Bu$_4$NI to give trimethylsilyl protected alcohol. TMS group is subsequently removed by treatment with TBAF.[55]

Lewis acid promoted ring opening reaction of 2-(2-phenylthio-cyclobutyl)oxiranes and oxetanes proceeds with (phenylthio)trimethylsilane to give allyl and homoallyl alcohols (eq 28).[56,57] Depending on the Lewis acid used, different regiochemistry is observed. Nucleophilic ring opening of the cyclopropane ring with (phenylthio)trimethylsilane has been reported for the cyclopropanated galactal derivative (eq 29).[58,59] Reaction takes place under TMSOTf catalysis and gives oxepane with good diastereo-selectivity.

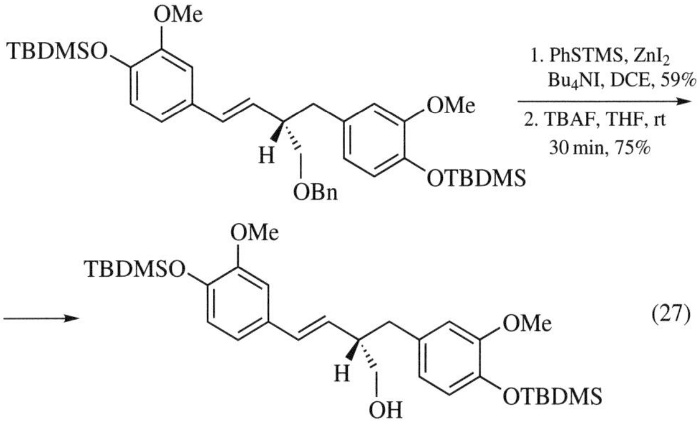

$$
\text{(27)}
$$

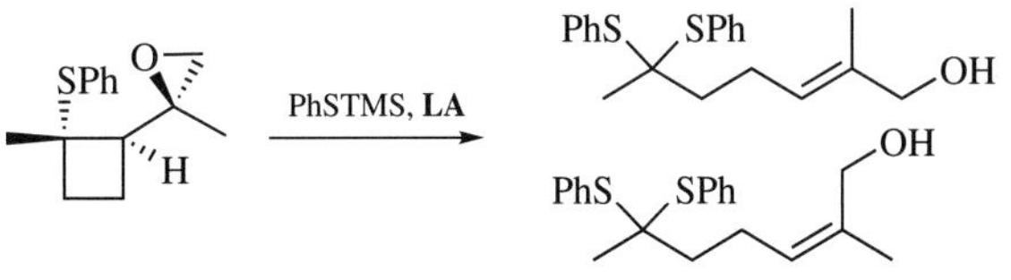

$$
\text{(28)}
$$

lewis acid:	yield, %	E:Z
TiCl$_2$(Oi-Pr)$_2$	77	97:3
EtAlCl$_2$	85	8:92

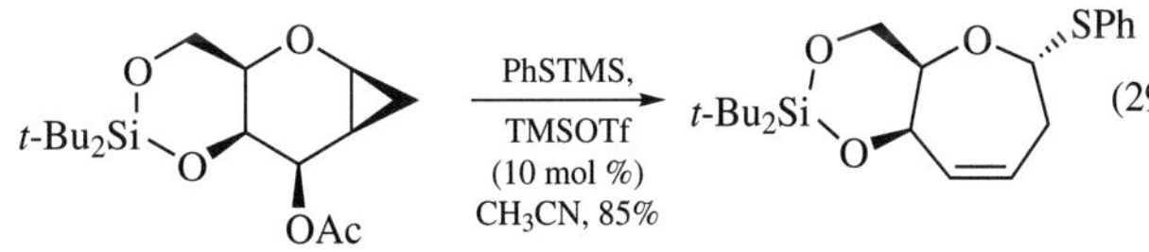

$$
\text{(29)}
$$

Conjugate Addition. Conjugate addition-aldol tandem reaction of α,β-unsaturated esters in the presence of catalytic amount of lithium phenylthiolate proceeds stereoselectively with the formation of alcohol (after protodesilylation). The aldol reaction of the lithium enolate with aldehyde takes place from the bottom face, *anti* to the phenylsulfenyl group in the thermodynamic enolate conformation (eq 30).[60]

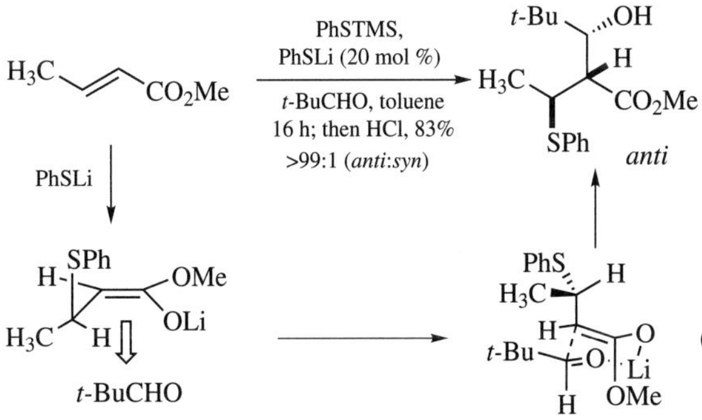

$$
\text{(30)}
$$

1. Evans, D. A.; Truesdale, L. K.; Grimm, K. G.; Nesbitt, S. L., *J. Am. Chem. Soc.* **1977**, *99*, 5009.

2. Kusche, A.; Hoffmann, R.; Munster, I.; Keiner, P.; Bruckner, R., *Tetrahedron Lett.* **1991**, *32*, 467.

3. Mukaiyama, T.; Takeda, T.; Atsumi, K., *Chem. Lett.* **1974**, 187.

4. Hanessian, S.; Guindon, Y., *Tetrahedron Lett.* **1980**, *21*, 2305.

5. Glass, R. M., *J. Organomet. Chem.* **1973**, *61*, 83.

6. Mori, I.; Bartlett, P. A.; Heathcock, C. H., *J. Org. Chem.* **1990**, *55*, 5966.

7. Takeuchi, Y.; Demachi, Y.; Yoshii, E., *Tetrahedron Lett.* **1979**, *20*, 1231.

8. Hanessian, S.; Guindon, Y., *Carbohydr. Res.* **1980**, *86*, C3.

9. Nicolaou, K. C.; Daines, R. A.; Ogawa, Y.; Chakraborty, T. K., *J. Am. Chem. Soc.* **1988**, *110*, 4696.

10. Guindon, Y.; Young, R. N.; Frenette, R., *Synth. Commun.* **1981**, *11*, 391.

11. Numata, T.; Togo, H.; Oae, S., *Chem. Lett.* **1979**, 329.

12. Soga, T.; Takenoshita, H.; Yamada, M.; Han, J. S.; Mukaiyama, T., *Bull. Chem. Soc. Jpn.* **1991**, *64*, 1108.

13. Degl'Innocenti; Ulivi, P.; Capperucci, A.; Mordini, A.; Reginato, G.; Ricci, A., *Synlett* **1992**, 499.

14. Dieter, R. K.; Pounds, S., *J. Org. Chem.* **1982**, *47*, 3174.

15. Kim, S.; Do, J. Y.; Kim, S. H.; Kim, D.-I., *J. Chem. Soc., Perkin Trans. 1* **1994**, 2357.

16. Sato, K.; Sasaki, M., *Org. Lett.* **2005**, *7*, 2441.

17. Inoue, M.; Miyazaki, K.; Uehara, H.; Maruyama, M.; Hirama, M., *Proc. Natl. Acad. Sci. USA* **2004**, *101*, 12013.

18. Jain, R. K.; Piskorz, C. F.; Matta, K. L., *Carbohydr. Res.* **1993**, *243*, 385.

19. Damager, I.; Olsen, C. E.; Moller, B. L.; Motawia, M. S., *Synthesis* **2002**, 418.

20. Johnston, J. N.; Paquette, L. A., *Tetrahedron Lett.* **1995**, *36*, 4341.

21. Paquette, L. A.; Barriault, L.; Pissarnitski, D.; Johnston, J. N., *J. Am. Chem. Soc.* **2000**, *122*, 619.

22. Wakao, M.; Fukase, K.; Kusumoto, S., *J. Org. Chem.* **2002**, *67*, 8182.

23. Vourloumis, D.; Winters, G. C.; Takahashi, M.; Simonsen, K. B.; Ayida, B. K.; Shandrick, S.; Zhao, Q.; Hermann, T., *ChemBioChem* **2003**, *4*, 879.

24. Buskas, T.; Garegg, P. J.; Konradsson, P.; Maloisel, J.-L., *Tetrahedron: Asymmetry* **1994**, *5*, 2187.

25. Cooksey, J.; Gunn, A.; Kocienski, P. J.; Kuhl, A.; Uppal, S.; Christopher, J. A.; Bell, R., *Org. Biomol. Chem.* **2004**, *2*, 1719.

26. Egusa, K.; Kusumoto, S.; Fukase, K., *Eur. J. Org. Chem.* **2003**, 3435.

27. Zhang, P.; Appleton, J.; Ling, C.-C.; Bundle, D. R., *Can. J. Chem.* **2002**, *80*, 1141.

28. Cipollone, A.; Berettoni, M.; Bigioni, M.; Binaschi, M.; Cermele, C.; Monteagudo, E.; Olivieri, L.; Palomba, D.; Animati, F.; Goso, C.; Maggi, C. A., *Bioorg. Med. Chem.* **2002**, *10*, 1459.

29. Dunkel, R.; Treu, J.; Martin, H.; Hoffmann, R., *Tetrahedron: Asymmetry* **1999**, *10*, 1539.

30. Vera-Ayoso, Y.; Borrachero, P.; Cabrera-Escribano, F.; Dianez, M. J.; Estrada, M. D.; Gomez-Guillen, M.; Lopez-Castro, A.; Perez-Garrido, S., *Tetrahedron: Asymmetry* **2001**, *12*, 2031.

31. Roeper, S.; Wartchow, R.; Hoffmann, H. M., *R. Org. Lett.* **2002**, *4*, 3179.

32. De Vicente, J.; Betzemeier, B.; Rychnovsky, S. D., *Org. Lett.* **2005**, *7*, 1853.

33. Oberthur, M.; Peters, S.; Kumar Das, S.; Lichtenthaler, F. W., *Carbohydr. Res.* **2002**, *337*, 2171.

34. Zhu, Y.-H.; Vogel, P., *Synlett* **2001**, 79.

35. Steunenberg, P.; Jeanneret, V.; Zhu, Y.-H.; Vogel, P., *Tetrahedron: Asymmetry* **2005**, *16*, 337.

36. Takahashi, S.; Kuzuhara, H.; Nakajima, M., *Tetrahedron* **2001**, *57*, 6915.

37. Arndt, S.; Hsieh-Wilson, L. C., *Org. Lett.* **2003**, *5*, 4179.

38. Lindberg, J.; Ohberg, L.; Garegg, P. J.; Konradsson, P., *Tetrahedron* **2002**, *58*, 1387.

39. Demchenko, A. V.; Wolfert, M. A.; Santhanam, B.; Moore, J. N.; Boons, G.-J., *J. Am. Chem. Soc.* **2003**, *125*, 6103.

40. Mukaiyama, T.; Ohno, T.; Nishimura, T.; Han, J. S.; Kobayashi, S., *Bull. Chem. Soc. Jpn.* **1991**, *64*, 2524.

41. Degl'Innocenti, A.; Ulivi, P.; Capperucci, A.; Mordini, A.; Reginato, G.; Ricci, A., *Synlett* **1992**, 499.

42. Degl'Innocenti, A.; Capperucci, A., *Eur. J. Org. Chem.* **2000**, 2171.

43. Tsay, S. C.; Lin, L. C.; Furth, P. A.; Shum, C. C.; King, D. B.; Yu, S. F.; Chen, B. L.; Hwu, J. R., *Synthesis* **1993**, 329.

44. Prakash, G. K. S.; Hoole, D.; Ha, D. S.; Wilkinson, J.; Olah, G. A., *ARKIVOC* **2002**, 50.

45. Trost, B. M.; Scanlan, T. S., *Tetrahedron Lett.* **1986**, *27*, 4141.

46. Ishiyama, T.; Mori, M.; Suzuki, A.; Miyaura, N., *J. Organomet. Chem.* **1996**, *525*, 225.

47. Tanabe, Y.; Mori, K.; Yoshida, Y., *J. Chem. Soc., Perkin Trans. 1* **1997**, 671.

48. Schneider, C., *Synlett* **2000**, 1840.

49. Tosaki, S.-Y.; Tsuji, R.; Ohshima, T.; Shibasaki, M., *J. Am. Chem. Soc.* **2005**, *127*, 2147.

50. Martin, G.; Sauleau, J.; David, M.; Sauleau, A.; Sinbandhit, S., *Can. J. Chem.* **1992**, *70*, 2190.

51. Forristal, I.; Lawson, K. R.; Rayner, C. M., *Tetrahedron Lett.* **1999**, *40*, 7015.

52. Mukaiyama, T.; Miyashita, M.; Shiina, I., *Chem. Lett.* **1992**, 1747.

53. Miyashita, M.; Shiina, I.; Miyoshi, S.; Mukaiyama, T., *Bull. Chem. Soc. Jpn.* **1993**, *66*, 1516.

54. Suzuki, T.; Matsumura, R.; Oku, K. I.; Taguchi, K.; Hagiwara, H.; Hoshi, T.; Ando, M., *Tetrahedron Lett.* **2001**, *42*, 65.

55. Shattuck, J. C.; Shreve, C. M.; Solomon, S. E., *Org. Lett.* **2001**, *3*, 3021.

56. Fujiwara, T.; Tsuruta, Y.; Takeda, T., *Tetrahedron Lett.* **1995**, *36*, 8435.

57. Fujiwara, T.; Sawabe, K.; Takeda, T., *Tetrahedron* **1997**, *53*, 8349.

58. Batchelor, R.; Hoberg, J. O., *Tetrahedron Lett.* **2003**, *44*, 9043.

59. Hoberg, J. O., *J. Org. Chem.* **1997**, *62*, 6615.

60. Ono, M.; Nishimura, K.; Nagaoka, Y.; Tomioka, K., *Tetrahedron Lett.* **1999**, *40*, 1509.

1-(Phenylthio)-1-(trimethylsilyl)-cyclopropane[1]

[74379-74-1] $C_{12}H_{18}SSi$ (MW 222.46)

InChI = 1S/C12H18SSi/c1-14(2,3)12(9-10-12)13-11-7-5-4-6-8-11/h4-8H,9-10H2,1-3H3

InChIKey = DEIZBLNWJYTTLE-UHFFFAOYSA-N

(Peterson alkenation;[1] cyclobutane synthesis[2] and cyclopentane[3] annulation via 1-lithio-1-(trimethylsilyl)cyclopropane)

Physical Data: bp 64 °C/0.1 mmHg; *d* 0.991 g cm^{-3}.

Form Supplied in: liquid; commercially available.

Analysis of Reagent Purity: FT-IR;[1c] also by NMR, IR, and mass spectrometry.[1b]

Preparative Methods: by reductive lithiation of 1,1-bis(phenyl-thio)cyclopropane followed by trimethylsilylation with chlorotrimethylsilane.[4,5]

Purification: by distillation.[1b]

Handling, Storage, and Precautions: use in a fume hood.

Peterson Alkenation. 1-Lithio-1-(trimethylsilyl)cyclopropane, derived from 1-phenylthio-1-(trimethylsilyl)cyclopropane by reductive lithiation with lithium 1-(dimethylamino)naphthalenide (LDMAN), condenses with aldehydes to give carbinols which are converted to alkylidenecyclopropanes under Peterson alkenation conditions (eq 1).[1]

Synthesis of Cyclobutanes. *m*-Chloroperbenzoic acid oxidation of 1-phenylthio-1-(trimethylsilyl)cyclopropane furnishes a sulfoxide which could undergo thermal sila-Pummerer rearrangement to give the masked cyclopropanone 1-(phenylthio)-1-trimethylsiloxy)cyclopropane (eq 2).[2a]

$$\text{(1)}$$

$$\text{(2)}$$

Alternative treatment of the sulfoxide with aldehydes in the presence of tetra-*n*-butylammonium fluoride gives the phenylsulfinylcarbinol and thence the phenylthiocarbinol following reduction. The two last-named compounds are intermediates in the synthesis of cyclobutanones and cyclobutenes, respectively (eqs 3 and 4).[2b,2c,2e]

$$\text{(3)}$$

$$\text{(4)}$$

Cyclopentane Annulation. Condensation of 1-lithio-1-(trimethylsilyl)cyclopropane with ketones gives vinylcyclopropane intermediates, which upon pyrolysis undergo vinylcyclopropane rearrangement to give annulated cyclopentene derivatives (eq 5).[3]

$$\text{(5)}$$

1. (a) Cohen, T.; Jung, S.-H.; Romberger, M. L.; McCullough, D. W., *Tetrahedron Lett.* **1988**, *29*, 25. (b) Cohen, T.; Sherbine, J. P.; Matz, J. R.; Hutchins, R. R.; McHenry, B. M.; Willey, P. R., *J. Am. Chem. Soc.* **1984**, *106*, 3245. (c) Pouchert, C. J. *Aldrich Library of FT-IR Spectra*; Aldrich: Milwaukee, WI, 1989; Vol 1, p 1669D.

2. (a) Bhupathy, M.; Cohen, T., *Tetrahedron Lett.* **1987** *41*, 4793. (b) Pohmakotr, M.; Sithikanchanakul, S., *Synth. Commun.* **1989**, *19*, 477.

(c) Hiroi, K.; Nakamura, H.; Anzai, T., *J. Am. Chem. Soc.* **1987**, *109*, 1249.
(d) Trost, B. M.; Keeley, D. E.; Arndt, H. C.; Rigby, J. H.; Bogdanowicz, M. J., *J. Am. Chem. Soc.* **1977**, *99*, 3080. (e) Trost, B. M.; Keeley, D. E.; Arndt, H. C.; Bogdanowicz, M. J., *J. Am. Chem. Soc.* **1977**, *99*, 3088.

3. Paquette, L. A.; Wells, G. J.; Horn, K. A.; Yan, T.-H., *Tetrahedron Lett.* **1982**, *23*, 263.

4. Cohen, T.; Matz, J. R., *Synth. Commun.* **1980**, *10*, 311.

5. Paquette, L. A.; Horn, K. A.; Wells, G. J., *Tetrahedron Lett.* **1982**, *23*, 259.

Uko E. Udodong
Eli Lilly Company, Indianapolis, IN, USA

1-Phenylthio-1-trimethylsilylethylene

[62762-20-3] $C_{11}H_{16}SSi$ (MW 208.43)

InChI = 1S/C11H16SSi/c1-10(13(2,3)4)12-11-8-6-5-7-9-11/h5-9H,1H2,2-4H3

InChIKey = VVPNCTFIKRGYRE-UHFFFAOYSA-N

(a reagent for C_2 homologation and a Michael acceptor)

Physical Data: colorless liquid; bp 130–134 °C/15 mmHg;[1] IR,[1,2] ^{1}H NMR,[1] and MS data[1] are available.

Solubility: very sol most organic solvents.

Preparative Methods: by lithiation of phenyl vinyl sulfide with lithium diisopropylamide, followed by silylation,[1,2] or the base-induced elimination of 2-chloro-1-trimethylsilyl-1-phenylthioethane,[2] or the lithiation of ethyl phenyl sulfoxide with LDA followed by silylation,[3] or the Pd-catalyzed coupling reactions of trimethylstannyl phenyl sulfide with 1-bromo-1-trimethylsilylethylene.[4]

Handling, Storage, and Precautions: use in a fume hood.

Addition of Organolithiums. Organolithiums undergo smooth additions to 1-phenylthio-1-trimethylsilylethylene at low temperature. The resulting adduct anions are stabilized by two anion-stabilizing substituents, phenylthio and trimethylsilyl. They can be quenched with a variety of electrophiles. Sulfines are formed by treatment with sulfur dioxide (eq 1),[5] phenyl 1-silylalkyl sulfides by hydrolytic quenching with NH_4Cl,[1] alkanals by oxidation with *m*-chloroperbenzoic acid and subsequent rearrangement reaction (eq 2),[6,7] phenyl silylalkyl sulfides by alkylation with a variety of alkylating agents,[8] and vinyl sulfides by condensation with aldehydes.[8]

$$\text{(1)}$$

$$\text{(2)}$$

Cyclopentenone Synthesis. When activated with Lewis acid catalysts or silver ion, 1-phenylthio-1-trimethylsilylethylene reacts with α,β-unsaturated acid chlorides at a low temperature in CH_2Cl_2.[2,9] The cation intermediates undergo a silicon-directed Nazarov cyclization to give 3-phenylthio-substituted cyclopentenone derivatives (eq 3). The phenylthio moiety at the β-position of cyclopentenones can be effectively utilized for the introduction of an alkyl substituent by addition/elimination.

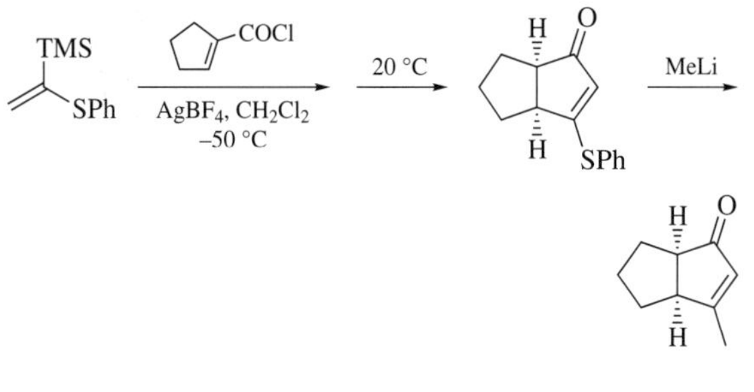

(3)

Cyclopropanation. Although the adduct anions formed by addition of organolithiums to 1-phenylthio-1-trimethylsilylethylene have high stability at low temperature, at higher temperatures (0 °C) in THF the anions may react with another molecule of 1-phenylthio-1-trimethylsilylethylene. The final anions undergo intramolecular alkylation under the reaction conditions to give cyclopropane derivatives in high yields (eq 4).[10] Employment of sulfur-stabilized anions as donor molecules is even more effective (eq 5).[11]

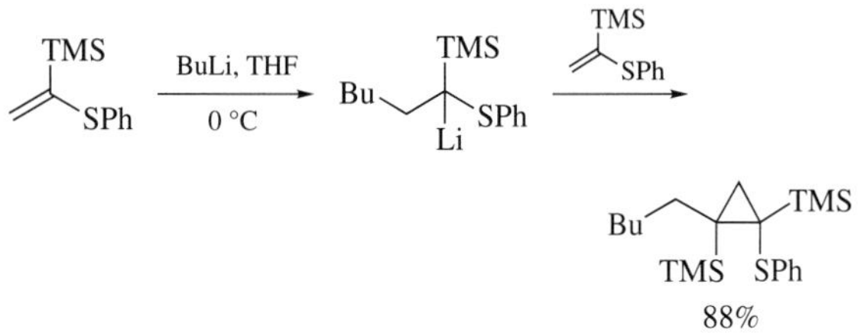

(4)

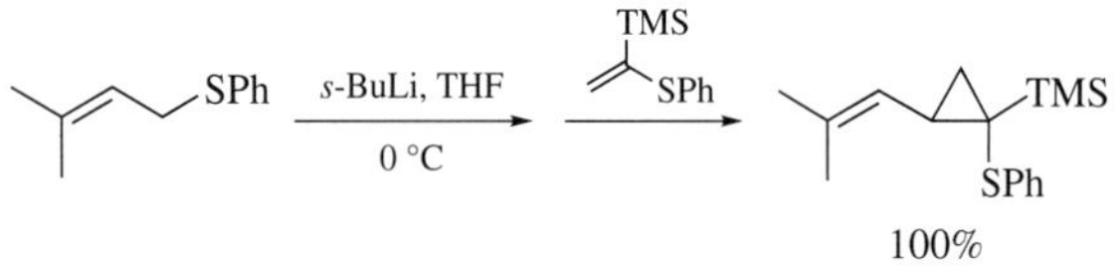

88%

(5)

100%

Cyclohexanone Annulation. The benzyllithium intermediates formed by metalation of *o*-toluamide derivatives with LDA at −78 °C react with 1-trimethylsilyl-1-phenylthioethylene. The resulting adduct anions undergo intramolecular cyclization on to the adjacent amide moiety to provide a convenient route for cyclohexanone annulation (eq 6).[12]

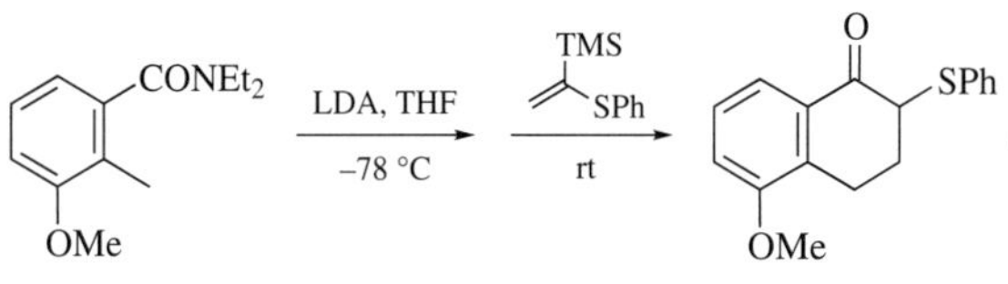

(6)

Analogous Reagents. 1-Phenylsulfinyl-1-trimethylsilylethylene (1)[2] and 1-phenylsulfonyl-1-trimethylsilylethylene (2) are used as synthetic analogs of 1-phenylthio-1-trimethylsilylethylene. The sulfoxide (1) is an electron-deficient alkene so it

is a highly reactive dienophile in Diels–Alder reactions. The cycloadduct with cyclopentadiene undergoes the sila-Pummerer rearrangement to a thioacetal, which is then hydrolyzed to the ketone (eq 7).[6] By this sequence, (1) becomes a useful synthetic equivalent of ketene. The sulfone (2) works as an excellent acceptor molecule for the Michael addition reactions using the aryllithium donors generated by the directed lithiation of *o*-chloroarenes. The adducts then undergo intramolecular cyclizations by the aid of metal amides to offer an excellent entry to benzocyclobutene derivatives.[13]

(7)

1. Ager, D. J., *J. Chem. Soc., Perkin Trans. 1* **1983**, 1131.

2. Cooke, F.; Moerck, R.; Schwindeman, J.; Magnus, P., *J. Org. Chem.* **1980**, *45*, 1046.

3. Miller, R. D.; Hässig, R., *Tetrahedron Lett.* **1984**, *25*, 5351.

4. Carpita, A.; Rossi, R.; Scamuzzi, B., *Tetrahedron Lett.* **1989**, *30*, 2699.

5. Van der Leij, M.; Zwanenburg, B., *Tetrahedron Lett.* **1978**, 3383.

6. Williams, R. V.; Lin, X., *J. Chem. Soc., Chem. Commun.* **1989**, 1872.

7. Ager, D. J., *Tetrahedron Lett.* **1981**, *22*, 587.

8. Ager, D. J., *J. Chem. Soc., Perkin Trans. 1* **1986**, 183.

9. Magnus, P.; Quagliato, D., *J. Org. Chem.* **1985**, *50*, 1621.

10. Kanemasa, S.; Kobayashi, H.; Tanaka, J.; Tsuge, O., *Bull. Chem. Soc. Jpn.* **1988**, *61*, 3957.

11. Cohen, T.; Sherbine, J. P.; Mendelson, S. A.; Myers, M., *Tetrahedron Lett.* **1985**, *26*, 2965.

12. Date, M.; Watanabe, M.; Furukawa, S., *Chem. Pharm. Bull.* **1990**, *38*, 902.

13. Iwao, M., *J. Org. Chem.* **1990**, *55*, 3622.

Shuji Kanemasa
Kyushu University, Kasuga, Japan

2-(Phenylthio)-2-(trimethylsilyl)propane[1]

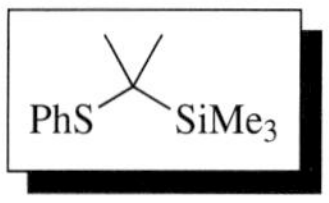

[89656-96-2] $C_{12}H_{20}SSi$ (MW 224.48)

InChI = 1S/C12H20SSi/c1-12(2,14(3,4)5)13-11-9-7-6-8-10-11/h6-10H,1-5H3

InChIKey = ZMDLIPFQILDDDR-UHFFFAOYSA-N

(Peterson alkenation[1] and preparation of ketones via 2-lithio-2-(trimethylsilyl)propane[2,3])

Physical Data: bp 70–80 °C/0.6 mmHg.

Preparative Method: by reductive lithiation of 2,2-bis(phenyl-thio)propane followed by trimethylsilylation with chlorotrimethylsilane.[2,3]

Purification: by flash chromatography and by Kugelrohr distillation.[3]

Handling, Storage, and Precautions: use in a fume hood.

Synthesis of Dienes via Peterson Alkenation. Reductive lithiation of 2-(phenylthio)-2-(trimethylsilyl)propane with lithium 1-(dimethylamino)naphthalenide (LDMAN) furnishes 2-lithio-2-(trimethylsilyl)propane, which is an intermediate for Peterson alkenation. The specific example in eq 1 features silicon-directed diene synthesis.[1]

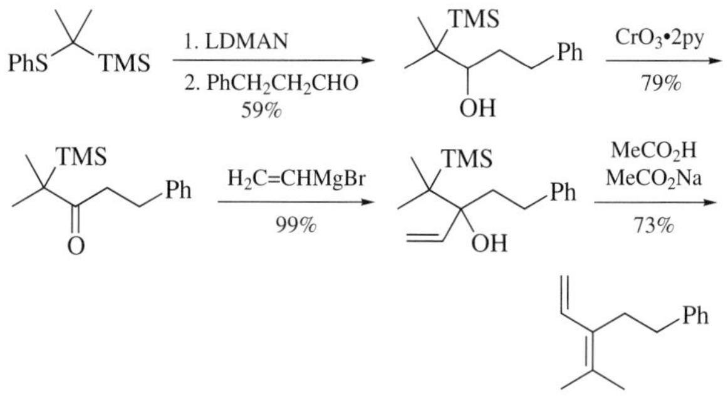

Synthesis of Ketones. *m*-chloroperbenzoic acid oxidizes 2-(phenylthio)-2-(trimethylsilyl)propane to the sulfone, which is then converted to the ketone via sila-Pummerer rearrangement (eq 2).[3]

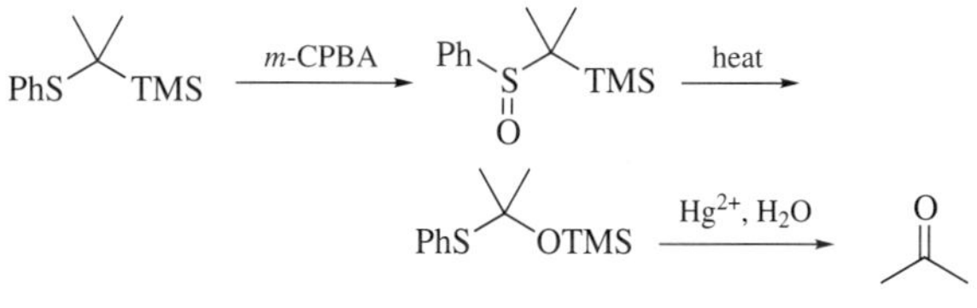

1. Brown, P. A.; Bonnert, R. V.; Jenkins, P. R.; Selim, M. R., *Tetrahedron Lett.* **1987**, *28*, 693.
2. Cohen, T.; Sherbine, J. P.; Matz, J. R.; Hutchins, R. R.; McHenry, B. M.; Willey, P. R., *J. Am. Chem. Soc.* **1984**, *106*, 3245.
3. Ager, D. J., *J. Chem. Soc., Perkin Trans. 1* **1986**, *183*, 195.

Uko E. Udodong

Eli Lilly Company, Indianapolis, IN, USA

Phenyl Trimethylsilyl Selenide

PhSeSiMe$_3$

[33861-17-5]   C$_9$H$_{14}$SeSi   (MW 229.28)

InChI = 1S/C9H14SeSi/c1-11(2,3)10-9-7-5-4-6-8-9/h4-8H, 1-3H3

InChIKey = MRRIPSZDPMMQEO-UHFFFAOYSA-N

(synthesis of benzeneselenol;[1a] introduction of the phenylseleno function into a variety of oxygen-containing compounds[2–6])

Physical Data: bp 92–93 °C/5 mmHg. ^{1}H NMR (CDCl$_3$) δ 0.35 (Me$_3$Si); ^{13}C NMR (CDCl$_3$) δ 1.49 (Me$_3$Si).

Solubility: sol benzene, Et$_2$O, THF, MeCN.

Preparative Methods: conveniently prepared by the reduction of diphenyl diselenide with sodium in THF, followed by silylation of thus formed PhSeNa with chlorotrimethylsilane.[1a,b] In a small scale reaction, silylation of PhSeLi, which can be prepared in situ from metallic selenium and phenyllithium in THF, is alternatively utilized.[1c]

Purification: distillation under reduced pressure.

Handling, Storage, and Precautions: although phenyl trimethylsilyl selenide isolated as a colorless liquid is somewhat sensitive to atmospheric moisture, it can be handled conveniently using syringe techniques and can be stored indefinitely under dry nitrogen (or argon). Use in a fume hood.

Reaction with Aldehydes and Ketones. Phenyl trimethylsilyl selenide does not react with carbonyl compounds at 25 °C or on heating. However, use of aluminum chloride as catalyst results in selective formation of selenoacetals, whereas the copresence of zinc halide in place of AlCl$_3$ leads to *O*-(trimethylsilyl) monoselenoacetals,[2a–d] which can be utilized as α-siloxy radical precursors (eq 1).[2e]

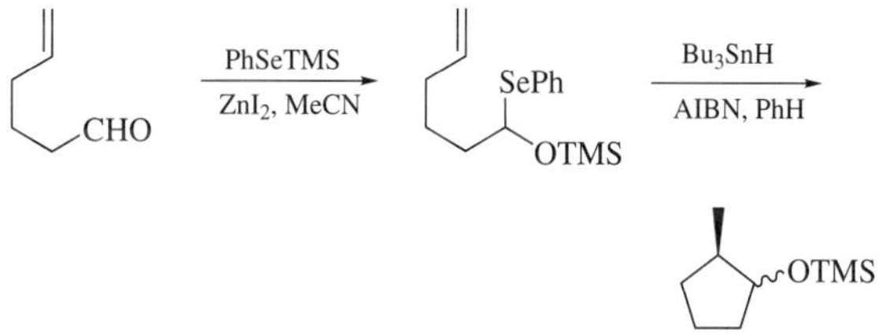

Reaction with Acetates and Lactones. In the presence of a catalytic amount of zinc iodide, alkyl acetates and lactones react with phenyl trimethylsilyl selenide to afford alkyl phenyl selenides and ω-(phenylseleno)carboxylic acids, respectively (eq 2).[3]

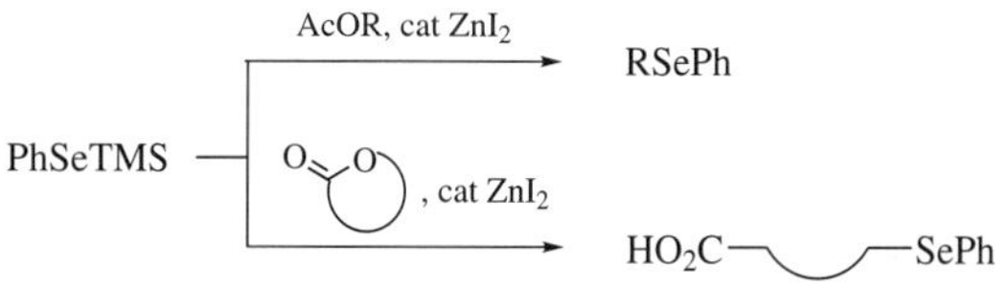

Reaction with α,β-Unsaturated Carbonyl Compounds. Conjugate addition of phenyl trimethylsilyl selenide to α,β-unsaturated aldehydes and ketones has been attained by using triphenylphosphine,[2c] zinc chloride,[2c] or trimethylsilyl trifluoromethanesulfonate[4] as the catalyst. Combination of this reaction with selenoxide elimination[5] provides a one-pot procedure for α-alkoxyalkylation of α,β-unsaturated ketones (eq 3).

Reaction with Cyclic Ethers. Ring opening of epoxides, oxetanes, and tetrahydrofurans with phenyl trimethylsilyl selenide takes place in the presence of ZnI_2 to provide β-, γ-, and δ-siloxyalkyl phenyl selenides, respectively, in good yields (eq 4).[6]

Miscellaneous. Introduction of a phenylseleno function into organic molecules has also been attained by the reaction of phenyl trimethylsilyl selenide with organic halides.[7] Palladium-catalyzed addition of $PhSeSiMe_3$ to arylacetylenes takes place regio- and stereoselectively to give β-silyl substituted vinylic selenides (eq 5).[8] $PhSeSiMe_3$ serves as a mild reducing agent for deoxygenation of sulfoxides, selenoxides, and telluroxides.[9]

Phenyl Trimethylsilyl Telluride. $PhTeSiMe_3$ is synthesized by the reaction of Me_3SiCl with $PhTeLi$,[10a] $PhTeNa$,[10b] or $PhTeMgBr$,[10c] and is purified by distillation (bp 77–79 °C/2 mmHg). It can be stored as a 2.5 M hexane solution in a Schlenk flask. Like $PhSeSiMe_3$, $PhTeSiMe_3$ reacts with cyclic ethers and lactones to give ring-opened products.[10b] Upon treatment with MeOH, benzenetellurol is formed in situ.[10d]

Related Reagents. Benzeneselenol; Bis(trimethylsilyl) Selenide.

---

1. (a) Miyoshi, N.; Ishii, H.; Kondo, K.; Murai, S.; Sonoda, N., *Synthesis* **1979**, 300. (b) Detty, M. R.; Seidler, M. D., *J. Org. Chem.* **1981**, *46*, 1283. (c) Drake, J. E.; Hemmings, R. T., *J. Chem. Soc., Dalton Trans.* **1976**, 1730.

2. (a) Dumont, W.; Krief, A., *Angew. Chem., Int. Ed. Engl.* **1977**, *16*, 540. (b) Clarembeau, M.; Cravador, A.; Dumont, W.; Hevesi, L.; Krief, A.; Lucchetti, J.; Van Ende, D., *Tetrahedron* **1985**, *41*, 4793. (c) Liotta, D.; Paty, P. B.; Johnston, J.; Zima, G., *Tetrahedron Lett.* **1978**, 5091. (d) Detty, M. R., *Tetrahedron Lett.* **1979**, 4189. (e) Keck, G. E.; Tafesh, A. M., *Synlett* **1990**, 257.

3. Miyoshi, N.; Ishii, H.; Murai, S.; Sonoda, N., *Chem. Lett.* **1979**, 873.

4. Suzuki, M.; Kawagishi, T.; Noyori, R., *Tetrahedron Lett.* **1981**, *22*, 1809.

5. Reich, H. J. *Oxidation in Organic Chemistry*; Trahanovsky, W., Ed.; Academic: New York, 1978; Part C, p 1.

6. (a) Miyoshi, N.; Kondo, K.; Murai, S.; Sonoda, N., *Chem. Lett.* **1979**, 909. (b) Miyoshi, N.; Hatayama, Y.; Ryu, I.; Kambe, N.; Murai, T.; Murai, S.; Sonoda, N., *Synthesis* **1988**, 175.

7. (a) Paquette, L. A.; Lagerwall, D. R.; Korth, H.-G., *J. Org. Chem.* **1992**, *57*, 5413. (b) Kato, S.; Yasui, E.; Terashima, K.; Ishihara, H.; Murai, T., *Bull. Chem. Soc. Jpn.* **1988**, *61*, 3931.

8. Ogawa, A.; Sonoda, N., *J. Synth. Org. Chem. Jpn.* **1993**, *51*, 815.

9. Detty, M. R., *J. Org. Chem.* **1979**, *44*, 4528.

10. (a) Drake, J. E.; Hemmings, R. T., *Inorg. Chem.* **1980**, *19*, 1879. (b) Sasaki, K.; Aso, Y.; Otsubo, T.; Ogura, F., *Tetrahedron Lett.* **1985**, *26*, 453. (c) Jones, C. H. W.; Sharma, R. D., *J. Organomet. Chem.* **1984**, *268*, 113. (d) Ohira, N.; Aso, Y.; Otsubo, T.; Ogura, F., *Chem. Lett.* **1984**, 853.

Akiya Ogawa
Osaka University, Osaka, Japan

Phosphine, Tris(trimethylsilyl)-

[15573-38-3] $C_9H_{27}PSi_3$ (MW 250.54)
InChI = 1S/C9H27PSi3/c1-11(2,3)10(12(4,5)6)13(7,8)9/h1-9H3
InChIKey = OUMZKMRZMVDEOF-UHFFFAOYSA-N

(user-friendly phosphorus source and alternative to phosphine gas, precursor of $(Me_3Si)_2PLi$, covalent synthon for the anion P^{3-})

Alternate Names: tris(trimethylsilyl)phosphane; $P(TMS)_3$.
Physical Data: mp 24 °C; bp 243–244 °C; d 0.863 g cm^{-3}; n_D^{20} 1.501–1.503.
Solubility: pentane, hexanes, methylene chloride, benzene, toluene, and acetonitrile.
Form Supplied in: clear colorless to yellow liquid (purity 95–98%); 10 wt% solution in hexanes; commercially available.
Preparative Methods: several preparative methods are known. These include the following: reaction of alkali metal phosphides ($NaPH_2$,[1] KPH_2,[2] Li_3P,[3] usually prepared by reaction of metal and phosphine gas or via metal alkyl derivative[4]) with chlorotrimethylsilane or fluorotrimethylsilane[2] in 1,2-dimethoxyethane or diethyl ether; reaction of sodium–potassium alloy with white or red[5] phosphorus in refluxing 1,2-dimethoxyethane for 24 h followed by addition of chlorotrimethylsilane and heating at reflux for 72 h (good stirring is necessary for high yield of product), evaporation of the solvent, and vacuum distillation (75% yield);[6,7] reaction of piperidinodichlorophosphine with lithium powder and chlorotrimethylsilane in refluxing tetrahydrofuran (71% yield);[5–9] reaction of phosphine with excess of trimethylsilyl triflate in the presence of a tertiary amine in an inert solvent (Et_2O) at low temperature (90% yield);[10] reaction of phosphorus trichloride, magnesium, and chlorotrimethylsilane (62% yield).[11] The last method is considered to be the most cost effective and also the safest approach.
Analysis of Reagent Purity: ^{31}P NMR (CD_2Cl_2): δ −250.00 ppm (s, 1P);[12] ^{31}P NMR (C_6D_6): δ −251.6 ppm (s, 1P);[5] ^1H NMR (C_6D_6): δ 0.06 ppm (d, 27H).[5]

---

*Article previously published in the electronic Encyclopedia of Reagents for Organic Synthesis as Tris(trimethylsilyl)phosphine DOI: 10.1002/047084289X.rn01332

Purification: vacuum distillation, 54–57 °C/0.5–1 mmHg;[7] 72 °C/1 mmHg.[9]

Handling, Storage, and Precautions: air sensitive; moisture sensitive; spontaneously flammable in air; handle and store under inert atmosphere; pyrophoric.

Introduction. Tris(trimethylsilyl)phosphine is a more stable analog of phosphine, but retains high reactivity due to the presence of the weak polar Si–P bond. It is nucleophilic and readily reacts with a range of electrophiles. Reaction with alkylating agents provides substituted phosphines, with acid chlorides phosphaalkenes can be obtained, and phosphabenzenes can be synthesized upon reaction with pyrylium salts. Its Lewis basic nature has also proved useful as it forms addition complexes with a range of Lewis acids, some of which undergo further reactions. It is stable in cold water, but at 100 °C undergoes reaction to produce hexamethyldisiloxane.[1] It can be oxidized with air[1] or nitrogen dioxide[4] to give the air-stable tris(trimethylsilyl)phosphate.

Alkylation, and Arylation and Related Reactions. Tris-(TMS)phosphine reacts with a range of alkylating reagents to form alkylphosphines. For example, the reaction of tris(TMS)phosphine with excess chloromethylamines gives tris(aminomethyl)-phosphines under mild conditions. Due to the high reactivity of the alkylating agent, and the increased nucleophilicity of the product compared to the substrate, tris(aminomethyl) derivatives are the sole products even if only 1 or 2 equiv of alkylating agent is used (eq 1).[9,13] In contrast, less activated alkylating agents such as N-chloromethylamides allow isolation of monoalkylated products (eq 2).[9,13] Similarly, tris(aminomethyl)phosphines can be synthesized by heating tris(TMS)phosphine with alkoxymethyldialkyl-lamines in the presence of catalytic methanol[9] or bis(dialkyl-amino)methanes in the presence of zinc chloride (eq 1).[9,14] Tris-(TMS)phosphine can be alkylated with an excess of chloromethyl-alkyl ethers in the presence of catalytic zinc chloride to give tris(alkoxymethyl) derivatives (eq 3).[9] Tris(trimethylsilyloxy-methyl)phosphine can be synthesized by reaction of tris(TMS)-phosphine with an excess of paraformaldehyde (eq 4).[9]

$$(Me_3Si)_3P \xrightarrow[87\%]{\substack{CH_2O \\ (3 \text{ equiv})}} Me_3SiO-P(OSiMe_3)(OSiMe_3) \quad (4)$$

An important reaction in this class involves the facile dialkyl-lation reaction of tris(TMS)phosphine with chiral cyclic sulfates to afford TMS-phospholanes in good yields (eq 5).[5] Such products are extensively used in the synthesis of chiral bisphospholane ligands for metal-catalyzed enantioselective hydrogenation reactions.[5,15] In another interesting example, 5-chloro-5H-dibenzo[a,d]cycloheptene reacts with tris(TMS)phosphine at room temperature to form 5-bis-(trimethylsilyl)phosphanyl-5H-dibenzo[a,d]cycloheptene in high yield (eq 6). The product was used for the preparation of dibenzo-1-phosphasemibullvalene.[16]

$$\text{(eq 5)} \quad R = Me: 62\%; \quad Et: 70\%$$

$$\text{(eq 6)} \quad \xrightarrow[\substack{toluene, rt \\ 2 \text{ days} \\ >90\%}]{(Me_3Si)_3P}$$

Tris(TMS)phosphine also undergoes arylation reactions with suitably activated aryl halides. Thus, tris-2-pyrimidinylphosphine can be obtained via fluoride-mediated trisarylation with 2-chloro-pyrimidine (eq 7).[17]

$$\text{(eq 7)} \quad \xrightarrow[\substack{18\text{-crown-6} \\ 51\%}]{(Me_3Si)_3P, KF}$$

Synthesis of Phosphabenzenes. Tris(TMS)phosphine is widely used for the preparation of phosphabenzenes (phosphi-nines, phosphorins).[18–20] and is a more user-friendly phosphorus source than toxic PH$_3$. Thus, pyrylium salts can be converted to phosphabenzenes upon treatment with tris(TMS)phosphine. Selected examples are given in eq 8. In contrast using phosphine for the same synthesis required elevated pressure.[21] The yields of this transformation depend on substituents in pyrylium ring[22] and the nature of anion.[23,24] However, with highly substituted pyrylium salts, the yields of phosphabenzenes can be relatively and reproducibly high.[22,24]

Reaction with Acid Chlorides and Related Reactions. The reaction of tris(TMS)phosphine with carboxylic acid chlorides to give phosphaalkenes is well established.[25–27] The reaction proceeds via initial formation of an acylphosphine species **1**, which usually undergoes a rapid [1,3]-silyl shift from phosphorus to oxygen with simultaneous formation of a P–C double bond (eq 9). The reactions ultimately furnish the phosphaalkenes **2** in good

$$R \overset{}{\diagup} R' \xrightarrow[71-87\%]{(Me_3Si)_3P} R' \overset{}{\diagdown} P \overset{}{\diagup} R' \quad (1)$$

R = Cl, R′= NEt$_2$, NPr$_2$, Ni-Bu$_2$, piperidinyl
R = R′ = NMe$_2$, NEt$_2$, NPr$_2$, Ni-Bu$_2$, piperidinyl
R = OMe, R′ = NPr$_2$, Ni-Bu$_2$, piperidinyl

$$R \overset{}{\diagup} Cl \xrightarrow{(Me_3Si)_3P} R \overset{}{\diagup} P(SiMe_3)_2 \quad (2)$$

$$R = \quad : 47\%; \quad : 38\%$$

$$RO \overset{}{\diagup} Cl \xrightarrow{(Me_3Si)_3P, ZnCl_2} RO \overset{}{\diagup} P \overset{}{\diagup} OR \quad (3)$$

R = Me, Et, Pr, Bu: 63–69%

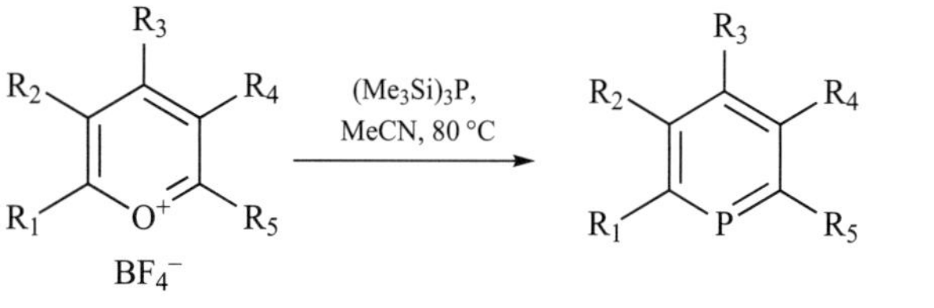

$$(8)$$

$R_1 = R_3 = Ph; R_2 = R_4 = H; R_5 = Ph, Me, 2\text{-MeOC}_6H_4; 3\text{-MeOC}_6H_4$
   $2\text{-naphthyl}: 46–81\%$
$R_1 = R_5 = Me; R_2 = R_4 = H; R_3 = Ph: 67\%$

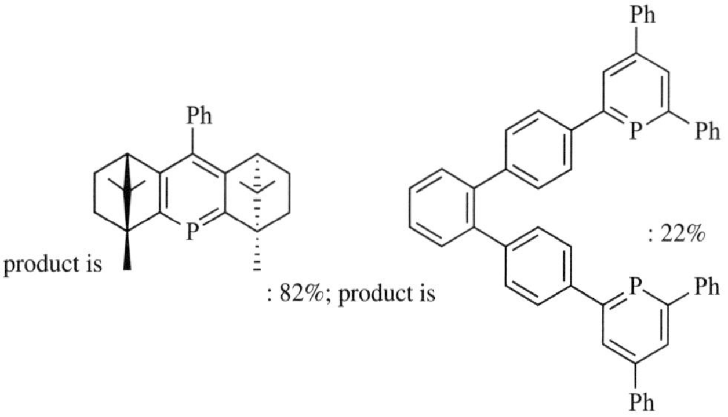

product is : 82%; product is : 22%

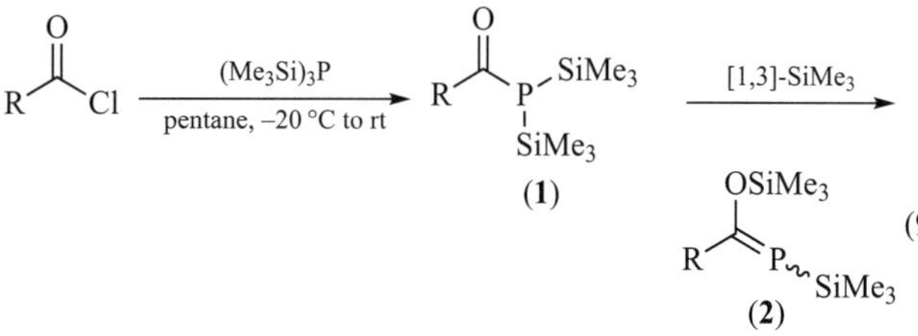

$$(9)$$

R = Me, Et, Pr, $n\text{-C}_{12}H_{24}$: 84–95% (2:1 - 4:1 Z:E); R = adamant-1-yl:
   67% (97:3 Z:E); cyclopropyl: 74% (19:81 Z:E);
2-methylcyclopropyl: 79% (18:82 Z:E); 1-methylcyclopropyl: 84%
(98:2 Z:E); 1-methyl-2,2-dichlorocyclopropyl: 94% (38:62 Z:E)

to high yields as mixtures of E- and Z-isomers, which can be readily differentiated by their different $J_{P,X}$ coupling constants in ^{31}P NMR spectra.[28] With more hindered acid chlorides, such as those derived from cyclopropane and adamantane carboxylic acids, the intermediate acylphosphines **1** are stable enough to allow complete characterization by ^{1}H, ^{13}C, and ^{31}P spectroscopy.[25,28]

The phosphaalkenes thus obtained are highly reactive substances and can be used in [3 + 2] cycloaddition reactions to give 1,2,4-oxazaphospholes[28] and 1,2,4-diazaphospholes,[28] in 'ene' reactions to furnish diphosphanes,[27] and in the preparation of phosphaalkynes (via elimination of hexamethyldisiloxane).[28,29] Interestingly, addition of CsF to the reaction between tris(TMS)-phosphine and an acyl chloride can provide phosphaalkynes in situ. These then undergo further reactions to provide a range of products whose distribution depends on the reaction times and ratios of reagents. One example is given in eq 10.[29]

Tris(TMS)phosphine can undergo a related reaction with isocyanates to furnish carbamoylphosphines and their imidate tautomers in a ratio of ~9:1. Such compounds are used in the synthesis of 1,2,4-azadiphospholes[30,31] and 1,3-azaphosphinines[32] (eq 11). Reaction with imidoyl chlorides can give 2-phosphaallylic cations,[33] which can be condensed with hydroxylamine to give 1,2,4-oxazaphospholes[33,34] and 1,2,4-diazaphospholes.[33–37] Selected examples are given in eq 12.

Tris(TMS)phosphine reacts with methyl chloroformate under mild conditions to form bis(trimethylsilyl)methoxycarbonylphosphine and tris(methoxycarbonyl)phosphine (or mixtures thereof) depending on the equivalents of methyl chloroformate used (eq 13).[9] In contrast to acylphosphines **1**, which readily undergo a phosphorus to oxygen 1,3-shift of a TMS group, compound **3** is stable and does not isomerize (eq 13).

Metalation and Subsequent Reactions.   Metalation of tris(TMS)phosphine by methyllithium gives lithium bis(TMS)-phosphide as a bis(THF) complex (eq 14).[6,7] The extent of reaction could be monitored by the ^{31}P NMR resonance signal of LiP(TMS)$_2$ that appeared at δ −302.4 ppm.[15]

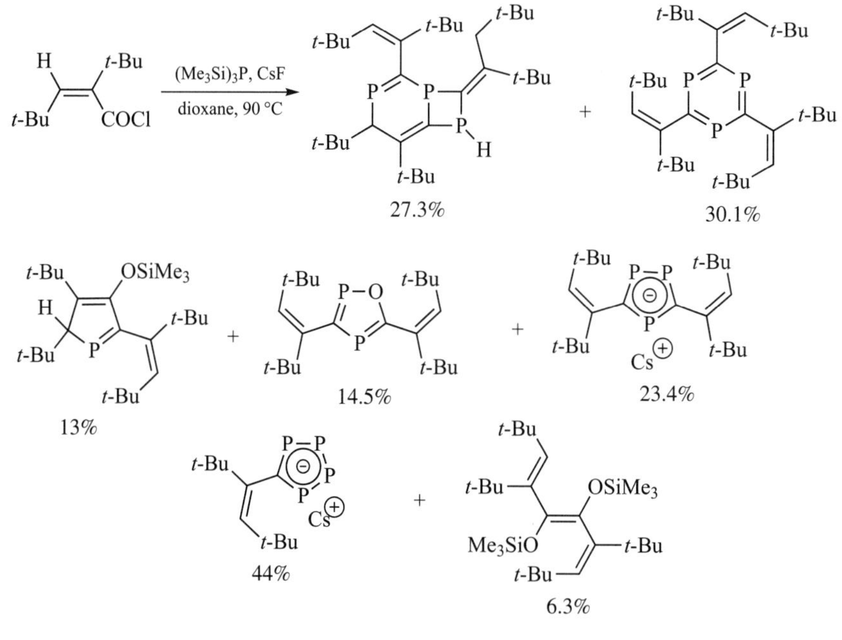

$$(10)$$

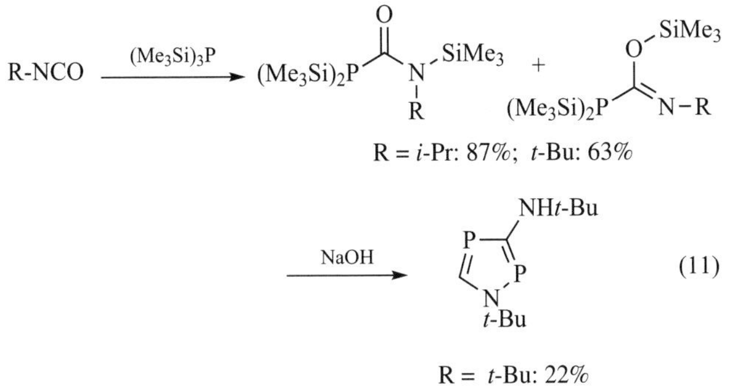

R-NCO $\xrightarrow{(Me_3Si)_3P}$ $(Me_3Si)_2P$—C(=O)—N(SiMe₃)(R) + $(Me_3Si)_2P$—C(OSiMe₃)=N—R

R = i-Pr: 87%; t-Bu: 63%

$\xrightarrow{NaOH}$ [structure with NHt-Bu, P, P, N, t-Bu]   (11)

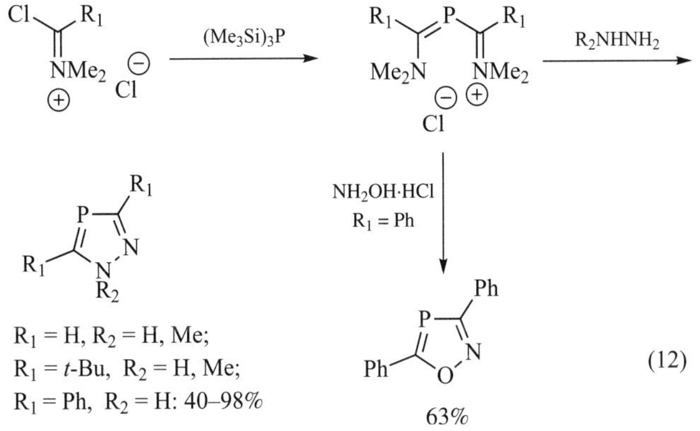

R = t-Bu: 22%

[structure: Cl, R₁, NMe₂, Cl⁻, ⊕] $\xrightarrow{(Me_3Si)_3P}$ [structure: R₁, P, R₁, Me₂N, NMe₂, Cl⁻ ⊕] $\xrightarrow{R_2NHNH_2}$

[triazole structure: R₁, P, N, R₁, N, R₂]

R₁ = H, R₂ = H, Me;
R₁ = t-Bu, R₂ = H, Me;
R₁ = Ph, R₂ = H: 40–98%

$\downarrow$ NH₂OH·HCl, R₁ = Ph

[structure: Ph, P, N, Ph, O]   (12)

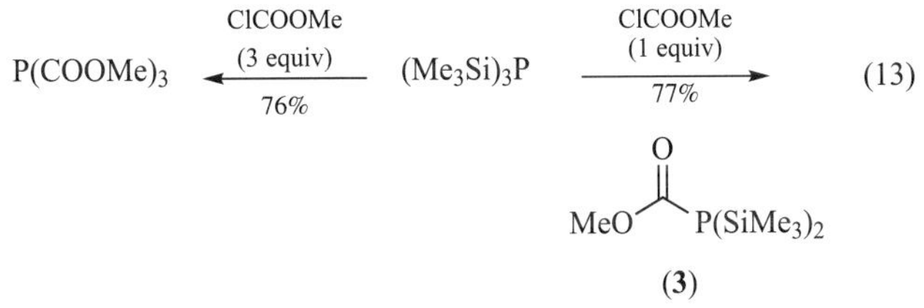

63%

$P(COOMe)_3$ $\xleftarrow[76\%]{\text{ClCOOMe (3 equiv)}}$ $(Me_3Si)_3P$ $\xrightarrow[77\%]{\text{ClCOOMe (1 equiv)}}$ (13)

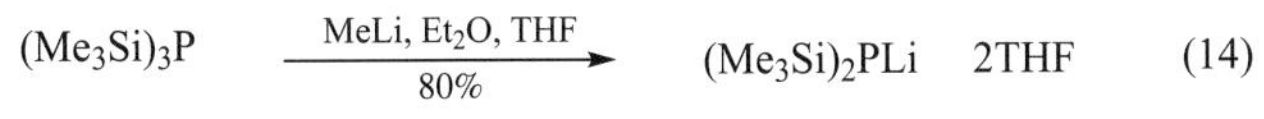

MeO—C(=O)—P(SiMe₃)₂

(3)

$(Me_3Si)_3P$ $\xrightarrow[80\%]{\text{MeLi, Et}_2\text{O, THF}}$ $(Me_3Si)_2PLi \cdot 2THF$ (14)

As one would expect, lithium bis(TMS)phosphide is a more reactive nucleophile than tris(TMS)phosphine itself and has been effectively used in analogous reactions as a P-centered anion. For example, alkylation with cyclic sulfates can be used to provide useful TMS-phospholanes via an intermediate open-chain adduct 4, which can alternatively undergo protodesilylation upon treatment with methanol to give primary phosphines (eq 15).[15]

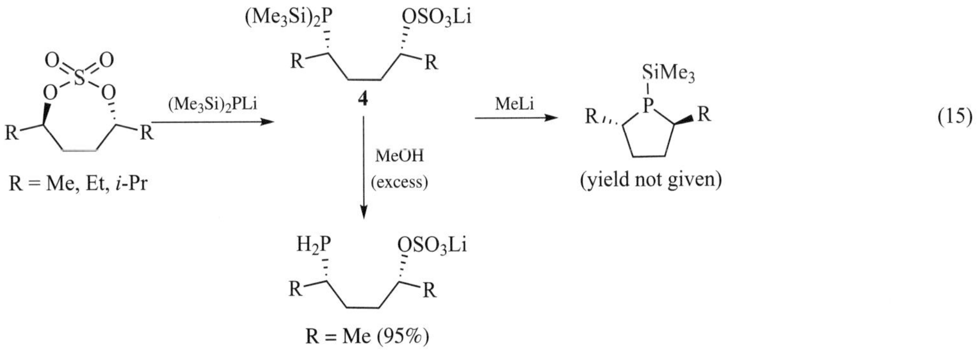

In another example, lithium bis(TMS)phosphide reacts smoothly with N,N-dialkylformamides under mild conditions to give substituted phosphaalkenes, whereas tris(TMS)phosphine does not react (eq 16).[9]

Similarly to tris(TMS)phosphine, reactions of lithium bis(TMS)-phosphide with acid chlorides are reported to give unstable acyl-phosphines 1, which rearrange to give phosphaalkenes 2. In some cases, the reactions perform better than those of the parent tris(TMS)phosphine. Thus, the reaction of lithium bis(TMS)-phosphide–THF complex with 1-adamantanoyl chloride affords the corresponding phosphaalkene 2 as a solid in 96% yield, compared to isolation as an oil in only 67% yield when tris(TMS)-phosphine was used.[28] A particularly interesting example involves the reaction of Z-2-t-butyl-4,4-dimethylpent-2-enoyl chloride that led to the stereoselective formation of 2,3-di-t-butyl-1-trimethylsilyl-4-trimethylsilyloxy-1,2-dihydrophosphete (eq 17). The mechanism involves formation of an acylphosphine, which undergoes a 1,3-TMS shift, affording a transient 1-phosphabuta-diene, which then undergoes electrocyclic ring closure to give the 1,2-dihydrophosphete.[12]

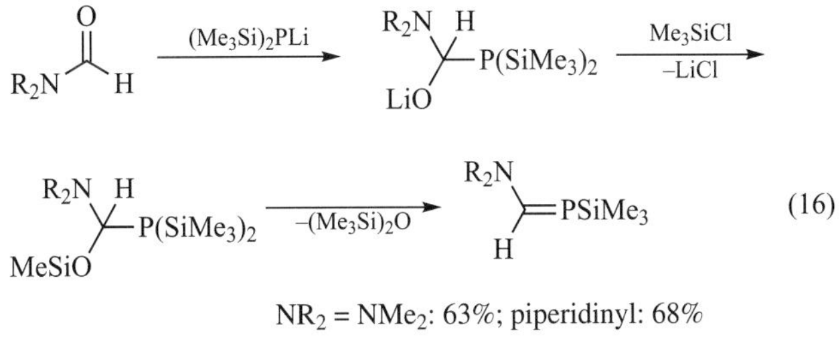

[structure: R₂N—CHO] $\xrightarrow{(Me_3Si)_2PLi}$ [structure: R₂N, H, LiO, P(SiMe₃)₂] $\xrightarrow[-LiCl]{Me_3SiCl}$

[structure: R₂N, H, MeSiO, P(SiMe₃)₂] $\xrightarrow{-(Me_3Si)_2O}$ [structure: R₂N, H, =PSiMe₃]   (16)

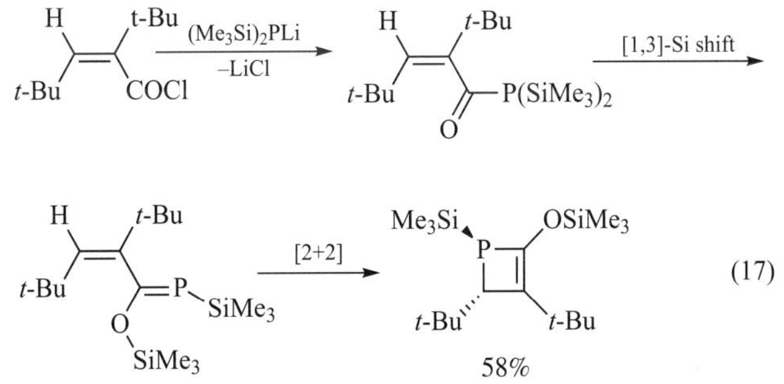

NR₂ = NMe₂: 63%; piperidinyl: 68%

[alkene structure: H, t-Bu, t-Bu, COCl] $\xrightarrow[-LiCl]{(Me_3Si)_2PLi}$ [structure: H, t-Bu, t-Bu, C(=O)—P(SiMe₃)₂] $\xrightarrow{[1,3]-\text{Si shift}}$

[structure: H, t-Bu, t-Bu, O(SiMe₃), P, SiMe₃] $\xrightarrow{[2+2]}$ [phosphete: Me₃Si, P, OSiMe₃, t-Bu, t-Bu]   (17)

58%

Metalation of tris(TMS)phosphine with butyllithium, followed by treatment with 2,3-di-t-butylcyclopropenone, and finally addition of excess TMSCl can provide the phosphatria-fulvene 5 in high yield, via a one-pot Peterson-type olefination

[cyclic sulfate structure: O₂S(O)(O), R, R, R = Me, Et, i-Pr] $\xrightarrow{(Me_3Si)_2PLi}$ [structure 4: (Me₃Si)₂P, OSO₃Li, R, R] $\xrightarrow{MeLi}$ [phospholane: SiMe₃, P, R, R]   (15)

4

(yield not given)

$\downarrow$ MeOH (excess)

[structure: H₂P, OSO₃Li, R, R]

R = Me (95%)

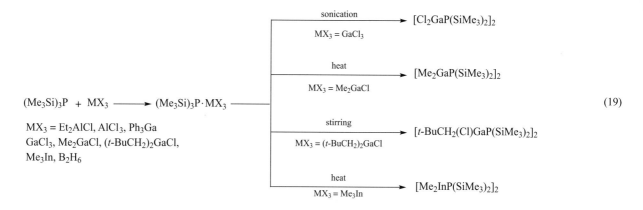

$$(\text{Me}_3\text{Si})_3\text{P} + \text{MX}_3 \longrightarrow (\text{Me}_3\text{Si})_3\text{P}\cdot\text{MX}_3 \tag{19}$$

$$\text{MX}_3 = \text{Et}_2\text{AlCl}, \text{AlCl}_3, \text{Ph}_3\text{Ga}$$
$$\text{GaCl}_3, \text{Me}_2\text{GaCl}, (t\text{-BuCH}_2)_2\text{GaCl},$$
$$\text{Me}_3\text{In}, \text{B}_2\text{H}_6$$

reaction (eq 18).[37] Phosphatriafulvenes synthesized by this method are free of contaminants and are used in the synthesis of 1,3-diphosphines by reaction with kinetically stabilized phosphaalkynes.[37] In other reactions, lithium bis(TMS)phosphide has been used in the synthesis of 1,2,4-thiadiphospholes[38,39] and 1,2,4-triphospholides.[40,41]

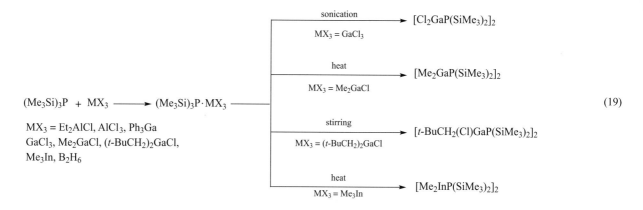

$$(\text{Me}_3\text{Si})_3\text{P} \xrightarrow[\text{90\%}]{\begin{array}{l}1.\ n\text{-BuLi}\\2.\\3.\ \text{TMSCl}\\-\text{LiCl}, -\text{TMS}_2\text{O}\end{array}} \tag{18}$$

Formation of Addition Complexes and Related Reactions. Tris(TMS)phosphine can form addition complexes with a number of Lewis acidic species. For example, reaction with diborane provides the crystalline adduct (TMS)$_3$P·BH$_3$ (eq 19).[1,4] In other examples, tris(TMS)phosphine forms isolable adducts with a range of aluminum, gallium, and indium species, such as Et$_2$AlCl,[42] iBu$_2$AlCl,[42] EtAlCl$_2$,[42] AlCl$_3$,[43] Me$_3$Al,[44] Ph$_3$Ga,[45] Ph$_2$GaCl,[45] GaCl$_3$,[46] (Me$_3$SiCH$_2$)$_3$Ga,[47] (t-BuCH$_2$)$_2$GaCl,[48] Me$_2$GaCl,[49] Me$_3$Ga,[44] (Me$_3$SiCH$_2$)$_3$In,[47] (Me$_3$SiCH$_2$)InCl$_2$,[50] (t-BuCH$_2$)$_2$ InCl,[51] and Me$_3$In,[52] which may then undergo further reactions involving 1,2-elimination or dehalosilylation (eq 19).[48]

In a more complex example, reaction of tris(TMS)phosphine and the tris(dichloroboryl)alkane **6** provides the interesting zwitterionic 1-phosphonium-2,4-diboretane-3-ide **7**, which is an unusual neutral monohomoaromatic (eq 20).

Tris(TMS)phosphine has been used for the preparation of group 13 and 15 (In, Ga, Bi) binary phosphides (GaP,[46] BiP,[54] and InP[55]) through initial reaction with the corresponding trichlorides followed by vacuum thermolysis (eq 21). The phosphides thus produced are extensively used for material science applications, such as the synthesis of phosphide semiconductors, and this approach to their synthesis is favored due to its easier handling and less elaborate laboratory workup procedures. Notably, the indium phosphide synthesized via this method is free from incorporated iodine.

The treatment of tetracarbonylnickel with tris(TMS)phosphine gives a quantitative yield of the stable complex tricarbonyl tris-(TMS)phosphine nickel(0), whose structure was supported by ^{31}P NMR studies (eq 22).[56]

Use as a Reducing Agent. Tris(TMS)phosphine can be used as a reducing agent for the selective removal of the two axial fluorine atoms in difluorotris(perfluoroalkyl)phosphoranes, to give the corresponding tris(perfluoroalkyl)phosphanes. Trimethylsilyl fluoride is produced as an easily removable by-product (eq 23).[57]

$$(\text{Me}_3\text{Si})_3\text{P} + \text{MCl}_3 \xrightarrow[\text{2. vacuum thermolysis}]{\text{1. PhMe, rt}} \text{MP} \tag{21}$$

$$\text{M = Ga, In, Bi}$$

$$\text{F}_2\text{P}[(\text{CF}_2)n\text{CF}_3]_3 \xrightarrow[-\text{Me}_3\text{SiF}]{(\text{Me}_3\text{Si})_3\text{P}} \text{P}[(\text{CF}_2)n\text{CF}_3]_3 \tag{23}$$
$$n = 1\text{--}4$$

1. Leffler, A. J.; Teach, E. G., *J. Am. Chem. Soc.* **1960**, *82*, 2710.

2. Bruker, A. B.; Balashova, L. D.; Soborovskii, L. Z., *Dokl. Akad. Nauk SSSR* **1960**, *135*, 843; *Chem. Abstr.* **1961**, *55*, 13301.

3. Parshall, G., W. U. S Pat 2,907,785 (1959) (E. I. du Pont de Nemours & Co).

4. Parshall, G. W.; Lindsey, R. V., Jr., *J. Am. Chem. Soc.* **1959**, *81*, 6273.

5. Holz, J.; Zayas, O.; Jiao, H.; Baumann, W.; Spannenberg, A.; Monsees, A.; Riermeier, T. H.; Almena, J.; Kadyrov, R.; Borner, A., *Chem. Eur. J.* **2006**, *12*, 5001.

6. Becker, G.; Schmidt, H.; Uhl, G.; Uhl, W., *Inorg. Synth.* **1990**, *27*, 243.

7. Askham, F. R.; Stanley, G. G.; Marques, E. C., *J. Am. Chem. Soc.* **1985**, *107*, 7423.

8. Niecke, E.; Westermann, H., *Synthesis* **1988**, *4*, 330.

9. Prishchenko, A. A.; Livantsov, M. V.; Novikova, O. P.; Livantsova, L. I.; Petrosyan, V. S., *Heteroatom. Chem.* **2010**, *21*, 441.

10. Uhlig, W.; Thust, U.; Tzschach, A.; Gomille, R. German Pat DD 274626A1 (1989) (Martin-Luther-Universitaet Halle-Wittenberg).

11. Schuman, H.; Rösch L., *Chem. Ber.* **1974**, *107*, 854.

12. Ionkin, A. S.; Marshall, W. J.; Fish, B. M.; Schiffhauer, M. F.; McEwen, C. N., *Chem. Commun.* **2008**, 5432.

13. Prischchenko, A. A.; Livantsov, M. V.: Pisarnitskii, D. A.; Petrosyan, V. S., *Zh. Obshch. Khim.* **1991**, *61*, 1016; *Chem. Abstr.* **1991**, *116*, 6649.

14. Prischchenko, A. A.; Livantsov, M. V.; Pisarnitskii, D. A.; Petrosyan, V. S., *Zh. Obshch. Khim.* **1990**, *60*, 460.

15. Burk, M. J.; Pizzano, A.; Martin, J. A.; Liable-Sands, L. M.; Rheingold, A. L., *Organometallics* **2000**, *19*, 250.

16. Geier, J.; Frison, G.; Grutzmacher, H., *Angew. Chem., Int. Ed.* **2003**, *42*, 3955.

17. Reetz, M. T.; Demuth, R.; Goddard, R., *Tetrahedron Lett.* **1998**, *39*, 7089.

18. Märkl, G., *Angew. Chem.* **1966**, *78*, 907; *Angew. Chem., Int. Ed. Engl.* **1966**, *5*, 846.

19. Ashe, A. J., III, *J. Am. Chem. Soc.* **1971**, *93*, 3293.

20. Muller, C.; Freixa, Z.; Lutz, M.; Spek, A. L.; Vogt, D.; Leeuwen, P. W. N. M., *Organometallics* **2008**, *27*, 834.

21. Breit, B.; Fuchs, E., *Synthesis* **2006**, *13*, 2121.

22. Bell, J. R.; Franken, A.; Garner, C. M., *Tetrahedron* **2009**, *65*, 9368.

23. Märkl, G.; Lieb, F.; Merz, A., *Angew. Chem., Int. Ed.* **1967**, *6*, 458.

24. DiMauro, E. F.; Kozlowski, M. C., *J. Chem. Soc., Perkin Trans. 1* **2002**, 439.

25. Kostitsyn, A. B.; Nefedov, O. M.; Heydt, H.; Regtiz, M., *Synthesis* **1994**, 161.

26. Slany, M.; Regtiz, M., *Synthesis* **1994**, 1262.

27. Marinetti, A.; Ricard, L.; Mathey, F., *Tetrahedron* **1993**, *49*, 10279.

28. Allspach, T.; Regitz, M.; Becker, G.; Becker, W., *Synthesis* **1986**, *1*, 31.

29. Ionkin, A. S.; Marshall, W. J.; Fish, B. M.; Schiffhauer, M. F.; Davidson, F.; McEwen, C. N., *Organometallics* **2009**, *28*, 2410.

30. Ionkin, A. S.; Ignat'eva, S. N.; Litvinov, I. A.; Naumov, V. A.; Arbuzov, B. A., *Heteroatom Chem.*, **1991**, *2*, 577.

31. Collier, S. J., *Sci. of Synth.*, **2004**, *13*, 717.

32. Appel, R.; Poppe, M., *Angew. Chem.* **1989**, *101*, 70; *Angew. Chem., Int. Ed. Engl.* **1989**, *28*, 53

33. Schmidpeter, A., *Phosphorus Sulfur Relat. Elem.* **1986**, *28*, 71.

34. Bansal, R. K.; Gupta, N., *Sci. of Synth.*, **2004**, *13*, 647.

35. Schmidpeter, A.; Willhalm, A., *Angew. Chem., Int. Ed. Engl.* **1984**, *23*, 903.

36. Bansal, R. K.; Gupta, N., *Sci. of Synth.*, **2004**, *13*, 689.

37. Hofmann, M. A.; Heydt, H.; Regtiz, M., *Synthesi* **2001**, *3*, 463.

38. Appel, R.; Moors, R., *Angew. Chem., Int. Ed. Engl.* **1986**, *25*, 567.

39. Collier, S. J., *Sci. of Synth.* **2004**, *13*, 659.

40. Callaghan, C.; Clentsmith, G. K. B.; Cloke, F. G. N.; Hitchcock, P. B.; Nixon, J. F.; Vickers, D. M., *Organometallics* **1999**, *18*, 793.

41. Bansal, R. K.; Gupta, N., *Sci. of Synth.*, **2004**, *13*, 729.

42. Wells, R. L.; McPhail, A. T.; Self, M. F.; Laske, J. A., *Organometallics* **1993**, *12*, 3333.

43. Wells, R. L.; McPhail, A. T.; Laske, J. A.; White, P. S., *Polyhedron* **1994**, *13*, 2737.

44. Wells, R. L.; McPhail, A. T.; Self, M. F.; Laske, J. A., *Organometallics* **1996**, *15*, 3980.

45. Wells, R. L.; Aubuchon, S. R.; Self, M. F.; Jasinski, J. P.; Woudenberg, R. C.; Butcher, R. J., *Organometallics* **1992**, *11*, 3370.

46. Wells, R. L.; Self, M. F.; McPhail, A. T.; Aubuchon, S. R.; Woudenberg, R.; Jasinski, J. P., *Organometallics* **1993**, *12*, 2832.

47. Wells, R. L.; Baldwin, R. A.; White, P. S., *Organometallics* **1995**, *14*, 2123.

48. Wells, R. L.; Baldwin, R. A.; White, P. S.; Pennington, W. T.; Rheingold, A. L.; Yap, G. P. A., *Organometallics* **1996**, *15*, 91.

49. Wiedmann, D.; Hausen, H.-D.; Weidlein, J., *Z. Anorg. Allg. Chem.* **1995**, *621*, 1351.

50. Wells, R. L.; McPhail, A. T.; Self, M. F., *Organometallics* **1992**, *11*, 221.

51. Wells, R. L.; McPhail, A. T.; Self, M. F., *Organometallics* **1993**, *12*, 3363.

52. Stuczynski, S. M.; Opila, R. L.; Marsh, P.; Brennan, J. G.; Steigerwald, M. L., *Chem. Mater.* **1991**, *3*, 379.

53. Sahin, Y.; Ziegler, A.; Happel, T.; Meyer, H.; Bayer, M. J.; Pritzkow, H.; Massa, W.; Hofmann, M.; Schleyer, P. V. R.; Siebert, W.; Berndt, A., *J. Organomet. Chem.* **2003**, *680*, 244.

54. Carmalt, C. J.; Cowley, A. H.; Hector, A. L.; Norman, N. C.; Parkin, I. P., *J. Chem. Soc., Chem. Commun.* **1994**, *17*, 1987.

55. Healy, M. D.; Laibinis, P. E.; Stupik, P. D.; Barron, A. R., *J. Chem. Soc., Chem. Commun.* **1989**, *6*, 359.

56. Schumann, H.; Stelzer, O., *Angew. Chem., Int. Ed. Engl.* **1967**, *6*, 701.

57. Kampa, J. J.; Nail, J. W.; Lagow, R. J., *Angew. Chem., Int. Ed. Engl.* **1995**, *34*, 1241.

Sergey A. Kosarev

AMRI Singapore Research Centre Pte Ltd, Singapore

Steven J. Collier

Codexis Laboratories Singapore Pte Ltd, Singapore

Polymethylhydrosiloxane

[9004-73-3] $C_9H_{30}O_4Si_5$

InChI = 1S/C9H30O4Si5/c1-14(10-15(2)12-17(4,5)6)11-16(3) 13-18(7,8)9/h14-16H,1-9H3

InChIKey = INRLRTZQQOALFL-UHFFFAOYSA-N

(reducing agent often used in conjunction with metal catalysts or nucleophilic activators)

Alternate Name: PMHS; methylhydrogensiloxane polymer; methylhydrosilicone homopolymer; 1,1,1,3,5,7,7,7-octamethyltetrasiloxane (PMHS-dimer) [16066-09-4]; polymethylhydrosiloxane trimethylsilyl terminated [63148-57-2] and [178873-19-3]; and poly(dimethysiloxane-*co*-methylhydrosiloxane) (96% wt methylhydrosiloxane monomer units) [63148-57-2] are sold as 'PMHS'.

Physical Data: colorless free flowing liquid; average molecular weight 1500-2200 g mol^{-1} (supplier dependent); effective mass per hydride of 60 g mol^{-1}; d = 1.006.

Solubility: most ethereal, chlorinated, or hydrocarbon solvents as well as EtOH, i-PrOH, warm DMF, and warm NMP; insoluble in MeOH, DMSO, acetonitrile, and water.

Preparative Methods: hydrolysis of methyldichlorosilane followed by heating (60–150 °C) the resultant mixture of cyclic silanes in the presence of hexamethyldisiloxane generates the linear polysiloxane.[1]

Handling, Storage, and Precaution: stable to air and moisture; incompatible with strong acids, bases, or oxidants (forms hydrogen upon decomposition); generally considered non-toxic, however thorough toxicity studies have not been performed; skin/eye contact and inhalation should be avoided.

General. Polymethylhydrosiloxane (PMHS) is an easily handled, inexpensive, non-toxic, and mild reducing agent. Although relatively inert towards organic functionality, PMHS can transfer its hydride to a variety of metal catalysts (including Sn, Ti, Zn, Cu, and Pd) which can then participate in a wide range of reductions. Alternatively, when made hypercoordinate by the action of fluoride or other nucleophiles, PMHS can act directly as a reducing agent. PMHS is attractive as a substitute for more expensive or hazardous silanes or siloxanes and as the stoichiometric reductant in catalytic organotin-mediated processes. Applications of PMHS in organic synthesis have been detailed in several reviews[2] including an excellent treatise by Professor Nicholas J. Lawrence and co-workers (Cardiff/UMIST).[2a]

Synthesis of Alkyltin Hydrides. Perhaps the most widely recognized use of PMHS is in the synthesis of trialkyltin hydrides. In 1967, Hayashi et al.[3] established the PMHS reduction of organotin oxides (via Si–H/Sn–O σ-bond metathesis) as a preparative route to trialkyl- and dialkyltin hydrides (eq 1). This method obviates the need for highly reactive reducing agents like LiAlH$_4$.[4] The method is also amenable to in situ generation and reaction of organotin reagents,[2a] with the order of reactivity for organotin oxides towards PMHS is Bu$_3$Sn(OEt)$_2$ > Bu$_3$SnOEt > (Bu$_3$Sn)$_2$O > Bu$_2$SnO > (Ph$_3$Sn)$_2$O > Bu$_3$SnOSiBu$_3$. Complete conversion of (Bu$_3$Sn)$_2$O to two equiv of Bu$_3$SnH typically requires elevated reaction temperatures (>80 °C), however Fu has shown that adding n-BuOH to the reaction mixture facilitates liberation of the second Bu$_3$SnH.[5]

$$(Bu_3Sn)_2O \xrightarrow[\text{neat, } \Delta]{\text{PMHS}} \begin{bmatrix} Bu_3SnOSiR_3 \\ + \\ Bu_3SnH \end{bmatrix} \xrightarrow[\text{neat, } \Delta]{\text{PMHS}} Bu_3SnH \quad (1)$$
$$87\%$$

PMHS alone does not reduce triorganotin halides[6] (or amides[7]) and thus the Hayashi method is not readily applied to the preparation of Me$_3$SnH or other organotin hydrides for which the corresponding organotin oxides are not commercially available. However, PMHS in combination with a fluoride source can effect the reduction of trialkyltin chlorides, bromides, and fluorides (eq 2).[8] The PMHS/fluoride/R$_3$Sn–X combination is also applicable to in situ generation and reaction of organotin hydrides.[2a,8,9,10,11]

$$Bu_3SnCl \xrightarrow[\substack{\text{Et}_2\text{O, rt} \\ \text{aq NaOH work-up}}]{\substack{\text{1.1 equiv. PMHS} \\ \text{2.2 equiv. aq KF}}} Bu_3SnH \quad (2)$$
$$82\%$$

Reductions of C–O Bonds. A wide range of catalytic systems employing PMHS can reduce carbonyl groups. PMHS/(Bu$_3$Sn)$_2$O does not efficiently reduce carbonyls, however slow addition of a ketone and PMHS to stoichiometric Bu$_2$SnO in toluene at 25 °C has been shown to be an efficient way of carrying out such reductions.[12]

Similar reductions can be made catalytic in tin by using dibutyltin dilaurate,[13] bis(dibutylacetoxytin) oxide (DBATO),[14] or polymer-supported organotins.[6] In some cases, these catalytic protocols have proven more successful than their stoichiometric counterparts (eq 3). However, enones suffer from competitive 1,4-reductions. Reductions using a chiral tin catalyst have recently

emerged (eq 4), although the observed enantioselectivities are poor to moderate (0-58% ee).[15]

$$(3)$$
$$81\%$$

$$(4)$$
$$95\%, 58\% \text{ ee}$$

Mimoun et al. has extensively studied zinc catalysts in combination with PMHS. In order to effect the catalytic cycle, a co-catalyst is necessary, usually LiAlH$_4$ or NaBH$_4$. The optimal [ZnH] combination appears to be 2-5 mol% Zn(2-ethylhexanoate)$_2$ (i.e. Zn(2-EH)$_2$), 2-5 mol% NaBH$_4$, and 1-2 equiv of PMHS.[16] Zinc-catalyst/PMHS allows for 1,2-reduction of saturated or α,β-unsaturated esters, aldehydes, and ketones (eq 5). It should be noted that in contrast to the organotin-mediated chemistry, these reactions initially afford the silyl ether, which is ultimately subjected to a separate hydrolytic work-up.

$$(5)$$
$$94\%$$

Zinc-mediated PMHS reductions have also been carried out enantioselectively.[17] Aromatic ketones afford alcohols in 64-81% ee (eq 6), while non-aromatic ketones are reduced less selectively (15-20% ee).

$$(6)$$
$$76\% \text{ ee}$$

The zinc catalyst can also reduce lactones to their corresponding lactols or diols by using one or two equiv of PMHS, respectively (eq 7). Terminal primary epoxides are opened to the corresponding secondary alcohols (eq 8), while more substituted epoxides are unaffected. Employing Pd(OAc)$_2$(PPh$_3$)$_2$ or Cu(2-EH)$_2$ can reverse the normal 1,2- over 1,4-selectivity of enone reductions.

$$(7)$$

$$Ph\text{–}O\text{–}\underset{}{\triangle} \quad \xrightarrow[\text{NaBH}_4 \text{ (2 mol \%)}, (i\text{-Pr})_2O]{\text{PMHS, Zn(2-EH)}_2 \text{ (2 mol \%)}} \quad Ph\text{–}O\text{–}\underset{\underset{90\%}{Me}}{\overset{OH}{|}} \qquad (8)$$

PMHS can also be used in conjunction with other copper catalysts to effect 1,4-reductions.[18] These reductions generate silyl enol ethers, which Lipshutz and co-workers have shown can be exploited in subsequent chemical events such as alkylations or condensations (eq 9). PMHS plus catalytic (PPh$_3$)CuH can also reduce saturated ketones and aldehydes in a 1,2-fashion.[19]

$$(9)$$

72–89%

Enantioselective 1,4-reductions of conjugated esters with a PMHS, CuCl, BINAP mixture (eq 10) are high yielding and allow asymmetric construction of β-keto stereogenic centers (80-92% ee).[20]

$$(10)$$

98%, 92% ee

For selective 1,4-reduction of α,β-unsaturated aldehydes PMHS can be used in combination with aq KF, catalytic Pd(0), and 10 mol% Bu$_3$SnH (eq 11). It is useful to note that no reduction of the α,β-carbonyl compound occurs in absence of the tin halide and that no reduction of the resultant aldehyde was observed.[8]

$$Ph\text{–}\!\!\!=\!\!\!\text{–}CHO \quad \xrightarrow[\text{(PPh}_3)_2\text{PdCl}_2 \text{ (0.8 mol \%)}]{\substack{\text{Bu}_3\text{SnCl (20 mol \%)} \\ \text{PMHS, aq KF} \\ \text{THF}}} \quad Ph\text{–}\!\!\!\backslash\!\!\!\text{–}CHO \qquad (11)$$

87%

Hypercoordinate silicates formed by reaction of PMHS with KF, TBAF,[21,22] or other nucleophiles,[23] can reduce ketones, aldehydes, and esters (eq 12).[24] With proper nucleophile choice, aldehydes can be reduced selectively over ketones, and ketones over esters (eq 13). Halides, nitriles, nitro groups, and olefins survive these conditions, but enones often undergo both 1,2- and 1,4-reductions. PMHS is worse than other silanes for performing enantioselective reductions with chiral fluoride sources, affording alcohols in only 9–36% ee.[25]

$$(12)$$

trans/cis
91:9

$$\underset{H}{\overset{O}{\|}}\text{–}(\ )_8\text{–}\underset{}{\overset{O}{\|}}\text{–}Me \quad \xrightarrow[\text{DMF, 100 °C}]{\text{PMHS, HCO}_2\text{K}} \quad HO\text{–}(\ )_8\text{–}\underset{}{\overset{O}{\|}}\text{–}Me \qquad (13)$$

63%

Buchwald and co-workers have developed a number of titanocene-based protocols involving catalytic Cp$_2$TiCl$_2$ activated by EtMgBr or n-BuLi. PMHS converts the resultant titanium species into a titanium hydride, which can reduce ketones and esters (eq 14).[26] Conjugate hydride addition is rarely seen with α,β-unsaturated esters, however 4-10% of the 1,4-reduction products has been observed during reductions of enones. Again, it is the silyl ether that is initially formed in these reactions. Employing Cp$_2$Ti(p-C$_6$H$_4$O)$_2$ or Cp$_2$TiF$_2$ as the catalyst allows for the conversion of lactones to lactols.[27] Both Reding[28] and Breeden[29] have also found that Ti(O-i-Pr)$_4$ can catalyze the reduction of esters to primary alcohols. Importantly, alkynes, bromides, chlorides, epoxides, and nitro groups are compatible with this chemistry.

$$Ph\text{–}\!\!\!\frown\!\!\!\text{–}CO_2Et \quad \xrightarrow[\substack{\text{2. PMHS} \\ \text{3. NaOH, H}_2\text{O}}]{\substack{\text{1. Cp}_2\text{TiCl}_2 \text{ (2 mol \%)} \\ \text{EtMgBr (4 mol \%), THF}}} \quad Ph\text{–}\!\!\!\frown\!\!\!\text{–}OH \qquad (14)$$

94%

Halterman and co-workers[30] and Buchwald et al.[26b] have used chiral titanocenes to asymmetrically reduce aryl and α,β-unsaturated ketones with relatively high enantioselectivity (82–97% ee) (eq 15). However, electron-withdrawing groups about the aryl ring appear to be problematic, as are saturated ketones.

$$(15)$$

(4–5 mol %)   95%, 96% ee

R = (R,R)-1,1'-binaphthyl-2,2'-diolate

In addition to carbonyl reductions, PMHS can also be used in deoxygenations. For example, PMHS, NaI, and TMSCl will generate benzyl iodides from benzaldehydes (eq 16). Unfortunately this solventless transformation does not work well with ketones or aliphatic aldehydes.[31] Fu and co-workers have shown that PMHS is instrumental in performing Barton-McCombie deoxygenations of thionocarbonates with only catalytic amounts of (Bu$_3$Sn)$_2$O (eq 17).[5] Interestingly, dithiocarbonates do not respond uniformly well to similar conditions.[32]

$$(16)$$

70–90%

$$(17)$$

70%

Reductions of C–N Bonds. An ethanolic mixture of PMHS and Pd/C will reduce oximes (eq 18)[33] to amines and reductively open aziridines (eq 19).[34] Either (a) PMHS plus catalytic *n*-butyltin tris(2-ethylhexanoate) (eq 20)[35] or (b) PMHS plus ZnCl$_2$ reduces imines.[36] The PMHS/DBATO combination reduces azides,[37] while PMHS/Ti(O-*i*-Pr)$_4$ can be applied to reductive aminations (eq 21).[38] Asymmetric imine reductions via chiral titanium complexes and PMHS are also viable, but very substrate dependent with nonaromatic imines working best (69–99% ee vs. 6–97% ee for aromatic imines).[39]

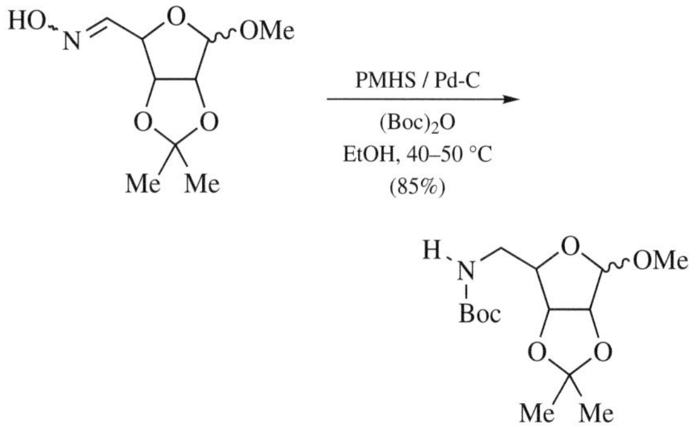

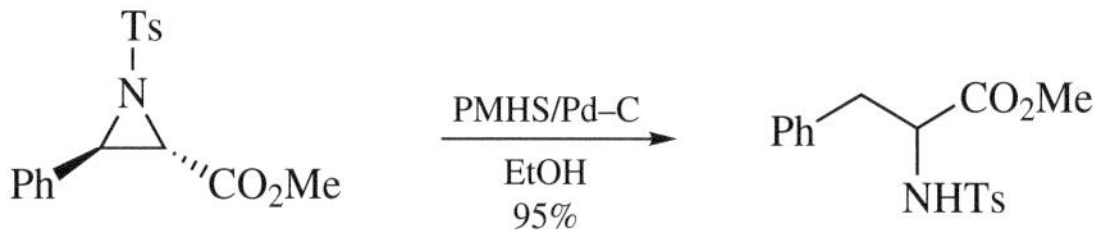

Reductions of C–X Bonds. Organotin hydrides generated in situ via PMHS reduction of Sn-O or Sn-X precursors easily reduce aromatic and aliphatic halides. These reactions can proceed either thermally or photochemically and have been used successfully to reduce geminal dihalides stepwise (eq 22).[12,40] Since organotin halides are the by-products of these reactions, the PMHS/fluoride/R$_3$Sn-X combination allows reactions to be carried out with catalytic amounts of tin.[8]

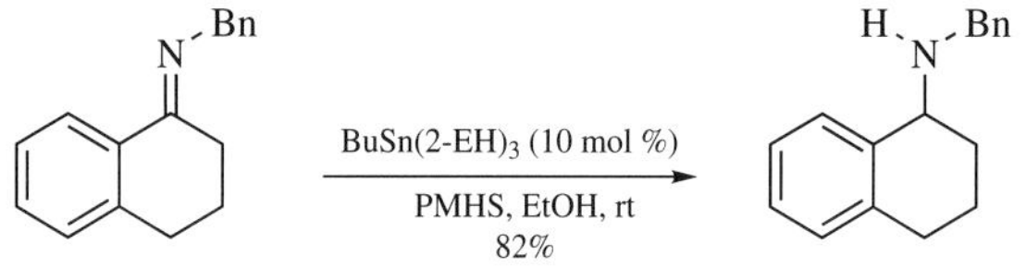

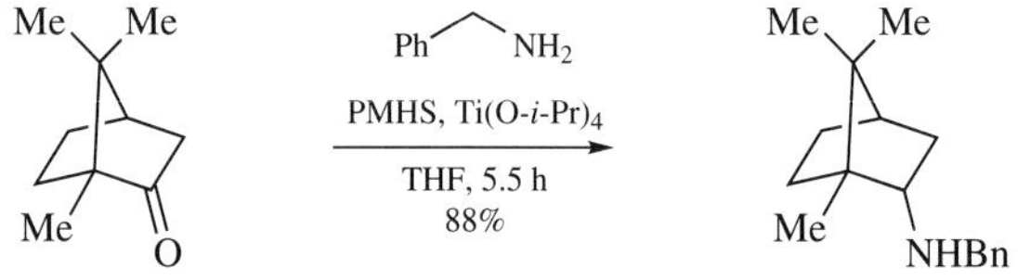

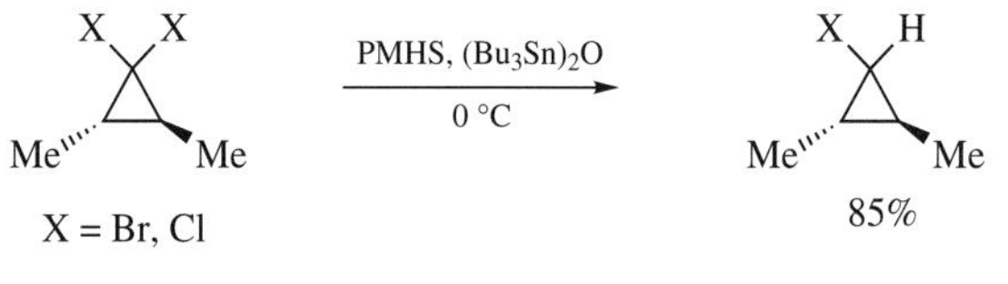

PMHS can be used directly as a hydride donor for the reduction of aryl and vinyl bromides or iodides along with α-halo ketones and acids provided the halide can be activated by Pd(0) (eq 23).[41]

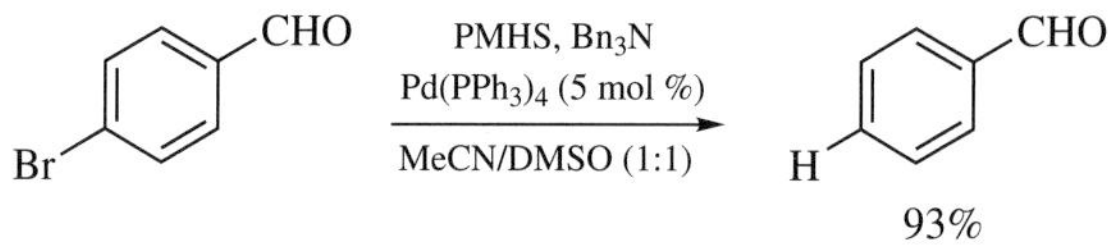

The PMHS/Sn and PMHS/Pd protocols tolerate a wide range of organic functionality including alcohols, alkenes, carbonyls, and nitro groups, thus complementing LiEt$_3$BH, catalytic hydrogenation, and other means of effecting halide reductions.

Finally, PMHS with sodium metal can be used to reduce aromatic chlorides (eq 24).[42]

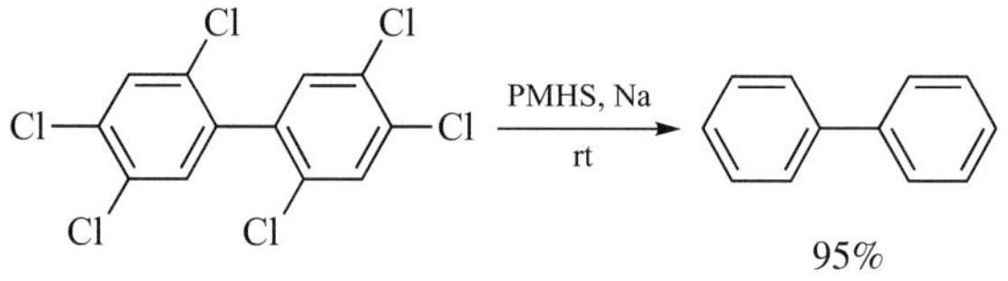

Reductions of P–O Bonds. Phosphine oxides can be efficiently reduced to phosphines[43] using stoichiometric amounts of Ti(O-*i*-Pr)$_4$ and PMHS at 50 °C (vs. 250 °C without Ti[44]). The reduction proceeds via a *syn* hydrotitanation and goes with retention of configuration when the phosphine oxide is chiral (eq 25).[45] The PMHS/Ti reagent combination has also proven amenable to reaction with polymer supported materials.[46]

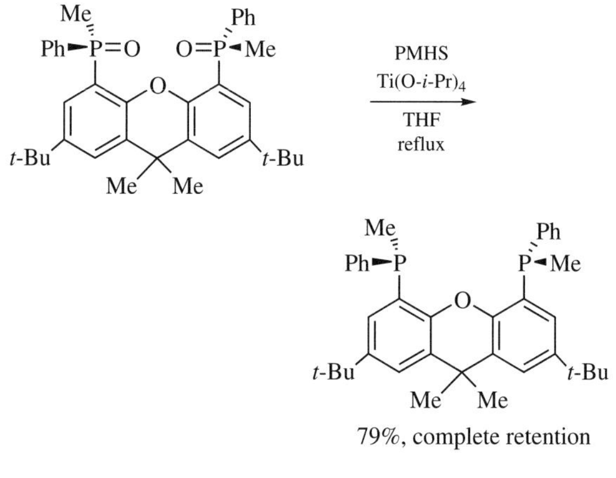

Miscellaneous Reductions. Carbon-sulfur bonds have been reduced using PMHS and (Bu$_3$Sn)$_2$O with AIBN (eq 26).[47] PMHS can also serve as a substitute for hydrogen in Pd/C catalytic hydrogenations of aromatic nitro groups and various alkenes including those of α,β-unsaturated ketones and esters (eq 27). It is useful to note that electron-rich *trans*-alkenes are not reduced.[14] PMHS and catalytic Pd(PPh$_3$)$_4$ are superior to standard hydrogenation conditions for the reduction of acyl fluorides to their corresponding aldehydes (eq 28).[48]

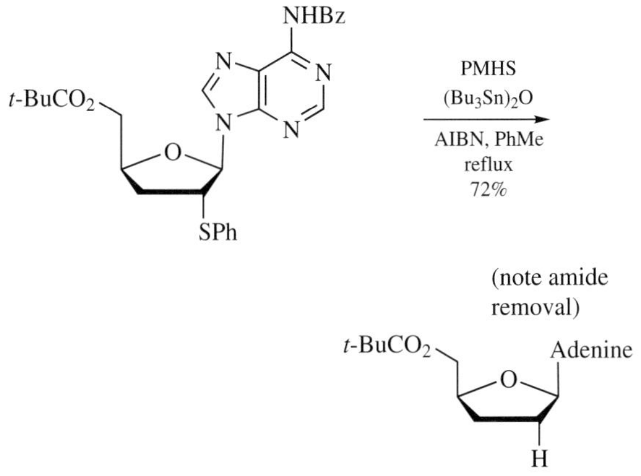

(note amide removal)

(26)

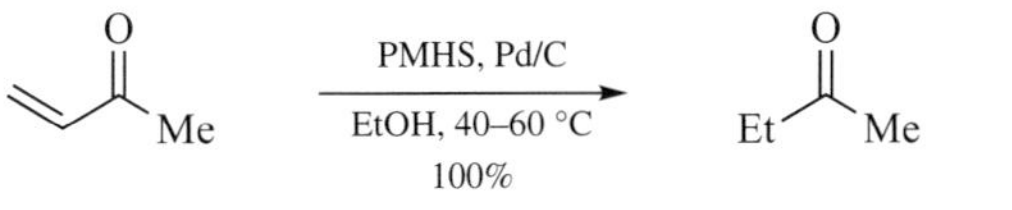

(27)

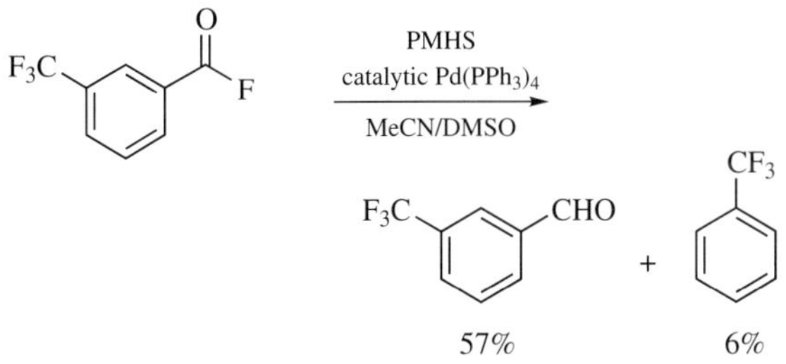

(28)

57%   6%

PMHS has also proven useful in the removal of various protective groups. For example, PMHS and catalytic Pd(PPh₃)₄ in the presence of ZnCl₂ allow for the selective cleavage of allylic ethers, amines, or esters in the presence of PMB, benzyl, TBS and other ethers (eq 29).[49]

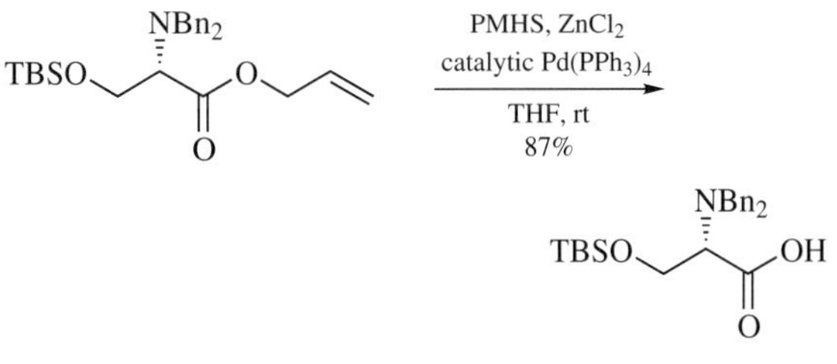

(29)

Benzylidene acetals of 1,2- or 1,3-diols can be opened with PMHS and AlCl₃ to generate a mono-benzylated alcohol, where the benzyl group ends up at the most sterically hindered alcohol (eq 30).[50] Azides, esters, and TBS-protected alcohols are among the functional groups that tolerate these conditions.

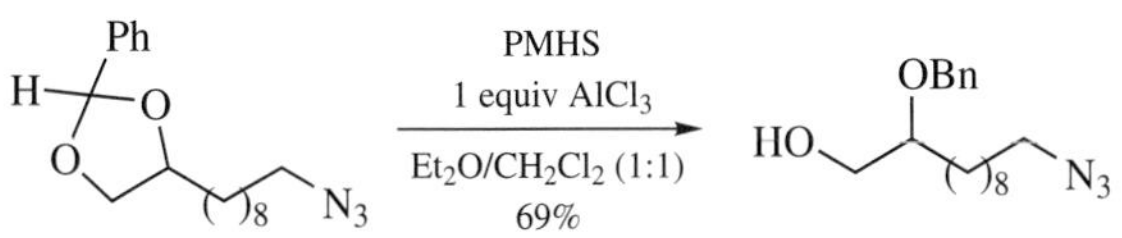

(30)

C–C Bond Forming and Related Reactions. In situ generated organotin hydrides can also be used in free radical C-C bond forming reactions. Terstiege and Maleczka have carried out such reactions with catalytic amounts of tin via their PMHS/fluoride/R₃Sn-X method for generating R₃SnH (eq 31).[8]

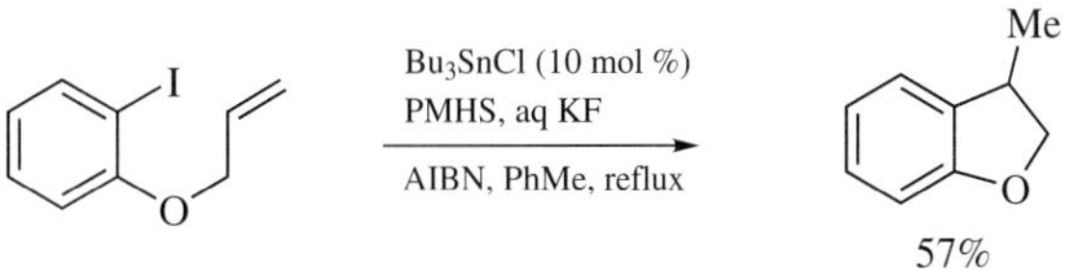

(31)

The same group has also used in situ generated R₃SnH in free radical and palladium-mediated hydrostannations of alkynes (eq 32).[9] This one-pot R₃SnH generation/Pd(0) hydrostannation protocol can also be coupled with an in situ Stille reaction. In such a sequence, PMHS can aid recycling of the tin by-product from the cross coupling, rendering the entire process catalytic in tin (eq 33).[10] Finally, in a testament to the thermal stability of PMHS, the stoichiometric R₃SnH generation/ hydrostannation/ Stille one-pot protocol can be carried out under microwave irradiation.[11]

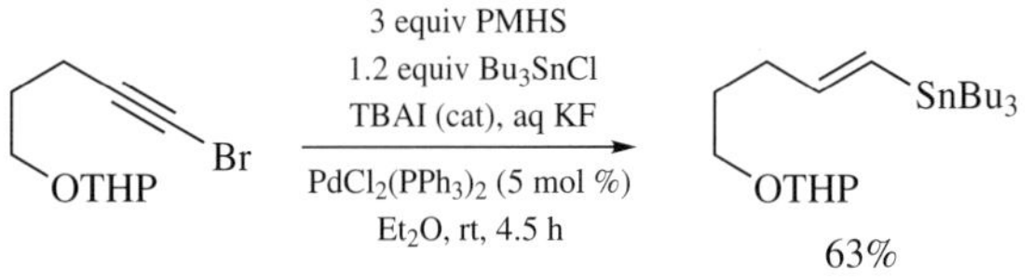

(32)

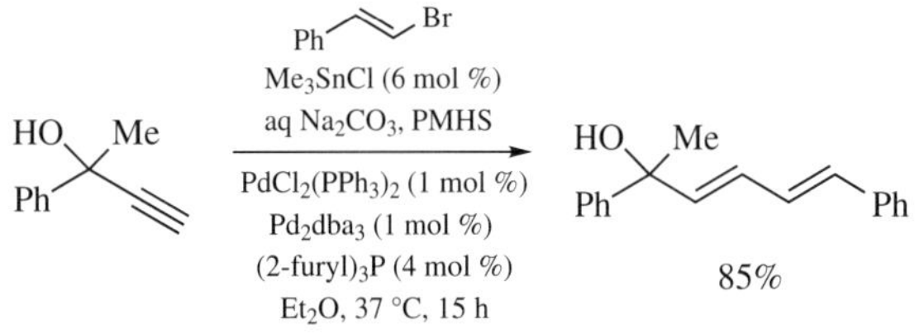

(33)

Several Friedel-Crafts processes have also exploited the reactivity of PMHS (eq 34),[51] and PMHS can serve as a substitute for R₃SnH in Pd(0)-mediated carbonylations (eq 35).[52]

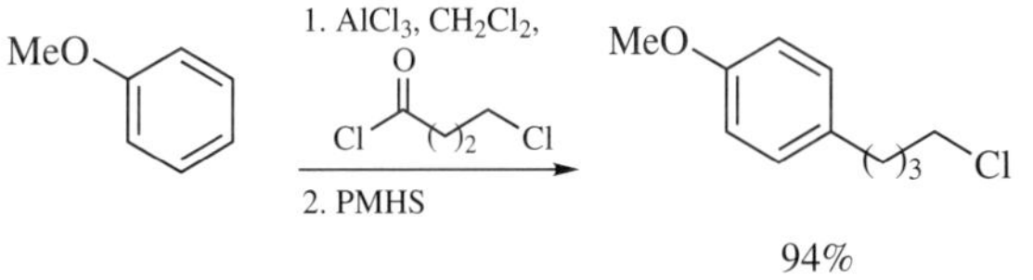

(34)

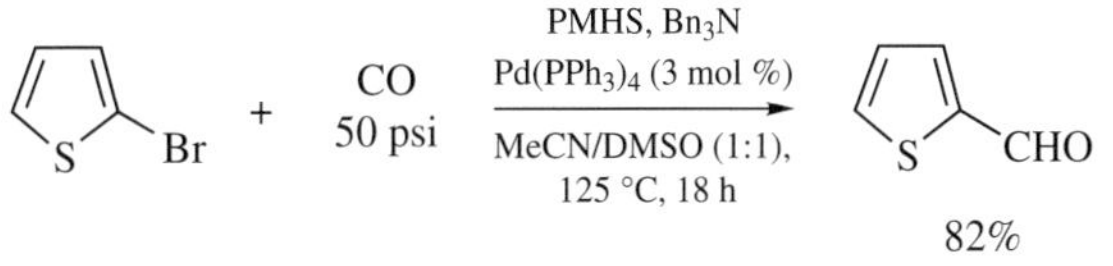

(35)

Final Notes. The polymeric nature of PMHS can make GC analysis of reaction mixtures difficult, however Lopez et al. found that substituting PMHS-dimer [16066-09-4] for PMHS facilitates such analysis with little effect on reactivity.[5] In terms of purification, the insoluble gels that can form during the reaction or after hydrolytic work-up can clog filter paper and glass frits. It is usually possible to remove this material by passing it over a wide pad of silica gel. Alternatively, we have found that freezing the crude reaction mixture in benzene over night leads to a more granular and filterable PMHS waste product.

1. Sauer, R.; Scheiber, W. J.; Brewer, S. D., *J. Am. Chem. Soc.* **1946**, *68*, 962.

2. (a) Lawrence, N. J.; Drew, M. D.; Bushell, S. M., *J. Chem. Soc., Perkin Trans. 1* **1999**, 3381. (b) Lipowitz, J.; Bowman, S. A., *Aldrichimica Acta*

1973, *6*, 1. (c) Fieser, L. F.; Fieser, M., in *Reagents for Organic Synthesis*; Wiley: New York, 1974, Vol. 4, p 393.

3. Hayashi, K.; Iyoda, J.; Shiihara I., *J. Organomet. Chem.* **1967**, *10*, 81.

4. Deleuze, H.; Maillard, B., *J. Organomet. Chem.* **1995**, *490*, C-14.

5. Lopez, R. M.; Hays, D. S.; Fu, G. C., *J. Am. Chem. Soc.* **1997**, *119*, 6949

6. Matlin, S. A.; Gandham, P. S., *J. Chem. Soc, Chem. Commun.* **1984**, 798.

7. Hays, D. S.; Fu, G. C., *J. Org. Chem.* **1997**, *62*, 7070.

8. Terstiege, I.; Maleczka, R. E, Jr, *J. Org. Chem.* **1999**, *64*, 342.

9. Maleczka, R. E., Jr; Terrell, L. R.; Clark, D. H.; Whitehead, S. L.; Gallagher, W. P.; Terstiege, I., *J. Org. Chem.* **1999**, *64*, 5958.

10. (a) Maleczka, R. E., Jr; Gallagher, W P.; Terstiege, I., *J. Am. Chem. Soc.* **2000**, *122*, 384. (b) Gallagher, W. P.; Terstiege, I.; Maleczka, R. E., Jr, *J. Am. Chem. Soc.* **2001**, *123*, 3194.

11. Maleczka, R. E., Jr; Lavis, J. M.; Clark, D. H.; Gallagher, W. P., *Org. Lett.* **2000**, *2*, 3655.

12. Grady, G. L.; Kuivila, H. G., *J. Org. Chem.* **1969**, *34*, 2014.

13. Nitzsche, S.; Wick, M., *Angew. Chem.* **1957**, *69*, 96.

14. Lipowitz, J.; Bowman, S. A., *J. Org. Chem.* **1973**, *38*, 162.

15. Lawrence, N. J.; Bushell, S. M., *Tetrahedron Lett.* **2000**, *41*, 4507.

16. Mimoun, H., *J. Org. Chem.* **1999**, *64*, 2582.

17. Mimoun, H.; de Saint Laumer, J.-Y.; Giannini, L.; Scopelliti, R.; Floriani, C., *J. Am. Chem. Soc.* **1999**, *121*, 6158.

18. Lipshutz, B. H.; Chrisman, W.; Noson, K.; Papa, P.; Sclafani, J. A.; Vivian, R. W.; Keith, J. M., *Tetrahedron* **2000**, *56*, 2779.

19. Lipshutz, B. H.; Chrisman, W.; Noson, K., *J. Organomet. Chem.* **2001**, *624*, 367.

20. Appella, D. H.; Moritani, Y.; Shintani, R.; Ferreira, E. M.; Buchwald, S. L., *J. Am. Chem. Soc.* **1999**, *121*, 9473.

21. Drew, M. D.; Lawrence, N. J.; Fontaine, D.; Sehkri, L.; Bowles, S. A.; Watson, W., *Synlett* **1997**, 989.

22. Chuit, C.; Corriu, R. J. P.; Perz, R.; Reyé, C., *Synthesis* **1982**, 981.

23. Chuit, C.; Corriu, R. J. P.; Reye, C.; Young, J. C., *Chem. Rev.* **1993**, *93*, 1371.

24. Kobayashi, Y.; Takahisa, E.; Nakano, M.; Watatani, K., *Tetrahedron* **1997**, *53*, 1627.

25. Drew, M. D.; Lawrence, N. J.; Watson, W.; Bowles, S. A., *Tetrahedron Lett.* **1997**, *38*, 5857.

26. (a) Barr, K. J.; Berk, S. C.; Buchwald, S. L., *J. Org. Chem.* **1994**, *59*, 4323. (b) Carter, M. B.; Schiøtt, B.; Gutiérrez, A.; Buchwald, S. L., *J. Am. Chem. Soc.* **1994**, *116*, 11667.

27. (a) Verdaguer, X.; Berk, S. C.; Buchwald, S. L., *J. Am. Chem. Soc.* **1995**, *117*, 12641. (b) Verdaguer, X.; Hansen, M. C.; Berk, S. C.; Buchwald, S. L., *J. Org. Chem.* **1997**, *62*, 8522.

28. Reding, M. T.; Buchwald, S. L., *J. Org. Chem.* **1995**, *60*, 7884.

29. Breeden, S. W.; Lawrence, N. J., *Synlett* **1994**, 833.

30. Halterman, R. L.; Ramsey, T. M.; Chen, Z., *J. Org. Chem.* **1994**, *59*, 2642.

31. Aizpurua, J. M.; Lecea, B.; Palomo, C., *Can. J. Chem.* **1986**, *64*, 2342.

32. Conway, R. J.; Nagel, J. P.; Stick, R. V.; Tilbrook, D. M. G., *Aust. J. Chem.* **1985**, *38*, 939.

33. Chandrasekhar, S.; Reddy, M. V.; Chandraiah, L., *Synlett* **2000**, *9*, 1351.

34. Chandrasekhar, S.; Ahmed, M., *Tetrahedron Lett.* **1999**, *40*, 9325.

35. Lopez, R. M.; Fu, G. C., *Tetrahedron* **1997**, *53*, 16349.

36. Chandrasekhar, S.; Reddy, M. V.; Chandraiah, L., *Synthetic Commun.* **1999**, *29*, 3981.

37. Hays, D. S.; Fu, G. C., *J. Org. Chem.* **1998**, *63*, 2796.

38. Chandrasekhar, S.; Reddy, Ch. R.; Ahmed, M., *Synlett* **2000**, *11*, 1655.

39. (a) Verdaguer, X.; Lange, U. E. W.; Buchwald, S. L., *Angew. Chem., Int. Ed. Engl.* **1998**, *37*, 1103. (b) Hansen, M. C.; Buchwald, S. L., *Tetrahedron Lett.* **1999**, *40*, 2033. (c) Hansen, M. C.; Buchwald, S. L., *Org. Lett.* **2000**, *2*, 713.

40. Grignon-Dubois, M.; Dunogues, J., *J. Organomet. Chem.* **1986**, *309*, 35.

41. Pri-Bar, I.; Buchman, O., *J. Org. Chem.* **1986**, *51*, 734.

42. (a) Hawari, J., *J. Organomet. Chem.* **1992**, *437*, 91. (b) *US Patent* 4 973 783, 27 Nov 1990.

43. (a) Warren, S.; Wyatt, P., *Tetrahedron Asymm* **1996**, *7*, 989. (b) Warren, S.; Wyatt, P., *J. Chem. Soc., Perkin Trans. 1* **1998**, 249. (c) Russel, M. G.; Warren, S., *Tetrahedron Lett.* **1998**, *39*, 7995. (d) Lawrence, N. J.; Muhammad, F., *Tetrahedron* **1998**, *54*, 15361. (e) Lawrence, N. J.; Beynek, H., *Synlett* **1998**, 497. (f) Ariffin, A.; Blake, A. J.; Li, W.-S.; Simpkins, N. S., *Synlett* **1997**, 1453. (g) Coumbe, T.; Lawrence, N. J.; Muhammad, F., *Tetrahedron Lett.* **1994**, *35*, 625.

44. (a) Fritzshe, H.; Hasserodt, U.; Korte, F., *Chem. Ber.* **1964**, *97*, 1988. (b) Fritzshe, H.; Hasserodt, U.; Korte, F., *Chem. Ber.* **1965**, *98*, 1681.

45. Hamada, Y.; Matsuura, F.; Oku, M.; Hatano, K.; Shioiri, T., *Tetrahedron Lett.* **1997**, *38*, 8961.

46. Sieber, F.; Wenworth, P., Jr; Janda, K. D., *Molecules* **2000**, *5*, 1018.

47. Kawakami, H.; Ebata, T.; Koseki, K.; Okano, K.; Matsumoto, K.; Matsushita, H., *Heterocycles* **1993**, *36*, 2765.

48. Braden, R.; Himmler, T., *J. Organomet. Chem.* **1989**, *367*, C12.

49. Chandrasekhar, S.; Reddy, Ch. R.; Rao, R. J., *Tetrahedron* **2001**, *57*, 3435.

50. Chandrasekhar, S.; Reddy, Y. R.; Reddy, Ch. R., *Chem. Lett.* **1998**, 1273.

51. Jaxa-Chamiec, A.; Shah, V. P.; Kruse, L. I., *J. Chem. Soc., Perkin Trans. 1* **1989**, 1705.

52. Pri-Bar, I.; Buchman, O., *J. Org. Chem.* **1984**, *49*, 4009.

Jérôme M. Lavis & Robert E. Maleczka Jr
Michigan State University, Michigan, USA

Potassium Hexamethyldisilazide

[40949-94-8] $C_6H_{18}KNSi_2$ (MW 199.53)

InChI = 1S/C6H18NSi2.K/c1-8(2,3)7-9(4,5)6;/h1-6H3;/q-1;+1

InChIKey = IUBQJLUDMLPAGT-UHFFFAOYSA-N

(sterically hindered base)

Alternate Name: KHMDS; potassium bis(trimethylsilyl)amide.

Solubility: soluble in THF, ether, benzene, toluene.[1]

Form Supplied in: commercially available as moisture-sensitive, tan powder, 95% pure, and 0.5 M solution in toluene.

Analysis of Reagent Purity: solid state structures of $[KN(SiMe_3)_2]_2$[3] and $[KN(SiMe_3)_2 \cdot 2 \text{ toluene}]_2$[4] have been determined by X-ray diffraction; solutions may be titrated using fluorene,[2] 2,2'-bipyridine,[5] and 4-phenylbenzylidene benzylamine[6] as indicators.

Preparative Methods: prepared and isolated by the procedure of Wannagat and Niederpruem.[1] A more convenient in situ generation from potassium hydride and hexamethyldisilane is described by Brown.[2]

Handling, Storage, and Precautions: the dry solid and solutions are inflammable and must be stored in the absence of moisture. These should be handled and stored under a nitrogen atmosphere. Use in a fume hood.

Original Commentary

Brett T. Watson

Bristol-Myers Squibb Pharmaceutical Research Institute, Wallingford, CT, USA

Use as a Sterically Hindered Base for Enolate Generation.
Potassium bis(trimethysilyl)amide, $KN(TMS)_2$, has been shown to be a good base for the formation of kinetic enolates from carbonyl groups bearing α-hydrogens.[7] For example, treatment of 2-methylcyclohexanone with $KN(TMS)_2$ at low temperature followed by trapping with triethylborane and iodomethane gave good selectivity for 2,6-dimethylcyclohexanone (eq 1). In comparison, the use of potassium hydride for this transformation gave good selectivity for 2,2-dimethylcyclohexanone, which is the product derived from the thermodynamic enolate (eq 2).[8]

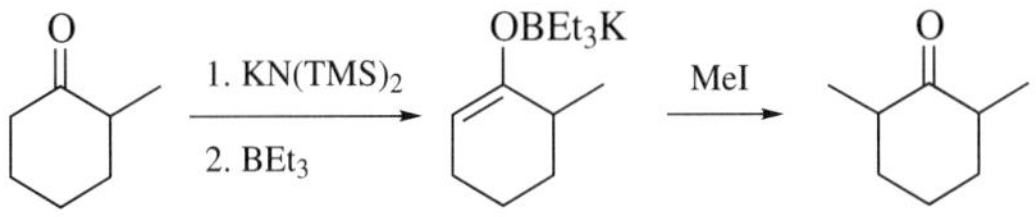

$$93\% \text{ 2,6-}$$
$$86\% \text{ yield} \tag{1}$$

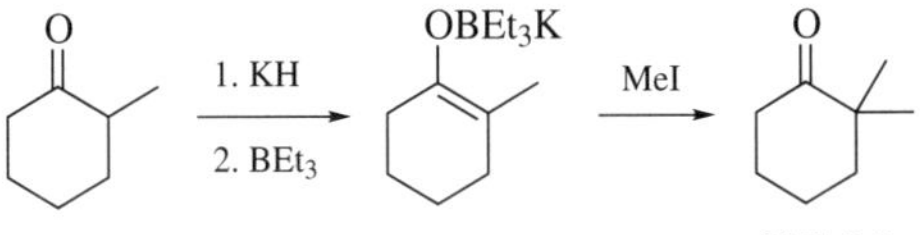

$$90\% \text{ 2,2-}$$
$$79\% \text{ yield} \tag{2}$$

This reagent has been shown to be a good base for the generation of highly reactive potassium enolates;[9] for example, treatment of various ketones and esters bearing α-hydrogens with $KN(TMS)_2$ followed by 2 equiv of N-F-saccharinsultam allowed isolation of the difluorinated product (eq 3).

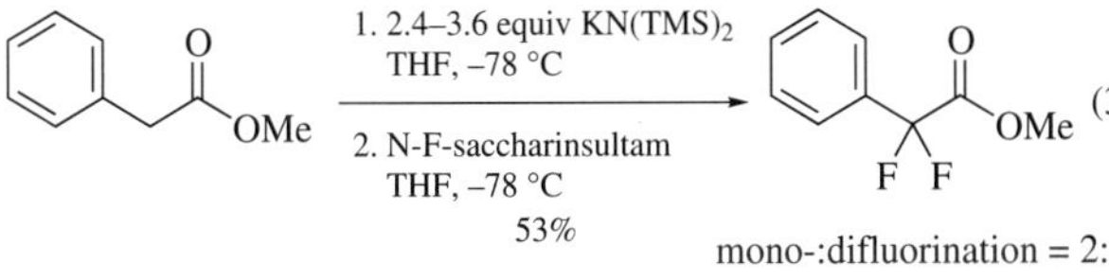

$$\text{mono-:difluorination} = 2:98$$

In a study on the electrophilic azide transfer to chiral enolates, Evans[10] found that the use of potassium bis(trimethylsilyl)amide was crucial for this process. The $KN(TMS)_2$ played a dual role in the reaction; as a base, it was used for the stereoselective generation of the (Z)-enolate (1). Reaction of this enolate with trisyl azide gave an intermediate triazene species (2) (eq 4). The potassium counterion from the $KN(TMS)_2$ used for enolate formation was important for the decomposition of the triazene to the desired azide. Use of other hindered bases such as lithium hexamethyldisilazide allowed preparation of the intermediate triazene; however, the lithium ion did not catalyze the decomposition of the triazene to the azide.[10a] This methodology has been utilized in the synthesis of cyclic tripeptides.[10b]

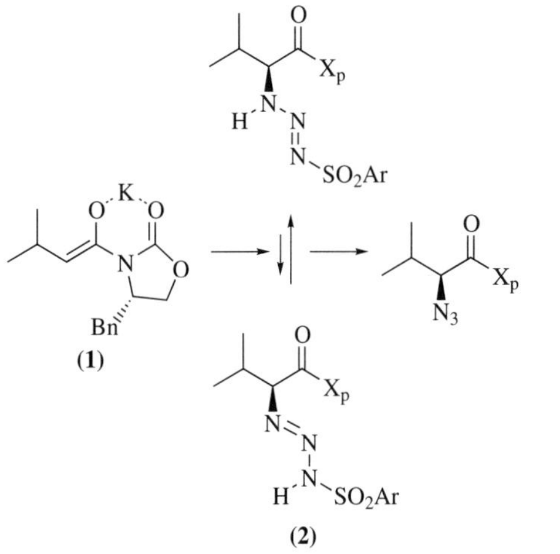

Treatment of carbonyl species bearing acidic α-hydrogens with potassium bis(trimethylsilyl)amide has also been shown to generate anions which, due to the larger, less coordinating potassium cation, allow the negative charge to be stabilized by other features in the molecule rather than as the potassium enolate. Treatment of 9-acetyl-*cis,cis,cis,cis*-cyclonona-1,3,5,7-tetraene with this reagent gave an anionic species which was characterized by spectroscopic methods to be more like the [9]-annulene anion than the nonafulvene enolate. In this case the negative charge is more fully stabilized by delocalization into the ring to form the aromatic species rather than as the potassium enolate. Use of the bis(trimethylsilyl)amide bearing the more strongly coordinating lithium cation led to an intermediate which appeared to be lithium nonafulvene enolate. Addition of chlorotrimethylsilane to each of these intermediates gave the same nonafulvenesilyl enol ether (eq 5).[11]

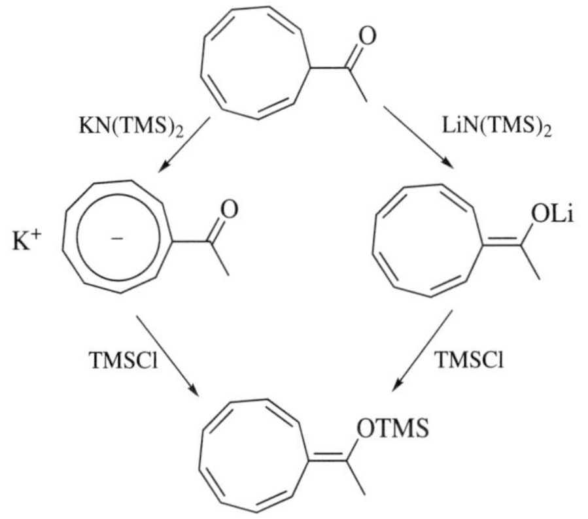

Selective Formation of Linear Conjugated Dienolates. Potassium bis(trimethylsilyl)amide has been shown to be an efficient base for the selective generation of linear-conjugated dienolates from α,β-unsaturated ketones.[12] As shown in eqs 6 and 7, treatment of both cyclic and acyclic α,β-unsaturated enones with $KN(TMS)_2$ in a solvent mixture of DMF/THF (2:1) followed by quenching with methyl chloroformate gave excellent selectivities for the products derived from the linear dienolate anion. In comparison, the use of lithium bases for this reaction gave products derived from the cross-conjugated dienolate anions. This methodology, however, did not work for 1-cyclohexenyl methyl ketone, in which case the product from the cross-conjugated dienolate anion was isolated exclusively (eq 8).

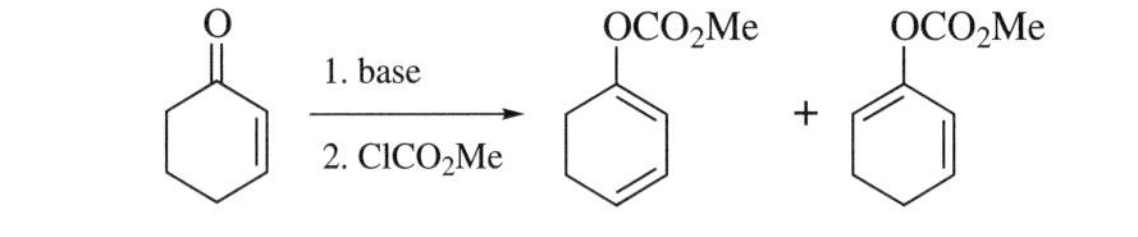

Base	Linear	Cross	Yield (%)
KN(TMS)$_2$	>99	–	34
LiN(TMS)$_2$	75	25	68
LDA	–	99	44

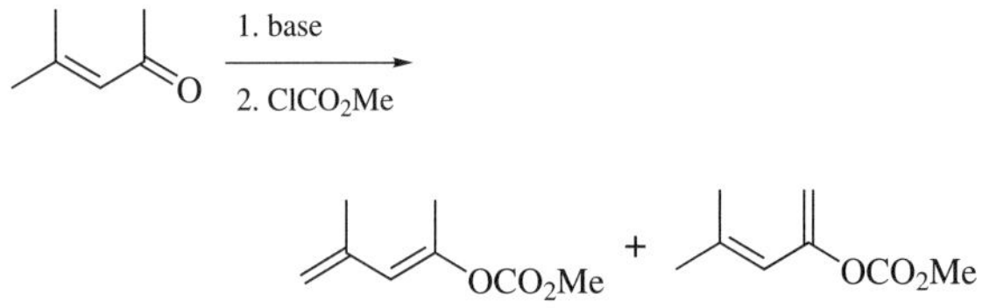

Base	Linear	Cross	Yield (%)
KN(TMS)$_2$	99	–	34
LDA	16	84	50

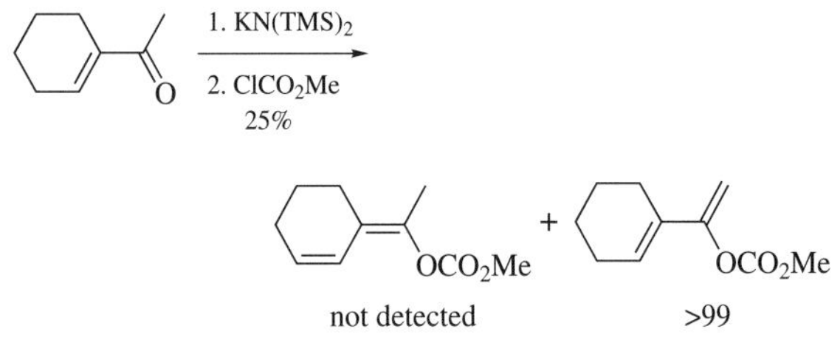

Stereoselective Generation of Alkyl (Z)-3-Alkenoates. Deconjugative isomerization of 2-alkenoates to 3-alkenoates occurs via γ-deprotonation of the α,β-unsaturated ester to form an intermediate dienolate anion. In most cases, the α-carbon is more reactive to protonation[13] and allows for the isolation of the 3-alkenoate. If the C-4 position bears a methyl group, this transformation is usually stereospecific, leading to the (Z)-3-alkenoate; however, when groups larger than a methyl occupy the C-4 position, the reaction becomes increasingly stereorandom.[13d] Potassium bis(trimethylsilyl)amide, however, was shown to be a good base for the stereoselective isomerization of 2,4-dimethyl-3-pentyl (E)-2-dodecenoate, which bears a long C-4 substituent, to the corresponding (Z)-3-dodecenoate (eq 9).[14a]

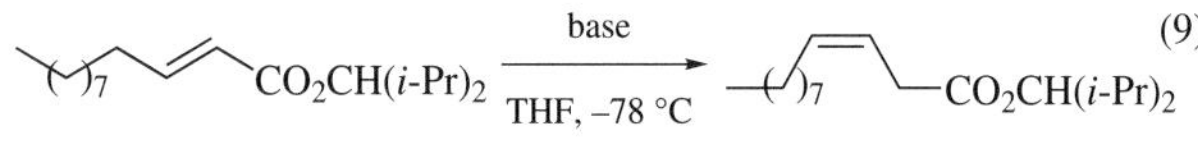

Base	(Z):(E)	Yield (%)
LDA/HMPA	84:16	85
KN(TMS)$_2$	97:3	64

This reagent has been used to stereoselectively prepare (Z)-3-alkenoate moieties for use in the syntheses of insect pheromones.[14]

Generation of α-Keto Acid Equivalents (Dianions of Glycolic Acid Thioacetals). Potassium bis(trimethylsilyl)amide was found to be the optimal reagent for the generation of the dianion of glycolic acid thioacetals. This reagent may be used to effect a

nucleophilic α-keto acid homologation. Treatment of the starting bis(ethylthio)acetic acid with KN(TMS)$_2$ proceeded to give the corresponding soluble dianionic species. This underwent alkylation with a variety of halides and tosylates (eq 10) and subsequent hydrolysis allowed isolation of the desired α-keto acids.[15]

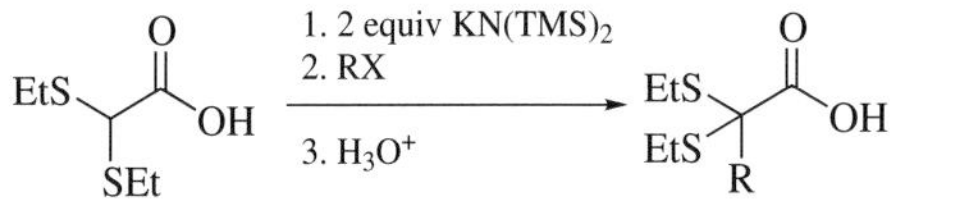

RX	Yield (%)
MeI	100
EtOTs	100
i-PrOTs	72
CyOTs	64
PhCH$_2$Cl	100

The dianion was also shown to undergo ring-opening reactions with epoxides and aziridines (eq 11).

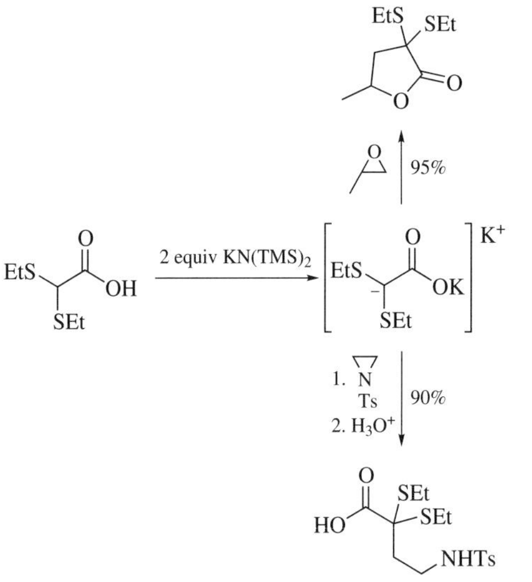

Generation of Ylides and Phosphonate Anions.

Ylides. In the Wittig reaction, lithium salt-free conditions have been shown to improve (Z/E) ratios of the alkenes which are prepared;[16] sodium hexamethyldisilazide has been shown to be a good base for generating these conditions. In a Wittig-based synthesis of (Z)-trisubstituted allylic alcohols, potassium bis(trimethylsilyl)amide was shown to be the reagent of choice for preparing the starting ylides.[17] These were allowed to react with protected α-hydroxy ketones and depending upon the substitution pattern of the ylide and/or the ketone, stereoselectivities ranging from good to excellent were achieved (eqs 12–14).

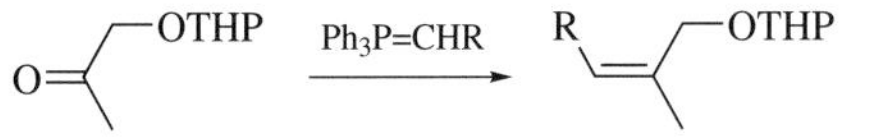

R	(Z):(E)	Yield (%)
n-Pr	60:1	87
i-Pr	6:1	45

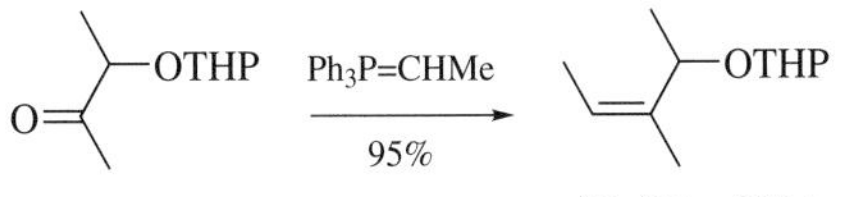

(13)

$(Z):(E) = 200:1$

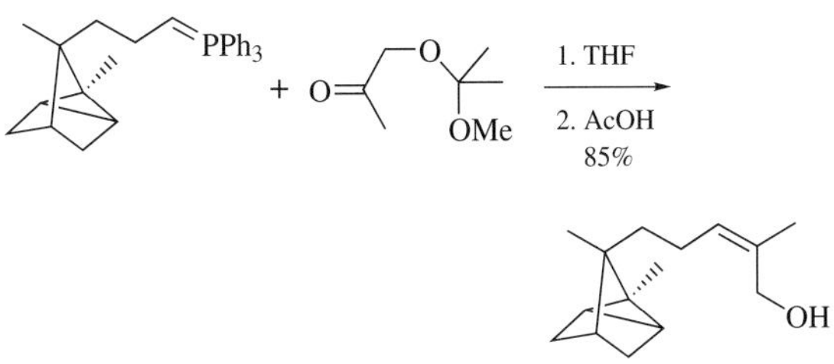

(14)

>99% stereoisomeric purity

(16)

(17)

(18)

95% *cis* + 5% *trans*

Phosphonates. In a Horner–Emmons-based synthesis of di- and trisubstituted (Z)-α,β-unsaturated esters, the strongly dissociated base system of potassium bis(trimethylsilyl)amide/18-crown-6 was used to prepare the desired phosphonate anions. This base system, coupled with highly electrophilic bis(trifluoroethyl)phosphono esters, gave phosphonate anions which, when allowed to react with aldehydes, gave excellent selectivity for the (Z)-α,β-unsaturated esters (eq 15).[18]

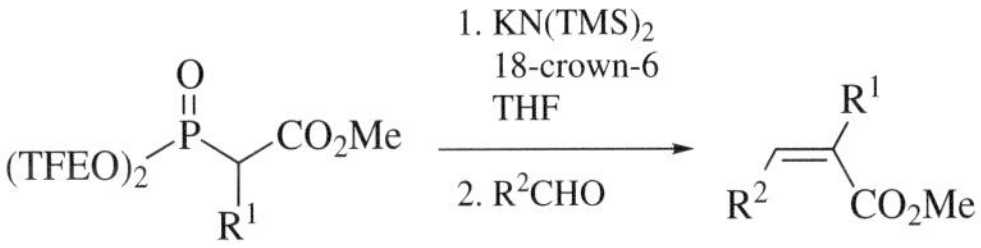

(15)

R^{2}CHO	R^1	$(Z):(E)$	Yield (%)
Me(CH$_2$)$_6$CHO	H	12:1	90
Me(CH$_2$)$_6$CHO	Me	46:1	88
Me(CH$_2$)$_2$CH=CHCHO	H	>50:1	87
Me(CH$_2$)$_2$CH=CHCHO	Me	>50:1	79
CyCHO	H	4:1	71
CyCHO	Me	>50:1	80
PhCHO	H	>50:1	95
PhCHO	Me	30:1	95

Intramolecular Cyclizations.

Haloacetal Cyclizations. Intramolecular closure of a carbanion onto an α-haloacetal has been shown to be a valuable method for the formation of carbocycles.[19a] Potassium bis(trimethylsilyl) amide was found to be the most useful base for the formation of the necessary carbanions. This methodology may be used for the formation of single carbocycles (eq 16), for annulation onto existing ring systems (eq 17), and for the formation of multiple ring systems in a single step (eq 18). In the case of annulations forming decalin or hydrindan systems, this ring closure proceeded to give largely the *cis*-fused bicycles (eq 17). For the reaction shown in eq 18, in which two rings are being formed, the stereochemistry of the ring closure was found to be dependent upon the counter ion of the bis(trimethylsilyl)amide; use of the potassium base allowed isolation of the *cis*-decalin system as the major product (95%), whereas use of lithium bis(trimethylsilyl)amide led to the isolation of the *trans*-decalin (95%).[19]

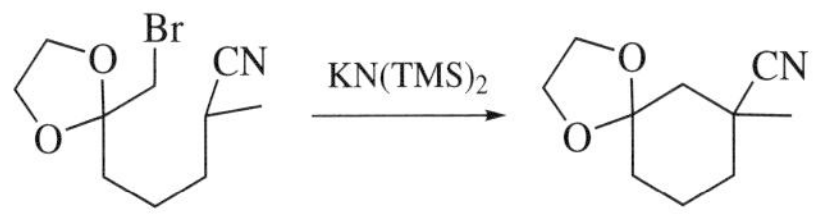

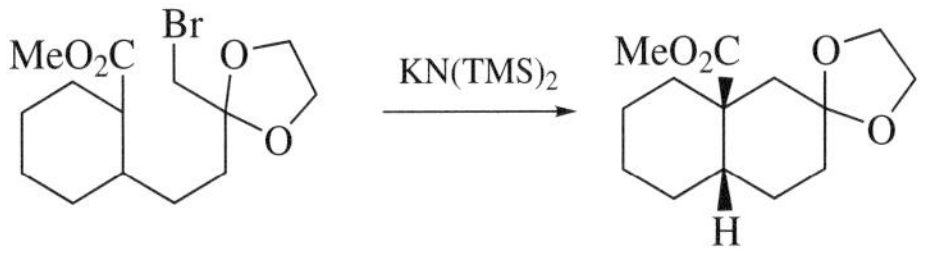

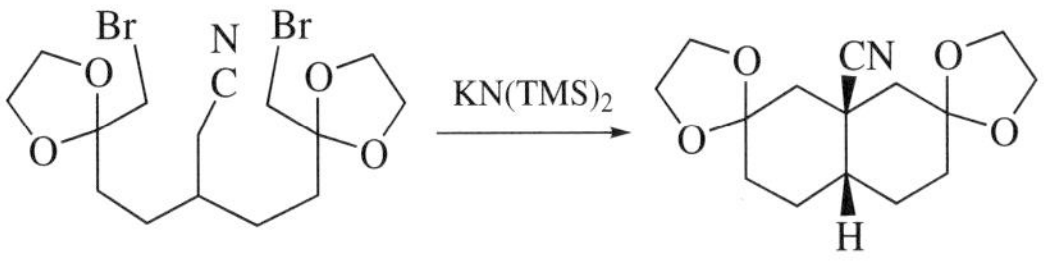

Intramolecular Lactonization. In a general method for the formation of 14- and 16-membered lactones via intramolecular alkylation,[20] potassium bis(trimethylsilyl) amide was shown to be a useful base for this transformation (eq 19).

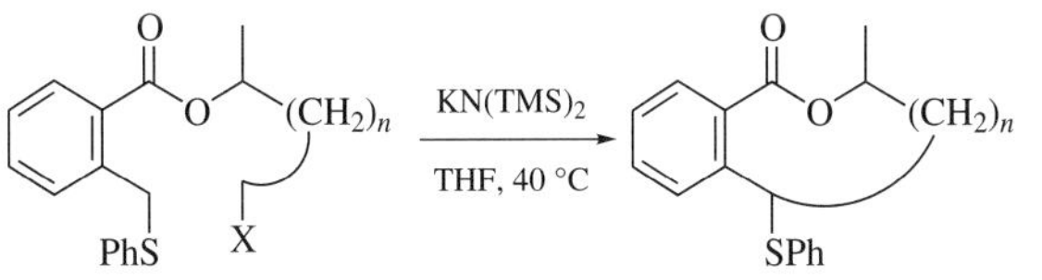

(19)

$n = 5, 0\%;\ 7, 75\%;\ 8, 71\%$

Intramolecular Rearrangement. Potassium bis(trimethylsilyl)amide was shown to be a good base for the generation of a diallylic anion which underwent a biogenetically inspired intramolecular cyclization, forming $(\pm)$-dictyopterene B (eq 20).[21]

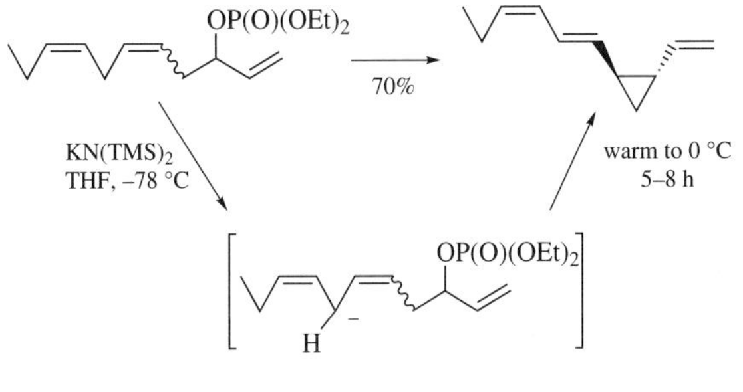

(20)

Synthesis of Vinyl Fluorides. Addition of potassium bis(trimethylsilyl)amide to β-fluoro-β-silyl alcohols was shown to selectively effect a Peterson-type alkenation reaction to form vinyl fluorides (eq 21).[22] Treatment of a primary β-fluoro-β-silyl alcohol with KN(TMS)$_2$ led cleanly to the terminal alkene. Use of a *syn*-substituted secondary alcohol led to the stereoselective formation of the (Z)-substituted alkene (eq 22); reaction of the *anti*-isomer, however, demonstrated no $(Z:E)$ selectivity.

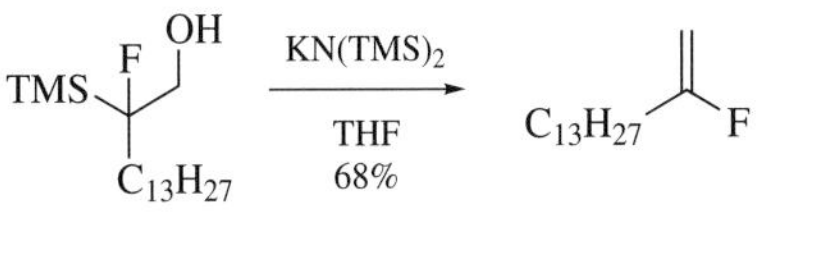

(21)

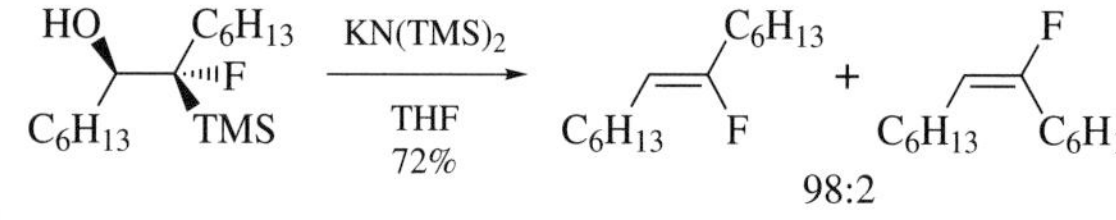

$$\text{(22)}$$

Oxyanionic Cope Rearrangement. Potassium bis(trimethylsilyl)amide/18-crown-6 was shown to be a convenient alternative to potassium hydride for the generation of anions for oxyanionic Cope rearrangements (eq 23).[23]

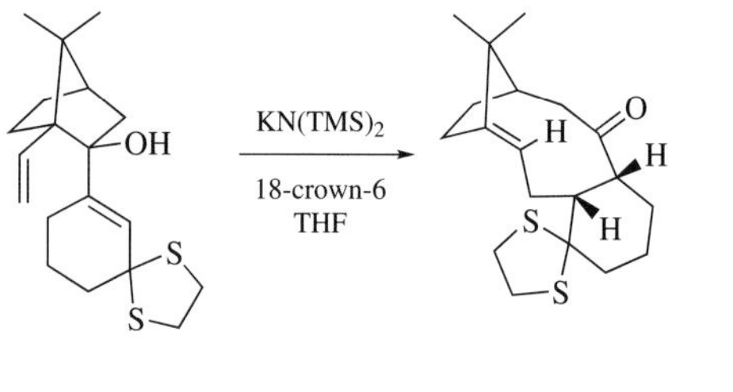

$$\text{(23)}$$

Stereoselective Synthesis of Functionalized Cyclopentenes. Potassium bis(trimethylsilyl)amide was shown to be an effective base for the base-induced ring contraction of thiocarbonyl Diels–Alder adducts (eq 24).[24] lithium diisopropylamide has also been shown to be equally effective for this transformation.

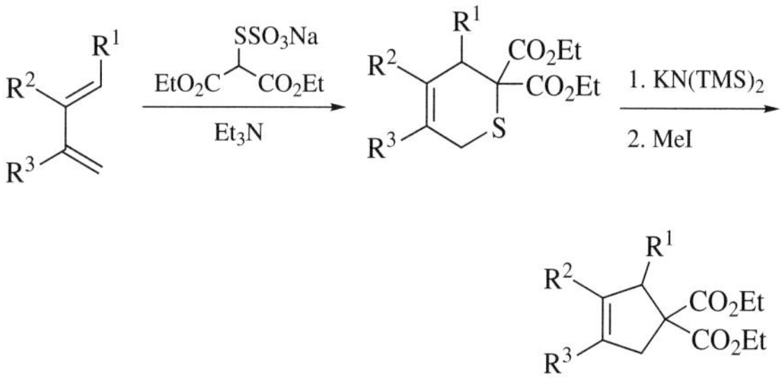

$$\text{(24)}$$

First Update

Professor Hélène Lebel
Université de Montréal, Montréal, Canada

Nitrogen Source: Synthesis of Amines. KHMDS has been exploited as a nitrogen source in a few systems. For instance, the ring opening of (trifluoromethyl)oxirane with KHMDS, followed by acidic hydrolysis produced the desired amino alcohol in 61% yield without any variation in the enantiomeric purity (eq 25).[25]

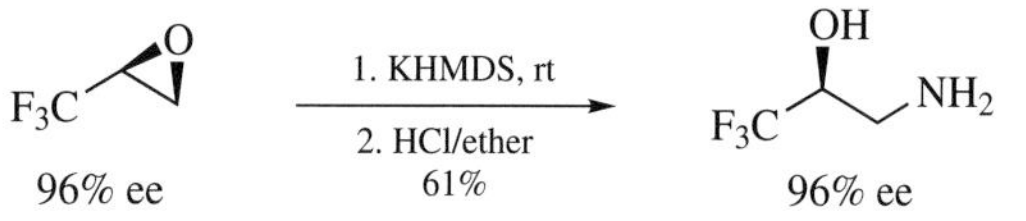

$$\text{(25)}$$

KHMDS was shown to be as effective as LHMDS[26] in the boron-assisted displacement of secondary chloride which led to the corresponding protected amine with inversion of configuration (eq 26).[27] The chemoselective displacement of the chloride is greatly facilitated by the α-boro substituent, which is known to cause a rate acceleration on the attack of nucleophiles of approximately two orders of magnitude relative to the primary halide.[28]

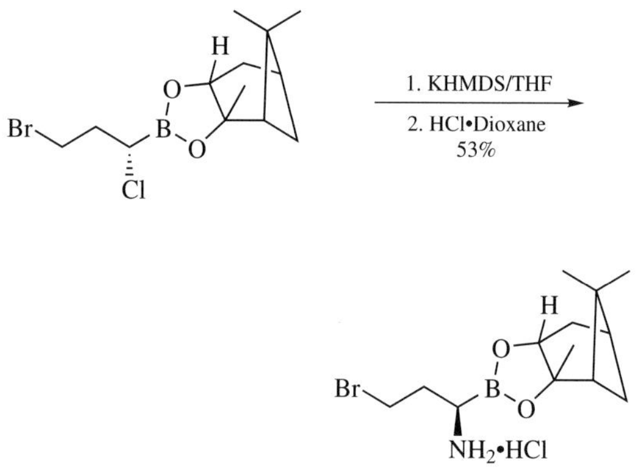

$$\text{(26)}$$

KHMDS was also used in the synthesis of aminocinnolines from trifluoromethylhydrazones (eq 27).[29] The deprotonation of the hydrazone is followed by the spontaneous loss of the fluoride anion to produce the corresponding difluoroalkene. Subsequent nucleophilic attack of KHMDS and cyclization with loss of the second fluoride leads to an intermediate that upon aromatization produces the desired product.

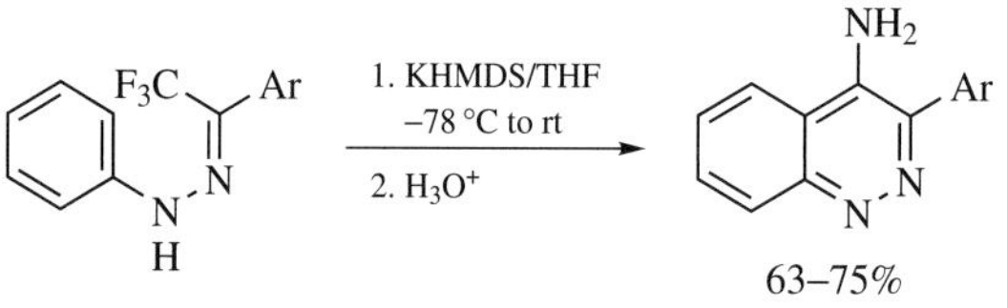

$$\text{(27)}$$

Both LHMDS and KHMDS have been successfully utilized as nitrogen sources in palladium(0) catalyzed aminations of allyl chlorides (eq 28). Up to 90% conversion and 55% isolated yield were obtained for the synthesis of allyl hexamethyldisilazide and no cyclization via deprotonation or significant hydrolysis during the work-up was observed.[30]

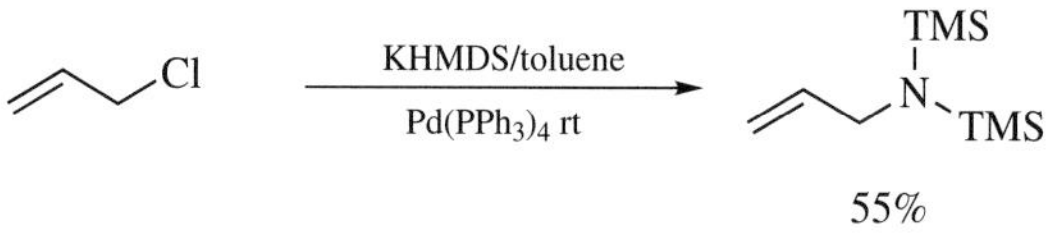

$$\text{(28)}$$

Deprotonation of Imidazolium and Pyrazolium Salts: Synthesis of N-Heterocyclic Carbene Ligands. Potassium bases, such as potassium t-butoxide[31] and KHMDS, have been selected for the deprotonation of imidazolium and pyrazolium salts to generate N-heterocyclic carbene ligands, which have emerged over the last decade as an important part of transition metal-catalyzed homogeneous catalysis.[32,33] For instance, 1,1'-methylene-3,3'-di-t-butyldiimidazole-2,2'-diylidene and 1,2-ethylene-3,3'-di-t-butyldiimidazole-2,2'-diylidene were prepared in good yields (ca. 60%) by deprotonation of the corresponding imidazolium dibromides with KHMDS in THF (eq 29).[34] It is important to note that this base leaves methylene and ethylene bridges intact.

KHMDS was also used in the preparation of pyridine- and phosphine-functionalized N-heterocyclic carbenes (eqs 30 and 31),[35] as well as for the synthesis of N-functionalized pincer bis-carbene ligands (eq 32).[36]

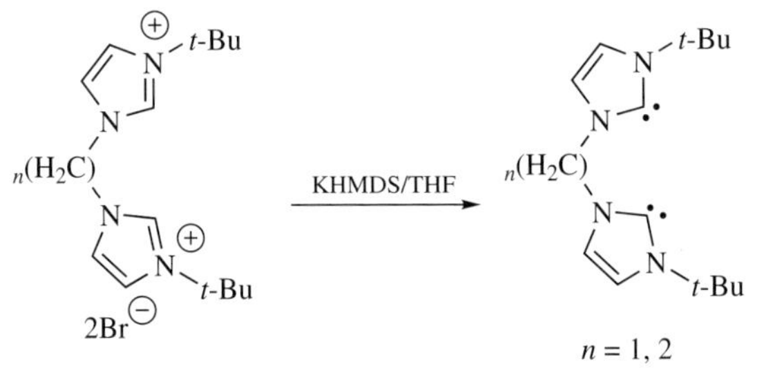

(29)

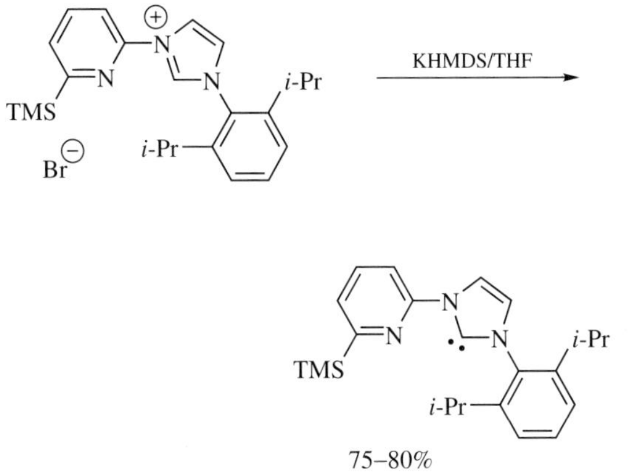

$n = 1, 2$

(30)

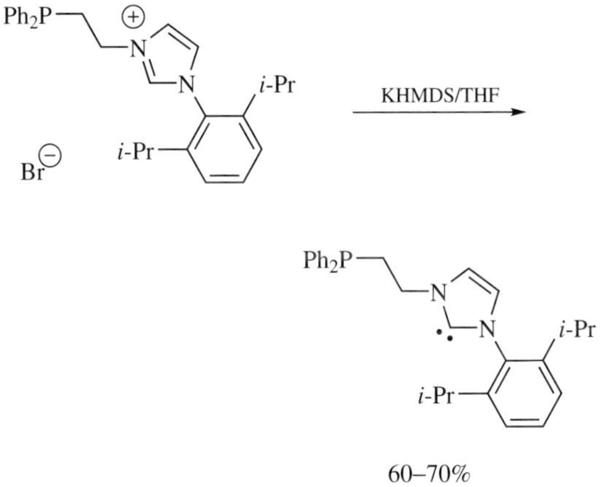

75–80%

(31)

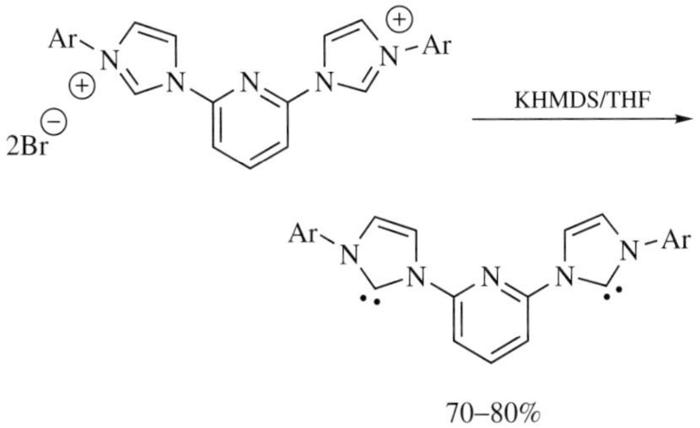

60–70%

(32)

70–80%

A series of chiral triazolium salts have been reacted with a base to form the corresponding chiral carbenes, which was shown to catalyze the Stetter reaction efficiently and to provide 1,4-dicarbonyl products in high yields and enantioselectivities (eq 33).[37] A survey of common bases identified KHMDS as providing an optimal balance between the yield and selectivity in this reaction. The reaction is sensitive to the nature of the Michael

acceptor: while electron deficient E-alkenes provided the desired product in good yields and enantioselectivities, no reaction was observed in the case of Z-alkenes.[38]

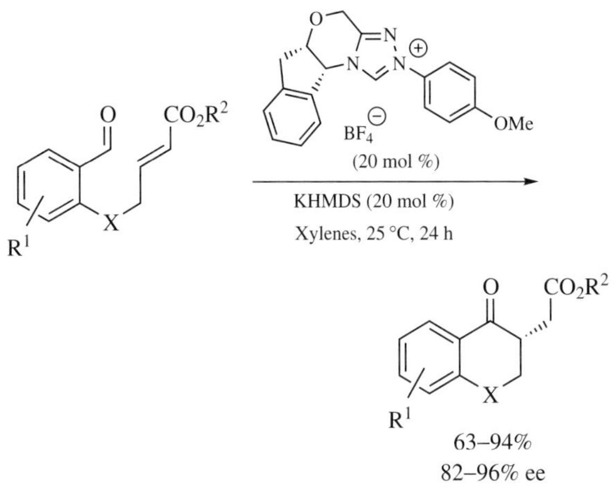

(33)

63–94%
82–96% ee

Palladium-Catalyzed α-Arylation of Carbonyl Compounds. A variety of bases have been used in the palladium-catalyzed α-arylation of carbonyl derivatives.[39,40] The pKa of the carbonyl moiety determines the choice of the base. The preferred bases for the α-arylation with ester derivatives are either NaHMDS (t-butyl propionate) or LiHMDS (t-butyl acetate); as KHMDS was reported to lead to lower yield because of competing hydrodehalogenation.[41–43] More sensitive substrates such as α-imino esters,[44] malonates, or cyanoesters[45] required the use of a milder base, as decomposition was observed with HMDS bases.

Both sodium t-butoxide and potassium hexamethyldisilazide were used in palladium-catalyzed ketone arylations.[46,47] While reactions involving electron-neutral or electron-rich aryl halides were more selective for monoarylation when KHMDS was used, sodium t-butoxide gave good selectivity with electron-poor aryl halides, without direct decomposition of the aryl halide. Bis(diphenylphosphino)ferrocene-ligated palladium complexes were active catalysts and led to the formation of a variety of substituted ketones in excellent yields (eq 34).

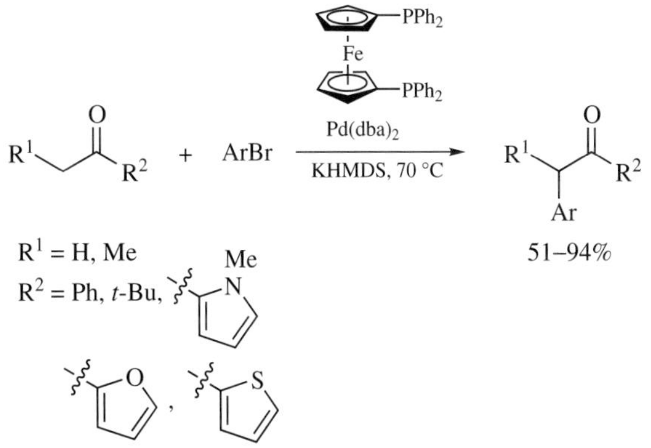

(34)

$R^1 = H, Me$
$R^2 = Ph, t$-Bu,

51–94%

$Ar = Ph, 2\text{-}MePh, 4\text{-}MeOPh$

More recently, the use of potassium phosphate has also been reported for the α-arylation of ketones with chloroarenes.[48]

KHMDS was the most effective base for the intermolecular arylation of N,N-dialkylamides (eq 35).[49] Higher yields were observed with KHMDS compared to LiTMP. Use of NaOt-Bu resulted in low conversion together with a high level of undesired side products, while LDA appeared to deactivate the system. Coupling of unfunctionalized and electron-rich aryl bromides with N,N-dimethylacetamide afforded α-aryl amides in moderate to good yields when the reaction was conducted with at least 2 equiv of KHMDS base. Diarylation of acetamides as well as hydro-dehalogenation of the aryl halides were side reactions that limited this process.

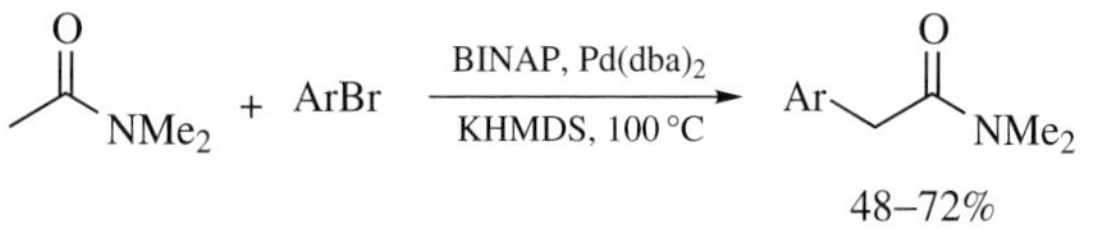

$$
\underset{}{\text{O}} + \text{ArBr} \xrightarrow[\text{KHMDS, 100 °C}]{\text{BINAP, Pd(dba)}_2} \text{Ar} \underset{\text{NMe}_2}{\overset{\text{O}}{}} \quad (35)
$$

Ar = 4-MePh, 2-MePh,
4-MeOPh, 2-naphthyl

Alkylation of Nitrile Derivatives. Tertiary benzylic nitriles were prepared from aryl fluorides and secondary nitrile anion.[50] In the presence of 4 equiv of nitrile and 1.5 equiv of a base, the nucleophilic aromatic substitution of fluoroarenes led to tertiary benzylic nitriles (eq 36). KHMDS was the best base for this reaction, as LiHMDS and NaHMDS provided lower yields. The desired product was not observed when Cs_2CO_3, LDA, or t-BuOK were used. With KHMDS, the reaction proceeded in high yields with a variety of substrates.

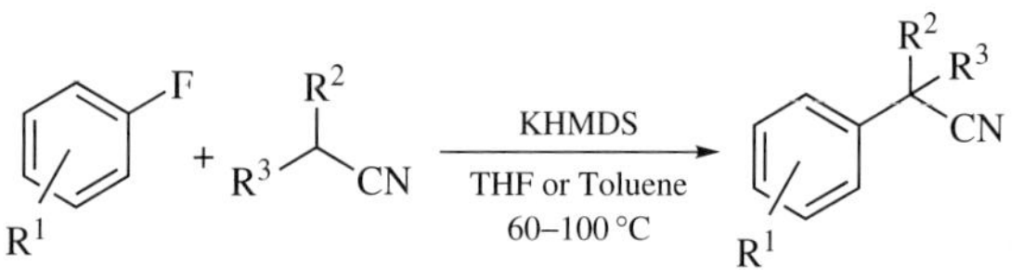

$$
\xrightarrow[\substack{\text{THF or Toluene} \\ 60–100 °C}]{\text{KHMDS}} \quad (36)
$$

66–95%

R^1 = 2-OMe, 3-OMe, 4-OMe, 3,5-OMe, 2-Cl, 4-Cl, H, 3-Me,
4-CN, 4-CF$_3$

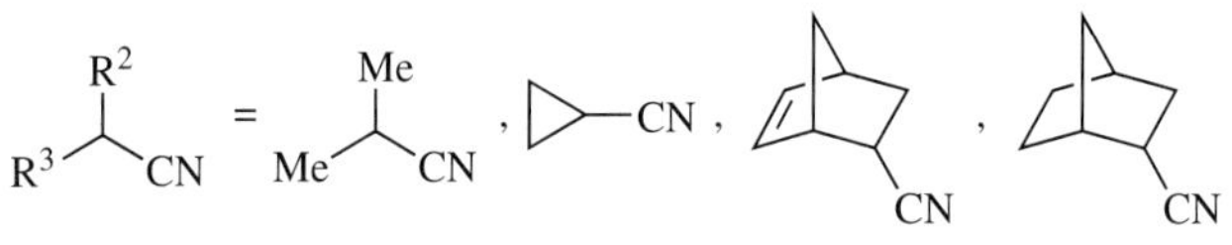

Treatment of readily available arylacetonitriles with KHMDS and subsequent alkylation with α,ω-dibromo or dichloroalkanes produces cycloalkyl adducts in good yields and short reaction time (eq 37).[51] In this process, the nitrile moiety serves as a masked aldehyde, which could be revealed upon reduction with DIBAL-H.

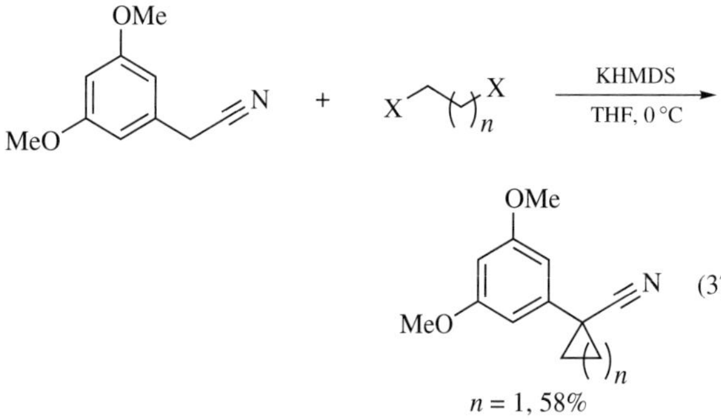

$$
\xrightarrow[\text{THF, 0 °C}]{\text{KHMDS}} \quad (37)
$$

n = 1, 58%
n = 2, 64%
n = 3, 88%
n = 4, 94%

Synthesis of Pyridines. The annulation of α-aryl carbonyl derivatives with vinamidinium hexafluorophosphate salts in the presence of a base gave access to the corresponding 3-arylpyridine.[52] While potassium t-butoxide was used with ketones, the annulation of aldehydes was performed with KHMDS (eq 38).

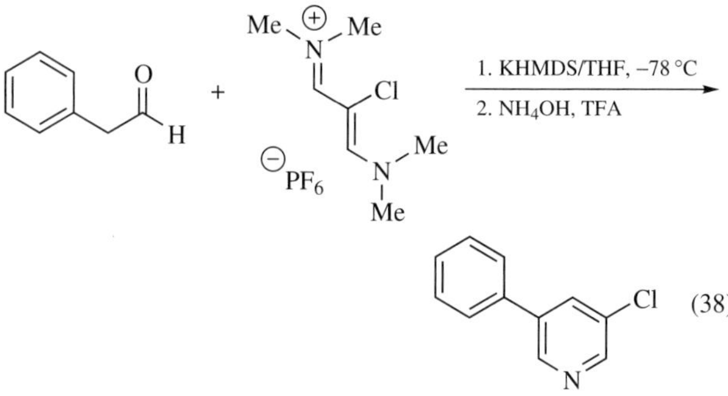

$$
\xrightarrow[\text{2. NH}_4\text{OH, TFA}]{\text{1. KHMDS/THF, −78 °C}} \quad (38)
$$

80%

Deprotonation of Oxazolines. New chiral oxazoline-sulfoxide ligands have recently been prepared and utilized as chiral ligands in copper(II)-catalyzed enantioselective Diels–Alder reactions.[53] These new ligands were prepared via sulfinylation of chiral 1,3-oxazoline to lead to unsubstituted oxazoline sulfoxide derivatives. While the first methylation was conducted using LDA and methyl iodide, the introduction of the second methyl was carried out using KHMDS and methyl iodide to afford the desired chiral ligand (eq 39).

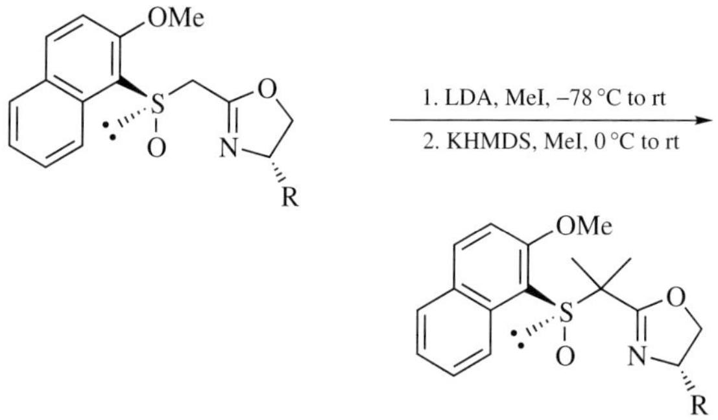

$$
\xrightarrow[\text{2. KHMDS, MeI, 0 °C to rt}]{\text{1. LDA, MeI, −78 °C to rt}} \quad (39)
$$

R = t-Bu, Ph, Bn

Deprotonation of Alkyne and Propargylic Derivatives. KHMDS is a strong enough base to deprotonate terminal alkynes. It has been particularly useful for the intramolecular condensation of alkyne anions with aldehydes. For instance, the final cyclization in the synthesis of a new bicyclic tetrahydropyridine system was carried out in the presence of KHMDS (eq 40).[54] The same reaction failed when using LDA in THF.

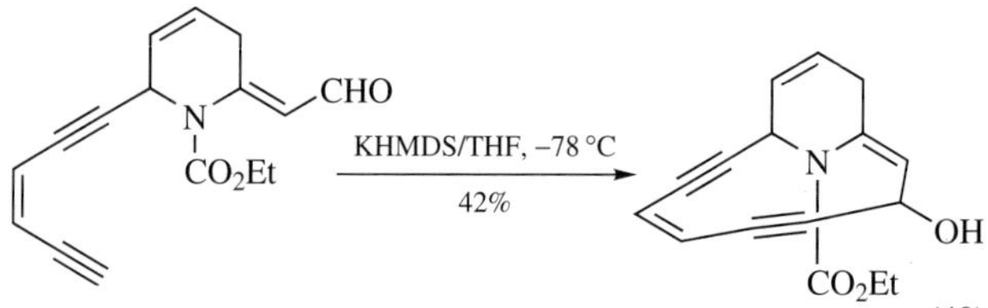

$$
\xrightarrow[42\%]{\text{KHMDS/THF, −78 °C}} \quad (40)
$$

In a tandem cyclization of propynyloxyethyl derivatives, KHMDS is as efficient as LDA or NaHMDS (eq 41).[55] This cyclization proceeded through the formation of an alkynyl metal, followed by a cyclization which led to a carbenic species that underwent an intramolecular C–H insertion.

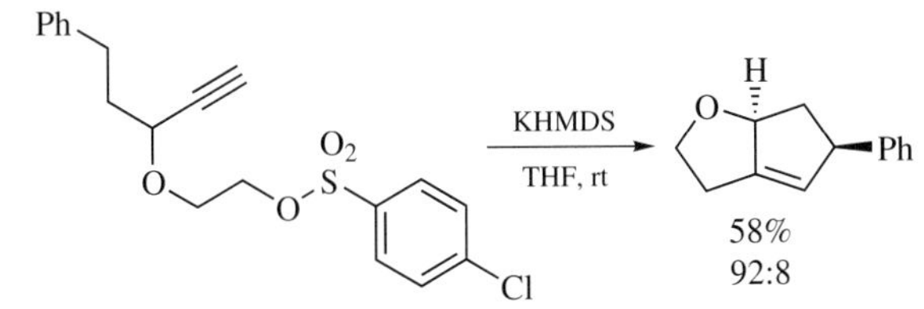

$$(41)$$

A convenient method for the preparation of 1-phenylthio-3-alken-1-ynes from 4-tetrahydropyranyloxy-1-phenylthio-2-alkynes in the presence of base was reported (eq 42).[56]

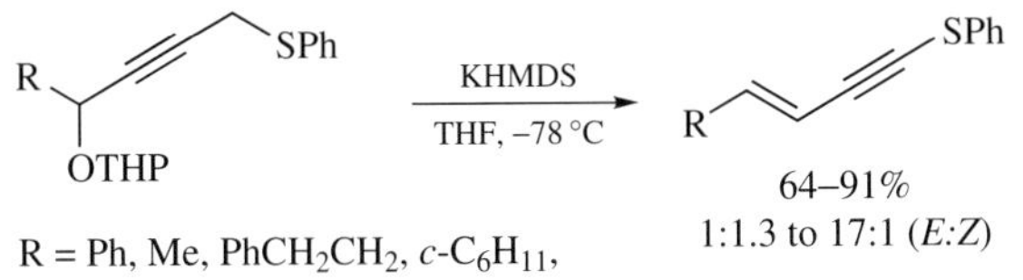

$$(42)$$

R = Ph, Me, PhCH$_2$CH$_2$, c-C$_6$H$_{11}$, 1-naphthyl, 9-anthryl

The use of KHMDS led mainly to the formation of the E-isomer, whereas the Z-isomer was formed as the major product in the presence of MeLi. The presence of the masked hydroxyl group at the 4-position of the substrate is essential, as treatment of corresponding 4-hydroxy-1-phenylthio-2-alkynes with KHMDS produced 4-hydroxy-1-phenylthio-1,2-alkadienes (eq 43).

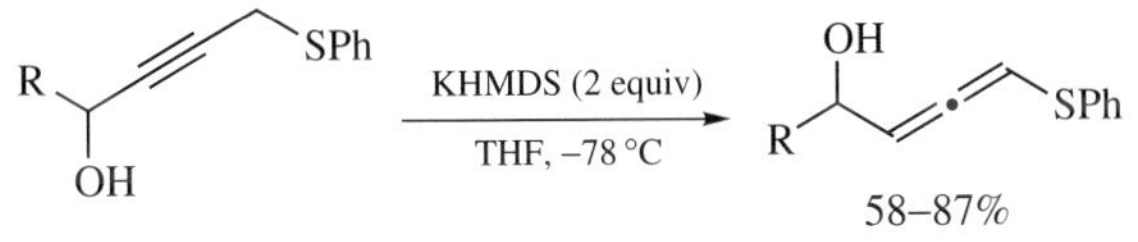

$$(43)$$

R = Ph, Me, PhCH$_2$CH$_2$

Cyclization of 1,1-Disubstituted Alkenes to Cyclopentenes. A general method for the cyclization of an unactivated 1,1-disubstituted alkene to the corresponding cyclopentene has been described.[57] Bromination followed by the addition of KHMDS in the same pot gave the vinyl bromide intermediate, which reacted further in situ to give the alkylidene carbene. This is then inserted in a 1,5-fashion into a C–H bond to yield the corresponding cyclopentene (eq 44). KHMDS proved to be superior in this process as compared to LiHMDS and NaHMDS.

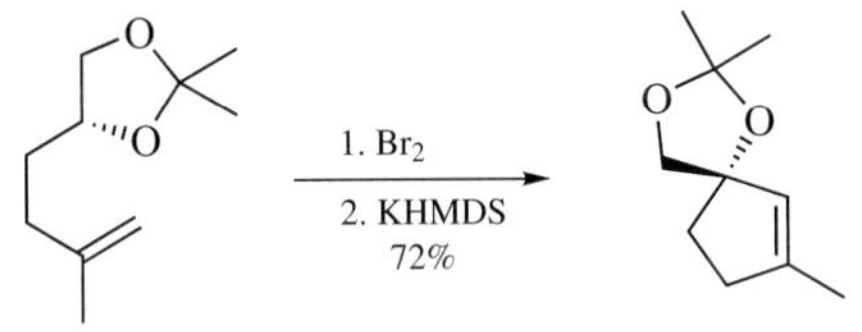

$$(44)$$

Darzens Reaction. KHMDS was used to deprotonate bromomethylketones to generate the corresponding bromoenolate that upon reaction with various carbonyl derivatives produced a variety of epoxides. This method has been particularly useful to synthesize epoxides derived from amino acids with high yields and

selectivity (eq 45).[58] Both lithium and sodium enolate produced lower yields and diastereoselectivities.

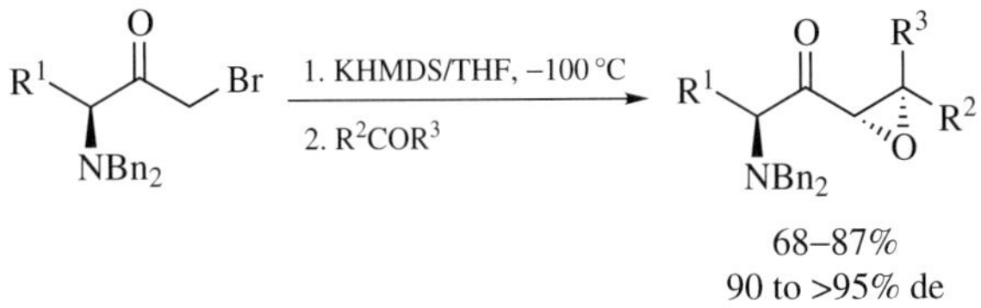

$$(45)$$

Deprotonation of Amino Acid Derivatives. The direct asymmetric α-methylation of α-amino acid derivatives was recently reported to proceed with good efficiency and enantioselectivity when using KHMDS in toluene (eq 46).[59] It was established that the chiral induction was based on the dynamic chirality of enolates, as the di-boc derivative and the cyclic acetal gave racemic products.[60]

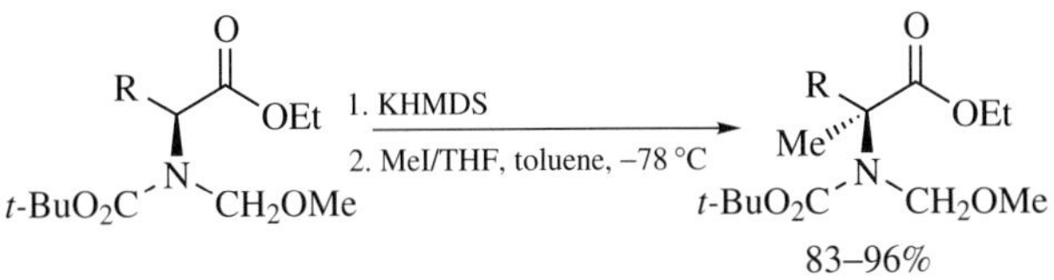

$$(46)$$

An intramolecular version of this process has been also reported, for which it was also shown that KHMDS provided superior results than LHDMS.[61] The corresponding cyclic amino acids could be recovered in 61–95% yield and with 94–98% enantiomeric excess.

Epoxidation with Alkylhydroperoxide. The epoxidation of quinone monoacetals in the presence of tritylhydroperoxide using KHMDS was shown to produce the desired epoxides in good yields (eq 47).[62,63]

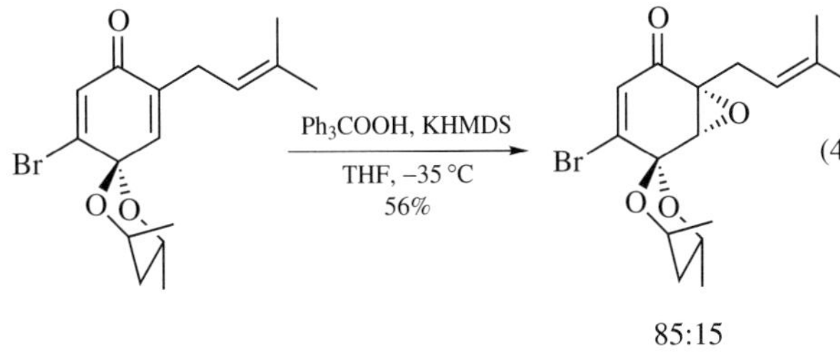

$$(47)$$

Alkylation of Phosphorus Derivatives. The formation of diethyl phosphite anions has been achieved successfully with KHMDS and LiHMDS.[64] The following reaction with acid chlorides is strongly influenced by the counter-ion, as only the rearrangement product was observed with the lithium base, whereas the desired hydroxybisphosphate was the major product when using the potassium base (eq 48).

A series of new phosphinooxazoline ligands[65] have been recently prepared and tested in the asymmetric Heck reaction.[66,67] Synthesis of the ligands involved the aromatic nucleophilic substitution of aryl fluorides with phosphide nucleophile generated from the corresponding phosphine and KHMDS (eq 49). The reaction proceeded in good yields, but proved to be more sluggish with electron-rich aryl fluorides and failed completely when the addition of electron-deficient phosphines was attempted.

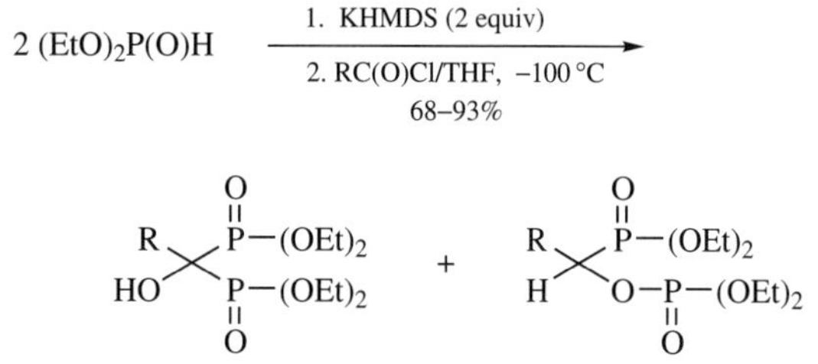

$$2 \ (EtO)_2P(O)H \xrightarrow[\substack{2. \ RC(O)Cl/THF, \ -100\,°C \\ 68–93\%}]{1. \ KHMDS \ (2 \ equiv)}$$

$$(48)$$

6–7:1

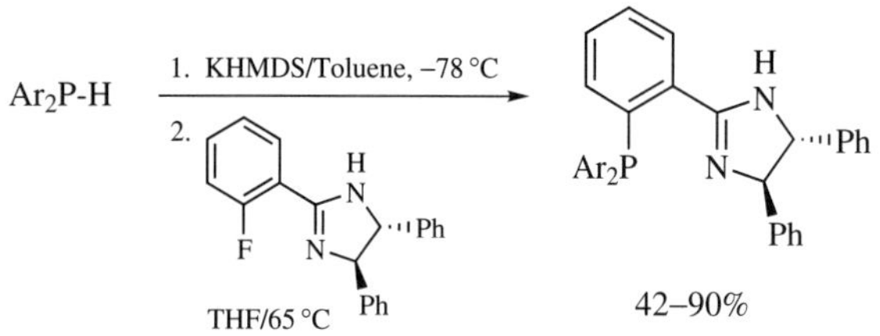

$$Ar_2P\text{-}H \xrightarrow[\substack{2.}]{1. \ KHMDS/Toluene, \ -78\,°C} \quad (49)$$

THF/65 °C

42–90%

Acetal Fission in Desymmetrization of 1,2-Diols. The asymmetric desymmetrization of cyclic *meso*-1,2-diols was accomplished via diastereoselective acetal fission. After acetalization of *meso*-1,2-diols with the C_2-symmetric bis-sulfoxide ketone, the resulting acetal was subjected to base-promoted acetal fission, followed by acetylation or benzylation to give the desymmetrized diol derivatives. Interestingly, the counter-cation of the base had a remarkable effect on the diastereoselectivity of the reaction. While LHMDS produce the desired compound with low diastereoselectivity, 90% and >96% diastereomeric excesses were obtained with NaHMDS and KHMDS, respectively. The best results were obtained using 3 equiv of KHMDS and 18-crown-6 in THF, which led to the formation of the desired acetate in 90% yield and 96% de (eq 50).[68]

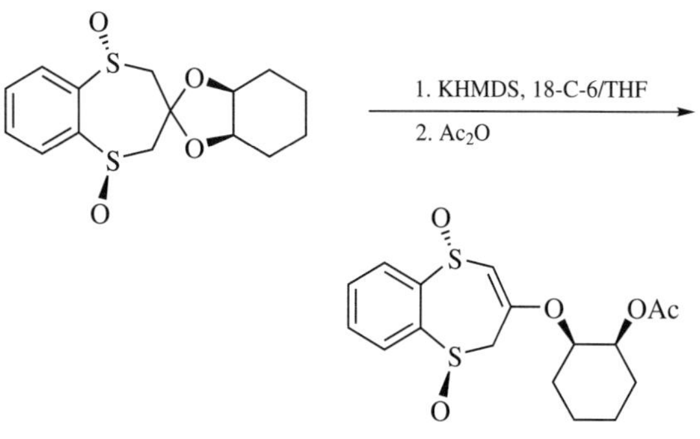

$$\xrightarrow[\substack{2. \ Ac_2O}]{1. \ KHMDS, \ 18\text{-}C\text{-}6/THF}$$

$$(50)$$

91%, 96% de

Generation of Sulfonium Ylides. Sulfonium ylides could be generated very efficiently by deprotonation of the corresponding sulfonium salt with a strong base, such as KHMDS. The formation of vinyloxiranes from sulfonium ylides and aldehydes has been recently investigated. It was shown that in the presence of lithium bromide and silylated diphenylsulfonium allylide, produced from the corresponding sulfonium salt and KHMDS, aldehydes led to the corresponding vinyl epoxide with high *cis*-selectivity (eq 51).[69]

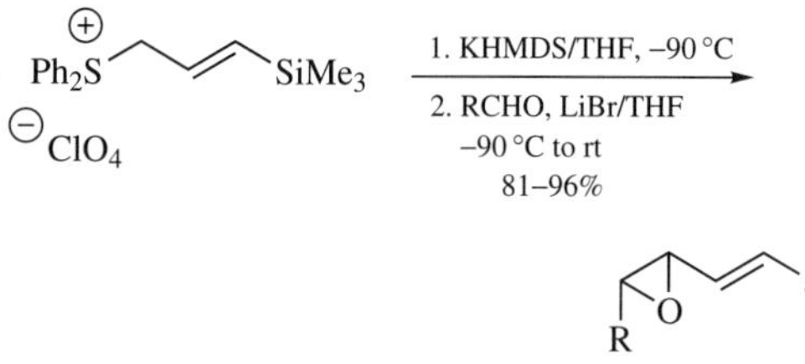

$$\xrightarrow[\substack{2. \ RCHO, \ LiBr/THF \\ -90\,°C \ to \ rt \\ 81–96\%}]{1. \ KHMDS/THF, \ -90\,°C}$$

$$(51)$$

80:20 to 94:6 (*cis:trans*)

More recently, an asymmetric version has been reported based on the use of chiral sulfonium salts.[70] In the presence of a base, the chiral sulfonium salt was reacted with various carbonyl derivatives to provide the corresponding *trans*-epoxide with excellent diastereomeric ratios and enantiomeric excesses. Potassium hydroxide and phosphazene bases are usually utilized for this process, although it has been shown that KHMDS is indeed equally effective (eq 52).

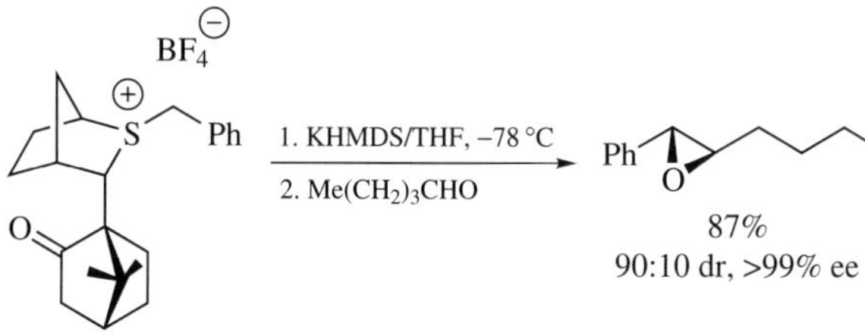

$$\xrightarrow[\substack{2. \ Me(CH_2)_3CHO}]{1. \ KHMDS/THF, \ -78\,°C}$$

$$(52)$$

87%
90:10 dr, >99% ee

The formation of metal carbenes from sulfonium ylides has been also recently disclosed.[71] Treatment of the sulfonium salt with KHMDS followed by the addition of tris(triphenylphosphine) ruthenium dichloride and phosphine exchange with tricyclohexylphosphine led to the formation of the Grubb's metathesis catalyst in 98% yield (eq 53). This process has been shown to be quite general and to work with various transition metals.

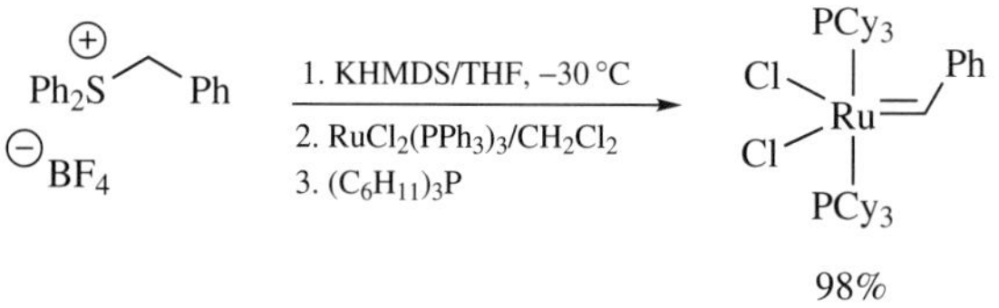

$$\xrightarrow[\substack{2. \ RuCl_2(PPh_3)_3/CH_2Cl_2 \\ 3. \ (C_6H_{11})_3P}]{1. \ KHMDS/THF, \ -30\,°C}$$

$$(53)$$

98%

Generation of Sulfur Ylides: Julia Olefination and Related Processes. Recently, the 'modified' Julia olefination, which employed certain heteroarylsulfones instead of the traditional phenylsulfones, has emerged as a powerful tool for alkene synthesis.[72] Although the reaction was first reported with LDA, bases such as LHMDS, NaHMDS, and KHMDS are now commonly used.[73] In addition, solvent as well as base counter-cation have been shown to markedly affect the stereochemical outcome of the olefination reaction. For instance, KHMDS was less selective than NaHMDS for the coupling between benzothiazoylsulfone (**1**) and cyclopropane carboxaldehyde (**2**) in toluene, furnishing a 3.7:1 ratio compared to a 10:1 ratio favoring the Z-isomer. However, both bases provided a 1:1 mixture of isomers when the reaction was run in DMF (eq 54).[74]

In a different system, KHMDS proved to be the most Z-selective for the formation of a triene from an allylic sulfone and a conjugated aldehyde (eq 55).

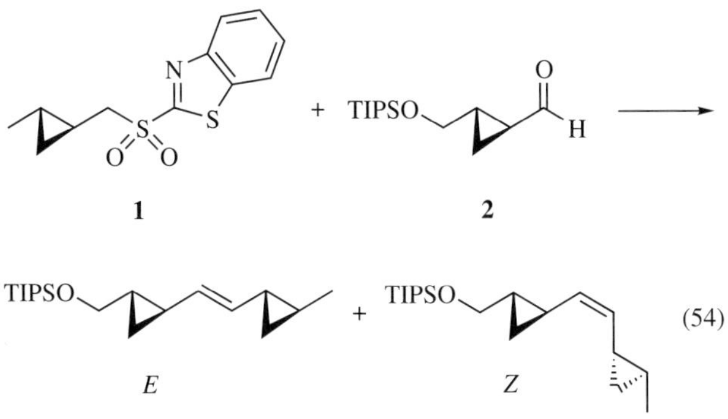

$$(54)$$

Conditions	E:Z
NaHMDS, THF	1.1:1
KHMDS, THF	1.2:1
NaHMDS, Toluene	1:10
KHMDS, Toluene	1:3.7

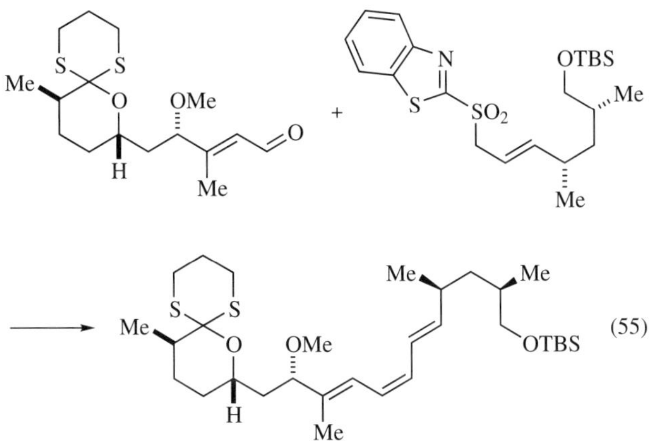

$$(55)$$

Conditions	E:Z
LiHMDS	1:2.4
NaHMDS	1:1.3
KHMDS	1:4.6

1-Phenyl-1*H*-tetrazol-5-yl sulfones were introduced as another alternative to phenylsulfones; these lead to alkenes in a one-pot procedure from aldehydes. Usually the highest selectivities are obtained when using KHMDS.[75] For instance, alkene **3** was obtained in 100% yield and an 84:16 *E:Z* ratio with NaHMDS, whereas the use of KHMDS led to the same product in 59% yield and a 99:1 *E:Z* ratio (eq 56).

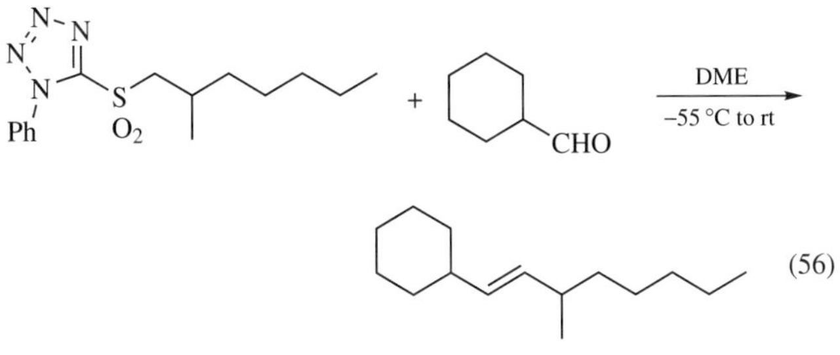

$$(56)$$

3

NaHMDS: 100%; 84:16 (*E:Z*)
KHMDS: 59%; 99:1 (*E:Z*)

Many examples of the use of this strategy with KHMDS has been reported to construct double bonds in the total synthesis of

natural products. For instance, the *E,E,E*-triene of thiazinotrienomycin E was constructed according to this process in 85% yield and with 10:1 selectivity (eq 57).[76]

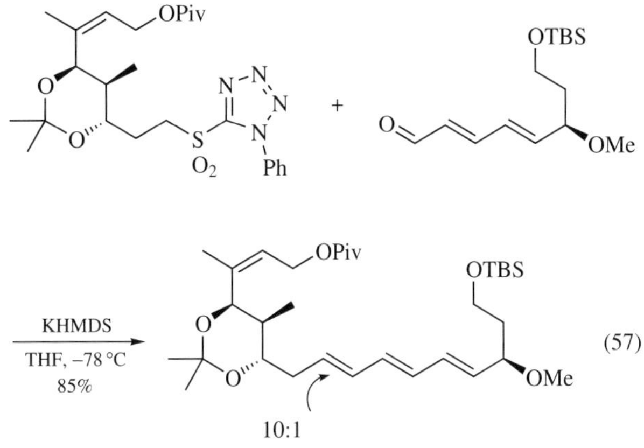

$$(57)$$

10:1

Disubstituted *E*-alkenes were also prepared according to this strategy as exemplified in the total synthesis of brefeldin A,[77] amphidinolide A,[78] and leucascandrolide A.[79]

High *Z*-selectivities were observed with benzylic and allylic 1-*t*-butyl-1*H*-tetrazol-5-yl sulfones when reacted with aliphatic aldehydes in the presence of KHMDS (eq 58).[80]

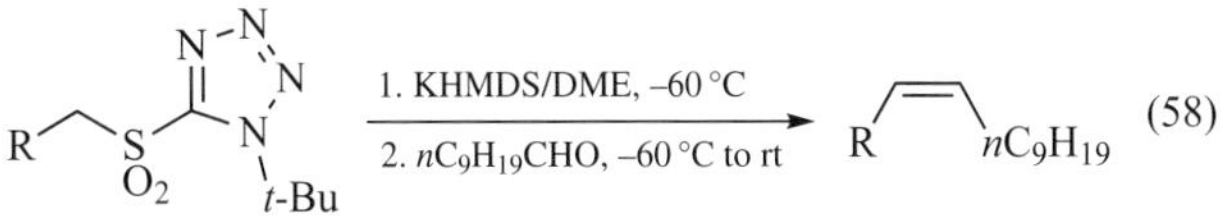

$$(58)$$

R = Ph, 95%; >99:1 (*E:Z*)
R = CH=CH₂, 60%; 96:4 (*E:Z*)

It was shown that *E,Z*-dienes could be synthesized from α,β-unsaturated aldehydes and 2-pyridylsulfones in good yields and selectivities.[81,82] NaHMDS and KHMDS were equally effective in promoting this condensation (eq 59).

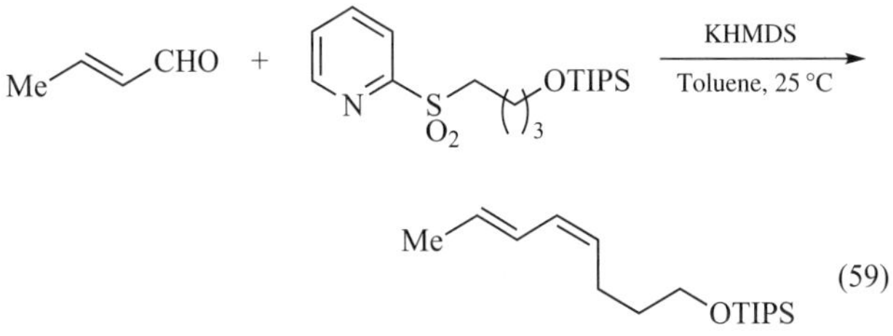

$$(59)$$

67%, 91:9 (*E,Z:E,E*)

Generation of Other Heteroatom Ylides. KHMDS has been also used in the generation of other heteroatom ylides, such as tellurium[83] and arsonium ylides.[84]

Second Update

Helena C. Malinakova & John C. Hershberger
University of Kansas, Lawrence, KS, USA

Base for α-Deprotonation of Carbonyl Compounds. Potassium hexamethyldisilazide, KN(TMS)₂ (KHMDS) has been identified as the optimum base for an efficient epimerization of the stereogenic C-2 carbon in the esters of phenyl kainic acid and in *cis*-prolinoglutamic esters.[85] Bases included in an initial survey, e.g., NaH, DBU, LHMDS, MeONa, and KOH afforded low yields of the epimerized products. Although promising results were observed with KH activated by the 18-crown-6 ether, the protocol suffered from poor reproducibility. In contrast, the treatment of *N*-Cbz-protected methyl *cis*-3-prolinoglutamate with solid KHMDS (2.5 equiv, 0 °C, THF, 6 h) resulted in 80% epimerization without the occurrence of undesired degradation processes (eq 60). The application of KHMDS (5 equiv, THF, 0 °C–rt, 16 h) to epimerization of *endo* (*cis*(C2–C3), *cis*(C3–C4)) *N*-Cbz-protected 4-phenyl methyl kainate afforded 95% epimerization accompanied by significant saponification, and subsequent treatment with diazomethane was required to recover the epimerized ester in a good yield (60–75%) (eq 61). In contrast, the *exo* (*cis*(C2–C3), *trans*(C3–C4)), 3-phenyl methyl kainate ester proved to be much more resistant to epimerization, which only proceeded to the extent of 40%.

$$(60)$$

80% epimerization

$$(61)$$

95% epimerization

In the synthesis of a second-generation anticancer taxoid ortataxel, deprotonation of 13-oxobaccatin III by KHMDS (THF–HMPA, 83:17) followed by electrophilic oxidation with (1*R*)-(10-camphorsulfonyl)oxaziridine afforded the C14-hydroxylated taxoid in 70% yield (eq 62).[86] An analogous oxidation mediated by *t*-BuOK (THF–DMPU, 83:17) afforded the product in 83% yield (eq 62). The 14β-OH epimer was formed exclusively, in accord with the approach of the oxidant from the less sterically hindered β-face. The performance of several bases was compared, indicating the order of reactivity *t*-BuOK > LDEA (lithium diethylamide) > KHMDS > NaHMDS, in THF/HMPA or DMPU solvent mixtures.

Oxalactims, for example 5*H*-methyl-2-phenyl-oxazol-4-one, were investigated as synthetically powerful synthons for the construction of α-hydroxy acids via α-deprotonation followed by molybdenum-catalyzed asymmetric allylation and a subsequent hydrolysis of the oxalactim rings. Among the series of the HMDS bases used in the Mo-catalyzed asymmetric allylation, LHMDS was found to be the base of choice providing better yields, regio- (branched/linear), diastereo- and enantioselectivity than the corresponding allylation reactions mediated by KHMDS (eq 63).[87]

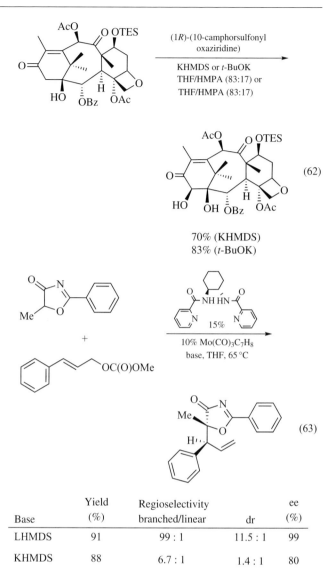

$$(62)$$

70% (KHMDS)
83% (*t*-BuOK)

$$(63)$$

Base	Yield (%)	Regioselectivity branched/linear	dr	ee (%)
LHMDS	91	99 : 1	11.5 : 1	99
KHMDS	88	6.7 : 1	1.4 : 1	80

Potassium hexamethyldisilazide, KHMDS, has been used successfully in the past for α-deprotonation of sulfoxides.[53,68] Aiming to probe the nature of the transition state in the reaction of α-metalated (*R*)-methyl *p*-tolyl sulfoxide with *N*-(benzylidene) aniline under kinetic conditions, the effect of the choice of the bases for the α-deprotonation on the reactivity and diastereoselectivity was studied, using LDA, LHMDS, NaHMDS, and KHMDS.[88] Best yields and diastereoselectivities (90%, 84:16) were achieved with LDA, whereas LHMDS showed diminished reactivity (50%) without a loss of diastereoselectivity (87:13). Improved yields and lower diastereoselectivities were recorded in reactions mediated by NaHMDS and KHMDS, providing the amine in 60% (dr 81:19) and 77% (dr 69:31) yields and diastereoselectivities, respectively (eq 64).

These results, along with quantum mechanical calculations, supported the conclusion that a six-membered chelated "flat-chair-like" transition state operated under the kinetic conditions. The diastereoselectivity of reactions between Li-salts of the (*R*)-methyl *p*-tolyl sulfoxide formed by α-deprotonation with *n*-BuLi, could be further enhanced by the addition of external C_2-symmetrical bidentate ligand (*R,R*)-1,2-*N,N*-bis(trifluoromethanesulfonyla-mino)cyclohexane in the form of the lithium *N,N*-dianion providing the amine in 80% yield with a diastereoselectivity of 99:1, apparently achieving a "match" between the chirality of the additive and the chirality of the sulfoxide reagent.

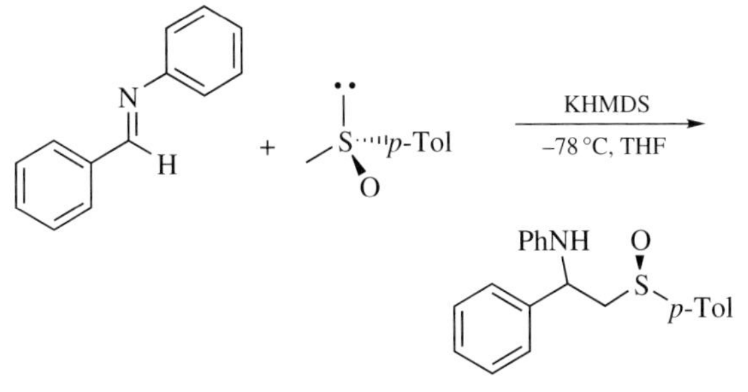

(64)

77% (69:31)

KHMDS has been used to effect α-deprotonation of O-silyl protected cyanohydrins derived from 2-p-tolylsulfinyl benzaldehyde followed by trapping of the C-nucleophile with diverse C-electrophiles, providing a powerful alternative approach to cyanohydrins of ketones.[89] The remote 1,4-asymmetric induction was equally effective for either epimer (diastereomer) of the O-TIPS protected cyanohydrin, and an equimolar mixture of the two epimers was employed. Both KHMDS and LHMDS bases provided the substituted cyanohydrins from reactions with highly reactive electrophiles (ClCOOMe and ClCOMe) in excellent yields and diastereoselectivities (dr > 98:2) (eq 65). The deprotonation induced by KHMDS led to more reactive nucleophiles, shortening the reaction times. Notably, in alkylations of Eschenmoser's salt, and benzyl and allyl bromides, the application of LHMDS instead of KHMDS improved the diastereoselectivity. The stereoselectivity of the alkylations mediated by KHMDS could be increased by the inclusion of the 18-crown-6 ether additive. To rationalize the observed trends, the deprotonation with KHMDS was proposed to give rise to pyramidal potassium C-bonded enolates, in contrast to planar lithium N-bonded enolates arising from the deprotonation with LHMDS. The steric discrimination of the enantiotopic faces would be achieved more efficiently with the planar N-enolates considering the different directions of the approach by the electrophiles. Thus, the 18-crown-6 ether additive serves to minimize the pyramidal shape of the potassium enolates, improving diastereoselectivity. The increase in the diastereoselectivity observed with highly reactive electrophiles (e.g., ClCOOMe) presumably reflects the involvement of an early transition state resembling the structures of the planar N-lithium enolates, or "planarized" potassium enolates complexed with the 18-crown-6 additive, effectively discriminating against one of the two enantiotopic pathways for the approach of the electrophiles.

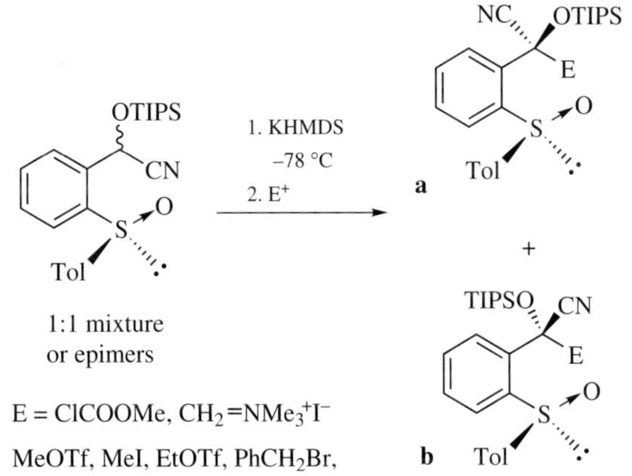

The deprotonation of a stereogenic carbon in chiral nonracemic substrates with KHMDS has been used as a first step of sequences relying on the memory of chirality for asymmetric synthesis of organic products.[90,91] The treatment of N-Boc-protected amino esters featuring a Michael acceptor group with KHMDS in DMF–THF (1:1) at $-78\,°C$ for 30 min provided trisubstituted pyrrolidines, piperidines and tetrahydroisoquinolines, and in good yields (65–74%) and diastereoselectivities (4:1 for pyrrolidine, and 1:0 for piperidines and tetrahydroisoquinoline) and excellent enantiomeric excesses (91–98% ee) (eqs 66 and 67).[90]

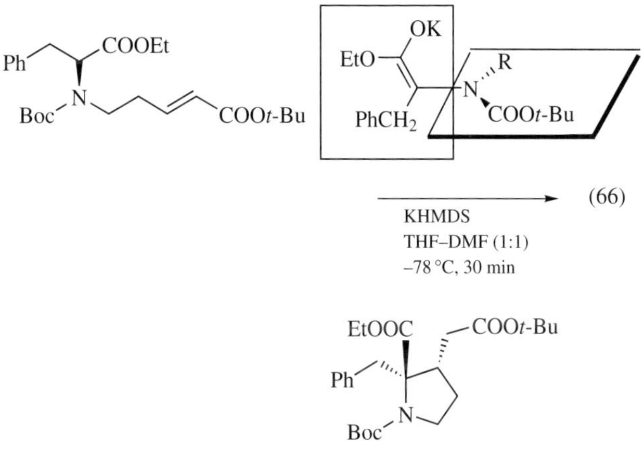

(66)

65%, 4:1 dr
(91% ee major diastereomer)

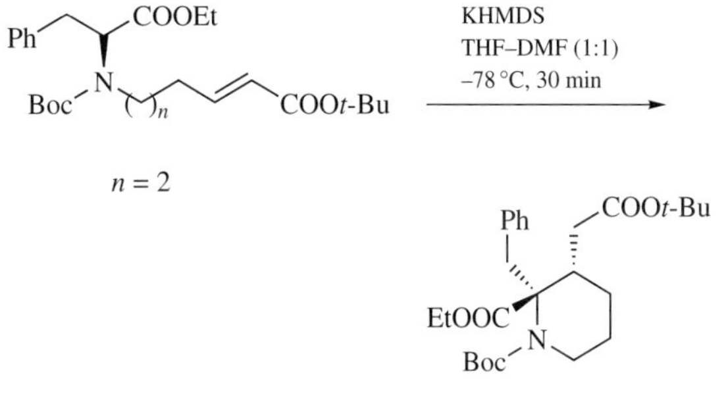

(67)

66% (97% ee)

(65)

E = ClCOOMe, CH$_2$=NMe$_3{}^+$I$^-$

MeOTf, MeI, EtOTf, PhCH$_2$Br,

CH$_2$=CHCH$_2$Br

73–85% yields
of major epimers

dr **a:b** = 98:2–70:30
or 30:70 (E = PhCH$_2$Br)

The choice of the N-protecting group (N-Boc) proved to be critical for achieving a high enantiomeric excess of the cyclization reaction. In contrast to KHMDS, lithium amide bases (LHMDS or LiTMP) did not afford detectable quantities of the anticipated heterocycles. The mechanism of asymmetry transfer was proposed to rely on the formation of axially chiral nonracemic enolate (eq 66) with a chiral C–N axis, the racemization barrier for which was found to be 16.0 kcal mol^{-1}.[90]

The deprotonation of C-3 substituted 1,4-benzodiazepin-2-ones with KHMDS afforded an enantiopure, conformationally chiral enolate with a relatively bulky $N-1$ substituent, which provided a sufficient barrier to enolate racemization.[91] In contrast, LHMDS proved to be ineffective in this transformation, and KHMDS served as an operationally simpler alternative to a mixed base consisting of LDA/n-BuLi. At $-100\,°C$ a *simultaneous* addition of KHMDS and an excess of benzyl bromide to the substrate provided the alkylated product in 75% ee (78% yield). However, the simultaneous addition could not be carried out with a more reactive benzyl iodide. Ultimately, the best result was achieved in a stepwise addition mode, when the deprotonation with KHMDS (20 min) at $-109\,°C$ (THF–HMPA) was followed by the addition of an excess of benzyl iodide affording the 3,3-disubstituted benzodiazepine in 97% ee (93% yield) (eq 68).[91]

$$(68)$$

93% (97% ee)

The utility of a series of different inorganic bases, including LHMDS, NaHMDS, KHMDS, t-BuONa, t-BuOK, t-BuOLi, LTMP, LDA, PhOLi, MeONa, and EtONa for the alkylation of cyclohexanone with benzyl bromide in a microreactor was studied.[92] The microreactor employed a field-induced electrokinetic flow acting as a pump. To achieve a sufficient electrokinetic mobilization, the inorganic bases had to be solubilized by the addition of stoichiometric quantities of the appropriate crown ethers. With a single exception (LHMDS with 12-crown-4 ether), the desired electrokinetic mobility was reached with all other bases under these conditions.

α-Deprotonation of aliphatic nitriles with NaHMDS or KHMDS bases operated as a key step in a mild and transition metal-free α-arylation of primary and secondary aliphatic nitriles with activated heteroaryl halides.[93] The α-nitrile carbanion had to be generated in the presence of the heteroaryl halide to minimize competing decomposition reactions. Notably, monoarylation occurred selectively with primary carbonitriles. A correlation between the choice of the metal ion in the silylamide bases and the leaving group in the heteroaryl halide was observed. The $S_N Ar$ reaction performed best when NaHMDS was used with bromide and chloride leaving groups, and KHMDS with substrates featuring a fluoride leaving group (eq 69).[93]

$$(69)$$

X = Cl, M = Na; yield: 61%
X = F, M = K; yield: 88%

Carbon nucleophiles can be used to construct carbon-transition metal bonds in stable organometallic complexes via an $S_N 2$ displacement at the metal center. For example, deprotonation at the α-position to the ester groups in the chiral nonracemic (ethoxycarbonylmethyl)-N-(trifluoromethanesulfonyl)-2-(aminophenyl)iodopalladium(II) complex followed by a rapid irreversible displacement of the iodide afforded new C(sp^3)–Pd bonds, yielding a diastereomerically enriched palladacycle with a Pd-bonded stereogenic carbon.[94] The ratio of atropisomers arising from a hindered rotation about the aryl–Pd bond in the aryliodopalladium complex proved to play an important role, and the results shown in eq 70 were achieved with an aza-aryliodopalladium complex featuring a 98:2 ratio of atropisomers. Apparently, the ring-closure reaction was faster than the interconversion of the atropisomers. Deprotonation with KHMDS followed by the displacement of the iodide afforded excellent yields (89% and 93%) of the palladacycle in good diasteroselectivities (60% and 72%) favoring the diastereomer with the (S) configuration at the Pd-bonded stereogenic carbon. The application of sterically hindered bases with tightly chelating cations proved to be important for achieving high diastereoselectivites. Consequently, deprotonation with LDA (THF, rt) afforded the palladacycle in 61% yield and 80% de. The optimum diastereoselectiviy was realized following the replacement of the iodide counterion with acetate utilizing AgOAc additive and t-BuOK base (THF, $-78\,°C$) yielding the palladacycle in 99% yield and 92% de (eq 70).[94]

98:2 atropisomer ratio

$$(70)$$

base: KHMDS; 93% yield (72% de)
base: AgOAc/t-BuOK; 99% yield (92% de)

Base for Deprotonation of Heteroatom–Hydrogen (X–H) Bonds. The deprotonation of heteroatom–hydrogen bonds with KHMDS constitutes a key step in diverse methods for the preparation of heterocycles.[95,96] A cyclization of 3,4-dihydro-2-pyridones derived from (S)-phenylglycinol featuring a β-enamino ester functionality and an internal hydroxyl group via an intramolecular Michael addition was achieved though the deprotonation of the hydroxyl group with KHMDS, NaHMDS, or LHMDS in THF at 0 °C.[95] In all cases, *cis*-diastereomers of the bicyclic lactams were formed, and only the configuration of the carbon adjacent to the ester group was variable. Reactions performed under kinetic conditions with 0.1 equiv of LHMDS (2 h) gave the best diastereoselectivity (>98:2 dr). However, the same diastereoselectivity (>98:2 dr) was also realized with KHMDS (0.1 equiv) when the reaction was stopped after just 1 h (eq 71).[95] In contrast, the application of 0.5 equiv of a strong base (KHMDS) induced the epimerization of the acidic proton providing almost 1:1 ratios of *cis/trans* diastereomers.

$$>80\% \ (>98\text{:}2 \ dr) \tag{71}$$

KHMDS was used to promote the condensation of an imidazole precursor with commercially available carbamates providing 1-substituted 7-PMB-protected xanthines in a single step.[96] KHMDS served to deprotonate the N–H bonds in both substrates, yielding a metal amide and an isocyanate via fragmentation of the metalated carbamate, which participated in a condensation reaction providing potassium salts of the xanthines. In most cases, the potassium salt precipitated directly from the crude reaction mixtures, and was easily N-allylated in a subsequent step under mild conditions. The choice of the base for the condensation reaction was optimized. Sodium bases (EtONa, MeONa, t-BuONa), and unhindered potassium bases (EtOK) gave low to moderate yields of the corresponding N-metal salts, whereas lithium bases (LHMDS, LDA) gave modest yields and required longer reaction times. A problematic base-induced degradation of the imidazole substrate could be minimized by using strong sterically hindered potassium bases (e.g., KHMDS, t-AmOK, t-BuOK) and by adding a solution of the base in diglyme slowly to the preheated (70–80 °C) and stirred diglyme solutions of substrates (eq 72).

In the preparation of a series of phenylalkylphosphonamidate derivatives of glutamic acid, KHMDS was employed to deprotonate the H–P bond in the dibenzyl phosphite, generating a P-nucleophile that reacted with phenylalkyl bromides to provide targeted phosphonates in good yields (eq 73).[97]

The deprotonation of N–H bonds in diverse oxazolidin-2-ones with KHMDS as the base followed by the treatment of the crude reaction mixtures with trimethylsilylethynyl iodonium triflate electrophiles afforded trimethylsilyl-terminated N-ethynyl oxazolidinones in 50–60% yields (eq 74).[98] Desilylation could be realized on the purified products or the crude reaction mixtures, and the alkynyl oxazolidinones were elaborated into novel stannyl enamines in the subsequent steps. In contrast, protocols employing n-BuLi in toluene, or Cs_2CO_3 in DMF gave yields lower than 20%. The procedure could be successfully applied to chiral oxazolidinones, since substitution at the C4 position of the oxazolidinones did not have a detrimental effect on reactivity.

Pyrrole carbinols were identified as new base-labile protecting groups for aldehydes.[99] The optimal bases for the preparation of a pyrrole carbinol via the reaction of metal pyrrolates with *iso*-butyraldehyde were sought. Alkali metal pyrrolates performed better than their alkaline earth counterparts. Lithium and sodium pyrrolates prepared via the deprotonation of pyrrole with n-BuLi, LHMDS, and NaHMDS afforded the corresponding pyrrole carbinol in 90%, 85% and 85% yields, respectively, whereas potassium pyrrolate obtained using KHMDS afforded the pyrrole carbinol in a lower yield (45%) (eq 75).[99]

Base	Yield
n-BuLi	90
LHMDS	85
NaHMDS	85
KHMDS	45

KHMDS could also be used for the deprotonation of an N–H bond present in a ligand of an aminoiridium complex, providing an amidoiridium complex required in a study elucidating the kinetic and thermodynamic barriers to interconversions of late metal amido hydride complexes and late metal amine complexes.[100] The treatment of an ammonia complex $(PCP)Ir(H)(Cl)(NH_3)$ with

KHMDS (2 equiv) in THF ($-78\,°C$) afforded the targeted amido complex $(PCP)Ir(H)(NH_2)$ in a quantitative yield accompanied by HMDS and KCl (eq 76).[100] The product was stable below $-10\,°C$, and could be characterized by NMR techniques.

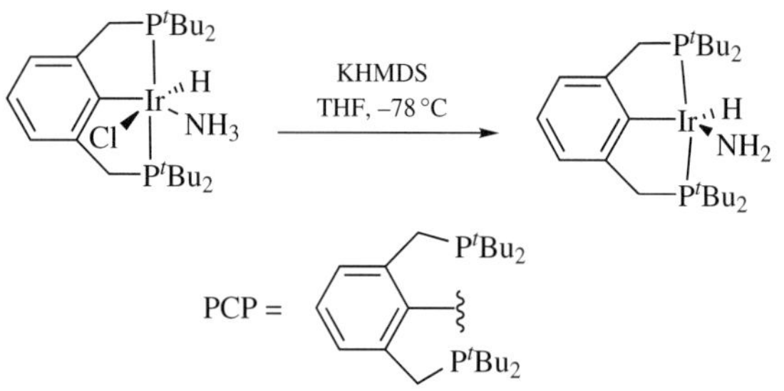

(76)

KHMDS has been used as the key reagent in different approaches to a stereocontrolled glycosidation and functionalization of carbohydrates.[101,102] The choice of the base and solvent for the deprotonation of the glycosidic O–H group in reactions of 2,3,4,6-tetra-O-benzyl-D-glucopyranose or 2,3,4,6-tetra-O-benzyl-D-mannopyranose with 2-chloro-6-nitrobenzothiazole was found to be a critical factor in controlling the α/β selectivity.[101] In the synthesis of 2,3,4,6-tetra-O-benzyl-D-mannopyranoside both stereoisomers (α or β) could be prepared selectively. The α-isomer was obtained by using KHMDS (THF, rt, 0.5 h) giving 90% yield of a 73:27 ratio of α:β stereoisomers (eq 77), whereas a preferential formation of the β-stereoisomer was realized using NaHMDS (THF, rt, 0.5 h) giving 92% yield of a 34:66 ratio of α:β stereoisomers (eq 78).

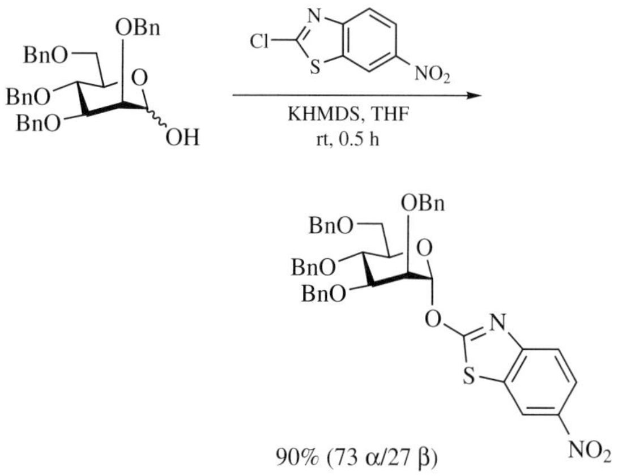

(77)

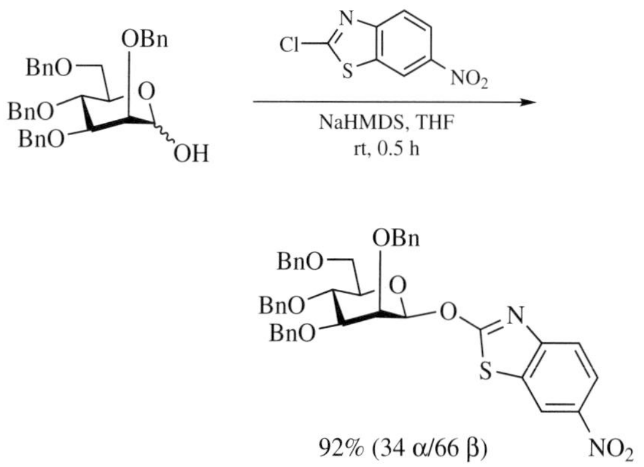

90% (73 α/27 β)

(78)

92% (34 α/66 β)

In contrast, a stereoselective preparation of the α-isomer of the 2,3,4,6-tetra-O-benzyl-D-glucopyranoside employed LHMDS

base (THF–DMF, $0\,°C$, 48 h) and gave a 92% yield of an 88:12 ratio of α:β stereoisomers.[101]

A conversion of bromohydrins featuring a glycosidic hydroxyl group of cyclic hemiacetals into epoxides mediated by different bases, for example KHMDS, followed by the epoxide opening by nucleophiles represents a useful strategy for a stereocontrolled functionalization of carbohydrates.[102,103] For example, an α-stereoselective coupling of N-acetyllactosamine-derived bromo-hydrins to sialic acid-derived thiols or thioacetates was realized via an *in situ* formation of an epoxide, which was effectively mediated by KHMDS at low temperatures (eqs 79 and 80).[103] The reaction temperature proved to be critical for achieving a good diastereoselectivity.

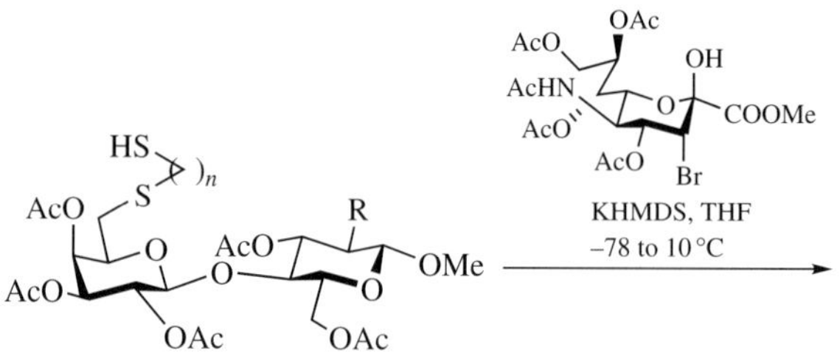

(79)

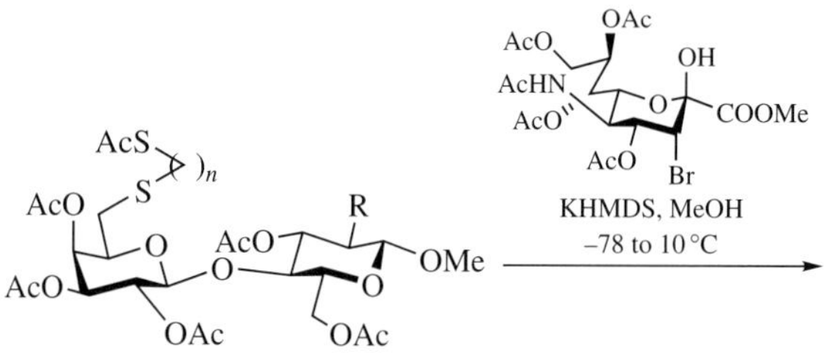

(80)

R = NHAc or OAc

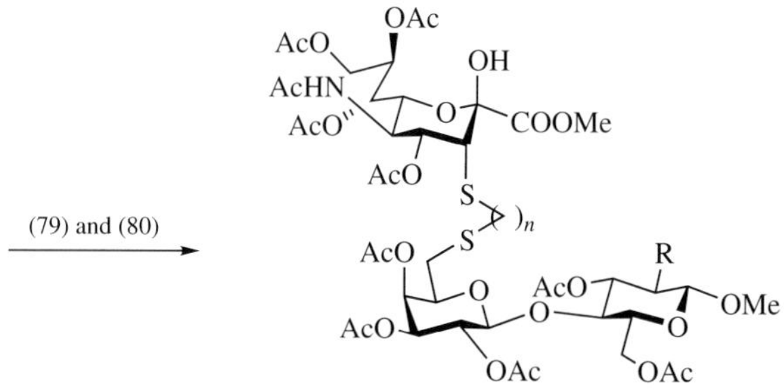

(79) and (80)

21–54% (R = NHAc, n = 1–5)
22–68% (R = OAc, n = 1–5)

Deprotonation of the N–H Bond in Sulfonamides: Catalyst in an Anionic Ring-opening Polymerization. The addition of 5 mol % each of N-benzyl methanesulfonamide and KHMDS to a solution of 2-n-decyl-N-mesylaziridine in DMF initiated a ring-opening polymerization providing a very low polydispersity polyamine polymer (eq 81).[104]

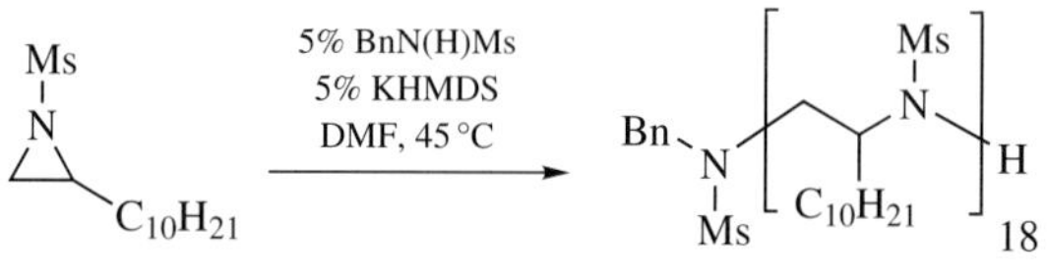

(81)

racemic

M_n = 4800, PDA = 1.06

Polymerization of racemic aziridines afforded soluble polymers, whereas the use of enantiopure aziridines provided sparingly soluble polymers. KHMDS serves to deprotonate the primary sulfonamide initiator, providing an amide anion, which then functions as a nucleophile to promote the ring-opening polymerization of the aziridine. The polymerization kinetics were shown to feature a first-order dependence on both the aziridine substrate and the anionic initiator.

Nucleophilic Nitrogen Source in Polymer Synthesis. KHMDS was used as a stoichiometric *N*-nucleophile to initiate an anionic polymerization of ethylene oxide, providing the α-amino-ω-hydroxyl-poly(ethyleneglycol) (PEG) (THF, 50 °C, 15 h) polymers with different molecular weights (2100, 4400, 7200) (eq 82).[105] The *N*-silyl groups were removed under acidic conditions, and terminating functionalities with structures of known pharmaceutical agents (sulfadiazine and chlorambucil) were attached to the amino and hydroxyl groups via carbamate or ester groups. The anticancer activity of the resulting polymers was studied.

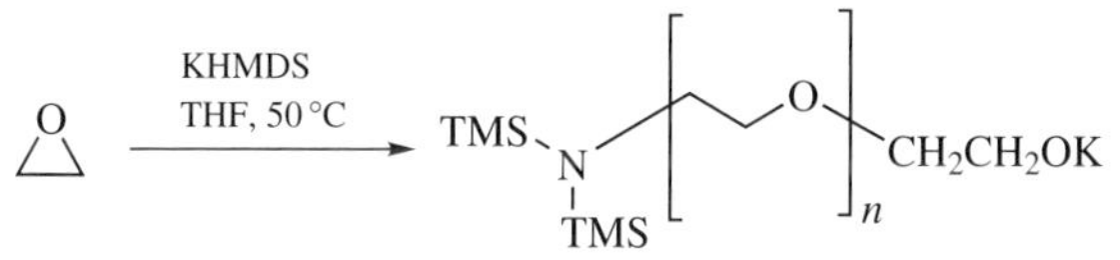

(82)

A method for terminal functionalization of polystyrene chains relying on platinum-catalyzed hydrosilylation of ω-silyl hydride functionalized polymers with bis(trimethylsilyl)allylamine was developed. KHMDS has been treated with allyl bromide in HMDS (0 °C, 1 h) to provide the requisite TMS-protected allylamine.[106]

Base-mediated Rearrangements of the Carbon Skeleton. KHMDS has been utilized to induce the [3,3] sigmatropic anionic oxy-Cope rearrangement with subsequent C-methylation, providing the ring-expanded α-methylated ketone in an excellent yield (81%) in a one step reaction (eq 83).[107]

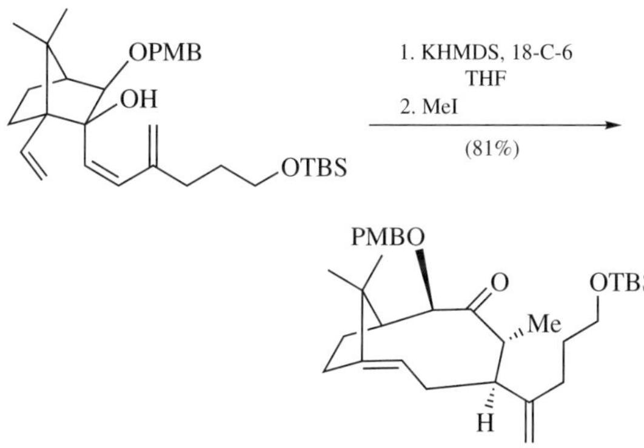

(83)

An attempted KHMDS-mediated oxygenation reaction of *cis*-C9–C10 taxol precursor with an excess (10 equiv) of KHMDS (THF, −78 °C) utilizing an excess of 18-crown-6 ether (30 equiv) afforded a hydroxylated product featuring a rearranged carbon skeleton corresponding to the ring system found in taxol. Apparently, the *cis*-C9–C10 ring system was prone to a facile α-ketol rearrangement, and the α-oxygenation occurred only after the skeletal rearrangement took place (eq 84).[107]

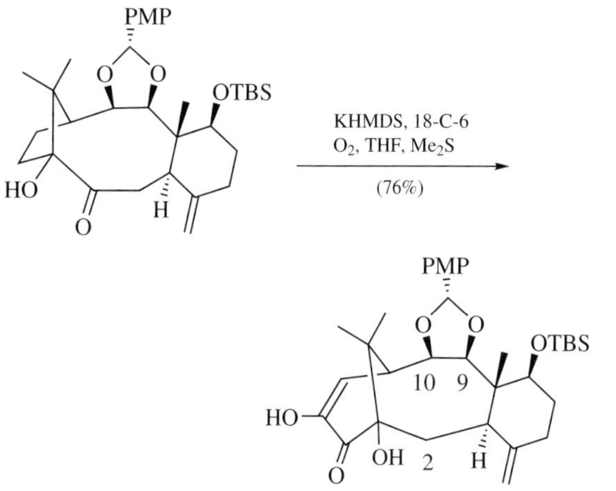

(84)

In contrast, an analogous KHMDS-mediated oxygenation reaction applied to the *trans*-C9–C10 PMP acetal-protected taxol precursor afforded the C-2 oxygenated product in a good yield. The oxidation was not accompanied by the α-ketol rearrangement of the *trans*-C9–C10 ring system. Apparently, the *trans*-locked PMP-acetal introduced a significant steric strain preventing the operation of the α-ketol rearrangement.

KHMDS has been studied as a reagent capable of inducing base-mediated rearrangements of 2-benzyloxycyclooctanone and its Δ^5-unsaturated congener.[108] Exploring the feasibility of an O $\rightarrow$ C 1,2-shift, 2-benzyloxycyclooctanone was treated with bases including NaH in DMF, LDA in THF, and KHMDS in THF. However, the O $\rightarrow$ C 1,2-shift failed to occur. In contrast, the skeletal rearrangement took place in 10 min at room temperature when the reactivity of KHMDS was increased by the addition of the 18-crown-6 ether. Under these conditions, both 2-benzyloxycyclooctanone and 2-benzyl-2-hydroxycyclooctanone afforded a ring-contracted 1-benzylcarbonyl-1-hydroxycycloheptanone in an identical excellent yield (91%) (eqs 85 and eq 86).[108]

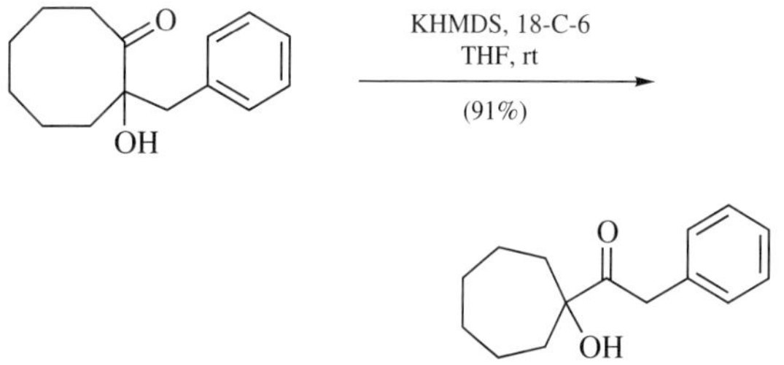

(85)

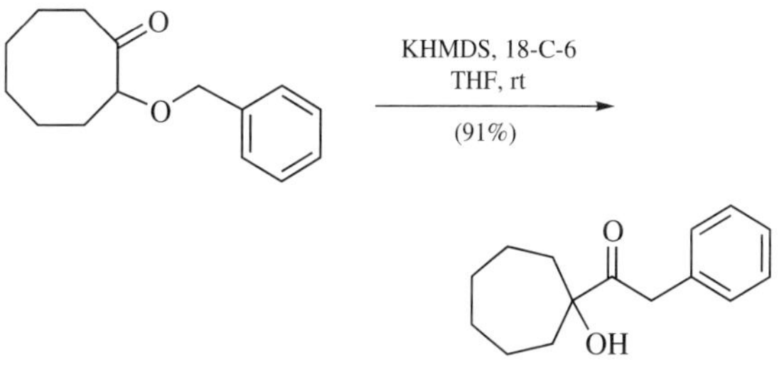

(86)

The rearrangement pathway was rationalized by a sequential O $\rightarrow$ C 1,2-shift (in the case of the 2-benzyloxycyclooctanone) followed by the α-ketol rearrangement. In contrast, analogous rearrangements of the Δ^5-unsaturated analogs under the same conditions required a minimum of 8 h, providing only modest yields of the rearranged product (39–41%). Notably, the

α-ketol rearrangement is a thermodynamically controlled process implying greater thermodynamic stability of the seven-membered isomers.

KHMDS in the presence or absence of the 18-crown-6 ether were used to induce an acyloin rearrangement of 1,3-di-*t*-butyldimethylsilyloxybicyclo[2.2.2]oct-5-en-2-ones to the 1,8-*t*-butyldimethylsilyloxybicyclo[3.2.1]oct-3-en-2-ones.[109] The bicyclo[2.2.2]octenone bearing a formyl (CHO) group afforded good yields of the rearranged bicyclo[3.2.1]octenone with *exo* stereochemistry at C-7 in reactions mediated by KHMDS (80%) or KH (66%). Interestingly, the addition of the 18-crown-6 to the reaction mediated by KHMDS gave only a low yield (35%) and a 1:1 ratio of C-7 *endo/exo* rearranged products (eq 87).[109] The bicyclo[2.2.2]octenone bearing a cyano (CN) group failed to proceed with KH base, whereas the reactions mediated by KHMDS without or with the addition of 18-crown-6 ether afforded 77% and 62% combined yields of C-7 *endo/exo* epimers, respectively (eq 88).[109]

66% (A) and 80% (B)

(87)

77% (B) and 62% (C)

(88)

In contrast, treatment of the bicyclo[2.2.2]octenone bearing a nitro (NO$_2$) group with KHMDS led only to the epimerization of the C-7 stereocenter in the substrate, whereas KH in THF or NaH in DMF induced the acyloin rearrangement in good yields.

The application of KHMDS was essential to realizing the Ireland–Claisen rearrangement of esters of allylic cyclohexenols.[110] Under the optimized conditions (KHMDS, toluene, −78 °C to reflux, TMSCl), the rearranged cyclohexene product was obtained in a high yield (76%, dr 3:2) (eq 89).[110] Other bases, including LDA, LHMDS or NaHMDS, gave lower yields, and the application of TIPSOTf and Hunig's base resulted in significant decomposition. However, the presence of Me or Br substituents at the C-2 carbon in the cyclohexenol ring of the substrates reduced the yields to less than 50%.

(89)

Base	Yield (%)	dr
KHMDS	76	3:2
NaHMDS	44	1.4:1
LHMDS	12	2.1:1

Base-mediated (Formal) Cycloadditions. KHMDS, as well as other bases, have been used in the synthesis of pyrroles via a formal [3 + 2] cycloaddition of metalated isocyanides with ethyl cyclopropylpropiolate.[111] The effect of the nature of the base on the ratio of the desired pyrrole product to the undesired ketone byproduct was investigated. The proposed mechanism consists of a formal cycloaddition of the metalated isocyanides across the triple bond, followed by a 1,5-hydrogen shift and protonation. The reaction between benzyl isocyanide and *t*-butyl cyclopropylpropiolate mediated by *t*-BuOK (2 h, rt) afforded a mixture of both the products in 96% combined yield and 1:1 ratio (eq 90).[111] Similar results were achieved in reactions mediated by KHMDS both at rt and −78 °C (93% and 86% combined yields of 1:1 mixtures) (eq 90). The replacement of KHMDS with NaHMDS and LiHMDS led to a nearly exclusive formation of the undesired ketone product. The content of the undesired ketone product appeared to increase when harder cations were used. Accordingly, the application of *t*-BuOCs (2 h, rt) afforded 91% combined yield of the two products in a 96:4 ratio favoring the heterocycle (eq 90).

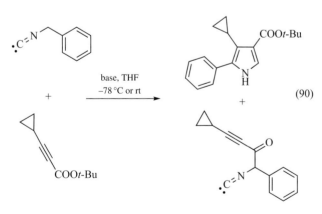

(90)

Base	Combined yield (%)	Pyrrole:ketone ratio
t-BuOK	96	51:49
KHMDS	93	47:53
NaHMDS	98	6:94
LHMDS	98	2:98
t-BuOCs	91	96:4

Several examples of the formal cycloaddition reaction mediated effectively by KHMDS were described (eq 91).

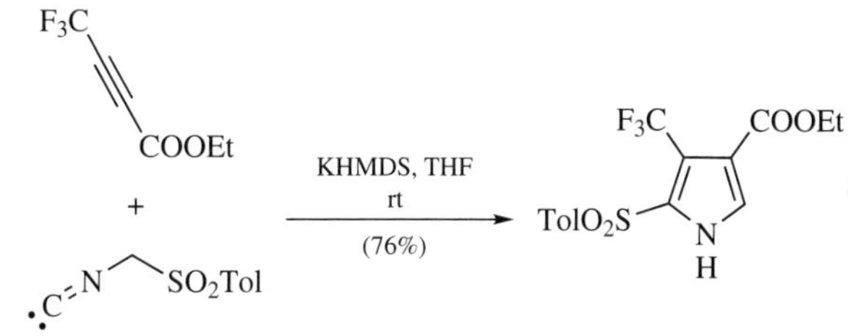

(91)

The effect of the base on the yield and diastereoselectivity of the 1,3-dipolar cycloaddition of azomethine ylides derived from iminoesters with (S)-2-p-tolylsulfinyl-2-cyclopentenone was studied.[112] With a variety of bases (e.g., AgOAc/DBU, AgOAc/TEA) the cycloaddition proceeded with complete regioselectivity and endo selectivity, but low diastereoselectivity (facial selectivity). The diastereoselectivity was dependent on the cation of the base, and the comparison of the performance of LHMDS, NaHMDS and KHMDS indicated that the best yields and diastereoselectivities could be achieved with the lithium cation, giving 91% yield of the diastereomer A and a 98/2 ratio of A/B diastereomers (eq 92). The lower rate and stereoselectivity of reactions involving sodium and potassium cations (eq 92) was rationalized by the operation of a stepwise addition/ring-closure mechanism, in which the cation acts as a tether between the ester oxygen and the oxygen of the p-tolylsulfinyl groups in the two substrates. In contrast to the lithium cation, sodium or potassium cations were unable to fulfill this role with comparable efficiency.

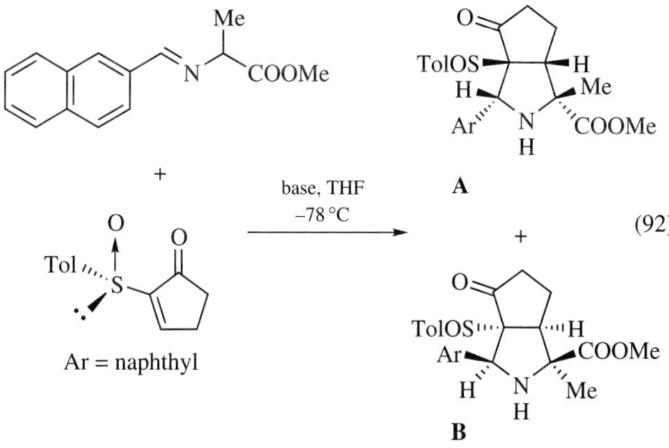

(92)

Base	Yield of A (%)	A:B ratio
LHMDS	91	98:2
NaHMDS	48	68:32
KHMDS	54	75:25

Base-mediated Olefination Reactions. A Peterson olefination reagent (t-BuO)Ph$_2$SiCH$_2$CN, which has shown good selectivity for the formation of (Z)-β-monosubstituted α,β-unsaturated cyanides was developed, and the effect of the cations in the bases and the substrate structures on the reaction yield and the (Z)-stereoselectivity was investigated.[113] The results of the olefinations of various *aromatic aldehydes* mediated by n-BuLi, KHMDS and NaHMDS revealed only minimal effects of the cation in the

base. In general, (Z)-selectivity exceeded 9:1, and good yields were obtained. For example, olefination of 2-pyridylcarboxaldehyde gave optimum results (88% yield, 92:8, Z:E ratio) when mediated by KHMDS (eq 93). In contrast, the olefinations of diverse *aliphatic aldehydes* with (t-BuO)Ph$_2$SiCH$_2$CN mediated by KHMDS proved to be sensitive to the steric bulk of substituents at the α-carbon of the aldehyde, giving a lower (Z)-selectivity. The (Z)-selectivity increased to good levels with only minimal decrease in the overall yields when KHMDS was replaced with NaHMDS and n-BuLi (eq 93).

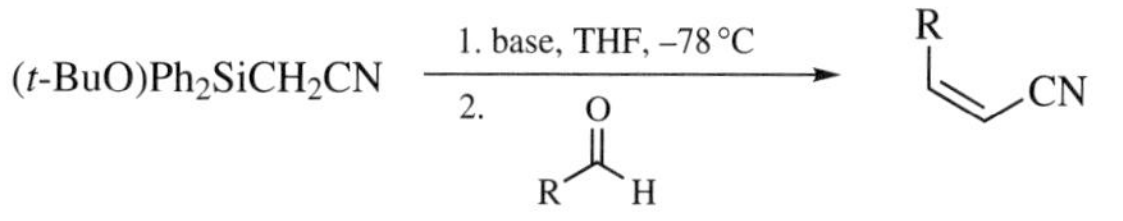

(93)

R	Base	Yield	Z:E ratio
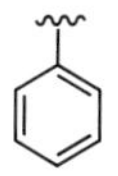	KHMDS	74	96:4
	NaHMDS	79	98:2
	n-BuLi	93	99:1
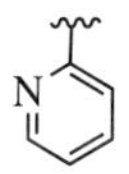	KHMDS	88	92:8
	NaHMDS	75	90:10
	n-BuLi/TMEDA	71	91:9
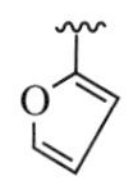	KHMDS	62	98:2
	NaHMDS	85	94:6
	n-BuLi	90	97:3

R	Base	Yield	Z:E ratio
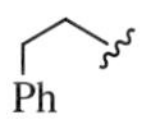	KHMDS	99	96:4
	NaHMDS	88	98:2
	n-BuLi	99	96:4
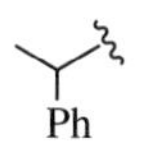	KHMDS	20	98:2
	NaHMDS	55	96:4
	n-BuLi/TMEDA	60	98:2
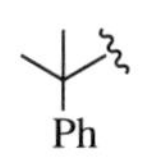	KHMDS	93	86:14
	NaHMDS	85	95:5
	n-BuLi	87	95:5

KHMDS has been used as a base in Wittig-type olefinations, which served as the key steps in the synthesis of all eight geometric isomers of methyl 2,4,6-decatrienoate desired as precursors for several pheromones.[114] Dienyl acetals (eq 94 and eq 95) were converted into the corresponding trienyl esters with *cis*-stereoselectivity under the modified Horner–Wadsworth–Emmons olefination conditions utilizing electrophilic phosphonate $(CF_3CH_2O)_2P(O)CH_2COOCH_3$ and a strongly dissociated base system. Thus, reactions mediated by KHMDS along with the 18-crown-6 ether provided the trienyl esters in 87% and 88% yields, and 95:5 and 93:7 Z:E ratios, respectively, after flash column chromatography (eqs 94 and 95). A partial isomerization of the 6-Z bond in the ester product (eq 95) occurred in the presence of the excess of the 18-crown-6 ether due to the increased polarity of the medium, accounting for the lower selectivity (93:7).

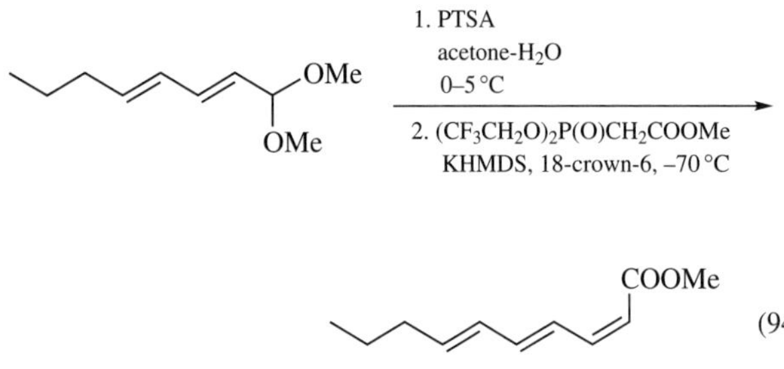

$$(94)$$

95:5 (2Z), 87% yield

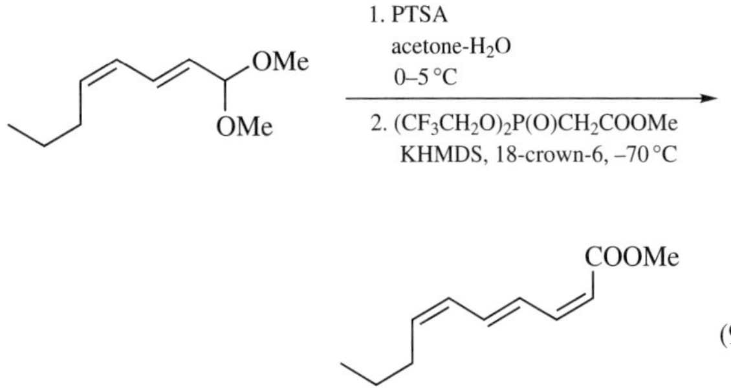

$$(95)$$

93:7 (2Z), 88% yield

Generation of Ylides. A cyclopropanation of alkylidene and arylidene malonates with arsonium ylides was developed, providing access to *trans*-2,3-disubstituted cyclopropane 1,1-dicarboxylic acids bearing a vinyl substituent.[115] The choice of bases for the reactions between ethyl propylidene malonate and arsonium cinnamylide, generated in situ by deprotonation of a corresponding arsonium bromide, was optimized. Experiments with LHMDS, NaHMDS and KHMDS revealed that the best yields and diastereoselectivities could be achieved with KHMDS (88% yield, *trans*:*cis* = 16:1) (eq 96). The application of LHMDS led to a decrease in both the yields and diastereoselectivities (30%, 1.5:1). The transformation proved to be generally applicable, allowing an efficient cyclopropanation of arylidene malonates (eq 97).

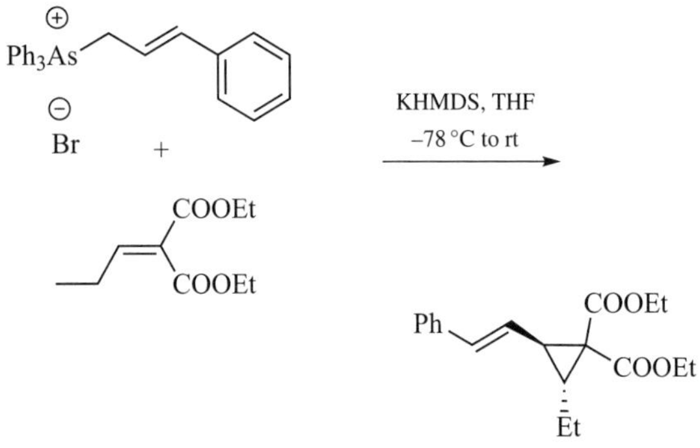

$$(96)$$

88% (*trans*/*cis* 16/1)

KHMDS was used as a base for the deprotonation of benzyltrimethylphosphonium iodide (KHMDS, 0.5 M in toluene, 4 h, 0 °C to rt) to afford the corresponding phosphorus ylide, which was reacted *in situ* with phenylnitrile or 2-furylnitrile, providing *N*-vinylic phosphazenes (eq 98).[116] A mechanistic path-

way involving the construction and subsequent ring-opening of an azaphosphetane (eq 98) was proposed to rationalize the reaction course. The *N*-vinylicphosphazenes were reacted with alkynyl esters to afford unsaturated amines or imines.

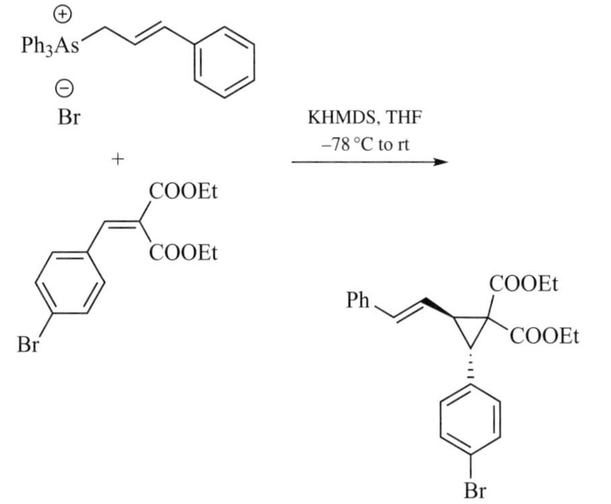

$$(97)$$

94% (*trans*/*cis* 27/1)

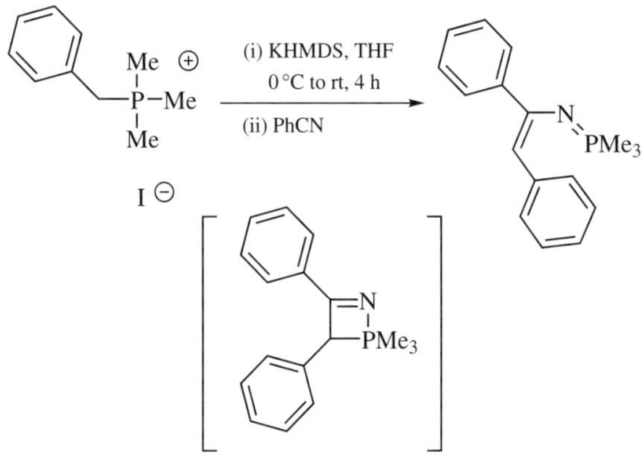

$$(98)$$

Base in Pd-catalyzed C–N Bond-forming Reactions. Suitable bases for successful cross-coupling of aromatic bromides bearing reactive functionalities (OH, AcNH, CH$_3$CO, NH$_2$CO) with amines via a catalytic system employing a new bicyclic triaminophosphine ligand along with Pd(OAc)$_2$ catalyst (toluene, 80 °C) were sought.[117] The cross-coupling reactions, which failed using weak bases, could be efficiently realized employing *t*-BuO-Na/LDA mixtures, LHMDS, KHMDS, or NaHMDS bases. The results for coupling of 3-hydroxyphenyl bromide to morpholine (toluene, 80 °C, 22 h) are shown in eq 99.

The reasons for the effectiveness of silylated amide bases might consist in an internal protection of the reactive functionalities, either via the formation of tightly associated lithium ion pairs following the deprotonation, or via the formation of silylated alcohols or silylated amides operating as protecting groups. Both pathways would prevent the alkoxide or amide coordination to palladium resulting in catalyst deactivation.

Base for C–H or N–H Deprotonation of Precursors to Carbene or Amide Ligands in the Preparation of Metal Complexes. KHMDS has been used as a base for the deprotonation of N–H or B–H bonds in imine-based ligands or trispyrazolyl borate-derived ligands to synthesize calcium complexes, which were used as initiators of the ring-opening polymerization of lactides.[118] A first example of a two-coordinate, neutral In[I] singlet carbene complex has been prepared via an in situ deprotonation of the imine ligand precursor with KHMDS in the presence of In[I]I to afford the In(I) complex in 36% unoptimized yield as pale yellow, thermally stable crystals.[119]

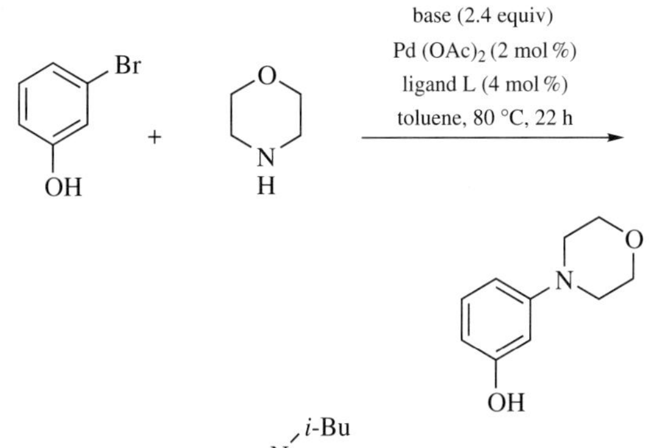

(99)

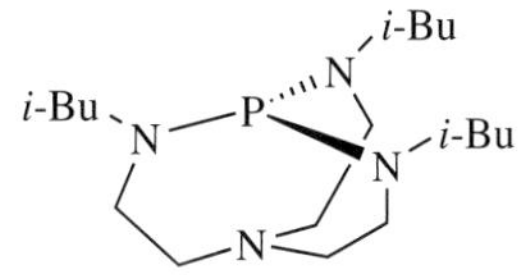

Base	Yield (%)	Base	Yield (%)
t-BuONa	Trace	LiHMDS	92
Cs_2CO_3	NR	KHMDS	85
t-BuONa/LDA	80	NaHMDS	88

A bidentate ferrocenyl-*N*-heterocyclic carbene ligand precursor in the form of the bis-imidazolinium salt was deprotonated in situ with KHMDS and reacted with [PdCl₂(cod)] to provide a bis(carbene)palladium(II) dichloride complex, which was structurally characterized (eq 100).[120]

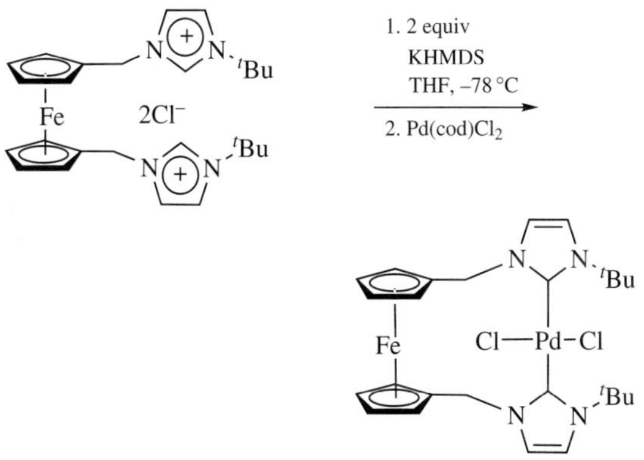

(100)

An in situ treatment of the iminium salt shown in eq 101 with KHMDS afforded a four-membered *N*-heterocyclic carbene ligand, which reacted with the ruthenium complex (PPh₃)Cl₂Ru= CH-*o*-O-*i*-PrC₆H₅ (60 °C, overnight) providing the ruthenium carbene complex in 30% yield as an air stable brown solid (eq 101), which was characterized by NMR and tested for its reactivity in ring closing metathesis.[121]

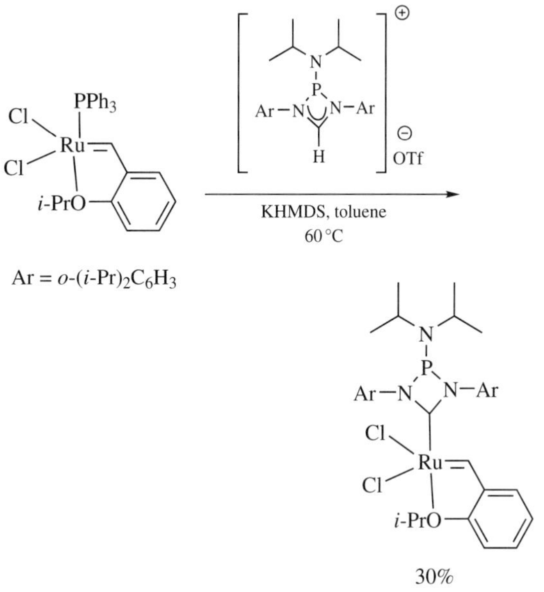

$Ar = o$-(i-Pr)₂C₆H₃

(101)

30%

Synergistic Effects in Metalation of Unactivated C–H Bonds. The metalation of toluene has been achieved using a bimetallic base KZn(HMDS)₃, providing the metalated benzyl product [KZn(HMDS)₂CH₂Ph], whereas an analogous metalation could not be achieved either with KHMDS or Zn(HMDS)₂.[122] A mixture of KHMDS and Zn(HMDS)₂ (10 mmol each) in toluene (20 mL) was stirred for 15 min at room temperature followed by vigorous heating for 5 min to afford the metalated benzyl product as a solid [KZn(HMDS)₂CH₂Ph]. Under more strenuous conditions the metalation could not be realized with either KHMDS or Zn(HMDS)₂, clearly demonstrating that the described deprotonative metalation is synergic by its origin. The synergic effect did not operate when Zn was replaced with Mg, and the synergy-driven deprotonation could be extended to *m*-xylene and mesitylene.

1. Wannagat, U.; Niederpruem, H., *Ber. Dtsch. Chem. Ges.* **1961**, *94*, 1540.

2. (a) Brown, C. A., *Synthesis* **1974**, 427. (b) Brown, C. A., *J. Org. Chem.* **1974**, *39*, 3913.

3. Tesh, K. F.; Hanusa, T. P.; Huffman, J. C., *Inorg. Chem.* **1990**, *29*, 1584.

4. Williard, P. G., *Acta Crystallogr.* **1988**, *C44*, 270.

5. Ireland, R. E.; Meissner, R. S., *J. Org. Chem.* **1991**, *56*, 4566.

6. Duhamel, L.; Plaquevent, J.-C., *J. Organomet. Chem.* **1993**, *448*, 1.

7. (a) Evans, D. A. in *Asymmetric Synthesis*; Morrison, J. D., Ed.; Academic: New York, 1984; Vol. 3, p 1. (b) Brown, C. A., *J. Org. Chem.* **1974**, *39*, 3913.

8. Negishi, E.; Chatterjee, S., *Tetrahedron Lett.* **1983**, *24*, 1341.

9. Differding, E.; Rueegg, G.; Lang, R. W., *Tetrahedron Lett.* **1991**, *32*, 1779.

10. (a) Evans, D. A.; Britton, T. C., *J. Am. Chem. Soc.* **1987**, *109*, 6881. (b) Evans, D. A.; Ellman, J. A., *J. Am. Chem. Soc.* **1989**, *111*, 1063.

11. (a) Boche, G.; Heidenhain, F., *Angew. Chem., Int. Ed. Engl.* **1978**, *17*, 283. (b) Boche, G.; Heidenhain, F.; Thiel, W.; Eiben, R., *Ber. Dtsch. Chem. Ges.* **1982**, *115*, 3167.

12. Kawanisi, M.; Itoh, Y.; Hieda, T.; Kozima, S.; Hitomi, T.; Kobayashi, K., *Chem. Lett.* **1985**, 647.

13. (a) Rathke, M. W.; Sullivan, D., *Tetrahedron Lett.* **1972**, 4249. (b) Herrman, J. L.; Kieczykowski, G. R.; Schlessinger, R. H., *Tetrahedron*

Lett. **1973**, 2433. (c) Krebs, E. P., *Helv. Chim. Acta* **1981**, *64*, 1023. (d) Kende, A. S.; Toder, B. H., *J. Org. Chem.* **1982**, *47*, 163. (e) Ikeda, Y.; Yamamoto, H., *Tetrahedron Lett.* **1984**, *25*, 5181.

14. (a) Ikeda, Y.; Ukai, J.; Ikeda, N.; Yamamoto, H., *Tetrahedron* **1987**, *43*, 743. (b) Chattopadhyay, A.; Mamdapur, V. R., *Synth. Commun.* **1990**, *20*, 2225.

15. (a) Bates, G. S., *Chem. Commun.* **1979**, 161. (b) Bates, G. S.; Ramaswamy, S., *Can. J. Chem.* **1980**, *58*, 716.

16. (a) Schlosser, M.; Christmann, K. F., *Justus Liebigs Ann. Chem.* **1967**, *708*, 1. (b) Schlosser, M., *Top. Stereochem.* **1970**, *5*, 1. (c) Schlosser, M.; Schaub, B.; de Oliveira-Neto, J.; Jeganathan, S., *Chimia* **1986**, *40*, 244. (d) Schaub, B.; Jeganathan, S.; Schlosser, M., *Chimia* **1986**, *40*, 246.

17. Sreekumar, C.; Darst, K. P.; Still, W. C., *J. Org. Chem.* **1980**, *45*, 4260.

18. Still, W. C.; Gennari, C., *Tetrahedron Lett.* **1983**, *24*, 4405.

19. (a) Stork, G.; Gardner, J. O.; Boeckman, R. K., Jr.; Parker, K. A., *J. Am. Chem. Soc.* **1973**, *95*, 2014. (b) Stork, G.; Boeckman, R. K., Jr., *J. Am. Chem. Soc.* **1973**, *95*, 2016.

20. Takahashi, T.; Kazuyuki, K.; Tsuji, J., *Tetrahedron Lett.* **1978**, 4917.

21. Abraham, W. D.; Cohen, T., *J. Am. Chem. Soc.* **1991**, *113*, 2313.

22. Shimizu, M.; Yoshioka, H., *Tetrahedron Lett.* **1989**, *30*, 967.

23. Paquette, L. A.; Pegg, N. A.; Toops, D.; Maynard, G. D.; Rogers, R. D., *J. Am. Chem. Soc.* **1990**, *112*, 277.

24. Larsen, S. D., *J. Am. Chem. Soc.* **1988**, *110*, 5932.

25. Brown, H. C.; Ramachandran, P. V.; Gong, B. Q.; Brown, H. C., *J. Org. Chem.* **1995**, *60*, 41.

26. Mantri, P.; Duffy, D. E.; Kettner, C. A., *J. Org. Chem.* **1996**, *61*, 5690.

27. Wityak, J.; Earl, R. A.; Abelman, M. M.; Bethel, Y. B.; Fisher, B. N.; Kauffman, G. S.; Kettner, C. A.; Ma, P.; McMillan, J. L.; Mersinger, L. J.; Pesti, J.; Pierce, M. E.; Rankin, F. W.; Chorvat, R. J.; Confalone, P. N., *J. Org. Chem.* **1995**, *60*, 3717.

28. Matteson, D. S.; Schaumberg, G. D., *J. Org. Chem.* **1966**, *31*, 726.

29. Kiselyov, A. S., *Tetrahedron Lett.* **1995**, *36*, 1383.

30. Bruning, J., *Tetrahedron Lett.* **1997**, *38*, 3187.

31. See for example: Grasa, G. A.; Guveli, T.; Singh, R.; Nolan, S. P., *J. Org. Chem.* **2003**, *68*, 2812.

32. Herrmann, W. A., *Angew. Chem. Int. Ed.* **2002**, *41*, 1290.

33. Yong, B. S.; Nolan, S. P., *Chemtracts: Org. Chem.* **2003**, *16*, 205.

34. Douthwaite, R. E.; Haussinger, D.; Green, M. L. H.; Silcock, P. J.; Gomes, P. T.; Martins, A. M.; Danopoulos, A. A., *Organometallics* **1999**, *18*, 4584.

35. Danopoulos, A. A.; Winston, S.; Gelbrich, T.; Hursthouse, M. B.; Tooze, R. P., *Chem. Commun.* **2002**, 482.

36. Danopoulos, A. A.; Winston, S.; Motherwell, W. B., *Chem. Commun.* **2002**, 1376.

37. Kerr, M. S.; de Alaniz, J. R.; Rovis, T., *J. Am. Chem. Soc.* **2002**, *124*, 10298.

38. Kerr, M. S.; Rovis, T., *Synlett* **2003**, 1934.

39. Culkin, D. A.; Hartwig, J. F., *Acc. Chem. Res.* **2003**, *36*, 234.

40. Miura, M.; Nomura, M., *Top. Curr. Chem.* **2002**, *219*, 211.

41. Lloyd-Jones, G. C., *Angew. Chem., Int. Ed.* **2002**, *41*, 953.

42. Lee, S.; Beare, N. A.; Hartwig, J. F., *J. Am. Chem. Soc.* **2001**, *123*, 8410.

43. Jorgensen, M.; Lee, S.; Liu, X. X.; Wolkowski, J. P.; Hartwig, J. F., *J. Am. Chem. Soc.* **2002**, *124*, 12557.

44. Gaertzen, O.; Buchwald, S. L., *J. Org. Chem.* **2002**, *67*, 465.

45. Beare, N. A.; Hartwig, J. F., *J. Org. Chem.* **2002**, *67*, 541.

46. Hamann, B. C.; Hartwig, J. F., *J. Am. Chem. Soc.* **1997**, *119*, 12382.

47. Fox, J. M.; Huang, X. H.; Chieffi, A.; Buchwald, S. L., *J. Am. Chem. Soc.* **2000**, *122*, 1360.

48. Ehrentraut, A.; Zapf, A.; Beller, M., *Adv. Synth. Catal.* **2002**, *344*, 209.

49. Shaughnessy, K. H.; Hamann, B. C.; Hartwig, J. F., *J. Org. Chem.* **1998**, *63*, 6546.

50. Caron, S.; Vazquez, E.; Wojcik, J. M., *J. Am. Chem. Soc.* **2000**, *122*, 712.

51. Papahatjis, D. P.; Nikas, S.; Tsotinis, A.; Vlachou, M.; Makriyannis, A., *Chem. Lett.* **2001**, 192.

52. Marcoux, J. F.; Corley, E. G.; Rossen, K.; Pye, P.; Wu, J.; Robbins, M. A.; Davies, I. W.; Larsen, R. D.; Reider, P. J., *Org. Lett.* **2000**, *2*, 2339.

53. Watanabe, K.; Hirasawa, T.; Hiroi, K., *Heterocycles* **2002**, *58*, 93.

54. Brana, M. F.; Moran, M.; Devega, M. J. P.; Pitaromero, I., *Tetrahedron Lett.* **1994**, *35*, 8655.

55. Harada, T.; Fujiwara, T.; Iwazaki, K.; Oku, A., *Org. Lett.* **2000**, *2*, 1855.

56. Ogawa, A.; Sakagami, K.; Shima, A.; Suzuki, H.; Komiya, S.; Katano, Y.; Mitsunobu, O., *Tetrahedron Lett.* **2002**, *43*, 6387.

57. Taber, D. F.; Christos, T. E.; Neubert, T. D.; Batra, D., *J. Org. Chem.* **1999**, *64*, 9673.

58. Barluenga, J.; Baragana, B.; Concellon, J. M.; Pinera-Nicolas, A.; Diaz, M. R.; Garcia-Granda, S., *J. Org. Chem.* **1999**, *64*, 5048.

59. Kawabata, T.; Suzuki, H.; Nagae, Y.; Fuji, K., *Angew. Chem. Int. Ed.* **2000**, *39*, 2155.

60. Kawabata, T.; Fuji, K., *Top. Stereo. Chem.* **2003**, *23*, 175.

61. Kawabata, T.; Kawakami, S.; Majumdar, S., *J. Am. Chem. Soc.* **2003**, *125*, 13012.

62. Corey, E. J.; Wu, L. I., *J. Am. Chem. Soc.* **1993**, *115*, 9327.

63. Li, C. M.; Johnson, R. P.; Porco, J. A., *J. Am. Chem. Soc.* **2003**, *125*, 5095.

64. Ruel, R.; Bouvier, J. P.; Young, R. N., *J. Org. Chem.* **1995**, *60*, 5209.

65. Helmchen, G.; Pfaltz, A., *Acc. Chem. Res.* **2000**, *33*, 336.

66. Busacca, C. A.; Grossbach, D.; So, R. C.; O'Brien, E. M.; Spinelli, E. M., *Org. Lett.* **2003**, *5*, 595.

67. Erratum: Busacca, C. A.; Grossbach, D.; So, R. C.; O'Brien, E. M.; Spinelli, E. M., *Org. Lett.* **2003**, *5*, 1595.

68. Maezaki, N.; Sakamoto, A.; Nagahashi, N.; Soejima, M.; Li, Y. X.; Imamura, T.; Kojima, N.; Ohishi, H.; Sakaguchi, K.; Iwata, C.; Tanaka, T., *J. Org. Chem.* **2000**, *65*, 3284.

69. Zhou, Y. G.; Li, A. H.; Hou, X. L.; Dai, L. X., *Chem. Commun.* **1996**, 1353.

70. Aggarwal, V. K.; Bae, I.; Lee, H. Y.; Richardson, J.; Williams, D. T., *Angew. Chem. Int. Ed.* **2003**, *42*, 3274.

71. Gandelman, M.; Rybtchinski, B.; Ashkenazi, N.; Gauvin, R. M.; Milstein, D., *J. Am. Chem. Soc.* **2001**, *123*, 5372.

72. Blakemore, P. R., *J. Chem. Soc., Perkin Trans. 1* **2002**, 2563.

73. Baudin, J. B.; Hareau, G.; Julia, S. A.; Lorne, R.; Ruel, O., *Bull. Soc. Chim. Fr.* **1993**, *130*, 856.

74. Charette, A. B.; Lebel, H., *J. Am. Chem. Soc.* **1996**, *118*, 10327.

75. Blakemore, P. R.; Cole, W. J.; Kocienski, P. J.; Morley, A., *Synlett* **1998**, 26.

76. Smith, A. B.; Wan, Z. H., *J. Org. Chem.* **2000**, *65*, 3738.

77. Trost, B. M.; Crawley, M. L., *J. Am. Chem. Soc.* **2002**, *124*, 9328.

78. Trost, B. M.; Chisholm, J. D.; Wrobleski, S. T.; Jung, M., *J. Am. Chem. Soc.* **2002**, *124*, 12420.

79. Fettes, A.; Carreira, E. M., *Angew. Chem. Int. Ed.* **2002**, *41*, 4098.

80. Kocienski, P. J.; Bell, A.; Blakemore, P. R., *Synlett* **2000**, 365.

81. Charette, A. B.; Berthelette, C.; St-Martin, D., *Tetrahedron Lett.* **2001**, *42*, 5149.

82. Erratum: Charette, A. B.; Berthelette, C.; St-Martin, D., *Tetrahedron Lett.* **2001**, *42*, 6619.

83. Tang, Y.; Ye, S.; Huang, Z. Z.; Huang, Y. Z., *Heteroatom Chem.* **2002**, *13*, 463.

84. Dai, W.-M.; Wu, A.; Wu, H., *Tetrahedron: Asymmetry* **2002**, *13*, 2187.

85. Klotz, P.; Mann, A., *Tetrahedron Lett.* **2003**, *44*, 1927.

86. Baldelli, E.; Battaglia, A.; Bombardelli, E.; Carenzi, G.; Fontana, G.; Gambini, A.; Gelmi, M. L.; Guerrini, A.; Pocar, D., *J. Org. Chem.* **2003**, *68*, 9773.

A list of General Abbreviations appears on the front Endpapers

87. Trost, B. M.; Dogra, K.; Franzini, M., *J. Am. Chem. Soc.* **2004**, *126*, 1944.

88. Pedersen, B.; Rein, T.; Sotofte, I.; Norrby, P. O.; Tanner, D., *Collect. Czech. Chem. Commun.* **2003**, *68*, 885.

89. García Ruano, J. L.; Martín-Castro, A. M.; Tato, F.; Pastor, C. J., *J. Org. Chem.* **2005**, *70*, 7346.

90. Kawabata, T.; Majumdar, S.; Tsubaki, K.; Monguchi, D., *Org. Biomol. Chem.* **2005**, *3*, 1609.

91. Carlier, P. R.; Lam, P. C. H.; DeGuzman, J. C.; Zhao, H., *Tetrahedron: Asymmetry* **2005**, *16*, 2998.

92. Wiles, C.; Watts, P.; Haswell, S. J.; Pombo-Villar, E., *Tetrahedron* **2005**, *61*, 10757.

93. Klapars, A.; Waldman, J. H.; Campos, K. R.; Jensen, M. S.; McLaughlin, M.; Chung, J. Y. L.; Cvetovich, R. J.; Chen, C., *J. Org. Chem.* **2005**, *70*, 10186.

94. Lu, G.; Malinakova, H. C., *J. Org. Chem.* **2004**, *69*, 4701.

95. Agami, C.; Dechoux, L.; Hebbe, S., *Tetrahedron Lett.* **2003**, *44*, 5311.

96. Zavialov, I. A.; Dahanukar, V. H.; Nguyen, H.; Orr, C.; Andrews, D. R., *Org. Lett.* **2004**, *6*, 2237.

97. Maung, J.; Mallari, J. P.; Girtsman, T. A.; Wu, L. Y.; Rowley, J. A.; Santiago, N. M.; Brunelle, A. N.; Berkman, C. E., *Bioorg. Med. Chem.* **2004**, *12*, 4969.

98. Naud, S.; Cintrat, J. C., *Synthesis* **2003**, 1391.

99. Dixon, D. J.; Scott, M. S.; Luckhurst, C. A., *Synlett* **2003**, 2317.

100. Kanzelberger, M.; Zhang, X.; Emge, T. J.; Goldman, A. S.; Zhao, J.; Incarvito, C.; Hartwig, J. F., *J. Am. Chem. Soc.* **2003**, *125*, 13644.

101. Hashihayata, T.; Mandai, H.; Mukaiyama, T., *Bull. Chem. Soc. Jpn.* **2004**, *77*, 169.

102. Kobayashi, S.; Takahashi, Y.; Komano, K.; Alizadeh, B. H.; Kawada, Y.; Oishi, T.; Tanaka, S. I.; Ogasawara, Y.; Sasaki, S. Y.; Hirama, M., *Tetrahedron* **2004**, *60*, 8375.

103. Hinou, H.; Sun, X. S.; Ito, Y., *J. Org. Chem.* **2003**, *68*, 5602.

104. Stewart, I. C.; Lee, C. C.; Bergman, R. G.; Toste, D. F., *J. Am. Chem. Soc.* **2005**, *127*, 17616.

105. Jia, Z.; Zhang, H.; Huang, J., *Bioorg. Med. Chem. Lett.* **2003**, *13*, 2531.

106. Quirk, R. P.; Kim, H.; Polce, M. J.; Wesdemiotis, C., *Macromolecules* **2005**, *38*, 7895.

107. Paquette, L. A.; Hofferberth, J. E., *J. Org. Chem.* **2003**, *68*, 2266.

108. Vilotijevic, I.; Yang, J.; Hilmey, D.; Paquette, L. A., *Synthesis* **2003**, 1872.

109. Katayama, S.; Yamauchi, M., *Chem. Pharm. Bull.* **2005**, *53*, 666.

110. Beaulieu, P.; Ogilvie, W. W., *Tetrahedron Lett.* **2003**, *44*, 8883.

111. Larionov, O. V.; de Meijere, A., *Angew. Chem., Int. Ed.* **2005**, *44*, 5664.

112. García Ruano, J. L.; Tito, A.; Peromingo, M. T., *J. Org. Chem.* **2003**, *68*, 10013.

113. Kojima, S.; Fukuzaki, T.; Yamakawa, A.; Murai, Y., *Org. Lett.* **2004**, *6*, 3917.

114. Khrimian, A., *Tetrahedron* **2005**, *61*, 3651.

115. Jiang, H.; Deng, X.; Sun, X.; Tang, Y.; Dai, L. X., *J. Org. Chem.* **2005**, *70*, 10202.

116. Palacios, F.; Alonso, C.; Pagalday, J.; Ochoa de Retana, A. M.; Rubiales, G., *Org. Biomol. Chem.* **2003**, *1*, 1112.

117. Urgaonkar, S.; Verkade, J. G., *Adv. Synth. Catal.* **2004**, *346*, 611.

118. Chisholm, M. H.; Gallucci, J. C.; Phomphrai, K., *Inorg. Chem.* **2004**, *43*, 6717.

119. Hill, M. S.; Hitchcock, P. B., *Chem. Commun.* **2004**, 1818.

120. Coleman, K. S.; Turberville, S.; Pascu, S. I.; Green, M. L. H., *J. Organomet. Chem.* **2005**, *690*, 653.

121. Despagnet-Ayoub, E.; Grubbs, R. H., *Organometallics* **2005**, *24*, 338.

122. Clegg, W.; Forbes, G. C.; Kennedy, A. R.; Mulvey, R. E.; Liddle, S. T., *Chem. Commun.* **2003**, 406.

(+)-9-(1*S*,2*S*-Pseudoephedrinyl)-(10*R*)-(trimethylsilyl)-9-borabicyclo[3.3.2]-decane[1]

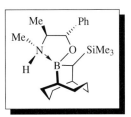

[848701-34-8]   C$_{22}$H$_{38}$BNOSi   (MW 371.44)

InChI = 1S/C22H38BNOSi/c1-17(24-2)21(18-11-7-6-8-12-18)25-23-20-15-9-13-19(14-10-16-20)22(23)26(3,4)5/h6-8,11-12,17,19-22,24H,9-10,13-16H2,1-5H3/t17-,19-,20+,21+,22?/m0/s1

InChIKey = ZOZYQBFBLCMBTA-FQXALJIWSA-N

(chiral pseudoephedrine borinic acid derivative for the preparation of reagents for the asymmetric hydroboration, allyl-, allenyl-, crotyl-, propargyl-, and alkynylation of aldehydes and aldimines through Grignard or organolithium procedures)

Physical Data: mp 130–134 °C, $[\alpha]_D^{22} = +60°$ ($c = 4.5$, CDCl$_3$).
Solubility: sol CHCl$_3$, CH$_2$Cl$_2$, THF.
Preparation: both enantiomers are available in pure form from the transesterification of racemic *B*-MeO-10-TMS-9-BBD (TMSCHN$_2$ plus *B*-MeO-9-BBN) with pseudoephedrine in MeCN; the sequential protocol utilizes both enantiomers of pseudoephedrine to provide both enantiomers (ca. 66% total yield) of the borinate complexes in a single procedure.
Analysis of Reagent Purity: in solution, both the open and closed forms of the complex can be observed through ^{13}C NMR by analyzing the TMS signal of the complex in CDCl$_3$ (δ 1.7 (open) and δ 0.7 (closed)). Even very minor amounts of the undesired diastereomer (δ 1.1) can be observed if present.[2] Recrystallization (MeCN/CH$_2$Cl$_2$ ~8:1 (0.5 M)) after complete dissolution at reflux (20–40 h) gives the pure crystalline enantiomer.
Form Supplied in: colorless crystalline solid (Scheme 1); both enantiomers are commercially available.

Scheme 1

Handling, Storage, and Precautions: crystalline solid that can be handled in the open atmosphere; best stored under dry nitrogen to preserve product purity.

Allylboration.[1]   The direct conversion of the complex **1** to **2** was accomplished by its reaction with AllMgBr in Et$_2$O at $-78\,°$C (eq 1). The stable allylboranes **2** (**2*R***: 856675-79-1; **2*S***: 856675-80-4) can be isolated (98%) in chemically and optically pure form through simple filtration through Celite under N$_2$ to remove the Mg^{2+} salts, followed by concentration in vacuo. Under N$_2$, **2** is stable for weeks at 25 °C. In practice, **2** need not be isolated, but rather quantitatively generated with the simple addition of AllMgBr in Et$_2$O to **1**. This admixture is directly used for the allylations, the Mg^{2+} salts not interfering with the rate of allylation.

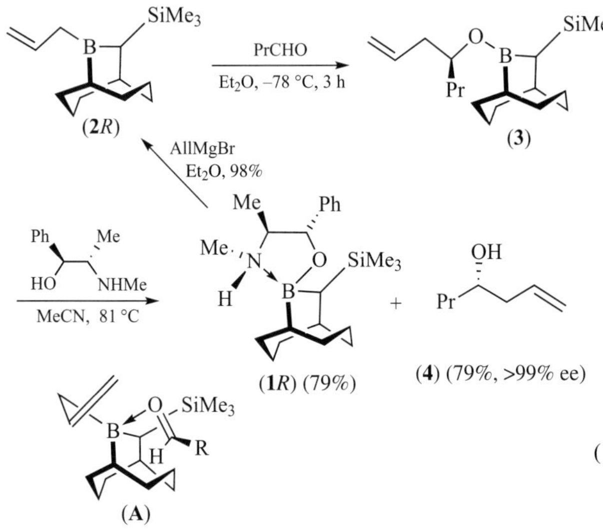

(1)

The asymmetric allylboration of representative aldehydes was conducted with either **2*R*** or **2*S*** in Et$_2$O (3 h, $-78\,°$C) with the homoallylic alcohols **4** being isolated in high yield (70–92%) and ee (96–99% ee) (eq 1). This process is quite general, being effective for alkyl, aryl, heteroaryl, and unsaturated aldehydes. The complex **1** is regenerated through the reaction protocol and isolated in 68–84% yield, thereby efficiently recycling the chiral borane moiety. Due to their rigid bicyclic ring structure, the reagents **2** are highly selective over a wide temperature range such that little diminution (2–8% ee) in optical purity is observed in the product alcohols for reactions conducted at +25 vs. $-78\,°$C. As can be represented by the pre-transition state model **A**, the preferred mode of attack is *anti* complexation of the aldehyde with the borane. The aldehyde bonds to boron on the side *cis* to the TMS group because the O is smaller than the sp^3 C of the allyl group. It is oriented down to permit a viable Zimmerman–Traxler chair-like transition state to be reached without interference from the TMS group. The boat side of the boat–chair ring conformation is favored on the TMS side of the ring making a small chiral pocket.

Even with challenging chiral substrates, **2** exhibits remarkable levels of reagent control as was illustrated in recent studies toward the synthesis of mycolactone (**7**) where the desired anti-Felkin alcohol **6** was produced as a single diastereomer from the 'mismatched' allylboration combination (eq 2).[3]

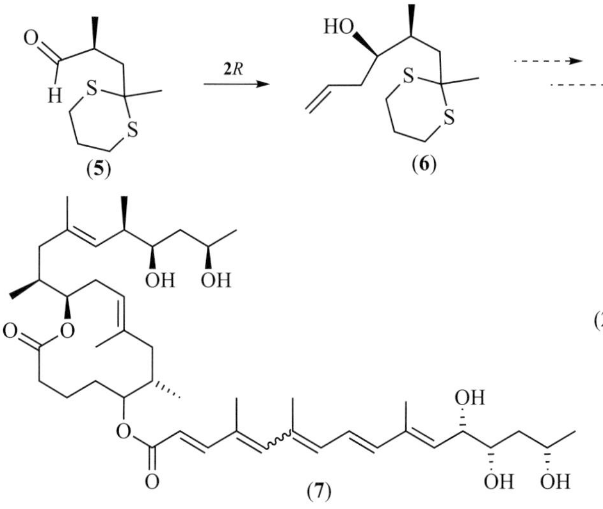

(2)

(7)

The allylation of *N*-H aldimines generated from the methanolysis of both *N*-TMS and *N*-Al(Bu-*i*)$_2$ aldimine precursors with **2** provides corresponding adducts **8** (^{11}B NMR δ 51). Treatment of **8** with the appropriate enantiomer of pseudoephedrine (PE, 10 h, MeCN, reflux) provides **1** (50–70%) and the desired homoallylic amines **9** (60–90%, 68–89% ee) from aliphatic, aryl, and heteroaryl aldimines (eq 3). An acidic workup (2 M HCl) was employed for the *N*-DIBAL imines to provide **9**.[4] This lowered selectivity compared to aldehydes is expected from the smaller size difference of NH vs. the allylic sp^3 C compared to O vs. this C.

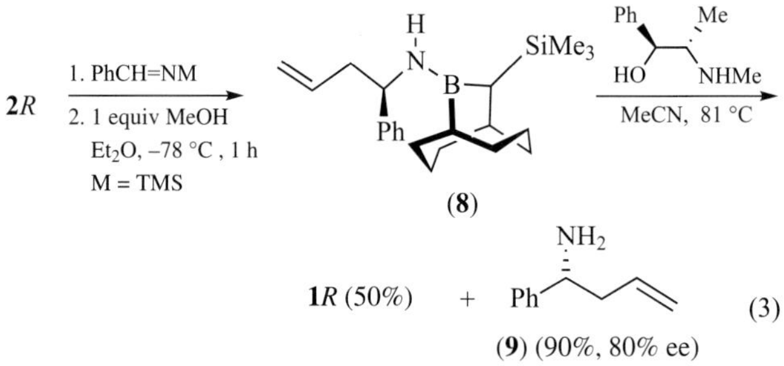

(3)

Methallylboration.[5]   The synthesis of the *B*-methallyl-10-TMS-9-BBD (**11** (**11*S***: 1001158-42-4; **11*R***: 1001158-41-3) is accomplished through a Grignard procedure employing the *B*-OMe precursor **10** (**10*S***: 856675-97-3; **10*R***:856675-96-2) (eq 4). The complex **1** is unreactive toward methallylmagnesium chloride requiring the preparation of **10** from **1** through a two-step procedure (85% overall). The homoallylic alcohols **12** are produced efficiently (69–89%) in high ee (94–99%). The allylation of *N*-H aldimines can be accomplished but in much lower ee (i.e., 1 example, 38%). The homoallylic alcohols **12** provide efficient access to the corresponding β-hydroxy ketones through ozonolysis.

Crotylboration.[1]   The preparation of the *Z*- and *E*-crotylboranes **14** (**14*S***: 856676-03-4; **14*R***: 56676-00-1) and **15** (**15*S***: 856675-99-5; **15*R***: 856675-02-3) in >98% isomeric purities, respectively, was accomplished through the deprotonation of 2-butenes with Schlosser's 'Superbase'. The borinate complexes **13**

are demethoxylated with TMSOTf. The modifications to previous routes to crotylboranes that were employed for **14** and **15** include (1) the use of THF solutions of KO(Bu-*t*) rather than the solid reagent, which dramatically reduces the time required for the deprotonation of *cis*-2-butene from 12 to 1 h and increases the geometric purity of **14** from 94 to 98%, (2) conducting the entire process at −78 °C, which results in the clean formation of **13** (¹¹B NMR δ 3.6) avoiding isomerization of the 'crotylpotassium' reagents and the double addition of the crotylmetallic (¹¹B NMR δ −9.6) to **10**, and (3) the use of TMSOTf as the Lewis acid, which avoids the unwanted decrotylation of **13** to regenerate **10** and cleanly produces **14** or **15** (95%) in high geometric purity (>98%) (eq 5). The reagents undergo slow isomerization to *E/Z* mixtures at room temperature, but are configurationally stable under the reaction conditions employed.

$$2R \xrightarrow[\substack{C_6H_{14} \\ 3\,h,\,reflux}]{MeOH} \textbf{(10R)} (87\%) \xrightarrow[\substack{THF,\,C_6H_{14} \\ 0\,°C,\,30\,min}]{CH_2=MeCH_2MgCl} \textbf{(11R)} (98\%)$$

$$\xrightarrow[\substack{1.\ i\text{-PrCHO} \\ -78\,°C,\,3\,h \\ 2.\ 1S,2S\text{-PE} \\ MeCN \\ reflux,\,12\,h}]{} \textbf{1R} (74\%) \ + \ \textbf{(12)} (75\%,\,97\%\ ee) \tag{4}$$

$$\xrightarrow[\substack{1.\ t\text{-BuOK/THF} \\ 2.\ n\text{-BuLi} \\ -78\,°C,\,1\,h \\ 3.\ \textbf{10},\,Et_2O \\ -78\,°C,\,15\,min}]{} \textbf{(13)}$$

$$\xrightarrow[-78\,°C,\,15\,min]{TMSOTf} \textbf{(14)} \ \ or \ \ \textbf{(15)} \tag{5}$$

Crotylboration with **14** and **15** has been used to prepare all four stereoisomers of β-methyl homoallylic alcohols in high dr (>98:2) and ee (94–99%) (eq 6). Both oxidative (alkaline H₂O₂) and nonoxidative (PE) workup procedures have been developed for these conversions.

$$\textbf{14R} \xrightarrow[\substack{2.\ workup}]{1.\ PrCHO,\,Et_2O \\ -78\,°C} \textbf{(16)} (94\%,\,>98:2\ syn/anti,\,96\%\ ee) \tag{6}$$

Allenylboration.[6] The synthesis of **17** (**17S**: 848617-98-1; **17R**: 848617-97-0) from **1** is accomplished in high yield (93%) through the Grignard reagent derived from propargyl bromide. Either enantiomer is isolable in pure form, being thermally stable to distillation (bp 88–90 °C, 0.1 mmHg). Asymmetric allenylboration of representative aldehydes with the storable **17** provides

18 cleanly (74–82%) in high ee (93–95%) with predictable stereochemistry. The slightly lowered selectivities with the allenyl- vs. allylborane systems can be attributed to the smaller size of sp² vs. sp³ C reducing the difference in effective size between this group and the aldehydic O. The procedure also regenerates **1** (80–85%) for its direct conversion back to **17** and facilitates the efficient recovery of the pseudoephedrine (90%). The net process is the synthetic equivalent of the asymmetric addition of allenylmagnesium bromide to aldehydes (eq 7).

$$\textbf{1R} \xrightarrow[\substack{Et_2O \\ 93\%}]{C_3H_3MgBr} \textbf{(17R)} \xrightarrow[\substack{Et_2O,\,-78\,°C,\,3\,h \\ 2.\ Ph,\,Me \\ HO\ NHMe \\ MeCN,\,81\,°C}]{1.\ 2\text{-furylCHO}} \textbf{1R} (80\%) \ + \ \textbf{(18)} (80\%,\,95\%\ ee) \tag{7}$$

Rapidly finding new synthetic applications, **17** was found to add to the Lewis acid-mediated ring-opening products derived from vinyl oxiranes (e.g., **19**) to produce homopropargylic alcohols (**20**, 81%, 91% ee) (eq 8).[7]

$$\textbf{(19)} \xrightarrow[\substack{THF,\,-78\,°C,\,8\,h \\ 2.\ H_2O_2,\,NaOH,\,14\,h,\,rt}]{1.\ \textbf{17R} \\ 10\,mol\,\%\ Sc(OTf)_3} \textbf{(20)} (81\%,\,91\%\ ee) \tag{8}$$

As part of his synthesis of antifungal and anticancer agent cruentaren A (**23**), Fürstner employed **17S** for the asymmetric propargylation of **21** providing **22** in high yield and diastereoselectivity (75%, >95% de) (eq 9).[8]

$$\textbf{(21)} \xrightarrow[Et_2O,\,-78\,°C]{17S} \textbf{(22)} (75\%,\,>95\%\ de) \tag{9}$$

(23)

In a truly novel organoborane conversion, the alkynylborane **24** (**24S**: 906650-63-3; **24R**: 906650-62-2) derivatives were found to provide a highly useful route to allenylboranes **25** (**25S**: 935677-61-5; **25R**: 935677-30-8) through a silyl-mediated 1,3-borotropic rearrangement that occurs with complete retention of configuration (eq 10).[2]

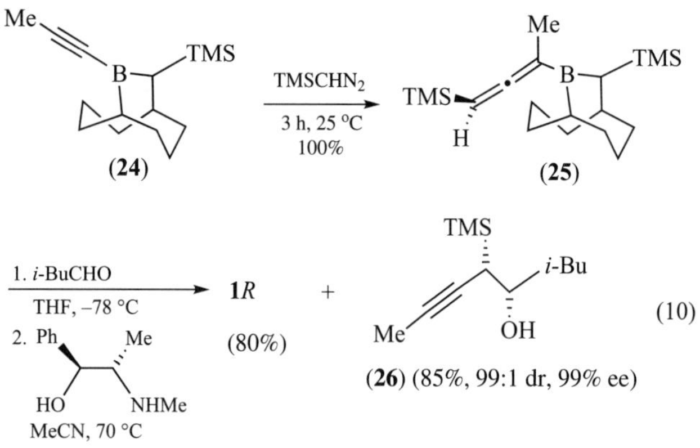

(10)

(26) (85%, 99:1 dr, 99% ee)

The smooth addition to aldehydes with **25** occurs at −78 °C to provide β-TMS homopropargylic alcohols (**26**, 80–96%) in extremely high dr (99:1) and ee (94–99%). When the enynylborane **27** (**28S**: 906650-65-5; **28R**: 906650-64-4) undergoes the TMSCH insertion and rearranges to **28** (**28S**: 929082-50-8; **28R**: 929082-62-2), allylboration is faster than allenylboration. Thus, **28** adds to benzaldehyde to form the 1,2,3-trienyl carbinol **29** in chemically and optically pure form (eq 11).

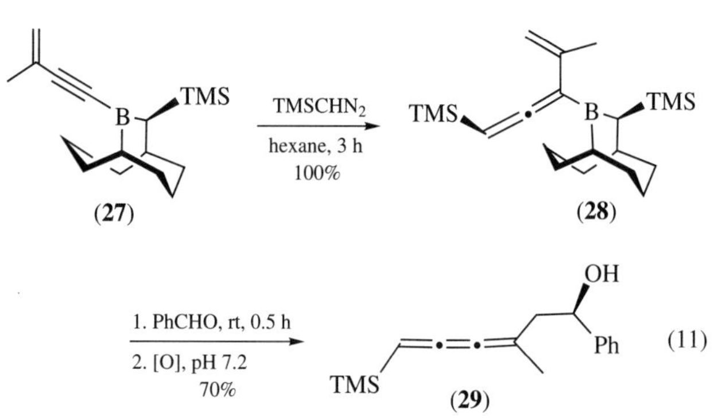

(11)

These fascinating allenylborane reagents (e.g., **30** (**30S**: 935677-63-7; **30R**: 935677-35-3)) undergo protiodeborylation to provide essentially optically pure allenylsilanes (**31** (e.g., **31S**: 935677-59-1), 55–94%) (eq 12).[9]

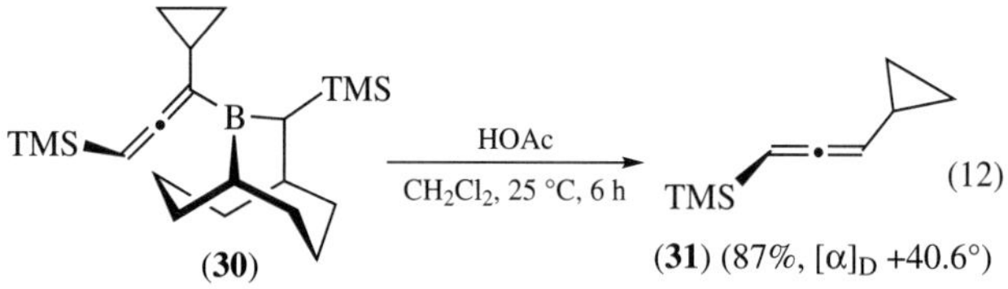

(12)

The additional α-substitution in allenylborane reagents such as **25**, **28**, and **30** compared to **17** enhances the enantioselectivity for the propargylation of aldehydes. However, this effect is particularly apparent in their additions to N-H aldimines, which produce homopropargylic amines in good yields (91%), but disappointing ee (from PhHC=NH, 73% ee). With **32** (**32S**: 935677-62-6; **32R**: 929082-37-1), the homopropargylic amine **33** is produced as essentially a single isomer (51–85%, >99 syn, 92–99% ee) (eq 13).[9]

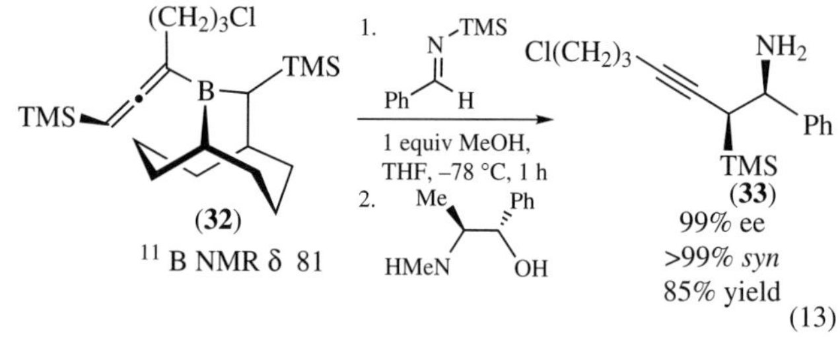

(13)

(33)
99% ee
>99% syn
85% yield

Propargylboration.[10]   The asymmetric propargylboration of aldehydes at −78 °C in <3 h with **34** (**34S**: 872359-60-9; **34R**: 872359-59-6) provides silylated α-allenyl carbinols **35** (60–87%) in high ee (94– >98% ee) (eq 14). The reagents **35** are easily prepared in both enantiomeric forms through a simple Grignard procedure and air-stable borinate complexes **1**. Interestingly, the ozonolysis of **35** proceeds smoothly through an acylsilane intermediate to give a TMS ester that is hydrolyzed to the α-hydroxy acid quantitatively with water.

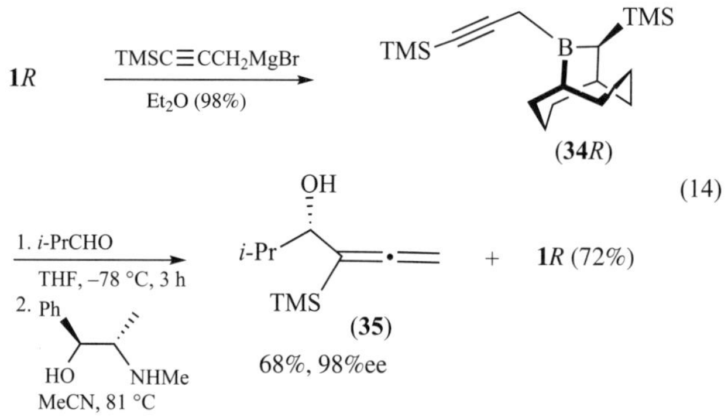

(14)

(35)
68%, 98%ee

Alkynylboration.[11]   The asymmetric synthesis of N-propargylamides through the Michael addition of alkynylboranes (e.g., **27**) to N-acylimines has been accomplished. The N-acetylimines provide the best substrates for the process exhibiting high selectivity (56–95% ee) with predictable stereochemistry (eq 15). In several cases such as **36** (formed in 82% ee), the propargyl amides crystallize in essentially pure form (97–99% ee). The process regenerates **1** for its direct conversion back to the alkynylborane and facilitates the efficient recovery of the pseudoephedrine.

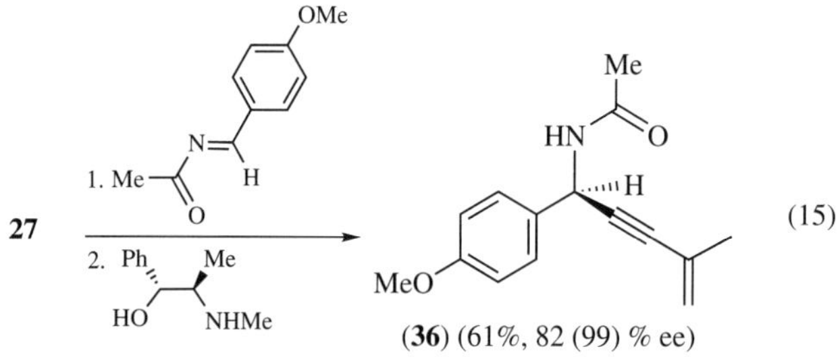

(15)

(36) (61%, 82 (99) % ee)

Asymmetric Hydroboration.[12–14]   Through the reaction of LiAlH₃(OEt) with **1**, the corresponding borohydrides **37** are generated quantitatively. Its reaction with Me₃SiCl provides the borane **38** (**38S**: 1042678-87-4; **38R**: 1042678-84-1) that exists in equilibrium with its dimer. Less reactive than the 10-Ph-9-BBD reagent, **38** exhibits the remarkable property that, in asymmetric hydroborations, it is more sensitive to differences in the substitution pattern at the beta position of the C=C relative to the boron placement rather than at the alpha position. This results in the hydroboration of *cis*-2-butene with the **38***R* reagent giving 2*S*-butanol (98%, 84% ee) while *trans*-2-butene gives 2*R*-butanol (95%, 95% ee). Thus, as can be seen in eq **16**, the asymmetric hydroboration of α-methylstyrene with **38***S* provides the 2*R*-phenyl-1-propanol in 83% yield and 66% ee. The observed product ee's are proportional to the differences in size between two beta groups. The preferred boat–chair conformation presents a relatively flat approach to the reagent's BH with only the TMS exerting a major steric influence. The observed product stereochemistry can be understood through these models (Scheme 2).

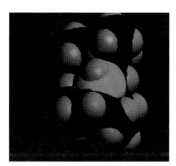

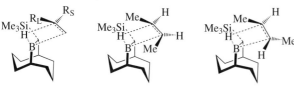

Scheme 2

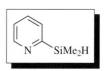

1*S* → LiAlH₃(OEt) → **(37)** → TMSCl → **(38)**

→ **(39)** → H₂O₂, base → **(40)** (83%, 66% ee)   (16)

1. Burgos, C. H.; Canales, E.; Matos, K.; Soderquist, J. A., *J. Am. Chem. Soc.* **2005**, *127*, 8044.
2. Canales, E.; González, A. Z.; Soderquist, J. A., *Angew. Chem., Int. Ed.* **2007**, *46*, 397.
3. Alexander, M. D.; Fontaine, S. D.; La Clair, J. J.; DiPasquale, A. G.; Rheingold, A. L.; Burkart, M. D., *Chem. Commun.* **2006**, 4602.
4. Hernández, E.; Canales, E.; González, E.; Soderquist, J. A., *Pure Appl. Chem.* **2006**, *7*, 1389.
5. Román, J. G.; Soderquist, J. A., *J. Org. Chem.* **2007**, *72*, 9772.
6. Lai, C.; Soderquist, J. A., *Org. Lett.* **2005**, *7*, 799.
7. Maddess, M. L.; Lautens, M., *Org. Lett.* **2005**, *7*, 3557.
8. Fürstner, A.; Bindl, M.; Jean, L., *Angew. Chem., Int. Ed.* **2007**, *46*, 9275.
9. González, A. Z.; Soderquist, J. A., *Org. Lett.* **2007**, *9*, 1081.
10. Hernández, E.; Soderquist, J. A., *Org. Lett.* **2005**, *7*, 5397.
11. González, A. Z.; Canales, E.; Soderquist, J. A., *Org. Lett.* **2006**, *8*, 3331.
12. González, A. Z.; Román, J. G.; González, E.; Martínez, J.; Medina, J. R.; Matos, K.; Soderquist, J. A., *J. Am. Chem. Soc.* **2008**, *130*, 9218.
13. Soderquist, J. A.; Matos, K.; Burgos, C. H.; Lai, C.; Vaquer, J.; Medina, J. R. In *Contemporary Boron Chemistry*; Davidson, M. G.; Hughes, A. K.; Marder, T. B.; Wade, K., Eds.; Royal Society of Chemistry: Cambridge, UK, 2000; p 472.
14. Soderquist, J. A.; Matos, K.; Burgos, C. H.; Lai, C.; Vaquer, J.; Medina, J. R.; Huang, S. D. In *Organoboranes for Syntheses, A. C. S. Symposium Series 783*; Ramachandran, P. V.; Brown, H. C., Eds.; American Chemical Society: Washington, DC, 2000; Chapter 13, p 176.

John A. Soderquist
University of Puerto Rico, Rio Piedras, Puerto Rico

2-Pyridyldimethylsilane[1]

[21032-48-4]   C₇H₁₁NSi   (MW 137.26)
InChI = 1S/C7H11NSi/c1-9(2)7-5-3-4-6-8-7/h3-6,9H,1-2H3
InChIKey = FTSXFFCALJHVGU-UHFFFAOYSA-N

(reagent used for metal-catalyzed hydrosilylation of alkynes and alkenes)

Physical Data: bp 68–72 °C/20 mm Hg.
Solubility: soluble in most common organic solvents.
Form Supplied in: colorless liquid.
Analysis of Reagent Purity: ¹H NMR (CDCl₃) δ 0.40 (d, J = 3.9 Hz, 6H), 4.46 (septet, J = 3.9 Hz, 1H), 7.22 (ddd, J = 7.5, 4.8, 1.2 Hz, 1H), 7.54 (ddd, J = 7.5, 1.8, 1.2 Hz, 1H), 7.60 (td, J = 7.5, 1.8 Hz, 1H), 8.77 (ddd, J = 4.8, 1.8, 1.2 Hz, 1H).
Preparative Methods: can be prepared by reaction of 2-pyridyllithium with chlorodimethylsilane in dry Et₂O at −78 °C under argon.
Purity: distillation.

Introduction. Hydrosilanes are widely used reagents in the metal-catalyzed hydrosilylation (addition of H–Si bond across carbon-carbon and carbon-heteroatom unsaturated bonds). There are several excellent reviews on this subject in the literature.[2] In addition to widely used hydrosilanes such as trichlorosilane, triethylsilane, and dimethylphenylsilane, 2-pyridyldimethylsilane has been recently introduced as a unique multifunctional hydrosilane in this chemistry.[1,3,4] Presumably because of the presence of catalyst-directing pyridyl group on silicon, 2-pyridyldimethylsilane exhibits greater reactivity in Rh-catalyzed hydrosilylation of alkenes than those of other related hydrosilanes. Moreover, the presence of the basic pyridyl group enables easy product separation and facile catalyst recovery through simple acid/base extraction.

Rh-Catalyzed Hydrosilylation of Alkenes. Treatment of 2-pyridyldimethylsilane with 1-alkenes in the presence of a catalytic amount of chlorotris(triphenylphosphine)rhodium gives terminally silylated products at room temperature with virtually complete regioselectivity (Table 1).[1] Internal alkenes are usually inert under these conditions. The preferred solvent is acetonitrile, but other common organic solvents such as toluene, dichloromethane, and tetrahydrofuran can also be used.

Table 1 Rh-catalyzed hydrosilylation of alkenes[1]

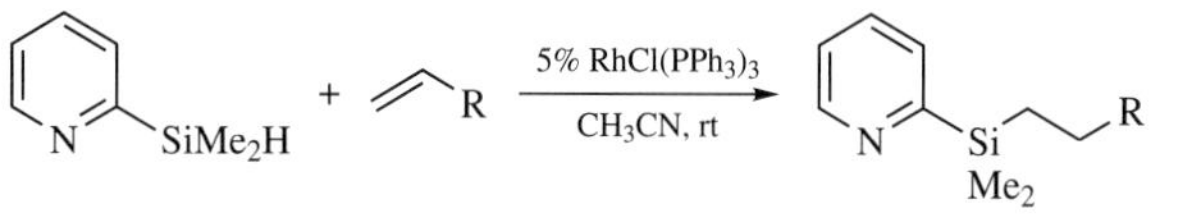

Entry	Alkene	Product	Yield (%)	Purity (%)
1	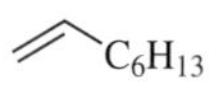 C₆H₁₃	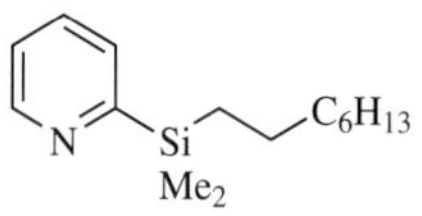 C₆H₁₃	86	>95
2	 Ph	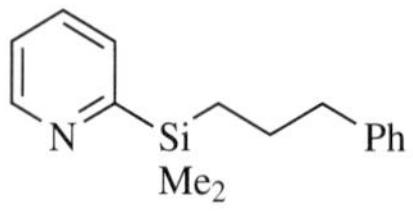 Ph	86	>95
3	 SiMe₃	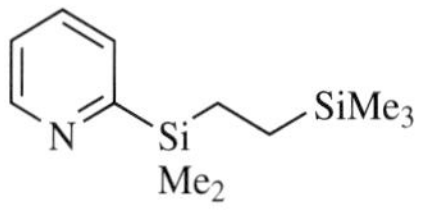 SiMe₃	93	>95
4	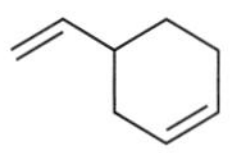	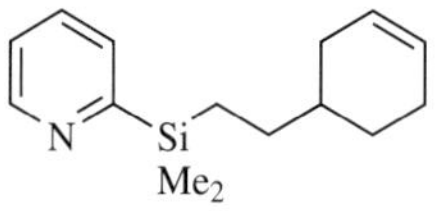	89	>95
5	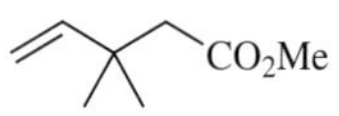 CO₂Me	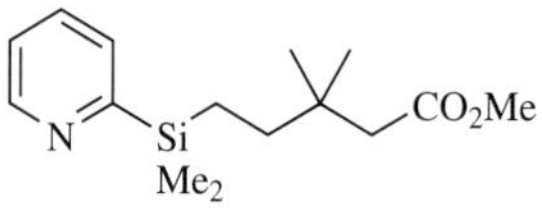 CO₂Me	83	>95
6	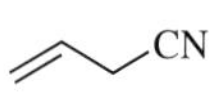 CN	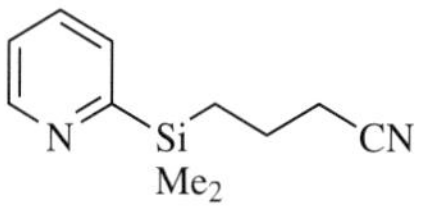 CN	84	>95

The reactivity of 2-pyridyldimethylsilane is much greater than that of structurally and electronically related 3-pyridyl-, 4-pyridyl-, and phenyl-dimethylsilanes in the Rh-catalyzed reaction shown in eq 1.[1] Coordination-induced facile alkene insertion into the Rh–Si bond has been suggested as an origin of this accelerating effect. Interestingly, however, the reactivity of 2-pyridyldimethylsilane is the lowest among the four hydrosilanes when a Pt complex is used as a catalyst.[1]

The additional advantage of using 2-pyridyldimethylsilane is that, by making use of the phase tag property of 2-pyridyldimeth-

Ar	Yield (%)
2-pyridyl	91
3-pyridyl	4
4-pyridyl	1
phenyl	1

ylsilyl group,[3,5,6] most hydrosilylation products can be isolated by simple acid/base extraction. Chromatographic purifications are usually not required for achieving high purity (>95%) of products (Table 1). Moreover, the rhodium catalyst can be recovered during the acid/base extraction and reused in further reactions with ease.[1]

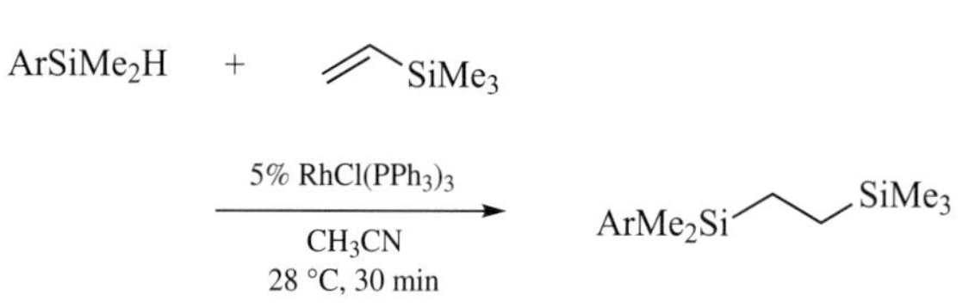

The resultant organo(2-pyridyl)silanes can be transformed into the corresponding alcohols by Tamao–Fleming-type oxidation[7] under the influence of hydrogen peroxide and potassium fluoride (eq 2).[8]

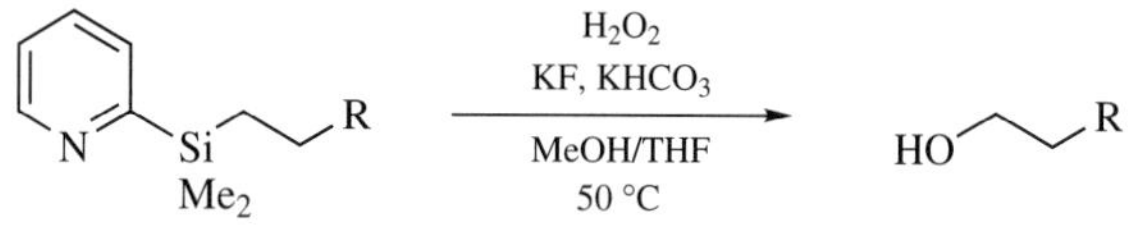

Table 2 Pt-catalyzed hydrosilylation of alkynes[1]

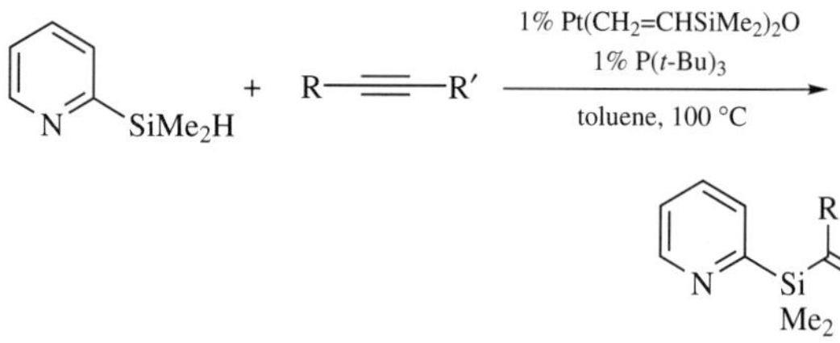

Entry	Alkyne	Product	Yield (%)	Regioselectivity (%)
1	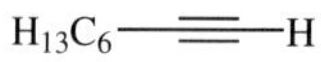 H₁₃C₆——H	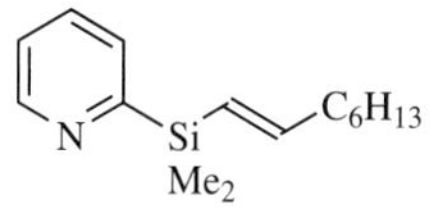 C₆H₁₃	89	>99
2	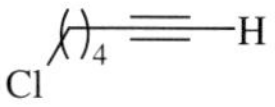 Cl(CH₂)₄——H	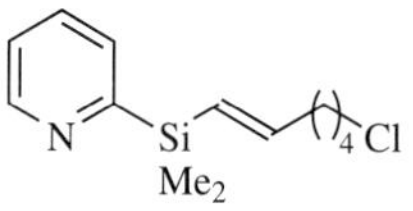 (CH₂)₄Cl	84	>99
3	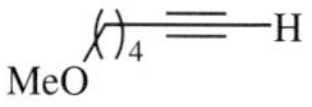 MeO(CH₂)₄——H	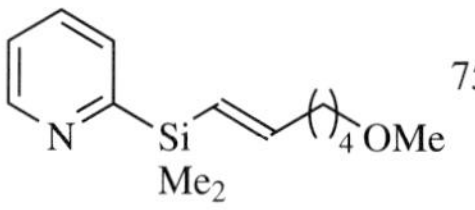 (CH₂)₄OMe	75	>99
4	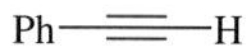 Ph——H	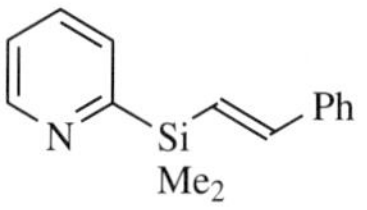 Ph	93	>99
5	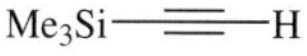 Me₃Si——H	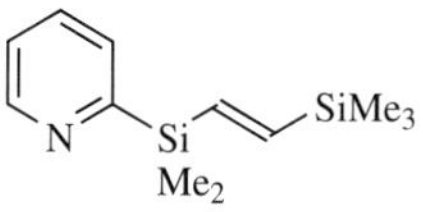 SiMe₃	63	90
6[a]	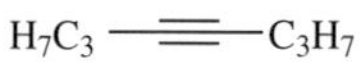 H₇C₃——C₃H₇	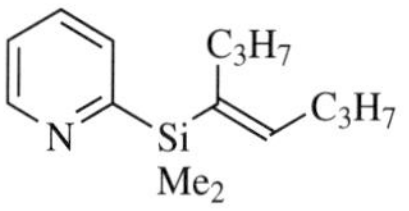 C₃H₇ / C₃H₇	85	-
7[a]	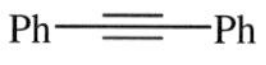 Ph——Ph	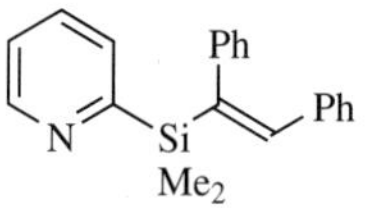 Ph / Ph	86	-

[a]PPh₃ is used instead of P(t-Bu)₃.

Pt-Catalyzed Hydrosilylation of Alkynes. In the catalytic hydrosilylation of alkynes using 2-pyridyldimethylsilane, a Pt complex is preferable over other transition metal complexes. Thus, the treatment of 2-pyridyldimethylsilane with alkynes in the presence of a catalytic amount of $Pt(CH_2=CHSiMe_2)_2O$ modified with $P(t-Bu)_3$ ligand[9] gives alkenyl(2-pyridyl)silanes regioselectively (Table 2).[1] The use of $P(t-Bu)_3$ is crucial for attaining high regioselectivity. For the hydrosilylation of symmetrical internal alkynes, where the regioselectivity is not an issue, the use of PPh_3 instead of $P(t-Bu)_3$ is preferable because it is cheaper and easier to handle (entries 6 and 7).

The resultant alkenyl(2-pyridyl)silanes can be subjected to further transformations. For example, the treatment of alkenyl (2-pyridyl)silanes with aryl iodides in the presence of tetrabutylammonium fluoride and palladium catalyst affords substituted olefins in good yields (eq 3).[10] This cross-coupling reaction is suggested to proceed through the intermediacy of a highly reactive alkenylsilanol.[11] The treatment of alkenyl(2-pyridyl)silanes with tetrabutylammonium fluoride effects simple protodesilylation (eq 4).[12] The reaction with acetyl chloride in the presence of aluminum chloride affords the corresponding α,β-unsaturated enones (eq 5).[13]

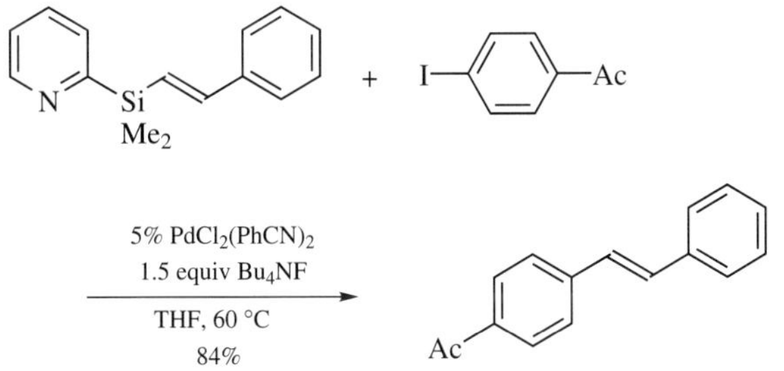

$$(3)$$

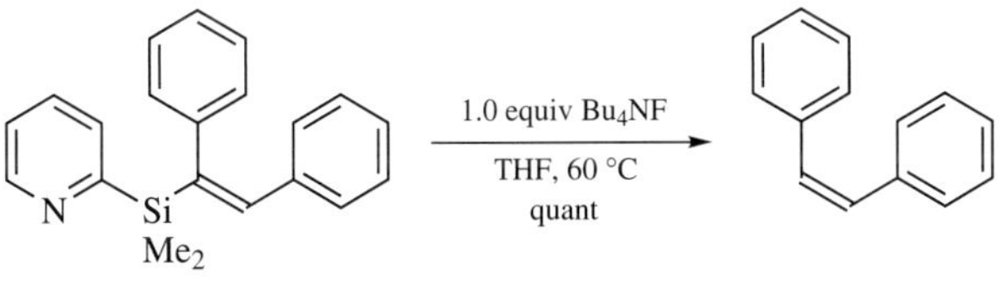

$$(4)$$

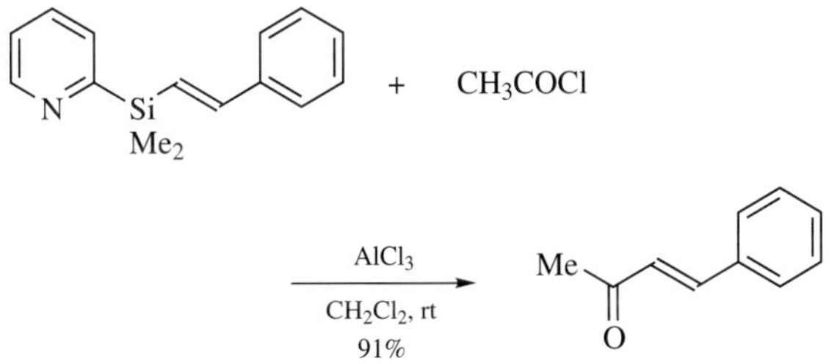

$$(5)$$

Other than replacing the 2-pyridyldimethylsilyl group by other functionalities, a further addition reaction across the C=C bond of alkenyl(2-pyridyl)silanes is feasible. The addition of Grignard reagents (carbomagnesation) to alkenyl(2-pyridyl)silane gives α-silyl organomagnesium compounds which can further react with various electrophiles (eq 6).[14] The Pd-catalyzed Mizoroki–Heck reaction (carbopalladation) of alkenyl(2-pyridyl)silane with

aryl iodide affords β,β-disubstituted vinylsilane in virtually complete stereoselectivity (eq 7).[12,15] These two addition reactions are believed to proceed through the agency of efficient pyridyl-to-metal coordination effects.[16]

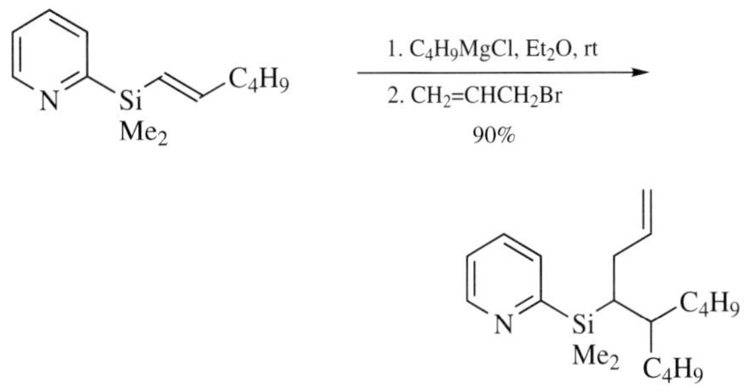

$$(6)$$

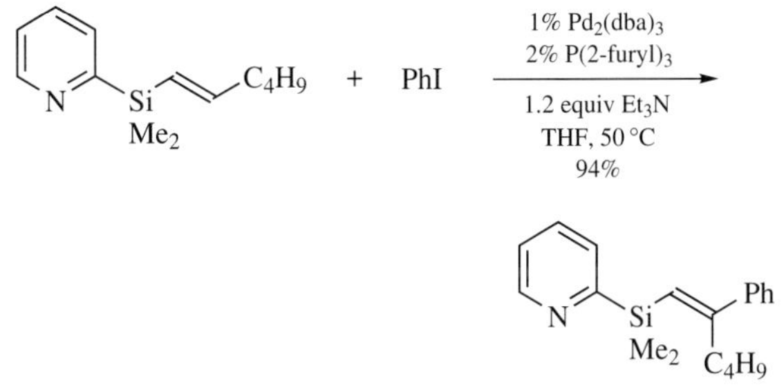

$$(7)$$

Related Reagents. Trichlorosilane; triethylsilane; dimethylphenylsilane; chlorotris(triphenylphosphine)rhodium(I); hydrogen peroxide; potassium fluoride; tetrabutylammonium fluoride; acetyl chloride; aluminum chloride.

1. Itami, K.; Mitsudo, K.; Nishino, A.; Yoshida, J., *J. Org. Chem.* **2002**, *67*, 2645.

2. (a) Marciniec, B., In *Comprehensive Handbook on Hydrosilylation*; Pergamon Press: Oxford, 1992. (b) Hiyama, T.; Kusumoto, T., In *Comprehensive Organic Synthesis*; Trost, B. M.; Fleming, I., Eds.; Pergamon Press: Oxford, 1991, Vol. 8, p 763.

3. Yoshida, J.; Itami, K.; Mitsudo, K.; Suga, S., *Tetrahedron Lett.* **1999**, *40*, 3403.

4. Itami, K.; Mitsudo, K.; Nishino, A.; Yoshida, J., *Chem. Lett.* **2001**, 1088.

5. Yoshida, J.; Itami, K., *J. Synth. Org. Chem. Jpn.* **2001**, *59*, 1086.

6. Yoshida, J.; Itami, K., *Chem. Rev.* **2002**, *102*, 3693.

7. (a) Tamao, K., In *Advances in Silicon Chemistry*; Larson, G. L., Ed.; JAI Press Inc., 1996, Vol. 3, p 1. (b) Fleming, I., *ChemTracts: Org. Chem.* **1996**, 1. (c) Jones, G. R.; Landais, Y., *Tetrahedron* **1996**, *52*, 7599.

8. Itami, K.; Mitsudo, K.; Yoshida, J., *J. Org. Chem.* **1999**, *64*, 8709.

9. (a) Chandra, G.; Lo, P. Y.; Hitchcock, P. B.; Lappert, M. F., *Organometallics* **1987**, *6*, 191. (b) Takahashi, K.; Minami, T.; Ohara, Y.; Hiyama, T., *Tetrahedron Lett.* **1993**, *34*, 8263. (c) Denmark, S. E.; Wang, Z., *Org. Lett.* **2001**, *3*, 1073. (d) Hosoi, K.; Nozaki, K.; Hiyama, T., *Chem. Lett.* **2002**, 138.

10. Itami, K.; Nokami, T.; Yoshida, J., *J. Am. Chem. Soc.* **2001**, *123*, 5600.

11. (a) Hirabayashi, K.; Mori, A.; Kawashima, J.; Suguro, M.; Nishihara, Y.; Hiyama, T., *J. Org. Chem.* **2000**, *65*, 5342. (b) Denmark, S. E.; Sweis, R. F., *Acc. Chem. Res.* **2002**, *35*, 835.

12. Itami, K.; Nokami, T.; Ishimura, Y.; Mitsudo, K.; Kamei, T.; Yoshida, J., *J. Am. Chem. Soc.* **2001**, *123*, 11577.

13. Itami, K.; Nokami, T.; Yoshida, J., *Tetrahedron* **2001**, *57*, 5045.

14. Itami, K.; Mitsudo, K.; Yoshida, J., *Angew. Chem., Int. Ed.* **2001**, *40*, 2337.

15. Itami, K.; Mitsudo, K.; Kamei, T.; Koike, T.; Nokami, T.; Yoshida, J., *J. Am. Chem. Soc.* **2000**, *122*, 12013.

16. Itami, K.; Mitsudo, K.; Nokami, T.; Kamei, T.; Koike, T.; Yoshida, J., *J. Organomet. Chem.* **2002**, *653*, 105.

Jun-ichi Yoshida & Kenichiro Itami
Kyoto University, Kyoto, Japan

S

Silane, 1,1'-(Chloromethylene)bis[1,1,1]-trimethyl-

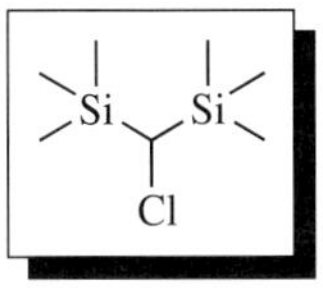

[5926-35-2]   C$_7$H$_{19}$ClSi$_2$ (MW 194.85)

InChI = 1S/C7H19ClSi2/c1-9(2,3)7(8)10(4,5)6/h7H,1-6H3

InChIKey = XNJGZHVYPBNLEB-UHFFFAOYSA-N

(versatile C$_1$ building block; broad application in Peterson olefination reactions; Grignard reagent participates readily in Kumada cross-coupling reactions with aryl and vinyl halides; useful in synthesis of methylenephosphine analogs; as sterically demanding ligand for a number of main group and transition metal complexes)

Alternate Names: chlorobis(trimethylsilyl)methane; bis(trimethylsilyl) methyl chloride; bis(trimethylsilyl)chloromethane; [chloro(trimethylsilyl)methyl]trimethylsilane.

Physical Data: bp 57–60 °C (15 mmHg); flash point 48 °C; d 0.892 g mL^{-1} (25 °C).

Solubility: soluble in common organic solvents (DMF, THF, Et$_2$O, CH$_2$Cl$_2$).

Form Supplied in: colorless liquid, commercially available.

Preparative Method: a convenient one-pot procedure has been developed by Kemp and Cowley in which trimethylsilylchloride reacts with dichloromethane in the presence of nBuLi at −110 °C to give bis(trimethylsilyl)dichloromethane.[1] This intermediate is sequentially treated with nBuLi and ethanol to yield chlorobis(trimethylsilyl)methane. A number of other syntheses have been described.[2–8]

Purification: fractionally distilled through a Vigreaux column.[1]

Handling, Storage, and Precautions: flammable. Chlorobis(trimethylsilyl)methane is irritating to the eyes, skin, and respiratory system and should only be handled in a well-ventilated fume hood. In addition, proper protective clothing, gloves, and eye/face protection should be worn when title compound is in use. In case of contact with eyes, rinse immediately with water and seek medical advice.

Peterson Olefination Reactions. McNulty and Das have shown that the treatment of chlorobis(trimethylsilyl)methane (**1**) with s-BuLi at low temperatures favors lithium–halogen exchange processes over deprotonation pathways (eq 1).[9] The resulting lithiated derivative **2** undergoes Peterson olefination reactions[10] with a number of benzaldehyde derivatives to produce the corresponding β-silyl styrene products with good E-selectivity in moderate to high yields. When aliphatic aldehydes are used as starting materials in these reactions, a significant loss of E/Z-olefin selectivity is observed.

A majority of reaction sequences rely on the use of nucleophilic (metalated) variants of **1**.[11] However, the electrophilic character of the sterically congested methine carbon has been successfully exploited via displacement reactions with sodium azide.[12] Ogata and Shimizu have also reported that **1** undergoes nucleophilic attack by 1,2,4-triazole in the presence of potassium carbonate to provide 1-[bis(trimethylsilyl)methyl]-1,2,4-triazole (**3**) (eq 2).[13] Subsequent introduction of TBAF results in desilylation of **3** to afford an α-silyl carbanion intermediate that provides 1-vinyl-1,2,4-triazole products upon condensations with aldehydes or ketones (eq 3). While high yields of the alkene products are generally obtained, mixtures of E/Z-isomers are observed in all cases.

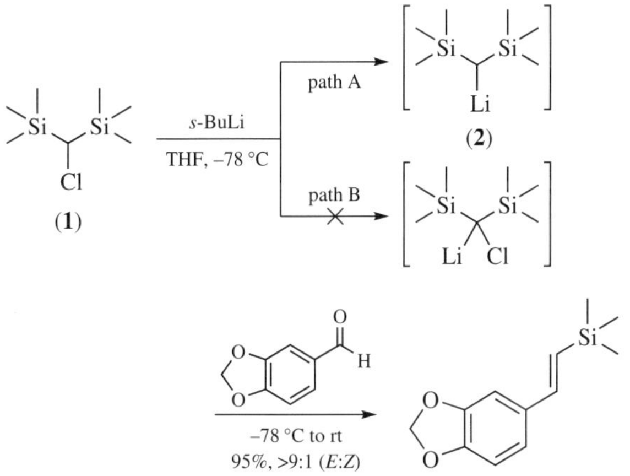

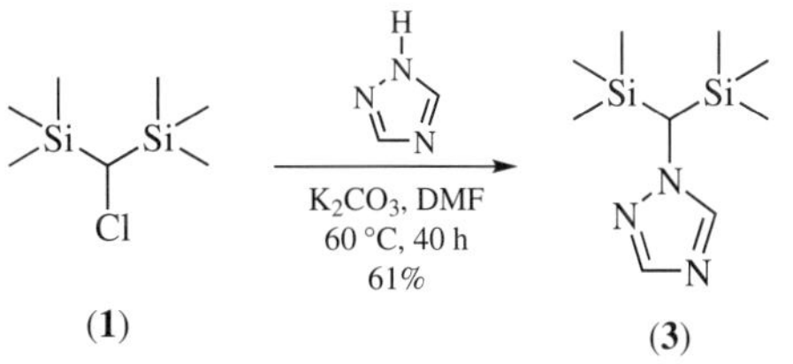

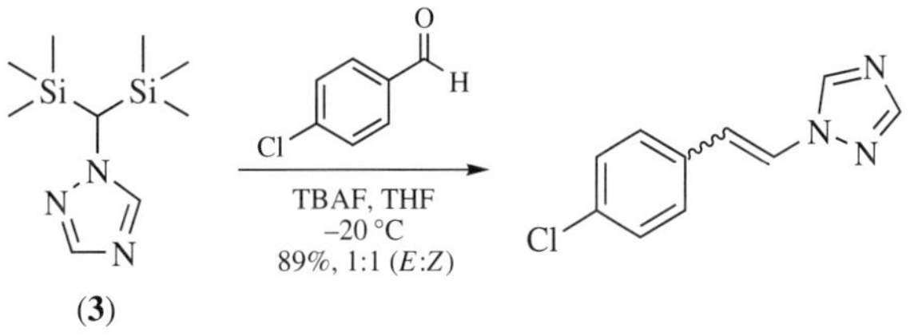

Kumada Cross-coupling Reactions. Yang and coworkers have shown that the Grignard reagent **4**, derived from chlorobis(trimethylsilyl)methane (**1**), undergoes a palladium-catalyzed Kumada cross-coupling reaction with p-bromostyrene (no yield provided, eq 4).[14] The product of this reaction readily participates in polymerization processes to afford silicon-containing homopolymers that are known to be highly resistant to etching by oxygen plasma and therefore useful in a number of materials-based applications.

Williams and coworkers have described a novel approach to bis-trimethylsilyl substituted allylating reagents via high-yielding Kumada cross-coupling reactions involving **4** and vinyl halides (eq 5).[15] These studies resulted in the synthesis of a versatile 3,3-bis(trimethylsilyl)-2-methyl-1-propene (**5**) reagent. In the presence of BF$_3$·OEt$_2$ and an aldehyde, silane **5** was shown to participate in ene reactions to provide products in which the bis-trimethylsilyl subunit is preserved (eq 6). Alternatively, when stronger Lewis acids (SnCl$_4$) are employed in CH$_2$Cl$_2$ solutions of aldehydes, a Sakurai S$_{E}'$ process gives rise to trisubstituted olefins bearing a synthetically useful vinylsilane functionality (eq 7). The methodology was examined with a range of aldehydes, and in all cases, the allylation products were exclusively obtained as (*E*) geometric isomers under these conditions.

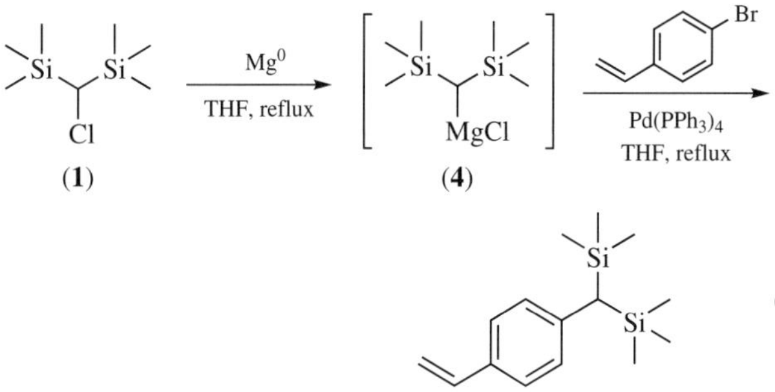

(4)

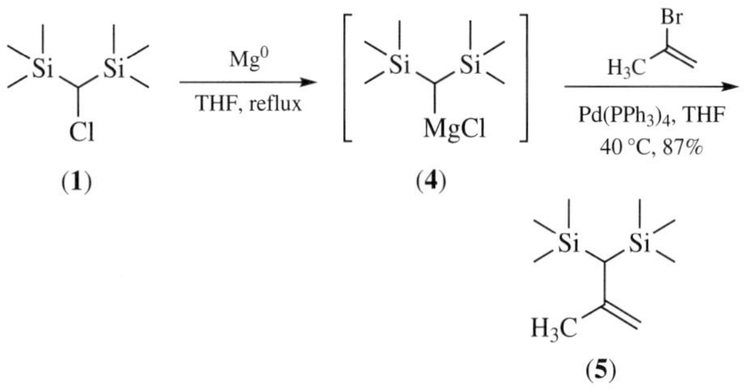

(5)

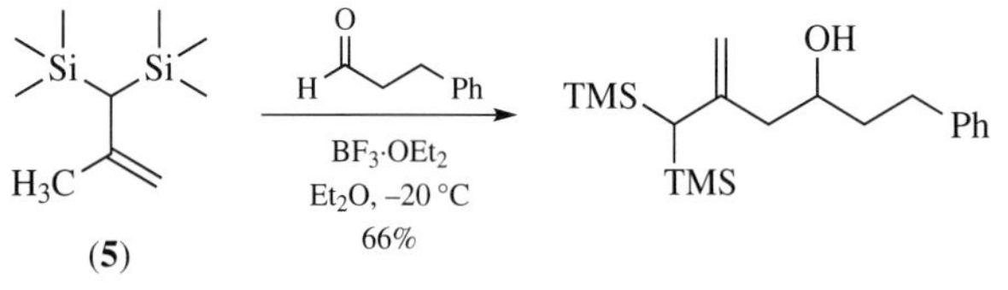

(6)

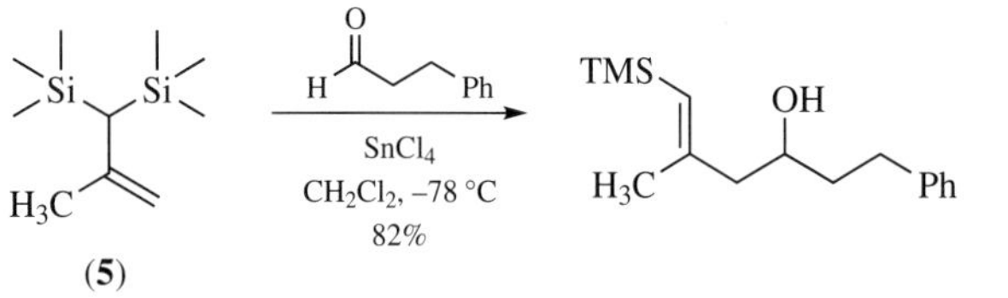

(7)

Synthesis of Methylenephosphine Derivatives. 'Low coordinate' phosphorous compounds, including methylenephosphines,[16–18] are of interest due to their unique reactivity. Neilson and coworkers have described a useful approach to such compounds, in which the incorporation of silyl substituents results in a highly polarized P–C double bond.[19] Grignard reagent **4** was utilized for reactions with MesPCl$_2$ to produce the mesitylchlorophosphine **6** (no yield given, eq 8) that was subjected to conditions for dehydrohalogenation in the presence of DBU (no yield given,

eq 9). Generally, the methylenephosphine product (**7**) displayed interesting reactivity due to the stabilization of cationic character at phosphorous via hyperconjugation by the presence of the silyl moieties. This β-silicon effect accounts for the propensity of **7** to undergo P-methylation in the presence MeLi (no yield given, eq 10). Exposure of the reactive anionic intermediate to MeI resulted in a second methylation of phosphorous, which provides a stable ylide product (**8**).

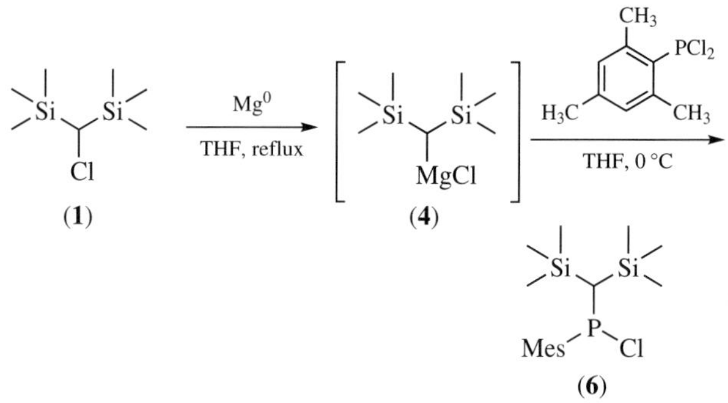

(8)

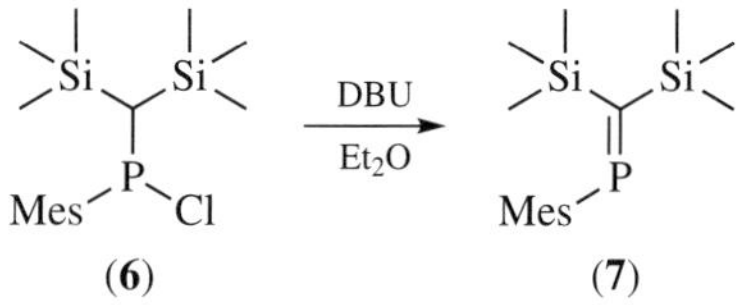

(9)

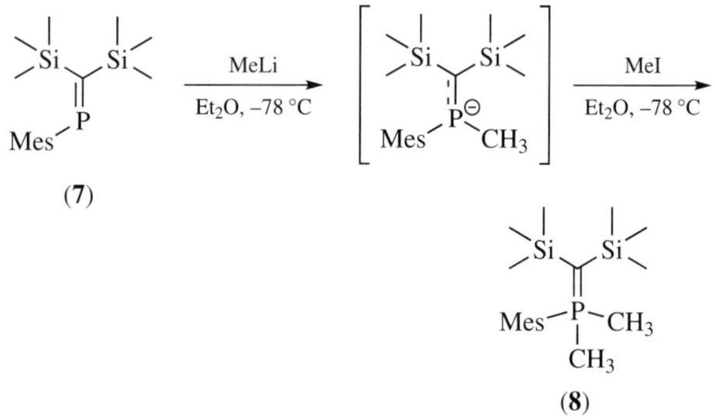

(10)

Participation as a Sterically Demanding Ligand. Chlorobis(trimethylsilyl)methane (**1**) has found application in the design and development of novel transition,[20,21] and main group[22–34] metal complexes due to the kinetic stability imparted by its steric bulk. The majority of these compounds are prepared via the metalation of **1** and the displacement of halides at activated metal centers. For example, Lappert and Goldwhite have shown that the lithio species **2** reacts with 1 or 0.5 equiv of PCl$_3$ to give mono- and disubstituted chlorophosphines, **9** and **10**, respectively (eq 11).[35] Irradiation of a mixture of **9** and **10** in the presence of an electron rich olefin resulted in the formation of the stable phosphinyl radical **11** (no yield given, eq 12), which displayed remarkable persistence ($t_{1/2} > 1$ year). The author attributed this behavior to the steric bulk provided by the sterically demanding ligand derived from **1**. Furthermore, Cowley and coworkers have shown that **11** is a useful ligand for the generation of new classes of organometallic radicals (no yield given, eq 13).[36]

$$(11)$$

$$(12)$$

$$(13)$$

(9) R = Cl (73%)
(10) R = CH(TMS)$_2$ (60%)

(9) R = Cl
(10) R = CH(TMS)$_2$

(11)

(11)

Related Reagents. Silane, 1,1'-(bromomethylene)bis[1,1,1]-trimethyl- [29955-12-2]; silane, 1,1'-(iodomethylene)bis[1,1,1]-trimethyl- [29954-87-8]; silane, (chloromethylene-*d*)bis[trimethyl-(9Cl)] [118043-92-8]; bis(trimethylsilyl)dichloromethane [15951-41-4]; bis(trimethylsilyl)methane [2117-28-4]; silane, (1-chloroethylidene)bis[trimethyl-(9Cl)] [27484-02-2]; silane, (1-bromoethylidene)bis[trimethyl-(9Cl)] [29954-86-7].

1. Kemp, R. A.; Cowley, A. H., *Inorg. Synth.* **1997**, *31*, 128.
2. Eisch, J. J.; Galle, J. E.; Piotrowski, A.; Tsai, M.-R., *J. Org. Chem.* **1982**, *47*, 5051.
3. Barton, T. J.; Hoekman, S. K., *J. Am. Chem. Soc.* **1980**, *102*, 1584.
4. Dimmel, D. D.; Wilkie, C. A.; Ramon, F., *J. Org. Chem.* **1972**, *37*, 2662.
5. Pons, P.; Biran, C.; Bordeau, M.; Dunogues, J., *J. Organomet. Chem.* **1988**, *358*, 31.
6. Kumada, M.; Ishikawa, M., *J. Organomet. Chem.* **1964**, *1*, 411.
7. Cowley, A. H.; Kemp, R. A., *Synth. React. Inorg. Met.-Org. Chem.* **1981**, *11*, 591.
8. Cook, M. A.; Eaborn, C.; Walton, D. R. M., *J. Organomet. Chem.* **1971**, *29*, 389.
9. McNulty, J.; Das, P., *Chem. Commun.* **2008**, 1244.
10. Groebel, B.; Seebach, D., *Angew. Chem., Int. Ed. Engl.* **1974**, *86*, 102.
11. Breunig, H. J.; Kanig, W.; Soltani-Neshan, A., *Polyhedron* **1983**, *2*, 291.
12. Palomo, C.; Aizpurua, J. M.; Garcia, J. M.; Legido, M., *J. Chem. Soc., Chem. Commun.* **1991**, 524.
13. Shimizu, S.; Ogata, M., *J. Org. Chem.* **1987**, *52*, 2314.
14. Fukukawa, K.-I.; Zhu, L.; Gopalan, P.; Ueda, M.; Yang, S., *Macromolecules* **2005**, *38*, 263.
15. Williams, D. R.; Morales-Ramos, A. I.; Williams, C. M., *Org. Lett.* **2006**, *8*, 4393.
16. Appel, R., *Inorg. Synth.* **1986**, *24*, 117.
17. Ford, R. R.; Neilson, R. H., *Polyhedron* **1986**, *5*, 643.
18. Thoraval, J. Y.; Nagai, W.; Yeung Lam Ko, Y. Y.; Carrie, R., *Tetrahedron Lett.* **1990**, *46*, 3859.
19. Xie, Z.-M.; Wisian-Neilson, P.; Neilson, R. H., *Organometallics* **1985**, *4*, 339.
20. Beswick, C. L.; Marks, T. J., *Organometallics* **1999**, *18*, 2410.
21. Beswick, C. L.; Marks, T. J., *J. Am. Chem. Soc.* **2000**, *122*, 10358.
22. Davidson, P. J.; Harris, D. H.; Lappert, M. F., *J. Chem. Soc., Dalton Trans.* **1976**, 2268.
23. Cotton, J. D.; Davidson, P. J.; Lappert, M. F., *J. Chem. Soc., Dalton Trans.* **1976**, 2275.
24. Gynane, M. J. S.; Lappert, M. F.; Carty, A. J.; Taylor, N. J., *J. Chem. Soc., Dalton Trans.* **1977**, 2009.
25. Cowley, A. H.; Lasch, J. G.; Norman, N. C.; Pakulski, M.; Whittlesey, B. R., *J. Chem. Soc., Chem. Commun.* **1983**, 881.
26. Breunig, H. J.; Kanig, W.; Soltani-Neshan, A., *Polyhedron* **1983**, *2*, 291.
27. Cowley, A. H.; Kilduff, J. E.; Lasch, J. G.; Mehrotra, S. K.; Norman, N. C.; Pakulski, M.; Whittlesey, B. R.; Atwood, J. L.; Hunter, W. E., *Inorg. Chem.* **1984**, *23*, 2582.
28. Fjeldberg, T.; Haaland, A.; Schilling, B. E. R.; Lappert, M. F.; Thorne, A. J., *J. Chem. Soc., Dalton Trans.* **1986**, 1551.
29. Althaus, H.; Breunig, H. J.; Rosler, R.; Lork, E., *Organometallics* **1999**, *18*, 328.
30. Matsumoto, T.; Tokitoh, N.; Okazaki, R., *J. Am. Chem. Soc.* **1999**, *121*, 8811.
31. Kano, N.; Shibata, K.; Tokitoh, N.; Okazaki, R., *Organometallics* **1999**, *18*, 2999.
32. Balazs, L.; Breunig, H. J.; Ghesner, I.; Lork, E., *J. Organomet. Chem.* **2002**, *648*, 33.
33. Saito, M.; Tokitoh, N.; Okazaki, R., *J. Am. Chem. Soc.* **2004**, *126*, 15572.
34. Balazs, L.; Breunig, H. J.; Silvestru, C.; Varga, R., *Z. Naturforsch.* **2005**, *60b*, 1321.
35. Gynane, M. J. S.; Hudson, A.; Lappert, M. F.; Power, P. P.; Goldwhite, H., *J. Chem. Soc., Dalton Trans.* **1980**, 2428.
36. Cowley, A. H.; Kemp, R. A.; Wilburn, J. C., *J. Am. Chem. Soc.* **1982**, *104*, 332.

David R. Williams & Martin J. Walsh
Indiana University, Bloomington, IN, USA

Silane, Chlorotri-2-propen-1-yl-*

(1)

[17865-20-2]   C$_9$H$_{15}$SiCl   (MW 186.76)

InChI = 1S/C9H15ClSi/c1-4-7-11(10,8-5-2)9-6-3/h4-6H,1-3,7-9H2

InChIKey = LMQNPINUXNVGGV-UHFFFAOYSA-N

(dendrimer capping agent)

Alternate Names: triallylchlorosilane, triallylsilyl chloride.
Physical Data: bp 97–98 °C (30 mmHg); *d* 0.9197 g cm^{-3}, n_D^{25} 1.4779.
Solubility: sol organics (benzene, Et$_2$O, CHCl$_3$, DCM, THF, DMF).
Form Supplied in: colorless liquid.
Analysis of Reagent Purity: ^{1}H NMR, ^{13}C NMR, GC, HPLC.
Preparative Methods: allylation of tetrachlorosilane via Grignard addition of allylmagnesium chloride,[1] treatment of triallylsilane with anhydrous CuCl$_2$ under basic conditions in

*Article previously published in the electronic Encyclopedia of Reagents for Organic Synthesis as Chloro tris-2-propen-1-yl- DOI: 10.1002/047084289X. rn01223

acetonitrile,[2] or via allylation of tetrachlorosilane with allyl-samarium bromide.[3]

Purification: fractional distillation under reduced pressure, bp 97–98 °C @ 30 mmHg.

Handling, Storage, and Precautions: toxic; use only in fume hood; reagent can be absorbed through skin; always wear gloves when handling this reagent.

Protection of Alcohols as Triallylsilyl Ethers. Typically, triallylsilyl chloride (**1**) is used to introduce a triallylsilyl group into a dendrimer, either as a capping group or as latent functionality. The most common method of forming triallylsilyl ethers involves the use of triallylsilyl chloride and a base (eq 1).[2,4] To date, only the protection of phenol derivatives have been reported.

$$\text{(2)} \xrightarrow[\substack{\text{Et}_3\text{N} \\ 69\%}]{\text{(allyl)}_3\text{SiCl}} \text{(3)} \tag{1}$$

Carbosilylation. Aryl triallylsilanes have been synthesized via various methods, including the lithium-mediated silyl group migration (eq 2) or homo-Brook rearrangement of (4-bromophenyl)triallylsilyl ether (**3**) to afford (4-triallylsilyl)phenol (**4**).[2] The introduction of the triallylsilyl moiety can also be achieved via lithium–halogen exchange (eq 3),[4,5] followed by treatment with triallylchlorosilane (**1**), or through quenching a Snieckus-type directed *ortho* metalation (eq 4) with triallylsilyl chloride (**1**).[6]

$$\text{(3)} \xrightarrow[\substack{2.\ \text{NH}_4\text{Cl} \\ 81\%}]{1.\ t\text{-BuLi, THF} \atop -78\ °\text{C}} \text{(4)} \tag{2}$$

$$\text{(5)} \xrightarrow[\substack{2.\ \text{(allyl)}_3\text{SiCl} \\ 46\%}]{1.\ t\text{-BuLi, Et}_2\text{O} \atop -78°\text{C}} \text{(6)} \tag{3}$$

$$\text{(7)} \xrightarrow[\substack{2.\ n\text{-BuLi} \\ 3.\ \text{(allyl)}_3\text{SiCl} \\ 46\%}]{1.\ \text{LiNMeCH}_2\text{CH}_2\text{NMe}_2 \atop \text{THF, }-78\ °\text{C}} \text{(8)} \tag{4}$$

1. (a) Sommer, L. H.; U. S. Pat. 2,512,390 (1950). (b) Scott, R. E.; Frisch, K. C., *J. Am. Chem. Soc.* **1951**, *73*, 2599. (c) Delumulle, L.; Van Der Kelen, G. P., *J. Mol. Struct.* **1981**, *70*, 207. (d) Henry, C.; Brook, M. A., *Tetrahedron* **1994**, *50*, 11379.

2. Gossage, R. A.; Muños-Martinez, E.; van Koten, G., *Tetrahedron Lett.* **1998**, *39*, 2397.

3. Li, Z.; Cao, X.; Lai, G.; Liu, J.; Ni, Y.; Wu, J.; Qiu, H., *J. Organomet. Chem.* **2006**, *691*, 4740.

4. Gossage, R. A.; Muños-Martinez, E.; Frey, H.; Burgath, A.; Lutz, M.; Spek, A. L.; van Koten, G., *Chem. Eur. J.* **1999**, *5*, 2191.

5. Deng, X.; Mayeux, A.; Cai, C., *J. Org. Chem.* **2002**, *67*, 5279.

6. Bashiardes, G.; Chaussebourg, V.; Laverdan, G.; Pornet, J., *Chem. Commun.* **2004**, 122.

Craig R. Smith
The Ohio State University, Columbus, OH, USA

Silane, 1,1'-(1-Propene-1,3-diyl)bis[1,1,1-trimethyl-(E), Silane, 1,1'-(1-Propene-1,3-diyl)bis[1,1,1-trimethyl-(Z), Silane, 1,1'-(1-Propene-1,3-diyl)bis[1,1,1-trimethyl-($E + Z$)*

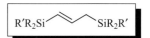

R = R' = Me,
[52152-48-4]   C$_9$H$_{22}$Si$_2$ (MW 186.44)
InChI = 1S/C9H22Si2/c1-10(2,3)8-7-9-11(4,5)6/h7-8H,9H2,
     1-6H3/b8-7+
InChIKey = SVOLIZOGHWXZQV-BQYQJAHWSA-N
R = Me, R' = *t*-Bu
[227612-82-0]   C$_{15}$H$_{34}$Si$_2$ (MW 270.60)
InChI = 1S/C15H34Si2/c1-14(2,3)16(7,8)12-11-13-17(9,10)15
     (4,5)6/h11-12H,13H2,1-10H3/b12-11+
InChIKey = SSUQULQUWRHVMP-VAWYXSNFSA-N
R = Me, R'= Ph
[34774-00-9]   C$_{19}$H$_{26}$Si$_2$ (MW 310.58)
InChI = 1S/C19H26Si2/c1-20(2,18-12-7-5-8-13-18)16-11-17-21
     (3,4)19-14-9-6-10-15-19/h5-16H,17H2,1-4H3/b16-11+
InChIKey = KKFWBUGWJKYYQH-LFIBNONCSA-N

(reagents used as an extension to conventional allylsilane chemistry)

Alternate Names: 1,1'-(1-propene-1,3-diyl)bis[1,1,1-trimethylsilane]; (1*E*)-1-propene-1,3-diylbis[trimethylsilane]; (*E*)-prop-1-ene-1,3-diylbis(trimethylsilane); (1*Z*)-1-propene-1,3-diylbis[trimethylsilane], CAS: 141291-16-9; Bis(*t*-butyldimethylsilyl) derivative: (1*E*)-1-propene-1,3-diylbis[(1,1-dimethylethyl)dimethylsilane]; Bis(dimethylphenyl) derivative: [[(1*E*)-3-dimethylphenylsilyl)-1-propen-1-yl]dimethylsilyl]-benzene; other derivatives are known but less widely used.

*Article previously published in the electronic Encyclopedia of Reagents for Organic Synthesis as 1,3-Bis(silyl)propenes. DOI: 10.1002/047084289X.rn01327.

Physical Data: oil.

Solubility: soluble in most organic solvents.

Preparative Methods: lithiation of 1,3-dichloropropene followed by treatment with a silyl chloride;[1] reaction of 1,3-dichloropropene with Mg^0 and a silyl chloride electrophile;[2] treatment of allylsilanes with BuLi/TMEDA and a silyl electrophile;[3] reaction of allyl alcohol with HMPA, Mg^0, and a Lewis acid catalyst;[4] cross-metathesis of allylsilanes;[5] reductive lithiation of allylthioethers.[6,7]

Handling, Storage, and Precautions: can be stored on the bench top for several months, but the bis(trimethylsilyl) derivative is somewhat volatile.

General. Traditional allylsilane chemistry has been utilized for the stereoselective formation of C–C bonds and exploits the ability of the C–Si bond to stabilize a β-carbocation through a hyperconjugative $\sigma \to p$ donation. The β-silicon effect is responsible for the observed regioselectivity and nucleophilicity of allylsilanes toward electrophiles.[8,9] 1,3-Bis(silyl)propenes offer an extension to this valuable chemistry, as they can often undergo typical allylsilane chemistry, while leaving a handle for subsequent functionalization. Some drawbacks of these compounds are the reduced nucleophilicity due to increased sterics compared to allylsilanes and the relative ease of protodesilylation. Nevertheless, these reagents have seen use in the stereoselective synthesis of C–C bonds. 1,3-Bis(silyl)propenes also play an important role in the stabilization of metal complexes, offering both insight into the reactivity of the parent allyl compounds and utility in polymerization reactions. These compounds have also been utilized for the construction of silacyclobutenes,[10] aromatic tetraboranes,[11] and phosphadiene ligands.[12]

C–C Bond Construction. The addition of 1,3-bis(silyl)propenes to a variety of electrophiles can be accomplished under Lewis acid catalysis or by using the corresponding 1,3-disilylallyl anion and this represents an important method for C–C bond construction. A single case of the addition to an α,β-unsaturated ketone is reported in the context of the perhydroazulenes[13] and photoaddition of 1,3-bis(trimethylsilyl)propene to phenyl pyrrolidinium perchlorate has been disclosed,[14] though these examples will not be discussed in detail.

Cycloadditions. The reaction of chlorosulfonyl isocyanate with 1,3-bis(silyl)propenes yields the corresponding azetidinones (eq 1).[15] It was found that the reaction is stereoselective with regard to the double bond geometry, as the E isomer yields exclusively the *trans* product and Z isomer gives the *cis* product. Unsymmetrical 1,3-bis(silyl)propenes have also been explored in this context as a route to 4-hydroxymethylazetidin-2-ones and eventual routes toward β-lactam antibiotics were envisaged.[16]

Me$_3$Si〜SiMe$_3$ → (CISO$_2$NCO, Na$_2$SO$_3$, NaHCO$_3$, CH$_2$Cl$_2$, –20 °C) → [azetidinone product]   51% yield   (1)

Lewis Acid-catalyzed Additions to Aldehydes. Somfai and coworkers realized the stereoselective synthesis of functionalized pyrrolidines via the [3 + 2]-annulation of *N*-Ts-α-amino aldehydes and 1,3-bis(silyl)propenes (eq 2).[17] A variety of symmetrical bis(silyl)propenes were screened in this reaction and it was found that silanes containing more electron-withdrawing functionalities, such as Cl or OEt, were unreactive, probably due to their decreased nucleophilicity.

PhMe$_2$Si〜SiMe$_2$Ph + [α-amino aldehyde, NHTs] → Lewis acid (LA), 33–69% yield → [intermediate] → [pyrrolidine product]

$R' = i$-Pr, Ph, Me, CH$_2$OTBS   (2)

Interestingly, in this reaction the 1,3-bis(silyl)propene functions solely as a 1,2-dipole equivalent with no migration of the silyl moiety and contrasts with the reactivity of allylsilanes that typically act as 1,3-dipoles with a concurrent 1,2-silyl migration.[18] Use of the Z-bissilane or the β-substituted amino aldehyde did not yield the desired annulated product and instead only the corresponding diene was isolated. This annulation reaction provides facile access toward polyhydroxylated pyrrolidines, as the silyl moieties can be readily transformed into hydroxyl groups via the Tamao–Fleming oxidation.[19–21] This sequence was applied to the synthesis of the monocyclic glycosidase inhibitor DGDP, as well as to (+)-alexine.[22]

Lewis Acid-catalyzed Additions to Glyoxylates. The vinylation of glyoxylates and glyoxamides using 1,3-bis(silyl)propenes was demonstrated by Somfai and coworkers (eq 3). This reaction proceeds via the nucleophilic addition of a 1,3-bis(silyl)propene onto the Lewis acid-activated carbonyl moiety, followed by a [1,3]-Brook rearrangement of the intermediate cation. Interestingly, the reaction was not sensitive to the steric bulk of the silane used and symmetrical bis(silyl)propenes with silyl moieties ranging from SiMe$_3$ to SiPh$_2$$t$Bu were competent in the reaction without significant alteration in yield.

R'$_3$Si〜SiR'$_3$ + [glyoxylate, R] → SnCl$_4$, CH$_2$Cl$_2$, –78 °C, 4 Å MS → 71–89% yield → [OSiR'$_3$ product]   (3)

R = NHPh, OEt

Lewis Acid-catalyzed Additions to Acid Chlorides. The Lewis acid-catalyzed addition of 1,3-bis(silyl)propenes to acid chlorides was explored by Tubul and Santelli in hopes of obtaining the corresponding cyclohexenones.[23] However, the reaction yielded divinylketones, presumably via a metal (Ti or Si) dienolate (eq 4). These products are known to undergo Nazarov cyclization. With a cyclic acid chloride, a chloroenone by-product was also observed.

Lewis Acid-catalyzed Addition to Epoxides. The Lewis acid-catalyzed addition of 1,3-bis(silyl)propenes to epoxides has been studied for both inter- and intramolecular reactions.[24] In the intermolecular reaction, monosubstituted epoxides give higher yields (66–53%) than bis-substituted epoxides (33–19%) (eq 5). It is proposed that the epoxide opens in a S_N1 fashion, followed by nucleophilic attack of the bis(silyl)propene on the resultant cation. This leads to a silicon-stabilized carbocation, which undergoes C→O silyl migration yielding the final alkene.

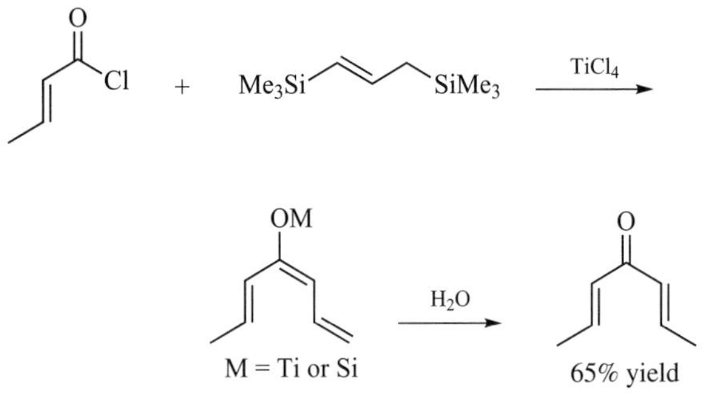

(4)

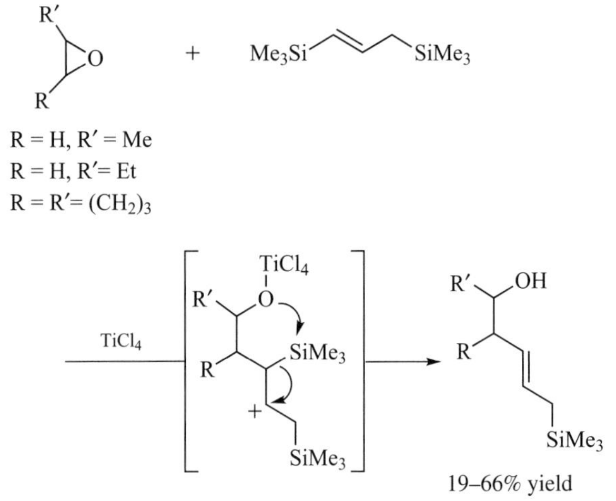

M = Ti or Si   65% yield

R = H, R′ = Me
R = H, R′ = Et
R = R′ = (CH₂)₃

19–66% yield   (5)

For the intramolecular case (eq 6), there is regioselective ring opening of the epoxide at the more substituted position. There are several competing pathways for this reaction and contraction of the cyclohexane ring reveals that the cationic charge is better stabilized by a β-silicon atom than in the intermediate tertiary carbocation. Several dienes were made using this methodology, although polymerization was a competing reaction.

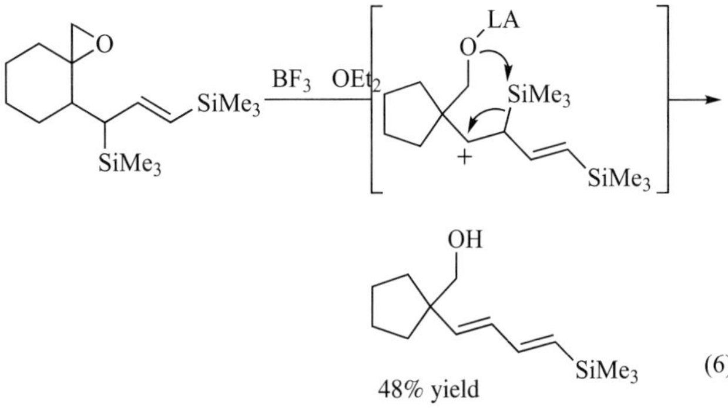

(6)

48% yield

Anionic Additions to Epoxides. The addition of 1,3-bis(silyl)-allyl anions to epoxides yields the corresponding substituted bis(silyl)propenes in good yields and moderate dr (eq 7).[25] It

was envisioned that deprotonation of the resulting alcohol would lead to a [1,4] C→O silyl shift, directly constructing the analogous vinyl cyclopropane from a net homo-Peterson reaction.[26] However, only desilylation was observed. Quenching the reaction with MeI revealed that the necessary [1,4] C→O silyl shift had indeed ensued as only methylation of the allylic positions was observed. However, despite this migration, the cyclization to the cyclopropane could not be realized.

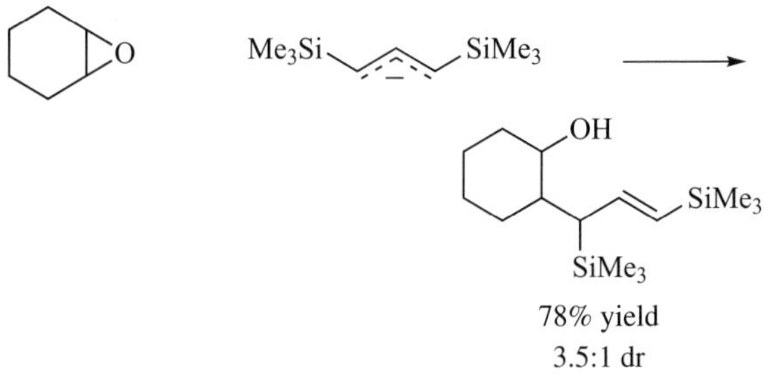

78% yield
3.5:1 dr   (7)

Derivatization of the alcohol to the tosylate, followed by treatment with Bu_4NF, however, did provide access to the desired functionalized vinyl cyclopropane in 61% yield (eq 8). Unfortunately, protodesilylation was seen as a side reaction.

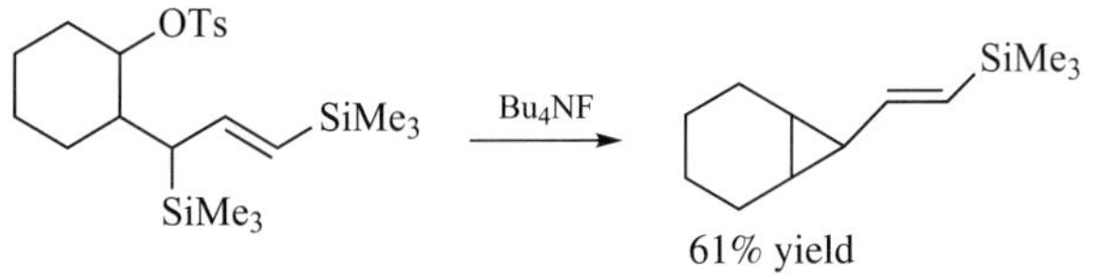

61% yield   (8)

This same strategy was also applied to tethered epoxides to yield 1-(2-vinylcyclopropyl)alkanols (eq 9).[27] In this sequence, the tosylate was replaced by an epoxide as the target of the intramolecular nucleophilic attack. This type of strategy was applied to the synthesis of dictyopterene A.

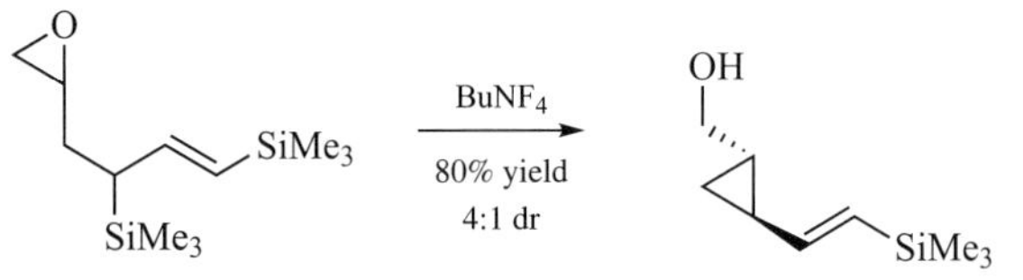

80% yield
4:1 dr   (9)

Schaumann and coworkers have also examined the addition of 1,3-bis(silyl)allyl anions to epoxides followed by epoxidation. The first addition of the bis(silyl) anion to the epoxide gave a 3:1 mixture of diastereomers differing at the stereocenter α to the silyl moiety, while the subsequent epoxidation gave the product as a single diastereomer albeit in low yield (31%). Treatment of the resulting epoxy alcohol with a Lewis acid then yielded *O*-Me δ-lactol regioselectively in a moderate 48% yield (eq 10).[28]

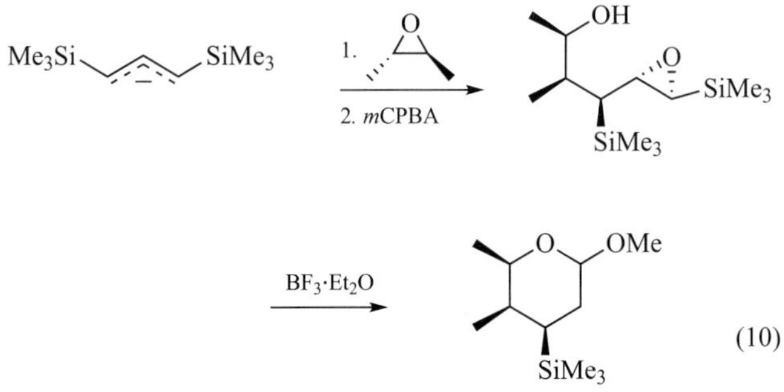

(10)

Anionic Additions to Aldehydes. The addition of 1,3-bis(silyl)-propenes to aldehydes and ketones to yield the vinyl silyl alcohol was explored.[29] This was done using TBAF and good to excellent yields were achieved (eq 11). A further extension of this work was the addition of the (1,3-bis(silyl)allyl)lithium to ketones and aldehydes (eq 12). In this reaction, the substituted silyl diene was isolated in moderate to good yields. These substrates were then explored as ligands for both iron and manganese complexes.

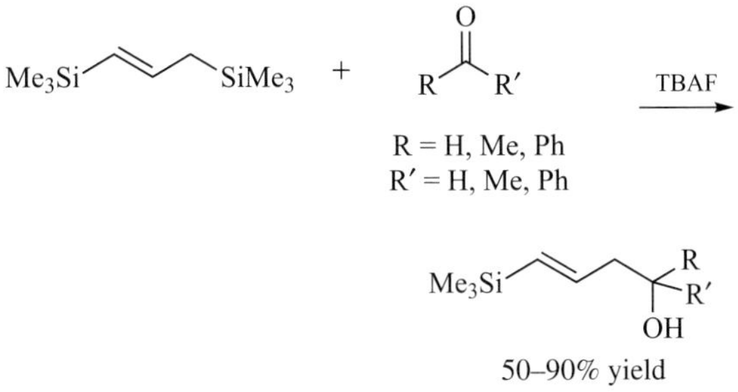

$$\text{50–90\% yield} \quad (11)$$

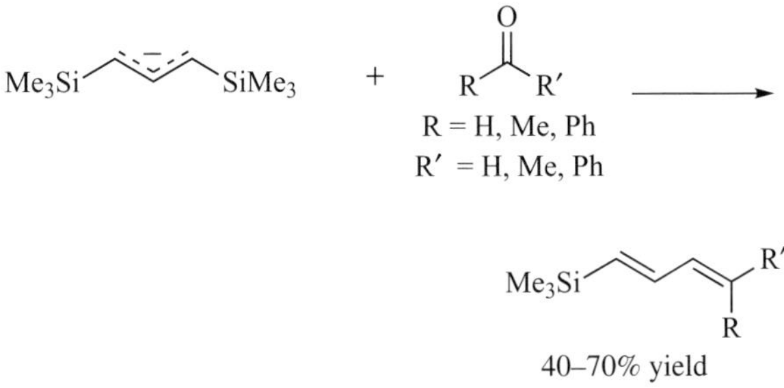

$$\text{40–70\% yield} \quad (12)$$

Removable Silicon Tethers. Cox and coworkers have explored the use of removable silicon tethers as a method of controlling the relative stereochemistry of the addition of allylsilane to aldehydes.[30,31] This strategy relies on the construction of unsymmetrical bis(silyl)propenes with a pendant aldehyde (eq 13). When the dimethylsilyl ether was used as the tether only the corresponding diene was observed, but by increasing the steric bulk of the silane to the diethylsilyl ether the construction of the oxasilacycle was possible. Use of the diisopropyl tether completely suppressed formation of the diene by-product; however, the 1,3-stereoinduction was eroded as four diastereomers were observed instead of two. Tamao–Fleming oxidation of the two diastereomers (2.7:1–9.7:1 dr) derived from the diethylsilyl ether tether yielded the stereodefined 1,2,4-triol in the case of the major

diastereomer and the diene in the case of the minor oxasilacycle. Steric hindrance in the transition state of the elimination reaction is proposed to be the origin of the preferential oxidation of the major diastereomer. Use of the Z-allylsilane derivative led to an enhanced 1,4-stereoinduction (7:1–21:1 dr) with similar yields.

Stereoselective Preparation of 2-Alkenyltrimethylsilanes. Epoxidation of 1,3-bis(trimethylsilyl)propene creates a versatile building block for the stereoselective construction of both *E*- and *Z*-2-alkenyltrimethylsilanes via a magnesium-mediated rearrangement to bis(trimethylsilyl)propanal.[32] It was found that treatment of the resulting aldehyde with a variety of Grignard reagents then leads to an intermediate alcohol species, which can be transformed stereoselectively into an alkene depending on the conditions chosen. This allows for the synthesis of valuable substituted allylsilanes for further transformations.[2]

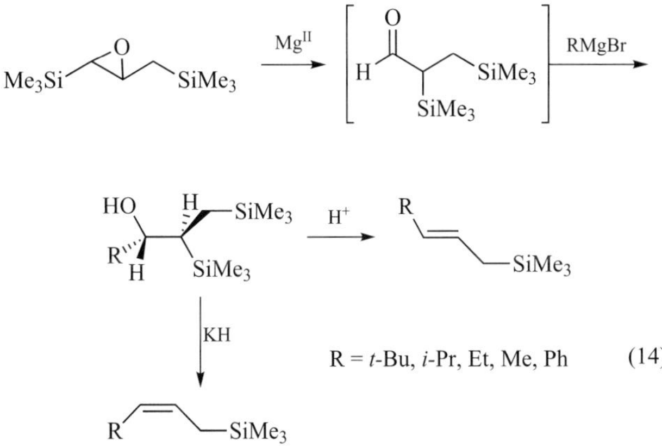

$$R = t\text{-Bu}, \ i\text{-Pr}, \ \text{Et}, \ \text{Me}, \ \text{Ph} \quad (14)$$

1,3-Bis(silyl)propenes and Metals. Allyl metal complexes play an important role in organometallic chemistry.[33,34] Often the parent allyl complexes are not thermally stable or discrete complexes. In order to gain further insight into these parent compounds, 1,3-bis(trimethylsilyl)propene or 1,3-bis(*t*-butyldimethylsilyl)propene ligands have been used as surrogates for the allyl group as the steric bulk of the silicon moieties stabilize the electron-deficient metal center.[35] Complexes of Th,[35] Cr,[36,37] Ni,[38] Ti,[39] and Zn[39] have been studied. These compounds have indeed shown improved thermal stability. It is believed that the restricted motion of the allyl ligands hinders decomposition, which allows for more in-depth studies of the nature of these compounds.

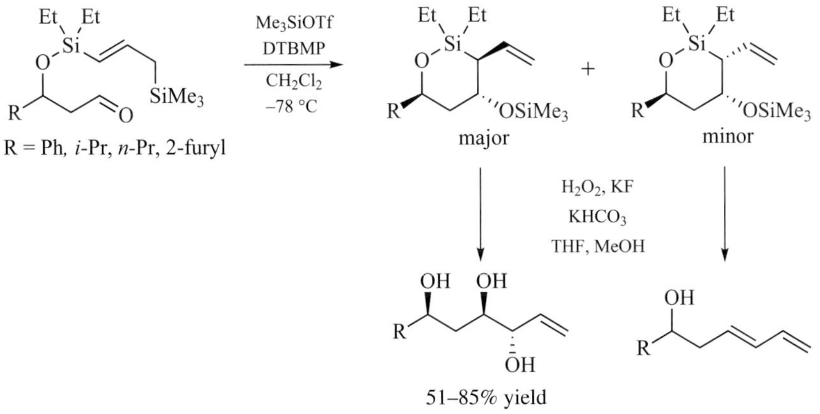

$$\text{51–85\% yield} \quad (13)$$

Several lanthanide metal complexes (Sm,[40,41] Nd,[41,42] Ce, Eu, Tb, and Yb[42]) have also been characterized in hopes of elucidating active catalysts for the polymerization of methyl methacrylate. However, it was found that the potassium 1,3-bis(trimethylsilyl)-allyl salt was more active (higher turnover frequency) for the polymerization of methyl methacrylate than the neutral lanthanide complexes mentioned above. This leads to the further characterization of several anionic complexes of heavy alkali metals and 1,3-bis(trimethylsilyl)propenes.[43]

The dynamics of a (1,3-disilylallyl)lithium and TMEDA complex has also been studied by NMR.[44] Isotopic perturbation of equilibrium revealed that the lithium salt is symmetrical and favors an *exo–exo* configuration. The corresponding allyl lithium salt was revealed to be unsymmetrical in similar studies, although allyl sodium and allyl potassium salts were found to be symmetrical.

1,3-Bis(silyl)propenes have also been briefly explored using Pd allyl chemistry,[45,46] as well as cross-metathesis.[5,47]

1. Guijarro, A.; Yus, M., *Tetrahedron* **1994**, *50*, 13269.

2. Shimizu, N.; Imazu, S.; Shibata, F.; Tsuno, Y., *Bull. Chem. Soc. Jpn.* **1991**, *64*, 1122.

3. Burns, G. T.; Barton, T. J., *J. Organomet. Chem.* **1981**, *216*, C5.

4. Biran, C.; Duffaut, N.; Dunogues, J.; Calas, R., *J. Organomet. Chem.* **1975**, *91*, 279.

5. Kakiuchi, F.; Yamada, A.; Chatani, N.; Murai, S.; Furukawa, N.; Seki, Y., *Organometallics* **1999**, *18*, 2033.

6. Streiff, S.; Ribeiro, N.; Désaubry, L., *J. Org. Chem.* **2004**, *69*, 7592.

7. Streiff, S.; Ribeiro, N.; Desaubry, L., *Chem. Commun.* **2004**, 346.

8. Masse, C. E.; Panek, J. S., *Chem. Rev.* **1995**, *95*, 1293.

9. Chabaud, L.; James, P.; Landais, Y., *Eur. J. Org. Chem.* **2004**, 3173.

10. Burns, G. T.; Barton, T. J., *J. Organomet. Chem.* **1981**, *216*, C5.

11. Präsang, C.; Sahin, Y.; Hofmann, M.; Geiseler, G.; Massa, W.; Berndt, A., *Eur. J. Inorg. Chem.* **2004**, 3063.

12. Boyd, B. A.; Thoma, R. J.; Watson, W. H.; Neilson, R. H., *Organometallics* **1988**, *7*, 572.

13. House, H. O.; Gaa, P. C.; Lee, J. H. C., Van Derveer, D., *J. Org. Chem.* **1983**, *48*, 1670.

14. Ohga, K.; Mariano, P. S., *J. Am. Chem. Soc.* **1982**, *104*, 617.

15. Nativi, C.; Perrotta, E.; Ricci, A.; Taddei, M., *Tetrahedron Lett.* **1991**, *32*, 2265.

16. Colvin, E. W.; Monteith, M., *J. Chem. Soc., Chem. Commun.* **1990**, 1230.

17. Restorp, P.; Fischer, A.; Somfai, P., *J. Am. Chem. Soc.* **2006**, *128*, 12646.

18. Restorp, P.; Dressel, M.; Somfai, P., *Synthesis* **2007**, 1576.

19. Fleming, I.; Henning, R.; Plaut, H., *J. Chem. Soc., Chem. Commun.* **1984**, 29.

20. Tamao, K.; Ishida, N., *J. Organomet. Chem.* **1984**, *269*, c37.

21. Jones, G. R.; Landais, Y., *Tetrahedron* **1996**, *52*, 7599.

22. Dressel, M.; Restorp, P.; Somfai, P., *Chem. Eur. J.* **2008**, *14*, 3072.

23. Tubul, A.; Santelli, M., *Tetrahedron* **1988**, *44*, 3975.

24. Nowak, A.; Bolte, O.; Schaumann, E., *Tetrahedron Lett.* **1998**, *39*, 529.

25. Schaumann, E.; Kirschning, A.; Narjes, F., *J. Org. Chem.* **1991**, *56*, 717.

26. Fleming, I.; Floyd, C. D., *J. Chem. Soc., Perkin Trans. 1* **1981**, 969.

27. Narjes, F.; Bolte, O.; Icheln, D.; Koenig, W. A.; Schaumann, E., *J. Org. Chem.* **1993**, *58*, 626.

28. Flörke, H., Schaumann, E., *Synthesis* **1996**, 647.

29. Corriu, R.; Escudie, N.; Guerin, C., *J. Organomet. Chem.* **1984**, *264*, 207.

30. Beignet, J.; Jervis, P. J.; Cox, L. R., *J. Org. Chem.* **2008**, *73*, 5462.

31. Ramalho, R.; Jervis, P. J.; Kariuki, B. M.; Humphries, A. C.; Cox, L. R., *J. Org. Chem.* **2008**, *73*, 1631.

32. Shimizu, N.; Shibata, F.; Tsuno, Y., *Chem. Lett.* **1985**, 1593.

33. Kazmaier, U.; Pohlman, M.; In *Metal-Catalyzed Cross-Coupling Reactions* 2nd ed.; de Meijere, A.; Diederich, F. Eds.; Wiley-VCH Verlag GmbH & Co. KGaA: Weinheim, 2004; Vol. 1, p 531.

34. Hegedus, L.; *Transition Metals in the Synthesis of Complex Organic Molecules*; University Science Books: Sausalito, CA, 1999; p 245.

35. Carlson, C. N.; Hanusa, T. P.; Brennessel, W. W., *J. Am. Chem. Soc.* **2004**, *126*, 10550.

36. Smith, J. D.; Hanusa, T. P.; Young, V. G., *J. Am. Chem. Soc.* **2001**, *123*, 6455.

37. Carlson, C. N.; Smith, J. D.; Hanusa, T. P.; Brennessel, W. W.; Young, V. G., *J. Organomet. Chem.* **2003**, *683*, 191.

38. Quisenberry, K. T.; Smith, J. D.; Voehler, M.; Stec, D. F.; Hanusa, T. P.; Brennessel, W. W., *J. Am. Chem. Soc.* **2005**, *127*, 4376.

39. Ray, B.; Neyroud, T. G.; Kapon, M.; Eichen, Y.; Eisen, M. S., *Organometallics* **2001**, *20*, 3044.

40. Ihara, E.; Koyama, K.; Yasuda, H.; Kanehisa, N.; Kai, Y., *J. Organomet. Chem.* **1999**, *574*, 40.

41. Woodman, T. J.; Schormann, M.; Hughes, D. L.; Bochmann, M., *Organometallics* **2003**, *22*, 3028.

42. Simpson, C. K.; White, R. E.; Carlson, C. N.; Wrobleski, D. A.; Kuehl, C. J.; Croce, T. A.; Steele, I. M.; Scott, B. L.; Young, V. G.; Hanusa, T. P.; Sattelberger, A. P.; John, K. D., *Organometallics* **2005**, *24*, 3685.

43. Quisenberry, K. T.; Gren, C. K.; White, R. E.; Hanusa, T. P.; Brennessel, W. W., *Organometallics* **2007**, *26*, 4354.

44. Fraenkel, G.; Chow, A.; Winchester, W. R., *J. Am. Chem. Soc.* **1990**, *112*, 1382.

45. Karabelas, K.; Westerlund, C.; Hallberg, A., *J. Org. Chem.* **1985**, *50*, 3896.

46. Corriu, R.; Masse, J., *J. Organomet. Chem.* **1973**, *57*, C5.

47. Pietraszuk, C.; Fischer, H., *Chem. Commun.* **2000**, 2463.

Brinton Seashore-Ludlow & Peter Somfai
Royal Institute of Technology, Stockholm, Sweden

Sodium Hexamethyldisilazide[1]

NaN(SiMe₃)₂

[1070-89-9] $C_6H_{18}NNaSi_2$ (MW 183.42)

InChI = 1S/C6H18NSi2.Na/c1-8(2,3)7-9(4,5)6;/h1-6H3;/q-1;+1

InChIKey = WRIKHQLVHPKCJU-UHFFFAOYSA-N

(useful as a sterically hindered base and as a nucleophile)

Alternate Names: NaHMDS; sodium bis(trimethylsilyl)amide.

Physical Data: mp 171–175 °C; bp 170 °C/2 mmHg.

Solubility: soluble in THF, ether, benzene, toluene.[1]

Form Supplied in: (a) off-white powder (95%); (b) solution in THF (1.0 M); (c) solution in toluene (0.6 M).

Analysis of Reagent Purity: THF solutions of the reagent may be titrated using 4-phenylbenzylidenebenzylamine as an indicator.[2]

Handling, Storage, and Precaution: the dry solid and solutions are flammable and must be stored in the absence of moisture. These should be handled and stored under a nitrogen atmosphere. Use in a fume hood.

Original Commentary

Brett T. Watson

Bristol-Myers Squibb Pharmaceutical Research Institute, Wallingford, CT, USA

Introduction. Sodium bis(trimethylsilyl)amide is a synthetically useful reagent in that it combines both high basicity[3] and nucleophilicity,[4] each of which may be exploited for useful organic transformations such as selective formation of enolates,[5] preparation of Wittig reagents,[6] formation of acyl anion equivalents,[7] and the generation of carbenoid species.[8] As a nucleophile, it has been used as a nitrogen source for the preparation of primary amines.[9,10]

Sterically Hindered Base for Enolate Formation. Like other metal dialkylamide bases, sodium bis(trimethylsilyl)amide is sufficiently basic to deprotonate carbonyl-activated carbon acids[5] and is sterically hindered, allowing good initial kinetic vs. thermodynamic deprotonation ratios.[11] The presence of the sodium counterion also allows for subsequent equilibration to the thermodynamically more stable enolate.[5f] More recently, this base has been used in the stereoselective generation of enolates for subsequent alkylation or oxidation in asymmetric syntheses.[12] As shown in eq 1, NaHMDS was used to selectively generate a (Z)-enolate; alkylation with iodomethane proceeded with excellent diastereoselectivity.[12a] In this case, use of the sodium enolate was preferred as it was more reactive than the corresponding lithium enolate at lower temperatures.

$$(1)$$

79%
99:1 diastereoselectivity

The reagent has been used for the enolization of carbonyl compounds in a number of syntheses.[13] For ketones and aldehydes which do not have enolizable protons, NaHMDS may be used to prepare the corresponding TMS-imine.[14]

Generation of Ylides for Wittig Reactions. In the Wittig reaction, salt-free conditions have been shown to improve (Z):(E) ratios of the alkenes which are prepared.[15] NaHMDS has been shown to be a good base for generating ylides under lithium-salt-free conditions.[6] It has been used in a number of syntheses to selectively prepare (Z)-alkenes.[16] Ylides generated under these conditions have been shown to undergo other ylide reactions such as C-acylations of thiolesters and inter- and intramolecular cyclization.[6] Although Wittig-based syntheses of vinyl halides exist,[17] NaHMDS has been shown to be the base of choice for the generation of iodomethylenetriphenylphosphorane for the stereo-

selective synthesis of (Z)-1-iodoalkenes from aldehydes and ketones (eq 2).[18]

$$(2)$$

96%
(Z):(E) = 62:1

61%

NaHMDS has been shown to be the necessary base for the generation of the ylide anion of sodium cyanotriphenylphosphoranylidenemethanide, which may be alkylated with various electrophiles and in turn used as an ylide to react with carbonyl compounds.[19] NaHMDS was used as the base of choice in a Horner–Emmons–Wadsworth-based synthesis of terminal conjugated enynes.[20]

Intramolecular Alkylation via Protected Cyanohydrins (Acyl Anion Equivalents). Although NaHMDS was not the base of choice for the generation of protected cyanohydrin acyl carbanion equivalents in the original references,[21] it has been shown to be an important reagent for intramolecular alkylation using this strategy (eqs 3 and 4).[7,22] The advantages of this reagent are (a) that it allows high yields of intramolecularly cyclized products with little intermolecular alkylation and (b) the carbanion produced in this manner acts only as a nucleophile without isomerization of double bonds α,β to the anion or other existing double bonds in the molecule. Small and medium rings as well as macrocycles[22a] have been reported using this methodology (eqs 3 and 4).

$$(3)$$

$$(4)$$

Generation of Carbenoid Species. Metal bis(trimethylsilyl) amides may be used to effect α-eliminations.[23] It is proposed that these nucleophilic agents undergo a hydrogen–metal exchange reaction with polyhalomethanes to give stable carbenoid species.[23b] NaHMDS has been used to generate carbenoid species which have been used in a one-step synthesis of monobromocyclopropanes (eqs 5 and 6).[23c,d] NaHMDS has been shown to give better yields than the corresponding lithium or potassium amides in this reaction.

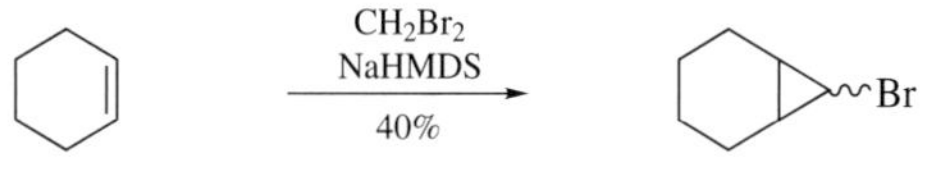

$$\text{(5)} \qquad cis{:}trans = 1.5{:}1$$

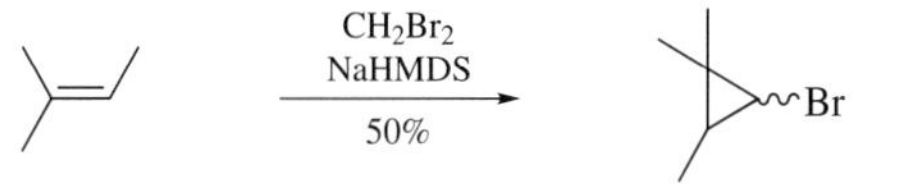

$$\text{(6)}$$

A similar study which evaluated the use of NaHMDS versus *n*-butyllithium for the generation of the active carbenoid species from 1,1-dichloroethane and subsequent reaction with alkenes, forming 1-chloro-1-methylcyclopropanes, suggested that the amide gave very similar results to those with *n*-butyllithium.[24]

In an initial report, the carbenoid species formed by the treatment of diiodomethane with NaHMDS was shown to react as a nucleophile, displacing primary halides and leading to a synthesis of 1,1-diiodoalkanes; this is formally a 1,1-diiodomethylene homologation (eq 7).[25] This methodology is limited in that electrophiles which contain functionality that allows facile E2 elimination (i.e. allyl) form a mixture of the desired 1,1-diiodo compound and the iododiene. In the case of allyl bromide, addition of 2 equiv of the sodium reagent allows isolation of the iododiene as the major product.

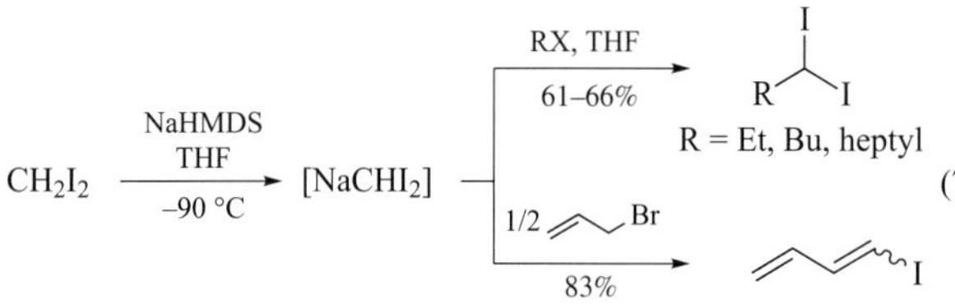

$$\text{(7)}$$

Synthesis of Primary Amines. The nucleophilic properties of this reagent may be utilized in the S_N2 displacement of primary alkyl bromides, iodides, and tosylates to form bis(trimethylsilyl) amines (**1**) (eq 8).[9a] HCl hydrolysis of (**1**) allows isolation of the corresponding hydrochloride salt of the amine, which may be readily separated from the byproduct, bis(trimethylsilyl) ether. In one example a secondary allylic bromide also underwent the conversion with good yield.

$$\text{RX + NaHMDS} \longrightarrow \text{R–N(TMS)}_2 \longrightarrow \text{RNH}_3\text{Cl + (TMS)}_2\text{O} \quad \text{(8)}$$

		(1)	(2)
R	X		
Me	I	75%	99%
Et	OTs	77%	98%
Br	Br	73%	100%
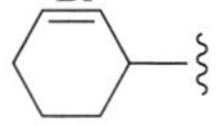 cyclohexenyl	Br	66%	97%

Aminomethylation. NaHMDS may be used as the nitrogen source in a general method for the addition of an aminomethyl group (eq 9).[10] The reagent is allowed to react with chloromethyl methyl ether, forming the intermediate aminoether. Addition of Grignard reagents to this compound allows the displacement of the

methoxy group, leaving the bis(trimethylsilyl)-protected amines. Acidic hydrolysis of these allows isolation of the hydrochloride salt of the corresponding amine in good yields.

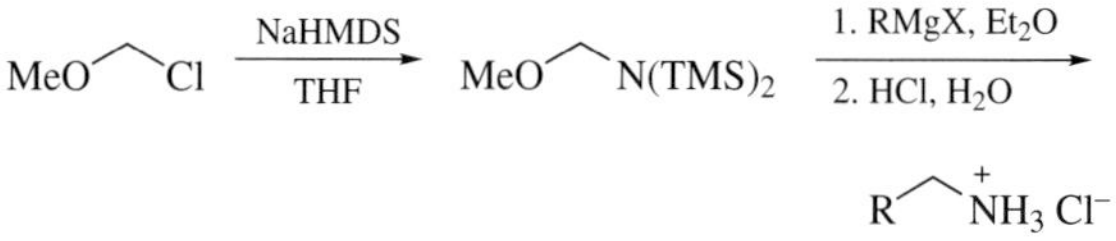

$$\text{(9)}$$

R = Me, 89%; allyl, 76%; Cy, 75%; Ph, 78%; propargyl, 66%

First Update

Professor Hélène Lebel
Université de Montréal, Montréal, Canada

Synthesis of Sulfinamides. The nucleophilic substitution of chiral enantiopure sulfinate esters with NaHMDS led to S-O bond breakage with inversion of configuration at the S atom, giving quantitative conversion to chiral sulfinamides (eq 10). Lithium amide in liquid ammonia at $-78\,^\circ$C could be also used, although in some cases such as in the synthesis of (*R*)-(+)-*p*-toluenesulfinamide higher ee's were observed when using NaHMDS.[26]

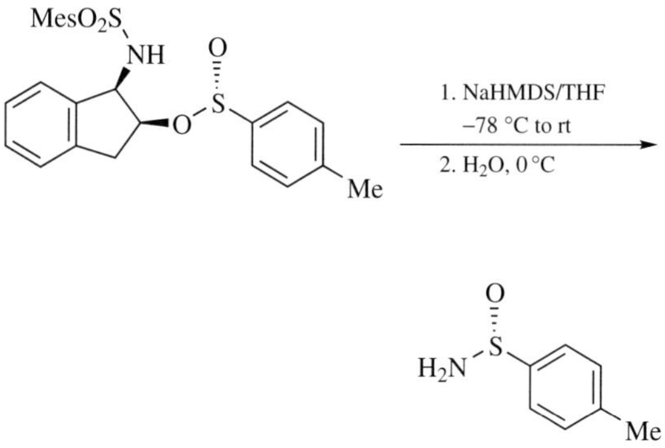

$$\text{(10)}$$

85%
99% ee

***N*-Alkylation of Aryl Amines.** Various aryl amines and aminopyridines have been reacted with di-*t*-butyldicarbonate in the presence of 2 equiv of NaHMDS in THF to lead to the corresponding Boc-protected amines in high yields.[27] A series of anilines were treated with 1.5 to 2.0 equiv of NaHMDS and reacted with the appropriate esters or lactones to give the corresponding *N*-aryl amides in 88–99% yields (eq 11).[28] No diminution of the reaction rate was observed with sterically hindered esters and the method could be applied to the preparation of *N*-aryl amides containing functional groups that are incompatible with strong bases such as *n*-BuLi. The reaction with alkyl amines was not efficient as the mechanism involved the formation of the corresponding sodium amide, and NaHMDS is strong enough to deprotonate aryl amines, but not alkyl amines.

Palladium-Catalyzed α-Arylation of Carbonyl Compounds and Nitriles. A variety of bases have been used in the palladium-catalyzed α-arylation of carbonyl derivatives.[29,30] The pKa of the carbonyl moiety determines the choice of the base. For instance,

α-arylation with ester derivatives requires the use of strong bases such as NaHMDS.[31] The best yields for the arylation of t-butyl acetate were reported with LHMDS, whereas the arylation of t-butyl propionate occurred in higher yields in the presence of NaHMDS (eq 12).[32]

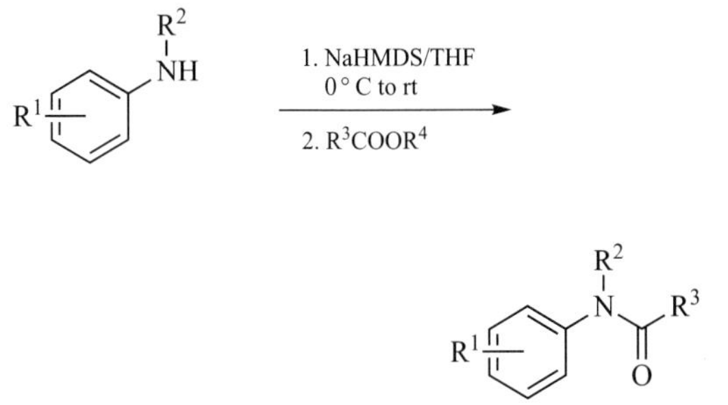

(11)

88–99%

R^1 = H, F, Br, I, OMe, CF$_3$, PhCO
R^2 = H, Me
R^3 = H, alkyl, aryl
R^4 = Me, Et, t-Bu

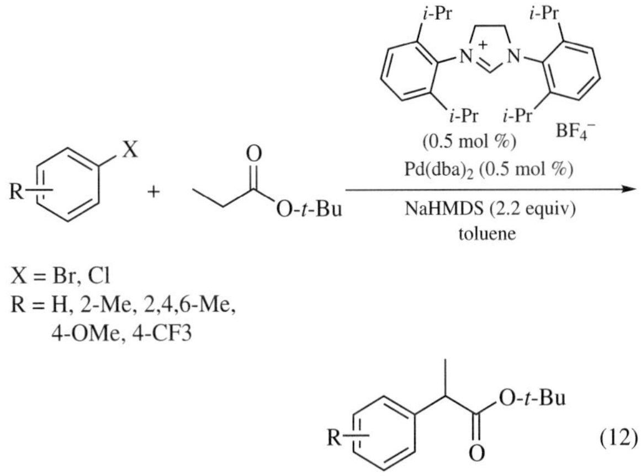

X = Br, Cl
R = H, 2-Me, 2,4,6-Me,
    4-OMe, 4-CF3

(12)

71–88%

It was shown later that LiNCy$_2$ could also be very effective for this reaction.[33] More sensitive substrates such as α-imino esters,[34] malonates, or cyanoesters[35] required the use of a milder base, as decomposition was observed with NaHMDS.

However, NaHMDS was used with success in nickel–BINAP-catalyzed enantioselective α-arylation of α-substituted γ-butyrolactones (eq 13). Coupled with an accelerating effect of Zn(II) salts, α-quaternization is achieved with high enantioselectivities and moderate to excellent yields.[36]

Sodium t-butoxide is usually used in palladium-catalyzed ketone arylations. However, this base was incompatible with aryl halides substituted with an electron-withdrawing group. Replacement by NaHMDS led to α-aryl ketones in good yields (eq 14).[37]

Finally, the α-arylation of nitrile derivatives was also reported using NaHMDS (eq 15).[38]

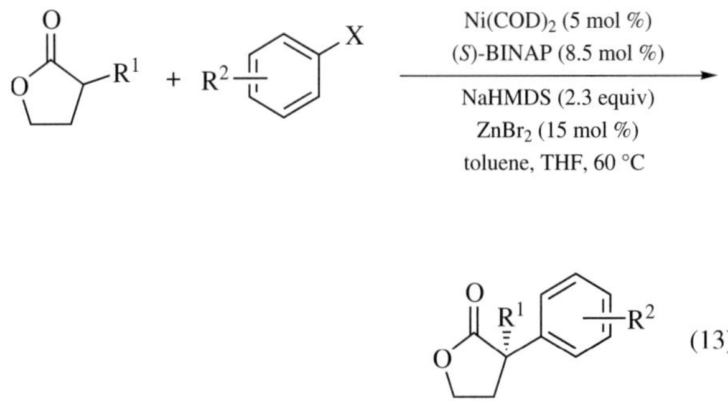

(13)

57–95%
90–99% ee

R^1 = Me, Bn, Allyl, n-Pr
X = Br, Cl
R^2 = H, 3-NMe$_2$, 4-OTBS, 4-t-Bu, 3-OMe, 4-OMe, 4-CF$_3$,
       2-napthyl, 3-CO$_2t$-Bu, 4-CO$_2t$-Bu

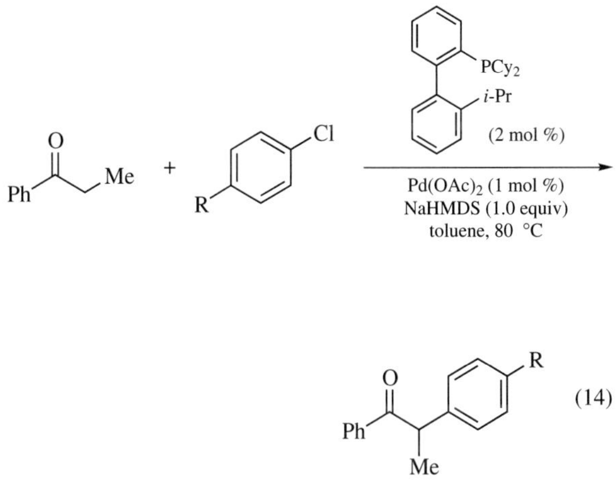

(14)

76–78%
R = CN or CO$_2$Me

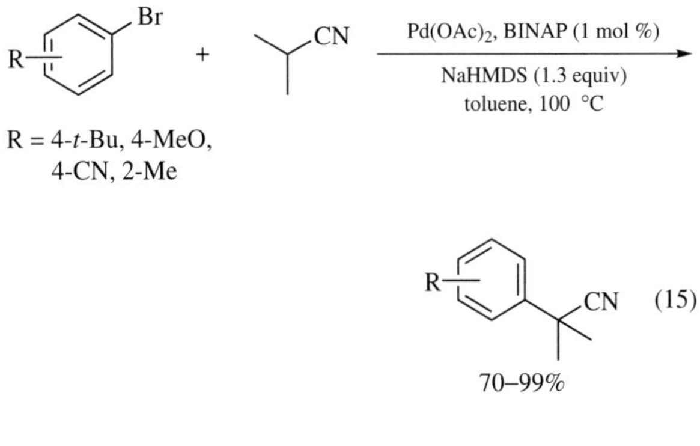

R = 4-t-Bu, 4-MeO,
    4-CN, 2-Me

(15)

70–99%

Transition Metal-Catalyzed Allylic Alkylation. Simple ketone enolates were found to be suitable nucleophiles for palladium-catalyzed allylic alkylations.[39] The palladium-catalyzed asymmetric allylic alkylation of α-aryl ketones will take place in the presence of NaHMDS and allyl acetate to produce the desired α,α-disubstituted ketone derivative in high yields and ee (eq 16).[40]

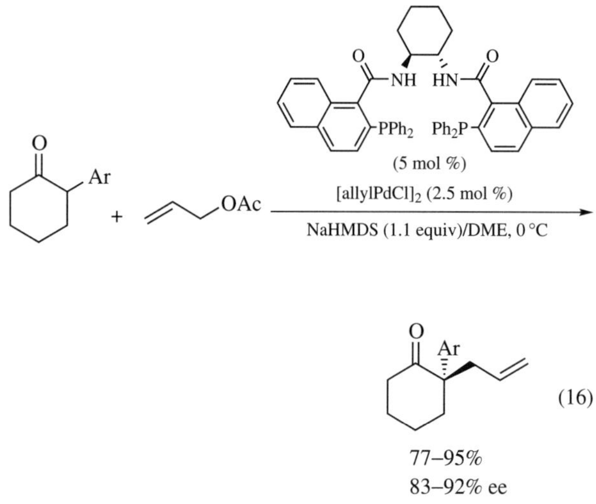

$$(16)$$

77–95%
83–92% ee

Synthesis of 1-Bromoalkynes. The dehydrobromination of 1,1-dibromoalkenes leading to 1-bromoalkynes was usually performed with NaHMDS (eq 17).[41] In comparison, stronger bases such as butyllithium led to the corresponding acetylide anion via a metal–halogen process.[42]

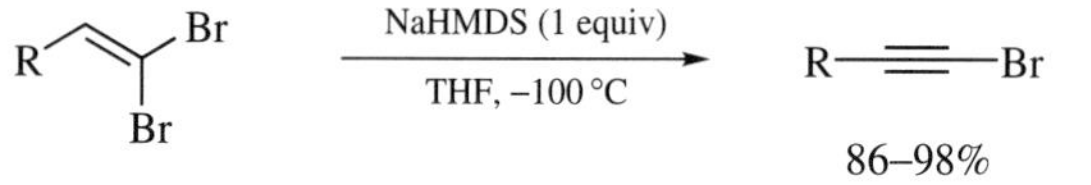

$$(17)$$

86–98%

Synthesis of Cyclopropene. When allyl chloride was dropped into a solution of NaHMDS in toluene at 110 °C, cyclopropene could be isolated in a trap/ampoule at −80 °C in 40% yield and >95% purity (eq 18).[43]

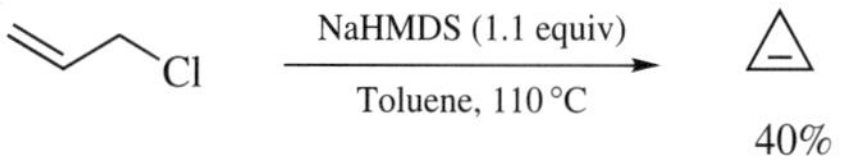

$$(18)$$

40%

Deprotection of Alkyl Aryl Ether. Both NaHMDS and LDA can act as efficient nucleophiles in O-demethylation reactions (eq 19).[44] The reaction is run in a sealed tube at 185 °C in a mixture of THF and 1,3-dimethyl-2-imidazolidinone (DMEU). This synthetic strategy is applicable to methoxyarenes, including benzene, naphthalene, anthracene, biphenyl, and pyridine, which bear either an electron-withdrawing or an electron-donating group, in addition to being efficient with benzodioxoles that lead to the corresponding catechols in excellent yields. As the activity of NaHMDS is lower than that of LDA, mono-O-demethylation of o-dimethoxybenzenes could be accomplished only with the former.

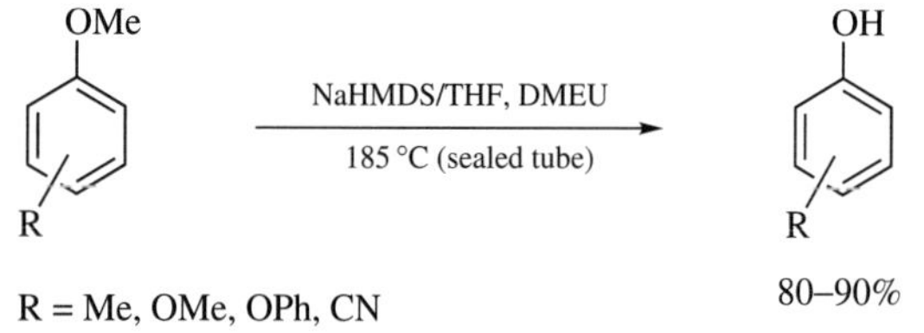

$$(19)$$

R = Me, OMe, OPh, CN

80–90%

Synthesis of Aromatic Nitriles from Esters. A one-flask method has been developed for the conversion of aromatic esters

to the corresponding nitriles by use of NaHMDS in a sealed tube at 185 °C in a mixture of THF and 1,3-dimethyl-2-imidazolidinone (DMEU) (eq 20).[45] The transformation proceeded with good to excellent yields. The synthetic strategy is only applicable to aromatic esters that bear an electron-donating substituent such as hydroxy or methoxy. In the latter case, competitive O-demethylation is observed, thus leading to a mixture of nitrile products. The reaction has been also applied to indole-3-carboxylate. However, simple unsubstituted methyl benzoate failed to give the desired product.

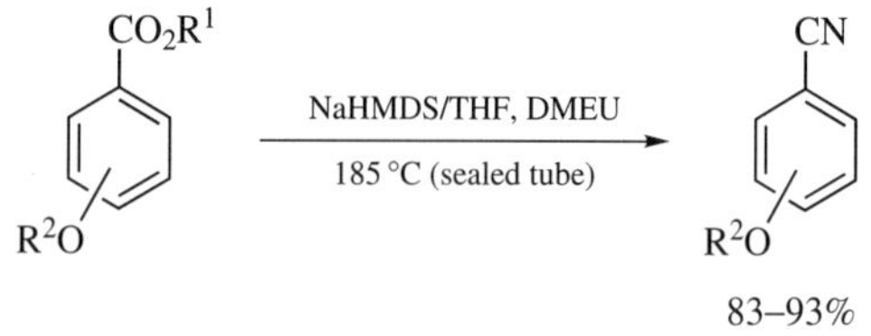

$$(20)$$

83–93%

$R^1 = Me, Et$
$R^2 = H, Me$

Deprotonation of Alkynes. NaHMDS is a base strong enough to deprotonate terminal alkynes. It has been particularly useful for the condensation of alkyne anions with aldehydes, both intermolecularly (eq 21)[46] and intramolecularly (eq 22).[47]

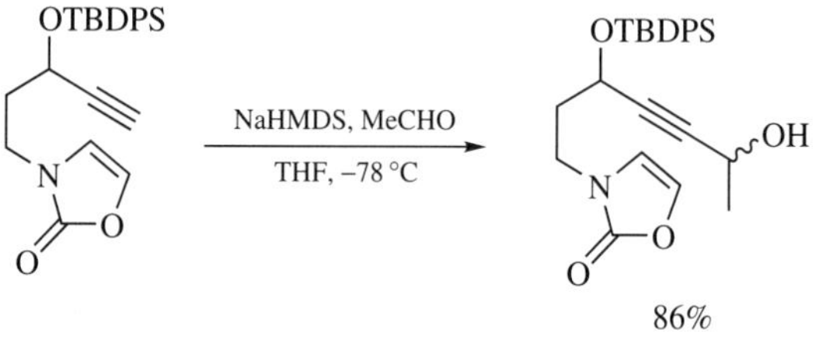

$$(21)$$

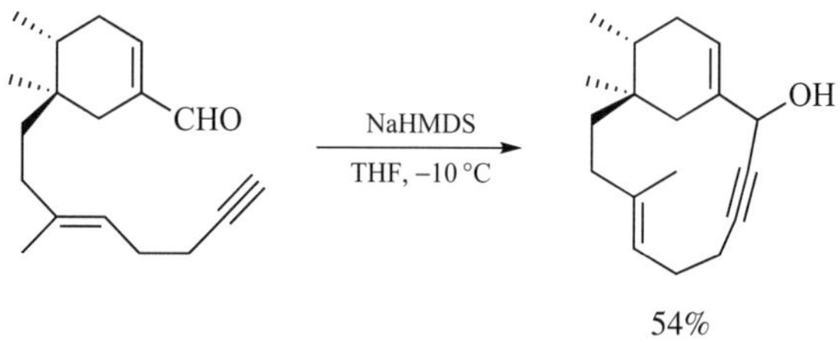

86%

$$(22)$$

54%

In a tandem cyclization of propynyloxyethyl derivatives, NaHMDS is as efficient as LDA or KHMDS (eq 23).[48] This cyclization proceeded through the formation of an alkyne metal, followed by a cyclization which led to a carbenic species that underwent an intramolecular C-H insertion.

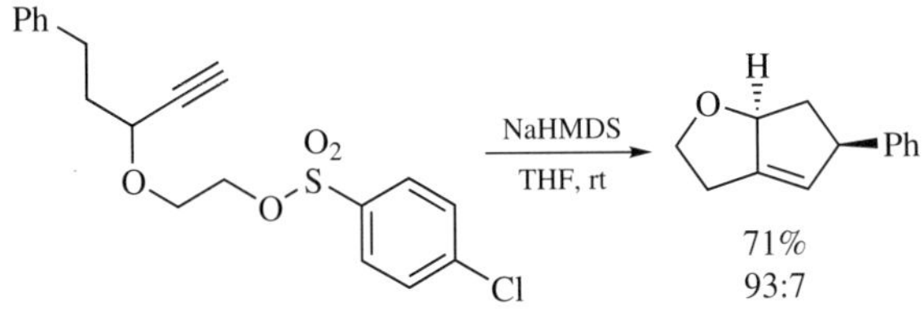

$$\text{(23)}$$

71%
93:7

Enantioselective Oxidation with (Camphorsulfonyl) oxaziridines. The synthesis of 2-substituted-2-hydroxy-1-tetralones from 2-substituted-1-tetralone sodium enolates, obtained from the corresponding ketones and NaHMDS, proceeded with high enantioselectivities in the presence of (+)-(2R,8aR*)-[(8,8-dichlorocamphoryl)sulfonyl]oxaziridine (eq 24).[49]

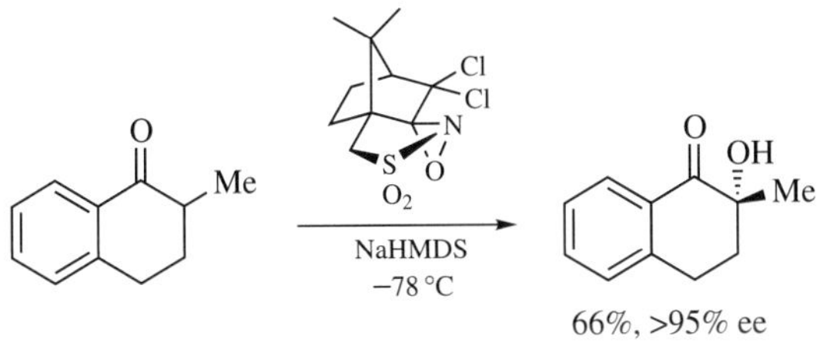

$$\text{(24)}$$

66%, >95% ee

A similar strategy was used to prepare α-hydroxy phosphonates and the highest selectivity was observed when NaHMDS was used as a base (eq 25).[50,51]

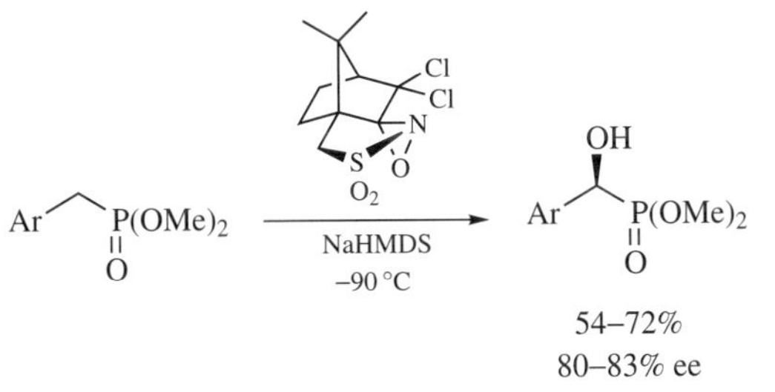

$$\text{(25)}$$

54–72%
80–83% ee

Deprotonation of Trimethylsilyldiazomethane. Treatment of trimethylsilyldiazomethane with NaHMDS led efficiently to the formation of the corresponding sodium ion. Ring opening of N-carboxylated β-lactams with this anion followed by photolytic Wolff rearrangement provided γ-lactams in very good yields (eq 26).[52]

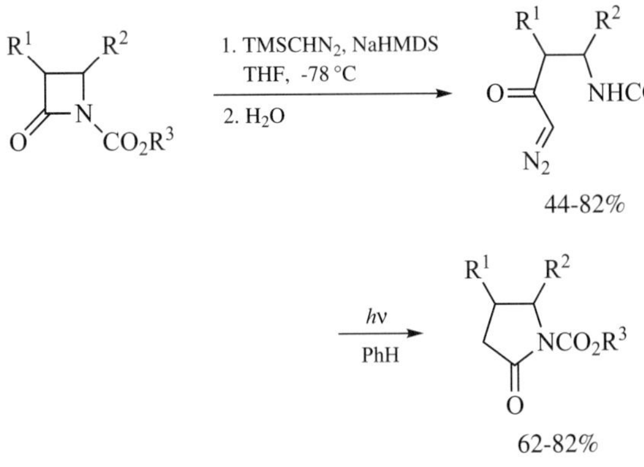

$$\text{(26)}$$

44–82%

62–82%

Deprotonation of Tosylhydrazones. The deprotonation of tosylhydrazones with NaHMDS provides the corresponding sodium salts very efficiently. The formation of allylic alcohols from sugar hydrazones was accomplished when NaHMDS was combined with LiAlH$_4$ (eq 27).[53]

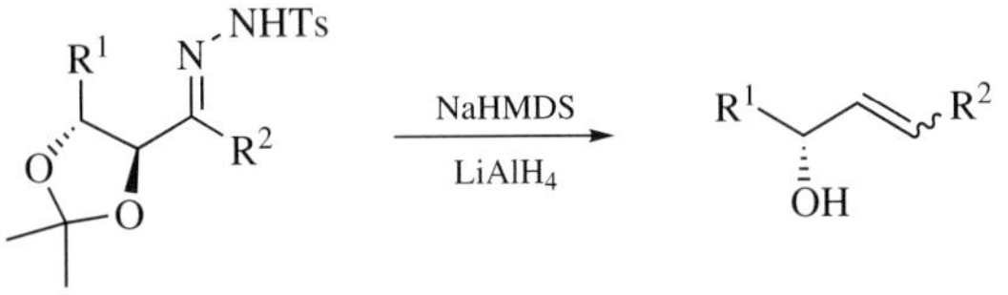

$$\text{(27)}$$

Tosylhydrazone salts could also be decomposed into the corresponding diazo compounds in the presence of a phase transfer catalyst. The addition of late transition metal complexes leads to the formation of metal carbenoid species which undergo various reactions such as cyclopropanation, aziridination, epoxidation, and C-H insertion.[54,55] While both cyclopropanation and aziridination work equally well in the presence of the sodium and lithium tosylhydrazone salts, it was established that the sodium salt provided higher yields and selectivities in the reaction with aldehydes which led to the formation of epoxides (eq 28).[56,57]

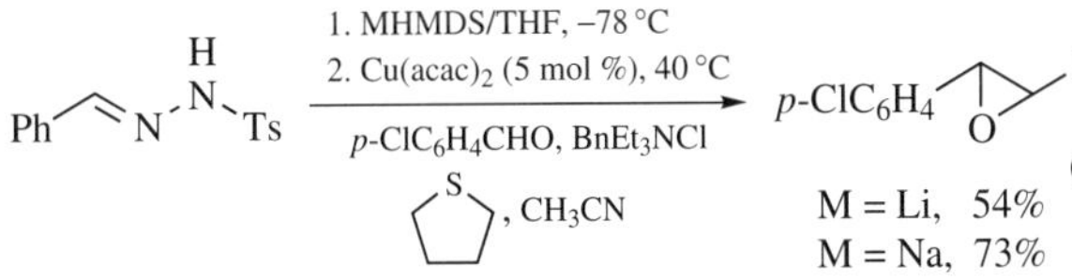

$$\text{(28)}$$

M = Li, 54%
M = Na, 73%

Most tosylhydrazone sodium salts could be usually isolated from the reaction of the corresponding tosylhydrazones with sodium methoxide. However, a number of more functionalized hydrazone sodium salts, such as those derived from alkenyl aryl sulfonylhydrazones, ketone sulfonylhydrazones, and trimethylsilylacrolein sulfonylhydrazones, could not be isolated and an in situ salt generation procedure using NaHMDS was developed. A mixture of the desired hydrazones and NaHMDS was stirred at −78 °C then concentrated down before the addition of rhodium acetate, tetrahydrothiophene, BnEt$_3$NCl, and the aldehyde to produce the required epoxide in high yields (eq 29).

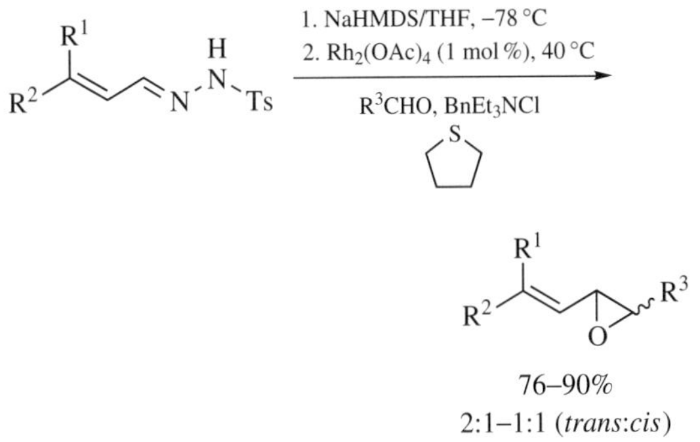

$$\text{(29)}$$

76–90%
2:1–1:1 (trans:cis)

Generation of Phosphorus Ylides and Phosphonate Anions. NaHMDS is the most utilized base for the deprotonation of a variety of phosphonium salts to generate the corresponding ylides, which then undergo Wittig reaction with a carbonyl compound. More recently, it was shown that such a base is compatible with a variety of other systems. For instance, it was shown that allenes and dienes could be prepared, respectively, from aromatic and alicyclic aldehydes when reacted with $(Me_2N)_3P=CH_2$ in the presence of 4 equiv of NaHMDS and titanium trichloride isopropoxide (eqs 30 and 31).[58,59]

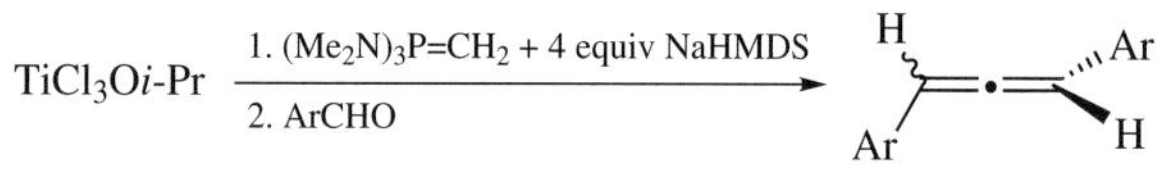

$$TiCl_3Oi\text{-}Pr \xrightarrow[\text{2. ArCHO}]{\text{1. }(Me_2N)_3P=CH_2 + 4 \text{ equiv NaHMDS}} \quad (30)$$

40–69%

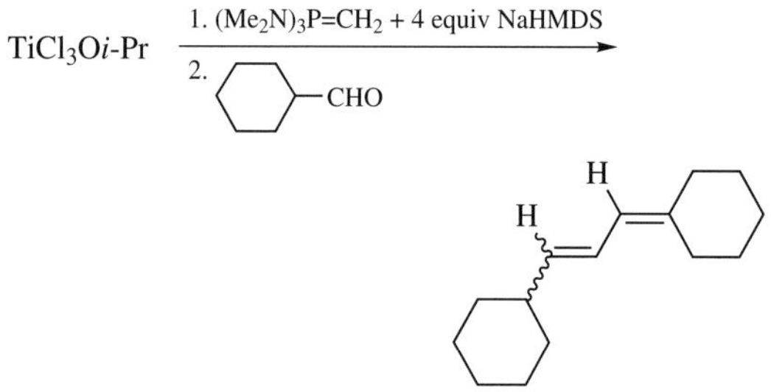

$$TiCl_3Oi\text{-}Pr \xrightarrow[\text{2.}]{\text{1. }(Me_2N)_3P=CH_2 + 4 \text{ equiv NaHMDS}} \quad (31)$$

85%
$E:Z = 7.9{:}1$

This strategy has also been applied to the one-pot double deoxygenation of simple alkyl- and polyether-tethered aromatic dialdehydes to give macrocyclic allenes in high yield without the need for slow-addition techniques.[60]

NaHMDS was also used for the generation of chloroallyl phosphonamide anion, which could further be added to oximes to yield a variety of cis-disubstituted N-alkoxy aziridines in enantiomerically pure form (eq 32).[61] Oxidative cleavage of the chiral auxiliary followed by derivatization of products led to the formation of enantiopure N-alkoxy aziridines.

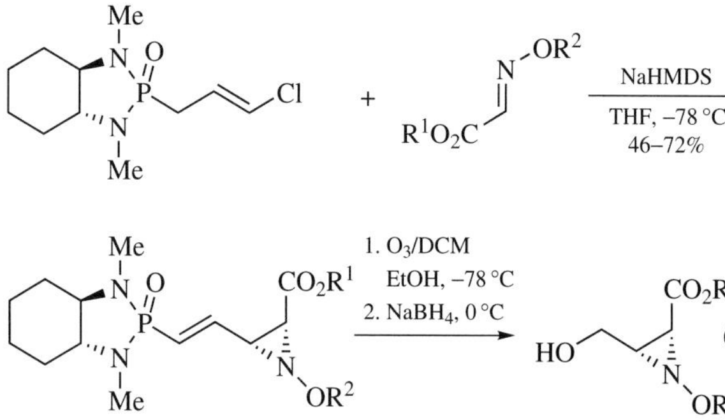

$$ \text{(32)} $$

74–94%

The preparation of Z-α,β-unsaturated amides by using Horner–Wadsworth–Emmons reagents, (diphenylphosphono)acetamides, has been recently reported.[62] High $Z:E$ ratios were observed with aromatic aldehydes and potassium t-butoxide, whereas the best Z-selectivities for aliphatic aldehydes were obtained when NaHMDS was used as base (eq 33).

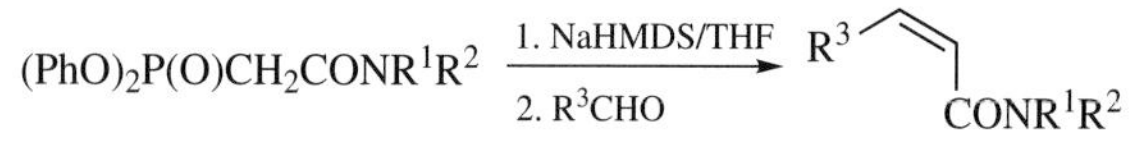

$$(PhO)_2P(O)CH_2CONR^1R^2 \xrightarrow[\text{2. }R^3CHO]{\text{1. NaHMDS/THF}} \quad (33)$$

90–94%
$Z:E = 85{:}15 \text{ to } 94{:}6$

The electrophilic fluorination of benzylic phosphonates has been reported and the optimal reaction conditions involved the use of 2 equiv of NaHMDS and 2.5 equiv of N-fluorobenzenesulfonimide.[63]

New semi-stabilized ylides such as $PhCH=P(MeNCH_2CH_2)_3N$ have been prepared from $RCH_2P(MeNCH_2CH_2)_3N$ employing various bases: NaHMDS, KHMDS, LDA, t-BuOK.[64] The reaction of such ylides with aldehydes provided alkenes in high yield with quantitative E-selectivity (eq 34). Although the E-selectivity is maintained despite the change in the metal ion of the ionic base, NaHMDS furnished usually the highest selectivities.

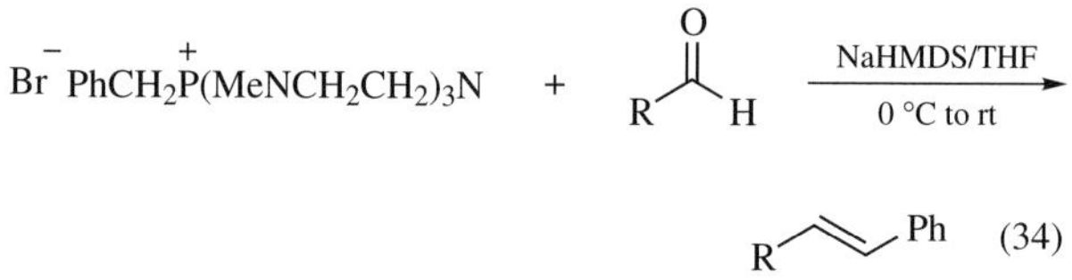

$$\bar{B}r \; PhCH_2\overset{+}{P}(MeNCH_2CH_2)_3N \;+\; \underset{R}{\overset{O}{\|}}{}_H \xrightarrow[\text{0 °C to rt}]{\text{NaHMDS/THF}} \quad (34)$$

86–94%
>99:1, $(E{:}Z)$

Generation of Sulfur Ylides: Julia Olefination and Related Reactions. Deprotonation at a position α to a sulfone group is effected using a strong lithium base, typically n-butyllithium, LDA, or LiHMDS, to give an anion that then reacts with an aldehyde or ketone leading to the corresponding β-hydroxy sulfone. There are, however, few examples that used NaHMDS as the base. For instance, in the synthesis of curacin A, condensation of the requisite sulfone and aldehyde in the presence of NaHMDS led to the corresponding hydroxy sulfone, which was further elaborated to the desired diene (eq 35).[65]

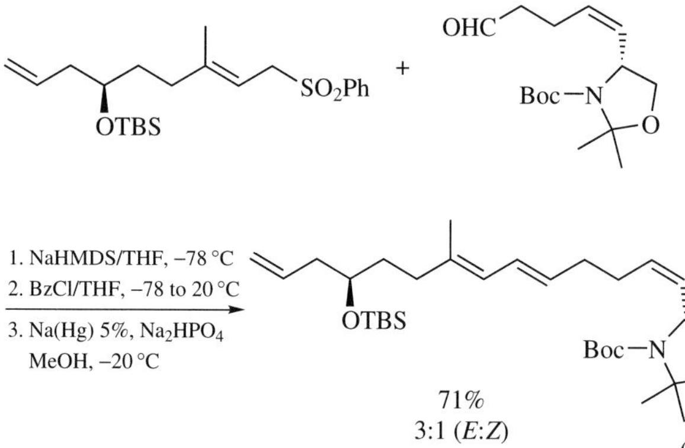

1. NaHMDS/THF, –78 °C
2. BzCl/THF, –78 to 20 °C
3. Na(Hg) 5%, Na_2HPO_4
   MeOH, –20 °C

71%
3:1 $(E{:}Z)$
$$ \text{(35)} $$

The β-hydroxy imidazolyl sulfone derivatives were readily prepared from imidazolyl sulfones and aldehydes in the presence of NaHMDS (eq 36).[66]

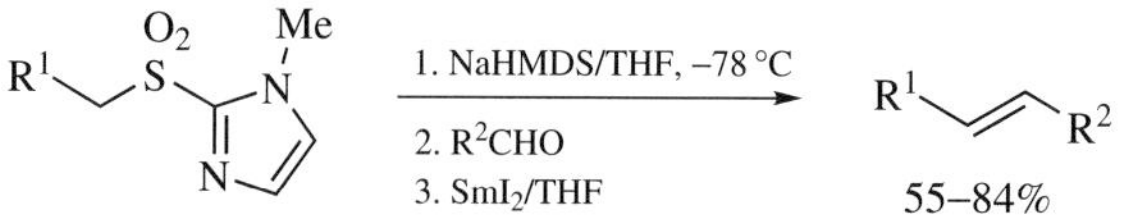

$$R^1 \xrightarrow[\text{2. }R^2CHO]{\substack{\text{1. NaHMDS/THF, –78 °C} \\ \text{3. SmI}_2\text{/THF}}} R^1 \diagdown R^2 \quad (36)$$

55–84%

The reductive elimination reaction of the β-hydroxy imidazolyl sulfone derivatives with sodium amalgam or samarium diiodide provided mainly the desired E-alkenes in good yields.

More recently, the 'modified' Julia olefination, which employed certain heteroarylsulfones instead of the traditional phenylsulfones, has emerged as a powerful tool for alkene synthesis.[67] Although the reaction was first reported with LDA, bases such as LHMDS, NaHMDS, and KHMDS are now commonly used.[68] In addition, solvent as well as base counterion have been shown to markedly affect the stereochemical outcome of the olefination reaction. For instance, it was shown that the *E*-isomer was major when benzothiazoylsulfone (**1**) and cyclopropane carboxaldehyde (**2**) were reacted with NaHMDS in DMF, whereas the *Z*-isomer was obtained in the presence of NaHMDS in either toluene or dichloromethane (eq 37).[69]

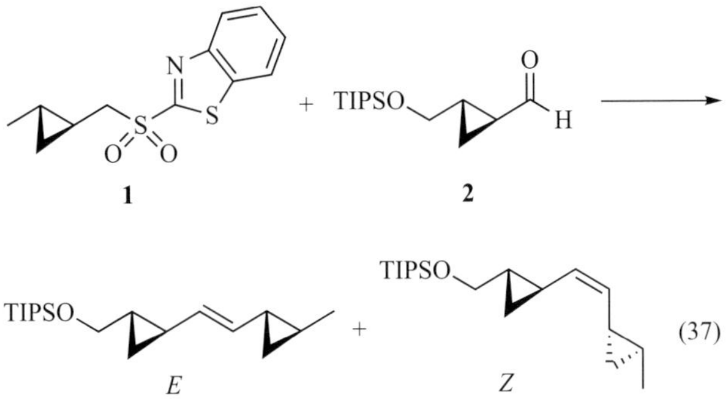

Conditions	*E:Z*
NaHMDS, DMF	3.5:1
NaHMDS, Toluene	1:10
NaHMDS, CH$_2$Cl$_2$	1:10

The former reaction conditions were used for the synthesis of *E*-polycyclopropane alkene (**3**), a precursor to the natural product U-106305 (eq 38).

A similar strategy was used for the final coupling in a synthesis of okadaic acid, which involves an olefination reaction (eq 39).[70]

More recently, the same reaction conditions have been used for the preparation of another cyclopropyl alkene fragment in the total synthesis of ambruticin S (eq 40).[71]

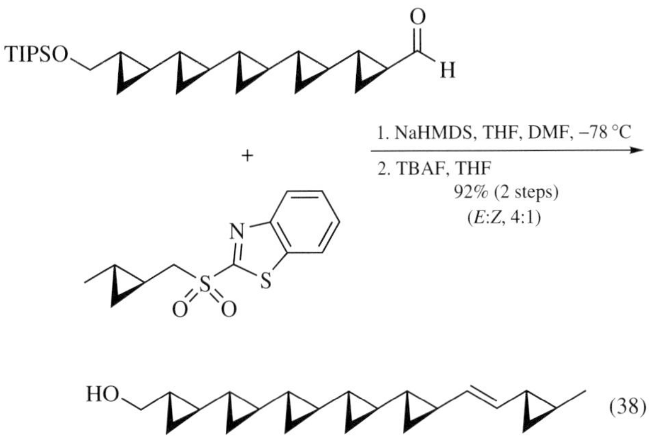

The synthesis of fluoroalkene derivatives has recently been disclosed using 2-(1-fluoroethyl)sulfonyl-1,3-benzothiazole in the presence of various aldehydes and ketones.[72,73] Either NaHMDS or *t*-BuOK could be used as a base.

1-Phenyl-1*H*-tetrazol-5-yl sulfones were introduced as another alternative to phenyl sulfones, which led to alkenes in a one-pot procedure from aldehydes. Usually the highest selectivities are obtained while using KHMDS, although higher yields are observed with sodium hexamethyldisilazide.[74] For instance, alkene **4** was obtained in 100% yield and an 84:16 *E:Z* ratio with NaHMDS, whereas the use of KHMDS led to the same product with 59% yield and a 99:1 *E:Z* ratio (eq 41).

More recently, this strategy has been used to construct the double bond of strictifolione.[75] The use of NaHMDS led to formation of a mixture (4:1) of isomeric *E*- and *Z*-alkenes in 34% yield (eq 42).

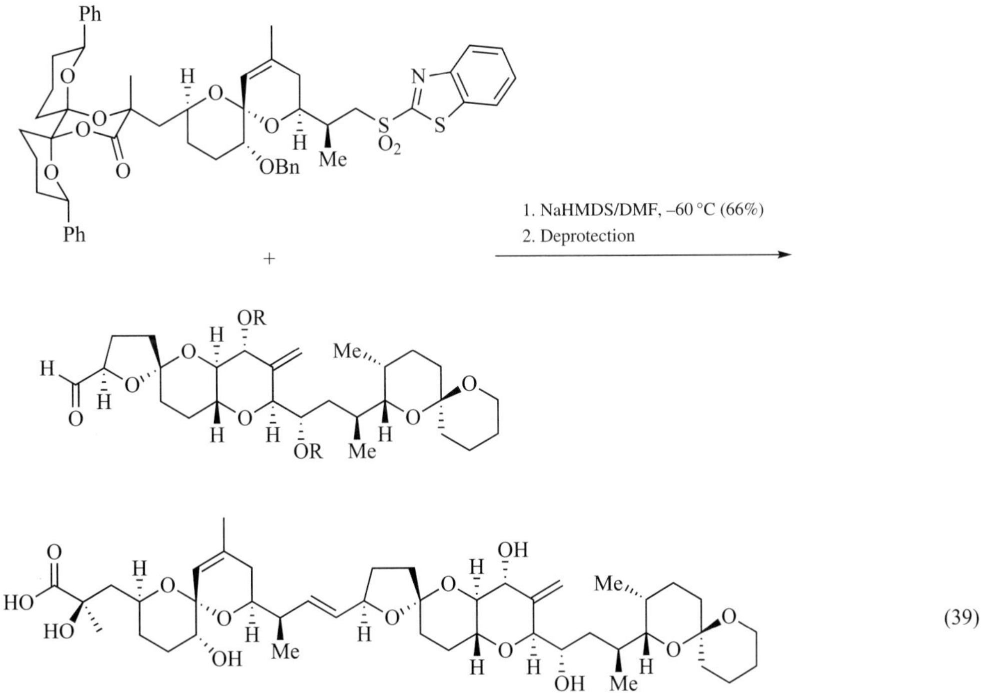

It was shown that E,Z-dienes could be synthesized from α,β-unsaturated aldehydes and 2-pyridyl sulfones with good yields and selectivities.[76,77] NaHMDS and KHMDS were equally effective in promoting this condensation (eq 43).

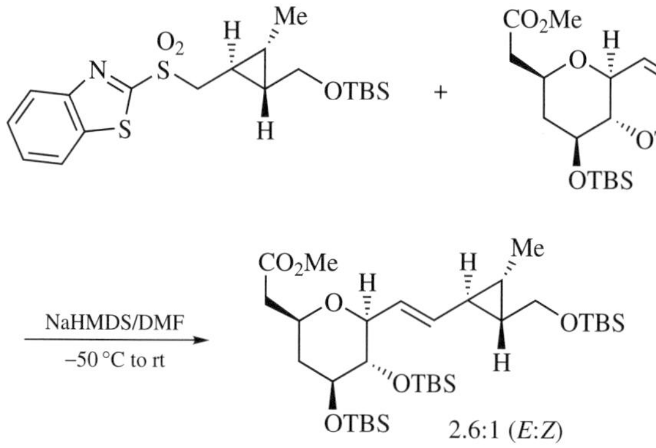

Finally, sulfonium ylides could be generated from the corresponding salt and NaHMDS. When reacted with imines, sulfonium ylides furnished the corresponding aziridines in high yields and selectivities (eq 44).[78]

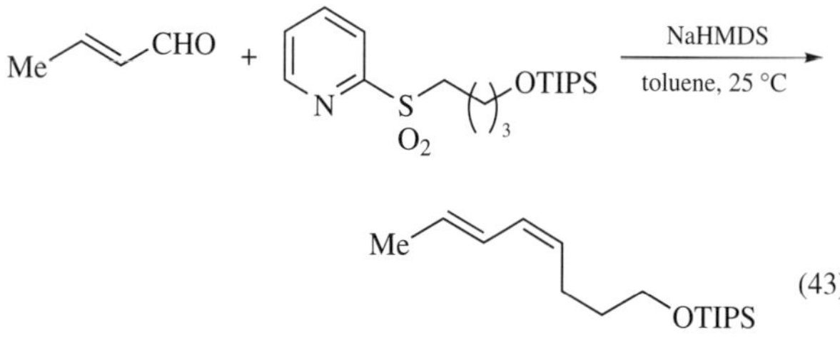

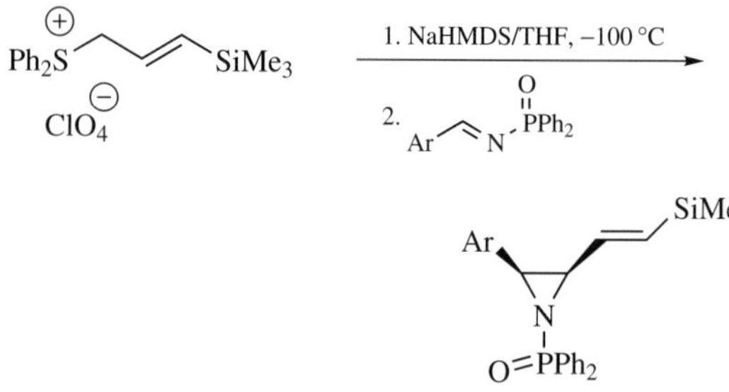

(40)

2.6:1 $(E:Z)$

(44)

72–94%
85:15 to 99:1 $(cis:trans)$

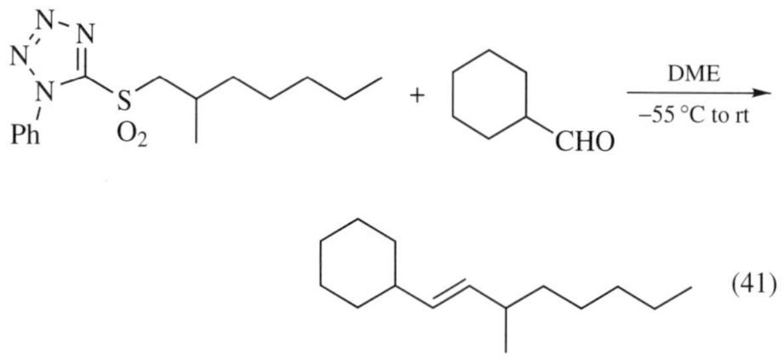

(41)

4

NaHMDS: 100%; 84:16 $(E:Z)$
KHMDS: 59%; 99:1 $(E:Z)$

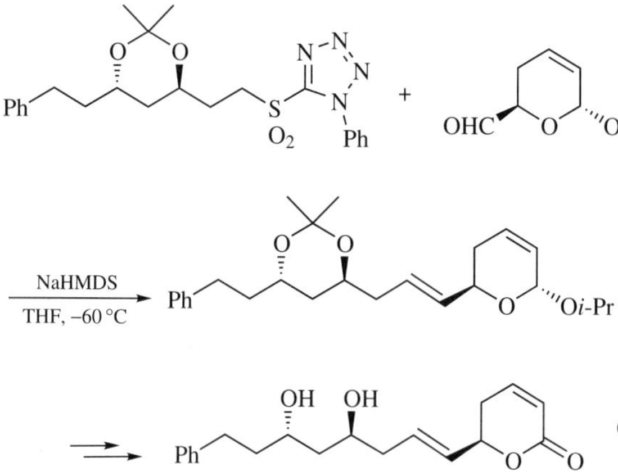

Generation of Bismuthonium Ylides. The use of NaHMDS as base to deprotonate bismuthonium salts has been reported. Higher yields for the subsequent condensation were usually obtained when using NaHMDS compared to potassium t-butoxide (eq 45).[79]

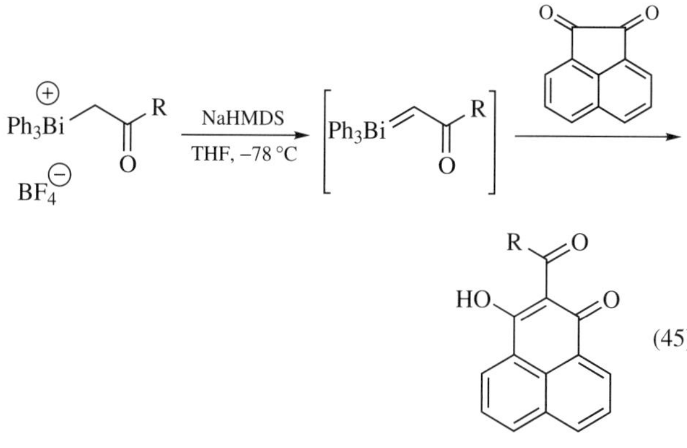

(45)

35–67%

(42)

Strictifolione

Related Reagents. Lithium hexamethyldisilazide; potassium hexamethyldisilazide.

(43)

54%, 91:9 $(E,Z:E,E)$

1. Wannagat, U.; Niederpruem, H., *Chem. Ber.* **1961**, *94*, 1540.
2. Duhamel, L.; Plaquevent, J. C., *J. Organomet. Chem.* **1993**, *448*, 1.
3. Barletta, G.; Chung, A. C.; Rios, C. B.; Jordan, F.; Schlegel, J. M., *J. Am. Chem. Soc.* **1990**, *112*, 8144.
4. (a) Capozzi, G.; Gori, L.; Menichetti, S., *Tetrahedron Lett.* **1990**, *31*, 6213. (b) Capozzi, G.; Gori, L.; Menichetti, S.; Nativi, C., *J. Chem. Soc., Perkin Trans. 1* **1992**, 1923.
5. (a) Evans, D. A., In *Asymmetric Synthesis*; Morrison, J. D., Ed.; Academic: New York, 1984; Vol 3, p 1. (b) Tanabe, M.; Crowe, D. F., *J. Chem. Soc., Chem. Common.* **1969**, 1498. (c) Barton, D. H. R.; Hesse, R. H.; Pechet, M. M.; Wiltshire, C., *J. Chem. Soc., Chem. Common.* **1972**, 1017. (d) Krüger, C. R.; Rochow, E., *J. Organomet. Chem.* **1964**, *1*, 476. (e) Krüger, C. R.; Rochow, E. G., *Angew. Chem., Int. Ed. Engl.* **1963**, *2*, 617. (f) Gaudemar, M.; Bellassoued, M., *Tetrahedron Lett.* **1989**, *30*, 2779.

6. Bestmann, H. J.; Stransky, W.; Vostrowsky, O., *Chem. Ber.* **1976**, *109*, 1694.

7. Stork, G.; Depezay, J. C.; d'Angelo, J., *Tetrahedron Lett.* **1975**, 389.

8. Martel, B.; Hiriart, J. M., *Synthesis* **1972**, 201.

9. (a) Bestmann, H. J.; Woelfel, G., *Chem. Ber.* **1984**, *117*, 1250. (b) Anteunis, M. J. O.; Callens, R. De Witte, M.; Reyniers, M. F.; Spiessens, L., *Bull. Soc. Chim. Belg.* **1987**, *96*, 545.

10. Bestmann, H. J.; Woelfel, G.; Mederer, K., *Synthesis* **1987**, 848.

11. Barton, D. H. R.; Hesse, R. H.; Tarzia, G.; Pechet, M. M., *Chem. Commun.* **1969**, 1497.

12. (a) Evans, D. A.; Ennis, M. D.; Mathre, D. J., *J. Am. Chem. Soc.* **1982**, *104*, 1737. (b) Evans, D. A.; Morrissey, M. M.; Dorow, R. L., *J. Am. Chem. Soc.* **1985**, *107*, 4346. (c) Davis, F. A.; Haque, M. S. Przeslawski, R. M., *J. Org. Chem.* **1989**, *54*, 2021.

13. (a) Schmidt, U.; Riedl, B., *J. Chem. Soc., Chem. Commun.* **1992**, 1186. (b) Glazer, E. A.; Koss, D. A.; Olson, J. A.; Ricketts, A. P.; Schaaf, T. K.; Wiscount, R. J., Jr., *J. Med. Chem.* **1992**, *35*, 1839.

14. Krueger, C.; Rochow, E. G.; Wannagat, U., *Chem. Ber.* **1963**, *96*, 2132.

15. (a) Schlosser, M.; Christmann, K. F., *Justus Liebigs Ann. Chem.* **1967**, *708*, 1. (b) Schlosser, M., *Top. Stereochem.* **1970**, *5*, 1. (c) Schlosser, M.; Schaub, B.; de Oliveira-Neto, J.; Jeganathan, S., *Chimia* **1986**, *40*, 244. (d) Schaub, B.; Jeganathan, S.; Schlosser, M., *Chimia* **1986**, *40*, 246.

16. (a) Corey, E. J.; Su, W., *Tetrahedron Lett.* **1990**, *31*, 3833. (b) Niwa, H.; Inagaki, H.; Yamada, K., *Tetrahedron Lett.* **1991**, *32*, 5127. (c) Chattopadhyay, A.; Mamdapur, V. R., *Synth. Commun.* **1990**, *20*, 2225. (d) Mueller, S.; Schmidt, R. R., *Helv. Chim. Acta* **1993**, *76*, 616.

17. (a) Miyano, S.; Izumi, Y.; Fuji, K.; Ohno, Y.; Hashimoto, H., *Bull. Chem. Soc. Jpn.* **1979**, *52*, 1197. (b) Smithers, R. H., *J. Org. Chem.* **1978**, *43*, 2833.

18. Stork, G.; Zhao, K., *Tetrahedron Lett.* **1989**, *30*, 2173.

19. Bestmann, H. J.; Schmidt, M., *Angew. Chem., Int. Ed. Engl.* **1987**, *26*, 79.

20. Gibson, A. W.; Humphrey, G. R.; Kennedy, D. J.; Wright, S. H. B., *Synthesis* **1991**, 414.

21. (a) Stork, G.; Maldonado, L., *J. Am. Chem. Soc.* **1971**, *93*, 5286. (b) Stork, G.; Maldonado, L., *J. Am. Chem. Soc.* **1974**, *96*, 5272.

22. (a) Takahashi, T.; Nagashima, T. Tsuji, J., *Tetrahedron Lett.* **1981**, 1359. (b) Takahashi, T.; Nemoto, H.; Tsuji, J., *Tetrahedron Lett.* **1983**, 2005.

23. (a) Martel, B.; Aly, E., *J. Organomet. Chem.* **1971**, *29*, 61. (b) Martel, B.; Hiriart, J. M., *Tetrahedron Lett.* **1971**, 2737. (c) Martel, B.; Hiriart, J. M., *Synthesis* **1972**, 201. (d) Martel, B.; Hiriart, J. M., *Angew. Chem., Int. Ed. Engl.* **1972**, *11*, 326.

24. Arora, S.; Binger, P., *Synthesis* **1974**, 801.

25. Charreau, P.; Julia, M.; Verpeaux, J. N., *Bull. Soc. Chem. Fr. Part 2* **1990**, *127*, 275.

26. Han, Z. X.; Krishnamurthy, D.; Grover, P.; Fang, Q. K.; Senanayake, C. H., *J. Am. Chem. Soc.* **2002**, *124*, 7880.

27. Kelly, T. A.; McNeil, D. W., *Tetrahedron Lett.* **1994**, *35*, 9003.

28. Wang, J. J.; Rosingana, M.; Discordia, R. P.; Soundararajan, N.; Polniaszek, R., *Synlett* **2001**, 1485.

29. Culkin, D. A.; Hartwig, J. F., *Acc. Chem. Res.* **2003**, *36*, 234.

30. Miura, M.; Nomura, M., *Top. Curr. Chem.* **2002**, *219*, 211.

31. Lloyd-Jones, G. C., *Angew. Chem. Int. Ed.* **2002**, *41*, 953.

32. Lee, S.; Beare, N. A.; Hartwig, J. F., *J. Am. Chem. Soc.* **2001**, *123*, 8410.

33. Jorgensen, M.; Lee, S.; Liu, X. X.; Wolkowski, J. P.; Hartwig, J. F., *J. Am. Chem. Soc.* **2002**, *124*, 12557.

34. Gaertzen, O.; Buchwald, S. L., *J. Org. Chem.* **2002**, *67*, 465.

35. Beare, N. A.; Hartwig, J. F., *J. Org. Chem.* **2002**, *67*, 541.

36. Spielvogel, D. J.; Buchwald, S. L., *J. Am. Chem. Soc.* **2002**, *124*, 3500.

37. Fox, J. M.; Huang, X. H.; Chieffi, A.; Buchwald, S. L., *J. Am. Chem. Soc.* **2000**, *122*, 1360.

38. Culkin, D. A.; Hartwig, J. F., *J. Am. Chem. Soc.* **2002**, *124*, 9330.

39. Kazmaier, U., *Curr. Org. Chem.* **2003**, *7*, 317.

40. Trost, B. M.; Schroeder, G. M.; Kristensen, J., *Angew. Chem. Int. Ed.* **2002**, *41*, 3492.

41. Grandjean, D.; Pale, P.; Chuche, J., *Tetrahedron Lett.* **1994**, *35*, 3529.

42. Corey, E. J.; Fuchs, P. L., *Tetrahedron Lett.* **1972**, 3769.

43. Binger, P.; Wedemann, P.; Brinker, U. H., *Org. Synth.* **2000**, *77*, 254.

44. Hwu, J. R.; Wong, F. F.; Huang, J. J.; Tsay, S. C., *J. Org. Chem.* **1997**, *62*, 4097.

45. Hwu, J. R.; Hsu, C. H.; Wong, F. F.; Chung, C. S.; Hakimelahi, G. H., *Synthesis* **1998**, 329.

46. Nomura, L.; Mukai, C., *Org. Lett.* **2002**, *4*, 4301.

47. Houghton, T. J.; Choi, S.; Rawal, V. H., *Org. Lett.* **2001**, *3*, 3615.

48. Harada, T.; Fujiwara, T.; Iwazaki, K.; Oku, A., *Org. Lett.* **2000**, *2*, 1855.

49. Chen, B. C.; Murphy, C. K.; Kumar, A.; Reddy, R. T.; Clark, C.; Zhou, P.; Lewis, B. M.; Gala, D.; Mergelsberg, I.; Scherer, D.; Buckley, J.; DiBenedetto, D.; Davis, F. A., *Org. Synth.* **1996**, *73*, 159.

50. Pogatchnik, D. M.; Wiemer, D. F., *Tetrahedron Lett.* **1997**, *38*, 3495.

51. Skropeta, D.; Schmidt, R. R., *Tetrahedron: Asymmetry* **2003**, *14*, 265.

52. Ha, D. C.; Kang, S.; Chung, C. M.; Lim, H. K., *Tetrahedron Lett.* **1998**, *39*, 7541.

53. Chandrasekhar, S.; Mohapatra, S.; Takhi, M., *Synlett* **1996**, 759.

54. Aggarwal, V. K.; Alonso, E.; Fang, G. Y.; Ferrar, M.; Hynd, G.; Porcelloni, M., *Angew. Chem. Int. Ed.* **2001**, *40*, 1433.

55. Aggarwal, V. K.; de Vicente, J.; Bonnert, R. V., *Org. Lett.* **2001**, *3*, 2785.

56. Aggarwal, V. K.; Alonso, E.; Fang, G. Y.; Ferrar, M.; Hynd, G.; Porcelloni, M., *Angew. Chem. Int. Ed.* **2001**, *40*, 1430.

57. Aggarwal, V. K.; Alonso, E.; Bae, I.; Hynd, G.; Lydon, K. M.; Palmer, M. J.; Patel, M.; Porcelloni, M.; Richardson, J.; Stenson, R. A.; Studley, J. R.; Vasse, J.-L., Winn, C. L., *J. Am. Chem. Soc.* **2003**, *125*, 10926.

58. Reynolds, K. A.; Dopico, P. G.; Sundermann, M. J.; Hughes, K. A.; Finn, M. G., *J. Org. Chem.* **1993**, *58*, 1298.

59. Reynolds, K. A.; Finn, M. G., *J. Org. Chem.* **1997**, *62*, 2574.

60. Brody, M. S.; Williams, R. M.; Finn, M. G., *J. Am. Chem. Soc.* **1997**, *119*, 3429.

61. Hanessian, S.; Cantin, L. D., *Tetrahedron Lett.* **2000**, *41*, 787.

62. Ando, K., *Synlett* **2001**, 1272.

63. Taylor, S. D.; Dinaut, A. N.; Thadani, A. N.; Huang, Z., *Tetrahedron Lett.* **1996**, *37*, 8089.

64. Wang, Z. G.; Zhang, G. T.; Guzei, I.; Verkade, J. G., *J. Org. Chem.* **2001**, *66*, 3521.

65. Aube, J.; Hoemann, M. Z.; Agrios, K. A.; Aube, J., *Tetrahedron* **1997**, *53*, 11087.

66. Kende, A. S.; Mendoza, J. S., *Tetrahedron Lett.* **1990**, *31*, 7105.

67. Blakemore, P. R., *J. Chem. Soc., Perkin Trans. 1* **2002**, 2563.

68. Baudin, J. B.; Hareau, G.; Julia, S. A.; Lorne, R.; Ruel, O., *Bull. Soc. Chim. Fr.* **1993**, *130*, 856.

69. Charette, A. B.; Lebel, H., *J. Am. Chem. Soc.* **1996**, *118*, 10327.

70. Ley, S. V.; Humphries, A. C.; Eick, H.; Downham, R.; Ross, A. R.; Boyce, R. J.; Pavey, J. B. J.; Pietruszka, J., *J. Chem. Soc., Perkin Trans. 1* **1998**, 3907.

71. Kirkland, T. A.; Colucci, J.; Geraci, L. S.; Marx, M. A.; Schneider, M.; Kaelin, D. E.; Martin, S. F., *J. Am. Chem. Soc.* **2001**, *123*, 12432.

72. EP Pat. 1209154 A (to Pazenok, S.; Demoute, J.-P.; Zard, S.; Lequeux, T.) (2002).

73. Chevrie, D.; Lequeux, T.; Demoute, J. P.; Pazenok, S., *Tetrahedron Lett.* **2003**, *44*, 8127.

74. Blakemore, P. R.; Cole, W. J.; Kocienski, P. J.; Morley, A., *Synlett* **1998**, 26.

75. Juliawaty, L. D.; Watanabe, Y.; Kitajima, M.; Achmad, S. A.; Takayama, H.; Aimi, N., *Tetrahedron Lett.* **2002**, *43*, 8657.

76. Charette, A. B.; Berthelette, C.; St-Martin, D., *Tetrahedron Lett.* **2001**, *42*, 5149.

77. Erratum: Charette, A. B.; Berthelette, C.; St-Martin, D., *Tetrahedron Lett.* **2001**, *42*, 6619.

78. Yang, X. F.; Mang, M. J.; Hou, X. L.; Dai, L. X., *J. Org. Chem.* **2002**, *67*, 8097.

79. Rahman, M. M.; Matano, Y.; Suzuki, H., *Synthesis* **1999**, 395.

T

Tetrabutylammonium Difluorotriphenylsilicate

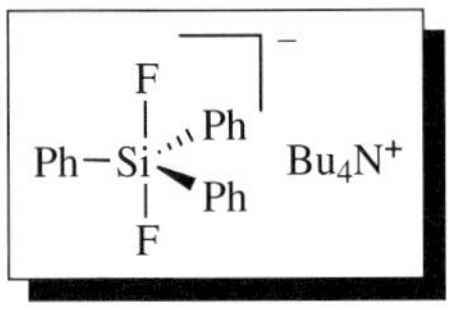

[163931-61-1] $C_{34}H_{51}F_2NSi$ (MW 539.88)

InChI = 1S/C18H15F2Si.C16H36N/c19-21(20,16-10-4-1-5-11-
       16,17-12-6-2-7-13-17)18-14-8-3-9-15-18;1-5-9-13-
       17(14-10-6-2,15-11-7-3)16-12-8-4/h1-15H;5-16H2,
       1-4H3/q-1;+1

InChIKey = RQBKGJOQACIQDG-UHFFFAOYSA-N

(non-hygroscopic, organic-soluble reagent used as a nucleophilic
fluoride source[1])

Alternate Name: TBAT.
Physical Data: mp 155–156 °C (EtOAc).
Solubility: soluble in most organic solvents.
Form Supplied in: colorless solid.
Preparative Methods: prepared by treatment of triphenylsilyl fluoride with tetrabutylammonium fluoride (TBAF),[1] or by reaction of triphenylsilane with tetrabutylammonium hydrogen difluoride.[2]
Purification: recrystallized from EtOAc or CH$_2$Cl$_2$/hexanes.
Handling, Storage, and Precautions: reagent is extremely sensitive to acid.

Original Commentary

Nigel S. Simpkins
University of Nottingham, Nottingham, UK

Nucleophilic Fluorination. TBAT was introduced by DeShong and co-workers as a non-hygroscopic substitute for the more widely used Bu$_4$NF (TBAF).[3] The initial report described the use of TBAT as a fluoride source for nucleophilic substitution reactions of a range of alkyl halides and sulfonates, and provided some comparisons with TBAF, which gave generally inferior results. The use of excess TBAT with sulfonate leaving groups gave some of the best results, for example, the conversion of triflate **1** into fluoride **2** with clean inversion of stereochemistry (eq 1).

In the reaction of the mesylate corresponding to **1**, the product fluoride was obtained as a mixture of diastereomers due to some epimerization. TBAT has also been employed in reactions of 6-chloropurine nucleosides, to give the corresponding fluorinated

derivatives,[4] and in a two-step ring expansion–nucleophilic addition sequence of ketones, which led in one case to a lactam with a fluoroethyl group (eq 2).[5]

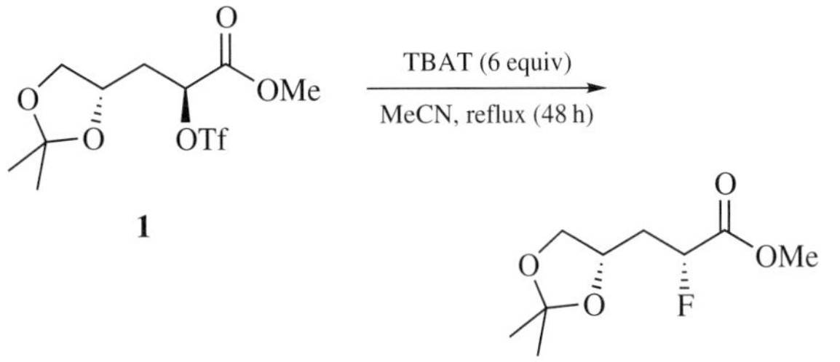

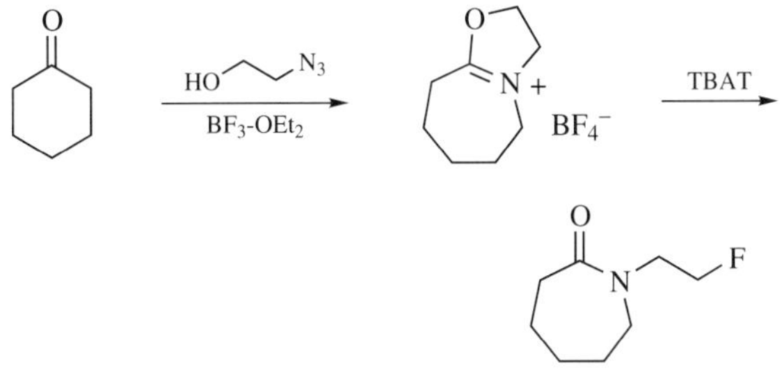

Allylation Reactions. In 1996, Pilcher and DeShong described the use of TBAT for the in situ generation of carbanions from various functionalized silane precursors, especially allylsilane (eqs 3 and 4).[6]

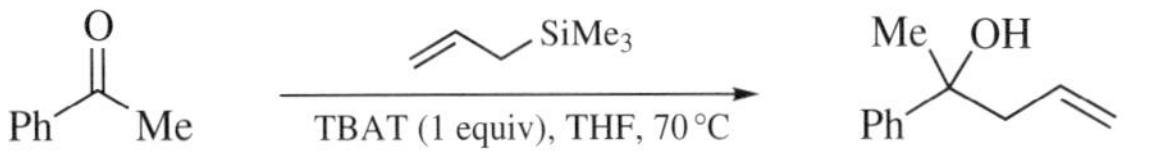

The fluoride mediated allylation of imines using allyltrimethylsilane is especially noteworthy since it was unprecedented until this report. A range of related reactions were described including alkylation and addition reactions of benzylic and acetylenic silanes and trimethylsilyldithiane. Under the conditions employed, phenyltrimethylsilane and vinyltrimethylsilane were unreactive as nucleophilic components, and epoxides did not participate as electrophiles. TBAT has been shown to be an effective initiator for an intramolecular allylsilane addition involving a dihydropyridone acceptor.[7] Asymmetric variants of allylations that employ TBAT have been described, involving either the use of chiral *N*-acylhydrazones as auxiliaries[8] or catalytic amounts of a *p*-Tol-BINAP–CuCl=TBAT combination.[9]

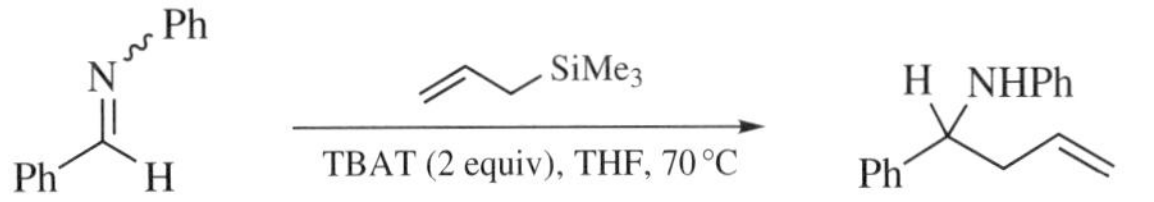

Trifluoromethylation.

The Olah research group has published several reports that describe the use of Me$_3$SiCF$_3$ in combination with appropriate fluoride sources, including TBAT, for trifluoromethylation of sulfonyl or sulfinyl imines.[10–12] The latter type of process can provide access to chiral trifluoromethylated amines of various types by use of non-racemic sulfinyl imines as starting materials, e.g., eq 5.[11]

Addition to imines of type **3** was found to be very highly diastereoselective and subsequent hydrolysis of intermediate **4** gave the final chiral amine salts of type **5**. Somewhat related trifluoromethylation of carbonyl compounds has been demonstrated using a combination of HCF_3 and $(Me_3Si)_3N$ and catalytic TBAT,[13] although in this case the fluoride serves to generate a base from the silylamine. This type of system was also used to trifluoromethylate disulfides.[13]

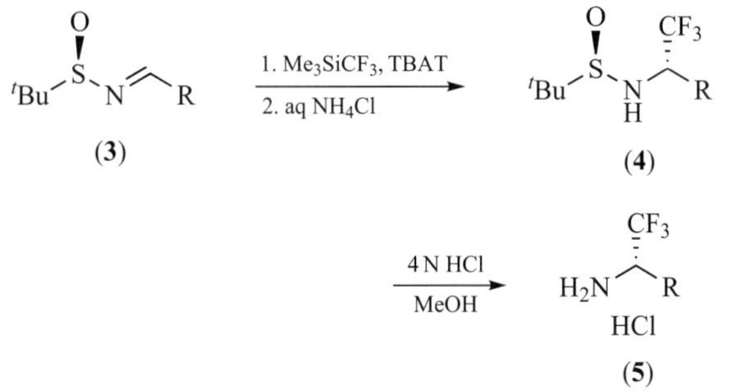

Palladium Mediated Couplings.   TBAT is an effective phenylating reagent for palladium catalyzed reactions involving cross-coupling with arenes, e.g., aryl iodides and triflates, and also with allylic alcohol derivatives via π-allyl complexes (eq 6 and eq 7).[14–16]

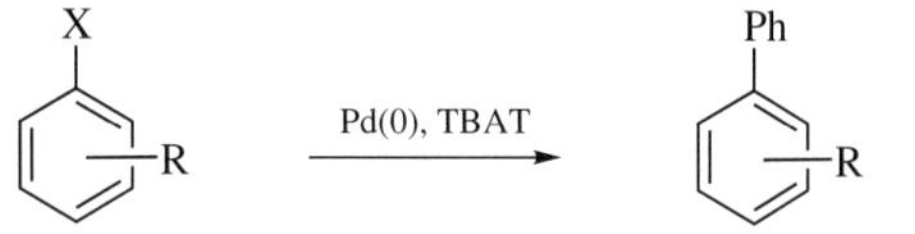

In the biaryl synthesis, the use of 2 equiv of TBAT proved optimal in terms of chemical yield and minimizing homocoupling of the arene partner.[14] Phenylation of allylic benzoates via π-allyl complexes, using either stannyl or silyl derivatives as reaction partners, proceeds by overall inversion, presumably by transmetallation and aryl transfer from palladium to carbon.[15] In a range of reactions, TBAT was shown to give generally superior results to $PhSnMe_3$ in terms of yields and levels of regiocontrol.[16] In studies of the regio- and enantiocontrol possible in allylic alkylations under palladium catalysis, TBAT was found to have a beneficial effect by promoting the π–σ–π interconversion process in the intermediate organopalladium species.[17,18]

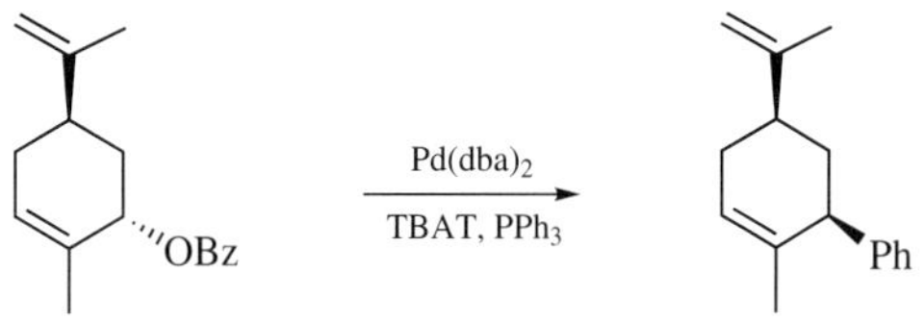

Mukaiyama Aldol and Related Processes.   The Carreira group has developed an asymmetric catalytic aldol reaction that involves addition of a silyl dienolate to an aldehyde partner in the presence of a chiral catalyst generated in situ from (S)-Tol-BINAP, $Cu(OTf)_2$, and TBAT, e.g., eq 8.[19,20]

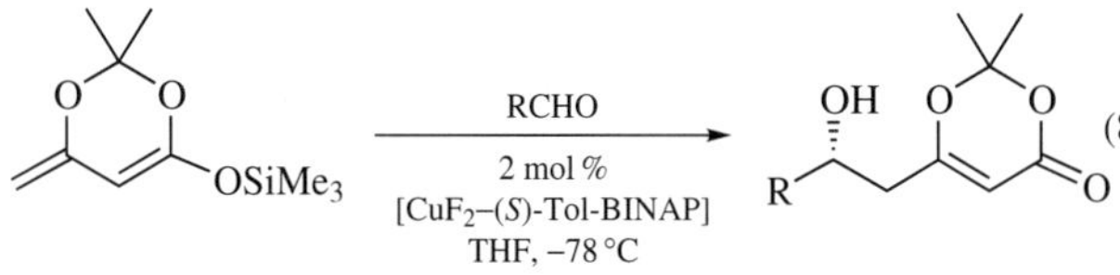

This process, which is thought to involve a metalloenolate intermediate, delivers the addition products in very good yield and enantioselectivities, and has been applied in the total synthesis of Leucascandrolide A.[21] TBAT has also been employed in regiocontrolled alkylations of enol silanes derived from cyclopentenones,[22] in bridgehead substitution of a ketone having a trimethylsilyl group at the bridgehead,[23] and in initiating homoenolate reactivity starting with a β-silylester.[24]

Miscellaneous.   An asymmetric Henry reaction involving combination of an aldehyde with a silyl nitronate, catalyzed by $Cu(OTf)_2$ in the presence of TBAT and a chiral BOX ligand, gives the nitroaldol products in moderate yield and stereoselectivity.[25] TBAT is effective in triggering desilylation of certain functionalized silyloximes, resulting in concomitant cyclization and nitrone formation (eq 9).[26]

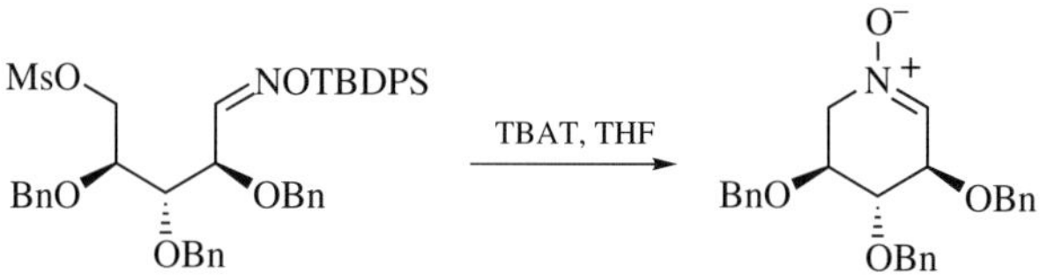

A range of densely functionalized five- and six-membered nitrones were formed in this way, and the method is proposed as suitable for the synthesis of the alkaloid laccarin. Under the influence of nucleophilic fluoride sources, including TBAT, a carbamoyl silane undergoes addition to ketones and aldehydes to give α-silyloxyamide products.[27] TBAT can be used in combination with Me_3SiN_3 to effect glycosyl azide formation,[28] and finally it can be used to effect self-condensation of some unsaturated carbonyl compounds, a process in which fluoride presumably acts as a base.[29]

First Update

Frederick E. Nytko, III & Philip DeShong
University of Maryland, College Park, MD, USA

Alkylation and Allylation.   The use of TBAT as a fluoride source for the in situ formation of pentacoordinate silicates from silanes of alkyl and allyl substituents, generating electron-rich carbon centers capable of acting as nucleophiles toward electron deficient moieties, has been well documented for over a decade. Recently, N-acylhydrazones have been shown to undergo highly diastereoselective allylations when treated with allylsilanes, TBAT, and indium(III) trifluoromethanesulfonate (eq 10).[30]

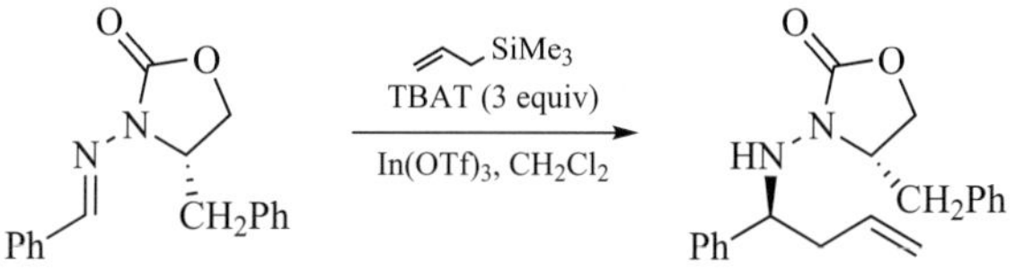

(58% yield, 94:6 dr)

High diastereoselectivity in the allylated product is controlled through chelation of the allylfluorosilicate to the Lewis acid associated with the chiral N-acylhydrazone, causing Si-face addition. Cleavage of the N–N bond produced homoallylic amines.

It has also been shown that bridgehead alkylsilanes of bridged ketones can undergo *ipso*-substitution when treated with TBAT in the presence of electrophiles at room temperature (eq 11).[31]

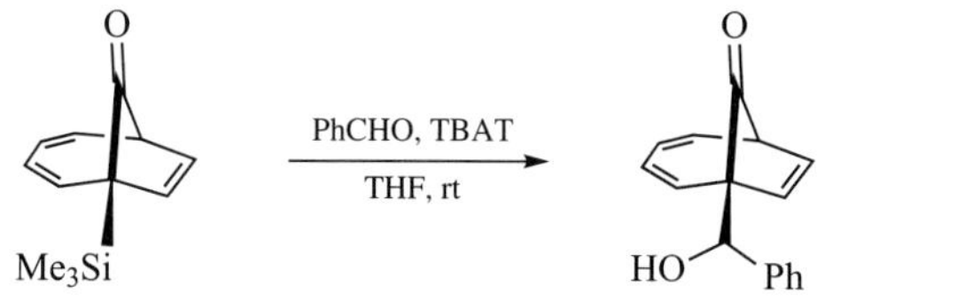

(72% yield) (11)

Similarly, in the presence of TBAT, tethered silanes have been shown to undergo fluoride-catalyzed, intramolecular cyclizations. *N*-Propargylsilane-substituted dihydropyridones provided the corresponding 1-vinylidene indolizidines as a single isomer under mild conditions (eq 12).[32] Attempts to cyclize the propargylic silanes with fluoride sources other than TBAT, or with Lewis acids, failed to provide product.

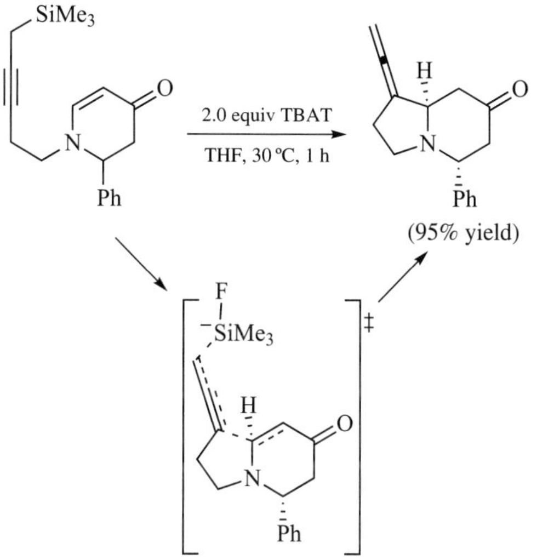

(95% yield) (12)

Palladium-Catalyzed Allylation and Cross-Coupling. The benefit of using TBAT over other sources of fluoride in asymmetric allylic alkylation is twofold; it not only provides fluoride to regenerate the enolate from the protected silyl enol ether but also the affinity of TBAT for the palladium of π-allyl (eq 13)[33] or π-aryl (eq 14)[34] complexes expedites π–σ–π equilibration, providing greater regio- and enantioselectivity of coupled products compared to cesium fluoride.[35]

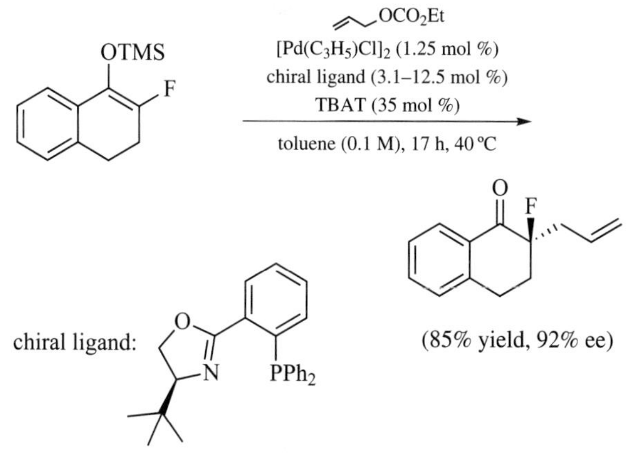

(85% yield, 92% ee) (13)

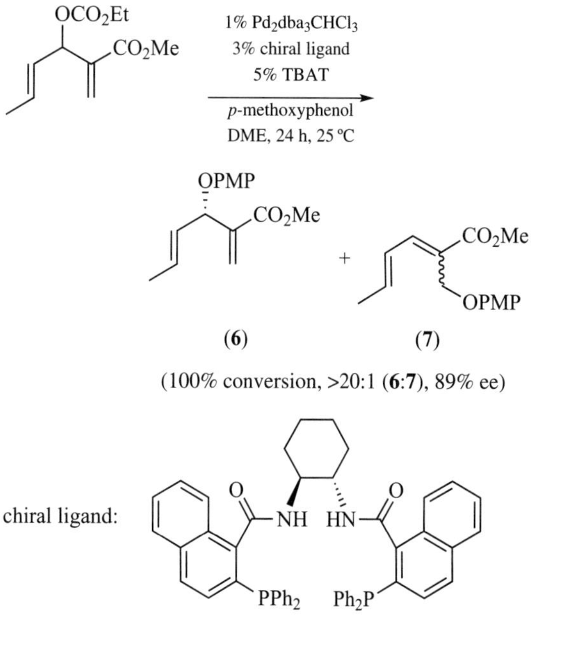

(100% conversion, >20:1 (**6:7**), 89% ee)

chiral ligand:

(*S*,*S*)-**LN**

Aryne Annulation. Treatment of *o*-(trimethylsilyl)aryl triflates with TBAT results in the formation of benzyne intermediates.[36] Fluoride-induced 1,2-eliminations of *o*-(trimethylsilyl)aryl triflates, which are readily prepared from phenols, have been shown to provide the respective aryne products, which, at room temperature, precipitously undergo cycloaddition in the presence of Diels–Alder adducts, including, but not limited to, tethered acyclic dienes, tethered enynes (eq 15),[36] azomethine imines (eq 16),[37] ene carbamates, and acetamidoacrylates (eq 17).[38]

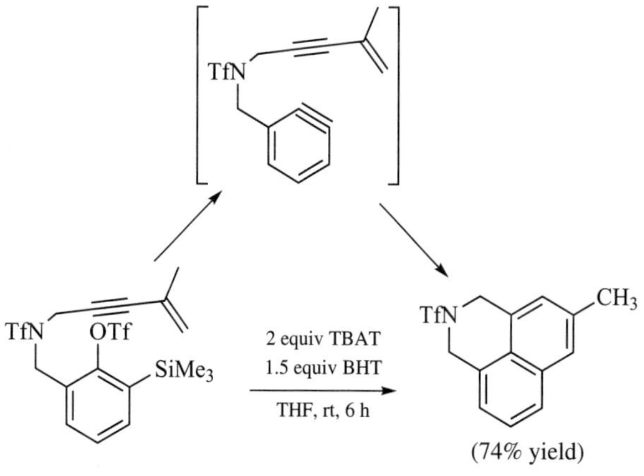

(74% yield) (15)

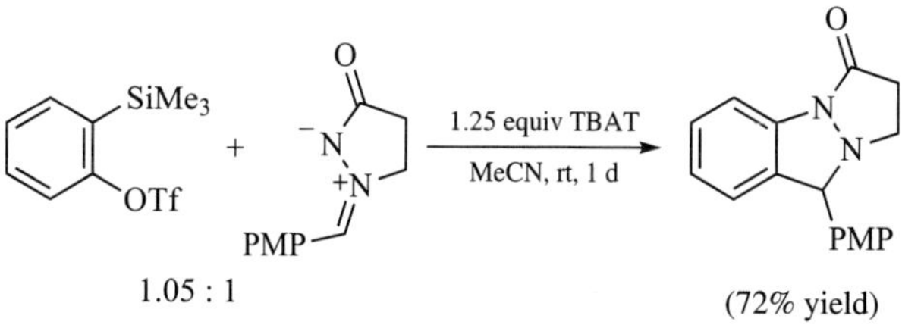

(72% yield) (16)

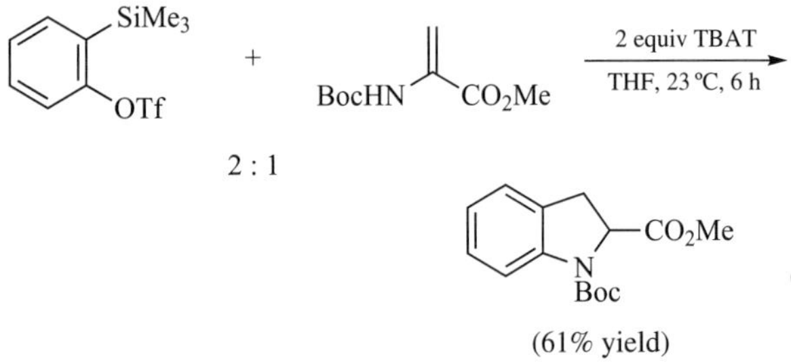

$$(17)$$

(61% yield)

For the tethered enyne system, the use of TBAT at low concentrations provides 74% yield of the fused cycloadduct product, outperforming both TBAF (24–32% yield) and CsF (43% yield).[36]

Fluoroalkylation.   The nucleophilic addition of fluorinated alkyl groups to electron deficient centers is a reaction of great interest to biological and medicinal chemists, given the high degree of lipophilicity imparted on the final product.[39] The addition of (trifluoromethyl)trimethylsilane across aldimines under treatment with TBAT was far superior (92% yield) to other fluorinating agents, including TMAF (72% yield) and CsF (25% yield) (eq 18).[40]

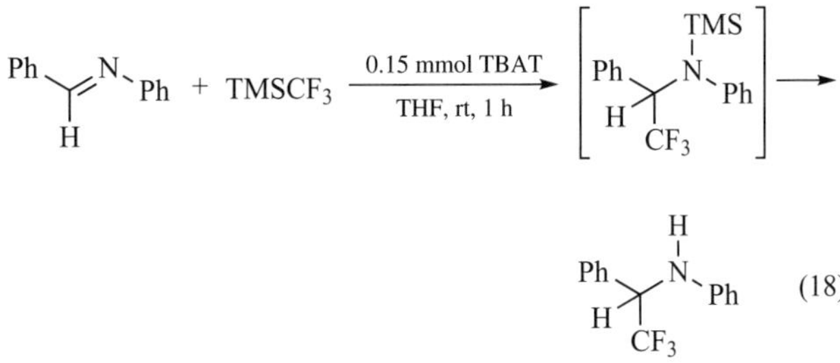

$$(18)$$

Depending on the relative amount of TBAT used, both tri- and difluoromethylated amines were formed. The latter product hypothetically arose from the association of excess TMS with the lone pair of the amine, followed by abstraction of TMS by excess fluoride, subsequently ejecting HF (see section on desilylative defluorination).

Similarly, through the use of [difluoro(phenylthio)methyl]trimethylsilane[41,42] or [difluoro(phenylseleno)methyl]trimethylsilane[43] treated with TBAT, in the presence of amides, imines, or aldehydes, the respective difluoromethylene phenylsulfide or difluoromethylene phenylselenide is produced (eq 19).

In the presence of tributyl tin reagents and the radical initiator AIBN, both sulfides and selenides undergo tin hydride-mediated radical cyclizations, providing gem-difluoromethylene products.

Cyanation.   The addition of nitrile functionalities to molecules is a viable route for the introduction of carbonyl derivatives. An example of this approach can be seen in the Lewis base-catalyzed cyanomethylation of benzaldehyde with (trimethylsilyl)propionitrile.[44] The Lewis base catalyst activates the Si–C bond of (trimethylsilyl)propionitrile, allowing the subsequent attack at the target aldehyde center, providing the substituted benzylic alcohol in good to excellent yields. However, upon treatment of the same starting materials with both Lewis base catalyst and TBAT, overall yields declined (eq 20).

It has also been shown that in the presence of 3-methylcyclohexenone, treatment of diethylaluminum cyanide with TBAT provides the cyanide conjugate addition product.[45] The enolate can be captured when the reaction is run in the presence of triflic anhydride, providing the triflate.

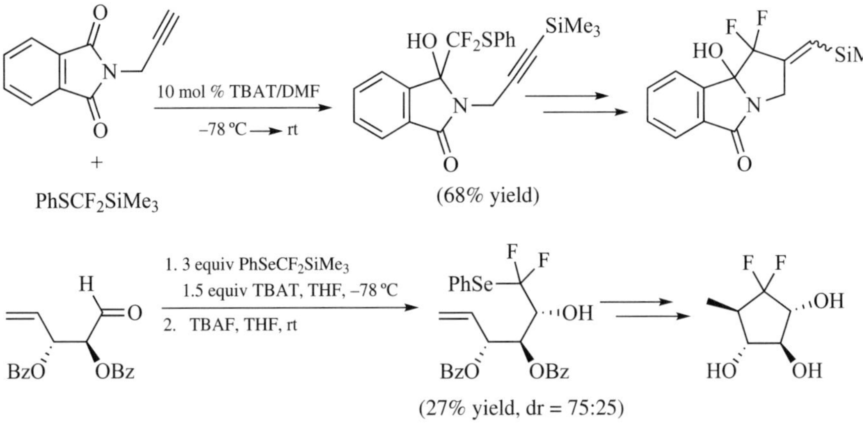

(68% yield)

$$(19)$$

(27% yield, dr = 75:25)

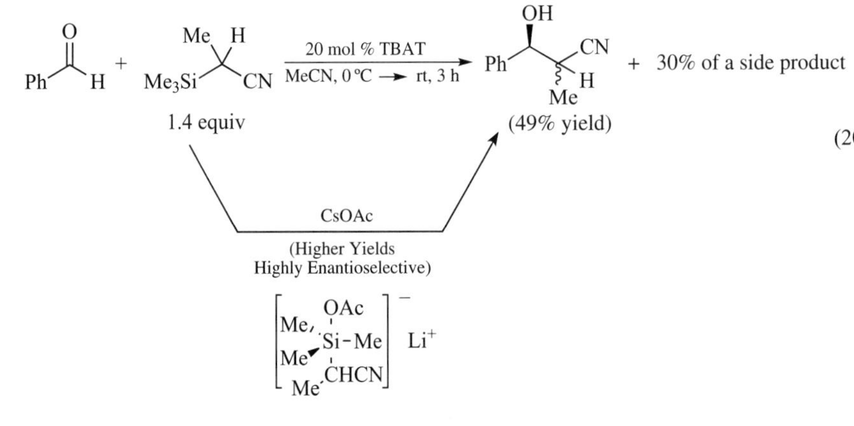

$$(20)$$

Mukaiyama Reaction of Aldehydes. An optically pure variant of Carreira's catalyst, [CuF(S)-Tol-BINAP], can be formed in situ from Cu(OTf)$_2$, chiral Tol-BINAP ligand, and TBAT. Reactions of silyl dienolates with aldehydes, in the presence of this catalyst, provide α,β-unsaturated lactones, under good regio-, diastereo-, and enantiocontrol, via the asymmetric vinylogous Mukaiyama reaction (eq 21).[46] By varying reaction conditions such as copper source, ligand, and dienolate structure, the overall product yield and the regio-, diastereo-, and enantiocontrol displayed in the products were greatly varied.

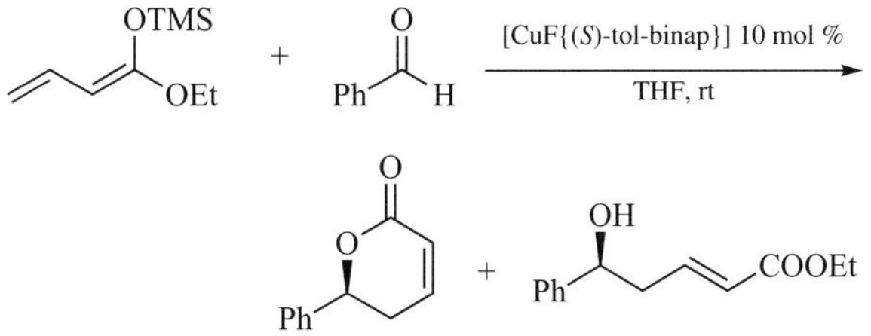

(up to 85% yield, 90:10 (**8:9**), 87% ee)

The availability for further derivatization at the conjugated double bond of the lactone proves an interesting approach to several complex natural products.

Desilylative Defluorination. The use of TBAT as a source of fluoride anion for desilylation has been previously discussed. By having the silyl functionality conjugated through a π-system to a trifluoromethyl group, the subsequent desilylation with TBAT will induce defluorination, forming the difluoromethylene moiety, while catalytically liberating fluoride ion to further support the reaction (eq 22).[47]

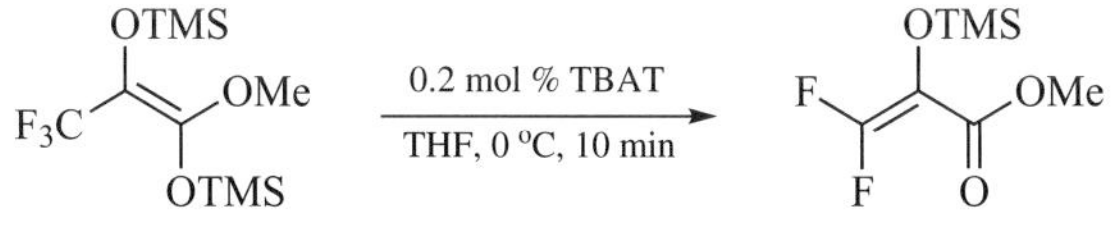

(95% yield)

Addition of different electrophiles to the difluoromethylene site allows access to a variety of β-substituted β,β-difluoropyruvate derivates.

Similarly, the inductively electron-withdrawing nature of a trifluoromethyl group attached to the π-electron system of an arene allows more facile breaking of the C-F bond. Treatment of 1-(3$'$-chlorophenyl)-1-trimethylsilyl-1,2,2,2-tetrafluoroethane with TBAT undergoes desilylative defluorination to provide 3$'$-chloro-1,2,2-trifluorostyrene in excellent yield (eq 23).[48]

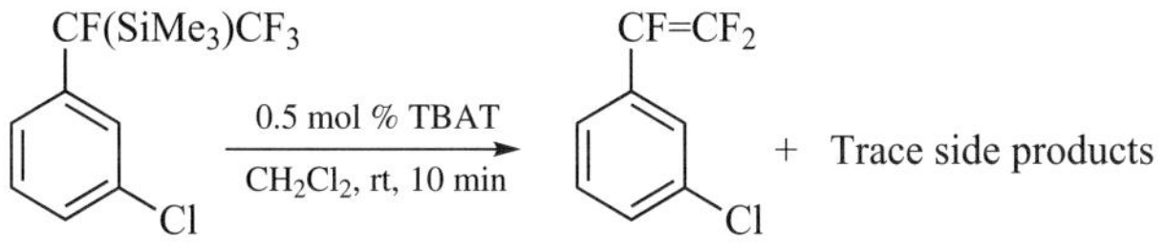

(96% yield)

(23)

Fluoride-Based Desilylation. Treatment of symmetrical O,O-acetals with triethylsilyl trifluoromethanesulfonate in the presence of 2,4,6-collidine yields weakly electrophilic collidinium

salts. The introduction of an azide functionality to the molecule via treatment with trimethylsilyl azide in the presence of TBAT proceeds rapidly and in good yield (eq 24).[49]

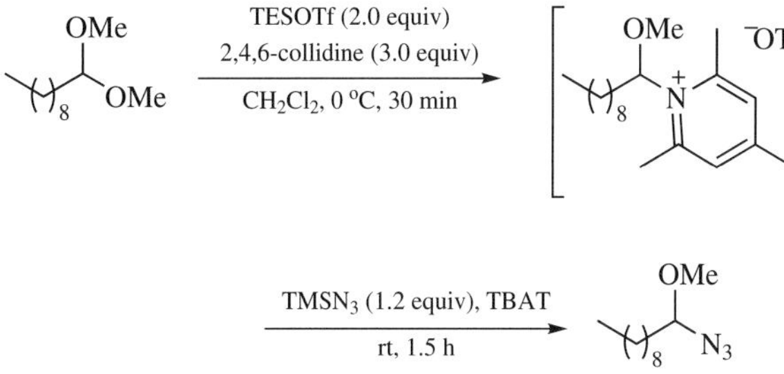

(24)

(77% yield)

In the presence of fluoride anion, electronically poor silyl enol ethers will undergo an "anti-Baldwin" 3-*exo*-dig cyclization to provide vinylidene cyclopropanes. The use of TBAT at room temperature provided product rapidly and in good yield (eq 25).[50]

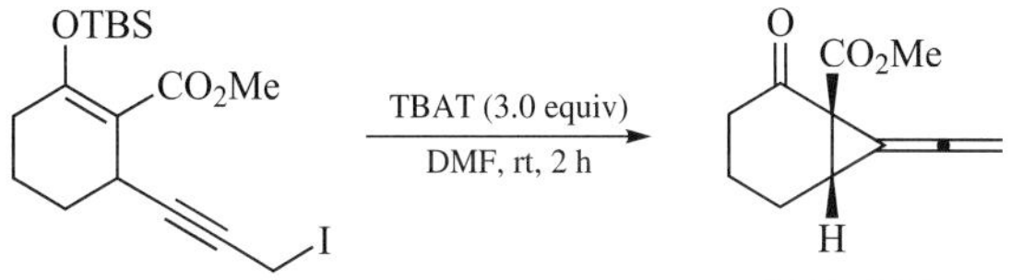

(25)

(82% yield)

Treatment of silylated thiazolium carbinols with a fluoride anion source resulted in the formation of an acyl anion equivalent, which proceeded to acylate various nitroalkenes, affording β-nitro ketones (eq 26).[51]

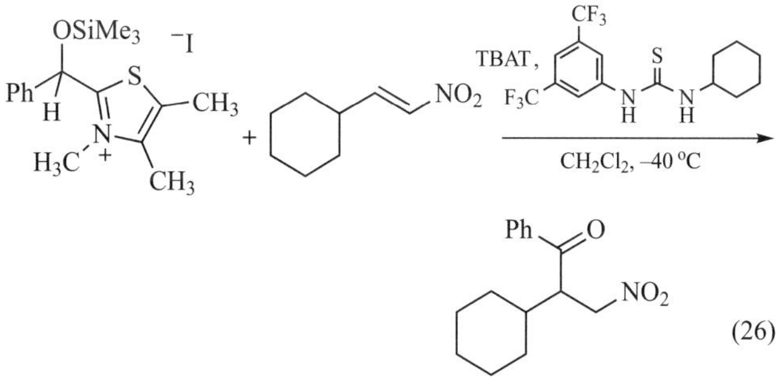

(26)

(66% yield)

The addition of thiourea derivatives provided improved yields, as well as the opportunity to affect the enantioselectivity of newly formed chiral centers through the use of optically pure derivatives.

Miscellaneous. Tetrabutylammonium salts have long been known to possess antibacterial abilities toward both Gram-positive and Gram-negative bacteria, with the hypothesized mode of action being association of the tetrabutylammonium cation with the bacterial cell wall, displacing cellular calcium and magnesium cations. Recently, by undertaking zone of inhibition tests on different TBA salts, activity was determined as a function of counterion.[52] TBAT displayed good activity toward *Staphylococcus epidermidis*, the representative Gram-positive bacteria; however, activity toward the Gram-negative representative, *Escherichia coli*, was poor.

Phase transfer-catalyzed fluorination of alkyl halides and sulfonates is efficiently achieved via treatment with solid potassium fluoride in the presence of triphenyltin fluoride cocatalyst. Similarly, the tetraalkylammonium salts of tin, germanium, and silicon have also been studied for their ability to promote phase transfer-catalyzed fluorination. Notably, TBAT was shown to be not nearly as efficient a cocatalyst under conditions analogous to those used for the tin compounds.[53]

1. Handy, C. J.; Lam, Y.-F.; DeShong, P., *J. Org. Chem.* **2000**, *65*, 3542.

2. Albanese, D.; Landini, D.; Penso, M., *Tetrahedron Lett.* **1995**, *36*, 8865.

3. Pilcher, A. S.; Ammon, H. L.; DeShong, P., *J. Am. Chem. Soc.* **1995**, *117*, 5166.

4. Gurvich, V.; Kim, H.-Y.; Hodge, R. P.; Harris, C. M.; Harris, T. M., *Nucleosides Nucleotides* **1999**, *18*, 2327.

5. Gracias, V.; Milligan, G. L.; Aubé, J., *J. Org. Chem.* **1996**, *61*, 10.

6. Pilcher, A. S.; DeShong, P., *J. Org. Chem.* **1996**, *61*, 6901.

7. Furman, B.; Dziedzic, M., *Tetrahedron Lett.* **2003**, *44*, 6629.

8. Friestad, G. K.; Ding, H., *Angew. Chem., Int. Ed.* **2001**, *40*, 4491.

9. Yamasaki, S.; Fujii, K.; Wada, R.; Kanai, M.; Shibasaki, M., *J. Am. Chem. Soc.* **2002**, *124*, 6536.

10. Prakash, G. K. S.; Mandal, M.; Olah, G. A., *Synlett* **2001**, 77.

11. Prakash, G. K. S.; Mandal, M.; Olah, G. A., *Angew. Chem., Int. Ed.* **2001**, *40*, 589.

12. Prakash, G. K. S.; Mandal, M.; Olah, G. A., *Org. Lett.* **2001**, *3*, 2847.

13. Large, S.; Roques, N.; Langlois, B. R., *J. Org. Chem.* **2000**, *65*, 8848.

14. Mowery, M. E.; DeShong, P., *J. Org. Chem.* **1999**, *64*, 3266.

15. Brescia, M.-R.; DeShong, P., *J. Org. Chem.* **1998**, *63*, 3156.

16. Hoke, M. E.; Brescia, M.-R.; Bogaczyk, S.; DeShong, P.; King, B. W.; Crimmins, M. T., *J. Org. Chem.* **2002**, *67*, 327.

17. Trost, B. M.; Toste, F. D., *J. Am. Chem. Soc.* **1998**, *121*, 4545.

18. Trost, B. M.; Jiang, C., *J. Am. Chem. Soc.* **2001**, *123*, 12907.

19. Kruger, J.; Carreira, E. M., *J. Am. Chem. Soc.* **1998**, *120*, 837.

20. Pagenkopf, B. L.; Kruger, J.; Stojanovic, A.; Carreira, E. M., *Angew. Chem., Int. Ed.* **1998**, *37*, 3124.

21. (a) Fettes, A.; Carreira, E. M., *Angew. Chem., Int. Ed.* **2002**, *41*, 4098. (b) Bluet, G.; Campagne, J.-M., *J. Org. Chem.* **2001**, *66*, 4293.

22. Yun, J.; Buchwald, S. L., *Org. Lett.* **2001**, *3*, 1129.

23. Blake, A. J.; Giblin, G. M. P.; Kirk, D. T.; Simpkins, N. S.; Wilson, C., *Chem. Commun.* **2001**, 2668.

24. DiMauro, E.; Fry, A. J., *Tetrahedron Lett.* **1999**, *40*, 7945.

25. Risgaard, T.; Gothelf, K. V.; Jørgensen, K. A., *Org. Biomol. Chem.* **2003**, *1*, 153.

26. Tamura, O.; Toyao, A.; Ishibashi, H., *Synlett* **2002**, 1344.

27. Cunico, R. F., *Tetrahedron Lett.* **2002**, *43*, 355.

28. Soli, E. D.; DeShong, P., *J. Org. Chem.* **1999**, *64*, 9724.

29. Xuan, J. X.; Fry, A. J., *Tetrahedron Lett.* **2001**, *42*, 3275.

30. Friestad, G. K.; Korapala, C. S.; Ding, H., *J. Org. Chem.* **2006**, *71*, 281.

31. Hayes, C. J.; Simpkins, N. S.; Kirk, D. T.; Mitchell, L.; Baudoux, J.; Blake, A. J.; Wilson, C., *J. Am. Chem. Soc.* **2009**, *131*, 8196.

32. Dziedzic, M.; Lipner, G.; Illangua, J. M.; Furman, B., *Tetrahedron* **2005**, *61*, 8641.

33. Belanger, E.; Cantin, K.; Messe, O.; Tremblay M.; Paquin, J. F., *J. Am. Chem. Soc.* **2007**, *129*, 1034.

34. Trost, B. M.; Brennan, M. K., *Org. Lett.* **2007**, *9*, 3961.

35. Trost, B. M.; Brennan, M. K., *Org. Lett.* **2006**, *8*, 2027.

36. Hayes, M. E.; Shinokubo, H.; Danheiser, R. L., *Org. Lett.* **2005**, *7*, 3917.

37. Shi, F.; Mancuso, R.; Larock, R. C., *Tetrahedron Lett.* **2009**, *50*, 4067.

38. Gilmore, C. D.; Allan, K. M.; Stoltz, B. M., *J. Am. Chem. Soc.* **2008**, *130*, 1558.

39. Roussel, S.; Billard, T.; Langlois, B. R.; Saint-James, L., *Chem. Eur. J.* **2005**, *11*, 939.

40. Prakash, G. K. S.; Mogi, R.; Olah, G. A., *Org. Lett.* **2006**, *8*, 3589.

41. Li, Y.; Hu, J., *Angew. Chem., Int. Ed.* **2007**, *46*, 2489.

42. Bootwicha, T.; Panichakul, D.; Kuhakarn, C.; Prabpai, S.; Kongsaeree, P.; Tuchinda, P.; Reutrakul, V.; Pohmakotr, M., *J. Org. Chem.* **2009**, *74*, 3798.

43. Fourriere, G.; Lalot, J.; Van Hijfte, N.; Quirion, J. C.; Leclerc, E., *Tetrahedron Lett.* **2009**, *50*, 7048.

44. Kawano, Y.; Kaneko, N.; Mukaiyama, T., *Chem. Lett.* **2005**, *34*, 1508.

45. Peese, K. M.; Gin, D. Y., *J. Am. Chem. Soc.* **2006**, *128*, 8734.

46. Bazan-Tejeda, B.; Bluet, G.; Broustal, G.; Campagne, J. M., *Chem.-Eur. J.* **2006**, *12*, 8358.

47. Takikawa, G.; Toma, K.; Uneyama, K., *Tetrahedron Lett.* **2006**, *47*, 6509.

48. Nakamura, Y.; Uneyama, K., *J. Org. Chem.* **2007**, *72*, 5894.

49. Fujioka, H.; Okitsu, T.; Ohnaka, T.; Li, R.; Kubo, O.; Okamoto, K.; Sawama, Y.; Kita, Y., *J. Org. Chem.* **2007**, *72*, 7898.

50. Campbell, M. J.; Pohlhaus, P. D.; Min, G.; Ohmatsu, K.; Johnson, J. S., *J. Am. Chem. Soc.* **2008**, *130*, 9180.

51. Mattson, A. E.; Zuhl, A. M.; Reynolds, T. E.; Scheidt, K. A., *J. Am. Chem. Soc.* **2006**, *128*, 4932.

52. Ingalsbe, M. L.; St. Denis, J. D.; McGahan, M. E.; Steiner, W. W.; Priefer, R., *Bioorg. Med. Chem. Lett.* **2009**, *19*, 4984.

53. Makosza, M.; Bujok, R., *J. Fluorine Chem.* **2005**, *126*, 209.

2,4,6,8-Tetraethenyl-2,4,6,8-tetramethyl-cyclotetrasiloxane

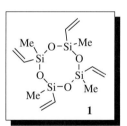

[2554-06-5] $C_{12}H_{24}O_4Si_4$ (MW 344.66)

InChI = 1S/C12H24O4Si4/c1-9-17(5)13-18(6,10-2)15-20(8, 12-4)16-19(7,11-3)14-17/h9-12H,1-4H2,5-8H3

InChIKey = VMAWODUEPLAHOE-UHFFFAOYSA-N

(reagent for the vinylation of aryl halides,[1] building block for carbosilane dendrimers[2])

Physical Data: bp 224–224.5 °C.

Solubility: soluble in most organic solvents.

Form Supplied in: clear oil.

Purification: distillation.

Handling, Storage, and Precautions: 2,4,6,8-Tetraethenyl-2,4, 6,8-tetramethylcyclotetrasiloxane is a stable oil that can be stored in a bottle without precaution for air and moisture exposure.

Introduction. 2,4,6,8-Tetraethenyl-2,4,6,8-tetramethylcyclotetrasiloxane (**1**) (D_4^V) and the trimer, 2,4,6-triethenyl-2,4,6-trimethylcyclotrisiloxane (**2**) (D_3^V) are the two commercially available cyclic polyvinylmethylsiloxanes.[3] D_4^V is used as a vinyl donor in the palladium-catalyzed vinylation of aryl halides and as a building block in polymer chemistry and dendrimer synthesis.

The palladium-catalyzed vinylation of substituted aryl halides using D_4^V is a general and mild method for the preparation of a variety of substituted styrenes or dienes. Each of the four vinyl groups on D_4^V is available for transfer, therefore, only 0.3–0.5 equiv are required for successful vinylation. Tetrabutyl-ammonium fluoride (TBAF) is used to activate D_4^V for in these reactions.

The title compound offers two sites for polymerization, both at the silicon center and on the vinyl group. These structural features make D_4^V attractive for polymer chemistry and are exploited in the preparation of thin films and related materials.[4,5] A related application of D_4^V involves the construction of carbosilane dendrimers by functionalization of the vinyl groups. Each generation of dendrimer is prepared by a two-step iterative process consisting of hydrosilylation followed by addition of allylmagnesium bromide. The silicon unit introduced by hydrosilylation serves as a branch point for further generations.

Whereas the vinyl groups of D_4^V are accessible for functionalization by hydroboration or hydrosilylation, they are inert to functionalization by cross-metathesis.[6] Alternatively, formal metathesis products can be obtained by the ruthenium-catalyzed silylative coupling reaction.[7] This method involves the combination of a vinyl silane and an olefin in the presence of a ruthenium catalyst, to provide an alkenylsilane (see eq 7). The application of this reaction to D_4^V provides substitution at each of the four vinyl groups, resulting in a cyclic tetraalkenyltetramethylcyclotetrasiloxane. The silylative coupling reaction of both D_4^V and D_3^V has been demonstrated with styrenes and enol ethers.[8]

Palladium-catalyzed Vinylation Reactions. Over the last decade, palladium-catalyzed cross-coupling reactions have emerged as the premier strategy for the construction of a bond between two unsaturated carbon centers. A vinyl group is one of the simplest unsaturated carbon donors, and a variety of vinylating reagents have been developed. These reagents include magnesium-,[9] boron-,[10] and tin-based[11] compounds, all of which react effectively with a range of aryl halides. Alternatively, silicon-based donors including vinyltrimethylsilane[12] and vinylsilacyclobutane[13] have been used to vinylate aryl iodides.

In the course of studies on organosilicon-based cross-coupling reactions, Denmark and co-workers found that vinyldisiloxanes and vinylpolysiloxanes are competent organometallic donors.[14] A series of commercially available di- and polyvinylsiloxanes are effective in the palladium-catalyzed cross-coupling reaction with 4'-iodoacetophenone (**5**, Table 1; eq 1).

2   **3**   **4**

Polysiloxanes **1–3** were each competent at effecting the transformation, providing excellent yields in short reaction times. D_4^V

was selected for further development on the basis of cost and efficiency of vinyl transfer (Table 1).[15] The optimized coupling process works well for a range of electron-poor (entries 1, 2, 4, 7, and 8, Table 2) and electron-rich (entries 3, and 6) aryl iodides and tolerates a wide range of functional groups (entries 1-5). The anisyl derivatives (entries 3 and 6) required additional amounts of D_4^V and TBAF to proceed to completion, because of a slow rate of coupling and competitive catalyst decomposition (eq 2).

Table 1 Optimization of the coupling of vinylpolysiloxanes with **5**

Entry	Vinylsiloxan (equiv)	Vinyl (equiv)	TBAF (equiv)	Yield (%)
1	**1** (0.3)	1.2	2.0	88
2	**2** (0.4)	1.2	2.0	85
3	**3** (0.4)	1.2	2.0	89
4	**4** (0.25)	1.5	2.0	51

Table 2 Scope of aryl iodides in vinylation reaction

Entry	Aryl	D_4^V (equiv)	TBAF (equiv)	Time (min)	Yield (%)
1	$(4\text{-MeCO})C_6H_4$	0.3	2.0	10	88
2	$(4\text{-EtO}_2C)C_6H_4$	0.3	2.0	10	85
3	$(4\text{-MeO})C_6H_4$	0.4	3.0	360	63
4	$(3\text{-O}_2N)C_6H_4$	0.3	2.0	10	87
5	$(3\text{-HOCH}_2)C_6H_4$	0.3	2.0	480	59
6	$(2\text{-MeO})C_6H_4$	0.4	3.0	24 h	72
7	$(2\text{-MeO}_2C)C_6H_4$	0.3	2.0	480	83
8	1-Naphthyl	0.3	2.0	180	64

The scope of electrophile used in the vinylation reaction can be extended to aryl bromides by employing 2-(di-t-butylphosphino)-biphenyl[16] as a ligand for palladium and raising the temperature to 50 °C (eq 3).[17] The optimized reaction conditions employ a slightly higher loading of D_4^V in comparison to the reaction with aryl iodides. The additional amount of D_4^V is needed to suppress a secondary Heck reaction that provides the symmetrical stilbene as a byproduct. A broad range of styrene derivatives can be prepared by this method as shown in Table 3.

Table 3 Scope of aryl bromides in vinylation reaction

Entry	Aryl	Time (h)	Yield (%)
1	$(4\text{-MeCO})C_6H_4$	3	91
2	$(4\text{-EtO}_2C)C_6H_4$	5	83
3	$(4\text{-MeO})C_6H_4$	10	86
4	$(4\text{-HOCH}_2)C_6H_4$	14	54
5	$(4\text{-AcNH})C_6H_4$	12	77
6	$(2\text{-EtO}_2C)C_6H_4$	5	86
7	$(2\text{-O}_2N)C_6H_4$	2	85
8	$(2\text{-Et})C_6H_4$	17	75
9	$(2\text{-MeO})C_6H_4$	20	80
10	$(2\text{-Me}_2N)C_6H_4$	24	78
11	1-Naphthyl	3	71
12	2-Naphthyl	3	81
13	$(2,4,6\text{-Me}_3)C_6H_2$	48	72
14	3-Quinolyl	3	89

In general, the reaction times are longer than the vinylation of the corresponding aryl iodides, although the yields are somewhat higher. The reaction shows high functional group compatibility (entries 1–5, 7, 9, and 14) and tolerates severe steric crowding at the site of vinylation (entries 8, 10, and 13). The electronic nature of the aryl bromide significantly influences the reaction time, as electron-poor substrates couple much faster than electron-rich substrates (entries 1 vs. 3, 7 vs. 10), but the yields are not significantly affected. Alkenyl bromides also serve as substrates under these conditions as demonstrated in the synthesis of a prenyl α-amino acid (eq 4).[18]

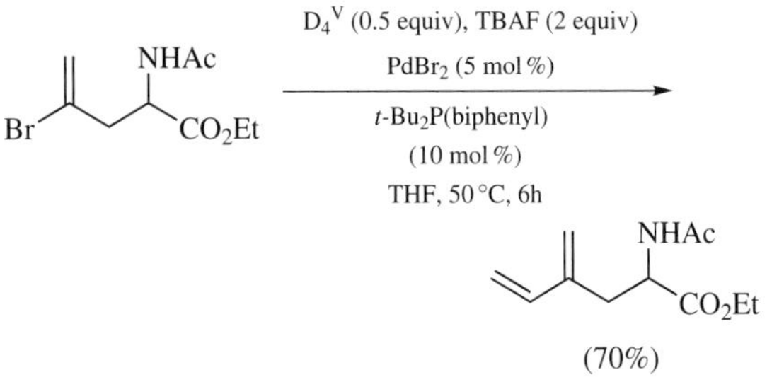

(4)

Carbosilane Dendrimer Synthesis. $D_4{}^V$ is also used as a building block for the preparation of carbosilane dendrimers.[19] The preparation of these dendrimers follows the method of van der Made, which involves alternating cycles of platinum-catalyzed hydrosilylation of a terminal alkene with a chlorosilane and alkylation of the resulting chlorosilane with allylmagnesium bromide.[20] The first step in the preparation of the dendrimeric structure involves the hydrosilylation of $D_4{}^V$ with dichloromethylsilane in the presence of platinum on carbon as the catalyst. This process introduces a methyldichlorosubstituent on each of the four vinyl groups to afford **7** in excellent yield (eq 5).

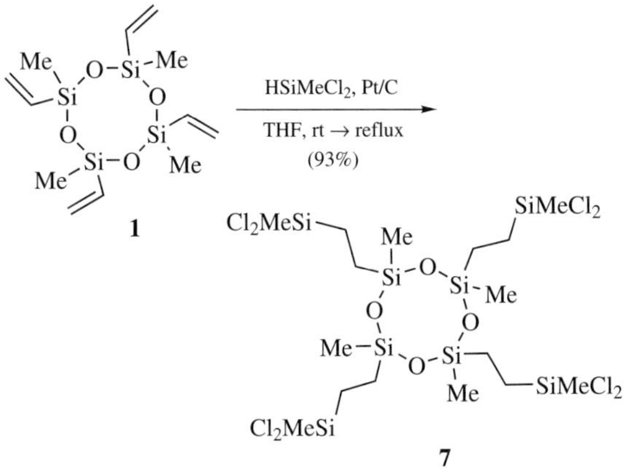

In the next step of the dendrimer synthesis, treatment of the dichloromethylsilyl-capped product with allylmagnesium bromide provides eight terminal alkenes and provides the first-generation product. Subsequent cycles of hydrosilylation and alkylation provide the second- and third-generation dendrimers in similar fashion. The third-generation dendrimers were ultimately capped with phenols, pendant allyl- or alkoxy-[4,21] groups, or phenylethylene units.[22]

(5)

Alternatively, the combination of $D_4{}^V$ and 9-BBN (and subsequent oxidation) provides the 2-hydroxyethyl-substituted cyclotetrasiloxane building block **8** with excellent regioselectivity (eq 6).[23] The terminal hydroxyl groups can be further functionalized by oxidation or silylation in the elaboration to alternative dendrimer structures.

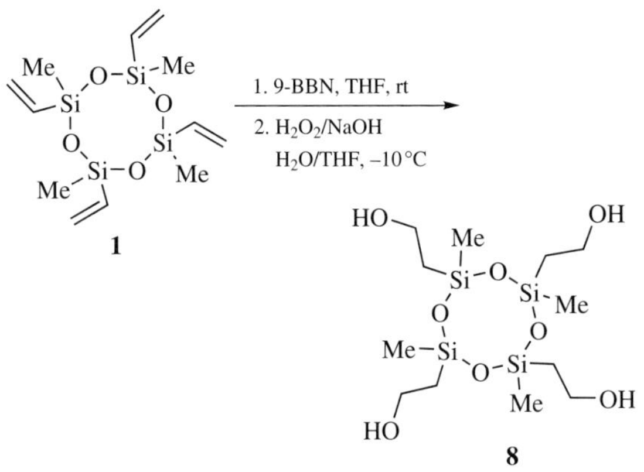

(6)

Ruthenium-catalyzed Silylative Functionalization. Alkenylsilanes generally do not participate in traditional cross-metathesis reactions.[6] Alternatively, the ruthenium-catalyzed silylative coupling reaction of vinylsilanes and olefins provides formal metathesis products (eq 7). Although the products are the same, the reactions proceed through a distinct mechanism, converting vinylsilanes into substituted alkenylsilanes.

(7)

Application of this reaction to cyclic vinylsiloxanes results in β-substitution at each of the vinyl groups on the cyclic siloxanes.[7] Enol ethers, silyl enol ethers, and styrene all effectively functionalize cyclic siloxanes in high yield. The combination of $D_4{}^V$ with styrene under silylative functionalization conditions provides tetrastyryltetramethylcyclotrisiloxane (**9**) in 92% yield with the phenyl subsituent exclusively at the β-carbon (eq 8), while the same reactants under standard ruthenium cross-metathesis conditions afford only monofunctionalization in 10% conversion. Moreover, the reaction of $D_4{}^V$ or $D_3{}^V$ and styrene affords only the E-isomer, whereas the combination of $D_3{}^V$ and an enol ether provides a mixture of the E- and Z-isomers.

styrene (4.4 equiv)
RuHCl(CO)(PCy$_3$)$_2$
(0.1 mol %)
Toluene, 80 °C, 18 h
(92%)

1

(8)

9

Miscellaneous Reactions. 2,4,6,8-Tetraethenyl-2,4,6,8-tetra-methylcyclotetrasiloxane acts as a chelating ligand for a bimetallic nickel(0) complex,[24] similar to Karstedt's catalyst.[25] The combination of D$_4^V$ and tetramethylcyclotetrasiloxane in the presence of Karstedt's catalyst cross-links the two tetramers to prepare a platinum-containing glass.[26] Additionally, Bennetau and Boileau have functionalized D$_4^V$ with C$_8$F$_{17}$I under ultrasonic irradiation conditions to provide tetra(perfluorooctyl)-substituted D$_4^V$ after reductive cleavage of the iodide with tributyltin hydride.[27]

---

1. Denmark, S. E.; Wang, Z., *J. Organomet. Chem.* **2001**, *624*, 372.
2. (a) Son, H.-J.; Han, W.-S.; Kim, H.; Kim, C.; Ko, J.; Lee, C.; Kang, S. O., *Organometallics* **2006**, *25*, 766. (b) Lim, C.; Park, J., *Synthesis* **1999**, 1804.
3. Kantor, S. W.; Osthoff, R. C.; Hurd, D. T., *J. Am. Chem. Soc.* **1955**, *77*, 1685.
4. Hammouch, S. O.; Beinert, G. J.; Herz, J. E., *Polymer* **1996**, *37*, 3353.
5. Kang, H.; Lee, J.; Park, J.; Lee, H. H., *Nanotechnology* **2006**, *17*, 197.
6. Pietraszuk, C.; Fischer, H., *J. Chem. Soc. Chem. Commun.* **2000**, 2463.
7. Marciniec, B., *Appl. Organomet. Chem.* **2000**, *14*, 527.
8. Itami, Y.; Marciniec, B.; Kubicki, M., *Organometallics* **2003**, *22*, 3717.
9. Bumagin, N. A.; Luzikova, E. V., *J. Organomet. Chem.* **1997**, *532*, 271.
10. (a) Kerins, F.; O'Shea, D. F., *J. Org. Chem.* **2002**, *67*, 4968. (b) Molander, G. A.; Brown, A. R., *J. Org. Chem.* **2006**, *71*, 9681.
11. McKean, D. R.; Parrinello, G.; Renaldo, A. F.; Stille, J. K., *J. Org. Chem.* **1987**, *52*, 422.
12. Hatanaka, Y.; Hiyama, T., *J. Org. Chem.* **1988**, *53*, 918.
13. Denmark, S. E.; Wang, Z., *Synthesis* **2000**, 999.
14. Denmark, S. E.; Wehrli, D.; Choi, J. Y., *Org. Lett.* **2000**, *2*, 2491.
15. **1**, Gelest cat #: SIT7900.0, $199/mol; **2**, Gelest cat #: SIT8737.0, $1198/mol; **3**, Gelest cat #: SIT8725.0, $1137/mol.
16. Aranyos, A.; Old, D. W.; Kiyomori, A.; Wolfe, J. P.; Sadighi, J. P.; Buchwald, S. L., *J. Am. Chem. Soc.* **1999**, *121*, 4369.
17. Denmark, S. E.; Butler, C. R., *Org. Lett.* **2006**, *8*, 63.
18. Chen, R.; Lee, V.; Adlington, R. M.; Baldwin, J. E., *Synthesis* **2007**, 113.
19. Kim, C.; An, K., *J. Organomet. Chem.* **1997**, *547*, 55.
20. van der Made, A. W.; van Leeuwen, P. W. N. M.; de Wilde, J. C.; Brandes, R. A. C., *Adv. Mater.* **1993**, *5*, 466.
21. Kim, C.; Kim, H., *J. Organomet. Chem.* **2003**, *673*, 77.
22. Kim, C.; Ryu, M., *J. Polym. Sci. Pt. A* **2000**, *38*, 764.
23. Kim, C.; Son, S.; Kim, B., *J. Organomet. Chem.* **1999**, *588*, 1.
24. Hitchcock, P. B.; Lappert, M. F.; Maciejewski, H., *J. Organomet. Chem.* **2000**, *605*, 221.
25. Karstedt, B. D. U.S. Pat. 3,419,593 (1968).
26. Stein, J.; Lewis, L. N.; Gao, Y.; Scott, R. A., *J. Am. Chem. Soc.* **1999**, *121*, 3693.
27. Beyout, E.; Babin, P.; Bennetau, B.; Dunogues, J.; Teyssie, D.; Boileau, S., *Tet. Lett.* **1995**, *36*, 1843.

Scott E. Denmark & Christopher R. Butler
University of Illinois, Urbana, IL, USA

3-Tetrahydropyranyloxy-1-trimethylsilyl-1-propyne[1]

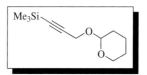

[36551-06-1]   C$_{11}$H$_{20}$O$_2$Si   (MW 212.40)
InChI = 1S/C11H20O2Si/c1-14(2,3)10-6-9-13-11-7-4-5-8-12-11
/h11H,4-5,7-9H2,1-3H3
InChIKey = SNFCEOGLSUQETE-UHFFFAOYSA-N

(metal derivative is used for 1-alkoxypropargylation;[2,3] used in synthesis of functionalized vinylsilanes[4])

Physical Data: bp 108 °C/11 mmHg; n_D^{20} 1.4630.[2]
Solubility: sol THF, ether, and most organic solvents.
Preparative Methods: from 2-propynol via tetrahydropyranyl ether formation, deprotonation, and trimethylsilylation;[2–6] from 3-trimethylsilylpropynol via tetrahydropyranyl ether formation.[4]
Handling, Storage, and Precautions: store in the absence of air and moisture.

Alkoxypropargylation. The ambident character of the propargylic anion, which may be in equilibrium with the allenic form, is responsible for its limited use in synthesis. In general the structure and reactivity of the ambident anion depend on the nature of the substrate, the counter cation, and the solvent.[7,8] There is also an *erythro–threo* stereoselectivity problem when alkylated propargylic anions react with aldehydes or unsymmetrical ketones. In contrast, the zinc and titanium reagents derived from the title compound possess the allenic structure and, upon reaction with aldehydes, lead almost exclusively to the β-acetylenic alcohol (eq 1), presumably by a chelate transition state (S$_E$i' process).[3,5] The reaction also leads preferentially to the *erythro* diastereomer. The stereoselectivity is highest with titanium as the metal and THF as the solvent.[6,9]

Whereas the titanium reagent is not reactive towards ketones, the zinc derivative has been found to react even with sterically encumbered ketones, such as 17-keto steroids, to yield the β-acetylenic alcohol in good yield (75%).[2]

Synthesis of Vinylsilanes. The title compound can be used in reductive alkylation sequences for the synthesis of functionalized vinylsilanes with high isomeric purity (eq 2). One sequence consists of a regioselective *cis*-hydroboration, transmetalation, and carbodemetalation.[4,10] In another sequence, diisobutyl-aluminum hydride reduction–bromination leads to brominated vinylsilanes,[11] which can be further transformed via halogen–metal exchange, followed by reaction with iodides,[12] or with aldehydes and ketones.[13]

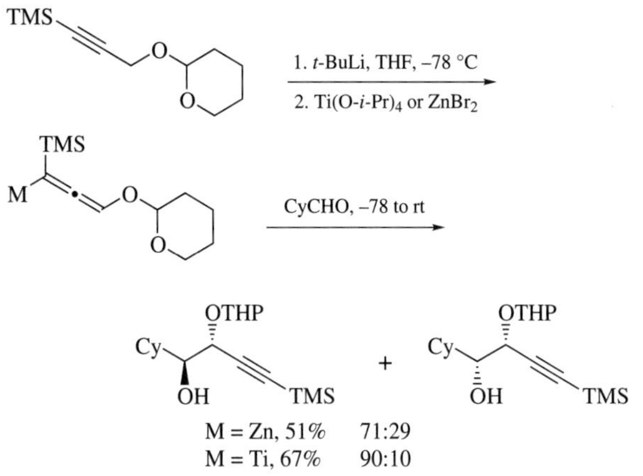

$$M = Zn, \ 51\% \quad 71:29$$
$$M = Ti, \ 67\% \quad 90:10$$

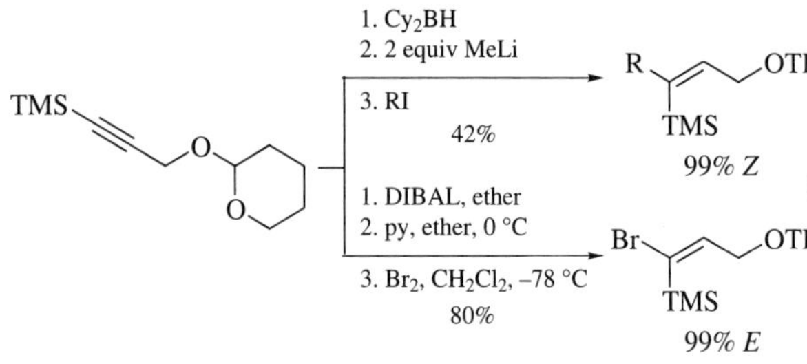

1. *Fieser & Fieser* **1986**, *12*, 353.
2. Chwastek, H.; Epsztein, R.; Le Goff, N., *Tetrahedron* **1973**, *29*, 883.
3. Ishiguro, M.; Ikeda, N.; Yamamoto, H., *J. Org. Chem.* **1982**, *47*, 2225.
4. Uchida, K.; Utimoto, K.; Nozaki, H., *Tetrahedron* **1977**, *33*, 2987.
5. Furuta, K.; Ishiguro, M.; Haruta, R.; Ikeda, N.; Yamamoto, H., *Bull. Chem. Soc. Jpn.* **1984**, *57*, 2768.
6. Hiraoka, H.; Furuta, K.; Ikeda, N.; Yamamoto, H., *Bull. Chem. Soc. Jpn.* **1984**, *57*, 2777.
7. (a) Klein, J. In *The Chemistry of the Carbon–Carbon Triple Bond*; Patai, S., Ed.; Wiley: New York, 1978; Part 1, p 343. (b) Moreau, J.-L. In *The Chemistry of Ketenes, Allenes and Related Compounds*; Patai, S., Ed.; Wiley: New York, 1980; p 363.
8. Suzuki, M.; Morita, Y.; Nayori, R., *J. Org. Chem.* **1990**, *55*, 441.
9. Parsons, P. J.; Willis, P. A.; Eyley, S. C., *J. Chem. Soc., Chem. Commun.* **1988**, 283.
10. Bell, V. L.; Giddings, P. J.; Holmes, A. B.; Mock, G. A.; Raphael, R. A., *J. Chem. Soc., Perkin Trans. 1* **1986**, 1515.
11. Zweifel, G.; Lewis, W., *J. Org. Chem.* **1978**, *43*, 2739.
12. Miller, R. B.; Al-Hassan, M. I., *Tetrahedron Lett.* **1983**, *24*, 2055.
13. Miller, R. B.; Al-Hassan, M. I., *J. Org. Chem.* **1983**, *48*, 4113.

Pierre J. De Clercq & Frank Nuyttens
Universiteit Gent, Gent, Belgium

1,1,3,3-Tetramethyldisilazane

$$(HMe_2Si)_2NH$$

[15933-59-2] $\quad C_4H_{15}NSi_2 \quad$ (MW 133.38)
InChI = 1S/C4H15NSi2/c1-6(2)5-7(3)4/h5-7H,1-4H3
InChIKey = NQCZAYQXPJEPDS-UHFFFAOYSA-N

(intramolecular hydrosilation of allyl alcohols, homoallyl alcohols, and homopropargyl alcohols for the regio- and stereoselective synthesis of polyols)

Physical Data: bp 99–100 °C; n_D^{20} 1.4040; d 0.752 g cm^{-3}.
Solubility: common organic solvents.
Form Supplied in: neat liquid.
Handling, Storage, and Precautions: moisture sensitive but no special precautions necessary for handling in the air for a short period of time. May be stored under nitrogen in a tightly capped bottle in a refrigerator. Use in a fume hood.

Intramolecular Hydrosilation.[1,2] Allyl and homoallyl alcohols are transformed into 1,3-diols in a highly regio- and stereoselective fashion via intramolecular hydrosilation followed by oxidative cleavage of the silicon–carbon bonds by hydrogen peroxide, which proceeds with complete retention of configuration at carbon (eqs 1 and 2).[3,4]

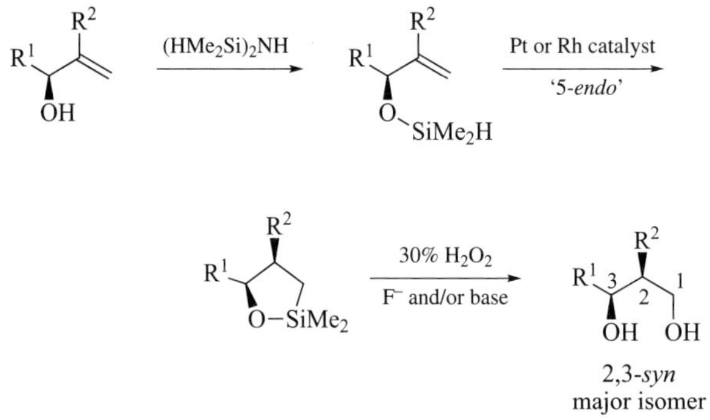

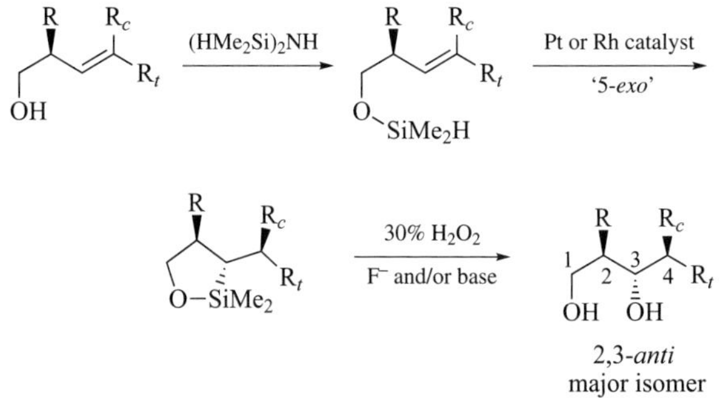

Thus, an alcohol is converted into a hydrodimethylsilyl ether by treatment with 1,1,3,3-tetramethyldisilazane in the absence or presence of a catalytic amount of ammonium chloride as a promoter. The ammonia produced in this reaction and the excess disilazane must be removed prior to the hydrosilation step. The intramolecular hydrosilation can be achieved by using a platinum or rhodium catalyst (≤1 mol %) such as acidic $H_2PtCl_6 \cdot 6H_2O$ in *i*-PrOH (Speier's catalyst) or THF, neutral $Pt[\{(CH_2{=}CH)Me_2Si\}_2O]_2$ in xylene, or chlorotris(triphenylphosphine)rhodium(I) (Wilkinson's catalyst). After the reaction, the catalyst, which causes a rapid decomposition of hydrogen peroxide in the subsequent oxidation step, must be removed by stirring the mixture with crystalline EDTA·2Na or activated carbon followed by filtration.[2] The oxidative cleavage of the organosilane intermediate usually proceeds smoothly at room temperature under the standard conditions shown in the following examples.

The reaction of terminal allyl alcohols proceeds in a 5-*endo* fashion to give five-membered ring compounds regioselectively and stereoselectively. Subsequent oxidation affords 2,3-*syn*-1,3-diols preferentially, regardless of the nature of the catalyst (eq 1).[5] The stereoselectivity increases with increased bulkiness of the allylic substituent R^1 and the nature of the alkene substituent R^2 (see below).[5] 5-*Exo* type cyclization occurs with homoallyl alcohols to form five-membered heterocycles and 1,3-diols after oxidation. Two chiral centers are produced in this reaction. The 2,3-relationship (*anti*) is controlled by the allylic substituent, while the 3,4-relationship is determined by the stereochemistry of the alkene; the hydrosilation occurs by *cis* addition of Si–H to the alkene (eq 2).[5b]

This methodology has been applied to the construction of a variety of stereoisomers of polypropionate skeletons, as exemplified by the formation of a tetraol from a symmetrical bis-allyl alcohol by a sequence involving three intramolecular hydrosilation–oxidation steps (eq 3).[5b]

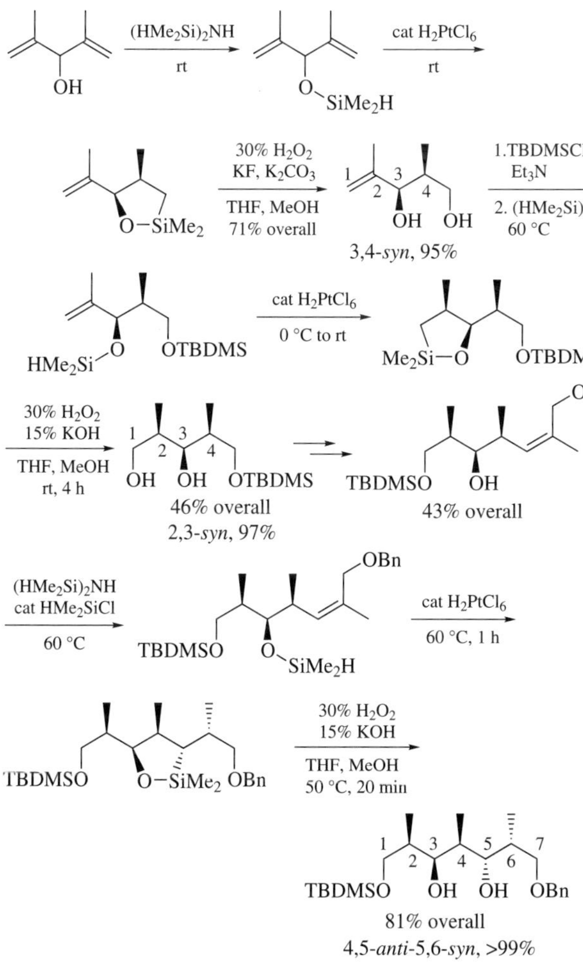

Extremely high stereoselectivity has been attained in the intramolecular hydrosilation of α-hydroxy enol ethers, leading to 1,2,3-triol derivatives with 2,3-*syn* stereochemistry (eq 4).[6] It has been pointed out, however, that the stereoselectivity largely depends upon the nature of the catalyst, Pt or Rh, and the presence or absence of the disilazane in the Pt case (eq 5).[7]

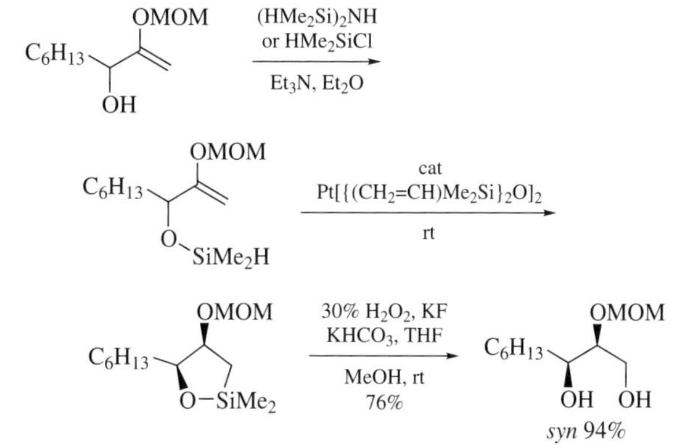

(4)

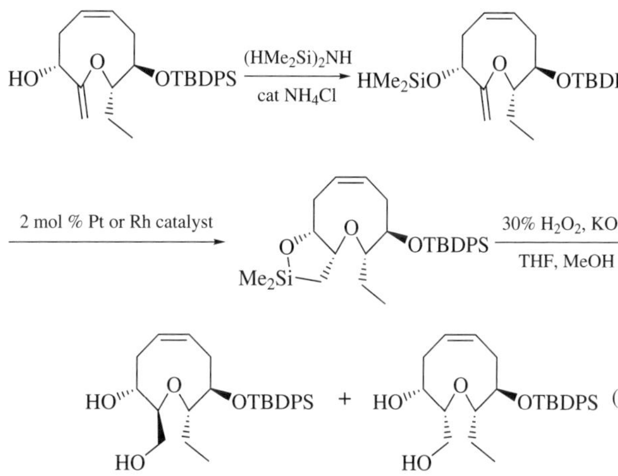

(5)

catalyst

$Pt[\{CH_2{=}CH)Me_2Si\}_2O]_2$, 72%   16:84
$Pt[\{CH_2{=}CH)Me_2Si\}_2O]_2$
+ $(HMe_2Si)_2NH$ (10 mol %), 84%   83:17
$Rh(acac)(1,5\text{-}cod)$, 61%   >95:<5

(3)

81% overall
4,5-*anti*-5,6-*syn*, >99%

The intramolecular hydrosilation of homopropargyl alcohols also proceeds in a 5-*exo* manner to form five-membered cyclic vinylsilanes exclusively. Subsequent oxidation affords a β-hydroxy ketone (eq 6).[8] The vinylsilane also undergoes a Pd-catalyzed cross-coupling reaction with aryl or alkenyl halides stereoselectively (eq 6).[9] The intramolecular hydrosilation thus provides an efficient methodology for the regio- and/or stereoselective functionalization and carbon–carbon bond formation of the alkyne moiety in homopropargyl alcohol.

While the five-membered cyclic alkoxysilane moiety has been found to tolerate a wide range of chemical transformations (OsO_4 oxidation, Swern oxidation, Horner–Emmons alkenation, protection conditions such as Tr/py^+ BF_4^- and MeI/NaH, and catalytic hydrogenation), the carbon–silicon bond can be cleaved efficiently by tetra-n-butylammonium fluoride in DMF (eq 7).[10]

C6H13—≡—OH $\xrightarrow{\text{(HMe}_2\text{Si)}_2\text{NH}}$ C6H13—≡—O—SiHMe2 $\xrightarrow[\text{rt to 60 °C}]{\text{cat H}_2\text{PtCl}_6}$

C6H13 —SiMe2 (cyclic with O) $\xrightarrow[\text{THF, MeOH, 60 °C}]{30\% \text{ H}_2\text{O}_2, \text{KHCO}_3, \text{KF}}$ C6H13—CH2—C(=O)—CH2—CH2—OH

72% overall   (6)

Br—CH=CH—Ph | $\xrightarrow[\text{P(OEt)}_3, \text{THF, 50 °C}]{\text{Bu}_4\text{NF} \atop \text{cat [PdCl}(\eta^3\text{-C}_3\text{H}_5)]_2}$

C6H13—C(=CH—CH=CH—Ph)—CH2CH2—OH

42% overall, (E,E) > 99%

OH—...—OBn (with Me) $\xrightarrow[\text{cat NH}_4\text{Cl}]{\text{(HMe}_2\text{Si)}_2\text{NH}}$ O—SiMe2H—...—OBn $\xrightarrow[\text{rt}]{\text{cat H}_2\text{PtCl}_6}$

quantitative

O—SiMe2 (cyclic)—...—OBn $\longrightarrow$ O—SiMe2 (cyclic)—...—CH=CH—C(=O)—N(Me)(OMe) $\xrightarrow[\substack{\text{acetone, H}_2\text{O} \\ 80\%}]{\substack{\text{cat OsO}_4 \\ \text{NMO}}}$

quantitative   70% overall

OH—CH2—[Me2Si—O cyclic]—...—C(=O)—N(Me)(OMe), OH OMe $\xrightarrow[\text{DMF, 50 °C}]{\text{Bu}_4\text{NF}}$ OTr—...—OH—...—C(=O)—N(Me)(OMe), OMe OMe   (7)

77% overall

The intramolecular hydrosilation of allyl alcohols containing an ester group at the terminal carbon proceeds in a 5-*endo* fashion to form the five-membered cyclic products with modest stereoselectivity. The silyl group α to the carbonyl group is readily cleaved by a fluoride ion in protic solvents (eq 8).[11]

OH—CH(iPr)—C(Me)=CH—CO2Me $\xrightarrow{\text{(HMe}_2\text{Si)}_2\text{NH}}$ O—SiMe2H—CH(iPr)—C(Me)=CH—CO2Me

$\xrightarrow[\substack{\text{ClCH}_2\text{CH}_2\text{Cl} \\ \text{rt to 55 °C}}]{\substack{\text{cat} \\ \text{Pt[\{(CH}_2=\text{CH)Me}_2\text{Si\}}_2\text{O]}_2}}$ O—SiMe2 (cyclic)—...CO2Me + O—SiMe2 (cyclic)—...CO2Me

$\xrightarrow[\substack{\text{MeOH, rt} \\ 74\%}]{\text{sat. KF}}$ lactone + lactone   (8)

77:23

Catalytic asymmetric intramolecular hydrosilation of allyl alcohols has been achieved by using chlorodiphenylsilane[12] or 1-chloro-1-silacyclohexane[13] as the silylating agent.

1. Tamao, K., *J. Synth. Org. Chem. Jpn.* **1988**, *46*, 861.

2. Tamao, K.; Nakagawa, Y.; Ito, Y., *Org. Synth.* **1995**, in press.

3. Tamao, K.; Ishida, N.; Tanaka, T.; Kumada, M., *Organometallics* **1983**, *2*, 1694.

4. Colvin, E. W., *Comprehensive Organic Synthesis* **1991**, *7*, 641.

5. (a) Tamao, K.; Tanaka, T.; Nakajima, T.; Sumiya, R.; Arai, H.; Ito, Y., *Tetrahedron Lett.* **1986**, *27*, 3377. (b) Tamao, K.; Nakajima, T.; Sumiya, R.; Arai, H.; Higuchi, N.; Ito, Y., *J. Am. Chem. Soc.* **1986**, *108*, 6090. (c) Anwar, S.; Davis, A. P., *Proc. R. Ir. Acad.* **1989**, *89B*, 71.

6. Tamao, K.; Nakagawa, Y.; Arai, H.; Higuchi, N.; Ito, Y., *J. Am. Chem. Soc.* **1988**, *110*, 3712.

7. (a) Curtis, N. R.; Holmes, A. B.; Looney, M. G., *Tetrahedron Lett.* **1992**, *33*, 671. (b) Curtis, N. R.; Holmes, A. B., *Tetrahedron Lett.* **1992**, *33*, 675.

8. Tamao, K.; Maeda, K.; Tanaka, T.; Ito, Y., *Tetrahedron Lett.* **1988**, *29*, 6955.

9. Tamao, K.; Kobayashi, K.; Ito, Y., *Tetrahedron Lett.* **1989**, *30*, 6051.

10. Hale, M. R.; Hoveyda, A. H., *J. Org. Chem.* **1992**, *57*, 1643.

11. Denmark, S. E.; Forbes, D. C., *Tetrahedron Lett.* **1992**, *33*, 5037.

12. Tamao, K.; Tohma, T.; Inui, N.; Nakayama, O.; Ito, Y., *Tetrahedron Lett.* **1990**, *31*, 7333.

13. Bergens, S. H.; Noheda, P.; Whelan, J.; Bosnich, B., *J. Am. Chem. Soc.* **1992**, *114*, 2121.

Kohei Tamao
Kyoto University, Kyoto, Japan

1,1,2,2-Tetraphenyldisilane

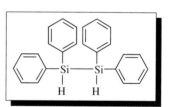

[16343-18-3] $C_{24}H_{22}Si_2$ (MW 366.61)

InChI = 1S/C24H22Si2/c1-5-13-21(14-6-1)25(22-15-7-2-8-16-22)26(23-17-9-3-10-18-23)24-19-11-4-12-20-24/h1-20,25-26H

InChIKey = LKWATTABEBNHPP-UHFFFAOYSA-N

(reagent used for radical reduction, intramolecular and intermolecular carbon–carbon bond formation, 1,2-elimination, etc., of halides, chalcogenides, and xanthates)

Physical Data: mp 79–80 °C, IR(KBr): 2120 cm^{-1}(SiH). ^{1}H-NMR (CDCl$_3$, TMS) $\delta = 5.19$ (2H, s, SiH), 7.26–7.30 (8H, m, Ph), 7.33–7.38 (4H, m, Ph), 7.45–7.48 (8H, m, Ph). ^{29}Si-NMR (CDCl$_3$, TMS) $\delta = -34.96$. X-ray crystal structure analysis: Si–Si (2.35 Å), Si–H (0.97 Å).

Solubility: soluble in chloroform and most organic solvents, less soluble in ethanol.

Preparative Methods: generally, 1,1,2,2-tetraphenyldisilane is prepared by the coupling reaction of diphenylsilane in the presence of diphenyltitanocene without solvent under heating conditions at 110 °C for 24 h.

Purity: 1,1,2,2-tetraphenyldisilane is purified by column chromatography on silicagel from the reaction mixture (eluent; chloroform:hexane = 1:7).

Handling, Storage, and Precaution: it is an air-stable solid.

Introduction. In synthetic organic chemistry, tributyltin hydride has been well used as an effective radical mediator.[1] However, it is well known and recognized that organotin compounds are highly toxic, and the complete removal of the tin residues from the reaction products is very troublesome.

As an alternative to tributyltin hydride, tris(trimethylsilyl)silane (TTMSS), which is a much less toxic reagent, was developed by Chatgilialoglu,[2] However, TTMSS is a moderately stable oil which is easily oxidized by molecular oxygen in air, and thus lacks stability under aerobic conditions, which limits its reactivity range. The Si–H bond dissociation energy and radical reactivity of pentamethyldisilane have been studied,[3] and the Si–H bond dissociation energy of pentamethyldisilane was measured to be 85.3 kcal/mol. It is still strong as compared with that of tributyltin hydride (74.0 kcal/mol) and TTMSS (79.0 kcal/mol). Therefore, pentamethyldisilane has low hydrogen-donating ability and can be used just for the reduction of halides (Cl, Br, I), selenides, and xanthates initiated by peroxides under heating conditions in toluene.

Now, 1,1,2,2-tetraphenyldisilane (TPDS) is more effective than pentamethyldisilane, since one phenyl group on the Si–H group probably reduces by 2 kcal/mol the Si–H bond dissociation energy. TPDS was first prepared by Gilman; however, only the reactivity on hydrolysis by hydroxide, ethoxide, and amine, and the addition to an N-N double bond have been studied,[4] TPDS is a stable crystalline solid; no decomposition was observed for 3 months under air at room temperature, and it is sufficiently stable not only in general organic solvents but also in ethanol.[5]

Reduction. TPDS showed good reactivity in the reduction of methyl 6-bromo-6-deoxy-α-D-glucopyranoside initiated by Et_3B in ethanol, as shown in eq 1. Only one of the two hydrogen atoms bonded to silicon atoms in TPDS participates in these reactions. The present reaction with TPDS initiated by Et_3B or AIBN can be used for the reduction of various alkyl bromides or iodides derived from steroids, sugars, nucleosides, and aromatic bromides in high yields.[5] Recently, it has been reported that the reduction of halides with diphenylsilane and phenylsilane initiated by peroxide proceeds effectively to give the corresponding reduction products in good yield,[6] However, under the present conditions, the reduction of halides with diphenylsilane initiated by AIBN gave only trace amounts of the reduction products, together with the starting halides. The difference in reactivity between TPDS and diphenylsilane is significant.

$$\text{(1)}$$

One advantage in the use of chalcogenides is that sugar chalcogenides are much more stable than sugar halides for storage and chemical treatment. Therefore, an effective radical reaction of chalcogenides with TPDS would be very convenient. Practically, adamantyl phenyl selenide and phenyl tridecyl selenide can be reduced effectively by the TPDS–AIBN method or by the TPDS–Et_3B method. Phenyl 2,3,4,6-tetra-O-acetyl-1-phenylseleno-β-D-galactopyranoside and ethyl 2-phenylseleno-3,4,6-tri-O-acetyl-β-D-glucopyranoside (eq 2) were reduced in high yields by both methods. Phenyl 2,3,5-tri-O-benzyl-1-phenylseleno-D-

ribofuranoside and the 2-phenylselenoadenosine derivative are also reduced by the TPDS–AIBN method in good yield.[7]

$$\text{(2)}$$

Deoxygenation reactions of hydroxy groups in carbohydrates, nucleosides, and antibiotics are very important. Today, the most effective and practical deoxygenation method of hydroxy groups is the Barton–McCombie reaction, which is the reaction of xanthates with tributyltin-hydride–AIBN or tributyltin-hydride–Et_3B;[8] or phenylsilane or diphenylsilane in the presence of peroxide or Et_3B;[9] trialkylsilane in the presence of peroxide–thiol (polarity reversal catalyst);[10] 5,10-dihydrosilanthrene–AIBN;[11] TTMSS–AIBN;[2] dibutylphosphineoxide–(AIBN or Et_3B);[12] or phosphineborane–AIBN.[13] The modified Barton–McCombie reactions have found widespread use in organic synthesis. TPDS–AIBN also showed good reactivity in the deoxygenative reduction of various xanthates derived from steroids, sugars, and nucleosides to give the corresponding reduction products (eq 3).[14] An imidazole thiocarbonyl group can be also used instead of a methyl dithiocarbonate group.

$$\text{(3)}$$

Reductive Addition. Radical addition to activated olefins with alkyl bromides to form a carbon–carbon bond can be also carried out in the presence of TPDS initiated by AIBN in ethanol. Other activated olefins such as diethyl vinylphosphonate and ethyl acrylate can be also used in the addition reaction. The reductive addition of 1-bromoadamantane to diethyl vinylphosphonate with TPDS initiated by AIBN to form the corresponding diethyl 2(1-adamantyl)ethylphosphonate in good yield is shown in eq 4.

The addition reaction with other alkyl bromides and xanthates to activated olefins such as phenyl vinyl sulfone and diethyl vinylphosphonate, with TPDS–AIBN, also gives the corresponding reductive addition products in moderate to good yields (40–84%).[5,14] Unfortunately, reductive addition of chalcogenides to activated olefin with the TPDS–AIBN method does not work effectively due to the lower reactivity of chalcogenides toward the silyl radical derived from TPDS, in comparison with the corresponding halides or xanthates.

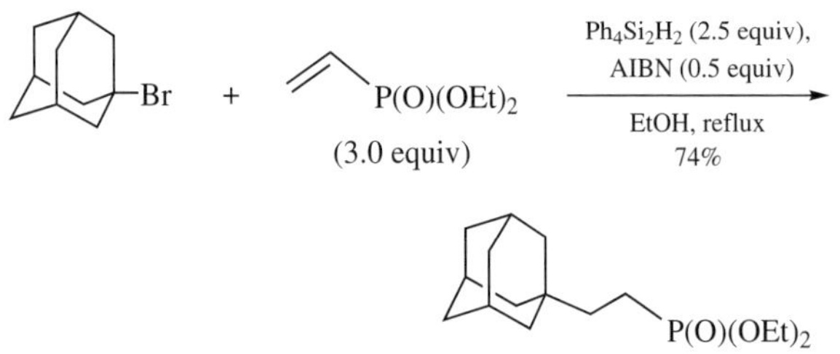

$$\text{(4)}$$

Alkylation of Heteroaromatics. Treatment of alkyl bromides, TPDS, and heteroaromatic bases, which are activated by protonation with trifluoroacetic acid in ethanol at reflux temperature, gives the corresponding alkylated heteroaromatic bases in moderate to good yields. Generally, the radical alkylation to heteroaromatic bases with secondary and tertiary alkyl bromides proceeds effectively to generate the corresponding alkylated products in good yields (55–93%), while the yields of alkylation products with primary alkyl bromides go down (~30%).[5] Alkyl bromides and iodides show the same reactivity, while alkyl chloride does not react at all. As an example, adamantylation of pyridine with 1-bromoadamantane and TPDS initiated by AIBN is shown in eq 5 to give the corresponding 2-adamantylpyridine and 4-adamantylpyridine in 79% yield (63:27).

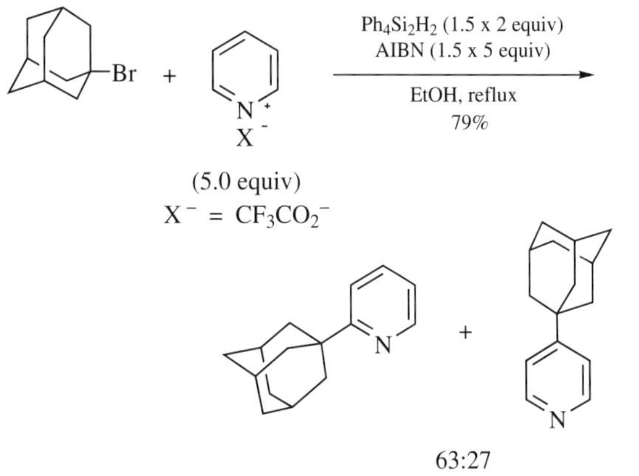

$$\text{(5)}$$

63:27

The present reactions proceed in ethanol instead of a toxic organic solvent such as benzene or toluene.

Cyclization. *5-exo-trig* and *6-exo-trig* radical cyclization involving multiple bonds is a powerful and versatile method for the construction of cyclic systems. Bicyclic sugars are especially interesting compounds because of their use as building blocks for the synthesis of natural products as well as their biological activities. Recently, bicyclic sugars have been prepared extensively through the radical cyclization of sugar halides with tributyltin hydride.[15] However, the concentration of tributyltin hydride in solution must be quite low to avoid simple reduction. As shown in eq 6, utilization of the TPDS–Et$_3$B method and the TPDS–AIBN method gives only the cyclization product in 84% yield and 78% yields, respectively, under the standard conditions.[16] In contrast, the tributyltinhydride–Et$_3$B method and the tributyltinhydride–AIBN method produce the bicyclic sugar

in 37% and 65% yields together with the direct reduction product in 44% and 32% yields, respectively.

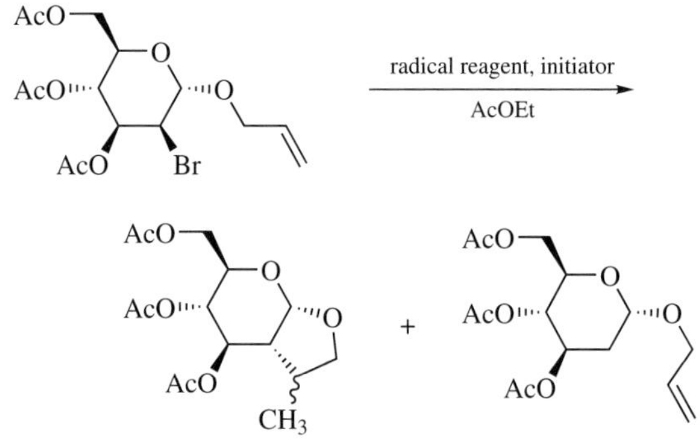

$$\text{(6)}$$

Reagent[a]	Initiator[b]	Yields	(%)
Ph$_4$Si$_2$H$_2$	Et$_3$B	84	0
	AIBN	78	0
Bu$_3$SnH	Et$_3$B	37	44
	AIBN	65	32

[a]Ph$_4$Si$_2$H$_2$ or Bn$_3$SnH (1.2 equiv).
[b]Et$_3$B (0.6 equiv) or AIBN (0.5 equiv), Et$_3$B : rt, AIBN : reflux.

In other sugar compounds, the corresponding bicyclic sugars can be obtained in good yields by both the TPDS–Et$_3$B method and the TPDS–AIBN method. These bicyclic sugars consist of a cis-fused ring derived from 5-*exo-trig* or 5-*exo-dig* closure.

1,2-Elimination. The formation of olefins from the *vic*-diols via 1,2-dixanthates is very important, especially after the discovery of anti-HIV nuclosides such as D4C (2′,3′-didehydro-2′,3′-dideoxycytidine) and D4T (2′,3′-didehydro-3′-deoxythymidine).[17] Treatment of adenosine 1,2-dixanthates with the TPDS–AIBN method under refluxing conditions in ethyl acetate generates the corresponding 2′,3′-dideoxygenated adenosine nucleoside in quite good yield as shown in eq 7.[18]

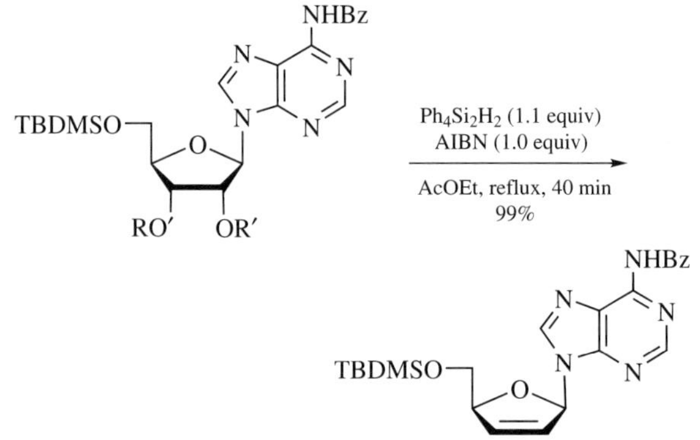

$$\text{(7)}$$

The reactions with TPDS initiated by Et$_3$B at room temperature give the corresponding olefins in poor yields, because the radical β-elimination reaction requires heating conditions.

Aryl Transfer. The biaryl skeleton is pharmacologically interesting and important as a building block for a large number of natural products, chiral ligands in asymmetric synthesis, polymers, and advanced materials. Treatment of various

N-methyl-N-(2-bromoaryl)arenesulfonamides with TPDS and AIBN under heating conditions produces the corresponding biaryl products in moderate yields through the intramolecular radical *ipso*-substitution (eq 8). TPDS is the most effective reagent from among diphenylsilane, tributyltin hydride, tris(trimethylsilyl)silane, and TPDS for 1,5-*ipso*-substitution on the sulfonamides.[19]

Ring Expansion. TPDS-mediated radical ring expansion of β-haloalkyl cyclic β-keto esters to form ring-expanded cyclic keto esters through the radical cyclization of the initially formed carbon radicals to the carbonyl group in an 3-*exo-trig* manner, followed by β-cleavage of the resulting bicyclic alkoxy radical intermediate, proceeds effectively (eq 9). TPDS produces the ring-expansion products in better yields than those obtained with tributyltin hydride,[20] though it showed almost the same reactivity as that of tris(trimethylsilyl)silane. Chain extension of acyclic bromoalkyl β-keto esters also proceeds effectively as shown in eq 10. Here again, TPDS showed almost the same reactivity as those with tris(trimethylsilyl)silane, and tributyltin hydride showed poor results.

Aryldiphenylsilane. Aryl halides are converted to aryldiphenylsilanes in moderate to good yields in the presence of TPDS

and CsF in DMPU (eq 11).[21] The reactivity of aryl halides is as follows: Br > I > Cl.

Related Reagents. Tributyltin hydride; tris(trimethylsilyl)silane.

1. (a) Giese, B., *Radicals in Organic Synthesis: Formation of Carbon-Carbon Bonds*; Pergamon: Oxford, 1986. (b) Fossey, J.; Lefort, D.; Sorba, J., *Free Radicals in Organic Chemistry*; Wiley: Chichester, 1995. (c) Togo, H., *Organic Chemistry of Free Radicals*; Kodansha: Tokyo, 2001.

2. (a) Chatgilialoglu, C., *Acc. Chem. Res.* **1992**, *25*, 188. (b) Chatgilialoglu, C., *Chem. Rev.* **1995**, *95*, 1229.

3. (a) Lusztyk, J.; Maillard, B.; Ingold, K. U., *J. Org. Chem.* **1986**, *51*, 2457. (b) Kanabus-Kaminska, J. M.; Hawari, J. A.; Griller, D.; Chatgilialoglu, C., *J. Am. Chem. Soc.* **1987**, *109*, 5267. (c) Ballestri, M.; Chatgilialoglu, C.; Guerra, M.; Guerrini, A.; Lucarini, M.; Seconi, G., *J. Chem. Soc., Perkin Trans 1* **1993**, 421.

4. (a) Steudel, W.; Gilman, H., *J. Am. Chem. Soc.* **1960**, *82*, 6129. (b) Winkler, H. J. S.; Gilman, H., *J. Org. Chem.* **1961**, *26*, 1265. (c) Schott Von, G.; Langecker, W., *Z. Anorg. Allg. Chem.* **1968**, *358*, 210. (d) Linke, Karl-H.; Göhausen, H. J., *Chem. Ber.* **1973**, *106*, 3438. (e) Lachance, N.; Gallant, M., *Tetrahedron Lett.* **1998**, *39*, 171.

5. (a) Yamazaki, O.; Togo, H.; Matsubayashi, S.; Yokoyama, M., *Tetrahedron Lett.* **1998**, *39*, 1921. (b) Yamazaki, O.; Togo, H.; Matsubayashi, S.; Yokoyama, M., *Tetrahedron* **1999**, *55*, 3735.

6. (a) Barton, D. H. R.; Jang, D. O.; Jaszberenyi, J. C., *Tetrahedron Lett.* **1991**, *32*, 2567; (b) Cole, S. J.; Kirwan, N.; Roberts, B. P.; Willis, C. R., *J. Chem. Soc., Perkin Trans 1* **1991**, 103. (c) Yamazaki, O.; Togo, H.; Nogami, G., *Bull. Chem. Soc. Jpn.* **1997**, *70*, 2519.

7. Yamazaki, O.; Togo, H.; Yokoyama, M., *J. Chem. Soc., Perkin Trans 1* **1999**, 2891.

8. (a) Barton, D. H. R.; McCombie, S. W., *J. Chem. Soc., Perkin Trans 1* **1975**, 1574. (b) Barton, D. H. R.; Motherwell, W. B.; Stange, A., *Synthesis* **1981**, 743. (c) Barton, D. H. R.; Hartwig, W.; Hay-Motherwell, R. S.; Motherwell, W. B.; Stange, A., *Tetrahedron Lett.* **1982**, *23*, 2019. (d) Barton, D. H. R.; Jang, D. O.; Jaszberenyi, J. C., *Tetrahedron Lett.* **1990**, *31*, 3991. (e) Barton, D. H. R.; Parekh, S. I.; Tse, C. L., *Tetrahedron Lett.* **1993**, *34*, 2733.

9. (a) Barton, D. H. R.; Jang, D. O.; Jaszberenyi, J. C., *Tetrahedron Lett.* **1990**, *31*, 4681. (b) Barton, D. H. R.; Jang, D. O.; Jaszberenyi, J. C., *Synlett* **1991**, 435.

10. (a) Cole, S. J.; Kirwan, J. N.; Roberts, B. P.; Willis, C. R., *J. Chem. Soc., Perkin Trans. 1* **1991**, 103. (b) Allen, R. P.; Roberts, B. P.; Willis, C. R., *Chem. Commun.* **1989**, 1387.

11. Gimisis, T.; Ballestri, M.; Ferreri, C.; Chatgilialoglu, C., *Tetrahedron Lett.* **1995**, *36*, 3897.

12. (a) Barton, D. H. R.; Jang, D. O.; Jaszberenyi, J. C., *Tetrahedron Lett.* **1992**, *33*, 5709. (b) Jang, D. O.; Cho, D. H.; Barton, D. H. R., *Synlett* **1998**, 39.

13. Barton, D. H. R.; Jacob, M., *Tetrahedron Lett.* **1998**, *39*, 1331.

14. Togo, H.; Matsubayashi, S.; Yamazaki, O.; Yokoyama, M., *J. Org. Chem.* **2000**, *65*, 2816.

15. (a) Audin, C.; Lancelin, J. M.; Beau, J. M., *Tetrahedron Lett.* **1988**, *29*, 3691. (b) De Mesmaeker, A.; Hoffmann, P.; Ernst, B., *Tetrahedron Lett.* **1988**, *29*, 6585. (c) Chapleur, Y.; Moufid, N. J., *Chem. Commun.* **1989**, 39. (d) De Mesmaeker, A.; Hoffmann, P.; Ernst, B., *Tetrahedron Lett.* **1989**, *30*, 57. (e) Ferrier, R. J.; Petersen, P. M., *Tetrahedron* **1990**, *46*, 1. (f) Lesueur, C.; Nouguier, R.; Bertrand, M. P.; Hoffmann, P.; De Mesmaeker, A., *Tetrahedron* **1994**, *50*, 5369.

16. Yamazaki, O.; Yamaguchi, K.; Yokoyama, M.; Togo, H., *J. Org. Chem.* **2000**, *65*, 5440.

17. (a) Lin, T. S.; Yang, J. H.; Liu, M. C.; Zhu, J. L., *Tetrahedron Lett.* **1990**, *31*, 3829. (b) Cosford, N. D. P.; Schinazi, R. F., *J. Org. Chem.* **1991**, *56*, 2161. (c) Barton, D. H. R.; Jang, D. O.; Jaszberenyi, J. C., *Tetrahedron* **1993**, *49*, 2793.

18. Togo, H.; Sugi, M.; Toyama K. C. R., *Acad. Sci. Paris, Chimie* **2001**, *4*, 539.

19. Ryokawa, A.; Togo, H., *Tetrahedron* **2001**, *57*, 5915.

20. Sugi, M.; Togo, H., *Tetrahedron* **2002**, *58*, 3171.

21. Lachance, N.; Gallant, M., *Tetrahedron Lett.* **1998**, *39*, 171.

Hideo Togo
Chiba University, Chiba, Japan

Trialkylsilyl Cyanide, Polymer-supported

[7677-24-9]P

(reagent used as a safer replacement for liquid TMS-CN, loading ca. 1 mmol/g)

Compatibility for Organic Reactions: swelling in DCM, THF, DMF, and most organic solvents. Not compatible with H_2O and alcohols as solvents.

Analysis of Reagent Purity: loading can be estimated in a qualitative fashion by IR (band at 2185 cm^{-1}) but also can be quantitatively determined by magic angle spinning NMR (^{1}H–MAS–NMR) or by elemental analysis (vide infra).

Preparative Methods: commercially available PS-DES resin (**1**)[1,2] is converted to the reactive chloride intermediate **3** following a published procedure (eq 1).[2,3] A flame-dried, 250-mL two-necked round-bottom flask equipped with an argon inlet and a bubbler was charged with a suspension of **1** (5 g, 4.8 mmol, loading 0.94 mmol/g) and 1,3-dichloro-5,5-dimethylhydantoin (**2**) (2.84 g, 14.4 mmol, 3 equiv) in dry CH_2Cl_2 (50 mL).[3] After 2 h stirring at rt with an orbital shaker, the resin was filtered and washed repeatedly under N_2 atmosphere with dry CH_2Cl_2 (5 × 100 mL) and dry THF (3 × 100 mL). The washed resin **3** (5 g, 4.8 mmol, loading 0.94 mmol/g) was immediately added to a flame-dried two-necked round-bottom flask equipped with an argon inlet and a reflux condenser carrying on top a bubbler. TMS-CN (**4**) (45 mL, 335 mmol, large excess) was added in a single portion, the reaction mixture was heated to 80 °C, and stirred at this temperature with an orbital shaker for 36 h. After

cooling to rt, filtering, and repeated washing under an N_2 atmosphere with dry CH_2Cl_2 (20 × 20 mL), the pure title reagent **5** was obtained (eq 2). Its loading was calculated as 0.92 mmol/g by elemental analysis (calculated for 0.94 mmol/g 1.31% N, found 1.28% N), corresponding to an overall 97.7% conversion.

Handling, Storage, and Precaution: the thoroughly washed resin **5** is perfectly safe and does not release toxic materials under any of the experimental conditions tested. A resin aliquot stored on a shelf at rt was submitted to IR analysis at regular times (7 days, 1 month and 4 months) and it consistently showed identical IR spectra, proving its stability. Storage under more controlled conditions (nitrogen atmosphere, −20 °C) should provide indefinitely stable samples.

Additions to Carbonyl Groups. TMS-CN is known to react rapidly and efficiently with carbonyl compounds to provide cyanohydrins.[4,5] The supported material **5** reacts with neat cinnamaldehyde (**6a**) under conditions already reported[6] to produce the corresponding cyanohydrin **7a** as a released material (eq 3) in 82% yield after chromatography (needed to remove excess reagents).[3]

The same reaction performed under different experimental conditions[7] yields the supported cyanohydrin **8a** (eq 4) by addition and simultaneous silyl transfer to the oxygen atom in >85% yield as determined by ^{1}H–MAS–NMR.[3]

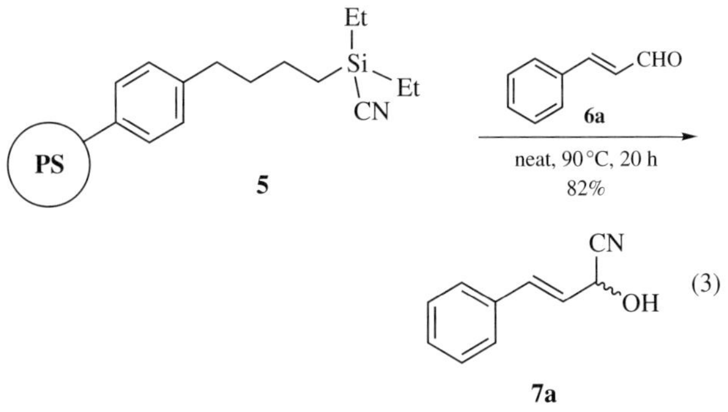

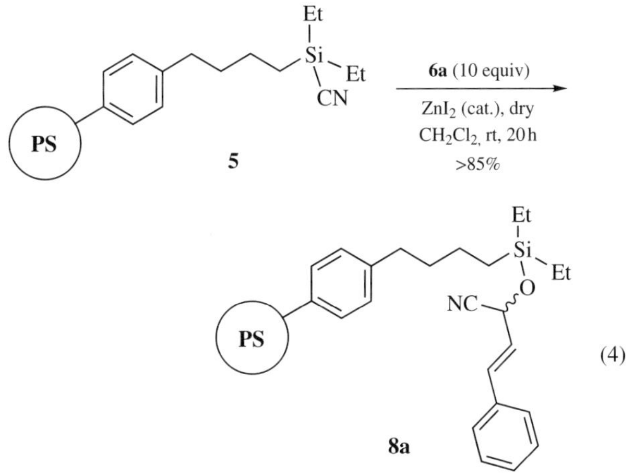

provided high yields of the expected free cyanohydrin (**7d,7f,7i,7j**). Resin **8h** yielded an equimolar mixture of **7h** and of the aldehyde **11h** after chromatography, while **8e** yielded only the aldehyde **11e** after chromatography due most likely to the strong electron-withdrawing character of its substituent.

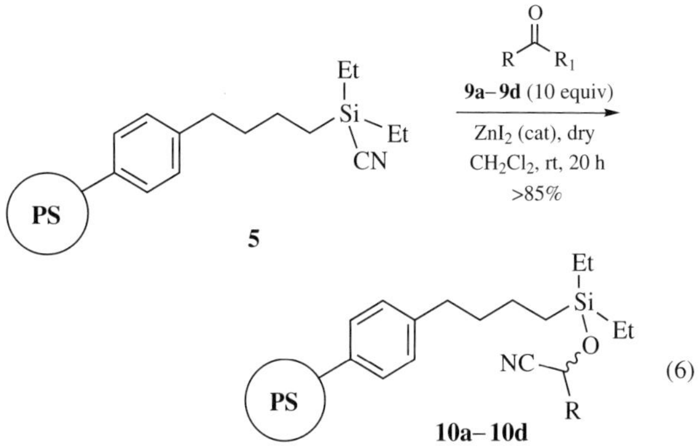

$$(3)$$

$$(4)$$

R = Me, R$_1$ = *p*-MeOPh (**10a**);

R = Et, R$_1$ = *p*-FPh (**10b**);

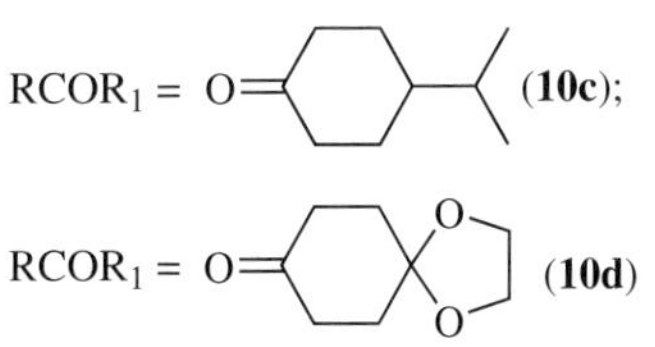

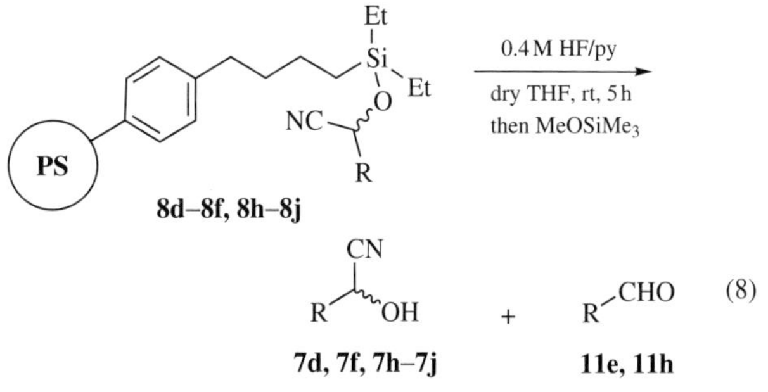

RCOR$_1$ = (**10c**);

RCOR$_1$ = (**10d**)

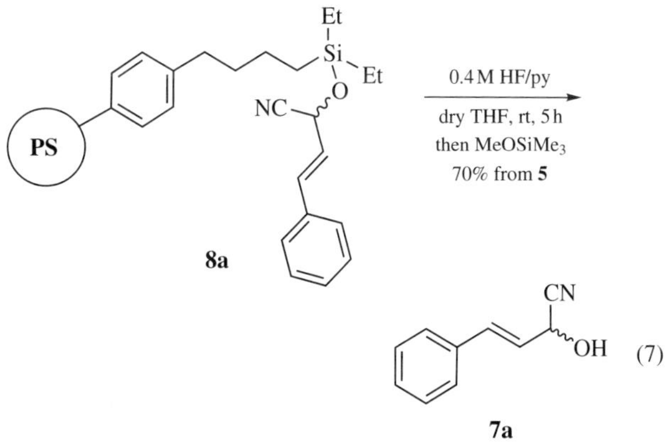

Resin **8a** shows extensive shelf stability based on IR and ^{1}H–MAS–NMR, showing no sign of degradation for up to 4 months. Several other aldehydes (eq 5) and ketones (eq 6) were reacted with **5** in order to expand the scope of the supported transformation.[3] The only failure was observed when nicotinaldehyde was used. Supported cyanohydrins **8b–8k** and **10a–10d** were obtained in > 85% yield, as estimated by ^{1}H–MAS–NMR analysis. Resin **8g** was also submitted to elemental analysis, and based on its bromine content a 96% conversion from **5** was determined.

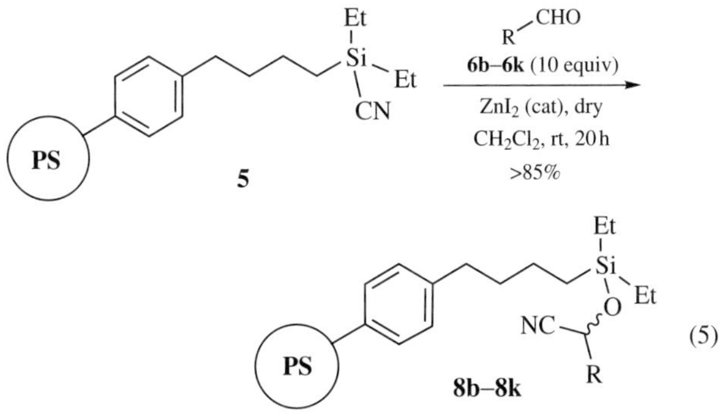

$$(5)$$

$$(6)$$

$$(7)$$

$$(8)$$

R = Me (**8b**), *t*-Bu (**8c**), Ph (**8d**), *p*-NO$_2$Ph (**8e**)
  p-MeOPh (**8f**), *p*-Br (**8g**), β-naphthyl (**8h**),
  2-thienyl (**8i**), 2-furyl (**8j**), 2-(1-MePyrrolyl) (**8k**)

Resin **8a** was reacted under optimized hydrolytic conditions to provide the free cyanohydrin **7a** by treatment with HF/pyridine complex, followed by addition of MeOSiMe$_3$ (as HF scavenger) and chromatography (eq 7).[3] Several other experimental protocols could be used, but either yields were lower or the experimental protocol was more complex. Some supported cyanohydrins were also treated under hydrolytic conditions (eq 8).[3] Four of them

7d: 98%; **7f**: 63%; **7h**: 50%;
7i: 89%; **7j**: 87%; **11e**: 95%; **11h**: 50%

Resin **8h** was also cleaved with acids in EtOH to provide the ethyl ester **12h** in solution (eq 9) in a satisfactory 63% overall yield.[3]

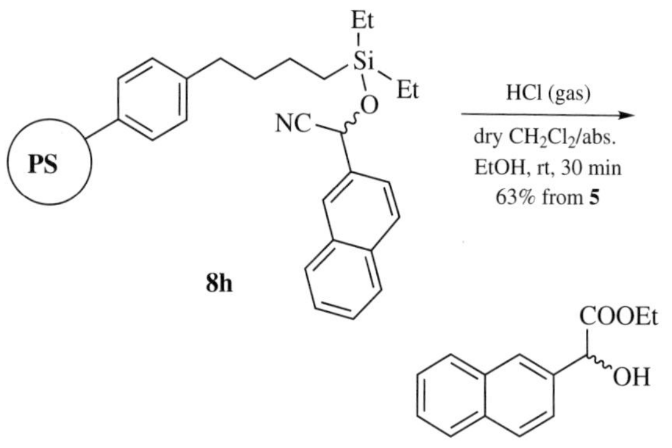

8h

$$\xrightarrow[\substack{\text{dry CH}_2\text{Cl}_2/\text{abs.}\\\text{EtOH, rt, 30 min}\\63\% \text{ from } \mathbf{5}}]{\text{HCl (gas)}}$$

(9)

12h

These alternative cleavage conditions were tested on a small library pool composed of equimolar quantities of supported cyanohydrins **8a–8k** (eq 10).[3] The cleavage mixture was collected and analyzed by ESI–MS which showed all expected molecular ions. Each of them was detected in similar intensity, thus indicating high yields in the cleavage, with the exception of α-hydroxyesters **12e**, **12i**, and **12j** whose intensity was significantly lower.

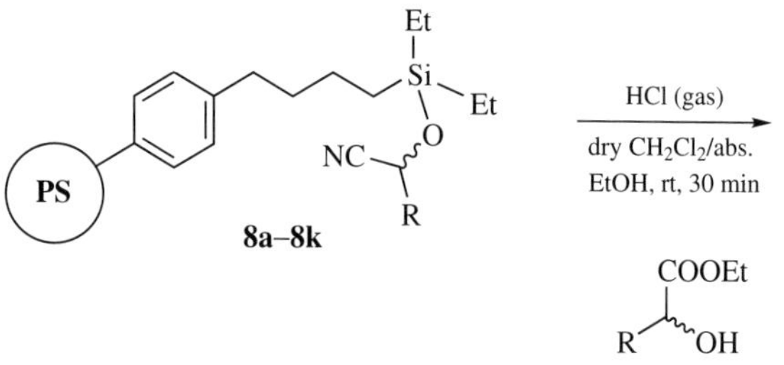

8a–8k

$$\xrightarrow[\substack{\text{dry CH}_2\text{Cl}_2/\text{abs.}\\\text{EtOH, rt, 30 min}}]{\text{HCl (gas)}}$$

(10)

12a–12k

12a–12d, 12f–12h, 12k: strong ESI–MS detection
12e, 12i, 12j: weak ESI–MS detection

Related Reagents. Polymer-supported diethyl silane; polymer-supported trimethylsilyl azide.

1. Hu, Y.; Porco, J. A., Jr; Labadie, J. W.; Gooding, O. W.; Trost, B. M., *J. Org. Chem.* **1998**, *63*, 4518.

2. Hu, Y.; Porco, J. A., Jr, *Tetrahedron Lett.* **1998**, *39*, 2711.

3. Missio, A.; Marchioro, C.; Rossi, T.; Panunzio, M.; Selva, S.; Seneci, P., *Biotechnol. Bioeng.* **2000**, *71*, 38.

4. Golinski, M.; Brock, C. P.; Watt, D. S., *J. Org. Chem.* **1993**, *58*, 159.

5. Gregory, R. J. H., *Chem. Rev.* **1999**, *99*, 3649.

6. Evans, D. A.; Truesdale, L. K.; Carroll, G. L., *J. Chem. Soc. Chem. Commun.* **1973**, 55.

7. Deuchert, K.; Hertenstein, U.; Huenig, S., *Synthesis* **1973**, 777.

Pierfausto Seneci
Nucleotide Analog Design AG, Munich, Germany

Tribenzylchlorosilane

$(PhCH_2)_3SiCl$

[18740-59-5] $C_{21}H_{21}ClSi$ (MW 336.96)
InChI = 1S/C21H21ClSi/c22-23(16-19-10-4-1-5-11-19,17-20-12-6-2-7-13-20)18-21-14-8-3-9-15-21/h1-15H,16-18H2
InChIKey = UTXPCJHKADAFBB-UHFFFAOYSA-N

(reagent for alcohol protection[5])

Physical Data: mp 141–142 °C;[1,2] a somewhat higher value (143–145 °C) is cited in the Aldrich catalog.

Solubility: sparingly sol cold petroleum ether; moderately sol hot petroleum ether; sol ether.

Form Supplied in: white crystalline solid; commercially available.

Preparative Methods: prepared by reaction of benzylmagnesium chloride with SiCl4 in ether;[3] separation from mono- and dibenzylated chlorosilane byproducts is then accomplished by fractional distillation (boiling range 300–360 °C/100 mmHg). Other preparations involve the chlorination of tribenzylsilane using Cl2[1] or an acyl chloride.[1,3] No yields are reported for any of the above preparative procedures.

Purification: can be recrystallized from petroleum ether.[2]

Handling, Storage, and Precautions: corrosive; should be handled with caution. It does not fume perceptibly in air,[2] but it is moisture sensitive and is slowly decomposed by water over several hours to give the corresponding silanol.[3]

Protection of Alcohols. Tribenzylchlorosilane is highly reactive towards nucleophilic displacement of chloride ion,[4] but it has found only limited application as a bulky silyl protecting group for alcohols. The protection of the allylic alcohol (**1**) as the tribenzylsilyl ether (**2**) (eq 1) allowed for high π-facial selectivity in a subsequent epoxidation reaction at the 10,11-double bond.[5] The tribenzylsilyl ether group can be efficiently deprotected to the free alcohol using AcOH/THF/H2O (3:1:1) at room temperature.[5]

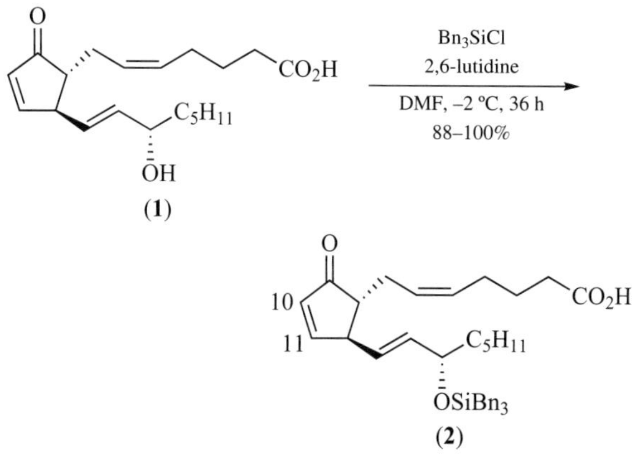

(1)

$$\xrightarrow[\substack{\text{2,6-lutidine}\\\text{DMF, }-2\,°\text{C, 36 h}\\88–100\%}]{\text{Bn}_3\text{SiCl}}$$

(1)

(2)

Silylation of Azomethine Anions. The only other synthetically useful reaction reported for tribenzylchlorosilane involves the *C*-silylation of azomethine anions. Deprotonation of the parent imine (**Butyllithium**) followed by *C*-trapping of the resulting azomethine anion affords the α'-silylimine in moderate yield (eq 2).[6]

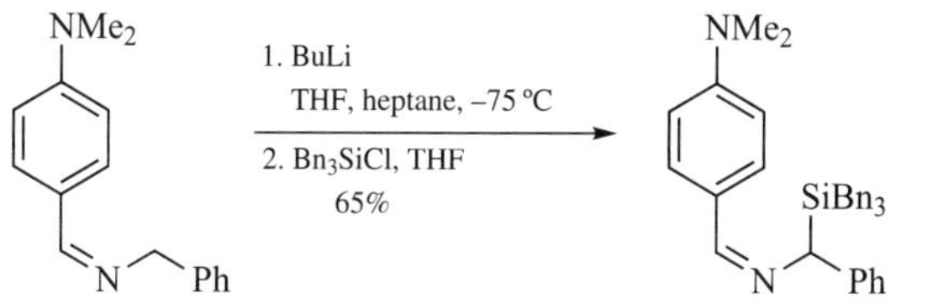

$$ (2) $$

Related Reagents. *t*-Butyldimethylchlorosilane; *t*-Butyldimethylsilyl Trifluoromethanesulfonate; Chlorotriethylsilane.

1. Jenkins, J. W.; Post, H. W., *J. Org. Chem.* **1950**, *15*, 556.
2. Robison, R.; Kipping, F. S., *J. Chem. Soc.* **1908**, 439.
3. Martin, G.; Kipping, S. F., *J. Chem. Soc.* **1909**, 302.
4. Grant, M. W.; Prince, R. H., *J. Chem. Soc. (C)* **1969**, 1138.
5. Corey, E. J.; Ensley, H. E., *J. Org. Chem.* **1973**, *38*, 3187.
6. Popowski, E.; Konzempel, K.; Schott, G., *Z. Chem.* **1974**, *14*, 289.

Paul Sampson
Kent State University, Kent, OH, USA

Tri(*t*-butoxy)silanethiol

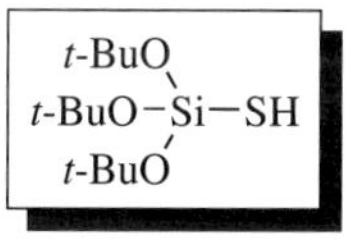

[690-52-8] $C_{12}H_{28}O_3SSi$ (MW 280.50)
InChI = 1S/C12H28O3SSi/c1-10(2,3)13-17(16,14-11(4,5)6)
        15-12(7,8)9/h16H,1-9H3
InChIKey = ZVUGYOCGLCLJAV-UHFFFAOYSA-N

(reagent used as an efficient hydrogen donor and catalyst in the context of polarity reversal catalysis in radical chain reactions)

Alternate Name: TBST.
Physical Data: bp 113–115 °C/35 mmHg;[1] 95 °C/15 mmHg.[2]
Solubility: soluble in common organic solvents.
Form Supplied: colorless liquid, not commercially available.
Analysis of Reagent Purity: the reagent may be checked by bp or analyzed by ^{1}H NMR spectroscopy.
Preparative Method: this reagent can be prepared by alcoholysis of silicon disulfide (SiS$_2$).[1,2] Representative procedure: powdered silicon disulfide (95% pure; 20.2 g, 0.21 mol) was charged into a 100 ml round-bottomed flask containing a robust stirrer bar and equipped with a reflux condenser. *t*-Butyl alcohol (60.0 g, 0.81 mol) was added and the mixture was stirred and heated under reflux under nitrogen for 72 h. The cooled reaction mixture was filtered through Celite to remove unreacted silicon disulfide and the filter cake was washed

with diethyl ether. Excess of alcohol and diethyl ether were removed from the filtrate by rotary evaporation and the residual oil was distilled under reduced pressure to give the silanethiol (18.1 g, 31%) as a colorless liquid, bp 95 °C/15 mmHg, 113–115 °C/35 mmHg).
Purification: this reagent may be purified by distillation.
Handling, Storage, and Precautions: thiols are known to be susceptible to autooxidation. This reagent may be toxic and may possess an unpleasant order. It should be kept under an inert atmosphere and handled with care.

Thiol-catalyzed Radical-chain Cyclization of Unsaturated Acetals and Thioacetals. When the unsaturated dioxolane **1** and a radical initiator, 2,2-di(*t*-butylperoxy)butane (DBPB), were heated at 125 °C in octane in the presence of tri(*t*-butoxy) silanethiol (TBST), the spirocyclic ketal **2** was formed cleanly and isolated in 92% yield (eq 1).[3] When the reaction was performed in the absence of TBST, compound **2** was not detected. TBST is believed to promote the generation of the 1,3-dioxolan-2-yl radical **3** by hydrogen-atom abstraction from **1** in a process termed polarity-reversal catalysis [4] (eq 2).

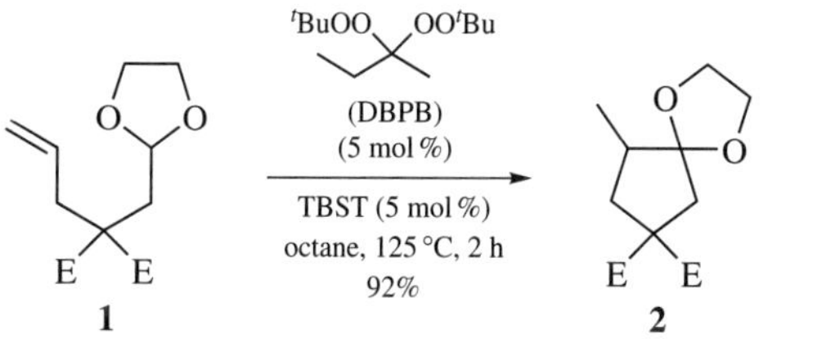

$$ (1) $$

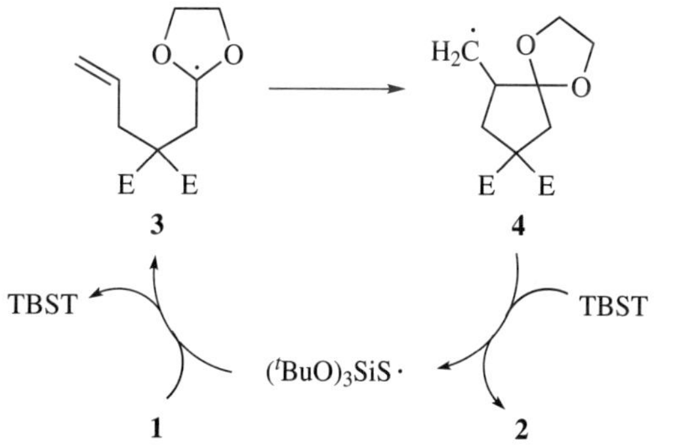

$$ (2) $$

In the same context, TBST effectively catalyzes the cyclization of the unsaturated 1,3-dithiane **5** to give the cyclized product **6** in excellent yield (eq 3).

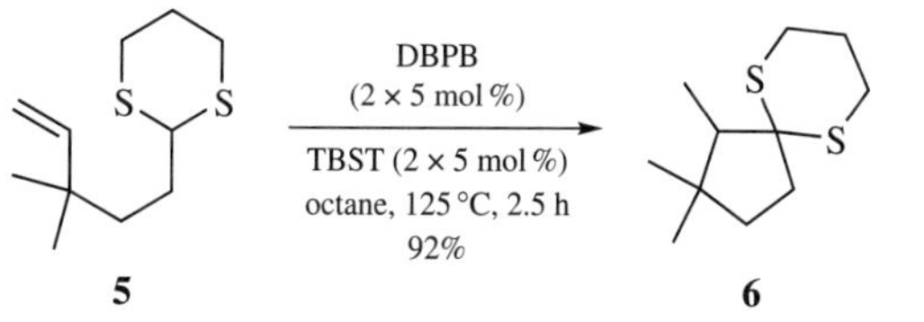

$$ (3) $$

Selective Radical-chain Epimerization at C–H Bonds. When heated to 125 °C in a hydrocarbon solvent in the presence of a thiol as the catalyst and DBPB as initiator, the *cis*-cyclic ketal **7** underwent selective epimerization at the C–H center α to oxygen to give the thermodynamically more stable *trans*-epimer **8**. Silanethiols were found to be more effective protic-polarity reversal catalysts than alkanethiols (eq 4).[5] Thus, in the presence of TBST and a small amount of collidine as the scavenger of adventitious acid formed under the reaction conditions, the epimerization of **7** proceeded smoothly to give **8** in 84% conversion after 1 h. Interestingly, when triphenylsilanethiol was used as the catalyst, the coadministration of collidine proved to be detrimental, resulting in a suppression of the isomerization. This is probably because this latter thiol is susceptible to nucleophilic attack by the base. The improved performance of TBST is therefore attributed to its stability toward nucleophilic substitution at the silicon center.

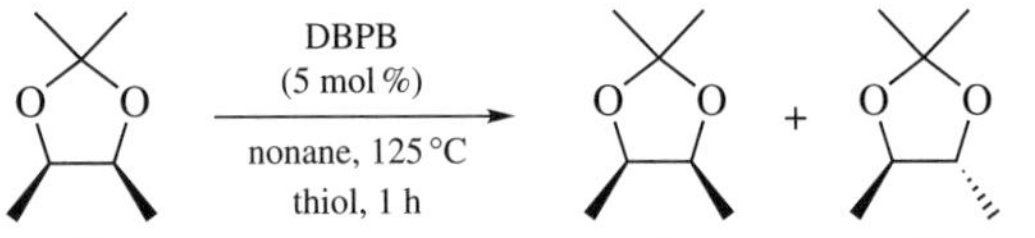

$$\text{(4)}$$

tert-C$_{12}$H$_{25}$SH	63:37
+ collidine (10 mol %)	
Ph$_3$SiSH	24:76
TBST (5 mol %)	16:84
+ collidine (10 mol %)	

In a sequence of derivatization, epimerization, and deprotection, the thermodynamically more stable *trans*-diol **9** was transformed to the less stable *cis*-isomer **10** through an efficient **11**→**12** epimerization (95% conversion) catalyzed by TBST (eq 5).[6]

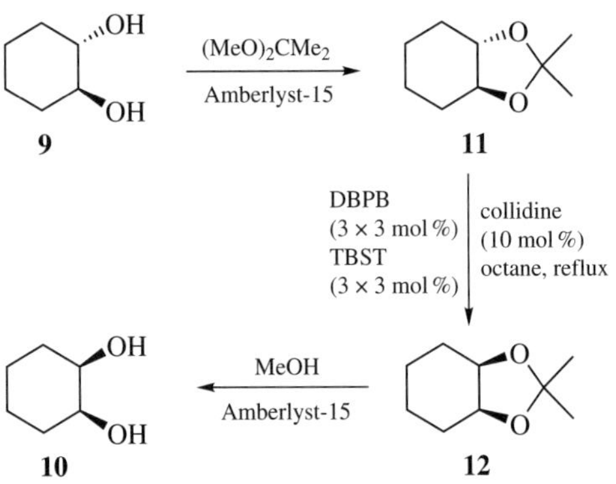

$$\text{(5)}$$

Cyclic ketals derived from carbohydrates also undergo synthetically useful thiol-catalyzed epimerization. For example, the α-D-mannofuranoside (**13**) was epimerized selectively at C-5 to give the β-L-gulofuranoside (**18**) in 30% conversion and 24% isolated yield (eq 6).[6]

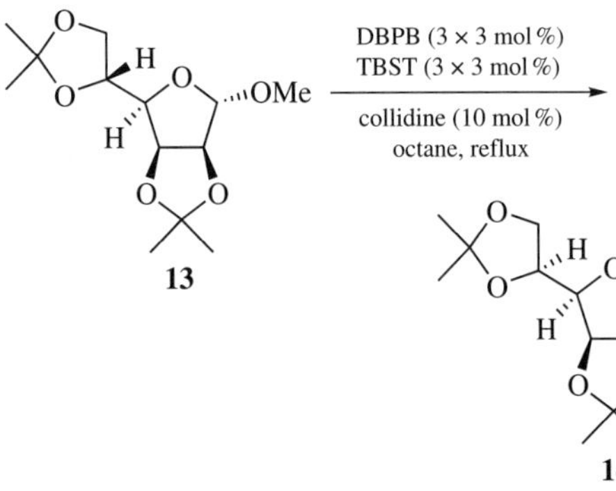

$$\text{(6)}$$

Radical-chain Deoxygenation of Tertiary Alcohols. As outlined by the general reaction scheme in eq 7, methoxymethyl (MOM) ethers derived from tertiary alcohols undergo TBST-mediated deoxygenative cleavage to form hydrocarbon products.[7] For example, when MOM ethers **15–17** were treated with DBPB in the presence of TBST and collidine in refluxing octane, the deoxygenation products **18–20** were isolated in high yields (eq 8).

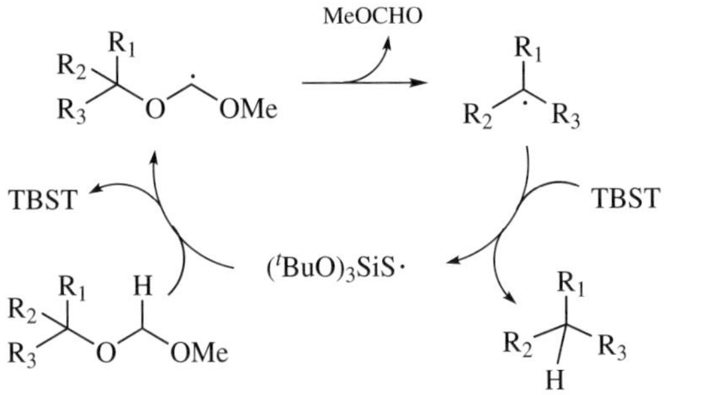

$$\text{(7)}$$

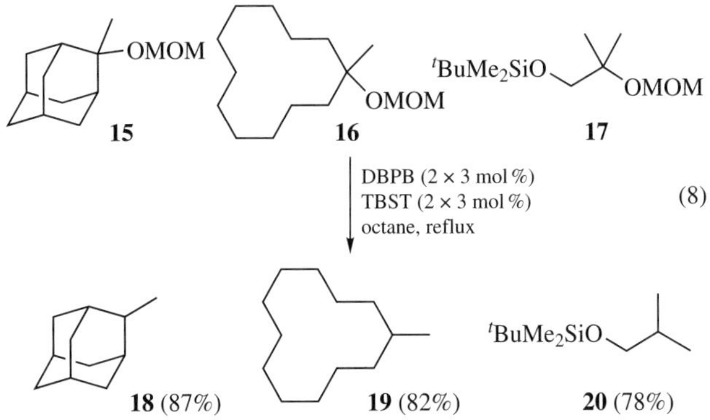

$$\text{(8)}$$

When the tertiary MOM ether **21**, which can be prepared from diacetone D-glucose, was subject to this polarity-reversal-catalysis protocol, the deoxygenated products **22** and **23** were isolated in excellent yield (90%) and good diastereoselectivity (91:9) (eq 9).[7]

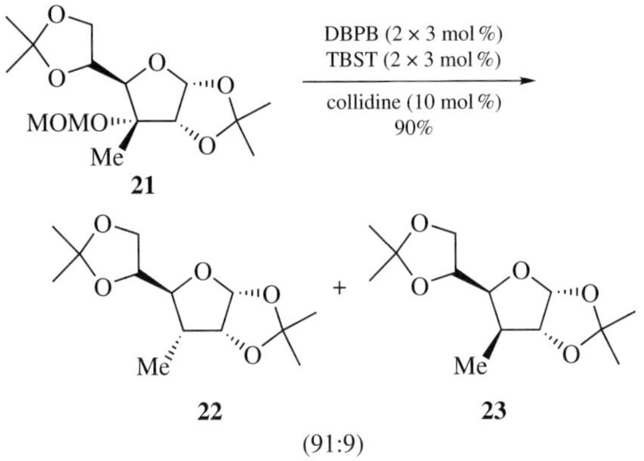

$$\text{(9)}$$

Radical-chain Redox Rearrangement of Cyclic Acetals. Benzylidene acetals of diols undergo radical-chain cleavage to give benzoate esters via the formation and subsequent selective fragmentation of a di(α-alkoxyl)benzyl radical.[8] Thus, cyclic acetal **24** was quantitatively converted to benzoate **25** under the catalysis of TBST (eq 10). Application to the carbohydrate derived cyclic acetal **26** led to the formation of the 6-deoxy glycoside **27** in high yield (eq 11).[8,9]

$$(10)$$

24   **25**

26

$$(11)$$

27

This chemistry has been successfully applied to the conversion of the tartaric acid derived acetal **28** to the ester derivative **30** of the unusual (R)-malic acid (eq 12).[8]

$$(12)$$

28

29   **30**

Other Applications.   TBST is widely used as a thiolate ligand for the synthesis of various metal thiolates as exemplified by the formation of Zn(II) thiolate **31** (eq 13)[10] and silver(II) complex **32** (eq 14).[11]

$$(13)$$

31

$$(14)$$

32

Related Reagents.   Triisopropylsilanethiol (TIPST); triphenyl-silanethiol (TPST).

1. Piekos, R.; Wojnowski, W., *Z. Anorg. Allg. Chem.* **1962**, *318*, 212.

2. Dang, H.-S.; Roberts, B. P.; Tocher, D. A., *J. Chem. Soc., Perkin Trans. 1* **2001**, 2452.

3. Dang, H.-S.; Roberts, B. P., *Tetrahedron Lett.* **1999**, *40*, 8929.

4. Roberts, B. P., *Chem. Soc. Rev.* **1999**, *28*, 25.

5. Dang, H.-S.; Roberts, B. P., *Tetrahedron Lett.* **1999**, *40*, 4271.

6. Dang, H.-S.; Roberts, B. P., *Tetrahedron Lett.* **2000**, *41*, 8595.

7. Dang, H.-S.; Franchi, P.; Roberts, B. P., *Chem. Commun.* **2000**, 499.

8. Roberts, B. P.; Smits, T. M., *Tetrahedron Lett.* **2001**, *42*, 137.

9. Dang, H.-S.; Roberts, B. P.; Sekhon, J.; Smits, T. M., *Org. Bimol. Chem.* **2003**, *1*, 1330.

10. Becker, B.; Radacki, K.; Wojnowski, W., *J. Organomet. Chem.* **1996**, *521*, 39.

11. Chojnacki, J.; Becker, B.; Konitz, A.; Potrzebowski, M.; Wojnowski, W., *J. Chem. Soc., Dalton Trans.* **1999**, 3063.

Qingwei Yao

Northern Illinois University, DeKalb, IL, USA

(E)-1-Tri-n-butylstannyl-2-trimethylsilylethylene[1,2]

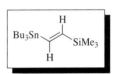

[58207-97-9] $C_{17}H_{38}SiSn$ (MW 389.35)

InChI = 1S/C5H11Si.3C4H9.Sn/c1-5-6(2,3)4;3*1-3-4-2;/h1,5H,2-4H3;3*1,3-4H2,2H3;

InChIKey = JDQLLFOVXMXKNW-UHFFFAOYSA-N

(used in the Pd^0-catalyzed cross coupling of vinyl and aryl halides or triflates;[1,2] a source of β-trimethylsilylvinyllithium[3])

Physical Data: bp 96–102 °C/0.5 mmHg.

Analysis of Reagent Purity: IR 1538 cm^{-1} (C=C); ^{1}H NMR (CDCl$_3$, CH$_2$Cl$_2$) δ 0.12 (9H, s), 0.95 (9H, t, J = 7 Hz), 1.41 (18H, m), 6.65 and 7.08 (2H, AB pattern, J = 23 Hz).

Solubility: sol ethers, DMF, hydrocarbon, aromatic solvents.

Preparative Method: bis(tri-n-butylstannyl)ethylene is treated successively with molar equivalents of n-butyllithium and chlorotrimethylsilane to give (E)-1-tri-n-butylstannyl-2-trimethylsilylethylene.[5]

Handling, Storage, and Precautions: most tin compounds are toxic and are readily absorbed through the skin.[4] Their preparation and use should be carried out at all times in a well-ventilated fume hood.

Pd^0-Catalyzed Cross Couplings.   In general, (E)-1-tri-n-butylstannyl-2-trimethylsilylethylene undergoes the same types of Pd^0-catalyzed cross-coupling reactions as other substituted vinyl-stannanes. It readily couples with vinyl halides in the presence of catalytic amounts of Pd^0, giving 1,3-dienes in high yields (eq 1).[6] Vinyl and aryl triflates can undergo the cross-coupling reactions with the organostannane reagent, yielding 1,3-dienes provided lithium chloride is also present (eq 2).[7,8]

When the coupling reactions of vinyl and aryl triflates are carried out in the presence of carbon monoxide and LiCl, good yields of the cross-coupled ketones are obtained (eqs 3 and 4).[9] This is a particularly attractive route to divinyl ketones which are important substrates in the Nazarov cyclization.

The reagent also couples predominantly 1,4 to acyclic vinyl epoxides to give allylic alcohols. The regioselectivity is controlled by the substitution pattern of the vinyloxirane. The 1,4 versus 1,2 selectivity could be further enhanced by the addition of water (10 equiv based on the vinyl epoxide) (eq 5).[10,11]

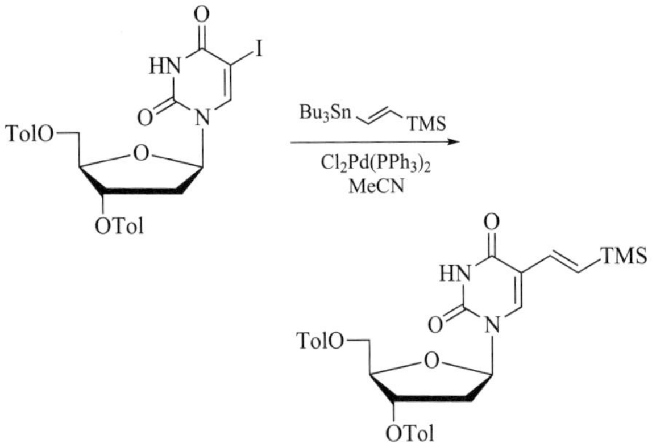

(1)

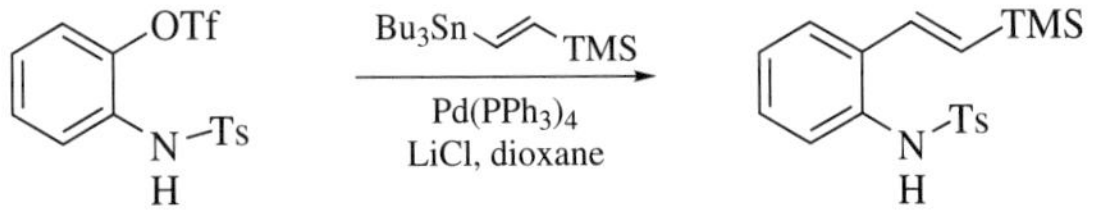

(2)

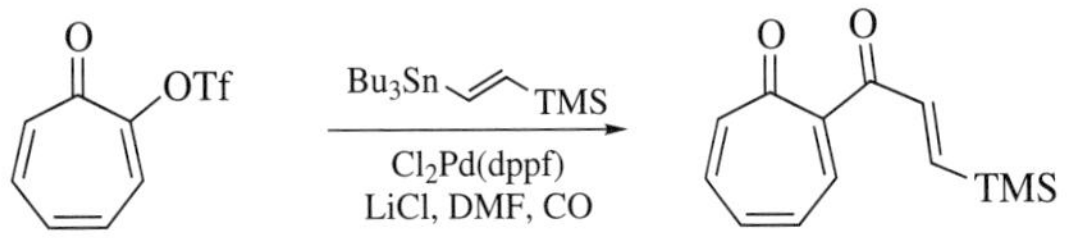

(3)

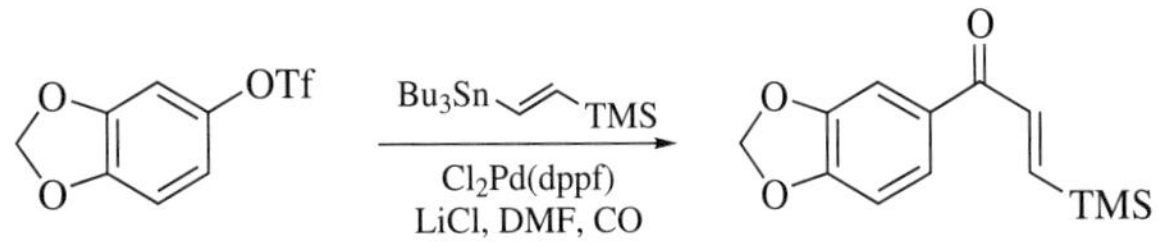

(4)

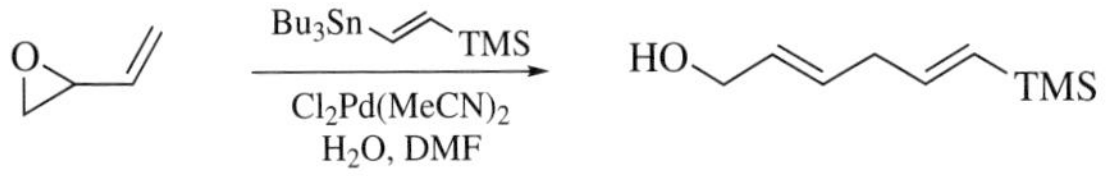

(5)

For additional discussion about palladium-catalyzed coupling reactions, see also those entries dealing with organopalladium catalysis (e.g. tetrakis(triphenylphosphine)palladium(0), tris-(dibenzylideneacetone)dipalladium–chloroform, and (E)-1-trimethylsilyl-2-trimethylstannylethylene.

Transmetalation Reactions. Transmetalation of trimethylsilyl-2-trimethylstannylethylene at low temperature with BuLi affords *trans*-β-trimethylsilyllithium in high yield, as evidenced by derivatization of 2-methyldimedone isobutyl ether (eq 6).[3,12–14]

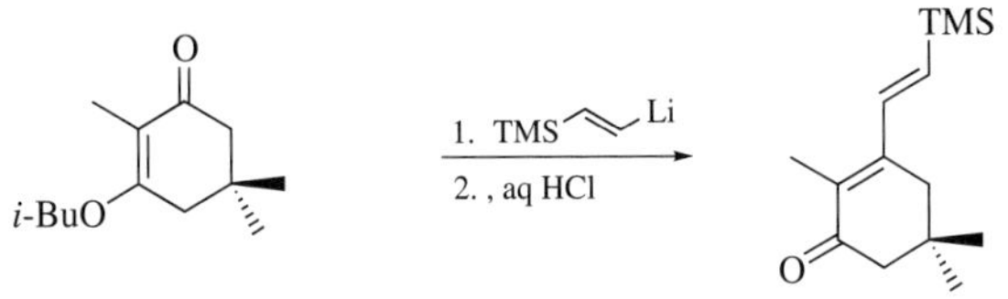

(6)

1. Stille, J. K., *Pure Appl. Chem.* **1985**, *57*, 1771.
2. Stille, J. K., *Angew. Chem., Int. Ed. Engl.* **1986**, *25*, 508.
3. Cunico, R. F.; Clayton, F. J., *J. Org. Chem.* **1976**, *41*, 1480.
4. Krigman, M. R.; Silverman, A. P., *Neurotoxicology* **1984**, *5*, 129.
5. Seyferth, D.; Vick, S. C., *J. Organomet. Chem.* **1978**, *144*, 1.

6. Crisp, G. T., *Synth. Commun.* **1989**, *19*, 2117.
7. Echavarren, A. M.; Stille, J. K., *J. Am. Chem. Soc.* **1987**, *109*, 5478.
8. Crisp, G. T.; Flynn, B. L., *Tetrahedron Lett.* **1990**, *31*, 1347.
9. Echavarren, A. M.; Stille, J. K., *J. Am. Chem. Soc.* **1988**, *110*, 1557.
10. Echavarren, A. M.; Tueting, D. R.; Stille, J. K., *J. Am. Chem. Soc.* **1988**, *110*, 4039.
11. Tueting, D. R.; Echavarren, A. M.; Stille, J. K., *Tetrahedron* **1989**, *45*, 979.
12. Seyferth, D.; Vick, S. C., *J. Organomet. Chem.* **1978**, *144*, 1.
13. Burke, S. D.; Murtiashaw, C. W.; Dike, M. S.; Strickland, S. M. S.; Saunders, J. O., *J. Org. Chem.* **1981**, *46*, 2400.
14. Padaw, A.; Eisenbarth, P.; Venkataramanan, M. K.; Wong, G. S. K., *J. Org. Chem.* **1987**, *52*, 2427.

Kevin J. Moriarty
Rhône-Poulenc Rorer, Collegeville, PA, USA

Trichloroiodosilane

[13465-85-5]   Cl₃ISi   (MW 261.34)

Cl_3ISi

InChI = 1S/Cl3ISi/c1-5(2,3)4
InChIKey = OIIOOEGXCZBZMX-UHFFFAOYSA-N

(selective ether-cleaving reagent;[1] also cleaves acetals;[1,2] mediates reaction of aromatic aldehydes with nitriles to yield secondary amides;[3] aromatic aldehydes and acrylonitrile give iminoaldehydes[4])

Physical Data: bp 114.5 °C/760 mmHg; mp <−60 °C.
Preparative Methods: prepared through halogen exchange by heating $SiCl_4$ and sodium iodide in methylene chloride–acetonitrile or toluene–acetonitrile.[1] It also has been prepared by the reaction of $SiCl_4$ and hydrogen iodide,[5] by heating $SiHCl_3$ with iodine,[5] and by heating silicon with ICl.[6]
Handling, Storage, and Precautions: Cl_3SiI is a moisture-sensitive material; it could be conveniently used in situ. Use in a fume hood.

Ether Cleavage. An important application of Cl_3SiI is in regiospecific and selective cleavage of alkyl–alkyl and alkyl–aryl ethers. Primary–secondary alkyl ethers are cleaved to yield primary alkyl iodide and secondary alcohol on workup. This cleavage pattern is opposite to that of boron halides like 9-bromo-9-borabicyclo[3.3.1]nonane[7] and boron tribromide,[8] which yield secondary bromide and primary alcohol (eq 1). Compared to iodotrimethylsilane, Cl_3SiI shows a greater degree of regiospecificity and slightly faster rate of reaction.[1]

Selective cleavage of diethers is possible with this reagent (eq 2).

Miscellaneous Transformations. Aromatic aldehydes and nitriles like acetonitrile or benzonitrile are converted by Cl_3SiI to secondary amides, through a redox reaction (eq 3). When acrylonitrile is used instead, the product is an imino aldehyde (eq 4).

The reactions with Cl₃SiI, TMSI, and 9-Br-9-BBN on cyclohexyl methyl ether:

$$\text{(1)}$$

100:0 (Cl₃SiI)

90:10 (TMSI)

0:100 (9-Br-9-BBN)

$$\text{(2)}$$

75% → 82%

$$\text{PhCHO} + \text{MeCN} + 2\,\text{Cl}_3\text{SiI} + 2\,\text{H}_2\text{O} \longrightarrow$$

$$\text{(3)}$$

$$+\ \text{I}_2 + 2\,\text{Cl}_3\text{SiOH}$$

$$\text{PhCHO} + \text{CN} \xrightarrow[\text{H}_2\text{O}]{\text{Cl}_3\text{SiI}}$$ (4)

1. Bhatt, M. V.; Elmorsy, S. S., *Synthesis* **1982**, 1048.
2. Elmorsy, S. S.; Bhatt, M. V.; Pelter, A., *Tetrahedron Lett.* **1992**, *33*, 1657.
3. Elmorsy, S. S.; Nour, M. A.; Kandeel, E. M.; Pelter, A., *Tetrahedron Lett.* **1991**, *32*, 1825.
4. Elmorsy, S. S.; Badaway, D. S.; Nour, M. A.; Pelter, A., *Tetrahedron Lett.* **1991**, *32*, 5421.
5. Besson, A., *C. R. Hebd. Seances Acad. Sci.* **1891**, *112*, 611/4.
6. Besson, A., *C. R. Hebd. Seances Acad. Sci.* **1891**, *112*, 1314/6.
7. Bhatt, M. V., *J. Organomet. Chem.* **1978**, *156*, 221.
8. Bhatt, M. V.; Kulkarni, S. U., *Synthesis* **1983**, 249.

M. Vivekananda Bhatt
Indian Institute of Science, Bangalore, India

Triethoxysilane

HSi(OEt)₃

[998-30-1]   C₆H₁₆O₃Si   (MW 164.31)

InChI = 1S/C6H16O3Si/c1-4-7-10(8-5-2)9-6-3/h10H,4-6H2,1-3H3

InChIKey = QQQSFSZALRVCSZ-UHFFFAOYSA-N

(useful reagent for the hydrosilylation of carbon–carbon multiple
bonds; reducing agent for carbonyl groups)

Physical Data: bp 134–135 °C; d 0.89 g cm⁻³.

Solubility: sol diethyl ether, THF, alkanes, aromatic and chlorinated solvents.

Form Supplied in: colorless liquid; widely available.

Handling, Storage, and Precautions: is moisture sensitive and
should be handled in a well-ventilated hood. Contact with the
eyes and skin should be avoided.

Original Commentary

Robert J. P. Corriu & Christian Guérin
Université Montpellier II, France

Hydrosilylation of Carbon–Carbon Multiple Bonds.
HSi(OEt)₃ is an efficient reagent for the hydrosilylation of alkenes,
alkynes, and conjugated dienes (eq 1).[1] The reaction is promoted by a variety of transition metal catalysts, including platinum, ruthenium, and rhodium. Chloroplatinic acid hexahydrate,
H₂PtCl₆·6H₂O, is by far the most widely used catalyst and has
been shown to be very efficient, particularly in the hydrosilylation
of alkenic substrates.[2]

$$+\ \text{HSi(OEt)}_3 \xrightarrow{\text{[Pt]}} \text{(EtO)}_3\text{Si} \qquad \text{(1)}$$

[Pt] = H₂PtCl₆·6H₂O in *t*-BuOH

The reaction is especially valuable for the introduction of functionalized groups at silicon,[3] as this cannot be achieved by simple
organometallic procedures (eq 1).[4] Furthermore, the reaction may
be quite regioselective and stereoselective, as in the case of the
hydrosilylation of ocimene (eq 2).[5]

$$\xrightarrow[\substack{100\ °\text{C, 24 h} \\ 98\%}]{\text{(PPh}_3\text{)}_3\text{RhCl}} \quad \text{Si(OEt)}_3 \quad + \quad \text{(EtO)}_3\text{Si} \qquad \text{(2)}$$

98:2

**Reduction of Functional Groups Catalyzed by Fluoride or
Alkoxide Salts.[6]** A general and powerful reduction method of
carbonyl compounds consists of using HSi(OEt)₃ activated by
potassium fluoride or cesium fluoride.[7] The reaction can proceed without solvent (eq 3)[7a] and is accelerated by polar solvents,
e.g. DMSO or DMF.[8] The reduction of aldehydes and ketones is
rapid at room temperature; also, esters are smoothly converted to
alcohols (eq 4).[7b]

$$\text{C}_6\text{H}_{13}\text{CHO} + \text{HSi(OEt)}_3 \xrightarrow[\text{25 °C, 4 h}]{\text{KF}} \xrightarrow{\text{H}_3\text{O}^+} \text{C}_6\text{H}_{13}\text{CH}_2\text{OH} \quad \text{(3)}$$

70%

$$+\ \text{HSi(OEt)}_3 \xrightarrow[\text{25 °C, 0.5 h}]{\text{CsF}} \xrightarrow{\text{H}_3\text{O}^+}$$

$$\text{OH} \qquad \text{(4)}$$

70%

The reaction is highly chemoselective: aldehydes and ketones possessing functional groups, such as carbon–carbon double bonds, or bromo, nitro, amide, and ester groups, are reduced selectively to the corresponding alcohols (eqs 5 and 6).[7a, 9]

$$
\text{Ph} \overset{O}{\underset{2}{\bigvee}} CO_2Me + HSi(OEt)_3 \xrightarrow[25\,°C,\,2.5\,h]{CsF} \xrightarrow{H_3O^+}
$$

$$
\text{Ph} \overset{O}{\underset{O}{\bigcirc}} \quad (5)
$$

85%

$$
\overset{O\ \ \ O}{\bigvee}_{NHPh} + HSi(OEt)_3 \xrightarrow[25\,°C,\,0.2\,h]{CsF} \xrightarrow{H_3O^+}
$$

$$
\overset{OH\ \ \ O}{\bigvee}_{NHPh} \quad (6)
$$

90%

Alternatively, solutions of $HSi(OEt)_3$ with lithium ethoxide or pinacolate may be used to convert aldehydes and ketones to alcohols.[10] Enantioselective reduction of prochiral ketones is achieved by use of a mixture of $HSi(OEt)_3$ and the dilithium salt of a chiral diol or amino alcohol.[11]

The reaction between carbonyl compounds and $HSi(OEt)_3$ is efficiently catalyzed by inorganic solid bases such as hydroxyapatite, $Ca_{10}(PO_4)_6(OH)_2$, at temperatures ranging from 25 to 90 °C.[12] Enones afford the corresponding 1,2-addition products.

$HSi(OEt)_3$ is easily converted into a more reactive form, namely a pentacoordinate hydridosilicate (eq 7).[13a]

$$
HSi(OEt)_3 + KOEt \xrightarrow[75\%]{\underset{25\,°C,\,2\,h}{THF}} K[HSi(OEt)_4] \quad (7)
$$

$K[HSi(OEt)_4]$ reduces aldehydes, ketones, and esters in the absence of added catalyst.[13] Treatment of nonenolizable amides yields the corresponding aldehydes. Reaction with 1 equiv of isocyanate in diethyl ether or THF leads quantitatively to the potassium imidate, which can be quenched in situ (eq 8).[13b] Furthermore, use of an excess of aryl isocyanate yields to the corresponding isocyanurate.[13b]

$$
\begin{array}{c}
R^{\diagdown N \diagdown C \diagup O} \\
+ \\
K[HSi(OEt)_4] \\
R = Ph\ or\ Cy
\end{array}
\xrightarrow[rt]{Et_2O\ or\ THF}
K^+[R\text{–}N=CH\text{–}O]^- + Si(OEt)_4
$$

$$
\xrightarrow{H^+} \quad \xrightarrow{MeI} \quad \xrightarrow{MeCOCl} \quad (8)
$$

$$
\underset{R=Ph,\,66\%}{R^{\diagdown N \diagdown CHO}_{H}} \quad \underset{R=Ph,\,70\%}{R^{\diagdown N \diagdown CHO}_{Me}} \quad \underset{R=Ph,\,70\%}{R^{\diagdown N \diagdown CHO}_{COMe}}
$$

Titanium-Catalyzed Reduction of Esters to Alcohols. $HSi(OEt)_3$ is used stoichiometrically in conjunction with a titanium-based catalytic system, conveniently prepared by the reaction of 2 equiv of n-butyllithium with dichlorobis(cyclopentadienyl)titanium (eq 9).[14a]

$$
5\%\ Cp_2TiCl_2 \xrightarrow[-78\,°C,\,15\,min]{10\%\ BuLi,\ THF} \xrightarrow[rt,\,0.5–2\,h]{\underset{2\ equiv\ HSi(OEt)_3}{Br(CH_2)_5CO_2Et}}
$$

$$
\underset{Br}{\bigvee}_5 OSi(OEt)_3 \xrightarrow{1N\ NaOH} \underset{Br}{\bigvee}_5 OH \quad (9)
$$

78%

The combination of a catalytic amount of titanium tetraisopropoxide, an inexpensive and air-stable liquid, and $HSi(OEt)_3$ also generates an effective and mild system for the conversion of esters into primary alcohols.[14b]

Selective reduction of a great variety of esters has been achieved with these methods. The procedures represent safer and convenient alternatives to those employing reducing agents such as diisobutylaluminum hydride and lithium aluminum hydride.

First Update

Karl A. Scheidt & Robert B. Lettan II
Northwestern University, Evanston, IL, USA

Hydrosilylation of Carbon–Carbon Multiple Bonds. $HSi(OEt)_3$ has been widely utilized in the hydrosilylation of unsaturated carbon–carbon bonds. Recent advancements have focused on the versatility of functional groups and isomer selectivity of the resulting alkene when using alkynes as substrates. The transition metal-catalyzed hydrosilylation of enamines has been extensively investigated. Seminal contributions from Skoda-Földes[15] and by Murai and Kato[16] demonstrate that the regioselectivity of these processes is strongly catalyst dependent (eq 10). However, in these cases the substitution on the silicon examined was limited to triethylsilane and phenyldimethylsilane. Further investigations by Sieburth have shown that a regioselective hydrosilylation of α-alkyl-α-aminosilanes can be accomplished in high yields in the presence of triethoxysilane and $Rh_2(OAc)_4$ (eq 11).[17] The use of alkyl- and aryl-substituted silanes for rhodium-catalyzed hydrosilylations provides high levels of regioselectivity, but in much lower yields compared to those obtained with triethoxysilane. The large difference in yield can be attributed to the increased reactivity of triethoxysilane due to the σ-electron withdrawing effects associated with the ethoxy substituents.[18]

$$
\underset{N}{\overset{O}{\bigcirc}} \xrightarrow[\underset{RhCl(PPh_3)_3}{PtCl_2}]{HSi(OEt)_3} \underset{N\ \ CH_3}{\overset{O\quad Si(OEt)_3}{\bigcirc}} + \underset{N}{\overset{O}{\bigcirc}} Si(OEt)_3 \quad (10)
$$

$$
0:100 \\
92:8
$$

$$
BocHN \overset{}{\diagdown} \underset{CH_3}{\diagup} \xrightarrow[\underset{toluene,\ reflux}{Rh_2(OAc)_4}]{HSi(OEt)_3} BocHN \underset{CH_3}{\overset{Si(OEt)_3}{\diagdown \diagup}} \quad (11)
$$

Stereoselective hydrosilylations of alkynes have been explored using a wide variety of rhodium and ruthenium catalysts.[19]

Similar to the results seen for the hydrosilylation of enamines (vide supra), the levels of regioselectivity with the hydrosilylation of terminal alkynes greatly depends on catalyst structure. In addition, catalyst ligand substitution has a direct effect on the stereoselectivity attained for the olefin products. Ozawa and Katayama have achieved the highly selective generation of (E)- and (Z)-alkenylsilanes through slight changes of the ligands on the ruthenium metal (eqs 12 and 13).[20] For example, the use of trialkylphosphine ligands on ruthenium promotes the generation of (Z)-alkenylsilanes. Trost has subsequently reported the use of Cp*Ru(MeCN)$_3$PF$_6$ to produce the Markovnikov hydrosilylation product (eq 14).[21] The vinylsilanes synthesized by these methods are particularly useful for palladium-catalyzed cross-coupling reactions,[22] as acceptors in conjugate additions,[23] as masked aldehydes and ketones through the Tamao–Fleming oxidation,[24] and as terminators for cation cyclizations.[25] Additionally, vinylsilanes are stable enough to undergo further synthetic manipulations, unlike their vinylborane counterparts.

$$\text{(12)}$$

$$\text{(13)}$$

$$\text{(14)}$$

The hydrosilylation of terminal alkynes disclosed by Trost can be applied to internal alkynes as well.[21] Remarkably, the (Z)-isomer is generated in this process, resulting from trans addition during hydrosilylation. The protodesilylation of these silylated products in the presence of copper(I) iodide and tetrabutylammonium fluoride (TBAF)[26] or silver(I) fluoride (eq 15)[27] leads to internal *trans*-olefins. This two-step method is a useful synthetic transformation to access (E)-alkenes from internal alkynes.[27a,28] In contrast, the chemoselective reduction of alkynes to the corresponding (Z)-alkenes is conventionally accomplished readily with Lindlar's catalyst.[29] The complementary process to afford (E)-olefins has proven much more difficult. Methods involving metal hydrides,[30] dissolving metal reductions,[31] and low-valent chromium salts[32] provide the desired chemical conversion, albeit with certain limitations. For example, functional substitution at the propargylic position (alcohols, amines, and carbonyl units) is often necessary to achieve selectivity in these transformations. Conversely, the hydrosilylation/protodesilylation protocol is a mild method for the reduction of alkynes to (E)-alkenes.

$$\text{(15)}$$

The hydrosilylation of unsaturated carbon–carbon bonds has also been extended to rhodium-catalyzed silylcarbocyclizations.[33] In the presence of Rh$_4$(CO)$_{12}$ and triethoxysilane, a rigid triyne backbone can undergo a silylcarbotricyclization cascade reaction to yield [5,6,5]-tricycles (eq 16). Similar to the results observed by Sieburth for the hydrosilylation of enamines,[17] the alkoxysilane functionality provides significant rate enhancement in comparison to silylcarbocyclizations using alkyl- and arylsilane reagents. The incorporation of carbonyl functionality as terminal electrophiles into these cyclizations has also been successful.[33b] Rhodium-catalyzed carbonylative silylcarbocyclizations proceed in the presence of carbon monoxide (10 atm) to incorporate a carbonyl unit, usually as the aldehyde. Both of these tandem addition/cyclization strategies produce functionalized carbocycles with simultaneous incorporation of silyl functionality as aryl- and vinylsilanes. These alkenylsilanes can then be exploited for further synthetic manipulations as discussed above.[22–25]

$$\text{(16)}$$

Hydrosilylation of Carbonyl Compounds and Imines. The regioselective synthesis of enol silyl ethers from the hydrosilylation of α,β-unsaturated carbonyl compounds can be accomplished using triethoxysilane (eq 17). Both rhodium[34] and platinum[35] complexes catalyze this transformation, which promotes highly regioselective 1,4-reduction versus 1,2-hydrosilylation of α,β-unsaturated carbonyl compounds.[36]

$$\text{(17)}$$

Similar to the results discussed for the silylcarbocyclizations of carbon–carbon multiple bonds, reductive cyclizations in the presence of carbonyl compounds are readily achieved. Crowe has developed a titanium-catalyzed procedure for the intramolecular reductive coupling of δ,ε-unsaturated carbonyl compounds in the presence of triethoxysilane (eq 18).[37] The electronic advantage of triethoxysilane is demonstrated by the lack of reductive coupling in the presence of less reactive silanes, such as triethylsilane and diphenylsilane. With this method, Mori has utilized nickel(0) catalysts to generate five- and six-membered carbocycles and pyrrolidine derivatives.[38] Furthermore, coordination of a chiral phosphine ligand to the nickel catalyst renders the reaction moderately enantioselective.[38f]

$$(18)$$

Asymmetric hydrosilylation of ketones[39] and ketoimines[40] has been demonstrated in the absence of transition metal catalysts. Using catalytic amounts of chiral-alkoxide Lewis bases such as binaphthol (BINOL), Kagan was able to facilitate the asymmetric reduction of ketones (eq 19).[39a] This process is believed to arise from activation of the triethoxysilane by mono-alkoxide addition to give an activated pentavalent intermediate, which can undergo coordination of an aldehyde. This highly ordered hexacoordinate transition state directs reduction in an asymmetric manner, with subsequent catalyst regeneration. Brook was able to facilitate a similar tactic for asymmetric reduction by employing histidine as a bi-dentate Lewis base activator of triethoxysilane.[39b] A similar chiral lithium-alkoxide-catalyzed asymmetric reduction of imines was demonstrated by Hosomi with the di-lithio salt of BINOL and trimethoxysilane.[40]

$$(19)$$

Hydrosilylation of Cyclopropanes. β-Silylated olefins are prepared via the hydrosilylation of methylenecyclopropanes in the presence of catalytic amounts of Rh(I) complexes (eq 20).[41] The methylenecyclopropanes used in the process are versatile substrates that have become readily available and have various applications in organic synthesis as building blocks.[42]

$$(20)$$

The hydrosilylation of methylenecyclopropanes is proposed to proceed via oxidative addition to the olefin, followed by rhodium migration across the strained cyclopropane ring, and eventual reductive elimination to give the silyl-substituted olefins. The process is compatible with aromatic and aliphatic substitution on the olefin and often requires heating. Additionally, cyclopropyl-substituted methylenecyclopropanes may be selectively silylated to give alkenes containing one, two, or three β-silylated olefin chains.

Silylation of Aryl and Alkenyl Halides. Aryl halides can be successfully silylated in the presence of palladium and rhodium catalysts.[43] The palladium(0)-catalyzed coupling reaction of triethoxysilane with aryl iodides and bromides was developed by Masuda and Murata (eq 21),[43a] and later improved on by DeShong.[43b] This is an efficient process in most cases, resulting in facile formation of arylsilanes. One limitation of this palladium(0)-catalyzed process is the decreased effectiveness in the presence of *ortho*-substituted and electron-deficient aryl halides. The successively developed rhodium(I)-catalyzed process by Murata and Masuda is broadly applicable to a wide range

of sterically demanding and electronically diverse aryl halides in high yield.[43c] The arylsilanes accessed with these cross-coupling methods are both valuable intermediates in organic synthesis[44] and useful cross-linking agents.[45]

$$(21)$$

In addition to the silylation of aryl halides, alkenyl iodides can also be silylated in the presence of a palladium(0) catalyst (eq 22).[46] This silylation proceeds stereoselectively with retention of the carbon–carbon bond stereochemistry, with neither the α- nor (Z)-isomer being produced. The byproduct that arises from this reaction is the saturated β-aryl triethoxysilane resulting from hydrosilylative reduction of the olefin (97:3 unsaturated:saturated). Interesting to note is the increased observance of this saturated byproduct in the presence of other silanes, including dimethoxymethylsilane (93:7), triethylsilane (87:13), dimethylphenylsilane (87:13) and triphenylsilane (90:10), once again demonstrating the specific advantage of triethoxysilane. The alkenylsilanes produced in this reaction are versatile intermediates that have been used effectively in other synthetic transformations.[47]

$$(22)$$

Reduction of Phosphine Oxides. In the presence of titanium(IV) and triethoxysilane, phosphine oxides can be reduced to the corresponding phosphines (eq 23).[48]

$$(23)$$

Drawing from the work conducted by Buchwald for the reduction of esters to alcohols,[14] triethoxysilane is used to promote catalyst turnover and not in the direct silylation of the reacting substrate. This approach is similar to other titanium hydride/phosphine oxide reduction strategies, including the use of lithium aluminum hydride/titanium(IV) chloride,[49] and the use of magnesium and cyclopentyldienyltitanium chloride.[50] In this particular case, the catalytic amount of titanium(IV) isopropoxide is proposed to react with triethoxysilane to produce a titanium hydride species and $Si(OR)_4$. The newly formed titanium hydride then proceeds to reduce the phosphine oxide with concomitant generation of bis-titanium oxide. The bis-titanium oxide then reacts with $Si(OR)_4$ to regenerate the titanium(IV) catalyst and $O[Si(OR)_3]_2$, thus completing the catalytic cycle.

Related Reagents. (2-Dimethylaminomethylphenyl)phenyl-silane; diphenylsilane–cesium fluoride; phenylsilane–cesium fluoride; triethylsilane; triphenylsilane.

1. (a) Lukevics, E.; Belyakova, Z. V.; Pomerantseva, M. G.; Voronkov, M. G., *J. Organomet. Chem. Libr.* **1977**, *17*, 1. (b) Speier, J. L., *Adv. Organomet. Chem.* **1979**, *17*, 407.

2. Marciniec, B.; Gulinski, J.; Urbaniak, W. *Comprehensive Handbook on Hydrosilylation*; Marciniec, B., Ed.; Pergamon: Oxford, 1991.

3. Ojima, I. In *The Chemistry of Organic Silicon Compounds*; Patai, S.; Rappoport, Z., Eds.; Wiley: Chichester, 1989; Part 2, Chapter 25, p 1479.

4. Plueddemann, E. P.; Fanger, G., *J. Am. Chem. Soc.* **1959**, *81*, 2632.

5. Ojima, I.; Kumagai, M., *J. Organomet. Chem.* **1978**, *157*, 359.

6. Chuit, C.; Corriu, R. J. P.; Reyé, C.; Young, J. C., *Chem. Rev.* **1993**, *93*, pp. 1371–1448.

7. (a) Boyer, J.; Corriu, R.; Perz, R.; Reyé, C., *Tetrahedron* **1981**, *37*, 2165. (b) Boyer, J.; Corriu, R.; Perz, R.; Poirier, M.; Reyé, C., *Synthesis* **1981**, 558.

8. Chuit, C.; Corriu, R.; Perz, R.; Reyé, C., *Synthesis* **1982**, 981.

9. Boyer, J.; Corriu, R.; Perz, R.; Reyé, C., *J. Chem. Soc., Chem. Commun.* **1981**, 121.

10. Hosomi, A.; Hayashida, H.; Kohra, S.; Tominaga, Y., *J. Chem. Soc., Chem. Commun.* **1986**, 1411.

11. Kohra, S.; Hayashida, H.; Tominaga, Y.; Hosomi, A., *Tetrahedron Lett.* **1988**, *29*(1), 89.

12. (a) Izumi, Y.; Nanami, H.; Higuchi, K.; Onaka, M., *Tetrahedron Lett.* **1991**, *32*, 4741. (b) Izumi, Y.; Onaka, M., *J. Mol. Catal.* **1992**, *74*, 35.

13. (a) Corriu, R.; Guerin, C.; Henner, B.; Wang, Q., *Organometallics* **1991**, *10*, 2297. (b) Corriu, R.; Guerin, C.; Henner, B.; Wang, Q., *Inorg. Chim. Acta* **1992**, *198–200*, 705.

14. (a) Berk, S. C.; Kreutzer, K. A.; Buchwald, S. L., *J. Am. Chem. Soc.* **1991**, *113*, 5093. (b) Berk, S. C.; Buchwald, S. L., *J. Org. Chem.* **1992**, *57*, 3751.

15. Skoda-Földes, R.; Kollár, L.; Heil, B., *J. Organomet. Chem.* **1991**, *408*, 297.

16. (a) Murai, T.; Oda, T.; Kimura, F.; Onishi, H.; Kanda, T.; Kato, S., *J. Chem. Soc., Chem. Commun.* **1994**, 2143. (b) Murai, T.; Kimura, F.; Tsutsui, K.; Hasegawa, K.; Kato, S., *Organometallics* **1998**, *17*, 926.

17. Hewitt, G. W.; Somers, J. J.; Sieburth, S.; McN. *Tetrahedron Lett.* **2000**, *41*, 10175.

18. (a) Ojima, I.; Hirai, K. In *Asymmetric Synthesis*, Morrison, J. D., Ed.; Academic Press: New York, 1985, Vol. 5, p. 103. (b) Ojima, I. In *The Chemistry of Organic Silicon Compounds*; Patai, S., Rappoport, Z.. Eds.; Wiley: New York, 1989, Vol. 2, p. 1479.

19. (a) Hiyama, T.; Kusumoto, T. In *Comprehensive Organic Synthesis*; Trost, B. M., Fleming, I., Eds; Pergamon Press: Oxford, 1991, Vol. 8, p. 763. (b) Reichl, J. A.; Berry, D. H., *Adv. Organomet. Chem.* **1993**, *43*, 197.

20. Katayama, H.; Taniguchi, K.; Kobayashi, M.; Sagawa, T.; Minami, T.; Ozawa, F., *J. Organomet. Chem.* **2002**, *645*, 192.

21. Trost, B. M.; Ball, Z. T., *J. Am. Chem. Soc.* **2001**, *123*, 12726.

22. (a) Hatanaka, Y.; Hiyama, T., *Synlett* **1991**, 845. (b) Mowery, M. E.; DeShong, P., *Org. Lett.* **1999**, *1*, 2137. (c) Denmark, S. E.; Neuville, L., *Org. Lett.* **2000**, *2*, 3221.

23. Bunlaksananusorn, T.; Rodriguez, A. L.; Knochel, P., *Chem. Commun.* **2001**, 745.

24. Tamao, K.; Kumada, M.; Maeda, K., *Tetrahedron Lett.* **1984**, *25*, 321.

25. Blumenkopf, T. A.; Overman, L. E., *Chem. Rev.* **1986**, *86*, 7.

26. (a) Trost, B. M.; Ball, Z. T.; Jöge, T., *J. Am. Chem. Soc.* **2002**, *124*, 7922. (b) Fürstner, A.; Radkowski, K., *Chem. Commun.* **2002**, 2182.

27. (a) Fürstner, A.; Radkowski, K., *Chem. Commun.* **2002**, 2182. (b) Lacombe, F.; Radkowski, K.; Günter, S.; Fürstner, A., *Tetrahedron* **2004**, *60*, 7315.

28. Marcinec, B. In *Comprehensive Handbook on Hydrosilylation*; Pergamon Press: Oxford, 1992.

29. Lindlar, H.; Dubuis, R. In *Organic Synthesis*; Wiley & Sons: New York, 1973; Collect. Vol. 5, p. 880.

30. (a) Jones, T. K.; Denmark, S. E. In *Organic Synthesis*; Wiley & Sons: New York, 1985; Collect. Vol. 7, p 524. (b) Tsuda, T.; Yoshida, T.; Kawamoto, T.; Saegusa, T., *J. Org. Chem.* **1987**, *52*, 1624.

31. (a) Chan, K. K.; Cohen, N.; Denoble, J. P.; Specian, A. C.; Saucy, G., *J. Org. Chem.* **1976**, *41* 3497. (b) Brandsma, L.; Nieuwenhuizen, W. F.; Zwikker, J. W.; Maeorg, U., *Eur. J. Org. Chem.* **1999**, *775*, 5.

32. (a) Castro, C. E.; Stephens, R. D., *J. Am. Chem. Soc.* **1964**, *86*, 4358. (b) Hanson, J. R., *Synthesis* **1974**, 1. (c) Smith, A. B.; Levenberg, P. A.; Suits, J. Z., *Synthesis* **1986**, 184. (d) Carreira, E. M.; Dubois, J., *J. Am. Chem. Soc.* **1995**, *117*, 8106.

33. (a) Ojima, I.; Vu, A. T.; McCullagh, J. V.; Kinoshita, A., *J. Am. Chem. Soc.* **1999**, *121*, 3230. (b) Ojima, I.; Vu, A. T.; Lee, S.-Y.; McCullagh, J. V.; Moralee, A. C.; Fujiwara, M.; Hoang, T. H., *J. Am. Chem. Soc.* **2002**, *124*, 9164.

34. (a) Ojima, I.; Kogure, T., *Organometallics* **1982**, *1*, 1390. (b) Revis, A.; Hilty, K. T., *J. Org. Chem.* **1990**, *55*, 2972. (c) Chan, T. H.; Zheng, G. Z., *Tetrahedron Lett.* **1993**, *34*, 3095.

35. (a) Speier, J. L.; Webster, J. A.; Barnes, G. H., *J. Am. Chem. Soc.* **1957**, *79*, 974. (b) Sadykh-Zade, S. I.; Petrov, A. D., *Zh. Obshch. Khim.* **1959**, *29*, 3194.

36. Ojima, I.; Donovan, R. J.; Cols, N., *Organometallics* **1991**, *10*, 2606.

37. Crowe, W. E.; Rachita, M. J., *J. Am. Chem. Soc.* **1995**, *117*, 6787.

38. (a) Sato, Y.; Takimoto, M.; Hayashi, K.; Katsuhara, T.; Takagi, K.; Mori, M., *J. Am. Chem. Soc.* **1994**, *116*, 9771. (b) Sato, Y.; Takimoto, M.; Mori, M., *Tetrahedron Lett.* **1996**, *37*, 887. (c) Sato, Y.; Takimoto, M.; Mori, M., *Synlett* **1997**, 734. (d) Sato, Y.; Saito, N.; Mori, M., *Tetrahedron Lett.* **1997**, *38*, 3931. (e) Sato, Y.; Saito, N.; Mori, M., *Tetrahedron* **1998**, *54*, 1153. (f) Sato, Y.; Saito, N.; Mori, M., *J. Org. Chem.* **2002**, *67*, 9310.

39. (a) Schiffers, R.; Kagan, H. B., *Synlett* **1997**, 1175. (b) Laronde, F. J.; Brook, M. A., *Tetrahedron Lett.* **1999**, *40*, 3507.

40. Nishikori, H.; Yoshihara, R.; Hosomi, A., *Synlett* **2003**, 561.

41. (a) Bessmertnykh, A. G.; Blinov, K. A.; Grishin, Y. K.; Donskaya, N. A.; Beletskaya, I. P., *Russ. J. Org. Chem.* **1996**, *32*, 1620. (b) Bessmertnykh, A. G.; Blinov, K. A.; Grishin, Y. K.; Donskaya, N. A.; Tveritinova, E. V.; Yur'eva, N. M.; Beletskaya, I. P., *J. Org. Chem.* **1997**, *62*, 6069. (c) Bessmertnykh, A. G.; Blinov, K. A.; Grishin, Y. K.; Donskaya, N. A.; Tveritinova, E. V.; Beletskaya, I. P., *Russ. J. Org. Chem.* **1998**, *34*, 799.

42. (a) Binger, P.; Buech, M., *Top. Curr. Chem.* **1987**, *135*, 77. (b) Trost, B. M., *Angew. Chem., Int. Ed. Engl.* **1986**, *25*, 1. (c) DeMeijere, A.; Wessjohann, L., *Synlett* **1990**, 20.

43. (a) Murata, M.; Suzuki, K.; Watanabe, S.; Masuda, Y., *J. Org. Chem.* **1997**, *62*, 8569. (b) Manoso, A. S.; DeShong, P., *J. Org. Chem.* **2001**, *66*, 7449. (c) Murata, M.; Ishikura, M.; Nagata, M.; Watanabe, S.; Masuda, Y., *Org. Lett.* **2002**, *4*, 1843.

44. (a) Hatanaka, Y.; Fukushima, S.; Hiyama, T., *Chem. Lett.* **1989**, *18*, 1171. (b) Hatanaka, Y.; Hiyama, T., *Chem. Lett.* **1989**, *18*, 2049. (c) Hatanaka, Y.; Hiyama, T., *Synlett* **1991**, *18*, 845.

45. (a) Baney, R. H.; Itoh, M.; Sakakibara, A.; Suzuki, T., *Chem. Rev.* **1995**, *95*, 1409. (b) Loy, D. A.; Shea, K. J., *Chem. Rev.* **1995**, *95*, 1431.

46. Murata, M.; Watanabe, S.; Masuda, Y., *Tetrahedron Lett.* **1999**, *40*, 9255.

47. Colvin, E. W. In *Comprehensive Organometallic Chemistry II*; Abel, E. W.; Stone, F. G. A.; Wilkinson, G., Eds.; Pergamon: Oxford, 1995; Vol. 11, p. 313.

48. Coumbe, T.; Lawrence, N. J.; Muhammad, F., *Tetrahedron Lett.* **1994**, *35*, 625.

49. Dzhemilev, U. M.; Yu. Gabaidullin, L.; Tolstikov, H. A.; Zeenova, L. M., *Chem. Abs.* **1980**, *93*, 25841.

50. Mathey, F.; Maillet, R., *Tetrahedron Lett.* **1980**, *21*, 2525.

Triethylsilane[1]

[617-86-7]   C$_6$H$_{16}$Si   (MW 116.31)
InChI = 1S/C6H16Si/c1-4-7(5-2)6-3/h7H,4-6H2,1-3H3
InChIKey = AQRLNPVMDITEJU-UHFFFAOYSA-N

(mild reducing agent for many functional groups)

Physical Data: mp $-156.9\,^\circ$C; bp $107.7\,^\circ$C; d 0.7309 g cm^{-3}.
Solubility: insol H$_2$O; sol hydrocarbons, halocarbons, ethers.
Form Supplied in: colorless liquid; widely available.
Purification: simple distillation, if needed.
Handling, Storage, and Precautions: triethylsilane is physically very similar to comparable hydrocarbons. It is a flammable, but not pyrophoric, liquid. As with all organosilicon hydrides, it is capable of releasing hydrogen gas upon storage, particularly in the presence of acids, bases, or fluoride-releasing salts. Proper precautions should be taken to vent possible hydrogen buildup when opening vessels in which triethylsilane is stored.

Original Commentary

James L. Fry
The University of Toledo, OH, USA

Introduction. Triethylsilane serves as an exemplar for organosilicon hydride behavior as a mild reducing agent. It is frequently chosen as a synthetic reagent because of its availability, convenient physical properties, and economy relative to other organosilicon hydrides which might otherwise be suitable for effecting specific chemical transformations.

Hydrosilylations. Addition of triethylsilane across multiple bonds occurs under the influence of a large number of metal catalysts.[2] Terminal alkynes undergo hydrosilylations easily with triethylsilane in the presence of platinum,[3] rhodium,[3a,4] ruthenium,[5] osmium,[6] or iridium[4] catalysts. For example, phenylacetylene can form three possible isomeric hydrosilylation products with triethylsilane; the (Z)-β-, the (E)-β-, and the α-products (eq 1). The (Z)-β-isomer is formed exclusively or preferentially with ruthenium[5] and some rhodium[4] catalysts, whereas the (E)-β-isomer is the major product formed with platinum[3] or iridium[4] catalysts. In the presence of a catalyst and carbon monoxide, terminal alkynes undergo silylcarbonylation reactions with triethylsilane to give (Z)- and (E)-β-silylacrylaldehydes.[7] Phenylacetylene gives an 82% yield of a mixture of the (Z)- and (E)-isomers in a 10:1 ratio when 0.3 mol % of dirhodium(II) tetrakis (perfluorobutyrate) catalyst is used under atmospheric pressure at 0 °C in dichloromethane (eq 2).[7d] Terminal alkenes react with triethylsilane in the presence of this catalyst to form either 'normal' anti-Markovnikov hydrosilylation products or allyl- or vinylsilanes, depending on whether the alkene is added to the silane or vice versa.[8] A mixture of 1-hexene and triethylsilane in the presence of 2 mol % of an iridium catalyst ([IrCl(CO)$_3$]$_n$) reacts under 50 atm of carbon monoxide to give a 50% yield of a mixture of

the (Z)- and (E)-enol silyl ether isomers in a 1:2 ratio (eq 3).[9] Hydrolysis yields the derived acylsilane quantitatively.[9]

$$Ph\!-\!\!\equiv\ +\ Et_3SiH\ \xrightarrow{\text{cat}}$$

$$\underset{Ph}{\overset{H}{\diagdown}}\!\!\!=\!\!\!\underset{SiEt_3}{\overset{H}{\diagup}}\ +\ \underset{Ph}{\overset{H}{\diagdown}}\!\!\!=\!\!\!\underset{H}{\overset{SiEt_3}{\diagup}}\ +\ \underset{Ph}{\overset{Et_3Si}{\diagdown}}\!\!\!=\!\!\!\underset{H}{\overset{H}{\diagup}}\quad(1)$$

$$Ph\!-\!\!\equiv\ +\ Et_3SiH\ +\ CO\ \xrightarrow{\text{cat}}$$

$$\underset{Ph}{\overset{OHC}{\diagdown}}\!\!\!=\!\!\!\underset{H}{\overset{SiEt_3}{\diagup}}\ +\ \underset{Ph}{\overset{OHC}{\diagdown}}\!\!\!=\!\!\!\underset{SiEt_3}{\overset{H}{\diagup}}\quad(2)$$

$$BuCH\!=\!CH_2\ +\ Et_3SiH\ +\ CO\ \xrightarrow{\text{cat}}$$

$$\underset{H}{\overset{Bu}{\diagdown}}\!\!\!=\!\!\!\underset{SiEt_3}{\overset{OSiEt_3}{\diagup}}\ +\ \underset{H}{\overset{Bu}{\diagdown}}\!\!\!=\!\!\!\underset{OSiEt_3}{\overset{SiEt_3}{\diagup}}\quad(3)$$

A number of metal complexes catalyze the hydrosilylation of various carbonyl compounds by triethylsilane.[10] Stereoselectivity is observed in the hydrosilylation of ketones[11] as in the reactions of 4-t-butylcyclohexanone and triethylsilane catalyzed by ruthenium,[12] chromium,[13] and rhodium[12,14] metal complexes (eq 4). Triethylsilane and chlorotris(triphenylphosphine)rhodium(I) catalyst effect the regioselective 1,4-hydrosilylation of α,β-unsaturated ketones and aldehydes.[15,16] Reduction of mesityl oxide in this manner results in a 95% yield of product that consists of 1,4- and 1,2-hydrosilylation isomers in a 99:1 ratio (eq 5). This is an exact complement to the use of phenylsilane, where the ratio of respective isomers is reversed to 1:99.[16]

$$\underset{t\text{-Bu}}{\diagdown}\!\!\diagup\!\!\overset{O}{\diagup}\ \xrightarrow[\text{cat}]{Et_3SiH}$$

$$\underset{t\text{-Bu}}{\diagdown}\!\!\overset{OSiEt_3}{\underset{H}{\diagup}}\ +\ \underset{t\text{-Bu}}{\diagdown}\!\!\overset{H}{\underset{OSiEt_3}{\diagup}}\quad(4)$$

(Ph$_3$P)$_3$RuCl$_2$, AgTFA, PhMe, Δ 5:95
Et$_4$N$^+$ [HCr$_2$(CO)$_{10}$]$^-$, DME, Δ 10:90
(Ph$_3$P)$_3$RhCl, PhMe, Δ 11:89
[Rh(η^3-C$_3$H$_5$){P(OMe)$_3$}$_3$], PhH   29:71

$$\underset{O}{\diagdown}\!\!\diagup\!\!\diagdown\ +\ Et_3SiH\ \xrightarrow{\text{cat}}\ \underset{OSiEt_3}{\diagdown}\!\!\diagup\!\!\diagdown\ +\ \underset{OSiEt_3}{\diagdown}\!\!\diagup\!\!\diagup\quad(5)$$
$$\qquad\qquad\qquad\qquad\qquad 99:1$$

Silane Alcoholysis. Triethylsilane reacts with alcohols in the presence of metal catalysts to give triethylsilyl ethers.[17] The use of dirhodium(II) perfluorobutyrate as a catalyst enables regioselective formation of monosilyl ethers from diols (eq 6).[17a]

$$\underset{Bu}{\overset{Bu}{\diagdown}}\underset{OH}{\overset{}{\diagup}}\!\!\diagup\!\!\diagdown\!\!\diagup OH\ \xrightarrow[\substack{Rh_2(pfb)_4\\92\%}]{Et_3SiH}\ \underset{Bu}{\overset{Bu}{\diagdown}}\underset{OH}{\overset{}{\diagup}}\!\!\diagup\!\!\diagdown\!\!\diagup OSiEt_3\ +\ H_2\quad(6)$$

Formation of Singlet Oxygen. Triethylsilane reacts with ozone at $-78\,^\circ$C in inert solvents to form triethylsilyl hydrotrioxide, which decomposes at slightly elevated temperatures to produce triethylsilanol and singlet oxygen. This is a convenient way to generate this species for use in organic synthesis.[18]

Reduction of Acyl Derivatives to Aldehydes. Aroyl chlorides and bromides give modest yields of aryl aldehydes when refluxed in diethyl ether with triethylsilane and aluminum chloride.[19] Better yields of both alkyl and aryl aldehydes are obtained from mixtures of acyl chlorides or bromides and triethylsilane by using a small amount of 10% palladium on carbon catalyst (eq 7).[20] This same combination of triethylsilane and catalyst can effect the reduction of ethyl thiol esters to aldehydes, even in sensitive polyfunctional compounds (eq 8).[21]

$$C_7H_{15}COCl + Et_3SiH \xrightarrow[83\%]{10\% \text{ Pd/C}} C_7H_{15}CHO \qquad (7)$$

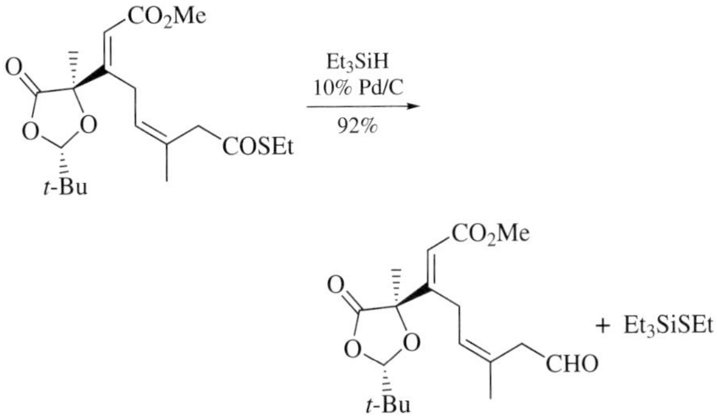

(8)

Radical Chain Reductions. Triethylsilane can replace toxic and difficult to remove organotin reagents for synthetic reductions under radical chain conditions. Although it is not as reactive as tri-n-butylstannane,[22] careful choice of initiator, solvent, and additives leads to effective reductions of alkyl halides,[23,24] alkyl sulfides,[23] and alcohol derivatives such as O-alkyl S-methyl dithiocarbonate (xanthate) and thionocarbonate esters.[22,23,25,26] Portionwise addition of 0.6 equiv of dibenzoyl peroxide to a refluxing triethylsilane solution of O-cholestan-3β-yl O'-(4-fluorophenyl) thionocarbonate gives a 93% yield of cholestane (eq 9).[22] The same method converts bis-xanthates of vic-diols into alkenes (eq 10).[22] Addition of a small amount of thiol such as t-dodecanethiol to serve as a 'polarity reversal catalyst'[24] with strong radical initiators in nonaromatic solvents also gives good results.[23,25] Treatment of ethyl 4-bromobutanoate with four equiv

(9)

of triethylsilane, two equiv of dilauroyl peroxide (DLP), and 2 mol % of t-dodecanethiol in refluxing cyclohexane for 1 hour yields ethyl butanoate in 97% yield (eq 11).[23]

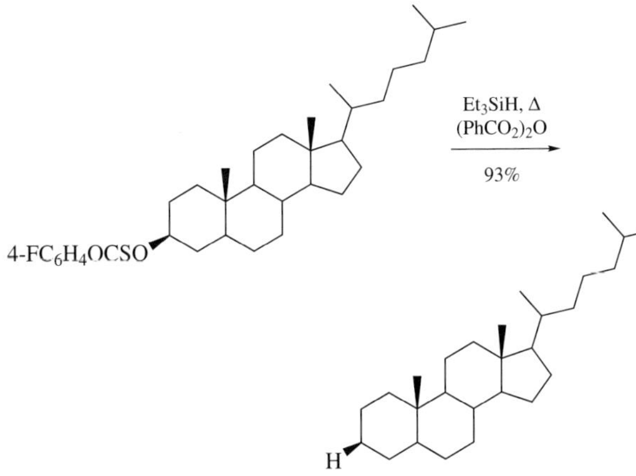

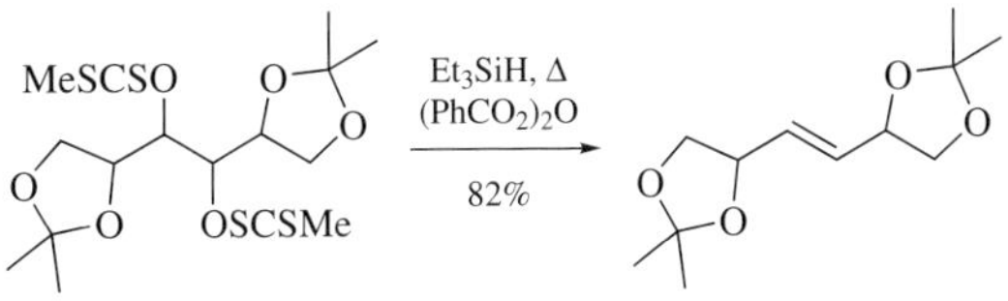

(10)

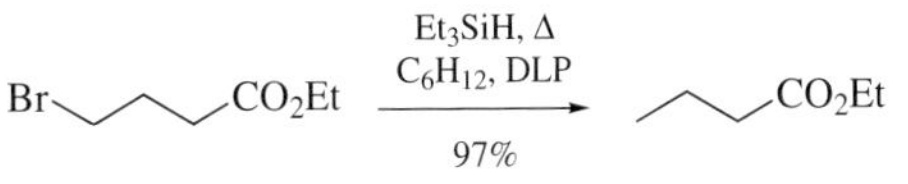

(11)

Ionic Hydrogenations and Reductive Substitutions. The polar nature of the Si–H bond enables triethylsilane to act as a hydride donor to electron-deficient centers. Combined with Brønsted or Lewis acids this forms the basis for many useful synthetic transformations.[27] Use of trifluoromethanesulfonic acid (triflic acid) at low temperatures enables even simple alkenes to be reduced to alkanes in high yields (eq 12).[28] Boron trifluoride monohydrate is effective in promoting the reduction of polycyclic aromatic compounds (eq 13).[29] Combined with thiols, it enables sulfides to be prepared directly from aldehydes and ketones (eq 14).[30] Combinations of triethylsilane with either trifluoroacetic acid/ammonium fluoride or pyridinium poly (hydrogen fluoride) (PPHF) are effective for the reductions of alkenes, alcohols, and ketones (eq 15).[31] Immobilized strong acids such as iron- or copper-exchanged montmorillonite K10[32] or the superacid nafion-H^{33} facilitate reductions of aldehydes and ketones[32] or of acetals[33] by increasing the ease of product separation (eq 16). Boron trifluoride and triethylsilane are an effective combination for the reduction of alcohols, aldehydes, ketones (eq 17),[34] and epoxides.[35] boron trifluoride etherate sometimes may be substituted for the free gas.[36]

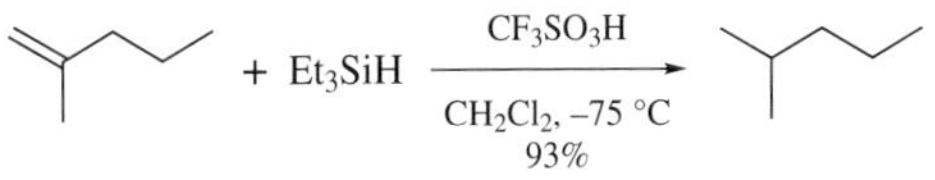

(12)

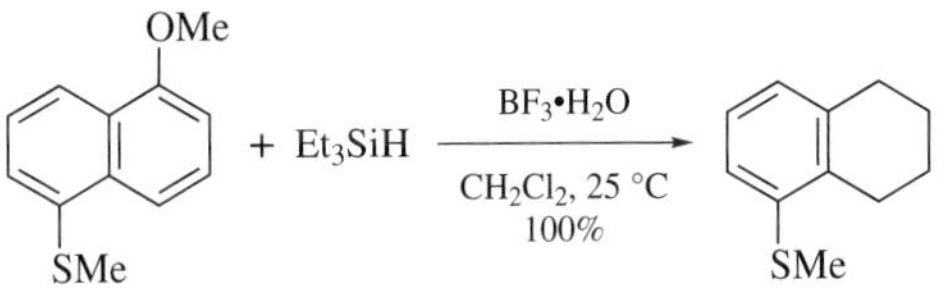

(13)

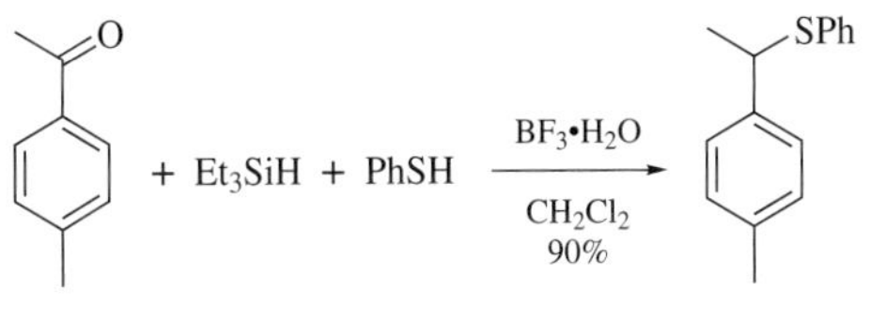

(14)

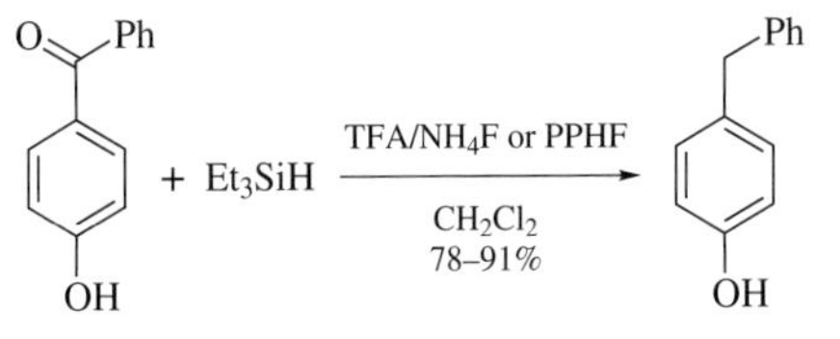

(15)

$$ \text{(16)} $$

$$ \text{(17)} $$

Triethylsilane in 3M ethereal lithium perchlorate solution effects the reduction of secondary allylic alcohols and acetates (eq 18).[37] The combination of triethylsilane and titanium(IV) chloride is a particularly effective reagent pair for the selective reduction of acetals.[38] Treatment of (±)-frontalin with this pair gives an 82% yield of tetrahydropyran products with a *cis:trans* ratio of 99:1 (eq 19).[38b] This exactly complements the 1:99 product ratio of the same products obtained with diisobutylaluminum hydride.[38b]

$$ \text{(18)} $$

$$ \text{(19)} $$

Triethylsilane and trityl salts[39] or trimethylsilyl trifluoromethanesulfonate[40] are effective for the reduction of various ketones and acetals, as are combinations of chlorotrimethylsilane and indium(III) chloride[41] and tin(II) bromide and acetyl bromide.[42] Isophthaldehyde undergoes reductive polycondensation to a polyether when treated with triethylsilane and triphenylmethyl perchlorate.[43]

Triethylsilane reduces nitrilium ions to aldimines,[44] diazonium ions to hydrocarbons,[45] and aids in the deprotection of amino acids.[46] With aluminum halides, it reduces alkyl halides to hydrocarbons.[47]

First Update

Ronald J. Rahaim Jr. & Robert E. Maleczka Jr.
Michigan State University, East Lansing, MI, USA

Additional Hydrosilylations. Hydrosilylations of terminal alkynes with triethylsilane (eq 1) have been improved in terms of their regio- and stereocontrol as well as in other aspects of their operation. Through the employment of Pt(DVDS),[48] Pt-catalyzed hydrosilylations of 1-alkynes[3] can now be performed at room temperature and in water with very high selectivity for the (E)-β-vinylsilanes (eq 20). It has also been shown that PtO₂ catalyzes the internal hydrosilylation of aryl alkynes under *ortho*-substituent regiocontrol (eq 21).[49] Strong preference for the (E)-β-vinylsilanes during the hydrosilylation of 1-alkynes has also been observed with cationic Ru-catalysts[50] and [RhCl(nbd)]₂/dppp,[51] the latter of which can also be employed in water (eq 20).

Hydrosilylations afford α-vinylsilanes when catalyzed by [CpRu(MeCN)₃]PF₆[52] (eq 20). (Z)-β-Vinylsilanes are similarly made under [RuCl₂(p-cymene)]₂ catalysis[53] or by the *trans* hydrosilylation of 1-alkynes under Lewis acid (AlCl₃)[54] catalysis.

$$ \text{(20)} $$

R = C₁₀H₂₁	[RhCl(nbd)]₂ + dppp	93:3:4
R = Ph	[Cp*Rh(BINAP)](SbF₆)₂	97:0:3
R = C₄H₉	Pt(DVDS)EP	100:0:0
R = C₄H₉	[RuCl₂(p-cymene)]₂	4:96:0
R = PhCH₂	AlCl₃	0:100:0
R = C₆H₁₂CO₂H	[CpRu(MeCN)₃]PF₆	2.5:2.5:95

$$ \text{(21)} $$

AlCl₃ can also promote the hydrosilylation of allenes and alkenes.[54] With regard to the hydrosilylation of alkenes; Rh-catalyzed reactions of Et₃SiH and methylenecyclopropanes provide a convenient route to homoallylic silanes (eq 22).[55]

$$ \text{(22)} $$

The hydrosilylation of carbonyl compounds with Et₃SiH (eq 4) has also been the subject of additional research. Owing to these efforts, carbonyls can now be directly converted to their triethylsilyl (TES) ethers with copper catalysts in the company of a bidentate phosphine[56] or N-heterocyclic carbene[57] ligand. Triethylsilyl ethers can also be made from carbonyl compounds and Et₃SiH in the presence of rhenium(V) oxo-complexes.[58]

Additional Silane Alcoholysis. The direct silylation of alcohols with triethylsilane (eq 26)[17] continues to be an interesting, if somewhat underused, method to TES protect alcohols. Recent works have demonstrated that this process is promoted by a number of catalysts including PdCl₂,[59] a Au(I) catalyst,[60] and the Lewis acid B(C₆F₆)₃[61] (eq 23).

$$ R{-}OH \xrightarrow[\text{catalyst}]{Et_3SiH} R{-}OSiEt_3 \quad \text{(23)} $$

R =	Catalyst	% Yield
1° and 2° aliphatic	PdCl₂	78–98
1°, 2°, 3° aliphatic or Ar	Au(I)	80–100
2°, 3° aliphatic or Ar	B(C₆F₅)₃	95–100

Additional Ionic Hydrogenation and Reductive Substitutions.

Nitrogen Containing Functional Group Reductions. As previously discussed, triethylsilane can donate its hydride to carbonyls and other functional groups (eqs 12–19).[27–43] A variety of transition metals have recently emerged as promoters of such reactions, especially for reductions of nitrogen containing moieties. For example, organic azides are efficiently transformed to their Boc-protected amines with catalytic palladium in the presence of di-t-butyl dicarbonate (eq 24).[62]

$$Ar-N\overset{H}{\underset{Me}{}} \xleftarrow[\substack{SnCl_4 \\ (0-91\%)}]{Et_3SiH} R-N_3 \xrightarrow[\substack{cat\ Pd,\ Boc_2O \\ (64-98\%)}]{Et_3SiH} R-N\overset{Boc}{\underset{H}{}} \quad (24)$$

$$R = ArCH_2 \qquad R = aryl,\ 1°,\ 2°\ aliphatics$$

Alternatively, if such azides bear a 1°-benzylic group they can be converted to N-methylanilines by reaction with Et_3SiH and $SnCl_4$.[63] Wilkinson's catalyst and Et_3SiH reduce aromatic nitro groups to their amines in moderate to good yields,[64] while the combination of $Pd(OAc)_2$ and Et_3SiH in a THF–water mixture reduces aliphatic nitro groups to the N-hydroxylamines (eq 25).[65]

$$R-NO_2 \xrightarrow[cat]{Et_3SiH} R-N\overset{H}{\underset{Y}{}}$$

R = aryl; cat = $Rh(PPh_3)_3Cl$;
Y = H (0–90%)

R = aliphatic; cat = $Pd(OAc)_2$;
Y = OH (31–89%)

$$(25)$$

Imines are reduced by triethylsilane to their amines when the proper Ir[66,67] or Ni[68] catalysts are employed. Non-metal-mediated reductions of C=N groups by Et_3SiH are also possible. Among these, the trifluorosulfonic acid promoted reductive amidation of aliphatic and aromatic aldehydes with Et_3SiH is an excellent way to mono N-alkylate aliphatic and aromatic amides, thioamides, carbamates, and ureas (eq 26).[69] It is also worth noting that trifluorosulfonic acid/Et_3SiH reduces acyl- and tosylhydrazones to hydrazines[70,71] and 2-aminopyrimidines to 2-amino-dihydro- or 2-aminotetrahydropyrimidines (eq 27).[72]

$$R^1\overset{X}{\underset{NH_2}{}} + R^2CHO \xrightarrow[TFA]{Et_3SiH} R^1\overset{X}{\underset{H}{N}}\overset{}{}R^2 \quad 63-97\% \quad (26)$$

X = O or S; R^1 = aliphatic, aryl, OR, $NR'R''$; R^2 = aliphatic, aryl

$$(27)$$

R^1 = H, CH_2, Me, Ph, SO_2Ar; R^2 = H, CH_2, Me;
R^3 = H, Me, C(O)Me, Ar, vinyl, Br

Reductive Etherifications and Acetal Reductions. Additional applications of triethylsilane in the reduction of C–O bonds also continue to surface. The Kusanov–Parnes dehydrative reduction[27] of hemiacetals and acetals with trifluorosulfonic acid/Et_3SiH has proven especially valuable. Under such conditions, 4,6-O-benzylidene acetal glucose derivatives can be asymmetrically deprotected to 6-O-benzyl-4-hydroxy derivatives (eq 28)[73] and thioketone

derivatives can be converted to *syn*-2,3-bisaryl (or heteroaryl) dihydrobenzoxanthins with excellent stereo- and chemoselectivity (eq 29).[74] Triethylsilane is also useful in a number of related acetal reductions, including those used for the formation of C-glycosides. For example, Et_3SiH reductively opens 1,3-dioxolan-4-ones to 2-alkoxy carboxylic acids when catalyzed by $TiCl_4$.[75] Furthermore, functionalized tetrahydrofurans are generated in good yield from 1,2-O-isopropylidenefuranose derivatives with boron trifluoride etherate and Et_3SiH (eq 30).[76] These same conditions lead to 1,4- or 1,5-anhydroalditols when applied to methyl furanosides or pyranosides.[77]

$$\xrightarrow[\substack{TFA \\ (80-95\%)}]{Et_3SiH}$$

R^1 = Ac, Bn; R^2 = OAc, OBn, NHAc

$$(28)$$

$$\xrightarrow[\substack{TFA \\ CH_2Cl_2 \\ (30-95\%)}]{Et_3SiH}$$

$$(29)$$

$$\xrightarrow[BF_3\cdot Et_2O]{Et_3SiH} \qquad 72-95\% \quad (30)$$

Novel syntheses of amino acids have also employed triethylsilane C–O bond cleavages of N,O-acetals. In this way, N-methylamino acid derivatives are isolated in high yields from the Fmoc or Cbz protected 5-oxazolidinone precursors, using TFA and Et_3SiH,[78,79] or with the Lewis acid $AlCl_3$ and Et_3SiH (eq 31).[80] A one-pot preparation of N-methyl-α-amino acid dipeptides can be accomplished from an oxazolidinone, amino acid, TFA, and Et_3SiH combination.[81]

$$P-N\overset{R}{\underset{O}{}}=O \xrightarrow[\substack{TFA\ or\ AlCl_3 \\ (27-100\%)}]{Et_3SiH} P-N\overset{R}{\underset{Me}{}}CO_2H \quad (31)$$

P = Fmoc or Cbz

Triethylsilane can also facilitate the high yielding reductive formation of dialkyl ethers from carbonyls and silyl ethers. For example, the combination of 4-bromobenzaldehyde, trimethylsilyl protected benzyl alcohol, and Et_3SiH in the presence of catalytic amounts of $FeCl_3$ will result in the reduction *and* benzylation of the carbonyl group (eq 32).[82] Similarly, $Cu(OTf)_2$ has been shown to aid Et_3SiH in the reductive etherification of variety of carbonyl compounds with *n*-octyl trimethylsilyl ether to give the alkyl ethers in moderate to good yields.[83] Likewise, TMSOTf catalyzes the conversion of tetrahydropyranyl ethers to benzyl ethers with Et_3SiH and benzaldehyde, and diphenylmethyl ethers with Et_3SiH and diphenylmethyl formate.[84] Symmetrical and unsymmetrical ethers are afforded in good yield from carbonyl compounds with silyl ethers (or alcohols) and Et_3SiH catalyzed by bismuth trihalide salts.[85] An intramolecular version of this procedure has been nicely applied to the construction of *cis*-2,6-di- and trisubstituted tetrahydropyrans.[86]

$$\text{(32)}$$

In a related process, triethylsilane plus $SnCl_4$ can expediently convert appropriately protected aldol products to fully protected 1,3-diols. Moreover, the synthesis of *syn*-1,3-ethylidene acetals from 1-(2-methoxyethoxy)ethyl-protected β-hydroxy ketones with $SnCl_4$ and Et_3SiH can occur with very high levels of diastereocontrol (eq 33).[87]

$$\text{(33)}$$

syn/anti: >200/1

Ether Cleavages. Triethylsilane and $B(C_6F_5)_3$ can also be used for the general cleavage of ether bonds to their corresponding triethylsilyl ether and hydrocarbon.[61b] This chemistry can selectively cleave differently substituted ethers (e.g., primary alkyl ethers cleave preferentially over secondary, tertiary, or aryl ether groups), but it should be noted that only a limited number of such examples have been reported. Furthermore, chemoselectivity can be an issue as $Et_3SiH/B(C_6F_5)_3$ can deoxygenate primary alcohols and acetals, as well as perform the aforementioned silane alcoholyses. Nonetheless, Et_3SiH and TFA are well suited for taking triphenylmethyl (trityl, Tr) protective groups off hydroxyls (eq 34),[73] aziridines,[88] or peptides[89] even when other acid-sensitive functional groups are present. Triethylsilane has also been employed in the deprotection of triphenylmethyl-protected nucleotides, but with dichloroacetic acid in dichloromethane.[90]

$$R-O-Tr \quad \xrightarrow[\text{2. 80\% aq HOAc/THF}]{\substack{\text{1. } Et_3SiH \\ \text{cat TESOTf, } CH_2Cl_2}} \quad R-OH \ + \ H-Tr \quad \text{(34)}$$
$$\text{(87–99\%)}$$

Ester Reductions and Miscellaneous Reductive Substitutions.
Triethylsilane can react with esters in a number of ways. Aliphatic esters and lactones are reduced to acyclic and cyclic ethers when treated with $TiCl_4$, TMSOTf, and Et_3SiH (eq 35).[91] Propargylic acetates, on the other hand, will undergo reductive cleavage of their C–O bonds when treated with catalytic amounts of indium(III) bromide and Et_3SiH.[92] Aryl and enol triflates are reduced when exposed to Et_3SiH and a Pd–phosphine complex[93] (eq 36), whereas aromatic and aliphatic iodides, bromides, and chlorides are dehalogenated with Et_3SiH and catalytic $PdCl_2$ (also see eq 11[23]). Curiously, Et_3SiH and $PdCl_2$ can also be used to make C–X bonds, as alcohols are converted to the corresponding halide with $PdCl_2$, Et_3SiH, and iodomethane, dibromomethane, or hexachloroethane (eq 37).[94] Likewise, lactones will undergo a ring-opening halosilylation with $PdCl_2$, Et_3SiH, and iodomethane, or allyl bromide, producing the triethylsilyl ω-iodo- or ω-bromoalkanoates.[95]

$$R^1 \underset{O}{\overset{O}{\|}} {O}^{-R^2} \quad \xrightarrow[\substack{\text{TMSOTf, } TiCl_4 \\ (40–89\%)}]{Et_3SiH} $$
$$R^1 \diagup O {\diagdown}^{R^2} \quad R^1 \ \& \ R^2 = \text{aliphatic} \quad \text{(35)}$$

$$R-X \quad \xrightarrow[\text{cat "Pd"}]{Et_3SiH} \quad R-H \quad \begin{array}{l} R = \text{aliphatic or aryl; } X = \text{I, Br, Cl} \\ (78–95\%) \\ R = \text{vinyl or aryl; } X = \text{OTf} \\ (41–100\%) \end{array} \quad \text{(36)}$$

$$Ph \diagup OH \quad \xrightarrow[\substack{RX, PdCl_2 \\ (94–98\%)}]{Et_3SiH} \quad Ph \diagup X \quad \begin{array}{l} RX = MeI \\ RX = CH_2Br_2 \\ RX = CCl_3CCl_3 \end{array} \quad \text{(37)}$$

Reductive Couplings and Cyclizations. As previously discussed, triethylsilane can react with both activated (eq 5)[15,16] and non-activated olefins (eq 12[28]). Recent developments in this area include the saturation of alkenes by Et_3SiH under catalysis by Grubb's 1st generation catalyst. A particularly elegant application of this chemistry is possible when ring closing metathesis (RCM) is kinetically favored. In such cases one can effect a one-pot ring closure/alkene reduction in good overall yield (eq 38).[96]

$$\text{(38)}$$
$$\begin{array}{l} n = 0, X = NTs \\ R = H \ (76\%) \\ n = 1, X = O \\ R = Ar \ (75\%) \end{array}$$

Alkenes, along with alkynes, allenes, or dienes, can also participate in triethylsilane promoted reductive couplings. Aldehydes, in particular, are good at coupling with the intermediates of nickel-catalyzed additions of Et_3SiH across alkenes, allenes, dienes, or alkynes (eq 39).[97] These reactions tend to be highly regioselective; as are the indium(III) bromide catalyzed reductive syn aldol between aldehydes, enones, and Et_3SiH (eq 40).[98] Finally, in the presence of ethylaluminum sesquichloride and Et_3SiH, alkylchloroformates participate in what have been termed Friedel–Crafts alkylations of alkenes.[99]

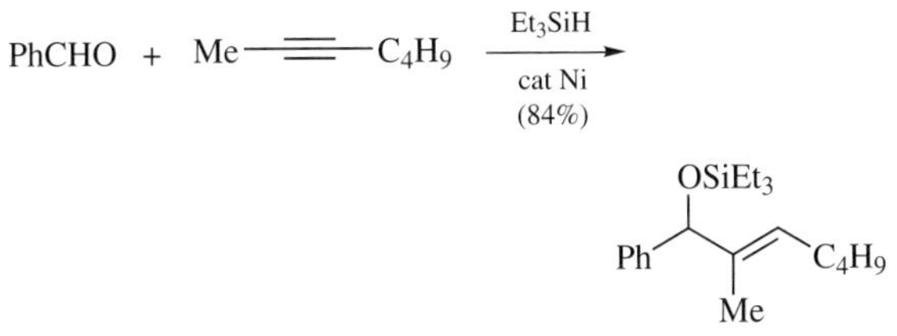

$$PhCHO + Me\!\!=\!\!\!=\!\!\!-C_4H_9 \xrightarrow[\substack{(84\%)}]{\substack{Et_3SiH \\ cat\ Ni}}$$

(39)

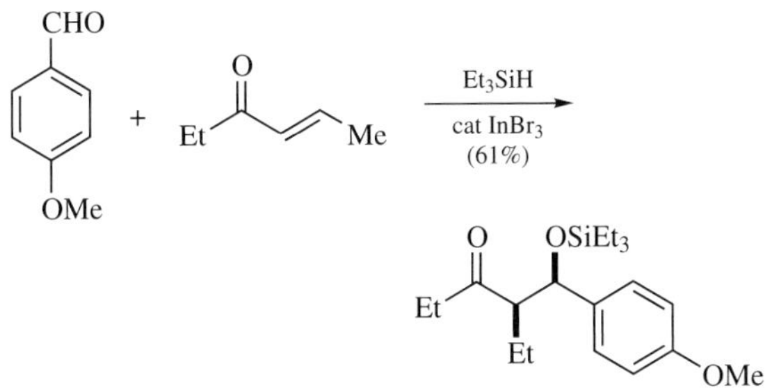

(40)

syn/anti >99/1

Imines can serve as electrophiles in similar processes. For example, tetrahydropyran or tetrahydrofuran containing amino acids are synthesized in good yield from a $TiCl_4$ catalyzed coupling of cyclic enol ethers, *N*-tosyl imino ester, and triethylsilane (eq 41).[100] Triethylsilane and a palladium-catalyst can prompt the cyclization–hydrosilylation of 1,6- and 1,7-dienes with good yields and moderate to high stereoselectivity (eq 42).[101] Cationic rhodium catalyzes a cyclization–hydrosilylation of 1,6-enynes,[102] whereas palladium catalyzes a regio- and stereoselective cycloreduction.[103] This reaction has also been applied to haloenynes[104] and bimetallic cobalt/rhodium nanoparticles in an atmosphere of carbon monoxide to effect a carbonylative silylcarbocyclization (eq 43).[105] 1,6-Diynes with cationic platinum, or cationic rhodium, in conjunction with Et_3SiH undergo a cyclization–hydrosilylation.[106,107] The combination of Et_3SiH, rhodium, carbon monoxide, and allenyl–carbonyl compounds yields *cis*-2-triethylsilylvinyl-cyclopentanols and cyclohexanols (eq 44).[108] For this reaction the investigators mention that Et_3SiH is superior to Ph_3SiH, Me_2PhSiH, and $(EtO)_3SiH$. This contrasts most other reports of hydrosilylations with Et_3SiH, where no particular advantage is either attributed or demonstrated for Et_3SiH over other silanes. Finally, reductive Nazarov cyclizations can also take place with Et_3SiH and a Lewis acid.[109]

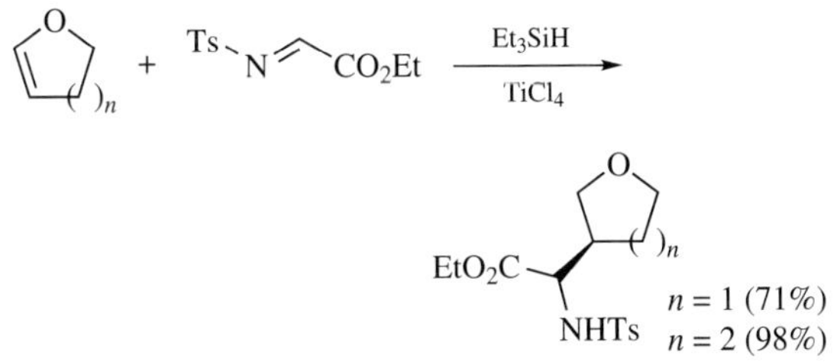

(41)

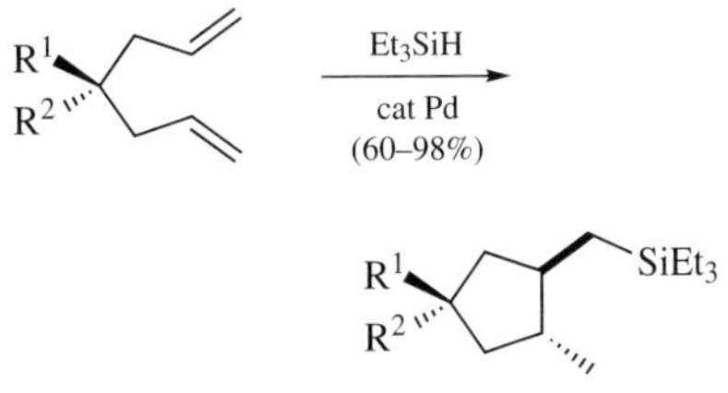

n = 1 (71%)
n = 2 (98%)

R^1 = ester
R^2 = H, Me, Ph, (42)
ester, CN,
SO_2Me

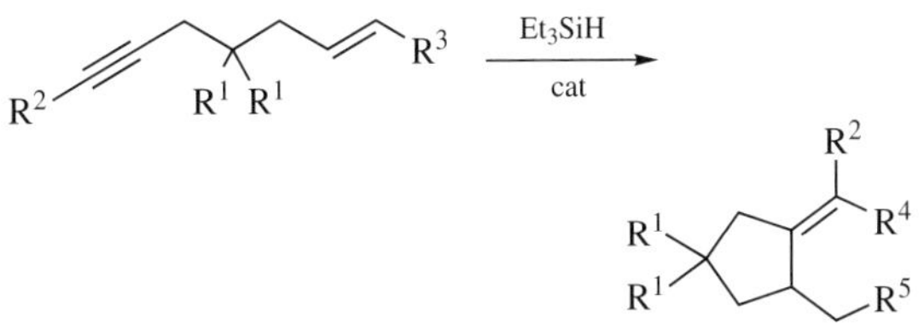

(43)

R^1	R^2	R^3	Catalyst	R^4	R^5	% Yield
CO_2Et	H	Ph	$Pd(dppe)Cl_2$	H	Ph	85
Me	Me	H	$[Rh(COD)_2]SbF_6$	$SiEt_3$	H	65
CO_2Me	H	H	$Co_2Rh_2 + CO$	$SiEt_3$	CHO	93

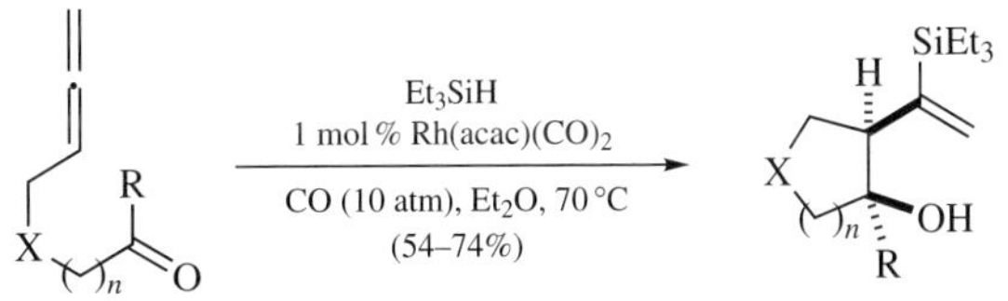

(44)

$X = C(CO_2Et)_2$, NTs, O; R = H, Me, Et; n = 1, 2

When reacting alkenes with triethylsilane it is necessary to keep in mind that the $PdCl_2/Et_3SiH$ combination also promotes the double bond isomerization of monosubstituted aliphatic olefins[110] and α-alkylidene cyclic carbonyl compounds are isomerized to α,β-unsaturated cyclic carbonyls with tris(triphenylphosphine) rhodium chloride.[111]

Aromatic Silylations. Aryltriethylsilanes are synthesized in moderate to good yield from electron-rich *meta*- and *para*-substituted aryl iodides, by $Pd(P\text{-}tBu)_3$ in the presence of K_3PO_4 and triethylsilane.[112] Platinum oxide in conjunction with sodium acetate and Et_3SiH silylates *meta* and *para* substituted aryl iodides and bromides that contain electron-withdrawing groups.[113] *ortho*-Triethylsilyl aromatics are accessed with $Ru_3(CO)_{12}$ using azoles, imines, pyridines, amides, and esters as directing groups; the system tolerates electron-donating and withdrawing groups (eq 45).[114] This method has also been applied to the silylation of benzylic C–H bonds.[115]

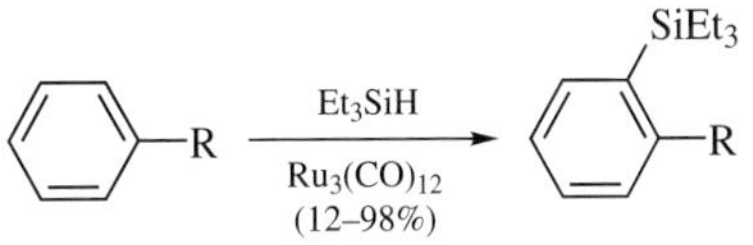

R = azole, imine,
ester, amide, (45)
pyridine,

Generation of Other Triethylsilyl Reagents, etc. Triethylsilane is also used in the synthesis of various other reagents for organic synthesis. Triethylsilyl cyanide, which is used for the silylcyanation of aldehydes and ketones, can be prepared from Et_3SiH and acetonitrile in the presence of catalytic amounts of $Cp(CO)_2FeMe$.[116] Bromotriethylsilane is prepared when Et_3SiH

reacts with copper(II) bromide and catalytic amounts of copper(I) iodide[117] or with PdCl$_2$ and allyl bromide.[118] Et$_3$SiH can also reduce Bu$_3$SnCl to Bu$_3$SnH, which when carried out in the presence of alkynes, allenes, or alkenes can undergo Lewis acid promoted hydrostannation reactions (eq 46).[119] This represents the first example of Lewis acid catalyzed hydrostannations with in situ generated tributyltin hydride. Significantly, Et$_3$SiH succeeded in this reaction where hydrosiloxanes failed. Lastly, Et$_3$SiH reacts with indium(III) chloride to generate dichloroindium hydride.[119]

$$R \underset{}{=\!\!\!=\!\!\!=} R^1 \xrightarrow[\substack{\text{toluene, 0 °C to rt} \\ (70–90\%)}]{\substack{\text{Et}_3\text{SiH, Bu}_3\text{SnCl} \\ 10 \text{ mol \% B(C}_6\text{F}_3)_3}} \underset{\substack{\\ R^1}}{\overset{\substack{R \\ \|}}{\underset{H}{\qquad}}}\!\!SnBu_3 \quad (46)$$

Z:E ~ 90:10

Related Reagents.   Phenylsilane–cesium fluoride; tri-n-butylstannane;   tricarbonylchloroiridium–diethyl(methyl) silane–carbon monoxide; triethylsilane–trifluoroacetic acid.

1. (a) Fleming, I. In *Comprehensive Organic Chemistry*; Barton, D.; Ollis, W. D., Eds.; Pergamon: New York, 1979; Vol. 3, p 541. (b) Colvin, E. *Silicon in Organic Synthesis*; Butterworths: Boston, 1981. (c) Weber, W. P. *Silicon Reagents for Organic Synthesis*; Springer: New York, 1983. (d) *The Chemistry of Organic Silicon Compounds*; Patai, S.; Rappoport, Z., Eds.; Wiley: New York, 1989. (e) Corey, J. Y. In *Advances in Silicon Chemistry*; Larson, G. L., Ed.; JAI: Greenwich, CT, 1991; Vol. 1, p 327.

2. (a) Lukevics, E., *Russ. Chem. Rev. (Engl. Transl.)* **1977**, *46*, 264. (b) Speier, J. L., *Adv. Organomet. Chem.* **1979**, *17*, 407. (c) Keinan, E., *Pure Appl. Chem.* **1989**, *61*, 1737.

3. (a) Doyle, M. P.; High, K. G.; Nesloney, C. L.; Clayton, T. W., Jr.; Lin, J., *Organometallics* **1991**, *10*, 1225. (b) Lewis, L. N.; Sy, K. G.; Bryant, G. L., Jr.; Donahue, P. E., *Organometallics* **1991**, *10*, 3750.

4. Kopylova, L. I.; Pukhnarevich, V. B.; Voronkov, M. G., *Zh. Obshch. Khim.* **1991**, *61*, 2418.

5. Esteruelas, M. A.; Herrero, J.; Oro, L. A., *Organometallics* **1993**, *12*, 2377.

6. Esteruelas, M. A.; Oro, L. A.; Valero, C., *Organometallics* **1991**, *10*, 462.

7. (a) Murai, S.; Sonada, N., *Angew. Chem., Int. Ed. Engl.* **1979**, *18*, 837. (b) Matsuda, I.; Ogiso, A.; Sato, S.; Izumi, Y., *J. Am. Chem. Soc.* **1989**, *111*, 2332. (c) Ojima, I.; Ingallina, P.; Donovan, R. J.; Clos, N., *Organometallics* **1991**, *10*, 38. (d) Doyle, M. P.; Shanklin, M. S., *Organometallics* **1993**, *12*, 11.

8. Doyle, M. P.; Devora, G. A.; Nefedov, A. O.; High, K. G., *Organometallics* **1992**, *11*, 549.

9. Chatani, N.; Ikeda, S.; Ohe, K.; Murai, S., *J. Am. Chem. Soc.* **1992**, *114*, 9710.

10. (a) Eaborn, C.; Odell, K.; Pidcock, A., *J. Organomet. Chem.* **1973**, *63*, 93. (b) Corriu, R. J. P.; Moreau, J. J. E., *J. Chem. Soc., Chem. Commun.* **1973**, 38. (c) Ojima, I.; Nihonyanagi, M.; Kogure, T.; Kumagai, M.; Horiuchi, S.; Nakatasugawa, K.; Nagai, Y., *J. Organomet. Chem.* **1975**, *94*, 449.

11. Ojima, I.; Nihonyanagi, M.; Nagai, Y., *Bull. Chem. Soc. Jpn.* **1972**, *45*, 3722.

12. Semmelhack, M. F.; Misra, R. N., *J. Org. Chem.* **1982**, *47*, 2469.

13. Fuchikami, T.; Ubukata, Y.; Tanaka, Y., *Tetrahedron Lett.* **1991**, *32*, 1199.

14. Bottrill, M.; Green, M., *J. Organomet. Chem.* **1976**, *111*, C6.

15. Ojima, I.; Kogure, T.; Nihonyanagi, M.; Nagai, Y., *Bull. Chem. Soc. Jpn.* **1972**, *45*, 3506.

16. Ojima, I.; Kogure, T., *Organometallics* **1982**, *1*, 1390.

17. (a) Doyle, M. P.; High, K. G.; Bagheri, V.; Pieters, R. J.; Lewis, P. J.; Pearson, M. M., *J. Org. Chem.* **1990**, *55*, 6082. (b) Zakharkin, L. I.; Zhigareva, G. G., *Izv. Akad. Nauk SSSR, Ser. Khim.* **1992**, 1284. (c) Barton, D. H. R.; Kelly, M. J., *Tetrahedron Lett.* **1992**, *33*, 5041.

18. Corey, E. J.; Mehrota, M. M.; Khan, A. U., *J. Am. Chem. Soc.* **1986**, *108*, 2472.

19. Jenkins, J. W.; Post, H. W., *J. Org. Chem.* **1950**, *15*, 556.

20. Citron, J. D., *J. Org. Chem.* **1969**, *34*, 1977.

21. Fukuyama, T.; Lin, S.-C.; Li, L., *J. Am. Chem. Soc.* **1990**, *112*, 7050.

22. Barton, D. H. R.; Jang, D. O.; Jaszberenyi, J. C., *Tetrahedron Lett.* **1991**, *32*, 7187: *Tetrahedron* **1993**, *49*, 2793.

23. Cole, S. J.; Kirwan, J. N.; Roberts, B. P.; Willis, C. R., *J. Chem. Soc., Perkin Trans. 1* **1991**, 103.

24. Allen, R. P.; Roberts, B. P.; Willis, C. R., *J. Chem. Soc., Chem. Commun.* **1989**, 1387.

25. Kirwin, J. N.; Roberts, B. P.; Willis, C. R., *Tetrahedron Lett.* **1990**, *31*, 5093.

26. Cf. Chatgilialoglu, C.; Ferreri, C.; Lucarini, M., *J. Org. Chem.* **1993**, *58*, 249.

27. (a) Kursanov, D. N.; Parnes, Z. N., *Russ. Chem. Rev. (Engl. Transl.)* **1969**, *38*, 812. (b) Kursanov, D. N.; Parnes, Z. N.; Loim, N. M., *Synthesis* **1974**, 633. (c) Nagai, Y., *Org. Prep. Proced. Int.* **1980**, *12*, 13. (d) Kursanov, D. N.; Parnes, Z. N.; Kalinkin, M. I.; Loim, N. M. *Ionic Hydrogenation and Related Reactions*; Harwood: Chur, Switzerland, **1985**.

28. Bullock, R. M.; Rappoli, B. J., *J. Chem. Soc., Chem. Commun.* **1989**, 1447.

29. (a) Larsen, J. W.; Chang, L. W., *J. Org. Chem.* **1979**, *44*, 1168. (b) Eckert-Maksic, M.; Margetic, D., *Energy Fuels* **1991**, *5*, 327. (c) Eckert-Maksic, M.; Margetic, D., *Energy Fuels* **1993**, *7*, 315.

30. Olah, G. A.; Wang, Q.; Trivedi, N. J.; Prakash, G. K. S., *Synthesis* **1992**, 465.

31. Olah, G. A.; Wang, Q.; Prakash, G. K. S., *Synlett* **1992**, 647.

32. Izumi, Y.; Nanami, H.; Higuchi, K.; Onaka, M., *Tetrahedron Lett.* **1991**, *32*, 4741.

33. Olah, G. A.; Yamato, T.; Iyer, P. S.; Prakash, G. K. S., *J. Org. Chem.* **1986**, *51*, 2826.

34. (a) Adlington, M. G.; Orfanopoulos, M.; Fry, J. L., *Tetrahedron Lett.* **1976**, 2955. (b) Fry, J. L.; Orfanopoulos, M.; Adlington, M. G.; Dittman, W. R., Jr.; Silverman, S. B., *J. Org. Chem.* **1978**, *43*, 374. (c) Fry, J. L.; Silverman, S. B.; Orfanopoulos, M., *Org. Synth.* **1981**, *60*, 108.

35. Fry, J. L.; Mraz, T. J., *Tetrahedron Lett.* **1979**, 849.

36. (a) Doyle, M. P.; West, C. T.; Donnelly, S. J.; McOsker, C. C., *J. Organomet. Chem.* **1976**, *117*, 129. (b) Dailey, O. D., Jr., *J. Org. Chem.* **1987**, *52*, 1984. (c) Krause, G. A.; Molina, M. T., *J. Org. Chem.* **1988**, *53*, 752. (d) Gil, J. F.; Ramón, D. J.; Yus, M., *Tetrahedron* **1993**, *49*, 4923.

37. Wustrow, D. J.; Smith, W. J., III; Wise, L. D., *Tetrahedron Lett.* **1994**, *35*, 61.

38. (a) Kotsuki, H.; Ushio, Y.; Kadota, I.; Ochi, M., *Chem. Lett.* **1988**, 927. (b) Ishihara, K.; Mori, A.; Yamamoto, H., *Tetrahedron* **1990**, *46*, 4595.

39. (a) Tsunoda, T.; Suzuki, M.; Noyori, R., *Tetrahedron Lett.* **1979**, 4679. (b) Kato, J.; Iwasawa, N.; Mukaiyama, T., *Chem. Lett.* **1985**, *6*, 743. (c) Kira, M.; Hino, T.; Sakurai, H., *Chem. Lett.* **1992**, 555.

40. (a) Bennek, J. A.; Gray, G. R., *J. Org. Chem.* **1987**, *52*, 892. (b) Sassaman, M. B.; Kotian, K. D.; Prakash, G. K. S.; Olah, G. A., *J. Org. Chem.* **1987**, *52*, 4314.

41. (a) Mukaiyama, T.; Ohno, T.; Nishimura, T.; Han, J. S.; Kobayashi, S., *Chem. Lett.* **1990**, 2239. (b) Mukaiyama, T.; Ohno, T.; Nishimura, T.; Han, J. S.; Kobayashi, S., *Bull. Chem. Soc. Jpn.* **1991**, *64*, 2524.

42. (a) Oriyama, T.; Iwanami, K.; Tsukamoto, K.; Ichimura, Y.; Koga, G., *Bull. Chem. Soc. Jpn.* **1991**, *64*, 1410. (b) Oriyama, T.; Ichimura, Y.; Koga, G., *Bull. Chem. Soc. Jpn.* **1991**, *64*, 2581.

43. Yokozawa, T.; Nakamura, F., *Makromol. Chem., Rapid Commun.* **1993**, *14*, 167.

44. (a) Fry, J. L., *J. Chem. Soc., Chem. Commun.* **1974**, 45. (b) Fry, J. L.; Ott, R. A., *J. Org. Chem.* **1981**, *46*, 602.

45. Nakayama, J.; Yoshida, M.; Simamura, O., *Tetrahedron* **1970**, *26*, 4609.

46. Mehta, A.; Jaouhari, R.; Benson, T. J.; Douglas, K. T., *Tetrahedron Lett.* **1992**, *33*, 5441.

47. (a) Doyle, M. P.; McOsker, C. C.; West, C. T., *J. Org. Chem.* **1976**, *41*, 1393. (b) Parnes, Z. N.; Romanova, V. S.; Vol'pin, M. E., *J. Org. Chem. USSR (Engl. Transl.)* **1988**, *24*, 254.

48. (a) Aneetha, H.; Wu, W.; Verkada, J. G., *Organometallics* **2005**, *24*, 2590. (b) Wu, W.; Li, C.-J., *Chem. Commun.* **2003**, 1668.

49. Hamze, A.; Provot, O.; Alami, M.; Brion, J.-D., *Org. Lett.* **2005**, *7*, 5625.

50. (a) Takeuchi, R.; Nitta, S.; Watanabe, D., *J. Chem. Soc. Chem. Commun.* **1994**, 1777. (b) Faller, J. W.; D'Alliessi, D. G., *Organometallics* **2002**, *21*, 1743.

51. Sato, A.; Kinoshita, H.; Shinokubo, H.; Oshima, K., *Org. Lett.* **2004**, *6*, 2217.

52. Trost, B. M.; Ball, Z. T., *J. Am. Chem. Soc.* **2005**, *127*, 17644.

53. Na, Y.; Chang, S., *Org. Lett.* **2000**, *2*, 1887.

54. (a) Sudo, T.; Asao, N.; Gevorgyan, V.; Yamamoto, Y., *J. Org. Chem.* **1999**, *64*, 2494. (b) Song, Y.-S.; Yoo, B. R.; Lee, G.-H.; Jung, I. N., *Organometallics* **1999**, *18*, 3109.

55. Bessmertnykh, A. G.; Blinov, K. A.; Grishin, Y. K.; Donskaya, N. A.; Tveritinova, E. V.; Yur'eva, N. M.; Beletskaya, I. P., *J. Org. Chem.* **1997**, *62*, 6069.

56. Lipshutz, B. H.; Caires, C. C.; Kuipers, P.; Chrisman, W., *Org. Lett.* **2003**, *5*, 3085.

57. Díez-González, S.; Kaur, H.; Zinn, F. K.; Stevens, E. D.; Nolan, S. P., *J. Org. Chem.* **2005**, *70*, 4784.

58. Ison, E. A.; Trivedi, E. R.; Corbin, R. A.; Abu-Omar, M. M., *J. Am. Chem. Soc.* **2005**, *127*, 15374.

59. Mirza-Aghayan, M.; Boukherroub, R.; Bolourtchian, M., *J. Organomet. Chem.* **2005**, *690*, 2372.

60. Ito, H.; Takagi, K.; Miyahara, T.; Sawamura, M., *Org. Lett.* **2005**, *7*, 3001.

61. (a) Blackwell, J. M.; Foster, K. L.; Beck, V. H.; Piers, W. E., *J. Org. Chem.* **1999**, *64*, 4887. (b) Gevorgyan, V.; Rubin, M.; Benson, S.; Liu, J.-X.; Yamamoto, Y., *J. Org. Chem.* **2000**, *65*, 6179.

62. Kotsuki, H.; Ohishi, T.; Araki, T., *Tetrahedron Lett.* **1997**, *38*, 2129.

63. Lopez, F. J.; Nitzan, D., *Tetrahedron Lett.* **1999**, *40*, 2071.

64. Brinkman, H. R.; Miles, W. H.; Hilborn, M. D.; Smith, M. C., *Synth. Commun.* **1996**, *26*, 973.

65. Rahaim, R. J., Jr.; Maleczka, R. E., Jr. *Org. Lett.* **2005**, *7*, 5087.

66. Field, L. D.; Messerle, B. A.; Rumble, S. L., *Eur. J. Org. Chem.* **2005**, 2881.

67. In-situ formed imine: Mizuta, T.; Sakaguchi, S.; Ishii, Y., *J. Org. Chem.* **2005**, *70*, 2195.

68. Vetter, A. H.; Berkessel, A., *Synthesis* **1995**, 419.

69. Dubé, D.; Scholte, A. A., *Tetrahedron Lett.* **1999**, *40*, 2295.

70. Wu, P.-L.; Peng, S.-Y.; Magrath, J., *Synthesis* **1995**, 435.

71. Wu, P.-L.; Peng, S.-Y.; Magrath, J., *Synthesis* **1996**, 249.

72. Baskaran, S.; Hanan, E.; Byun, D.; Shen, W., *Tetrahedron Lett.* **2004**, *45*, 2107.

73. Imagawa, H.; Tsuchihashi, T.; Singh, R. K.; Yamamoto, H.; Sugihara, T.; Nishizawa, M., *Org. Lett.* **2003**, *5*, 153.

74. Kim, S.; Wu, J. Y.; Chen, H. Y.; DiNinno, F., *Org. Lett.* **2003**, *5*, 685.

75. Winneroski, L. L.; Xu, Y., *J. Org. Chem.* **2004**, *69*, 4948.

76. Ewing, G. J.; Robins, M. J., *Org. Lett.* **1999**, *1*, 635.

77. (a) Rolf, D.; Gray, G. R., *J. Am. Chem. Soc.* **1982**, *104*, 3539. (b) Rolf, D.; Bennek, J. A.; Gray, G. R., *J. Carbohydr. Chem.* **1983**, *2*, 373. (c) Bennek, J. A.; Gray, G. R., *J. Org. Chem.* **1987**, *52*, 892.

78. Luke, R. W. A.; Boyce, P. G. T.; Dorling, E. K., *Tetrahedron Lett.* **1996**, *37*, 263.

79. Aurelio, L.; Brownlee, R. T. C.; Hughes, A. B., *Org. Lett.* **2002**, *4*, 3767.

80. Zhang, S.; Govender, T.; Norstrom, T.; Arvidsson, P. I., *J. Org. Chem.* **2005**, *70*, 6918.

81. Dorow, R. L.; Gingrich, D. E., *Tetrahedron Lett.* **1999**, *40*, 467.

82. (a) Iwanami, K.; Seo, H.; Tobita, Y.; Oriyama, T., *Synthesis* **2005**, 183. (b) Iwanami, K.; Yano, K.; Oriyama, T., *Synthesis* **2005**, 2669.

83. Yang, W.-C.; Lu, X.-A.; Kulkarni, S. S.; Hung, S.-C., *Tetrahedron Lett.* **2003**, *44*, 7837.

84. Suzuki, T.; Kobayashi, K.; Noda, K.; Oriyama, T., *Synth. Commun.* **2001**, *31*, 2761.

85. (a) Wada, M.; Nagayama, S.; Mizutani, K.; Hiroi, R.; Miyoshi, N., *Chem. Lett.* **2002**, 248. (b) Komatsu, N.; Ishida, J.; Suzuki, H., *Tetrahedron Lett.* **1997**, *38*, 7219. (c) Bajwa, J. S.; Jiang, X.; Slade, J.; Prasad, K.; Repic, O.; Blacklock, T. J., *Tetrahedron Lett.* **2002**, *43*, 6709.

86. Evans, P. A.; Cui, J.; Gharpure, S. J.; Hinkle, R. J., *J. Am. Chem. Soc.* **2003**, *125*, 11456.

87. Cullen, A. J.; Sammakia, T., *Org. Lett.* **2004**, *6*, 3143.

88. Vedejs, E.; Klapars, A.; Warner, D. L.; Weiss, A. H., *J. Org. Chem.* **2001**, *66*, 7542.

89. Kadereit, D.; Deck, P.; Heinemann, I.; Waldmann, H., *Chem. Eur. J.* **2001**, *7*, 1184.

90. Ravikumar, V. T.; Krotz, A. H.; Cole, D. L., *Tetrahedron Lett.* **1995**, *36*, 6587.

91. Yato, M.; Homma, K.; Ishida, A., *Tetrahedron* **2001**, *57*, 5353.

92. Sakai, N.; Hirasawa, M.; Konakahara, T., *Tetrahedron Lett.* **2005**, *46*, 6407.

93. Kotsuki, H.; Datta, P. K.; Hayakawa, H.; Suenaga, H., *Synthesis* **1995**, 1348.

94. Ferreri, C.; Costantino, C.; Chatgilialoglu, C.; Boukherroub, R.; Manuel, G., *J. Organomet. Chem.* **1998**, *554*, 135.

95. Iwata, A.; Ohshita, J.; Tang, H.; Kunai, A.; Yamamoto, Y.; Matui, C., *J. Org. Chem.* **2002**, *67*, 3927.

96. Menozzi, C.; Dalko, P. I.; Cossy, J., *Synlett* **2005**, 2449.

97. (a) Montgomery, J., *Angew. Chem., Int. Ed.* **2004**, *43*, 3890. (b) Mahandru, G. M.; Liu, G.; Montgomery, J., *J. Am. Chem. Soc.* **2004**, *126*, 3698. (c) Ng, S.-S.; Jamison, T. F., *J. Am. Chem. Soc.* **2005**, *127*, 7320. (d) Knapp-Reed, B.; Mahandru, G. M.; Montgomery, J., *J. Am. Chem. Soc.* **2005**, *127*, 13156.

98. Shibata, I.; Kato, H.; Ishida, T.; Yasuda, M.; Baba, A., *Angew. Chem., Int. Ed.* **2004**, *43*, 711.

99. Biermann, U.; Metzger, J. O., *Angew. Chem., Int. Ed.* **1999**, *38*, 3675.

100. Ghosh, A. K.; Xu, C.-X.; Kulkarni, S. S.; Wink, D., *Org. Lett.* **2005**, *7*, 7.

101. (a) Widenhoefer, R. A.; Stengone, C. N., *J. Org. Chem.* **1999**, *64*, 8681. (b) Widenhoefer, R. A.; Perch, N. S., *Org. Lett.* **1999**, *1*, 1103. (c) Perch, N. S.; Pei, T.; Widenhoefer, R. A., *J. Org. Chem.* **2000**, *65*, 3836. (d) Wang, X.; Chakrapani, H.; Stengone, C. N.; Widenhoefer, R. A., *J. Org. Chem.* **2001**, *66*, 1755.

102. Chakrapani, H.; Liu, C.; Widenhoefer, R. A., *Org. Lett.* **2003**, *5*, 157.

103. Oh, C. H.; Jung, H. H.; Sung, H. R.; Kim, J. D., *Tetrahedron* **2001**, *57*, 1723.

104. Oh, C. H.; Park, S. J., *Tetrahedron Lett.* **2003**, *44*, 3785.

105. Park, K. H.; Jung, I. G.; Kim, S. Y.; Chung, Y. K., *Org. Lett.* **2003**, *5*, 4967.

106. Wang, X.; Chakrapani, H.; Madine, J. W.; Keyerleber, M. A.; Widenhoefer, R. A., *J. Org. Chem.* **2002**, *67*, 2778.

107. Liu, C.; Widenhoefer, R. A., *Organometallics* **2002**, *21*, 5666.

108. Kang, S.-K.; Hong, Y.-T.; Leen, J.-H.; Kim, W.-Y.; Lee, I.; Yu, C.-M., *Org. Lett.* **2003**, *5*, 2813.

109. (a) Giese, S.; West, F. G., *Tetrahedron Lett.* **1998**, *39*, 8393–8396. (b) Giese, S.; West F. G., *Tetrahedron* **2000**, *56*, 10221.

110. Miraz-Aghayan, M.; Boukherroub, R.; Bolourtchian, M.; Hoseini, M.; Tabar-Hydar, K., *J. Organomet. Chem.* **2003**, *678*, 1.

111. Tanaka, M.; Mitsuhashi, H.; Maruno, M.; Wakamatsu, T., *Chem. Lett.* **1994**, 1455.

112. Yamanoi, Y., *J. Org. Chem.* **2005**, *70*, 9607.

113. Hamze, A.; Provot, O.; Alami, M.; Brion, J.-D., *Org. Lett.* **2006**, *8*, 931.

114. Kakiuchi, F.; Matsumoto, M.; Tsuchiya, K.; Igi, K.; Hayamizu, T.; Chatani, N.; Murai, S., *J. Organomet. Chem.* **2003**, *686*, 134.

115. Kakiuchi, F.; Tsuchiya, K.; Matsumoto, M.; Mizushima, E.; Chatani, N., *J. Am. Chem. Soc.* **2004**, *126*, 12792.

116. Itazaki, M.; Nakazawa, H., *Chem. Lett.* **2005**, *34*, 1054.

117. Kunai, A.; Ochi, T.; Iwata, A.; Ohshita, J., *Chem. Lett.* **2001**, 1228.

118. Gevorgyan, V.; Liu, J.-X.; Yamamoto, Y., *Chem. Commun.* **1998**, 37.

119. Hayashi, N.; Shibata, I.; Baba, A., *Org. Lett.* **2004**, *6*, 4981.

Triethylsilyl Hydrotrioxide

$$Et_3SiOOOH$$

[101631-06-5] $C_6H_{16}O_3Si$ (MW 164.31)

InChI = 1S/C6H16O3Si/c1-4-10(5-2,6-3)9-8-7/h7H,4-6H2,1-3H3

InChIKey = NTBHQWQOMYCCJI-UHFFFAOYSA-N

(alkene oxidation, dioxetane formation[1])

Preparative Methods: addition of triethylsilane (2 equiv) to a cold (−78 °C), saturated methylene chloride solution of ozone (ca. 0.04 M) results in discharge of the color with formation of triethylsilyl hydrotrioxide within 45 s.[2]

Handling, Storage, and Precautions: is prepared in solution immediately before use. In methylene chloride at −78 °C, the half-life of the reagent has been estimated to be a few minutes.[2] Precautions appropriate for reactions which utilize or generate peroxides should be used.

Alkene Oxidation Reactions. The reaction of triethylsilane with ozone and use of the intermediate triethylsilyl hydrotrioxide in oxidation reactions have been described. Initial oxidation reactions reported included the formation of the 9,10-endoperoxide from 9,10-dimethylanthracene (eq 1), and an allylic hydroperoxide from 2,3-dimethyl-2-butene (eq 2). Researchers also observed a near-IR emission from triethylsilyl hydrotrioxide as it decomposed at −60 °C, consistent with generation of singlet oxygen.[2] Other workers have characterized triethylsilyl hydrotrioxide by NMR spectroscopy and and measured the kinetics of its decomposition in deuterated acetone.[3]

$$(1)$$

$$(2)$$

Subsequent work examined the use of triethylsilyl hydrotrioxide in alkene oxidations and found it to be effective in oxidative cleavage of unactivated terminal and internal alkenes into carbonyl fragments, and in formation of 1,2-dioxetanes from vinyl aromatics and vinyl ethers.[1,4–6] The results of this research indicated that the reaction proceeds via direct interaction of the alkene and triethylsilyl hydrotrioxide without the intermediacy of singlet oxygen. Allylbenzene reacted with triethylsilyl hydrotrioxide followed by lithium aluminum hydride reduction to give 2-phenylethanol (eq 3). The same sequence of reactions gave 1-nonanol and 1,9-nonanediol in 64% and 74% yield, respectively, from methyl oleate (eq 4).[1] It has been noted in the cleavage of alkenes by triethylsilyl hydrotrioxide that the presence of an ether or ester function in the molecule results in improved yields, even when the oxygen function is remote from the double bond. Addition of diethyl ether or ethyl acetate does not improve the reactions of unsaturated hydrocarbons.[5]

$$(3)$$

$$(4)$$

In the case of an enol ether derivative (eq 5), reaction with triethylsilyl hydrotrioxide produced an intermediate dioxetane which was cleaved to give 3-phenylpropionaldehyde upon warming. In contrast, reaction of the same enol ether with photochemically generated singlet oxygen proceeded via an ene pathway. After hydrolysis, this gave an α,β-unsaturated aldehyde in 37% yield.[1]

$$(5)$$

Reaction of triethylsilyl hydrotrioxide with a keto vinyl ether and subsequent rearrangement of the dioxetane intermediate resulted in formation of the 1,2,4-trioxane product in 58% overall yield (eq 6).[5] This model system for the naturally occurring antimalarial artemisin was similarly prepared in 48% yield using singlet oxygen.[7] Both singlet oxygen and triethylsilyl hydrotrioxide have been used in the synthesis of a regiospecifically oxygen-18 labeled model system.[6]

$$(6)$$

1. Posner, G. H.; Webb, K. S.; Nelson, W. M.; Kishimoto, T.; Seliger, H. H., *J. Org. Chem.* **1989**, *54*, 3252.

2. Corey, E. J.; Mehrotra, M. M.; Khan, A. U., *J. Am. Chem. Soc.* **1986**, *108*, 2472.

3. Koenig, M.; Barrau, J.; Hamida, N. B., *J. Organomet. Chem.* **1988**, *356*, 133.

4. Posner, G. H.; Weitzberg, M.; Nelson, W. M.; Murr, B. L.; Seliger, H. H., *J. Am. Chem. Soc.* **1987**, *109*, 278.

5. Posner, G. H.; Oh, C. H.; Milhous, W. K., *Tetrahedron Lett.* **1991**, *32*, 4235.

6. Posner, G. H.; Oh, C. H., *J. Am. Chem. Soc.* **1992**, *114*, 8328.

7. Jefford, L. W.; Verlarde, J.; Bernardinelli, G., *Tetrahedron Lett.* **1989**, *30*, 4485.

Charles G. Caldwell

Merck Research Laboratories, Rahway, NJ, USA

3-Triethylsilyloxy-1,4-pentadiene

[62418-65-9] $C_{11}H_{22}OSi$ (MW 198.42)

InChI = 1S/C11H22OSi/c1-6-11(7-2)12-13(8-3,9-4)10-5/h6-7, 11H,1-2,8-10H2,3-5H3

InChIKey = RCMWXQHEDHJRTI-UHFFFAOYSA-N

(a masked vinyl ketone synthon;[7,8] a precursor for either the diene or dienophile in the intramolecular Diels–Alder reaction[2])

Physical Data: bp 140 °C; n_D^{20} 1.4390.

Solubility: sol pentane, benzene, THF, ether, CH_2Cl_2.

Analysis of Reagent Purity: NMR, GC (4.56 min on 1 mm 5% OV 225 on Chromosorb W at 108 °C) or TLC: R_f = 0.15 (pentane) or 0.6 (benzene).

Form Supplied in: commercially available.

Preparative Method: by silylation of 3-hydroxy-1,4-pentadiene with chlorotriethylsilane.[1,2] Other relatives of the title compound are available by silylation with chlorotrimethylsilane[2] or *t*-butyldimethylchlorosilane[3] 3-Trimethylsilyloxy-1,4-pentadiene is usually prepared and used in situ. Storage of 3-trimethylsilyl-1,4-pentadiene at −30 °C is recommended.[2] The dianion of the parent 3-hydroxy-1,4-pentadiene can also be prepared.[4]

Handling, Storage, and Precautions: use in a fume hood.

General Discussion.   3-Triethylsilyloxy-1,4-pentadiene may be used to prepare the useful 3-triethylsilyloxypentadienyl anion via metalation with *sec*-butyllithium (eq 1).[1,2]

$$\text{(1)}$$

Reaction of the 3-triethylsilyloxypentadienyl anion with electrophiles is usually regioselective for the γ-adduct (eq 2 and Table 1).[2–7] This reaction provides a direct route to substituted dienes, as well as to vinyl ketones after hydrolysis. A representative procedure is as follows: a solution of 3-triethylsiloxy-1,4-pentadiene in THF was cooled to −78 °C and a solution of *s*-butyllithium in cyclohexane (1.1 equiv) was added dropwise over 30 min. The electrophile dissolved in THF was then added dropwise to the often yellow pentadienyl anion solution. The reaction mixture was stirred for 15 min and then quenched with saturated NH_4Cl solution and the product extracted with ether.

$$\text{(2)}$$

Table 1 Regioselective reactions of 3-Triethylsilyloxy-1,4-pentadiene with typical electrophiles

Entry	Electrophile (E^+)	Yield (%)	Ratio γ to α product	Ref.
1	H_2O	87	100:0	2
2	TMSCl	89	100:0	2
3	MeCl	85	100:0	2
4	*i*-PrI	85	23:77	2
5	$PhCH_2Cl$	84	49:51	2
6	CH_2O	84	83:17	2
7	cyclohexanone	77	96:4	2
8	methallyl bromide	63	4:1	5
9	$PhSiMe_2Cl$	75	100:0	7

A wide variety of electrophiles may be used and the selectivity is based on kinetics. In some cases the products have been thermally interconverted (cf. entry 8, Table 1).[5,6] Because of polarity differences, the isomeric products usually can be separated by silica gel chromatography.

Elaboration of the alkylation products to give vinyl ketones may be accomplished with potassium fluoride in methanol at −10 to 0 °C or in *i*-PrOH at 25 °C (eq 3). For example, preparation of β-hydroxylethyl vinyl ketone synthons has been accomplished using a $PhMe_2Si$ unit as a masked hydroxyl group (eq 3).[7,8]

$$\text{(3)}$$

The method has been used to prepare intermediates for intramolecular Diels–Alder reactions (eqs 4 and 5).[2–6] The title reagent may provide ether the diene for the intramolecular Diels–Alder (eq 4)[4] or, after hydrolysis, the dienophile (eq 5).[3]

$$\text{(4)}$$

$$\text{(5)}$$

1. Oppolzer, W.; Snowden, R. L., *Tetrahedron Lett.* **1976**, 4187.

2. Oppolzer, W.; Snowden, R. L.; Simmons, D. P., *Helv. Chim. Acta* **1981**, *64*, 2002.

3. Wilson, S. R.; Jacob, L., *J. Org. Chem.* **1992**, *57*, 4380.

4. Wang, W.-B.; Roskamp, E. J., *Tetrahedron Lett.* **1992**, 7631.

5. Shea, K. J.; Wise, S., *Tetrahedron Lett.* **1979**, 1011.

6. Shea, K. J.; Wise, S.; Burke, L. D.; Davis, P. D.; Gilman, J. W.; Greeley, A. C., *J. Am. Chem. Soc.* **1982**, *104*, 5708.

7. Kim, S.; Emeric, G.; Fuchs, P. L., *J. Org. Chem.* **1992**, *57*, 7362.

8. Fleming, I., *Pure Appl. Chem.* **1988**, *60*, 71.

Stephen R. Wilson

New York University, NY, USA

(α-Triethylsilyl)propionaldehyde Cyclohexylimine

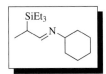

[119711-55-6] $C_{15}H_{31}NSi$ (MW 253.56)

InChI = 1S/C15H31NSi/c1-5-17(6-2,7-3)14(4)13-16-15-11-9-8-10-12-15/h13-15H,5-12H2,1-4H3

InChIKey = DWZBXMFEWTUGRV-UHFFFAOYSA-N

(reagent used for the selective formation of (*E*)-α-methyl-α,β-unsaturated aldehydes[1])

Physical Data: bp 84–86 °C/0.1 mmHg.

Solubility: sol THF, sol ether.

Analysis of Reagent Purity: ^{1}H NMR (300 MHz, CDCl3) δ 7.65 (d, *J* = 6.7, 1H); 2.87 (tt, *J* = 10.6 and 4.1 Hz, 1H); 2.03 (pentet, *J* = 7.0 Hz, 1H); 1.82–1.20 (m, 10H); 1.16 (d, *J* = 7.0 Hz, 3H); 0.96 (t, *J* = 7.8 Hz, 9H); 0.60 (q, *J* = 7.9 Hz, 6H).[1]

Preparative Method: prepared from propionaldehyde *N*-cyclohexylimine[1,2] by treatment with lithium diisopropylamide in THF at −78 °C for 30 min followed by the addition of chlorotriethylsilane and gradually warming to 0 °C over a 3–4 h period.[2,3]

Purification: by fractional distillation.[1]

Handling, Storage, and Precautions: use in a well-ventilated fume hood.

Formation of (*E*)-α-Methyl-α,β-unsaturated Aldehydes.
The anions of α-silyl aldimines have been used to prepare unsaturated aldehydes by means of Peterson alkenation reactions.[4,5] The anions of α-trimethylsilyl- or α-triethylsilylpropionaldehyde *t*-butylimine are easily prepared by treatment of the silyl aldimine with LDA or *s*-butyllithium (eq 1).[4]

$$
\begin{array}{c}
\underset{\textbf{(1) R = Me}}{\underset{\textbf{(2) R = Et}}{}} \quad \xrightarrow[\text{THF, 0 °C}]{\text{LDA or }s\text{-BuLi}} \quad \underset{\textbf{(3) R = Me}}{\underset{\textbf{(4) R = Et}}{}}
\end{array} \quad (1)
$$

Condensation of the α-lithio species with aldehydes and ketones conveniently yields α,β-unsaturated aldehydes following hydrolysis of the intermediate unsaturated imines (eqs 2 and 3).[3,4]

$$(2)$$

90%

$$(3)$$

77%

While the trimethyl and triethylsilyl aldimines (**1**) and (**2**) showed similar behavior in producing α,β-unsaturated aldehydes, the ability to isolate the triethylsilyl reagent (**2**) without contamination by the *N*-silyl enamine isomer makes it the preferred reagent.[3] Additionally, the replacement of the *t*-butylimine with a cyclohexylimine affords a reagent (**5**) which is easily purified by fractional distillation.[1] The anion of α-triethylsilylpropionaldehyde cyclohexylimine (**5**) is formed by the addition of *s*-BuLi in THF at −78 °C (eq 4).[1] This anion condenses with aldehydes to produce α-methyl-α,β-unsaturated aldehydes in good yield after hydrolysis (eq 4).

The ratio of (*E*)- to (*Z*)-alkene products is dependent on the workup conditions. Addition of the α-silylaldimine anion to a carbonyl species gives a mixture of (*E*)- and (*Z*)-α-β-unsaturated imines, which upon hydrolysis with aqueous oxalic acid affords moderately selective formation of (*E*)-α,β-unsaturated aldehydes (eqs 4 and 5). Treatment of the mixture of imines with anhydrous

(**5**)

$$(4)$$

91%, (*E*):(*Z*) = 100:1
[(*E*):(*Z*) = 2:1 with (CO$_2$H)$_2$ workup]

(*Z*)-imine   (*E*)-imine

$$(5)$$

84%, (*E*):(*Z*) = 100:1
[85%, (*E*):(*Z*) = 2:1 with (CO$_2$H)$_2$ workup]

trifluoroacetic acid in THF for a short period of time prior to aqueous hydrolysis serves to isomerize the imines and produce the *trans*-aldehyde with high selectivity (eqs 4 and 5).[1] The anhydrous trifluoroacetic acid workup protocol thus provides a highly selective method for the preparation of (E)-α-methyl-α,β-unsaturated aldehydes.[6]

1. Desmond, R.; Mills, S. G.; Volante, R. P.; Shinkai, I., *Tetrahedron Lett.* **1988**, *29*, 3895.

2. Campbell, K. N.; Sommers, A. H.; Campbell, B. K., *J. Am. Chem. Soc.* **1944**, *66*, 82.

3. Schlessinger, R. H.; Poss, M. A.; Richardson, S.; Lin, P., *Tetrahedron Lett.* **1985**, *26*, 2391.

4. Corey, E. J.; Enders, D.; Bock, M. G., *Tetrahedron Lett.* **1976**, 7.

5. Peterson, D. J., *J. Org. Chem.* **1968**, *33*, 780.

6. (a) Corey, E. J.; Huang, H. C., *Tetrahedron Lett.* **1989**, *30*, 5235. (b) Jones, T. K.; Reamer, R. A.; Desmond, R.; Mills, S. G., *J. Am. Chem. Soc.* **1990**, *112*, 2998. (c) Martin, S. F.; Dodge, J. A.; Burgess, L. E.; Hartmann, M., *J. Org. Chem.* **1992**, *57*, 1070.

John M. McGill
Eli Lilly Company, Lafayette, IN USA

Triethylsilyl Trifluoromethanesulfonate[1]

$$Et_3SiOSO_2CF_3$$

[79271-56-0] $C_7H_{15}F_3O_3SSi$ (MW 264.38)
InChI = 1S/C7H15F3O3SSi/c1-4-15(5-2,6-3)13-14(11,12)7
     (8,9)10/h4-6H2,1-3H3
InChIKey = STMPXDBGVJZCEX-UHFFFAOYSA-N

(potent silylating agent;[2–4] Lewis acid catalyst)

Physical Data: 85–86 °C/12 mmHg; *d* 1.169 g cm^{-3}.
Solubility: readily sol hydrocarbons, dialkyl ethers, halogenated solvents. CH$_2$Cl$_2$ is employed most commonly. Reactions in 1,2-dichloroethane proceed faster than those in CCl$_4$ or Et$_2$O. Protic solvents and THF react with trialkylsilyl triflates and are therefore not suitable.
Form Supplied in: neat colorless liquid.
Preparative Method: can be prepared by reacting chlorotriethylsilane with trifluoromethanesulfonic acid followed by distillation.[1]
Handling, Storage, and Precautions: trialkylsilyl triflates are generally corrosive and moisture sensitive. Appropriate precautions should be taken to ensure that the reagent is handled and stored under rigorously anhydrous conditions.

Reactive Silylating Agent. Triethylsilyl ethers are generally more stable towards hydrolysis than are trimethylsilyl ethers, and consequently the Et$_3$Si moiety has gained increasing use as a protecting group for alcohols. However, since it is often difficult to silylate sterically hindered hydroxyl groups using Et$_3$SiCl, triethylsilyl perchlorate and triethylsilyl triflate (Et$_3$SiOTf) were introduced to overcome this problem. Although both reagents are

much more potent than Et$_3$SiCl, the triflate is considered to be safer than the perchlorate and is more convenient because it is commercially available.

Secondary and tertiary alcohols can be silylated under mild conditions using Et$_3$SiOTf and 2,6-lutidine or a trialkylamine as a proton scavenger (eqs 1 and 2).[2–4]

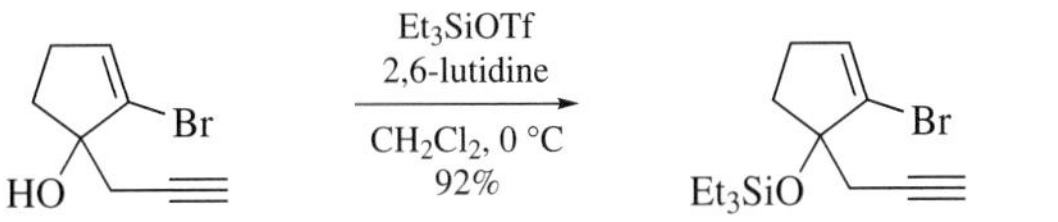

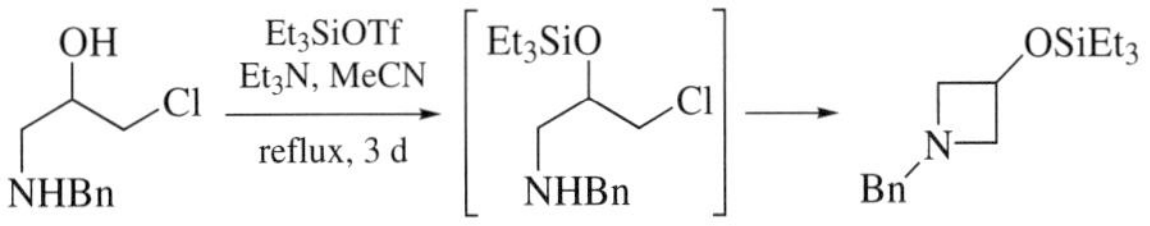

Silyl enol ethers can be conveniently prepared by treatment of ketones with Et$_3$SiOTf and triethylamine (eq 3).[1,5] Similarly, esters can be converted to silyl ketene acetals; however, large amounts of the *C*-silylated product may also be isolated (eq 4).[1,6,7] Equilibration of *O*- and *C*-silylated products may occur in the presence of catalytic amounts of Et$_3$SiOTf.[1]

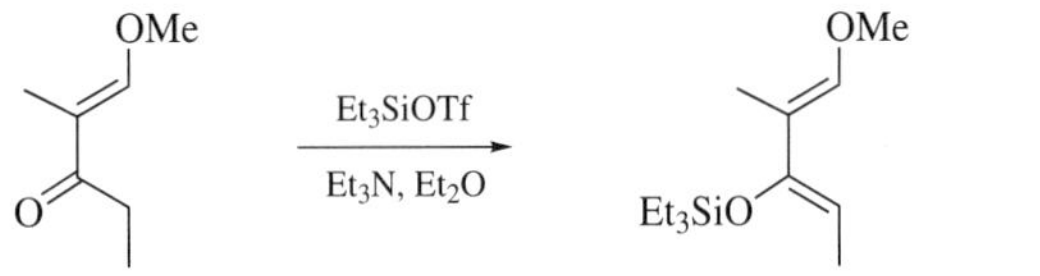

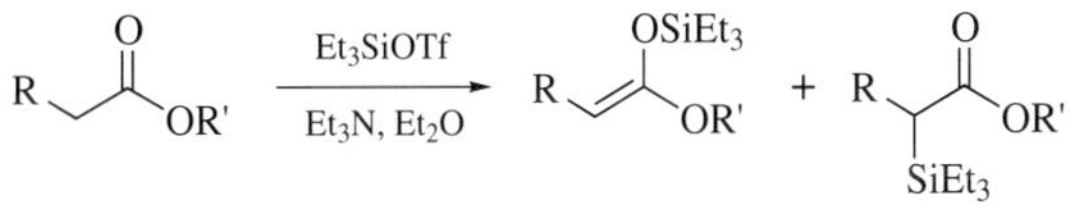

α-Diazo esters react with Et$_3$SiOTf and Hünig's base (diisopropylethylamine) to give exclusively α-silyl-α-diazo esters, which can be converted to other silylated compounds by loss of dinitrogen (eq 5).[8]

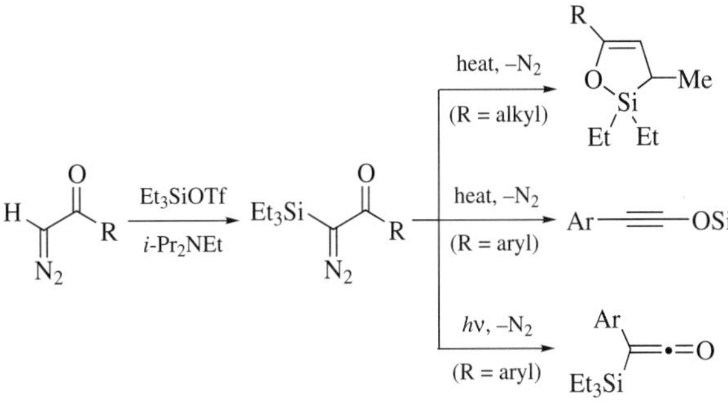

Nitroalkanes have been shown to give α-siloxy *O*-silyloximes upon reaction with two molar equivalents of the silylating agent (eq 6).[9] The reaction is believed to proceed via a nitrogen-to-carbon 1,3-silyloxy rearrangement.

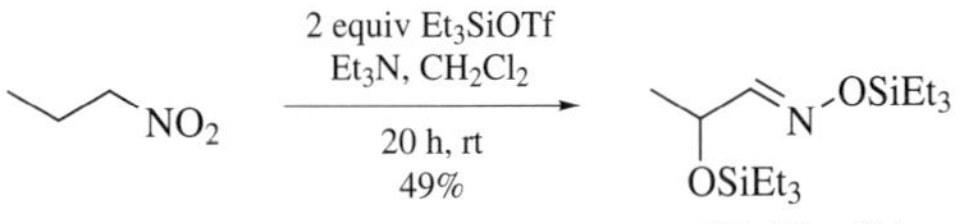

Linderman and Ghannam have utilized Et$_3$SiOTf to trap the alkoxide formed from the addition of stannyl anion to aldehydes (eq 7).[10] Upon treatment with excess *n*-butyllithium, these adducts undergo a reverse Brook rearrangement to afford α-hydroxysilanes.

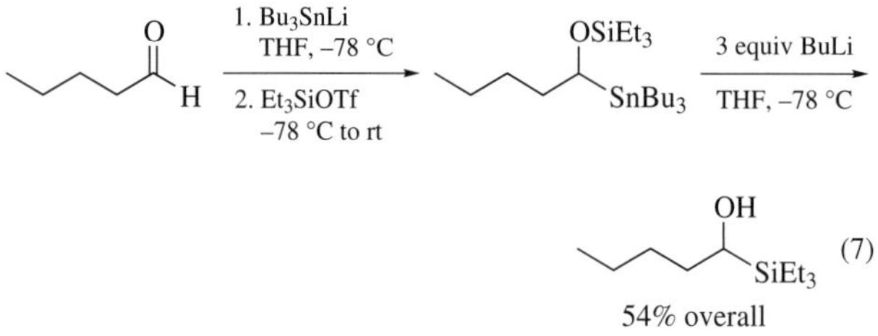

(7)

54% overall

Lewis Acid Catalyst. Jefford has reported that condensation reactions of 2-trimethylsiloxyfuran with aldehydes can be catalyzed by Et₃SiOTf to give mainly the *threo* addition product (eq 8).[11] Conversely, the *erythro* adduct is favored when fluoride ion is used as the catalyst.

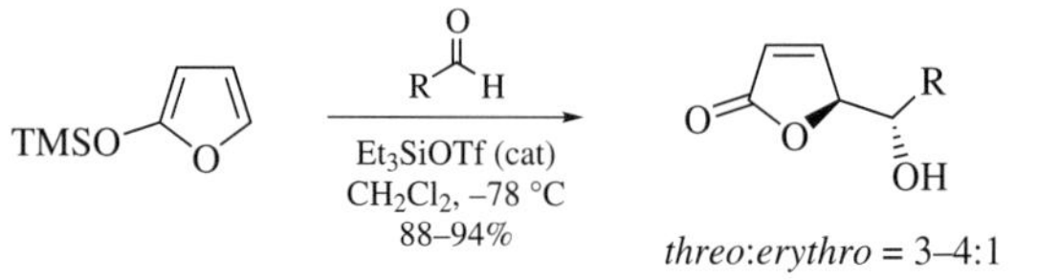

(8)

threo:erythro = 3–4:1

Fraser-Reid has reported that an equimolar mixture of N-iodosuccinimide and Et₃SiOTf efficiently promotes the glycosylation of hindered glycoside donors with n-pentenyl glycoside acceptors (eq 9).[12]

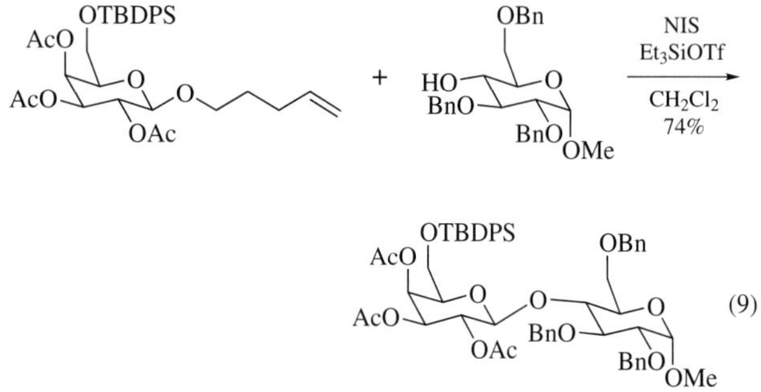

(9)

Related Reagents. Chlorotriethylsilane; Triethylsilyl Perchlorate; Trimethylsilyl Trifluoromethanesulfonate.

1. Emde, H.; Domsch, D.; Feger, H.; Frick, U.; Götz, A.; Hergott, H. H.; Hofmann, K.; Kober, W.; Krägeloh, K.; Oesterle, T.; Steppan, W.; West, W.; Simchen, G., *Synthesis* **1982**, 1.
2. Fujiwara, K.; Sakai, H.; Hirama, M., *J. Org. Chem.* **1991**, *56*, 1688.
3. Higgins, R. H., *J. Heterocycl. Chem.* **1987**, *24*, 1489.
4. McCarthy, P. A.; Kageyama, M., *J. Org. Chem.* **1987**, *52*, 4681.
5. Danishefsky, S.; Harvey, D. F., *J. Am. Chem. Soc.* **1985**, *107*, 6647.
6. Emde, H.; Götz, A.; Hofmann, K.; Simchen, G., *Liebigs Ann. Chem.* **1981**, 1643.
7. Emde, H.; Simchen, G., *Liebigs Ann. Chem.* **1983**, 816.
8. (a) Maas, G.; Brückmann, R., *J. Org. Chem.* **1985**, *50*, 2801; (b) Brückmann, R.; Maas, G., *Chem. Ber.* **1987**, *120*, 635.
9. (a) Ferger, H.; Simchen, G., *Liebigs Ann. Chem.* **1986**, 428; (b) Feger, H.; Simchen, G., *Synthesis* **1981**, 378.
10. (a) Linderman, R. J.; Ghannam, A., *J. Am. Chem. Soc.* **1990**, *112*, 2392;(b) Linderman, R. J.; Ghannam, A., *J. Org. Chem.* **1988**, *53*, 2878.
11. Jefford, C. W.; Jaggi, D.; Boukouvalas, J., *Tetrahedron Lett.* **1987**, *28*, 4037.
12. Konradsson, P.; Udodong, U. E.; Fraser-Reid, B., *Tetrahedron Lett.* **1990**, *31*, 4313.

Edward Turos
State University of New York at Buffalo, NY, USA

Trimethylsilyl Trifluoromethanesulfonate[1]

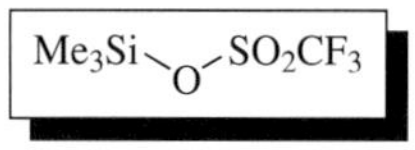

[27607-77-8]   C₄H₉F₃O₃SSi   (MW 222.29)
InChI = 1S/C4H9F3O3SSi/c1-12(2,3)10-11(8,9)4(5,6)7/
    h1-3H3
InChIKey = FTVLMFQEYACZNP-UHFFFAOYSA-N

Alternate Name: TMSOTf.
Physical Data: bp 45–47 °C/17 mmHg, 39–40 °C/12 mmHg; d 1.225 g cm⁻³.
Solubility: sol aliphatic and aromatic hydrocarbons, haloalkanes, ethers.
Form Supplied in: colorless liquid; commercially available.
Preparative Methods: may be prepared by a variety of methods.[2]
Handling, Storage, and Precautions: flammable; corrosive; very hygroscopic.

Original Commentary

Joseph Sweeney & Gemma Perkins
University of Bristol, Bristol, UK

Silylation. TMSOTf is widely used in the conversion of carbonyl compounds to their enol ethers. The conversion is some 10^9 faster with TMSOTf/triethylamine than with chlorotrimethylsilane (eqs 1–3).[3–5]

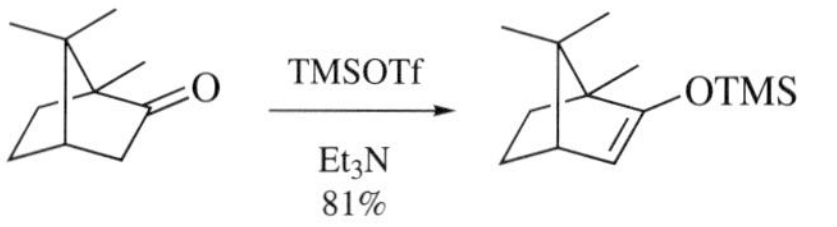

(1)

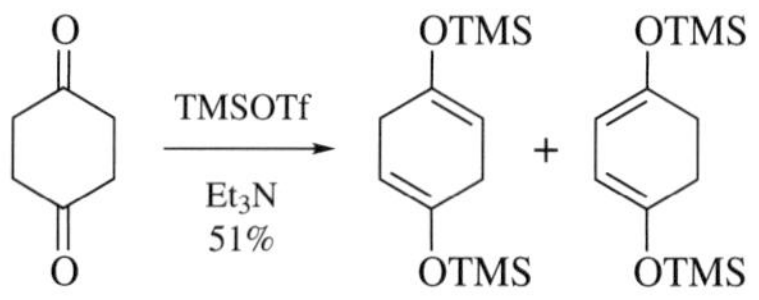

(2)

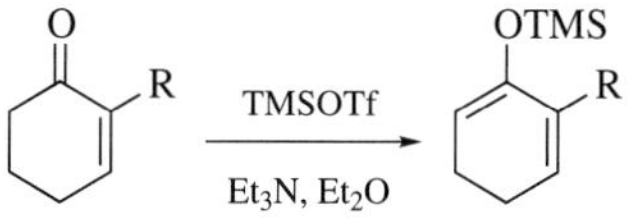

(3)

Dicarbonyl compounds are converted to the corresponding bis-enol ethers; this method is an improvement over the previous two-step method (eq 4).[6]

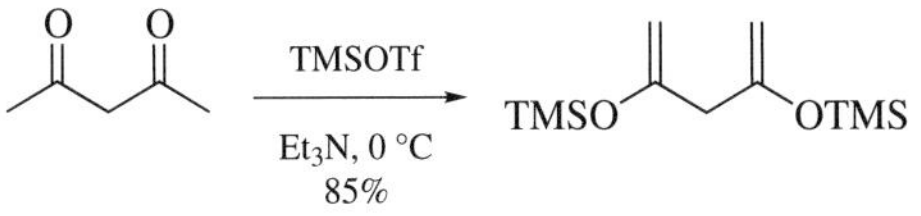

(4)

In general, TMSOTf has a tendency to C-silylation which is seen most clearly in the reaction of esters, where C-silylation dominates over O-silylation. The exact ratio of products obtained depends on the ester structure[7] (eq 5).[8] Nitriles undergo C-silylation; primary nitriles may undergo C,C-disilylation.[9]

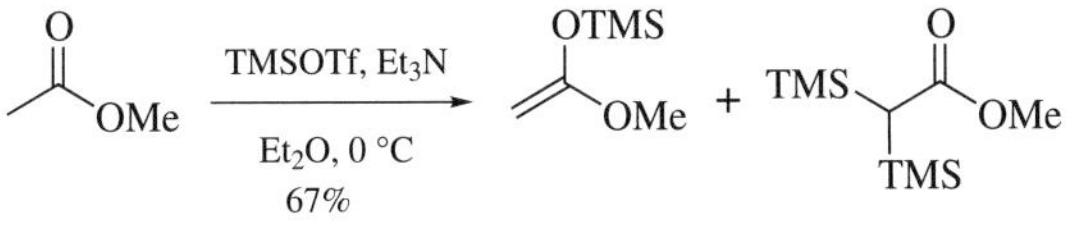

(5)

84:16

TMS enol ethers may be prepared by rearrangement of α-ketosilanes in the presence of catalytic TMSOTf (eq 6).[10,11]

(6)

Enhanced regioselectivity is obtained when trimethylsilyl enol ethers are prepared by treatment of α-trimethylsilyl ketones with catalytic TMSOTf (eq 7).[12]

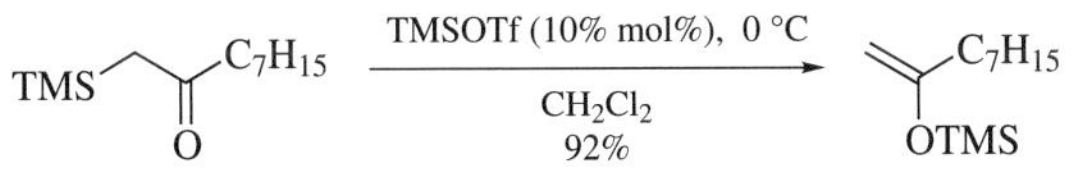

(7)

The reaction of imines with TMSOTf in the presence of Et_3N gives N-silylenamines.[13]

Ethers do not react, but epoxides are cleaved to give silyl ethers of allylic alcohols in the presence of TMSOTf and 1,8-diazabicyclo[5.4.0]undec-7-ene; The regiochemistry of the reaction is dependent on the structure of the epoxide (eq 8).[14]

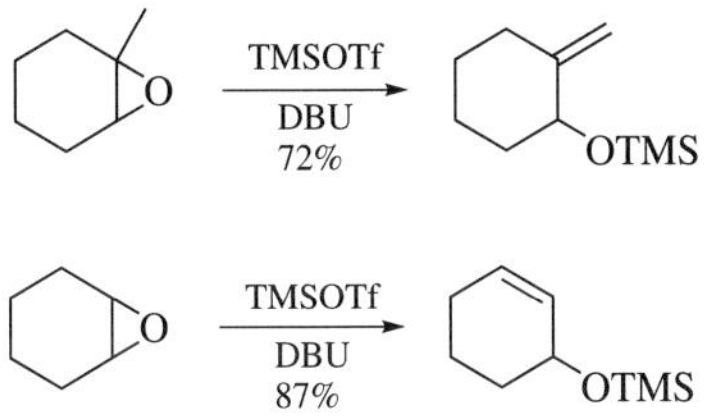

(8)

Indoles and pyrroles undergo efficient C-silylation with TMSOTf (eq 9).[15]

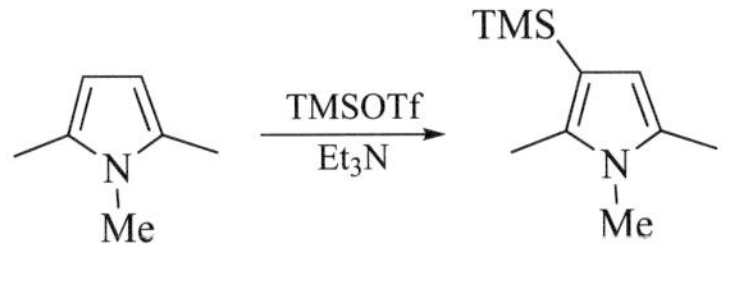

(9)

t-Butyl esters are dealkylatively silylated to give TMS esters by TMSOTf; benzyl esters are inert under the same conditions.[16]

Imines formed from unsaturated amines and α-carbonyl esters undergo ene reactions in the presence of TMSOTf to form cyclic amino acids.[17]

Carbonyl Activation. 1,3-Dioxolanation of conjugated enals is facilitated by TMSOTf in the presence of 1,2-bis(trimethylsilyl-oxy)ethane. In particular, highly selective protection of sterically differentiated ketones is possible (eq 10).[18] Selective protection of ketones in the presence of enals is also facilitated (eq 11).[19]

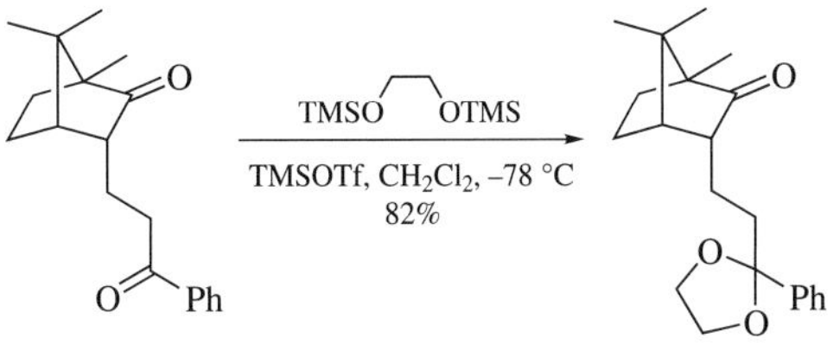

(10)

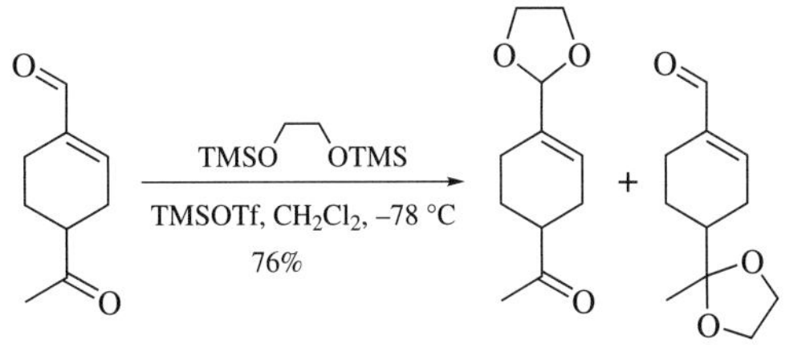

(11)

1:27

The similar reaction of 2-alkyl-1,3-disilyloxypropanes with chiral ketones is highly selective and has been used to prepare spiroacetal starting materials for an asymmetric synthesis of α-tocopherol subunits (eq 12).[20]

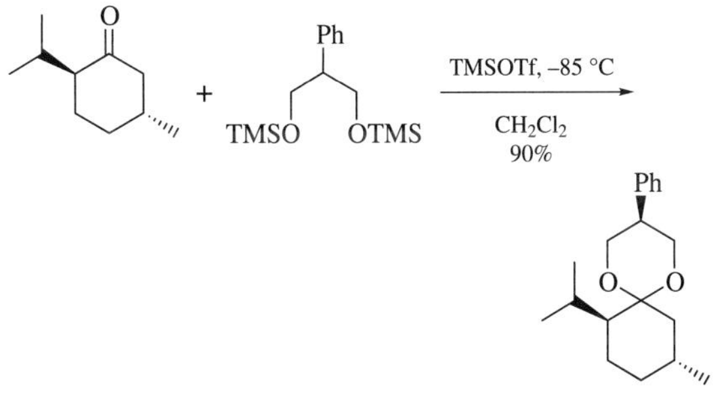

(12)

The preparation of spiro-fused dioxolanes (useful as chiral gly-colic enolate equivalents) also employs TMSOTf (eq 13).[21]

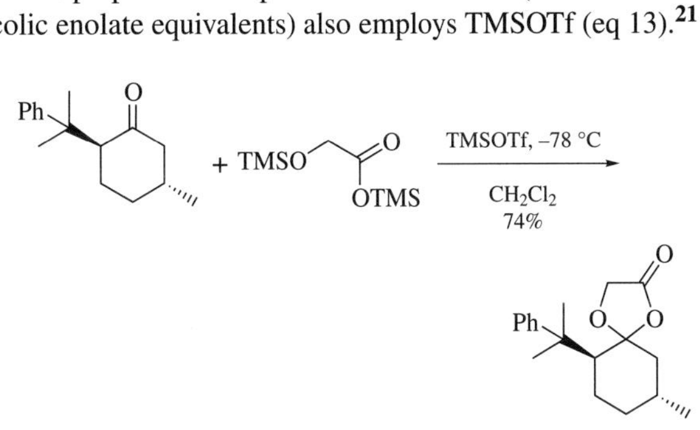

(13)

~ 1:1 mixture

TMSOTf mediates a stereoselective aldol-type condensation of silyl enol ethers and acetals (or orthoesters). The nonbasic reaction conditions are extremely mild. TMSOTf catalyzes many aldol-type reactions; in particular, the reaction of relatively non-nucleophilic enol derivatives with carbonyl compounds is facile in the presence of the silyl triflate. The activation of acetals was first reported by Noyori and has since been widely employed (eq 14).[22,23]

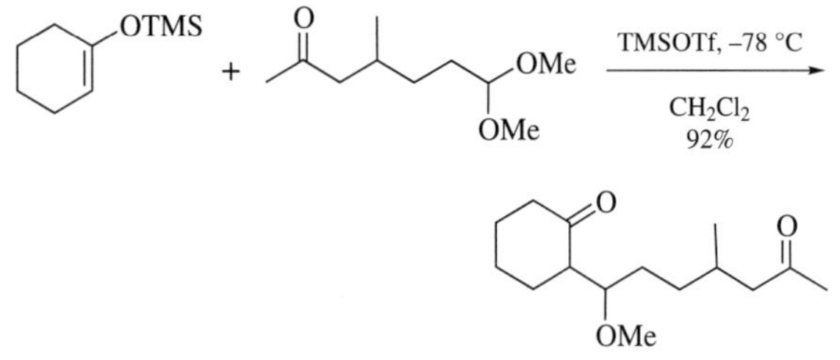

(14)

In an extension to this work, TMSOTf catalyzes the first step of a [3 + 2] annulation sequence which allows facile synthesis of fused cyclopentanes possessing bridgehead hydroxy groups (eq 15).[24]

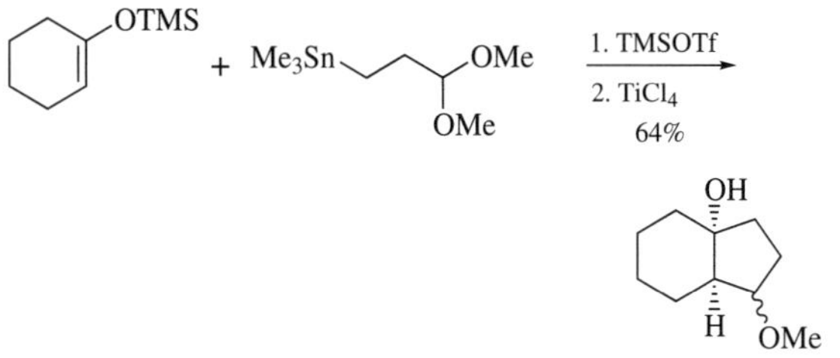

(15)

The use of TMSOTf in aldol reactions of silyl enol ethers and ketene acetals with aldehydes is ubiquitous. Many refinements of the basic reaction have appeared. An example is shown in eq 16.[25]

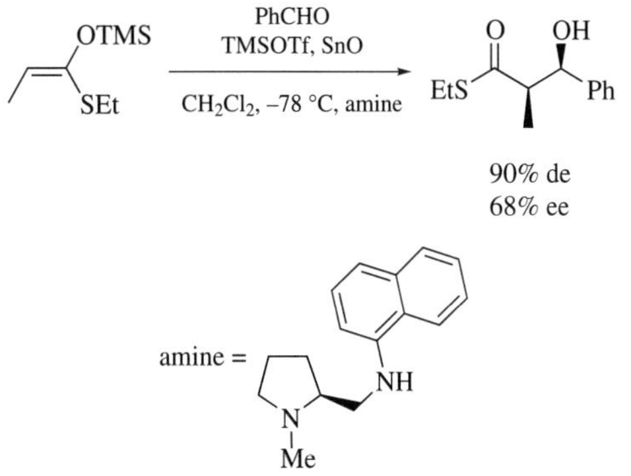

(16)

90% de
68% ee

amine =

The use of TMSOTf in the reaction of silyl ketene acetals with imines offers an improvement over other methods (such as Ti^{IV}- or Zn^{II}-mediated processes) in that truly catalytic amounts of activator may be used (eq 17);[26] this reaction may be used as the crucial step in a general synthesis of 3-(1'-hydroxyethyl)-2-azetidinones (eq 18).[27]

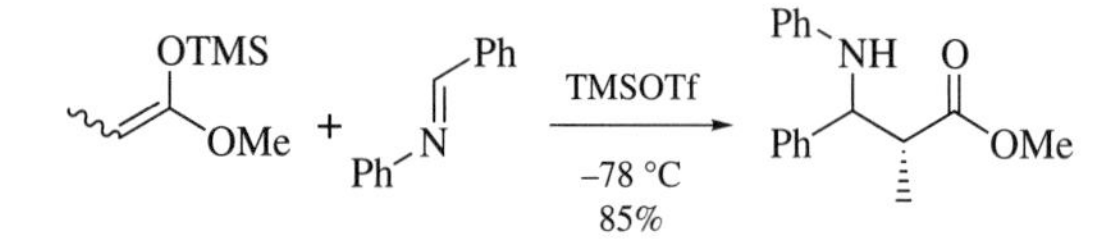

(17)

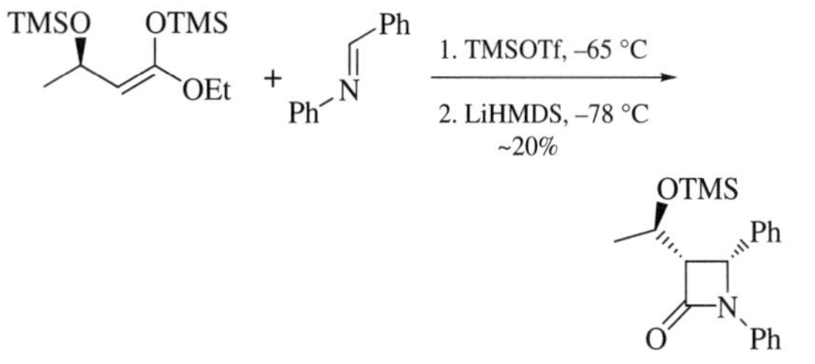

(18)

Stereoselective cyclization of α,β-unsaturated enamide esters is induced by TMSOTf and has been used as a route to quinolizidines and indolizidines (eq 19).[28]

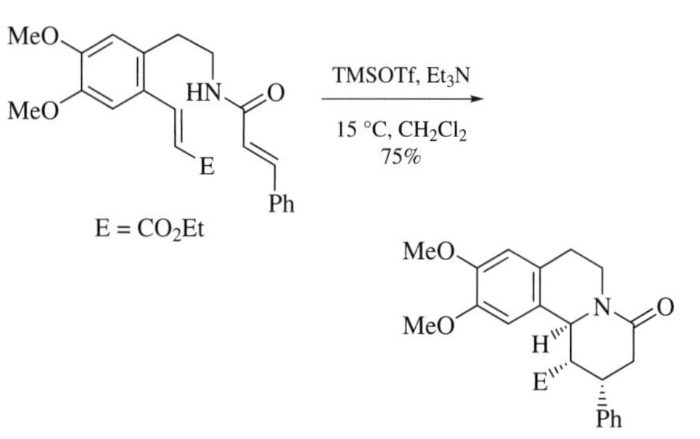

E = CO$_2$Et

(19)

The formation of nitrones by reaction of aldehydes and ketones with N-methyl-N,O-bis(trimethylsilyl)hydroxylamine is accelerated when TMSOTf is used as a catalyst; the acceleration is particularly pronounced when the carbonyl group is under a strong electronic influence (eq 20).[29]

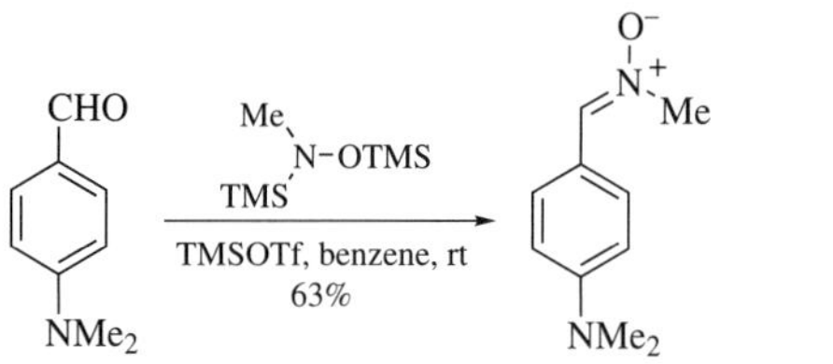

(20)

β-Stannylcyclohexanones undergo a stereoselective ring contraction when treated with TMSOTf at low temperature. When other Lewis acids were employed, a mixture of ring-contracted and protiodestannylated products was obtained (eq 21).[30]

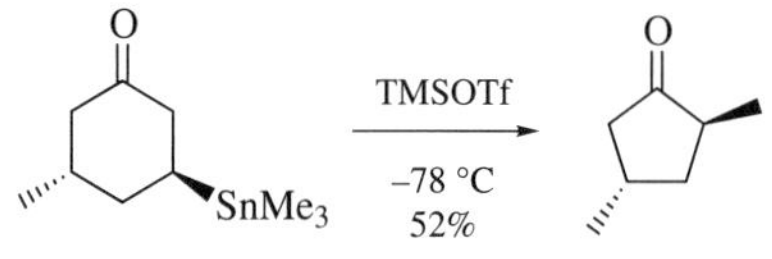

(21)

The often difficult conjugate addition of alkynyl organometallic reagents to enones is greatly facilitated by TMSOTf. In particular, alkynyl zinc reagents (normally unreactive with α,β-unsaturated carbonyl compounds) add in good yield (eq 22).[31] The proportion of 1,4-addition depends on the substitution pattern of the substrate.

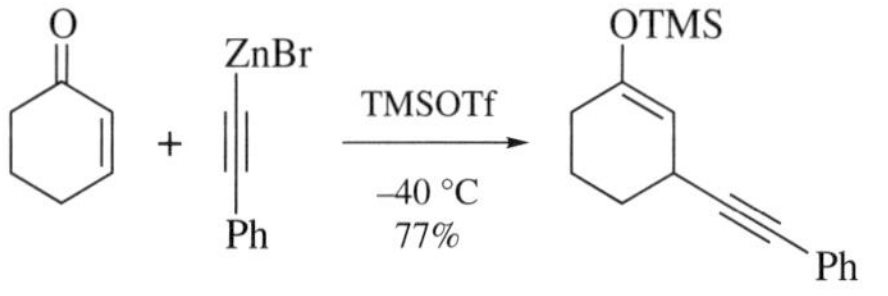

(22)

The 1,4-addition of phosphines to enones in the presence of TMSOTf gives β-phosphonium silyl enol ethers, which may be deprotonated and alkylated in situ (eq 23).[32]

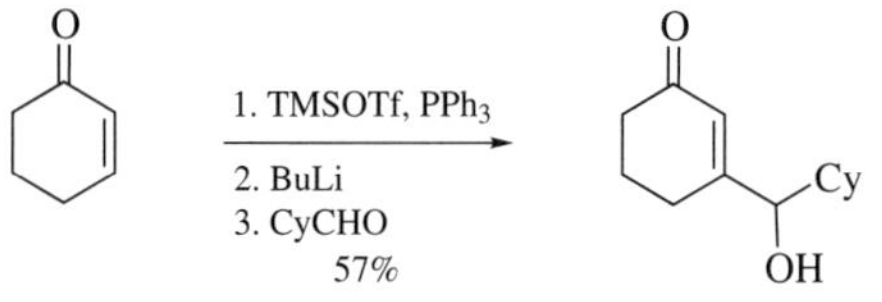

(23)

Miscellaneous. Methyl glucopyranosides and glycopyranosyl chlorides undergo allylation with allylsilanes under TMSOTf catalysis to give predominantly α-allylated carbohydrate analogs (eq 24).[33]

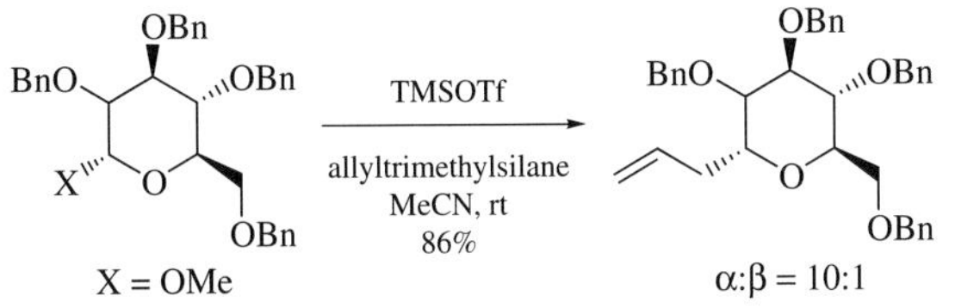

$$\text{(24)}$$

X = OMe   α:β = 10:1

Glycosidation is a reaction of massive importance and widespread employment. TMSOTf activates many selective glycosidation reactions (eq 25).[34]

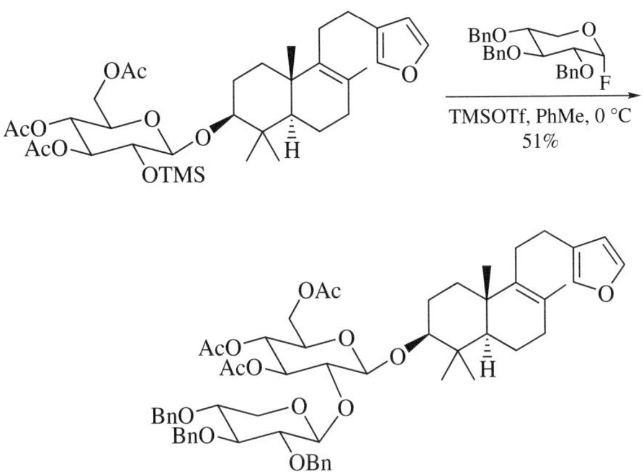

$$\text{(25)}$$

TMSOTf activation for coupling of 1-O-acylated glycosyl donors has been employed in a synthesis of avermectin disaccharides (eq 26).[35]

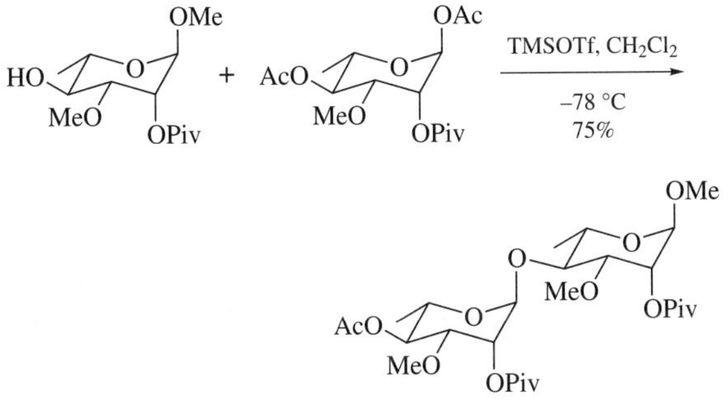

$$\text{(26)}$$

Similar activation is efficient in couplings with trichloroimidates[36] and O-silylated sugars.[37,38]

2-Substituted Δ^3-piperidines may be prepared by the reaction of 4-hydroxy-1,2,3,4-tetrahydropyridines with a variety of carbon and heteronucleophiles in the presence of TMSOTf (eqs 27 and 28).[39]

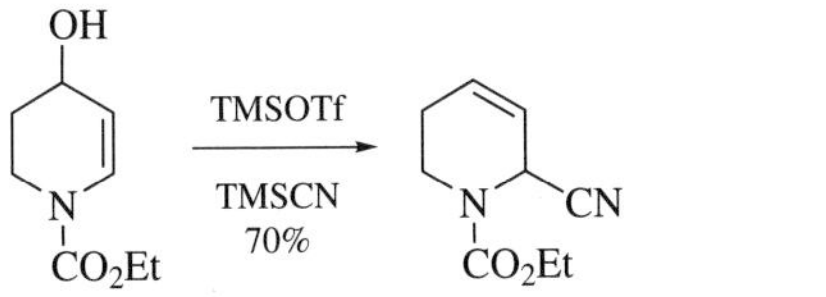

$$\text{(27)}$$

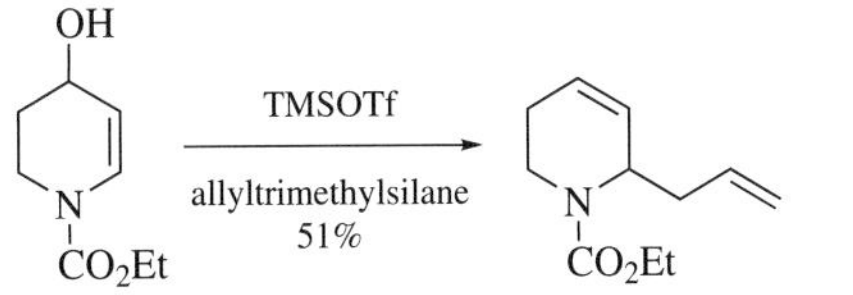

$$\text{(28)}$$

Iodolactamization is facilitated by the sequential reaction of unsaturated amides with TMSOTf and iodine (eq 29).[40]

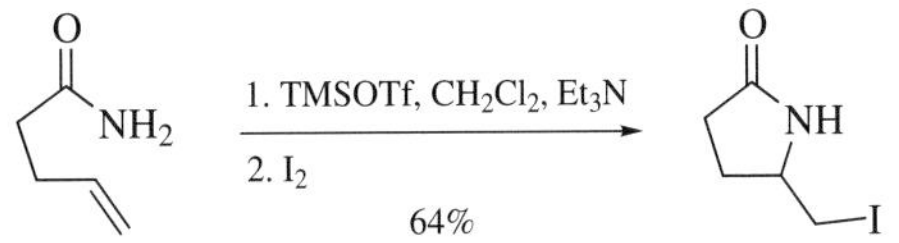

$$\text{(29)}$$

By use of a silicon-directed Beckmann fragmentation, cyclic (E)-β-trimethylsilylketoxime acetates are cleaved in high yield in the presence of catalytic TMSOTf to give the corresponding unsaturated nitriles. Regio- and stereocontrol are complete (eq 30).[41]

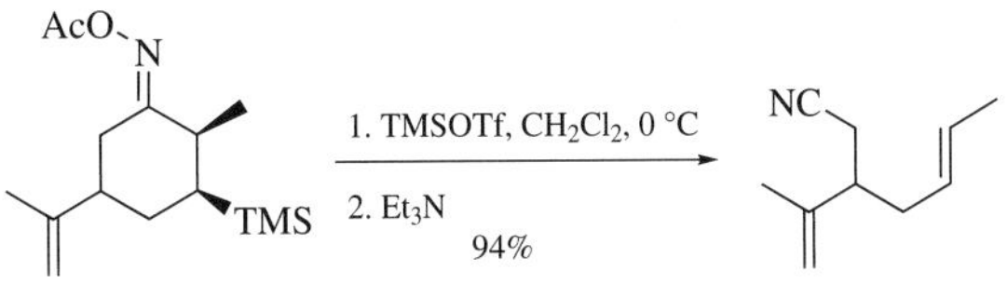

$$\text{(30)}$$

A general route to enol ethers is provided by the reaction of acetals with TMSOTf in the presence of a hindered base (eq 31).[42] The method is efficient for dioxolanes and noncyclic acetals.

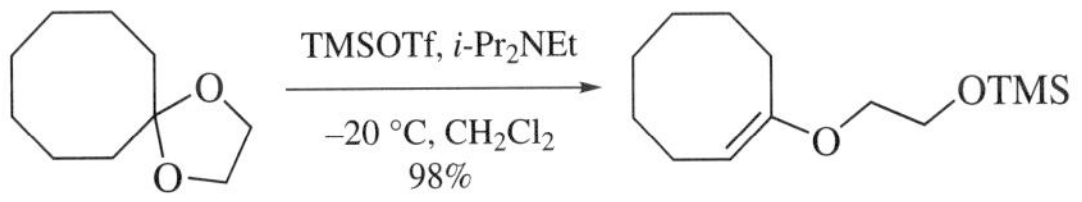

$$\text{(31)}$$

α-Halo sulfoxides are converted to α-halovinyl sulfides by reaction with excess TMSOTf (eq 32),[43] while α-cyano- and α-alkoxycarbonyl sulfoxides undergo a similar reaction (eq 33).[44] TMSOTf is reported as much superior to iodotrimethylsilane in these reactions.

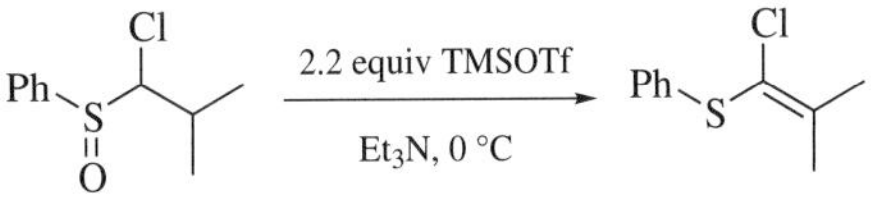

$$\text{(32)}$$

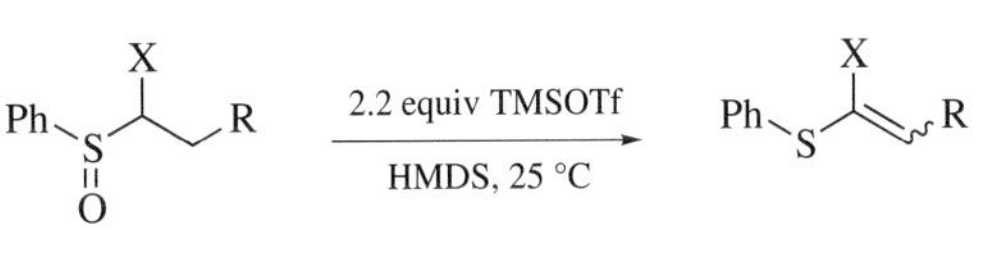

$$\text{(33)}$$

X = CN or CO$_2$R′

First Update

Enrique Aguilar & Manuel A. Fernández-Rodríguez
Universidad de Oviedo, Asturias, Spain

The introduction of TMSOTf as a highly electrophilic silylating reagent led to its widespread use and there are a very large number of reactions which employ the reagent in either stoichiometric or catalytic amounts.

***O*-Silylation.**   The formation of TMS ethers can be achieved by reacting the requisite alcohol with TMSOTf and an amine (triethylamine, pyridine, or 2,6-lutidine) in dichloromethane (eq 34).[45]

$$ \text{(34)} $$

The combination of TMSOTf and Et$_3$N in dichloromethane (DCM) allows the direct conversion of *p*-methoxybenzyl ethers into silyl-protected alcohols, thus affording an expedient way to replace the benzyl ether-type protective group with the silyl ether-type one (eq 35).[46]

$$ \text{(35)} $$

The preparation of silyl enol ethers from carbonyl compounds represents one of the major uses of TMSOTf. Recently, the stereochemistry and regiospecificity of such transformation has been addressed for aldehydes[47] and α-(*N*-alkoxycarbonylamino) ketones,[48] respectively. On the other hand, enantiopure silyl enol ethers can be formed by addition of TMSOTf to zinc enolates, which are obtained from the copper-catalyzed enantioselective conjugate addition of dialkylzinc reagents to cyclic (eq 36) and acyclic enones.[49]

$$ \text{(36)} $$

ligand =

The addition of dimethyl sulphide to α,β-unsaturated carbonyl compounds in the presence of TMSOTf generates highly reactive 3-trialkylsilyloxyalk-2-enylenesulfonium salts, which permits the introduction of a wide variety of nucleophiles at the β-position as well as α-functionalization.[50]

Bis-silylation of imides in both oxygen atoms results in the formation of cyclic 2-azadienes (eq 37).[51]

$$ \text{(37)} $$

Silylated 2-oxidienes and bis-silylated 2,3-dioxidienes are prepared from α,β-unsaturated ketones[52] and 1,2-dicarbonyl compounds,[53] respectively; in the latter case, the reagent employed for

the first silylation is trimethylsilyl bromide while TMSOTf is required for the second silylation. 1,3-Bis-silyloxydienes are prepared by direct silylation of 1,3-dicarbonyl compounds with TMSOTf/Et$_3$N.[54] They also can be obtained as mixtures of *Z/E*-isomers by lithium reductive cleavage of isoxazoles followed by the slow addition of an excess of TMSOTf/Et$_3$N (eq 38).[55]

$$ \text{(38)} $$

The formation of stable *N,O*-acetal TMS ethers, which are excellent precursors of *N*-acyliminium ions, is easily achieved by DIBAL reduction of *N*-acylamides followed by in situ protection with TMSOTf/pyridine (eq 39).[56] 2,6-Lutidine has also been used as base.[57] The DIBAL reduction-TMSOTf/pyridine silylation sequence has also been applied to the formation of monosilyl *O,O*-acetal from esters.[58]

$$ \text{(39)} $$

Silylated α,β-unsaturated oximes, which can be easily desilylated to enoximes, can be readily prepared by treatment of aliphatic nitro compounds with an excess of TMSOTf in the presence of Et$_3$N, provided that an electron-withdrawing group is located at the β-position (eq 40).[59]

$$ \text{(40)} $$

The in situ formation of silyl enol ether intermediates appears to be a valuable strategy to accomplish further transformations. That is what happens for the palladium(II) acetate-mediated preparation of indenones from indanones (eq 41),[60] or for diastereoselective enol silane coupling reactions (eq 42).[61]

$$ \text{(41)} $$

$$ \text{(42)} $$

A Claisen rearrangement of allylic esters is also effected by TMSOTf in the presence of triethylamine; when chiral esters are employed the transfer of chirality is, however, quite low. Higher degrees of 1,4-asymmetric transmission are reached if bulkier silyl triflates and bulkier amines are employed (eq 43).[62]

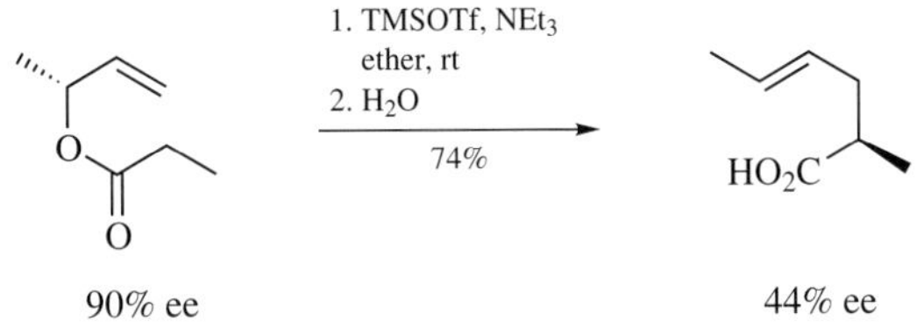

(43)

90% ee

44% ee

***C*-Silylation.** Depending on the reaction conditions, secondary amides can be either *C*-silylated or *N*-silylated. Kinetic *C*-silylation of *N*-methylacetamide can be quantitatively achieved by treatment with excess TMSOTf (3 equiv) and Et$_3$N (3 equiv) at 0 °C for 2 min, followed by aqueous work-up (eq 44).[63]

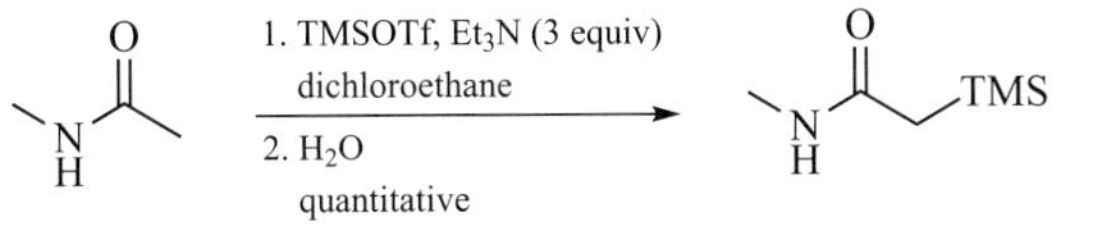

(44)

The silylation of aromatic[64] or heteroaromatic compounds can be performed via magnesiated or lithiated intermediates, which can be accessed by hydrogen-metal or halogen-metal atom exchange. Thus, C-4 silylation of imidazoles is achieved from 4-iodo imidazoles (eq 45),[65] while C-2 silylated oxazoles are prepared by addition of TMSOTf to the lithiated oxazole (eq 46); interestingly, for this reaction, the employment of TMSCl as electrophile resulted in the formation of the *O*-silylation product.[66] Pyrazole *N*-oxides and 1,2,3-triazole *N*-oxides are also *C*-silylated on the ring and at exocyclic α-positions in high yields by a one-pot procedure, which is initiated by *O*-silylation followed by deprotonation, and subsequently terminated by silylation of the generated anion.[67]

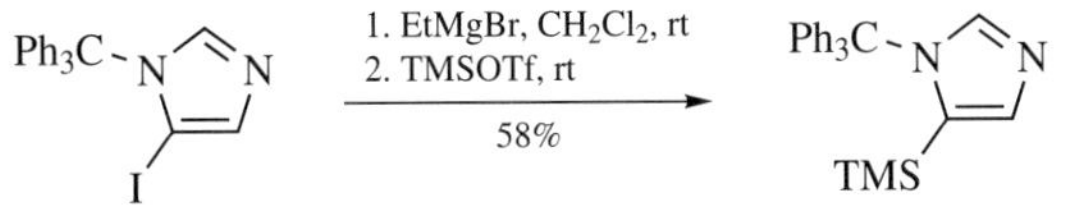

(45)

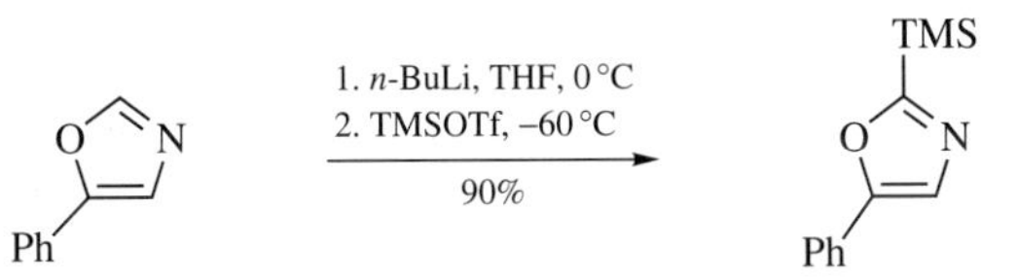

(46)

The enantioselective *C*-silylation of allylic substrates such as *N*-(*t*-butoxycarbonyl)-*N*-(*p*-methoxyphenyl)allylamines[68] or 1,3-diphenylpropene[69] is accomplished with butyllithium in the presence of (−)-sparteine, followed by the addition of TMSOTf (eq 47). The same procedure allows the asymmetric deprotonation-substitution of arenetricarbonyl(0) complexes,[70] while chiral bis(oxazolines) have been the ligands of choice to perform such transformation with aryl benzyl sulfides;[71] in these reactions, different yields and enantioselectivities are reached if trimethylsilyl chloride is used as silylating reagent, although there is a substrate dependence and no definite rules can be established.

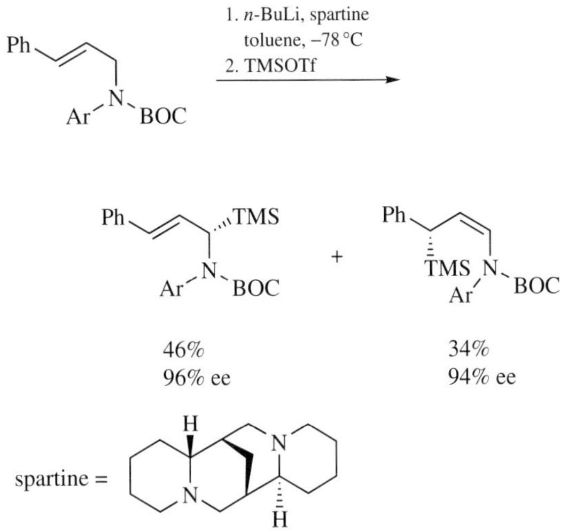

(47)

46%   34%
96% ee   94% ee

spartine =

1,1-Dibromo-4-methyl-3-(trimethylsilyloxy)pentane is cleanly silylated with TMSOTf through a preformed carbenoid species (eq 48). On the other hand, silylation can be carried out with TMS-imidazole albeit in lower yields, while the treatment with trimethylsilyl chloride is unsuccessful.[72]

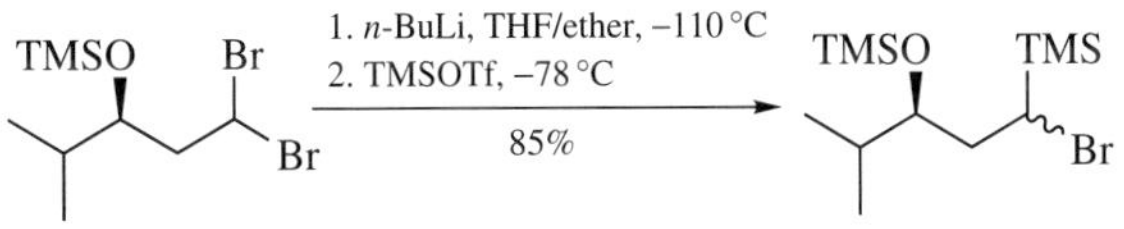

(48)

C-Silylated cyclohexyl diazo esters can be prepared from the appropriate diazo acetates by treatment with TMSOTf and ethyl diisopropylamine in ether at −78 °C (eq 49).[73]

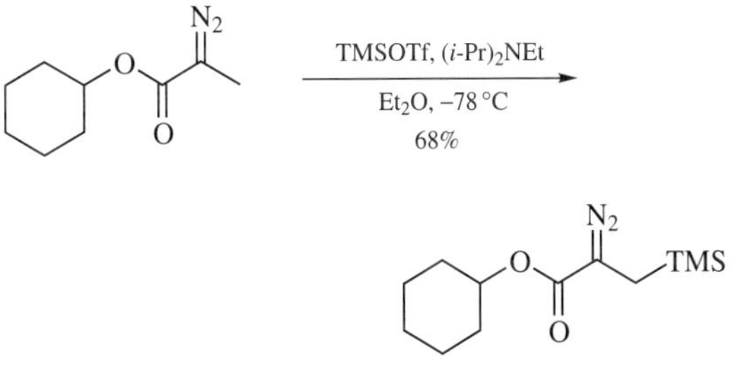

(49)

N-Silylation. The *N*-bis-silylation of α-amino acids with TMSOTf is only effective for glycine (eq 50); for other α-amino acids *N*-mono-silylation prevails because the larger size of the carbon chain at the α-position hinders bis-silylation.[74]

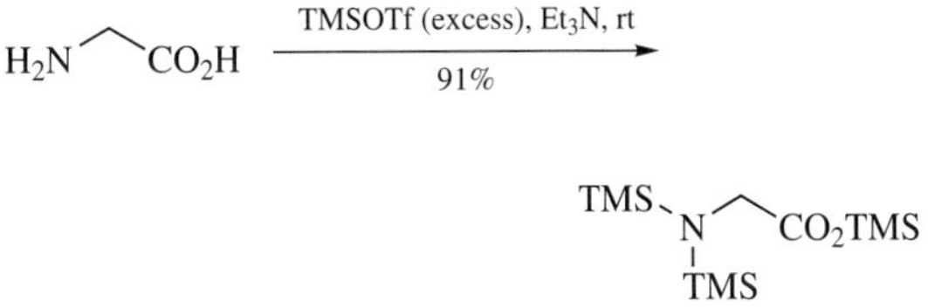

(50)

TMSOTf is the reagent of choice to transform imines into *N*-silyl iminium salts, which have been postulated to be the reactive intermediates in several reactions (eq 51).[75]

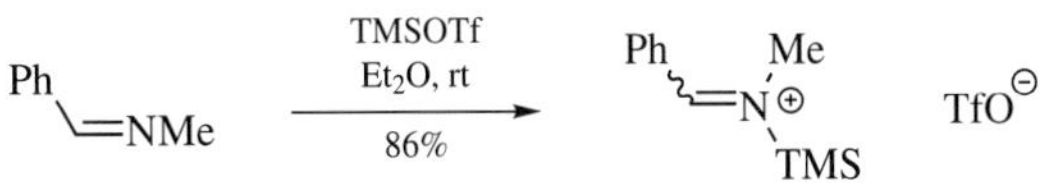

(51)

Secondary amides[76] and *O*-aryl *N*-isopropylcarbamates[77] can be converted in situ into their *N*-silylated derivatives which, although not isolated, are stable for further transformations (eq 52).

$$1. \text{TMSOTf, TMEDA, Et}_2\text{O}$$
$$2. \text{s-BuLi, TMEDA, } -78\,^\circ\text{C}$$
$$3. \text{EX, } -78\,^\circ\text{C}$$
$$4. \text{MeOH}$$
$$63\text{-}94\%$$

(52)

***C,O*-Bis-silylation.**   Bis-silylation of α,β-unsaturated carbonyl compounds can be achieved by palladium-TMSOTf-catalyzed addition of disilanes to enones, enals, or aromatic aldehydes via an η^3-silyloxyallylpalladium intermediate.[78] The scope of the reaction is wide and it generally leads to 1,4-bis-silylated compounds, although for cinnamaldehyde, it gives a mixture of 1,4- and 1,2-addition products, and 1,2-bis-silylated adducts are obtained for aromatic aldehydes (eq 53).

$$\text{Pd(OAc)}_2 \text{ (10 mol \%)}$$
$$\text{TMSOTf (10 mol \%)}$$
$$\text{C}_6\text{D}_6, \text{ rt}$$
$$40\text{-}99\%$$

(53)

Si-Si = PhMe$_2$Si-SiMe$_2$Ph or Me$_3$Si-SiMe$_3$

Carbonyl Activation.   TMSOTf frequently acts as a Lewis acid and it is able to activate several functional groups (the carbonyl group, the acetal unit, the nitrone moiety,...) thus facilitating different kinds of reactions.

In this regard, it has promoted the direct conversion of carbonyl compounds into *O,S*-acetals by reaction of aldehydes with 1 equiv each of Me$_3$SSiPh and a silyl ether in CH$_2$Cl$_2$ at low temperature; the mentioned compounds are isolated as major reaction products (eq 54).[79]

$$\text{R}^1\text{CHO} + \text{Me}_3\text{SiOR}^2 + \text{Me}_3\text{SiSPh} \xrightarrow[37\text{-}93\%]{\substack{4\text{-}200 \text{ mol \% TMSOTf} \\ \text{CH}_2\text{Cl}_2, -78\,^\circ\text{C}}}$$

(54)

1,3-Dioxolanation is a very common strategy for the protection of carbonyl groups. In this regard, an efficient dioxolanation of ketones in the presence of aldehydes is achieved by treating the dicarbonyl compound with TMSOTf and dimethyl sulphide in dichloromethane (eq 55).[80] On the other hand, both aldehydes and ketones can be protected as 4-trimethylsilyl-1,3-dioxolanes by using catalytic amounts of TMSOTf; this protective group

can be selectively cleaved to regenerate the carbonyl compound in the presence of a 1,3-dioxolane with either LiBF$_4$ or HF in acetonitrile.[81]

$$1. \text{TMSOTf, Me}_2\text{S, DCM, } -78\,^\circ\text{C}$$
$$2. \text{TMSOTf, TMSO} \diagup\diagdown \text{OTMS}$$
$$3. \text{aq K}_2\text{CO}_3$$
$$80\%$$

(55)

The role of the ligand has been found to be crucial in the silyl Lewis acid Mukaiyama aldol reaction, which opens interesting applications for synthetic organic chemistry. When TMSOTf induces the reaction, the silyl group of TMSOTf remains in the product and that of the silyl enol ether becomes the catalyst for the next catalytic cycle; however, if the reaction is promoted by TMSNTf$_2$, the silyl group of the catalyst is not released from -NTf$_2$ and that of the silyl enol ether intermolecularly transfers to the product (eq 56).[82]

$$\text{PhCHO} + \underset{\text{Ph}}{\overset{\text{OTBDMS}}{\diagup\diagdown}} \xrightarrow[\text{Et}_2\text{O, } -100\,^\circ\text{C}]{\text{catalyst (1 equiv)}}$$

(56)

catalyst			
TMSOTf	24%	99	1
TMSNTf$_2$	>99%	1	>99

The direct preparation of esters from aldehydes has been reported but different conditions are required depending on the nature of the aldehyde. The aldehydes are treated with the corresponding silyl ether followed by addition of AIBN and NBS in CCl$_4$; for aromatic aldehydes, the presence of TMSOTf is required to complete the radical oxidation to the ester; for aliphatic aldehydes, the acetals are isolated under the above-mentioned reaction conditions (eq 57), while the radical oxidation to the esters only takes place in the absence of TMSOTf.[83]

The condensation of a carbonyl compound with an isonitrile usually takes place in the presence of a carboxylic acid to give carboxamides (Passerini reaction) and plays a central role in combinatorial chemistry. Activation of the carbonyl group can be achieved by TMSOTf (eq 58), although higher yields can be obtained with the system Zn(OTf)$_2$/TMSCl.[84]

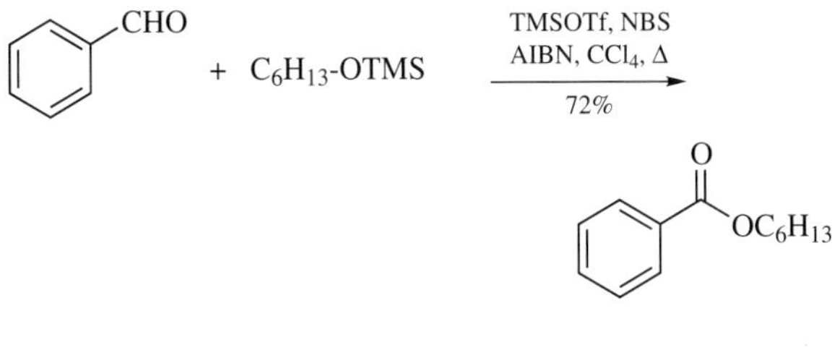

$$ \text{(57)} $$

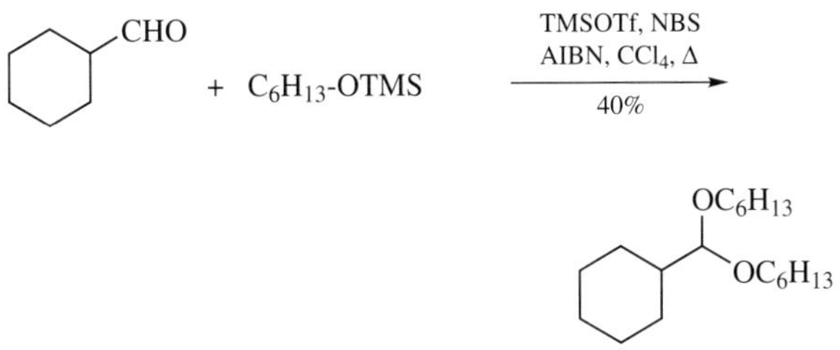

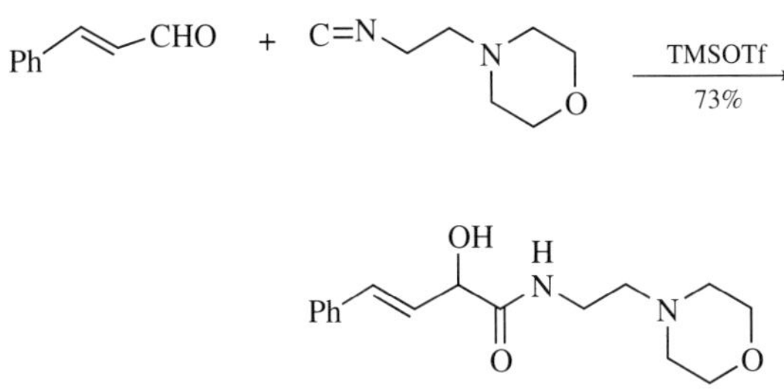

$$ \text{(58)} $$

A previously known thermal ring-contraction, which follows an initial 1,3-dipolar cycloaddition between azides and cyclic enones, can be smoothly promoted by TMSOTf to give enaminones (eq 59).[85]

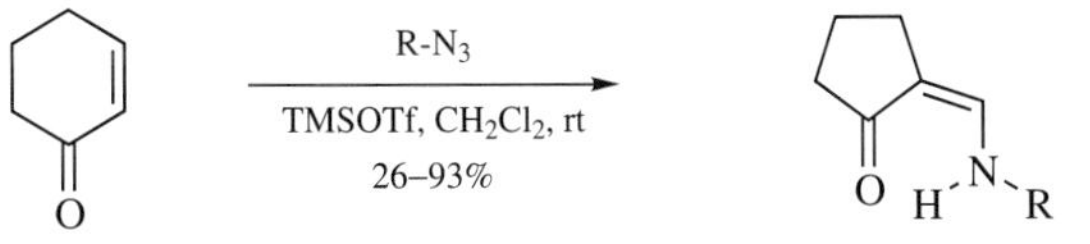

$$ \text{(59)} $$

Z/E ratio = 1:1 to 2:1

Also, 3-stannyl cyclohexanones experience a ring contraction to a 2-methylcyclopentanone upon treatment with TMSOTf in dichloromethane (eq 60). This transformation has been used as a key step for the synthesis of (+)-β-cuparenone.[86]

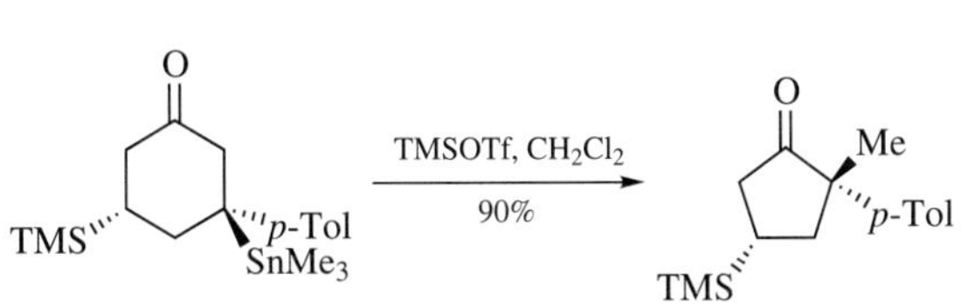

$$ \text{(60)} $$

The addition of bis(trimethylsilyl)formamide to methyl pyruvate is readily accomplished in the presence of a catalytic amount of TMSOTf (eq 61).[87]

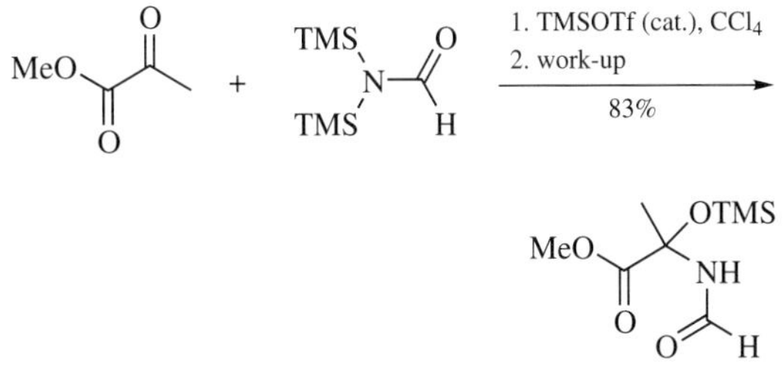

$$ \text{(61)} $$

Similarly, pyridine addition to aldehydes is also promoted by TMSOTf in a three-component reaction to form [1-(trimethylsiloxy)alkyl]pyridinium salts (eq 62), which may act as group transfer reagents or as precursors for the analogous phosphonium salts.[88]

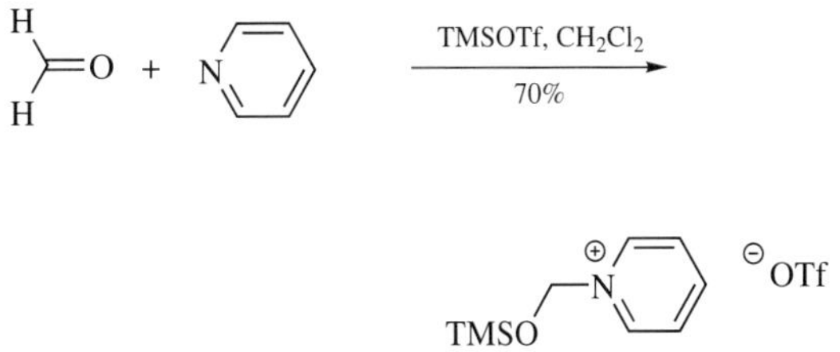

$$ \text{(62)} $$

The allylation of aldehydes can be achieved in the presence of ketones in a highly chemoselective fashion by preferential in situ conversion of aldehydes into 1-silyloxysulfonium salts, by treatment with dimethyl sulphide and TMSOTf and subsequent displacement with allyltributylstannane (eq 63).[89]

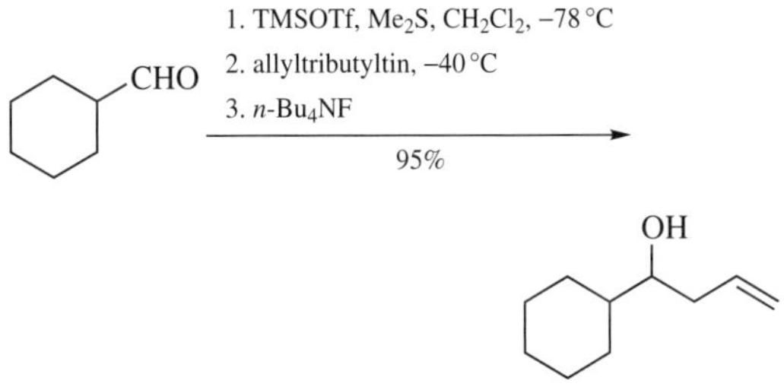

$$ \text{(63)} $$

TMSOTf also promotes an intramolecular addition of allylsilanes to aldehydes (Sakurai reaction), when used with 2,6-di-t-butyl-4-methylpyridine[90] (2,6-DTBMP) (eq 64), and ketones.[91]

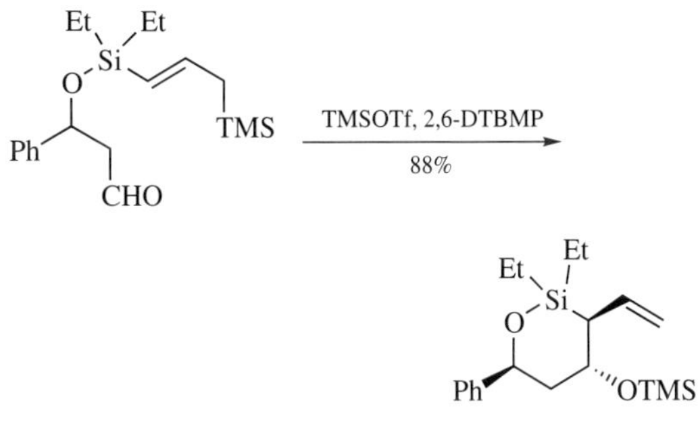

$$ \text{(64)} $$

dr = 3.9:1

An intermolecular version of the Sakurai reaction has been developed. It proceeds at temperatures as low as −78 °C, using substoichiometric amounts of the Lewis acid, to form five- or six-membered oxygenated heterocycles from cyclic allylsiloxanes and aldehydes through a chairlike transition state (eq 65).

Acetals may also act as electrophiles in this kind of reaction.[92] β-Borylallylsilanes also undergo nucleophilic allylation to lead to the preparation of functionalized alkenylboranes, which may participate in further transformations.[93]

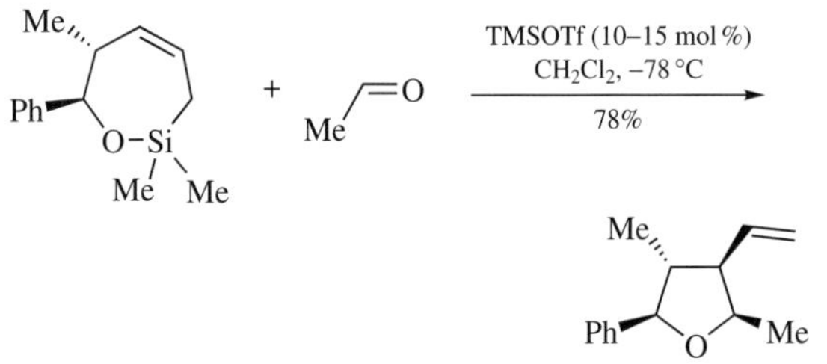

$$(65)$$

The Lewis acid-mediated ring-opening reaction of arylcarbonyl activated cyclopropanes with arylaldehydes is successfully performed with 1 equiv of TMSOTf in refluxing 1,2-dichloroethane to form α-substituted α,β-enones in good yields (eq 66).[94]

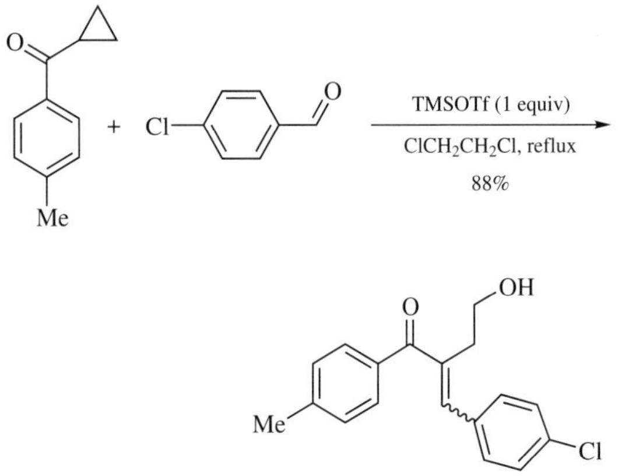

$$(66)$$

Z:E = 24:76

An improving effect of TMSOTf (and other trialkylsilyl triflates) in the enantioselectivity has been disclosed for the Cu-catalyzed 1,4-addition of Me₃Al to a 4,4-disubstituted cyclohexa-2,5-dienone, using a chiral oxazoline as ligand.[95] The presence of TMSOTf also helps to increase the yield for the conjugate addition of copper acetylides to α,β-unsaturated ketones and aldehydes; however, trimethylsilyl iodide has been more effective while other trimethylsilyl halides or BF₃ inhibit the addition.[96]

The conjugate addition of organozinc reagents to α,β-enones is readily mediated by TMSOTf (eq 67), thus avoiding previous requirements of transmetallation to organocopper compounds or the use of nickel complexes as catalysts.[97]

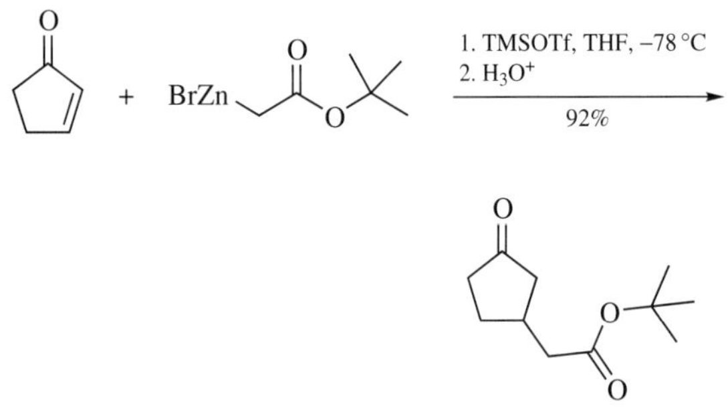

$$(67)$$

An asymmetric intramolecular Michael-aldol reaction which leads to nonracemic tricyclic cyclobutanes is performed by using TMSOTf and bis[(R)-1-phenylethyl]amine as chiral amine,

but only moderate enantioselectivities are reached (eq 68).[98] A similar reaction sequence can also be carried out with TMSOTf and HMDS as base, with (−)-8-phenylmenthol as the chiral auxiliary; however, the iodotrimethylsilane-HMDS system is more efficient in terms of yield and diastereoselectivity.[99] The combination Et₃N/TMSOTf (or some other trialkylsilyl triflates) has been used to accomplish an intramolecular Michael reaction, which was the key step for the synthesis of sesquiterpene (±)-ricciocarpin A.[100]

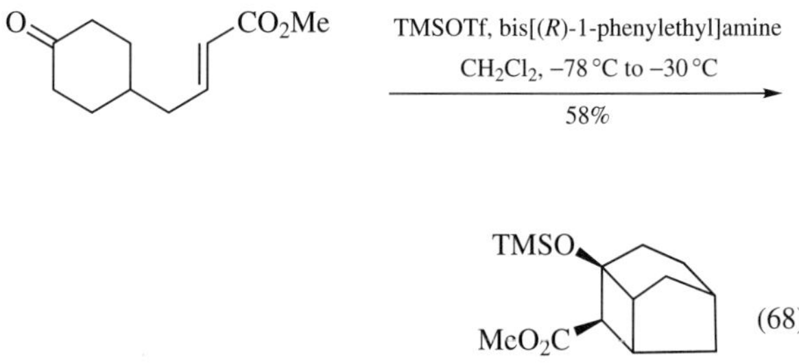

$$(68)$$

ee = 23%

On the other hand, β-functionalization of α,β-unsaturated lactones and esters is readily achieved through a phosphoniosilylation process: the addition of an aldehyde to the generated ylide followed by the fast addition of TMSOTf gave optimum results for the hydroxyalkylation reaction (eq 69). In the absence of TMSOTf alkylated products are obtained instead of the hydroxyalkylated ones.[101]

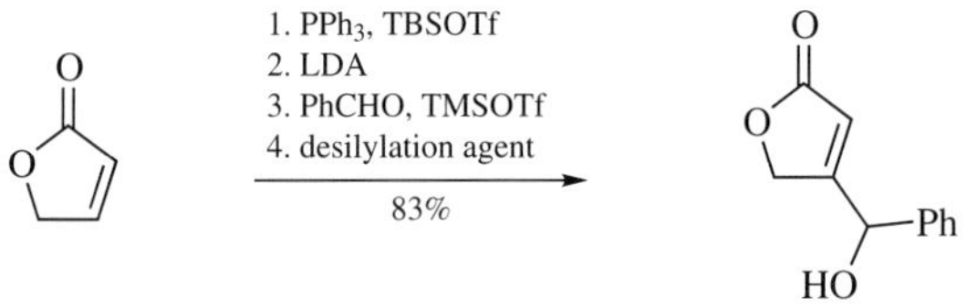

$$(69)$$

The combination of Me₃Al and TMSOTf allows the conversion of a carbonyl into a geminal dimethyl functionality (eq 70).[102] Other methylating reagents such as MeTiCl₃, Me₂TiCl₂, Me₂Zn, or Me₃Al itself were not successful for this transformation.

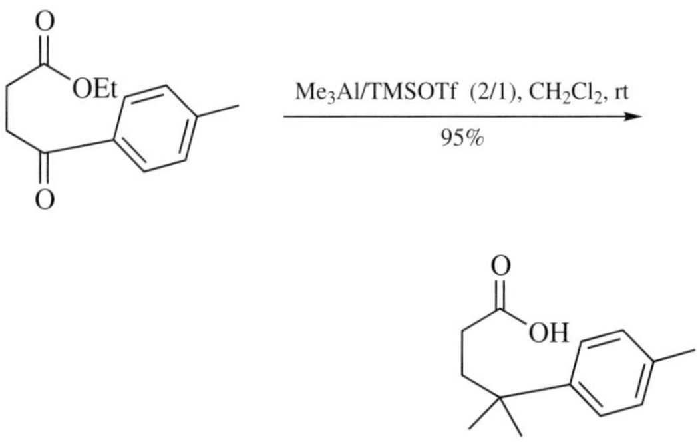

$$(70)$$

The coupling reaction between an electron-deficient alkene and an aldehyde (Baylis-Hillman reaction) usually requires a catalyst/catalytic system (typically, a tertiary amine and a Lewis acid) to be successful. The base catalyst is not necessary when pyridine-2-carboxaldehyde is employed as electrophile; it is enough with the activation effected by stoichiometric amounts of TMSOTf for the reaction to proceed to give indolizidine derivatives (eq 71).[103]

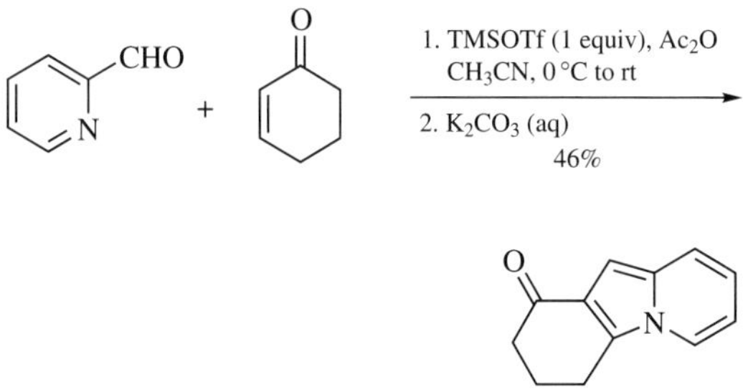

(71)

This reaction can be promoted by the use of dimethyl sulphide as a catalyst (Chalcogeno-Baylis-Hillman reaction) in the presence of stoichiometric amounts of TMSOTf; in the case of the reaction between methyl vinyl ketone and p-nitrobenzaldehyde a 2:1 adduct was isolated (eq 72), with incorporation of two units of ketone into the final product.[104]

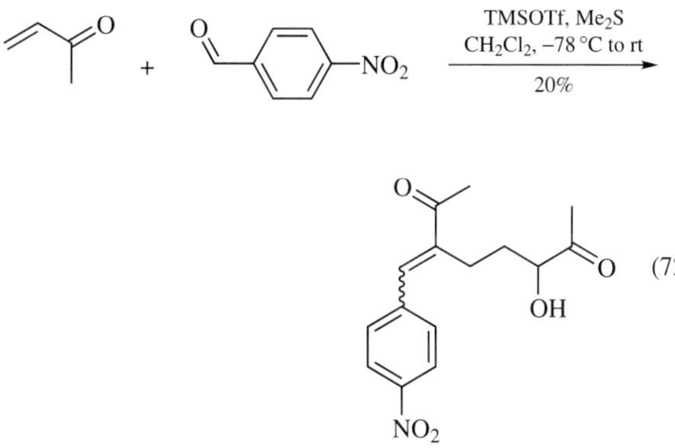

(72)

A synthesis of cross-conjugated 2-cyclopenten-1-ones from dialkenyl ketones is readily induced by TMSOTf (eq 73). A strong fluorine-directing effect has been observed for such Nazarov-type cyclization, as mixtures of products have been observed for nonfluorinated dialkenyl ketones. The addition of 1,1,1,3,3,3-hexafluoro-2-propanol (HFIP) as a cosolvent dramatically accelerates the cyclization. Other acids such as $BF_3 \cdot OEt_2$, $FeCl_3$, polyphosphoric acid, or TiOH are less effective while neither TMSI nor TMSOMe promote this cyclization at all in CH_2Cl_2.[105] 3-Ethoxycarbonyltetrahydro-γ-pyrones also undergo such Nazarov-type cyclization.[106]

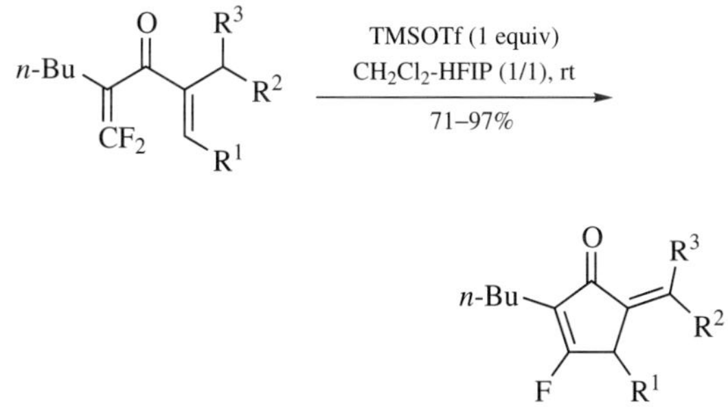

(73)

A catalytic amount of TMSOTf is enough to catalyze the [4 + 3] cycloaddition between 1,4-dicarbonyl compounds and bis(trimethylsilyl) enol ethers; the process is highly chemo- and regioselective (eq 74).[107] When 1,5-diketones are employed, the title compound is regarded as the ideal catalyst to achieve [5 + 3]

cyclizations with the bis(trimethylsilyl)enol ether of methyl acetoacetate.[108]

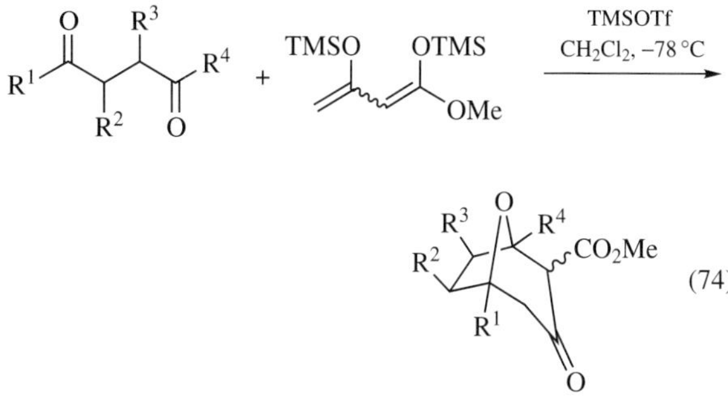

(74)

In a highly polar medium such as 3.0 M lithium perchlorate-ethyl acetate, TMSOTf is an effective reagent for promoting intermolecular[109] or intramolecular[110] all-carbon cationic [5 + 2] cycloaddition reactions (eq 75).

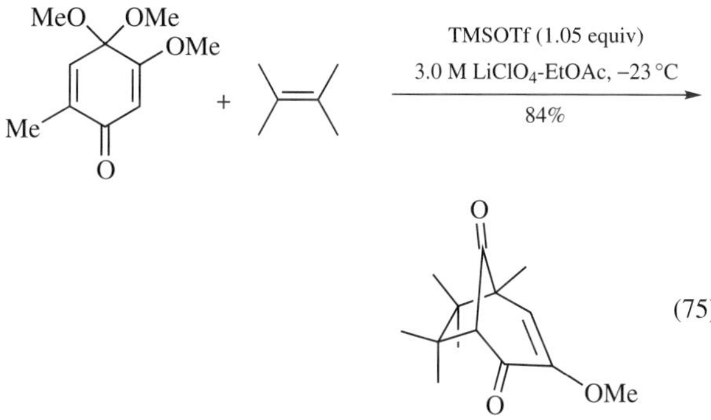

(75)

TMSOTf also interacts with carbonyl[111] or imino[112] groups promoting intermolecular hetero-Diels-Alder reactions with efficient control of the stereochemistry (eq 76).

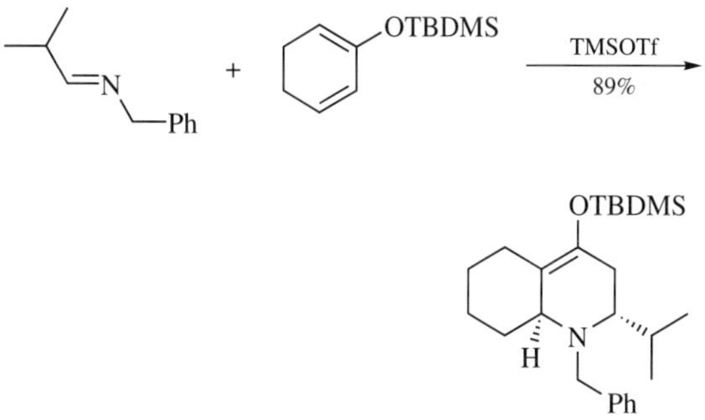

(76)

Acetal Activation. TMSOTf acts as a catalyst for the addition of several nucleophiles (allylsilanes, allylstannanes, silyl enol ethers, trimethylsilyl cyanide) toward N,O-acetals, which were obtained following a DIBAL reduction-silylation sequence from acyl amides, by promoting the formation of the corresponding acyliminium ions (eq 77).[56a] In such reactions, boron trifluoride etherate gives usually slightly higher yields. TMSOTf also acts as a catalyst for the analogous addition of silyl enol ethers to O,O-acetals.[58]

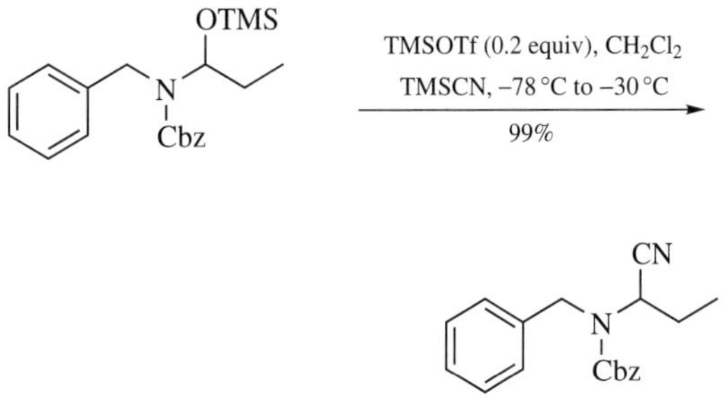

$$(77)$$

$$(80)$$

In other reactions that proceed via an acyliminium ion, *O*-vinyl *N,O*-acetals rearrange smoothly to β-(*N*-acylamino)aldehydes at 0 °C in CH$_2$Cl$_2$ in the presence of TMSOTf with moderate to high diastereoselectivities (eq 78).[113] However, TMSOTf failed to promote aprotic alkyne-iminium cyclizations, which are readily enhanced by TMSCl, TMSBr, or SiCl$_4$.[114] On the other hand, TMSOTf assists in the addition of enols to heteroaromatic imines or hydroxyaminal intermediates.[115]

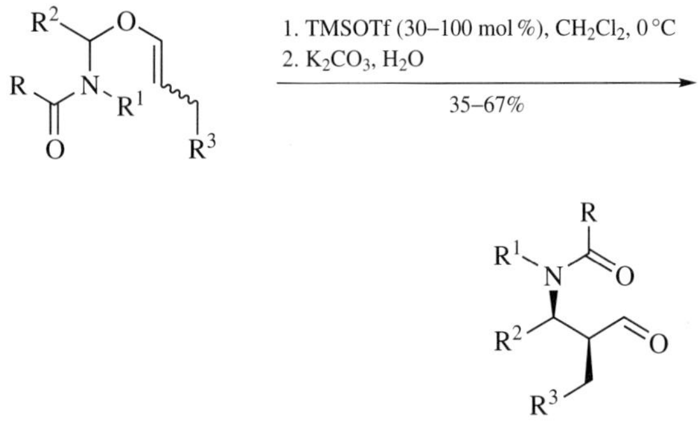

$$(78)$$

dr = 20–96%

The tropane alkaloid skeleton can be accessed in one pot via domino ene-type reactions of acetone silyl enol ether, the first one of them being intermolecular, with catalytic use of TMSOTf (eq 79).[116] Alternatively, asymmetric tropinones can be reached by cyclization of 1,3-bis-silyl enol ethers with acyl-iminium triflates.[117]

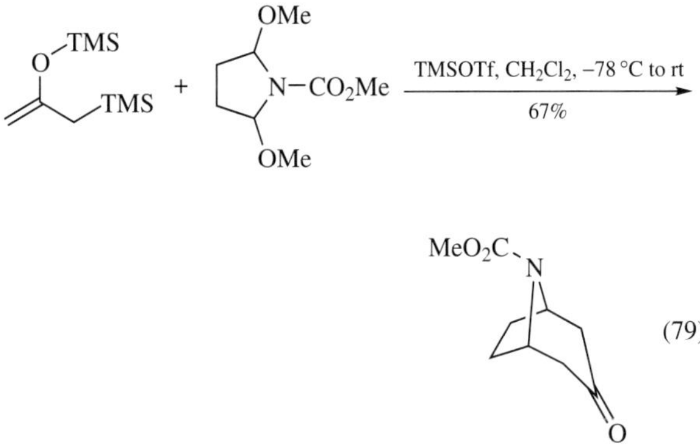

$$(79)$$

TMSOTf is effective in both stoichiometric and catalytic modes to promote the [4 + 3] cycloaddition of dienes to 1,1-dimethoxy-acetone-derived silyl enol ethers (eq 80). SnBr$_4$, SnCl$_4$, TiCl$_4$, SbCl$_5$, BCl$_3$, and Ph$_3$C$^+$BF$_4^-$ gave satisfactory results when employed in stoichiometric amounts, while the cycloaddition does not take place for Sn(OAc)$_4$, SnCl$_2$, ZnBr$_2$, or AgBF$_4$.[118]

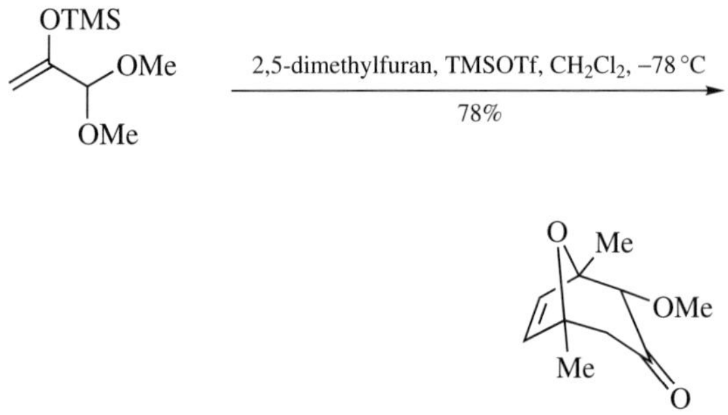

A previously described procedure, reported to be efficient for dioxolanes and noncyclic acetals,[42] has been generalized for both cyclic and acyclic acetals of ketones and aldehydes,[119] and it has been successfully applied to the opening of 1,3-dioxanes to enol ethers (eq 81).[120] However, aromatization has occurred for 4-(*N*-benzoylamino)-4-phenyl-2,5-cyclohexadienone dimethyl ketal (eq 82).[121] This method has also allowed the development of 1-methylcyclopropyl ether as a protecting group for the hydroxyl moiety which is prepared regioselectively from terminal acetonides.[122]

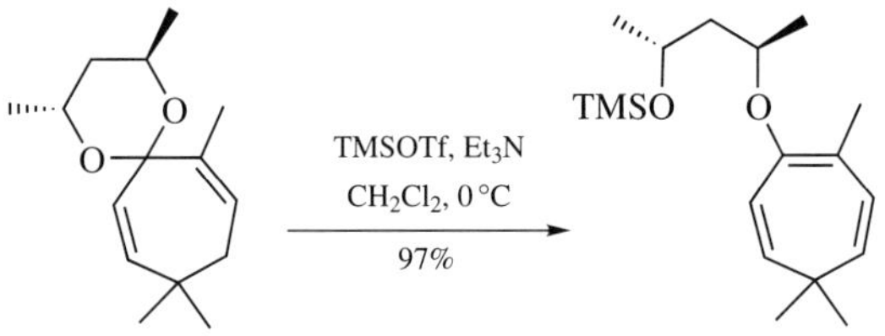

$$(81)$$

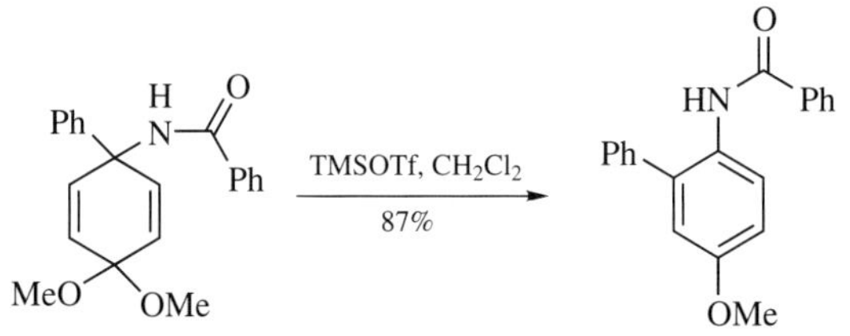

$$(82)$$

The allylation of acetals is also promoted by TMSOTf. In fact, TMSOTf is the only viable Lewis acid for the intermolecular allylation which is achieved employing allylborates, generated in situ from triallylborane, as the allyl source, and butyllithium in THF at −78 °C (eq 83).[123] The intramolecular allylation of mixed silyl-substituted acetals reaches high diastereoselectivities when promoted by TMSOTf (eq 84), although the diastereoselectivities are even higher and very good yields are reached when the intramolecular allylation reaction is promoted by boron trifluoride etherate without prior synthesis of the mixed acetal.[124] However, depending on the structure of the mixed acetal, TiCl$_4$ provides higher diastereoselectivities while the intramolecular allylation with TMSOTf as Lewis acid proceeds without stereoselectivity.[125] Thioacetals are suitable substrates for the TMSOTf-promoted addition of allylmetals.[126] On the other hand TiCl$_4$, SnCl$_4$, or Et$_2$AlCl provide higher yields although slightly lower diastereoselectivities than TMSOTf for the addition of other nucleophiles such as silyl enol ethers to trialkylsilyl-substituted acyclic-mixed acetals.[127]

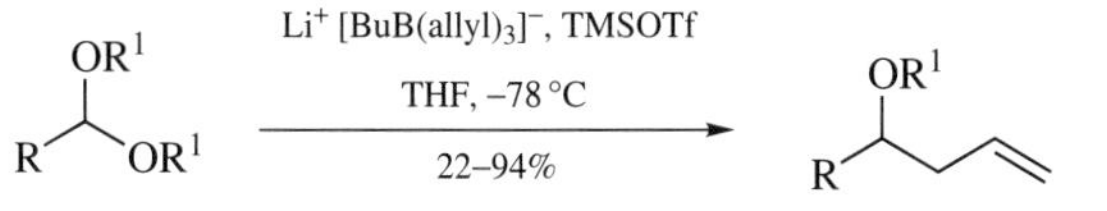

$$ \text{(83)} $$

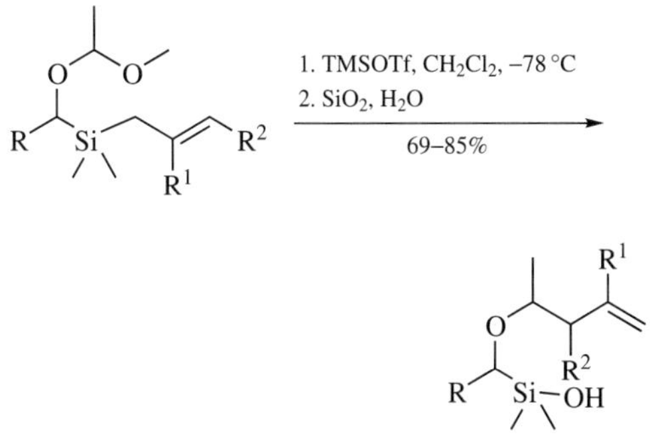

$$ \text{(84)} $$

diastereoselectivity = 1.6–34:1

$$ \text{(87)} $$

The reduction of ketals to protected secondary alcohols is readily accomplished in high yield with borane-dimethyl sulphide upon activation with TMSOTf at −78 °C (eq 85). Other Lewis acids require higher temperatures, and 1 equiv of TMSOTf is essential for complete conversion of the ketal.[128] The solvent has a decisive role for controlling the site selectivity in the case of unsymmetrical diols.[129]

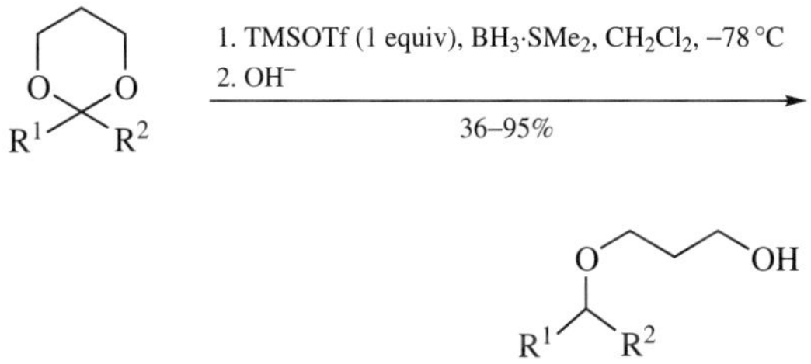

$$ \text{(85)} $$

TMSOTf also plays an important role in sugar chemistry. In fact it often mediates in the formation of the glycosidic bond, either alone[130] or in conjunction with trimethylsilyl chloride,[131] N-bromosuccinimide (NBS),[132] or N-iodosuccinimide (NIS).[133] In such reactions, different moieties have acted as leaving groups[134] in the glycosyl donor and O-, N-, or C-nucleophiles[135] have been employed (eq 86). The addition of hydroxyl unprotected sugar units to the exo-anomeric double bond of ketene dithioacetals, which act as glycosyl donors, is also promoted by TMSOTf.[136]

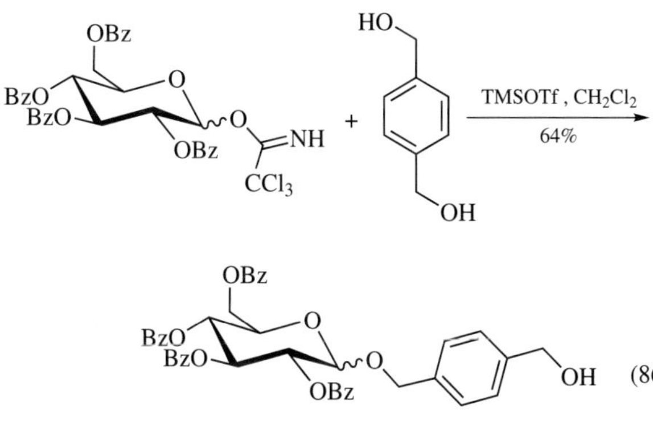

$$ \text{(86)} $$

A TMSOTf-induced intramolecular cyclization of silyl enol ethers also takes place with orthoesters (eq 87).[137]

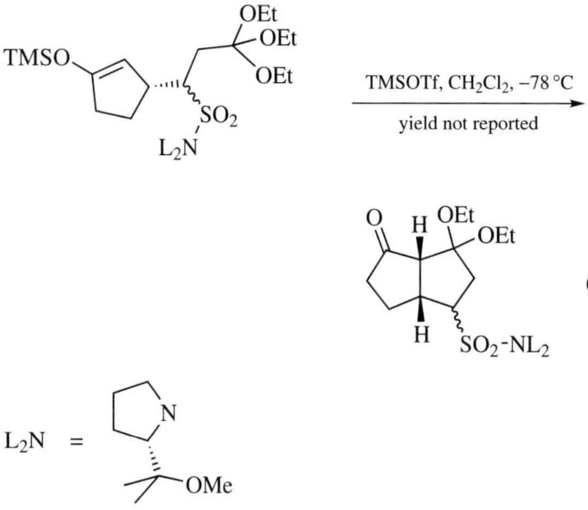

Nitrone Activation. The nucleophilic addition to aldonitrones depends on the nature of the metal involved and the presence/absence of an activator. Thus, allylsilanes add to aromatic nitrones to give homoallylhydroxylamines in excellent yields (eq 88);[138] other Lewis acids were unsuccessful with TiCl$_4$ as the only exception, although it produced just a 32% yield.

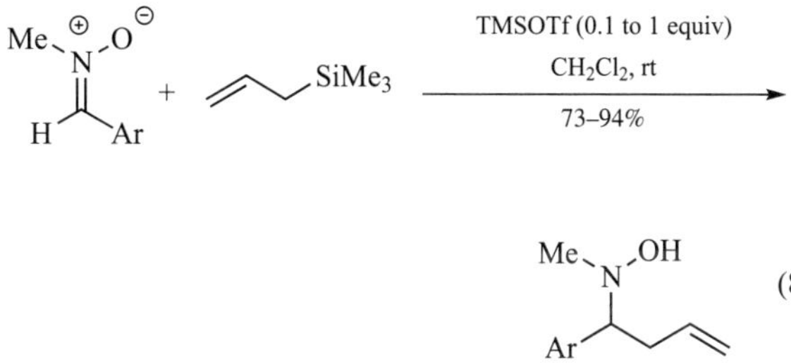

$$ \text{(88)} $$

The [3 + 2] dipolar cycloaddition of nonaromatic nitrones with allyltrimethylsilane, which requires temperatures >100 °C without a catalyst, is carried out at temperatures ≤20 °C in the presence of TMSOTf (eq 89), with moderate to good yields and low to moderate diastereoselectivities.[139]

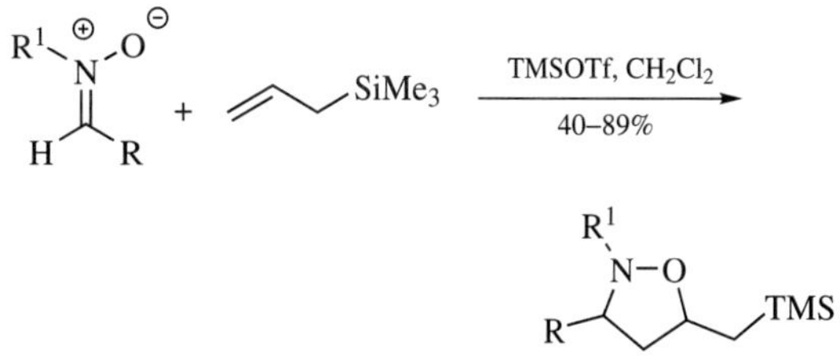

$$ \text{(89)} $$

cis/trans = 61/39 to 15/85

On the other hand, allylstannanes add to nitrones in the presence of TMSOTf to give O-silylated homoallylic hydroxylamines in high yields; when the crude reaction mixture is quenched with NIS, 5-iodomethylisoxazolines are formed in excellent yields (eq 90).[140]

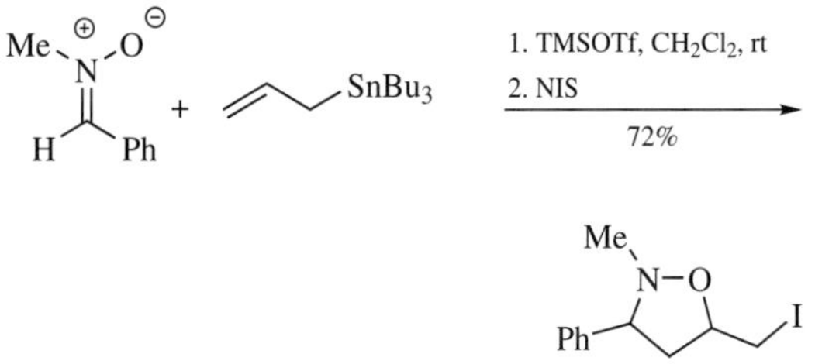

$$cis/trans = 55/45 \quad (90)$$

The addition of silyl ketene acetals to chiral aldonitrones requires the use of Lewis acids as activating reagents. Whereas activation with TMSOTf followed by a treatment with hydrofluoric acid-pyridine leads to the *syn*-adducts of isoxazolidin-5-ones (eq 91), the use of diethylaluminium chloride or boron trifluoride etherate leads to *anti*-compounds.[141]

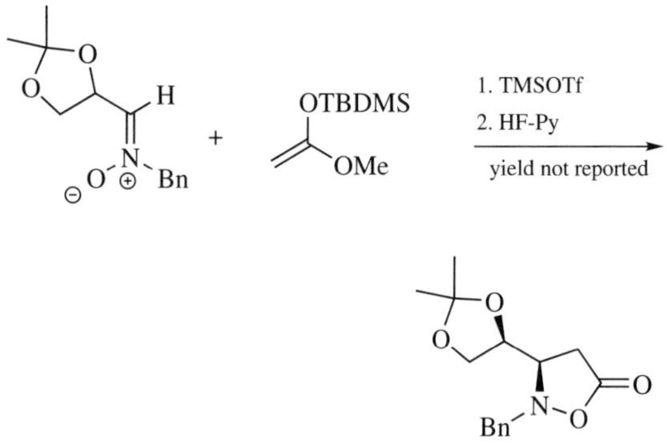

Epoxide Ring Opening. One-pot alkylation-*O*-silylation reactions of epoxides take place in excellent yields when the epoxides are treated with trialkylaluminum-TMSOTf and Et₃N (1.5 equiv of each one) in methylene chloride at −50 °C (eq 92). Other trialkylsilyl triflates also undergo this reaction, which is stereospecific: *anti*-compounds are obtained from *trans*-epoxides while *cis*-epoxides yield *syn*-adducts.[142]

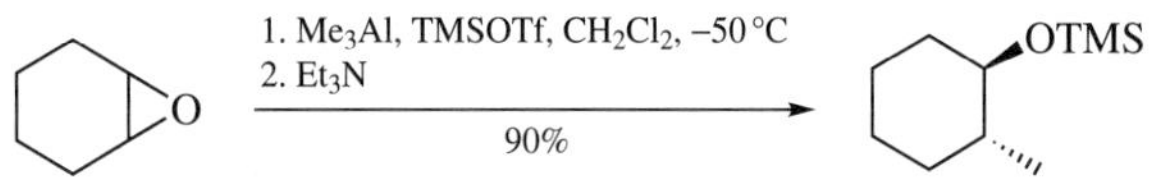

Chiral aryl epoxides are converted into hydroxy-protected chiral α-hydroxy aryl ketones with complete retention of the chiral center and good regioselectivity by treatment with a combination of DMSO and TMSOTf (eq 93). Other reagent combinations such as DMSO/BF₃·OEt₂ or DMSO/TfOH lead to lower yields of the desired alcohols.[143]

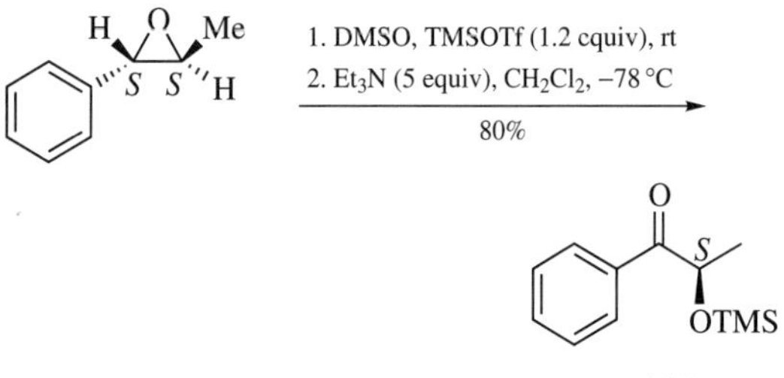

The TMSOTf-induced intramolecular reaction of allylstannanes with epoxides causes a regio- and stereoselective cyclization to five- or six-membered carbocycles (eq 94), depending on the substituents of the starting material. TiCl₄ also gives high yields and slightly higher selectivities; catalytic amounts of Lewis acids are insufficient for completion of the reactions.[144]

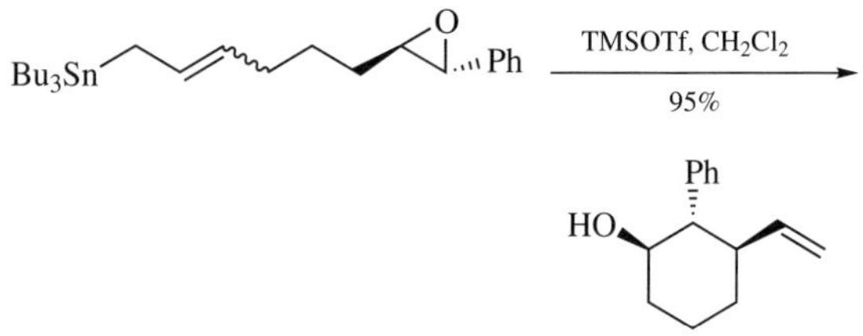

Silylated aldol adducts can be reached by using a nonaldol rearrangement promoted by treatment of bulky epoxy bis-silyl ethers with TMSOTf/*i*-Pr₂NEt in methylene chloride at −50 °C (eq 95).[145] Bulky mesylated epoxy silyl ethers also undergo this transformation; however, a silyl triflate-promoted Payne rearrangement was observed as a side reaction depending on the stereochemistry of the starting epoxide.[146]

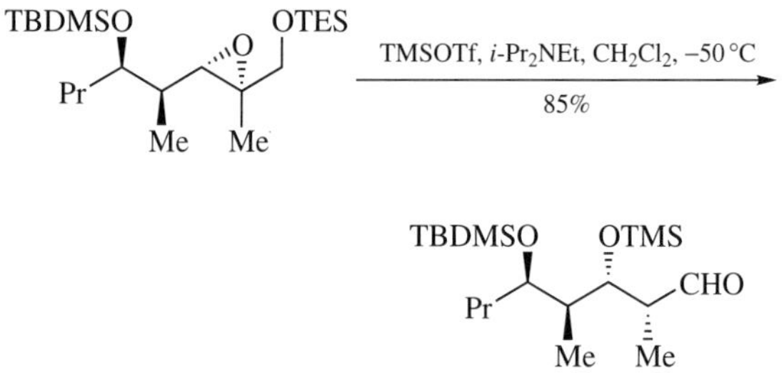

Cleavage of Protecting Groups. THP ethers of primary, secondary, and phenolic alcohols can be conveniently deprotected at room temperature by treatment with 1.2 equiv of TMSOTf in methylene chloride (eq 96).[147] Deprotection of *t*-butyldimethylsilyl ethers in the presence of a *t*-butyldiphenylsilyl ether has been smoothly accomplished using TMSOTf at −78 °C, as one of the steps of the total synthesis of marine macrolide ulapualide A.[148]

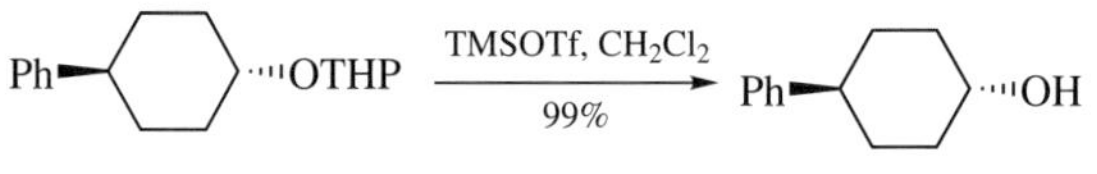

TMSOTf has been reported to be a selective and efficient regent for the cleavage of *t*-butyl esters in the presence of *t*-butyl ethers, under virtually neutral conditions (eq 97), while iodotrimethysilane effects the deprotection of both species. No racemization was observed when this procedure was applied to the synthesis of *t*-butoxy amino acids.[149]

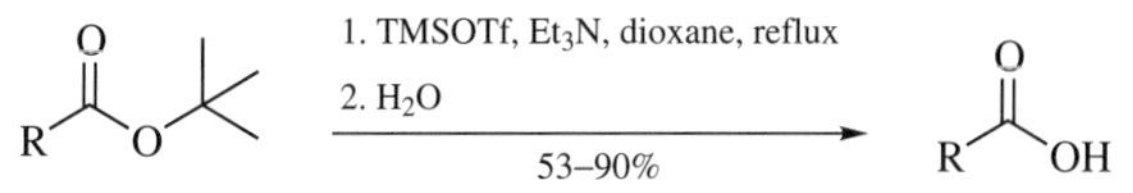

6-Substituted isopropyl-protected guaiacols undergo transprotection in one step when treated with TMSOTf in anhydrous acetonitrile (eq 98); very high yields are obtained for this transformation if the substituent at position 6 is bulky and electronwithdrawing. However, mono- and disubstituted isopropyl phenyl ethers undergo Friedel-Crafts acetylation rather than transprotection.[150]

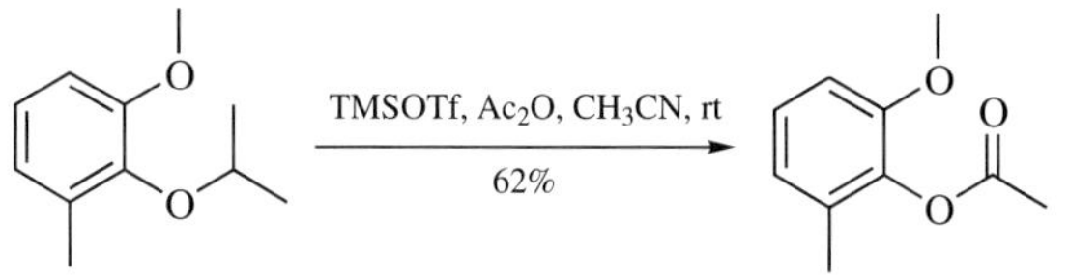

(98)

The system TMSOTf/2,6-lutidine has been proved to be highly effective for the deprotection of cyclic and acyclic acetals from aldehydes (eq 99); on the other hand, high chemoselectivity has been achieved when using triethylsilyl triflate/2,6-lutidine, which can deprotect acetals in the presence of ketals.[151]

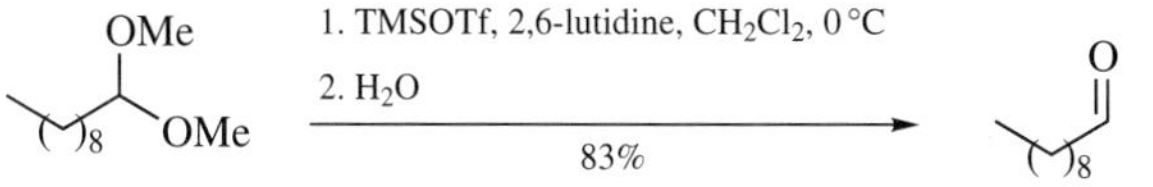

(99)

The same combination of reagents (TMSOTf/2,6-lutidine) has been employed to deprotect N-t-butoxycarbonyl groups from substrates in the solid phase synthesis of several peptides, without cleaving the substrates from the support as would occur in the case of using TFA with TFA-sensitive resins (such as Rink's amide resin) (eq 100). This method has great potential for the solid-phase synthesis of small molecule libraries.[152] Such a reagent combination had been previously developed for solution phase reactions of nonpeptidic substrates.[153]

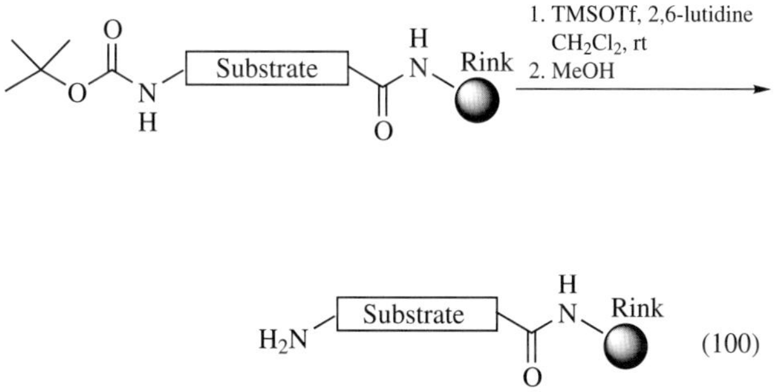

(100)

A combination of thioanisole and TMSOTf was used to nucleophilically release phenols from a polystyrene resin (eq 101).[154]

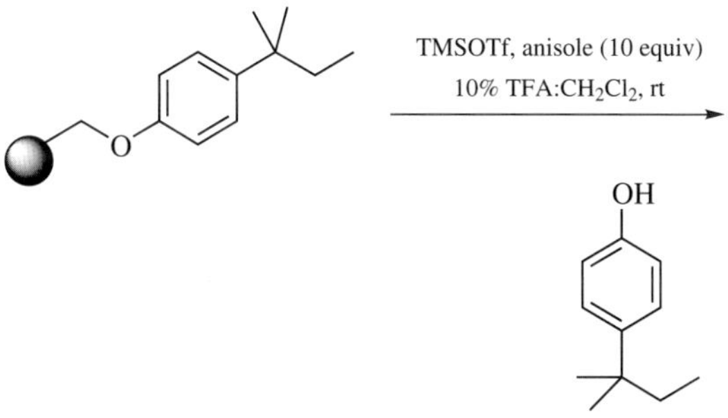

(101)

Hypervalent Iodine Chemistry. The formation of hypervalent iodine complexes is often promoted by TMSOTf. Thus, TMS-alkynes can be transformed with iodosobenzene and TMSOTf into phenyl iodonium triflates in moderate to good yields, which may be subsequently converted into alkyneamides (eq 102).[155]

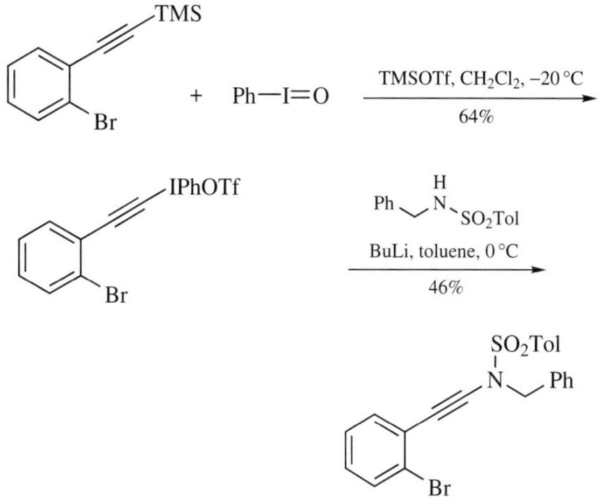

(102)

The TMSOTf-mediated reaction of benziodoxoles with aromatic nitrogenated heterocycles[156] or azides[157] permits the preparation of hypervalent iodine complexes bearing nitrogen ligands (eq 103) as well as several other unsymmetrical tricoordinate iodinanes.[158] TMSOTf has found widespread use as a reagent for the preparation of fluorinated hypervalent iodine compounds.[159]

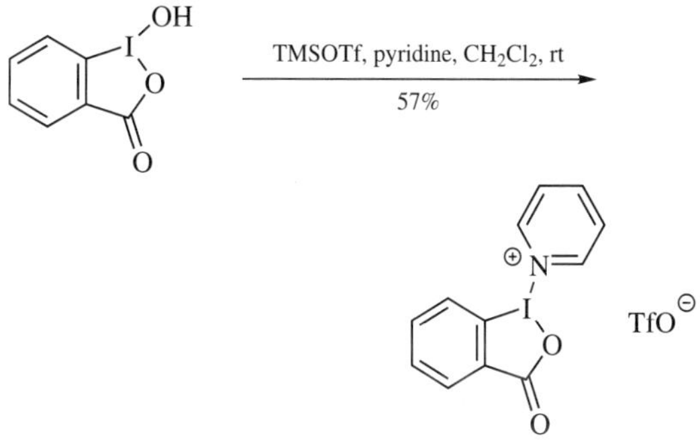

(103)

1-Alkynylbenziodoxoles are selectively formed in high preparative yield by the sequential addition of trimethylsilylacetylenes and pyridine to benzoiodoxole triflates,[160] which are prepared in situ from 2-iodosylbenzoic acid and TMSOTf (eq 104).

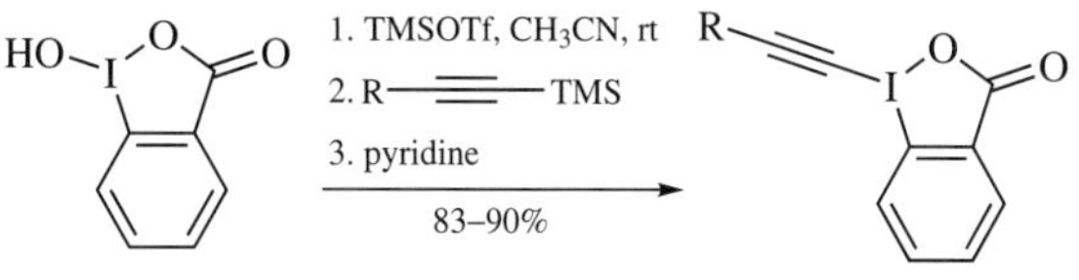

(104)

On the other hand, TMSOTf activates acetoxybenziodazoles to afford 3-iminobenziodoxoles upon reaction with amides and alcohols via an acid-catalyzed rearrangement (eq 105).[161]

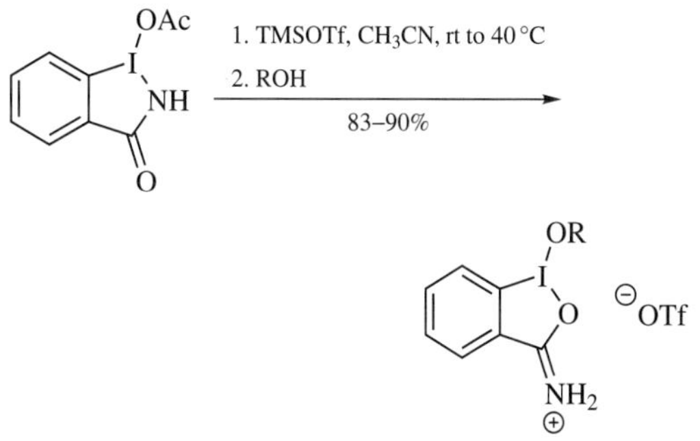

(105)

The formation of cationic iodonium macrocycles such as rhomboids, squares, or pentagons can be carried out in high yields by reaction of hypervalent iodoarenes with TMS-substituted aromatic compounds in the presence of TMSOTf (eq 106).[162]

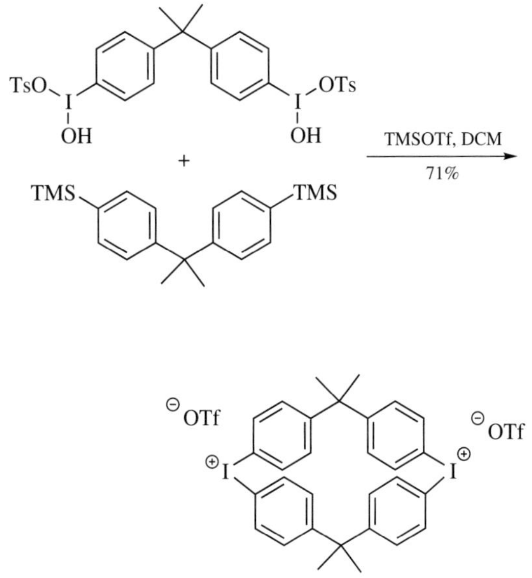

(106)

A combination of a hypervalent iodine(III) reagent [phenyliodine(III) bis(trifluoroacetate)] and TMSOTf has shown great efficiency for the synthesis of pyrroloiminoquinones from 3-(azidoethyl)indole derivatives (eq 107).[163]

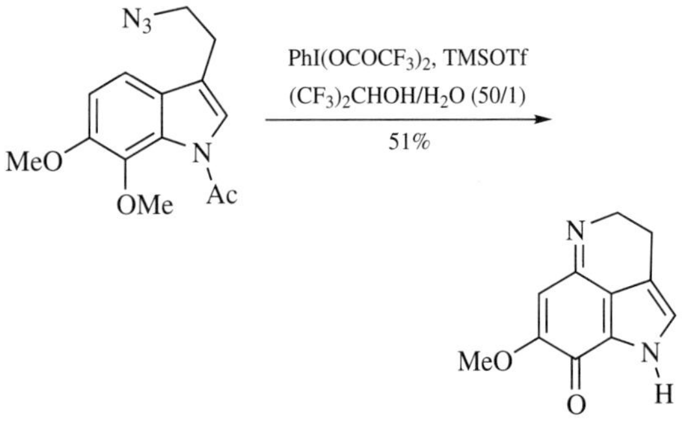

(107)

Miscellaneous. TMSOTf has been also the Lewis acid of choice (2 equiv) to promote the olefination reaction to give α-cyano α,β-unsaturated aldehydes from 2-cyano-3-ethoxy-2-en-1-ols (eq 108), which takes place at −78 °C.[164]

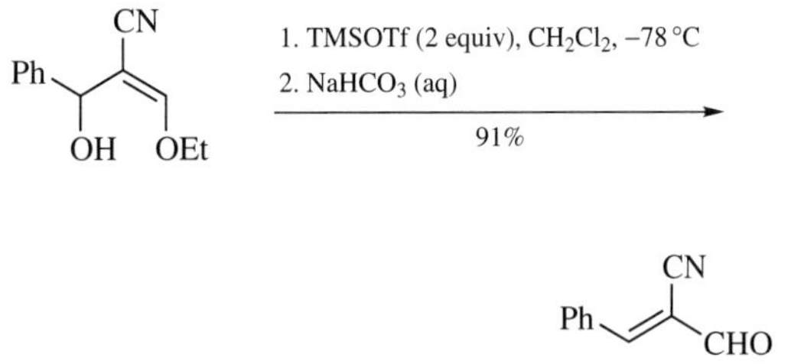

(108)

The reaction of pyridine with chiral acyl chlorides mediated by TMSOTf results in the formation of chiral N-acylpyridinium ions which can be trapped with organometallic reagents to form N-acyldihydropyridines or N-acyldihydropyridones in a diastereoselective manner (eq 109).[165] Catalytic TMSOTf also partakes in the acylation of cyclic bis(trimethylsiloxy)-1,3-dienes, which are synthons of 1,3-dicarbonyl dianions.[54]

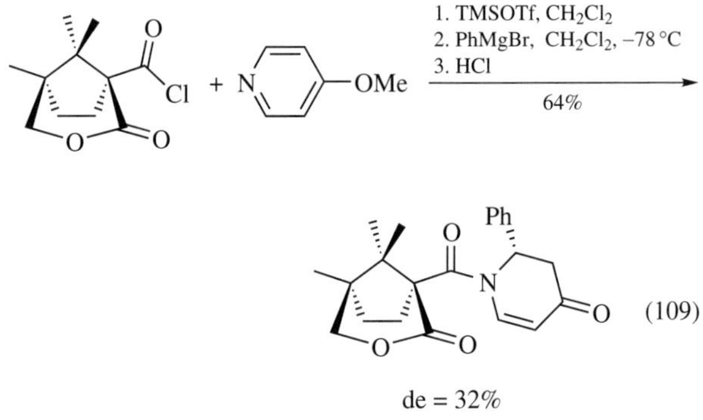

(109)

de = 32%

The reaction of 1,1-bis(trimethylsiloxy)ketene acetals with oxalyl chloride results in a straightforward one-pot synthesis of a variety of 3-hydroxymaleic anhydrides (eq 110).[166]

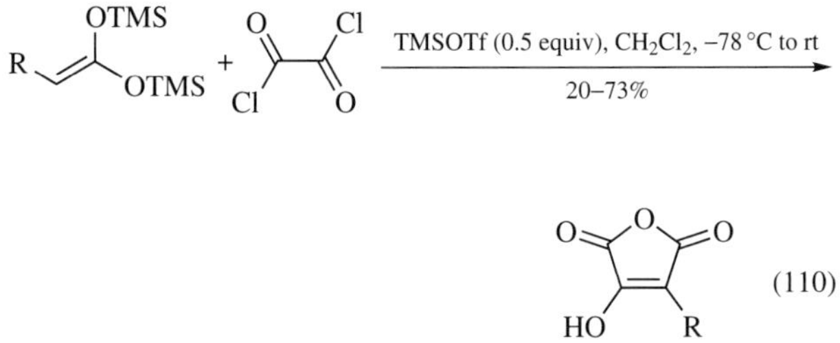

(110)

TMSOTf gives the best yields, among the tested Lewis acids (ZnBr$_2$, TiCl$_4$, SnCl$_4$), for the synthesis of functionalized 1,2,3,4-β-tetrahydrocarbolines from enamido ketones,[167] through the formation of N-acyliminium ions as intermediates (eq 111).

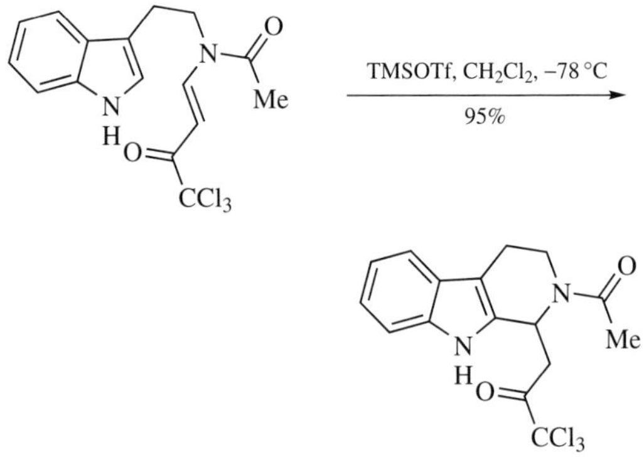

(111)

4-Trimethylsilyloxy-1-benzothiopyrylium triflate and 4-trimethylsilyloxyquinolinium triflate, which can be easily generated by TMSOTf addition to 4H-1-benzothiopyran-4-one[168] and 4-quinolone,[169] respectively, behave as Michael acceptors as they undergo the in situ TMSOTf-promoted addition of C-nucleophiles such as allylstannanes and silyl enol ethers, respectively. In a similar manner, 4-trimethylsilyloxypyrylium triflate, easily accessible in situ from pyran-4-one, undergoes a domino reaction sequence with 2 equiv of a silyloxybuta-1,3-diene leading to the formation of tetrahydro-2H-chromenes as sole products with high yields (eq 112).[170]

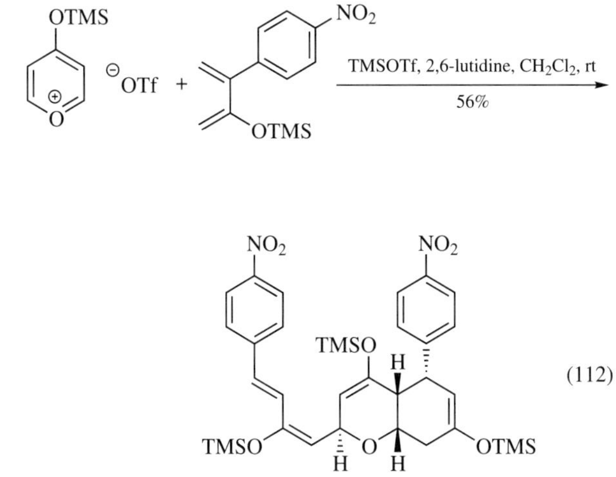

(112)

TMSOTf promotes the Diels-Alder cycloaddition of *p*-tolyl vinyl sulfoxide with furan or cyclopentadiene;[171] the reaction is slower and less stereoselective than when induced by ethyl Meerwein's reagent (Et$_3$OBF$_4$). It also promotes the [4 + 2] cycloaddition between 2,2-dimethoxyethyl acrylate and several dienes with high yields and stereoselectivities (eq 113);[172] TiCl$_4$ and Ph$_3$CClO$_4$ also give satisfactory results in such processes.

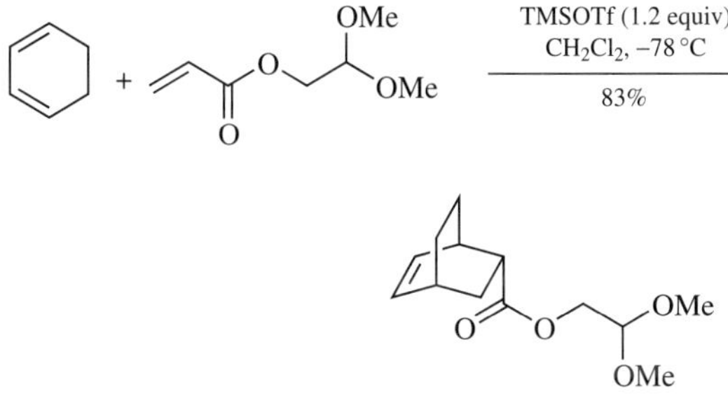

(113)

endo:exo = >99:1

Just catalytic amounts of the title reagent are required to efficiently mediate in the conversion of Diels-Alder adducts of 1-methoxy-3-[(trimethylsilyl)oxy]-1,3-butadiene (Danishefsky's diene) to cyclohexenones (eq 114).[173] Dilute HCl or concentrated acid lead to a greater percentage of side products.

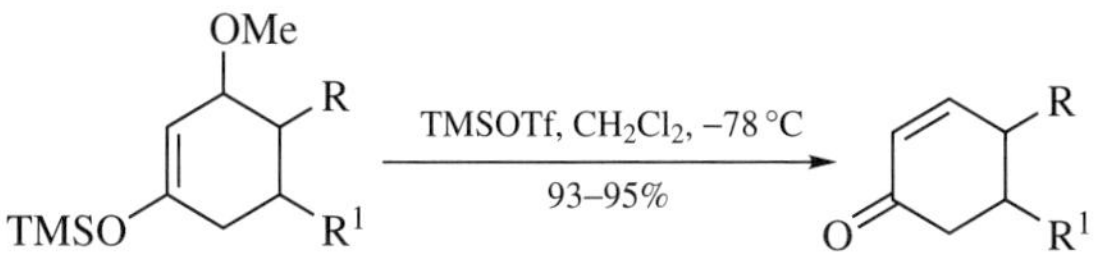

(114)

The addition of Lewis acids also improves the yield for a Beckmann fragmentation followed by carbon-carbon bond formation through reaction of oxime acetates with organoaluminum reagents (eq 115). ZnCl$_2$ is generally more effective than TMSOTf for this transformation.[174]

The mixture of 1,3-dibromo-5,5-dimethylhydantoin (DBDMH)-TMSOTf in methylene chloride has proved to be more reactive than DBDMH alone for the aromatic bromination of arenes.[175]

On the contrary, debromination of α-bromo carboxylic acid derivatives can be achieved in high yield by the combination of triphenylphosphine and TMSOTf (eq 116). Other Lewis acids such as germanium(IV) chloride proved to be even more efficient for such transformation.[176]

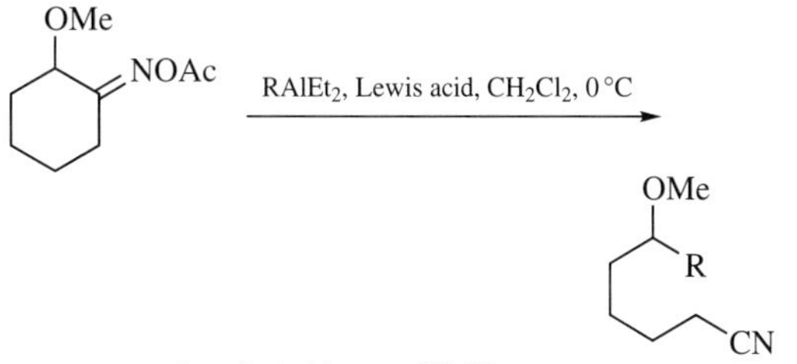

(115)

Lewis Acid	Yields
TMSOTf	39–62%
ZnCl$_2$	35–91%

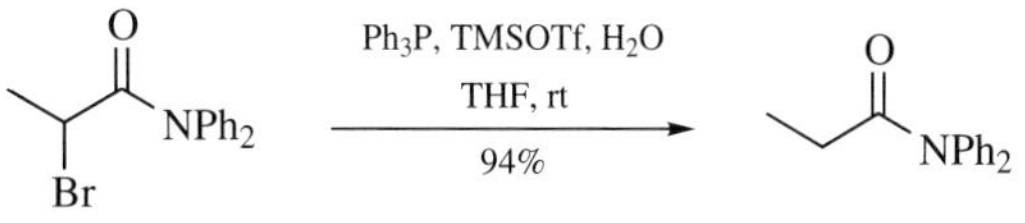

(116)

The generation of thionium ions from sulfoxides bearing an α-hydrogen can be carried out by using TMSOTf-Et$_3$N as initiator to afford Pummerer reaction products (eq 117).[177]

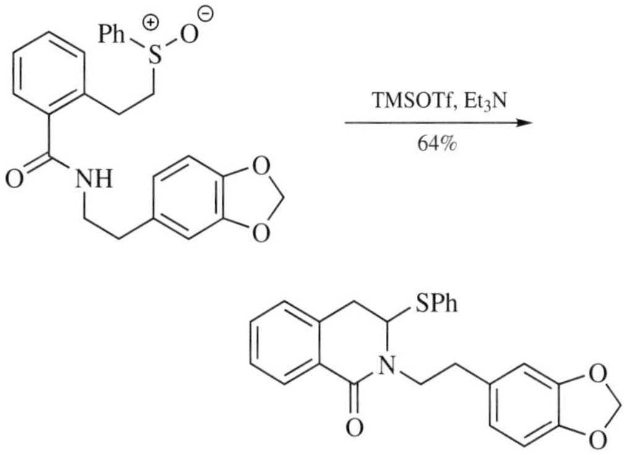

(117)

Several coupling reactions have been activated by the addition of TMSOTf.

It has been pointed out that the triflate ion increases the electrophilic character of *N*-acyl quaternary salts of imidazoles and accelerates its coupling reaction to silyl enol ethers.[178] In fact, in the absence of this Lewis acid, the yield of the coupling product diminishes (eq 118). Thiazoles, benzothiazoles, and benzoimidazoles react in a similar manner.[179]

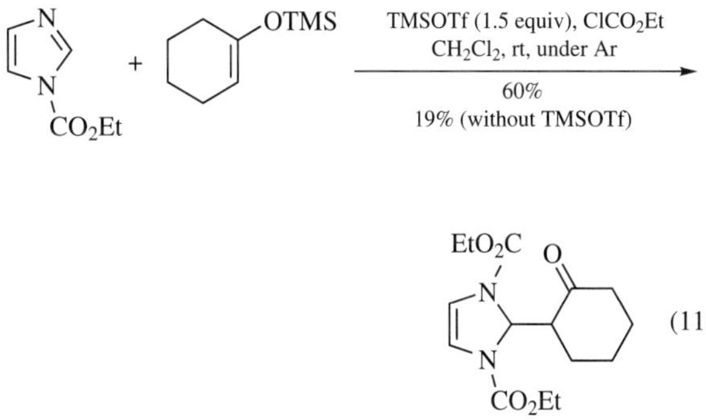

(118)

The nucleophilic attack of silyl enol glycine derivative to a dienyl acetate organometallic complex proceeds readily and regiospecifically at −78 °C in the presence of TMSOTf to afford a dienyl glycine derivative in excellent yield (eq 119).[180]

The nucleophilic substitution of α-acetoxyhydrazones by silyl enol ethers takes place with high chemical yields and moderate diastereomeric excesses. Among the several Lewis acids tested (AsCl$_3$, AlCl$_3$, TiCl$_4$, ZnCl$_2$,…), only BF$_3$·OEt$_2$ provides a slightly higher diastereoselectivity than TMSOTf.[181]

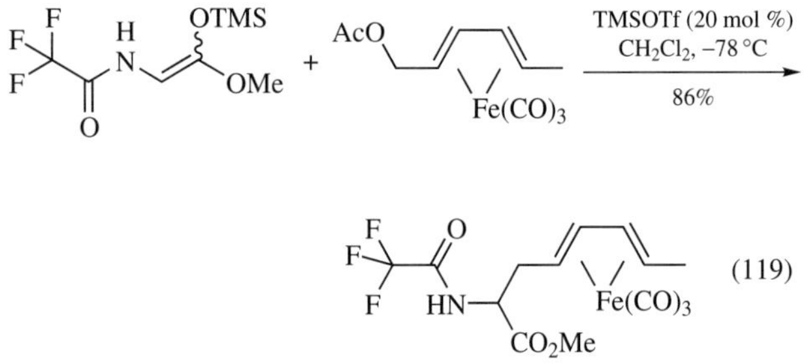

$$(119)$$

α-Propargyl carbocations, which can be generated by treatment of (arene)Cr(CO)$_3$-substituted propargyl acetates with TMSOTf,[182] react with a variety of *C*-, *S*-, or *N*-nucleophiles to give good yields of the corresponding propargyl derivatives. If the α-propargyl carbocations are prepared from arene-substituted propargyl silyl ethers,[183] the nature of the products (allenyl or propargyl compounds) depends mainly on the substituents of the starting material, although the nucleophile also may exert some influence; for example, diphenyl-substituted propargyl silyl ethers usually lead to the formation of allenyl derivatives. α,β-Unsaturated carbonyl compounds are obtained when these substituted propargyl silyl ethers are treated with TMSOTf followed by the addition of water (eq 120).[183]

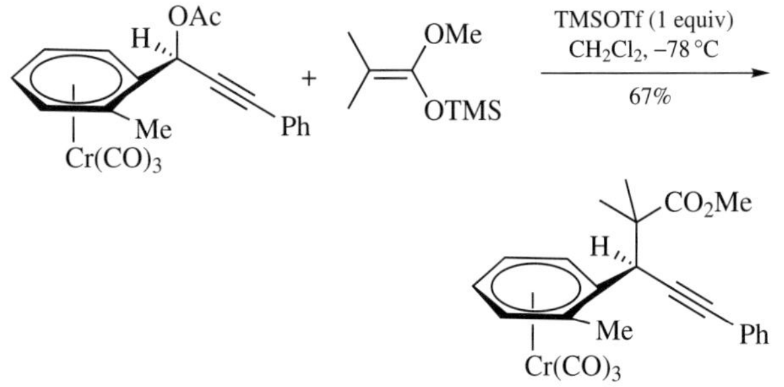

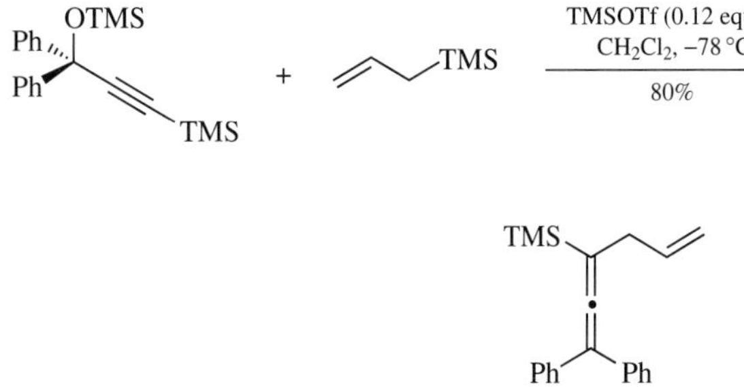

$$(120)$$

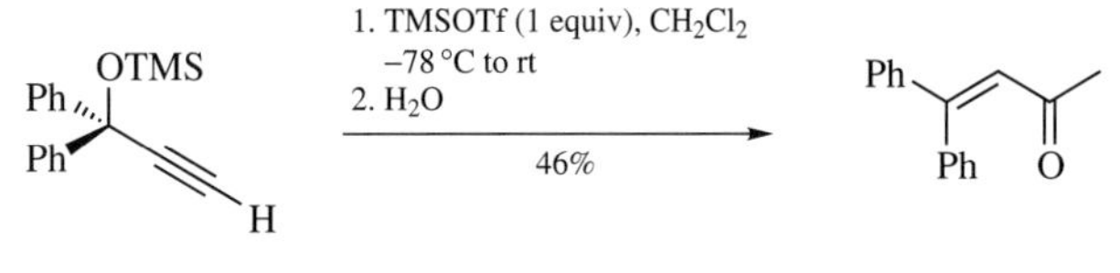

A TMSOTf-induced stereospecific cationic *syn*-[1,2] silyl shift occurs with retention of the stereochemistry at the migrating terminus (eq 121).[184]

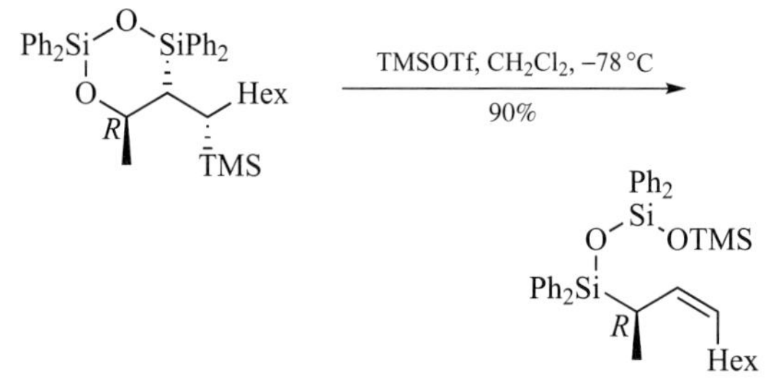

$$(121)$$

Iron carbene complexes bearing chirality at the carbene ligand can be generated from optically pure bimetallic (chromium-iron) complexes by the addition of 1 equiv of TMSOTf in the presence of an olefin, which in situ undergoes an asymmetric cyclopropanation. Excellent ee's are obtained when the reaction is carried out with *gem*-disubstituted olefins (eq 122).[185]

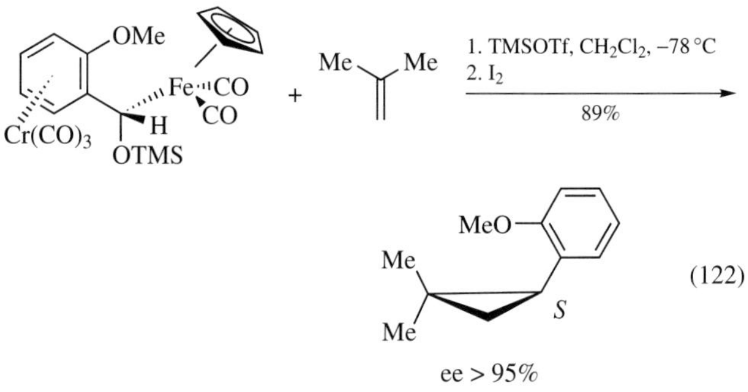

$$(122)$$

ee > 95%

Benzylic silanes bearing an electron-donating group experience an oxidative carbon-silicon bond cleavage selectively with oxovanadium(V) compounds to permit an intermolecular carbon-carbon bond formation; the addition of TMSOTf resulted in a more facile coupling (eq 123).[186]

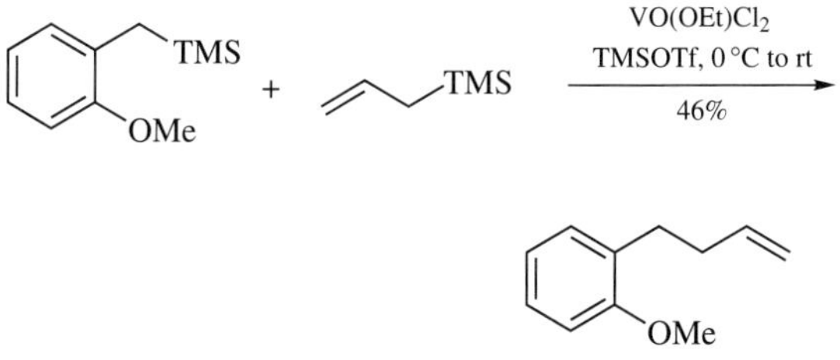

$$(123)$$

TMSOTf may act as a Lewis acid promoter for the chiral oxovanadium complex-catalyzed oxidative coupling of 2-naphthols. In the enantioselective version, chlorotrimethylsilane affords higher enantiomeric excesses (eq 124).[187]

(124)

ee = 42%
ee = 48% (with TMSCl)

catalyst =

1. Reviews: (a) Emde, H.; Domsch, D.; Feger, H.; Frick, U.; Götz, H. H.; Hofmann, K.; Kober, W.; Krägeloh, K.; Oesterle, T.; Steppan, W.; West, W.; Simchen, G., *Synthesis* **1982**, 1. (b) Noyori, R.; Murata, S.; Suzuki, M., *Tetrahedron* **1981**, *37*, 3899. (c) Stang, P. J.; White, M. R., *Aldrichim. Acta* **1983**, *16*, 15. Preparation: (d) Olah, G. H.; Husain, A.; Gupta, B. G. B.; Salem, G. F.; Narang, S. C., *J. Org. Chem.* **1981**, *46*, 5212. (e) Morita, T.; Okamoto, Y.; Sakurai, H., *Synthesis* **1981**, 745. (f) Demuth, M.; Mikhail, G., *Synthesis* **1982**, 827. (g) Ballester, M.; Palomo, A. L., *Synthesis* **1983**, 571. (h) Demuth, M.; Mikhail, G., *Tetrahedron* **1983**, *39*, 991. (i) Aizpurua, J. M.; Palomo, C., *Synthesis* **1985**, 206.

2. Simchen, G.; Kober, W., *Synthesis* **1976**, 259.

3. Hergott, H. H.; Simchen, G., *Liebigs Ann. Chem.* **1980**, 1718.

4. Simchen, G.; Kober, W., *Synthesis* **1976**, 259.

5. Emde, H.; Götz, A.; Hofmann, K.; Simchen, G., *Liebigs Ann. Chem.* **1981**, 1643.

6. Krägeloh, K.; Simchen, G., *Synthesis* **1981**, 30.

7. Emde, H.; Simchen, G., *Liebigs Ann. Chem.* **1983**, 816.

8. Emde, H.; Simchen, G., *Synthesis* **1977**, 636.

9. Emde, H.; Simchen, G., *Synthesis* **1977**, 867.

10. Yamamoto, Y.; Ohdoi, K.; Nakatani, M.; Akiba, K., *Chem. Lett.* **1984**, 1967.

11. Emde, H.; Götz, A.; Hofmann, K.; Simchen, G., *Liebigs Ann. Chem.* **1981**, 1643.

12. Matsuda, I.; Sato, S.; Hattori, M.; Izumi, Y., *Tetrahedron Lett.* **1985**, *26*, 3215.

13. Ahlbrecht, H.; Düber, E. O., *Synthesis* **1980**, 630.

14. Murata, S.; Suzuki, M.; Noyori, R., *J. Am. Chem. Soc.* **1980**, *102*, 2738.

15. Frick, U.; Simchen, G., *Synthesis* **1984**, 929.

16. Borgulya, J.; Bernauer, K., *Synthesis* **1980**, 545.

17. Tietze, L. F.; Bratz, M., *Synthesis* **1989**, 439.

18. Hwu, J. R.; Wetzel, J. M., *J. Org. Chem.* **1985**, *50*, 3946.

19. Hwu, J. R.; Robl, J. A., *J. Org. Chem.* **1987**, *52*, 188.

20. Harada, T.; Hayashiya, T.; Wada, I.; Iwa-ake, N.; Oku, A., *J. Am. Chem. Soc.* **1987**, *109*, 527.

21. Pearson, W. H.; Cheng, M-C., *J. Am. Chem. Soc.* **1986**, *51*, 3746.

22. Murata, S.; Suzuki, M.; Noyori, R., *J. Am. Chem. Soc.* **1980**, *102*, 3248.

23. Murata, S.; Suzuki, M.; Noyori, R., *Tetrahedron* **1988**, *44*, 4259.

24. Lee, T. V.; Richardson, K. A., *Tetrahedron Lett.* **1985**, *26*, 3629.

25. Mukaiyama, T.; Uchiro, H.; Kobayashi, S., *Chem. Lett.* **1990**, 1147.

26. Guanti, G.; Narisano, E.; Banfi, L., *Tetrahedron Lett.* **1987**, *28*, 4331.

27. Guanti, G.; Narisano, E.; Banfi, L., *Tetrahedron Lett.* **1987**, *28*, 4335.

28. Ihara, M.; Tsuruta, M.; Fukumoto, K.; Kametani, T., *J. Chem. Soc., Chem. Commun* **1985**, 1159.

29. Robl, J. A.; Hwu, J. R., *J. Org. Chem.* **1985**, *50*, 5913.

30. Sato, T.; Watanabe, T.; Hayata, T.; Tsukui, T., *J. Chem. Soc., Chem. Commun.* **1989**, 153.

31. Kim, S.; Lee, J. M., *Tetrahedron Lett.* **1990**, *31*, 7627.

32. Kim, S.; Lee, P. H., *Tetrahedron Lett.* **1988**, *29*, 5413.

33. Hosomi, A.; Sakata, Y.; Sakurai, H., *Tetrahedron Lett.* **1984**, *25*, 2383.

34. Yamada, H.; Nishizawa, M., *Tetrahedron* **1992**, 3021.

35. Rainer, H.; Scharf, H.-D.; Runsink, J., *Liebigs Ann. Chem.* **1992**, 103.

36. Schmidt, R. R., *Angew. Chem., Int. Ed. Engl.* **1986**, *25*, 212.

37. Tietze, L.-F.; Fischer, R.; Guder, H.-J., *Tetrahedron Lett.* **1982**, *23*, 4661.

38. Mukaiyama, T.; Matsubara, K., *Chem. Lett.* **1992**, 1041.

39. Kozikowski, A. P.; Park, P., *J. Org. Chem.* **1984**, *49*, 1674.

40. Knapp, S.; Rodriques, K. E., *Tetrahedron Lett.* **1985**, *26*, 1803.

41. Nishiyama, H.; Sakuta, K.; Osaka, N.; Itoh, K., *Tetrahedron Lett.* **1983**, *24*, 4021.

42. Gassman, P. G.; Burns, S. J., *J. Org. Chem.* **1988**, *53*, 5574.

43. Miller, R. D.; Hässig, R., *Synth. Commun.* **1984**, *14*, 1285.

44. Miller, R. D.; Hässig, R., *Tetrahedron Lett.* **1985**, *26*, 2395.

45. (a) Marigo, M.; Wabnitz, T. C.; Fielenbach, D.; Jorgensen, K. A., *Angew. Chem., Int. Ed.* **2005**, *44*, 794. (b) Lilly, M. J.; Sherburn, M. S., *Chem. Commun.* **1997**, 967.

46. Oriyama, T.; Yatabe, K.; Kawada, Y.; Koga, G., *Synlett* **1995**, 45.

47. Guha, S. K.; Shibayama, A.; Abe, D.; Sakaguchi, M.; Ukaji, Y.; Inomata, K., *Bull. Chem. Soc. Jpn.* **2004**, *77*, 2147.

48. Rossi, L.; Pecunioso, A., *Tetrahedron Lett.* **1994**, *35*, 5285.

49. Knopf, O.; Alexakis, A., *Org. Lett.* **2002**, *4*, 3835.

50. Kim, S.; Park, J. H.; Kim, Y. G.; Lee, J. M., *J. Chem. Soc., Chem. Commun.* **1993**, 1188.

51. Ghosez, L.; Bayard, P.; Nshimyumukiza, P.; Gouverneur, V.; Sainte, F.; Beaudegnies, R.; Rivera, M.; Frisque-Hesbain, A.-M.; Wynants, C., *Tetrahedron* **1995**, *51*, 11021.

52. (a) Plenio, H.; Aberle, C., *Chem. Commun.* **1996**, 2123. (b) Trost, B. M.; Urabe, H., *Tetrahedron Lett.* **1990**, *31*, 615.

53. Rivera, J. M.; Rebek, J., Jr., *J. Am. Chem. Soc.* **2000**, *122*, 7811.

54. Langer, P.; Schneider, T., *Synlett* **2000**, 497.

55. Barbero, A.; Pulido, F. J., *Synthesis* **2004**, 401.

56. Suh, Y.-G.; Shin, D.-Y.; Jung, J.-K.; Kim, S.-H., *Chem. Commun.* **2002**, 1064.

57. Wybrow, R. A. J.; Edwards, A. S.; Stevenson, N. G.; Adams, H.; Johnstone, C.; Harrity, J. P. A., *Tetrahedron* **2004**, *60*, 8869.

58. (a) Kanwar, S.; Trehan, S., *Tetrahedron Lett.* **2005**, *46*, 1329. (b) Kiyooka, S.-i.; Shirouchi, M.; Kaneko, Y., *Tetrahedron Lett.* **1993**, *34*, 1491.

59. Danilenko, V.; Tishkov, A. A.; Ioffe, S. L.; Lyapkalo, I. M.; Strelenko, Y. A.; Tartakovsky, V. A., *Synthesis* **2002**, 635.

60. Hauser, F. M.; Zhou, M.; Sun, Y., *Synth. Commun.* **2001**, *31*, 77.

61. Miller, S. J.; Bayne, C. D., *J. Org. Chem.* **1997**, *62*, 5680.

62. Kobayashi, M.; Masumoto, K.; Nakai, E.-i; Nakai, T., *Tetrahedron Lett.* **1996**, *37*, 3005.

63. Werner, R. M.; Barwick, M.; Davis, J. T., *Tetrahedron Lett.* **1995**, *36*, 7395.

64. (a) Slagt, M. Q.; Rodríguez, G.; Grutters, M. M. P.; Gebbink, R. J. M. K.; Klopper, W.; Jenneskens, L. W.; Lutz, M.; Spek, A. L.; van Koten, G., *Chem. Eur. J.* **2004**, *10*, 1331. (b) Motoyama, Y.; Koga, Y.; Kobayashi, K.; Aoki, K.; Nishiyama, H., *Chem. Eur. J.* **2002**, *8*, 2968.

65. Turner, R. M.; Ley, S. V.; Lindell, S. D., *Synlett* **1993**, 748.

66. Miller, R. A.; Smith, R. M.; Karady, S.; Reamer, R. A., *Tetrahedron Lett.* **2002**, *43*, 935.

67. Begtrup, M.; Vedsø, P., *J. Chem. Soc., Perkin Trans. 1* **1993**, 625.

68. Weisenburger, G. A.; Faibish, N. C.; Pippel, D. J.; Beak, P., *J. Am. Chem. Soc.* **1999**, *121*, 9522 and references cited therein.

69. Marr, F.; Fröhlich, R.; Hoppe, D., *Tetrahedron: Asymmetry* **2002**, *13*, 2587.

70. Wilhelm, R.; Sebhat, I. K.; White, A. J. P.; Williams, D. J.; Widdowson, D. A., *Tetrahedron: Asymmetry* **2000**, *11*, 5003.

71. Nakamura, S.; Nakagawa, R.; Watanabe, Y.; Toru, T., *J. Am. Chem. Soc.* **2000**, *122*, 11340.

72. Hoffmann, R. W.; Bewersdorf, M.; Krüger, M.; Mikolaiski, W.; Stürmer, R., *Chem. Ber.* **1991**, *124*, 1243.

73. Müller, P.; Lacrampe, F.; Bernardinelli, G., *Tetrahedron: Asymmetry* **2003**, *14*, 1503.

74. Cavelier-Frontin, F.; Jacquier, R.; Paladino, J.; Verducci, J., *Tetrahedron* **1991**, *47*, 9807.

75. Grumbach, H.-J.; Arend, M.; Risch, N., *Synthesis* **1996**, 883.

76. Esker, J. L.; Newcomb, M., *Tetrahedron Lett.* **1992**, *33*, 5913.

77. Kauch, M.; Hoppe, D., *Can. J. Chem.* **2001**, *79*, 1736.

78. Ogoshi, S.; Tomiyasu, S.; Morita, M.; Kurosawa, H., *J. Am. Chem. Soc.* **2002**, *124*, 11598.

79. (a) Hoffmann, R.; Brückner, R., *Chem. Ber.* **1992**, *125*, 1471. (b) Kusche, A.; Hoffmann, R.; Münster, I.; Keiner, P.; Brückner, R., *Tetrahedron Lett.* **1991**, *32*, 467.

80. Kim, S.; Kim. Y. G.; Kim, D.-i., *Tetrahedron Lett.* **1992**, *33*, 2565.

81. Lillie, B. M.; Avery, M. A., *Tetrahedron Lett.* **1994**, *35*, 969.

82. Ishihara, K.; Hiraiwa, Y.; Yamamoto, H., *Chem. Commun.* **2002**, 1564.

83. Markó, I. E.; Mekhalfia, A.; Ollis, W. D., *Synlett* **1990**, 347.

84. Xia, Q.; Ganem, B., *Org. Lett.* **2002**, *4*, 1631.

85. Reddy, D. S.; Judd, W. R.; Aubé, J., *Org. Lett.* **2003**, *5*, 3899.

86. Sato, T.; Hayashi, M.; Hayata, T., *Tetrahedron* **1992**, *48*, 4099.

87. Roos, E. C.; López, M. C.; Brook, M. A.; Hiemstra, H.; Speckamp, W. N.; Kaptein, B.; Kamphuis, J.; Schoemaker, H. E., *J. Org. Chem.* **1993**, *58*, 3259.

88. Anders, E.; Hertlein, K.; Stankowiak, A.; Irmer, E., *Synthesis* **1992**, 577.

89. Kim, S.; Kim, S. H., *Tetrahedron Lett.* **1995**, *36*, 3723.

90. Beignet, J.; Cox, L. R., *Org. Lett.* **2003**, *5*, 4231.

91. Roux, M.-C.; Wartski, L.; Nierlich, M.; Lance, M., *Tetrahedron* **1996**, *52*, 10083.

92. (a) Miles, S. M.; Marsden, S. P.; Leatherbarrow, R. J.; Coates, W. J., *J. Org. Chem.* **2004**, *69*, 6874. (b) Meyer, C.; Cossy, J., *Tetrahedron Lett.* **1997**, *38*, 7861.

93. Suginome, M.; Ohmori, Y.; Ito, Y., *J. Am. Chem. Soc.* **2001**, *123*, 4601.

94. Shi, M.; Yang, Y.-H.; Xu, B., *Tetrahedron* **2005**, *61*, 1983.

95. Takemoto, Y.; Kuraoka, S.; Hamaue, N.; Iwata, C., *Tetrahedron: Asymmetry* **1996**, *7*, 993.

96. Eriksson, M.; Iliefski, T.; Nilsson, M.; Olsson, T., *J. Org. Chem.* **1997**, *62*, 182.

97. Kim, S.; Han, W. S.; Lee, J. M., *Bull. Korean Chem. Soc.* **1992**, *13*, 466.

98. Takasu, K.; Misawa, K.; Yamada, M.; Furuta, Y.; Taniguchi, T.; Ihara, M., *Chem. Commun.* **2000**, 1739.

99. Takasu, K.; Ueno, M.; Ihara, M., *J. Org. Chem.* **2001**, *66*, 4667.

100. Ihara, M.; Suzuki, S.; Taniguchi, N.; Fukumoto, K., *J. Chem. Soc., Chem. Commun.* **1993**, 755.

101. Jung, S. H.; Kim, J. H., *Bull. Korean Chem. Soc.* **2002**, *23*, 1375.

102. Kim, C. U.; Misco, P. F.; Luh, B. Y.; Mansuri, M. M., *Tetrahedron Lett.* **1994**, *35*, 3017.

103. Basavaiah, D.; Rao, A. J., *Chem. Commun.* **2003**, 604.

104. Kataoka, T.; Iwama, T.; Tsujiyama, S.-i., Iwamura, T.; Watanabe, S.-i., *Tetrahedron* **1998**, *54*, 11813.

105. Ichikawa, J.; Miyazaki, S.; Fujiwara, M.; Minami, T., *J. Org. Chem.* **1995**, *60*, 2320.

106. Andrews, J. F. P.; Regan, A. C., *Tetrahedron Lett.* **1991**, *32*, 7731.

107. Molander, G. A.; Cameron, K. O., *J. Org. Chem.* **1991**, *56*, 2617.

108. (a) Molander, G. A.; Cameron, K. O., *J. Org. Chem.* **1993**, *58*, 5931. (b) Molander, G. A.; Cameron, K. O., *J. Am. Chem. Soc.* **1993**, *115*, 830.

109. Collins, J. L.; Grieco, P. A.; Walker, J. K., *Tetrahedron Lett.* **1997**, *38*, 1321.

110. Grieco, P. A.; Walker, J. K., *Tetrahedron* **1997**, *53*, 8975.

111. Tietze, L. F.; Schneider, C., *Synlett* **1992**, 755.

112. Akiba, K.-y.; Motoshima, T.; Ishimaru, K.; Yabuta, K.; Hirota, H.; Yamamoto, Y., *Synlett* **1993**, 657.

113. Arenz, T.; Frauenrath, H.; Raabe, G.; Zorn, M., *Liebigs Ann. Chem.* **1994**, 931.

114. Murata, Y.; Overman, L. E., *Heterocycles* **1996**, *42*, 549.

115. Moutou, J. L.; Schmitt, M.; Wermuth, C. G.; Bourguignon, J. J., *Tetrahedron Lett.* **1994**, *35*, 6883.

116. Mikami, K.; Ohmura, H., *Chem. Commun.* **2002**, 2626.

117. Albrecht, U.; Armbrust, H.; Langer, P., *Synlett* **2004**, 143.

118. Murray, D. H.; Albizati, K. F., *Tetrahedron Lett.* **1990**, *29*, 4109.

119. Gassman, P. G.; Burns, S. J.; Pfister, K. B., *J. Org. Chem.* **1993**, *58*, 1449.

120. Futagawa, T.; Nishiyama, N.; Tai, A.; Okuyama, T.; Sugimura, T., *Tetrahedron* **2002**, *58*, 9279.

121. Swenton, J. S.; Bonke, B. R.; Clark, W. M.; Chen, C.-P.; Martin, K. V., *J. Org. Chem.* **1990**, *55*, 2027.

122. Rychnovsky, S. D.; Kim, J., *Tetrahedron Lett.* **1991**, *32*, 7219.

123. Hunter, R.; Michael, J. P.; Tomlinson, G. D., *Tetrahedron* **1994**, *50*, 871.

124. Linderman, R. J.; Chen, K., *J. Org. Chem.* **1996**, *61*, 2441.

125. Fujita, K.; Inoue, A.; Shinokubo, H.; Oshima, K., *Org. Lett.* **1999**, *1*, 917.

126. Sato, T.; Otera, J.; Nozaki, H., *J. Org. Chem.* **1990**, *55*, 6116.

127. Linderman, R. J.; Anklekar, T. V., *J. Org. Chem.* **1992**, *57*, 5078.

128. Hunter, R.; Bartels, B.; Michael, J. P., *Tetrahedron Lett.* **1991**, *32*, 1095.

129. Hunter, R.; Bartels, B., *J. Chem. Soc., Perkin Trans. 1* **1991**, 2887.

130. (a) Roush, W. R.; Bennett, C. E.; Roberts, S. E., *J. Org. Chem.* **2001**, *66*, 6389. (b) Kovensky, J.; Sinaÿ, P., *Eur. J. Org. Chem.* **2000**, 3523. (c) Iimori, T.; Kobayashi, H.; Hashimoto, S.-i.; Ikegami, S., *Heterocycles* **1996**, *42*, 485.

131. Yadav, V.; Chu, C. K.; Rais, R. H.; Al Safarjalani, O. N.; Guarcello, V.; Naguib, F. N. M.; el Kouni, M. H., *J. Med. Chem.* **2004**, *47*, 1987.

132. Qin, Z.-H.; Li, H.; Cai, M.-S.; Li, Z.-J., *Carbohydr. Res.* **2002**, *337*, 31.

133. Crich, D.; de la Mora, M.; Vinod, A. U., *J. Org. Chem.* **2003**, *68*, 8142.

134. (a) Tanimori, S.; Ohta, T.; Kirihata, M., *Bioorg. Med. Chem. Lett.* **2002**, *12*, 1135. (b) Jenkins, D. J.; Marwood, R. D.; Potter, B. V. L., *Chem. Commun.* **1997**, 449.

135. (a) Pradeepkumar, P. I.; Amirkhanov, N. V.; Chattopadhyaya, J., *Org. Biomol. Chem.* **2003**, *1*, 81. (b) Ali, I. A. I.; Abdel-Rahman, A. A.-H.; El Ashry, E. S. H.; Schmidt, R. R., *Synthesis* **2003**, 1065.

136. Mlynarski, J.; Banaszek, A., *Tetrahedron: Asymmetry* **2000**, *11*, 3737.

137. Huart, C.; Ghosez, L., *Angew. Chem., Int. Ed. Engl.* **1997**, *36*, 634.

138. Wuts, P. G. M.; Jung, Y.-W., *J. Org. Chem.* **1988**, *53*, 1957.

139. Dhavale, D. D.; Trombini, C., *Heterocycles* **1992**, *34*, 2253.

140. Gianotti, M.; Lombardo, M.; Trombini, C., *Tetrahedron Lett.* **1998**, *39*, 1643.

141. Merino, P.; Mates, J. A., *ARKIVOC* **2000**, *xi*, 12 and references cited therein.

142. Shanmugam, P.; Miyashita, M., *Org. Lett.* **2003**, *5*, 3265.

143. (a) Gala, D.; DiBenedetto, D. J., *Tetrahedron Lett.* **1994**, *35*, 8299. (b) Raubo, P.; Wicha, J., *J. Org. Chem.* **1994**, *59*, 4355.

144. Yoshitake, M.; Yamamoto, M.; Kohmoto, S.; Yamada, K., *J. Chem. Soc., Perkin Trans. 1* **1990**, 1226.

145. Jung, M. E.; Hoffmann, B.; Rausch, B.; Contreras, J.-M., *Org. Lett.* **2003**, *5*, 3159 and references cited therein.

146. Jung, M. E.; van der Heuvel, A., *Tetrahedron Lett.* **2002**, *43*, 8169.

147. Oriyama, T.; Yatabe, K.; Sugawara, S.; Machiguchi, Y.; Koga, G., *Synlett* **1996**, 523.

148. Chattopadhyay, S. K.; Pattenden, G., *Tetrahedron Lett.* **1998**, *39*, 6095.

149. Trzeciak, A.; Bannwarth, W., *Synthesis* **1996**, 1433.

150. Williams, C. M.; Mander, L. N., *Tetrahedron Lett.* **2004**, *45*, 667.

151. Fujioka, H.; Sawama, Y.; Murata, N.; Okitsu, T.; Kubo, O.; Matsuda, S.; Kita, Y., *J. Am. Chem. Soc.* **2004**, *126*, 11800.

152. Zhang, A.; Russell, D. H.; Zhu, J.; Burgess, K., *Tetrahedron Lett.* **1998**, *39*, 7439.

153. Sakaitani, M.; Ohfune, Y., *J. Org. Chem.* **1990**, *55*, 870.

154. Todd, M. H.; Abell, C., *Tetrahedron Lett.* **2000**, *41*, 8183.

155. Klein, M.; König, B., *Tetrahedron* **2004**, *60*, 1087.

156. Zhdankin, V. V.; Koposov, A. Y.; Yashin, N. V., *Tetrahedron Lett.* **2002**, *43*, 5735.

157. Zhdankin, V. V.; Kuehl, C. J.; Krasutsky, A. P.; Formaneck, M. S.; Bolz, J. T., *Tetrahedron Lett.* **1994**, *35*, 9677.

158. Zhdankin, V. V.; Crittell, C. M.; Stang, P. J.; Zefirov, N. S., *Tetrahedron Lett.* **1990**, *31*, 4821.

159. Zhdankin, V. V.; Kuehl, C. J.; Simonsen, A. J., *J. Org. Chem.* **1996**, *61*, 8272.

160. Zhdankin, V. V.; Kuehl, C. J.; Krasutsky, A. P.; Bolz, J. T.; Simonsen, A. J., *J. Org. Chem.* **1996**, *61*, 6547.

161. Zhdankin, V. V.; Arbit, R. M.; McSherry, M.; Mismash, B.; Young, V. G., Jr. *J. Am. Chem. Soc.* **1997**, *119*, 7408.

162. Radhakrishnan, U.; Stang, P. J., *J. Org. Chem.* **2003**, *68*, 9209.

163. Kita, Y.; Watanabe, H.; Egi, M.; Saiki, T.; Fukuoka, Y.; Tohma, H., *J. Chem. Soc., Perkin Trans. 1* **1998**, 635 and references cited therein.

164. Yoshimatsu, M.; Yamaguchi, S.; Matsubara, Y., *J. Chem. Soc., Perkin Trans. 1* **2001**, 2560.

165. (a) Hoesl, C. E.; Maurus, M.; Pabel, J.; Polborn, K.; Wanner, K. T., *Tetrahedron* **2002**, *58*, 6757. (b) Pabel, J.; Hösl, C. E.; Maurus, M.; Ege, M.; Wanner, K. T., *J. Org. Chem.* **2000**, *65*, 9272.

166. Ullah, E.; Langer, P., *Synlett* **2004**, 2782.

167. Tietze, L. F.; Wichmann, J., *Liebigs Ann. Chem.* **1992**, 1063.

168. (a) Beifuss, U.; Tietze, M.; Gehm, H., *Synlett* **1996**, 182. (b) Beifuss, U.; Gehm, H.; Noltemeyer, M.; Schmidt, H.-G., *Angew. Chem., Int. Ed.* **1995**, *34*, 647.

169. Beifuss, U.; Schniske, U.; Feder, G., *Tetrahedron* **2001**, *57*, 1005.

170. Beifuss, U.; Goldenstein, K.; Döring, F.; Lehmann, C.; Noltemeyer, M., *Angew. Chem., Int. Ed.* **2001**, *40*, 568.

171. Ronan, B.; Kagan, H. B., *Tetrahedron: Asymmetry* **1991**, *2*, 75.

172. Hashimoto, Y.; Saigo, K.; Machida, S.; Hasegawa, M., *Tetrahedron Lett.* **1990**, *39*, 5625.

173. Vorndam, P. E., *J. Org. Chem.* **1990**, *55*, 3693.

174. Fujioka, H.; Yamanaka, T.; Takuma, K.; Miyazaki, M.; Kita, Y., *J. Chem. Soc., Chem. Commun.* **1991**, 533.

175. Chassaing, C.; Haudrechy, A.; Langlois, Y., *Tetrahedron Lett.* **1997**, *38*, 4415.

176. Kagoshima, H.; Hashimoto, Y.; Oguro, D.; Kutsuna, T.; Saigo, K., *Tetrahedron Lett.* **1998**, *39*, 1203.

177. (a) Padwa, A.; Waterson, A. G., *Tetrahedron Lett.* **1998**, *39*, 8585. (b) Padwa, A.; Hennig, R.; Kappe, C. O.; Reger, T. S., *J. Org. Chem.* **1998**, *63*, 1144. (c) Padwa, A.; Gunn, D. E., Jr.; Osterhout, M. H., *Synthesis* **1997**, 1353.

178. Itoh, T.; Miyazaki, M.; Nagata, K.; Ohsawa, A., *Tetrahedron* **2000**, *56*, 4383.

179. Itoh, T.; Miyazaki, M.; Nagata, K.; Matsuya, Y.; Ohsawa, A., *Heterocycles* **1999**, *50*, 667.

180. Kalita, B.; Nicholas, K. M., *Tetrahedron* **2004**, *60*, 10771.

181. Enders, D.; Maaβen, R.; Runsink, J., *Tetrahedron: Asymmetry* **1998**, *9*, 2155.

182. Müller, T. J. J.; Netz, A., *Tetrahedron Lett.* **1999**, *40*, 3145 and references cited therein.

183. Ishikawa, T.; Okano, M.; Aikawa, T.; Saito, S., *J. Org. Chem.* **2001**, *66*, 4635.

184. Suginome, M.; Takama, A.; Ito, Y., *J. Am. Chem. Soc.* **1998**, *120*, 1930.

185. (a) Wang, Q.; Mayer, M. F.; Brennan, C.; Yang, F.; Hossain, M. M.; Grubisha, D. S.; Bennet, D., *Tetrahedron* **2000**, *56*, 4881. (b) Vargas, R. M.; Theys, R. D.; Hossain, M. M., *J. Am. Chem. Soc.* **1992**, *114*, 777.

186. Hirao, T.; Fujii, T.; Ohshiro, Y., *Tetrahedron Lett.* **1994**, *35*, 8005.

187. Chu, C.-Y.; Hwang, D.-R.; Wang, S.-K.; Uang, B.-J., *Chem. Commun.* **2001**, 980.

Trifluoromethanesulfonic Acid, Chlorobis (1,1-dimethylethyl)silyl Ester

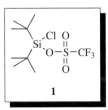

1

[134436-16-1] $C_9H_{18}ClF_3O_3SSi$ (MW 326.83)

InChI = 1S/C9H18ClF3O3SSi/c1-7(2,3)18(10,8(4,5)6)16-17(14,15)9(11,12)13/h1-6H3

InChIKey = KFJSFTNXTRLUFF-UHFFFAOYSA-N

(reagent used for the preparation of unsymmetrically substituted silaketals *t*-Bu$_2$Si(OR^1)(OR^2) and other silicon compounds of type *t*-Bu$_2$SiXY)

Physical Data: bp 60 °C/0.3 mm Hg.

Solubility: soluble in most dry organic solvents, e.g., CH$_2$Cl$_2$, THF, Et$_2$O, and DMF.

Form Supplied in: colorless liquid.

Analysis of Reagent Purity: ^{1}H NMR.

Preparative Methods: trifluoromethanesulfonic acid chlorobis (1,1-dimethylethyl)silyl ester (**1**) can be prepared (eq 1) on a medium scale as follows:[1] Di-*t*-butylchlorosilane (25.0 g, 140 mmol) was stirred at −15 °C in a flask equipped with a dropping funnel and thermometer. Triflic acid (24 mL, 40.6 g, 270 mmol) was added at a rate sufficient to maintain the temperature below −15 °C. After complete addition, the cooling bath was removed and the reaction mixture was stirred at 25 °C overnight. The crude reaction mixture consisted of a *t*-Bu$_2$Si(Cl)OTf (^{1}H NMR: δ = 1.19) and *t*-Bu$_2$Si(H)OTf (δ = 1.12) in a 91:9 ratio. The crude product was purified by distillation using a spinning band column yielding 24.0 g (53%) of **1**. Under the chosen reaction conditions, no subsequent displacement of chloride to produce *t*-Bu$_2$Si(OTf)$_2$ was observed.

This substitution process is slow and would require elevated temperatures.

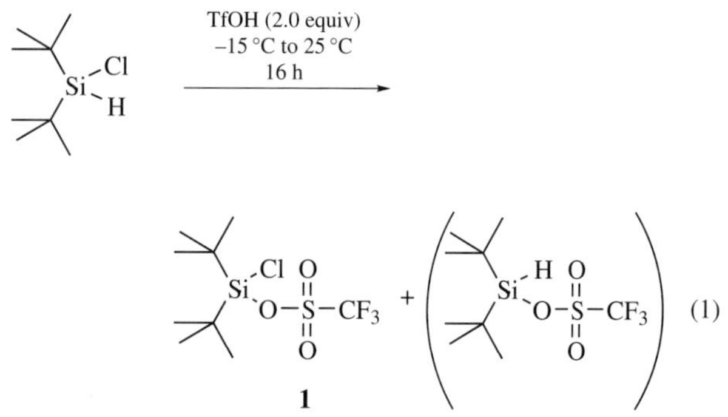

$$(1)$$

1

Purity: distillation using a spinning band column.

Handling, Storage, and Precaution: store under dry and inert atmosphere.

Unsymmetrical Silaketal Formation. Silicon tethers[2–4] are very commonly used to convert intermolecular into intramolecular reactions. Many examples for the use of such temporary tethers are known, for e.g., in intramolecular Diels–Alder reaction, 1,3-dipolar cycloadditions, radical cyclization, and metathetic ring closures. Using the stepwise reaction of R_2SiCl_2 or $R_2Si(OTf)_2$ with two different nucleophiles can lead to significant amounts of symmetrically substituted silanes as by-products. Therefore, a controlled stepwise introduction of two different nucleophiles is preferable. Since the triflate anion is a much better leaving group than chloride, selective introduction of one nucleophile or stepwise introduction of two nucleophiles is possible using **1** (eq 2).

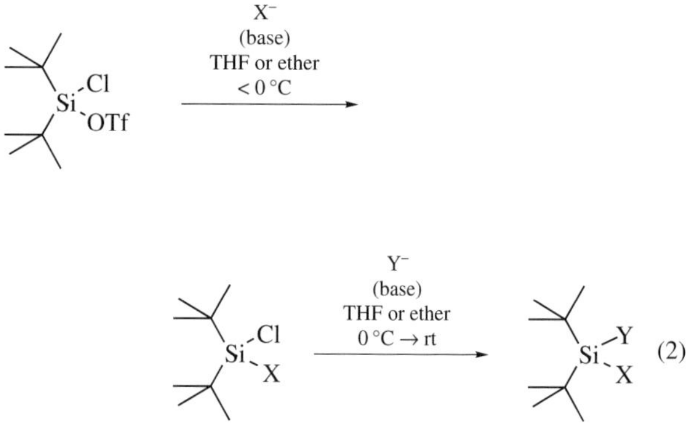

$$(2)$$

Usually, the first nucleophile (X^-) is introduced at a low temperature (-78 to $0\,^{\circ}C$) leading to the substituted chlorosilane. At room temperature or above, the second replacement step occurs to give the disubstituted silane (t-Bu_2SiXY). When protic nucleophiles are employed, bases such as triethylamine, ethyldiisopropylamine, and pyridine are required. With alcohols or alkoxides as nucleophiles, silaketals [t-$Bu_2Si(OR^1)(OR^2)$] are obtained. The cyclic silaketals resulting from an intramolecular reaction between R^1 and R^2 can be subsequently cleaved to yield diols using 3 equiv of $Bu_4N^+F^-$ in THF at $40\,^{\circ}C$.[1]

Reaction of **1** with ethyl 4-hydroxycrotonate (**2**) using pyridine as a base gives the mono-functionalized product **3** that is stable under the conditions used for chromatography. Reaction of chlorosilyl compound **3** with a preformed lithium enolate yields the desired unsymmetrical silaketal **4**, which undergoes intramolecular [4 + 2]-cycloaddition upon heating (eq 3).[1]

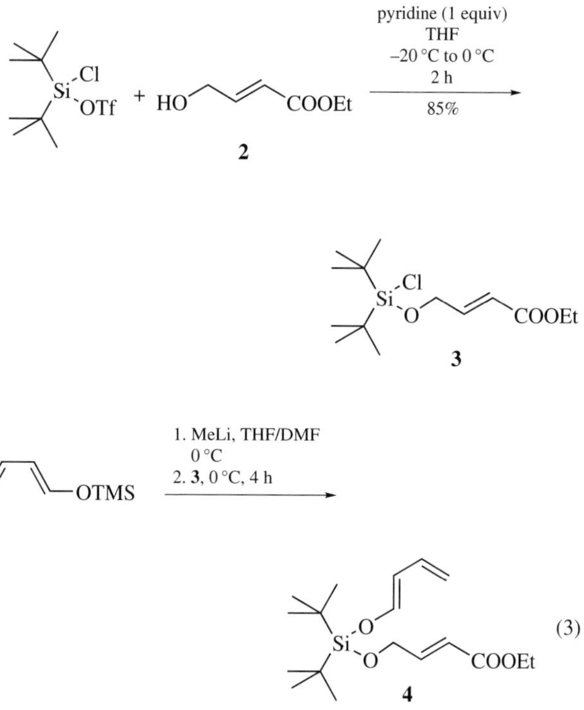

$$(3)$$

4

A similar approach is used for the synthesis of silaketal **6** as a key intermediate in a total synthesis of (+)-castanospermine and related indolizidine and pyrrolizidine alkaloids (eq 4).[5] First, the triflate moiety in **1** is replaced by reaction with lithium butadienolate generated in situ from the reaction of 2,5-dihydrofuran and n-BuLi. Treatment of the monosubstituted intermediate with potassium nitroacetaldehyde in a chloroform–acetonitrile mixture yields the desired sila-tethered compound **6**, which is then submitted to a Lewis acid-assisted tandem [4 + 2]/[3 + 2]-cycloaddition.

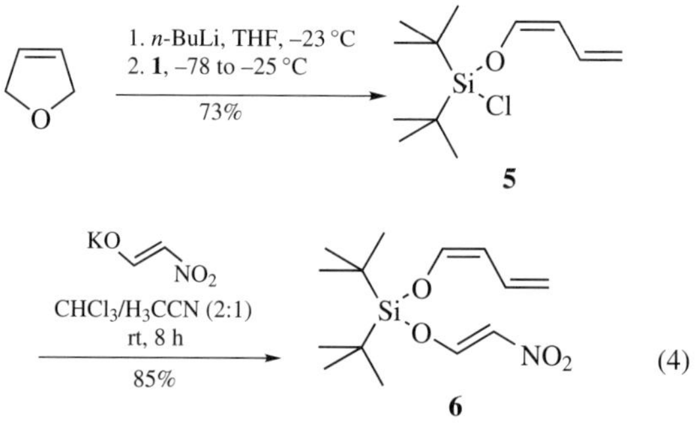

$$(4)$$

6

The synthesis of α-silyl-α-diazoacetates containing silicon–heteroatom bonds is also possible by a stepwise reaction of **1** with different nucleophiles (eqs 5 and 6).[6] In the first step, the reaction of alkyl diazoacetates with **1** in the presence of Hünig's base gives rise to (chlorosilyl)diazoacetates, e.g., **7** (eq 5).

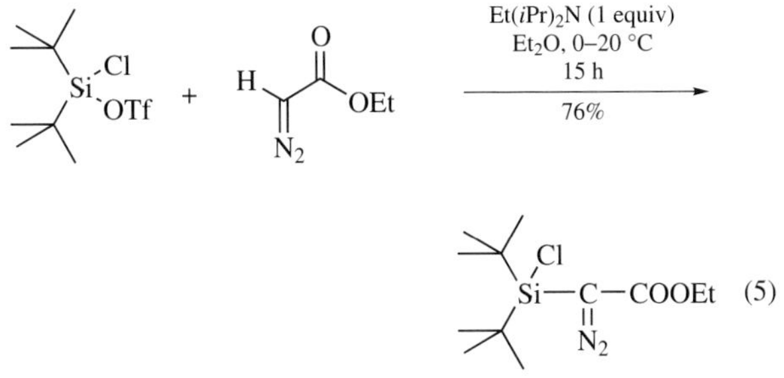

$$ (5) $$

In the second step (eq 6), diazoacetates such as **7** can be transformed smoothly into the thermally quite stable azidosilyl-, isocyanatosilyl-, and isothiocyanatosilyl-diazoacetates (**8–10**) by reaction with the appropriate pseudohalide.

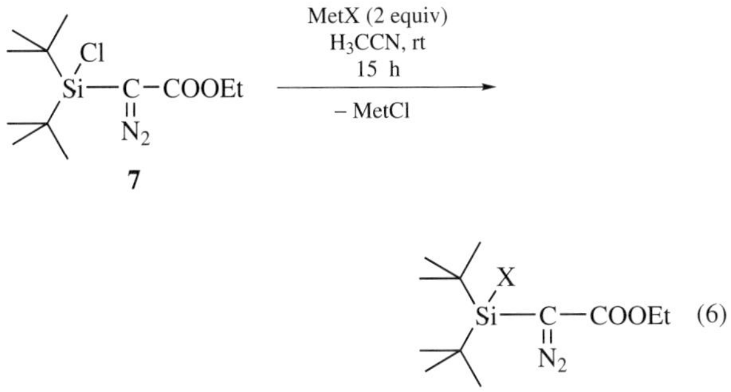

$$ (6) $$

MetX	Yield (%)	Product
NaN$_3$	64	**8**
KOCN	60	**9**
KSCN	60	**10**

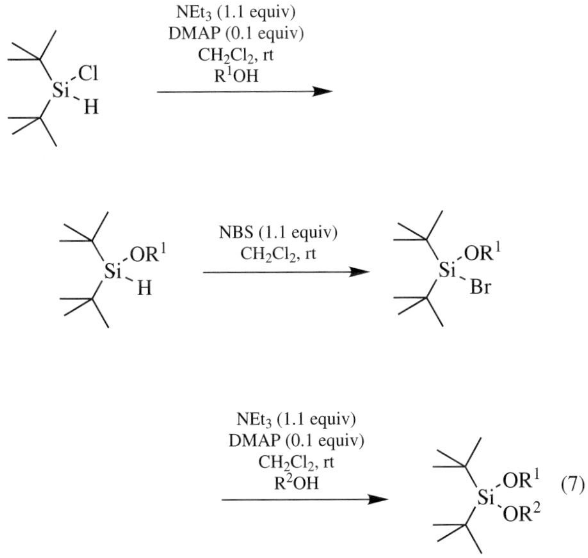

$$ (7) $$

Alternative Methodologies. Subsequent treatment of t-Bu$_2$SiCl$_2$ or t-Bu$_2$Si(OTf)$_2$ with nucleophiles under the appropriate conditions[7,8] is possible in selected cases. Alternatively,

a stepwise functionalization of t-Bu$_2$Si(H)Cl (eq 7) has been introduced recently.[9]

1. Gillard, J. W.; Fortin, R.; Grimm, E. L.; Maillard, M.; Tjepkema, M.; Bernstein, M. A.; Glaser, R., *Tetrahedron Lett.* **1991**, *31*, 1145.
2. Bols, M.; Skrydstrup, T., *Chem. Rev.* **1995**, *95*, 1253.
3. Fensterbank, L.; Malacria, M.; Sieburth, S. M., *Synthesis* **1997**, 813.
4. Gauthier, Jr, D. R.; Zandi, K. S.; Shea, K. J., *Tetrahedron* **1998**, *54*, 2289.
5. Denmark, S. E.; Martinborough, E. A., *J. Am. Chem. Soc.* **1999**, *121*, 3046.
6. Maas, G.; Bender, S., *Synthesis* **1999**, 1175.
7. Kablean, S. N.; Marsden, S. P.; Craig, A. M., *Tetrahedron Lett.* **1998**, *39*, 5109.
8. Fronda, A.; Krebs, F.; Daucher, B.; Werle, T.; Maas, G., *J. Organomet. Chem.* **1992**, *424*, 253.
9. Petit, M.; Chouraqui, G.; Aubert, C.; Malacria, M., *Org. Lett.* **2003**, *5*, 2037.

Gerhard Maas & Jürgen Schatz
University of Ulm, Ulm, Germany

Trifluoromethyltrimethylsilane

Me$_3$SiCF$_3$

[81290-20-2]   C$_4$H$_9$F$_3$Si   (MW 142.22)

InChI = 1S/C4H9F3Si/c1-8(2,3)4(5,6)7/h1-3H3

InChIKey = MWKJTNBSKNUMFN-UHFFFAOYSA-N

(nucleophilic trifluoromethylating agent;[1,2] difluorocarbene precursor[1])

Physical Data: bp 54–55 °C; d^{20} 0.963 g cm^{-3}.

Solubility: sol THF, ether, CH$_2$Cl$_2$.

Form Supplied in: colorless liquid; commercially available.

Preparative Methods: is prepared[2,6] based on the original procedure of Ruppert,[7] by the reaction of chlorotrimethylsilane with the complex of trifluoromethyl bromide and hexaethylphosphorous triamide in benzonitrile (eq 1). Other less convenient procedures for its preparation are also reported.[8,9]

$$ \text{TMSCl} + \text{CF}_3\text{Br} \xrightarrow[\substack{\text{benzonitrile} \\ 75\%}]{(\text{Et}_2\text{N})_3\text{P}} \text{TMSCF}_3 \quad (1) $$

Further preparative methods are also based on the reaction of CF$_3$Br or CF$_3$I with TMSCl,[6,7] consisting in electrochemical[24–26] or aluminum-mediated reductive procedures.[27] The need for a more convenient method, avoiding the use of CF$_3$Br because of its ozone depletion potential, has led to the development of a new protocol that employs a magnesium-mediated cleavage of phenyltrifluoromethylsulfoxide[28] (eq 2).

$$ \text{PhSOCF}_3 + \text{TMSCl} \xrightarrow[\substack{\text{DMF} \\ 81\%}]{\text{Mg}} \text{TMSCF}_3 \quad (2) $$

Handling, Storage, and Precautions: is an acid-, base-, and moisture-sensitive compound and should be stored under anhydrous conditions in a refrigerator. Use in a fume hood.

Original Commentary

George. A. Olah, G. K. Surya Prakash, Qi Wang & Xing-Ya Li
University of Southern California, Los Angeles, CA, USA

Introduction. TMSCF$_3$ is a valuable reagent for trifluoromethylation of electrophilic substrates under nucleophilic catalysis or initiation.[1,2] Homologous trifluoromethyltrialkylsilanes have also been used for the same purpose.[3-5]

TMSCF$_3$ reacts as a trifluoromethide equivalent with a wide variety of electrophilic substrates such as carbonyls, sulfonyl fluorides, sulfoxides, deactivated aromatics, sulfur dioxide, and alkenes, etc.

Reactions with Carbonyl Compounds. TMSCF$_3$ reacts with aldehydes in the presence of a catalytic amount of tetra-*n*-butylammonium fluoride (TBAF) in THF to form the corresponding trifluoromethylated carbinols in good to excellent yields following aqueous hydrolysis of the silyl ethers (eq 3).[2,3,6,10] The reaction also works very well for ketones under the same conditions, with the exception of extremely hindered ones such as 1,7,7-trimethylbicyclo[2.2.1]heptan-2-one, di-1-adamantyl ketone, and fenchone. The reaction has been characterized as a fluorideinduced autocatalytic reaction.[3,10] Other initiators such as tris(dimethylamino)sulfonium difluorotrimethylsilicate (TASF), potassium fluoride, Ph$_3$SnF$_2$$^-$, and RO$^-$ can also be used for these reactions. For the reactions of TMSCF$_3$ and perfluorinated ketones and pentafluorobenzaldehyde, excess of KF is needed.[16,17]

$$\text{(3)}$$

60–90%

TMSCF$_3$ has been used in the preparation of polycyclic aromatic carcinogens, the key step being the addition of TMSCF$_3$ to the carbonyl group (eq 4).[11]

$$\text{(4)}$$

A series of tripeptides containing the trifluoromethyl group has been prepared as potent inhibitors of human leukocyte elastase (HLE) by using TMSCF$_3$.[12] TMSCF$_3$ has also been used to prepare trifluoromethyl analogs of L-fucose and 6-deoxy-D-ribose.[13] Gassman et al. have prepared 1-trifluoromethylindene starting from 1-indanone.[14]

1,2-Diketones such as benzil give only the monoadduct. On the other hand, the highly enolizable cyclohexane-1,3-dione did not give the addition product with TMSCF$_3$.[15]

A series of α,β-conjugated enones and ynones react with TMSCF$_3$ to give predominant 1,2-addition products (eq 5).[3,15]

$$\text{(5)}$$

50–90%

Simple unactivated esters do not react with TMSCF$_3$.[3] However, activated esters such as trifluoroacetic acid esters do react.[3] Cyclic esters, i.e. lactones, react with TMSCF$_3$ to give the corresponding adducts.[3] An efficient and simple synthesis of trifluoropyruvic acid monohydrate has been developed starting from di-*t*-butyl oxalate.[18]

Direct trifluoromethylation of α-keto esters give Mosher's acid derivatives (eq 6).[19]

$$\text{(6)}$$

Acyl halides such as benzoyl chloride react with TMSCF$_3$ to give a mixture of the trifluoroacetophenone and hexafluorocumyl alcohol in the presence of equimolar TBAF.[3] Cyclic anhydrides react readily with TMSCF$_3$; however, a stochiometric amount of TBAF is required. Acyclic anhydrides react less cleanly.[3]

Simple amides, such as benzamide and acetamide, do not react with TMSCF$_3$ even with molar quantities of TBAF.[3] However, an activated amide carbonyl such as in *N*-trifluoroacetyl-piperidine reacts to give an adduct which upon subsequent hydrolysis gives hexafluoroacetone trihydrate.[15] Imidazolidinetriones react with TMSCF$_3$ to give 5-trifluoromethyl-5-hydroxyimidazolidine-2,4-diones upon aqueous acid workup.[20] Imides such as *N*-methylsuccinimide react smoothly to afford the hemiaminal adducts (eq 7).[15]

$$\text{(7)}$$

Sulfur Derivatives. Kirchmer and Patel have reported[17] the preparation of trifluoromethylsulfinyl fluorides, sulfonyl fluorides, sulfoxides, and sulfuranes using TMSCF$_3$ in the presence of a catalytic amount of KF.

TMSCF$_3$ reacts with dimethyl sulfoxide in the presence of catalytic amount of TBAF to give Me$_2$CF$_3$SOTMS.[17] Arylsulfonyl fluorides react with TMSCF$_3$ to give the corresponding trifluoromethyl sulfones.[1,21]

Sulfur dioxide reacts with TMSCF$_3$ in the presence of TMSONa to give the sodium trifluoromethyl sulfinate. The sulfinate has been further oxidized to trifluoromethanesulfonic acid in 30% overall yield.[1]

Nitroso Group. Nitrosobenzene reacts with TMSCF$_3$ to afford the *O*-silylated trifluoromethylated hydroxylamine quantitatively.[1]

Aromatic Compounds. Nucleophilic trifluoromethylation of aromatic compounds containing nitro, fluoro, and trifluoromethyl groups as substituents using TMSCF$_3$ has been investigated.[22,15] Yagupolskii et al. have reported that TMSCF$_3$–TASF (1:1) reacts with 1,2,4,5-tetrakis(trifluoromethyl)benzene at −30 °C to give the stable carbanion salt (eq 8).[23]

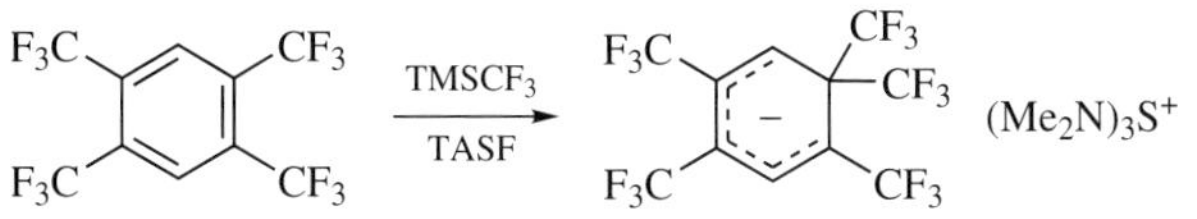

$$\text{(Me}_2\text{N)}_3\text{S}^+ \quad (8)$$

Phosphorus Compounds.   Treatment of (BuO)$_2$P(O)F with TMSCF$_3$ and a catalytic amount of KF gave (BuO)$_2$P(O)CF$_3$ in 93% isolated yield.[1]

Generation of Difluorocarbene.   Treatment of TMSCF$_3$ with an anhydrous fluoride source such as TASF in THF results in the generation of singlet difluorocarbene. In the presence of an acceptor such as tetramethylethylene, the corresponding adduct can be isolated.[1]

First Update

María Sánchez-Roselló & Carlos del Pozo Losada
Universidad de Valencia, Valencia, Spain

José Luis Aceña
Centro de Investigación Principe Felipe, Valencia, Spain

General   Reactivity.   Trifluoromethyltrimethylsilane (TMSCF$_3$, Ruppert–Prakash reagent[7,10]) has been extensively used for introducing CF$_3$ groups into a variety of organic and organometallic electrophiles.[29–31] Its high reactivity toward trifluoromethylation processes is likely a consequence of the weakness of the Si–CF$_3$ bond.[32] Most reactions of TMSCF$_3$ take place by fluoride anion activation through pentavalent silicon species which are believed to act as the reactive intermediates in the CF$_3$ group transfer. Evidence of these intermediates has been demonstrated by NMR spectroscopy[33–35] as well as by isolation and X-ray analysis of a tris(dimethylamino)sulfonium salt[36] (eq 9).

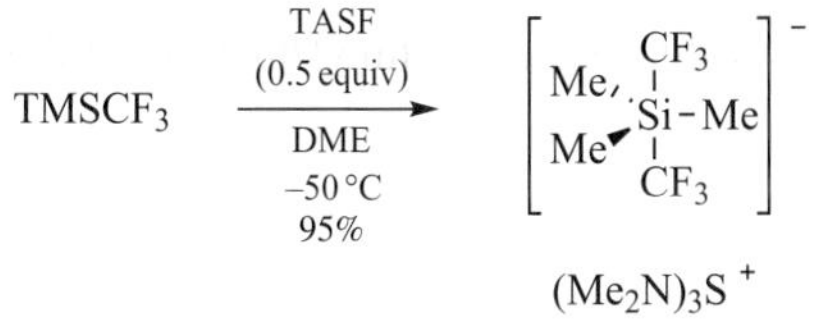

Reactions with Carbonyl Compounds and Related Derivatives.   Since the original procedure for the trifluoromethylation of carbonyl compounds with TMSCF$_3$ was reported[10] (employing catalytic TBAF as initiator in THF as solvent), several improved protocols have appeared over the years. Thus, removal of residual water by pre-drying the TBAF solution over activated 4 Å molecular sieves, in combination with non-polar aprotic solvents, made possible the trifluoromethylation of less reactive substrates such as hindered ketones and aliphatic or aromatic esters[37] (eq 10). Other anhydrous sources of fluoride anions have been successfully applied, including CsF,[38] tetramethylammonium fluoride

(TMAF), tetrabutylammonium difluorotriphenylsilicate (TBAT) or the mixed salt 1:1 CsF–CsOH.[39] Furthermore, the reaction can benefit from the use of ionic liquids as reaction media.[40] Finally, a polymer-supported fluoride initiator has also been described, thus improving the ease of purification of the products.[41]

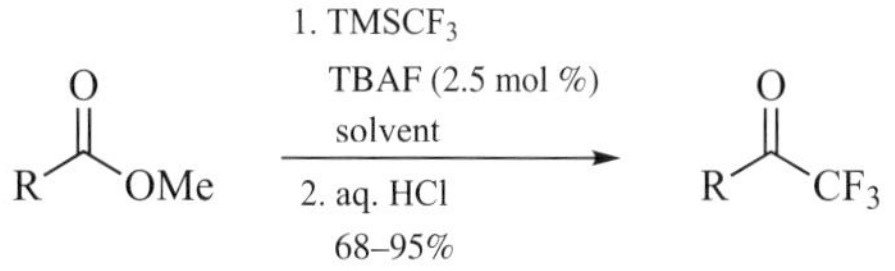

R = alkyl, alkenyl, aryl
solvent: CH$_2$Cl$_2$, toluene, pentane

The introduction of two CF$_3$ groups with TMSCF$_3$ was reported using anhydrides or activated esters as precursors[42] (eq 11).

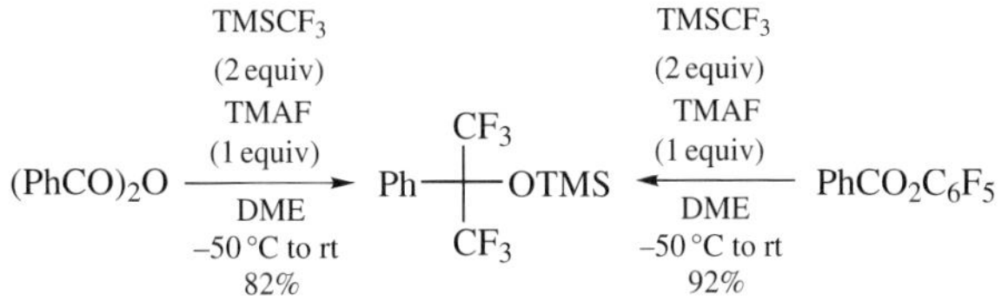

Non-fluoride initiators have been employed satisfactorily as well, including Lewis bases (amines,[43] amine N-oxides,[44] carbonates and phosphates,[45] LiOAc,[46,47] t-Bu$_3$P[48]), Lewis acids,[49] N-heterocyclic carbenes,[50,51] or even without initiator in DMSO as solvent.[52] The addition of TMSCF$_3$ to carbonyl compounds in the presence of a chiral initiator allows the enantioselective preparation of trifluoromethyl alcohols.[53] For this purpose, quaternary ammonium fluorides derived from cinchona alkaloids (**1**) have been employed, affording moderate (up to 51% ee)[54] to high enantioselectivities (up to 92% ee).[55] Also, the corresponding bromides were used in combination with an external fluoride source (KF or TMAF, up to 94% ee)[56,57] or with disodium (R)-binaphtholate (up to 71% ee),[58] or simply a cinchonidine-derived ammonium phenoxide (up to 87% ee).[59,60] Moreover, the use of a chiral TASF derivative (**2**) has also been reported (up to 52% ee).[61]

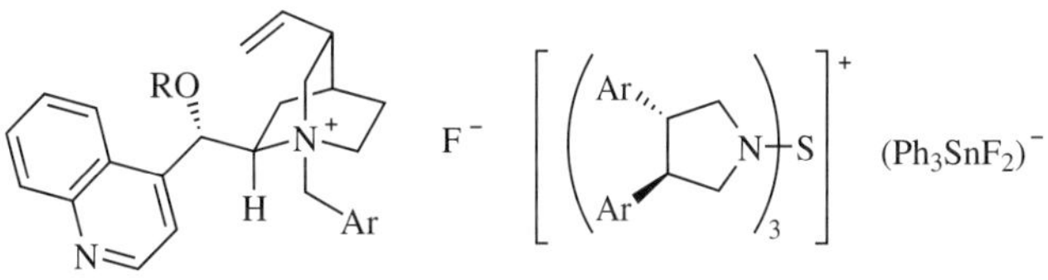

(1)   (2)

Other routes to optically active CF$_3$-containing molecules are based on the diastereoselective addition of TMSCF$_3$ to chiral substrates, including aldehydes,[62,63] ketones[64–66] (eq 12), α-ketoesters[67] or lactones.[68–70]

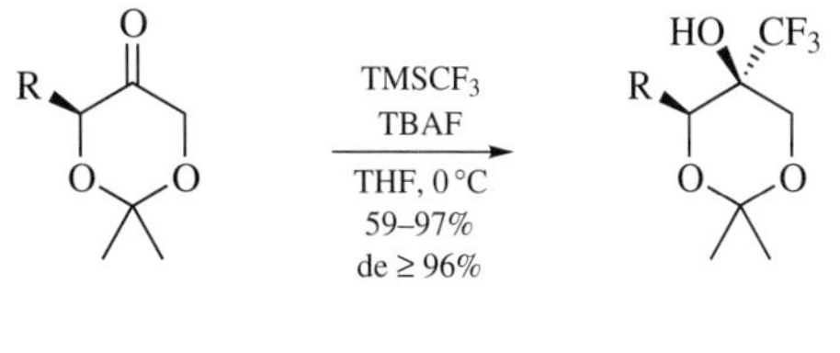

Imines are much less reactive than carbonyl compounds toward TMSCF$_3$, and the process proceeded mostly in the case of highly activated C=N bonds, using azirines,[71] polyfluorinated imines[72,73] and N-sulfonyl-[74,75] or N-sulfinylimines[76–79] as substrates. In the latter case, highly diastereoselective trifluoromethylations were achieved from chiral N-sulfinylimines (eq 13). Nitrones are another source of activated C=N bonds, suitable for the introduction of CF$_3$ groups with TMSCF$_3$.[80,81]

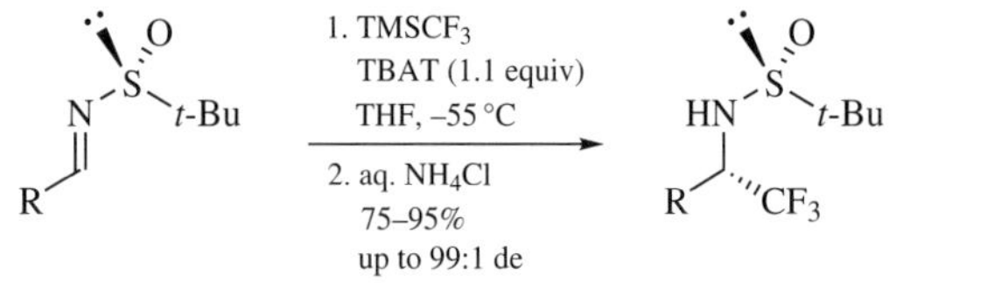

(13)

In some cases, less activated N-aryl imines were transformed into the corresponding trifluoromethylamines by reaction with TMSCF$_3$ in the presence of CsF and TMS-imidazole,[82] or with fluoride initiators such as TBAT or TMAF[83] (eq 14). However, in the latter example higher amounts of fluoride caused the loss of HF and therefore difluoromethylamines were obtained after reduction with NaBH$_4$. When a stoichiometric amount of TMAF was employed, tetramethylammonium amides were formed and further treated with a variety of electrophiles to produce N-substituted trifluoromethylamines.[84] N-Methyl imines derived from salicylaldehydes reacted with TMSCF$_3$ after forming an iminium cation[85] (eq 15).

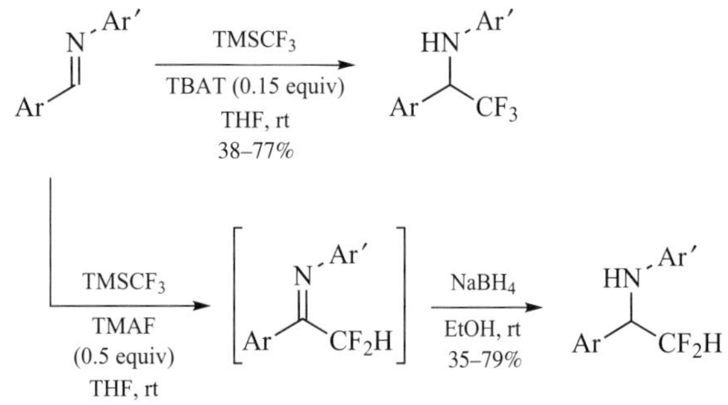

(14)

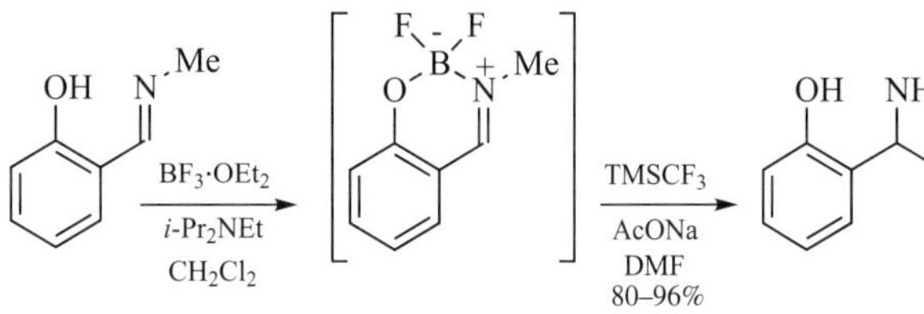

(15)

Difluoroenoxysilanes can be obtained from acylsilanes through a Brook rearrangement induced by TMSCF$_3$ and a weakly nucleophilic fluoride source (tetrabutylammonium difluorotriphenylstannate, DFTPS)[86,87] (eq 16). The resulting compounds can be isolated or reacted in a one-pot procedure with a variety of electrophiles, including glycosyl[88,89] and allyl[90,91] donors, Michael acceptors,[92] aldehydes[93] and imines.[94] The use of bis(acylsilanes) as substrates led to a subsequent cyclization reaction[95] (eq 17). A similar process has been observed in 1-arylketophosphonates, whereas the alkyl counterparts led to 1-alkyl-2,2,2-trifluoro-1-trimethylsilyloxyethylphosphonates[96] (eq 18).

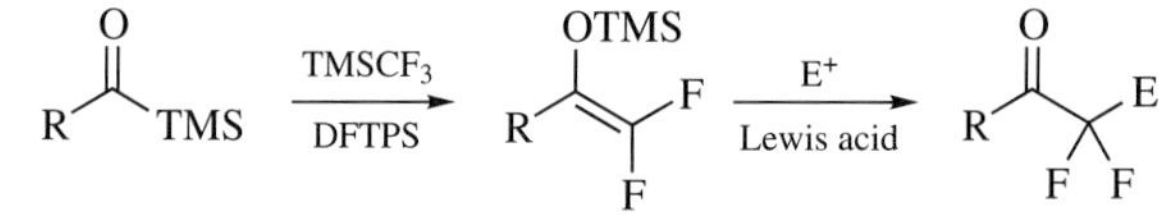

(16)

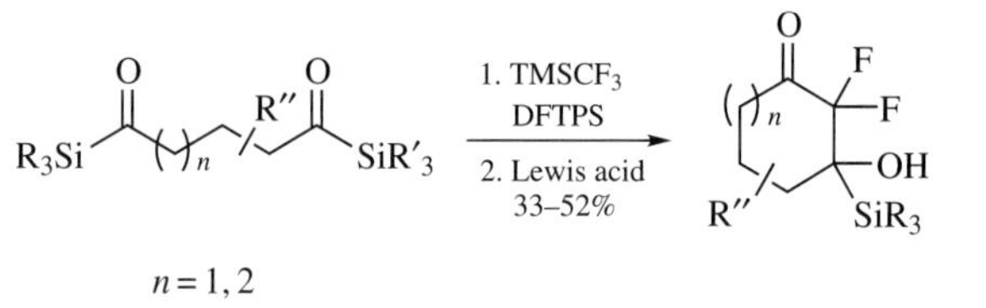

(17)

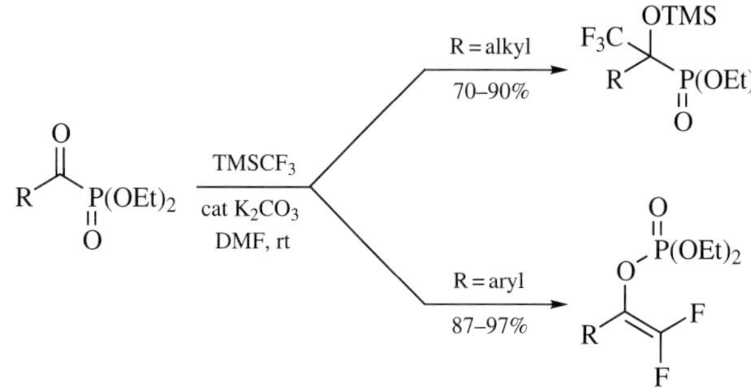

(18)

Aromatic thioketones can react with TMSCF$_3$ affording mixtures of compounds arising from both thiophilic and carbophilic attack.[97,98]

Additions to Heteroallenes. TMSCF$_3$ and a fluoride anion reacted with SO$_2$, CO$_2$, CS$_2$ and COS to produce trifluoromethanesulfinate, trifluoroacetate, trifluoromethylthiocarboxylate and trifluoromethyldithiocarboxylate salts, respectively.[99,100] Addition of TMSCF$_3$ to isocyanates and isothiocyanates gave N-substituted trifluoroacetamides and trifluorothioacetamides, respectively[101] (eq 19), whereas N-substituted sulfinylamines yielded the corresponding trifluoromethanesulfinylamides[102] (eq 20).

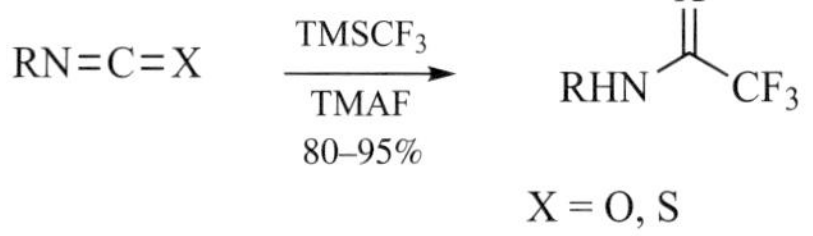

(19)

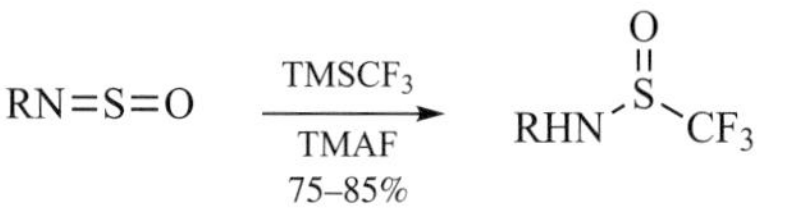

(20)

Alkyl Halides and Triflates. n-Alkyl bromides and iodides have been trifluoromethylated with TMSCF$_3$ in the presence of a stoichiometric amount of CsF and 15-crown-5 in DME as solvent, whereas the corresponding chlorides were unreactive.[103] Alkyl triflates have also been transformed into CF$_3$-substituted alkanes with TMSCF$_3$ and TMAF.[104] In contrast, aryl triflates experience nucleophilic attack on the sulfur atom to give hypervalent sulfur compounds[105] (eq 21).

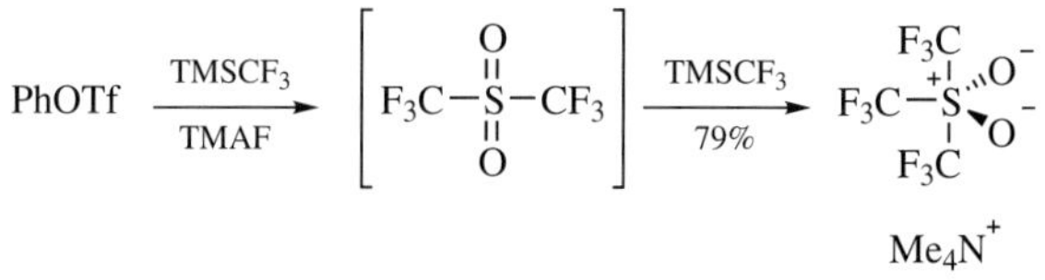

(21)

1,4-Additions to Conjugated Systems. The reaction of TMSCF$_3$ with α,β-unsaturated aldehydes or ketones under fluoride activation affords almost exclusively the 1,2-addition products.[10,106–108] Moreover, the 1,4-addition pathway may also participate in some cases,[109] and occurs predominantly in 2-perfluoroalkyl chromones and related systems[110–112] (eq 22). A 1,6-addition process in a suitably substituted substrate has also been described.[113] A more general method for the 1,4-addition of CF$_3$ groups to different enones consisted in the prior complexation of the carbonyl with a bulky Lewis acid [aluminum tris(2,6-diphenylphenoxide)].[114]

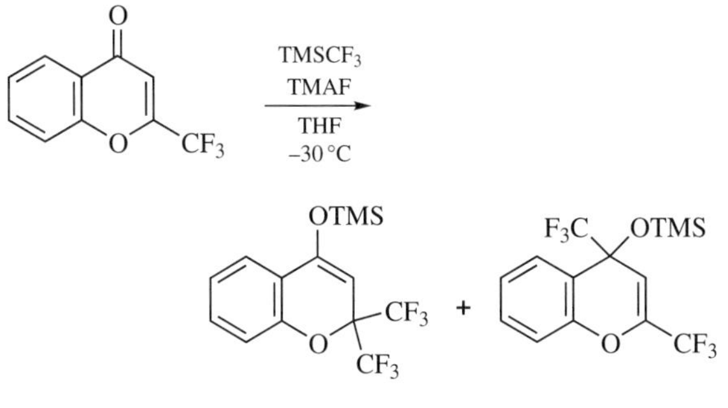

(92:8)

Aromatic Compounds. TMSCF$_3$ was used to convert haloarenes into trifluoromethyl arenes by an aromatic nucleophilic substitution process, employing a trifluoromethylcopper intermediate[5] prepared in situ with KF/CuI,[115,116] whereas the reaction of halo- or nitroarenes without copper reagents only afforded low yields of isomeric substitution products.[117,118] However, nitroarenes containing other electron-withdrawing groups were trifluoromethylated ortho or para to the nitro group, and subsequent oxidation with dimethyldioxirane (DMD) led to CF$_3$-substituted phenols[119] (eq 23). N-Alkyl salts of azinium systems were treated with TMSCF$_3$ in the presence of KF and catalytic Ph$_3$SnF to give trifluoromethyl dihydroazines which could be reoxidized to produce CF$_3$-substituted azines[120] (eq 24).

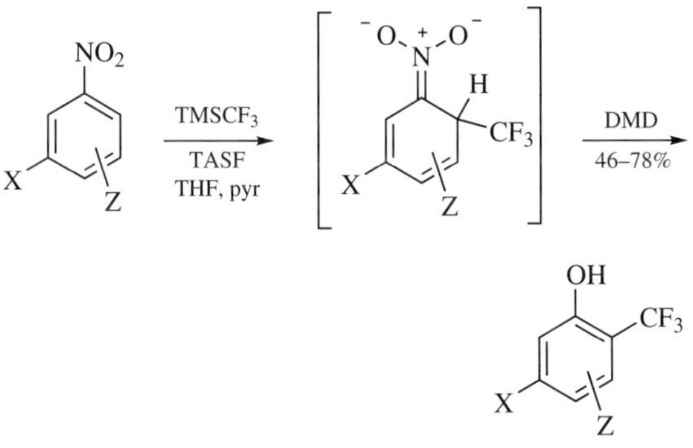

+ para isomers

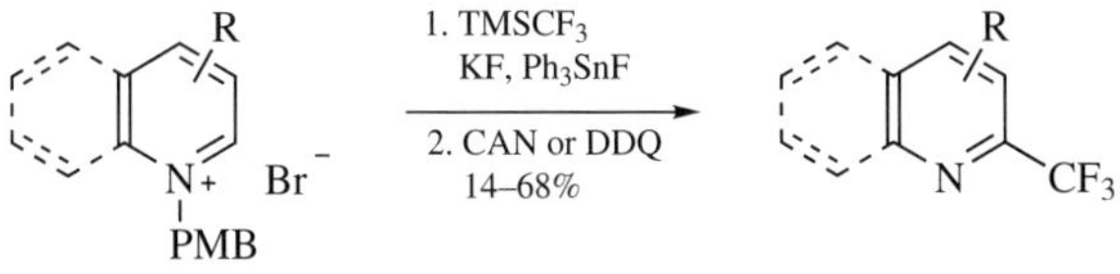

Iodine Compounds. Hypervalent iodine compounds containing a CF$_3$ group have been prepared by a chlorine-OAc exchange and further reaction with TMSCF$_3$ and TBAT in a one-pot procedure[121,122] (eq 25).

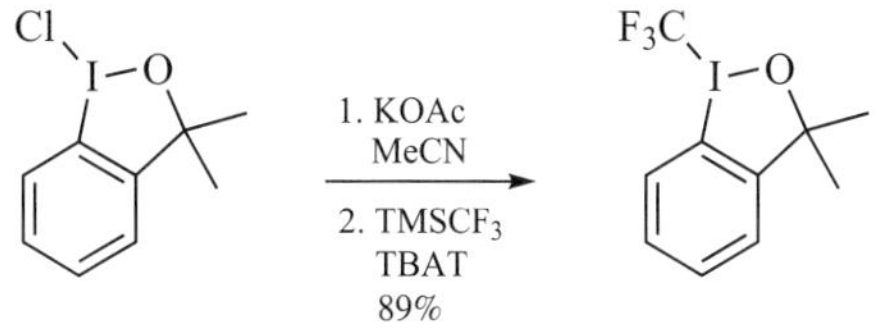

Sulfur, Selenium and Tellurium Derivatives. TMSCF$_3$ reacts with elemental sulfur, selenium or tellurium in the presence of a fluoride anion, leading to trifluoromethylthiolates, seleniates or tellurates, respectively.[123–125] Aryltrifluoromethylsulfides, sulfoxides and sulfones were obtained from the corresponding arylsulfenyl, sulfinyl and sulfonyl halides, respectively, by treatment with TMSCF$_3$ and TASF.[126] The reaction of disulfides and diselenides with TMSCF$_3$ and TBAF gave trifluoromethyl thioethers and selenoethers, respectively[127] (eq 26). The same procedure can also be carried out from thiocyanates and selenocyanates.[128]

$$R-X-X-R \xrightarrow[\substack{THF, 0\,^\circ C \\ 25-80\%}]{\substack{TMSCF_3 \\ TBAF}} R-X-CF_3 \quad (26)$$

X = S, Se

R = alkyl, aryl

Phosphorus Compounds. TMSCF$_3$ has been used to attach CF$_3$ groups to both P^{III} and P^V compounds. Thus, a number of tricoordinate P^{III}-CF$_3$ compounds have been prepared from P-F,[129] P-CN[130] or P-OAr[131–133] precursors. Hexafluorocyclotriphosphazene reacted with TMSCF$_3$ in the presence of catalytic CsF to afford hexakis(trifluoromethyl)cyclotriphosphazene in 90% yield[134] (eq 27).

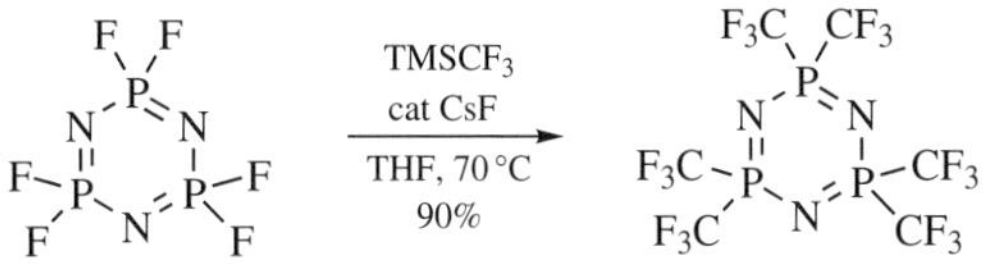

Boron Compounds. KCF$_3$BF$_3$ has been prepared in 85% yield by reaction of trimethoxyborane with TMSCF$_3$ and KF and subsequent treatment of the KCF$_3$B(OMe)$_3$ intermediate with aqueous HF.[135] Other trifluoromethyltrialkoxyborate salts were obtained in analogous manner and transformed into trifluoromethylboronic esters by reaction with suitable electrophiles.[136]

Tin Compounds. Bis(tributyltin) oxide has been transformed into n-Bu$_3$SnCF$_3$ in quantitative yield by reaction with TMSCF$_3$ and TBAF.[137]

Organometallic Compounds. Several trifluoromethyl organometallic complexes have been prepared by halogen-CF$_3$ exchange mediated by TMSCF$_3$. Those include titanium(IV),[138] rhodium(I)[139] and palladium(II) complexes[140–142] (eq 28). In contrast, the corresponding ruthenium(II) and osmium(II) compounds isomerized to difluorocarbene complexes[143,144] (eq 29). Trifluoromethylcopper(I) and gold(I) or (III) compounds were obtained from the corresponding alkoxy- or aryloxymetal complexes.[145]

$$ (28) $$

$$ (29) $$

L = t-Bu$_2$MeP

Silylation Reactions. TMSCF$_3$ and a fluoride anion have been used to protect terminal alkynes effectively with a TMS group.[146,147] This is due to the high nucleophilicity of the trifluoromethide anion, which abstracts a proton from the acetylene thus forming HCF$_3$ gas. In analogous manner, tertiary alcohols have been protected as their corresponding trimethylsilyl ethers.[148,149]

Related Reagents. trifluoromethylcopper(I); trifluoroiodomethane; (trifluoromethyl)dibenzothiophenium triflate and tetrafluoroborate; (trifluoromethyl)dibenzoselenophenium triflate and tetrafluoroborate.

1. Prakash, G. K. S. In *Synthetic Fluorine Chemistry*; Olah, G. A.; Chambers, R. D.; Prakash, G. K. S., Eds.; Wiley: Chichester, 1992, Chapter 10.

2. Bosmans, J. P., *Janssen Chim. Acta* **1992**, *10*, 22 (*Chem. Abstr.* **1992**, *117*, 7213q).

3. Krishnamurti, R.; Bellew, D. R.; Prakash, G. K. S., *J. Org. Chem.* **1991**, *56*, 984.

4. Stahly, G. P.; Bell, D. R., *J. Org. Chem.* **1989**, *54*, 2873.

5. Urata, H.; Fuchikami, T., *Tetrahedron Lett.* **1991**, *32*, 91.

6. Ramaiah, P.; Krishnamurti, R.; Prakash, G. K. S., *Org. Synth.* **1995**, *72*, 232.

7. Ruppert, I.; Schlich, K.; Volbach, W., *Tetrahedron Lett.* **1984**, *25*, 2195.

8. Pawelke, G., *J. Fluorine Chem.* **1989**, *42*, 429.

9. Eaborn, C.; Griffiths, R. W.; Pidcock, A., *J. Organomet. Chem.* **1982**, *225*, 331.

10. Prakash, G. K. S.; Krishnamurti, R.; Olah, G. A., *J. Am. Chem. Soc.* **1989**, *111*, 393.

11. Coombs, M. M.; Zepik, H. H., *J. Chem. Soc., Chem. Commun.* **1992**, 1376.

12. Skiles, J. W.; Fuchs, V.; Miao, C.; Sorcek, R.; Grozinger, K. G.; Mauldin, S. C.; Vitous, J.; Mui, P. W.; Jacober, S.; Chow, G.; Matteo, M.; Skoog, M.; Weldon, S. M.; Possanza, G.; Keirns, J.; Letts, G.; Rosenthal, A. S., *J. Med. Chem.* **1992**, *35*, 641.

13. Bansal, R. C.; Dean, B.; Hakomori, S-I. Toyokuni, T., *J. Chem. Soc., Chem. Commun.* **1991**, 796.

14. Gassman, P. G.; Ray, J. A.; Wenthold, P. G.; Mickelson, J. W., *J. Org. Chem.* **1991**, *56*, 5143.

15. Kantamneni, S. Ph. D Dissertation, University of Southern California, July 1993.

16. Kotun, S. P.; Anderson, J. D. O.; DesMarteau, D. D., *J. Org. Chem.* **1992**, *57*, 1124.

17. Patel, N. R.; Kirchmeier, R. L., *Inorg. Chem.* **1992**, *31*, 2537.

18. Broicher, V.; Geffken, D., *Tetrahedron Lett.* **1989**, *30*, 5243.

19. Ramaiah, P.; Prakash, G. K. S., *Synlett* **1991**, 643.

20. Broicher, V.; Geffken, D., *Arch. Pharm. (Weinheim, Ger.)* **1990**, *323*, 929.

21. Kolomeitsev, A. A.; Movchun, V. N.; Kondratenko, N. V.; Yagupolskii, Yu. L., *Synthesis* **1990**, 1151.

22. Bardin, V. V.; Kolomeitsev, A. A.; Furin, G. G.; Yagupolskii, Yu. L., *Izv. Akad. Nauk SSSR, Ser. Khim.* **1990**, 1693 (*Chem. Abstr.* **1991**, *115*, 279 503c).

23. Kolomeitsev, A. A.; Movchun, V. N.; Yagupolskii, Yu. L., *Tetrahedron Lett.* **1992**, *33*, 6191.

24. Aymard, F.; Nedelec, J. -Y.; Perichon, J., *Tetrahedron Lett.* **1994**, *35*, 8623.

25. Prakash, G. K. S.; Deffieux, D.; Yudin, A. K.; Olah, G. A., *Synlett* **1994**, 1057.

26. Martynov, B. I.; Stepanov, A. A., *J. Fluorine Chem.* **1997**, *85*, 127.

27. Grobe, J.; Hegge, J., *Synlett* **1995**, 641.

28. Prakash, G. K. S.; Hu, J.; Olah, G. A., *J. Org. Chem.* **2003**, *68*, 4457.

29. Prakash, G. K. S.; Yudin, A. K., *Chem. Rev.* **1997**, *97*, 757.

30. Singh, R. P.; Shreeve, J. M., *Tetrahedron* **2000**, *56*, 7613.

31. Prakash, G. K. S.; Mandal, M., *J. Fluorine Chem.* **2001**, *112*, 123.

32. Klatte, K.; Christen, D.; Merke, I.; Stahl, W.; Oberhammer, H., *J. Phys. Chem. A* **2005**, *109*, 8438.

33. Adams, D. J.; Clark, J. H.; Hansen, L. B.; Sanders, V. C.; Tavener, S. J., *J. Fluorine Chem.* **1998**, *92*, 123.

34. Maggiarosa, N.; Tyrra, W.; Naumann, D.; Kirij, N. V.; Yagupolskii, Yu. L., *Angew. Chem., Int. Ed.* **1999**, *38*, 2252.

35. Tyrra, W.; Kremlev, M. M.; Naumann, D.; Scherer, H.; Schmidt, H.; Hoge, B.; Pantenburg, I.; Yagupolskii, Yu. L., *Chem. Eur. J.* **2005**, *11*, 6514.

36. Kolomeitsev, A.; Bissky, G.; Lork, E.; Movchun, V.; Rusanov, E.; Kirsch, P.; Röschenthaler, G.-V., *Chem. Commun.* **1999**, 1017.

37. Wiedemann, J.; Heiner, T.; Mloston, G.; Prakash, G. K. S.; Olah, G. A., *Angew. Chem., Int. Ed.* **1998**, *37*, 820.

38. Singh, R. P.; Cao, G.; Kirchmeier, R. L.; Shreeve, J. M., *J. Org. Chem.* **1999**, *64*, 2873.

39. Busch-Petersen, J.; Bo, X.; Corey, E. J., *Tetrahedron Lett.* **1999**, *40*, 2065.

40. Kim, J.; Shreeve, J. M., *Org. Biomol. Chem.* **2004**, *2*, 2728.

41. Baxendale, I. R.; Ley, S. V.; Lumeras, W.; Nesi, M., *Comb. Chem. High Throughput Screening* **2002**, *5*, 197.

42. Babadzhanova, L. A.; Kirij, N. V.; Yagupolskii, Yu. L.; Tyrra, W.; Naumann, D., *Tetrahedron* **2005**, *61*, 1813.

43. Hagiwara, T.; Kobayashi, T.; Fuchikami, T., *Main Group Chem.* **1997**, *2*, 13.

44. Prakash, G. K. S.; Mandal, M.; Panja, C.; Mathew, T.; Olah, G. A., *J. Fluorine Chem.* **2003**, *123*, 61.

45. Prakash, G. K. S.; Panja, C.; Vaghoo, H.; Surampudi, V.; Kultyshev, R.; Mandal, M.; Rasul, G.; Mathew, T.; Olah, G. A., *J. Org. Chem.* **2006**, *71*, 6806.

46. Mukaiyama, T.; Kawano, Y.; Fujisawa, H., *Chem. Lett.* **2005**, *34*, 88.

47. Kawano, Y.; Kaneko, N.; Mukaiyama, T., *Bull. Chem. Soc. Jpn.* **2006**, *79*, 1133.

48. Mizuta, S.; Shibata, N.; Sato, T.; Fujimoto, H.; Nakamura, S.; Toru, T., *Synlett* **2006**, 267.

49. Mizuta, S.; Shibata, N.; Ogawa, S.; Fujimoto, H.; Nakamura, S.; Toru, T., *Chem. Commun.* **2006**, 2575.

50. Song, J. J.; Tan, Z.; Reeves, J. T.; Gallou, F.; Yee, N. K.; Senanayake, C. H., *Org. Lett.* **2005**, *7*, 2193.

51. Bjerre, J.; Hauch Fenger, T.; Marinescu, L. G.; Bols, M., *Eur. J. Org. Chem.* **2007**, 704.

52. Iwanami, K.; Oriyama, T., *Synlett* **2006**, 112.

53. Ma, J. -A.; Cahard, D., *Chem. Rev.* **2004**, *104*, 6119.

54. Iseki, K.; Nagai, T.; Kobayashi, Y., *Tetrahedron Lett.* **1994**, *35*, 3137.

55. Caron, S.; Do, N. M.; Arpin, P.; Larivée, A., *Synthesis* **2003**, 1693.

56. Mizuta, S.; Shibata, N.; Hibino, M.; Nagano, S.; Nakamura, S.; Toru, T., *Tetrahedron* **2007**, *63*, 8521.

57. Mizuta, S.; Shibata, N.; Akiti, S.; Fujimoto, H.; Nakamura, S.; Toru, T., *Org. Lett.* **2007**, *9*, 3707.

58. Zhao, H.; Qin, B.; Liu, X.; Feng, X., *Tetrahedron* **2007**, *63*, 6822.

59. Nagao, H.; Yamane, Y.; Mukaiyama, T., *Chem. Lett.* **2007**, *36*, 666.

60. Nagao, H.; Kawano, Y.; Mukaiyama, T., *Bull. Chem. Soc. Jpn.* **2007**, *80*, 2406.

61. Kuroki, Y.; Iseki, K., *Tetrahedron Lett.* **1999**, *40*, 8231.

62. Sugimoto, H.; Nakamura, S.; Shibata, Y.; Shibata, N.; Toru, T., *Tetrahedron Lett.* **2006**, *47*, 1337.

63. Pedrosa, R.; Sayalero, S.; Vicente, M., *Tetrahedron* **2006**, *62*, 10400.

64. Pedrosa, R.; Sayalero, S.; Vicente, M.; Maestro, A., *J. Org. Chem.* **2006**, *71*, 2177.

65. Massicot, F.; Monnier-Bennoit, N.; Deka, N.; Plantier-Royon, R.; Portella, C., *J. Org. Chem.* **2007**, *72*, 1174.

66. Enders, D.; Herriger, C., *Eur. J. Org. Chem.* **2007**, 1085.

67. Song, J. J.; Tan, Z.; Xu, J.; Reeves, J. T.; Yee, N. K.; Ramdas, R.; Gallou, F.; Kuzmich, K.; DeLattre, L.; Lee, H.; Feng, X.; Senanayake, C. H., *J. Org. Chem.* **2007**, *72*, 292.

68. Walter, M. W.; Adlington, R. M.; Baldwin, J. E.; Schofield, C. J., *J. Org. Chem.* **1998**, *63*, 5179.

69. Magueur, G.; Crousse, B.; Ourévitch, M.; Bonnet-Delpon, D.; Bégué, J.-P., *J. Fluorine Chem.* **2006**, *127*, 637.

70. Fustero, S.; Albert, L.; Aceña, J. L.; Sanz-Cervera, J. F.; Asensio, A., *Org. Lett.* **2008**, *10*, 605.

71. Félix, C. P.; Khatimi, N.; Laurent, A. J., *Tetrahedron Lett.* **1994**, *35*, 3303.

72. Banks, R. E.; Besheesh, M. K.; Lawrence, N. J.; Tovell, D. J., *J. Fluorine Chem.* **1999**, *97*, 79.

73. Petrov, V. A., *Tetrahedron Lett.* **2000**, *41*, 6959.

74. Prakash, G. K. S.; Mandal, M.; Olah, G. A., *Synlett* **2001**, 77.

75. Kawano, Y.; Fujisawa, H.; Mukaiyama, T., *Chem. Lett.* **2005**, *34*, 422.

76. Prakash, G. K. S.; Mandal, M.; Olah, G. A., *Angew. Chem., Int. Ed.* **2001**, *40*, 589.

77. Prakash, G. K. S.; Mandal, M.; Olah, G. A., *Org. Lett.* **2001**, *3*, 2847.

78. Prakash, G. K. S.; Mandal, M., *J. Am. Chem. Soc.* **2002**, *124*, 6538.

79. Kawano, Y.; Mukaiyama, T., *Chem. Lett.* **2005**, *34*, 894.

80. Nelson, D. W.; Easley, R. A.; Pintea, B. N. V., *Tetrahedron Lett.* **1999**, *40*, 25.

81. Nelson, D. W.; Owens, J.; Hiraldo, D., *J. Org. Chem.* **2001**, *66*, 2572.

82. Blazejewski, J. -C.; Anselmi, E.; Wilmshurst, M. P., *Tetrahedron Lett.* **1999**, *40*, 5475.

83. Prakash, G. K. S.; Mogi, R.; Olah, G. A., *Org. Lett.* **2006**, *8*, 3589.

84. Kirij, N. V.; Babadzhanova, L. A.; Movchun, V. N.; Yagupolskii, Yu. L.; Tyrra, W.; Naumann, D.; Fischer, H. T. M.; Scherer, H., *J. Fluorine Chem.* **2008**, *129*, 14.

85. Dilman, A. D.; Arkhipov, D. E.; Levin, V. V.; Belyakov, P. A.; Korlyukov, A. A.; Struchkova, M. I.; Tartakovsky, V. A., *J. Org. Chem.* **2007**, *72*, 8604.

86. Brigaud, T.; Doussot, P.; Portella, C., *J. Chem. Soc., Chem. Commun.* **1994**, 2117.

87. Portella, C.; Brigaud, T.; Lefebvre, O.; Plantier-Royon, R., *J. Fluorine Chem.* **2000**, *101*, 193.

88. Brigaud, T.; Lefebvre, O.; Plantier-Royon, R.; Portella, C., *Tetrahedron Lett.* **1996**, *37*, 6115.

89. Berber, H.; Brigaud, T.; Lefebvre, O.; Plantier-Royon, R.; Portella, C., *Chem. Eur. J.* **2001**, *7*, 903.

90. Lefebvre, O.; Brigaud, T.; Portella, C., *Tetrahedron* **1999**, *55*, 7233.

91. Lefebvre, O.; Brigaud, T.; Portella, C., *J. Org. Chem.* **2001**, *66*, 4348.

92. Lefebvre, O.; Brigaud, T.; Portella, C., *Tetrahedron* **1998**, *54*, 5939.

93. Lefebvre, O.; Brigaud, T.; Portella, C., *J. Org. Chem.* **2001**, *66*, 1941.

94. Jonet, S.; Cherouvrier, F.; Brigaud, T.; Portella, C., *Eur. J. Org. Chem.* **2005**, 4304.

95. Saleur, D.; Bouillon, J. -P.; Portella, C., *J. Org. Chem.* **2001**, *66*, 4543.

96. Demir, A. S.; Eymur, S., *J. Org. Chem.* **2007**, *72*, 8527.

97. Mlostoń, G.; Prakash, G. K. S.; Olah, G. A.; Heimgartner, H., *Helv. Chim. Acta* **2002**, *85*, 1644.

98. Large-Radix, S.; Billard, T.; Langlois, B. R., *J. Fluorine Chem.* **2003**, *124*, 147.

99. Singh, R. P.; Shreeve, J. M., *Chem. Commun.* **2002**, 1818.

100. Babadzhanova, L. A.; Kirij, N. V.; Yagupolskii, Yu. L., *J. Fluorine Chem.* **2004**, *125*, 1095.

101. Kirij, N. V.; Yagupolskii, Yu. L.; Petukh, N. V.; Tyrra, W.; Naumann, D., *Tetrahedron Lett.* **2001**, *42*, 8181.

102. Yagupolskii, Yu. L.; Kirij, N. V.; Shevchenko, A. V.; Tyrra, W.; Naumann, D., *Tetrahedron Lett.* **2002**, *43*, 3029.

103. Tyrra, W.; Naumann, D.; Quadt, S.; Buslei, S.; Yagupolskii, Yu. L.; Kremlev, M. M., *J. Fluorine Chem.* **2007**, *128*, 813.

104. Sevenard, D. V.; Kirsch, P.; Röschenthaler, G. -V.; Movchun, V. N.; Kolomeitsev, A. A., *Synlett* **2001**, 379.

105. Sevenard, D. V.; Kolomeitsev, A. A.; Hoge, B.; Lork, E.; Röschenthaler, G.-V., *J. Am. Chem. Soc.* **2003**, *125*, 12366.

106. Singh, R. P.; Kirchmeier, R. L.; Shreeve, J. A., *Org. Lett.* **1999**, *1*, 1047.

107. Prakash, G. K. S.; Tongco, E. C.; Mathew, T.; Vankar, Y. D.; Olah, G. A., *J. Fluorine Chem.* **2000**, *101*, 199.

108. Prakash, G. K. S.; Mandal, M.; Schweizer, S.; Petasis, N. A.; Olah, G. A., *Org. Lett.* **2000**, *2*, 3173.

109. Cleve, A.; Klar, U.; Schwede, W., *J. Fluorine Chem.* **2005**, *126*, 217.

110. Sosnovskikh, V. Ya.; Sevenard, D. V.; Usachev, B. I.; Röschenthaler, G.-V., *Tetrahedron Lett.* **2003**, *44*, 2097.

111. Sosnovskikh, V. Ya.; Usachev, B. I.; Sevenard, D. V.; Röschenthaler, G.-V., *J. Org. Chem.* **2003**, *68*, 7747.

112. Sosnovskikh, V. Ya.; Usachev, B. I.; Sevenard, D. V.; Röschenthaler, G.-V., *J. Fluorine Chem.* **2005**, *126*, 779.

113. Sosnovskikh, V. Ya.; Usachev, B. I.; Permyakov, M. N.; Sevenard, D. V.; Röschenthaler, G.-V., *Russ. Chem. Bull.* **2006**, *55*, 1687.

114. Sevenard, D. V.; Sosnovskikh, V. Ya.; Kolomeitsev, A. A.; Königsmann, M. H.; Röschenthaler, G.-V., *Tetrahedron Lett.* **2003**, *44*, 7623.

115. Cottet, F.; Schlosser, M., *Eur. J. Org. Chem.* **2002**, 327.

116. Cottet, F.; Marull, M.; Lefebvre, O.; Schlosser, M., *Eur. J. Org. Chem.* **2003**, 1559.

117. Adams, D. J.; Clark, J. H.; Hansen, L. B.; Sanders, V. C.; Tavener, S. J., *J. Chem. Soc., Perkin Trans. 1* **1998**, 3081.

118. Adams, D. J.; Clark, J. H.; Heath, P. A.; Hansen, L. B.; Sanders, V. C.; Tavener, S. J., *J. Fluorine Chem.* **2000**, *101*, 187.

119. Surowiec, M.; Makosza, M., *Tetrahedron* **2004**, *60*, 5019.

120. Loska, R.; Majcher, M.; Makosza, M., *J. Org. Chem.* **2007**, *72*, 5574.

121. Eisenberger, P.; Gischig, S.; Togni, A., *Chem. Eur. J.* **2006**, *12*, 2579.

122. Kieltsch, I.; Eisenberger, P.; Togni, A., *Angew. Chem., Int. Ed.* **2007**, *46*, 754.

123. Tyrra, W.; Naumann, D.; Hoge, B.; Yagupolskii, Yu. L., *J. Fluorine Chem.* **2003**, *119*, 101.

124. Tyrra, W.; Naumann, D.; Yagupolskii, Yu. L., *J. Fluorine Chem.* **2003**, *123*, 183.

125. Tyrra, W.; Kirij, N. V.; Naumann, D.; Yagupolskii, Yu. L., *J. Fluorine Chem.* **2004**, *125*, 1437.

126. Movchun, V. N.; Kolomeitsev, A. A.; Yagupolskii, Yu. L., *J. Fluorine Chem.* **1995**, *70*, 255.

127. Billard, T.; Langlois, B. R., *Tetrahedron Lett.* **1996**, *37*, 6865.

128. Billard, T.; Large, S.; Langlois, B. R., *Tetrahedron Lett.* **1997**, *38*, 65.

129. Tworowska, I.; Dąbkowski, W.; Michalski, J., *Angew. Chem., Int. Ed.* **2001**, *40*, 2898.

130. Panne, P.; Naumann, D.; Hoge, B., *J. Fluorine Chem.* **2001**, *112*, 283.

131. Murphy-Jolly, M. B.; Lewis, L. C.; Caffyn, A. J. M., *Chem. Commun.* **2005**, 4479.

132. Adams, J. J.; Lau, A.; Arulsamy, N.; Roddick, D. M., *Inorg. Chem.* **2007**, *46*, 11328.

133. Pavlenko, N. V.; Babadzhanova, L. A.; Gerus, I. I.; Yagupolskii, Yu. L.; Tyrra, W.; Naumann, D., *Eur. J. Inorg. Chem.* **2007**, 1501.

134. Singh, R. P.; Vij, A.; Kirchmeier, R. L.; Shreeve, J. M., *Inorg. Chem.* **2000**, *39*, 375.

135. Molander, G. A.; Hoag, B. P., *Organometallics* **2003**, *22*, 3313.

136. Kolomeitsev, A. A.; Kadyrov, A. A.; Szczepkowska-Sztolcman, J.; Milewska, M.; Koroniak, H.; Bissky, G.; Barten, J. A.; Röschenthaler, G.-V., *Tetrahedron Lett.* **2003**, *44*, 8273.

137. Prakash, G. K. S.; Yudin, A. K.; Deffieux, D.; Olah, G. A., *Synlett* **1996**, 151.

138. Taw, F. L.; Scott, B. L.; Kiplinger, J. L., *J. Am. Chem. Soc.* **2003**, *125*, 14712.

139. Vicente, J.; Gil-Rubio, J.; Guerrero-Leal, J.; Bautista, D., *Organometallics* **2004**, *23*, 4871.

140. Culkin, D. A.; Hartwig, J. F., *Organometallics* **2004**, *23*, 3398.

141. Grushin, V. V.; Marshall, W. J., *J. Am. Chem. Soc.* **2006**, *128*, 4632.

142. Grushin, V. V.; Marshall, W. J., *J. Am. Chem. Soc.* **2006**, *128*, 12644.

143. Huang, D.; Caulton, K. G., *J. Am. Chem. Soc.* **1997**, *119*, 3185.

144. Huang, D.; Koren, P. R.; Folting, K.; Davidson, E. R.; Caulton, K. G., *J. Am. Chem. Soc.* **2000**, *122*, 8916.

145. Usui, Y.; Noma, J.; Hirano, M.; Komiya, S., *Inorg. Chim. Acta* **2000**, *309*, 151.

146. Ishizaki, M.; Hoshino, O., *Tetrahedron* **2000**, *56*, 8813.

147. Yoshimatsu, M.; Kuribayashi, M., *J. Chem. Soc., Perkin Trans. 1* **2001**, 1256.

148. Odinokov, V. N.; Savchenko, R. G.; Nazmeeva, S. R.; Galyautdinov, I. V.; Khalilov, L. M., *Russ. Chem. Bull.* **2002**, *51*, 1937.

149. Odinokov, V. N.; Savchenko, R. G.; Nazmeeva, S. R.; Galyautdinov, I. V., *Russ. Chem. Bull.* **2002**, *51*, 1963.

Triisopropylsilanethiol

[156275-96-6] $C_9H_{22}SSi$ (MW 190.42)

InChI = 1S/C9H22SSi/c1-7(2)11(10,8(3)4)9(5)6/h7-10H,1-6H3

InChIKey = CPKHFNMJTBLKLK-UHFFFAOYSA-N

(convenient synthetic equivalent of H_2S for the synthesis of alkanethiols and unsymmetrical dialkyl sulfides; used as polarity-reversal catalyst in radical reactions)

Alternate Name: HSTIPS, TIPST.
Physical Data: bp 70–73 °C (2 mmHg); *d* 0.887 g cm^{-3}.
Solubility: soluble in most organic solvents.

Form Supplied in: liquid; commercially available.
Analysis of Reagent Purity: IR (neat) 2948, 2870, 2558, 1465, 1386, 1369, 882 cm^{-1}; ^{1}H NMR (CDCl$_3$) δ −0.56 (s, 1H), 1.07 (m, 21H); ^{13}C NMR (CDCl$_3$) δ 13.4, 18.2
Purification: distillation or liquid chromatography (Al$_2$O$_3$ or SiO$_2$).
Handling, Storage, and Precautions: stable to aqueous work-up. Irritating to eyes, respiratory system, and skin.

Use as a Nucleophile: A Synthetic Equivalent of H_2S.
HSTIPS is prepared in quantitative yield (98%) from the reaction of LiSH (readily obtained by reacting H_2S with *n*-BuLi in THF) with TIPSCl at −78 °C. HSTIPS is a convenient synthetic equivalent of H_2S for the synthesis of alkanethiols and unsymmetrical dialkyl sulfides (eq 1).[1]

$$H_2S \xrightarrow[\text{2. TIPSCl}]{\text{1. }n\text{-BuLi}} HSTIPS \xrightarrow[\text{2. R-X}]{\text{1. KH}}$$

RSTIPS → (H$_2$O) → RSH

$$RSTIPS \xrightarrow[\text{or TBAF/THF}]{\text{CsF/DMF}} \qquad (1)$$

RSTIPS → (R′-X) → RSR′

The potassium salt, isolated as a crystalline material (mp > 250 °C, recrystallized from dry toluene) reacts with a variety of primary and secondary alkylating agents (bromides, tosylates) in THF at 25 °C to give the corresponding RSTIPS derivatives in good yields (73–95%).

The TIPS group imparts a unique stability to silyl sulfides. They present a remarkable resistance toward hydrolysis. The TIPS group can be removed under mild conditions. Thus, treatment with CsF/DMF or TBAF/THF at room temperature followed by hydrolysis affords thiols. Alternatively, the intermediate thiolate can be alkylated in situ with an alkyl halide such as allyl bromide to provide the corresponding sulfide (eq 2).

$$\text{PhCH}_2\text{STIPS} \xrightarrow{\text{1. TBAF/THF}} \begin{cases} \text{2. H}_2\text{O} \quad (73\%) \rightarrow \text{PhCH}_2\text{SH} \\ \text{2. C}_3\text{H}_5\text{Br} \quad (76\%) \rightarrow \text{PhCH}_2\text{S-allyl} \end{cases} \qquad (2)$$

The intermediate RSTIPS derivative in eq 3, prepared by treating the corresponding chloride in the presence of TBAI with NaSTIPS, has been alkylated in 94% yield with bromoacetonitrile.[2]

The completely regio- and stereospecific ring-opening of epoxides can be achieved with HSTIPS/DBU (1 equiv of each reagent with respect to the epoxide).[3] The process is governed by steric interactions, the selectivity is far superior to the use of triphenylsilanethiol/Et$_3$N.[4] This is also the case for styrene oxide, which undergoes exclusive ring-opening from the less-hindered position. Furthermore, accompanying the ring-opening, the TIPS group migrates to give the *O*-TIPS-protected β-hydroxyalkylthiols exclusively (the related process is inefficient with triphenylsily-thiolates[4]). This reaction allows chiral synthons to be built from chiral epoxides (eq 4).[3]

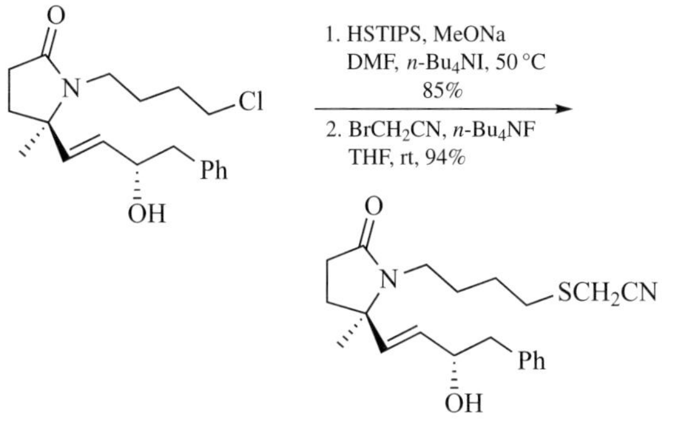

(3)

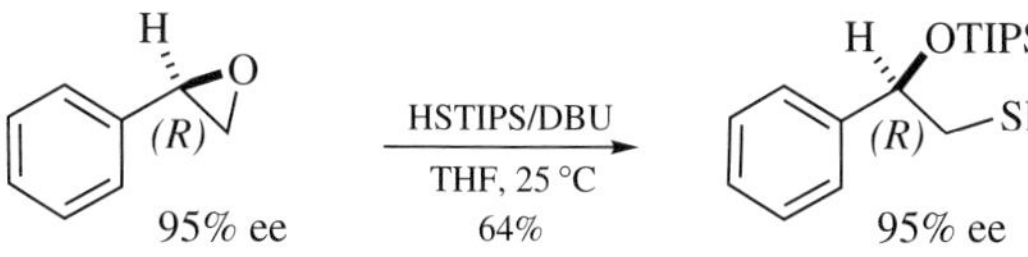

(4)

As shown in eq 5, ring-opening of 1,2-epoxybutane with KSTIPS results in the formation of the corresponding O-TIPS-protected potassium thiolate which is readily alkylated or acylated.

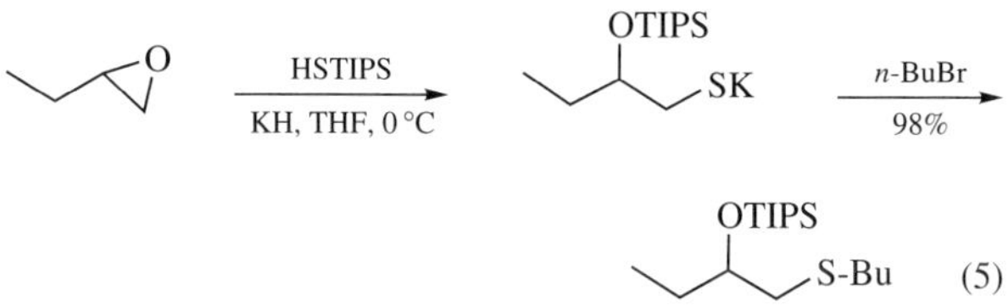

(5)

The Mitsunobu reaction (PPh$_3$/DEAD) in the presence of HSTIPS has been used to convert the enediyne primary alcohol in eq 6 to the protected sulfide, which is readily converted upon treatment with CsF/DMF and PhthNSSMe to the corresponding disulfide.[5]

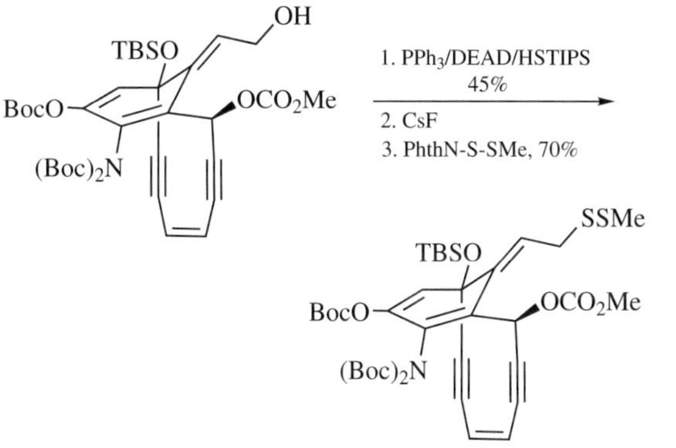

(6)

Synthesis of 9-BBN-derived Alkyl and Aryl Boranes. (TIPS)S-9-BBN is easily available from 9-BBN-H and HSTIPS (1:1, 135–150 °C (12 h), 85%, bp 143–146 °C, 0.3 Torr). It provides a useful entry, through its reaction with organolithiums or Grignards, to R-9-BBN derivatives which cannot be prepared via hydroboration (eq 7).[6] With the exception of vinyl-MgBr, the formation of a 1:1 adduct proceeds readily. Simple heating of this adduct under vacuum liberates the corresponding borane,

which distills in pure form from the mixture. Acidification of the residue with aq HCl regenerates HSTIPS contaminated with a minor amount of HOTIPS ($\sim$ 3%). The unique ability of the STIPS group to stabilize the intermediate adduct is critical for the success of the procedure.

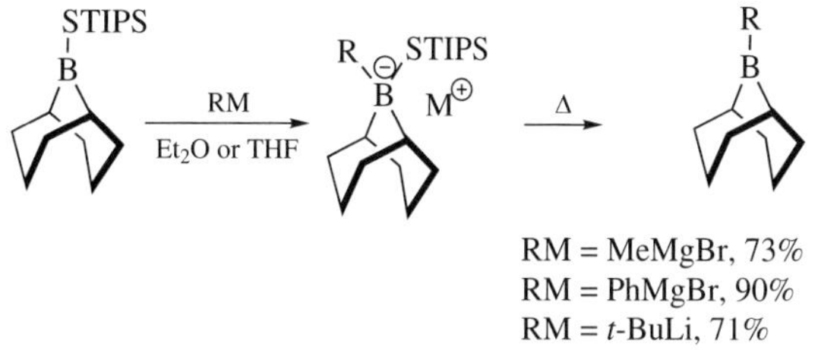

(7)

RM = MeMgBr, 73%
RM = PhMgBr, 90%
RM = t-BuLi, 71%

Palladium-catalyzed Cross-coupling Reactions. The Pd[0]-catalyzed cross-coupling of vinyl and aryl halides with KSTIPS affords the corresponding silyl sulfides, which can be used either to prepare thiols or sulfides.[7] The reaction proceeds with complete retention of configuration in the case of vinyl derivatives (eq 8). The yields are good to excellent (61–93%) when the reaction is carried out on a small scale (2 mmol).

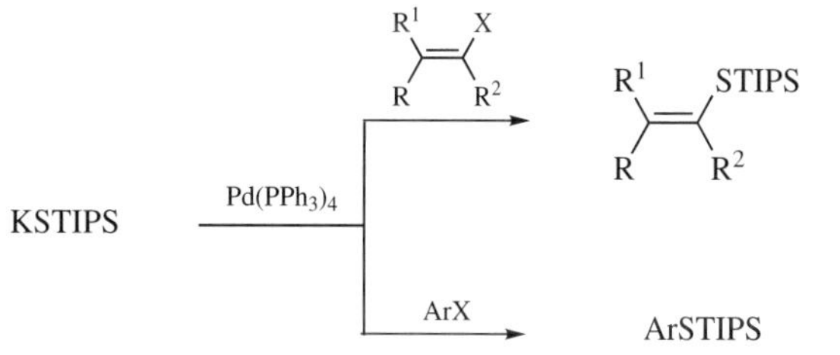

(8)

As exemplified in eq 9, the reaction of the appropriate aryl halide with KSTIPS, followed by treatment with CsF and subsequent hydrolysis, affords 1-methyl-5-indolethiol.[8]

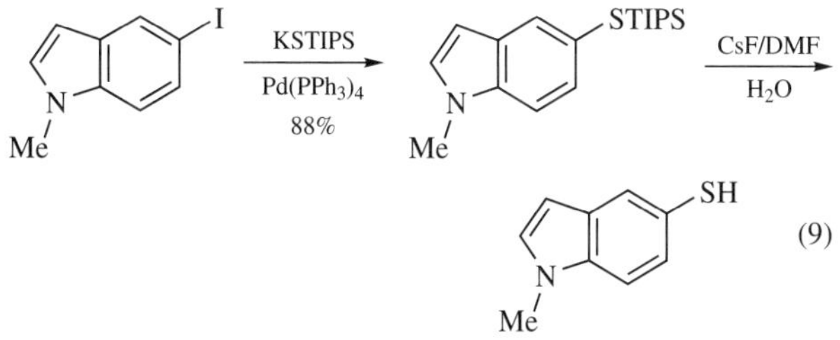

(9)

Conditions using Pd(OAc)$_2$/1,1′-bis(diisopropylphosphino)-ferrocene (DiPPF) as the catalyst are also very efficient for the coupling of aryl bromides and chlorides with a variety of aliphatic thiols including HSTIPS (eq 10).[9]

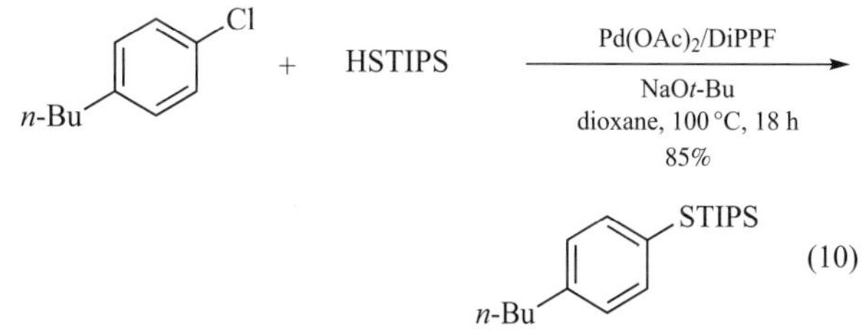

(10)

Bis(triisopropylsilyl) disulfide ((TIPSS)$_2$) is easily prepared as a bright yellow solid (mp 38 °C) from HSTIPS via the oxidation of its sodium salt (eq 11).[10,11]

$$\text{HSTIPS} \quad \xrightarrow[\text{2. I}_2]{\text{1. NaH}} \quad \text{(TIPSS)}_2 \quad (11)$$

This disulfide offers a straightforward entry to Z-1,2-bis-alkyl-sulfanyl-alkenes, which are not available in satisfying yields from aliphatic disulfides. As shown in eq 12, the protected disulfide first adds to the triple bond via a cross-coupling reaction catalyzed by Pd^0. The adduct is converted through deprotection with TBAF to the bis-vinyl thiolate, which reacts with methyl iodide to give the corresponding bis-alkylsulfide (eq 12).[10]

$$R^1 \!\!\!\equiv\!\!\! R^2 \xrightarrow[\text{Pd(PPh}_3)_4/80\,^\circ\text{C}]{\text{(TIPSS)}_2} \left[\begin{matrix} \text{STIPS} \\ R^1 \diagdown \diagup \text{STIPS} \\ R^2 \end{matrix} \right] \xrightarrow[\text{Me-I}]{\text{TBAF}}$$

$$\begin{matrix} \text{SMe} \\ R^1 \diagdown \diagup \text{SMe} \\ R^2 \end{matrix} \quad (12)$$

$$20\text{--}94\%$$

The coupling reaction is performed under N_2, at 80 or 90 °C, in a benzene or a toluene solution of the alkyne (1.0 M) in the presence of a slight excess of (TIPSS)$_2$ and a catalytic amount of Pd(PPh$_3$)$_4$ (5 mol %). The intermediate protected disulfides may be isolated. They can be stored for months at room temperature. Deprotection is effected with TBAF (or CsOAc) in the presence of an excess of an alkylating agent. Terminal alkynes, with the exception of phenylacetylene and t-butylacetylene, give high yields (70–94%). Yields are sensitive to the substitution at the propargylic carbon. Disubstituted alkynes are less reactive. The reaction has been extended to functional alkynes.[11] Unactivated halides, epoxides, acyl chlorides, thiophosgene, or phenyl chlorothionoformates (eq 13) all function efficiently as electrophiles in this reaction.

$$\begin{matrix} \text{STIPS} \\ \diagup \diagdown \text{STIPS} \end{matrix} \xrightarrow[\text{TBAF, 0 °C, toluene}]{\underset{\text{Cl}}{\overset{\text{O}}{\|}}\text{C}\text{SPh}} \quad \text{(product)} \quad (13)$$

$$75\%$$

Use as Ligand in Ziegler-Natta Polymerization of Ethylene in Solution. A medium pressure process for the polymerization of ethylene has been developed in the presence of a catalytic system involving a monocyclopentadienyl titanium species, containing TIPSthiolate as the heteroligand and two activable ligands (2Cl or 2Me), associated with an ionic activator such as triphenylcarbenium tetrakis(pentafluorophenyl)borate.[12]

Use in Radical Reactions. The radical reduction of alkyl halides by silanes is promoted by thiols (eq 14). According to Roberts, the thiol acts as a polarity-reversal catalyst.[13] The direct atom transfer in eq 15 is replaced by the sequence of the more rapid reactions in eqs 16 and 17.

$$\text{RHal} + R_3\text{SiH} \longrightarrow \text{RH} + R_3\text{SiHal} \quad (14)$$

$$R^\cdot + R_3\text{SiH} \longrightarrow \text{RH} + R_3\text{Si}^\cdot \quad (15)$$

$$R^\cdot + \text{XSH} \longrightarrow \text{RH} + \text{XS}^\cdot \quad (16)$$

$$\text{XS}^\cdot + R_3\text{SiH} \rightleftharpoons \text{XSH} + R_3\text{Si}^\cdot \quad (17)$$

Intriguingly, the reduction of xanthates by triphenylsilane (TPS), in the presence of TBHN (t-butylhyponitrite) as initiator at 60 °C, does not require any additional thiol catalyst (eq 18). In all likelihood, triphenylsilanethiol (HSTPS) is formed in the reaction medium via the in situ decomposition of TPSS(C=O)SMe. The latter gives rise to carbonyl sulfide according to eq 19. A subsequent radical chain mechanism explains the formation of the silanethiol that would then serve as the active catalyst. The validity of the proposed mechanism is attested by the synthesis of a series of silanethiols, among which HSTIPS, according to eqs 20–22.[14] HSTIPS is obtained in 74% yield from the reaction under pressure of carbonyl sulfide with triisopropylsilane in solution in dioxane (0.5 M) at 60 °C in the presence of TBHN (5 mol %).

$$\text{ROC(=S)SMe} + \text{Ph}_3\text{SiH} \longrightarrow \text{RH} + \text{Ph}_3\text{SiSC(=O)SMe} \quad (18)$$

$$\text{Ph}_3\text{SiSC(=O)SMe} \longrightarrow \text{Ph}_3\text{SiSMe} + \text{COS} \quad (19)$$

$$(i\text{-Pr})_3\text{Si}^\cdot + \text{COS} \longrightarrow (i\text{-Pr})_3\text{SiS}\dot{\text{C}}\text{=O} \quad (20)$$

$$(i\text{-Pr})_3\text{SiS}\dot{\text{C}}\text{=O} \longrightarrow \text{CO} + (i\text{-Pr})_3\text{SiS}^\cdot \quad (21)$$

$$(i\text{-Pr})_3\text{SiH} + (i\text{-Pr})_3\text{SiS}^\cdot \longrightarrow (i\text{-Pr})_3\text{SiSH} + (i\text{-Pr})_3\text{Si}^\cdot \quad (22)$$

The rate constant for hydrogen abstraction from t-butyl(methyl)phenylsilane by TIPSS$^\cdot$ is $2.1\text{--}2.8 \times 10^5$ M^{-1} s^{-1} at 60 °C. The S–H bond in silanethiol is stronger than in alkanethiols by ca. 4 kJ mol^{-1}.[15]

The use of silanethiols as polarity-reversal catalysts is highly recommended in the addition of nucleophilic carbon-centered radicals to double bonds.[13] This is especially so for the radical-chain addition of aldehydes to alkenes. In this particular reaction, triorganosilanethiols are very efficient catalysts for the addition of butanal to both electron-rich isopropenyl acetate, or to electron-deficient ethyl crotonate. The results obtained with HSTIPS are given in eqs 23 and 24.[16]

$$n\text{-BuCHO} + \begin{matrix}\diagup\diagdown \\ \text{AcO}\end{matrix} \xrightarrow[\substack{\text{dioxane, 60 °C, TBHN (5 mol %)} \\ 76\%}]{\text{HSTIPS (10 mol %)}}$$

$$\underset{n\text{-Bu}}{\overset{\text{O} \quad\quad \text{OAc}}{\diagdown\diagup\diagdown\diagup}} \quad (23)$$

$$n\text{-BuCHO} + \begin{matrix}\diagup\diagdown \\ \text{CO}_2\text{Et}\end{matrix} \xrightarrow[\substack{\text{dioxane, 60 °C, TBHN (5 mol %)} \\ 98\%}]{\text{HSTIPS (10 mol %)}}$$

$$\underset{n\text{-Bu}}{\overset{\text{O}}{\diagdown\diagup\diagdown\diagup}}\text{CO}_2\text{Et} \quad (24)$$

Very good yields are also obtained for the HSTIPS-catalyzed cyclization of (S)-(−)-citronellal in dioxane at 60 °C (eq 25);[16] 10 mol % of HSTIPS is added in two portions, and a 58/42 mixture of menthone and isomenthone is isolated.

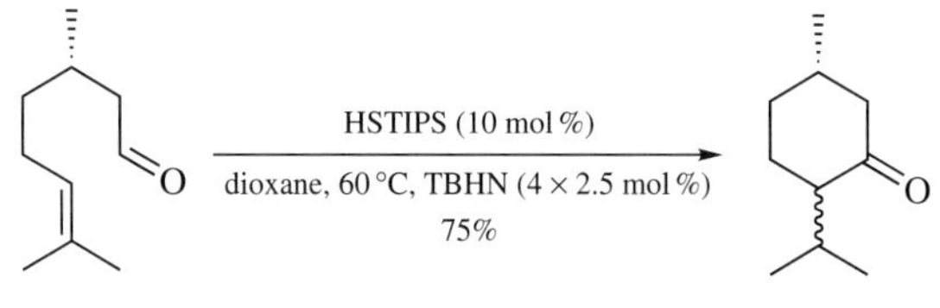

$$(25)$$

md : 58/42

The ring-opening of 2-phenyl-1,3-dioxan-2-yl radicals is also catalyzed by HSTIPS in conjunction with di-*t*-butyl peroxide (DTPB) as initiator,[17] when benzoate esters are formed. The regioselectivity of the reaction is controlled by kinetic factors. A strong preference for the β-scission of the primary C–O bond (91:9) is observed for the *trans*-fused-benzylidene acetal in eq 26.

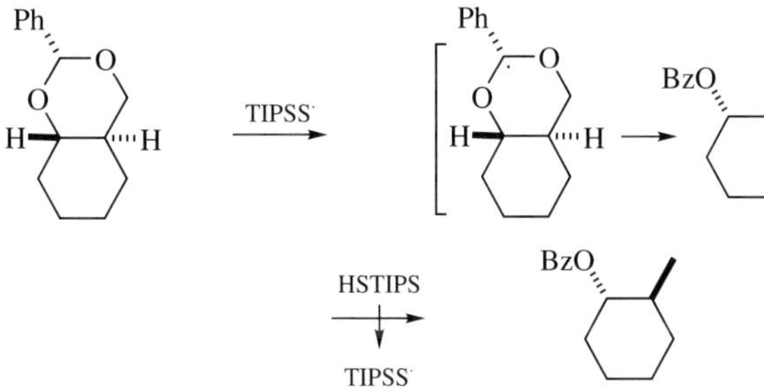

$$(26)$$

91%

(91:9 mixture of isomers)

In contrast, ring-opening of the *cis*-fused analog demonstrates only a slight preference for the formation of the secondary carbon-centered radical (51:49) from the intermediate dioxanyl radical (eq 27).

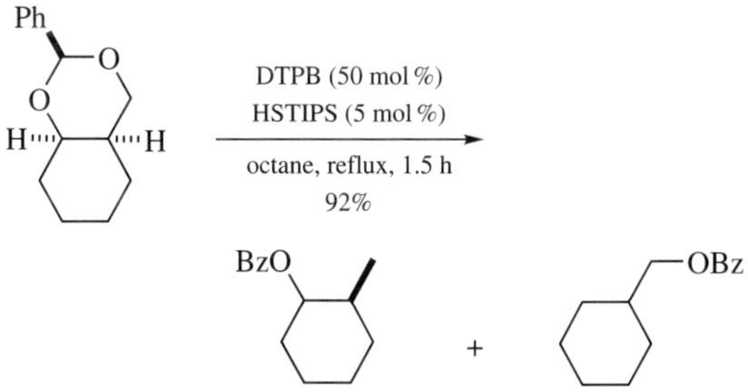

$$(27)$$

49:51

HSTIPS and tri-*t*-butoxysilanethiol are equally effective and robust acidic polarity-reversal catalysts since they are both remarkably stable to hydrolysis. The HSTIPS catalyst is preferred owing to the fact that it is commercially available. The reaction is conducted under inert atmosphere, at reflux in dry octane, in the presence of 5 mol % of the catalyst. It is to be noted that, when DTPB is used, 50 mol % of the initiator is necessary.

HSTIPS has been used to catalyze the addition of silyl radicals to camptothecin.[18] The best results were obtained in dioxane. At 105 °C, the 7-silyl product predominates over the 12-silyl isomer that can be readily separated by flash chromatography on silica gel (CH$_2$Cl$_2$ followed by 5% acetone in CH$_2$Cl$_2$). The detailed mechanism is not clear, in particular, that of the oxidative rearomatization step; the reversibility of the addition of the silyl radical has not been demonstrated.

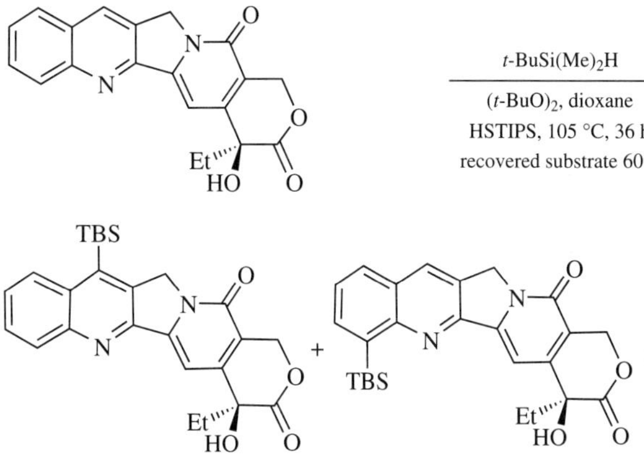

$$(28)$$

23%   11%

As exemplified in eq 28 with *t*-butyldimethylsilane, a series of silanes have been added. Despite the relatively low conversion which is close to 40%, this reaction is the key step in the semisynthesis of silatecans—an important class of lipophilic campothecin analog—which is shorter and higher yielding than their total synthesis through the radical annulation approach.[19]

1. Miranda, E. I.; Diaz, M. J.; Rosado, I.; Soderquist, J. A., *Tetrahedron Lett.* **1994**, *35*, 3221.
2. Billot, X.; Chateauneuf, A.; Chauret, N.; Denis, D.; Greig, G.; Mathieu, M.-C.; Metters, K. M.; Slipetz, D. M.; Young, R. N., *Bioorg. Med. Chem. Lett.* **2003**, *13*, 1129.
3. de Pomar, J. C. J.; Soderquist, J. A., *Tetrahedron Lett.* **1998**, *39*, 4409.
4. Brittain, J.; Gareau, Y., *Tetrahedron Lett.* **1993**, *34*, 3363.
5. Magnus, P.; Miknis, G. F.; Press, N. J.; Grandjean, D.; Taylor, G. M.; Harling, J., *J. Am. Chem. Soc.* **1997**, *119*, 6739.
6. Soderquist, J. A.; de Pomar, J. C. J., *Tetrahedron Lett.* **2000**, *41*, 3537.
7. Rane, A. M.; Miranda, E. I.; Soderquist, J. A., *Tetrahedron Lett.* **1994**, *35*, 3225.
8. Winn, M.; Reilly, E. B.; Liu, G.; Huth, J. R.; Jae, H.-S.; Freeman, J.; Pei, Z.; Xin, Z.; Lynch, J.; Kester, J.; von Geldern, T. W.; Leitza, S.; De Vries, P.; Dickinson, R.; Mussatto, D.; Okasinski, G. F., *J. Med. Chem.* **2001**, *44*, 4393.
9. Murata, M.; Buchwald, S. L., *Tetrahedron* **2004**, *60*, 7397.
10. Gareau, Y.; Orellana, A., *Synlett* **1997**, 803.
11. Gareau, Y.; Tremblay, M.; Gauvreau, D.; Juteau, H., *Tetrahedron* **2001**, *57*, 5739.
12. Gao, X.; Wang, Q.; Von Haken Spence, R. E.; Brown, S. J.; Zoricak, P., *PCT Int. Application*, WO 9940130 (1999) .
13. Roberts, B. P., *Chem. Soc. Rev.* **1999**, *28*, 25.
14. Cai, Y.; Roberts, B. P., *Tetrahedron Lett.* **2001**, *42*, 763.
15. Cai, Y.; Roberts, B. P., *J. Chem. Soc., Perkin Trans. 2* **2002**, 1858.
16. Dang, H.-S.; Roberts, B. P., *J. Chem. Soc., Perkin Trans. 1* **1998**, 67.
17. Cai, Y.; Dang, H.-S, Roberts, B. P., *J. Chem. Soc., Perkin Trans. 1* **2002**, 2449.
18. Du, W.; Kaskar, B.; Blumberg, P.; Subramanian, P.-K.; Curran, D. P., *Bioorg. Med. Chem.* **2003**, *11*, 451.
19. Josien, H.; Bom, D.; Curran, D. P.; Zheng, Y.-H.; Chou, T. C., *Bioorg. Med. Chem. Lett.* **1997**, *7*, 3189.

Laurence Feray & Michèle P. Bertrand
Université Paul Cézanne: Aix-Marseille III, Marseille, France

(Triisopropylsilyl)acetylene

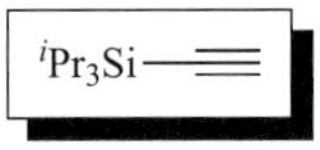

[89343-06-6] $C_{11}H_{22}Si$ (MW 182.38)

InChI = 1S/C11H22Si/c1-8-12(9(2)3,10(4)5)11(6)7/h1,9-11H,2-7H3

InChIKey = KZGWPHUWNWRTEP-UHFFFAOYSA-N

(silyl-protected acetylene often used in transition metal-catalyzed C–C bond formation reactions)

Alternate Name: TIPS-acetylene.

Physical Data: colorless clear liquid; bp 50–52 °C at 0.8 hPa (0.6 mmHg), flash point 56 °C (closed cup); $d = 0.813$.

Solubility: soluble in most organic solvents; unknown water solubility.

Handling, Storage, and Precautions: stable to air and moisture; incompatible with strong acids, bases, or oxidants; flammable vapor and liquid; skin/eye contact and inhalation should be avoided.

General. (Triisopropylsilyl)acetylene (TIPS-acetylene) is an easily handled and inexpensive monoprotected acetylene used as an attractive substitute for trimethylsilylacetylene (TMS-acetylene). The bulkier silyl protecting group of TIPS-acetylene provides stability in a wider range of reaction conditions than TMS-acetylene. Its higher boiling point also provides better handling and safety than TMS-acetylene (bp 87–88 °C at 12 hPa (9 mmHg)). The general utility of TIPS-acetylene is often highlighted in the transition metal-catalyzed C–C bond formations.

Transition Metal-catalyzed Coupling Reactions. TIPS-acetylene has been widely used for transition metal-catalyzed coupling reactions with a variety of aryl-X and vinyl-X (X = I, Br, Cl, and OTf) compounds as useful electrophiles in the presence of proper cocatalysts or additives, leading to the corresponding ethynyl-functionalized derivatives.

Treatment of TIPS-acetylene with aryl (eq 1)[1] and vinyl halides (eqs 2 and 3)[2] and triflates (eq 4)[3] and an appropriate Pd(0) catalyst (Sonogashira coupling) in the presence of copper(I) as cocatalyst affords arylalkynes and enynes in good yield.

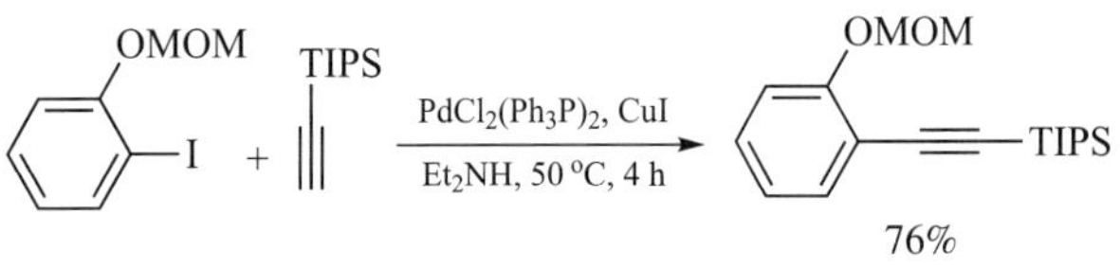

$$\text{(1)}$$

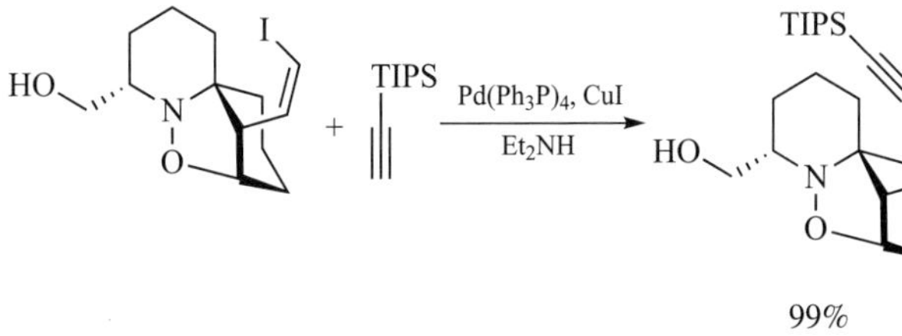

76%

$$\text{(2)}$$

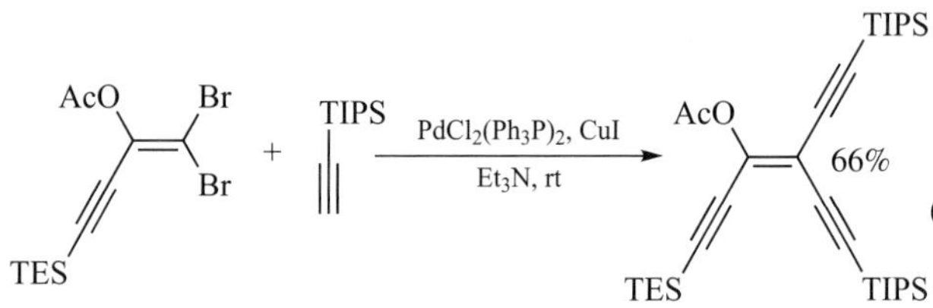

99%

TIPS-acetylene can also be used in conjunction with an Fe(III) catalyst to effect cross-coupling reactions. Nakamura and coworkers have achieved a new Fe(III)-catalyzed enyne cross-coupling of alkynyl Grignard reagent with alkenyl bromides and triflates in the presence of LiBr as a crucial additive to provide the corresponding conjugated enyne in high yield (eq 5).[4] It is useful to note that this new protocol does not require expensive transition metals and ligands.

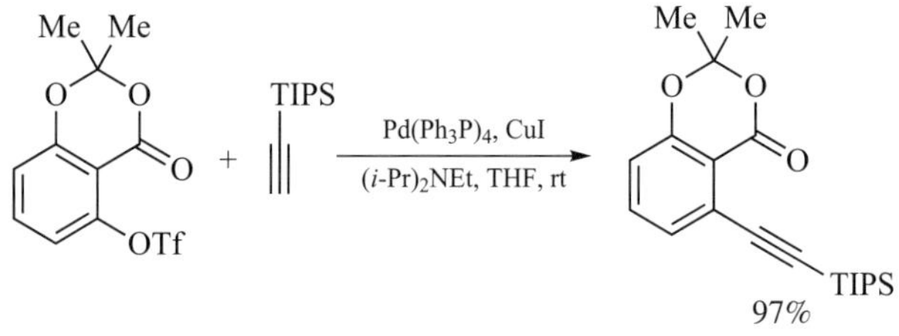

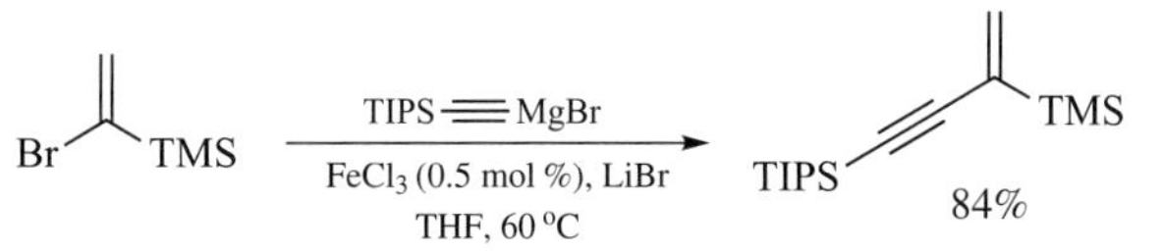

$$\text{(3)}$$

66%

$$\text{(4)}$$

97%

$$\text{(5)}$$

84%

The difference in reactivities of aryl halide (aryl-X, X = I, Br, Cl) bonds can allow a selective coupling protocol for specific aryl halides (eq 6),[5] leading to orthogonal functionalization of aryl halides.

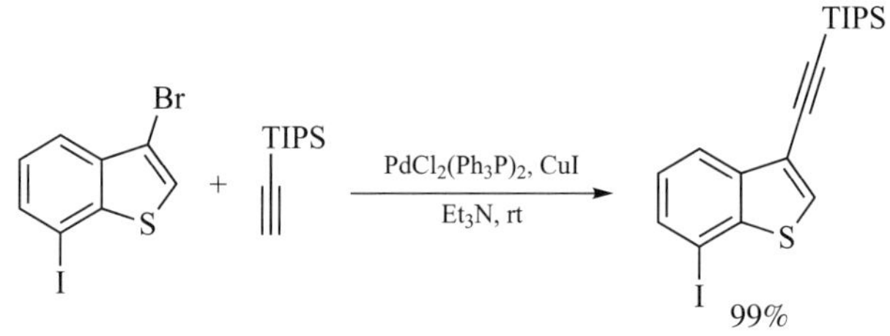

$$\text{(6)}$$

99%

In general, the electrophilic coupling partners for Sonogashira reactions are limited to aryl and vinyl halides and triflates due to the propensity of unproductive β-H-elimination of metal-alkyl intermediates that are derived from β-H-containing alkyl halides through oxidative coupling. Hu and coworkers have developed the first Ni-based method for Sonogashira coupling of nonactivated alkyl halides to generate highly functionalized substituted alkynes in comparable yield by reactions of alkyl halides with acetylide anions (eq 7).[6]

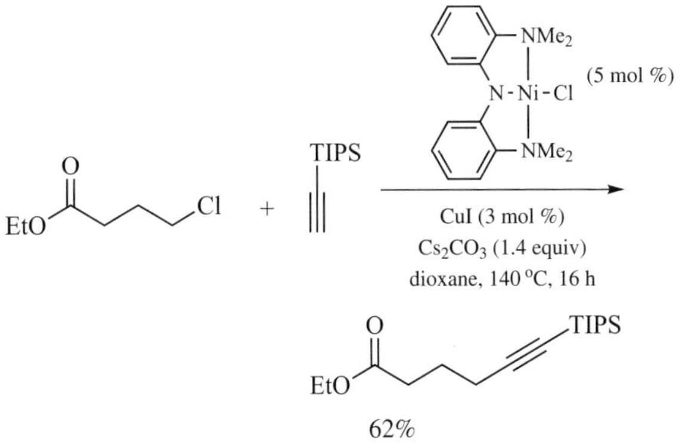

(7)

62%

Reaction of TIPS-acetylide with Electrophiles. Lithium and zinc (triisopropylsilyl)acetylide are easily generated from TIPS-acetylene through the deprotonation by *n*-butyllithium[7] and Et_2Zn[8] at low temperature, respectively. The corresponding Grignard reagent can be prepared from (triisopropylsilyl)acetylide and methylmagnesium bromide.[9] A cerate derivative has also been described.[10] All of these acetylides react with aldehydes (eq 8)[7a–c] and ketones (eq 9),[7d,e,10] esters (eq 10),[7f] amides (eq 11),[7g] and imines (eq 12)[9] to generate alcohols, ketones, and amines, respectively.

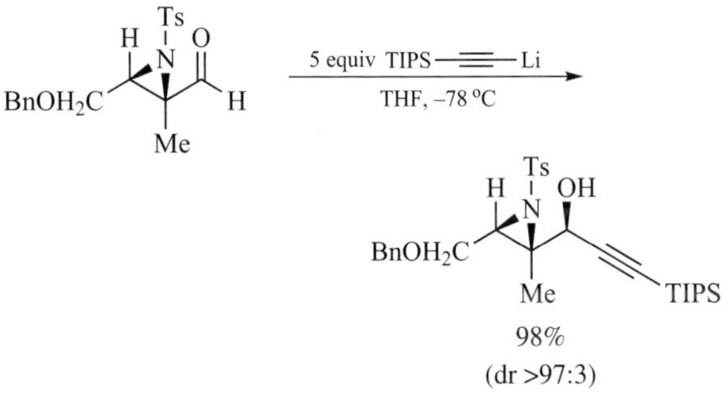

(8)

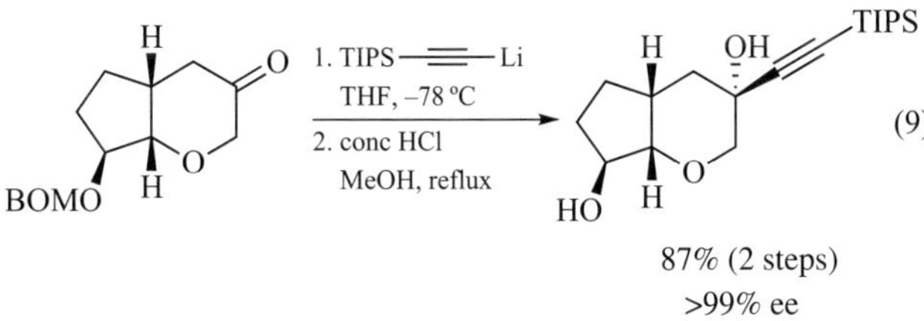

98%
(dr >97:3)

(9)

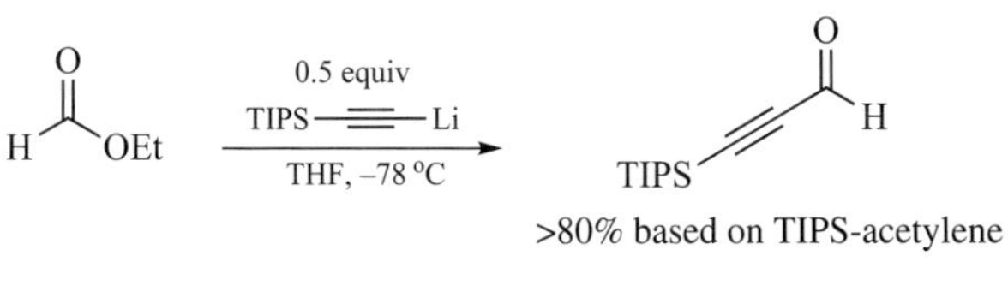

87% (2 steps)
>99% ee

(10)

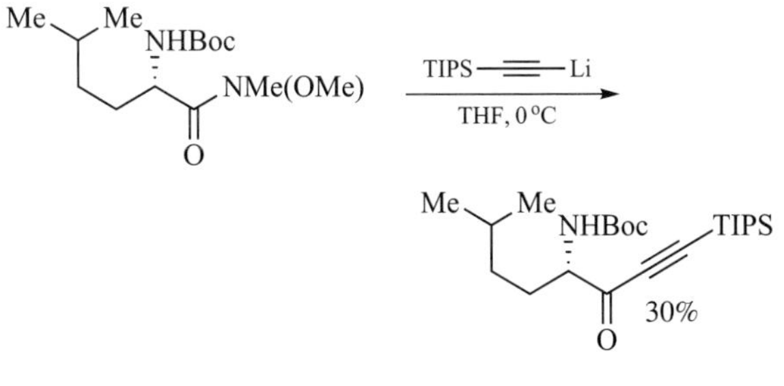

>80% based on TIPS-acetylene

(11)

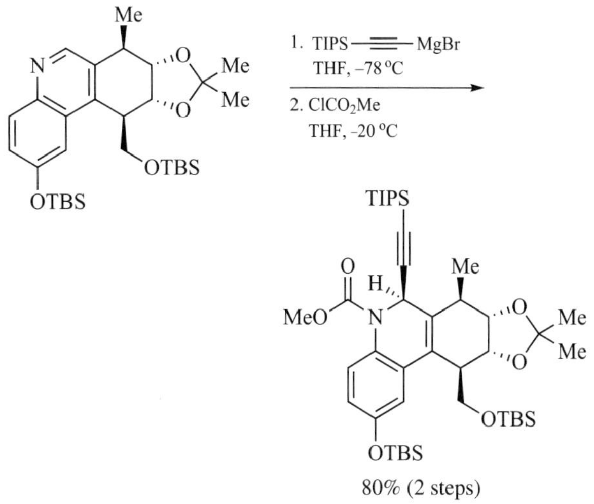

30%

(12)

80% (2 steps)

An enantioselective 1,2-addition of zinc acetylide to aryl aldehydes employing a catalytic amount of chiral BINOL ligand in combination with $Ti(Oi\text{-}Pr)_4$ is high yielding and allows asymmetric construction of chiral propargylic alcohols with excellent stereocontrol (eqs 13 and 14).[8a,b]

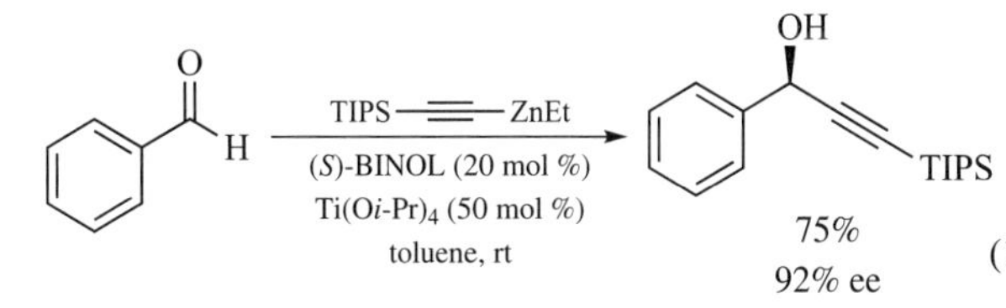

(13)

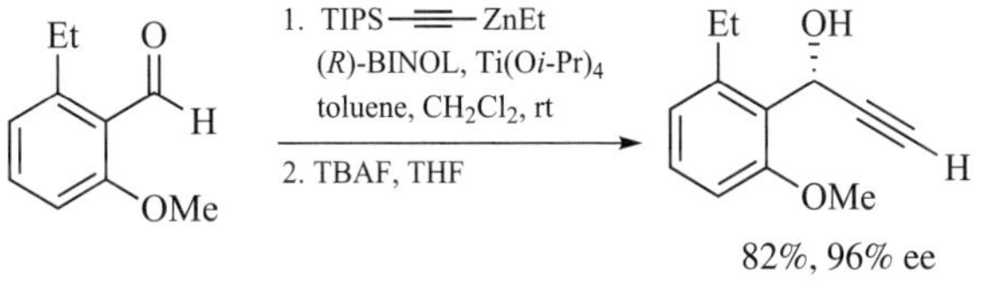

75%
92% ee

(14)

82%, 96% ee

Synthesis of Polyynes. Conjugated diyne and polyyne units, which exhibit unusual electrochemical, optical, and structural properties, can be efficiently constructed from 1-haloalkynes via homologation by one acetylene unit with TIPS-acetylene through a transition metal-catalyzed cross-coupling reaction,[11,12] which allows an access to a new class of polyyne framework.

The copper-catalyzed cross-coupling of TIPS-acetylene with bis iodo 1,6-heptadiyne generated a triisopropylsilyl-capped tetrayne in good yield (eq 15).[11]

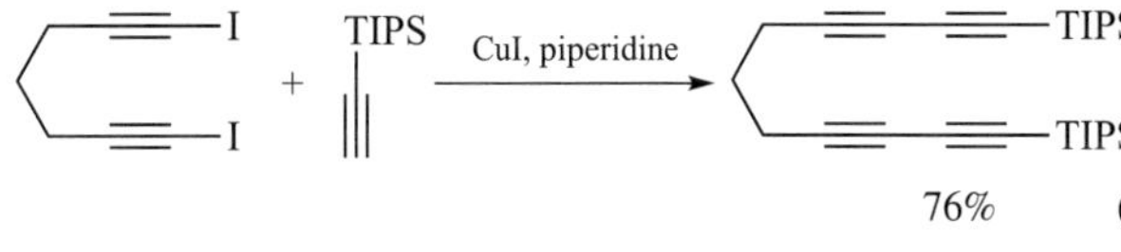

(15)

76%

Kim and coworkers have developed a novel iterative strategy for the synthesis of unsymmetrically substituted polyynes from bromoalkynes through palladium-catalyzed cross-coupling with TIPS-acetylene and a subsequent in situ one-pot AgF-mediated desilylative bromination in good yield (eq 16).[12a] This strategy enables an efficient preparation of different polyyne-containing sphingoid base analogs (eq 17) as well as a facile solid-phase synthetic pathway to generate a library of natural product-like polyynes difficult to obtain by other methods.[12b,c]

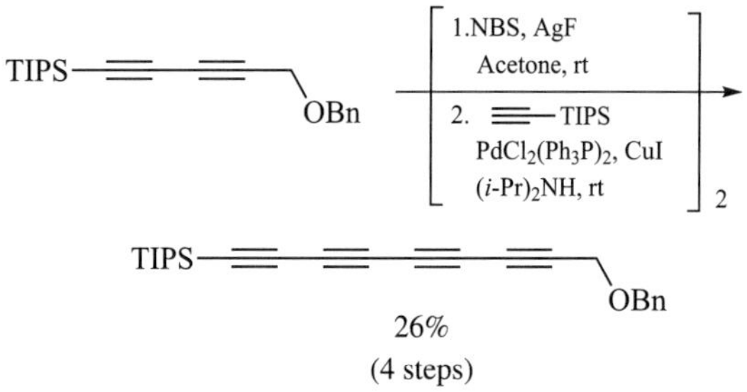

$$(16)$$

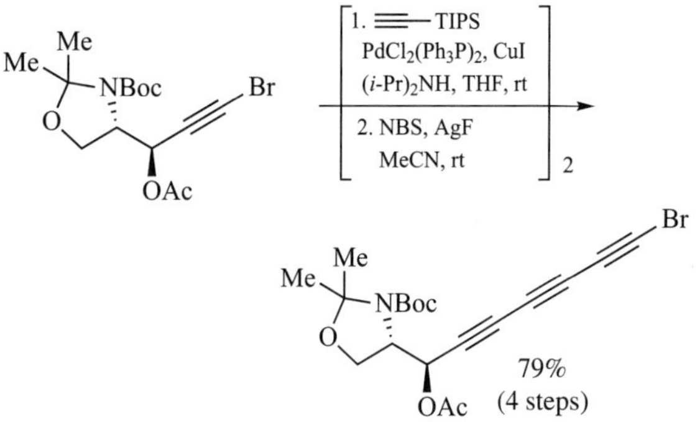

26%
(4 steps)

$$(17)$$

79%
(4 steps)

$$(20)$$

85%
(A/B = 91:9)

Transition Metal-catalyzed Cross-addition of TIPS-acetylene to Alkynes. Transition metal-catalyzed coupling reaction of a bulky silyl-substituted terminal alkyne through the addition of the C–H bond to carbon–carbon triple bonds can lead to a highly selective formation of enynes.[13]

Inoue and coworkers reported that dinuclear and mononuclear palladium complexes possessing N,N'-bis[2-(diphenylphosphino)phenyl]amidate (DPFAM) as a multidentate ligand were effective as catalysts for the selective cross-addition of TIPS-acetylene to a variety of internal and terminal alkynes (eq 18).[13a]

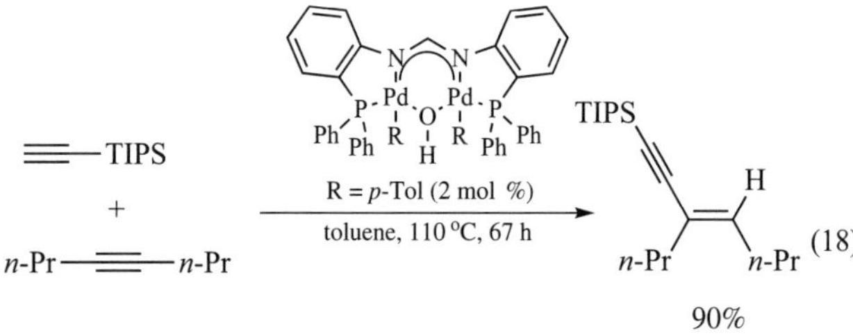

$$(18)$$

90%

Nickel-catalyzed three-component cross-trimerization has also been carried out in a highly chemo-, regio-, and stereoselective fashion. The 1:2 cross-trimerization involving TIPS-acetylene (eq 19)[12b] and two internal alkynes and the 1:1:1 three-component cross-trimerization (eq 20)[12c] of three distinct alkynes in the presence of [Ni(cod)$_2$] and a phosphine ligand afford the corresponding TIPS-substituted 1,3-diene-5-ynes in good yield with a high selectivity.

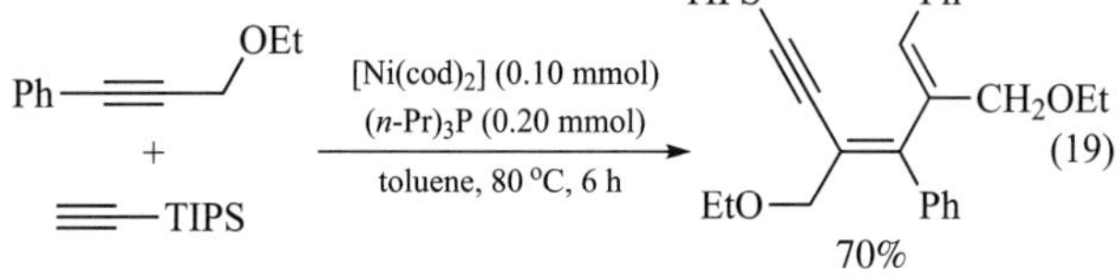

$$(19)$$

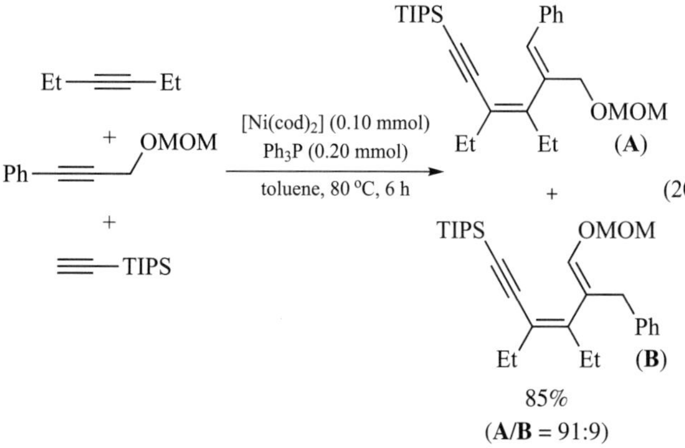

70%

Hydroalkynylation and Its Application. Shirakura and Suginome demonstrated TIPS-acetylene's utility for regioselective hydroalkynylations in combination of Ni(0) catalysts and bulky phosphine ligands on various alkene systems including activated alkenes, 1,3-dienes[14] (eq 21), and methylene cyclopropanes[15] (eq 22), or inactivated alkenes[16] such as styrenes (eq 23) using mild reaction conditions.

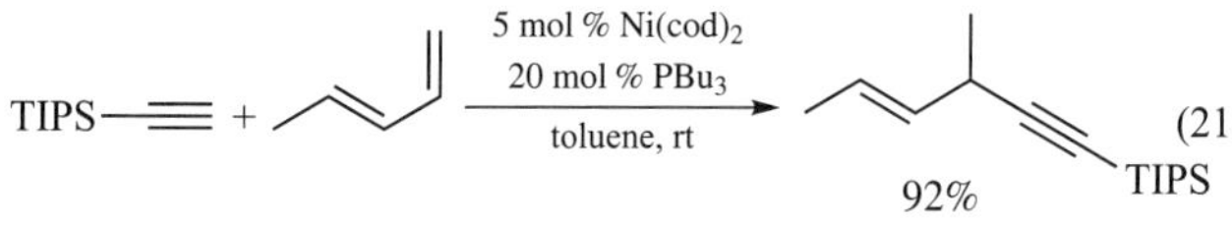

$$(21)$$

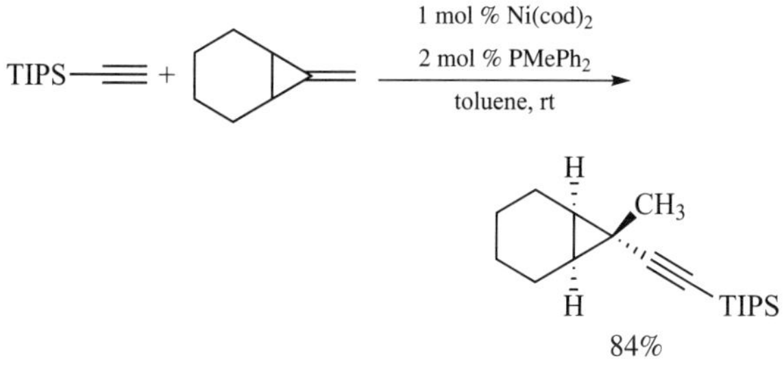

92%

$$(22)$$

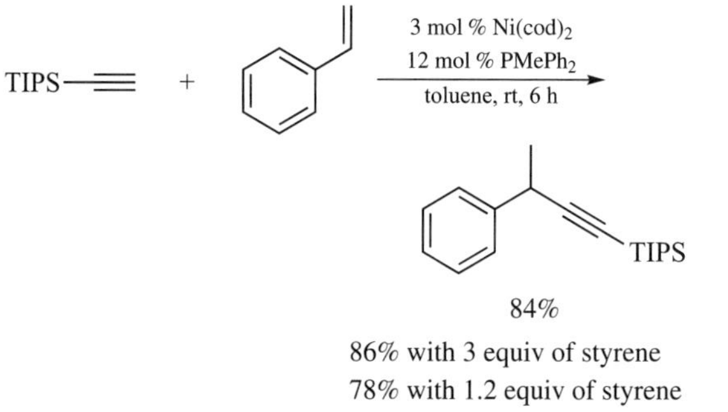

84%

$$(23)$$

84%
86% with 3 equiv of styrene
78% with 1.2 equiv of styrene

For a three-component reaction, Ogata and Fukuzawa demonstrated that one molecule of TIPS-acetylene was hydroalkynylated to a second molecule of TIPS-acetylene via Ni(0) catalysis and the adduct was nicely intercepted by norbornenes to furnish a series of elaborated norbornanes (eq 24).[17]

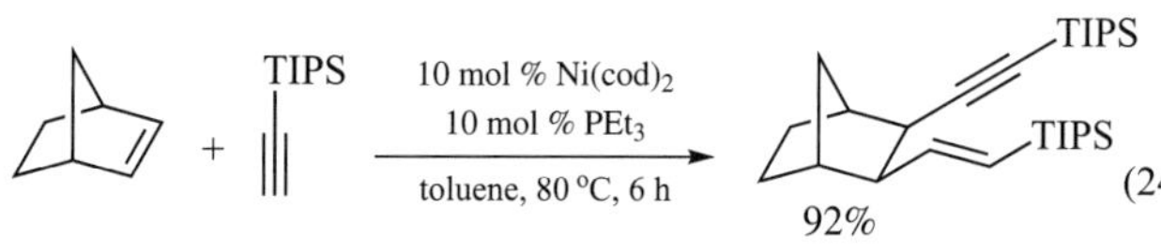

$$(24)$$

92%

Direct Alkynylation. Ni-catalyzed direct alkynylation on azoles has been achieved without assistance of halogen-mediated activation by using oxygen as oxidant in the catalytic cycle. TIPS-acetylene was added to a variety of nonhalogenated azoles to produce the Sonogashira coupling products (eq 25).[18]

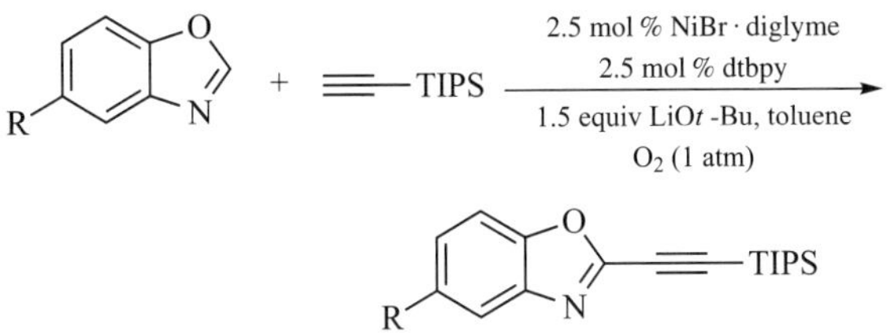

$$R = Me; 49\% (80 \,°C)$$
$$Ph; 57\% (50 \,°C)$$
$$Cl; 61\% (80 \,°C)$$

(25)

For direct alkynylation with TIPS-acetylene, a hypervalent iodine TIPS-acetylene reagent can also be used as demonstrated in Waser's recent works. Alkynylation on C-3 of indole was catalyzed by AuCl (eq 26)[19] and the Wacker cyclization's intermediate was intercepted by the hypervalent iodine TIPS-acetylene (eq 27).[20]

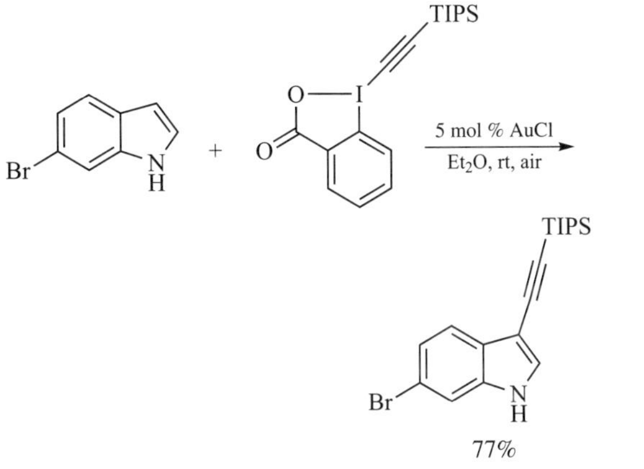

(26)

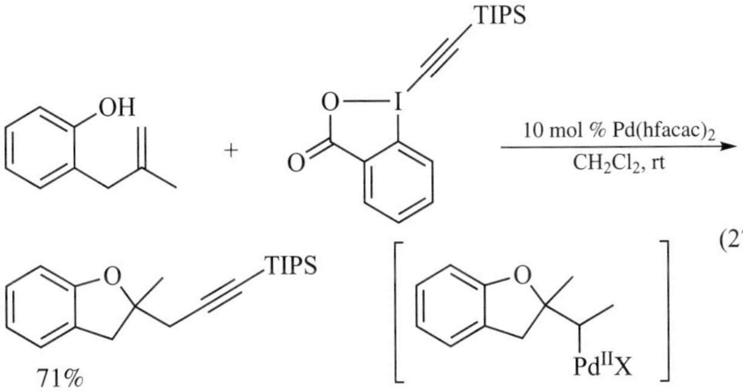

77%

(27)

71%

Conjugate Addition. Catalytic asymmetric conjugate addition of acetylenes to α,β-unsaturated carbonyl compounds is one of the most challenging tasks in organic chemistry and TIPS-acetylene has been used as a useful tool reagent to confirm the viability of the developed protocols. Rh(I)-catalyzed conjugate alkynylation has been extensively developed by Hayashi and coworkers. Asymmetric conjugate addition of TIPS-acetylene was accomplished via a Rh(I) catalysis (eq 28) and the effects of the alkyl substituents on the silylacetylenes were examined for their

efficiency in conjugate addition to 1-phenyl-2-buten-1-one. TIPS-acetylene was the most favorable silylacetylene due to its slow rate of unwanted dimerization (eq 29).[21]

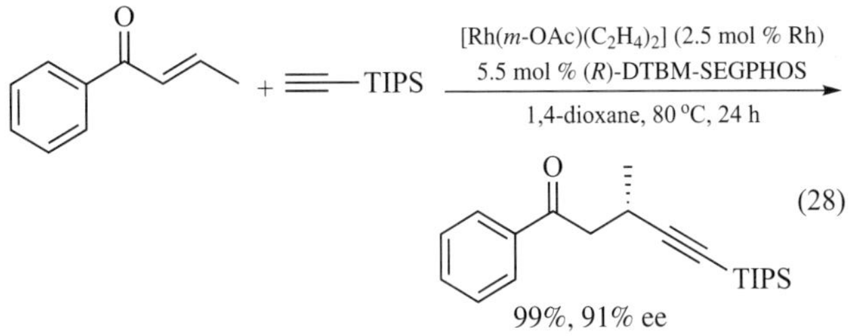

(28)

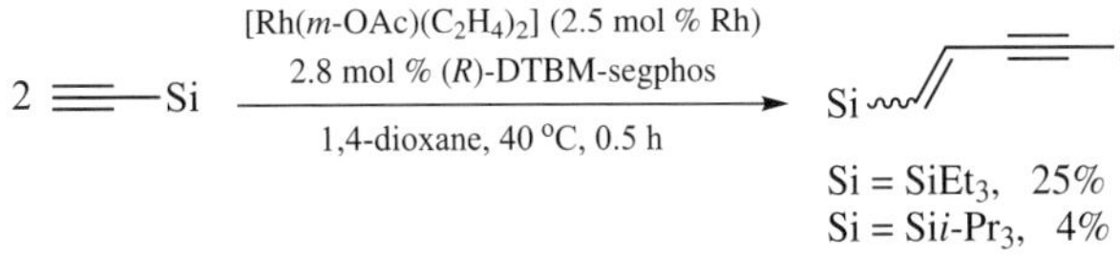

99%, 91% ee

(29)

Si = SiEt₃, 25%
Si = Sii-Pr₃, 4%

The dimerization of TIPS-acetylene in the Rh-catalyzed conjugate addition can be further minimized by employing the TIPS-acetylene elaborated with an alkynylsilanol group and an improved reaction protocol (eq 30).[22] In the same report, this new protocol was applied to conjugate alkynylation of cyclic enones (eq 31).

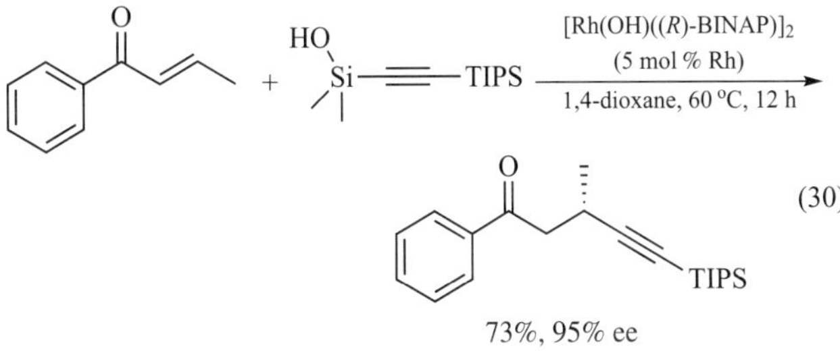

(30)

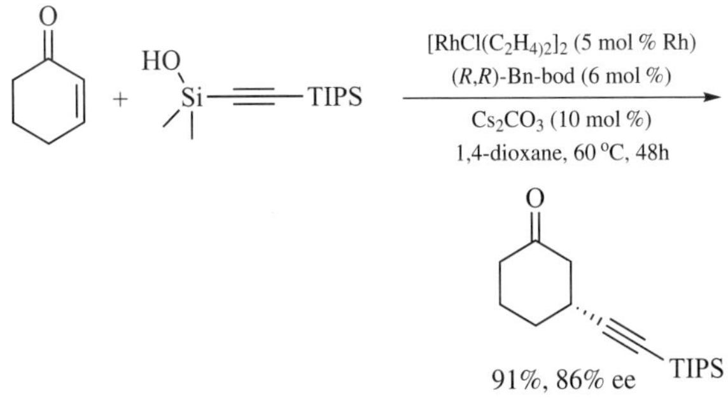

73%, 95% ee

(31)

91%, 86% ee

In a different context, Nishimura, Hayashi and coworkers demonstrated an asymmetric ring-opening alkynylation of aza-benzonorbornadienes using TIPS-acetylene, which was superior to other acetylenes in yield and stereoselectivity (eq 32).[23]

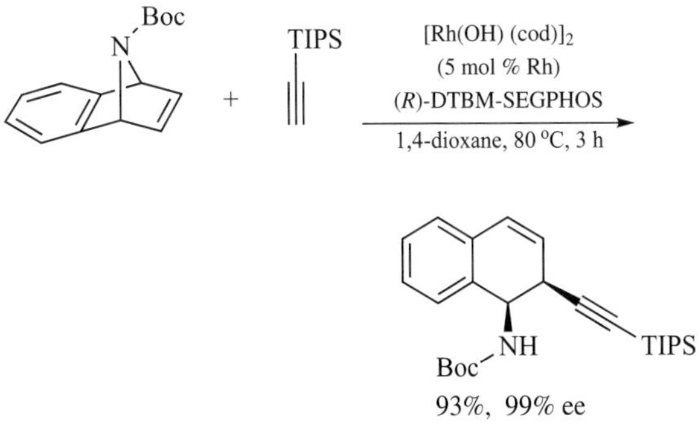

(32)

93%, 99% ee

Cycloaddition. A cobalt-catalyzed system has delivered highly efficient and regioselective Diels–Alder cycloadditions between trialkylsilylacetylenes and 1,3-dienes. TIPS-acetylene was equally effective as other silylacetylenes (TMS-, TES-, TPS-) for the Diels–Alder reaction with isoprene (eq 33).[24]

$$
\begin{array}{c}
\text{1. catalyst A or B} \\
\text{20 mol \% Zn powder} \\
\text{20 mol \% ZnI}_2\text{, CH}_2\text{Cl}_2\text{, rt, 12 h} \\
\text{2. DDQ}
\end{array}
$$

88%
(catalyst A adduct)

91%
(catalyst B adduct)

catalyst A = 10 mol % CoBr₂(mesityl-pyridin-2-yl-methyleneamine)
20 mol % Fe powder
catalyst B = 10 mol % CoBr₂(1,2-diphenylphosphinoethane)

(33)

Ring-opening Reactions. In combination with a Lewis acid, lithium TIPS-acetylide reacts effectively with epoxides to form the desired adducts (eq 34).[25]

$$
\text{BF}_3 \cdot \text{THF, THF} \quad -78\,^\circ\text{C}
$$

(34)

94%

1. (a) Giese, M. W.; Moser, W. H., *Org. Lett.* **2008**, *10*, 4215. (b) Merkul, E.; Boersch, C.; Frank, W.; Müller, T. J. J., *Org. Lett.* **2009**, *11*, 2269. (c) Jin, M.-J.; Lee, D.-H., *Angew. Chem., Int. Ed.* **2010**, *49*, 1119.
2. (a) Macdonald, J. M.; Horsely, H. T.; Ryan, J. H.; Saubern, S.; Holmes, A. B., *Org. Lett.* **2008**, *10*, 4227. (b) Rankin, T.; Tykwinski, R. R., *Org. Lett.* **2003**, *5*, 213.
3. Molander, G. A.; Dehmel, F., *J. Am. Chem. Soc.* **2004**, *126*, 10313.
4. Hatakeyama, T.; Yoshimoto, Y.; Gabriel, T.; Nakamura, M., *Org. Lett.* **2008**, *10*, 5341.
5. Ie, Y.; Hirose, T.; Aso, Y., *J. Mater. Chem.* **2009**, *19*, 8169.
6. Vechorkin, O.; Barmaz, D.; Proust, V.; Hu, X., *J. Am . Chem. Soc.* **2009**, *131*, 12078.
7. (a) Schomaker, J. M.; Geiser, A. R.; Huang, R.; Borhan, B., *J. Am. Chem. Soc.* **2007**, *129*, 3794. (b) Kim, H.; Baker, J. B.; Lee, S.-U.; Park, Y.; Bolduc, K. L.; Park, H.-B.; Dickens, M. G.; Lee, D.-S.; Kim, Y.; Kim, S. H.; Hong, J., *J. Am. Chem. Soc.* **2009**, *131*, 3192. (c) Boukouvalas, J.; Wang, J.-X., *Org. Lett.* **2008**, *10*, 3397. (d) Miyazaki, T.; Yokoshima, S.; Simizu, S.; Osada, H.; Tokuyama, H.; Fukuyama, T., *Org. Lett.* **2007**, *9*, 4737. (e) Chun, D.; Cheng, Y.; Wudl, F., *Angew. Chem., Int. Ed.* **2008**, *47*, 8380. (f) Marti, C.; Carreira, E. M., *J. Am. Chem. Soc.* **2005**, *127*, 11505. (g) Alemany, C.; Bach, J.; Farràs, J.; Garcia, J., *Org. Lett.* **1999**, *1*, 1831.
8. (a) Moore, D.; Pu, L., *Org. Lett.* **2002**, *4*, 1855. (b) Boukouvalas, J.; Pouliot, M.; Robichaud, J.; MacNeil, S.; Snieckus, V., *Org. Lett.* **2006**, *8*, 3597.
9. Shair, M.; Yoon, T. Y.; Mosny, K. K.; Chou, T. C.; Danishefsky, S. J., *J. Am. Chem. Soc.* **1996**, *118*, 9509.
10. Sumi, S.; Matsumoto, K.; Tokuyama, H.; Fukuyama, T., *Org. Lett.* **2003**, *5*, 1891.
11. Matano, Y.; Nakashima, M.; Imahori, H., *Angew. Chem., Int. Ed.* **2009**, *48*, 4002.
12. (a) Kim, S.; Kim, S.; Lee, T.; Ko, H.; Kim, D., *Org. Lett.* **2004**, *6*, 3601. (b) Kim, S.; Lee, Y. M.; Kang, H. R.; Cho, J.; Lee, T.; Kim, D., *Org. Lett.* **2007**, *9*, 2127. (c) Lee, S.; Lee, T. M.; Lee, Y.; Kim, D.; Kim, S., *Angew. Chem., Int. Ed.* **2007**, *46*, 8422.
13. (a) Tsukada, N.; Ninomiya, S.; Aoyama, Y.; Inoue, Y., *Org. Lett.* **2007**, *9*, 2919. (b) Ogata, K.; Murayama, H.; Sugasawa, J.; Suzuki, N.; Fukuzawa, S.-I., *J. Am. Chem. Soc.* **2009**, *131*, 3176. (c) Ogata, K.; Sugasawa, J.; Fukuzawa, S.-I., *Angew. Chem., Int. Ed.* **2009**, *48*, 6078.
14. Shirakura, M.; Suginome, M., *J. Am. Chem. Soc.* **2008**, *130*, 5410.
15. Shirakura, M.; Suginome, M., *J. Am. Chem. Soc.* **2009**, *131*, 5060.
16. Shirakura, M.; Suginome, M., *Org. Lett.* **2009**, *11*, 523.
17. Ogata, K.; Sugasawa, J.; Atsuumi, Y.; Fukuzawa, S., *Org. Lett.* **2010**, *12*, 148.
18. Matsuyama, N.; Kitahara, M.; Hirano, K.; Satoh, T.; Miura, M., *Org. Lett.* **2010**, *12*, 2358.
19. Brand, J. P.; Charpentier, J.; Waser, J., *Angew. Chem., Int. Ed.* **2009**, *48*, 9346.
20. Nicolai, S.; Erard, S.; Gonzalez, D. F.; Waser, J., *Org. Lett.* **2010**, *12*, 384.
21. Nishimura, T.; Guo, X.-X.; Uchiyama, N.; Katoh, T.; Hayashi, T., *J. Am. Chem. Soc.* **2008**, *130*, 1576.
22. Nishimura, T.; Tokuji, S.; Sawano, T.; Hayashi, T., *Org. Lett.* **2009**, *11*, 3222.
23. Nishimura, T.; Tsurumaki, E.; Kawamoto, T.; Hayashi, T., *Org. Lett.* **2008**, *10*, 4057.
24. Hilt, G.; Janikowski, J., *Org. Lett.* **2009**, *11*, 773.
25. Gaunt, M. J.; Hook, D. F.; Tanner, H. R.; Ley, S. V., *Org. Lett.* **2003**, *5*, 4815.

Dongjoo Lee
Dankuk University, Cheonan, Korea

Young-Shin Kwak
Korea Institute of Bioscience and Biotechnology, Ochang, Korea

Sanghee Kim
Seoul National University, Seoul, Korea

Triisopropylsilyl Chloride

i-Pr₃SiCl

[13154-24-0]   C₉H₂₁ClSi   (MW 192.84)
InChI = 1S/C9H21ClSi/c1-7(2)11(10,8(3)4)9(5)6/h7-9H,1-6H3
InChIKey = KQIADDMXRMTWHZ-UHFFFAOYSA-N

(hydroxy protecting group;[1,2] formation of triisopropylsilyl ynol ethers;[3] *N*-protection of pyrroles;[4,5] prevents chelation with Grignard reagents[6])

Alternative Name: TIPSCl; chlorotriisopropylsilane.
Physical Data: bp 198 °C/739 mmHg; *d* 0.901 g cm⁻³.
Solubility: sol THF, DMF, CH₂Cl₂.
Form Supplied in: clear, colorless liquid; commercially available (99% purity).
Analysis of Reagent Purity: bp; NMR.
Purification: distillation under reduced pressure.

Handling, Storage, and Precautions: moisture sensitive; therefore should be stored under an inert atmosphere; corrosive; use in a fume hood.

Original Commentary

Ellen M. Leahy
Affymax Research Institute , Palo Alto, CA, USA

Hydroxy Protecting Group. Several hindered triorganosilyl protecting groups have been developed to mask the hydroxy functionality. Although the *t*-butyldiphenylsilyl (TBDPS) and *t*-butyldimethylsilyl (TBDMS) groups are the most widely used, the triisopropylsilyl group (TIPS) has several properties which make it particularly attractive for use in a multi-step synthesis.[1,2]

Introduction of the TIPS group is most frequently accomplished using TIPSCl and imidazole in DMF,[1] although several other methods exist, including using TIPSCl and 4-dimethylaminopyridine in CH_2Cl_2. It is possible to silylate hydroxy groups selectively in different steric environments. For example, primary alcohols can be silylated in the presence of secondary alcohols (eq 1)[7] and less hindered secondary alcohols can be protected in the presence of more hindered ones (eq 2).[8]

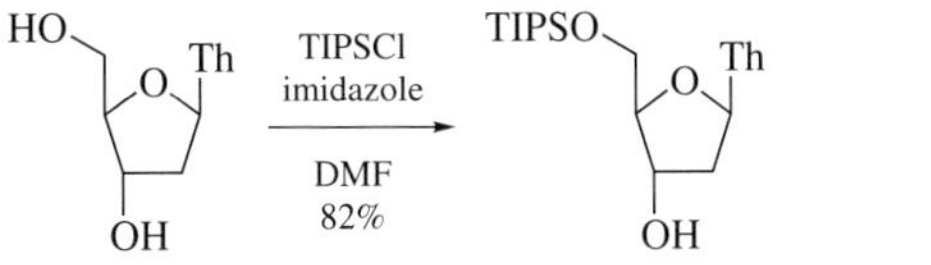

$$(1)$$

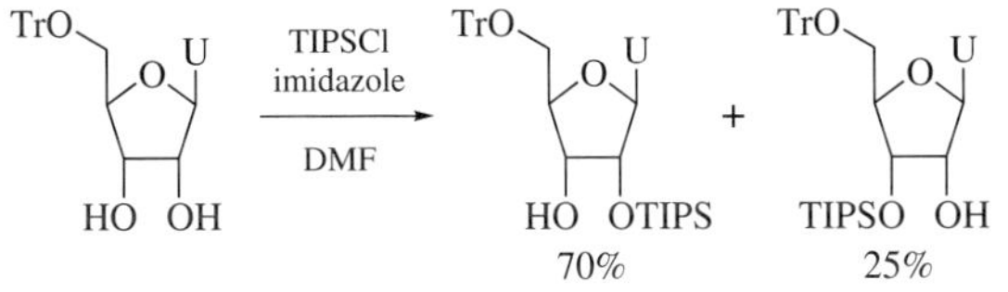

$$(2)$$

A comparison of the stability of different trialkylsilyl groups has shown that TIPS ethers are more stable than TBDMS ethers but less stable than TBDPS ethers toward acid hydrolysis.[2] The rate difference is large enough that a TBDMS group can be removed in the presence of a TIPS group (eq 3).[7] Under basic hydrolysis, TIPS ethers are more stable than TBDMS or TBDPS ethers.[2] The cleavage of TIPS ethers can also be accomplished by using tetra-*n*-butylammonium fluoride (TBAF) in THF at rt.[1] This is most convenient if only one silyl protecting group is present or if all of those present can be removed in one synthetic step.

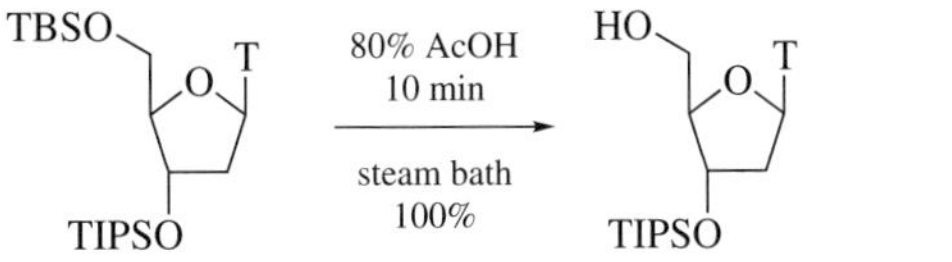

$$(3)$$

An additional feature of TIPS ethers is that they are volatile enough to make them amenable to GC and MS analysis. In fact, MS fragmentation patterns have been used to discern the structures of isomeric nucleosides.[9]

Formation of Silyl Ynol Ethers. Esters can also be transformed into triisopropylsilyl ynol ethers.[3] The ester is first converted to the ynolate anion, followed by treatment with TIPSCl to

furnish the TIPS ynol ether (eq 4). This method has even proven successful with lactones as starting materials (eq 5).

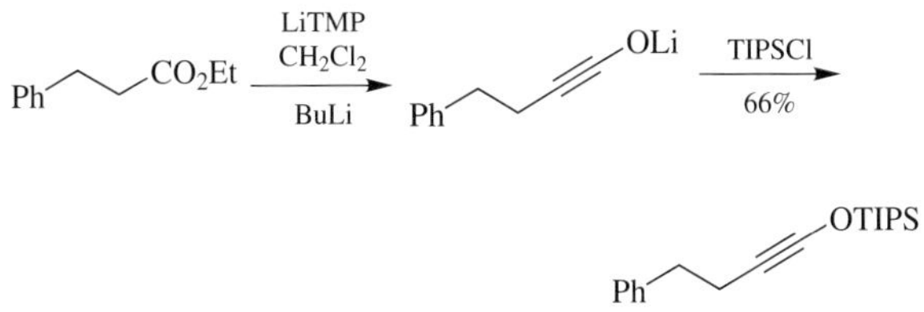

$$(4)$$

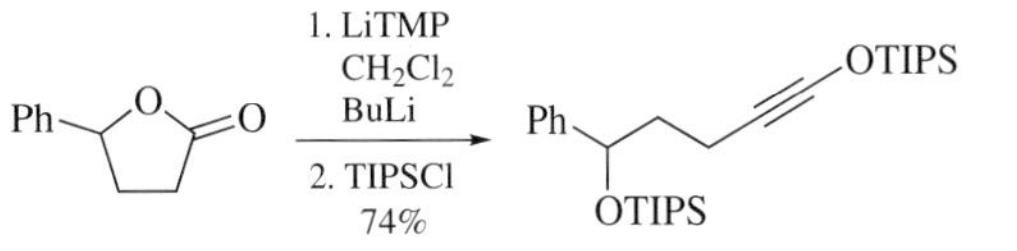

$$(5)$$

N-Protection of Pyrroles. Pyrroles typically undergo electrophilic substitution at the α-(2)-position, but when protected as *N*-triisopropylsilylpyrroles, substitution occurs exclusively at the β-(3)-position (eq 6).[4] It has been shown that 3-bromo-1-(triisopropylsilyl)pyrrole undergoes rapid halogen–metal exchange with *n*-butyllithium to generate the 3-lithiopyrrole, which can be trapped by electrophiles to provide the silylated 3-substituted pyrrole (eq 7).[5] As expected, the silyl group can be removed with TBAF to furnish the 3-substituted pyrrole. It should also be mentioned that the *N*-TIPS group also enhances the stability of some pyrroles. For example, 3-bromopyrrole is very unstable, but the *N*-silylated derivative is stable for an indefinite period of time.[5]

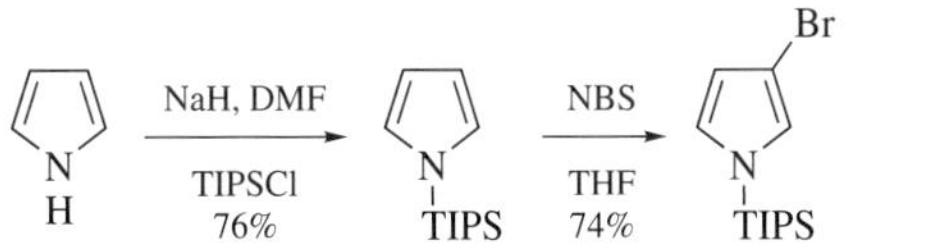

$$(6)$$

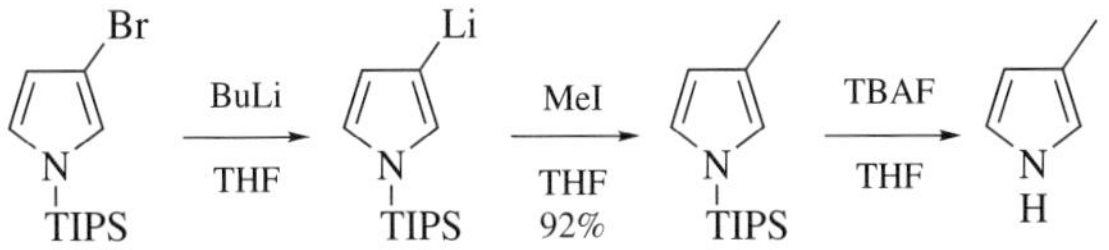

$$(7)$$

Prevention of Chelation in Grignard Reactions. The bulky TIPS protecting group has proven extremely effective in preventing competing chelation of α- and β-oxygen functionalities during Grignard reactions (eq 8).[6] It was determined that this effect is steric and not electronic since TMS and TBDMS ethers did not affect selectivity.

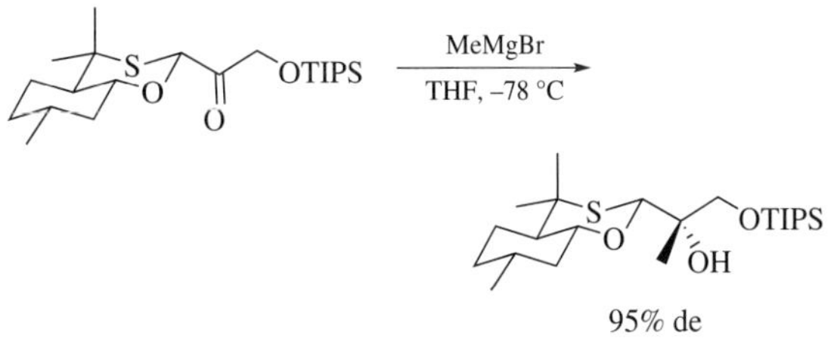

$$(8)$$

First Update

R. David Crouch
Dickinson College, Carlisle, PA, USA

Protection of Hydroxyl Groups. Triisopropylsilyl chloride (TIPSCl) continues to be widely used as the means of introducing the TIPS protecting groups onto alcohols.[10] The steric bulk of the three isopropyl groups on the silicon atom[11] provides considerable stability under acidic and basic conditions and allows selective removal of smaller, more labile silyl groups in the presence of TIPS ethers.[12,13]

Silylation with TIPSCl is most typically carried out using imidazole and an organic solvent[10] in a reaction that is so reliable that few alternatives have been described. However, a solvent-free microwave-mediated version of this reaction has been described in which excess TIPSCl and imidazole were mixed with alcohol and irradiated for up to 2 min (eq 9).[14] This reaction also allows for selective protection of 1° alcohols in the presence of 2° alcohols.

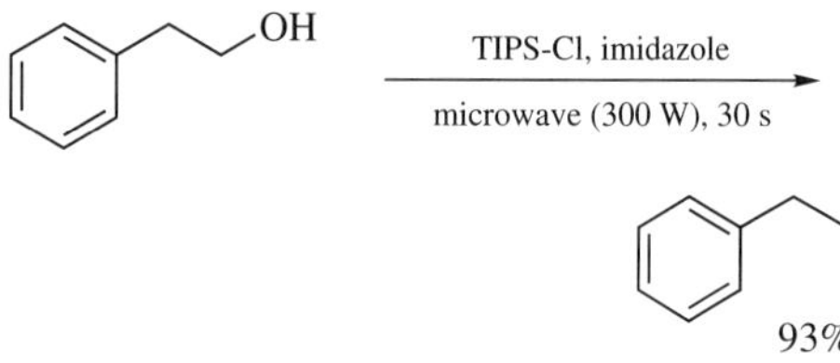

$$\text{(9)}$$

TIPSCl has also been used to prepare O-triisopropylsilylbenzamide (or, TIPS-Si-BEZA) (eq 10), which is especially effective for the conversion of sterically crowded 3° alcohols into TIPS ethers (eq 11).[15]

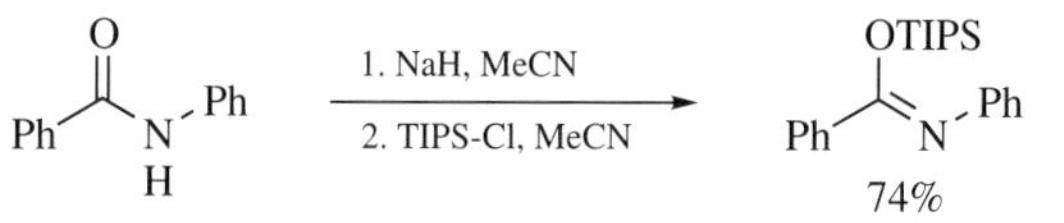

$$\text{(10)}$$

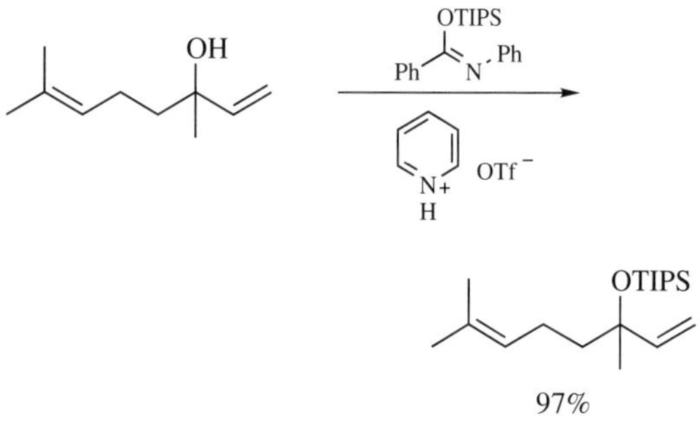

$$\text{(11)}$$

Secondary alcohols have been shown to undergo kinetic silylation in the presence of TIPSCl and chiral guanidines which function as superbases.[16] For example, treatment of 1,2,3,4-tetrahydro-1-naphthol yielded the enantiomerically-enriched TIPS ether in 15% yield but 70% ee (eq 12).[16]

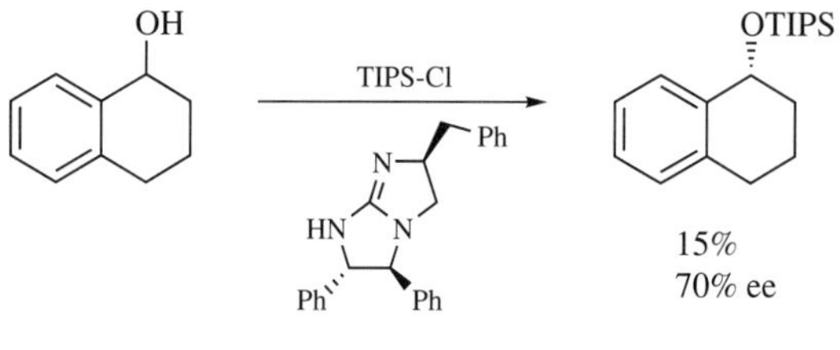

$$\text{(12)}$$

Microwave irradiation of a solvent-free mixture of menthol with excess TIPSCl and imidazole produced TIPS-protected menthol in reasonable yields.[17]

Protection of Terminal Alkynes. Reaction of a terminal alkyne with BuLi[18,19] or a Grignard reagent[20] followed by TIPSCl produces TIPS-protected alkynes, providing an unusually robust silylated terminal alkyne. If more than one terminal alkyne is present, the use of TIPS to protect one of the terminal alkynes and a more labile group to protect the other allows the subsequent deprotection of one terminal alkyne in the presence of another.[10,12,13] For example, a differentially protected enediyne can undergo selective protiodesilylation under mild conditions (eq 13).[21]

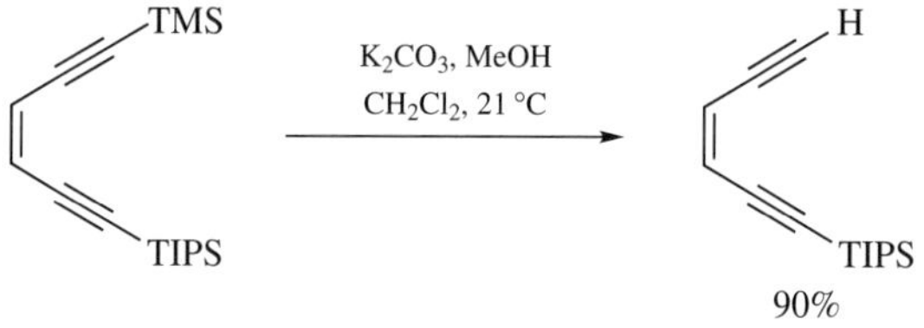

$$\text{(13)}$$

TIPSCl was used as the electrophile in trapping of the dianion of methyl 2-butynoate (eq 14).[22] When the more labile TMSCl was used, rapid hydrolysis of the silyl ketene acetal occurred, forming methyl 4-trimethylsilyl-3-butynoate.

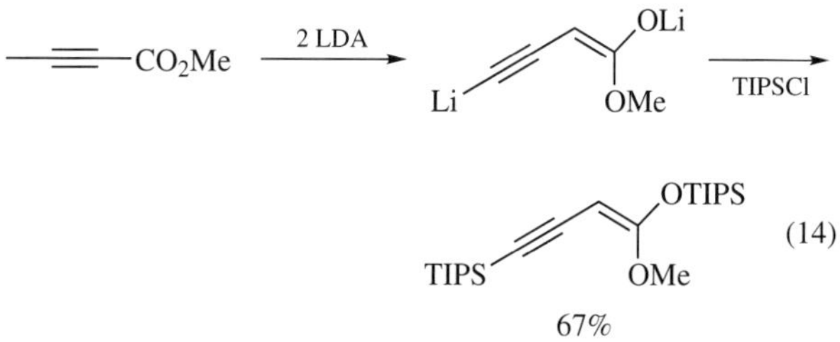

$$\text{(14)}$$

Protection of Pyrroles, Indoles, and Other Heterocycles. TIPS protection of the nitrogen in pyrrole and indole has a significant effect on electrophilic substitution in the aromatic ring system. TIPS-protected pyrrole can be produced by simply stirring pyrrole with TIPSCl while warming from −78 °C to room temperature (eq 15).[23] The resultant TIPS-pyrrole undergoes bromination with NBS to form 2,3,4-tribromopyrrole, after TBAF-mediated removal of the TIPS group (eq 15).[23]

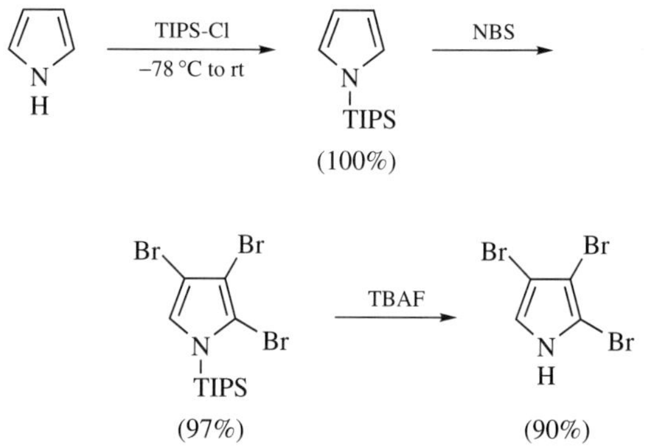

$$\text{(15)}$$

When unprotected pyrrole is treated with NBS, the major product is 2,3,5-tribromopyrrole.[23] The tendency for the TIPS group to direct halogen-metal exchange at C-3 position[5] in protected indoles has also been exploited in the solid phase synthesis of lamellarins.[24]

TIPS groups have been used in different ways to prevent undesired C-2 lithiation of indoles and benzothiophenes. The introduction of a TIPS group onto 4-bromo-1H-pyrrolo[2,3b]pyridine permitted metalation to occur exclusively at C-4, affording the 4-fluoro product (eq 16).[25] The TIPS group was removed with TBAF. Without TIPS, transmetalation at C-2 was observed.

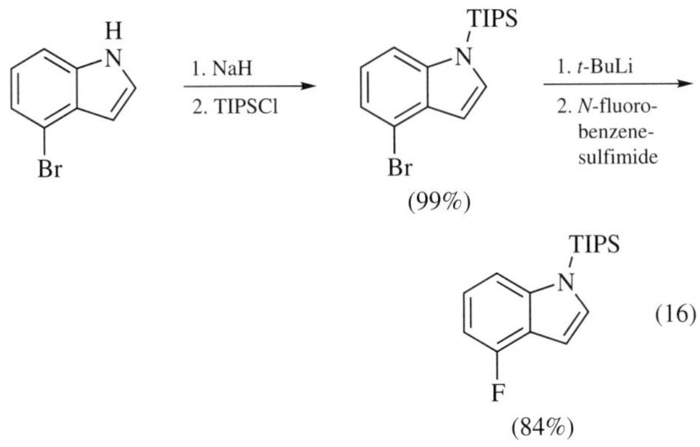

$$\text{(16)}$$

In a somewhat different approach, TIPS was used to protect the C-2 carbon of benzothiophene by treatment with LDA followed by TIPSCl (eq 17).[26] Subsequent metalation with BuLi in hexane/TMEDA produced the desired 7-lithio-2-triisopropylsilyl-benzothiophene which reacted with a Boc-protected piperidone to afford the 7-substituted benzothiophene. The TIPS group was removed by treatment with TFA in CH$_2$Cl$_2$.

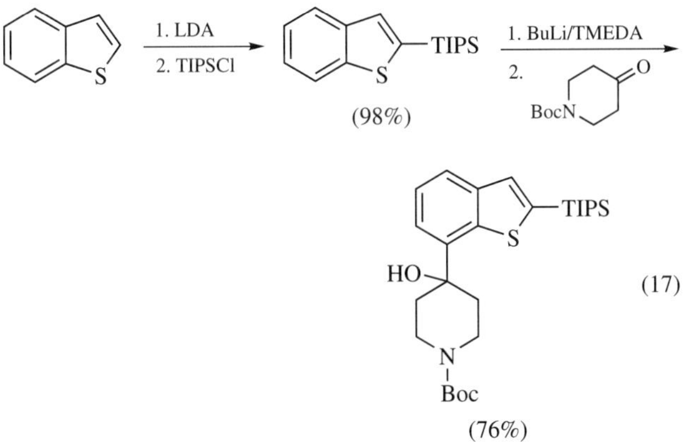

$$\text{(17)}$$

When 2,5-dibromo-3,4-di-t-butylthiophene was metalated with t-BuLi and quenched with TIPSCl, the steric bulk of the resulting tetrasubstituted thiophene was such that the ring adopted a nonplanar geometry.[27]

Preparation of "Supersilylating" Agents. TIPSCl reacts with boron tris(trifluoromethanesulfonate) or B(OTf)$_3$ in CH$_2$Cl$_2$ to form a so-called supersilylating agent with 1:1 stoichiometry (eq 18).[28]

$$i\text{-Pr}_3\text{Si-Cl} \ + \ \text{B(OTf)}_3 \ \xrightarrow{\text{CH}_2\text{Cl}_2} \ i\text{-Pr}_3\text{SiB(OTf)}_3\text{Cl} \quad \text{(18)}$$

An analogous reagent prepared from TIPSOTf and B(OTf)$_3$ serves as an efficient catalyst for the addition of allyltrimethyl-

silane to aldehydes and provides good control of Cram-type selectivity in Mukaiyama aldol reactions.[29] TIPS-B(OTf)$_3$Cl gives similar results (eq 19).[30]

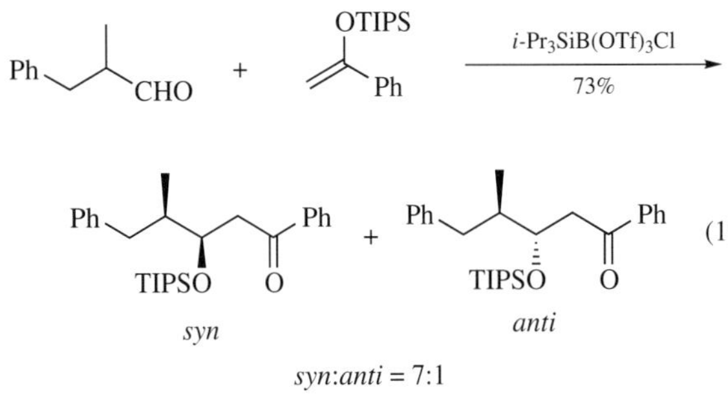

$$\text{(19)}$$

$$syn{:}anti = 7{:}1$$

Reactions with Hydrazine. TIPSCl reacts with an excess of anhydrous hydrazine to form triisopropylsilylhydrazine which can be used to prepare TIPS hydrazones (eq 20).[31] TIPS hydrazones are unusually stable in air but readily form asymmetric azines upon treatment with TBAF and a slight excess of ketone or aldehyde (eq 20).[31]

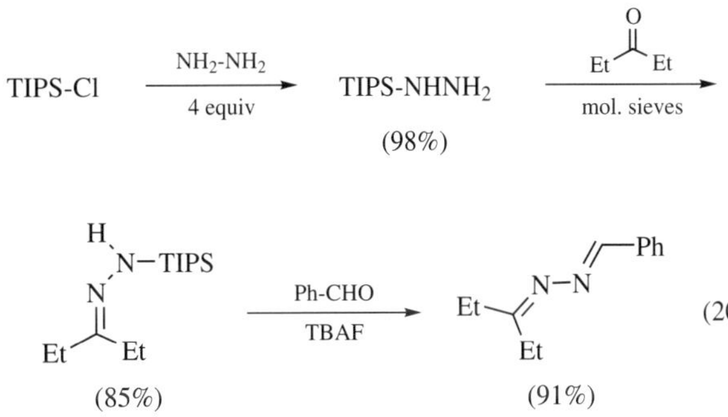

$$\text{TIPS-Cl} \xrightarrow[\text{4 equiv}]{\text{NH}_2\text{-NH}_2} \text{TIPS-NHNH}_2 \quad (98\%)$$

$$\text{(20)}$$

By contrast, treatment of anhydrous hydrazine with 2 equiv of t-BuLi followed by the addition of 2 equiv of TIPSCl leads to the formation of 1,2-bis(triisopropylsilyl)hydrazine (eq 21).[32] This compound, in the presence of catalytic Sc(OTf)$_3$ and an aldehyde, forms the analogous TIPS hydrazone (eq 22).[32]

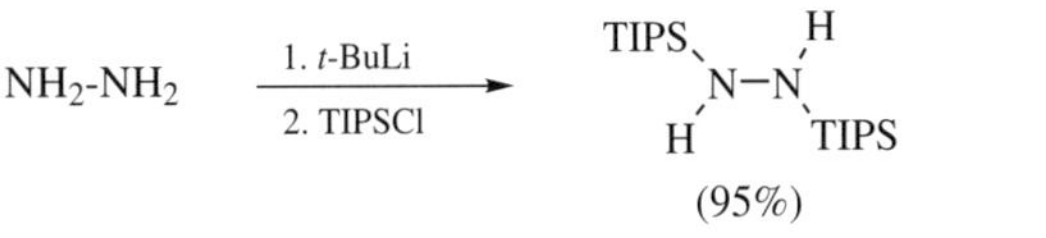

$$\text{NH}_2\text{-NH}_2 \xrightarrow[\text{2. TIPSCl}]{\text{1. } t\text{-BuLi}} \quad \text{(21)}$$

$$(95\%)$$

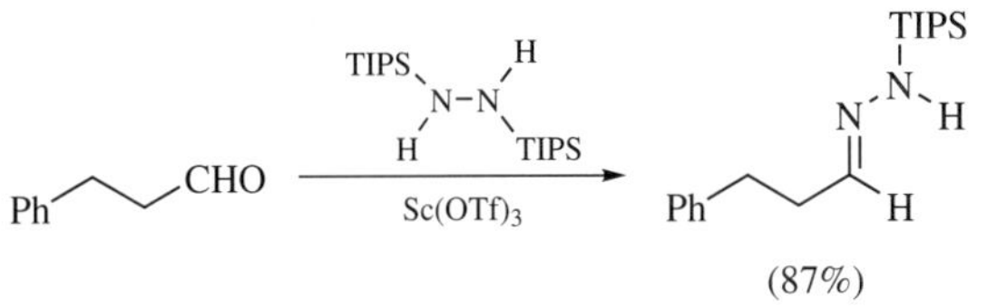

$$\text{(22)}$$

$$(87\%)$$

Reactions with Other Nucleophiles. Treatment of TIPSCl with aq NH$_3$ produced triisopropylsilanol which reacted with KOH to form a hindered base that can be used to produce elimination products without significant competition from substitution pathways (eq 23).[33]

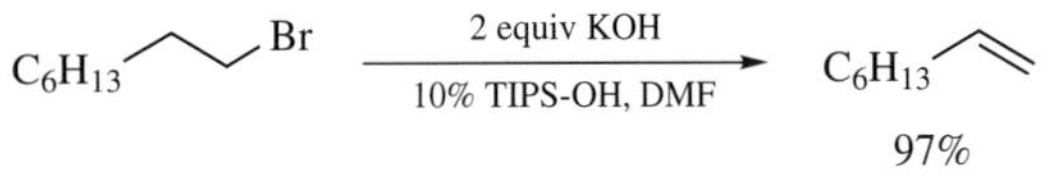

$$\text{C}_6\text{H}_{13}\text{---Br} \xrightarrow[\text{10\% TIPS-OH, DMF}]{\text{2 equiv KOH}} \text{C}_6\text{H}_{13} \quad \text{(23)}$$

$$97\%$$

Alkylation of H_2S with alkyl halides often leads to the formation of a mixture of alkanethiols, disulfides, and elimination products. But, when the lithium hydrosulfide was treated with TIPSCl (eq 24), the resulting triisopropylsilanethiol had one site blocked, allowing clean monoalkylation with a variety of electrophiles, including allylic, benzylic, and 1° or 2° halides and tosylates (eq 25).[34]

$$H_2S \xrightarrow[\text{2. TIPSCl}]{\text{1. BuLi}} \text{TIPS-SH} \qquad (24)$$
$$98\%$$

$$\text{TIPS-SH} \xrightarrow{\text{KH}} \text{TIPS-S}^- \text{ K}^+ \xrightarrow{\text{Ph} \diagup\diagdown \text{Cl}} $$
$$95\%$$

$$\text{Ph} \diagup\diagdown \text{S-TIPS} \qquad (25)$$
$$83\%$$

Removal of the TIPS group with TBAF or CsF in the presence of H_2O led to the formation of thiol.[34] Unsymmetrical sulfides can be formed when a suitable alkylating agent was used to trap the thiolate intermediate after addition of the fluoride source (eq 26).[34]

$$\text{Ph} \diagup\diagdown \text{S-TIPS} \xrightarrow[\text{2. } \diagup\diagdown\text{Br}]{\text{1. TBAF}} \text{Ph} \diagup\diagdown \text{S} \diagup\diagdown\diagup \qquad (26)$$
$$76\%$$

TIPSCl reacted with 2-lithio-1,3-dithiane to form a precursor to formyltriisopropylsilane, which was unmasked in a two-step sequence (eq 27).[35]

$$\underset{\text{H}\quad\text{TIPS}}{\overset{\text{S}\quad\text{S}}{\diagdown\diagup}} \xrightarrow[\text{MeOH}]{\text{HgCl}_2/\text{HgO}} \underset{\text{H}\quad\text{TIPS}}{\overset{\text{MeO}\quad\text{OMe}}{\diagdown\diagup}} \xrightarrow[\text{H}_2\text{O, MeCN}]{\text{LiBF}_4}$$
$$89\%$$

$$\underset{\text{H}\quad\text{TIPS}}{\overset{\text{O}}{\diagdown\!\!\parallel\!\!\diagup}} \qquad (27)$$
$$91\%$$

Although formyltriisopropylsilane ignites in air, the hydrazone derivative was isolable and formyltriisopropylsilane was used in situ in Wittig and aldol reactions.

Related Reagents. *t*-Butyldimethylchlorosilane; *t*-butyldiphenylchlorosilane; triisopropylsilyl; trifluoromethanesulfonate.

1. Green, T. W.; Wuts, P. G. M., *Protective Groups in Organic Synthesis*, 2nd ed.; Wiley: New York, 1991; p 74.
2. Cunico, R. F.; Bedell, L., *J. Org. Chem.* **1980**, *45*, 4797.
3. Kowalski, C. J.; Lal, G. S.; Haque, M. S., *J. Am. Chem. Soc.* **1986**, *108*, 7127.
4. Muchowski, J. M.; Solas, D. R., *Tetrahedron Lett.* **1983**, *24*, 3455.
5. Kozikowski, A. P.; Cheng, X.-M., *J. Org. Chem.* **1984**, *49*, 3239.
6. Frye, S. V.; Eliel, E. L., *Tetrahedron Lett.* **1986**, *27*, 3223.
7. Ogilvie, K. K.; Thompson, E. A.; Quilliam, M. A.; Westmore, J. B., *Tetrahedron Lett.* **1974**, 2865.
8. Ogilvie, K. K.; Sadana, K. L.; Thompson, E. A.; Quilliam, M. A.; Westmore, J. B., *Tetrahedron Lett.* **1974**, 2861.
9. Ogilvie, K. K.; Beaucage, S. L.; Entwistle, D. W.; Thompson, E. A.; Quilliam, M. A.; Westmore, J. B., *J. Carbohydr., Nucleosides, Nucleotides* **1976**, 197.
10. Greene, T. W.; Wuts, P. G. M., *Protective Groups in Organic Synthesis*; 3rd ed.; Wiley: New York, 1999; p 123.
11. Rucker, C., *Chem. Rev.* **1995**, *95*, 1009.
12. Nelson, T. D.; Crouch, R. D., *Synthesis* **1996**, 1031.
13. Crouch, R. D., *Tetrahedron* **2004**, *60*, 5833.
14. Khalafi-Nezhad, A.; Alamdari, R. F.; Zekri, N., *Tetrahedron* **2000**, *56*, 7503.
15. Misaki, T.; Kurihara, M.; Tanabe, Y., *Chem. Commun.* **2001**, 2478.
16. Isobe, T.; Fukuda, K.; Araki, Y.; Ishikawa, T., *Chem. Commun.* **2001**, 243.
17. Hatano, B.; Toyota, S.; Toda, F., *Green Chemistry* **2001**, *3*, 140.
18. Ogoshi, S.; Nishiguchi, S.; Tsutsumi, K.; Kurosawa, H., *J. Org. Chem.* **1995**, *60*, 4650.
19. Hlavaty, J.; Kavan, L.; Sticha, M., *J. Chem. Soc., Perkin Trans. 1* **2002**, 705.
20. Soderquist, J. A.; Colberg, J. C.; Del Valle, L., *J. Am. Chem. Soc.* **1989**, *111*, 4873.
21. Lu, Y. F.; Harwig, C. W.; Fallis, A. G., *J. Org. Chem.* **1993**, *58*, 4202.
22. Lepore, S. D.; He, Y.; Damisse, P., *J. Org. Chem.* **2004**, *69*, 9171.
23. John, E. A.; Pollet, P.; Gelbaum, L.; Kubanek, J., *J. Nat. Prod.* **2004**, *67*, 1929.
24. Marfil, M.; Albericio, F.; Alvarez, M., *Tetrahedron* **2004**, *60*, 8659.
25. Thibault, C.; L'Heureux, A.; Bhide, R. S.; Ruel, R., *Org. Lett.* **2003**, *5*, 5023.
26. Hansen, M. M.; Clayton, M. T.; Godfrey, A. G.; Grutsch, J. L., Jr.; Keast, S. S.; Kohlman, D. T.; McSpadden, A. R.; Pedersen, S. W.; Ward, J. A.; Xu, Y. C., *Synlett* **2004**, 1351.
27. Nakayama, J.; Yu, T.; Sugihara, Y.; Ishii, A.; Kumakura, S., *Heterocycles* **1997**, *45*, 1267.
28. Davis, A. P.; Muir, J. E.; Plunkett, S. J., *Tetrahedron Lett.* **1996**, *37*, 9401.
29. Davis, A. P.; Plunkett, S. J., *J. Chem. Soc., Chem. Commun.* **1995**, 2173.
30. Davis, A. P.; Plunkett, S. J.; Muir, J. E., *Chem. Commun.* **1998**, 1797.
31. Justo de Pomar, J. C.; Soderquist, J. A., *Tetrahedron Lett.* **2000**, *41*, 3285.
32. Furrow, M. E.; Myers, A. G., *J. Am. Chem. Soc.* **2004**, *126*, 5436.
33. Soderquist, J. A.; Vaquer, J.; Diaz, M. J.; Rane, A. N.; Bordwell, F. G.; Zhang, S., *Tetrahedron Lett.* **1996**, *37*, 2561.
34. Miranda, E. I.; Diaz, M. J.; Rosado, I.; Soderquist, J. A., *Tetrahedron Lett.* **1994**, *35*, 3221.
35. Soderquist, J. A.; Miranda, E. I., *J. Am. Chem. Soc.* **1992**, *114*, 10078.

Triisopropylsilylethynyl Triflone

$$\underset{\diagup}{\overset{\diagdown}{\text{Si}}}-\text{C}\equiv\text{C}-\overset{\overset{\text{O}}{\parallel}}{\underset{\underset{\text{O}}{\parallel}}{\text{S}}}-\text{CF}_3 \qquad \text{or} \qquad i\text{-Pr}_3\text{Si}-\text{C}\equiv\text{C}-\text{SO}_2\text{CF}_3$$

[196789-82-9] $C_{12}H_{21}F_3O_2SSi$ (MW 314.44)

InChI = 1S/C12H21F3O2SSi/c1-9(2)19(10(3)4,11(5)6)8-7-18(16,17)12(13,14)15/h9-11H,1-6H3

InChIKey = ROAJHHMTMBSXBG-UHFFFAOYSA-N

(reagent for radical alkynylation of C–H or C–I bonds)

Alternate Name: TIPS-acetylene triflone or TIPS-ethynyl triflone.
Solubility: most aprotic organic solvents.
Form Supplied in: colorless liquid not commercially available.
Analysis of Reagent Purity: reagent must be made fresh.

Handling, Storage, and Precaution: sensitive to free radical sources.

Only two recent papers[1,2] have described this reagent, which is used for free radical initiated alkynylation. In the first paper, the direct alkynylation of C–H bonds proceeds as summarized in eq 1 by the facile C–H bond abstraction by the very electrophilic trifluoromethyl radical **4** generated by cleavage of the CF_3SO_2 radical (**7**) and loss of SO_2. The alkyl radical **5** so generated reacts with the triflone reagent **2** to form the vinyl radical (**6**) which in turn eliminates the SO_2CF_3 radical (**7**) to propagate the chain and afford the attached alkynyl group in **3**.

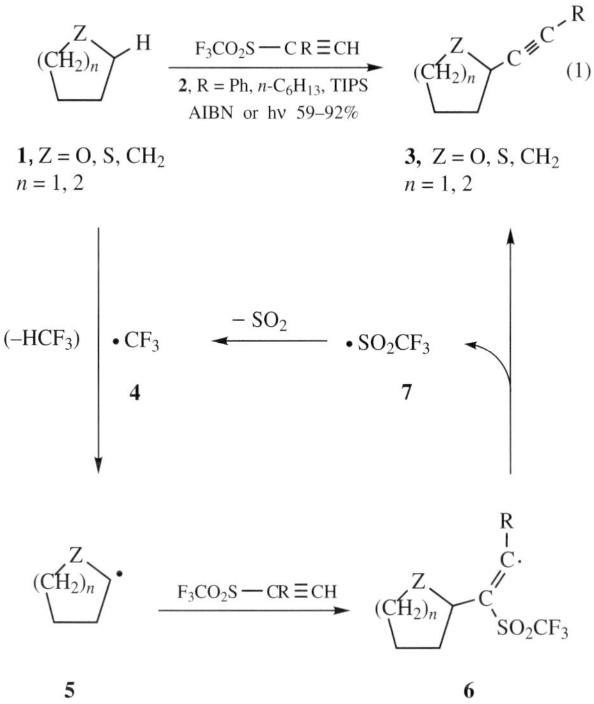

1, Z = O, S, CH_2
n = 1, 2

3, Z = O, S, CH_2
n = 1, 2

5

6

The TIPS variant in eq 1 is used not only neat on the cyclic C–H substrates THF, tetrahydrothiophene, and cyclohexane, but also in acetonitrile with adamantane, which substituted exclusively at the tertiary C–H (50% yield). Alkynylation was also successful with distal functionality, i.e. R = $(CH_2)_2OSiR_3$ and $(CH_2)_3Cl$, but the triflone could not be formed with ether functionality closer to the acetylene.

Previously, the alkynylation had been reported[3,4] with other attached hydrocarbon groups (R = Ph and *n*-hexyl in eq 1). The presence of a second acetylene group in the hydrocarbon group R was successful only with four methylenes, but not less, separating the two acetylenes. However, with enough separation the outer acetylene succeeded with only H-substitution to yield **8** in eq 2 and the product could be carried through triflation and a second alkynylation to form **9**.

The necessary silyl triflone reagents proved difficult or impossible to make with less hindered silanes (TMS or TBDMS) by the reaction of the silyl-acetylene anion (*n*-BuLi/Et_2O/$-78\,^{\circ}C$) with Tf_2O. It was this triflation which failed with the proximal ethers and substituted acetylenes above.

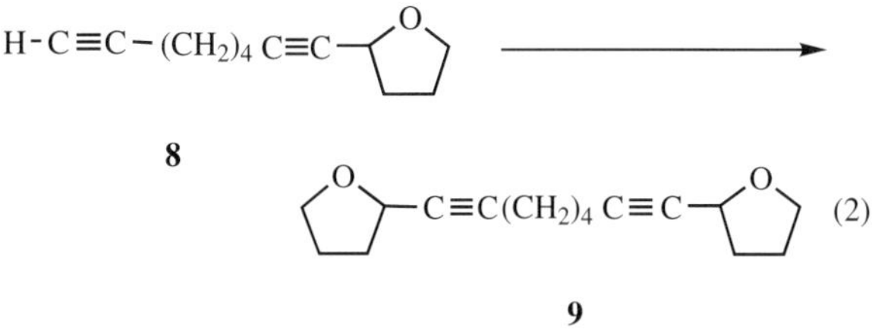

8

9

In the second paper,[2] the reaction was extended to alkynylate C–I bonds photolytically by the added intermediacy of hexabutyldistannane to generate the radical from the iodide. Bromides were inert in this alkynylation, suggesting that this differential reactivity of the two halogens should prove advantageous in synthetic applications.

The reaction outlined in eq 3, was conducted photolytically in benzene solution, with the benzene presumably scavenging the trifluoromethyl radical, as it is scavenged in eq 2. A dozen examples of iodides were successful in yields generally over 60% and with retention of configuration. These room-temperature examples show the reaction to be compatible with such diverse functionality as free hydroxyl, ester, amide, thiazole, and potential β-elimination substrates, and succeeded with primary, secondary, and tertiary iodides.

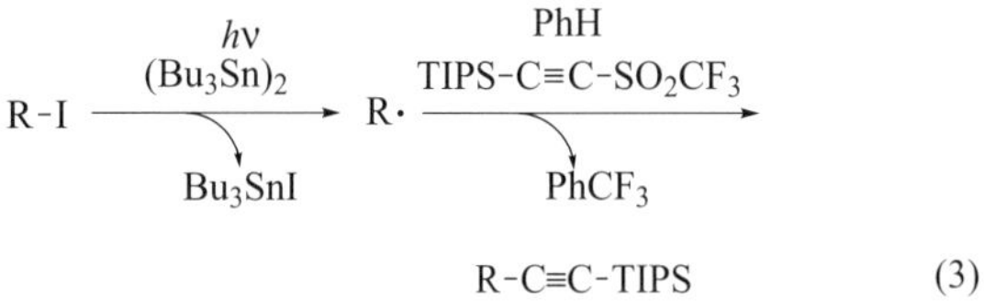

The special value of the TIPS group lies in its easy removal (TBAF/$25\,^{\circ}C$/2 h) to make the acetylene available for further substitution. Furthermore, the present availability of the TIPS-protected acetylene triflone should make possible a further exploration of the addition and cycloaddition reactions of the acetylene activated by the strong electron-withdrawing power of the triflone group.

1. Xiang, J.; Jiang, W.; Fuchs, P. L., *Tetrahedron Lett.* **1997**, *38*, 6635.
2. Xiang, J.; Fuchs, P. L., *Tetrahedron Lett.* **1998**, *39*, 8597.
3. Gong, J.; Fuchs, P. L., *J. Am. Chem. Soc.* **1996**, *118*, 4486.
4. Xiang, J.; Fuchs, P. L., *Tetrahedron Lett.* **1996**, *37*, 5269.

James B. Hendrickson
Brandeis University, Waltham, MA, USA

Triisopropylsilyl Trifluoromethanesulfonate[1]

$(i\text{-Pr})_3SiOSO_2CF_3$

[80522-42-5] $C_{10}H_{21}F_3O_3SSi$ (MW 306.46)
InChI = 1S/C10H21F3O3SSi/c1-7(2)18(8(3)4,9(5)6)16-17(14,15)
10(11,12)13/h7-9H,1-6H3
InChIKey = LHJCZOXMCGQVDQ-UHFFFAOYSA-N

(highly reactive silylating agent and Lewis acid capable of converting primary and secondary alcohols[1] to the corresponding triisopropylsilyl ethers and converting ketones[1] and lactones[2] into their enol silyl ethers; protection of terminal alkynes;[3] promoting conjugate addition of alkynylzinc compounds to α,β-enones;[4] preparation of (triisopropylsilyl)diazomethane[5])

Alternate Name: TIPS triflate.

Physical Data: colorless oil; bp 83–87 °C/1.7 mmHg; d 1.173 $g\,cm^{-3}$.

Solubility: sol most organic solvents, such as pentane, CH_2Cl_2, etc.

Form Supplied in: liquid; widely available.

Analysis of Reagent Purity: ^1H NMR (CDCl$_3$) δ 1.6–1.05 (m). although the reagent is commercially available from various vendors, there have been recent observations about *n*-propyldiisopropylsilyl triflate contaminating TIPSOTf. The presence of this impurity is observed easily by proton NMR.[6]

Preparative Methods: to 38.2 g (0.242 mol) of triisopropylsilane at 0 °C under argon is added 23.8 mL (0.266 mol) of trifluoromethanesulfonic acid dropwise. The solution is stirred at 22 °C for 16 h, at which time no further hydrogen gas evolves (removed through a bubbler). The resulting product is distilled through a 30-cm vacuum jacketed Vigreux column under reduced pressure: 71.7 g (97% yield) of TIPS triflate; bp 83–87 °C/1.7 mmHg.[1a]

Handling, Storage, and Precautions: store under argon at 0 °C; unpleasant odor; reacts rapidly with water and other protic solvents.

Original Commentary

Duy H. Hua & Jinshan Chen
Kansas State University, Manhattan, KS, USA

Silylation of Alcohols.[1]   Primary and secondary alcohols are silylated by TIPS triflate in excellent yields. Treatment of 1-phenylethanol with 1.3 equiv of TIPS triflate and 2.5 equiv of 2,6-lutidine in CH_2Cl_2 at 0 °C for 2 h provides a 98% yield of α-(triisopropylsilyloxy)ethylbenzene (eq 1).[1a]

$$ ROH \xrightarrow[CH_2Cl_2]{(i\text{-}Pr)_3SiOTf,\ 2,6\text{-lutidine}} ROSi(i\text{-}Pr)_3 \qquad (1) $$

Formations of Enol Silyl Ethers.[1,2]   Aldehydes, ketones, and lactones are readily converted into the corresponding enol TIPS ethers. Reactions of cycloalkanones with 1.1 equiv of TIPS triflate and 1.5 equiv of triethylamine in benzene at 23 °C for 1 h gives >98% yields of the corresponding enol silyl ethers (eq 2).[1a]

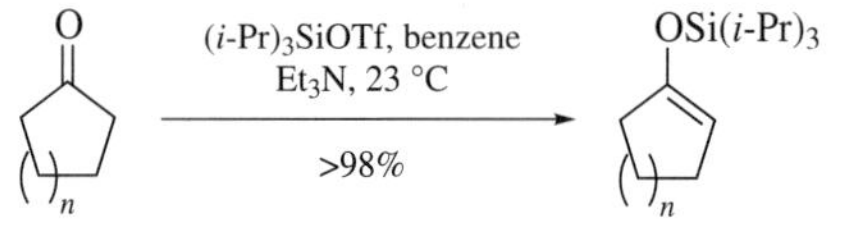

$$ (2) $$

Silylation of 2-morpholinones with 1.2 equiv of TIPS triflate and 1.5 equiv of triethylamine in CDCl$_3$ for 2 min provides silyl

ketene acetals, which upon standing at rt undergo Claisen rearrangement to afford pipecolic esters (eq 3).[2]

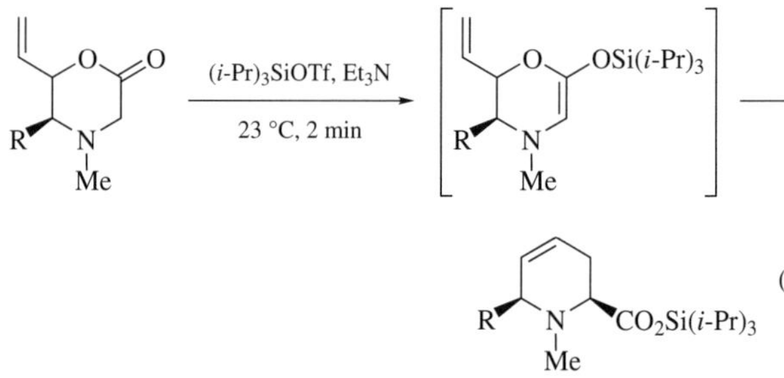

$$ (3) $$

R = Me, 64%
R = Bn, 74%
R = *i*-Pr, 70%

Alkynyltriisopropylsilanes.[3]   The acidic alkynic H can be protected with TIPS triflate. Silylation of 1-lithiopropyne with 1 equiv of TIPS triflate in ether at −40 to 0 °C gives an 87% yield of 1-TIPS-propyne (eq 4).[3a] 1,3-Bis(triisopropylsilyl)propyne, derived from treatment of 1-TIPS-propyne with *n*-butyllithium in THF at −20 °C for 15 min followed by TIPS triflate at −78 to −40 °C (eq 5), is lithiated with *n*-BuLi in THF at −20 °C for 15 min and then allowed to react with aldehydes (eq 6). The *cis*-enynes are isolated in high yields.[3a] TIPS-propargylmagnesium bromide together with copper(I) iodide has been used in the displacement of the mesylate derivative of farnesol.[3b]

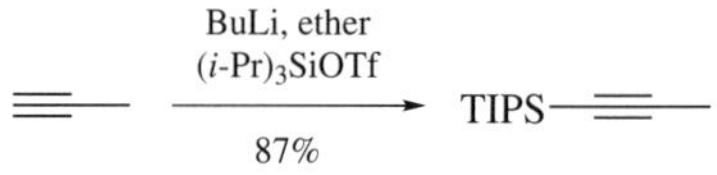

$$ (4) $$

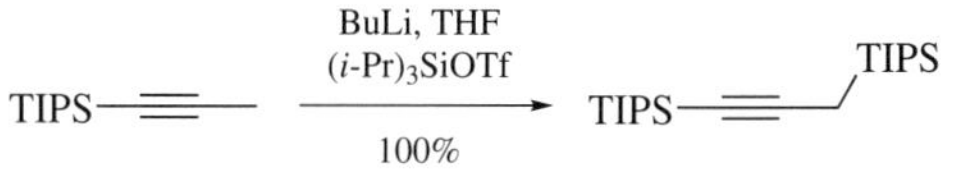

$$ (5) $$

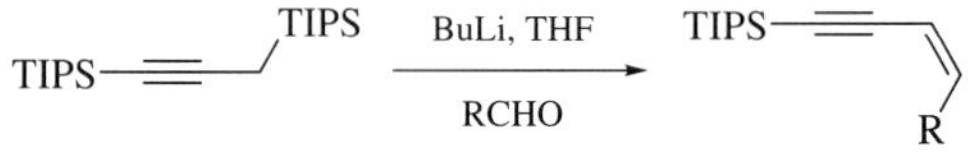

$$ (6) $$

$(Z):(E) = 20:1$

R = Cy, 71%
R = C_6H_{13}, 57%
R = *t*-Bu, 79%

Conjugate Addition of Alkynylzinc Bromides.[4]   Alkynylzinc bromides undergo conjugate addition with α,β-unsaturated ketones in the presence of *t*-butyldimethylsilyl trifluoromethanesulfonate or TIPS triflate (trimethylsilyl trifluoromethanesulfonate is also effective) in ether–THF at −40 °C to give the corresponding 1,4-adducts (54–96% yields). A representative example is illustrated in eq 7.

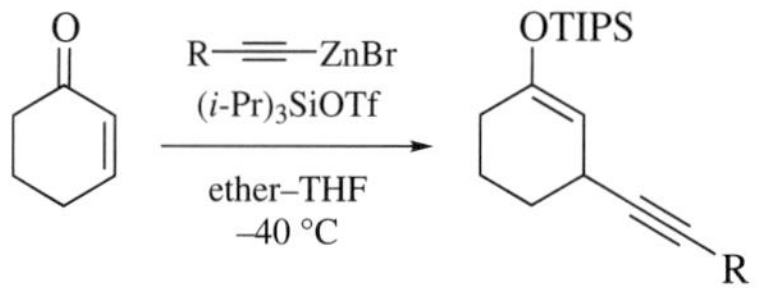

$$ (7) $$

(Triisopropylsilyl)diazomethane.[5]   Silylation of diazomethane with TIPS triflate and diisopropylethylamine in ether at

$-20\,°C$ to $25\,°C$ gives a 45% yield of (triisopropylsilyl)diazomethane (eq 8).[5] This silylated diazomethane is used to prepare stable silyl-substituted nitrilimines.[5]

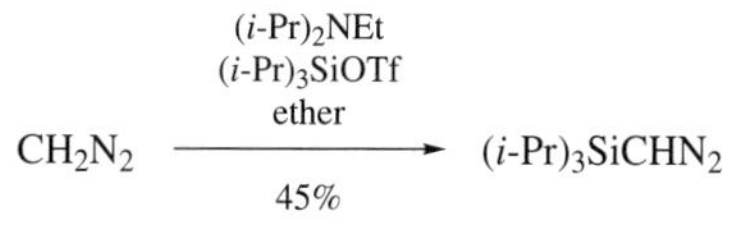

$$CH_2N_2 \xrightarrow[\substack{\text{ether} \\ 45\%}]{\substack{(i\text{-Pr})_2NEt \\ (i\text{-Pr})_3SiOTf}} (i\text{-Pr})_3SiCHN_2 \qquad (8)$$

First Update

Ross Miller

Merck Research Laboratories, Rahway, NJ, USA

Some highlights on the utility of TIPSOTf published since 1995 are shown below.[7]

Triisopropylsiloxycarbonyl (Tsoc) and BIPSOP Protecting Groups for Amines.[8,9] Two new useful nitrogen protecting groups that rely on the stability of the TIPS group have been developed. In both cases, TIPSOTf was relied on for the formation of the actual amine protecting group. BIPSOP (*N*-substituted-2,5-bis[(triisopropylsilyl)oxy]pyrroles) were formed by reaction of primary amines with succinic anhydride followed by conversion to the bis-siloxypyrrole with TIPSOTf (eq 9). BIPSOP was most useful for the protection of amines under strongly basic reactions. Removal of this protecting group was achieved by treatment with dilute acid followed by hydrazine treatment.

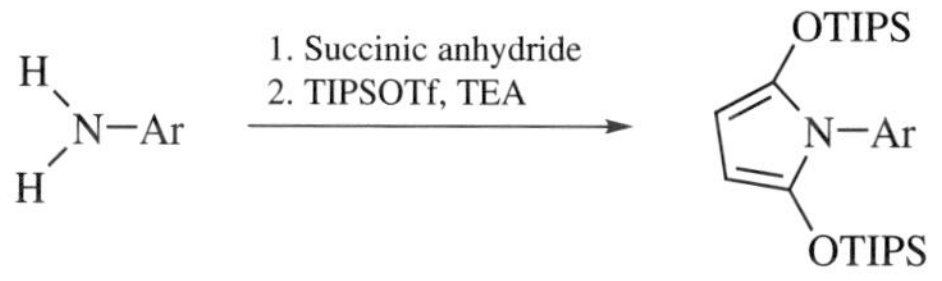

$$\underset{H}{\overset{H}{\underset{}{}}}N{-}Ar \xrightarrow[]{\substack{\text{1. Succinic anhydride} \\ \text{2. TIPSOTf, TEA}}} \underset{\text{OTIPS}}{\overset{\text{OTIPS}}{\underset{}{}}}N{-}Ar \qquad (9)$$

The Tsoc group was developed for both primary and secondary amine protection and shown to be stable to acid, base, and hydrogenation conditions (eq 10). Removal was achieved with fluoride treatment.

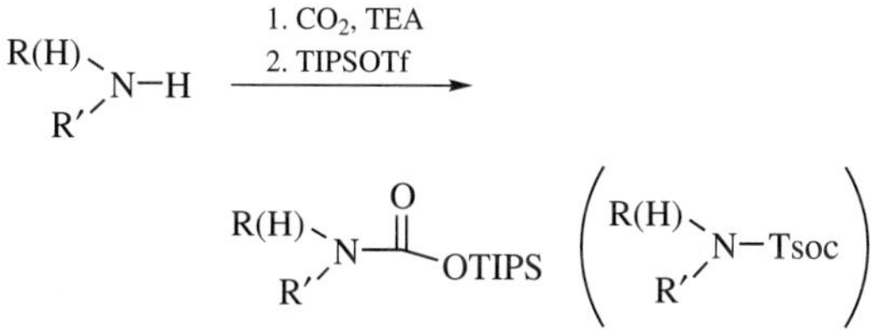

$$\underset{R'}{\overset{R(H)}{\underset{}{}}}N{-}H \xrightarrow[]{\substack{\text{1. } CO_2,\ TEA \\ \text{2. TIPSOTf}}}$$
$$\underset{R'}{\overset{R(H)}{\underset{}{}}}N\overset{O}{\underset{}{}}\text{OTIPS} \left(\underset{R'}{\overset{R(H)}{\underset{}{}}}N{-}\text{Tsoc}\right) \qquad (10)$$

TIPS Protection for Oxazoles/Regioselective Trapping of C-2 Oxazole Anions. Deprotonation of oxazoles occurs selectively at the C-2 position, and reaction with either TIPS triflate or TIPS chloride gave completely opposite C-silylation or O-silylation selectivity (eq 11).[10] The same remarkable trend was observed for the other, less hindered silyl chloride and -triflates; however, the stability of those products did not allow facile isolation and characterization. The C-2 TIPS oxazole derivative was shown to be stable to strongly basic reactions that allowed further oxazole functionalization (eq 12).[11] Removal of the TIPS group was then achieved with dilute acid.

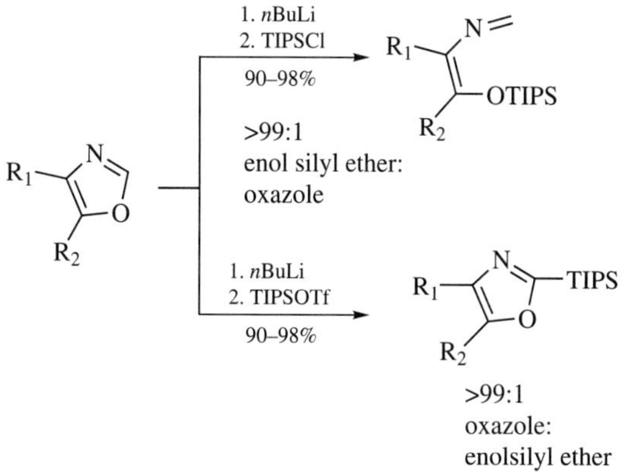

$$(11)$$

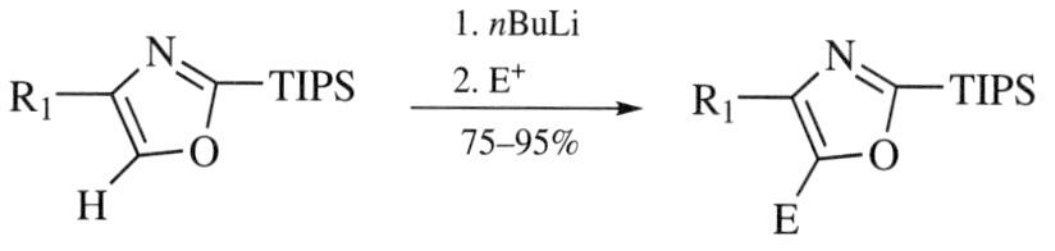

>99:1
enol silyl ether:
oxazole

>99:1
oxazole:
enolsilyl ether

$$(12)$$

Benzannulation Using Triisopropylsilyl Vinyl Ketenes. Addition of lithium ynolates generated from ethynylsilyl ethers with silyl vinyl ketenes gave highly substituted benzenes (eq 13).[12] TIPS protection, in contrast to TBDMS, allowed preparation and purification by distillation and silica gel chromatography of these sensitive starting materials.

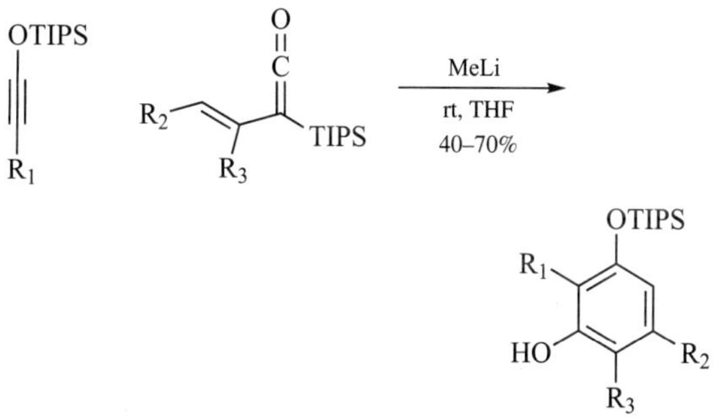

$$(13)$$

Cyclization of 1-Silyloxy-1,5-diynes. A $HNTf_2$-promoted 5-*endodig* cyclization of 1-siloxy-1,5-diynes has been shown to give highly selective β-halo enones by abstraction of a halogen from halocarbons in good yields (eq 14).[13]

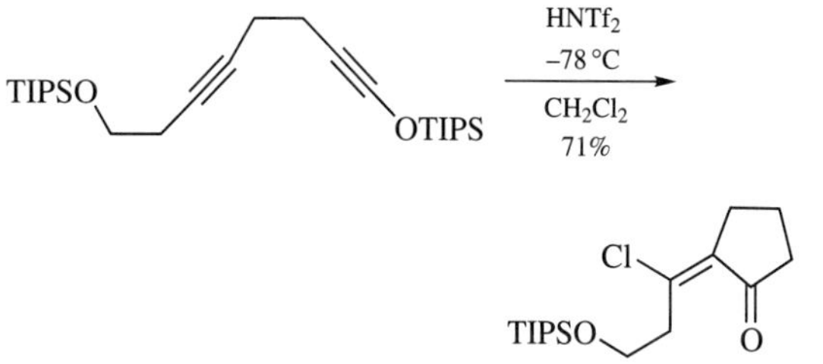

$$(14)$$

Desymmetrization of Tartaric Acid Esters. Selective reductions of the esters of tartaric acid using TIPS as the controlling element allowed the desymmetrization of tartaric acid and facile access to enantiopure alcohols and diol esters (eqs 15 and 16). Formation of the TIPS protected tartaric acids was achieved in high yields using TIPSOTf as a requirement.[14]

$$\text{HO} \quad \text{OTIPS}$$
$$i\text{-PrO} \overset{}{\underset{\overset{\parallel}{O}}{\diagup}} \underset{\overset{\parallel}{O}}{\diagdown} O i\text{-Pr} \xrightarrow{\begin{array}{c} \text{NaBH}_4 \\ \text{BH}_3\cdot\text{DMS} \end{array}}$$

$$\underset{\text{OH}}{\overset{\text{HO} \quad \text{OTIPS}}{\diagup}} \underset{\overset{\parallel}{O}}{} O i\text{-Pr} \quad + \quad i\text{-PrO} \underset{\overset{\parallel}{O}}{} \overset{\text{HO} \quad \text{OTIPS}}{\diagdown} \underset{\text{OH}}{} \quad (15)$$

$$>99:1$$

$$\underset{\overset{\parallel}{O}}{\text{MeO}} \overset{\text{HO} \quad \text{OTIPS}}{\diagup} \underset{\overset{\parallel}{O}}{} \text{OMe} \xrightarrow{\begin{array}{c} \text{NaBH}_4 \\ \text{LiCl, EtOH} \end{array}} \underset{\text{OH} \quad \text{OH}}{\overset{\text{HO} \quad \text{OTIPS}}{\diagup}} \quad (16)$$

Related Reagent.   Triisopropylsilyl Chloride.

1. (a) Corey, E. J.; Cho, H.; Rucker, C.; Hua, D. H., *Tetrahedron Lett.* **1981**, *22*, 3455. (b) For a review of trialkylsilyltriflates: Emde, H.; Domsch, D.; Feger, H.; Frick, U.; Gotz, A.; Hergott, H. H.; Hofmann, K.; Kober, W.; Krageloh, K.; Oesterle, T.; Steppan, W.; West, W.; Simchen, G., *Synthesis* **1982**, 1.

2. Angle, S. R.; Breitenbucker, J. G.; Arnaiz, D. O., *J. Org. Chem.* **1992**, *57*, 5947.

3. (a) Corey, E. J.; Rucker, C., *Tetrahedron Lett.* **1982**, *23*, 719. (b) Marshall, J. A.; Andersen, M. W., *J. Org. Chem.* **1992**, *57*, 2766.

4. Kim, S.; Lee, J. M., *Tetrahedron Lett.* **1990**, *31*, 7627.

5. Castan, F.; Baceiredo, A.; Bigg, D.; Bertrand, G., *J. Org. Chem.* **1991**, *56*, 1801.

6. Barden, D. J.; Fleming, I., *J. Chem. Soc., Chem. Commun.* **2001**, 2366.

7. A comprehensive review of all aspects of this reagent (770 references) up to 1995 has been published. See Rucker, C., *Chem. Rev.* **1995**, *95*, 1009.

8. Lipshutz, B. M.; Papa, P.; Keith, J. M., *J. Org. Chem.* **1999**, *64*, 3792.

9. Martin, S. F.; Limberakis, C., *Tetrahedron Lett.* **1997**, *38*, 2617.

10. Miller, R. A.; Smith, R. M.; Karady, S.; Reamer, R. A., *Tetrahedron Lett.* **2002**, *43*, 935.

11. Miller, R. A.; Smith, R. M.; Marcune, B., *J. Org. Chem.* **2005**, *70*, 9074.

12. Austin, W. F.; Zhang, Y.; Danheiser, R., *Org. Lett.* **2005**, *7*, 3905.

13. Sun, J.; Kozmin, S. A., *J. Am. Chem. Soc.* **2005**, *127*, 13512.

14. McNulty, J.; Mao, J., *Tetrahedron Lett.* **2002**, *43*, 3857.

Trimethylsilylacetic Acid[1]

$$\text{Me}_3\text{Si}\diagup\diagdown\text{CO}_2\text{H}$$

[2345-38-2]   C$_5$H$_{12}$O$_2$Si   (MW 132.26)

InChI = 1S/C5H12O2Si/c1-8(2,3)4-5(6)7/h4H2,1-3H3,(H,6,7)

InChIKey = JDMMZVAKMAONFU-UHFFFAOYSA-N

(preparation of α,β-unsaturated carboxylic acids and α-trimethylsilylbutyrolactones[2])

Physical Data: mp 40 °C.

Solubility: sol ethereal solvents.

Preparative Method: obtained from the reaction of trimethylsilylmethylmagnesium chloride with carbon dioxide;[3] can also

be prepared by reaction of acetic acid with 2 equiv of lithium diisopropylamide, followed by chlorotrimethylsilane and hydrolysis.[4]

Handling, Storage, and Precautions: prone to isomerization and should not be stored for extensive periods.

Anionic Reactions.   In a manner similar to α-silyl esters, trimethylsilylacetic acid (**1**) can be used to prepare α,β-unsaturated carboxylic acids from carbonyl compounds by a Peterson alkenation reaction (eq 1).[1] However, the dianion of acid (**1**) is required.[2] The yields are generally inferior to analogous reactions with α-silyl esters.[5]

$$\underset{(\mathbf{1})}{\text{TMS}\diagup\diagdown\text{CO}_2\text{H}} \xrightarrow[\text{2. R}^1\text{R}^2\text{CO}]{\text{1. 2.2 equiv LDA, THF, 0 °C}} \underset{\text{R}^2}{\overset{\text{R}^1}{\diagdown}}\diagup\text{CO}_2\text{H} \quad (1)$$

The dianion of acid (**1**) also reacts with alkyl halides to afford substituted α-trimethylsilylcarboxylic acids.[2] With epoxides, the dianion provides γ-hydroxy acids that cyclize to α-silylbutyrolactones (eq 2).[2] The resultant α-silyl lactones (**2**) are precursors to a wide variety of functionalized derivatives. The direct conversion of a butyrolactone to its α-silyl analog is complicated by the competition of the *O*-silylation pathway.

$$\underset{(\mathbf{1})}{\text{TMS}\diagup\diagdown\text{CO}_2\text{H}} \xrightarrow[\underset{\text{O}}{\overset{\text{R}}{\triangle}}]{\text{1. 2.2 equiv LDA, THF, 0 °C}} \underset{\underset{(\mathbf{2})}{\text{TMS}}}{\overset{\text{R}}{\diagdown}} \quad (2)$$

Derivatives.   Trimethylsilylacetyl thiolesters, which have been used as precursors to β-lactams, are prepared by reaction of the appropriate thiol with the acid chloride derivative of (**1**).[6] Acid (**1**) has also been used as a precursor to 1-diazo-3-trimethylsilylacetone, which provides cyclopropyl trimethylsilylmethyl ketones upon reaction with alkenes, through reaction of the acid chloride or a mixed anhydride derivatives of the acid (**1**) with diazomethane.[4]

Silylacetic acids undergo a thermal rearrangement to afford the silyl ester (eq 3). The same migration of the silyl group is observed at lower temperatures when base catalysis is employed.[7]

$$\underset{\text{R}_3\text{Si}}{\overset{\text{R}^1\ \text{R}^2}{\diagdown}}\diagup\text{CO}_2\text{H} \xrightarrow[\text{or base (e.g. Et}_3\text{N})]{\Delta} \underset{\text{R}^1}{\overset{\text{R}^2}{\diagdown}}\diagup\text{CO}_2\text{SiR}_3 \quad (3)$$

Related Reagents.   *t*-Butyl α-Lithiobis(trimethylsilyl)acetate; *t*-Butyl Trimethylsilylacetate; Dilithioacetate; Ethyl Lithio(trimethylsilyl)acetate; Ethyl Trimethylsilylacetate.

1. Ager, D. J., *Org. React.* **1990**, *38*, 1.

2. Grieco, P. A.; Wang, C.-L. J.; Burke, S. D., *J. Chem. Soc., Chem. Commun.* **1975**, 537.

3. Sommer, L. H.; Gold, J. R.; Goldberg, G. M.; Marans, N. S., *J. Am. Chem. Soc.* **1949**, *71*, 1509.

4. Tsuge, O.; Kanemasa, S.; Suzuki, T.; Matsuda, K., *Bull. Chem. Soc. Jpn.* **1986**, *59*, 2851.

5. Crimmin, M. J.; O'Hanlon, P. J.; Rogers, N. H., *J. Chem. Soc., Perkin Trans. 1* **1985**, 541.

6. Tajima, Y.; Yoshida, A.; Takeda, N.; Oida, S., *Tetrahedron Lett.* **1985**, *26*, 673; Lucast, D. H.; Wemple, J., *Tetrahedron Lett.* **1977**, 1103.

7. Brook, A. G.; Anderson, D. G.; Duff, J. M., *J. Am. Chem. Soc.* **1968**, *90*, 3876.

David J. Ager
The NutraSweet Company, Mount Prospect, IL, USA

Trimethylsilylacetone[1]

[5908-40-7] $C_6H_{14}OSi$ (MW 130.29)

InChI = 1S/C6H14OSi/c1-6(7)5-8(2,3)4/h5H2,1-4H3

InChIKey = NNWLMTLHOGRONX-UHFFFAOYSA-N

(reacts with Grignard or organolithium reagents to provide, after elimination, 2-substituted 2-propenes;[2] in combination with fluoride ion, gives chemistry of the corresponding enolate ion;[3] can direct electrophilic substitution to either side of the carbonyl group[3b,4])

Physical Data: clear liquid; bp 74 °C/96 mmHg; d 0.8274 g cm^{-3} (26 °C), n_D^{26} 1.4188.

Preparative Method: the best procedure for the preparation of trimethylsilylacetone is the reaction of trimethylsilylmethylmagnesium chloride or bromide with acetic anhydride.[5] Other useful preparative methods for α-trimethylsilyl ketones involve the reaction of trimethylsilylmethylmagnesium chloride with acid chlorides,[6] or with aldehydes followed by oxidation of the resulting β-hydroxysilane to the ketone.[2e] When the trialkylsilyl group is very large, particularly the triisopropylsilyl group, a rearrangement from the silyl enol ether to the α-silyl ketone is possible.[7]

Handling, Storage, and Precautions: α-trimethylsilyl ketones thermally rearrange to the more stable silyl enol ethers at temperatures above approximately 100 °C and normally within 1 h at 140 °C or above; they hydrolyze to the parent ketone upon treatment with acid or base and are difficult to purify by chromatography due to desilylation; use in a fume hood.

Preparation of 2-substituted α-Alkenes. The reaction of trimethylsilylacetone with Grignard and organolithium reagents has been employed in the synthesis of 2-substituted propenes via the formation of a β-hydroxysilane, which is then subjected to a Peterson elimination (eq 1).[2] It was also used to prepare β,γ-unsaturated esters, nitriles, and amides (eq 2).[2b]

$$\text{TMS}\underset{}{\overset{O}{\|}}\ \xrightarrow[\text{ether}]{\text{BuLi}}\ \text{TMS}\overset{OH}{\underset{Bu}{|}}\ \xrightarrow[\text{NaOMe}]{\text{HOAc}}\ \overset{Bu}{=}\qquad(1)$$

$$\text{TMS}\underset{R}{\overset{O}{\|}}\ \xrightarrow{\text{LiCH}_2\text{Y}}\ \underset{R}{=}\!\!\text{CH}_2\text{Y}\qquad(2)$$

Y = CO$_2$-t-Bu, CN, CONMe$_2$

The reaction of acid chlorides with 2 equiv of a trimethylsilylmethyl organometallic reagent, where the metal can be chloromagnesium, lithium, or, best, dichlorocerium, proved to be an excellent direct synthesis of allylsilanes via the intermediacy of an α-trimethylsilyl ketone (eq 3).[2d] By reaction of α-trimethylsilyl ketones with vinylmagnesium bromide, one can prepare 2-substituted 1,3-butadienes (eq 4).[2c,2f]

$$\text{C}_9\text{H}_{19}\underset{Cl}{\overset{O}{\|}}\ \xrightarrow{\text{2 equiv TMSCH}_2\text{CeCl}_2}\ \text{C}_9\text{H}_{19}\diagup\!\!\diagdown\text{TMS}\qquad(3)$$

$$\xrightarrow[\text{2. AcOH, NaOAc}]{\text{1. H}_2\text{C=CHMgBr}}\qquad(4)$$

OTBDMS   OTBDMS

Reaction with Electrophiles. The electrophilic substitution of α-silyl ketones gives the same products as those derived from the corresponding silyl enol ether. Thus reaction with bromine or thionyl chloride gives the α-bromo (or chloro) ketone, with the halogen replacing the trimethylsilyl group (eq 5).[4a,b]

$$\text{TMS}\underset{}{\overset{O}{\|}}\!\!\diagdown\!\!\!<\ \xrightarrow[\text{or SOCl}_2]{\text{Br}_2, \text{CCl}_4}\ (\text{Cl})\text{Br}\underset{}{\overset{O}{\|}}\!\!\diagdown\!\!\!<\qquad(5)$$

The Lewis acid-catalyzed reaction with aldehydes results in aldol condensation to give the β-hydroxy ketone with substitution on the side of the trimethylsilyl group as would be the case with the silyl enol ether. One potential advantage of this reaction is the ability to control the regioselectivity through the synthesis of the α-silyl ketone (eq 6).[3b]

$$\text{TMS}\underset{Et}{\overset{O}{\|}}\ \xrightarrow[\text{PhCHO}]{\text{TiCl}_4}\ \text{Ph}\overset{OH}{\underset{Et}{|}}\!\!\overset{O}{\underset{}{\|}}\qquad(6)$$

threo:erythro mixture

Electrophilic reaction with acetals is also possible, as shown with the optically active acetal (eq 7), which gives very high diastereoselectivity.[8] This has been employed in an approach to aklavinone[9] and *cis*-2-substituted 6-methylpiperidines.[4e]

$$\ +\ \text{TMS}\underset{}{\overset{O}{\|}}\ \xrightarrow[\text{CH}_2\text{Cl}_2]{\text{TiCl}_4}\qquad(7)$$

<3% of other diastereomer

Reaction with Fluoride Ion. The reaction of α-trimethylsilyl ketones with fluoride ion results in attack of the fluoride ion on the trimethylsilyl group and formation of the corresponding enolate ion. This represents an 'in situ' method for regioselective generation of the enolate ion in the presence of the electrophile (eqs 8 and 9). This can be especially advantageous when the more substituted enolate ion is desired.[3]

(8)

(9)

Reaction with Strong Bases. Trimethylsilylmethyl ketones that do not carry another α-substituent can be regioselectively deprotonated on the side bearing the trimethylsilyl group (eq 10). This can lead to α,β-unsaturated ketones when the resulting enolate reacts with aldehydes (eq 11).[10]

major   minor

(10)

(11)

The enolate chemistry of α-substituted α-trimethylsilyl ketones can be directed to either side of the ketone. Thus deprotonation with LDA occurs away from the trimethylsilyl side for steric reasons (eq 12), whereas Lewis acid-catalyzed reactions occur on the trimethylsilyl side of the molecule (eq 13).[10]

(12)

(13)

Related Reagents. *t*-Butyl Trimethylsilylacetate; *N,N*-Dimethyl-2-(trimethylsilyl)acetamide; Ethyl 2-(Methyldiphenylsilyl)propanoate; Ethyl Trimethylsilylacetate; Trimethylsilylacetic Acid.

1. For a review of α-silyl ketones and α-silyl carbonyl compounds consult: (a) Brook, A. G., *Acc. Chem. Res.* **1974**, *7*, 77. (b) Larson, G. L., *Pure Appl. Chem.* **1990**, *62*, 2021.
2. (a) Hudrlik, P. F.; Peterson, D., *Tetrahedron Lett.* **1972**, 1785. (b) Ruden, R. A.; Gaffney, B. L., *Synth. Commun.* **1975**, *5*, 15. (c) Brown, P. A.; Bonnert, R. V.; Jenkins, P. R.; Selim, M. R., *Tetrahedron Lett.* **1987**, *28*,

693. (d) Anderson, M. B.; Fuchs, P. L., *Synth. Commun.* **1987**, *17*, 621. (e) Hudrlik, P. F.; Peterson, D., *J. Am. Chem. Soc.* **1975**, *97*, 1464. (f) Brown, P. A.; Jenkins, P. R.; Fawcett, J.; Russell, D. R., *J. Chem. Soc., Chem. Commun.* **1984**, 253.
3. (a) Fiorenza, M.; Mordini, A.; Papaleo, S.; Pastorelli, S.; Ricci, A., *Tetrahedron Lett.* **1985**, *26*, 787. (b) Inoue, T.; Sato, T.; Kuwajima, I., *J. Org. Chem.* **1984**, *49*, 4671. (c) Paquette, L. A.; Blankenship, C.; Wells, G. J., *J. Am. Chem. Soc.* **1984**, *106*, 6442.
4. (a) Sato, S.; Matsuda, I.; Izumi, Y., *Tetrahedron Lett.* **1985**, *26*, 1527. (b) Benneche, T.; Christiansen, M. L.; Undheim, K., *Acta Chem. Scand.* **1986**, *B40*, 700. (c) Kuwajima, I.; Inoue, T.; Sato, T., *Tetrahedron Lett.* **1978**, 4887. (d) Pellon, P.; Hamelin, J., *Tetrahedron Lett.* **1986**, *27*, 5611. (e) Ryckman, D. M.; Stevens, R. V., *J. Org. Chem.* **1987**, *52*, 4274. (f) McNamara, J. M.; Kishi, Y., *J. Am. Chem. Soc.* **1982**, *104*, 7371. (g) Trost, B. M.; Schneider, S., *J. Am. Chem. Soc.* **1989**, *111*, 4430.
5. Hauser, C. R.; Hance, C. R., *J. Am. Chem. Soc.* **1952**, *74*, 5091.
6. Chan, T. H.; Chang, E.; Vinokur, E., *Tetrahedron Lett.* **1970**, 1137.
7. Corey, E. J.; Rücker, C., *Tetrahedron Lett.* **1984**, *25*, 4345.
8. Johnson, W. S.; Edington, C.; Elliot, J. D.; Silverman, I. R., *J. Am. Chem. Soc.* **1984**, *106*, 7588.
9. Pearlman, B. A.; McNamara, J. M.; Hasan, I.; Hatakeyama, S.; Sekizaki, H.; Kishi, Y., *J. Am. Chem. Soc.* **1981**, *103*, 4248.
10. Matsuda, I.; Okada, H.; Sato, S.; Izumi, Y., *Tetrahedron Lett.* **1984**, *25*, 3879.

Gerald L. Larson
Hüls America, Piscataway, NJ, USA

Trimethylsilylacetonitrile

[18293-53-3]   C$_5$H$_{11}$NSi   (MW 113.26)

InChI = 1S/C5H11NSi/c1-7(2,3)5-4-6/h5H2,1-3H3

InChIKey = WJKMUGYQMMXILX-UHFFFAOYSA-N

(α-carbanion is a useful alternative to LiCH$_2$CN for nucleophilic cyanomethylation; (*Z*)-selective Peterson alkenation)

Physical Data: bp 84–85 °C/54 mmHg,[1] 66 °C/35 mmHg,[2] 65–70 °C/20 mmHg;[3] *d* 0.827 g cm^{-3}.

Solubility: sol most common organic solvents.

Form Supplied in: liquid; commercially available.

Preparative Methods: from chlorotrimethylsilane, zinc, and XCH$_2$CN, (X = Cl, 61% yield; X = Br, 81% yield).[3]

Handling, Storage, and Precaution: is an organic cyanide and should be handled with due care in a fume hood.

Original Commentary

David Watt & Miroslaw Golinski
University of Kentucky, Lexington, KY, USA

Metalation. Exposure to lithium diisopropylamide or *n*-butyllithium in ether[3] or THF at −78 °C generates lithiotrimethylsilylacetonitrile *[70980-14-2]*, which is stable at −78 to −20 °C.[3–5]

Peterson Alkenation (Cyanomethylenation). The anion, typically lithiotrimethylsilylacetonitrile, undergoes Peterson alkenations with aldehydes to furnish α,β-unsaturated nitriles, principally as the (Z) isomer (eq 1),[6–9] in contrast to the Horner–Emmons–Wittig condensation, which gives predominantly the (E) isomer.[9] Use of the boronate-stabilized anion of trimethylsilylacetonitrile, especially in the presence of HMPA, increases the (Z) selectivity (eq 1) but no effect is observed in the case of benzaldehyde.[6,7] Acyclic α-chlorocarbonyl compounds[8] exhibit poor (E/Z) stereoselectivity. Other trialkylsilylacetonitriles (e.g. Ph_3SiCH_2CN, t-$BuMe_2SiCH_2CN$)[6,7] exhibit similar (E/Z) stereoselectivity, but higher homologs of trimethylsilylacetonitrile (e.g. $Me_2PhSiCH(Me)CN$, $Me_2PhSiCH(Et)CN$) give 1:1 (E/Z) mixtures.[10]

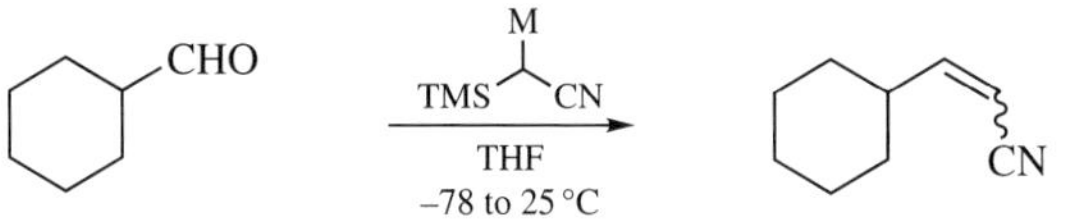

$$M = Li, 82\%, (Z):(E) = 7:1$$
$$M = B(O\text{-}i\text{-}Pr)_2, 90\%, (Z):(E) = 12:1$$

Michael Addition. The reaction of α,β-unsaturated aldehydes[3,6] and ketones[11,12] with trimethylsilylacetonitrile proceeds with 1,2- and 1,4-regioselectivity, respectively. A synthesis of (+)-sesbanimide[11,13] highlights the utility of trimethylsilylacetonitrile in a Michael addition to an α,β-unsaturated ester: lithiotrimethylacetonitrile gives exclusively the 1,4-addition product, whereas lithioacetonitrile gives mainly the 1,2-addition product. Desilylation of the former adduct results in the product of overall 1,4-addition of lithioacetonitrile.

Reactions with Other Electrophiles. The anion of trimethylsilylacetonitrile reacts with allylic, propargylic, and benzylic bromides,[14] with some alkyl bromides and iodides,[5,14] with epoxides to afford γ-(trimethylsilyloxy)nitriles,[3] and with Me_3GeCl,[15] Me_3SiCl,[16] and $(Me_3SiO)_2$[17] to afford the expected α-substitution products. Trimethylsilylacetonitrile also reacts with $(n\text{-}Bu)_3SnOMe$[18] to afford $(n\text{-}Bu)_3SnCH_2CN$.

Cyanomethylation. Trimethylsilylacetonitrile is a useful synthetic equivalent for the anion of acetonitrile in condensation reactions with carbonyl compounds[19–21] and glycosyl fluorides,[22] leading to β-(trimethylsilyloxy)nitriles (eq 2)[21] and cyanomethyl glycosides, respectively. However, the addition of trimethylsilylacetonitrile to hindered ketones using cyanide catalysis fails;[20b] enolizable ketones may form silyl enol ethers;[19,20b] and the β-(trimethylsilyloxy)nitriles derived from cyclic ketones undergo spontaneous elimination to give α,β-unsaturated nitriles.[19]

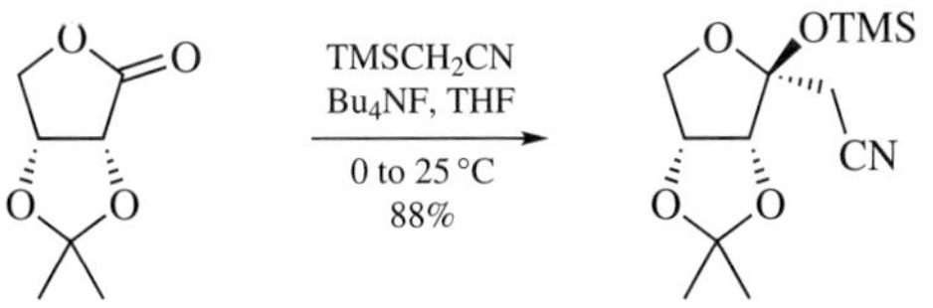

First Update

María Ribagorda
Universidad Autónoma de Madrid, Madrid, Spain

Peterson Alkenation (Cyanomethylenation). Two significant improvements have been developed in the Peterson olefination reaction with α-cyano carbanions and carbonyl compounds. The first involves a catalytic base procedure using nonmetallic organic phosphazene t-Bu-P4 superbase. The corresponding enenitriles were obtained from trimethylsilylacetonitrile (TMSCH$_2$CN) and the corresponding carbonyl compound in the presence of t-Bu-P4 (10 mol %) in good yields, and variable E/Z diastereoselectivities.[23] For example, acetophenone gave the corresponding alkene in 63% yield as an E/Z mixture (89:11) (eq 3), whereas chalcone afforded an E:Z mixture (50:50) of the resultant diene in 80% yield (eq 4).

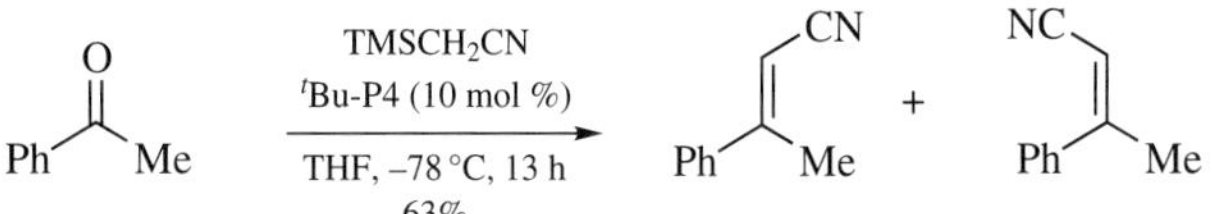

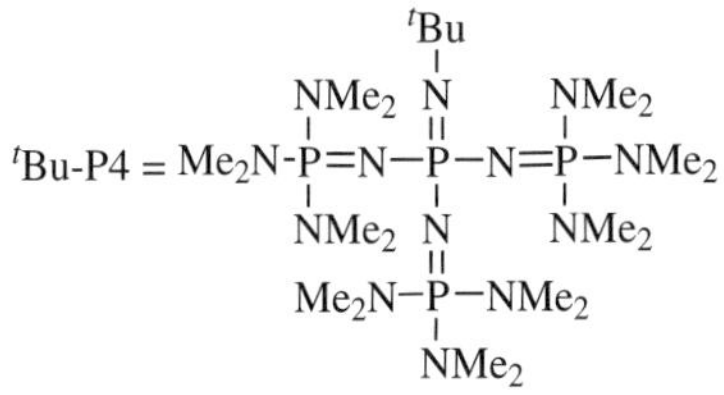

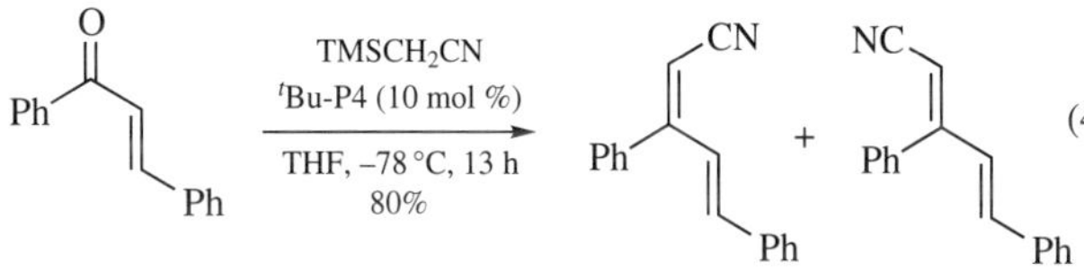

Phosphazene t-Bu-P4 superbase also catalyzed cyanomethylenation of formanilides with trimethylsilylacetonitrile, affording the corresponding enaminoester with moderate to good yields. The procedure is applicable to N-methyl, N-benzyl, or N-allyl formanilides (eqs 5–7).

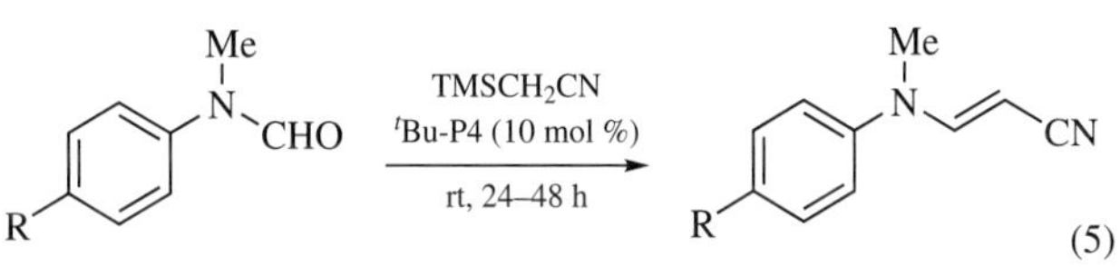

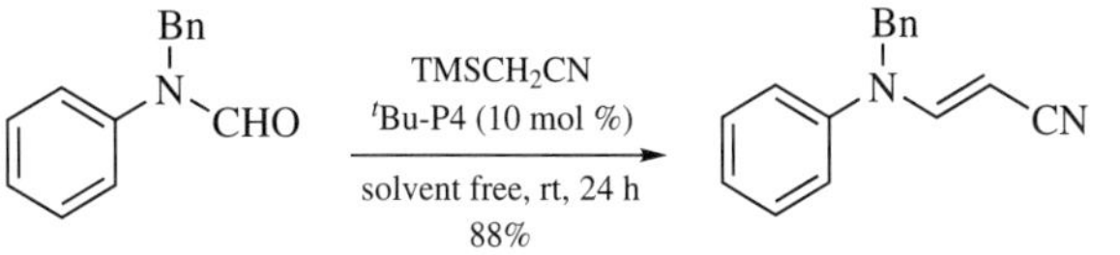

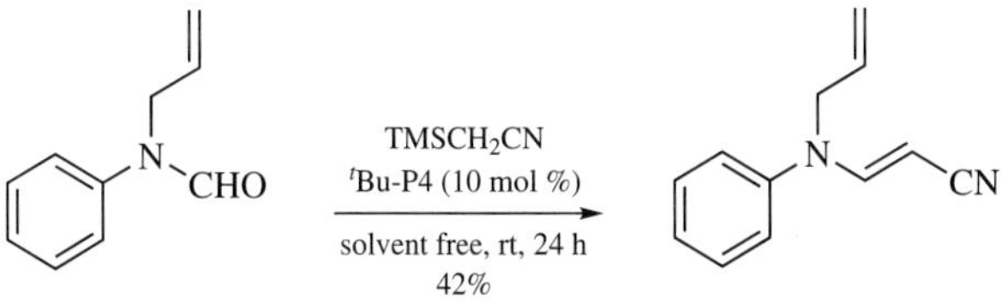

The second improvement developed in the cyanomethylenation reaction concerns the use of a modified *t*-butoxydiphenyl substituted silylacetonitrile, which provides excellent *Z* diastereoselectivities of the final alkene. Thus, $({}^tBuO)Ph_2SiCH_2TMS$ reacts upon base treatment with several aromatic and heteroaromatic aldehydes, leading to the corresponding α,β-unsaturated cyanides in good yields and *Z:E* selectivities from 92:8 to >98:2 (eqs 8–10).[24] Better results were reported using *n*-BuLi, potassium hexamethyldisilazinamide (KHMDS), or sodium hexamethyldisilazinamide (NaHMDS) as base at $-78\,°C$. Aliphatic aldehydes (eq 11) and formyl derivatives with quaternary α-carbons (eq 12) were successfully transformed to the corresponding *Z* olefin. Conjugated aliphatic aldehydes also furnished the alkenylation reaction, rendering conjugated systems of defined geometry (eq 13).

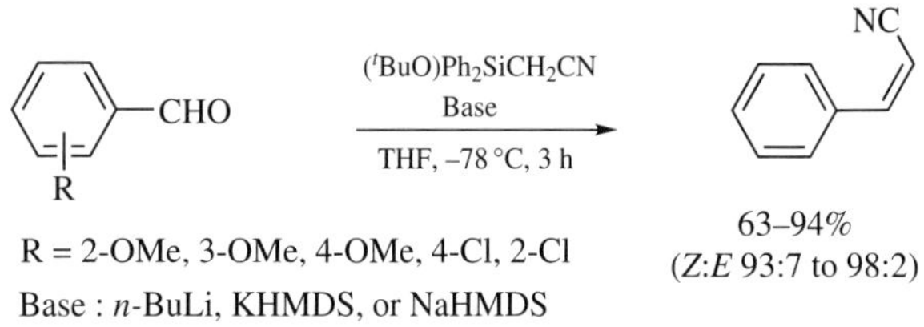

$$
\text{(8)}
$$

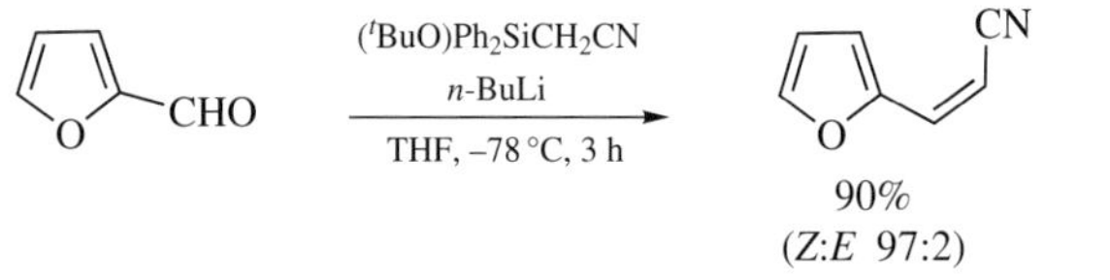

R = 2-OMe, 3-OMe, 4-OMe, 4-Cl, 2-Cl

Base : *n*-BuLi, KHMDS, or NaHMDS

63–94%
(*Z:E* 93:7 to 98:2)

90%
(*Z:E* 97:2)

$$
\text{(9)}
$$

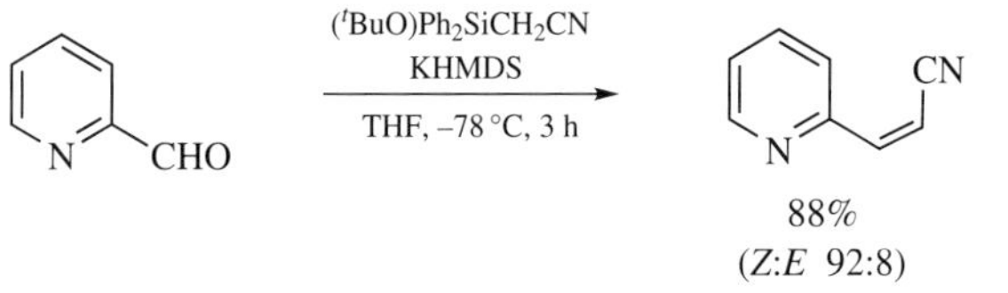

88%
(*Z:E* 92:8)

$$
\text{(10)}
$$

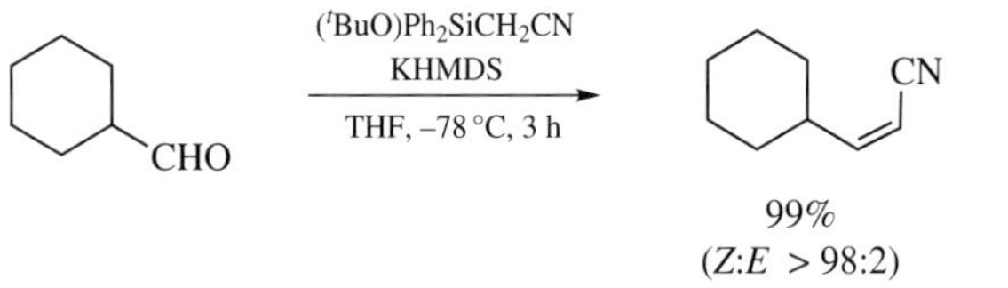

99%
(*Z:E* > 98:2)

$$
\text{(11)}
$$

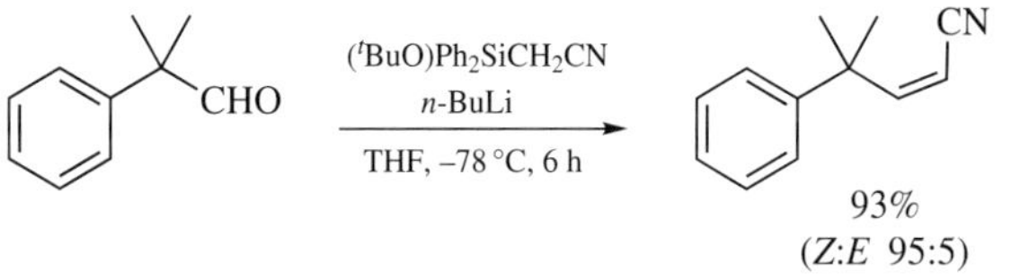

93%
(*Z:E* 95:5)

$$
\text{(12)}
$$

85%
(*Z:E* 94:6)

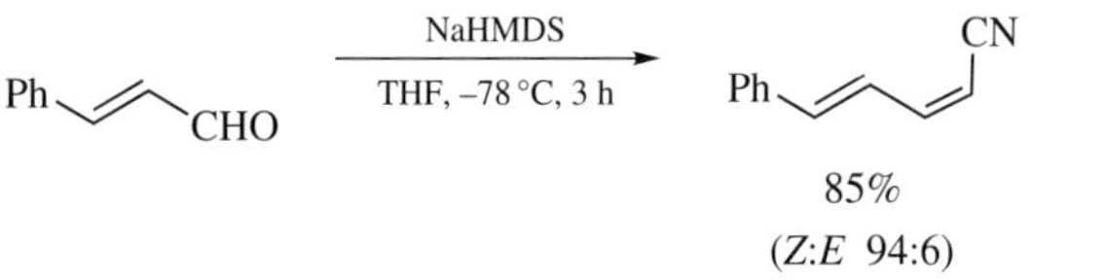

$$
\text{(13)}
$$

Some other examples of Peterson olefination using $TMSCH_2CN$ have been reported to date. For instance, this reaction

has been used in the initial construction of the D ring of the stemodane diterpenoid skeleton of Maritimol, where the addition of the diisopropyl boronate anion of trimethylsilylacetonitrile to a functionalized aldehyde exclusively provides the desired enenitrile as a 6:1 isomeric *E/Z* mixture in 79% yield (eq 14).[25]

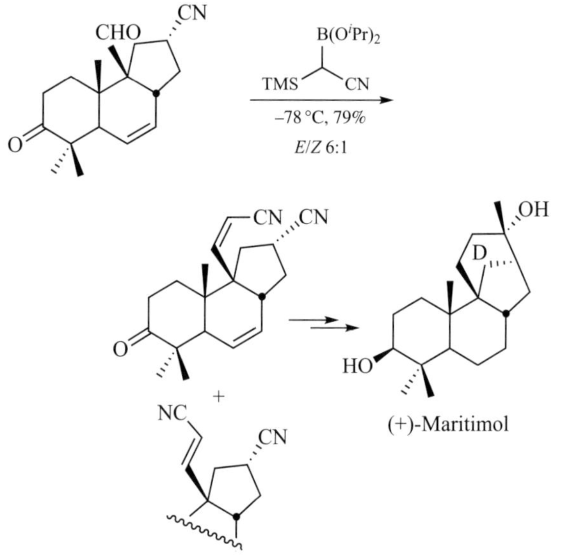

$$
\text{(14)}
$$

A double Peterson alkenylation reaction has been successfully conducted using two equivalents of the lithium anion of trimethylsilylacetonitrile, generated with *n*-BuLi, and the 5,5-dithianenonanedial shown in equation eq 15, affording the corresponding (Z,Z)-α,β-unsaturated dinitrile isomer in 73% yield. The residual material was assigned to the (Z,E)-stereoisomer.[26]

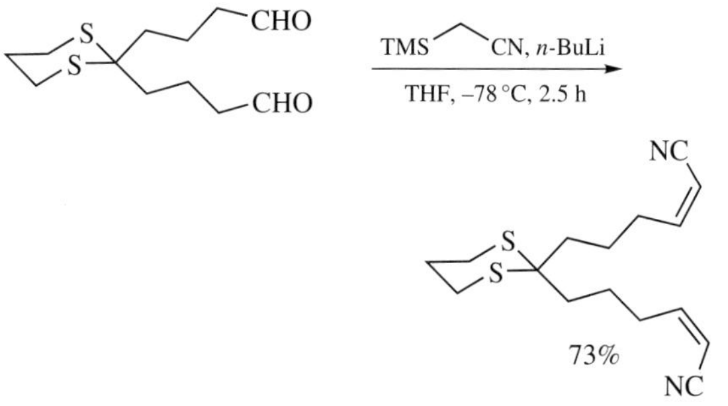

$$
\text{(15)}
$$

73%

Reactions with Other Electrophiles. The reaction of epibromohydrin with an excess of the lithium anion of trimethylsilylacetonitrile, generated with *n*-BuLi, gave trimethylsilyl substituted cyclopropane in good yield and excellent diastereoselectivity. The reaction was performed in the presence of lithium perchlorate as a Lewis acid in THF (eq 16).[27]

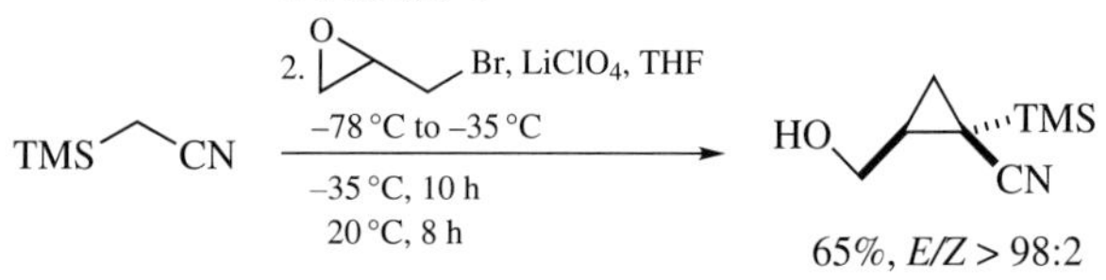

$$
\text{(16)}
$$

65%, *E/Z* > 98:2

Reaction of the lithium anion of trimethylsilylacetonitrile with *N*-benzyl-2-(bromomethyl)aziridines, followed by trimethylsilyl group cleavage by means of an aqueous NaOH workup, gave 2-cyanoethylaziridines in 65% yield (eq 17).[28]

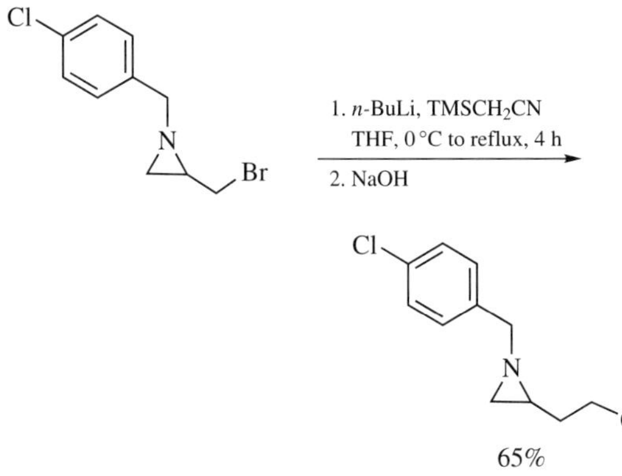

$$(17)$$

65%

Trimethylsilylacetonitrile regioselectively reacts with methyl-quinolinium or isoquinolinium iodides in the presence of a fluoride source, albeit in moderate yields.[29] Cesium fluoride provides better yields than potassium fluoride. *N*-Methylquinolinium iodides exclusively afford the 2-cyanomethyl substituted 1,2-dihydro-*N*-methylquinolines (eq 18), and *N*-methyl-isoquinolinium iodide gave the 1-cyanomethyl compound (eq 19). In the case of 2-methyl substituted *N*-methylquinolinium iodide, an 88:12 mixture of the C-2 and C-4 regioisomers were obtained in 21 and 11% yields, respectively. The 2,3-benzomethylquinolinium or isoquinolinium iodides regios-electively yielded cyanomethylated derivatives at C-4 and C-6, respectively (eqs 20 and 21).

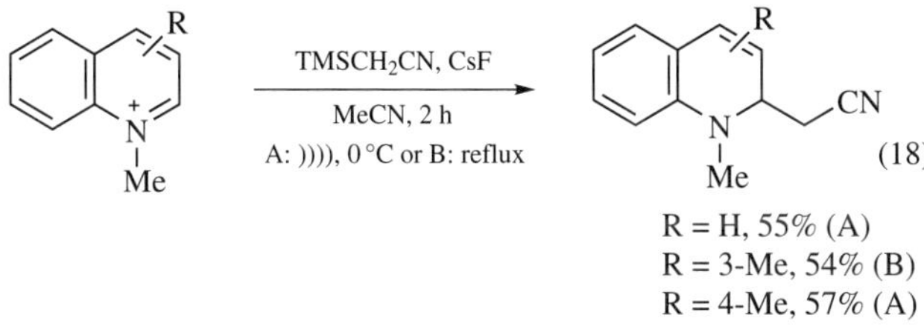

$$(18)$$

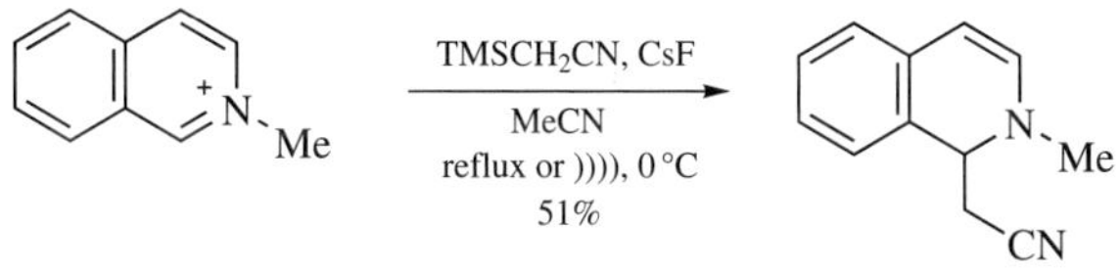

R = H, 55% (A)
R = 3-Me, 54% (B)
R = 4-Me, 57% (A)

$$(19)$$

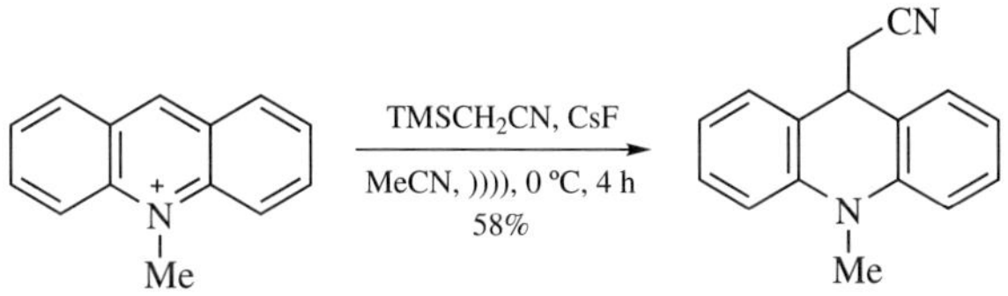

$$(20)$$

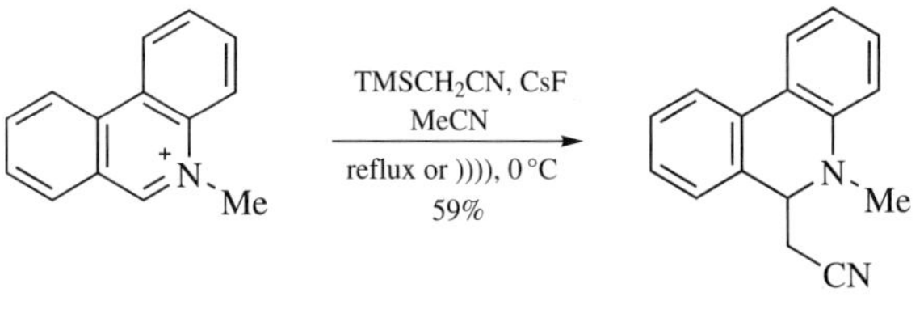

$$(21)$$

Cyanomethylation. Efficient catalytic cyanomethylation reactions of carbonyl compounds with TMSCH$_2$CN have been reported in the presence of catalytic copper complex (CuF·3PPh$_3$·2EtOH) and a stoichiometric amount of (EtO)$_3$SiF. The reaction is followed by a desilylation step, yielding the

corresponding β-hydroxy nitriles in good yields.[30] Dramatic improvement on the reactivity was observed when trimethylsilyl-acetonitrile was slowly added to the reaction mixture. The reaction is generally applicable to a wide range of aromatic aldehydes and aryl methyl ketones (eqs 22 and 23). Under these conditions cyanomethylation successfully proceeds with ketones and alde-hydes having enolizable groups in good to excellent yields (eq 24). Catalytic enantioselective cyanomethylation of acetophenone using chiral (*S*)-tol-BINAP diphosphine as copper ligands gave the corresponding β-hydroxy nitrile with moderate yield (48%) and enantioselectivitity (49% ee) (eq 25).

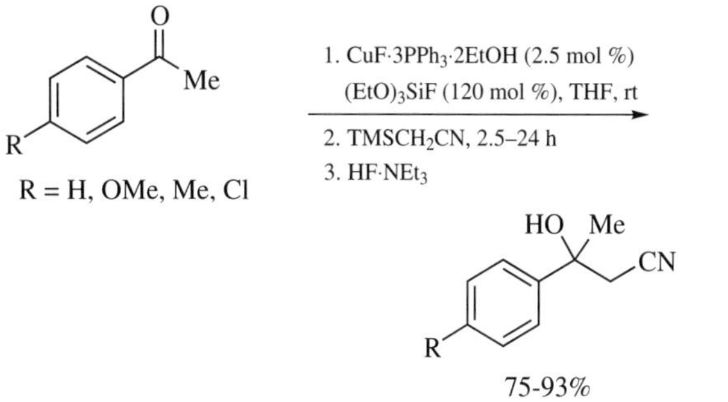

$$(22)$$

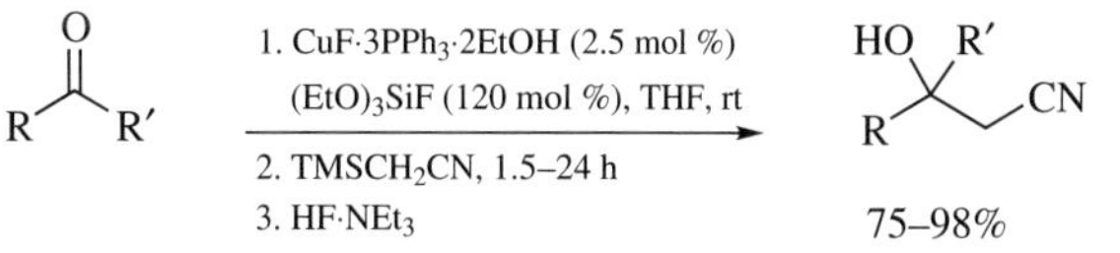

75-93%

$$(23)$$

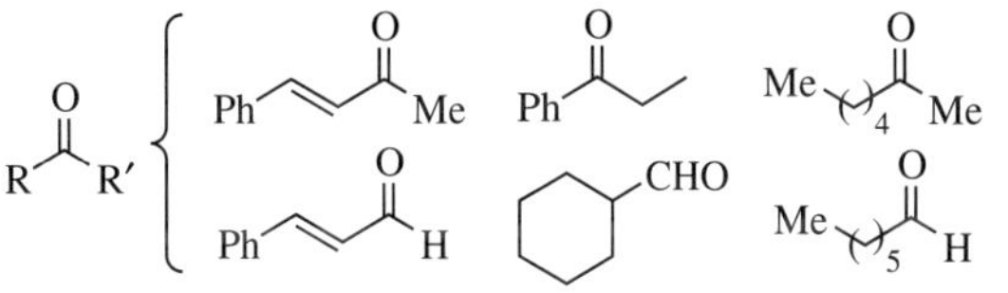

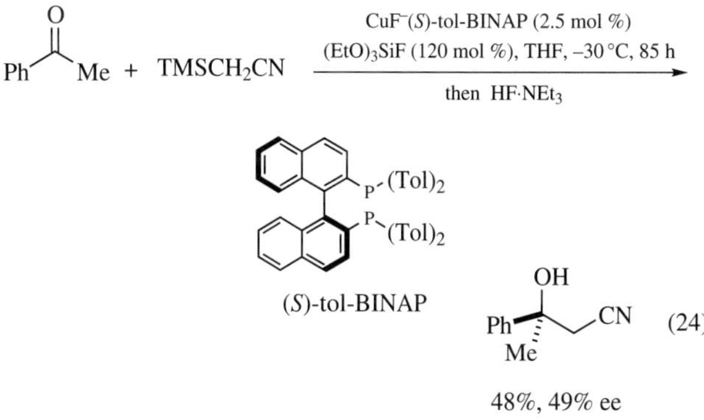

75-98%

(*S*)-tol-BINAP

$$(24)$$

48%, 49% ee

Cyanomethylations of aldehydes and ketones with TMSCH$_2$CN have also been reported in the presence of catalytic lithium acetate (10 mol %) as a Lewis base catalyst. The reaction was performed in a dimethylformamide (DMF) solution at 0 °C, giving rise, after an acid workup, to the corresponding β-hydroxy nitriles in very good yields.[31] Sodium, potassium, cesium, or ammonium acetates behave similarly. The reaction successfully proceeds with sev-eral aromatic aldehydes, with electron-withdrawing and -releasing groups (eq 25), heteroaromatic aldehydes (eq 26), base sensitive aldehydes, and cinnamaldehyde (eq 27), affording exclusively the 1,2 addition product. Specifically, the reaction can be applied to nonenolizable ketones such as benzophenone, phenyl trifluoro-methyl ketone, or chalcone (eq 28). However, competitive silyl enol ether formation occurs with ketones having α-protons, decreasing the yield of the desired β-hydroxy nitrile.

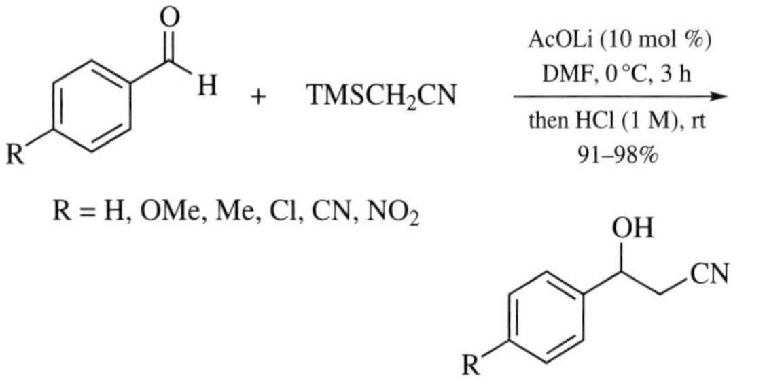

(25)

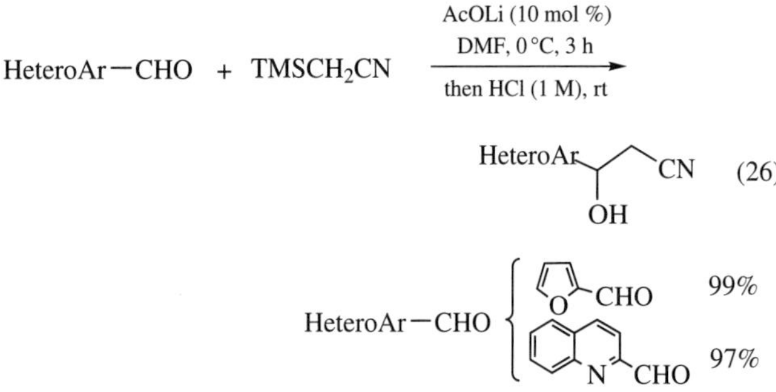

R = H, OMe, Me, Cl, CN, NO$_2$

(26)

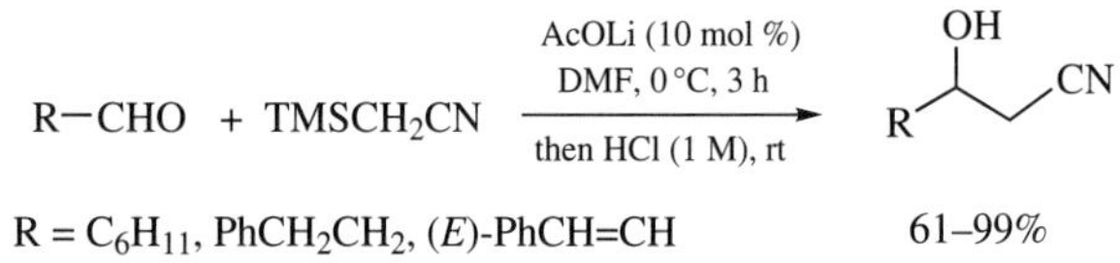

(27)

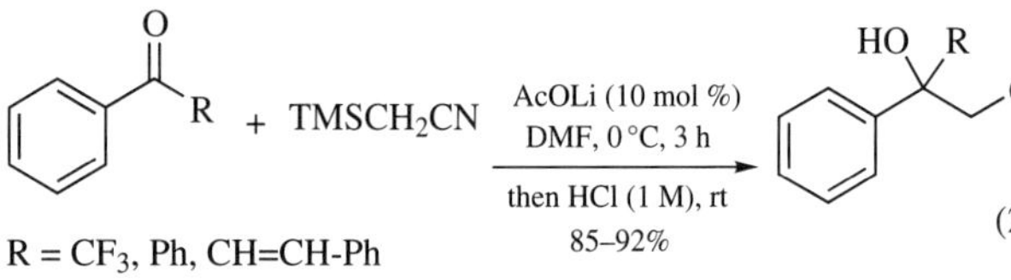

R = C$_6$H$_{11}$, PhCH$_2$CH$_2$, (E)-PhCH=CH   61–99%

(28)

R = CF$_3$, Ph, CH=CH-Ph   85–92%

Under these reaction conditions TMSCH$_2$CN also reacts with several N-tosylarylaldimines, having electron-withdrawing or -donating groups, giving rise to β-amino nitrile compounds in good yields (eq 29).[32] t-Butylaldimine affords the corresponding β-amino nitrile. However, the reaction cannot be performed with aliphatic aldimines having enolizable protons because the competitive abstraction of the α-proton takes place.

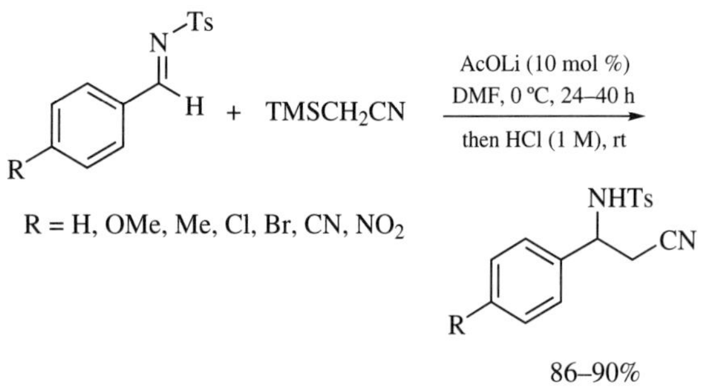

R = H, OMe, Me, Cl, Br, CN, NO$_2$

(29)

86–90%

An improvement [in terms of yield (81%) and amount of supported catalysts] in the synthesis of 3-phenyl-3-(trimethylsilyl-oxy)propionitrile from TMSCH$_2$CN and benzaldehyde has been

reported. The reaction is catalyzed by KF/Al$_2$O$_3$ (eq 30).[33] The reaction solvent is crucial to minimize the elimination product (α,β-unsaturated nitrile).

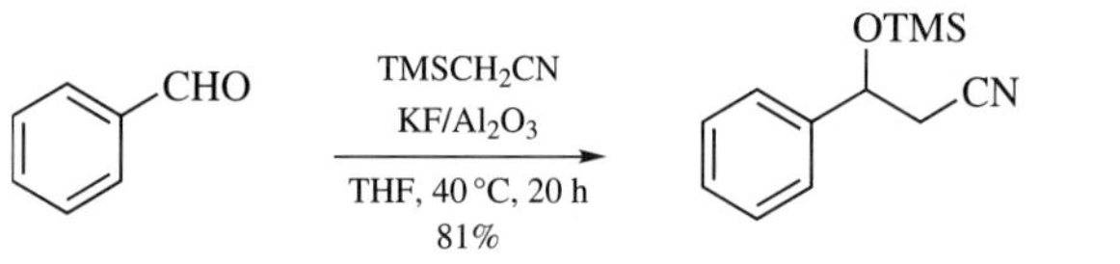

(30)

Metal-Catalyzed Reactions. Palladium-catalyzed cross-coupling reactions between aryl bromides and trimethylsilylace-tonitrile proceed in good to excellent yields.[34] The carbocyanation was conducted using Pd$_2$(dba)$_3$ (dba = dibenzylidenacetone) as a catalyst, and the reaction was performed using a phosphine ligand and ZnF$_2$ as additives in refluxing DMF. Changing the nature of the fluoride salt has a significant effect on the reactivity, since none of the desired reaction product was observed in its absence and other fluoride additives such as CsF or KF exclusively produced the diarylation product. The coupling process occurs in polar solvents [DMF or N-methylpyrrolidinone (NMP)] in high yields while nonpolar solvents inhibit the reaction. The process is generally applicable to an array of aryl bromides incorporating a variety of functional groups. Some representative examples are shown in eqs 31 and 32. Different phosphine ligands were used depending upon the nature of the aryl bromide. Thus, better results for cross-coupling with unhindered aryl bromides were achieved using Xantphos bidentate phosphine ligand. Coupling reactions of ortho-substituted aryl bromides were conducted with P^{t}Bu$_3$ as the palladium ligand and PhP(^{t}Bu)$_2$ was used in the nitrile monoarylation reaction of electron-rich 4-(N,N-dimethylamino) bromobenzene.

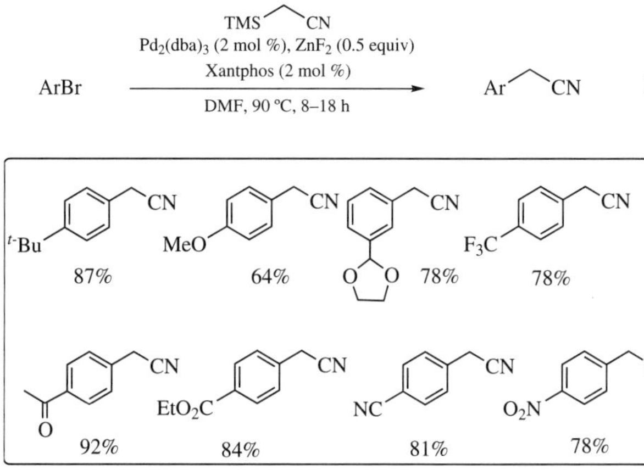

(31)

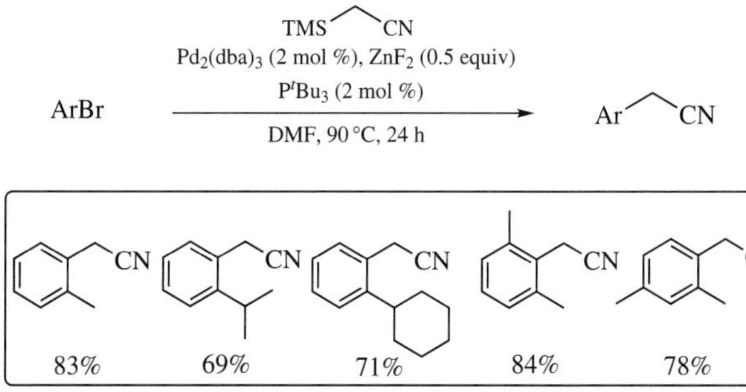

(32)

The nickel-catalyzed arylcyanation of alkynes with aryl cyanides can be conducted in the presence of a Lewis acid. Some alkyl nitriles also take part in the carbocyanation reaction. In particular, reaction of trimethylsilylacetonitrile with 4-octyne in the presence of Ni(cod)$_2$, 2-Mes-C$_6$H$_4$-PCy$_2$, and AlMe$_2$Cl in toluene at reflux, gave the desired allylsilane, albeit in moderate yield (29%) (eq 33).[35]

$$Pr\text{---}\!\!\equiv\!\!\text{---}Pr \xrightarrow[\text{Toluene, 80°C, 17 h}]{\substack{\text{TMS}\diagup\diagdown\text{CN} \\ \text{Ni(cod)}_2\ (5\ \text{mol \%}),\ \text{AlMe}_2\text{Cl(20 mol \%)} \\ \text{2-Mes-C}_6\text{H}_4\text{-PCy}_2\ (10\ \text{mol \%})}}$$

$$
\begin{array}{c}
\text{TMS}\diagdown\qquad\diagup\text{CN} \\
\diagup\qquad\diagdown \\
\text{Pr}\qquad\text{Pr}
\end{array}
\qquad (33)
$$

29%

Rhodium-catalyzed transannulation of 7-halo pyridotriazole with trimethylsilylacetonitrile leads to the generation of imidazopyridine as depicted in equation eq 34.[36] The mechanism proposed involves generation of a Rh-carbenoid, which upon reaction with TMSCH$_2$CN leads to the corresponding imidazopyridine in 70% yield.

$$
\xrightarrow[\substack{\text{Toluene, 60°C} \\ 70\%}]{\substack{\text{TMSCH}_2\text{CN} \\ \text{Rh}_2(\text{OAc})_4\ (1\ \text{mol \%})}}
\qquad (34)
$$

Related Reagents. *t*-Butyl Trimethylsilylacetate; *N,N*-Dimethyl-2-(trimethylsilyl)acetamide; Ethyl Lithio(trimethylsilyl)-acetate; Ethyl 2-(Methyldiphenylsilyl)propanoate; Ethyl Trimethylsilylacetate; Lithioacetonitrile

1. Prober, M., *J. Am. Chem. Soc.* **1955**, *77*, 3224.

2. Ekouya, A.; Dunogues, J.; Duffaut, N.; Calas, R., *J. Organomet. Chem.* **1978**, *148*, 225 (*Chem. Abstr.* **1978**, *88*, 190 961b).

3. Matsuda, I.; Murata, S.; Ishii, Y., *J. Chem. Soc., Perkin Trans. 1* **1979**, 26.

4. Kanemoto, N.; Inoue, S.; Sato, Y., *Synth. Commun.* **1987**, *17*, 1273.

5. Wells, G. J.; Yan, T.-H.; Paquette, L. A., *J. Org. Chem.* **1984**, *49*, 3604.

6. Haruta, R.; Ishiguro, M.; Furuta, K.; Mori, A.; Ikeda, N.; Yamamoto, H., *Chem. Lett.* **1982**, 1093.

7. Furuta, K.; Ishiguro, M.; Haruta, R.; Ikeda, N.; Yamamoto, H., *Bull. Chem. Soc. Jpn.* **1984**, *57*, 2768.

8. Mauzé, B.; Miginiac, L., *Synth. Commun.* **1990**, *20*, 2251.

9. Zimmerman, H. E.; Klun, R. T., *Tetrahedron* **1978**, *34*, 1775.

10. Ojima, I.; Kumagai, M., *Tetrahedron Lett.* **1974**, *15*, 4005.

11. Tomioka, K.; Koga, K., *Tetrahedron Lett.* **1984**, *25*, 1599.

12. Paquette, L. A.; Friedrich, D.; Pinard, E.; Williams, J. P.; St. Laurent, D.; Roden, B. A., *J. Am. Chem. Soc.* **1993**, *115*, 4377.

13. Tomioka, K.; Hagiwara, A.; Koga, K., *Tetrahedron Lett.* **1988**, *29*, 3095.

14. Mauzé, B.; Miginiac, L., *J. Organomet. Chem.* **1991**, *411*, 69 (*Chem. Abstr.* **1991**, *115*, 136 175m).

15. Inoue, S.; Sato, Y., *Organometallics* **1986**, *5*, 1197.

16. Palomo, C.; Aizpurua, J. M.; García, J. M.; Ganboa, I.; Cossio, F. P.; Lecea, B.; Lopez, C., *J. Org. Chem.* **1990**, *55*, 2498.

17. Dembech, P.; Guerrini, A.; Ricci, A.; Seconi, G.; Taddei, M., *Tetrahedron* **1990**, *46*, 2999.

18. Nair, V.; Turner, G. A.; Buenger, G. S.; Chamberlain, S. D., *J. Org. Chem.* **1988**, *53*, 3051.

19. Palomo, C.; Aizpurua, J. M.; López, M. C.; Lecea, B., *J. Chem. Soc., Perkin Trans. 1* **1989**, 1692.

20. (a) Gostevskii, B. A.; Kruglaya, O. A.; Albanov, A. I.; Vyazankin, N. S., *J. Org. Chem. USSR (Engl. Transl.)* **1979**, *15*, 983. (b) Gostevskii, B. A.; Kruglaya, O. A.; Albanov, A. I.; Vyazankin, N. S., *J. Organomet. Chem.* **1980**, *187*, 157.

21. Csuk, R.; Glänzer, B. I., *J. Carbohydr. Chem.* **1990**, *9*, 809.

22. Nicolaou, K. C.; Dolle, R. E.; Chucholowski, A.; Randall, J. L., *J. Chem. Soc., Chem. Commun.* **1984**, 1153.

23. Kobayashi, K.; Ueno, M.; Yoshinor, K., *Chem. Commun.* **2006**, 3128.

24. Kojima, S.; Fukuzaki, T.; Yamawaka, A.; Murai, Y., *Org. Lett.* **2004**, *6*, 3917.

25. Toró, A.; Nowak, P.; Deslongchamps, P., *J. Am. Chem. Soc.* **2000**, *122*, 4526.

26. Stockman, R. A., *Tetrahedron Lett.* **2000**, *41*, 9163.

27. Langer, P.; Freifeld, I., *Org. Lett.* **2001**, *3*, 3903.

28. D'hooghe, M.; Vervisch, K.; De Kimpe, N., *J. Org. Chem.* **2007**, *72*, 7329.

29. Diaba, F.; Houerou, C. L.; Grignon-Dubois, M.; Gerval, P., *J. Org. Chem.* **2000**, *65*, 907.

30. Suto, Y.; Kumagai, N.; Matsunaga, S.; Kanai, M.; Shibasaki, M., *Org. Lett.* **2003**, *5*, 3147.

31. Kawano, Y.; Kaneko, N.; Mukaiyama, T., *Chem. Lett.* **2005**, *34*, 1508.

32. Kawano, Y.; Kaneko, N.; Mukaiyama, T., *Chem. Lett.* **2005**, *34*, 1134.

33. Kawanami, Y.; Yuasa, H.; Toriyama, F.; Yoshida, S.; Baba, T., *Catal. Commun.* **2003**, *4*, 455.

34. Wu, L.; Hartwig, J. F., *J. Am. Chem. Soc.* **2007**, *129*, 2428.

35. Nakao, Y.; Yada, A.; Ebata, S.; Hiyama, T., *J. Am. Chem. Soc.* **2005**, *127*, 15824.

36. Chuprakov, S.; Hwang, F. W.; Gevorgyan, V., *Angew. Chem., Int. Ed.* **2007**, *46*, 4757.

Trimethylsilylacetylene

[1066-54-2]   C$_5$H$_{10}$Si   (MW 98.24)

InChI = 1S/C5H10Si/c1-5-6(2,3)4/h1H,2-4H3

InChIKey = CWMFRHBXRUITQE-UHFFFAOYSA-N

(ethynylation by palladium(0)-catalyzed coupling/condensation with aryl and vinyl halides[1] and triflates,[2] or by nucleophilic attack of the corresponding acetylide on electrophilic centers;[3,4] reacts with alkyl iodides,[5] tin hydrides,[6] and dichloroketene[7] in a regioselective and stereoselective manner)

Alternative Name: trimethylsilylethyne; TMSA.

Physical Data: bp 53 °C; *d* 0.695 g cm^{-3}.

Solubility: sol all organic solvents.

Form Supplied in: colorless transparent liquid; supplied in ampules.

Handling, Storage, and Precautions: once transferred from the ampule to a sample bottle, it can be stored for long periods without loss of purity and material if stored cold. It is a flammable liquid classified as an irritant. Use in a fume hood.

Original Commentary

Godson C. Nwokogu
Hampton University, VA, USA

Ethynylations. All ethynylation processes involve two steps: (a) coupling of the trimethylsilylethynyl group to the substrate either by a palladium(0)-catalyzed reaction or by nucleophilic attack of the derived acetylide on an electrophilic center; and (b) replacement of the trimethylsilyl group with a proton. Although 2-methylbut-3-yn-2-ol can be used for ethynylation and is a much cheaper reagent than TMSA, the advantage of TMSA is in the mild conditions needed for removal of the trimethylsilyl group. Deprotection of the 2-methyl-2-hydroxybutynyl group requires heating in toluene at >70 °C with NaH[8] or NaOH,[9] whereas replacement of the trimethylsilyl group occurs at room temperature and can be effected with dilute aqueous methanol solutions of NaOH or KOH,[10] with LiOH in aqueous THF,[11] or with mild bases such as K_2CO_3[12] or Na_2CO_3[13] in MeOH and KF in aqueous DMF.[14] The yields usually range from good to high.

Palladium(0)-Catalyzed Coupling Reactions. Treatment of the title reagent with vinyl (eq 1)[1b] and aryl (eq 2)[1c] halides and triflates (eqs 3 and 4)[2c,d] and an appropriate Pd catalyst affords vinylalkynes in good yield. For halides, the order of reactivity is I > Br > Cl.[15] Vinyl chlorides undergo this reaction, and one-step diethynylation of dichloroethylenes can be achieved in good yield (eq 5).[16] Aromatic chlorides, however, undergo ethynylation with terminal alkynes only if there is a strong electron-withdrawing group, such as nitro, on the ring.[15]

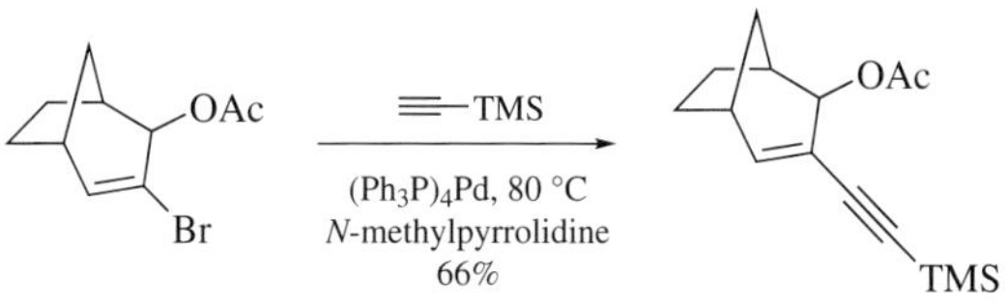

(1)

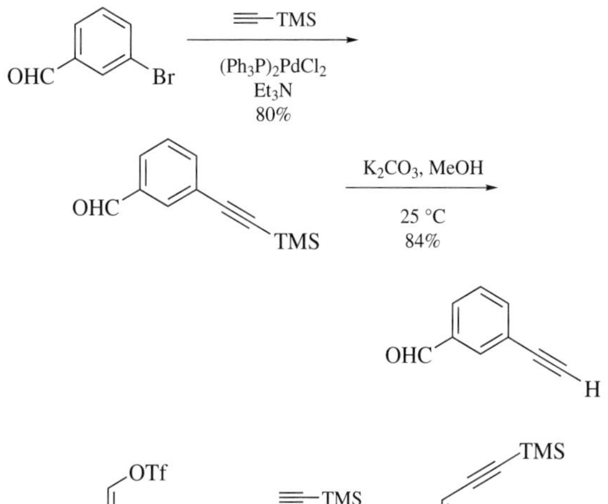

(2)

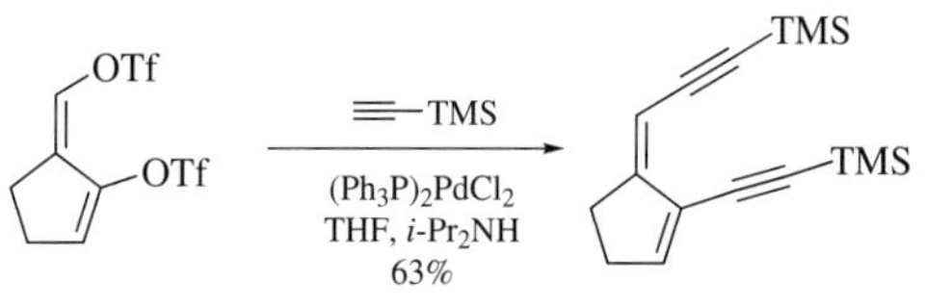

(3)

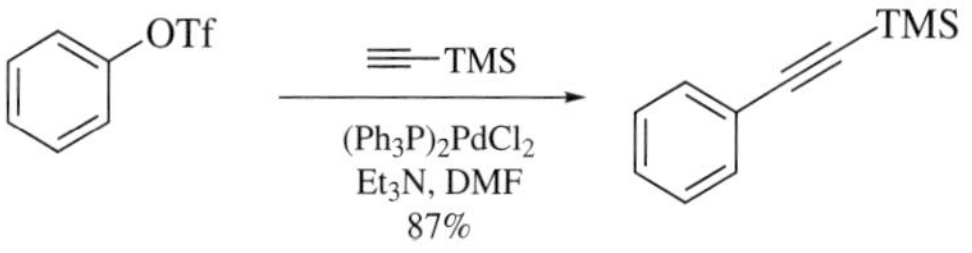

(4)

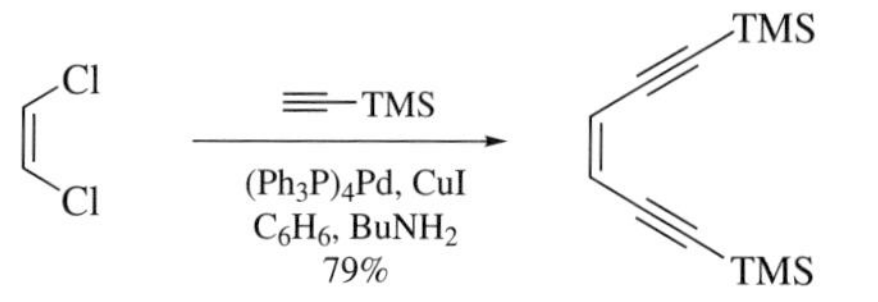

(5)

Heteroaromatic and heterocyclic vinyl halides and triflates can also be effectively ethynylated using TMSA (eqs 6 and 7).[17] The product heterocycles can be elaborated into polycyclic compounds (eq 8).[18]

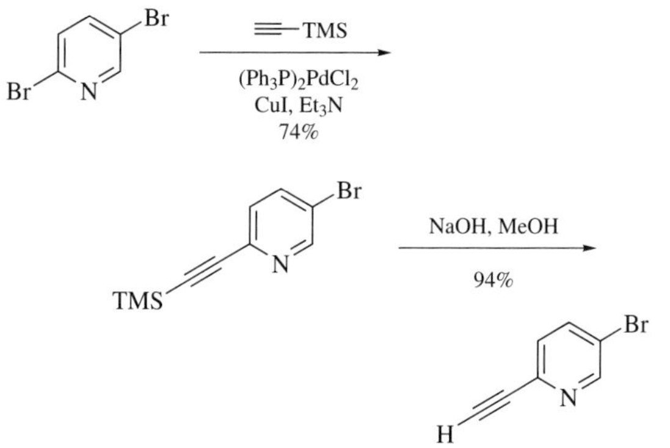

(6)

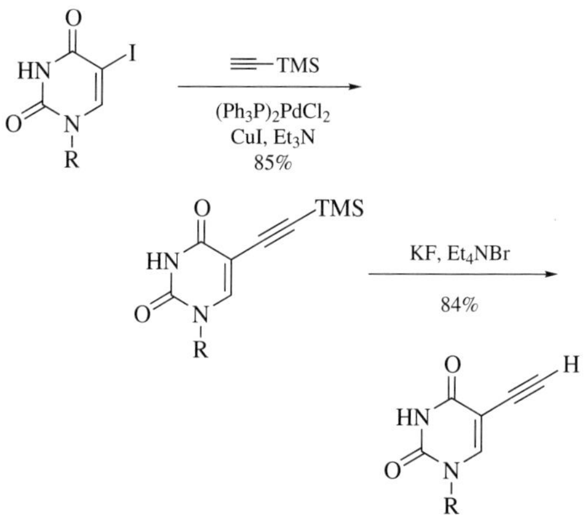

(7)

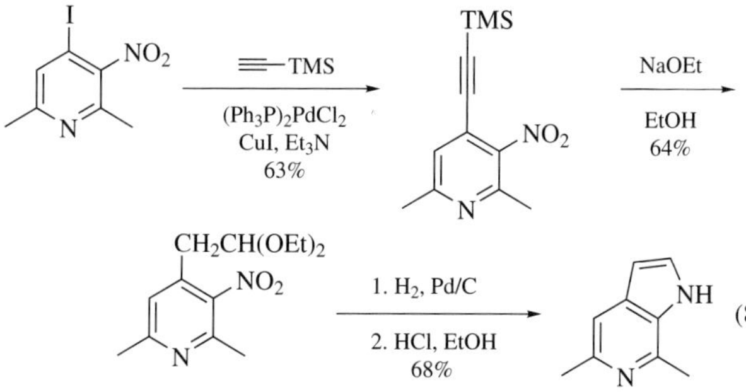

(8)

Aromatic compounds with nitro groups *ortho* to the alkyne have been converted into indoles (eq 9).[19] Alkynyl ketones have been prepared from either the corresponding acyl halides[20] or by carbonylative ethynylation[21] of vinyl halides and triflates with TMSA.

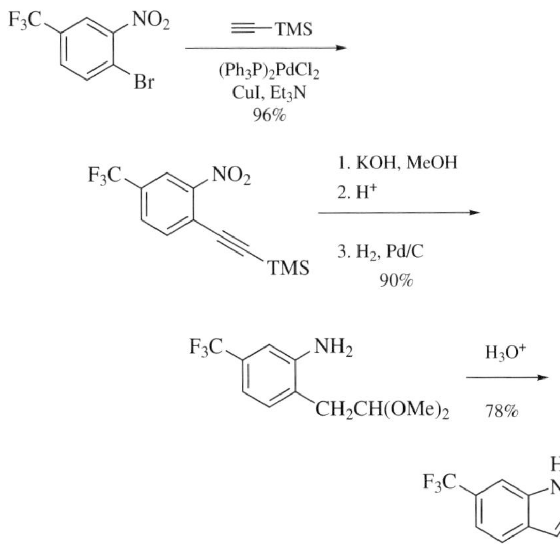

$$\text{(9)}$$

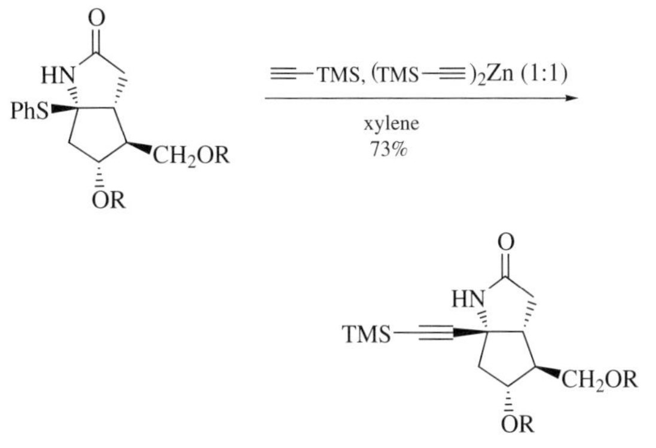

Generally, the palladium(0)-catalyzed reaction requires a base to deprotonate the terminal alkyne. Often, the solvent for the reaction is an amine which also serves as base. Alternatively, the amine can be used in slightly more than a stoichiometric amount in nonbasic solvents such as THF,[2c] benzene,[16b] or DMF.[1b,22] Sodium alkoxides[22] and acetate[2a] have also been used as bases. Various Pd[II] and Pd[0] complexes are effective for the coupling reaction. In one case, Pd[0]-catalyzed coupling of TMSA with a vinyl halide gives appreciable amounts (40–45%) of a fulvene instead of the expected enyne when the reaction is conducted using $(MeCN)_2PdCl_2$ in the absence of CuI (eq 10).[23] With appropriate choice of reaction conditions, however, it is possible to reduce or totally eliminate this fulvene formation (eq 10).

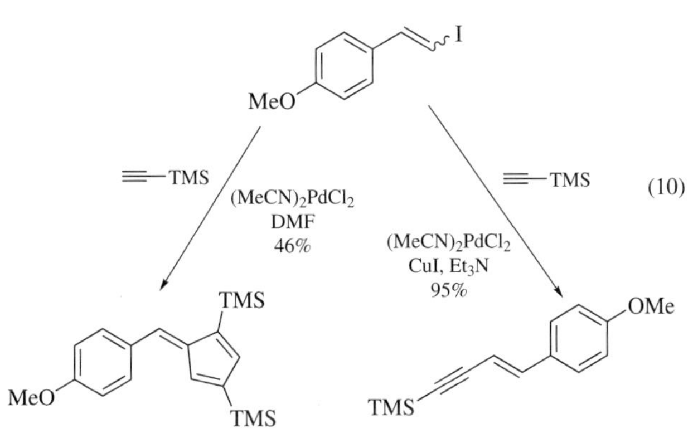

$$\text{(10)}$$

Reaction of Trimethylsilylacetylides with Electrophiles. lithium (trimethylsilyl)acetylide is easily generated from TMSA and *n*-butyllithium[4c,3b] or methyllithium[24] at low temperatures. The corresponding Grignard reagent can be prepared from TMSA and ethylmagnesium bromide. The zinc chloride derivative can be generated by transmetalation of the lithium acetylide.[4a] A cerate derivative has also been described.[25] All of these acetylides add to the carbonyl group of ketones[26] and aldehydes[3,24] to generate alcohols. The tertiary sulfide shown in eq 11 reacts smoothly with zinc acetylides without elimination, a complication that occurs with the more basic Li and Mg acetylides.[4a]

$$\text{(11)}$$

Radical-Initiated and Transition Metal-Catalyzed Additions. Some radical and transition metal-catalyzed additions to TMSA are unique when compared with additions to other terminals alkynes, because they show remarkable regioselectivity and/or stereoselectivity. The regioselectivity of a metal-catalyzed addition may be complementary to that of a radical-initiated process (eq 12). For example, rhodium[27] and molybdenum[28] complex-catalyzed additions of trialkyltin or triaryltin hydrides to TMSA give mainly the 1,1-disubstituted ethylenes, whereas radical hydrostannylation through sonication[29] or triethylborane[6] initiation gives the 1,2-adducts with the (*E*)-isomers predominating. Other terminal alkynes undergo radical or metal-catalyzed hydrostannylation with either poorer or reverse selectivity.

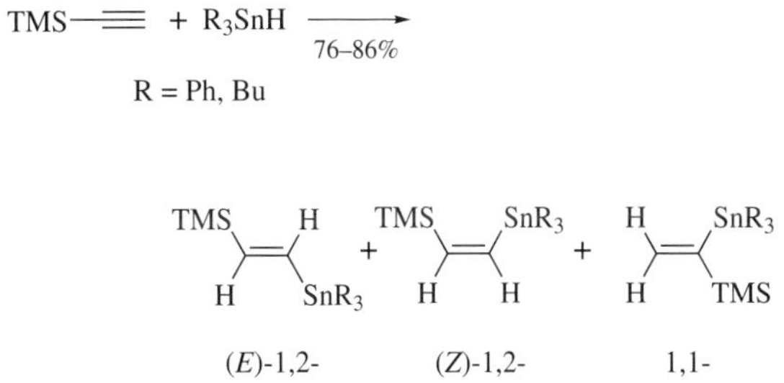

$$\text{(12)}$$

Et$_3$B-initiated radical addition of various alkyl iodides to TMSA occurs with high regioselectivity, giving (*Z*)-1-iodo-1-trimethylsilyl-2-alkylethylenes stereospecifically (eq 13).[5]

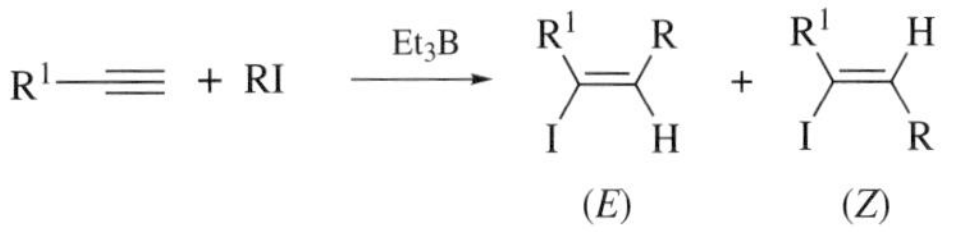

$$\text{(13)}$$

$$R^1 = TMS, R = i\text{-}Pr, 79\% \quad 0:100$$
$$R^1 = Ph, R = i\text{-}Pr, 81\% \quad 21:79$$

Cycloaddition Reactions. Unlike [2 + 2] additions involving other alkynes, TMSA adds to dichloroketene[7,30] and keteniminium salts[31] in a highly regioselective manner to give cyclobutanones (eqs 14 and 15). The regiochemistry in this cycloaddition is opposite to that predicted from the electronic effects of the trimethylsilyl group and has been explained using MO considerations.[30] The cycloaddition of TMSA to Fischer carbene complexes to provide naphthoquinones has also been reported (eq 16).[32]

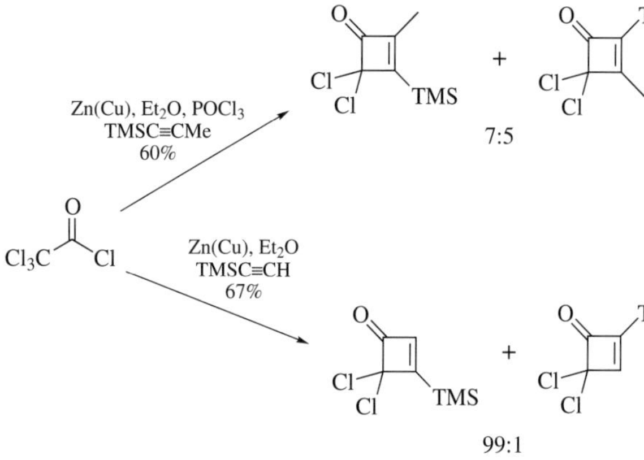

$$\text{(14)}$$

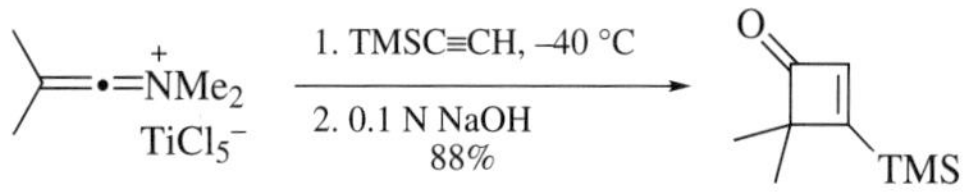

$$\text{(15)}$$

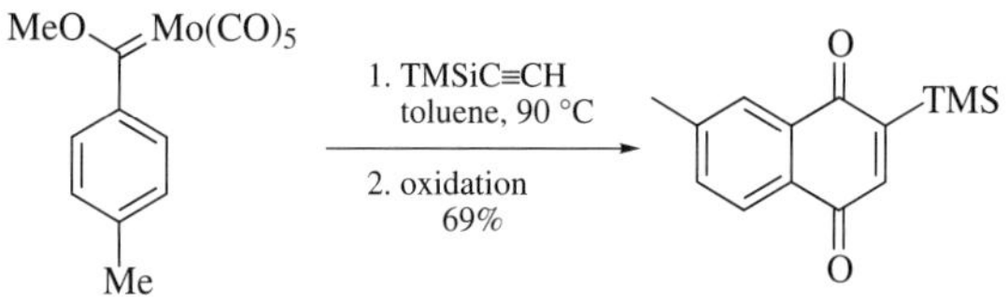

$$\text{(16)}$$

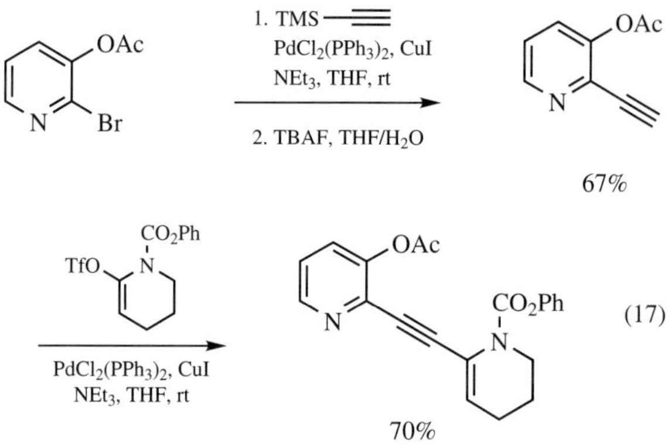

$$\text{(17)}$$

A useful alternative under neutral conditions is the coupling reaction of alkynylsilanes with aryl- (or alkenyl-) triflates using a CuCl/Pd[0] cocatalyst system (eq 18).[35] For the mechanistic pathway the authors presume that a transmetalation from silicon to copper takes place as the keystep (CuI being less effective in the transmetalation), allowing the synthesis of symmetrical and unsymmetrical arylalkynes.

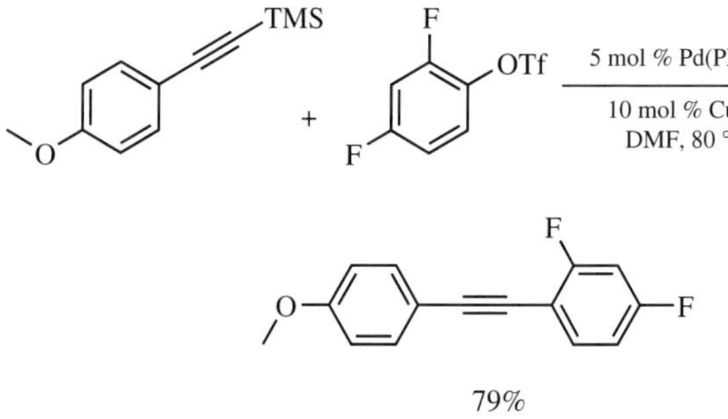

$$\text{(18)}$$

An efficient one-pot synthesis of symmetrical and unsymmetrical bisarylethynes under modified Sonogashira conditions has also been reported. Here the initial aryl-alkynylsilane from the first coupling step is used directly in the second coupling (eq 19).[36]

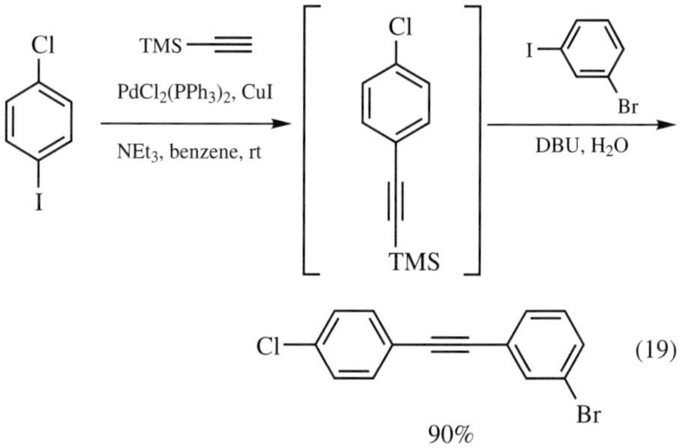

$$\text{(19)}$$

First Update

Saskia Zemolka
Grünenthal GmbH, Aachen, Germany

Florian Dehmel
Altana Pharma AG, Konstanz, Germany

Ethynylations. Trimethylsilylacetylene (TMSA) is one of the most versatile building blocks in organic synthesis to introduce a *C*2 moiety. These ethynylations with TMSA proceed via the terminal end of TMSA, yielding protected acetylenes with an easily removable TMS-protecting group. The deprotection can be accomplished with mild bases or fluoride source (in many cases TBAF) to obtain another reactive terminal acetylene.[33]

Pd[0]-catalyzed Coupling Reactions. TMSA is widely used in Pd-catalyzed cross coupling reactions enabling the introduction of an ethyne moiety onto an aryl- (or alkenyl-) -halide or -triflate. Among the various coupling protocols, the Sonogashira reaction (i.e., the Pd[0]-catalyzed coupling in the presence of a catalytic amount of CuI and an amine base) has taken the most prominent role.

By the combination of two consecutive Sonogashira coupling reactions, the synthesis of internal alkynes, such as bisaryl-alkynes and arylalkenyl-alkynes, can be achieved. Because the terminal silyl group is inert to further coupling under the standard Sonogashira conditions, it must be removed prior to the second coupling step by common methods (e.g., TBAF or basic conditions; eq 17).[34]

The Sonogashira reaction with TMSA could also be successfully accomplished in the coupling with different heteroaryles such as pyrroles,[37] thiophenes,[38] pyrimidines,[39] purines,[40] and quinazolines.[41]

Very often the heteroaryl-alkynylsilanes derived from a Sonogashira coupling with TMSA are further converted into more sophisticated ring systems by subsequent cyclization reactions.

Those tandem processes involve, for example, the preparation of simple indoles (eq 20),[42] fused indoles (eq 21)[43]), benzofurans,[44] and benzothiophenes (eq 22).[45]

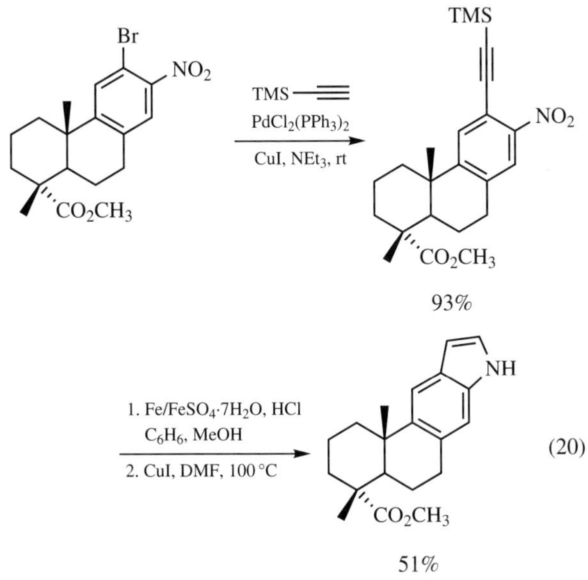

(20)

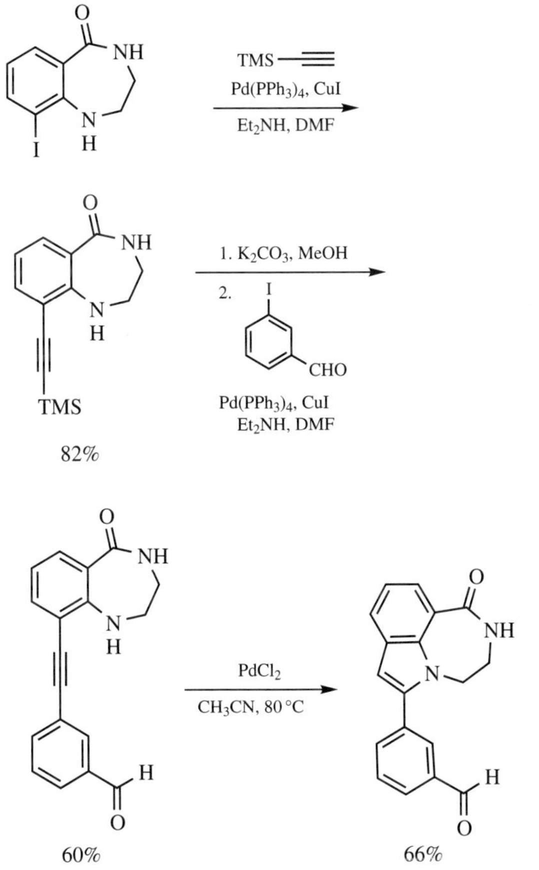

(21)

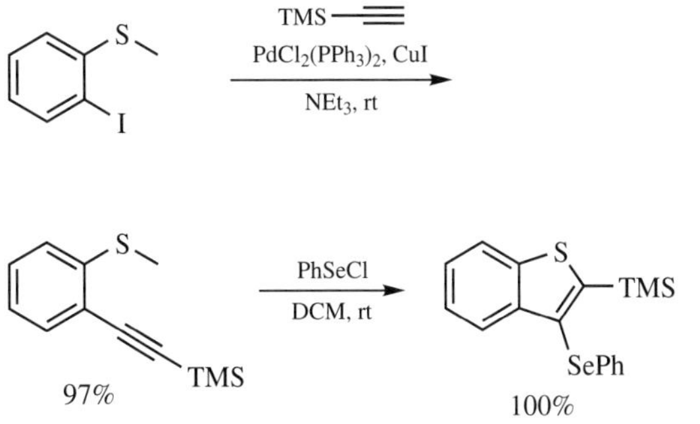

(22)

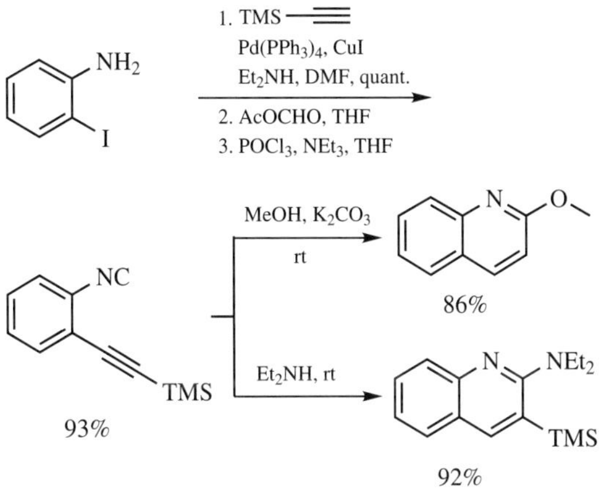

(23)

Also widely used is the combination of a Sonogashira reaction followed by a Pauson-Khand cyclization furnishing annulated cyclopentenones (eq 24).[47]

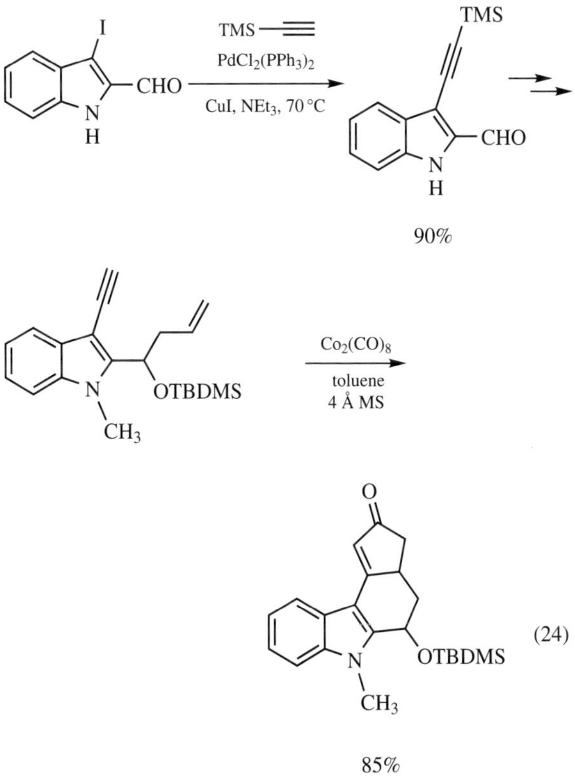

(24)

A similar reaction sequence, which involves a Sonogashira reaction with TMSA followed by a nucleophile triggered 6-*endo*-cyclization, provides access to quinolines (eq 23).[46]

By switching to a heterocumulenic system in place of the olefin moiety, pyrroloindolones can be synthesized (eq 25).[48]

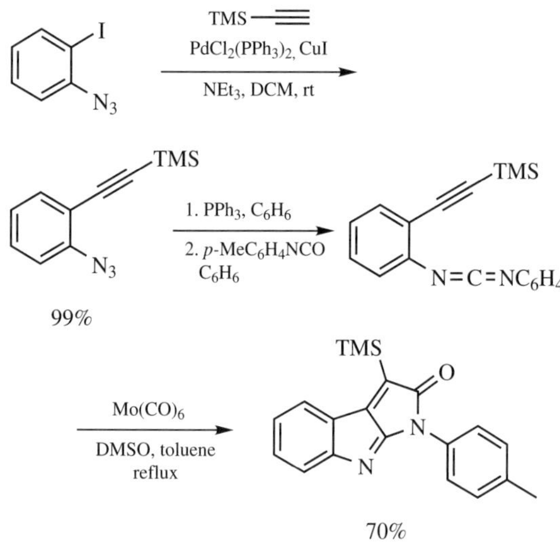

(25)

70%

Recently, the coupling of TMSA with heteroaryl carboxylic acid chlorides furnishing TMS-ynones has been reported. Such TMS-ynones resemble versatile synthetic equivalents of β-keto-aldehydes,[49,50] which can be further elaborated into 2,4-disubstituted pyrimidines (eq 26).

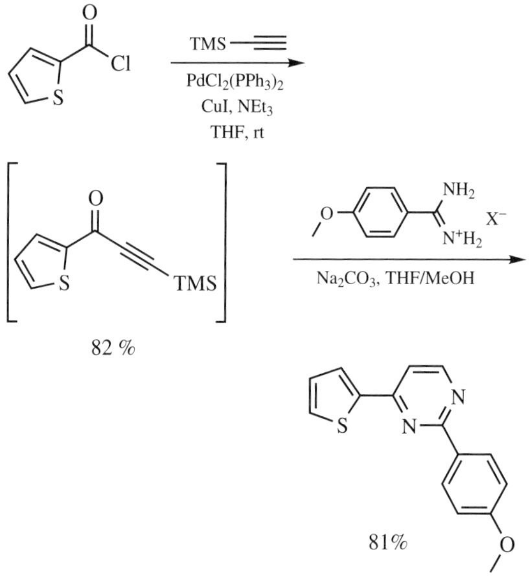

(26)

81%

Even though the Sonogashira reaction is a very versatile reaction carried out under mild reaction conditions, which gives rise to a wide range of terminal and internal acetylenes, it suffers from a major limitation: often the reaction is accompanied by the formation of considerable amounts of dimeric by-products which arise from homocoupling (Hay coupling or Glaser coupling) of the terminal copper-acetylide when exposed to air or an oxidant. These coupling products are not only undesirable from an economic point of view (e.g., in the case of an expensive alkyne), but are also difficult to separate from the desired products. In order to diminish the homocoupling process several approaches have been employed. One possibility is to keep the effective concentration of free acetylene at a minimum, so it is continuously generated in situ by the slow addition of TBAF to the reaction mixture containing

the TMS-masked acetylene.[53] In another approach the reaction is carried out under a dilute hydrogen atmosphere, limiting the occurrence of homomeric side-products to traces of oxygen in the reaction mixture.[52]

A copper- and amine-free version of the Sonogashira reaction has also recently been achieved by employing air-stable aminophosphine ligands in the presence of an inorganic base, thus excluding the formation of Glaser coupling products.[53]

Besides the widely used Sonogashira reaction with TMSA (utilizing in situ prepared copper-acetylides), the introduction of the TMSA moiety via Negishi-type coupling reactions of the corresponding zinc acetylides,[54] Suzuki-Miyaura reactions using air-stable potassium alkynyltrifluoroborates,[55] cross-coupling reactions employing indium acetylides,[56,57] and sodium tetraacetylidealuminates[58] have been reported.

Reaction of Trimethylsilylacetylene/Acetylide with Electrophiles. Deprotonation of TMSA with n-BuLi or Grignard reagents produces nucleophilic acetylides, which can react with various electrophilic carbon centers such as carbonyls, alkyl halides, or epoxides.

The products of the nucleophilic addition to carbonyl groups are valuable precursors for further transformations such as intramolecular Pauson-Khand reactions (eq 27),[59] ring expansions (eq 28),[60] rearrangements affording silylated allenes (eq 29),[61] or reductions and halogenation reactions.[62]

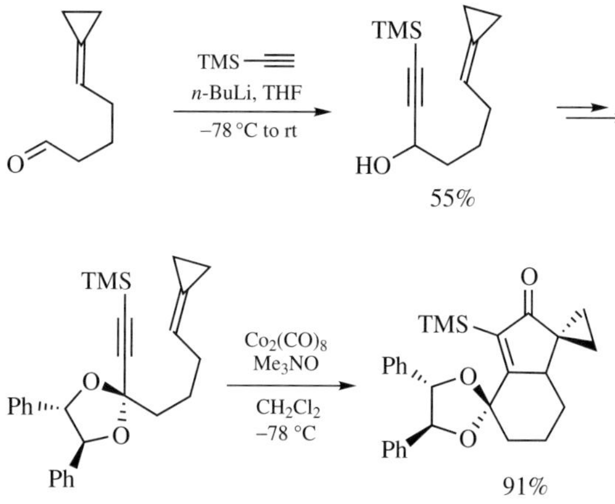

(27)

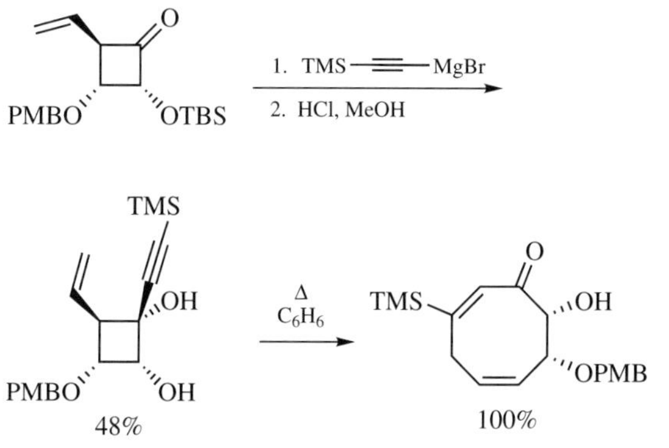

91%

(28)

100%

The magnesium acetylide of TMSA can also react by nucleophilic addition to activated pyrimidines yielding alkynylated dihydropyrimidinones (eq 30).[63]

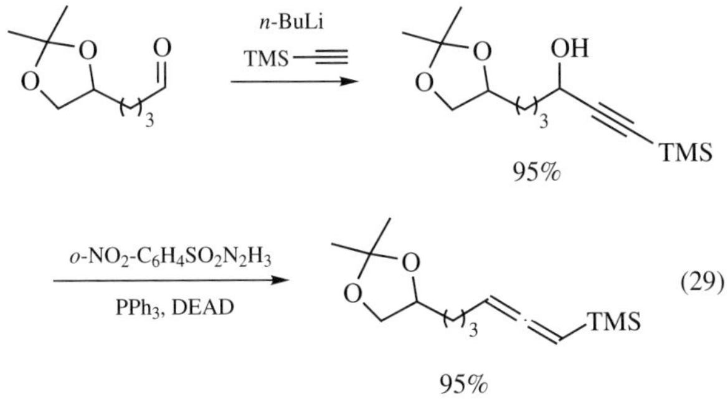

TMSA can also undergo Lewis acid- or transition metal-promoted additions (e.g., with Zn(OTf)$_2$,[64] AuBr$_3$,[65] InBr$_3$[66] or [Ir(COD)]$_2$[67]) to carbonyl centers or imines providing access to propargylic alcohols or amines.

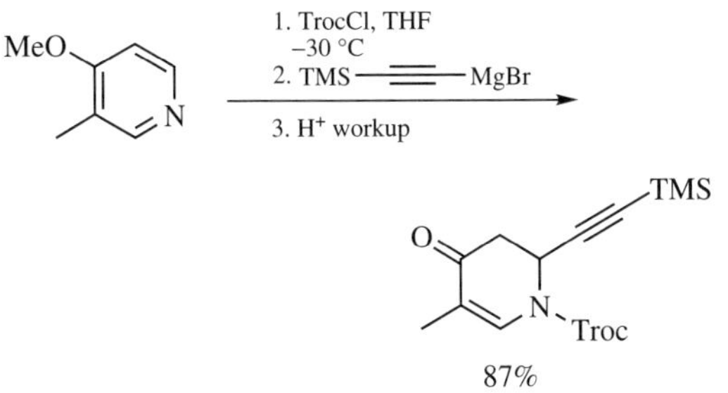

Enantioselective formation of secondary propargyl alcohols by the addition of TMSA to the carbonyl group of aldehydes have been reported employing Zn(OTf)$_2$ and N-methylephedrine (eq 31).[68,69]

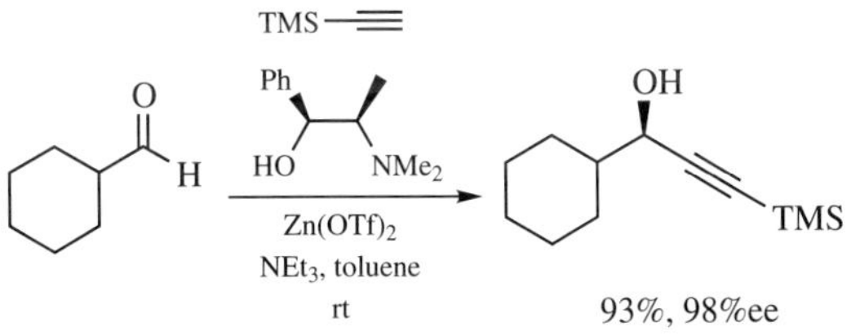

Furthermore, the diastereoselective addition of TMSA employing Et$_2$Zn, Ti(Oi-Pr)$_4$, and a chiral ligand has been reported.[70]

Chiral propargylic amines are also accessible when TMSA is added to imines[71,72] or enamines[73] (eq 32) in the presence of Zr(Oi-Pr)$_4$ or CuBr and chiral ligands.

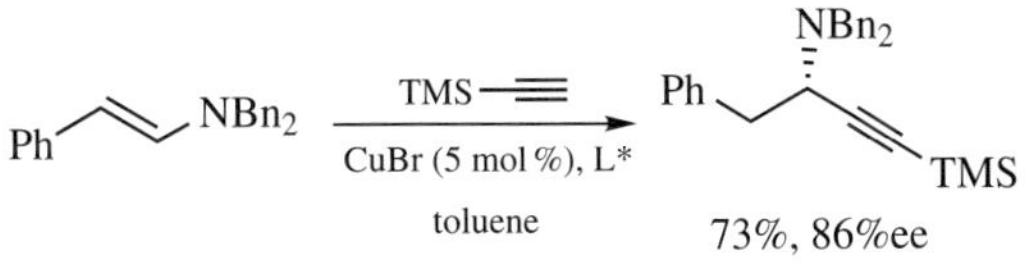

The Zn(OTf)$_2$-mediated diastereoselective addition of TMSA to nitrones leads to optically pure N-hydroxylamines.[74]

The low basicity of deprotonated TMSA allows S$_N$2-type reactions with primary alkyl halides without a competing elimination reaction. In the given example,[75] the TMS protection is removed under Lewis acidic conditions (TiCl$_4$) and the resulting terminal alkyne is reacted in situ with an allyl pivaloate (eq 33).

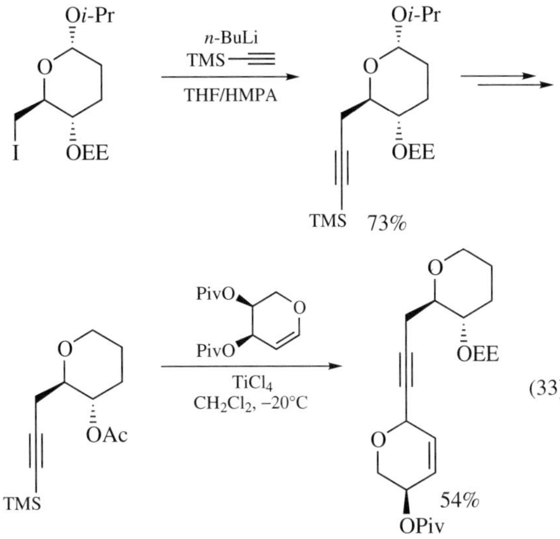

If the TMSA-derived acetylide is added to a ketone bearing a leaving group in the α-position, trisubstituted epoxides can be obtained (eq 34).[76]

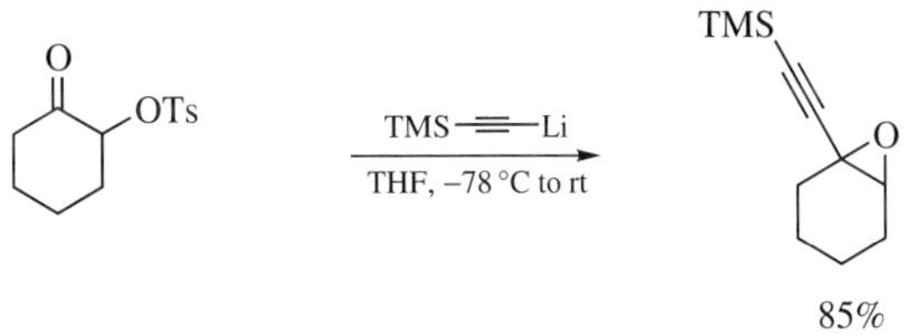

Because of the soft nature of TMS-acetylides, the nucleophilic ring opening of epoxides can be accomplished representing an efficient method for the preparation of homopropargylic alcohols (here exemplified by a sequence from the total synthesis of Epothilone A by Danishefsky) (eq 35).[77] In this synthesis, the TMS group again serves as more than just a protecting group for the acetylene but represents the precursor for an iodine, which is subsequently converted into a (Z)-iodo-alkene.

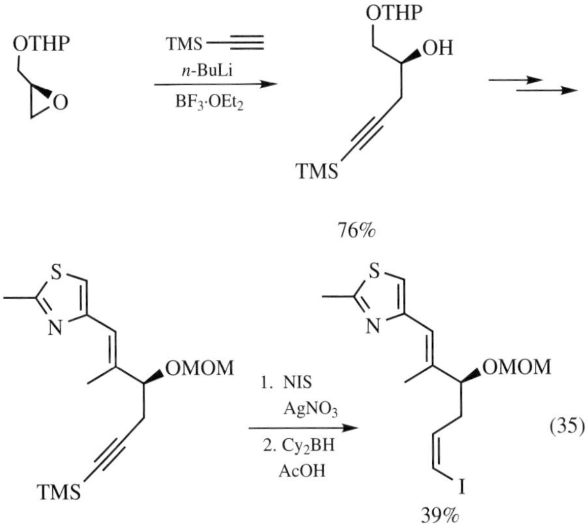

Cycloaddition Reactions. TMSA can serve as a $C2$-unit in the assembly of hetero-aromatic and -aliphatic ring systems. Four-membered rings can be easily prepared by the cobalt-catalyzed [2 + 2] cycloaddition onto alkenes (eq 36).[78]

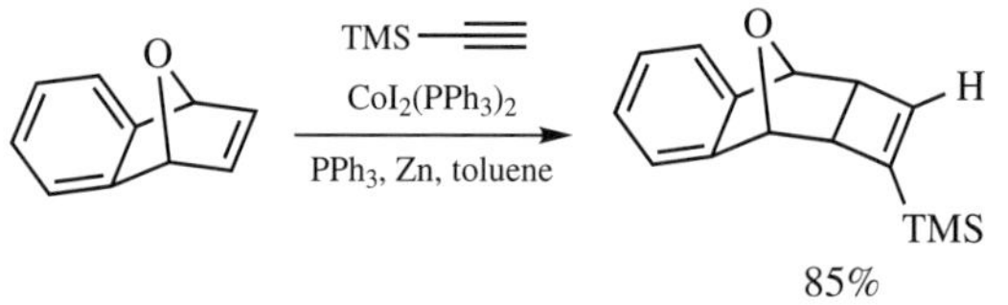

$$(36)$$

85%

$$(40)$$

45%

Another cobalt-mediated cyclization reaction is the Pauson Khand reaction (PKR). In a [2 + 2 + 1] fashion, the reaction of TMSA with alkenes delivers regioselectively silylated cyclopentenones. Following the trend typically observed in the PKR, the bulky TMS-group is always positioned α to the carbonyl group (eqs 37 and 38).[79,80]

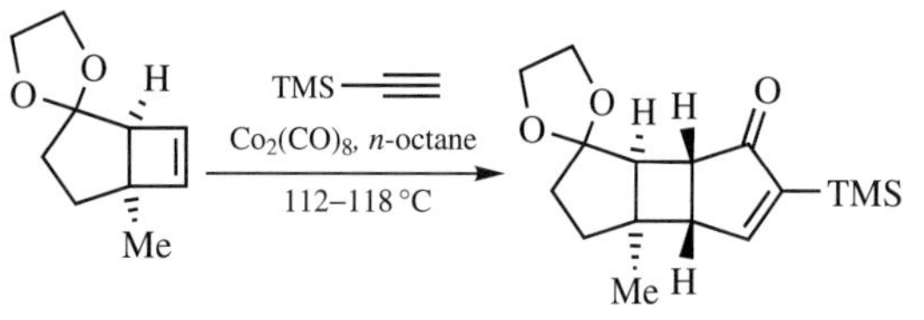

$$(37)$$

86%

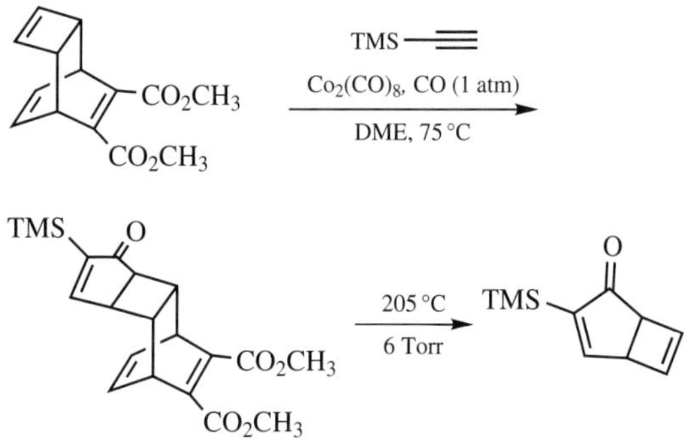

86%   85%   (38)

Various five-membered heterocycles such as triazoles (eq 39)[81] and isoxazoles (eq 40)[82] are also accessible from TMSA through [2 + 3] dipolar cycloaddition reactions. Because of the steric demand of the TMS-group, these reactions proceed regiospecifically.

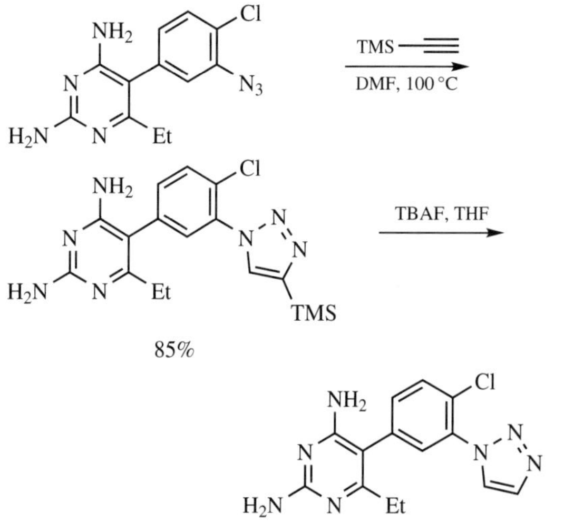

85%

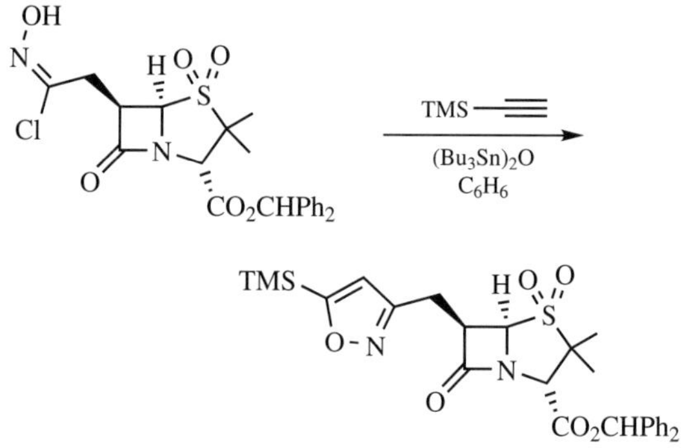

$$(39)$$

74%

TMSA may also be used for the construction of six-membered rings. In an intermolecular Reppe-type [2+2+2] cyclotrimerization catalyzed by cationic rhodium complexes substituted arenes (eq 41) are generated in a highly chemo- and regioselective fashion.[83]

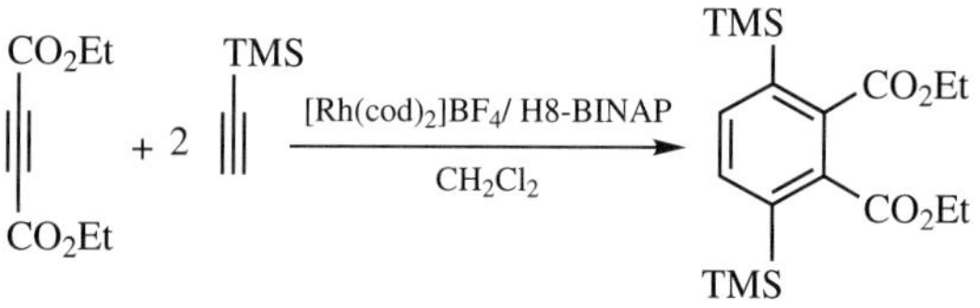

$$(41)$$

57%

Similar regioselectivities are observed when a titanium-calixarene complex was used as the catalyst.[84]

A remarkable example for the chemo- and regioselectivity observed during a [2 + 2 + 2] reaction represents the intermolecular cyclotrimerization between TMSA on one hand, another terminal alkyne on the other hand, and an enone[85] mediated by a binary Ni/Al catalytic system (eq 42).

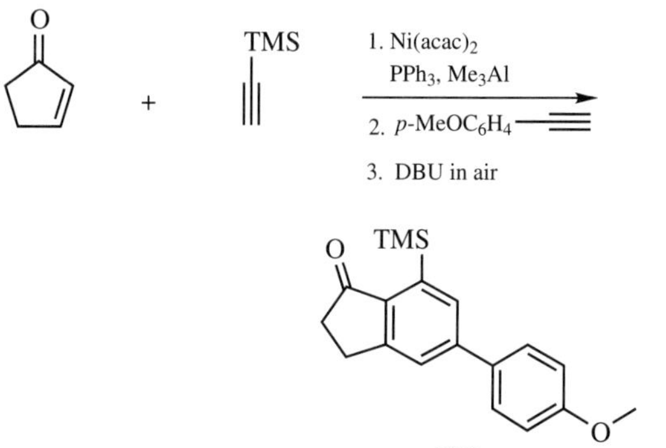

$$(42)$$

60%

Furthermore, substituting a nitrile for the alkyne as the cyclization partner provides access to silylated pyridines (eq 43).[86]

$$\text{(43)} \quad 65\%$$

$$\text{(46)} \quad 90\%$$

1,4-Dihydrobenzenes are obtained when TMSA is used as the dienophile in homo Diels-Alder reactions under Co^{II}-[87] or Rh^{I}-[88] catalysis.

TMSA serves furthermore as a reagent in the [3 + 2 + 1] benzannulation with Fischer carbene complexes (Dötz reaction) furnishing silylated naphthoquinones in good yields (eq 44).[89]

$$\text{(44)} \quad 61\%$$

Syntheses of naphthalenes derived from a [4 + 2] benzannulation reaction with TMSA have also been reported.[90]

A Rh^{I}-catalyzed tandem one-pot coupling cyclization reaction with TMSA and trifluoroacetimidoyl chlorides delivers 2-trifluoromethylated quinolines with good regioselectivity (eq 45).[91]

$$\text{(45)} \quad 82\% \quad (95:5)$$

TMSA can also be utilized in [5 + 2] cycloadditions to assemble cycloheptadienes (eq 46).[92]

Other cycloaddition reactions with TMSA forming seven-[93] or eight-membered rings[94] have been described.

Further Transformations. The ethynylated products derived from the introduction of the TMSA-moiety are valuable precursors for further transformations.

The easily removable TMS-protecting group provides access to a free alkyne. The mild conditions for the deprotection allow the discrimination between the TMS group and other acetylene protecting groups (e.g., TIPS or TBDMS) and therefore the selective generation of monofunctionalized enediynes (eq 47).[95,96]

$$\text{(47)} \quad 62\% \quad 95\%$$

It is also possible to remove the TMS-group in situ for subsequent conversions. CuCl in 1,3-dimethyl-2-imidazolidinone (DMI) is an effective combination for the direct conversion of TMS-acetylenes and acid halides into ynones (eq 48).[97] Other reagents such as CsOH:CsF (1:1)[98] or InBr$_3$[99] show similar reactivity.

$$\text{(48)} \quad 77\%$$

Moreover, the TMS group may serve as a precursor for synthetically valuable alkynyl halides (eq 49).[100] The alkynyl halides can be directly used in further coupling reactions (e.g., Nozaki-Hiyama-Kishi reactions) or can be selectively reduced to yield (Z)-alkenyl halides (eq 50).[101]

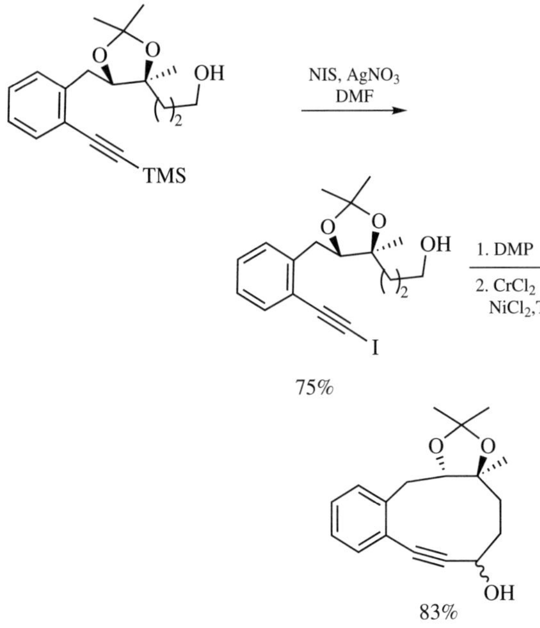

The corresponding (E)-alkenyl halides (eq 51)[102] as well as α-bromostyrenes[103] are also readily accessible from the TMSA-ethynylated products.

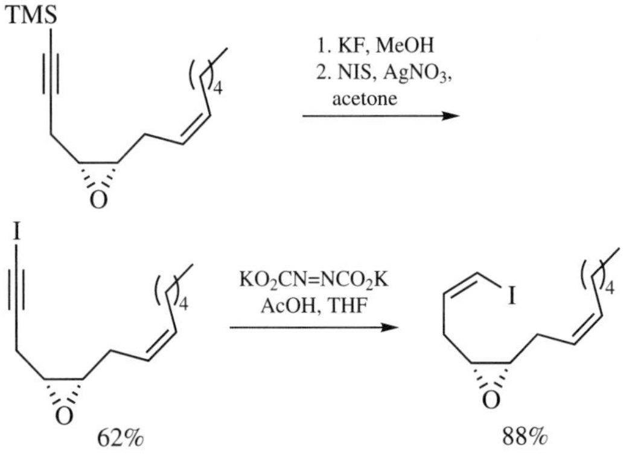

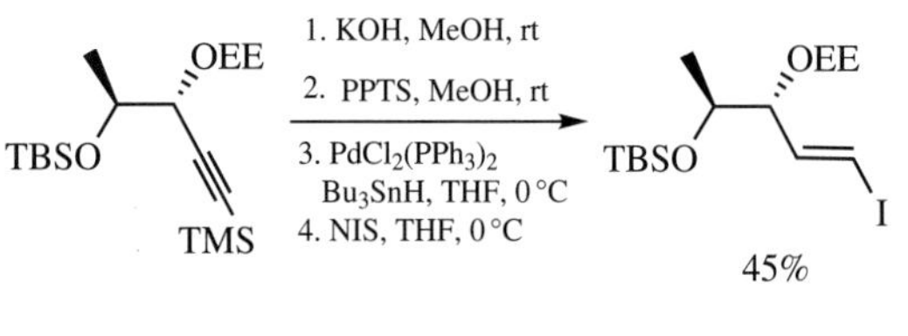

The alkyne moiety of the coupling products between TMSA and aryl halides is also of great synthetic value, as these compounds are interesting substrates for the preparation of aryl acetic acids (eq 52)[104] or acetophenones (eq 53).[105]

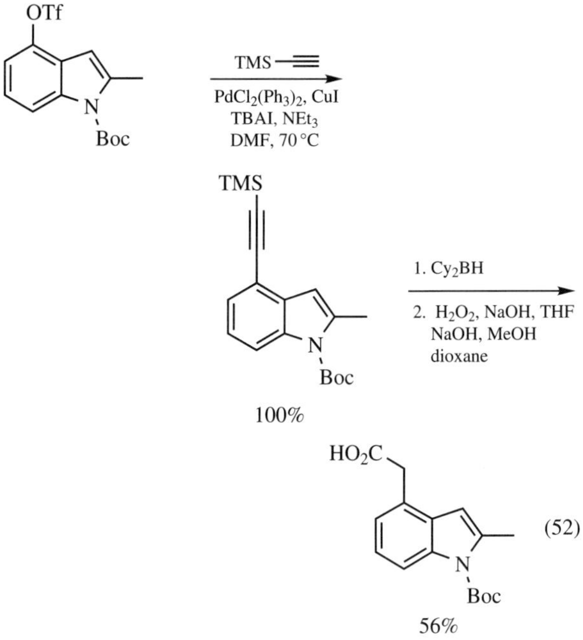

Related Reagents. Triisopropylacetylene; t-butyldimethyl-silylacetylene; 2-methyl-but-3-yn-2-ol.

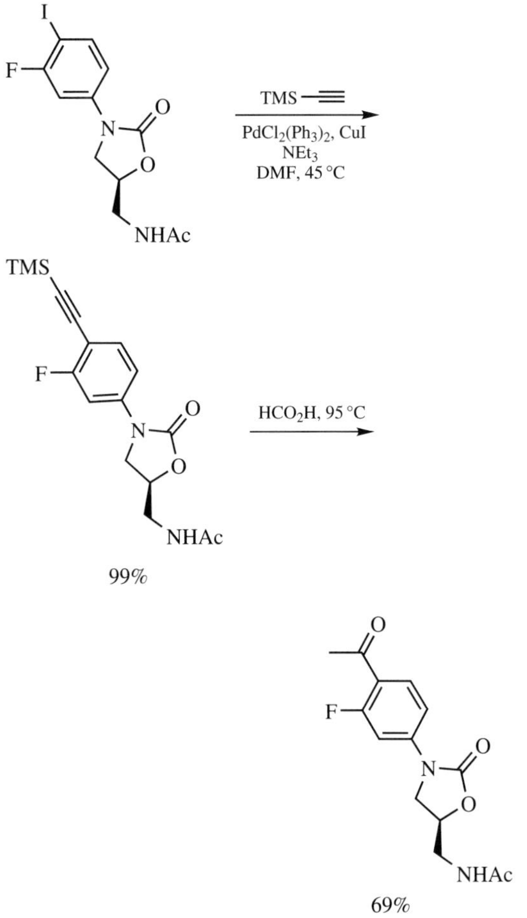

1. (a) Jeffery-Luong, T.; Linstrumelle, G., *Synthesis* **1983**, *32*. (b) Nwokogu, G., *J. Org. Chem.* **1985**, *50*, 3900. (c) Austin, W. B.; Bilow, N.; Kellaghan, W. J.; Lau, K. S. Y., *J. Org. Chem.* **1981**, *46*, 2280.

(d) Myers, A. G.; Alauddin, M. M.; Fuhry, M. A. M.; Dragovich, P. S.; Finney, N. S.; Harrington, P. M., *Tetrahedron Lett.* **1989**, *50*, 6997. (e) Konno, S.; Fujihara, S.; Yamanaka, S., *Heterocycles* **1984**, *22*, 2245. (f) Feldman, K. S., *Tetrahedron Lett.* **1982**, *23*; 3031.

2. (a) Cacchi, S., *Synthesis* **1986**, 320. (b) Suffert, J.; Bruckner, R., *Tetrahedron Lett.* **1991**, *32*, 1453. (c) Bruckner, R.; Scheuplein, S. W.; Suffert, J., *Tetrahedron Lett.* **1991**, *32*, 1449. (d) Chen, Q.-Y., Yang, Z.-Y., *Tetrahedron Lett.* **1986**, *27*, 1171.

3. (a) Kuroda, S.; Katsuki, T.; Yamaguchi, M., *Tetrahedron Lett.* **1987**, *28*, 803. (b) Kitano, Y.; Matsumoto, T.; Sato, F., *Tetrahedron* **1988**, *44*, 4073.

4. (a) Mori, S.; Iwakura, H.; Takechi, S., *Tetrahedron Lett.* **1988**, *29*, 5391. (b) Baumeler, A.; Brade, W.; Eugster, C. H., *Helv. Chim. Acta* **1990**, *73*, 700. (c) White, J. D.; Somers, T. C.; Reddy, G. N., *J. Am. Chem. Soc.* **1986**, *108*, 5352.

5. Ichinose, Y.; Matsunaga, S.; Fugami, K.; Oshima, K.; Utimoto, K., *Tetrahedron Lett.* **1989**, *30*, 3155.

6. Nozaki, K.; Oshima, K.; Utimoto, K., *J. Am. Chem. Soc.* **1987**, *109*, 2547.

7. Hassner, A.; Dillon, J. L., Jr., *J. Org. Chem.* **1983**, *48*, 3382.

8. Havens, S. J.; Hergenrother, P. M., *J. Org. Chem.* **1985**, *50*, 1763.

9. Sabourin, E. T.; Onopchenko, A., *J. Org. Chem.* **1983**, *48*, 5135.

10. (a) Bakthavachalam, V.; d'Alarco, M.; Leonard, N. J., *J. Org. Chem.* **1984**, *49*, 289. (b) Jensen, B. J.; Hergenrother, P. M., *J. Polym. Sci., Polym. Chem. Ed* **1985**, *23*, 2233.

11. Magnus, P.; Annoura, H.; Harling, J., *J. Org. Chem.* **1990**, *55*, 1709.

12. Havens, S. J.; Hergenrother, P. M., *J. Polym. Sci., Polym. Chem. Ed.* **1984**, *22*, 3011.

13. Jensen, B. J.; Hergenrother, P. M.; Nwokogu, G., *J. Macromol. Sci.–Pure Appl. Chem.* **1993**, *A30*, 449.

14. Semmelhack, M. F.; Neu, T.; Foubelo, F., *Tetrahedron Lett.* **1992**, *33*, 3277.

15. Fitton, P.; Rick, A. E., *J. Organomet. Chem.* **1971**, *28*, 287.

16. (a) Ratovelomanana, V.; Hammoud, A.; Linstrumelle, G., *Tetrahedron Lett.* **1987**, *28*, 1649. (b) Vollhardt, K. P. C.; Winn, L. S., *Tetrahedron Lett.* **1985**, *26*, 709.

17. (a) Tilley, J. W.; Zawoiski, S., *J. Org. Chem.* **1988**, *53*, 386. (b) Robins, M. J.; Barr, P. J., *J. Org. Chem.* **1983**, *48*, 1854. (c) Konno, S.; Fujimura, S.; Yamanaka, H., *Heterocycles* **1984**, *22*, 2245.

18. Sakamoto, T.; Kondo, Y.; Yamanaka, H., *Heterocycles* **1984**, *22*, 1347.

19. (a) Tischler, A. N.; Lanza, T. J., *Tetrahedron Lett.* **1986**, *27*, 1653. (b) Sakamoto, T.; Kondo, Y.; Yamanaka, H., *Heterocycles* **1986**, *24*, 31.

20. Logue, M. W.; Teng, K., *J. Org. Chem.* **1982**, *47*, 2549.

21. Ciattini, P. G.; Morera, E.; Ortar, G., *Tetrahedron Lett.* **1991**, *32*, 6449.

22. Cassar, L., *J. Organomet. Chem.* **1975**, *93*, 253.

23. Lee, G. C. M.; Tobias, B.; Holmes, J. M.; Harcourt, D. A.; Garst, M. E., *J. Am. Chem. Soc.* **1990**, *112*, 9330.

24. Wenkert, E.; Leftin, M. H.; Michelotti, E. L., *J. Org. Chem.* **1985**, *50*, 1122.

25. Tamura, Y.; Sasho, M.; Ohe, H.; Akai, S.; Kita, Y., *Tetrahedron Lett.* **1985**, *26*, 1549.

26. (a) Thies, R. W., Daruwala, K. P., *J. Org. Chem.* **1987**, *52*, 3798. (b) Kiesewetter, D. O.; Katzenellenbogen, J. A.; Kilbourn, M. R.; Welch, M. J., *J. Org. Chem.* **1984**, *49*, 4900.

27. Kikukawa, K.; Umekawa, H.; Wada, F.; Matsuda, T., *Chem. Lett.* **1988**, *5*, 881.

28. Zhang, X. H.; Guibe, F.; Balavoine, G., *J. Org. Chem.* **1990**, *55*, 1857.

29. Nakamura, E.; Machii, D.; Inubushi, T., *J. Am. Chem. Soc.* **1989**, *111*, 6849.

30. Danheiser, R. L.; Sard, H., *Tetrahedron Lett.* **1982**, *24*, 23.

31. Schmidt, C.; Sahraoui-Taleb, S.; Differding, E.; Dehasse-De Lombaert, C. G.; Ghosez, L., *Tetrahedron Lett.* **1984**, *25*, 5043.

32. Dotz, H. K.; Larbig, H., *J. Organomet. Chem.* **1991**, *405*, C38.

33. Greene, T. W.; Wuts P. G. M. *Protective Groups in Organic Synthesis*, 3rd Ed. Wiley: New York, 1999; p 654.

34. Lindström S.; Ripa L.; Hallberg A., *Org. Lett.* **2000**, *2*, 2291.

35. Nishihara, Y.; Ikegashira, K.; Hirabayashi, K.; Ando, J.-i.; Mori, A.; Hiyama, T., *J. Org. Chem.* **2000**, *65*, 1780.

36. Mio, M. J.; Kopel, L. C.; Braun, J. B.; Gadzikwa, T. L.; Hull, K. L.; Brisbois, R. G.; Markworth, C. J.; Grieco, P. A., *Org. Lett.* **2002**, *4*, 3199.

37. (a) Cho, D. H.; Lee, J. H.; Kim, B. H., *J. Org. Chem.* **1999**, *64*, 8048. (b) Liu, J.-H.; Chan, H.-W.; Wong, H. N. C., *J. Org. Chem.* **2000**, *65*, 3274. (c) Liu, J.-H.; Yang, Q.-C.; Mak, T. C. W.; Wong, H. N. C., *J. Org. Chem.* **2000**, *65*, 3587. (d) Tu, B.; Ghosh, B.; Lightner, D. A., *J. Org. Chem.* **2003**, *68*, 8950. (e) Tu, B.; Lightner, D. A., *J. Heterocycl. Chem.* **2003**, *40*, 707.

38. Ringenbach, C.; De Nicola, A.; Ziessel, R., *J. Org. Chem.* **2003**, *68*, 4708.

39. Petricci, E.; Radi, M.; Corelli, F.; Botta, M., *Tetrahedron Lett.* **2003**, *44*, 9181.

40. Lang, P.; Magnin, G.; Mathis, G.; Burger, A.; Biellmann, J.-F., *J. Org. Chem.* **2000**, *65*, 7825.

41. Vaidya, C. M.; Wright, J. E.; Rosowsky, A., *J. Med. Chem.* **2002**, *45*, 1690.

42. Fonseca, T.; Gignate, B.; Marques, M. M.; Gilchrist, T. L.; De Clercq, E., *Bioorg. Med. Chem.* **2004**, *12*, 103.

43. Tikhe, G. J.; Webber, S. E.; Hostomsky, Z.; Maegley, K. A.; Ekkers, A.; Li, J.; Yu, X.-H., Almassy, R. J.; Kumpf, R. A.; Boritzki, T. J.; Zhang, C.; Calabrese, C. R.; Curtin, N. J.; Kyle, S.; Thomas, H. D.; Wang, L.-Z., Calvert, A. H.; Golding, B. T.; Griffin, R. J.; Newell, D. R., *J. Med. Chem.* **2004**, *47*, 5467.

44. Wyatt, P. G.; Allen, M. J.; Chilcott, J.; Gardner, C. J.; Livermore, D. G.; Mordaunt, J. E.; Nerozzi, F.; Patel, M.; Perren, M. J.; Weingarten, G. G.; Shabbir, S.; Woollard, P. M.; Zhou, P., *Bioorg. Med. Chem. Lett.* **2002**, *12*, 1405.

45. Yue, D.; Larock, R. C., *J. Org. Chem.* **2002**, *67*, 1905.

46. Suginome, M.; Fukuda, T.; Ito, Y., *Org. Lett.* **1999**, *1*, 1977.

47. Pérez-Serano, L.; Dominguez, G.; Pérez-Castells, J., *J. Org. Chem.* **2004**, *69*, 5413.

48. Saito, T.; Shiotani, M.; Otani, T.; Hasaba, S., *Heterocycles* **2003**, *60*, 1045.

49. Karpov, A. S.; Müller, T. J. J., *Org. Lett.* **2003**, *5*, 3451.

50. Karpov, A. S.; Müller, T. J. J., *Synthesis* **2003**, 2815.

51. Cosford, N. D. P.; Tehrani, L.; Roppe, J.; Schweiger, E.; Smith, N. D.; Anderson, J.; Bristow, L.; Brodkin, J.; Jiang, X.; McDonald, I.; Rao, S.; Washburn, M.; Varney, M. A., *J. Med. Chem.* **2003**, *46*, 204.

52. Elangovan, A.; Wang, Y.-H.; Ho, T.-I., *Org. Lett.* **2003**, *5*, 1841.

53. Cheng, J.; Sun, Y.; Wang, F.; Guo, M.; Xu, J.-H.; Pan, Y.; Zhang, Z., *J. Org. Chem.* **2004**, *69*, 5428.

54. Anastasia, L.; Negishi, E.-I., *Org. Lett.* **2001**, *3*, 3111.

55. Molander, G. A.; Katona, B. W.; Machrouhi, F., *J. Org. Chem.* **2002**, *67*, 8416.

56. Pérez, I.; Sestelo, J. P.; Sarandeses, L. A., *J. Am. Chem. Soc.* **2001**, *123*, 4155.

57. Sakai, N.; Annaka, K.; Konakahara, T., *Org. Lett.* **2004**, *6*, 1527.

58. Gelman, D.; Tsvelikhovsky, D.; Molander, G. A.; Blum, J., *J. Org. Chem.* **2002**, *67*, 6287.

59. Krafft, M. E.; Boñaga, L. V. R.; Felts, A. S.; Hirosawa, C.; Kerrigan, S., *J. Org. Chem.* **2003**, *68*, 6039.

60. Paquette, L. A.; Kim, I. H.; Cunière, N., *Org. Lett.* **2003**, *5*, 221.

61. McCaleb, K. L.; Halcomb, R. L., *Org. Lett.* **2000**, *2*, 2631.

62. Sawad, D.; Kanai, M.; Shibasaki, M., *J. Am. Chem. Soc.* **2000**, *122*, 10521.

63. Xie, C.; Runnegar, M. T.; Snider, B. B., *J. Am. Chem. Soc.* **2000**, *122*, 5017.

64. Fischer, C.; Carreira, E. M., *Org. Lett.* **2004**, *6*, 1497.

65. Wie, C.; Li, C.-J., *J. Am. Chem. Soc.* **2003**, *125*, 9584.

66. Sakai, N.; Hirasawa, M.; Konakahara, T., *Tetrahedron Lett.* **2003**, *44*, 4171.

67. Fischer, C.; Carreira, E. M., *Org. Lett.* **2001**, *3*, 4319.

68. Frantz, D. E.; Fässler, R.; Carreira, E. M., *J. Am. Chem. Soc.* **2000**, *122*, 1806.

69. Trost, B. M.; Ameriks, M. K., *Org. Lett.* **2004**, *6*, 1745.

70. Marshall, J. A.; Bourbeau, M. P., *Org. Lett.* **2003**, *5*, 3197.

71. Traverse, J. F.; Hoveyda, A. H.; Snapper, M. L., *Org. Lett.* **2003**, *5*, 3273.

72. Knöpfel, T. F.; Aschwanden, P.; Ichikawa, T.; Wanatabe, T.; Carreira, E. M., *Angew. Chem., Int. Ed.* **2004**, *43*, 5971.

73. Koradin, C.; Polborn, K.; Knochel, P., *Angew. Chem., Int. Ed.* **2002**, *41*, 2535.

74. Fässler, R.; Frantz, D. E.; Oetiker, J.; Carreira, E. M., *Angew. Chem., Int. Ed.* **2002**, *41*, 3054.

75. Hosokawa, S.; Isobe, M., *J. Org. Chem.* **1999**, *64*, 37.

76. Nicolaou, K. C.; Montangnon, T.; Ulve, T.; Baran, P. S.; Zhong, Y.-L.; Sarabia, F., *J. Am. Chem. Soc.* **2002**, *124*, 5718.

77. Meng, D.; Bertinato, P.; Balog, A.; Su, D.-S.; Kamenecka, T.; Sorensen, E. J.; Danishefsky, S. J., *J. Am. Chem. Soc.* **1997**, *119*, 10073.

78. Chao, K. C.; Rayabarapu, D. K.; Wang, C.-C.; Cheng, C.-H., *J. Org. Chem.* **2001**, *66*, 8804.

79. Kowlczyk, B. A.; Smith, T. C.; Dauben, W. G., *J. Org. Chem.* **1998**, *63*, 1379.

80. Gibson, S.; Mainolfi, N.; Kalindjian, S. B.; Wright, P. T., *Angew. Chem., Int. Ed.* **2004**, *43*, 5680.

81. Chan, D. C. M.; Laughton, C. A.; Queener, S. F.; Stevens, M. F. G., *Bioorg. Med. Chem.* **2002**, *10*, 3001.

82. Sandanayaka, V. P.; Yang, Y., *Org. Lett.* **2000**, *2*, 3087.

83. Tanaka, K.; Shirasaka, K., *Org. Lett.* **2003**, *5*, 4697.

84. Ozerov, O. V.; Patrick, B. O.; Ladipo, F. T., *J. Am. Chem. Soc.* **2000**, *122*, 6423.

85. Mori, N.; Ikeda, S.-I.; Sato, Y., *J. Am. Chem. Soc.* **1999**, *121*, 2722.

86. Suzuki, D.; Tanaka, R.; Urabe, H.; Sato, F., *J. Am. Chem. Soc.* **2002**, *124*, 3518.

87. Hilt, G.; du Mesnil, F.-X., *Tetrahedron Lett.* **2000**, *41*, 6757.

88. Paik, S.-J.; Son, S. U.; Chung, Y. K., *Org. Lett.* **1999**, *1*, 2045.

89. Pulley, S. R.; Sen, S.; Vorogushin, A.; Swanson, E., *Org. Lett.* **1999**, *1*, 1721.

90. Asao, N.; Nogami, T.; Lee, S.; Yamamoto, Y., *J. Am. Chem. Soc.* **2003**, *125*, 10921.

91. Amii, H.; Kishikawa, Y.; Uneyama, K., *Org. Lett.* **2001**, *3*, 1109.

92. Wender, P. A.; Barzilay, C. M.; Dyckman, A. J., *J. Am. Chem. Soc.* **2001**, *123*, 179.

93. Barluenga, J.; Barrio, P.; Lopez, L. A.; Tomas, M.; Garcia-Granda, S.; Alvarez-Rua, C., *Angew. Chem., Int. Ed.* **2003**, *42*, 3008.

94. Barluenga, J.; Aznar, F.; Palomero, M. A., *Angew. Chem., Int. Ed.* **2000**, *39*, 4346.

95. Lu, Y.-F.; Harwig, C. W.; Fallis, A. G., *J. Org. Chem.* **1993**, *58*, 4202.

96. Molander, G. A.; Dehmel, F., *J. Am. Chem. Soc.* **2004**, *126*, 10313.

97. Ito, H.; Arimoto, K.; Sensui, H.-o.; Hosomi, A., *Tetrahedron Lett.* **1997**, *38*, 3977.

98. Busch-Petersen, J.; Bo, Y.; Corey, E. J., *Tetrahedron Lett.* **1999**, *40*, 2065.

99. Yadav, J. S.; Reddy, B. V. S.; Reddy, M. S.; Parimala, G., *Synthesis* **2003**, 2390.

100. Sandoval, C.; Redero, E.; Mateos-Timoneda, M. A.; Bermejo, F. A., *Tetrahedron Lett.* **2002**, *43*, 6521.

101. Falck, J. R.; Kumar, P. S.; Reddy, Y. K.; Zou, G.; Capdevilla, J. H., *Tetrahedron Lett.* **2001**, *42*, 7211.

102. Takahashi, T.; Kusaka, S.-I.; Doi, T.; Sunazuka, T.; Omura, S., *Angew. Chem., Int. Ed.* **2003**, *42*, 5230.

103. Takemura, I.; Imura, K.; Matsumoto, T.; Suzuki, K., *Org. Lett.* **2004**, *6*, 2503.

104. Torisu, K.; Kobayashi, K.; Iwahashi, M.; Nakai, Y.; Onoda, T.; Nagase, T.; Sugimoto, I.; Okada, Y.; Matsumoto, R.; Nanbu, F.; Ohuchida, S.; Nakai, H.; Toda, M., *Bioorg. Med. Chem.* **2004**, *12*, 4685.

105. Lee, C. S.; Allwine, D. A.; Barbachyn, M. R.; Grega, K. C.; Dolak, L. A.; Ford, C. W.; Jensen, R. M.; Seest, E. P.; Hamel, J. C.; Schaadt, R. D.; Stapert, D.; Yagi, B. H.; Zurenko, G. E.; Genin, M. J., *Bioorg. Med. Chem.* **2001**, *9*, 3243.

(Trimethylsilyl)allene[1]

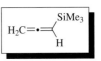

[14657-22-8]   C$_6$H$_{12}$Si   (MW 112.27)

InChI = 1S/C6H12Si/c1-5-6-7(2,3)4/h6H,1H2,2-4H3

InChIKey = SWOGHYPZXWCLBU-UHFFFAOYSA-N

(propargylic anion equivalent;[2–4] substrate for ene reaction[5])

Alternate Name: trimethyl-1,2-propadienylsilane.

Physical Data: bp 90–93 °C.

Solubility: sol CH$_2$Cl$_2$, benzene, most organic solvents.

Form Supplied in: colorless liquid; not commercially available.

Analysis of Reagent Purity: IR (tf) 2955, 2900, 1935, 1250, 1210, 1055, 840, 800, 750, and 690 cm^{-1}; ^{1}H NMR (60 MHz, CDCl$_3$) δ 0.15 (s, 9H), 4.27 (d, 2H, *J* = 7.7), 4.88 (dd, 1H, *J* = 6.6, 7.7).[4b]

Preparative Method: two methods have been reported for the preparation of (trimethylsilyl)allene [(TMS)allene] (**1**). Reductive deoxygenation of the tosylhydrazone derivative affords the title compound in 51% yield.[4b] The tosylhydrazone is readily prepared from the corresponding aldehyde, which in turn is accessed by formylation of (trimethylsilyl)ethynylmagnesium bromide with DMF (eq 1).[6] (TMS)allene has also been prepared by flash vacuum pyrolysis of methyl (trimethylsilyl)-propargyl ether, which is obtained from silylation of methyl propargyl ether (eq 2).[7]

$$\text{OHC}\!\!-\!\!\!\equiv\!\!\!-\text{TMS} \xrightarrow[\substack{\text{reflux, 3.5 h} \\ 94\%}]{\substack{1.1 \text{ equiv TsNHNH}_2 \\ \text{Et}_2\text{O}}}$$

$$\text{TsHN}\!-\!\text{N}\diagdown\!\!\!\equiv\!\!\!-\text{TMS} \xrightarrow[\substack{\text{DMF–sulfolane–pentane} \\ 51\%}]{\substack{8 \text{ equiv NaBH}_3\text{CN} \\ \text{pH 1–2, 40–60 °C, 3 d}}} \!\!=\!\!\bullet\!\!=\!\!\diagup^{\text{TMS}} \quad (1)$$
(1)

$$\text{MeO}\diagdown\!\!\!\equiv \xrightarrow[\substack{2.\text{ TMSCl}}]{\substack{1.\text{ BuLi, Et}_2\text{O}}} \text{MeO}\diagdown\!\!\!\equiv\!\!\!-\text{TMS} \xrightarrow[\substack{0.01 \text{ mmHg} \\ 72\%}]{\substack{640 °C}}$$

$$=\!\!\bullet\!\!=\!\!\diagup^{\text{TMS}} \quad (2)$$
(1)

Purification: distillation at atmospheric pressure.[4b]
Handling, Storage, and Precautions: stable indefinitely when stored under nitrogen in the refrigerator.

Propargylic Anion Equivalent. (TMS)allene reacts with electrophiles at the C-3 position in an S_E2' process analogous to electrophilic substitution reactions of allyl- and propargylsilanes.[8] For example, upon treatment with trimethylsilyl chlorosulfonate or sulfur trioxide–1,4-dioxane, (TMS)allene yields silyl esters of sulfonic acids (eq 3).[2] (TMS)allene undergoes conjugate addition with α,β-unsaturated acyl cyanides to yield δ,ε-acetylenic acyl cyanides.[3]

$$=\!\!=\!\!\bullet\!\!\diagup^{\text{TMS}} \xrightarrow[\substack{\text{rt, 1h} \\ 80\%}]{\substack{\text{ClSO}_3\text{TMS or dioxane}\bullet\text{SO}_3 \\ \text{CH}_2\text{Cl}_2}} =\!\!=\!\!\diagup^{\text{SO}_2\text{TMS}} \quad (3)$$
$$(1)$$

Particularly important is the reaction of (TMS)allene with aldehydes and ketones, which provides a convenient route to secondary and tertiary homopropargylic alcohols, respectively. Treatment of (1) (1.1–1.5 equiv) with a mixture of carbonyl compound and 1.1–1.5 equiv of titanium(IV) chloride in methylene chloride produces mixtures of homopropargylic alcohols and (trimethylsilyl)vinyl chlorides. Exposure of the crude reaction product to 2.5 equiv of potassium fluoride in DMSO[9] then furnishes the desired homopropargylic alcohols (eq 4).[4]

$$=\!\!=\!\!\bullet\!\!\diagup^{\text{TMS}} \xrightarrow[\substack{\text{TiCl}_4 \\ \text{CH}_2\text{Cl}_2}]{\substack{\text{O} \\ R^1\!\!\diagdown\!\!\diagup\!\! R^2}} \begin{matrix} R^1 \\ R^2 \end{matrix}\!\!\diagup^{\text{OH}}\!\!\diagdown\!\!\diagup^{\text{Cl}}_{\text{TMS}} \xrightarrow[\text{rt}]{\text{KF, DMSO}}$$
$$(1)$$

$$\begin{matrix} R^1 \\ R^2 \end{matrix}\!\!\diagup^{\text{OH}}\!\!\diagup\!\!\equiv \quad (4)$$

Carbonyl compound	% Yield
PhCH$_2$CH$_2$CHO	84
PhCH$_2$COMe	72
Cyclohexanone	89

Several alternative reagents function as propargylic anion equivalents.[10] Homopropargylic alcohols can be prepared by the addition of allenylboronate esters to aldehydes (though not to ketones).[11] Brown has recently reported that 9-allenyl-9-BBN reacts with aldehydes and ketones to produce homopropargylic alcohols in good yields.[12] Allenyllithium unfortunately combines with carbonyl compounds to afford mixtures of allenic and acetylenic alcohols.[13] Allenylmagnesium bromide reacts with carbonyl compounds to furnish homopropargylic alcohols in poor yields; in some cases, mixtures of allenic and alkynic products are obtained.[12] Diallenyltin dibromide, which is generated by reaction of propargyl bromide with tin metal in the presence of aluminum, reacts with both aldehydes and ketones to afford homopropargylic alcohols in good yield.[14] Alternatively, triisopropylsilyl derivatives of these alkynes can be prepared by the reaction of the lithium derivative of 1-(triisopropylsilyl)propyne with ketones and aldehydes by using the procedure described by Corey and Rucker.[15] Trimethylsilyl analogs of these alkynes have been prepared using the Reformatsky reagent derived from trimethylsilylpropargyl bromide: zinc-mediated condensation of trimethylsilylpropargyl

bromide with aldehydes and ketones delivers homopropargylic alcohols (eq 5).[16]

$$\underset{\text{BrZn}}{\diagup}\!\!=\!\!=\!\!-\text{TMS} \xrightarrow[\text{0 °C, 1 h}]{\text{PhCHO, THF}} \underset{\text{Ph}}{\diagup}\!\!\diagup^{\text{OH}}\!\!\diagup^{\text{TMS}}\!\!=\!\! \quad (5)$$

Synthesis of Silylacetylenes via Ene Reaction. (TMS)allene undergoes ene reactions with 4-phenyl-1,2,4-triazoline-3,5-dione and other reactive enophiles to give silylacetylenes.[5]

Related Reagents. Allenyllithium; 1-Methyl-1-(trimethylsilyl)allene; Allyltrimethylsilane; Propargylmagnesium Bromide; 3-Trimethylsilyl-1-propyne.

---

1. Review: Panek, J. S., *Comprehensive Organic Synthesis* **1991**, *2*, 579.
2. Bourgeois, P.; Calas, R.; Merault, G., *J. Organomet. Chem.* **1977**, *141*, 23.
3. (a) Jellal, A.; Santelli, M., *Tetrahedron Lett.* **1980**, *21*, 4487. (b) Santelli, M.; Abed, D. E.; Jellal, A., *J. Org. Chem.* **1986**, *51*, 1199.
4. (a) Danheiser, R. L.; Carini, D. J., *J. Org. Chem.* **1980**, *45*, 3925. (b) Danheiser, R. L.; Carini, D. J.; Fink, D. M.; Basak, A., *Tetrahedron* **1983**, *39*, 935. (c) Danheiser, R. L.; Carini, D. J.; Kwasigroch, D. A., *J. Org. Chem.* **1986**, *51*, 3870.
5. Laporterie, A.; Dubac, J.; Manuel, G.; Deleris, G.; Kowalski, J.; Dunogues, J.; Calas, R., *Tetrahedron* **1978**, *34*, 2669.
6. Komarov, N. V.; Yarosh, O. G.; Astaf'eva, L. N., *J. Gen. Chem. USSR (Engl. Transl.)* **1966**, *36*, 920.
7. Hopf, H.; Naujoks, E., *Tetrahedron Lett.* **1988**, *29*, 609.
8. See: ref. 1; Fleming, I., *Comprehensive Organic Synthesis* **1991**, *2*, 563; Fleming, K.; Dunogues, J.; Smithers, R., *Org. React.* **1989**, *37*, 57.
9. Cunico, R. F.; Dexheimer, E. M., *J. Am. Chem. Soc.* **1972**, *94*, 2868.
10. For reviews of the chemistry of propargylic anion equivalents, see: (a) Yamamoto, H., *Comprehensive Organic Synthesis* **1991**, *2*, 81. (b) Epsztein, R. In *Comprehensive Carbanion Chemistry*; Buncel, E., Durst, T., Eds.; Elsevier: Amsterdam, 1984; Part B, p 107. (c) Moreau, J.-L. In *The Chemistry of Ketenes, Allenes, and Related Compounds*; Patai, S., Ed.; Wiley: New York, 1978; p 343.
11. (a) Favre, E.; Gaudemar, M., *J. Organomet. Chem.* **1974**, *76*, 297. (b) Favre, E.; Gaudemar, M., *J. Organomet. Chem.* **1974**, *76*, 305. (c) Haruta, R.; Ishiguro, M.; Ikeda, N.; Yamamoto, H., *J. Am. Chem. Soc.* **1982**, *104*, 7667.
12. Brown, H. C.; Khire, U. R.; Racherla, U. S., *Tetrahedron Lett.* **1993**, *34*, 15, and references therein.
13. For example, see: Clinet, J.-C.; Linstrumelle, G., *Synthesis* **1981**, 875.
14. Nokami, J.; Tamaoka, T.; Okawara, R., *Chem. Lett.* **1984**, 1939.
15. Corey, E. J.; Rucker, C., *Tetrahedron Lett.* **1982**, *23*, 719.
16. Daniels, R. G.; Paquette, L. A., *Tetrahedron Lett.* **1981**, *22*, 1579.

Katherine L. Lee & Rick L. Danheiser
Massachusetts Institute of Technology, Cambridge, MA, USA

Trimethylsilylallyllithium[1]

[67965-38-2]   C₆H₁₃LiSi   (MW 120.22)

$C_6H_{13}LiSi$ (MW 120.22)

InChI = 1S/C6H13Si.Li/c1-5-6-7(2,3)4;/h5-6H,1H2,2-4H3;
InChIKey = UAIXLBKOPNNMJY-UHFFFAOYSA-N

(addition to ketones and aldehydes;[2] synthesis of dienes with defined stereochemistry;[3] reaction with alkyl halides,[4] epoxides,[5]
and other electrophiles[6])

Alternate Name: [1-(trimethylsilyl)-2-propenyl]lithium.
Physical Data: ¹H and ¹³C NMR solution studies of the TMEDA
   complex in THF[7a] and the X-ray structure of the TMEDA com-
   plex of a closely related reagent have been reported.[7b]
Solubility: sol THF, ether.
Preparative Method: prepared in situ, under an inert atmosphere,
   by reaction of allyltrimethylsilane with *n*-butyllithium or with
   s-butyllithium.[1,7a]
Handling, Storage, and Precaution: solutions are inflammable.
   Preclude contact of the solutions with air and moisture.

Addition to Ketones and Aldehydes.[2] Trimethylsilylallyllithium (**1**) adds to most aldehydes and ketones to give γ-(*E*)-
vinylsilanes (eq 1) and only very small amounts of α-condensation
products or Peterson elimination products.[2] The stereoselectivity
has been investigated by NMR studies carried out in solution,
which show that (**1**) exists exclusively in the *exo* conformation.[7]

However, other types of silylallyllithium compounds with
internal complexing capabilities are suspected to exist in the
endo conformation under certain conditions. For example,
(dialkylaminomethyl)dimethylsilylallyllithiums give the usual γ-
(*E*)-vinylsilane products with reasonable selectivity when they
react with aldehydes and ketones in toluene. However, if a small
amount of 1,2-dimethoxyethane is added to the reagent, the stereoselectivity of the reaction is reversed, giving the γ-(*Z*)-vinylsilane
as the major product (eq 2).[8]

These γ-condensation products are useful in organic synthesis.
For example, the vinylsilane portion of the molecule can be used as
an internal nucleophile in cyclization reactions, as demonstrated
in the synthesis of substituted aromatic compounds (eq 3).[9] They
can also be used as lactone precursors (eq 4), as demonstrated in
the synthesis of the steroidal 17-spiro-γ-lactone (**2**).[2]

The reactivity of the double bond in these hydroxyvinylsilane
products is affected by the trimethylsilyl group. The reaction
of (**3**) with *N*-bromosuccinimide leads to the formation of a β-
bromooxetane at 0 °C or of a diene product at 40 °C (eq 5). The
expected formation of the tetrahydrofuran product has not been
observed.[2]

Synthesis of Dienes with Defined Stereochemistry.[3] The
normal γ-regioselectivity in the reaction of (**1**) with ketones and
aldehydes can be modified by the use of other metal counter-ions.
Thus α-substitution products can be obtained selectively if magnesium bromide or other Lewis acids are added to the reaction
media (eq 6). The resulting alcohols are also diastereomerically
enhanced. These observations are explained by the formation of a
six-membered ring transition state (**4**) where the metal ion M⁺ coordinates with the carbonyl oxygen. Other reagents used to control
the regio- and stereoselectivity of the reactions of (**1**) with aldehydes are shown in (eq 7).[2c,3c–g] The alcohols are useful for the

stereospecific formation of dienes by the Peterson alkenation reaction (eq 6).[3a] The same logical approach has been used to obtain (E)- or (Z)-4-aryl (or alkyl)-(E)-1-(trimethylsilyl)-1,3-butadienes from the deprotonation of 1,3-bis(trimethylsilyl)propene, an α-trimethylsilylated equivalent of (1), and reaction with benzaldehyde (eq 8).[3b]

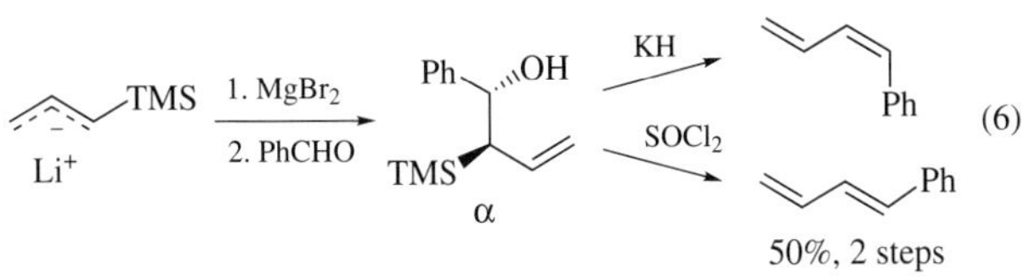

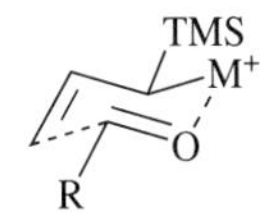

(6)

50%, 2 steps

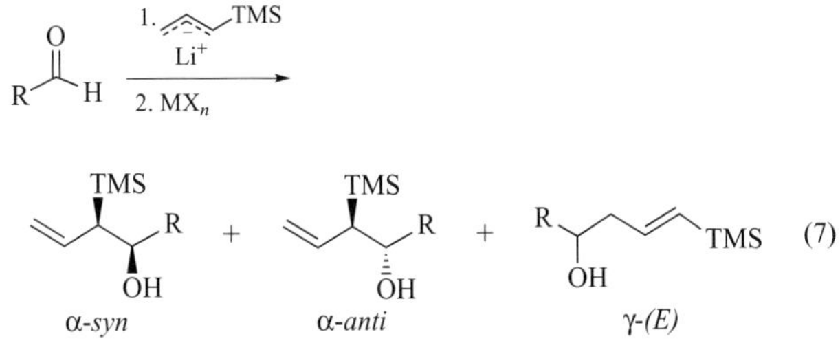

(4)

(7)

MX$_n$	RCHO	α-syn	α-anti	γ-(E)	Combined yield (%)
ZnEt$_2$	Benzaldehyde	100*	–	–	74[3c]
BR$_2$Cl	Isobutyraldehyde	–	100	–	60[3d]
CuCN	Cinnamaldehyde	–	–	100	75[2c]
AlEt$_3$	Benzaldehyde	37	43	2	51[3d,e]
AlEt$_2$Cl	Benzaldehyde	4	88	2	66[3d,e]
SnBu$_3$ + BF$_3$	Benzaldehyde	100*	–	–	77[3d]
TiCp$_2$Cl	Benzaldehyde	–	100*	–	95[3f]
Ti(O-i-Pr)$_4$	Benzaldehyde	–	>99	<1	80–83[3g]

*The ratios were inferred from the stereochemistry of the dienes obtained after Peterson eliminations of the α-condensation products

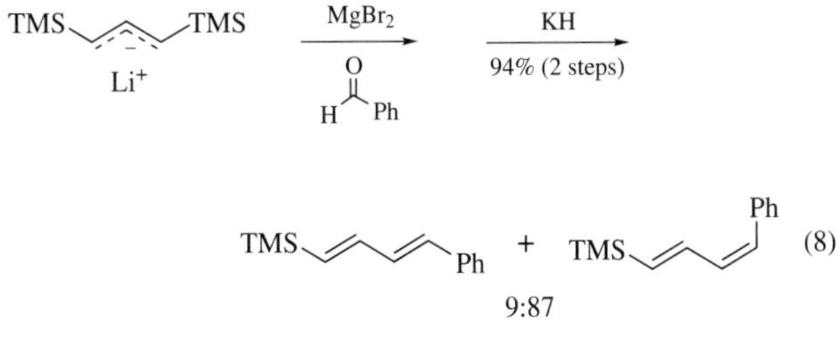

9:87

(8)

Reaction with Alkyl Halides, Epoxides, and Other Electrophiles.[4–6] Trimethylsilylallyllithium reacts with alkyl halides at low temperature, giving mixtures of α- and γ-(E) products in ratios that vary with the electrophile and the reaction conditions (eq 9).[4] In these reactions the γ-(E)-alkylation product generally predominates. For example, when trimethylsilylallyllithium is formed by deprotonation of allyltrimethylsilane with n-butyllithium in the presence of potassium t-butoxide in THF, the (E)-vinylsilane alkylation product is obtained in better then

85% yield (eq 10). This has been exploited in the synthesis of (Z)-9-tricosene epoxide, the sex pheromone of the gypsy moth.[4a]

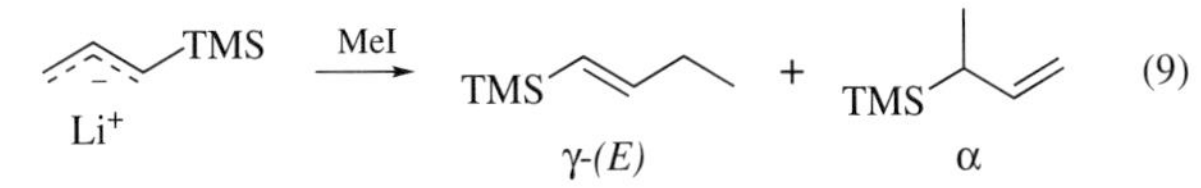

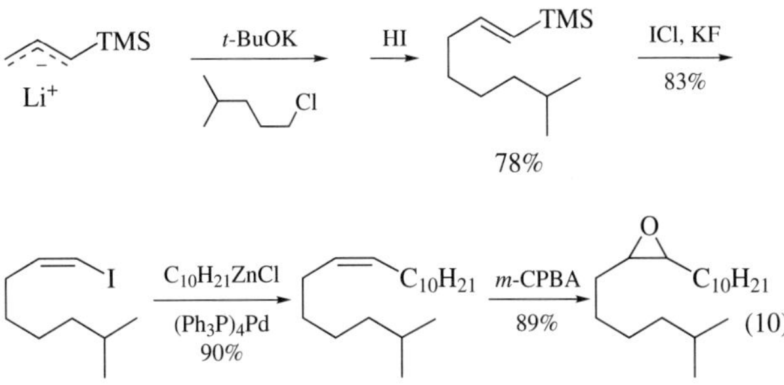

(9)

(10)

This γ-regioselectivity can be further improved by the use of bulky substituents attached to the silicon atom.[4a,b] In recent years, attempts have been made at developing reagents closely related to trimethylsilylallyllithium that will permit α-alkylation in high regio- and diastereoselectivity. In some instances, significant changes in regioselection can be achieved once the silicon is substituted with alkoxy or aminomethyl groups.[4c,4e] In particular, the use of chiral, lithium-chelating, amino groups attached to silicon as shown in eq 11), has been successfully applied to the regio- and diastereoselective alkylation of allylsilanes.[4c,e] This same approach gives even better results in the synthesis of chiral propargylic alcohols (eq 12).[4d]

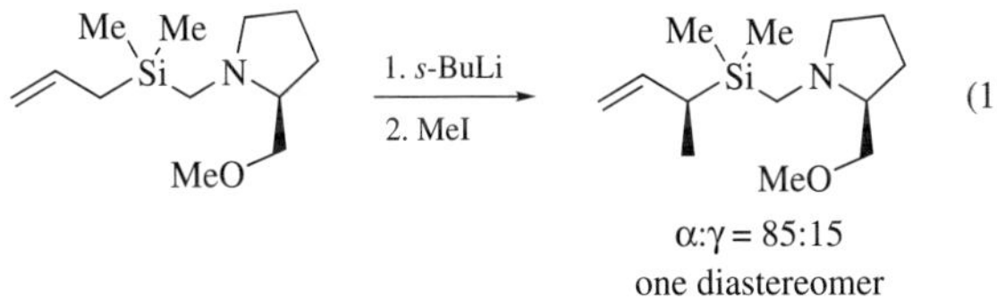

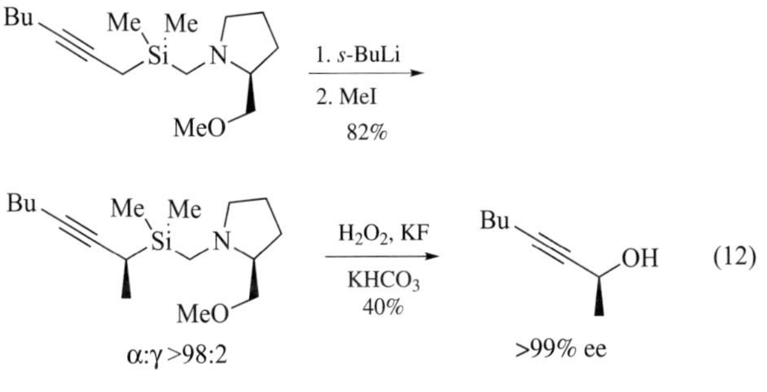

(11)

α:γ = 85:15
one diastereomer

(12)

α:γ >98:2   >99% ee

Trimethylsilylallyllithium reacts with epoxides (at −40 °C in THF for 2 h) to give alcohols in good yields. Highest α-selectivity is obtained with smaller monosubstituted epoxides; larger trisubstituted epoxides give γ-products. In unsymmetrical epoxides, attack is on the less hindered side[5a] with the notable exception of trimethylsilyl-substituted ones, where the attack occurs predominantly on the carbon bearing the trimethylsilyl group. This approach has been used to synthesize 1-silyl-substituted 1,4-dienes (eq 13).[5b] Trimethylsilylallyllithium also reacts with many other electrophiles including imines,[6a] carbon dioxide,[6b] substituted naphthalene,[6c] selenocyanogen,[6d] and dimethylformamide.[2a]

$$R^1 = R^2 = R^3 = H \qquad 10{:}1$$
$$R^1 = R^2 = Me, R^3 = H \qquad 1{:}2$$
$$R^1, R^2 = \text{-(CH}_2)_4\text{-}, R^3 = Me \qquad 1{:}10$$

Related Reagents. The literature contains numerous examples of reagents having different substitutions on the silicon moiety. The introduction of bulky and/or aromatic groups on the silicon atom, for example in (**5**) (R = Ph, p-tolyl, vinyl, t-Bu),[6b] (**6**),[6b] and (**7**) (R = OMe, Ph, Et, i-Pr),[2a, 4a, 6b, 10] has been used to modify the reactivity of the allylic anion. The influence exerted by these groups is mediated through their steric and/or electronic interactions with the allylic anion and with its counter ion. It is not surprising then that the presence of chelating amino or ether functions on the silicon atom, for example in (**5**) [R = NEt$_2$, N(i-Pr)$_2$, NH(CH$_2$CH$_2$NMe$_2$), pyrrolidino, CH$_2$(methoxymethyl-pyrrolidine), CH$_2$N(CH$_2$CH$_2$OMe)$_2$], also modifies the reactivity of the allyllithium attached to it.[4c, 11] The silylallyllithium reagents are normally obtained by deprotonation of the corresponding allylsilanes. However, γ-deprotonation of vinylsilanes can also be used for the same purposes, as is the case when either (**8**) or (**9**) are used to give the corresponding silylallyllithium reagent.[12] Other reagents (**10**)–(**17**) having heteroatom substituents on the allyl group have also been used.[3b, 7b, 13–19]

A few carbon-substituted reagents (**18**)–(**20**) have also been prepared.[4b, 6b, 20] Of these, reagent (**18**) has been used in the synthesis of frontalin, a natural product. The last three examples (**21**)–(**23**) show interesting steric and electronic effects on the regio- and stereoselectivity of the alkylation and condensation reactions.[4e, 21, 22]

Related Reagents. Allyllithium Crotyllithium 9-[1-(Trimethylsilyl)-2(E)-butenyl]-9-borabicyclo[3.3.1]nonane 5-Trimethylsilyl-1,3-pentadiene

1. For reviews, see; Colvin, E. W. *Silicon in Organic Synthesis*; Butterworths: London, 1981; p 118. Weber, W. P. *Silicon Reagents for Organic Synthesis*; Springer: Berlin, 1983; p 199, and references cited therein.

2. (a) Corriu, R. J. P.; Masse, J.; Samate, D., *J. Organomet. Chem.* **1975**, *93*, 71. (b) Ehlinger, E.; Magnus, P., *Tetrahedron Lett.* **1980**, *21*, 11; Ehlinger, E.; Magnus, P., *J. Am. Chem. Soc.* **1980**, *102*, 5004, and references cited therein. (c) Corriu, R. J. P.; Guerin, C.; M'Boula, J., *Tetrahedron Lett.* **1981**, *22*, 2985.

3. (a) Lau, P. K. W.; Chan, T. H., *Tetrahedron Lett.* **1978**, *19*, 2383. (b) Chan, T. H.; Li, J. S., *J. Chem. Soc., Chem. Commun.* **1982**, 969; Carter, M. J.; Fleming, I., *J. Chem. Soc., Chem. Commun.* **1976**, 679; Carter, M. J.; Fleming, I.; Percival, A., *J. Chem. Soc., Perkin Trans. 1* **1981**, 2415; Corriu, R.; Escudie, N.; Guerin, C., *J. Organomet. Chem.* **1984**, *264*, 207. (c) Wakamatsu, K.; Oshima, K.; Utimoto, K., *Chem. Lett.* **1987**, 2029. (d) Yamamoto, Y.; Saito, Y.; Maruyama, K., *J. Chem. Soc., Chem. Commun.* **1982**, 1326; Yamamoto, Y.; Saito, Y.; Maruyama, K., *J. Organomet. Chem.* **1985**, *292*, 311, and references cited therein; Hoffmann, R. W.; Brinkmann, H.; Frenking, G., *Chem. Ber.* **1990**, *123*, 2387; Tsai. D. J. S.; Matteson, D. S., *Tetrahedron Lett.* **1981**, *22*, 2751. (e) Yamamoto, Y.; Yatagai, H.; Saito, Y.; Maruyama, K., *J. Org. Chem.* **1984**, *49*, 1096; Naruta, Y.; Uno, H.; Maruyama, K., *Chem. Lett.* **1982**, 961. (f) Sato, F.; Suzuki, Y.; Sato, M., *Tetrahedron Lett.* **1982**, *23*, 4589. (g) Reetz, M. T.; Wenderoth, B., *Tetrahedron Lett.* **1982**, *23*, 5259; Reetz, M. T.; Steinbach, R.; Westermann, J.; Peter, R.; Wenderoth, B., *Chem. Ber.* **1985**, *118*, 1441; Ikeda, Y.; Yamamoto, H., *Bull. Chem. Soc. Jpn.* **1986**, *59*, 657.

4. (a) Chan, T. H.; Koumaglo, K., *J. Organomet. Chem.* **1985**, *285*, 109; Koumaglo, K.; Chan, T. H., *Tetrahedron Lett.* **1984**, *25*, 717. (b) Li, L. H.; Wang, D.; Chan, T. H., *Tetrahedron Lett.* **1991**, *32*, 2879; Chan, T. H.; Chen, L. M.; Wang, D.; Li, L. H., *Can. J. Chem.* **1993**, *71*, 60. (c) Horvath, R. F.; Chan, T. H., *J. Org. Chem.* **1989**, *54*, 317; Lamothe, S.; Cook, K. L.; Chan, T. H., *Can. J. Chem.* **1992**, *70*, 1733. (d) Hartley, R. C.; Lamothe, S.; Chan, T. H., *Tetrahedron Lett.* **1993**, *34*, 1449. (e) Lamothe, S.; Chan, T. H., *Tetrahedron Lett.* **1991**, *32*, 1847.

5. (a) Schaumann, E.; Kirschning, A., *Tetrahedron Lett.* **1988**, *29*, 4281. (b) Schaumann, E.; Kirschning, A., *J. Chem. Soc., Perkin Trans. 1* **1990**, 419; Kirschning, A.; Narjes, F.; Schaumann, E., *Liebigs Ann. Chem.* **1991**, 933. See also ref. 1.

6. (a) Guyot, B.; Pornet, J.; Miginiac, L., *Synth. Commun.* **1990**, *20*, 2409. (b) Uno, H., *Bull. Chem. Soc. Jpn.* **1986**, *59*, 2471; Naruta, Y.; Uno, H.; Maruyama, K., *Chem. Lett.* **1982**, 961. (c) Gant, T. G.; Meyers, A. I., *J. Am. Chem. Soc.* **1992**, *114*, 1010. (d) Meinke, P. T.; Krafft, G. A.; Guram, A., *J. Org. Chem.* **1988**, *53*, 3632.

7. (a) Fraenkel, G.; Chow, A.; Winchester, W. R., *J. Am. Chem. Soc.* **1990**, *112*, 2582, and references cited therein. (b) Boche, G.; Fraenkel, G.; Cabral, J.; Harms, K.; van Eikema Hommes, N. J. R.; Lohrenz, J.; Marsch, M.; von Ragué Schleyer, P., *J. Am. Chem. Soc.* **1992**, *114*, 1562.

8. Chan, T. H.; Labrecque, D., *Tetrahedron Lett.* **1992**, *33*, 7997. Similar results are obtained with boronate complexes; see: Tsai, D. J. S.; Matteson, D. S., *Organometallics* **1983**, *2*, 236.

9. Tius, M. A., *Tetrahedron Lett.* **1981**, *22*, 3335.

10. Muchowski, J. M.; Naef, R.; Maddox, M. L., *Tetrahedron Lett.* **1985**, *26*, 5375.

11. Tamao, K.; Nakajo, E.; Ito, Y., *Tetrahedron* **1988**, *44*, 3997; see also Refs. 4b and 4c.

12. Wakamatsu, K.; Oshima, K.; Utimoto, K., *Chem. Lett.* **1987**, 2029.

13. Seyferth, D.; Mammarella, R. E., *J. Organomet. Chem.* **1978**, *156*, 279.

14. Ikeda, Y.; Furuta, K.; Meguriya, N.; Ikeda, N.; Yamamoto, H., *J. Am. Chem. Soc.* **1982**, *104*, 7663; Kyler, K. S.; Netzel, M. A.; Arseniyadis, S.; Watt, D. S., *J. Org. Chem.* **1983**, *48*, 383; Furuta, K.; Ikeda, Y.; Meguriya, N.; Ikeda, N.; Yamamoto, H., *Bull. Chem. Soc. Jpn.* **1984**, *57*, 2781.

15. Murai, A.; Abiko, A.; Shimada, N.; Masamune, T., *Tetrahedron Lett.* **1984**, *25*, 4951; Marsch, M.; Harms, K.; Zschage, O.; Hoppe, D.; Boche, G., *Angew. Chem., Int. Ed. Engl.* **1991**, *30*, 321.

16. Ukai, J.; Ikeda, Y.; Ikeda, N.; Yamamoto, H., *Tetrahedron Lett.* **1984**, *25*, 5173; **1984**, *25*, 5177.

17. Trost, B. M.; Self, C. R., *J. Am. Chem. Soc.* **1983**, *105*, 5942.

18. Reich, H. J.; Clark, M. C.; Willis, W. W., Jr., *J. Org. Chem.* **1982**, *47*, 1618.

19. Degl'Innocenti, A.; Ulivi, P.; Capperucci, A.; Reginato, G.; Mordini, A.; Ricci, A., *Synlett* **1992**, 883.

20. Mordini, A.; Palio, G.; Ricci, A.; Taddei, M., *Tetrahedron Lett.* **1988**, *29*, 4991.

21. Sternberg, E.; Binger, P., *Tetrahedron Lett.* **1985**, *26*, 301.

22. Yasuda, H.; Nishi, T.; Miyanaga, S.; Nakamura, A., *Organometallics* **1985**, *4*, 359.

Denis Labrecque & Tak-Hang Chan
McGill University, Montreal, Montreal, Quebec, Canada

(*E*)-1-Trimethylsilyl-1,3-butadiene[1]

[71504-26-2]   C$_7$H$_{14}$Si   (MW 126.30)

InChI = 1S/C7H14Si/c1-5-6-7-8(2,3)4/h5-7H,1H2,2-4H3/b7-6+

InChIKey = YFSJFUCGCAQAJA-VOTSOKGWSA-N

(diene for Diels–Alder reactions,[2,3] creating allylsilanes that can be converted to a variety of functionalized cyclohexenes with a shift of the double bond from its original position in the cycloaddition;[3] synthesis of a chiral allylborane;[4] synthesis of allylsilanes[5,6] and trienes[7])

Physical Data: bp 110–115 °C, 70–74 °C/210 mmHg.

Solubility: freely sol all organic solvents.

Analysis of Reagent Purity: NMR: δ (CDCl$_3$) 6.6–4.9 (5H, m), 0.02 (9H, s).[3] UV: λ$_{max}$ (EtOH) 231 nm (ε 22 000).[2]

Preparative Method: prepared geometrically pure by Wittig[2] or Peterson[3] reactions on (*E*)-3-trimethylsilylpropenal,[8] by pyrolysis of silylated 3-sulfolenes,[9] and by coupling (*E*)-2-trimethylsilylvinyl bromide with lithium divinylcuprate.[10] It can also be prepared geometrically impure, but largely (*E*), from allyltrimethylsilane by a vinylogous Ramberg–Bäcklund

reaction,[11] and from 1,3-butadienyl-1-magnesium chloride and chlorotrimethylsilane.[4]

Handling, Storage, and Precautions: reasonably stable in the absence of air and radical initiators. It should be kept in a stoppered flask or sealed ampule in the presence of hydroquinone. Its toxicity is unknown, but the presence of a trimethylsilyl group is generally benign.

Diels–Alder Reactions. (*E*)-1-Trimethylsilyl-1,3-butadiene undergoes Diels–Alder reactions with the usual dienophiles, with unimpaired *endo* stereoselectivity, but at a somewhat slower rate than 1,3-butadiene itself (eq 1).[2,3] The regioselectivity with unsymmetrical dienophiles is poor (eq 2).[2,3] The adducts are allylsilanes, which react with electrophiles in the usual way (eqs 3 and 4) to give cyclohexenes with a double bond shifted from the original position.[3]

Because of the low level of regiocontrol imparted by the silyl group, other substituents on 1-silylated dienes can impart better regiocontrol (eq 5).[3]

Trimethylsilylallylboranes. The diene is used as a starting material for the synthesis of the chiral allylborane (**1**) or its enantiomer (eq 6).[4]

Allylsilanes. Reduction of the diene with lithium and *t*-butanol in liquid ammonia gives crotyltrimethylsilane,[5] and reaction with Grignard reagents gives longer chain (*E*)-allylsilanes (eq 7).[6]

$$TMS \diagup\diagdown\diagup\diagdown \xrightarrow[\text{THF, } -15\,°C]{\text{RMgBr}} TMS \diagup\diagdown\diagup\diagdown R \qquad (7)$$

Triene Synthesis. Like other vinylsilanes, 1-trimethylsilyl-1,3-butadiene can be used in a Hiyama coupling reaction with vinyl iodides (eq 8).[7]

$$\diagup\diagdown\diagup\diagdown TMS + I\diagup\diagdown\diagup\diagdown\diagup\diagdown \xrightarrow[\text{(EtO)}_3\text{P, THF, 50\,°C}]{\text{TASF, } (\eta^3\text{-C}_3\text{H}_5\text{PdCl})_2}{78\%}$$

$$\diagup\diagdown\diagup\diagdown\diagup\diagdown\diagup\diagdown\diagup\diagdown \qquad (8)$$

1. Luh, T.-Y.; Wong, K.-T., *Synthesis* **1993**, 349.
2. Jung, M. E.; Gaede, B., *Tetrahedron* **1979**, *35*, 621.
3. Carter, M. J.; Fleming, I.; Percival, A., *J. Chem. Soc., Perkin Trans. 1* **1981**, 2415.
4. Short, R. P; Masamune, S., *J. Am. Chem. Soc.* **1989**, *111*, 1892.
5. Koshutina, L. L.; Koshutin, V. I., *Zh. Obshch. Khim.* **1978**, *48*, 932 (*Chem. Abstr.* **1978**, *89*, 43 592z).
6. Koshutin, V. I.; *Zh. Obshch. Khim.* **1978**, *48*, 1665 (*Chem. Abstr.* **1978**, *89*, 163 645r).
7. Hatanaka, Y.; Hiyama, T., *J. Org. Chem.* **1988**, *53*, 918.
8. Jones, T. K.; Denmark, S. E., *Org. Synth., Coll. Vol.* **1990**, *7*, 524, and ref. 3.
9. (a) Bloch, R.; Abecassis, J., *Tetrahedron Lett.* **1983**, *24*, 1247. (b) Chou, T.; Tso, H.-H.; Tao, Y.-T.; Lin, L. C., *J. Org. Chem.* **1987**, *52*, 244.
10. Koshutin, V. I.; Nazarenko, N. P., *Zh. Obshch. Khim.* **1982**, *52*, 2376 (*Chem. Abstr.* **1983**, *98*, 89 445t).
11. Block, E.; Aslam, M.; Eswarakrishnan, V.; Gebreyes, K.; Hutchinson, J.; Iyer, R.; Laffitte, J.-A.; Wall, A., *J. Am. Chem. Soc.* **1986**, *108*, 4568.

Ian Fleming
Cambridge University, Cambridge, UK

3-Trimethylsilyl-3-buten-2-one

[43209-86-5] $C_7H_{14}OSi$ (MW 142.30)

InChI = 1S/C7H14OSi/c1-6(8)7(2)9(3,4)5/h2H2,1,3-5H3

InChIKey = SDWXIGKTVZPGPG-UHFFFAOYSA-N

(methyl vinyl ketone homolog[1] useful as a Michael acceptor in annulation reactions,[2,3] γ,δ-unsaturated ketone formation,[4] and Peterson condensations;[5] precursor to a stabilized Diels–Alder diene[6])

Physical Data: bp 72 °C/50 mmHg.

Analysis of Reagent Purity: ^{1}H NMR (CCl4): δ 6.53 (d, *J* = 2 Hz, 1H, C*H*), 6.18 (d, *J* = 2 Hz, 1H, C*H*), 2.23 (s, 3H, C*H*3), 0.14 [s, 9H, Si(C*H*3)3]. Analysis by gas chromatography on a 1.85m 3% silicon gum rubber (SE-30) column at 25 °C gives a single peak.

Preparative Method: prepared by the reaction of (1-bromovinyl)trimethylsilane with acetaldehyde.[1] An alternative synthesis from acrolein has been published.[7]

Handling, Storage, and Precaution:

shows no tendency to deteriorate when stored under an argon atmosphere at $-20\,°C$.

Annulation. 3-Trimethylsilyl-3-buten-2-one has been employed as a methyl vinyl ketone homolog in an improved method for the annulation of ketones.[2] Based on work by Stork and Ganem,[3] who employed 3-triethylsilyl-3-buten-2-one as a Michael acceptor in the Robinson annulation reaction, Suzuki and co-workers[8] prepared a functionalized bicyclic ketone via a silyl enol ether as shown in eq 1. In general, the annulation of 2-alkylcycloketones with methyl vinyl ketone and its homologs produces rather poor yields of the desired cyclized products.

$$(1)$$

The conjugate addition of enolate anions to activated 3-trimethylsilyl-3-buten-2-one helped solve another long-standing problem in organic synthesis by permitting the annulation reaction to be carried out in aprotic solvents under conditions where enolate equilibration is avoided. The annulation of thermodynamically unstable lithium enolates with MVK, where equilibration to the more stable enolate occurs prior to Michael addition, often yields a mixture of structural isomers.[9] For example, Boeckman successfully employed 3-trimethylsilyl-3-buten-2-one in a Robinson annulation sequence (eq 2). Thus treatment of cyclohexenone with lithium dimethylcuprate in diethyl ether and then with 3-trimethylsilyl-3-buten-2-one gives the desired Michael adduct, which is converted into the functionalized octalone in 52% overall yield.[10]

$$(2)$$

Boeckman originally postulated that the enolate copper bond played a part in the success of this sequence. In 1974, however, Boeckman[11] and Stork and Singh[12] reported simultaneously that lithium enolates can be used under aprotic conditions for regio-specific annulation with 3-trimethylsilyl-3-buten-2-one (eq 3).

The most general annulation procedure is to first generate an enolate by lithium–ammonia reduction of an enone in the presence of *t*-butanol; the enolate is then trapped as the trimethylsilyl enol ether, which can be examined spectroscopically to establish homogeneity. The enolate is then regenerated with methyllithium in dry glyme or diethyl ether at −78 °C prior to the addition of the α-silylated vinyl ketone. In this case, only traces of the linear tricyclic ketone are formed. The trimethylsilyl group helps stabilize the intermediate carbanion formed by conjugate addition, relative to that of the starting ketone, thus facilitating the forward reaction and at the same time discouraging equilibration of the enolate of the starting ketone.

Michael Condensation. In 1988, Hagiwara and co-workers extended this annulation technology to include the two-fold Michael reaction of kinetic enolates derived from 1-acetylcyclohexenes (eq 4).[13] The kinetic enolates of 1-acetylcyclohexenes are generated from the corresponding trimethylsilyl enol ethers by treatment with MeLi in THF prior to the addition of 3-trimethylsilyl-3-buten-2-one. This one-pot annulation produces the desired decalone as a single isomer in 39% yield.

1-Alkenyldialkoxyboranes also react with 3-trimethylsilyl-3-buten-2-one in the presence of boron trifluoride etherate through a facile 1,4-addition to give γ,δ-unsaturated ketones in good yields with high regio- and stereoselectivity (eq 5). Without BF$_3$ etherate this reaction does not occur and other Lewis acids, such as AlCl$_3$, TiCl$_4$, SnCl$_4$, and ZnCl$_2$, are less effective.[4]

Michael–Peterson Condensation. 3-Trimethylsilyl-3-buten-2-one also undergoes smooth Michael addition with Grignard reagents (R = Me, *n*-Pr, *i*-Pr, *t*-Bu, Ph), generating magnesium enolates which are then trapped with benzaldehyde to give (*E*)- and (*Z*)-enone isomers after Peterson condensation (eq 6).[5] For example, treatment of the α-silyl vinyl ketone with methylmagnesium iodide followed by reaction with benzaldehyde yields a 7:1

mixture of (*E*)- and (*Z*)-isomers of 3-ethyl-4-phenyl-3-buten-2-one in 45% yield. (*E*)-Alkenes become the major products under thermodynamic control when the condensation with benzaldehyde is carried out at room temperature in diethyl ether. (*Z*)-Isomers are favored as kinetically controlled products at −78 °C in THF.

Diene Adduct. 3-Trimethylsilyl-3-buten-2-one undergoes a Shapiro reaction to give the 2,3-bis(trimethylsilyl)buta-1,3-diene in 32% yield (eq 7). This compound undergoes Diels–Alder reactions with a number of dienophiles (e.g. maleic anhydride, benzoquinone) in benzene at 60 °C to give silylated cycloaddition adducts.[6]

1. Boeckman, R. K., Jr.; Blum, D. M.; Ganem, B.; Halvey, N., *Org. Synth., Coll. Vol.* **1988**, *6*, 1033.

2. (a) Shishido, K.; Hiroya, K.; Fukumoto, K.; Kametani, T., *J. Chem. Soc., Perkin Trans. 1* **1986**, 837. (b) Bonnert, R. V.; Jenkins, P. R., *J. Chem. Soc., Chem. Commun.* **1987**, 6. (c) Murai, A.; Tanimoto, N.; Sakamoto, N.; Masamune, T., *J. Am. Chem. Soc.* **1988**, *110*, 1985. (d) Rigby, J. H.; Kierkus, P. C.; Head, D., *Tetrahedron Lett.* **1989**, *30*, 5073. (e) Paquette, L. A.; Sauer, D. R.; Cleary, D. G.; Kinsella, M. A.; Blackwell, C. M.; Anderson, L. G., *J. Am. Chem. Soc.* **1992**, *114*, 7375.

3. Stork, G.; Ganem, B., *J. Am. Chem. Soc.* **1973**, *95*, 6152.

4. Hara, S.; Hyuga, S.; Aoyama, M.; Sato, M.; Suzuki, A., *Tetrahedron Lett.* **1990**, *31*, 247.

5. Tanaka, J.; Kobayashi, H.; Kanemasa, S.; Tsuge, O., *Bull. Chem. Soc. Jpn.* **1989**, *62*, 1193.

6. Garratt, P. J.; Tsotinis, A., *Tetrahedron Lett.* **1986**, *27*, 2761.

7. Okumoto, H.; Tsuji, J., *Synth. Commun.* **1982**, *12*, 1015.

8. Suzuki, T.; Sato, E.; Unno, K.; Kametani, T., *J. Chem. Soc., Chem. Commun.* **1988**, 724.

9. Marshall, J. A.; Fanta, W. I., *J. Org. Chem.* **1964**, *29*, 2501, and references cited therein.

10. (a) Boeckman, R. K., Jr., *J. Am. Chem. Soc.* **1973**, *95*, 6867. (b) Boeckman, R. K. Jr.; Blum, D. M.; Ganem, B., *Org. Synth., Coll. Vol.* **1988**, *6*, 666. (c) See also: Takahashi, T.; Naito, Y.; Tsuji, J., *J. Am. Chem. Soc.* **1981**, *103*, 5261.

11. Boeckman, R. K., Jr., *J. Am. Chem. Soc.* **1974**, *96*, 6179.

12. Stork, G.; Singh, J., *J. Am. Chem. Soc.* **1974**, *96*, 6181.

13. Hagiwara, H.; Akama, T.; Okano, A.; Uda, H., *Chem. Lett.* **1988**, 1793.

Bradley B. Brown
University of California, Berkeley, CA, USA

9-[1-(Trimethylsilyl)-2(E)-butenyl]-9-borabicyclo[3.3.1]nonane

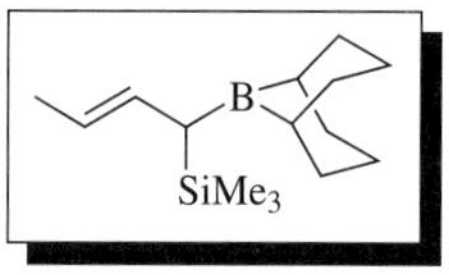

[100701-75-5] $C_{15}H_{29}BSi$ (MW 248.34)

InChI = 1S/C15H29BSi/c1-5-8-15(17(2,3)4)16-13-9-6-10-14(16)12-7-11-13/h5,8,13-15H,6-7,9-12H2,1-4H3/b8-5+/t13-,14+,15?

InChIKey = CTEZPJQQPSDWEJ-QDHMSOTKSA-N

(*anti*-(Z)-selective allylation reagent for aldehydes and pyruvates;[2] stereoselective synthesis of (Z)-2-butenylsilane[1])

Alternate Names: 9-trimethylsilylcrotyl-9-borabicyclo[3.3.1]-nonane.

Physical Data: bp 85–87 °C/0.03 mmHg.

Solubility: sol most organic solvents.

Preparative Method: prepared as described by Yatagai et al.[1]

Handling, Storage, and Precautions: the neat liquid must be handled and stored in the absence of air. May be kept as CH_2Cl_2 or THF solutions under a N_2 atmosphere. Use in a fume hood.

***Anti*-(Z)-Selective Crotylation of Aldehydes.** The reaction of α-trimethylsilyl-substituted crotyl-9-BBN with aldehydes in the presence of pyridine affords *anti*-homoallyl alcohols having (Z) double bond geometry (eq 1).[2] The use of 1 or 2 equiv of pyridine is essential to obtain high stereoselectivity. Without pyridine, a complex mixture of products, coupled both at the α- and γ-position of the crotyl reagent, are obtained. Perhaps pyridine coordinates the boron atom to form the corresponding ate complex, which prevents allylic rearrangement of the crotylboron. The reaction with aldehydes proceeds through a six-membered chair transition state in which the trimethylsilyl group adopts an axial position, giving (Z) geometry of the double bond in the homoallyl alcohol products. The *anti*-(Z)-alkenylsilanes thus obtained are useful for the further elaboration of complex molecules. For example, epoxidation with *m*-chloroperbenzoic acid proceeds with high diastereoselectivity to give the corresponding epoxides in good yield (eq 2).

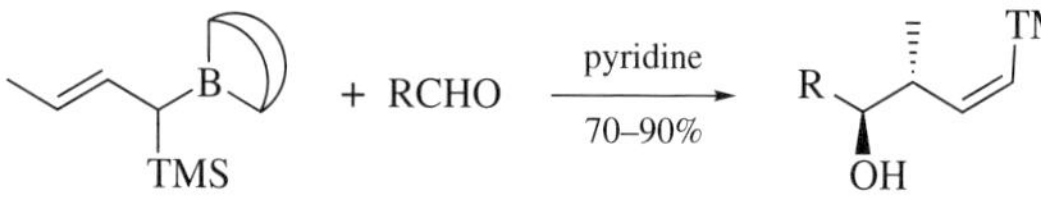

R = Ph, Bu, PhCH=CH

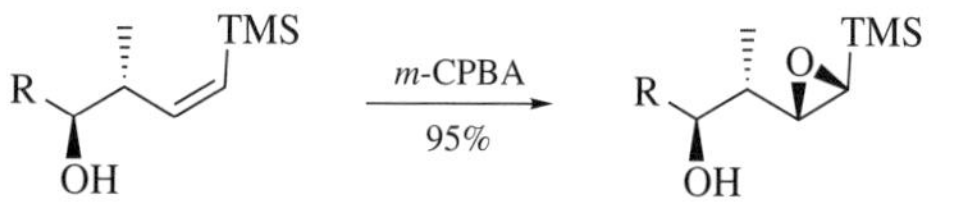

The reaction of methyl pyruvate with α-silyl-substituted crotyl-9-BBN gives the *anti*-(Z)-adduct as a single diastereoisomer in 60% yield along with the α-adduct (35%) (eq 3).[3] *n*-Butyllithium is used instead of pyridine for converting the crotylboron reagent

to the corresponding ate complex. The *anti*-(Z)-adduct can be converted into *cis*-crobarbatic acid methyl ester via the procedure[4] of Magnus (eq 4). The reaction of simple crotyl-9-BBN with pyruvates produces the *anti*-isomers either predominantly or exclusively. The *anti* diastereoselectivity can be improved by increasing the steric bulk of the ester groups; *anti/syn* = 73:27 in the case of the methyl ester, whereas the ratio becomes 100:0 in the case of the 2,6-di-*t*-butyl-4-methylphenyl ester. However, as mentioned above, the α-silyl-substituted crotyl-9-BBN produces a 100:0 selectivity even in the case of the methyl ester.

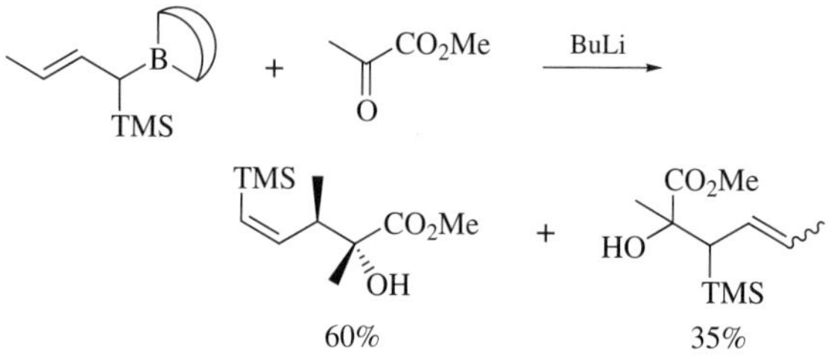

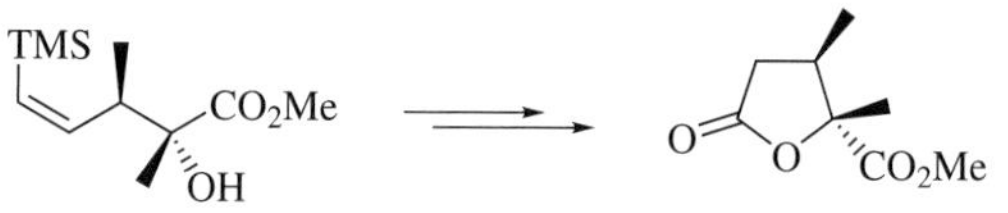

α-Trimethylstannylallyl-9-BBN, prepared by a procedure similar to the synthesis of α-trimethylsilylallyl-9-BBN, also exhibits *anti*-(Z) diastereoselectivity in reactions with aldehydes and pyruvates, as observed in the case of the silyl reagent.

Synthesis of (Z)-2-Alkenylsilanes. Protonolysis of 2-alkenyl-1-trimethylsilyl-9-BBN affords (Z)-2-alkenylsilanes with high stereoselectivity (eq 5).[1] Similarly, protonolysis of 2-alkenyl-1-trimethylstannyl-9-BBN gives (Z)-2-alkenylstannanes.

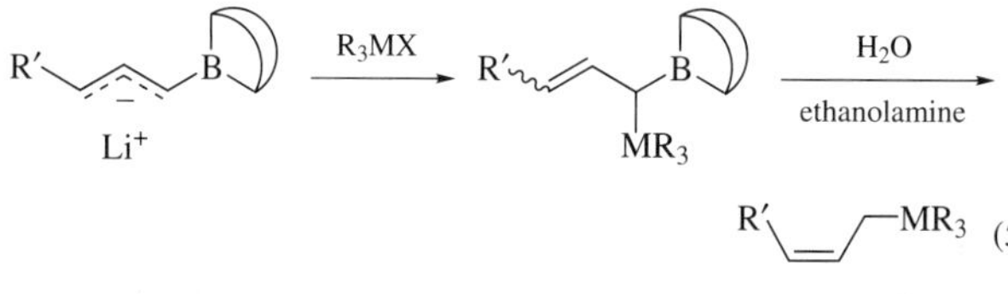

$R_3M = Me_3Si, Me_3Sn, Bu_3Sn$

Related Reagent. *B*-Allyl-9-borabicyclo[3.3.1]nonane.

1. Yatagai, H.; Yamamoto, Y.; Maruyama, K., *J. Am. Chem. Soc.* **1980**, *102*, 4548.

2. Yamamoto, Y.; Yatagai, H.; Maruyama, K., *J. Am. Chem. Soc.* **1981**, *103*, 3229.

3. (a) Yamamoto, Y.; Komatsu, T.; Maruyama, K., *J. Chem. Soc., Chem. Commun.* **1983**, 191. (b) Yamamoto, Y.; Maruyama, K.; Komatsu, T.; Ito, W., *J. Org. Chem.* **1986**, *51*, 886.

4. Ehlinger, E.; Magnus, P., *J. Am. Chem. Soc.* **1980**, *102*, 5004.

Yoshinori Yamamoto
Tohoku University, Sendai, Japan

Trimethylsilylcopper[1]

$$\boxed{\text{Me}_3\text{SiCu}}$$

[91899-54-6] C_3H_9CuSi (MW 136.76)

InChI = 1S/C3H9Si.Cu/c1-4(2)3;/h1-3H3;

InChIKey = LEJJRFFBNLQENA-UHFFFAOYSA-N

(nucleophilic silylmetal reagent useful for the regioselective preparation of allyltrimethylsilanes from allylic halides,[2–4] allylic sulfonates,[4] and allylic phosphates[5])

Preparative Method: prepared in situ by adding 1 equiv of copper(I) iodide dissolved in dimethyl sulfide (2.5 M solution) to a vigorously stirred solution of 1 equiv of trimethylsilyllithium[6] (prepared from hexamethyldisilane and methyllithium–lithium bromide complex) in HMPA at 0–5 °C under argon.[2] The reagent is used immediately in silylation reactions.

Handling, Storage, and Precautions: prepared and used under dry air-free conditions (argon atmosphere, dry solvents, etc.). Since reactions with this reagent employ HMPA (a carcinogen) as solvent, suitable precautions should be taken. Use in a fume hood.

Silylcopper Reagents.[1] Trimethylsilylcopper (TMSCu) is one of the many silylcopper reagents which were developed in the 1980s and used in the synthesis of allylsilanes from various allylic substrates. Other well-studied silylcopper reagents include $(Me_3Si)_2CuLi$,[7] $(Me_3Si)Cu(CN)Li$,[8] $(PhMe_2Si)_2CuLi$,[7,9–11] and $(PhMe_2Si)_2Cu(CN)Li_2$.[12–14] In all cases, these reagents are prepared in situ from 1 equiv of a silyllithium reagent (Me_3SiLi or $PhMe_2SiLi$) by reaction with 0.5–1.0 equiv of copper(I) iodide or copper(I) cyanide.

Synthesis of Allylsilanes. TMSCu reacts with a variety of allylic substrates to form allylsilanes stereoselectively and in good to excellent yields. Thus the reagent reacts readily with primary allylic halides,[2–4] sulfonates from secondary and tertiary alcohols,[4] and primary and secondary allylic phosphates[5] to yield a variety of allylsilanes. Representative examples of these reactions are shown in eqs 1–3. This methodology seems particularly attractive for the synthesis of 3-trimethylsilyl-1-alkenes which are presently unavailable

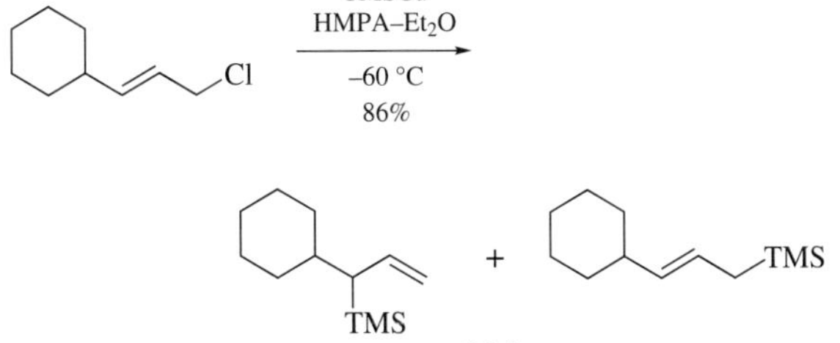

$$\text{(1)}$$

$$92:8$$

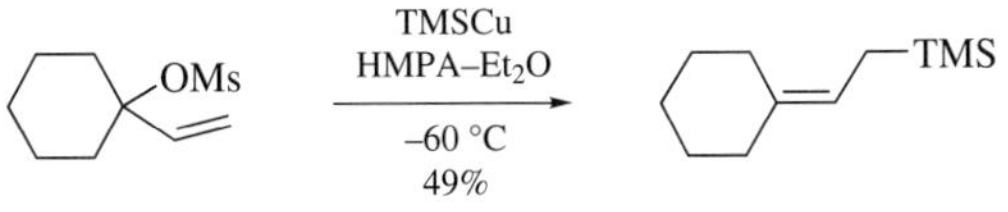

$$\text{(2)}$$

by other methods. For example, (*E*)-8-*t*-butyldiphenylsilyloxy-2,6-dimethyl-3-trimethylsilyl-1,6-octadiene, a key intermediate in Corey's synthesis of tricyclohexaprenol, has been prepared from a primary allylic chloride using this methodology (eq 4).[15] Moreover, this method has been used in the preparation of 2,3-bis(trimethylsilyl)alk-1-enes (eq 5)[4] and (2-bromoallyl)trimethylsilane (eq 6),[2] a useful intermediate in organic synthesis.[8]

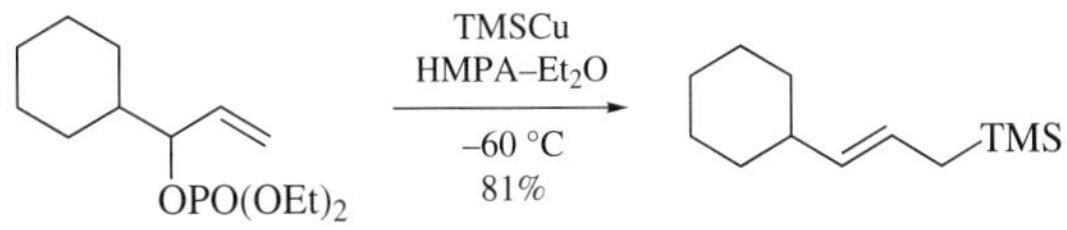

$$\text{(3)}$$

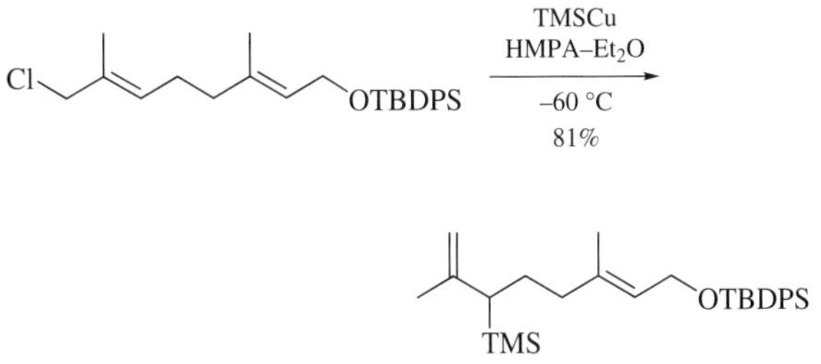

$$\text{(4)}$$

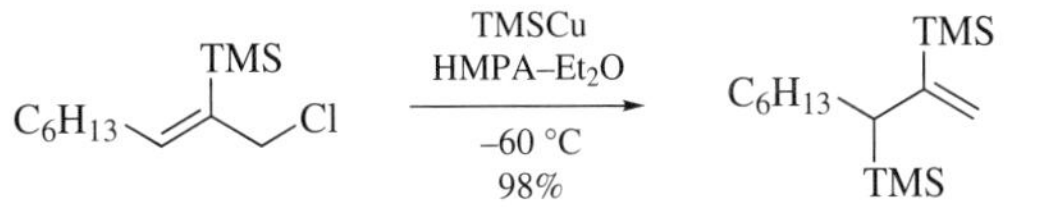

$$\text{(5)}$$

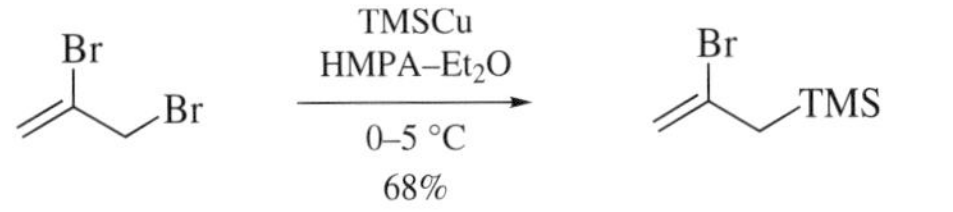

$$\text{(6)}$$

The reactions of TMSCu are in marked contrast to the reactions of trimethylsilyllithium with similar substrates.[3,4] Whereas treatment of a 1-chloro-2-alkene with TMSCu affords a 3-trimethylsilyl-1-alkene, reaction of TMSLi with these same starting materials yields terminal (*E*)-allylsilanes where the stereochemistry of the double bond is retained in the product (eq 7). Thus a single allylic halide yields either of two regioisomers by proper choice of reaction conditions. With allylic phosphates, TMSCu affords as major product the allylsilane resulting from attack by what is formally an S_N2'-like reaction. With TMSLi, however, the major product is always the isomeric allylsilane having a more substituted internal double bond (eq 8).

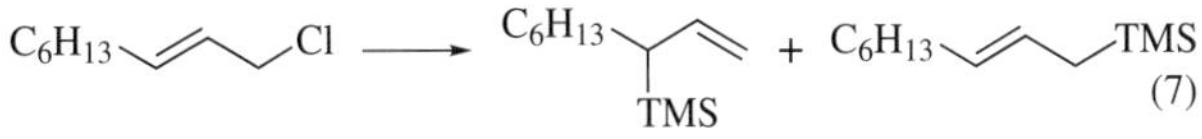

$$\text{(7)}$$

TMSCu 87%
TMSLi 78%
98:2
0:100

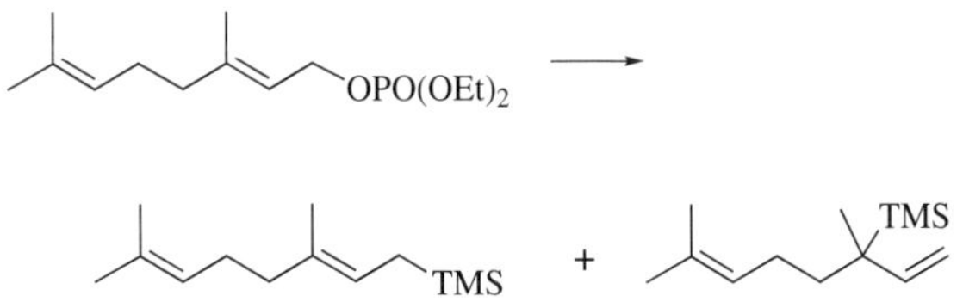

$$\text{(8)}$$

TMSCu 92%
TMSLi 69%
25:75
100:0

Other silylcuprate reagents have also been used to prepare allylsilanes from allylic halides,[10] acetates,[11,13] and urethanes.[13c,14,16] Moreover, other silylmetal reagents have been used in similar preparations of allylsilanes. For example, silylaluminum reagents have been used to prepare allylsilanes from allylic acetates[17] and allylic phosphates,[18] and silylmanganese reagents react with allylic sulfides and allylic ethers.[19] One advantage that the present methodology has over many other methods is that it utilizes a *trimethyl*silylmetal reagent to afford an allyl*trimethyl*silane. The silyl byproducts obtained in the preparation and reaction of these allylsilanes are volatile and easily separable from the desired product. Many of the other existing methods yield allylsilanes where one or more of the alkyl groups on the silicon is replaced by phenyl.

Related Reagents. Lithium Cyano(dimethylphenylsilyl)-cuprate; Trimethylsilyllithium.

1. (a) Sarkar, T. K., *Synthesis* **1990**, 969. (b) Colvin, E. W. *Silicon Reagents in Organic Synthesis*; Academic: New York, 1988; pp 27, 51. (c) Colvin, E. W. *Silicon in Organic Synthesis*; Butterworths: Boston, 1981; p 134.

2. Smith, J. G.; Quinn, N. R.; Viswanathan, M., *Synth. Commun.* **1983**, *13*, 1.

3. Smith, J. G.; Quinn, N. R.; Viswanathan, M., *Synth. Commun.* **1983**, *13*, 773.

4. Smith, J. G.; Drozda, S. E.; Petraglia, S. P.; Quinn, N. R.; Rice, E. M.; Taylor, B. S.; Viswanathan, M., *J. Org. Chem.* **1984**, *49*, 4112.

5. Smith, J. G.; Henke, S. L.; Mohler, E. M.; Morgan, L.; Rajan, N. I., *Synth. Commun.* **1991**, *21*, 1999.

6. Still, W. C., *J. Org. Chem.* **1976**, *41*, 3063.

7. Ager, D. J.; Fleming, I., *J. Chem. Soc., Chem. Commun.* **1978**, 177.

8. Trost, B. M.; Chan, D. M. T., *J. Am. Chem. Soc.* **1982**, *104*, 3733.

9. Ager, D. J.; Fleming, I.; Patel, S. K., *J. Chem. Soc., Perkin Trans. 1* **1981**, 2520.

10. (a) Laycock, B.; Maynard, I.; Wickham, G.; Kitching, W., *Aust. J. Chem.* **1988**, *41*, 693. (b) Laycock, B.; Kitching, W.; Wickham, G., *Tetrahedron Lett.* **1983**, *24*, 5785.

11. Fleming, I.; Marchi, D., Jr. *Synthesis* **1981**, 560.

12. Fleming, I.; Newton, T. W.; Roessler, F., *J. Chem. Soc., Perkin Trans. 1* **1981**, 2527.

13. (a) Fleming, I.; Terrett, N. K., *J. Organomet. Chem.* **1984**, *264*, 99. (b) Fleming, I.; Terrett, N. K., *Tetrahedron Lett.* **1983**, *24*, 4151. (c) Fleming, I.; Thomas, A. P., *J. Chem. Soc., Chem. Commun.* **1985**, 411.

14. Fleming, I.; Thomas, A. P., *J. Chem. Soc., Chem. Commun.* **1986**, 1456.

15. Corey, E. J.; Burk, R. M., *Tetrahedron Lett.* **1987**, *28*, 6413.

16. Fleming, I.; Newton, T. W., *J. Chem. Soc., Perkin Trans. 1* **1984**, 1805.

17. Trost, B. M.; Yoshida, J.; Lautens, M., *J. Am. Chem. Soc.* **1983**, *105*, 4494.

18. Okuda, Y.; Sato, M.; Oshima, K.; Nozaki, H., *Tetrahedron Lett.* **1983**, *24*, 2015.

19. (a) Fugami, K.; Oshima, K.; Utimoto, K.; Nozaki, H., *Tetrahedron Lett.* **1986**, *27*, 2161. (b) Fugami, K.; Hibino, J.; Nakatsukasa, S.; Matsubara, S.; Oshima, K.; Utimoto, K.; Nozaki, H., *Tetrahedron* **1988**, *44*, 4277.

Janice Gorzynski Smith
Mount Holyoke College, South Hadley, MA, USA

Trimethylsilyldiazomethane[1]

[18107-18-1]   C4H10N2Si   (MW 114.25)
InChI = 1S/C4H10N2Si/c1-7(2,3)4-6-5/h4H,1-3H3
InChIKey = ONDSBJMLAHVLMI-UHFFFAOYSA-N

(one-carbon homologation reagent; stable, safe substitute for diazomethane; [C–N–N] 1,3-dipole for the preparation of azoles[1])

Physical Data: bp 96 °C/775 mmHg; n_D^{25} 1.4362.[2]
Solubility: sol most organic solvents; insol H2O.
Form Supplied in: commercially available as 2 M and 10 w/w% solutions in hexane, and 10 w/w% solution in CH2Cl2; also available as 2 M solutions in diethyl ether or hexanes.
Analysis of Reagent Purity: concentration in hexane is determined by ^{1}H NMR.[3]
Preparative Methods: prepared by the diazo-transfer reaction of trimethylsilylmethylmagnesium chloride with diphenyl phosphorazidate (DPPA) (eq 1).[3]

$$ \text{TMS} \diagdown \text{Cl} \xrightarrow{\text{Mg}} \text{TMS} \diagup \text{MgCl} \xrightarrow{(\text{PhO})_2\text{P(O)N}_3} \text{TMS} \diagdown_{\text{N}_2} \quad (1) $$

Handling, Storage, and Precautions: should be protected from light.

Original Commentary

Takayuki Shioiri & Toyohiko Aoyama
Nagoya City University, Japan

One-Carbon Homologation. Along with its lithium salt, which is easily prepared by lithiation of trimethylsilyldiazomethane (TMSCHN2) with *n*-butyllithium, TMSCHN2 behaves in a similar way to diazomethane as a one-carbon homologation reagent. TMSCHN2 is acylated with aromatic acid chlorides in the presence of triethylamine to give α-trimethylsilyl diazo ketones. In the acylation with aliphatic acid chlorides, the use of 2 equiv of TMSCHN2 without triethylamine is recommended. The crude diazo ketones undergo thermal Wolff rearrangement to give the homologated carboxylic acid derivatives (eqs 2 and 3).[4]

$$ \text{(eq 2)} $$

with conditions:
1. TMSCHN2, Et3N
2. PhNH2, 180 °C, 2,4,6-trimethylpyridine
80%

$$ \text{(eq 3)} $$

with conditions:
1. 2 equiv TMSCHN2
2. PhCH2OH, 180 °C, 2,4,6-trimethylpyridine
77%

Various ketones react with TMSCHN2 in the presence of boron trifluoride etherate to give the chain or ring homologated ketones (eqs 4–6).[5] The bulky trimethylsilyl group of TMSCHN2 allows for regioselective methylene insertion (eq 5). Homologation of aliphatic and alicyclic aldehydes with TMSCHN$_2$ in the presence of magnesium bromide smoothly gives methyl ketones after acidic hydrolysis of the initially formed β-keto silanes (eq 7).[6]

$$PhCOCH_2Ph \xrightarrow[\substack{CH_2Cl_2, -15\ ^\circ C, 1\ h \\ 74\%}]{TMSCHN_2,\ BF_3\cdot Et_2O} PhCOCH_2CH_2Ph \quad (4)$$

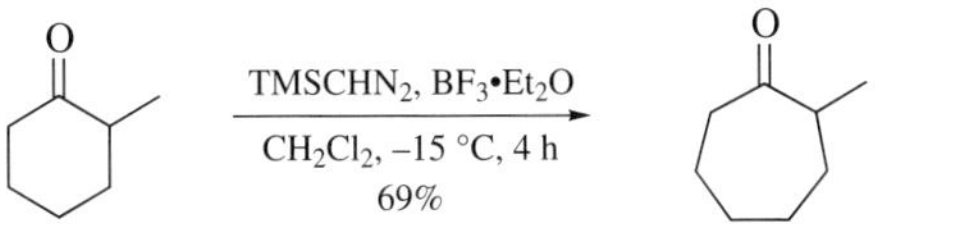

(5)

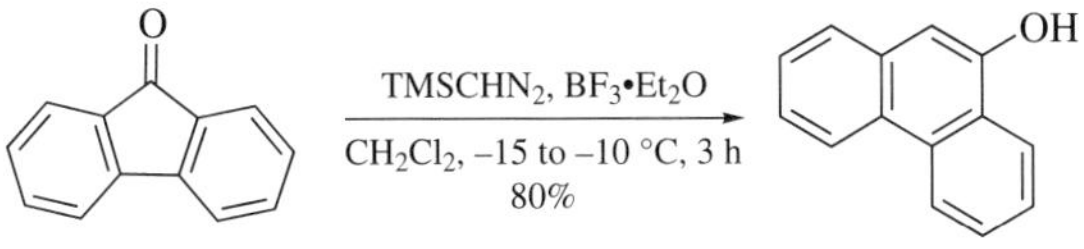

(6)

$$t\text{-BuCHO} \xrightarrow[\substack{2.\ 10\%\ aq\ HCl \\ 89\%}]{1.\ TMSCHN_2,\ MgBr_2} t\text{-BuCOMe} \quad (7)$$

O-Methylation of carboxylic acids, phenols, enols, and alcohols can be accomplished with TMSCHN$_2$ under different reaction conditions. TMSCHN$_2$ instantaneously reacts with carboxylic acids in benzene in the presence of methanol at room temperature to give methyl esters in nearly quantitative yields (eq 8).[7] This method is useful for quantitative gas chromatographic analysis of fatty acids. Similarly, O-methylation of phenols and enols with TMSCHN$_2$ can be accomplished, but requires the use of diisopropylethylamine (eqs 9 and 10).[8] Although methanol is recommended in these O-methylation reactions, methanol is not the methylating agent. Various alcohols also undergo O-methylation with TMSCHN$_2$ in the presence of 42% aq. tetrafluoroboric acid, smoothly giving methyl ethers (eq 11).[9]

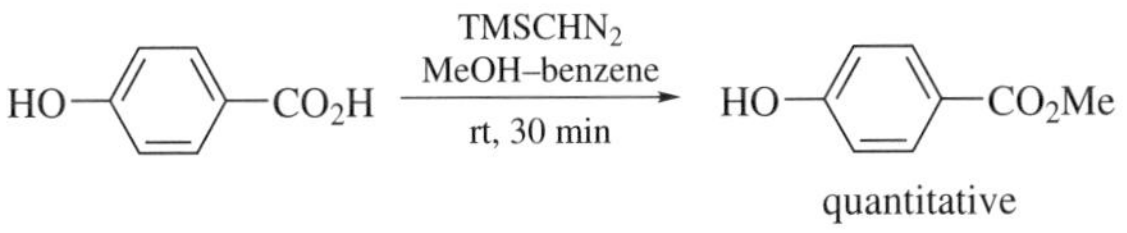

(8)

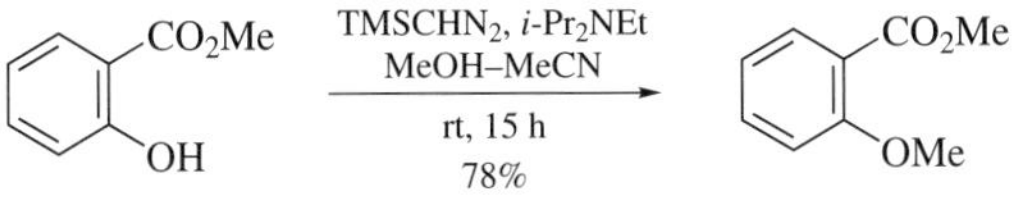

(9)

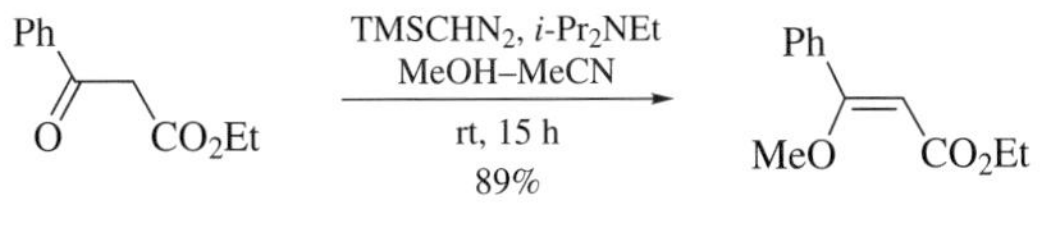

(10)

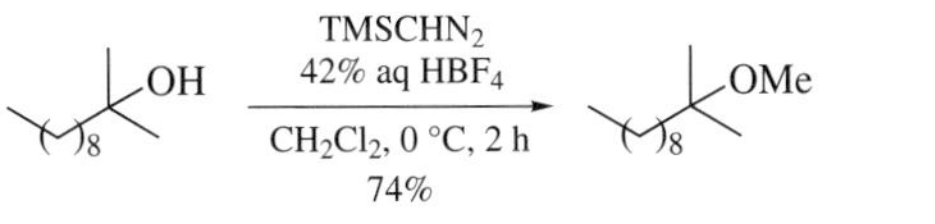

(11)

Alkylation of the lithium salt of TMSCHN$_2$ (TMSC(Li)N$_2$) gives α-trimethylsilyl diazoalkanes which are useful for the preparation of vinylsilanes and acylsilanes. Decomposition of α-trimethylsilyl diazoalkanes in the presence of a catalytic amount of copper(I) chloride gives mainly (E)-vinylsilanes (eq 12),[10] while replacement of CuCl with rhodium(II) pivalate affords (Z)-vinylsilanes as the major products (eq 12).[11] Oxidation of

α-trimethylsilyl diazoalkanes with m-chloroperbenzoic acid in a two-phase system of benzene and phosphate buffer (pH 7.6) affords acylsilanes (α-keto silanes) (eq 12).[12]

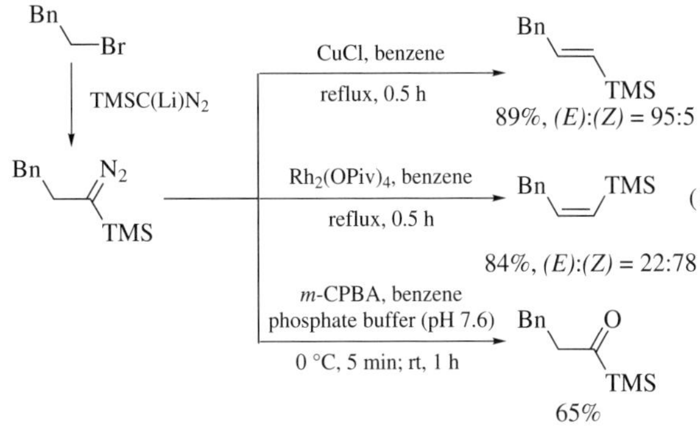

(12)

(E)-β-Trimethylsilylstyrenes are formed by reaction of alkanesulfonyl chlorides with TMSCHN$_2$ in the presence of triethylamine (eq 13).[13] TMSC(Li)N$_2$ reacts with carbonyl compounds to give α-diazo-β-hydroxy silanes which readily decompose to give α,β-epoxy silanes (eq 14).[14] However, benzophenone gives diphenylacetylene under similar reaction conditions (eq 15).[15]

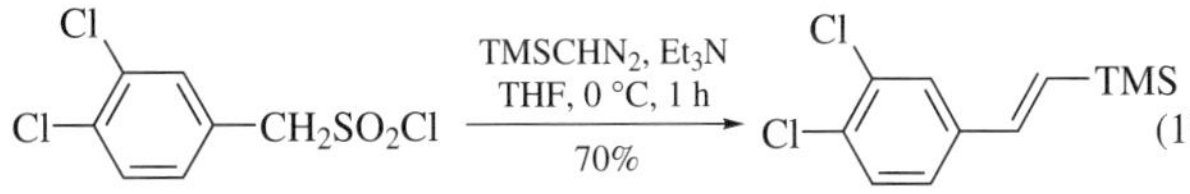

(13)

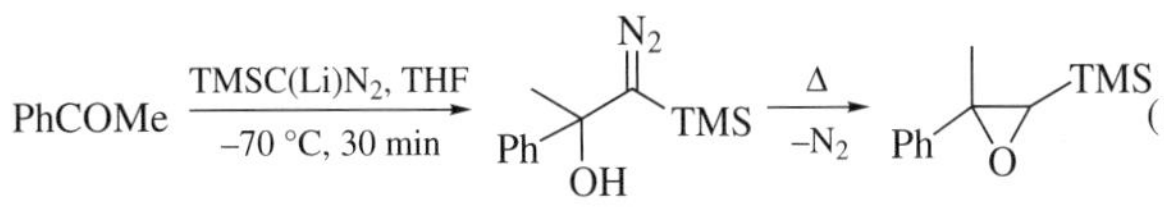

(14)

$$PhCOPh \xrightarrow[\substack{Et_2O,\ rt,\ 2\ h \\ 80\%}]{TMSC(Li)N_2} Ph\text{---}\!\!\equiv\!\!\text{---}Ph \quad (15)$$

Silylcyclopropanes are formed by reaction of alkenes with TMSCHN$_2$ in the presence of either palladium(II) chloride or CuCl depending upon the substrate (eqs 16 and 17).[16] Silylcyclopropanones are also formed by reaction with trialkylsilyl and germyl ketenes (eq 18).[17]

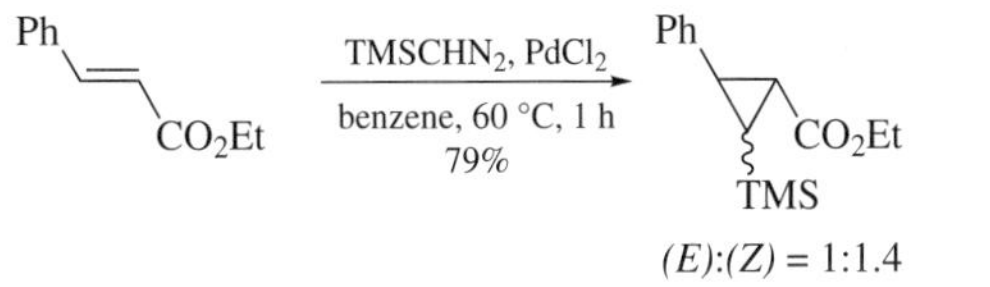

(16)

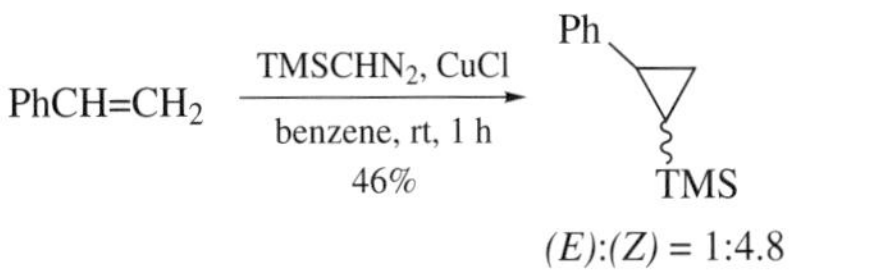

(E):$(Z) = 1$:1.4

(17)

(E):$(Z) = 1$:4.8

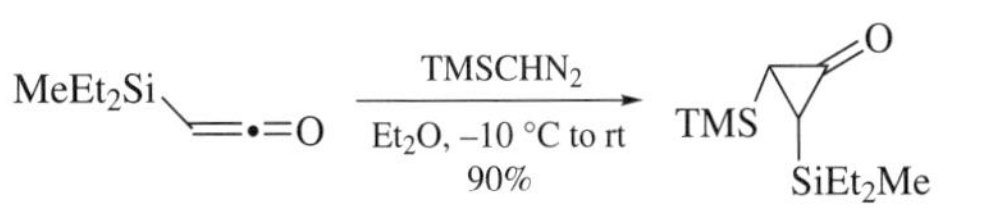

(18)

[C–N–N] Azole Synthon. TMSCHN$_2$, mainly as its lithium salt, TMSC(Li)N$_2$, behaves like a 1,3-dipole for the preparation of [C–N–N] azoles. The reaction mode is similar to that of diazomethane but not in the same fashion. TMSC(Li)N$_2$ (2 equiv) reacts with carboxylic esters to give 2-substituted 5-trimethylsilyl-tetrazoles (eq 19).[18] Treatment of thiono and dithio esters with TMSC(Li)N$_2$ followed by direct workup with aqueous methanol gives 5-substituted 1,2,3-thiadiazoles (eq 20).[19] While reaction of di-t-butyl thioketone with TMSCHN$_2$ produces the episulfide with evolution of nitrogen (eq 21),[20] its reaction with TMSC(Li)N$_2$ leads to removal of one t-butyl group to give the 1,2,3-thiadiazole (eq 21).[20]

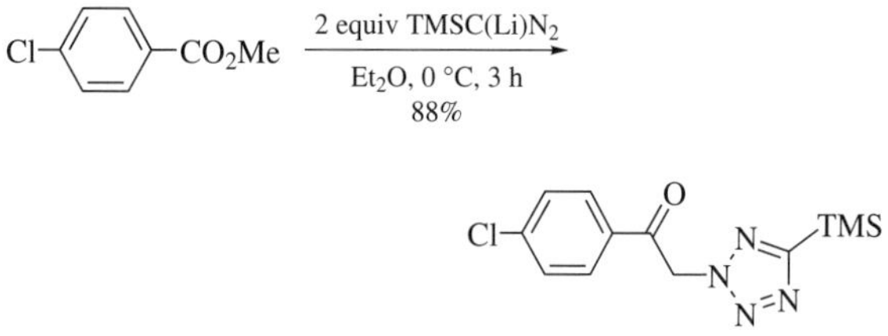

$$ (19) $$

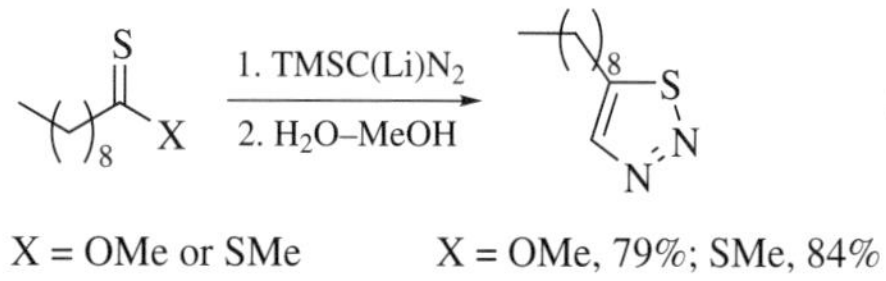

$$ (20) $$

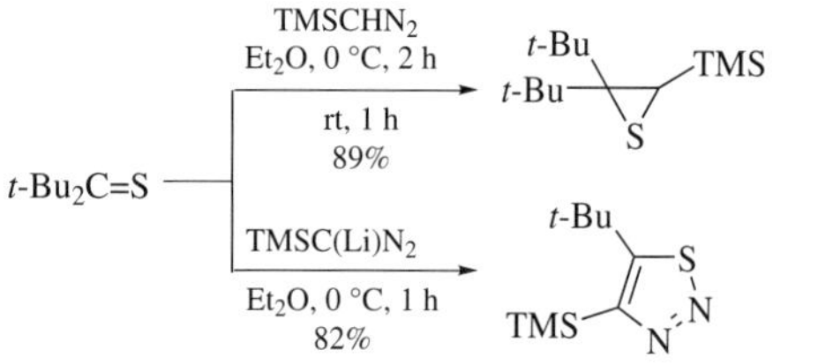

X = OMe or SMe   X = OMe, 79%; SMe, 84%

$$ (21) $$

TMSCHN$_2$ reacts with activated nitriles only, such as cyanogen halides, to give 1,2,3-triazoles.[21] In contrast with this, TMSC(Li)N$_2$ smoothly reacts with various nitriles including aromatic, heteroaromatic, and aliphatic nitriles, giving 4-substituted 5-tri-methylsilyl-1,2,3-triazoles (eq 22).[22] However, reaction of α,β-unsaturated nitriles with TMSC(Li)N$_2$ in Et$_2$O affords 3(or 5)-trimethylsilylpyrazoles, in which the nitrile group acts as a leaving group (eq 23).[23] Although α,β-unsaturated nitriles bearing bulky substituents at the α- and/or β-positions of the nitrile group undergo reaction with TMSC(Li)N$_2$ to give pyrazoles, significant amounts of 1,2,3-triazoles are also formed. Changing the reaction solvent from Et$_2$O to THF allows for predominant formation of pyrazoles (eq 24).[23] Complete exclusion of the formation of 1,2,3-triazoles can be achieved when the nitrile group is replaced by a phenylsulfonyl species.[24] Thus reaction of α,β-unsaturated sulfones with TMSC(Li)N$_2$ affords pyrazoles in excellent yields (eq 25). The geometry of the double bond of α,β-unsaturated sulfones is not critical in the reaction. When both a cyano and a sulfonyl group are present as a leaving group, elimination of the sulfonyl group occurs preferentially (eq 26).[24] The trimethylsilyl group attached to the heteroaromatic products is easily removed with 10% aq. KOH in EtOH or HCl–KF.

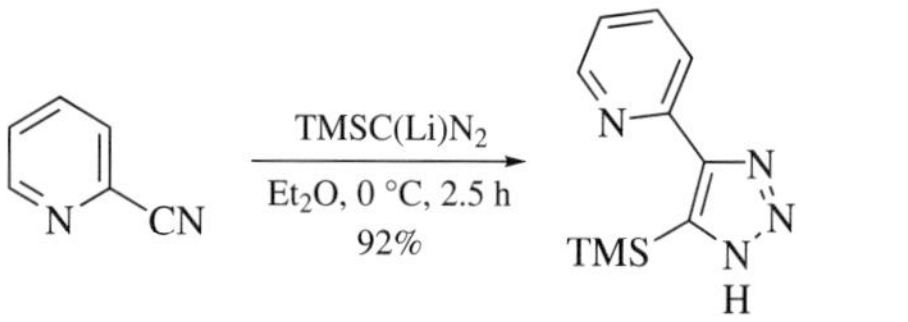

$$ (22) $$

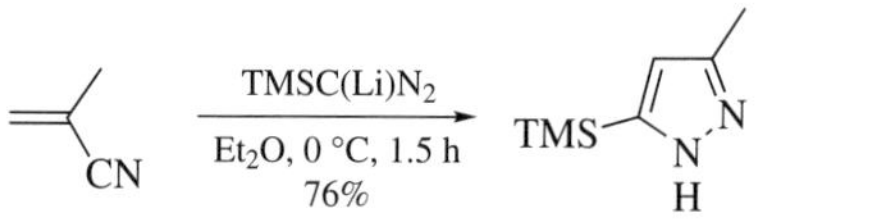

$$ (23) $$

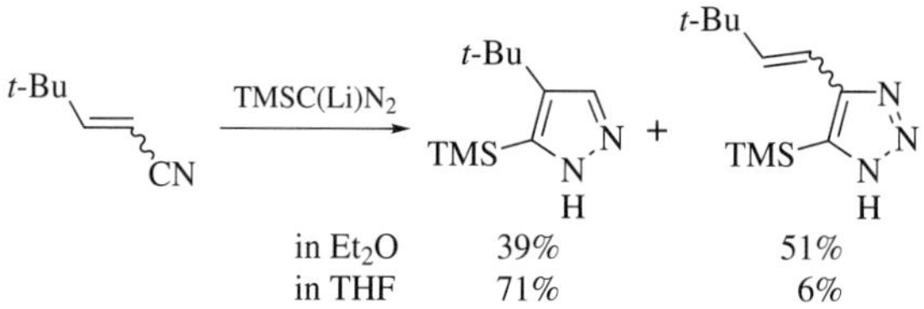

	in Et$_2$O	in THF
	39%	51%
	71%	6%

$$ (24) $$

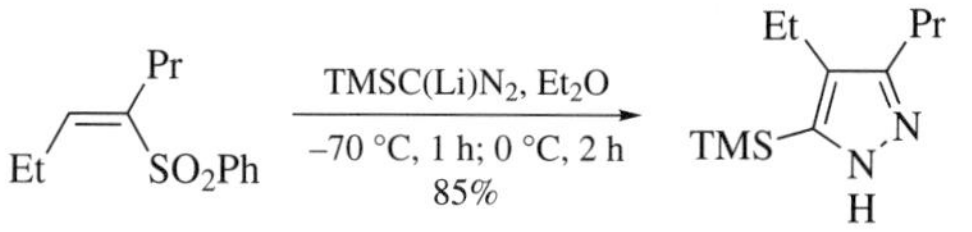

$$ (25) $$

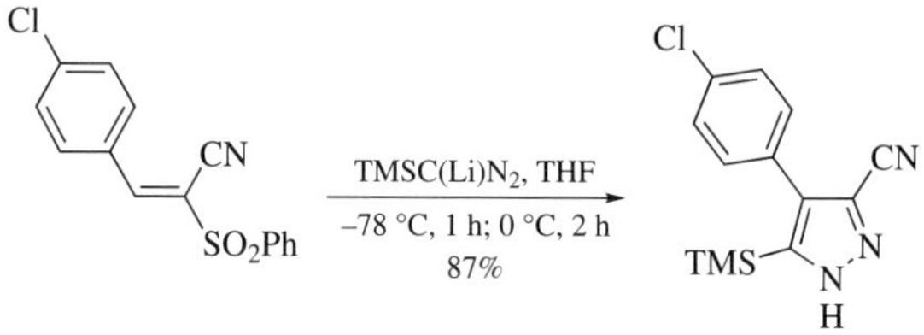

$$ (26) $$

Various 1,2,3-triazoles can be prepared by reaction of TMSC(Li)N$_2$ with various heterocumulenes. Reaction of isocyanates with TMSC(Li)N$_2$ gives 5-hydroxy-1,2,3-triazoles (eq 27).[25] It has been clearly demonstrated that the reaction proceeds by a stepwise process and not by a concerted 1,3-dipolar cycloaddition mechanism. Isothiocyanates also react with TMSC(Li)N$_2$ in THF to give lithium 1,2,3-triazole-5-thiolates which are treated in situ with alkyl halides to furnish 1-substituted 4-trimethylsilyl-5-alkylthio-1,2,3-triazoles in excellent yields (eq 28).[26] However, changing the reaction solvent from THF to Et$_2$O causes a dramatic solvent effect. Thus treatment of isothiocyanates with TMSC(Li)N$_2$ in Et$_2$O affords 2-amino-1,3,4-thiadiazoles in good yields (eq 28).[27] Reaction of ketenimines with TMSC(Li)N$_2$ smoothly proceeds to give 1,5-disubstituted 4-trimethylsilyl-1,2,3-triazoles in high yields (eq 29).[28] Ketenimines bearing an electron-withdrawing group at one position of the carbon–carbon double bond react with TMSC(Li)N$_2$ to give 4-aminopyrazoles as the major products (eq 30).[29]

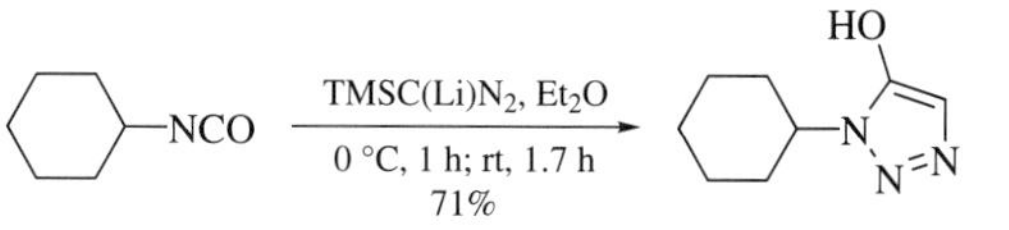

$$ (27) $$

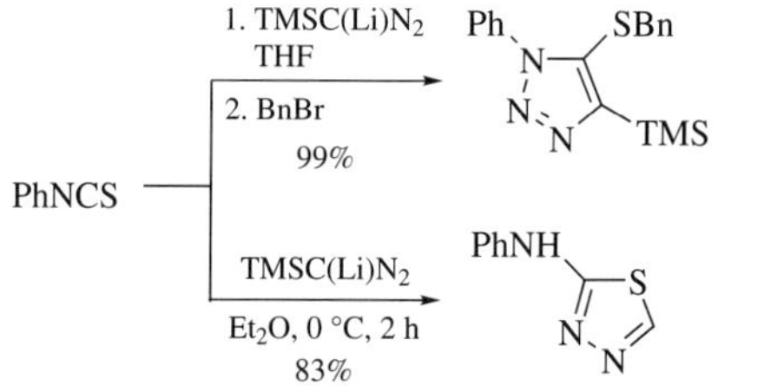

$$ (28) $$

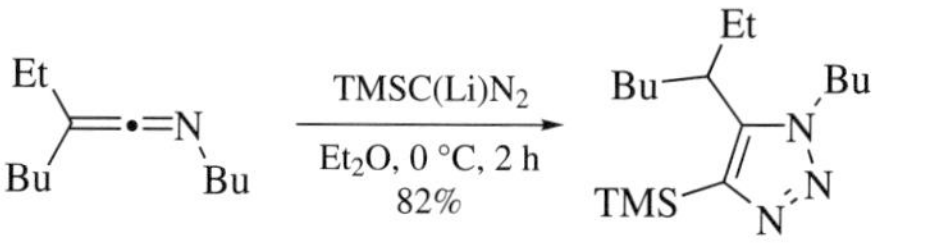

$$ (29) $$

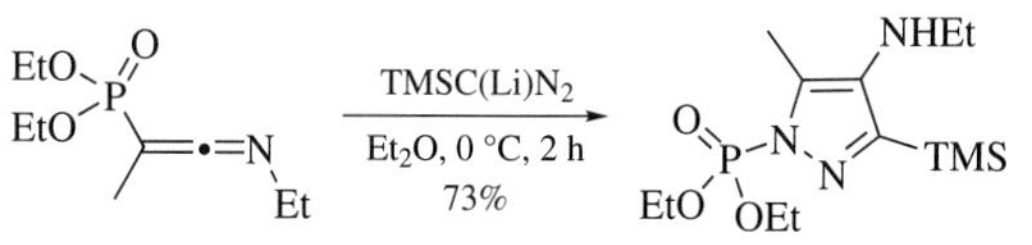

$$\text{(30)} \qquad 73\%$$

Pyrazoles are formed by reaction of TMSCHN$_2$ or TMSC(Li)N$_2$ with some alkynes (eqs 31 and 32)[24,30] and quinones (eq 33).[31] Some miscellaneous examples of the reactivity of TMSCHN$_2$ or its lithium salt are shown in eqs 34–36.[20,31,32]

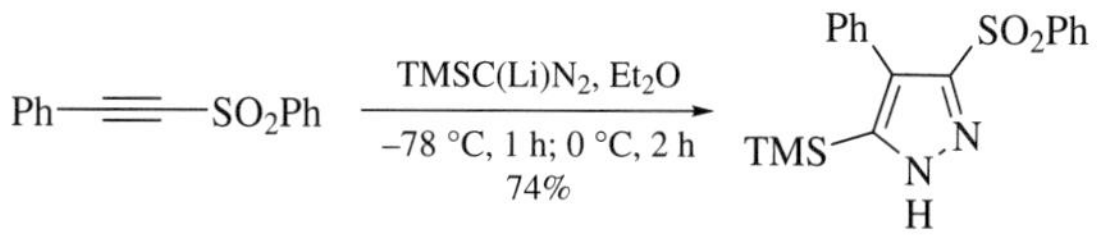

$$\text{(31)} \qquad 74\%$$

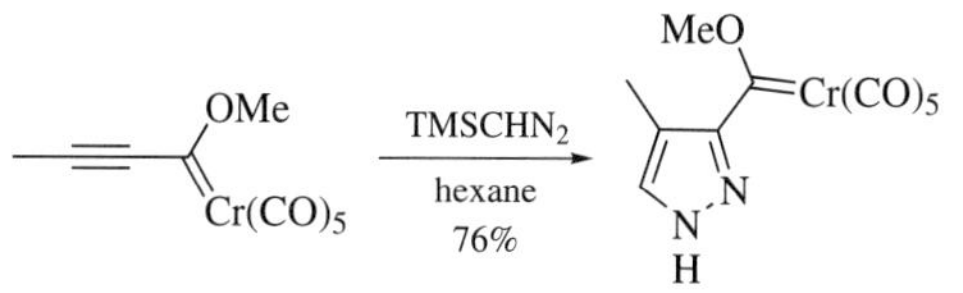

$$\text{(32)} \qquad 76\%$$

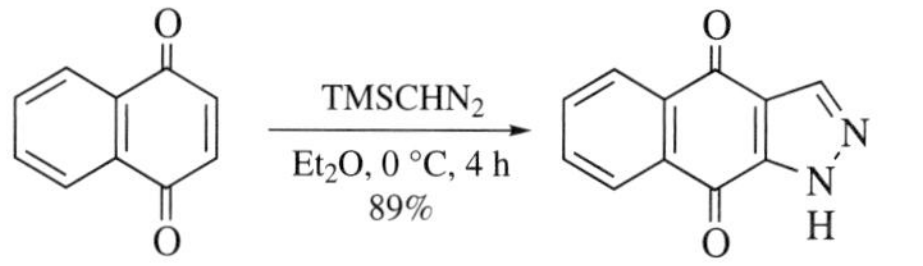

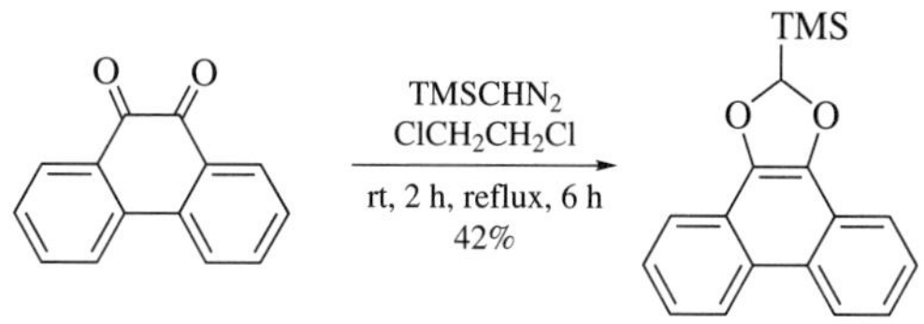

$$\text{(33)} \qquad 89\%$$

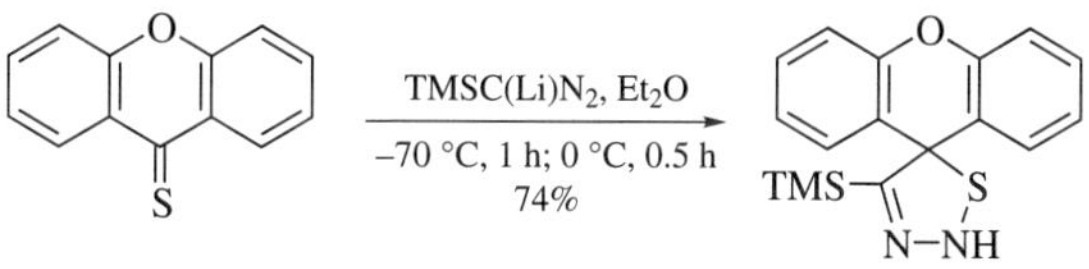

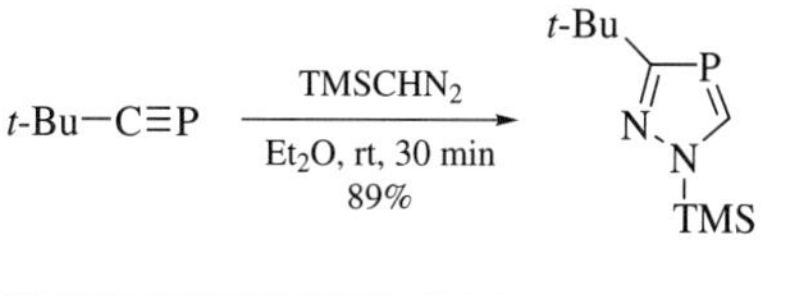

$$\text{(34)} \qquad 42\%$$

$$\text{(35)} \qquad 74\%$$

$$\text{(36)} \qquad 89\%$$

First Update

Timothy Snowden

The University of Alabama, Tuscaloosa, AL, USA

Trimethylsilyldiazomethane has been shown to have a nucleophilicity between that of silyl enol ethers and enamines in dichloromethane. Trimethylsilyldiazomethane is 1.5 orders of magnitude less nucleophilic than diazomethane and 4 orders of magnitude more nucleophilic than ethyl diazoacetate.[33] A crystal structure of lithium trimethylsilyldiazomethane, TMSC(Li)N$_2$, has also been obtained.[34]

One-Carbon Homologation. When acid chlorides are unavailable or problematic, the Arndt-Eistert-type homologation of carboxylic acid derivatives is possible by adding TMSCHN$_2$ to a

mixed anhydride or to a solution of carboxylic acid and DCC, followed by Wolff rearrangement of the intermediate diazoketone (eq 37).[35] The use of DCC as the activating agent limits the reaction to a maximum 50% yield because of the intermediate symmetric anhydride formed. A two-step ring expansion of *N*-carboxylated-β-lactams to the corresponding γ-lactams is affected by treating the former with NaHMDS and TMSCHN$_2$, followed by photo-Wolff rearrangement of the ring-opened diazoketone. The reaction occurs in moderate overall yield without epimerization of resident stereocenters (eq 38).[36]

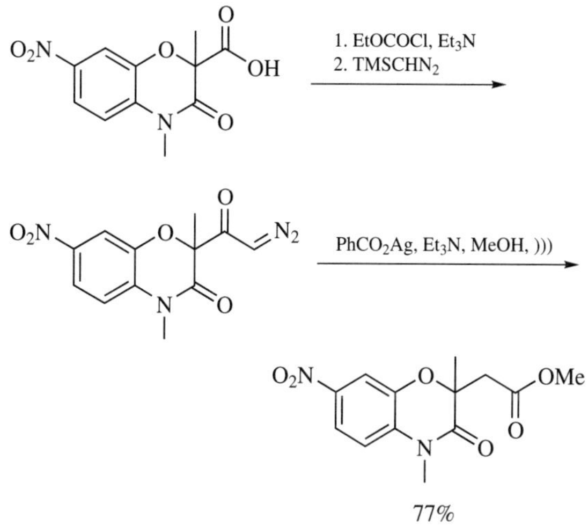

$$\text{(37)} \qquad 77\%$$

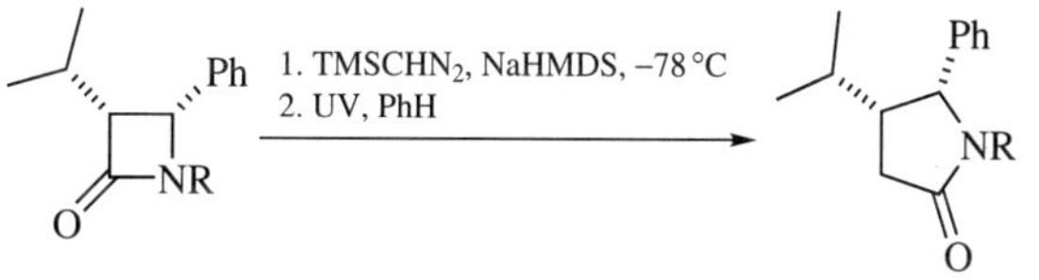

$$\text{(38)} \qquad \begin{array}{l} R = \text{Cbz (66\%)} \\ R = \text{CO}_2\text{Et (66\%)} \end{array}$$

One-carbon homologation of aldehydes and ketones using trimethylaluminum or methylaluminum bis(2,6-di-*t*-butyl-4-methylphenoxide) (MAD) has been reported to be more regioselective and higher yielding than homologation using boron trifluoride etherate in some cases (eqs 39 and 40).[37] Organoaluminum reagents also promote the direct conversion of aliphatic, alicyclic, and aromatic aldehydes to homologous methyl ketones using TMSCHN$_2$ (eq 41).[38] The complementary conversion of dialkyl ketones to homologated aldehydes is possible by reacting TMSC(Li)N$_2$ and lithium diisopropylamine with the ketone (eq 42). This is a particularly singular transformation, although yields (16–84%) are highly substrate dependent and conditions are unsuitable for base-sensitive substrates.[39]

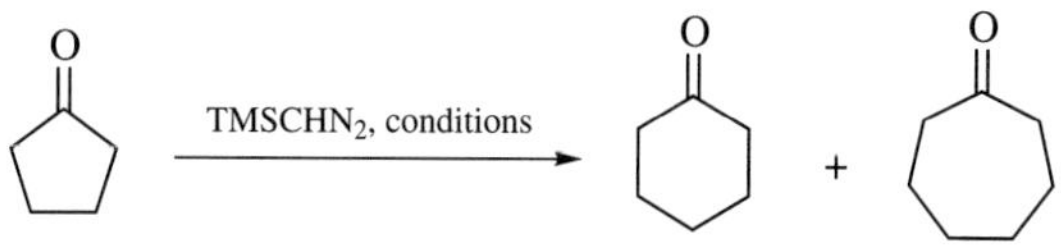

$$\text{(39)}$$

Conditions	Yield
Me$_3$Al, CH$_2$Cl$_2$, −20 °C	68% (96:2)
BF$_3$·Et$_2$O, CH$_2$Cl$_2$, −20 °C	35% (64:23)

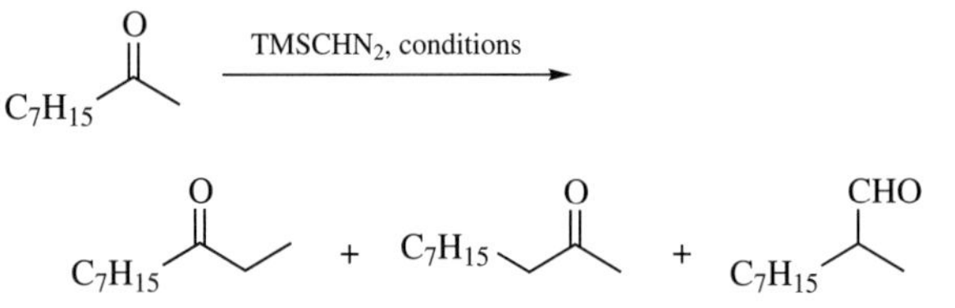

(40)

Conditions	Yield
MAD, CH$_2$Cl$_2$, −78 °C	75% (85:15:0)
Me$_3$Al, CH$_2$Cl$_2$, −20 °C	92% (31:36:33)
BF$_3$·Et$_2$O, CH$_2$Cl$_2$, −20 °C	87% (51:40:9)

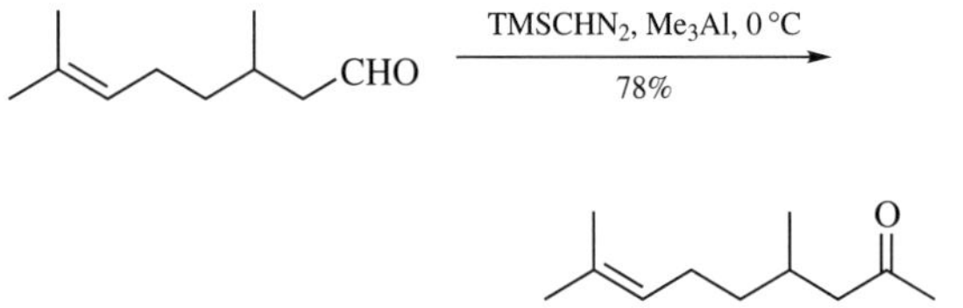

(41)

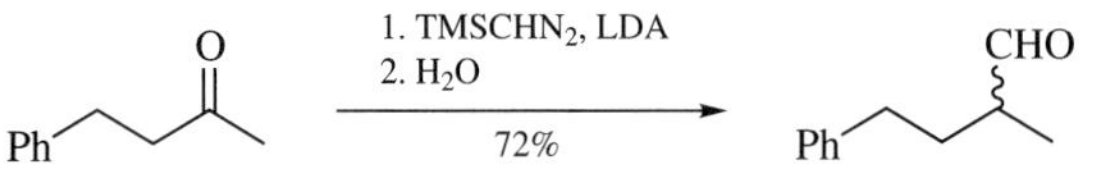

(42)

Alkenylation. Trimethylsilyldiazomethane is a useful reagent for the methylenation of carbonyl compounds in the presence of catalytic ruthenium[40] or rhodium complexes. Wilkinson's catalyst along with stoichiometric TMSCHN$_2$, 2-propanol, and triphenylphosphine rapidly methylenates a variety of sensitive ketones and aldehydes under nonbasic conditions in high yield (eqs 43–45).[41] The use of 2-trimethylsilanylethyl 4-diphenylphosphanylbenzoate (DPPBE)[42] in place of triphenylphosphine simplifies alkene purification without sacrificing overall yield.[41b] The TMSCHN$_2$ and Wilkinson's catalyst system is superior to standard Wittig conditions not only for methylenation of enolizable substrates but also for enhanced chemoselectivity in the methylenation of keto-aldehydes (eq 46).[41a,43]

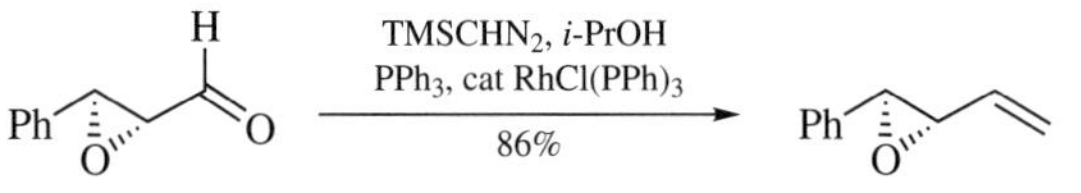

(43)

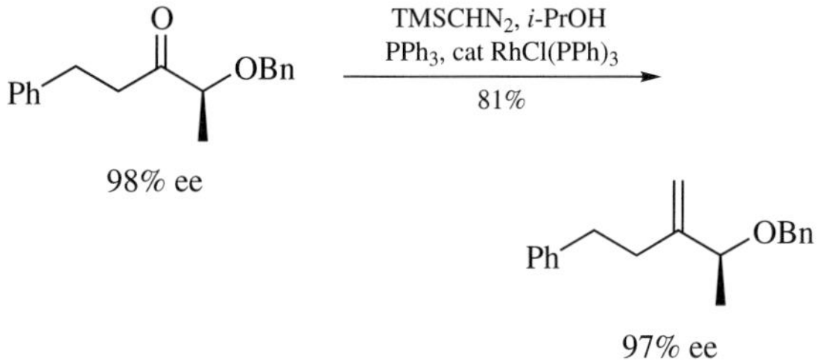

(44)

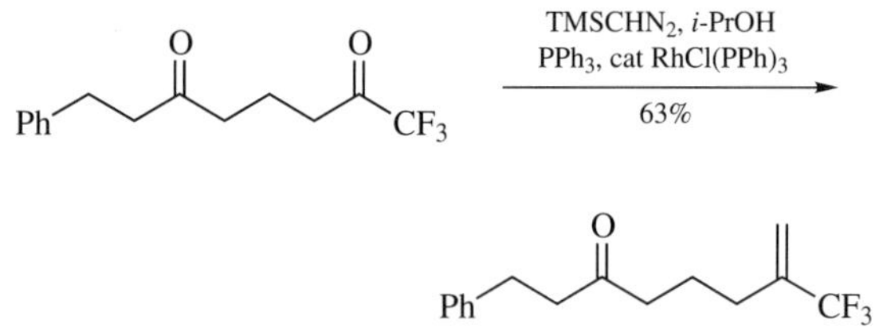

97% ee

(45)

Substrates suitable for the Colvin rearrangement[15] have been extended to enolizable aryl alkyl ketones and both aromatic and aliphatic aldehydes using TMSC(Li)N$_2$ (eq 47). These conditions are reported to be superior to those employing dimethyl diazomethylphosphonate (DAMP) with regard to reaction times and range of permissible substrates. However, the reaction is not suitable for dialkyl ketones (see eq 42).[44]

The reaction of 2 equiv of TMSCHN$_2$ with an alkyne in the presence of catalytic RuCl(cod)Cp* affords 1,4-bis(trimethyl-silyl)buta-1,3-dienes in moderate to excellent yield (eq 48).[45] The resulting terminal vinylsilanes may be readily converted to a variety of functionalities. This ruthenium catalyst also facilitates enyne metathesis with TMSCHN$_2$ to generate bicyclo[3.1.0] hexane derivatives bearing a heteroatom in the cyclopentane ring (eq 49). The reactivity of the enynes is greater with propargyl sulfonamides than with the corresponding ethers.[46] A related cascade metathesis reaction involving the Ni(cod)$_2$-catalyzed [4 + 2 + 1] cycloaddition of TMSCHN$_2$ and a dienyne offers an efficient approach to unsaturated bicyclo[5.3.0]decane systems (eq 50).[47]

Cyclopropane and Aziridine Formation. Aliphatic ketones react with TMSC(Li)N$_2$ in DME to generate an intermediate singlet alkylidene carbene. The carbene reacts with a large excess of alkene to afford alkylidenyl cyclopropanes in yields ranging from <30% to 69% (eq 51).[48]

Improvements in the stereoselective preparation of silylcyclo-propanes[16] (eqs 16 and 17) from alkenes and TMSCHN$_2$ have been reported based upon the judicious choice of catalytic metal and ligand. The meso-tetra-p-tolylporphyrin iron(II) complex, (TTP)Fe, and TMSCHN$_2$ convert styrene to the corresponding

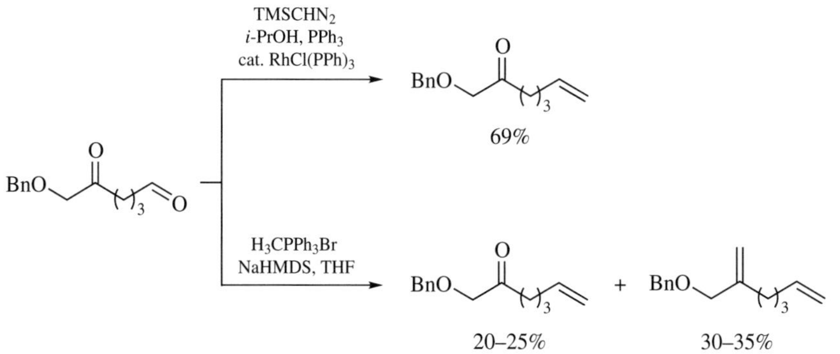

(46)

trans-1-phenyl-2-trimethylsilylcyclopropane in 89% yield with 13:1 diastereoselectivity.[49] The selection of [Cu(CH₃CN)₄]PF₆ and an appropriate bis(oxazoline) ligand provides highly diastereo- and enantioselective cyclopropanation of styrene and its derivatives, albeit with compromised yield compared to other protocols.[50]

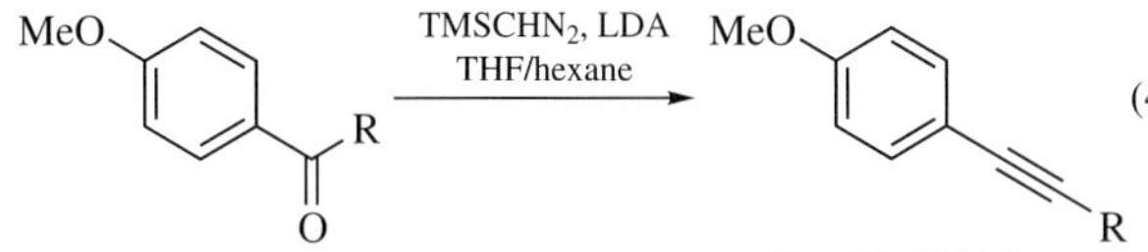

(47)

R = Me (82%)
R = H (86%)

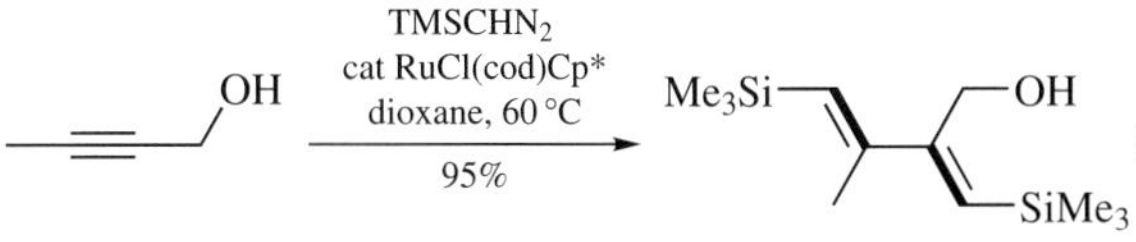

(48)

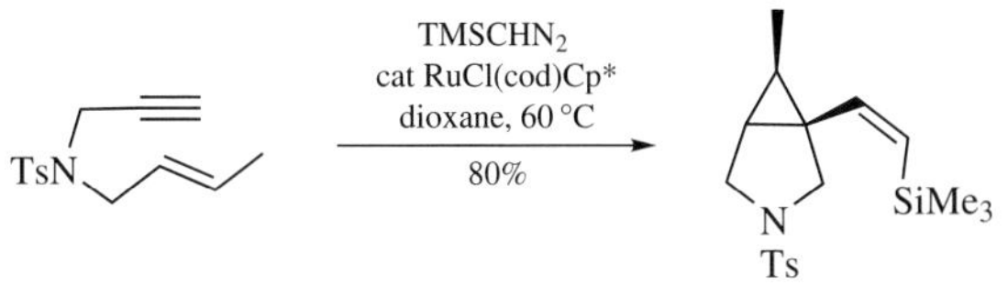

95%

(49)

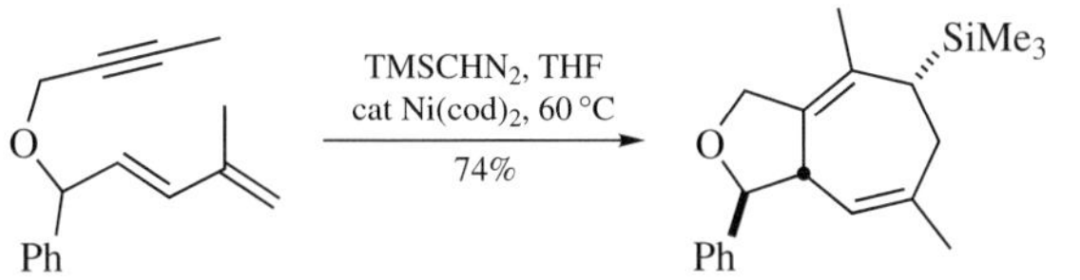

80%

(50)

74%

>95:5 dr

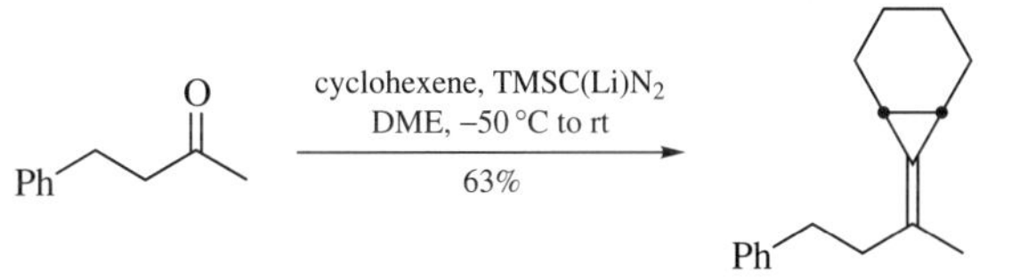

(51)

63%

Substituted *N*-sulfonyltrimethylsilylaziridines are formed when *N*-sulfonylaldimines are treated with TMSCHN₂. The reaction is highly diastereoselective, preferentially affording *cis*-products in good yields,[51] although electron releasing substituents attached to the *N*-sulfonylaldimine decrease the efficiency (eq 52).[52] The *C*-silylaziridines are useful precursors to a variety of products created by desilylation and trapping with electrophiles. This process proceeds with complete retention of stereochemistry. Alternatively, nucleophilic ring opening proceeds with excellent regiocontrol, occurring only at the silicon-bearing carbon atom (eq 53).[53] Substituted *N*-sulfonyltrimethylsilylaziridines are also formed with moderate enantioselectivity by treating *N*-tosyl α-imino esters with TMSCHN₂ and a catalytic BINAP-copper(I) complex to give the corresponding *cis*-aziridine or a bis(oxazoline)-copper(I) complex to create the analogous *trans*-diastereomer.[54]

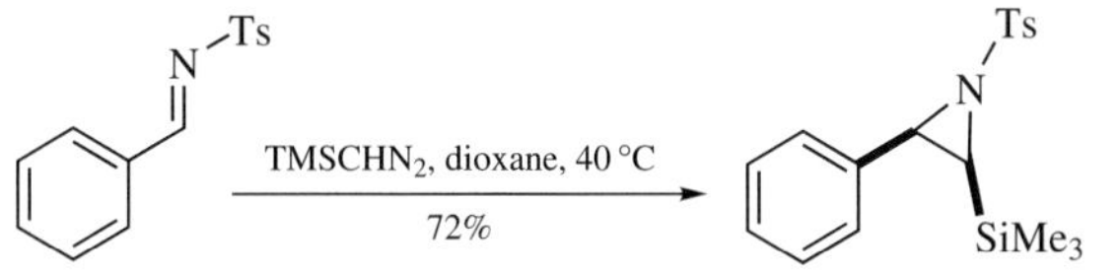

(52)

95:5 dr

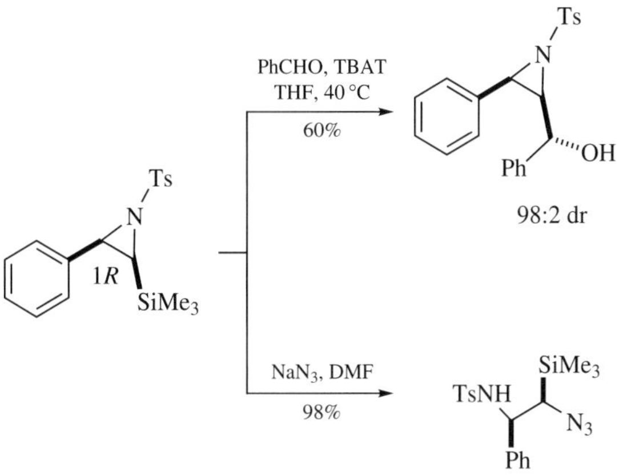

(53)

TBAT = tetrabutylammonium triphenyldifluorosilicate

Insertion Reactions. Trimethylsilyldiazomethane undergoes net insertion between the B–C bond of borinate and boronate esters. Thus, olefin hydroboration, followed by treatment with TMSCHN₂, oxidation, and desilylation offers a method for hydroxymethylation of alkenes in fair to moderate yield (eq 54).[55] Stable, chiral allenylboranes (**1R** and **1S**) are prepared by reaction of TMSCHN₂ with *B*-MeO-9-BBN followed by resolution with pseudoephedrine and reaction with allenyl magnesium bromide (eq 55). Compounds **1R** and **1S** are useful for the asymmetric allenylboration of aldehydes with predictable absolute stereochemistry. The precursors to **1** are then readily recovered during work-up.[56]

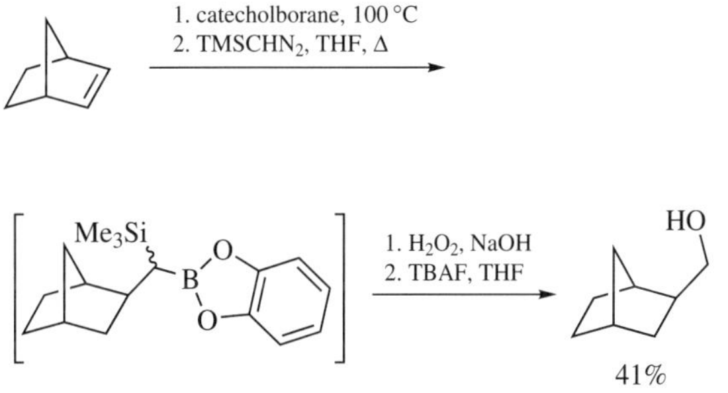

(54)

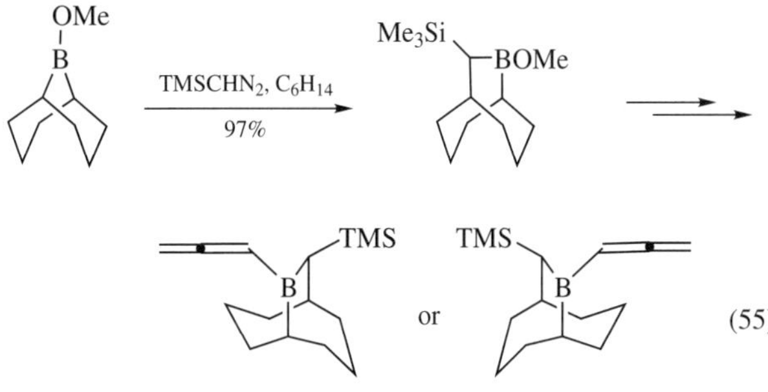

41%

97%

(55)

1R   **1S**

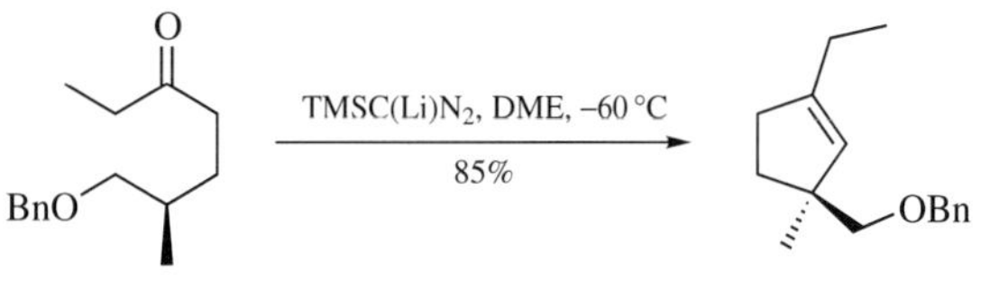

(56)

85%

Lithium trimethylsilyldiazomethane has proved particularly useful in the conversion of ketones into alkylidene carbenes, vide supra, that readily undergo 1,5 C–H insertion reactions to afford cyclopentenes (eq 56).[57] Yields are generally good and the chemoselectivity of C–H insertion is predictable. The C–H insertion of the singlet carbene into heteroatom-bearing stereocenters proceeds with retention of stereochemistry (eqs 57 and 58).[58] Reaction with acetals affords spiroketals (eq 59)[59] or 2-cyclopentenones after acetal hydrolysis (eq 60).[60]

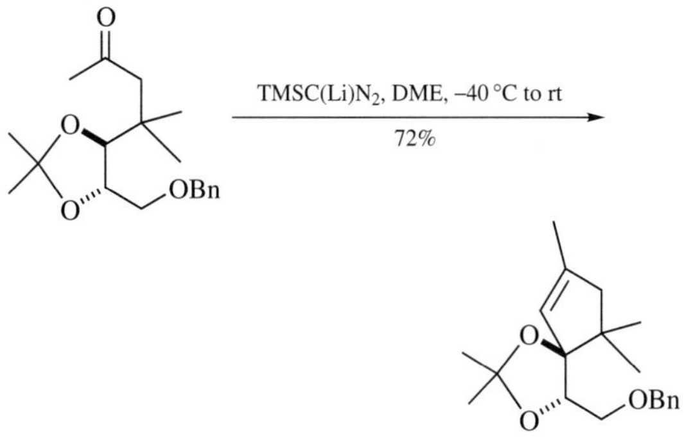

(57)

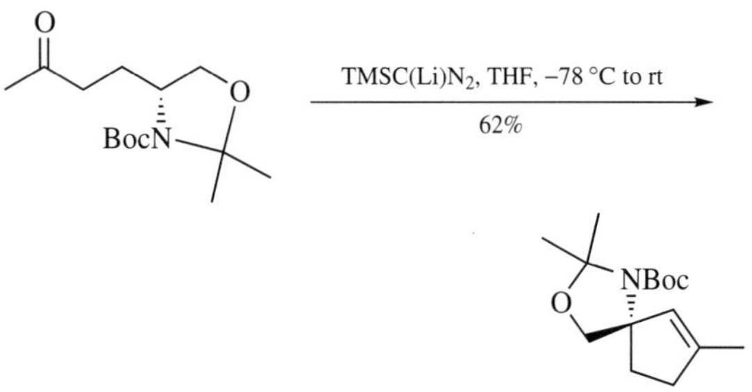

(58)

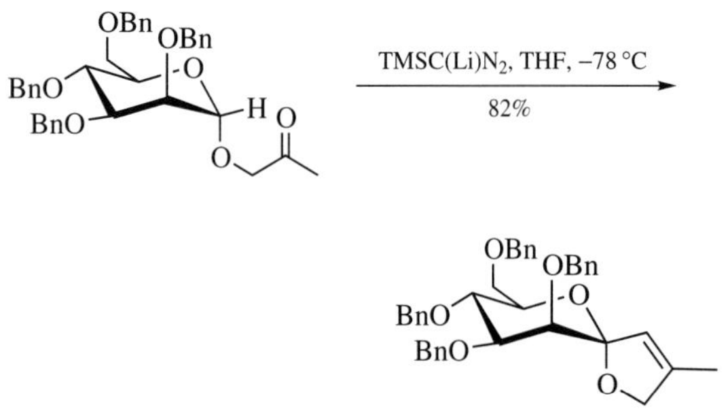

(59)

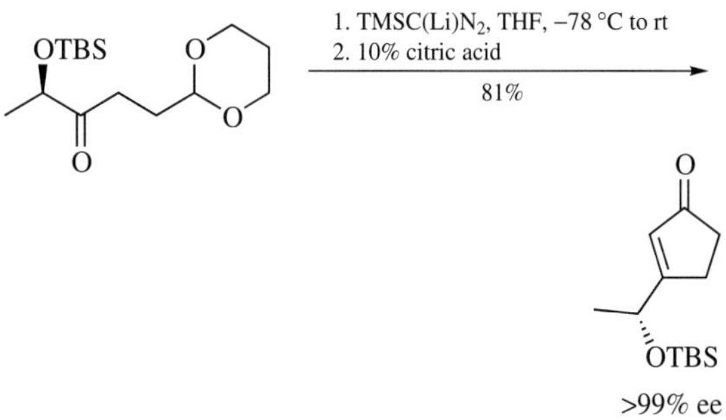

(60)

>99% ee

The intermediate carbenes generated from ketones and TMSC-(Li)N₂ have also been shown to undergo 1,5 O–Si and N–H insertion to produce 5-trimethylsilyl-2,3-dihydrofurans[61] and 2-pyrrolines or 3-substituted indoles, respectively (eqs 61 and 62).[62] Azulene 1-carboxylic esters may be generated in moderate yield by the reaction of TMSC(Li)N₂ with alkyl 4-aryl-2-oxobutanoates, followed by oxidation with manganese dioxide.[63]

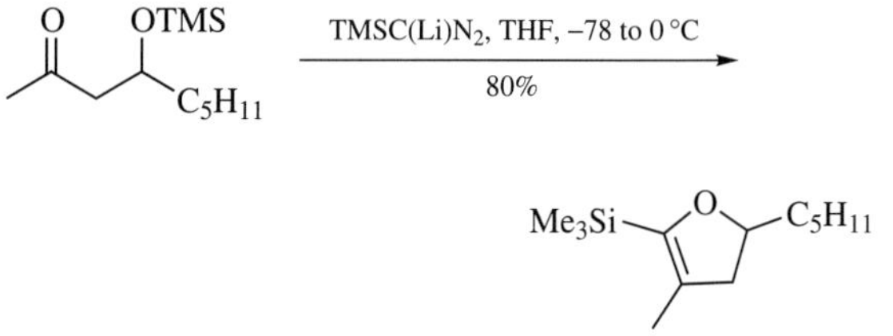

(61)

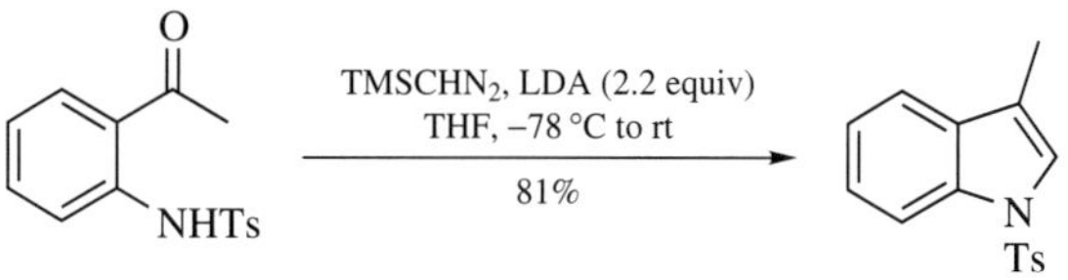

(62)

Ketenylation Reactions. Lithium silylethynolates are conveniently prepared by introducing carbon monoxide to TMSC(Li)N₂. The resulting ynolate, when mixed with trimethylaluminum, adds to epoxides to give 2-trimethylsilyl-γ-lactones with excellent regioselectivity and stereospecificity (eq 63).[64] Silylethynolates also add to aziridines, without Lewis acid activation, leading to 2-trimethylsilyl-γ-lactams. Introduction of excess aldehyde during lactam preparation affords α-alkylidene lactams in good yields via Peterson olefination (eq 64).[65]

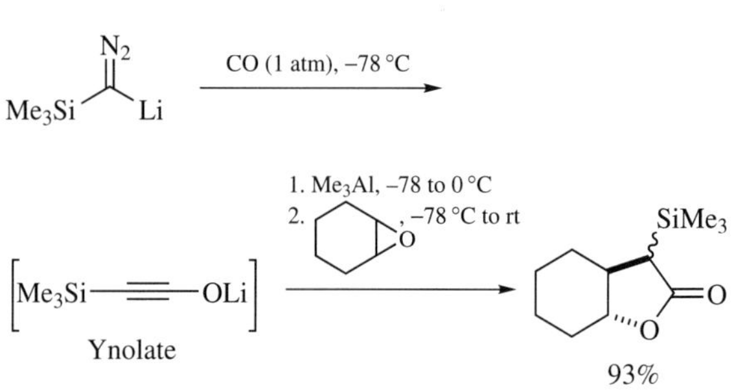

(63)

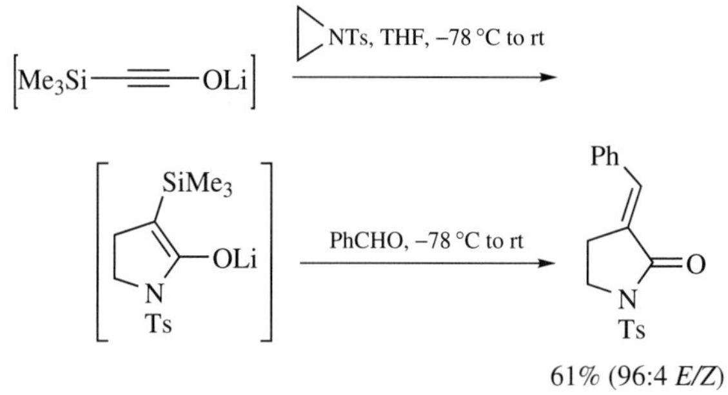

93%

(64)

61% (96:4 E/Z)

Preparation of Homoallylic Sulfides. When allyl sulfides are treated with TMSCHN₂ and a transition metal catalyst, an allylsulfonium ylide is formed that rapidly undergoes a diastereoselective [2,3] sigmatropic rearrangement in high yield (eq 65). Trimethylsilyldiazomethane is superior to ethyl diazoacetate for the transformation and rhodium(II) acetate is the catalyst of choice.[66] The homoallylic α-silyl sulfides thus obtained are suitable substrates for subsequent Peterson olefination or conversion to homoallylic aldehydes. Propargyl sulfides are also suitable reactants with TMSCHN₂ and catalytic iron(II) to generate allenyl α-silyl sulfides.[67]

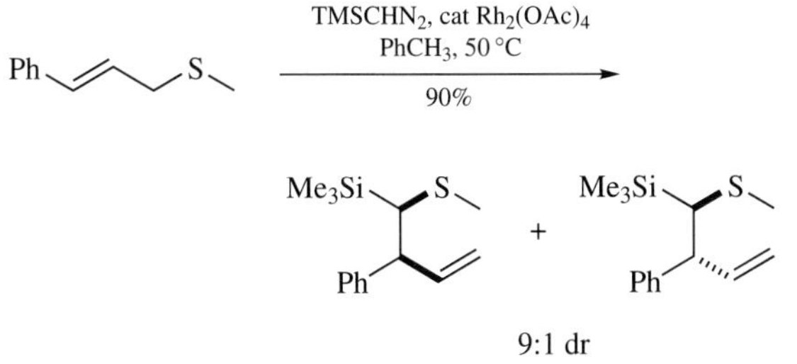

$$Ph\diagup\diagdown S\diagdown \xrightarrow[90\%]{\substack{\text{TMSCHN}_2,\ \text{cat Rh}_2(\text{OAc})_4 \\ \text{PhCH}_3,\ 50\,^{\circ}\text{C}}}$$

(65)

9:1 dr

Pericyclic Reactions. Trimethylsilyldiazomethane reacts with chiral acrylates to create optically active Δ^1-pyrazolines via regioselective asymmetric [3 + 2] cycloaddition. Subsequent protodesilylation affords Δ^2-pyrazolines in good yields (eq 66). This procedure offers a convenient route to azaprolines.[68]

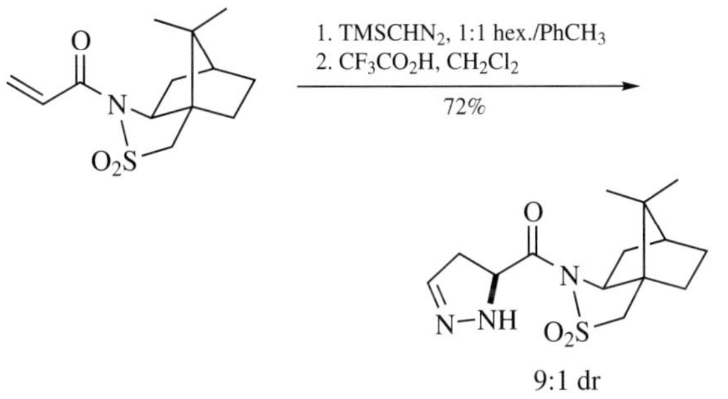

(66)

9:1 dr

Di- and trisubstituted furans may also be prepared by reacting TMSCHN$_2$ with an acyl isocyanate. The intermediate 4-trimethylsilyloxy oxazole readily undergoes Diels-Alder cycloaddition in situ with a suitable dienophile to construct the product (eq 67).[69]

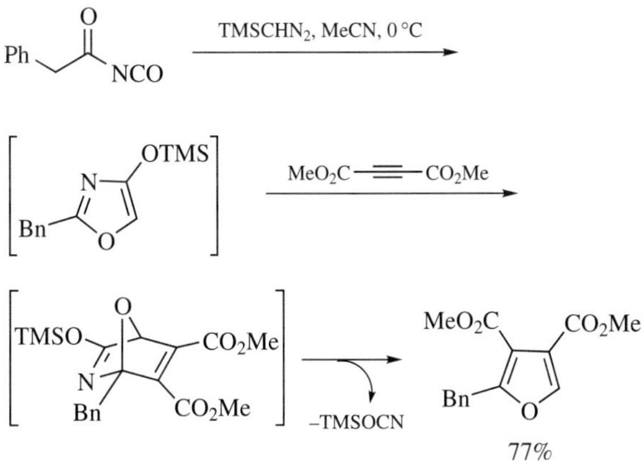

(67)

77%

Trimethylsilyldiazomethane smoothly reacts with (trialkylsilyl)vinylketenes, ultimately prepared from diazoketones, in a net [4 + 1] annulation process to afford 2-trialkylsilylcyclopentenones in good to excellent yields (eq 68).[70] The products are readily desilylated by treatment with methanesulfonic acid in methanol. An analogous reaction with (trialkylsilyl)arylketenes gives trialkylsilyl-2-indanone derivatives (eq 69).[71]

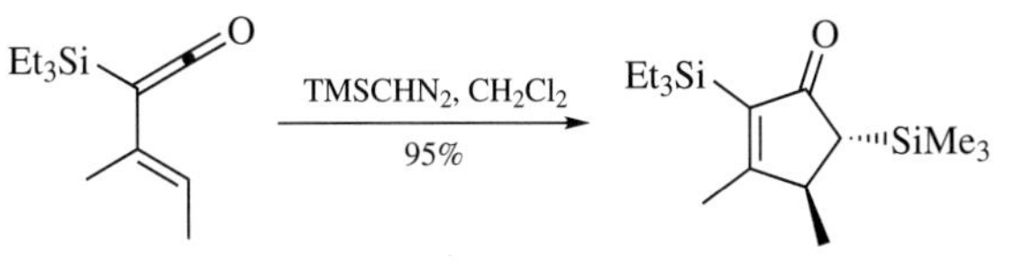

(68)

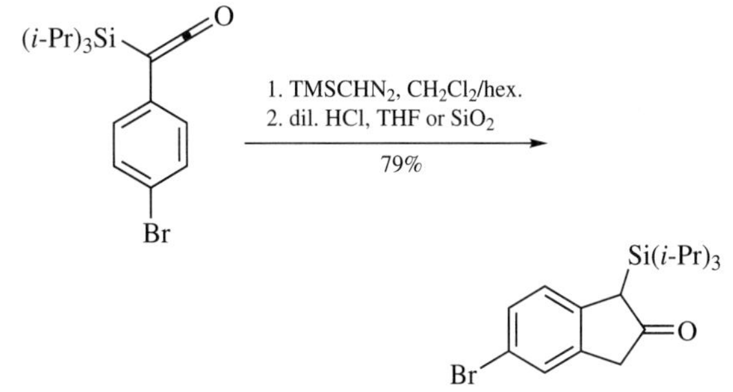

95%

(69)

Miscellaneous Reactions. Trimethylsilyldiazomethane converts acid- and base-sensitive maleic anhydride derivatives into the corresponding bis(methyl esters) (eq 70).[72] Terminal silyl enol ethers are conveniently prepared from aldehydes by first treating the carbonyl compound with TMSC(Li)N$_2$, followed sequentially by methanol and Rh$_2$(OAc)$_4$ (eq 71). The method works well with base-sensitive substrates and is superior to the attempted regioselective deprotonation/O-silylation of the corresponding methyl ketone.[73]

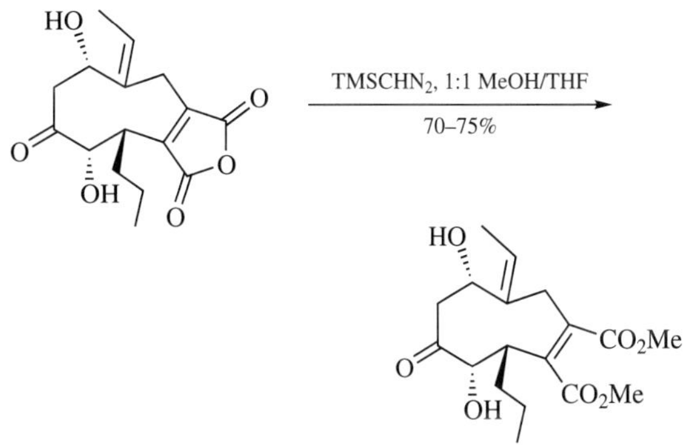

(70)

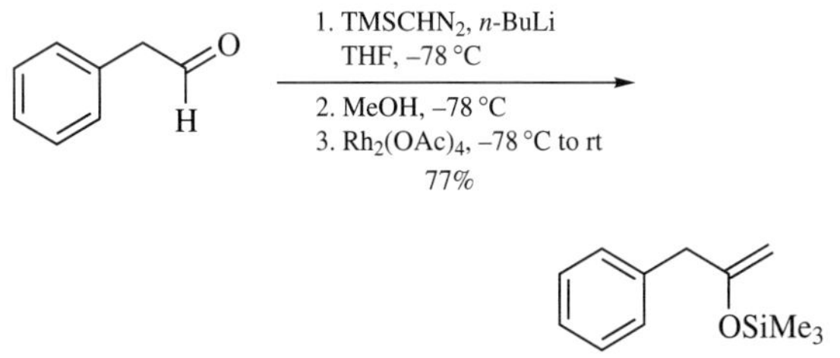

(71)

1. (a) Shioiri, T.; Aoyama, T., *J. Synth. Org. Chem. Jpn* **1986**, *44*, 149 (*Chem. Abstr.* **1986**, *104*, 168 525q). (b) Aoyama, T., *Yakugaku Zasshi* **1991**, *111*, 570 (*Chem. Abstr.* **1992**, *116*, 58 332q). (c) Anderson, R.; Anderson, S. B. In *Advances in Silicon Chemistry*; Larson, G. L., Ed.; JAI: Greenwich, CT, 1991; Vol. 1, p 303. (d) Shioiri, T.; Aoyama, T. In *Advances in the Use of Synthons in Organic Chemistry*; Dondoni, A., Ed.; JAI: London, 1993; Vol. 1, p 51.

2. Seyferth, D.; Menzel, H.; Dow, A. W.; Flood, T. C., *J. Organomet. Chem.* **1972**, *44*, 279.

3. Shioiri, T.; Aoyama, T.; Mori, S., *Org. Synth.* **1990**, *68*, 1.

4. Aoyama, T.; Shioiri, T., *Chem. Pharm. Bull.* **1981**, *29*, 3249.

5. Hashimoto, N.; Aoyama, T.; Shioiri, T., *Chem. Pharm. Bull.* **1982**, *30*, 119.

6. Aoyama, T.; Shioiri, T., *Synthesis* **1988**, 228.

7. Hashimoto, N.; Aoyama, T.; Shioiri, T., *Chem. Pharm. Bull.* **1981**, *29*, 1475.

8. Aoyama, T.; Terasawa, S.; Sudo, K.; Shioiri, T., *Chem. Pharm. Bull.* **1984**, *32*, 3759.

9. Aoyama, T.; Shioiri, T., *Tetrahedron Lett.* **1990**, *31*, 5507.

10. Aoyama, T.; Shioiri, T., *Tetrahedron Lett.* **1988**, *29*, 6295.

11. Aoyama, T.; Shioiri, T., *Chem. Pharm. Bull.* **1989**, *37*, 2261.

12. Aoyama, T.; Shioiri, T., *Tetrahedron Lett.* **1986**, *27*, 2005.

13. Aoyama, T.; Toyama, S.; Tamaki, N.; Shioiri, T., *Chem. Pharm. Bull.* **1983**, *31*, 2957.

14. Schöllkopf, U.; Scholz, H.-U., *Synthesis* **1976**, 271.

15. Colvin, E. W.; Hamill, B. J., *J. Chem. Soc., Perkin Trans. 1* **1977**, 869.

16. Aoyama, T.; Iwamoto, Y.; Nishigaki, S.; Shioiri, T., *Chem. Pharm. Bull.* **1989**, *37*, 253.

17. Zaitseva, G. S.; Lutsenko, I. F.; Kisin, A. V.; Baukov, Y. I.; Lorberth, J., *J. Organomet. Chem.* **1988**, *345*, 253.

18. Aoyama, T.; Shioiri, T., *Chem. Pharm. Bull.* **1982**, *30*, 3450.

19. Aoyama, T.; Iwamoto, Y.; Shioiri, T., *Heterocycles* **1986**, *24*, 589.

20. Shioiri, T.; Iwamoto, Y.; Aoyama, T., *Heterocycles* **1987**, *26*, 1467.

21. Crossman, J. M.; Haszeldine, R. N.; Tipping, A. E., *J. Chem. Soc., Dalton Trans.* **1973**, 483.

22. Aoyama, T.; Sudo, K.; Shioiri, T., *Chem. Pharm. Bull.* **1982**, *30*, 3849.

23. Aoyama, T; Inoue, S.; Shioiri, T., *Tetrahedron Lett.* **1984**, *25*, 433.

24. Asaki, T.; Aoyama, T.; Shioiri, T., *Heterocycles* **1988**, *27*, 343.

25. Aoyama, T.; Kabeya, M.; Fukushima, A.; Shioiri, T., *Heterocycles* **1985**, *23*, 2363.

26. Aoyama, T.; Kabeya, M.; Shioiri, T., *Heterocycles* **1985**, *23*, 2371.

27. Aoyama, T.; Kabeya, M.; Fukushima, A.; Shioiri, T., *Heterocycles* **1985**, *23*, 2367.

28. Aoyama, T.; Katsuta, S.; Shioiri, T., *Heterocycles* **1989**, *28*, 133.

29. Aoyama, T.; Nakano, T.; Marumo, K.; Uno, Y.; Shioiri, T., *Synthesis* **1991**, 1163.

30. Chan, K. S.; Wulff, W. D., *J. Am. Chem. Soc.* **1986**, *108*, 5229.

31. Aoyama, T.; Nakano, T.; Nishigaki, S.; Shioiri, T., *Heterocycles* **1990**, *30*, 375.

32. Rösch, W.; Hees, U.; Regitz, M., *Chem. Ber.* **1987**, *120*, 1645.

33. Bug, T.; Hartnagel, M.; Schlierf, C.; Mayr, H., *Chem. Eur. J.* **2003**, *9*, 4068.

34. Feeder, N.; Hendy, M. A.; Raithby, P. R.; Snaith, R.; Wheatley, A. E. H., *Eur. J. Org. Chem.* **1998**, 861.

35. Cesar, J.; Dolenc, M. S., *Tetrahedron Lett.* **2001**, *42*, 7099.

36. Ha, D.-C.; Kang, S.; Chung, C.-M.; Lim, H.-K., *Tetrahedron Lett.* **1998**, *39*, 7541.

37. (a) Maruoka, K.; Concepcion, A. B.; Yamamoto, H., *J. Org. Chem.* **1994**, *59*, 4725. (b) Maruoka, K.; Concepcion, A. B.; Yamamoto, H., *Synthesis* **1994**, 1283.

38. Maruoka, K.; Concepcion, A. B.; Yamamoto, H., *Synlett* **1994**, 521.

39. Miwa, K.; Aoyama, T.; Shioiri, T., *Synlett* **1994**, 109.

40. Lebel, H.; Paquet, V., *Organometallics* **2004**, *23*, 1187.

41. (a) Lebel, H.; Guay, D.; Paquet, V.; Huard, K., *Org. Lett.* **2004**, *6*, 3047. (b) Lebel, H.; Paquet, V., *J. Am. Chem. Soc.* **2004**, *126*, 320.

42. Yoakim, C.; Guse, I.; O'Meara, J. A.; Thavonekham, B., *Synlett* **2003**, 473.

43. Lebel, H.; Paquet, V.; Proulx, C., *Angew. Chem., Int. Ed.* **2001**, *40*, 2887.

44. Miwa, K.; Aoyama, T.; Shioiri, T., *Synlett* **1994**, 107.

45. Le Paih, J.; Derien, S.; Oezdemir, I.; Dixneuf, P. H., *J. Am. Chem. Soc.* **2000**, *122*, 7400.

46. Monnier, F.; Castillo, D.; Derien, S.; Toupet, L.; Dixneuf, P. H., *Angew. Chem., Int. Ed.* **2003**, *42*, 5474.

47. Yike, N.; Montgomery, J., *J. Am. Chem. Soc.* **2004**, *126*, 11162.

48. Sakai, A.; Aoyama, T.; Shioiri, T., *Tetrahedron* **1999**, *55*, 3687.

49. Hamaker, C. G.; Mirafzal, G. A.; Woo, L. K., *Organometallics* **2001**, *20*, 5171.

50. France, M. B.; Milojevich, A. K.; Stitt, T. A.; Kim, A. J., *Tetrahedron Lett.* **2003**, *44*, 9287.

51. Hori, R.; Aoyama, T.; Shioiri, T., *Tetrahedron Lett.* **2000**, *41*, 9455.

52. Aggarwal, V. K.; Ferrara, M., *Org. Lett.* **2000**, *2*, 4107.

53. Aggarwal, V. K.; Alonso, E.; Ferrara, M.; Spey, S. E., *J. Org. Chem.* **2002**, *67*, 2335.

54. Juhl, K.; Hazell, R. G.; Jorgensen, K. A., *J. Chem. Soc., Perkin Trans. 1* **1999**, 2293.

55. Goddard, J.-P.; Le Gall, T.; Mioskowski, C., *Org. Lett.* **2000**, *2*, 1455.

56. Lai, C.; Soderquist, J. A., *Org. Lett.* **2005**, *7*, 799.

57. (a) Ohira, S.; Moritani, T., *J. Chem. Soc., Chem. Commun.* **1992**, 721. (b) Taber, D. F.; Meagley, R. P., *Tetrahedron Lett.* **1994**, *35*, 7909.

58. (a) Taber, D. F.; Walter, R.; Meagley, R. P., *J. Org. Chem.* **1994**, *59*, 6014. (b) Gabaitsekgosi, R.; Hayes, C. J., *Tetrahedron Lett.* **1999**, *40*, 7713.

59. Wardrop, D. J.; Zhang, W.; Fritz, J., *Org. Lett.* **2002**, *4*, 489.

60. Sakai, A.; Aoyama, T.; Shioiri, T., *Tetrahedron Lett.* **2000**, *41*, 6859.

61. Miwa, K.; Aoyama, T.; Shioiri, T., *Synlett* **1994**, 461.

62. (a) Yagi, T.; Aoyama, T.; Shioiri, T., *Synlett* **1997**, 1063. (b) Miyagi, T.; Hari, Y.; Aoyama, T., *Tetrahedron Lett.* **2004**, *45*, 6303.

63. Hari, Y.; Tanaka, S.; Takuma, Y.; Aoyama, T., *Synlett* **2003**, 2151.

64. Kai, K.; Iwamoto, K.; Chatani, N.; Murai, S., *J. Am. Chem. Soc.* **1996**, *118*, 7634.

65. Iwamoto, K.; Kojima, M.; Chatani, N.; Murai, S., *J. Org. Chem.* **2001**, *66*, 169.

66. (a) Carter, D. S.; Van Vranken, D. L., *Tetrahedron Lett.* **1999**, *40*, 1617. (b) Aggarwal, V. K.; Ferrara, M.; Hainz, R.; Spey, S. E., *Tetrahedron Lett.* **1999**, *40*, 8923.

67. Prabharasuth, R.; Van Vranken, D. L., *J. Org. Chem.* **2001**, *66*, 5256.

68. Mish, M. R.; Guerra, F. M.; Carreira, E. M., *J. Am. Chem. Soc.* **1997**, *119*, 8379.

69. Hari, Y.; Iguchi, T.; Aoyama, T., *Synthesis* **2004**, 1359.

70. Loebach, J. L.; Bennett, D. M.; Danheiser, R. L., *J. Am. Chem. Soc.* **1998**, *120*, 9690.

71. Dalton, A. M.; Zhang, Y.; Davie, C. P.; Danheiser, R. L., *Org. Lett.* **2002**, *4*, 2465.

72. Fields, S. C.; Dent, W. H., III; Green, F. R., III; Tromiczak, E. G., *Tetrahedron Lett.* **1996**, *37*, 1967.

73. Aggarwal, V. K.; Sheldon, C. G.; MacDonald, G. J.; Martin, W. P., *J. Am. Chem. Soc.* **2002**, *124*, 10300.

Trimethylsilyldiethylamine[1]

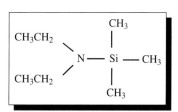

[996-50-9]   C$_7$H$_{19}$NSi   (MW 145.36)

InChI = 1S/C7H19NSi/c1-6-8(7-2)9(3,4)5/h6-7H2,1-5H3

InChIKey = JOOMLFKONHCLCJ-UHFFFAOYSA-N

(reagent used as an electrophilic trimethylsilyl source for cleavage of cyclic ethers,[2] esters, oxazolidines, and cyclic acetals; also as a nucleophilic source of the diethylamino group[3])

Alternate Names: *N*-(trimethylsilyl)diethylamine, diethyl(trimethylsilyl)amine, TMSDEA.

Physical Data: bp 127–129 °C; *d* 0.767 g cm^{-3}; n_{20}^D 1.4110.

Solubility: soluble in toluene, pentane, hexane, acetonitrile, ether, THF, and acetone.

Form Supplied in: liquid; commercially available.

Analysis of Reagent Purity: NMR, GC.

Preparative Methods: can be prepared by the reaction of diethylamine with chlorotrimethylsilane,[4,18] hexamethyldisilazane,[5] or cyanotrimethylsilane.[6]

Purity: can be purified by distillation.

Handling, Storage, and Precaution: stable indefinitely when maintained in a dry environment; moisture-sensitive reagent.

Original Commentary

Harold W. Pinnick
Bucknell University, Lewisburg, PA, USA

Trimethylsilyl Group Transfer. The major use of this reagent has been in the protection of functional groups (particularly hydroxyl) as the trimethylsilyl derivatives. This is illustrated in work on brefeldin A (eq 1)[7] and in the synthesis of prostaglandin E$_2$ methyl ester (eq 2), where selective protection of two hydroxyl groups occurred.[2] In addition, axial alcohols are converted into trimethylsilyl ethers much faster than equatorial alcohols.[8] Another multifunctional molecule which has been protected is the alcohol produced by cleaving the enone shown in eq 3.[9]

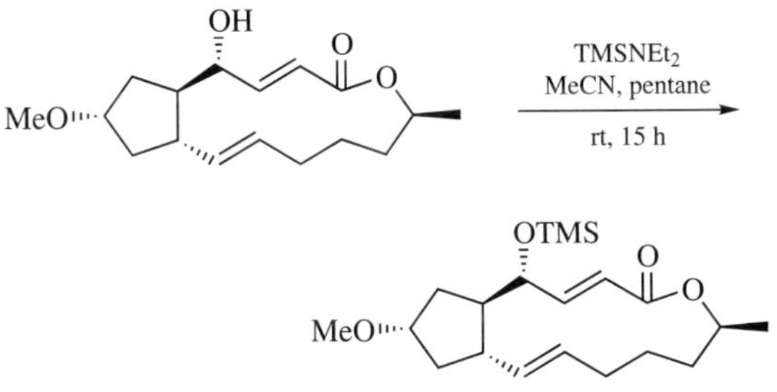

(1)

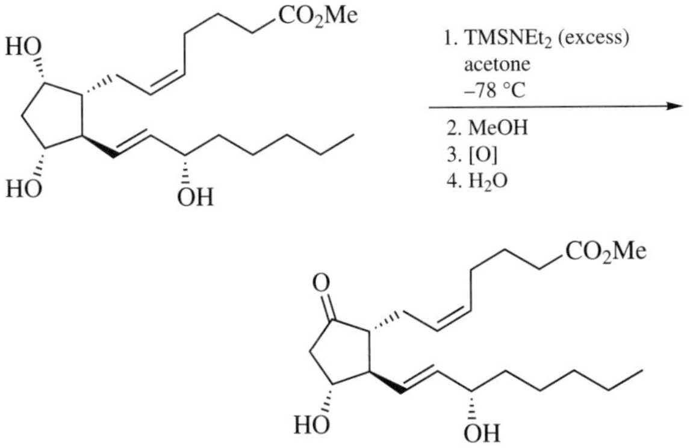

(2)

Unsaturated γ-butyrolactones are converted into 2-silyloxyfurans,[10] while the corresponding *N*-methyllactams give trimethylsilyloxypyrroles.[11] Amines can be converted into trimethylsilyl derivatives, as in the case of putrescine which forms the *N,N'*-bis(trimethylsilyl) compound in 95% yield.[12] Carboxylic acids can be silylated, as in the case of amino acid hydrochloride salts in DMF.[13] The hydroxyl group of acid-sensitive compounds,

such as that in eq 4, can be protected so that conversion to the phosphoryl chloride moiety gives trimethylsilyl chloride instead of HCl as a byproduct, thus enabling the other groups to survive.[14]

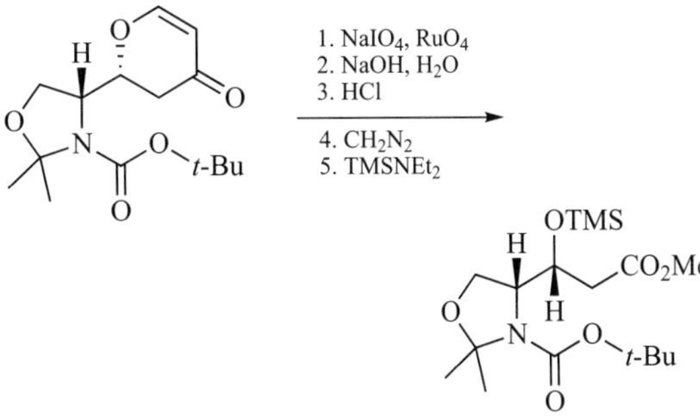

(3)

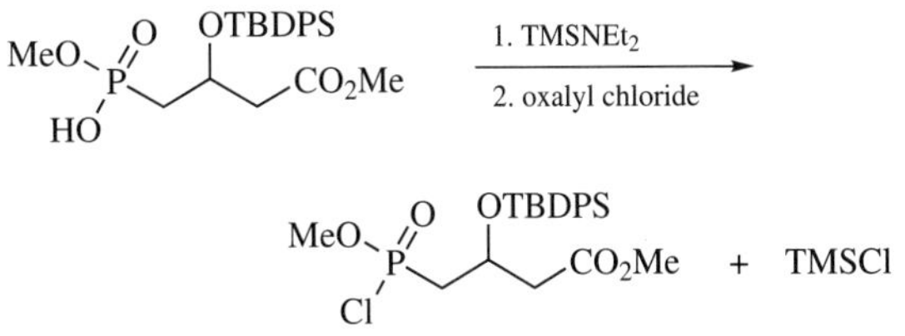

(4)

Diethylamino Group Transfer. The diethylamino group can be transferred to electrophilic compounds. *Oxalyl chloride* is converted into the monoamide with 1 equiv of TMSNEt$_2$ at -23 °C and the bisamide with excess silylamine at room temperature;[2] benzeneselenenyl chloride gives the selenenamide.[15] Unsaturated lactones give β-diethylamino lactones with TMSNEt$_2$ and a trace of diethylamine.[10] Conjugated enones give β-amino enol silyl ethers (eq 5),[16] while 1-triphenylsilylpropynone gives simple conjugate addition of diethylamine.[17]

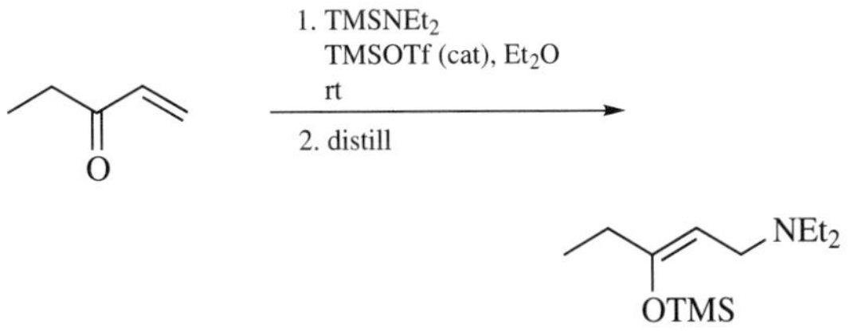

(5)

First Update

Dmitriy A. Bondar
The Ohio State University, Columbus, OH, USA

Trimethylsilyldiethylamine as Halosilane Equivalent. Trimethylsilyldiethylamine has recently found applications in preparation (in situ) of difficult-to-handle and highly moisture-sensitive iodosilanes and bromosilanes. This method has been used for a variety of synthetic transformations, including cleavage of cyclic ethers, cyclic acetals, and oxazolidines. The conversion of esters to carboxylic acids may also be accomplished with trimethylsilyldiethylamine as shown in the following examples. It may be suggested that other reactions of iodosilane and bromosilane can be carried out by means of trimethylsilyldiethylamine.

A 1:2 mixture of trimethylsilyldiethylamine and methyl iodide acts as a synthetic equivalent of trimethylsilyl iodide in reactions with cyclic ethers to afford ring-opened α-iodo-ω-siloxyalkanes in good yield (eq 6).[19]

$$\text{(6)}$$

Similarly, a 1:2 mixture of trimethylsilyldiethylamine and allyl bromide will afford cleavage of tetrahydrofuran in toluene at 80–90 °C for 12 h (eq 7).

$$\text{(7)}$$

Treatment of methyl *p*-methylbenzoate with 1.2 equiv of trimethylsilyldiethylamine and 1.3 equiv of methyl iodide in toluene, followed by gradual heating to 90 °C over a period of 4 h and additional stirring at 100–110 °C provides the corresponding trimethyl silyl ester. Subsequent hydrolysis affords *p*-methylbenzoic acid in 85% yield based on 90% conversion (reaction monitored by GC) (eq 8).[20] Benzoic esters substituted with an electron-donating substituent such as methyl or methoxy groups are more readily dealkylated in contrast with electron-deficient esters, such as with chlorine atom substituent, or aliphatic carboxylic acid esters.

$$\text{(8)}$$

Cyclic ketals react with trimethylsilyldiethylamine and 2 equiv of methyl iodide in toluene at 80 °C for 12 h to afford ring-opened 2-(trimethylsiloxy)ethyl enol ethers (eq 9).[21]

$$\text{(9)}$$

On the other hand, those with diaryl substituents on the ketal ring provide deprotected ketones in high yield (eq 10). Iodosiloxy alkanes are also obtained as co-products. This method can be used for deprotection of aromatic ketals under neutral conditions.[21]

$$\text{(10)}$$

In addition, 1,3-oxazolidines react under similar conditions with the 1:2 mixture to provide siloxyethylimines (or enamines) via C-O bond cleavage (eq 11).

$$\text{(11)}$$

In analogous reactions, trimethylsilyldiethylamine has been used for the ring opening of oxiranes.[22,23]

Silylation of Alcohols. Trimethylsilyl ethers can be prepared in good yield by reacting alcohols with trimethylsilyldiethylamine.[2,7]

$$\text{(12)}$$

As illustrated during the synthesis of erythronolide A, a trimethylsilyl ether can be successfully formed under mild conditions by stirring the alcohol with trimethylsilyldiethylamine in THF (eq 13).[25]

$$\text{(13)}$$

A noteworthy observation is that axial alcohols are transformed into silyl ethers significantly faster than equatorial alcohols.[8] The addition of trimethylsilyldiethylamine to an equimolar amount of 1-methylpyrrol-2(5*H*)-one under mild conditions gives 1-methyl-2-(trimethylsiloxy)pyrrole (eq 14).[11]

70%

Mild Generation of Phosphonochloridates. The phosphonochloridates are used in the synthesis of phosphorous-containing inhibitors. A new method has been developed using trimethylsilyldiethylamine protocol which enables the synthesis of a phosphonochloridate from a phosphonic acid monoalkyl ester under mild conditions, allowing acid-sensitive protecting groups to remain intact.[14] The hydroxyl group of an acid-sensitive t-butyldiphenylsilyl (TBDPS) protected phosphonic acid can be silylated with trimethylsilyldiethylamine prior to treatment with oxalyl chloride. Because, TMSCl is generated rather than HCl as a by-product, the TBDPS protecting group stays intact during this transformation (eq 15).

Silylation of Amines. While there are several effective methods for preparation of N-(trimethylsilyl)amines, only a few successful procedures for N,N-bis(trimethylsilyl)alkylamines synthesis have been reported. If silylation of ethylamine is attempted with trimethylsilyl chloride, only 13% of N,N-bis(trimethylsilyl)ethylamine is obtained in addition to the major product N-(trimethylsilyl)ethylamine. However, N,N-bis(trimethylsilyl)amines can be prepared by silylation of monosilylamines with trimethylsilyldiethylamine in the presence of catalytic ammonium sulfate or ammonium chloride.[26,27] Another convenient method for the preparation of N,N-bis(trimethylsilyl)alkylamines has recently been reported.[28] There trimethylsilyldiethylamine has been found to be effective in the conversion of primary amines, especially aromatic systems, and their monotrimethylsilyl derivatives into the corresponding bistrimethylsilyl products in high yields. For example, when isopropyl amine (or aniline) is refluxed with 1.1 equiv of trimethylsilyldiethylamine and 1.15 equiv of methyl iodide in toluene for 4 h, bis(trimethylsilyl)isopropylamine [or bis(trimethylsilyl)aniline] can be obtained in 99% yield (or 88% yield based on GC) (eqs 16 and 17).

99%

88%

Diethylamino Group Transfer. Trimethylsilyldiethylamine can be used for the convenient synthesis of amides that retain the N,N-diethyl moiety in their structures. Propionyl chloride is converted into the amide with 2 equiv of trimethylsilyldiethylamine in hexanes at ambient temperature in 5 h. Oxalyl chloride gives monoamide when treated with 1 equiv of trimethylsilyldiethylamine at $-23\,°C$, and gives bisamide with excess of the reagent at room temperature.[3] Trimethylsilyldiethylamine has been successfully used for the conversion of phenylselenyl chloride into the phenylselenenamide.[15]

Conjugate Addition. Conjugate addition of trimethylsilyl diethylamine to enones gives β-amino silyl enol ethers.[16] Thus, in the presence of catalytic amounts of trimethylsilyl triflate, trimethylsilyldiethylamine adds to ethyl vinyl ketone in a 1,4-fashion. The reaction requires 50 mol % excess of trimethylsilyldiethylamine, and proceeds in ether at room temperature under nitrogen. The product N,N-diethyl-3-[(trimethylsilyl)oxy]-(E)-2-pentamine is distilled out under reduced pressure in 45% yield (eq 18). The major product is the E-enol ether containing approximately 5% of other trimethylsilyl ether isomers.

46%

Aminoalkylation of Aldehydes. Iminium salts are important intermediates in organic synthesis. These salts can be produced in situ by the reaction of trimethylsilyldiethylamine and a concentrated ethereal solution of $LiClO_4$ with a variety of aliphatic or aromatic aldehydes.[29] Subsequent addition of various nucleophiles to the resulting iminium species gives rise to diverse aminoalkylation products (eq 19).

Trimethylsilyldiethylamine is a viable alternative to trimethylsilyl ketene acetals or halo esters for the preparation of N,N-dialkylamino esters. A convenient one-pot method for the preparation of these compounds has been recently reported.[30] The new route is based on aminoalkylation of aldehydes in the presence of (trimethysilyl)dialkylamine and a functionalized zinc reagent [$BrZn(CH_2)_nCO_2R$]. The protocol requires that an aldehyde be treated with a eightfold excess of 5 M ethereal $LiClO_4$. Trimethylsilyldiethylamine is added and the mixture is stirred for an additional 30 min. Subsequent addition to the freshly prepared bromoalkylzinc ester affords *N,N*-dialkylamino esters in moderate to good yields (Table 1).

Table 1 Preparation of *N,N*-dialkylamino esters in moderate to good yields using trimethylsilyldiethylamine and organozinc reagents

$$RCHO \xrightarrow[\text{LiClO}_4]{\text{Me}_3\text{SiNEt}_2} \xrightarrow{\text{RZnX}} \text{Product}$$

RCHO	RZnX	Product	Yield (%)
Ph⌒CHO	BrZnCH₂CO₂Me	Ph—(NEt₂)—CO₂Me	67
(thiophene)CHO	BrZnCH₂CO₂Me	(thiophene)—(NEt₂)—CO₂Me	73
PhCHO	BrZnCH₂CO₂Me	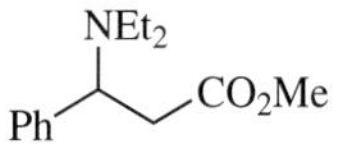 Ph—(NEt₂)—CO₂Me	74
i-PrCHO	BrZnCH₂CO₂Me	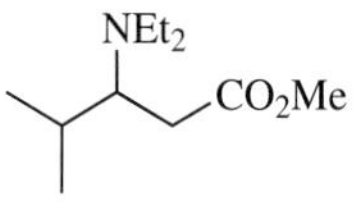 (i-Pr)—(NEt₂)—CO₂Me	52

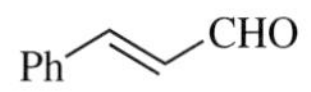

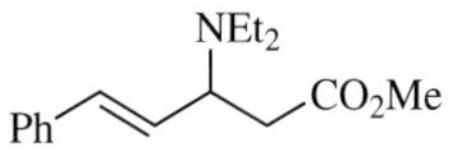

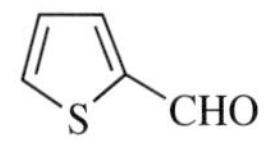

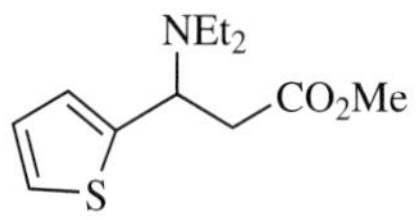

This reaction can be extended further to include other functionalized organozinc reagents. The products are derivatized amines as shown below (Table 2).

Table 2 Preparation of derivatized amines using trimethylsilyldiethylamine and functionalized organozinc reagents

$$RCHO \xrightarrow[\text{LiClO}_4]{\text{Me}_3\text{SiNEt}_2} \xrightarrow{\text{RZnX}} \text{Product}$$

RCHO	RZnX	Product	Yield (%)
i-PrCHO	ClZnC≡CSiMe₃	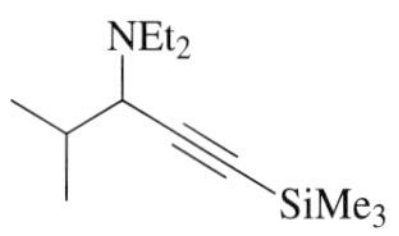 (NEt₂)...SiMe₃	82
i-PrCHO	BrZnCH₂CH=CH₂	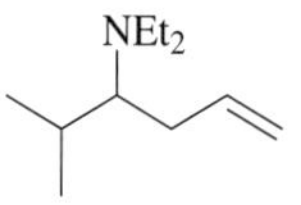 (NEt₂)...	62

The choice of nucleophiles is not limited to organozinc compounds. β-(Diethylamino) sulfoxides (eq 20) and β-(diethylamino) sulfones (eq 21) have been successfully prepared

in a one-pot LiClO₄-promoted condensation of aldehydes with trimethylsilyldiethylamine followed by addition of lithium salt of sulfoxides or sulfones, respectively.[31] In the case of sulfoxides, minor diastereoselectivity is observed. Changes in temperature and solvent do not have any significant effect on the diastereoselectivity.

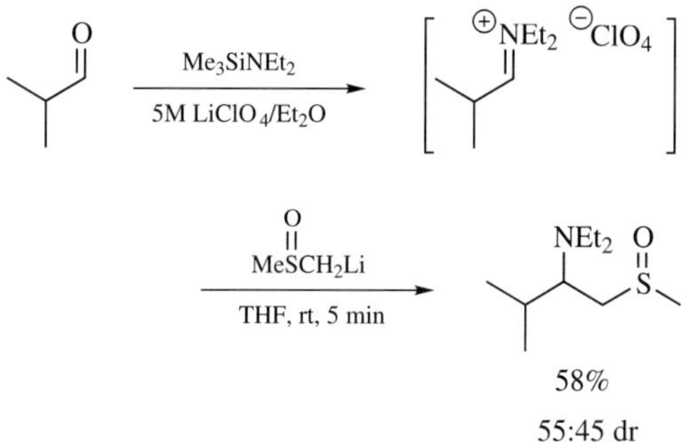

$$(20)$$

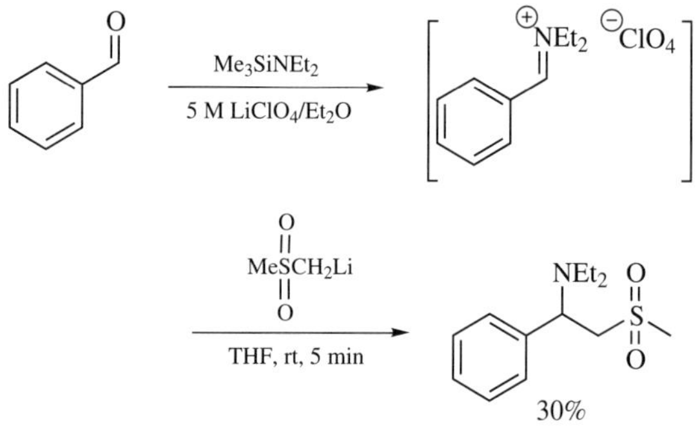

58%

55:45 dr

$$(21)$$

30%

Enamine and Silyl Enol Ether Preparation. Both enamines and silyl enol ethers are useful intermediates in organic synthesis. Aldehydes and ketones can be transformed into enamines via treatment with 2–3 equiv of trimethylsilyldiethylamine in the presence of a trace of *p*-toluenesulfonic acid.[32] The presence of excess silylamine obviates the normal requirements for water removal by azeotropic distillation or by addition of an insoluble drying agent. No solvent or heating is required for the transformation, and the reaction normally proceeds at room temperature. However, heating the reaction mixture allows enamine synthesis in equally high yield without an acid catalyst.

The same conditions applied to cyclic ketones fused with an aromatic core affords unexpected results. Silyl enol ethers are the primary products observed, and steric hindrance of the *peri*-H has been presumed to obstruct the enamine formation between the nitrogen and the carbonyl carbon, allowing for reversal of silylamine reactivity. Under these conditions, silicon attacks the oxygen electrophilically to give enol ether in good yield.[33]

Nevertheless, silyl enol ethers can be conveniently prepared by the reaction of ketones with trimethylsilyldiethylamine and methyl iodide. The process involves heating a mixture of 1.2 equiv of trimethyl silyldiethylamine and 1.3 equiv of methyl iodide at 60 °C for 1 h, and subsequently ketone is added dropwise (eqs 22 and 23).[34]

$$(22)$$

89%

$$(23)$$

91%

The combination of trimethylsilyldiethylamine and methyl iodide transforms both the cyclic and the acyclic ketones into silyl enol ethers in high yield, with favored formation of the more thermodynamically stable isomer.

Mediation of 1,4-Conjugate Additions. The high lability of the formyl group under basic and acidic conditions makes it difficult to cleanly generate enolates or enols. In order to arrest the pronounced reactivity under the said conditions, the formyl group is usually transformed into a stable synthon. Trimethylsilyldiethylamine can be utilized to solve such problems. A novel protocol for direct 1,4-addition of naked aldehydes to electron-deficient double bonds employs a catalytic amount of trimethylsilyldiethylamine.[35] For example, addition of but-3-en-2-one to a stirred solution of decanal and 0.5 equiv of trimethylsilyldiethylamine under nitrogen atmosphere yields the keto aldehyde in 67% yield (eq 24). The same conditions convert benzaldehyde and ethyl vinyl ketone into 2-benzyl-5-oxoheptanal in 68% yield (eq 25).

67%   (24)

68%   (25)

Trimethylsilyldiethylamine-promoted intramolecular 1,4-conjugate addition has also been reported. Reaction of an aldehydic enone, such as that in eq 26, with trimethylsilyldiethylamine in acetonitrile at room temperature provides a cyclic product in 73% yield with 7:3 diastereoselectivity.

$$(26)$$

73%

1. For a general review of compounds containing silicon–nitrogen bonds, see: Fessenden, R.; Fessenden, J. S., *Chem. Rev.* **1961**, *61*, 361.

2. Yankee, E. W.; Lin, C. H.; Fried, J., *Chem. Commun.* **1972**, 1120.

3. Bowser, J. R.; Williams, P. J.; Kurz, K., *J. Org. Chem.* **1983**, *48*, 4111.

4. Sauer, R. O.; Hasek, R. H., *J. Am. Chem. Soc.* **1946**, *68*, 241.

5. Langer, S. H.; Connell, S.; Wender, I., *J. Org. Chem.* **1958**, *23*, 50.

6. Mai, K.; Patil, G., *J. Org. Chem.* **1986**, *51*, 3545.

7. LeDrian, C.; Greene, A. E., *J. Am. Chem. Soc.* **1982**, *104*, 5473.

8. Weisz, I.; Felfoldi, K.; Kovacs, K., *Acta Chim. Acad. Sci. Hung.* **1968**, *58*, 189.

9. Garner, P., *Tetrahedron Lett.* **1984**, *25*, 5855.

10. Fiorenza, M.; Ricci, A.; Romanelli, M. N.; Taddei, M.; Dembech, P.; Seconi, G., *Heterocycles* **1982**, *19*, 2327.

11. Fiorenza, M.; Reginato, G.; Ricci, A.; Taddei, M.; Dembech, P., *J. Org. Chem.* **1984**, *49*, 551.

12. Taddei, M.; Tempesti, F., *Synth. Commun.* **1985**, *15*, 1019.

13. Rogozhin, S. V.; Davidovich, Y. A.; Yurtanov, A. I., *Synthesis* **1975**, 113.

14. Robl, J. A.; Duncan, L. A.; Pluscec, J.; Karanewshy, D. S.; Gordon, E. M.; Ciosek, C. P., Jr; Rich, L. C.; Dehmel, V. C.; Slusarchyk, D. A.; Harrity, T. W.; Obrien, K. A., *J. Med. Chem.* **1991**, *34*, 2804.

15. Back, T. G.; Kerr, R. G., *Can. J. Chem.* **1986**, *64*, 308.

16. Pratt, N. E.; Albizati, K. F., *J. Org. Chem.* **1990**, *55*, 770.

17. degl'Innocenti, A.; Capperucci, A.; Reginato, G.; Mordini, A.; Ricci, A., *Tetrahedron Lett.* **1992**, *33*, 1507.

18. Pike, R. A.; Schank, R. L., *J. Org. Chem.* **1962**, *27*, 2190.

19. Ohshita, J.; Iwata, A.; Kanetani, F.; Kunai, A.; Yamamoto, Y.; Matui, C., *J. Org. Chem.* **1999**, *64*, 8024.

20. Yamamoto, Y.; Shimizu, H.; Hamada, Y., *J. Organomet. Chem.* **1996**, *509*, 119.

21. Iwata, A.; Tang, H.; Kunai, A.; Ohshita, J.; Yamamoto, Y.; Matui, C., *J. Org. Chem.* **2002**, *67*, 5170.

22. Papini, A.; Ricci, A.; Taddei, M.; Seconi, G.; Dembech, P., *J. Chem. Soc., Perkin Trans. 1* **1984**, 2261.

23. Yamamoto, Y.; Shimizu, H.; Matui, C.; Chinda, M., *Main Group Chem.* **1996**, *1*, 409.

24. Evans, D. A.; Bartroli, J., *Tetrahedron Lett.* **1982**, *23*, 807.

25. Stork, G.; Rychnovsky, S. D., *J. Am. Chem. Soc.* **1987**, *109*, 1564.

26. Speier, J. L.; Zimmerman, R.; Webster, J., *J. Am. Chem. Soc.* **1956**, *78*, 2278.

27. Hils, J.; Hagen, V.; Ludwig, H.; Ruehlmann, K., *Chem. Ber.* **1966**, *99*, 776.

28. Hamada, Y.; Yamamoto, Y.; Shimizu, H., *J. Organomet. Chem.* **1996**, *510*, 1.

29. Saidi, M. R.; Heydari, A.; Ipaktschi, J., *Chem. Ber.* **1994**, *127*, 1761.

30. Saidi, M. R.; Khalaji, H. R.; Ipaktschi, J., *J. Chem. Soc., Perkin Trans. 1* **1997**, 1983.

31. Naimi-Jamal, M. R.; Ipaktschi, J.; Saidi, M. R., *Euro. J. Org. Chem.* **2000**, 1735.

32. Comi, R.; Franck, R. W.; Reitano, M.; Weinreb, S. M., *Tetrahedron, Lett.* **1973**, 3107.

33. Hellberg, L. H.; Juarez, A., *Tetrahedron Lett.* **1974**, 3553.

34. Yamamoto, Y.; Matui, C., *Organometallics* **1997**, *16*, 2204.

35. Hagiwara, H.; Komatsubara, N.; Ono, H.; Okabe, T.; Hoshi, T.; Suzuki, T.; Ando, M.; Kato, M., *J. Chem. Soc., Perkin Trans. 1* **2001**, 316.

2-Trimethylsilyl-1,3-dithiane

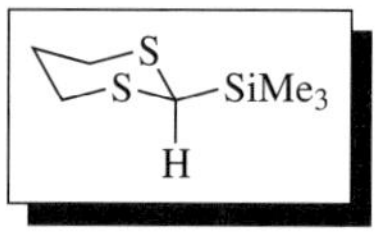

[13411-42-2]   C7H16S2Si   (MW 192.46)

InChI = 1S/C7H16S2Si/c1-10(2,3)7-8-5-4-6-9-7/h7H,4-6H2,1-3H3

InChIKey = BTTUMVHWIAXYPJ-UHFFFAOYSA-N

(precursor of dithioketene acetals[1] which are good substrates for both cationic[2] and anionic[3] cyclization processes; the anion reacts with unsaturated ketones and aldehydes to give vinyl dithioketene acetals that can be used as dienes[4] or for the preparation of substituted α,β-unsaturated alkyl ketones;[5] alkylation of the anion provides a general synthesis of acylsilanes[6,7])

Physical Data: bp 54–55 °C/0.17 mmHg.

Solubility: insol H_2O; sol organic solvents.

Form Supplied in: commercially available.

Preparative Method: prepared by alkylation of 2-lithio-1,3-dithiane with chlorotrimethylsilane (eq 1).[1]

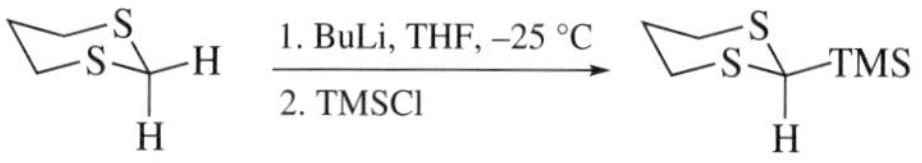

Handling, Storage, and Precautions: use in a fume hood.

Original Commentary

John W. Benbow

Lehigh University, Bethlehem, PA, USA

2-Lithio-2-trimethylsilyl-1,3-dithiane. This reagent (**1**) is generated from the title compound by treatment with *n*-butyllithium at −78 °C (eq 2)[1] and is the species utilized in many of the following transformations.

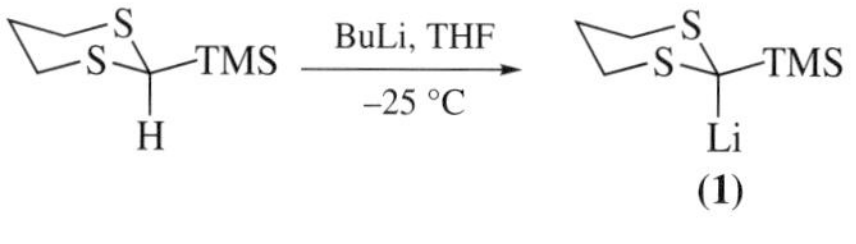

Thioketene Acetals. Anion (**1**) reacts with aldehydes and ketones to provide the corresponding thioketene acetals;[1] aryl and unsaturated aldehydes and ketones are good substrates for this reaction (eqs 3 and 4). Enolizable alkyl ketones also react to provide thioketene acetals (eq 5). Alternative methods for the preparation of these cyclic thioketene acetals involve the use of phosphonate derivatives,[8] mixed zinc–titanium organometallic reagents,[9] and *N,N*-dimethyl thioamides.[10] The phosphonate reagents are more nucleophilic than (**1**) and are superior when competitive deprotonation is a problem.

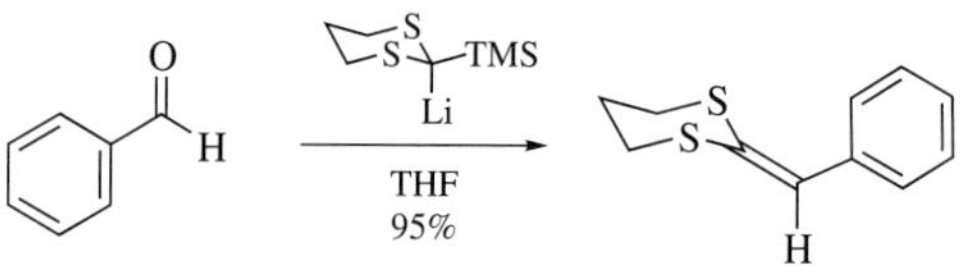

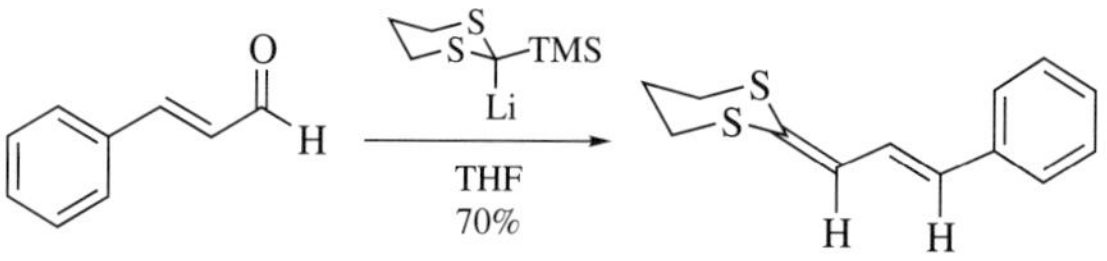

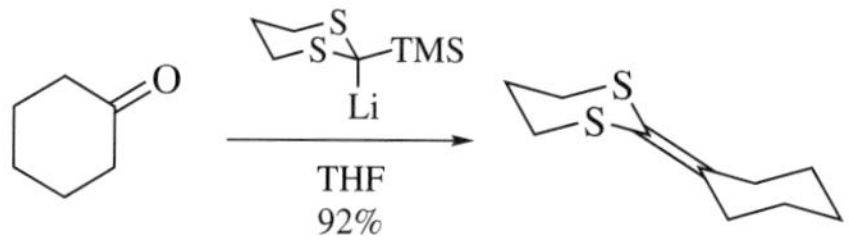

A variation of this method has also been developed for cases in which reactions with (**1**) give poor results (eq 6). Thus the carbonyl compound is treated with 2-lithio-1,3-dithiane followed by TMSCl to generate the silyl ether. Subsequent addition of a second equivalent of *n*-butyllithium effects alkenation, affording the thioketene acetal in good yield.[11]

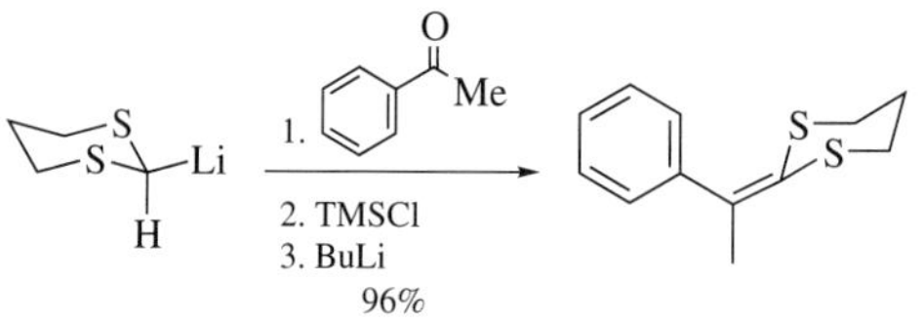

Cationic and Anionic Cyclizations. The thioketene acetals derived from the reaction of anion (**1**) have found several applications in cyclization methodology. The thioketene acetals can be used as either electrophiles (eq 7)[2] or nucleophiles (eq 8)[3] in a cyclization process which depends on the experimental conditions.

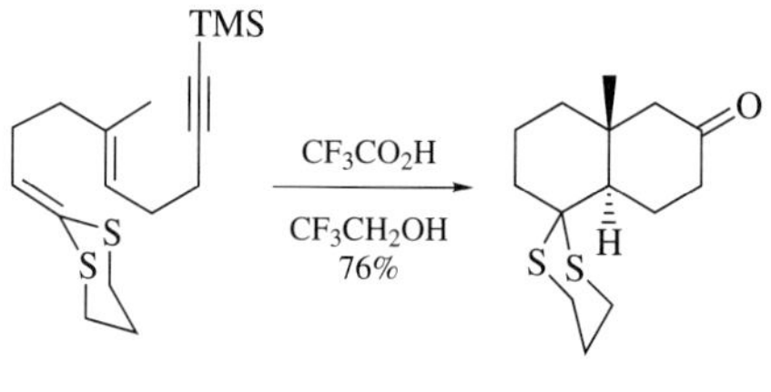

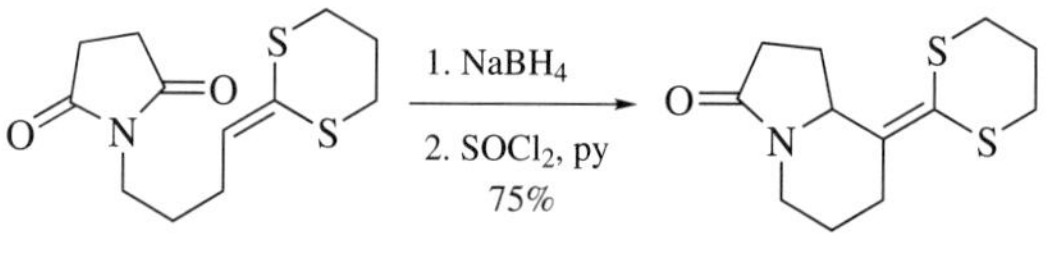

Substituted trimethylsilyldithianes can be converted into nucleophiles via fluoride ion desilylation and used in anionic cyclization strategies.[12] This anion not only undergoes condensations with carbonyls (eq 9), but it is also an effective Michael donor (eq 10).

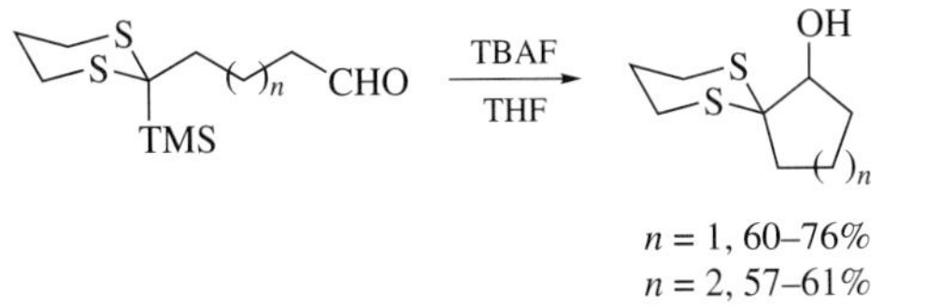

(9)

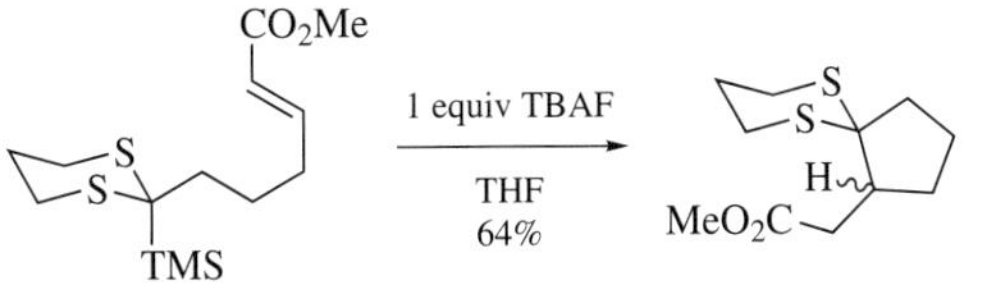

n = 1, 60–76%
n = 2, 57–61%

(10)

(14)

(15)

Unsaturated Dithioketene Acetals.

The reaction of anion (**1**) with various α,β-unsaturated aldehydes and ketones occurs in a 1,2-fashion and provides vinyl thioketene acetals that are acceptable dienes for Diels–Alder reactions (eq 11);[4] dienes of this type are difficult to access using other protocols. This method is also convenient for the introduction of a protected carbonyl group.

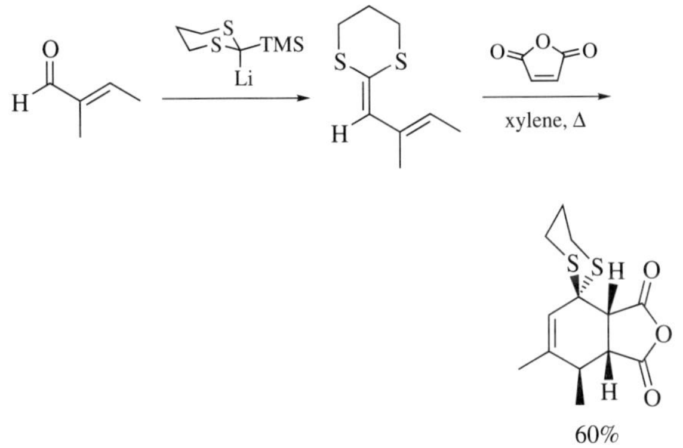

Acylsilanes.[14]

The reaction of anion (**1**) with alkyl halides generates a functionalized dithiane which, when hydrolyzed under mild conditions, provides the corresponding acylsilane (eq 16).[2] More highly substituted acylsilanes have been accessed by initial formation of the dithiane from the aldehyde, followed by deprotonation and silylation with TMSCl (eq 17). Acylsilanes have been used as sterically hindered aldehyde equivalents for regiocontrol in addition reactions (eq 18)[15] and also as precursors for silyl-substituted vinyl triflates (eq 19).[16]

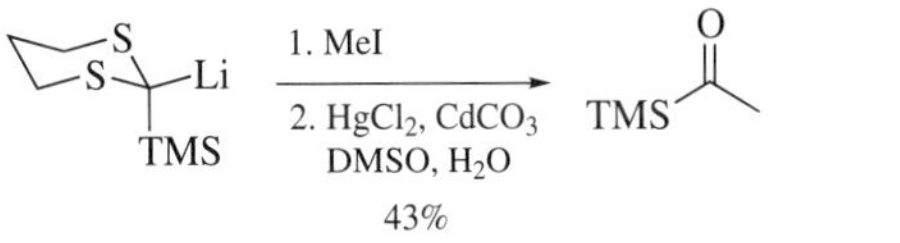

(11)

60%

(16)

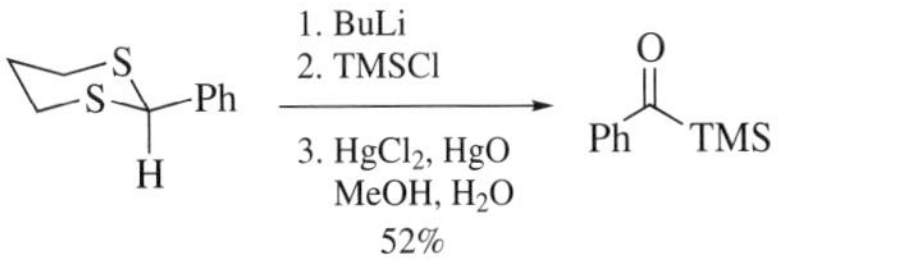

43%

(17)

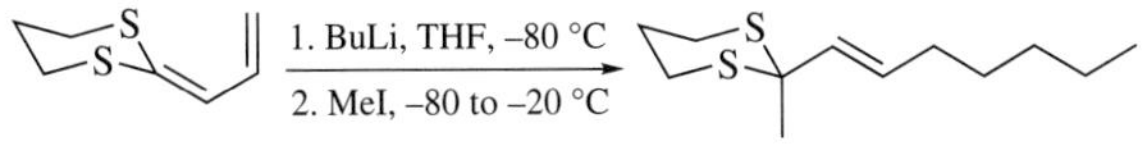

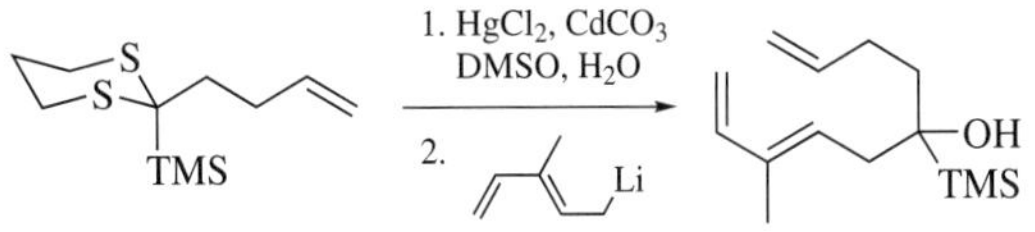

52%

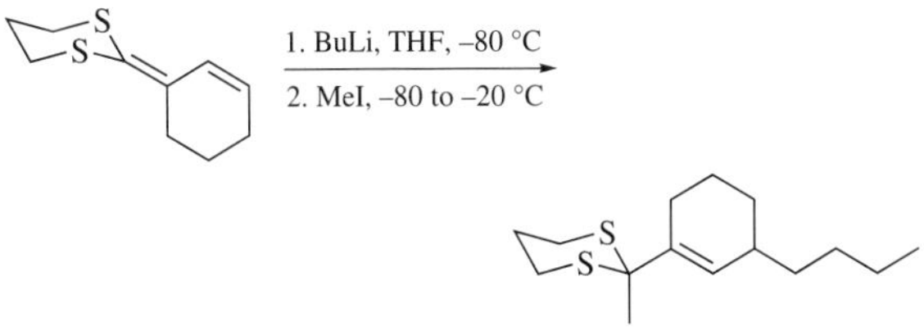

Treatment of acrolein and cyclohexenone with 2-lithio-2-trimethylsilyl-1,3-dithiane provides the corresponding unsaturated dithioketene acetals (eq 12).[5a] Subsequent exposure to *n*-butyllithium followed by MeI produces protected forms of α,β-unsaturated ketones in almost quantitative yield; anion addition occurs distal to the dithiane and alkylation takes place at the dithiane stabilized anion (eqs 13 and 14).[5b] Another interesting application is the use of suitably substituted 2-aryl-2-trimethylsilyl-1,3-dithianes as precursors for *o*-quinodimethanes (eq 15).[13]

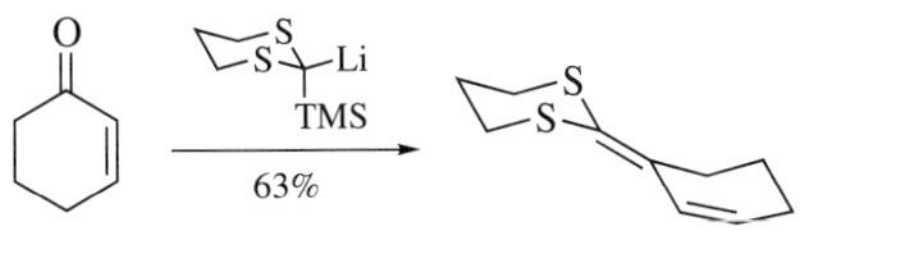

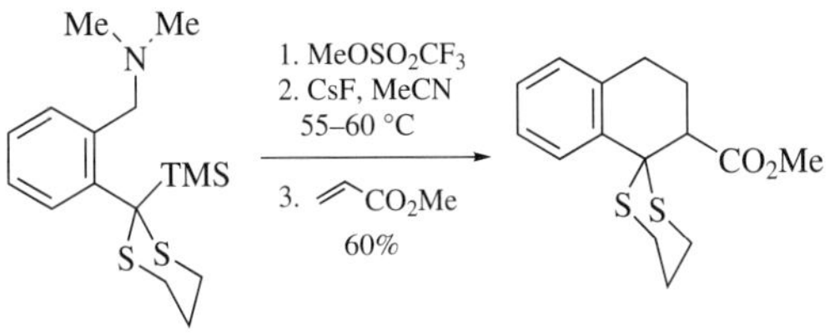

(18)

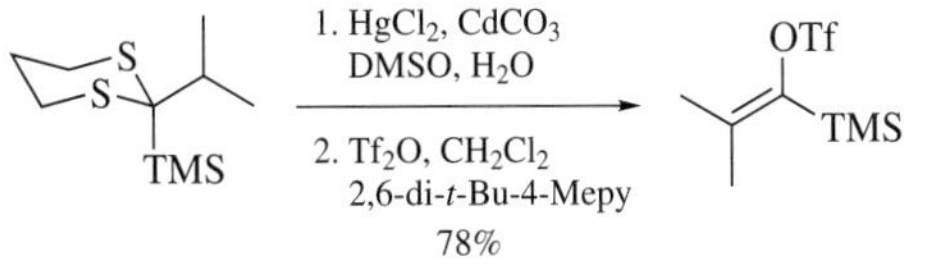

(19)

78%

(12)

63%

There are several alternative methods for the formation of acylsilanes; these include the reaction of silyl cuprates with acid halides,[17] palladium-assisted coupling of acyl halides with hexamethyldisilane,[18] the reaction of thiopyridyl esters with tris(trimethylsilyl)aluminum in the presence of Cu[I] salts,[19] and the hydroboration of trimethylsilyl-substituted alkynes.[20] While all of these procedures are complementary, the method involving the title compound provides an easy, inexpensive route to acylsilanes.

(13)

First Update

Megan A. Foley & Amos B. Smith, III
University of Pennsylvania, Philadelphia, PA, USA

One Carbon Homologation of Aldehydes via a Peterson Olefination. 2-Trimethylsilyl-1,3-dithianes have been widely used in the formation of thioketene acetals. This reaction is the first of a two-step procedure that results in the one carbon homologation of an aldehyde or ketone via a 2-lithio-2-TMS-1,3-dithiane-mediated Peterson olefination, followed by hydrolysis of the dithiane to give the corresponding ester (eq 20).[21]

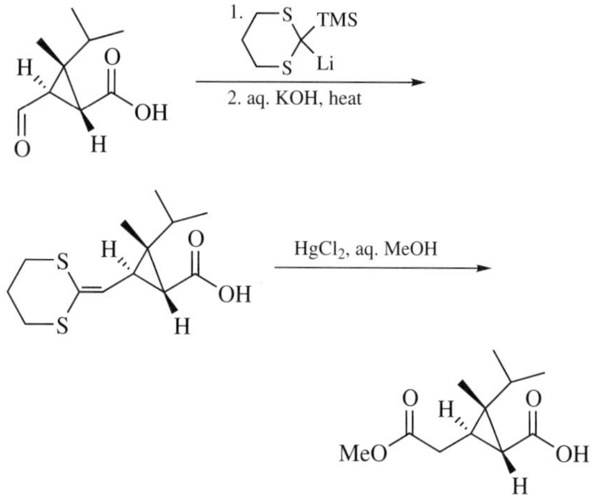

(20)

This two-step procedure has also been employed in a ring expansion reaction to convert a γ-lactone to a δ-lactone (eq 21). [22]

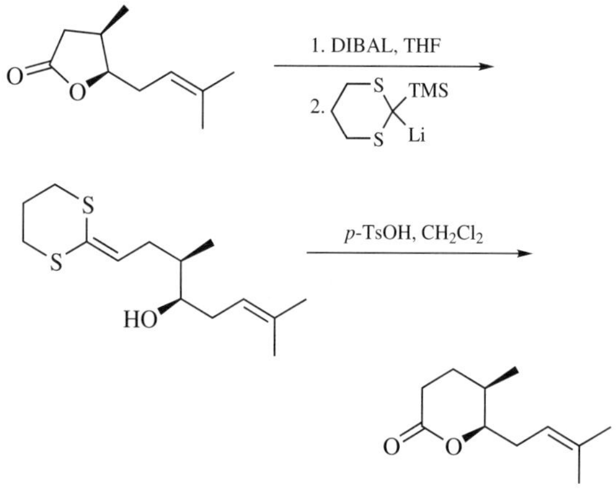

(21)

The one carbon homologue can be obtained in the aldehyde oxidation state employing a three-step procedure involving a 2-lithio-2-TMS-1,3-dithiane-mediated Peterson olefination, followed in turn by reduction of the resulting double bond and dithiane removal. This sequence was employed by Just and Manoharan furnishing the desired aldehyde as a single diastereomer (eq 22).[23]

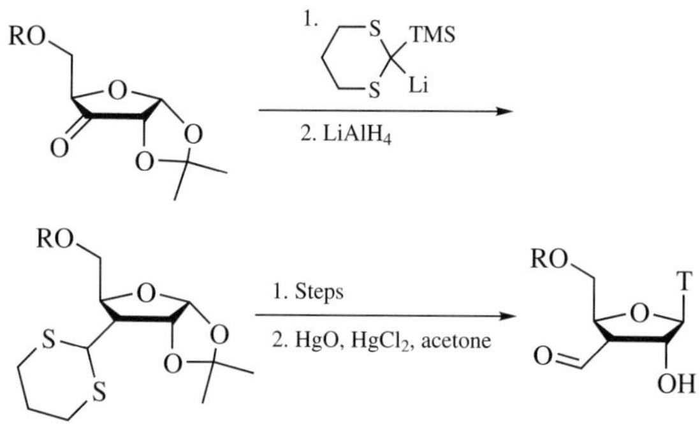

(22)

Cyclopropane Formation via a Homo-Peterson Reaction. In 1989, Schaumann and Friese disclosed a synthesis of cyclopropanes from oxirane and 2-lithio-2-TMS-1,3-dithiane employing a homo-Peterson reaction (eq 23).[24] Use of thiophenol as the anion stabilizing group led to cyclopropanes via an alternate reaction pathway.

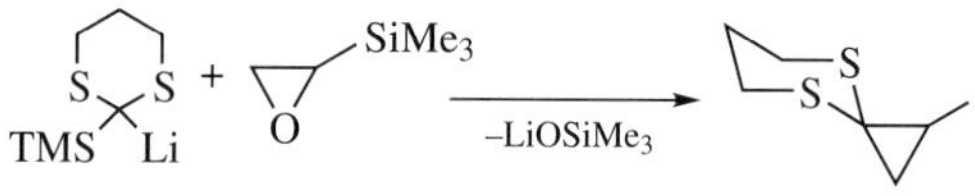

(23)

Silicon-Induced Domino Reactions. In a series of papers, Schaumann et al. disclosed the synthesis of various cyclic compounds utilizing a silicon-induced reaction cascade. Treatment of an epoxyalkyl tosylate with 2-lithio-2-TMS-1,3-dithiane leads to the corresponding lithium alkoxide. Subsequent 1,4-Brook rearrangement and displacement of the tosylates affords cyclopentanols in good yield (eq 24).[25,26]

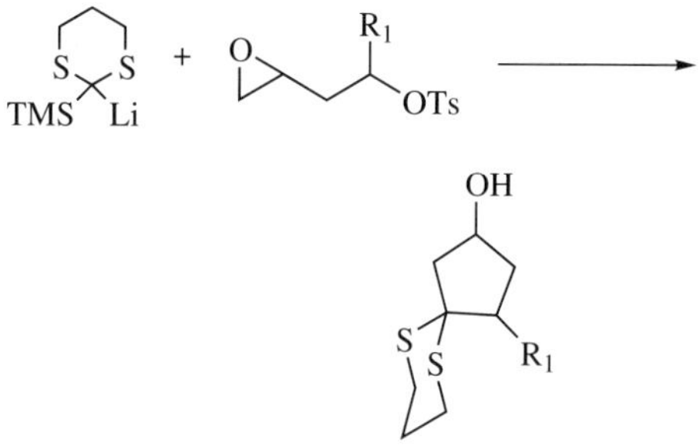

(24)

(a) R$_1$ = H   (a) R$_1$ = H (65%)
(b) R$_1$ = Me   (b) R$_1$ = Me (63%)
                              (a) 1:1 mixture of diastereomers

Additionally, Schaumann demonstrated that this approach can be used in the synthesis of pyrrolidine- and piperidine-2,3-diones. Treatment of bromoalkyl isocyanates with 2-lithio-2-TMS-1,3-dithiane provides, after Brook rearrangement and nucleophilic displacement of bromide, the corresponding lactams with moderate to good yields. Protection of the nitrogen followed by a two-step dithiane hydrolysis affords pyrrolidine- and piperidine-2,3-diones (eq 25).[27]

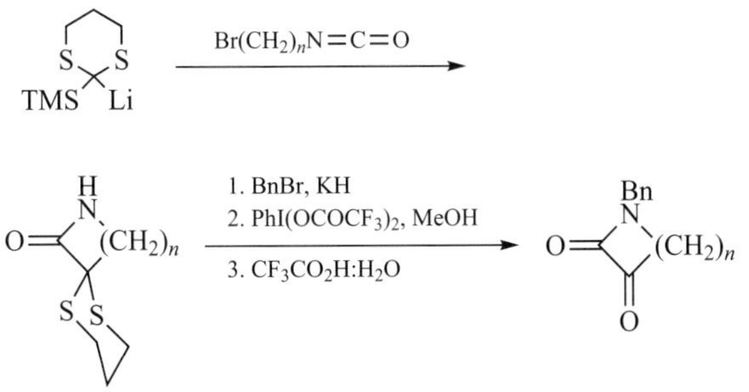

(25)

In 1999, Schaumann and co-workers extended the scope of this reaction to include bisepoxides as the electrophile component of the domino reaction.[28] Initial studies were carried out using both 2-lithio-2-TMS-1,3-dithiane and the lithium anion of dimethylsilylthioacetal. However, the dimethylsilylthioacetal proved to be more reactive and was used in subsequent experiments.

Subsequently Le Merrer and co-workers employed this transformation to generate iminosugar analogs from 1,2:5,6-dianhydro-3,4-O-methylethylidene-L-iditol.[29] Use of the parent 2-lithio-1,3-dithiane led to formation of a bis-dithiane resulting from opening of each epoxide by a separate dithiane anion. However, upon use of 2-lithio-2-TMS-1,3-dithiane a Brook rearrangement took place, followed by ring opening of the second epoxide in both a 7-endo-tet and a 6-exo-tet process to furnish a mixture of two products (eq 26). The yield of this reaction suffered due to loss of the TMS protecting group. The remainder of the studies were completed with the more robust 2-t-butyldimethylsilyl-1,3-dithiane.

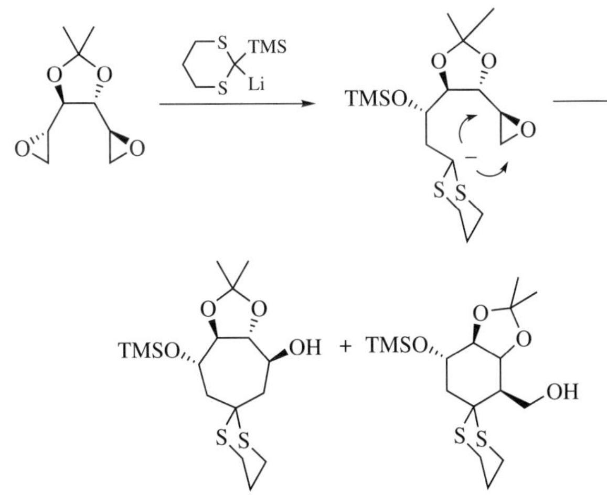

(26)

Multicomponent Linchpin Reactions. 2-Lithio-2-TMS-1,3-dithiane has also been employed as a bi-directional nucleophile or linchpin in epoxide opening reactions where both openings occur in an intermolecular fashion. The dithiane anion first attacks the less hindered carbon of the epoxide then, in the presence of THF and 12-crown-4, 1,4-Brook rearrangement occurs with transfer of the TMS group to reveal a new dithiane anion that can react with an additional equivalent of epoxide (eq 27). This reaction, first reported by Tietze et al. in 1994, suffered from a lack of control over the rearrangement step, thus limiting the use to the formation of symmetric 1,5-diols.[30]

In 1997 Smith and Boldi disclosed a similar reaction utilizing a solvent controlled Brook rearrangement to furnish unsymmetric

adducts.[31] In these reactions the initial addition is carried out in ether and the Brook rearrangement is triggered by the addition of a polar solvent, typically HMPA. This protocol, unlike that of Tietze, permitted the union of different epoxides. However, due to the aforementioned lability of the resulting TMS ether, most of the subsequent work employed 2-TBDPS-1,3-dithiane. In general, excellent yields of multicomponent coupling products were obtained with the TBS dithiane congener.

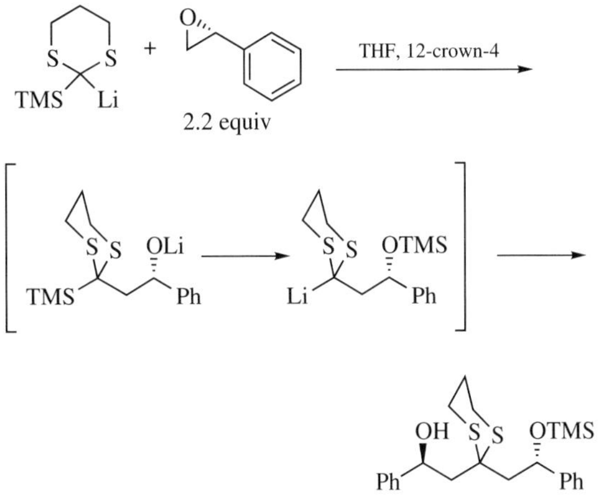

(27)

S_N2 Versus S_N2' Reaction of 2-TMS-1,3-Dithiane With Vinyl Epoxides. In 2002 Smith et al. demonstrated that 2-lithio-2-TMS-1,3-dithiane reacts with vinyl epoxides in an S_N2 fashion exclusively, while other, larger silyl dithianes afford mixtures of S_N2 and S_N2' products.[32] Particularly useful, the large 2-lithio-2-triisopropyl-1,3-dithiane provides solely the S_N2' product. Furthermore, it was found that trans epoxides furnish syn products and cis epoxides produced anti products, albeit in modest yields (eq 28). However, under the reaction conditions (THF, HMPA) the TMS group underwent a 1,4-Brook rearrangement to afford a 1:1 mixture of the anticipated homoallylic alcohol and the rearranged silyl ether.

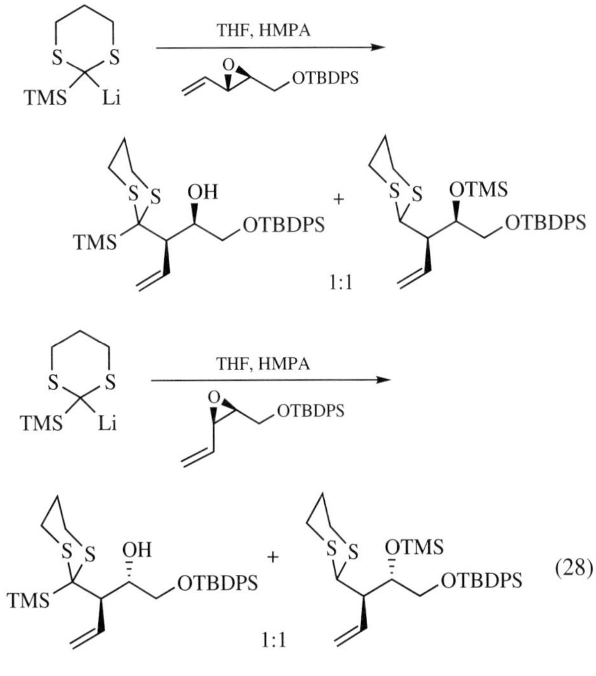

(28)

To control the timing of the Brook rearrangement, the reactions were performed in ether without HMPA. In these cases, the only product observed resulted from S_N2' attack on the vinyl epoxide (eq 29).[32]

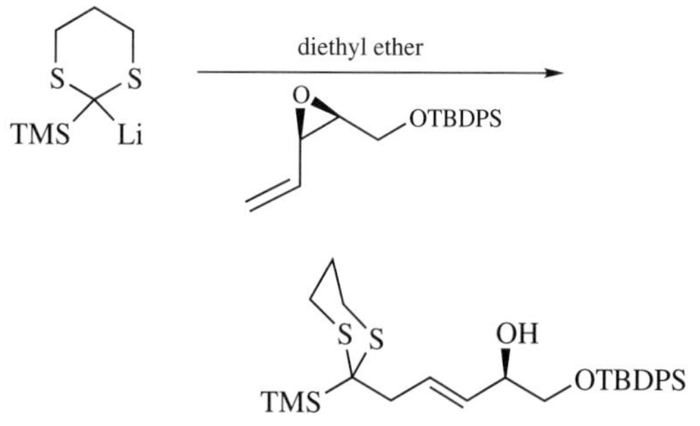

(29)

Formation of Bis(acylsilanes). 2-TMS-1,3-dithiane has also been used in the generation of acylsilanes. Recently 1,4-, 1,5-, 1,6- and 1,7-bis(acylsilanes) have been reported and used to produce silyl-substituted heterocycles. For example, Bouillon and Portella reported, in 1997, the formation of 1,5- and 1,6-bis(acylsilanes) employing 2-lithio-2-TMS-1,3-dithiane and epichlorohydrin and diepoxide, respectively, as electrophiles (eqs 30 and 31).[33] In the latter case the acylsilane was not isolated, as upon treatment with CaCO$_3$ and I$_2$ the resulting acylsilane underwent cyclization to furnish the 2,5-bis-trimethylsilylfuran.

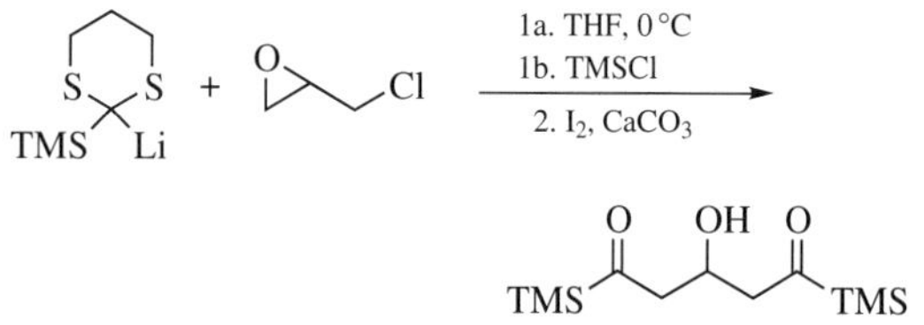

(30)

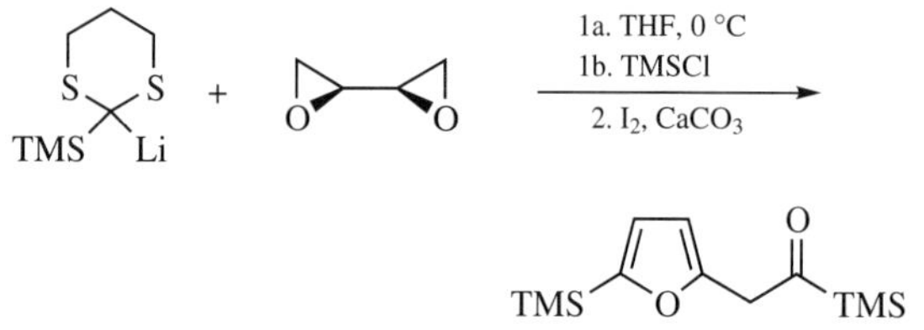

(31)

Portella also reported the formation of non-hydroxy-containing 1,4- and 1,5-bis(acylsilanes). For example, treatment of 2-TMS-1,3-dithiane with n-butyllithium followed by exposure to a variety of 1,2-bis(triflates) provided symmetric and unsymmetric 1,4-bisdithianes in modest to good yield.[34] Treatment with mercury salts or MeI revealed the bis(acylsilane), which upon exposure to PTSA cyclized to afford the corresponding 2,5-bis-trimethylsilylfuran (eq 32).

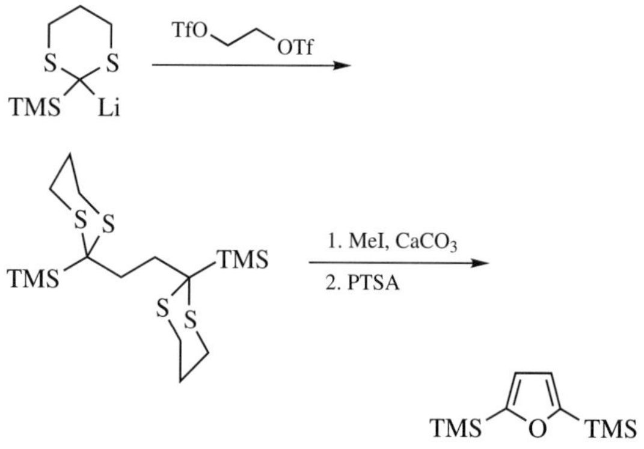

(32)

Formation of 1,5-bis(acylsilanes) was achieved via exposure of 2-lithio-2-TMS-1,3-dithiane to 1,3-dihalopropanes. Dithiane removal followed by treatment with PTSA then provides the 2,6-bis-trimethylsilyl-4H-pyran (eq 33).[35]

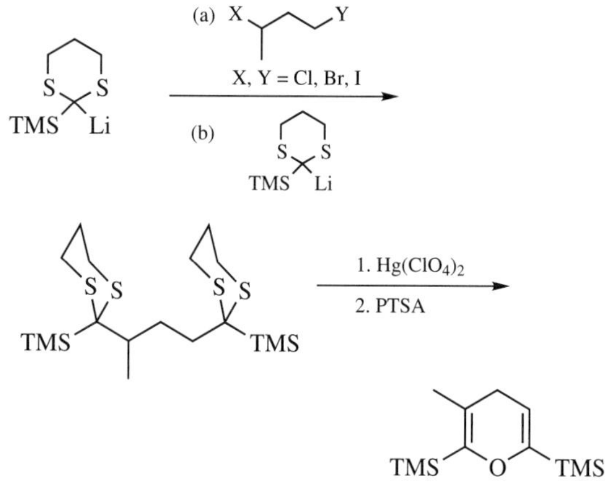

(33)

The scope of this reaction was subsequently extended to the formation of 1,6- and 1,7-bis(acylsilanes) through the reaction of 2-lithio-2-TMS-1,3-dithiane with various dihaloalkanes followed by removal of the dithianes (eq 34).[36]

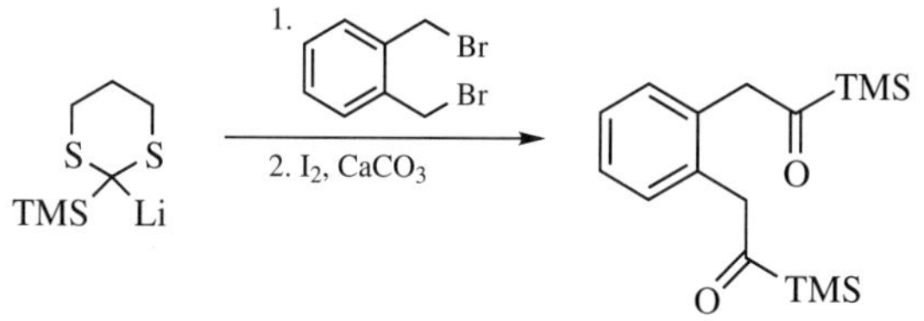

(34)

1. (a) Carey, F. A.; Court, A. S., *J. Org. Chem.* **1972**, *37*, 1926. (b) Seebach, D.; Grobel, B.-T.; Beck, A. K.; Braun, M.; Geiss, K.-H., *Angew. Chem., Int. Ed. Engl.* **1972**, *11*, 443.

2. Brinkmeyer, R. S., *Tetrahedron Lett.* **1979**, *20*, 207.

3. Chamberlin, A. R.; Chung, J. Y. L., *Tetrahedron Lett.* **1982**, *23*, 2619.

4. Carey, F. A.; Court, A. S., *J. Org. Chem.* **1972**, *37*, 4474.

5. (a) Seebach, D.; Kolb, M.; Grobel, B.-T., *Chem. Ber.* **1973**, *106*, 2277. (b) Seebach, D.; Kolb, M.; Grobel, B.-T., *Angew. Chem., Int. Ed. Engl.* **1973**, *12*, 69.

6. Brook, A. G.; Duff, J. M.; Jones, P. F.; Davis, N. R., *J. Am. Chem. Soc.* **1967**, *89*, 431.

7. Corey, E. J.; Seebach, D.; Freedman, R., *J. Am. Chem. Soc.* **1967**, *89*, 434.

8. (a) Mikolajczyk, M.; Grzejszczak, S.; Zatorski, A.; Mlotkowska, B.; Gross, H.; Costisella, B., *Tetrahedron* **1978**, *34*, 3081. (b) Mikolajczyk, M.; Balczewski, P., *Tetrahedron* **1992**, *48*, 8697.

9. Takai, K.; Fujimura, O.; Kataoka, Y.; Utimoto, K., *Tetrahedron Lett.* **1989**, *30*, 211.

10. Harada, T.; Tamaru, Y.; Yoshida, Z., *Tetrahedron Lett.* **1979**, *20*, 3525.

11. Chamchaang, W.; Prankprakma, V.; Tarnchompoo, B.; Thebtaranonth, C.; Thebtaranonth, Y., *Synthesis* **1982**, 579.

12. (a) Andersen, N. H.; McCrae, D. A.; Grotjahn, D. B.; Gabhe, S. Y.; Theodore, L. J.; Ippolito, R. M.; Sarkar, T. K., *Tetrahedron* **1981**, *37*, 4069. (b) Grotjahn, D. B.; Andersen, N. H., *J. Chem. Soc., Chem. Commun.* **1981**, 306.

13. Ito, Y.; Nakajo, E.; Sho, K.; Saegusa, T., *Synthesis* **1985**, 698.

14. For a review see: Bulman Page, P. C.; Klair, S. S.; Rosenthal, S., *Chem. Soc. Rev.* **1990**, *19*, 147.

15. Wilson, S. R.; Hague, M. S.; Misra, R. N., *J. Org. Chem.* **1982**, *47*, 747.

16. (a) Stang, P. J.; Fox, D. P., *J. Org. Chem.* **1977**, *42*, 1667. (b) Fox, D. P.; Bjork, J. A.; Stang, P. J., *J. Org. Chem.* **1983**, *48*, 3994.

17. Capperucci, A.; degl'Innocenti, A.; Faggi, C.; Ricci, A., *J. Org. Chem.* **1988**, *53*, 3612.

18. Yamamoto, K.; Suzuki, S.; Tsuji, J., *Tetrahedron Lett.* **1980**, *21*, 1653.

19. Nakada, M.; Nakamura, S.-I.; Kobayashi, S.; Ohno, M., *Tetrahedron Lett.* **1991**, *32*, 4929.

20. (a) Hassner, A.; Soderquist, J. A., *J. Organomet. Chem.* **1977**, *131*, C1. (b) Miller, J. A.; Zweifel, G., *Synthesis* **1981**, 288.

21. Reichelt, A.; Gaul, C.; Frey, R. R.; Kennedy, A.; Martin, S. F., *J. Org. Chem.* **2002**, *62*, 4062.

22. Honda, T.; Ishikawa, F.; Yamane, S.-I., *Heterocycles* **2000**, *52*, 313.

23. Prhavc, P.; Just, G.; Bhat, B.; Cook, P. D.; Manoharan, M., *Tetrahedron Lett.* **2000**, *41*, 9967.

24. Schaumann, E.; Friese, C., *Tetrahedron Lett.* **1989**, *30*, 7033.

25. Fischer, M.-R.; Kirschning, A.; Michel, T.; Schaumann, E., *Angew. Chem., Int. Ed. Engl.* **1994**, *33*, 217.

26. Michel, T.; Kirschning, A.; Beier, C.; Bräuer, N.; Schaumann, E.; Adiwidjaja, G., *Liebigs Ann.* **1996**, 1811.

27. Jung, A.; Koch, O.; Ries, M.; Schaumann, E., *Synlett* **2000**, 92.

28. Bräuer, N.; Dreeßen, S.; Schaumann, E., *Tetrahedron Lett.* **1999**, *40*, 2921.

29. Gravier-Pelletier, C.; Maton, W.; Dintinger, T.; Tellier, C.; Le, Merrer, Y., *Tetrahedron* **2003**, *59*, 8705.

30. Tietze, L. F.; Geissler, J. A.; Jakobi, G., *Synlett* **1994**, 511.

31. Smith, III, A. B.; Boldi, A. M., *J. Am. Chem. Soc.* **1997**, *119*, 6925.

32. Smith, III, A. B.; Pitram, S. M.; Gaunt, M. J.; Kozmin, S. A., *J. Am. Chem. Soc.* **2002**, *124*, 14516.

33. Bouillon, J.-P.; Portella, C., *Tetrahedron Lett.* **1997**, *37*, 6595.

34. Saleur, D.; Bouillon, J.-P.; Portella, C., *Tetrahedron Lett.* **2000**, *41*, 321.

35. Bouillon, J.-P.; Saleur, D.; Portella, C., *Synthesis* **2000**, 843.

36. Bouillon, J.-P.; Portella, C., *Eur. J. Org. Chem.* **1999**, 1571.

2-(Trimethylsilyl)ethanesulfonamide

[125486-96-6] $C_5H_{15}NO_2SSi$ (MW 181.33)

InChI = 1S/C5H15NO2SSi/c1-10(2,3)5-4-9(6,7)8/h4-5H2,1-3H3,(H2,6,7,8)

InChIKey = MZASHBBAFBWNFL-UHFFFAOYSA-N

(source of nucleophilic nitrogen that can react in manifold ways with electrophiles prior to removal of SES group with fluoride ion)

Alternate Name: β-(trimethylsilyl)ethanesulfonamide, SES-NH$_2$.

Physical Data: mp 85–86 °C.

Solubility: soluble in benzene, chloroform, dichloromethane, ether, methanol and toluene.

Preparative Methods: prepared by reaction of 2-(trimethylsilyl) ethanesulfonyl chloride (accessible from the commercially available sodium salt of the corresponding sulfonic acid or starting from vinyl trimethylsilane) with gaseous ammonia (eq 1).[1]

$$Me_3Si\diagup\diagdown \xrightarrow[\substack{PhCO_3t\text{-}Bu \\ MeOH, \Delta}]{NaHSO_3} Me_3Si\diagup\diagdown SO_3Na \xrightarrow{PCl_5/CCl_4}$$

$$Me_3Si\diagup\diagdown SO_2Cl \xrightarrow[86\%]{NH_3(g)} Me_3Si\diagup\diagdown SO_2NH_2 \quad (1)$$

Purity: recrystallized twice from ether by cooling solutions to −20 °C.

Handling, Storage, and Precaution: shelf-stable solid; avoid contact with sources of fluoride ion; toxicity unknown.

Preparation of *N*-Sulfonylimines. The title sulfonamide has been used to generate *N*-sulfinyl-2-(trimethylsilyl)ethanesulfonamide (SES-NSO),[1c,2] which reacts with aldehydes in the presence of BF$_3$•Et$_2$O to form, after loss of SO$_2$, *N*-sulfonylimine derivatives. These can be trapped with 2,3-dimethylbuta-1,3-diene in Diels–Alder cycloaddition reactions (eq 2).[2] The SES group can then be removed using a fluoride ion source.[1a,2]

SES-NSO also reacts with ketones in the presence of Lewis acids and the resulting *N*-sulfonylimines can be reduced to the corresponding sulfonamides with NaBH$_3$CN (eq 3). Such a reaction sequence has been employed in a synthesis of the antitumor antibiotic (−)-bactobolin.[3]

N-Sulfonylaldimines produced by condensation of SES-NH$_2$ with various aldehydes are converted into enantiomerically enriched aziridines upon reaction with certain sulfur ylides.[4] The same types of aldimines undergo [3 + 2] cycloaddition reactions with 2,3-butadienoates in the presence of triphenylphosphine to give 2,5-dihydropyrroles,[5] while that derived from SES-NSO and ethyl glyoxylate reacts efficiently with chiral sulfonimidoyl substituted bis(allyl)titanium complexes to give β-alkyl-γ,δ-unsaturated α-amino acid derivatives.[6]

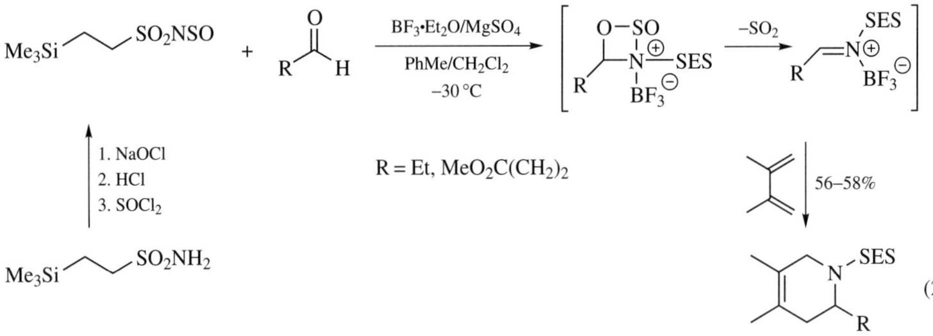

(2)

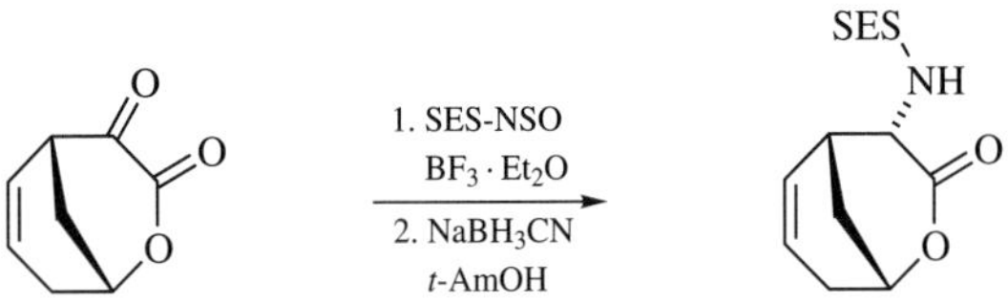

(3)

Mitsunobu Reactions. The readily obtained *N*-Boc derivative of SES-NH$_2$ is sufficiently acidic that it can participate effectively in Mitsunobu reactions with alcohols. Both *N*-protecting groups can be efficiently removed from the product amine derivative under different conditions (tetra-*n*-butyl ammonium fluoride or trifluoroacetic acid for SES or Boc, respectively) and without racemization.[7]

Azaglycosylation Chemistry. Reaction of SES-NH$_2$ with bromine in a 4% aqueous NaOH solution affords the corresponding *N,N*-dibromosulfonamide (SES-NBr$_2$) which has been employed in azaglycosylation of glucals (eq 4).[8] SES-NH$_2$ itself can be used directly for this same purpose in the presence of iodonium di-*sym*-collidine perchlorate.[8–10]

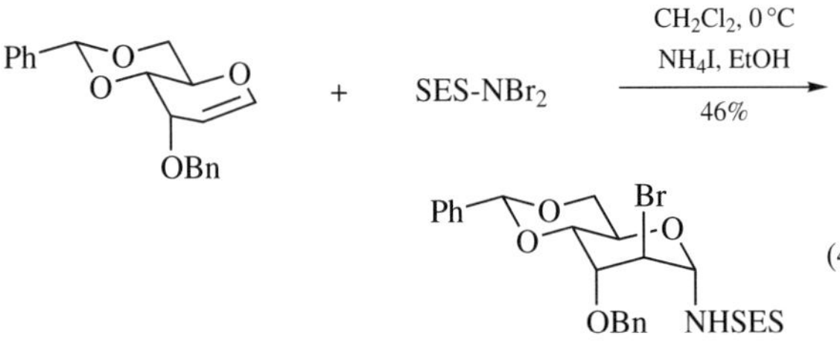

(4)

Facile anomerization of the products of such reactions has been noted.[11]

Preparation of Sulfodiimides. The sulfodiimide derived from SES-NH$_2$ has been shown to engage in a diastereofacially selective Sharpless-Kresze type ene reaction with a ring-fused cyclopentene to give an allylically aminated product (eq 5) that has been exploited in a total synthesis of the marine alkaloid agelastatin A.[1c,12]

Preparation of (*N*-SES-imino)phenyliodinane. Reaction of SES-NH$_2$ with PhI(OAc)$_2$ in the presence of methanolic potassium hydroxide affords (*N*-SES-imino)phenyliodinane that reacts with olefins in the presence of copper (I or II) salts to give aziridines (eq 6).[13] A one-pot variant on this procedure wherein the (*N*-SES-imino)phenyliodinane is generated *in situ* has been reported.[14] The SES-group associated with the product aziridine can be readily removed without accompanying cleavage of the three-membered ring.[13]

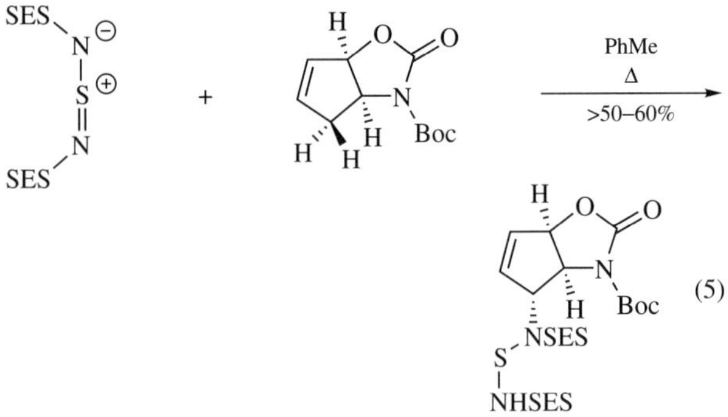

(5)

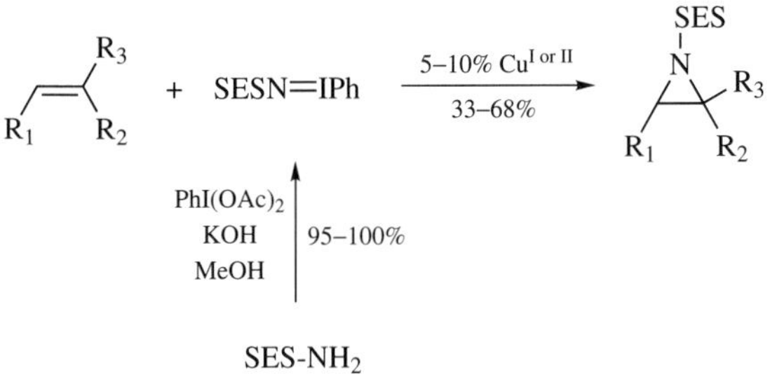

(6)

Miscellaneous Reactions. SES-NH$_2$ has been used in the synthesis of a spheroidal porphyrin. Thus, treatment of a tetrabromoporphyrin with SES-NH$_2$ in the presence of Cs$_2$CO$_3$ resulted in the production of an *N*-linked bis-porphyrin in 10% yield.[15]

1. (a) Weinreb, S. M.; Demko, D. M.; Lessen, T. A.; Demers, J. P., *Tetrahedron Lett.* **1986**, *27*, 2099. (b) Weinreb, S. M.; Chase, C. E.; Wipf, P.; Venkatraman, S., *Org. Synth.* **1998**, *75*, 161. (c) Stien, D.; Anderson, G. T.; Chase, C. E.; Koh, Y.; Weinreb, S. M., *J. Am. Chem. Soc.* **1999**, *121*, 9574.

2. Sisko, J.; Weinreb, S. M., *Tetrahedron Lett.* **1989**, *30*, 3037.

3. Garigipati, R. S.; Tschaen, D. M.; Weinreb, S. M., *J. Am. Chem. Soc.* **1990**, *112*, 3475.

4. Aggarwal, V. K.; Thompson, A.; Jones, R. V. H.; Standen, M. C. H., *J. Org. Chem.* **1996**, *61*, 8368.

5. Xu, Z.; Lu, X., *J. Org. Chem.* **1998**, *63*, 5031.

6. Schleusner, M.; Gais, H.-J.; Koep, S.; Raabe, G., *J. Am. Chem. Soc.* **2002**, *124*, 7789.

7. Campbell, J. A.; Hart, D. J., *J. Org. Chem.* **1993**, *58*, 2900.

8. Griffith, D. A.; Danishefsky, S. J., *J. Am. Chem. Soc.* **1996**, *118*, 9526.

9. Ritzeler, O.; Henning, L.; Findeisen, M.; Welzel, P.; Müller, D., *Tetrahedron* **1997**, *53*, 1665.

10. Wang, Z.-G.; Zhang, X.; Visser, M.; Live, D.; Zatorski, A.; Iserloh, U.; Lloyd, K. O.; Danishefsky, S. J., *Angew. Chem. Int. Ed. Engl.* **2001**, *40*, 1728.

11. Owens, J. O.; Yeung, B. K. S.; Hill, D. C.; Petillo, P. A., *J. Org. Chem.* **2001**, *66*, 1484.

12. Anderson, G. T.; Chase, C. E.; Koh, Y.; Stien, D.; Weinreb, S. M.; Shang, M., *J. Org. Chem.* **1998**, *63*, 7594.

13. Dauban, P.; Dodd, R. H., *J. Org. Chem.* **1999**, *64*, 5304.

14. Dauban, P.; Sanière, L.; Tarrade, A.; Dodd, R. H., *J. Am. Chem. Soc.* **2001**, *123*, 7707.

15. Zhang, H.-Y.; Yu, J.-Q.; Bruice, T. C., *Tetrahedron* **1994**, *50*, 11339.

Martin G. Banwell & Jens Renner

The Australian National University, Canberra, Australian Capital Territory, Australia

β-Trimethylsilylethanesulfonyl Chloride

[106018-85-3] $C_5H_{13}ClO_2SSi$ (MW 200.79)

InChI = 1S/C5H13ClO2SSi/c1-10(2,3)5-4-9(6,7)8/h4-5H2,1-3H3

InChIKey = BLPMCIWDCRGIJV-UHFFFAOYSA-N

(protection of primary and secondary amines as their sulfonamides, which are cleaved by fluoride ion[1])

Alternate Name: SESCl.

Physical Data: bp 60 °C/0.1 mmHg; yellow oil.

Solubility: sol most common organic solvents.

Preparative Methods: can be most conveniently synthesized from commercially available vinyltrimethylsilane (**1**) (eq 1).[1] Radical addition of sodium bisulfite to the vinyl group catalyzed by *t*-butyl perbenzoate yields the sulfonate salt (**2**) which can be directly converted to SESCl (**3**) with phosphorus(V) chloride. The chloride (**3**) can then be purified by distillation. The intermediate sulfonate salt (**2**) is commercially available. The chloride (**3**) can also be prepared in 62% yield from the salt (**2**) using sulfuryl chloride and triphenylphosphine (eq 2).[2] A less convenient procedure to synthesize SESCl (**3**) using β-trimethylsilylethylmagnesium chloride (**4**) and sulfuryl chloride has also been developed (eq 3).[1]

Handling, Storage, and Precautions: stable liquid that can be stored at room temperature for weeks. Prone to hydrolysis.

Original Commentary

Steven M. Weinreb & Janet L. Ralbovsky

The Pennsylvania State University, University Park, PA, USA

Protection of Amines. Sulfonamides are among the most stable of amine protecting groups and it is this stability that detracts from their utility, since harsh reaction conditions are often needed for their removal. The advantage of using the β-trimethylsilylethanesulfonyl (SES) protecting group is that the sulfonamide (**5**) can be easily cleaved to regenerate the parent amine (eq 4)[1] in good yields with either cesium fluoride (2–3 equiv) in DMF at 95 °C for 9–40 h, or tetra-*n*-butylammonium fluoride trihydrate (3 equiv) in refluxing MeCN. The main disadvantage of the latter procedure is an occasional difficulty in separating tetrabutylammonium salts from some amines.

The sulfonamides can be prepared from a wide variety of primary and secondary amines using sulfonyl chloride (**3**) in DMF containing triethylamine. For aromatic and heterocyclic amines, sodium hydride is the preferred base. The sulfonamides are generally quite stable and are untouched by refluxing TFA, 6 M HCl in refluxing THF, 1 M TBAF in refluxing THF, LiBF₄ in refluxing MeCN, BF₃·OEt₂, and 40% HF in ethanol. Table 1 lists a few examples of various amines and their protection as SES sulfonamides and cleavage with CsF in DMF at 95 °C.[1]

SESCl in Synthesis.[3] The SES group has been used successfully in the synthesis of glycosides (eq 5).[4] Reaction of (**6**) with SES-sulfonamide and iodonium di-sym-collidine perchlorate provides the iodo sulfonamide (**7**) in 82% yield. Treatment of (**7**) with (benzyloxy)tributylstannane in the presence of silver(I) trifluoromethanesulfonate provides the β-benzyl glycoside (**8**). Fluoride treatment of (**8**) removes both the silyl ether and the SES group, giving the amino alcohol (**9**).

The smooth removal of the SES protecting group from pyrrole and pyrrole-containing peptides demonstrates the synthetic potential of this protecting group in heterocyclic chemistry (eq 6).[5] The SES group of (**10**) is removed with TBAF·3H₂O in DMF at room temperature to yield (**11**). Other protecting groups (i.e. mesylate, triflate) cannot be removed without destruction of the substrates.

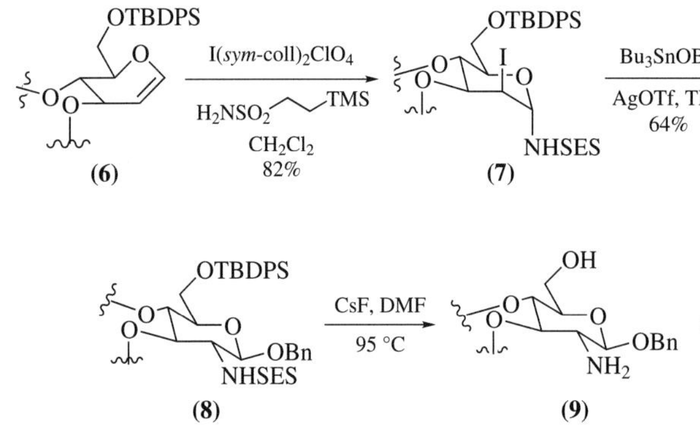

(5)

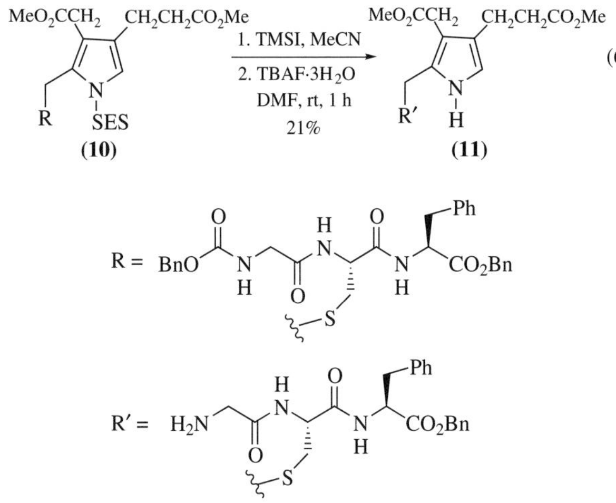

(6)

R =

R' =

The *N*-SES group can be incorporated by treating an aldehyde with *N*-sulfinyl-β-trimethylsilylethanesulfonamide (SESNSO) (13), which can be made by treating the sulfonamide (12) with thionyl chloride and a catalytic amount of *N,N*-dichloro-*p*-toluenesulfonamide (eq 7) (see also *N*-sulfinyl-*p*-toluenesulfonamide.[6] The *N*-sulfonyl imine can be used in situ in a number of reactions. For example, the *N*-sulfonyl imine from aldehyde (14) reacts with 2,3-dimethylbutadiene (eq 8)[7] to give the Diels–Alder adduct (15). Treatment of (15) with fluoride ion affords the bicyclic lactam (16). Also, the *N*-sulfonyl imine derived from isobutyraldehyde and (13) reacts with vinylmagnesium bromide to provide the allylic SES-sulfonamide (17) in 65% yield (eq 9).[8]

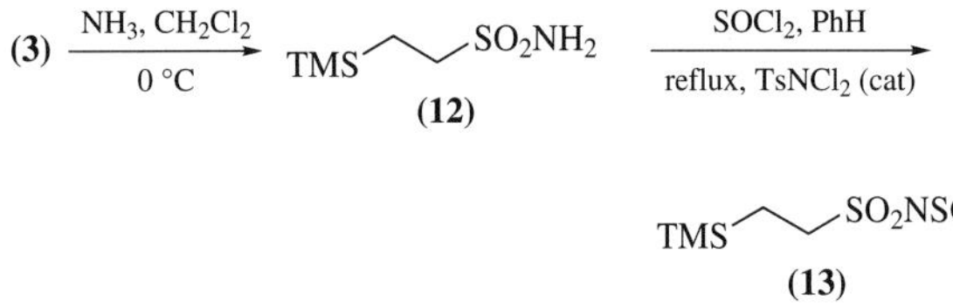

$$(3) \xrightarrow[0\ °C]{NH_3,\ CH_2Cl_2} TMS\diagup SO_2NH_2 \xrightarrow[reflux,\ TsNCl_2\ (cat)]{SOCl_2,\ PhH}$$

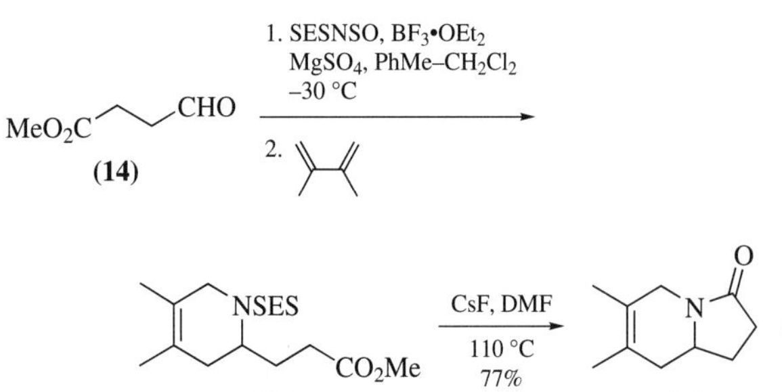

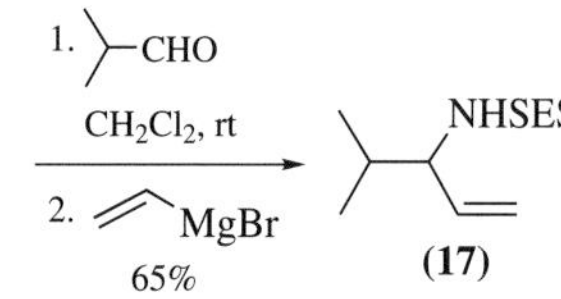

(12)

$$TMS\diagup SO_2NSO \quad (7)$$

(13)

(8)

(15)   (16)

(9)

In a total synthesis of the antitumor antibiotic (−)-bactobolin (18), the choice of protecting group on the nitrogen was crucial (eq 10).[6] Unlike other protecting groups, the SES group is compatible with a wide variety of transformations and reagents, and is easily removed at the end of the synthesis.

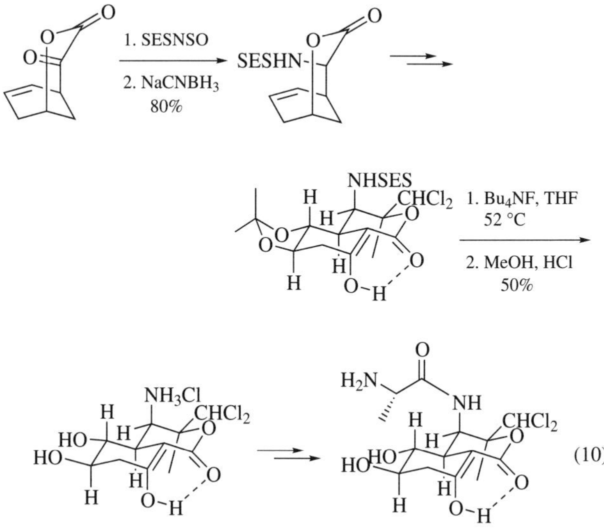

(10)

(18)

First Update

Valérie Declerck, Patrice Ribière, Jean Martinez & Frédéric Lamaty
Université Montpelier, Montpellier, France

Protection of an Amine. A simple and general synthesis of *N*-SES amino acids[9,10] starts from the corresponding salt using a temporary trimethylsilane protection of the carboxylic acid function, followed by reaction with SES-Cl[11] (eq 11 and Table 2).

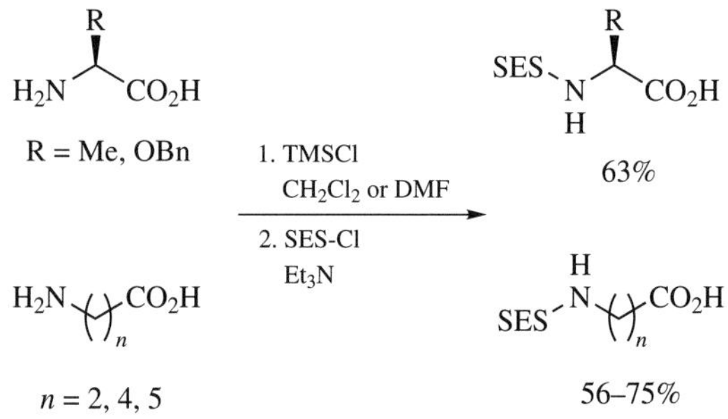

(11)

These SES-protected amino acids were shown to react in peptide synthesis[10,12] like the usual Boc, Z, or Fmoc amino acids (eq 12). No racemization was detected in the peptide product.

Table 2 Synthesis of SES-protected amino acids

Amino acid	Yield (%)
H-Ala-OH	63
H-Pro-OH	85
H-β-Ala-OH	56
5-Aminovaleric acid	72
6-Aminocaproic acid	75

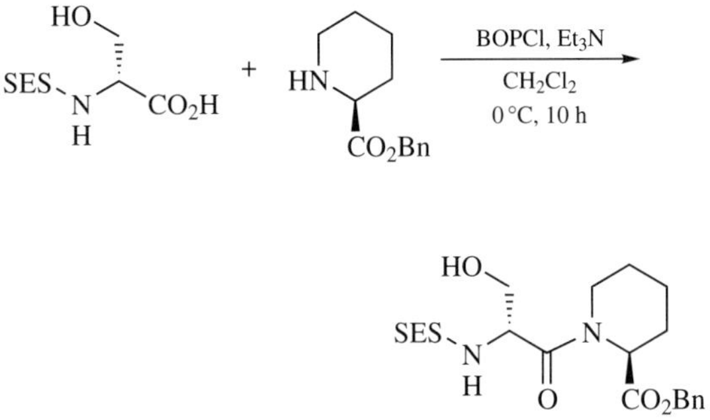

$$(12)$$

82%

The amino function of the side chain of ornithine has been protected as described or directly in aqueous medium[13] without protecting the carboxylic acid function (eq 13). The protected ornithine has been used in the synthesis of Ramoplanin,[13–15] a lipoglycodepsipeptide with potent antibacterial activity, and in the synthesis of the Chlorofusin cyclic peptide.[16]

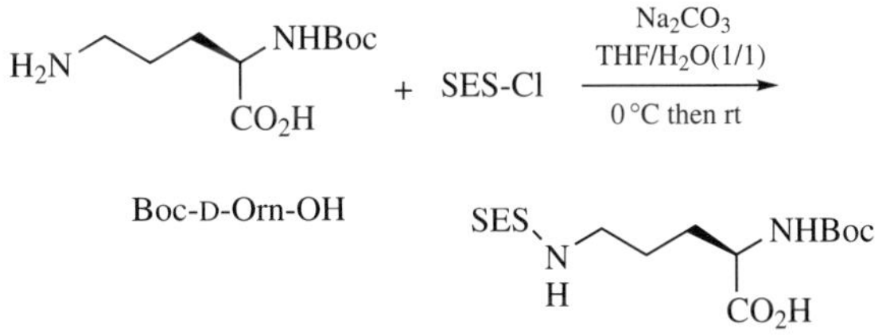

$$(13)$$

***N*-Alkylation.** Few examples of the Mitsunobu reaction have been described for SES-protected primary amines, however, an intramolecular reaction was used to prepare a perhydro-1,4-diazepin-2-one[10,11] (eq 14), which is a putative mimic of a peptide γ-turn with an ethylene link.

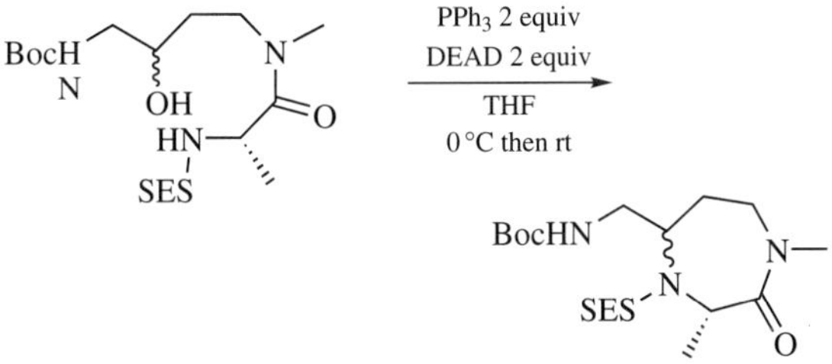

$$(14)$$

50%

Intermolecular Mitsunobu reaction was used in the synthesis of a 1,4-oxazepine (eq 15).[17]

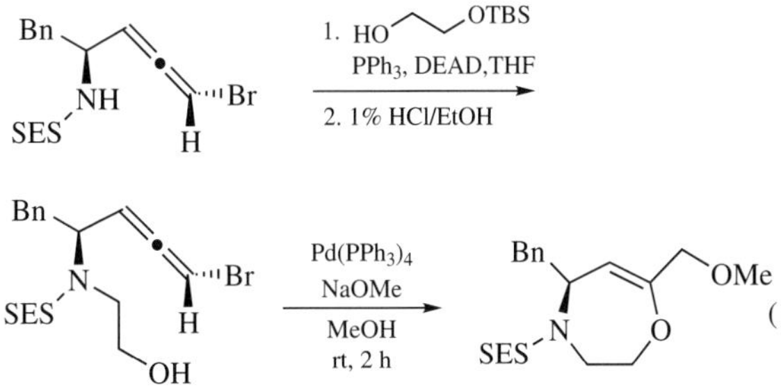

58%   78%

$$(15)$$

Deprotection Conditions. While cesium fluoride (CsF) in DMF and tetrabutylammonium fluoride (TBAF) in THF are the most common reaction conditions for deprotection, other reagents include hydrofluoric acid for deprotection of depsipeptides[12,13,18] and tris(dimethylamino)sulfonium difluorotrimethylsilicate $((Me_2N)_3S^+$ $Me_3SiF_2^-$, TASF) for deprotection of aziridines.[19,20]

The orthogonality of the SES protecting group has been illustrated for the synthesis of Homocaldopentamine,[21] where five different nitrogen protecting groups (Boc, allyl, TcBoc, azide, SES) are selectively deprotected (TFA, Pd(PPh_3)_4/NDBMA, Zn_{dust}/AcOH, PPh_3/H_2O, CsF/DMF, respectively).

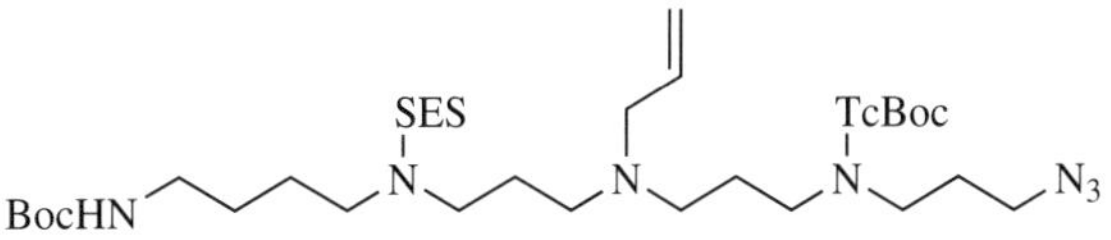

Formation of 2-Trimethylsilylethanesulfonylimines. The 2-trimethylsilylethanesulfonylimines (**20**) can be obtained by direct condensation between an aldehyde and SES-NH_2 (**19**) (obtained from reaction of SES-Cl and NH_3) in the presence of a Lewis acid using Dean-Stark azeotropic distillation (eq 16).[22,23]

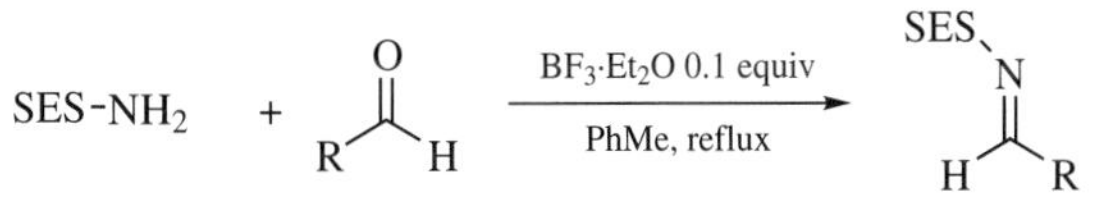

$$(16)$$

19   **20**
                                        57–82%

R- = Ph-, 4-Cl-C_6H_4-, 4-Me-C_6H_4-,
    (*E*)-PhCH=CH-, *t*-Bu-, *c*-C_6H_{11}-, *n*-Bu-

The reaction of 2,3-butadienoate with SES imines catalyzed by triphenylphosphine provides 2,5-dihydro-pyrrole-3-carboxylates (eq 17) with an excellent yield and a high regioselectivity.[23] Triphenylphosphine reacts with the allene to form an allyl carbanion, which is trapped by the SES imine in a [3 + 2] cycloaddition.

Similar pyrrolines have been obtained by sequential *aza*-Baylis-Hillman/alkylation/ring-closing metathesis reactions.[24] The reaction of SES-NH_2 (**19**) in the three-component *aza*-Baylis-Hillman reaction produces various unsaturated β-amino esters. This reaction, which corresponds to α-coupling of methyl acrylate with an activated imine, suggests in situ formation of the sulfonylimine prior to coupling with methyl acrylate (eq 18). Alkylation of the β-amino esters followed by ring-closing metathesis yields the pyrrolines. Dehydrodesulfinylation in the presence of TBAF[23] or *t*-BuOK[24] gives the corresponding pyrrole (eq 19).

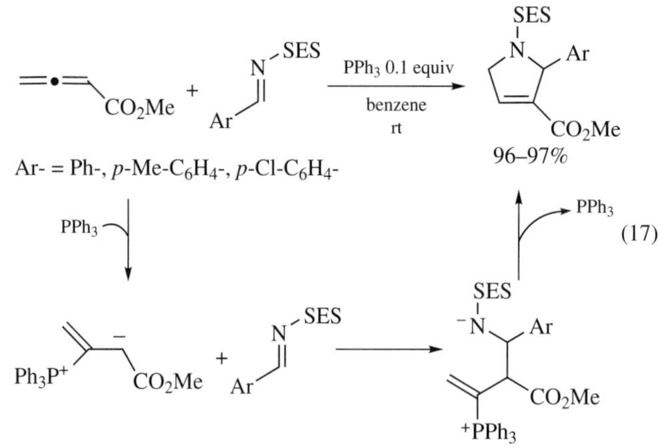

Ar- = Ph-, p-Me-C$_6$H$_4$-, p-Cl-C$_6$H$_4$-

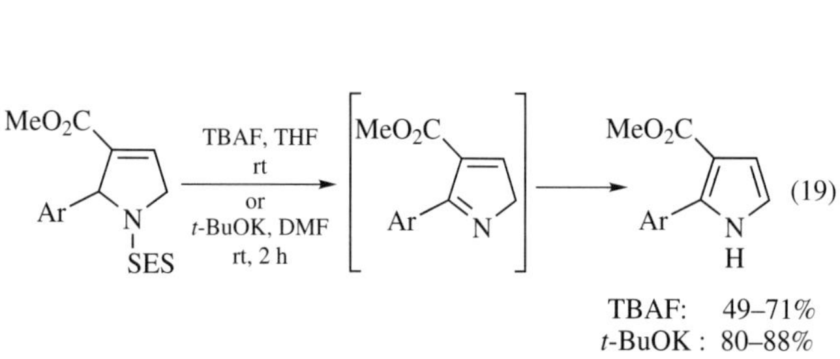

(17)

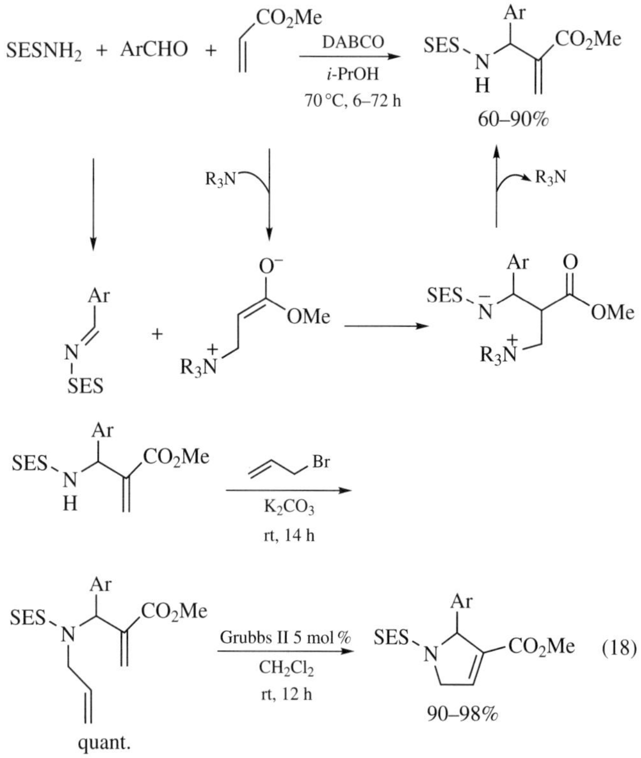

R- = Ph-, p-Cl-C$_6$H$_4$-, 2-furyl-

98–99%
ee = 90–93%

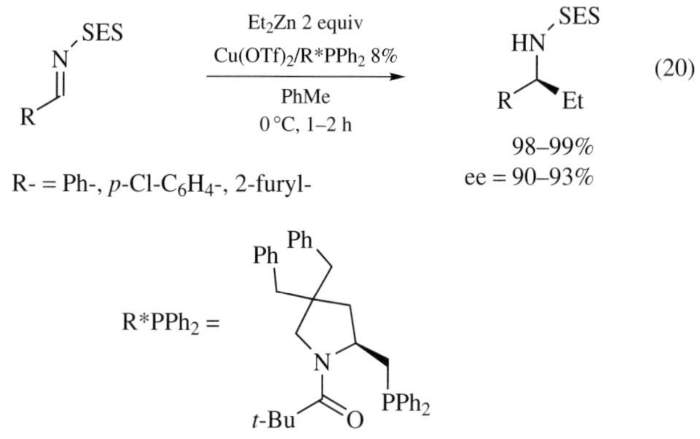

(20)

A catalytic and enantioselective alkylation of N,O-acetals,[26] protected by the SES group, is an interesting method to prepare various chiral α-amino esters from a common precursor, the SES hydroxyglycine (**21**), which is easier to form and more resistant to hydrolysis than the corresponding imine (eq 21). A first equivalent of enol silyl ether reacts with N,O-acetal (**21**) to form the imine intermediate in situ. Eventually, a second equivalent of enol silyl ether is necessary to perform the alkylation of the Cu(I) SES imine complex.

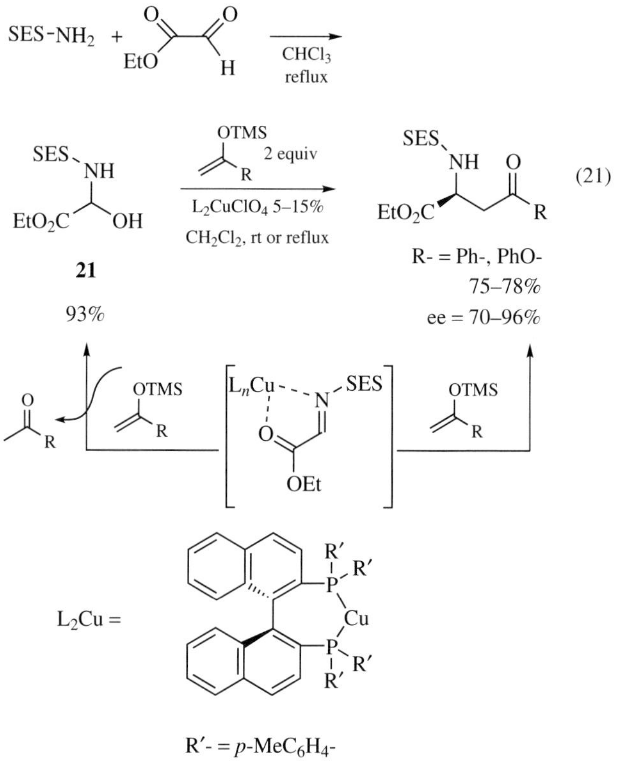

SESNH$_2$ + ArCHO + ... 60–90%

R- = Ph-, PhO-
75–78%
ee = 70–96%

(21)

(18)

90–98%

quant.

TBAF: 49–71%
t-BuOK: 80–88%

(19)

21

93%

L$_2$Cu =

R′- = p-MeC$_6$H$_4$-

The highly electrophilic character of the SES imine imparts good reactivity towards many "soft" organometallic compounds. Organozincs were added to several SES imines in the presence of a catalytic amount of copper(II) triflate and an enantiopure amidophosphine to deliver chiral sulfonamides in very good yields and enantiomeric excesses (eq 20).[25]

A direct access to the synthesis of dihydroquinolines (eq 22) was developed.[27] The vinyl quinone mono- or di-SES-imide, accessible by oxidation of the corresponding aminophenol or phenylenediamine, undergoes a thermal 6π-electrocyclization in the presence of a polar nonprotic additive. Dehydrodesulfinylation of the SES group will lead to the quinoline. The vinyl quinone monoimide substrate can also provide the protected indole by photochemical cyclization.

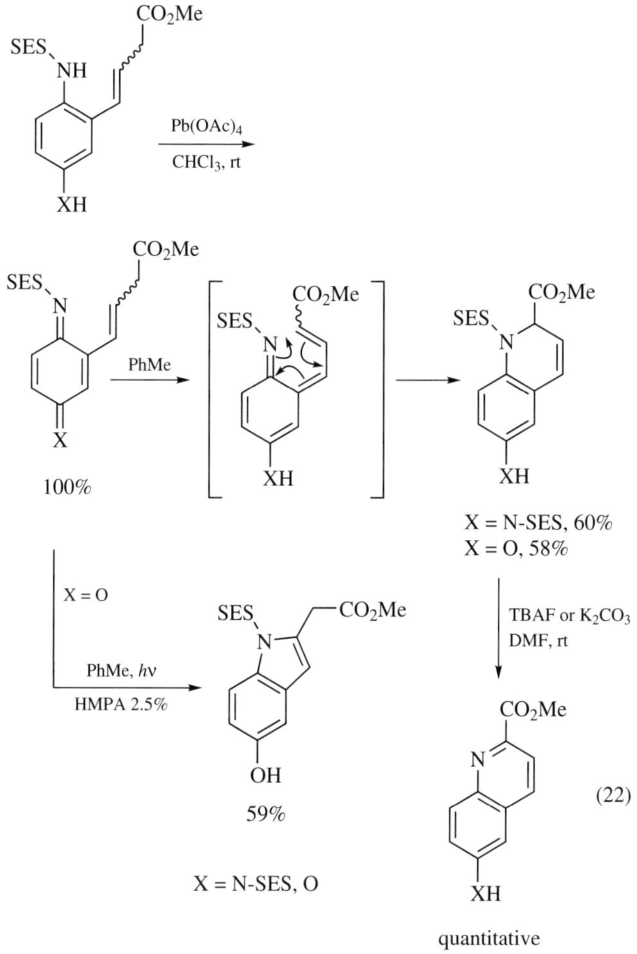

100%

X = N-SES, 60%
X = O, 58%

X = O

PhMe, *hv*
HMPA 2.5%

59%

X = N-SES, O

TBAF or K_2CO_3
DMF, rt

quantitative   (22)

Table 3 Synthesis of chiral aziridines

R^1	R^2	Sulfide	Yield (%)	ee (%)	*cis/trans*	Ref.
Ph	Ph	0.2 equiv	47	95	1/3	22
(*E*)-PhCH=CH	Ph	0.2 equiv	62	93	1/5	22
C_6H_{11}	Ph	1 equiv	72	89	1/1	22

and iodobenzene diacetate in a quantitative yield as a stable water-sensitive solid, which can be stored under argon at $-20\,^{\circ}C$. This product is a nitrene precursor that reacts with olefins in the presence of a copper catalyst to provide the corresponding aziridines. The reaction was performed on various olefins (terminal, electrons-rich, electron-poor, or cyclic) in moderate yields (Table 4).

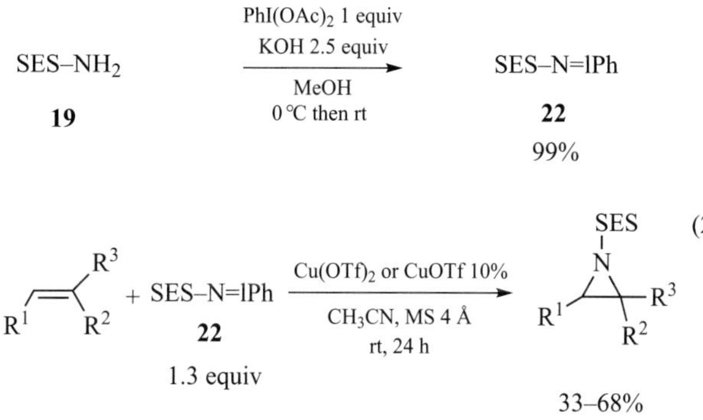

The 2-trimethylsilylethanesulfonylimines can also react with a carbene donor to give chiral aziridines,[22] a class of organic compounds valued as substrates for nucleophilic ring-opening reactions (eq 23 and Table 3). The strategy involves the addition of a catalytic amount of a chiral sulfur ylide to a sulfonylimine. The sulfur ylide is generated by nucleophilic attack of a chiral sulfide on a carbenoid formed from a diazo compound and a metal salt. The use of an electron-withdrawing group on the imine nitrogen, such as the SES group, inactivates the nonstereoselective direct reaction between the metallocarbene and the imine, but also increases the rate of addition of the sulfur ylide to the imine. The metal salt used is usually $Rh_2(OAc)_4$, but $Cu(acac)_2$ can also be employed with a slight decrease in both yield and enantiomeric excess.

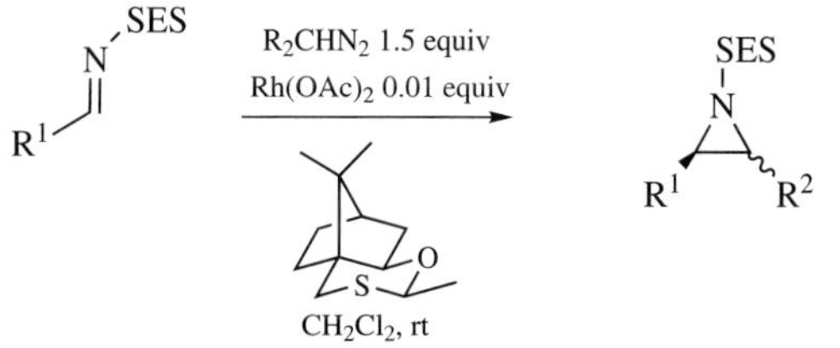

Table 4 Synthesis of SES-protected aziridines

Substrate	Catalyst	Yield (%)
Ph⌒⟍	$Cu(OTf)_2$	58
	CuOTf	68
Ph⌒⌒⟍	$Cu(OTf)_2$	40
Ph⌒⟍CO_2Me	$Cu(OTf)_2$	37
	CuOTf	39
⟍CO_2Me	$Cu(OTf)_2$	49
⟍CO_2Me	$Cu(OTf)_2$	52
	CuOTf	60
⟍CO_2Me	$Cu(OTf)_2$	47
CO_2t-Bu, Ph	$Cu(OTf)_2$	48
⬡	$Cu(OTf)_2$	34
⬡	$Cu(OTf)_2$	55
	CuOTf	67
⌒⌒⟍	$Cu(OTf)_2$	33
	CuOTf	43

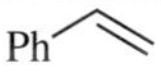

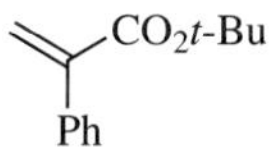

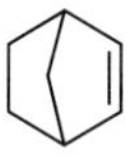

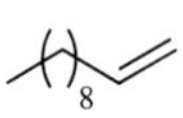

The (*N*-SES-Imino) Phenyliodinane or SESN=IPh Reagent. A new iminoiodinane reagent for the Evans copper-catalyzed aziridination was developed (eq 24).[19] This [*N*-(alkylsulfonyl)-imino]phenyliodinane (**22**) is synthesized from $SES-NH_2$ (**19**)

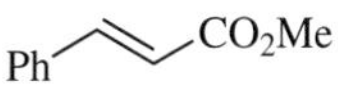

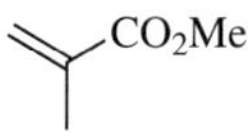

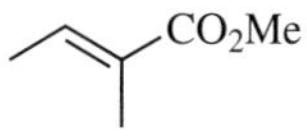

To avoid the difficulty with handling the iminoiodinane (**22**), a one-pot version of this reaction was developed.[28] The compound (**22**) was formed in situ from the easily accessible iodosylbenzene (PhI=O) and $SES-NH_2$ (**19**) (eq 25). Yields are very close to those obtained with iminoiodinane (**22**) (Table 5) and, in spite of the presence of PhI=O, which is a powerful oxygen donor, the rate of epoxidation is very low (lower than 10% in the absence of SES-NH₂). This method greatly simplifies the copper-catalyzed aziridination of olefins.

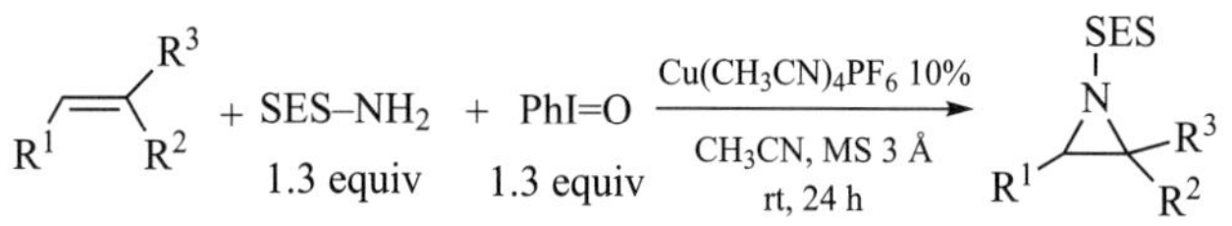

(25)

Table 5 One-pot preparation of aziridines

Substrate	Yield (%)
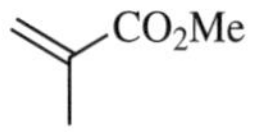	43
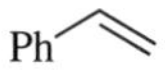	68
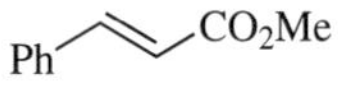	53
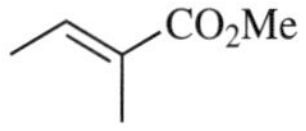	48
	46
	63

Synthesis of chiral SES aziridines was also performed by using a chiral nitridomanganese complex activated by silver salt and SES-Cl (**3**) (eq 26 and Table 6).[20] Like that for the sulfonyl-iminoiodinane process, this reaction proceeds via nitrogen atom transfer.

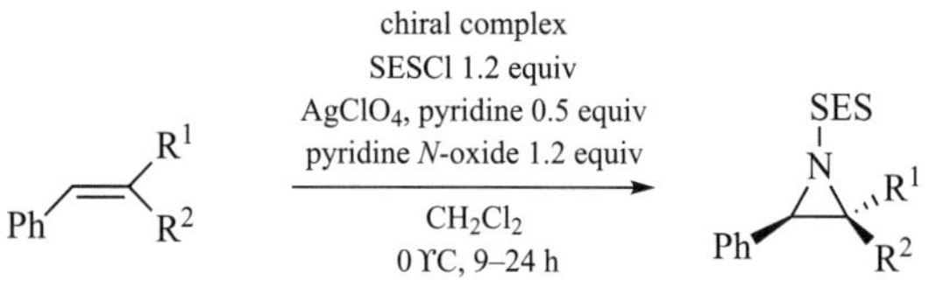

(26)

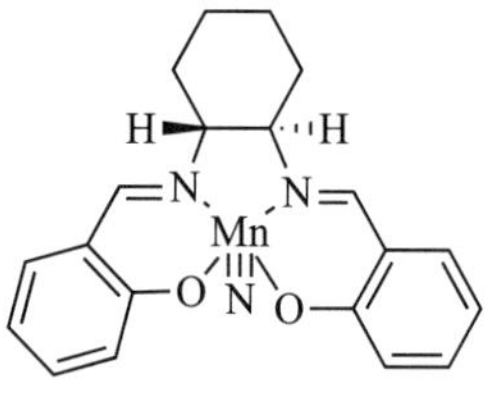

chiral complex

SES-N₃. SES-N₃ is generated in situ from SES-Cl (**3**) and NaN₃. This reagent acts as a nitrene precursor to provide a SES-protected monosulfoximine (eq 27).[29]

Table 6 Synthesis of chiral aziridines

R¹	R²	Yields (%)	ee (%)
H	H	60	40
Me	H	70	83
n-Pr	H	50	85
i-Pr	H	62	90
c-C₆H₁₁	H	62	93
H	Me	37	24

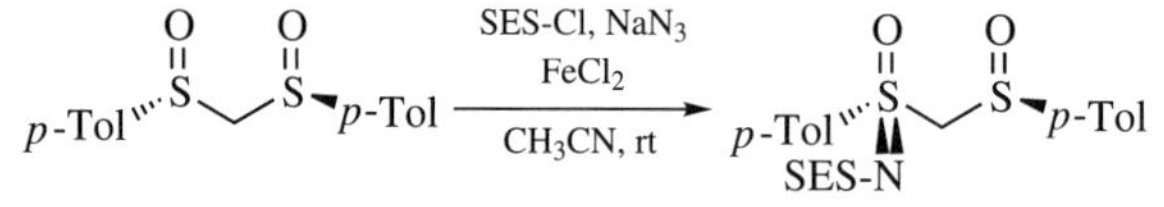

(27)

66%

SES-N₃ was used to promote the *N*-alkylation of an amide precursor for the synthesis of Epoxomicin.[30] The reaction proceeds via a thiatriazoline intermediate formed by a [2 + 3] cycloaddition or by a stepwise diazo transfer-like mechanism. This intermediate then decomposes stepwise or by a *retro*-[2 + 3] reaction to yield the amide (eq 28).

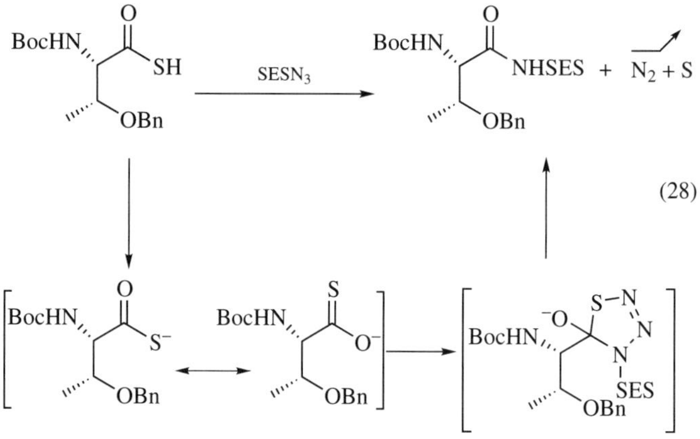

(28)

1. Weinreb, S. M.; Demko, D. M.; Lessen, T. A., *Tetrahedron Lett.* **1986**, *27*, 2099.

2. Huang, J.; Widlanski, T. S., *Tetrahedron Lett.* **1992**, *33*, 2657.

3. For the use of SESCl in the synthesis of *t*-butyl [[2-(trimethylsilyl)ethyl]sulfonyl]carbamate, a useful reagent in Mitsunobu reactions, see *N*-(*t*-butoxycarbonyl)-*p*-toluenesulfonamide; Campbell, J. A.; Hart, D. J., *J. Org. Chem.* **1993**, *58*, 2900.

4. Danishefsky, S. J.; Koseki, K.; Griffith, D. A.; Gervay, J.; Peterson, J. M.; McDonald, F. E.; Oriyama, T., *J. Am. Chem. Soc.* **1992**, *114*, 8331.

5. Miller, A. D.; Leeper, F. J.; Battersby, A. R., *J. Chem. Soc., Perkin Trans. 1* **1989**, 1943.

6. Garigipati, R. S.; Tschaen, D. M.; Weinreb, S. M., *J. Am. Chem. Soc.* **1990**, *112*, 3475.

7. Sisko, J.; Weinreb, S. M., *Tetrahedron Lett.* **1989**, *30*, 3037.

8. Sisko, J.; Weinreb, S. M., *J. Org. Chem.* **1990**, *55*, 393.

9. Nouvet, A.; Binard, M.; Lamaty, F.; Lazaro, R., *Lett. Pept. Sci.* **1999**, *6*, 239.

10. Nouvet, A.; Binard, M.; Lamaty, F.; Martinez, J.; Lazaro, R., *Tetrahedron* **1999**, *55*, 4685.

11. Chambert, S.; Désiré, J.; Décout, J.-L., *Synthesis* **2002**, 2319.

12. Boger, D. L.; Chen, J.-H.; Saionz, K. W., *J. Am. Chem. Soc.* **1996**, *118*, 1629.

13. Jiang, W.; Wanner, J.; Lee, R. J.; Bounaud, P.-Y.; Boger, D. L., *J. Am. Chem. Soc.* **2003**, *125*, 1877.

14. Shin, D.; Rew, Y.; Boger, D. L., *Proc. Natl. Acad. Sci. U. S. A.* **2004**, *101*, 11977.

15. Rew, Y.; Shin, D.; Hwang, I.; Boger, D. L., *J. Am. Chem. Soc.* **2004**, *126*, 1041.

16. Desai, P.; Pfeiffer, S. S.; Boger, D. L., *Org. Lett.* **2003**, *5*, 5047.

17. Ohno, H.; Hamaguchi, H.; Ohata, M.; Kosaka, S.; Tanaka, T., *J. Am. Chem. Soc.* **2004**, *126*, 8744.

18. Boger, D. L.; Ledeboer, M. W.; Kume, M.; Searcey, M.; Jin, Q., *J. Am. Chem. Soc.* **1999**, *121*, 11375.

19. Dauban, P.; Dodd, R. H., *J. Org. Chem.* **1999**, *64*, 5304.

20. Nishimura, M.; Minakata, S.; Takahashi, T.; Oderaotoshi, Y.; Komatsu, M., *J. Org. Chem.* **2002**, *67*, 2101.

21. Pak, J. K.; Hesse, M., *J. Org. Chem.* **1998**, *63*, 8200.

22. Aggarwal, V. K.; Ferrara, M.; O'Brien, C. J.; Thompson, A.; Jones, R. V. H.; Fieldhouse, R., *J. Chem. Soc., Perkin Trans. 1* **2001**, 1635.

23. Xu, Z.; Lu, X., *J. Org. Chem.* **1998**, *63*, 5031.

24. Declerck, V.; Ribière, P.; Martinez, J.; Lamaty, F., *J. Org. Chem.* **2004**, *69*, 8372.

25. Fujihara, H.; Nagai, K.; Tomioka, K., *J. Am. Chem. Soc.* **2000**, *122*, 12055.

26. Ferraris, D.; Dudding, T.; Young, B.; Drury, W. J., III; Lectka, T., *J. Org. Chem.* **1999**, *64*, 2168.

27. Parker, K. A.; Mindt, T. L., *Org. Lett.* **2002**, *4*, 4265.

28. Dauban, P.; Saniere, L.; Tarrade, A.; Dodd, R. H., *J. Am. Chem. Soc.* **2001**, *123*, 7707.

29. Reggelin, M.; Weinberger, H.; Spohr, V., *Adv. Synth. Catal.* **2004**, *346*, 1295.

30. Katukojvala, S.; Barlett, K. N.; Lotesta, S. D.; Williams, L. J., *J. Am. Chem. Soc.* **2004**, *126*, 15348.

[*N*-(2-(Trimethylsilyl)ethanesulfonyl)-imino]phenyliodane

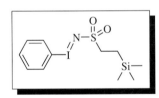

[236122-13-7] $C_{11}H_{18}INO_2SSi$ (MW 383.32)

InChI = 1S/C11H18INO2SSi/c1-17(2,3)10-9-16(14,15)13-12-11-7-5-4-6-8-11/h4-8H,9-10H2,1-3H3

InChIKey = PDQYQRCBDUNXBI-UHFFFAOYSA-N

(iodine(III) reagent used as a nitrogen atom source in the transition-metal catalyzed aziridination of olefins or in the sulfoximination of sulfoxides)[1,2]

Alternate Name: phenyl{[(2-(trimethylsilyl)ethyl)sulfonyl]-amino} iodonium, inner salt.

Physical Data: mp 84–85.5 °C.[3]

Solubility: slightly soluble in CH_2Cl_2, $CHCl_3$ (slow decomposition observed).

Form Supplied in: pale yellow solid. Typical impurities are SesNH₂ and PhI.

Analysis of Reagent Purity: ^{1}H NMR analysis of iminoiodane purity is unreliable since it is unstable and only slightly soluble in CDCl₃. Iodometric titration is possible. Another test of purity could be the use of the reagent in a standard aziridination applied to a five-molar excess of styrene.[4]

Preparative Methods: the protocol is adapted from the original synthesis of I-N ylides.[5] SesNH₂ (prepared by reaction of Ses-Cl[6] with concentrated aqueous ammonia at 0 °C in CH₃CN), KOH pellets (2.5 equiv), and PhI(OAc)₂ (1.0 equiv) are successively added to anhydrous methanol at 0 °C. After 3 h of stirring from 0 °C to rt, an expeditive work-up, described below, is followed. The mixture is diluted at 0 °C with freshly distilled CH_2Cl_2 and washed with ice water. After separation, the organic phase is dried over MgSO₄ and evaporated to dryness at room temperature. More reliable and practical now is the in situ generation of PhINSes from PhIO and SesNH₂.[7] This procedure is strongly recommended since the isolation of iminoiodinanes is particularly troublesome.

Handling, Storage, and Precaution: for optimal results, the reagent should be stored under argon at −20 °C. Irritating to skin. As with any hypervalent iodine reagent, caution is required while heating.

Copper-Catalyzed Aziridination of Olefins. In the presence of a catalytic quantity of copper salts and activated 3 or 4 Å molecular sieves in acetonitrile, the title reagent (PhI=NSes) reacts with a wide range of alkenes, the stoichiometrically limiting component, to give the corresponding aziridines in 35–60% yield (eq 1).[3] Slightly better results are obtained when the iminoiodane is generated in situ from iodosylbenzene and SesNH₂,[7] while copper(I) salts, typically copper(I) trifluoromethanesulfonate and tetrakis(acetonitrile)copper(I) hexafluorophosphate, afford higher yields. The reactivity of PhI=NSes is thus comparable to that of [*N*-(*p*-toluenesulfonyl)imino]phenyliodane (PhI=NTs).[4]

$$R^1\underset{R^2}{\overset{R^3}{=}} \quad \xrightarrow[\substack{\text{Molecular sieves, CH}_3\text{CN} \\ 35\text{--}60\%}]{\substack{1.2\text{--}1.3 \text{ equiv PhI}=\text{NSes} \\ 10 \text{ mol }\% \text{ Cu}^{\text{I or II}}}} \quad R^1\overset{H}{\underset{R^2}{\bigtriangleup}}^{\text{NSes}}_{\text{'''}R^3} \quad (1)$$

1.0 equiv

The resulting *N*-(Ses)aziridines are valuable synthetic intermediates. The electron-withdrawing character of the Ses-group allows ring opening of the aziridines by nucleophiles under mild conditions (eq 2),[3,8] as is the case with activated *N*-(Ns)aziridines prepared from [*N*-(*p*-nitrobenzenesulfonyl)imino]phenyliodane (PhI=NNs)[9] or *N*-(Ts)aziridines.

$$\text{SesN}\overset{}{\underset{}{\bigtriangleup}}\text{—OTr} \quad \xrightarrow[\substack{\text{dioxane, 90 °C} \\ 89\text{--}95\%}]{3 \text{ equiv BnOH, 3 equiv NaH}} \quad \text{BnO}\overset{\text{NHSes}}{\underset{}{\diagup}}\text{OTr} \quad (2)$$

More significantly, the *N*-(Ses)aziridines can be deprotected without concomitant opening of the three-membered ring (eq 3)[3,10] by use of tris(dimethylamino)sulfonium difluorotrimethylsilicate (TASF), a fluoride anion source that is soluble in polar organic solvents at room temperature. Such clean deprotection also occurs with *N*-(Ts)aziridines by using reductive conditions, namely magnesium in methanol[11] or sodium naphthalenide.[12]

The direct copper-catalyzed iodosyl-mediated nitrogen transfer to olefins compares with the parent rhodium-catalyzed process that is made possible by the combination of iodosylbenzene diacetate, magnesium oxide, and sulfamates.[13] Other recent promising nitrene transfer methods involve the bromine-catalyzed aziridination of olefins using chloramine-T[14] and the direct electrochemical aziridination with *N*-aminophthalimide.[15]

Sulfoximination of Sulfoxides. Copper(I)- or copper(II)-catalyzed sulfoximination of sulfoxides[16,17] with PhI=NSes[18] leads to the corresponding *N*-(Ses)sulfoximines (eq 4) which could be cleanly deprotected to the free sulfoximines by use of tetrabutylammonium fluoride.

Related Reagents. [*N*-(*p*-Toluenesulfonyl)imino]phenyl iodane; [*N*-(*p*-nitrobenzenesulfonyl)imino]phenyliodane.

---

1. Müller, P., In *Advances in Catalytic Processes*; Doyle, M. P., ed.; JAI Press: Greenwich, CT, 1997, p 113.

2. Zhdankin, V. V.; Stang, P. J., *Chem Rev.* **2002**, *102*, 2523.

3. Dauban, P.; Dodd, R. H., *J. Org. Chem.* **1999**, *64*, 5304.

4. Evans, D. A.; Faul, M. M.; Bilodeau, M. T., *J. Am. Chem. Soc.* **1994**, *116*, 2742.

5. Yamada, Y.; Yamamoto, T.; Okawara, M., *Chem. Lett.* **1975**, 361.

6. Weinreb, S. M.; Chase, C. E.; Wipf, P.; Venkatraman, S., *Org. Synth.* **1998**, *75*, 161.

7. Dauban, P.; Sanière, L.; Tarrade, A.; Dodd, R. H., *J. Am. Chem. Soc.* **2001**, *123*, 7707.

8. Wipf, P.; Venkatraman, S.; Miller, C. P., *Tetrahedron Lett.* **1995**, *36*, 3639.

9. Södergren, M. J.; Alonso, D. A.; Bedekar, A. V.; Andersson, P. G., *Tetrahedron Lett.* **1997**, *38*, 6897.

10. Di Chenna, P. H.; Dauban, P.; Ghini, A.; Baggio, R.; Garland, M. T.; Burton, G.; Dodd, R. H., *Tetrahedron* **2003**, *59*, 1009.

11. Alonso, D. A.; Andersson, P. G., *J. Org. Chem.* **1998**, *63*, 9455.

12. Bergmeier, S. C.; Seth, P. P., *Tetrahedron Lett.* **1999**, *40*, 6181.

13. Guthikonda, K.; Du Bois, J., *J. Am. Chem. Soc.* **2002**, *124*, 13672.

14. Jeong, J. U.; Tao, B.; Sagasser, I.; Henniges, H.; Sharpless, K. B., *J. Am. Chem. Soc.* **1998**, *120*, 6844.

15. Siu, T.; Yudin, A. K., *J. Am. Chem. Soc.* **2002**, *124*, 530.

16. Bolm, C.; Muniz, K.; Aguilar, N.; Kesselgruber, M.; Raabe, G., *Synthesis* **1999**, 1251.

17. Lacôte, E.; Amatore, M.; Fensterbank, L.; Malacria, M., *Synlett* **2002**, 116.

18. Cren, S.; Kinahan, T. C.; Skinner, C. L.; Tye, H., *Tetrahedron Lett.* **2002**, *43*, 2749.

Philippe Dauban & Robert H. Dodd
Institut de Chimie des Substances Naturelles, CNRS, France

2-(Trimethylsilyl)ethanethiol[1]

[18143-30-1] $C_5H_{14}SSi$ (MW 134.32)
InChI = 1S/C5H14SSi/c1-7(2,3)5-4-6/h6H,4-5H2,1-3H3
InChIKey = BCOLNMGFOWHFNI-UHFFFAOYSA-N

(reagent used as a monoprotected sulfide dianion equivalent for the two-step installation of a divalent sulfur atom into organic molecules)

Physical Data: bp 144–146 °C [52–54 °C (25 Torr)]; d 0.850.
Solubility: soluble in MeOH, CH$_2$Cl$_2$, THF, and most organic solvents.
Form Supplied in: liquid from major research chemical suppliers at a cost of ca. US$60/g (7.4 mmol) of 97% pure material.
Analysis of Reagent Purity: GC.
Preparative Methods: the high commercial cost of 2-(trimethylsilyl)ethanethiol necessitates methods for its lab scale synthesis. The radical addition of thiolacetic acid to vinyltrimethylsilane occurs thermally to afford a 9:1 ratio of regioisomers, the 1- and 2-(trimethylsilyl)ethyl thiolesters of acetic acid (eq 1).[2,3] Photochemical initiation may improve the ratio of regioisomers obtained.[4] Purification of the thiolacetate by careful fractional distillation should be performed prior to conversion to the thiol in order to ensure high purity. The conversion to thiol is best performed with LAH[3] or methanolic K$_2$CO$_3$,[2] but ammonolysis may also be suitable.[5] Saponification with KOH in water/ethanol is slightly less efficient.[4] Direct radical addition of liquid H$_2$S to vinyl trimethylsilane has been demonstrated, but the procedure is low-yielding and inconvenient.[4]

Purity: distillation.
Handling, Storage, and Precaution: the thiol possesses a strong unpleasant aroma that combines the characteristic olfactory features of thiol and low molecular weight silane. As a thiol, the reagent is incompatible with oxygen, base, and common oxidants. Distillation should be carried out with care since

uncontrollable frothing may occur under reduced pressure, even with vigorous stirring.

Thiol Substitution Reactions. Vital to the usefulness of 2-(trimethylsilyl)ethanethiol is that it behaves as a typical thiol. Base mediated reactions that proceed as expected include alkylation[6-9] and nucleophilic ring openings.[3,8,10,11] Moreover, addition-eliminations on unsaturated systems,[12,13] Michael additions,[3] and *cis*-additions to triple bonds (eq 2) are all routine.[14,15] To access 2-(trimethylsilyl)ethyl thioethers, a useful procedure is hydrolysis of 2-(trimethylsilyl)ethyl thioacetate and alkylation of the thiolate in a single pot.[16,17]

2-(Trimethylsilyl)ethanethiol will also participate in aromatic[18] and heteroaromatic displacements[10] of halogen atoms. A recent example shows that 2-(trimethylsilyl)ethanethiol is an alternative for the commonly employed potassium ethyl xanthate[19,20] as a latent thiol in nitrogen displacement chemistry of aromatic diazonium ions (eq 3). The low yield occurs in the substitution step, and is believed to be due to the sensitive doubly brominated substrate.[21]

Reactions at acid centers to create S-2-(trimethylsilyl) ethanethiolesters are also common. Carboxylic S-thiolesters are formed in high yield by the DCC/DMAP mediated coupling of 2-(trimethylsilyl)ethanethiol and carboxylic acids[3] or by thiol substitution on a carboxylic acid chloride.[21] S-2-(Trimethylsilyl) ethyl p-toluenethiolsulfonate is formed by treating 2-(trimethylsilyl)ethanethiol with tosyl bromide (eq 4).[22] The product is a useful electrophile for carbon nucleophiles,[3,23,24] allowing the introduction of the 2-(trimethylsilyl)ethylthio unit by an alternative mechanism. Independent of the thiol, aryl and alkyl 2-trimethylsilylethyl thioethers may be prepared by the radical addition of the appropriate arene- or alkanethiol to vinyl trimethylsilane in a reaction comparable to that of eq 1.[7,16,17,25]

The true value of the products of such 2-(trimethylsilyl)-ethylthio installation reactions arises by way of manipulation and usually removal of the 2-(trimethylsilyl)ethyl group. Such reactions, which appear to be independent of the means of introduction

of the 2-(trimethylsilyl)ethylthio group, are the focus of the ensuing sections in this article.

Introduction of a Divalent Sulfur Equivalent. One function where 2-(trimethylsilyl)ethanethiol exhibits its value is as a mononucleophilic sulfide equivalent where the sulfur acts as a nucleophile only once. Hence the reagent displays the nucleophilic strength of a typical alkanethiol to afford a stable product. Liberation of the 2-(trimethylsilyl)ethyl group with alkali metal fluorides and 18-crown-6 or tetralkylammonium fluorides converts the sulfide moiety to a thiolate (eq 5).[13] Innocuous trimethylsilyl fluoride and ethylene are believed to be the by-products of fluoride treatment.

In order for the 2-(trimethylsilyl)ethylthio unit to lose its protecting group upon treatment with the fluoride, the sulfur must be attached to an unsaturated carbon.[3,10,12,13,15,18,21,23,25] Hence the fluoride mediated sulfur deprotection is feasible for 2-(trimethylsilyl)ethylthio substituted (het)arenes,[18,21,25] alkenes,[13,15] alkynes,[23] and acid derivatives such as carboxylic[3,21] and selenothiophosphinic acid salts.[12] In thiolate form, the substrates have value for the formation of self-assembled monolayers[23] or as metal complexing agents.[12] Simple addition of acid to the thiolate to give a stable thiol characterizes 2-(trimethylsilyl)ethanethiol as a simple $M^+(HS)^-$ equivalent that is only capable of a single substitution reaction.[18]

Alkylation of the thiolate consequently defines the reagent as a doubly nucleophilic sulfur, taking the form of M_2S (eq 6).[23,25] However, until deprotection is instigated, one valency of the thiol remains protected with the 2-(trimethylsilyl)ethyl substituent, a group tolerant of numerous other chemical transformations.[25]

The double addition of two 2-(trimethylsilyl)ethanethiols and hence placement of two 2-(trimethylsilyl)ethylthio groups close to one another in a molecule offers a protocol for use of 2-(trimethylsilyl)ethanethiol as a $M^{+-}S-S^- M^+$ equivalent. This chemistry was developed while pursuing the formation of the central 1,2-dithiin ring system of the thiarubrine natural products. The general premise requires the double *cis*-addition of the thiol to the opposite ends of a conjugated diyne, a process that brings the sulfur atoms to the requisite proximity (eq 2). Fluoride induced deprotection offers two thiolates that can be connected using either aq $K_3Fe(CN)_6$, I_2 or I_2/KI as oxidants. Although yields tend to be low to fair, the protocol has led to the formation of the 1,2-dithiin ring of the antibiotics Thiarubrine A (eq 7)[14] and Thiarubrine C.[15]

Thiarubrine A

Overall ring formation can be achieved using other thiol protecting reagents. The use of benzyl thiol requires reducing metal conditions for sulfur dealkylation,[26,27] conditions that may not always be inert to the conjugated acetylene or other substituents that are harbored on the 3- and 6-positions of some thiarubrines.[14] *t*-Butanethiol has also proved viable for the preparation of the 1,2-dithiin skeleton. Cleavage of the *t*-Bu groups and disulfide bond formation require a mild oxidant such as NBS[28] or involve a less inviting three-step protocol beginning with sulfenyl chloride exposure.[27,29] Base induced liberation of acetyl[30,31] and 2-cyanoethyl[32] sulfur protecting groups has also been utilized for the formation of the 1,2-dithiin ring. The latter is reportedly useful for larger scale preparations, particularly when the 3,6-groups have been appropriately adapted.

Of all the options, the use of 2-(trimethylsilyl)ethanethiol is possibly the most expensive. Nevertheless its versatility and enduring popularity as a $M^{+-}S-S^- M^+$ equivalent is demonstrated by the preparation of a 1,2-benzodithiolanone for studies modeling the bioactivity of leinamycin (eq 8).[21] Moreover, the use of fluoride as a deprotecting agent is unique and often innocuous to other functionalities. Divorced of 1,2-dithiin or -dithiolanone chemistry, an isolated example of direct conversion of an alkyl 2-(trimethylsilyl)ethyl sulfide to the dialkyl disulfide using BrCN (70%) has been reported.[8]

When the 2-(trimethylsilyl)ethyl component is installed on an alkyl substrate, simple fluoride treatment fails to effect trimethylsilylethyl removal.[3] To employ a 2-(trimethylsilyl)ethyl alkyl thioether as a protected alkylthiol, the sulfur must first be functionalized with an electrophilic agent and once achieved, 2-(trimethylsilyl)ethyl loss from the intermediate sulfonium ion is facilitated.

Conversion of 2-(trimethylsilyl)ethyl alkyl thioethers to alkanethiols was first realized by treatment of the thioethers with

$MeSS^+Me_2BF_4^-$ and MeSSMe to give a methyl alkyl disulfide, which can be transformed into the thiol using Bu_3P in $MeOH/H_2O$ (eq 9).[3]

Direct alkylation has been achieved in special circumstances. Thus 2-(trimethylsilyl)ethyl sulfides, obtainable from alkylation of 2-(trimethylsilyl)ethanethiol or by radical addition of thiols to vinyl trimethylsilane, react with α-glycosyl bromide with the assistance of Ag(I) activation to offer a general synthesis of β-thioglycosides. Yields generally vary (34–99%), and the β:α ratio can be quite high (eq 10).[7]

β (shown):α = 99:1

Methods for the electrophilic modification of the 2-(trimethylsilyl)ethyl group to other groups are known. For example, 2-(trimethylsilyl)ethyl sulfides are a proven source of thiolesters: the simple treatment of the sulfide with a carboxyl chloride and a Ag(I) salt provides the thiolester, often in high yield.[33] Formation of SCN groups from SCH_2CH_2TMS units occurs through the treatment of the latter with BrCN. The protocol positions the thiocyanate group into a variety of substrates including nucleosides (eq 11).[8]

Source of Sulfur Acid Derivatives. Removal of the trimethylsilylethyl group can also occur when the sulfur is in a higher oxidation state. Such a process will liberate a sulfur acid derivative and following this theme, two major areas of focus exist in the literature.

Alkenyl 2-(trimethylsilyl)ethyl sulfones, which are readily prepared from 2-(trimethylsilyl)ethanethiol by way of Wittig chemistry and oxidation, provide a source of 1-alkenesulfinate salts.[6]

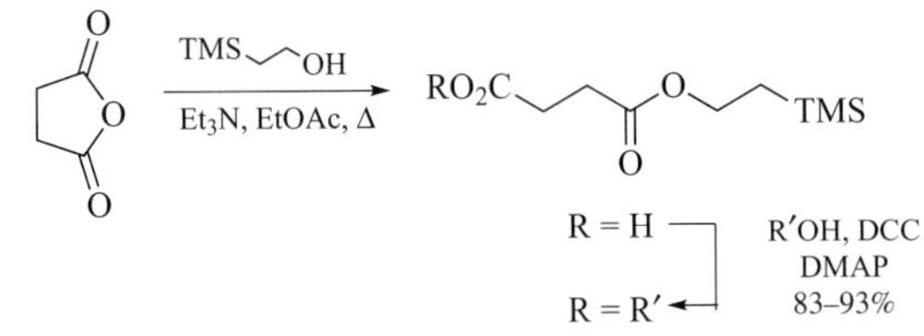

Transesterification of methyl esters to 2-(trimethylsilyl)ethyl esters under mild and neutral conditions takes place in the presence of titanium tetraisopropoxide (eqs 6–8).[11] Deprotection of the 2-(trimethylsilyl)ethyl ester in the presence of an O-TBDMS protected secondary hydroxyl group has been achieved (eq 9).[11b] An alternative method for the transesterification uses 1,8-diazabicyclo[5.4.0]undec-7-ene/lithium bromide and 2-(trimethylsilyl) ethanol.[12]

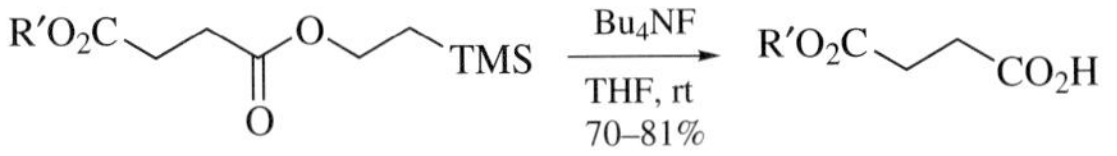

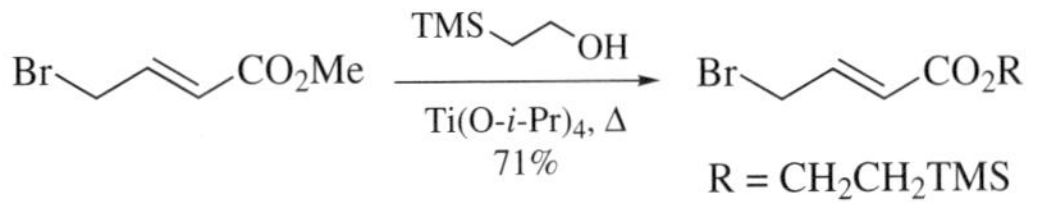

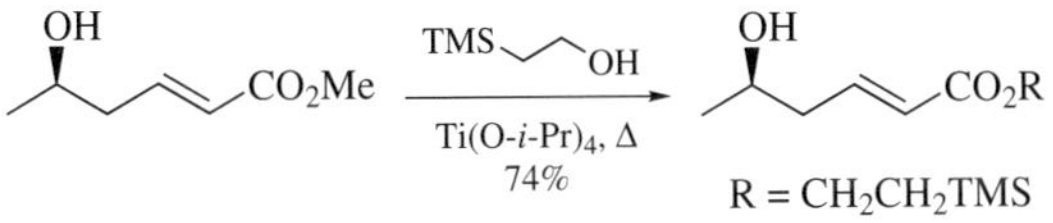

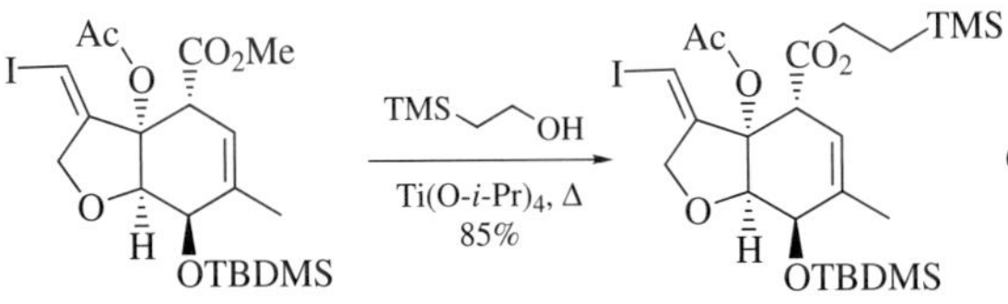

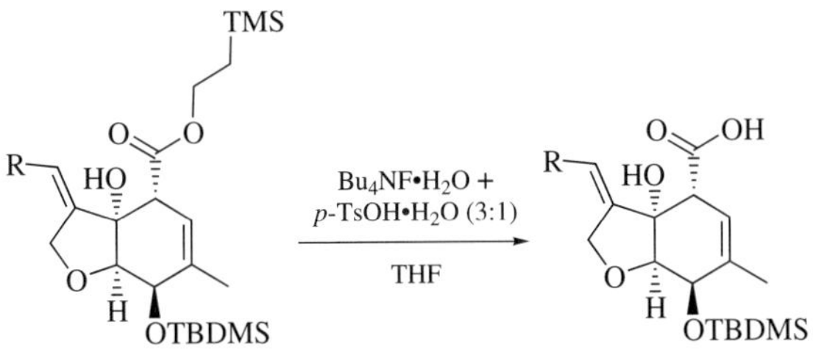

Methods for the protection of pyranosides and furanosides as 2-(trimethylsilyl)ethyl glycosides (eq 12) and deprotection using dry lithium tetrafluoroborate in MeCN have been developed (eq 13).[14]

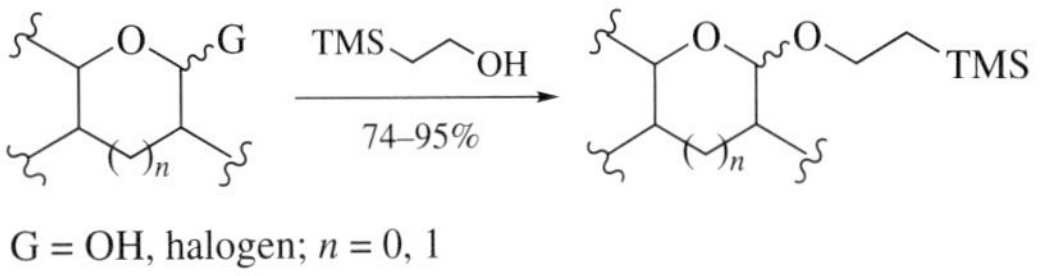

2-(Trimethylsilyl)ethanol has been used as a protecting group in phosphate monoester synthesis and involves the use of 2-(trimethylsilyl)ethyl dichlorophosphite (eq 14).[15] Bis[2-(trimethylsilyl)ethyl] N,N-diisopropylphosphoramidite has been prepared from dichloro(diisopropylamino)phosphine and 2-(trimethylsilyl) ethanol and used as a phosphitylating agent in the synthesis of phosphotyrosine containing peptides (eq 15).[16]

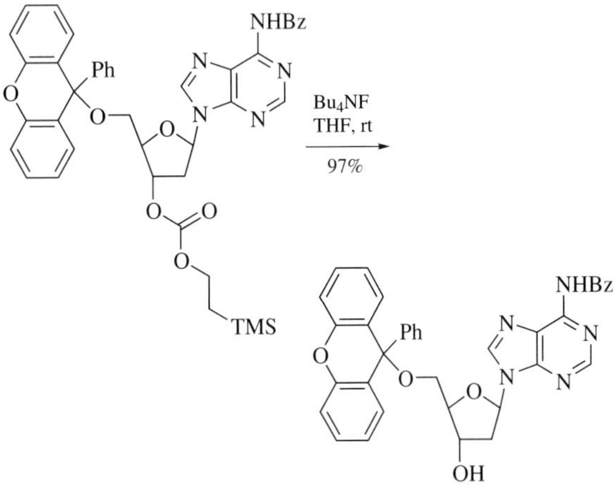

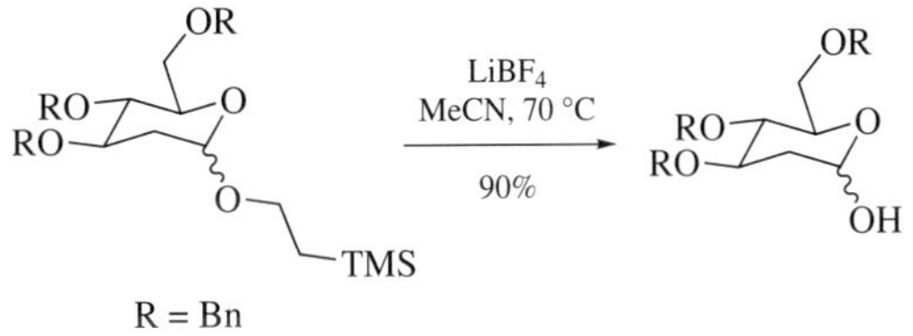

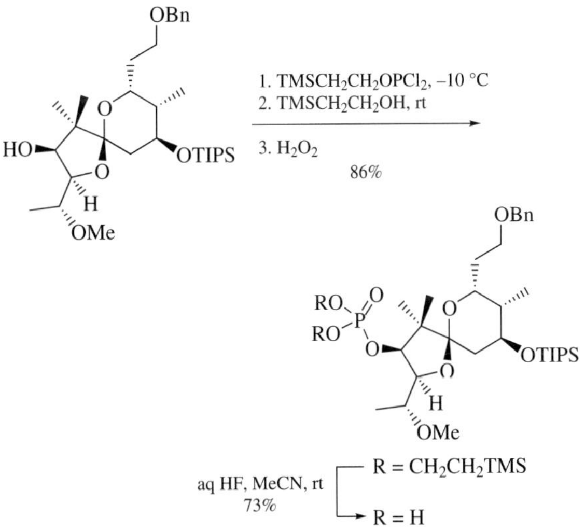

Protection of hydroxyl groups as 2-(trimethylsilyl)ethyl ethers[13a] and 2-(trimethylsilyl)ethyl carbonates[13b] (eq 10) has been utilized as illustrated below. The 2-(trimethylsilyl)ethyl carbonate group can be cleaved under mild conditions using TBAF in dry THF (eq 11).[13b]

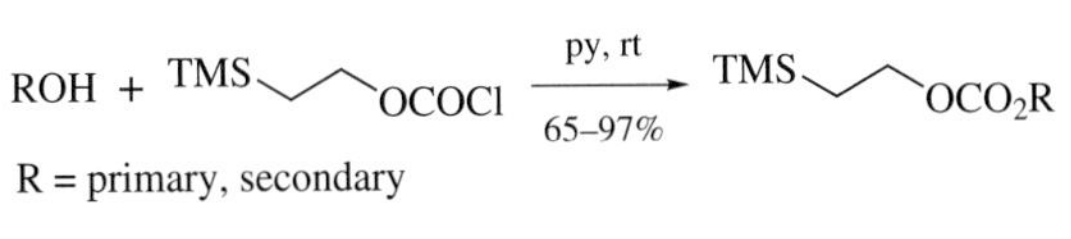

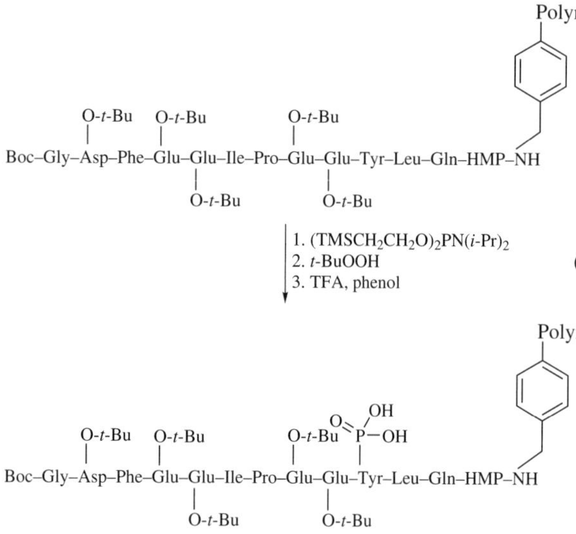

$$(15)$$

2-(Trimethylsilyl)ethoxycarbonyl (Teoc) groups have been used to protect amine functionalities (eq 16).[17] Using a mixture of tetra-n-butylammonium chloride and KF·2H$_2$O deprotects the Teoc group.[18] N-Debenzylation and concurrent protection as N-Teoc results when tertiary N-benzylamines are treated with 2-(trimethylsilyl)ethyl chloroformate.[19]

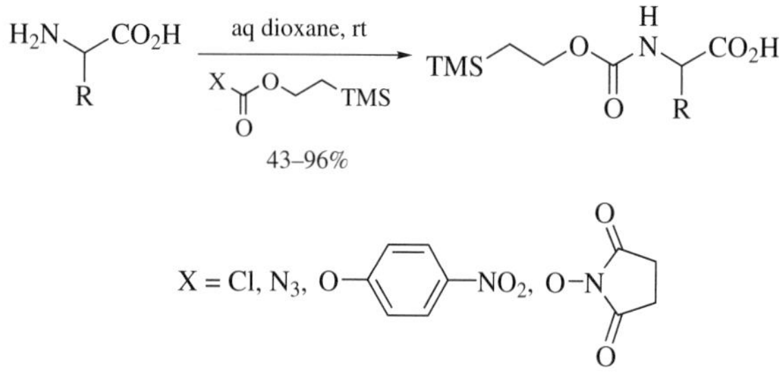

$$(16)$$

A new method for the synthesis of imidazolones involves the replacement of the C-2 nitro group of N-protected dinitroimidazoles by nucleophilic addition of the sodium salt of 2-(trimethylsilyl)ethanol (eq 17).[20]

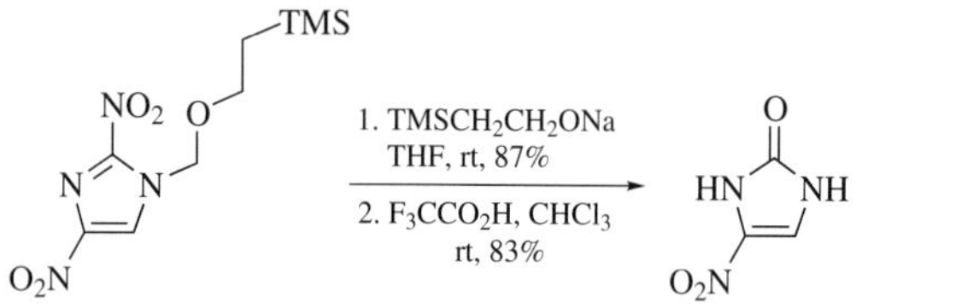

$$(17)$$

Reaction of 2-(trimethylsilyl)ethyl benzenesulfenate with halides in the presence of TBAF yields phenyl sulfoxides (eq 18).[21] 2-(Trimethylsilyl)ethyl benzenesulfenate is prepared by the reaction of benzenesulfenyl chloride and the lithium salt of 2-(trimethylsilyl)ethanol.

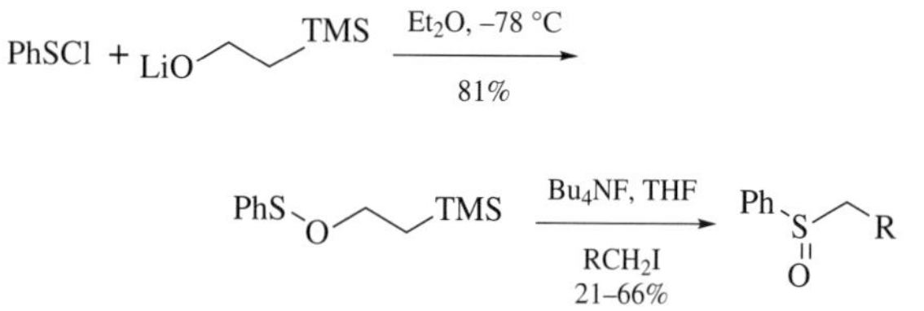

$$(18)$$

First Update

Christiane Yoakim

Boehringer Ingelheim (Canada) Ltd., Laval, Quebec, Canada

Phenol and Acid Protection. The synthesis of trimethylsilylethyl ethers and esters is readily achieved under a number of conditions. Their ease of preparation and stability under a wide variety of reaction conditions, combined with the fact that deprotection can be achieved under almost neutral conditions, have enlarged the scope of this protecting group. Protection of phenols and acids is easily achieved under Mitsunobu conditions as shown in eqs 19 and 20. Upon completion of the synthesis the phenol (eq 19)[22] was deprotected using cesium fluoride in DMF at 160 °C in 89% yield. As well, en route to the synthesis of Daurichromeric acid, the trimethylsilylethyl ester (eq 20)[23] was cleaved with TBAF in THF with a yield of 94%.

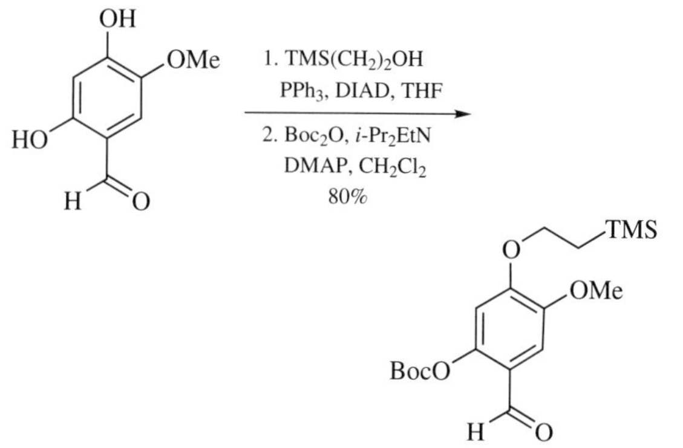

$$(19)$$

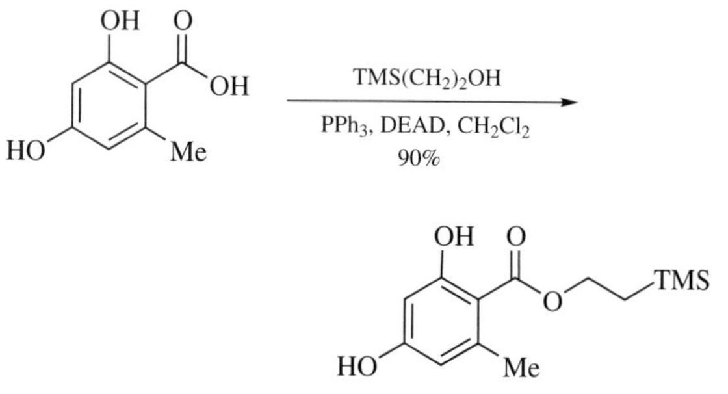

$$(20)$$

The trimethylsilylethyl ester functionality has been utilized in the design of a new phosphine reagent for the Mitsunobu reaction. Treatment of 4-diphenylphosphanyl-benzoic acid with 2-(trimethylsilyl)ethanol under dehydrating conditions gave the corresponding ester in excellent yield (eq 21).[24] It was demonstrated that 4-diphenylphosphanyl-benzoic acid 2-trimethylsilanyl-ethyl ester (DPPBE) is an efficient reagent for the preparation of esters, ethers and phthalimides. The corresponding phosphine oxide is easily removed after cleavage of the ester by washing with aqueous base. This practical procedure was efficiently utilized for the synthesis of complex molecules such as macrocyclic NS3 protease inhibitors of the hepatitis C virus.[25] In addition; DPPBE was used as alternative to triphenylphosphine in the rhodium-catalyzed methylenation of aldehydes using trimethylsilyldiazomethane.[26]

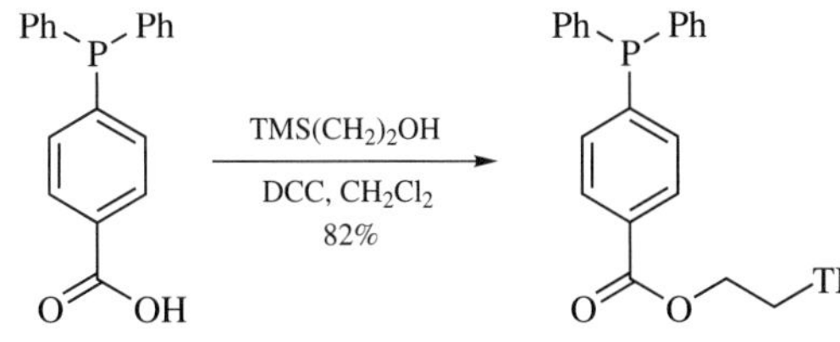

(21)

DPPBE

Aromatic fluorides can be converted into trimethyl (2-phenoxyethyl)silanes. In particular, aromatic fluorides bearing moderate to strong electron-withdrawing groups in the *ortho-* and *para*-position gave modest to good yields (eq 22).[27]

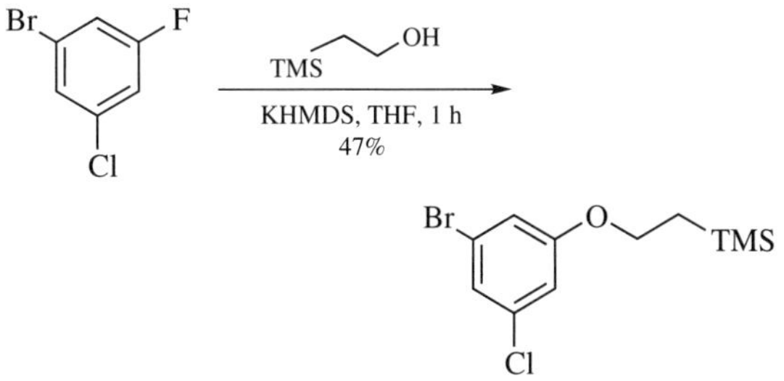

(22)

Alcohol Protection. The choice of alcohol protection is crucial in the total synthesis of natural products. Protection of propargyl alcohol was achieved via alkylation of 2-(trimethylsilyl)ethanol with propargyl bromide in presence of sodium hydride in THF at 0 °C to room temperature (eq 23).[28] This protected alcohol is a useful 3-carbon unit synthon.

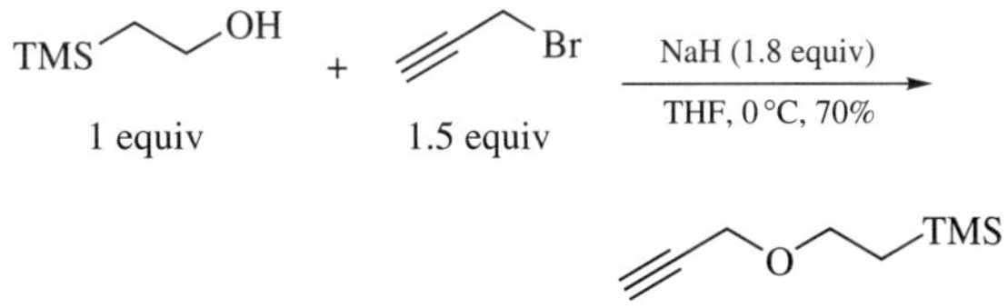

(23)

2-(Trimethylsilyl)ethyl vinyl ether was developed as a reagent for protecting alcohols under non-basic conditions. The reagent was prepared by treatment of a solution 2-(trimethylsilyl)ethanol in ethyl vinyl ether with a catalytic amount of mercuric trifluoroacetate and triethylamine (eq 24).[29]

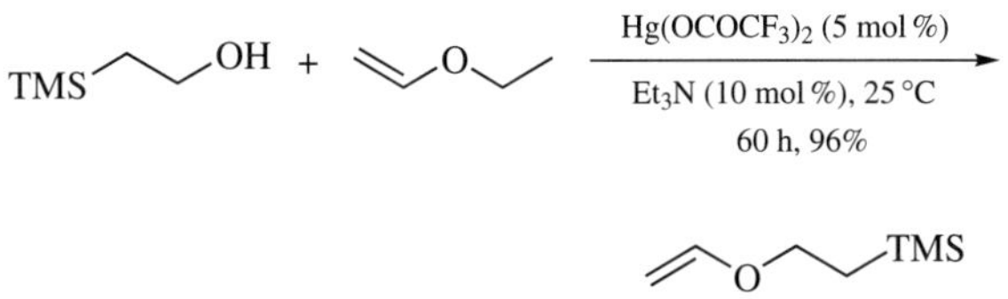

(24)

As exemplified in eq 25, protection of the secondary alcohol was carried out under mild acidic conditions such as pyridinium *p*-toluenesulfonate in methylene chloride. Removal of the 1-[(2-trimethylsilyl)ethoxy]ethyl group can be achieved using a fluoride source such as TBAF.[29]

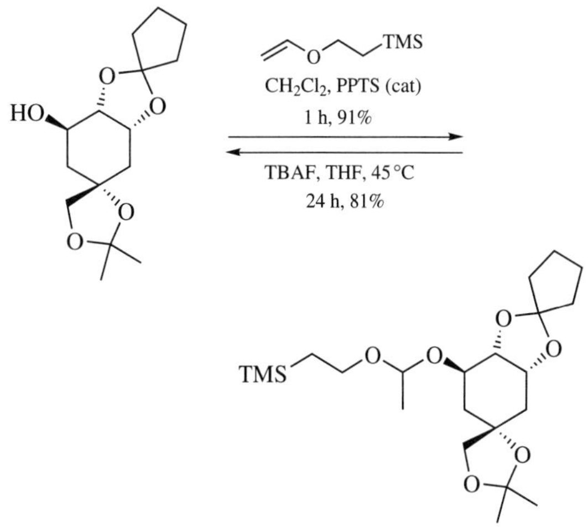

(25)

Hemiacetal Protection. 2-(Trimethylsilyl)ethanol is also used to protect hemiacetals as shown in eq 26. The hemiacetal was submitted to azeotropic distillation of water in the presence of a catalytic amount of acid. Although the reaction time is rather long the yield based on recovered starting material is good. After completion of the synthesis deprotection was accomplished with trifluoroacetic acid in dichloromethane.[30]

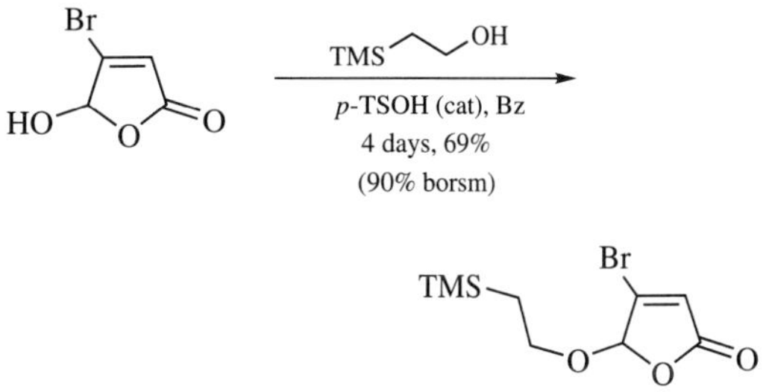

(26)

Protection of hemiacetals could also be achieved by addition of a drying agent such as magnesium sulfate (eq 27) to give an excellent yield over the two steps in a 3.5:1 mixture.[31]

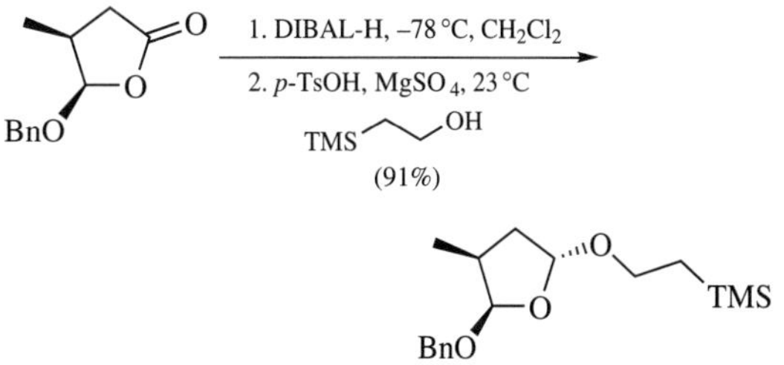

(27)

Amine Protection. Carbamates are frequently used in organic synthesis for the protection of amines. Recently a number of new procedures have been developed to introduce the 2-trimethylsilylethoxycarbonyl group (Teoc). Conversion of mesitylamine into mesityl isocyanate by reaction with di-*t*-butyldicarbonate in presence of DMAP, followed by addition of 2-(trimethylsilyl)ethanol, gave the corresponding *N*-mesitylcarbamate in good yield along with 2% of the *t*-butyl *N*-mesitylcarbamate (eq 28).[32]

Reaction of secondary amines with carbonyldiimidazole followed by activation with methyl iodide provided the carbamoyl imidazolium salts. Nucleophilic displacement using the sodium salt of 2-(trimethylsilyl)ethanol provided the corresponding carbamate in moderate yield (eq 29).[33]

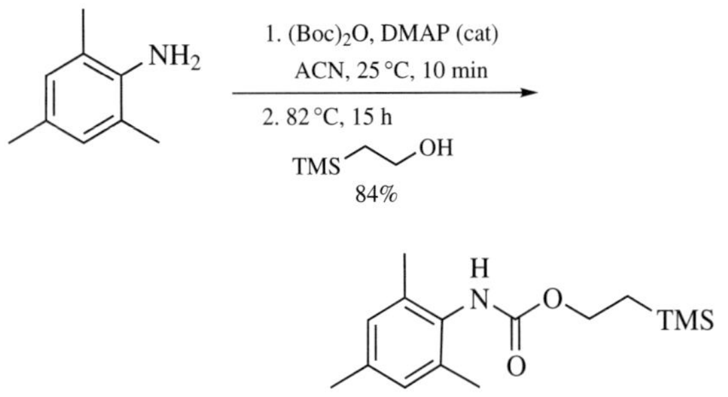

(28)

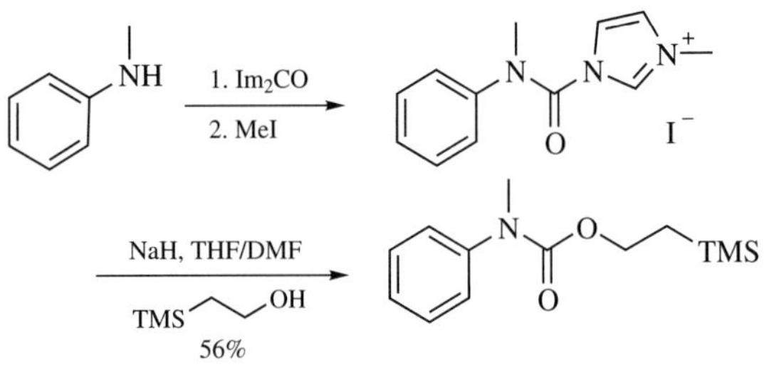

(29)

Trans-protection of N-benzyl piperidine is also possible using in situ generation of 2-(trimethylsilyl)ethyl chloroformate in the presence of a large excess of potassium carbonate. Upon completion of the synthetic sequence, deprotection was carried out using a fluoride source such as cesium fluoride in DMF at 90 °C (eq 30).[34]

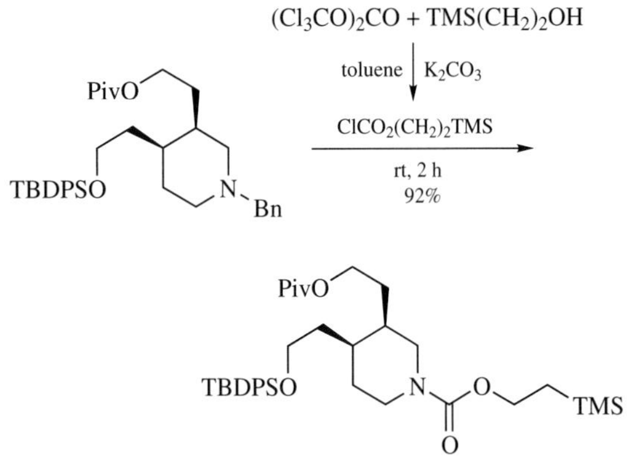

(30)

Cyclopropanes containing alpha amino acids are important intermediates in medicinal chemistry. The racemic "syn" ethyl-substituted cyclopropane amino ester was prepared via Curtius rearrangement to generate the Teoc derivative followed by fluoride-induced deprotection (eq 31).[35]

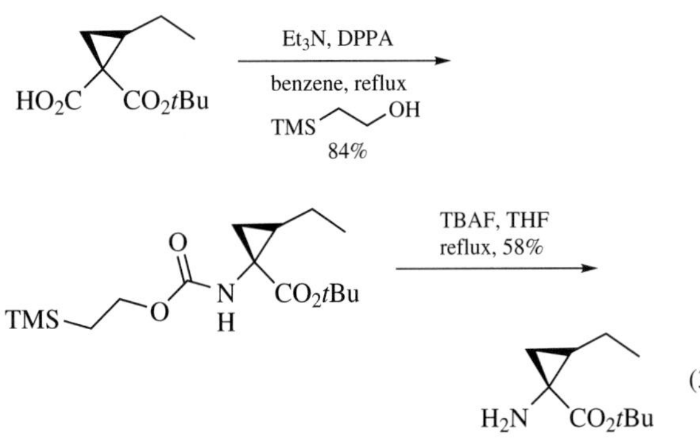

(31)

Enol Ether Synthesis. A 2-(trimethylsilyl)ethyl-based dithioorthoformate was prepared via copper(II) bromide-promoted oxidative coupling of bis(phenylthio)-methyltributylstannane. The dithioorthoformate thus prepared is treated with a titanocene(II) reagent followed by the addition of ketones or esters to promote alkoxymethylidenation as exemplified in eq 32.[36]

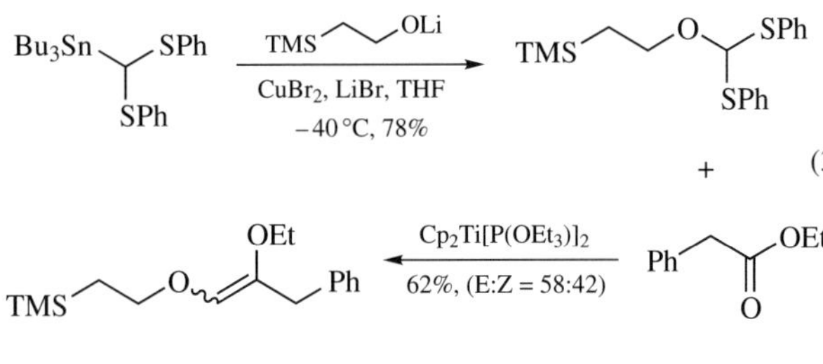

(32)

Carbohydrate Chemistry. The 2-(trimethylsilyl)ethyl group plays an important role in the protection of the anomeric center of carbohydrates. 2-Acetamido-β-D-glucopyranosides are converted to the corresponding 2-(trimethylsilyl)ethyl β-glycosides in the presence of a catalytic amount of camphorsulfonic acid (CSA) with azeotropic removal of acetic acid in excellent yield (eq 33).[37]

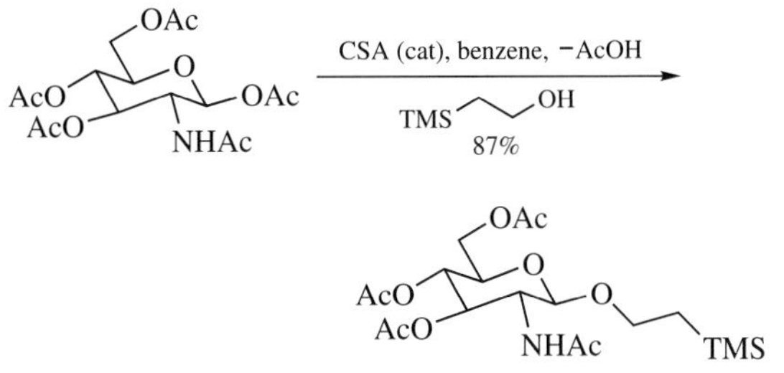

(33)

2-Nitro thioglycosides are activated with NIS/TMSOTf in the presence of the alcohol to give the desired glycoside in excellent selectivity and yields (eq 34).[38] A similar protocol was developed for the α-D-mannopyranosides using a silver trifluoromethane-sulfonate (AgOTf) mediated Koenigs–Knorr reaction to give the α-2-(trimethylsilyl)ethyl mannopyranoside (eq 35).[39]

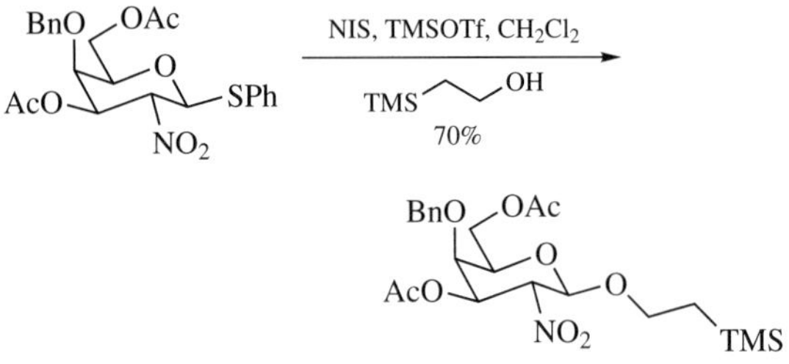

(34)

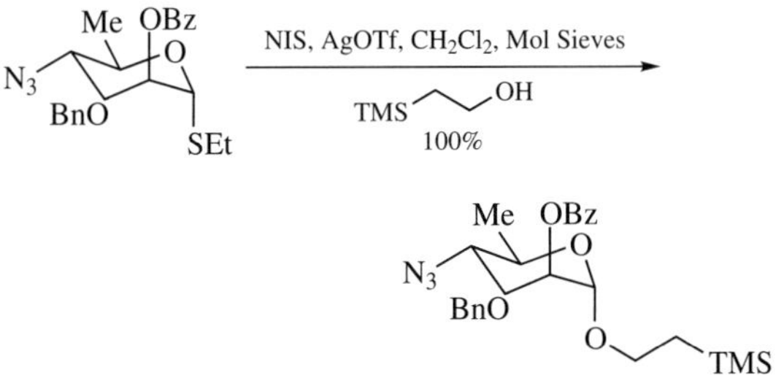

(35)

1. Fessenden, R. J.; Fessenden, J. S., *J. Org. Chem.* **1967**, *32*, 3535.

2. Gerlach, H., *Helv. Chim. Acta* **1977**, *60*, 3039.

3. Jansson, K.; Ahlfors, S.; Frejd, T.; Kihlberg, J.; Magnusson, G.; Dahmen, J.; Noori, G.; Stenvall, K., *J. Org. Chem.* **1988**, *53*, 5629.

4. Rosowsky, A.; Wright, J. E., *J. Org. Chem.* **1983**, *48*, 1539.

5. Soderquist, J. A.; Hassner, A., *J. Organomet. Chem.* **1978**, *156*, C12.

6. Soderquist, J. A.; Brown, H. C., *J. Org. Chem.* **1980**, *45*, 3571.

7. Soderquist, J. A.; Thompson, K. L., *J. Organomet. Chem.* **1978**, *159*, 237.

8. Mancini, M. L.; Honek, J. F., *Tetrahedron Lett.* **1982**, *23*, 3249.

9. From acids:(a) Sieber, P., *Helv. Chim. Acta* **1977**, *60*, 2711. (b) Brook, M. A.; Chan, T. H., *Synthesis* **1983**, 201. (c) White, J. D.; Jayasinghe, L. R., *Tetrahedron Lett.* **1988**, *29*, 2139. From an acid chloride: see Ref. 2. From an anhydride: Vedejs, E.; Larsen, S. D., *J. Am. Chem. Soc.* **1984**, *106*, 3030. For cleavage: see Ref. 9a, and Forsch, R. A.; Rosowsky, A., *J. Org. Chem.* **1984**, *49*, 1305.

10. Pouzar, V.; Drasar, P.; Cerny, I.; Havel, M., *Synth. Commun.* **1984**, *14*, 501.

11. (a) Seebach, D.; Hungerbühler, E.; Naef, R.; Schnurrenberger, P.; Weidmann, B.; Züger, M., *Synthesis* **1982**, 138. (b) Férézou, J. P.; Julia, M.; Liu, L. W.; Pancrazi, A., *Synlett* **1991**, 618.

12. Seebach, D.; Thaler, A.; Blaser, D.; Ko, S. Y., *Helv. Chim. Acta* **1991**, *74*, 1102.

13. (a) Burke, S. D.; Pacofsky, G. J.; Piscopio, A. D., *Tetrahedron Lett.* **1986**, *27*, 3345. (b) Gioeli, C.; Balgobin, N.; Josephson, S.; Chattopadhyaya, J. B., *Tetrahedron Lett.* **1981**, *22*, 969.

14. Lipshutz, B. H.; Pegram, J. J.; Morey, M. C., *Tetrahedron Lett.* **1981**, *22*, 4603; also see Ref. 3.

15. Sawabe, A.; Filla, S. A.; Masamune, S., *Tetrahedron Lett.* **1992**, *33*, 7685.

16. Chao, H.-G.; Bernatowicz, M. S.; Klimas, C. E.; Matsueda, G. R., *Tetrahedron Lett.* **1993**, *34*, 3377.

17. (a) Shute, R. E.; Rich, D. H., *Synthesis* **1987**, 346. (b) Ref. 4. (c) Carpino, L. A.; Tsao, J.-H., *J. Chem. Soc., Chem. Commun.* **1978**, 358.

18. Carpino, L. A.; Sau, A. C., *J. Chem. Soc., Chem. Commun.* **1979**, 514.

19. Campbell, A. L.; Pilipauskas, D. R.; Khanna, I. K.; Rhodes, R. A., *Tetrahedron Lett.* **1987**, *28*, 2331.

20. Marlin, J. E.; Killpack, M. O., *Heterocycles* **1992**, *34*, 1385.

21. Oida, T.; Ohnishi, A.; Shimamaki, T.; Hayashi, Y.; Tanimoto, S., *Bull. Chem. Soc. Jpn.* **1991**, *64*, 702.

22. Selenski, C.; Pettus, T. R. R., *J. Org. Chem.* **2004**, *69*, 9196.

23. Kang, Y.; Mei, Y.; Du, Y.; Jin, Z., *Org. Lett.* **2003**, *5*, 4481.

24. Yoakim, C.; Guse, I.; O'Meara, J. A.; Thavonekham, B., *Synlett* **2003**, 473.

25. Llinàs-Brunet, M.; Bailey, M. D.; Bolger, G.; Brochu, C.; Faucher, A.-M.; Ferland, J.-M.; Garneau, M.; Ghiro, E.; Gorys, V.; Grand-Maître, C.; Halmos, T.; Lapeyre-Paquette, N.; Liard, F.; Poirier, M.; Rhéaume, M.; Tsantrizos, Y. S.; Lamarre, D., *J. Med. Chem.* **2004**, *47*, 1605.

26. Lebel, H.; Paquet, V., *J. Am. Chem. Soc.* **2004**, *126*, 320.

27. Grecian, S. A.; Hadida, S.; Warren, S. D., *Tetrahedron Lett.* **2005**, *46*, 4683.

28. Schelessinger, R. H.; Gillman, K. W., *Tetrahedron Lett.* **1999**, *40*, 1257.

29. Wu, J.; Shull, B. K.; Koreeda, M., *Tetrahedron Lett.* **1996**, *37*, 3647.

30. Coombs, J.; Lattmann, E.; Hoffmann, H. M. R., *Synthesis* **1998**, 1367.

31. Ghosh, A. K.; Liu, C., *J. Am. Chem. Soc.* **2003**, *125*, 2374.

32. Knölker, H.-J.; Braxmeier, T., *Tetrahedron Lett.* **1996**, *37*, 5861.

33. Batey, R. A.; Yoshina-Ishii, C.; Taylor, S. D.; Santhakumar, V., *Tetrahedron Lett.* **1999**, *40*, 2669.

34. Igarashi, J.; Kobayashi, Y., *Tetrahedron Lett.* **2005**, *46*, 6381.

35. Rancourt, J.; Cameron, D. R.; Gorys, V.; Lamarre, D.; Poirier, M.; Thibeault, D.; LLinàs-Brunet, M., *J. Med. Chem.* **2004**, *47*, 2511.

36. Takeda, T.; Sato, K.; Tsubouchi, A., *Synthesis* **2004**, 1457.

37. Sowell, C. G.; Livesay, M. T.; Johnson, D. A., *Tetrahedron Lett.* **1996**, *37*, 609.

38. Barroca, N.; Schmidt, R. R., *Org. Lett.* **2004**, *6*, 1551.

39. Saksena, R.; Zhang, J.; Kováč, P., *J. Carbohydr. Chem.* **2002**, *21*, 453.

2-(Trimethylsilyl)ethoxycarbonyl-imidazole

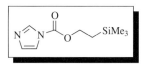

[81616-20-8] $C_9H_{16}N_2O_2Si$ (MW 212.36)

InChI = 1S/C9H16N2O2Si/c1-14(2,3)7-6-13-9(12)11-5-4-10-8-11/h4-5,8H,6-7H2,1-3H3

InChIKey = IHNKPDZWSVHJCU-UHFFFAOYSA-N

(introduction of the trimethylsilylethoxycarbonyl protecting group[1])

Physical Data: mp 29–30.5 °C;[1] bp 99–100 °C/1 mmHg.[3]

Solubility: insol H_2O; sol benzene, ether, THF, chloroform.

Form Supplied in: not commercially available.

Analysis of Reagent Purity: NMR.[1]

Preparative Method: from 2-(trimethylsilyl)ethanol and N,N'-carbonyldiimidazole.[1]

Purification: silica gel chromatography.[1]

Handling, Storage, and Precautions: stable crystalline solid; possibly moisture sensitive.

Protecting Group Formation. 2-(Trimethylsilyl)ethoxy-carbonylimidazole (**1**) is a useful reagent for the introduction of the 2-trimethylsilylethoxycarbonyl (Teoc) alcohol protecting group. For example, (**1**) has been used to protect the secondary hydroxyl group of verrucarol (**2**) as the Teoc derivative (eq 1).[1] Note that the apparently low yield (54%) of this reaction is due to the presence of a second reactive hydroxyl group in (**2**). The yield based on consumed (**2**) is a more respectable 81%. Although other examples of the use of (**1**) have not yet been reported, the reagent appears to have significant potential for the protection of alcohols and amines.

Alternative reagents for the introduction of the Teoc group include 2-(trimethylsilyl)ethyl chloroformate (**3**; X = Cl),[2] trimethylsilylethyl azidoformate (**3**; X = N₃),[2] trimethylsilylethyl methoxyvinyl carbonate (**4**),[3] trimethylsilylethyl 4-nitrophenyl carbonate (**5**),[4] and the aminoxy carbonates (**6**).[7] Of these, only the 4-nitrophenyl carbonate (**5**) is commercially available. The carbonate (**5**) has been used to block NH groups in amino acids[4,5] and indoles.[6]

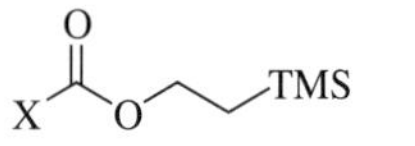

(3) 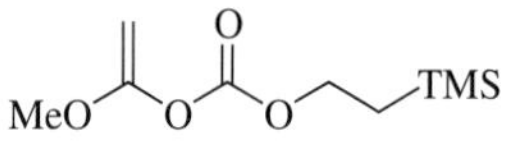   **(4)**

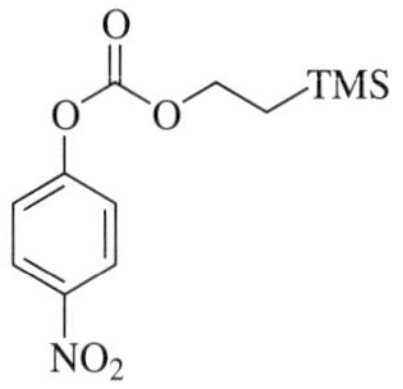

(5) 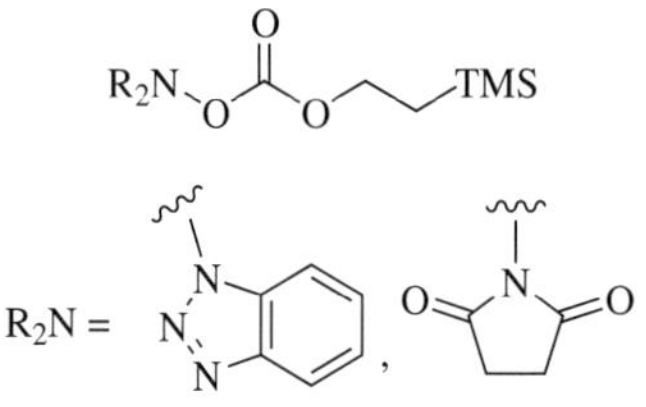   **(6)**

$$R_2N =$$

Although (**1**) is not commercially available, it is a stable, crystalline solid which is readily prepared in excellent yield by condensation of commercially available 2-trimethylsilylethanol and carbonyldiimidazole (eq 2).[1] This procedure avoids the use of potentially dangerous phosgene gas which is required for the preparation of the other reagents. Alternatively, (**1**) may be prepared in a somewhat less convenient manner from (**4**).[3] The stability, easy preparation, and reactivity of (**1**) combine to make it an excellent reagent for the introduction of the Teoc protecting group.

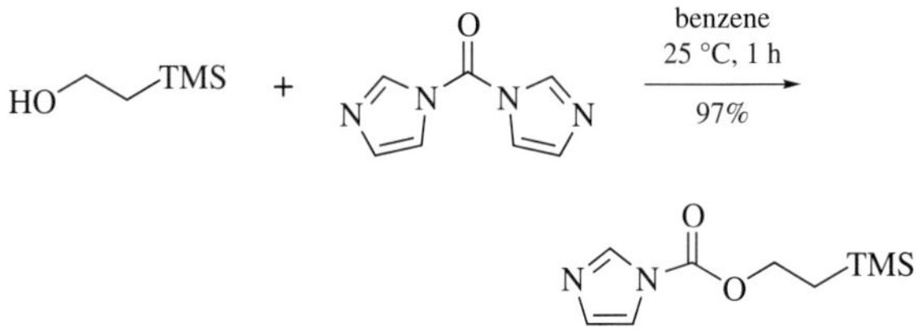

(**1**)

1. Roush, W. R.; Blizzard, T. A., *J. Org. Chem.* **1984**, *49*, 4332.
2. Carpino, L. A.; Tsao, J.-H., *J. Chem. Soc., Chem. Commun.* **1978**, 358.
3. Kita, Y.; Haruta, J.; Yasuda, H.; Fukunaga, K.; Shirouchi, Y.; Tamura, Y., *J. Org. Chem.* **1982**, *47*, 2697.
4. Wuensch, E.; Moroder, L., *Hoppe-Seyler's Z. Physiol. Chem.* **1981**, *362*, 1289 (*Chem. Abstr.* **1982**, *96*, 7063p).
5. Rosowsky, A.; Wright, J. E., *J. Org. Chem.* **1983**, *48*, 1539.
6. Dhanak, D.; Reese, C. B., *J. Chem. Soc., Perkin Trans. 1* **1986**, 2181.
7. Shute, R. E.; Rich, D. H., *Synthesis* **1987**, 346.

Timothy A. Blizzard
Merck Research Laboratories, Rahway, NJ, USA

2-(Trimethylsilyl)ethoxymethyl Chloride[1]

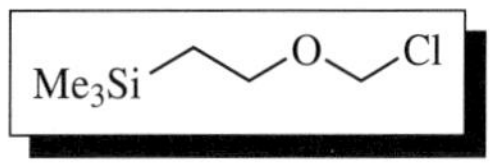

[76513-69-4] $C_6H_{15}ClOSi$ (MW 166.75)
InChI = 1S/C6H15ClOSi/c1-9(2,3)5-4-8-6-7/h4-6H2,1-3H3
InChIKey = BPXKZEMBEZGUAH-UHFFFAOYSA-N

(protection of alcohols,[1,2] secondary aryl amines,[3] and imidazole, indole, and pyrrole nitrogens;[4–6] electrophilic formaldehyde equivalent;[7] acyl anion equivalent[8])

Alternative Name: SEM-Cl.
Physical Data: bp 57–59 °C/8 mmHg; *d* 0.942 g cm^{-3}.
Solubility: sol most organic solvents (pentane, CH_2Cl_2, Et_2O, THF, DMF, DMPU, HMPA).
Form Supplied in: liquid; commercially available 90–95% pure; HCl is typical impurity.
Analysis of Reagent Purity: GC.
Preparative Method: several syntheses have been reported.[2,9–12]
Handling, Storage, and Precautions: water sensitive; corrosive; should be stored in a glass container under an inert atmosphere; a lachrymator; flammable (fp 46 °C).

Original Commentary

Jill Earley
Indiana University, Bloomington, IN, USA

Reagent for the Protection of Alcohols. Lipshutz and coworkers introduced the use of SEM-Cl for the protection of primary, secondary, and tertiary alcohols (eq 1).[2] SEM-Cl is now widely employed in organic synthesis for the protection of hydroxyl functionalities (eqs 2 and 3).[13,14]

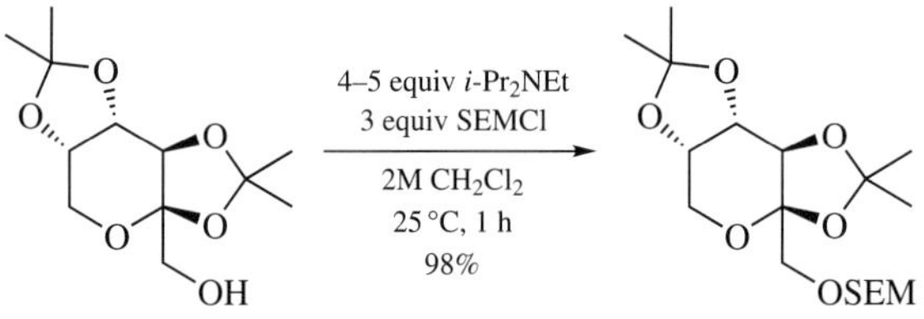

(1)

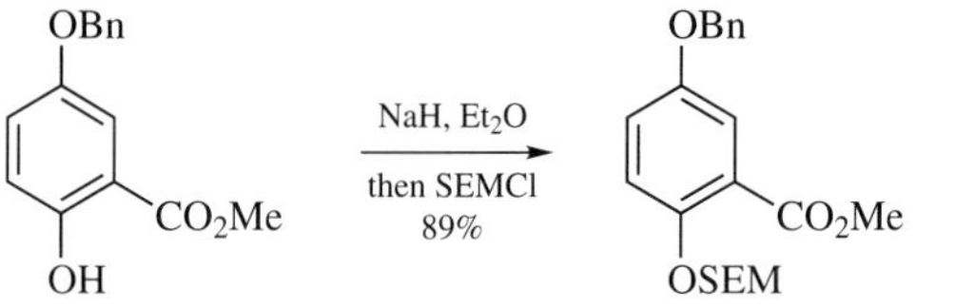

(2)

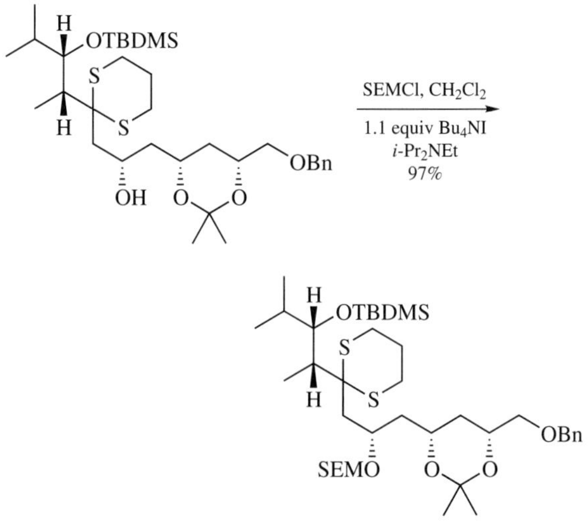

(3)

The resulting SEM ethers are stable under a variety of conditions. Most SEM ethers are cleaved with a fluoride anion source (eqs 4 and 5),[2,15–17] although this fragmentation is generally much less facile than fluoride-induced cleavage of silyloxy bonds. Therefore the selective deprotection of other silyl ethers in the presence of SEM ethers is possible (eq 6).[15] Vigorous conditions with anhydrous fluoride ion are required for the cleavage of some tertiary SEM ethers (eq 7).[18,19]

bromide (eq 8).[21] SEM, MOM, and MEM phenolic protective groups are removed with diphosphorus tetraiodide (eq 9).[22]

$$ (4) $$

$$ (5) $$

$$ (6) $$

$$ R = SEM $$

$$ R = H \quad (7) $$

The stability of SEM ethers has necessitated the development of a variety of other deprotection methods. SEM protective groups are stable to the acidic conditions used to hydrolyze THP, TBDMS, and MOM ethers (AcOH, H_2O, THF, $45\,°C$),[2] but can be removed under strongly acidic conditions with trifluoroacetic acid[20] SEM, MTM, and MOM ethers are selectively cleaved in the presence of MEM, TBDMS, and benzyl ethers with magnesium

$$ (8) $$

$$ (9) $$

SEM-Cl has also proven useful in carbohydrate synthesis where the resulting SEM ethers can be cleaved under less acidic conditions than those required for the cleavage of MOM and MEM ethers (eq 10).[23]

$$ (10) $$

Reagent for the Protection of Acids. Carboxylic acids can be protected as SEM esters (eq 11).[24,25] Cleavage of the SEM esters occurs with refluxing methanol[24] or magnesium bromide (eq 12).[25,26]

$$ (11) $$

$$ (12) $$

Reagent for the Protection of Secondary Amines. SEM-Cl can be used to protect secondary aromatic amines (eq 13),[3] and is an ideal reagent for the protection of imidazoles, indoles, and pyrroles (eqs 14–16).[4–6,27] Many functional groups are compatible with the introduction and cleavage of SEM amines, and the SEM substituent is unusually stable to further functionalization of the molecule.[4–6,27] Recently, a one-pot method for protection and alkylation of imidazoles employing *n*-butyllithium and SEM-Cl has been developed (eq 17).[28–30]

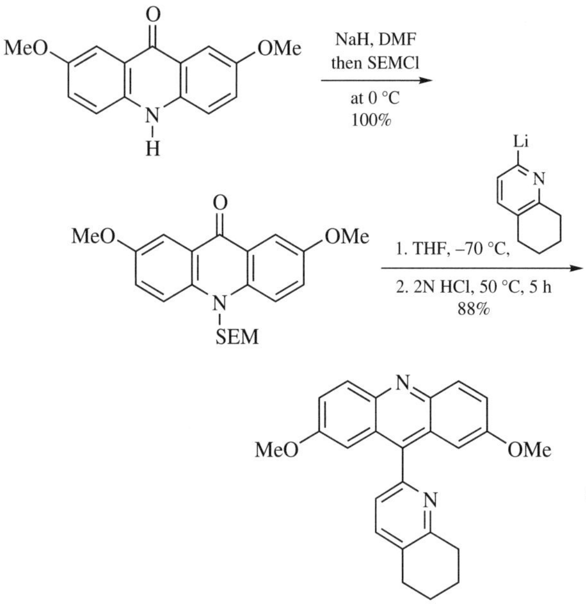

(13)

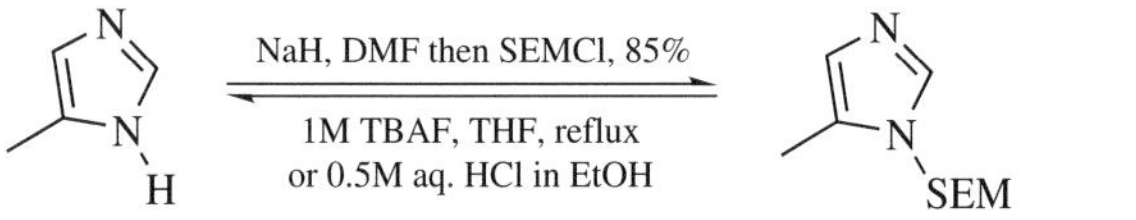

(14)

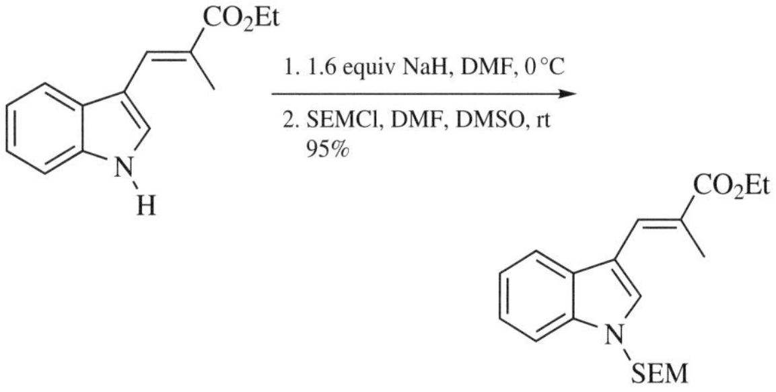

(15)

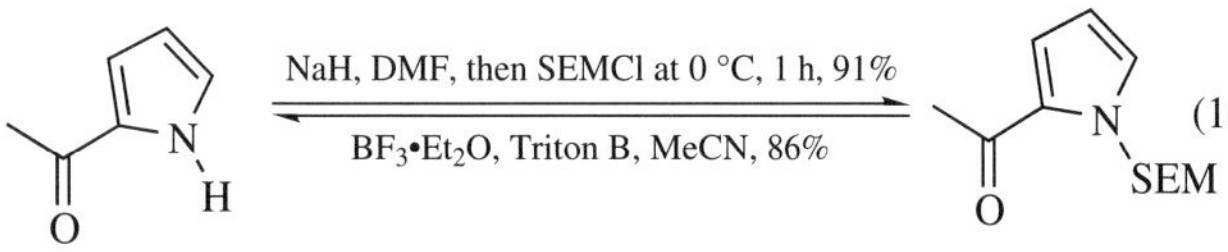

(16)

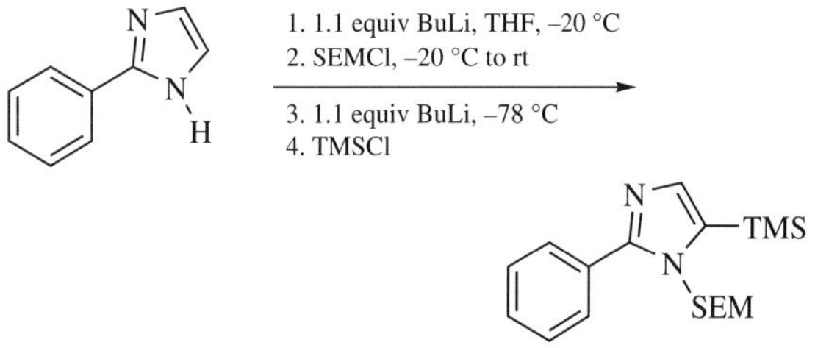

(17)

One-Carbon Homologations. SEM-Cl has functioned as an alternative to the use of formaldehyde (eq 18).[7] The enolate of the lactone (**1**) undergoes *C*-alkylation with SEM-Cl, and the resulting SEM substituent can be subsequently treated with tri-fluoroacetic acid to afford the alcohol (**2**).

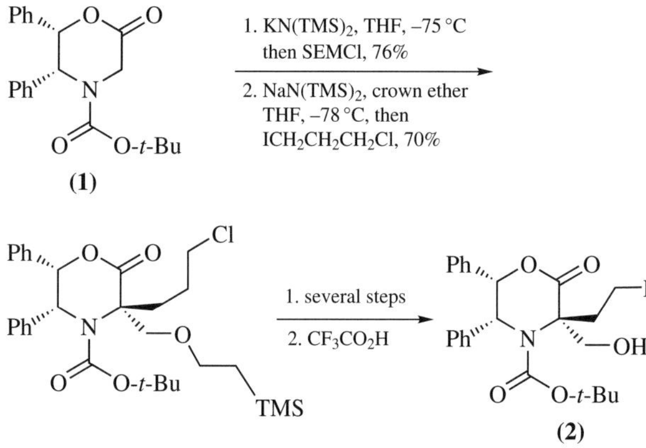

(18)

SEM-Cl has been transformed to a formaldehyde carbanion equivalent.[8] SEM-Cl is converted to the ylide upon treatment with triphenylphosphine and sodium hydride. This ylide reacts with a variety of aldehydes and ketones affording enol ethers (eq 19).[8] Hydrolysis with 5% aqueous HF gives the corresponding aldehyde.

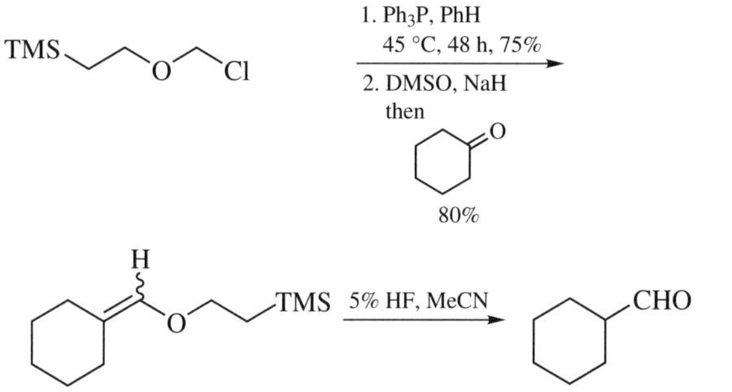

(19)

First Update

Christopher M. Adams

Novartis Institutes for Biomedical Research, Cambridge, MA, USA

Reagent for the Protection of Alcohols. SEM-Cl continues to be used most commonly to protect alcohols as their corresponding 2-(trimethylsilyl)ethoxymethyl ethers. Generally, the afore-mentioned conditions of reacting an alcohol with SEM-Cl in the presence of a trialkylamine base in dichloromethane remain the primary means of installation.[2] However, a few new methods have been disclosed that offer some distinct advantages. Blass and co-workers have reported the protection of phenols, which employs potassium fluoride on alumina (KF/Al$_2$O$_3$) in place of a trial-kylamine base.[31] The advantage of this method is that it alleviates the need for an aqueous work-up, and in most cases the products can be isolated in high yield by simply filtering away the alumina followed by solvent removal (eq 20).

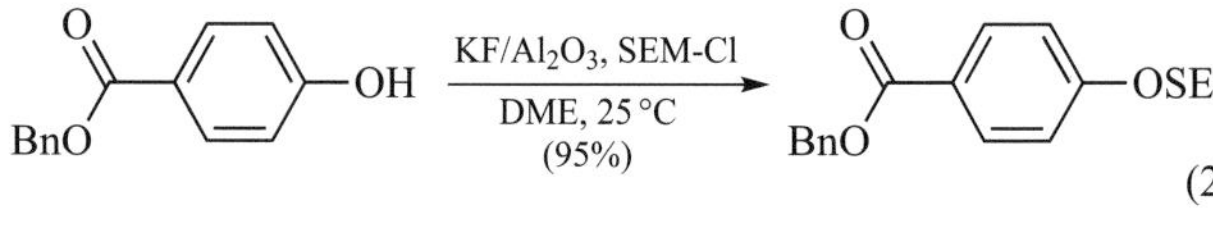

(20)

In addition, Halcomb and Freeman-Cook have shown that replacing the standard trialkylamine base with 2,6-di-*t*-butylpyridine can permit the selective protection of a primary hydroxyl over secondary and tertiary hydroxyls (eq 21).[32]

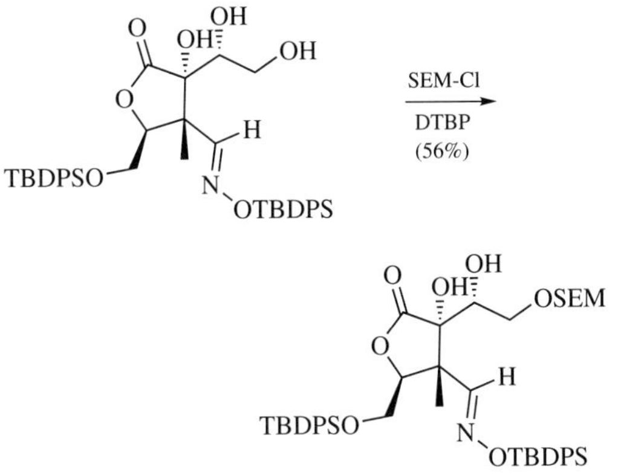

The continued use of SEM-Cl for the protection of hydroxyls in a wide array of substrates has necessitated the development of several new deprotection strategies. In addition to the originally disclosed nucleophilic fluoride deprotection,[2] the use of HF as an aqueous solution[33] or buffered with pyridine[34] has also gained prominence. The use of Lewis acids with or without the use of a thiol scavenger has also gained widespread adoption. Notably, $BF_3 \cdot OEt_2$,[35,36] ZnI_2,[37] $ZnBr_2$,[38] and $LiBF_4$,[39] have all been used with success. Hoffman and Vakalopoulos have also extended the use $MgBr_2$.[21,38] They have shown that the use of nitromethane as a cosolvent alleviates the need for thiol additives and permits the selective deprotection of SEM ethers in the presence of silyl ethers (eq 22).[38]

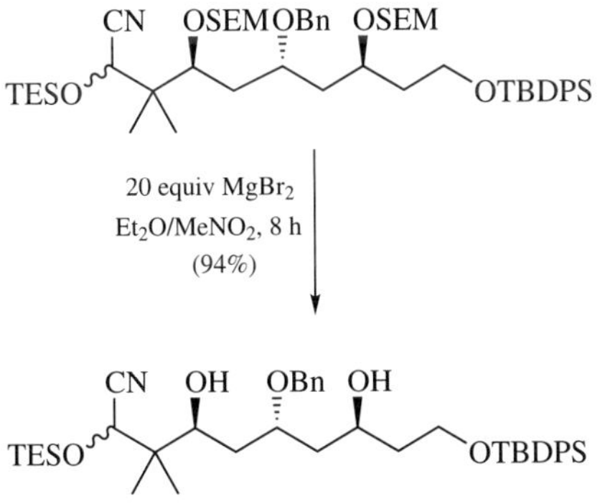

Other recently developed methods for SEM ether deprotection include the use of $CBr_4/MeOH$ either at reflux or in the presence of UV light[40] and $I_2/MeOH$ at 50 °C.[41]

Reagent for the Protection of Secondary Amines. SEM-Cl has also maintained its prominence in the area of heterocyclic nitrogen protection. Of particular note is the use of SEM-Cl for the chemoselective protection of uracil[42] and indazole[43] derivatives (eqs 23 and 24, respectively).

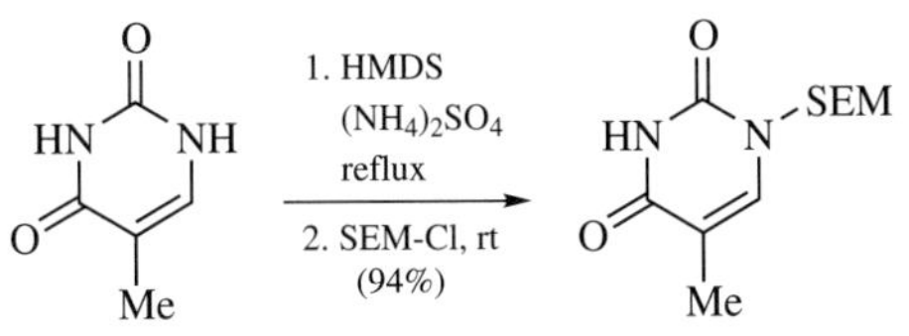

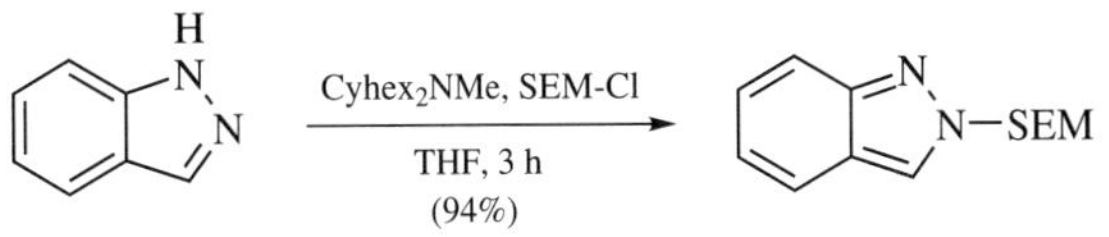

SEM-Cl has also found use in the protection of the nitrogen of amides[44] (eq 25) and sulfonamides.[45]

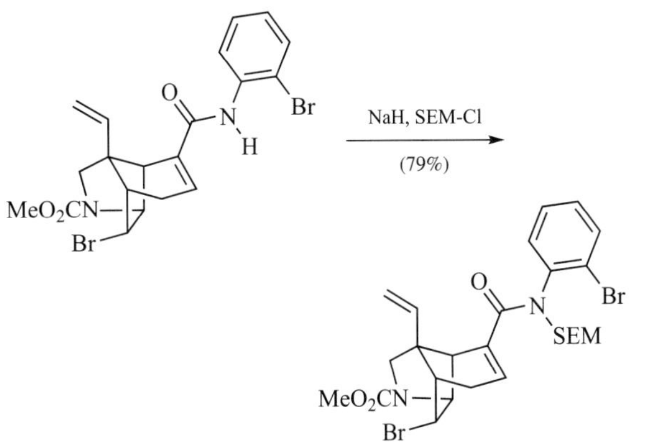

One-carbon Homologations. SEM-Cl can be used as a formaldehyde equivalent, which upon alkylation affords, directly, a protected hydroxyl. SEM-Cl does not suffer from some of the handling liabilities of formaldehyde (e.g., the need for cracking or the use of aqueous solutions) and as such has shown specific promise in the area of asymmetric alkylations. Asymmetric alkylations of enolates have been accomplished via the employment of chiral auxiliaries including oxazolidinones (eq 26)[46] and RAMP/SAMP hydrazones.[47,48] Furthermore, Schultz and co-workers have utilized SEM-Cl to trap chiral enolates derived from Birch reductions (eq 27).[49]

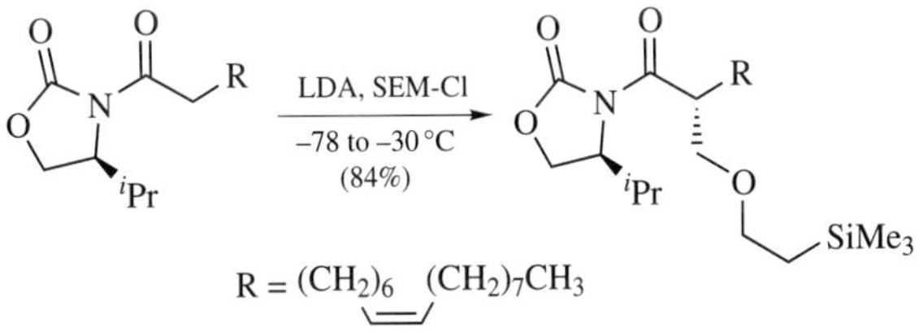

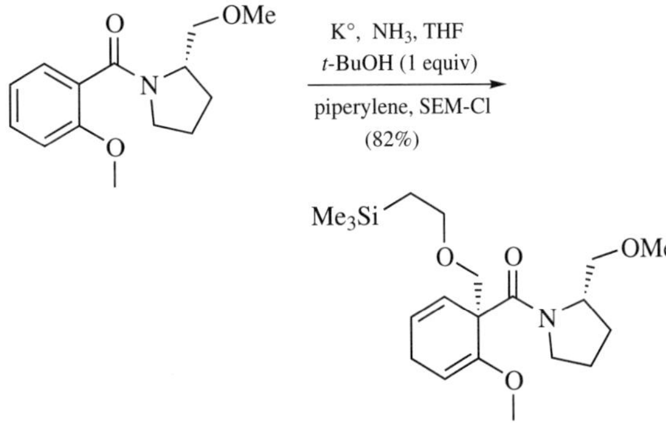

As described above, SEM-Cl has found use as a formaldehyde carbanion equivalent[8] via reaction with triphenylphosphine to produce a Wittig salt. However, it is no longer necessary to generate the Wittig salt from SEM-Cl now that 2-(trimethylsilyl)ethoxymethyl-triphenylphosphonium chloride (CAS# 82495-75-8) is commercially available.

Cyclizations. In addition to the use of SEM-Cl as a means of one-carbon homologation, it has also been employed in the generation of acetals capable of undergoing a variety of cyclization events. Alkylation of allylic alcohols with SEM-Cl permits haloacetalization to afford di- and trisubstituted 1,2-dioxanes (eq 28).[50] Alternatively, homoallylic alcohols permit Prins-type cyclizations (eq 29) to furnish 2,4,5-trisubstituted tetrahydropyrans.[51] Transition metals can also be employed to promote the formation of substituted benzofurans (eq 30).[52]

1. Greene, T. W.; Wuts, P. G. M. *Protective Groups in Organic Synthesis*, 2nd ed.; Wiley: New York, 1991.

2. Lipshutz, B. H.; Pegram, J. J., *Tetrahedron Lett.* **1980**, *21*, 3343.

3. Zeng, Z.; Zimmerman, S. C., *Tetrahedron Lett.* **1988**, *29*, 5123.

4. Whitten, J. P.; Matthews, D. P.; McCarthy, J. R., *J. Org. Chem.* **1986**, *51*, 1891.

5. Ley, S. V.; Smith, S. C.; Woodward, P. R., *Tetrahedron* **1992**, *48*, 1145.

6. Muchowski, J. M.; Solas, D. R., *J. Org. Chem.* **1984**, *49*, 203.

7. Baldwin, J. E.; Lee, V.; Schofield, C. J., *Synlett* **1992**, 249.

8. Schönauer, K.; Zbiral, E., *Tetrahedron Lett.* **1983**, *24*, 573.

9. Sonderquist, J. A.; Hassner, A., *J. Organomet. Chem.* **1978**, *156*, C12.

10. Sonderquist, J. A.; Thompson, K. L., *J. Organomet. Chem.* **1978**, *159*, 237.

11. Gerlach, H., *Helv. Chim. Acta* **1977**, *60*, 3039.

12. Fessenden, R. J.; Fessenden, J. S., *J. Org. Chem.* **1967**, *32*, 3535.

13. Kotecha, N. R.; Ley, S. V.; Mantegani, S., *Synlett* **1992**, 395.

14. Lipshutz, B. H.; Moretti, R.; Crow, R., *Tetrahedron Lett.* **1989**, *30*, 15.

15. Williams, D. R.; Jass, P. A.; Tse, H.-L. A.; Gaston, R. D., *J. Am. Chem. Soc.* **1990**, *112*, 4552.

16. Shull, B. K.; Koreeda, M., *J. Org. Chem.* **1990**, *55*, 99.

17. Ireland, R. E.; Varney, M. D., *J. Org. Chem.* **1986**, *51*, 635.

18. Kan, T.; Hashimoto, M.; Yanagiya, M.; Shirahama, H., *Tetrahedron Lett.* **1988**, *29*, 5417.

19. Lipshutz, B. H.; Miller, T. A., *Tetrahedron Lett.* **1989**, *30*, 7149.

20. Schlessinger, R. H.; Poss, M. A.; Richardson, S., *J. Am. Chem. Soc.* **1986**, *108*, 3112.

21. Kim, S.; Kee, I. S.; Park, Y. H.; Park, J. H., *Synlett* **1991**, 183.

22. Saimoto, H.; Kusano, Y.; Hiyama, T., *Tetrahedron Lett.* **1986**, *27*, 1607.

23. Pinto, B. M.; Buiting, M. M. W.; Reimer, K. B., *J. Org. Chem.* **1990**, *55*, 2177.

24. Logusch, E. W., *Tetrahedron Lett.* **1984**, *25*, 4195.

25. Kim, S; Park, Y. H.; Kee, I. S., *Tetrahedron Lett.* **1991**, *32*, 3099.

26. Salomon, C. J.; Mata, E. G.; Mascaretti, O. A., *Tetrahedron* **1993**, *49*, 3691.

27. Lipshutz, B. H.; Vaccaro, W.; Huff, B., *Tetrahedron Lett.* **1986**, *27*, 4095.

28. Lipshutz, B. H.; Huff, B.; Hagen, W., *Tetrahedron Lett.* **1988**, *29*, 3411.

29. Demuth, T. P. Jr.; Lever, D. C.; Gorgos, L. M.; Hogan, C. M.; Chu, J., *J. Org. Chem.* **1992**, *57*, 2963.

30. For a similar method with pyrroles, see Ref. 13.

31. Blass, B. E.; Harris, C. L.; Portlock, D. E., *Tetrahedron Lett.* **2001**, *42*, 1611.

32. Freeman-Cook, K. D.; Halcomb, R. L., *J. Org. Chem.* **2000**, *65*, 6153.

33. Vadalà, A.; Finzi, P. V.; Zanoni, G.; Vidari, G., *Eur. J. Org. Chem.* **2003**, 642.

34. Nakamura T.; Shiozaki, M., *Tetrahedron* **2002**, *58*, 8779.

35. Takano, S.; Murakami, T.; Samizu, K.; Ogasawara, K., *Heterocycles* **1994**, *39*, 67.

36. Marshall, J. A.; Hinkle, K. W., *Tetrahedron Lett.* **1998**, *39*, 1303.

37. Sugimoto, M.; Suzuki, T.; Hagiwara, H.; Hoshi, T., *Tetrahedron Lett.* **2007**, *48*, 1109.

38. Vakalopoulos, A.; Hoffman, H. M. R., *Org. Lett.* **2000**, *2*, 1447.

39. Ng, R. A.; Guan, J.; Alford, V. C., Jr.; Lanter, J. C.; Allan, G. F.; Sbriscia, T.; Linton, O.; Lundeen, S. G.; Sui, Z., *Bioorg. Med. Chem. Lett.* **2007**, *17*, 784.

40. Chen, M.-Y.; Lee, A. S.-Y., *J. Org. Chem.* **2002**, *67*, 1384.

41. Keith, J. M., *Tetrahedron Lett.* **2004**, *45*, 2739.

42. Arias, L.; Guzmán, S.; Jaime-Figueroa, S.; Morgans, D. J., Jr.; Pahilla, F.; Pérez-Medrano, A.; Quintero, C.; Romero, M.; Sandoval, L., *Synlett* **1997**, 1233.

43. Luo, G.; Chen, L.; Dubowchik, G., *J. Org. Chem.* **2006**, *71*, 5392.

44. Madin, A.; O'Connell, C. J.; Oh, T.; Old, D. W.; Overman, L. E.; Sharp, M. J., *J. Am. Chem. Soc.* **2005**, *127*, 18054.

45. Davis, F. A.; Srirajan, V., *J. Org. Chem.* **2000**, *65*, 3248.

46. Eichelberger, U.; Neundorf, I.; Hennig, L.; Findeisen, M.; Giesa, S.; Müller, D.; Welzel, P., *Tetrahedron* **2002**, *58*, 545.

47. Enders, D.; Hundertmark, T.; Lampe, C.; Jegelka, U.; Scharfbillig, I., *Eur. J. Org. Chem.* **1998**, 2839.

48. Enders, D.; Gries, J., *Synthesis* **2005**, *20*, 3508.

49. Khim, S.-K.; Dai, M.; Zhang, X.; Chen, L.; Pettus, L.; Thakkar, K.; Schultz, A. G., *J. Org. Chem.* **2004**, *69*. 7728.

50. Clausen, R. P.; Bols, M., *J. Org. Chem.* **2000**, *65*, 2797.

51. Al-Mutairi, E. H.; Crosby, S. R.; Darzi, J.; Harding, J. R.; Hughes, R. A.; King, C. D.; Simpson, T. J.; Smith, R. W.; Willis, C. L., *Chem. Commun.* **2001**, 835.

52. Fürstner, A.; Davies, P. W., *J. Am. Chem. Soc.* **2005**, *127*, 15024.

[2-(Trimethylsilyl)ethoxymethyl]-triphenylphosphonium chloride*

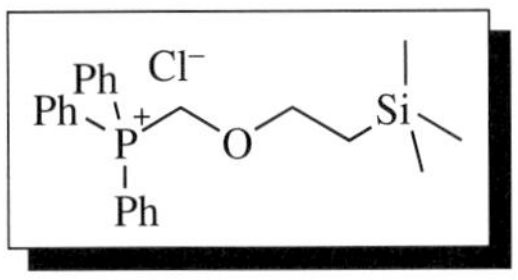

[82495-75-8] $C_{24}H_{30}ClOPSi$ (MW 429.01)

InChI = 1S/C24H30ClOPSi/c1-28(2,3)20-19-26-21-27(25,22-13-7-4-8-14-22,23-15-9-5-10-16-23)24-17-11-6-12-18-24/h4-18H,19-21H2,1-3H3

InChIKey = FXUYLZMROPGYNP-UHFFFAOYSA-N

(reagent undergoes deprotonation and ylide formation permitting the construction of enol ethers from aldehydes and ketones)

Physical Data: mp 165–169 °C.

Solubility: sol H_2O and dichloromethane; sparingly soluble in THF.

Form Supplied in: white crystalline solid; widely available.

Preparative Method: involves reaction of triphenylphosphine with 2-(trimethylsilyl)ethoxymethyl chloride in benzene at gentle reflux for 18 h. Upon cooling, the resulting phosphonium salt can be recovered by filtration and purified by triturating with diethyl ether in 94% yield.[1]

Handling, Storage, and Precautions: may absorb moisture. Bottles should be kept tightly sealed and stored under nitrogen or argon in a dry place. This compound is suspected to be an irritant to the skin and eyes.

Enol Ether Formation. Triphenyl[[2-(trimethylsilyl)ethoxy]methyl]-phosphonium chloride is used almost exclusively in Wittig-type reactions with aldehydes and ketones. The reagent was first described by Fuchs and coworkers for the preparation of an electron-rich diene, suitable for elaboration via Diels–Alder reaction (eq 1).[1] In the Fuchs example, the ylide is generated by

reaction with potassium *t*-butoxide in THF and undergoes smooth reaction with an α,β-unsaturated ketone. The resulting alkoxy diene is formed as a separable 7:1 *ZE/ZZ* mixture in 84% yield. Further elaboration afforded the Diels–Alder precursor, which cyclized in high yield.

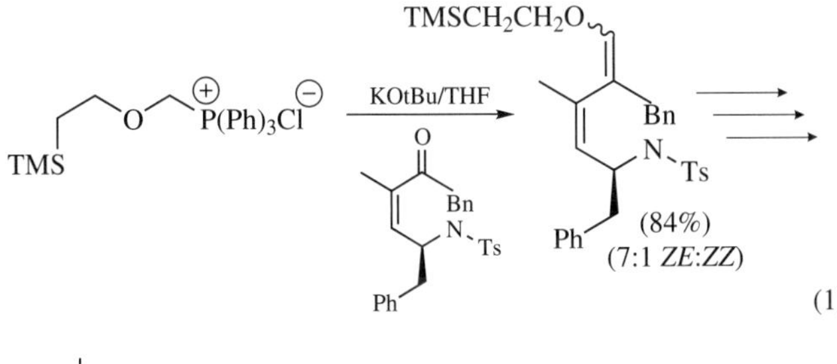

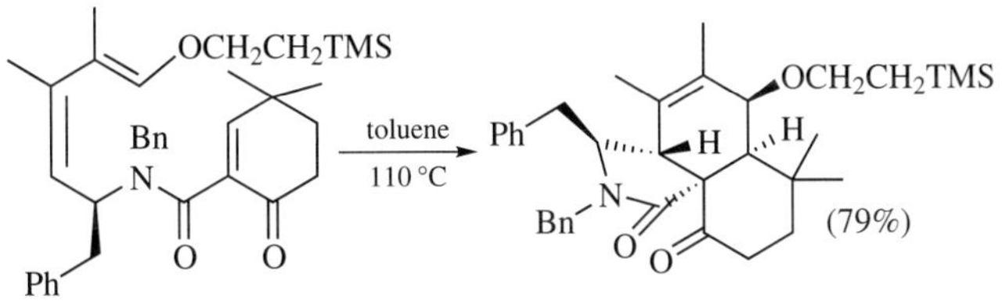

Subsequent to the Fuchs disclosure, Zbiral and Schönauer further explored the uses of triphenyl[[2-(trimethylsilyl)ethoxy]-methyl]-phosphonium chloride.[2,3] Various aldehydes and ketones were transformed to the corresponding enol ethers, which upon hydrolysis afforded aldehydes extended by one carbon (eqs 2 and 3).

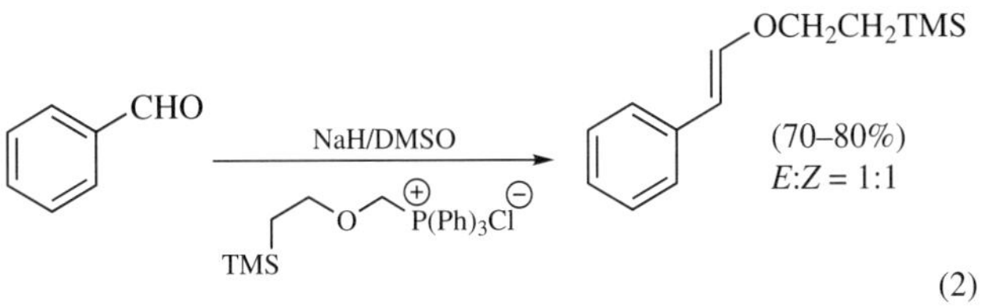

(2)

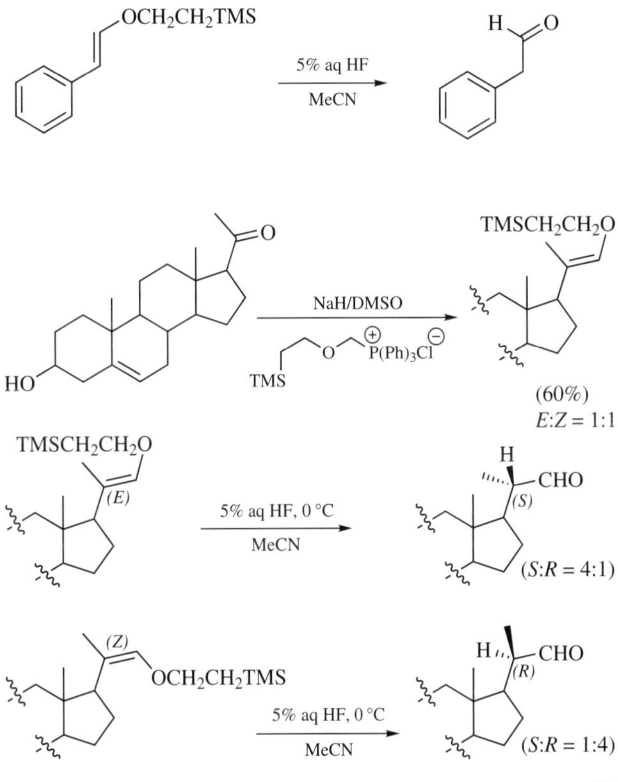

(3)

The initial olefination was accomplished using the Wittig olefination conditions developed by Corey (NaH/DMSO)[4] to afford *E/Z* mixtures of enol ethers. The enol ethers undergo hydrolysis to the corresponding aldehydes upon treatment with 5% aqueous HF in acetonitrile. Modest degrees of stereoselectivity were observed. The ability of fluoride to mediate the hydrolysis of the (trimethylsilyl)ethoxymethyl moiety may provide opportunities not available when (methoxymethyl)triphenylphosphonium salts are employed. However, it should be noted that Zbiral and Schönauer reported that hydrolysis could not be realized with the use of tetrabutylammonium formate or triethylamine · 2HF.

(1)

---

*Article previously published in the electronic Encyclopedia of Reagents for Organic Synthesis as Triphenyl[2-(trimethylsilyl)ethoxy]methyl]-phosphonium Chloride. DOI: 10.1002/047084289X. rn01122.

More recently, triphenyl[[2-(trimethylsilyl)ethoxy]methyl]-phosphonium chloride has found use in homologating a variety of architecturally complex aldehydes and ketones (eqs 4–7).[5–8] Highlights include the chemoselective olefination of an aldehyde over a ketone (eq 5),[6] the olefination of a sterically congested α-epoxy aldehyde (eq 6),[7] and the homologation of a particularly challenging ketone by Overman and coworkers (eq 7).[8]

1. Pyne, S. G.; Hensel, M. J.; Fuchs, P. L., *J. Am. Chem. Soc.* **1982**, *104*, 5179.

2. Schönauer, K.; Zbiral, E., *Tetrahedron Lett.* **1983**, *24*, 573.

3. Schönauer, K.; Zbiral, E., *Liebigs Ann. Chem.* **1983**, 1031.

4. Greenwald, R.; Chaykovsky, M.; Corey, E. J., *J. Org. Chem.* **1963**, *28*, 1128.

5. Horner, J. H.; Tanaka, N.; Newcomb, M., *J. Am. Chem. Soc.* **1998**, *120*, 10379.

6. Paquette, L.; Zeng, Q.; Wang, H.-L.; Shih, T.-L., *Eur. J. Org. Chem.* **2000**, 2187.

7. Kuba, M.; Furuichi, N.; Katsumura, S., *Chem. Lett.* **2002**, 1248.

8. Becker, M. H.; Chua, P.; Downham, R.; Douglas, C. J.; Garg, N. K.; Hiebert, S.; Jaroch, S.; Matsuoka, R. T.; Middleton, J. A.; Ng, F. W.; Overman, L. E., *J. Am. Chem. Soc.* **2007**, *129*, 11987.

Christopher M. Adams
Novartis Institutes for Biomedical Research, Inc.,
Cambridge, MA, USA

1. TMSCH$_2$CH$_2$OCH$_2$P$^+$(Ph)$_3$Cl$^-$
   LDA, THF, 0 °C (62%)
2. CF$_3$CO$_2$H, CH$_2$Cl$_2$ (28%)

$$\qquad (4)$$

TMSCH$_2$CH$_2$OCH$_2$P$^+$(Ph)$_3$Cl$^-$
KHMDS, Et$_2$O, 0 °C (44%)

$(E{:}Z = 1{:}3)$ $\qquad (5)$

1. TMSCH$_2$CH$_2$OCH$_2$P$^+$(Ph)$_3$Cl$^-$
   KOtBu, Et$_2$O, 0 °C
2. 1N HCl aq, THF, rt, 5 min
3. MnO$_2$, ether, rt (49%, 3 steps)

$$\qquad (6)$$

1. TMSCH$_2$CH$_2$OCH$_2$P$^+$(Ph)$_3$Cl$^-$
   KHMDS, THF, −78 °C to rt
2. 5% aq HF, MeCN (91%, 2 steps)

$$\qquad (7)$$

2-(Trimethylsilyl)ethyl Chloroformate

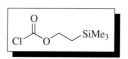

[20160-60-5]　　　　C$_6$H$_{13}$ClO$_2$Si　　　　(MW 180.73)
InChI = 1S/C6H13ClO2Si/c1-10(2,3)5-4-9-6(7)8/h4-5H2,1-3H3
InChIKey = BTEQQLFQAPLTLI-UHFFFAOYSA-N

(introduction of the trimethylsilylethoxycarbonyl (Teoc) protecting group for amines[1–3] or alcohols;[4,5] activation of tertiary amines for dealkylation[6] or ring opening;[7–10] starting material for alternative reagents for introduction of the trimethylsilylethoxycarbonyl protecting group[11])

Physical Data: bp 34 °C/1 mmHg;[6] bp 42–43 °C/4 mmHg;[12] *d* 0.9944 g cm^{-3}.[1]

Solubility: insol H$_2$O; sol benzene, ether, THF, chloroform.

Form Supplied in: not commercially available.

Analysis of Reagent Purity: NMR.

Preparative Method: prepared from 2-(trimethylsilyl)ethanol and phosgene.[6,11,12]

Purification: distillation.

Handling, Storage, and Precautions: moisture sensitive; stable for three months under nitrogen at 0 °C;[6] probably toxic; handle in a fume hood.

Protection of Amines.[1] 2-(Trimethylsilyl)ethyl chloroformate (**1**) is a useful reagent for the introduction of the trimethylsilylethoxycarbonyl (Teoc) protecting group for amines.[1–3] For example, *p*-chloroaniline reacts with (**1**) to form the expected trimethylsilylethyl carbamate (eq 1).[1] The Teoc protecting group is stable to hydrogenation (H$_2$, palladium on carbon) and mild acid or base. It can be removed with fluoride ion or with a strong acid (e.g. trifluoroacetic acid). It is possible to selectively cleave a *t*-butyldimethylsilyl

(TBDMS) ether in the presence of a trimethylsilylethyl carbamate (eq 2).[3] For a discussion of alternative reagents for the introduction of the Teoc group, see 2-(trimethylsilyl)-ethoxycarbonylimidazole.

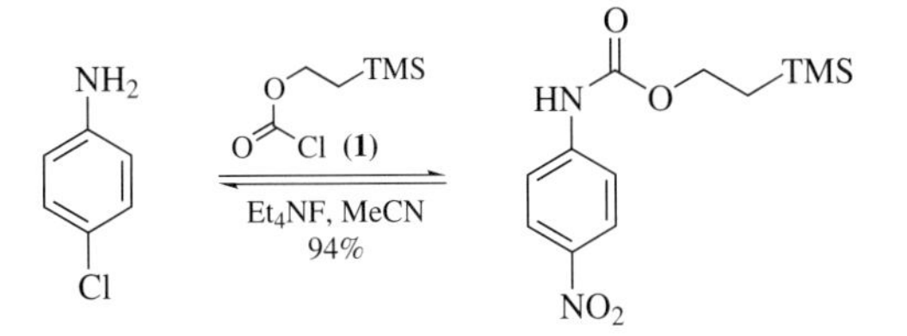

(1)

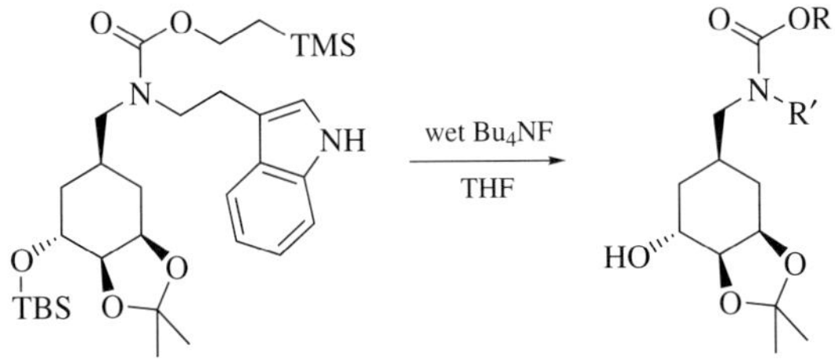

(2)

Protection of Alcohols.[4]

2-(Trimethylsilyl)ethyl chloroformate (1) is also useful for the introduction of the trimethylsilylethoxycarbonyl (Teoc) protecting group for alcohols.[4,5] The Teoc protecting group is stable to mild acid but it is base sensitive. It can be cleaved with fluoride ion or with zinc chloride or zinc bromide (eq 3).[4]

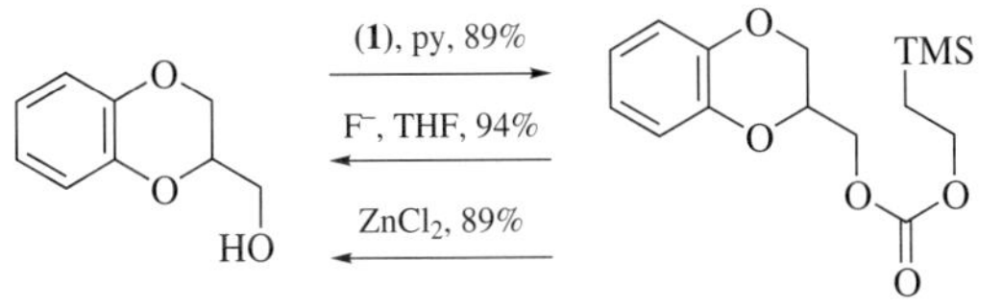

(3)

Dealkylation or Ring Opening of Tertiary Amines.[6–10]

Under certain conditions, 2-(trimethylsilyl)ethyl chloroformate (1) is reactive enough to acylate a tertiary amine. The resulting quaternary ammonium species is a sufficiently good leaving group for one of the carbon–nitrogen bonds to be cleaved by reaction with a suitable nucleophile. This process has been used to debenzylate tertiary amines (eq 4).[6] Subsequent removal of the Teoc group (see above) to afford the free amine results in a net debenzylation of the tertiary amine. Demethylation of tertiary amines is also possible but yields are generally lower.[6] Alternatively, for cyclic tertiary amines, acylation of the nitrogen followed by reaction with a nucleophile can result in a ring-opening reaction (eq 5).[7] In a strained ring system (e.g. an aziridine), acylation of the nitrogen followed by reaction with a nucleophile can result in ring opening even for secondary amines (eq 6).[8]

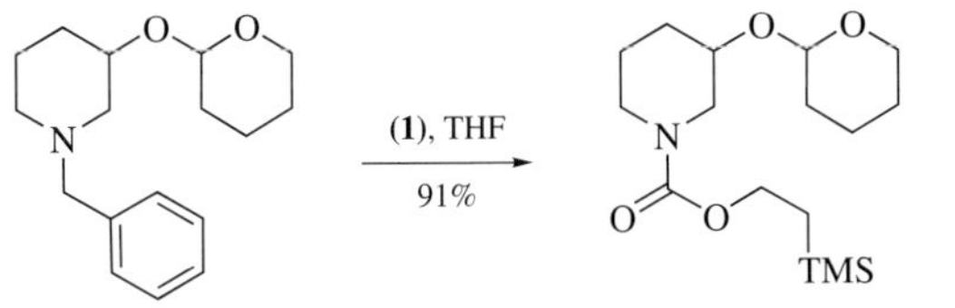

(4)

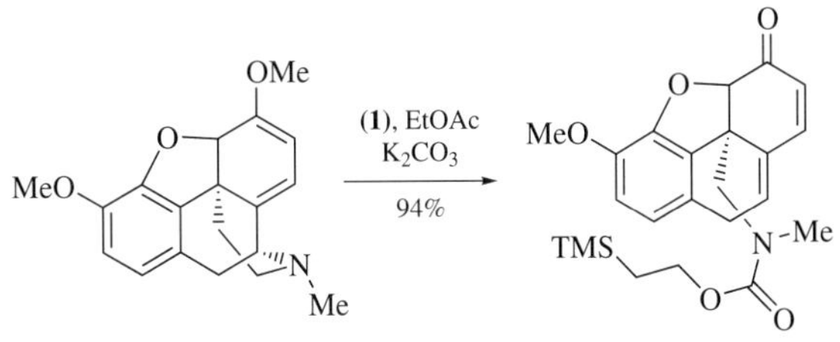

(5)

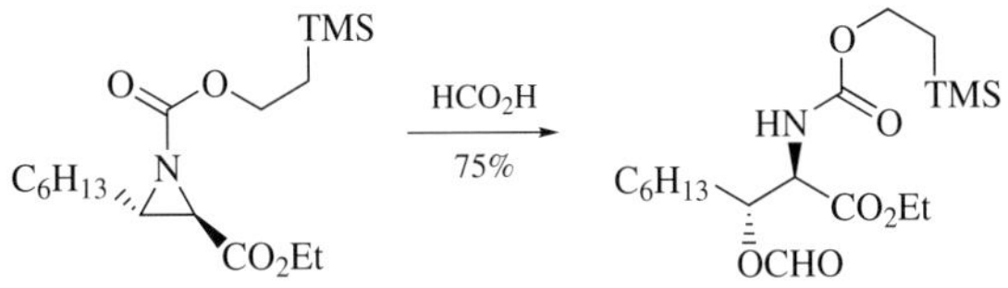

(6)

Synthesis of Alternative Teoc Reagents.[11]

The chloroformate (1) has been used as a starting material for the preparation of alternative reagents for the introduction of the Teoc group (eq 7).[11] For a discussion of additional alternative reagents for the introduction of the Teoc group, see 2-(trimethylsilyl)ethoxycarbonylimidazole.

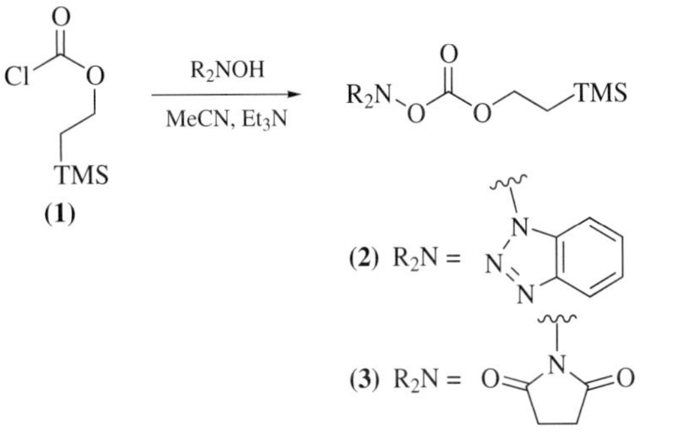

(7)

(2) $R_2N =$

(3) $R_2N =$

1. Carpino, L. A.; Tsao, J.-H., *J. Chem. Soc., Chem. Commun.* **1978**, 358.
2. Trost, B. M.; Cossy, J., *J. Am. Chem. Soc.* **1982**, *104*, 6881.
3. Trost, B. M.; Romero, A. G., *J. Org. Chem.* **1986**, *51*, 2332.
4. Gioeli, C.; Balgobin, N.; Josephson, S.; Chattopadhyaya, J. B., *Tetrahedron Lett.* **1981**, *22*, 969.
5. Schuda, P. F.; Ammon, H. L.; Heimann, M. R.; Bhattacharjee, S., *J. Org. Chem.* **1982**, *47*, 3434.
6. Campbell, A. L.; Pilipauskas, D. R.; Khanna, I. K.; Rhodes, R. A., *Tetrahedron Lett.* **1987**, *28*, 2331.
7. Toth, J. E.; Fuchs, P. L., *J. Org. Chem.* **1986**, *51*, 2594.
8. Legters, J.; Willems, J. G. H.; Thijs, L.; Zwanenburg, B., *Recl. Trav. Chim. Pays-Bas* **1992**, *111*, 59.
9. Kim, G.; Chu-Moyer, M. Y.; Danishefsky, S. J., *J. Am. Chem. Soc.* **1990**, *112*, 2003.
10. Kim, G.; Chu-Moyer, M. Y.; Danishefsky, S. J.; Schulte, G. K., *J. Am. Chem. Soc.* **1993**, *115*, 30.
11. Shute, R. E.; Rich, D. H., *Synthesis* **1987**, 346.
12. Kozyukov, V. P.; Sheludyakov, V. D.; Mironov, V. F., *J. Gen. Chem. USSR (Engl. Transl.)* **1968**, *38*, 1133.

Timothy A. Blizzard
Merck Research Laboratories, Rahway, NJ, USA

2-(Trimethylsilyl)ethyl 3-Nitro-1*H*-1,2,4-triazole-1-carboxylate[1]

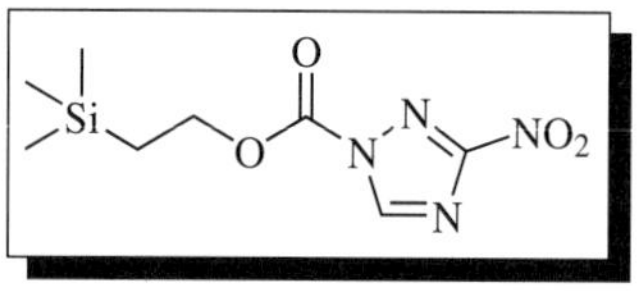

[1001067-09-9] $C_8H_{14}N_4O_4Si$ (MW 258.31)

InChI = 1S/C8H14N4O4Si/c1-17(2,3)5-4-16-8(13)11-6-9-7
(10-11)12(14)15/h6H,4-5H2,1-3H3

InChIKey = BHTJYIDOVYPNKR-UHFFFAOYSA-N

(reagent used for the introduction of 2-(trimethylsilyl)ethoxy-
carbonyl (Teoc) protecting group in amines and alcohols)

Alternate Name: Teoc-NT.

Physical Data: mp (dec.) 85–86 °C.

Solubility: sol organics (CH_2Cl_2, THF, CH_3CN, EtOAc); reacts
with amines.

Form Supplied in: light yellow crystalline solid; widely available.

Analysis of Reagent Purity: ^{1}H NMR (500 MHz, CD_3CN) δ 8.97
(1H, s), 4.67–4.60 (2H, m), 1.27–1.20 (2H, m), 0.08 (9H, s);
^{13}C NMR (125 MHz, CD_3CN) δ 164.0, 148.8, 147.2, 70.8, 17.9,
−1.7; IR (neat) ν 3137, 2958, 1777, 1564, 1514, 1290, 1230,
1170, 1015 cm^{-1}; crystallographic data are available as supplementary no. CCDC 675623.

Handling, Storage, and Precautions: Teoc-NT may release CO_2
gas upon prolonged storage; therefore, the bottle should be
opened carefully. To prevent decomposition, store under an inert gas atmosphere and refrigerate.

General Informations.   2-(Trimethylsilyl)ethoxycarbonyl
(Teoc) group is one of the most useful protecting groups for
amines, which can be cleaved by fluoride ion under neutral
conditions.[2] However, 2-(trimethylsilyl)ethyl chloroformate
(Teoc-Cl),[3,4] which is generally used for the introduction of the
Teoc protecting group in amines, is quite unstable and easy to
decompose; therefore, distillation is recommended at regular
intervals.[2,5] On the other hand, Teoc-NT is a highly stable,
nonhygroscopic crystalline material and can be stored for long
periods without decomposition.[1,6] Generally, carbamate formation proceeds quickly under neutral conditions, and highly pure
carbamates are obtained without tedious column chromatographic
purification. Furthermore, the coproduct 3-nitro-1,2,4-triazole
(NT) can be recycled.

Reactions with Amines and Aminoalcohols.   Reactions of
Teoc-NT with amines and aminoalcohols are summarized in
Table 1.[1] Primary and secondary amines react rapidly with
1 equiv of Teoc-NT in CH_2Cl_2 at room temperature, precipitating
3-nitro-1,2,4-triazole (NT)[7,8] as by-product. NT is easily separated by simple filtration or aqueous workup with 5% $NaHCO_3$,
and highly pure (>99%) acylation products are obtained without
column chromatographic purification. Aminoalcohols chemoselectively react with Teoc-NT to give the corresponding carbamates
in highly pure form. The hydroxyl groups of aminoalcohols do not
react with Teoc-NT, even employing excess Teoc-NT.[1] In general,
aromatic amines are less reactive compared to aliphatic amines,
and the reaction of aniline with Teoc-NT is slow. However,
in these cases, addition of base (Et_3N) is effective and gives the
desired carbamate in high yield.

Table 1

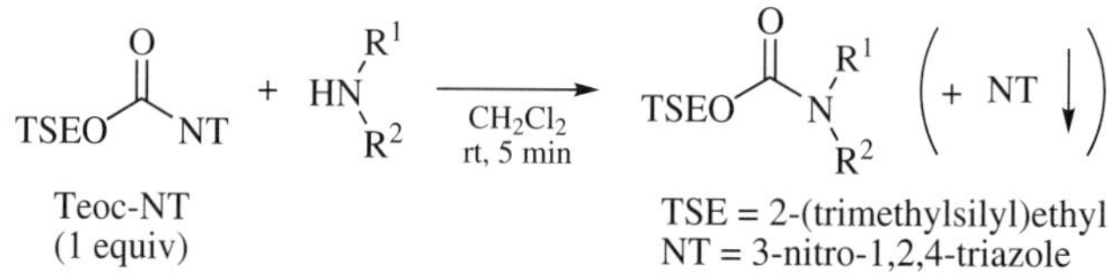

Entry	Amine	Yield (%)	Remarks
1	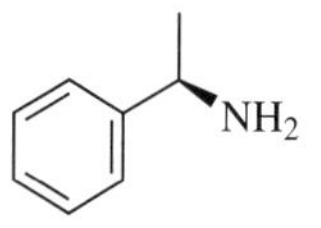	95	Solvent: CH_2Cl_2–5% $NaHCO_3$ (1:1, v/v)
2	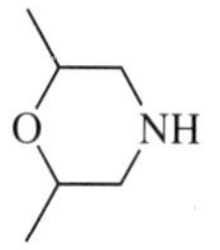	quant	–
3	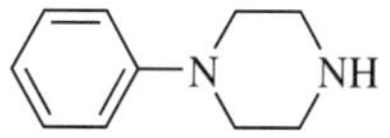	quant	–
4	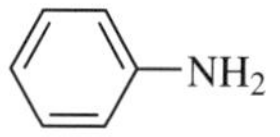	94	60 min, Et_3N (5 min)
5	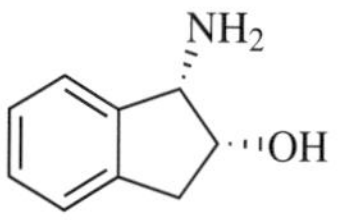	96	Solvent: CH_2Cl_2–5% $NAHCO_3$ (1:1, v/v)
6		quant	30 min

A list of General Abbreviations appears on the front Endpapers

Amine Protection for Nucleosides. Teoc is expected to be a useful protecting group for the exocyclic amino group of nucleobases, since it is a fluoride ion labile protecting group. However, introduction of the Teoc group to guanine base is not accomplished by using Teoc-Cl,[9] partially due to the instability of the reagent.[5] On the other hand, Teoc-NT smoothly reacts with amino group of nucleobases to afford protected nucleoside derivatives in high yield after column chromatographic purification (Table 2).[1]

Table 2

Entry	Base	X	Time (h)	Yield (%)	Remarks
1	Cytosine	2	1	quant	–
2	Adenine	4	12	98	–
3	Guanine	4	14	80	Et$_3$N(10 equiv)

Reaction with Alcohols. Teoc-NT does not react with alcohols in the absence of base; however, reaction with benzyl alcohol in the presence of Et$_3$N proceeds quickly to give highly pure (>99%) carbonate in 92% yield after aqueous workup with 5% NaHCO$_3$ (eq 1).[6] Even the secondary alcohol ((+)-menthol) reacts with Teoc-NT to afford the corresponding carbonate in 97% yield after column chromatography (eq 2).[6]

$$\text{(1)}$$

$$\text{(2)}$$

Related NT Reagents. Similarly to Teoc-NT, other commonly used acylating reagents can be converted to NT reagents (Table 3).[7] Each of these reagents is a stable crystalline material and can be used in a similar manner as Teoc-NT.

Table 3

Entry	R	Abbreviation	mp (dec.) (°C)	Yield (%)
1	BnO-	Z-NT	112–113	95
2	Cl$_3$CCH$_2$-	Troc-NT	145–146	92
3	9-Fluorenylmethory	Fmoc-NT	170–171	95
4	t-BuO-	Boc-NT	77–78	55
5	PhO-	Px-NT	157–158	87
6	p-MeO-C$_6$H$_4$-	Ani-NT	108–109	92
7	PhOCH$_2$-	Pac-NT	133–134	91
8	CH$_3$(CH$_2$)$_{14}$-	Pal-NT	62–63	95

1. Shimizu, M.; Sodeoka, M., *Org. Lett.* **2007**, *9*, 5231.

2. Wuts, P. G. M.; Greene, T. W., *Greene's Protecting Groups in Organic Synthesis*, 4th ed.; Wiley: New Jersey, **2007**, 719.

3. Kozyukov, V. P.; Sheludyakov, V. D.; Mironov, V. F., *J. Gen. Chem. USSR* **1968**, *38*, 1133.

4. Carpino, L. A.; Tsao, J.-H.; Ringsdorf, H.; Fell, E.; Hettrich, G., *Chem. Commun.* **1978**, 358.

5. Campbell, A. L.; Pilipauskas, D. R.; Khanna, I. K.; Rhodes, R. A., *Tetrahedron Lett.* **1987**, *28*, 2331.

6. Shimizu, M.; Sodeoka, M., *Heterocycles* **2008**, *76*, 1301.

7. *Physical Methods in Heterocyclic Chemistry*, Katritsky, A. R., ed; Academic Press: New York, 1963, Vol. 1.

8. Brown, E. J., *Aust. J. Chem.* **1969**, *22*, 2251.

9. Sekine, M.; Tobe, M.; Nagayama, T.; Wada, T., *Lett. Org. Chem.* **2004**, *1*, 179.

Mamoru Shimizu
Chiralgen Ltd., Kashiwa, Japan

(Trimethylsilyl)ethynylcopper(I)[1]

[53210-13-2]   C$_5$H$_9$CuSi   (MW 160.78)
InChI = 1S/C5H9Si.Cu/c1-5-6(2,3)4;/h2-4H3;
InChIKey = DNXSGHQBAIKZLI-UHFFFAOYSA-N

(coupling agent for introduction of the (trimethylsilyl)ethynyl group;[2] preparation of mixed cuprates[3])

Alternate Names: copper(I) trimethylsilylacetylide.
Physical Data: orange-red unstable solid.
Solubility: sol THF, THF–HMPA mixtures; insol ether.
Preparative Method: by treatment of trimethylsilylacetylene with copper(I) t-butoxide,[2] by addition of CuBr·SMe$_2$ to a THF solution of lithium (trimethylsilyl)acetylide,[4] or by treatment of trimethylsilylacetylene with triethyl phosphite and copper(I) chloride.[5]

Handling, Storage, and Precautions: best generated in situ and used immediately.[2] THF solutions are unstable above $-20\,^{\circ}\text{C}$.[2] The solid is unstable, and all such copper(I) acetylides are potentially explosive. Use in a fume hood.

Coupling Agent. Copper trimethylsilylacetylide is a versatile coupling agent for introduction of the trimethylsilylethynyl group to activated molecules. Due to the instability of the reagent, the Castro coupling reaction cannot be attempted, but additions to acyl chlorides,[2] activated amines,[6] and terminal (methoxy)iodoallenes have been reported.[7] This last reaction proceeds via an S_N2' mechanism, and a similar addition has been observed with α-acetoxy alkynes (eq 1) giving allenic alkynes in high yield.[8] Coupling between copper TMS-acetylide and 1,2-dichloroalkenes has also been demonstrated.[9,10] Coupling of the copper acetylide (generated in situ) to give the biologically important enediyne framework can be achieved using palladium catalysis (eq 2).[10]

R = Br, 33%; OAc, 98%; OSOMe, 98%

Mixed Organocuprates. The trimethylsilylethynyl group acts as a nontransferable dummy ligand when copper trimethylsilylacetylide is converted to a mixed alkyl or vinyl cuprate.[3,11,12] In one report, utilization of a mixed cuprate derived from a terminally lithiated allene allows chemoselective conjugate delivery of the allenic moiety to an α,β-unsaturated system (eq 3), which gives a functionalized and versatile enone system on hydrolysis.[3] The effects of chlorotrimethylsilane on such conjugate additions have been studied.[11] On using a mixed cuprate derived from combination with *n*-butyllithium in the presence of TMSCl, smooth and efficient 1,4-butylation ensues (eq 4).[11] The stoichiometry of the TMSCl employed has a marked effect on the reaction profile. Mixed magnesioorganocuprates have also been employed in conjugate addition reactions.[12] The highest yields of the corresponding 1,4-addition products are recovered using a stoichiometry of 3:1 Grignard reagent–copper TMS-acetylide.[12]

Copper Homoenolates. Magnesium homoenolates, typically generated by addition of vinylmagnesium bromide to an acyltrimethylsilane, undergo conjugate addition to enones (eq 5) in the presence of copper trimethylsilylacetylide, giving easy access to 1,6-diketones.[4,13] The reaction presumably proceeds via a mixed copper homoenolate, with the trimethylsilylethynyl group acting as a nontransferable ligand.

Related Reagents. 3-Trimethylsilyl-2-propynylcopper(I).

1. (a) Normant, J. F., *Synthesis* **1972**, 63. (b) Sladkov, A. M.; Gol'ding, I. R., *Russ. Chem. Rev. (Engl. Transl.)* **1979**, *48*, 868. (c) Lipshutz, B. H.; Sengupta, S., *Org. React.* **1992**, *41*, 135. (d) *Fieser & Fieser* **1977**, *6*, 148.
2. Logue, M. W.; Moore, G. L., *J. Org. Chem.* **1975**, *40*, 131.
3. Matsuoka, R.; Horiguchi, Y.; Kuwajima, I., *Tetrahedron Lett.* **1987**, *28*, 1299.
4. Enda, J.; Kuwajima, I., *J. Am. Chem. Soc.* **1985**, *107*, 5495.
5. Shostakovskii, M. F.; Polyakova, L. A.; Vasil'eva, L. V.; Polyakov, A. I., *Zh. Org. Khim.* **1966**, *2*, 1899.
6. Boche, G.; Bernheim, M.; Niessner, M., *Angew. Chem., Int. Ed. Engl.* **1983**, *22*, 53.
7. Oostveen, J. M.; Westmijze, H.; Vermeer, P., *J. Org. Chem.* **1980**, *45*, 1158.
8. Ruitenberg, K.; Kleijn, H.; Westmijze, H.; Meijer, J.; Vermeer, P., *Recl. Trav. Chim. Pays-Bas* **1982**, *101*, 405.
9. Rubin, Y.; Diederich, F., *J. Am. Chem. Soc.* **1989**, *111*, 6870.
10. Vollhardt, K. P. C.; Winn, L. S., *Tetrahedron Lett.* **1985**, *26*, 709.
11. Sakata, H.; Kuwajima, I., *Tetrahedron Lett.* **1987**, *28*, 5719.
12. Drouin, J.; Rousseau, G., *J. Organomet. Chem.* **1985**, *289*, 223.
13. Enda, J.; Matsutani, T.; Kuwajima, I., *Tetrahedron Lett.* **1984**, *25*, 5307.

Graham B. Jones & Brant J. Chapman
Clemson University, Clemson, SC, USA

Trimethylsilyl Fluorosulfonyldifluoroacetate

$$\text{FSO}_2\text{CF}_2\text{CO}_2\text{Si(CH}_3)_3$$

[120801-75-4] $C_5H_9F_3O_4SSi$ (MW 218.14)

InChI = 1S/C5H9F3O4SSi/c1-14(2,3)12-4(9)5(6,7)13(8,10)11/h1-3H3

InChIKey = XHVSCKNABCCCAC-UHFFFAOYSA-N

(difluorocarbene precursor)

Physical Data:[1,2] 73–74 $^{\circ}$C at 35mm Hg.
Solubility:[1,2] soluble in ether, toluene, and diglyme.
Form Supplied in: colorless liquid.

Preparative Methods: [1,2] prepared by treatment of $FO_2SCF_2CO_2H$ with $(CH_3)_3SiCl$.[1,2]

Purity: 3–4% of residual acid ($FO_2SCF_2CO_2H$) may be removed by treatment with NEt_3 if necessary.

Handling, Storage, and Precaution: [3] very susceptible to hydrolysis, best to use freshly prepared in reactions and to avoid all contact with skin.

Synthesis of *gem*-Difluorocyclopropanes. Trimethylsilyl fluorosulfonyldifluoroacetate is slowly added to a mixture of catalytic amounts of NaF, olefin, and solvent at a moderate temperature under N_2 for several hours to give the *gem*-difluorocyclopropanes. The reactions proceed cleanly and with unprecedented efficiency, as shown in eq 1.[1,2]

$$FSO_2CF_2CO_2SiMe_3 \quad + \quad \text{(1.5 equiv)} \quad \xrightarrow[\substack{105\,°C,\,\sim 2h \\ \text{neat or toluene}}]{NaF\,(0.012\;equiv)} \quad (1)$$

R = n-C_6H_{13} (74%), CH_2O_2CPh (78%), $CH_2CH_2O_2CPh$ (89%), $(CH_2)_3O_2CPh$ (97%), $CO_2C_4H_9$ (73%)

Notably, the reagent can be successfully applied to quite unreactive alkenes such as acrylate esters (73%). The factors, e.g., solvent, fluoride source, reaction temperature, and rate of addition influencing the yields have been investigated in some detail.[3] The superiority of the reagent over $ClCF_2CO_2Na$ is demonstrated in the synthesis of tris-*gem*-difluorocyclopropane eq 2.[4]

Method A: $ClCF_2CO_2Na$(50 equiv), diglyme, 180 °C, yield=54%
B: $FSO_2CF_2CO_2TMS$(10 equiv), NaF(0.012eq), 105 °C, Yield=72%

Thus, the desired product was obtained in 72% yield using 10 equiv of this reagent (method B) at 105 °C, while in 54% yield employing 50 eq of $ClCF_2CO_2Na$ at 180 °C.

The reagent does not react with α,β-unsaturated aldehydes and ketones under the reaction conditions used (i.e., in the presence of 10 mol% of NaF in diglyme at 120 °C) due to the stronger electron-deficiency of enone carbon-carbon double bond (versus acrylate esters). In order to overcome the difficulty, a protection and deprotection of enone carbonyl group is an alternative choice.[5]

Treatment of 1,3 dioxolane of α,β-unsaturated aromatic aldehydes and ketones with the reagent (30 equiv) in diglyme in the presence of 1 mol% NaF at 110–130 °C for 2h gives the corresponding [1 + 2] adducts in moderate to good yields. Deprotection of the adducts is achieved by treatment with saturated oxalic acid in dioxane at 110 °C for 6h.

X = Cl, CH_3O, CF_3, CH_3
R = H, CH_3, C_6H_5, $4\text{-}CH_3C_6H_4$
    $4\text{-}CH_3OC_6H_4$

The normal α-aryl-*gem*-difluorocyclopropyl ketones are obtained by acidic hydrolysis of α,β-unsaturated ketones except for those compounds having an electron-donating group (CH_3O, CH_3) on the phenyl ring. 1-Aryl-2-fluorofurans are the only isolated products derived from the hydrolysis of α-aryl-*gem*-difluorocyclopropylacetals (R=H) except the compounds possessing a strong electron-withdrawing group (CF_3) at *para* position of the phenyl ring. As for the α-aryl-*gem*-difluorocyclopropylmethyl ketals (R=CH_3), the ratio of normal products to furan derivatives obtained is dependent on the electronic properties of substrates on the phenyl ring; i.e., for an electron-donating group, furan derivatives are the sole products, whereas for an electron-withdrawing group, the normal ketone is dominant, if not exclusively produced.

These results might be rationalized as follows: The carbocation of aldehydes (R=H) and methyl ketones (R=CH_3) generated undergoes an intramolecular shift easily with simultaneous collapse of carbon-carbon bond opposite to *gem*-difluoromethyl followed by an attack of end oxygen on the resultant carbonium ion (Y=H) to yield β-fluorofuran derivatives after HF elimination (Path **a**). An alternative mechanism is that cyclopropane ring opening occurs during the protonation with the direct formation of a new cation (Y=CH_2CH_2OH) (Path **b**) (eq 4).

Synthesis of gem-Difluorocyclopropenes. It is known that *gem*-difluorocyclopropenes are unstable (decompose slowly at 25 °C).[6] When phenylacetylene is treated with the reagent in the presence of 10 mol% NaF in diglyme at 120 °C for 1h, the desired product (65%) turns black quickly and finally loses fluorine on standing when exposed to air at room temperature. Nevertheless, the compound could be trapped by bromine to give a rather stable adduct (eq 5).[7]

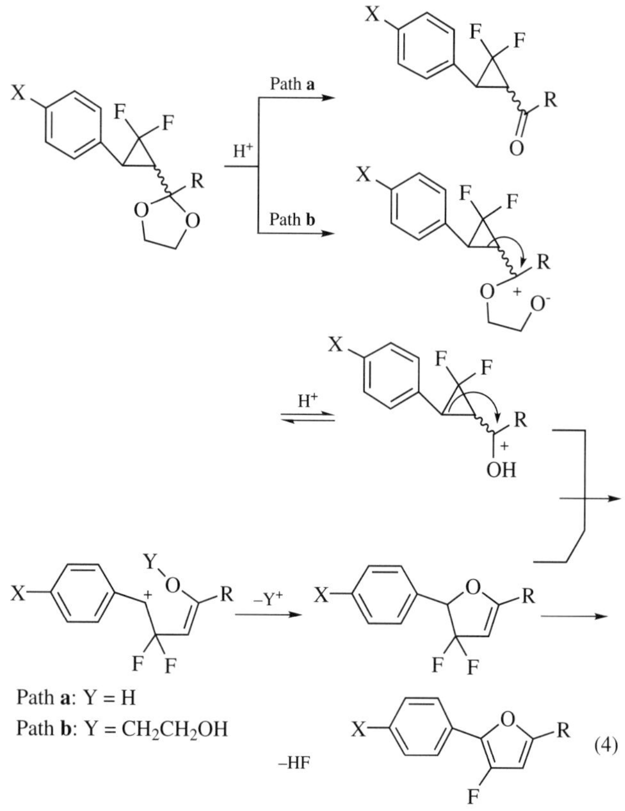

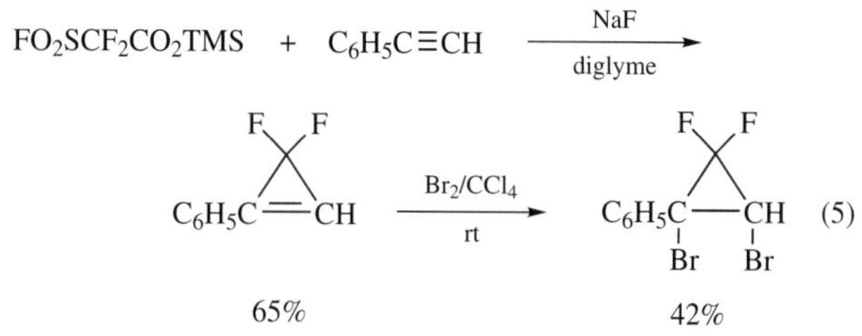

Path a: Y = H
Path b: Y = CH$_2$CH$_2$OH

(4)

$$FO_2SCF_2CO_2TMS + C_6H_5C \equiv CH \xrightarrow[\text{diglyme}]{NaF}$$

65%   → Br$_2$/CCl$_4$, rt → (5)   42%

On the basis of these results, *gem*-difluoroiodocyclopropenes are expected to be more stable than the noniodinated ones, if 1-iodoalkynes are utilized as substrates. This is indeed the case. Heating iodoalkynes and this reagent in diglyme in the presence of 10 mol% NaF at 110–124 °C for 0.5h gives 3,3-difluoro 1-iodocyclopropenes in 69–82% yield. As expected, the products are rather stable. They can be stored at room temperature without decomposition for several hours and much longer in a solvent at refrigerator temperatures. These new iodides can be trifluoromethylated and functionized via the Heck reaction (eq 6).[7]

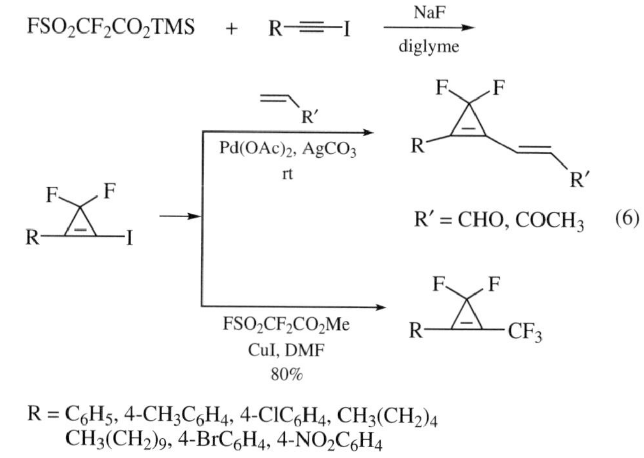

$$FSO_2CF_2CO_2TMS + R \equiv I \xrightarrow[\text{diglyme}]{NaF}$$

Pd(OAc)$_2$, AgCO$_3$, rt

R′ = CHO, COCH$_3$ (6)

FSO$_2$CF$_2$CO$_2$Me
CuI, DMF
80%

R = C$_6$H$_5$, 4-CH$_3$C$_6$H$_4$, 4-ClC$_6$H$_4$, CH$_3$(CH$_2$)$_4$
CH$_3$(CH$_2$)$_9$, 4-BrC$_6$H$_4$, 4-NO$_2$C$_6$H$_4$

Related Reagents.[3] Other trialkylsilyl fluorosulfonyldifluoroacetates can be prepared and have been found to have similar reactivities.

1. Tian, F.; Kruge, V.; Bautista, O.; Duan, J.-X., Li, A. R.; Dolbier Jr, W. R.; Chen, Q.-Y., *Org. Lett.* **2000**, *2*, 563.
2. Dolbier Jr., W. R.; Tian, F.; Duan, J.-X.; Chen, Q.-Y., *Org. Synth.* **2003**, *80*, 172.
3. Dolbier Jr., W. R.; Tian, F.; Duan, J.-X.; Li, A.-R.; Ait-Mohand, S., et al., *J. Fluorine Chem.* **2004**, *125*, 459.
4. Itoh, T.; Ishida, N.; Mitsukura, K.; Uneyama, K., *J. Fluorine Chem.* **2001**, *112*, 63.
5. Xu, W.; Chen, Q.-Y., *Org. Biomol. Chem.* **2003**, *1*, 1151.
6. Bessard, Y.; Schlosser, M., *Tetrahedron* **1991**, *47*, 7323.
7. Xu, W.; Chen. Q.-Y., *J. Org. Chem.* **2002**, *67*, 9421.

Qing-Yun Chen
*Shanghai Institute of Organic Chemistry,
Chinese Academy of Sciences, Shanghai, China*

N-(Trimethylsilyl)imidazole[1]

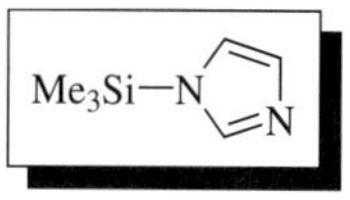

[18156-74-6]   C$_6$H$_{12}$N$_2$Si   (MW 140.29)
InChI = 1S/C6H12N2Si/c1-9(2,3)8-5-4-7-6-8/h4-6H,1-3H3
InChIKey = YKFRUJSEPGHZFJ-UHFFFAOYSA-N

(silylating agent for alcohols[1] and 1,3-dicarbonyl compounds;[2] reaction with esters to give imidazolides; preparation of O-trimethylsilyl monothioacetals; aromatization of the A-ring of steroids)

Physical Data: bp 93–94 °C/14 mmHg; fp 5 °C; *d* 0.956 g cm^{-3}.
Form Supplied in: liquid.
Handling, Storage, and Precautions: flammable. Harmful by inhalation, in contact with skin, and if swallowed. Possible carcinogen. Use in a fume hood. When using, wear protective gear and clothing. Handle and store under N$_2$. Store in a cool dry place.

Original Commentary

Janet Wisniewski Grissom & Gamini Gunawardena
University of Utah, Salt Lake City, UT, USA

General Discussion. *N*-(Trimethylsilyl)imidazole is a powerful silylating agent for alcohols (eq 1).[2] 1,3-Dicarbonyl compounds, although usually existing mostly in the enol form, are not efficiently silylated by standard methods. *N*-(Trimethylsilyl)-imidazole has proved to be an ideal reagent for this transformation (eq 2).[3]

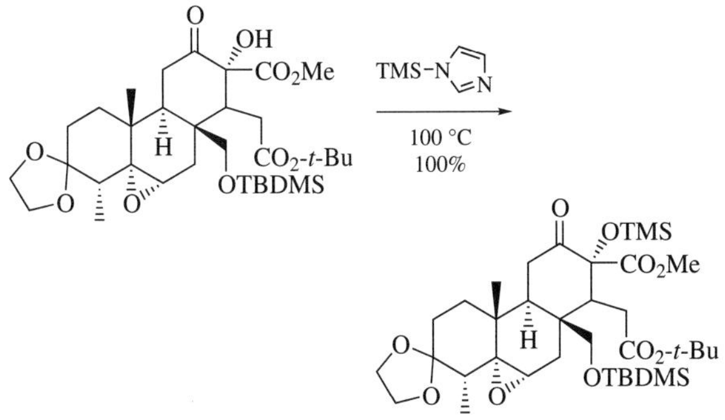

$$(1)$$

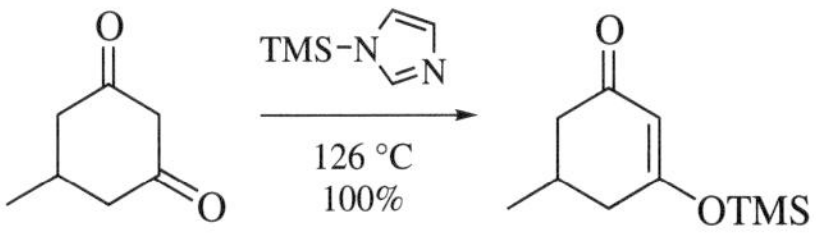

$$(2)$$

N-(Trimethylsilyl)imidazole reacts with phenyl and 2,2,2-trifluoroethyl esters under base catalysis to give acylimidazolides (eq 3).[4] These imidazolides show reactivity similar to acyl chlorides and therefore are useful in synthesis.[5]

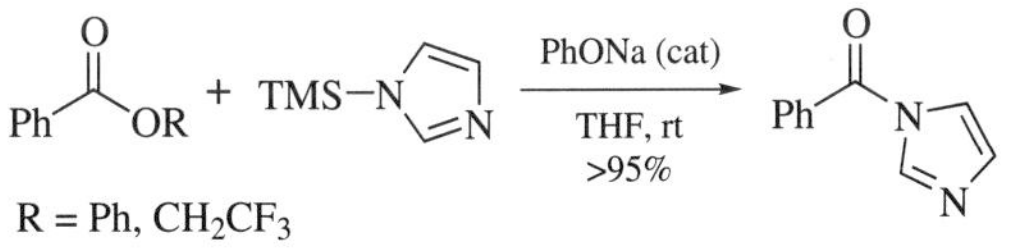

$$(3)$$

R = Ph, CH$_2$CF$_3$

N-(Trimethylsilyl)imidazole is used in the preparation of *O*-trimethylsilyl monothioacetals directly from corresponding carbonyl compounds and thiols (eq 4).[6] The basicity of the medium due to the imidazole generated during the reaction effectively prevents dithioacetalation.

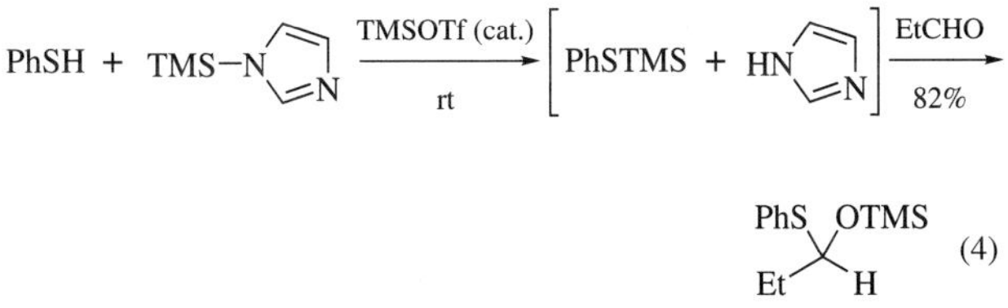

$$(4)$$

Aromatization of the A-ring of steroid systems has been achieved by the use of *N*-(trimethylsilyl)imidazole (eq 5).[7]

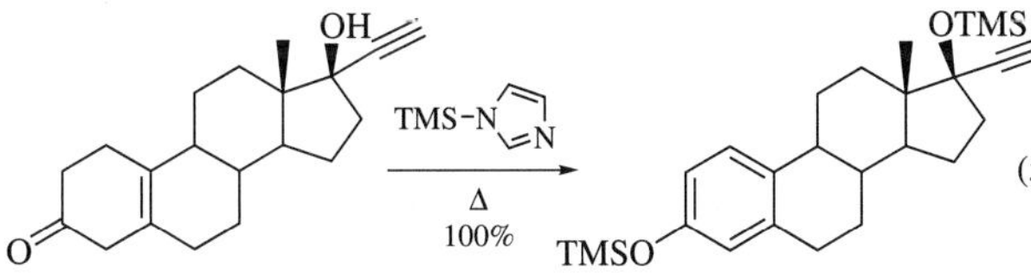

$$(5)$$

First Update

Thanasis Gimisis & Crina Cismaş
University of Athens, Athens, Greece

The use of *N*-(trimethylsilyl)imidazole (TMSIM) as a silylating reagent for alcohols has been already demonstrated and intensively applied in organic synthesis. Additionally, *N*-(trimethylsilyl)imidazole can be a good precursor for imidazole functionalized organic compounds.

Hydroxyl Silylation Reactions. *N*-(Trimethylsilyl)imidazole has been effective in the protection of a hindered allylic alcohol (eq 6), where common TBS protecting reagents were unreactive (TBS-Cl, TBS-imidazole, *N*-TBS-*N*-methyltrifluoro acetamide) or caused decomposition (TBS-OTf).[8]

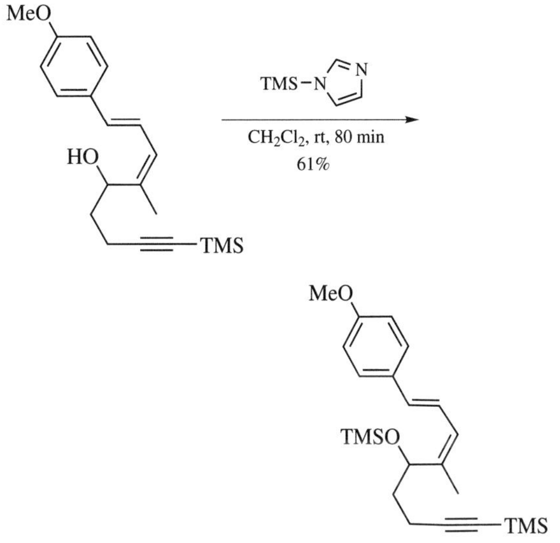

$$(6)$$

It has been shown that addition of catalytic amounts of TBAF (0.02 equiv.) to *N*-(trimethylsilyl)imidazole promotes the quantitative silylation of alcohols in complex polyhydroxy compounds under milder conditions. The reaction proved to be highly chemoselective, with other functional groups like ketones, enones, epoxides, amines or carboxylic acids being unreactive under these conditions. The silylation of a complex derivative that could not be achieved without TBAF was effected in 1 h (eq 7).[9]

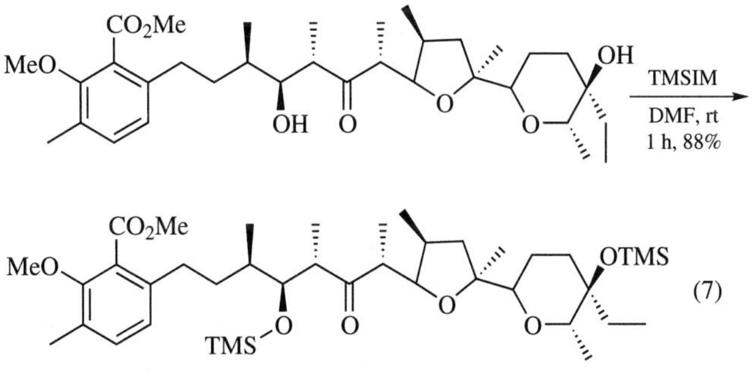

$$(7)$$

In a series of studies directed toward the development of new trifluoromethylating reagents, *N*-(trimethylsilyl)imidazole has been used for the protection of intermediate hemiaminals (eq 8)[10] and amino alcohols (eq 9).[11]

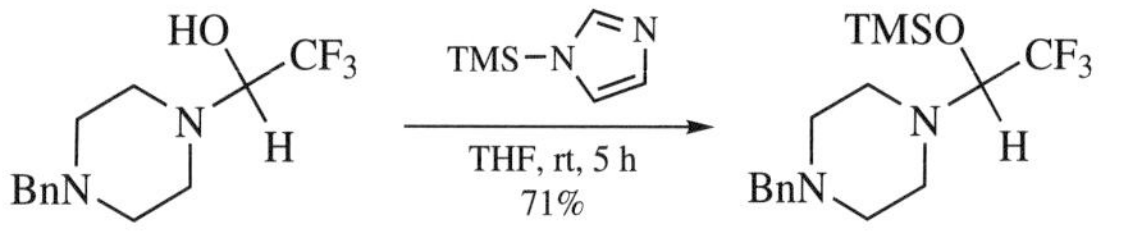

$$ (8) $$

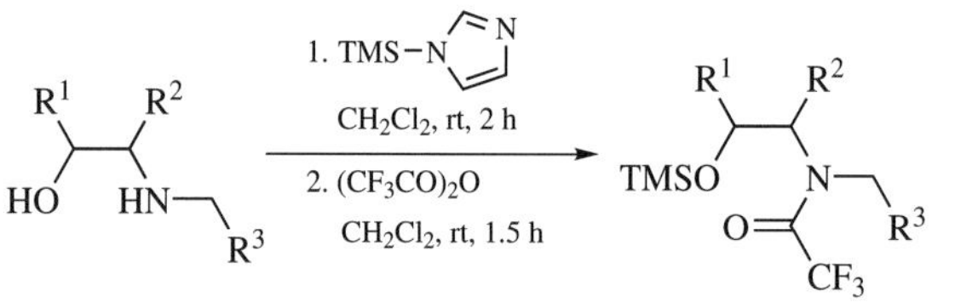

$$ (9) $$

R¹	H	H	Ph	H	H	Ph
R²	H	H	H	Ph	Ph	Me
R³	H	Ph	H	Ph	*n*-Bu	H

N-Alkylhydroxylamines can also be silylated by *N*-(trimethylsilyl)imidazole in the presence of a suitable base (eq 10). Oxidative coupling of *N*-alkyl-*O*-(trimethylsilyl)hydroxylamines with cyanocuprates bearing aromatic and heteroaromatic transferable ligands leads to the formation of *N*-alkyl aromatic and heteroaromatic secondary amines (eq 11).[12]

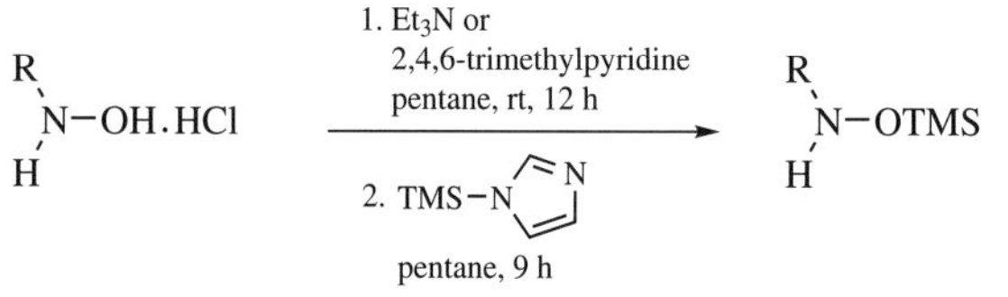

$$ (10) $$

R = Me, *i*-Pr, *t*-Bu

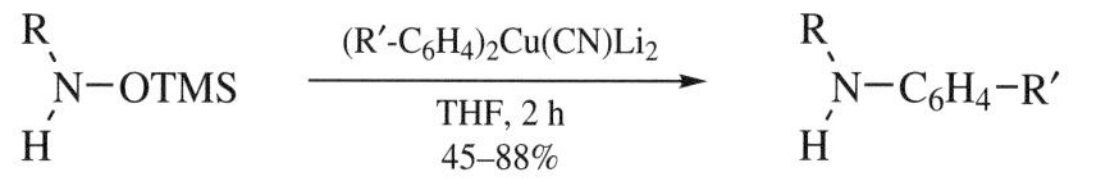

$$ (11) $$

for R = Me, R′ = H, *p*-Me, *m*-OMe, *p*-OMe
for R = *i*-Pr, R′ = H, *p*-Me, *p*-F, *p*-OMe
for R = *t*-Bu, R′ = H, *p*-Me, *m*-OMe, *p*-F

Trimethylsilylated *N*-alkylhydroxylamines have also been used for the conversion of catecholborane esters into amines with retention of configuration.[13]

Alkyl esters can be converted to monosilyl acetals, useful for the synthesis of different homoallylic alcohols or ethers,[14] via reduction and trapping with *N*-(trimethylsilyl)imidazole (eq 12). Trapping of the reactive aluminum intermediate was less effective with TMS-OTf and ineffective with chlorosilanes. It was also proposed that monosilyl acetals can be a better alternative to the commonly used dialkyl acetals for the protection of carbonyl compounds.

A growing interest has surged in the application of *N*-(trimethylsilyl)imidazole for the derivatization of various reaction mixtures prior to GC/MS analysis. Silylation under optimal conditions (TMSIM:pyrrolidine = 1:2), provided the corresponding fatty acid pyrrolidides that were directly analyzed by GC–MS.[15] The analysis of the aminoglycoside antibiotics kanamicin sulfate and gentamicin sulfate was performed on capillary GC–MS with the hydroxyl groups of the analytes derivatized with

N-(trimethylsilyl)imidazole.[16] The derivatization was also successfully applied to the accurate detection of a diastereomeric process impurity of commercial formoterol fumarate,[17] to the identification and quantification of estrogens in ground water and swine lagoon samples,[18] and to the analysis of dihydrostreptomycin.[19] It was used to establish the diastereomeric ratio of aldol products in a 1,2-asymmetric induction study of enolate additions to heteroatom-substituted aldehydes.[20]

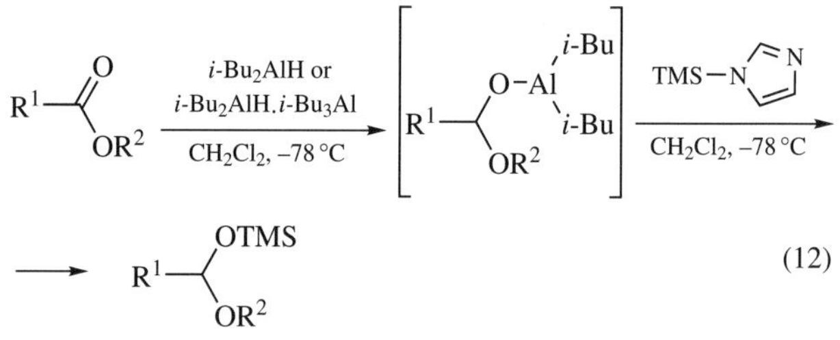

$$ (12) $$

for R² = Me, R¹ = *n*-heptyl, *n*-Pr
for R² = Et, R¹ = *n*-heptyl, *i*-Bu, –CH(Me)OMe, –CH(Ph)OMe, –CH(Me)N=CPh₂

It has been shown that trimethylsilylation of the silanol groups of Ti-substituted mesoporous molecular sieves (Ti-MCM-41), a catalyst of olefin epoxidation reactions in aqueous H_2O_2, improve the catalytic activity and the selectivity toward epoxide formation. *N*-(Trimethylsilyl)imidazole was the most effective in improving the hydrophobicity and catalytic performance.[21]

The phenolic function is also a good substrate for TMS silylation with TMSIM, demonstrated by the selective formation of the cone isomer of the *O*-silylated derivative of a hexahomotriazacalix[3]arene.[22]

Silyl Aminal Formation Reactions. The related *N*-(*t*-butyldimethylsilyl)imidazole reagent has been used for the protection of an aldehyde function.[23,24] Although stable towards reductants and Grignard reagents,[23] the formed *N*,*O*-silyl aminal has exhibited the same reactivity toward organolithium reagents as the parent compound.[24] This reactivity has been attributed to the acidity of the 2-H which leads to a novel retro-[1,4] Brook rearrangement (eq 13). Consequently, 2-Me substitution of the imidazolyl moiety leads to *N*,*O*-acetals stable in the presence of organolithium reagents.

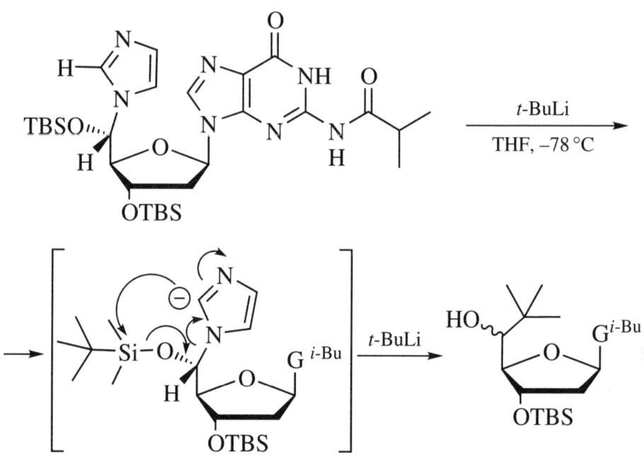

$$ (13) $$

In a similar fashion, *N*-(trimethylsilyl)imidazole has been used for the temporary protection of an aldehyde function prior to the reduction of an ester function in a complex structure (eq 14).[25] The *N,O* acetal was used without further purification and the parent aldehyde was easily recovered during the work-up after the ester reduction.

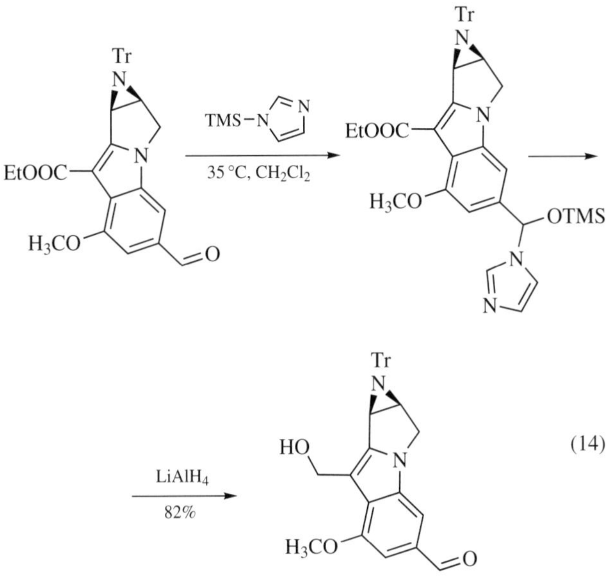

(14)

These findings suggest that the stability of imidazolyl acetals can be adjusted by the choice of protecting silyl imidazole reagent.

Nitrogen Silylation Reactions. *N*-(Trimethylsilyl)imidazole can serve as an electrophilic trap for the highly reactive intermediate formed in the reaction of *N*-aryl imines with trifluoromethyltrimethylsilane under fluoride catalysis (eq 15).[26] In the absence of *N*-(trimethylsilyl)imidazole, the reactive intermediate decomposes to difluorocarbene, shifting the equilibrium toward the starting imine derivative. Chlorotrimethylsilane or *N*-methyl-*N*-trimethylsilylacetamide were not able to promote the formation of monotrifluoromethylated amine, with the former leading to almost complete recovery of the starting trifluoromethyltrimethylsilane while the latter induced decomposition of the imine.

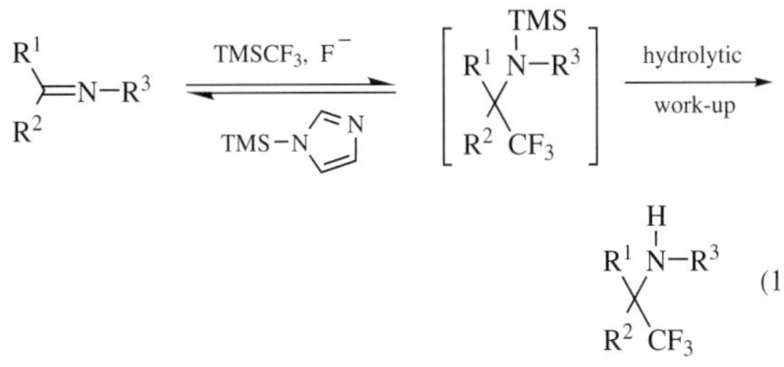

(15)

R^1	Ph	Ph	Ph	Ph	Ph	2-Naphthyl	2-Furyl	3-MeOC$_6$H$_4$
R^2	H	H	CF$_3$	Me	H	H	H	H
R^3	Ph	3-CF$_3$C$_6$H$_4$	Ph	Ph	PhCH$_2$	Ph	Ph	Ph

Acyl Imidazole Formation. *N*-(Trimethylsilyl)imidazole has proven to be a useful reagent in the synthesis of 1-acylimidazoles and amides under mild conditions. Aroyl chlorides react rapidly to form 1-aroylimidazoles which, after removal of the generated chlorotrimethylsilane, were used without further purification

as coupling reagents in Wittig reactions with acceptor-stabilized phosphoranes.[27]

Free carboxylic acids are not reactive toward *N*-(trimethylsilyl)-imidazole. It has been recently shown that bis- and especially tris(imidazol-1-yl)silane compounds reacted readily with free carboxylic acids to form 1-acylimidazoles in good to very good yields.[28] The imidazol-1-yl silane derivatives are readily available by the *trans*-silylation reaction between *N*-(trimethylsilyl)imidazole and chlorosilanes (eq 16).

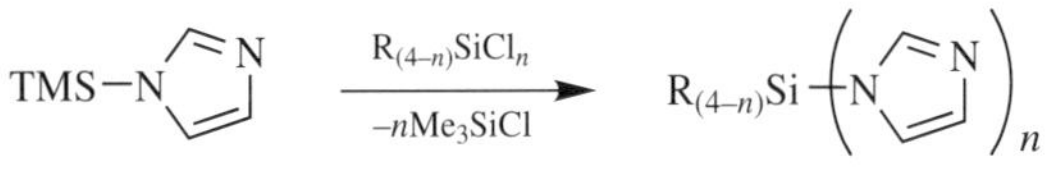

R = Me, Bu

This method was applied in the condensation of free carboxylic acids with amines, and the corresponding carboxamides were obtained in good to high yields. The low solubility of tetrakis-(imidazol-1-yl)silane was circumvented by using the 2-Me-imidazol-1-yl analogue, in which case the only side product of the reaction was silica [(SiO$_2$)$_n$], which is insoluble and can be removed easily by filtration (eq 17).

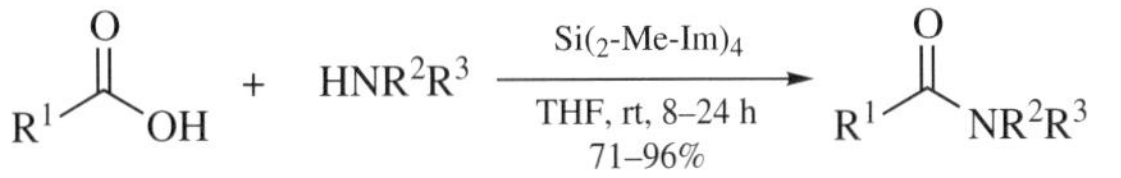

R^1	Ph	Ph(CH$_2$)$_2$	Ph(CH$_2$)$_2$	Ph(CH$_2$)$_2$	Ph(CH$_2$)$_2$
R^2	Ph(CH$_2$)$_3$	Ph(CH$_2$)$_3$	PhCHMe	Ph	PhCH$_2$
R^3	H	H	H	H	Me

Michael Addition Reactions. *N*-(Trimethylsilyl)imidazole reacted smoothly with acetylenic silyl ketones and the Michael addition adduct (*β*-imidazolyl acylsilane) was obtained in good yield.[29] The reaction is regio- and stereospecific, with no traces of the 1,2-addition product or *cis* isomer being observed (eq 18).

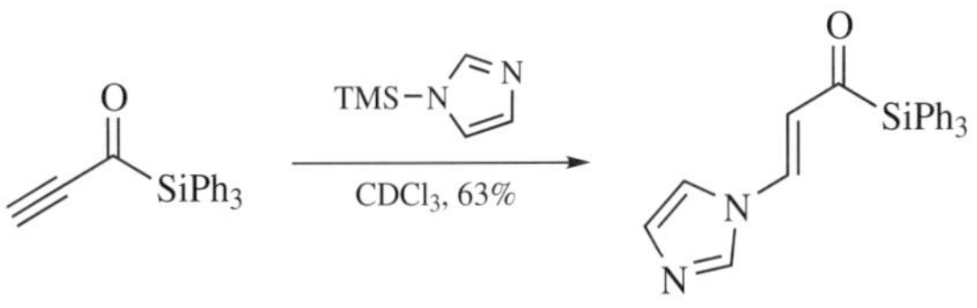

(18)

The coupling of *N*-(trimethylsilyl)imidazole with *N,N*-bis(silyloxy)enamines leads to the formation of *α*-imidazolyl-substituted oximes bearing functional groups at different positions relative to the hydroxyimino group. The reaction is fast, high yielding, and stereoselective providing the *E*-isomer. The unprotected imidazole is not suitable for effective *C,N*-coupling, since only 5% of the corresponding oxime derivatives were obtained (eq 19).[30]

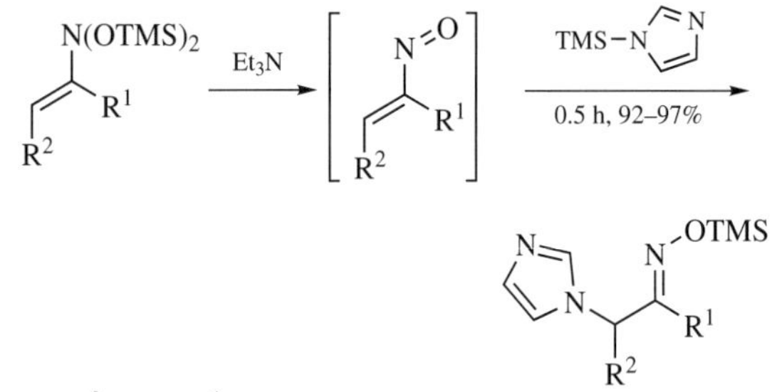

(19)

for R² = Me, R¹ = H, COOMe

for R² = H, R¹ = Me

Substitution Reactions. 3-(1-*t*-Butyldimethylsilyloxy)-ethyl-4-phenylsulfinylazetidin-2-one has been quantitatively and stereospecifically converted to *trans*-4-imidazo-3-(1-*t*-butyldi-methylsilyloxy)ethylazetidin-2-one using N-(trimethylsilyl)imidazole as a nucleophile (eq 20).[31]

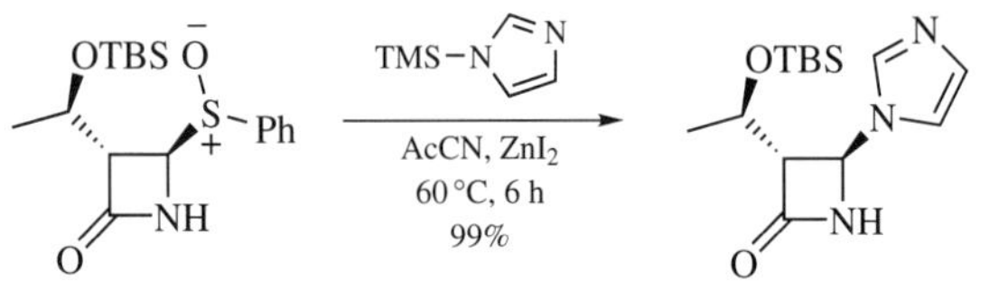

(20)

N-(Trimethylsilyl)imidazole has proven to be a good nucleophile for the ring opening of epoxides,[32] but a moderate one for the ring opening of either optically pure N-Cbz-L-serine-β-lactone[33] or thiiranium ion intermediates.[34,35]

The benzofuran epoxide ring opening (eq 21) proceeded rapidly, and the *cis* isomer of the imidazole adduct was formed exclusively. Intramolecular hydrogen bonding between the hydroxyl function and imidazole was proposed for this unexpected stereospecificity.[32]

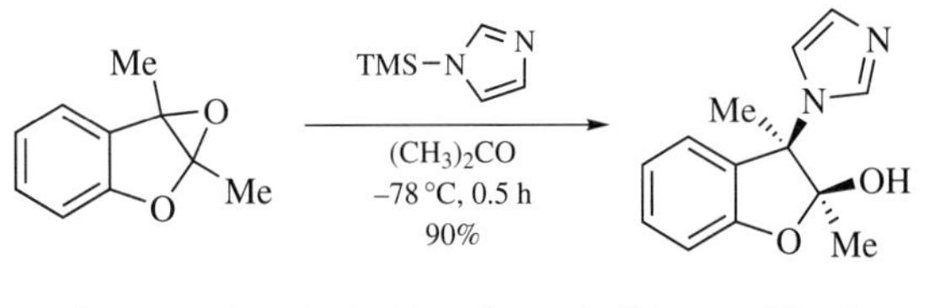

(21)

N,N'-Disubstituted imidazolium halides, utilized as ionic liquids, are readily available in a one-step procedure using N-(trimethylsilyl)imidazole in combination with alkyl- or aryl halides. Important features of this synthetic approach are that the synthesis is performed under anhydrous conditions, and all side products are readily removed.[36] Under this procedure, a wide range of ionic liquids with different counter anions have become available.[37,38] Polycondensation reactions in the above ionic liquids have led to the formation of high molecular weight polyimides, polynaphthoyleneimides or poly[naphthoylene-bis(benzimidazole)]s in the absence of any other catalyst.

Phosphoroimidazolidate Formation. N-(Trimethylsilyl)-imidazole has become an important reagent in nucleotide chemistry. In this respect, the formation of phosphate derivatives is almost quantitative, with the side products being easily removable volatile compounds.

Methyl phosphoroimidazolidate was generated oxidatively by treatment of methyl H-phosphonate with N-(trimethylsilyl)imidazole in the presence of CCl₄ (eq 22). It was demonstrated to be

the most reactive methoxyphosphoryl donor for the synthesis of CH₃pppG (eq 23). Other reagents such as methyl phosphoroimidazolidate or piperididate gave low yields in the above reaction.[39]

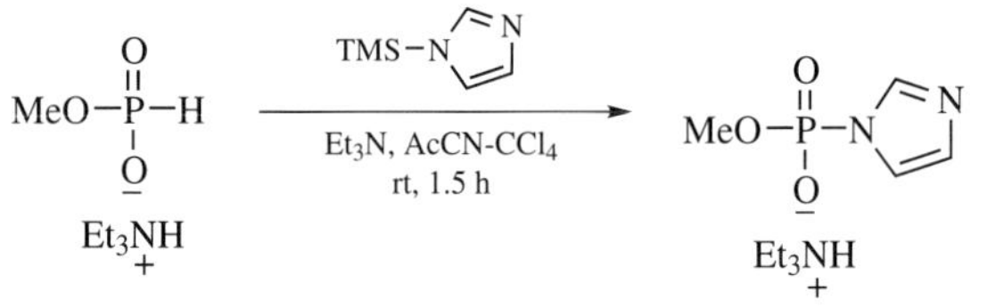

(22)

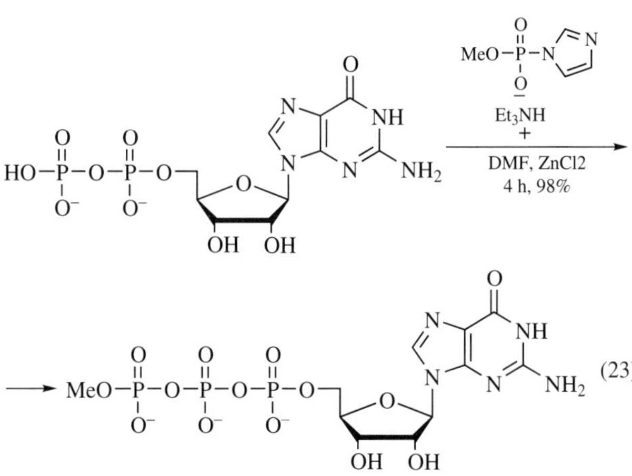

(23)

The same method was applied to oxidative 5'-phosphoroimidazolidate formation in 5'-H-phosphonate derivatives.[40,41]

2',3'-Dideoxy nucleoside analogues, containing a single hydroxyl function, can be directly converted into 5'-triphosphates by phosphorylating the 5'-hydroxyl group with tris-imidazolylphosphate (TIP) (eq 24). The crystalline TIP reagent was prepared by precipitation from a concentrated reaction mixture containing N-(trimethylsilyl)imidazole and phosphoryl chloride in anhydrous toluene.[42]

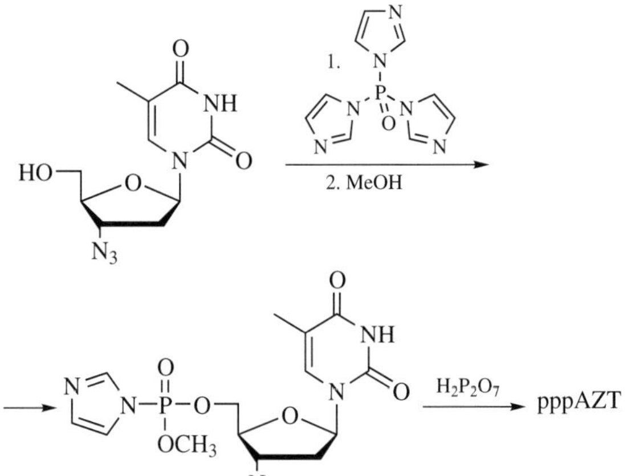

(24)

N-(Trimethylsilyl)imidazole has also been used for the activation of the phosphorus group at the 2' or 3' position of nucleosides in oligonucleotide synthesis via the phosphoroimidazolidate or thioamidite methods.[43,44]

Other Uses. A large number of macrocyclic imidazolylboranes were synthesized via a macrocyclization reaction of N-(trimethylsilyl)imidazole with haloboranes under high dilution conditions.[45,46]

It has been demonstrated that *N*-(trimethylsilyl)imidazole is able to transform transition metal complexes containing a metal–Cl bond to the corresponding metal–imidazole analogues. This transformation has been reported for transition metal complexes of Ni(II), Pd(II), Pt(II),[47] Ru(II), Ru(III),[48] Te(IV)[49] and Ti(IV).[50]

1. For reviews on silylations, see: (a) Birkofer, L.; Ritter, A., *Angew. Chem., Int. Ed. Engl.* **1965**, *4*, 417. (b) Klebe, J. F., *Acc. Chem. Res.* **1970**, *3*, 299. (c) Colvin, E. W., *Chem. Soc. Rev.* **1978**, *7*, 15. (d) Cooper, B. E., *Chem. Ind. (London)* **1978**, 794. (e) Fleming, I., *Chem. Soc. Rev.* **1981**, *10*, 83. (f) Lalonde, M.; Chan, T. H., *Synthesis* **1985**, 817.

2. Kerwin, S. M.; Paul, A. G.; Heathcock, C. H., *J. Org. Chem.* **1987**, *52*, 1686.

3. Zhou, Y. F.; Huang, N. Z., *Synth. Commun.* **1982**, *12*, 795.

4. Bates, G. S.; Diakur, J.; Masamune, S., *Tetrahedron Lett.* **1976**, 4423.

5. Staab, H. A., *Angew. Chem., Int. Ed. Engl.* **1962**, *1*, 351.

6. Sassaman, M. B.; Surya Prakash, G. K.; Olah, G. A., *Synthesis* **1990**, 104.

7. Kulkarni, S.; Abdel-Baky, S.; Le Quesne, P. W.; Vouros, P., *Steroids* **1989**, *53*, 131.

8. Wender, P. A.; Smith, T. E., *Tetrahedron* **1998**, *54*, 1255.

9. Tanabe, Y.; Murakami, M.; Kitaichi, K.; Yoshida, Y., *Tetrahedron Lett.* **1994**, *35*, 8409.

10. Billard, T.; Langlois, B. R.; Blond, G., *Tetrahedron Lett.* **2000**, *41*, 8777.

11. Joubert, J.; Roussel, S.; Christophe, C.; Billard, T.; Langlois, B. R.; Vidal, T., *Angew. Chem., Int. Ed.* **2003**, *42*, 3133.

12. Bernardi, P.; Dembech, P.; Fabbri, G.; Ricci, A.; Seconi, G., *J. Org. Chem.* **1999**, *64*, 641.

13. Knight, F. I.; Brown, J. M.; Lazzari, D.; Ricci, A.; Blacker, A. J., *Tetrahedron* **1997**, *53*, 11411.

14. Sames, D.; Liu, Y.; DeYoung, L.; Polt, R., *J. Org. Chem.* **1995**, *60*, 2153.

15. Vetter, W.; Walther, W., *J. Chromatogr.* **1990**, *513*, 405.

16. Preu, M.; Guyot, D.; Petz, M., *J. Chromatogr. A* **1998**, *818*, 95.

17. Akapo, S. O.; Wegner, M.; Mamangun, A.; McCrea, C.; Asif, M.; Dussex, J.-C., *J. Chromatogr. A* **2004**, *1045*, 211.

18. Fine, D. D.; Breidenbach, G. P.; Price, T. L.; Hutchins, S. R., *J. Chromatogr. A* **2003**, *1017*, 167.

19. Preu, M.; Petz, M., *J. Chromatogr. A* **1999**, *840*, 81.

20. Evans, D. A.; Siska, S. J.; Cee, V. J., *Angew. Chem., Int. Ed.* **2003**, *42*, 1761.

21. Bu, J.; Rhee, H.-K., *Catal. Lett.* **2000**, *66*, 245.

22. Chirakul, P.; Hampton, P. D.; Duesler, E. N., *Tetrahedron Lett.* **1998**, *39*, 5473.

23. Quan, L. G.; Cha, J. K., *Synlett* **2001**, 1925

24. Gimisis, T.; Arsenyan, P.; Georganakis, D.; Leondiadis, L., *Synlett* **2003**, 1451.

25. Kim, M.; Vedejs, E., *J. Org. Chem.* **2004**, *69*, 7262.

26. Blazejewski, J.-C.; Anselmi, E.; Wilmshurst, M. P., *Tetrahedron Lett.* **1999**, *40*, 5475.

27. Bohlmann, R.; Strehlke, P., *Tetrahedron Lett.* **1996**, *37*, 7249.

28. Tozawa, T.; Yamane, Y.; Mukaiyama, T., *Chem. Lett.* **2005**, *34*, 734.

29. Degl'Innocenti, A.; Capperucci, A.; Reginato, G.; Mordini, A.; Ricci, A., *Tetrahedron Lett.* **1992**, *33*, 1507.

30. Lesiv, A. V.; Ioffe, S. L.; Strelenko, Y. A.; Tartakovsky, V. A., *Helv. Chim. Acta* **2002**, *85*, 3489.

31. Kita, Y.; Shibata, N.; Yoshida, N.; Tohjo, T., *Tetrahedron Lett.* **1991**, *32*, 2375.

32. Adam, W.; Hadjiarapoglou, L.; Mosandl, T.; Saha-Moeller, C. R.; Wild, D., *J. Am. Chem. Soc.* **1991**, *113*, 8005.

33. Ratemi, E. S.; Vederas, J. C., *Tetrahedron Lett.* **1994**, *35*, 7605.

34. Gill, D. M.; Pegg, N. A.; Rayner, C. M., *J. Chem. Soc. Perkin Trans. 1* **1993**, 1371.

35. Gill, D. M.; Pegg, N. A.; Rayner, C. M., *Tetrahedron* **1996**, *52*, 3609.

36. Harlow, K. J.; Hill, A. F.; Welton, T., *Synthesis* **1996**, 697.

37. Zhao, D.; Fei, Z.; Ohlin, C. A.; Laurenczy, G.; Dyson, P. J., *Chem. Commun.* **2004**, 2500.

38. Vygodskii, Y. S.; Lozinskaya, E. I.; Shaplov, A. S.; Lyssenko, K. A.; Antipin, M. Y.; Urman, Y. G., *Polymer* **2004**, *45*, 5031.

39. Kadokura, M.; Wada, T.; Urashima, C.; Sekine, M., *Tetrahedron Lett.* **1997**, *38*, 8359.

40. Sekine, M.; Ushioda, M.; Wada, T.; Seio, K., *Tetrahedron Lett.* **2003**, *44*, 1703.

41. Sekine, M.; Aoyagi, M.; Ushioda, M.; Ohkubo, A.; Seio, K., *J. Org. Chem.* **2005**, *70*, 8400.

42. Vinogradov, S. V.; Kohli, E.; Zeman, A. D., *Mol. Pharm.* **2005**, *2*, 449.

43. Beigelman, L.; Matulic-Adamic, J.; Haeberli, P.; Usman, N.; Dong, B.; Silverman, R. H.; Khamnei, S.; Torrence, P. F., *Nucleic Acids Res.* **1995**, *23*, 3989.

44. Kitamura, A.; Horie Y.; Yoshida T., *Chem. Lett.* **2000**, *29*, 1134.

45. Weiss, A.; Pritzkow, H.; Siebert, W., *Angew. Chem., Int. Ed.* **2000**, *39*, 547.

46. Weiss, A.; Barba, V.; Pritzkow, H.; Siebert, W., *J. Organomet. Chem.* **2003**, *680*, 294.

47. Siddiqi, Z. A.; Qidwai, S. N.; Mathew, V. J.; Khan, A. A., *Synth. React. Inorg. Met.-Org. Chem.* **1993**, *23*, 1735.

48. Siddiqi, Z. A.; Qidwai, S. N.; Mathew, V. J., *Synth. React. Inorg. Met.-Org. Chem.* **1993**, *23*, 709.

49. Shakir, M.; Islam, K. S.; Hasan, S. S.; Jahan, N., *Phosphor. Sulfur Silicon Relat. Elem.* **1998**, *136*, 603.

50. Hensen, K.; Lemke, A.; Naether, C., *Z. Anorg. Allg. Chem.* **1997**, *623*, 1973.

Trimethylsilylketene[1]

[4071-85-6] $C_5H_{10}OSi$ (MW 114.24)

InChI = 1S/C5H10OSi/c1-7(2,3)5-4-6/h5H,1-3H3

InChIKey = GYUIBRZEAWWIHU-UHFFFAOYSA-N

(reactive acylating agent for amines and alcohols;[2] building block for synthesis of coumarins;[3] synthesis of α-silyl ketones via the addition of organocerium reagents;[4] treatment with stabilized ylides forms trimethylsilyl-substituted allenes;[2a] cycloaddition with aldehydes affords β-lactones;[5] forms small rings with diazomethane;[6] treatment with *n*-BuLi forms a ketene enolate[7])

Physical Data: [2a,7] bp 81–82 °C; *d* 0.80 g cm^{-3}.

Solubility: sol CH_2Cl_2, $CHCl_3$, CCl_4, THF, diethyl ether, and most standard organic solvents; reacts with alcoholic and amine solvents.

Form Supplied in: colorless oil; not commercially available.

Analysis of Reagent Purity: [8] IR (CCl_4) 2130 cm^{-1}; ^{1}H NMR (60 MHz, CCl_4) δ 1.65 (s, 1 H), 0.12 (s, 9 H) ppm; ^{13}C NMR (50.3 MHz, $CDCl_3$) δ 179.4, 0.5, $-$0.2 ppm.

Preparative Methods: most often prepared (eq 1) by pyrolysis of ethoxy(trimethylsilyl)acetylene at 120 °C (100 mmol scale, 65% yield).[2a] Recently, pyrolysis of *t*-butoxy(trimethylsilyl) acetylene has been shown to be a convenient alternative for the preparation of trimethylsilylketene (**1**). Thermal decomposition of *t*-butoxy(trimethylsilyl)acetylene causes elimination of 2-methylpropene slowly at temperatures as low as 50 °C and instantaneously at 100–110 °C (30 mmol scale, 63% yield).[8] The main advantage of this method is that it is possible to generate trimethylsilylketene in the presence of nucleophiles, leading to in situ trimethylsilylacetylation (eq 2). Increased shielding of the triple bond prevents problems such as polymerization and nucleophilic attack that occur when the ketene is generated in situ from (trimethylsilyl)ethoxyacetylene. Trimethylsilylketene can also be prepared (eq 3) via the dehydration of commercially available trimethylsilylacetic acid with 1,3-dicyclohexylcarbodiimide (DCC) in the presence of a catalytic amount of triethylamine (100 mmol scale, 63%).[9] Other typical methods used for ketene generation such as dehydrohalogenation of the acyl chloride[10] and pyrolysis of the anhydride[10b,11] have been applied to the preparation of (**1**); however, both methods afford low yields.

$$\text{(1)}$$

$$\text{(2)}$$

$$\text{(3)}$$

There have been no significant developments in the methods used to prepare trimethylsilylketene (TMSK). However, Black et al. have published slight modifications.[30] to the original preparation by Ruden,[2a] which primarily deals with accessing ethoxyacetylene.

Purification: purified by distillation at 82 °C/760 mmHg.

Handling, Storage, and Precautions: unusually stable for an aldoketene with respect to dimerization and decomposition. Samples stored neat under nitrogen at room temperature show no noticeable decomposition after several months.

Original Commentary

Jennifer L. Loebach & Rick L. Danheiser
Massachusetts Institute of Technology, Cambridge, MA, USA

Trimethylsilylacetylation of Alcohols and Amines. Trimethylsilylketene (**1**) has been used as a potent acylating agent for amines and alcohols (eq 4).[2a] It reacts almost instantaneously

with hindered amines to form α-silyl amides in quantitative yield. Reaction with alcohols such as *t*-butanol is much slower (CCl_4, 48 h, rt, 80% yield). However, boron trifluoride etherate strongly catalyzes α-silylacetate formation. Hindered tertiary alcohols that cannot be acylated by standard reagents such as benzoyl chloride, acetyl chloride, and acetic anhydride (even in the presence of DMAP) can be acylated by (**1**). Desilylation can be effected using potassium fluoride in methanol to produce the acetate directly (eq 5).[12] When the addition of alcohols to (**1**) is catalyzed by zinc halides, a high degree of functionality can be tolerated in the substrate, including carbonyl, acetal, thioacetyl, epoxy, and alkenic groups.[2b] In contrast, $BF_3 \cdot Et_2O$ catalysis results in partial product desilylation with alcohols containing carbonyl groups and also causes cleavage of acetal groups. Other reactions used to prepare α-silyl acetates including the Reformatsky reaction,[13] *C*-silylation of esters,[14] and silyl migration from acylsilanes[15] are incompatible with many active functional groups. Functionalized α-silyl acetates serve as useful precursors to butenolides (eq 6).[2b]

$$\text{(4)}$$

$$\text{(5)}$$

$$\text{(6)}$$

Synthesis of Coumarins via Cyclization–Elimination. Reaction of (**1**) with *o*-acylphenols affords coumarins in high yields in a one-pot reaction (eq 7).[3] The reaction is applicable to a variety of functionalized aromatic systems. Coumarins form more readily in somewhat higher yield using this procedure as compared to a similar method involving the use of a cumulated phosphorus ylide reagent, ketenylidenetriphenylphosphorane.[16] Workup and product isolation is easier in view of the triphenylphosphine oxide byproduct formed in the ylide reaction. Other methods used to convert *o*-acylphenols to coumarins include approaches based on Knovenagel[17] and Pechmann[18] reactions. The method described above involves milder conditions and often provides coumarins in higher yields.

(7)

(11)

One-pot Formation of α-Silyl Ketones. α-Silyl ketones can be prepared in a one-pot synthesis by the addition of organocerium reagents to (**1**) followed by subsequent quenching with aqueous ammonium chloride or alkyl halides (eq 8).[4] The organocerium reagents can be easily prepared from cerium(III) chloride and an alkyl- or aryllithium. Organolithium reagents are not suitable for this reaction because proton abstraction is preferred resulting only in complicated reaction mixtures of products.[7] A variety of other methods have been reported for the synthesis of α-silyl ketones, including methods based on 1,3-O to C silyl migration from lithiated silyl enol ethers,[19] other types of migration from silicon-containing compounds,[20] isomerization of silicon-containing allyl alcohols,[21] and Si–H insertion reactions of diazo ketones,[22] as well as classical preparations involving carboxylic acid derivatives[23] and the oxidation of β-hydroxy silanes.[21,23] However, most of these methods require multistep procedures and are less convenient.

(8)

Preparation of Trimethylsilyl-Substituted Allenes. Treatment of (**1**) with a stabilized phosphorus ylide, (ethoxycarbonylmethylene)triphenylphosphorane, affords the silyl substituted allenic ester in 85% yield (eq 9).[2a] These alkenation reactions only occur with stabilized phosphorus ylides; unstabilized ylides reportedly form complex mixtures with trimethylsilylketene.

(9)

Preparation of β-Lactones. Reaction of (**1**) with saturated aldehydes in the presence of BF$_3$·Et$_2$O results in mixtures of both cis- and trans-2-oxetanones (eq 10). Ketones are reported not to undergo cycloadditions with (**1**).[5c,d] Cycloadditions have been described with both saturated and α,β-unsaturated aldehydes. Distillation is reported to promote a 1,3-shift of the organosilyl group accompanied by ring opening to yield the trimethylsilyl dienoate ester.[5a]

(10)

cis:trans = 60:40

Recently, a method for the stereoselective preparation of the cis-substituted β-lactones using catalytic methylaluminum bis(4-bromo-2,6-di-t-butylphenoxide) has been described (eq 11).[5e] This method is reported to work well for alkyl, aryl, and unsaturated aldehydes. Stoichiometric amounts of catalyst lead to desilylation followed by ring opening to afford (Z)-alkenoic acids.

Reaction with Diazomethane to Form Silylated Cyclopropanes and Cyclobutanones. The reaction of (**1**) and diazomethane results in a mixture of products. Treatment of equimolar amounts of (**1**) with diazomethane at $-130\,^\circ$C leads to (trimethylsilyl)cyclopropanone in moderate yield (eq 12).[6] The product obtained can then react with a second equivalent of diazomethane upon warming to $-78\,^\circ$C, resulting in ring expansion to a mixture of 2- and 3-(trimethylsilyl)cyclobutanones. Alternatively, these isomeric products may be obtained directly with 2 equiv of diazomethane. Treatment of the isomeric (trimethylsilyl)cyclobutanone mixture with methanol makes it possible to obtain pure 3-substituted isomer in 84% yield.[24] This 3-(trimethylsilyl)cyclobutanone derivative can also be formed by a more elaborate route via the regioselective [2 + 2] addition of dichloroketene to trimethylsilylacetylene followed by hydrogenation and reductive removal of the two chlorine atoms.[25] trimethylsilyldiazomethane has also been reported to react with (**1**) to form bis-silyl substituted cyclopropanones.[26]

(12)

40:60

Formation of the Ketene Enolate. Addition of (**1**) to a solution of n-butyllithium at $-100\,^\circ$C presumably forms the ketene enolate. Quenching with chlorotrimethylsilane then affords the bis-silylketene in high yield (eq 13).[7] Attempts to trap the enolate with other electrophiles have been unsuccessful. Other methods[27] affording bis(trimethylsilyl)ketene have been reported including a very similar method involving deprotonation of (**1**) with triethylamine and quenching with trimethylsilyl trifluoromethanesulfonate.[28] However, the yield for this reaction is much lower (48%). Other reactions report bis(trimethylsilyl)ketene as a byproduct.[29]

(13)

First Update

Jan A. R. Adams

ARIAD Pharmaceuticals, Inc., Cambridge, MA, USA

Preparation of Trimethylsilyl-substituted Allenes. Kita et al. have explored the addition of stabilized Wittig reagents to TMSK. Their more recent efforts[31] have led to minor technical improvements with regard to TMSK. Notably, they have demonstrated that use of *t*-butyldimethylsilylketene permits reaction with a much wider range of Wittig-type ylides.

Preparation of β-Lactones. Recently, there has been significant interest in the use of TMSK for the synthesis of β-lactones. Rapid advances have been made in this area since the early work of Yamamoto illustrated that methylaluminum bis(4-bromo-2,6-di-*t*-butylphenoxide) can catalyze the stereoselective formation of β-lactones from TMSK and saturated aldehydes.[5e,32] Simpler Lewis acids, such as EtAlCl$_2$[33] and MgBr$_2$·Et$_2$O,[34] have successfully been employed with a chiral aldehyde to effect the chelation-controlled stereoselective synthesis of β-lactones (eq 14).[34,35]

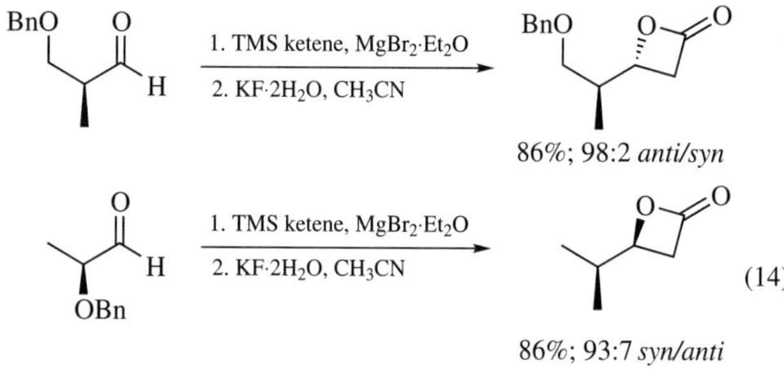

$$86\%; 98:2 \ anti/syn$$

$$86\%; 93:7 \ syn/anti \tag{14}$$

Shortly thereafter, Kocienski and coworkers reported the first catalytic enantioselective β-lactone synthesis utilizing TMSK in the presence of chiral methylaluminoimidazolines (eq 15).[36]

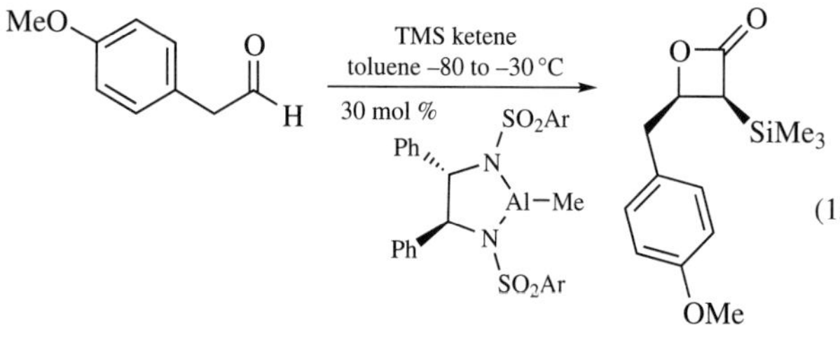

$$\tag{15}$$

Numerous metals and ligands can afford high levels of enantioselectivity for the [2 + 2] cycloaddition of TMSK to aldehydes, gloxylates, and ketoesters. These include Ti-TADDOL,[37] Al-triamine-ligands,[38] dirhodium(II)-carboxamidates[39] and Cu(II)-bisoxazolines (eq 16).[40]

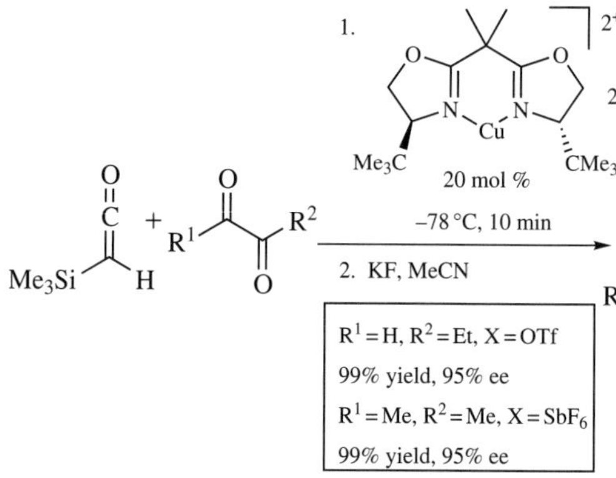

In addition to advances in enantioselective synthesis, efforts have been made to expand the scope of this reaction. Black et al. have shown that the use of BF$_3$·Et$_2$O and extended reaction times converts saturated and unsaturated aldehydes and ketones directly to α,β-unsaturated carboxylic acids in good yields (eq 17).[41]

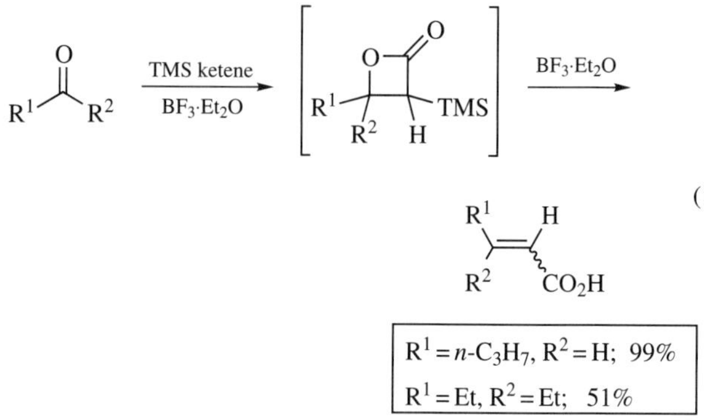

$$\tag{16}$$

R^1 = H, R^2 = Et, X = OTf
99% yield, 95% ee
R^1 = Me, R^2 = Me, X = SbF$_6$
99% yield, 95% ee

$$\tag{17}$$

R^1 = *n*-C$_3$H$_7$, R^2 = H; 99%
R^1 = Et, R^2 = Et; 51%

Furthermore, there has been recent progress extending the [2 + 2] cycloaddition to imines, thus providing access to β-lactams (eq 18).[42]

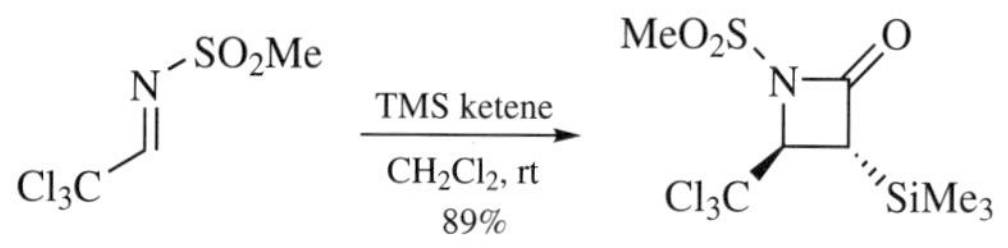

$$\tag{18}$$

Formation of Ketene Enolate. The formation and trapping of enolates *directly* from TMSK has shown little progress over the past several years. As previously reported, TMSK enolates can be trapped with chlorotrimethylsilane to afford the bis-silylketene.[7] However, trapping the enolate with the sterically more encumbered *t*-butyldimethylsilyl chloride furnishes the corresponding silyl ynol ether.[43] Although the direct trapping of TMSK enolates appears to be limited in scope, silyl ynolate anions can react with a variety of carbon electrophiles.[44] However, in this instance the use of TMSK is avoided since the silyl ynolates are derived from trimethylsilyldiazomethane, via lithiation, followed by addition to carbon monoxide, and loss of nitrogen gas.

Synthesis of Heterocycles. TMSK has also gained prominence as a useful reagent for accessing various heterocycles. TMSK reacts readily with α,*N*-diarylnitrones to afford *N*-alkyloxindoles, which can subsequently be converted to oxindoles upon hydrolysis (eq 19).[45] The analogous ring expanded 3,3a-dihydro-3-trimethylsilyl-1-azaazulen-2(1H)-ones can also be accessed via an [8 + 2] cycloaddition of TMSK with 8-azaheptafulvenes.[46]

(19)

(24)

TMSK can also react with in situ generated aza-Wittig reagents to furnish azeto[2,1-*b*]quinazolines[47] and isoquinolines (eq 20).[48]

(20)

TMSK has also been employed in the synthesis of pyranones via a [4 + 2] cycloaddition with an electron-rich 1,3-diene (eq 21).[49] Alternatively, 2 equiv of TMSK can be used to react with an enamine to afford pyranones or the corresponding resorcinols.[50] 1,3-Diaza-1,3-dienes have also been reacted with TMSK in an analogous way to furnish 4(3H)-pyrimidinones.[51]

(21)

2-Pyridones can be synthesized via the reaction of TMSK with acyl isocyanates, followed by the treatment with electron-deficient acetylenes (eq 22).[52] Alternatively, bicyclic 2-pyridinones can be accessed by replacing the electron-deficient acetylenes with cyclic enamines.[53]

(22)

TMSK can also be used in [3 + 2] cycloadditions to generate pyrroloisoquinolines[54] and pyrazolo[5,1-*a*]isoquinolines,[55] from isoquinolinium methylides and *N*-tosyliminoisoquinolinium ylides, respectively (eqs 23 and 24).

(23)

1. Shchukouskaya, L. L.; Pal'chik, R. I.; Lazarev, A. N., *Dokl. Akad. Nauk SSSR* **1965**, *164*, 357.

2. (a) Ruden, R. A., *J. Org. Chem.* **1974**, *39*, 3607. (b) Kita, Y.; Sekihachi, J.; Hayashi, Y-Z.; Da, Y.; Yamamoto, M.; Akai, S., *J. Org. Chem.* **1990**, *55*, 1108.

3. Taylor, R. T.; Cassell, R. A., *Synthesis* **1982**, 672.

4. Kita, Y.; Matsuda, S.; Kitagaki, S.; Tsuzuki, Y.; Akai, S., *Synlett* **1991**, 401.

5. (a) Brady, W. T.; Saidi, K., *J. Org. Chem.* **1979**, *44*, 733. (b) Mead, K. T.; Yang, H-L., *Tetrahedron Lett.* **1989**, *30*, 6829. (c) Zaitseva, G. S.; Vasil'eva, L. I.; Vinokurova, N. G.; Safronova, O. A.; Baukov, Y. I., *Zh. Obshch. Khim.* **1978**, *48*, 1363. (d) Zaitseva, G. S.; Vinokurova, N. G.; Baukov, Y. I., *Zh. Obshch. Khim.* **1975**, *45*, 1398. (e) Maruoka, K.; Concepcion, A. B.; Yamamoto, H., *Synlett* **1992**, 31.

6. (a) Zaitseva, G. S.; Bogdanova, G. S.; Baukov, Y. I.; Lutsenko, I. F., *Zh. Obshch. Khim.* **1978**, *48*, 131. (b) Brady, W. T.; Cheng, T. C., *J. Organomet. Chem.* **1977**, *137*, 287. (c) Zaitseva, G. S.; Bogdanova, G. S.; Baukov, Y. I.; Lutsenko, I. F., *J. Organomet. Chem.* **1976**, *121*, C21. (d) Zaitseva, G. S.; Krylova, G. S.; Perelygina, O. P.; Baukov, Y. I.; Lutsenko, I. F., *Zh. Obshch. Khim.* **1981**, *51*, 2252.

7. Woodbury, R. P.; Long, N. R.; Rathke, M. W., *J. Org. Chem.* **1978**, *43*, 376.

8. Valenti, E.; Pericas, M. A.; Serratosa, F., *J. Org. Chem.* **1990**, *55*, 395.

9. Olah, G. A.; Wu, A.; Farooq, O., *Synthesis* **1989**, 568.

10. (a) Lutsenko, I. F.; Baukov, Y. I.; Kostyuk, A. S.; Savelyeva, N. I.; Krysina, V. K., *J. Organomet. Chem.* **1969**, *17*, 241. (b) Kostyuk, A. S.; Boyadzhan, Zh. G.; Zaitseva, G. S.; Sergeev, V. N.; Savel'eva, N. I.; Baukov, Y. I.; Lutsenko, I. F., *Zh. Obshch. Khim.* **1979**, *49*, 1543.

11. Kostyuk, A. S.; Dudukina, O. V.; Burlachenko, G. S.; Baukov, Y. I.; Lutsenko, I. F., *Zh. Obshch. Khim.* **1969**, *39*, 467.

12. (a) Danheiser, R. L. In *Strategy and Tactics in Organic Synthesis*, Lindberg, T., Ed.; Academic: Orlando, Florida, 1984; Vol. 1, Chapter 2. (b) Danheiser, R. L. Ph. D. Thesis, Harvard University, 1978.

13. Fessenden, R. J.; Fessenden, J. S., *J. Org. Chem.* **1967**, *32*, 3535.

14. (a) Rathke, M. W.; Sullivan, D. F., *Synth. Commun.* **1973**, *3*, 67. (b) Larson, G. L.; Fuentes, L. M., *J. Am. Chem. Soc.* **1981**, *103*, 2418. (c) Larson, G. L.; Cruz de Maldonado, V.; Fuentes, L. M.; Torres, L. E., *J. Org. Chem.* **1988**, *53*, 633. (d) Emde, H.; Simchen, G., *Synthesis* **1977**, 867; *Liebigs Ann. Chem.* **1983**, 816.

15. Kuwajima, I.; Matsumoto, K.; Inoue, T., *Chem. Lett.* **1979**, 41.

16. Bestmann, H. J.; Schmid, G.; Sandmeier, D., *Angew. Chem.* **1976**, *88*, 92; *Angew. Chem., Int. Ed. Engl.* **1976**, *15*, 115.

17. Kaufman, K. D., *J. Org. Chem.* **1961**, *26*, 117.

18. Sethna, S.; Phadke, R., *Org. React.* **1953**, *7*, 1.

19. (a) Kuwajima, I.; Takeda, R., *Tetrahedron Lett.* **1981**, *22*, 2381. (b) Corey, E. J.; Rucker, C., *Tetrahedron Lett.* **1984**, *25*, 4345. (c) Sampson, P.; Hammond, G. B.; Wiemer, D. F., *J. Org. Chem.* **1986**, *51*, 4342.

20. (a) Sato, T.; Abe, T.; Kuwajima, I., *Tetrahedron Lett.* **1978**, 259. (b) Obayashi, M.; Utimoto, K.; Nozaki, H., *Bull. Chem. Soc. Jpn.* **1979**, *52*, 2646. (c) Cunico, R. F., *Tetrahedron Lett.* **1986**, *27*, 4269. (d) Cunico, R. F.; Kuan, C. P., *J. Org. Chem.* **1990**, *55*, 4634.

21. Sato, S.; Matsuda, I.; Izumi, Y., *J. Organomet. Chem.* **1988**, *344*, 71.

22. Bagheri, V.; Doyle, M. P.; Taunton, J.; Claxton, E. E., *J. Org. Chem.* **1988**, *53*, 6158.

23. (a) McNamara, J. M.; Kishi, Y., *J. Am. Chem. Soc.* **1982**, *104*, 7371. (b) Hudrlik, P. F.; Peterson, D., *J. Am. Chem. Soc.* **1975**, *97*, 1464. (c) Lutsenko, I. F.; Baukov, Y. I.; Dudukina, O. V.; Kramarova, E. N., *J. Organomet. Chem.* **1968**, *11*, 35. (d) Seitz, D. E.; Zapata, A., *Synthesis* **1981**, 557. (e) Lee, T. V.; Channon, J. A.; Cregg, C.; Porter, J. R.; Roden, F. S.; Yeoh, H. T-L., *Tetrahedron* **1989**, *45*, 5877.

24. Tarakanova, A. V.; Baranova, S. V.; Boganov, A. M.; Zefirov, N. S., *Zh. Org. Khim.* **1986**, *22*, 1095.

25. Hassner, A.; Dillon, J. L., *J. Org. Chem.* **1983**, *48*, 3382.

26. (a) Fedorenko, E. N.; Zaitseva, G. S.; Baukov, Y. I.; Lutsenko, I. F., *Zh. Obshch. Khim.* **1986**, *56*, 2431. (b) Zaitseva, G. S.; Kisin, A. N.; Fedorenko, E. N.; Nosova, V. M.; Livantsova, L. I.; Baukov, Y. I., *Zh. Obshch. Khim.* **1987**, *57*, 2049. (c) Zaitseva, G. S.; Lutsenko, I. F.; Kisin, A. V.; Baukov, Y. I.; Lorberth, J., *J. Organomet. Chem.* **1988**, *345*, 253.

27. (a) Sullivan, D. F.; Woodbury, R. P.; Rathke, M. W., *J. Org. Chem.* **1977**, *42*, 2038. (b) Uhlig, W.; Tzschach, A., *Z. Chem.* **1988**, *28*, 409.

28. Efimova, I. V.; Kazankova, M. A.; Lutsenko, I. F., *Zh. Obshch. Khim.* **1985**, *55*, 1647.

29. (a) Maas, G.; Schneider, K.; Ando, W., *J. Chem. Soc., Chem. Commun.* **1988**, 72,(b) Groh, B. L.; Magrum, G. R.; Barton, T. J., *J. Am. Chem. Soc.* **1987**, *109*, 7568.

30. Black, H. T.; Farrel, J. R.; Probst, D. A.; Zotz, M. C., *Synth. Commun.* **2002**, *32*, 2083.

31. Kita, Y.; Tsusuki, Y.; Kitagaki, S.; Akai, S., *Chem. Pharm. Bull.* **1994**, *42*, 233.

32. Concepcion, A. B.; Maruoka, K.; Yamamoto, H., *Tetrahedron* **1995**, *51*, 4011.

33. Pommier, A.; Pons, J.-M.; Kocienski, P. J.; Wong, L., *Synthesis* **1994**, 1294.

34. Zemribo, R.; Romo, D., *Tetrahedron Lett.* **1995**, *36*, 4159.

35. Singelton, D. A.; Wang, Y.; Yang, H. W.; Romo, D., *Angew. Chem., Int. Ed.* **2002**, *41*, 1572.

36. Dymock, B. W.; Kocienski, P. J.; Pons, J.-M., *J. Chem. Soc., Chem. Commun.* **1996**, 105.

37. (a) Yang, H. W.; Romo, D., *Tetrahedron Lett.* **1998**, *39*, 2877. (b) Romo, D.; Harrison, P. M. H.; Jenkins, S. I.; Riddoch, R. W.; Park, K.; Yang, H. W.; Zhao, C.; Wright, G. D., *Bioorg. Med. Chem.* **1998**, *6*, 1255.

38. Nelson, S. G.; Wan, Z., *Org. Lett.* **2000**, *2*, 1883.

39. Forslund, R. E.; Cain, J.; Colyer, J.; Doyle, M. P., *Adv. Synth. Catal.* **2005**, *347*, 87.

40. Evans, D. A.; Janey, J. M., *Org. Lett.* **2001**, *3*, 2125.

41. Black, T. H.; Zhang, Y.; Huang, J., *Synth. Commun.* **1995**, *25*, 15.

42. Pelotier, B.; Rajzmann, M.; Pons, J.-M.; Campomanes, P.; López, R.; Sordo, T. L., *Eur. J. Org. Chem.* **2005**, 2599.

43. Akai, S.; Kitagaki, S.; Naka, T.; Yamamoto, K.; Tsuzuki, Y.; Matsumoto, K.; Kita, Y., *J. Chem. Soc., Perkin Trans. 1* **1996**, 1704.

44. Iwamoto, K.; Kojima, M.; Chatani, N.; Murai, S., *J. Org. Chem.* **2001**, *66*, 169.

45. Takaoka, K.; Aoyama, T.; Shioiri, T., *Tetrahedron Lett.* **1999**, *40*, 3017.

46. Takaoka, K.; Aoyama, T.; Shioiri, T., *Heterocycles* **2001**, *54*, 209.

47. Cossío, F. P.; Arrieta, A., *J. Org. Chem.* **2000**, *65*, 3663.

48. Molina, P.; Vidal, A.; Tovar, F., *Synthesis* **1997**, 963.

49. Ito, T.; Aoyama, T.; Shioiri, T., *Tetrahedron Lett.* **1993**, *34*, 6583.

50. Takaoka, K.; Aoyama, T.; Shioiri, T., *Synlett* **1994**, 1005.

51. Arai, S.; Sakurai, T.; Asakura, H.; Fuma, S.-y.; Shioiri, T.; Aoyama, T., *Heterocycles* **2001**, *55*, 2283.

52. Takaoka, K.; Aoyama, T.; Shioiri, T., *Tetrahedron Lett.* **1996**, *37*, 4973.

53. Takaoka, K.; Aoyama, T.; Shioiri, T., *Tetrahedron Lett.* **1996**, *37*, 4977.

54. Kobayashi, M.; Tanabe, M.; Kondo, K.; Aoyama, T., *Tetrahedron Lett.* **2006**, *47*, 1469.

55. Kobayashi, M.; Kondo, K.; Aoyama, T., *Tetrahedron Lett.* **2007**, *48*, 7019.

Trimethylsilyllithium

[18000-27-6] C$_3$H$_9$SiLi (MW 80.15)

InChI = 1S/C3H9Si.Li/c1-4(2)3;/h1-3H3;

InChIKey = DTMCOMZUESDMSN-UHFFFAOYSA-N

(synthesis of β-silyl ketones;[1] homologation of sterically hindered ketones;[2] inversion of alkene stereochemistry[3])

Preparative Method: by the reaction of hexamethyldisilane and methyllithium in HMPA.

Handling, Storage, and Precautions: should be used immediately after preparation. The red HMPA solution of trimethylsilyllithium is sensitive to air and moisture. HMPA is carcinogenic and should be used only in a well-ventilated hood. Contact with eyes and skin should be avoided.

Original Commentary

Russell J. Linderman
North Carolina State University, Raleigh, NC, USA

General Discussion. Triorganosilyl alkali metal species in which at least one substituent on silicon is an aryl group were prepared by cleavage of the corresponding hexaorganodisilanes by alkali metals in the 1950s; however, the hexaalkyldisilanes are inert to metal cleavage.[4] The generation of trimethylsilyl alkali metal species has more recently been accomplished by the reaction of hexamethyldisilane and potassium methoxide,[3] sodium methoxide,[5] or methyllithium[1] in HMPA. trimethylsilylpotassium (TMSK) and trimethylsilylsodium have also been obtained by the reaction of hexamethyldisilane and KH or NaH in HMPA.[6] The most useful method for the generation of trimethylsilyllithium (TMSLi) is the procedure reported by Still.[1] A deep red solution of TMSLi is obtained upon reaction of hexamethyldisilane with methyllithium in HMPA at 0 °C in 15 min (eq 1); however, further reaction leading to a disilane anion is possible (eq 2).[7] The solid state structure of TMSLi has been determined by single crystal X-ray analysis as a hexameric species.[8]

$$\text{TMS-TMS} \xrightarrow[\text{HMPA, 0 °C}]{\text{MeLi}} \text{TMSLi} + \text{Me}_4\text{Si} \qquad (1)$$

$$\text{TMS-TMS} + \text{TMSLi} \longrightarrow \text{TMS-SiMe}_2\text{Li} + \text{Me}_4\text{Si} \qquad (2)$$

TMSLi adds to cyclohexenone in THF–HMPA to give exclusively the 1,4-addition product (eq 3).[1] The intermediate enolate may be stereoselectively alkylated at carbon with alkyl halides, or *O*-silylated with chlorotrimethylsilane to provide the enol ether. Nucleophilic 1,4-addition to cyclohexenones is quite stereoselective. Reaction of 5-methylcyclohex-2-enone with TMSLi occurs predominantly by axial attack, resulting in a 92:8 ratio of axial to equatorial products.[9] TMSLi also undergoes diastereoselective nucleophilic addition to 1-naphthyloxazolines (eq 4);[10] however, the addition of the silyl nucleophile is not as selective as the addition of simple alkyllithium reagents. The diminished diastereoselection is presumably due to the presence of HMPA required in the generation of TMSLi.

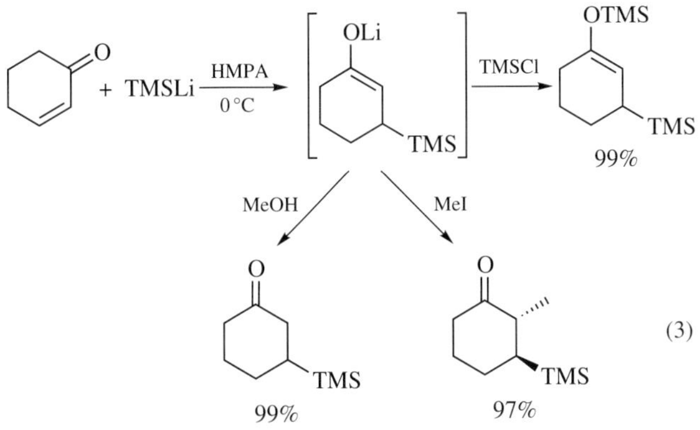

(3)

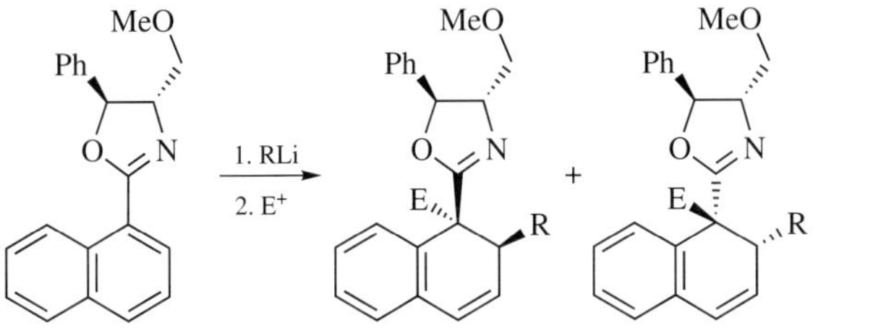

(4)

R = Bu, E = MeCO₂Cl   94:6
R = TMS, E⁺ = MeI   40:60

The β-silyl ketones obtained by 1,4-addition of TMSLi have been employed in the synthesis of alkenic acids by means of a silicon directed Baeyer–Villager oxidation (eq 5).[11] Peracid oxidation of the β-silyl ketone results in formation of the unstable seven-membered ring lactone which undergoes acid-catalyzed ring opening with concomitant stereoselective 1,2-migration of the trimethylsilyl group to provide the stable six-membered ring lactone. Saponification followed by stereoselective alkene generation, either via *anti*-elimination using boron trifluoride etherate or *syn*-elimination using potassium hydride, leads to the *trans*- or *cis*-alkene, respectively. The directed Baeyer–Villager approach has been applied to the stereoselective synthesis of *exo*- and *endo*-brevicomin.[12] The regioselectivity of the β-silyl-directed oxidation is not completely controlled. Quaternary carbons migrate preferentially over a less substituted β-silyl alkyl group.

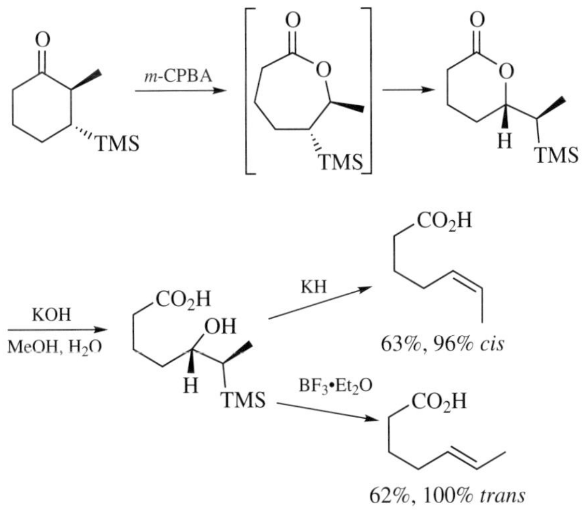

(5)

63%, 96% *cis*

62%, 100% *trans*

TMSLi also undergoes direct nucleophilic 1,2-addition to saturated ketones and aldehydes. A novel route for the homologation of sterically hindered ketones which involves the addition of TMSLi to the α-trimethylsilyloxy aldehyde derived from the cyanohydrin of a ketone has been developed.[2] The initial alkoxide undergoes an intramolecular oxygen-to-oxygen silyl migration, leading to a β-hydroxy silane (eq 6). Treatment of the β-hydroxy silane with excess base induces alkene formation by a Peterson alkenation reaction. Hydrolysis of the silyl enol ether product then provides the homologated aldehyde. Reduction of the aldehyde to the hydroxymethyl group provides a key intermediate in the synthesis of aphidicolin (eq 6),[2] stemodin, and stemodinone.[13]

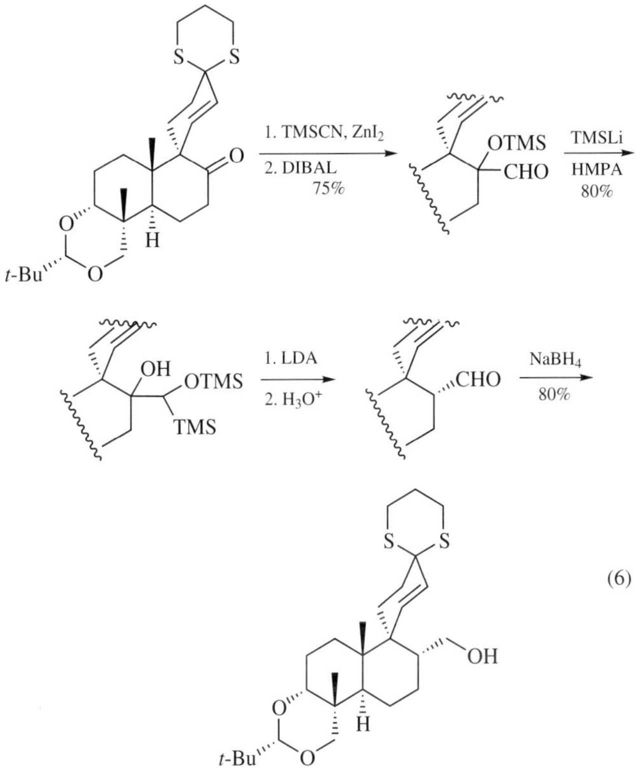

(6)

Stereoselective nucleophilic addition of TMSLi to 2-methylcyclohexanone provides an intermediate for the study of the stereochemistry of the aliphatic Brook rearrangement (eq 7).[14] The reverse Brook rearrangement has provided a method for the synthesis of a variety of (α-hydroxyalkyl)trialkylsilanes which does not require the difficult preparation of trialkylsilyl anions.[15] The method involves nucleophilic addition of readily available tri-*n*-butylstannyllithium to a carbonyl, *O*-silylation, and subsequent O-to-C migration of the silyl group induced by Sn–Li transmetalation (eq 8). The yields of (α-hydroxyalkyl)trimethylsilanes obtained by this procedure are frequently greater than the yields obtained by the direct addition of TMSLi. Direct addition of TMSLi[16] or TMSK[3] to epoxides has also been reported. Due to the stereospecific nature of the Peterson alkenation reaction, the epoxidation and TMSK deoxygenation of an alkene can result in stereospecific inversion of the double bond stereochemistry (eq 9). The epoxidation–deoxygenation sequence has been employed as a method to protect an alkene during catalytic hydrogenation.[16] Allylsilanes may be prepared by the S_N2 displacement of allyl chlorides.[17] The reaction appears to be a direct nucleophilic displacement and does not involve electron transfer processes. TMSLi can also undergo transmetalation to a variety of other organometallic species such as trimethylsilylcopper. Vinylsilanes can be obtained by the reaction of an alkyne with TMSLi in the presence of Mn^{II} and methylmagnesium chloride.[18] TMSLi has not been used as extensively as the more readily accessible dimethylphenylsilyllithium for generation of other silyl organometallic reagents.[4,19]

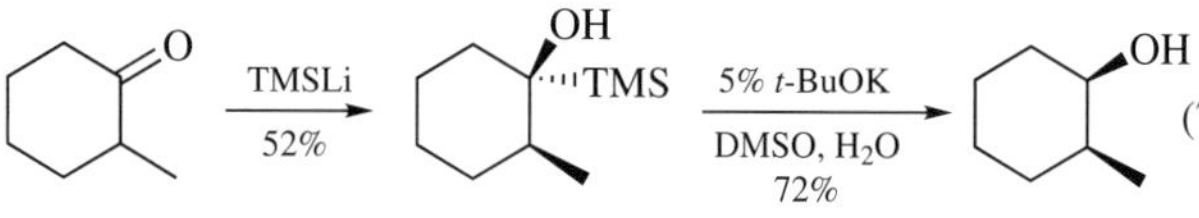

$$ (7) $$

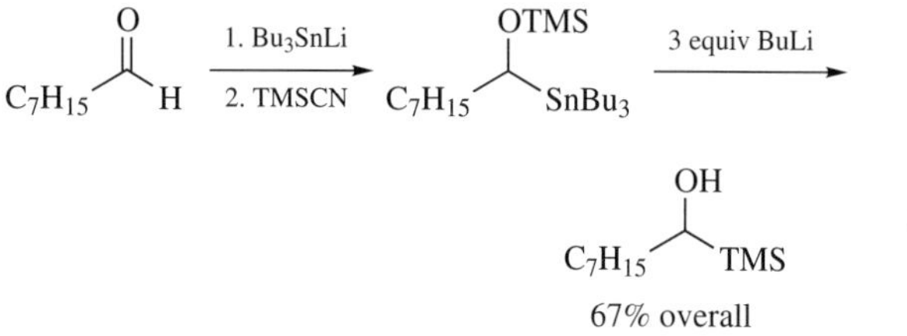

$$ (8) $$

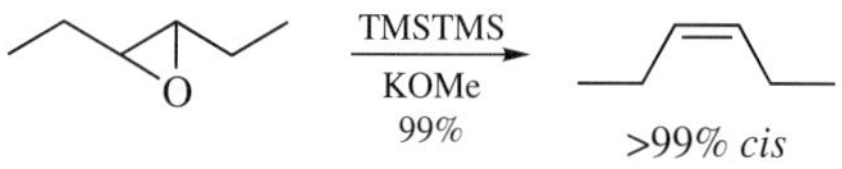

$$ (9) $$

First Update

Nikola Stiasni & Martin Hiersemann
Technische Universität Dortmund, Dortmund, Germany

The conversion of nitrones to imines could be achieved using TMSLi in good yields (82–88%).[20] Similarly, pyridine *N*-oxides containing electron-donating groups (R = Me, OMe) were deoxygenated to give the corresponding *N*-pyridines in 73–86% yields (eq 10). Quinoline *N*-oxide, isoquinoline *N*-oxide, and 2,2′-bipyridine *N,N′*-dioxide could be successfully deoxygenated to furnish the corresponding heterocycles in 81–84% yields.[20]

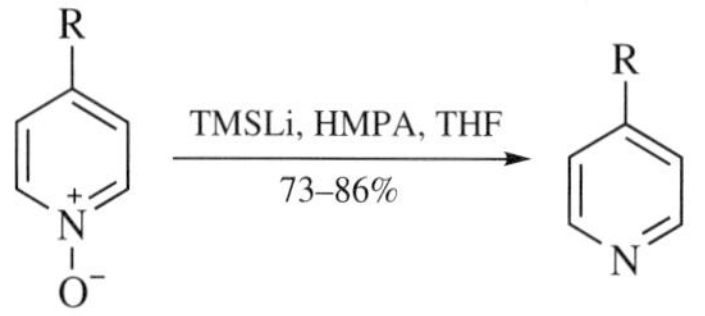

$$ (10) $$

1-Bromo-(*Z*)-1-alkenylboronate esters reacted with TMSLi to give 1-trimethylsilyl-(*E*)-1-alkenylboronate esters, which upon oxidation with sodium acetate and hydrogen peroxide yielded the corresponding alkyl trimethylsilyl ketones in 72–85% yields (eq 11).[21]

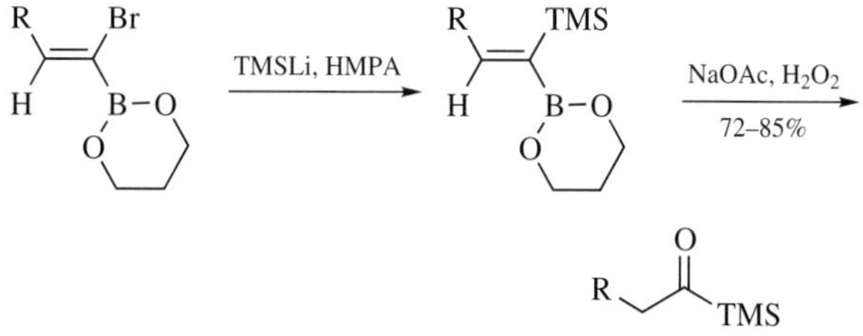

$$ (11) $$

In the synthesis of rabelomycin, pentamethyldisilyllithium was generated from TMSLi and 1 equiv of hexamethyldisilane, and

added to the α,β-unsaturated ketone (**1**) to give exclusively 1,4-addition product (**2**) in 58% yield (eq 12).[22] The pentamethyldisilyl group was exploited to mask the tertiary hydroxy group[23] that was prone to elimination under basic conditions and to direct a regioselective deprotonation, since it destabilizes anions β- to the silicon atom. At the end of synthesis, the Si–Si bond was cleaved by $AlCl_3$ to yield a silanol, which was then oxidized to a hydroxy group on treatment with KF/H_2O_2.[22]

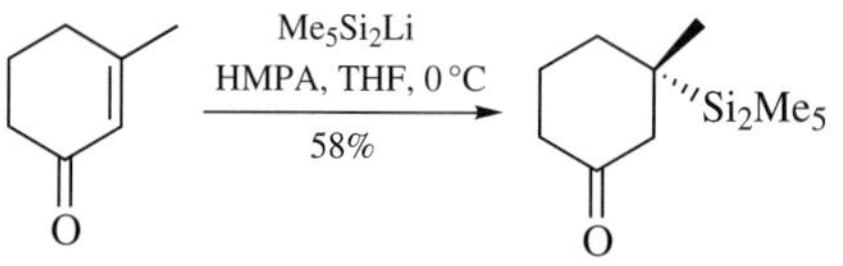

$$ (12) $$

In the synthetic study toward garsubellin A, the C–C double bond of the α,β-unsaturated ketone (**3**) was masked by the introduction of the pentamethyldisilyl group and easily regenerated at the end of the synthesis by exposing β-silyl ketone (**4**) to a Tamao-type oxidation followed by β-elimination (eq 13).[24]

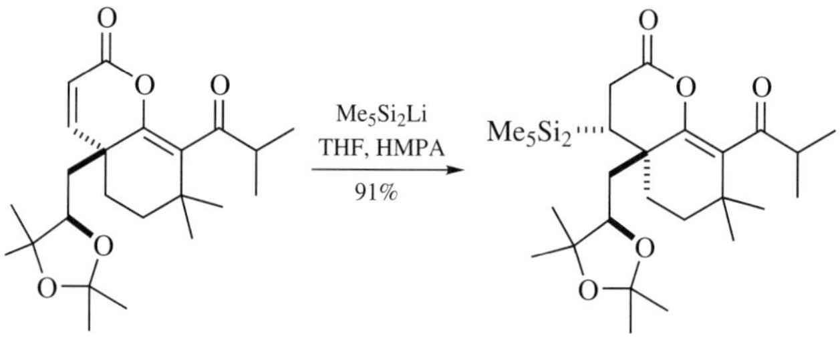

$$ (13) $$

TMSLi addition to Michael acceptor (**5**), followed by an aldol reaction, provided the β-hydroxy ester (**6**) in 72% yield over two steps, which could be easily converted to the corresponding allyltrimethylsilane in two additional steps (eq 14).[25]

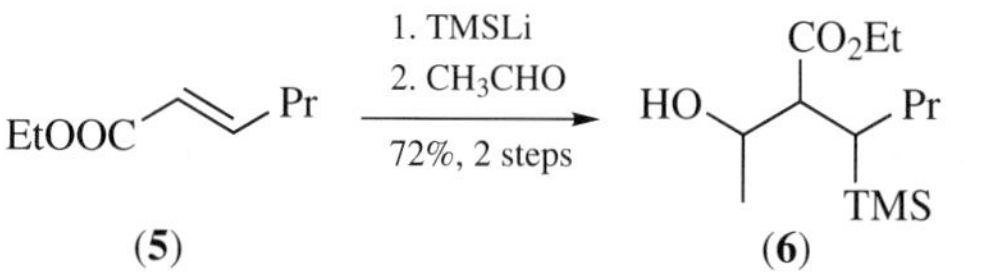

$$ (14) $$

Allyltrimethylsilane (**8**) could be synthesized in 86% yield by reacting TMSLi with allyl chloride (**7**) (eq 15).[26]

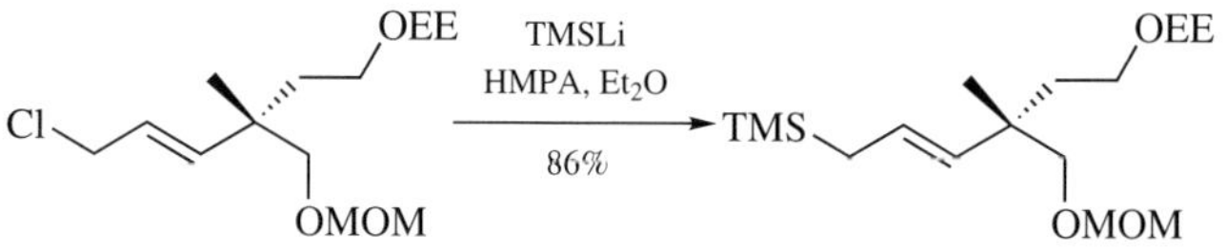

$$ (15) $$

Benzyl silyl ether (**9**) reacted with a slight excess of TMSLi to give the corresponding ketone (**10**) in 90% yield (eq 16).[27]

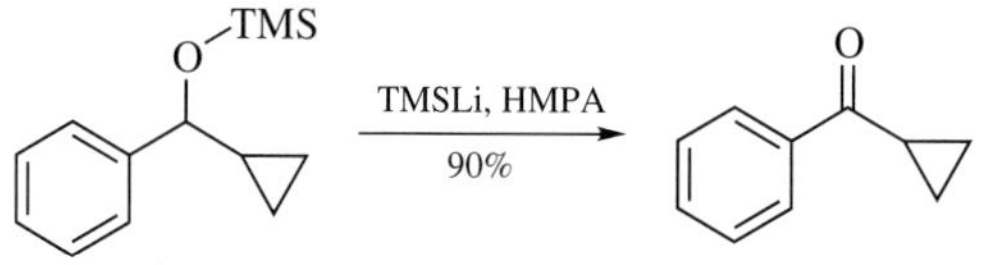

$$ (16) $$

In the total synthesis of dysidiolide, the α,β-unsaturated ketone (**11**) was treated with TMSLi to give exclusively the axial β-TMS ketone (**12**) in 64% yield (eq 17). The introduction of the trimethylsilyl group initiates a Brook rearrangement later on in the synthesis, which established the fully substituted bicyclic core of dysidiolide.[28]

Related Reagents. Dimethyl(phenyl)silane; Dimethylphenylsilyllithium; Hexamethyldisilane; Trimethylsilyl cuprate; Trimethylsilylpotassium.

1. Still, W. C., *J. Org. Chem.* **1976**, *41*, 3063.
2. Corey, E. J.; Tius, M. A.; Das, J., *J. Am. Chem. Soc.* **1980**, *102*, 1742.
3. Dervan, P. B.; Shippey, M. A., *J. Am. Chem. Soc.* **1976**, *98*, 1265.
4. Gilman, H.; Lichtenwalter, G. D., *J. Am. Chem. Soc.* **1958**, *80*, 608.
5. Sakurai, H.; Okada, A.; Kira, M.; Yonezawa, K., *Tetrahedron Lett.* **1971**, 1511.
6. Corriu, R. J. P.; Guerin, C., *J. Chem. Soc., Chem. Commun.* **1980**, 168.
7. Hudrlik, P. F.; Waugh, M. A.; Hudrlik, A. M., *J. Organomet. Chem.* **1984**, *271*, 69.
8. Ilsley, W. H.; Schaaf, T. F.; Glick, M. D.; Oliver, J. P., *J. Am. Chem. Soc.* **1980**, *102*, 3769.
9. Wickham, G.; Olszowy, H. A.; Kitching, W., *J. Org. Chem.* **1982**, *47*, 3788.
10. Barner, B. A.; Meyers, A. I., *J. Am. Chem. Soc.* **1984**, *106*, 1865.
11. Hudrlik, P. F.; Hudrlik, A. M.; Nagendrappa, G.; Yimenu, T.; Zellers, E. T.; Chin, E., *J. Am. Chem. Soc.* **1980**, *102*, 6894.
12. Hudrlik, P. F.; Hudrlik, A. M.; Yimenu, T.; Waugh, M. A.; Nagendrappa, G., *Tetrahedron* **1988**, *44*, 3791.
13. Corey, E. J.; Tius, M. A.; Das, J., *J. Am. Chem. Soc.* **1980**, *102*, 7612.
14. Hudrlik, P. F.; Hudrlik, A. M.; Kulkarni, A. K., *J. Am. Chem. Soc.* **1982**, *104*, 6809.
15. Linderman, R. J.; Ghannam, A., *J. Am. Chem. Soc.* **1990**, *112*, 2392.
16. Oliver, J. E.; Schwarz, M.; Klun, J. A.; Lusby, W. R.; Waters, R. A., *Tetrahedron Lett.* **1993**, *34*, 1593.
17. Smith, J. G.; Drozda, S. E.; Petraglia, S. P.; Quinn, N. R.; Rice, E. M.; Taylor, B. S.; Viswanathan, M., *J. Org. Chem.* **1984**, *49*, 4112.
18. Hibino, J.; Nakatsukasa, S.; Fugami, K.; Matsubara, S.; Oshima, K.; Nozaki, H., *J. Am. Chem. Soc.* **1985**, *107*, 6416.
19. Fleming, I.; Newton, T. W.; Roessler, F., *J. Chem. Soc., Perkin Trans. 1* **1981**, 2527.
20. Hwi, J. R.; Tseng, W. N.; Patel, H. V.; Wong, F. F.; Horng, D.-N.; Liaw, B. R.; Lin, L. C., *J. Org. Chem.* **1999**, *64*, 2211.
21. Bhat, N. G.; Tamm, A.; Gorena, A., *Synlett* **2004**, 297.
22. Krohn, K.; Khanbabaee, K., *Angew. Chem., Int. Ed. Engl.* **1994**, *33*, 99.
23. Suginome, M.; Matsunaga, S.; Ito, Y., *Synlett* **1995**, 941.
24. Usuda, H.; Kanai, M.; Shibasaki, M., *Org. Lett.* **2002**, *4*, 859.
25. Hirose, T.; Sunazuka, T.; Shirahata, T.; Yamamoto, D.; Harigaya, Y.; Kuwajima, I.; Omura, S., *Org. Lett.* **2002**, *4*, 501.
26. Noguchi, N.; Nakada, M., *Org. Lett.* **2006**, *8*, 2039.
27. Hwu, J. R.; Tsay, S.-C.; Wang, N.; Hakimelahi, G. H., *Organometallics* **1994**, *13*, 2461.
28. Corey, E. J.; Roberts, B. E., *J. Am. Chem. Soc.* **1997**, *119*, 12425.

Trimethylsilyl Methanenitronate[1]

($R^1 = R^2 = H$, $R_3Si = TMS$)
[51146-35-1] $C_4H_{11}NO_2Si$ (MW 133.25)
InChI = 1S/C4H11NO2Si/c1-5(6)7-8(2,3)4/h1H2,2-4H3
InChIKey = VNEHTFSOCPDKQH-UHFFFAOYSA-N
($R^1 = C_5H_{11}$, $R^2 = H$, $R_3Si = TBDMS$)
[75157-17-4] $C_{12}H_{27}NO_2Si$ (MW 245.49)
InChI = 1S/C12H27NO2Si/c1-7-8-9-10-11-13(14)15-16(5,6)12(2,3)4/h11H,7-10H2,1-6H3
InChIKey = ZUUIVXNUQSEOEW-UHFFFAOYSA-N
($R^1R^2 = (CH_2)_5$, $R_3Si = TBDMS$)
[75157-19-6] $C_{12}H_{25}NO_2Si$ (MW 243.47)
InChI = 1S/C12H25NO2Si/c1-12(2,3)16(4,5)15-13(14)11-9-7-6-8-10-11/h6-10H2,1-5H3
InChIKey = FOCLRGXYPZCUTN-UHFFFAOYSA-N

(react with alkenes in a 1,3-dipolar cycloaddition reaction;[1c,3–5] undergo Bu_4NF-mediated diastereoselective carbonyl addition to aldehydes;[6–12] react with alkyllithium reagents to give oximes;[13] oxidative coupling leads to 1,2-dinitro alkanes;[14,15] cross coupling with silyl enol ethers or enamines gives β-nitro carbonyl derivatives;[15] conversion of thiocarbonyl to carbonyl groups;[1c,16] can be converted into carbonyl compounds (cf. Nef reaction)[17])

Alternate Name: [(trimethylsilyl)-*aci*-nitro]methane.
Physical Data: $R^1 = C_5H_{11}$, $R^2 = H$, $R_3Si = TBDMS$: bp 80–90 °C/0.02 mmHg. $R^1R^2 = (CH_2)_5$, $R_3Si = TBDMS$: bp 150 °C/0.01 mmHg. A more complete list of silyl nitronates is given by Torssell.[1c]
Solubility: sol pentane and in all nonprotic common organic solvents.
Preparative Methods: a large number of silylation conditions can be applied to primary or secondary nitroalkanes,[1,2] including: R_3SiCl/Et_3N (or Ag^+ or Li_2S), R_3SiOTf, LDA/R_3SiCl, R_3SiCl/DBU,[17a] silylated amides, etc. The first reports were published by Ioffe, Tartakovskii, and their colleagues in the early 1970s.[1] The silyl nitronates are isolated by nonaqueous workup and purified by bulb-to-bulb distillation, with the TBDMS derivatives being much more thermally stable than the TMS derivatives.[2] From crystal structure analyses and NMR studies it is concluded that the silyl group migrates rapidly from one nitronate oxygen to the other and that the more stable configuration of silyl nitronates derived from primary nitroalkanes is (*E*).[2,8]
Handling, Storage, and Precautions: although there are indications that some trimethylsilyl nitronates are thermally unstable,[1c] there have been no reports of violent decompositions. Silyl nitronates are, of course, extremely sensitive to moisture, and they are more resistant to base than to acid. All silyl nitronates should be kept under an inert atmosphere and stored in a freezer.

Reactions of Silyl Nitronates with C–C Bond Formation.
Silyl nitronates are synthetically equivalent to nitrile oxides in
[3 + 2] cycloadditions. The [3 + 2] adducts shown in eq 1 lose tri-
alkylsilanol very readily, with formation of Δ^2-isoxazolines.[1c,3–5]
Silyl nitronates are somewhat less reactive than nitrile oxides,
which is not a disadvantage in intramolecular cycloadditions.[3]
The reaction is also applicable to the CF_3-substituted silyl nitronate ($R^1 = CF_3$, $R^2 = H$).[18] Depending upon the method of
reduction, either the amino alcohol (**1**) or its epimer (**2**) can be
obtained with a diastereoselectivity of ca. 4:1. When the silyl nitronate is derived from a secondary nitroalkane, no silanol elimination can occur; the corresponding isoxazolidines undergo a
rearrangement to nitroso silyl ethers such as (**3**)[1d,19] The isoxazolidines derived from primary nitroalkanes are not only precursors to amino alcohols but also to β-hydroxy ketones. Thus the
nitrile oxide/silyl nitronate [3 + 2] cycloaddition route constitutes
an alternative access to aldols.[1c,20,21] This method becomes especially attractive when rendered enantioselective. Addition of a
silyl nitronate from a primary nitroalkane to a chiral acrylamide
(such as 10,2-camphorsultam,[4] trans-2,5-dimethylpyrrolidine,[22]
or Kemp–Rebek acid derivatives[5]), silanol elimination, and reductive removal of the auxiliary gives 3-substituted Δ^2-isoxazoline-
5-methanols in either enantiomeric form (eq 2).

$$R^2 = \text{alkyl, vinyl, phenyl, COR, CONR}_2, \text{CO}_2R$$

The second most important synthetic application of silyl nitronates in C–C bond-forming reactions is their fluoride-mediated
addition to aldehydes.[6–12] Silyl nitronates from secondary nitroalkanes lead to free nitro aldols such as (**4**),[8] while those from
primary nitro alkanes give silylated products. In contrast to the
classical Henry reaction, the silyl variant is highly diastereoselective with aldehydes, furnishing erythro-O-silylated nitro aldols
(e.g. **5**).[9] It is important that the reaction temperature does not
rise above 0 °C, otherwise threo/erythro equilibration takes place.
The same erythro-nitro aldol derivatives are available by diaste-
oselective protonation of silyloxy nitronates (eq 3) (usually the
dr is >20:1), while the nonsilylated threo-epimers ($R^3 = H$, dr =
7:3–20:1) are formed by kinetic protonation of lithioxy lithio nitronates in THF/DMPU (eq 4).[9] Other recent modifications of the
nitroaldol addition using titanium nitronates[23] or ClSiR$_3$ in situ[24]
are less selective. It should also be mentioned that there are recent
reports[25] about the enantioselective addition of nitromethane to
aldehydes in the presence of rare earth binaphthol complexes.

Reactions of Silyl Nitronates with Strong Base.[13]   With 2
equiv of an alkyllithium, the nitronates from primary nitroalkanes
give oximes with the newly introduced alkyl group attached to the
oxime carbon (eq 5). The analogous reaction of silyl nitronates
from secondary nitroalkanes produces oximes in which chain extension has occurred in the α-position (eq 6). These reactions take
place when alkyllithium is added to 0.1 molar silyl nitronate in
THF at dry-ice temperature, with subsequent warming to room
temperature before aqueous workup. Probably a nitrile oxide is
the intermediate in the first case and a nitroso alkene in the second
case. Finally, oxidative cross couplings of silyl nitronates with silyl
enol ethers, ketene acetals, or enamines produce β-nitro carbonyl
compounds (eq 7) or, by HNO_2 elimination, α,β-unsaturated
ketones and esters.[15]

Functionalization Reactions of Silyl Nitronates.   Silyl nitronates can be used for a number of transformations in which
the carbon skeleton is not changed. Thus they are intermediates
en route from nitroalkanes to ketones (the transform[26] of the Nef
reaction). Peroxy acid treatment converts silyl nitronates, which
would not survive the classical conditions of the Nef reaction, to
ketones[17a] (eq 8). Aldehydes can be obtained analogously, using
stannyl nitronates.[17b]

Silyl nitronates can also be further silylated to the interesting *N,N*-bis(silyloxy)enamines (eq 9).[27] In contrast to the *N,N*-bis(lithioxy)enamines, the double bond in the bis(silyloxy)enamines appears to have electrophilic rather than nucleophilic reactivity. With primary and secondary amines, α-amino oximes are produced (eq 9)[27] in a kind of S_N' substitution, followed by hydrolytic desilylation. In this manner, the bis(silyloxy)enamine is reacting as a nitroso alkene.

Conversion of Thioketones to Ketones.[16] Thioketones generated by a Norrish-type photofragmentation of a sulfenyl acetophenone are trapped in situ by [3 + 2] dipolar cycloaddition with a silyl nitronate (eq 10). Fluoride treatment of the resulting heterocycle produces the ketone.[16b] This transformation is compatible with a variety of functional groups and has been used as part of a synthetic manipulation in which an α-acyl cyclic thioether is converted stereoselectively, with ring enlargement, to a ketolactone (methynolide synthesis).[16a]

Silyl Nitronate Reactivity Pattern. As illustrated by the examples described above, silyl nitronates provide a^1 and d^1 acyl and aminoalkyl synthons (**6** and **7**), as well as a^2 α-carbonyl and aminoalkyl synthetic building blocks (**8** and **9**).[1e, 28]

Related Reagents. Lithium α-Lithiomethanenitronate; Nitroethane; Nitromethane; 1-Nitropropane; Phenylsulfonylnitromethane.

---

1. (a) Colvin, E. W. *Silicon in Organic Synthesis*; Butterworths: London, 1981. (b) Colvin, E. W. In *The Chemistry of the Metal–Carbon Bond*; Hartley, F. R., Ed.; Wiley: Chichester, 1987; Vol. 4, Chapter 6, p 539. (c) Torssell, K. B. G. *Nitrile Oxides, Nitrones, and Nitronates in Organic Synthesis*; VCH: Weinheim, 1988. (d) Döpp, D.; Döpp, H., *Methoden Org. Chem. (Houben-Weyl)* **1990**, *E14b*, 780. (e) Seebach, D.; Colvin, E. W.; Lehr, F.; Weller, T., *Chimia* **1979**, *33*, 1.

2. Colvin, E. W.; Beck, A. K.; Bastani, B.; Seebach, D.; Kai, Y.; Dunitz, J. D., *Helv. Chim. Acta* **1980**, *63*, 697.

3. Dehaen, W.; Hassner, A., *Tetrahedron Lett.* **1990**, *31*, 743.

4. Kim, B. H.; Lee, J. Y., *Tetrahedron: Asymmetry* **1991**, *2*, 1359.

5. Stack, J. A.; Heffner, T. A.; Geib, S. J.; Curran, D. P., *Tetrahedron* **1993**, *49*, 995.

6. Colvin, E. W.; Seebach, D., *J. Chem. Soc., Chem. Commun.* **1978**, 689.

7. Seebach, D.; Beck, A. K.; Lehr, F.; Weller, T.; Colvin, E. W., *Angew. Chem., Int. Ed. Engl.* **1981**, *20*, 397.

8. Colvin, E. W.; Beck, A. K.; Seebach, D., *Helv. Chim. Acta* **1981**, *64*, 2264.

9. Seebach, D.; Beck, A. K.; Mukhopadhyay, T.; Thomas, E., *Helv. Chim. Acta* **1982**, *65*, 1101.

10. Öhrlein, R.; Jäger, V., *Tetrahedron Lett.* **1988**, *29*, 6083.

11. Martin, O. R.; Khamis, F. E.; El-Shenawy, H. A.; Rao, S. P., *Tetrahedron Lett.* **1989**, *30*, 6139.

12. Martin, O. R.; Khamis, F. E.; Rao, S. P., *Tetrahedron Lett.* **1989**, *30*, 6143.

13. Colvin, E. W.; Robertson, A. D.; Seebach, D.; Beck, A. K., *J. Chem. Soc., Chem. Commun.* **1981**, 952.

14. Kai, Y.; Knochel, P.; Kwiatkowski, S.; Dunitz, J. D.; Oth, J. F. M.; Seebach, D.; Kalinowski, H.-O., *Helv. Chim. Acta* **1982**, *65*, 137. In this paper a procedure for the coupling of lithio nitronates with Pb(OAc)$_4$ is given; silyl nitronates can be coupled in the same way.

15. Narasaka, K.; Iwakura, K.; Okauchi, T., *Chem. Lett.* **1991**, 423.

16. (a) Vedejs, E.; Buchanan, R. A.; Watanabe, Y., *J. Am. Chem. Soc.* **1989**, *111*, 8430. (b) Vedejs, E.; Perry, D. A., *J. Org. Chem.* **1984**, *49*, 573.

17. (a) Aizpurua, J. M.; Oiarbide, M.; Palomo, C., *Tetrahedron Lett.* **1987**, *28*, 5361. (b) Aizpurua, J. M.; Oiarbide, M.; Palomo, C., *Tetrahedron Lett.* **1987**, *28*, 5365.

18. Originally, we had problems reproducing the preparation of F$_3$CCH=N(O)OTBDMS (Beck, A. K.; Seebach, D., *Chem. Ber.* **1991**, *124*, 2897; *Chem. Abstr.* **1992**, *116*, 40 553c); using Torsell's procedure we are able to prepare this silyl nitronate: Marti, R. E.; Heiner, J.; Seebach, D., *Liebigs Ann. Chem.* **1995**, in press.

19. Mukerji, S. K.; Torssell, K. B. G., *Acta Chem. Scand.* **1981**, *B35*, 643.

20. Curran, D. P. In *Advances in Cycloaddition*; Curran, D. P., Ed.; JAI: Greenwich, CT, 1988; Vol. 1, p 129.

21. Jäger, V.; Müller, I.; Schohe, R.; Frey, M.; Ehrler, R.; Häfele, B.; Schröter, D., *Lect. Heterocycl. Chem.* **1985**, *8*, 79.

22. Whitesell, J. K., *Chem. Rev.* **1989**, *89*, 1581.

23. Barrett, A. G. M.; Robyr, C.; Spilling, C. D., *J. Org. Chem.* **1989**, *54*, 1233.

24. Fernández, R.; Gasch, C.; Gómez-Sánchez, A.; Vílchez, J. E., *Tetrahedron Lett.* **1991**, *32*, 3225.

25. (a) Sasai, H.; Suzuki, T.; Itoh, N.; Arai, S.; Shibasaki, M., *Tetrahedron Lett.* **1993**, *34*, 2657. (b) Sasai, H.; Itoh, N.; Suzuki, T.; Shibasaki, M., *Tetrahedron Lett.* **1993**, *34*, 855. (c) Sasai, H.; Suzuki, T.; Itoh, N.; Shibasaki, M., *Tetrahedron Lett.* **1993**, *34*, 851. (d) Sasai, H.; Suzuki, T.; Arai, S.; Arai, T.; Shibasaki, M., *J. Am. Chem. Soc.* **1992**, *114*, 4418. (e) Sasai, H.; Suzuki, T.; Itoh, N.; Tanaka, K.; Date, T.; Okamura, K.; Shibasaki, M., *J. Am. Chem. Soc.* **1993**, *115*, 10 372.

26. Corey, E. J.; Cheng, X.-M. *The Logic of Chemical Synthesis*; Wiley: New York, 1989.

27. Feger, H.; Simchen, G., *Liebigs Ann. Chem.* **1986**, 1456 (*Chem. Abstr.* **1987**, *106*, 33 161p).

28. Seebach, D., *Angew. Chem., Int. Ed. Engl.* **1979**, *18*, 239.

Albert K. Beck & Dieter Seebach
Eidgenössische Technische Hochschule Zürich, Switzerland

1-Trimethylsilyl-1-methoxyallene[1]

(X = H)
[77129-88-5]   C$_7$H$_{14}$OSi   (MW 142.30)
InChI = 1S/C7H14OSi/c1-6-7(8-2)9(3,4)5/h1H2,2-5H3
InChIKey = KXBFNKYGTOEYOY-UHFFFAOYSA-N
(X = Li)
[82200-98-4]   C$_7$H$_{13}$LiOSi   (MW 148.23)
InChI = 1S/C7H13OSi.Li/c1-6-7(8-2)9(3,4)5;/h1H,2-5H3;
InChIKey = JOQKHAYPRNCRIB-UHFFFAOYSA-N

(precursor to acylsilanes, 2-trimethylsilylfurans, and dihydrofurans)

Physical Data: bp 40–43 °C/28 mmHg.[2]
Solubility: sol THF, ether.
Preparative Method: from methoxyallene by deprotonation with *n*-butyllithium and quenching the allenyl anion with chlorotrimethylsilane.[3]
Handling, Storage, and Precautions: best stored at −20 °C or below, over anhydrous K$_2$CO$_3$. Hydrolysis, desilylation, or polymerization take place in the presence of acid.

α,β-Unsaturated Acylsilanes. Deprotonation at C-3 of 1-trimethylsilyl-1-methoxyallene (**1**) takes place with *n*-BuLi in THF. Trapping the lithioallene with an alkyl halide, followed by hydrolysis of the methyl enol ether with trifluoroacetic acid, leads to the (*E*)-acylsilane (eq 1).[3] Fluorodesilylation with tetra-*n*-butylammonium fluoride prior to the hydrolysis step produces the corresponding α,β-unsaturated aldehyde.

$$
\begin{array}{ccc}
\text{(1)} & \xrightarrow[\substack{\text{2. RX} \\ R = Bu, 82\%}]{\text{1. BuLi, THF}} & \xrightarrow[\substack{R = Bu, 76\%}]{\text{TFA}}
\end{array}
\qquad (1)
$$

Furan and Dihydrofuran Synthesis. Trapping the 3-lithio derivative of (**1**) with decanal leads to the allenic alcohol (eq 2). Aqueous acid under carefully controlled conditions produces 2-trimethylsilyl-5-nonylfuran.[2] The addition product of the 3-lithio

derivative of (**1**) with cyclopentanone undergoes acid-catalyzed conversion to the butenolide (**2**) (eq 3). This unusual transformation apparently takes place through formal loss of Me$_3$Si$^-$ from a cyclic intermediate.[2]

$$
\xrightarrow[\text{0 °C, 20 min}]{\text{HClO}_4, \text{ aq THF}} \qquad (2)
$$

25% overall from (**1**)

$$
\xrightarrow[70\%]{\text{H}_3\text{O}^+} \qquad (3)
$$

(**2**)

Deprotonation of (**1**) with lithium diisopropylamide takes a different course (eq 4). The initially formed allenyl anion isomerizes to the acetylide, which is trapped with ketones or aldehydes to produce propargyl alcohols (**3**).[4] The isomerization is postulated to take place through proton transfer steps mediated by diisopropylamine. This is consistent with the observation that no such isomerization takes place with alkyllithium reagents. The propargyl alcohols (**3**) are converted to methoxydihydrofurans (**4**) with catalytic potassium hydride in DMSO.[4]

$$
\text{(1)} \xrightarrow[\substack{\text{then} \\ \text{10 min, 0 °C}}]{\substack{\text{1 equiv LDA} \\ \text{Et}_2\text{O, 0 °C, 1 h;}}} \text{(3)} \xrightarrow[\text{DMSO, rt}]{\text{cat KH}} \text{(4)} \qquad (4)
$$

R^1	R^2	(**3**)	(**4**)
-(CH$_2$)$_5$-		89%	71%
-(CH$_2$)$_6$-		86%	77%
C$_8$H$_{17}$	H	83%	56%

Related Silanes. 1-Trimethylsilyl-1-(ethoxyethoxy)allene (**5**) undergoes efficient reaction with electrophiles to produce substituted unsaturated acylsilanes (eq 5).[5] The reaction products can be converted to diverse 2-silyloxy-1,3-butadienes. Exposure of (**5**) to boron trifluoride etherate leads to rearrangement in which an oxonium ion is intercepted at C-2 of the allene (eq 5).[6] This rearrangement proceeds via an intermolecular pathway. Acylstannanes can also be prepared through this method.

Epoxidation of *t*-butyldimethylsilyl-1-(ethoxyethoxy)allene (**6**) with *m*-chloroperbenzoic acid leads to an α-keto acylsilane,[6] presumably through an allene oxide intermediate (eq 6). Deprotonation of (**6**) at C-3 and trapping of the anion with selenium, followed by iodomethane, produces an allenyl selenide. The reaction of this material with peracid follows a different course, leading to an acetylenic acylsilane, presumably via [2,3]-sigmatropic rearrangement of an allenyl selenide (eq 6).[7]

Physical Data: bp 43 °C/43 mmHg; IR ν_{max} 2100 cm^{-1}; ^{1}H NMR (CDCl$_3$) δ = 0.12 (s, 9H), 2.75 (s, 2H); MS *m/z* 129, 73.

Solubility: sol organic solvents; insol water.

Form Supplied in: colorless liquid.

Preparative Method: obtained in high yield by reaction of sodium azide and (chloromethyl)trimethylsilane in DMF[1a] or HMPA[1b] at 80 °C.

Handling, Storage, and Precautions: useful, safe substitute for methyl azide, bp 20 °C, which is explosive. Trimethylsilylmethyl azide is stable to temperatures of at least 120 °C and can be stored in a refrigerator for more than 6 months.

Amination. Trimethylsilylmethyl azide reacts with aryl Grignard reagents at room temperature to give aniline derivatives in 70–95% yield (eq 1).[1a] An intermediate of the reaction is probably a triazene, which is hydrolyzed by neutral water to yield an amine. Amination of Grignard reagent with other azides (sulfonyl azide and thiomethyl azide) is known, but relatively drastic conditions are required to decompose the intermediate.[2] The azide also reacts with aryllithium compounds to afford amines (eq 2), but yields are generally lower.[1a,3]

$$ArX \xrightarrow{Mg} ArMgX \xrightarrow[\text{2. H}_2\text{O}]{\text{1. TMSCH}_2\text{N}_3} ArNH_2 \quad (1)$$

$$\begin{matrix} ArH \\ ArX \end{matrix} \xrightarrow{BuLi} ArLi \xrightarrow[\text{2. H}_2\text{O}]{\text{1. TMSCH}_2\text{N}_3} ArNH_2 \quad (2)$$

Azomethine Ylide Precursor. The azide is readily converted into a precursor of azomethine ylides (eq 3).[4,5] Reactions of the ylides with carbon–carbon multiple bonds and carbonyl group are used in a synthesis of pyrroline and oxazole derivatives (eq 4).[6–8]

Silylmethyl-Substituted Heterocumulenes. (Trimethylsilylmethyl)iminotriphenylphosphorane is obtained from a one-pot reaction of the azide and triphenylphosphine (eq 3). The reactions of the phosphorane with carbon dioxide or carbon disulfide (eq 5) give trimethylsilylmethyl isocyanate or trimethylsilylmethyl isothiocyanate in 68 and 94% yields, respectively.[4]

E$^+$	E	Yield
H$_2$SO$_4$	H	65%
PhSeCl	PhSe	73%
SO$_2$Cl$_2$	Cl	65%

1. (a) Huche, M., *Bull. Soc. Claim. Fr., Part 2* **1978**, 313. (b) Zimmer, R., *Synthesis* **1993**, 165. (c) Schuster, H. F.; Coppola, G. M. *Allenes in Organic Synthesis*; Wiley: New York, 1984; p 215. (c) Ricci, A.; degl'Innocenti, A., *Synthesis* **1989**, 647.

2. Pappalardo, P.; Ehlinger, E.; Magnus, P., *Tetrahedron Lett.* **1982**, *23*, 309.

3. Clinet, J.-C.; Linstrumelle, G., *Tetrahedron Lett.* **1980**, *21*, 3987.

4. Kuwajima, I.; Sugahara, S.; Enda, J., *Tetrahedron Lett.* **1983**, *24*, 1061.

5. Reich, H. J.; Kelly, M. J.; Olson, R. E.; Holtan, R. C., *Tetrahedron* **1983**, *39*, 949.

6. Ricci, A.; degl'Innocenti, A.; Capperucci, A.; Faggi, C.; Seconi, G.; Favaretto, L., *Synlett* **1990**, 471.

7. Reich, H. J.; Kelly, M. J., *J. Am. Chem. Soc.* **1982**, *104*, 1119.

Marcus A. Tius
University of Hawaii, Honolulu, HI, USA

Trimethylsilylmethyl Azide[1]

[87576-94-1]   C$_4$H$_{11}$N$_3$Si   (MW 129.27)

InChI = 1S/C4H11N3Si/c1-8(2,3)4-6-7-5/h4H2,1-3H3

InChIKey = HKBUFTCADGLKAS-UHFFFAOYSA-N

(amination of organometallic compounds; preparation of heterocyclic compounds; precursor of azomethine ylide)

$$\text{TMSCH}_2\text{N=PPh}_3 \quad \overset{\text{CO}_2}{\underset{68\%}{\nearrow}} \quad \text{TMSCH}_2\text{N=C=O}$$
$$\overset{\text{CS}_2}{\underset{94\%}{\searrow}} \quad \text{TMSCH}_2\text{N=C=S} \qquad (5)$$

[3 + 2] Cycloaddition. The azide reacts with alkynes (eq 6),[1b] phosphaalkynes (eq 7),[9] and isothiocyanates (eq 8)[10] to give [3 + 2] cycloaddition products, which can be used as synthons for the preparation of a large variety of other heterocycles.

$$\text{TMSCH}_2\text{N}_3 \; + \; \text{R}\!\!-\!\!\equiv\!\!-\!\!\text{R}' \quad \xrightarrow{99-100\%} \qquad (6)$$

$$\text{TMSCH}_2\text{N}_3 \; + \; t\text{-Bu}\!\!-\!\!\equiv\!\!\text{P} \quad \xrightarrow{96\%} \qquad (7)$$

$$\text{TMSCH}_2\text{N}_3 \; + \; \text{Ar}-\text{N}=\!\!\bullet\!\!=\text{S} \quad \xrightarrow{65\%} \qquad (8)$$

Miscellaneous. The azide is used for the preparation of other heterocyclic compounds. Typical examples are the formations of disilaaziridine (eq 9)[11] and *s*-triazine (eq 10)[12] derivatives.

$$\text{TMSCH}_2\text{N}_3 \; + \; \text{Mes}_2\text{Si=SiMes}_2 \; \longrightarrow$$

$$\qquad + \qquad (9)$$

$$\text{TMSCH}_2\text{N}_3 \; + \; \text{RCOCl} \quad \xrightarrow[73\%]{\text{F}^-} \qquad (10)$$

$$\text{R} = p\text{-MeC}_6\text{H}_4$$

1. (a) Nishiyama, K.; Tanaka, N., *J. Chem. Soc., Chem. Commun.* **1983**, 1322. (b) Tsuge, O.; Kanemasa, S.; Matsuda, K., *Chem. Lett.* **1983**, 1131.
2. Smith, P. A. S.; Rowe, C. D.; Bruner, L. B., *J. Org. Chem.* **1969**, *34*, 3430; Trost, B. M.; Pearson, W. H., *J. Am. Chem. Soc.* **1981**, *103*, 2483.
3. Okazaki, R.; Unno, M.; Inamoto, N., *Chem. Lett.* **1987**, 2293.
4. Tsuge, O.; Kanemasa, S.; Matsuda, K., *J. Org. Chem.* **1984**, *49*, 2688.
5. Letellier, M.; McPhee, D. J.; Griller, D., *Synth. Commun.* **1988**, *18*, 1975.
6. Anderson, W. K.; Dabrah, T. T., *Synth. Commun.* **1986**, *16*, 559.
7. Padwa, A.; Gasdaska, J. R.; Haffmanns, G.; Rebello, H., *J. Org. Chem.* **1987**, *52*, 1027.
8. Anderson, W. K.; Kinder, F. R. Jr., *J. Heterocycl. Chem.* **1990**, *27*, 975.
9. Roesch, W.; Facklam, T.; Regitz, M., *Tetrahedron* **1987**, *43*, 3247.
10. L'Abbe, G.; Brems, P.; Albrecht, E., *J. Heterocycl. Chem.* **1990**, *27*, 1059.
11. Gillette, G. R.; West, R., *J. Organomet. Chem.* **1990**, *394*, 45.
12. Nishiyama, K.; Mikuni, H.; Harada, M., *Bull. Chem. Soc. Jpn.* **1985**, *58*, 3381.

Kozaburo Nishiyama
Tokai University, Shizuoka, Japan

1-[(Trimethylsilyl)methyl]-1*H*-benzotriazole

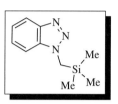

[122296-00-8] $C_{10}H_{15}N_3Si$ (MW 205.34)

InChI = 1S/C10H15N3Si/c1-14(2,3)8-13-10-7-5-4-6-9(10)11-12-13/h4-7H,8H2,1-3H3

InChIKey = VRGRHLJMRVGQCS-UHFFFAOYSA-N

(preparation of ketones, carboxylic acids, and fused heterocycles)

Physical Data: mp 55–56 °C.

Solubility: well soluble in most organic solvents.

Form Supplied in: colorless needles.

Analysis of Reagent Purity: ^{1}H NMR (CDCl$_3$) δ 7.80–7.60 (m, 1H), 7.20–6.80 (m, 3H), 4.00 (s, 2H), 0.20 (s, 9H); ^{13}C NMR (CDCl$_3$) δ 145.4, 133.6, 126.5, 123.4, 119.5, 109.4, 38.7, −2.1.

Preparative Methods: 1-[(trimethylsilyl)methyl]-1*H*-benzotriazole is readily prepared in a reaction of sodium benzotriazolide with chloromethyltrimethylsilane.[1]

Purification: recrystallization from hexanes.

Handling, Storage, and Precautions: the material is stable in ambient atmosphere, but it is slowly decomposed by water, especially in the presence of acids, bases, or fluoride anions.

Preparation of Ketones. 1-[(Trimethylsilyl)methyl]benzotriazole reacts readily with acyl chlorides to provide the corresponding (benzotriazol-1-yl)methyl ketones in good yields. One example of such reactions is given in eq 1, and similar results are reported for reactions with benzoyl, acetyl, phenylacetyl, and other acyl chlorides.[1] As shown by an example in eq 2, the benzotriazolyl moiety in (benzotriazol-1-yl)methyl ketones is easily removed by reduction with zinc in acetic acid to provide the corresponding methyl ketones. To prepare higher ketones, lithiated 1-[(trimethylsilyl)methyl]benzotriazole is alkylated first and then subjected to the regular reactions with acyl chlorides and zinc. Thus, in a reaction of 1-[(trimethylsilyl)methyl]benzotriazole with *n*-BuLi followed by benzyl bromide, 1-[1-(trimethylsilyl)-2-phenylethyl]benzotriazole is obtained in 81% yield. Subsequent treatment of this product with 4-methylbenzoyl chloride and then with zinc in acetic acid provides 4-methylphenyl 2-phenylethyl ketone in 60% yield (eq 3). Although ketones of the type R^1–CO–CH_2R^2 can be prepared directly in reactions of acid chlorides R^1COCl with appropriate Grignard reagents,[2–4] assembling of the R^2CH_2 group from two fragments in this method may be the best synthetic option when the corresponding Grignard reagents are not easily available.

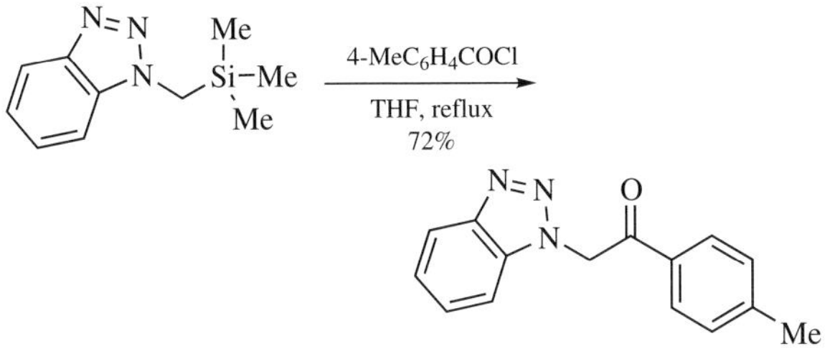

(1)

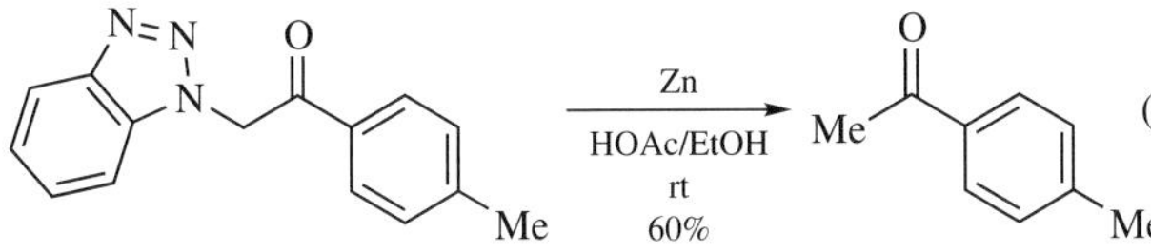

(2)

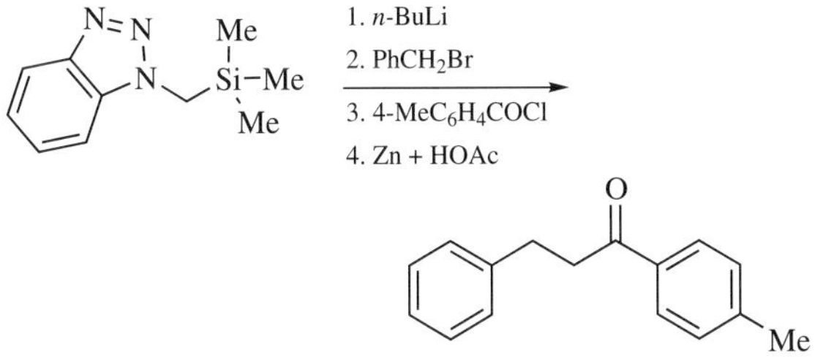

(3)

One-carbon Homologation of Carboxylic Acids. 1-[(Trimethylsilyl)methyl]benzotriazole converts benzoyl chlorides to the corresponding (benzotriazol-1-yl)methyl aryl ketones in high yields (see eq 1). Treatment with triflic anhydride and 2,6-lutidine in CH_2Cl_2 converts these ketones into their enolate triflates in 83–95% yields (eq 4). In the subsequent steps, the triflates are treated with sodium methoxide and then with ethanolic HCl to afford ethyl esters of the corresponding arylacetic acids in 89–98% yields (eq 5). The proposed reaction mechanism involves elimination of trifluoromethanesulfonic acid with sodium methoxide and final alcoholysis of the obtained 1-(arylethynyl)benzotriazole intermediates with ethanolic HCl.[5,6] A comparable classical method for one-carbon homologation of carboxylic acids, the Arndt–Eistert reaction, involves difficult to handle diazomethane and α-diazoketones.[7,8]

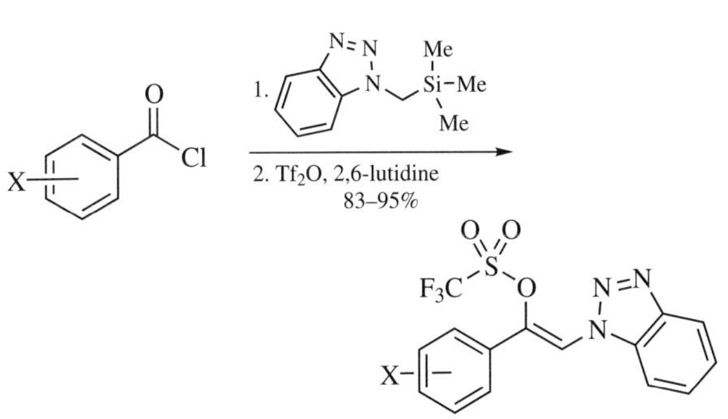

(4)

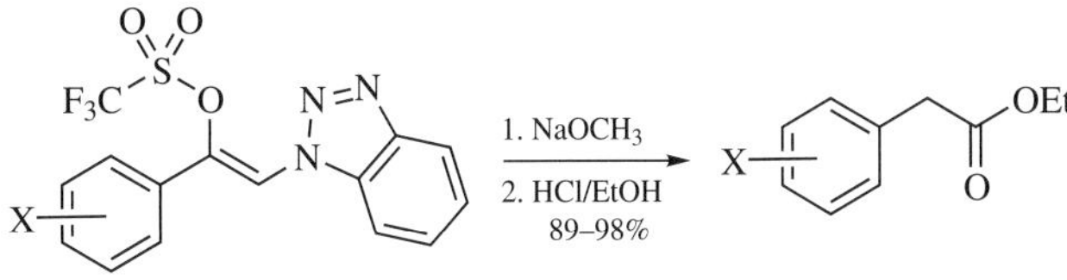

X = H, 4-Cl, 4-MeO, 2-Me, 3-Me, or 4-Me   (5)

The above procedure does not work well with aliphatic carboxylic acids. In this case, aliphatic analogs of triflates from eq 4 are efficiently converted to 1-(1-alkyn-1-yl)benzotriazoles by treatment with 10% NaOH (eq 6). In the following step, the alkynyl derivatives are treated with *p*-toluenesulfonic acid monohydrate (in acetonitrile at 65 °C) to generate the corresponding enol tosylates, which are subsequently hydrolyzed with TBAF and 1 N HCl to furnish the desired acids (eq 7). The lowest yield (28%) is reported for R = $PhCH_2CH_2$, and all other yields fall into the range of 48–68%.[6]

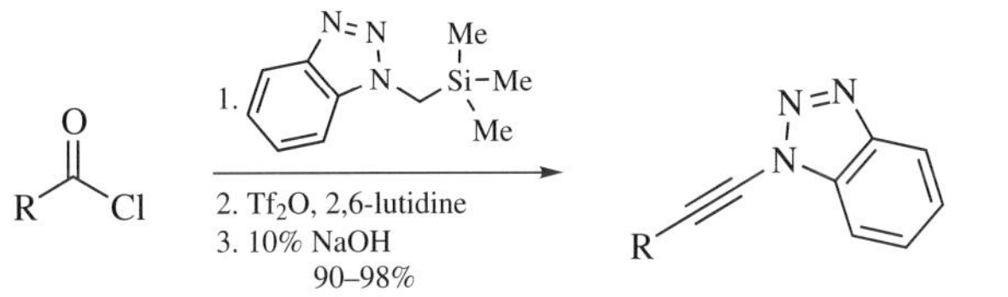

(6)

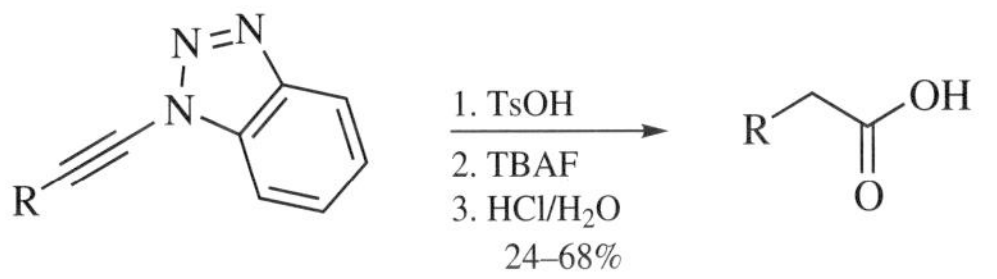

(7)

R = Me, $CH_3(CH_2)_4$, $CH_3(CH_2)_6$, $(CH_3)_3CCH_2$, $(CH_3)_3CCH_2CH(CH_3)CH_2$, or $PhCH_2CH_2$

Synthesis of Fused Heterocycles. Use of chloroacetyl chloride in the reaction with 1-[(trimethylsilyl)methyl]benzotriazole opens new possibilities. Thus, the 1-(benzotriazol-1-yl)-3-chloroacetone obtained in the first step is further reacted with 2-mercaptobenzothiophene to give 1-(benzotriazol-1-yl)-3-[(benzothiophen-2-yl)thio]acetone in a practically quantitative yield (eq 8). Cyclization of the obtained acetone derivative with $ZnCl_2$ readily provides 3-[(benzotriazol-1-yl)methyl][2,3-*b*]benzothiophene (eq 9). The structure can be further modified by substitution of the benzotriazolyl moiety with other groups. Preparation of several other fused thiophene systems, starting from other aryl mercaptans, is also reported.[9]

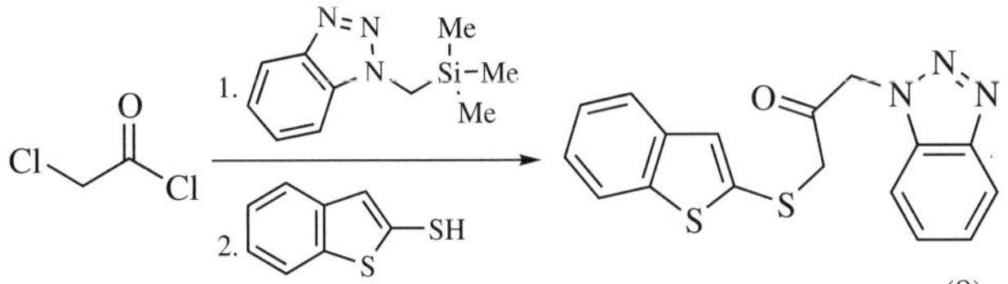

(8)

$$(9)$$

1-(Benzotriazol-1-yl)-3-chloroacetone, obtained from a reaction of 1-[(trimethylsilyl)methyl]benzotriazole with chloroacetyl chloride, is a convenient precursor for various heterocycles. For example, in its reaction with monosubstituted thioureas, 2-amino-4-[(benzotriazol-1-yl)methyl]thiazoles are obtained in good yields (eq 10).[10] This reaction is similar to the general synthesis of 4-substituted 2-aminothiazoles from thioureas and halogeno-methyl ketones;[11–13] however, in this case, the (benzotriazol-1-yl)methyl substituent allows further transformations. Thus, in reactions with chalcones, 2-amino-5,7-diarylbenzothiazoles are obtained in 59–74% yields (eq 11). Analogous condensations of 1-(benzotriazol-1-yl)-3-chloroacetone with 2-methylpyridines and 2-aminopyridines produce corresponding 2-[(benzotriazol-1-yl)methyl]indolizines and 2-[(benzotriazol-1-yl)methyl]imidazo-[1,2-a]pyridines, respectively.[10]

$$(10)$$

$$(11)$$

R^1 = H, Ph, 4-MeOC$_6$H$_4$, 4-ClC$_6$H$_4$, 4-O$_2$NC$_6$H$_4$, 2-ClC$_6$H$_4$, or 1-naphthyl
R^2 = Ph, 4-MeOC$_6$H$_4$, 4-ClC$_6$H$_4$, or 4-O$_2$NC$_6$H$_4$
R^3 = Ph or 4-MeC$_6$H$_4$

1. Katritzky, A. R.; Lam, J. N., *Heteroatom. Chem.* **1990**, *1*, 21.

2. Sato, F.; Inoue, M.; Oguro, K.; Sato, M., *Tetrahedron Lett.* **1979**, *20*, 4303.

3. Fiandanese, V.; Marchese, G.; Martina, V.; Ronzini, L., *Tetrahedron Lett.* **1984**, *25*, 4805.

4. Cardellicchio, C.; Fiandanese, V.; Marchese, G.; Ronzini, L., *Tetrahedron Lett.* **1987**, *28*, 2053.

5. Katritzky, A. R.; Zhang, S.; Fang, Y., *Org. Lett.* **2000**, *2*, 3789.

6. Katritzky, A. R.; Zhang, S.; Hussein, A. H. M.; Fang, Y.; Steel, P. J., *J. Org. Chem.* **2001**, *66*, 5606.

7. Winum, J. Y.; Kamal, M.; Leydet, A.; Roque, J. P.; Montero, J. L., *Tetrahedron Lett.* **1996**, *37*, 1781.

8. Aller, E.; Molina, P.; Lorenzo, A., *Synlett* **2000**, *4*, 526.

9. Katritzky, A. R.; Vvedensky, V. Y.; Tymoshenko, D. O., *J. Chem. Soc., Perkin Trans. 1* **2001**, 2483.

10. Katritzky, A. R.; Tymoshenko, D. O.; Monteux, D.; Vvedensky, V.; Nikonov, G.; Cooper, C. B.; Deshpande, M., *J. Org. Chem.* **2000**, *64*, 8059.

11. Qian, C. Y.; Jin, Z. T.; Yin, B. Z.; Imafuku, K., *J. Heterocycl. Chem.* **1989**, *26*, 601.

12. Tanaka, K.; Nomura, K.; Oda, H.; Yoshida, S.; Mitsuhashi, K., *J. Heterocycl. Chem.* **1991**, *28*, 907.

13. South, M. S., *J. Heterocycl. Chem.* **1991**, *28*, 1003.

Stanislaw Rachwal
Cortex Pharmaceuticals, Irvine, CA, USA

2-Trimethylsilylmethyl-1,3-butadiene[1]

[70901-64-3]   C$_8$H$_{16}$Si   (MW 140.33)
InChI = 1S/C8H16Si/c1-6-8(2)7-9(3,4)5/h6H,1-2,7H2,3-5H3
InChIKey = JGVXFONXFYPFAK-UHFFFAOYSA-N

(isoprenylation reagent;[1] Diels–Alder diene[1])

Physical Data: bp 69–70 °C/80 mmHg.
Analysis of Reagent Purity: ^{1}H NMR.
Purification: by distillation.
Solubility: sol common organic solvents.
Preparative Methods: prepared most conveniently by the coupling reaction of the Grignard reagent prepared from (chloromethyl)trimethylsilane with 2-chloro-1,3-butadiene (chloroprene) in the presence of a catalytic amount of Ni[Ph$_2$P-(CH$_2$)$_3$PPh$_2$]Cl$_2$ (91% yield) (eq 1).[2] Alternatively, it can be prepared by the reaction of the same Grignard reagent with al-lenylmethyl phosphate in 57–73% yield (eq 2).[3] Direct metala-tion of isoprene followed by the reaction with chlorotrimethyl-silane gives 2-trimethylsilylmethyl-1,3-butadiene in low yield.[4] Thermal isomerization of 1-trimethylsilylmethyl-cyclobutene to 2-trimethylsilylmethyl-1,3-butadiene has also been reported.[5]

$$TMSCH_2MgCl + \quad \xrightarrow{cat} \quad (1)$$

cat = Ni[PPh$_2$(CH$_2$)$_3$PPh$_2$]Cl$_2$

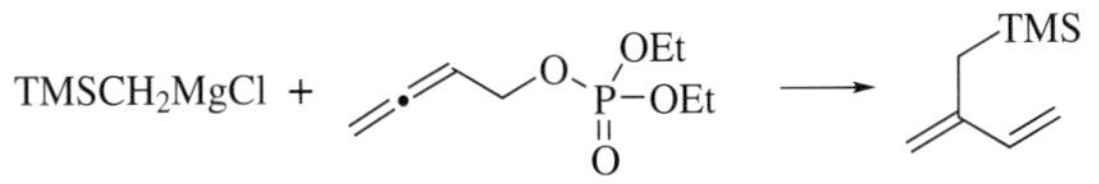

(2)

Handling, Storage, and Precautions: can be stored in a glass bottle under nitrogen.

Isoprenylation. Like other allylic silanes, 2-trimethylsilylmethyl-1,3-butadiene reacts with various electrophilic species such as acetals, acid chlorides, and carbonyl compounds with the aid of a Lewis acid to give the corresponding isoprenylated compounds (eqs 3–6).

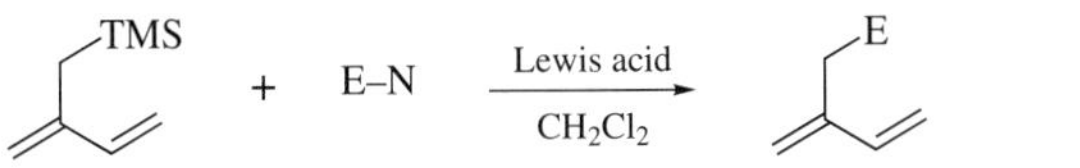

(3)

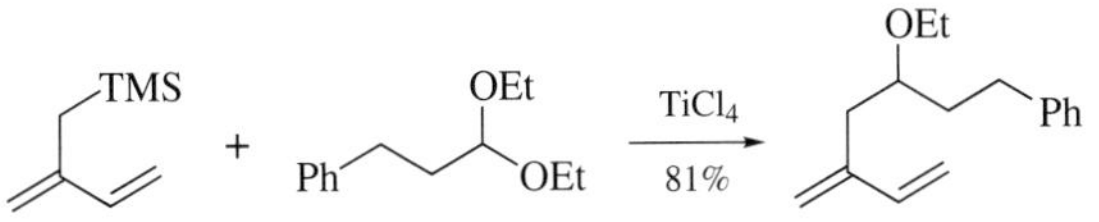

(4)

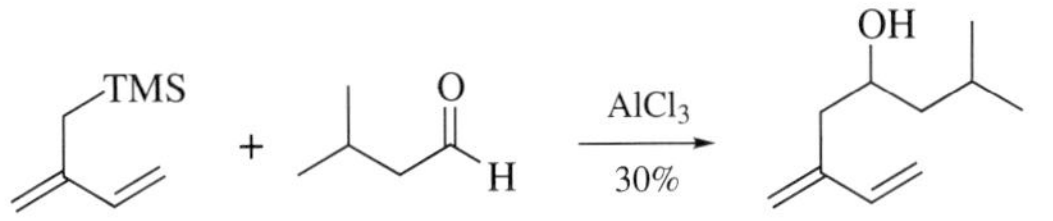

(5)

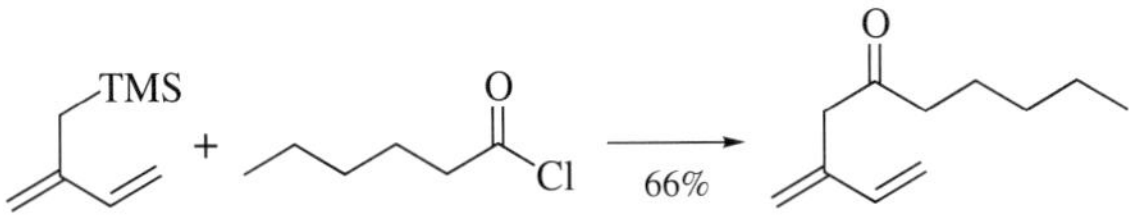

(6)

The direct reaction of 2-trimethylsilylmethyl-1,3-butadiene with isovaleraldehyde gives (±)-ipsenol (2-methyl-6-methylene-7-octen-4-ol) in rather low yield (30%) (eq 5). However, ipsenol is obtained in 62% overall yield by the reaction of 2-trimethylsilylmethyl-1,3-butadiene with isovaleryl chloride, followed by reduction with diisobutylaluminum hydride (eq 7). Similarly, (±)-ipsdienol (2-methyl-6-methylene-2,7-octadien-4-ol) is obtained by reduction of myrcenone, prepared by the reaction of 2-trimethylsilylmethyl-1,3-butadiene with 3,3-dimethylacryloyl chloride, in 75% overall yield (eq 8).

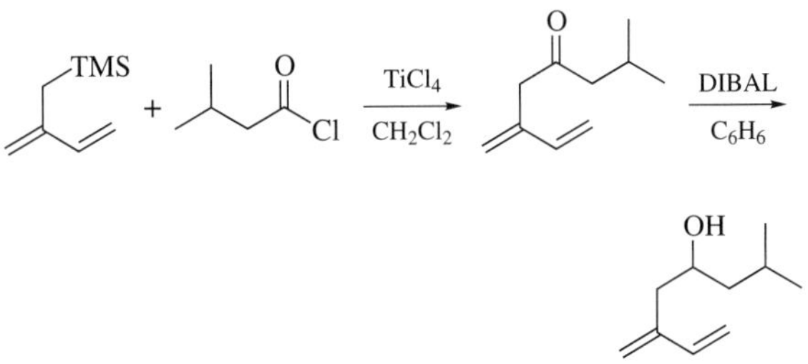

(7)

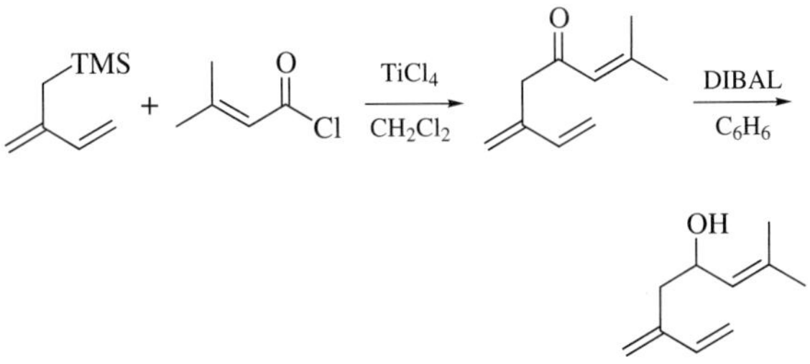

(8)

The methyl ether of ipsenol can be prepared by the reaction of 2-trimethylsilylmethyl-1,3-butadiene and isovaleraldehyde

dimethyl acetal with iodotrimethylsilane catalysis (90%) (eq 9)[6] and from methyl 1-chloro-3-methylbutyl ether (72%).[7] Isoprenylation of carbonyl compounds with 2-trimethylsilylmethyl-1,3-butadiene initiated by a catalytic amount of tetra-*n*-butylammonium fluoride (TBAF) (eq 10) is the most convenient route to ipsenol and ipsdienol.[8] Reactions with γ-vinylbutyrolactone[9] and *N*-alkylmethyleneiminium salt[10] are also reported.

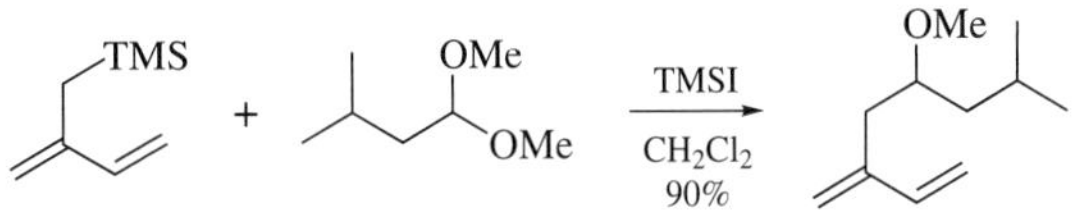

(9)

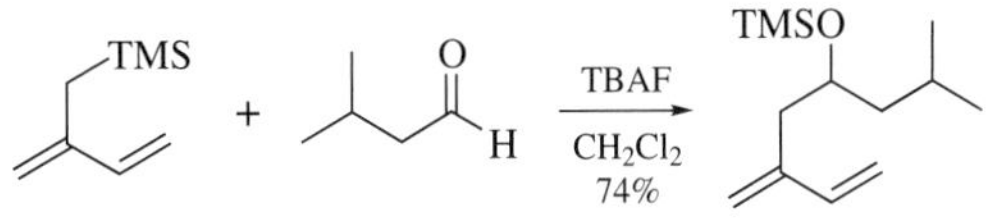

(10)

Diels–Alder Reactions. 2-Trimethylsilylmethyl-1,2-butadiene and 2-trimethylstannylmethyl-1,3-butadiene undergo facile cycloaddition with dienophiles (eq 11). The reactions of 2-trimethylsilylmethyl- and 2-trimethylstannylmethyl-1,3-butadiene with unsymmetrical dienophiles give the so-called *para* product predominantly, and the selectivity is much higher in the reactions with 2-trimethylsilylmethyl- and 2-trimethylstannylmethyl-1,3-butadiene compared to reactions with isoprene. The *para/meta* ratios in the reactions with methyl acrylate (eq 12) are: 70/30 (X = H), 84/16 (X = SiMe₃), and 91/9 (X = SnMe₃).[11]

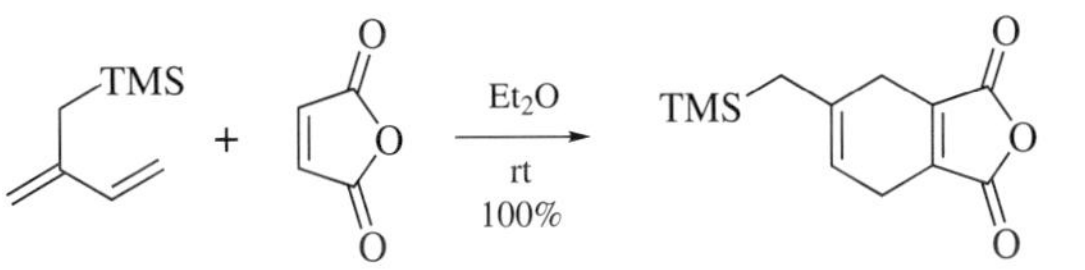

(11)

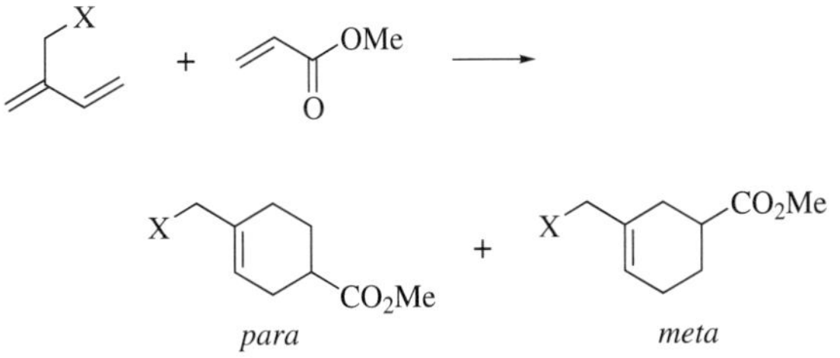

(12)

2-Trimethylsilylmethyl-1,3-butadiene undergoes highly regioselective aluminum chloride-catalyzed Diels–Alder reactions with dienophiles such as acrolein and methyl vinyl ketone in which the '*para*' isomers are obtained almost exclusively (eqs 13 and 14). The adducts are converted readily to a variety of naturally occurring mono- and sesquiterpenes (eq 15).[12]

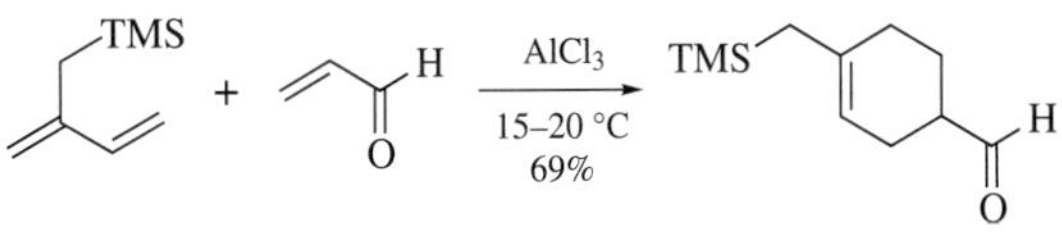

(13)

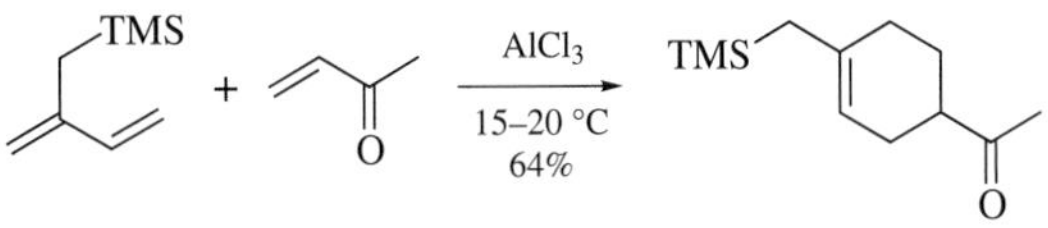

(14)

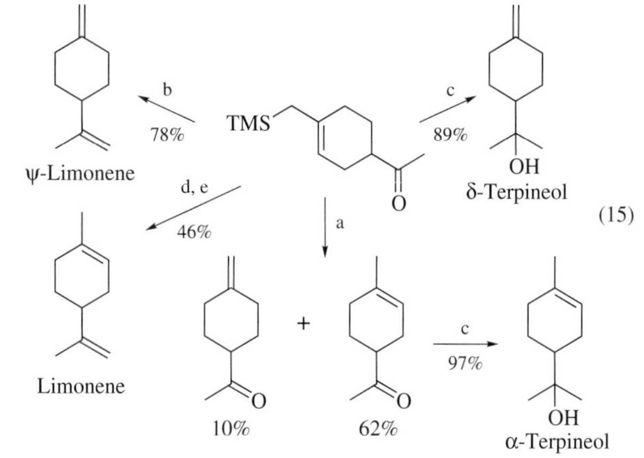

(15)

a: KF, DMSO, 120 °C, 12 h
b: 1. TMSCH₂MgCl, Et₂O, 35 °C, 2 h. 2. MeCOCl, MeOH, 0 °C, 15 min
c: 1. MeMgBr, Et₂O, 35 °C, 2 h. 2. HCl, MeOH, rt, 20 min
d: TMSCH₂MgCl, Et₂O, 35 °C, 2 h
e: CsF, DMSO, 130 °C, 3 h

7.   Sakurai, H.; Sakata, Y.; Hosomi, A., *Chem. Lett.* **1983**, 409.

8.   Hosomi, A.; Araki, Y.; Sakurai, H., *J. Org. Chem.* **1983**, *48*, 3122.

9.   Kawashima, M.; Fujisawa, T., *Bull. Chem. Soc. Jpn.* **1988**, *61*, 4051.

10. Larsen, S. D.; Grieco, P. A.; Fobare, W. F., *J. Am. Chem. Soc.* **1986**, *108*, 3512.

11. Hosomi, A.; Saito, M.; Sakurai, H., *Tetrahedron Lett.* **1980**, *21*, 355.

12. Hosomi, A.; Iguchi, H.; Sasaki, J.; Sakurai, H., *Tetrahedron Lett.* **1982**, *23*, 551.

13. Kasatkin, A. N.; Kulak, A. N.; Tolstikov, G. A., *J. Organomet. Chem.* **1988**, *346*, 23.

14. Pankayatselvan, R.; Nicholas, K. M., *J. Organomet. Chem.* **1990**, *384*, 361.

15. Ding, Y.-X.; Weber, W. P., *Macromolecules* **1988**, *21*, 2672.

Hideki Sakurai
Tohoku University, Sendai, Japan

Formation of π-Allylic Complex Followed by Acyldemetalation. 2-Trimethylsilylmethyl-1,3-butadiene forms a titanium(III) complex by the reaction with dichlorobis-(cyclopentadienyl)titanium and *n*-PrMgBr. The complex reacts with carboxylic acid chlorides RCOCl (R = alkyl, alkenyl) to give β,γ-unsaturated ketones (eq 16).[13]

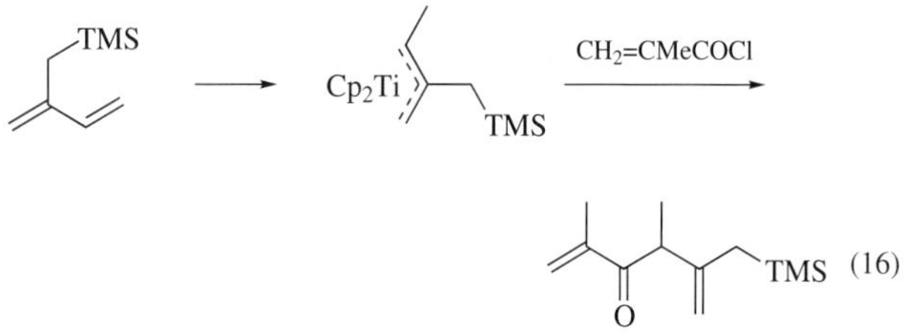

(16)

The regioselectivity of nucleophilic additions to the Co(CO)₃BF₄ complex has also been examined.[14] Ziegler–Natta polymerization of 2-trimethylsilylmethyl-1,3-butadiene catalyzed by triethylaluminum and titanium(IV) chloride gives predominantly *cis*-1,4-polymer. However, anionic polymerization yields a polymer whose microstructure is composed of *cis*-1,4-, *trans*-1,4-, and 3,4-units.[15]

Related Reagents.   Allyltrimethylsilane; 2-Trimethylstannyl-methyl-1,3-butadiene.

1.   (a) Fieser, M.; Danheiser, R. L.; Roush, W., *Fieser & Fieser* **1981**, *9*, 493. (b) Fieser, M., *Fieser & Fieser* **1982**, *10*, 432. (c) Fieser, M., *Fieser & Fieser* **1984**, *11*, 580. (d) Fieser, M., *Fieser & Fieser* **1986**, *12*, 24, 539. (e) Sakurai, H., *Pure Appl. Chem.* **1982**, *54*, 1. (f) Sakurai, H.; Hosomi, A.; Saito, M.; Sasaki, K.; Iguchi, H.; Sasaki, J.; Araki, Y., *Tetrahedron* **1983**, *39*, 883.

2.   Hosomi, A.; Saito, M.; Sakurai, H., *Tetrahedron Lett.* **1979**, 429.

3.   Djahanbini, D.; Cazes, B.; Gore, J., *Tetrahedron* **1985**, *41*, 867.

4.   Klusener, P. A. A.; Tip, L.; Brandsma, L., *Tetrahedron* **1991**, *47*, 2041.

5.   Wilson, S. R.; Philips, L. R.; Natalie, K., *J. Am. Chem. Soc.* **1979**, *101*, 3340.

6.   Sakurai, H.; Sasaki, K.; Hosomi, A., *Tetrahedron Lett.* **1981**, *22*, 745.

3-[(Trimethylsilyl)methyl]-3-butenoic Acid Methyl Ester

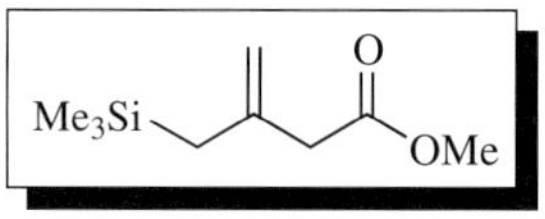

[70639-89-3]   C₉H₁₈O₂Si   (MW 186.11)

InChI = 1S/C9H18O2Si/c1-8(6-9(10)11-2)7-12(3,4)5/h1,6-7H2, 2-5H3

InChIKey = DPBRXLFGMUIBMC-UHFFFAOYSA-N

(isoprenoid acid synthon which reacts readily with a variety of electrophiles)

Physical Data: pale-yellow oil; bp for precursor acid 53 °C/0.01 Torr.[1,2]

Solubility: soluble in all common organic solvents.

Analysis of Reagent Purity: the precursor acid has been characterized by ¹H NMR and IR spectroscopies.[1,2]

Preparative Methods: readily prepared by reaction of diketene with trimethylsilylmethylmagnesium chloride in the presence of a Ni(II) catalyst,[1,2] followed by esterification of the resulting acid using diazomethane.[3]

Purity: distillation at reduced pressure.

Handling, Storage, and Precaution: should be stored at or below room temperature away from acids and other electrophilic species.

Deprotonation and Alkylation Reactions.   Reaction of the title compound with LDA and trapping of the resulting ester enolate with alkyl halides leads to mixtures of the α- and γ-alkylated products (eq 1), the ratio of which can be strongly influenced by the presence or absence of CuI.[4] Related studies have been carried out on the dianion derived from the precursor acid,[1,2,5] and one product of such processes exploited in the synthesis of the racemic modification of the diterpene trixagol.[1]

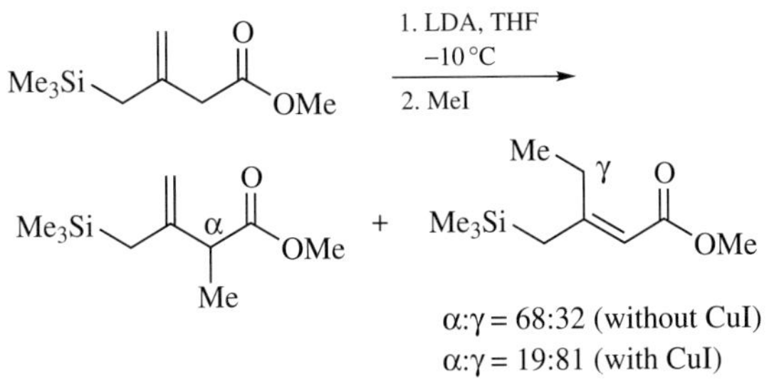

$$\alpha{:}\gamma = 68{:}32 \text{ (without CuI)}$$
$$\alpha{:}\gamma = 19{:}81 \text{ (with CuI)}$$
(1)

$$\alpha{:}\beta = 95{:}5$$

(3)

(4)

Reactions with Electrophiles. The allyltrimethylsilyl moiety associated with the title compound reacts readily with electrophiles, and this process has been exploited in the synthesis of optically active carbacephems. Thus, following on from earlier work,[6] 4-acetoxyazetidin-2-one was treated (eq 2) with 3-[(trimethylsilyl)methyl]-3-butenoic acid methyl ester in the presence of boron trifluoride diethyl etherate to give the corresponding allylated azetidin-2-one.[7,8] The readily generated and racemic N-hydroxymethylated derivative was then subject to a lipase-mediated kinetic resolution using vinyl acetate as an irreversible acyl-transfer agent. The ensuing mixture of alcohol and acetate was separated chromatographically with the latter product being obtained in >95% ee. Further manipulations of such resolved materials provided a variety of carbacephems.

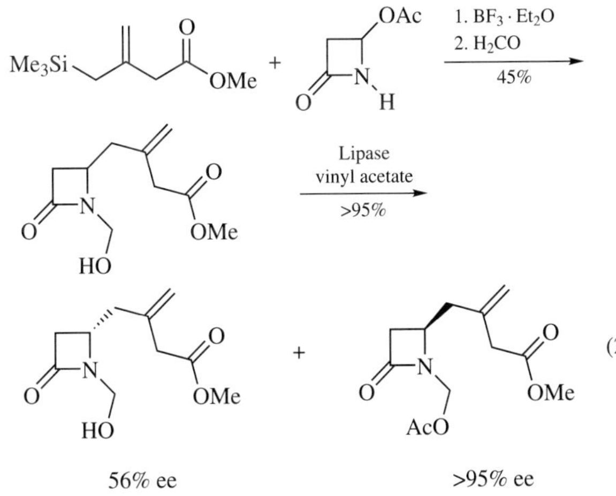

56% ee   >95% ee
(2)

Oxocarbenium ions generated from the pyranose forms of certain protected aldohexoses react with the title reagent to give α-C-glycosyl isoprenoid compounds (eq 3) in varying yields.[3,9] Such adducts have been manipulated to provide C-glycosides incorporating, *inter alia*, a butenolide residue.

Oxocarbenium ions derived from squaric acid derivatives have also been intercepted with 3-[(trimethylsilyl)methyl]-3-butenoic acid methyl ester.[10,11] The ensuing adducts are converted into bicyclo[3.2.0]heptenones (eq 4) upon thermolysis. Related chemistry has been employed in the trapping of a thiacarbenium ion annulated to a C_{60} core.[12]

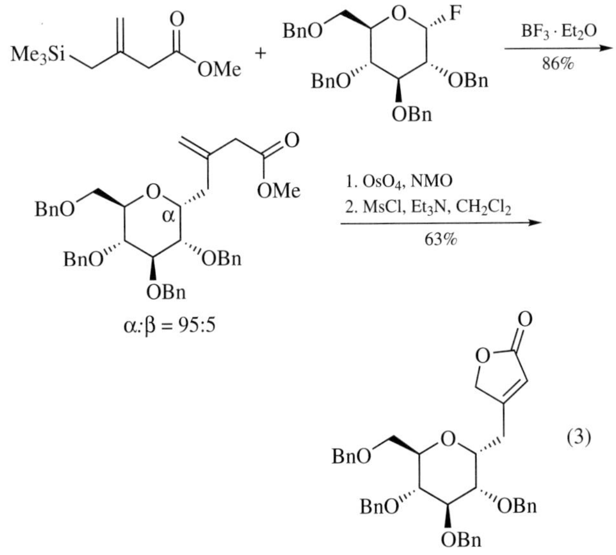

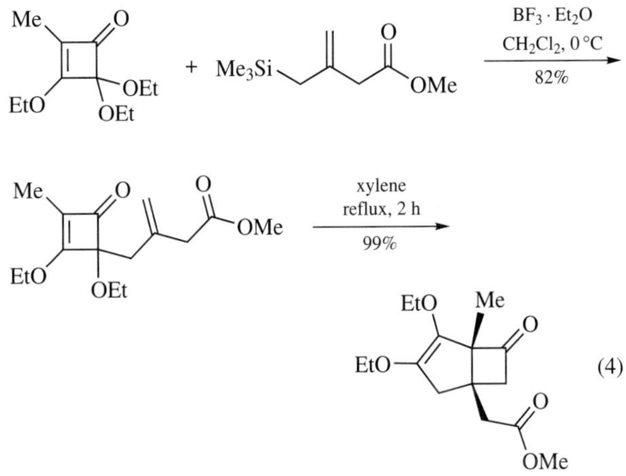

The reaction of 3-[(trimethylsilyl)methyl]-3-butenoic acid methyl ester with various simple nitriles in the presence of BCl_3 has afforded varying mixtures of pyridones, pyridines, and acylation products (eq 5).[13]

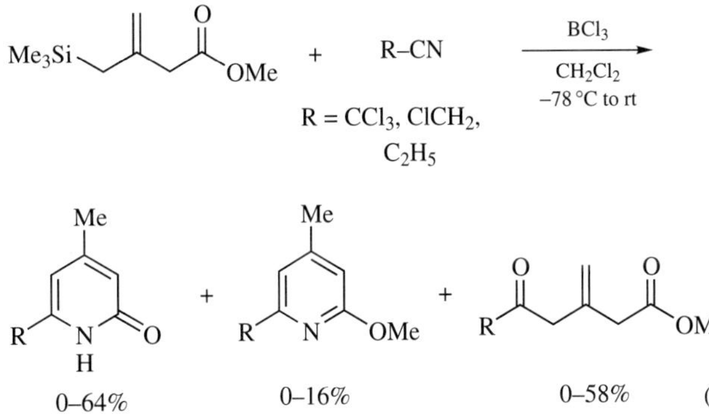

0–64%   0–16%   0–58%   (5)

1. Armstrong, R. J.; Weiler, L., *Can. J. Chem.* **1983**, *61*, 2530.

2. Itoh, K.; Yogo, T.; Ishii, Y., *Chem. Lett.* **1977**, 103.

3. Jégou, A.; Pacheco, C.; Veyrières, A., *Synlett* **1998**, 81.

4. Nishiyama, H.; Itagaki, K.; Takahashi, K.; Itoh, K., *Tetrahedron Lett.* **1981**, *22*, 1691.

5. Itoh, K.; Fukui, M.; Kurachi, Y., *J. Chem. Soc., Chem. Commun.* **1977**, 500.

6. Aratani, M.; Sawada, K.; Hashimoto, M., *Tetrahedron Lett.* **1982**, *23*, 3921.

7. Oumoch, S.; Rousseau, G., *Bioorg. Med. Chem. Lett.* **1994**, *4*, 2841.

8. Oumoch, S.; Rousseau, G., *Bull. Soc. Chim. Fr.* **1996**, *133*, 997.

9. Jégou, A.; Pacheco, C.; Veyrières, A., *Tetrahedron* **1998**, *54*, 14779.

10. Yamamoto, Y.; Ohno, M.; Eguchi, S., *Chem. Lett.* **1995**, 525.

11. Yamamoto, Y.; Ohno, M.; Eguchi, S., *Bull. Chem. Soc. Jpn.* **1996**, *69*, 1353.

12. Ishida, H.; Itoh, K.; Ohno, M., *Tetrahedron* **2001**, *57*, 1737.

13. Hamana, H.; Sugasawa, T., *Chem. Lett.* **1985**, 921–924.

Martin G. Banwell & Brian D. Kelly
The Australian National University, Canberra, ACT, Australia

Trimethylsilylmethyllithium[1]

[1822-00-0] $C_4H_{11}LiSi$ (MW 94.18)

InChI = 1S/C4H11Si.Li/c1-5(2,3)4;/h1H2,2-4H3;

InChIKey = RJPBNMJOQNIBQW-UHFFFAOYSA-N

(methylenation of carbonyl compounds;[1b] reacts with carboxylic acid derivatives to provide α-silyl ketones;[1d,2] synthesis of allylsilanes[1k,3])

Physical Data: mp 112 °C.

Solubility: sol ethereal solvents; reacts with protic solvents.

Preparative Methods: obtained by reaction of (chloromethyl)-trimethylsilane with lithium metal in an inert solvent.[4] It is also available by displacement of a heteroatom group based on sulfur,[5] silicon,[6] or tin.[7]

Handling, Storage, and Precautions: reacts with protic solvents. It should be prepared and handled in inert solvents under an atmosphere of dry nitrogen or argon.

Original Commentary

David J. Ager
The NutraSweet Company, Mount Prospect, IL, USA

Peterson Alkenation. Trimethylsilylmethyllithium (**1**) provides an alternative to a Wittig approach for the preparation of methylene compounds from carbonyl precursors.[1b,c,7] In some cases the use of (**1**) is superior to the Wittig approach.[8] Condensation of (**1**) with a carbonyl compound results in the formation of a β-hydroxysilane. Elimination to the alkene can be accomplished by use of acidic or basic conditions (eq 1).[1a,b,9] acetyl chloride or thionyl chloride can also be used to accomplish this elimination.[10] A wide variety of aldehydes and ketones have been used as substrates in this reaction.[1b] The use of cerium(III) chloride has been advocated with reagent (**1**) to favor nucleophilic addition with enolizable carbonyl compounds. The use of the lithium agent (**1**) gives superior yields compared to the use of trimethylsilylmethyl-magnesium chloride with cerium.[11]

The carbonyl compound can also contain additional functionality.[1b,12] Thus, treatment of an α,β-epoxy ketone with excess lithium reagent (**1**) provides the allyl alcohol (**2**) (eq 2).[13] The use of an α-phenyl selenoaldehyde as electrophile allows either an allyl selenide or a β-silyl aldehyde to be obtained, depending upon the reaction conditions used with the hydroxysilane (eq 3).[14] With α,β-unsaturated ketones, the lithium reagent (**1**) adds in the 1,2-sense; the Grignard analog can provide 1,4-addition.[15] The cuprate derived from (**1**) undergoes the expected reactions for this class of compounds, such as 1,4-addition.[16]

$$\text{TMS}\diagup\text{Li} \quad \xrightarrow{R^1R^2CO} \quad \text{TMS}\diagdown\underset{R^1}{\overset{OH}{\diagup}}R^2 \quad \xrightarrow[\text{or base}]{\text{acid}} \quad \overset{R^1}{\underset{R^2}{\diagup}} \qquad (1)$$
$$(1)$$

$$\text{(eq 2)} \qquad (2)$$

$$\underset{PhSe}{\overset{R^1\ R^2}{\diagup}}\text{CHO} \quad \xrightarrow[Et_2O, -78\ °C]{(1)} \quad \underset{PhSe}{\overset{R^1\ R^2}{\diagup}}\underset{OH}{\diagdown}\text{TMS} \longrightarrow$$

$$\xrightarrow[CH_2Cl_2]{SnCl_2} \quad \underset{R^2}{\overset{R^1}{\diagup}}\diagdown\text{SePh}$$

$$\xrightarrow[CH_2Cl_2]{AgNO_3/celite} \quad \underset{PhSe}{\overset{R^1\ R^2}{\diagup}}\diagdown\text{TMS} \qquad (3)$$

Reaction with Carboxylic Acid Derivatives. Reaction of (**1**) with carboxylic acid derivatives provides β-silyl ketones (**3**) (eq 4). The reaction yield is very dependent upon the presence of α-hydrogens in the substrate; lower yields are obtained when deprotonation can occur at this center.[2] The reaction occurs with esters, lactones, acid chlorides, and the parent carboxylic acids.[7,17] The resultant β-silyl ketones can be desilylated by simple hydrolysis,[2] or used as a substrate in a Peterson alkenation approach to enones.[1b] The use of a cerium reagent with the acid chloride has been advocated for the preparation of allylsilanes.[18]

$$\underset{X = Cl\ or\ OR^1}{\overset{O}{\underset{R}{\diagup}}\diagup X} \quad \xrightarrow{(1)} \quad \underset{(3)}{\overset{O}{\underset{R}{\diagup}}\diagup\text{TMS}} \quad \xrightarrow[HO^-\ or\ F^-]{H_3O^+} \quad \overset{O}{\underset{R}{\diagup}} \qquad (4)$$

Other Electrophiles. A mixed higher order cuprate derived from (**1**) allows for the stereoselective preparation of allylsilanes through an epoxide ring opening (eq 5); use of a lower order cuprate results in low yields and mixtures of products.[3]

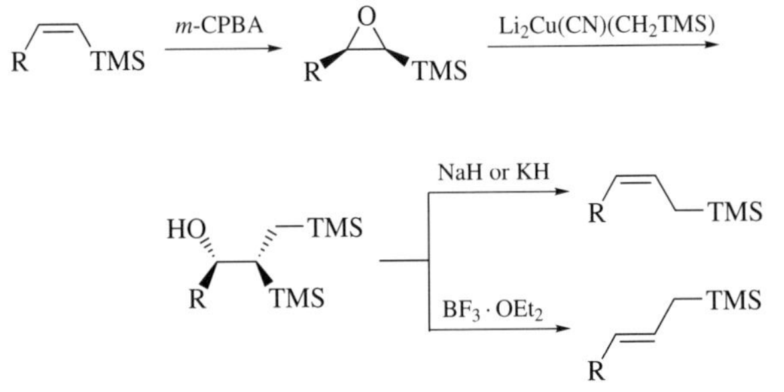

(5)

The organolithium reagent (**1**) also reacts with a wide variety of other electrophiles, including silyl chlorides to provide bis(silyl)methane derivatives,[4a,19] and nitriles to provide β-silyl amines after in situ reduction of the intermediate imine derivative.[20] α-Silyl epoxides are opened to provide the substituted vinylsilane.[21] Reaction of (**1**) with arenesulfonyl fluorides provides α-silyl sulfones,[16c] key intermediates for the preparation of vinyl sulfones.[5b,c,22] Reaction of the lithium reagent (**1**) with aluminum chloride followed by a vinyl triflate results in formation of an allylsilane (eq 6).[23] With carbon monoxide as an electrophile with (**1**), a rapid entry to acylsilanes, or silyl enol ethers by subsequent reaction with chlorotrimethylsilane, can be realized.[24] The alkyllithium (**1**) is the precursor to a wide variety of organometallic compounds that contain the trimethylsilylmethyl ligand.[25]

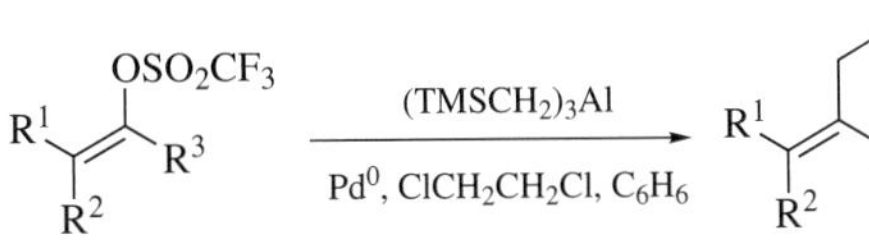

(6)

Analogs. The displacement of a heteroatom to introduce the lithium allows for a wide range of substituted analogs of the reagent (**1**) to be prepared.[1b,5,26] The presence of an aromatic group on the same carbon atom as the silyl moiety allows for direct deprotonation by butyllithium to form the lithium reagent.[27] Higher alkyl analogs of (**1**) are also available by the addition of an alkyllithium to a vinylsilane.[1a,10,27,28] All of these higher homologs react with carbonyl compounds and other electrophiles, as expected.[1b,d]

First Update

David Ager
DSM, Raleigh, NC, USA

Peterson Alkenylation. The Peterson alkenylation reaction continues to be the major use of trimethylsilylmethyllithium (**1**) and converts a carbonyl group to a methylene (eq 1).[29] Some enantioselectivity has been seen for the addition of **1** to benzaldehyde in the presence of chiral ligands.[30] The ketone substrates can contain functional groups that are compatible with the basic conditions.[31] The use of cerium(III) chloride to promote nucleophilic addition of **1** to an enolizable carbonyl compound has continued to prove advantageous.[32]

For the reaction of nonenolizable ketones with **1**, the use of diethylaluminum chloride has been advocated as a method to cause the elimination of water to form a vinylsilane (eq 7).[33]

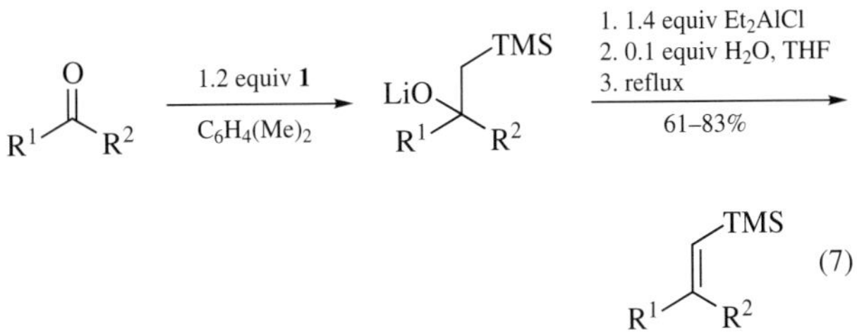

(7)

Reaction with Carboxylic Acid Derivatives. The reaction of a methyl ester to provide a methyl ketone (eq 4) has been used in total syntheses of epothilones.[34] A Weinreb amide has also been used successfully with **1** to give a methyl ketone in high yield (eq 4; X = NMeOMe).[35]

Other Electrophiles. The alkyllithium (**1**) adds to an imine. This approach has been used to prepare a linker for solid-phase peptide synthesis (eq 8).[36]

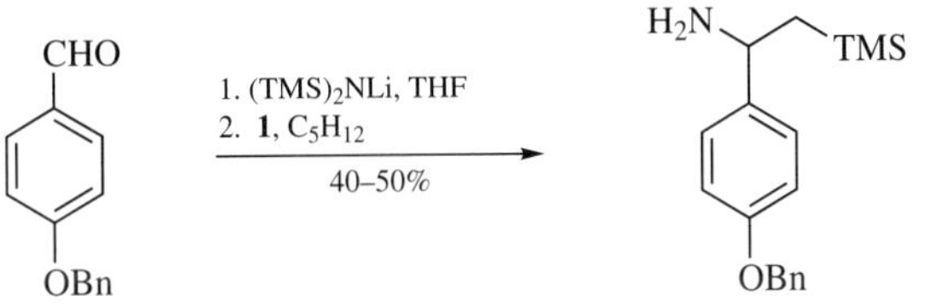

(8)

As with other alkyllithiums, **1** can react with epoxides in an alkylative deoxygenation reaction to provide an allyl alcohol. At least 2 equiv of the alkyllithium are required as ring opening occurs from the α-lithioepoxide, whose formation requires the first equivalent of the alkyllithium to act as a base.[37] With functionalized epoxides such as **4**, subsequent reactions can also occur as illustrated in eq 9.[38] An analogous reaction is observed with 3,4-epoxytetrahydrofurans to give 1,2-diols.[39]

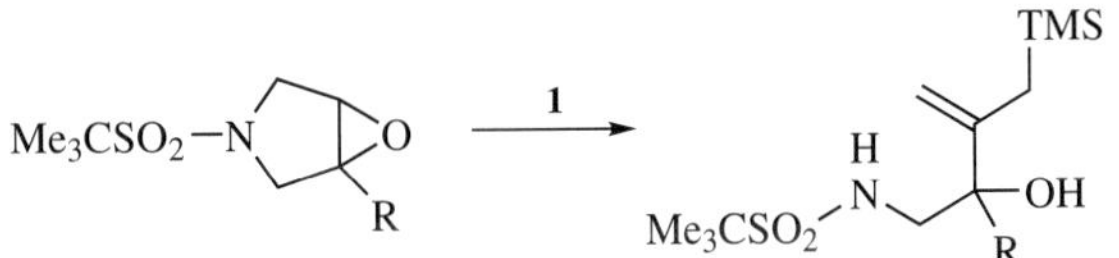

(9)

The alkyllithium (**1**) reacts with allylic carbamates in the presence of copper(I) to afford the substituted product (eq 10).[40]

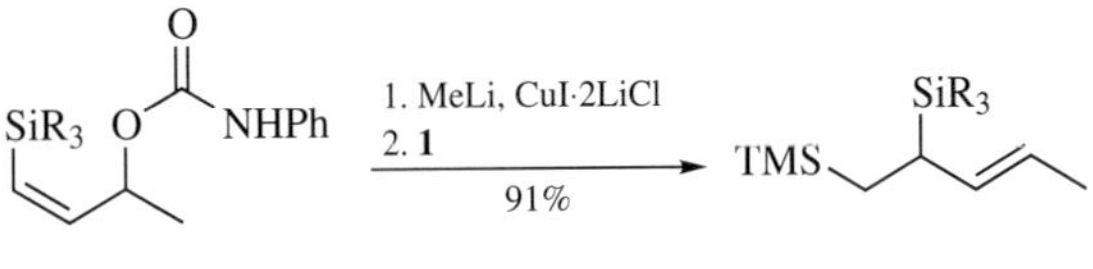

(10)

β-Ketosilanes can be accessed by reaction of **1** with (Z)-1-bromo-1-alkenylboronate esters (eq 11).[41]

$$R^1 \text{–} Br \text{ on } B\text{–}O \quad \xrightarrow{1} \quad R^1 \text{...} B\text{–}O \text{ ...TMS} \quad \xrightarrow[\text{H}_2\text{O}_2]{\text{NaOAc}}$$

$$R^1 \underset{\text{TMS}}{\overset{O}{\big\Vert}} \qquad (11)$$

Related Reagents. Bis(trimethylsilyl)methane; (diisopropoxymethylsilyl)methylmagnesium chloride; trimethylsilylmethylmagnesium chloride; (trimethylstannylmethyl)lithium.

1. (a) Chan, T.-H., *Acc. Chem. Res.* **1977**, *10*, 442. (b) Ager, D. J., *Org. React.* **1990**, *38*, 1. (c) Ager, D. J., *Synthesis* **1984**, 384. (d) Fleming, I. In *Comprehensive Organic Chemistry*; Barton, D. H. R.; Ollis, W. D., Eds.; Pergamon: Oxford, 1979; Vol. 3; p 541. (e) Weber, W. P., *Silicon Reagents for Organic Synthesis–Concepts in Organic Chemistry*; Springer: New York, 1983; Vol. 14. (f) Colvin, E. W., *Silicon in Organic Synthesis*; Butterworths: London, 1981. (g) Colvin, E. W., *Chem. Soc. Rev.* **1978**, *7*, 15. (h) Magnus, P., *Aldrichim. Acta* **1980**, *13*, 43. (i) Magnus, P. D.; Sarkar, T.; Djuric, S. In *Comprehensive Organometallic Chemistry*; Wilkinson, G.; Stone, F. G. A.; Abel, E. W., Eds.; Pergamon: Oxford, 1982; Vol. 7; p 515. (j) Birkofer, L.; Stuhl, O., *Top. Curr. Chem.* **1980**, *88*, 33. (k) Chan, T. H.; Fleming, I., *Synthesis* **1979**, 761.

2. Demuth, M., *Helv. Chim. Acta* **1978**, *61*, 3136.

3. Soderquist, J. A.; Santiago, B., *Tetrahedron Lett.* **1989**, *30*, 5693.

4. Sommer, L. H.; Mitch, F. A.; Goldberg, G. M., *J. Am. Chem. Soc.* **1949**, *71*, 2746; Connolly, J. W.; Urry, G., *Inorg. Chem.* **1963**, *2*, 645.

5. (a) Cohen, T.; Sherbine, J. P.; Matz, J. R.; Hutchins, R. R.; McHenry, B. M.; Willey, P. R., *J. Am. Chem. Soc.* **1984**, *106*, 3245. (b) Ager, D. J., *J. Chem. Soc., Perkin Trans. 1* **1986**, 183. (c) Ager, D. J., *J. Chem. Soc., Perkin Trans. 1* **1986**, 195.

6. Sakurai, H.; Nishiwaki, K.; Kira, M., *Tetrahedron Lett.* **1973**, 4193.

7. Seitz, D. E.; Zapata, A., *Synthesis* **1981**, 557.

8. Jung, M. E.; Hudspeth, J. P., *J. Am. Chem. Soc.* **1980**, *102*, 2463.

9. Hudrlik, P. F.; Peterson, D., *J. Am. Chem. Soc.* **1975**, *97*, 1464; Hudrlik, P. F.; Peterson, D., *Tetrahedron Lett.* **1974**, 1133; Olah, G. A.; Reddy, V. P.; Prakash, G. K. S., *Synthesis* **1991**, 29.

10. Chan, T. H.; Chang, E., *J. Org. Chem.* **1974**, *39*, 3264.

11. Johnson, C. R.; Tait, B. D., *J. Org. Chem.* **1987**, *52*, 281.

12. Davis, F. A.; Kumar, A., *Tetrahedron Lett.* **1991**, *32*, 7671; Wilson, S. R.; Venkatesan, A. M.; Augelli-Szafran, C. E.; Yasmin, A., *Tetrahedron Lett.* **1991**, *32*, 2339; Evans, J. M.; Kallmerten, J., *Synlett* **1992**, 269.

13. Sato, T.; Kikuchi, T.; Sootome, N.; Murayama, E., *Tetrahedron Lett.* **1985**, *26*, 2205.

14. Nishiyama, H.; Kitajima, T.; Yamamoto, A.; Itoh, K., *J. Chem. Soc., Chem. Commun.* **1982**, 1232; Nishiyama, H.; Itagaki, K.; Osaka, N.; Itoh, K., *Tetrahedron Lett.* **1982**, *23*, 4103.

15. Taylor, R. T.; Galloway, J. G., *J. Organomet. Chem.* **1981**, *220*, 295.

16. (a) Horiguchi, Y.; Kataoka, Y.; Kuwajima, I., *Tetrahedron Lett.* **1989**, *30*, 3327. (b) Klaver, W. J.; Moolenaar, M. J.; Hiemstra, H.; Speckamp, W. N., *Tetrahedron* **1988**, *44*, 3805. (c) Frye, L. L.; Sullivan, E. L.; Cusak, K. P.; Funaro, J. M., *J. Org. Chem.* **1992**, *57*, 697.

17. Ruden, R. A.; Gaffney, B. L., *Synth. Commun.* **1975**, *5*, 15; Wagner, J.; Vogel, P., *Tetrahedron* **1991**, *47*, 9641; Glänzer, B. I.; Csuk, R., *Carbohydr. Res.* **1991**, *220*, 79.

18. Anderson, M. B.; Fuchs, P. L., *Synth. Commun.* **1987**, *17*, 621.

19. Peterson, D. J., *J. Organomet. Chem.* **1967**, *9*, 373; Barton, T. J.; Hoekman, S. K., *J. Am. Chem. Soc.* **1980**, *102*, 1584.

20. Cunico, R. F., *J. Org. Chem.* **1990**, *55*, 4474.

21. Santiago, B.; Lopez, C.; Soderquist, J. A., *Tetrahedron Lett.* **1991**, *32*, 3457.

22. Jones, P. S.; Ley, S. V.; Simpkins, N. S.; Whittle, A. J., *Tetrahedron* **1986**, *42*, 6519.

23. Saulnier, M. G.; Kadow, J. F.; Tun, M. M.; Langley, D. R.; Vyas, D. M., *J. Am. Chem. Soc.* **1989**, *111*, 8320.

24. Murai, S.; Ryu, I.; Iriguchi, J.; Sonoda, N., *J. Am. Chem. Soc.* **1984**, *106*, 2440.

25. Armitage, D. A. In *Comprehensive Organometallic Chemistry*; Wilkinson, G.; Stone, F. G. A.; Abel, E. W., Eds.; Pergamon: Oxford, 1982; Vol. 2; p 1.

26. Dumont, W.; Krief, A., *Angew. Chem., Int. Ed. Engl.* **1976**, *15*, 161; Barrett, A. G. M.; Hill, J. M.; Wallace, E. M.; Flygare, J. A., *Synlett* **1991**, 764; Barrett, A. G. M.; Hill, J. M.; Wallace, E. M., *J. Org. Chem.* **1992**, *57*, 386.

27. Peterson, D. J., *J. Org. Chem.* **1968**, *33*, 780; Brook, A. G.; Duff, J. M.; Anderson, D. G., *Can. J. Chem.* **1970**, *48*, 561; Chan, T. H.; Chang, E.; Vinokur, E., *Tetrahedron Lett.* **1970**, 1137.

28. Tamao, K.; Kanatani, R.; Kumada, M., *Tetrahedron Lett.* **1984**, *25*, 1905.

29. van Staden, L. F.; Gravestock, D.; Ager, D. J., *Chem. Soc. Rev.* **2002**, *31*, 195.

30. Schön, M.; Naef, R., *Tetrahedron: Asymmetry* **1999**, *10*, 169

31. Mukai, C.; Yamashita, H.; Ichiryu, T.; Hanaoka, M., *Tetrahedron* **2000**, *56*, 2203

32. Donkervoort, J. G.; Gordon, A. R.; Johnstone, C.; Kerr, W. J.; Lange, U., *Tetrahedron* **1996**, *52*, 7391.

33. Kwan, M. L.; Battiste, M. A., *Tetrahedron Lett.* **2002**, *43*, 8765; Kwan, M. L.; Yeung, C. W.; Breno, K. L.; Doxsee, K. M., *Tetrahedron Lett.* **2001**, *42*, 1411.

34. Mulzer, J.; Mantoulidis, A.; Öhler, E., *J. Org. Chem.* **2000**, *65*, 7456; Storer, R. I.; Takemoto, T.; Jackson, P. S.; Brown, D. S.; Baxendale, I. R.; Ley, S. V., *Chem. Euro. J.* **2004**, *10*, 2529.

35. Yang, C.; Yasuda, N., *Bioorg. Med. Chem. Lett.* **1998**, *8*, 255.

36. Chao, H.-G.; Bernatowicz, M. S.; Matsueda, G. R., *J. Org. Chem.* **1993**, *58*, 2640.

37. Crandall, J. K.; Apparu, M., *Org. React.* **1983**, *29*, 345.

38. Hodgson, D. M.; Miles, T. J.; Witherington, J., *Tetrahedron* **2003**, *59*, 9729; Hogdson, D. M.; Maxwell, C. R.; Miles, T. J.; Paruch, E.; Stent, M. A. H.; Matthews, I. R.; Wilson, F. X.; Witherington, J., *Angew. Chem., Int. Ed. Engl.* **2002**, *41*, 4313; Hodgson, D. M.; Paruch, E., *Tetrahedron* **2004**, *60*, 5185; Hodgson, D. M.; Miles, T. J.; Witherington, J., *Synlett* **2000**, 310.

39. Hodgson, D. M.; Stent, M. A. H.; Wilson, F. X., *Org. Lett.* **2001**, *3*, 3401.

40. Smitrovich, J. H.; Woerpel, K. A., *J. Magn. Reson.* **2000**, *65*, 1601; *J. Am. Chem. Soc.* **1998**, *120*, 12998.

41. Bhat, N. G.; Martinez, C.; De Los Santos, J., *Tetrahedron Lett.* **2000**, *41*, 6541.

Trimethylsilylmethylmagnesium Chloride[1]

Me₃Si⌃MgCl

[13170-43-9]   C₄H₁₁ClMgSi   (MW 147.00)

InChI = 1S/C4H11Si.ClH.Mg/c1-5(2,3)4;;/h1H2,2-4H3;1H;/q;;+1/p-1

InChIKey = NAQATMJWCJCHOZ-UHFFFAOYSA-M

(methylenation of carbonyl compounds;[1b,2] provides a variety of methods to prepare allylsilanes[1k])

Solubility: sol ethereal solvents; reacts with protic solvents.
Preparative Method: from (chloromethyl)trimethylsilane and magnesium in an ethereal solvent.[3,4]
Handling, Storage, and Precautions: this Grignard reagent reacts with protic solvents.

Original Commentary

David J. Ager
The NutraSweet Company, Mount Prospect, IL, USA

Peterson Alkenation. Trimethylsilylmethylmagnesium chloride (**1**) reacts with carbonyl compounds to give β-hydroxysilanes (**2**).[2,4,5] These silanes can then be eliminated to provide an alkene under acidic or basic conditions, such as with sodium hydride or potassium hydride (eq 1).[1a,1b,5,6] The elimination can also be accomplished by acetyl chloride or thionyl chloride.[7] For the introduction of *exo*-methylene groups, reagent (**1**) has been found to be superior to a Wittig approach;[8] the silicon reagent reacts rapidly and the byproduct is simple to remove.[1b]

This methodology has found application in the carbohydrate field for homologation of a saccharide,[9] as other functional groups can be tolerated.[1b,10] The resultant alkene can be functionalized in a wide variety of ways.[11] The use of paraformaldehyde as electrophile provides a simple method to 2-(trimethylsilyl)ethanol.[12]

Many of the uses of the Grignard reagent (**1**) are complementary to those of trimethylsilylmethyllithium, although the cerium reagent derived from the lithium analog provides higher yields with enolizable aldehydes and ketones.[13]

With α,β-unsaturated carbonyl compounds the Grignard reagent reacts in a 1,2-manner,[14] although 1,4-addition can be observed in certain cases.[15] The resultant β-hydroxysilane from a 1,2-addition can be isomerized to the β-ketosilane with a rhodium catalyst.[16] In the presence of copper(I), the Grignard reagent (**1**) reacts in a 1,4-manner with α,β-unsaturated carbonyl compounds.[15,17]

The use of substituted carbonyl compounds allows for the formation of functionalized alkenes; for example, α,β-epoxy ketones afford the monoepoxide of a diene.[18] The successive treatment of α-chlorocarbonyl compounds with (**1**) and lithium powder provides a regioselective entry to allylsilanes (eq 2).[19]

Reagent (**1**) also reacts with imines that, in turn, can be generated in situ.[20]

Reaction with Carboxylic Acid Derivatives. In addition to carbonyl compounds, reagent (**1**) also reacts with carboxylic acid derivatives.[4] Thus lactones provide hydroxy allylsilanes (**3**) (eq 3).[21]

Reaction of excess (**1**) with esters provides the tertiary alcohol in an analogous manner,[1d] and subsequent elimination provides the allylsilane.[22] The addition of the second equivalent of (**1**), however, is dependent on the steric requirements of the intermediate β-silyl ketone.[23] The addition of chlorotrimethylsilane to the reaction mixture has been advocated as higher yields of the resultant allylsilane are obtained.[24] The use of cerium(III) chloride with Grignard reagent (**1**) promotes nucleophilic attack on esters, and the allylsilanes can be obtained in high yield (eq 4).[25] The yields seem to be higher than for the analogous reaction between an acid chloride and the cerium reagent prepared from trimethylsilylmethyllithium.[26] 1,2-Addition is observed with α,β-unsaturated esters.[25] Other functional groups, such as acetals, thioacetals, halogens, hydroxy, acetates, and sulfides, can be incorporated at the α-position of the ester group without detrimental effects.[27]

Reaction with diketene in the presence of a nickel catalyst gives 3-(trimethylsilylmethyl)but-3-enoic acid (**4**) (eq 5).[28]

The Grignard reagent (**1**) does react with acid chlorides to provide ketones after hydrolytic workup.[2] The use of a copper(I) catalyst allows isolation of β-silyl ketones (eq 6).[29]

β-Silyl ketones are hydrolytically unstable and can be converted to the desilylated ketone by simple acid or base treatment,[30] or used in a Peterson alkenation reaction to provide enones.[1b,29a] They are also precursors to silyl enol ethers by rearrangement.[29c,31] Reaction of the β-silyl ketone with a vinyl Grignard reagent provides a rapid entry to 2-substituted 1,3-dienes by a Peterson protocol.[32]

Reaction of (**1**) with carbon dioxide provides trimethylsilylacetic acid,[33] while treatment of (**1**) with ethyl chloroformate gives ethyl trimethylsilylacetate.[34] The use of ethyl oxalyl chloride as substrate for (**1**) provides a simple preparation of ethyl 2-(trimethylsilylmethyl)propenoate.[35] Condensation of (**1**) with benzonitrile resulted in isolation of acetophenone (64%) and desoxybenzoin (45%).[4]

Reaction with Alkyl Halides. The Grignard reagent (**1**) can be alkylated by allyl halides to afford the homoallylsilane.[4,36] The use of a nickel(II) catalyst facilitates coupling of the Grignard reagent (**1**) to vinyl halides[37] and aryl halides, triflates and O-carbamates[38] and provides a useful method for the preparation of allyl- and benzylsilanes (eq 7). Palladium catalysis is also effective to couple (**1**) with vinyl halides.[37b,39]

$$R^1 \diagup\!\!\!\!\!\diagdown Br \xrightarrow[\text{NiCl}_2(\text{dppp})]{(1)} R^1 \diagup\!\!\!\!\!\diagdown\!\!\!\!\!\diagdown TMS \qquad (7)$$

Reaction with Sulfur Compounds. In a similar coupling reaction to those of alkyl halides, (**1**) reacts in the presence of a nickel catalyst with allylic dithioacetals to yield 1-(trimethylsilyl) butadienes (eq 8).[40]

$$\underset{R^2}{\underset{|}{R^1}}\diagup\!\!\!\!\!\diagdown\!\!\!\!\!\overset{SR^3}{\underset{SR^3}{\diagdown}} \xrightarrow[\text{Et}_2\text{O, C}_6\text{H}_6]{(1), \text{NiCl}_2(\text{PPh}_3)} R^1\diagup\!\!\!\!\!\overset{}{\underset{R^2}{\diagdown}}\!\!\!\!\!\diagup\!\!\!\!\!\diagdown TMS \quad (8)$$

Dithioacetals derived from alkyl aryl ketones react with the Grignard reagent (**1**) in the presence of a nickel catalyst.[41,42] The use of an orthothioester as substrate in place of a thioacetal provides 1,3-bis(trimethylsilyl)propenes as a mixture of the (E)- and (Z)-isomers.[43] With α-oxoketene dithioacetals (**5**), the Grignard reagent (**1**) reacts, in the presence of copper(I) iodide, to provide 1-trimethylsilyloxy-3-thiadienes (eq 9).[44]

$$\underset{R^2}{\underset{|}{\overset{\text{O}}{R^1}}}\!\!\!\!\overset{\text{SMe}}{\underset{\text{SMe}}{}} \xrightarrow[\text{Et}_2\text{O}]{(1), \text{CuI}} \left[R^1\overset{\text{O}}{\underset{R^2}{}}\!\!\!\!\overset{\text{TMS}}{\underset{\text{SMe}}{}} \right] \longrightarrow$$

$$\underset{R^2}{\overset{\text{TMSO}}{R^1}}\!\!\!\!\overset{}{\underset{\text{SMe}}{}} \qquad (9)$$

Reaction with Other Electrophiles. Reaction occurs between (**1**) and epoxides to yield γ-hydroxysilanes.[45] Michael addition is observed with nitroalkenes; the silyl group can then promote a Nef reaction to afford β-silyl ketones (eq 10).[46]

$$\underset{\text{NO}_2}{\overset{\text{CH}_3}{C_4H_9}}\!\!\!\! \xrightarrow[\text{2. 10\% H}_2\text{SO}_4]{\text{1. (1), THF}} \underset{\text{O}}{\overset{\text{TMSCH}_2}{C_4H_9}}\!\!\!\!\overset{\text{CH}_3}{} \quad (10)$$
$$52\%$$

With aromatic nitro compounds, nucleophilic addition of (**1**) occurs on the aromatic nucleus (eq 11).[47]

$$\underset{X}{\overset{\text{NO}_2}{\bigcirc}} \xrightarrow{(1), \text{THF}} \underset{X}{\overset{\text{NO}_2}{\bigcirc}}\!\!\!\!\text{TMS} \qquad (11)$$

A nickel catalyst allows reaction between (**1**) and an enol phosphate,[48] silyl enol ether,[49] or substituted dihydrofurans[50] and dihydropyrans[51] to afford allylsilanes. Additional functionality can be tolerated in the substrate.[52]

Coupling of (**1**) with propargylic tosylates, mesylates, or acetates in the presence of copper(I) leads to α-trimethylsilylallenes,[53] while the use of a propargyl alcohol substrate leads to the substituted allylsilane (eq 12).[54]

$$\equiv\!\!\!\!\diagup\!\!\!\!\!\diagdown\text{OH} \xrightarrow[\substack{\text{2. pH 9}\\64\%}]{\substack{\text{1. (1), }s\text{-BuMgCl}\\\text{CuCl, Et}_2\text{O}}} \text{TMS}\diagup\!\!\!\!\!\overset{}{\diagdown}\!\!\!\!\!\diagup\!\!\!\!\!\diagdown\text{OH} \quad (12)$$

With alkynes, copper-catalyzed addition of the Grignard reagent (**1**) provides the allylsilane.[55] The intermediate vinylcopper reagent can also be trapped with electrophiles.[56]

The reagent (**1**) also reacts with wide variety of electrophiles, including carbon dioxide,[33] cyanogen,[57] silyl chlorides,[3,58] metal halides,[59] and phosphorus(III) chloride[59a] With methyl phenyl sulfinate, reaction of (**1**) leads to the formation of (phenylthio)(trimethylsiloxy)methane, a protected form of formaldehyde, by a sila-Pummerer rearrangement.[60]

The Grignard reagent (**1**) is the precursor of numerous organometallic compounds that contain the trimethylsilylmethyl ligand.[61]

First Update

Hideki Yorimitsu & Koichiro Oshima
Kyoto University, Kyoto, Japan

Peterson Alkenation. The Peterson alkenation is as useful and reliable as the Wittig reaction. The former is often better than the latter with respect to the workup procedure. Recent development of the use of trimethylsilylmethylmagnesium chloride in the Peterson alkenation is directed toward the syntheses of complex molecules of biological interest.

The alkenations of aldehydes and ketones are applied to the synthesis of key intermediates for carbapenem antibiotics (eq 13),[62] naturally occurring sesquiterpenoids,[63] a taxol analog,[64] a β-glucosidase inhibitor,[65] and so on.[66]

Lactols, masked aldehydes, undergo the alkenation to yield hydroxyalkenes.[67] For instance, the synthesis of Hemibrevetoxin B is achieved by using ring-closing metathesis as a key step, the vinyl group for which is prepared by the alkenation (eq 14).[67a]

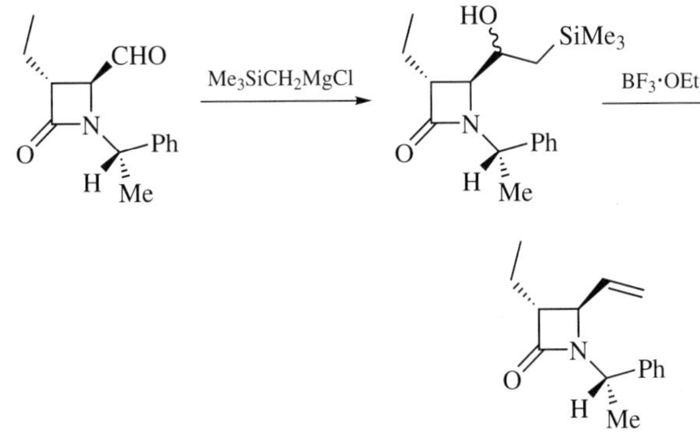

(13)

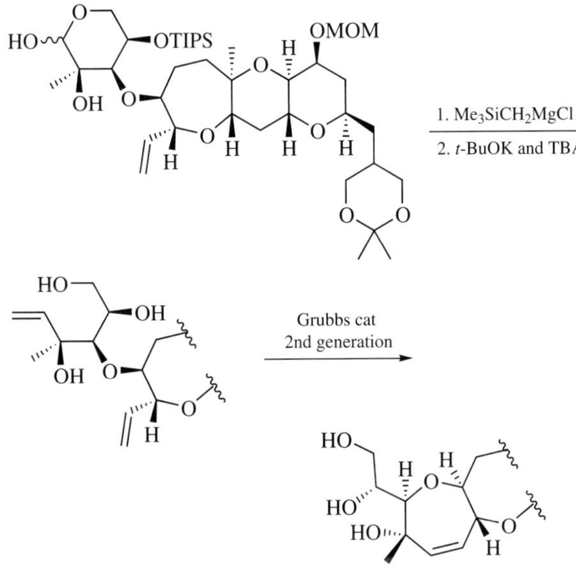

(16)

(+)-alkaloid 241D

A fluorine-containing allylsilane is prepared from methyl chloro-difluoroacetate, wherein no cerium salt is necessary (eq 17).[71]

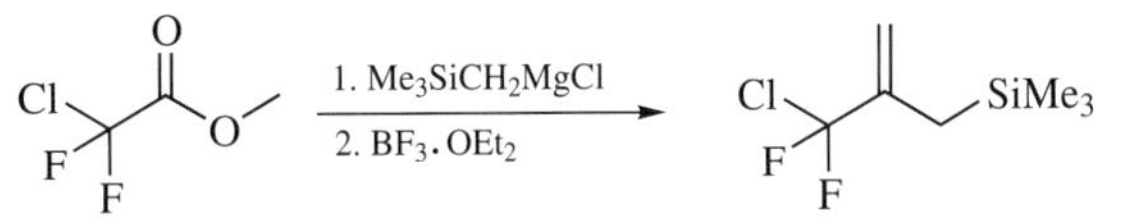

(17)

Cross-coupling Reactions. Trimethylsilylmethylmagnesium chloride reacts with organic halides (or pseudohalides), especially aryl and alkenyl halides, in the presence of transition metal catalysts. The reactions directly provide allylic or benzylic trimethylsilanes of significant synthetic use.

Alkenyl phosphates,[72] prepared from the corresponding ketones, as well as alkenyl halides[73] undergo nickel-catalyzed cross-coupling reactions to yield allylsilanes (eq 18). The reactions are stereospecific.

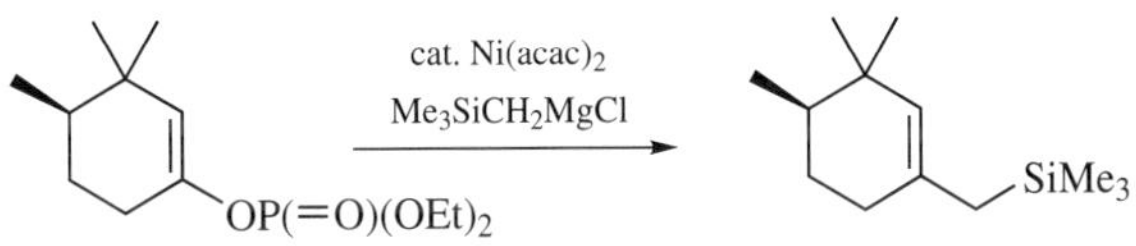

(18)

Alkenyl iodides, bromides or triflates are the coupling partners when palladium catalysts are employed (eq 19).[74] The coupling reaction is highly chemoselective. The allylsilanes formed are available to conduct the Hosomi–Sakurai allylation, leading to natural products.

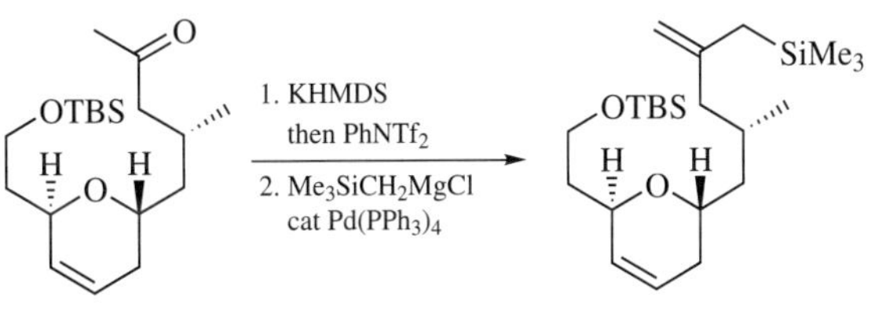

(19)

Alkenyl or aryl selenides react with trimethylsilylmethyl-magnesium chloride under either nickel or palladium catalysis (eq 20).[75a]

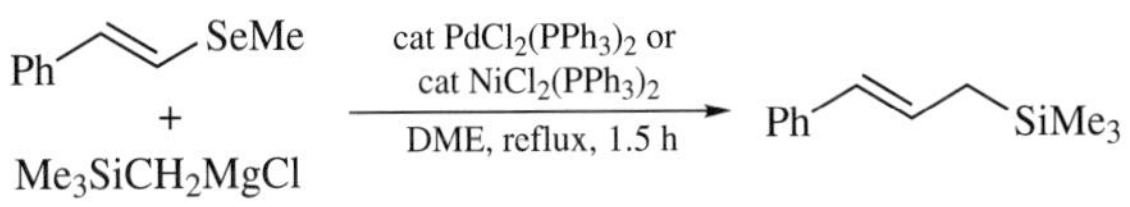

(20)

The cross-coupling reaction of 2-perfluorooctyl-substituted iodoethane provides the corresponding cross-coupling product in high yield, without suffering from slow oxidative addition of the

(14)

Combined with the Sharpless dihydroxylation reaction, the Peterson alkenation offers a conversion of an oxo moiety to a chiral 1,2-dihydroxyethyl group.[68] It is worth noting that the Peterson alkenation technology is superior to the Wittig reaction in the synthesis of vinylfuran (eq 15).[68a]

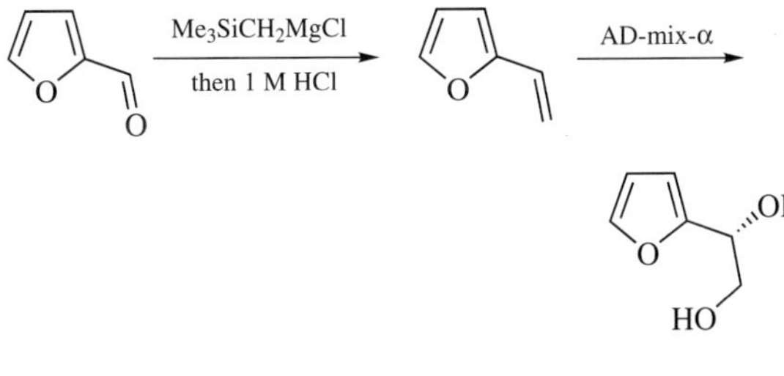

(15)

Cerium(III) chloride mediates nucleophilic attack of Me₃SiCH₂MgCl to esters to form allylsilanes. The reaction allows the syntheses of allylsilanes having a functionalized alkyl chain at the β position. As a representative, the reactions of chiral β-hydroxy- and β-amino ester with Me₃SiCH₂MgCl in the presence of CeCl₃ provide chiral allylsilanes, which are applicable to the synthesis of natural products (eq 16).[69,70]

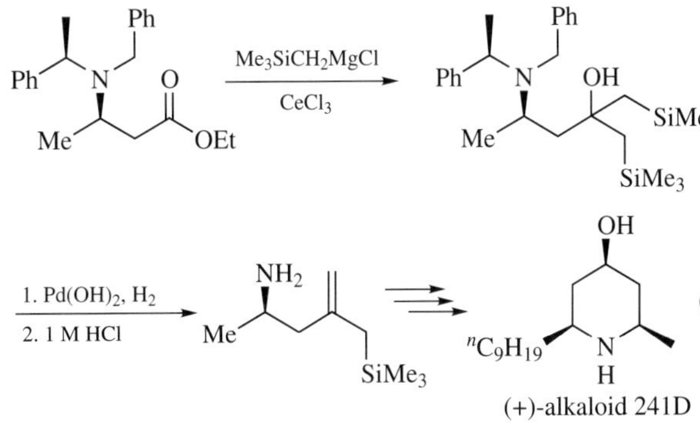

iodide and β-hydride elimination from the alkylcopper intermediate (eq 21).[75b]

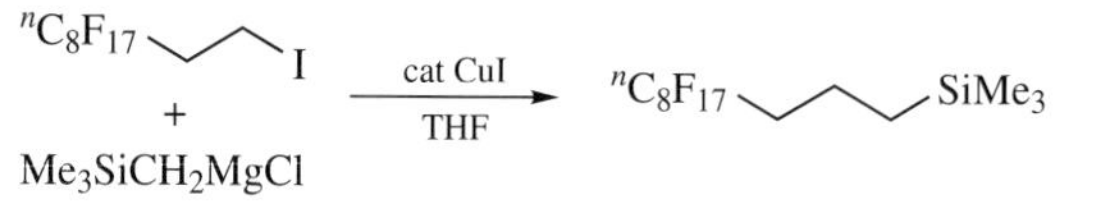

$$^{n}C_8F_{17}\diagdown\diagup I + Me_3SiCH_2MgCl \xrightarrow[\text{THF}]{\text{cat CuI}} {}^{n}C_8F_{17}\diagdown\diagup\diagdown SiMe_3 \quad (21)$$

The copper-mediated S_N2' reaction of (E)-3-t-butyldimethylsilyl-1-methyl-2-propenyl carbamate with trimethylsilylmethylmagnesium chloride furnishes α-trimethylsilylmethyl-substituted (Z)-crotylsilane regio- and stereoselectively (eq 22).[76] On the other hand, a similar reaction of the corresponding (Z) carbamate with trimethylsilylmethyllithium exhibits the opposite stereoselectivity (eq 23).

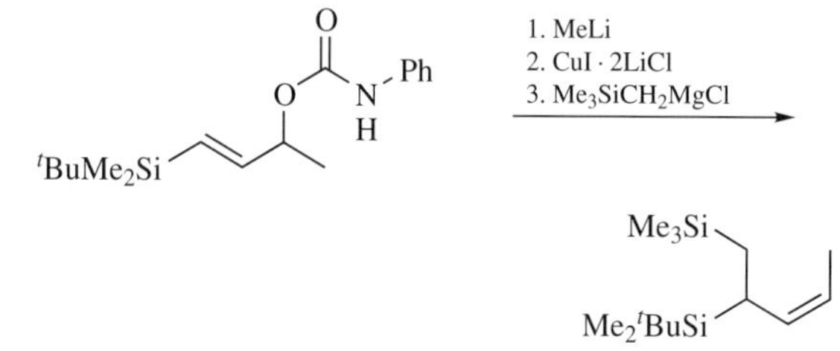

$$ (22) $$

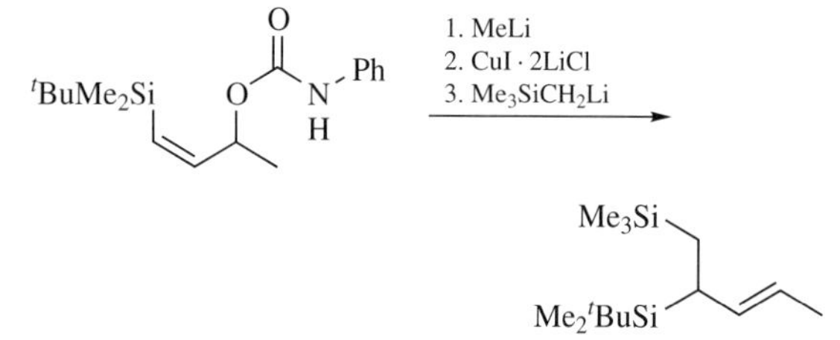

$$ (23) $$

The S_N2' reaction of propargyl tosylates with trimethylsilylmethylmagnesium chloride in the presence of a stoichiometric amount of CuCN·2LiCl permits the large-scale synthesis of allenylsilanes (eq 24).[77]

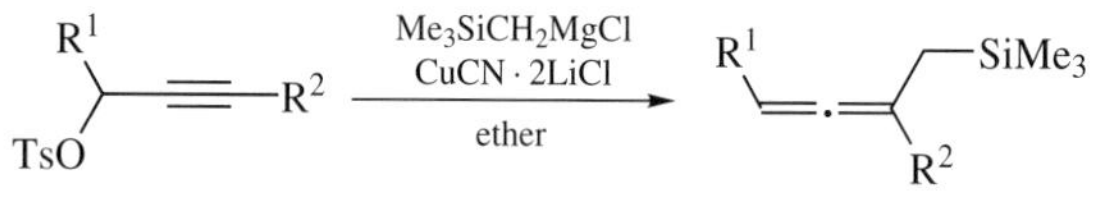

$$ (24) $$

Cobalt catalysts also promote the cross-coupling reaction of aryl halides, alkenyl halides, and allyl methyl ethers to yield arylmethyl-, allyl-, and homoallyltrimethylsilanes, respectively (eq 25).[78]

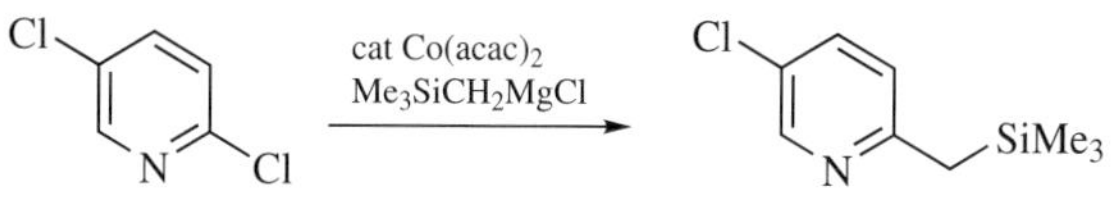

$$ (25) $$

The cobalt-mediated reaction of diphenylmethyl(dibromomethyl)silanes with trimethylsilylmethylmagnesium chloride yields 1,2-disilylethenes in a regio- and stereoselective fashion (eq 26).[79] The reaction is applicable to the synthesis of 1,1,2-trisilylethenes starting from $(R_3Si)_2CBr_2$. Manganese[80] and copper[81] salts also mediate the reaction of gem-dibromo compounds.

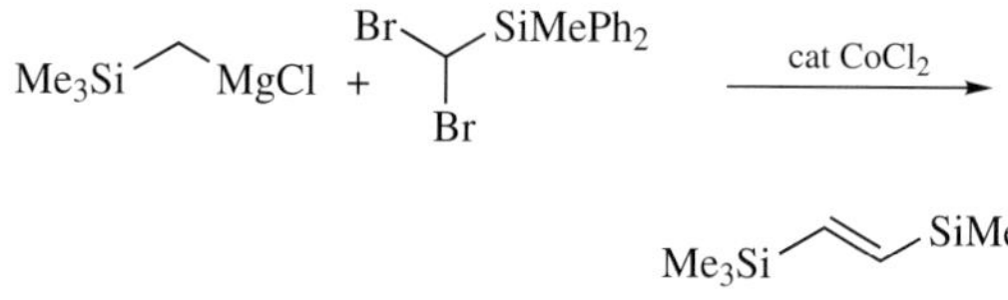

$$ (26) $$

Carbometalation. Treatment of allenyl sulfones with $(Me_3SiCH_2)_2CuMgCl$, prepared from Me_3SiCH_2MgCl and CuI, provides useful synthetic intermediates, β-(trimethylsilylmethyl) allyl sulfones, in good yield (eq 27).[82] Copper-catalyzed addition of Me_3SiCH_2ZnX (Me_3SiCH_2MgCl and $ZnBr_2$) to alkynyl sulfoxides affords syn-adducts exclusively, whereas a similar addition reaction of Me_3SiCH_2MgCl resulted in poor stereoselectivity (eq 28).[83] Me_3SiCH_2MgCl also adds to C_{60} to afford $C_{60}(CH_2SiMe_3)H$ and $C_{60}(CH_2SiMe_3)_2$ in THF and toluene, respectively.[84]

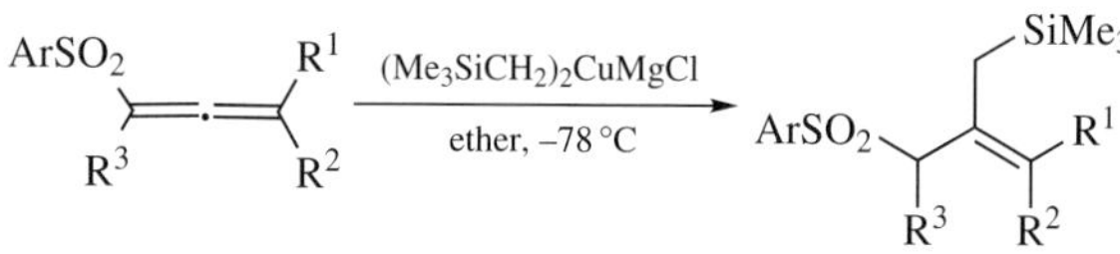

$$ (27) $$

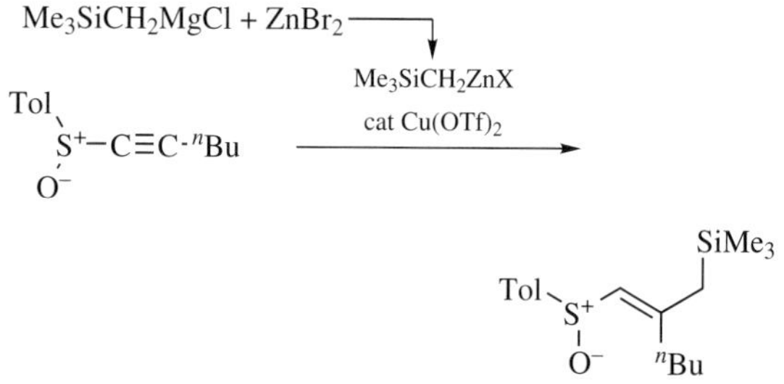

$$ (28) $$

Due to their electron deficiency, aromatic nitro compounds are susceptible to the addition of Me_3SiCH_2MgCl (eq 29).[85] The adducts, magnesium nitronates, recover their aromaticity upon mild oxidation.

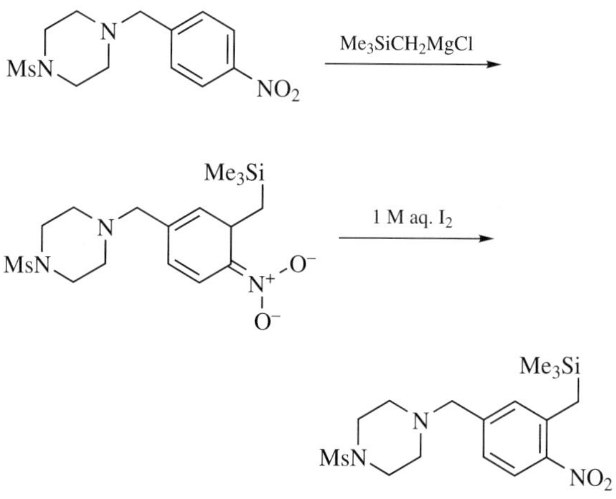

$$ (29) $$

As a Generator of Alkyl Radicals from Alkyl Halides in the Presence of Cobalt Catalysts. A conceptually novel use of Me_3SiCH_2MgCl recently emerged in cobalt-catalyzed Heck-type transformations.[86] The palladium-catalyzed Heck reaction always employs aryl- or alkenyl halides. In contrast, alkyl halides are not available for use as starting material, due to predominant β-hydride elimination from the corresponding alkylpalladium intermediate. New cobalt–Me$_3$SiCH$_2$MgCl systems now realize the involvement of alkyl halides in the Heck-type alkylation of styrenes, by taking advantage of the generation of an alkyl radical from alkyl halides via a single electron transfer process.

A mixture of bromocyclohexane and styrene is treated with Me$_3$SiCH$_2$MgCl in the presence of a catalytic amount of [CoCl$_2$(dpph)] (dpph = 1,6-bis(diphenylphosphino)hexane) in ether to yield β-cyclohexylstyrene in high yield (eq 30).[86a] Additional results are shown in Table 1. Not only secondary alkyl bromides but also primary and tertiary bromides can participate in the reaction. Notably, alkyl chlorides, usually inert in transition metal-mediated reactions, are suitable substrates as well (entries 6–8). Aromatic ester and amide moieties can survive under the reaction conditions (entries 14, 15). The reaction of an optically pure bromo acetal having a (+)-isomenthoxy group provides the styrylated chiral acetal, which will enjoy a variety of transformations (eq 31).[86b] Based on a detailed mechanistic investigation, the reaction proceeds in the following sequence (eq 32): (1) single electron transfer from an electron-rich cobalt complex to alkyl halide to generate the corresponding alkyl radical, (2) radical addition onto styrene to form a benzylic radical, (3) capture of the benzylic radical by a cobalt complex, (4) β-hydride elimination from the benzylcobalt intermediate.

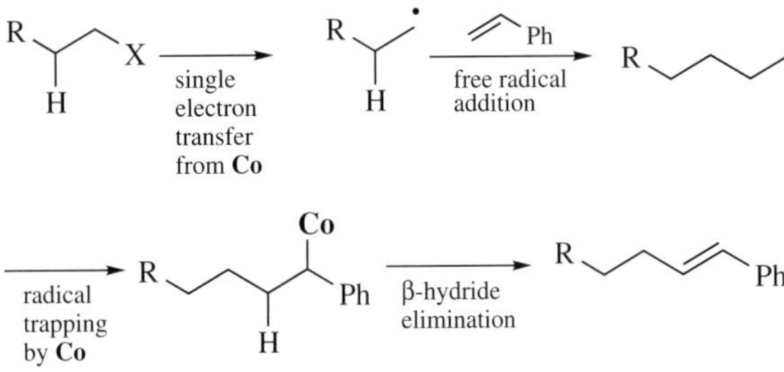

(32)

The cobalt-catalyzed reaction is applicable to intramolecular cyclization of 6-halo-1-hexene derivatives (Table 2).[86c] In the intramolecular version, dppb, 1,4-bis(diphenylphosphino)butane, is the best ligand, and a higher temperature is necessary. The same system also effects the conversion of aryl iodides having an olefinic moiety at a proper position into the cyclized product (eq 33).

Table 2 Cobalt-catalyzed intramolecular cyclization of 6-halo-1-hexene derivatives

Entry	Substrate	Product	Yield (%)
1	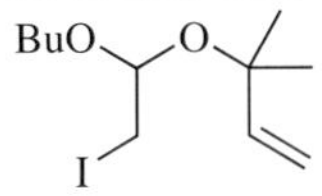	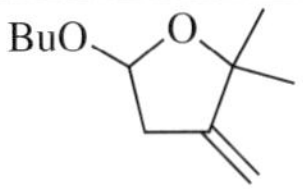	79
2	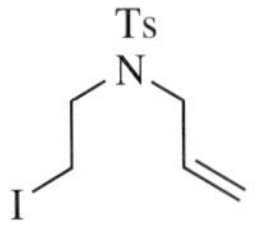	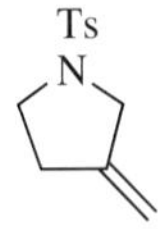	91
3	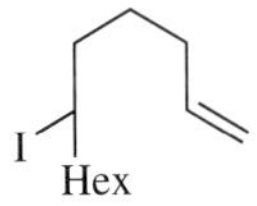	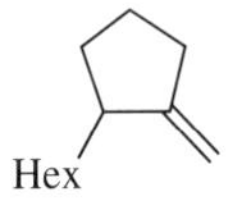	82

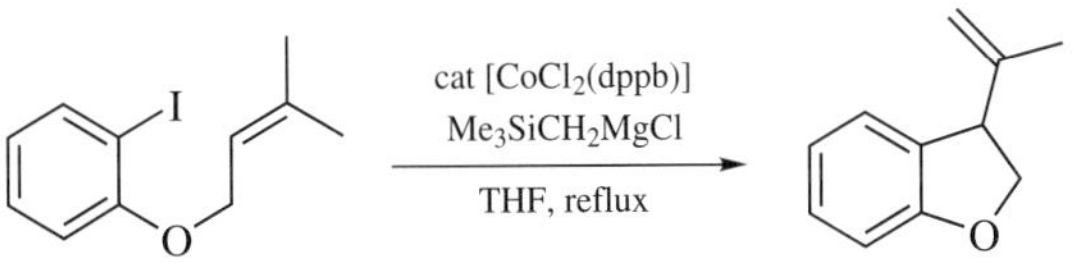

(33)

The reaction of epoxides with styrene provides homocinnamyl alcohols in good yield (eq 34).[86d] The reaction involves ring-opening of the epoxide to yield the magnesium alkoxide of a *vic*-halohydrin, which then undergoes the cobalt-catalyzed styrylation. Use of trimethylsilylmethylmagnesium bromide and cobalt(II) bromide, instead of chlorides, facilitates the reaction of epoxide since the more nucleophilic bromide efficiently attacks the epoxide. Conversion of tosylaziridines affords homocinnamyl amines.

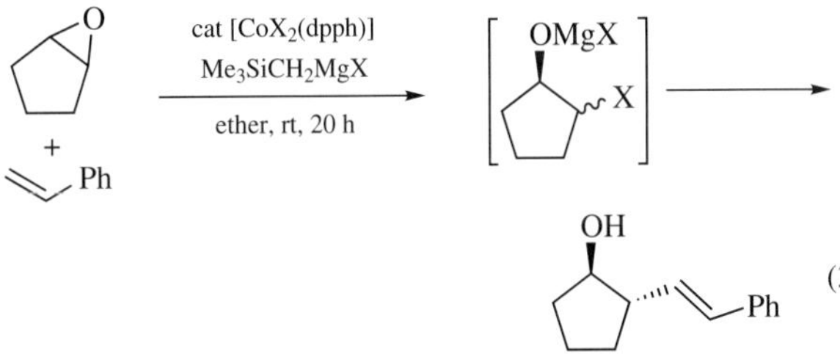

(34)

X = Cl, 62%; Br, 71%

Table 1 Cobalt-catalyzed trimethylsilylmethylmagnesium-promoted styrylation of alkyl halides

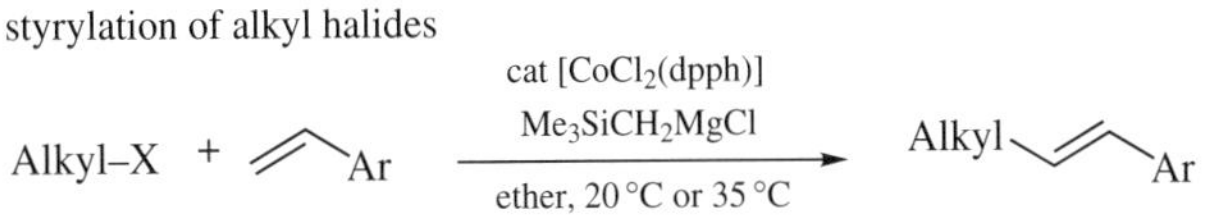

Entry	Alkyl–X	Ar	Yield (%)
1	$^nC_6H_{13}CH(CH_3)Br$	Ph	73
2	$^nC_{12}H_{25}Br$	Ph	76
3	1-Adamantyl–Br	Ph	87
4	$^tC_4H_9Br$	Ph	67
5	$^nC_{12}H_{25}I$	Ph	57
6	$^nC_{12}H_{25}Cl$	Ph	74
7	1-Adamantyl–Cl	Ph	90
8	$^cC_6H_{11}Cl$	Ph	84
9	$^cC_6H_{11}Br$	C_6H_4-p-Me	87
10	$^cC_6H_{11}Br$	C_6H_4-p-Cl	85
11	$^cC_6H_{11}Br$	C_6H_4-m-Cl	82
12	$^cC_6H_{11}Br$	C_6H_4-o-Cl	85
13	$^cC_6H_{11}Br$	C_6H_4-p-OMe	82
14	$^cC_6H_{11}Br$	C_6H_4-m-CON(CH$_2$Ph)$_2$	95
15	$^cC_6H_{11}Br$	C_6H_4-m-COO^{t}C$_4$H$_9$	66

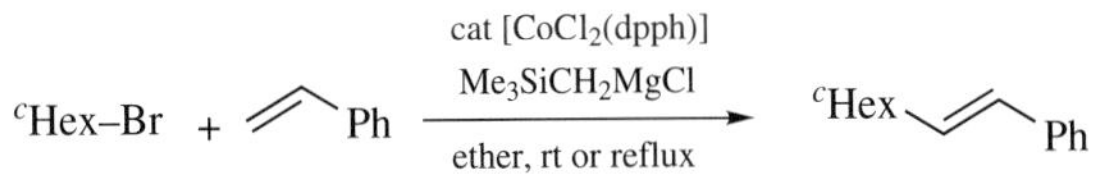

(30)

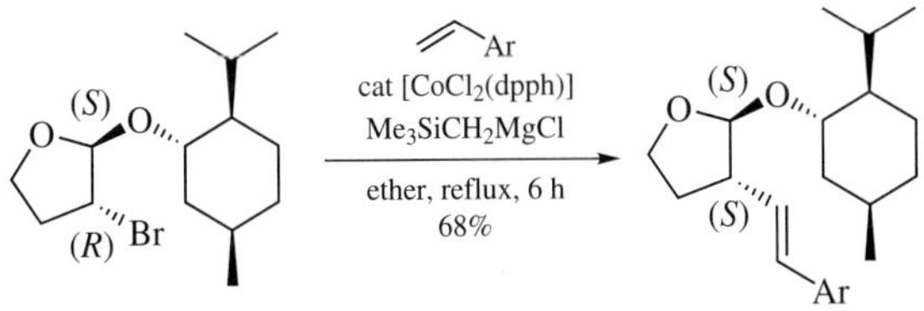

(31)

When 1,3-diene is the radical acceptor, a three-component coupling reaction of alkyl halide, 1,3-diene, and trimethylsilylmethylmagnesium chloride takes place to afford homoallylsilane.[86e] The final step of the three-component coupling is reductive elimination from allyl(trimethylsilylmethyl)cobalt that predominates over conceivable β-hydride elimination.

$$\text{(35)}$$

Miscellaneous Reactions. α-Nitrocycloalkanones undergo nucleophilic addition of trimethylsilylmethylmagnesium chloride to yield (ω-nitro-β-ketoalkyl)trimethylsilanes by ring-opening (eq 36).[87]

$$\text{(36)}$$

Terminally unsaturated nitriles are obtained upon treatment of nitroalkenes with trimethylsilylmethylmagnesium chloride followed by phosphorus trichloride (eq 37).[88]

$$\text{(37)}$$

In special cases, the Grignard reagent serves to deprotect acetals, wherein chelation plays a key role to direct the deprotection (eq 38).[89]

$$\text{(38)}$$

Aerobic oxidation of trimethylsilylmethylmagnesium chloride in the presence of α-methylstyrene results in addition of trimethylsilylmethyl radical to the carbon–carbon double bond followed by oxygenation, yielding 2-phenyl-4-trimethylsilyl-2-butanol (eq 39).[90]

$$\text{(39)}$$

Related Reagents. (Diisopropoxymethylsilyl)methylmagnesium chloride; trimethylsilylmethylpotassium; (trimethylstannylmethyl)lithium; trimethylsilylmethyllithium; Grubbs catalyst; AD-mix or osmium(VIII) tetraoxide; cerium(III) chloride; nickel(II) acetylacetonate; cobalt(II) chloride.

---

1. (a) Chan, T.-H., *Acc. Chem. Res.* **1977**, *10*, 442. (b) Ager, D. J., *Org. React.* **1990**, *38*, 1. (c) Ager, D. J., *Synthesis* **1984**, 384. (d) Fleming, I. In *Comprehensive Organic Chemistry*; Barton, D. H. R.; Ollis, W. D., Eds.; Pergamon: Oxford, 1979; Vol. 3, p 541. (e) Weber, W. P. *Silicon Reagents for Organic Synthesis–Concepts in Organic Chemistry*; Springer: New York, 1983; Vol. 14. (f) Colvin, E. W. *Silicon in Organic Synthesis*; Butterworths: London, 1981. (g) Colvin, E. W., *Chem. Soc. Rev.* **1978**, *7*, 15. (h) Magnus, P., *Aldrichim. Acta* **1980**, *13*, 43. (i) Magnus, P. D.; Sarkar, T.; Djuric, S. In *Comprehensive Organometallic Chemistry*; Wikinson, G.; Stone, F. G. A.; Abel, E. W., Eds.; Pergamon: Oxford, 1982; Vol. 7, p 515. (j) Birkofer, L.; Stuhl, O., *Top. Curr. Chem.* **1980**, *88*, 33. (k) Chan, T. H.; Fleming, I., *Synthesis* **1979**, 761.

2. Chan, T. H.; Chang, E.; Vinokur, E., *Tetrahedron Lett.* **1970**, 1137.

3. Sommer, L. H.; Goldberg, G. M.; Gold, J.; Whitmore, F. C., *J. Am. Chem. Soc.* **1947**, *69*, 980.

4. Hauser, C. R.; Hance, C. R., *J. Am. Chem. Soc.* **1952**, *74*, 5091.

5. Peterson, D. J., *J. Org. Chem.* **1968**, *33*, 780.

6. (a) Hudrlik, P. F.; Peterson, D., *J. Am. Chem. Soc.* **1975**, *97*, 1464. (b) Hudrlik, P. F.; Peterson, D., *Tetrahedron Lett.* **1974**, 1133.

7. Chan, T. H.; Chang, E., *J. Org. Chem.* **1974**, *39*, 3264.

8. (a) Boeckman, R. K. Jr.; Silver, S. M., *Tetrahedron Lett.* **1973**, 3497. (b) Akiyama, T.; Ohnari, M.; Shima, H.; Ozaki, S., *Synlett* **1991**, 831.

9. (a) Jones, K.; Wood, W. W., *J. Chem. Soc., Perkin Trans. 1* **1987**, 537. (b) Ferrier, R. J.; Stütz, A. E., *Carbohydr. Res.* **1990**, *205*, 283. (c) Udodong, U. E.; Fraser-Reid, B., *J. Org. Chem.* **1988**, *53*, 2131. (d) Udodong, U. E.; Fraser-Reid, B., *J. Org. Chem.* **1989**, *54*, 2103.

10. Lin, J.; Nikaido, M. M.; Clark, G., *J. Org. Chem.* **1987**, *52*, 3745.

11. (a) Glänzer, B. I.; Györgydeák, Z.; Bernet, B.; Vasella, A., *Helv. Chim. Acta* **1991**, *74*, 343. (b) Ager, D. J.; East, M. B., *Tetrahedron* **1993**, *49*, 5683.

12. Mancini, M.; Honek, J. F., *Tetrahedron Lett.* **1982**, *23*, 3249.

13. Johnson, C. R.; Tait, B. D., *J. Org. Chem.* **1987**, *52*, 281.

14. (a) Carter, M. J.; Fleming, I., *J. Chem. Soc., Chem. Commun.* **1976**, 679. (b) Pillot, J.-P.; Dunoguès, J.; Calas, R., *J. Chem. Res. (S)* **1977**, 268.

15. Taylor, R. T.; Galloway, J. G., *J. Organomet. Chem.* **1981**, *220*, 295.

16. (a) Sato, S.; Okada, H.; Matsuda, I.; Izumi, Y., *Tetrahedron Lett.* **1984**, *25*, 769. (b) Matsuda, I.; Okada, H.; Sato, S.; Izumi, Y., *Tetrahedron Lett.* **1984**, *25*, 3879.

17. (a) Hatanaka, Y.; Kuwajima, I., *J. Org. Chem.* **1986**, *51*, 1932. (b) Horiguchi, Y.; Kataoka, Y.; Kuwajima, I., *Tetrahedron Lett.* **1989**, *30*, 3327. (c) Fujiwara, T.; Suda, A.; Takeda, T., *Chem. Lett.* **1991**, 1619.

18. Kitahara, T.; Kurata, H.; Mori, K., *Tetrahedron* **1988**, *44*, 4339.

19. Barluenga, J.; Fernández-Simón, J. L.; Concellón, J. M.; Yus, M., *Tetrahedron Lett.* **1989**, *30*, 5927.

20. Sisko, J.; Weinreb, S. M., *J. Org. Chem.* **1990**, *55*, 393.

21. Ochiai, M.; Fujita, E.; Arimoto, M.; Yamaguchi, H., *Chem. Pharm. Bull.* **1985**, *33*, 989.

22. de Raadt, A.; Stütz, A. E., *Carbohydr. Res.* **1991**, *220*, 101.

23. Fleming, I.; Pearce, A., *J. Chem. Soc., Perkin Trans. 1* **1981**, 251.

24. Box, V. G. S.; Brown, D. P., *Heterocycles* **1991**, *32*, 1273.

25. Narayanan, B. A.; Bunelle, W. H., *Tetrahedron Lett.* **1987**, *28*, 6261.

26. (a) Anderson, M. B.; Fuchs, P. L., *Synth. Commun.* **1987**, *17*, 621. (b) Hojo, M.; Ohsumi, K.; Hosomi, A., *Tetrahedron Lett.* **1992**, *33*, 5981.

27. (a) Lee, T. V.; Channon, J. A.; Cregg, C.; Porter, J. R.; Roden, F. S.; Yeoh, H. T.-L., *Tetrahedron* **1989**, *45*, 5877. (b) Calò, V.; Lopez, L.; Pesce, G., *J. Organomet. Chem.* **1988**, *353*, 405.

28. (a) Itoh, K.; Fukui, M.; Kurachi, Y., *J. Chem. Soc., Chem. Commun.* **1977**, 500. (b) Itoh, K.; Yogo, T.; Ishii, Y., *Chem. Lett.* **1977**, 103. (c) Lee, T. V.; Boucher, R. J.; Rockell, C. J. M., *Tetrahedron Lett.* **1988**, *29*, 689.

29. (a) Fürstner, A.; Kollegger, G.; Weidmann, H., *J. Organomet. Chem.* **1991**, *414*, 295. (b) Pearlman, B. A.; McNamara, J. M.; Hasan, I.; Hatakeyama, S.; Sekizaki, H.; Kishi, Y., *J. Am. Chem. Soc.* **1981**, *103*, 4248. (c) Yamamoto, Y.; Ohdoi, K.; Nakatani, M.; Akiba, K.-y., *Chem. Lett.* **1984**, 1967. (d) Koerwitz, F. L.; Hammond, G. B.; Wiemer, D. F., *J. Org. Chem.* **1989**, *54*, 738.

30. Whitmore, F. C.; Sommer, L. H.; Gold, J.; Van Strien, R. E., *J. Am. Chem. Soc.* **1947**, *69*, 1551.

31. Larson, G. L.; Montes de López-Cepero, I.; Torres, L. E., *Tetrahedron Lett.* **1984**, *25*, 1673.

32. Brown, P. A.; Bonnert, R. V.; Jenkins, P. R.; Lawrence, N. J.; Selim, M. R., *J. Chem. Soc., Perkin Trans. 1* **1991**, 1893.

33. Sommer, L. H.; Gold, J. R.; Goldberg, G. M.; Marans, N. S., *J. Am. Chem. Soc.* **1949**, *71*, 1509.

34. Gold, J. R.; Sommer, L. H.; Whitmore, F. C., *J. Am. Chem. Soc.* **1948**, *70*, 2874.

35. Haider, A., *Synthesis* **1985**, 271.

36. Hosomi, A.; Masunari, T.; Tominaga, Y.; Yanagi, T.; Hojo, M., *Tetrahedron Lett.* **1990**, *31*, 6201.

37. (a) Hayashi, T.; Kabeta, K.; Hamachi, I.; Kumada, M., *Tetrahedron Lett.* **1983**, *24*, 2865. (b) Negishi, E.-i.; Luo, F.-T.; Rand, C. L., *Tetrahedron Lett.* **1982**, *23*, 27. (c) Kitano, Y.; Matsumoto, T.; Wakasa, T.; Okamoto, S.; Shimazaki, T.; Kobayashi, Y.; Sato, F.; Miyaji, K.; Arai, K., *Tetrahedron Lett.* **1987**, *28*, 6351.

38. (a) Tamao, K.; Sumitani, K.; Kiso, Y.; Zembayashi, M.; Fujioka, A.; Kodama, S.-i.; Nakajima, I.; Minato, A.; Kumada, M., *Bull. Chem. Soc. Jpn.* **1976**, *49*, 1958. (b) Sengupta, S.; Leite, M.; Raslan, D. S.; Quesnelle, C.; Snieckus, V., *J. Org. Chem.* **1992**, *57*, 4066.

39. Andreini, B. P.; Carpita, A.; Rossi, R.; Scamuzzi, B., *Tetrahedron* **1989**, *45*, 5621.

40. Ni, Z.-J.; Luh, T.-Y., *J. Org. Chem.* **1988**, *53*, 5582.

41. (a) Ni, Z.-J.; Luh, T.-Y., *J. Chem. Soc., Chem. Commun.* **1988**, 1011. (b) Wong, K.-T.; Ni, Z.-J.; Luh, T.-Y., *J. Chem. Soc., Perkin Trans. 1* **1991**, 3113. (c) Cheng, W.-L.; Luh, T.-Y., *J. Org. Chem.* **1992**, *57*, 3516. (d) Wong, K.-T.; Luh, T.-Y., *J. Chem. Soc., Chem. Commun.* **1992**, 564. (e) Wong, K.-T.; Luh, T.-Y., *J. Am. Chem. Soc.* **1992**, *114*, 7308.

42. (a) Ni, Z.-J.; Luh, T.-Y., *J. Org. Chem.* **1988**, *53*, 2129. (b) Ni, Z.-J.; Yang, P.-F.; Ng, D. K. P.; Tzeng, Y.-L.; Luh, T.-Y., *J. Am. Chem. Soc.* **1990**, *112*, 9356.

43. Tzeng, Y.-L.; Luh, T.-Y.; Fang, J.-M., *J. Chem. Soc., Chem. Commun.* **1990**, 399.

44. Tominaga, Y.; Kamio, C.; Hosomi, A., *Chem. Lett.* **1989**, 1761.

45. (a) Sommer, L. H.; Van Strien, R. E.; Whitmore, F. C., *J. Am. Chem. Soc.* **1949**, *71*, 3056. (b) Fleming, I.; Loreto, M. A.; Wallace, I. H. M.; Michael, J. P., *J. Chem. Soc., Perkin Trans. 1* **1986**, 349. (c) Camp Schuda, A. D.; Mazzocchi, P. H.; Fritz, G.; Morgan, T., *Synthesis* **1986**, 309.

46. Hwu, J. R.; Gilbert, B. A., *J. Am. Chem. Soc.* **1991**, *113*, 5917.

47. (a) Bartoli, G.; Bosco, M.; Dalpozzo, R.; Todesco, P. E., *J. Org. Chem.* **1986**, *51*, 3694. (b) Bartoli, G.; Bosco, M.; Dalpozzo, R.; Todesco, P. E., *J. Chem. Soc., Chem. Commun.* **1988**, 807. (c) Bartoli, G.; Bosco, M.; Dal Pozzo, R.; Petrini, M., *Tetrahedron* **1987**, *43*, 4221.

48. (a) Hayashi, T.; Fujiwa, T.; Okamata, Y.; Katsuro, Y.; Kumada, M., *Synthesis* **1981**, 1001. (b) Danishefsky, S. J.; Mantlo, N., *J. Am. Chem. Soc.* **1988**, *110*, 8129.

49. Hayashi, T.; Katsuro, Y.; Kumada, M., *Tetrahedron Lett.* **1980**, *21*, 3915.

50. Wadman, S.; Whitby, R.; Yeates, C.; Kocieński, P.; Cooper, K., *J. Chem. Soc., Chem. Commun.* **1987**, 241.

51. Kocieński, P.; Dixon, N. J.; Wadman, S., *Tetrahedron Lett.* **1988**, *29*, 2353.

52. Pettersson, L.; Frejd, T.; Magnusson, G., *Tetrahedron Lett.* **1987**, *28*, 2753.

53. (a) Westmijze, H.; Vermeer, P., *Synthesis* **1979**, 390. (b) Montury, M.; Psaume, B.; Goré, J., *Tetrahedron Lett.* **1980**, *21*, 163. (c) Trost, B. M.; Urabe, H., *J. Am. Chem. Soc.* **1990**, *112*, 4982.

54. (a) Kleijn, H.; Vermeer, P., *J. Org. Chem.* **1985**, *50*, 5143. (b) Trost, B. M.; Matelich, M. C., *Synthesis* **1992**, 151. (c) Trost, B. M.; Matelich, M. C., *J. Am. Chem. Soc.* **1991**, *113*, 9007.

55. Foulon, J. P.; Bourgain-Commerçon, M.; Normant, J. F., *Tetrahedron* **1986**, *42*, 1389.

56. Foulon, J. P.; Bourgain-Commerçon, M.; Normant, J. F., *Tetrahedron* **1986**, *42*, 1399.

57. Prober, M., *J. Am. Chem. Soc.* **1955**, *77*, 3224.

58. Sommer, L. H.; Mitch, F. A.; Goldberg, G. M., *J. Am. Chem. Soc.* **1949**, *71*, 2746.

59. (a) Seyferth, D.; Freyer, W., *J. Org. Chem.* **1961**, *26*, 2604. (b) Westerhausen, M.; Rademacher, B.; Poll, W., *J. Organomet. Chem.* **1991**, *421*, 175.

60. (a) Brook, A. G.; Anderson, D. G., *Can. J. Chem.* **1968**, *46*, 2115. (b) Ager, D. J., *Chem. Soc. Rev.* **1982**, *11*, 493.

61. Armitage, D. A. In *Comprehensive Organometallic Chemistry*; Wilkinson, G.; Stone, F. G. A.; Abel, E. W., Eds.; Pergamon: Oxford, 1982; Vol. 2, p 1.

62. Ishibashi, H.; Kameoka, C.; Kodama, K.; Ikeda, M., *Tetrahedron* **1996**, *52*, 489.

63. Takao, K.; Tsujita, T.; Hara, M.; Tadano, K., *J. Org. Chem.* **2002**, *67*, 6690.

64. Vázquez, A.; Williams, R. M., *J. Org. Chem.* **2000**, *65*, 7865.

65. Akiyama, T.; Shima, H.; Ohnari, M.; Okazaki, T.; Ozaki, S., *Bull. Chem. Soc. Jpn.* **1993**, *66*, 3760.

66. (a) Guevel, A. C.; Hart, D. J., *J. Org. Chem.* **1996**, *61*, 473. (b) Rossiter, S., *Tetrahedron Lett.* **2002**, *43*, 4671. (c) Shiina, J.; Nishiyama, S., *Tetrahedron* **2003**, *59*, 6039. (d) Valla, A.; Laurent, A.; Prat, V.; Andriamialisoa, Z.; Cartier, D.; Giraud, M.; Labia, R.; Potier, P., *Tetrahedron Lett.* **2001**, *42*, 4795.

67. (a) Zakarian, A.; Batch, A.; Holton, R. A., *J. Am. Chem. Soc.* **2003**, *125*, 7822. (b) Fürstner, A.; Baumgartner, J., *Tetrahedron* **1993**, *49*, 8541. (c) Storer, R. I.; Takemoto, T.; Jackson, P. S.; Brown, D. S.; Baxendale, I. R.; Ley, S. V., *Chem. Eur. J.* **2004**, *10*, 2529. (d) Compostella, F.; Franchini, L.; De Libero, G.; Palmisano, G.; Ronchetti, F.; Panza, L., *Tetrahedron* **2002**, *58*, 8703.

68. (a) Harris, J. M.; Keranen, M. D.; O'Doherty, G. A., *J. Org. Chem.* **1999**, *64*, 2982. (b) Haukaas, M. H.; O'Doherty, G. A., *Org. Lett.* **2002**, *4*, 1771. (c) Paquette, L. A.; Efremov, I., *J. Am. Chem. Soc.* **2001**, *123*, 4492. (d) Tietze, L. F.; Görlitzer, J., *Synlett* **1997**, 1049.

69. (a) Diaz, L. C.; Giacomini, R., *Tetrahedron Lett.* **1998**, *39*, 5343. (b) Bardot, V.; Gelas-Mialhe, Y.; Gramain, J. C.; Remuson, R., *Tetrahedron Asymm.* **1997**, *8*, 1111. (c) Monfray, J.; Gelas-Mialhe, Y.; Gramain, J. C.; Remuson, R., *Tetrahedron Lett.* **2003**, *44*, 5785. (d) Monfray, J.; Gelas-Mialhe, Y.; Gramain, J. C.; Remuson, R., *Tetrahedron: Asymmetry* **2005**, *16*, 1025.

70. (a) Clark, J. S.; Dossetter, A. G.; Blake, A. J.; Li, W. S.; Whittingham, W. G., *Chem. Commun.* **1999**, 749. (b) Chakraborty, T. K.; Das, S., *Tetrahedron Lett.* **2001**, *42*, 3387. (c) Tanino, K.; Shimizu, T.; Miyama, M.; Kuwajima, I., *J. Am. Chem. Soc.* **2000**, *122*, 6116.

71. Ishihara, T.; Miwatashi, S.; Kuroboshi, M.; Utimoto, K., *Tetrahedron Lett. 32*, 1069.

72. (a) Monti, H.; Audran, G.; Monti, J. P.; Leandri, G., *J. Org. Chem.* **1996**, *61*, 6021. (b) Laval, G.; Audran, G.; Sanchez, S.; Monti, H., *Tetrahedron: Asymmetry* **1999**, *10*, 1927. (c) Nicolaou, K. C.; Shi, G. Q.; Namoto, K.; Bernal, F., *Chem. Commun.* **1998**, 1757.

73. (a) Organ, M. G.; Murray, A. P., *J. Org. Chem.* **1997**, *62*, 1523. (b) Tsukazaki, M.; Snieckus, V., *Tetrahedron Lett.* **1993**, *34*, 411. (c) Uenishi, J.; Kawahama, R.; Izaki, Y.; Yonemitsu, O., *Tetrahedron* **2000**, *56*, 3493.

74. (a) Mulzer, J.; Hanbauer, M., *Tetrahedron Lett.* **2002**, *43*, 3381. (b) Williams, D. R.; Plummer, S. V.; Patnaik, S., *Angew. Chem., Int. Ed.* **2003**, *42*, 3934. (c) Fish, P. V., *Tetrahedron Lett.* **1994**, *35*, 7181. (d) Pérez-Balado, C.; Markó, I. E., *Tetrahedron Lett.* **2005**, *46*, 4887. (e) Yakura, T.; Muramatsu, W.; Uenishi, J., *Chem. Pharm. Bull.* **2005**, *53*, 989. (f) Uenishi, J.; Ohmi, M., *Angew. Chem., Int. Ed.* **2005**, *44*, 2756. (g) Uenishi, J.; Matsui, K., *Tetrahedron Lett.* **2001**, *42*, 4353.

75. (a) Hevesi, L.; Hermans, B.; Allard, C., *Tetrahedron Lett.* **1994**, *35*, 6729. (b) Shimizu, R.; Yoneda, E.; Fuchikami, T., *Tetrahedron Lett.* **1996**, *37*, 5557.

76. (a) Smitrovich, J. H.; Woerpel, K. A., *J. Am. Chem. Soc.* **1998**, *120*, 12998. (b) Smitrovich, J. H.; Woerpel, K. A., *J. Org. Chem.* **2000**, *65*, 1601. (c) Smitrovich, J. H.; Woerpel, K. A., *Synthesis* **2002**, 2778.

77. Mentink, G.; van Maarseveen, J. H.; Hiemstra, H., *Org. Lett.* **2002**, *4*, 3497.

78. (a) Mizutani, K.; Yorimitsu, H.; Oshima, K., *Chem. Lett.* **2004**, *33*, 832. (b) Ohmiya, H.; Mizutani, K.; Yorimitsu, H.; Oshima, K., *Tetrahedron* **2006**, *62*, 1410. (c) Ohmiya, H.; Yorimitsu, H.; Oshima, K., *Chem. Lett.* **2004**, *33*, 1240. (d) Kamachi, T.; Kuno, A.; Matsuno, C.; Okamoto, S., *Tetrahedron Lett.* **2004**, *45*, 4677.

79. Ohmiya, H.; Yorimitsu, H.; Oshima, K., *Angew. Chem., Int. Ed.* **2005**, *44*, 3488.

80. Kakiya, H.; Shinokubo, H.; Oshima, K., *Bull. Chem. Soc. Jpn.* **2000**, *73*, 2139.

81. Inoue, A.; Kondo, J.; Shinokubo, H.; Oshima, K., *J. Am. Chem. Soc.* **2001**, *123*, 11109.

82. Harmata, M.; Herron, B. F., *Synthesis* **1992**, 202.

83. Maezaki, N.; Sawamoto, H.; Suzuki, T.; Yoshigami, R.; Tanaka, T., *J. Org. Chem.* **2004**, *69*, 8387.

84. Nagashima, H.; Terasaki, H.; Kimura, E.; Nakajima, K.; Itoh, K., *J. Org. Chem.* **1994**, *59*, 1246.

85. (a) Wong, A.; Kuethe, J. T.; Davies, I. W.; Hughes, D. L., *J. Org. Chem.* **2004**, *69*, 7761. (b) Kuethe, J. T.; Wong, A.; Qu, C.; Smitrovich, J.; Davies, I. W.; Hughes, D. L., *J. Org. Chem.* **2005**, *70*, 2555.

86. (a) Ikeda, Y.; Nakamura, T.; Yorimitsu, H.; Oshima, K., *J. Am. Chem. Soc.* **2002**, *124*, 6514. (b) Affo, W.; Ohmiya, H.; Fujioka, T.; Ikeda, Y.; Nakamura, T.; Yorimitsu, H.; Oshima, K.; Imamura, Y.; Mizuta, T.; Miyoshi, K., unpublished results. (c) Fujioka, T.; Nakamura, T.; Yorimitsu, H.; Oshima, K., *Org. Lett.* **2002**, *4*, 2257. (d) Ikeda, Y.; Yorimitsu, H.; Shinokubo, H.; Oshima, K., *Adv. Synth. Catal.* **2004**, *346*, 1631. (e) Mizutani, K.; Shinokubo, H.; Oshima, K., *Org. Lett.* **2003**, *5*, 3959.

87. Ballini, R.; Bartoli, G.; Giovannini, R.; Marcantoni, E.; Petrini, M., *Tetrahedron Lett.* **1993**, *34*, 3301.

88. Tso, H. H.; Gilbert, B. A.; Hwu, J. R., *J. Chem. Soc., Chem. Commun.* **1993**, 669.

89. Chiang, C. C.; Chen, Y. H.; Hsieh, Y. T.; Luh, T. Y., *J. Org. Chem.* **2000**, *65*, 4694.

90. Nobe, Y.; Arayama, K.; Urabe, H., *J. Am. Chem. Soc.* **2005**, *127*, 18006.

Trimethylsilylmethylpotassium[1]

(**1**; M = K)
[53127-82-5] $C_4H_{11}KSi$ (MW 126.34)
InChI = 1S/C4H11Si.K/c1-5(2,3)4;/h1H2,2-4H3;
InChIKey = HLDWFMKKBGSCEZ-UHFFFAOYSA-N
(**2**; M = Na)
[53127-81-4] $C_4H_{11}NaSi$ (MW 110.23)
InChI = 1S/C4H11Si.Na/c1-5(2,3)4;/h1H2,2-4H3;
InChIKey = YMMDCARGHHHAMC-UHFFFAOYSA-N

(can act as a strong base with benzylic and allylic compounds[2])

Preparative Methods: trimethylsilylmethylpotassium (**1**) and sodium (**2**) are prepared by reaction of bis(trimethylsilyl)-mercury with the appropriate metal as a suspension in cyclohexane.[2,3] Trimethylsilylmethylsodium can also be prepared by reaction of bis(trimethylsilyl)methane with sodium methoxide in HMPA.[4]

Handling, Storage, and Precautions: both of these organometallic reagents should be prepared just prior to use and not stored for extended periods. All reactions with these reagents should be performed under an inert atmosphere.

Base Reactions. Both metal reagents (**1**) and (**2**) act as bases. The tricyclic ether (**3**) is readily deprotonated by the hindered base (**1**), but pentylsodium or butylpotassium proved superior in other cases.[2] However, reagent (**1**) provides an alternative to the *n*-butyllithium–potassium-butoxide system (eq 1).[1]

$$\text{(3)} \xrightarrow[58\%]{\substack{1.\ \text{TMSCH}_2\text{K (1)} \\ 2.\ \text{TMSCl}}} \text{TMS} \qquad (1)$$

Both reagents (**1**) and (**2**) react with benzylic compounds to afford benzyl anions. The reaction can accommodate alkyl or silyl groups on the benzylic position or as substituents on the aromatic ring (eq 2).[2,3]

$$\text{ArCH}_2\text{R} \xrightarrow[c\text{-C}_6\text{H}_{12}\text{ or THF}]{\text{TMSCH}_2\text{M}} \text{ArCHR}^-\text{M}^+ \qquad (2)$$

$$M = K\ (\mathbf{1})\ \text{or Na}\ (\mathbf{2})$$

The use of trimethylsilylmethylpotassium (**1**) as a base also allows deprotonation of allylic positions. The allyl carbanions produced by the action of substrate with (**1**) react with ethylene oxide to afford the 4-en-1-ol product; unsymmetrical allylic anions can result in a mixture of products, although reaction is favored at the least substituted center (eq 3).[5]

$$\xrightarrow[\text{2. ethylene oxide}]{1.\ (\mathbf{1}),\ \text{THF}}$$

OH   37%   +   HO   28%   (3)

The potassium reagent (1) will deprotonate simple alkenes to give allylic anions.[2] Subsequent reaction with a chloroborane results in the allylic borane, which, in turn, can be converted to an allyl alcohol (eq 4).[6]

$$\text{[structure]} \xrightarrow[\substack{3.\ H_2O_2,\ NaOH \\ 46\%}]{\substack{1.\ (1) \\ 2.\ Et_2BCl}} \text{[structure]—OH} \qquad (4)$$

Pentadienyl-type organometallic reagents are available by reaction of (1) with homoconjugated and conjugated dienes, without significant polymerization. The regioselectivity of the metalation is dependent on the structure of the substrate, 'U-shaped' delocalized systems usually being preferred.[7] However, the factors affecting the regioselectivity are subtle, as illustrated by reaction at the 14-position of 7-dehydrocholesterol (4) (eq 5).[8]

$$\text{[steroid structure (4), HO—]} \xrightarrow[\substack{4.\ H^+ \\ 5.\ CH_2N_2 \\ 43\%}]{\substack{1.\ KH,\ THF \\ 2.\ (1) \\ 3.\ CO_2}} \text{[steroid structure, HO—, CO_2Me]} \qquad (5)$$

(4)   dr = 2:1

Peterson Alkenation. Trimethylsilylmethylsodium (2) is prepared in situ by the action of sodium methoxide on bis(trimethylsilyl)methane (5),[4] and reacts with nonenolizable carbonyl compounds to provide alkenes by a Peterson alkenation reaction (eq 6). Alternative procedures are available that circumvent the use of HMPA, and allow reaction with enolizable carbonyl compounds.[9]

$$(TMS)_2CH_2 \xrightarrow[HMPA]{NaOMe} TMSCH_2Na \xrightarrow{Ph_2CO} H_2C=CPh_2 \qquad (6)$$
$$(5) \qquad\qquad\qquad (2) \qquad\qquad 53\%$$

Related Reagents. Trimethylsilylmethyllithium; Trimethylsilylmethylmagnesium Chloride.

1. Schlosser, M., *Angew. Chem., Int. Ed. Engl.* **1974**, *13*, 701.
2. Hartmann, J.; Schlosser, M., *Helv. Chim. Acta* **1976**, *59*, 453.
3. Hart, A. J.; O'Brien, D. H.; Russell, C. R., *J. Organomet. Chem.* **1974**, *72*, C19.
4. Sakurai, H.; Nishiwaki, K.-i.; Kira, M., *Tetrahedron Lett.* **1973**, 4193.
5. Hartmann, J.; Schlosser, M., *Synthesis* **1975**, 328.
6. Zaidlewicz, M., *J. Organomet. Chem.* **1985**, *293*, 139.
7. (a) Schlosser, M.; Rauchschwalbe, G., *J. Am. Chem. Soc.* **1978**, *100*, 3258. (b) Bosshardt, H.; Schlosser, M., *Helv. Chim. Acta* **1980**, *63*, 2393. (c) Schlosser, M.; Bosshardt, H.; Walde, A.; Stähle, M., *Angew. Chem., Int. Ed. Engl.* **1980**, *19*, 303.
8. Moret, E.; Schlosser, M., *Tetrahedron Lett.* **1984**, *25*, 1449.
9. (a) Ager, D. J., *Org. React.* **1990**, *38*, 1. (b) Johnson, C. R.; Tait, B. D., *J. Org. Chem.* **1987**, *52*, 281.

David J. Ager
The NutraSweet Company, Mount Prospect, IL, USA

2-Trimethylsilylmethyl-2-propen-1-ol[1]

[81302-80-9]   C$_7$H$_{16}$OSi   (MW 144.32)

InChI = 1S/C7H16OSi/c1-7(5-8)6-9(2,3)4/h8H,1,5-6H2,2-4H3

InChIKey = BWNLJXBYSUHUET-UHFFFAOYSA-N

(derivatives function as trimethylenemethane (TMM) precursors and undergo cyclopentannulation reactions; methylenecyclopentanes; 2-acetoxymethyl-3-trimethylsilylpropene;[2] [3 + 2] annulation;[3] three-carbon condensative expansion;[4] cyclocontraction–spiroannulation[5])

Alternate Names: 2-(hydroxymethyl)-3-trimethylsilylpropene; 2-(hydroxymethyl)-2-propenyltrimethylsilane; (hydroxymethyl)allyl trimethylsilane; 2-(trimethylsilylmethyl)allyl alcohol.

Physical Data: bp 54–56 °C/2 mmHg; n_D^{20} 1.454; *d* 0.861 g cm^{-3}; fp 71 °C.

Solubility: insol water; sol all organic liquids.

Form Supplied in: liquid, 90% pure; remainder of the mixture (approx. 10%) is 3-trimethylsilyl-2-methyl-2-propen-1-ol or 3-trimethylsilylmethyl-2-propen-1-ol.

Preparative Methods: prepared in four steps from methallyl alcohol. The synthesis begins with the metalation with *n*-butyllithium (2 equiv) producing the dianion which is bis-silylated by trapping with chlorotrimethylsilane to generate the allylsilane. The TMS ether is subsequently removed by hydrolysis (eq 1). The primary allylic alcohol serves as a precursor to more functionalized reagents.

$$\text{[structure OH]} \xrightarrow[\substack{2.\ TMSCl}]{\substack{1.\ 2\ equiv\ BuLi,\ TMEDA,\ Et_2O}} \text{[structure OTMS, TMS]} \xrightarrow{H_3O^+}$$

$$\text{[structure OH, TMS]} \qquad (1)$$

Useful derivatives of the alcohol are the primary allylic chloride, mesylate, and acetate. These are easily prepared in a short number of steps as shown (eq 2) or by functionalization of the primary allylic alcohol.

$$\text{[structure Cl, Cl]} \xrightarrow[\substack{Et_3N,\ CuI}]{\substack{1.25\ equiv\ Cl_3SiH}} \text{[structure SiCl_3, Cl]} \xrightarrow{MeMgCl} \text{[structure TMS, Cl]} \qquad (2)$$

Purification: by distillation under reduced pressure.

Handling, Storage, and Precautions: flammable liquid; irritant; store at 0–4 °C.

Introduction. The title allylsilane reagent has been employed as a conjunctive reagent which is considered to be a synthetic equivalent of a zwitterionic, bifunctional compound possessing a nucleophilic allylic anion synthon and an electrophilic cation synthon in the same molecule (1).

conjunctive reagent   synthetic equivalent

(1)

[3 + 2] Annulations. The primary allylic halide 2-chloromethyl-3-trimethylsilyl-1-propene or the mesylate derived from the corresponding alcohol behaves as a bifunctional reagent possessing both nucleophilic and electrophilic reaction centers. For example, they have been shown to participate in [3 + 2] carbon annulations for the construction of methylenecyclopentanes. The reaction proceeds through a Lewis acid-promoted conjugate addition (Sakurai reaction) followed by an internal alkylation of the derived ketone enolate (eq 3).

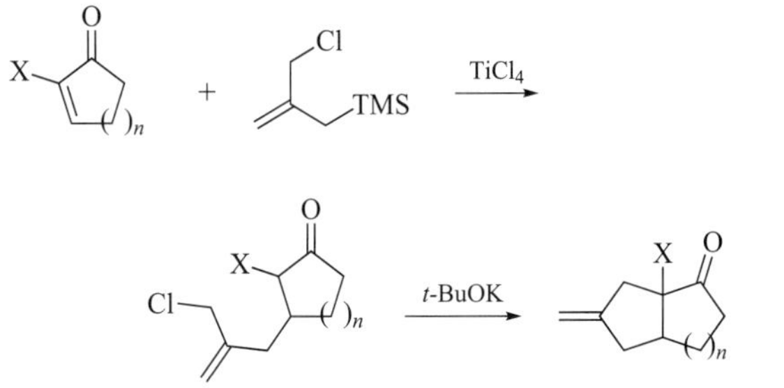

n	X	Yield (%)	Yield (%)
2	H	80	60
1	H	8	–
2	SPh	80	79
1	SPh	63	73

In a related study, the annulation is combined with a ring expansion reaction. The reaction utilizes a β-keto sulfone as the two-carbon component along with an allylic mesylate as the bifunctional reagent. The alkylation/cyclization process is outlined in eq 4. The subsequent ring expansion appears to be a general process except for cases involving fused six- and seven-membered ring systems (eq 5).

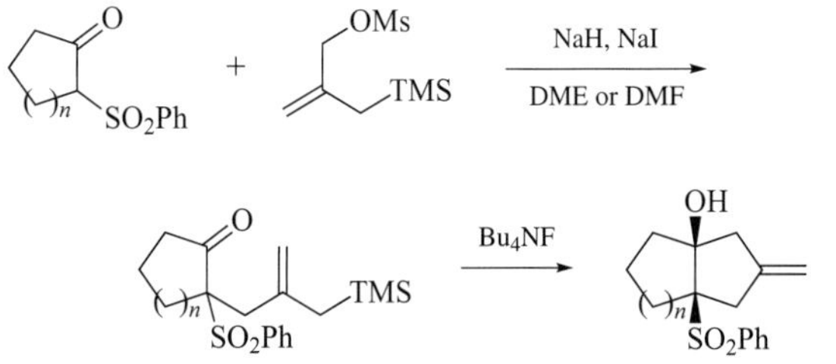

n	Alkylation yield (%)	Cyclization yield (%)
1	81	94
2	71	79
3	80	79

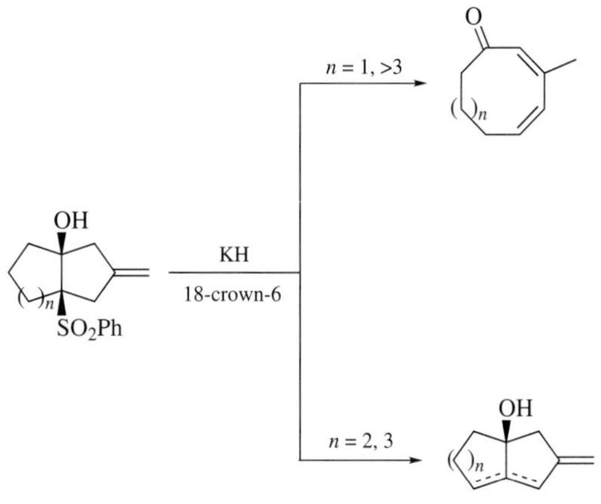

Methylenecyclopentane Annulation. The acylated derivative 2-acetyloxymethyl-3-trimethylsilylpropene is known to be an effective three-carbon component in cyclopentane annulations. It has been demonstrated that 2-acetoxymethylallyltrimethylsilane adds to a variety of electron-deficient alkenes in the presence of a catalytic amount of tetrakis(triphenylphosphine) palladium(0) and 1,2-bis(diphenylphosphino)ethane to produce methylene-cyclopentanes (eqs 6–8).

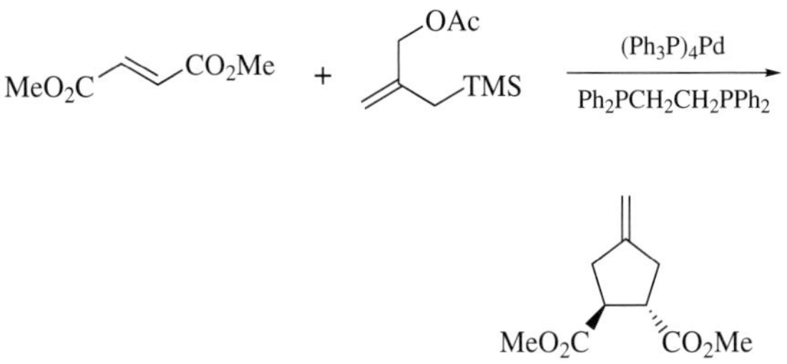

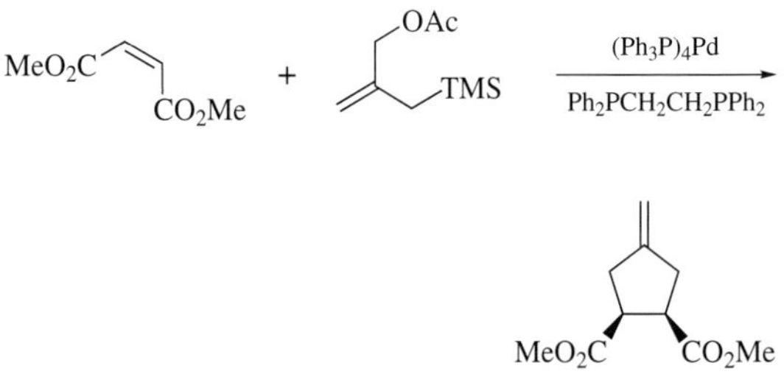

cis:trans = 1.3:1

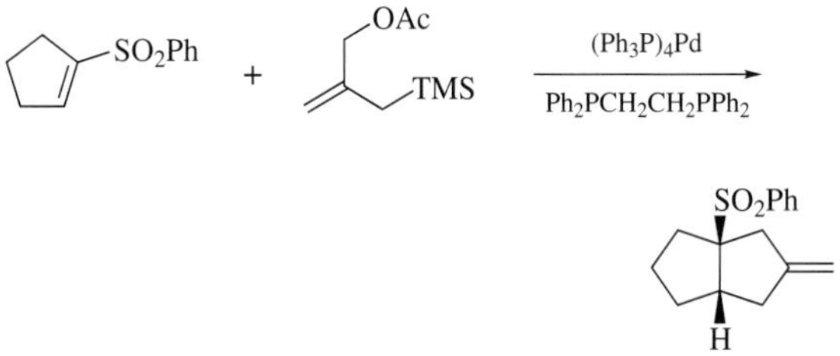

Cyclocontraction–Spiroannulation. The mesylate derivative of 2-hydroxymethyl-3-trimethylsilylpropene participates in an interesting annulation reaction involving a β-keto sulfone as the two-carbon component (eq 9).

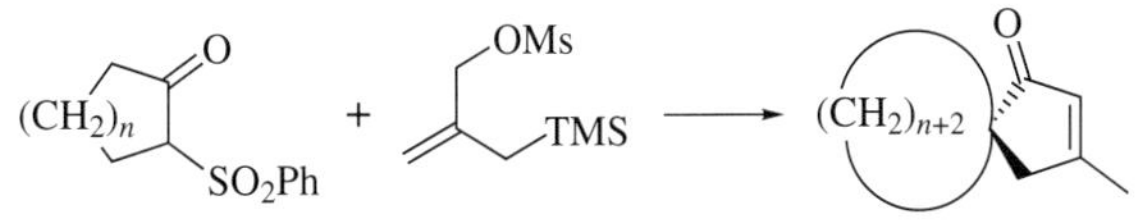

The reaction is thought to involve a pinacol-type rearrangement in which the phenylsulfonyl group serves as a leaving group in the presence of a Lewis acid (eq 10).

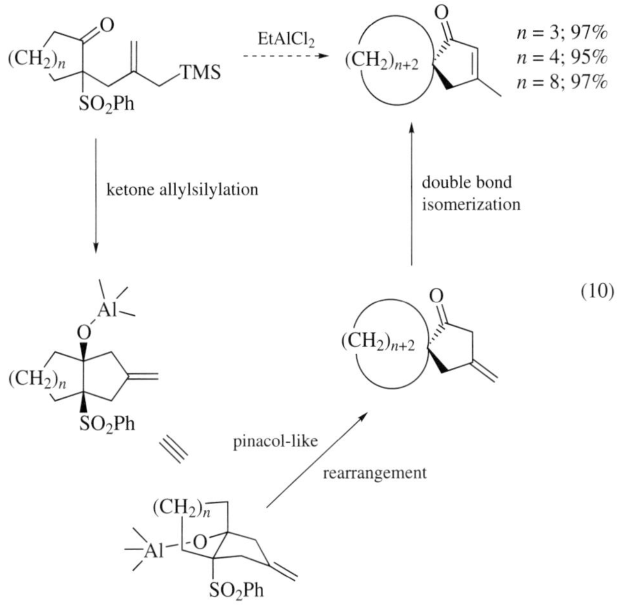

(10)

Related Reagents. 3-Acetoxy-2-trimethylsilylmethyl-1-propene 1; 3-Iodo-2-trimethylsilylmethyl-1-propene.

1. Trost, B. M.; Chan, D. M. T.; Nanninga, T. N., *Org. Synth.* **1984**, *62*, 58.
2. Trost, B. M.; Chan, D. M. T., *J. Am. Chem. Soc.* **1979**, *101*, 6429.
3. Knapp, S.; O'Connor, U.; Mobilio, D., *Tetrahedron Lett.* **1980**, *21*, 4557.
4. Trost, B. M.; Vincent, J. E., *J. Am. Chem. Soc.* **1980**, *102*, 5680.
5. Trost, B. M.; Adams, B. R., *J. Am. Chem. Soc.* **1983**, *105*, 4849.

James S. Panek & Pier F. Cirillo
Boston University, MA, USA

1-Trimethylsilyloxy-1,3-butadiene

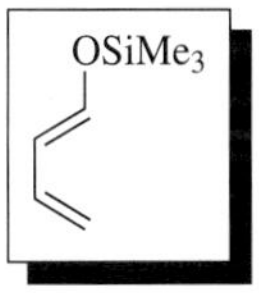

[6651-43-0]   C$_7$H$_{14}$OSi   (MW 142.30)

InChIKey = 1S/C7H14OSi/c1-5-6-7-8-9(2,3)4/h5-7H,1H2,
    2-4H3

InChIKey = UQGOYQLRRBTVFM-UHFFFAOYSA-N

(easily prepared[1] reactive diene for Diels–Alder reactions and other cycloadditions;[2–21] reactive silyl enol ether for aldol and Michael reactions,[22–28] and electrophilic additions[29–32])

Physical Data: bp 131 °C, bp 49.5 °C/25 mmHg; *d* 0.811 g cm^{-3}.
Solubility: sol most standard organic solvents.
Form Supplied in: liquid commercially available (98% purity) as an approximately 85:15 mixture of (*E*)- and (*Z*)-isomers.
Preparative Method: can be prepared easily.[1]
Handling, Storage, and Precautions: is a flammable liquid and is moisture sensitive.

Diels–Alder Reactions and Other Cycloadditions. Although commercially available, 1-trimethylsilyloxy-1,3-butadiene (**1**) can be easily prepared by silylation of crotonaldehyde.[1] It has often been used as a reactive diene in Diels–Alder reactions. For example, reaction with dimethyl acetylenedicarboxylate (**2**) affords the cyclohexadiene diester (**3**) in 68% yield. This initial Diels–Alder adduct can be converted into two different aromatic products by the proper choice of conditions: namely, thermal elimination affords the phthalate (**4**), while oxidation produces the phenol (**5**), both in good yield (eq 1).[2] Reaction with methyl 3-nitroacrylate (**6**) followed by hydrolysis of the initial adduct (**7**) and elimination of the *β*-nitro group leads to the cyclohexadienol (**8**) (eq 2).[3] Many other Diels–Alder reactions of this type have been carried out using (**1**), as shown in Table 1.[4] In general, the *endo* adduct is favored, especially at lower temperatures.

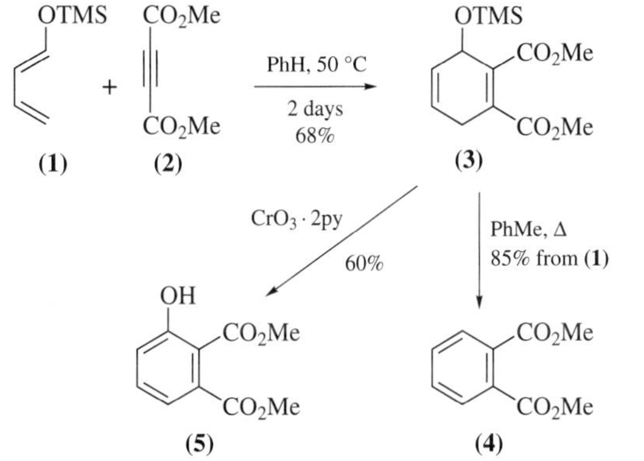

(1)

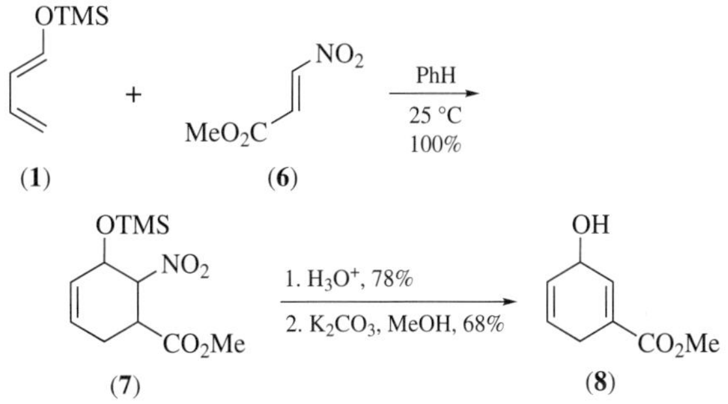

(2)

Cyclic enones, lactones, and lactams have also been used often as dienophiles in Diels–Alder reactions with (**1**), again giving mainly the *endo* adduct,[5] e.g. (**12**) reacted with (**1**) to give (**13**) as the major product in 74% yield (eq 3).[5a] As mentioned earlier, the initial adducts are often oxidized (either directly or after hydrolysis of the silyl ether) with Jones reagent to the corresponding enone,[6] e.g. (**14**) to give (**15**) (eq 4).[6a] This corresponds to the annulation of a cyclohexenone unit onto an existing enone or unsaturated lactone unit and has been used often in synthesis, e.g. in the preparation of aureolic acid derivatives such as (17) from (16) and (1) (eq 5).[6c]

Table 1 Diels–Alder reactions of 1-trimethylsilyl-1,3-butadiene (1)

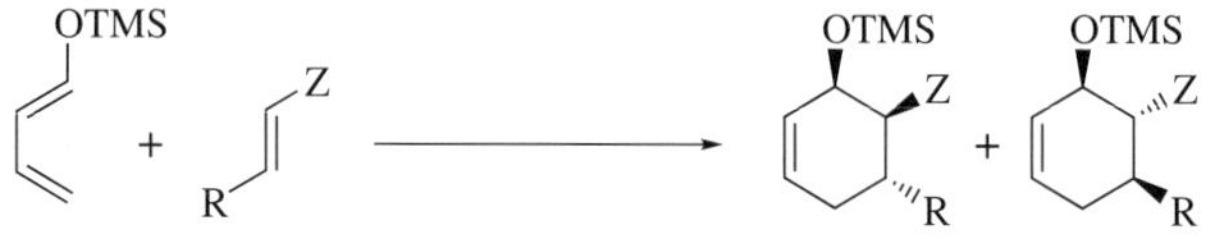

Dienophile	Conditions	Yield	Ratio (10):(11)
Z = CHO, R = H	50 °C, 48 h	88%	11:1[4a]
Z = CHO, R = H	130 °C, 7 h	53%	86:14[4b]
Z = CHO, R = Me	150 °C, 7 h	69%	65:35[4b]
Z = CHO, R = CO$_2$Me	40 °C, 48 h	>90%	83:17[4c]
Z = R = COPh	80 °C, 5 h	93%	no stereochem. given[4d]
Z = NO$_2$, R = TMS	100 °C, 37 h	83%	100:0[4e]
Z = NO$_2$, R = OCOPh	25 °C	90%	6:1[4f]

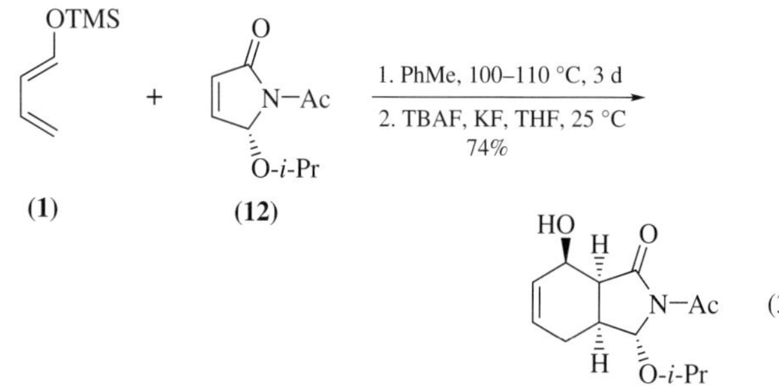

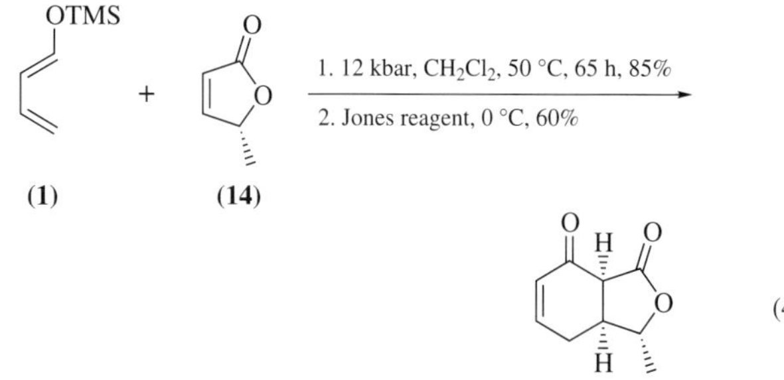

There are some exceptions to the preference for *endo* stereochemistry, especially with unsaturated sulfones.[7] A curious and useful reversal of the stereochemical preference has been reported, namely Diels–Alder cycloaddition of the pyrazolecarboxylate (18) with (1) followed by photochemical elimination of nitrogen gave mainly the *endo* adduct (19n) (eq 6), whereas cycloaddition of 1 with the cyclopropenecarboxylate (20) gave nearly exclusively the *exo* adduct (19x) (eq 7).[8]

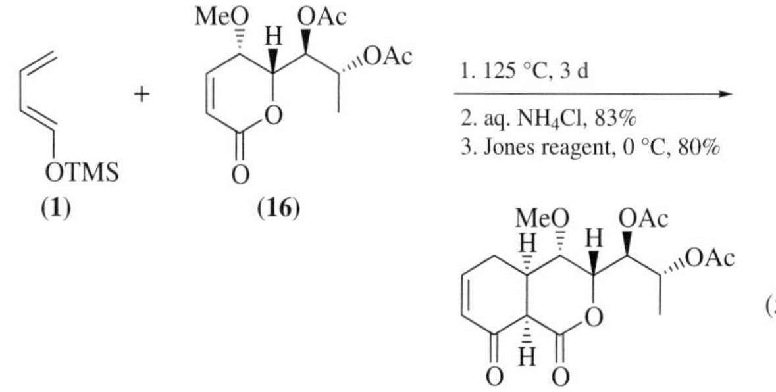

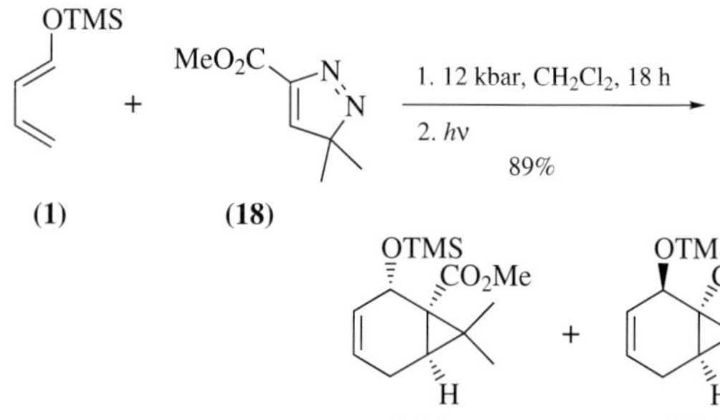

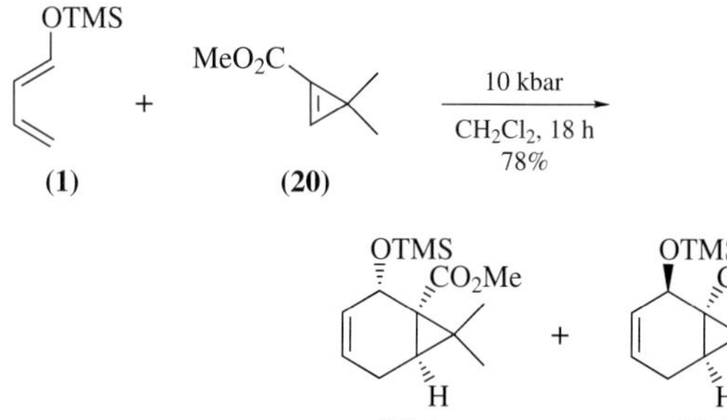

Stereocontrol with acyclic dienophiles can also be high,[9] e.g. (21) giving only (22) (eq 8),[9a] although the relative stereochemistry of the adjacent allylic center can cause nearly stereorandom addition as well.[9a] The versatility of the initial adducts has been evidenced most clearly in the synthesis of anthraquinones and their derivatives. The adducts of (1) with various quinones and substituted quinones have been transformed into simple aromatics,[10] phenols,[11] e.g. (23) gave (24) (eq 9),[11a] cyclohexenones,[12] e.g. (25) gave (26) (eq 10),[12a] and allylic alcohols,[13] all in excellent yields.

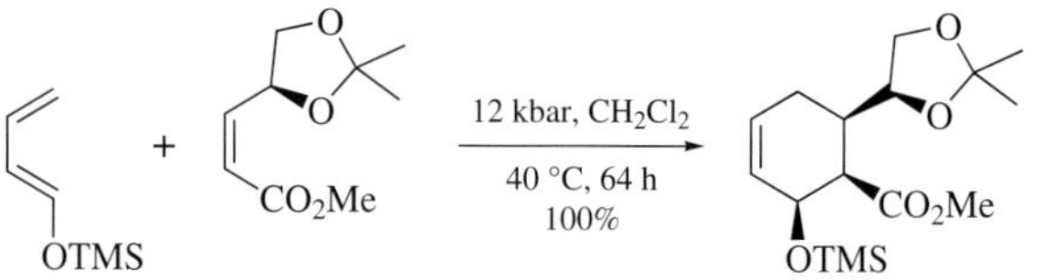

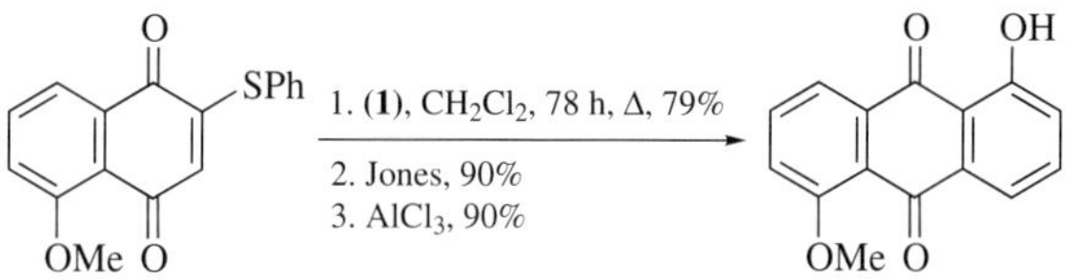

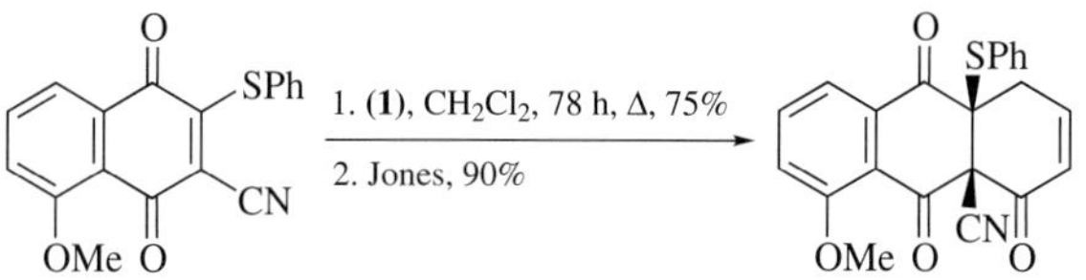

The reaction can be carried out with excellent enantiocontrol (generally 80% or better) using juglone (27) and a catalyst prepared from trimethyl borate and a tartaramide to give (28) (eq 11).[13a] Finally, with metal complexes of tropene and tropone systems, one can obtain either normal [4 + 2] cycloadditions[14a] or novel [6 + 4] cycloadditions, e.g. (29) giving (30) (eq 12).[14b,c] Other [4 + 2] cycloadditions have also been reported, using as dienophiles aldehydes[15] to give pyran derivatives such as (31), vinyl chlorides,[16] singlet oxygen,[17] nitrosoalkanes[18] to give compounds such as (32), and phosphaalkenes (eq 13).[19] Several [2 + 2] cyclizations are known, namely carbene and ketene additions to (1), all of which occur at the unsubstituted double bond.[20] One clever use of (1) in synthesis is the tandem [2 + 2]–Cope process which converts (33) into (34) (eq 14).[20d] Finally, a nickel(0)-catalyzed [4 + 4] cyclization of (1) gives the *trans*-3,4-bis-(silyloxy)-1,5-cyclooctadiene (35) in excellent yield (eq 15).[21]

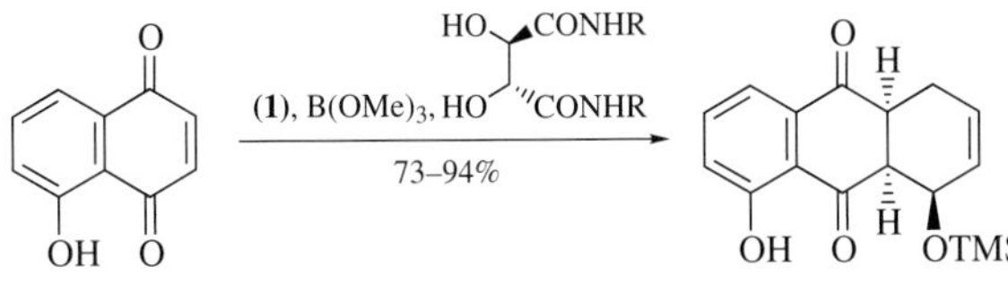

(11)

(27)   (28) ~80% ee

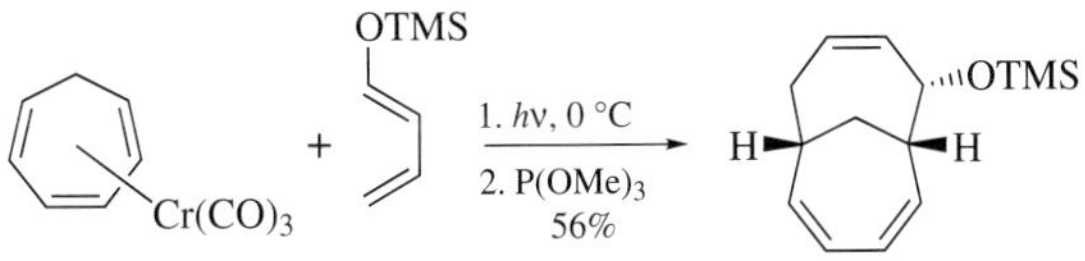

(12)

(29)   (1)   (30)

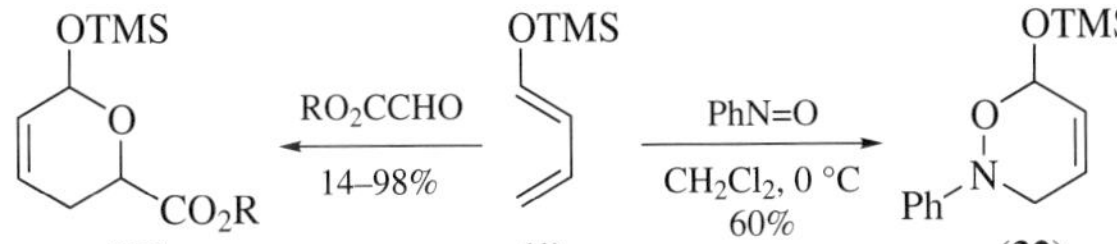

(13)

(31)   (1)   (32)
trans/cis mixture

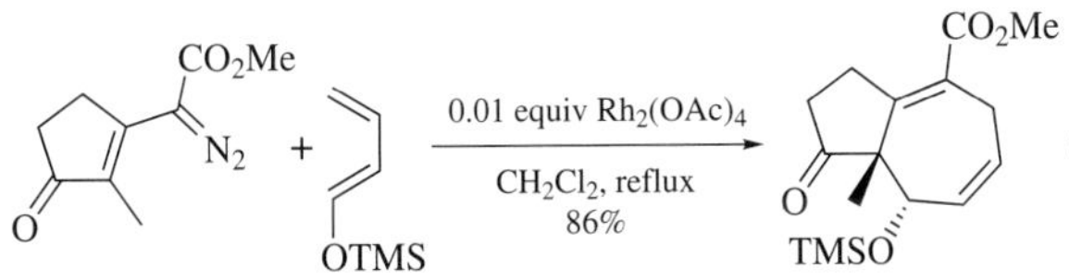

(14)

(33)   (1)   (34)

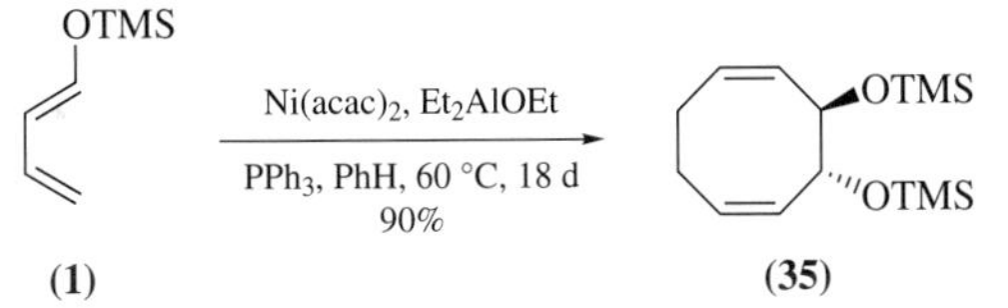

(15)

(1)   (35)

Aldol Condensations. The nucleophilicity of the γ-carbon atom in 1-trimethylsilyloxy-1,3-butadiene (1) allows Lewis acid-catalyzed aldol condensations to be carried out, with (1) acting as the equivalent of the enolate of crotonaldehyde.[22–26] The products usually have the (E) stereochemistry about the newly formed double bond. Orthoesters are good electrophiles;[22,1d] when orthoformate is used as the electrophile, the (E)-monoacetal of glutacondialdehyde (36) is formed in good yield.[22e] This was then used in a very short synthesis of the antiviral agent AZT (37) (eq 16).[22e] 1,3-Dithienium salts and 2-alkoxydithiolanes have also been used to produce the dithioacetals corresponding to (36).[23] Simple acetals and α-chloro ethers can be used as the electrophiles with (1)

to generate initially δ-alkoxy-α,β-unsaturated aldehydes and then the doubly unsaturated aldehydes after treatment with base.[24,1c] A clever approach to the synthesis of indole from pyrrole involves the condensation of (1) with the endoperoxides derived from N-alkoxycarbonylpyrrole as shown in eq 17.[25a] An approach to the piperidine alkaloids uses similar chemistry.[25b,c]

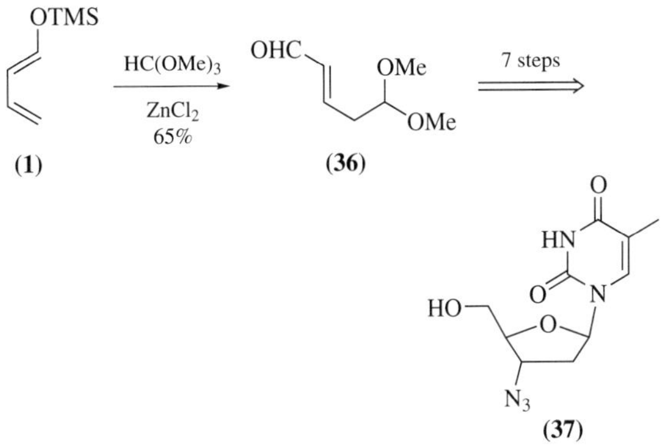

(16)

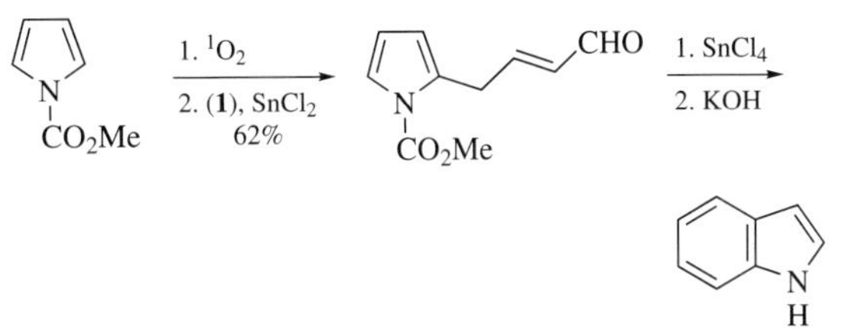

(1)   (36)   (37)

(17)

Finally, a conceptually similar reaction involves the addition of (1) to an acyl imminium salt to give an intermediate (38) (eq 18) which was then used in an intramolecular Diels–Alder approach to the heteroyohimboid alkaloids.[26] A Michael adduct is formed when (1) is allowed to react with 2,2-bis(phenylsulfonyl)styrene, probably as the result of a Diels–Alder reaction in which the adduct reverses to a zwitterion and internally deprotonates.[27] Also the lithium enolate derived from (1) has been added in a Michael fashion to α,β-unsaturated ketones to give, after silylation, the same adducts as the direct Diels–Alder reaction but at much lower temperatures.[28]

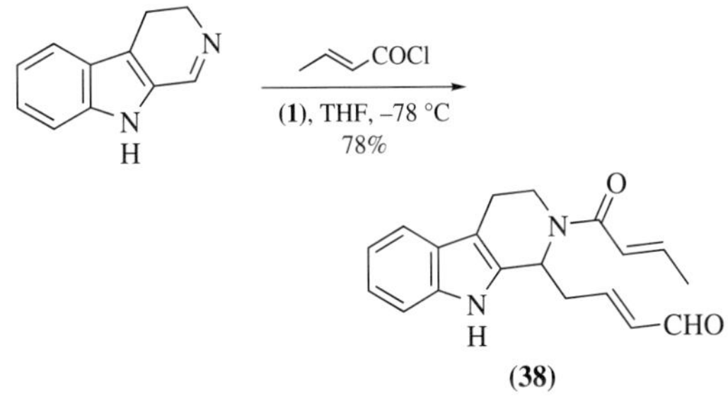

(18)

(38)

Electrophilic Additions. Finally, various electrophiles (other than the carbon species in aldol and other condensations) have been added to (1) in good yield, namely: benzenesulfenyl chloride,[29] bromine or N-bromosuccinimide,[30] and t-butyl hypochlorite[31] There are also several reports of the reaction of (1) with various transition metal electrophiles which generate either a silyloxydiene bound to the metal[32] or a crotonaldehyde unit bound to the metal.[33]

Related Reagents. 1-Acetoxy-1,3-butadiene; 2-Methoxy-1,3-butadiene; 2-Methoxy-3-phenylthio-1,3-butadiene; 1-Methoxy-3-trimethylsilyloxy-1,3-butadiene 1; 2-Trimethyl-silyloxy-1,3-butadiene.

---

1. (a) Belg. Patent 670 769, 1966 (*Chem. Abstr.* **1966**, *65*, 5487d). (b) Cazeau, P.; Frainnet, E., *Bull. Soc. Claim. Fr., Part 2* **1972**, 1658. (c) Ishida, A.; Mukaiyama, T., *Bull. Chem. Soc. Jpn.* **1977**, *50*, 1161. (d) Makin, S. M.; Kruglikova, R. I.; Popova, T. P.; Chernyshev, A. I., *J. Org. Chem. USSR (Engl. Transl.)* **1982**, *18*, 834. (e) Makin, S. M.; Kruglikova, R. I.; Shavrygina, O. A.; Chernyshev, A. I.; Popova, T. P.; Tung, N. F., *J. Org. Chem. USSR (Engl. Transl.)* **1982**, *18*, 250. (f) Cazeau, P.; Duboudin, F.; Moulines, F.; Babot, O.; Dunoguès, J., *Tetrahedron* **1987**, *43*, 2089. (g) Iqbal, J.; Khan, M. A., *Synth. Commun.* **1989**, *19*, 515.

2. Yamamoto, K.; Suzuki, S.; Tsuji, J., *Chem. Lett.* **1978**, 649.

3. Danishefsky, S. J.; Prisbylla, M. P.; Hiner, S., *J. Am. Chem. Soc.* **1978**, *100*, 2918.

4. (a) Kurth, M. J.; Brown, E. G.; Hendra, E.; Hope, H., *J. Org. Chem.* **1985**, *50*, 1115. (b) Makin, S. M.; Tung, N. F.; Shavrygina, O. A.; Arshava, B. M.; Romanova, I. A., *J. Org. Chem. USSR (Engl. Transl.)* **1983**, *19*, 640. (c) Kakushima, M.; Scott, D. E., *Can. J. Chem.* **1979**, *57*, 1399. (d) Oida, T.; Tanimoto, S.; Sugimoto, O.; Okano, M., *Synthesis* **1980**, 131. (e) Padwa, A.; MacDonald, J. G., *J. Org. Chem.* **1983**, *48*, 3189. (f) Kraus, G. A.; Thurston, J.; Thomas, P. J.; Jacobson, R. A.; Su, Y., *Tetrahedron Lett.* **1988**, *29*, 1879.

5. (a) Koot, W.-J.; Hiemstra, H.; Speckamp, W. N., *J. Org. Chem.* **1992**, *57*, 1059. (b) Alonso, D.; Font, J.; Ortuño, R. M.; d'Angelo, J.; Guingant, A.; Bois, C., *Tetrahedron* **1991**, *47*, 5895. (c) Tsuda, Y.; Horiguchi, Y.; Sano, T., *Heterocycles* **1976**, *4*, 1355.

6. (a) Ortuño, R. M.; Guingant, A.; d'Angelo, J., *Tetrahedron Lett.* **1988**, *29*, 6989. (b) Moriarty, K. J.; Shen, C.-C.; Paquette, L. A., *Synlett* **1990**, *5*, 263. (c) Franck, R. W.; Subramanian, C. S.; John, T. V.; Blount, J. F., *Tetrahedron Lett.* **1984**, *25*, 2439. (d) Kallmerten, J., *Tetrahedron Lett.* **1984**, *25*, 2843. (e) Card, P. J., *J. Org. Chem.* **1982**, *47*, 2169.

7. (a) Cossu, S.; Delogu, G.; De Lucchi, O.; Fabbri, D.; Licini, G., *Angew. Chem., Int. Ed. Engl.* **1989**, *28*, 766. (b) Hayakawa, K.; Nishiyama, H.; Kanematsu, K., *J. Org. Chem.* **1985**, *50*, 512.

8. Rigby, J. H.; Kierkus, P. C., *J. Am. Chem. Soc.* **1989**, *111*, 4125.

9. (a) Casas, R.; Parella, T.; Branchadell, V.; Oliva, A.; Ortuño, R. M., *Tetrahedron* **1992**, *48*, 2659. (b) Serrano, J. A.; Cáceres, L. E.; Román, E., *J. Chem. Soc., Perkin Trans. 1* **1992**, 941.

10. (a) Echavarren, A.; Prados, P.; Fariña, F., *Tetrahedron* **1984**, *40*, 4561. (b) Kraus, G. A.; Molina, M. T.; Walling, J. A., *J. Chem. Soc., Chem. Commun.* **1986**, 1568. (c) Lee, J.; Snyder, J. K., *J. Org. Chem.* **1990**, *55*, 4995. (d) Lee, J.; Tang, J.; Snyder, J. K., *Tetrahedron Lett.* **1987**, *28*, 3427.

11. (a) Laugraud, S.; Guingant, A.; Chassagrand, C.; d'Angelo, J., *J. Org. Chem.* **1988**, *53*, 1557. (b) Kraus, G. A.; Chen, L., *J. Org. Chem.* **1991**, *56*, 5098. (c) McKenzie, T. C.; Hassen, W.; Macdonald, J. F., *Tetrahedron Lett.* **1987**, *28*, 5435. (d) Cameron, D. W.; Feutrill, G. I.; Gibson, C. L.; Read, R. W., *Tetrahedron Lett.* **1985**, *26*, 3887.

12. (a) Kraus, G. A.; Walling, J. A., *Tetrahedron Lett.* **1986**, *27*, 1873. (b) Laugraud, S.; Guingant, A.; d'Angelo, J., *Tetrahedron Lett.* **1989**, *30*, 83.

13. (a) Maruoka, K.; Sakurai, M.; Fujiwara, J.; Yamamoto, H., *Tetrahedron Lett.* **1986**, *27*, 4895. (b) Kraus, G. A.; Taschner, M. J., *J. Org. Chem.* **1980**, *45*, 1174. (c) Carretero, J. C.; Cuevas, J. C.; Echavarren, A.; Fariña, F.; Prados, P., *J. Chem. Res. (S)* **1984**, 6.

14. (a) Rigby, J. H.; Ogbu, C. O., *Tetrahedron Lett.* **1990**, *31*, 3385. (b) Rigby, J. H.; Ateeq, H. S.; Charles, N. R.; Cuisiat, S. V.; Ferguson, M. D.; Henshilwood, J. A.; Krueger, A. C.; Ogbu, C. O.; Short, K. M.; Heeg, M. J., *J. Am. Chem. Soc.* **1993**, *115*, 1382. (c) Rigby, J. H.; Ateeq, H. S., *J. Am. Chem. Soc.* **1990**, *112*, 6442.

15. (a) Cervinka, O.; Svatos, A.; Trska, P.; Pech, P., *Collect. Czech. Chem. Commun.* **1990**, *55*, 230. (b) Achmatowicz, O., Jr.; Bialecka-Florjanczyk, E., *Tetrahedron* **1990**, *46*, 5317. (c) Bélanger, J.; Landry, N. L.; Paré, J. R. J.; Jankowski, K., *J. Org. Chem.* **1982**, *47*, 3649.

16. (a) South, M. S.; Liebeskind, L. S., *J. Org. Chem.* **1982**, *47*, 3815. (b) Seitz, G.; v. Gemmern, R., *Synthesis* **1987**, 953.

17. Clennan, E. L.; L'Esperance, R. P., *Tetrahedron Lett.* **1983**, *24*, 4291.

18. McClure, K. F.; Danishefsky, S. J., *J. Org. Chem.* **1991**, *56*, 850.

19. Märkl, G.; Kallmünzer, A., *Tetrahedron Lett.* **1989**, *30*, 5245.

20. (a) Wenkert, E.; Goodwin, T. E.; Ranu, B. C., *J. Org. Chem.* **1977**, *42*, 2137. (b) Tomilov, Yu. V.; Kostitsyn, A. B.; Shulishov, E. V.; Nefedov, O. M., *Synthesis* **1990**, 246. (c) Brady, W. T.; Lloyd, R. M., *J. Org. Chem.* **1981**, *46*, 1322. (d) Cantrell, W. R., Jr.; Davies, H. M. L., *J. Org. Chem.* **1991**, *56*, 723.

21. (a) Tenaglia, A.; Brun, P.; Waegele, B., *J. Organomet. Chem.* **1985**, *285*, 343. (b) Brun, P.; Tenaglia, A.; Waegele, B., *Tetrahedron Lett.* **1983**, *24*, 385.

22. (a) Makin, S. M.; Raifel'd, Yu. E.; Zil'berg, L. L.; Arshava, B. M., *J. Org. Chem. USSR (Engl. Transl.)* **1984**, *20*, 189. (b) Akgün, E.; Pindur, U., *Synthesis* **1984**, 227. (c) Akgün, E.; Pindur, U., *Monatsh. Chem.* **1984**, *115*, 587. (d) Duhamel, L.; Ple, G.; Ramondenc, Y., *Tetrahedron Lett.* **1989**, *30*, 7377. (e) Jung, M. E.; Gardiner, J. M., *J. Org. Chem.* **1991**, *56*, 2614.

23. (a) Paterson, I.; Price, L. G., *Tetrahedron Lett.* **1981**, *22*, 2833. (b) Hatanaka, K.; Tanimoto, S.; Sugimoto, T.; Okano, M., *Tetrahedron Lett.* **1981**, *22*, 3243.

24. (a) Makin, S. M.; Kruglikova, R. I.; Popova, T. P., *J. Org. Chem. USSR (Engl. Transl.)* **1982**, *18*, 1001. (b) Makin, S. M.; Shavrygina, O. A.; Kruglikova, R. I.; Mikerin, I. E.; Tung, N. F.; Ermakova, G. A., *J. Org. Chem. USSR (Engl. Transl.)* **1983**, *19*, 1994. (c) Makin, S. M.; Dymshakova, G. M.; Granenkina, L. S., *J. Org. Chem. USSR (Engl. Transl.)* **1986**, *22*, 256. (d) Ishida, A.; Mukaiyama, T., *Bull. Chem. Soc. Jpn.* **1978**, *51*, 2077. (e) Ishida, A.; Mukaiyama, T., *Chem. Lett.* **1977**, 467. (f) Ibragimov, M. A.; Lazareva, M. I.; Smit, W. A., *Synthesis* **1985**, 880.

25. (a) Natsume, M.; Muratake, H., *Tetrahedron Lett.* **1979**, 3477. (b) Natsume, M.; Muratake, H., *Heterocycles* **1980**, *14*, 615. (c) Natsume, M.; Muratake, H., *Heterocycles* **1981**, *16*, 973.

26. Martin, S. A.; Benage, B.; Geraci, L. S.; Hunter, J. E.; Mortimore, M., *J. Am. Chem. Soc.* **1991**, *113*, 6161.

27. De Lucchi, O.; Fabbri, D.; Lucchini, V., *Tetrahedron* **1992**, *48*, 1485.

28. Kraus, G. A.; Sugimoto, H., *Tetrahedron Lett.* **1977**, 3929.

29. Fleming, I.; Goldhill, J.; Paterson, I., *Tetrahedron Lett.* **1979**, 3205.

30. Duhamel, L.; Guillemont, J.; Le Gallic, Y.; Plé, G.; Poirier, J.-M.; Ramondenc, Y.; Chabardes, P., *Tetrahedron Lett.* **1990**, *31*, 3129.

31. Decor, J. P. Ger. Patent 2 708 281, 1977 (*Chem. Abstr.* **1977**, *87*, 200 795w).

32. Tolstikov, G. A.; Miftakhov, M. S.; Menakov, Yu. B.; Lomakina, S. I., *Zh. Obshch. Khim.* **1976**, *46*, 2630.

33. (a) Benyunes, S. A.; Day, J. P.; Green, M.; Al-Saadoon, A. W.; Waring, T. L., *Angew. Chem., Int. Ed. Engl.* **1990**, *29*, 1416. (b) Benyunes, S. A.; Green, M.; Grimshire, M. J., *Organometallics* **1989**, *8*, 2268.

Michael E. Jung
University of California, Los Angeles, CA, USA

2-Trimethylsilyloxy-1,3-butadiene

[38053-91-7]   C$_7$H$_{14}$OSi   (MW 142.30)

InChI = 1S/C7H14OSi/c1-6-7(2)8-9(3,4)5/h6H,1-2H2,3-5H3

InChIKey = JOAPBVRQZQYKMS-UHFFFAOYSA-N

(easily prepared[1] reactive diene for Diels–Alder reactions[2–40] and other cycloadditions;[41–44] reactive silyl enol ether for aldol, Michael reactions, and electrophilic additions[45–46])

Physical Data: bp 50–55 °C/50 mmHg; d 0.811 g cm^{-3}.

Solubility: sol most standard organic solvents.

Form Supplied in: liquid available commercially.

Preparative Method: can be prepared easily from methyl vinyl ketone.[1]

Handling, Storage, and Precautions: is a flammable liquid and is moisture sensitive.

Diels–Alder Reactions. Although commercially available, 2-trimethylsilyloxy-1,3-butadiene (**1**) can be easily prepared by silylation of methyl vinyl ketone.[1] It has been often used as a reactive diene in Diels–Alder reactions. In nearly all cases with a strongly activated dienophile, the trimethylsilyloxy group ends up 1,4 to the activating group in the adduct. Several representative examples are shown in Table 1.[1–14] The initial Diels–Alder adduct is a silyl enol ether and as such can be converted into several different products by simple silyl enol ether chemistry, e.g. the adduct (**3**) from the reaction of (**1**) with dimethyl fumarate (**2**) (eq 1) can be hydrolyzed to the cyclohexanone (**4**) in either acid or base, converted into the α-bromo or α-hydroxy ketone (**5**) or (**6**) on treatment with bromine or *N*-bromosuccinimide and *m*-chloroperbenzoic acid, respectively, and finally via a Mukaiyama-type aldol (PhCHO/titanium(IV) chloride) to the enone (**7**).[1b,2]

Alkynic dienophiles also work well, e.g. ethyl propiolate and (**1**) produce the expected cyclohexadiene ester in 77% yield.[15] In addition to simple dienophiles, a number of allenic dienophiles have also been utilized, as shown in Table 2.[16–19] Cyclic dienophiles, enones, lactams, etc., are often reacted with (**1**) to give fused bicyclic systems, but generally the yields are only in the 25–45% range,[20–24] e.g. reaction of (**1**) with (**13**) gives (**14**) in 27% yield (eq 2).[20b]

Table 1 Diels–Alder reactions of 2-trimethylsilyl-1,3-butadiene (**1**)

Z	R^1	R^2	R^3	Conditions	Yield (%)	Ratio	Ref.
CN	CN	CN	CN	PhH, 25 °C	90	100:0	1f
CO$_2$Me	H	H	H	toluene, 110 °C, 18 h	35	100:0	1b, 2
COMe	H	H	H	toluene, 110 °C, 18 h	60	100:0	1b, 2
CO$_2$Et	H	CO$_2$Et	H	toluene, 110 °C, 18 h	77	100:0	1b, 2
CO$_2$Et	H	H	CO$_2$Et	toluene, 110 °C, 18 h	39	100:0	1b, 2
CO$_2$Et	H	H	CO$_2$Et	toluene, 140 °C, 7 h	95	100:0	1c
CHO	H	H	H	toluene, 110 °C, 24 h	81^b	100:0	3
CO$_2$Me	Me	H	H	xylene, 140 °C, 48 h	68	100:0	4
CHO	Me	H	H	toluene, 110 °C, 45 h	54	100:0	5
NO$_2$	H	CO$_2$Me	H	PhH, 25 °C, 42 h	100 (72)^a	100:0	6
NO$_2$	H	OCOPh	H	–	95	100:0	7
CO$_2$Me	H	CO$_2$-*t*-Bu	H	CH$_2$Cl$_2$, –20 °C, MAD, 50 h	48	99:1	8
SO$_2$Ph	SO$_2$Ph	H	H	CH$_2$Cl$_2$, 20 °C, 24 h	100 (79)^a	100:0	9
SO$_2$Ph	SO$_2$Ph	Me	H	LiClO$_4$, Et$_2$O, 50 °C, 24 h	95	100:0	10
SO$_2$Ph	SO$_2$Ph	Ph	H	LiClO$_4$, Et$_2$O, 50 °C, 24 h	90	100:0	10
SO$_2$Ph	H	CF$_3$	H	toluene, 110 °C, 4 h	89 (68)^a	66:23	11
CF$_3$	H	H	H	neat, 150 °C, 72 h	17	76:24	12
	CO$_2$CMe$_2$O$_2$C	(CH$_2$)$_2$CX$_2$(CH$_2$)$_2$		PhH, 80 °C, 22 h	97	100:0	13
	COCR=CRO	H	H	toluene, 110 °C, 12 h	(53)^a	100:0	14

a Yield of cyclohexanone after hydrolysis. b Product is 93% pure.

Table 2 Allenic dienophiles in Diels–Alder reactions with (1)

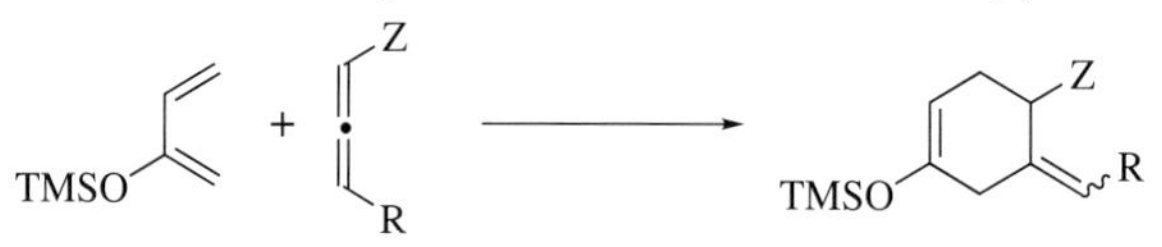

Z	R	Conditions	Yield (%)	Ref.
CO_2TMS	CO_2TMS	PhH, 82 °C, 20 h	80	16
CO_2Et	CO_2Et	PhH, 82 °C, 14 h	97	17
SO_2Ph	H	160 °C, 3 h	79	18
CO_2TMS	$CH_2OTBDMS$	PhMe, 110 °C, 24 h	62	19

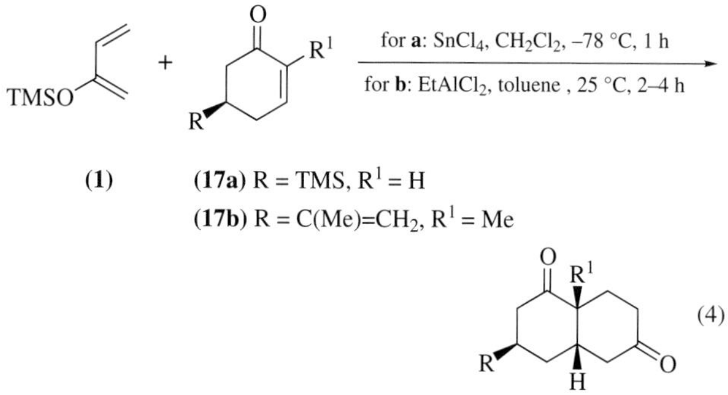

More reactive dienophiles, e.g. β-nitroenones,[21] enediones,[22] 1,2-disulfonylethylene,[23] etc., give higher yields, as do Lewis acids,[24] e.g. zinc bromide (eq 3),[24a] and the use of high pressure.[25] Stereofacial differentiation of (1) with cyclic enones,[26] lactones,[27] and lactams[28] is generally excellent, e.g. (1) reacts with (17a,b) to give (18a,b) with high selectivity in good yield (eq 4).[26] Reaction of (1) with quinones has often been used[29] in syntheses of anthraquinones and their derivatives, e.g. to prepare the A ring of the anthracyclines.[29b–g] Double Michael reactions[30] have been used quite often instead of Diels–Alder reactions to provide the same products, often in higher yields. The preparation of the azabicyclic ketone (20) is an interesting example of a tandem Diels–Alder reaction–aldol-type condensation (eq 5).[31] An example of an aldol reaction followed by a Diels–Alder is also known.[32]

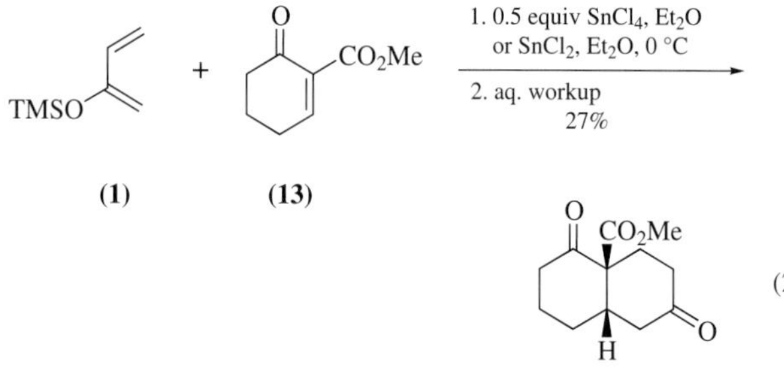

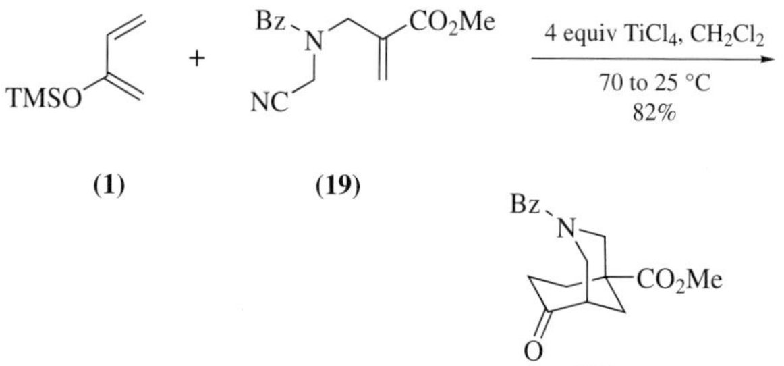

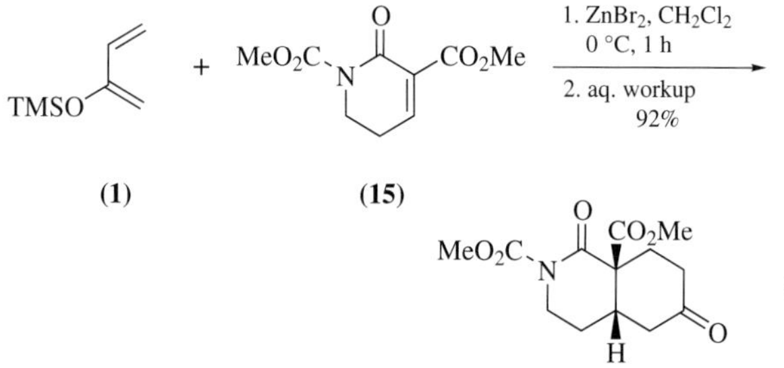

Finally, there is one example of (1) acting as the dienophile, namely in the Diels–Alder reaction with α-nitrosostyrene, where 6-vinyl-6-silyloxy-4,5-dihydrooxazine is formed in 65% yield.[33]

Heterodienophiles. Many heterodienophiles have been reacted with (1).[34–37] Mesoxalate, glyoxalate, and hexafluoroacetone all give the 4-pyranone silyl ether derivatives in fair yield.[34] The corresponding thiomesoxalate affords the 3-thiopyranone-4,4-diester in good yield,[35a,b] as does an α-oxosulfine,[35c] while alkyl thioformates yield the 4-thiopyranones.[35d] By far the largest group of heterodienophiles used are imine derivatives.[36,37] Activated imines, e.g. diacyl imines,[36] afford the 4-piperidones, while aryl-fused dihydropyridines, e.g. (21), furnish fused 4-piperidones, e.g. (22), in good yield when reacted with (1) in the presence of a Lewis acid (eq 6).[37]

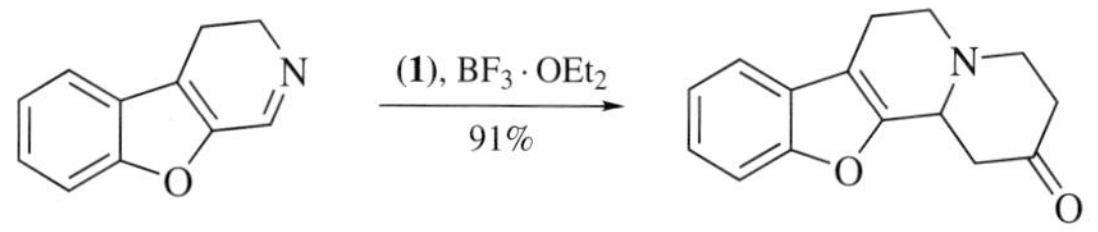

Finally, several imines give products of an aldol-type condensation rather than Diels–Alder reaction when reacted with (1) in the presence of Lewis acids.[38] Phosphaalkenes react with (1) to give ultimately the aromatic phosphinine[39] and dimethyl azodicarboxylate affords the expected [4 + 2] adduct in 83% yield.[40]

[2 + 2] and Other Cycloadditions, Carbene Additions. Several [2 + 2] cyclizations of (1) with various alkenes are known.[41] Quite often these vinyl cyclobutyl silyl ethers can be thermally rearranged to the normal Diels–Alder product,[16,19,41] e.g. (23) giving (25) via (24) (eq 7).[41a] These [2 + 2] adducts can also be transformed via palladium catalysis into α-methylenecyclopentanones[41d,e] in good yield, e.g. (26) gives (28) via (27) (eq 8).[41d] Metal complexes of 1,3,5-cycloheptatriene react with (1) to give a novel [6 + 4] cycloaddition, e.g. (29) giving (30) (eq 9).[42]

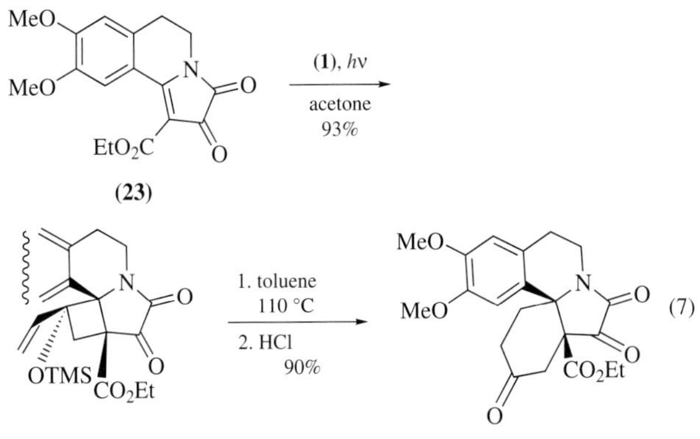

(7)

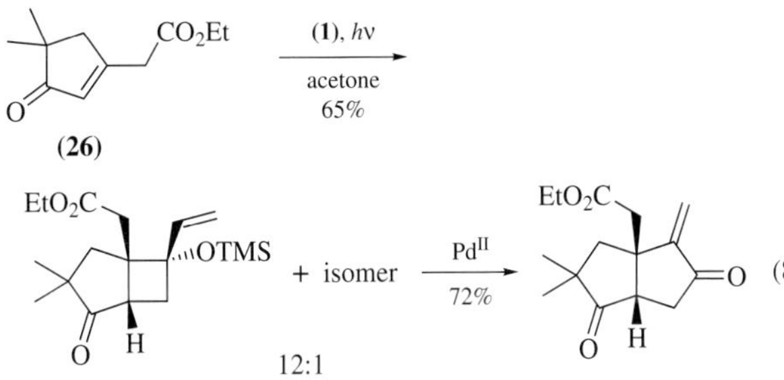

(8)

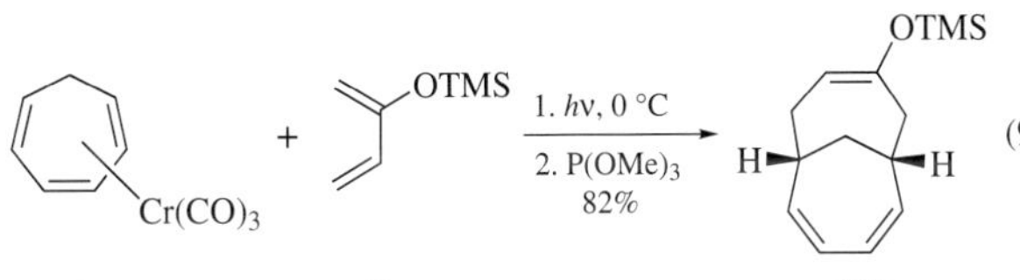

(9)

Even [3 + 2] cycloadditions occur when cyclopropyl ketones are treated with (1) under photolytic or Lewis acid conditions to produce cyclopentane systems.[43] Cyclopropanations of (1) are well known, in all cases adding to the more electron-rich silyloxy alkene.[1a,44] The vinylcyclopropanol silyl ethers can be converted into a number of different products, e.g. cyclopentanones, 2-methylcyclobutanones, vinyl ketones, and β-alkoxy ketones (eq 10). One clever use of (1) in synthesis is the tandem [2 + 2]–Cope process which converts (34) into (35) (eq 11).[44i]

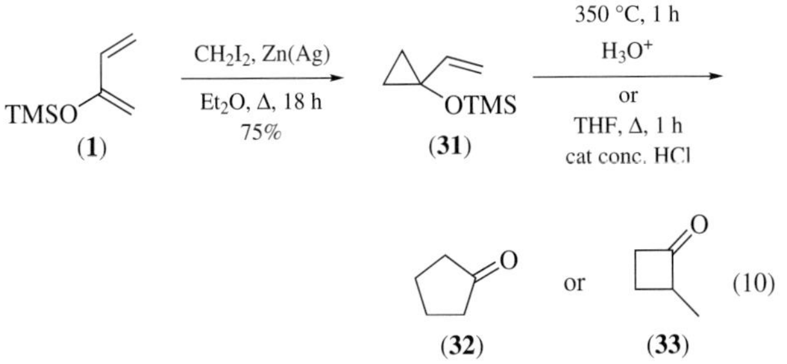

(10)

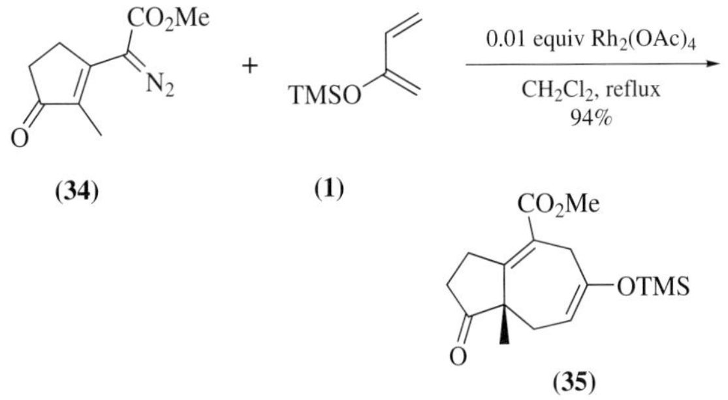

(11)

Aldol Condensations and Electrophilic Additions. 2-Trimethylsilyloxybutadiene (1) can act as the enolate of methyl vinyl ketone (MVK) towards reactive carbon atoms in an aldol-like process to give the products of overall addition of MVK to reactive centers.[45] Various strong electrophiles have been added to (1), e.g. bromine, NBS, m-CPBA, alkylating agents, and acylating agents, usually reacting at the carbon of the silyl enol ether to give α-bromo, α-silyloxy, or α-alkyl ketones in good yields.[46] With acylating agents, however, the O-acylated products are obtained.[46e]

Related Reagents. 1-Acetoxy-1,3-butadiene; 2-Methoxy-1,3-butadiene; 1-Methoxy-3-trimethylsilyloxy-1,3-butadiene; 1-Trimethylsilyloxy-1,3-butadiene.

1. (a) Girard, C.; Amice, P.; Barnier, J. P.; Conia, J. M., *Tetrahedron Lett.* **1974**, 3329. (b) Jung, M. E.; McCombs, C. A., *Tetrahedron Lett.* **1976**, 2935. (c) Jung, M. E.; McCombs, C. A., *Org. Synth., Coll. Vol.* **1988**, *6*, 445. (d) Cazeau, P.; Moulines, F.; LaPorte, O.; Duboudin, F., *J. Organomet. Chem.* **1980**, *201*, C9. (e) Cazeau, P.; Duboudin, F.; Moulines, F.; Babot, O.; Dunoguès, J., *Tetrahedron* **1987**, *43*, 2089. (f) Cazeau, P.; Frainnet, E., *Bull. Soc. Claim. Fr., Part 2* **1972**, 1658.

2. Jung, M. E.; McCombs, C. A.; Takeda, Y.; Pan, Y.-G., *J. Am. Chem. Soc.* **1981**, *103*, 6677.

3. Yin, T.-K.; Lee, J. G.; Borden, W. T., *J. Org. Chem.* **1985**, *50*, 531.

4. Baldwin, J. E.; Broline, B. M., *J. Org. Chem.* **1982**, *47*, 1385.

5. Rigby, J. H.; Kotnis, A. S., *Tetrahedron Lett.* **1987**, *28*, 4943.

6. Danishefsky, S. J.; Prisbylla, M. P.; Hiner, S., *J. Am. Chem. Soc.* **1978**, *100*, 2918.

7. Kraus, G. A.; Thurston, J.; Thomas, P. J.; Jacobson, R. A.; Su, Y., *Tetrahedron Lett.* **1988**, *29*, 1879.

8. Maruoka, K.; Saito, S.; Yamamoto, H., *J. Am. Chem. Soc.* **1992**, *114*, 1089.

9. Rao, Y. K.; Nagarajan, M., *Synthesis* **1984**, 757.

10. De Lucchi, O.; Fabbri, D.; Lucchini, V., *Tetrahedron* **1992**, *48*, 1485.

11. Taguchi, T.; Hosoda, A.; Tomizawa, G.; Kawara, A.; Masuo, T.; Suda, Y.; Nakajima, M.; Kobayashi, Y., *Chem. Pharm. Bull.* **1987**, *35*, 909.

12. Ojima, I.; Yatabe, M.; Fuchikami, T., *J. Org. Chem.* **1982**, *47*, 2051.

13. Bell, V. L.; Holmes, A. B.; Hsu, S.-Y.; Mock, G. A.; Raphael, R. A., *J. Chem. Soc., Perkin Trans. 1* **1986**, 1502.

14. Ko, B.-S.; Oritani, T.; Yamashita, K., *Agric. Biol. Chem.* **1990**, *54*, 2199.

15. Ackland, D. A.; Pinhey, J. T., *J. Chem. Soc., Perkin Trans. 1* **1987**, 2689.

16. Jung, M. E.; Node, M.; Pfluger, R.; Lyster, M. A.; Lowe, J. A., III, *J. Org. Chem.* **1982**, *47*, 1150.

17. Kozikowski, A. P.; Schmiesing, R., *Synth. Commun.* **1978**, *8*, 363.

18. Hayakawa, K.; Nishiyama, H.; Kanematsu, K., *J. Org. Chem.* **1985**, *50*, 512.

19. Jung, M. E.; Lowe, J. A., III; Lyster, M. A.; Node, M.; Pfluger, R.; Brown, R. W., *Tetrahedron* **1984**, *40*, 4751.

20. (a) Liu, H.-J.; Ngooi, T. K., *Synth. Commun.* **1982**, *12*, 715. (b) Liu, H.-J.; Ngooi, T. K.; Browne, E. N. C., *Can. J. Chem.* **1988**, *66*, 3143. (c) Kraus, G. A.; Gottschalk, P., *J. Org. Chem.* **1984**, *49*, 1153. (d) Ghosh, S.; Saha, S., *Tetrahedron Lett.* **1985**, *26*, 5325. (e) Ghosh, S.; Saha Roy, S.; Saha, G., *Tetrahedron* **1988**, *44*, 6235. (f) Sano, T.; Toda, J.; Kashiwaba, N.; Ohshima, T.; Tsuda, Y., *Chem. Pharm. Bull.* **1987**, *35*, 479. (g) Rigby, J. H.; Ogbu, C. O., *Tetrahedron Lett.* **1990**, *31*, 3385. (h) Seitz, G.; v Gemmern, R., *Synthesis* **1987**, 953. (i) Wulff, W. D.; Yang, D. C., *J. Am. Chem. Soc.* **1984**, *106*, 7565.

21. Corey, E. J.; Estreicher, H., *Tetrahedron Lett.* **1981**, *22*, 603.

22. Danishefsky, S.; Kahn, M., *Tetrahedron Lett.* **1981**, *22*, 489.

23. De Lucchi, O.; Fabbri, D.; Cossu, S.; Valle, G., *J. Org. Chem.* **1991**, *56*, 1888.

24. (a) de Oliveira Imbroisi, D.; Simpkins, N. S., *J. Chem. Soc., Perkin Trans. 1* **1991**, 1815. (b) de Oliveira Imbroisi, D.; Simpkins, N. S., *Tetrahedron Lett.* **1989**, *30*, 4309. (c) Fujiwara, T.; Ohsaka, T.; Inoue, T.; Takeda, T., *Tetrahedron Lett.* **1988**, *29*, 6283.

25. (a) Laugraud, S.; Guingant, A.; d'Angelo, J., *Tetrahedron Lett.* **1989**, *30*, 83. (b) Branchadell, V.; Sodupe, M.; Ortuño, R. M.; Oliva, A.; Gomez-Pardo, D.; Guingant, A.; d'Angelo, J., *J. Org. Chem.* **1991**, *56*, 4135.

26. (a) Asaoka, M.; Nishimura, K.; Takei, H., *Bull. Chem. Soc. Jpn.* **1990**, *63*, 407. (b) Haaksma, A. A.; Jansen, B. J. M.; de Groot, A., *Tetrahedron* **1992**, *48*, 3121.

27. de Jong, J. C.; van Bolhuis, F.; Feringa, B. L., *Tetrahedron: Asymmetry* **1991**, *2*, 1247.

28. (a) Koot, W.-J.; Hiemstra, H.; Speckamp, W. N., *J. Org. Chem.* **1992**, *57*, 1059. (b) Baldwin, S. W.; Greenspan, P.; Alaimo, C.; McPhail, A. T., *Tetrahedron Lett.* **1991**, *32*, 5877.

29. (a) Boisvert, L.; Brassard, P., *J. Org. Chem.* **1988**, *53*, 4052. (b) Tamura, Y.; Wada, A.; Sasho, M.; Fukunaga, K.; Maeda, H.; Kita, Y., *J. Org. Chem.* **1982**, *47*, 4376. (c) Tamura, Y.; Sasho, M.; Akai, S.; Wada, A.; Kita, Y., *Tetrahedron* **1984**, *40*, 4539. (d) Carretero, J. C.; Cuevas, J. C.; Echavarren, A.; Fariña, F.; Prados, P., *J. Chem. Res. (S)* **1984**, 6. (e) Cameron, D. W.; Feutrill, G. I.; Griffiths, P. G.; Merrett, B. K., *Tetrahedron Lett.* **1986**, *27*, 2421. (f) Preston, P. N.; Winwick, T.; Morley, J. O., *J. Chem. Soc., Perkin Trans. 1* **1985**, 39. (g) Laugraud, S.; Guingant, A.; d'Angelo, J., *Tetrahedron Lett.* **1992**, *33*, 1289.

30. (a) Mukaiyama, T.; Sagawa, Y.; Kobayashi, S., *Chem. Lett.* **1986**, 1821. (b) Inokuchi, T.; Kurokawa, Y.; Kusumoto, M.; Tanigawa, S.; Takagishi, S.; Torii, S., *Bull. Chem. Soc. Jpn.* **1989**, *62*, 3739. (c) see also ref 20c.

31. Yang, T.-K.; Hung, S.-M.; Lee, D.-S.; Hong, A.-W.; Cheng, C.-C., *Tetrahedron Lett.* **1989**, *30*, 4973.

32. Simpkins, N. S., *Tetrahedron* **1991**, *47*, 323.

33. Hippeli, C.; Reissig, H.-U., *Synthesis* **1987**, 77.

34. (a) Bélanger, J.; Landry, N. L.; Paré, J. R. J.; Jankowski, K., *J. Org. Chem.* **1982**, *47*, 3649. (b) Kirmse, W.; Mrotzeck, U., *Ber. Dtsch. Chem. Ges./Chem. Ber.* **1988**, *121*, 485. (c) Ishihara, T.; Shinjo, H.; Inoue, Y.; Ando, T., *J. Fluorine Chem.* **1983**, *22*, 1.

35. (a) Larsen, S. D., *J. Am. Chem. Soc.* **1988**, *110*, 5932. (b) Kirby, G. W.; McGregor, W. M., *J. Chem. Soc., Perkin Trans. 1* **1990**, 3175. (c) Rewunkel, J. G. M.; Zwanenberg, B., *Recl. Trav. Chim. Pays-Bas* **1990**, *101*, 190. (d) Herczegh, P.; Zsély, M.; Bognár, R., *Tetrahedron Lett.* **1986**, *27*, 1509.

36. (a) Jung, M. E.; Shishido, K.; Light, L.; Davis, L., *Tetrahedron Lett.* **1981**, *22*, 4607. (b) Hamley, P.; Holmes, A. B.; Kee, A.; Ladduwahetty, T.; Smith, D. F., *Synlett* **1991**, 29.

37. (a) Vacca, J. P., *Tetrahedron Lett.* **1985**, *26*, 1277. (b) Ryan, K. M.; Reamer, R. A.; Volante, R. P.; Shinkai, I., *Tetrahedron Lett.* **1987**, *28*, 2103. (c) Huff, J. R.; Baldwin, J. J.; de Solms, S. J.; Guare, J. P.; Hunt, C. A.; Randall, W. C.; Sanders, W. S.; Smith, S. J.; Vacca, J. P.; Zrada, M. M., *J. Med. Chem.* **1988**, *31*, 641.

38. (a) Ueda, Y.; Maynard, S. C., *Tetrahedron Lett.* **1985**, *26*, 6309. (b) Schrader, T.; Steglich, W., *Synthesis* **1990**, 1153. (c) Hartmann, P.; Obrecht, J.-P., *Synth. Commun.* **1988**, *18*, 553.

39. (a) Märkl, G.; Kallmünzer, A., *Tetrahedron Lett.* **1989**, *30*, 5245. (b) Markovskiii, L. N.; Romanenko, V. D.; Kachkovskaya, L. S., *J. Gen. Chem. USSR (Engl. Transl.)* **1985**, *55*, 2488.

40. Vartanyan, R. S.; Gyul'budagyan, A.; Khanamiryan, A. Kh.; Mkrtumyan, E. N.; Kazaryan, Zh. V., *Khim. Geterotsikl. Soedin.* **1991**, 783.

41. (a) Sano, T.; Toda, J.; Ohshima, T.; Tsuda, Y., *Chem. Pharm. Bull.* **1992**, *40*, 873. (b) Sano, T.; Toda, J.; Horiguchi, Y.; Imafuku, K.; Tsuda, Y., *Heterocycles* **1981**, *16*, 1463. (c) Sano, T.; Toda, J.; Tsuda, Y.; Yamaguchi, K.; Sakai, S.-I., *Chem. Pharm. Bull.* **1984**, *32*, 3255. (d) Demuth, M.; Pandey, B.; Wietfeld, B.; Said, H.; Viader, J., *Helv. Chim. Acta* **1988**, *71*, 1392. (e) de Almeida-Barbosa, L.-C.; Mann, J., *J. Chem. Soc., Perkin Trans. 1* **1992**, 337. (f) Dolbier, W. R., Jr.; Piedrahita, C.; Houk, K. N.; Strozier, R. W.; Gandour, R. W., *Tetrahedron Lett.* **1978**, 2231. (g) Hassner, A.; Naisdorf, S., *Tetrahedron Lett.* **1986**, *27*, 6389.

42. (a) Rigby, J. H.; Ateeq, H. S., *J. Am. Chem. Soc.* **1990**, *112*, 6442. (b) Rigby, J. H.; Ateeq, H. S.; Charles, N. R.; Cuisiat, S. V.; Ferguson, M. D.; Henshilwood, J. A.; Krueger, A. C.; Ogbu, C. O.; Short, K. M.; Heeg, M. J., *J. Am. Chem. Soc.* **1993**, *115*, 1382.

43. (a) Demuth, M.; Wietfeld, B.; Pandey, B.; Schaffner, K., *Angew. Chem., Int. Ed. Engl.* **1985**, *24*, 763. (b) Komatsu, M.; Suehiro, I.; Horiguchi, Y.; Kuwajima, I., *Synlett* **1991**, 771.

44. (a) Conia, J. M., *J. Chem. Res. (S)* **1978**, 182. (b) Girard, C.; Conia, J. M., *Tetrahedron Lett.* **1974**, 3333. (c) Olofson, R. A.; Lotts, K. D.; Barber, G. N., *Tetrahedron Lett.* **1976**, 3779. (d) Kunkel, E.; Reichelt, I.; Reissig, H.-U., *Liebigs Ann. Chem.* **1984**, 512. (e) Zschiesche, R.; Reissig, H.-U., *Liebigs Ann. Chem.* **1987**, 387. (f) Kuehne, M.; Pitner, J. B., *J. Org. Chem.* **1989**, *54*, 4553. (g) Shi, G.; Xu, Y., *J. Org. Chem.* **1990**, *55*, 3383. (h) de Meijere, A.; Schulz, T.-J.; Kostikov, R. R.; Graupner, F.; Murr, T.; Beifeldt, T., *Synthesis* **1991**, 547. (i) Cantrell, W. R., Jr.; Davies, H. M. L., *J. Org. Chem.* **1991**, *56*, 723.

45. (a) Natsume, M.; Muratake, H., *Tetrahedron Lett.* **1979**, 3477. (b) Conde-Frieboes, K.; Hoppe, D., *Synlett* **1990**, 99.

46. (a) Lüönd, R. M.; Cuomo, J.; Neier, R. W., *J. Org. Chem.* **1992**, *57*, 5005. (b) Herman, T.; Carlson, R., *Tetrahedron Lett.* **1989**, *30*, 3657. (c) Pennanen, S. I., *Synth. Commun.* **1985**, *15*, 865, 1063. (d) Takeda, K.; Ayabe, A.; Kawashima, H.; Harigaya, Y., *Tetrahedron Lett.* **1992**, *33*, 951. (e) Olofson, R. A.; Cuomo, J., *Tetrahedron Lett.* **1980**, *21*, 819.

Michael E. Jung
University of California, Los Angeles, CA, USA

2-Trimethylsilyloxyfuran[1]

[61550-02-5]   C$_7$H$_{12}$O$_2$Si   (MW 156.28)

InChI = 1S/C7H12O2Si/c1-10(2,3)9-7-5-4-6-8-7/h4-6H,1-3H3

InChIKey = ILBCDLOQMRDXLN-UHFFFAOYSA-N

(provides 5-substituted 2(5H)-furanones by alkylation,[2] aldolization,[3] and conjugate addition;[4] transforms quinones into furo-[3,2-b]benzofurans;[5] useful for the four-carbon elongation of sugars[6])

Alternate Name: TMSOF.

Physical Data: bp 44–46 °C/17 mmHg; *d* 0.93 g mL^{-1}.

Solubility: sol most organic solvents, e.g. CH$_2$Cl$_2$, Et$_2$O, benzene, THF, MeCN.

Form Supplied in: colorless liquid; commercially available (98% pure) but expensive.

Preparative Method: accessible by silylation of 2(5*H*)-fura-none,[1c,5b] which is obtained at very low cost by oxidation of furfural.

Handling, Storage, and Precautions: flammable liquid; sensitive to moisture. To avoid hydrolysis, it should be kept under Ar at −18 °C or below. For best results, reactions should be performed under strictly anhydrous conditions in aprotic solvents. Use in a fume hood.

Original Commentary

John Boukouvalas
Université Laval, Québec, Canada

Synthetic Applications. In recent years, 2-trimethylsilyloxy-furan has emerged as a keystone for the rapid assembly of a wide variety of 5-substituted 2(5*H*)-furanones. This is readily achieved by reaction with an electrophile and desilylation. Electrophilic attack usually occurs exclusively at C-5;[3–10] products arising by C-3 attack are observed only rarely.[2c,11] Consequently, this electron-rich furan is viewed as a stable, readily accessible synthon for the γ-anion of 2(5*H*)-furanone (**1**).

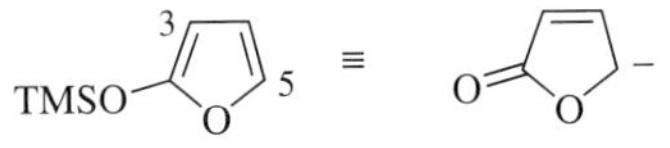

$$\textbf{(1)}$$

Carbon–Heteroatom Bond Formation. Bromination of 2-trimethylsilyloxyfurans affords the corresponding 5-bromo-2(5*H*)-furanones in quantitative yields (eq 1).[1a] Similarly, reaction with lead(II) acetate provides 5-acetoxy-2(5*H*)-furanones (eq 2).[8] Oxidation can also be effected by using iodosylbenzene in conjunction with a nucleophile, such as *p*-toluenesulfonic acid, and boron trifluoride etherate (eq 3).[9] Replacement of TsOH by azidotrimethylsilane in eq 3 provides 5-azido-2(5*H*)-furanone.[9]

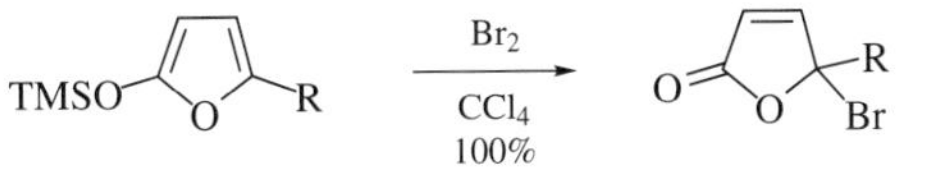

$$\text{(1)}$$

R = H or Me

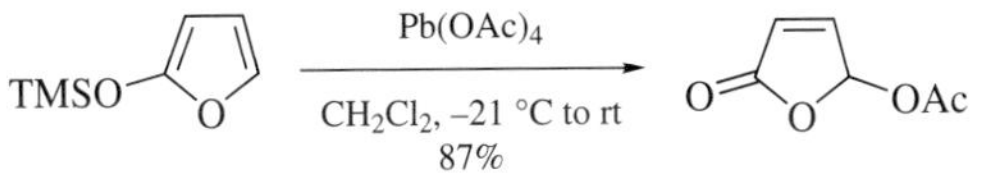

$$\text{(2)}$$

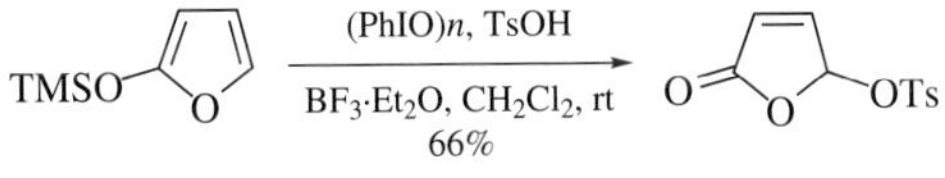

$$\text{(3)}$$

Alkylation. Upon activation with silver(I) trifluoroacetate, primary alkyl iodides and allylic bromides are readily transformed into 5-alkyl- and 5-allyl-2(5*H*)-furanones, respectively (eq 4).[2] Significantly, typical S_N1-unreactive halides such as ethyl iodoacetate work equally well.[2b] This method has served as a basis for short syntheses of several natural products,[13] including eldanolide[2a] and the Whisky lactones.[2c]

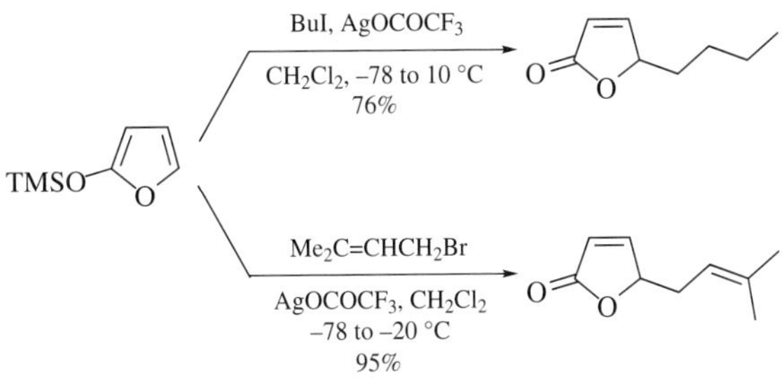

$$\text{(4)}$$

Alternatively, 5-allyl-2(5*H*)-furanones can be obtained from TMSOF and allylic acetates using lithium perchlorate in ether.[13] 5-Propargyl-2(5*H*)-furanones are prepared by coupling with hexacarbonyldicobalt complexes of propargylium cations (eq 5).[10] Reaction of the intermediate complex with cerium(IV) ammonium nitrate removes the cobalt. Fluoroalkylation is achieved by using bis(fluoroalkanoyl) peroxides in 1,1,2-trichloro-1,2,2-trifluoroethane (Freon 113) (eq 6).[14] This process is believed to involve combination of a fluoroalkyl radical with the furan-derived radical cation.

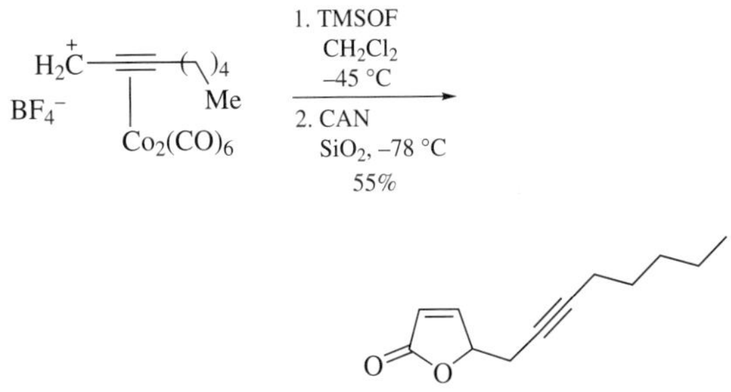

$$\text{(5)}$$

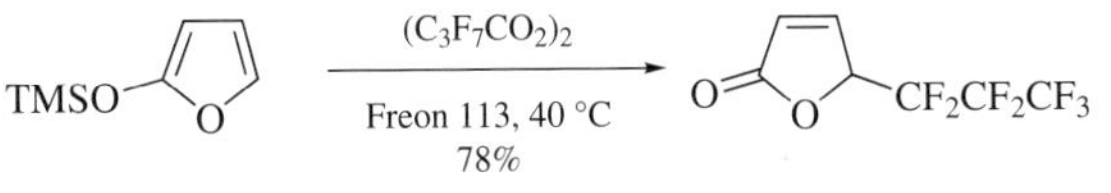

$$\text{(6)}$$

Aldol-Type Reactions. The facile condensation of 2-trialkyl-silyloxyfurans with carbonyl compounds[3] and their derivatives[7a] has spawned many synthetic applications.[15–19] Aldol products derived from aldehydes, ketones, or acetals can be readily transformed into 4-ylidenebutenolides.[3a] Dehydration of 5-(1′-hydroxyalkyl)-2(5*H*)-furanones is achieved by using acetic anhydride and 4-pyrrolidinopyridine (PPY) in triethylamine (eq 7).[3a]

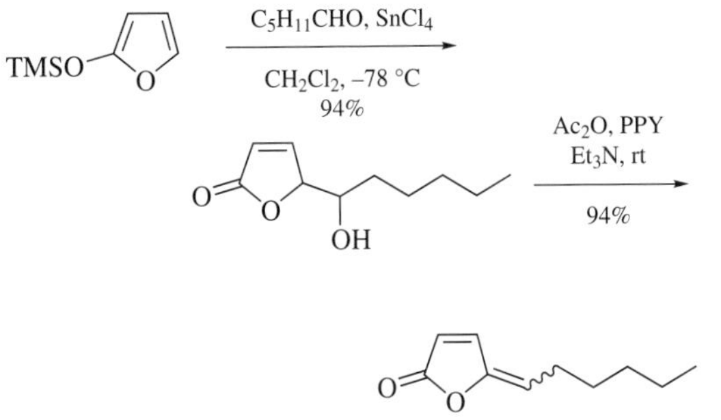

$$\text{(7)}$$

$$(E){:}(Z) = 1{:}1$$

The diastereoselectivity of the reaction of TMSOF with aldehydes can be controlled by the choice of catalyst. Lewis acids such as triethylsilyl trifluoromethanesulfonate (TESOTf), BF₃·OEt₂, and tin(IV) chloride favor *syn*-selectivity, while the use of fluoride ion confers *anti*-selectivity (eq 8).[3b] In both situations, open transition states are involved.[3b] Impressive diastereofacial preference is encountered in the aldol reaction of TMSOF with chiral aldehydes (eq 9).[6a] This chemistry is particularly useful for the four-carbon elongation of sugars.[6]

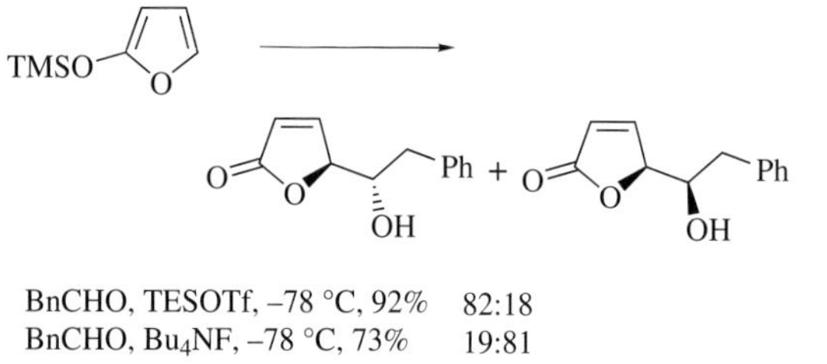

(8)

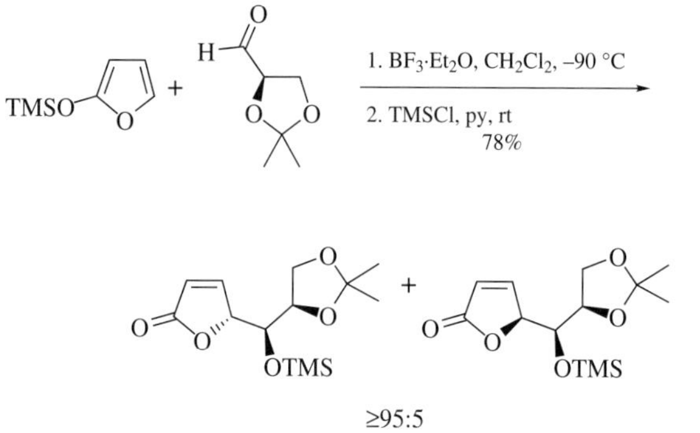

BnCHO, TESOTf, –78 °C, 92%   82:18
BnCHO, Bu₄NF, –78 °C, 73%   19:81

(9)

≥95:5

Ketones,[16] orthoamides,[17] and *N*-acyl-*N,O*-acetals[18] also undergo diastereoselective condensation with TMSOF in the presence of Lewis acids. Thus 1-benzyloxycarbonyl-2-ethoxypyrrolidine leads preferentially to the *syn* product (eq 10).[19]

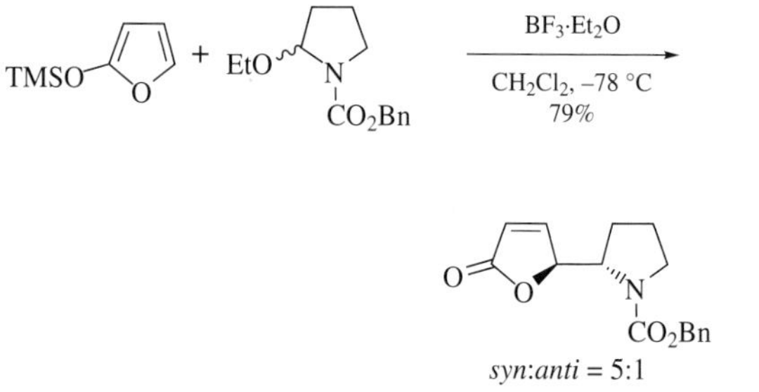

(10)

syn:anti = 5:1

Conjugate Addition. In comparison to aldol condensation, the conjugate addition of 2-trialkylsilyloxyfurans to enones has received only scant attention.[8,20] Yet some interesting applications have already surfaced. Especially noteworthy is the high level of stereochemical discrimination in the few recorded cases.[21] For example, full diastereoselectivity is observed in the fluoride ion-induced addition of TMSOF to a chalcone (eq 11).[4]

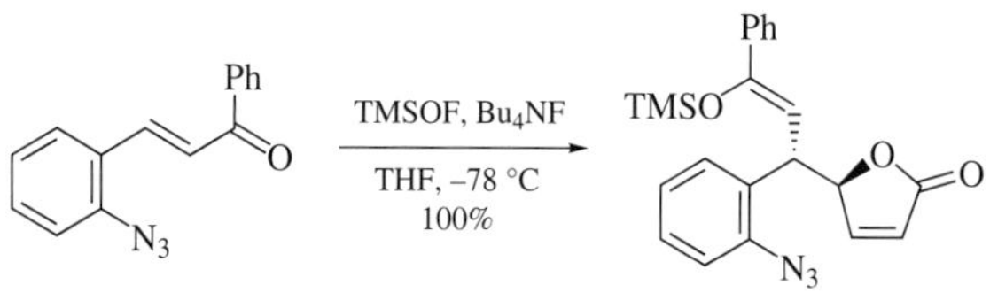

(11)

Activated quinones undergo uncatalyzed reaction with TMSOF to produce furo[3,2-*b*]benzofurans (eq 12).[22] Furofuran annulation is explicable by initial conjugate addition of the furan, subsequent aromatization, and eventual intramolecular conjugate

addition of the resulting hydroxy group onto the butenolide.[5] This protocol is of relevance to the synthesis of pyranonaphthoquinone antibiotics.[23]

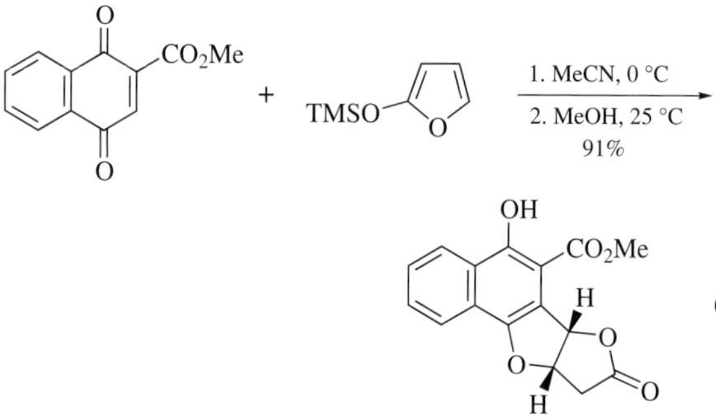

(12)

Diels–Alder Reaction. In view of the propensity of furans to undergo [4 + 2] cycloadditions,[24] 2-trialkylsilyloxyfurans are potentially useful for the rapid construction of various phenol derivatives.[25] Although TMSOF performs poorly in Diels–Alder reactions, some of its more substituted analogs are highly effective.[26] An illustration is provided by the short synthesis of a naturally occurring hydroxyphthalide (**2**) via cycloaddition of 3,5-dimethyl-2-(trimethylsilyloxy)furan with maleic anhydride (eq 13).[26] Unlike TMSOF, the more stable 2-pivaloyloxyfuran is strongly recommended for similar transformations.[27]

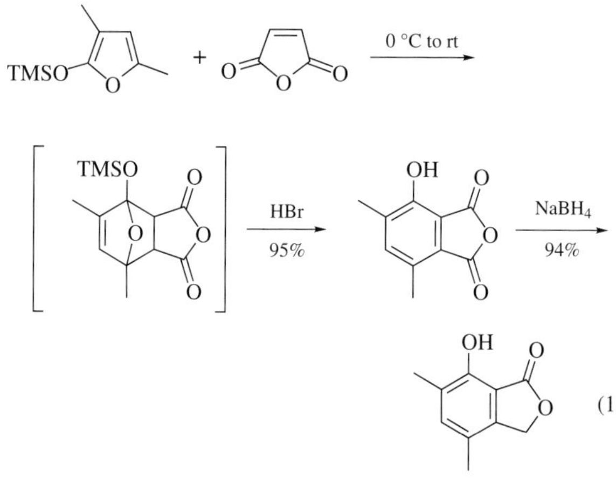

(13)

(**2**)

First Update

Jonathan Sperry & Margaret A. Brimble
University of Auckland, Auckland, New Zealand

Carbon-heteroatom bond formation. The γ-thiocyanation of 2-trimethylsilyloxyfuran (TMSOF) is achieved using the reagent combination of (dichloroiodo)benzene and lead(II) thiocyanate (eq 14).[31]

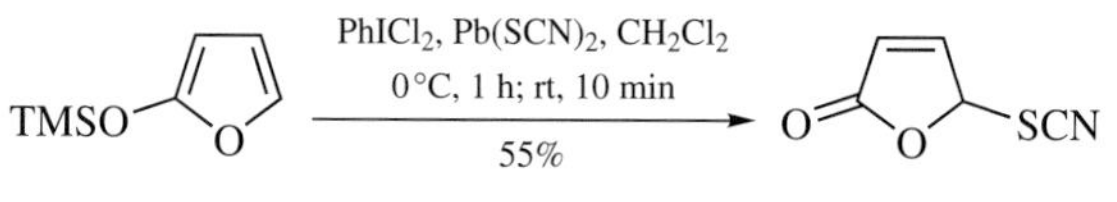

(14)

Alkylation. The Lewis acid promoted stereoselective $C5$ alkylation of TMSOF is possible with acetoxytetrahydrofurans,[32a–d] β-lactam acetates,[33] 1,4-dioxane-2-one acetates,[34] dihydroartemisenin acetate,[35] and homochiral ortho-ester and oxazolidine derivatives.[36] Lithium cobalt-bis-dicarbollide is an effective catalyst for the alkylation of TMSOF with isophorol acetate (eq 15).[37]

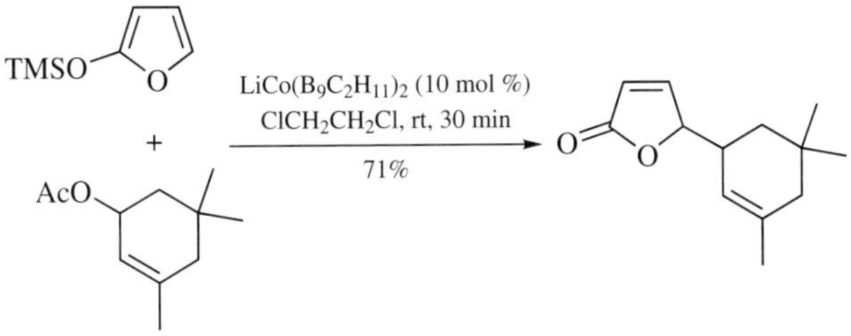

(15)

Lewis acid catalyzed regioselective ring opening of a [2.2.1] bicyclic acetal with TMSOF occurs, furnishing a 2,5-disubstituted tetrahydrofuran adduct as a mixture of *erythro* and *threo* isomers (eq 16).[38]

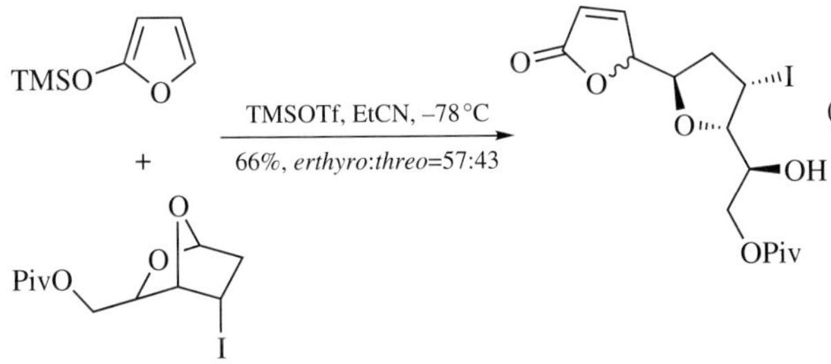

(16)

TMSOF acts as the final component in a *p*-toluenesulfonyl chloride mediated coupling of two alkyl vinyl ether units, providing a general procedure for the synthesis of a polyfunctional γ-butenolide, albeit as a mixture of four diastereomers. The reaction proceeds via ring expansion of an episulfonium ion (ESI) followed by ring opening of the resulting thiophanium ion (TSI) with TMSOF (eq 17).[39]

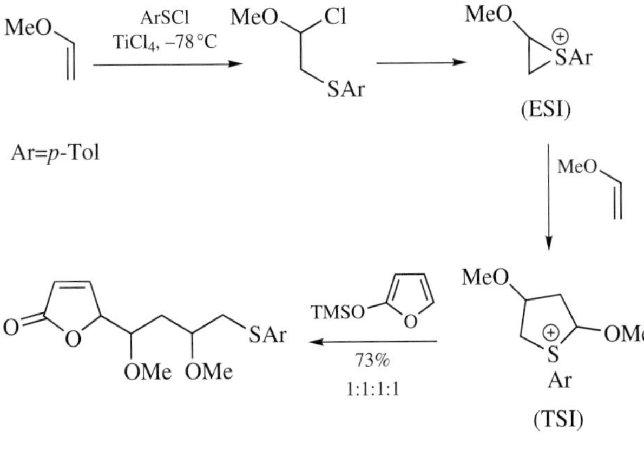

(17)

Allylic alcohols derived from 1,4-dioxene can be used to alkylate TMSOF in the presence of lithium perchlorate. The reaction proceeds through an oxocarbenium ion, with one example exhibiting an equal amount of side product arising from the alkylation at the tertiary carbon (eq 18).[40]

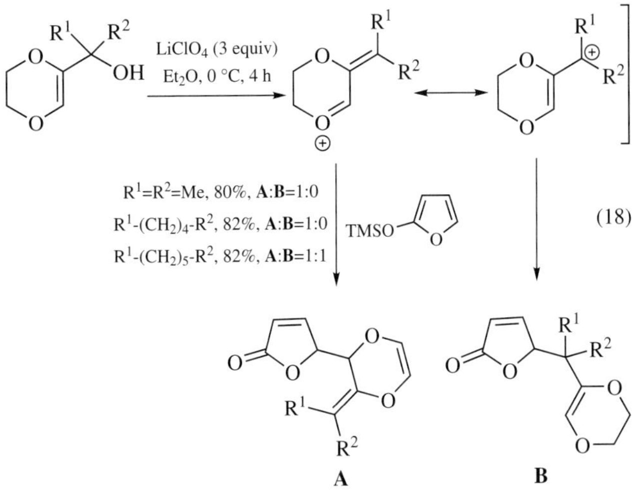

$R^1=R^2=Me$, 80%, **A:B**=1:0
R^1-$(CH_2)_4$-R^2, 82%, **A:B**=1:0
R^1-$(CH_2)_5$-R^2, 82%, **A:B**=1:1

(18)

A diastereoselective construction of γ-butenolides can be achieved via an organocatalytic allylic alkylation of Morita-Baylis-Hillman acetates with TMSOF in the presence of triphenylphosphine. In all cases involving methyl vinyl ketone derived allylic acetates (R^1=Me), a small amount of a regioisomer **B** is formed, an observation that is greatly reduced with the use of methyl acrylate (R^1=OMe) derived allylic acetates (eq 19, Table 1). Preliminary studies on controlling the absolute stereochemical outcome of the reaction have shown promising results using the chiral auxiliary 8-phenylmenthol (R^1).[41]

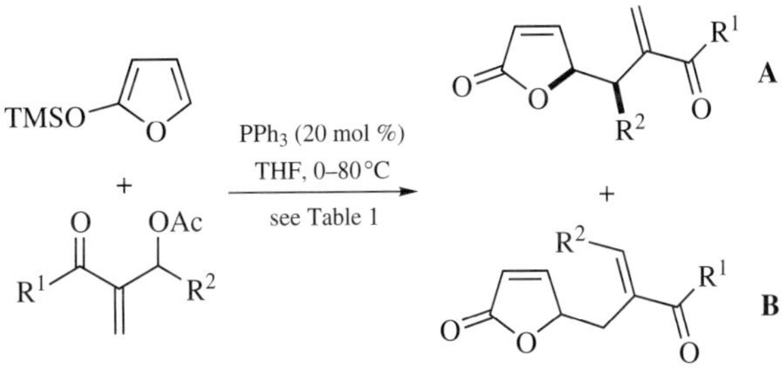

(19)

Table 1

R^1	R^2	Yield A(%)	dr A	Yield B(%)
Me	4-O_2N-C_6H_4	88	>95:5	9
Me	Ph	80	>95:5	5
Me	^{i}Pr	63	>95:5	7
Me	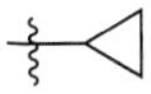 cyclopropyl	80	20:1	5
Me	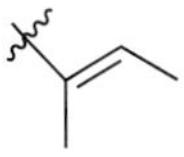 2-methyl-1-propenyl	45	>95:5	10
Me	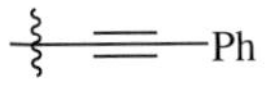 —≡—Ph	88	24:1	5
OMe	4-O_2N-C_6H_4	84	>95:5	6
OMe	Ph	86	>95:5	1
OMe	^{i}Pr	67	2.8:1	0
OMe	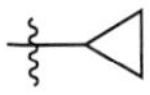 cyclopropyl	83	3.5:1	0
OMe	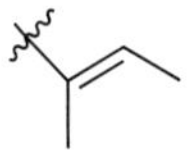 2-methyl-1-propenyl	62	>95:5	0
OMe	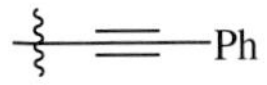 —≡—Ph	94	>95:5	0

The zinc dichloride promoted condensation of TMSOF with various *ortho*-esters followed by base induced cyclization of the resulting adducts provides a convenient two step route to functionalized spirocyclic butenolides (eq 20). Various other silyloxyfurans can also be used in this process.[42]

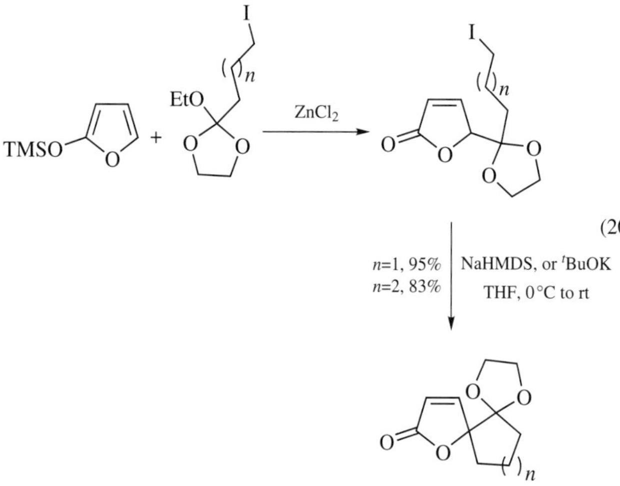

$$(20)$$

$$n=1, 95\%$$
$$n=2, 83\%$$
NaHMDS, or ^{t}BuOK
THF, 0 °C to rt

Alkenylation/Arylation.   The palladium-catalyzed coupling of TMSOF with hypervalent iodonium salts under mild aqueous conditions provides a route to γ-substituted β,γ-unsaturated butyrolactones in excellent yield (eq 21).[43]

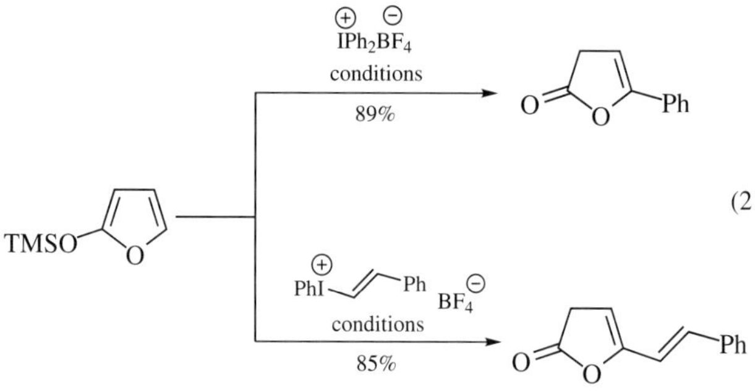

$$(21)$$

conditions—Pd(OAc)$_2$ (2 mol %), DME-H$_2$O (4:1), 30 °C, 30 min

The palladium-catalyzed arylation of TMSOF with triarylantimony diacetates provides a facile route to 5-arylcyclopenten-2-ones in excellent yield (eq 22).[44]

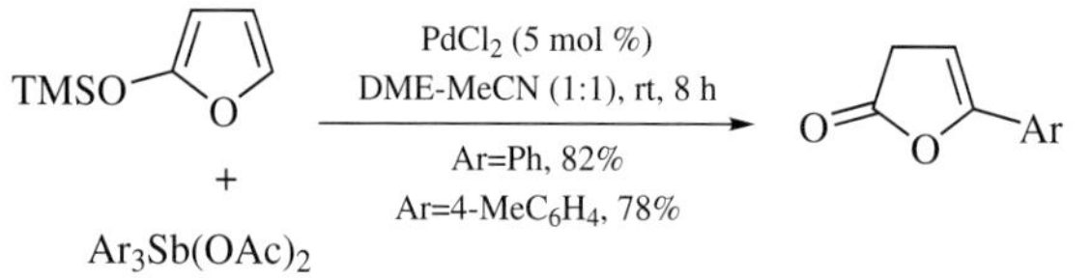

$$(22)$$

Aldol-type Reactions.   The aldol condensation of TMSOF with carbonyl compounds and their derivatives has given rise to a vast array of synthetic applications with a variety of conditions used to catalyze the reaction. The absence of chiral catalysts often leads to lack of selectivity with a mixture of stereoisomers being obtained.[45a–e] Dehydration of the products obtained from these aldol reactions can be achieved using potassium acetate in acetic acid furnishing a mixture of *E*- and *Z*-trienes, key intermediates for

the total synthesis and structural confirmation of goniobutenolides A and B (eq 23).[46] *E*-selectivity for similar dehydration products has been observed using pyridine and triflic anhydride or Burgess' reagent, with increased *Z*-selectivity observed when mesyl chloride is employed for the dehydration step.[47] The use of Burgess' reagent to enhance *E*-selectivity is observed during a total synthesis of (+)-pyrenolide D.[48]

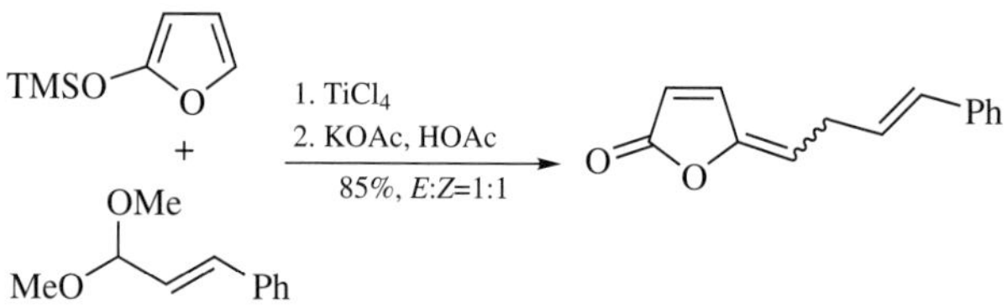

$$(23)$$

Mukaiyama aldol coupling of acyclic *O,S*-acetals with TMSOF provides a key step in the enantioselective syntheses of goniobutenolides A and B (eq 24). Elimination of benzenethiol is accomplished with silver fluoride and pyridine.[49] The Mukaiyama aldol reaction between TMSOF and mucaholic acid represents the first reaction of this type involving the latter reagent.[50]

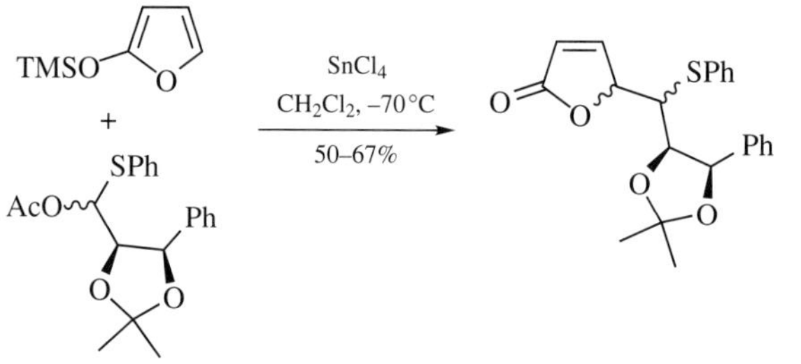

$$(24)$$

Various homochiral γ-substituted butenolides are accessible via highly diastereoselective homologations of TMSOF with simple aldehyde or imine precursors. Employing such butenolides as chiral templates is exploited during the synthesis of higher-carbon sugars,[51a,b] *C*-glycosyl α-amino acids,[52a,b] azasugars,[53a–c] enantiomeric tetrahydroxyquinolizidines,[54] the total syntheses of (+)-2,8,8a-tri-*epi*-swainsonine, (−)-1-*epi*-swainsonine,[55] and (+)-goniofurfurone.[56a,b] A highly stereoselective nucleophilic addition of TMSOF to *N*-glyoxyloyl (2*R*)-bornane-10,2-sultam provides a substituted butenolide in excellent yield with a high level of diastereoselectivity (eq 25).[57]

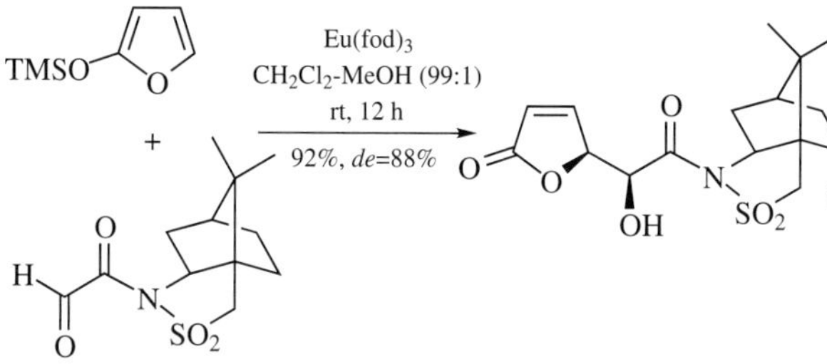

$$(25)$$

The Lewis acid promoted (typically boron trifluoride) condensation of cyclic iminium ions with TMSOF provides a facile route to a variety of substituted butenolide products. In this manner, the Lewis acid promoted condensation of pyrrolidine derived iminium ions with TMSOF provides the desired *threo*-product as the major isomer (eq 26).[58] This step is pivotal in the synthesis of a tachykinin NK-2 receptor antagonist,[59] the C1-C32 fragment of aza-solamin[60] and various biologically active indolizidinones.[61]

A variety of substituted variants of TMSOF have also been used in this method.[62a,b] One step decarboxylation-oxidation-alkylation methods involving TMSOF and an iminium ion intermediate generates a *threo*-pyrrolidine-furanone as the major product, with (diacetoxyiodo)benzene (DIB) being the oxidant of choice. Increased yields are obtained when *N*-acyl groups are employed on the pyrrolidine ring (eq 27).[63a,b] TMSOF can also undergo successful Lewis acid promoted addition to iminium ions generated from five to seven membered ring *N,O*-acetal TMS ethers. Decreased diastereoselectivity is observed with increased ring size (eq 28).[64] The Lewis acid-free reaction of TMSOF with a variety of preformed iminium salts does not furnish the expected aminoalkylated products, rather a series of *γ*-arylidenebutenolides arising from spontaneous deamination are formed (eq 29).[65] Benzotriazole precursors are also successfully used as acyliminium ion precursors in the reaction with TMSOF,[66] while an eco-friendly solvent-free method for iminium ion generation has also been reported.[67] The addition of TMSOF to acyclic imines provides a key step in the formal synthesis of thymine polyoxin C.[68] TMSOF has also been seen to undergo addition to acyclic fluorinated aldimines with excellent diastereocontrol.[69] Products arising from an imine addition side reaction are observed during an attempted aza Diels-Alder reaction between TMSOF and *N*-diethylphosphoryl imines.[70]

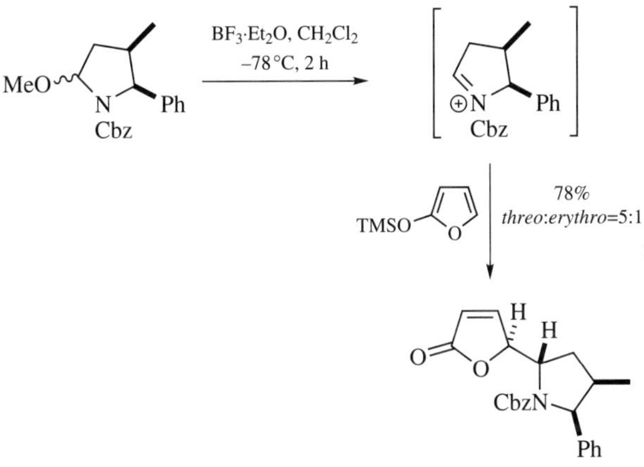

enantioselectivity can be achieved by portionwise addition of aldehyde and TMSOF to the preformed complex in diethyl ether. The addition of various chiral activators has been shown to increase yield, diastereo- and enantioselectivity even further.[71a–c] This method was particularly useful during the structural confirmation of *iso*-cladospolide B as it allowed stereoselective syntheses of all four possible diastereomers in enantiopure form.[72] A contribution toward the total synthesis of caribenolide also incorporates this optimized method.[73] Cationic (*R,R*)-Cr(Salen) complexes catalyze the enantioselective addition of TMSOF to aldehydes furnishing 5-substituted butenolides, although diastereoselectivity is modest. The high enantioselectivity is dependent upon the addition of water and secondary alcohols to the reaction.[74a,b]

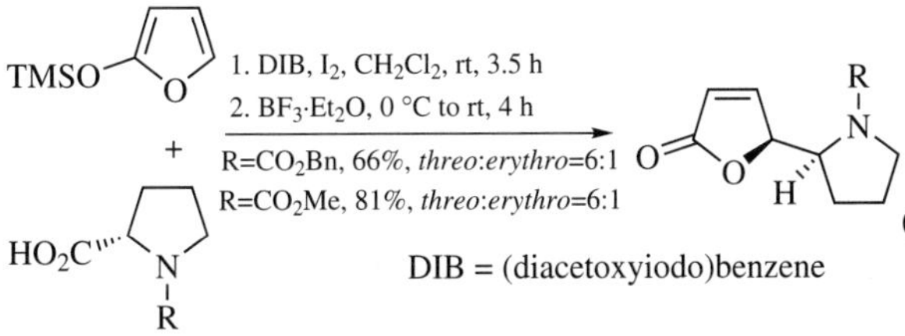

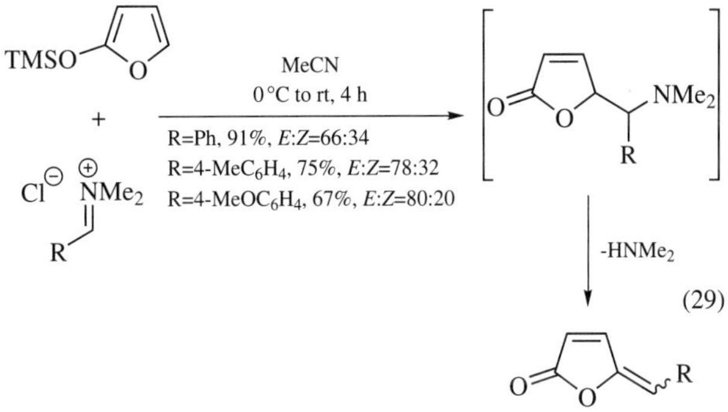

$$(29)$$

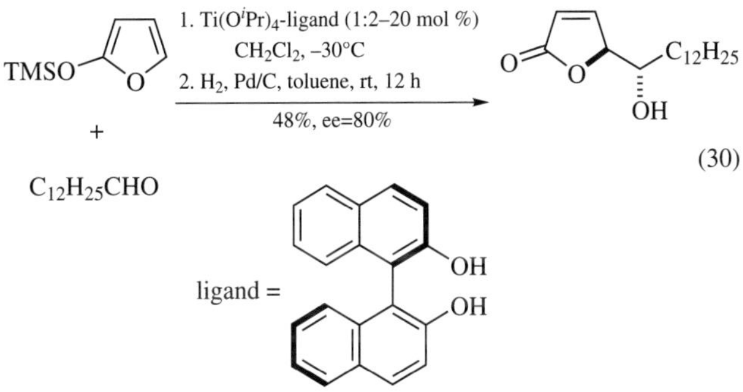

$$(26)$$

$$(27)$$

DIB = (diacetoxyiodo)benzene

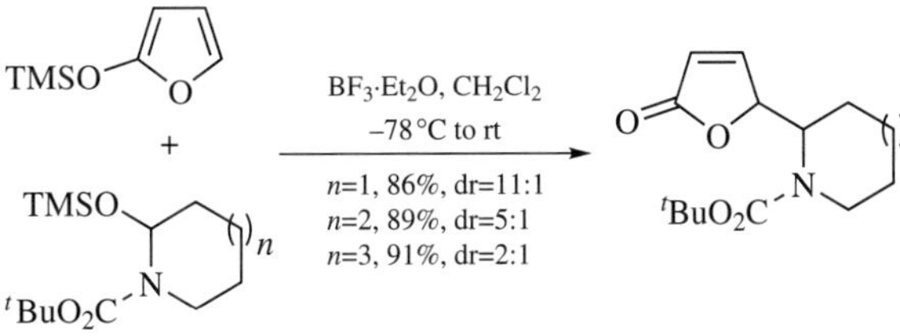

$$(28)$$

The first example of TMSOF participating in an enantioselective Mukaiyama aldol reaction involves the addition of a binol-Ti(IV) complex. The method is exemplified by the asymmetric aldol reaction between TMSOF and tridecanal, a key step during the total synthesis of (+)-muricatacin (eq 30). An improvement in

$$(30)$$

C_2-symmetric bis(oxazolinyl)pyridine (pybox)-Cu(II)-complexes have been shown to catalyze the enantioselective Mukaiyama aldol reaction between (benzyloxy)acetaldehyde and TMSOF in excellent yield, diastereo- and enantioselectivity (eq 31).[75]

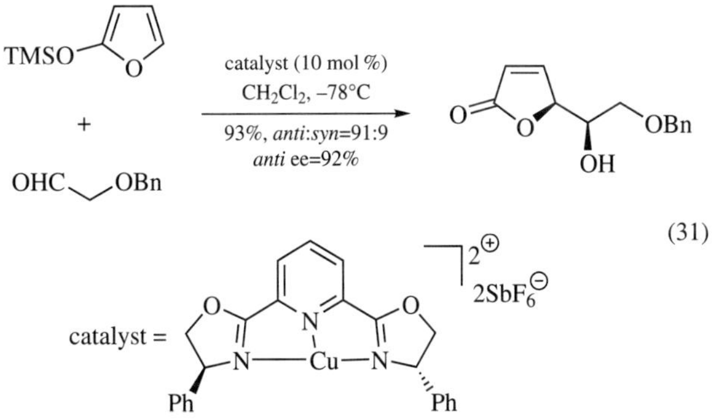

$$(31)$$

Conjugate Addition (1,4-Addition, Mukaiyama–Michael). The Michael addition of TMSOF to activated 1,4-benzoquinones provides an efficient entry into the furobenzofuran (eq 32) and furonaphthofuran ring system,[76] a pivotal step in the construction of various pyranonaphthoquinone antibiotics. The stereochemistry of the bridgehead protons can be controlled by use of a quinone bearing a chiral auxiliary.[77a–c] These conjugate addition reactions have been used to great effect in the synthesis of various monomeric[78a–e] and dimeric[79a,b] pyranonaphthoquinone antibiotics. Heterocyclic cage compounds can be stereochemically synthesized by the conjugate addition of TMSOF to enantiopure [(S)R]-[(p-tolylsulfinyl)methyl]-p-quinols in the presence of TBAF.[80]

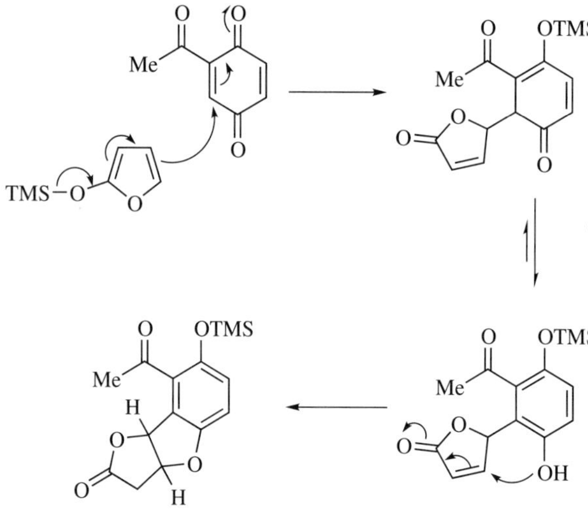

$$(32)$$

The Michael addition of TMSOF to oxazolidinone substituted enones provides a basis for an asymmetric variant of this reaction that has attracted considerable interest. The Lewis acid promoted Michael addition of TMSOF to 3-[(E)-2-butenoyl]-1,3-oxazolidin-2-one has been examined, with Cu(II)-bis(oxazoline) complexes exhibiting excellent enantioselectivity with moderate to good anti-selectivity. The addition of hexafluoroisopropanol is essential for the excellent enantioselectivity observed (eq 33). The anti-selectivity increases somewhat by changing the catalytic system to (ScOTf)$_3$-3,3'-bis(diethylaminomethyl)-1,1'-bi-2-naphthol, but with a significant decrease in enantioselectivity. 3-Methyl-TMSOF is also used in this reaction.[81a,b] This method is prevalent in a synthesis of trans-whisky lactone.[82] Catalysis of the same reaction with a R,R-diPh-pybox ligand in the presence of various trivalent (Sc, La, Eu, Yb) and tetravalent (Ce) cations confers total anti-selectivity, quantitative yield, and excellent enantioselectivity (eq 34), with related lanthanide-based pybox ligands showing similarly excellent stereocontrol.[83a,b] Binaphthyldiimine-Ni(II) complexes are also viable catalysts for this reaction, furnishing the Michael products with excellent enantioselectivity and anti-selectivity.[84]

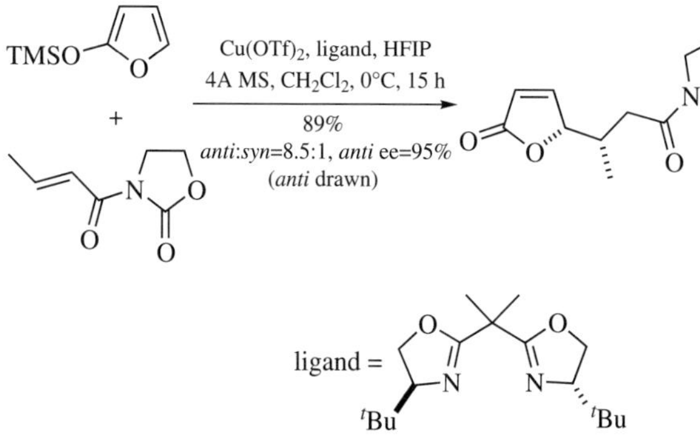

$$(33)$$

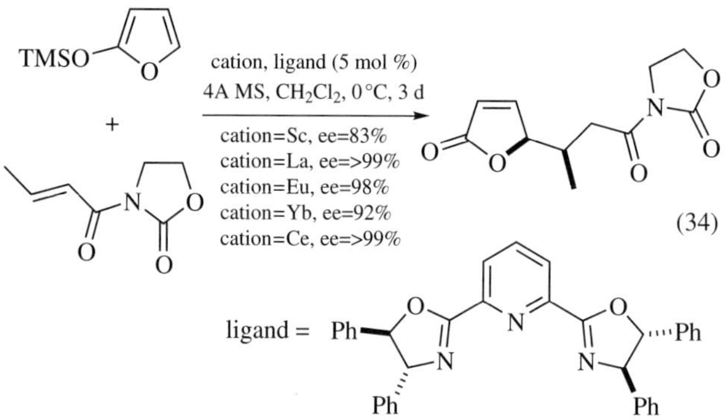

$$(34)$$

The first report of an enantioselective organocatalytic Mukaiyama–Michael reaction involves the use of an amine catalyst during the 1,4-addition of TMSOF to the electron deficient olefin (E)-but-2-enal. The catalyst is employed as its dinitrobenzoic acid (DNBA) salt (eq 35). Several other substituted TMSOFs are used in this reaction with similar levels of success.[85] A formal enantioselective synthesis of (+)-compactin incorporates this method.[86]

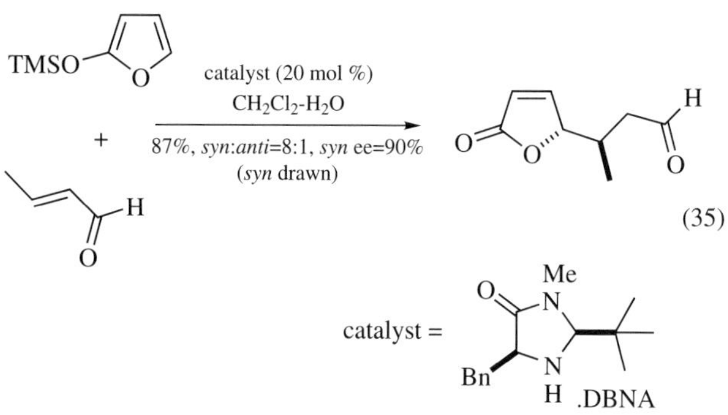

$$(35)$$

An asymmetric uncatalyzed Mukaiyama–Michael reaction between TMSOF and enantiopure tungsten carbene complexes provides rapid asymmetric access to complex γ-butenolides with excellent anti and face selectivity (eq 36).[87]

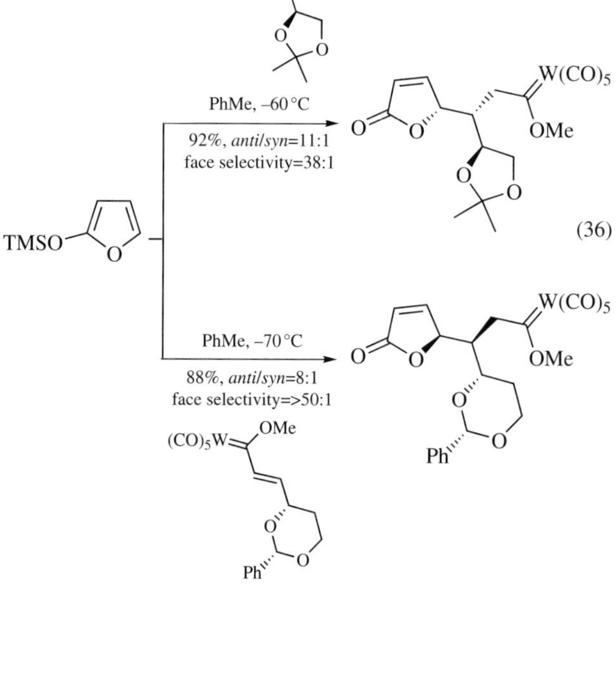

$$(36)$$

Cycloadditions. The electron donating nature of the trimethylsilyloxy group in TMSOF promotes regiospecific ring opening of the 7-oxa bridge of the Diels-Alder cycloadduct obtained with hexafluoro-2-butyne, furnishing 2,3-bis(trifluoromethyl)-4-(trimethylsilyl)phenol in a single step (eq 37).[88]

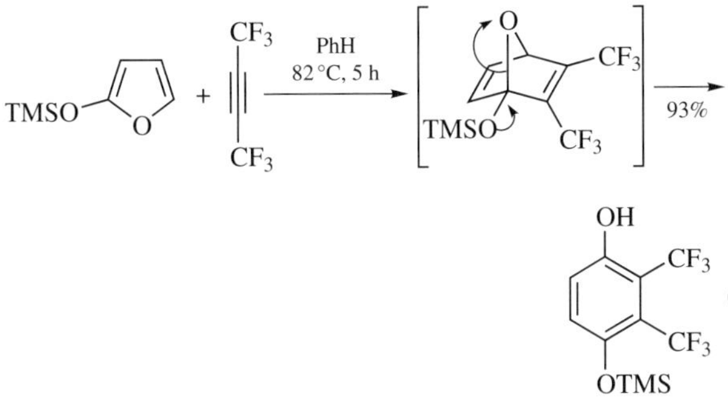

Trimethylsilyl trifluoromethanesulfonate (TMSOTF) promotes the formal [1+3]-dipolar cycloaddition of nitrones with TMSOF, providing a simple route to tetrahydrofuro[2,3-d]isoxazol(2H)-ones after fluoride ion mediated cyclization of the initially formed butenolides. Depending upon the substituents on the nitrone, complementary stereochemical courses of the cycloaddition can be achieved (eq 38, Table 2).[89] The scope of the reaction has been broadened by the use of several Lewis acid catalysts and chiral nitrones derived from lactaldehyde. As with previous examples, the choice of a benzyl substituent on the nitrogen drives the formation of the *anti*-product.[90] The synthetic versatility of this method is demonstrated by a high yielding conversion of the tetrahydrofuro[2,3-d]isoxazol(2H)-ones into various 5-amino-2,5-dideoxy-heptono-1,5-lactams by straightforward hydrogenation over Pearlman's catalyst observed during an expeditious synthesis of *rac*-fagomine.[91a,b] The use of 5-membered cyclic nitrones derived from (R,R)-tartaric acid followed by hydrogenation provides the basis for a short synthesis of a partially protected 1,2,7,8-indolizidinetetrol.[92] The stereochemical outcome of the cycloaddition reaction between TMSOF and N-gulosylnitrone is controlled by the presence of the N-chiral auxiliary on the nitrone, providing a key step during the successful synthesis of the C-terminal amino acid component of nikkomycin Bz.[93]

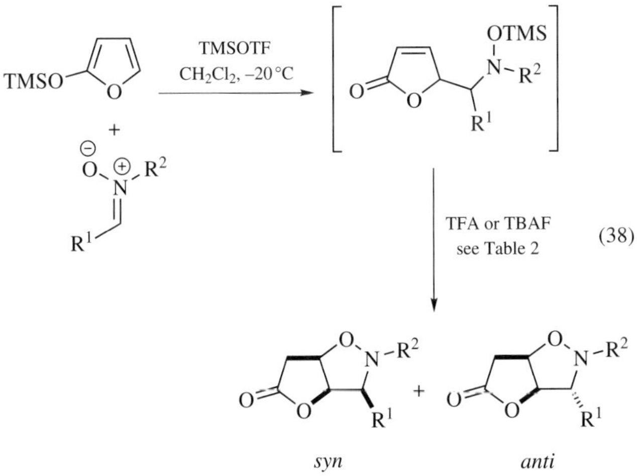

The 1,3-dimethyloxyallyl cation, generated in situ by treatment of 2,4-dibromo-3-pentenone with copper and sodium iodide,

Table 2

R^1	R^2	Yield (%)	syn:anti
Ph	Me	96	84:16
2-Thienyl	Me	84	96:4
1-Naphthyl	Me	85	77:23
^{i}Pr	Me	55	45:55
^{i}Bu	Me	83	56:44
Et	Bn	65	23:77
^{i}Pr	Bn	80	12:88
^{t}Bu	Bn	46	29:71
Ph	Bn	90	82:18
2-Thienyl	Bn	48	95:5

undergoes an *endo*-selective [4 + 3]-cycloaddition reaction with TMSOF generating a C-1 functionalized 8-oxabicyclo[3.2.1]-6-octen-3-one in excellent yield (eq 39).[94a,b]

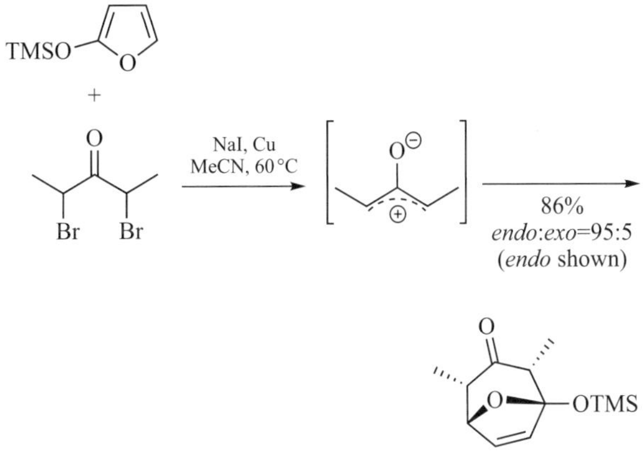

Ring Opening. Oxidation of TMSOF with dimethyldioxirane (DMDO) followed by ring opening of the resulting hydroxybutenolide with a Wittig reagent furnishes exclusively the (2Z, 4E) monoacid product (eq 40).[95] Ring opening of TMSOF is also achieved by treatment with aryl diazoacetates under rhodium(II)-catalysis, affording the corresponding 6-aryl-6-oxo-(2Z, 4E)-hexadienoic acids (eq 41).[96]

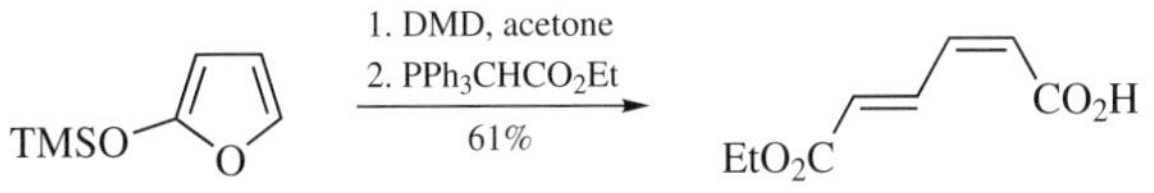

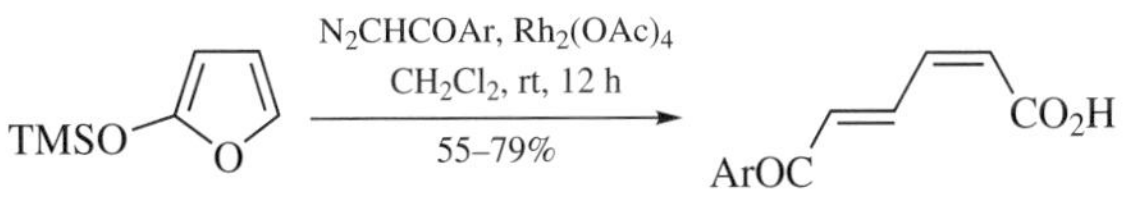

Ar=Ph, 2-thienyl, 4-MeOC$_6$H$_4$, 2-naphthyl

Related Reagents. Several other 2-trialkylsilyloxyfurans have served as building blocks for the synthesis of natural products[28a,b] and analogs.[28c] The commercially available (4-methoxy-2-(trimethylsilyloxy)furan (3) is especially useful for preparing 5-substituted methyl tetronates.[29] 5-Ethylthio-2-(trimethylsilyloxy)furan (4)[21a] has been employed for the synthesis of mitomycins.[21] The easily prepared 2-(t-butyldimethylsilyloxy)-3-methylfuran (5)[12a] is the reagent of choice for constructing furanosesquiterpenes[12] and spirocyclic alkaloids of the *Ergot* family.[30]

Other related reagents are γ-butyrolactone, dihydro-5-(hydroxymethyl)-2(3H)-furanone, 2,5-dihydro-2,5-dimethoxyfuran, furan, 5-lithio-2,3-dihydrofuran, and β-vinyl-α,β-butenolide.

---

1. (a) Yoshii, E.; Koizumi, T.; Kitatsuji, E.; Kawazoe, T.; Kaneko, T., *Heterocycles* **1976**, *4*, 1663. (b) Brownbridge, P., *Synthesis* **1983**, 85. (c) Jefford, C. W.; Jaggi, D.; Sledeski, A. W.; Boukouvalas, J. In *Studies in Natural Products Chemistry*; Atta-ur-Rahman, Ed.; Elsevier: Amsterdam, 1989; Vol 3, p 157.

2. (a) Jefford, C. W.; Sledeski, A. W.; Boukouvalas, J., *Tetrahedron Lett.* **1987**, *28*, 949. (b) Jefford, C. W.; Sledeski, A. W.; Boukouvalas, J., *J. Chem. Soc., Chem. Commun.* **1988**, 364. (c) Jefford, C. W.; Sledeski, A. W.; Boukouvalas, J., *Helv. Chim. Acta* **1989**, *72*, 1362.

3. (a) Asaoka, M.; Yanagida, N.; Ishibashi, K.; Takei, H., *Tetrahedron Lett.* **1981**, *22*, 4269. (b) Jefford, C. W.; Jaggi, D.; Boukouvalas, J., *Tetrahedron Lett.* **1987**, *28*, 4037.

4. Fukuyama, T.; Goto, S., *Tetrahedron Lett.* **1989**, *30*, 6491.

5. (a) Brimble, M. A.; Gibson, J. J.; Baker, R.; Brimble, M. T.; Kee, A. A.; O'Mahony, M. J., *Tetrahedron Lett.* **1987**, *28*, 4891. (b) Brimble, M. A.; Brimble, M. T.; Gibson, J. J., *J. Chem. Soc., Perkin Trans. 1* **1989**, 179.

6. (a) Casiraghi, G.; Colombo, L.; Rassu, G.; Spanu, P., *J. Org. Chem.* **1991**, *56*, 2135. (b) Casiraghi, G.; Pinna, L.; Rassu, G.; Spanu, P.; Ulgheri, F., *Tetrahedron: Asymmetry* **1993**, *4*, 681.

7. (a) Asaoka, M.; Sugimura, N.; Takei, H., *Bull. Chem. Soc. Jpn.* **1979**, *52*, 1953. (b) Fiorenza, M.; Ricci, A.; Romanelli, M. N.; Taddei, M.; Dembech, P.; Seconi, G., *Heterocycles* **1982**, *19*, 2327. (c) Kubota, T.; Iijima, M.; Tanaka, T., *Tetrahedron Lett.* **1992**, *33*, 1351.

8. Asaoka, M.; Yanagida, N.; Sugimura, N.; Takei, H., *Bull. Chem. Soc. Jpn.* **1980**, *53*, 1061.

9. Moriarty, R. M.; Vaid, R. K.; Hopkins, T. E.; Vaid, B. K.; Tuncay, A., *Tetrahedron Lett.* **1989**, *30*, 3019.

10. Stuart, J. G.; Nicholas, K. M., *Heterocycles* **1991**, *32*, 949.

11. (a) Brown, D. W.; Campbell, M. M.; Taylor, A. P.; Zhang, X., *Tetrahedron Lett.* **1987**, *28*, 985. (b) Loreto, M. A.; Pellacani, L.; Tardella, P. A., *Tetrahedron Lett.* **1989**, *30*, 5025.

12. (a) Jefford, C. W.; Sledeski, A. W.; Rossier, J.-C.; Boukouvalas, J., *Tetrahedron Lett.* **1990**, *31*, 5741. (b) Jefford, C. W.; Huang, P.-Z.; Rossier, J.-C.; Sledeski, A. W.; Boukouvalas, J., *Synlett* **1990**, 745.

13. Pearson, W. H.; Schkeryantz, J. M., *J. Org. Chem.* **1992**, *57*, 2986.

14. Yoshida, M.; Imai, R.; Komatsu, Y.; Morinaga, Y.; Kamigata, N.; Iyoda, M., *J. Chem. Soc., Perkin Trans. 1* **1993**, 501.

15. (a) Asaoka, M.; Yanagida, N.; Takei, H., *Tetrahedron Lett.* **1980**, *21*, 4611. (b) Rassu, G.; Pinna, L.; Spanu, P.; Culeddu, N.; Casiraghi, G.; Fava, G. G.; Ferrari, M. B.; Pelosi, G., *Tetrahedron* **1992**, *48*, 727. (c) Fukuyama, T.; Xu, L.; Goto, S., *J. Am. Chem. Soc.* **1992**, *114*, 383.

16. Jefford, C. W.; Jaggi, D.; Bernardinelli, G.; Boukouvalas, J., *Tetrahedron Lett.* **1987**, *28*, 4041.

17. Bernardi, A.; Cardani, S.; Carugo, O.; Colombo, L.; Scolastico, C.; Villa, R., *Tetrahedron Lett.* **1990**, *31*, 2779.

18. Harding, K. E.; Coleman, M. T.; Liu, L. T., *Tetrahedron Lett.* **1991**, *32*, 3795.

19. Martin, S. F.; Corbett, J. W., *Synthesis* **1992**, 55.

20. (a) Aso, M.; Hayakawa, K.; Kanematsu, K., *J. Org. Chem.* **1989**, *54*, 5597. (b) Boivin, J.; Crépon, E.; Zard, S. Z., *Tetrahedron Lett.* **1991**, *32*, 199.

21. (a) Fukuyama, T.; Yang, L., *J. Am. Chem. Soc.* **1987**, *109*, 7881. (b) Fukuyama, T.; Yang, L., *J. Am. Chem. Soc.* **1989**, *111*, 8303.

22. Brimble, M. A.; Spicer, J. A., *Aust. J. Chem.* **1991**, *44*, 197.

23. (a) Brimble, M. A.; Stuart, S. J., *J. Chem. Soc., Perkin Trans. 1* **1990**, 881. (b) Brimble, M. A.; Nairn, M. R., *J. Chem. Soc., Perkin Trans. 1* **1992**, 579. (c) Brimble, M. A.; Lynds, S. M., *J. Chem. Soc., Perkin Trans. 1* **1994**, 493.

24. Wong, H. N. C.; Ng, T.-K.; Wong, T.-Y.; Xing, Y. D., *Heterocycles* **1984**, *22*, 875.

25. (a) Garver, L. C.; van Tamelen, E. E., *J. Am. Chem. Soc.* **1982**, *104*, 867. See also: (b) Gravel, D.; Deziel, R.; Brisse, F.; Hechler, L., *Can. J. Chem.* **1981**, *59*, 2997. (c) Hanessian, S.; Beaulieu, P.; Dubé, D., *Tetrahedron Lett.* **1986**, *27*, 5071.

26. Asaoka, M.; Miyake, K.; Takei, H., *Chem. Lett.* **1977**, 167.

27. Näsman, J.-A. H., *Synthesis* **1985**, 788.

28. (a) Boukouvalas, J.; Maltais, F., *Tetrahedron Lett.* **1994**, *35*, 5769. (b) Boukouvalas, J.; Maltais, F.; Lachance, N., *Tetrahedron Lett.* **1994**, *35*, 7897. (c) Kawada, K.; Kitagawa, O.; Taguchi, T.; Hanzawa, Y.; Kobayashi, Y.; Itaka, Y., *Chem. Pharm. Bull.* **1985**, *33*, 4216.

29. (a) Pelter, A.; Al-Bayati, R.; Lewis, W., *Tetrahedron Lett.* **1982**, *23*, 353. (b) Pelter, A.; Al-Bayati, R. I. H.; Ayoub, M. T.; Lewis, W.; Pardasani, P.; Hansel, R., *J. Chem. Soc., Perkin Trans. 1* **1987**, 717. (c) Pelter, A.; Ward, R. S.; Sitir, A., *Tetrahedron: Asymmetry* **1995**, *5*, 1745.

30. Martin, S. F.; Liras, S., *J. Am. Chem. Soc.* **1993**, *115*, 10450.

31. Prakash, O.; Harpreet, K.; Hitesh, B.; Neena, R.; Singh, S. P.; Moriarty, R. M., *J. Org. Chem.* **2001**, *66*, 2019.

32. (a) Figadere, B.; Chaboche, C.; Peyrat, J. F.; Cave, A., *Tetrahedron Lett.* **1993**, *34*, 8093. (b) Figadere, B.; Peyrat, J. F.; Cave, A., *J. Org. Chem.* **1997**, *62*, 3428. (c) Hanessian, S.; Grillo, T. A., *J. Org. Chem.* **1998**, *63*, 1049. (d) Hanessian, S.; Giroux, S.; Buffat, M., *Org. Lett.* **2005**, *7*, 3989.

33. Hanessian, S.; Reddy, G. B., *Bioorg. Med. Chem. Lett.* **1994**, *4*, 2285.

34. Fujioka, H.; Matsunaga, N.; Kitagawa, H.; Nagatomi, Y.; Kondo, M.; Kita, Y., *Tetrahedron: Asymmetry* **1995**, *6*, 2117.

35. Ma, J.; Katz, E.; Kyle, D. E.; Ziffer, H., *J. Med. Chem.* **2000**, *43*, 4228.

36. Pelter, A.; Ward, R. S.; Sirit, A., *Tetrahedron: Asymmetry* **1994**, *5*, 1745.

37. Grieco, P. A.; DuBay, W. J.; Todd, L. J., *Tetrahedron Lett.* **1996**, *37*, 8707.

38. Lemee, L.; Jegou, A.; Veyrieres, A., *Tetrahedron Lett.* **1999**, *40*, 2761.

39. Lazareva, M. I.; Kryschenko, Y. K.; Hayford, A.; Lovdahl, M.; Caple, R.; Smit, W. A., *Tetrahedron Lett.* **1998**, *39*, 1083.

40. Hanna, I.; Ricard, L., *Tetrahedron Lett.* **1999**, *40*, 863.

41. Cho, C.-W.; Krische, M. J., *Angew. Chem., Int. Ed.* **2004**, *116*, 6857; *Angew. Chem., Int. Ed.* **2004**, *43*, 6689.

42. Maulide, N.; Marko, I. E., *Org. Lett.* **2006**, *8*, 3705.

43. Kang, S.-K.; Yamaguchi, T.; Ho, P.-S.; Kim, W.-Y.; Yoon, S.-K., *Tetrahedron Lett.* **1997**, *38*, 1947.

44. Kang, S.-K.; Ryu, H.-C.; Hong, Y.-T., *J. Chem. Soc., Perkin. Trans. 1* **2000**, *20*, 3350.

45. (a) Holstein Wagner, S.; Lundt, I., *J. Chem. Soc., Perkin. Trans. 1* **2001**, *8*, 780. (b) Li, Y.; Snyder, L. B.; Langley, D. R., *Bioorg. Med. Chem. Lett.* **2003**, *13*, 3261. (c) Seong-Chol, B.; Yong, K.; Mi-Young, K.; Byun-Zun, A., *Arch. Pharm. Res.* **2004**, *27*, 485. (d) Takao, K.; Yasui, H.; Yamamoto, S.; Sasaki, D.; Kawasaki, S.; Watanabe, G.; Tadano, K., *J. Org. Chem.* **2004**, *69*, 8789. (e) Acocella, M. R.; De Rosa, M.; Massa, A.; Palombi, L.; Villano, R.; Scettri, A., *Tetrahedron* **2005**, *61*, 4091.

46. Xu, D.; Sharpless, B. K., *Tetrahedron Lett.* **1994**, *35*, 4685.

47. von der Ohe, F.; Bruckner, R., *New. J. Chem.* **2000**, *24*, 659.

48. Engstrom, K. M.; Mendoza, M. R.; Navarro-Villalobos, M.; Gin, D. Y., *Angew. Chem., Int. Ed.* **2001**, *113*, 1162; *Angew. Chem., Int. Ed.* **2001**, *40*, 1128.

49. Ko, S. Y.; Lerpiniere, J., *Tetrahedron Lett.* **1995**, *36*, 2101.

50. Angell, P.; Zhang, J.; Belmont, D.; Curran, T.; Davidson, J. G., *Tetrahedron Lett.* **2005**, *46*, 2029.

51. (a) Casiraghi, G.; Colombo, L.; Rassu, G.; Spanu, P., *J. Org. Chem.* **1990**, *55*, 2565. (b) Rassu, G.; Spanu, P.; Casiraghi, G.; Pinna, L., *Tetrahedron* **1991**, *47*, 8025.

52. (a) Mashima, K.; Matsumura, Y.; Kusano, K.; Kumobayashi, H.; Sayo, N.; Hori, Y.; Ishizaki, T.; Akutagawa, S.; Takaya, H., *J. Chem. Soc., Chem. Commun.* **1991**, 609. (b) Casiraghi, G.; Colombo, L.; Rassu, G.; Spanu, P., *J. Org. Chem.* **1991**, *56*, 6523.

53. (a) Rassu, G.; Pinna, L.; Spanu, P.; Culeddu, N.; Casiraghi, G.; Gasparri, F.; Belicchi, F. M.; Pelosi, G., *Tetrahedron* **1992**, *48*, 727. (b) Casiraghi, G.; Rassu, G.; Spanu, P.; Pinna, L., *J. Org. Chem.* **1992**, *57*, 3760. (c) Rassu, G.; Casiraghi, G.; Spanu, P.; Pinna, L.; Fava, G. G.; Belichi, F. M.; Pelosi, G., *Tetrahedron: Asymmetry* **1992**, *3*, 1035.

54. Rassu, G.; Casiraghi, G.; Pinna, L.; Spanu, P.; Ulgheri, F.; Cornia, M.; Zanardi, F., *Tetrahedron* **1993**, *49*, 6627.

55. Casiraghi, G.; Rassu, G.; Spanu, P.; Pinna, L.; Ulgheri, F., *J. Org. Chem.* **1993**, *58*, 3397.

56. (a) Mukia, C.; Kim, I. J.; Hanaoka, M., *Tetrahedron Lett.* **1993**, *34*, 6081. (b) Mukia, C.; Hirai, S.; Kim, I. J.; Kido, M.; Hanaoka, M., *Tetrahedron* **1996**, *52*, 6547.

57. Bauer, T., *Tetrahedron: Asymmetry* **1996**, *7*, 981.

58. Hanessian, S.; Raghavan, S., *Bioorg. Med. Chem. Lett.* **1994**, *4*, 1697.

59. Hanessian, S.; McNaughton-Smith, G., *Bioorg. Med. Chem. Lett.* **1996**, *6*, 1567.

60. Pichon, M.; Hocquemiller, R.; Figadere, B., *Tetrahedron Lett.* **1999**, *40*, 8567.

61. Hanessian, S.; Therrien, E.; Granberg, K.; Nilsson, I., *Bioorg. Med. Chem. Lett.* **2002**, *12*, 2907.

62. (a) Morimoto, Y.; Nishida, K.; Hayashi, Y.; Shirahama, H., *Tetrahedron Lett.* **1993**, *34*, 5773. (b) Martin, S. F.; Barr, K. J., *J. Am. Chem. Soc.* **1996**, *118*, 3299.

63. (a) Boto, A.; Hernandez, R.; Suarez, E., *Tetrahedron Lett.* **2000**; *41*, 2899. (b) Boto, A.; Hernandez, R.; Suarez, E., *J. Org. Chem.* **2000**; *65*, 4930.

64. Suh, Y.-G.; Kim, S.-H.; Jung, J.-K.; Shin, D.-Y., *Tetrahedron Lett.* **2002**, *43*, 3165.

65. Piper, S.; Risch, N., *Arkivoc.* **2003**, *1*, 86.

66. Meester, W. J. N.; van Maarseveen, J. H.; Kirchsteiger, K.; Hermkens, P. H. H.; Schoemaker, H. E.; Hiemstra, H.; Rutjes, F. P. J. T., *Arkivoc.* **2004**, *2*, 122.

67. Tranchant, M.-J.; Moine, C.; Ben Othman, R.; Bousquet, T.; Othman, M.; Dalla, V., *Tetrahedron Lett.* **2006**, *47*, 4477.

68. Harding, K. E.; Southard, J. M., *Tetrahedron: Asymmetry* **2005**, *16*, 1845.

69. Spanedda, M. V.; Ourevitch, M.; Crousse, B.; Begue, J.-P.; Bonnet-Delphon, D., *Tetrahedron Lett.* **2004**, *45*, 5023.

70. Di Bari, L.; Guillarme, S.; Hermitage, S.; Howard, J. A. K.; Jay, D. A.; Pescitelli, G.; Whiting, A.; Yufit, D. S., *Synlett* **2004**, 708.

71. (a) Szlosek, M.; Franck, X.; Figadere, B.; Cave, A., *J. Org. Chem.* **1998**, *63*, 5169. (b) Szlosek, M.; Peyrat, J.-F.; Chaboche, C.; Franck, X.; Hocquemiller, R.; Figadere, B., *New J. Chem.* **2000**, *24*, 337. (c) Szlosek, M.; Figadere, B., *Angew. Chem., Int. Ed.* **2000**, *112*, 1869; *Angew. Chem., Int. Ed.* **2000**, *39*, 1799.

72. Franck, X.; Vaz Araujo, M. E.; Jullian, J.-C.; Hocquemiller, R.; Figadere, B., *Tetrahedron Lett.* **2001**, *42*, 2801.

73. Jalce, G.; Franck, X.; Seon-Meniel, B.; Hocquemiller, R.; Figadere, B., *Tetrahedron Lett.* **2006**, *47*, 5905.

74. (a) Matsuoka, Y.; Irie, R.; Katsuki, T., *Chem. Lett.* **2003**, *32*, 584. (b) Onitsuka, S.; Matsuoka, Y.; Irie, R.; Katsuki, T., *Chem. Lett.* **2003**, *32*, 974.

75. Evans, D. A.; Kozlowski, M. C.; Murry, J. A.; Burgey, C. S.; Campos, K. R.; Connell, B. T.; Staples, R. J., *J. Am. Chem. Soc.* **1999**, *121*, 669.

76. Brimble, M. A.; Elliot, R. J. R., *Tetrahedron* **1997**, *53*, 7715.

77. (a) Brimble, M. A.; Duncalf, L. J.; Reid, D. C. W., *Tetrahedron: Asymmetry* **1995**, *6*, 263. (b) Brimble, M. A.; McEwan, J. F.; Turner, P., *Tetrahedron: Asymmetry* **1998**, *9*, 1257. (c) Carreno, M. C.; Ruano,

J. L. G.; Urbano, A.; Remor, C. Z.; Arroyo, Y., *Tetrahedron: Asymmetry* **1999**, *10*, 4357.

78. (a) Brimble, M. A.; Phythian, S. J.; Prabaharan, H., *J. Chem. Soc., Perkin. Trans. 1* **1995**, *22*, 2855. (b) Brimble, M. A.; Duncalf, L. J.; Phythian, S. J., *Tetrahedron Lett.* **1995**, *36*, 9209. (c) Brimble, M. A.; Brenstrum, T. J., *J. Chem. Soc., Perkin. Trans. 1* **2001**, *14*, 1624. (d) Brimble, M. A.; Davey, R. M.; McLeod, M. D., *Synlett* **2002**, 1318. (e) Brimble, M. A.; Davey, R. M.; McLeod, M. D.; Murphy, M., *Org. Biomol. Chem.* **2003** 1690.

79. (a) Brimble, M. A.; Neville, D.; Duncalf, L. J., *Tetrahedron Lett.* **1998**, *39*, 5647. (b) Brimble, M. A.; Lai, M. Y. H., *Org. Biomol. Chem.* **2003**, 4227.

80. Carreno, M. C.; Luzon, C. G.; Ribagorda, M., *Chem. Eur. J.* **2002**, *8*, 208.

81. (a) Kitajima, H.; Katsuki, T., *Synlett* **1997**, 568. (b) Kitajima, H.; Katsuji, I.; Katsuki, T., *Tetrahedron* **1997**, *53*, 17015.

82. Nishikori, H.; Ito, K.; Katsuki, T., *Tetrahedron: Asymmetry* **1998**, *9*, 1165.

83. (a) Desimoni, G.; Faita, G.; Filippone, S.; Mella, M.; Zampori, M. G.; Zema, M., *Tetrahedron* **2001**, *57*, 10203. (b) Desimoni, G.; Faita, G.; Guala, M.; Laurenti, A.; Mella, M., *Chem. Eur. J.* **2005**, *11*, 3816.

84. Suga, H.; Kitamura, T.; Kakehi, A.; Baba, T., *J. Chem. Soc., Chem. Commun.* **2004**, 1414.

85. Brown, S. P.; Goodwin, N. C.; MacMillan, D. W. C., *J. Am. Chem. Soc.* **2003**, *125*, 1192.

86. Robichaud, J.; Tremblay, F., *Org. Lett.* **2006**, *8*, 597.

87. Barluenga, J.; de Prado, A.; Santamaria, J.; Tomàs, M., *Angew. Chem.* **2005**, *117*, 6741; *Angew. Chem., Int. Ed.* **2005**, *44*, 6583.

88. Zhu, G.-D.; Staeger, M. A.; Boyd, S. A., *Org Lett.* **2000**, *2*, 3345.

89. Camiletti, C.; Poletti, L.; Trombini, C., *J. Org. Chem.* **1994**, *59*, 6843.

90. Castellari, C.; Lombardo, M.; Pietropaolo, G.; Trombini, C., *Tetrahedron: Asymmetry* **1996**, *7*, 1059.

91. (a) Degiorgis, F.; Lombardo, M.; Trombini, C., *Tetrahedron* **1997**, *53*, 11721. (b) Degiorgis, F.; Lombardo, M.; Trombini, C., *Synthesis* **1997**, 1243.

92. Lombardo, M.; Trombini, C., *Tetrahedron* **1999**, *56*, 323.

93. Mita, N.; Tamura, O.; Ishibashi, H.; Sakamoto, M., *Org. Lett.* **2002**, *4*, 1111.

94. (a) Montaña, A. M.; Ribes, S.; Grima, P. M.; García, F., *Chem. Lett.* **1997**, *26*, 847. (b) Montaña, A. M.; Ribes, S.; Grima, P. M.; García, F.; Solans, X.; Font-Bardia, M., *Tetrahedron* **1997**, *53*, 11669.

95. Adger, B. J.; Barrett, C.; Brennan, J.; McGuigan, P.; McKervey, M. A.; Tarbit, B., *J. Chem. Soc., Chem. Commun.* **1993**, 1220.

96. Shieh, P. C.; Ong, C. W., *Tetrahedron* **2001**, *57*, 7303.

(Z)-2-(Trimethylsilyloxy)vinyllithium

[78108-48-2]   C$_5$H$_{11}$LiOSi   (MW 122.19)

InChI = 1S/C5H11OSi.Li/c1-5-6-7(2,3)4;/h1,5H,2-4H3;

InChIKey = INZLCJPRKXVKAB-UHFFFAOYSA-N

(vinylogation of aldehydes and ketones under mild conditions; preparation of β-methoxy aldehydes and β-silyl silyl enol ethers)

Solubility: sol Et$_2$O, THF.

Preparative Methods: bromine–lithium exchange starting from (Z)-2-bromo-1-(trimethylsilyloxy)ethylene in dry Et$_2$O

(0.15 M) and *t*-Butyllithium at −70 °C under inert atmosphere.[1] One equivalent of *t*-BuLi is sufficient if (Z)-(trimethylsilyloxy)-vinyllithium is condensed with reactive electrophilic reagents which are added 10–20 min after the end of the addition of *t*-BuLi.[1,2] In other cases, two equivalents of *t*-BuLi are needed, the second equivalent is used to destroy *t*-BuBr formed during the exchange[3] and the electrophilic reagents are added after the end of addition of *t*-BuLi.

Handling, Storage, and Precautions: when prepared with 2 equiv of *t*-BuLi the reagent is stable in Et₂O over 20 h at −70 °C; 1 from −60 °C it begins to decompose with formation of acetylene.[4]

General Discussion. (Z)-(Trimethylsilyloxy)vinyllithium (**1**) has essentially been used for a one-pot vinylogation of carbonyl compounds. Condensation of (**1**) with aldehydes and ketones (**2**) occurs readily at low temperature (−70 °C); then treatment of the reaction mixture with an acidic solution in mild conditions (1 N hydrochloric acid, 0 °C to rt) produces the α,β-unsaturated aldehydes (**3**), without double bond migration. The (E) isomers are obtained alone (from aldehydic compounds) or very predominantly (from ketonic compounds). The intermediate adducts, γ-hydroxy enol ethers (**4**), can be isolated by slightly basic mild hydrolysis (eq 1).[1]

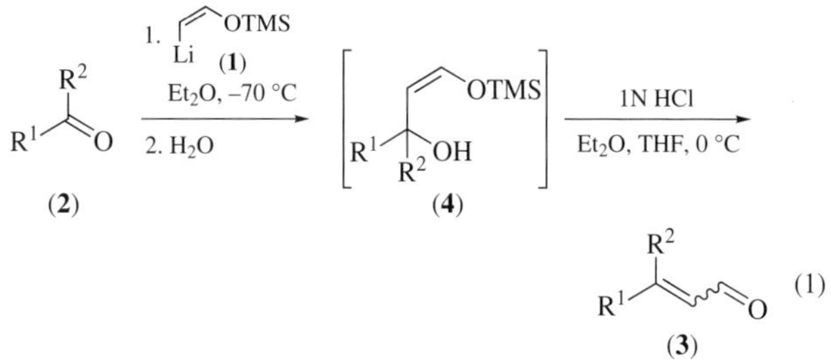

As condensation with α,β-unsaturated compounds occurs exclusively in a 1,2-fashion,[1,2] this reagent has been used in terpene synthesis. A synthesis of retinal (**3b**) in three steps from β-ionone (**2a**) (via β-ionylideneacetaldehyde and C₁₈ ketone) was reported using reagent **1** for the first and third steps (eq 2).[5]

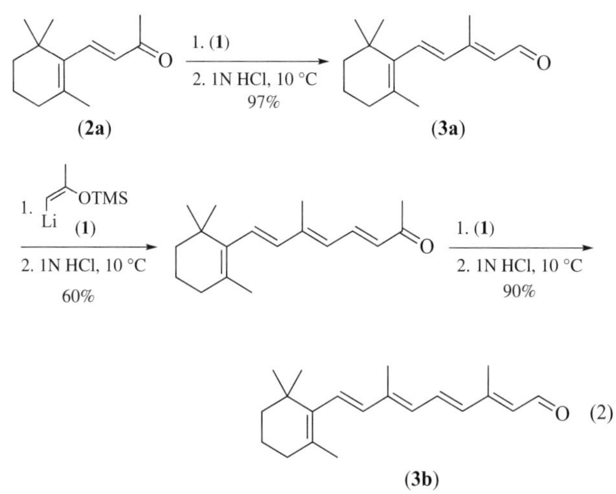

For the step (**2a**) → (**3a**), reagent (**1**) competes favorably with other classical vinylogation reagents: formylmethylenetriphenylphosphorane (0%);[6] lithium cyclohexylvinylamide (2 steps, 34%);[7] lithio salt of 5,6-dihydro-2,4,4,6-tetramethyl-1,3(4H)-oxazine (3 steps, 54%);[8] lithium *t*-butyl-2-trimethylsilylvinylamide (1 step, 88%)[9] (see Duhamel et al.[2] for other comparative studies).

Thus (Z)-(trimethylsilyloxy)vinyllithium (**1**) is an excellent reagent for the preparation of building blocks such as (**4**), (**5a**), and (**5b**) (eq 3), which are useful for short convergent syntheses of terpenoid compounds (eqs 4 and 5).[10]

Condensation of acetylacetaldehyde dimethyl acetal with reagent (**1**) leads to the δ-aldolacetal (**3c**) in 90% yield, which has been transformed by classical methods into bromoacetal (**4**) and bromo enol ethers (**5a,5b**) (eq 3).[10]

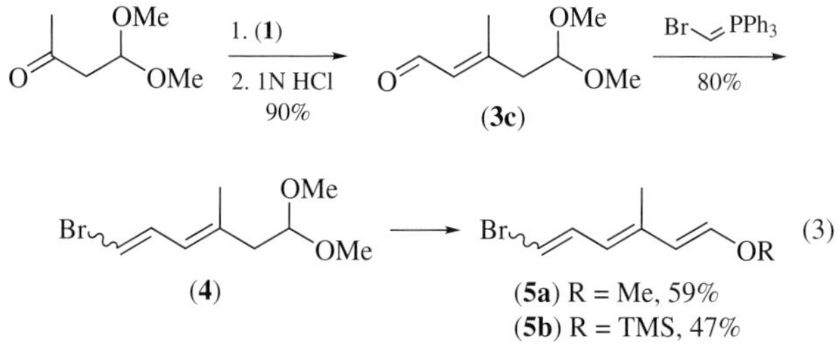

For the first time a one-step synthesis of retinal (**3b**) from β-ionone (**2a**) was reported, using reagents (**5a**) and (**5b**) (eq 4),[10] whereas two-step syntheses of dehydrocitral from acetone (60%), *pseudo*-retinal from *pseudo*-ionone (61%), and retinal from β-ionone (55%) were described using reagent (**4**).[10]

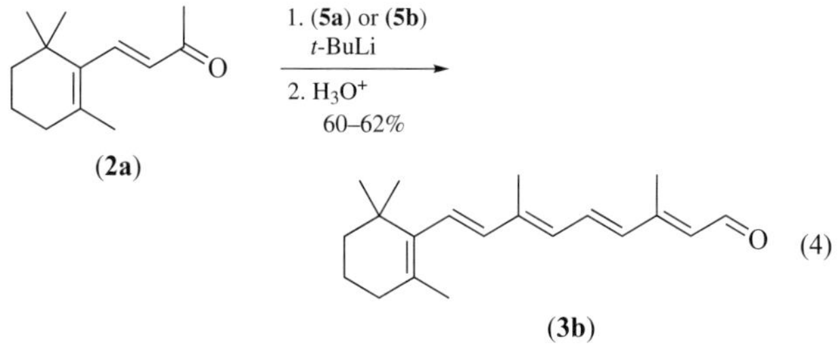

A short synthesis of phytol (**6**) from 6-methyl-5-hepten-2-one was described using reagents (**1**) and (**4**). Condensation of 6-methyl-5-hepten-2-one with (**4**) yielded the hydroxy acetal (**7**), which was converted in one pot into *pseudo*-phytone (**8**). Catalytic hydrogenation of (**8**) gave phytone, which was transformed into phytol (**6**) after condensation with (**1**) followed by reduction (eq 5).[11]

The adduct of (**1**) with carbonyl compounds can be trapped by methyl fluorosulfonate, leading to γ-methoxy enol ethers (**9**) precursors of β-methoxy aldehydes (**10**) (eq 6).[12]

The reaction schemes on the left side of the page show:

1. (4)
2. 1N HCl
80%

(7) →
1. 1N HCl, acetone, reflux
2. 1N NaOH, rt, 1 h
70%

(8) →
H₂, Pd/C
85%

→
1. (1)
2. LiAlH₄
76%

(6) (5)

$R^1\text{CHO}$
1. (1), Et₂O
2. MeSO₃F
−70 °C to rt
55–79%
→ (9) $R^1 = Ph, C_5H_{11}$
BuLi, THF
−20 °C
93%
→ (10) $R^1 = C_5H_{11}$ (6)

1. (a) Duhamel, L.; Tombret, F., *J. Org. Chem.* **1981**, *46*, 3741. (b) Duhamel, L.; Tombret, F.; Mollier, Y., *J. Organomet. Chem.* **1985**, *280*, 1.
2. Duhamel, L.; Plé, G.; Contreras, B., *Org. Prep. Proceed. Int.* **1986**, *18*, 219.
3. Neumann, H.; Seebach, D., *Chem. Ber.* **1978**, *111*, 2785.
4. Baudrillard, V.; Plé, G.; Davoust, D., *J. Org. Chem.* **1995**, *60*, 1473.
5. Duhamel, L.; Duhamel, P.; Lecouvé, J. P., *J. Chem. Res. (S)* **1986**, 34.
6. Trippett, S.; Walker, D. M., *Chem. Ind. (London)* **1960**, 202.
7. Wittig, G.; Reiff, H., *Angew. Chem., Int. Ed. Engl.* **1968**, *7*, 7.
8. Meyers, A. I.; Nabeya, A.; Adickes, H. W.; Politzer, I. R.; Fitzpatrick, J. M.; Malone, G. R., *J. Am. Chem. Soc.* **1969**, *91*, 764.
9. Akita, H., *Chem. Pharm. Bull.* **1983**, *31*, 1796.
10. Duhamel, L.; Duhamel, L.; Lecouvé, J. P., *Tetrahedron* **1987**, *43*, 4349.
11. Duhamel, L.; Ancel, J. E., *J. Chem. Res. (S)* **1990**, 154.
12. Rivière, P. Personal communication.
13. Duhamel, L.; Gralak, J.; Ngono, B., *J. Organomet. Chem.* **1994**, *464*, C11.
14. Duhamel, L.; Gralak, J.; Ngono, B., *J. Organomet. Chem.* **1989**, *363*, C4.

Lucette Duhamel
University of Rouen, Mont-Saint-Aignan, France

5-Trimethylsilyl-1,3-pentadiene

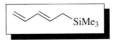

[72952-73-9]   C₈H₁₆Si   (MW 140.33)

InChI = 1S/C8H16Si/c1-5-6-7-8-9(2,3)4/h5-7H,1,8H2,2-4H3
InChIKey = ZFPBFBJPENLNKQ-UHFFFAOYSA-N

(useful pentadienylation reagent;[1] an allyl silane;[7] a precursor for 1-trimethylsilyldienes and intramolecular Diels–Alder substrates[2])

Alternate Name: 2,4-pentadienyltrimethylsilane.
Physical Data: bp 41 °C/36 mmHg or 105 °C/760 mmHg; n_D^{20} 1.4570.
Solubility: sol hexane, benzene, CH₂Cl₂, ether, THF.
Analysis of Reagent Purity: by NMR, GC (8.4 min on 3 mm × 3 M 5% Apiezon L on Chromosorb W at 80 °C) or TLC: 5% AgNO₃/silica gel (4:1 hexane/benzene).
Preparative Method: may be prepared in 85% yield by the reaction of pentadienyllithium with chlorotrimethylsilane in THF at room temperature.[1-3] Generally the product is ≈95% (*E*) isomer as judged by IR (strong band at 1000 cm⁻¹) or NMR (*trans* coupling constant of 15.2 Hz for proton at C-3, 5.92 ppm).
Handling, Storage, and Precautions: stable liquid; the toxicological properties of this reagent are unknown. It should be handled in a well ventilated hood.

Depending on the experimental conditions, reaction of (**1**) with *t*-butyldimethylsilyl trifluoromethanesulfonate leads either to the expected β-trimethysilyl enol ether (**11**), or to its isomer (**12**) formed by a 1,3-migration of the trimethylsilyl group from the oxygen atom to the carbon atom (**1** → **13**). Reaction with *t*-butyldimethylchlorosilane always gives (**12**).[13] Trimethylsilyl enol ether (**11**) is easily hydrolyzed to 2-(*t*-butyldimethylsilyl)-acetaldehyde (**14**) (eq 7).[14]

(1) →
(13)
TBDMSOTf, THF
−70 to 0 °C, 2 min
67%
→ (11)
78%
TBDMSOTf, THF
−70 to 0 °C, 90 min
→ (12) (7)
68%
1N HCl
Et₂O, rt
→ (14)

The study of the ¹H NMR spectra of (**1**) in Et₂O-*d*₁₀ at −70 °C is in agreement with the proposed structure; the isomeric enolate (**13**) is not normally detected (eq 7), but in THF at −70 °C it slowly forms.[4]

General Discussion. 2,4-Pentadienyltrimethylsilane is used as a stable (storable) nucleophilic substitute for pentadienyllithium. This compound reacts with Lewis acid (boron trifluoride etherate, iodotrimethylsilane, titanium(IV) chloride) activated electrophiles in the same manner as allylsilanes (eq 1). Generally only ε-substitution is observed (cf. Table 1).[1,4,5,7]

Table 1 Reaction of $H_2C=CHCH=CHCH_2MMe_3$ with typical electrophiles

Entry	Electrophile (E^+)	M	Lewis acid	Product	Yield (%)	Ref
1	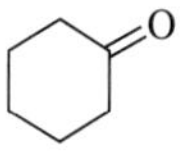	Si	TiCl$_4$	ε-adduct	64	1
2	PhCH(OMe)$_2$	Si	BF$_3\cdot$OEt$_2$	ε-adduct	85	4,5
3	PhCHO	Si	BF$_3\cdot$OEt$_2$	ε-adduct	16	5
4	t-BuCOCl	Si	BF$_3\cdot$OEt$_2$	34% ε-adduct, 25% γ-adduct	-	5
5	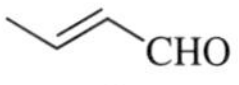 CHO	Si	BF$_3\cdot$OEt$_2$	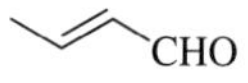 CHO	40	1,7
6	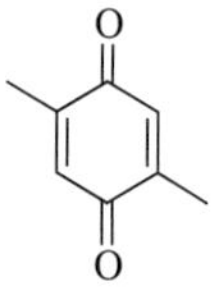	Sn	BF$_3\cdot$OEt$_2$	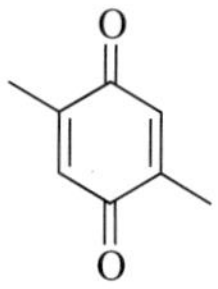	70	7
7	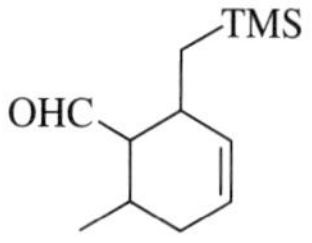	Si	BF$_3\cdot$OEt$_2$	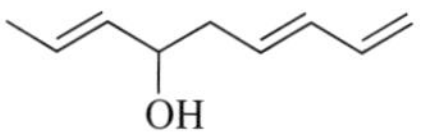	91	7
8	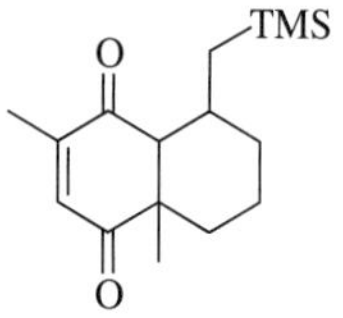	Sn	BF$_3\cdot$OEt$_2$	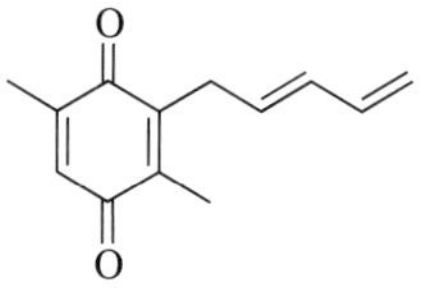	85[a]	7

[a] After reoxidation to the quinone.

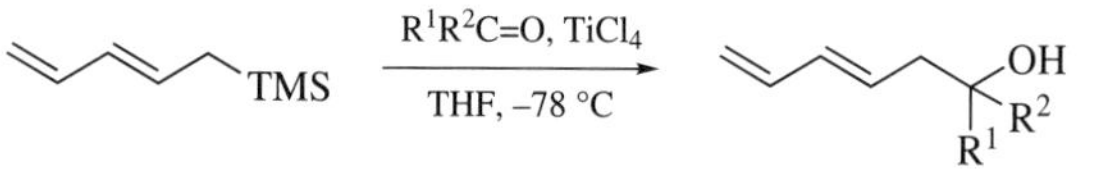

$$\text{TMS} \xrightarrow[\text{THF, }-78\,°C]{R^1R^2C=O,\ TiCl_4} \quad \text{OH} \quad (1)$$

Other electrophiles which have been used include acetals,[4] acid chlorides,[5] and heteroatom-based electrophiles. A useful varient of the title compound is the more reactive 2,4-pentadienyltrimethylstannane (Table 1, entries 6 and 8).[6,7] Lewis acid-catalyzed conjugate addition reactions have been observed with the tin analog,[7,8] but are not observed with 2,4-pentadienyltrimethylsilane due to a competing Diels–Alder reaction (Table 1, entries 5 and 7).

In a typical reaction of 2,4-pentadienyltrimethylsilane with electrophiles, a carbonyl compound was dissolved in CH$_2$Cl$_2$ and cooled to $-40\,°C$ and 0.3 equiv TiCl$_4$ (or BF$_3\cdot$OEt$_2$) was added. After 5 min, 0.66 equiv 2,4-pentadienyltrimethylsilane was added and the mixture warmed to $5–25\,°C$ over a 5 min to 2 h period (depending on the reactivity of the electrophile). After quenching with NaHCO$_3$ the product was extracted with ether.

2,4-Pentadienyltrimethylsilane may also be deprotonated adjacent to silicon with lithium diisopropylamide, forming the 1-(Me$_3$Si)-pentadienyl anion.[2] This anion reacts with many electrophiles to give ε-substituted dienes (eq 2) which are ideal candidates for protodesilyation or intramolecular Diels–Alder reactions.

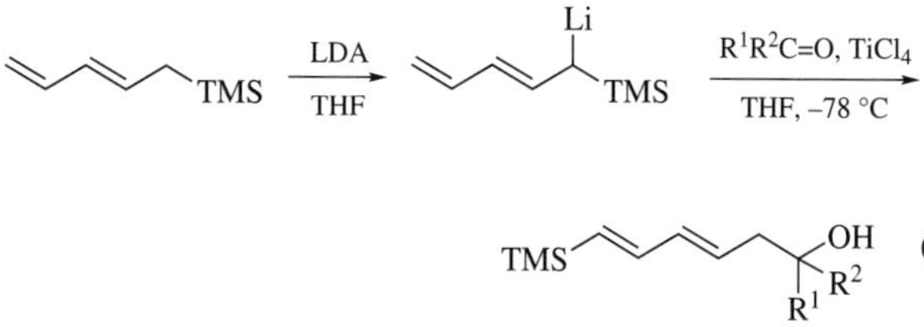

The reaction of the 1-(Me$_3$Si)-pentadienyl anion with cyclohexanone gives a 59/41 mixture of γ- and ε-regioisomers.

1. Seyferth, D.; Pornet, J., *J. Org. Chem.* **1980**, *45*, 1722.
2. Oppolzer, W.; Burford, S. C.; Marazza, F., *Helv. Chim. Acta* **1980**, *63*, 555.
3. Seyferth, D.; Pornet, J.; Weinstein, R. M., *Organometallics* **1982**, *1*, 1651.
4. Sakurai, H.; Sasaki, K.; Hosomi, A., *Tetrahedron Lett.* **1981**, 745.
5. Hosomi, A.; Saito, M.; Sakurai, H., *Tetrahedron Lett.* **1980**, 3783.
6. Jones, M.; Kitching, W., *J. Organomet. Chem.* **1983**, *247*, C5.
7. Naruta, Y.; Nagai, N.; Arita, Y.; Maruyama, K., *Chem. Lett.* **1983**, 1683.
8. Nishigaichi, Y.; Fujimoto, M.; Takuwa, A., *J. Chem. Soc., Perkin Trans. 1* **1992**, 2581.

Stephen R. Wilson
New York University, NY, USA

2-(Trimethylsilyl)phenyl Triflate

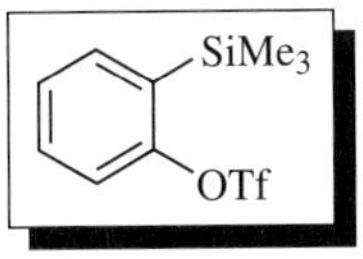

[88284-48-4] $C_{10}H_{13}F_3O_3SSi$ (MW 298.35)

InChI = 1S/C10H13F3O3SSi/c1-18(2,3)9-7-5-4-6-8(9)16-17
        (14,15)10(11,12)13/h4-7H,1-3H3

InChIKey = XBHPFCIWRHJDCP-UHFFFAOYSA-N

(reagent used as an aryne precursor in a variety of reactions)

Physical Data: bp 70 °C/2 mmHg; d 1.229 g mL^{-1}
Form Supplied in: liquid; widely available.
Handling, Storage, and Precautions: stable liquid. Incompatible with fluoride, hydroxide, or alkoxides, especially upon warming.

Kobayashi and co-workers introduced 2-(trimethylsilyl)phenyl triflate as an aryne precursor subject to benzyne formation without the need of a strong base.[1] Fluoride-induced desilylation and rapid elimination of the sulfonate provide efficient access to benzyne in acetonitrile at room temperature (eq 1). Other solvents such as THF, acetone, dichloromethane, DME, and toluene may be used, but such conditions may require heating for benzyne formation and frequently afford diminished yields. Preparation of the benzyne intermediate is even possible in protic media, albeit with decreased efficiency and limited applicability in subsequent reaction steps.[1]

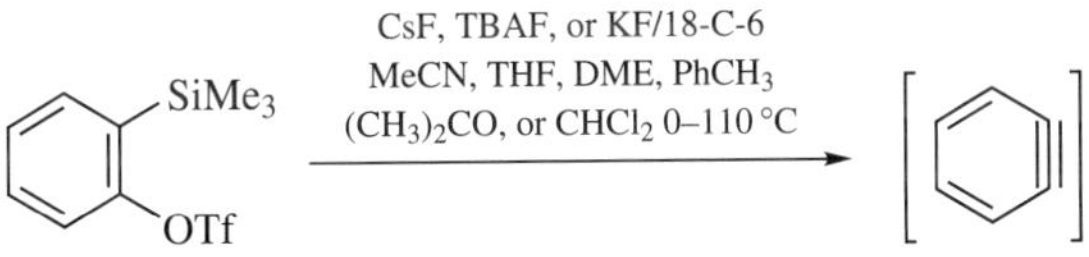

(1)

Various fluoride sources are employed for benzyne generation. Cesium fluoride or TBAF are most commonly used; however, the combination of potassium fluoride and 18-crown-6 also works well. The use of excess fluoride, typically 2 to 4 equiv, tends to provide superior results.

Polycyclic Arenes. Trimethylsilylphenyl triflates have found widespread use as benzyne precursors in the preparation of polycyclic arenes. Triphenylenes are made via palladium-catalyzed [2 + 2 + 2] cyclotrimerization of arynes,[2] by palladium-catalyzed annulation of the aryne and 2-halobiaryls,[3] or via carbopalladation/carbocyclization of arynes with substituted iodobenzenes (eq 2).[4] All three approaches furnish substituted triphenylenes in high reported yields. Substituted phenanthrenes or naphthalenes are similarly prepared using 2-(trimethylsilyl) phenyl triflate as an aryne precursor. Treatment of the reagent with fluoride, a palladium catalyst, and a deactivated alkene,[5] allyl halide,[6] or internal alkyne[7] provides 9-substituted or 9,10-disubstituted phenanthrenes in moderate to good yields (eq 3). Substituted naphthalenes are created by reacting the aryne with alternate combinations of the listed reagents using specific palladium catalysts.[7,8] Aryl naphthalene lignans are attainable through a Pd-catalyzed [2 + 2 + 2] cocyclization of the aryne and diynes.

This route has been used to synthesize lignan-containing natural products.[9] Extended fused polycyclic arenes are also accessible using related approaches.[10]

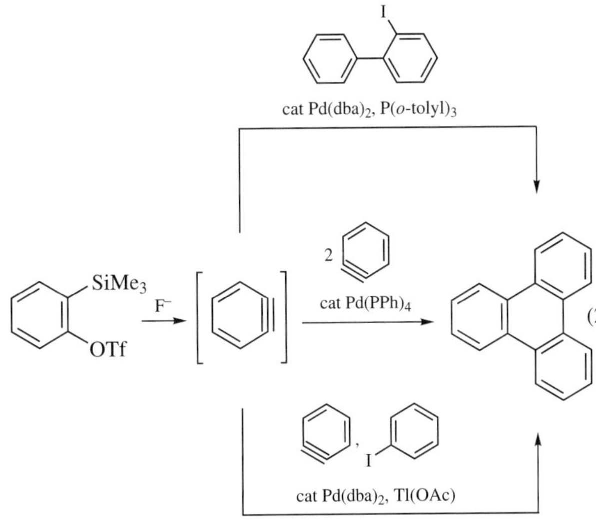

(2)

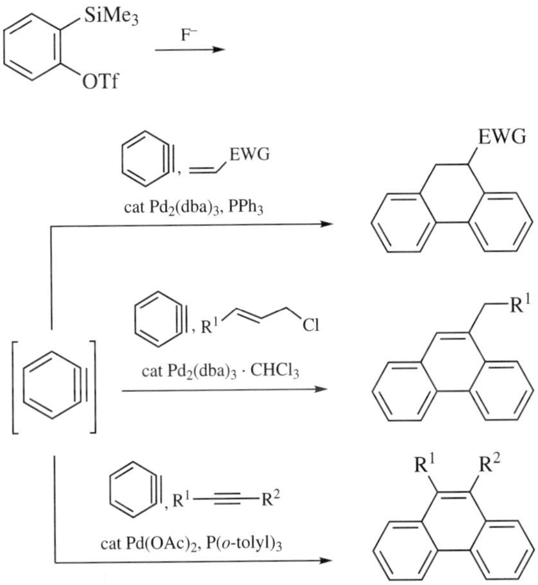

(3)

The palladium-catalyzed annulation of arynes, derived from 2-(trimethylsilyl)phenyl triflate, by 2-halobenzaldehydes provides variably substituted fluoren-9-ones in good yields (eq 4).[11] Catalytic dicobalt octacarbonyl under CO pressure furnishes anthraquinones in the presence of the aryne derived from 2-(trimethylsilyl)phenyl triflate.[12]

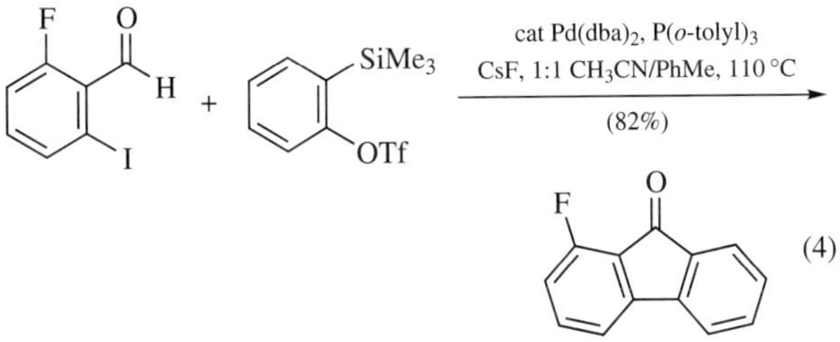

(4)

Heteroatom Arylation. Benzynes have a proven history as superior electrophiles in heteroatom arylation reactions. As such, aryne generation from 2-(trimethylsilyl)phenyl triflates offers particularly convenient access to anilines, anisoles, thioanisoles, etc. at ambient temperature using easily administered fluoride sources. The N-arylation of amines, azaarenes, or sulfonamides proceeds in uniformly excellent yields via this approach. Phenols and arenecarboxylic acids are also suitable nucleophiles, providing biaryl ethers or phenyl esters (eq 5).[13] Trifluoromethylamides and sulfinamides serve as both nucleophiles and acylating agents under these conditions to afford 2-trifluoroacylanilines or 2-trifluoromethylsulfoxyanilines, respectively (eq 6).[14] Similarly, cyclic ureas provide rapid access to simple benzodiazepines (eq 7) and benzodiazocines.

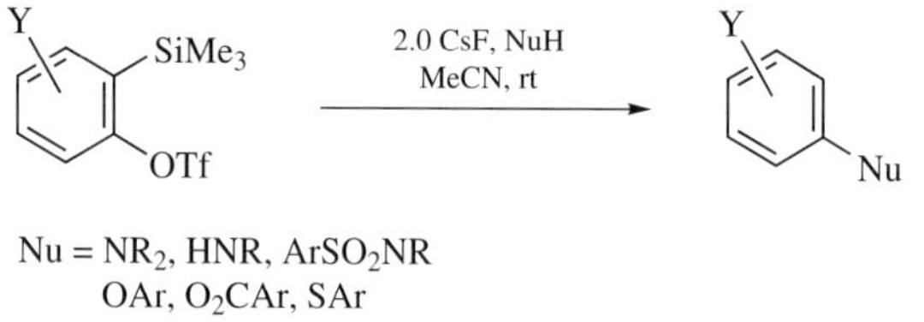

(5)

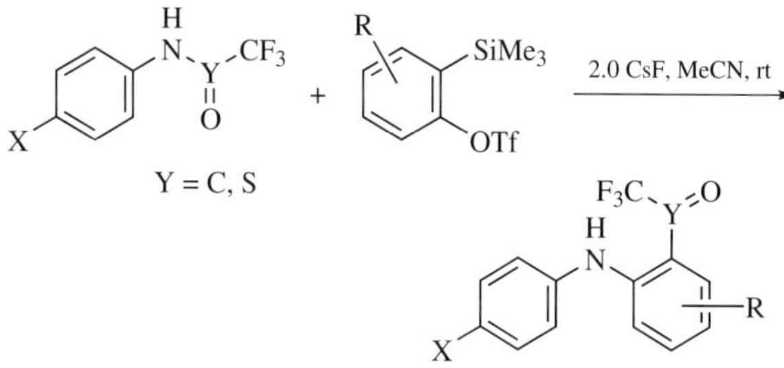

Nu = NR₂, HNR, ArSO₂NR
OAr, O₂CAr, SAr

(6)

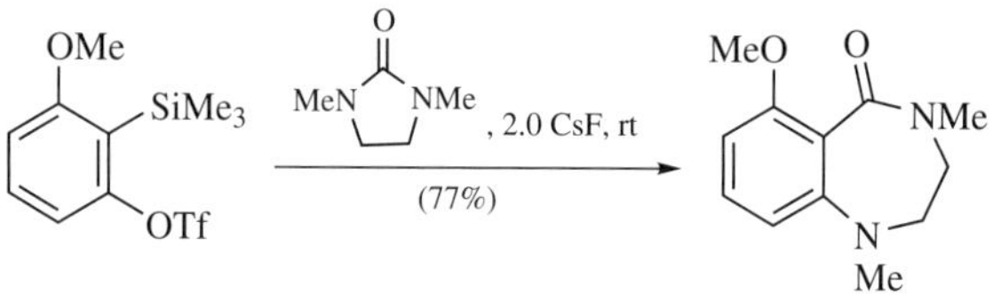

(7)

Acyclic ureas give 2-aminobenzamides in moderate yields via an analogous benzyne arylation/acylation pathway.[15] Heating azaarenes with 2-(trimethylsilyl)phenyl triflate and CsF in a cyanocarbon solvent furnishes 2-cyanomethyl-N-aryl azacycles in good yields. The nitriles must possess a hydrogen at the 2-position and are deprotonated by the zwitterionic arylation intermediate prior to addition by the azaarenium moiety (Reissert-type reaction).[16] Palladium-catalyzed disilylation[17] or distannylation[18] of the aryne is possible by the treatment of various 2-(trimethylsilyl) phenyl triflates with KF/18-crown-6, t-octyl isocyanide ligand, and a disilane or distannane. This method is also amenable to the preparation of the 2,2'-heterosubstituted biphenyls via employment of an alternate Pd⁰-ligand.[19]

Heteroarenes and Benzannulated Heterocycles. The heteroatom arylation methods involving 2-(trimethylsilyl)phenyl triflates have been extended to the preparation of heterocycles. Acridones, xanthones, and thioxanthones are readily obtained by reacting methyl 2-aminobenzoate, methyl salicylate, or methyl thiosalicylate, respectively, with the aryne generated from 2-(trimethylsilyl)phenyl triflate (eq 8).[20] The aryne is also used to prepare various benzannulated heterocycles via a three-component coupling process where at least one other component is an imine or carbonyl compound. This procedure provides simple entry to racemic substituted benzoiminofurans,[21] 2-iminoisoindolines,[22]

9-arylxanthenes,[23] or benzoxazinones.[24] Carbazoles or dibenzofurans are attained in high yields via the coupling of 2-iodoanilines or 2-iodophenols with 2-silylphenyl triflates followed by intramolecular palladium-catalyzed cyclization (eq 9).[25]

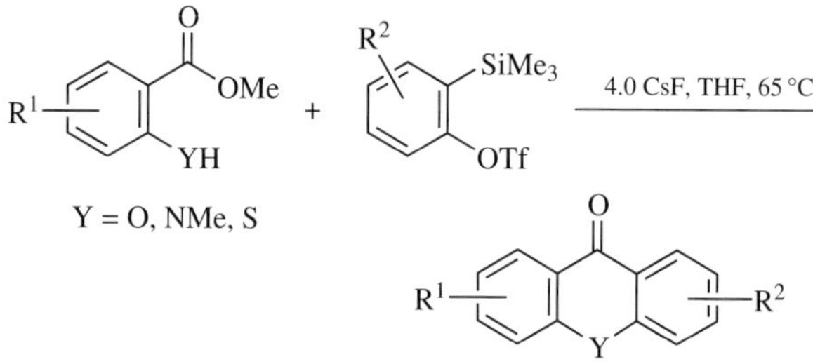

(8)

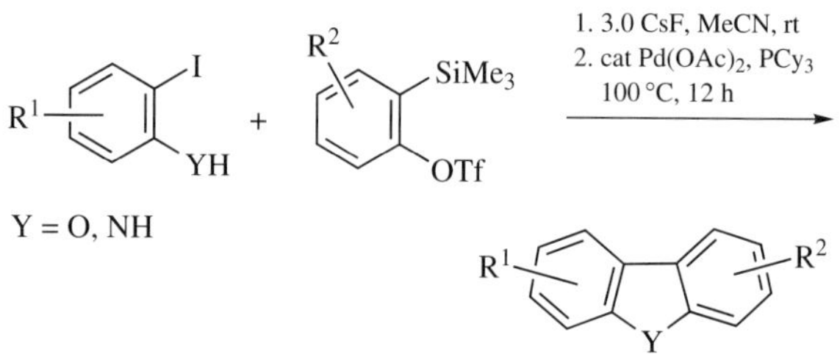

(9)

Carbon Arylations. Active methylene compounds are susceptible to the arylation/acylation reactions described above. Treatment of β-diketones, β-ketoesters, or dialkyl malonates with a 2-(trimethylsilyl)phenyl triflate and fluoride affords the aryl alkylation–acylation products in moderate to high yields (eq 10).[26] β-Ketonitriles and 2-cyanoformates are also compatible with this process, furnishing 2-cyanomethylphenones or alkyl benzoates, respectively.[27] A transition-metal-free regioselective coupling between pyridine N-oxides and arynes gives 3-(2-hydroxyaryl)-pyridines in good to excellent yields. This approach avoids the low yields and regioisomeric mixtures commonly observed in the preparation of such heterocycles.[28] The benzyne generated from 2-(trimethylsilyl)phenyl triflate may also be copolymerized with pyridine to give alternating o-phenylene and 2,3-dihydropyridine containing copolymers.[29]

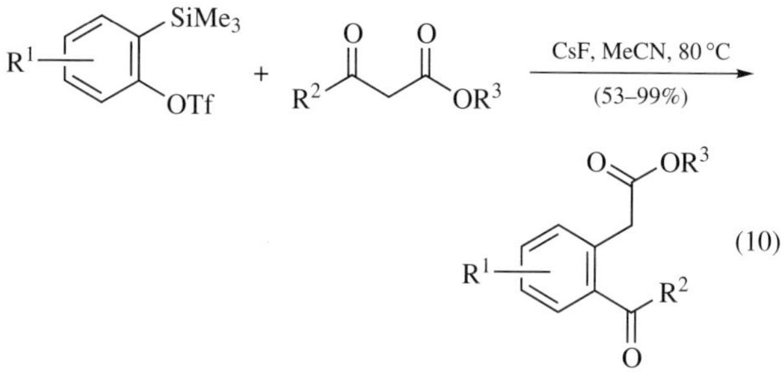

(10)

A three component coupling reaction involving carbopalladation of the aryne followed by a Heck coupling with t-butyl acrylate affords ortho-substituted cinnamic acids in good yields (eq 11).[30] An ene reaction between the aryne, generated in THF at room temperature, and an alkyne creates allenylbenzenes in moderate yields.[31] The reaction of π-allylpalladium species with the benzyne created from 2-(trimethylsilyl)phenyl triflate provides access to several types of products in multicomponent coupling reactions. For example, bis(π-allylpalladium) reagents afford 1,2-diallylated benzenes in moderate yields,[32] while allylation followed by Stille coupling with alkynylstannanes[33] or Suzuki–Miyaura coupling with arylboronic acids[34] yields 1-allyl-2-alkynylbenzenes or 2-allylbiphenyls, respectively (eq 12).

(11)

(12)

$$Y = CH_2CH=CH_2, Ar, C\equiv CR^2$$

The benzyne created from 2-(trimethylsilyl)phenyl triflate was used in a highly diastereoselective aryne Diels–Alder reaction with a diene bearing Oppolzer's sultam. This approach to *cis*-functionalized 1,4-dihydronaphthalenes was reportedly the first aryne Diels–Alder reaction to provide enantioenriched cyclo-adducts.[35] An unusual route to β-aminoketones involves the treatment of 2-(trimethylsilyl)phenyl triflate with TBAF and an asymmetric vinyldihydropyridone. The resultant aryne Diels–Alder cycloadduct undergoes aromatization/elimination to create the *N*-acyl-β-aminoketone (eq 13).[36] This method was featured in a multistep synthesis of an unnatural α-amino acid.

(13)

Related Reagents. (Phenyl)-[*o*-(trimethylsilyl)phenyl]iodonium triflate; 2-diazoniobenzenecarboxylate.

1. Himeshima, Y.; Sonoda, T.; Kobayashi, H., *Chem. Lett.* **1983**, 1211.
2. Peña, D.; Escadero, S.; Pérez, D.; Guitián, E., *Angew. Chem., Int. Ed.* **1998**, *37*, 2659.
3. Liu, Z.; Larock, R. C., *J. Org. Chem.* **2007**, *72*, 223.
4. Thatai Jayanth, T.; Cheng, C.-H., *Chem. Commun.* **2006**, 894.
5. Quintana, I.; Boersma, A. J.; Peña, D.; Pérez, D.; Guitián, E., *Org. Lett.* **2006**, *8*, 3347.

6. Yoshikawa, E.; Radhakrishnan, K.V.; Yamamoto, Y., *J. Am. Chem. Soc.* **2000**, *122*, 7280.
7. (a) Peña, D.; Pérez, D.; Guitián, E.; Castedo, L., *J. Org. Chem.* **2000**, *65*, 6944. (b) Radhakrishnan, K. V.; Yoshikawa, E.; Yamamoto, Y., *Tetrahedron Lett.* **1999**, *40*, 7533.
8. Hsieh, J.-C.; Cheng, C.-H., *Chem. Commun.* **2005**, 2459.
9. Sato, Y.; Tamura, T.; Mori, M., *Angew. Chem., Int. Ed.* **2004**, *43*, 2436.
10. (a) Thatai Jayanth, T.; Jeganmohan, M.; Cheng, C.-H., *J. Org. Chem.* **2004**, *69*, 8445. (b) Romero, C.; Peña, D.; Pérez, D.; Guitián, E., *Chem. Eur. J.* **2006**, *12*, 5677.
11. Zhang, X.; Larock, R. C., *Org. Lett.* **2005**, *7*, 3973.
12. Chatani, N.; Kamitani, A.; Oshita, M.; Fukumoto, Y.; Murai, S., *J. Am. Chem. Soc.* **2001**, *123*, 12686.
13. Liu, Z.; Larock, R. C., *J. Org. Chem.* **2006**, *71*, 3198.
14. Liu, Z.; Larock, R. C., *J. Am. Chem. Soc.* **2005**, *127*, 13112.
15. Yoshida, H.; Shirakawa, E.; Honda, Y.; Hiyama, T., *Angew. Chem., Int. Ed.* **2002**, *41*, 3247.
16. Jeganmohan, M.; Cheng, C.-H., *Chem. Commun.* **2006**, 2454.
17. Yoshida, H.; Ikadai, J.; Shudo, M.; Ohshita, J.; Kunai, A., *Organometallics* **2005**, *24*, 156.
18. Yoshida, H.; Tanino, K.; Ohshita, J.; Kunai, A., *Angew. Chem., Int. Ed.* **2004**, *43*, 5052.
19. Yoshida, H.; Tanino, K.; Ohshita, J.; Kunai, A., *Chem. Commun.* **2005**, 5678.
20. Zhao, J.; Larock, R. C., *J. Org. Chem.* **2007**, *72*, 583.
21. Yoshida, H.; Fukushima, H.; Ohshita, J.; Kunai, A., *Angew. Chem., Int. Ed.* **2004**, *43*, 3935.
22. Yoshida, H.; Fukushima, H.; Ohshita, J.; Kunai, A., *Tetrahedron Lett.* **2004**, *45*, 8659.
23. Yoshida, H.; Watanabe, M.; Fukushima, H.; Ohshita, J.; Kunai, A., *Org. Lett.* **2004**, *6*, 4049.
24. Yoshida, H.; Fukushima, H.; Ohshita, J.; Kunai, A., *J. Am. Chem. Soc.* **2006**, *128*, 11040.
25. Liu, Z.; Larock, R. C., *Tetrahedron* **2007**, *63*, 347.
26. (a) Tambar, U. K.; Stoltz, B. M., *J. Am. Chem. Soc.* **2005**, *127*, 5340. (b) Yoshida, H.; Watanabe, M.; Ohshita, J.; Kunai, A., *Chem. Commun.* **2005**, 3292.
27. Yoshida, H.; Watanabe, M.; Ohshita, J.; Kunai, A., *Tetrahedron Lett.* **2005**, *46*, 6729.
28. Raminelli, C.; Liu, Z.; Larock, R. C., *J. Org. Chem.* **2006**, *71*, 4689.
29. Ihara, E.; Kurokawa, A.; Koda, T.; Muraki, T.; Itoh, T.; Inoue, K., *Macromolecules* **2005**, *38*, 2167.
30. Henderson, J. L.; Edwards, A. S.; Greaney, M. F., *J. Am. Chem. Soc.* **2006**, *128*, 7426.
31. Thatai Jayanth, T.; Masilamani, J.; Cheng, M.-J.; Chu, S.-Y.; Cheng, C.-H., *J. Am. Chem. Soc.* **2006**, *128*, 2232.
32. Yoshikawa, E.; Radhakrishnan, K. V.; Yamamoto, Y., *Tetrahedron Lett.* **2000**, *41*, 729.
33. Jeganmohan, M.; Cheng, C.-H., *Org. Lett.* **2004**, *6*, 2821.
34. Thatai Jayanth, T.; Jeganmohan, M.; Cheng, C.-H., *Org. Lett.* **2005**, *7*, 2921.
35. Dockendorff, C.; Sahli, S.; Olsen, M.; Milhau, L.; Lautens, M., *J. Am. Chem. Soc.* **2005**, *127*, 15028.
36. Comins, D. L.; Kuethe, J. T.; Miller, T. M.; Février, F. C.; Brooks, C. A., *J. Org. Chem.* **2005**, *70*, 5221.

Timothy S. Snowden
The University of Alabama, Tuscaloosa, AL, USA

Trimethylsilylpotassium

[56859-17-7] C$_3$H$_9$KSi (MW 112.31)

InChIKey = 1S/C3H9Si.K/c1-4(2)3;/h1-3H3;

InChIKey = LKVHNWCEAZYJRS-UHFFFAOYSA-N

(highly reactive nucleophile and one-electron transfer reagent;[1-3] deoxygenates substituted epoxides with inversion of stereochemistry[4,5])

Solubility: sol DME, THF, HMPA, hexane.

Analysis of Reagent Purity: ^{13}C NMR (HMPA) δ 4.2.[6]

Preparative Methods: prepared quantitatively by the cleavage of the silicon–silicon bond of hexamethyldisilane with potassium hydride or potassium alkoxides (KOMe or potassium *t*-butoxide) (eqs 1 and 2).[1,2,7-9] It can also be conveniently prepared by deprotonation of Me$_3$SiH with KH (eq 3).[2,3] These procedures circumvent the need to handle volatile organomercury compounds that previous routes required.[10]

$$\text{TMS-TMS} \xrightarrow[\text{or DME, 18-crown-6}]{\substack{\text{KH} \\ \text{DME or DME–HMPA}}} \text{TMSK + TMSH} \quad (1)$$

$$\text{TMS-TMS} \xrightarrow[\text{DME, 18-crown-6}]{\text{KOMe}} \text{TMSK + TMSH} \quad (2)$$

$$\text{TMSH} \xrightarrow[\text{DME–HMPA}]{\text{KH}} \text{TMSK} \quad (3)$$

Handling, Storage, and Precautions: solutions of Me$_3$SiK are air and moisture sensitive and must be handled under inert atmospheres. The toxicity of Me$_3$SiK is not currently known; however, the preparations employing hexamethylphosphoric triamide and 18-crown-6 necessitate handling these solutions in a fume hood using chemically resistant gloves.

Nucleophilic and Electron Transfer Reactions. Me$_3$SiK is a highly reactive nucleophile and one-electron transfer reagent. It readily couples with halides and enones via S$_N$2 and 1,4-additions (eqs 4 and 5). When 1-bromohex-5-ene and benzophenone are treated with Me$_3$SiK, electron transfer induced cyclization and dimerization occur (eqs 6 and 7). The course of these reactions can be substantially influenced by the nature of the solvent, resulting in enhancement of either nucleophilic substitution or electron transfer.[1-3]

$$\text{(eq 4)}$$

$$\text{(eq 5)}$$

$$\text{(eq 6)}$$

Aryl halides can be converted in one step to aryl trimethylsilanes in good yields with Me$_3$SiK (eq 8).[11] These are particularly useful intermediates for directing electrophilic substitutions on carbon.[12]

$$\text{(eq 8)}$$

Deoxygenation of Epoxides with Inversion. Me$_3$SiK stereospecifically deoxygenates substituted epoxides with inversion of stereochemistry. Reactions of *cis*- and *trans*-epoxides with Me$_3$SiSiMe$_3$ and KOMe give the *trans*- and *cis*-alkenes, respectively (eqs 9 and 10).[4,5]

$$\text{(eq 9)}$$

$$\text{(eq 10)}$$

---

1. Sakurai, H.; Kondo, F., *J. Organomet. Chem.* **1975**, *92*, C46.
2. Corriu, R. J. P.; Guérin, C., *J. Chem. Soc., Chem. Commun.* **1980**, 168.
3. Corriu, R. J. P.; Guérin, C.; Kolani, B., *Bull. Soc. Claim. Fr.* **1985**, *5*, 973.
4. Dervan, P. B.; Shippey, M. A., *J. Am. Chem. Soc.* **1976**, *98*, 1265.
5. Koreeda, M.; Koizumi, N.; Teicher, B. A., *J. Chem. Soc., Chem. Commun.* **1976**, 1035.
6. Olah, G. A.; Hunadi, R. J., *J. Am. Chem. Soc.* **1980**, *102*, 6989.
7. Sakurai, H.; Kira, M.; Umino, H., *Chem. Lett.* **1977**, *11*, 1265.
8. Banik, G. M.; Silverman, R. B., *J. Am. Chem. Soc.* **1990**, *112*, 4499.
9. Buncel, E.; Venkatachalam, T. K.; Edlund, U., *J. Organomet. Chem.* **1992**, *437*, 85.
10. Hengge, E.; Holtschmidt, N., *J. Organomet. Chem.* **1968**, *12*, P5.
11. Shippey, M. A.; Dervan, P. B., *J. Org. Chem.* **1977**, *42*, 2654.
12. Eaborn, C., *J. Organomet. Chem.* **1975**, *100*, 43.

Kevin J. Moriarty

Rhône-Poulenc Rorer, Collegeville, PA, USA

2-Trimethylsilyl-1,3-propanediol

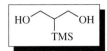

[189066-36-2]   C$_6$H$_{16}$O$_2$Si   (148.2)

InChI = 1S/C6H16O2Si/c1-9(2,3)6(4-7)5-8/h6-8H,4-5H2,1-3H3
InChIKey = BGMVXIDWBZLIRZ-UHFFFAOYSA-N

(reagent for the protection of ketones and aldehydes under mild conditions[1])

Alternate Name: cyclo-SEM.
Physical Data: mp 36–38 °C; bp 110–130 °C/2 mmHg.
Solubility: soluble in organic solvents, insoluble in water.
Form Supplied in: white, crystalline solid.
Analysis of Reagent Purity: ^{1}H NMR, capillary GC.
Preparative Methods: preparation (eq 1) is performed by lithium–halogen exchange of 1-(bromovinyl)trimethylsilane (**1**),[2] with 2.1 equiv of *t*-BuLi in diethyl ether at −78 °C for 0.5 h followed by cold cannulation into a slurry of paraformaldehyde in diethyl ether and warming to room temperature over 5 h.[3] The crude 2-trimethylsilyl-2-propenol (**2**) is purified, after water work-up and extraction, by column chromatography (5% EtOAc/petroleum ether) followed by bulb-to-bulb distillation (110–114 °C/60 mmHg). Treatment of **2** with 2.1 equiv of freshly prepared thexylborane[4] at −10 °C followed by 18 h at room temperature generates the boronate ester which is subsequently hydrolyzed by treatment with two equiv of methanol followed by a standard H$_2$O$_2$/NaOH work-up. Column chromatography on base-treated silica gel with 50% EtOAc/petroleum ether yields 2-trimethylsilyl-1,3-propanediol (**3**), as a white crystalline solid.

TMS—C(=CH$_2$)—Br
1

$$\xrightarrow[\text{2. (HCHO)n, Et}_2\text{O} \\ -78\,°\text{C to rt, 5 h}]{\text{1. } t\text{-BuLi, Et}_2\text{O} \\ -78\,°\text{C, 0.5 h}}$$

TMS—C(=CH$_2$)—CH$_2$OH
2

$$\xrightarrow[\text{2. NaOH, 30\% H}_2\text{O}_2]{\text{1. thx-BH}_2\text{, THF, 18 h, rt}}$$

HO—CH$_2$—CH(TMS)—CH$_2$OH   (1)

3

Purity: column chromatography using silica gel as the stationary phase and a mobile phase consisting of 50% EtOAc/petroleum ether will provide reagent of excellent quality. Recrystallization from either petroleum ether or hexanes, or bulb-to-bulb distillation (110–130 °C/2 mmHg) may also be used to purify the cyclo-SEM diol (**3**).
Handling, Storage, and Precaution: store in a closed container. No noticeable degradation after 12 months on the bench top.

Protection of Ketones and Aldehydes. The protection of ketones and aldehydes (eqs 2–6) may be accomplished upon treatment with 2–5 equiv of cyclo-SEM diol **3** and 0.25 equiv of an acid source, either camphorsulfonic acid or *p*-toluenesulfonic acid

in methylene chloride. Activated, powdered molecular sieves are used as a water scavenger and methylene chloride is the solvent of choice, although toluene may be used instead. Reactions take between 2 and 36 h, depending on the amount of diol **3** used, the reaction concentration, and the extent of steric hindrance around the carbonyl group.

$$\xrightarrow[\text{4 Å molecular sieves, 2 h, rt} \\ \text{92–98\%}]{\text{HO}\diagdown\text{OH (TMS) 2 equiv} \\ \text{0.25 equiv CSA, CH}_2\text{Cl}_2,}$$ (2)

$$\xrightarrow[\text{4 Å molecular sieves, 10 h, rt} \\ \text{93\%}]{\text{HO}\diagdown\text{OH (TMS) 5 equiv} \\ \text{0.25 equiv CSA, CH}_2\text{Cl}_2,}$$ (3)

$$\xrightarrow[\text{4 Å molecular sieves, 18 h, rt} \\ \text{81\%}]{\text{HO}\diagdown\text{OH (TMS) 4 equiv} \\ \text{0.25 equiv CSA, CH}_2\text{Cl}_2,}$$ (4)

$$\xrightarrow[\text{4 Å molecular sieves, 20 h, rt} \\ \text{89\%}]{\text{HO}\diagdown\text{OH (TMS) 5 equiv} \\ \text{0.25 equiv CSA, CH}_2\text{Cl}_2,}$$ (5)

$$\xrightarrow[\text{4 Å molecular sieves, 20 h, rt} \\ \text{87\%}]{\text{HO}\diagdown\text{OH (TMS) 1.5 equiv} \\ \text{0.25 equiv CSA, CH}_2\text{Cl}_2,}$$ (6)

The conditions for the formation of cyclo-SEM ketals are considerably milder than are those for the corresponding 1,3-dioxanes or 1,2-dioxolanes, which often require larger excess of diol and extended reflux times in benzene or toluene.[5] Attempts at installation of a cyclo-SEM ketal under refluxing toluene or benzene conditions were unsuccessful, leading to decomposition of cyclo-SEM diol **3**.

Saturated ketones and aldehydes react quickly and in high yields, even with α-substitution. Electronically deactivated ketones (cyclohexenone, acetophenone) and aldehydes (benzaldehyde), on the other hand, react poorly. Ketones and aldehydes with steric hindrance require a larger excess of *cyclo*-SEM diol to proceed to satisfactory levels of completion (eqs 4 and 5). Excess *cyclo*-SEM diol **3** may be recovered during product isolation by column chromatography, eluting with 50% EtOAc/petroleum ether. Recovery of 80–90% of the unreacted *cyclo*-SEM diol **3** is typical.

Unsymmetrical ketones and aldehydes protected as *cyclo*-SEM ketals or acetals form two distinct *cis*- and *trans*-isomers (eq 7). The two isomers are individually separable and are uniquely identifiable by ^{1}H and ^{13}C NMR. This added complexity limits the utility of *cyclo*-SEM as a protecting group.

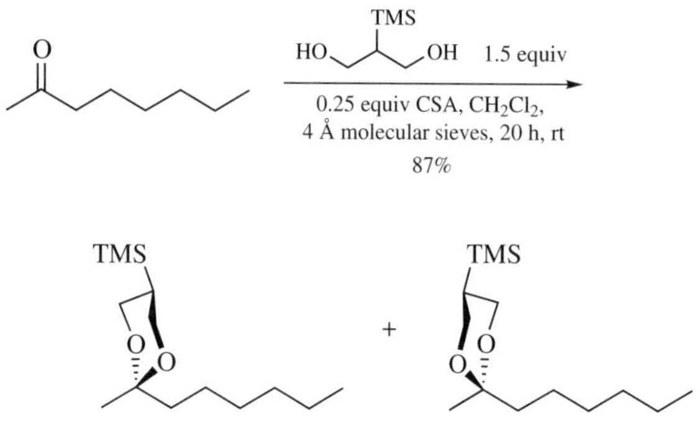

$$(7)$$

trans   *cis*

Deprotection of *cyclo*-SEM Ketals and Acetals. Removal of the *cyclo*-SEM ketal is performed by heating at reflux in the presence of one equiv of LiBF$_4$ in THF (eqs 8–12).[1] Use of a larger excess of LiBF$_4$ decreases the deprotection time, and acetonitrile may be substituted for THF. These mild, non-reductive, non-oxidative conditions of removal provide significant advantage over other carbonyl protecting groups.[5]

$$(8)$$

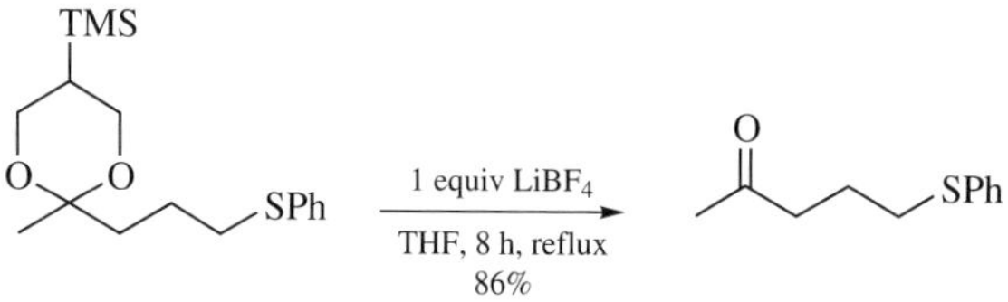

$$(9)$$

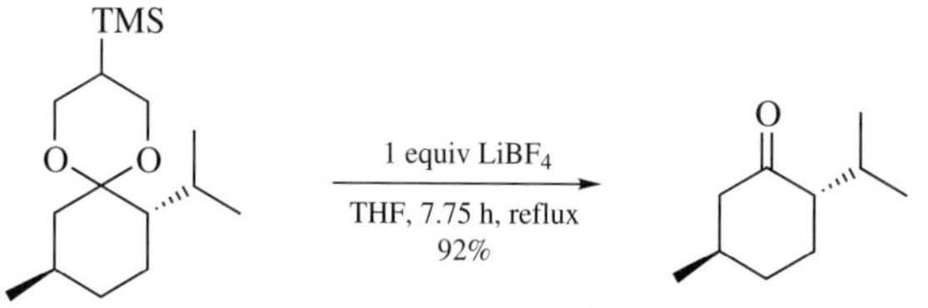

$$(10)$$

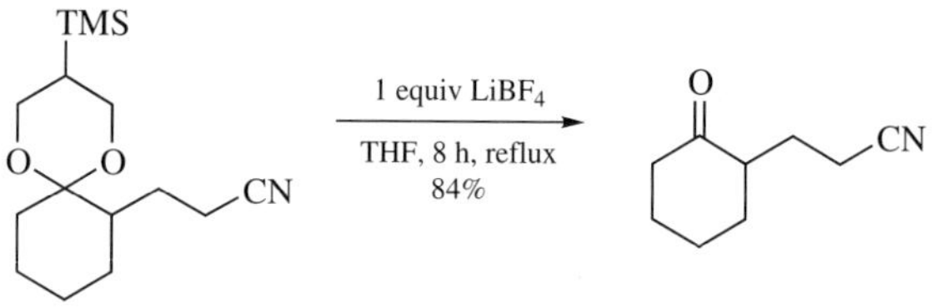

$$(11)$$

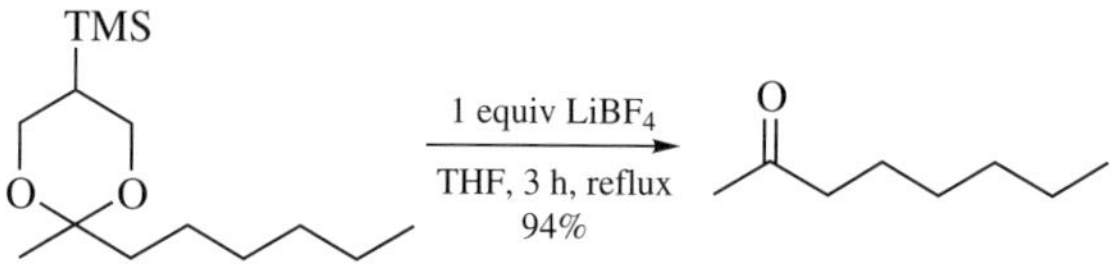

$$(12)$$

LiBF$_4$ removes *cyclo*-SEM ketals in refluxing THF (eq 13) at rates that exceed those observed for 1,3-dioxanes (eq 14) or 1,3-dioxolanes (eq 15) under these conditions, thereby allowing selective deprotection of *cyclo*-SEM ketals.

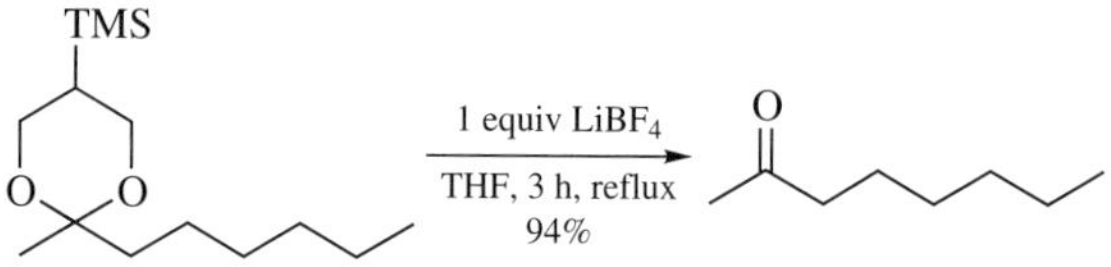

$$(13)$$

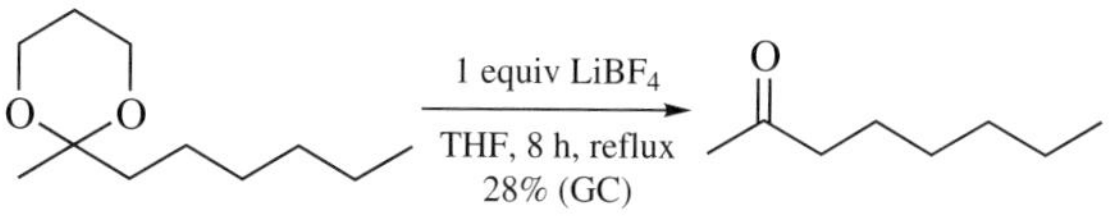

$$(14)$$

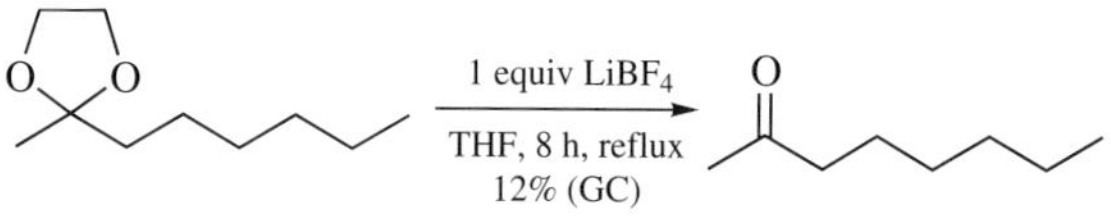

$$(15)$$

TBAF fails to remove the *cyclo*-SEM ketal and facilitates protiodesilylation, returning a 1,3-dioxlane.

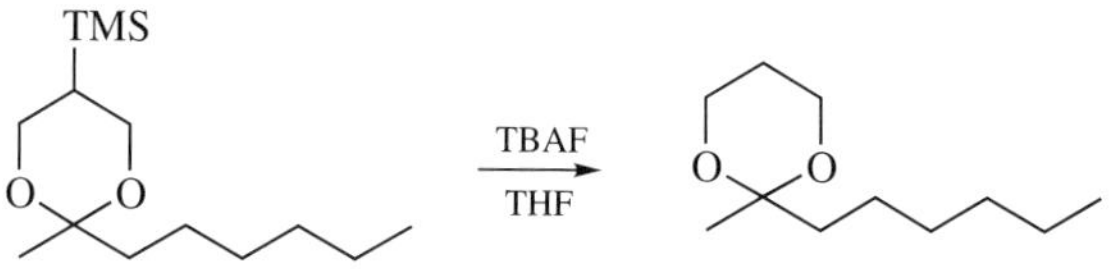

$$(16)$$

Related Reagents. 1,3-Propanediol; 1,2-ethanediol; 2,3-bis (trimethylsilyloxy)trimethylsilylpropane.[6]

1. Lipshutz, B. H.; Mollard, P.; Lindsley, C. L.; Chang, V., *Tetrahedron Lett.* **1997**, *38*, 1873.

2. (a) Boeckmann, R. K., Jr; Blum, D. M.; Ganem, B.; Haley, N., *Org. Syn. Coll. Vol. VI*, **1988**, 1033. (b) Boeckmann, R. K.; Blum, D. M.; Ganem, B.; Halvey, N., *Org. Synth.* **1978**, *58*, 152.

3. Overman, L. E.; Renhowe, P. A., *J. Org. Chem.* **1994**, *59*, 4138.

4. Brown, H. C.; Negishi, E. J., *Am. Chem. Soc.* **1972**, *94*, 3567.

5. Greene, T. W.; Wuts, P. G. M., *Protective Groups in Organic Synthesis*, 2nd edn. Chichester: Wiley, 1991, pp 119, 175.

6. Lillie, B. M.; Avery, M. A., *Tetrahedron Lett.* **1994**, *35*, 969.

Bruce H. Lipshutz & Paul Mollard
University of California, Santa Barbara, California, USA

3-Trimethylsilyl-2-propen-1-yl Acetate

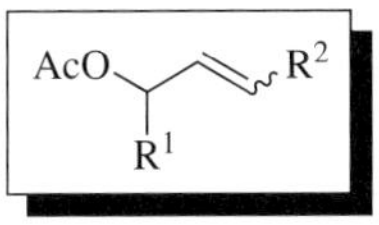

1; R^1 = H, R^2 = (E)-SiMe$_3$)

[86422-21-1]   C$_8$H$_{16}$O$_2$Si   (MW 172.33)

InChI = 1S/C8H16O2Si/c1-8(9)10-6-5-7-11(2,3)4/h5,7H,6H2,
    1-4H3/b7-5+

InChIKey = KWDYUHNACUTFNF-FNORWQNLSA-N

2; R^1 = H, R^2 = (Z)-SiMe$_3$)

[86422-22-2]

InChI = 1S/C8H16O2Si/c1-8(9)10-6-5-7-11(2,3)4/h5,7H,6H2,
    1-4H3/b7-5-

InChIKey = KWDYUHNACUTFNF-ALCCZGGFSA-N

(E + Z)

[80401-14-5]

InChI = 1S/C8H16O2Si/c1-8(9)10-6-5-7-11(2,3)4/h5,7H,6H2,
    1-4H3

InChIKey = KWDYUHNACUTFNF-UHFFFAOYSA-N

3; R^1 = SiMe$_3$, R^2 = H)

[80401-11-2]

InChI = 1S/C8H16O2Si/c1-6-8(10-7(2)9)11(3,4)5/h6,8H,1H2,
    2-5H3

InChIKey = IAMQLXWPWSUFPI-UHFFFAOYSA-N

(preparation of vinylsilanes via the corresponding allylmetal complexes;[1] oxygenated allylsilane reagents for nucleophilic addition and substitution reactions[2])

Preparative Methods: (i) from 3-trimethylsilyl-2-propyn-1-ol by reduction (sodium bis(2-methoxyethoxy)aluminum hydride (Red-Al), 70%)[3] and acetylation (acetyl chloride, pyridine, 78%);[4] (ii) from 3-trimethylsilyl-2-propyn-1-ol by reduction (P-2 raney nickel, H$_2$, 86%)[5] and acetylation; (iii) from allyloxytrimethylsilane by a metalation–rearrangement sequence (*t*-butyllithium, 90%)[2a,6] and acetylation (acetic anhydride, triethylamine, 76%).[2a]

Handling, Storage, and Precautions: use in a fume hood.

Preparation of Vinylsilanes. Both (E)- and (Z)-3-trimethylsilyl-2-propen-1-yl acetates (**1**) and (**2**), together with their allylic isomer (**3**) (1-trimethylsilyl-2-propen-1-yl acetate) are useful reagents for the formation of vinylsilanes by way of transition metal-catalyzed allylic alkylations. Reaction of the π-allylpalladium intermediate generated from (**1**)–(**3**) with several types of nucleophiles has been studied.[4,7] In general, these alkylations exhibit high regioselectivity, with reaction occurring at the position γ to silicon. diethyl malonate sodium enolate reacts with either (**2**) or (**3**) in the presence of Pd^0 to give alkylated products in good yields and favoring the (Z)-vinylsilane isomer (eq 1).[1] Cyclohexanone enamine participates in the alkylation reaction, giving the (E)-vinylsilane as the sole product (eq 2).[1] Although lithium enolates are problematic, addition of tri-*n*-butyltin trifluoroacetate allows product formation in good yield (eq 3).[8]

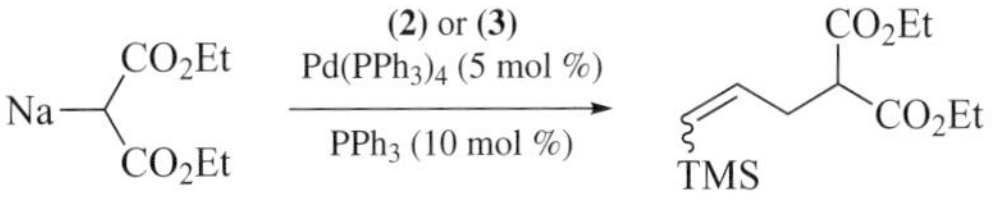

Reagent	Yield (%)	(E):(Z)
(**2**)	72	22:78
(**3**)	65	27:73

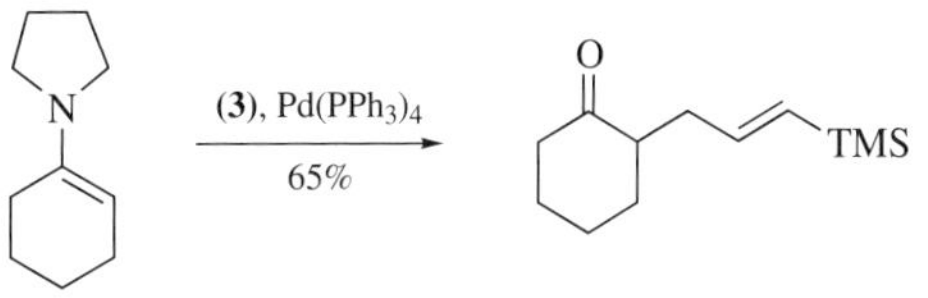

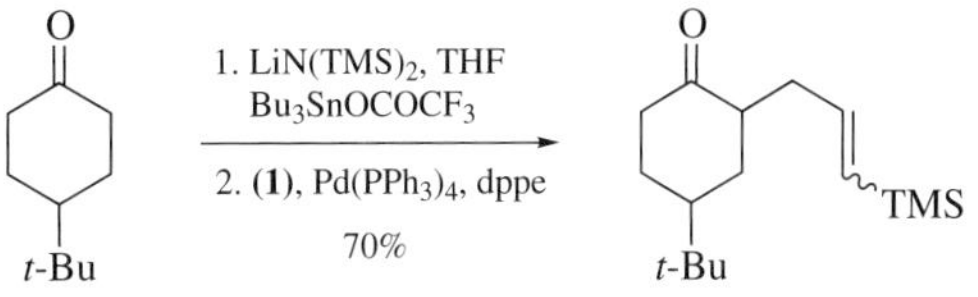

Other metals besides palladium are also effective.[9] Reaction of the cationic allyltetracarbonyliron complex derived from (**1**) or (**2**) with silyl enol ethers, O-silyl ketene acetals, or allylstannanes, followed by oxidative decomplexation, gives the vinylsilane products.[10] The process was shown to occur with near complete retention of stereochemistry (cf. eqs 4 and 5).[10]

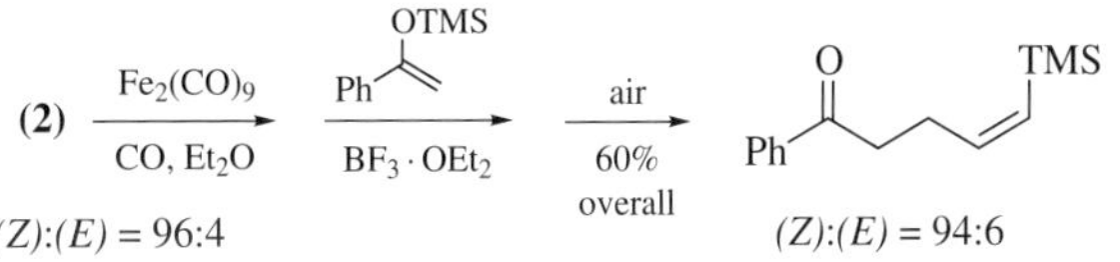

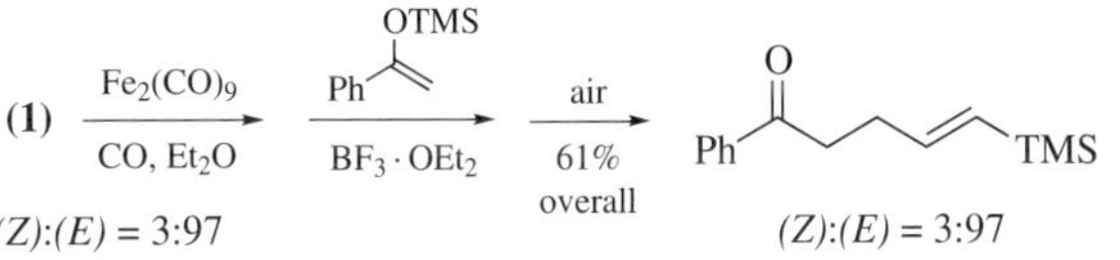

Nucleophilic Addition and Substitution Reactions. Allylsilane (**3**) has been shown to undergo conjugate addition to enones in the presence of tetra-*n*-butylammonium fluoride (eq 6).[2a] The reaction demonstrates high regioselectivity, as no products arising from 1,2-addition to the enone or attack at the γ-position of the allylsilane were isolated. Stereoselective C-glycosidation can be effected by reaction of allylsilane (**3**) with D-mannopyranoside derivatives in the presence of boron trifluoride etherate (eq 7).[2b] The α-C-glycoside arises from axial addition to the pyranoside oxonium ion.

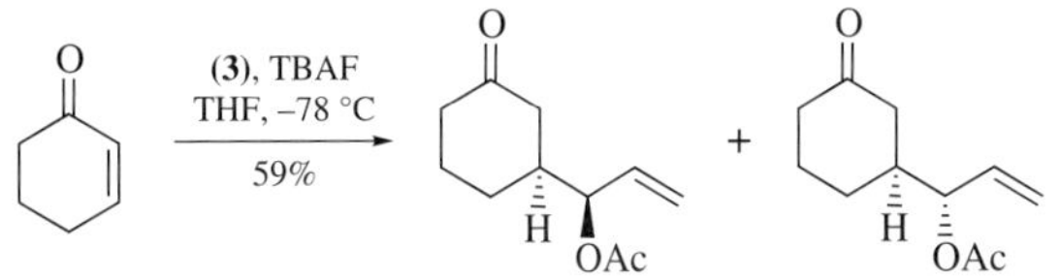

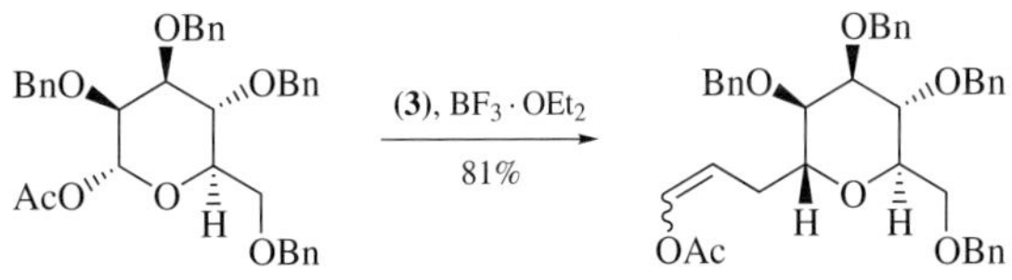

Related Reagents. Allyltrimethylsilane; 3-Bromo-1-trimethylsilyl-1-propene; 1-Pyrrolidino-1-cyclohexene; Tetrakis-(triphenylphosphine)palladium(0); Trimethylsilylallyllithium.

1. Hirao, T.; Enda, J.; Ohshiro, Y.; Agawa, T., *Tetrahedron Lett.* **1981**, *22*, 3079.

2. (a) Panek, J. S.; Sparks, M. A., *Tetrahedron Lett.* **1987**, *28*, 4649. (b) Panek, J. S.; Sparks, M. A., *J. Org. Chem.* **1989**, *54*, 2034. (c) Aicher, T. D.; Buszek, K. R.; Fang, F. G.; Forsyth, C. J.; Jung, S. H.; Kishi, Y.; Scola, P. M., *Tetrahedron Lett.* **1992**, *33*, 1549.

3. Jones, T. K.; Denmark, S. E., *Org. Synth., Coll. Vol.* **1990**, *7*, 524.

4. Trost, B. M.; Self, C. R., *J. Am. Chem. Soc.* **1983**, *105*, 5942.

5. Hiemstra, H.; Klaver, W. J.; Speckamp, W. N., *Recl. Trav. Chim. Pays-Bas* **1986**, *105*, 299.

6. Danheiser, R. L.; Fink, D. M.; Okano, K.; Tsai, Y.; Szczepanski, S. W., *Org. Synth.* **1988**, *66*, 14.

7. Mori, M.; Isono, N.; Kaneta, N.; Shibasaki, M., *J. Org. Chem.* **1993**, *58*, 2972.

8. Trost, B. M.; Self, C. R., *J. Org. Chem.* **1984**, *49*, 468.

9. For use of Mo(CO)$_6$, see: Trost, B. M.; Lautens, M., *Organometallics* **1983**, *2*, 1687.

10. Gajda, C.; Green, J. R., *Synlett* **1992**, 973.

David L. Clark
University of California, Berkeley, CA, USA

3-Trimethylsilyl-1-propyne

[13361-64-3]   C$_6$H$_{12}$Si   (MW 112.27)

InChI = 1S/C6H12Si/c1-5-6-7(2,3)4/h1H,6H2,2-4H3

InChIKey = ULYLMHUHFUQKOE-UHFFFAOYSA-N

(propargylation agent; allenylation agent; can give addition reactions on the triple bond)

Alternate Name: propargyltrimethylsilane.

Physical Data: bp 91–93 °C/760 mmHg; 40 °C/140 mmHg; d 0.753 g mL^{-1}; n_D^{20} 1.4140. IR (cm^{-1}, neat): 3320 s, 2120 m (C≡CH); 1250 s, 845 s, 755 w (Si–C). ^{1}H NMR (CCl$_4$, δ ppm): 0.11 (s, 9H, SiMe$_3$); 1.41 (d, *J* 2.5 Hz, 2H, CH$_2$); 1.66 (t, *J* 2.5 Hz, 1H, CH).[1-4]

Solubility: sol ether, THF, CH$_2$Cl$_2$.

Form Supplied in: colorless liquid, commercially available.

Analysis of Reagent Purity: IR and NMR.

Preparative Methods: the main method of preparation is shown in eq 1.[1-3] The title reagent can be obtained by other methods,[4] but in poor yields.

$$\text{=\!\!\!=\!\!\!\diagdown\!\!^{Br}} + \text{Mg} \xrightarrow[20\ ^\circ\text{C}]{\text{ether}} \left[\text{=\!\!=\!\!•\!\!\diagdown\!\!^{MgBr}}\right] \xrightarrow[-5\text{ to }-20\ ^\circ\text{C}]{\text{TMSCl}}$$

$$\text{=\!\!\!=\!\!\!\diagdown\!\!^{TMS}} + \text{=\!\!=\!\!•\!\!\diagdown\!\!^{TMS}} \qquad (1)$$

$$90{:}10$$

Purification: by distillation.

Handling, Storage, and Precautions: usually stored in a refrigerator. The reagent should be handled in a well-ventilated hood.

Original Commentary

Léone Miginiac
University of Poitiers, Poitiers, France

Reactivity. 3-Trimethylsilyl-1-propyne is the simplest of the propargylic silanes, which are convenient reagents in organic synthesis.[5,6]

Propargylation Reactions. 3-Trimethylsilyl-1-propyne is metalated easily by using an organomagnesium or a lithium compound, leading to the corresponding alkynic organometallic reagent which can react with many electrophiles.

Alkylation Reactions.[7-9] These reactions may be obtained with various halides, a number of them having other functionalities (eqs 2 and 3).

$$\text{=\!\!=\!\!\diagdown\!\!^{TMS}} \xrightarrow[\substack{\text{2. TMSCl, }-30\text{ to }20\ ^\circ\text{C} \\ 47\%}]{\text{1. BuLi, hexane, }-30\text{ to }0\ ^\circ\text{C}} \text{TMS\!=\!\!=\!\!\diagdown\!\!^{TMS}} \quad (2)$$

$$\text{=\!\!=\!\!\diagdown\!\!^{TMS}} \xrightarrow[\substack{\text{2. ICH}_2\text{TMS} \\ 60\%}]{\text{1. BuLi, hexane, THF}} \text{TMS\!\!\diagup\!\!=\!\!=\!\!\diagdown\!\!^{TMS}} \quad (3)$$

Reactions with Epoxides.[10,11] These reactions allow preparation of β-alkynic alcohols (eq 4); the use of *n*-butyllithium, followed by diethylaluminum chloride, may increase the yield.

$$\text{=\!\!=\!\!\diagdown\!\!^{TMS}} \xrightarrow[\substack{\text{2. } \triangledown\!\!\!\!_O\text{, THF, }-4\text{ to }20\ ^\circ\text{C} \\ 51\%}]{\text{1. EtMgBr, THF, }20\ ^\circ\text{C}} \text{TMS\!\!\diagup\!\!=\!\!=\!\!\diagup\!\!\diagdown\!\!^{OH}} \quad (4)$$

Reactions with Aldehydes and Ketones.[12-16] These are exemplified by eqs 5 and 6. Under the conditions of eq 6, the major product results from further reaction between the initial alcoholate and the excess formaldehyde. Secondary and tertiary alcohols are also obtained easily (eq 7). The yields are best when *n*-BuLi is used as the base and when HMPA is used as a cosolvent in the reaction with the ketone.

$$\text{=\!\!=\!\!\diagdown\!\!^{TMS}} \xrightarrow[\substack{\text{2. (CH}_2\text{O)}_n\text{, }-78\text{ to }20\ ^\circ\text{C} \\ 76\%}]{\text{1. BuLi, hexane, THF, }-78\ ^\circ\text{C}} \text{TMS\!\!\diagup\!\!=\!\!=\!\!\diagup\!\!\diagdown\!\!^{OH}} \quad (5)$$

$$\text{=\!\!=\!\!\diagdown\!\!^{TMS}} \xrightarrow[\substack{\text{2. (CH}_2\text{O)}_n\text{, }0\text{ to }20\ ^\circ\text{C} \\ \text{3. TMSCl, }0\text{ to }20\ ^\circ\text{C}}]{\text{1. BuLi, hexane, THF, }-30\ ^\circ\text{C}}$$

$$\underset{13\%}{\text{TMS\!\!\diagup\!\!\overset{OTMS}{=\!\!=\!\!\diagup\!\!\diagdown}}} + \underset{42\%}{\text{TMS\!\!\diagup\!\!=\!\!=\!\!\diagup\!\!\diagdown\!\!^{O}\!\!\diagdown\!\!^{OTMS}}} \quad (6)$$

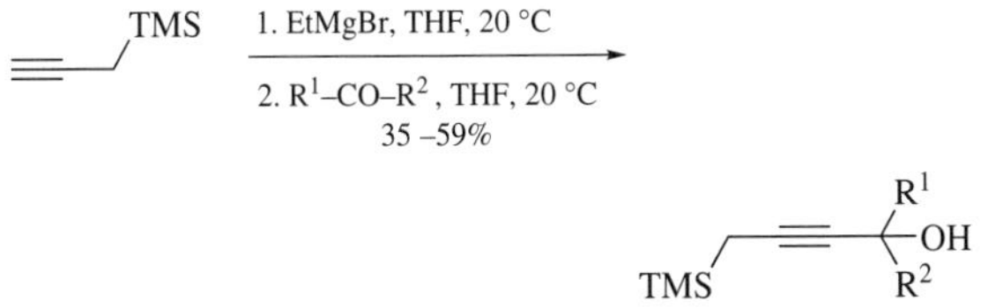

Reactions with Chloroformates.[17]   These are exemplified by eq 8.

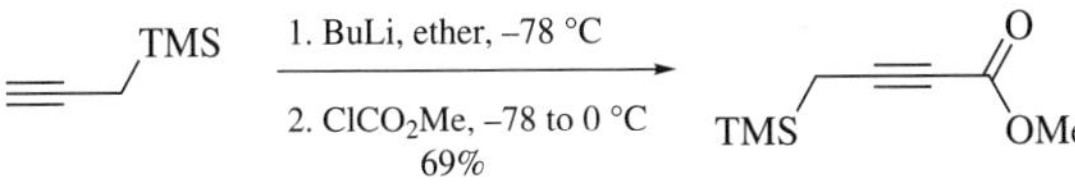

$$\text{TMS} \xrightarrow[\substack{2.\ \text{ClCO}_2\text{Me, }-78\text{ to }0\ ^\circ\text{C} \\ 69\%}]{1.\ \text{BuLi, ether, }-78\ ^\circ\text{C}} \quad (8)$$

Reactions with Polyoxymethylene and Amines[13] (Mannich Reaction).   These are exemplified by eq 9.

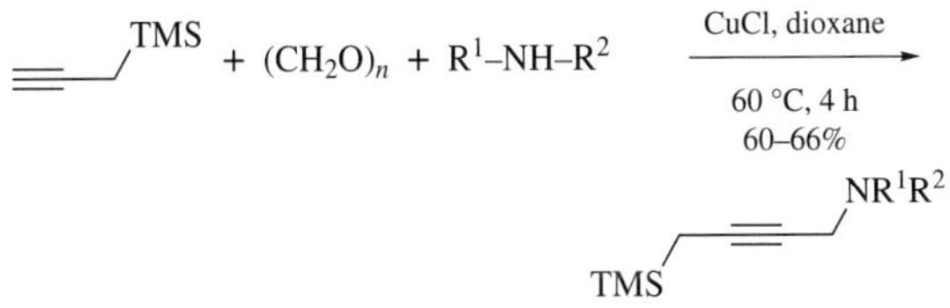

$$\text{TMS} + (\text{CH}_2\text{O})_n + \text{R}^1\text{–NH–R}^2 \xrightarrow[\substack{60\ ^\circ\text{C, }4\text{ h} \\ 60\text{–}66\%}]{\text{CuCl, dioxane}} \quad (9)$$

Allenylation Reactions.   The propargyltrimethylsilanes undergo Lewis acid-catalyzed reactions with electrophiles, with regiospecific rearrangement, to give substituted allenes (eq 10).[18–21] The catalysts usually employed are aluminum chloride, boron trifluoride etherate, and titanium(IV) chloride.

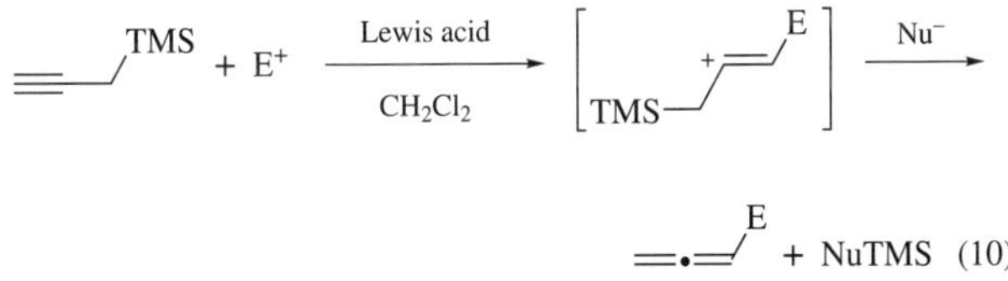

$$\text{TMS} + \text{E}^+ \xrightarrow[\text{CH}_2\text{Cl}_2]{\text{Lewis acid}} \left[\text{TMS} \overset{+}{\underset{}{}} \text{E} \right] \xrightarrow{\text{Nu}^-}$$

$$ \overset{\text{E}}{=\!\bullet\!=} + \text{NuTMS} \quad (10)$$

Substitution Reactions.   Few examples have been described. Generally, the reaction occurs with regiospecific rearrangement. Substitution of halides.[18] is shown in eq 11, and of OH, OSiMe₃, and OCOR groups[19,20] in eq 12.

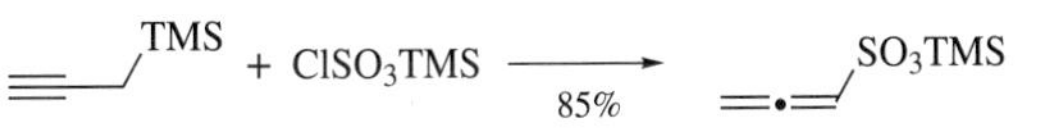

$$\text{TMS} + \text{ClSO}_3\text{TMS} \xrightarrow{85\%} =\!\bullet\!=\!\!-\text{SO}_3\text{TMS} \quad (11)$$

$$(\text{OC})_3\text{Cr} + \text{TMS} \xrightarrow[\text{CH}_2\text{Cl}_2]{\text{BF}_3\cdot\text{OEt}_2} \quad (12)$$

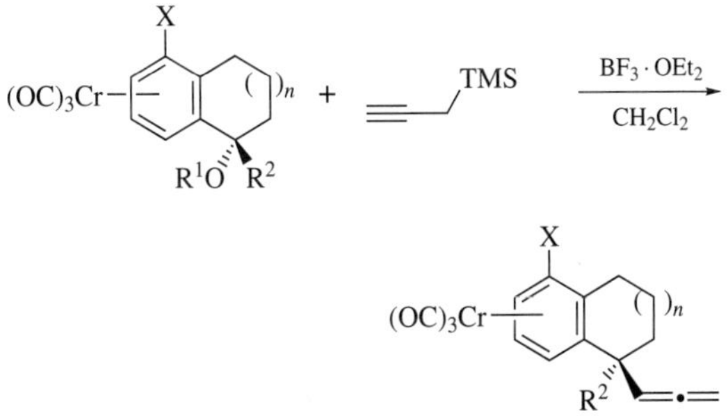

Ortho-propynyliodoarenes are obtained by reductive iodono-Claisen rearrangement of allenyliodinanes (eq 13).[21] When both

ortho positions and the *para* position of aryliodinanes are occupied, *ipso*-propynylarenes are formed (eq 14).

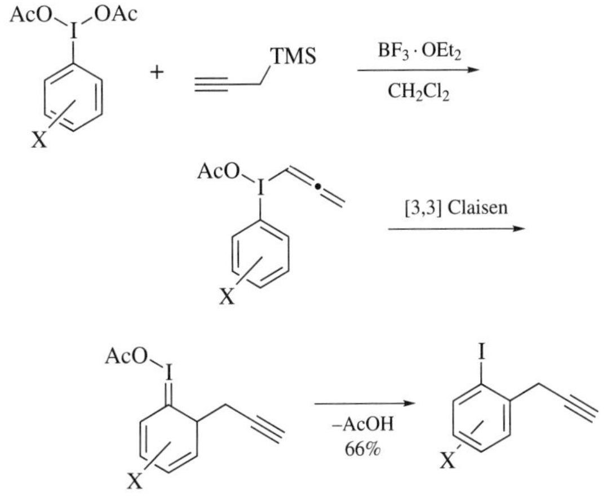

$$[3,3] \text{ Claisen}$$

$$\xrightarrow[66\%]{-\text{AcOH}} \quad (13)$$

$$\text{TMS} + \xrightarrow[\substack{\text{CH}_2\text{Cl}_2 \\ 57\%}]{\text{BF}_3\cdot\text{OEt}_2} \quad (14)$$

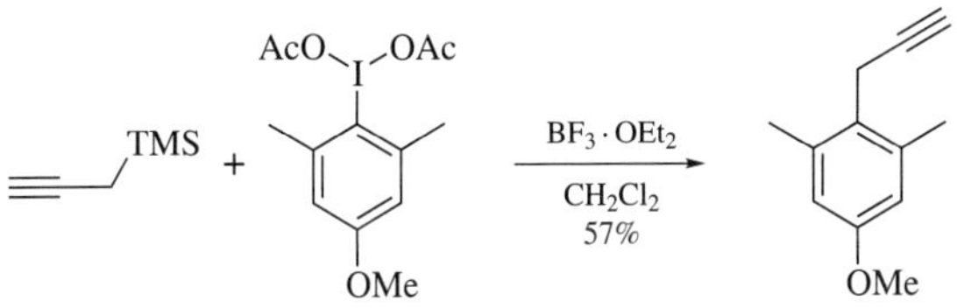

Reactions with Acetals.[22,23] and Hemiacetals[24–26]   These are exemplified by eqs 15 and 16. The synthesis of *C*-glycosides bearing an allenyl group has been accomplished[25,26] by use of 3-trimethylsilyl-1-propyne.

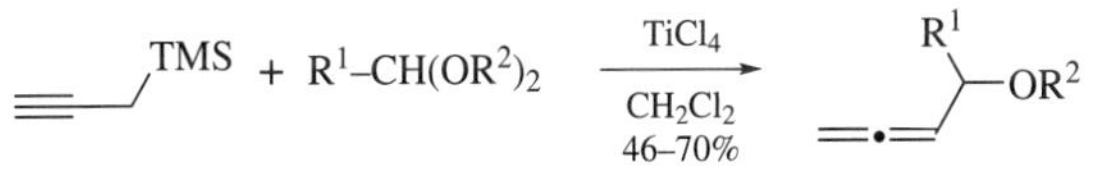

$$\text{TMS} + \text{R}^1\text{–CH(OR}^2)_2 \xrightarrow[\substack{\text{CH}_2\text{Cl}_2 \\ 46\text{–}70\%}]{\text{TiCl}_4} \quad (15)$$

$$\text{TMS} + \xrightarrow[\substack{\text{CH}_2\text{Cl}_2,\ -78\text{ to }20\ ^\circ\text{C} \\ 92\%}]{\text{BF}_3\cdot\text{OEt}_2} \quad (16)$$

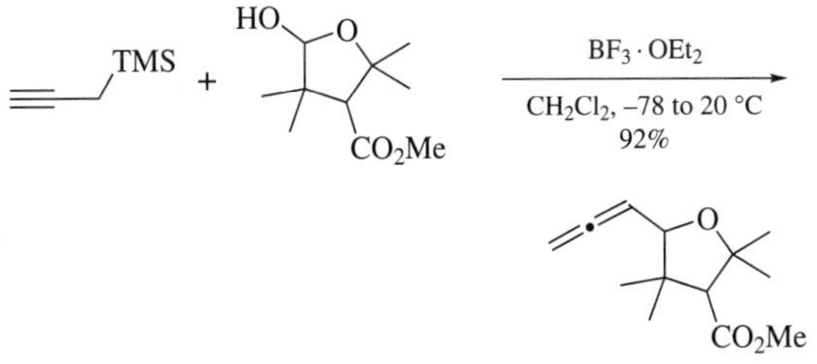

Reactions with Aldehydes and Ketones.[22,27–29]   These reactions generally lead to α-allenyl alcohols (eq 17). With titanium(IV) chloride as catalyst, the major product is a chloroprenic derivative formed from the α-allenyl *O*-silylated alcohol (eq 18).[22] The yields are good with aliphatic aldehydes bearing a primary or a secondary group and with activated ketones; they are poorer with aliphatic ketones (23–25%). With two equivalents of an aldehyde, the formation of a heterocyclic compound may be observed (eq 19).[28] By using tetra-*n*-butylammonium fluoride as catalyst, α-allenyl alcohols are easily obtained[29] with aliphatic and aromatic aldehydes (eq 20).

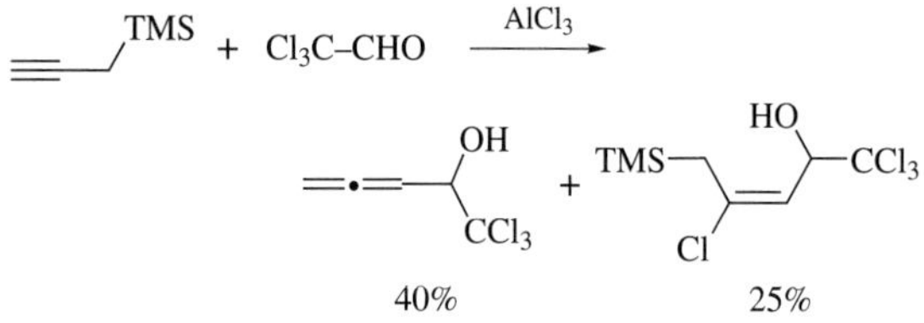

(17)

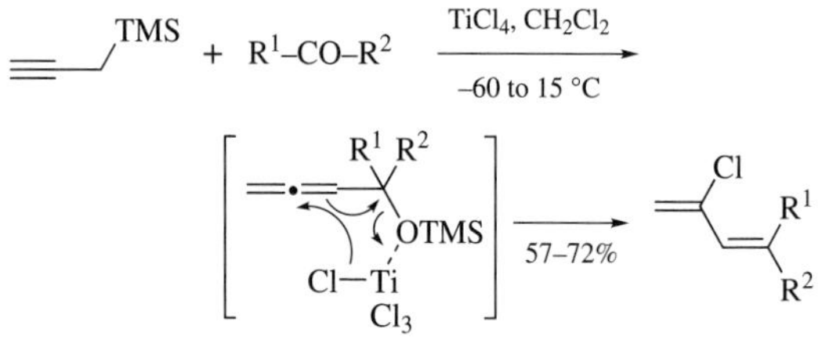

40%   25%

(18)   57–72%

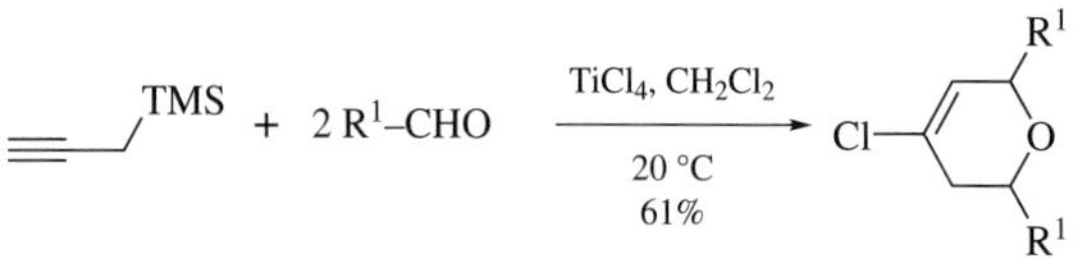

(19)   61%

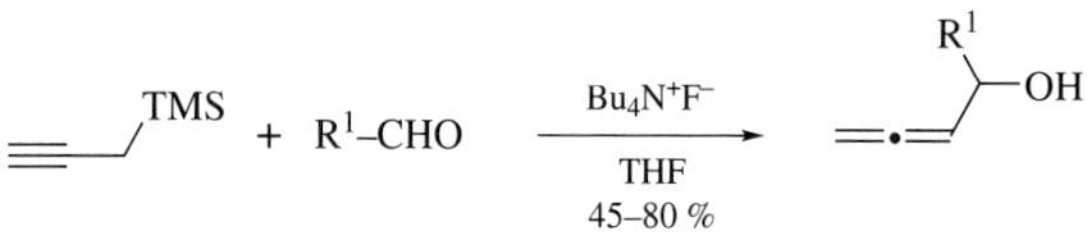

(20)

Reactions with in situ Generated Iminium and α-Acyl Iminium Ions.[30–32] These reactions allow the preparation of α-allenyl amines and amides (eqs 21 and 22). The photoinduced addition reaction leads to a mixture of two products (eq 23).[32]

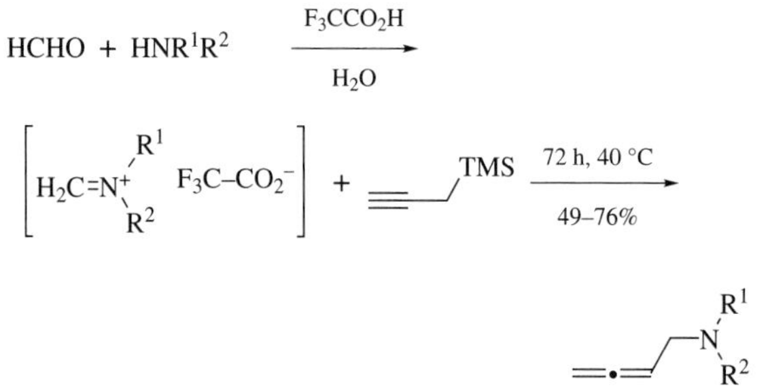

(21)

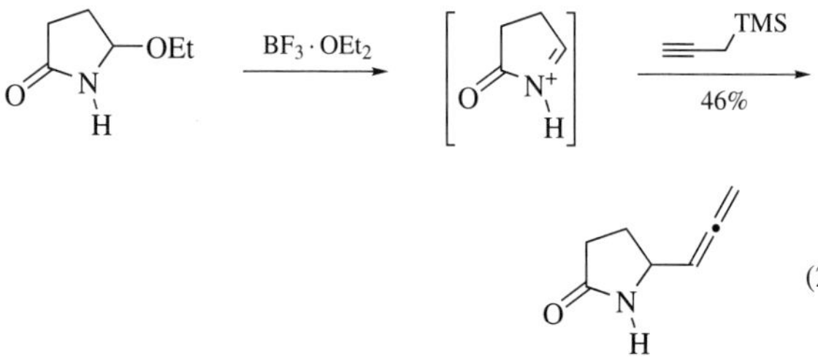

(22)

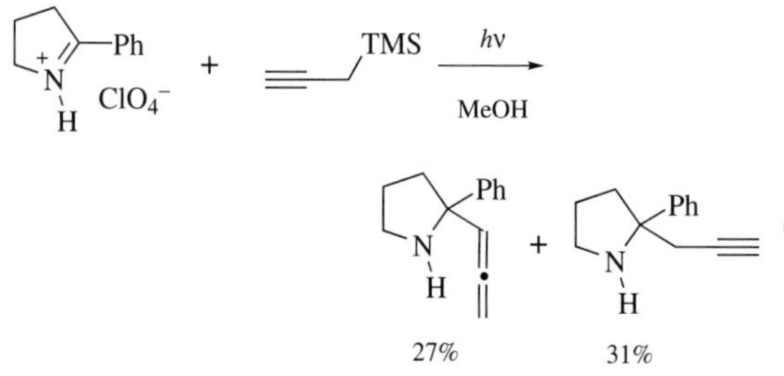

(23)

27%   31%

Reactions with Acid Chlorides.[33] These reactions lead to α-allenyl ketones (eq 24).

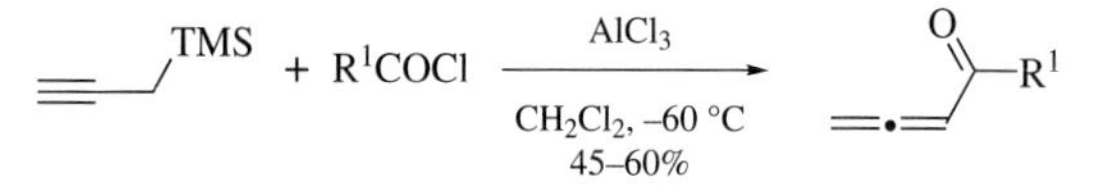

(24)   45–60%

Reactions with Conjugated Heteroatomic Systems.[3a, 34] 3-Trimethylsilyl-1-propyne seems unable to undergo 1,4-addition to monofunctional conjugated carbonyl derivatives such as $R'CH=CH–CO–R''$, $R'CH=CH–CO_2R''$, or $R'CH=CH–CN$, but readily undergoes 1,4-addition reactions with alkylidenemalonates and analogs (eq 25)[3a] and with conjugated acyl cyanides (eq 26).[34] Similar results are obtained with alkylideneacetylacetates and cyanoacetates (45–65%).[3a]

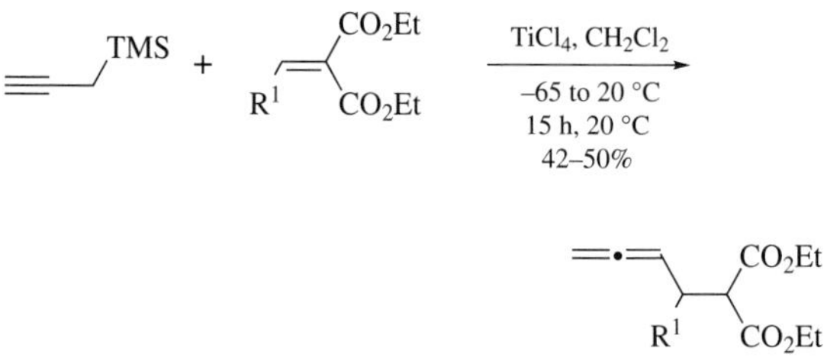

(25)

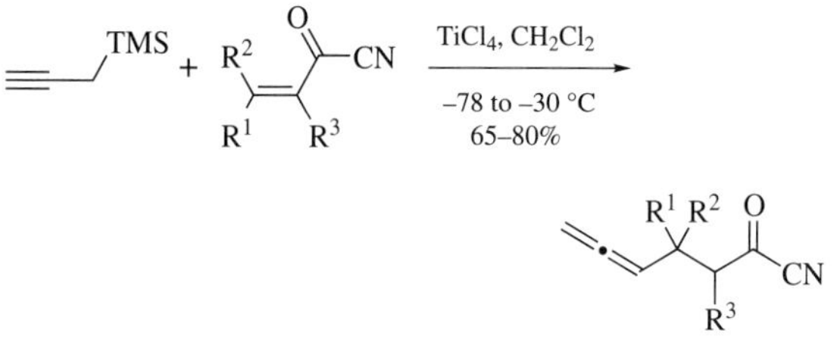

(26)

Reactions of the Triple Bond.

Preparation of Complexes with Transition Metals.[5b, 35] 3-Trimethylsilyl-1-propyne reacts with octacarbonyldicobalt to give the complex shown in eq 27.[35] This complex can react with alkenes to lead to 2-trimethylsilylmethylcyclopentenones (28–30%).

(27)

Preparation of a Bis-silylated Conjugated Enyne.[36] This is shown in eq 28.

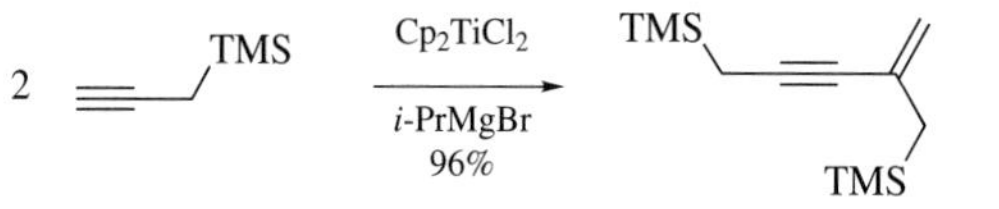

(28)   96%

First Update

Sape S. Kinderman & Henk Hiemstra
University of Amsterdam, Amsterdam, The Netherlands

Propargylation Reactions.

Alkylations. 3-Trimethylsilyl-1-propyne gets deprotonated twice using a strong base, after which it can be added to an electrophile (eq 29).[37]

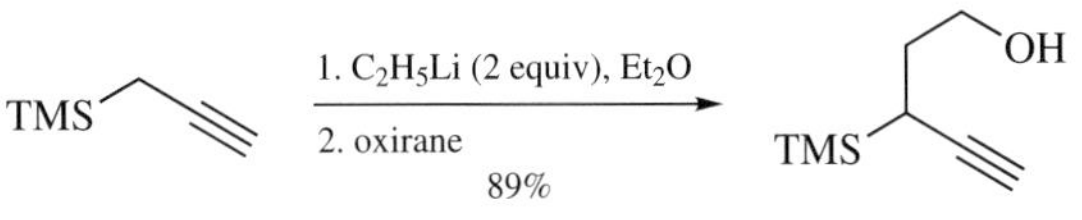
TMS
1. C_2H_5Li (2 equiv), Et_2O
2. oxirane
89%
OH
TMS

(29)

Metal Alkynilide Additions to Aldehydes, Ketones, Imines, and Nitrones. 3-Trimethylsilyl-1-propyne is transformed into the corresponding metal alkynilide by treatment with a catalytic amount of an amine base and a metal triflate. These metal alkynilides give an addition reaction, for example, to nitrones (eq 30).[38]

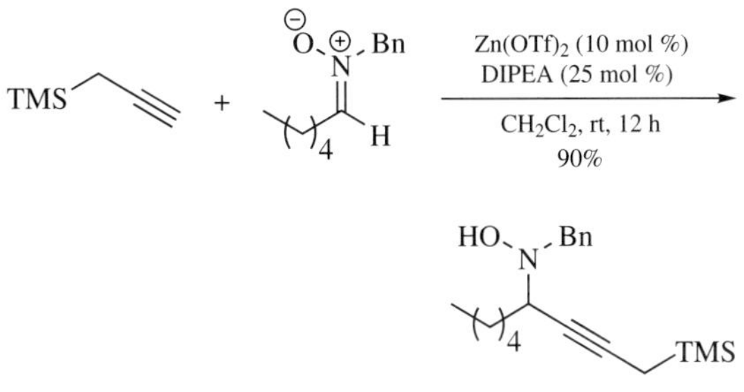
TMS
+
Bn
H
4
$Zn(OTf)_2$ (10 mol %)
DIPEA (25 mol %)
CH_2Cl_2, rt, 12 h
90%
HO
Bn
N
TMS

(30)

Using silver triflate, the reagent adds to imino esters without an amine base (eq 31).[39]

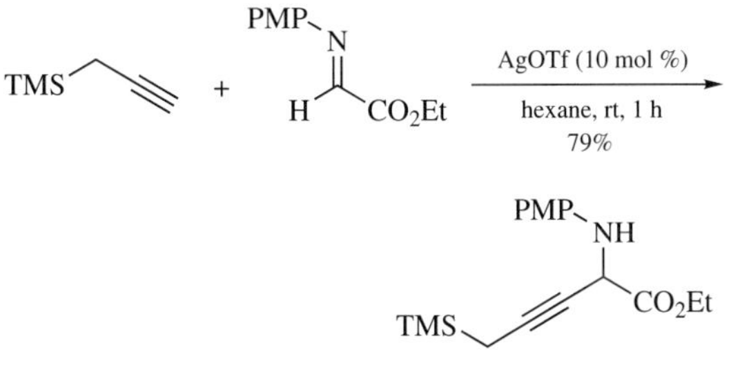
TMS
+
PMP
N
H
CO_2Et
AgOTf (10 mol %)
hexane, rt, 1 h
79%
PMP
NH
CO_2Et
TMS

(31)

Effective enantioselective versions of these additions have been developed, exemplified by eqs 32[40] and 33.[41]

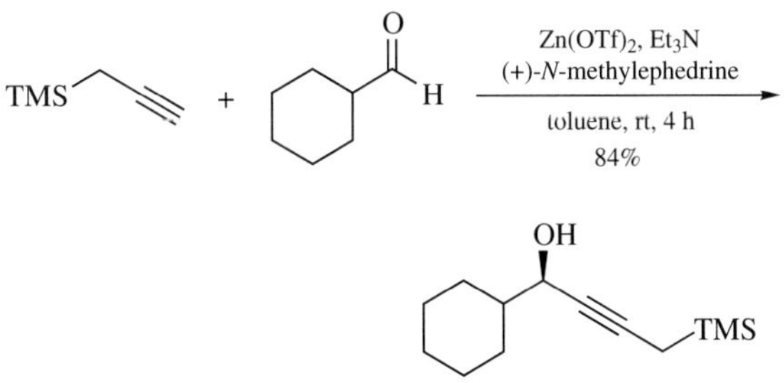
TMS
+
O
H
$Zn(OTf)_2$, Et_3N
(+)-N-methylephedrine
toluene, rt, 4 h
84%
OH
TMS

(32)

98% ee

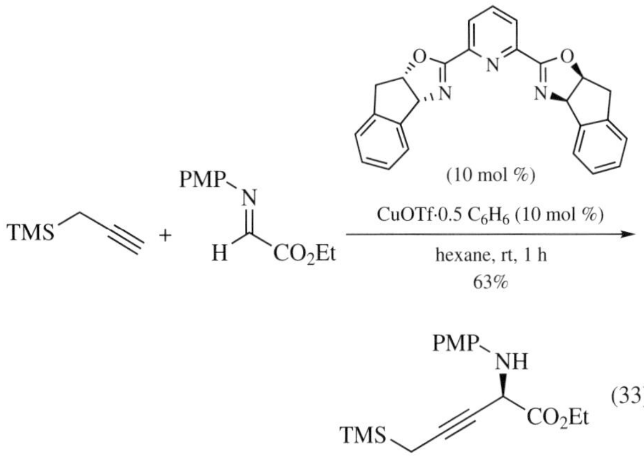
(10 mol %)
TMS
+
PMP
N
H
CO_2Et
CuOTf·0.5 C_6H_6 (10 mol %)
hexane, rt, 1 h
63%
PMP
NH
CO_2Et
TMS

(33)

77% ee

Reactions of the Triple Bond.

Preparation of Complexes. Upon reaction with $AlH_3{\cdot}NMe_3$ a carba–alane complex is formed (eq 34).[42]

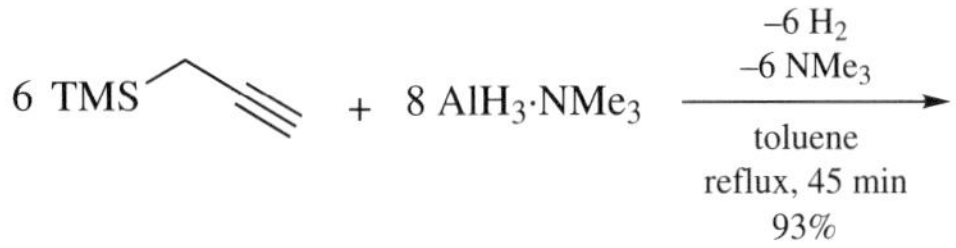
6 TMS
+ 8 $AlH_3{\cdot}NMe_3$
$-6\ H_2$
$-6\ NMe_3$
toluene
reflux, 45 min
93%

$[(AlH)_6(AlNMe_3)_2(CCH_2CH_2SiMe_3)_6]$ (34)

The triple bond of 3-trimethylsilyl-1-propyne can undergo various types of carbometallation reactions, generating very useful reactive metal alkenylides.

Hydrozirconation and Zirconium-catalyzed Carboalumination. Hydrozirconation of 3-trimethylsilyl-1-propyne using Schwartz's reagent gives the alkenylzirconocene that can be further reacted to provide 1,3-dienes after Lewis-acid-catalyzed addition to aldehydes (eq 35),[43] vinyl sulfones from reaction with sulfonyl chlorides (eq 36),[44] and allylsilanes upon transmetallation with CuBr followed by 1,4-addition (eq 37).[45]

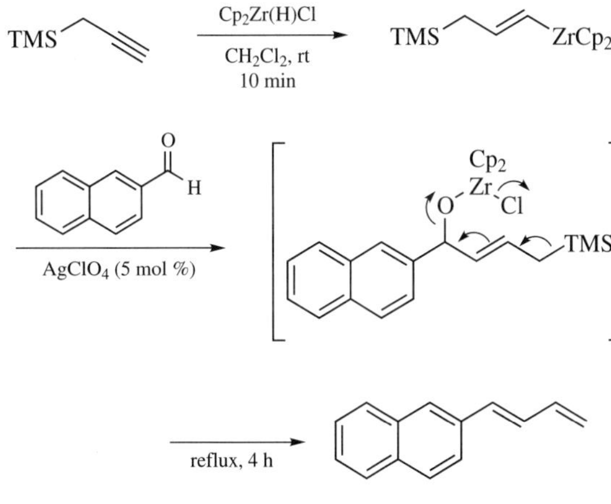
TMS
$Cp_2Zr(H)Cl$
CH_2Cl_2, rt
10 min
TMS
$ZrCp_2Cl$
O
H
$AgClO_4$ (5 mol %)
Cp_2
Zr
O
Cl
TMS
reflux, 4 h

(35)

87%, 96:4 (*E*/*Z*)

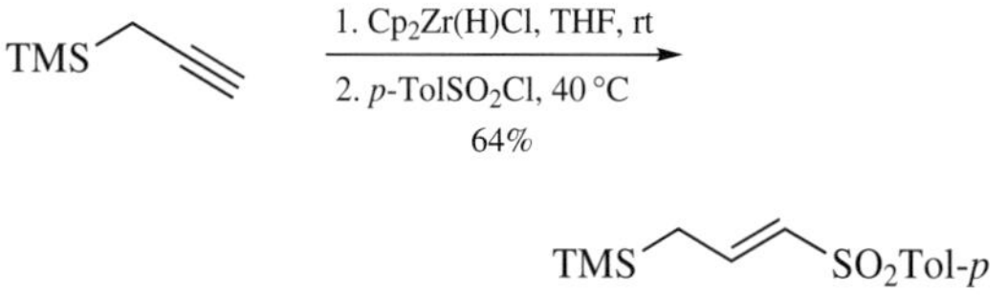

TMS —≡ $\xrightarrow[\substack{2.\ p\text{-}TolSO_2Cl,\ 40\,°C \\ 64\%}]{1.\ Cp_2Zr(H)Cl,\ THF,\ rt}$

TMS ⌢⌢ SO₂Tol-*p*   (36)

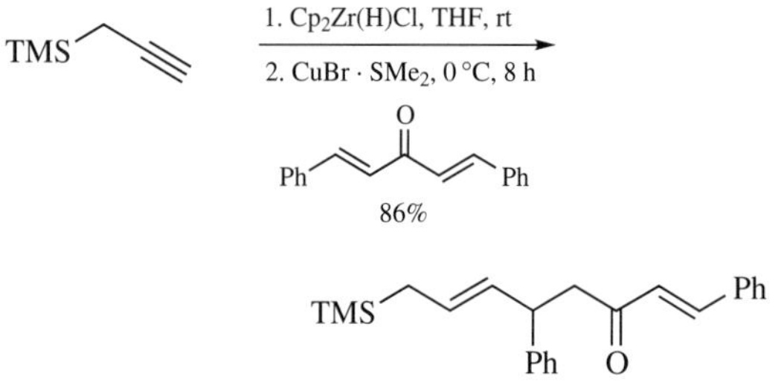

TMS —≡ $\xrightarrow[\substack{2.\ CuBr \cdot SMe_2,\ 0\,°C,\ 8\ h}]{1.\ Cp_2Zr(H)Cl,\ THF,\ rt}$

Ph ⌢⌢ Ph   (O)   86%

TMS ⌢⌢ Ph   (37)

3-Trimethylsilyl-1-propyne undergoes zirconium-catalyzed carboalumination. The intermediate aluminum compound can react with different electrophiles (eq 38).[46–48]

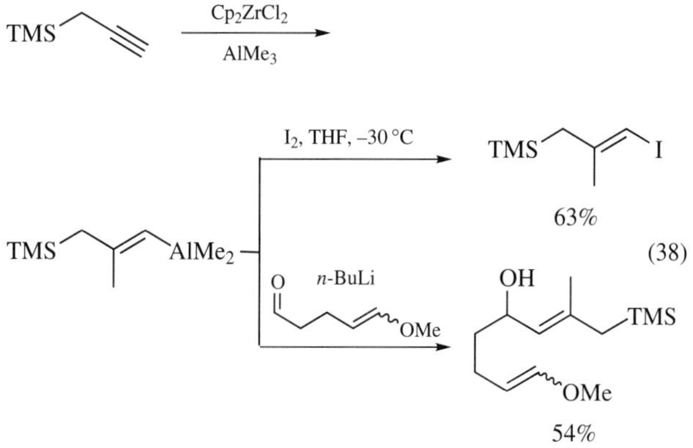

TMS —≡ $\xrightarrow[AlMe_3]{Cp_2ZrCl_2}$

TMS ⌢ AlMe₂

$\xrightarrow{I_2,\ THF,\ -30\,°C}$ TMS ⌢ I   63%   (38)

$\xrightarrow[\text{OMe}]{O\quad n\text{-BuLi}}$ TMS ... OMe   OH   54%

Silyl-cupration. Stoichiometric reaction of 3-trimethylsilyl-1-propyne with a silyl-cuprate reagent gives, after quenching with a proton source at −78 °C, a bis-silylated propene (eq 39).[49]

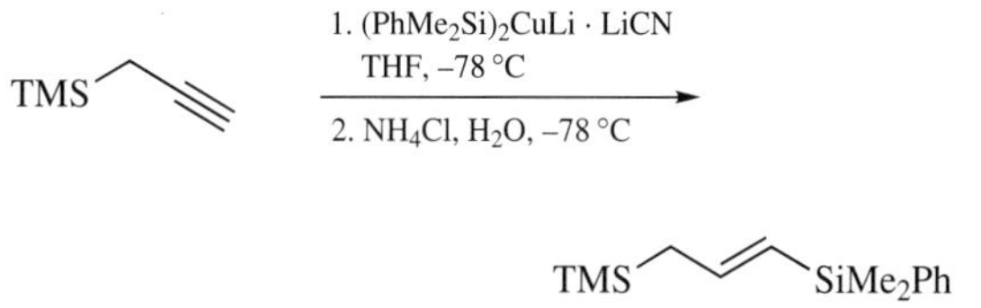

TMS —≡ $\xrightarrow[\substack{2.\ NH_4Cl,\ H_2O,\ -78\,°C}]{\substack{1.\ (PhMe_2Si)_2CuLi \cdot LiCN \\ THF,\ -78\,°C}}$

TMS ⌢⌢ SiMe₂Ph   (39)

Rhodium-catalyzed Additions. The title compound has been a test reagent in several rhodium-catalyzed additions to the triple bond. For example, a rhodium-catalyzed silylcarbamoylation in the presence of a secondary amine afforded the silylated product in reasonable yield (eq 40).[50] In the presence of a CO atmosphere and with 5% rhodium on alumina as catalyst, stereoselective hydrosilylation has been reported (eq 41).[51]

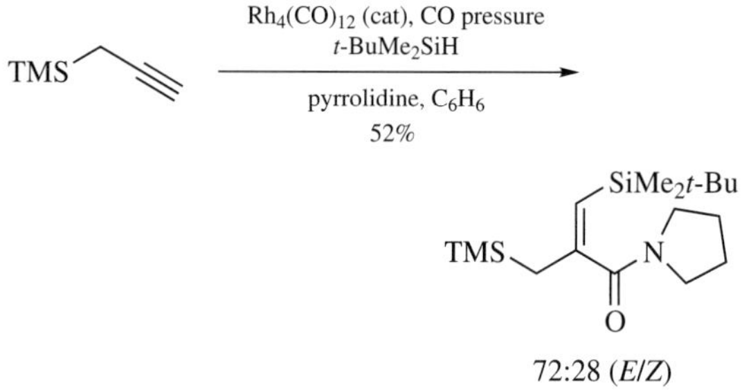

TMS —≡ $\xrightarrow[\substack{pyrrolidine,\ C_6H_6 \\ 52\%}]{\substack{Rh_4(CO)_{12}\ (cat),\ CO\ pressure \\ t\text{-}BuMe_2SiH}}$

SiMe₂*t*-Bu
TMS ⌢ N   O   (40)

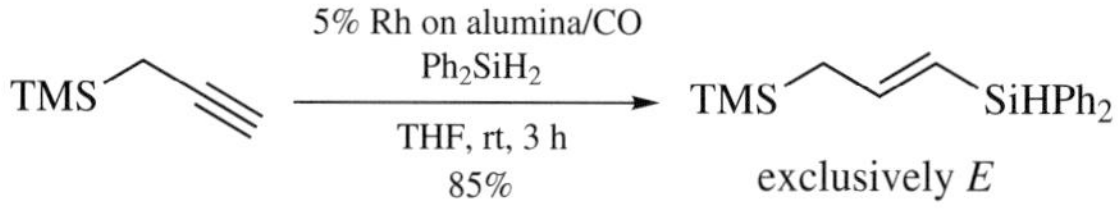

72:28 (*E/Z*)

TMS —≡ $\xrightarrow[\substack{THF,\ rt,\ 3\ h \\ 85\%}]{\substack{5\%\ Rh\ on\ alumina/CO \\ Ph_2SiH_2}}$ TMS ⌢⌢ SiHPh₂   (41)

exclusively *E*

A rhodium-catalyzed hydrophosphinylation was developed to give the corresponding alkenylphosphine oxide (eq 42).[52]

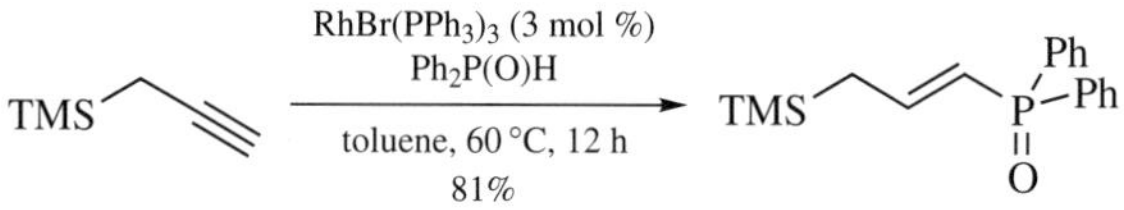

TMS —≡ $\xrightarrow[\substack{toluene,\ 60\,°C,\ 12\ h \\ 81\%}]{\substack{RhBr(PPh_3)_3\ (3\ mol\ \%) \\ Ph_2P(O)H}}$

TMS ⌢⌢ P—Ph   Ph   O   (42)

Nickel-catalyzed Carbostannylation. When 3-trimethylsilyl-1-propyne is subjected to a 1,2-diene in the presence of an alkynyl-stannane and a nickel catalyst, a carbostannylation takes place with high *Z*-selectivity under the specified conditions. Alternative conditions provided the *E*-isomer as the major product (eq 43).[53]

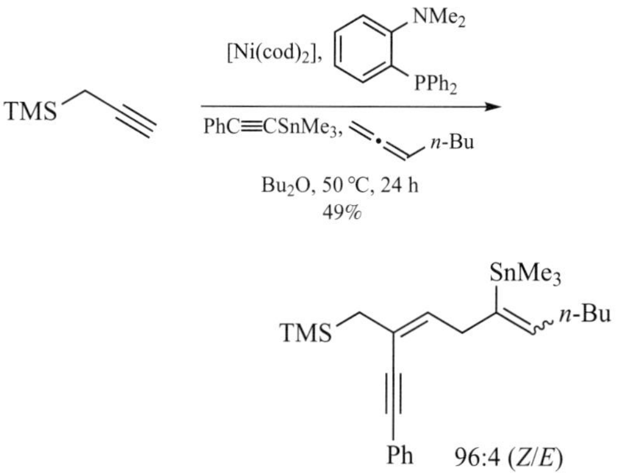

TMS —≡ $\xrightarrow[\substack{Bu_2O,\ 50\,°C,\ 24\ h \\ 49\%}]{\substack{[Ni(cod)_2],\ \text{(NMe}_2/PPh_2) \\ PhC\equiv CSnMe_3,\ \text{•}=\text{•}\ n\text{-Bu}}}$

SnMe₃
TMS ⌢ ⌢ *n*-Bu
Ph   96:4 (*Z/E*)   (43)

Miscellaneous Triple Bond Additions. An example has been published in which a Fischer carbene chromium complex adds to 3-trimethylsilyl-1-propyne, to give the Dötz benzannulation product in 62% yield (eq 44).[54]

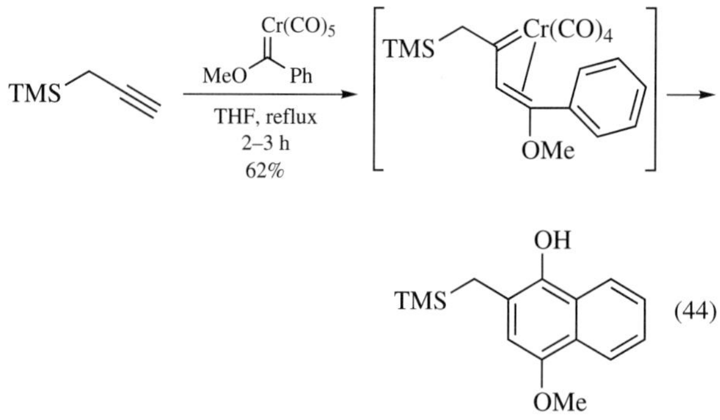

TMS —≡ $\xrightarrow[\substack{THF,\ reflux \\ 2–3\ h \\ 62\%}]{\substack{Cr(CO)_5 \\ MeO\ /\ Ph}}$

[TMS ⌢ Cr(CO)₄ / Ph / OMe] →

OH
TMS ⌢ / OMe   (44)

In a similar fashion, the intermediate chromium complex can react further to give a substituted cyclopentadiene in moderate yield (eq 45).[55]

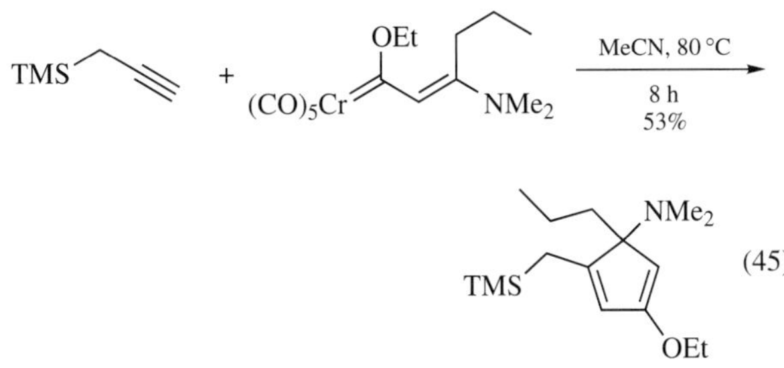

(45)

The nickel-catalyzed synthesis of 1,3-dienes has been reported utilizing the reactivity of 3-trimethylsilyl-1-propyne (eq 46).[56]

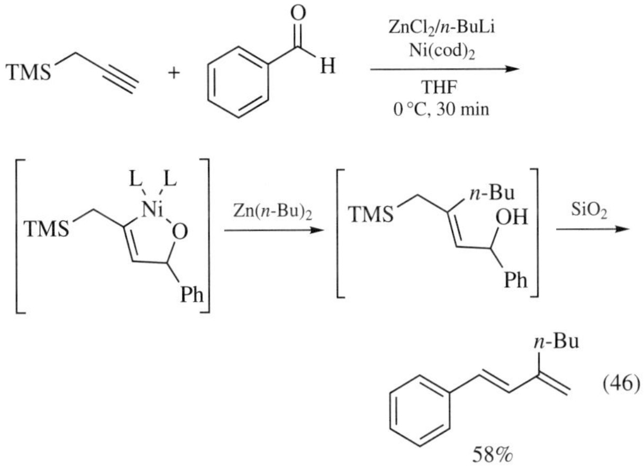

(46)

58%

3-Trimethylsilyl-1-propyne reacts regioselectively with an in situ generated sulfenic acid to afford a mixture of diastereoisomeric sulfoxides (eq 47).[57,58]

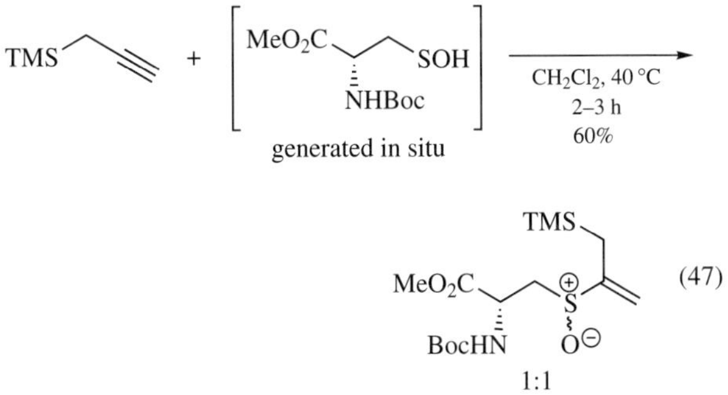

(47)

1:1

Palladium-catalyzed Cross Couplings. 3-Trimethylsilyl-1-propyne has been used in several cases to introduce an alkyne onto an aromatic ring system through a palladium-catalyzed cross coupling with the corresponding aryl halides[59–62] or to generate conjugated enynes using an alkenyl iodide or -triflate.[63,64] For each process an example is given (eqs 48[59] and 49[64]).

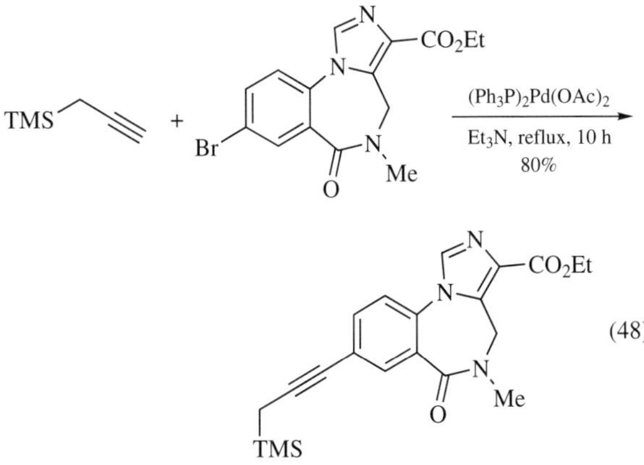

(48)

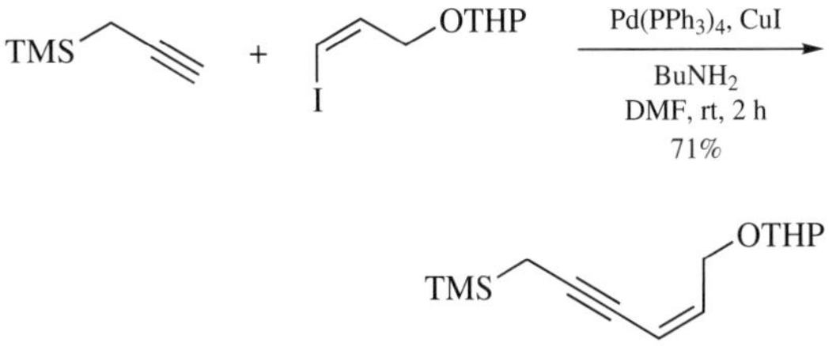

(49)

Ene Reaction. An ene reaction of 3-trimethylsilyl-1-propyne is possible with iminium ions when a non-nucleophilic counterion is used. The ene product salts can be hydrolyzed to yield the secondary amines or reduced to afford the corresponding tertiary amines (eq 50).[65]

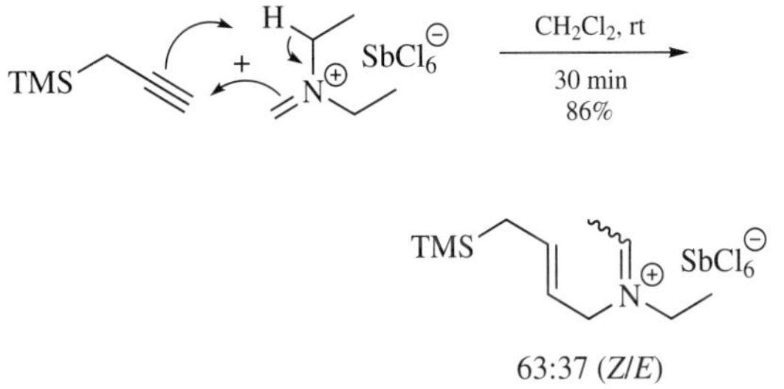

(50)

63:37 (Z/E)

Cycloadditions. 3-Trimethylsilyl-1-propyne is a useful reagent in several types of cycloadditions. Among them are [6 + 2] cycloadditions, homo Diels–Alder additions, and 1,3-dipolar cycloadditions. For example, a cobalt(I)-catalyzed [6 + 2] cycloaddition with cycloheptatriene in 1,2-dichloroethane affords the product in good yield (eq 51).[66] In a similar reaction with cyclooctatetraene, a mixture of compounds is obtained.[67]

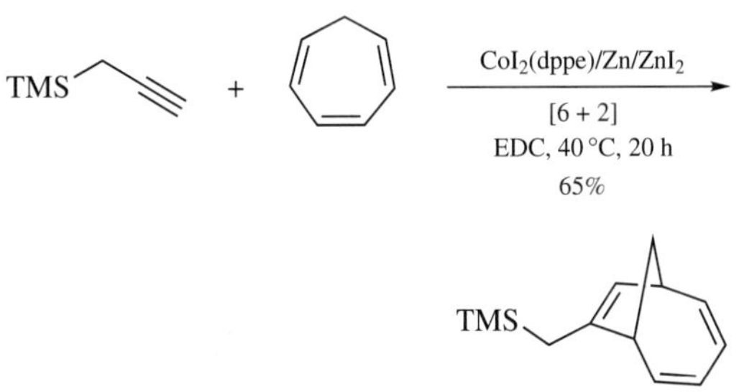

(51)

The cobalt-catalyzed $[2\pi + 2\pi + 2\pi]$ homo Diels–Alder of 3-trimethylsilyl-1-propyne with norbornadiene delivers in good yield the expected multicyclic product (eq 52).[68]

(52)

3-Trimethylsilyl-1-propyne reacts as the dipolarophile with an in situ generated cyclic carbonyl ylide to give two cycloadduct products in an 86:14 ratio (eq 53).[69]

86:14
(53)

The title reagent reacts with in situ generated azido-4-methyl-benzene as the 1,3-dipole at 70 °C in the presence of copper(I) as catalyst (eq 54).[70]

(54)

3-Trimethylsilyl-1-propyne is utilized as a crucial reagent in the synthesis of Feist's esters. Thus, it reacts with ethyl diazoacetate in the presence of a chiral rhodium catalyst. This induces an asymmetric [2 + 1] cycloaddition in reasonable yield with high ee (eq 55).[71] Desilylation and attack of the allylsilane onto carbon dioxide gives the monoester of Feist's acid. Esterification affords the diester of Feist's acid in almost enantiopure form.

(55)
98% ee

Related Reagents. (Trimethylsilyl)allene.

A list of General Abbreviations appears on the front Endpapers

1. Masson, J. C.; Le Quan, M.; Cadiot, P., *Bull. Soc. Claim. Fr., Part 3* **1967**, 777.
2. Slutsky, J.; Kwart, H., *J. Am. Chem. Soc.* **1973**, *95*, 8678.
3. (a) Pornet, J.; Kolani, N'B.; Mesnard, D.; Miginiac, L.; Jaworski, K., *J. Organomet. Chem.* **1982**, *236*, 177. (b) Damour, D. Ph.D. Thesis, University of Poitiers (France), 1987.
4. (a) Bourgeois, P.; Mérault, G., *J. Organomet. Chem.* **1972**, *39*, C44. (b) Mérault, G.; Bourgeois, P.; Dunoguès, J., *C. R. Hebd. Seances Acad. Sci., Ser. C* **1972**, *274*, 1857.
5. (a) Weber, W. P. *Silicon Reagents for Organic Synthesis*; Springer: New York, 1983: p 136. (b) Weber, W. P. *Silicon Reagents for Organic Synthesis*; Springer: New York, 1983: p 179.
6. Dunoguès, J., *Actualité Chim.* **1986**, *3*, 11.
7. Pornet, J.; Mesnard, D.; Miginiac, L., *Tetrahedron Lett.* **1982**, *23*, 4083.
8. Pornet, J.; Kolani, N'B.; Miginiac, L., *Tetrahedron Lett.* **1981**, *22*, 3609.
9. Schinzer, D.; Solyom, S.; Becker, M., *Tetrahedron Lett.* **1985**, *26*, 1831.
10. (a) Hiemstra, H.; Sno, M. H. A. M.; Vijn, R. J.; Speckamp, W. N., *J. Org. Chem.* **1985**, *50*, 4014. (b) Hiemstra, H.; Klaver, W. J.; Speckamp, W. N., *J. Org. Chem.* **1984**, *49*, 1149.
11. Pornet, J.; Damour, D.; Miginiac, L., *Tetrahedron* **1986**, *42*, 2017.
12. Pornet, J.; Randrianoélina, B.; Miginiac, L., *Tetrahedron Lett.* **1984**, *25*, 651.
13. Damour, D.; Pornet, J.; Miginiac, L., *J. Organomet. Chem.* **1988**, *349*, 43.
14. Mastalerz, H., *J. Org. Chem.* **1984**, *49*, 4092.
15. Pornet, J.; Damour, D.; Randrianoélina, B.; Miginiac, L., *Tetrahedron* **1986**, *42*, 2501.
16. Nativi, C.; Taddéi, M.; Mann, A., *Tetrahedron Lett.* **1987**, *28*, 347.
17. Hojo, M.; Tomita, K.; Hosomi, A., *Tetrahedron Lett.* **1993**, *34*, 485.
18. Bourgeois, P.; Mérault, G., *C. R. Hebd. Seances Acad. Sci., Ser. C* **1971**, *273*, 714.
19. Ohno, M.; Matsuoka, S.; Eguchi, S., *J. Org. Chem.* **1986**, *51*, 4553.
20. Uemura, M.; Kobayashi, T.; Hayashi, Y., *Synthesis* **1986**, 386.
21. (a) Ochiai, M.; Ito, T.; Takaoka, Y.; Masaki, Y., *J. Am. Chem. Soc.* **1991**, *113*, 1319. (b) Ochiai, M.; Ito, T.; Masaki, Y., *J. Chem. Soc., Chem. Commun.* **1992**, 15. (c) Ochiai, M.; Ito, T.; Shiro, M., *J. Chem. Soc., Chem. Commun.* **1993**, 218.
22. Pornet, J., *Tetrahedron Lett.* **1981**, *22*, 453.
23. Pornet, J.; Miginiac, L.; Jaworski, K.; Randrianoélina, B., *Organometallics* **1985**, *4*, 333.
24. Brückner, C.; Holzinger, H.; Reissig, H-U., *J. Org. Chem.* **1988**, *53*, 2450.
25. Bertozzi, C. R.; Bednarski, M. D., *Tetrahedron Lett.* **1992**, *33*, 3109.
26. Babirad, S. A.; Wang, Y.; Kishi, Y., *J. Org. Chem.* **1987**, *52*, 1370.
27. Deleris, G.; Dunoguès, J.; Calas, R., *J. Organomet. Chem.* **1975**, *93*, 43.
28. Coppi, L.; Ricci, A.; Taddéi, M., *Tetrahedron Lett.* **1987**, *28*, 973.
29. Pornet, J., *Tetrahedron Lett.* **1981**, *22*, 455.
30. Damour, D.; Pornet, J.; Randrianoélina, B.; Miginiac, L., *J. Organomet. Chem.* **1990**, *396*, 289.
31. (a) Hiemstra, H.; Fortgens, H. P.; Speckamp, W. N., *Tetrahedron Lett.* **1984**, *25*, 3115. (b) Esch, P. M.; Hiemstra, H.; Speckamp, W. N., *Tetrahedron Lett.* **1988**, *29*, 367.
32. Haddaway, K.; Somekawa, K.; Fleming, P.; Tossell, J. A.; Mariano, P. S., *J. Org. Chem.* **1987**, *52*, 4239.
33. Pillot, J. P.; Bennetau, B.; Dunoguès, J.; Calas, R., *Tetrahedron Lett.* **1981**, *22*, 3401.
34. (a) Santelli, M.; El Abed, D.; Jellal, A., *J. Org. Chem.* **1986**, *51*, 1199. (b) Jellal, A.; Santelli, M., *Tetrahedron Lett.* **1980**, *21*, 4487.
35. Billington, D. C.; Kerr, W. J.; Pauson, P. L., *J. Organomet. Chem.* **1988**, *341*, 181.
36. Akita, M.; Yasuda, H.; Nakamura, A., *Bull. Chem. Soc. Jpn.* **1984**, *57*, 480.

37. Pornet, J.; Aubert, P.; Randrianoélina, B.; Miginiac, L., *J. Organomet. Chem.* **1994**, *481*, 217.

38. Frantz, D. E.; Fässler, R.; Carreira, E. M., *J. Am. Chem. Soc.* **1999**, *121*, 11245.

39. Ji, J.-X.; Au-Yeung, T. T.-L.; Wu, J.; Yip, C. W.; Chan, A. S. C., *Adv. Synth. Catal.* **2004**, *346*, 42.

40. Frantz, D. E.; Fässler, R.; Carreira, E. M.; *J. Am. Chem. Soc.* **2000**, *122*, 1806.

41. Ji, J.-X.; Wu, J.; Chan, A. S. C., *Proc. Natl. Acad. Sci. USA* **2005**, *102*, 11196.

42. Stasch, A.; Ferbinteanu, M.; Prust, J.; Zheng, W.; Cimpoesu, F.; Roesky, H. W.; Magull, J.; Schmidt, H.-G., Noltemeyer, M., *J. Am. Chem. Soc.* **2002**, *124*, 5441.

43. Suzuki, K.; Hasegawa, T.; Imai, T.; Maeta, H.; Ohba, S., *Tetrahedron* **1995**, *51*, 4483.

44. Duan, D.-H.; Huang, X., *Synlett* **1999**, 317.

45. Huang, X.; Pi, J., *Synlett* **2003**, 481.

46. Negishi, E.; Luo, F.-T.; Rand, C. L., *Tetrahedron Lett.* **1982**, *23*, 27.

47. Rayner, C. M.; Astles, P. C.; Paquette, L. A., *J. Am. Chem. Soc.* **1992**, *114*, 3926.

48. Frey, D. A.; Reddy, S. H. K.; Moeller, K. D., *J. Org. Chem.* **1999**, *64*, 2805.

49. Angeles Cubillo de Dios, M.; Fleming, I.; Friedhoff, W.; Woode, P. D. W., *J. Organomet. Chem.* **2001**, *624*, 69.

50. Matsuda, I.; Takeuchi, K.; Itoh, K., *Tetrahedron Lett.* **1999**, *40*, 2553.

51. JoongLee, S.; KyeuPark, M.; HeeHan, B., *Silicon Chemistry* **2002**, *1*, 41.

52. Han, L-B.; Zhao, C-Q.; Tanaka, M., *J. Org. Chem.* **2001**, *66*, 5929.

53. Shirakawa, E.; Yamamoto, Y.; Nakao, Y.; Oda, S.; Tsuchimoto, T.; Hiyama, T., *Angew. Chem., Int. Ed.* **2004**, *43*, 3448.

54. Patel, P. P.; Zhu, Y.; Zhang, L.; Herndon, J. W., *J. Organomet. Chem.* **2004**, *689*, 3379.

55. Flynn, B. L.; Funke, F. J.; Silveira, C. C.; de Meijere, A., *Synlett* **1995**, 1007.

56. Qi, X.; Montgomery, J., *J. Org. Chem.* **1999**, *64*, 9310.

57. Aversa, M. C.; Barattucci, A.; Bonaccorsi, P.; Giannetto, P., *J. Org. Chem.* **2005**, *70*, 1986.

58. Aversa, M. C.; Barattucci, A.; Bonaccorsi, P.; Giannetto, P., *Phosphorus, Sulfur Silicon Relat. Elem.* **2005**, *180*, 1203.

59. Liu, R.; Hu, R. J.; Zhang, P.; Skolnick, P.; Cook, J. M., *J. Med. Chem.* **1996**, *39*, 1928.

60. Seela, F.; Zulauf, M., *Synthesis* **1996**, 726.

61. Kottysch, T.; Ahlborn, C.; Brotzel, F.; Richert, C., *Chem. Eur. J.* **2004**, *10*, 4017.

62. Yao, T.; Campo, M. A.; Larock, R. C.; *J. Org. Chem.* **2005**, *70*, 3511.

63. Suffert, J.; Eggers, A.; Scheuplein, S. W.; Brückner, R., *Tetrahedron Lett.* **1993**, *34*, 4177.

64. Kabbara, J.; Hoffmann, C.; Schinzer, D., *Synthesis* **1995**, 299.

65. Ofial, A. R.; Mayr, H., *Angew. Chem., Int. Ed. Engl.* **1997**, *36*, 143.

66. Achard, M.; Tenaglia, A.; Buono, G., *Org. Lett.* **2005**, *7*, 2353.

67. Achard, M.; Mosrin, M.; Tenaglia, A.; Buono, G., *J. Org. Chem.* **2006**, *71*, 2907.

68. Lautens, M.; Tam, W.; Lautens, J. C.; Edwards, L. G.; Crudden, C. M.; Smith, A. C., *J. Am. Chem. Soc.* **1995**, *117*, 6863.

69. Hodgson, D. M.; Le Strat, F.; Avery, T. D.; Donohue, A. C.; Brückl, T., *J. Org. Chem.* **2004**, *69*, 8796.

70. Andersen, J.; Bolvig, S.; Liang, X., *Synlett* **2005**, 2941.

71. Wheatherhead-Kloster, R. A.; Corey, E. J., *Org. Lett.* **2006**, *8*, 171.

3-Trimethylsilyl-2-propynylcopper(I)[1]

[55630-32-5]   C$_6$H$_{11}$CuSi   (MW 174.81)

InChI = 1S/C6H11Si.Cu/c1-5-6-7(2,3)4;/h1H2,2-4H3;

InChIKey = VKHYHOFWVNJNQN-UHFFFAOYSA-N

(agent for the introduction of the trimethylsilylpropynyl group via 1,6-conjugate additions[2] and vinyl substitutions;[3] preparation of lithium bis(trimethylsilylpropynyl)cuprate[4])

Alternate Name: trimethylsilylpropargylcopper

Physical Data: brown heterogenous solution in ether.

Solubility: sol pyridine, THF; slightly sol ether.

Analysis of Reagent Purity: addition of aqueous ammonium chloride gives >95% conversion to 1-trimethylsilylpropyne.[2]

Preparative Method: a solution of 1-lithio-3-trimethylsilyl-1-propyne containing *N,N,N',N'*-tetramethylethylenediamine (TMEDA) in ether[2] or THF is added to a slurry of copper(I) iodide in ether at −78 °C, then warmed to −10 °C and used immediately.[3]

Handling, Storage, and Precautions: dry copper(I) acetylides are potentially explosive. Use in a fume hood.

Coupling Reactions. Trimethylsilylpropynylcopper adds to conjugated esters in a 1,6-manner if facilitated, giving mixtures of products dependent on the nature of the system.[2] Addition to an unsubstituted (*E*)-dienoic ester gave an 8:2 mixture of allene:propynyl-substituted adducts (eq 1).

$$\text{(eq 1)}$$

80%   20%   (1)

Best conversion yields (80%) are obtained by using 3.5 equiv of the copper reagent; use of 1.2 equiv results in a lower yield (42%) but an unaltered ratio.[2] If a δ-substituted dienoic ester is employed, the ratio reverses (eq 2), showing marked sensitivity to steric effects in this region;[2] for best conversion to products (70%), 3.5 equiv of copper reagent are employed.[2]

$$\text{(eq 2)}$$

18%   82%   (2)

Finally, if a β,γ-disubstituted dienoic ester is used, selectivity reverses in favor of the allenic adduct (eq 3), pointing to the δ-carbon as the single most important dictator of steric control in this reaction (in this sequence, only 1.2 equiv of cuprate were employed, accounting for the low (35%) overall conversion to products). Results obtained using *t*-butyldimethylsilylpropynylcopper paralleled those indicated above. As expected, however, the ratios of products showed sensitivity towards this more sterically bulky coupling agent, with a 95:5 ratio of allene:alkyne products in the addition to the unsubstituted dienoate (eq 1) and a 97:3 ratio of alkyne:allene in the addition to the δ-substituted system (eq 2).[2] Vinyl substitution has been reported, using an iodohexene and 2 equiv of trimethylsilylpropynylcopper; the corresponding adduct was obtained in 80% yield (eq 4).[3] Using 1 equiv of copper acetylide, however, the yield drops to 63%. A mechanism has been proposed to account for this coupling, and is assumed to involve a four-centered transition state assembly.[3]

$$\text{(3)}$$

85%   15%

$$\text{(4)}$$

0–5 °C, pyridine, 2 h
80%

Lithioorganocuprates. Lithium bis(trimethylsilylpropynyl)-copper, prepared by addition of lithium trimethylsilylpropyne to copper(I) iodide, has been used as a nucleophilic source of the trimethylsilylpropynyl group (eq 5).[4] Chemoselective 1,2-addition resulted in a 90% isolated yield of the desired secondary alkynyl alcohol (eq 5); use of the corresponding lithium or bromomagnesium acetylides resulted in substantial recovery of allenic products.[4]

$$\text{(5)}$$

THF, –78 °C
90%

Related Reagents. 1-Hexynylcopper(I); 1-Pentynylcopper(I); (Trimethylsilyl)ethynylcopper(I).

1. (a) Normant, J. F., *Synthesis* **1972**, 63. (b) Sladkov, A. M.; Gol'ding, I. R., *Russ. Chem. Rev. (Engl. Transl.)* **1979**, *48*, 868. (c) Lipshutz, B. H.; Sengupta, S., *Org. React.* **1992**, *41*, 135. (d) *Fieser & Fieser* **1977**, *6*, 638.

2. Ganem, B., *Tetrahedron Lett.* **1974**, 4467.

3. Commercon, A.; Normant, J.; Villieras, J., *J. Organomet. Chem.* **1975**, *93*, 415.

4. Vedejs, E.; Dent, W. H.; Gapinski, D. M.; McClure, C. K., *J. Am. Chem. Soc.* **1987**, *109*, 5437.

Graham B. Jones & Robert S. Huber
Clemson University, SC, USA

2-(Trimethylsilyl)thiazole[1]

[79265-30-8] $C_6H_{11}NSSi$ (MW 157.34)

InChI = 1S/C6H11NSSi/c1-9(2,3)6-7-4-5-8-6/h4-5H,1-3H3
InChIKey = VJCHUDDPWPQOLH-UHFFFAOYSA-N

(useful formyl anion equivalent;[1] effective one-carbon homologating reagent of alkoxy aldehydes, amino aldehydes, and dialdoses[2])

Physical Data: bp 51–53 °C/10 mmHg; d 0.992 g mL^{-1}; n_D 1.4980.
Solubility: insol H_2O; very sol CH_2Cl_2, diethyl ether, THF.
Form Supplied in: liquid; widely available.
Analysis of Reagent Purity: ^{1}H and ^{13}C NMR.
Purification: distillation under reduced pressure.
Preparative Method: although 2-(trimethylsilyl)thiazole (**1**) is commercially available, it can be easily prepared on a multigram scale from 2-bromothiazole, *n*-butyllithium, and chlorotrimethylsilane as shown in eq 1.[3]

$$\text{(1)}$$

1. BuLi
2. TMSCl

(1)

Handling, Storage, and Precautions: is stable upon storage under an inert atmosphere in a refrigerator. Nevertheless, it should be freshly distilled for best results. Irritating to the eyes, respiratory system, and skin. Wear suitable protective clothing, gloves and eye/face protection. Use in a fume hood.

Original Commentary

Alessandro Dondoni
University of Ferrara, Ferrara, Italy

Pedro Merino
University of Zaragoza, Zaragoza, Spain

General Discussion. 2-(Trimethylsilyl)thiazole (**1**) reacts with various C-electrophiles such as azolium halides (eq 2),[4] ketenes (eq 3),[5,6] acyl chlorides (eq 4),[6,7] ketones, and aldehydes (eq 5)[6–8] under mild conditions to give the corresponding 2-substituted thiazoles in very good isolated yields. No catalysts are required in these carbodesilylation reactions.

$$\text{(2)}$$

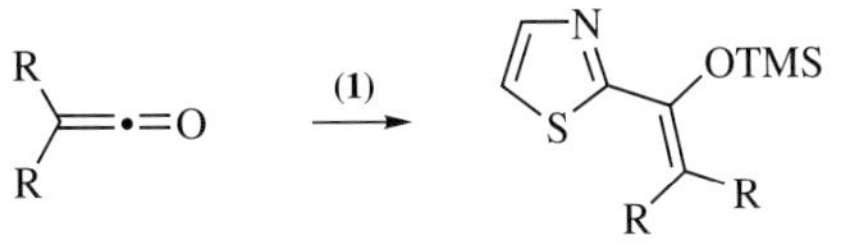

$$(3)$$

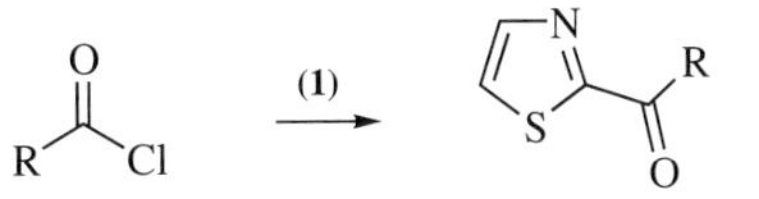

$$(4)$$

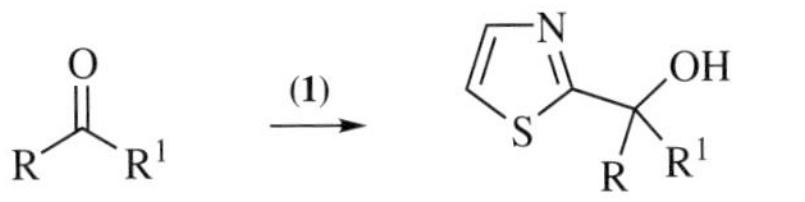

$$(5)$$

R = H, CF$_3$

The general mechanism via an *N*-thiazolium-2-ylide, earlier proposed for all these reactions,[6,7] has found substantial support in the identification of spirodioxolane intermediates in the reaction of (**1**) with aldehydes.[9]

The reaction of (**1**) with chiral alkoxy aldehydes led to the discovery of a new methodology for the one-carbon chain elongation of these compounds.[2a,b] The methodology proved to be capable of iteration and therefore suitable for the stereoselective synthesis of polyalkoxy aldehydes (carbohydrate-like materials).

The homologation of D-glyceraldehyde acetonide (**2**) serves to illustrate the method.[3] This consists of two sequential operations: first, addition of 2-(trimethylsilyl)thiazole (**1**) to the aldehyde (**2**) followed by in situ desilylation and protection as the *O*-benzyl derivative (**3**) (eq 6); secondly, aldehyde release in the resultant adduct[10] by a three-step sequence involving methylation of the thiazole ring to the *N*-methylthiazolium salt; reduction to thiazolidine; hydrolysis to α-benzyloxy aldehyde (**4**) (eq 7).

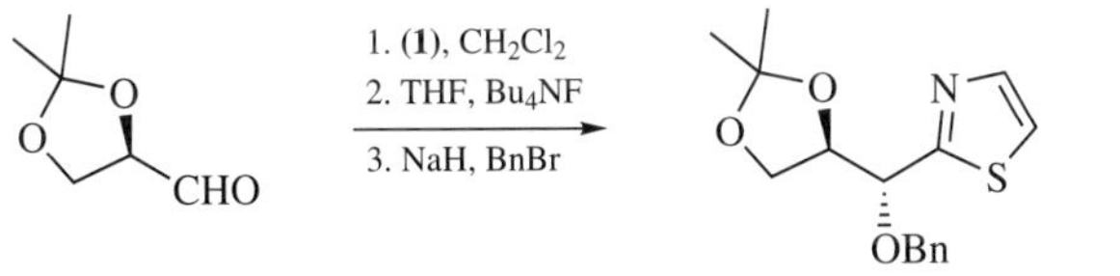

$$(6)$$

ds 95%

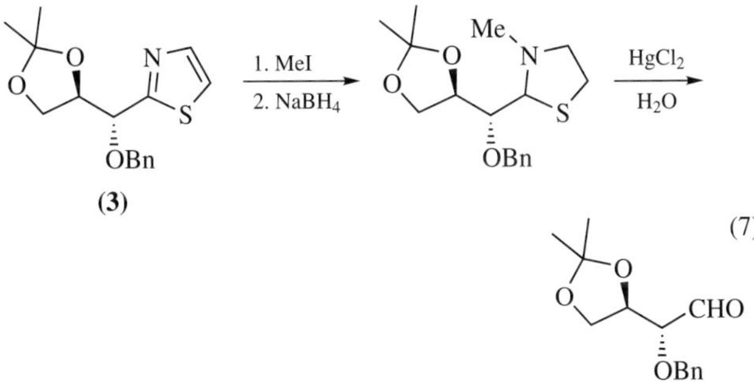

$$(7)$$

Hence, (**1**) serves as a synthetic equivalent for the formyl anion synthon which adds to the aldehyde in a stereoselective manner, creating a new chiral hydroxymethylene center. The repetition of the first and second operational sequences above over six consecutive cycles with high levels of diastereoselectivity and chemical yields in each cycle provided a series of protected D-aldoses[2c] having up to nine carbons in the chain (eq 8).

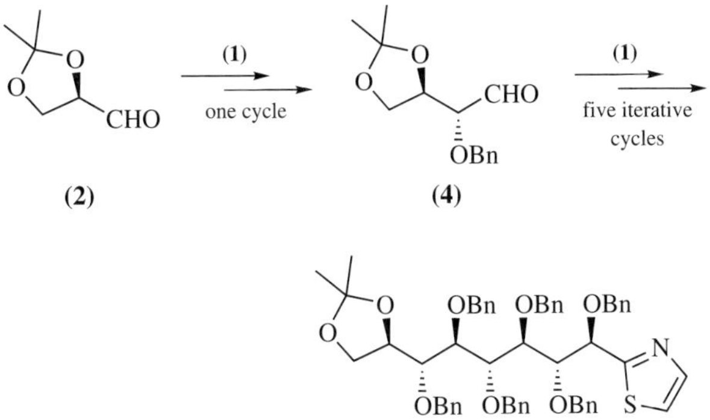

$$(8)$$

The utility of this approach to long-chain sugars (thiazole route to carbohydrates) is illustrated by the conversion of protected L-threose (**5**) and dialdogalactopyranose (**6**) into higher homologs (eqs 9 and 10).[2c,d] The extension of this methodology to other dialdoses has been reported.[11]

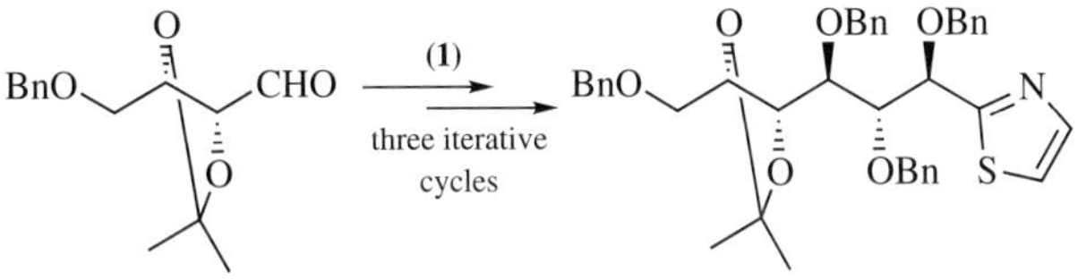

$$(9)$$

$$(10)$$

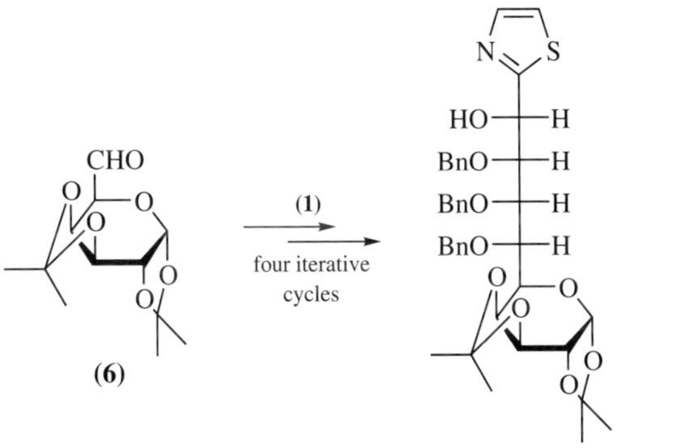

This technique, in combination with an inversion of configuration of the carbon adjacent to the thiazole ring, has been employed[2e] for the chain extension of D-glyceraldehyde acetonide (**2**) into all possible tetrose and pentose homologs (eq 11).

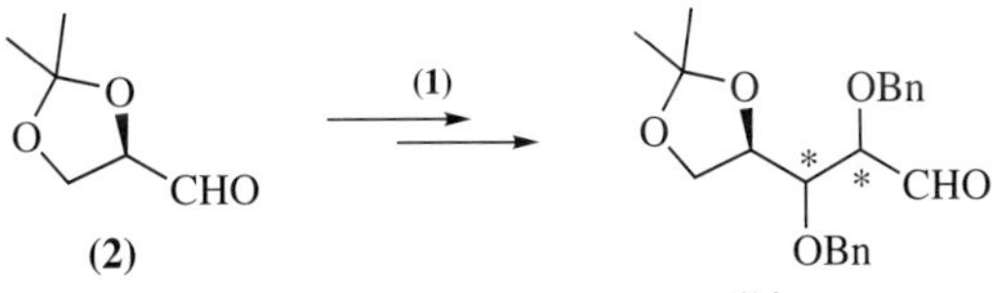

$$(11)$$

all isomers

The 2-(trimethylsilyl)thiazole-based one-carbon extension strategy was also used for the construction of amino tetroses as well as amino pentoses[12] using α-amino aldehydes as chiral educts. An interesting finding was that the reactions of (**1**) with the *N*-diprotected α-amino aldehydes were *anti* selective (eq 12), whereas those with the *N*-monoprotected α-amino aldehydes were *syn* selective (eq 13).

These opposite diastereoselectivities were explained by assuming a Felkin–Anh–Houk model (*anti* addition) and a proton-bridged Cram cyclic model (*syn* addition) for the case of differentially protected α-amino aldehydes (**7**) and (**8**). The resulting

1,2-amino alcohols proved to be convenient precursors to L-amino sugars and sphingosines.[12]

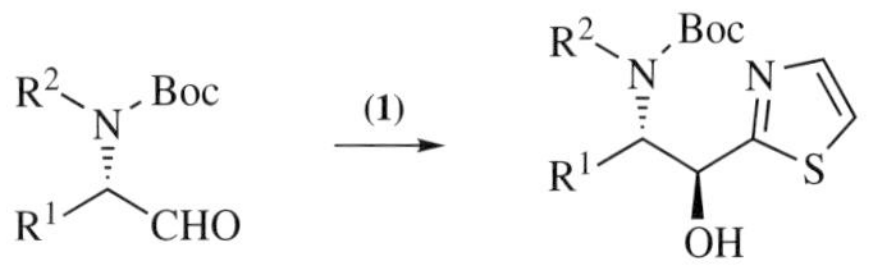

(12)

ds = 85–92%

(13)

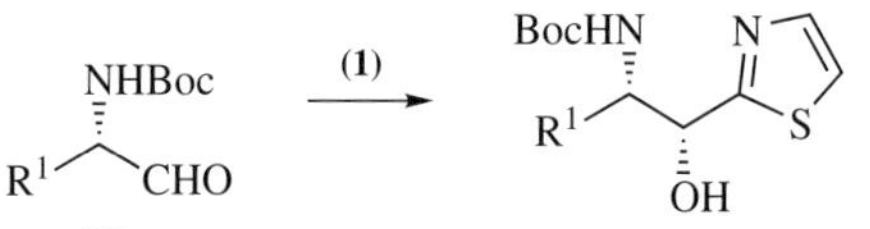

ds = 80%

This strategy has been also employed for the synthesis of a dipeptide mimic,[13] as illustrated in eq 14.

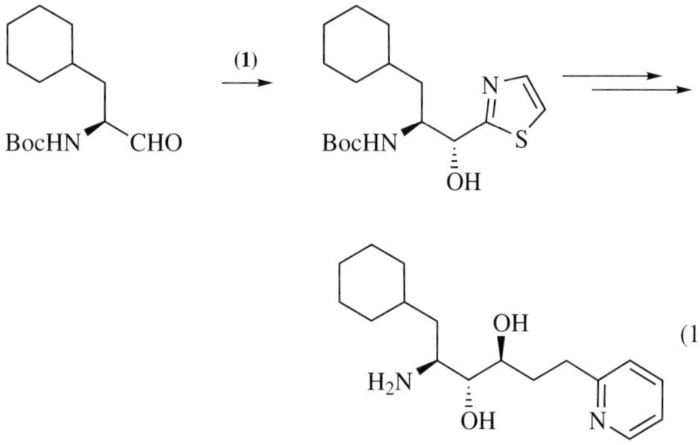

(14)

In summary, 2-(trimethylsilyl)thiazole (1) appears to be a useful formyl anion equivalent.[14] The advantages over other precursors to the formyl anion can be found in the stability of the thiazole ring to a wide range of reaction conditions and its ready conversion into the formyl group under mild and neutral conditions. Few of the many formyl anion equivalents[14] have been demonstrated to be capable of producing labile α-alkoxy aldehydes without racemization.

First Update

Francisco Sánchez-Sancho, María M. Zarzuelo,
& María Garranzo

PharmaMar S.A., Colmenar Viejo, Spain

Homologation of Aldehydes. The use of the thiazole ring as a convenient precursor of the formyl group (*thiazole aldehyde synthesis*) has been widely documented by Dondoni and coworkers.[15,16] Especially 2-(trimethylsilyl)thiazole (2-TST; 1),[17,18] known as Dondoni's reagent, has shown its synthetic usefulness and versatility in the preparation of a variety of biologically active compounds.[19] The utilization of 2-TST (1) as an effective formyl anion equivalent in the homologation of chiral aldehydes is well established. Dondoni's one-carbon homologation procedure is one of the most successful protocols described to date for the homologation of aldehydes. The mild and simple experimental conditions enhance the interest and synthetic potential of this efficient and reproducible process. The process consists of

a three-step reaction sequence involving a 2-TST 1 addition to the aldehyde, hydroxyl protection, and aldehyde unmasking. In addition, the highly stereoselective asymmetric induction obtained with α-alkoxy and α-amino aldehydes has been used in a number of synthetic approaches for the preparation of enantiomerically pure products with biological interest, such as the pseudopeptide microbial agent AI-77-B (eq 15)[20,21] or the natural cerebroside 9 (eq 16),[22] both starting from formyloxazolidines.

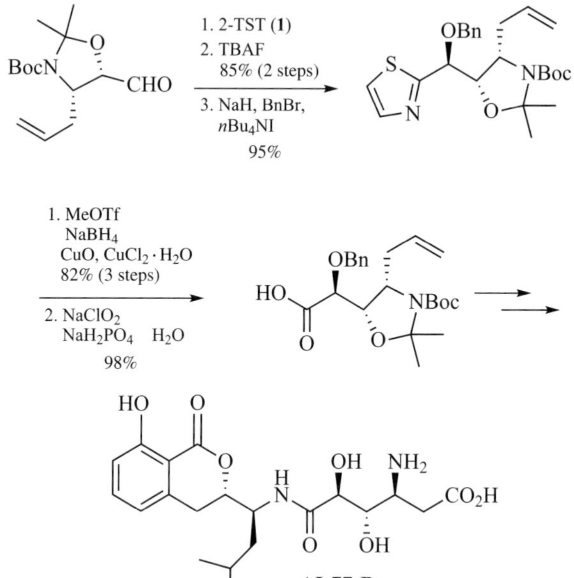

(15)

AI-77-B

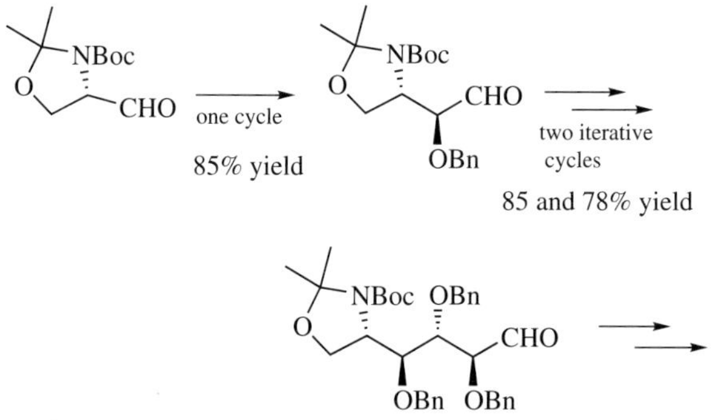

cycle=

1. 2-TST (1)

2. NaH, BnBr

3. MeI, NaBH$_4$, Hg^{2+}, H$_2$O

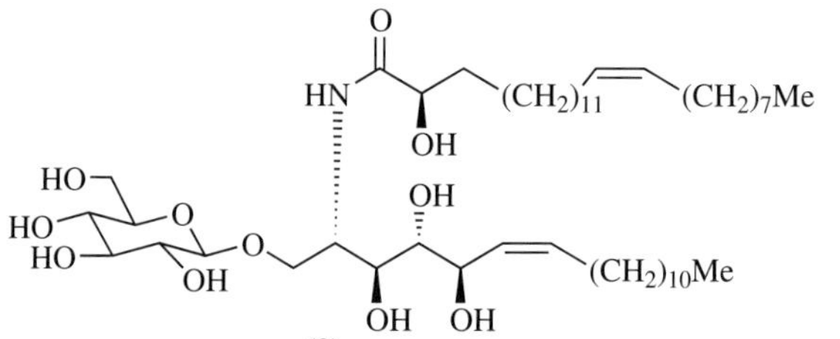

(16)

(9)

The thiazole-to-formyl conversion entails a three-step reaction sequence: *N*-methylation to an *N*-methylthiazolium salt, then hydride reduction to thiazolidine, and finally metal-ion-assisted hydrolysis to aldehyde. This latter step can be performed with CuCl$_2$–CuO (introduced by Dondoni and coworkers)[10] or AgNO$_3$

(originally used by Corey and Boger),[23] as alternatives to the classical hydrolysis with the harmful $HgCl_2$. Moreover, these conditions are compatible with commonly used hydroxyl and amino-protective groups as well as with a variety of functional groups.

The one-carbon homologation method of protected monosaccharides, employing **1** as formyl anion equivalent, has been used for the multigram scale synthesis of rare hexoses, and the disaccharide subunit of Bleomycin A_2 from L-xylose (eq 17).[24,25]

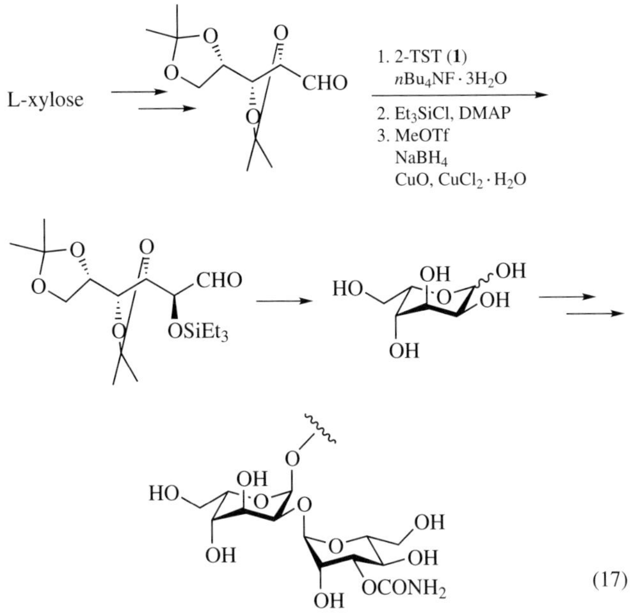

L-xylose

disaccharide subunit of Bleomycin A_2 (17)

The method was applied to the construction of the fragment GH in the total stereoselective synthesis of the carbohydrate Everninomicin 13,348-1 (eq 18).[26] Although the addition of 2-TST (**1**) led to a lack of selectivity (1:1 mixture of diastereoisomers). The wrong diastereoisomer was recycled by an oxidation/reduction process.

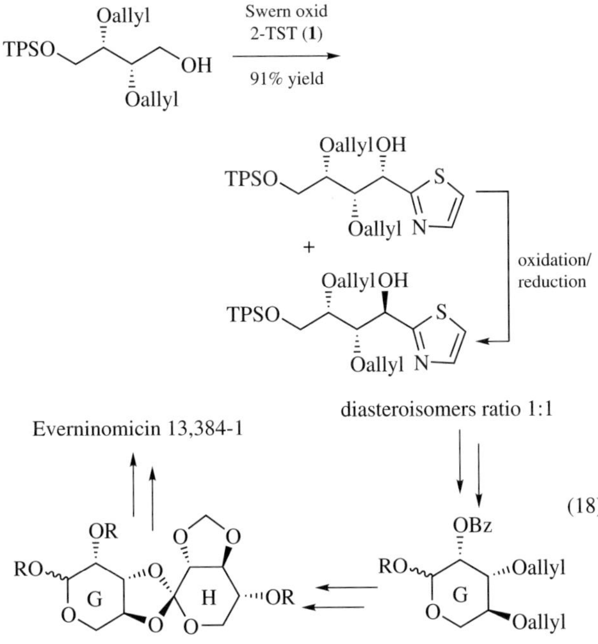

Everninomicin 13,384-1

diasteroisomers ratio 1:1 (18)

Usually the Dondoni method gives the *anti*-1,2-diol. However, the *syn*-1,2-diol has also been obtained.[27,28] The *syn* selectivity was applied, for example, to the first stereo-controlled synthesis of the C1–C13 fragment of Nystatin A_1 (eq 19).[29]

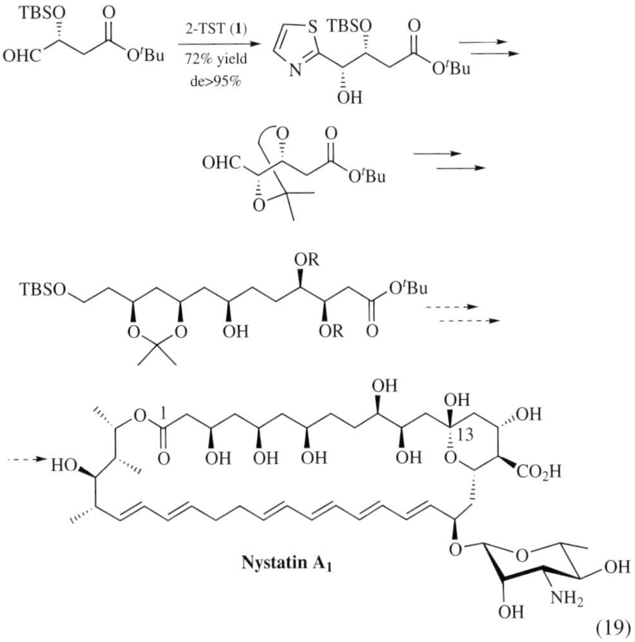

Nystatin A_1 (19)

The mechanism of the aldehyde homologation is predicted to occur through a multistep process that consists of the concerted formation of an *N*-thiazolium 2-ylide intermediate (**10**), followed by nucleophilic addition of this intermediate to another aldehyde molecule (**11**). Loss of the first aldehyde gives the addition product (eq 20).[15] The observation of the spiro(thiazoline-2,4′-dioxolane) (**12**) by NMR in a kinetic study[9] provides additional evidence for this mechanism. The conversion of compound **12** into the final product should proceed through the loss of two molecules of aldehyde and the transfer of the silyl group to the oxygen atom in the C-2-alkyl moiety.

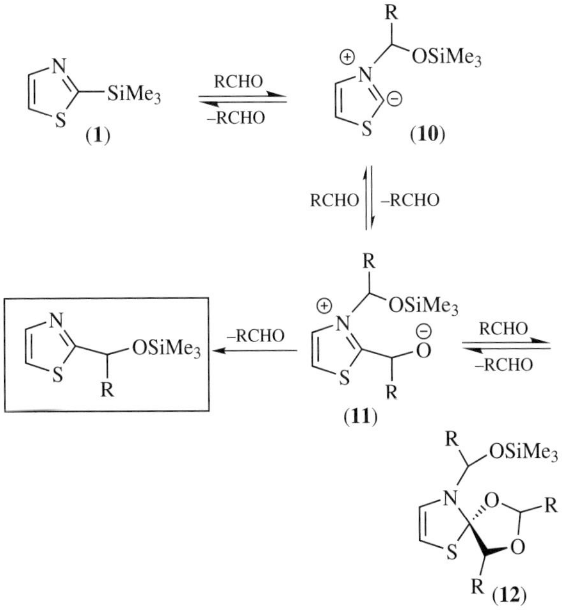

(20)

This mechanistic proposal is in agreement with the results of a theoretical study based on *ab initio* calculations,[30] which also showed that the N-thiazolium 2-ylide (**10**) was formed via concerted nitrogen alkylation and silyl group migration from C-2 to the oxygen atom in the N-alkyl moiety.

The hypothesis of the N-thiazolium-2-ylide intermediate (**10**) as a key intermediate is consistent with the fact that electrophilic aldehydes can promote the addition of 2-TST (**1**) to electrophilic ketones, in particular to α,α'-dialkoxy ketones.[31] The effect of several aldehydes on the rate of the reaction of 2-TST (**1**) with two different ketones (**13** and **14**), at different temperatures and at different concentrations of the ketones and aldehydes, was studied. It was found that electron-poor aldehydes, particularly 2-fluorobenzaldehyde, enhance the rate of addition of 2-TST (**1**) to those studied ketones (eqs 21 and 22).

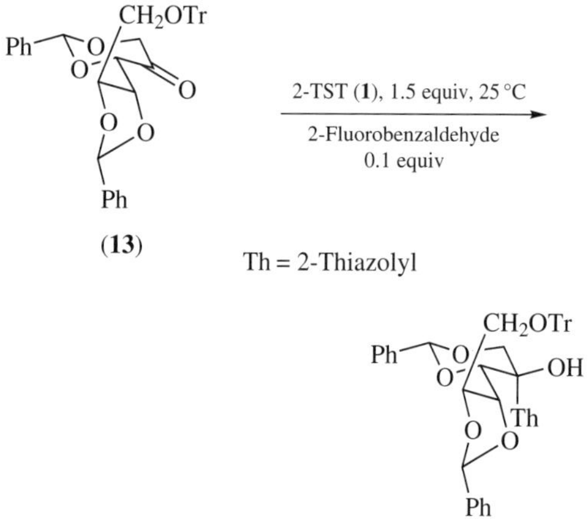

(13)

Th = 2-Thiazolyl

(21)

100% conversion after 24 h
(12% conversion without aldehyde)

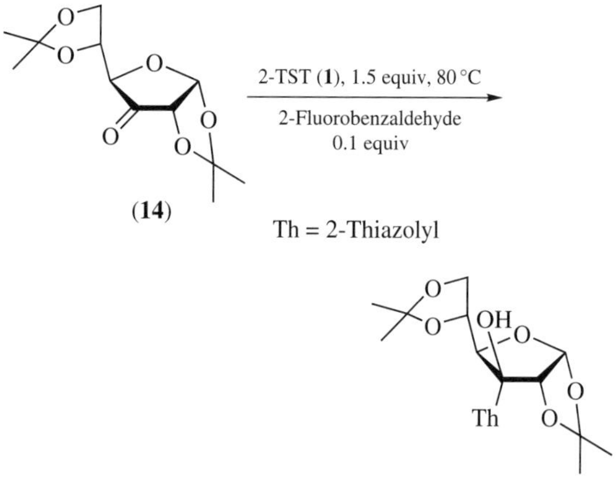

(14)

Th = 2-Thiazolyl

(22)

100% conversion after 19 h
(41 h without aldehyde)

Another example of the thiazole-based one-carbon homologation is the conversion of 4-oxoazetidine-2-carbaldehydes (**15**) into α-alkoxy β-lactam acetaldehydes (**16**).[32] The final acetaldehyde derivatives (**16**) were obtained as single isomers in moderate yields (eq 23). The observed *syn* diastereoselectivity for the 2-TST (**1**) addition step is explained by the authors by invoking the Felkin–Anh model, similar to related addition processes described previously by the same research group.

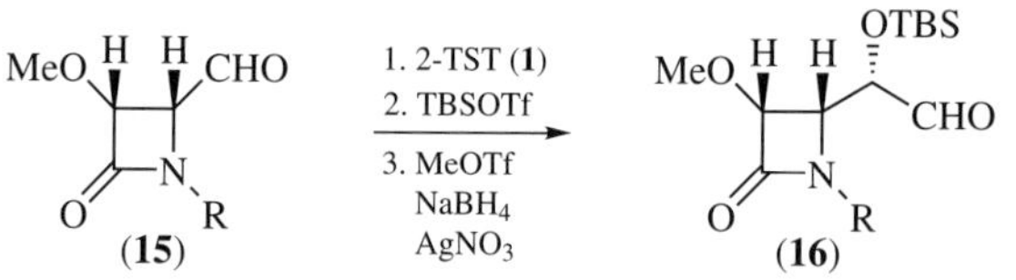

(23)

R = allyl, benzyl   52–56%

Remarkably, the addition of 2-TST (**1**) to N-aryl-4-oxoazetidine-2-carbaldehydes (**17**) led predominantly to enantiopure α-alkoxy-γ-keto acid derivatives (**18**) via a novel N1–C4 bond cleavage of the β-lactam nucleus (eq 24).[33] The same procedure was also applied recently to the asymmetric synthesis of conformationally constrained spirocyclic and tertiary α-alkoxy-γ-keto amides.[34]

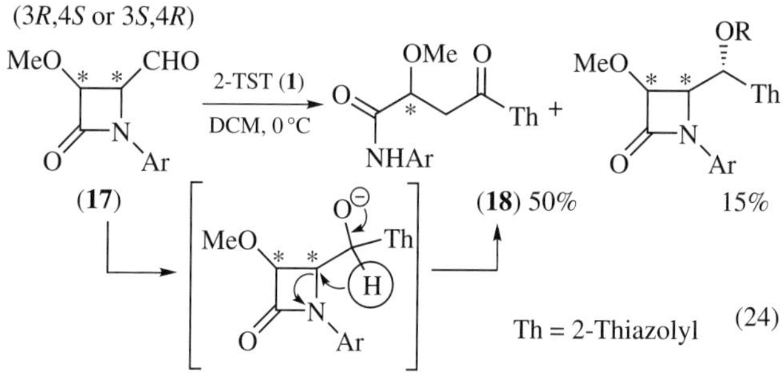

(24)

The proposed mechanism to account for the ring cleavage involves a 1,2-migration of hydrogen in the intermediate alkoxide with associated cleavage of the 2-azetidinone ring.

Metalation and Reactivity with Electrophiles. The reaction of 2-(trimethylsilyl)thiazole (**1**) with carbon electrophiles such as aldehydes, ketones, ketenes, carboxylic acid chlorides, and azaaryl cations has attracted considerable attention.[15–19] In this series, Nagasaki and coworkers reported the use of trimethylsilyl heteroarenes as the heteroarenyl carbanion donors in the electrophilic cyanation and described, for example, the electrophilic cyanation of 2-TST (**1**) with p-toluenesulfonyl cyanide in the absence of solvent (eq 25).[35]

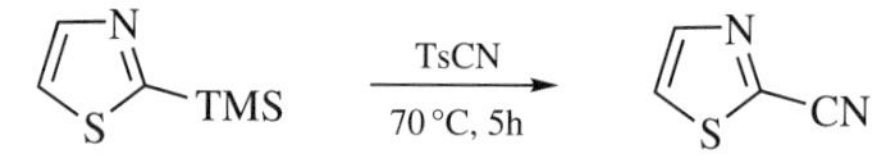

(25)

(1)   73%

However, 2-TST (**1**) turned out to be unreactive toward nitrones as electrophiles. To overcome this difficulty and accomplish the homologation of aldehydes to α-amino aldehydes, Dondoni and coworkers reported the use of C-2 metalated thiazoles, instead of 2-TST (**1**), and nitrones as derivatives of the aldehydes (eq 26).[36]

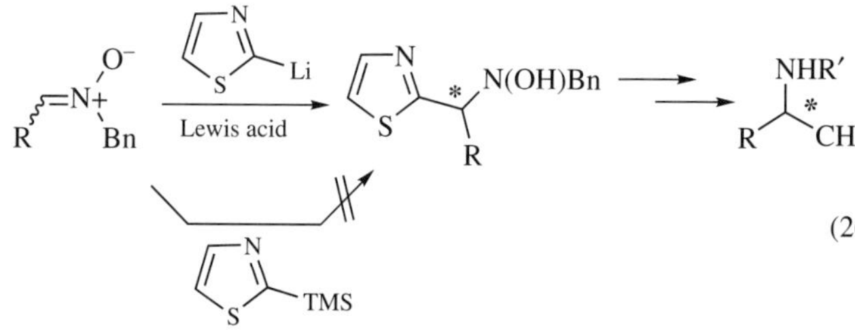

(26)

(1)

Although the reactivity with different carbon electrophiles has been widely documented, there are not many examples of synthetically useful carbon–sulfur bond formation at the thiazole carbons. For instance, the regio- and chemoselective cleavage of the carbon–silicon bond of 2-TST (1) by trifluoromethylsulfenyl chloride was reported to take place at −78 °C. The expected 2-(trifluoromethylthio)thiazole was obtained with satisfactory yields (eq 27).[37]

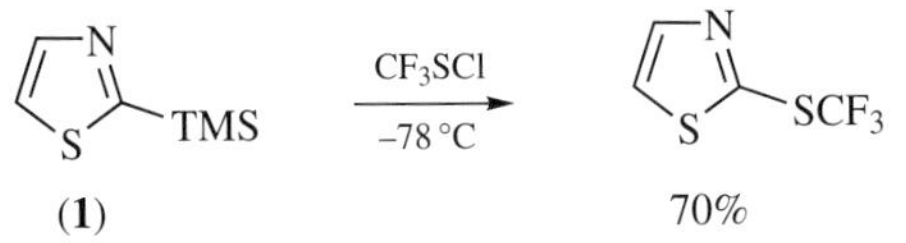

$$ (27) $$

Additionally, 2-TST (1) has been employed as a starting material for the preparation of thiazoles substituted at C-2 and C-5. Typically, the trimethylsilyl group operates as a protecting group of the C-2 position of the thiazole ring, which can be eliminated in the last step to obtain C-5-substituted thiazoles or transformed into another functional group to obtain finally C-2 or C-5 disubstituted thiazoles. For example, in the synthesis of several analogs of retinoic acid,[38] 2-TST (1) was carboxylated in the 5-position (n-BuLi, THF, −78 °C, EtO$_2$CCl) and the 2-trimethylsilyl group was iodinated (I$_2$, THF) to give ethyl 2-iodo-5-thiazolecarboxylate (19) This compound was further elaborated to obtain a retinoic acid analog (eq 28).

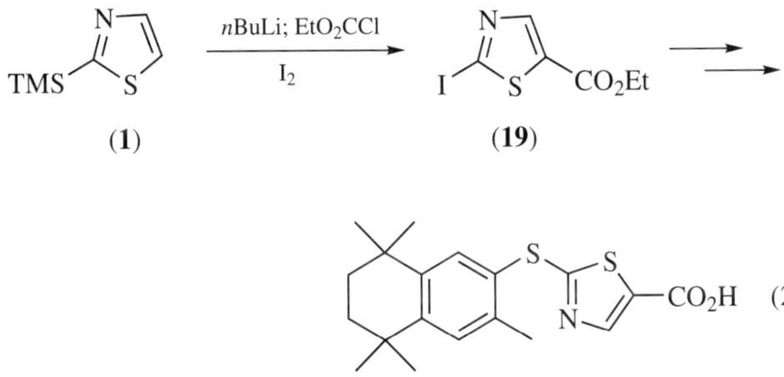

$$ (28) $$

The regioselective C-5 lithiation of 2-TST (1) followed by transmetalation with zinc chloride and palladium(0)-catalyzed Negishi cross-coupling was employed to prepare 5-aryl- and 5-heteroaryl-substituted thiazoles (20) in a one-pot procedure (eq 29).[39] The lithiation and transmetalation were performed in diethyl ether because of the instability of 5-lithio-2-trimethylsilylthiazole in THF at −78 °C. The 2-trimethylsilyl group of the cross-coupled products was cleaved quantitatively under acidic conditions.

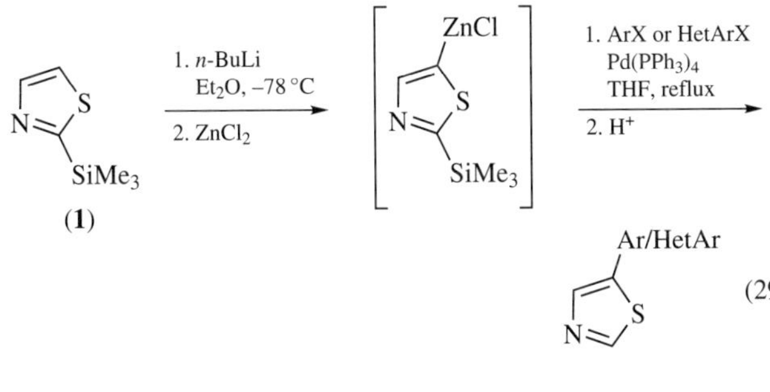

$$ (29) \ (20) \ 26\text{–}88\% $$

Similarly, the regioselective C-5 lithiation of 2-TST (1) followed by transmetalation with magnesium bromide afforded the corresponding heterocyclic Grignard reagent.[40] This compound was then reacted with an activated pyroglutamate derivative (21) in a 1,4-addition–elimination reaction to yield exclusively the intermediate 22 with *E* stereochemistry. Further elaboration of this compound through lactam ring hydrolysis and subsequent acid treatment afforded the corresponding glutamic acid derivative (eq 30).

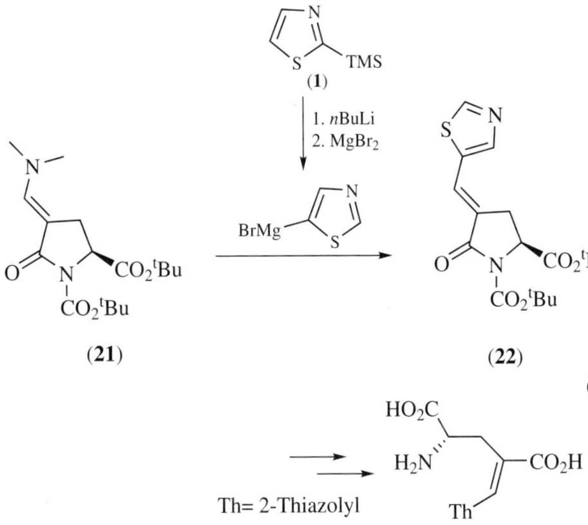

$$ (30) $$

Th= 2-Thiazolyl

Cross-coupling Reactions. Several metalated and halogenated thiazole species have been used for cross-coupling reactions, especially under Stille and Suzuki conditions.[41–43] However, there are not many examples of the direct use of 2-(trimethylsilyl)thiazole (1) in cross-coupling reactions with aryl or heteroaryl halides or triflates.

Hosomi and coworkers described the cross-coupling reaction between arylsilanes or heteroarylsilanes and aryl halides mediated by a copper(I) salt.[44] In particular, the cross-coupling reaction of 2-TST (1) with iodobenzene under fluoride- and palladium-free conditions (eq 31) was examined employing different solvents and copper(I) salts (Table 1). The yield of the product was increased in polar solvents such as 1,3-dimethyl-2-imidazolidinone (DMI). In addition, it was found that the reaction was best promoted by CuOC$_6$F$_5$ because of the strong affinity of pentafluorophenoxide ion to the silicon atom, resulting in an accelerated transfer of the thiazole group to copper.

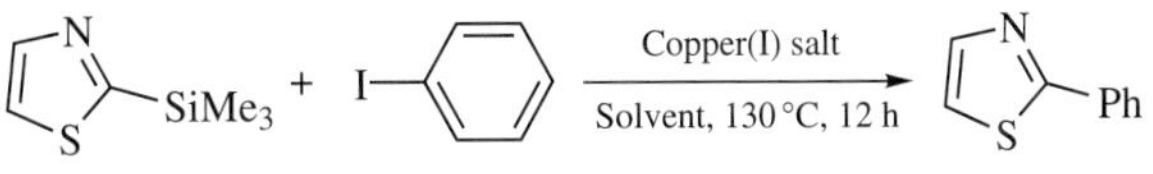

$$ (31) $$

This provides a new method for the coupling of C(sp2)-C(sp2) centers under mild conditions. Furthermore, the absence of a fluoride ion as an activator of the organosilane represents a significant improvement of the synthetic utility over other existing coupling methods because in this case, no desilylation of silyl ethers occurs during the process. For example, 4-(triisopropylsilyloxy)-iodobenzene (24) effectively reacted with 2-TST (1) in the presence of copper(I) iodide and sodium pentafluorophenoxide to afford the product (25) in good yield without desilylation (eq 32).

Table 1 Copper(I) salt mediated cross coupling of 2-TST (**1**) with iodobenzene*

Copper(I) salt	Solvent	Yield of 23 (%)
CuCl	Toluene	4
CuCl	Diglyme	48
CuCl	DMI	76
$CuOC_6F_5$	*DMI*	*93*
CuBr	DMI	53
CuI	DMI	53
CuCN	DMI	12

*Selected results from reference 44.

$$(32)$$

$$(25)\ 87\%$$

Alternatively, the palladium-catalyzed cross-coupling reaction of 2-TST (**1**) with biaryl triflates has also been described without any fluoride anion source.[45] The cross-coupling reaction of BINOL mono(triflate) with 2-TST (**1**) was performed using $Pd(OAc)_2$ and 1,4-bis(diphenylphosphino)butane (dppb) in the presence of potassium carbonate and lithium chloride in DMF at 140 °C for 17 h. The reaction proceeded efficiently to afford the corresponding monothiazole (**26**) in 71% yield (eq 33).

$$(33)$$

$$(26)\ 71\%$$

The same reaction conditions were employed with BINOL bis(triflate) with an excess amount of 2-TST (10 equiv) to obtain the bis(thiazole). However, only a 31% yield of the desired product was obtained, together with a 6% yield of the monothiazole.

In conclusion, 2-(trimethylsilyl)thiazole (**1**) has demonstrated extensively its utility as a formyl anion equivalent, especially in the homologation of aldehydes, where 2-TST (**1**) displays several advantages over the existing methods.[46] Its particular properties, such as high chemical stability, ease of handling and high reactivity under different conditions, together with the mild and simple experimental conditions and the high stereoselectivity obtained in the addition reactions, account for the great synthetic usefulness of this reagent. Moreover, the thiazole-to-formyl conversion is simple, reliable, and compatible with commonly used protective groups and stereogenic centers.

Additional reactivity of 2-TST (**1**) such as metalation, reaction with different electrophiles, and cross-coupling reactions expands the applicability of this reagent to the synthesis of substituted thiazoles.

Related Reagents. (4a*R*)-(4aα,7α,8aβ)-Hexahydro-4,4,7-trimethyl-4*H*-1,3-benzoxathiin; 2-Lithio-1,3-dithiane; 1,1,3,3-Tetramethylbutyl Isocyanide; Benzothiazole; Bis(phenylthio)-methane; *N,N*-Diethylaminoacetonitrile; Diethyl Morpholino-methylphosphonate; 1,3-Dithiane; 2-(2,6-Dimethylpiperidino) acetonitrile; Cyanotrimethylsilane; Samarium(II)iodide-1,3-dioxolane; (Phenylthiomethyl)trimethylsilane.

1. For some leading references on the use of 2-(trimethylsilyl)thiazole, see: (a) Dondoni, A., *Pure Appl. Chem.* **1990**, *62*, 643. (b) Dondoni, A. In *Modern Synthetic Methods*; Scheffold, R., Ed.; Helvetica Chimica Acta: Basel, 1992. (c) Dondoni, A. In *New Aspects of Organic Chemistry II*; Yoshida, Z.; Ohshino, Y., Eds.; VCH: Weinheim, 1992.

2. (a) Dondoni, A.; Fogagnolo, M.; Medici, A.; Pedrini, P., *Tetrahedron Lett.* **1985**, *26*, 5477. (b) Dondoni, A.; Fantin, G.; Fogagnolo, M.; Medici, A., *Angew. Chem., Int. Ed. Engl.* **1986**, *25*, 835. (c) Dondoni, A.; Fantin, G.; Fogagnolo, M.; Medici, A.; Pedrini, P., *J. Org. Chem.* **1989**, *54*, 693. (d) Dondoni, A.; Fantin, G.; Fogagnolo, M.; Medici, A., *Tetrahedron* **1987**, *43*, 3533. (e) Dondoni, A.; Orduna, J.; Merino, P., *Synthesis* **1992**, 201.

3. Dondoni, A.; Merino, P., *Org. Synth.* **1993**, *72*, 21.

4. Dondoni, A.; Dall'Occo, T.; Galliani, G.; Mastellari, A.; Medici, A., *Tetrahedron Lett.* **1984**, *25*, 3637.

5. Medici, A.; Pedrini, P.; Dondoni, A., *J. Chem. Soc., Chem. Commun.* **1981**, 655.

6. Dondoni, A.; Fantin, G.; Fogagnolo, M.; Medici, A.; Pedrini, P., *J. Org. Chem.* **1988**, *53*, 1748.

7. Medici, A.; Fantin, G.; Fogagnolo, M.; Dondoni, A., *Tetrahedron Lett.* **1983**, *24*, 2901.

8. Dondoni, A.; Boscarato, A.; Formaglio, P.; Bégué, J.-P.; Benayoud, F., *Synthesis* **1995**, 654.

9. Dondoni, A.; Douglas, A. W.; Shinkai, I., *J. Org. Chem.* **1993**, *58*, 3196.

10. For a modification of the original procedure, see: Dondoni, A.; Marra, A.; Perrone, D., *J. Org. Chem.* **1993**, *58*, 275.

11. (a) Maillard, Ph.; Huel, C.; Momenteau, M., *Tetrahedron Lett.* **1992**, *33*, 8081. (b) Khare, N. K.; Sood, R. K.; Aspinall, G. O., *Can. J. Chem.* **1994**, *72*, 237.

12. (a) Dondoni, A.; Fantin, G.; Fogagnolo, M.; Medici, A., *J. Chem. Soc., Chem. Commun.* **1988**, 10. (b) Dondoni, A.; Fantin, G.; Fogagnolo, M.; Pedrini, P., *J. Org. Chem.* **1990**, *55*, 1439.

13. Wagner, A.; Mollath, M., *Tetrahedron Lett.* **1993**, *34*, 619.

14. (a) Dondoni, A.; Colombo, L. In *Advances in the Use of Synthons in Organic Chemistry*; Dondoni A., Ed.; JAI Press: London, 1993: (b) Ager, D. J. In *Umpoled Synthons*; Hase, T. A.; Ed.; Wiley: New York, 1987; Chapter 2.

15. Dondoni, A., *Synthesis* **1998**, 1681.

16. Dondoni, A.; Marra, A., *Chem. Rev.* **2004**, *104*, 2557.

17. Dondoni, A.; Merino, P., *Org. Synth., Coll. Vol. 9* **1998**, 52; Vol. 72, **1995**, 21.

18. Dondoni, A.; Perrone, D., *Org. Synth., Coll. Vol. 10* **2004**, 140; Vol. 77 **2000**, 78.

19. Dondoni, A.; Perrone, D., *Aldrichimica Acta* **1997**, *30*, 35.

20. Ghosh, A. K.; Bischoff, A.; Cappiello, J., *Org. Lett.* **2001**, *3*, 2677.

21. Ghosh, A. K.; Bischoff, A.; Cappiello, J., *Eur. J. Org. Chem.* **2003**, 821.

22. Cateni, F.; Zacchigna, M.; Zilic, J.; Di Luca, G., *Helv. Chim. Acta* **2007**, *90*, 282.

23. Corey, E. J.; Boger, D. L., *Tetrahedron Lett.* **1978**, *19*, 5.

24. Dondoni, A.; Marra, A.; Massi, A., *Carbohydrate Lett.* **1997**, *2*, 367.

25. Dondoni, A.; Marra, A.; Massi, A., *J. Org. Chem.* **1997**, *62*, 6261.

26. Nicolaou, K. C.; Mitchell, H. J.; Fylaktakidou, K. C.; Rodríguez, R. M.; Suzuki, H., *Chem. Eur. J.* **2000**, *6*, 3116.

27. Dondoni, A.; Perrone, D.; Merino, P., *J. Org. Chem.* **1995**, *60* 8080.

28. Dondoni, A.; Orduna, J.; Merino, P., *Synthesis* **1992**, 201.

29. Solladié, G.; Wilb, N.; Bauder, C., *J. Org. Chem.* **1999**, *64* 5447.

30. Wu, Y.-D.; Lee, J. K.; Houk, K. N.; Dondoni, A., *J. Org. Chem.* **1996**, *61*, 1922.

31. Carcano, M.; Vasella, A., *Helv. Chim. Acta* **1998**, *81*, 889.

32. Alcaide, B.; Almendros, P.; Redondo, M. C.; Ruiz, M. P., *J. Org. Chem.* **2005**, *70*, 8890.

33. Alcaide, B.; Almendros, P.; Redondo, M. C., *Org. Lett.* **2004**, *6*, 1765.

34. Alcaide, B.; Almendros, P.; Redondo, M. C., *Eur. J. Org. Chem.* **2007**, 3707.

35. Nagasaki, I.; Suzuki, Y.; Iwamoto, K.; Higashino, T.; Miyashita, A., *Heterocycles* **1997**, *46*, 443.

36. Dondoni, A.; Franco, S.; Junquera, F.; Merchán, F. L.; Merino, P.; Tejero, T.; Bertolasi, V., *Chem. Eur. J.* **1995**, *1*, 505.

37. Munavalli, S.; Rohrbaugh, D. K.; Rossman, D. I.; Durst, H. D.; Dondoni, A., *Phosphorus, Sulfur Silicon Relat. Elem.* **2002**, *177*, 2465.

38. Beard, R. L.; Colon, D. F.; Song, T. K.; Davies, P. J. A.; Kochhar, D. M.; Chandraratna, R. A. S., *J. Med. Chem.* **1996**, *39*, 3556.

39. Jensen, J.; Skjæbæk, N.; Vedsø, P., *Synthesis* **2001**, 128.

40. Valgeirsson, J.; Christensen, J. K.; Kristensen, A. S.; Pickering, D. S.; Nielsen, B.; Fischer, C. H.; Bräuner-Osborne, H.; Nielsen, E.Ø.; Krogsgaard-Larsena, P.; Madsena, U., *Bioorg. Med. Chem.* **2003**, *11*, 4341.

41. Sharp, M.; Snieckus, V., *Tetrahedron Lett.* **1985**, *26*, 5997.

42. Dondoni, A.; Fogagnolo, M.; Medici, A.; Negrini, E., *Synthesis* **1987**, 185.

43. Gronowitz, S.; Peters, D., *Heterocycles* **1990**, *30*, 645.

44. Ito, H.; Sensui, H.; Arimoto, K.; Miura, K.; Hosomi, A., *Chem. Lett.* **1997**, *7*, 639.

45. Hara, O.; Nakamura, T.; Sato, F.; Makino, K.; Hamada, Y., *Heterocycles* **2006**, *68*, 1.

46. For a comparison of 2-TST (**1**) with alternative reagents, see: Dondoni, A.; Marra, A.; In *Preparative Carbohydrate Chemistry*; Hanessian, S., Ed.; Marcel Dekker, Inc., CRC Press: New York, 1997; Chapter 9.

Trimethylsilyltributylstannane[1]

$$\boxed{Me_3SiSnBu_3}$$

[17955-46-3] $C_{15}H_{36}SiSn$ (MW 383.32)

InChI = 1S/3C4H9.C3H9Si.Sn/c3*1-3-4-2;1-4(2)3;/h3*1,3-4H2,2H3;1-3H3;

InChIKey = MZUUBYSIDDKUKE-UHFFFAOYSA-N

(used for the vicinal bis-functionalization of alkynes, dienes, allenes, activated olefins, 1,1-addition to isonitriles, for generation of metal-free aryl and vinyl anions, for generation of benzynes and quinonedimethanes; stereoselective bis-functionalization-cyclization of diynes, eneynes, bis-allenes, alleneynes, and allene-aldehydes; silylation of allylic, propargylic derivatives, and aryl bromides; formation of silyl and stannyl cuprates)

Physical Data: colorless liquid. bp 88 °C (0.2 mmHg).
Solubility: soluble in THF, ether, CH_2Cl_2, MeOH, hexane, insoluble in water.

Preparative Method: by treating freshly distilled tri-*n*-butyl-stannane with LDA followed by quenching with chlorotri-methylsilane.[2a]
Purification: the crude mixture can be purified by distillation through an efficient Vigreux column or by chromatography on silica gel using hexane as eluent.
Handling, Storage, and Precautions: tin compounds can be toxic. The reagent is moderately air- and moisture-stable and it can be stored at room temperature for months. Because of its volatility this compound should be handled in an efficient fume hood, and appropriate care must be taken to avoid inhalation hazards.

Introduction. Trimethylsilyltributylstannane is one example of a group of accessible reagents with varying alkyl/aryl groups on silicon and tin. These reagents sometimes show significant differences in reactivity, depending on the alkyl/aryl groups present and the relevant references for individual applications should be consulted for details and the optimum choice of alkyl/aryl groups.

Addition to Alkynes. Reactions of trimethylsilyltributylstannane ($Me_3SiSnBu_3$) with a variety of terminal alkynes affords synthetically useful *vic*-silyl(stannyl)alkenes (eq 1).[2–4] The reaction is typically catalyzed by Pd(0)-complexes in neat mixtures of reagents or in solvents such as THF, benzene, or toluene at temperatures in the range of 25–80 °C. The reaction occurs with complete regio- and stereoselectivity with exclusive *syn*-addition giving terminal vinylsilanes. Exceptions are noted in the reactions of acetylene and trimethylsilylacetylene where mixtures of *cis*- and *trans*-alkenes are obtained. The range of functional groups tolerated in this reaction includes hydroxyl, 3°-amino, cyano, chloro, ether, β-lactam, epoxide, olefin, internal acetylene, and ester moieties. Some propargyl-substituted acetylenes lead to diminished yields. Sterically hindered *t*-butylacetylene undergoes a very sluggish reaction with $(Ph_3P)_4Pd(0)$ as catalyst, but reacts readily in the presence of a *t*-isonitrile ligand.[4] Propargyl bromides and acetates give multiple side products including substitution products.[5] Propargylic alcohols give acceptable yields with $Pd_2(dba)_3 \cdot CHCl_3/Ph_3P$ as catalyst at reflux in THF.[3e] The Pd(0)-catalyzed 1,2-additions of silylstannanes to 1-alkoxyacetylenes, the regioselectivity of addition depends on the catalyst and the nature of substituents on Si and Sn (eqs 2 and 3).[3c,4a] Some internal alkynes with an ester group attached to the triple bond also undergo the reaction with trimethylsilyltributylstannane upon prolonged heating to give exclusively the *Z-vic*-silylstannylalkene (eq 4).[3b] However, depending on the substitution, the regioselectivity and, in some cases, the stereoselectivity of double bond are diminished (eq 5).[6]

$$Me_3SiSnBu_3 + R{\equiv\!\equiv}H \xrightarrow[60-70\,°C]{cat\ Pd(0)} \underset{R}{\overset{Bu_3Sn}{\diagdown}}{=}\underset{H}{\overset{SiMe_3}{\diagup}} \quad (1)$$

R = alkyl, aryl, $NC(CH_2)_3$, $Cl(CH_2)_3$, $HO(CH_2)_3$, EtO_2C, $(EtO)_2CH$, $PhMeNCH_2$, $R'C(OH)(H)–$

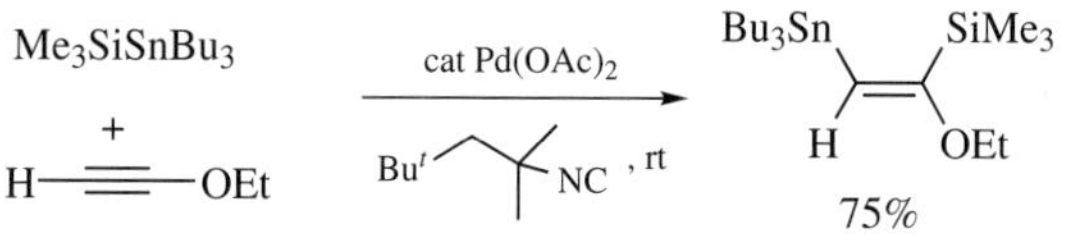

Equation (2):

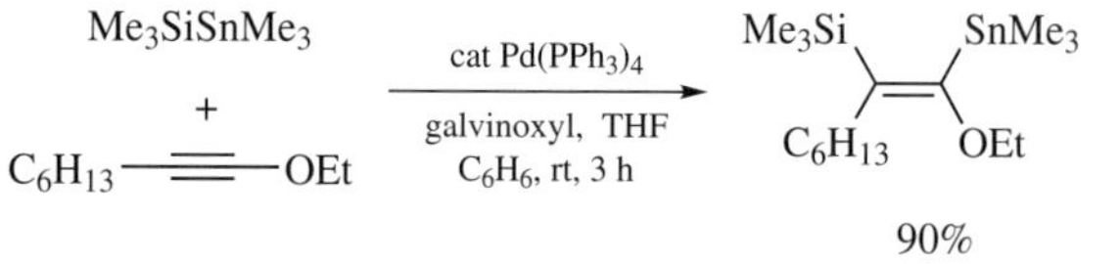

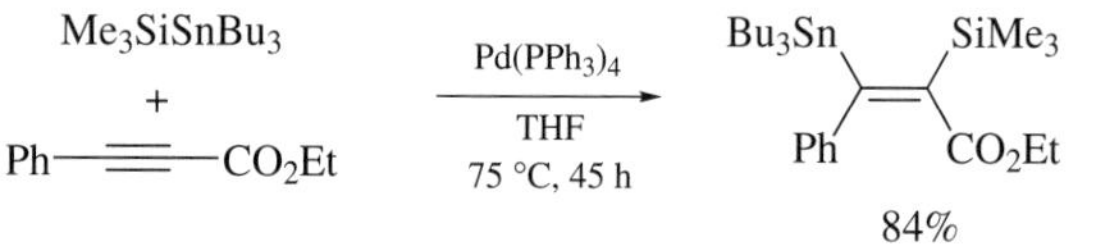

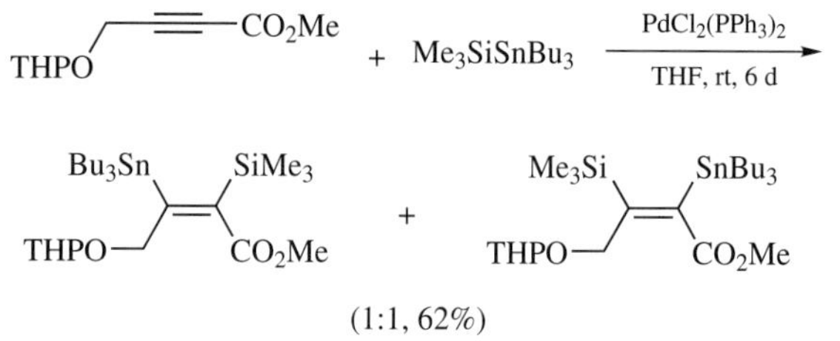

$$Me_3SiSnBu_3 + H\!\!\equiv\!\!-OEt \xrightarrow[\text{Bu}^t\text{NC, rt}]{\text{cat Pd(OAc)}_2} \underset{H}{Bu_3Sn}\!\!=\!\!\underset{OEt}{SiMe_3} \quad 75\%$$

Equation (3):

$$Me_3SiSnMe_3 + C_6H_{13}\!\!\equiv\!\!-OEt \xrightarrow[\text{galvinoxyl, THF}]{\text{cat Pd(PPh}_3)_4}{C_6H_6, \text{rt, 3 h}} \underset{C_6H_{13}}{Me_3Si}\!\!=\!\!\underset{OEt}{SnMe_3} \quad 90\%$$

Equation (4):

$$Me_3SiSnBu_3 + Ph\!\!\equiv\!\!-CO_2Et \xrightarrow[\text{75 °C, 45 h}]{\text{Pd(PPh}_3)_4}{\text{THF}} \underset{Ph}{Bu_3Sn}\!\!=\!\!\underset{CO_2Et}{SiMe_3} \quad 84\%$$

Equation (5):

THPO–≡–CO₂Me + Me₃SiSnBu₃ →[PdCl₂(PPh₃)₂ / THF, rt, 6 d] Bu₃Sn/SiMe₃ THPO/CO₂Me + Me₃Si/SnBu₃ THPO/CO₂Me (1:1, 62%)

Equation (8):

$$Z = Bu_3Sn \xrightarrow[94\%]{I_2, CH_2Cl_2} Z = I$$

Equation (9):

1. Cp₂TiCl (1.1–1.2 equiv) THF, rt, 0.5 h 2. H⁺, H₂O (88% → 78%)

Equation (10):

1. R-X, BnPd(Cl)(PPh₃)₂ CuI, DMF 2. CF₃CO₂H/THF/H₂O

R = allyl, benzyl, vinyl, acyl

Synthetic Applications of β-Trialkylsilylvinylstannanes.

β-Trialkylsilylvinyllithium intermediates generated from the silylstannane adducts of acetylenes can be trapped with electrophiles with retention of configuration when R ≠ Ar (eq 6).[3a,b] Lewis acid (AlCl₃/CH₂Cl₂)-catalyzed reactions of the adducts with acid chlorides give acylation products,[3a] while Stille coupling with organic halides can be used to prepare various other β-silylvinyl derivatives (eq 7).[3b,c] 1,2-Silylstannyl olefins can be converted into β-silylvinyl bromides and iodides, which, in appropriate substrates can be further used for annulation via Heck or radical reactions (eq 8).[3d] The remarkable functional group compatibility of silylstannylation allows the preparation and cyclization of suitable epoxystannyl olefin precursors for Ti(III)-mediated[7] radical cyclization (eq 9).[3d] Highly substituted acylsilanes can be prepared from 1-alkoxyacetylenes by silylstannylation followed by further reactions of the stannyl moiety (Stille coupling, Sn-halogen exchange), and finally hydrolysis of the enol ether (eq 10).[4a] Acylsilanes can also be prepared from acid chlorides and silylstannanes in the presence of [(allyl)PdCl]₂ and triethylphosphite under CO.[8]

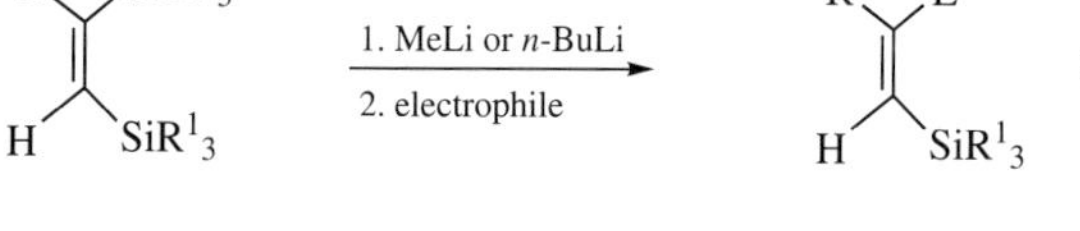

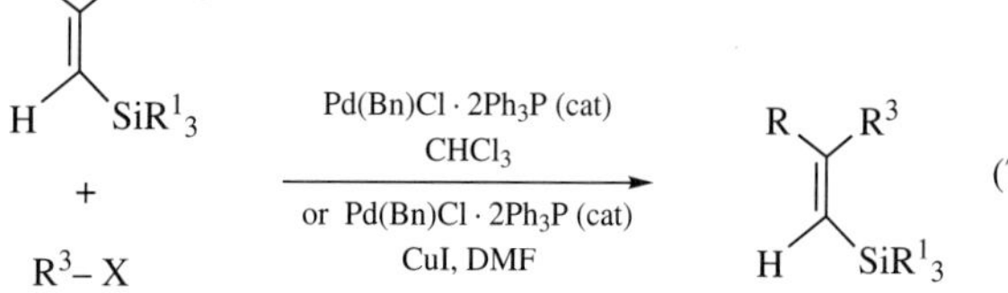

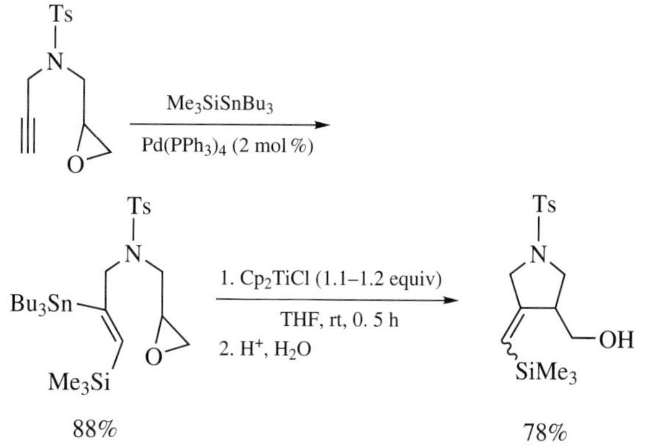

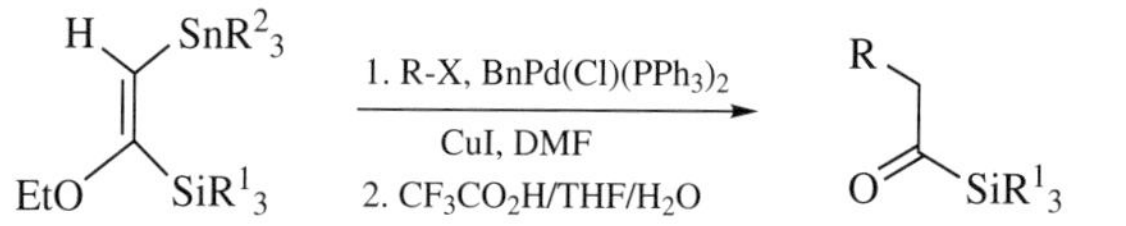

Addition to Isonitriles.

Silylstannanes undergo 1,1-addition to isonitriles giving synthetically useful silylstannyl iminomethanes (eq 11). These intermediates are readily converted into the corresponding iminomethyl-litium, -copper, or -cuprate reagents which undergo the expected 1,2- or 1,4-additions with appropriate electrophiles.[9]

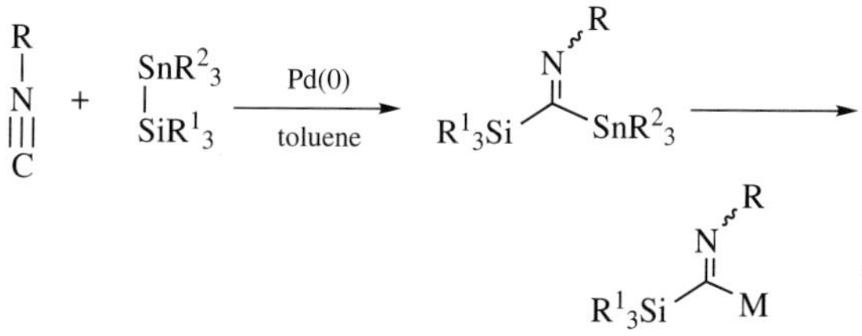

Equation (6):

1. MeLi or n-BuLi 2. electrophile

E = H, Me, -C(O)H, HOC(H)R

Equation (11):

M = Li, Cu (ligand)

Equation (7):

Pd(Bn)Cl · 2Ph₃P (cat) CHCl₃ or Pd(Bn)Cl · 2Ph₃P (cat) CuI, DMF

R³–X

R³ = allyl, cinnamyl, benzyl, Ph, RC(O)-

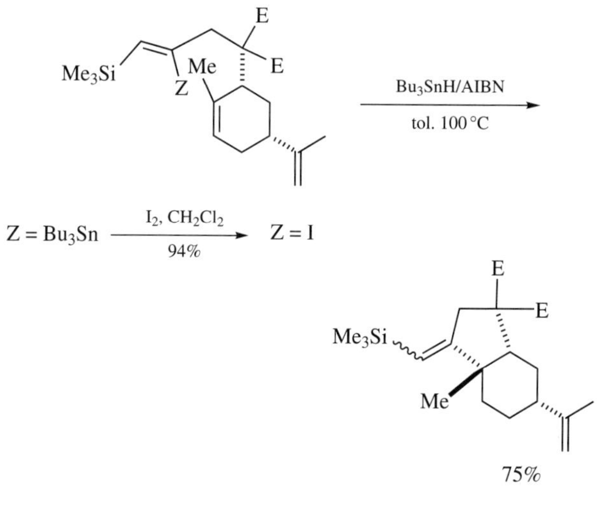

Addition to Alkenes and 1,3-Dienes.

Silylstannanes add to selected alkenes under Pd-catalysis. The applicable alkenes include ethylene, strained bicyclic alkenes,[10a] cyclopropylidenecyclopropane,[10b] or cyclopropenes[10c] (eq 12).

$$Me_3Si\text{-}SnBu_3 + \left\{ \begin{array}{l} \text{(12)} \end{array} \right.$$

Addition of silylstannanes to 1,3-dienes in the presence of $Pt(CO)_2(PPh_3)_2$ leads to *trans*-1,4-silastannyl-2-butene as a single regio- and stereoisomer (eq 13).[11a] However, in the 1,4-carbosilylation of 1,3-dienes with acid halides, disilanes give better yields than silylstannanes.[11b]

$$\text{(13)}$$
$$84\%$$

Addition to Activated Double Bonds. Trimethylsilyltributylstannane can also be activated with a catalytic amount of "naked" cyanide (KCN/18-crown-6, Bu_4NCN, *tris*-diethylaminosulfonium [TAS] cyanide) or potassium *t*-butoxide,[2] or with a stoichiometric amount of $BnEt_3NCl$[12a] to produce the stannyl anion, which undergoes 1,4-addition to α,β-unsaturated compounds (eq 14). Aldehydes, but not ketones, undergo 1,2-additions.[13] The reaction of silylstannanes with methyl propiolate leads to β,β-distannyl esters which, after reduction, can be lithiated to generate useful 3-carbon elongation synthons (eq 15).[12] Silylstannanes can be used to prepare higher order stannyl and silylcuprates, which undergo classical substitution and addition reactions leading to C–Si and C–Sn bonds.[14]

$$\text{(14)}$$

$$\text{(15)}$$
$$85\% \ (E/Z = 35/1)$$

Addition to Allenes. In a reaction that is highly dependent on the ligands employed, with Pd-catalyzed addition of silylstannanes to allenes shows high regio- and stereoselectivity under the optimum conditions (eqs 16 and 17).[15a–d] Three-component

coupling can be achieved by the reaction of an allene, a silylstannane, and an aryl halide (eq 18).[15e]

$$\text{(16)}$$
$$91\%$$

$$\text{(17)}$$
$$82\%$$

$$\text{(18)}$$
$$92\%$$

Generation and Reactions of Metal-free Stannyl Anions from $Me_3SiSnBu_3$ and Highly Dissociated Fluorides. Aryl, vinyl, and allyl anions are easily generated by reactions of the appropriate precursor with $Me_3SiSnBu_3$ and a reactive fluoride source. Presumably, the reaction proceeds through a highly dissociated tributylstannyl anion. The examples (eqs 19–22)[1c, 16] illustrate the use of various halide sources for the synthesis of these reactive intermediates.

$$\text{(19)}$$
$$86\%$$

$$BnO(CH_2)_3 \equiv H \quad \text{(20)}$$

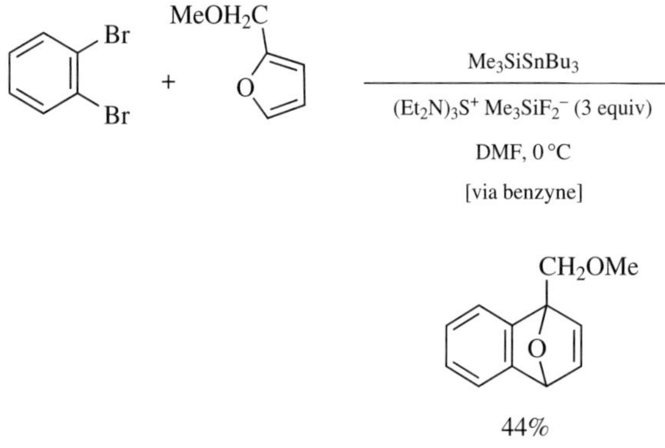

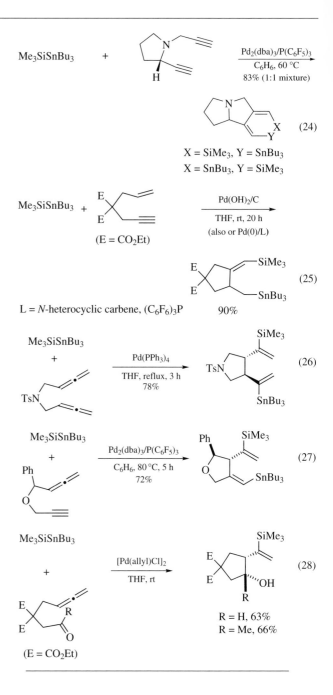

Reactions with Propargylic and Allylic Derivatives, α-Halo-ketones, and Arylbromides. Propargylic chlorides, mesylates, and epoxides react with structurally related silylstannanes in the presence of Pd pincer complexes giving propargyl- and allenyl-stannanes and silanes in high yields.[5] Conversion of aromatic allyl acetates to allylsilanes can also be accomplished by silylstannane reagents in the presence of Pd(0) and LiCl.[17a] Reaction of Me₃SiSnBu₃ with α-haloketones gives silylenol ethers in moderate yields under catalysis by PdCl₂/P(OMe)₃.[17b] Likewise, aryl bromides react with silylstannanes to give the corresponding trialkylsilylarene (not stannane) derivatives.[17c]

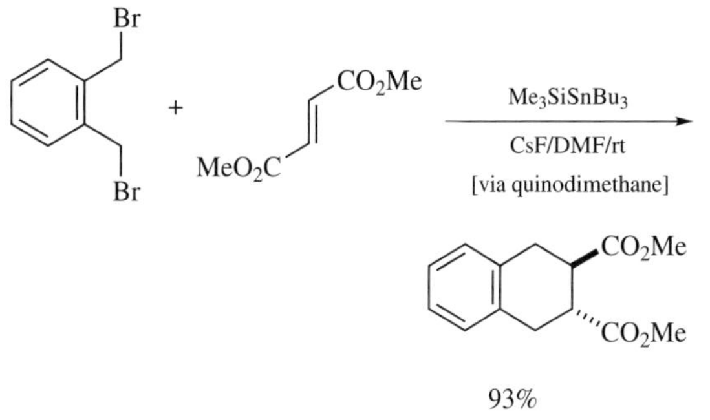

Cyclization Reactions Involving Me₃SiSnBu₃ Additions. When silylstannanes add to a substrate containing two C–C π-bonds, cyclization occurs concurrent with addition of silyl and stannyl groups. Cyclizations are reported for diynes (eqs 23 and 24),[18] enynes (eq 25),[3d,19] bisallenes (eq 26),[20] and allenynes (eq 27).[15d,21] The diyne adducts, (Z,Z)-1,2-bisalkylidenecyclo-alkanes are chiral but undergo facile helical enantiomerization at room temperature.[18b] Allene-aldehydes and allene-ketones are also viable substrates for cyclizations leading to stereoselective formation of cis-2-alkenylcyclopentanols (eq 28).[22] The reactivity of the initial silylpalladation process appears to decrease in the order: allene, alkyne, and alkene. The corresponding 1,5-bisdienes gave poor selectivity when the reaction was carried out with the only viable catalyst, PdCl₂(COD).[23]

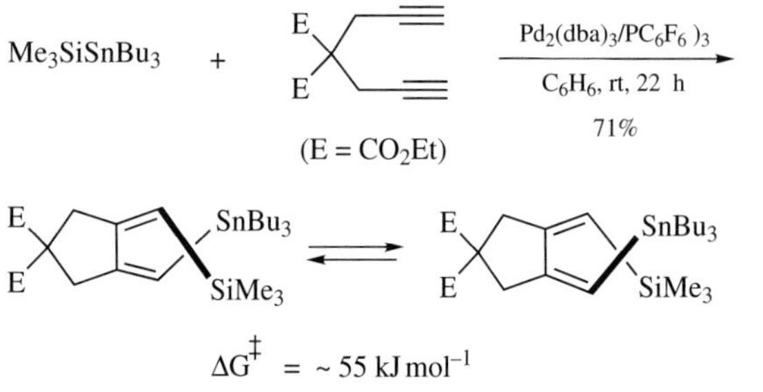

$$\text{Me}_3\text{SiSnBu}_3 \quad + \quad \overset{E}{\underset{E}{\diagup}}\diagdown\diagdown \xrightarrow[\text{C}_6\text{H}_6,\text{rt},22\ \text{h}]{\text{Pd}_2(\text{dba})_3/\text{PC}_6\text{F}_6)_3}$$

(E = CO₂Et)

71%

$$\overset{E}{\underset{E}{}} \quad \rightleftharpoons \quad \overset{E}{\underset{E}{}} \quad (23)$$

ΔG‡ = ~ 55 kJ mol⁻¹

1. (a) Hemeon, I.; Singer, R. D., *Science of Synthesis* **2002**, *4*, 205. (b) Suginome, M.; Ito, Y., *Chem. Rev.* **2000**, *100*, 3221. (c) Mori, M.; Isono, N.; Wakamatsu, H., *Synlett* **1999**, 269.

2. (a) Chenard, B. L.; Laganis, E. D.; Davidson, F.; RajanBabu, T. V., *J. Org. Chem.* **1985**, *50*, 3666. (b) First reference to this compound in the literature: Tamborski, C.; Ford, F. E.; Soloski, E. J., *J. Org. Chem.* **1963**, *28*, 237.

3. (a) Chenard, B. L.; Van Zyl, C. M., *J. Org. Chem.* **1986**, *51*, 3561. (b) Mitchell, T. N.; Wickenkamp, R.; Amamria, A.; Dicke, A.; Schneider, U., *J. Org. Chem.* **1987**, *52*, 4868. (c) Casson, S.; Kocienski, P.; Reid, G.; Smith, N.; Street, J. M.; Webster, M., *Synthesis* **1994**, 1301. (d) Apte, S.; Radetich, B.; Shin, S.; RajanBabu, T. V., *Org. Lett.* **2004**, *6*, 4053. (e) Nielsen, T. E.; Le Quement, S.; Tanner, D., *Synthesis* **2004**, 1381.

4. Use of more active ligands: (a) t-octylisonitrile: Murakami, M.; Amii, H.; Takizawa, N.; Ito, Y., *Organometallics* **1993**, *12*, 4223. (b) For the use of tricyclohexylphosphine: see ref. 3d. (c) For a theoretical study see: Hada, M.; Tanaka, Y.; Ito, M.; Murakami, M.; Amii, H.; Ito, Y.; Nakatsuji, H., *J. Am. Chem. Soc.* **1994**, *116*, 8754.

5. Kjellgren, J.; Sunden, H.; Szabo, K. J., *J. Am. Chem. Soc.* **2005**, *127*, 1787.

6. Mabon, R.; Richecoeur, A. M. E.; Sweeney, J. B., *J. Org. Chem.* **1999**, *64*, 328.

7. RajanBabu, T. V.; Nugent, W. A., *J. Am. Chem. Soc.* **1994**, *116*, 986.

8. Geng, F.; Maleczka, R. E., Jr., *Tetrahedron Lett.* **1999**, *40*, 3113.

9. (a) Ito, Y.; Bando, T.; Matsuura, T.; Ishikawa, M., *J. Chem. Soc., Chem. Commun.* **1986**, 980. (b) Ito, Y.; Matsuura, T.; Murakami, M., *J. Am. Chem. Soc.* **1987**, *109*, 7888. (c) Murakami, M.; Matsuura, T.; Ito, Y., *Tetrahedron Lett.* **1988**, *29*, 355.

10. (a) Obora, Y.; Tsuji, Y.; Asayama, M.; Kawamura, T., *Organometallics* **1993**, *12*, 4697. (b) Pohlmann, T.; de Meijere, A., *Org. Lett.* **2000**, *2*, 3877. (c) Rubina, M.; Rubin, M.; Gevorgyan, V., *J. Am. Chem. Soc.* **2002**, *124*, 11566.

11. (a) Tsuji, Y.; Obora, Y., *J. Am. Chem. Soc.* **1991**, *113*, 9368. (b) Obora, Y.; Tsuji, Y.; Kawamura, T., *J. Am. Chem. Soc.* **1995**, *117*, 9814.

12. (a) Isono, N.; Mori, M., *J. Org. Chem.* **1998**, *63*, 1773. (b) For addition elimination to give β-stannylenones, see: Liebeskind, L. S.; Stone, G. B.; Zhang, S., *J. Org. Chem.* **1994**, *59*, 7917.

13. Bhatt, R. K.; Ye, J.; Falck, J. R., *Tetrahedron Lett.* **1994**, *35*, 4081.

14. Lipshutz, B. H.; Reuter, D. C.; Ellsworth, E. L., *J. Org. Chem.* **1989**, *54*, 4975.

15. (a) Mitchell, T. N.; Killing, H.; Dicke, R.; Wickenkamp, R., *J. Chem. Soc., Chem. Commun.* **1985**, 354. (b) Mitchell, T. N.; Schneider, U., *J. Organomet. Chem.* **1991**, *407*, 319. (c) Jeganmohan, M.; Shanmugasundaram, M.; Chang, K.-J.; Cheng, C.-H., *Chem. Commun.* **2002**, 2552. (d) Kumareswaran, R.; Shin, S.; Gallou, I.; RajanBabu, T. V., *J. Org. Chem.* **2004**, *69*, 7157. (e) Wu, M.-Y.; Yang, F.-Y.; Cheng, C.-H., *J. Org. Chem.* **1999**, *64*, 2471.

16. (a) Mori, M.; Isono, N.; Kaneta, N.; Shibasaki, M., *J. Org. Chem.* **1993**, *58*, 2972. (b) Sato, H.; Isono, N.; Okamura, K.; Date, T.; Mori, M., *Tetrahedron Lett.* **1994**, *35*, 2035. (c) Sato, H.; Isono, N.; Miyoshi, I.; Mori, M., *Tetrahedron* **1996**, *52*, 8143.

17. (a) Tsuji, Y.; Kajita, S.; Isobe, S.; Funato, M., *J. Org. Chem.* **1993**, *58*, 3607. (b) Kosugi, M.; Ohya, T.; Migita, T., *Bull. Chem. Soc. Jpn.* **1983**, *56*, 3539. (c) Azizian, H.; Eaborn, C.; Pidcock, A., *J. Organomet. Chem.* **1981**, *215*, 49.

18. (a) Greau, S.; Radetich, B.; RajanBabu, T. V., *J. Am. Chem. Soc.* **2000**, *122*, 8579. (b) Warren, S.; Chow, A.; Fraenkel, G.; RajanBabu, T. V., *J. Am. Chem. Soc.* **2003**, *125*, 15402.

19. (a) Sato, Y.; Imakuni, N.; Hirose, T.; Wakamatsu, H.; Mori, M., *J. Organomet. Chem.* **2003**, *687*, 392. (b) Lautens, M.; Mancuso, J., *Synlett* **2002**, 394. (c) ref. 3d

20. Kang, S.-K.; Baik, T.-G.; Kulak, A. N.; Ha, Y.-H.; Lim, Y.; Park, J., *J. Am. Chem. Soc.* **2000**, *122*, 11529.

21. Shin, S.; RajanBabu, T. V., *J. Am. Chem. Soc.* **2001**, *123*, 8416.

22. (a) Kang, S.-K.; Ha, Y.-H.; Ko, B.-S.; Lim, Y.; Jung, J., *Angew. Chem., Int. Ed.* **2002**, *41*, 343. (b) Kumareswaran, R.; Gallucci, J.; RajanBabu, T. V., *J. Org. Chem.* **2004**, *69*, 9151. (c) For a diene-aldehyde cyclization see: Sato, Y.; Saito, N.; Mori, M., *Chem. Lett.* **2002**, 18.

23. Obora, Y.; Tsuji, Y.; Kakehi, T.; Kobayashi, M.; Shinkai, Y.; Ebihara, M.; Kawamura, T., *Chem. Soc., Perkin Trans.* **1995**, 599.

T. V. (Babu) RajanBabu
The Ohio State University, Columbus, OH, USA

Seunghoon Shin
Hanyang University, Seoul, Korea

3-Trimethylsilyl-3-trimethylsilyloxy-1-propene[1]

[66662-17-7] $C_9H_{22}OSi_2$ (MW 202.49)

InChI = 1S/C9H22OSi2/c1-8-9(11(2,3)4)10-12(5,6)7/h8-9H, 1H2,2-7H3

InChIKey = YIXCNTGLARMYPD-UHFFFAOYSA-N

(homoenolate anion[2] and homoenolate dianion[3] equivalents; synthesis of acylsilanes)

Physical Data: bp 62 °C/35 mmHg; n_D^{20} 1.4228.

Preparative Methods: allyl trimethylsilyl ether is lithiated with *s*-butyllithium in THF–HMPA to give a rapidly equilibrating mixture of a carbanion and a silylallyloxy anion (eq 1). Silylation of the mixture with chlorotrimethylsilane affords 3-trimethylsilyl-3-trimethylsilyloxy-1-propene (**1**) exclusively. Several other derivatives with bulky substituent(s) on the silyloxy group were also prepared.[2]

Handling, Storage, and Precautions: use in a fume hood.

$$\text{\small OTMS} \xrightarrow[\text{THF–HMPA}]{s\text{-BuLi}}$$

$$\left[\begin{array}{c} \text{Li} \\ \text{OTMS} \end{array} \rightleftharpoons \begin{array}{c} \text{TMS} \\ \text{OLi} \end{array} \right] \xrightarrow{R^1_2R^2\text{SiCl}} \begin{array}{c} \text{TMS} \\ \text{OSiR}^1_2R^2 \end{array} \quad (1)$$

$R^1 = R^2 = \text{Me}, 80\%$; $R^1 = R^2 = \text{Et}, 72\%$; $R^1 = \text{Me}, R^2 = t\text{-Bu}, 76\%$

Homoenolate Anion Equivalent. The reaction of (**1**) with acid chlorides in the presence of titanium(IV) chloride affords the corresponding γ-keto aldehydes after hydrolysis (eq 2).[2] *O*-Acylation to form the 1-silyl allyl ester (17–30%) competes with the S_E' alkylation of the double bond. Bulkier silyloxy derivatives give γ-keto aldehydes in higher yield.

$$\begin{array}{c} \text{TMS} \\ \text{OSiR}^1_2R^2 \end{array} + \ i\text{-PrCOCl} \xrightarrow[\text{2. H}_2\text{O}]{\text{1. TiCl}_4, \text{CH}_2\text{Cl}_2} $$

$$ \quad (2)$$

$R^1 = R^2 = \text{Me}$	20%	30%
$R^1 = R^2 = \text{Et}$	43%	17%
$R^1 = \text{Me}, R^2 = t\text{-Bu}$	45%	–

Homoenolate Dianion Equivalent. The intermediate in the reaction is believed to be an enol silane and/or enol titanate which gives the expected product with carbon electrophiles such as acetals (eq 3).[3]

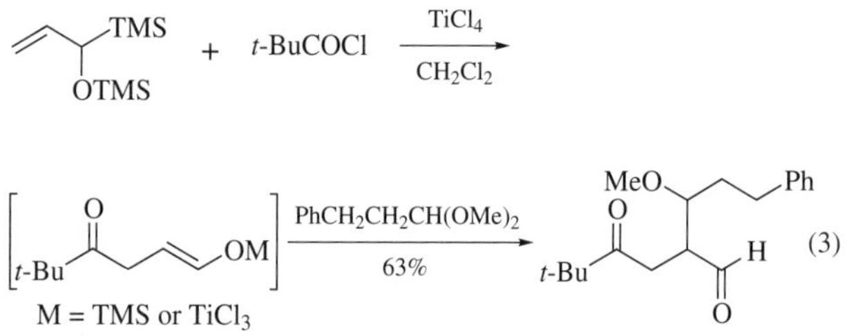

$$
\begin{array}{c}
\left[t\text{-Bu} \underset{O}{\overset{O}{\|}} \diagdown\diagup OM \right] \xrightarrow[63\%]{PhCH_2CH_2CH(OMe)_2} t\text{-Bu} \cdots H \quad (3) \\
M = TMS \text{ or } TiCl_3
\end{array}
$$

Synthesis of Acylsilanes. The title reagent (**1**) and other related derivatives undergo sequential isomerization to α-silylalkenyl ethers upon treatment with palladium on carbon which subsequently give acylsilanes by hydrolysis (eq 4).[4]

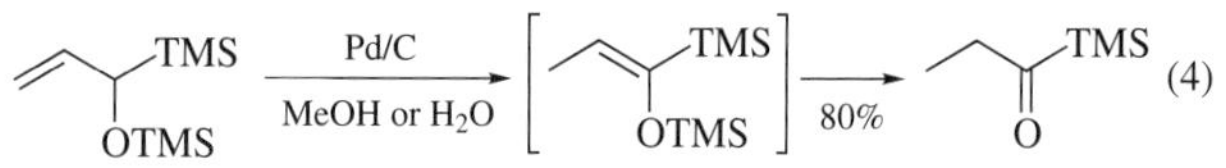

$$
\diagup\diagdown_{OTMS}^{TMS} \xrightarrow[MeOH \text{ or } H_2O]{Pd/C} \left[\diagup\diagdown_{OTMS}^{TMS} \right] \xrightarrow{80\%} \diagup\underset{O}{\overset{}{\diagdown}}^{TMS} \quad (4)
$$

Related Reagents. Allyltrimethylsilane; 1-Ethoxy-1-(trimethylsilyloxy)cyclopropane; 1-Methoxyallyllithium; Trimethylsilylallyllithium.

---

1. Fieser, M., *Fieser & Fieser* **1980**, *8*, 485.
2. Hosomi, A.; Hashimoto, H.; Sakurai, H., *J. Org. Chem.* **1978**, *43*, 2551.
3. Hosomi, A.; Hashimoto, H.; Kobayashi, H.; Sakurai, H., *Chem. Lett.* **1979**, 245.
4. Hosomi, A.; Hashimoto, H.; Sakurai, H., *J. Organomet. Chem.* **1979**, *175*, C1.

Hideki Sakaurai
Tohoku University, Sendai, Japan

(*E*)-1-Trimethylsilyl-2-trimethylstannylethylene[1,2]

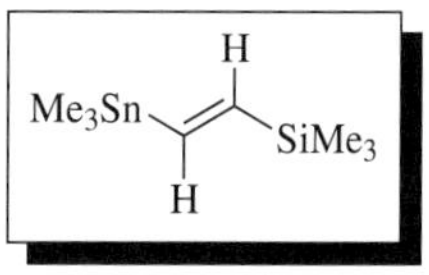

[65801-56-1]*   C$_8$H$_{20}$SiSn   (MW 263.08)

InChI = 1S/C5H11Si.3CH3.Sn/c1-5-6(2,3)4;;;;/h1,5H,2-4H3;3*1H3;

InChIKey = PJDCAXMJPGJRMP-UHFFFAOYSA-N

(reagent for Pd^0-catalyzed cross-coupling reactions with aryl and vinyl halides or triflates;[1,2] source of β-trimethylsilylvinyllithium[3])

Physical Data: bp 22 °C/0.04 mmHg; n_D 1.4723.
Solubility: sol ether, DMF, hydrocarbon, aromatic solvents.
Analysis of Reagent Purity: ^{1}H NMR (CCl$_4$/CH$_2$Cl$_2$) δ 0.12 (s, 9H, Me$_3$Si), 0.17 (s, 9H, Me$_3$Sn), 6.30–7.20 (AB pattern, J_{AB} = 22 Hz, 2H, vinyl).
Preparative Method: trans-1,2-bis(tri-*n*-butylstannyl)ethylene is treated successively with *n*-butyllithium, chlorotrimethylsilane, *n*-BuLi, and chlorotrimethylstannane to give (*E*)-trimethylsilyl-2-trimethylstannylethylene in good yield.[3]

Handling, Storage, and Precautions: organotin reagents are extremely toxic and readily absorb through the skin.[4] The reagent should be handled at all times in an efficient fume hood, wearing the appropriate personal safety equipment (i.e. gloves, safety eyewear).

* CAS number corrected 15th October 2001.

Palladium-Catalyzed Coupling Reactions. Vinyl and aromatic halides undergo palladium-catalyzed cross-coupling reactions with trimethylsilyl-2-trimethylstannylethylene to give substituted 1,3-dienes and styrenes, respectively (eqs 1 and 2).[5,6] Vinyl triflates also undergo cross-coupling reactions to give 1,3-dienes, provided that lithium chloride is added to the reaction (eq 3).[7,8]

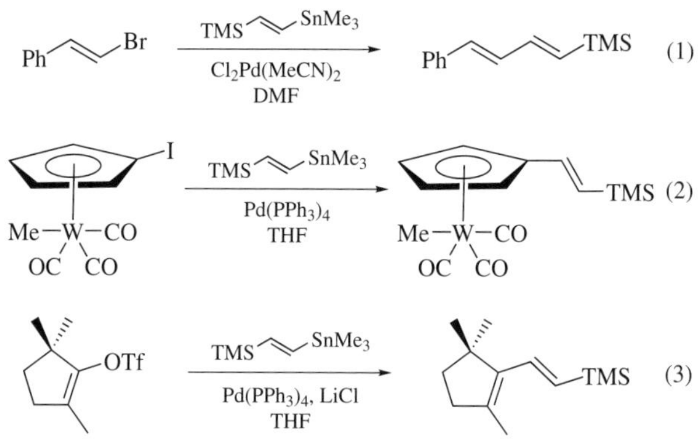

The reactions employ mild conditions, are regioselective and stereospecific, and give high yields of products. In addition, tedious protection–deprotection steps are not necessary because the reactions tolerate a wide variety of functional groups.[5,6]

When the coupling of vinyl triflates is carried out in the presence of carbon monoxide and lithium chloride, carbon monoxide insertion occurs, yielding unsymmetrical divinyl ketones (eqs 4 and 5).[9,10] This is a particularly useful procedure for the formation of intermediates for the Nazarov cyclization reaction.[11]

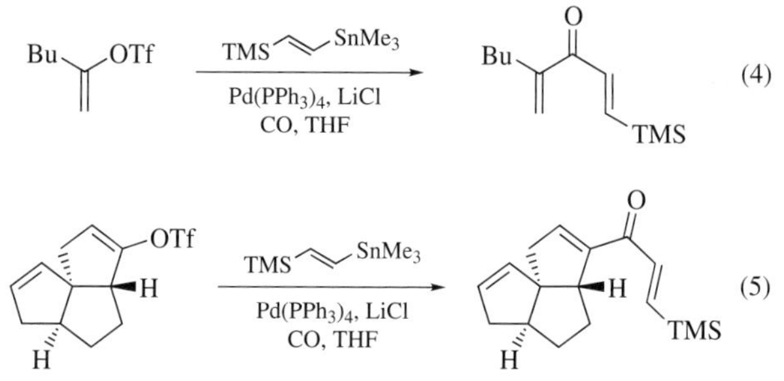

For additional discussion about palladium-catalyzed coupling reactions, see also those entries dealing with organopalladium complexes (e.g. tetrakis(triphenylphosphine)palladium(0), tris(dibenzylideneacetone)dipalladium–chloroform, and/or (*E*)-1-tri-*n*-butylstannyl-2-trimethylsilylethylene).

Transmetalation Reactions. Transmetalation of trimethylsilyl-2-trimethylstannylethylene at low temperature with *n*-BuLi affords (*E*)-2-(trimethylsilyl)vinyllithium in high yield (eq 6).[3]

$$Me_3Sn\diagdown\diagup TMS \xrightarrow{BuLi} Li\diagdown\diagup TMS \qquad (6)$$

1. Stille, J. K., *Angew. Chem., Int. Ed. Engl.* **1986**, *25*, 508.

2. Scott, W. J.; McMurry, J. E., *Acc. Chem. Res.* **1988**, *21*, 47.

3. Seyferth, D.; Vick, S. C., *J. Organomet. Chem.* **1978**, *144*, 1.

4. Krigman, M. R.; Silverman, A. P., *Neurotoxicology* **1984**, *5*, 129.

5. Stille, J. K.; Groh, B. L., *J. Am. Chem. Soc.* **1987**, *109*, 813.

6. Brehm, E. C.; Stille, J. K.; Meyers, A. I., *Organometallics* **1992**, *11*, 938.

7. Scott, W. J.; Crisp, G. T.; Stille, J. K., *J. Am. Chem. Soc.* **1984**, *106*, 4630.

8. Scott, W. J.; Stille, J. K., *J. Am. Chem. Soc.* **1986**, *108*, 3033.

9. Crisp, G. T.; Scott, W. J.; Stille, J. K., *J. Am. Chem. Soc.* **1984**, *106*, 7500.

10. Cheney, D. L.; Paquette, L. A., *J. Org. Chem.* **1989**, *54*, 3334.

11. Santelli-Rouvier, C.; Santelli, M., *Synthesis* **1983**, 429.

Kevin J. Moriarty

Rhône-Poulenc Rorer, Collegeville, PA, USA

(Trimethylsilyl)vinylketene

[75232-81-4]   C$_7$H$_{12}$OSi   (MW 140.28)

InChI = 1S/C7H12OSi/c1-5-7(6-8)9(2,3)4/h5H,1H2,2-4H3

InChIKey = JRRCBKRLXRMIJR-UHFFFAOYSA-N

(reactive diene for [4 + 2] cycloadditions[1])

Alternate Name: 2-(trimethylsilyl)-1,3-butadien-1-one.

Physical Data: bp 49.0–50.5 °C/1.6 mmHg.

Solubility: sol CH$_2$Cl$_2$, CHCl$_3$, CCl$_4$, toluene, benzene, hexane, most organic solvents.

Form Supplied in: yellow-green liquid; not commercially available.

Analysis of Reagent Purity: IR (CDCl$_3$) 2085, 1610 cm^{-1}; ^{1}H NMR (CDCl$_3$) δ 0.25 (s, 9 H), 4.82 (dd, $J = 1$, 10 Hz, 1 H), 4.88 (dd, $J = 1$, 17 Hz, 1 H), 5.92 (dd, $J = 10$, 17 Hz, 1 H) ppm; ^{13}C NMR (CDCl$_3$) δ −1.0 (q), 22.3 (s), 111.6 (t), 125.1 (d), 183.7 (s) ppm.

Preparative Methods: conveniently prepared as outlined in eq 1. Treatment of 1-(trimethylsilyl)propyne (**2**)[2] with 1.1 equiv of diisobutylaluminum hydride (25 °C, 21 h) and 1.1 equiv of methyllithium (0 °C, 0.5 h) in ether–hexane,[2] followed by reaction of the resulting vinylalanate with anhydrous carbon dioxide,[3] yields (Z)-2-(trimethylsilyl)-2-butenoic acid (**3**) in 68% yield. Exposure of the potassium salt of this acid (**4**) to 1.1 equiv of oxalyl chloride in pentane containing a catalytic amount of *N,N*-dimethylformamide (0–25 °C, 1.5 h) then produced a mixture of the acid chloride (**5**) and its geometric isomer, which was dehydrohalogenated without further purification. A solution of (**5**) in pentane was added dropwise over 1–2 h to a solution of 0.9 equiv of triethylamine in pentane

at 25 °C, and the resulting mixture was heated at reflux for 15–24 h and then filtered with the aid of pentane. Solvent was evaporated at −50 °C (0.5 mmHg), and the residue was distilled at 25 °C (1 mmHg) and then again at 5 mmHg into a receiver cooled at −78 °C. In this manner a yellow-green liquid of (trimethylsilyl)vinylketene (**1**) was obtained in 39–50% overall yield (from **3**).

$$\equiv-TMS \longrightarrow \underset{\underset{(3)\ X=OH}{\underset{(4)\ X=O^-K^+}{(5)\ X=Cl}}}{\overset{TMS}{\underset{COX}{H}\diagup}} \longrightarrow TMS\diagup\diagdown{}^{=O} \qquad (1)$$

(**2**)   (**3**) X = OH   (**1**)
                 (**4**) X = O$^-$ K$^+$
                 (**5**) X = Cl

Handling, Storage, and Precautions: purified vinylketenes can be stored under nitrogen in solution at 0 °C without appreciable decomposition for 1–2 weeks.

Reactive Enophile in [4 + 2] Cycloadditions. Vinylketenes are not effective as dienes in Diels–Alder reactions because they undergo only [2 + 2] cycloaddition with alkenes, as predicted by frontier molecular orbital theory. However, silylketenes exhibit dramatically different properties from those found for most ketenes.[4] (Trimethylsilyl)vinylketene (**1**) is a relatively stable isolable compound which does not enter into typical [2 + 2] cycloadditions with electron-rich alkenes. Instead, (**1**) participates in Diels–Alder reactions with a variety of alkenic and alkynic dienophiles. The directing effect of the carbonyl group dominates in controlling the regiochemical course of cycloadditions using this diene. For example, reaction of (**1**) with methyl propiolate produced methyl 3-(trimethylsilyl)salicylate with the expected regiochemical orientation. Protodesilylation of this adduct with trifluoroacetic acid in chloroform (25 °C, 24 h) afforded methyl salicylate in 78% yield (eq 2).

$$\underset{CO_2Me}{|||} \xrightarrow[\underset{45\%}{}]{(1),\ toluene \\ 95\ °C,\ 63\ h} \underset{TMS}{\overset{OH}{\bigcirc}}CO_2Me \xrightarrow[\underset{78\%}{}]{TFA,\ CHCl_3 \\ 25\ °C,\ 24\ h}$$

$$\underset{}{\overset{OH}{\bigcirc}}CO_2Me \qquad (2)$$

Diels–Alder addition of (**1**) to alkenic dienophiles furnishes cyclohexenone derivatives (eqs 3 and 4). Addition of (**1**) to naphthoquinone afforded a mixture of several cycloadducts which could be oxidized to a single anthraquinone (eq 5).

$$\underset{EtO_2C}{\overset{CO_2Et}{\diagup}} \xrightarrow[\underset{62\%}{}]{(1),\ toluene \\ 95\ °C,\ 38\ h} TMS\overset{O}{\bigcirc}\underset{{}^{\prime\prime\prime}CO_2Et}{CO_2Et} \qquad (3)$$

$$\underset{O}{\overset{O}{\bigcirc}}N-Ph \xrightarrow[\underset{74\%}{}]{(1),\ CHCl_3 \\ 40\ °C,\ 24\ h} TMS\overset{O\ H\ O}{\bigcirc}N-Ph \qquad (4)$$

$$\text{(5)}$$

The reactivity of (trimethylsilyl)vinylketene compares favorably to previously reported vinylketene equivalents. Both vinylketene acetals[5] and thioacetals[6] have been reported to undergo [4+2] cycloadditions in good yield. However, due to the limited reactivity of hindered (Z)-dienes, only highly electrophilic dienophiles react. Specially activated vinylketene thioacetal substrates have been made with an electron-donating group at the 3-position of the diene to promote reaction with weaker dienophiles. Unfortunately, due to their strong nucleophilic character, these dienes were found to participate in a stepwise Michael-type process. Concerted [4+2] cycloadditions were only seen with a few dienophiles.[6c] Vinylketenimines are shown to be isolable by distillation, are stable at room temperature, and undergo [4+2] cycloadditions in moderate to good yield. Again, reaction only occurs with electron-deficient dienophiles.[7]

A general route to substituted (silyl)vinylketenes has been developed and several substituted ketenes have been prepared. The synthesis of 1-(1-cyclohexenyl)-2-triisopropylsilylketene is representative of this general approach (eq 6).[10]

$$\text{(6)}$$

Thus 'detrifluoroacetylative' diazo transfer[8] furnished the diazo ketone in 80% yield, which was converted to the silyl diazo ketone in 75% yield by the method of Maas.[9] The key photolysis step was effected by irradiating a degassed 0.1 M solution of the silyl diazo ketone in hexane in a Vycor tube using a low-pressure mercury lamp (300 nm) for 2 h. Concentration and chromatographic purification furnished the desired (silyl)vinylketene in 73% yield.[10]

1. Danheiser, R. L.; Sard, H., *J. Org. Chem.* **1980**, *45*, 4810.
2. (a) Eisch, J. J.; Damasevitz, G. A., *J. Org. Chem.* **1976**, *41*, 2214. (b) Uchida, K.; Utimoto, K.; Nozaki, H., *J. Org. Chem.* **1976**, *41*, 2215.
3. Zweifel, G.; Steele, R. B., *J. Am. Chem. Soc.* **1967**, *89*, 2754.
4. Allen, A. D.; Tidwell, T. T., *Tetrahedron Lett.* **1991**, *32*, 847.
5. (a) McElvain, S. M.; Morris, L. R., *J. Am. Chem. Soc.* **1952**, *74*, 2657. (b) Banville, J.; Grandmaison, J. L.; Lang, G.; Brassard, P., *Can. J. Chem.* **1974**, *52*, 80. (c) Banville, J.; Brassard, P., *J. Chem. Soc., Perkin Trans. 1* **1976**, 1852. (d) Banville, J.; Brassard, P., *J. Org. Chem.* **1976**, *41*, 3018. (e) Grandmaison, J. L.; Brassard, P., *Tetrahedron* **1977**, *33*, 2047. (f) Grandmaison, J. L.; Brassard, P., *J. Org. Chem.* **1978**, *43*, 1435. (g) Roberge, G.; Brassard, P., *J. Chem. Soc., Perkin Trans. 1* **1978**, 1041. (h) Gompper, R.; Sobotta, R., *Tetrahedron Lett.* **1979**, *11*, 921.
6. (a) Carey, F. A.; Court, A. S., *J. Org. Chem.* **1972**, *37*, 1926, 4474. (b) Kelly, T. R.; Goerner, R. N.; Gillard, J. W.; Prazak, B. K., *Tetrahedron Lett.* **1976**, *43*, 3869. (c) Danishefsky, S.; McKee, R.; Singh, R. K., *J. Org. Chem.* **1976**, *41*, 2934.

7. (a) Sonveaux, E.; Ghosez, L., *J. Am. Chem. Soc.* **1973**, *95*, 5417. (b) Differding, E.; Vandevelde, O.; Roekens, B.; Van, T. T.; Ghosez, L., *Tetrahedron Lett.* **1987**, *28*, 397.
8. Danheiser, R. L.; Miller, R. F.; Brisbois, R. G.; Park, S. Z., *J. Org. Chem.* **1990**, *55*, 1959.
9. Maas, G.; Bruckman, R., *J. Org. Chem.* **1985**, *50*, 2801.
10. Danheiser, R. L.; Loebach, J. L. unpublished results.

Jennifer L. Loebach & Rick L. Danheiser
Massachusetts Institute of Technology, Cambridge, MA, USA

1-(Trimethylsilyl)vinyllithium

[51666-94-5]   C₅H₁₁LiSi   (MW 106.19)
InChI = 1S/C5H11Si.Li/c1-5-6(2,3)4;/h1H2,2-4H3;
InChIKey = XMMHXPNYMBBQRC-UHFFFAOYSA-N

(adds 1,2 to ketones and aldehydes;[1] 1,4-additions to enones via the cuprates;[2] diene synthesis;[3] formation of vinylsilanes[4])

Alternate Names: 1-lithio-1-(trimethylsilyl)ethylene; α-(trimethylsilyl)vinyllithium.
Solubility: sol THF, ether.
Preparative Method: prepared in situ under an inert atmosphere by lithium–halogen exchange of (1-bromovinyl)trimethylsilane with *t*-butyllithium.
Handling, Storage, and Precautions: solutions are inflammable; it is critical to preclude contact with air and moisture.

Addition to Ketones and Aldehydes.[1] α-(Trimethylsilyl)vinyllithium **1** adds to ketones and aldehydes. The adducts do not undergo the expected Peterson elimination reaction when treated with sodium hydride or potassium hydride, but the corresponding allene can still be obtained if the alcohol is converted to the chloride followed by fluoride-catalyzed β-elimination (eq 1).[5a,b] Other uses have also been made of the allyl chlorides made in this manner: they react with cuprates,[5c] or they can be oxidized to the epoxides to serve as precursors to allene oxides.[5d–i]

$$\text{(1)}$$

Under certain conditions (KH in HMPA), products from the homo-Brook rearrangement are observed (eq 2).[6] This is therefore a good method of desilylation of β-hydroxyvinylsilanes to the corresponding vinyl alcohols. Other conditions for desilylation were found to give better yields (eq 3).[7] Since α-(trimethylsilyl)-vinyllithium was found to add to chiral aldehydes with better diastereoselectivity than vinyllithium, the overall sequence is an improvement for chelation controlled 1,2-asymmetric induction.[7a] In the sequence of eq 3, diastereoselectivity varied as the protecting group R' was changed; it was 20:1 for R'-benzyl. This illustrates the usefulness of α-(trimethylsilyl)vinyllithium in stereoselective polyol synthesis.[7]

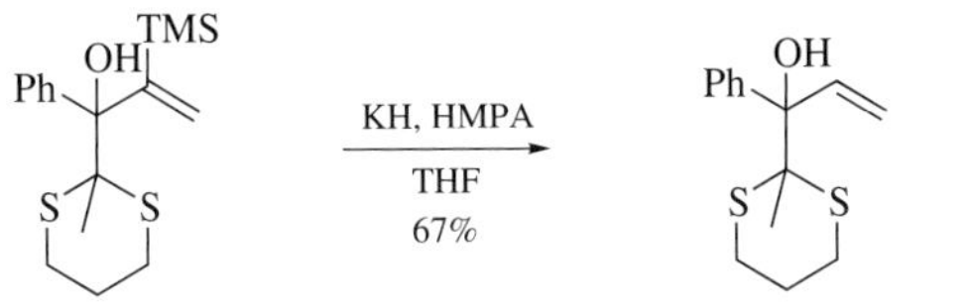

$$(2)$$

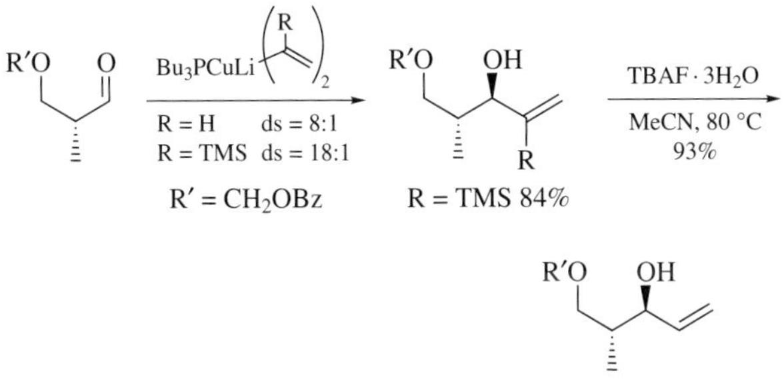

$$(3)$$

The alcohol adducts from α-(trimethylsilyl)vinyllithium and aldehydes have found many uses in organic synthesis. The ketones obtained by oxidation (eq 4) are especially good Michael acceptors and have been used in a modified Robinson annulation reaction for the construction of cyclohexenones.[8] Cyclopropanation leads to cyclopropylsilane adducts, which can be converted into a variety of cyclopentenes (eq 5).[9] Halogenation followed by stannylation gives synthons (eq 6) useful for making substituted vinylsilanes through radical reactions.[10]

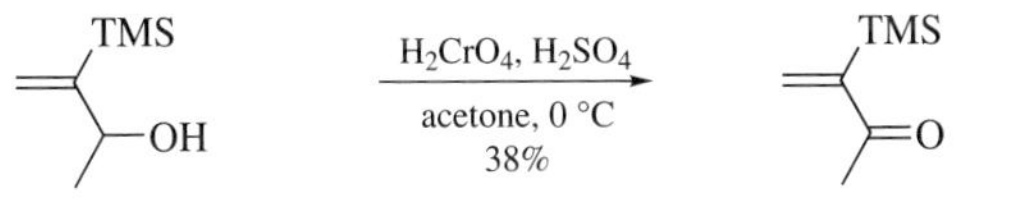

$$(4)$$

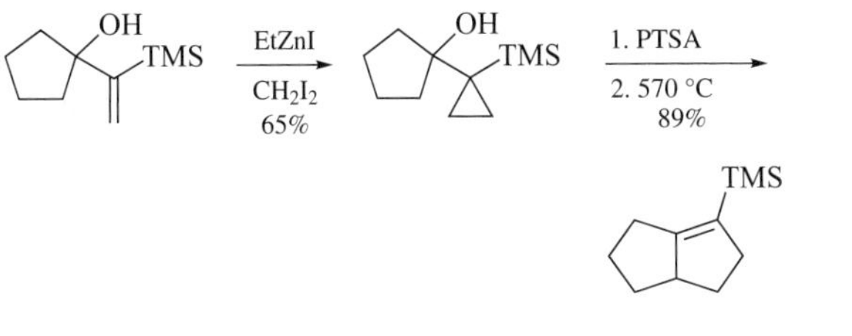

$$(5)$$

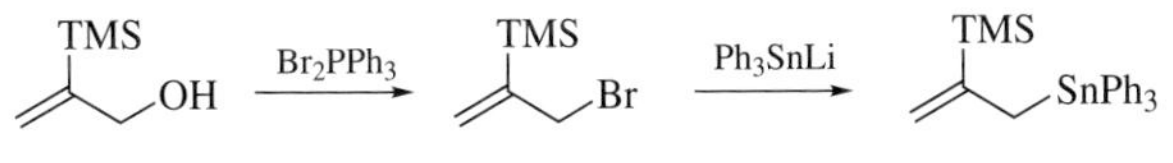

Rearrangements of alcohols bearing an epoxide functionality give ketones that can be stereoselectively reduced. The new stereocenter formed is of opposite configuration when the trimethylsilyl group 'R' is present on the vinyl moiety (eq 7).[11b,c] Thus α-(trimethylsilyl)vinyllithium can be viewed as a large vinyl anion equivalent (e.g. eq 3) which complements vinylmagnesium and lithium chemistry.[11] These reactions were elegantly used in the total syntheses of avenaciolide, isoavenaciolide,[11b,c] and (+)- and (−)-eldanolide.[11f]

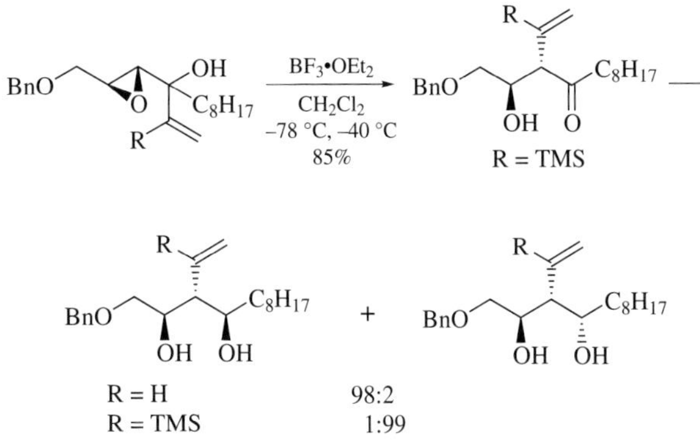

$$(7)$$

$$
\begin{array}{cc}
R = H & 98:2 \\
R = TMS & 1:99
\end{array}
$$

1,4-Additions to Enones via the Cuprates.[2] α-(Trimethylsilyl)vinyllithium undergoes Michael addition to enones, in good yields, in the presence of copper salts (eq 8).[2] The vinyl group can be oxidized to the epoxide, which is then converted to the acyl group.[2,12] The overall transformation is the addition of an acyl anion equivalent. α-(Trimethylsilyl)vinyllithium compares favorably with other acyl anion equivalents, particularly with the vinyl ether analogs which do not add well to hindered enones.[13]

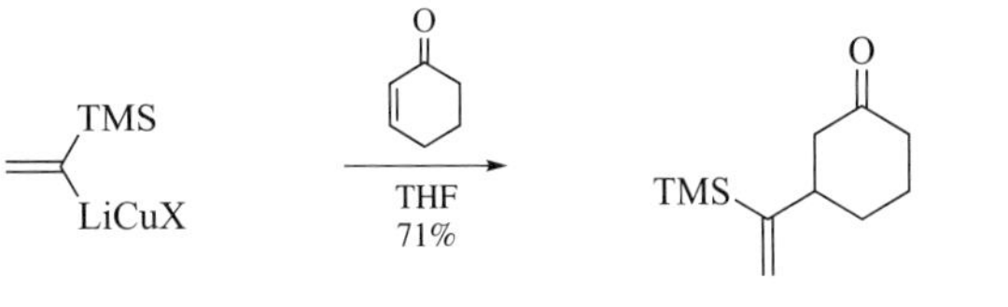

$$(8)$$

Diene Synthesis.[3] α-(Trimethylsilyl)vinyllithium dimerizes in the presence of copper(I) iodide. This reaction was first observed as a side product during the formation of organo-cuprates,[3a] then developed for the synthesis of 2,3-bis(trimethylsilyl)buta-1,3-diene (eq 9).[3b]

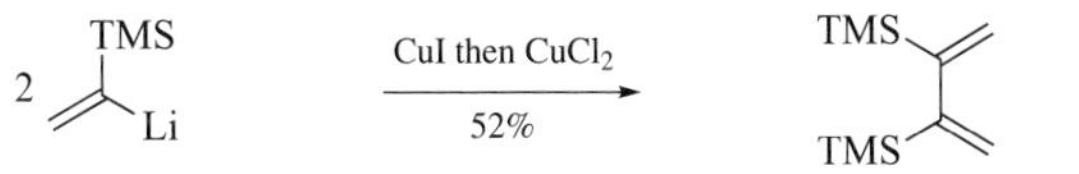

$$(9)$$

Reactions with Other Electrophiles.[4] α-(Trimethylsilyl)-vinyllithium reacts with ethylene oxide to give an intermediate which has been used to set up an impressive eight-membered ring cyclization for the synthesis of (−)-laurenyne (eq 10).[4a] Here, vinyltrimethylsilane was preferred over the simple vinyl moiety because of its greater reactivity.

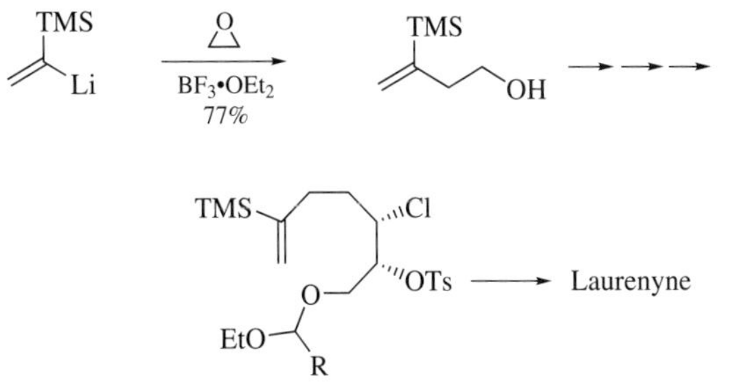

$$(10)$$

The reaction of this anion with carbon monoxide leads to the formation of an interesting cyclopropyl product (eq 11).[4c] Other electrophiles, for example selenium cyanide,[4d] iodine,[4f] chlorosilanes,[4g] and imines,[4h] react with α-(trimethylsilyl)vinyllithium to give the corresponding adducts.

$$(11)$$

Related Reagents. Other types of 1-lithio-1-vinylsilanes have been made, mainly through the lithium–halogen exchange reaction described before. For example, one or all of the methyl groups on silicon can be substituted by phenyls. The compounds (**2**)[4c] and (**3**)[14] have properties similar to α-(trimethylsilyl)vinyllithium.

Monosubstitution in the 2-position on the vinyl moiety has been extensively exemplified in the literature. The stereochemistry of the double bond implies two possible types of reagents. For example, the 2-(Z)-monosubstituted (**4**,[15] **5**[11d]) and the 2-(E)-monosubstituted (**6**,[16a] **7**[16b]) 1-lithiovinylsilanes have been treated with carbonyl and alkyl electrophiles.

(**4**) R = C₆H₁₃   (**6**) R = i-Pr
(**5**) R = PhCH₂OCH₂   (**7**) R = Ph

The stereochemistry of the 1-lithiovinylsilanes is easily lost, as exemplified in eq 12 where (**9**), prepared by a transmetalation reaction, subsequently equilibrates to a mixture containing (**9**) and (**10**).[15,16b]

$$(12)$$

One other related reagent includes the species (**11**),[17] obtained from silylcupration of the corresponding lithium acetylide. This is yet another way to obtain 1-lithiovinylsilanes substituted in the β-position.

See also lithium (trimethylsilyl)acetylide, (E)-2-(trimethylsilyl)vinyllithium, and vinyllithium.

1. Gröebel, B. T.; Seebach, D., *Angew. Chem., Int. Ed. Engl.* **1974**, *86*, 102.
2. Boeckman, Jr., R. K.; Bruza, K. J., *Tetrahedron Lett.* **1974**, 3365.
3. (a) Huynh, C.; Linstrumelle, G., *Tetrahedron Lett.* **1979**, 1073. (b) Garratt, P. J.; Tsotinis, A., *Tetrahedron Lett.* **1986**, *27*, 2761.
4. (a) Blumenkopf, T. A.; Bratz, M.; Castaneda, A.; Look, G. C.; Overman, L. E.; Rodriguez, D.; Thompson, A. S., *J. Am. Chem. Soc.* **1990**, *112*, 4386. (b) Blumenkopf, T. A.; Look, G. C.; Overman, L. E., *J. Am. Chem. Soc.* **1990**, *112*, 4399. (c) Ryu, I.; Hayama, Y.; Hirai, A.; Sonoda, N.; Orita, A.; Ohe, K.; Murai, S., *J. Am. Chem. Soc.* **1990**, *112*, 7061. (d) Meinke, P. T.; Krafft, G. A.; Guram, A., *J. Org. Chem.* **1988**, *53*, 3632. (e) Karabelas, K.; Anders, H., *J. Org. Chem.* **1988**, *53*, 4909. (f) Jones, P. R.; Bates, T. F., *J. Am. Chem. Soc.* **1987**, *109*, 913. (g) Cunico, R. F.; Kuan, C. P., *J. Org. Chem.* **1992**, *57*, 3331. (h) Claremon, D. A.; Lumma, P. K.; Phillips, B. T., *J. Am. Chem. Soc.* **1986**, *108*, 8265.
5. (a) Chan, T. H.; Mychajlowskij, W.; Ong, B. S.; Harpp D. N., *J. Organomet. Chem.* **1976**, *107*, C1. (b) Mychajlowski, W.; Chan, T. H., *Tetrahedron Lett.* **1976**, *17*, 4439. (c) Amouroux, R.; Chan, T. H., *Tetrahedron Lett.* **1978**, *19*, 4453. (d) Chan, T. H.; Ong B. S., *J. Org. Chem.* **1978**, *43*, 2994. (e) Chan, T. H.; Ong, B. S.; Mychajlowskij, W., *Tetrahedron Lett.* **1976**, *17*, 3253. (f) Chan, T. H.; Ong, B. S., *Tetrahedron Lett.* **1976**, *17*, 3257. (g) Chan, T. H.; Lau, P. K. W.; Li, M. P., *Tetrahedron Lett.* **1976**, 2667. (h) Chan, T. H.; Li, M. P.; Mychajlowskij, W.; Harpp, D. N., *Tetrahedron Lett.* **1974**, *14*, 3511. (i) Ong, B. S.; Chan T. H., *Heterocycles* **1977**, *7*, 913.
6. Wilson, S. R.; Georgiadis, G. M., *J. Org. Chem.* **1983**, *48*, 4143.
7. (a) Burke, S. D.; Piscopio, A. D.; Marron, B. E.; Matulenko, M. A.; Pan, G., *Tetrahedron Lett.* **1991**, *32*, 857. (b) Chan, T. H.; Mychajlowskij, W., *Tetrahedron Lett.* **1974**, 3479. (c) Sato, F.; Tanaka, Y.; Sato, M., *J. Chem. Soc., Chem. Commun.* **1983**, 165. (d) Sato, F.; Kobayashi, Y.; Takahashi, O.; Chiba, T.; Takeda, Y.; Kusakabe, M., *J. Chem. Soc., Chem. Commun.* **1985**, 1636.
8. (a) Zhou, W.; Wei, G., *Synthesis* **1990**, *9*, 822. (b) Boeckman, Jr., R. K., *Tetrahedron* **1983**, *39*, 925. (c) ApSimon, J. W.; Sequin, R. P.; Huber, C. P., *Can. J. Chem.* **1982**, *60*, 509. (d) Boeckman, Jr., R. K.; Blum, D. M.; Arthur, S. D., *J. Am. Chem. Soc.* **1979**, *101*, 5060.
9. (a) Wells, G. J.; Yan, T. H.; Paquette, L. A., *J. Org. Chem.* **1984**, *49*, 3604. (b) Paquette, L. A.; Wells, G. J.; Horn, K. A.; Yan T. H., *Tetrahedron* **1983**, *39*, 913. (c) Boeckman, Jr., R. K.; Blum, D. M.; Ganem, D. M.; Halvey, N., *Org. Synth.* **1978**, *58*, 152. and references cited therein.
10. Lee, E.; Yu, S. G.; Hur, C. U.; Yang, S. M., *Tetrahedron Lett.* **1988**, *29*, 6969.
11. (a) Shimazaki, M.; Hara, H.; Suzuki, K.; Tsuchihashi, G., *Tetrahedron Lett.* **1987**, *28*, 5891. (b) Suzuki, K.; Shimazaki, M.; Tsuchihashi, G., *Tetrahedron Lett.* **1986**, *27*, 6233. (c) Suzuki, K.; Miyazawa, M.; Shimazaki, M.; Tsuchihashi, G., *Tetrahedron Lett.* **1986**, *27*, 6237. (d) Suzuki, K.; Ohkuma, T.; Miyazawa, M.; Tsuchihashi, G., *Tetrahedron Lett.* **1986**, *27*, 373. (e) Maruoka, K.; Hasegawa, M.; Yamamoto, H.; Suzuki, K.; Shimazaki, M.; Tsuchihashi, G., *J. Am. Chem. Soc.* **1986**, *108*, 3827. (f) Suzuki, K.; Tsuchihashi, G., *Tetrahedron Lett.* **1985**, *26*, 861.
12. Gröbel, B. T.; Seebach, D., *Angew. Chem., Int. Ed. Engl.* **1974**, *13*, 83.
13. Boeckman, Jr., R. K.; Bruza, K. J., *J. Org. Chem.* **1979**, *44*, 4781.
14. Brook, A. G.; Duff, J. M., *Can. J. Chem.* **1973**, *51*, 2024.
15. Negishi, E.; Takahashi, T., *J. Am. Chem. Soc.* **1986**, *108*, 3402, and references cited therein.
16. (a) Khripach, V. A.; Zhabinskiy, V. N.; Olkhovick, V. K., *Tetrahedron Lett.* **1990**, *31*, 4937. (b) Mitchell, T. N.; Reimann, W., *J. Organomet. Chem.* **1985**, *281*, 163 and references cited therein.
17. Fleming, I.; Newton, T. W.; Roessler, F., *J. Chem. Soc., Perkin Trans. 1* **1981**, 2527.

Denis Labrecque & Tak-Hang Chan
McGill University, Montreal, Quebec, Canada

(E)-2-(Trimethylsilyl)vinyllithium

[55339-31-6]   C$_5$H$_{11}$LiSi   (MW 106.19)
InChI = 1S/C5H11Si.Li/c1-5-6(2,3)4;/h1,5H,2-4H3;
InChIKey = CIJZBCRRKSOQOF-UHFFFAOYSA-N

(for introduction of the β-(trimethylsilyl)vinyl moiety; the resulting alkenylsilanes undergo useful electrophilic substitutions of the TMS group[1])

Form Supplied in: not available commercially; prepared in situ in either THF or diethyl ether.

Preparative Methods: attempts to produce synthetically useful quantities of (E)-2-(trimethylsilyl)vinyllithium (**1**) by reaction of lithium metal with (2-bromovinyl)trimethylsilane (**2**) failed because of concomitant metalation–elimination processes involving (**1**) and (**2**).[2] However, the transmetalation process between (**2**) and excess *t*-butyllithium has been effectively used to provide (**1**) (eq 1).[3] An alternative route to (**1**) employs the transmetalation of a 1-(trimethylsilyl)-2-stannylethylene with organolithium reagents (eqs 2 and 3).[4,5]

$$\text{TMS}\diagup\diagdown_{\text{Br}} \xrightarrow[-78 \text{ to } 0\,°C]{t\text{-BuLi, Et}_2\text{O}} \text{TMS}\diagup\diagdown_{\text{Li}} \quad (1)$$
$$\text{(2)} \qquad\qquad\qquad\qquad \text{(1)}$$

$$\text{TMS}\diagup\diagdown_{\text{SnPh}_3} \xrightarrow[25\,°C]{\text{PhLi, Et}_2\text{O}} \text{TMS}\diagup\diagdown_{\text{Li}} + \text{Ph}_4\text{Sn (insoluble)} \quad (2)$$
$$\qquad\qquad\qquad\qquad\qquad\qquad \text{(1)}$$

$$\text{TMS}\diagup\diagdown_{\text{SnBu}_3} \xrightarrow[-78 \text{ to } -30\,°C]{\text{BuLi, THF}} \text{TMS}\diagup\diagdown_{\text{Li}} + \text{Bu}_4\text{Sn} \quad (3)$$
$$\qquad\qquad\qquad\qquad\qquad \text{(1)}$$

Handling, Storage, and Precautions: must be prepared and transferred under inert gas (Ar, N$_2$) to exclude oxygen and moisture.

Substitution Reactions. Alkylating agents containing primary alkyl or methyl groups react with (**1**) in synthetically useful yields (eqs 4 and 5).[4,5] Reaction of (**1**) with 1,2-dibromoethane affords (**2**),[4] while the use of metal or metalloidal halides generally results in high yields of coupling products (eqs 6–9).[4–7]

$$\text{TMS}\diagup\diagdown_{\text{Li}} \xrightarrow[\substack{\text{THF} \\ 81\%}]{\text{BuBr}} \text{TMS}\diagup\diagdown_{\text{Bu}} \quad (4)$$
$$\text{(1)}$$

$$\text{TMS}\diagup\diagdown_{\text{Li}} \xrightarrow[\substack{\text{Et}_2\text{O} \\ 76\%}]{(\text{MeO})_2\text{SO}_2} \text{TMS}\diagup\diagdown \quad (5)$$
$$\text{(1)}$$

$$\text{TMS}\diagup\diagdown_{\text{Li}} \xrightarrow[\substack{\text{THF} \\ 90\%}]{\text{TMSCl}} \text{TMS}\diagup\diagdown_{\text{TMS}} \quad (6)$$
$$\text{(1)}$$

$$\text{TMS}\diagup\diagdown_{\text{Li}} \xrightarrow[\substack{\text{Et}_2\text{O} \\ 88\%}]{\text{Me}_3\text{SnCl}} \text{TMS}\diagup\diagdown_{\text{SnMe}_3} \quad (7)$$
$$\text{(1)}$$

$$\text{TMS}\diagup\diagdown_{\text{Li}} \xrightarrow[\substack{\text{THF} \\ 94\%}]{\text{TBDPSCl}} \text{TMS}\diagup\diagdown_{\text{TBDPS}} \quad (8)$$
$$\text{(1)}$$

$$\text{TMS}\diagup\diagdown_{\text{Li}} \xrightarrow[\substack{\text{THF} \\ 32\%}]{\text{CpFe(CO)}_2\text{I}} \text{TMS}\diagup\diagdown_{\text{Fe(CO)}_2\text{Cp}} \quad (9)$$
$$\text{(1)}$$

An important subset of the reaction with metal halides is the conversion of (**1**) into more discriminating reagents. Thus the reaction of (**1**) with copper(I) iodide forms a diorganocuprate (**3**) which adds in a 1,4-fashion to conjugated enones (eqs 10 and 11).[3a,b,8] In this same vein, (**2**) was transformed into an alkenylzinc reagent which displayed selectivity between adjacent aldehyde and ester functionalities (eq 12).[9]

$$\left(\text{TMS}\diagup\diagdown\right)_2\text{CuLi}_2\text{I} \xrightarrow[\substack{\text{Et}_2\text{O} \\ 82\%}]{} \text{TMS}\diagup\diagdown \quad (10)$$
$$\text{(3)}$$

$$\left(\text{TMS}\diagup\diagdown\right)_2\text{CuLi}_2\text{I} \xrightarrow[\substack{\text{Et}_2\text{O} \\ 53\%}]{} \quad (11)$$
$$\text{(3)}$$

$$\text{TMS}\diagup\diagdown_{\text{Br}} \xrightarrow{\substack{1.\ t\text{-BuLi, Et}_2\text{O, }-78\,°C \\ 2.\ \text{ZnCl}_2\text{, THF, }-20\,°C \\ 3.\ \text{CHOCO}_2\text{Bu}}} \text{TMS}\diagup\diagdown_{\text{CO}_2\text{Bu}} \quad (12)$$
$$\text{(2)}$$

Addition Reactions. The behavior of (**1**) with carbonyl-containing compounds is completely straightforward, with aldehydes (eqs 13 and 14),[10,11] ketones (eqs 15 and 16),[12,13] and carbon dioxide (eq 17)[4] leading cleanly to expected products. Sato and co-workers have observed (**1**) to display considerable chemoselectivity in favor of an aldehyde functionality when presented with an ester as the alternative (eq 18).[13]

$$\text{OHC}\diagdown\diagup_{\text{Ph}} \xrightarrow[\substack{\text{THF} \\ 84\%}]{(1)} \text{TMS}\diagup\diagdown\diagdown_{\text{Ph}} \quad (13)$$

$$\text{OHC}\diagdown\diagup^{\text{O}}\diagdown_{\text{Ph}} \xrightarrow[\substack{\text{THF} \\ >80\%}]{(1)} \text{TMS}\diagup\diagdown\diagdown^{\text{O}}\diagdown_{\text{Ph}} \quad (14)$$

$$ \xrightarrow[\substack{\text{Et}_2\text{O} \\ 88\%}]{(1)} \quad (15)$$

$$ \xrightarrow{\substack{1.\ (1)\text{, THF} \\ 2.\ \text{H}_3\text{O}^+ \\ 93\%}} \quad (16)$$

$$\text{TMS}\diagup\diagdown_{\text{Li}} \xrightarrow{\substack{1.\ \text{CO}_2\text{, THF} \\ 2.\ \text{H}_3\text{O}^+ \\ 86\%}} \text{TMS}\diagup\diagdown_{\text{CO}_2\text{H}} \quad (17)$$
$$\text{(1)}$$

$$\text{OHC}\diagdown\diagup\diagdown_{\text{CO}_2\text{Me}} \xrightarrow[\substack{\text{THF} \\ 56\%}]{(1)} \text{TMS}\diagup\diagdown\diagdown\diagdown_{\text{CO}_2\text{Me}} \quad (18)$$

Related Reagents. (2-Bromovinyl)trimethylsilane; (*E*)-1-Lithio-2-tributylstannylethylene; (*E*)-1-Tri-*n*-butylstannyl-2-trimethylsilylethylene.

1. Alkenylsilane chemistry reviews: (a) Fleming, I.; Dunogues, J.; Smithers, R., *Org. React.* **1989**, *37*, 57. (b) Blumenkopf, T. A.; Overman, L. E., *Chem. Rev.* **1986**, *86*, 857. (c) Colvin, E. W. *Silicon in Organic Synthesis*; Butterworths: London, 1981.

2. (a) Husk, G. R.; Velitchko, A. M., *J. Organomet. Chem.* **1973**, *49*, 85. (b) Also see attempts at the direct lithiation of vinyltrimethylsilane: Khotimskii, V. S.; Bryantseva, I. S.; Durgar'yan, S. G.; Petrovskii, P. V., *Izv. Akad. Nauk SSSR, Ser. Khim.* **1984**, 470.

3. (a) Boeckman, Jr., R. K.; Bruza, K. J., *Tetrahedron Lett.* **1974**, 3365. (b) Boeckman, Jr., R. K.; Bruza, K. J., *J. Org. Chem.* **1979**, *44*, 4781. (c) Miller, S. A.; Gadwood, R. C., *J. Org. Chem.* **1988**, *53*, 2214.

4. (a) Cunico, R. F.; Clayton, F. J., *J. Org. Chem.* **1976**, *41*, 1480. (b) The related reagent, (*E*)-2-(tri-*n*-butylstannyl)vinyllithium, has been similarly prepared: Corey, E. J.; Wollenberg, R. H., *J. Am. Chem. Soc.* **1974**, *96*, 5581.

5. Seyferth, D.; Vick, S. C., *J. Organomet. Chem.* **1978**, *144*, 1.

6. Doyle, M. M.; Jackson, W. R.; Perlmutter, P., *Aust. J. Chem.* **1989**, *42*, 1907.

7. Landrum, B. E.; Lay, Jr., J. O.; Allison, N. T., *Organometallics* **1988**, *7*, 787.

8. Belmont, D. T.; Paquette, L. A., *J. Org. Chem.* **1985**, *50*, 4102.

9. Trost, B. M.; Self, C. R., *J. Am. Chem. Soc.* **1983**, *105*, 5942.

10. Jenkins, P. R.; Gut, R.; Wetter, H.; Eschenmoser, A., *Helv. Chim. Acta* **1979**, *62*, 1922.

11. Kusakabe, M.; Kato, H.; Sato, F., *Chem. Lett.* **1987**, 2163.

12. Burke, S. D.; Murtiashaw, C. W.; Dike, M. S.; Strickland, S. M. S.; Saunders, J. O., *J. Org. Chem.* **1981**, *46*, 2400.

13. Kobayashi, Y.; Shimazaki, T.; Sato, F., *Tetrahedron Lett.* **1987**, *28*, 5849.

Robert F. Cunico
Northern Illinois University, DeKalb, IL, USA

1-(Trimethylsilyl)vinylmagnesium Bromide–Copper(I) Iodide[1]

[49750-22-3]   C$_5$H$_{11}$BrMgSi   (MW 203.46)
InChI = 1S/C5H11Si.BrH.Mg/c1-5-6(2,3)4;;/h1H2,2-4H3;1H;/
    q;;+1/p-1
InChIKey = UIRQXGNUQMSHFO-UHFFFAOYSA-M
[7681-65-4]   CuI   (MW 190.45)
InChI = 1S/Cu.HI/h;1H/q+1;/p-1
InChIKey = LSXDOTMGLUJQCM-UHFFFAOYSA-M

(reacts with epoxides to give homoallylic alcohols useful for the synthesis of α-methylenebutyrolactones;[2] can function as a synthon for epoxides;[3,4] adds to D-glyceraldehyde acetonide to give epoxy alcohols and D-threose derivatives[3,4])

Preparative Methods: prepared in situ under anhydrous conditions by reacting 1-(trimethylsilyl)vinylmagnesium bromide (**1**) with copper(I) iodide (**2**) in various molar ratios.

A list of General Abbreviations appears on the front Endpapers

Handling, Storage, and Precautions: use in a fume hood.

α-Methylenebutyrolactone Synthesis. Epoxides react with (**1**) in the presence of catalytic amounts of (**2**) to give homoallylic alcohols (**3**) (eq 1) which were used in a synthesis of α-methylene lactones (**4**) (eq 2).[2] In the absence of (**2**) the reaction fails to give the desired homoallylic alcohols (**3**).

$$R^3 \overset{R^2}{\underset{O}{\diagdown}} R^1 \xrightarrow[71–85\%]{\text{(1), 0.05 equiv (2)}} R^3 \underset{OH\ TMS}{\overset{R^1}{\underset{R^2}{\diagdown}}} \quad (1)$$

(**3**)

$$R^3 \underset{OH\ TMS}{\overset{R^1}{\underset{R^2}{\diagdown}}} \xrightarrow[\text{2. NaOMe}]{\text{1. Br}_2} R^3 \underset{OH\ Br}{\overset{R^1}{\underset{R^2}{\diagdown}}} \xrightarrow[50–75\%]{\text{Ni(CO)}_4, \text{KOAc}}$$

(**3**)

$$\quad (2)$$

(**4**)

Addition to D-Glyceraldehyde Acetonide: Preparation of Epoxy Alcohols and D-Threose Derivatives. 2,3-*O*-isopropylideneglyceraldehyde (**5**) reacts with (**1**) in the presence of excess (**2**) to give the *syn* addition product (**6**) with high diastereoselectivity (>98:2 *syn:anti*, eq 3).[3,4] Sharpless epoxidation of (**6**) followed by protodesilylation gave epoxy alcohol (**7**) (eq 4).[3] Alternatively, (**6**) can be converted to the D-threose derivative (**8**) by a sequence of three steps: protodesilylation, protection, and ozonolysis.[4] Replacement of copper(I) cyanide and methyllithium for CuI in this reaction gives predominantly the *anti* addition product in 87% yield.

$$\xrightarrow[86\%]{\substack{1.5 \text{ equiv (1)} \\ 1.7 \text{ equiv (2)} \\ \text{THF·Me}_2\text{S}}} \quad (3)$$

(**5**)   (**6**)

$$\xrightarrow[\substack{2.\ t\text{-BuOK} \\ \text{Bu}_4\text{NF}}]{\substack{1.\ t\text{-BuOOH} \\ \text{VO(acac)}_2}} \quad (7)$$

(**6**)

$$\xrightarrow[66–77\%]{\substack{1.\ \text{NaH, HMPA} \\ 2.\ \text{protection} \\ 3.\ \text{O}_3, \text{Me}_2\text{S}}} \quad (4)$$

(**8**)

Related Reagents. 2-(3-Trimethylsilyl-1-propenyl)magnesium Bromide–Copper(I) Iodide; Vinylmagnesium Bromide; Vinylmagnesium Bromide–Copper(I) Iodide.

1. For a recent review of organocopper reagents, see: Lipshutz, B. H.; Sengupta, S., *Org. React.* **1992**, *41*, 135.
2. Matsuda, I., *Chem. Lett.* **1978**, 773.
3. Sato, F.; Kobayashi, Y.; Takahashi, O.; Chiba, T.; Takeda, Y.; Kusakabe, M., *J. Chem. Soc., Chem. Commun.* **1985**, 1636.
4. Kusakabe, M.; Sato, F., *Chem. Lett.* **1986**, 1473.

Michael N. Greco
The R. W. Johnson Pharmaceutical Research Institute,
Spring House, PA, USA

Trimethyl{2-[(tributylstannyl)methyl]-2-propenyl}silane

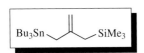

[164662-96-8] $C_{19}H_{42}SiSn$ (MW 417.33)

InChI = 1S/C7H15Si.3C4H9.Sn/c1-7(2)6-8(3,4)5;3*1-3-4-2;/h1
-2,6H2,3-5H3;3*1,3-4H2,2H3;

InChIKey = FPFPBKGQPNENSZ-UHFFFAOYSA-N

(reagent serves as an isobutene dianion equivalent)

Physical Data: bp 132–135 °C/0.40 Torr.[1]

Solubility: soluble in benzene, DME, ether, hexanes, methylene chloride, and THF.

Analysis of Reagent Purity: the compound has been characterized by ^{1}H and ^{13}C NMR spectroscopies as well as by IR and HRMS.[2]

Preparative Methods: most conveniently prepared by reacting tributylstannyllithium with commercially available 2-chloromethyl-3-trimethylsilyl-1-propene,[1,2] although methods involving reaction of the dithiocarbamate or allylic sulfide derived from 2-trimethylsilylmethylprop-2-en-ol with Bu$_3$SnH and AIBN have been reported.[3]

Purification: the reagent should be purified by distillation as it decomposes during chromatography over silica gel.

Handling, Storage, and Precautions: the reagent can be stored in a refrigerator for up to 8 weeks without appreciable decomposition. Reagent kept for a longer time should be redistilled before use.

Original Commentary

Martin G. Banwell & Brian D. Kelly
The Australian National University, Canberra, ACT, Australia

Selective Electrophilic Addition Reactions. The difference in reactivity between the allylstannane and allylsilane functionalities incorporated within this bismetalloid allows for selective reaction with electrophiles (eq 1).[3] The more nucleophilic allylstannane moiety reacts preferentially with electrophiles such as acid chlorides and aldehydes under thermal conditions, thereby providing functionalized allylsilanes (eq 2).[4] Related processes have been induced photochemically.[5] The allylstannane function-

ality also reacts selectively with electrophiles in the presence of a weak Lewis acid such as Et$_3$Al, but with stronger ones, e.g. BF$_3$ · OEt$_2$, a reversal of reactivity is observed (eq 3).[3]

A convenient two-step synthesis of a variety of *cis*- or *trans*-disubstituted pyrans from two distinct aldehydes and the title reagent has been reported (eq 4).[6] The initial step involves allylation of the first aldehyde with the stannane moiety using the mild and chiral Lewis acid 1,1′-bi-2,2′-naphtholtitanium tetraisopropoxide (BITIP) as catalyst. The resulting enantiomerically enriched (92–96% ee) hydroxyallylsilane is then treated with the second aldehyde in the presence of TMSOTf to give the target pyran. A similar stepwise manipulation of the two carbon-metalloid bonds in the title reagent has been used in a one-pot synthesis of *cis*-2,6-disubstituted-4-methylenepiperidines (eq 5).[7]

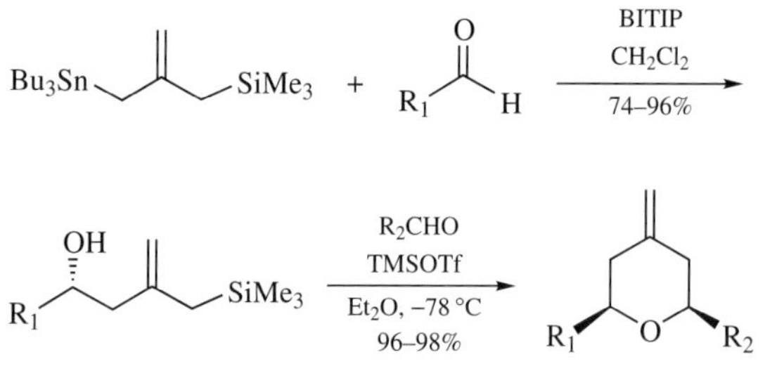

$$(4)$$

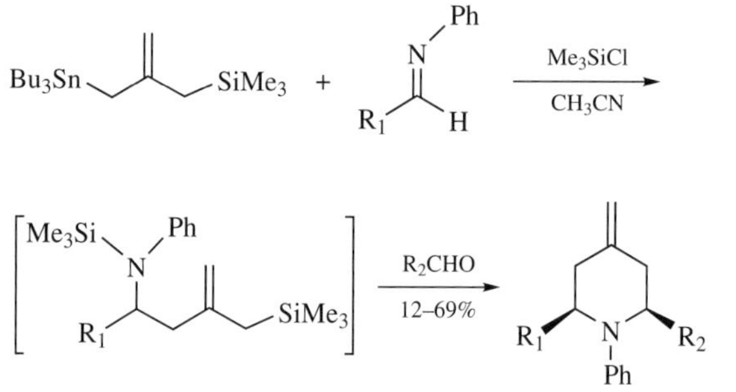

$$(5)$$

$$(8)$$

[4 + 3] Cycloaddition Reactions. The title reagent reacts with hexacarbonyldicobalt complexes of, for example, 1,4-diethoxy-2-butyne in the presence of 5 equiv of $BF_3 \cdot OEt_2$ to produce cycloheptyne cobalt complexes (eq 6).[8] Related chemistry[9] has been employed in the construction of an open-chain equivalent of such complexes.

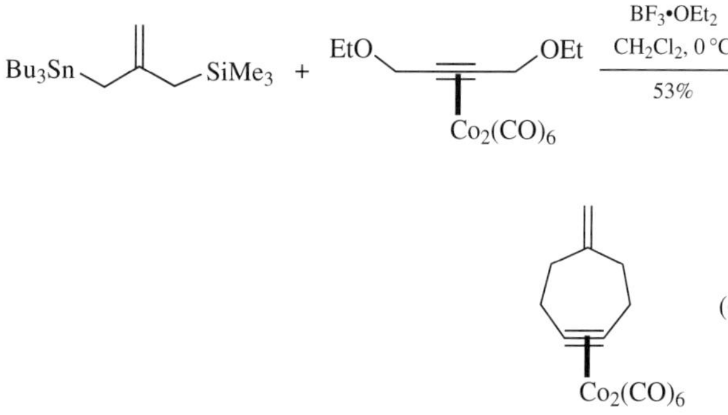

$$(6)$$

Radical Allylation Reactions. Trimethyl{2-[(tributylstannyl) methyl]-2-propenyl}silane reacts with organic halides under photochemical conditions (irradiation with a medium pressure mercury lamp through a pyrex filter) at temperatures below 20 °C to give various allylsilanes (eq 7).[2,10] The ease of the reaction was found to be halide dependent with those affording more electrophilic radicals generally behaving most effectively.

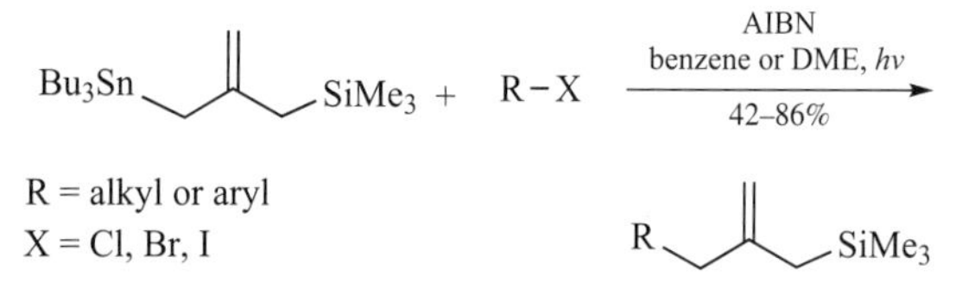

R = alkyl or aryl
X = Cl, Br, I

$$(7)$$

***trans*-Metallation Processes.** Treatment of the title reagent with *n*-BuLi and trapping of the resulting organolithium with an epoxy-tosylate has been used in the construction of compounds relevant to the studies on the biosynthesis of sterols (eq 8).[11]

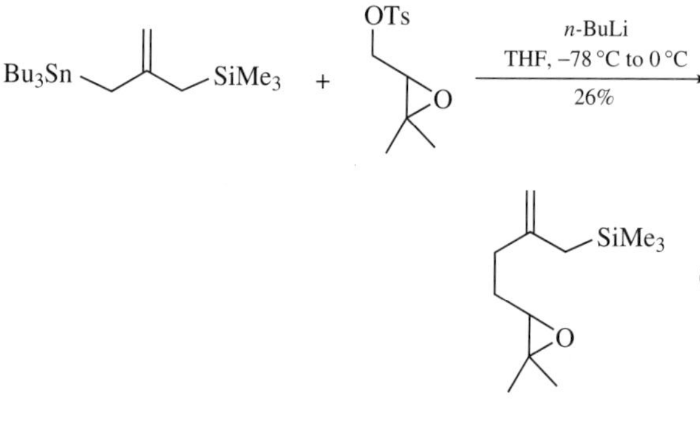

First Update

Tanya A. Bradford & Martin G. Banwell
The Australian National University, Canberra, ACT, Australia

Preparative Methods. A new method for the preparation of the title reagent has been reported.[12] This involves the sequential replacement, using *trans*-metallation procedures, of the two methylselenyl residues within 3-(methylseleno)-2,2-[(methylseleno)methyl]-1-propene by a trimethylsilyl and then a tributylstannyl moiety. The reagent is obtained in 52% overall yield.

Selective Electrophilic Addition Reactions. The diastereoselective reaction of trimethyl{2-[(tributylstannyl)methyl]-2-propenyl}silane with a β-oxygenated and chiral, nonracemic aldehyde (eq 9) has been used in the first step of a convergent synthesis of the tetrahydropyran fragment associated with a series of 'C7 diversified' bryostatin analogs.[13] When a large excess of the promoter derived from (*R*)-BINOL and $Ti(Oi$-$Pr)_4$ was employed, the same reagent has been shown to add in a highly diastereoselective 1,2-manner to an α,β-unsaturated aldehyde incorporating remote stereogenic centers.[14] The ensuing allylsilane was engaged in a TMSOTf-catalyzed intermolecular Sakurai reaction with an aliphatic aldehyde to give a *cis*-2,6-disubstituted tetrahydropyran embodying the tricyclic core of bryostatin.

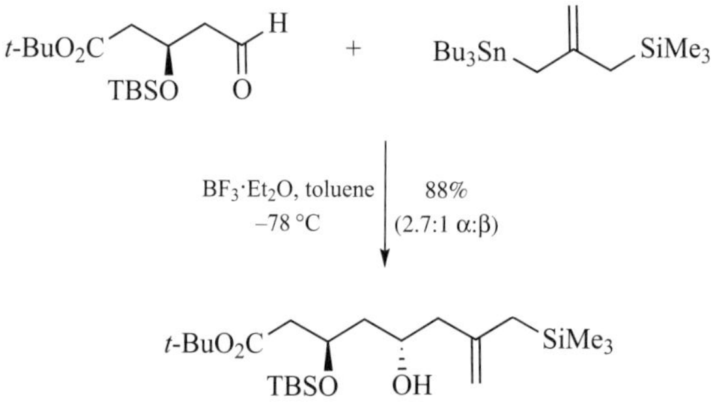

$$(9)$$

New protocols for the catalytic asymmetric monoallylation of aldehydes using the title reagent continue to be developed. For example, it has been shown (eq 10) that a range of aromatic and aliphatic aldehydes engage, in the presence of 5 mol% {[(*R*)-BINOL]-$Ti^{IV}[OCH(CF_3)_2]_2$}, in an enantioselective addition reaction with trimethyl{2-[(tributylstannyl)methyl]-2-propenyl}silane to give the expected addition products in ≥90% ee and 54–94% chemical yield.[15]

The products of such processes have been manipulated in a variety of ways. Thus, the derived *O*-TMS ethers undergo TMSNTf$_2$-promoted conversion into *cis*-2,6-disubstituted tetrahydropyrans, acetals, or spiroacetals. In certain cases, 2,6-bridged tetrahydropyrans can be formed in the presence of Sn(OTf)$_2$. Reactions analogous to those shown in eq 10 have been used during the development of total syntheses of the polyketide (+)-dactylolide[16] as well as the alkaloids (−)-adaline and (−)-euphococcinine.[17]

$$\text{(10)}$$

R = Ph, PhCH$_2$CH$_2$, *n*-C$_6$H$_{13}$, PhCH=CH, EtO$_2$CCH$_2$CH$_2$CH$_2$
CH$_3$COCH$_2$CH$_2$CH$_2$, BuCOCH$_2$(CH$_2$)$_2$CH$_2$, or BuCOCH$_2$CH$_2$CH$_2$

trans-Metallation Processes. The allyllithium obtained by treating trimethyl{2-[(tributylstannyl)methyl]-2-propenyl}silane with methyllithium has been added to cyclic imines (generated in situ via a Beckmann rearrangement), thus affording α-allylated amines (eq 11). On treatment with a mixture of TFA and formaldehyde, such products are converted, via a tandem Mannich/aza-Cope rearrangement reaction sequence, into the corresponding 1-azabicyclo[*n*.4.0]alkane.[18]

$$\text{(11)}$$

1. Kang, K.-T.; U, J. S.; Park, D. K.; Kim, J. G.; Kim, W. J., *Bull. Korean Chem. Soc.* **1995**, *16*, 464.
2. Clive, D. L. J.; Paul, C. C.; Wang, Z., *J. Org. Chem.* **1997**, *62*, 7028.
3. Majetich, G.; Nishidie, H.; Zhang, Y., *J. Chem. Soc., Perkin Trans. 1* **1995**, 453.
4. Kang, K.-T.; Sung, T. M.; Kim, J. K.; Kwon, Y. M., *Synth. Commun.* **1997**, *27*, 1173.
5. Takuwa, A.; Saito, H.; Nishigaichi, Y., *Chem. Commun.* **1999**, 1963.
6. Keck, G. E.; Covel, J. A.; Schiff, T.; Yu, T., *Org. Lett.* **2002**, *4*, 1189.
7. Kang, K.-T.; Kim, E. H.; Kim, W. J.; Song, N. S.; Shin, J. K.; Cho, B. Y., *Synlett* **1998**, 921.
8. Patel, M. M.; Green, J. R., *Chem. Commun.* **1999**, 509.
9. Green, J. R., *Chem. Commun.* **1998**, 1751.
10. Kang, K.-T.; Hwang, S. S.; Kwak, W. Y.; Yoon, U. C., *Bull. Korean Chem. Soc.* **1999**, *20*, 801.
11. Corey, E. J.; Staas, D. D., *J. Am. Chem. Soc.* **1998**, *120*, 3526.
12. Driesschaert, B.; Leroy, B., *Synlett* **2006**, 2148.
13. Wender, P. A.; Verma, V. A., *Org. Lett.* **2008**, *10*, 3331.
14. Keck, G. E.; Truong, A. P., *Org. Lett.* **2005**, *7*, 2153.
15. Yu, C.-M.; Lee, J.-Y.; So, B.; Hong, J., *Angew. Chem., Int. Ed.* **2002**, *41*, 161.
16. Sanchez, C. C.; Keck, G. E., *Org. Lett.* **2005**, *7*, 3053.
17. Lee, B.; Kwon, J.; Yu, C.-M., *Synlett* **2009**, 1498.
18. Kang, K.-T.; Sung, T. M.; Jung, H. C.; Lee, J. G., *Bull. Korean Chem. Soc.* **2008**, *29*, 1669.

Triphenylsilane

Ph$_3$SiH

[789-25-3]   C$_{18}$H$_{16}$Si   (MW 260.43)
InChI = 1S/C18H16Si/c1-4-10-16(11-5-1)19(17-12-6-2-7-13-17)18-14-8-3-9-15-18/h1-15,19H
InChIKey = AKQNYQDSIDKVJZ-UHFFFAOYSA-N

(reducing agent for esters, xanthates, and polychloroalkanes; protecting group for alcohols)

Physical Data: mp 47 °C; bp 152 °C/2 mmHg.
Solubility: sol most organic solvents.
Form Supplied in: white solid; commercially available.
Handling, Storage, and Precautions: this reagent is stable in air. Because the toxicogical properties are unknown, it should be handled in a well-ventilated fume hood. Contact with the eyes and skin should be avoided.

Original Commentary

Hiroshi Sano
Gunma University, Kiryu, Japan

Deoxygenations. Ph$_3$SiH is a useful reducing reagent under radical conditions. Esters are reduced to hydrocarbons at 140 °C by Ph$_3$SiH in the presence of a radical generator, 1,1-Di-*t*-butyl Peroxide (DTBP) (eq 1).[1]

$$\text{(1)}$$

The best results are obtained with acetate esters compared with other esters, such as isobutyrate, pivalate, or benzoate. Acetates derived from primary, secondary, and tertiary alcohols are all reduced in high yield. Other silanes such as tripropylsilane and diphenylmethylsilane are not as effective, and other radical generators such as AIBN and dibenzoyl peroxide are not suitable for the deoxygenations.

This reaction can be applied to deoxygenation of carbohydrates. Thus both *O*-acetylfuranoses and -pyranoses can be converted to the corresponding deoxy sugars (eqs 2 and 3).[2]

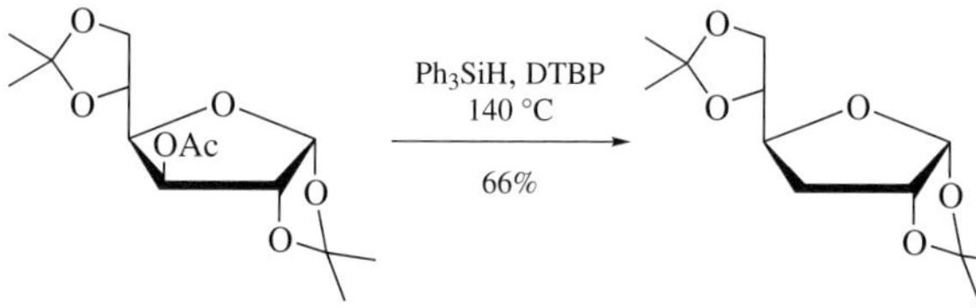

$$\text{Ph}_3\text{SiH, DTBP} \quad 140\,°C \quad 66\%$$

(2)

$$\text{Ph}_3\text{SiH, DTBP} \quad 140\,°C \quad 70\%$$

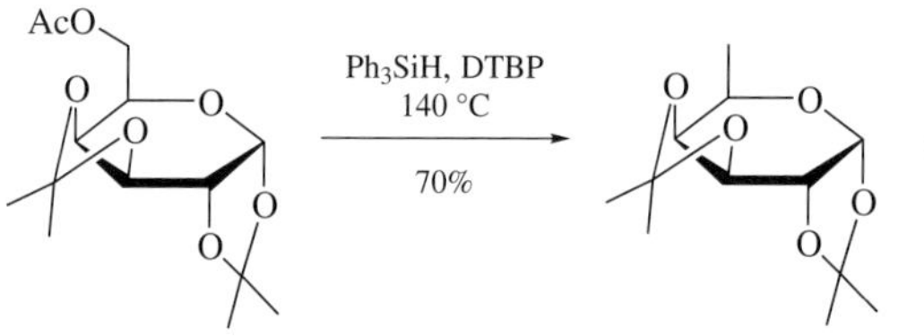

(3)

Xanthates (eq 4)[3] and perhaloalkanes[4] are also reduced by Ph_3SiH.

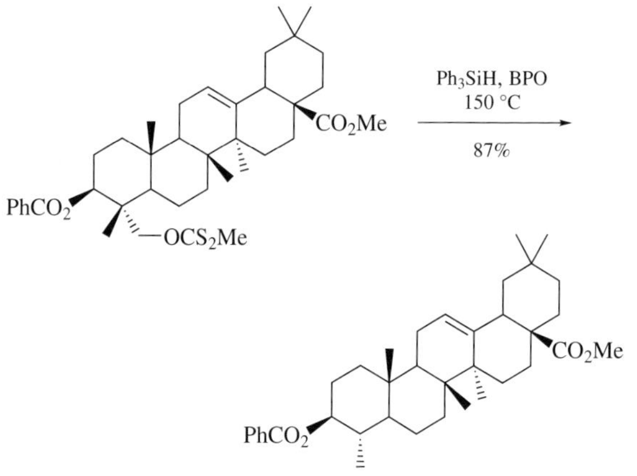

$$\text{Ph}_3\text{SiH, BPO} \quad 150\,°C \quad 87\%$$

(4)

Ph_3SiH in combination with trifluoroacetic acid has been used for ionic deoxygenation of tertiary alcohols (eq 5).[5]

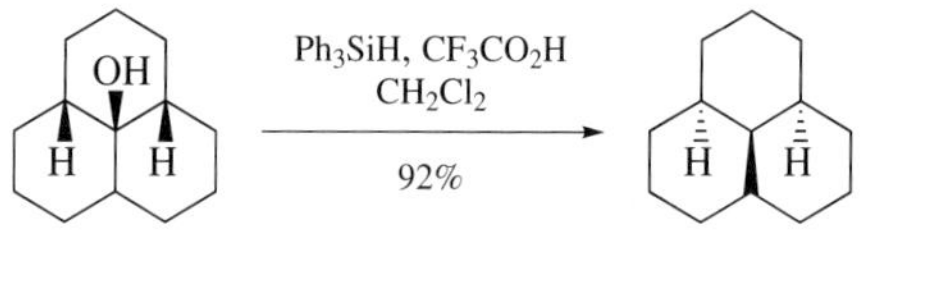

$$\text{Ph}_3\text{SiH, CF}_3\text{CO}_2\text{H} \quad \text{CH}_2\text{Cl}_2 \quad 92\%$$

(5)

Protecting Group. Triphenylsilane is used in certain cases for the preparation of triphenylsilyl ethers, which serve as alcohol protecting groups.[6] The triphenylsilyl group is considerably more stable (about 400 times) than the TMS group toward acidic hydrolysis.[7]

First Update

Nicole M. Torres & Robert E. Maleczka Jr.
Michigan State University, East Lansing, MI, USA

Introduction. In the years following the original EROS article on triphenylsilane, many new and improved uses of this reagent as a reducing agent for a variety of functional groups have appeared in the literature. Perhaps most prevalent among these recent studies have been investigations of triphenylsilane as a reagent for the transition metal-mediated hydrosilylation of alkenes, alkynes, and carbonyls. This update shall discuss these new findings, paying particular attention to instances where triphenylsilane is unique among other silanes for such functional group transformations.

Deoxygenations. Deoxygenations using Ph_3SiH and a radical initiator (i.e., DTBP) were shown in the previous EROS article to proceed for esters, xanthates, and perhaloalkanes (eqs 1–4) at elevated temperatures ($\sim$140 °C) in moderate yields. It has since been shown that DTBP-initiated deoxygenations of *N*-phenylthioxocarbamate-derived aliphatic alcohols can be achieved in excellent yield. Moreover, Et_3B facilitates silane reductions at lower temperatures, and thus the identical deoxygenation was achieved in 20 min at rt (eq 6).[8,9] This procedure has been extended to a variety of xanthates and thiocarbonates using Ph_3SiH and AIBN.[10] Likewise, decarbonylative reductions of a 2-pyridylmethyl ester can be achieved with $\text{Ru}_3(\text{CO})_{12}$ and Ph_3SiH as the hydrogenation reagent (eq 7).[11]

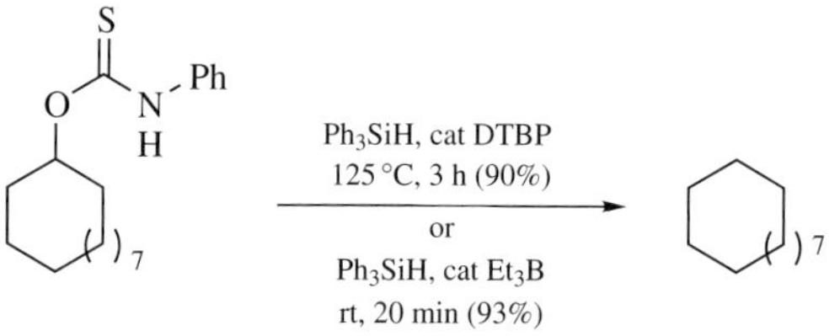

$$\text{Ph}_3\text{SiH, cat DTBP} \quad 125\,°C, 3\,h\,(90\%)$$
or
$$\text{Ph}_3\text{SiH, cat Et}_3\text{B} \quad \text{rt, 20 min}\,(93\%)$$

(6)

$$\text{Ph}_3\text{SiH} \quad \text{cat Ru}_3(\text{CO})_{12} \quad (67\%)$$

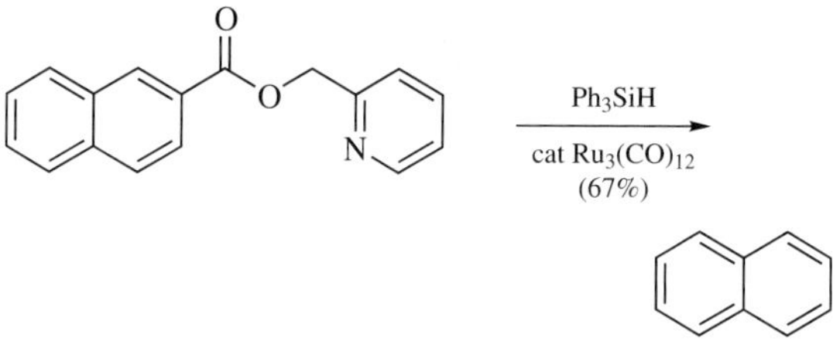

(7)

Hydrosilylations.[12] The addition of a triorganosilane across multiple bonds can be aided by a number of catalysts. The most commonly used triorganosilane is triethylsilane and the reader in referred to the EROS articles on Et_3SiH for a more in depth analysis of the various catalyst systems employed for hydrosilylation reactions. This update shall detail some advancements not discussed in the aforementioned article and areas in which Ph_3SiH has proven to be a more (or less) appropriate triorganosilane. One general advantage of Ph_3SiH and other arylsilanes in C–Si bond forming reactions is the opportunity for subsequent Tamao–Fleming oxidation to afford the corresponding alcohols.[13]

Alkenes.

Radical Chain Hydrosilylations. Thiols such as *t*-dodecanethiol have been shown to catalyze the radical-chain addition of Ph_3SiH to terminal alkenes when initiated by di-*t*-butyl hyponitrite (TBHN).[14] Slow catalyst addition was necessary to afford the product in good yield with Ph_3SiH, whereas methyl thioglycolate ($\text{MeO}_2\text{CCH}_2\text{SH}$) and triphenylsilanethiol (TPST) have been shown to be more effective catalysts, therefore, negating controlled thiol addition, especially with the use of triphenylsilane as opposed to other trialkylsilanes (eq 8). These products were formed as a racemic mixture but carbohydrate-derived homochiral thiol catalysts may be used to achieve enantioselective (76% ee) additions (eq 9).[15]

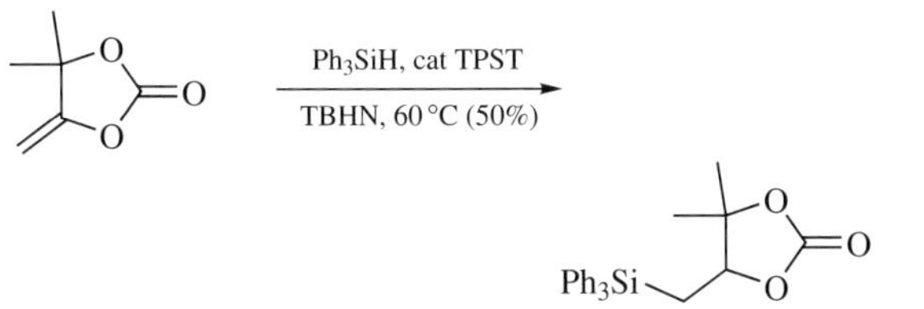

$$\text{(8)}$$

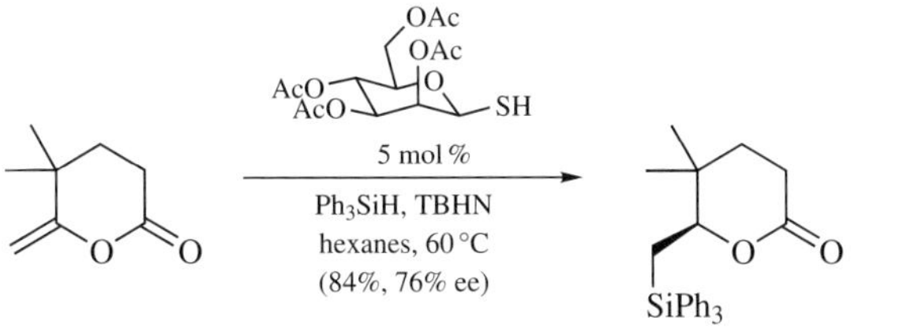

$$\text{(9)}$$

Metal-catalyzed Hydrosilylations. Various metals, particularly Pd, Pt, and Rh, have also been shown to catalyze alkene hydrosilylations.[16] The 1,4-addition of Ph$_3$SiH across 1,3-dienes can give rise to up to four different allylic silane isomers (eq 10).[17] Oxygen has been shown to have an affect on the regioselectivity, favoring the formation of the "head-product" (compound **II**). When *E*- or *Z*-isomers were possible, only the *Z*-isomer was observed. With silanes other than Ph$_3$SiH, reaction under an inert atmosphere favors the "tail-product" (compound **I**), but with Ph$_3$SiH a mixture of products was observed.[17] *N*-Hydroxyphthalimide (NHPI)catalyzed coupling of alkenes with Ph$_3$SiH and Co(II)-catalyst in O$_2$ provides the hydroxysilylation product at rt (eq 11).[18] In contrast, use of Et$_3$SiH required a reaction temperature of 60 °C to afford the hydroxysilylation product and even then lower yields were observed. Disparate reactivity between Ph$_3$SiH and Et$_3$SiH has also been noted during the hydrosilylation of vinylcyclopropanes, where the former reagent allows reaction *without* opening of the cyclopropane.[19]

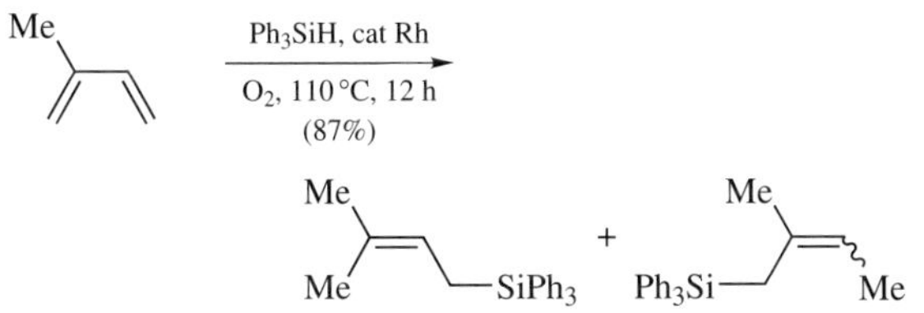

$$\text{(10)}$$

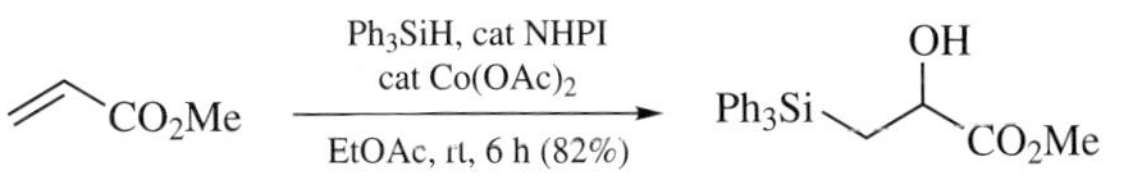

$$\text{(11)}$$

Lewis Acid-assisted Hydrosilylations. To date the only Lewis acid-mediated hydrosilylation utilizing arylsilanes has been achieved with assistance from B(C$_6$F$_5$)$_3$ (eq 12).[20]

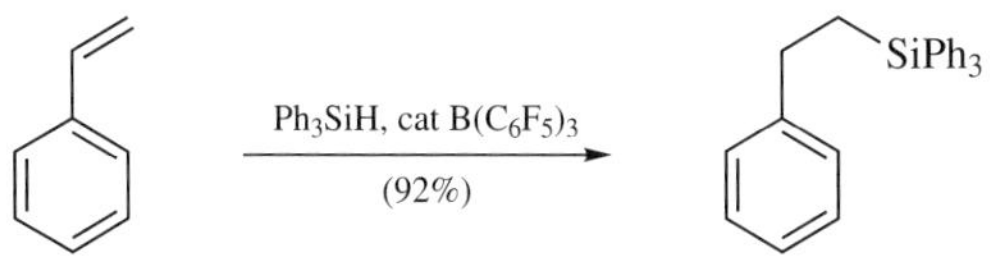

$$\text{(12)}$$

Alkynes.

Metal-catalyzed Hydrosilylations. Many factors such as the type of catalyst employed, substrate, hydrosilylating agent, and solvent can impact the ability of these reactions to afford the β-(*E*)-, β-(*Z*)-, and α-substituted vinylsilanes preferentially. High selectivity for the β-(*E*)-vinylsilane can be achieved through the use of Pt(DVDS)/P(*i*-BuNCH$_2$CH$_2$)$_3$N (DVDS = [(H$_2$C=CH) Me$_2$Si]$_2$O) and exhibits high functional group tolerance including ethers, halides, cyano groups, esters, and alcohols (eq 13).[21] Notably, free amines do not require protection to avoid alkyne polymerization as is needed with cationic [Rh(COD)$_2$]BF$_4$ catalysis.[22] β-(*E*)-Vinylsilanes have also been selectively achieved through utilization of various Pt[23,24] (via photoactivation), Pd,[25] and Ir[26] catalysts.

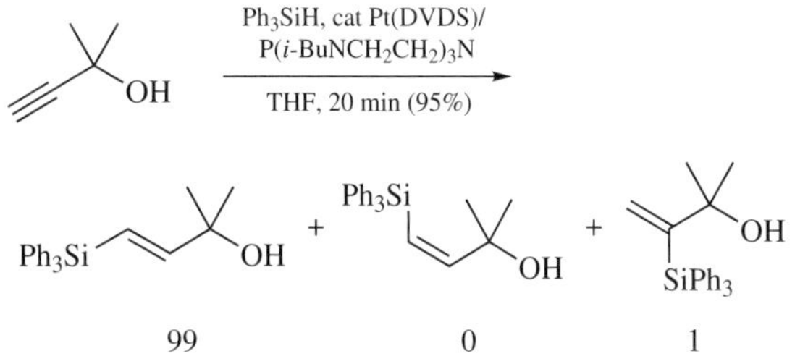

$$\text{(13)}$$

Neutral Rh catalysts allow access to the β-(*Z*)-vinylsilanes.[27] Ru[28] and Rh[29] *N*-heterocyclic carbenes and EtAlCl$_2$[30] Lewis acid catalysis have also shown β-(*Z*) selectivity. α-Substituted vinylsilanes can be obtained when Ru carbene complexes are used.[31]

Hydrosilylation of carbonyls with Ph$_3$SiH can provide direct access to the corresponding triphenylsilyl ethers via a Rh[29,32] complex. Side products of silyl enol ethers and alcohols are occasionally observed. Aromatic aldehydes, ketones, and esters are hydrosilylated by a borane[33,34] catalyst (eq 14). Thioketones and imines may also be hydrosilylated to the corresponding heteroatom-silylated products using B(C$_6$F$_5$)$_3$[35] or an ytterbium[36] catalyst, respectively. α,β-Unsaturated carbonyls have been selectively converted to their triphenylsilyl enol ethers upon treatment with Ph$_3$SiH and Karstedt's catalyst.[37] It should be noted that triethylsilane has been more widely studied for the hydrosilylation of carbonyls, and it would appear that some, but certainly not all, of the methods used for that reagent can be successfully applied to Ph$_3$SiH.

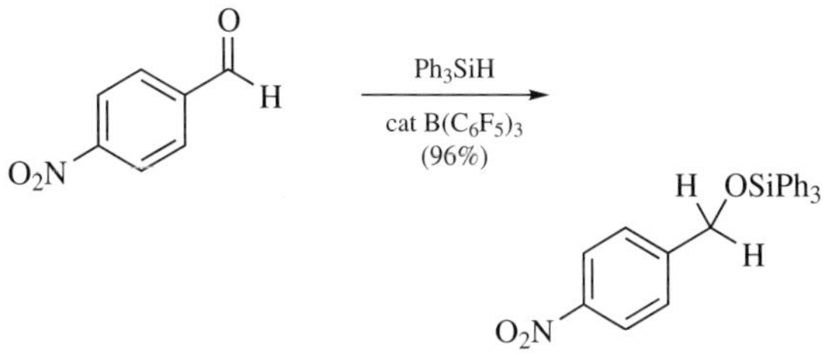

$$\text{(14)}$$

Silylformylation of alkynals with Ph$_3$SiH proceeds in a *syn*-selective fashion upon treatment with Rh(acac)(CO)$_2$ under

10 atm of CO (eq 15).[38] Ph$_3$SiH has proven to be a more efficient silane as compared with Et$_3$SiH although longer reaction times are required. The Pd-catalyzed carbocyclization of tetraenes proceeds rapidly with Ph$_3$SiH, exhibiting similar reactivity to Et$_3$SiH (eq 16).[39] Tandem hydrosilylation-intramolecular aldol rections afforded substituted cycloalkanols in 42% yield with 1.5:1 *cis: trans* ratio (eq 17).[40] Here, Et$_3$SiH appears to be a more efficient silane, affording the requisite product in 81% yield and higher *cis*-selectivity (3:1).

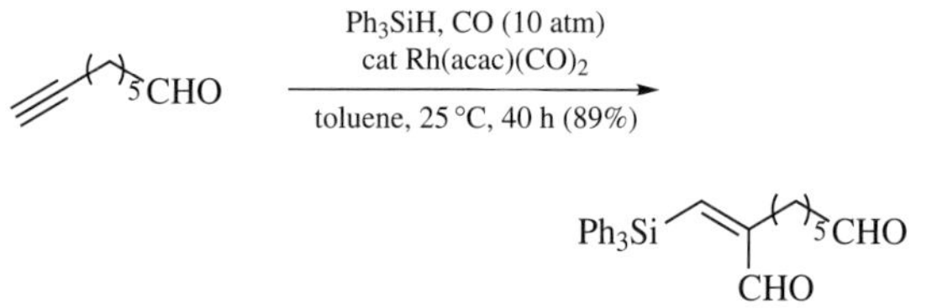

$$Ph_3Si \diagdown \diagup_5 CHO \quad (15)$$

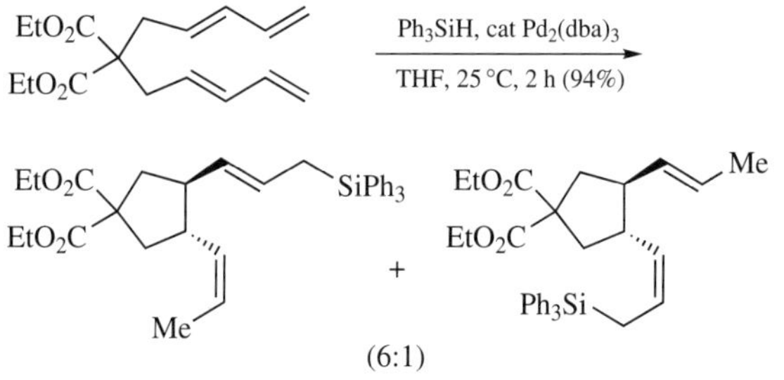

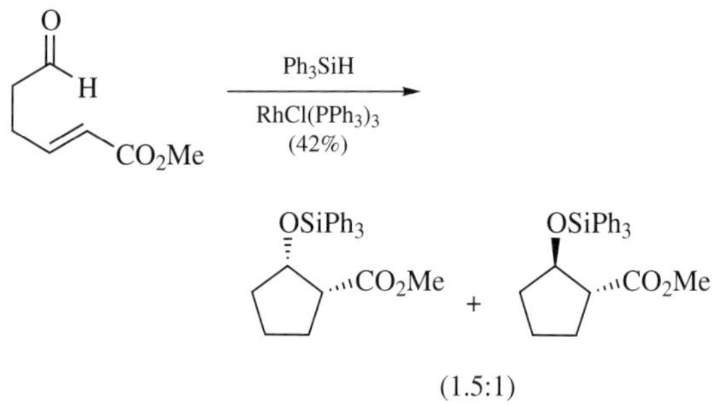

(6:1)

(1.5:1)

Reductions.

Heteroatom Reductions. Treatment of alkyl halides with a thiol-Ph$_3$SiH couple promotes dehalogenation (eq 18).[41] A defluorination using catalytic Os-hydrides as the hydride donor and stoichiometric Ph$_3$SiH to regenerate the Os-hydride has been achieved.[42] Reductions of 1° alcohols to the corresponding deoxygenated products proceed after treatment with catalytic B(C$_6$F$_5$)$_3$ and 3 equiv of silane. When 2° and 3° alcohols are used, the triphenylsilyl ethers are obtained. Ph$_2$CHOH and Ph$_3$COH afforded the deoxygenated products with only 1.1 equiv of silane. The authors reported the results for the reaction with Et$_3$SiH and stated that Ph$_3$SiH produced similar results.[43]

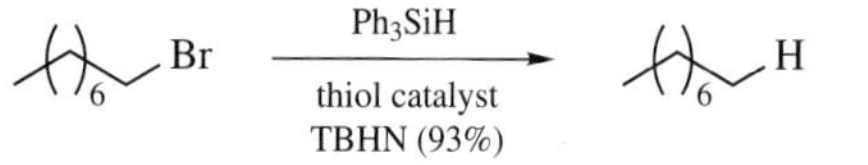

Multiple Bond Reductions. Imines bearing aryl substitution on C and N may be reduced upon treatment with Ph$_3$SiH and MoO$_2$Cl$_2$ (eq 19).[44] Best results were obtained when electron withdrawing substituents were present on the aryl rings.

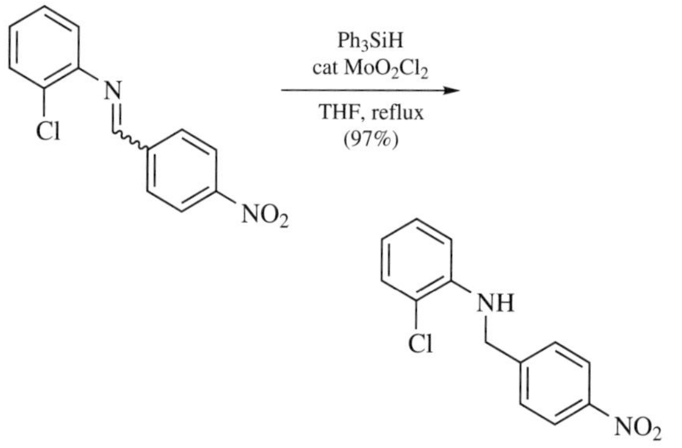

Alkene Hydrogenations.
Treatment of olefins with Zr catalysts effected their saturation in 50% yield along with 50% dehydrogenative silylation as calculated by GC (eq 20).[45] Ru carbene complexes also afforded the reduced product in the presence of Ph$_3$SiH. Reduction of an olefin via radical-chain reductive carboxyalkylation proceeded with Ph$_3$SiH and a homochiral thiol catalyst upon TBHN initiation (eq 21).[46] The olefin of vinylstannanes could be reduced without any protodestannylated product produced (eq 22).[47]

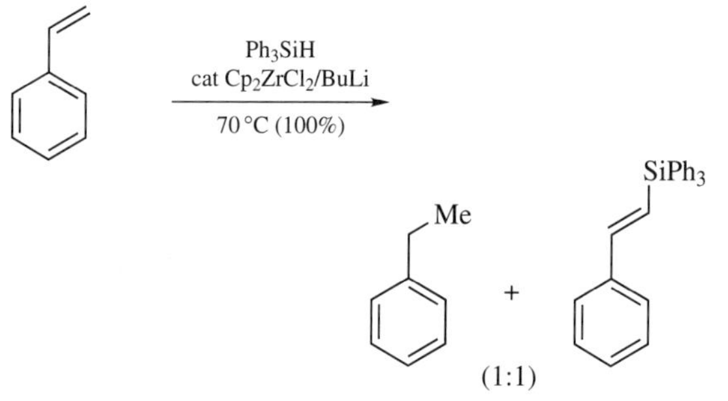

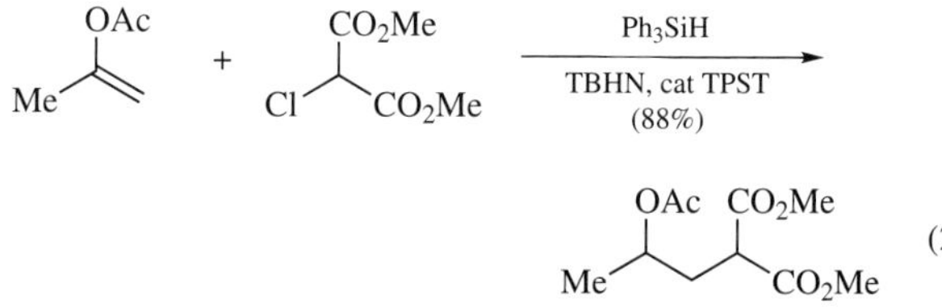

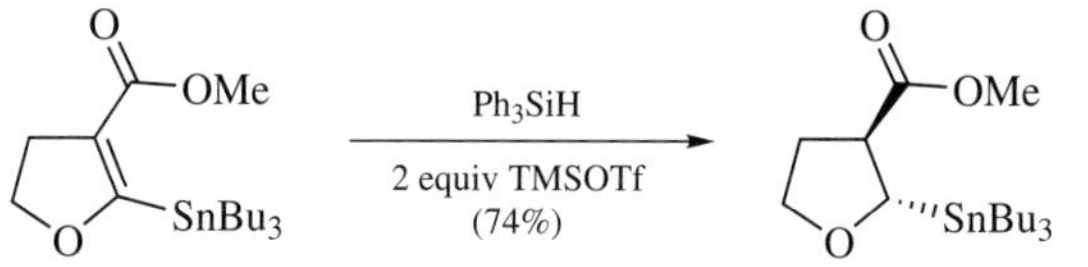

(1:1)

Silylations.

Si–O Bond Forming Reactions.

Alcoholysis. One route to silyl-protected alcohols is by dehydrogenative silylation in the presence of triorganosilanes. Conversion of alcohols to the corresponding triphenylsilyl ethers have been mediated by transition metal catalysts including Pd,[48] Cu,[49,50] and Au[51] in excellent yields. These catalysts exhibit high selectivity and functional group tolerance. A wide range of functionalized 1°, 2°, and 3° alcohols (alkenes, alkynes, halides, cyanos, ketones, and esters) and functionalized phenols were silylated upon treatment with catalytic B(C$_6$F$_5$)$_3$ and Ph$_3$SiH in excellent yields.[52] KOH/18-crown-6 can serve as a simple and economical combination for the silylation of 1° and 2° alcohols. For this method, phenylhydrosilanes, such as Ph$_3$SiH, were superior to Me$_3$SiH (eq 23).[53]

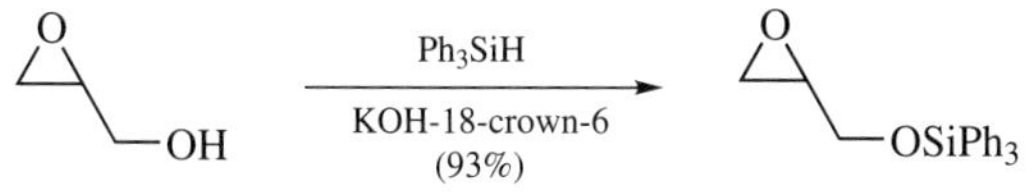

(23)

Si–C Bond Forming Reactions.

Halogen/Si Exchange. Aryl- and vinyl iodides have been treated with Pd(t-Bu$_3$P)$_2$[54] (eq 24) and Pd$_2$(dba)$_3$·CHCl$_3$,[55] respectively, to afford the triphenylsilyl-transferred products in moderate to excellent yields.

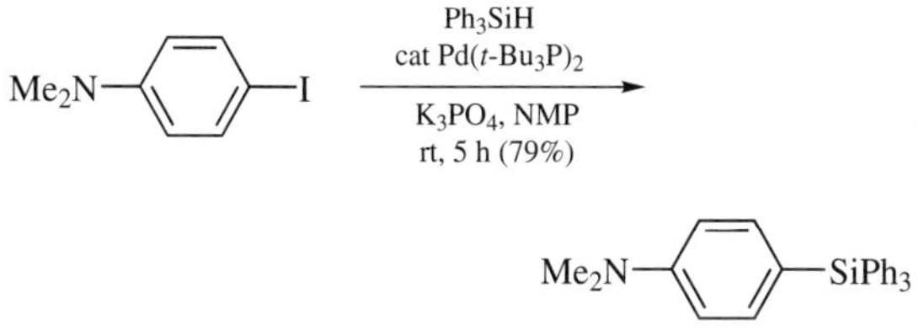

(24)

Dehydrogenative Silylation. Ytterbium–imine complexes have catalyzed the dehydrogenative silylation of terminal alkynes (eq 25).[56]

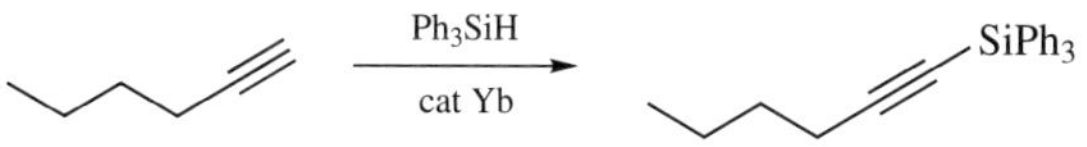

(25)

Carbene Insertion. Carbenoid insertion into silanes of various metals including Cu[57] (eq 26), Rh[58,59] (derived from diazoesters), and Cr[60,61] carbenoids can proceed in good to excellent yields. When chiral ligands on the metal or chiral auxiliaries (eq 27)[61] were implemented, good to excellent diastereocontrol was possible (46:1 dr).

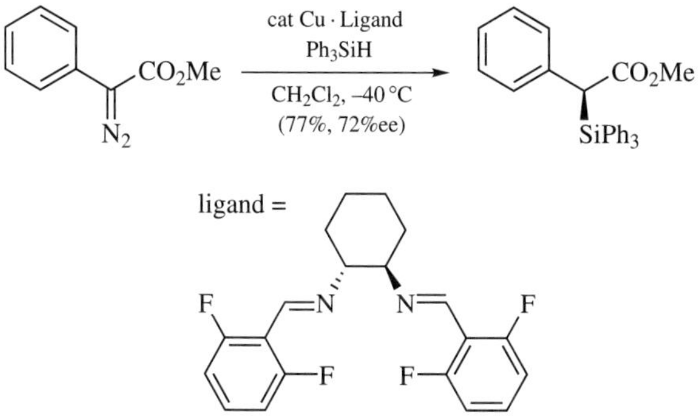

(26)

Via Benzylic C–H Activation. Silylation by nitrogen atom-directed activation of benzylic sp^3 C–H bonds has been reported employing triphenylsilane, catalytic Ru$_3$(CO)$_{12}$, and norbornene as a hydrogen acceptor (eq 28). No bissilylation products were detected with Ph$_3$SiH, as opposed to with Et$_3$SiH, where the disilylated product was major.[62]

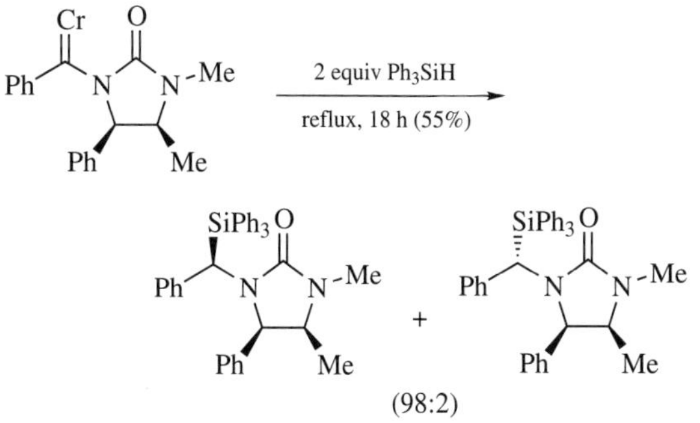

(27)

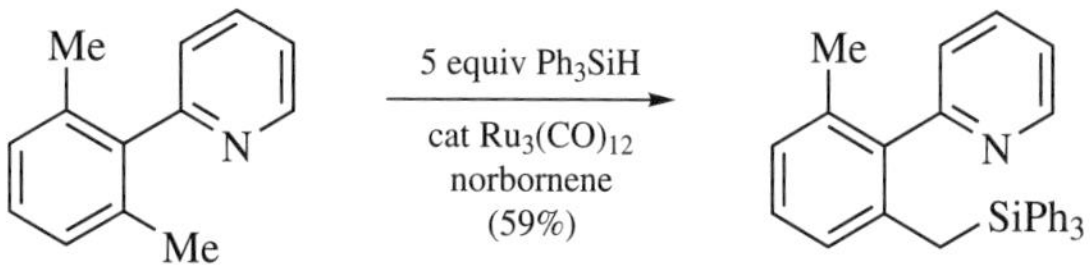

(28)

Miscellaneous.

Cyclizations. Lewis acid-catalyzed cyclizations of epoxy ketones with Ph$_3$SiH may afford cyclic ketals and substituted tetrahydrofuran or pyran rings depending on the Lewis acid and substrate (eq 29).[63]

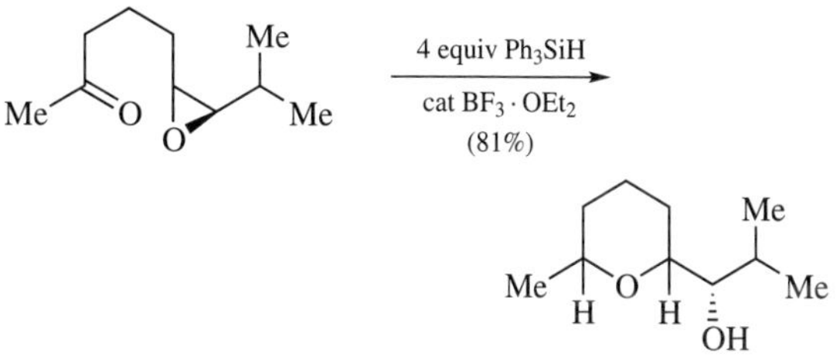

(29)

An asymmetric cyclization/hydrosilylation of a ω-formyl-1,3-diene has been catalyzed by Ni(0) in the presence of Ph$_3$SiH to afford the five-membered carbocycle. The same reaction with other silanes has been applied to the construction of six-membered carbocycles and pyrrolidines. However, the operative mechanism has been shown to be silane dependent and thus care should be taken before extrapolating those results to any proposed reaction schemes with Ph$_3$SiH.[64] Rh-catalyzed cyclization/hydrosilylation of enynes with Ph$_3$SiH afforded the silylcarbocyclization product, but for these reactions other silanes [e.g., Me$_2$PhSiH, (MeO)$_3$SiH, etc.] provided better results.[65] Cyclization/hydrosilylation of 1,6-dienes with Ph$_3$SiH have been demonstrated with a cationic Pd-catalyst (eq 30).[66]

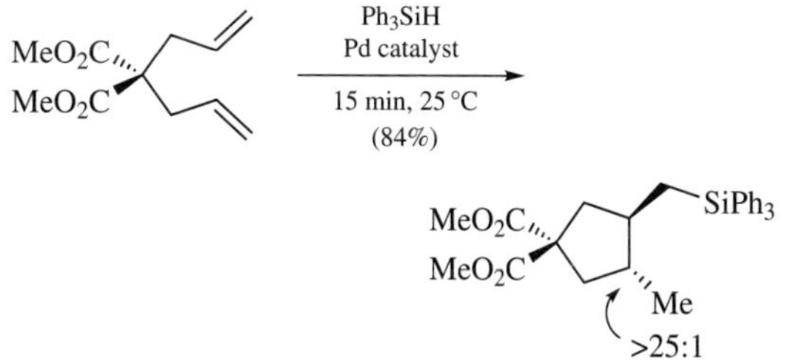

(30)

Hydroborations. Ph$_3$SiH can convert tri- and dihaloboranes to the corresponding di- and monohaloboranes, which can then be used in the in situ hydroboration of alkenes and alkynes. Notably, these reactions may be run from $-78\,^{\circ}\mathrm{C}$ to rt, thereby negating the usual need for elevated temperatures ($\sim100\,^{\circ}\mathrm{C}$) (eq 31).[67] Oxygen functionalities were typically not tolerated by this method.

$$\underset{3}{\bigwedge} \quad \xrightarrow[\text{BBr}_3,\ \text{rt}]{\text{Ph}_3\text{SiH}} \quad \underset{3}{\bigwedge}\text{B}\bigwedge_3 \quad (31)$$
$$\text{Br}$$

Cross-coupling Reactions. Triphenyl(difluoro)silicates arise from the reaction of Ph₃SiH and quaternary onium hydrogendifluoride. These species can then be subject to cross-coupling phenylation reactions under Pd-catalysis. Usefully, no homocoupled product formation was observed under these conditions (eq 32).[68]

$$\text{Ph}_3\text{SiH} \quad \xrightarrow{\text{Bu}_4\text{N}^+\ \text{HF}_2^-} \quad \overset{\ominus}{\underset{\text{F}}{\overset{\text{F}}{\text{Ph}-\text{Si}}}}\overset{\text{Ph}}{\underset{\text{Ph}}{}} \ \text{Bu}_4\text{N}^+ \quad \xrightarrow[\substack{p\text{-nitro-}\\ \text{iodobenzene}\\ (89\%)}]{\text{cat Pd}}$$

$$\overset{\text{Ph}}{\underset{\text{NO}_2}{\bigcirc}} \quad (32)$$

Related Reagents. 1,4-Bis(diphenylhydrosilyl)benzene; Triethylsilane Acid.

1. Sano, H.; Ogata, M.; Migita, T., *Chem. Lett.* **1986**, 77.

2. Sano, H.; Takeda, T.; Migita, T., *Synthesis* **1988**, 402.

3. Barton, D. H. R.; Blundell, P.; Dorchak, J.; Jang, D. O.; Jaszberenyi, J. C., *Tetrahedron* **1991**, *47*, 8969.

4. (a) Nagai, Y.; Yamazaki, K.; Shiojima, I.; Kobori, N.; Hayashi, M., *J. Organomet. Chem.* **1967**, *9*, 21. (b) Sommer, L. H.; Ulland, L. A., *J. Am. Chem. Soc.* **1972**, *94*, 3803. (c) Lesage, M.; Simões, J. A. M.; Griller, D., *J. Org. Chem.* **1990**, *55*, 5413.

5. Carey, F. A.; Tremper, H. S., *J. Org. Chem.* **1971**, *36*, 758.

6. Sommer, L. H. *Stereochemistry, Mechanism and Silicon; An Introduction to the Dynamic Stereochemistry and Reaction Mechanisms of Silicon Centers*; McGraw-Hill: New York, 1965; p 126.

7. (a) Lukevics, E.; Dzintara, M., *J. Organomet. Chem.* **1984**, *271*, 307. (b) Horner, L.; Mathias, J., *J. Organomet. Chem.* **1985**, *282*, 175.

8. Oba, M.; Nishiyama, K., *Tetrahedron* **1994**, *50*, 10193.

9. Oba, M.; Nishiyama, K., *Synthesis* **1994**, 624.

10. Barton, D. H. R.; Jang, D. O.; Jaszberenyi, J. C., *Tetrahedron* **1993**, *49*, 2793.

11. Tatamidani, H.; Yokota, K.; Kakiuchi, F.; Chatani, N., *J. Org. Chem.* **2004**, *69*, 5615.

12. Trost, B. M.; Ball, Z. T., *Synthesis* **2005**, 853.

13. Jones, G. R.; Landais, Y., *Tetrahedron* **1996**, *52*, 7599.

14. Dang, H. S.; Roberts, B. P., *Tetrahedron Lett.* **1995**, *36*, 2875.

15. Haque, M. B.; Roberts, B. P.; Tocher, D. A., *J. Chem. Soc., Perkin Trans. 1* **1998**, 2881.

16. Ojima, I., *The Chemistry of Organic Silicon Compounds*; Wiley: New York, **1989**.

17. Gustafsson, M.; Frejd, T., *J. Chem. Soc., Perkin Trans. 1* **2002**, 102.

18. Tayama, O.; Iwahama, T.; Sakaguchi, S.; Ishii, Y., *Eur. J. Org. Chem.* **2003**, 2286.

19. Itazaki, M.; Nishihara, Y.; Osakada, K., *J. Org. Chem.* **2002**, *67*, 6889.

20. Rubin, M.; Schwier, T.; Gevorgyan, N., *J. Org. Chem.* **2002**, *67*, 1936.

21. Aneetha, H.; Wu, W.; Verkade, J. G., *Organometallics* **2005**, *24*, 2590.

22. Takeuchi, R.; Ebata, I., *Organometallics* **1997**, *16*, 3707.

23. Wang, F.; Neckers, D. C., *J. Organomet. Chem* **2003**, *665*, 1.

24. Lewis, F. D.; Salvi, G. D., *Inorg. Chem.* **1995**, *34*, 3182.

25. Motoda, D.; Shinokubo, H.; Oshima, K., *Synlett* **2002**, 1529.

26. Tanke, R. S.; Crabtree, R. H., *J. Am. Chem. Soc.* **1990**, *112*, 7984.

27. Imlinger, N.; Wurst, K.; Buchmeiser, M. R., *Monatsh. Chem.* **2005**, *136*, 47.

28. Maifeld, S. V.; Tran, M. N.; Lee, D., *Tetrahedron Lett.* **2005**, *46*, 105.

29. Imlinger, N.; Wurst, K.; Buchmeiser, M. R., *J. Organomet. Chem* **2005**, *690*, 4433.

30. Sudo, T.; Asao, N.; Gevorgyan, V.; Yamamoto, Y., *J. Org. Chem.* **1999**, *64*, 2494.

31. Menozzi, C.; Dalko, P. I.; Cossy, J., *J. Org. Chem.* **2005**, *70*, 10717.

32. Reyes, C.; Prock, A.; Giering, W. P., *J. Organomet. Chem* **2003**, *671*, 13.

33. Parks, D. J.; Blackwell, J. M.; Piers, W. E., *J. Org. Chem.* **2000**, *65*, 3090.

34. Parks, D. J.; Piers, W. E., *J. Am. Chem. Soc.* **1996**, *118*, 9440.

35. Harrison, D. J.; McDonald, R.; Rosenberg, L., *Organometallics* **2005**, *24*, 1398.

36. Takaki, K.; Kamata, T.; Miura, Y.; Shishido, T.; Takehira, K., *J. Org. Chem.* **1999**, *64*, 3891.

37. Johnson, C. R.; Raheja, R. K., *J. Org. Chem.* **1994**, *59*, 2287.

38. Ojima, I.; Tzamarioudaki, M.; Tsai, C. Y., *J. Am. Chem. Soc.* **1994**, *116*, 3643.

39. Takacs, J. M.; Chandramouli, S., *Organometallics* **1990**, *9*, 2877.

40. Emiabata-Smith, D.; McKillop, A.; Mills, C.; Motherwell, W. B.; Whitehead, A. J., *Synlett* **2001**, 1302.

41. Allen, R. P.; Roberts, B. P.; Willis, C. R., *J. Chem. Soc., Chem. Commun.* **1989**, 1387.

42. Renkema, K. B.; Werner-Zwanziger, U.; Pagel, M. D.; Caulton, K. G., *J. Mol. Catal.; A* **2004**, *224*, 125.

43. Gevorgyan, V.; Rubin, M.; Benson, S.; Liu, J. X.; Yamamoto, Y., *J. Org. Chem.* **2000**, *65*, 6179.

44. Fernandes, A. C.; Romão, C. C., *Tetrahedron Lett.* **2005**, *46*, 8881.

45. Kesti, M. R.; Waymouth, R. M., *Organometallics* **1992**, *11*, 1095.

46. Dang, H. S.; Kim, K. M.; Roberts, B. P., *Chem. Commun.* **1998**, 1413.

47. Zhao, Y. K.; Quayle, P.; Kuo, E. A., *Tetrahedron Lett.* **1994**, *35*, 4179.

48. Purkayshtha, A.; Baruah, J. B., *J. Mol. Catal. A* **2003**, *198*, 47.

49. Purkayashtha, A.; Baruah, J. B., *Silicon Chemistry* **2002**, 1.

50. Ito, H.; Watanabe, A.; Sawamura, M., *Org. Lett.* **2005**, *7*, 1869.

51. Ito, H.; Takagi, K.; Miyahara, T.; Sawamura, M., *Org. Lett.* **2005**, *7*, 3001.

52. Blackwell, J. M.; Foster, K. L.; Beck, V. H.; Piers, W. E., *J. Org. Chem.* **1999**, *64*, 4887.

53. Le Bideau, F.; Coradin, T.; Henique, J.; Samuel, E., *Chem. Commun.* **2001**, 1408.

54. Yamanoi, Y., *J. Org. Chem.* **2005**, *70*, 9607.

55. Murata, M.; Watanabe, S.; Masuda, Y., *Tetrahedron Lett.* **1999**, *40*, 9255.

56. Takaki, K.; Kurioka, M.; Kamata, T.; Takehira, K.; Makioka, Y.; Fujiwara, Y., *J. Org. Chem.* **1998**, *63*, 9265.

57. Dakin, L. A.; Schaus, S. E.; Jacobsen, E. N.; Panek, J. S., *Tetrahedron Lett.* **1998**, *39*, 8947.

58. Kitagaki, S.; Kinoshita, M.; Takeba, M.; Anada, M.; Hashimoto, S., *Tetrahedron: Asymmetry* **2000**, *11*, 3855.

59. Landais, Y.; Planchenault, D., *Tetrahedron* **1997**, *53*, 2855.

60. Mak, C. C.; Tse, M. K.; Chan, K. S., *J. Org. Chem.* **1994**, *59*, 3585.

61. Parisi, M.; Solo, A.; Wulff, W. D.; Guzei, I. A.; Rheingold, A. L., *Organometallics* **1998**, *17*, 3696.

62. Kakiuchi, F.; Tsuchiya, K.; Matsumoto, M.; Mizushima, E.; Chatani, N., *J. Am. Chem. Soc.* **2004**, *126*, 12792.

63. Fotsch, C. H.; Chamberlin, A. R., *J. Org. Chem.* **1991**, *56*, 4141.

64. Sato, Y.; Saito, N.; Mori, M., *J. Org. Chem.* **2002**, *67*, 9310.

65. Ojima, I.; Vu, A. T.; Lee, S. Y.; McCullagh, J. V.; Moralee, A. C.; Fujiwara, M.; Hoang, T. H., *J. Am. Chem. Soc.* **2002**, *124*, 9164.

66. Widenhoefer, R. A.; Stengone, C. N., *J. Org. Chem.* **1999**, *64*, 8681.

67. Soundararajan, R.; Matteson, D. S., *Organometallics* **1995**, *14*, 4157.

68. Penso, M.; Albanese, D.; Landini, D.; Lupi, V., *J. Mol. Catal.; A* **2003**, *204*, 177.

Tris(dimethylamino)sulfonium Difluorotrimethylsilicate[1]

$$(R_2N)_3S^+Me_3SiF_2^-$$

(R = Me)

[59218-87-0] $C_9H_{27}F_2N_3SSi$ (MW 275.55)

InChI = 1S/C6H18N3S.C3H9F2Si/c1-7(2)10(8(3)4)9(5)6;1-6(2,3,4)5/h1-6H3;1-3H3/q+1;-1

InChIKey = JMGVTLYEFSBAGJ-UHFFFAOYSA-N

(R = Et)

[59201-86-4] $C_{15}H_{39}F_2N_3SSi$ (MW 359.73)

InChI = 1S/C12H30N3S.C3H9F2Si/c1-7-13(8-2)16(14(9-3)10-4)15(11-5)12-6;1-6(2,3,4)5/h7-12H2,1-6H3;1-3H3/q+1;-1

InChIKey = WRVWLZUYPSYYEW-UHFFFAOYSA-N

(anhydrous fluoride ion source; synthesis of C–F compounds by nucleophilic displacement of sulfonates;[3] promoter for electrophilic reactions of silyl enolates of ketones and esters;[4-6] source of sulfonium cation capable of stabilizing or imparting high nucleophilic reactivity to other anions;[4-8a,10a] activator of vinylsilanes in Pd-catalyzed cross-coupling reactions;[7] also used for generation[8] and reactions[10] of α- and β-halo carbanions; hydrosilylation[11] and cyanomethylation[14] of ketones)

Alternate Name: TASF.

Physical Data: R = Me, mp 98–101 °C; R = Et, mp 90–95 °C.

Solubility: R = Me, sol MeCN, pyridine, benzonitrile; partially sol THF. R = Et, sol THF, MeCN. Both react slowly with MeCN.

Form Supplied in: R = Me, white crystalline solid, ~90% pure; major impurity is tris(dimethylamino)sulfonium bifluoride [$(Me_2N)_3S^+$ HF_2^-].

Analysis of Reagent Purity: mp; ^{19}F NMR δ (at 200 MHz, CFCl$_3$ standard) TASMe$_3$SiF$_2$ (CD$_3$CN) δ −60.3; TASHF$_2$ −145.8 (d, J_{HF} = 120 Hz).

Preparative Methods: the methyl derivative is prepared by the reaction of dimethylaminotrimethylsilane and sulfur tetrafluoride at −70 °C to rt in ether; the precipitated solid is filtered off.[1a] The ethyl derivative is best prepared by the reaction of N,N-diethylaminosulfur trifluoride (DAST) and diethylaminotrimethylsilane.[1b,11b]

Handling, Storage, and Precautions: because of the extreme hygroscopic nature of this compound, it is best handled in a dry box or a polyethylene glove bag filled with high purity nitrogen. Use in a fume hood.

Original Commentary

T. V. (Babu) RajanBabu
The Ohio State University, Columbus, OH, USA

William J. Middleton & Victor J. Tortorelli
Ursinus College, Collegeville, PA, USA

Introduction. The acronym TASF has been used to refer to both $(Me_2N)_3S^+F_2Me_3Si^-$ and $(Et_2N)_3S^+F_2Me_3Si^-$. To eliminate confusion, the sulfonium salt containing dimethylamino groups is referred to as TASF(Me), and the sulfonium salt containing diethylamino groups is referred to as TASF(Et). Reactions of both reagents are similar. Since both of these salts can be prepared in a rigorously anhydrous state, they have an advantage over quaternary ammonium fluorides which usually contain some water. TASF(Me) has a slight advantage over TASF(Et) in that it is highly crystalline and easier to prepare in a high state of purity, whereas TASF(Et) has an advantage over TASF(Me) in that it has greater solubility in organic solvents. The tris(dialkylamino)sulfonium cation is often referred to by the acronym TAS.

TASF is a source of organic soluble fluoride ion[2] with a bulky noncoordinating counter ion (eq 1).[9]

$$TAS^+ Me_3SiF_2^- \rightleftharpoons Me_3SiF + TAS^+ + F^- \quad (1)$$

Fluoride Ion Source in Nucleophilic Displacements. TASF(Me) can be used to prepare fluorides from halides[1b] and sulfonates[3] under relatively mild conditions (eq 2).

$$\qquad (2)$$

Generation of Enolates and Enolate Surrogates from Enol Silanes. Enol silanes react with TASF(Et) to give highly reactive 'naked enolates' which have been characterized by NMR and electrochemical measurements.[4] These enolates, generated in situ, can be regioselectively alkylated without complications from polyalkylation or rearrangements of the alkylating agent (eq 3).

$$\qquad (3)$$

In the presence of excess fluorotrimethylsilane, TASF(Et) catalyzes aldol reactions of silyl enol ethers and aldehydes.[4] The stereochemical course of the reaction (*syn* selectivity, independent of the enol geometry, (Z)- or (E)-**1**, eq 4), has been interpreted as arising from an extended transition state in which steric and charge repulsions are minimized.

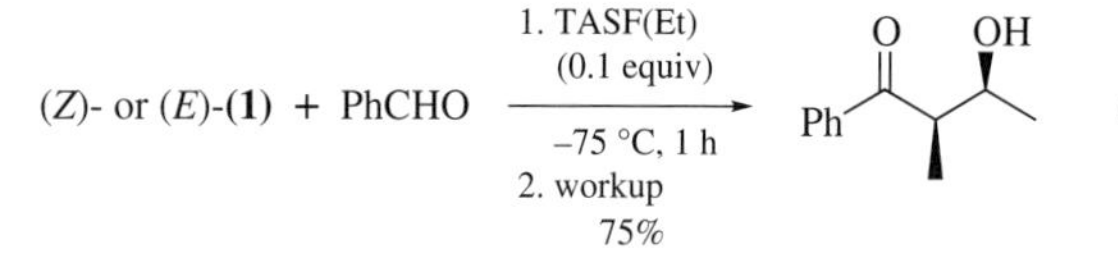

$$(Z)\text{- or }(E)\text{-}(\mathbf{1}) + PhCHO \xrightarrow[\substack{2.\ workup \\ 75\%}]{\substack{1.\ TASF(Et) \\ (0.1\ equiv) \\ -75\ ^\circ C,\ 1\ h}} \quad (4)$$

Very potent carbon nucleophiles formally equivalent to ester enolates are generated by the interaction of TASF(Me) with unhindered trialkylsilyl ketene acetals. In contrast to lithium enolates, these TAS enolates add 1,4 (nonstereoselectively) to α,β-unsaturated ketones. These adducts can be alkylated in situ to form two new C–C bonds in one pot, or they can be hydrolyzed to give 1,5-dicarbonyl compounds (eq 5).[5a]

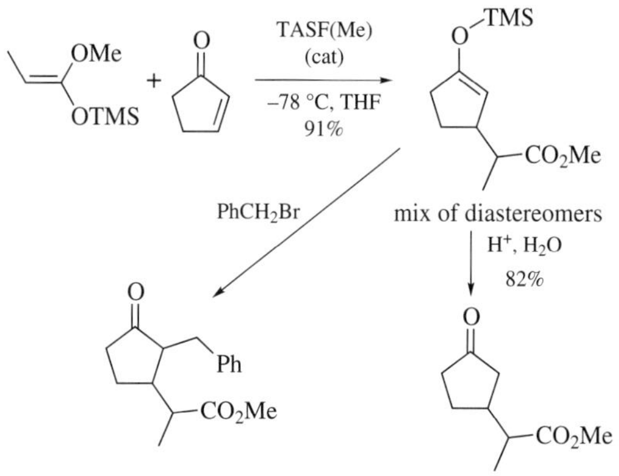

$$(5)$$

Conjugated esters undergo sequential additions to form polymers (group transfer polymerization).[5b] The molecular weight and the end group functionality of the polymer can be controlled by this method. Mechanistic studies indicate an associative intramolecular silicon transfer process via (2), with concomitant C–C bond formation during the polymer growth (eq 6).

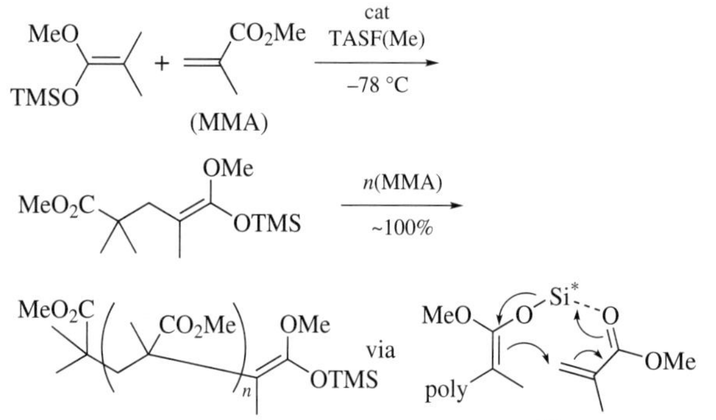

$$(6)$$

(2)

$$Si^* = SiR_3F$$

poly = polymer chain

Silyl enol ethers and ketene silyl acetals add to aromatic nitro compounds in the presence of TASF(Me) to give intermediate dihydro aromatic nitronates which can be oxidized with bromine or 2,3-dichloro-5,6-dicyano-1,4-benzoquinone to give α-nitroaryl carbonyl compounds;[6a] the latter are precursors for indoles and oxindoles.[6b] The reaction is widely applicable to alkyl-, halo-, and alkoxy-substituted aromatic nitro compounds, including heterocyclic and polynuclear derivatives (eq 7).

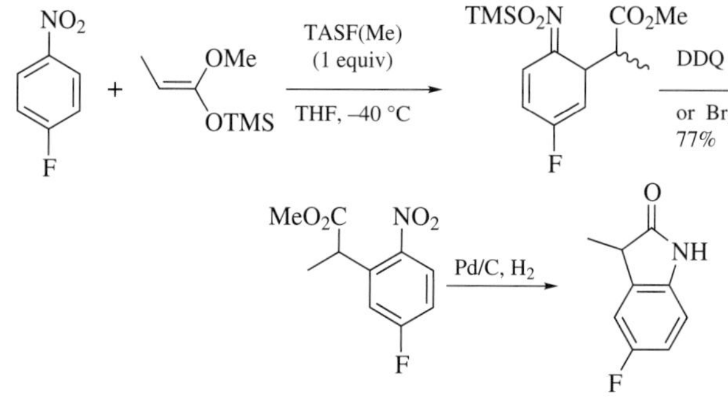

A list of General Abbreviations appears on the front Endpapers

$$(7)$$

Cross-Coupling Reactions. TASF(Et) activates vinyl-, alkynyl-, and allylsilanes in the Pd-mediated cross-coupling with vinyl and aryl iodides and bromides.[7] As illustrated in eqs 8–10, the reaction is stereospecific and chemoselective. This cross-coupling protocol is remarkably tolerant towards a variety of other functional groups such as carbonyl, amino, hydroxy, and nitro. Vinylsilanes can be synthesized from hexamethyldisilane and vinyl iodides in the presence of TASF(Et) (eq 10) via cleavage of a Si–Si bond.[7a] Aryl iodides can also be synthesized by this method. TASF is superior to tetra-n-butylammonium fluoride for these reactions. In the absence of a vinylsilane reagent, one of the methyl groups from the difluorotrimethylsilicate is substituted for the halide (eq 11).[7d]

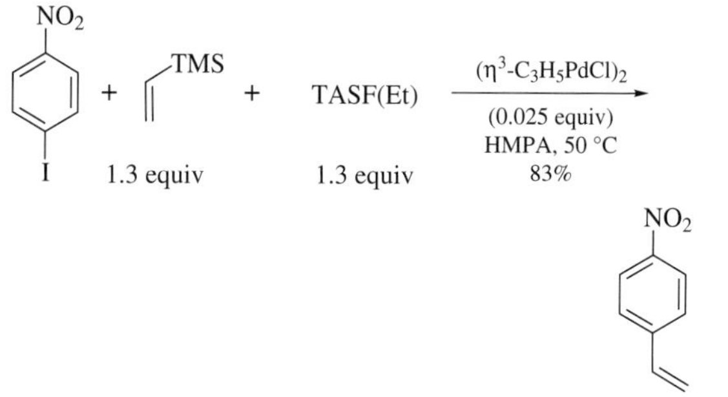

$$(8)$$

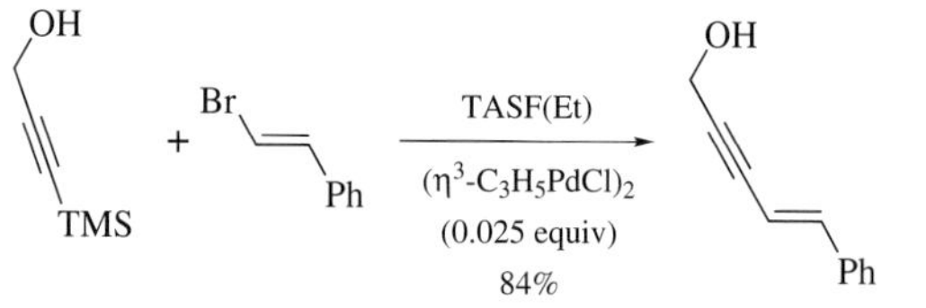

$$(9)$$

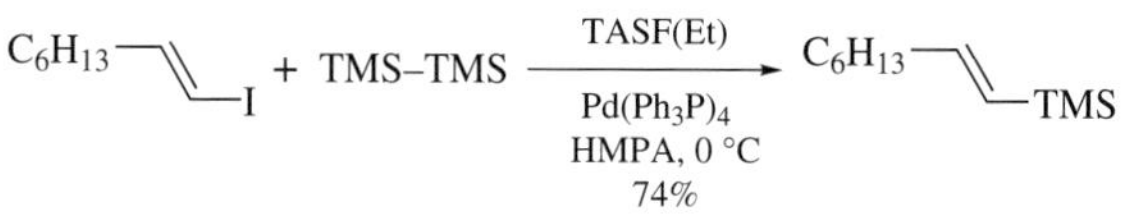

$$(10)$$

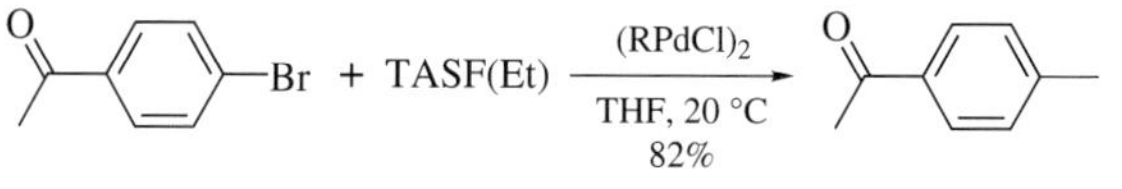

$$(11)$$

Generation and Reactions of Unusual Carbanions. Both α- and β-halo carbanions are generally labile species and their generation and reactions require extremely low temperatures. TASF(Me) has been used to prepare several stable and isolable perfluorinated carbanions (eq 12)[8a] or alkoxides.[8b] As compared to the corresponding metal salts, the TAS$^+$ counterion has little coordination to the fluorines of the anion,[9] and this presumably slows the decomposition of the TAS salts to carbenoids or alkenes.

Addition of α- and β-halo carbanions to carbonyl compounds may be achieved by the in situ generation of these species by the reaction of TASF(Et) with the corresponding silylated derivatives (eq 13).[10]

$$\text{TASF(Me)} + \underset{F}{\overset{F_3C}{\diagdown}}\!\!\!\!\underset{F}{\overset{CF_3}{\diagup}} \xrightarrow[94\%]{\substack{\text{THF} \\ 0\,°C\ \text{to rt}}} (CF_3)_3C^-\,TAS^+ + \text{TMSF} \quad (12)$$
$$\text{mp } 170\ °C$$

$$\underset{Ph}{\overset{}{\diagdown}}\!\!\!\text{CHO} + \text{TMSCCl}_3 \xrightarrow[\substack{2.\ H^+ \\ 75\%}]{\substack{1.\ 0.1\ \text{equiv TASF(Et)} \\ \text{THF, } 0\,°C,\ 4\ h}} \underset{Ph}{\overset{}{\diagdown}}\!\!\!\underset{OH}{\overset{}{\diagup}}\!\!\!CCl_3 \quad (13)$$
$$87\%\ syn$$

Other Applications. TASF(Et) catalyzes the addition of dimethyl(phenyl)silane to α-alkoxy, -acyl or -amido ketones to give the corresponding *anti* aldols (eq 14).[11] This complements the acid-catalyzed reduction which gives the *syn* isomer.

$$\underset{Ph}{\overset{O\quad O}{\diagup\!\!\diagdown}}\!\!\!N\!\!\bigcirc + \text{TASF(Et)} + \text{PhMe}_2\text{SiH} \xrightarrow[\substack{0\,°C,\ 12\ h \\ 77\%}]{\text{THF, HMPA}}$$

$$\underset{Ph}{\overset{OH\quad O}{\diagup\!\!\diagdown}}\!\!\!N\!\!\bigcirc \quad (14)$$
$$anti{:}syn = 99{:}1$$

Dimethyl(phenyl)silane will reduce aldehydes and ketones to hydroxyl compounds under very mild conditions in the presence of a catalytic amount of TASF(Et).[11c]

Aryl or vinyl anions can be generated by the reaction of the corresponding vinyl iodide with Bu_3Sn anion, which in turn is produced from TASF(Et) and $Bu_3SnSiMe_3$. With an appropriately placed carbonyl group, an intramolecular cyclization ensues (eq 15).[12]

$$\text{(structure)} + \text{TMS–SnBu}_3 \xrightarrow[\substack{\text{DMF, 1 h} \\ -30\,°C \\ 93\%}]{\text{TASF(Et)}} \text{(structure)} \quad (15)$$

A useful variation of the Peterson alkenation relies on the generation of α-silyl carbanions from geminal disilyl compounds containing an additional stabilizing group (CO_2R, SPh, SO_2Ph, OMe, CN, Ph) at the α-carbon.[13] A related reaction is the cyanomethylation of ketones and aldehydes with trimethylsilylacetonitrile in the presence of TASF(Me).[14] TASF(Me) was found to be the best fluoride ion source for the synthesis of aryl trifluoromethyl sulfones from the corresponding sulfonyl fluorides and trifluoromethyltrimethylsilane or Me_3SnCF_3 (eq 16).[15]

$$\text{PhSO}_2\text{F} + \text{TMSCF}_3 \xrightarrow[\substack{\text{THF, } 20\,°C \\ 96\%}]{\text{TASF(Me)}} \text{PhSO}_2\text{CF}_3 \quad (16)$$

The carbanions formed by scission of a C–Si bond with TASF can also be oxygenated. Benzylic trimethylsilyl groups can be converted to hydroxyl groups in 20–95% yield by reaction with TASF(Me) in the presence of oxygen and trimethyl phosphite (eq 17).[16] No other source of fluoride ion was found that could replace TASF.

$$\underset{Ar}{\overset{R}{\diagdown}}\!\!\!\text{TMS} \xrightarrow{\text{TASF(Me)}} \underset{Ar}{\overset{R}{\diagdown}}\!\!\!- TAS^+ \xrightarrow{O_2}$$

$$\underset{Ar}{\overset{R}{\diagdown}}\!\!\!O-O^-\,TAS^+ \xrightarrow[20-95\%]{P(OMe)_3} \underset{Ar}{\overset{R}{\diagdown}}\!\!\!OH \quad (17)$$

First Update

Myriam Roy & Richard E. Taylor
University of Notre Dame, Notre Dame, IN, USA

Introduction. The relative energy of a Si–X bond has been shown to be proportional to the electronegativity of X.[17,18] Thus, fluorine, the most electronegative element, has a high affinity for silicon. Moreover, in contrast to carbon, silicon is able to temporarily increase its covalency through accessibility of vacant low energy 3d orbitals. The generation of a hypervalent species lengthens the silicon bonds and renders the silicon atom more accessible. Although the Si–X bond is stronger than the corresponding C–X, the Si–X bond is kinetically more reactive. The increased attraction of nucleophiles to the greater partial positive charge of the silicon is one potential explanation. Thus, tris(dimethylamino)sulfonium difluorotrimethyl silicate (TASF), a pentacoordinated silicate, is a ready source of fluoride ion.

Numerous fluoride sources are readily available. However, it is clear that counterion and solvent can substantially affect its reactivity. TASF is now known to be a mild source of highly nucleophilic "naked" fluoride due to the high degree of dissociation.[9,19] Much like tetrabutylammonium fluoride (TBAF), it has the advantage of being soluble in organic solvents in contrast to alkaline metal salts such as CsF and KF. Moreover, TASF is much less hydroscopic and can be accessed in its anhydrous form.[20,21] These properties make TASF an excellent reagent for the generation of alkyl fluorides as well as the activation of silicon necessary for deprotection of silyl ethers, participation in organometallic cross-coupling reactions, and the generation of carbon and tin-based anionic species.

Sensitive Deprotections. Since the previous review, TASF has emerged as a good alternative to TBAF for the mild deprotection of silyl ethers of acid- and/or base-sensitive compounds.[22] Side reactions, commonly observed during deprotection using TBAF, include complete decomposition (encountered during the synthesis of amphidinolides[23] and aurisides[24]), acyl migration[20,25,26] (encountered during the synthesis of papulacandins[20]), elimination (encountered during the synthesis of bafiloycin A1[27] and dolabriferol[28]), and olefin isomerization (encountered during the synthesis of apicularen A analogs[29]). In each case, the use of anhydrous TASF provided the desired intermediate in high yield without undesired product formation. One particular striking comparison of TASF and TBAF is shown in eqs 18 and 19.[25] The difference in reactivity between TBAF and TASF can be explained by the relative basicity of the two reagents. Although TBAF is less expensive, it always contains a non-negligible amount of water and anhydrous preparations are inefficient.[20,21,30] Thus, tetrabutylammonium hydroxide is likely responsible for most of the side reactions observed.

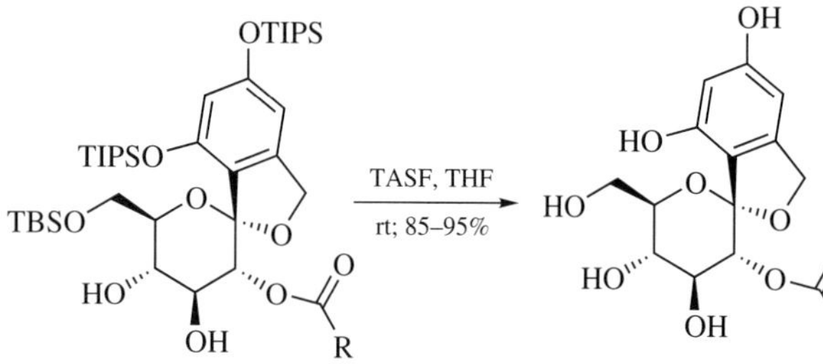

(18)

TIPS = triisopropylsilyl
TBS = *t*-butyldimethylsilyl

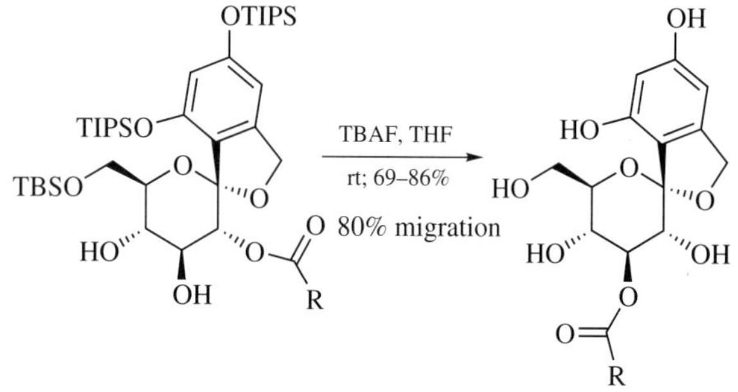

(19)

80% migration

In one case, in particular, Roush observed the elimination of a silyl ether even with an anhydrous preparation of TASF. However, inclusion of several equivalents of water attenuated the basicity of the fluoride and produced the desired product in good yield without competitive elimination.[27] Anhydrous TASF does foster similar side reactions with some highly sensitive substrates, such as those found during the synthesis of bafilomycin A1,[27] (−)-osmundalactone,[26] and (−)-muricatacin.[26] While HF•pyridine and HF•Et$_3$N are alternative reagents for base sensitive substrates, reactivity problems are often encountered, leading to very slow or incomplete reactions, or even no reaction. However, TASF in a solution of DMF/water (2–3 equiv of water relative to the amount of TASF) appears to be even milder than the anhydrous reagent, as shown in the synthesis of bafilomycin A1.[27]

In addition to fluoride source and solvent, one must also consider the effect of functionality present in the substrate. Specifically positioned unprotected hydroxyl groups can act as internal acids and bases under certain conditions. In the case of bafilomycin A1, the alcohol at the C18 position is precisely positioned to deprotonate C21 in a six-membered ring transition state (eqs 20 and 21). The basicity of TASF (and TBAF) might activate this internal base leading only to the elimination by-product. The addition of water seems to buffer the basicity of TASF, preventing alcohol deprotonation, and providing the desired unprotected hemiketal.

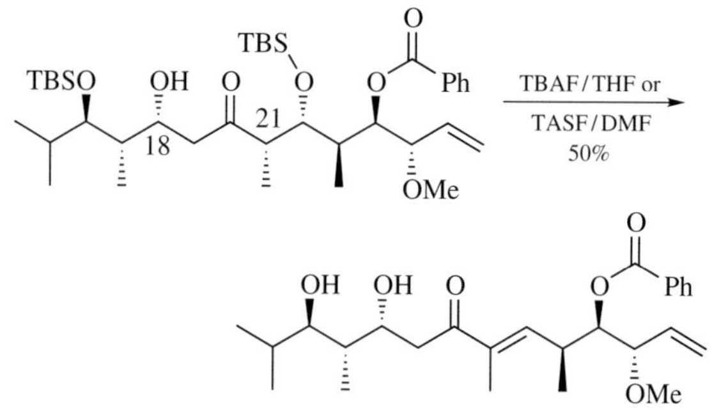

(20)

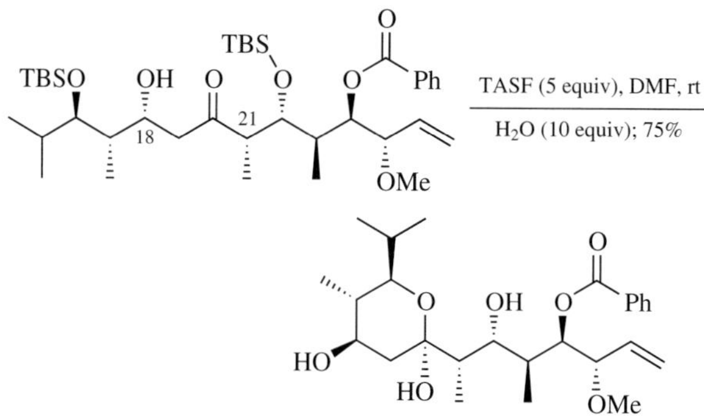

(21)

Acid-sensitive compounds are also efficiently deprotected (Rhizoxin D).[31] In this case, however, the reaction should be conducted in a nalgene vial because TASF reacts with glass to produce HF. TASF exhibits a fairly good selectivity for primary TBDPS over secondary TBS[22,32,33] or TIPS (eq 22).[31,34] On the contrary, in the case of a molecule containing a primary TBS and a secondary TBDPS, the use of 1.2 equiv of TASF led to a mixture of products due to the competitive reactivity of both protecting groups.[22] Secondary TBS and TES are also not well differentiated.[22] The use of a large excess of reagent (5 equiv) leads to global desilylation.

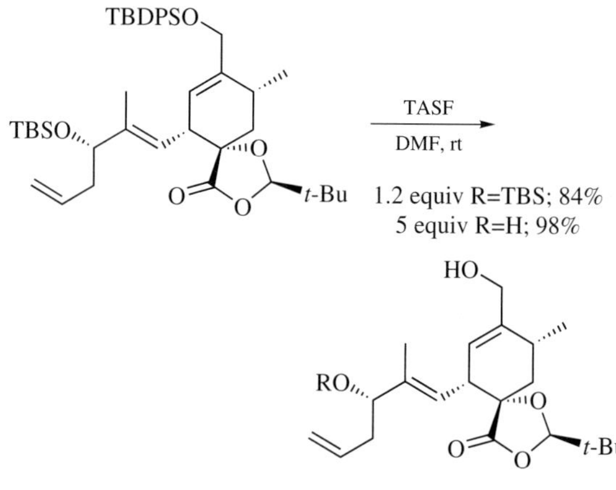

(22)

TBDPS = *t*-butyldiphenylsilyl

Interestingly, during the synthesis of (+)-concanamycin F, the allylic and the secondary DEIPS were deprotected selectively over the secondary and the vinylic TES groups even with a large excess of TASF (eq 23).[35]

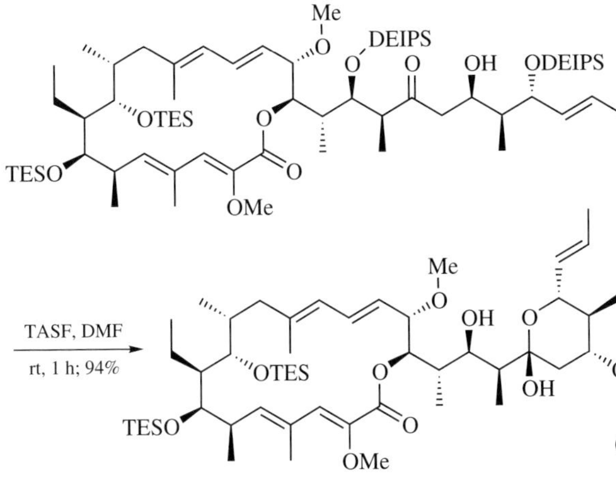

$$\text{(23)}$$

DEIPS = diethylisopropylsilyl
TES = triethylsilyl

Generation of Enolates and Enolate Surrogates from Enol Silanes. Silyl enol ethers in the presence of TASF generate TAS-enolates that undergo highly diastereoselective aldol and Michael reactions.[36] The selectivity observed for Michael reactions is opposite to that obtained using mercury diiodide as a Lewis acid promoter,[37,38] thus allowing access to both geometries (eq 24). Tandem reactions provide cyclized products, also in high diastereoselectivity.[39–41]

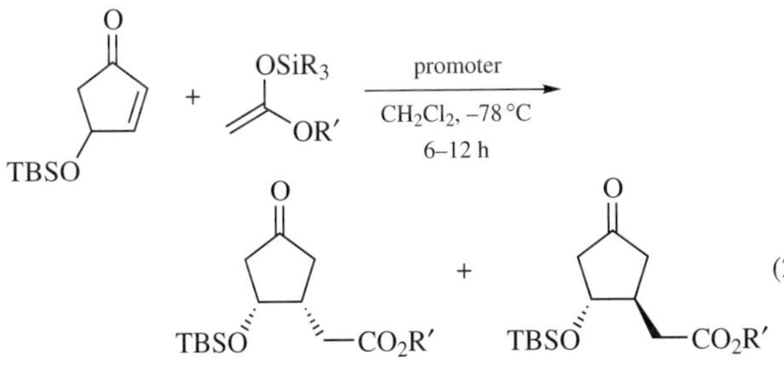

$$\text{(24)}$$

| R = Me, R′ = Et | TASF | 0 : 100 | 87% |
| R = R′ = Et | HgI$_2$ (0.1 equiv) | 95 : 5 | 74% |

An aldol mechanism was investigated by NMR monitoring.[36] The sterically large TAS-enolate attacks the less-hindered face accounting for the observed diastereoselectivity. Some hindered TAS-enolates have shown basic properties that can cause dimerization of monosaccharide lactones.[41] Cyclic enol ethers were also used to generate bicyclic lactams[42] (eq 25) or β-keto sulfoxides[43] (eq 26). TASF appears to be a more potent source of fluoride than TBAF or BTAF, providing higher yields.

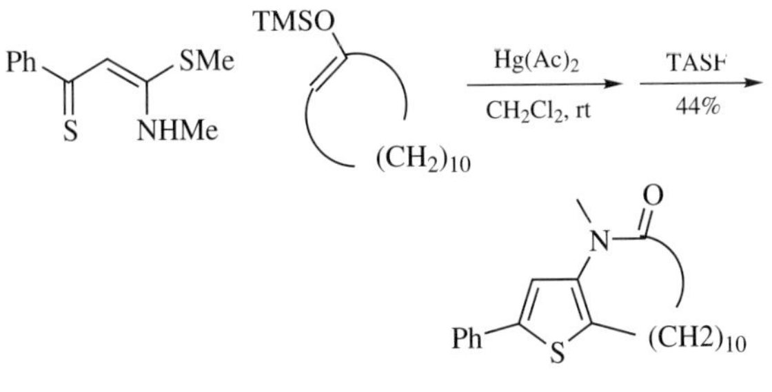

$$\text{(25)}$$

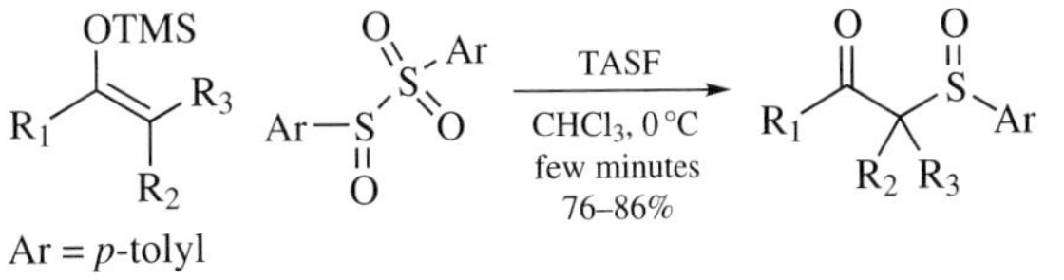

$$\text{(26)}$$

Ar = p-tolyl

Cross-coupling Reactions. The use of activated silicon species in cross-coupling reactions appears to be a practical alternative to toxic organotin reagents. However, the poor nucleophilicity of silicon toward organopalladium intermediates is an important drawback that still needs to be addressed. The quantitative addition of fluoride greatly accelerates the rate-determining transmetallation step. One proposed hypothesis of such for behavior is the formation of a pentacoordinated silicate of increased nucleophilicity toward organopalladium species. Recently, the transmetallation step was investigated independently of the oxidative addition and reductive elimination through the production of a strained intermediate[44,45] (eq 27).

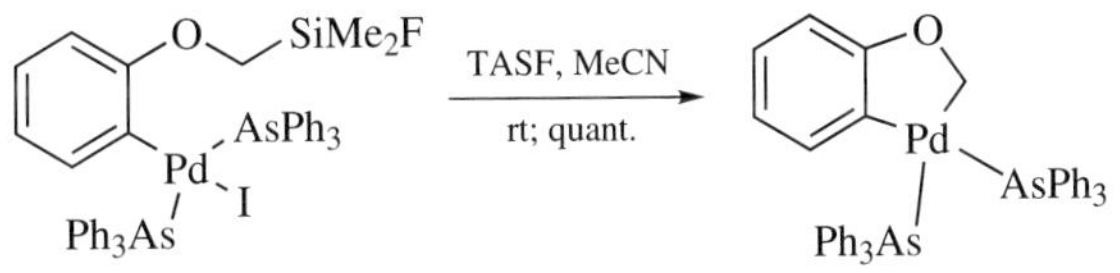

$$\text{(27)}$$

The difference in reactivity between the stannane and the silane was also advantageously exploited in a cross-coupling reaction (eq 28). Very good selectivity was obtained and the mechanism was investigated.[46–50]

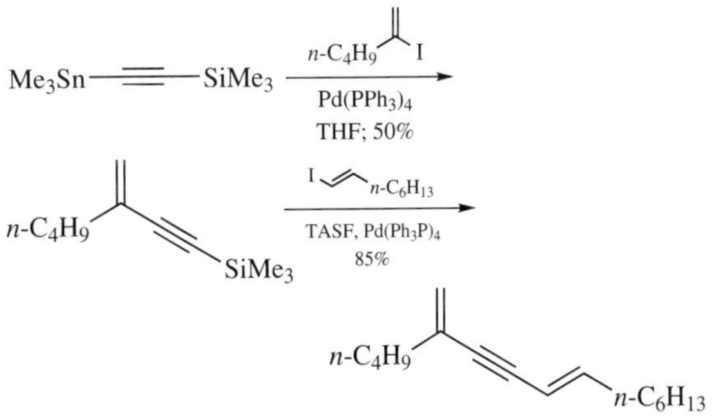

$$\text{(28)}$$

Cycloadditions and the Generation of Unusual Carbanions. Dienes or dipoles generated in situ from the cleavage of a TMS group efficiently undergo cycloaddition with aromatic aldehydes. While a net [3 + 2] cycloaddition occurs with a stoichiometric amount of fluoride[51] (eq 29), the hetero Diels–Alder of benzocyclobutane proceeds with only a catalytic amount of TASF[52] (eq 30). These are particularly mild conditions to produce the reactive o quinodimethane usually generated thermally.

In situ cleavage of C–Si bonds also generates carbon nucleophiles that can add to carbonyls and aromatic rings,[53] or displace leaving groups on both aromatic rings[54] and sulfurs.[55] An interesting example is the TASF-mediated coupling reaction of polymeric allyltrimethylsilanes with aldehydes.[56] The reaction was far superior to Lewis acid promoted reactivity which induced undesired cross-linking.

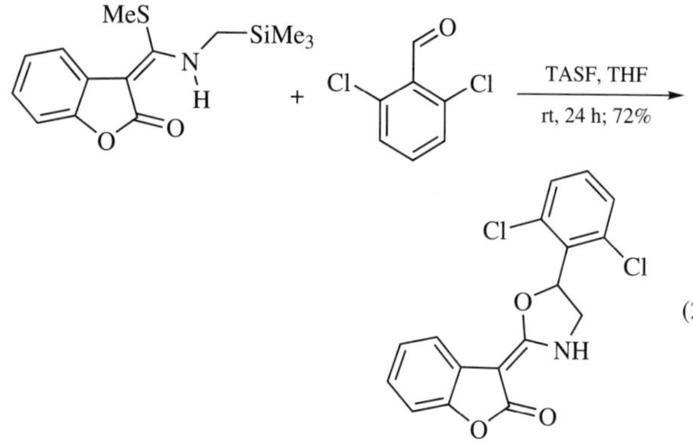

$$(29)$$

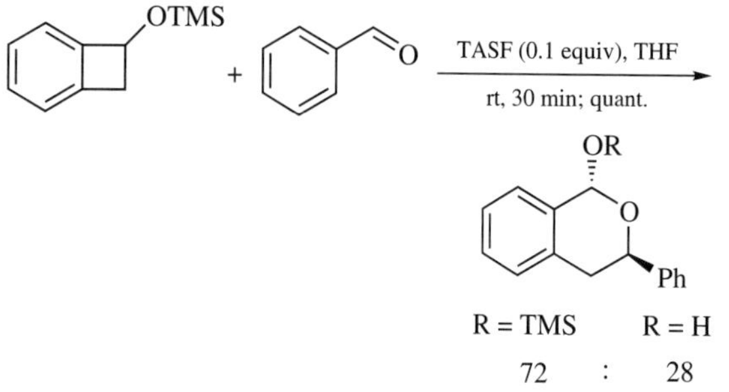

$$(30)$$

$$R = TMS \quad\quad R = H$$
$$72 \quad : \quad 28$$

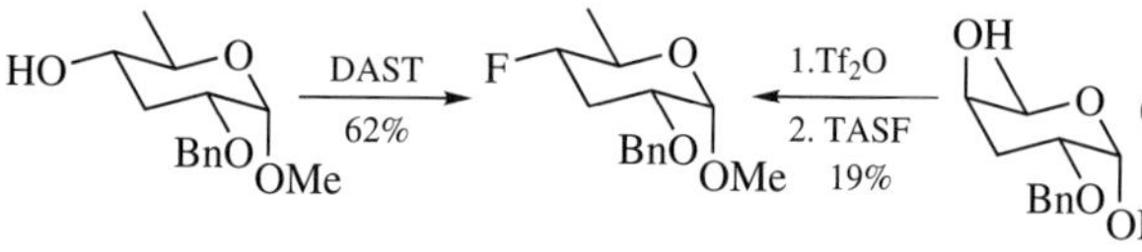

$$(33)$$

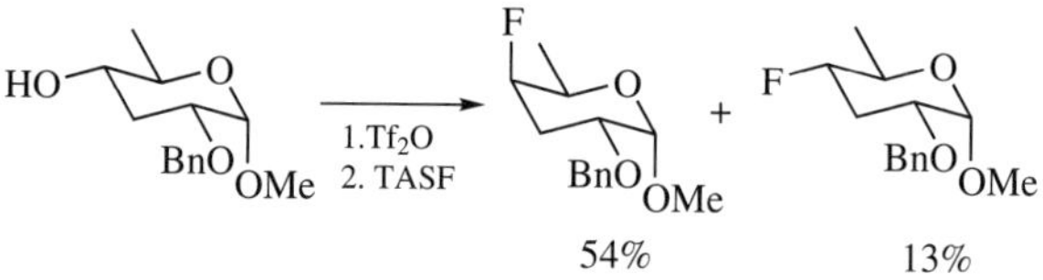

$$(34)$$

The fluoride ion from TASF also adds to aldehydes[66] or acid fluorides[67] to provide fluorinated or perfluorinated ethers (eqs 35 and 36).

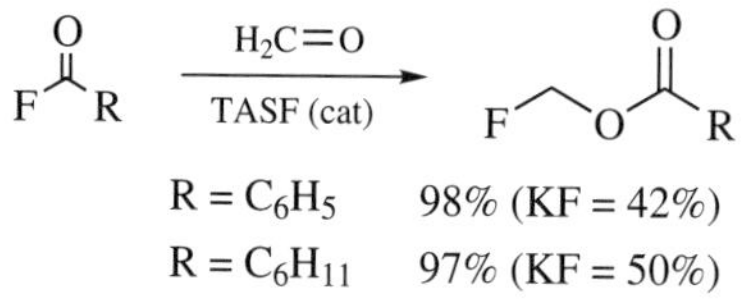

$$R = C_6H_5 \quad\quad 98\% \; (KF = 42\%)$$
$$R = C_6H_{11} \quad\quad 97\% \; (KF = 50\%)$$

$$(35)$$

$$(36)$$

Stannyl Anions. Stannyl anions, generated in situ from Sn–Si bond cleavage, readily undergo halogen metal exchange with aryl halides.[57-61] The resulting anions have been shown to condense with carbonyls (eq 31) or generate highly reactive intermediates like carbenes (eq 32), benzynes, o-quinodimethanes, or cumulenes through 1,1-, 1,2-, or 1,4-eliminations. These mild conditions allow easy access to synthetically useful reactive intermediates.

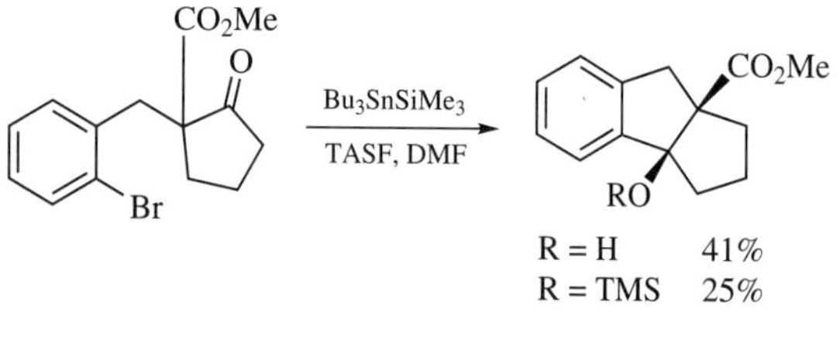

$$(31)$$

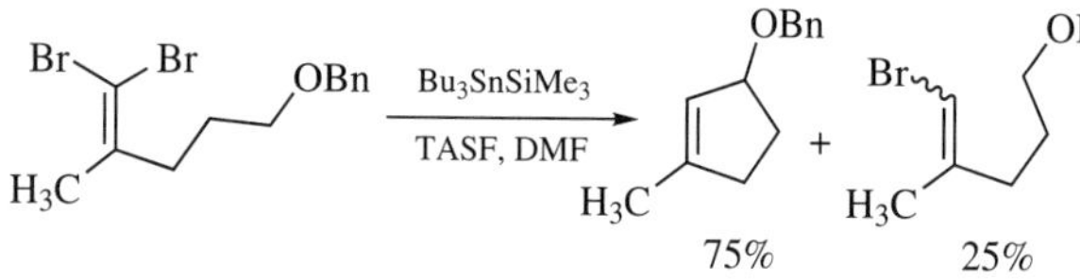

$$R = H \quad\quad 41\%$$
$$R = TMS \quad\quad 25\%$$

$$(32)$$

$$75\% \quad\quad 25\%$$

Nucleophilicity of the Fluoride. Although TASF has been valuable for the direct incorporation for fluoride into organic substrates, the nucleophilicity of the fluoride from TASF has not been exclusively exploited for silicon-based reactions. Recent examples of substitution reactions have been published since the previous article, mainly for sugar derivatives.[62-64] Unfortunately, the reactions are typically low yielding and elimination byproducts are often observed. Despite these complications, TASF is often considered to be the mildest reagent for such reactions. In one case, the source of fluoride (DAST or TASF) gave stereocomplementary products (eqs 33 and 34).[65]

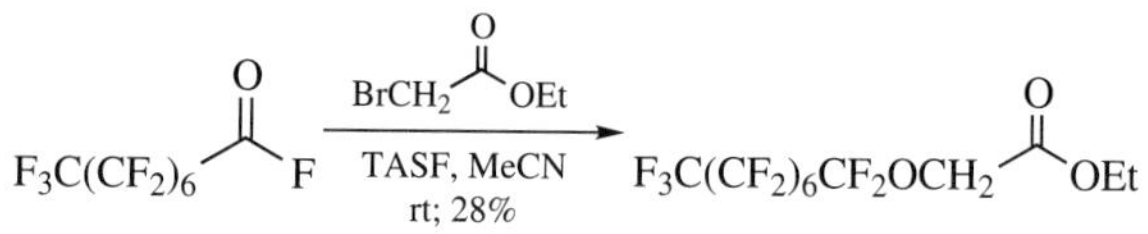

1. (a) Middleton, W. J., Org. Synth. **1985**, 64, 221. (b) Middleton, W. J. U. S. Patent 3 940 402 **1976** (Chem. Abstr. **1976**, 85, 6388j). See also Ref. 11(b).

2. For a review of applications of fluoride ion in organic synthesis, see: Clark, J. H., Chem. Rev. **1980**, 80, 429.

3. (a) Card, P. J.; Hitz, W. D., J. Am. Chem. Soc. **1984**, 106, 5348. (b) Doboszewski, B.; Hay, G. W.; Szarek, W. A., Can. J. Chem. **1987**, 65, 412.

4. (a) Noyori, R.; Nishida, I.; Sakata, J., J. Am. Chem. Soc. **1983**, 105, 1598. (b) Noyori, R.; Nishida, I.; Sakata, J., Tetrahedron Lett. **1981**, 22, 3993.

5. (a) RajanBabu, T. V., J. Org. Chem. **1984**, 49, 2083. (b) Webster, O. W.; Hertler, W. R.; Sogah, D. Y.; Farnham, W. B.; RajanBabu, T. V., J. Am. Chem. Soc. **1983**, 105, 5706.

6. (a) RajanBabu, T. V.; Reddy, G. S.; Fukunaga, T., J. Am. Chem. Soc. **1985**, 107, 5473. (b) RajanBabu, T. V.; Chenard, B. L.; Petti, M. A., J. Org. Chem. **1986**, 51, 1704.

7. (a) Hatanaka, Y.; Hiyama, T., Tetrahedron Lett. **1987**, 28, 4715. (b) Hatanaka, Y.; Hiyama, T., J. Org. Chem. **1988**, 53, 918. (c) Hatanaka, Y.; Fukushima, S.; Hiyama, T., Heterocycles **1990**, 30, 303. (d) Hatanaka, Y.; Hiyama, T., Tetrahedron Lett. **1988**, 29, 97.

8. (a) Smart, B. E.; Middleton, W. J.; Farnham, W. B., J. Am. Chem. Soc. **1986**, 108, 4905. (b) Farnham, W. B.; Smart, B. E.; Middleton, W. J.; Calabrese, J. C.; Dixon, D. A., J. Am. Chem. Soc. **1985**, 107, 4565.

9. For a discussion of structural aspects of TAS salts, see: (a) Farnham, W. B.; Dixon, D. A.; Middleton, W. J.; Calabrese, J. C.; Harlow, R. L.; Whitney, J. F.; Jones, G. A.; Guggenberger, L. J., J. Am. Chem. Soc. **1987**, 109, 476. (b) Dixon, D. A.; Farnham, W. B.; Heilemann, W.; Mews, R.; Noltemeyer, M., Heteroatom Chem. **1993**, 4, 287.

10. (a) Fujita, M.; Hiyama, T., J. Am. Chem. Soc. **1985**, 107, 4085. (b) Hiyama, T.; Obayashi, M.; Sawahata, M., Tetrahedron Lett. **1983**, 24, 4113. See also: de Jesus, M. A.; Prieto, J. A.; del Valle, L.; Larson, G. L., Synth. Commun. **1987**, 17, 1047.

11. (a) Fujita, M.; Hiyama, T., J. Org. Chem. **1988**, 53, 5405. (b) Fujita, M.; Hiyama, T., Org. Synth. **1990**, 69, 44. (c) Fujita, M; Hiyama, T., Tetrahedron Lett. **1987**, 28, 2263.

12. Mori, M.; Isono, N.; Kaneta, N.; Shibasaki, M., *J. Org. Chem.* **1993**, *58*, 2972.

13. Palomo, C.; Aizpurua, J. M.; García, J. M.; Ganboa, I.; Cossio, F. P.; Lecea, B.; López, C., *J. Org. Chem.* **1990**, *55*, 2498. Also see: Padwa, A.; Chen, Y.-Y.; Dent, W.; Nimmesgern, H., *J. Org. Chem.* **1985**, *50*, 4006.

14. Palomo, C.; Aizpurua, J. M.; López, M. C.; Lecea, B., *J. Chem. Soc., Perkin Trans. 1* **1989**, 1692.

15. Kolomeitsev, A. A.; Movchun, V. N.; Kondratenko, N. V.; Yagupolski, Yu. L., *Synthesis* **1990**, 1151.

16. Vedejs, E.; Pribish, J. R., *J. Org. Chem.* **1988**, *53*, 1593.

17. Colvin, E. W. *Silicon Reagents in Organic Synthesis*; Academic Press Inc.: San Diego, CA, **1988**.

18. Manoso, A. S., PhD dissertation, *University of Maryland*, 2004, and cited references.

19. Wrackmeyer, B.; Gerstmann, S.; Herberhold, M.; Webb, G. A.; Kurosu, H., *Magn. Reson. Chem.* **1994**, *32*, 492.

20. Barrett, A. G. M.; Peña, M.; Willardsen, J. A., *J. Org. Chem.* **1996**, *61*, 1082.

21. Pilcher, A. S.; DeShong, P., *J. Org. Chem.* **1996**, *61*, 6901.

22. Scheidt, K. A.; Chen, H.; Follows, B. C.; Chemler, S. R.; Scott Coffey, D.; Roush, W. R., *J. Org. Chem.* **1998**, *63*, 6436.

23. Aissa, C.; Riveiros, R.; Ragot, J.; Fürstner, A., *J. Am. Chem. Soc.* **2003**, *125*, 15512.

24. Paterson, I.; Florence, G. J.; Heimann, A. C.; Mackay, A. C., *Angew. Chem., Int. Ed.* **2005**, *44*, 1130.

25. Hamdouchi, C.; Jaramillo, C.; Lopez-Prados, J.; Rubio, A., *Tetrahedron Lett.* **2002**, *43*, 3875.

26. Carda, M.; Rodríguez, S.; González, F.; Castillo, E.; Villanueva, A. J.; Marco, J. A., *Eur. J. Org. Chem.* **2002**, 2649.

27. Scheidt, K. A.; Bannister, T. D.; Tasaka, A.; Wendt, M. D.; Savall, B. M.; Fegley, G. J.; Roush, W. R., *J. Am. Chem. Soc* **2002**, *124*, 6981.

28. Lister, T.; Perkins, M. V., *Org. Lett.* **2006**, *8*, 1827.

29. Petri, A. F.; Sasse, F.; Maier, M. E., *Eur. J. Org. Chem.* **2005**, 1865.

30. Denmark, S. E.; Sweis, R. F., *Acc. Chem. Res.* **2002**, *35*, 835.

31. White, J. D.; Blakemore, P. R.; Green, N. J.; Hauser, E. B.; Holoboski, M. A.; Keown, L. E.; Nylund Kolz, C. S.; Phillips, B. W., *J. Org. Chem.* **2002**, *67*, 7750.

32. Aoyagi, S.; Hirashima, S.; Saito, K.; Kibayashi, C., *J. Org. Chem.* **2002**, *67*, 5517.

33. Hirashima, S.; Aoyagi, S.; Kibayashi, C., *J. Am. Chem. Soc* **1999**, *121*, 9873.

34. N'Zoutani, M.-A.; Lensen, N.; Pancrazi, A.; Ardisson, J., *Synlett* **2005**, 489.

35. Paterson, I.; Doughty, V. A.; McLeod, M. D.; Trieselmann, T., *Angew. Chem., Int. Ed.* **2000**, *39*, 1308.

36. Csuk, R.; Schaade, M., *Tetrahedron* **1994**, *50*, 3333.

37. Otera, J.; Fujita, Y.; Fukuzumi, S.; Hirai, K.-I.; Gu, J.-H.; Nakai, T., *Tetrahedron Lett.* **1995**, *36*, 95.

38. Danishefsky, S. J.; Cabal, M. P.; Chow, K., *J. Am. Chem. Soc.* **1989**, *111*, 3456.

39. Klimko, P. G.; Singleton, D. A., *Synthesis* **1994**, 979.

40. Klimko, P. G.; Singleton, D. A., *J. Org. Chem.* **1992**, *57*, 1733.

41. Csuk, R.; Schaade, M.; Schmidt, A., *Tetrahedron* **1994**, *50*, 11885.

42. Lee, J. S.; Lee, D. J.; Kim, B. S.; Kim, K., *J. Chem. Soc., Perkin Trans. 1* **2001**, *21*, 2774.

43. Caputo, R.; Ferreri, C.; Longobardo, L.; Palumbo, G.; Pedatella, S., *Synth. Commun.* **1993**, *23*, 1515.

44. Mateo, C.; Fernández-Rivas, C.; Echavarren, A. M.; Cárdenas, D. J., *Organometallics* **1997**, *16*, 1997.

45. Mateo, C.; Fernández-Rivas, C.; Echavarren, A. M.; Cárdenas, D. J., *Organometallics* **1998**, *17*, 3661.

46. Hatanaka, Y.; Hiyama, T., *Synlett* **1991**, 845.

47. Hatanaka, Y.; Goda, K.-J.; Hiyama, T., *Tetrahedron Lett.* **1994**, *35*, 1279.

48. Hatanaka, Y.; Ebina, Y.; Hiyama, T., *J. Am. Chem. Soc* **1991**, *113*, 7075.

49. Denmark, S. E.; Amishiro, N., *J. Org. Chem.* **2003**, *68*, 6997.

50. Wan, D.; Quing, F.-L., *J. Fluorine Chem.* **1999**, *94*, 105.

51. Tominaga, Y.; Takada, S.; Kohra, S., *Heterocycles* **1994**, *39*, 15.

52. Chino, K.; Takada, T.; Endo, T., *Synth. Commun.* **1996**, *26*, 2145.

53. Surowiec, M.; Makosza, M., *Tetrahedron* **2004**, *60*, 5019.

54. Omotowa, B. A.; Shreeve, J. M., *Organometallics* **2000**, *19*, 2664.

55. Movchun, V. N.; Kolomeitsev, A. A.; Yagupolskii, Y. L., *J. Fluorine Chem.* **1995**, *70*, 255.

56. Sanda, F.; Murata, J.; Endo, T., *Macromolecules* **1997**, *30*, 160.

57. Mori, M.; Isono, N.; Kaneta, N.; Shibasaki, M., *J. Org. Chem.* **1993**, *58*, 2972.

58. Honda, T.; Mori, M., *Chem. Lett.* **1994**, 1013.

59. Sato, H.; Isono, N.; Miyoshi, I.; Mori, M., *Tetrahedron* **1996**, *52*, 8143.

60. Sato, H.; Isono, N.; Okamura, K.; Date, T.; Mori, M., *Tetrahedron Lett.* **1994**, *35*, 2035.

61. Mori, M.; Isono, N.; Wakamatsu, H., *Synlett* **1999**, 269.

62. Di, J.; Rajanikanth, B.; Szarek, W. A., *J. Chem. Soc., Perkin Trans. 1* **1992**, 2151.

63. Mori, Y.; Morishima, N., *Bull. Chem. Soc. Jpn.* **1994**, *67*, 236.

64. Mulard, L. A.; Kováč, P.; Glaudemans, C. P. J., *Carbohydr. Res.* **1994**, *259*, 117.

65. Mori, Y.; Morishima, N., *Chem. Pharm. Bull.* **1991**, *39*, 1088.

66. Schlosser, M.; Limat, D., *Tetrahedron* **1995**, *50*, 5807.

67. Haywood, L.; Mc Kee, S.; Middleton, W. J., *J. Fluorine Chem.* **1991**, *51*, 419.

Tris(trimethylsilyl)aluminum[1]

[65343-66-0]   C$_9$H$_{27}$AlSi$_3$ (MW 246.61)

InChI = 1S/3C3H9Si.Al/c3*1-4(2)3;/h3*1-3H3;

InChIKey = PMTNABPGINFFGB-UHFFFAOYSA-N

(·OEt$_2$)

[75441-10-0]   C$_{13}$H$_{37}$AlOSi$_3$ (MW 320.75)

InChI = 1S/C4H10O.3C3H9Si.Al/c1-3-5-4-2;3*1-4(2)3;/h3-4H2,1-2H3;3*1-3H3;

InChIKey = DAICGAZEJMWRAW-UHFFFAOYSA-N

(nucleophilic silylation;[2,3] silylalumination of terminal alkynes;[4] preparation of allyl-, aryl-, and alkenylsilanes[5,6])

Physical Data: mp 60 °C (dec).

Solubility: sol ethers, pentane, benzene; reacts violently with protic solvents.

Form Supplied in: white needles; not commercially available.

Preparative Methods: to aluminum powder and granular aluminum in ether is added mercury(0) and chlorotrimethylsilane followed by lithium wire. After the reaction is complete, the solution is filtered and ether is removed. The solid residue is stirred with pentane, which effects removal of insoluble Li(Me$_3$Si)$_4$Al. Partial concentration and cooling of the pentane solution gives (Me$_3$Si)$_3$Al·OEt$_2$ as white needles.[1b, 2c]

Handling, Storage, and Precautions: ignites spontaneously in air. The reagent can be conveniently handled as a stock solution in pentane, which can be stored under an inert atmosphere in a freezer. The molarity can be determined by titration.[2c, 6]

Carbonyl Additions. The reagent is a useful nucleophilic trimethylsilyl transfer agent.[2a] Enones and enals undergo 1,2-addition in ether at room temperature, while 1,4-addition occurs at $-78\,^\circ$C (eq 1).[2b] Ester carbonyls, which are amenable to attack only in boiling ether in the presence of aluminum chloride, give α,α-bis(trimethylsilyl)alkanols.[2a]

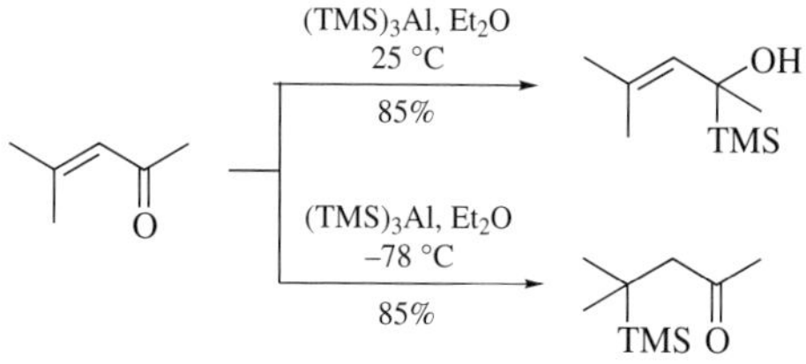

(1)

Functionalized acylsilanes are accessible by the reaction of $(Me_3Si)_3Al \cdot OEt_2$ with (S)-2-pyridyl esters in the presence of stoichiometric copper(I) cyanide (eq 2).[3]

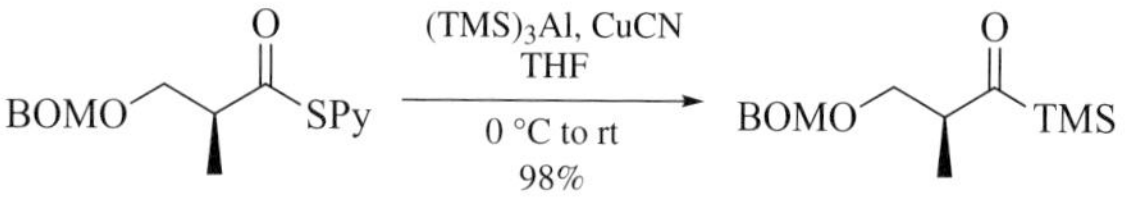

Silylalumination of Alkynes. Silylalumination of 1-alkynes provides a convenient entry to vinylsilane derivatives in a regio- and stereoselective manner (eq 3).[4]

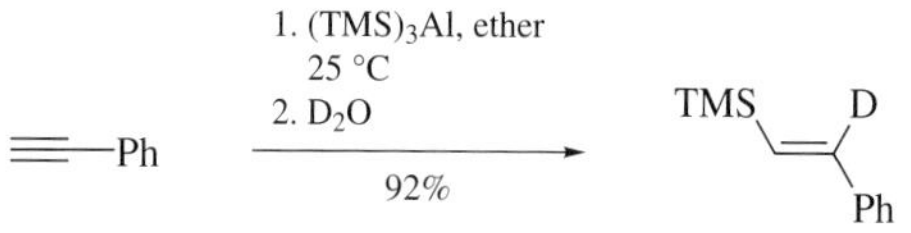

Transition Metal Promoted Coupling Reactions. In the presence of a Ni catalyst, the reagent undergoes cross-coupling reactions with aryl bromides (eq 4).[5] Alkenyl iodides are stereospecifically converted to alkenylsilanes in the presence of a Pd catalyst (eq 5).

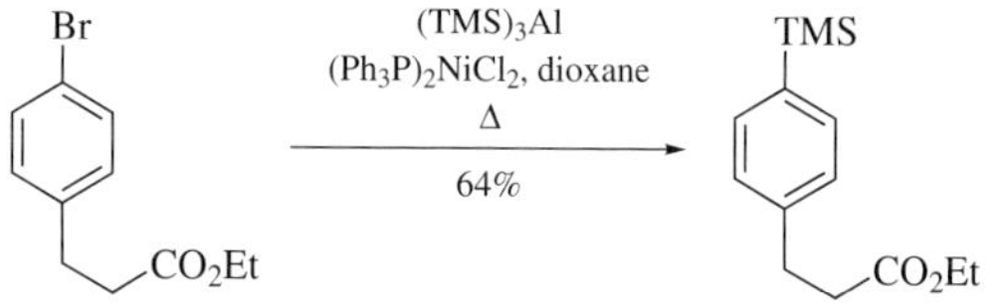

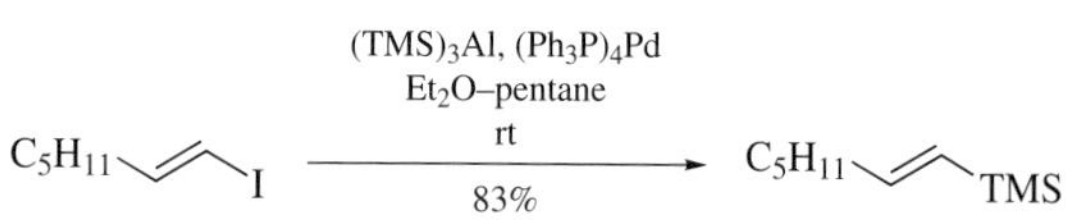

Allylic acetates are converted to allylsilanes in the presence of Pd[0] complexes or hexacarbonylmolybdenum. The regioselectivity depends on the catalyst. The Mo[0] catalyst leads to products with the silyl group attached to the less-hindered end of the allyl group. The regioselectivity of the Pd[0]-catalyzed reaction depends on the solvent and, in particular, on the ligands. The two catalysts also differ with respect to stereochemistry (eq 6).[6]

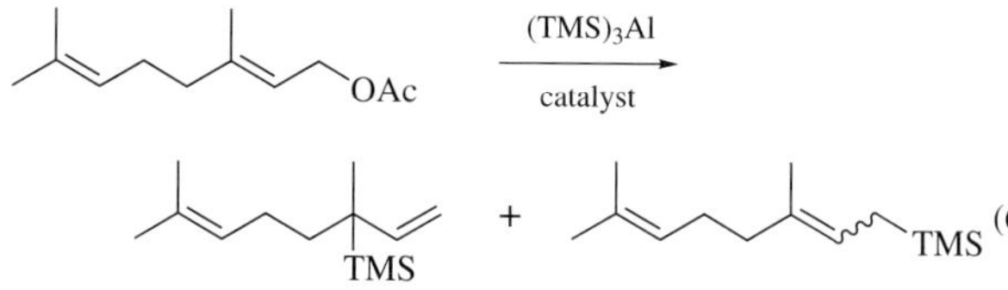

| (Ph$_3$P)$_4$Pd, THF, 25 °C | 86% | 61: 39 |
| Mo(CO)$_6$, toluene, reflux | 72% | 0:100 |

1. (a) Rösch, L., *Angew. Chem., Int. Ed. Engl.* **1977**, *16*, 480. (b) Rösch, L.; Altnau, G., *J. Organomet. Chem.* **1980**, *195*, 47. (c) Rösch, L.; Altnau, G.; Krüger, C.; Tsay, Y.-H., *Z. Naturforsch., Tell B* **1983**, *38B*, 34.

2. (a) Rösch, L.; Altnau, G.; Otto, W. H., *Angew. Chem., Int. Ed. Engl.* **1981**, *20*, 581. (b) Altnau, G.; Rösch, L.; Jas, G., *Tetrahedron Lett.* **1983**, *24*, 45. (c) Avery, M. A.; Chong, W. K. M.; Jennings-White, C., *J. Am. Chem. Soc.* **1992**, *114*, 974.

3. Nakada, M.; Nakamura, S.; Kobayashi, S.; Ohno, M., *Tetrahedron Lett.* **1991**, *32*, 4929.

4. Altnau, G.; Rösch, L.; Bohlmann, F.; Lonitz, M., *Tetrahedron Lett.* **1980**, *21*, 4069.

5. Trost, B. M.; Yoshida, J., *Tetrahedron Lett.* **1983**, *24*, 4895.

6. Trost, B. M.; Yoshida, J.; Lautens, M., *J. Am. Chem. Soc.* **1983**, *105*, 4494.

Keisuke Suzuki & Tetsuya Nagasawa
Keio University, Yokohama, Japan

Tris(trimethylsilyl)methane[1]

$$(Me_3Si)_3CH$$

[1068-69-5]   C$_{10}$H$_{28}$Si$_3$ (MW 232.65)

InChI = 1S/C10H28Si3/c1-11(2,3)10(12(4,5)6)13(7,8)9/h10H,1-9H3

InChIKey = BNZSPXKCIAAEJK-UHFFFAOYSA-N

(precursor of $(Me_3Si)_3CLi$, useful to introduce bulky tris(trimethylsilyl)methyl (trisyl) group; Peterson alkenation reagent)

Physical Data: bp 219 °C, 93–95 °C/13.5 mmHg; d 0.836 g cm^{-3}; n_D^{20} 1.4657.

Solubility: sol common organic solvents.

Analysis of Reagent Purity: ^{1}H NMR.

Purification: by distillation.

Preparative Methods: the reaction of chlorotrimethylsilane, lithium metal, and chloroform in THF gives tris(trimethylsilyl)methane.[2]

Handling, Storage, and Precautions: can be stored in a glass bottle.

Synthesis of Tris(trimethylsilyl)methyl Bromide.[3] Tris(trimethylsilyl)methyl bromide (trisyl bromide) can be prepared in 75% yield by photochemical bromination of neat tris(trimethylsilyl)methane at 180–190 °C in less than 5 h (eq 1).

$$(TMS)_3CH + Br_2 \xrightarrow[75\%]{\substack{h\nu \\ 180-190\,°C}} (TMS)_3CBr + HBr \quad (1)$$

Trisyl bromide is very reactive toward lithium, magnesium, and phenyllithium to give the corresponding organometallic reagents. These reagents are useful in the Peterson alkenation reaction.

Metalation. Tris(trimethylsilyl)methyllithium can be prepared by metalation of tris(trimethylsilyl)methane by methyllithium in a mixture of diethyl ether and THF (eq 2).[4] An improved procedure is described as follows. To a solution of tris(trimethylsilyl)methane in THF is added a solution of MeLi (10% excess) in diethyl ether with stirring under nitrogen. The diethyl ether is distilled off and the residual solution is boiled under reflux under nitrogen. After 2 h, about 95% of tris(trimethylsilyl)methane is converted to tris(trimethylsilyl)methyllithium. Tris(trimethylsilyl)methyllithium is important for the introduction of $(Me_3Si)_3C$ (trisyl) groups, and for use in Peterson alkenations (eq 3).

$$(TMS)_3CH + MeLi \xrightarrow{Et_2O-THF} (TMS)_3CLi \quad (2)$$

$$(TMS)_3CLi + CH_2O \xrightarrow{70\%} (TMS)_2C=CH_2 \quad (3)$$

Formation of Bis(trimethylsilyl)methyllithium. The carbon–silicon bond of tris(trimethylsilyl)methane can be cleaved by lithium methoxide in HMPA to form bis(trimethylsilyl)methyllithium, which subsequently reacts with carbonyl compounds to form alkenes (eq 4).[5]

$$(TMS)_3CH + LiOMe \xrightarrow{HMPA} (TMS)_2CHLi \xrightarrow{Ph_2CO} Ph_2C=CHTMS \quad (4)$$
$$51\%$$

Similar cleavage of the Si–C bond occurs with fluoride ion. The in situ reaction with carbonyl compounds gives the corresponding alkenes. Thus the reaction of tris(trimethylsilyl)methane with nonenolizable carbonyl compounds in the presence of tris(dimethylamino)sulfonium difluorotrimethylsilicate (TASF) gives the corresponding alkenes in excellent yield (eq 5).[6]

$$(TMS)_3CH + XC_6H_4CHO \xrightarrow[THF]{TASF} XC_6H_4CH=CHTMS \quad (5)$$
$$X = H, 94\%$$
$$X = p\text{-MeO}, 94\%$$

Related Reagent. Bis(trimethylsilyl)methane.

1. *Fieser & Fieser* **1975**, *5*, 617.
2. Merker, R. L.; Scott, M. J., *J. Organomet. Chem.* **1965**, *4*, 98.
3. Smith, C. L.; James, L. M.; Sibley, K. L., *Organometallics* **1992**, *11*, 2938.
4. Cook, M. A.; Eaborn, C.; Jukes, A. E.; Walton, D. R. M., *J. Organomet. Chem.* **1970**, *24*, 529.
5. Sakurai, H.; Nishiwaki, K.; Kira, M., *Tetrahedron Lett.* **1973**, 4193.
6. Palomo, C.; Aizpurua, J. M.; García, J. M.; Ganboa, I.; Cossio, F. P.; Lecea, B.; López, C., *J. Org. Chem.* **1990**, *55*, 2498.

Hideki Sakurai
Tohoku University, Sendai, Japan

Tris(trimethylsilyl)silane[1]

$(Me_3Si)_3SiH$

[1873-77-4] $C_9H_{28}Si_4$ (MW 248.73)
InChI = 1S/C9H28Si4/c1-11(2,3)10(12(4,5)6)13(7,8)9/h10H,
    1-9H3
InChIKey = SQMFULTZZQBFBM-UHFFFAOYSA-N

(mediator of radical reactions;[1,2] nontoxic substitute for tri-*n*-butylstannane in radical reactions; slower hydrogen donor than tri-*n*-butylstannane[3])

Alternatives names: TTMSS, $(TMS)_3SiH$.
Physical Data: bp 82–84 °C/12 mmHg; *d* 0.806 g cm^{-3}; n_D^{20} 1.489.
Solubility: sol pentane, ether, toluene, THF; modestly sol acetone, acetonitrile; insol H_2O; decomposes rapidly in methanol and other alcohols.
Form Supplied in: colorless liquid; commercially available.
Preparative Method: easy to synthesize.[4]
Handling, Storage, and Precautions: is slightly sensitive to oxygen and should be stored under nitrogen.[5] It showed no toxicity in several biological test systems.[6]

Original Commentary

Bernd Giese & Joachim Dickhaut
University of Basel, Basel, Switzerland

Functional Group Reductions. Tris(trimethylsilyl)silane is an effective radical reducing agent for organic halides, selenides, xanthates, isocyanides,[2] and acid chlorides (Table 1).[7] The reactions are carried out at 75–90 °C in toluene in the presence of a radical initiator, i.e. azobisisobutyronitrile Chromatographic workup affords the products. The silicon-containing byproducts are easily separated. The silane can also be used catalytically when sodium borohydride is employed as the coreductant.[8] If a halide (bromide or iodide) is treated under photochemical initiation conditions with an excess of sodium borohydride and a small amount of tris(trimethylsilyl)silane or its corresponding halide, the silane is continously regenerated from the silyl halide.

Iodides and bromides are reduced by tris(trimethylsilyl)silane to the corresponding hydrocarbons in high yield after a short reaction time (0.5 h). From tertiary to secondary and primary chlorides the reduction becomes increasingly difficult. A longer reaction time and periodic addition of initiator is required. Photochemical initiation can be used and is quite efficient.[9] Tris (trimethylsilyl)silane is superior to tri-*n*-butylstannane in

Table 1 Reduction of Several Organic Compounds by Tris(trimethylsilyl) silane[2,7]

RX	Yield RH (%)	RX	Yield RH (%)
 ⬡–Cl	82	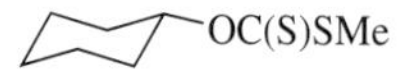 ⬡–OC(S)SMe	86
 [norbornyl]–Br	90	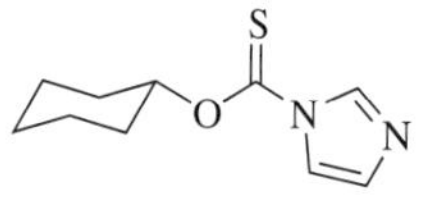 ⬡–O–C(=S)–imidazole	95
 ⬡–I	97	 ⬡–SePh	99
 ⬡–NC	95	 ⬡–COCl	92

replacing an isocyanide group by hydrogen. The reaction with tin hydride requires high temperatures (boiling xylene for primary isocyanides) and periodic addition of initiator. Using the silane, primary, secondary, and tertiary isocyanides are reduced at 80 °C in high yields. The reduction of selenides by tris(trimethylsilyl) silane proceeds with high yields; however, the corresponding reaction of sulfides is inefficient.

Acyl chlorides are converted by tris(trimethylsilyl)silane to the corresponding hydrocarbons. Tertiary and secondary acid chlorides react at 80 °C, while the reduction of primary derivatives requires higher temperatures.[7] The radical deoxygenation of hydroxyl groups is carried out by conversion of the alcohol to a thionocarbonate, which can be reduced by tris(trimethylsilyl) silane (eq 1). This very mild method is especially useful in natural product synthesis. It has been utilized for the deoxygenation of lanosterol (eq 2)[6] and the dideoxygenation of 1,6-anhydro-D-glucose (eq 3).[10]

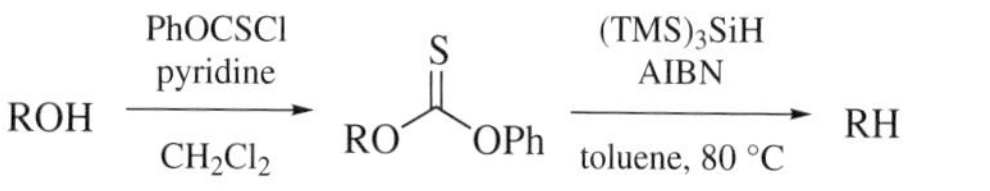

$$ROH \xrightarrow[\substack{\text{PhOCSCl} \\ \text{pyridine} \\ CH_2Cl_2}]{} RO\overset{S}{\underset{}{\diagdown}}OPh \xrightarrow[\substack{\text{toluene, 80 °C}}]{\substack{(TMS)_3SiH \\ AIBN}} RH \quad (1)$$

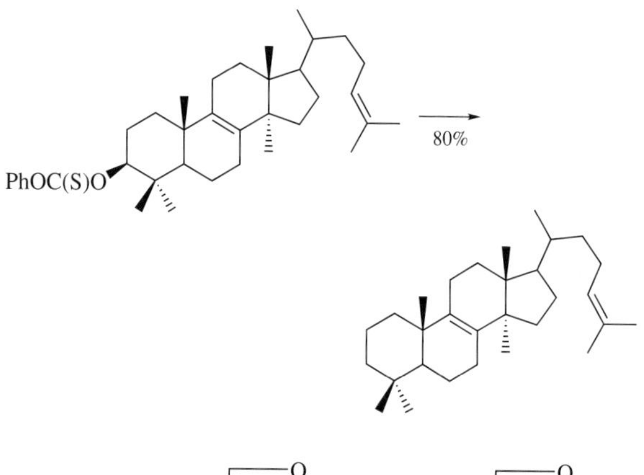

(2)

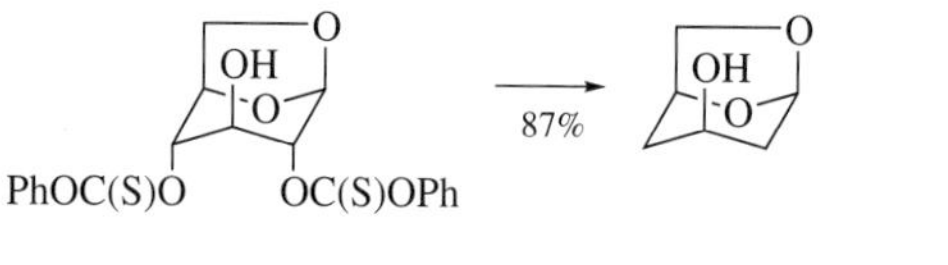

80%

(3)

87%

Radical deoxygenation of the *cis*-unsaturated fatty acid derivative with tris(trimethylsilyl)silane gives methyl triacont-21-*trans*-enoate together with the saturated compound (eq 4). If the reaction is carried out with tri-*n*-butyltin hydride, the configuration remains unchanged.[11]

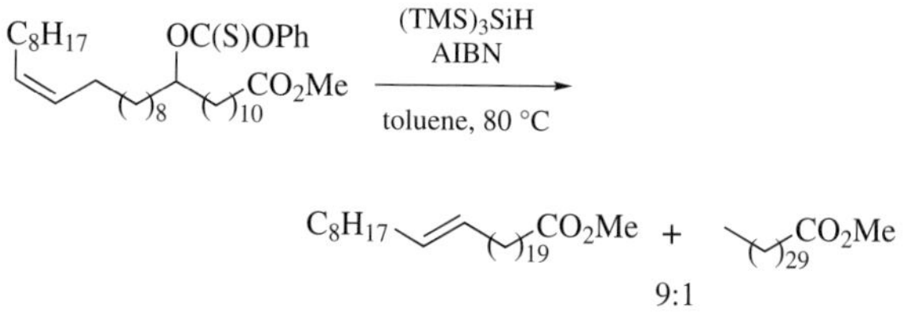

(4)

9:1

Hydrosilylation of Double Bonds. Tris(trimethylsilyl)silane is capable of radical hydrosilylation of dialkyl ketones,[12] alkenes,[12,13] and alkynes.[13] Hydrosilylation of alkenes yields the anti-Markovnikov products with high regio- and good diastereoselectivity (eq 5). By using a chiral alkene, complete stereocontrol can be achieved (eq 6).[14] The silyl group can be converted to a hydroxyl group by Tamao oxidation.[13]

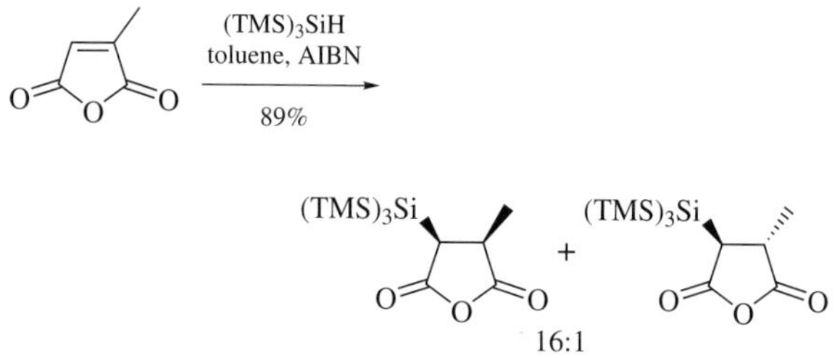

89%

(5)

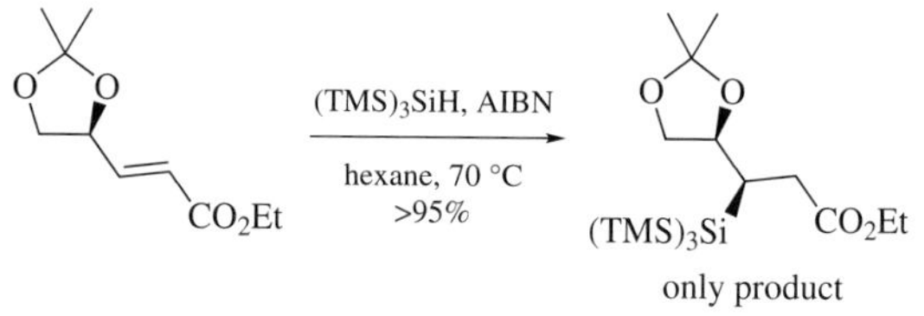

16:1

(6)

>95%

only product

Monosubstituted alkynes give alkenes in high yield and stereoselectivity. The formation of (*E*)- or (*Z*)-alkenes depends on the steric demand of the substituents (eq 7). 1,2-Disubstituted phenylalkynes are attacked exclusively *β* to the phenylated alkyne carbon atom.[13] The silyl moiety can be replaced by a bromine atom with overall retention of configuration (eq 8).[13]

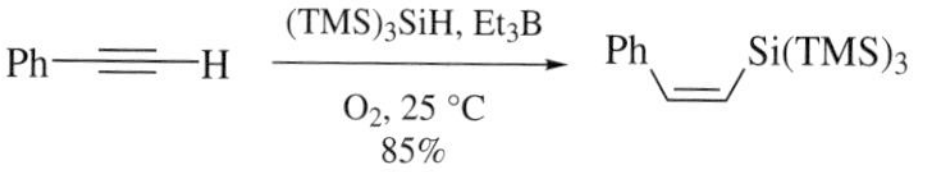

$$Ph\text{—}\equiv\text{—}H \xrightarrow[\substack{O_2, 25 °C \\ 85\%}]{(TMS)_3SiH, Et_3B} Ph\diagdown\diagup Si(TMS)_3 \quad (7)$$

$$Ph\diagdown\diagup Si(TMS)_3 \xrightarrow[\substack{-78 °C \\ 75\%}]{Br_2, CH_2Cl_2} Ph\diagdown\diagup Br \quad (8)$$

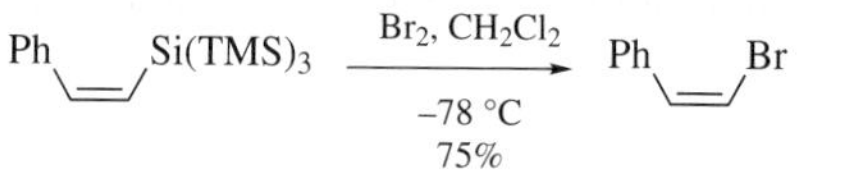

The hydrosilylation of ketones is in general slower than the corresponding reaction of alkenes and alkynes. In the case of sterically hindered ketones, a catalytic amount of a thiol is necessary to carry out the reaction.[15] The resulting silyl ethers can be easily desilylated by standard procedures. With 4-*t*-butylcyclohexanone the *trans* isomer is formed as the main product (eq 9). The hydrosilylation of a ketone bearing a chiral center in the adjacent position yields mainly the Felkin–Anh product (eq 10).[15]

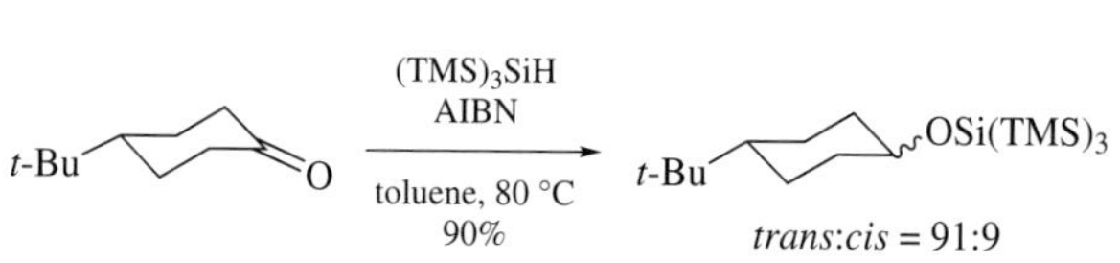

(9)

90%

trans:*cis* = 91:9

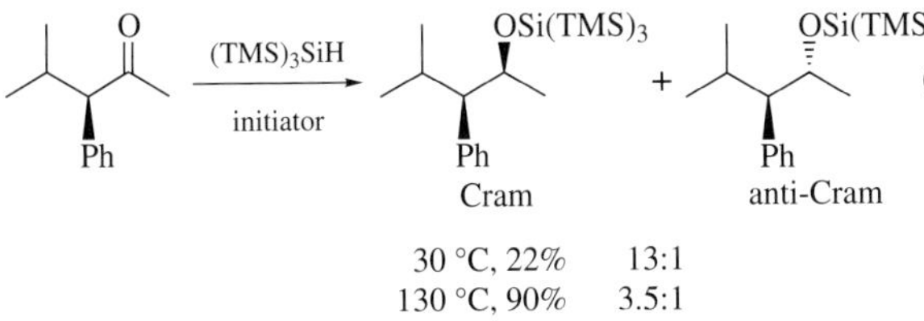

$$ (10) $$

30 °C, 22%	13:1	
130 °C, 90%	3.5:1	

Intramolecular Reactions. Tris(trimethylsilyl)silane is an effective mediator of radical cyclizations.[16] In addition to halides and selenides, secondary isocyanides can be used as precursors for intramolecular C–C bond formation,[17] which is impossible using the tin hydride (eq 11). Selective cleavage of the carbon–sulfur bond of a 1,3-dithiolane, 1,3-dithiane,[18] 1,3-oxathiolane, or 1,3-thiazolidine[19] derivative is an efficient process to generate carbon-centered radicals, which can undergo cyclization (eq 12).

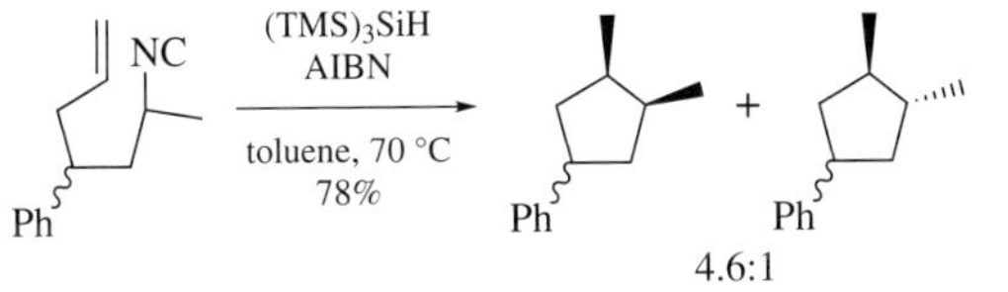

$$ (11) $$

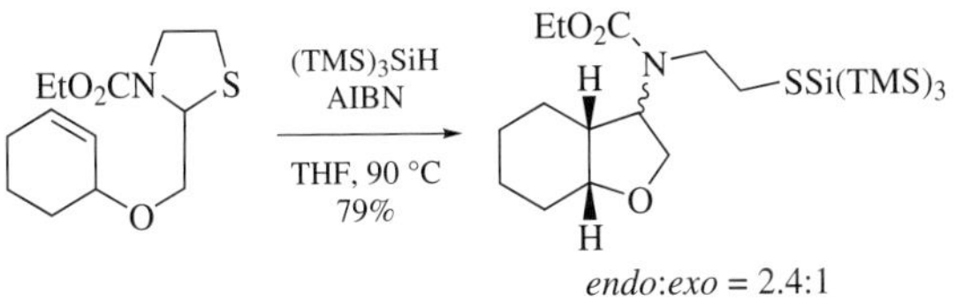

4.6:1

$$ (12) $$

endo:exo = 2.4:1

2-Benzylseleno-1-(2-iodophenyl)ethanol reacts smoothly with tris(trimethylsilyl)silane to give benzo[*b*]selenophene (eq 13).[20] A similar homolytic substitution reaction at the silicon atom yields a sila bicycle.[21]

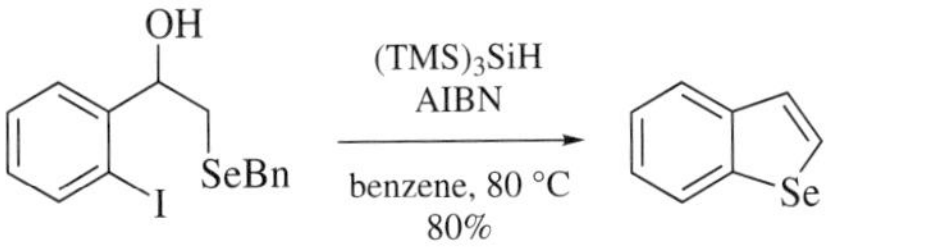

$$ (13) $$

The silane is superior to the tin reagent in the radical rearrangement of glycosyl halides to 2-deoxy sugars (eq 14).[16] Aromatization of the A-ring of 9,10-secosteroids can be achieved by a mild, radical-induced fragmentation reaction of 3-oxo-1,4-diene steroids (eq 15).[22]

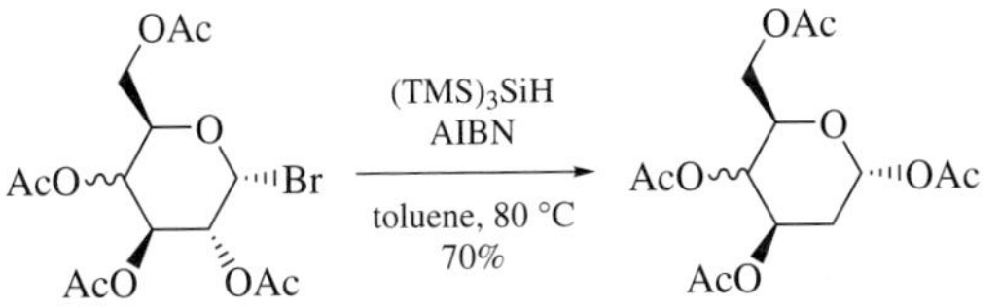

$$ (14) $$

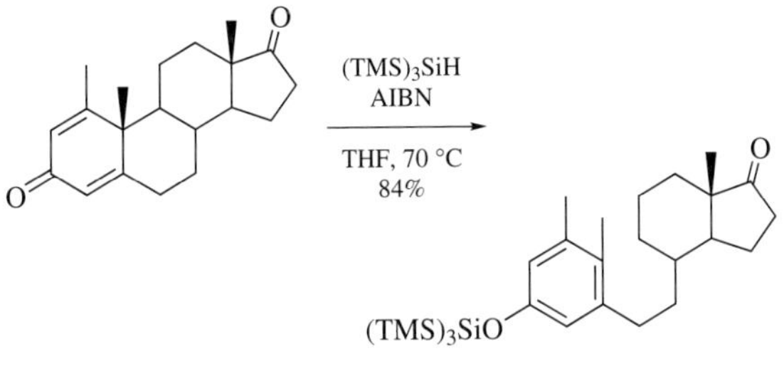

$$ (15) $$

Intermolecular Reactions. Radical carbon–carbon bond formation can be carried out with tris(trimethylsilyl)silane.[16] Again, it is possible to use isocyanides as precursors (eqs 16 and 17).[17]

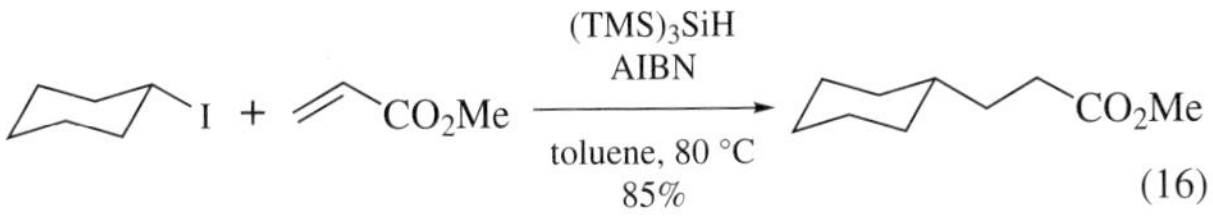

$$ (16) $$

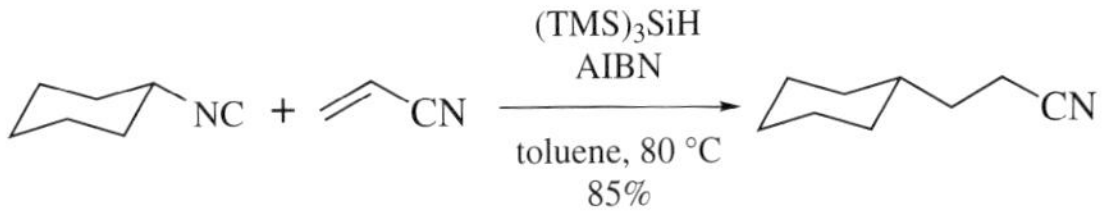

$$ (17) $$

Nonradical Reactions. Tris(trimethylsilyl)silane reacts with carbenium ions to form a silicenium ion.[23] In this case, tris(trimethylsilyl)silane is only slightly more reactive than trimethylsilane. The reaction of the silane with methyl diazoacetate in the presence of copper catalyst gives the α-silyl ester (eq 18).[24]

$$ (TMS)_3SiH + N_2CHCO_2Me \xrightarrow[90\,°C]{Cu} (TMS)_3SiCH_2CO_2Me \quad (18) $$

First Update

Chryssostomos Chatgilialoglu
ISOF, Consiglio Nazionale delle Ricerche, Bologna, Italy

Functional Group Reductions. The procedures for reductive removal of functional groups by (TMS)$_3$SiH are numerous and have recently been summarized in a book.[25] Some examples of the extension of these reactions in the area of nucleoside chemistry are worth mentioning, since the choice of this reagent on the route to biologically active compounds also includes the absence of cell toxicity compared to tin hydride. For example, the reductions of bromide **1** and chloride **2** are achieved in 94% and 92% yields, respectively, at 80 °C using AIBN as the radical initiator.[26,27] Similarly, deoxygenations in the 2′ position via the O-thioxocarbamate **3** and O-arylthiocarbonate **4** are realized in 97% and 91% yields, respectively, under similar conditions.[28,29]

(TMS)$_3$SiH is not soluble in water and does not suffer from any significant reaction with water at 100 °C for a few hours. Taking advantage of this observation, the reduction of different organic halides, bromonucleosides among them, was successfully carried out in yields ranging from 75% to quantitative, using (TMS)$_3$SiH in a heterogeneous system with water as solvent (eq 19).[30] This procedure, employing 2-mercaptoethanol as the catalyst and the hydrophobic 1,1′-azobis(cyclohexanecarbonitrile) (ACCN) as the initiator, illustrates that (TMS)$_3$SiH can be the radical-based reducing agent of choice in aqueous medium with additional benefits, such as ease of purification and environmental compatibility.

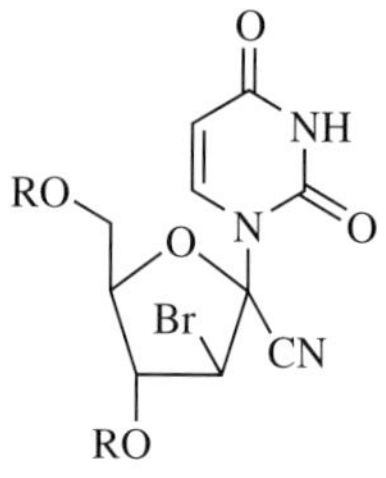

1

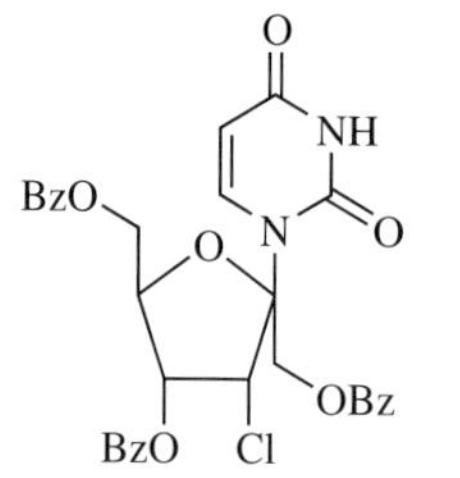

2

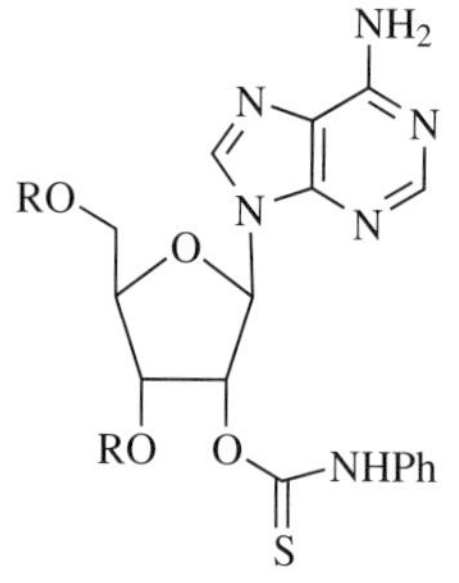

3

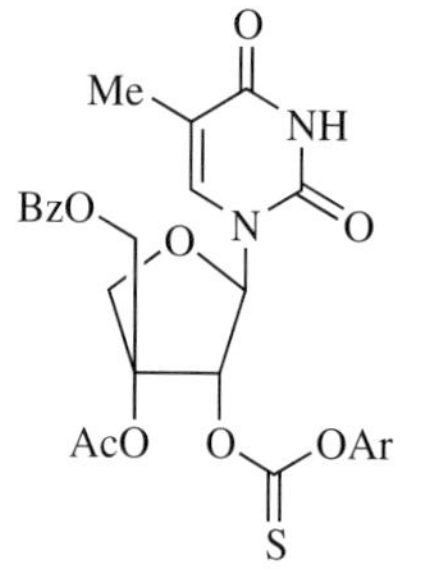

4

$$RX + (TMS)_3SiH \xrightarrow[\substack{ACCN \\ HOCH_2CH_2SH \\ H_2O}]{} RH \quad (19)$$

>90%

Reductive decarboxylation can be carried out by $(TMS)_3SiH$, and it is used as the key step to construct the chiral *cis*-cyclopropane structure in compounds designed as antidopaminergic agents (eq 20).[31] The high observed *cis* selectivity is due to hydrogen abstraction from the sterically demanding $(TMS)_3SiH$, which occurs from the less-hindered side of the intermediate cyclopropyl radical.

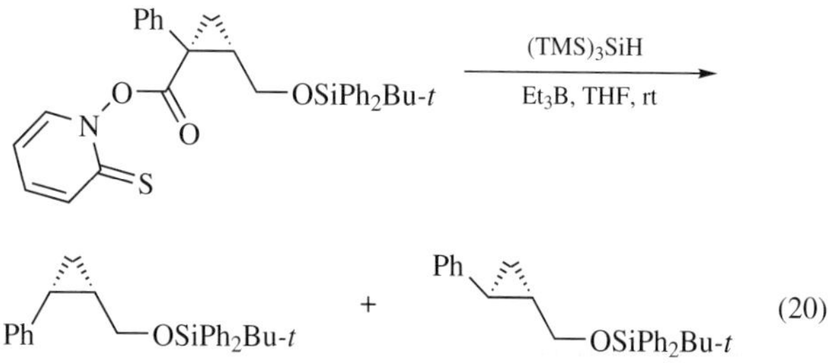

$$(20)$$

74% (*cis/trans* = 10:1)

Replacement of a pyridinium moiety by hydrogen with the formation of 3-fluoro-2-aminopyridine derivatives are achieved in good yields with $(TMS)_3SiH$ under standard radical chain conditions (eq 21).[32]

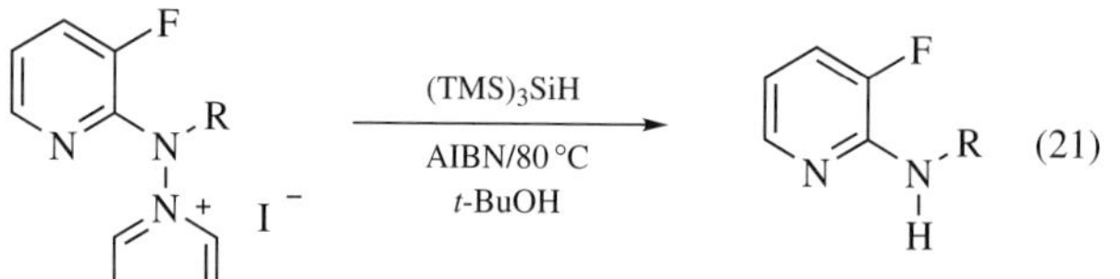

$$(21)$$

R = Me, Et

70–73%

$(TMS)_3SiH$ reacts with phosphine sulfides and phosphine selenides under free radical conditions to give the corresponding

phosphines or, after treatment with BH_3–THF, the corresponding phosphine–borane complex in good to excellent yields.[33] Stereochemical studies on P-chiral phosphine sulphides showed that these reductions proceed with retention of configuration.

Z/E Isomerization. $(TMS)_3Si^\bullet$ radicals are found to add to a variety of double bonds reversibly and therefore, to isomerize alkenes.[34] An example based on the isomerization of the 1,5,9-cyclododecatriene (**5**) is shown (eq 22). The final isomeric composition of 78:20:2 for (E,E,E)-**5**:(Z,E,E)-**5**:(Z,Z,E)-**5**, which is independent of the starting isomer or isomeric mixture, is reached in 5 h with a yield of about 80% by using $(TMS)_3SiH/t$-BuOOBu-$t/143\,°C$.

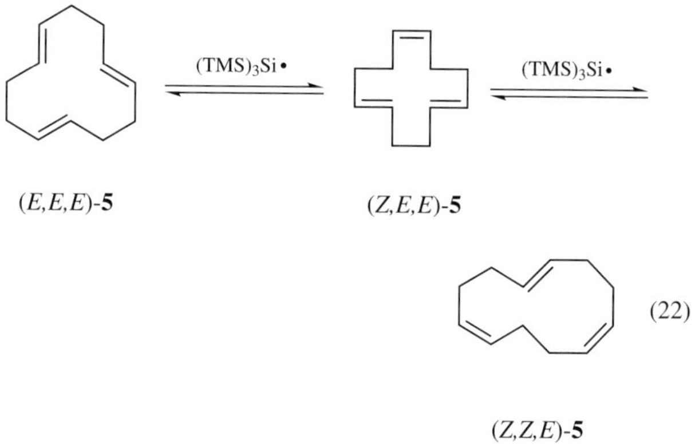

$$(22)$$

(E,E,E)-**5** (Z,E,E)-**5**

(Z,Z,E)-**5**

Hydrosilylation. The addition of $(TMS)_3SiH$ to a number of mono acetylenes has been studied in some detail.[13,35] These radical-based reactions are highly regioselective (*anti*-Markovnikov) and give terminal $(TMS)_3Si$-substituted alkenes in good yields (eq 7). These reactions proceed well also without solvent (eq 23).[36] On the other hand, reaction in the presence of $AlCl_3$ in CH_2Cl_2 gave exclusively *gem*-disubstituted olefin (eq 24).[36] The presence of Lewis acids shift the reaction mechanism from radical to ionic, affording a complementary regioselectivity.

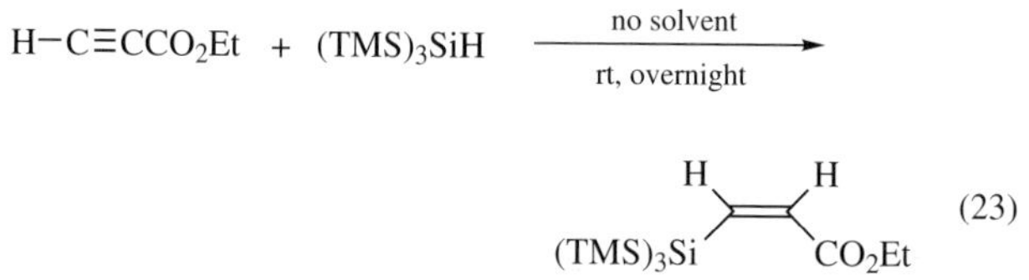

$$H-C{\equiv}CCO_2Et + (TMS)_3SiH \xrightarrow[\text{rt, overnight}]{\text{no solvent}}$$

$$(23)$$

92%

$$H-C{\equiv}CCO_2Et + (TMS)_3SiH \xrightarrow[\substack{CH_2Cl_2 \\ 0\,°C, 4\,h}]{AlCl_3}$$

$$(24)$$

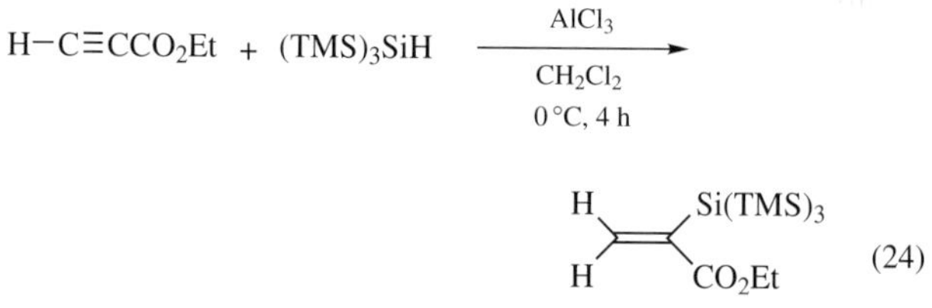

62%

Intramolecular Reactions. Consecutive radical reactions mediated by $(TMS)_3SiH$ have been discussed in some detail.[25] In the construction of carbocycles, five-membered ring formation has been used for preparing fused cyclic compounds, such as functionalized diquinanes.[37] The reaction of **6** with $(TMS)_3SiH$

furnished the expected product **7** in 80% yield in an $\alpha{:}\beta$ ratio of 82:18 as the result of a kinetically controlled reaction (eq 25).

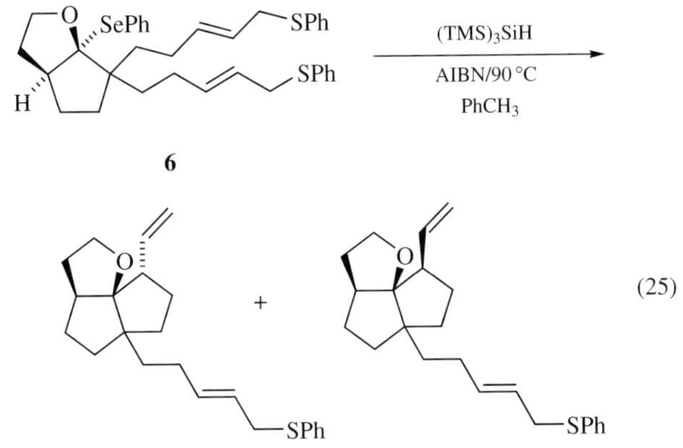

$$(25)$$

7, 80% ($\alpha{:}\beta$ = 82:18)

Complex skeletons such as triquinanes could be prepared by a key silyl radical addition to conjugated dienes to form allylic type radicals with subsequent intramolecular addition to C=C double bonds. The exposure of **8** to (TMS)$_3$SiH and AIBN at 80 °C gave the triquinane **9** in an unoptimized yield of 51% (eq 26).[38]

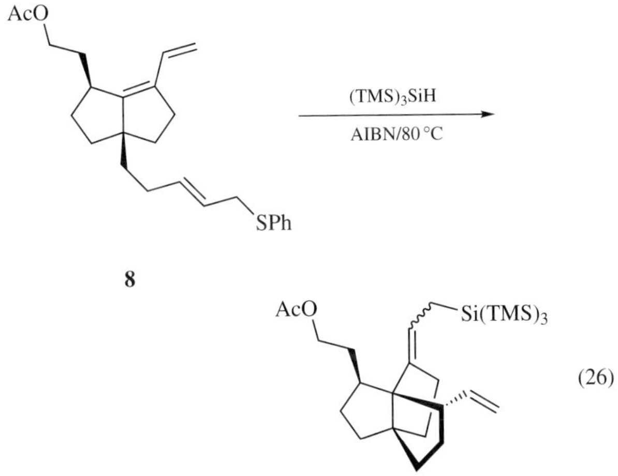

$$(26)$$

9, 51%

As a strategy for the construction of cyclic ethers, the radical cyclization of β-alkoxyacrylates was used for the preparation of *cis*-2,5-disubstituted tetrahydrofurans and *cis*-2,6-disubstituted tetrahydropyrans. An example is given with the β-alkoxymethacrylate **10** as precursor of the optically active benzyl ether of (+)-methyl nonactate, exclusively formed as the threo product (eq 27).[39]

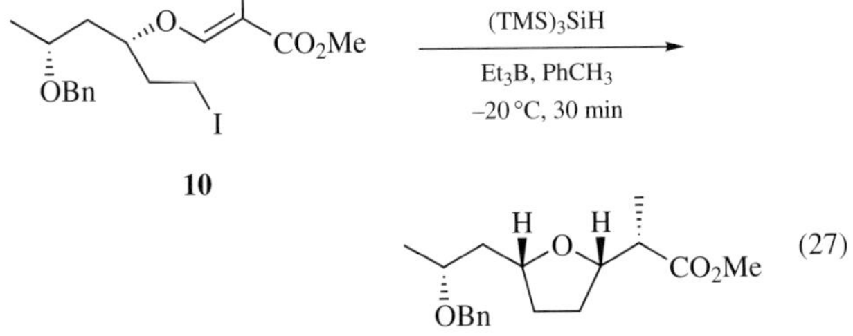

$$(27)$$

90%, de > 96%

Using (TMS)$_3$SiH as the mediator, phenylseleno esters can be conveniently used as a precursor of acyl radicals. An example is the key step for the enantioselective synthesis of the nonisoprenoid sesquiterpene (−)-kumausallene, obtained by radical cyclization at low temperature in a 32:1 mixture in favor of the 2,5-*cis* diastereoisomer (eq 28).[40]

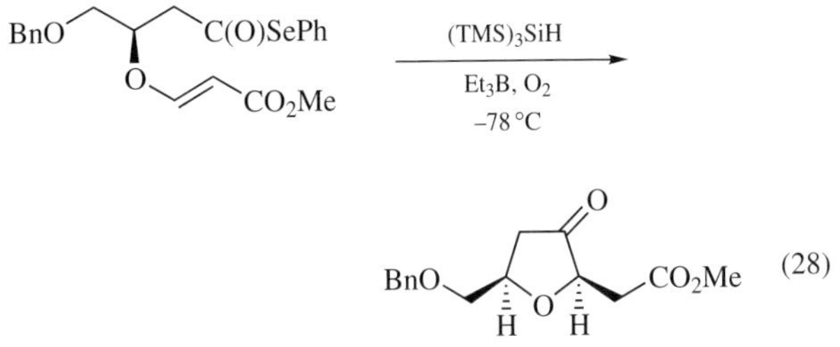

$$(28)$$

92%, *cis:trans* = 32:1

Another example of acyl radical cyclization is given in eq 29.[41] The careful choice of the configuration of the double bond combined with conformational features of the preexisting ring in the starting material can improve the poor diastereoselectivity of 6-*exo-trig* cyclizations.

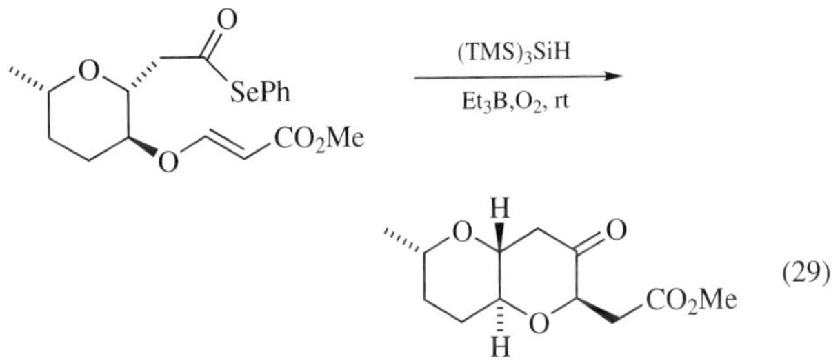

$$(29)$$

91%, ds > 95:5

Radical cyclization to triple bonds is used as the key step for the synthesis of oxygen heterocycles.[42] It can be combined with Lewis acid complexation with aluminum tris(2,6-diphenyl phenoxide) (ATPH), which assists the radical cyclization.[43] The β-iodo ether **11** can be complexed with 2 equiv of ATPH to achieve a relevant template effect, facilitating the subsequent radical intramolecular addition and orienting the (TMS)$_3$SiH approach from one face (eq 30). The resulting quantitatively formed cyclization products show a preferential Z configuration.

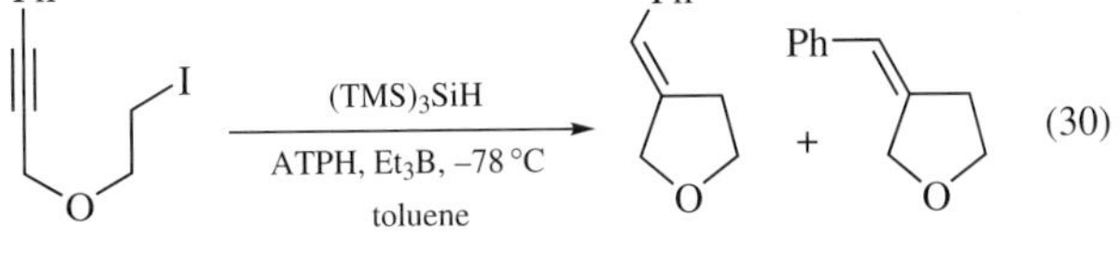

$$(30)$$

11

(E) (Z)

99%, $E{:}Z$ = <1:>99

In alkaloid synthesis, cyclization strategies have been mediated by (TMS)$_3$SiH. An example is given by the total synthesis of (−)-slaframine starting from a phenylseleno derivative (eq 31).[44]

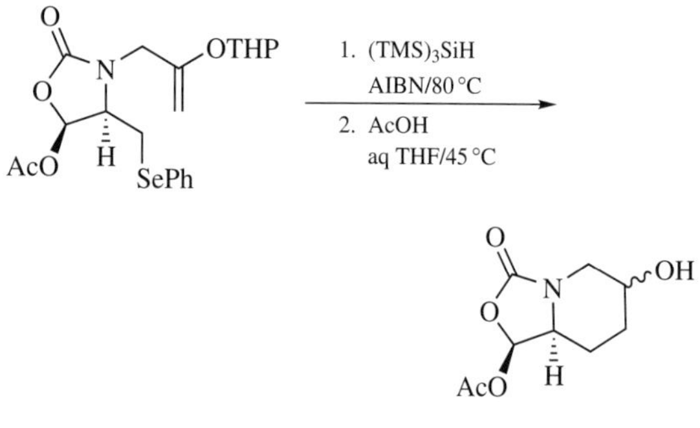

$$76\%,\ \alpha{:}\beta = 7{:}1 \tag{31}$$

The affinity of $(TMS)_3SiH$ toward azides allows this functionality to be used as a radical acceptor.[45] An example is given in eq 32 where the amine product was tosylated before work-up.

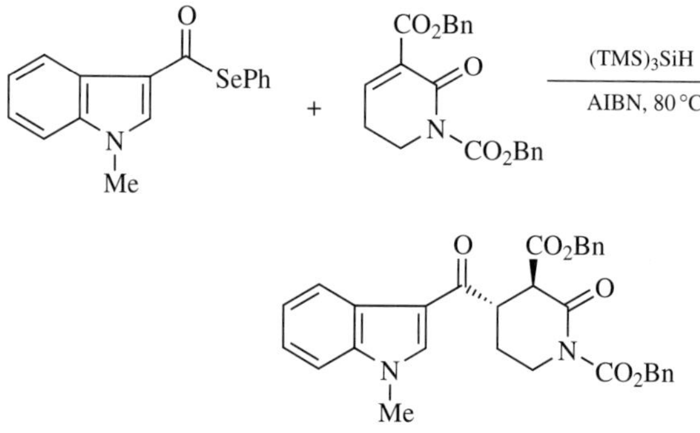

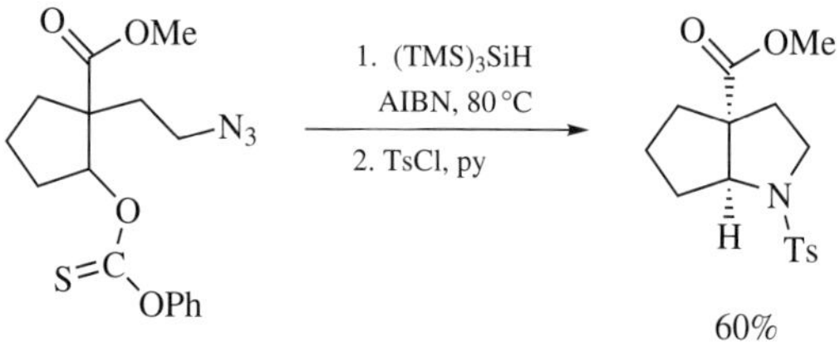

$$60\% \tag{32}$$

Intermolecular Reactions. The intermolecular C–C bond formation mediated by $(TMS)_3SiH$ has been the subject of several synthetically useful investigations. The effect of the bulky $(TMS)_3SiH$ can be appreciated in the example of β- or γ-substituted α-methylenebutyrolactones with n-BuI (eq 33).[46] The formation of α,β- or α,γ-disubstituted lactones was obtained in good yield and diastereoselection when one of the substituents is a phenyl ring.

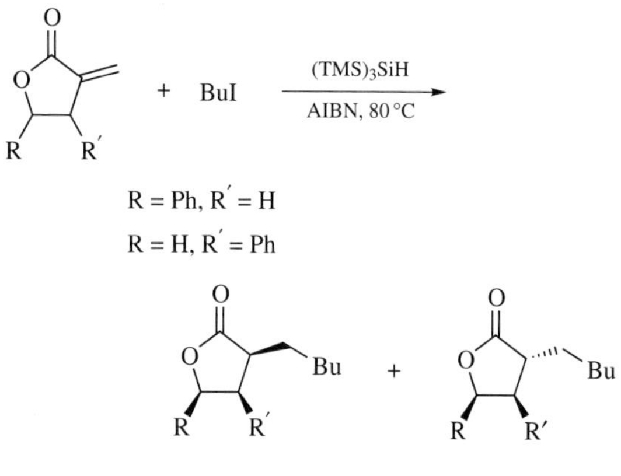

R = Ph, R' = H

R = H, R' = Ph

$$\tag{33}$$

60%, *cis*:*trans* = 98:2

60%, *cis*:*trans* = 94:6

The use of $(TMS)_3SiH$ with acyl selenides can also lead to new C–C bond formation, as shown with the α,β-unsaturated lactam ester (eq 34). The resulting ketone can be envisaged as potentially useful for the synthesis of 2-acylindole alkaloids.[47] Both the effects of H-donating ability and steric hindrance given by the silicon hydride are evident.

$$\tag{34}$$

72%

The synthesis of *N*-alkoxylamines is relevant since this is a class of initiators in 'living' radical polymerization. A method for C–O bond formation has been designed using $(TMS)_3SiH$, which consists of the trapping of alkyl radicals generated in situ from alkyl bromides or iodides by stable nitroxide radicals (eq 35).[48]

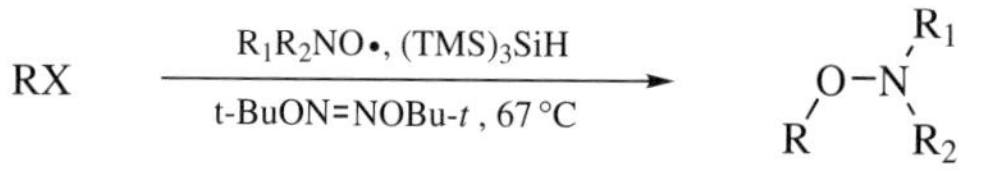

$$RX \xrightarrow[t\text{-BuON}=NO\text{Bu-}t,\ 67\,°C]{R_1R_2NO\bullet,\ (TMS)_3SiH} \quad R{-}O{-}N\begin{smallmatrix}R_1\\R_2\end{smallmatrix} \tag{35}$$

Examples of this intermolecular C–P bond formation by means of radical phosphonation[49] and phosphination[50] have been achieved by reaction of aryl halides with trialkyl phosphites and chlorodiphenylphosphine, respectively, in the presence of $(TMS)_3SiH$ under standard radical conditions. The phosphonation reaction (eq 36) works well either under UV irradiation at room temperature or in refluxing toluene. The radical phosphination (eq 37) required pyridine in boiling benzene for 20 h. Phosphinated products were handled as phosphine sulfides.

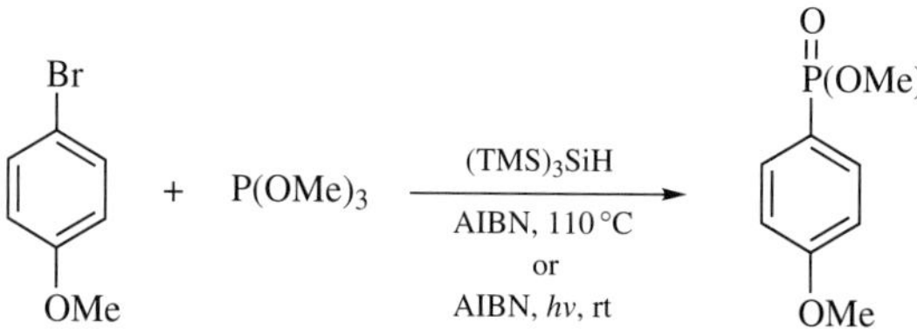

$$\tag{36}$$

90–93%

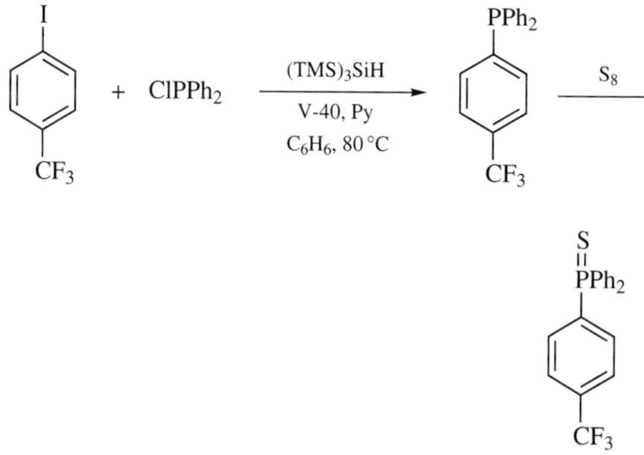

$$\tag{37}$$

78%

Tandem and Cascade Radical Reactions. An efficient carbonylation procedure can be achieved by a three-component coupling reaction mediated by $(TMS)_3SiH$ (eqs 38 and 39). It proceeds by the addition of an alkyl- or vinyl radical onto carbon monoxide with formation of an acyl radical intermediate, which can further react with electron deficient olefins to lead to the poly-functionalized compounds.[51]

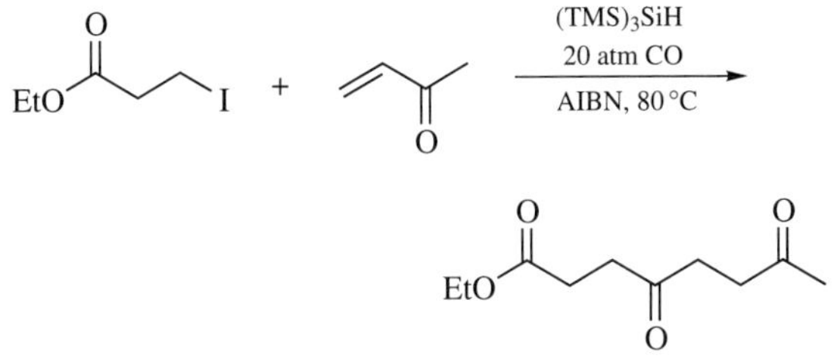

EtO + [structure] $\xrightarrow[\text{AIBN, 80 °C}]{\substack{(TMS)_3SiH \\ 20\ atm\ CO}}$

[structure]   (38)

45%

[structure] + [structure]CN $\xrightarrow[\text{AIBN, 80 °C}]{\substack{(TMS)_3SiH \\ 20\ atm\ CO}}$

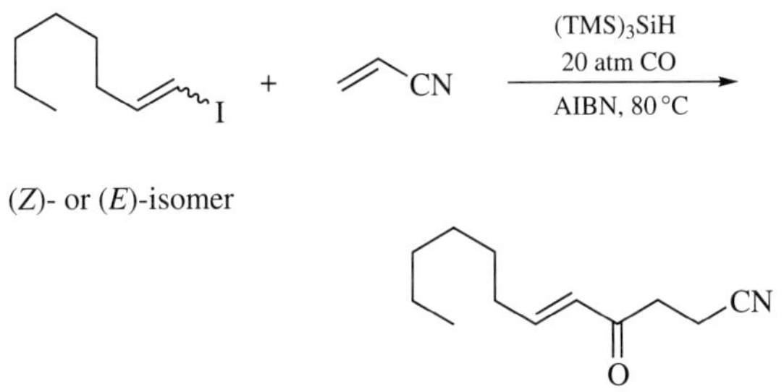

(Z)- or (E)-isomer

[structure]   (39)

(E)-isomer, 50%

The field of alkaloid synthesis via tandem cyclizations has favored the application of $(TMS)_3SiH$ over other radical-based reagents, due to its very low toxicity and chemoselectivity. For example, cyclization of the iodo aryl azide **12** mediated by $(TMS)_3$ SiH under standard experimental conditions, produced the N-Si(TMS)₃ protected alkaloid **13**, which after washing with dilute acid afforded the amine **14** in an overall 83% yield from **12** (eq 40).[52] The formation of the labile N-Si(TMS)₃ bond was thought to arise from the reaction of the product amine **14** with the by-product (TMS)₃SiI. The skeletons of (±)-horsfiline, (±)-aspidospermidine and (±)-vindoline have been achieved by this route.[52]

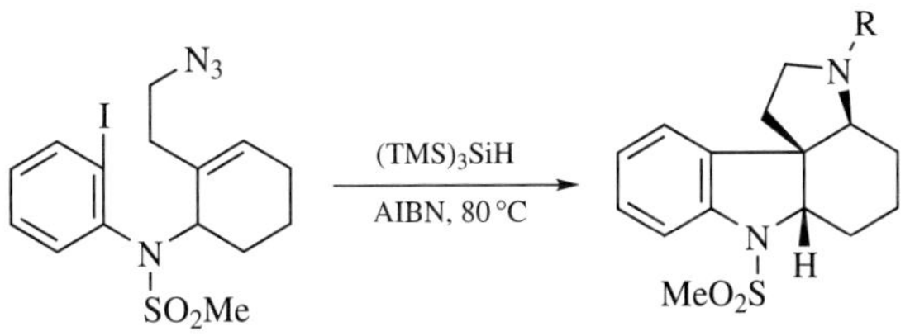

[structure]

12

$\xrightarrow[\text{AIBN, 80 °C}]{(TMS)_3SiH}$

[structure]   (40)

13 R = Si(TMS)₃
14 R = H, 83%

Another effective radical cascade strategy started from bro-momethyldimethylsilyl propargyl ethers.[53] The synthesis of functionalized cyclopentanone precursor **15** is achieved as a single diastereomer, starting from the reduction of the bromo derivative in the presence of $(TMS)_3SiH$ (eq 41). When differ-ent substituents are used in the skeleton, as in compound **16**, a completely different reaction pattern results (eq 42).

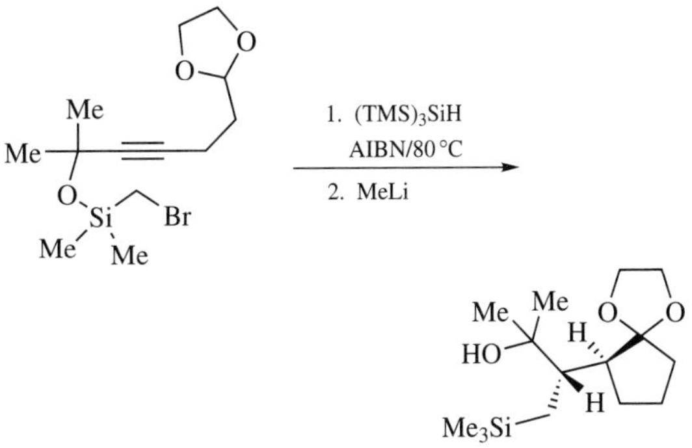

[structure]

$\xrightarrow{\substack{1.\ (TMS)_3SiH \\ AIBN/80\ °C \\ 2.\ MeLi}}$

[structure]   (41)

15, 92%

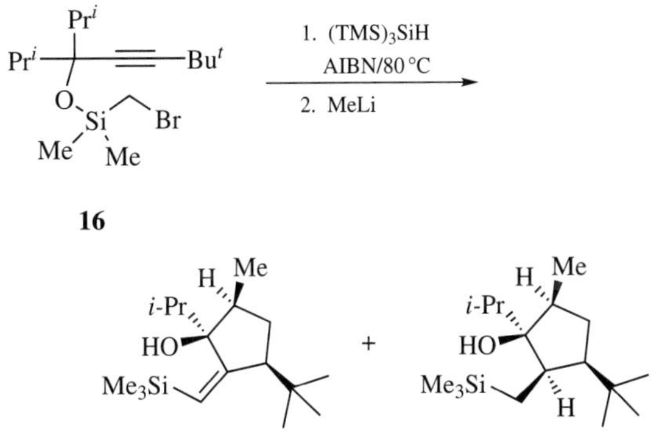

[structure]

16

$\xrightarrow{\substack{1.\ (TMS)_3SiH \\ AIBN/80\ °C \\ 2.\ MeLi}}$

[structure] + [structure]   (42)

17   **18**

95%, **17**:**18** = 90:10

An interesting radical carboxyarylation approach used in a rad-ical cascade represents the key step in the total synthesis of sev-eral biologically important natural products.[54] The thiocarbonate derivatives **19** (R = Me or TBS) react with 1.1 equiv of $(TMS)_3SiH$ in refluxing benzene and in the presence of AIBN (0.4 equiv added over 6 h) as radical initiator to produce compound **20** in 44% yield (eq 43).

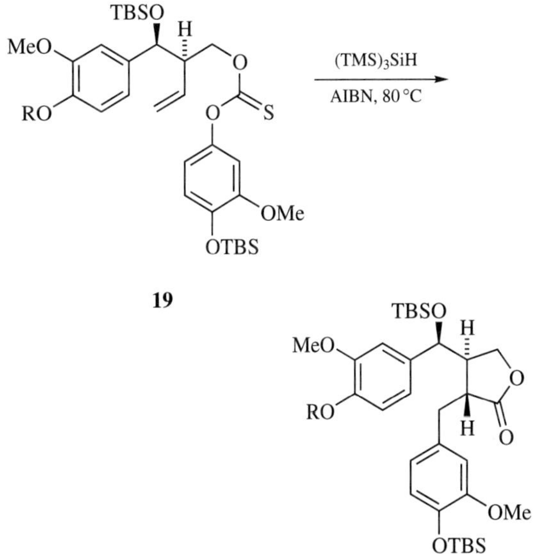

[structure]

19

$\xrightarrow[\text{AIBN, 80 °C}]{(TMS)_3SiH}$

[structure]   (43)

20, 44%

1. Chatgilialoglu, C., *Acc. Chem. Res.* **1992**, *25*, 188.

2. Ballestri, M.; Chatgilialoglu, C.; Clark, K. B.; Griller, D.; Giese, B.; Kopping, B., *J. Org. Chem.* **1991**, *56*, 678.

3. Chatgilialoglu, C.; Dickhaut, J.; Giese, B., *J. Org. Chem.* **1991**, *56*, 6399.

4. Dickhaut, J.; Giese, B., *Org. Synth.* **1991**, *70*, 164.

5. Chatgilialoglu, C.; Guarini, A.; Guerrini, A.; Seconi, G., *J. Org. Chem.* **1992**, *57*, 2207.

6. Schummer, D.; Höfle, G., *Synlett* **1990**, 705.

7. Ballestri, M.; Chatgilialoglu, C.; Cardi, N.; Sommazzi, A., *Tetrahedron Lett.* **1992**, *33*, 1787.

8. Lesage, M.; Chatgilialoglu, C.; Griller, D., *Tetrahedron Lett.* **1989**, *30*, 2733.

9. Chatgilialoglu, C.; Griller, D.; Lesage, M., *J. Org. Chem.* **1988**, *53*, 3641.

10. Barton, D. H. R.; Jang, D. O.; Jaszberenyi, J. C., *Tetrahedron Lett.* **1992**, *33*, 6629.

11. Johnson, D. W.; Poulos, A., *Tetrahedron Lett.* **1992**, *33*, 2045.

12. Kulicke, K. J.; Giese, B., *Synlett* **1990**, 91.

13. Kopping, B.; Chatgilialoglu, C.; Zehnder, M.; Giese, B., *J. Org. Chem.* **1992**, *57*, 3994.

14. Smadja, W.; Zahouily, M.; Malacria, M., *Tetrahedron Lett.* **1992**, *33*, 5511.

15. Giese, B.; Damm, W.; Dickhaut, J.; Wetterich, F.; Sun, S.; Curran, D. P., *Tetrahedron Lett.* **1991**, *32*, 6097.

16. Giese, B.; Kopping, B.; Chatgilialoglu, C., *Tetrahedron Lett.* **1989**, *30*, 681.

17. Chatgilialoglu, C.; Giese, B.; Kopping, B., *Tetrahedron Lett.* **1990**, *31*, 6013.

18. Arya, P.; Samson, C.; Lesage, M.; Griller, D., *J. Org. Chem.* **1990**, *55*, 6248.

19. Arya, P.; Lesage, M.; Wayner, D. D. M., *Tetrahedron Lett.* **1991**, *32*, 2853. Arya, P.; Wayner, D. D. M., *Tetrahedron Lett.* **1991**, *32*, 6265.

20. Schiesser, C. H.; Sutej, K., *Tetrahedron Lett.* **1992**, *33*, 5137.

21. Kulicke, K. J.; Chatgilialoglu, C.; Kopping, B.; Giese, B., *Helv. Chim. Acta* **1992**, *75*, 935.

22. Künzer, H.; Sauer, G.; Wiechert, R., *Tetrahedron Lett.* **1991**, *32*, 7247.

23. Mayr, H.; Basso, N.; Hagen, G., *J. Am. Chem. Soc.* **1992**, *114*, 3060.

24. Watanabe, H.; Nakano, T.; Araki, K.-I.; Matsumoto, H.; Nagai, Y., *J. Organomet. Chem.* **1974**, *69*, 389.

25. Chatgilialoglu, C. *Organosilanes in Radical Chemistry*; Wiley: Chichester, UK, 2004.

26. Chatgilialoglu, C.; Gimisis, T., *Chem. Commun.* **1998**, 1249.

27. Chatgilialoglu, C.; Costantino, C.; Ferreri, C.; Gimisis, T.; Romagnoli, A.; Romeo, R., *Nucleosides Nucleotides* **1999**, *18*, 637.

28. Oba, M.; Nishiyama, K., *Tetrahedron* **1994**, *50*, 10193.

29. Hammerschmidt, F.; Öhler, E.; Polsterer, J.-P.; Zbiral, E.; Balzarini, J.; DeClercq, E., *Liebigs Ann.* **1995**, 551.

30. Postigo, A.; Ferreri, C.; Navacchia, M. L.; Chatgilialoglu, C., *Synlett* **2005**, 2854.

31. Yamaguchi, K.; Kazuta, Y.; Abe, H.; Matsuda, A.; Shuto, S., *J. Org. Chem.* **2003**, *68*, 9255.

32. García de Viedma, A.; Martínez-Barrasa, V.; Burgos, C.; Luisa Izquiedro, M.; Alvarez-Builla, J., *J. Org. Chem.* **1999**, *64*, 1007.

33. Romeo, R.; Wozniak, L. A.; Chatgilialoglu, C., *Tetrahedron Lett.* **2000**, *41*, 9899.

34. Chatgilialoglu, C.; Ballestri, M.; Ferreri, C.; Vecchi, D., *J. Org. Chem.* **1995**, *60*, 3826.

35. Miura, K.; Oshima, K.; Utimoto, K., *Bull. Chem. Soc. Jpn.* **1993**, *66*, 2356.

36. Liu, Y.; Yamazaki, S.; Yamabe, S., *J. Org. Chem.* **2005**, *70*, 556.

37. Usui, S.; Paquette, L. A., *Tetrahedron Lett.* **1999**, *40*, 3495.

38. Paquette, L. A.; Usui, S., *Tetrahedron Lett.* **1999**, *40*, 3499.

39. Lee, E.; Choi, S. J., *Org. Lett.* **1999**, *1*, 1127.

40. Evans, P. A.; Murthy, V. S.; Roseman, J. D.; Rheingold, A. L., *Angew. Chem., Int. Ed.* **1999**, *38*, 3175.

41. Evans, P. A.; Roseman, J. D.; Garber, L. T., *J. Org. Chem.* **1996**, *61*, 4880.

42. Sasaki, K.; Kondo, Y.; Maruoka, K., *Angew. Chem., Int. Ed.* **2001**, *40*, 411.

43. Ooi, T.; Hokke, Y.; Maruoka, K., *Angew. Chem., Int. Ed. Engl.* **1997**, *36*, 1181.

44. Knapp, S.; Gibson, F. S., *J. Org. Chem.* **1992**, *57*, 4802.

45. Kim, S., *Pure Appl. Chem.* **1996**, *68*, 623.

46. Urabe, H.; Kobayashi, K.; Sato, F., *J. Chem. Soc. Chem. Commun.* **1995**, 1043.

47. Bennasar, M.-L.; Roca, T.; Griera, R.; Bassa, M.; Bosch, J., *J. Org. Chem.* **2002**, *67*, 6268.

48. Braslau, R.; Tsimelzon, A.; Gewandter, J., *Org. Lett.* **2004**, *6*, 2233.

49. Jiao, X.-Y.; Bentrude, W. G., *J. Org. Chem.* **2003**, *68*, 3303.

50. Sato, A.; Yorimitsu, H.; Oshima, K., *J. Am. Chem. Soc.* **2006**, *128*, 4240.

51. Ryu, I.; Sonoda, N., *Angew. Chem., Int. Ed. Engl.* **1996**, *35*, 1050.

52. Kizil, M.; Patro, B.; Callaghan, O.; Murphy, J. A.; Hursthouse, M. B.; Hibbs, D., *J. Org. Chem.* **1999**, *64*, 7856. Zhou, S.; Bommezijn, S.; Murphy, J. A., *Org. Lett.* **2002**, *4*, 443. Lizos, D. E.; Murphy, J. A., *Org. Biomol. Chem.* **2003**, *1*, 117.

53. Bogen, S.; Journet, M.; Malacria, M., *Synlett* **1994**, 958. Bogen, S.; Malacria, M., *J. Am. Chem. Soc.* **1996**, *118*, 3992. Bogen, S.; Gulea, M.; Fensterbank, L.; Malacria, M., *J. Org. Chem.* **1999**, *64*, 4920.

54. Reynolds, A. J.; Scott, A. J.; Turner, C. I.; Sherburn, M. S., *J. Am. Chem. Soc.* **2003**, *125*, 12108. Fischer, J.; Reynolds, A. J.; Sharp, L. A.; Sherburn, M. S., *Org. Lett.* **2004**, *6*, 1345.

Vinyltrimethylsilane[1]

[754-05-2] C$_5$H$_{12}$Si (MW 100.26)

InChI = 1S/C5H12Si/c1-5-6(2,3)4/h5H,1H2,2-4H3

InChIKey = GCSJLQSCSDMKTP-UHFFFAOYSA-N

(ethylene equivalent in electrophilic substitution reactions; precursor to 3-trimethylsilyl-3-buten-2-one, a methyl vinyl ketone surrogate for Robinson annulations; homologation of aldehydes to α,β-unsaturated aldehydes)

Physical Data: bp 55–57 °C; d_4^{20} 0.691 g cm^{-3}; n_D^{20} 1.391.

Solubility: sol most common organic solvents (THF, Et$_2$O, benzene, CH$_2$Cl$_2$, etc.)

Form Supplied in: commercially available from several suppliers.

Analysis of Reagent Purity: characterized by ^{1}H, ^{13}C, and ^{29}Si NMR,[2] and IR.[3] ^{1}H NMR (500 MHz) (CDCl$_3$): δ 0.08 (9H, s), 5.67 (1H, dd, J_{trans} = 20.3 MHz, J_{gem} = 3.8 Hz), 5.93 (1H, dd, J_{cis} = 14.7 Hz, J_{gem} = 3.8 Hz), 6.17 (1H, dd, J_{trans} = 20.3 Hz, J_{cis} = 14.7 Hz) ppm; ^{13}C NMR (125 MHz) (CDCl$_3$): δ −1.62, 130.88, 140.27 ppm; ^{29}Si NMR (19.9 MHz) ((MeO)$_4$Si): δ − 71.70 ppm; IR (CCl$_4$): ν 1594 cm^{-1}.

Preparative Method: prepared in 67–91% yield from vinylmagnesium bromide and chlorotrimethylsilane in THF.[4,5]

Purification: fractional distillation at 1 atm using an efficient Vigreux column. Some difficulty in removing trace amounts of THF (bp 67 °C) has been reported.

Handling, Storage, and Precautions: highly flammable; flash point −34 °C; hygroscopic. The reagent should be used in a well ventilated hood. Contact with the eyes and skin should be avoided.

Original Commentary

Glenn J. Fegley

Indiana University, Bloomington, IN, USA

Synthesis of Vinyl Aryl Sulfides. Vinyltrimethylsilane functions as an ethylene equivalent in electrophilic substitution reactions with sulfenyl chlorides (eq 1).[6] Reaction with various arylsulfenyl chlorides provides the stable adducts (**1**) in high yield. Treatment of (**1**) with fluoride ion furnishes vinyl aryl sulfides (**2**) in high yield.

Synthesis of 3-Triethylsilyl-3-buten-2-one. Vinyltriethylsilane can be elaborated in five steps to yield 3-triethylsilyl-3-buten-2-one (**4**) (eq 2), a methyl vinyl ketone surrogate useful in Robinson annulation reactions (eq 3).[7] The analogous annulation reaction using methyl vinyl ketone gives poor yields.

$$\text{TMS} \xrightarrow[\substack{CH_2Cl_2 \\ 87–95\%}]{ArSCl} \underset{\substack{\quad \\ Cl}}{\overset{ArS\quad TMS}{\diagup\diagdown}} \xrightarrow[\substack{KF\cdot 2H_2O,\ DMSO \\ 90–100\%}]{KF,\ THF\ or} \overset{ArS}{\diagup} \quad (1)$$

$$(\mathbf{1}) \qquad\qquad (\mathbf{2})$$

Ar = 2,4-dinitrophenyl, 2-nitrophenyl, phenyl, 4-chlorophenyl, 4-methylphenyl

$$\text{SiEt}_3 \xrightarrow[\substack{\text{2. Et}_2\text{NH}}]{\text{1. Br}_2} \underset{\text{SiEt}_3}{\overset{\text{Br}}{\diagup\diagdown}} \xrightarrow[\substack{\text{3. Jones}\\ \text{oxidation}}]{\substack{\text{1. Mg}^0 \\ \text{2. MeCHO}}} \overset{O}{\underset{\text{SiEt}_3}{\diagdown}} \quad (2)$$

$$(\mathbf{3}) \qquad\qquad (\mathbf{4})$$

$$\underset{\text{SiOEt}_3}{\overset{O}{\diagdown}} + \underset{}{\overset{\text{OLi}}{\bigcirc}} \xrightarrow[\text{MeOH}]{5\%\ \text{NaOMe}} \underset{O}{\overset{}{\bigcirc\bigcirc}} \quad (3)$$

$$(\mathbf{4}) \qquad\qquad\qquad \text{64\% from (3)}$$

Synthesis of α,β-Unsaturated Primary Amides. Vinylsilanes react with chlorosulfonyl isocyanate without the aid of Lewis acid catalysts to form β-lactams, e.g. (**5**), which subsequently undergo hydrolysis to the corresponding *trans*-α,β-unsaturated primary amides, e.g. (**6**) (eq 4).[8]

$$\text{Ph}\diagup\diagdown\text{TMS} + \underset{Cl}{\overset{O\ \ O}{\diagup S\diagdown}}\text{NCO} \xrightarrow{92\%} \left[\underset{\substack{N \\ SO_2Cl}}{Ph\overset{TMS}{\square}O} \right] \xrightarrow[63\%]{HCl}$$

$$(\mathbf{5})$$

$$\text{Ph}\diagup\diagdown\text{CONH}_2 \quad (4)$$

$$(\mathbf{6})$$

Synthesis of α,β-Unsaturated Aldehydes. Vinyltrimethylsilane undergoes [3 + 2] cycloaddition reactions with the aldehyde-derived nitrone (**7**) to provide the corresponding (trimethylsilyl)isoxazolidine adduct (**8**) in 84% yield (eq 5).[9] Treatment of (**8**) with aqueous HF furnishes the α,β-unsaturated aldehyde (**9**) in 95% yield. This methodology is general for the homologation of aldehydes and serves as an alternative to the traditional Wittig-type alkenations.

Synthesis of Bicyclopentenones. The title reagent serves as an ethylene equivalent in aliphatic Friedel–Crafts acylation reactions involving cyclic α,β-unsaturated acid chlorides. For example, the acid chloride (**10**) reacts with vinyltrimethylsilane in the presence of tin(IV) chloride to give the divinyl ketone intermediate (**11**), which then undergoes a Nazarov cyclization, thus producing the bicyclopentenone (**12**) in 46% overall yield (eq 6).[6,10]

Avoid Skin Contact with All Reagents

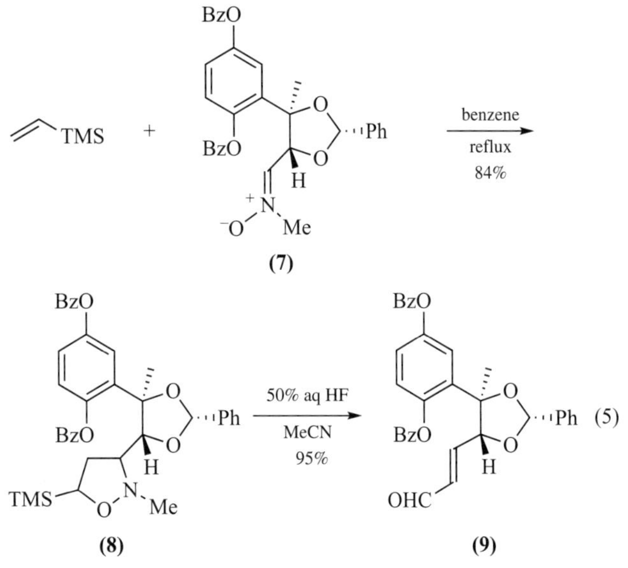

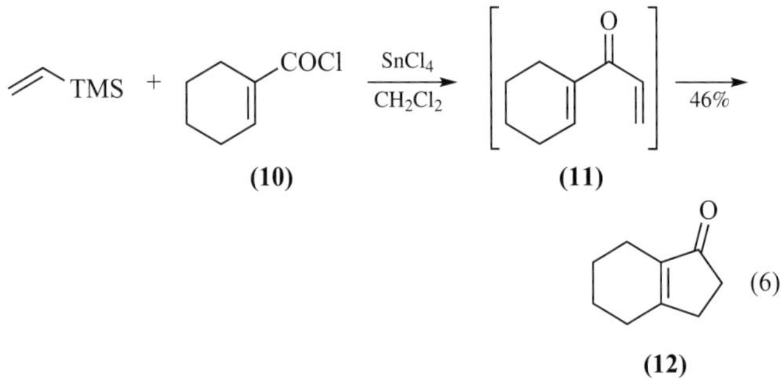

Synthesis of 1-Chlorocyclopropene. Vinyltrimethylsilane serves as a useful precursor to 1-chlorocyclopropene (**14**).[13] Reaction of the vinylsilane with phenyl(trichloromethyl)mercury[12] in refluxing benzene followed by fluoride-induced elimination of TMSCl from the cyclopropane (**13**) furnishes 1-chlorocyclopropene in good yield, as evidenced in the subsequent Diels–Alder reaction with 1,3-diphenylisobenzylfuran (eq 7). A previous synthesis[13] of 1-chlorocyclopropene gave only 5–10% yield.

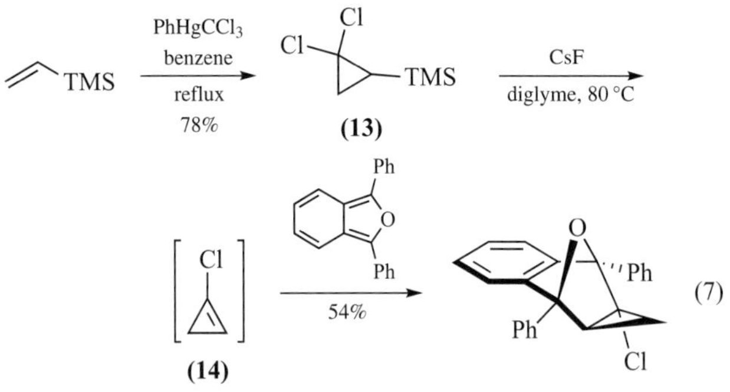

Radical Addition Reactions of Vinyltrimethylsilane. Reaction of the title reagent with benzenesulfonyl chloride under radical addition conditions[14] provides (*E*)-1-phenylsulfonyl-2-trimethylsilylethylene (**15**),[15] an acetylene equivalent in Diels–Alder reactions (eq 8). In similar fashion, thiophenol adds to vinyltrimethylsilane regioselectively.[16] Oxidation of the thiophe-nol adduct provides the sulfone (**16**) (eq 9), which serves as a latent exomethylene unit in the enantioselective synthesis of (1*R*)-[(methylenecyclopropyl)acetyl]-CoA (**17**).[17] The sulfone (**16**) is also used in the high yield preparation of allylsilanes, vinyl sulfones, and 2-(benzenesulfonyl)allyl alcohol derivatives.[18]

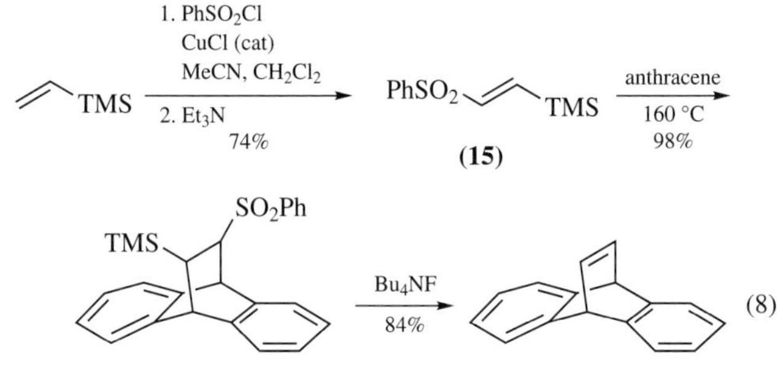

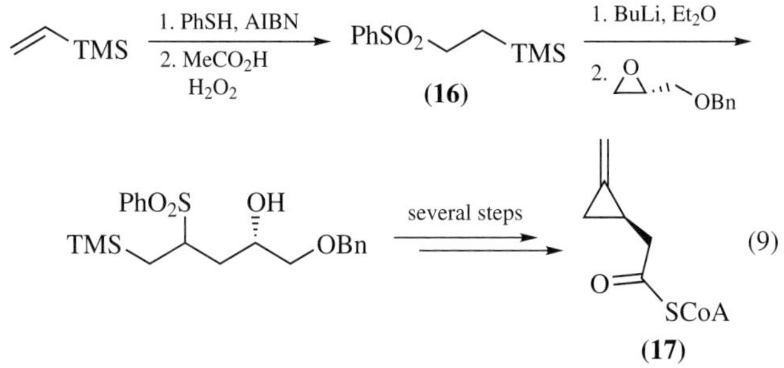

Reaction of Vinyltrimethylsilane with 2-Azaallylanions. Reaction of vinyltrimethylsilane with the nonstabilized 2-azaallyl anion (**19**), generated in situ from the (2-azaallyl)stannane (**18**) (eq 10), produces, after quenching with MeI, the pyrrolidine (**20**) as a single diastereomer in 77% yield (eq 11).[21] Some evidence suggests that the cycloaddition is stepwise and that the 'W-conformation' of the anion predominates in the cycloaddition sequence.

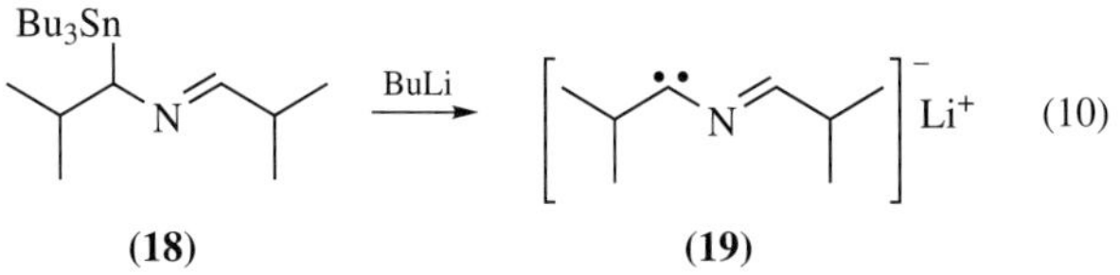

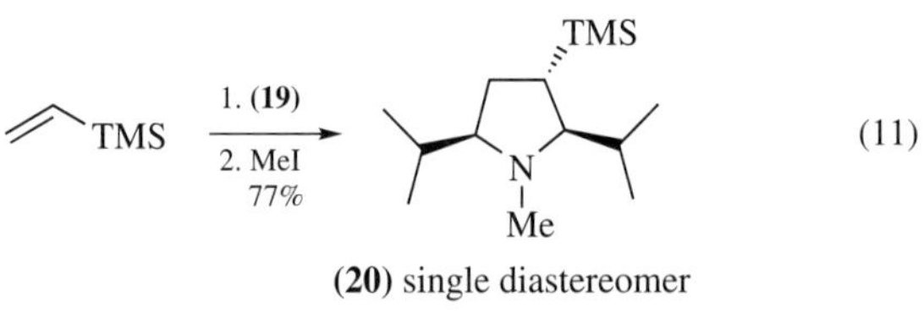

Regioselective Hydroesterification of Vinyltrimethylsilane. Vinyltrimethylsilane undergoes highly regioselective hydroesterification reactions to furnish either ethyl 3-(trimethylsilyl)-propionate (**21**) or ethyl 2-(trimethylsilyl)propionate (**22**), depending on the choice of catalyst and reaction conditions (eq 12).[20] Highly regioselective (>96% *β*-selective) hydroformylations of vinylsilanes containing bulky alkyl and aryl substituents on silicon have been achieved using carbonylhydridotris (triphenylphosphine)rhodium(I) as catalyst.[21]

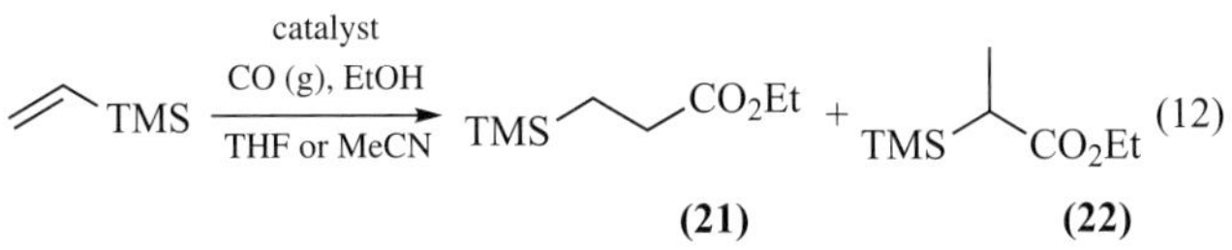

$$\text{(12)}$$

Catalyst	Yield (21)/(22)	Ratio (21):(22)	$P_{initial}(CO)$
PdCl$_2$(PPh$_3$)$_2$ (1–2 mol %)	94%	98:2	60 kg cm^{-2}
Co(CO)$_8$ (2–4 mol %)	74%	1:99	70 kg cm^{-2}

First Update

Gerald L. Larson

Gelest, Inc., Morrisville, PA, USA

3 + 2 Cycloaddition of Vinyltrimethylsilane with Nitrones.

Vinyltrimethylsilane reacts with nitrones to provide a 5-(trimethylsilyl)isoxazolidine in good yield as illustrated with the nitrone from the protected galactose (23) (eq 12). Treatment of the silylated isoxazolidine, (24) with acetyl chloride results in cleavage to the β-amino aldehyde with loss of the trimethylsilyl group (eq 13).[22] This reaction sequence was successfully applied to the synthesis of C$_7$ and C$_8$ aminodialdoses.

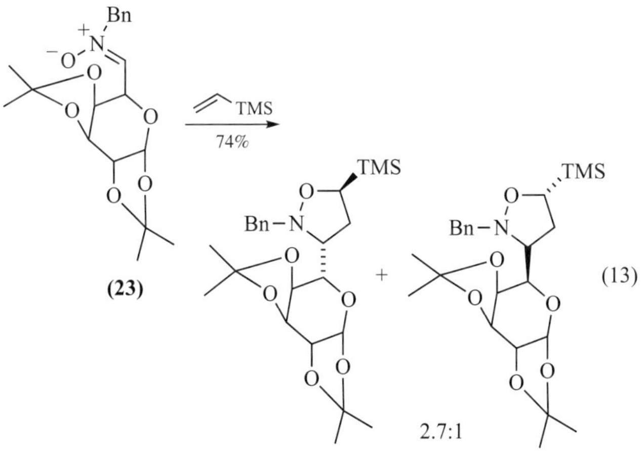

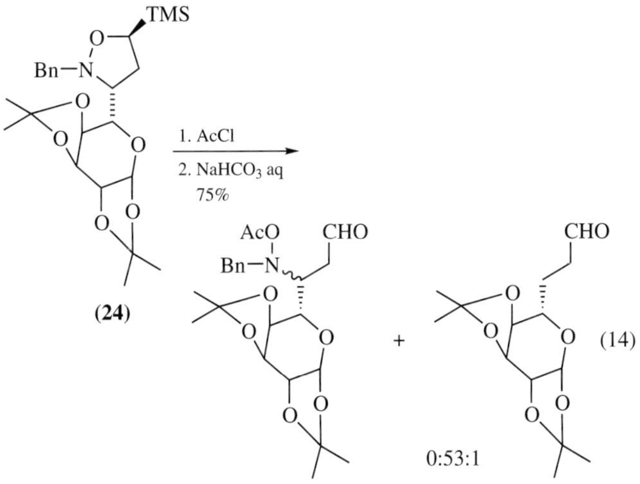

$$\text{(13)}$$

2.7:1

$$\text{(14)}$$

0:53:1

Addition of α-Iodo-α,α-Difluoroketones to Vinyltrimethylsilane.

The addition of α-iodo-α,α-difluoroketones to vinyltrimethylsilane occurs with catalysis by copper metal, palladium(0), or UV.[23,24] The resulting α-iodosilane reacts with ammonium hydroxide to form the 3,3-difluoro-5-trimethylsilyl-1-pyrrolines, which upon treatment with potassium fluoride gives the corresponding 2-substituted-3-fluoropyrrole (eq 15). This use of vinyltrimethylsilane adds to the standard approach, which only works with electron poor olefins.

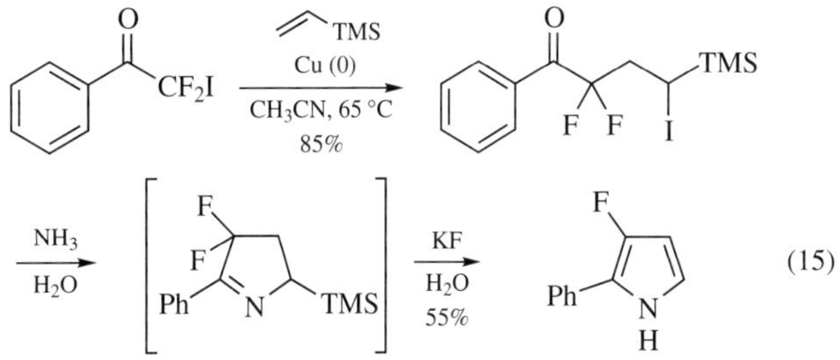

$$\text{(15)}$$

Addition of α-Iodosulfones to Vinyltrimethylsilane.

The addition of an α-iodosulfone to vinyltrimethylsilane yields a γ-trimethylsilyl-γ-iodosulfone, which upon treatment with the strong base, sodium bis(trimethylsilylamide), gives the β-trimethylsilylcyclopropyl sulfone. Elimination with n-butyllithium provides the 1-trimethylsilylcyclopropene (eq 16).[25]

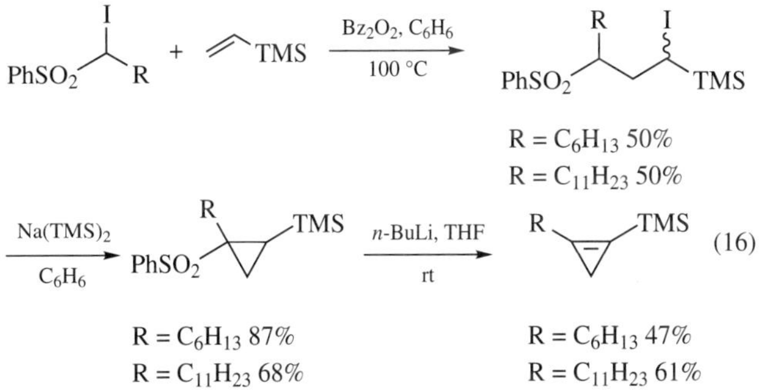

R = C$_6$H$_{13}$ 50%
R = C$_{11}$H$_{23}$ 50%

$$\text{(16)}$$

R = C$_6$H$_{13}$ 87%
R = C$_{11}$H$_{23}$ 68%

R = C$_6$H$_{13}$ 47%
R = C$_{11}$H$_{23}$ 61%

Titanium-mediated Formation of Trimethylsilylcyclopropanols, β-Trimethylsilyl Ketones and Cyclopropenes.

Vinyltrimethylsilane reacts with chlorotriisopropoxytitanium in the presence of isopropylmagnesium chloride to form the interesting titanium(II) titanacycle (25) (eq 17), which reacts with esters to form β-trimethylsilylcyclopropanols (eq 18). These can be converted to the β-trimethylsilyl ketone (eq 19) or the cyclopropene (eq 20).[26]

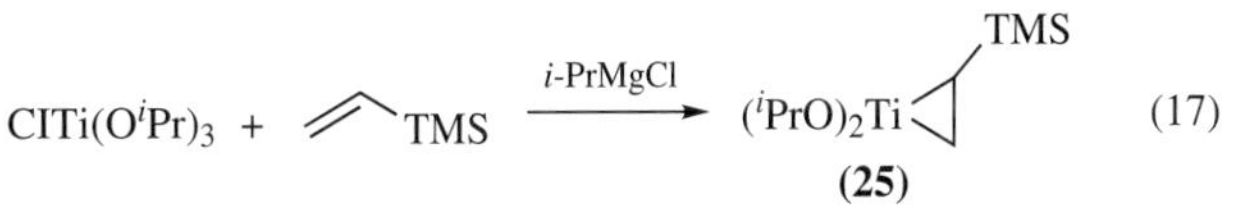

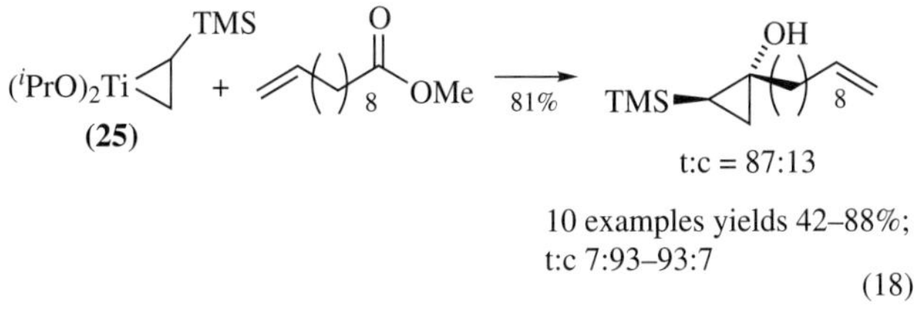

$$\text{(17)}$$

t:c = 87:13

10 examples yields 42–88%;
t:c 7:93–93:7

$$\text{(18)}$$

(19)

(20)

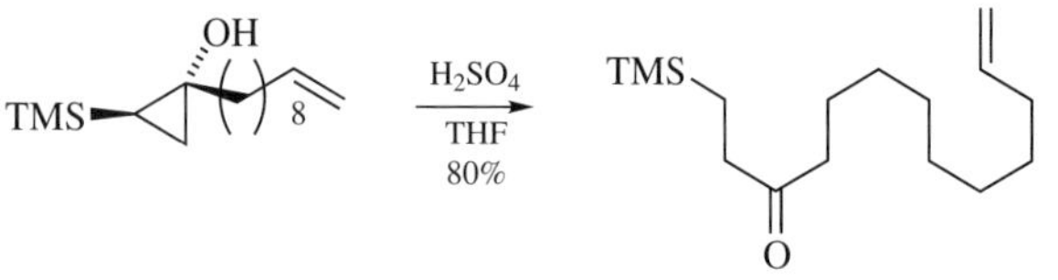

Hydrogen Acceptor Catalyst in the Conversion of Alcohols to Hydrogenated Wittig Adducts. The ruthenium carbene catalyst (**26**) readily loses hydrogen to vinyltrimethylsilane to form the complex (**27**), which in turn readily accepts hydrogen from a primary alcohol reforming (**26**) and an aldehyde. When this is carried out in the presence of a Wittig reagent, the aldehyde formed is converted to the unsaturated ester, which is then reduced to the saturated ester by (**26**) resulting in the overall conversion of a primary alcohol to a chain-extended ester (eq 21).[27]

(**26**) (**27**)

(21)

R = Bn; 80%
R = Me; 78%
R = *t*-Bu: 70%

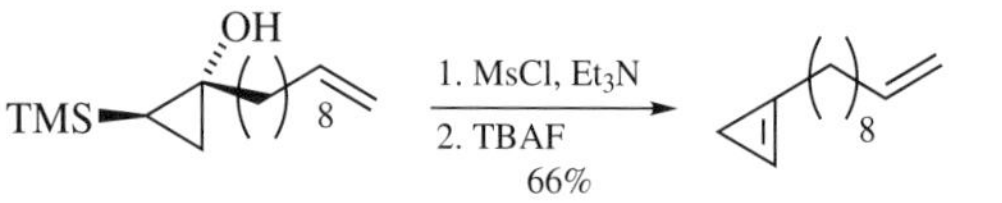

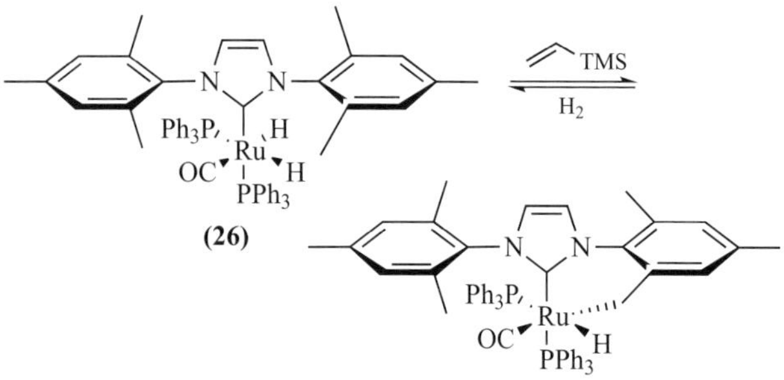

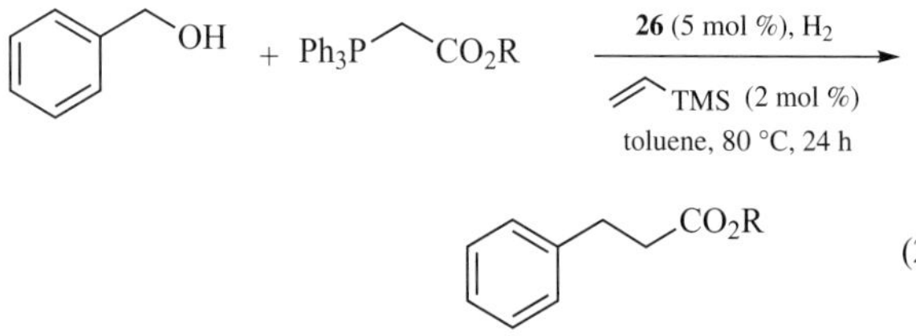

Formation of 2-Trimethylsilylaziridines. The photochemical or thermal reaction of vinyltrimethylsilane with ethyl azidoformate or phenyl azide provides the trimethylsilylaziridine in modest yields (eq 22). Reduction of the formyl amide gives the parent trimethylsilylaziridine (eq 23).[28] Trimethylsilylaziridines were also formed from the reaction of α-lithiochloromethyltrimethylsilane with imines, but this reaction only works with the imines of aldehydes. The reaction of bromotriazide, formed in situ from the reaction of bromine with sodium azide, with vinylsilanes forms silylaziridines as well. This reaction was not carried out with vinyltrimethylsilane. However, vinyltriphenylsilane gave a 50% yield of 2-triphenylsilylaziridine under these conditions.

(22)

R = Me; 25%
R = Et; 64%

(23)

UV; 50%
55 °C; 61%

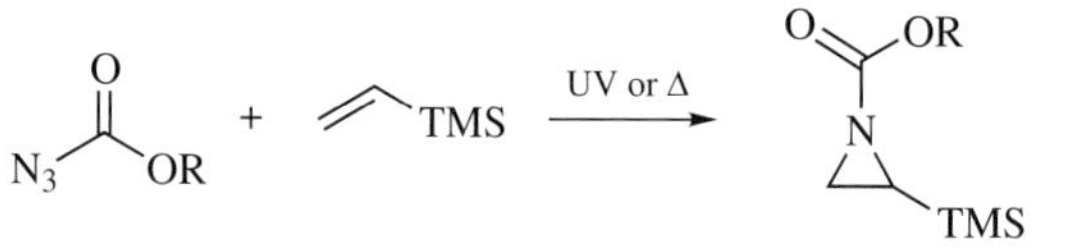

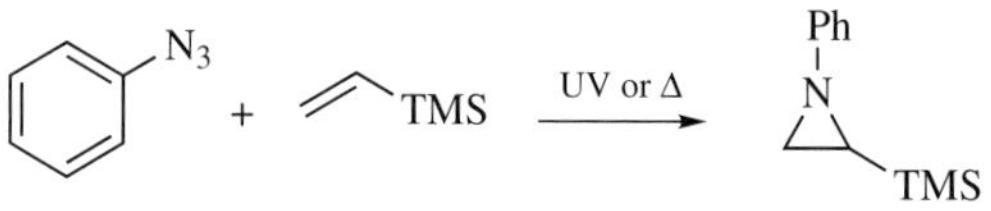

Formation of Sulfonyl Chlorides. Vinyltrimethylsilane undergoes the free radical addition of thiols to form the 2-thioethyltrimethylsilane. Oxidation to the sulfoxide and reaction with sulfuryl chloride provides 2-chloroethyltrimethylsilane and the sulfonyl chloride in good yield (eq 24).[29] The reaction sequence works well for the preparation of alkyl, aryl, and unsaturated sulfonyl chlorides.

12 examples; 80–98% 12 examples; 59–100% (24)

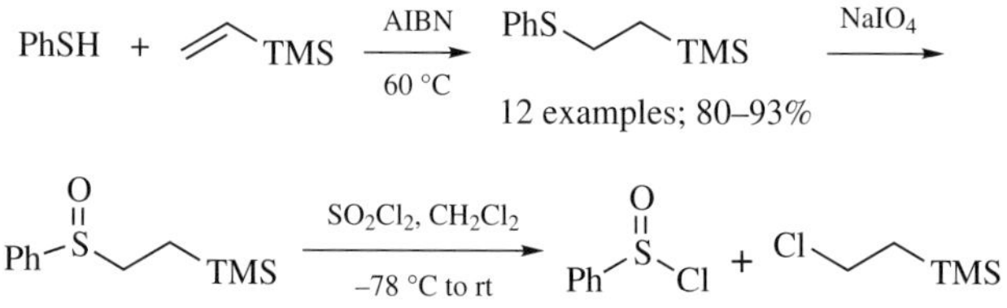

Improved Synthesis of 2-Trimethylsilylethylsulfonyl Chloride. Vinyltrimethylsilane is converted in high yield in a straightforward fashion to 2-trimethylsilylethylsulfonyl chloride, an excellent organosilyl group for the protection of primary amines and in the formation of 2-trimethylsilylsulfonamides. The use of a 10-fold amount of DMF catalyst over that previously employed[30] increases the yield by more than 10% (eq 25).[31]

(25)

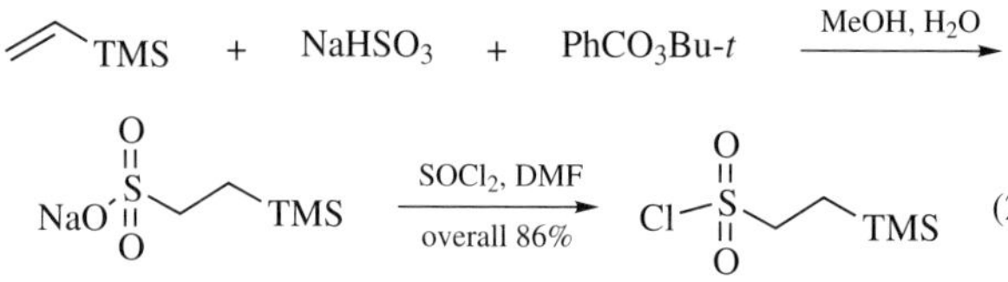

Formation of an Iron Carbonyl Trienone Complex. Vinyltrimethylsilane can be used in the well-known conversion of acid chlorides to enones in the synthesis of the iron trienone complex, (**28**) (eq 26).[32]

Synthesis of 2-Phenyl-2-Trimethylsilylethanol. Vinyltrimethylsilane is converted to the epoxytrimethylsilane, which is converted to 2-phenyl-2-trimethylsilylethanol. This is used to protect carboxylic acids as the 2-trimethylsilyl ester. Deprotection is readily carried out under the influence of TBAF. The reagent is used to protect amino acids and peptides and in the automated synthesis of peptides and glycopeptides (eq 27).[33–35]

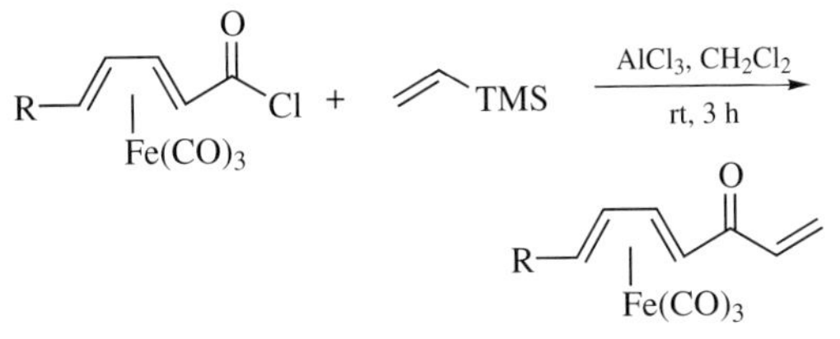

(26)

(28)

R = Me; 50%
R = Ph; 48%
R = C$_3$H$_7$; 28%

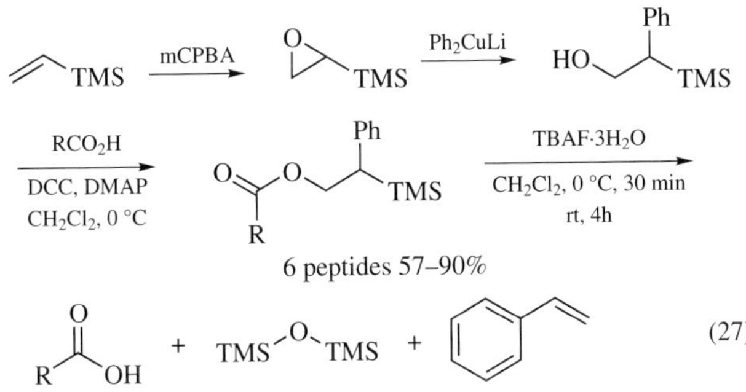

6 peptides 57–90%

(27)

Synthesis of 2-Vinylanilines. Vinyltrimethylsilane provides the vinyl group for a transformation of *N*-phenylsulfonimidoyl chloride to 2-alkenylanilines.[36] Thus, the reaction under the influence of a Lewis acid results in the formation of the trimethylsilylated cyclic benzothiazine (**29**) as a mixture of diastereoisomers accompanied by the desilylated analog (**30**) (eq 28). The silylated (**29**) can be alkylated and then the double bond introduced through a fluoride-induced, ring-opening elimination of the β-trimethylsilylsulfonamide group to give the 2-alkenylsulfonamide, which is converted to the aniline by standard protocol (eq 29).

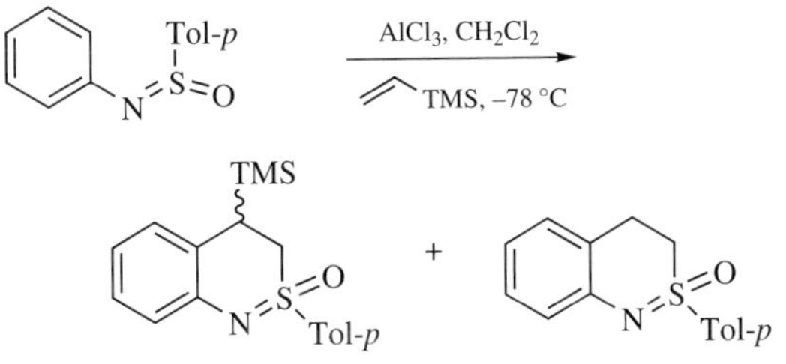

(28)

(29) 37%   **(30)** 30%

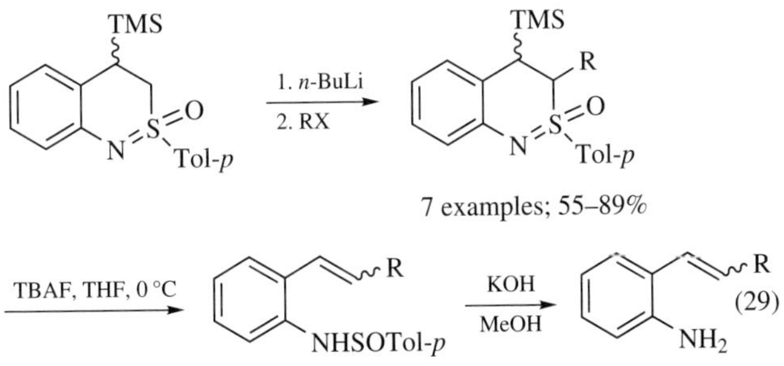

7 examples; 55–89%

(29)

Decarbonylative Vinylation of an Aromatic Ester. Vinyltrimethylsilane provides the vinyl group for the trimethylsilyl-influenced formation of a styrene from 4-nitrophenyl

4-cyanobenzoate, thus providing an alternative to the failure of ethylene to bring about this transformation (eq 30).[37]

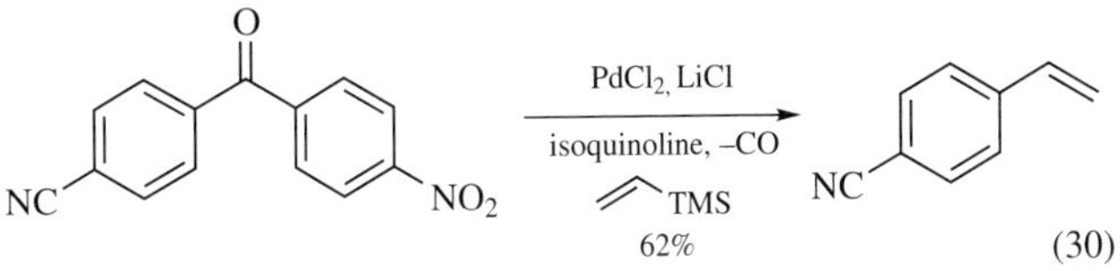

(30)

Asymmetric Epoxidation of Vinyltrimethylsilane. Vinyltrimethylsilane is converted to a predominant enantiomer of undetermined configuration by the ααββ-TAPP {ααββ-tetrakis(aminophenyl)porphyrin}-catalyzed epoxidation with iodosylbenzene (eq 31).[38] α,β-Epoxysilanes have been shown to have excellent synthetic utility.[39]

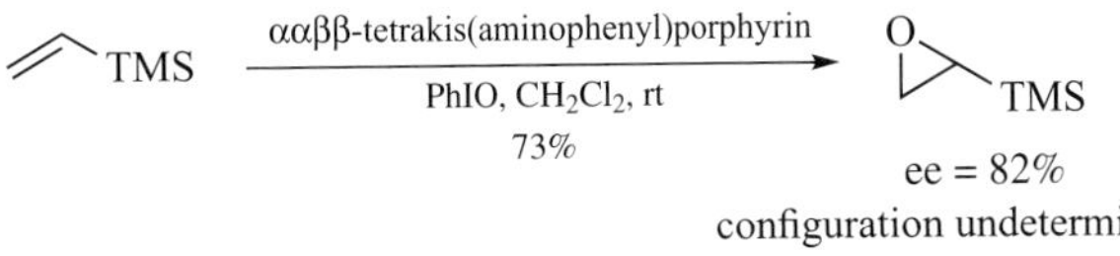

(31)

ee = 82%
configuration undetermined

Direct Silylation of Heteroarylcarbonyl Compounds. Under ruthenium catalysis vinyltrimethylsilane reacts to ortho silylate heteroaryl carbonyl compounds directly in good yields (eqs 32 and 33).[40] The reaction only works with heteroaromatic systems. The resulting aryltrimethylsilanes can be used to introduce electrophiles regioselectively through electrophilic desilylation. The reaction also works with vinyltriethoxysilane, opening the possibility of silicon-based cross-coupling reactions.

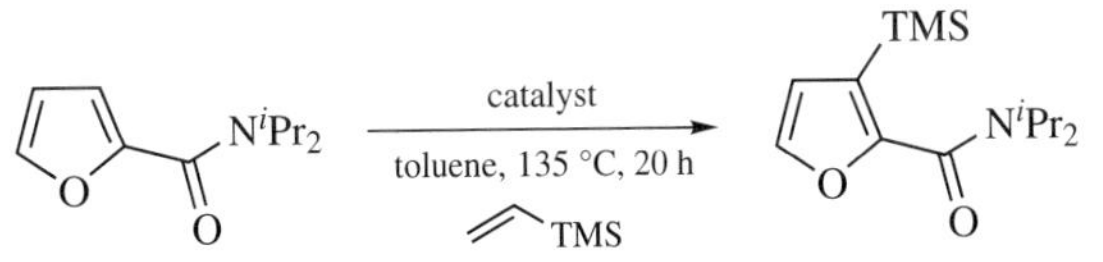

(32)

Ru$_3$(CO)$_{12}$, 88%
RuHCl(CO)(PPh$_3$)$_3$, 87%

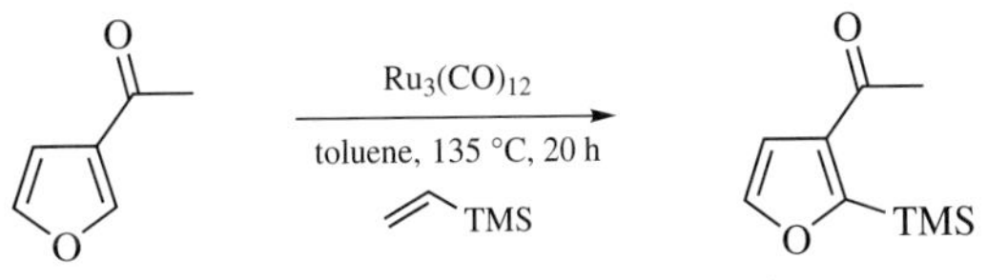

(33)

64%

Trimethylsilylation of Vinylboronates. Vinylboronate esters react with vinyltrimethylsilane under ruthenium catalysis to give the 1-trimethylsilyl-1-boronylethylene (eq 34). Such a species could be used in Suzuki-type cross coupling reactions to prepare various α-substituted vinyltrimethylsilanes.[41]

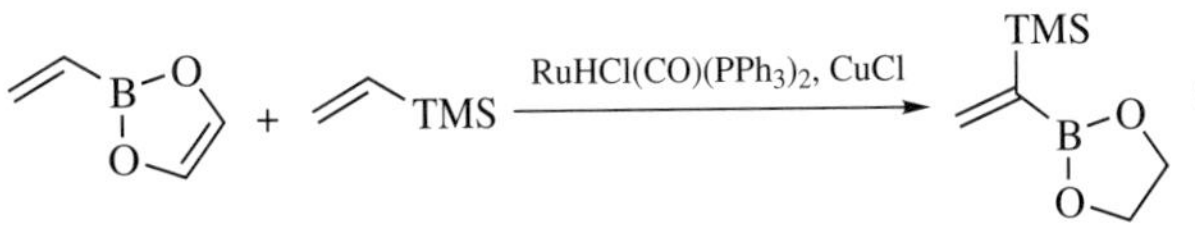

(34)

Preparation of Vinyldibromoborane. Vinyltrimethylsilane reacts with boron tribromide to give vinyldibromoborane and bromotrimethylsilane. The highly dienophilic vinyldibromoborane was not isolated but reacted directly with 2,3-dimethyl-1,3-butadiene and oxidized to give (31) in 74% overall yield (eq 35).[42]

$$BBr_3 + \diagup\!\!\!\!\diagup TMS \xrightarrow[\text{hexanes}]{75\,°C, 1\,h} \left[\diagup\!\!\!\!\diagup BBr_2 \right] \longrightarrow$$

$$\left[\text{BBr}_2 \right] \xrightarrow[\text{NaOH}]{H_2O_2 \atop 74\% \text{ overall}} \text{OH} \quad (35)$$

(31)

Trimethylsilylation of Alcohols. Vinyltrimethylsilane can be used to trimethylsilylate alcohols in good yields in the presence of Wilkinson's catalyst (eqs 36 and 37). Though an interesting reaction, it cannot compete with the numerous other existing methods for preparing trimethylsilyl ethers.[43]

$$\text{PhCH}_2\text{OH} \xrightarrow[\text{toluene, 100 °C, 2 h} \atop 93\%]{\diagup\!\!\!\!\diagup TMS \atop (PPh_3)_3RhCl} \text{PhCH}_2\text{OTMS} \quad (36)$$

$$\text{HO}\diagdown\!\!\!\diagup\text{OH} \atop \text{OH} \xrightarrow[\text{toluene, 100 °C, 2 h} \atop 88\%]{\diagup\!\!\!\!\diagup TMS \atop (PPh_3)_3RhCl} \text{TMSO}\diagdown\!\!\!\diagup\text{OTMS} \atop \text{OTMS} \quad (37)$$

β-Trimethylvinylation and Vinylation of Aryl Iodides. Vinyltrimethylsilane reacts with aryl iodides to either form *trans*-β-trimethylsilylstyrenes or styrenes (eqs 38 and 39).[44]

$$\text{PhI} + \diagup\!\!\!\!\diagup TMS \xrightarrow[\text{DMF, MS-4 Å, rt, 24 h} \atop 94\%]{\text{Bu}_4\text{NOAc, Pd(OAc)}_2} \text{Ph}\diagup\!\!\!\!\diagdown\!\!\!\!\diagup TMS \quad (38)$$
5 examples 85–94%

$$\text{PhI} + \diagup\!\!\!\!\diagup TMS \xrightarrow[\text{toluene, rt, 24 h} \atop 86\%]{\text{Bu}_4\text{NCl, Pd(dba)}_2, \text{KF}} \text{Ph}\diagup\!\!\!\!\diagdown\!\!\!\!\diagup TMS \quad (39)$$
5 examples 78–86%

Reaction of Vinyltrimethylsilane with 1,3-Butadienes. The cobalt-catalyzed reaction of vinyltrimethylsilane with 2,3-dimethyl-1,3-butadiene provides the 2-trimethylsilyl-1,4-butadiene (32) in excellent yield (eq 40).[45] The reaction with 2-methyl-1,3-butadiene gives both possible regioisomers in a ratio of 11.5:1 favoring reaction at the least substituted double bond of the diene (eq 41).

$$\diagup\!\!\!\diagdown + \diagup\!\!\!\!\diagup TMS \xrightarrow[\text{Bu}_4\text{NBH}_4, \text{CH}_2\text{Cl}_2, \text{rt} \atop 90\%]{\text{Co(dppe)Br}_2, \text{ZnI}_2} \diagup\!\!\!\!\diagup TMS \quad (40)$$

(32)

$$\diagup\!\!\!\diagdown + \diagup\!\!\!\!\diagup TMS \xrightarrow[\text{Bu}_4\text{NBH}_4, \text{CH}_2\text{Cl}_2, \text{rt} \atop 82\%]{\text{Co(dppe)Br}_2, \text{ZnI}_2}$$

$$\diagup\!\!\!\!\diagup TMS + \diagup\!\!\!\!\diagup TMS \quad (41)$$

Silylformylation of Vinyltrimethylsilane. The ethyldimethylsilylformylation of vinyltrimethylsilane leads to the interesting and potentially useful allylsilane (33), which is also the silyl enol ether of an acylsilane and a vinylsilane (eq 42).[46] In a similar fashion, the silylformylation employing phenyldimethylsilane gives a slightly higher yield and higher E to Z ratio along with the (Z)-silyl enol ether (34) (eq 43).[47] With the catalyst, [Ir(cod)(PCy₃)(OTMS)] (35) is formed exclusively in 92% yield in 24 h.

$$\diagup\!\!\!\!\diagup TMS \xrightarrow[\text{C}_6\text{H}_6, 140\,°C, 48\,h \atop 73\%]{\text{EtMe}_2\text{SiH, CO} \atop \text{IrCl(CO)}_3}$$

$$\text{TMS}\diagdown\!\!\!\diagup\overset{\text{OSiMe}_2\text{Et}}{\underset{\text{SiMe}_2\text{Et}}{|}} + \text{TMS}\diagdown\!\!\!\diagup\text{OSiMe}_2\text{Et} \quad (42)$$

(33) E:Z=73:27   (34)

$$\diagup\!\!\!\!\diagup TMS \xrightarrow[\text{80 °C, 60 h}]{\text{PhMe}_2\text{SiH, CO} \atop [\{\text{Ir}(\mu\text{-OTMS})(\text{cod})_2\}]}$$

$$\text{TMS}\diagdown\!\!\!\diagup\overset{\text{OSiMe}_2\text{Ph}}{\underset{\text{SiMe}_2\text{Ph}}{|}} + \text{TMS}\diagdown\!\!\!\diagup\text{OSiMe}_2\text{Ph} \quad (43)$$

82%; E:Z=93.7   (35) 14%

2-Trimethylsilylacetaldehyde. Hydroformylation of vinyltrimethylsilane under Rh(I) catalysis results in a 20% yield of a mixture of 2-trimethylsilylpropionaldehyde and 3-trimethylsilylpropionaldehyde (eq 44).[48]

$$\diagup\!\!\!\!\diagup TMS \xrightarrow[\text{100 °C, 1.5 h} \atop 20\%]{\text{Rh(cod)BPh}_4, \text{H}_2, \text{CO}} \underset{\text{TMS}}{\overset{O}{\diagup\!\!\!\!\diagdown}}H + \text{TMS}\diagdown\!\!\!\diagup\overset{O}{\diagdown}H \quad (44)$$

70:30

2-Trimethylsilylacetaldehyde Enol Ethers. The cross metathesis of vinyltrimethylsilane and enol ethers results in the formation of 2-trimethylsilyl enol ethers of acetaldehyde as a mixture of E and Z isomers in good yields (eqs 45 and 46).[49]

$$\diagup\!\!\!\!\diagup TMS + \diagup\!\!\!\!\diagup OEt \xrightarrow{\text{RhClH(CO)(PPh}_3)_2} \text{TMS}\diagdown\!\!\!\diagup\text{OEt} \quad (45)$$

$$\diagup\!\!\!\!\diagup TMS + \diagup\!\!\!\!\diagup OTMS \xrightarrow{\text{RhClH(CO)(PPh}_3)_2} \text{TMS}\diagdown\!\!\!\diagup\text{OTMS} \quad (46)$$

1. (a) Hudrlik, P. F. *New Applications of Organometallic Reagents in Organic Synthesis*; Seyferth, D., Ed.; Elsevier: Amsterdam, 1976; p 127. (b) Chan, T-H., *Acc. Chem. Res.* **1977**, *10*, 442. (c) Chan, T. H.; Fleming, I., *Synthesis* **1977**, 761. (d) Colvin, E. W., *Chem. Soc. Rev.* **1978**, *7*, 15. (e) Colvin, E. W. *Silicon in Organic Synthesis*; Butterworths: London, 1981; p 44. (f) Ager, D. J., *Chem. Soc. Rev.* **1982**, *11*, 493. (g) Weber, W. P. *Silicon Reagents for Organic Synthesis*; Springer: Berlin, 1983. (h) Parnes, Z. N.; Bolestova, G. I., *Synthesis* **1984**, 991. (i) Colvin, E. W. *Silicon Reagents in Organic Synthesis*; Academic: San Diego, 1988; p 7. (j) Fleming, I.; Donogues, J.; Smithers, R., *Org. React.* **1989**, *37*, 57.

2. Scholl, R. L.; Maciel, G. E.; Musker, W. K., *J. Am. Chem. Soc.* **1972**, *94*, 6376.

3. Katritzky, A. R.; Pinzelli, R. F.; Sinnott, M. V.; Topsom, R. D., *J. Am. Chem. Soc.* **1970**, *92*, 6861.

4. For preparations of the title reagent, see: (a) Nagel, R.; Post, H. W., *J. Org. Chem.* **1952**, *17*, 1379. (b) Rosenberg, S. D.; Walburn, J. J.; Stankovich, T. D.; Balint, A. E.; Ramsden, H. E., *J. Org. Chem.* **1957**, *22*, 1200. (c) Ottolenghi, A.; Fridkin, M.; Zilka, A., *Can. J. Chem.* **1963**, *41*, 2977. (d) Boeckman, R. K., Jr.; Blum, D. M.; Ganem, B.; Halvey, N., *Org. Synth., Coll. Vol.* **1988**, *6*, 1033.

5. For the preparation of vinyltrimethylsilane analogs, see Ref. 1.

6. (a) Cooke, F.; Moerck, R.; Schwindeman, J.; Magnus, P., *J. Org. Chem.* **1980**, *45*, 1046. (b) *Fieser & Fieser* **1982**, *10*, 444.

7. Stork, G.; Ganem, B., *J. Am. Chem. Soc.* **1973**, *95*, 6152.

8. Barton, T. J.; Rogido, R. J., *J. Org. Chem.* **1975**, *40*, 582.

9. (a) DeShong, P.; Li, W.; Kennington, J. W., Jr.; Ammon, H. L.; Leginus, J. M., *J. Org. Chem.* **1991**, *56*, 1364. (b) DeShong, P.; Leginus, J. M., *J. Org. Chem.* **1984**, *49*, 3421.

10. (a) Cooke, F.; Schwindeman, J.; Magnus, P., *Tetrahedron Lett.* **1979**, 1995. (b) *Fieser & Fieser* **1981**, *9*, 498.

11. Chan, T. H.; Massuda, D., *Tetrahedron Lett.* **1975**, 3383.

12. Nesmeyanov, A. N.; Fiedlina, R. K.; Velchko, F. K., *Dokl. Akad. Nauk SSSR* **1957**, *114*, 557.

13. Breslow, R.; Ryan, G.; Groves, J. T., *J. Am. Chem. Soc.* **1970**, *92*, 988.

14. (a) Pillot, J-P.; Dunogues, J.; Calas, R., *Synthesis* **1977**, 469. (b) *Fieser & Fieser* **1984**, *11*, 41.

15. Paquette, L. A.; Williams, R. V., *Tetrahedron Lett.* **1981**, *22*, 4643.

16. (a) Hsaio, C-N.; Shechter, H., *Tetrahedron Lett.* **1982**, *23*, 1963. (b) Hsaio, C-N.; Shecter, H., *J. Org. Chem.* **1988**, *53*, 2688.

17. Lai, M-T.; Oh, E.; Shih, Y.; Liu, H-W., *J. Org. Chem.* **1992**, *57*, 2471.

18. Hsiao, C-N.; Shecter, H., *Tetrahedron Lett.* **1982**, *23*, 3455.

19. Pearson, W. H.; Postich, M. J., *J. Org. Chem.* **1992**, *57*, 6354.

20. (a) Takeuchi, R.; Ishii, N.; Sugiura, M.; Sato, N., *J. Org. Chem.* **1992**, *57*, 4189. (b) Takeuchi, R.; Ishii, N.; Sato, N., *J. Chem. Soc., Chem. Commun.* **1991**, 1247.

21. Doyle, M. M.; Jackson, W. R.; Perlmutter, P., *Tetrahedron Lett.* **1989**, *30*, 233.

22. Borrachero, P.; Cabrera-Escribano, F.; Diánez, M. J.; Estrada, M. D.; Gómez-Guillen, M.; Castro, A. L.; Pérez-Garrido, S.; Torres, M. I., *Tetrahedron: Asymmetry* **2002**, *13*, 2025.

23. Kwak, K. C.; Oh, H.; Chung, D.-S.; Lee, Y.-H.; Baik, S.-H.; Kim, J. S.; Chai, K. C., *J. Korean Chem. Soc.* **2001**, *45*, 100.

24. Qui, Z.-M.; Burton, D. J., *Tetrahedron Lett.* **1995**, *36*, 5119.

25. Jankowski, P.; Masnyk, M.; Wicha, J., *Synlett* **1995**, 866.

26. Mizojiri, R.; Urabe, H.; Sato, F., *Tetrahedron Lett.* **1999**, *40*, 2557.

27. Edwards, M. G.; Jazzar, R. F. R.; Paine, B. M.; Shermer, D. J.; Whittlesey, M. K.; Williams, J. M. J.; Edney, D. D., *Chem. Commun.* **2004**, 90.

28. Bassindale, A. R.; Kyle, P. A.; Soobramanien, M.-C.; Taylor, P. G., *J. Chem. Soc., Perkin Trans 1* **2000**, 1173.

29. Schwan, A. I.; Strickler, R. R.; Dunn-Dufault, R.; Brillon, D., *Eur. J. Org. Chem.* **2001**, 1643.

30. Weinreb, S. M.; Chase, C. E.; Wipf, P., *Org. Synth.* **1997**, *75*, 161.

31. Parker, L. L.; Gowans, N. D.; Jones, S. W.; Robins, D. J., *Tetrahedron* **2003**, *59*, 10165.

32. Nakanishi, S.; Kumeta, K.; Takata, T., *Inorg. Chim. Acta* **1999**, *291*, 231.

33. Wagner, M.; Heiner, S.; Kunz, H., *Synlett* **2000**, 1753.

34. Wagner, M.; Kunz, H., *Synlett* **2000**, 400.

35. Wagner, M.; Kunz, H., *Angew. Chem., Int. Ed.* **2002**, *41*, 317.

36. Harmata, M.; Kahraman, M.; Jones, D. E.; Pavri, N.; Weatherwax, S. E., *Tetrahedron* **1998**, *54*, 9995.

37. Gooßen, L. J.; Paetzold, J., *Angew. Chem., Int. Ed.* **2002**, *41*, 1237.

38. Collman, J. P.; Wang, Z.; Straumanis, A.; Quelquejeu, M., *J. Am. Chem. Soc.* **1999**, *121*, 460.

39. Hudrlik, P. F.; Hudrlik, A. M. In *Advances in Silicon Chemistry*; Larson, G. L., Ed.; JAI Press (now Elsevier): Greenwich, CT, 1993.

40. Kakiuchi, P.; Matsumoto, M.; Sonoda, M.; Fukuyama, T.; Chatani, N.; Murai, S., *Chem. Lett.* **2000**, 750.

41. Jankowska, M.; Marciniec, B.; Pietraszuk, C.; Cytarska, J.; Zaidlewicz, M., *Tetrahedron Lett.* **2004**, *45*, 6615.

42. Singleton, D. A.; Leung, S.-W., *J. Organomet. Chem.* **1997**, *544*, 157.

43. Park, J.-W.; Chang, H.-J.; Jun, C.-H., *Synlett* **2006**, 771.

44. Jeffrey, T., *Tetrahedron Lett.* **1999**, *40*, 1673.

45. Hilt, G.; Lüers, S., *Synthesis* **2002**, 609.

46. Chatani, N.; Ikeda, S.; Ohe, K.; Murai, S., *J. Am. Chem. Soc.* **1992**, *114*, 9710.

47. Kawnacki, I.; Marciniec, B.; Szubert, K.; Kubicki, M., *Organometallics* **2005**, *24*, 6179.

48. Crudden, C. M.; Alper, H., *J. Org. Chem.* **1994**, *59*, 3091.

49. Marciniec, B.; Kujawa, M.; Pietraszuk, C., *Organometallics* **2000**, *18*, 1677.

List of Contributors

Jennifer L. Loebach	*Massachusetts Institute of Technology, Cambridge, MA, USA*	
	• Trimethylsilylketene	645
	• (Trimethylsilyl)vinylketene	725
Tyler R. Long	*University of Iowa, Iowa City, IA, USA*	
	• Polydimethylsiloxane Thimbles	403
Andreas Luxenburger	*Industrial Research Limited, Lower Hutt, New Zealand*	
	• Methyldiphenylchlorosilane	381
Gerhard Maas	*University of Ulm, Ulm, Germany*	
	• Trifluoromethanesulfonic Acid, Chlorobis (1,1-dimethylethyl)silyl Ester	537
Max Malacria	*Université Pierre et Marie Curie, Paris, France*	
	• (Bromomethyl)chlorodimethylsilane	85
Robert E. Maleczka Jr	*Michigan State University, East Lansing, MI, USA*	
	• Polymethylhydrosiloxane	427
	• Triethylsilane	506
	• Triphenylsilane	733
Helena C. Malinakova	*University of Kansas, Lawrence, KS, USA*	
	• Potassium Hexamethyldisilazide	432
Sven Mangelinckx	*Ghent University, Ghent, Belgium*	
	• Dimethylthexylsilyl Trifluoromethanesulfonate	265
Joseph P. Marino	*The University of Michigan, Ann Arbor, MI, USA*	
	• Dilithium Cyanobis(dimethylphenylsilyl)-cuprate	248
	• Lithium Cyano(dimethylphenylsilyl)cuprate	353
Michael J. Martinelli	*Lilly Research Laboratories, Indianapolis, IN, USA*	
	• Bromotrimethylsilane	92
	• Iodotrimethylsilane	325
Jean Martinez	*Université Montpelier, Montpellier, France*	
	• β-Trimethylsilylethanesulfonyl Chloride	611
Mark A. Matulenko	*University of Wisconsin at Madison, Madison, WI, USA*	
	• Bis(trimethylsilyl) Sulfide	83
Christopher G. McDaniel	*The Ohio State University, Columbus, OH, USA*	
	• Ammonium Hexafluorosilicate	25
John M. McGill	*Eli Lilly and Company, Lafayette, IN, USA*	
	• (α-Triethylsilyl)propionaldehyde Cyclohexylimine	516
Andrew L. McIver	*North Carolina State University, Raleigh, NC, USA*	
	• Chlorodimethylvinylsilane	157
Jeffrey A. McKinney	*Zeneca Pharmaceuticals, Wilmington, DE, USA*	
	• Phenylthiobis(trimethylsilyl)methane	411
Pedro Merino	*University of Zaragoza, Zaragoza, Spain*	
	• 2-(Trimethylsilyl)thiazole	712
William J. Middleton	*Ursinus College, Collegeville, PA, USA*	
	• Tris(dimethylamino)sulfonium Difluorotrimethylsilicate	739
Léone Miginiac	*University of Poitiers, Poitiers, France*	
	• 3-Trimethylsilyl-1-propyne	704
Ross Miller	*Merck Research Laboratories, Rahway, NJ, USA*	
	• Triisopropylsilyl Trifluoromethanesulfonate	559
Paul Mollard	*University of California, Santa Barbara, CA, USA*	
	• 2-Trimethylsilyl-1,3-propanediol	701

Reagent Formula Index

Subject Index

NOTES: **Bold** entries refer to main article titles, *see* refers to alternative names and acronyms, *see also* refers to related reagents